Brief Contents

Help students learn physics by doing physics

Dear Colleague,

Welcome to the second edition of our textbook *College Physics: Explore and Apply* and its supporting materials (Mastering™ Physics, the *Active Learning Guide* (ALG), and our *Instructor's Guide*)—a coherent learning system that helps students **learn physics by doing physics**!

Experiments, experiments...   Instead of being presented physics as a static set of established concepts and mathematical relations, students develop their own ideas just as physicists do: they *explore* and analyze **observational experiments**, identify patterns in the data, and propose explanations for the patterns. They then design **testing experiments** whose outcomes either confirm or contradict their explanations. Once tested, students *apply* explanations and relations for practical purposes and to problem solving.

A physics tool kit   To build problem-solving skills and confidence, students master proven visual tools (representations such as motion diagrams and energy bar charts) that serve as bridges between words and abstract mathematics and that form the basis of our overarching problem-solving strategy. Our unique and varied problems and activities promote 21st-century competences such as evaluation and communication and reinforce our practical approach with photo, video, and data analysis and real-life situations.

A flexible learning system   Students can work collaboratively on ALG activities in class (lectures, labs, and problem-solving sessions) and then read the textbook at home and solve end-of-chapter problems, or they can read the text and do the activities using Mastering Physics at home, then come to class and discuss their ideas. However they study, students will see physics as a living thing, a process in which they can participate as equal partners.

Why a new edition?   With a wealth of feedback from users of the first edition, our own ongoing experience and that of a gifted new co-author, and changes in the world in general and in education in particular, we embarked on this second edition in order to refine and strengthen our experiential learning system. Experiments are more focused and effective, our multiple-representation approach is expanded, topics have been added or moved to provide more flexibility, the writing, layout, and design are streamlined, and all the support materials are more tightly correlated to our approach and topics.

Working on this new edition has been hard work, but has enriched our lives as we've explored new ideas and applications. We hope that using our textbook will enrich the lives of your students!

Eugenia Etkina

Gorazd Planinsic

Alan Van Heuvelen

"This book made me think deeper and understand better."

—student at *Horry Georgetown Technical College*

PEARSON

A unique and active learning approach promotes deep and lasting

UPDATED! Observational Experiment Tables and Testing Experiment Tables: Students must make observations, analyze data, identify patterns, test hypotheses, and predict outcomes. Redesigned for clarity in the second edition, these tables encourage students to explore science through active discovery and critical thinking, constructing robust conceptual understanding.

NEW! Digitally Enhanced Experiment Tables now include embedded videos in the Pearson eText for an interactive experience. Accompanying questions are available in Mastering Physics to build skills essential to success in physics.

conceptual understanding of physics and the scientific process

TESTING EXPERIMENT TABLE 4.6   Testing the independence of horizontal and vertical motions

VIDEO TET 4.6

Testing experiment	Prediction	Outcome
At time zero, ball 1 is dropped. Simultaneously, ball 2 is shot horizontally when a compressed spring is released. Which ball hits the surface first?	Both balls start with zero initial vertical speed; thus their vertical motions are identical. Since we think that the vertical motion is independent of the horizontal motion, we predict that they will land at the same time.	When we try the experiment, the balls do land at the same time.

Conclusion

The outcome supports the hypothesis that horizontal and vertical motions are independent of each other.

> *"I like that the experiment tables... explain in detail why every step was important."*
>
> —student at *Mission College*

EXPANDED! Experiment videos and photos created by the authors enhance the active learning approach. Approximately 150 photos and 40 videos have been added to the textbook, as well as embedded in the Pearson eText, and scores more in the *Active Learning Guide* (ALG).

FIGURE 2.2 Long-exposure photographs of a moving cart with a blinking LED.

A wealth of practical and consistent guidance, examples, and opportunities

A four-step problem-solving approach in worked examples

consistently uses multiple representations to teach students how to solve complex physics problems. Students follow the steps of **Sketch & Translate, Simplify & Diagram, Represent Mathematically, Solve & Evaluate** to translate a problem statement into the language of physics, sketch and diagram the problem, represent it mathematically, solve the problem, and evaluate the result.

Physics Tool Boxes

Physics Tool Boxes focus on a particular skill, such as drawing a motion diagram, force diagram, or work-energy bar chart, to help students master the key tools they will need to utilize throughout the course to analyze physics processes and solve problems, bridging real phenomena and mathematics.

> *"It made me excited to learn physics! It has a systematic and easy-to-understand method for solving problems."*
>
> —student at *State University of West Georgia*

PHYSICS TOOL BOX 3.1 — Constructing a force diagram

1. Sketch the situation (a rock sinking into sand).

2. Circle the system (the rock).

3. Identify external interactions:
- The sand pushes up on the rock.
- Earth pulls down on the rock.

4. Place a dot at the side of the sketch, representing the system.

5. Draw force arrows to represent the external interactions.

6. Label the forces with a subscript with two elements.

56. * A frog jumps at an angle 30° above the horizontal. The origin of the coordinate system is at the point where the frog leaves the ground. Complete Table P4.55 by drawing check marks in the cells that correctly connect the quantities in the first column that describe the motion of the frog and the descriptions of what is happening to these quantities while the frog is moving. Consider the frog as a point-like object and assume that the resistive force exerted by the air is negligible.

TABLE P4.55

Physical quantity	Remains constant	Is changing	Increases only	Decreases only	Increases, then decreases	Decreases, then increases
x-coordinate magnitude						
y-coordinate magnitude						
Direction of velocity						
Magnitude of velocity						
Direction of acceleration						
Magnitude of acceleration						

29. * Your friend Devin has to solve the following problem: "You have a spring with spring constant k. You compress it by distance x and use it to shoot a steel ball of mass m into a sponge of mass M. After the collision, the ball and the sponge move a distance s along a rough surface and stop (see Figure P7.29). The coefficient of friction between the sponge and the surface is μ. Derive an expression that shows how the distance s depends on relevant physical quantities."

FIGURE P7.29

Devin derived the following equation:

$$s = \frac{kmx^2}{2(m + M)^2 g\mu}$$

Without deriving it, evaluate the equation that [...] reasonable? How do you know?

59. * Jeff and Natalie notice that a rubber balloon, which is first in a warm room, shrinks when they take it into the garden on a cold winter day. They propose two different explanations for the observed phenomenon: (a) the balloon is slowly leaking; (b) the balloon shrinks due to decreased temperature while the pressure in the balloon remains constant (isobaric compression). In order to test their proposed explanations, Jeff and Natalie perform three consecutive experiments: they measure the volume of the balloon and the temperature of the air near the balloon (1) in the room, (2) in the garden, and (3) again in the room. Their measurements, including uncertainties, are presented in the table below.

Exp. #	Location	Temperature	Volume of the balloon
1	Room	26.2 °C ± 0.1 °C	7500 cm³ ± 400 cm³
2	Garden	−15.3 °C ± 0.1 °C	6400 cm³ ± 400 cm³
3	Room	26.2 °C ± 0.1 °C	7300 cm³ ± 400 cm³

Based on the data, can Jeff and Natalie reject any of their hypotheses? Explain. Make sure you include uncertainties in your answer.

Pedagogically driven design and content changes

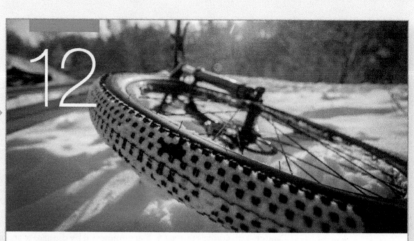

NEW! A fresh and modern design with a more transparent hierarchy of features and navigation structure, as well as an engaging chapter opener page and streamlined chapter summary, result in a more user-friendly resource, both for learning and for reference.

12

Gases

- Why does a plastic bottle left in a car overnight look crushed on a chilly morning?
- How hard is air pushing on your body?
- How long can the Sun shine?

When you inflate the tires of your bicycle in a warm basement in winter, they tend to look a bit flat when you take the bike outside. The same thing happens to a basketball—you need to pump it up before playing outside on a cold day. An empty plastic bottle left in a car looks crushed on a chilly morning. What do all those phenomena have in common, and how do we explain them?

BE SURE YOU KNOW HOW TO:
- Draw force diagrams (Section 3.1).
- Use Newton's second and third laws to analyze interactions of objects (Section 3.7 and 3.8).
- Use the impulse-momentum principle (Section 6.3).

IN CHAPTER 11, we learned that sound propagates due to the compression and decompression of air. But what exactly is being compressed? To answer this question and the ones above, we need to investigate what makes up a gas and how certain properties of gases can change.

352

---

308   **CHAPTER 10**   Vibrational Motion

Summary

Vibrational motion is the repetitive movement of an object back and forth about an **equilibrium position**. This vibration is due to the **restoring force** exerted by another object that tends to return the first object to its equilibrium position. An object's maximum displacement from equilibrium is the **amplitude** A of the vibration. **Period** T is the time interval for one complete vibration, and **frequency** f is the number of complete vibrations per second (in hertz). The frequency is the inverse of the period. (Section 10.1)

Object at end of spr

$F_{\text{Restoring}}$

$T = 2\pi$

Simple pendulum:

$F_{\text{Restoring}\,x} =$

$T = 2\pi \sqrt{}$

Simple harmonic motion is a mathematical model of vibrational motion when position x, velocity v, and acceleration a of the vibrating object change as sine or cosine functions with time. (Section 10.2)

$x = A \cos\left(\dfrac{2\pi}{T} t\right)$ Eq. (10.2)

$v_x = -\left(\dfrac{2\pi}{T}\right) A \sin\left(\dfrac{2\pi}{T} t\right)$ Eq. (10.3)

$a_x = -\left(\dfrac{2\pi}{T}\right)^2 A \cos\left(\dfrac{2\pi}{T} t\right)$ Eq. (10.4)

The **energy of a spring-object system** vibrating horizontally converts continuously from elastic potential energy when at the extreme positions to maximum kinetic energy when passing through the equilibrium position to a combination of energy types at other positions. (Section 10.3)

Object at end of spring: $x = \pm A$ $x = 0$ Other x
$U_s = K = K + U_s$

$E = \frac{1}{2} k x^2 + \frac{1}{2} m v^2$
$E = \frac{1}{2} k A^2$
$= \frac{1}{2} m v_{\max}^2$ Eq. (10.9)

The **energy of a pendulum-Earth system** converts continuously from gravitational potential energy when it is at the maximum height of a swing to kinetic energy when it is passing through the lowest point in the swing to a combination of energy types at other positions. (Section 10.5)

Simple pendulum: $\pm A$ 0 At other places
$U_g = K = K + U_g$

$E = mgy + \frac{1}{2} m v^2$
$E = mgy_{\max}$
$= \frac{1}{2} m v_{\max}^2$

Resonant energy transfer occurs when the frequency of the variable external force driving the oscillations is close to the **natural frequency** f_0 of the vibrating system. (Section 10.8)

REVISED! Streamlined text, layout, and figures throughout the book enhance the focus on central themes and topics, eliminating extraneous detail, resulting in **over 150 fewer pages** than the first edition and allowing students to study more efficiently.

enhance ease of use for students and instructors alike

FIGURE 19.14 A green LED. The electric circuit in (b) is used to collect the I-versus-ΔV data plotted in (c).

(a) One lead is longer than the other.

(b)

Green LED

+ −

Variable emf

(c)

Long lead connected to −, short to + ("wrong" direction)

Long lead connected to +, short to − ("right" direction)

$\vec{v}_{Po\,f}$

$v_{Po\,f} = +4.99 \times 10^5 \text{ m/s}$

$\vec{v}_{Rn\,i}$

$36.7°$

$\phi = ?$

$v_{Rn\,fx} = +4.0 \times 10^5 \text{ m/s}$

$\vec{v}_{\alpha f} = ?$

FIGURE 26.11 World lines for two objects and two light beams drawn on a spacetime diagram.

World line for light traveling left at speed c.

World line for light traveling right at speed c.

t (y)

$\frac{2}{5}c$ $\frac{3}{5}c$

$-c$

c

Future

Present

Past

x (ly)

NEW, REVISED, and EXPANDED! **Topics** include capacitors, AC circuits, LEDs, friction, 2-D collisions, energy, bar charts for rotational momentum and nuclear energy, ideal gas processes, thermodynamic engines, semiconductors, velocity selectors, and spacetime diagrams in special relativity.

$I_i\omega_i$

$I_f\omega_f$

$L_{Ai} + L_{Bi} + \Sigma\tau\Delta t = L_{Af} + L_{Bf}$

NEW! Integration of vector arithmetic into early chapters helps students develop vector-related skills in the context of learning physics. **Earlier placement of waves and oscillations** allows instructors to teach these topics with mechanics if preferred. Coverage with optics is also possible.

A flexible learning system adapts to any method of instruction

REVISED! The **Active Learning Guide** aligns with the textbook's chapters and supplements the knowledge-building approach of the textbook with activities that provide opportunities for further observation, testing, sketching, and analysis as well as collaboration, scientific reasoning, and argumentation. The *Active Learning Guide* can be used in class for individual or group work or assigned as homework and is now better integrated with the text. Now available via download in the Mastering Instructor Resource Center and customizable in print form via Pearson Collections.

> *"It is much easier to understand a concept when you can see it in action, and not just read it."*
>
> —student at San Antonio College

The **Instructor's Guide** provides key pedagogical principles of the textbook and elaborates on the implementation of the methodology used in the textbook, providing guidance on how to integrate the approach into your course.

and provides tools for easy implementation

NEW! Ready-to-Go Teaching Modules created for and by instructors make use of teaching tools for before, during, and after class, including new ideas for in-class activities. The modules incorporate the best that the text, Mastering Physics, and Learning Catalytics have to offer and guide instructors through using these resources in the most effective way. The modules can be accessed through the Instructor Resources area of Mastering Physics and as pre-built, customizable assignments.

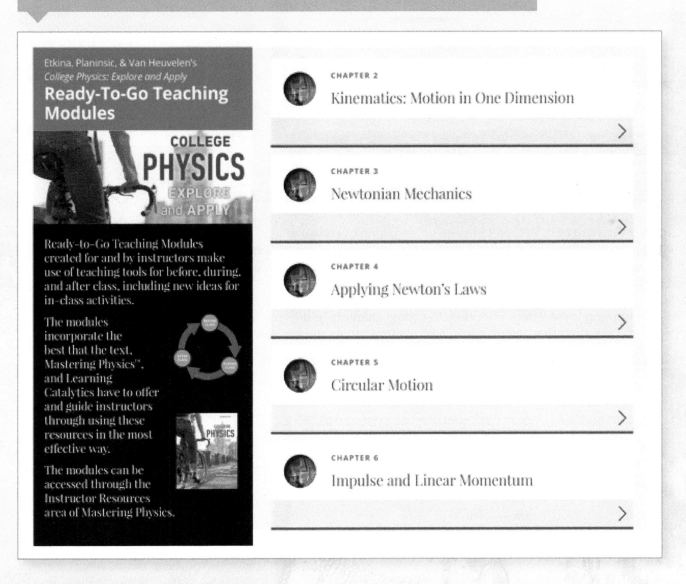

Etkina, Planinsic, & Van Heuvelen's
College Physics: Explore and Apply

Ready-To-Go Teaching Modules

COLLEGE
PHYSICS
EXPLORE and APPLY

Ready-to-Go Teaching Modules created for and by instructors make use of teaching tools for before, during, and after class, including new ideas for in-class activities.

The modules incorporate the best that the text, Mastering Physics™, and Learning Catalytics have to offer and guide instructors through using these resources in the most effective way.

The modules can be accessed through the Instructor Resources area of Mastering Physics.

CHAPTER 2
Kinematics: Motion in One Dimension

CHAPTER 3
Newtonian Mechanics

CHAPTER 4
Applying Newton's Laws

CHAPTER 5
Circular Motion

CHAPTER 6
Impulse and Linear Momentum

Mastering Physics

Build a basic understanding of physics principles and math skills

NEW! The Physics Primer relies on videos, hints, and feedback to refresh students' math skills in the context of physics and prepare them for success in the course. These tutorials can be assigned before the course begins as well as throughout the course as just-in-time remediation. The primer ensures students practice and maintain their math skills, while tying together mathematical operations and physics analysis.

Interactive Animated Videos provide an engaging overview of key topics with embedded assessment to help students check their understanding and to help professors identify areas of confusion. Note that these videos are not tied to the textbook and therefore do not use the language, symbols, and conceptual approaches of the book and ALG. The authors therefore recommend assigning these videos after class to expose students to different terminology and notation that they may come across from other sources.

Dynamic Study Modules (DSMs) help students study effectively on their own by continuously assessing their activity and performance in real time and adapting to their level of understanding. The content focuses on definitions, units, and the key relationships for topics across all of mechanics and electricity and magnetism.

www.MasteringPhysics.com

Show connections between physics and the real world as students learn to apply physics concepts via enhanced media

NEW! Direct Measurement Videos are short videos that show real situations of physical phenomena. Grids, rulers, and frame counters appear as overlays, helping students to make precise measurements of quantities such as position and time. Students then apply these quantities along with physics concepts to solve problems and answer questions about the motion of the objects in the video.

NEW! End-of-chapter problem types and 15% new questions and problems include multiple choice with multiple correct answers, find-a-pattern in data presented in a video or a table, ranking tasks, evaluate statements/claims/explanations/measuring procedures, evaluate solutions, design a device or a procedure that meets given criteria, and linearization problems. End-of-chapter problems have undergone careful analysis using Mastering Physics usage data to provide fine-tuned difficulty ratings and to produce a more varied, useful, and robust set of end-of-chapter problems.

Give students fingertip access to interactive tools

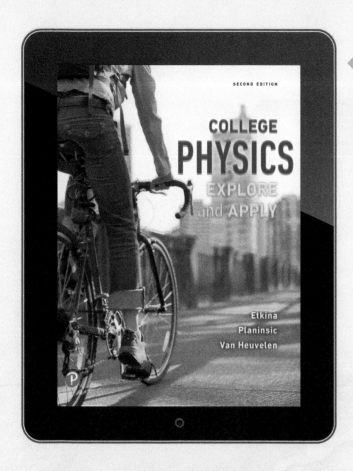

NEW! Pearson eText, optimized for mobile, seamlessly integrates videos such as the Observational Experiment Tables and other rich media with the text and gives students access to their textbook anytime, anywhere. Pearson eText is available with Mastering Physics when packaged with new books or as an upgrade students can purchase online.

Learning Catalytics™ helps generate class discussion, customize lectures, and promote peer-to-peer learning with real-time analytics. Learning Catalytics acts as a student response tool that uses students' smartphones, tablets, or laptops to engage them in more interactive tasks and thinking.

- **NEW!** Upload a full PowerPoint® deck for easy creation of slide questions.
- **NEW!** Team names are no longer case sensitive.
- Help your students develop critical thinking skills.
- Monitor responses to find out where your students are struggling.
- Rely on real-time data to adjust your teaching strategy.
- Automatically group students for discussion, teamwork, and peer-to-peer learning.

COLLEGE
PHYSICS

EXPLORE and APPLY

second edition

COLLEGE
PHYSICS
EXPLORE
and APPLY

Eugenia Etkina
RUTGERS UNIVERSITY

Gorazd Planinsic
UNIVERSITY OF LJUBLJANA

Alan Van Heuvelen
RUTGERS UNIVERSITY

 Pearson

New York, NY

Couseware Portfolio Management, Director: Jeanne Zalesky
Courseware Portfolio Manager: Darien Estes
Managing Producer: Kristen Flathman
Content Producer: Tiffany Mok
Courseware Director, Content Development: Jennifer Hart
Courseware Senior Analyst, Content Development: Alice
    Houston, Ph.D.
Senior Content Developer: David Hoogewerff
Courseware Editorial Assistant: Kristen Stephens and Leslie Lee
Rich Media Content Producer: Dustin Hennessey
Full-Service Vendor: Cenveo® Publisher Services
Full-Service Vendor Project Manager: Susan McNally,
    Cenveo® Publisher Services
Copyeditor: Joanna Dinsmore

Compositor: Cenveo® Publisher Services
Design Manager: Mark Ong, Side By Side Studios
Interior Designer: Lisa Buckley
Cover Designer: Lisa Buckley
Illustrators: Jim Atherton, Cenveo® Publisher Services
Rights & Permissions Project Manager: Kathleen Zander,
    Cenveo® Publisher Services
Rights & Permissions Management: Ben Ferrini
Photo Researcher: Karin Kipp, Cenveo® Publisher Services
Manufacturing Buyer: Stacey Weinberger
Director of Product Marketing: Allison Rona
Product Marketing Manager: Elizabeth Ellsworth Bell
Cover Photo Credit: Kari Medig/Aurora/Getty Images

Library of Congress Cataloging-in-Publication Data is on file with the Library of Congress.

www.pearson.com

ISBN 10: 0-134-60182-3 **ISBN 13:** 978-0-134-60182-3 (Student Edition)
ISBN 10: 0-134-68330-7 **ISBN 13:** 978-0-134-68330-0 (NASTA)

About the Authors

EUGENIA ETKINA is a Distinguished Professor at Rutgers, the State University of New Jersey. She holds a PhD in physics education from Moscow State Pedagogical University and has more than 35 years of experience teaching physics. She is a recipient of the 2014 Millikan Medal, awarded to educators who have made significant contributions to teaching physics, and is a fellow of the AAPT. Professor Etkina designed and now coordinates one of the largest programs in physics teacher preparation in the United States, conducts professional development for high school and university physics instructors, and participates in reforms to the undergraduate physics courses. In 1993 she developed a system in which students learn physics using processes that mirror scientific practice. That system, called *Investigative Science Learning Environment* (ISLE), serves as the basis for this textbook. Since 2000, Professor Etkina has conducted over 100 workshops for physics instructors, and she co-authored the first edition of *College Physics* and the *Active Learning Guide*. Professor Etkina is a dedicated teacher and an active researcher who has published over 60 peer-refereed articles.

GORAZD PLANINSIC is a Professor of Physics at the University of Ljubljana, Slovenia. He has a PhD in physics from the University of Ljubljana. Since 2000 he has led the Physics Education program, which prepares almost all high school physics teachers in the country of Slovenia. He started his career in MRI physics and later switched to physics education research. During the last 10 years, his work has mostly focused on the research of new experiments and how to use them more productively in teaching and learning physics. He is co-founder of the Slovenian hands-on science center House of Experiments. Professor Planinsic is co-author of more than 80 peer-refereed research articles and more than 20 popular science articles, and is the author of a university textbook for future physics teachers. In 2013 he received the Science Communicator of the Year award from the Slovenian Science Foundation.

ALAN VAN HEUVELEN holds a PhD in physics from the University of Colorado. He has been a pioneer in physics education research for several decades. He taught physics for 28 years at New Mexico State University, where he developed active learning materials including the *Active Learning Problem Sheets* (the *ALPS Kits*) and the *ActivPhysics* multimedia product. Materials such as these have improved student achievement on standardized qualitative and problem-solving tests. In 1993 he joined Ohio State University to help develop a physics education research group. He moved to Rutgers University in 2000 and retired in 2008. For his contributions to national physics education reform, he won the 1999 AAPT Millikan Medal and was selected a fellow of the American Physical Society. Over the span of his career he has led over 100 workshops on physics education reform. He worked with Professor Etkina in the development of the *Investigative Science Learning Environment* (ISLE) and co-authored the first edition of *College Physics* and the *Active Learning Guide*.

Preface

To the student

College Physics: Explore and Apply is more than just a book. It's a learning *companion*. As a companion, the book won't just tell you about physics; it will act as a guide to help you build physics ideas using methods similar to those that practicing scientists use to construct knowledge. The ideas that you build will be *yours*, not just a copy of someone else's ideas. As a result, the ideas of physics will be much easier for you to use when you need them: to succeed in your physics course, to obtain a good score on exams such as the MCAT, and to apply to everyday life.

Although few, if any, textbooks can honestly claim to be a pleasure to read, *College Physics: Explore and Apply* is designed to make the process interesting and engaging. The physics you learn in this book will help you understand many real-world phenomena, from why giant cruise ships are able to float to how telescopes work. The cover of the book communicates its spirit: you learn physics by exploring the natural world and applying it in your everyday life.

A great deal of research has been done over the past few decades on how students learn. We, as teachers and researchers, have been active participants in investigating the challenges students face in learning physics. We've developed unique strategies that have proven effective in helping students think like physicists. These strategies are grounded in *active learning with your peers*—deliberate, purposeful action on your part to learn something new. For learning to happen, one needs to talk to others, share ideas, listen, explain, and argue. It is in these deliberations that new knowledge is born. Learning is not passively memorizing so that you can repeat it later. When you learn actively, you engage with the material and—most importantly—share your ideas with others. You relate it to what you already know and benefit from the knowledge of your peers. You think about the material in as many different ways as you can. You ask yourself questions such as "Why does this make sense?" and "Under what circumstances does this not apply?" Skills developed during this process will be the most valuable in your future, no matter what profession you choose.

This book (your learning companion) includes many tools to support the active learning process: each problem-solving tool, worked example, observational experiment table, testing experiment table, review question, and end-of-chapter question and problem is designed to help you build your understanding of physics. To get the most out of these tools and the course, stay actively engaged in the process of developing ideas and applying them; form a learning group with your peers and try to work on the material together. When things get challenging, don't give up.

At this point you should turn to Chapter 1, Introducing Physics, and begin reading. That's where you'll learn the details of the approach that the book uses, what physics is, and how to be successful in the physics course you are taking.

To the instructor

Welcome to the second edition of *College Physics: Explore and Apply* and its supporting materials (Mastering™ Physics, the *Active Learning Guide* (ALG), and the *Instructor's Guide*), a coherent learning system that helps our students learn physics as an ongoing process rather than a static set of established concepts and mathematical relations. It is based on a framework known as ISLE (the *Investigative Science Learning Environment*). This framework originated in the work of Eugenia Etkina in the early 1990s. She designed a logical progression of student learning of physics that mirrors the processes in which physicists engage while constructing and applying knowledge. This progression was enriched in the early 2000s when Alan Van Heuvelen added his multiple representation approach. While logical flow represents a path for thinking, multiple representations are thinking tools. Since 2001, when ISLE curriculum development began, tens of thousands of students have been exposed to it as hundreds of instructors used the materials produced by the authors and their collaborators. Research on students learning physics through ISLE has shown that these students not only master the content of physics, but also become expert problem solvers, can design and evaluate their own experiments, communicate, and most importantly see physics as a process based on evidence as opposed to a set of rules that come from the book.

Experiments, experiments... The main feature of this system is that students practice developing physics concepts by following steps similar to those physicists use when developing and applying knowledge. The first introduction to a concept or a relation happens when students observe simple experiments (called **observational experiments**). Students learn to analyze these experiments, find patterns (either qualitative or quantitative) in the data, and develop multiple explanations for the patterns or quantitative relations. They then learn how to test the explanations and relations in new **testing experiments**. Sometimes the outcomes of the experiments might cause us to reject the explanations; often, they help us keep them. Students see how scientific ideas develop from evidence and are tested by evidence, and how evidence sometimes causes us to reject the proposed explanations. Finally, students learn how tested explanations and relations are applied for practical purposes and in problem solving. This is the process behind the subtitle of the book.

Explore and apply To help students explore and apply physics, we introduce them to tools: physics-specific representations, such as motion and force diagrams, momentum and energy bar charts, ray diagrams, and so forth. These representations serve as bridges between words and abstract mathematics. Research shows that students who use representations other than mathematics to solve problems are much more successful than those who just look for equations. We use a representations-based problem-solving strategy that helps students approach problem solving without fear and eventually develop not only problem-solving skills, but also confidence. The textbook and ALG introduce a whole library of novel problems and activities that help students develop competencies necessary for success in the 21st century: argumentation, evaluation, estimation, and communication. We use photo and video analysis, real-time data, and real-life situations to pose problems.

A flexible learning system There are multiple ways to use our learning system. Students can work collaboratively on ALG activities in class (lectures, labs, and problem-solving sessions) and then read the textbook and solve end-of-chapter problems at home, or they can first read the text and do the activities using Mastering Physics at home, then come to class and discuss their ideas. However they study, students will see physics as a living thing, a process in which they can participate as equal partners.

The key pedagogical principles of this book are described in detail in the first chapter of the *Instructor's Guide* that accompanies *College Physics*—please read that chapter. It elaborates on the implementation of the methodology that we use in this book and provides guidance on how to integrate the approach into your course.

While our philosophy informs *College Physics*, you need not fully subscribe to it to use this textbook. We've organized the book to fit the structure of most algebra-based physics courses: we begin with kinematics and Newton's laws, then move on to conserved quantities, statics, vibrations and waves, gases, fluids, thermodynamics, electricity and magnetism, optics, and finally modern physics. The structure of each chapter will work with any method of instruction. You can assign all of the innovative experimental tables and end-of-chapter problems, or only a few. The text provides thorough treatment of fundamental principles, supplementing this coverage with experimental evidence, new representations, an effective approach to problem solving, and interesting and motivating examples.

New to this edition

There were three main reasons behind the revisions in this second edition. (1) Users provided lots of feedback and we wanted to respond to it. (2) We (the authors) grew and changed, and learned more about how to help students learn, and our team changed—we have a new co-author, who is an expert in educational physics experiments and in the development of physics problems. (3) Finally, we wanted to respond to changes in the world (new physics discoveries, new technology, new skills required in the workplace) and to changes in education (the Next Generation Science Standards, reforms in the AP and MCAT exams). Our first edition was already well aligned with educational reforms, but the second edition strengthens this alignment even further.

We have therefore made the following global changes to the textbook, in addition to myriad smaller changes to individual chapters and elements:

- **An enhanced experiential approach**, with more experiment videos and photos (all created by the authors) and an updated and more focused and effective set of experiment tables, strengthens and improves the core foundation of the first edition. Approximately 150 photos and 40 videos have been added to the textbook, and even more to the ALG.

- **An expanded introductory chapter** (now Chapter 1) gives students a more detailed explanation of "How to use this book" to ensure they get the most out of the chapter features, use them actively, and learn how to think critically.

- **Integration of vector arithmetic in early chapters** allows students to develop vector-related skills in the context of learning physics, rather than its placement in an appendix in the first edition.

- **Earlier placement of waves and oscillations** allows instructors to teach these topics with mechanics if preferred. Coverage with optics is also still possible.

- **Significant new coverage of capacitors, AC circuits, and LEDs** (LEDs now permeate the whole book) expand the real-world and up-to-date applications of electricity.

- **Other new, revised, or expanded topics** include friction, 2-D collisions, energy, bar charts for rotational momentum and nuclear energy, ideal gas processes, thermodynamic engines, semiconductors, velocity selectors, and spacetime diagrams in special relativity.

- **Applications are integrated throughout each chapter,** rather than being grouped in the "Putting it all together" sections of the first edition, in order to optimize student engagement.

- **Problem-solving guidance is strengthened** by the careful revision of many Problem-Solving Strategy boxes and the review of each chapter's set of worked examples. The first edition Reasoning Skill boxes are renamed Physics Tool Boxes to better reflect their role; many have been significantly revised.

- **Streamlined text, layout, and figures** throughout the book enhance the focus on central themes and topics, eliminating extraneous detail. The second edition has over 150 fewer pages than the first edition, and the art program is updated with over 450 pieces of new or significantly revised art.

- **21st-century skills incorporated into many new worked examples and end-of-chapter problems** include data analysis, evaluation, and argumentation. Roughly 15% of all end-of-chapter questions and problems are new.

- **Careful analysis of Mastering Physics usage data** provides fine-tuned difficulty ratings and a more varied, useful, and robust set of end-of-chapter problems.

- **A fresh and modern design** provides a more transparent hierarchy of features and navigation structure, as well as an engaging chapter-opening page and streamlined chapter summary.

- **A significantly revised** *Active Learning Guide* is better integrated with the textbook, following the section sequence, and emphasizes collaboration, scientific reasoning, and argumentation.

All of the above sounds like a lot of work—and it was! But it was also lots of fun: we took photos of juice bottles sinking in the snow, we chased flying airplanes and running water striders, we drove cars with coffee cups on dashboards. Most exciting was our trip to a garbage plant to study and photograph the operation of an eddy current waste separator. Working on this new edition has enriched our lives, and we hope that using our textbook will enrich the lives of your students!

Instructor supplements

TIP All of the following materials are available for download on the Mastering Physics Instructor Resources page.

The **Instructor's Guide** (ISBN 0-134-89031-0), written by Eugenia Etkina, Gorazd Planinsic, David Brookes, and Alan Van Heuvelen, walks you through the innovative approaches they take to teaching physics. Each chapter of the *Instructor's Guide* contains a roadmap for assigning chapter content, *Active Learning Guide* assignments, homework, and videos of the experiments. In addition, the authors call out common pitfalls to mastering physics concepts and describe techniques that will help your students identify and overcome their misconceptions. Tips include how to manage the complex vocabulary of physics, when to use classroom response tools, and how to organize lab, lecture, and small-group learning time. Drawing from their extensive experience as teachers and researchers, the authors give you the support you need to make *College Physics* work for you.

The **Active Learning Guide** workbook (ISBN 0-134-60549-7) by Eugenia Etkina, David Brookes, Gorazd Planinsic, and Alan Van Heuvelen consists of carefully crafted cycles of in-class activities that provide an opportunity for students to conduct observational experiments, find patterns, develop explanations, and conduct the testing experiments for those explanations described in the textbook before they read it. These learning cycles are interspersed with "pivotal" activities that serve different purposes: (a) to introduce and familiarize students with new representational techniques, (b) to give students practice with representational techniques, (c) to directly address ideas that we *know* students struggle with (the goal is to encourage that struggle so that students reach a resolution either through their own discussion or by the instructor giving a "time for telling" lecture at the end of the activity), and (d) to provide scaffolding for students to work through an example or a passage in the textbook. The ALG also contains multiple experiments that can be used in labs. Whether the activities are assigned or not, students can always use this workbook to reinforce the concepts they have read about in the text, to practice applying the concepts to real-world scenarios, or to work with sketches, diagrams, and graphs that help them visualize the physics. The ALG is downloadable to share with your class; you may also talk to your sales representative about printing a custom version for your students.

The **Instructor Resource Materials** (ISBN 0-134-87386-6) on the Mastering Physics Instructor Resources page provide invaluable and easy-to-use resources for your class, organized by textbook chapter. The contents include a comprehensive library of all figures, photos, tables, and summaries from the textbook in JPEG and PowerPoint formats. A set of editable **Lecture Outlines**, **Open-Ended Questions**, and **Classroom Response System "Clicker" Questions** in PowerPoint will engage your students in class. Also included among the Instructor Resource Materials are the **Test Bank**, **Instructor Solutions Manual**, **Active Learning Guide**, **Active Learning Guide Solutions Manual**, and **Instructor Guide**.

Mastering™ Physics is the leading online homework, tutorial, and assessment platform designed to improve results by engaging students with powerful content. All Mastering resources, content, and tools are easy for both students and instructors to access in one convenient location. Instructors ensure that students arrive ready to learn by assigning educationally effective content before class and encourage critical thinking and retention with in-class resources such as Learning Catalytics™. Students can master concepts after class through traditional and adaptive homework assignments that provide hints and answer-specific feedback. The Mastering gradebook records scores for all automatically graded assignments in one place, while diagnostic tools give instructors access to rich data to assess student understanding and misconceptions.

New for the second edition of this book, Mastering Physics includes activities for students to do before coming to class, as an alternative to working through the *Active Learning Guide* activities prior to reading the textbook. These activities focus students' attention on observational experiments, helping them learn to identify patterns in the data, and on testing experiments, helping them learn how to make a prediction of an outcome of an experiment using an idea being tested, not personal intuition. Both skills are very important in science, but are very difficult to develop.

The significantly revised **Instructor's Solutions Manual**, provided as PDFs and editable Word files, gives complete solutions to all end-of-chapter questions and problems using the textbook's problem-solving strategy.

The **Test Bank**, which has also been significantly revised, contains more than 2000 high-quality problems, with a range of multiple-choice, true/false, short-answer, and regular homework-type questions. Test files are provided in TestGen® (an easy-to-use, fully networkable program for creating and editing quizzes and exams), as well as PDF and Word format.

Student supplements

 Physics experiment videos, accessed via the eText, with a smartphone through this QR code, at https://goo.gl/s2MerO, or online in the Mastering Physics Study Area, accompany most of the Observational and Testing Experiment Tables, as well as other discussions and problems in the textbook and in the ALG. Students can observe the exact experiment described in the text.

VIDEO

The **Pearson eText,** optimized for mobile, seamlessly integrates videos and other rich media with the text and gives students access to their textbook anytime, anywhere.

- The Pearson eText mobile app offers offline access and can be downloaded for most iOS and Android phones/tablets from the Apple App Store or Google Play
- Accessible (screen-reader ready)
- Configurable reading settings, including resizable type and night reading mode
- Instructor and student note-taking, highlighting, bookmarking, and search

The **Student Solutions Manual** (ISBN 0-134-88014-5) gives complete solutions to select odd-numbered end-of-chapter questions and problems using the textbook's problem-solving strategy.

In addition to content assigned by the instructor and this text's accompanying experiment videos, **Mastering™ Physics** also provides a wealth of self-study resources:

- **Dynamic Study Modules** assess student performance and activity in real time. They use data and analytics that personalize content to target each student's particular strengths and weaknesses. DSMs can be accessed from any computer, tablet, or smart phone.
- **PhET simulations** from the University of Colorado, Boulder are provided in the Mastering Physics Study Area to allow students to explore key concepts by interacting with these research-based simulations.
- **24/7 access to online tutors*** enables students to work one-on-one, in real time, with a tutor using an interactive whiteboard. Tutors will guide them through solving their problems using a problem-solving-based teaching style to help them learn underlying concepts. In this way, students will be better prepared to handle future assignments on their own.

Acknowledgments

We wish to thank the many people who helped us create this textbook and its supporting materials. First and foremost, we want to thank our team at Pearson Higher Education, especially Jeanne Zalesky, who believed that the book deserved a second edition; Alice Houston, who provided careful, constructive, creative, enriching, and always positive feedback on every aspect of the book and the ALG; Darien Estes, who fearlessly made pivotal decisions that made the new edition much better; Susan McNally, who tirelessly shepherded the book through all stages of production; and David Hoogewerff, who oversaw the Mastering Physics component of the program. Tiffany Mok and Leslie Lee oversaw the new edition of the *Active Learning Guide* and other supplements. Special thanks to Jim Smith and Cathy Murphy who helped shape the first edition of the book. We also want to thank Adam Black for believing in the future of the project.

Although Michael Gentile is not a co-author on the second edition, this work would be impossible without him; he contributed a huge amount to the first edition and provided continuous support for us when we were working on the second edition. No words will describe how grateful we are to have Paul Bunson on our team. Paul helped us with the end-of-chapter problem revisions and Mastering Physics and ALG activities, and provided many helpful suggestions, particularly on rotational mechanics, fluids, relativity, and quantum optics. In addition, he was the first to adopt the textbook even before the first edition was officially printed and since then has remained a vivid advocate and supporter of ISLE. We are indebted to Charlie Hibbard, who checked and rechecked every fact and calculation in the text. Brett Kraabel prepared detailed solutions for every end-of-chapter problem for the *Instructor's Solutions Manual*. We also want to thank all of the reviewers, in particular Jeremy Hohertz, who put their time and energy to providing thoughtful, constructive, and supportive feedback. We thank Matt Blackman for adding excellent problems to the Test Bank, Katerina Visnjic for her support of ISLE and the idea to expand energy bar charts to nuclear physics, and Mikhail Kagan for timely feedback. Our special thanks go to Lane Seeley for his thoughtful review of the energy chapter, which led to its deep revision. We thank Diane Jammula and Jay Pravin Kumar, who not only became avid supporters and users of ISLE but also helped create instructor resources for the second edition. We thank Ales Mohoric and Sergej Faletic for their suggestions on problems.

Our infinite thanks go to Xueli Zou, the first adopter of ISLE, and to Suzanne Brahmia, who came up with the Investigative Science Learning Environment acronym "ISLE" and was and is an effective user and tireless advocate of the ISLE learning strategy. Suzanne's ideas about relating physics and mathematics are reflected in many sections of the book. We are indebted to David Brookes, another tireless ISLE developer, whose research shaped the language we use. We thank all of Eugenia's students who are now physics teachers for providing feedback and ideas and using the book with their students.

We have been very lucky to belong to the physics teaching community. Ideas of many people in the field contributed to our understanding of how people learn physics and what approaches work best. These people include Arnold Arons, Fred Reif, Jill Larkin, Lillian McDermott, David Hestenes, Joe Redish, Stamatis Vokos, Jim Minstrell, David Maloney, Fred Goldberg, David Hammer, Andy Elby, Noah Finkelstein, David Meltzer, David Rosengrant, Anna Karelina, Sahana Murthy, Maria Ruibal-Villasenhor, Aaron Warren, Tom Okuma, Curt Hieggelke, and Paul D'Alessandris. We thank all of them and many others.

Personal notes from the authors

We wish to thank Valentin Etkin (Eugenia's father), an experimental physicist whose ideas gave rise to the ISLE philosophy many years ago, Inna Vishnyatskaya (Eugenia's mother), who never lost faith in the success of our book, and Dimitry and Alexander Gershenson (Eugenia's sons), who provided encouragement to Eugenia over the years. While teaching Alan how to play violin, Alan's uncle Harold Van Heuvelen provided an instructional system very different from that of traditional physics teaching. In Harold's system, many individual abilities (skills) were developed with instant feedback and combined over time to address the process of playing a complex piece of music. We tried to integrate this system into our ISLE physics learning system.

—Eugenia Etkina, Gorazd Planinsic, and Alan Van Heuvelen

*Please note that tutoring is available in selected Mastering products, and in those products you are eligible for one tutoring session of up to 30 minutes duration with your course. Additional hours can be purchased at reasonable rates.

Reviewers and classroom testers of the first and second editions

Ricardo Alarcon
Arizona State University

Eric Anderson
University of Maryland, Baltimore County

James Andrews
Youngstown State University

Aurelian Balan
Delta College

David Balogh
Fresno City College

Linda Barton
Rochester Institute of Technology

Ian Beatty
University of North Carolina at Greensboro

Robert Beichner
North Carolina State University, Raleigh

Aniket Bhattacharya
University of Central Florida

Luca Bombelli
University of Mississippi

Scott Bonham
Western Kentucky University

Gerald Brezina
San Antonio College

Paul Bunson
Lane Community College

Debra Burris
University of Central Arkansas

Hauke Busch
Georgia College & State University

Rebecca Butler
Pittsburg State University

Paul Camp
Spelman College

Amy Campbell
Georgia Gwinnett College

Kapila Castoldi
Oakland University

Juan Catala
Miami Dade College North

Colston Chandler
University of New Mexico

Soumitra Chattopadhyay
Georgia Highlands College

Betsy Chesnutt
Itawamba Community College

Chris Coffin
Oregon State University

Lawrence Coleman
University of California, Davis

Michael Crivello
San Diego Mesa College

Elain Dahl
Vincennes University

Jared Daily
Mid Plains Community College

Danielle Dalafave
The College of New Jersey

Charles De Leone
California State University, San Marcos

Carlos Delgado
Community College of Southern Nevada

Christos Deligkaris
Drury University

Dedra Demaree
Oregon State University

Karim Diff
Santa Fe Community College

Kathy Dimiduk
Cornell University

Diana Driscoll
Case Western Reserve University

Raymond Duplessis
Delgado Community College

Taner Edis
Truman State University

Bruce Emerson
Central Oregon Community College

Davene Eyres
North Seattle Community College

Xiaojuan Fan
Marshall University

Nail Fazleev
University of Texas at Arlington

Gerald Feldman
George Washington University

Frank Ferrone
Drexel University

Jane Flood
Muhlenberg College

Lewis Ford
Texas A&M University

Tom Foster
Southern Illinois University, Edwardsville

Joseph Ganem
Loyola University Maryland

Brian Geislinger
Gadsden State Community College

Richard Gelderman
Western Kentucky University

Lipika Ghosh
Virginia State University

Anne Gillis
Butler Community College

Martin Goldman
University of Colorado, Boulder

Greg Gowens
University of West Georgia

Michael Graf
Boston College

Alan Grafe
University of Michigan, Flint

Sigrid Greene
Alice Lloyd College

Elena Gregg
Oral Roberts University

Recine Gregg
Fordham University

John Gruber
San Jose State University

Arnold Guerra
Orange Coast College

Edwin Hach III
Rochester Institute of Technology

Steve Hagen
University of Florida, Gainesville

Thomas Hemmick
State University of New York, Stony Brook

Scott Hildreth
Chabot College

Zvonimir Hlousek
California State University, Long Beach

Jeremy Hohertz
Elon University

Mark Hollabaugh
Normandale Community College

Klaus Honscheid
Ohio State University

Kevin Hope
University of Montevallo

Joey Huston
Michigan State University

Richard Ignace
East Tennessee State University

Doug Ingram
Texas Christian University

George Irwin
Lamar University

Darrin Johnson
University of Minnesota

Adam Johnston
Weber State University

Mikhail Kagan
Pennsylvania State University

David Kaplan
Southern Illinois University, Edwardsville

James Kawamoto
Mission College

Julia Kennefick
University of Arkansas

Casey King
Horry-Georgetown Technical College

Patrick Koehn
Eastern Michigan University

Victor Kriss
Lewis-Clark State College

Peter Lanagan
College of Southern Nevada, Henderson

Albert Lee
California State University, Los Angeles

Todd Leif
Cloud County Community College

Eugene Levin
Iowa State University

Jenni Light
Lewis-Clark State College

Curtis Link
Montana Tech of The University of Montana

Donald Lofland
West Valley College

Susannah Lomant
Georgia Perimeter College

Rafael Lopez-Mobilia
University of Texas at San Antonio

Kingshuk Majumdar
Berea College

Gary Malek
Johnson County Community College

Eric Mandell
Bowling Green State University

Lyle Marschand
Northern Illinois University

Eric Martell
Millikin University

Donald Mathewson
Kwantlen Polytechnic University

Mark Matlin
Bryn Mawr College

Dave McGraw
University of Louisiana, Monroe

Ralph McGrew
Broome Community College

Timothy McKay
University of Michigan

David Meltzer
Arizona State University

William Miles
East Central Community College

Rabindra Mohapatra
University of Maryland, College Park

Enrique Moreno
Northeastern University

Joe Musser
Stephen F. Austin State University

Charles Nickles
University of Massachusetts, Dartmouth

Gregor Novak
United States Air Force Academy

Gonzalo Ordonez
Butler University

John Ostendorf
Vincennes University

Philip Patterson
Southern Polytechnic State University

Euguenia Peterson
Richard J. Daley College

Jeff Phillips
Loyola Marymount University

Francis Pichanick
University of Massachusetts, Amherst

Alberto Pinkas
New Jersey City College

Dmitri Popov
State University of New York, Brockport

Matthew Powell
West Virginia University

Stephanie Rafferty
Community College of Baltimore County, Essex

Roberto Ramos
Indiana Wesleyan University

Greg Recine
Fordham University

Edward Redish
University of Maryland, College Park

Lawrence Rees
Brigham Young University

Lou Reinisch
Jacksonville State University

Andrew Richter
Valparaiso University

Joshua Ridley
Murray State University

Melodi Rodrigue
University of Nevada, Reno

Charles Rogers
Southwestern Oklahoma State University

David Rosengrant
Kennesaw State University

Alvin Rosenthal
Western Michigan University

Lawrence Rowan
University of North Carolina at Chapel Hill

Roy Rubins
University of Texas at Arlington

Otto Sankey
Arizona State University

Rolf Schimmrigk
Indiana University, South Bend

Brian Schuft
North Carolina Agricultural and Technical State University

Sara Schultz
Montana State University, Moorhead

Bruce Schumm
University of California, Santa Cruz

David Schuster
Western Michigan State University

Lane Seeley
Seattle Pacific University

Bart Sheinberg
Houston Community College, Northwest College

Carmen Shepard
Southwestern Illinois College

Douglas Sherman
San Jose State University

Chandralekha Singh
University of Pittsburgh

David Snoke
University of Pittsburgh

David Sokoloff
University of Oregon

Nickolas Solomey
Wichita State University

Mark Stone
Northwest Vista College

Bernhard Stumpf
University of Idaho

Steven Sweeney
DeSales University

Tatsu Takeuchi
Virginia Polytechnic Institute and State University

Julie Talbot
University of West Georgia

Colin Terry
Ventura College

Beth Ann Thacker
Texas Tech University

James Thomas
University of New Mexico

John Thompson
University of Maine, Orono

Som Tyagi
Drexel University

David Ulrich
Portland Community College

Eswara P. Venugopal
University of Detroit, Mercy

James Vesenka
University of New England

Melissa Vigil
Marquette University

William Waggoner
San Antonio College

Kendra Wallis
Eastfield College

Jing Wang
Eastern Kentucky University

Tiffany Watkins
Boise State University

Laura Weinkauf
Jacksonville State University

William Weisberger
State University of New York, Stony Brook

John Wernegreen
Eastern Kentucky University

Daniel Whitmire
University of Louisiana at Lafayette

Luc Wille
Florida Atlantic University

Roy Wilson
Mississippi Gulf Coast Community College Perkinston

Brian Woodahl
Indiana University-Purdue University Indianapolis

Gary Wysin
Kansas State University, Manhattan

Chadwick Young
Nicholls State University

Jiang Yu
Fitchburg State College

Real-World Applications

Contents

Introducing Physics

In everyday life, a model of something (such as a model airplane or a model train) is usually a smaller, simpler, or idealized version of the original. An architect creates a model to show a building's essential elements and context. Physicists do something similar, but it might surprise you to hear that in physics, a marble is a very useful model of an airplane, a car, or the Moon. Read on and you will learn why.

- What are physics models?
- How is the word "law" used differently in physics than in the legal system?
- How do we solve physics problems such as determining the minimum runway length needed for an airplane?

WHETHER YOU'VE STUDIED ANY physics before or not, it's helpful to take a step back and consider what physics is about and how physicists think about things. You'll find that learning to analyze problems like a physicist will help you not only in this course, but also in others (and in life in general). This book is designed to help you do this, and this chapter will give you an overview of how to use this book to your best advantage.

FIGURE 1.1 Archaeologists applied principles from physics to determine that this skeleton of *Australopithecus afarensis*, nicknamed "Lucy," lived about 3.2 million years ago.

1.1 What is physics?

Physics is a fundamental experimental science encompassing subjects such as motion, waves, light, electricity, magnetism, atoms, and nuclei. Knowing physics allows you to understand many aspects of the world, from why bending over to lift a heavy load can injure your back to why Earth's climate is changing. Physics explains the very small—atoms and subatomic particles—and the very large—planets, galaxies, and black holes.

In each chapter of this textbook, we will apply our knowledge of physics to other fields of science and technology such as biology, medicine, geology, astronomy, architecture, engineering, agriculture, and anthropology. For instance, you will learn about techniques used by archeologists to determine the age of bones (Figure 1.1), about electron microscopes and airport metal detectors, and why high blood pressure indicates problems with the circulatory system.

In this book we will concentrate not only on developing an understanding of the important basic laws of physics but also on the processes that physicists employ to discover and use these laws. The processes (among many) include:

- Collecting and analyzing experimental data.
- Making explanations and experimentally testing them.
- Creating different representations (pictures, graphs, bar charts, etc.) of physical processes.
- Finding mathematical relations—mathematical models—between different variables.
- Testing those relations in new experiments.

The search for rules

Physicists search for general rules, or **laws**, that bring understanding to the chaotic behavior of our surroundings. In physics the word *law* means a causal mathematical relation between variables inferred from the data or through some reasoning process. Causal relations show how change in one quantity affects the change in another quantity, but they do not explain why such causation occurs. The laws, once discovered, often seem obvious, yet their discovery usually requires years of experimentation and theorizing. Despite being called "laws," they are temporary in the sense that new information often leads to their modification, revision, and, in some cases, abandonment.

For example, in 200 B.C. Apollonius of Perga watched the Sun and the stars moving in arcs across the sky and adopted the concept that Earth occupied the center of a revolving universe. Three hundred years later, Ptolemy developed a detailed model to explain the complicated motion of the planets in that Earth-centered universe. Ptolemy's model, which predicted with surprising accuracy the changing positions of the planets, was accepted for the next 1400 years. However, as the quality of observations improved, discrepancies between the predictions of Ptolemy's model and the real positions of the planets became bigger and bigger. A new model was needed. Copernicus, who studied astronomy around the time that Columbus sailed to America, developed a model of motion for the heavenly bodies in which the Sun resided at the center of the universe while Earth and the other planets moved in orbits around it. More than 100 years later the model was revised by Johannes Kepler and later supported by careful experiments by Galileo Galilei. Finally, 50 years after Galileo's death, Isaac Newton formulated three simple laws of motion and the universal law of gravitation, which together provided a successful explanation for the orbital motion of Earth and the other planets. These laws also allowed us to predict the positions of new planets, which at the time were not yet known. Newton's work turned the heliocentric model into the theory of gravitation. For nearly 300 years Newtonian theory went unaltered until Albert Einstein made several profound improvements to our understanding of motion and gravitation at the beginning of the 20th century.

Newton's inspiration provided not only the basic resolution of the 1800-year-old problem of the motion of the planets but also a general framework for analyzing the mechanical properties of nature (**Figure 1.2**). Newton's simple laws give us the understanding needed to guide rockets to the Moon, to build skyscrapers, and to lift heavy objects safely without injury.

It is difficult to appreciate the great struggles our predecessors endured as they developed an understanding that now seems routine. Today, similar struggles occur in most branches of science, though the questions being investigated have changed. How does the brain work? What causes Earth's magnetism? What is the nature of the pulsating sources of X-ray radiation in our galaxy? Is the recently discovered accelerated expansion of the universe really caused by a mysterious "dark energy," or is our interpretation of the observations of distant supernovae that revealed the acceleration incomplete?

The processes for devising and using new models

Physics is an experimental science. To answer questions, physicists do not just think and dream in their offices but constantly engage in experimental investigations. Physicists use special measuring devices to observe phenomena (natural and planned), describe their observations (carefully record them using words, numbers, graphs, etc.), find repeating features called patterns (for example, the distance traveled by a falling object is directly proportional to the square of the time in flight), and then try to explain these patterns. By doing this, physicists describe and answer the questions of "why" or "how" the phenomena happened and then deduce quantitative rules called mathematical models that explain the phenomena.

However, a deduced explanation or a mathematical model is not automatically accepted as true. Every model needs to undergo careful testing. When physicists test a model, they use the model to predict the outcomes of new experiments. As long as there is no experiment whose outcome is inconsistent with predictions made using the model, it is not disproved. However, a new experiment could be devised tomorrow whose outcome is not consistent with the prediction made using the model. The point is that there is no way to "prove" a model once and for all. At best, the model just hasn't been disproven yet.

A simple example will help you understand some processes that physicists follow when they study the world. Imagine that you walk into the house of your acquaintance Bob and see 10 tennis rackets of different quality and sizes. This is an **observational experiment**. During an observational experiment a scientist collects data that seem important. Sometimes it is an accidental or unplanned experiment. The scientist has no prior expectation of the outcome. In this case the number of tennis rackets and their quality and sizes represent the data. Having so many tennis rackets seems unusual to you, so you try to explain the data you collected (or, in other words, to explain why Bob has so many rackets) by devising several hypotheses. A **hypothesis** is an explanation that usually is based on some mechanism that is behind what is going on, or it can be a mathematical model describing the phenomenon. One hypothesis is that Bob has lots of children and they all play tennis. A second hypothesis is that Bob makes his living by fixing tennis rackets. A third hypothesis is that he is a thief who steals tennis rackets.

How do you decide which hypothesis is correct? You may reason: if Bob has many children who play tennis, and I walk around the house checking the sizes of clothes that I find, then I will find clothes of different sizes. Checking the clothing sizes is a new experiment, called a **testing experiment**. A testing experiment is different from an observational experiment. In a testing experiment, a specific hypothesis is being "put on trial." This hypothesis is used to construct a clear expectation of the outcome of the experiment. This clear expectation (based on the hypothesis being tested) is called a **prediction**. So you conduct the testing experiment by walking around the house checking the closets. You do find clothes of different sizes. This is the **outcome** of your testing experiment. Does it mean for absolute certain that Bob has the rackets

FIGURE 1.2 Thanks to Newton, we can explain the motion of the Moon. We can also build skyscrapers.

TIP Notice the difference between a hypothesis and a prediction. A *hypothesis* is an idea that explains why or how something that you observe happens. A *prediction* is a statement of what should happen in a particular experiment if the hypothesis being tested were true. The prediction is based on the hypothesis and cannot be made without a specific experiment in mind.

because all of his children play tennis? No; he could still be a racket repairman or a thief. Therefore, if the outcome of the testing experiment matches the prediction based on your hypothesis, you cannot say that you proved the hypothesis. All you can say is that you failed to disprove it. However, if you walk around the house and do not find any children's clothes, you can say with more confidence that the number of rackets in the house is not due to Bob having lots of children who play tennis. Still, this conclusion would only be valid if you made an **assumption**: Bob's children live in the house and wear clothes of different sizes. Generally, in order to reject a hypothesis you need to check the additional assumptions you made and determine if they are reasonable.

Imagine you have rejected the first hypothesis (you didn't find any children's clothes). Next you wish to test the hypothesis that Bob is a thief. This is your reasoning: *If* Bob is a thief (the hypothesis), *and* I walk around the house checking every drawer (the testing experiment), *then* I will not find any receipts for the tennis rackets (the prediction). You perform the experiment and you find no receipts. Does it mean that Bob is a thief? He might just be a disorganized father of many children or a busy repairman. However, if you find all of the receipts, you can say with more confidence that he is not a thief (but he could still be a repairman). Thus it is possible to disprove (rule out) a hypothesis, but it is not possible to prove it once and for all. The process that you went through to create and test your hypotheses is depicted in **Figure 1.3**. At the end of your investigation you might be left with a hypothesis that you failed to disprove. As a physicist you would now have some confidence in this hypothesis and start using it for practical applications.

Using this textbook you will learn physics by following a process similar to that described above. The section "How to use this book to learn physics" at the end of this chapter will explain how to master this process.

Physicists use words and the language of mathematics to express ideas about the world. But they also represent these ideas and the world itself in other ways—sketches, diagrams, and even cut-out paper models (James Watson made a paper model of DNA when trying to determine its structure). In physics, however, the ultimate goal is to understand the mechanisms behind physical phenomena and to devise mathematical models that allow for quantitative predictions of new phenomena. Thus, a big part of physics is identifying measurable properties of the phenomena (such as mass, speed, and force), collecting quantitative data, finding the patterns in that data, and creating mathematical models representing relations between the quantities.

FIGURE 1.3 Science is a cyclical process for creating and testing knowledge.

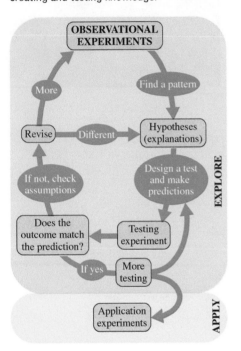

How will learning physics change your interactions with the world?

Even if you do not plan on becoming a professional physicist, learning physics can change the way you think about the world. For example, why do you feel cold when you wear wet clothes? Why is it safe to sit in a car during a lightning storm? Knowing physics will also help you understand what underlies many important technologies. How can a GPS know your present position and guide you to a distant location? How do power plants generate electric energy?

Studying physics is also a way to acquire the processes of knowledge construction, which will help you make decisions based on evidence rather than on personal opinions. When you hear an advertisement for a shampoo claim it will make your hair 97.5% stronger, you will ask: How do they know this? Did this number come from an experiment? If it did, was it tested? What assumptions did they make? Did they control for food consumed, exercise, air quality, etc.? Understanding physics will help you differentiate between actual evidence and unsubstantiated claims. For instance, before you accept a claim, you might ask about the data supporting the claim, what experiments were used to test the idea, and what assumptions were made. Thinking critically about the messages you hear will change the way you make decisions as a consumer and a citizen.

1.2 Modeling

You already met the word "model" in this chapter. In this section we will learn more about this term. Physicists study how the complex world works. To start the study of some aspect of that world, they often begin with a simplified version. Consider how you move your body when you walk. Your back foot on the pavement lifts and swings forward, only to stop for a short time when it again lands on the pavement, now ahead of you. Your arms swing back and forth. The trunk of your body moves forward steadily. Your head also moves forward but bobs up and down slightly. It would be very difficult to start our study of motion by analyzing all these complicated parts and movements. Thus, physicists create in their minds simplified representations, called **models**, of physical phenomena and then think of the phenomena in terms of those models. Physicists begin with very simple models and then add complexity as needed to investigate more detailed aspects of the phenomena.

A simplified object

To simplify real objects, physicists often neglect both the dimensions of objects (their sizes) and their structures (the different parts) and instead regard them as single **point-like objects** (Figure 1.4).

Is modeling a real object as a point-like object a good idea? Imagine a 100-m race. The winner is the first person to get a body part across the finish line. The judge needs to observe the movement of all body parts (or a photo of the parts) across a very small distance near the finish line to determine who had the fastest time. Here, that very small distance near the finish line is small compared to the size of the human body. This is a situation where modeling the runners as point-like objects is not reasonable. However, if you are interested in how long it takes an average person to run 100 m, then the movement of different body parts is not as important, since 100 m is much larger than the size of a runner. In this case, the runners can be modeled as point-like objects. Even though we are talking about the same situation (a 100-m race), the aspect of the situation that interests us determines how we choose to model the runners.

Consider an airplane landing on a runway (Figure 1.5a). We want to determine how long it takes for it to stop. Since all of its parts move together, the part we study does not matter. In that case it is reasonable to model the airplane as a point-like object. However, if we want to build a series of gates for planes to unload passengers (Figure 1.5b), then we need to consider the motion of the different parts of the airplane. For example, there must be enough room for an airplane to turn while maneuvering into and out of the gate. In this case the airplane cannot be modeled as a point-like object.

> **Point-like object** A point-like object is a simplified representation of a real object. As a rule of thumb, you can model a real object as a point-like object when one of the following two conditions is met: (a) when all of its parts move in the same way, or (b) when the object is much smaller than the other relevant lengths in the situation. The same object can be modeled as a point-like object in some situations but not in others.

Modeling

The process that we followed to decide when a real object could be considered a point-like object is an example of what is called **modeling**. The modeling of objects is the first step that physicists use when they study natural phenomena. In addition to simplifying the objects that they study, scientists simplify the interactions between objects and also the processes that are occurring in the real world. Then they add complexity as their understanding grows. Galileo Galilei is believed to be the first scientist to consciously model a phenomenon. In his studies of falling objects in the early 17th century, he chose to simplify the real phenomenon by ignoring the interactions of the falling objects with the air.

FIGURE 1.4 Physicists often model complex structures as point-like objects.

We can ignore the chassis flexing and the driver moving and model this racecar as a point-like object in order to study its motion.

FIGURE 1.5 An airplane can be considered a point-like object (a) when landing, but not (b) when parking.

(a)

(b)

> **Modeling** A model is a simplified representation of an object, a system (a group of objects), an interaction, or a process. A scientist creating the model decides which features to include and which to neglect.

As you see from above, there are different kinds of models. Those important for physics are mathematical models of processes. Models allow us to predict future phenomena.

1.3 Physical quantities

To describe physical phenomena quantitatively and model them mathematically, physicists construct **physical quantities**: features or characteristics of phenomena that can be determined experimentally, directly or indirectly. Determining the value of a physical quantity means comparing the characteristic to an assigned **unit** (a chosen standard).

Units of measure

Physicists describe physical quantities using the **SI system**, or *Le Système international d'unités*, whose origin goes back to the 1790s when King Louis XVI of France created a special commission to invent a new metric system of units. For example, in the SI system length is measured in meters. One meter is approximately the distance from your nose to the tip of the fingers of your outstretched arm. A long step is about one meter. Other units of length are related to the meter by powers of 10 using **prefixes** (milli, kilo, nano, ...). These prefixes relate smaller or bigger versions of the same unit to the basic unit. For example, 1 millimeter is 0.001 meter; 1 kilometer is 1000 meters. The prefixes are used when a measured quantity is much smaller or much larger than the basic unit. If the distance is much larger than 1 m, you might want to use the kilometer (10^3 m) instead. The most common prefixes and the powers of 10 to which they correspond are given in Table 1.1. In addition to the unit of length, the SI system has six other basic units, summarized in Table 1.2.

TABLE 1.1 Power of 10 prefixes

Prefix	Abbreviation	Value
Tera	T	10^{12}
Giga	G	10^9
Mega	M	10^6
Kilo	k	10^3
Hecto	h	10^2
Deka	da	10^1
Deci	d	10^{-1}
Centi	c	10^{-2}
Milli	m	10^{-3}
Micro	μ	10^{-6}
Nano	n	10^{-9}
Pico	p	10^{-12}
Femto	f	10^{-15}

TABLE 1.2 Basic SI physical quantities and their units

Physical quantity	Unit name and symbol	Physical description (approximate value)
Time	Second, s	One second is the time it takes for the heart to beat once.
Length	Meter, m	One meter is the length of one stride.
Mass	Kilogram, kg	One kilogram is the mass of 1 liter of water.
Electric current	Ampere, A	One ampere is the electric current through super bright white LEDs, such as those used in some street lamps.
Temperature	Kelvin, K	One kelvin degree is the same as 1 degree on the Celsius scale or about 2 degrees on the Fahrenheit scale. Half a kelvin (about 1 degree Fahrenheit) is the smallest temperature difference that the average person can detect by touching (with your hands).
Amount of matter	Mole, mol	One mole of anything contains 6×10^{23} of the units; one mole of oranges contains 6×10^{23} oranges.
Intensity of light	Candela, cd	One candela is the intensity of light produced by a relatively large candle at a distance of 1 m.

Table 1.2 provides a "feel" for some of the units but does not say exactly how each unit is defined. More careful definitions are important in order that measurements made by scientists in different parts of the world are consistent. However, to understand the precise definitions of these units, one needs to know more physics. We will learn how each unit is precisely defined when we investigate the concepts on which the definition is based.

Measuring instruments

Physicists use a measuring instrument to compare the quantity of interest with a standardized unit. Each measuring instrument is calibrated so that it reads in multiples of that unit. Some examples of measuring instruments are a thermometer to measure temperature (calibrated in degrees Celsius or degrees Fahrenheit), a watch to measure time intervals (calibrated in seconds), and a meter stick to measure the height of an object (calibrated in millimeters). We can now summarize these ideas about physical quantities and their units.

Physical quantity A physical quantity is a feature or characteristic of a physical phenomenon that can be measured in some unit. A measuring instrument is used to make a quantitative comparison of this characteristic with a unit of measure. Examples of physical quantities are your height, your body temperature, the speed of your car, and the temperature of air or water.

Significant digits

When we measure a physical quantity, the instrument we use and the circumstances under which we measure it determine how precisely we know the value of that quantity. Imagine that you wear a pedometer (a device that measures the number of steps that you take) and wish to determine the number of steps on average that you take per minute. You start walking at 3 p.m. according to your analog watch and stop when it shows 3:26, and see that the pedometer shows 2254 steps. You divide 2254 by 26 using your calculator, and it says 86.692307692307692. If you accept this number, it means that you know the number of steps per minute within plus or minus 0.000000000000001 steps/min. If you accept the number 86.69, it means that you know the number of steps to within 0.01 steps/min. If you accept the number 90, it means that you know the number of steps within 10 steps/min. Which answer should you use?

To answer this question, let's first focus on the measurements. Although it seems that you walked for 26 min, you could have walked for as few as 25 min or for as many as 27 min depending on whether you started a little after 3 p.m. or finished a little after 3:26. The number 26 does not give us enough information to know the time more precisely than that. The time measurement 26 min has two **significant digits**, or two numbers that carry meaning contributing to the precision of the result. The pedometer measurement 2254 has four significant digits. Should the result of dividing the number of steps by the amount of time you walked have two or four significant digits? If we accept four, it means that the number of steps per minute is known more precisely than the time measurement in minutes. This does not make sense. The number of significant digits in the final answer should be the same as the number of significant digits of the quantity used in the calculation that has the *smallest number of significant digits*. Thus, in our example, the average number of steps per minute should be 87. We rounded the result given by the calculator to two significant digits. This example shows that when we divide (or multiply) one term by another value, the number of significant digits in the result cannot be greater than the number of the significant digits in the term that has the fewest—in our case, two digits in the time. However, as we add or subtract values, the answer cannot have more decimal places than the term with fewest decimal places. For example, 30.517 s + 0.3 s = 30.8 s.

FIGURE 1.6 The precision of an instrument is determined by one-half of its smallest division. The smallest division of this measuring tape is 1 mm, so the precision is 0.5 mm.

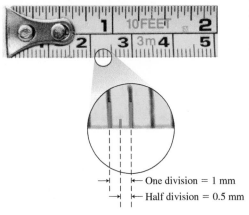

→| |← One division = 1 mm
→| |← Half division = 0.5 mm

Another issue with significant digits arises when a quantity is reported with no decimal points. For example, how many significant digits does 6500 have—two or four? This is where scientific notation helps. **Scientific notation** means writing numbers in terms of their power of 10. For example, we can write 6500 as 6.5×10^3. This means that the 6500 actually has two significant digits: 6 and 5. If we write 6500 as 6.50×10^3, it means 6500 has three significant digits: 6, 5, and 0. The number 6.50 is more precise than the number 6.5, because it means that you are confident in the number to the hundredths place. Scientific notation provides a compact way of writing large and small numbers and also allows us to indicate unambiguously the number of significant digits a quantity has.

Because we use instruments to measure quantities, we need to connect our knowledge of significant digits to the measurements. When we measure a quantity, we talk about precision. The **precision** of the value of a physical quantity is determined by one of two cases.

1. If the quantity is measured by a single instrument, its precision depends on the instrument used to measure it. **Figure 1.6** shows how the design of the measuring instrument determines its precision.

2. If the quantity is calculated from other measured quantities, then its precision depends on the least precise instrument out of all the instruments used to measure a quantity used in the calculation.

1.4 Making rough estimates

Sometimes it is useful to make a rough estimate of a physical quantity to help assess a situation or to make a decision. To do this, we use our personal knowledge or experience to get an approximate numerical value for an unknown quantity. Often the goal of the rough estimate is to determine the *order of magnitude* of the quantity—is it tens, hundreds, or thousands of the relevant units? Estimating is an extremely valuable skill not only in science but also in everyday life (**Figure 1.7**).

FIGURE 1.7 In everyday life, rough estimates are often sufficient.

For example, suppose we want to estimate the amount of drinking water that needs to be taken on a passenger airplane for a 4-h flight. We don't know exactly how many passengers are on the plane, but perhaps we assume that there are about 30 rows of seats with six seats per row. These data allow us to estimate the number of passengers: $30 \times 6 = 180$. After adding the crew and rounding the number, we can estimate the number of passengers to be 200. How many cups of water will each passenger drink? Some may drink six, but some only one. It seems reasonable to estimate that each person will drink about a cup of water every hour during the flight. Therefore, during a 4-h flight, 200 people will drink about 800 cups of water. How much water is in 1 cup? Again, if we don't know exactly, we can estimate it to be about 200 mL (a can of Coke holds about 350 mL). Our final estimation for the amount of water needed on the airplane is therefore 160,000 mL, or 160 L. You could argue that the airplane might be larger (with, say, 40 rows of eight seats) and that if the flight is during the daytime, people might drink more water (say six glasses per person). If you repeat the calculation using these new numbers, you will get about 380 L of water. This number is larger than our first estimate, but the order of magnitude is the same—hundreds of liters.

1.5 Vector and scalar quantities

There are two general types of physical quantities—those that contain information about magnitude as well as direction and those that contain magnitude information only. Physical quantities that do not contain information about direction are called **scalar quantities** and are written using *italic* symbols (*m*, *T*, etc.). Mass is a scalar quantity, as is temperature (**Figure 1.8**). To manipulate scalar quantities, you use stand-ard arithmetic and algebra rules—you add, subtract, multiply, and divide scalars as though they were ordinary numbers.

Physical quantities that contain information about magnitude and direction are called **vector quantities** and are represented by italic symbols with an arrow on top ($\vec{F}$, $\vec{v}$, etc.). The little arrow on top of the symbol always points to the right. The ac-tual direction of the vector quantity is shown in a diagram. For example, force is a physical quantity with both magnitude and direction (direction is very important if you are trying to hammer a nail into the wall). When you push a door, your push can be represented with a force arrow on a diagram; the stronger you push, the longer that arrow must be. The direction of the push is represented by the direction of that arrow (**Figure 1.9**). The arrow's direction indicates the direction of the vector, and the arrow's relative length indicates the vector's magnitude. The methods for manipulating vector quantities (adding and subtracting them as well as multiplying a vector quantity by a scalar quantity) will be introduced as needed in the following chapters.

1.6 How to use this book to learn physics

The goals of this textbook are to help you construct understanding of some of the most important ideas in physics, learn to use physics knowledge to analyze physical phe-nomena, and develop the general process skills that scientists use in the practice of science. One such skill is learning from a scientific text. Thus, by learning to work with your textbook efficiently and productively, you will not only learn physics but also de-velop textbook reading skills that will be helpful in any other science subject you study. To take advantage of all the tools that this textbook has to offer, we suggest that you read the text below and then, after you work through the first few chapters, come back to this section and read it again.

The most important strategy that will help you learn better is called **interrogation**. Interrogation means continually asking yourself the same question when reading the text. This question is so important that we put it in the box below:

> ## Why is this true?

Make sure that you ask yourself this question as often as possible so that eventually it becomes a habit. Out of all the strategies that are recommended for reading comprehension, this is the one that is directly connected to better learning outcomes. For example, in Section 1.2 you read the following sentence: "The modeling of objects is the first step that physicists use when they study natural phenomena." Ask yourself, "Why is this true?" Possibly because real-life phenomena are too complicated to be investigated in detail—it is much easier to describe the motion of a runner if you consider her to be a point-like object than to take into account the details of her arms, legs, hair, etc. Thus, simplifying, (or as physicists call it, modeling) is the first step that we take. By just stopping and interrogating yourself as often as possible about what is written in the book you will be able to understand and remember this information better.

Textbook features

This textbook has several features that repeat in every chapter. Recognizing these fea-tures and using them effectively not only will help you learn physics and shorten the time that you spend doing so, but also will help you develop good reading habits.

FIGURE 1.8 Temperature is a scalar quantity; it has magnitude, but not direction. However, it can be positive or negative.

FIGURE 1.9 The force that your hand exerts on a door is a vector quantity represented by an arrow.

$\vec{F}_{\text{Hand on Door}}$

Chapter-opening page and connection to real life

The large photo at the beginning of each chapter shows an aspect of your life or the world around you that can be explained using the physics in the chapter. The three questions in the margin below the photo also help you connect chapter material to real life or some exciting applications. The questions are all answered in the chapter, often in a quantitative way. For example, in Chapter 3, the question "How do seat belts and air bags save lives?" is answered in the last section, "Seat belts and air bags." When you start working with a new chapter, read the questions and the short explanation below the photo to get motivated about the material in the chapter. People learn and remember new knowledge much better if they are motivated to learn!

BE SURE YOU KNOW HOW TO:
- Draw a motion diagram for a moving object (Section 2.2).
- Determine the direction of acceleration using a motion diagram (Section 2.7).
- Add vectors graphically and determine their components (Sections 2.3, 2.4, and 2.6).

Be sure you know how to

At the bottom of each chapter-opening page is a bulleted list referring to sections from earlier chapters. This list highlights concepts and skills that are necessary to be successful in learning the upcoming new material. Consider the bullets as cues identifying the most important ideas in the previous chapters and where to find them. For example, in Chapter 3, one bullet says "Determine the direction of acceleration using a motion diagram (Section 2.7)." Ask yourself whether you know how to do that, and if you do not, review Section 2.7.

Physics Tool Boxes

These boxes show you the sequence of steps necessary to master a specific physics skill, such as drawing a force diagram or a bar chart. These skills are tools you will use to analyze physics processes and solve problems. They serve as bridges between real phenomena and mathematics (math is also one of the tools, but we treat it separately). If you learn to use these tools effectively, problem solving will be much easier. When you are learning a new tool, redraw all the steps in your notebook and, after that, work through the Conceptual Exercise that follows a Physics Tool Box. This process will help you practice the steps that you just learned.

Observational Experiment Tables

These tables help you learn how to observe simple experiments carefully and look for patterns. The goal of these tables is to show you where physics concepts and mathematical models come from. To find the patterns, you will need to use the tools that you learn in the Physics Tool Boxes. When working with an Observational Experiment Table, first watch the video of the experiment. Then read the description of the experiment in the left column of the table and try to analyze it without looking at the rest of the table. After you are done, compare your analysis to the analysis in the right-hand column(s) of the table and see if you agree with the described patterns. The experiment videos launch directly from the eText pages. They are also linked within

the Study Area of Mastering™ Physics, via this QR code, and at https://goo.gl /s2MerO. We recommend that you bookmark this page and use it when you work with the book.

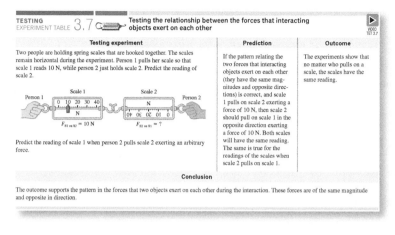

TESTING EXPERIMENT TABLE 3.7 — Testing the relationship between the forces that interacting objects exert on each other

Testing experiment	Prediction	Outcome
Two people are holding spring scales that are hooked together. The scales remain horizontal during the experiment. Person 1 pulls her scale so that scale 1 reads 10 N, while person 2 just holds scale 2. Predict the reading of scale 2.	If the pattern relating the two forces that interacting objects exert on each other (they have the same magnitudes and opposite directions) is correct, and scale 1 pulls on scale 2 exerting a force of 10 N, then scale 2 should pull on scale 1 in the opposite direction exerting a force of 10 N. Both scales will have the same reading. The same is true for the readings of the scales when scale 2 pulls on scale 1.	The experiments show that no matter who pulls on a scale, the scales have the same reading.

Predict the reading of scale 1 when person 2 pulls scale 2 exerting an arbitrary force.

Conclusion

The outcome supports the pattern in the forces that two objects exert on each other during the interaction. These forces are of the same magnitude and opposite in direction.

TIP Notice that the "vector sum of the forces" mentioned in the definition at right does not mean the sum of their magnitudes. Vectors are not added as numbers; their directions affect the magnitude of the vector sum.

Tips

These short comments within the text encourage the use of particular strategies, point your attention to important details that you might otherwise miss, or warn you about specific difficulties that many students experience.

REVIEW QUESTION 3.6 Newton's second law says that the acceleration of an object is inversely proportional to its mass. However, the acceleration with which all objects fall in the absence of air is the same. How can this be?

Review Questions

At the end of each section is a Review Question. The goal of these questions is to help you learn how to read a scientific text and interrogate it in a way that is similar to how physicists interrogate claims. This skill is a major component of critical thinking. After you read the question, first try to form your opinion about the answer and only after that reread the section and try to find evidence that will support or refute your opinion. Ask your instructor for the answers to Review Questions.

Problem solving

This textbook teaches you how to approach physics problem solving in a very systematic way. If you consistently follow the problem-solving strategy demonstrated in worked examples, we guarantee that by the end of the course you will be an expert problem solver. Every example has four problem-solving steps that we suggest you use when solving *any* physics problem. Start by reading the problem at least three times to completely understand the problem statement. Next, try to visualize the problem situation and draw a sketch. This is the beginning of the first step:

Sketch and translate

"Translate" means converting the problem's everyday language into the language of physical quantities. It is helpful to use the strategy "circle, box, underline" when trying to translate. *Circle* what is given and can be easily translated (for example, the given mass of a crash test dummy, 60 kg), *box* what needs to be found (such as the average force that an air bag and seat belt exert on the dummy), and *underline* the words that need interpretation for the translation (the word "stops" tells us that the final speed of the dummy is 0 m/s). Notice that sketching the process and representing it in different ways are important steps in solving any problem.

Testing Experiment Tables

These tables help you learn how to test patterns or explanations (many of them will be the patterns that emerged from the Observational Experiment Tables). Your procedure with these tables is different. First, read the description of the experiment in the left column, then try to use the pattern or explanation being tested to predict the outcome of the experiment. Only then should you watch the video or read about the outcome in the far right column.

■ Summary

Chapter summaries

The summary at the end of each chapter helps you reflect on what you learned. Compare different representations of the same idea and try to find anything you might have missed or misunderstood. Do you understand the connections? Can you write the mathematics illustrated by a diagram? This process is the best strategy to make sure you did not miss anything.

EXAMPLE 3.9 Force exerted by air bag on driver during collision

A 60-kg crash test dummy moving at 13.4 m/s (30 mi/h) stops during a collision in a distance of 0.65 m. Estimate the magnitude of the average force that the air bag and seat belt exert on the dummy.

Sketch and translate We sketch and label the situation as shown below, choosing the crash test dummy as the system. The positive x-direction is in the direction of motion, and the origin is at the position of the dummy at the start of the collision.

Simplify and diagram

Here you will make necessary simplifications and choose a productive physics representation (from your Tool Box). Determine how you can simplify the situation. How will you model the object of interest? What interactions can you neglect? You will learn a new representation in nearly every chapter; thus the choice of a productive one is key to solving the problem.

Simplify and diagram We model the dummy D as a point-like object and assume that the primary force exerted on the dummy while stopping is due to the air bag and seat belt's $\vec{F}_{\text{A on D}}$, shown in the force diagram. We can ignore the downward gravitational force that Earth exerts on the dummy $\vec{F}_{\text{E on D}}$ and the upward force that the car seat exerts on the dummy $\vec{F}_{\text{S on D}}$ since they add to zero and do not contribute to the acceleration.

Represent mathematically

In this step you will use the physics representation from the step above to construct the mathematical representation of the same situation. No numbers are needed here, only symbols of physical quantities. The goal of this step is to work with the mathematical representations to isolate the variable that you need to solve for. By representing the situation using these multiple methods and learning to translate from one type of representation to another, you will start giving meaning to the abstract symbols used in the mathematical description of the situation.

Represent mathematically To determine the dummy's acceleration, we use kinematics:

$$a_x = \frac{v_x^2 - v_{0x}^2}{2(x - x_0)}$$

Once we have the dummy's acceleration, we apply the x-component form of Newton's second law to find the force exerted by the air bag and seat belts on the dummy:

$$a_x = \frac{F_{\text{A on D }x}}{m_{\text{D}}} = \frac{-F_{\text{A on D}}}{m_{\text{D}}} = -\frac{F_{\text{A on D}}}{m_{\text{D}}}$$

$$\Rightarrow F_{\text{A on D}} = -m_{\text{D}}a_x$$

Solve and evaluate

Finally, you will plug the values of the physical quantities into the equation found in the step above (do not forget the units!) and solve for the variable you boxed in the first step. When you obtain a number, your job is not done—you need to evaluate your answer, or ask yourself whether your answer is reasonable. You will learn three major strategies for performing this evaluation in this book.

Solve and evaluate We know that $v_{0x} = +13.4 \text{ m/s}$ and $v_x = 0$ (the dummy has stopped). The initial position of the dummy is $x_0 = 0$ and the final position is $x = 0.65 \text{ m}$. The acceleration of the dummy while in contact with the air bag and seat belt is

$$a_x = \frac{0^2 - (13.4 \text{ m/s})^2}{2(0.65 \text{ m} - 0 \text{ m})} = -138 \text{ m/s}^2$$

Thus, the *magnitude* of the average force exerted by the air bag and seat belt on the dummy is

$$F_{\text{A on D}} = -(60 \text{ kg})(-138 \text{ m/s}^2) = 8300 \text{ N}$$

This force $\left[8300 \text{ N}(1 \text{ lb}/4.45 \text{ N}) = 1900 \text{ lb} \right]$ is almost 1 ton. Is this estimate reasonable? The magnitude is large, but experiments with crash test dummies in the real world are consistent with a force this large in magnitude, a very survivable collision.

Try to solve the example problems in the chapters by using this four-step strategy without looking at the solution. After finishing, compare your solution to the one described in the book. Then do the "Try it yourself" part of the example problem and compare your answer to the book's answer. If you are still having trouble, use the same strategy to actively solve other example problems in the text or problems from the Active Learning Guide (if you are using that companion book). Uncover the solutions to these worked examples only after you have tried to complete the problem on your own. Then follow the same process on the homework problems assigned by your instructor.

In addition to the full worked examples, the book provides Conceptual Exercises and Quantitative Exercises. These have fewer steps: "Sketch and translate" and "Simplify and diagram" in the Conceptual Exercises, and "Represent mathematically" and "Solve and evaluate" in the Quantitative Exercises.

Summary

We are confident that this book will act as a useful companion in your study of physics and that you will take from the course not just the knowledge of physics but also an understanding of the process of science that will help you in all your scientific endeavors. Learning physics through the approach used in this book builds a deeper understanding of physics concepts and an improved ability to solve difficult problems compared to traditional learning methods. In addition, you will learn to reason scientifically and be able to transfer those reasoning skills to many other aspects of your life.

Kinematics: Motion in One Dimension

When you drive, you are supposed to follow the three-second tailgating rule. When the car in front of you passes a sign at the side of the road, your car should be far enough behind it that it takes you 3 s to reach the same sign. In this chapter, we will learn the physics behind the three-second rule.

SCIENTISTS OFTEN ASK questions about things that most people accept as "just the way it is." For example, in the northern hemisphere, there are more hours of daylight in June than in December, whereas in the southern hemisphere, the opposite is true. Scientists want explanations for such simple phenomena. In this chapter, we will learn to describe a phenomenon that we encounter every day but rarely question—motion.

- What is a safe following distance between your car and the car in front of you?
- Can you be moving and not moving at the same time?
- Why do physicists say that an object thrown upward is falling?

BE SURE YOU KNOW HOW TO:

- Decide when an object can be modeled as a point-like object (Section 1.2).
- Use significant digits in calculations (Section 1.3).

2.1 What is motion?

Imagine you are a passenger in a car. You see your friend standing on the sidewalk and wave at her, turning your head to look at her while driving past her. She too turns her head as she waves back. We turn our heads to follow a moving object, so it is not surprising that your friend turns her head—you are in a moving car. But why do *you* turn your head? After all, your friend is standing still on the sidewalk. However, from your perspective, she is moving away from you. This example helps us understand a very important idea: **different observers see the same motion differently**. In other words, motion is relative to the *observer*, the person who is describing the motion. So who is moving in this scenario, you or your friend? The answer is that it depends on who is the observer. The driver of your car, for example, sees you (a passenger in the car) as not moving.

This example shows us that when describing motion we need to specify both the object whose motion we are describing (the **object of interest**) and the person who is doing the describing (the **observer**).

> **Motion** is a change in an object's position relative to a given observer during a certain time interval. Without identifying the observer, it is impossible to say whether the object of interest moved. Physicists say *motion is relative,* meaning that the motion of any object of interest depends on the point of view of the observer.

FIGURE 2.1 Motion is relative. Two observers explain the motion of the Sun relative to Earth differently.

(a)

An observer on Earth sees the Sun move in an arc across the sky.

(b)

An observer in a spaceship sees the person on Earth as rotating under a stationary Sun.

What makes the idea of relative motion confusing at first is that people intuitively use Earth as the object of reference—the object with respect to which they describe motion. If an object does not move with respect to Earth, many people would say that the object is not moving. That is why it took scientists thousands of years to understand why Earth has days and nights. An observer on Earth uses Earth as the object of reference and sees the Sun moving in an arc across the sky (**Figure 2.1a**). An observer on a distant spaceship, however, sees Earth rotating on its axis so that different parts of its surface face the Sun at different times (Figure 2.1b).

Reference frames

Specifying the observer before describing the motion of an object of interest is an extremely important part of constructing what physicists call a **reference frame**. **A reference frame includes an object of reference, a coordinate system with a scale for measuring distances, and a clock to measure time.** If the object of reference is large and cannot be considered a point-like object, it is important to specify where on the object of reference the origin of the coordinate system is placed. For example, if you want to describe the motion of a bicyclist and choose your object of reference to be Earth, you place the origin of the coordinate system at Earth's surface, not its center.

Modeling motion

When we model objects, we make simplified assumptions in order to analyze complicated situations. Just as we simplified an object to model it as a point-like object in Chapter 1, we can also simplify a process involving motion. What is the simplest way an object can move?

Imagine that you haven't ridden a bike in a while. You probably wouldn't start by attempting a turn; instead, you would ride in a straight line. This kind of motion is called **linear motion** or **one-dimensional motion**.

Linear motion is a model of motion that assumes that an object, considered as a point-like object, moves along a straight line.

For example, suppose we want to model a car's motion along a straight stretch of highway. We can assume that the car is a point-like object (it is small compared to the length of the highway) and the motion is linear motion (the highway is long and straight).

REVIEW QUESTION 2.1 What does the statement "Motion is relative" mean?

2.2 A conceptual description of motion

To describe linear motion more precisely, we start by devising a visual representation. Consider Observational Experiment Table 2.1, in which a bowling ball rolls on a smooth floor.

OBSERVATIONAL EXPERIMENT TABLE 2.1 Using dots to represent motion

 VIDEO OET 2.1

Observational experiment	Analysis
Experiment 1. You push a bowling ball (the object of interest) and let it roll on a smooth linoleum floor. Each second, you place a beanbag beside the bowling ball. The beanbags are evenly spaced.	The dots in this diagram represent the positions of the beanbags you placed each second as the bowling ball slowly rolled on the floor. • • • •
Experiment 2. You repeat Experiment 1, but you push the ball harder before you let it roll. The beanbags are farther apart but are still evenly spaced.	The dots in this diagram represent the positions of the beanbags, which are still evenly spaced but separated by a greater distance than the bags in Experiment 1. • • •
Experiment 3. You push the bowling ball and let it roll on a carpeted floor instead of a linoleum floor. The distance between the beanbags decreases as the ball rolls.	The dots in this diagram represent the decreasing distance between beanbags as the ball rolls on the carpet. • • • •
Experiment 4. You roll the ball on the linoleum floor and gently and continually push on it with a board. The distance between the beanbags increases as the pushed ball rolls.	The dots in this diagram represent the increasing distance between beanbags as the ball is continually pushed across the linoleum floor. • • • •

Pattern

- The spacing of the dots allows us to visualize the motion of the object of interest.
- When the object travels without speeding up or slowing down, the dots are evenly spaced.
- When the object slows down, the dots get closer together.
- When the object moves faster and faster, the dots get farther apart.

Motion diagrams

In the experiments in Table 2.1, we used beanbags as an approximate record of where the ball was located as time passed to help us visualize the motion of the ball. Let's now use blinking LEDs to help us more precisely visualize motion, this time of

FIGURE 2.2 Long-exposure photographs of a moving cart with a blinking LED.

(a)

(b)

(c)

(d)

(e)

a cart. **Figure 2.2a** shows a cart with an attached blinking LED (on for 31 ms, then off for 31 ms). Figure 2.2b is a long-exposure photo of the cart as it is moving on a horizontal surface at steady speed, similar to the situation in Experiment 1 of Table 2.1. In Figure 2.2c, the cart is moving faster. Next, in Figure 2.2d, the moving cart is being pushed in the direction opposite to its motion until it stops. In Figure 2.2e, the cart is being pushed from rest and moves faster and faster.

As you can see, the light traces in the four photos clearly illustrate the different types of motion of the cart. The same-length light traces in Figure 2.2b indicate that the cart moved the same distance each time the LED light was on (that is, was traveling at a steady speed), while the longer traces of Figure 2.2c show that the cart covered a longer distance during each light blink. In Figure 2.2d, the traces get shorter as the cart was slowing down and turn into a point when the cart stopped. In Figure 2.2e, the traces get longer as the cart was speeding up.

These LED photos help us construct a new representation of motion that has more information than the dot diagrams: this representation is called a **motion diagram**. We draw dots to represent the positions of the cart and, instead of the traces of light, we draw arrows to indicate the cart's direction of motion. The length of each arrow shows how fast the cart was moving at the location of that dot. These arrows are called **velocity arrows**.

Figure 2.3 shows different types of motion represented with dots and velocity arrows. Notice the small arrow above each letter v (for velocity) in the figures. The arrow indicates that this characteristic of motion has a direction as well as a magnitude. Remember from Chapter 1 that such quantities are called *vector quantities*.

FIGURE 2.3 Motion diagrams represent the types of motion shown in Table 2.1.

(a) The dots represent the positions of the ball at regular times.

The velocity arrows represent how fast the ball is moving and its direction.

$0\,s\;\vec{v}\quad 1\,s\;\vec{v}\quad 2\,s\;\vec{v}\quad 3\,s\;\vec{v}$

(b) This ball is moving at a constant but faster speed than the ball in (a).

$0\,s\quad \vec{v}\qquad 1\,s\quad \vec{v}\qquad 2\,s\quad \vec{v}$

(c) The ball is slowing down.

$\vec{v}\qquad\quad \vec{v}\quad \vec{v}$

(d) The ball is speeding up.

$\vec{v}\;\vec{v}\quad \vec{v}\qquad \vec{v}$

Velocity change arrows

In Experiment 4 in Table 2.1 the bowling ball was moving increasingly faster while being pushed. The velocity arrows in the motion diagram in Figure 2.3d thus are increasingly longer. We can represent this change by drawing a **velocity change ($\Delta\vec{v}$) arrow**. The Δ (delta) means a change in whatever quantity follows the Δ, a change in $\vec{v}$ in this case. The $\Delta\vec{v}$ doesn't tell us the exact increase or decrease in the velocity; it only indicates a qualitative difference between the velocities at two adjacent points in the diagram.

In **Figure 2.4a** we have redrawn the diagram shown in Figure 2.3d. For illustration purposes only, we number the $\vec{v}$ arrows for each position: $\vec{v}_1$, $\vec{v}_2$, $\vec{v}_3$, etc. To draw the velocity change arrow as the ball moves from position 2 to position 3 in Figure 2.4a, we place the arrow $\vec{v}_3$ directly above the arrow $\vec{v}_2$, as shown. The $\vec{v}_3$ arrow is longer than the $\vec{v}_2$ arrow. This tells us that the object was moving faster at position 3 than at position 2. To visualize the change in velocity, we need to think about how arrow $\vec{v}_2$ can be turned into $\vec{v}_3$. We can do it by placing the tail of a velocity change arrow $\Delta\vec{v}_{23}$ at the head of $\vec{v}_2$ so that the head of $\Delta\vec{v}_{23}$ makes $\vec{v}_2 + \Delta\vec{v}_{23}$ the same length as $\vec{v}_3$ (Figure 2.4b). Since they are the same length and point in the same direction, the two vectors $\vec{v}_2 + \Delta\vec{v}_{23}$ and $\vec{v}_3$ are equal:

$$\vec{v}_2 + \Delta\vec{v}_{23} = \vec{v}_3$$

FIGURE 2.4 Determining the magnitude and direction of the velocity change arrow in a motion diagram.

(a) The ball is speeding up.

$\vec{v}_1\quad \vec{v}_2\qquad\quad \vec{v}_3\qquad\qquad \vec{v}_4$

$\vec{v}_3$

How can we represent the change in velocity from 2 to 3?

$\vec{v}_2$

(b)

$\vec{v}_3$

$\vec{v}_2 + \Delta\vec{v}_{23} = \vec{v}_3$

$\vec{v}_2\quad \Delta\vec{v}_{23}\qquad \Delta\vec{v}_{23} = \vec{v}_3 - \vec{v}_2$

Add the $\Delta\vec{v}_{23}$ arrow to $\vec{v}_2$ to get $\vec{v}_3$.

Note that if we move $\vec{v}_2$ to the other side of the equation, then

$$\Delta \vec{v}_{23} = \vec{v}_3 - \vec{v}_2$$

Thus, $\Delta \vec{v}_{23}$ is the difference of the third velocity arrow and the second velocity arrow—the change in velocity between position 2 and position 3.

Making a complete motion diagram

We now place the $\Delta \vec{v}$ arrows above and between the dots in our diagrams where the velocity change occurred (see **Figure 2.5a**). The dots in these more detailed motion diagrams indicate the object's position at equal time intervals; velocity arrows and velocity change arrows are also included. **A $\Delta \vec{v}$ arrow points in the same direction as the $\vec{v}$ arrows when the object is speeding up; the $\Delta \vec{v}$ arrow points in the opposite direction of the $\vec{v}$ arrows when the object is slowing down.** When velocity changes by the same amount during each consecutive time interval, the $\Delta \vec{v}$ arrows for each interval are the same length. In such cases we need only one $\Delta \vec{v}$ arrow for the entire motion diagram (see Figure 2.5b).

Physics Tool Box 2.1 summarizes the procedure for constructing a motion diagram. Notice that in the experiment represented in this diagram, the object is moving from right to left and slowing down.

FIGURE 2.5 Two complete motion diagrams, including position dots, $\vec{v}$ arrows, and $\Delta \vec{v}$ arrows.

(a) A motion diagram showing the velocity change for each consecutive position change

(b) When the velocity change is constant from time interval to time interval, we need only one $\Delta \vec{v}$ for the diagram.

PHYSICS TOOL BOX 2.1 ⟩ **Constructing a motion diagram**

1. Draw dots to represent the position of the object with respect to the observer at equal time intervals.

2. Point velocity arrows in the direction of motion and draw their relative lengths to indicate approximately how fast the object is moving.

3. Draw a velocity change arrow to indicate how the velocity arrows are changing between adjacent positions.

Conceptual Exercise 2.1 will help you practice representing motion with motion diagrams. When working on Conceptual Exercises, first visualize the situation, then draw a sketch, and finally construct a physics representation (in this case, a motion diagram).

TIP ▸ When drawing a motion diagram, always specify the position of the observer.

CONCEPTUAL EXERCISE 2.1 ⟩ **Driving in the city**

A car at rest at a traffic light starts moving faster and faster when the light turns green. The car reaches the speed limit in 4 s, continues at the speed limit for 3 s, then, approaching a second, red, traffic light, slows down and stops in 2 s. At the light, the car waits for 1 s until it turns green. Meanwhile, a cyclist approaching the first green light keeps moving without slowing down or speeding up. She reaches the second stoplight just as it turns green. Draw a motion diagram for the car and another for the bicycle as seen by an observer on the ground. If you place one diagram below the other, it will be easier to compare them.

Sketch and translate Visualize the motion for the car and for the bicycle as seen by the observer on the ground. The car and the bicycle are our objects of interest.

The motion of the car has four distinct parts:

1. starting at rest and moving faster and faster for 4 s;
2. moving at a constant rate for 3 s;
3. slowing down to a stop for 2 s; and
4. sitting at rest for 1 s.

(CONTINUED)

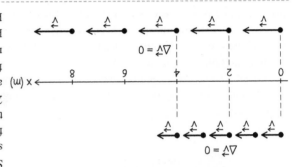

The bicycle moves at a constant rate with respect to the ground for the entire time.

Simplify and diagram We model the car and the bicycle as point-like objects (dots). In each motion diagram, there are 11 dots, one for each second of time (including one for time 0). The last two dots for the car are on top of each other since the car is at rest from time = 9 s to time = 10 s. The dots for the bicycle are evenly spaced.

Try it yourself Two bowling balls are rolling along a linoleum floor. One of them is moving twice as fast as the other. At time 0, they are next to each other on the floor. Construct motion diagrams for each ball during a time interval of 4 s, as seen by an observer on the ground. Indicate on the diagrams the locations at which the balls were next to each other at the same time. What are the possible mistakes that a student could make when doing this exercise?

Answer See the figure at left. The balls are side by side only at time 0—the first dot for each ball. It looks like they are side by side when at the 2-m position, but the slower ball is at the 2-m position at 2 s and the faster ball is there at 1 s. Similar reasoning applies for the 4-m positions—the balls reach that point at different times.

2.3 Operations with vectors

Physics Tool Box 2.1 tells you to determine the direction of a $\Delta\vec{v}$ arrow on a motion diagram. To do this, you need to know either how to add vectors (the operation $\vec{v}_2 + \Delta\vec{v}_{23} = \vec{v}_3$ in Figure 2.4) or how to subtract them (the operation $\Delta\vec{v}_{23} = \vec{v}_3 - \vec{v}_2$ in Figure 2.4). Here we discuss these operations in more detail. To start, we first consider more simple vector operations. Carefully examine **Figure 2.6**, which shows eight arrows of different lengths and directions representing vectors. The length of each arrow is the vector's magnitude, in any units you choose, and the orientation of each vector is determined by the direction of the arrow. Each vector has a *tail* (the point where it originates) and a *head* (the tip of the arrow). Which arrows represent the same vector? You see that arrows 1, 3, and 4 have the same length and the same direction: they therefore represent the same vector, $\vec{A}$. Arrows 2, 5, and 7 represent a different vector, vector $\vec{B}$. From this exercise we see that moving the vector parallel to itself will not change the vector. What about arrows 6 and 8? They have the same length as $\vec{A}$ and $\vec{B}$, respectively, but they point in the opposite directions. In this case, we say that vector 6 is $-\vec{A}$, the *negative* of vector $\vec{A}$, and vector 8 is $-\vec{B}$. The minus sign is used to mean that **$-\vec{A}$ has the same length as $\vec{A}$ but points in the opposite direction.**

Now that we know how to move vectors around and draw the negative of a vector, we can explore how to add and subtract vectors. What rules of vector addition and subtraction can you infer by analyzing the examples shown in Figure 2.6b (addition, $\vec{A} + \vec{B}$) and Figure 2.6c (subtraction, $\vec{A} - \vec{B}$)? Spend a few moments thinking about this before reading the following explanation of the rules.

FIGURE 2.6 Using arrows to represent vectors.

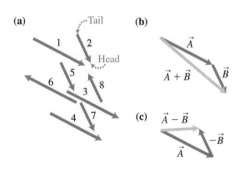

Adding vectors

Suppose we want to graphically add the two vectors $\vec{K}$ and $\vec{L}$ shown in **Figure 2.7a**. We can move a vector from one location to another as long as we do not change its magnitude or direction; therefore, we can place the tail of vector $\vec{L}$ at the head of $\vec{K}$, as shown in Figure 2.7b, without changing $\vec{L}$. Next, we draw another vector $\vec{R}$ from the tail of $\vec{K}$ to the head of $\vec{L}$, as shown in Figure 2.7c, to represent the result of the addition of the vectors $\vec{K}$ and $\vec{L}$. We can write the *resultant vector* (vector sum) as a mathematical equation: $\vec{R} = \vec{K} + \vec{L}$. As you can see in Figure 2.7c, the magnitude of the vector sum is not in general equal to the sum of the magnitudes of $\vec{K}$ and $\vec{L}$. However, there are two special cases, as Figure 2.7d shows:

- When two vectors are parallel and point in the same direction, the addition is much easier: the magnitude of the resultant vector equals the sum of the magnitudes of the vectors being added.

- When two vectors are parallel but point in opposite directions (called antiparallel), the magnitude of the resultant vector equals the absolute value of the difference in the magnitudes of the vectors being added.

If you need to add more vectors, follow the same procedure of "tail to head" for all of them, and then draw the resultant vector. For accurate results in graphical addition, draw carefully using a ruler and a protractor.

FIGURE 2.7 Graphical vector addition.

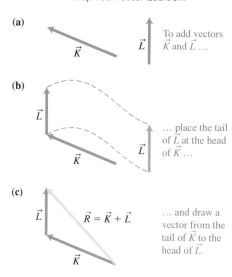

(d) Adding parallel and antiparallel vectors

FIGURE 2.8 Graphical vector subtraction.

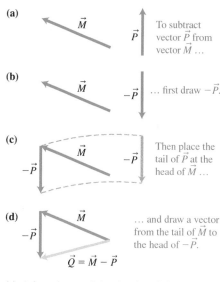

(e) Subtracting parallel and antiparallel vectors

Subtracting vectors

Suppose we now want to subtract vector $\vec{P}$ from vector $\vec{M}$, shown in **Figure 2.8a**. You can interpret the operation of subtracting vector $\vec{P}$ from vector $\vec{M}$ as the *addition* of vector $-\vec{P}$ and vector $\vec{M}$. The vector equation is $\vec{Q} = \vec{M} - \vec{P} = \vec{M} + (-\vec{P})$. As Figure 2.8 shows, we therefore need to draw the vector $-\vec{P}$ first, and then proceed with graphical vector addition of $-\vec{P}$ and $\vec{M}$. Figure 2.8e shows the subtraction of parallel and antiparallel vectors. This part is relevant to determining the direction and magnitude of the $\Delta\vec{v}$ vector on a motion diagram.

CONCEPTUAL EXERCISE 2.2 **Multiple ways to build vector $\vec{X}$**

Vector $\vec{X}$ is 3 cm long and points west. (a) Draw three different combinations of two vectors whose addition gives vector $\vec{X}$. One of the combinations should have antiparallel vectors. (b) Draw three different combinations of two vectors whose subtraction gives vector $\vec{X}$. One of the combinations should have parallel vectors.

Sketch and translate (a) We draw vector $\vec{X}$ and think of possible combinations of two vectors that we can add to get $\vec{X}$. Using the method for graphical addition, we know that the sum $\vec{X}$ must span from the tail of one vector to the head of the other when the two vectors are placed head to tail. The figure below shows three possible combinations. $\vec{H}$ and $\vec{K}$ are antiparallel vectors whose sum is $\vec{X}$.

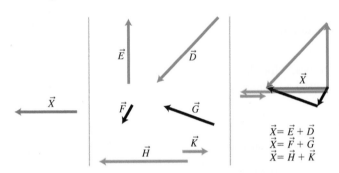

$$\vec{X} = \vec{E} + \vec{D}$$
$$\vec{X} = \vec{F} + \vec{G}$$
$$\vec{X} = \vec{H} + \vec{K}$$

(b) If the subtraction of one vector from another gives $\vec{X}$, then we know that the first vector plus $\vec{X}$ gives the second vector. We can therefore draw any vector for the first vector and then add $\vec{X}$ to it to get the second vector. The figure below shows three possible combinations. $\vec{L}$ and $\vec{M}$ are parallel vectors whose difference is $\vec{X}$.

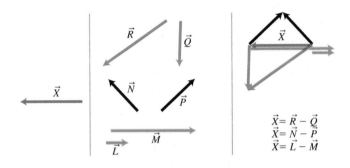

$$\vec{X} = \vec{R} - \vec{Q}$$
$$\vec{X} = \vec{N} - \vec{P}$$
$$\vec{X} = \vec{L} - \vec{M}$$

Try it yourself Examine Figure 2.4. Where in this figure do we use vector addition and where do we use vector subtraction?

Answer The velocity arrow $\vec{v}_3$ is a vector that is the sum of the two vectors $\vec{v}_2$ and $\Delta\vec{v}_{23}$: $\vec{v}_3 = \vec{v}_2 + \Delta\vec{v}_{23}$; the velocity change arrow is a vector $\Delta\vec{v}_{23}$ that is the result of subtraction of vector $\vec{v}_2$ from vector $\vec{v}_3$: $\Delta\vec{v}_{23} = \vec{v}_3 - \vec{v}_2$.

Multiplying a vector by a scalar

In addition to vector addition and subtraction, there is another operation with vectors that you will need when studying physics. It involves multiplying a vector by a positive or a negative number (a scalar quantity)—for example, finding vector $2\vec{C}$ when vector $\vec{C}$ is known. **Figure 2.9a** shows the result of this operation. Vector $2\vec{C}$ is twice as long as $\vec{C}$, or equal to the sum of two $\vec{C}$ vectors, and points in the same direction. Vector $3\vec{C}$ is three times as long, or equal to the sum of three $\vec{C}$ vectors, and points in the same direction. Figure 2.9b shows the result of dividing vector $\vec{D}$ by 2 and by 3. We see that the result is equal to multiplying vector $\vec{D}$ by $1/2$ and $1/3$, respectively (the vector is one-half or one-third of the length of vector $\vec{D}$). Finally, if you need to multiply vector $\vec{C}$ by a negative number, for example -2 or -3, you need to first draw vector $-\vec{C}$, as we learned above, and then multiply it by 2 or 3 (Figure 2.9c). In general, **when you multiply a vector by a scalar, the result is a vector parallel or antiparallel to the original vector whose magnitude equals the product of the magnitude of the original vector and the magnitude of the scalar.** [Note that if the scalar in this operation has units, the resulting vector has units that are the product of the original vector units and the units of the scalar.]

FIGURE 2.9 Multiplying a vector by a scalar.

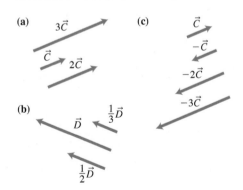

REVIEW QUESTION 2.3 Eugenia says that to find the $\Delta\vec{v}$ arrow on a motion diagram you need to subtract vectors. Alan says that you need to add vectors. Both are right. How can this be? Explain and show examples.

2.4 Quantities for describing motion

A motion diagram helps represent motion qualitatively. To analyze situations more precisely, for example, to determine how far a car will travel after the brakes are applied, we need to describe motion quantitatively. In this section, we devise some of the quantities we need to describe linear motion.

Time and time interval

People use the word "time" to mean the reading on a clock and how long a process takes. Physicists distinguish between these two meanings with different terms:

> **Time and time interval** *Time* (clock reading) t is the reading on a clock or some other time-measuring instrument. *Time interval* $(t_2 - t_1)$ or Δt is the difference between two times. In the SI system (metric units), the unit of both time and time interval is the second. Other units are minutes, hours, days, and years. Time and time interval are both scalar quantities.

Position, displacement, distance, and path length

We also need to precisely define four quantities that describe the location and motion of an object:

> **Position, displacement, distance, and path length** The *position* of an object is its location with respect to a particular coordinate system (usually indicated by x or y). The *displacement* of an object, usually indicated by $\vec{d}$, is a vector that starts from an object's initial position and ends at its final position. The magnitude (length) of the displacement vector is called the *distance d*. The *path length l* is how far the object moved as it traveled from its initial position to its final position. Imagine laying a string along the path the object took. The length of the string is the path length.

Figure 2.10a shows a car's initial position x_i at initial time t_i. The car first backs up (moving in the negative direction) toward the origin of the coordinate system at $x = 0$. The car stops and then moves in the positive x-direction to its final position x_f. Notice that the *initial position* and the *origin* of a coordinate system are not necessarily the same point! The displacement $\vec{d}$ for the whole trip is a vector that points from the starting position at x_i to the final position at x_f (Figure 2.10b). The distance for the trip is the magnitude of the displacement (always a positive value). The path length l is the distance from x_i to 0 plus the distance from 0 to x_f (Figure 2.10c). Note that the path length does not equal the distance.

Scalar component of displacement for motion along one axis

To describe linear motion quantitatively we first specify a reference frame. For simplicity we can point one coordinate axis either parallel or antiparallel to the object's direction of motion. For linear motion, we need only one coordinate axis to describe the object's changing position. In the example of the car trip, at the initial time t_i the car is at position $x(t_i)$, and at the final time t_f the car is at $x(t_f)$ (Figure 2.11). The notation $x(t)$

FIGURE 2.10 Position, displacement, distance, and path length for a short car trip.

(a) Positions x_i and x_f

(b) Displacement $\vec{d}$ and distance d

$$d = |x_f - x_i|$$

(c) Path length l

$$l = |0 - x_i| + |x_f - 0|$$

> **TIP** Sometimes we use the subscripts 1, 2, and 3 for times and the corresponding positions to communicate a sequence of different and distinguishable stages in any process, and sometimes we use the subscripts i (initial) and f (final) to communicate the sequence. Subscripts i and f mean the beginning and end of our observations, not the beginning and end of the process.

FIGURE 2.11 Indicating an object's position at a particular time t, for example, $x(t)$.

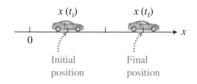

means the position x is a function of clock reading t (spoken "x of t"), not x multiplied by t. When we need to note a specific value of position x at a specific clock reading t_1, instead of writing $x(t_1)$, we will write x_1, keeping in mind that 1 is a shortcut for time t_1. The same applies to x_i, x_f, etc. The vector that points from the initial position x_i to the final position x_f is the **displacement**.

The quantity that we determine through the operation $x_f - x_i$ is called the **x-scalar component of the displacement** vector and is abbreviated d_x (usually we drop the term "scalar" and just call this the **x-component of the displacement**). **Figure 2.12a** shows that the initial position of person A is $x_{Ai} = +3.0$ m and the final position is $x_{Af} = +5.0$ m; thus the x-component of the person's displacement is

$$d_{Ax} = x_{Af} - x_{Ai} = (+5.0 \text{ m}) - (+3.0 \text{ m}) = +2.0 \text{ m}$$

The displacement is positive since the person moved in the positive x-direction. In Figure 2.12b, person B moved in the negative direction from the initial position of $+5.0$ m to the final position of $+3.0$ m; thus the x-component of displacement of the person is negative:

$$d_{Bx} = x_{Bf} - x_{Bi} = (+3.0 \text{ m}) - (+5.0 \text{ m}) = -2.0 \text{ m}$$

Distance is always positive, as it equals the absolute value of the displacement $|x_f - x_i|$. In the example above, the displacements for A and B are different, but the distances are both $+2.0$ m.

Significant digits

Note that in Figure 2.12 the positions were written as $+3.0$ m and $+5.0$ m. Could we have written them instead as $+3$ m and $+5$ m, or as $+3.00$ m and $+5.00$ m? The thickness of a human body from the back to the front is about 0.2 m (20 cm). Thus, we should be able to measure the person's location at one instant of time to within about 0.1 m but not to 0.01 m (1 cm). Thus, the locations can reasonably be given as $+3.0$ m, which implies an accuracy of ± 0.1 m. (For more on significant digits, see Chapter 1.)

REVIEW QUESTION 2.4 Jade went hiking between two camps that were separated by about 10 km. She hiked approximately 16 km to get from one camp to the other. Translate 10 km and 16 km into the language of physical quantities.

2.5 Representing motion with data tables and graphs

So far, we have learned how to represent linear motion with motion diagrams. In this section, we will learn to represent linear motion with data tables and graphs.

Imagine your friend (the object of interest) walking across the front of your classroom. To record her position, you drop a beanbag on the floor at her position each second (**Figure 2.13a**). The floor is the object of reference. The origin of the coordinate system is 1.00 m from the first beanbag, and the position axis points in the direction of motion. **Table 2.2** shows each bag's position and the corresponding clock reading. Do you see a pattern in the table's data? One way to determine if there is a pattern is to plot the data on a graph (Figure 2.13b). This graph is called a **kinematics position-versus-time graph**. In physics, the word **kinematics** means description of motion. Kinematics graphs contain more precise information about an object's motion than motion diagrams can.

FIGURE 2.12 The x-component of displacement is (a) positive; (b) negative.

(a)   Positive displacement when the person moves in the positive direction

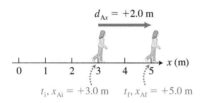

$t_i, x_{Ai} = +3.0$ m $t_f, x_{Af} = +5.0$ m

(b)   Negative displacement when the person moves in the negative direction

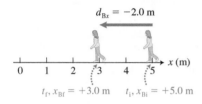

$t_f, x_{Bf} = +3.0$ m $t_i, x_{Bi} = +5.0$ m

FIGURE 2.13 Constructing a kinematics position-versus-time graph.

(a)

(b)

(c)

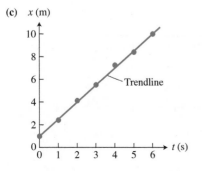

Time t is usually considered to be the independent variable, as time progresses even if there is no motion, so the horizontal axis will be the t-axis. Position x is the dependent variable (position changes with time), so the vertical axis will be the x-axis.

Plot the data in each row in Table 2.2 on the axes. Each point on the horizontal axis represents a time (clock reading). Each point on the vertical axis represents the position of a beanbag. When we draw lines through these points and perpendicular to the axes, they intersect at a single location—a dot on the graph that simultaneously represents a time and the corresponding position of the object. This dot *is not* a location in real space but rather a representation of the position of the beanbag at a specific time.

Is there a trend in the locations of the dots on the graph? We see that the position increases as the time increases. This makes sense. We can draw a smooth best-fit curve that passes as close as possible to the data points—a **trendline** (Figure 2.13c). It looks like a straight line in this particular case—the position is linearly dependent on time.

Correspondence between a motion diagram and position-versus-time graph

To understand how graphs relate to motion diagrams, consider the motion represented by the data in Table 2.2 and in Figure 2.13c. **Figure 2.14** shows a modified motion diagram for the data in Table 2.2 (the dot times are shown and the $\Delta \vec{v}$ arrows have been removed for simplicity) and the corresponding position-versus-time graph. The position of each dot on the motion diagram corresponds to a point on the position axis. The graph line combines the information about the position of an object and the clock reading when this position occurred. Note, for example, that the $t = 4.0$ s dot on the motion diagram at position $x = 7.26$ m is at 7.26 m on the position axis. The corresponding dot on the graph is at the intersection of the vertical line passing through 4.0 s and the horizontal line passing though 7.26 m.

The role of a reference frame

Always keep in mind that representations of motion (motion diagrams, tables, kinematics graphs, equations, etc.) depend on the reference frame chosen. Let's look at the representations of the motion of a cyclist using two different reference frames.

TABLE 2.2 Time-position data for linear motion

Clock reading (time)	Position
$t_0 = 0.0$ s	$x_0 = 1.00$ m
$t_1 = 1.0$ s	$x_1 = 2.42$ m
$t_2 = 2.0$ s	$x_2 = 4.13$ m
$t_3 = 3.0$ s	$x_3 = 5.52$ m
$t_4 = 4.0$ s	$x_4 = 7.26$ m
$t_5 = 5.0$ s	$x_5 = 8.41$ m
$t_6 = 6.0$ s	$x_6 = 10.00$ m

TIP ▸ The quantity that appears on the vertical axis of a graph can represent the position of an object whose actual position is changing along any axis. The position on the vertical axis does not mean the object is moving in the vertical direction.

FIGURE 2.14 Correspondence between a motion diagram and the position-versus-time graph.

CONCEPTUAL EXERCISE 2.3 ▸ **Effect of reference frame on motion description**

Two observers each use different reference frames to record the changing position of a bicycle rider. Both reference frames use Earth as the object of reference, but the origins of the coordinate systems and the directions of the x-axes are different. The data for the cyclist's trip are presented in Table 2.3 for observer 1 and in Table 2.4 for observer 2. Sketch a motion diagram and a position-versus-time graph for the motion when using each reference frame.

TABLE 2.3 Time-position data for cyclist when using reference frame 1

Clock reading (time)	Position
$t_0 = 0.0$ s	$x_0 = 40.0$ m
$t_1 = 1.0$ s	$x_1 = 30.0$ m
$t_2 = 2.0$ s	$x_2 = 20.0$ m
$t_3 = 3.0$ s	$x_3 = 10.0$ m
$t_4 = 4.0$ s	$x_4 = 0.0$ m

TABLE 2.4 Time-position data for cyclist when using reference frame 2

Clock reading (time)	Position
$t_0 = 0.0$ s	$x_0 = 0.0$ m
$t_1 = 1.0$ s	$x_1 = 10.0$ m
$t_2 = 2.0$ s	$x_2 = 20.0$ m
$t_3 = 3.0$ s	$x_3 = 30.0$ m
$t_4 = 4.0$ s	$x_4 = 40.0$ m

(CONTINUED)

Sketch and translate According to Table 2.3, the observer in reference frame 1 sees the cyclist (the object of interest) at time $t_0 = 0.0$ s at position $x_0 = 40.0$ m and at $t_4 = 4.0$ s at position $x_4 = 0.0$ m. Thus, the cyclist is moving in the negative direction relative to the coordinate axis in reference frame 1. Meanwhile, according to Table 2.4, the observer in reference frame 2 sees the cyclist at time $t_0 = 0.0$ s at position $x_0 = 0.0$ m and at time $t_4 = 4.0$ s at position $x_4 = 40.0$ m. Thus, the cyclist is moving in the positive direction relative to reference frame 2.

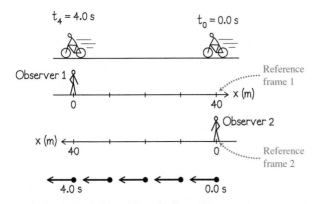

Simplify and diagram Since the size of the cyclist is small compared to the distance he is traveling, we can represent him as a point-like object. The motion diagram for the cyclist is the same for both observers, as they are using the same object of reference. Using the data in the tables, we plot kinematics position-versus-time graphs for each observer at top right. Although the graph for observer 1 looks very different from the graph for observer 2, they represent the same motion. The graphs look different because the reference frames are different.

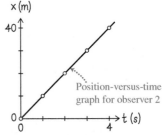

Try it yourself A third observer recorded the values (in **Table 2.5**) for the time and position of the same cyclist. Describe the reference frame of this observer.

TABLE 2.5 Data collected by the third observer

Clock reading (time)	Position
$t_0 = 0.0$ s	$x_0 = 0.0$ m
$t_1 = 1.0$ s	$x_1 = -20.0$ m
$t_2 = 2.0$ s	$x_2 = -40.0$ m
$t_3 = 3.0$ s	$x_3 = -60.0$ m
$t_4 = 4.0$ s	$x_4 = -80.0$ m

Answer The point of reference could be another cyclist moving in the opposite positive direction from the direction in which the first cyclist is traveling, with each covering the same distance relative to the ground during the same time interval.

REVIEW QUESTION 2.5 A position-versus-time graph representing a moving object is shown in **Figure 2.15**. What are the positions of the object at clock readings 2.6 s and 5.8 s?

FIGURE 2.15 A position-versus-time graph representing a moving object.

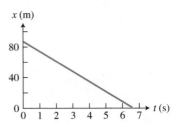

FIGURE 2.16 Positions of cars A and B at 0 s and 5.0 s.

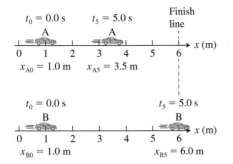

2.6 Constant velocity linear motion

In the last section we devised a graphical representation of motion. Here we will connect graphs to mathematical representations using the example of two motorized toy cars racing toward a finish line. At time 0 they are next to each other, but car B is moving faster than car A and reaches the finish line first (**Figure 2.16**). The data that we collect are shown in Observational Experiment **Table 2.6**. Earth is the object of reference. The origin of the coordinate system (the point of reference) is 1.0 m to the left of the position of the cars at $t_0 = 0$. The positive x-direction points right in the direction of the cars' motions. Now let's use the data to find a pattern.

OBSERVATIONAL
EXPERIMENT TABLE 2.6   Graphing the motion of cars

| Observational experiment | | Analysis |

Observational experiment

Data for car A

$t_0 = 0.0$ s	$x_0 = 1.0$ m
$t_1 = 1.0$ s	$x_1 = 1.4$ m
$t_2 = 2.0$ s	$x_2 = 1.9$ m
$t_3 = 3.0$ s	$x_3 = 2.5$ m
$t_4 = 4.0$ s	$x_4 = 2.9$ m
$t_5 = 5.0$ s	$x_5 = 3.5$ m

Data for car B

$t_0 = 0.0$ s	$x_0 = 1.0$ m
$t_1 = 1.0$ s	$x_1 = 1.9$ m
$t_2 = 2.0$ s	$x_2 = 3.0$ m
$t_3 = 3.0$ s	$x_3 = 3.9$ m
$t_4 = 4.0$ s	$x_4 = 5.0$ m
$t_5 = 5.0$ s	$x_5 = 6.0$ m

Analysis

We graph the data with the goal of finding a pattern. The trendlines for both cars are straight lines. The line for car B has a bigger angle with the time axis than the line for car A.

Pattern

It looks like a straight line is the simplest reasonable choice for the best-fit curve in both cases (the data points do not have to be exactly on the line).

In Table 2.6, the slope of the line representing the motion of car B is greater than the slope of the line representing the motion of car A. What is the physical meaning of this slope? In mathematics the value of a dependent variable is usually written as y and depends on the value of an independent variable, usually written as x. A function $y(x) = f(x)$ is an operation that one needs to do to x as an input to have y as the output. For a straight line, the function $y(x)$ is $y(x) = mx + b$, where m is the slope and b is the y-intercept—the value of the y when $x = 0$.

In the case of the cars, the independent variable is time t and the dependent variable is position x. The equation of a straight line becomes $x(t) = mt + b$, where b is the x-intercept of the line, and m is the slope of the line. The x-intercept is the x-position when $t = 0$, also called the initial position of the car x_0. Both cars started at the same location: $x_{A0} = x_{B0} = 1.0$ m.

To find the **slope** m of a straight line, we can choose *any* two points on the line and divide the change in the vertical quantity (Δx in this case) by the change in the horizontal quantity (Δt in this case):

$$m = \frac{x_2 - x_1}{t_2 - t_1} = \frac{\Delta x}{\Delta t}$$

For example, for car A the slope of the line is

$$m_A = \frac{3.5 \text{ m} - 1.0 \text{ m}}{5.0 \text{ s} - 0.0 \text{ s}} = +0.5 \text{ m/s}$$

The slope of the line for car B is

$$m_B = \frac{6.0 \text{ m} - 1.0 \text{ m}}{5.0 \text{ s} - 0.0 \text{ s}} = +1.0 \text{ m/s}$$

Now we have all the information we need to write mathematical equations that describe the motion of each of the two cars:

Car A: $x_A = (+0.5 \text{ m/s})t + (1.0 \text{ m})$

Car B: $x_B = (+1.0 \text{ m/s})t + (1.0 \text{ m})$

Notice that the units of the slope are meters per second. The slope indicates how the object's position changes with respect to time. The slope of the line contains more information than just how fast the car is going. It also tells us the direction of motion relative to the coordinate axis.

FIGURE 2.17 The sign of the slope indicates the direction of motion.

(a)

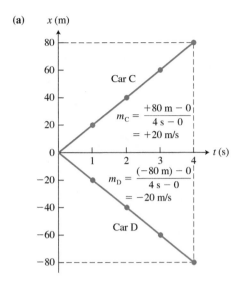

(b)

The position-versus-time graph for car D has negative slope.	The position-versus-time graph for car C has positive slope.

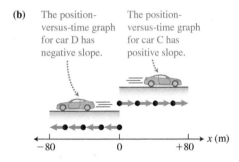

TIP Equation (2.1) allows you to use *any* change in position divided by the time interval during which that change occurred to obtain the same number—as long as the position-versus-time graph is a straight line (the object is moving at constant velocity). Later in the chapter, you will learn how to modify this equation for cases in which the velocity is not constant.

Consider the motions represented graphically in **Figure 2.17a**. The slope of the position-versus-time graph for car C is +20 m/s, but the slope of the position-versus-time graph for car D is −20 m/s. What is the significance of the minus sign? Car C is moving in the positive direction, but car D is moving in the negative direction (check the motion diagrams in Figure 2.17b). The magnitudes of the slopes of their position-versus-time graphs are the same, but the signs are different. Thus, in addition to the information about how fast the car is traveling (its **speed**), the slope tells in what direction it is traveling. Together, speed and direction are called **velocity**, and this is what the slope of a position-versus-time graph represents. You are already familiar with the term "velocity arrow" used on motion diagrams. Now you have a formal definition for velocity as a physical quantity.

> **Velocity and speed for constant velocity linear motion** For constant velocity linear motion, the component of velocity v_x along the axis of motion can be found as the slope of the position-versus-time graph or the ratio of the component of the displacement of an object $x_2 - x_1$ during *any* time interval $t_2 - t_1$:
>
> $$v_x = \frac{x_2 - x_1}{t_2 - t_1} = \frac{\Delta x}{\Delta t} \tag{2.1}$$
>
> where the subscript x next to the v symbol indicates the direction of the axis and t_1 and t_2 are the two clock readings when the positions x_1 and x_2 were measured. Examples of units of velocity are m/s, km/h, and mi/h (which is often written as mph). *Speed* is the magnitude of the velocity and is always a positive number.

Note that velocity is a vector quantity. In vector form, motion at constant velocity is $\vec{v} = \vec{d}/\Delta t$, where $\vec{d}$ is the displacement vector and Δt is the time interval. Here we divide a vector by a scalar. You learned this operation in Section 2.3. The velocity vector has the same direction as the displacement vector (direction of motion), but the magnitude is different as well the units. The magnitude of the velocity vector is speed. The displacement vector has units of length (m), and the time interval has units of time (s). Thus the velocity vector has the units of m/s. Since it is difficult to operate mathematically with vectors, we will work with components. When the velocity vector points in the positive direction of the x-axis, the x-component of velocity is equal to the magnitude of the velocity vector. When the velocity vector points in the negative direction of the axis, the x-component of velocity is equal to the magnitude of the velocity vector with the negative sign.

Equation of motion for constant velocity linear motion

We can rearrange Eq. (2.1) into a form that allows us to determine the position of an object at time t_2 knowing only its position at time t_1 and the x-component of its velocity: $x_2 = x_1 + v_x(t_2 - t_1)$. If we apply this equation for time 0 ($t_0 = 0$) when the initial position is x_0, then the position x at any later position time t can be written as follows:

> **Position equation for constant velocity linear motion**
>
> $$x = x_0 + v_x t \tag{2.2}$$
>
> where x is the function $x(t)$, position x_0 is the position of the object at time $t_0 = 0$ with respect to a particular reference frame, and the (constant) x-component of the velocity of the object v_x is the slope of the position-versus-time graph.

Below you see a new type of task—a Quantitative Exercise. Quantitative Exercises include two steps of the problem-solving process: "Represent mathematically" and "Solve and evaluate." Their purpose is to help you practice using new equations right away.

QUANTITATIVE EXERCISE 2.4 **A cyclist**

In Conceptual Exercise 2.3, you constructed graphs for the motion of a cyclist using two different reference frames. Now construct mathematical representations (equations) for the cyclist's motion for each of the two graphs. Do the equations indicate the same position for the cyclist at time $t = 6.0$ s?

Represent mathematically The cyclist moves at constant velocity; thus the general mathematical description of his motion is $x = x_0 + v_x t$, where

$$v_x = \frac{x_2 - x_1}{t_2 - t_1}$$

Solve and evaluate Using the graph for reference frame 1 in Conceptual Exercise 2.3, we see that the cyclist's initial position is $x_0 = +40$ m. The velocity along the x-axis (the slope of the graph) is

$$v_x = \frac{0 \text{ m} - 40 \text{ m}}{4 \text{ s} - 0 \text{ s}} = -10 \text{ m/s}$$

The minus sign indicates that the velocity points in the negative x-direction (toward the left) relative to that axis. The motion of the bike with respect to reference frame 1 is described by the equation

$$x = x_0 + v_x t = 40 \text{ m} + (-10 \text{ m/s})t$$

Using the graph for reference frame 2, we see that the cyclist's initial position is $x_0 = 0$ m. The x-component of the velocity along the axis of motion is

$$v_x = \frac{40 \text{ m} - 0 \text{ m}}{4 \text{ s} - 0 \text{ s}} = +10 \text{ m/s}$$

The positive sign indicates that the velocity points in the positive x-direction (toward the left). The motion of the bike relative to reference frame 2 is described by the equation

$$x = x_0 + v_x t = 0 \text{ m} + (10 \text{ m/s})t$$

The position of the cyclist at time $t_1 = 6$ s with respect to reference frame 1 is

$$x = 40 \text{ m} + (-10 \text{ m/s})(6 \text{ s}) = -20 \text{ m}$$

With respect to reference frame 2:

$$x = 0 \text{ m} + (+10 \text{ m/s})(6 \text{ s}) = +60 \text{ m}$$

How can the position of the cyclist be both -20 m and $+60$ m? Remember that the description of motion of an object depends on the reference frame. If you put a dot on coordinate axis 1 at the -20 m position and a dot on coordinate axis 2 at the $+60$ m position, you find that the dots are in fact at the same location, even though that location corresponds to a different position in each reference frame. Both descriptions of the motion are correct and consistent, but each one is meaningful only with respect to the corresponding reference frame.

Reference frame 1

Reference frame 2

Try it yourself Use the data for the motion of the cyclist as seen by the third observer in the "Try it yourself" part of Conceptual Exercise 2.3 to write the equation of motion. Why is the magnitude of the cyclist's velocity different than the 10 m/s in the example above?

Answer $x = (0 \text{ m}) + (-20 \text{ m/s})t$. The observer is moving with respect to Earth at the same speed in the direction opposite to the cyclist.

Below you see another new type of task—a worked example. The worked examples include all four steps of the problem-solving strategy we use in this book. (See Chapter 1 for descriptions of these steps.)

EXAMPLE 2.5 **You chase your sister**

Your young sister is running at 2.0 m/s toward a mud puddle that is 6.0 m in front of her. You are 10.0 m behind her running at 5.0 m/s to catch her before she jumps into the mud. Will she need a bath?

Sketch and translate We start by drawing a sketch of what is happening. Your sister and you are two objects of interest. Next, we choose a reference frame with Earth as the object of reference. The origin of the coordinate system is your initial position and the positive direction is toward the right, in the direction that you both run, as shown at right. Notice the subscripts for the given quantities. The first subscript

indicates the object, and the second indicates the time. For example, x_{Y0} means your position at the zero clock reading.

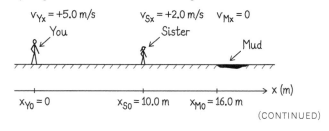

(CONTINUED)

We can now mathematically describe the positions and velocities of you and your sister at the beginning of the process. The initial clock reading is zero at the moment that you are at the origin. Note that you were both running before time zero; this just happened to be the time when we started analyzing the process. We want to know the time when you and your sister are at the same position. This will be the position where you catch up to her.

Simplify and diagram We assume that you and your sister are point-like objects. To sketch graphs of the motions, find your sister's position at 1 s by multiplying her speed by 1 s and adding it to her initial position. Do this for 2 and 3 s as well. Plot these values on a graph for the corresponding clock readings (1 s, 2 s, 3 s, etc.) and draw a line that extends through these points. Repeat this process for yourself.

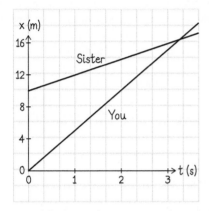

Represent mathematically Use Eq. (2.2) to construct mathematical representations of motion. The form of the equation is the same for both $(x = x_0 + v_x t)$; however, the values for the initial positions and the components of the velocities along the axis for your sister, $x_S = x_{S0} + v_{Sx}t$, and for you, $x_Y = x_{Y0} + v_{Yx}t$, are different. Now we can use the values of the quantities from the sketch.

Sister: $x_S = (10.0 \text{ m}) + (2.0 \text{ m/s})t$
You: $x_Y = (0.0 \text{ m}) + (5.0 \text{ m/s})t$

From the graphs, we see that the distance between you and your sister is shrinking with time. Do the equations tell the same story? For example, at time $t = 2.0$ s, your sister is at position

$$x_S(2 \text{ s}) = (10.0 \text{ m}) + (2.0 \text{ m/s})(2.0 \text{ s}) = 14.0 \text{ m}$$

and you are at

$$x_Y(2 \text{ s}) = (0.0 \text{ m})+(5.0 \text{ m/s})(2.0 \text{ s}) = 10.0 \text{ m}$$

You are catching up to your sister.

Solve and evaluate The time t at which the two of you are at the same position can be found by setting $x_S(t) = x_Y(t)$:

$$(10.0 \text{ m}) + (2.0 \text{ m/s})t = (0.0 \text{ m}) + (5.0 \text{ m/s})t$$

Rearrange the above to determine the time t when you are both at the same position:

$$(2.0 \text{ m/s})t - (5.0 \text{ m/s})t = (0.0 \text{ m}) - (10.0 \text{ m})$$
$$(-3.0 \text{ m/s})t = -(10.0 \text{ m})$$
$$t = 3.33333333 \text{ s}$$

The 3.33333333 s number produced by our calculator has many more significant digits than the givens. Should we round it to have the same number of significant digits as the given quantities? The rule of thumb is that if it is the final result, you need to round this number to 3.3, as the answer cannot be more precise than the given information. However, we do not round the result of an intermediate calculation. We use the result as is to calculate the next quantity needed to get the final answer and then round the final result.

Sister: $x_S(t) = (10.0 \text{ m}) + (2.0 \text{ m/s})(3.33333333 \text{ s})$
$= 16.7 \text{ m}$

You: $x_Y(t) = (0.0 \text{ m}) + (5.0 \text{ m/s})(3.33333333 \text{ s})$
$= 16.7 \text{ m}$

Note that if you used the rounded number 3.3 s, you would get 16.6 m for your sister and 16.5 m for you. These numbers are slightly smaller than the result calculated above. However, for our purposes it does not matter, as the goal of this example was to decide if you could catch your sister before she reaches the puddle. Since you caught her at a position of about 16.7 m, with the uncertainty of about 0.1 m, this position is slightly greater than the 16.0-m distance to the puddle. Therefore, your sister reaches the puddle before you and will indeed need a bath. This answer seems consistent with the graphical representation of the motion shown above. Note that we used the symbol t to denote the specific clock reading when you catch your sister. Often, the same symbol is used to denote both a variable and some specific value. Always ask yourself what a symbol means when you read it.

Try it yourself Describe the problem situation using a reference frame with your sister (not Earth) as the object and point of reference and the positive direction pointing toward the puddle.

Answer

With respect to this reference frame, your sister is at position 0 and at rest; you are initially at −10.0 m and moving toward your sister with velocity +3.0 m/s; and the mud puddle is initially at +6.0 m and moving toward your sister with velocity −2.0 m/s.

TIP   In the reference frame we chose in Example 2.5, the positions of you and your sister are always positive, as are the components of your velocities. Also, your initial position is zero. Thus the calculations are the easiest. Often the description of the motion of object(s) will be simplest in one particular reference frame. There is not one general rule for finding such a frame; however, it is helpful to consider which frame will result in positive values for the most quantities.

Graphing velocity

So far, we have learned to make position-versus-time graphs. We could also construct a graph of an object's velocity as a function of time. Consider Example 2.5, in which you chase your sister. Again we will use Earth as the object of reference. You are moving at a constant velocity whose x-component is $v_x = +5.0$ m/s. For your sister, the x-component of velocity is $+2.0$ m/s. We place clock readings on the horizontal axis and the x-components of your and your sister's velocities on the vertical axis; then we plot points for these velocities at each time (see **Figure 2.18a**). The best-fit curve for each person is a horizontal straight line, which makes sense since neither velocity is changing. For you, the equation of the best-fit line is $v_{Yx}(t) = +5.0$ m/s, and for your sister, it is $v_{Sx}(t) = +2.0$ m/s, where $v_x(t)$ represents the x-component of velocity as a function of time.

If instead we choose your sister as the object of reference, her velocity with respect to herself is zero, so the best-fit curve is again a horizontal line but at a value of 0.0 m/s instead of $+2.0$ m/s; your velocity is $+3.0$ m/s (see Figure 2.18b); and the mud's velocity is -2.0 m/s. The minus sign indicates that from your sister's point of view, the mud is moving in the negative direction toward her at speed $+2.0$ m/s.

Finding displacement from a velocity graph

We have just learned to construct a velocity-versus-time graph. Can we get anything more out of such graphs besides being able to represent velocity graphically? As you know, for constant velocity linear motion, the position of an object changes with time according to $x = x_0 + v_x t$. Rearranging this equation a bit, we get $x - x_0 = v_x t$, where t is the time interval during which the motion occurred, assuming that the motion started at zero clock reading $(t_0 = 0)$. The left side is the displacement of the object from time zero to time t. Now look at the right side: v_x is the vertical height of the velocity-versus-time graph line and t is the horizontal width from time zero to time t (see **Figure 2.19**). We can interpret the right side as the shaded area between the velocity-versus-time graph line and the time axis. In equation $x - x_0 = v_x t$, this area (the right side) equals the displacement of the object from time zero to time t on the left side of the equation. Here the displacement is a positive number because we chose an example with positive velocity.

Let's extend this reasoning to more general cases: an object initially at position x_1 at time t_1 and later at position x_2 at time t_2 and moving in either the positive or negative direction.

> **Displacement is the area between a velocity-versus-time graph line and the time axis** For motion with constant velocity, the magnitude of the displacement $x_2 - x_1$ (the distance traveled) of an object during a time interval from t_1 to t_2 is the area between a velocity-versus-time graph line and the time axis between those two clock readings. The displacement is the area with a plus sign when the velocity is positive and the area with a negative sign when velocity is negative.

FIGURE 2.18 Velocity-versus-time graphs.

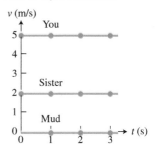

(a) v-versus-t graph lines with Earth as object of reference

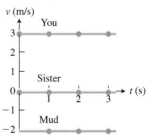

(b) v-versus-t graph lines with Sister as object of reference

TIP ▸ Notice that a horizontal line on a position-versus-time graph means that the object is at rest (the position is constant with time). A horizontal line on a velocity-versus-time graph means that the object is moving at constant velocity (its velocity does not change with time).

FIGURE 2.19 Using the v_x-versus-t graph to determine displacement $x - x_0$.

An object's displacement $x - x_0$ between $t_0 = 0$ and time interval $t - 0$ is the area between the v_x-versus-t curve and the t axis.

Displacement of you and your sister

Use the velocity-versus-time graphs shown in Figure 2.18a for you and your sister (see Example 2.5) to find your displacements with respect to Earth for the time interval from 0 to 3.0 s. The velocity components relative to Earth are $+2.0$ m/s for your sister and $+5.0$ m/s for you.

Represent mathematically For constant velocity motion, the object's displacement $x - x_0$ is the area of a rectangle whose vertical side equals the object's velocity v_x and whose horizontal side equals the time interval $t - t_0 = t - 0$ during which the motion occurred.

Solve and evaluate The displacement of your sister heading toward the mud puddle during the 3.0-s time interval is the product of her velocity and the time interval:

$$d_S = (x - x_0)_S = (+2.0 \text{ m/s})(3.0 \text{ s}) = +6.0 \text{ m}$$

(CONTINUED)

She was originally at position 10.0 m, so she is now at position $(10.0 + 6.0)$ m $= +16.0$ m. Your displacement during that same 3.0-s time interval is the product of your velocity and the time interval:

$$d_Y = (x - x_0)_Y = (+5.0 \text{ m/s})(3.0 \text{ s}) = +15.0 \text{ m}$$

You were originally at 0.0 m, so you are now at $+15.0$ m, 1.0 m behind your sister.

Try it yourself Determine the magnitudes of the displacements of you and your sister from time 0 to time 2.0 s and your positions at that time. Your initial position is zero and your sister's is 10 m.

Answer

Your sister's values are $d_S = 4.0$ m and $x_S = 14.0$ m, and your values are $d_Y = 10.0$ m and $x_Y = 10.0$ m. She is 4.0 m ahead of you at that time.

REVIEW QUESTION 2.6 Why is the following statement true? "Displacement is equal to the area between a velocity-versus-time graph line and the time axis with a positive or negative sign."

FIGURE 2.20 Velocity-versus-time graph for motion with changing velocity.

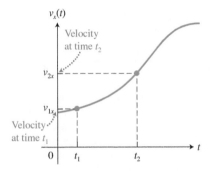

TIP Sometimes we use average speed instead of average velocity. To find the average speed divide the total path length by the time of travel. If the object is not changing its direction of motion, the average speed is equal to the magnitude of the average velocity.

FIGURE 2.21 Dropping a metal ball onto a foam block from different heights.

2.7 Motion at constant acceleration

In the last section, the function $v_x(t)$ was a horizontal line on the velocity-versus-time graph because the velocity was constant. How would the graph look if the velocity were changing? One example of such a graph is shown in **Figure 2.20**. A point on the curve indicates the velocity of the object shown on the vertical axis at a particular time shown on the horizontal axis. In this case, the velocity is continually changing and is positive.

Instantaneous velocity and average velocity

The velocity of an object at a particular time is called the **instantaneous velocity**. Figure 2.20 shows a velocity-versus-time graph for motion with continually changing instantaneous velocity. When an object's velocity is changing, we cannot use Eq. (2.1) to determine its instantaneous velocity using any time interval and a corresponding position change, because the ratio

$$v_x = \frac{x_2 - x_1}{t_2 - t_1} = \frac{\Delta x}{\Delta t}$$

is not the same for different time intervals the way it was when the object was moving at constant velocity. We can still use this equation to determine the **average velocity**, which is the ratio of the change in position and the time interval during which this change occurred. For motion at constant velocity, the instantaneous and average velocity are equal; for motion with changing velocity, they are not. However, if the two points chosen are very close together, Δx and Δt are both very small, but the ratio $\Delta x / \Delta t$ is not necessarily large or small. In fact, within a vanishingly small time interval, we can use $\Delta x / \Delta t$ as the definition of the x-component of instantaneous velocity.

When an object moves with changing velocity, its velocity can change quickly or slowly. To characterize the rate at which the velocity of an object is changing, we need a new physical quantity.

Acceleration

Let's start with a simple experiment. We take a metal ball and drop it onto a foam block that deforms when touched. The photos in **Figure 2.21** show that the dent in the block becomes deeper when the ball is dropped from a larger height. Can this result be due to a change in the velocity of the ball as it is falling? That is, does a longer fall mean a greater velocity and thus a bigger dent? We need to collect data to answer this question. The first two rows of Table 2.7 show position and time data for the falling ball:

TABLE 2.7 Position and time data for a falling ball

t (s)	0.000	0.100	0.200	0.300	0.400	0.500
y (m)	0.000	0.049	0.196	0.441	0.784	1.225
t^* (s)		0.050	0.150	0.250	0.350	0.450
v_{av} (m/s)		0.49	1.47	2.45	3.43	4.41

From the table you can see that the distance that the ball travels during each successive time interval increases; Figure 2.22 shows the position of the ball every 0.100 s. If the distance that the ball travels during each successive interval increases, the velocity must increase, too. To determine the average velocity during each time interval, we calculate the displacement of the ball between consecutive times and then divide by the time interval. For example, the average velocity between 0.100 s and 0.200 s is $(0.196 \text{ m} - 0.049 \text{ m})/(0.200 \text{ s} - 0.100 \text{ s}) = 1.47$ m/s; it is positive because we pointed the axis down. We can think of this average velocity as the instantaneous velocity that the ball had in the middle of the time interval—in this case, the 0.150 clock reading. We can now put these calculated velocities v_{av} and the associated clock readings t^* in the middle of each time interval in the two bottom rows of the table. These data now allow us to plot a velocity-versus-time graph (Figure 2.23). Notice that the best-fit curve for these data is a straight line. This occurs when the *velocity of the object increases or decreases at a constant rate*, that is, by the same amount during the same time interval. In our case, the velocity increases by 9.8 m/s every second in the downward direction.

FIGURE 2.23 A velocity-versus-time graph for a falling ball.

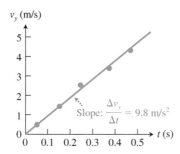

The physical quantity that characterizes how fast instantaneous velocity is changing is equal to the slope of the velocity-versus-time graph and is called **acceleration** $\vec{a}$. The vector symbol above this quantity indicates that it has direction, communicated by the sign of the slope of the velocity-versus-time graph for one-dimensional motion. Similar to velocity, the acceleration of an object moving in a straight line along an x-coordinate axis is defined as

$$a_x = \frac{v_{2x} - v_{1x}}{t_2 - t_1} = \frac{\Delta v_x}{\Delta t}$$

We see that the units of acceleration based on the definition of acceleration are $\frac{\text{m/s}}{\text{s}} = \frac{\text{m}}{\text{s} \cdot \text{s}} = \frac{\text{m}}{\text{s}^2}$.

In the case of a small falling ball, we found that its acceleration is about 9.8 m/s^2, which means that its velocity increases by 9.8 m/s every second.

To better understand acceleration, think of a cyclist and a car riding along a road at increasing speeds (Figures 2.24a and b, on the next page). The accelerations of both objects are constant (as seen from the analysis of the slopes of their velocity-versus-time graphs) but not the same; the car has a greater acceleration. If the car were slowing down,

FIGURE 2.22 The position of a falling ball every 0.100 s.

FIGURE 2.24 Velocity-versus-time graph when the velocity is changing at a constant rate.

(a) A moving cyclist

The velocity in the positive direction v_x is increasing at a constant rate.

(b) A moving car

The velocity in the positive direction is increasing at a greater rate than in (a).

(c) A car slowing down

The velocity is in the positive direction ($v_x > 0$ and is above the t axis). The acceleration is negative ($a_x < 0$) since v_x is decreasing and its $v_x(t)$ graph has a negative slope.

(d) A car speeding up

The velocity is in the negative direction (below the t axis). The acceleration is negative ($a_x < 0$) since v_x is increasing in the negative direction and its $v_x(t)$ graph has a negative slope.

its velocity-versus-time graph would instead have a negative slope, which corresponds to a decreasing speed and a negative acceleration, as v_{2x} is smaller than v_{1x} (Figure 2.24c). However, it is not necessary for an object moving with negative acceleration to be slowing down. Examine Figure 2.24d. The car has a negative acceleration and it is speeding up! This happens when the object is moving in the negative direction and has a negative component of velocity, but its speed in the negative direction is increasing in magnitude.

Because velocity is a vector quantity and the acceleration shows how quickly the velocity changes as time progresses, acceleration is also a vector quantity. We can define acceleration in a more general way. The **average acceleration** of an object during a time interval is the following:

$$\vec{a} = \frac{\vec{v}_2 - \vec{v}_1}{t_2 - t_1} = \frac{\Delta \vec{v}}{\Delta t}$$

To determine the acceleration, we need to determine the velocity change vector $\Delta \vec{v} = \vec{v}_2 - \vec{v}_1$. This equation involves the subtraction of vectors, but you can think of it in terms of addition by rearranging it to be $\vec{v}_1 + \Delta \vec{v} = \vec{v}_2$. Note that $\Delta \vec{v}$ is the vector that we add to $\vec{v}_1$ to get $\vec{v}_2$ (**Figure 2.25**). We did this when making motion diagrams in Section 2.3, only then we were not concerned with the exact lengths of the vectors. The acceleration vector $\vec{a} = \Delta \vec{v}/\Delta t$ is in the same direction as the velocity change vector $\Delta \vec{v}$, as the time interval Δt is a scalar quantity. The units of the acceleration vector are determined by dividing the units of the velocity change vector by the units of time.

Acceleration An object's average acceleration during a time interval Δt is the change in its velocity $\Delta \vec{v}$ divided by that time interval:

$$\vec{a} = \frac{\vec{v}_2 - \vec{v}_1}{t_2 - t_1} = \frac{\Delta \vec{v}}{\Delta t} \tag{2.3}$$

If Δt is very small, then the acceleration given by this equation is the **instantaneous acceleration** of the object. For one-dimensional motion, the component of the average acceleration along a particular axis (for example, for the x-axis) is

$$a_x = \frac{v_{2x} - v_{1x}}{t_2 - t_1} = \frac{\Delta v_x}{\Delta t} \tag{2.4}$$

The units of acceleration are $(m/s)/s = m/s^2$.

Note that if an object has an acceleration of $+6 \text{ m/s}^2$, it means that its velocity changes by $+6$ m/s in 1 s, or by $+12$ m/s in 2 s $[(+12 \text{ m/s})/(2 \text{ s}) = +6 \text{ m/s}^2]$.

It is possible for an object to have a zero velocity and nonzero acceleration—for example, at the moment when an object starts moving from rest. An object can also have a nonzero velocity and zero acceleration—for example, an object moving at constant velocity. Note that the acceleration of an object depends on the observer. For example, a car is accelerating for an observer on the ground but is not accelerating for the driver of that car.

FIGURE 2.25 How to determine the change in velocity $\Delta \vec{v}$.

We add $\Delta \vec{v}$ to $\vec{v}_1$ to get $\vec{v}_2$.

TIP If it is difficult for you to think about velocity and acceleration in abstract terms, try calculating the acceleration for simple integer velocities.

Determining velocity change from acceleration

If, at $t_0 = 0$ for linear motion, the x-component of the velocity of some object is v_{0x} and its acceleration a_x is constant, then its velocity v_x at a later time t can be determined by substituting these quantities into Eq. (2.4):

$$a_x = \frac{v_x - v_{0x}}{t - 0}$$

Rearranging, we get an expression for the changing instantaneous velocity of the object as a function of time:

$$v_x = v_{0x} + a_x t \qquad (2.5)$$

Here the value of the velocity at any clock reading is the instantaneous velocity. For one-dimensional motion, the directions of the vector components a_x, v_x, and v_{0x} are indicated by their signs relative to the axis of motion—positive if in the positive x-direction and negative if in the negative x-direction.

EXAMPLE 2.7   Bicycle ride

Suppose that you are sitting on a bench watching someone riding a bicycle on a flat, straight road. In a 2.0-s time interval, the velocity of the bicycle changes from -4.0 m/s to -7.0 m/s. Describe the motion of the bicycle as fully as possible.

Sketch and translate We can sketch the process as shown below. The bicycle (the object of interest) is moving in the negative direction with respect to the chosen reference frame. The components of the bicycle's velocity along the axis of motion are negative: $v_{0x} = -4.0$ m/s at time $t_0 = 0.0$ s and $v_x = -7.0$ m/s at $t = 2.0$ s. The speed of the bicycle (the magnitude of its velocity) increases. It is moving faster in the negative direction. We can determine the acceleration of the bicycle and describe the changes in its velocity.

Simplify and diagram Our sketch of the motion diagram for the bicycle is shown below. Note that $\Delta \vec{v}$ points in the negative x-direction.

Represent mathematically We apply Eq. (2.4) to determine the acceleration:

$$a_x = \frac{v_x - v_{0x}}{t - t_0}$$

Solve and evaluate Substituting the given velocities and times, we get

$$a_x = \frac{(-7.0\,\text{m/s}) - (-4.0\,\text{m/s})}{2.0\,\text{s} - 0.0\,\text{s}} = -1.5\ \text{m/s}^2$$

The bicycle's x-component of velocity at time 0 was -4.0 m/s. Its velocity was changing by -1.5 m/s each second. So 1 s later, its velocity was

$$v_{1s\,x} = v_{0x} + \Delta v_x$$
$$= (-4.0\ \text{m/s}) + (-1.5\ \text{m/s}) = -5.5\ \text{m/s}$$

During the second 1-s time interval, the velocity changed by another -1.5 m/s and was then $(-5.5\ \text{m/s}) + (-1.5\ \text{m/s}) = (-7.0\ \text{m/s})$. In this example, the bicycle was speeding up by 1.5 m/s each second in the negative direction.

Try it yourself A car's acceleration is -3.0 m/s^2. At time 0 its velocity is $+14$ m/s. What happens to the velocity of the car? What is its velocity after 3 s?

Answer

The car's velocity in the positive x-direction is decreasing. After 3 s, $v_x = 5.0$ m/s.

TIP   It is possible for an object to have a positive acceleration and be slowing down as well as to have a negative acceleration and be speeding up. When an object is speeding up, the acceleration vector is in the same direction as the velocity vector, and the velocity and acceleration components along the same axis have the same sign. When an object is slowing down, the acceleration is in the opposite direction relative to the velocity; their components have opposite signs.

Free fall

Let's return to the example that led us to the "discovery" of acceleration: dropping a metal ball onto a foam block (Figure 2.21). Galileo Galilei (1564–1642) did experiments similar to ours in which he dropped objects from different heights and examined the resulting dents in sand. However, he also noticed that objects of different shapes fall differently—for example, a piece of paper falls more slowly than a heavy brick. He hypothesized that this difference was due to air resistance. *If* his hypothesis were correct, *then* in a vacuum (space devoid of air), all objects would fall with the same acceleration. Indeed, in a large vacuum chamber, when a light feather and a heavy apple fall, they are always next to each other, meaning that they have the same velocity and acceleration at every instant (**Figure 2.26**). You can perform this testing experiment,

FIGURE 2.26 A feather and an apple fall side by side in a vacuum.

FIGURE 2.27 Motion for an upward thrown ball.

(a)

(b)

(c)

(d)

too: take a heavy ball and a tennis ball, drop them at the same time from a height of about 2 m, and videotape the process. You will see them moving side by side!

Earlier in this section we found that the acceleration of the falling metal ball was about 9.8 m/s². At the same location, without air resistance, all falling objects will have this same acceleration; however, at another location on Earth, the value of acceleration might be slightly different. We will learn why in Chapter 5.

If we videotape a small object thrown upward and then use the data to construct a velocity-versus-time graph, we find that its acceleration is still the same at all clock readings. For an upward-pointing axis, the object's acceleration is -9.8 m/s² on the way up, -9.8 m/s² on the way down, and even -9.8 m/s² at the instant when the object is momentarily at rest at the highest point of its motion. At all times during the object's flight, its velocity is changing at a rate of -9.8 m/s each second. A motion diagram and the graphs representing the position-versus-time, velocity-versus-time, and acceleration-versus-time are shown in **Figure 2.27**. The positive direction is up. Notice that when the object is at its maximum height on the position-versus-time graph, the velocity is instantaneously zero (the slope of the position-versus-time graph is zero). The acceleration is never zero, even at the moment when the velocity of the object is zero.

It might be tempting to think that at the instant an object is not moving, its acceleration must be zero. This is only true for an object that is at rest and remains at rest. In the case of an object thrown upward, if its acceleration at the top of the flight is zero, it would never descend (it would remain at rest at its highest point).

The motion described above is called a **free fall**. Free fall is an example of a simplified model of motion in which we assume that there is no air or any other medium affecting the motion of a dropped or thrown (upward or downward) object. All objects in free fall have the same acceleration. We will learn the explanation for this phenomenon in Chapter 3. Note that physicists use the term free fall even when the object is moving up.

REVIEW QUESTION 2.7 (a) Give an example in which an object with negative acceleration is speeding up. (b) Give an example in which an object with positive acceleration is slowing down. (c) Give an example in which an object with zero velocity is accelerating.

2.8 Displacement of an object moving at constant acceleration

Equation (2.2) allows us to predict the location of an object moving at constant velocity at any time if we know its velocity and initial position. Can we construct a similar equation for an object moving at constant acceleration? Suppose that at time zero an object is passing the mark x_0 on the axis of motion traveling at velocity v_0 and acceleration a_x. Where will the object be after the time interval $t - 0 = t$? To answer this question, we can use the concept of average velocity. If we know the object's average velocity and initial position, and we assume that the object is moving at this velocity during the time interval of interest, we can predict the position at any time using Eq. (2.2),
$x = x_0 + v_{av\,x}t$.

The average velocity is

$$v_{av\,x} = \frac{v_x + v_{0x}}{2}$$

where, according to Eq. (2.5), instantaneous velocity is $v_x = v_{0x} + a_x t$. Thus we can write

$$v_{av\,x} = \frac{v_x + v_{0x}}{2} = \frac{v_{0x} + a_x t + v_{0x}}{2} = \frac{2v_{0x} + a_x t}{2}$$

Therefore, the position of the accelerating object at any time is

$$x = x_0 + v_{avx}t = x_0 + \left(\frac{2v_{0x} + a_x t}{2}\right)t$$

$$\Rightarrow x = x_0 + v_{0x}t + \tfrac{1}{2}a_x t^2$$

(The symbol $\Rightarrow$ means that this equation follows from the previous equation.)

Does the above result make sense? Consider a limiting case, for example, when the object is traveling at a constant velocity (when $a_x = 0$). In this case the equation should reduce to the result from our investigation of linear motion with constant velocity ($x = x_0 + v_{x0}t$). It does. We can also check the units of each term in this equation for consistency (when terms in an equation are added or subtracted, each of those terms must have the same units). Each term has units of meters, so the units also check.

> **Position of an object during linear motion with constant acceleration** For any initial position x_0 at clock reading $t_0 = 0$, we can determine the position x of an object at any later time t, provided we also know the initial velocity v_{0x} of the object and its constant acceleration a_x:
>
> $$x = x_0 + v_{0x}t + \tfrac{1}{2}a_x t^2 \qquad (2.6)$$

Since t^2 appears in Eq. (2.6), the position-versus-time graph for this motion is not a straight line but a parabola (a parabola is a graph for a quadratic function) (**Figure 2.28a**). Unlike the position-versus-time graph for constant velocity motion where $\Delta x/\Delta t$ is the same for any time interval, the position-versus-time graph for accelerated motion is not a straight line; it does not have a constant slope. At different times, the change in position Δx has a different value for the same time interval Δt (Figure 2.28b). The line tangent to the position-versus-time graph at a particular time has a slope $\Delta x/\Delta t$ that equals the velocity v_x of the object at that time. The slopes of the tangent lines at different times for the graph in Figure 2.28b differ—they are greater when the position changes more during the same time interval.

The next example involves vertical motion. Thus, we will use a vertical y-coordinate axis and apply the equations of motion.

TIP You can only add or subtract physical quantities that have the same units. The result has the same units as those two quantities. When you multiply or divide quantities, you make a new physical quantity whose units are the result of multiplication or division of the original quantities.

FIGURE 2.28 (a) A position-versus-time graph for constant acceleration motion. (b) The instantaneous velocity v_x is the slope of the position-versus-time graph at a particular clock reading.

EXAMPLE 2.8   Estimating acceleration of a falling person

A woman jumps off a large boulder. She is moving at a speed of about 5 m/s when she reaches the ground. *Estimate* her acceleration during landing. Indicate any other quantities you use in the estimate.

Sketch and translate To solve the problem, we assume that the woman's knees bend while landing so that her body travels 0.4 m closer to the ground compared to the standing upright position. That is, 0.4 m is the stopping distance for the main part of her body. At right we have sketched the initial and final situations during the landing. We choose the central part of her body as the object of interest. We use a vertical reference with a y-axis pointing down and the final location of the central part of her body as the origin of the coordinate axis. The initial values of her motion at the time $t_0 = 0$ when she first touches the ground are $y_0 = -0.4$ m (the central part of her body is 0.4 m in the negative direction above her final position) and $v_{0y} = +5$ m/s

(she is moving downward, the positive direction relative to the y-axis). The final values at some unknown time t at the instant she stops are $y = 0$ and $v_y = 0$. Note: It is important to have a coordinate axis and the initial and final values with appropriate signs.

(CONTINUED)

Simplify and diagram The motion diagram at right represents her motion while stopping, assuming constant acceleration. We cannot model the woman as a point-like object in this situation, so we focus on the motion of her midsection.

Represent mathematically The challenge in representing this situation mathematically is that there are two unknowns: the magnitude of her acceleration a_y (the unknown we wish to determine) and the time interval t between when she first contacts the ground and when she comes to rest. However, both Eqs. (2.5) and (2.6) describe linear motion with constant acceleration and have a_y and t in them. Since we have two equations and two unknowns, we can handle this challenge by solving Eq. (2.5) for the time t

$$t = \frac{v_y - v_{0y}}{a_y}$$

and substitute it into rearranged Eq. (2.6):

$$y = y_0 + v_{0y}t + \tfrac{1}{2}a_y t^2$$

The result is:

$$y - y_0 = v_{0y}\left(\frac{v_y - v_{0y}}{a_y}\right) + \frac{a_y(v_y - v_{0y})^2}{2a_y^2}$$

Using algebra, we can simplify the above equation:

$$\Rightarrow 2a_y(y - y_0) = 2v_{0y}(v_y - v_{0y}) + (v_y - v_{0y})^2$$
$$\Rightarrow 2a_y(y - y_0) = (2v_{0y}v_y - 2v_{0y}^2)$$
$$+ (v_y^2 - 2v_y v_{0y} + v_{0y}^2)$$
$$\Rightarrow 2a_y(y - y_0) = v_y^2 - v_{0y}^2$$
$$\Rightarrow a_y = \frac{v_y^2 - v_{0y}^2}{2(y - y_0)}$$

Solve and evaluate Now we can use the previous equation to find her acceleration:

$$a_y = \frac{v_y^2 - v_{0y}^2}{2(y - y_0)}$$
$$= \frac{(0)^2 - (5 \text{ m/s})^2}{2[0 - (-0.4 \text{ m})]} = -31 \text{ m/s}^2 \approx -30 \text{ m/s}^2$$

Is the answer reasonable? The sign is negative. This means that the acceleration points upward, as does the velocity change arrow in the motion diagram. This is correct. The units for acceleration are correct. We cannot judge yet if the magnitude is reasonable. We will learn later that it is. The answer has one significant digit, as it should—the same as the information given in the problem statement.

- -

Try it yourself Using the expression

$$a_y = \frac{v_y^2 - v_{0y}^2}{2(y - y_0)}$$

decide how the acceleration would change if (a) the stopping distance doubles and (b) the initial speed doubles. Note that the final velocity is $v_y = 0$.

Answer

(a) a_y would be half the magnitude; (b) a_y would be four times the magnitude.

In the "Represent mathematically" step above, we developed a new mathematical relation that is useful when you do not know the time of travel. You do not need to memorize it, but it can come in handy when you solve problems.

> **A useful equation for linear motion with constant acceleration** for situations in which you do not know the time interval during which the changes in position and velocity occurred:
>
> $$2a_x(x - x_0) = v_x^2 - v_{0x}^2 \tag{2.7}$$

Figure 2.29 shows a different way to derive Eq. (2.6): using the velocity-versus-time graph. Examine the four parts of the figure carefully. You can use the idea of the displacement as the area between the constant velocity graph and the time axis as the starting point for this derivation (Figure 2.29a). Breaking the time axis into small intervals allows us to consider the velocity for those time intervals as constant and to apply the method we used in Figure 2.19 (Figure 2.29b). The result is that the displacement is the area of a trapezium (Figure 2.29c), which can be broken up into two easy-to-calculate areas (Figure 2.29d).

(a)

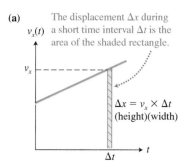

The displacement Δx during a short time interval Δt is the area of the shaded rectangle.

$\Delta x = v_x \times \Delta t$
(height)(width)

(b)

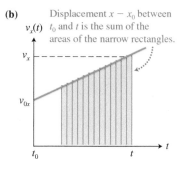

Displacement $x - x_0$ between t_0 and t is the sum of the areas of the narrow rectangles.

FIGURE 2.29 How to determine the displacement of an object moving at constant acceleration using its velocity-versus-time graph.

(c)

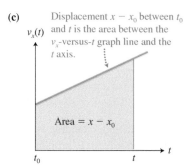

Displacement $x - x_0$ between t_0 and t is the area between the v_x-versus-t graph line and the t axis.

Area $= x - x_0$

(d)

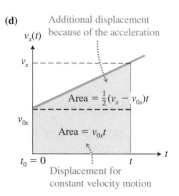

Additional displacement because of the acceleration

Area $= \frac{1}{2}(v_x - v_{0x})t$

Area $= v_{0x}t$

Displacement for constant velocity motion

REVIEW QUESTION 2.8 Explain qualitatively, without algebra, why the displacement of an object moving at constant acceleration is proportional to time squared, not to time to the power of 1 as it is for motion at constant velocity.

2.9 Skills for analyzing situations involving motion

To help analyze physical processes involving motion, we will represent processes in multiple ways: the words in the problem statement, a sketch, one or more diagrams, possibly a graph, and a mathematical description. Different representations have to agree with each other; in other words, they need to be consistent.

Motion at constant velocity

EXAMPLE 2.9   Two walking friends

You stand on a sidewalk and observe two friends walking at constant velocity. At time 0 Jim is 4.0 m east of you and walking away from you at speed 2.0 m/s. Also at time 0, Sarah is 10.0 m east of you and walking toward you at speed 1.5 m/s. Represent their motions with an initial sketch, with motion diagrams, and mathematically.

Sketch and translate We choose Earth as the object of reference with your position as the reference point. The positive direction points to the east. We have two objects of interest here: Jim and Sarah. Jim's initial position is $x_0 = +4.0$ m and his constant velocity is $v_x = +2.0$ m/s. Sarah's initial position is $X_0 = +10.0$ m and her

constant velocity is $V_x = -1.5$ m/s (the velocity is negative since she is moving westward). In our sketch we are using capital letters to represent Sarah and lowercase letters to represent Jim.

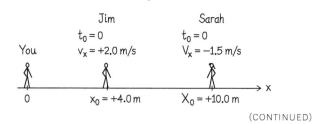

(CONTINUED)

Simplify and diagram We can model both friends as point-like objects since the distances they move are somewhat greater than their own sizes. The motion diagrams below represent their motions.

Represent mathematically Now construct equations to represent Jim's and Sarah's motion:

Jim: $x = +4.0 \text{ m} + (2.0 \text{ m/s})t$

Sarah: $X = +10.0 \text{ m} + (-1.5 \text{ m/s})t$

Solve and evaluate We were not asked to solve for any quantity. We will do it in the "Try it yourself" exercise.

Try it yourself Determine the time when Jim and Sarah are at the same position, and where that position is.

Answer

They are at the same position when $t = 1.7$ s and when $x = X = 7.4$ m.

Equation Jeopardy problems

Learning to read the mathematical language of physics with understanding is an important skill. To help develop this skill, this text includes Jeopardy-style problems. In this type of problem, you have to work backwards: you are given one or more equations and are asked to use them to construct a consistent sketch of a process. You then convert the sketch into a diagram of a process that is consistent with the equations and sketch. Finally, you invent a word problem that the equations could be used to solve. Note that there are often many possible word problems for a particular mathematical description.

CONCEPTUAL EXERCISE 2.10   Equation Jeopardy

The following equation describes an object's motion:

$$x = (5.0 \text{ m}) + (-3.0 \text{ m/s})t$$

Construct a sketch, a motion diagram, kinematics graphs, and a verbal description of a situation that is consistent with this equation. There are many possible situations that the equation describes equally well.

Sketch and translate This equation looks like a specific example of our general equation for the linear motion of an object with constant velocity: $x = x_0 + v_x t$. The minus sign in front of the 3.0 m/s indicates that the object is moving in the negative x-direction. At time 0, the object is located at position $x_0 = +5.0$ m with respect to some chosen object of reference and *is already moving*. Let's imagine that this chosen object of reference is a running person (the observer) and the equation represents the motion of a person (the object of interest) sitting on a bench as seen by the runner. The sketch below illustrates this possible scenario. The positive axis points from the observer (the runner) toward the person on the bench, and at time 0 the person sitting on it is 5.0 m in front of the runner and coming closer to the runner as time elapses.

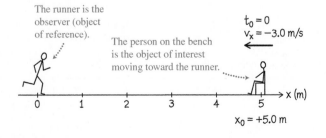

Simplify and diagram Model the object of interest as a point-like object. A motion diagram for the situation is shown below. The equal spacing of the dots and the equal lengths of velocity arrows both indicate that the object of interest is moving at constant velocity with respect to the observer.

$$\Delta \vec{v} = 0$$

Motion diagram for bench relative to runner

Position-versus-time and velocity-versus-time kinematics graphs of the process are shown below. The position-versus-time graph has a constant -3.0 m/s slope and a $+5.0$ m intercept with the vertical (position) axis. The velocity-versus-time graph has a constant value (-3.0 m/s) and a zero slope (the velocity is not changing). The following verbal description describes this particular process: A jogger sees a person on

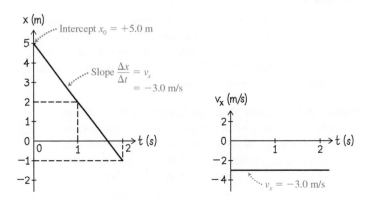

a bench in the park 5.0 m in front of him. The bench is approaching at a speed of 3.0 m/s as seen in the jogger's reference frame. The direction pointing from the jogger to the bench is positive.

Try it yourself Suppose we switch the roles of observer and object of reference. Now the person on the bench is the object of reference and observes the runner. We choose to describe the process by the same equation as in the example:

$$x = (5.0 \text{ m}) + (-3.0 \text{ m/s})t$$

Construct an initial sketch and a motion diagram that are consistent with the equation and with the new observer and new object of reference.

Answer

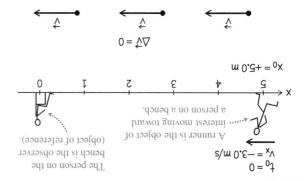

An initial sketch for this process and a consistent motion diagram are shown below.

The person on the bench is the observer (object of reference).

A runner is the object of interest moving toward a person on a bench.

$x_0 = +5.0$ m
$v_x = -3.0$ m/s
$t_0 = 0$

Motion diagram for runner relative to person on bench

Motion at constant nonzero acceleration

Now let's apply some representation techniques to linear motion with constant (nonzero) acceleration.

EXAMPLE 2.11   Equation Jeopardy

A process is represented mathematically by the following equation:

$$x = (-60 \text{ m}) + (10 \text{ m/s})t + (1.0 \text{ m/s}^2)t^2$$

Use the equation to construct an initial sketch, a motion diagram, and a verbal description of a process that is consistent with this equation.

Sketch and translate The above equation appears to be an application of Eq. (2.6), which we constructed to describe linear motion with constant acceleration, if we assume that the 1.0 m/s² in front of t^2 is the result of dividing 2.0 m/s² by 2:

$$x = x_0 + v_{0x}t + \tfrac{1}{2}a_xt^2$$
$$= (-60 \text{ m}) + (10 \text{ m/s})t + \tfrac{1}{2}(2.0 \text{ m/s}^2)t^2$$

It looks like the initial position of the object of interest is $x_0 = -60$ m, its initial velocity is $v_{0x} = +10$ m/s, and its acceleration is $a_x = +2.0$ m/s². Let's imagine that this equation describes the motion of a car on a straight highway passing a van in which you, the observer, are riding. The car is 60 m behind you and moving 10 m/s faster than your van. The car speeds up at a rate of 2.0 m/s² with respect to the van. The object of reference is you in the van; the positive direction is the direction in which the car and van are moving. A sketch of the situation follows.

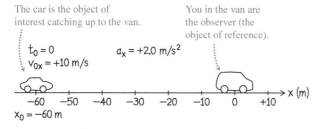

The car is the object of interest catching up to the van.

You in the van are the observer (the object of reference).

$t_0 = 0$
$v_{0x} = +10$ m/s
$a_x = +2.0$ m/s²
$x_0 = -60$ m

Simplify and diagram The car can be considered a point-like object—much smaller than the dimensions of the path it travels. The

car's velocity and acceleration are both positive. Thus, the car's velocity in the positive x-direction is increasing as it moves toward the van (toward the origin). Below is a motion diagram for the car's motion as seen from the van. The successive dots in the diagram are spaced increasingly farther apart as the velocity increases; the velocity arrows are drawn increasingly longer. The velocity change arrow (and the acceleration) point in the positive x-direction, that is, in same direction as the velocity arrows.

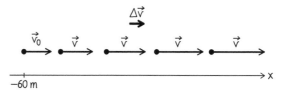

Represent mathematically The mathematical representation of the situation appears at the start of the Equation Jeopardy example.

Solve and evaluate To evaluate what we have done, we can check the consistency (agreement) of the different representations. For example, we can check whether the initial position and velocity in the equation, the sketch, and the motion diagram are consistent. In this case, they are.

Try it yourself Describe a different scenario for the same mathematical representation.

Answer This mathematical representation could also describe the motion of a cyclist moving on a straight path as seen by a person standing on a sidewalk 60 m in front of the cyclist. The positive direction is in the direction the cyclist is traveling. When the person starts observing the cyclist, she is moving at an initial velocity of $v_{0x} = +10$ m/s and speeding up with acceleration $a_x = +2.0$ m/s².

The problem-solving strategy in this text uses multiple representations, an approach that was found to be extremely useful in helping solve physics problems. The complete problem-solving strategy for a particular type of problem will be presented along with an example at the side. Below, the problem-solving steps for kinematics (motion) problems are presented on the left side and illustrated on the right side. The content of this example is also very important: it shows why it is necessary to remain a safe distance behind the car in front of you when driving.

PROBLEM-SOLVING STRATEGY 2.1 > Kinematics

EXAMPLE 2.12 **An accident involving tailgating**

A car follows about two car lengths (10.0 m) behind a van. At first, both vehicles are traveling at a conservative speed of 25 m/s (56 mi/h). The driver of the van suddenly slams on the brakes to avoid an accident, slowing down at 9.0 m/s^2. The car driver's reaction time is 0.80 s and the car's maximum acceleration while slowing down is also 9.0 m/s^2. Will the car be able to stop before hitting the van?

Sketch and translate

- Sketch the situation described in the problem. Choose the object of interest.

- Include an object of reference and a coordinate system. Indicate the origin and the positive direction.

- Label the sketch with relevant information.

- Identify the unknown that you need to find. Label it with a question mark on the sketch.

Below we represent this situation for each vehicle (we have two objects of interest). We use capital letters to indicate quantities referring to the van and lowercase letters for quantities referring to the car. We use the coordinate system shown with the origin of the coordinates at the initial position of the car's front bumper. The positive direction is in the direction of motion.

The process starts when the van starts braking. The van moves at constant negative acceleration throughout the entire process. We separate the motion of the car into two parts: (1) its motion before the driver applies the brakes (constant positive velocity) and (2) its motion after the driver starts braking (constant negative acceleration).

Simplify and diagram

- Decide how you will model each moving object (for example, as a point-like object).

- Can you model the motion as constant velocity or constant acceleration?

- Draw motion diagrams and kinematics graphs if needed.

We model each vehicle as a point-like object, but since we are trying to determine if they collide, we need to be more specific about their positions. The position of the car is the position of its front bumper. The position of the van is the position of its rear bumper. We look at the motion of each vehicle separately. If the car's final position is greater than the van's final position, then a collision has occurred at some point during their motion. Assume that the vehicles have constant acceleration so that we can apply our model of motion with constant acceleration. A velocity-versus-time graph line for each vehicle is shown at right.

Represent mathematically

• Use the sketch(es), motion diagram(s), and kinematics graph(s) to construct a mathematical representation (equations) of the process. Be sure to consider the sign of each quantity.

Equation (2.7) can be used to determine the distance the van travels while stopping:

$$2A_x(X - X_0) = V_x^2 - V_{0x}^2 \Rightarrow X = \frac{V_x^2 - V_{0x}^2}{2A_x} + X_0$$

The car Part 1 Since the car is initially traveling at constant velocity, we use Eq. (2.2). The subscript 0 indicates the moment the driver sees the van start slowing down. The subscript 1 indicates the moment the car driver starts braking.

$$x_1 = x_0 + v_{0x}t_1$$

The car Part 2 After applying the brakes, the car has an acceleration of -9.0 m/s^2. The subscript 2 indicates the moment the car stops moving. We represent this part of the motion using Eq. (2.7):

$$2a_x(x_2 - x_1) = v_{2x}^2 - v_{1x}^2$$

$$\Rightarrow x_2 = \frac{v_{2x}^2 - v_{1x}^2}{2a_x} + x_1 \Rightarrow x_2 = \frac{v_{2x}^2 - v_{1x}^2}{2a_x} + (x_0 + v_{0x}t_1)$$

The last step came from inserting the result from Part 1 for x_1.

- -

Solve and evaluate

• Solve the equations to find the answer to the question you are investigating.

• Evaluate the results to see if they are reasonable. Check the units and decide if the calculated quantities have reasonable values (sign, magnitude). Check limiting cases: examine whether the final equation leads to a reasonable result if one of the quantities is zero or infinity. This strategy applies when you derive a new equation while solving a problem.

The van's initial velocity is $V_0 = +25 \text{ m/s}$, its final velocity is $V_x = 0$, and its acceleration is $A = -9.0 \text{ m/s}^2$. Its initial position is two car lengths in front of the front of the car, so $X_0 = 2 \times 5.0 \text{ m} = 10 \text{ m}$. The final position of the van is

$$X = \frac{V_x^2 - V_{0x}^2}{2A_x} + X_0 = \frac{0^2 - (25 \text{ m/s})^2}{2(-9.0 \text{ m/s}^2)} + 10 \text{ m} = 45 \text{ m}$$

The car's initial position is $x_0 = 0$, its initial velocity is $v_{0x} = 25 \text{ m/s}$, its final velocity is $v_{2x} = 0$, and its acceleration when braking is $a_x = -9.0 \text{ m/s}^2$. The car's final position is

$$x_2 = \frac{v_{2x}^2 - v_{1x}^2}{2a_x} + (x_0 + v_{0x}t_1) = \frac{0^2 - (25 \text{ m/s})^2}{2(-9.0 \text{ m/s}^2)} + [0 \text{ m} + (25 \text{ m/s})(0.8 \text{ s})] = 55 \text{ m}$$

The car will stop about 10 m beyond where the van will stop. There will be a collision between the two vehicles.

This analysis illustrates why tailgating is such a big problem. The car traveled at a 25-m/s constant velocity during the relatively short 0.80-s reaction time. During the same 0.80 s, the van's velocity decreased by $(0.80 \text{ s})(-9.0 \text{ m/s}^2) = 7.2 \text{ m/s}$ from 25 m/s to about 18 m/s. So the van was moving somewhat slower than the car when the car finally started to brake. Since they were both slowing down at about the same rate, the tailgating vehicle's velocity was always greater than that of the vehicle in front until they hit.

- -

Try it yourself Two cars, one behind the other, are traveling at 30 m/s (67 mi/h). The front car hits the brakes and slows down at the rate of 10 m/s². The driver of the second car has a 1.0-s reaction time. The front car's speed has decreased to 20 m/s during that 1.0 s. The rear car traveling at 30 m/s starts braking, slowing down at the same rate of 10 m/s². How far behind the front car should the rear car be so it does not hit the front car?

Answer   The rear car should be at least 30 m behind the front car.

- -

REVIEW QUESTION 2.9 A car's motion with respect to the ground is described by the following function:

$$x = (-48 \text{ m}) + (12 \text{ m/s})t + (-2.0 \text{ m/s}^2)t^2$$

Mike says that its original position is (-48 m) and its acceleration is (-2.0 m/s^2). Do you agree? If yes, explain why; if not, explain how to correct his answer.

Summary

A **reference frame** consists of an object of reference, a point of reference on that object, a coordinate system whose origin is at the point of reference, and a clock. (Section 2.1)

Object of reference

Clock

Object of interest

0

Point of reference (origin)

Coordinate axis

x

Time or clock reading t (a scalar quantity) is the reading on a clock or another time measuring instrument. (Section 2.4)

t

Time: t

Time interval Δt (a scalar quantity) is the difference of two times. (Section 2.4)

t_1

t_2

Time interval:

$$\Delta t = t_2 - t_1$$

Position x (a scalar quantity) is the location of an object relative to the chosen origin. (Section 2.4)

Displacement $\vec{d}$ is a vector drawn from the initial position of an object to its final position. The x-component of the displacement d_x is the change in position of the object along the x-axis. (Section 2.4)

Distance d (a scalar quantity) is the magnitude of the displacement and is always positive. (Section 2.4)

Path length l is the length of a string laid along the path the object took. (Section 2.4)

0 x_0 x

$\vec{d}$

0 x_0 x

$d_x = x - x_0 > 0$ if $\vec{d}$ points in the positive direction of x-axis.

$d_x = x - x_0 < 0$ if $\vec{d}$ points in the negative direction.

$$d = |x - x_0|$$

Velocity $\vec{v}$ (a vector quantity) is the ratio of the displacement of an object during a time interval to that time interval. The velocity is *instantaneous* if the time interval is very small and *average* if the time interval is longer. (Section 2.6)

Speed v (a scalar quantity) is the magnitude of the velocity. (Section 2.6)

t_1

t_2

0 x_1 x_2 x

For constant velocity linear motion,

$$\vec{v} = \vec{d}/\Delta t$$

$$v_x = \frac{\Delta x}{\Delta t} = \frac{x_2 - x_1}{t_2 - t_1} \qquad \text{Eq. (2.1)}$$

$$v = \left| \frac{\Delta x}{\Delta t} \right|$$

Acceleration $\vec{a}$ (a vector quantity) is the ratio of the change in an object's velocity $\Delta \vec{v}$ during a time interval Δt to the time interval. The acceleration is *instantaneous* if the time interval is very small and *average* if the time interval is longer. (Section 2.7)

t_0

t

v_{0x}

v_x

0

x

$$a_x = \frac{v_{2x} - v_{1x}}{t_2 - t_1} = \frac{\Delta v_x}{\Delta t} \qquad \text{Eq. (2.4)}$$

(rearranged)

$$a_x = \frac{\Delta v_x}{\Delta t} = \frac{v_x - v_{0x}}{t - t_0} \qquad \text{Eq. (2.4)}$$

Motion with constant velocity or constant acceleration can be represented with a sketch, a motion diagram, kinematics graphs, and mathematically. (Sections 2.6—2.9)

$\Delta \vec{v} = 0$

$\vec{v}$ $\vec{v}$

$\Delta \vec{v}$

$\vec{v}$ $\vec{v}$ $\vec{v}$

x

t

v

t

$$v_x = v_{0x} + a_x t \qquad \text{Eq. (2.5)}$$

$$x = x_0 + v_{0x}t + \tfrac{1}{2}a_x t^2 \qquad \text{Eq. (2.6)}$$

$$2a_x(x - x_0) = v_x^2 - v_{0x}^2 \qquad \text{Eq. (2.7)}$$

(rearranged)

$$x - x_0 = \frac{v_x^2 - v_{0x}^2}{2a_x} \qquad \text{Eq. (2.7)}$$

Questions

Multiple Choice Questions

1. Match the general elements of physics knowledge (left) with the appropriate examples (right).

Model of a process	Free fall
Model of an object	Acceleration
Physical quantity	Rolling ball
Physical phenomenon	Point-like object

   (a) Model of a process—Acceleration; Model of an object—Point-like object; Physical quantity—Free fall; Physical phenomenon—Rolling ball.
   (b) Model of a process—Rolling ball; Model of an object—Point-like object; Physical quantity—Acceleration; Physical phenomenon—Free fall.
   (c) Model of a process—Free fall; Model of an object—Point-like object; Physical quantity—Acceleration; Physical phenomenon—Rolling ball.

2. Which group of quantities below consists only of scalar quantities?
   (a) Average speed, displacement, time interval
   (b) Average speed, path length, clock reading
   (c) Temperature, acceleration, position

3. Which of the following are examples of time interval?
   (1) I woke up at 7 a.m. (2) The lesson lasted 45 minutes. (3) Svetlana was born on November 26. (4) An astronaut orbited Earth in 4 hours.
   (a) 1, 2, 3, and 4   (b) 2 and 4   (c) 2
   (d) 4   (e) 3

4. A student said, "The displacement between my dorm and the lecture hall is 1 kilometer." Is he using the correct physical quantity for the information provided? What should he have called the 1 kilometer?
   (a) Distance   (b) Path length   (c) Position
   (d) Both a and b are correct.

5. An object moves so that its position depends on time as $x = +12$ m $- (4$ m/s$)t + (1$ m/s$^2)t^2$. Which statement below is not true?
   (a) The object is accelerating.
   (b) The speed of the object is always decreasing.
   (c) The object first moves in the negative direction and then in the positive direction.
   (d) The acceleration of the object is $+2$ m/s^2.
   (e) The object stops for an instant at 2.0 s.

6. Choose the correct approximate velocity-versus-time graph for the following hypothetical motion: a car moves at constant velocity, and then slows to a stop and without a pause moves in the opposite direction with the same acceleration (Figure Q2.6).

   FIGURE Q2.6

7. Figure Q2.7b shows the position-versus-time graph for a small red dot that was moving along the x-axis from point C at $t = 0$, as shown in Figure Q2.7a. Select the two answers that correctly describe the motion of the dot.

   FIGURE Q2.7

   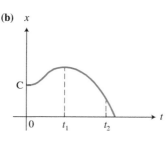

   (a) At t_1 the velocity of the dot was zero.
   (b) At t_1 the speed of the dot was greatest.
   (c) At t_1 the dot was at position D.
   (d) At t_2 the dot was at position A.
   (e) At t_2 the velocity of the dot was positive.

8. Oliver takes two identical marbles and drops the first one from a certain height. A short time later, he drops the second one from the same height. How does the distance between the marbles change while they are falling to the ground? Choose the best answer and explanation.
   (a) The distance increases because the first marble moves faster than the second marble at every instant of time.
   (b) The distance increases because the first marble moves with greater acceleration than the second marble.
   (c) The distance stays the same because both marbles are falling with the same acceleration.
   (d) The distance stays the same because the velocities of the marbles are equal at every instant of time.

9. Your car is traveling west at 12 m/s. A stoplight (the origin of the coordinate axis) to the west of you turns yellow when you are 20 m from the edge of the intersection (see Figure Q2.9). You apply the brakes and your car's speed decreases. Your car stops before it reaches the stoplight. What are the signs for the components of kinematics quantities?

   FIGURE Q2.9

   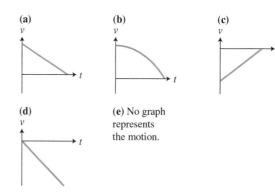

   $t_0 = 0$ s
   $x_0 = $ __20 m
   $v_{0x} = $ __12 m/s
   $a_x = $ __6.0 m/s^2

	x_0	v_{0x}	a_x
(a)	+	−	−
(b)	+	−	+
(c)	+	+	−
(d)	+	+	+
(e)	−	−	+

10. Which velocity-versus-time graph in Figure Q2.10 best describes the motion of the car in the previous problem (see Figure Q2.9) as it approaches the stoplight?

    FIGURE Q2.10

    (a)   (b)   (c)

    (d)   (e) No graph represents the motion.

11. Azra wants to determine the average speed of the high-speed train that operates between Paris and Lyon. Before she boards the train, she measures the length of the train L and the distance d between the electric line poles that are placed along the tracks. Which of the following procedures should she use?
    (a) Divide the distance between the adjacent poles by 10 times the time needed to pass 10 poles.
    (b) Divide the length of the train by $\frac{1}{10}$ of the time needed to pass 10 poles.
    (c) Count how many poles the train passed in 10 s and divide this number by $d \cdot 10$ s.
    (d) Count how many poles the train passed in 10 s and multiply this number by $\frac{d}{10\ s}$.

12. A sandbag hangs from a rope attached to a rising hot air balloon. The rope connecting the bag to the balloon is cut. How will two observers see the motion of the sandbag? Observer 1 is in the hot air balloon and observer 2 is on the ground?
    (a) Both 1 and 2 will see it go down.
    (b) 1 will see it go down and 2 will see it go up.
    (c) 1 will see it go down and 2 will see it go up and then down.

13. An apple falls from a tree. It hits the ground at a speed of about 5.0 m/s. What is the approximate height of the tree?
    (a) 2.5 m   (b) 1.3 m   (c) 10.0 m   (d) 2.4 m

14. You have two small metal balls. You drop the first ball and throw the other one in the downward direction. Choose the statements that are not correct.
    (a) The second ball will spend less time in flight.
    (b) The first ball will have a slower final speed when it reaches the ground.
    (c) The second ball will have larger acceleration.
    (d) Both balls will have the same acceleration.

15. Which of the graphs in **Figure Q2.15** represent the motion of an object that starts from rest and then, after undergoing some motion, returns to its initial position? Multiple answers could be correct.

FIGURE Q2.15

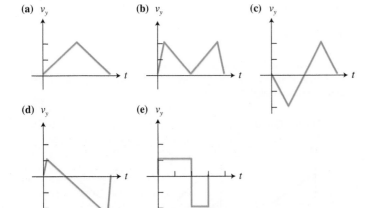

(a) v_y (b) v_y (c) v_y

(d) v_y (e) v_y

16. You throw a small ball upward and notice the time it takes to come back. If you then throw the same ball so that it takes twice as much time to come back, what is true about the motion of the ball the second time?
    (a) Its initial speed was twice the speed in the first experiment.
    (b) It traveled an upward distance that is twice the distance of the original toss.
    (c) It had twice as much acceleration on the way up as it did the first time.
    (d) The ball stopped at the highest point and had zero acceleration at that point.

Conceptual Questions

17. **Figure Q2.17** shows vectors $\vec{E}$, $\vec{F}$, and $\vec{G}$. Draw the following vectors using rules for graphical addition and subtraction:
    (a) $\vec{E} + \vec{F}$,   (b) $\vec{E} - \vec{F}$,
    (c) $\vec{E} + \vec{G}$,   (d) $-\vec{G} + \vec{F}$,
    (e) $\vec{E} + \vec{F} + \vec{G}$, and   (f) $\vec{E} - \vec{F} + \vec{G}$.

18. Peter is cycling along an 800-m straight stretch of a track. His speed is 13 m/s. Choose all of the graphical representations of motion from **Figure Q2.18** that correctly describe Peter's motion.

FIGURE Q2.17

FIGURE Q2.18

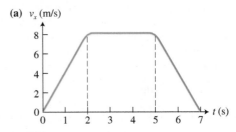

(a) x (b) x (c) x

(d) x (e) v (f) v

19. In what reasonable ways can you represent or describe the motion of a car traveling from one stoplight to the next? Construct each representation for the moving car.

20. What is the difference between speed and velocity? Between path length and distance? Between distance and displacement? Give an example of each.

21. What physical quantities do we use to describe motion? What does each quantity characterize? What are their SI units?

22. Devise stories describing each of the motions shown in each of the graphs in **Figure Q2.22**. Specify the object of reference.

FIGURE Q2.22

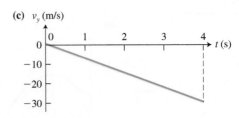

(a) v_x (m/s)

(b) v_y (m/s)

(c) v_y (m/s)

23. For each of the position-versus-time graphs in **Figure Q2.23**, draw
     velocity-versus-time graphs and acceleration-versus-time graphs.

FIGURE Q2.23

(a)

(b)

24. **Figure Q2.24** shows velocity-versus-
     time graphs for two objects, A and B.
     Draw motion diagrams that correspond
     to the motion of these two objects.

FIGURE Q2.24

25. Can an object have a nonzero velocity
     and zero acceleration? If so, give an
     example.
26. Can an object at one instant of time have
     zero velocity and nonzero acceleration?
     If so, give an example.
27. Your little sister has a battery-powered
     toy truck. When the truck is moving, how can you determine whether it
     has constant velocity, constant speed, constant acceleration, or changing
     acceleration? Explain in detail.
28. You throw a ball upward. Your friend says that at the top of its flight the
     ball has zero velocity and zero acceleration. Do you agree or disagree?
     If you agree, explain why. If you disagree, how would you convince your
     friend of your opinion?

Problems

Below, **BIO** indicates a problem with a biological or medical focus. Problems
labeled **EST** ask you to estimate the answer to a quantitative problem rather
than derive a specific answer. Asterisks indicate the level of difficulty of the
problem. Problems with no * are considered to be the least difficult. A single *
marks moderately difficult problems. Two ** indicate more difficult problems.

2.2 A conceptual description of motion

1. A car starts at rest from a stoplight and speeds up. It then moves at constant
    speed for a while. Then it slows down until reaching the next stoplight. Repre-
    sent the motion with a motion diagram as seen by the observer on the ground.
2. * You are an observer on the ground. (a) Draw two motion diagrams
    representing the motions of two runners moving at the same constant
    speeds in opposite directions toward you. Runner 1, coming from the east,
    reaches you in 5 s, and runner 2 reaches you in 3 s. (b) Draw a motion
    diagram for the second runner as seen by the first runner.
3. * A car is moving at constant speed on a highway. A second car catches up
    and passes the first car 5 s after it starts to speed up. Represent the situation
    with a motion diagram. Specify the observer with respect to whom you
    drew the diagram.
4. * A hat falls off a man's head and lands in the snow. Draw a motion
    diagram representing the motion of the hat as seen by the man.

2.3 Operations with vectors

5. **Figure P2.5** shows
    several displacement
    vectors that are all in
    the (x, y) plane. (a)
    List all displacement
    vectors with a scalar
    component equal to
    -2 units. (b) List all
    displacement vectors
    with a scalar compo-
    nent equal to 0 units.
    (c) List all displace-
    ment vectors that
    represent the distance
    traveled of 2 units.
    (d) Find three vectors
    that are related as
    $\vec{X} + \vec{Y} = \vec{Z}$ and three that are related as $\vec{X} - \vec{Y} = \vec{Z}$.

FIGURE P2.5

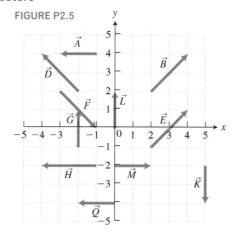

6. **Figure P2.6** shows an incomplete motion diagram for an object. (a) For
    each pair of adjacent velocities, draw a corresponding $\Delta\vec{v}$ vector. Use the
    grid to determine their lengths accurately. (b) Explain how to construct a
    $\Delta\vec{v}$ vector by using the rule for addition of vectors and how to do it using
    the rule for subtraction of vectors.

FIGURE P2.6

2.4 and 2.5 Quantities for describing motion and Repre-
senting motion with data tables and graphs

7. * You drive 100 km east, do some sightseeing, and then turn around and
    drive 50 km west, where you stop for lunch. (a) Represent your trip with a
    displacement vector. Choose an object of reference and coordinate axis so
    that the scalar component of this vector is (b) positive; (c) negative; (d) zero.
8. * Choose an object of reference and a set of coordinate axes associated with
    it. Show how two people can start and end their trips at different locations
    but still have the same displacement vectors in this reference frame.
9. The scalar x-component of a displacement vector for a trip is -70 km.
    Represent the trip using a coordinate axis and an object of reference. Then
    change the axis so that the displacement component becomes $+70$ km.
10. * You recorded your position with respect to the front door of your house
    as you walked to the mailbox. Examine the data presented in **Table 2.8** and
    answer the following questions: (a) What instruments might you have used
    to collect data? (b) Represent your motion using a position-versus-time
    graph. (c) Tell the story of your motion in words. (d) Show on the graph the
    displacement, distance, and path length.

TABLE 2.8

t (s)	1	2	3	4	5	6	7	8	9
x (steps)	2	4	9	13	18	20	16	11	9

2.6 Constant velocity linear motion

11. * You need to determine the time interval (in seconds) needed for light to pass an atomic nucleus. What information do you need? How will you use it? What simplifying assumptions about the objects and processes do you need to make? What approximately is that time interval?

12. A speedometer reads 65 mi/h. (a) Use as many different units as possible to represent the speed of the car. (b) If the speedometer reads 100 km/h, what is the car's speed in mi/h?

13. Convert the following record speeds so that they are in mi/h, km/h, and m/s. (a) Australian dragonfly—36 mi/h; (b) the diving peregrine falcon—349 km/h; and (c) the Lockheed SR-71 jet aircraft—980 m/s (about three times the speed of sound).

14. **EST** **Hair growth speed** Estimate the rate that your hair grows in meters per second. Indicate any assumptions you made.

15. * **EST** A kidnapped banker looking through a slit in a van window counts her heartbeats and observes that two highway exits pass in 80 heartbeats. She knows that the distance between the exits is 1.6 km (1 mile). (a) Estimate the van's speed. (b) Choose and describe a reference frame and draw a position-versus-time graph for the van.

16. * Some computer scanners scan documents by moving the scanner head with a constant speed across the document. A scanner can scan up to 30 pages per minute using an automatic paper feed tray. Estimate the maximum speed of the scanner head when scanning a standard letter sheet (21.59 cm × 27.94 cm). Indicate any assumptions that you made.

17. * **Equation Jeopardy** Two observers observe two different moving objects. However, they describe their motions mathematically with the same equation: $x(t) = 10 \text{ km} - (4 \text{ km/h})t$. (a) Write two short stories about these two motions. Specify where each observer is and what she is doing. What is happening to the moving object at $t = 0$? (b) Use significant digits to determine the interval within which the initial position is known.

18. * Your friend's pedometer shows that he took 17,000 steps in 2.50 h during a hike. Determine everything you can about the hike. What assumptions did you make? How certain are you in your answer? How would the answer change if the time were given as 2.5 h instead of 2.50 h?

19. During a hike, two friends were caught in a thunderstorm. Four seconds after seeing lightning from a distant cloud, they heard thunder. How far away was the cloud (in kilometers)? Write your answer as an interval using significant digits as your guide. Sound travels in air at about 340 m/s.

20. Light travels at a speed of 3.0×10^8 m/s in a vacuum. The approximate distance between Earth and the Sun is 150×10^6 km. How long does it take light to travel from the Sun to Earth? What are the margins within which you know the answer?

21. Proxima Centauri is 4.22 ± 0.01 light-years from Earth. Determine the length of 1 light-year and convert the distance to the star into meters. What is the uncertainty in the answer?

22. * Spaceships traveling to other planets in the solar system move at an average speed of 1.1×10^4 m/s. It took Voyager about 12 years to reach the orbit of Uranus. What can you learn about the solar system using these data? What assumption did you make? How did this assumption affect the results?

23. ** **Figure P2.23** shows a velocity-versus-time graph for the bicycle trips of two friends with respect to the parking lot where they started. (a) Determine their displacements in 20 s. (b) If Xena's position at time zero is 0 and Gabriele's position is 60 m, what time interval is needed for Xena to catch Gabriele? (c) Use the information from (b) to write the function $x(t)$ for Gabriele with respect to Xena.

24. * Table 2.9 shows position and time data for your walk along a straight path. (a) Tell everything you can about the walk. Specify the object of reference. (b) Draw a motion diagram, draw a graph $x(t)$, and write a function $x(t)$ that is consistent with the data and the chosen reference frame.

TABLE 2.9

Time (s)	Position (m)
0	80
10	40
20	0
30	−40
40	−80
50	−120

TABLE 2.10

Time (s)	Position (m)
0	−200
10	−120
20	−40
30	40
40	120
50	200

25. * Table 2.10 shows position and time data for your friend's bicycle ride along a straight bike path. (a) Tell everything you can about his ride. Specify the observer. (b) Draw a motion diagram, draw a graph $x(t)$, and write a function $x(t)$ that is consistent with the ride.

26. * You are walking to your physics class at speed 1.0 m/s with respect to the ground. Your friend leaves 2.0 min after you and is walking at speed 1.3 m/s in the same direction. How fast is she walking with respect to you? How far does your friend travel before she catches up with you? Indicate the uncertainty in your answers. Describe any assumptions that you made.

27. * Gabriele enters an east–west straight bike path at the 3.0-km mark and rides west at a constant speed of 8.0 m/s. At the same time, Xena rides east from the 1.0-km mark at a constant speed of 6.0 m/s. (a) Write functions $x(t)$ that describe their positions as a function of time with respect to Earth. (b) Where do they meet each other? In how many different ways can you solve this problem? (c) Write a function $x(t)$ that describes Xena's motion with respect to Gabriele.

28. * Jim is driving his car at 32 m/s (72 mi/h) along a highway where the speed limit is 25 m/s (55 mi/h). A highway patrol car observes him pass and quickly reaches a speed of 36 m/s. At that point, Jim is 300 m ahead of the patrol car. How far does the patrol car travel before catching Jim?

29. * You hike two-thirds of the way to the top of a hill at a speed of 3.0 mi/h and run the final third at a speed of 6.0 mi/h. What was your average speed?

30. * Olympic champion swimmer Michael Phelps swam at an average speed of 2.01 m/s during the first half of the time needed to complete a race. What was his average swimming speed during the second half of the race if he tied the record, which was at an average speed of 2.05 m/s?

31. * A car makes a 100-km trip. It travels the first 50 km at an average speed of 50 km/h. How fast must it travel the second 50 km so that its average speed is 100 km/h?

32. * Jane and Bob see each other when 100 m apart. They are moving toward each other, Jane at constant speed 4.0 m/s and Bob at constant speed 3.0 m/s with respect to the ground. What can you determine about this situation using these data?

33. * The graph in **Figure P2.33** represents four different motions. (a) Write a function $x(t)$ for each motion. (b) Use the information in the graph to determine as many quantities related to the motion of these objects as possible. (c) Act out these motions with two friends. (Hint: Think of what each object was doing at $t = 0$.)

FIGURE P2.33

FIGURE P2.23

2.7 and 2.8 Motion at constant acceleration and Displacement of an object moving at constant acceleration

34. A car starts from rest and reaches the speed of 10 m/s in 30 s. What can you determine about the motion of the car using this information?

35. A truck is traveling east at +16 m/s. (a) The driver sees that the road is empty and accelerates at +1.0 m/s^2 for 5.0 s. What can you determine about the truck's motion using these data? (b) The driver then sees a red light ahead and decelerates at -2.0 m/s^2 for 3.0 s. What can you determine about the truck's motion using these data? (c) Determine the values of the quantities you listed in (a) and (b).

36. **Bumper car collision** On a bumper car ride, friends smash their cars into each other (head-on), and each has a speed *change* of 3.2 m/s. If the magnitudes of acceleration of each car during the collision averaged 28 m/s^2, determine the time interval needed to stop and the stopping distance for each car while colliding. Specify your reference frame.

37. A bus leaves an intersection accelerating at +2.0 m/s^2. Where is the bus after 5.0 s? What assumption did you make? If this assumption is not valid, would the bus be closer or farther away from the intersection compared to your original answer? Explain.

38. A jogger is running at +4.0 m/s when a bus passes her. The bus is accelerating from +16.0 m/s to +20.0 m/s in 8.0 s. The jogger speeds up with the same acceleration. What can you determine about the jogger's motion using these data?

39. * The motion of a person as seen by another person is described by the equation $v = -3.0$ m/s $+ (0.5$ m/s$^2)t$. (a) Represent this motion with a motion diagram and position-, velocity-, and acceleration-versus-time graphs. (b) Say everything you can about this motion and describe what happens to the person when his speed becomes zero.

40. While cycling at a speed of 10 m/s, a cyclist starts going downhill with an acceleration of magnitude 1.2 m/s^2. The descent takes 10.0 s. What can you determine about the cyclist's motion using these data? What assumptions did you make?

41. * **EST** To his surprise, Daniel found that an egg did not break when he accidentally dropped it from a height of 0.40 m onto his floor, covered with 2.0-cm-thick carpet. Estimate the minimum acceleration of the egg while it was slowing down after touching the carpet. Indicate any assumptions you made.

FIGURE P2.41

42. **BIO** **Squid propulsion** *Lolliguncula brevis* squid use a form of jet propulsion to swim—they eject water out of jets that can point in different directions, allowing them to change direction quickly. When swimming at a speed of 0.15 m/s or greater, they can accelerate at 1.2 m/s^2. (a) Determine the time interval needed for a squid to increase its speed from 0.15 m/s to 0.45 m/s. (b) What other questions can you answer using the data?

43. **Dragster record on the desert** In 1977, Kitty O'Neil drove a hydrogen peroxide–powered rocket dragster for a record time interval (3.22 s) and final speed (663 km/h) on a 402-m-long Mojave Desert track. Determine her average acceleration during the race and the acceleration while stopping (it took about 20 s to stop). What assumptions did you make?

44. * Imagine that a sprinter accelerates from rest to a maximum speed of 10.8 m/s in 1.8 s. In what time interval will he finish the 100-m race if he keeps his speed constant at 10.8 m/s for the last part of the race? What assumptions did you make?

45. ** Two runners are running next to each other when one decides to speed up at constant acceleration a. The second runner notices the acceleration after a short time interval Δt when the distance between the runners is d. The second runner accelerates at the same acceleration. Represent their motions with a motion diagram and position-versus-time graph (both graph lines on the same set of axes). Use any of the representations to predict what will happen to the distance between the runners—will it stay d, increase, or decrease? Assume that the runners continue to have the same acceleration for the duration of the problem.

46. * **Meteorite hits car** In 1992, a 14-kg meteorite struck a car in Peekskill, NY, leaving a 20-cm-deep dent in the trunk. (a) If the meteorite was moving at 500 m/s before striking the car, what was the magnitude of its acceleration while stopping? Indicate any assumptions you made. (b) What other questions can you answer using the data in the problem?

47. **BIO** **Froghopper jump** A spittlebug called the froghopper (*Philaenus spumarius*) is believed to be the best jumper in the animal world. It pushes off with muscular rear legs for 0.0010 s, reaching a speed of 4.0 m/s. Determine its acceleration during this launch and the distance that the froghopper moves while its legs are pushing.

48. **Tennis serve** The fastest server in women's tennis is Sabine Lisicki, who recorded a serve of 131 mi/h (211 km/h) in 2014. If her racket pushed on the ball for a distance of 0.10 m, what was the average acceleration of the ball during her serve? What was the time interval for the racket-ball contact?

49. * **Shot from a cannon** In 1998, David "Cannonball" Smith set the distance record for being shot from a cannon (56.64 m). During a launch in the cannon's barrel, his speed increased from zero to 80 km/h in 0.40 s. While he was being stopped by the catching net, his speed decreased from 80 km/h to zero with an average acceleration of 180 m/s^2. What can you determine about Smith's flight using this information?

50. **Col. John Stapp's final sled run** Col. John Stapp led the U.S. Air Force Aero Medical Laboratory's research into the effects of higher accelerations. On Stapp's final sled run, the sled reached a speed of 282.5 m/s (632 mi/h) and then stopped with the aid of water brakes in 1.4 s. Stapp was barely conscious and lost his vision for several days but recovered. Determine his acceleration while stopping and the distance he traveled while stopping.

51. * Sprinter Usain Bolt reached a maximum speed of 11.2 m/s in 2.0 s while running the 100-m dash. (a) What was his acceleration? (b) What distance did he travel during this first 2.0 s of the race? (c) What assumptions did you make? (d) What time interval was needed to complete the race, assuming that he ran the last part of the race at his maximum speed? (e) What is the total time for the race? How certain are you of the number you calculated?

52. ** Imagine that Usain Bolt can reach his maximum speed in 1.7 s. What should be his maximum speed in order to tie the 19.19-s record for the 200-m dash?

53. * A bus is moving at a speed of 36 km/h. How far from a bus stop should the bus start to slow down so that the passengers feel comfortable (a comfortable acceleration is 1.2 m/s^2)?

54. * **EST** You want to estimate how fast your car accelerates. What information can you collect to answer this question? What assumptions do you need to make to do the calculation using the information?

55. * In your car, you covered 2.0 m during the first 1.0 s, 4.0 m during the second 1.0 s, 6.0 m during the third 1.0 s, and so forth. Was this motion at constant acceleration? Explain.

56. (a) Determine the acceleration of a car in which the velocity changes from -10 m/s to -20 m/s in 4.0 s. (b) Determine the car's acceleration if its velocity changes from -20 m/s to -18 m/s in 2.0 s. (c) Explain why the sign of the acceleration is different in (a) and (b).

57. You accidentally drop an eraser out the window of an apartment 15 m above the ground. (a) How long will it take for the eraser to reach the ground? (b) What speed will it have just before it reaches the ground? (c) If you multiply the time interval answer from (a) and the speed answer from (b), why is the result much more than 15 m?

58. * What is the average speed of the eraser in the previous problem from the instant it is released to the instant it reaches the ground?

59. You throw a tennis ball straight upward. The initial speed is about 12 m/s. Say everything you can about the motion of the ball. Is 12 m/s a realistic speed for an object that you can throw with your hands?

60. While skydiving, your parachute opens and you slow from 50.0 m/s to 8.0 m/s in 0.80 s. Determine the distance you fall while the parachute is opening. Some people faint if they experience acceleration greater than 5g (5 times 9.8 m/s^2). Will you feel faint? Explain and discuss simplifying assumptions inherent in your explanation.

61. * After landing from your skydiving experience, you are so excited that you throw your helmet upward. The helmet rises 5.0 m above your hands. What was the initial speed of the helmet when it left your hands? How long was it moving from the time it left your hands until it returned?

62. * You are standing on the rim of a canyon. You drop a rock and in 7.0 s hear the sound of it hitting the bottom. How deep is the canyon? What assumptions did you make? Examine how each assumption affects the answer. Does it lead to a larger or smaller depth than the calculated depth? (The speed of sound in air is about 340 m/s.)

63. * You are doing an experiment to determine your reaction time. Your friend holds a ruler. You place your fingers near the sides of the lower part of the ruler without touching it. The friend drops the ruler without warning you. You catch the ruler after it falls 12.0 cm. What was your reaction time?

64. **EST** **Cliff divers** Divers in Acapulco fall 36 m from a cliff into the water. Estimate their speed when they enter the water and the time interval needed to reach the water. What assumption did you make? Does this assumption make the calculated speed larger or smaller than actual speed?

65. * Galileo dropped a light rock and a heavy rock from the Leaning Tower of Pisa, which is about 55 m high. Suppose that Galileo dropped one rock 0.50 s before the second rock. With what initial velocity should he drop the second rock so that it reaches the ground at the same time as the first rock?

66. * A person holding a lunch bag is moving upward in a hot air balloon at a constant speed of 7.0 m/s. When the balloon is 24 m above the ground, she accidentally releases the bag. What is the speed of the bag just before it reaches the ground?

67. * A parachutist falling vertically at a constant speed of 10 m/s drops a penknife when 20 m above the ground. What is the speed of the knife just before it reaches the ground?

2.9 Skills for analyzing situations involving motion

68. A diagram representing the motion of two cars is shown in **Figure P2.68**. The number near each dot indicates the clock reading in seconds when the car passes that location. (a) Indicate times when the cars have the same speed. (b) Indicate times when they have the same position.

FIGURE P2.68

Car 1

Car 2

69. Use the velocity-versus-time graph lines in **Figure P2.69** to determine the change in the position of each car from 0 s to 60 s. Represent the motion of each car mathematically as a function $x(t)$. Their initial positions are A (200 m) and B (−200 m).

FIGURE P2.69

70. * An object moves so that its position changes in the following way: $x = 10$ m − $(4$ m/s$)t$. (a) Describe all of the known quantities for this motion. (b) Invent a story for the motion. (c) Draw a position-versus-time graph, and use the graph to determine when the object reaches the origin of the reference frame. (e) Act out the motion.

71. * While babysitting their younger brother, Chrisso and Devin are playing with toys. They notice that the squishy Piglet slows down in a repeatable way when they push it along the smooth wooden floor. They propose a hypothesis that the toy slows down

with a constant acceleration, which does not depend on the toy's initial velocity. For each of five different initial speeds, they measure the distance traveled by the toy from the time they stop pushing it to the time the toy stops moving, and they measure the corresponding time interval. Their data are presented below. Do the data support their hypothesis? Explain. If yes, determine the average acceleration of Piglet and the maximum speed with which Chrisso and Devin push Piglet.

Experiment #	Distance (m)	Time (s)
1	0.96	0.65
2	2.84	1.12
3	1.72	0.87
4	2.53	1.05
5	0.62	0.53

72. ** An object moves so that its position changes in the following way: $x(t) = -100$ m $+ (30$ m/s$)t + (3.0$ m/s$^2)t^2$. (a) What kind of motion is this (constant velocity, constant acceleration, or changing acceleration)? (b) Describe all of the known quantities for this motion. (c) Invent a story for the motion. (d) Draw a velocity-versus-time graph, and use it to determine when the object stops. (e) Use equations to determine when and where it stops. Did you get the same answer using graphs and equations?

73. * The positions of objects A and B with respect to Earth depend on time as follows: $x(t)_A = 10.0$ m $- (4.0$ m/s$)t$; $x(t)_B = -12$ m $+ (6$ m/s$)t$. Represent their motions on a motion diagram and graphically (position-versus-time and velocity-versus-time graphs). Use the graphical representations to find where and when they will meet. Confirm the result with mathematics.

74. * Two cars on a straight road at time zero are beside each other. The first car, traveling at speed 30 m/s, is passing the second car, which is traveling at 24 m/s. Seeing a cow on the road ahead, the driver of each car starts to slow down at 6.0 m/s^2. Represent the motions of the cars mathematically and on a velocity-versus-time graph from the point of view of a pedestrian. Where is each car when it stops?

75. * Oliver drops a tennis ball from a certain height above a concrete floor. **Figure P2.75** shows the velocity-versus-time graph of the ball's motion from the moment the ball is released to the moment the ball reaches its maximum height after bouncing up from the floor. (a) How is the y-axis directed: up or down? (b) Determine the initial height from which the ball is released and the final height to which the ball bounces. (c) Determine the average speed of the ball during the downward motion.

FIGURE P2.75

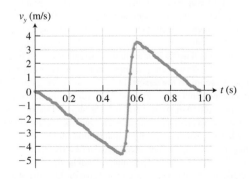

76. * **BIO** **EST** **Water striders**
Water striders are insects
that propel themselves on the
surface of ponds by creating
vortices in the water shed by
their driving legs. The velocity-versus-time graph of a
17-mm-long water strider
that moved in a straight line
was created from a video
(**Figure P2.76**). The insect
started from rest, sped up by
taking two strides, and then

slowed down until it stopped. Estimate (a) the maximum speed (in m/s),
(b) the maximum acceleration (in m/s^2), and (c) the total displacement
(in m) of the water strider. Note that the velocity on the graph is given in
units of length of water strider body per second.

FIGURE P2.76

77. You are traveling in your car at 20 m/s a distance of 20 m behind a car
traveling at the same speed. The driver of the other car slams on the brakes
to stop for a pedestrian who is crossing the street. Will you hit the car?
Your reaction time is 0.60 s. The maximum acceleration of each car is
9.0 m/s^2.

78. * You are driving a car behind another car. Both cars are moving at speed
80 km/h. What minimum distance behind the car in front should you drive
so that you do not crash into the car's rear end if the driver of that car slams
on the brakes? Indicate any assumptions you made.

79. * A driver with a 0.80-s reaction time applies the brakes, causing the car to
have 7.0-m/s^2 acceleration opposite the direction of motion. If the car is
initially traveling at 21 m/s, how far does the car travel during the reaction
time? How far does the car travel after the brakes are applied and while
skidding to a stop?

80. ** Some people in a hotel are dropping water balloons from their open
window onto the ground below. The balloons take 0.15 s to pass your
1.6-m-tall window. Where should security look for the raucous hotel
guests? Indicate any assumptions that you made in your solution.

81. ** **BIO** **EST** **Avoiding injury from hockey puck** Hockey players wear
protective helmets with facemasks. Why? Because the bone in the upper
part of the cheek (the zygomatic bone) can fracture if the acceleration
of a hockey puck due to its interaction with the bone exceeds 900g for
a time lasting 6.0 ms or longer. Suppose a player was not wearing a
facemask. Is it likely that the acceleration of a hockey puck when hitting
the bone would exceed these numbers? Use some reasonable numbers of
your choice and estimate the puck's acceleration if hitting an unprotected
zygomatic bone.

82. ** **EST** A bottle rocket burns for 1.6 s. After it stops burning, it continues
moving up to a maximum height of 80 m above the place where it stopped
burning. Estimate the acceleration of the rocket during launch. Indicate any
assumptions made during your solution. Examine their effect.

83. * **Data from state driver's manual** The state driver's manual lists the reaction
distances, braking distances, and total stopping distances for automobiles
traveling at different initial speeds (Table 2.11). Use the data to determine the
driver's reaction time interval and the acceleration of the automobile while
braking. The numbers assume dry surfaces for passenger vehicles.

TABLE 2.11 Data from driver's manual

Speed (mi/h)	Reaction distance (m)	Braking distance (m)	Total stopping distance (m)
20	7	7	14
40	13	32	45
60	20	91	111

84. ** **EST** Estimate the time interval needed to pass a semi-trailer truck on
a highway. If you are on a two-lane highway, how far away from you must
an approaching car be in order for you to safely pass the truck without
colliding with the oncoming traffic? Indicate any assumptions used in your
estimate.

85. * Car A is heading east at 30 m/s and Car B is heading west at 20 m/s.
Suddenly, as they approach each other, they see a one-way bridge ahead.
They are 100 m apart when they each apply the brakes. Car A's speed
decreases at 7.0 m/s each second and car B decreases at 9.0 m/s each
second. Do the cars collide? Provide two solutions: one using equations
and one using graphs.

Reading Passage Problems

BIO **Head injuries in sports** A research group at Dartmouth College has developed a Head Impact Telemetry (HIT) System that can be used to collect data
about head accelerations during impacts on the playing field. The researchers
observed 249,613 impacts from 423 football players at nine colleges and high
schools and collected collision data from participants in other sports. The accelerations during most head impacts ($>89\%$) in helmeted sports caused head
accelerations less than a magnitude of 400 m/s^2. However, a total of 11 concussions were diagnosed in players whose impacts caused accelerations between
600 and 1800 m/s^2, with most of the 11 over 1000 m/s^2.

86. Suppose that the magnitude of the head velocity change was 10 m/s.
Which time interval for the collision would be closest to producing a possible concussion with an acceleration of 1000 m/s^2?
(a) 1 s   (b) 0.1 s   (c) 10^{-2} s
(d) 10^{-3} s   (e) 10^{-4} s

87. Using numbers from the previous problem, which answer is closest to the
average speed of the head while stopping?
(a) 50 m/s   (b) 10 m/s   (c) 5 m/s
(d) 0.5 m/s   (e) 0.1 m/s

88. Suppose the average speed while stopping was 4 m/s (not necessarily the
correct value) and the collision lasted 0.01 s. Which answer is closest to the
head's stopping distance (the distance it moves while stopping)?
(a) 0.04 m   (b) 0.4 m   (c) 4 m
(d) 0.02 m   (e) 0.004 m

89. Use Eq. (2.7) and the numbers from Problem 86 to determine which
stopping distance is closest to that which would lead to a 1000 m/s^2 head
acceleration.
(a) 0.005 m   (b) 0.5 m   (c) 0.1 m
(d) 0.01 m   (e) 0.05 m

90. Choose the changes in the head impacts that would *reduce* the acceleration
during the impact.
1. A shorter impact time interval
2. A longer impact time interval
3. A shorter stopping distance
4. A longer stopping distance
5. A smaller initial speed
6. A larger initial speed
(a) 1, 4, 6   (b) 1, 3, 5   (c) 1, 4, 5
(d) 2, 4, 5   (e) 2, 4, 6

Automatic sliding doors   The first automatic sliding doors were described by Hero of Alexandria almost 2000 years ago. The doors were moved by hanging containers that were filled with water. Modern sliding doors open or close automatically. They are equipped with sensors that detect the proximity of a person and an electronic circuit that processes the signals from the sensors and drives the electromotor-based system that moves the doors. The sensors typically emit pulses of infrared light or ultrasound and detect the reflected pulses. By measuring the delay between emitted and received pulses, the system can determine the distance to the object from which the pulse was reflected. The whole system must be carefully designed to ensure safe and accurate functioning. Designers of such doors take into account several variables such as typical walking speeds of people and their dimensions.

Let's try to learn more about automatic sliding doors by analyzing the motion of a single-side automatic sliding door when a person is walking through the door. Figure 2.30 shows the position-versus-time graph of the motion of the edge of the door (marked with a red cross in the photo) from the moment the door starts opening to when the door is closed while a person walks toward and through the door. The doors are adjusted to start opening when a person is 2.0 m away.

FIGURE 2.30

91. How long does it take for the door to fully open?
    (a) 1.5 s   (b) 3 s   (c) 5.5 s   (d) 11 s
92. How long does it take for the door to close after it is opened?
    (a) 11 s   (b) 8 s   (c) 5.5 s   (d) 3 s
93. A person is walking at constant speed of 1.15 m/s toward and through the sliding door. How far from the door is the person when the door starts closing?
    (a) 2.3 m   (b) 4.3 m   (c) 6.3 m   (d) 8.3 m
94. What is the average opening speed of the door?
    (a) 0.1 m/s   (b) 0.3 m/s   (c) 0.6 m/s   (d) 3.0 m/s
95. What is the maximum speed of the door?
    (a) 0.1 m/s   (b) 0.3 m/s   (c) 0.6 m/s   (d) 3.0 m/s
96. A 50-cm-wide person is walking toward the door. What is the maximum walking speed of the person that will allow her to pass through the door without hitting it (assume the person aims for the opening)?
    (a) 0.6 m/s   (b) 1.2 m/s   (c) 1.7 m/s   (d) 2.5 m/s

Newtonian Mechanics

Seat belts and air bags save about 250,000 lives worldwide every year because they significantly reduce the risk of injury (belts alone by about 40% and belts with air bags by about 54%). How do seat belts and air bags provide this protection?

IN THE LAST CHAPTER, we learned to *describe* motion—for example, to determine a car's acceleration when it stops abruptly during a collision. However, we did not discuss the causes of the acceleration. In this chapter, we will learn *why* an object has a particular acceleration. This knowledge will help us *explain* the motion of many objects: cars, car passengers, elevators, skydivers, and even rockets.

- How do seat belts and air bags save lives?

- If you stand on a bathroom scale in a moving elevator, does its reading change?

- Can a parachutist survive a fall if the parachute does not open?

BE SURE YOU KNOW HOW TO:

- Draw a motion diagram for a moving object (Section 2.2).

- Determine the direction of acceleration using a motion diagram (Section 2.7).

- Add vectors graphically and determine their components (Sections 2.3, 2.4, and 2.6).

3.1 Describing and representing interactions

What causes objects to accelerate or maintain a constant velocity? Consider a simple experiment—standing on Rollerblades on a horizontal floor. No matter how hard you swing your arms or legs you cannot start moving by yourself; you need to either push off the floor or have someone push or pull you. Physicists say that the floor or the other person *interacts* with you, thus changing your motion. Objects can interact directly, when they touch each other, or at a distance, as when a magnet attracts or repels another magnet without touching it.

Choosing a system in a sketch of a process

FIGURE 3.1 A sketch of a car skidding to avoid a collision with a van.

We learned (in Chapter 2) that the first step in analyzing any process is sketching it. **Figure 3.1** shows a sketch of a car skidding to avoid a collision with a van. In this and later chapters we choose one particular object for detailed analysis. We call this object the **system**. All other objects that are not part of the system can interact with it (touch it, pull it, or push it) and are in the system's **environment**. Interactions between the system and objects in the environment are called **external interactions**.

> **System** A system is the object or group of objects that we choose to analyze. Everything outside that system is called its environment and consists of objects that might interact with the system (touch, push, or pull it) and affect its motion through external interactions.

FIGURE 3.2 Representing external interactions (forces exerted on a system).

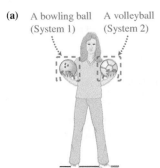

(a) A bowling ball (System 1)   A volleyball (System 2)

On the sketch, we draw a boundary around the system to emphasize the system choice (shown by the red dashed line around the skidding car in Figure 3.1). Sometimes, a single system object has parts—like the wheels on the car and its axles. The parts interact with each other. Since both parts are in the system, these are called **internal interactions**. In this chapter we will model an object like a car as a point-like object and ignore such internal interactions.

Representing interactions

External interactions affect the motion or lack of motion of a system. Consider holding a bowling ball in one hand and a volleyball in the other. Each ball is considered as a system (**Figure 3.2a**). What objects interact with each ball? Your hand pushes up hard to keep the bowling ball steady and much less hard to keep the volleyball steady (Figure 3.2b). We use an arrow to represent the upward push exerted by each hand on one of the balls. Notice that the arrow for the interaction of the hand with the bowling ball is longer than that for the hand with the volleyball.

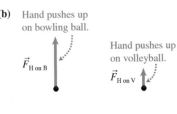

(b) Hand pushes up on bowling ball.

$\vec{F}_{\text{H on B}}$

Hand pushes up on volleyball.

$\vec{F}_{\text{H on V}}$

The arrow represents a "force" that is exerted by the hand on the ball. **A force is a physical quantity that characterizes how hard and in what direction an external object pushes or pulls on the system.** The symbol for force is $\vec{F}$ with a subscript that identifies both the external object that exerts the force and the system on which the force is exerted. For example, the hand pushing on the bowling ball is represented as $\vec{F}_{\text{H on B}}$. The force that the hand exerts on the volleyball is $\vec{F}_{\text{H on V}}$. The arrow above the symbol indicates that force is a vector quantity with a magnitude *and* direction. The SI unit of a force is the *newton* (N). When you hold a 100-g ball, you exert an upward force that is a little less than 1.0 N.

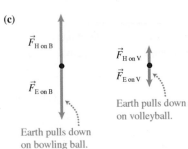

(c)

$\vec{F}_{\text{H on B}}$

$\vec{F}_{\text{E on B}}$

$\vec{F}_{\text{H on V}}$

$\vec{F}_{\text{E on V}}$

Earth pulls down on volleyball.

Earth pulls down on bowling ball.

Do any other objects exert forces on the balls? Intuitively, we know that something must be pulling down to balance the upward force your hands exert on the balls. The concept of **gravity** represents the interaction of planet Earth with the ball. Earth pulls downward on an object toward Earth's center. Because of this interaction we need to include a second arrow representing the force that Earth exerts on the ball $\vec{F}_{\text{E on B}}$ (see Figure 3.2c). Intuitively, we know that the two arrows for each ball should be of the same length, since the balls are not moving.

Do other objects besides the hands and Earth interact with the balls? Air surrounds everything close to Earth. Does it push down or up on the balls? Let's hypothesize that air pushes down. If so, then our hands have to push up harder on the balls to balance the combined effect of the downward push of the air and the downward pull of Earth. Let's test the hypothesis that air pushes down on the balls.

Testing a hypothesis

To test a hypothesis in science means to first accept it as a true statement (even if we disagree with it); then design an experiment whose outcome we can predict using this hypothesis (a testing experiment); then compare the outcome of the experiment and the prediction; and, finally, make a preliminary judgment about the hypothesis. If the outcome matches the prediction, we can say that the hypothesis has not been disproved by the experiment. When this happens, our confidence in the hypothesis increases. If the outcome and prediction are inconsistent, we need to reconsider the hypothesis and possibly reject it.

To test the hypothesis that air exerts a downward force on objects, we attach an empty closed water bottle to a spring and let it hang; the spring stretches (**Figure 3.3a**). Next we place the bottle and spring inside a large jar that is connected to a vacuum pump and pump the air out of the inside of the jar. We predict that *if* the air inside the jar pushes down on the bottle (the hypothesis), *then* when we pump the air out of the jar, it should be easier to support the bottle—the spring should stretch less (the prediction that follows from the hypothesis).

When we do the experiment, the outcome does not match the prediction— the spring actually stretches *slightly more* when the air is pumped out of the jar (Figure 3.3b). Evidently the air does not push down on the bottle; instead, it helps support the bottle by exerting an upward force on it. This outcome is surprising. If you study fluids, you will learn the mechanism by which air pushes up on objects.

Reflect

Let's reflect on what we have done here. We formulated an initial hypothesis—air *pushes down* on objects. Then we designed an experiment whose outcome we could predict using the hypothesis—the bottle on a spring in a vacuum jar. We used the hypothesis to make a prediction of the outcome of the testing experiment—the spring should stretch less in a vacuum. We then performed the experiment and found that something completely different happened. We revised our hypothesis—air *pushes up* slightly on objects. Note that air's upward push on the bottle is very small. Therefore, in many situations, and in all situations in this chapter, the effects of air can be ignored.

Drawing force diagrams

A **force diagram** (sometimes called a *free-body diagram*) represents the forces that objects in a system's environment exert on it (see Figure 3.2c). We represent the system by a dot to show that we model it as a point-like object. Arrows represent the forces. Unlike a motion diagram, a force diagram does not show us how a process changes with time; it shows us only the forces at a single instant. For processes in which no motion occurs, this makes no difference. But when motion does occur, we need to know if the force diagram is changing as the system moves.

Consider a rock dropped from above and sinking into sand, making a small crater. We construct a force diagram for the instant shortly after the rock touches the sand but before it completely stops moving.

FIGURE 3.3 A testing experiment to determine the effect of air on the bottle.

(a)

We attach a bottle to a spring. The spring stretches.

(b)

When the air is removed, the spring stretches farther. The air must have been exerting a small upward force on the bottle.

Vacuum pump

Constructing a force diagram

| 1. Sketch the situation (a rock sinking into sand). | 2. Circle the system (the rock). | 3. Identify external interactions:
• The sand pushes up on the rock.
• Earth pulls down on the rock. | 4. Place a dot at the side of the sketch, representing the system. |

5. Draw force arrows to represent the external interactions.

6. Label the forces with a subscript with two elements.

$\vec{F}_{\text{S on R}}$

$\vec{F}_{\text{E on R}}$

TIP Remember that on the force diagram, you only draw forces exerted *on* the system. Do not draw forces that the system exerts on other objects! For example, the rock exerts a force on the sand, but we do not include this force in the force diagram since the sand is not part of the system.

Notice that the upward-pointing arrow representing the force exerted by the sand on the rock is longer than the downward-pointing arrow representing the force exerted by Earth on the rock. The difference in lengths reflects the difference in the magnitudes of the forces.

Force diagram for a book

Book A sits on a table with book B on top of it. Construct a force diagram for book A.

Sketch and translate We sketch the situation below. We choose book A as the system. Notice that the dashed line around book A passes between the table and book A, and between book B and book A. It's important to be precise in the way you draw this line so that the separation between the system and the environment is clear. In this example, Earth, the table, and book B are external environmental objects that exert forces on book A.

Book B, table, and Earth are in the environment.

Earth

Simplify and diagram Draw a force diagram for book A, which is represented by a dot. Two objects in the environment touch book A. The table pushes up on the bottom surface of the book, exerting a force $\vec{F}_{\text{T on A}}$, and book B pushes down on the top surface of book A, exerting

force $\vec{F}_{\text{B on A}}$. In addition, Earth exerts a downward force on book A $\vec{F}_{\text{E on A}}$. Intuitively, we feel that these forces should somehow cancel each other. The diagram shows this by carefully representing the length of the forces. Note that if you add the two downward forces, their combined length is the same as the upward force. Below we show how to find a resultant force vector when several forces are exerted on an object.

$\vec{F}_{\text{T on A}}$

$\vec{F}_{\text{E on A}}$ $\vec{F}_{\text{B on A}}$

Try it yourself Construct a force diagram for book A after another book C is placed on top of book B.

Answer

The same three objects interact with book A. Earth exerts the same downward force on book A ($\vec{F}_{\text{E on A}}$). C does not directly touch A and exerts no force on A. However, C does push down on B, so B exerts a greater force on A ($\vec{F}_{\text{B on A}}$). Because the downward force of B on A is greater, the table exerts a greater upward force on book A ($\vec{F}_{\text{T on A}}$).

$\vec{F}_{\text{B on A}}$

$\vec{F}_{\text{E on A}}$

$\vec{F}_{\text{T on A}}$

Contact and noncontact forces

You can see in the above Conceptual Exercise that some of the forces exerted on book A are exerted by the objects directly touching the book—the table and book B. But Earth does not need to touch the book to exert a force on it. The force that Earth as a whole planet exerts on every object is an example of a gravitational force. Forces that require the interacting objects to be touching are called **contact forces**, while forces that do not require touching are called **noncontact forces**. The force a rope pulling on a crate exerts (sometimes called a *tension force*) is another example of a contact force; the force of a magnet on another magnet (a magnetic force) is another example of a noncontact force.

REVIEW QUESTION 3.1 How do we determine how many forces to draw on a force diagram?

3.2 Adding and measuring forces

Usually, more than one environmental object exerts a force on a system. How can we add them to find the total force? In this chapter we will restrict our attention to forces that are exerted and point along one axis. Consider the process of lifting a suitcase.

Adding force vectors graphically

You lift a suitcase straight up (**Figure 3.4a**). Earth pulls down on the suitcase, exerting a force of magnitude $F_{\text{E on S}} = 100$ N, and you exert an upward force of magnitude $F_{\text{Y on S}} = 150$ N (Figure 3.4b). What are the magnitude and direction of the total force exerted on the suitcase? The net effect of the two forces exerted along a vertical axis is the same as a 50-N force pointed straight up. Why? Remember that force is a vector. (We discussed the rules for vector addition in Chapter 2.) To add two vectors to find $\Sigma\vec{F} = \vec{F}_{\text{E on S}} + \vec{F}_{\text{Y on S}}$, we place them head to tail (see Figure 3.4c) and draw the vector that goes from the tail of the first vector to the head of the second vector. This new vector is the sum vector, or the resultant vector. In the case of forces it is called the **net force**.

To simplify operations with vectors, we will use components in our calculations. In this chapter we will only analyze situations when all forces are exerted along one axis (x or y) or when forces along one direction add to zero and all the remaining forces are along the other (perpendicular) direction. To find the component of a force along a particular direction, we first need to identify the positive direction, then compare the direction of the force with the identified positive direction. Forces that point in the positive direction of an axis have components equal to their magnitudes; forces that point in the negative direction have components equal to their magnitudes with a negative sign (see Sections 2.4 and 2.6 to review components of a vector). In our example of lifting a suitcase, the y-axis points up. The y-component of the force that you exert on the suitcase is positive and equal to $F_{\text{Y on S } y} = 150$ N, and the y-component of the force that Earth exerts on the suitcase is $F_{\text{E on S } y} = -100$ N. The sum of the y-components is $\Sigma F_y = 150$ N $+ (-100) = 50$ N. This is the y-component of the sum of the forces if the y-axis points up. If the positive direction is down, the sum of the forces has a y-component $\Sigma F_y = -50$ N.

If several external objects in the environment exert forces on the system, we still use vector addition to find the *sum* of the forces exerted on the system:

$$\Sigma\vec{F}_{\text{on S}} = \vec{F}_{1 \text{ on S}} + \vec{F}_{2 \text{ on S}} + \cdots + \vec{F}_{n \text{ on S}} \tag{3.1}$$

Measuring force magnitudes

The simplest device to measure forces is a spring scale (**Figure 3.5a**, on the next page); we will demonstrate this with a simple experiment. Imagine that a light plastic bag hangs from a light spring. The spring is not stretched. We then place one golf ball into the bag and observe that the spring stretches to a new length. We add a second ball and

FIGURE 3.4 The sum of the forces (the net force) exerted on the suitcase.

(a)

Sketch the situation and choose a system.

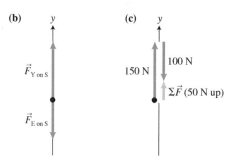

(b)

Draw a force diagram for the system showing the external forces exerted on the system.

(c)

Place the arrows head to tail. The sum of the forces goes from the tail of the first arrow to the head of the last arrow.

FIGURE 3.5 Using a spring scale to measure forces.

(a)

(b)

(c)

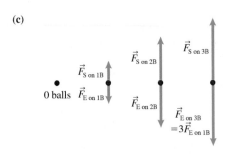

observe that the spring stretches twice as far. We add a third ball and observe that the spring stretches three times as far. Figure 3.5b shows the four situations; for each one, notice the change in the length of the spring. You can see that the spring stretches the same amount for each ball, and that it stretches three times more for three balls than for one ball.

Next, we choose the bag with the golf balls as the system and analyze the forces exerted in each case. We assume that Earth exerts the same force on each ball $\vec{F}_{\text{E on 1B}}$ that is independent of the presence of other balls. Thus, the total force exerted by Earth on the three-ball system is three times greater than the force exerted on the one-ball system. Figure 3.5c shows force diagrams for each case. Because the bag does not fall, the spring in each case must exert a force on the system $\vec{F}_{\text{S on \#B}}$ that is equal in magnitude and opposite in direction to the force that Earth exerts on the system $\vec{F}_{\text{E on \#B}}$ so that the sum of the forces exerted on the system with a number (#) of golf balls is zero.

This experiment provides us with one method to measure an unknown force that an object exerts on a system. We calibrate a spring in terms of some standard force, such as Earth's pull on one or more golf balls. Then if some unknown force is exerted on a system, we can use the spring to exert a balancing force on that system. The unknown force is equal in magnitude and opposite in direction to the force exerted by the spring. In this case, we would be measuring force in units equal to Earth's pull on a golf ball. We could use any spring to balance a known standard force (1 N, or approximately the force that Earth exerts on a 100-g object) and then calibrate this spring in newtons by placing marks at equal distances as you pull on its end with increasing force. We thus build a spring scale—the simplest instrument to measure forces.

TIP   In physics, force is a physical quantity that characterizes the direction and magnitude of an interaction between two objects. **For a force to exist, there must be two objects that interact,** just as a hug requires the interaction of two people. Force does not reside in an object. However, the idea that force resides in an object remains very strong; people say, "The truck's force caused a lot of damage to the telephone pole." We will be careful in this book to always identify the two interacting objects when speaking about any force. Remember, if you are considering a force that is exerted on a moving object but cannot find another object that interacts with it, then you are thinking of something else, not force.

REVIEW QUESTION 3.2 A book bag hanging from a spring scale is partially supported by a platform scale. The platform scale reads about 36 N and the spring scale reads about 28 N. Draw a sketch of the situation, construct a force diagram, and use them to find the magnitude of the force that Earth exerts on the bag.

3.3 Conceptual relationship between force and motion

When we drew a force diagram for a ball held by a person, we intuitively drew the forces exerted on the ball as being equal in magnitude. What if the person catches a ball falling from above? Would she still need to exert a force on the ball that has a magnitude equal to that Earth exerts on the ball? In other words, is there a relationship between the forces that are exerted on an object and the way the object moves?

Observational Experiment Table 3.1 helps us find out whether there is a pattern between the motion diagram and the force diagram for a moving object.

OBSERVATIONAL
EXPERIMENT TABLE 3.1   How are motion and forces related?

	Analysis	
Observational experiment	**Motion diagram**	**Force diagrams for first and third positions**
Experiment 1. A bowling ball B rolls on a very hard, smooth surface S without slowing down.	$\Delta\vec{v} = 0$	
Experiment 2. A ruler R lightly pushes the rolling bowling ball opposite the ball's direction of motion. The ball continues to move in the same direction, but slows down.	$\Delta\vec{v}$	
Experiment 3. A ruler R lightly pushes the rolling bowling ball in the direction of its motion. The ball speeds up.	$\Delta\vec{v}$	

Pattern

- In all the experiments, the vertical forces add to zero and cancel each other. We consider only forces exerted on the ball in the horizontal direction.
- In the first experiment, the sum of the forces exerted on the ball is zero; the ball's velocity remains constant.
- In the second and third experiments, when the ruler pushes the ball, the velocity change arrow ($\Delta\vec{v}$ arrow) points in the same direction as the sum of the forces.

Summary: The $\Delta\vec{v}$ arrow in all experiments is in the same direction as the sum of the forces. Notice that there is no pattern relating the *direction* of the velocity $\vec{v}$ to the direction of the sum of the forces. In Experiment 2, the velocity and the sum of the forces are in opposite directions, but in Experiment 3, they are in the same direction.

In each of the experiments in Table 3.1, the $\Delta\vec{v}$ arrow for a system and the sum of the forces $\Sigma\vec{F}$ that external objects exert on the system are in the same direction. In addition, we often observe that the $\vec{v}$ arrow for a system (the direction the object is moving) is in the same direction as the sum of the forces exerted on it. For example, a grocery cart moves in the direction the shopper pushes it and a soccer ball moves in the direction the player kicks it. We should test both ideas.

Testing possible relationships between force and motion

We have two possible ideas that relate motion and force:

1. An object's *velocity* $\vec{v}$ always points in the direction of the sum of the forces $\Sigma\vec{F}$ that other objects exert on it.
2. An object's *velocity change* $\Delta\vec{v}$ always points in the direction of the sum of the forces $\Sigma\vec{F}$ that other objects exert on it.

To test these two relationships, we use each to predict the outcome of the experiments in Testing Experiment Table 3.2. Then we perform the experiments and compare the outcomes with the predictions. From this comparison, we determine whether we can reject one or both of the relationships.

TESTING
EXPERIMENT TABLE 3.2 Testing velocity and the sum of the forces $\Sigma \vec{F}$

Testing experiment	Prediction	Outcome
Experiment 1. Tanya, wearing a blinking LED on her belt, coasts to the right on Rollerblades. Her friend pulls her back lightly on a rope. If you took a long-exposure photo of Tanya moving, predict what the LED traces will look like after her friend starts pulling the rope.	**Prediction based on idea 1:** The sum of the forces exerted on Tanya points to the left. Thus Tanya's velocity should immediately change from right to left. The LED traces should reverse direction the moment her friend starts pulling back. **Prediction based on idea 2:** Because the sum of the forces points toward the left, the $\Delta \vec{v}$ arrow should point left. Tanya should continue moving to the right, slowing down until she stops. The traces should continue in the same direction as before, only they should get shorter and shorter.	Tanya slows down and eventually stops. The traces are shown below.
Experiment 2. You throw a ball (with an attached LED) upward. If you took a long-exposure photo of the moving ball, predict what the LED traces will look like.	**Prediction based on idea 1:** The only force exerted on the ball in flight points down. Thus the ball should immediately begin moving downward after you release it. The LED traces should not appear above the height at which the ball was released. **Prediction based on idea 2:** Because the only force exerted on the ball in flight points down, the $\Delta \vec{v}$ arrow should point down, too. The ball should slow down until it stops, and then it will start speeding up moving down. The traces continue upward but will become shorter until the ball stops. Then they increase in length in the opposite direction.	The ball moves up at decreasing speed and then reverses direction and starts moving downward. The LED traces are shown at right.

Conclusion

- All outcomes contradict the predictions based on idea 1—we can reject it.
- All outcomes are consistent with the predictions based on idea 2. This does not necessarily mean it is true, but it does mean our confidence in the idea increases.

Recall that the $\Delta \vec{v}$ arrow in a motion diagram is in the same direction as the object's acceleration $\vec{a}$. Thus, based on this idea and these testing experiments, we can now accept idea 2 with greater confidence.

Relating forces and motion The velocity change ($\Delta \vec{v}$) arrow in an object's motion diagram (and its acceleration $\vec{a}$) point in the same direction as the sum of the forces that other objects exert on it. If the sum of the forces points in the same direction as the object's velocity, the object speeds up; if the sum points in the opposite direction, it slows down. If the sum of the forces is zero, the object continues with no change in velocity.

FIGURE 3.6 Force diagram for a ball being caught.

Reflect

How did we devise a relationship that allows us to explain why objects slow down, speed up, or continue at constant velocity? We first observed simple experiments and analyzed them with motion diagrams and force diagrams. We then tested two possible relationships between the objects' motion and the sum of all forces that other objects exerted on it. The above relationship emerged from this analysis and testing.

The relationship between forces and change in motion also applies to the situation of catching a dropped ball. **Figure 3.6** shows a ball with an attached blinking LED being dropped and caught and the corresponding force diagram for one instant during the catch.

CONCEPTUAL EXERCISE 3.2 ▶ **Diagram Jeopardy**

The force diagram shown here describes the forces that external objects (the surface and Earth) exert on a woman (in this scenario, the force diagram does not change with time). Describe three different types of motion of the woman that are consistent with the force diagram.

Sketch and translate Two equal-magnitude, oppositely directed forces are being exerted on the woman ($\Sigma \vec{F} = 0$). Thus, a motion diagram for the woman must have a zero velocity change ($\Delta \vec{v} = 0$).

Simplify and diagram Three possible motions consistent with this idea are shown below.

1. She stands at rest on a horizontal surface.

2. She glides at constant velocity on Rollerblades on a smooth horizontal surface.

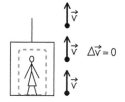

3. She stands on the floor of an elevator that moves up or down at constant velocity.

Note that in all three of the above, the velocity change arrow is zero. This is consistent with the sum of the forces being zero.

Try it yourself Suppose that the elevator described above was moving up at decreasing speed instead of at constant speed. How then would the force diagram be different?

Answer

A velocity change ($\Delta \vec{v}$) arrow for the woman's motion would now point down, opposite the direction of her velocity. Thus, the sum of the forces $\Sigma \vec{F}$ that other objects exert on her must also point down. This means that the magnitude of the upward force $\vec{F}_{\text{S on W}}$ that the elevator floor (surface) exerts on her must now be less than the magnitude of the downward force $\vec{F}_{\text{E on W}}$ that Earth exerts on her ($\vec{F}_{\text{S on W}} < \vec{F}_{\text{E on W}}$).

REVIEW QUESTION 3.3 An elevator in a tall office building moves downward at constant speed. How does the magnitude of the upward force exerted by the cable on the elevator $\vec{F}_{\text{C on El}}$ compare to the magnitude of the downward force exerted by Earth on the elevator $\vec{F}_{\text{E on El}}$? Explain your reasoning.

3.4 Inertial reference frames and Newton's first law

Our description of the motion of an object depends on the observer's reference frame. However, in this chapter we have tacitly assumed that all observers were standing on Earth's surface. Are there any observers who will see a chosen object moving with changing velocity even though the sum of the forces exerted on the object appears to be zero?

Inertial reference frames

In **Table 3.3**, we consider two different observers analyzing the same situation.

Two observers watch the same coffee mug

VIDEO
OET 3.3

Observational experiment	Analysis done by each observer
Experiment 1. Observer 1 is slouched down in the passenger seat of a car and cannot see outside the car. Suddenly, he observes a coffee mug (M) sliding toward him from the dashboard (D).	Observer 1 creates a motion diagram and a force diagram for the mug as he observes it. On the motion diagram, increasingly longer $\vec{v}$ arrows indicate that the mug's speed changes from zero to nonzero as seen by observer 1 even though no external object is exerting a force on it in that direction.
Experiment 2. Observer 2 stands on the ground beside the car. She observes that the car starts moving forward at increasing speed and that the mug remains stationary with respect to her.	Observer 2 creates a motion diagram and force diagram for the mug as she observes it. There are no $\vec{v}$ or $\Delta\vec{v}$ arrows on the diagram, and the mug is at rest relative to her.

Pattern

Observer 1: The forces exerted on the mug by Earth and by the dashboard surface add to zero. But the velocity of the mug increases as it slides off the dashboard. This is inconsistent with the rule relating the sum of the forces and the change in velocity.
Observer 2: The forces exerted on the mug by Earth and by the dashboard surface add to zero. Thus the velocity of the mug should not change, and it does not. This is consistent with the rule relating the sum of the forces and the change in velocity.

VIDEO
3.1

Observer 2 in Table 3.3 can account for what is happening using the rule relating the sum of the forces and changing velocity, but observer 1 cannot. For observer 1, the mug's velocity changes for no apparent reason.

Similarly, in the video in the margin, you see the balls fly inside the box for no reason. For the observer inside the box, Newton's laws cannot explain their behavior. Can you think of some other observers who would be able to explain the balls' motion?

It appears that the applicability of the relationship between the force and motion diagram depends on the reference frame of the observer. Observers who *can* explain the behavior of the mug (observer 2) and the balls by using the relationship between the sum of the forces and changing velocity are said to be observers in **inertial reference frames**. Those who *cannot* explain the behavior of the mug (observer 1) and the balls using this relationship are said to be observers in **noninertial reference frames**. Any observer who accelerates with respect to Earth is a noninertial reference frame observer.

Observers in inertial reference frames can explain the changes in velocity of objects by considering the forces exerted on them by other objects. Observers in noninertial reference frames cannot. From now on, we will always analyze phenomena from the point of view of observers in inertial reference frames. This idea is summarized by Isaac Newton's first law.

Newton's first law of motion For an observer in an inertial reference frame, when no other objects exert forces on a system or when the forces exerted on the system add to zero, the system continues moving at constant velocity (including remaining at rest).

TIP It is important to note that in physics some mathematical terms and symbols have several different meanings. Zero could mean a balance or an absence. For example, when we say that two forces add to zero, we mean that they balance the effects of each other; when we say an object moves with zero acceleration, we mean that the acceleration is absent. A negative sign in front of a quantity could mean direction or subtraction. Every time you meet a zero or a negative sign, stop and determine how it is being used in that equation.

Physicists have analyzed the motion of thousands of objects from the point of view of observers in inertial reference frames and found no contradictions to the rule. Newton's first law of motion limits the reference frames with respect to which the other laws that you will learn in this chapter are valid—these other laws work only for the observers in inertial reference frames. In this and following chapters we will assume that observers of events are not accelerating.

REVIEW QUESTION 3.4 What is the main difference between inertial and noninertial reference frames? Give an example.

Isaac Newton. Isaac Newton (1643–1727) invented differential and integral calculus, formulated the law of universal gravitation, developed a new theory of light, and put together the ideas for his three laws of motion.

3.5 Newton's second law

Our conceptual analyses in Section 3.3 allowed us to deduce a qualitative relationship between forces and changes in an object's motion. In this section we will learn how to predict the magnitude of an object's acceleration if we know the forces exerted on it. The experiments in Observational Experiment Table 3.4 will help us construct this quantitative relationship.

OBSERVATIONAL EXPERIMENT TABLE 3.4 Forces and resulting acceleration

VIDEO
OET 3.4

Observational experiment	Analysis

Experiment 1. A cart starts at rest on a smooth horizontal track. A spring scale continuously exerts one unit of force in the positive direction. The experiment is repeated four more times. Each time, the force probe exerts one additional unit of force on the cart (up to three units). We record the value of the force, and use a motion detector on the track to record the cart's speed and acceleration.

Using this information, we create velocity-versus-time and acceleration-versus-time graphs for two of the five different magnitudes of force. Note that the greater the force, the greater the acceleration.

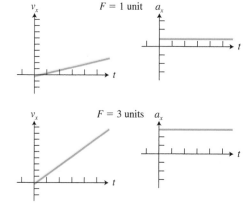

(CONTINUED)

Observational experiment	Analysis

Experiment 2. We repeat the same five experiments, only this time the cart is moving in the positive direction, and the probe pulls back on the cart in the negative direction so that the cart slows down.

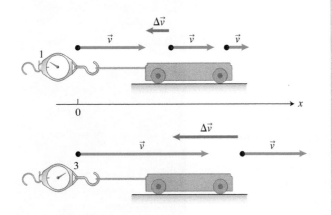

We create velocity-versus-time and acceleration-versus-time graphs for the cart when forces of two different magnitudes oppose the cart's motion.

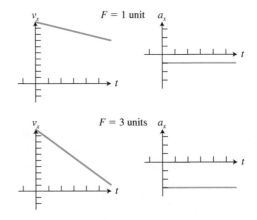

Pattern

- When the sum of the forces exerted on the cart is constant, its acceleration is constant—the cart's speed increases at a constant rate.
- When we plot acceleration versus force using the five positive and five negative values of the force, we obtain the graph at the right. The eleventh point is (0,0), which we know from previous experiments.
- The acceleration is directly proportional to the force exerted by the spring scale (in this case, it is the sum of all forces) and points in the direction of the force.

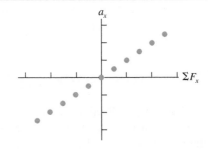

The outcome of these experiments expressed mathematically is as follows:

$$\vec{a} \propto \Sigma\vec{F} \tag{3.2}$$

where $\Sigma\vec{F}$ is the sum of all the forces that other objects exert on the system (not an additional force), and $\vec{a}$ is the system's acceleration. The symbol $\propto$ means "is proportional to." In other words, if the sum of the forces doubles, then the acceleration doubles. When the sum of the forces is zero, the acceleration is zero. When the sum of the forces exerted on an object is constant, the object's resulting acceleration (not velocity) is constant.

Mass, another physical quantity

Do other physical quantities affect acceleration? You know from experience that it is easy to pull an empty cardboard box across the floor, but it is much harder when it is full of books. The amount of matter being pulled must affect the acceleration.

Let's perform another experiment to find the quantitative effect of the amount of *matter* being pulled. We use a spring scale to pull one cart, then two carts stacked on top of each other, and then three and four carts on top of each other. In each case, the spring scale exerts the same force on the carts, regardless of how many carts are being pulled. The experiment is summarized in Observational Experiment Table 3.5.

OBSERVATIONAL
EXPERIMENT TABLE 3.5   Amount of matter and acceleration

Observational experiment	Analysis

Observational experiment

We pull the indicated number of stacked carts using an identical pulling force and measure the acceleration with a motion detector.

Number n of carts	Acceleration (m/s^2)
1	1.00
2	0.49
3	0.34
4	0.25

Motion detector   Carts   Spring scale

Analysis

We graph the acceleration versus number of carts for constant pulling force.

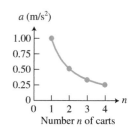

From the graph, we see that increasing the number of carts decreases the acceleration. To check whether this relationship is inversely proportional, we plot a versus $\frac{1}{n}$.

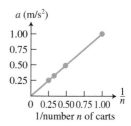

Pattern

Since the graph a versus $\dfrac{1}{n}$ is a straight line, we conclude that a is inversely proportional to n, which we write as $a \propto \dfrac{1}{n}$.

From the pattern observed in Table 3.5, we conclude that *the greater the amount of matter being pulled, the smaller the object's acceleration when the same force is exerted on it.* This property of an object, which affects its acceleration, is called **mass**.

To measure the mass of an object quantitatively, we first define a standard unit of mass. The choice for the unit of mass is arbitrary, but after the unit has been chosen, the masses of all other objects can be determined from this unit. The SI standard of mass is the kilogram (kg). A quart of milk has a mass of about 1 kg. Suppose, for example, that you exert a constant pulling force on a 1.0-kg object (and that all other forces exerted on this object add to zero), and you measure its acceleration. You then exert the same pulling force on another object of unknown mass. Your measurement indicates that it has half the acceleration of the standard 1.0-kg object. Thus, its mass is twice the standard mass (2.0 kg). This method is not practical for everyday use. Later we will learn another method of measuring the mass of an object, a method that is simple enough to use in everyday life.

Our experiments indicate that when the same force is exerted on two objects, the one with the greater mass will have a smaller acceleration. Mathematically:

$$a \propto \frac{1}{m} \tag{3.3}$$

Mass Mass m characterizes the amount of matter in an object. When the same unbalanced force is exerted on two objects, the object with greater mass has a smaller acceleration. The SI unit of mass is the kilogram (kg). Mass is a scalar quantity, and masses add as scalars.

Sum of the forces, mass and acceleration

We have found that the acceleration $\vec{a}$ of a system is proportional to the vector sum of the forces $\Sigma\vec{F}$ exerted on it by other objects [Eq. (3.2)] and inversely proportional to the mass m of the system [Eq. (3.3)]. We can combine these two proportionalities into a single equation on the next page.

$$\vec{a}_S \propto \frac{\Sigma \vec{F}_{\text{on S}}}{m_S} \tag{3.4}$$

Rearrange the above to get $m_S \vec{a}_S \propto \Sigma \vec{F}_{\text{on S}}$. We can turn this into an equation if we choose the unit of force to be $\text{kg} \cdot \text{m/s}^2$. Because force is such a ubiquitous quantity, physicists have given the force unit a special name called a **newton** (N). A force of 1 newton (1 N) causes an object with a mass of 1 kg to accelerate at 1 m/s^2.

$$1 \text{ N} = 1 \text{ kg} \cdot \text{m/s}^2 \tag{3.5}$$

TIP A force of one pound can also be defined (1 lb equals 4.45 N). When the sum of the forces exerted on an object of a unit of mass (called a slug in the Imperial system of units) is equal to 1 lb, the acceleration of the object is 1 foot/s^2. A slug is a relatively massive unit: 1 slug = 14.95 kg.

Equation (3.4), rewritten with an equality sign instead of a proportional sign, is called **Newton's second law**.

TIP Notice that the "vector sum of the forces" mentioned in the definition at right does not mean the sum of their magnitudes. Vectors are not added as numbers; their directions affect the magnitude of the vector sum.

Newton's second law The acceleration $\vec{a}_S$ of a system is proportional to the vector sum of all forces being exerted on the system and inversely proportional to the mass m of the system:

$$\vec{a}_S = \frac{\Sigma \vec{F}_{\text{on S}}}{m_S} = \frac{\vec{F}_{1 \text{ on S}} + \vec{F}_{2 \text{ on S}} + \cdots}{m_S} \tag{3.6}$$

Here the subscripts 1 and 2 stand for the objects exerting forces on the system.
The acceleration of the system points in the same direction as the vector sum of the forces.

Does this new equation make sense? For example, does it work in extreme cases? First, imagine an object with an infinitely large mass. According to the law, it will have zero acceleration for any process in which the sum of the forces exerted on it is finite:

$$\vec{a}_S = \frac{\Sigma \vec{F}_{\text{on S}}}{\infty} = 0$$

This seems reasonable, as an infinitely massive object would not change motion due to finite forces exerted on it. On the other hand, an object with a zero mass will have an infinitely large acceleration when a finite magnitude force is exerted on it:

$$\vec{a}_S = \frac{\Sigma \vec{F}_{\text{on S}}}{0} = \infty$$

Both extreme cases make sense. Newton's second law is a so-called *cause-effect relationship*. The right side of the equation (the sum of the forces being exerted divided by the mass of the system) is the cause of the effect (the acceleration) on the left side.

TIP In the cause-effect relationship, the quantities on the right side of the equation can be varied independently. In our case, these are the sum of the forces and the mass. Each can be changed independently of the other to affect acceleration. In the operational definition, the quantities on the right side of the equation cannot be varied independently. The velocity change depends on the time interval during which it occurs.

$$\underset{\text{Effect}}{\cdots\cdots\rightarrow} \vec{a}_S = \frac{\Sigma \vec{F}_{\text{on S}}}{m_S} \underset{\text{Cause}}{\leftarrow\cdots\cdots}$$

On the other hand, $\vec{a} = \Delta \vec{v}/\Delta t$ is called an *operational definition* of acceleration. It tells us how to determine the quantity acceleration but does not tell us *why* it has a particular value. For example, suppose that an elevator's speed changes from 2 m/s to 5 m/s in 3 s as it moves vertically along a straight line in the positive *y*-direction. The elevator's acceleration (using the definition of acceleration) is

$$a_y = \frac{5 \text{ m/s} - 2 \text{ m/s}}{3 \text{ s}} = +1 \text{ m/s}^2$$

This operational definition does not tell you the reason for the acceleration. If you know that the mass of the elevator is 500 kg and that Earth exerts a 5000-N downward force on the accelerating elevator and the cable exerts a 5500-N upward force on it, then using the cause-effect relationship of Newton's second law:

$$\frac{5500\ \text{N} + (-5000\ \text{N})}{500\ \text{kg}} = +1\ \text{m/s}^2$$

Thus, you obtain the same number using two different methods—one from kinematics (the part of physics that *describes* motion) and the other from dynamics (the part of physics that *explains* motion).

Force components used for forces along one axis

You can use the component form of Newton's second law for the *x*- and *y*-directions instead of the vector equation [Eq. (3.6)]:

$$a_{Sx} = \frac{\Sigma F_{\text{on S }x}}{m_S} \qquad\qquad (3.7x)$$

and

$$a_{Sy} = \frac{\Sigma F_{\text{on S }y}}{m_S} \qquad\qquad (3.7y)$$

The next example shows you how to apply Newton's second law in component form to solve problems.

EXAMPLE 3.3 **Lifting a suitcase**

Earth exerts an approximately 100-N force on a 10-kg suitcase. Suppose you exert an upward 120-N force on the suitcase. If the suitcase starts at rest, how fast is it traveling after lifting for 0.50 s?

Sketch and translate First, we sketch the initial and final states of the process, choosing the suitcase as the system. The sketch helps us visualize the process and also brings together all the known information, letting our brains focus on other aspects of solving the problem. One common aspect of problems like this is the use of a two-step strategy. Here, we use Newton's second law to determine the acceleration of the suitcase and then use kinematics to determine the suitcase's speed after lifting for 0.50 s.

Simplify and diagram Next, we construct a force diagram for the suitcase while being lifted. The *y*-components of the forces exerted on the suitcase are your upward pull on the suitcase $F_{\text{Y on S }y} = +F_{\text{Y on S}} = +120\ \text{N}$ and Earth's downward pull on the suitcase $F_{\text{E on S }y} = -F_{\text{E on S}} = -100\ \text{N}$. Because the upward force is larger, the suitcase will have an upward acceleration $\vec{a}$.

Represent mathematically Since all the forces are along the *y*-axis, we apply the *y*-component form of Newton's second law to determine

the suitcase's acceleration (notice how the subscripts in the equation below change from step to step):

$$a_{Sy} = \frac{\Sigma F_{\text{on S }y}}{m_S} = \frac{F_{\text{Y on S }y} + F_{\text{E on S }y}}{m_S}$$
$$= \frac{(+F_{\text{Y on S}}) + (-F_{\text{E on S}})}{m_S} = \frac{F_{\text{Y on S}} - F_{\text{E on S}}}{m_S}$$

After using Newton's second law to determine the acceleration of the suitcase, we then use kinematics to determine the suitcase's speed after traveling upward for 0.50 s:

$$v_y = v_{0y} + a_y t$$

The initial velocity is $v_{0y} = 0$.

Solve and evaluate Now substitute the known information in the Newton's second law *y*-component equation above to find the acceleration of the suitcase:

$$a_{Sy} = \frac{F_{\text{Y on S}} - F_{\text{E on S}}}{m_S} = \frac{120\ \text{N} - 100\ \text{N}}{10\ \text{kg}} = +2.0\ \text{m/s}^2$$

Insert this and other known information into the kinematics equation to find the vertical velocity of the suitcase after lifting for 0.50 s:

$$v_y = v_{0y} + a_y t = 0 + (+2.0\ \text{m/s}^2)(0.50\ \text{s}) = +1.0\ \text{m/s}$$

The unit for time is correct and the magnitude is reasonable.

Try it yourself How far up did you pull the suitcase during this 0.50 s?

Answer

The average speed while lifting it was $(0 + 1.0\ \text{m/s})/2 = 0.50\ \text{m/s}$. Thus you lifted the suitcase $y - y_0 = (0.50\ \text{m/s})(0.50\ \text{s}) = 0.25\ \text{m}$.

REVIEW QUESTION 3.5 Your friend says that $m\vec{a}$ is a force exerted on an object and it should be represented on the force diagram. Do you agree or disagree with him? Explain your answer.

3.6 Gravitational force law

In the last example, we were given the mass of a suitcase (10 kg) and the approximate magnitude of the gravitational force that Earth exerts on it (100 N). Is it possible to determine the magnitude of this force by just knowing the mass of the suitcase?

Imagine that we evacuate all the air from a 3.0-m-long Plexiglas tube, place a motion sensor at the top, and drop objects of various sizes, shapes, and compositions through the tube (as was done with a feather and an apple in Figure 2.26). The measurements taken by the motion sensor reveal that all objects fall straight down with the same acceleration, approximately 9.8 m/s². Earth is the only object that exerts the force on the falling object $\vec{F}_{\text{E on O}}$ during the entire flight. If we choose the positive y-axis pointing down and apply the y-component form of Newton's second law, we get

$$a_{Oy} = \frac{1}{m_O}F_{\text{E on O} y} = \frac{1}{m_O}(+F_{\text{E on O}}) = +\frac{F_{\text{E on O}}}{m_O}$$

Every object dropped in our experiment had the same free-fall acceleration, $g = 9.8$ m/s², even those with very different masses (such as a ping-pong ball and a lead ball). Thus, the gravitational force that Earth exerts on each object must be proportional to its mass so that the mass cancels when we calculate the acceleration. Earth must exert a force on a 10-kg object that is 10 times greater than that on a 1-kg object:

$$a_{Oy} = \frac{F_{\text{E on O}}}{m_O} = g$$

This reasoning leaves just one possibility for the magnitude of the force that Earth exerts on an object:

$$F_{\text{E on O}} = m_O g = m_O(9.8 \text{ m/s}^2)$$

The ratio of the force and the mass is a constant for all objects—the so-called gravitational coefficient g, already familiar to us as free-fall acceleration.

> **Gravitational force** The magnitude of the gravitational force that Earth exerts on any object $F_{\text{E on O}}$ when on or near its surface equals the product of the object's mass m and the coefficient g:
>
> $$F_{\text{E on O}} = m_O g \qquad (3.8)$$
>
> where $g = 9.8$ m/s² $= 9.8$ N/kg on or near Earth's surface. This force points toward the center of Earth.

TIP Equation (3.8) is a cause-effect relation because m_O and g can be varied independently.

The value of the free-fall acceleration g in the above gravitational force relation [Eq. (3.8)] does not mean that the object is actually falling. The g is just used to determine the magnitude of the gravitational force exerted on the object by Earth whether the object is falling, sitting at rest on a table, or moving down a waterslide. To avoid confusion, we will use $g = 9.8$ N/kg rather than 9.8 m/s² when calculating the gravitational force. Sometimes, when we do not need high precision in the calculation, we will even use 10 N/kg.

We will learn in the chapter on circular motion (Chapter 5) that the gravitational coefficient g at a particular point depends on the mass of Earth and on how far this point is from the center of Earth. On Mars or the Moon, the gravitational coefficient depends on the mass of Mars or the Moon, respectively. The gravitational coefficient is 1.6 N/kg on the Moon and 3.7 N/kg on Mars. You could throw a ball upward higher on the Moon since g is smaller there, resulting in a smaller force exerted downward on the ball.

TIP The *weight* of an object on a planet is the force that the planet exerts on the object. We will not use the term "weight of an object" because it implies that weight is a property of the object rather than an interaction between two objects.

REVIEW QUESTION 3.6 Newton's second law says that the acceleration of an object is inversely proportional to its mass. However, the acceleration with which all objects fall in the absence of air is the same. How can this be?

3.7 Skills for applying Newton's second law for one-dimensional processes

In this section we will develop a strategy that can be used whenever a process involves force and motion. We will introduce the strategy by applying it to the 2006 skydive of diving champion Michael Holmes. After more than 7000 successful jumps, Holmes jumped from an airplane 3700 m above Lake Taupo in New Zealand. His main parachute failed to open, and his backup chute became tangled in its cords. The partially opened backup parachute slowed his descent to about 36 m/s (80 mi/h) as he reached a 2-m-high thicket of wild shrubbery. Holmes survived.

TIP In all problems, unless specifically noted, we will assume that the observer is on the ground and is not accelerating.

PROBLEM-SOLVING STRATEGY 3.1   Applying Newton's laws for one-dimensional processes

EXAMPLE 3.4 Holmes's skydive

Michael Holmes (70 kg) was moving downward at 36 m/s (80 mi/h) and was stopped by 2.0-m-high shrubbery and the ground. Estimate the average force exerted by the shrubbery and ground on his body while stopping his fall.

Sketch and translate

- Sketch the process.
- Choose the system.
- Choose a coordinate system.
- Label the sketch with everything you know about the situation.
- Identify the unknown that you need to find. Label it with a question mark on the sketch.

We sketch the process, choosing Holmes as the system (H). We want to know the average force that the shrubbery and ground (S-G) exert on him from when he first touches the shrubbery to the instant when he stops. We choose the y-axis pointing up and the origin at the ground where Holmes comes to rest. We use kinematics to find his acceleration while stopping and Newton's second law to find the average force that the shrubbery and ground exerted on him while stopping him.

$m_H = 70 \text{ kg}$

Initial
$t_0 = 0$
$v_0 = -36 \text{ m/s}$
$y_0 = +2.0 \text{ m}$

Final
$v = 0$
$F_{\text{S-G on H}} = ?$
$y = 0$

(CONTINUED)

Simplify and diagram

- Make appropriate simplifying assumptions about the process. For example, can you neglect the size of the system? Can you assume that forces or acceleration is constant?

- Then represent the process with a motion diagram and/or force diagram(s). Make sure the diagrams are consistent with each other.

Represent mathematically

- Convert these qualitative representations into quantitative mathematical descriptions of the situation using kinematics equations and Newton's second law for motion along the axis. Determine the signs for the force components in the equations. Add the force components (with either positive or negative signs) to find the sum of the forces.

Solve and evaluate

- Substitute the known values into the mathematical expressions and solve for the unknowns.

- Finally, evaluate your work to see if it is reasonable (check units, limiting cases, and whether the answer has a reasonable magnitude). Check whether all representations—mathematical, pictorial, and graphical—are consistent with each other.

We model Holmes as a point-like object and assume that the forces being exerted on him are constant so that they lead to a constant acceleration. A motion diagram for his motion while stopping is shown along with the corresponding force diagram. To draw the force diagram we first identify the objects interacting with Holmes as he slows down: the shrubbery and ground (combined as one interaction) and Earth. The shrubbery and ground exert an upward force $\vec{F}_{\text{S-G on H}}$ on Holmes. Earth exerts a downward gravitational force $\vec{F}_{\text{E on H}}$. The force diagram is the same for all points of the motion diagram because the acceleration is constant. On the force diagram the arrow for $\vec{F}_{\text{S-G on H}}$ must be longer to match the motion diagram, which shows the velocity change arrow pointing up.

The y-component of Holmes's average acceleration is

$$a_y = \frac{v_y^2 - v_{0y}^2}{2(y - y_0)}$$

The y-component of Newton's second law with the positive y-direction up is

$$a_y = \frac{\Sigma F_{\text{on H } y}}{m_{\text{H}}}$$

The y-component of the force exerted by the shrubbery-ground on Holmes is $F_{\text{S-G on H } y} = +F_{\text{S-G on H}}$, and the y-component of the force exerted by Earth is $F_{\text{E on H } y} = -F_{\text{E on H}} = -m_{\text{H}}g$. Therefore,

$$a_y = \frac{F_{\text{S-G on H } y} + F_{\text{E on H } y}}{m_{\text{H}}} = \frac{(+F_{\text{S-G on H}}) + (-F_{\text{E on H}})}{m_{\text{H}}} = \frac{+F_{\text{S-G on H}} - m_{\text{H}}g}{m_{\text{H}}}$$

$$\Rightarrow F_{\text{S-G on H}} = m_{\text{H}}a_y + m_{\text{H}}g$$

Holmes's average acceleration was

$$a_y = \frac{0^2 - (-36 \text{ m/s})^2}{2(0 - 2.0 \text{ m})} = +324 \text{ m/s}^2$$

Holmes's initial velocity is negative, since he is moving in the negative direction. His initial position is +2.0 m at the top of the shrubbery, and his final position is zero at the ground. His velocity in the negative direction is decreasing, which means the velocity change and the acceleration both point in the opposite direction (positive). The average magnitude of the force exerted by the shrubbery and ground on Holmes is

$$F_{\text{S-G on H}} = m_{\text{H}}a_y + m_{\text{H}}g = (70 \text{ kg})(324 \text{ m/s}^2) + (70 \text{ kg})(9.8 \text{ N/kg})$$

$$= 22{,}680 \text{ kg} \cdot \text{m/s}^2 + 686 \text{ N} = 23{,}366 \text{ N} = 23{,}000 \text{ N}$$

The force has a magnitude greater than the force exerted by Earth—thus the results are consistent with the force diagram and motion diagram. The magnitude is huge and the units are correct. A limiting case for zero acceleration gives us a correct prediction—the force exerted on Holmes by the shrubbery and ground equals the force exerted by Earth.

Try it yourself   What process involving forces can be described by the equation $50 \text{ kg} \times 2 \text{ m/s}^2 = 50 \text{ kg} \times 9.8 \text{ N/kg} + (-390 \text{ N})$?

Answer   A 50-kg person is landing on a thick cushion that exerts a 390-N upward force on her but cannot slow her down. She continues to accelerate down, but her acceleration is less than g (only 2 m/s²). The vertical axis points down.

An elevator ride standing on a bathroom scale

In Example 3.5, we consider a much less dangerous process, one you could try the next time you ride an elevator. When you stand on a bathroom scale, the scale reading indicates how hard you are pushing on the scale. The force that it exerts on you balances the downward force that Earth exerts on you, resulting in your zero acceleration. What will the scale read if you stand on it in a moving elevator?

EXAMPLE 3.5   Elevator ride

You stand on a bathroom scale in an elevator as it makes a trip from the first floor to the tenth floor of a hotel. Your mass is 50 kg. When you stand on the scale in the stationary elevator, it reads 490 N (110 lb). What will the scale read (a) early in the trip while the elevator's upward acceleration is 1.0 m/s^2, (b) while the elevator moves up at a constant speed of 4.0 m/s, and (c) when the elevator slows to a stop with a downward acceleration of 1.0 m/s^2 magnitude?

Sketch and translate We sketch the situation as shown at right, choosing you as the system. The coordinate axis points upward with its origin at the first floor of the elevator shaft. Your mass is $m_Y = 50$ kg, the magnitude of the force that Earth exerts on you is $F_{E \text{ on } Y} = m_Y g = 490$ N, and your acceleration is (a) $a_y = +1.0$ m/s^2 (the upward velocity is increasing);
(b) $a_y = 0$ (v is a constant 4.0 m/s upward); and (c) $a_y = -1.0$ m/s^2 (the upward velocity is decreasing, so the acceleration points in the opposite, negative direction).

Simplify and diagram We model you as a point-like object and represent you as a dot in both the motion and force diagrams, shown for each part of the trip in Figures a, b, and c. On the diagrams, E represents Earth, Y is you, and S is the scale. The magnitude of the downward force that Earth exerts does not change (it equals $m_Y \vec{g}$, and neither m_Y nor $\vec{g}$ changes). Notice that the force diagrams and motion diagrams are consistent with each other for each part of the trip. The length of the arrows representing the force that the scale exerts on you changes from one case to the next so that the sum of the forces points in the same direction as your velocity change arrow.

Represent mathematically The motion and the forces are entirely along the vertical y-axis. Thus, we use the vertical y-component form of Newton's second law [Eq. (3.7y)] to analyze the process. There are two forces exerted on you (the system) so there will be two vertical y-component forces on the right side of the equation: the y-component of the force that Earth exerts on you, $F_{E \text{ on } Y\,y} = -m_Y g$, and the y-component of the force that the scale exerts on you, $F_{S \text{ on } Y\,y} = +F_{S \text{ on } Y}$:

$$a_{Yy} = \frac{\Sigma F_y}{m_Y} = \frac{F_{E \text{ on } Y\,y} + F_{S \text{ on } Y\,y}}{m_Y} = \frac{-m_Y g + F_{S \text{ on } Y}}{m_Y}$$

Multiplying both sides by m_Y, we get $a_{Yy} m_Y = -m_Y g + F_{S \text{ on } Y}$. We can now move $-m_Y g$ to the left side: $m_Y a_{Yy} + m_Y g = F_{S \text{ on } Y}$, or

$$F_{S \text{ on } Y} = m_Y a_{Yy} + m_Y g = m_Y a_{Yy} + 490 \text{ N}$$

(a) $\Delta \vec{v}$ and net force point up. The upward velocity is increasing.

(b) $\Delta \vec{v}$ and net force are zero. The velocity is constant.

(c) $\Delta \vec{v}$ and net force point down. The upward velocity is decreasing.

Remember that $m_Y g = 490$ N is the magnitude of the force that Earth exerts on you. The expression for $F_{S \text{ on } Y}$ gives the magnitude of the force that the scale exerts on you.

Solve and evaluate We can now use the last equation to predict the scale reading for the three parts of the trip.

(a) Early in the trip, the elevator is speeding up, and its acceleration is $a_{Yy} = +1.0$ m/s^2. During that time interval, the force exerted by the scale on you should be

$$F_{S \text{ on } Y} = m_Y a_{Yy} + 490 \text{ N}$$

$$= (50 \text{ kg})(+1.0 \text{ m/s}^2) + 490 \text{ N} = 540 \text{ N}$$

(b) In the middle of the trip, when the elevator moves at constant velocity, your acceleration is zero and the scale should read

$$F_{S \text{ on } Y} = m_Y a_{Yy} + 490 \text{ N}$$

$$= (50 \text{ kg})(0 \text{ m/s}^2) + 490 \text{ N} = 490 \text{ N}$$

(CONTINUED)

(c) When the elevator is slowing down near the end of the trip, its acceleration points downward and is $a_y = -1.0 \text{ m/s}^2$. Then the force exerted by the scale on you should be

$$F_{\text{S on Y}} = m_Y \, a_{Yy} + 490 \text{ N}$$

$$= (50 \text{ kg})(-1.0 \text{ m/s}^2) + 490 \text{ N} = 440 \text{ N}$$

When the elevator is at rest or moving at constant speed, the scale reading equals the magnitude of the force that Earth exerts on you. When the elevator accelerates upward, the scale reads more. When it accelerates downward, even if you are moving upward, the scale reads less. What is also important is that the motion and force diagrams in Figures a, b, and c are consistent with each other and the force diagrams are consistent with the predicted scale readings—an important consistency check of the motion diagrams, force diagrams, and math.

Try it yourself You are standing on the scale in the elevator. The scale first reads 490 N, then 440 N, then 490 N again, and finally 540 N. What could possibly be happening to the elevator?

Answer The elevator is at rest, then moves down with downward acceleration of 1 m/s², then moves at constant velocity, and finally starts to slow down with acceleration of 1 m/s² pointing upward.

It might not seem very important to know what a scale reads in an elevator as it moves, but if we consider the cable supporting the elevator, the value of such calculations becomes apparent. We deduce that the force that the cable exerts on the elevator when the elevator is accelerating upward is greater than that when the elevator is at rest or moving with constant velocity. This means that the cable must be strong enough to withstand a magnitude of force of at least $mg + ma$, where m is the mass of the elevator and its maximum load, and a is the magnitude of the elevator's maximum vertical acceleration.

REVIEW QUESTION 3.7 Three friends argue about the type of information a bathroom scale reports: Eugenia says that it reads the force that Earth exerts on a person, Alan says that it reads the sum of the forces exerted on the person by Earth and the scale, and Mike says that the scale reads the force that the person exerts on the scale. Who do you think is correct? Why?

3.8 Forces come in pairs: Newton's third law

So far, we have analyzed a system's acceleration due to the forces exerted on it by external objects. What effect does the system have on these external objects? To help answer this question, we observe the interaction of two objects and analyze what happens to each of them. Suppose, while wearing Rollerblades, you push abruptly on a wheeled cart that is loaded with a heavy box. If you and the cart are on a hard smooth floor, the cart starts to move (it accelerates), and you also start to move and accelerate in the opposite direction. Evidently, you exerted a force on the cart and the cart exerted a force on you. Since the accelerations were in opposite directions, the forces must point in opposite directions. Let's consider more quantitatively the effect of such a mutual interaction between two objects.

Consider the experiments in Observational Experiment Table 3.6. Two dynamics carts roll freely on a smooth track before colliding. We mount force probes on each cart in order to measure the forces that each cart exerts on the other while colliding. A motion sensor on each end of the track records the initial velocity of each cart before the collision.

Forces that two dynamics carts exert on each other

VIDEO
OET 3.6

Observational experiment	Analysis
Experiment 1. Two carts of different masses move toward each other on a level track. A motion detector indicates their speed before the collision, and force probes record the forces exerted by each cart on the other. Before the collision: $m_1 = 1.0 \text{ kg}, v_{1x} = +2 \text{ m/s}$ $m_2 = 0.5 \text{ kg}, v_{2x} = -2 \text{ m/s}$ Force probes	Because both carts changed velocities due to the collision, they must have exerted forces on each other. The computer recordings from the force probes show that the forces that the carts exert on each other vary with time and at each time have the same magnitude and point in opposite directions. Cart 1 exerts a force on cart 2 toward the right, and cart 2 exerts a force on cart 1 toward the left. 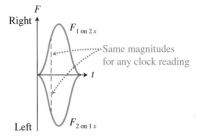
Experiment 2. Cart masses and velocities before collision: $m_1 = 1.0 \text{ kg}, v_{1x} = 0 \text{ m/s (at rest)}$ $m_2 = 0.5 \text{ kg}, v_{2x} = -1 \text{ m/s}$ 	Although the forces that the carts exert on each other are smaller than in the first experiment, the magnitudes of the forces at each time are still the same.
Experiment 3. Cart masses and velocities before collision: $m_1 = 1.0 \text{ kg}, v_{1x} = +2 \text{ m/s}$ $m_2 = 0.5 \text{ kg}, v_{2x} = -1 \text{ m/s}$ 	The same analysis applies.

Pattern

In each experiment, independent of the masses and velocities of the carts before the collisions, at every instant during the collision the force that cart 1 exerted on cart 2 $\vec{F}_{1 \text{ on } 2}$ had the same magnitude as but pointed in the opposite direction from the force that cart 2 exerted on cart 1 $\vec{F}_{2 \text{ on } 1}$.

The cart collisions in Table 3.6 indicate that the force that one cart exerts on another is equal in magnitude and opposite in direction to the force that the second cart exerts on the first.

$$\vec{F}_{1 \text{ on } 2} = -\vec{F}_{2 \text{ on } 1}$$

Will the pattern that we found allow us to correctly predict the results of a new experiment?

TESTING
EXPERIMENT TABLE 3.7   Testing the relationship between the forces that interacting objects exert on each other

VIDEO
TET 3.7

Testing experiment	Prediction	Outcome
Two people are holding spring scales that are hooked together. The scales remain horizontal during the experiment. Person 1 pulls her scale so that scale 1 reads 10 N, while person 2 just holds scale 2. Predict the reading of scale 2. Predict the reading of scale 1 when person 2 pulls scale 2 exerting an arbitrary force.	If the pattern relating the two forces that interacting objects exert on each other (they have the same magnitudes and opposite directions) is correct, and scale 1 pulls on scale 2 exerting a force of 10 N, then scale 2 should pull on scale 1 in the opposite direction exerting a force of 10 N. Both scales will have the same reading. The same is true for the readings of the scales when scale 2 pulls on scale 1.	The experiments show that no matter who pulls on a scale, the scales have the same reading.

Conclusion

The outcome supports the pattern in the forces that two objects exert on each other during the interaction. These forces are of the same magnitude and opposite in direction.

To be convinced of the validity of this conclusion, we need many more testing experiments. So far, physicists have found no experiments involving the dynamics of everyday processes in which interacting objects exert forces on each other that are not of equal magnitude and oppositely directed. This relationship between the forces is called **Newton's third law**.

> **Newton's third law** When two objects interact, object 1 exerts a force on object 2. Object 2 in turn exerts an equal-magnitude, oppositely directed force on object 1:
>
> $$\vec{F}_{1 \text{ on } 2} = -\vec{F}_{2 \text{ on } 1} \tag{3.9}$$
>
> Note that these forces are exerted on different objects and *cannot* be added to find the sum of the forces exerted on one object.

It seems counterintuitive that two interacting objects always exert forces of the same magnitude on each other. Imagine a game of ping-pong. A paddle hits the ball and the ball flies rapidly toward the other side of the table. However, the paddle moves forward with little change in motion. How is it possible that the light ball exerted a force on the paddle of the same magnitude as the force the paddle exerted on the ball?

To resolve this apparent contradiction, think about the masses of the interacting objects and their corresponding accelerations. If the forces are the same, the object with larger mass has a smaller magnitude acceleration than the object with smaller mass:

$$a_{\text{paddle}} = \frac{F_{\text{ball on paddle}}}{m_{\text{paddle}}} \quad \text{and} \quad a_{\text{ball}} = \frac{F_{\text{paddle on ball}}}{m_{\text{ball}}}$$

Because the mass of the ball is so small, the same force leads to a large change in velocity. The paddle's mass, on the other hand, is much larger. Thus, the same magnitude force leads to an almost zero velocity change. We observe the velocity change and incorrectly associate that alone with the force exerted on the object.

TIP Remember that the forces in Newton's third law are exerted on two different objects. This means that the two forces will never show up on the same force diagram, and they should not be added together to find the sum of the forces. You have to choose the system and consider only the forces exerted on *it*!

CONCEPTUAL EXERCISE 3.6   **A book on a table**

A book sits on a tabletop. Identify the forces exerted on the book by other objects. Then, for each of these forces, identify the force that the book exerts on another object. Explain why the book is not accelerating.

Sketch and translate Draw a sketch of the situation and choose the book as the system.

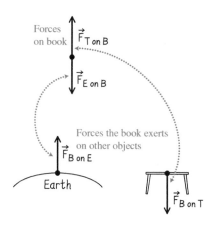

Simplify and diagram Assume that the tabletop is perfectly horizontal and model the book as a point-like object. A force diagram for the book is shown at right. Earth exerts a downward gravitational force on the book $\vec{F}_{E \text{ on } B}$, and the table exerts an upward force on the book $\vec{F}_{T \text{ on } B}$. Newton's second law explains why the book is not accelerating: the forces exerted on it by other objects are balanced and add to zero.

The subscripts on each force identify the two objects involved in the interaction. The Newton's third law pair force will have its subscripts reversed. For example, Earth exerts a downward gravitational force on the book ($\vec{F}_{E \text{ on } B}$). According to Newton's third law, the book must exert an equal-magnitude upward gravitational force on Earth ($\vec{F}_{B \text{ on } E} = -\vec{F}_{E \text{ on } B}$), as shown at right. The table exerts an upward contact force on the book ($\vec{F}_{T \text{ on } B}$), so the book must exert an equal-magnitude downward contact force on the table ($\vec{F}_{B \text{ on } T} = -\vec{F}_{T \text{ on } B}$).

Try it yourself A horse pulls on a sled that is stuck in snow and not moving. Your friend Chris says this happens because the horse exerts on the sled the same magnitude force that the sled exerts on the horse. Since the sum of the forces is zero, there is no acceleration. What is wrong with Chris's reasoning?

Answer Chris added the forces exerted on two different objects and did not consider all forces exerted on the sled. If you choose the sled as the system, then the horse pulls forward on the sled, and the snow exerts a backward, resistive force. If these two horizontal forces happen to be of the same magnitude, they add to zero, and the sled does not accelerate horizontally. If, on the other hand, we choose the horse as the system, the ground exerts a forward force on the horse's hooves (since the horse is exerting a force backward on the ground), and the sled pulls back on the horse. If those forces have the same magnitude, the net horizontal force is again zero, and the horse does not accelerate.

EXAMPLE 3.7   **Hairdryer on a scale**

Hairdryers contain a small propeller that pushes air away from the dryer through a nozzle. You place a hairdryer on a scale with the nozzle pointing up, and it reads 0.99 lb. When you turn the hairdryer on, so that the hairdryer is pushing the air upward, the reading of the scale increases to 1.09 lb. Explain the change in the reading qualitatively and quantitatively.

Sketch and translate We sketch the process (a) before the hairdryer is turned on and (b) after it is turned on. The system is the hairdryer (with the propeller). We point the positive y-axis down. We know that the force that the scale exerts on the hairdryer when it is off is $F_{S \text{ on } HDoff} = 0.99 \text{ lb} = 4.40 \text{ N}$. When the hairdryer is on, the force the scale exerts on it is $F_{S \text{ on } HDon} = 1.09 \text{ lb} = 4.85 \text{ N}$. We need to qualitatively explain the increase of the reading and then explain the quantitative increase of 0.45 N.

(CONTINUED)

Simplify and diagram After the hairdryer is turned on, the scale moves downward a little, but very quickly the situation stabilizes. Thus we will assume that the turned-on hairdryer is not moving with respect to the scale; the scale is stationary after it reaches the reading of 4.85 N. Therefore, the acceleration of the system is zero. There are two objects that interact with the dryer: Earth and the scale. We draw a force diagram for the system (a) when the hairdryer is off and (b) when it is on. The force that Earth exerts on the dryer is the same in both experiments. If diagram (a) is correct, this means that diagram (b) needs an additional downward force of 0.45 N to explain the increase in the reading. What object is exerting this force? Remember that when the dryer is on, the propeller inside is rotating. It pushes the air up, exerting a force $F_{\text{HDon on A}}$; therefore, according to Newton's third law, the air should push the propeller and consequently the whole hairdryer down. Accordingly, we add another force to the diagram in (b), $F_{\text{A on HDon}}$, so that the sum of the forces is still zero. We can now determine the magnitude of this force using the reading of the scale.

Represent mathematically Using the force diagram in (b) we can write

$$a_y = \frac{\Sigma F_y}{m_{\text{HD}}} = \frac{F_{\text{E on HDon }y} + F_{\text{S on HDon }y} + F_{\text{A on HDon }y}}{m_{\text{HD}}}$$

$$= \frac{F_{\text{E on HDon}} + (-F_{\text{S on HDon}}) + F_{\text{A on HDon}}}{m_{\text{HD}}} = 0$$

$$\Rightarrow F_{\text{A on HDon}} = F_{\text{S on HDon}} + (-F_{\text{E on HDon}})$$

Solve and evaluate Using the last equation from the previous step, we find the force that the air exerts on the hairdryer:

$$F_{\text{A on HDon}} = F_{\text{S on HDon}} + (-F_{\text{E on HDon}}) = 4.85\text{ N} - 4.45\text{ N} = 0.45\text{ N}$$

The result makes sense. The magnitude is smaller than the force Earth exerts on the hairdryer but not insignificant. Such an increase in supporting force can explain an observation you may have made while watering plants with a hose. If the flow of water is strong, you can feel the expelled water pushing back on the hose. The same principle explains the behavior of a rotating sprinkler.

Try it yourself An 80-kg basketball player pushes off a gym floor exerting a force that is 4 times greater than the force Earth exerts on him and that the push lasts 0.10 s (unrealistically short). Estimate how fast he is moving when he leaves the floor?

Answer

About 3 m/s. He would move faster if he took longer to push off while exerting the same force.

EXAMPLE 3.8　Learning to linearize data

Alex is investigating the motion of a battery-powered fan attached to a low-friction cart (a fan cart) that is moving on a horizontal track. As the fan blades rotate, they exert a force on the air, and the air exerts an equal and opposite force on the blades, making the cart move (an analogy for the fan cart is a hairdryer on wheels). Using a motion detector, Alex finds that the cart moves with constant acceleration. He also measures how the acceleration of the cart a_x depends on the mass of the objects that he adds to the cart (m_{added}). His measurements are shown in the table below.

m_{added} (kg)	a_x (m/s^2)
0.10	0.25
0.20	0.21
0.30	0.19
0.40	0.17
0.50	0.15

(a) Draw a force diagram for the cart and use it to explain why the cart moves at constant acceleration for fixed added mass. (b) Two physical quantities that are not listed in the table affect the motion of the cart. Determine these two quantities using the data above. (Hint: Rearrange the mathematical expression for the acceleration of the cart to obtain linear dependence on added mass and then plot the graph using the data in the table.)

Sketch and translate The cart is moving to the left on a horizontal surface. The acceleration of the cart is also to the left due to the fan pushing air to the right (the fan works in a similar way to the hairdryer that we investigated in the previous example). We know the masses of the additional objects on the cart and the respective accelerations. We also know that the acceleration of any system is affected by the sum of the forces exerted on it and the system's mass. Because the mass of the cart (together with the fan) contributes to the system's mass, we hypothesize that the quantities that we need to determine are the sum of the forces and the mass of the cart.

Simplify and diagram We assume that the fan cart can be modeled as a point-like object and that the track is smooth, so we do not need to worry about friction. The only objects interacting with the cart (C) are Earth (E), the track (T), and the air (A). The motion diagram and the force diagram for the cart are shown in the figures below and at right. From the force diagram, we see that the forces exerted on the cart by Earth and the track add to zero, and the only force that is causing the acceleration of the cart to the left is the force exerted by the air.

Represent mathematically We write Newton's second law in component form along the axis of motion:

$$a_x = \frac{F_{\text{A on C}}}{m_{\text{added}} + m_\text{C}}$$

Since m_{added} is changed in each experiment (it's an independent variable), it is useful to rearrange the equation so that it is a linear function with linear dependence on m_{added}:

$$\frac{1}{a_x} = \frac{m_{\text{added}} + m_\text{C}}{F_{\text{A on C}}} = \frac{m_{\text{added}}}{F_{\text{A on C}}} + \frac{m_\text{C}}{F_{\text{A on C}}}$$

and to plot $1/a_x$ against m_{added} (see graph at right). The function for $1/a_x$ is a linear function of the form $mx + b$; thus $m = 1/F_{\text{A on C}}$ is the slope of the graph and the intercept is $b = m_\text{C}/F_{\text{A on C}}$. (Do not confuse m for the slope with the symbol m for mass.)

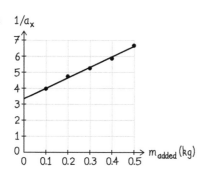

Solve and evaluate The intercept of the graph with the vertical axis is 1 divided by the acceleration the cart would have had if the added mass was zero. Thus $m_\text{C}/F_{\text{A on C}} = 3.35$ kg/N.

We can find the slope of the graph as Δy divided by Δx. For example, when $\Delta y = 6 - 3.35$ and $\Delta x = 0.4 - 0$, this ratio is $2.65/0.4 = 6.63$. The slope of the graph is thus

$$\frac{1}{F_{\text{A on C}}} = \frac{\Delta y}{\Delta x} = 6.63 \, \frac{1}{\text{N}}$$

Therefore, $F_{\text{A on C}} = 0.15$ N and $m_\text{C} = 0.51$ kg.

- -

Try it yourself How would the graph change (consider intercept and slope) if you placed a second fan on the same cart and repeat the experiment?

Answer The slope of the graph is inversely proportional to the force exerted by the air on the cart; therefore the slope will be cut in half. The intercept is inversely proportional to the force that the air exerts on the cart and is directly proportional to the mass of the cart and fans, therefore there is not enough data to determine it. But if we assume that the mass of the fan is much smaller than the mass of the cart, then the intercept will be between the original one and half of its value.

REVIEW QUESTION 3.8 Is the following sentence true? When you hold a heavy object in your hands, you exert the same magnitude force on the object as the object exerts on you but in the opposite direction, and because these forces add to zero, the object stays at rest.

3.9 Seat belts and air bags

At the beginning of the chapter we posed a question about air bags. How do they save lives? We now have all the physics needed to investigate this question. Consider **Figure 3.7**. An air bag is like a balloon that is packed in a small box in the steering wheel or the passenger side dashboard. Air bags are designed to deploy when a car has a negative acceleration of magnitude $10g$ [$10(9.8 \text{ m/s}^2) = 98 \text{ m/s}^2 \approx 100 \text{ m/s}^2$] or more. When a car has such a rapid decrease in speed, the bag inflates with nitrogen gas in about 0.04 s and forms a cushion for the occupant's chest and head. The bag has two important effects:

1. It spreads the force that stops the person over a larger area of the body.
2. It increases the stopping distance, and consequently the stopping time interval, thus reducing the average force stopping the occupant.

Why is spreading the stopping force over a larger area of the body an advantage? If a person uses only a seat belt, his head is not belted to the seat and tends to continue moving forward during a collision, even though his chest and waist are restrained. To stop the head without an air bag, the neck must exert considerable force on the head. This can cause a dangerous stretching of the spinal cord and muscles of the neck, a phenomenon known as "whiplash." The air bag exerts a more uniform force across the upper body and head and helps all parts to stop together.

FIGURE 3.7 An air bag stops a crash test dummy during a collision.

How does the air bag increase the stopping distance? Suppose a test car is traveling at a constant speed of 13.4 m/s (30 mi/h) until it collides head-on with a concrete wall. The front of the car crumples about 0.65 m. A crash test dummy is rigidly attached to the car's seat and is further protected by the rapidly inflating air bag. The dummy also travels about 0.65 m before coming to rest. Without an air bag or a seat belt, the dummy would continue to move forward at the initial velocity of the car. It would then crash into the steering wheel or windshield of the stopped car and stop in a distance much less than 0.65 m—like flying into a rigid wall. The smaller the stopping distance, the greater the acceleration, and therefore the greater the force that is exerted on the dummy. Let's estimate the average force exerted by the air bag on the body during a collision.

EXAMPLE 3.9   **Force exerted by air bag on driver during collision**

A 60-kg crash test dummy moving at 13.4 m/s (30 mi/h) stops during a collision in a distance of 0.65 m. Estimate the magnitude of the average force that the air bag and seat belt exert on the dummy.

Sketch and translate We sketch and label the situation as shown below, choosing the crash test dummy as the system. The positive x-direction is in the direction of motion, and the origin is at the position of the dummy at the start of the collision.

$t_0 = 0$ $F_{A \text{ on } D} = ?$ t
$x_0 = 0$ $x = 0.65$ m
$v_{0x} = 13.4$ m/s $v_x = 0$

$m_D = 60$ kg

Simplify and diagram We model the dummy D as a point-like object and assume that the primary force exerted on the dummy while stopping is due to the air bag and seat belt's $\vec{F}_{A \text{ on } D}$, shown in the force diagram. We can ignore the downward gravitational force that Earth exerts on the dummy $\vec{F}_{E \text{ on } D}$ and the upward force that the car seat exerts on the dummy $\vec{F}_{S \text{ on } D}$ since they add to zero and do not contribute to the acceleration.

Represent mathematically To determine the dummy's acceleration, we use kinematics:

$$a_x = \frac{v_x^2 - v_{0x}^2}{2(x - x_0)}$$

Once we have the dummy's acceleration, we apply the x-component form of Newton's second law to find the force exerted by the air bag and seat belts on the dummy:

$$a_x = \frac{F_{A \text{ on } D\, x}}{m_D} = \frac{-F_{A \text{ on } D}}{m_D} = -\frac{F_{A \text{ on } D}}{m_D}$$

$$\Rightarrow F_{A \text{ on } D} = -m_D a_x$$

Solve and evaluate We know that $v_{0x} = +13.4$ m/s and $v_x = 0$ (the dummy has stopped). The initial position of the dummy is $x_0 = 0$ and the final position is $x = 0.65$ m. The acceleration of the dummy while in contact with the air bag and seat belt is

$$a_x = \frac{0^2 - (13.4 \text{ m/s})^2}{2(0.65 \text{ m} - 0 \text{ m})} = -138 \text{ m/s}^2$$

Thus, the magnitude of the average force exerted by the air bag and seat belt on the dummy is

$$F_{A \text{ on } D} = -(60 \text{ kg})(-138 \text{ m/s}^2) = 8300 \text{ N}$$

This force $[8300 \text{ N}(1 \text{ lb}/4.45 \text{ N}) = 1900 \text{ lb}]$ is almost 1 ton. Is this estimate reasonable? The magnitude is large, but experiments with crash test dummies in the real world are consistent with a force this large in magnitude, a very survivable collision.

- -

Try it yourself Find the acceleration of the dummy and the magnitude of the average force needed to stop the dummy if it is not belted, has no air bag, and stops in 0.1 m when hitting a hard surface.

Answer -900 m/s² and 54,000 N.

REVIEW QUESTION 3.9 Explain how an air bag and seat belt reduce the force exerted on the driver of a car during a collision.

Summary

A **system** is circled in a sketch of a process. Other objects that interact with the system are called the **environment**. (Sections 3.1 and 3.5)

System: elevator

Environment: cable and Earth

The **force** that one object exerts on another characterizes an interaction between the two objects (a pull or a push). The unit of force is the newton (N); $1\ N = (1\ kg)(1\ m/s^2)$. (Sections 3.1 and 3.5)

A **force diagram** represents the forces that external objects exert on the system. The arrows in the diagram point in the directions of the forces, and their lengths indicate the relative magnitudes of the forces. (Sections 3.1 and 3.2)

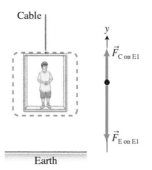

$\vec{F}_{\text{E on O}}$

Interacting objects

A force is denoted by the symbol $\vec{F}$ with two subscripts indicating the object that is exerting the force and the system object.

Newton's first law of motion If no other objects exert forces on a system or if the forces exerted on the system add to zero, then the system continues moving at constant velocity (as seen by observers in inertial reference frames). (Section 3.4)

Newton's second law The acceleration a_S of a system is proportional to the sum of the forces that other objects exert on the system and inversely proportional to its mass m. (Section 3.5)

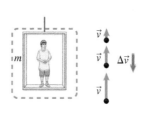

$$\vec{a}_S = \frac{\Sigma \vec{F}_{\text{on S}}}{m_S}$$

$$= \frac{\vec{F}_{1\text{ on S}} + \vec{F}_{2\text{ on S}} + \cdots}{m_S} \qquad \text{Eq. (3.6)}$$

Newton's third law Two objects exert equal-magnitude and opposite direction forces of the same type on each other. (Section 3.8)

$$\vec{F}_{1\text{ on }2} = -\vec{F}_{2\text{ on }1} \qquad \text{Eq. (3.9)}$$

The **gravitational force** $\vec{F}_{\text{E on O}}$ that Earth exerts on an object of mass m when on or near Earth's surface depends on the gravitational coefficient g of Earth. If on or near another planet or the Moon, the gravitational coefficient near those objects is different. (Section 3.6)

Magnitude

$$F_{\text{E on O}} = m_O g \qquad \text{Eq. (3.8)}$$

where

$$g = 9.8\ m/s^2 = 9.8\ N/kg$$

when the object is on or near Earth's surface.

Questions

Multiple Choice Questions

1. An upward-moving elevator slows to a stop as it approaches the top floor. Which answer below best describes the relative magnitudes of the upward force that the cable exerts on the elevator $\vec{F}_{\text{C on El}}$ and the downward gravitational force that Earth exerts on the elevator $\vec{F}_{\text{E on El}}$?
   (a) $F_{\text{C on El}} > F_{\text{E on El}}$ (b) $F_{\text{C on El}} = F_{\text{E on El}}$
   (c) $F_{\text{C on El}} < F_{\text{E on El}}$
   (d) Not enough information is given to answer the question.

2. You apply the brakes of your car abruptly and your book starts sliding off the front seat. Three observers explain this differently. Observer A says that the book continued moving and the car accelerated from underneath it. Observer B says that the car pushed forward on the book. Observer C says that she must be in a noninertial reference frame because the book started moving without any extra objects interacting with it. Which of the observers is correct?
   (a) A   (b) B
   (c) C   (d) A and C
   (e) All of the observers

3. Which of the statements below explains why a child lurches forward in a stroller when you abruptly stop the stroller?
   (a) The child does not lurch forward but instead continues her motion.
   (b) Your pull on the stroller causes the child to move in the opposite direction.
   (c) Newton's third law

4. Which observers can explain the phenomenon of whiplash, which occurs when a car stops abruptly, using Newton's laws?
   (a) The driver of the car   (b) A passenger in the car
   (c) An observer on the sidewalk beside the car and road

5. Which vector quantities describing a moving object are always in the same direction?
   (a) Velocity and acceleration
   (b) Velocity and the sum of the forces
   (c) Acceleration and the sum of the forces
   (d) Acceleration and force
   (e) Both b and c are correct.

6. You have probably observed that magnets attract objects made of certain metals, such as steel. You tie a steel paperclip to a tabletop with string and bring a strong magnet above it so that the paperclip is positioned as shown in Figure Q3.6. Which answer correctly compares the magnitude of the force exerted by the magnet on the paperclip to the magnitude of the force exerted by the paperclip on the magnet?
   (a) $F_{\text{magnet on clip}} > F_{\text{clip on magnet}}$
   (b) $F_{\text{magnet on clip}} < F_{\text{clip on magnet}}$
   (c) $F_{\text{magnet on clip}} = F_{\text{clip on magnet}}$

FIGURE Q3.6

Magnet

7. Which of the following velocity-versus-time graphs represent the motion of the object for which $\Sigma F_y > 0$? Choose all answers that are correct.

FIGURE Q3.7

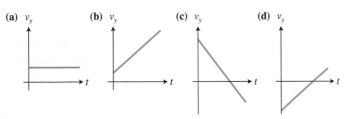

(a) v_y (b) v_y (c) v_y (d) v_y

8. A book sits on a tabletop. What force is the Newton's third law pair to the force that Earth exerts on the book? Choose the correct answer with the best explanation.
   (a) The force that the table exerts on the book because it is equal and opposite in direction to the force that Earth exerts on the book
   (b) The force that the table exerts on the book because the table and the book are touching each other
   (c) The force that the table exerts on the book because it describes the same interaction
   (d) The force that the book exerts on Earth because it describes the same interaction
   (e) The force that the book exerts on Earth because it is equal and opposite in direction to the force that Earth exerts on the book

9. A spaceship moves in outer space. What happens to its motion if there are no external forces exerted on it? If there is a constant force exerted on it in the direction of its motion? If something exerts a force opposite its motion?
   (a) It keeps moving; it speeds up with constant acceleration; it slows down with constant acceleration.
   (b) It slows down; it moves with constant velocity; it slows down.
   (c) It slows down; it moves with constant velocity; it stops instantly.

10. A 0.10-kg apple falls on Earth, whose mass is about 6×10^{24} kg. Which is true of the gravitational force that Earth exerts on the apple?
    (a) It is bigger than the force that the apple exerts on Earth by almost 25 orders of magnitude.
    (b) It is the same magnitude.
    (c) We do not know the magnitude of the force the apple exerts on Earth.

11. A man stands on a scale and holds a heavy object in his hands. What happens to the scale reading if the man quickly lifts the object upward and then stops lifting it?
    (a) The reading increases, returns briefly to the reading when standing stationary, then decreases.
    (b) The reading decreases, returns briefly to the reading when standing stationary, then increases.
    (c) Nothing, since the mass of the person with the object remains the same. Thus the reading does not change.

12. You stand on a bathroom scale in a moving elevator. What happens to the scale reading if the cable holding the elevator suddenly breaks?
    (a) The reading will increase.
    (b) The reading will not change.
    (c) The reading will decrease a little.
    (d) The reading will drop to 0 instantly.

13. A person pushes a 10-kg crate, exerting a 200-N force on it, but the crate's acceleration is only 5 m/s². Explain.
    (a) The crate pushes back on the person, thus the total force is reduced.
    (b) There are other forces exerted on the crate so that the total force is reduced.
    (c) Not enough information is given to answer the question.

14. Two small balls of the same material, one of mass m and the other of mass $2m$, are dropped simultaneously from the Leaning Tower of Pisa. On which ball does Earth exert a bigger force?
    (a) On the $2m$ ball
    (b) On the m ball
    (c) Earth exerts the same force on both balls because they fall with the same acceleration.

15. A box full of lead and a box of the same size full of feathers are floating inside a spaceship that has left the solar system. Choose equipment that you can use to compare their masses.
    (a) A balance scale
    (b) A digital scale
    (c) A watch with a second hand and a meter stick

16. **Figure Q3.16** shows an unlabeled force diagram for a moving object that is missing one force. The length of the sides of the square grid corresponds to a force magnitude of 1 N. Which additional force has to be exerted on the object so that the object (I) moves at a constant speed, (II) accelerates downward, (III) accelerates upward, and (IV) moves down? For each case, choose all answers from (a) to (f) that are correct.

FIGURE Q3.16

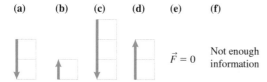

(a)   (b)   (c)   (d)   (e)   (f)

$\vec{F} = 0$ Not enough information

17. A person jumps from a wall and lands stiff-legged. Which statement best explains why the person is less likely to be injured when landing on soft sand than on concrete?
    (a) The concrete can exert a greater force than the sand.
    (b) The person sinks into the sand, increasing the stopping distance.
    (c) The upward acceleration of the person in the sand is less than on concrete, thus the force that the sand exerts on the person is less.
    (d) b and c
    (e) a, b, and c

18. A 3000-kg spaceship is moving away from a space station at a constant speed of 3 m/s. The astronaut in the spaceship decides to return to the space station by switching on engines that expel fuel so that the sum of the forces exerted on the spaceship by the expelled fuel points toward the space station. What is the magnitude of the minimum force needed to bring the spaceship back to the space station?
    (a) 9000 N
    (b) 1000 N
    (c) Any force larger than zero
    (d) The spaceship will keep moving away from the space station no matter how large the force.
    (e) Not enough information is given to answer the question.

Conceptual Questions

19. **Figure Q3.19** is a velocity-versus-time graph for the vertical motion of an object. Choose the correct combination (a, b, or c) of an acceleration-versus-time graph and a force-versus-time graph for the object.

FIGURE Q3.19

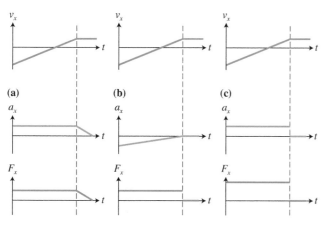

20. Explain the purpose of crumple zones, that is, the front of a car that collapses during a collision.
21. Explain why when landing on a firm surface after a fall you should not land with stiff legs.
22. A small car bumps into a large truck. Compare the forces that the truck exerts on the car and the car exerts on the truck if before the collision (a) the truck was stationary and the car was moving; (b) the car and the truck were moving in opposite directions; (c) the car and the truck were moving in the same direction.
23. You are pulling a sled. Compare the forces that you exert on the sled and the sled exerts on you if you (a) move at constant velocity; (b) speed up; (c) slow down.
24. You stand on a bathroom scale in a moving elevator. The elevator is moving up at increasing speed. The acceleration is constant. Draw three consecutive force diagrams for you.
25. You are holding a 100-g apple. (a) What is the force that you exert on the apple? (b) What is the force that the apple exerts on you? Support your answer with a motion diagram and a force diagram.
26. You throw a 100-g apple upward. (a) While the apple is still in your hand (we'll call this period "the throw"), is the force that you exert on the apple more than, less than, or the same as the force that you exert on the apple when you are holding it at rest? (b) Support your answer with a motion diagram for the throw and a force diagram for one instant during the throw. (c) Is the force that the apple exerts on you at this instant more than, less than, or the same as the force that you exert on the apple when you are holding it at rest? Explain your answer.
27. After having been thrown upward, a 100-g apple falls back into your hand and you catch it. (a) While you are catching it, is the force that you exert on the apple more than, less than, or the same as the force that you exert on the apple when you are holding it at rest? (b) Support your answer with a motion diagram and a force diagram for one instant of the catch. (c) Is the force that the apple exerts on you at this instant more than, less than, or the same as the force that you exert on the apple when you are holding it at rest? Explain your answer.

Problems

Below, **BIO** indicates a problem with a biological or medical focus. Problems labeled **EST** ask you to estimate the answer to a quantitative problem rather than derive a specific answer. Asterisks indicate the level of difficulty of the problem. Problems with no * are considered to be the least difficult. A single * marks moderately difficult problems. Two ** indicate more difficult problems.

3.1 and 3.2 Describing and representing interactions and Adding and measuring forces

1. * In **Figure P3.1** you see unlabeled force diagrams for balls in different situations. Match the diagrams with the following descriptions. (1) A ball is moving upward after it leaves your hand. (2) You hold a ball in your hand. (3) A ball is falling down. (4) You are throwing a ball (still in your hand) straight up. (5) You are lifting a ball at a constant pace. Explain your choices. Label the forces on the diagrams.

FIGURE P3.1

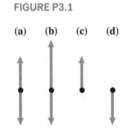

(a) (b) (c) (d)

2. Draw a force diagram (a) for a bag hanging at rest from a spring; (b) for the same bag sitting on a table; and (c) for the same bag that you start to lift so it moves up faster and faster.

3. For each of the following situations, draw the forces exerted on the moving object and identify the other object causing each force. (a) You pull a wagon along a level floor using a rope oriented horizontally. (b) A bus moving on a horizontal road slows down in order to stop. (c) You lift your overnight bag into the overhead compartment on an airplane.

4. You hang a book bag on a spring scale and place the bag on a platform scale so that the platform scale reads 25.7 N and the spring scale reads 17.6 N. (a) Draw a force diagram for the book bag to represent the situation. (b) What is the magnitude of the force that Earth exerts on the bag?

3.3 Conceptual relationship between force and motion

5. A block of dry ice slides at constant velocity along a smooth, horizontal surface. (a) Construct a motion diagram. (b) Draw position- and velocity-versus-time graphs. (c) Construct a force diagram for the block for three instants represented by dots on the motion diagram. Are the diagrams consistent with each other?

6. * You throw a ball upward. (a) Draw a motion diagram and two force diagrams for the ball on its way up and another motion diagram and two force diagrams for the ball on its way down. (b) Represent the motion of the ball with a position-versus-time graph and velocity-versus-time graph.

7. A string pulls horizontally on a cart so that it moves at increasing speed along a smooth horizontal surface. When the cart is moving medium-fast, the pulling is stopped abruptly. (a) Describe in words what happens to the cart's motion when the pulling stops. (b) Illustrate your description with motion diagrams, force diagrams, and position-versus-time and velocity-versus-time graphs. Indicate on the graphs when the pulling stopped. What assumptions did you make?

8. * Solving the previous problem, your friend says that after the string stops pulling, the cart starts slowing down. (a) Explain why your friend would think this way. (b) Do you agree with him? Explain your opinion. (c) Explain how you can design an experiment to test his idea.

9. * A string pulls horizontally on a cart so that it moves at increasing speed along a smooth horizontal surface. When the cart is moving medium-fast, the magnitude of the pulling force is reduced to half its former magnitude. (a) Describe what happens to the cart's motion after the reduction in the string pulling. (b) Illustrate your description with motion diagrams, force diagrams, and position-versus-time and velocity-versus-time graphs.

10. A block of dry ice slides at a constant velocity on a smooth horizontal surface. A second block of dry ice slides twice as fast on the same surface (at a higher constant velocity). Compare the resultant forces exerted on each block. Explain your reasoning.

11. Three motion diagrams for a moving elevator are shown in **Figure P3.11**. Construct two force diagrams (for two consecutive moments) for the elevator for *each* motion diagram. Be sure that the lengths of the force arrows are the appropriate relative lengths and that there is consistency between the force diagrams and the motion diagrams. What assumptions did you make?

FIGURE P3.11

12. * A student holds a thin aluminum pie pan horizontally 2 m above the ground and releases it. Using a motion detector, she obtains the graph shown in **Figure P3.12**. Based on her measurements, (a) draw force diagrams for the pie pan at times 0.05 s, 0.3 s, and 0.7 s, and (b) estimate the distance that the pan travels once it reaches constant speed.

FIGURE P3.12

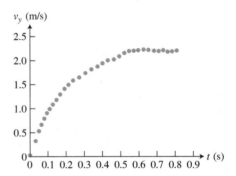

13. * Figures P3.11a, b, and c show three motion diagrams for an elevator moving downward. (a) For each diagram, say everything you can about the elevator's motion. (b) Draw a force diagram for each motion diagram. (c) Could you draw a different motion diagram for each force diagram? Explain how that is possible.

3.4 Inertial reference frames and Newton's first law

14. * A train traveling from New York to Philadelphia is passing a station. A ball is sitting on the floor of the train not moving with respect to the train. (a) Draw a force diagram and a motion diagram for the ball as seen by the observers on the train and on the platform. (b) The ball now starts accelerating forward relative to the floor. Draw force and motion diagrams for the ball as seen by the observers on the train and on the platform. Which of the observers can use Newton's first law to explain the ball's acceleration? Explain.

15. * Explain the phenomenon of whiplash from two points of view: that of an observer on the ground and an observer in the car.

3.5 Newton's second law

16. * An astronaut exerts a 100-N force pushing a beam into place on the International Space Station. The beam accelerates at 0.10 m/s². Determine the mass of the beam. What is the percent uncertainty in your answer?

17. Four people participate in a rope competition. Two of them pull the rope right, exerting forces of magnitude 330 N and 380 N. The other two pull left, exerting forces of magnitude 300 N and 400 N. What is the sum of the forces exerted on the rope?

18. * **Shot put throw** During a practice shot put throw, the 7.0-kg shot left world champion C. J. Hunter's hand at speed 13 m/s. While making the throw, his hand pushed the shot a distance of 1.7 m. Describe all the physical quantities you can determine using this information. Describe the assumptions you need to make to determine them.

19. * You know the sum of the forces $\Sigma \vec{F}$ exerted on an object of mass m during Δt seconds. The object is at rest at the beginning of the time interval. List three physical quantities that you can determine about that object's motion using this information. Then explain how you will determine them.

20. * You record the displacement of an object as a constant force is exerted on it. (a) If the time interval during which the force is exerted doubles, how does the object's displacement change? Indicate all the assumptions that you made. (b) Explain how your answer changes if one of the assumptions is not valid.

21. * **Equation Jeopardy 1** Invent a problem for which the following equation can be a solution:

$$(40 \text{ kg})a_x = 200 \text{ N} - 40 \text{ N}$$

3.6 and 3.7 Gravitational force law and Skills for applying Newton's second law for one-dimensional processes

22. * **Equation Jeopardy 2** Describe in words a problem for which the following equation is a solution and draw a force diagram that is consistent with the equation (specify the direction of the axis):

$$3.0 \text{ m/s}^2 \times 3.0 \text{ kg} = +29.4 \text{ N} - F_{R \text{ on } O}$$

23. **Equation Jeopardy 3** Describe in words a problem for which the following equation is a solution and draw a force diagram that is consistent with the equation (specify the direction of the axis):

$$\frac{0.8 \text{ m/s} - 1.2 \text{ m/s}}{1.6 \text{ s}} = \frac{\Sigma F_x}{50 \text{ kg}}$$

24. * **Equation Jeopardy 4** Describe in words a problem for which the following equation is a solution and draw a force diagram that is consistent with the equation (specify the direction of the axis):

$$2.0 \text{ m/s}^2 = \frac{196 \text{ N} - F_{P \text{ on } O}}{20 \text{ kg}}$$

25. * **Spider-Man** Spider-Man holds the bottom of an elevator with one hand. With his other hand, he holds a spider cord attached to a 50-kg box of explosives at the bottom of the cord. Determine the force that the cord exerts on the box if (a) the elevator is at rest; (b) the elevator accelerates up at 2.0 m/s²; (c) the upward-moving elevator's speed decreases at a rate of 2.0 m/s²; and (d) the elevator falls freely.

26. ** Matt is wearing Rollerblades. Beth pushes him along a hallway with a large spring, keeping the spring compressed and consequently the force that the spring exerts on Matt constant at all times. They conduct several experiments in which Matt starts from rest and travels 12.0 m while carrying objects of different mass in his backpack, recording the time interval for each trip. Their data are shown in the table at right.

Added mass (kg)	Time interval (s)
0	19.0
3.0	19.4
6.0	19.9
9.0	20.3
12.0	20.8
15.0	21.2

(a) Draw a force diagram for Matt and use it to explain why he is moving with a constant acceleration. (b) Two physical quantities that are not listed in the table also affect Matt's motion. Determine these two quantities using the data above. (Hint: This problem requires linearization. See Example 3.8 for help.)

27. * **Stuntwoman** The downward acceleration of a 60-kg stuntwoman near the end of a fall from a very high building is 7.0 m/s². What resistive force does the air exert on her body at that point?

28. **EST** Estimate the average force that a baseball pitcher's hand exerts on a 0.145-kg baseball as he throws a 40 m/s (90 mi/h) pitch. Indicate all of the assumptions you made.

29. * **Super Hornet jet take-off** A 2.1×10^4-kg F-18 Super Hornet jet airplane (see **Figure P3.29**) goes from zero to 265 km/h in 90 m during takeoff from the flight deck of the USS Nimitz aircraft carrier. What physical quantities can you determine using this information? Make a list and determine the values of three of them.

FIGURE P3.29

30. **Lunar Lander** The Lunar Lander of mass 2.0×10^4 kg made the last 150 m of its trip to the Moon's surface in 120 s, descending at approximately constant speed. The *Handbook of Lunar Pilots* indicates that the gravitational constant on the Moon is 1.633 N/kg. Using these quantities, what can you learn about the Lunar Lander's motion?

31. Aisha throws a 0.3-kg ball upward. Frances, standing on a balcony above Aisha, catches the ball by exerting a 1-N downward force on the ball. (a) Draw a motion diagram and a force diagram for the ball during the time interval when Frances is catching it. (b) Determine the acceleration of the ball.

32. * Students Lucia, Isabel, and Austin are investigating how snow stops a dropped 500-g lemon juice bottle. In particular, they are interested in how the force exerted by the snow depends on the age of the snow. They take high-speed videos of the bottle while it sinks into the snow, taking their first set of measurements 4 days after fresh snowfall and the second set of measurements 2 days later. After analyzing the videos frame by frame (see photo), they plot a graph that shows how the velocity of the bottle from the moment the bottle touches the snow changes for both types of snow (**Figure P3.32**).

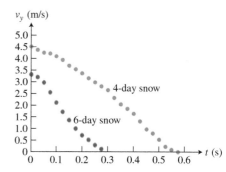

FIGURE P3.32

They each explain their results as follows:

Lucia: The 6-day snow exerts a larger force on the bottle because it stops the bottle in a shorter time.

Isabel: The time taken to stop the bottle does not say much about the force. The 6-day snow exerts a larger force on the bottle because the slope of its $v_y(t)$ graph is steeper.

Austin: We cannot compare the forces exerted by the snow because the initial velocities are different. We need to repeat the experiments and make sure we always drop the bottle from the same height.

Explain how each student reached her/his conclusion and decide who (if anyone) is correct. Indicate any assumptions that you have made.

33. * **Astronaut** Karen Nyberg, a 60-kg astronaut, sits on a bathroom scale in a rocket that is taking off vertically with an acceleration of $3g$. What does the scale read?

34. * A 0.10-kg apple falls off a tree branch that is 2.0 m above the grass. The apple sinks 0.060 m into the grass while stopping. Determine the force that the grass exerts on the apple while stopping it. Indicate any assumptions you made.

35. ** An 80-kg fireman slides 5.0 m down a fire pole. He holds the pole, which exerts a 500-N steady resistive force on the fireman. At the bottom he slows to a stop in 0.40 m by bending his knees. What can you determine using this information? Determine it.

3.8 Forces come in pairs: Newton's third law

36. * Earth exerts a 1.0-N gravitational force on an apple as it falls toward the ground. (a) What force does the apple exert on Earth? (b) Compare the accelerations of the apple and Earth due to these forces. The mass of the apple is about 100 g and the mass of Earth is about 6×10^{24} kg.

37. * You push a bowling ball down the lane toward the pins. Draw force diagrams for the ball (a) just before you let it go; (b) when the ball is rolling (for two clock readings); (c) as the ball is hitting a bowling pin. (d) For each force exerted on the ball in parts (a)–(c), draw the Newton's third law force beside the force diagram, and indicate the object on which these third law forces are exerted.

38. * **EST** (a) A 50-kg skater initially at rest throws a 4-kg medicine ball horizontally. Describe what happens to the skater and to the ball. (b) Estimate the acceleration of the ball during the throw and of the skater using a reasonable value for the force that a skater can exert on the medicine ball. (c) The skater moving to the right catches the ball moving to the left. After the catch, both objects move to the right. Draw force diagrams for the skater and for the ball while the ball is being caught.

39. ** **EST** Basketball player LeBron James can jump vertically over 0.9 m. Estimate the force that he exerts on the surface of the basketball court as he jumps. (a) Compare this force with the force that the surface exerts on James. Describe all assumptions used in your estimate and state how each assumption affects the result. (b) Repeat the problem looking at the time interval when he is landing back on the floor.

40. * **EST** The Scottish Tug of War Association contests involve eight-person teams pulling on a rope in opposite directions. Estimate the magnitude of the force that the rope exerts on each team. Indicate any assumptions you made and include a force diagram for a short section of the rope.

41. Consider the experiment described in Question 3.6 (Figure Q3.6). (a) Draw force diagrams for the magnet and for the paperclip. (b) Which of the forces that you have drawn are pairs according to Newton's third law? Assume all objects are at rest. (c) If the mass of the magnet is 0.300 kg and the force exerted by the hand on the magnet is 3.18 N, what is the magnitude of the force exerted by the paperclip on the magnet? (d) Can you determine the mass of the paperclip based on these data? Explain.

42. * **EST** A friend drops a 0.625-kg basketball from 2 m above you. (a) Estimate the smallest force that the basketball exerts on your hands while you are catching it. (b) How does it compare to the force that you have to exert on the ball to catch it? Use force diagrams to support your answer.

3.9 Seat belts and air bags

43. **Car safety** The National Transportation Safety Board indicates that a person in a car crash has a reasonable chance of survival if his or her acceleration is less than 300 m/s². (a) What magnitude force would cause this acceleration in such a collision? (b) What stopping distance is needed if the initial speed before the collision is 20 m/s (72 km/h, or 45 mi/h)? (c) Indicate any assumptions you made.

44. * A 70-kg person in a moving car stops during a car collision in a distance of 0.60 m. The stopping force that the air bag exerts on the person is 8000 N. Name at least three physical quantities describing the person's motion that you can determine using this information, and then determine them.

General Problems

45. * **BIO** **EST** **Left ventricle pumping** The lower left chamber of the heart (the left ventricle) pumps blood into the aorta. According to biophysical studies, a left ventricular contraction lasts about 0.20 s and pumps 88 g of blood. This blood starts at rest and after 0.20 s is moving through the aorta at about 2 m/s. (a) Estimate the force exerted on the blood by the left ventricle. (b) What is the percent uncertainty in your answer? (c) What assumptions did you make? Did the assumptions increase or decrease the calculated value of the force compared to the actual value?

46. ** **EST** **Acorn hits deck** You are sitting on a deck of your house surrounded by oak trees. You hear the sound of an acorn hitting the deck. You wonder if an acorn will do much damage if instead of the deck it hits your head. Make appropriate estimations and assumptions and provide a reasonable answer.

47. ** **EST** **Olympic dive** During a practice dive, Olympic diver Fu Mingxia reached a maximum height of 5.0 m above the water. She came to rest 0.40 s after hitting the water. Estimate the average force that the water exerted on her while stopping her.

48. * Komila knows that an egg breaks if it falls onto a concrete floor from a height of 0.4 m. She finds, however, that an egg does not break when it falls from the same height onto a floor that is covered with a 2.0-cm-thick carpet. Try to explain the outcome of her experiments. Draw a force diagram for the egg just before it stops. Based on your explanation, predict what the thickness of the carpet should be for an egg to survive a drop from a height of 1.0 m. Indicate any assumptions you have made.

49. ** **EST** You are doing squats on a bathroom scale. You decide to push off the scale and jump up. Estimate the reading as you push off and as you land. Indicate any assumptions you made.

50. ** **EST** Estimate the horizontal speed of the runner shown in Figure P3.50 at the instant she leaves contact with the starting blocks. Indicate any assumptions you made.

FIGURE P3.50

51. ** **EST** Estimate the maximum acceleration of Earth if all people got together and jumped up simultaneously.

52. ** **EST** Estimate how much Earth would move during the jump described in Problem 51.

Reading Passage Problems

Col. John Stapp crash tests From 1946 through 1958, Col. John Stapp headed the U.S. Air Force Aero Medical Laboratory's studies of the human body's ability to tolerate high accelerations during plane crashes. Conventional wisdom at the time indicated that a plane's negative acceleration should not exceed 180 m/s² (18 times gravitational acceleration, or $18g$). Stapp and his colleagues built a 700-kg "Gee Whiz" rocket sled, track, and stopping pistons to measure human tolerance to high acceleration. Starting in June 1949, Stapp and other live subjects rode the sled. In one of Stapp's rides, the sled started at rest and 360 m later was traveling at speed 67 m/s when its braking system was applied, stopping the sled in 6.0 m. He had demonstrated that $18g$ was not a limit for human deceleration.

53. In an early practice run while the rocket sled was stopping, a passenger dummy broke its restraining device and the window of the rocket sled and stopped after skidding down the track. What physics principle best explains this outcome?
    (a) Newton's first law
    (b) Newton's second law
    (c) Newton's third law
    (d) The first and second laws
    (e) All three laws

54. What is Stapp's 67 m/s speed in miles per hour?
    (a) 30 mi/h   (b) 40 mi/h   (c) 100 mi/h
    (d) 120 mi/h   (e) 150 mi/h
55. What is the magnitude of the acceleration of Stapp and his sled as their speed increased from zero to 67 m/s?
    (a) 5 m/s^2 (b) 6 m/s^2 (c) 10 m/s^2
    (d) 12 m/s^2 (e) 14 m/s^2
56. What is the magnitude of the acceleration of Stapp and his sled as their speed decreased from 67 m/s to zero?
    (a) 12g (b) 19g (c) 26g
    (d) 38g (e) 48g
57. What is the average force exerted by the restraining system on 80-kg Stapp while his speed decreased from 67 m/s to zero in a distance of 6.0 m?
    (a) 10,000 N   (b) 20,000 N   (c) 30,000 N
    (d) 40,000 N   (e) 50,000 N
58. What is the time interval for Stapp and his sled to stop as their speed decreased from 67 m/s to zero?
    (a) 0.09 s   (b) 0.18 s   (c) 0.34 s
    (d) 5.4 s   (e) 10.8 s

Using proportions A proportion is defined as an equality between two ratios; for instance, $a/b = c/d$. Proportions can be used to determine the expected change in one quantity when another quantity changes. Suppose, for example, that the speed of a car doubles. By what factor does the stopping distance of the car change? Proportions can also be used to answer everyday questions, such as whether a large container or a small container of a product is a better buy on a cost-per-unit-mass basis.

Suppose that a small pizza costs a certain amount. How much should a larger pizza of the same thickness cost? If the cost depends on the amount of ingredients used, then the cost should increase in proportion to the pizza's area and not in proportion to its diameter:

$$\text{Cost} = k(\text{Area}) = k(\pi r^2) \qquad (3.10)$$

where r is the radius of the pizza and k is a constant that depends on the price of the ingredients per unit area. If the area of the pizza doubles, the cost should double, but k remains unchanged.

Let us rearrange Eq. (3.10) so the two variable quantities (cost and radius) are on the right side of the equation and the constants are on the left:

$$k\pi = \frac{\text{Cost}}{r^2}$$

This equation should apply to any size pizza. If r increases, the cost should increase so that the ratio Cost/r^2 remains constant. Thus, we can write a proportion for pizzas of different sizes:

$$k\pi = \frac{\text{Cost}}{r^2} = \frac{\text{Cost}'}{r'^2}$$

For example, if a 3.5-in.-radius pizza costs $4.00, then a 5.0-in. radius pizza should cost

$$\text{Cost}' = \frac{r'^2}{r^2}\text{Cost} = \frac{(5.0 \text{ in})^2}{(3.5 \text{ in})^2}(\$4.00) = \$8.20$$

This process can be used for most equations relating two quantities that change while all other quantities remain constant.

59. The downward distance d that an object falls in a time interval t if starting at rest is $d = \frac{1}{2}at^2$. On the Moon, a rock falls 10.0 m in 3.50 s. How far will the object fall in 5.00 s, assuming the same acceleration?
    (a) 14.3 m   (b) 20.4 m   (c) 4.90 m
    (d) 7.00 m   (e) 10.0 m
60. The downward distance d that an object falls in a time interval t if starting at rest is $d = \frac{1}{2}at^2$. On the Moon, a rock falls 10.0 m in 3.50 s. What time interval t is needed for it to fall 15.0 m, assuming the same acceleration?
    (a) 2.33 s   (b) 2.86 s   (c) 3.50 s
    (d) 4.29 s   (e) 5.25 s
61. A car's braking distance d (the distance it travels if rolling to a stop after the brakes are applied) depends on its initial speed v_0, the maximum friction force $\vec{f}_{\text{R on C}}$ exerted by the road on the car, and the car's mass m according to the equation

$$\frac{2f_{s\,\text{max}}}{m}d = v_0^2$$

Suppose the braking distance for a particular car and road surface is 26 m when the initial speed is 18 m/s. What is the braking distance when traveling at 27 m/s?
    (a) 59 m   (b) 39 m   (c) 26 m
    (d) 17 m   (e) 12 m
62. You decide to open a pizza parlor. The ingredients require that you charge $4.50 for a 7.0-in.-diameter pizza. How large should you make a pizza whose price is $10.00, assuming the cost is based entirely on the cost of ingredients?
    (a) 1.4 in.   (b) 3.1 in.   (c) 7.0 in.
    (d) 10 in.   (e) 16 in.
63. A circular wool quilt of 1.2 m diameter costs $200. What should the price of a 1.6-m-diameter quilt be if it is to have the same cost per unit area?
    (a) $110   (b) $150   (c) $270   (d) $360

Applying Newton's Laws

- How does knowing physics help human cannonballs safely perform their tricks?

- Who would win a race down a waterslide: an adult or a small child?

- How does friction help us walk?

On March 10, 2011, David Smith, Jr., set the record for the distance traveled by a human cannonball—59 m, improving the record set by his father by more than 2 m. A human cannonball is a performer who is ejected from a cannon by a compressed spring or compressed air and lands on either a horizontal net or an inflated bag. How does the designer of the performance know where to put the net or bag? We'll learn how to make that determination in this chapter.

BE SURE YOU KNOW HOW TO:

- Use trigonometric functions (Mathematics appendix).

- Use motion diagrams and mathematical equations to describe motion (Sections 2.2, 2.6, and 2.7).

- Identify a system, construct a force diagram for it, and use the force diagram to apply Newton's second law (Sections 3.1, 3.5, and 3.7).

WHEN WE FIRST studied force and motion dynamics processes (in Chapter 3), we considered processes in which the forces exerted on an object by other objects were mainly along one axis—the axis along which motion occurred. These processes are common, but most everyday life processes involve forces that are *not* all directed only along the axis of motion. In this chapter we will learn to apply Newton's laws to those more complex processes.

4.1 Vectors in two dimensions and force components

To apply Newton's second law to situations in which the force vectors do not all point along the coordinate axes, we need to learn how to break forces into their *vector components* (also called *component vectors*). We start by using a vector quantity with which we are already very familiar—displacement $\vec{d}$, a vector that starts from an object's initial position and ends at its final position.

Finding vector components

Suppose you want to take a trip and can reach your final destination in two different ways (see **Figure 4.1**). Route 1 is a direct path that is represented by a displacement vector $\vec{C}$. Its tail represents your initial location and its head represents your final destination. Route 2 goes along two roads, represented by displacement vectors $\vec{A}$ and $\vec{B}$, that are perpendicular to each other. You first travel along $\vec{A}$, then along $\vec{B}$, and end at the same final destination.

Route 2 is an example of how to add displacement vectors graphically (you learned vector addition in Chapter 2). Because the initial and final positions are the same for either route, we say that $\vec{C} = \vec{A} + \vec{B}$. Any vector (for example, $\vec{C}$) can be replaced by two perpendicular vectors (for example, $\vec{A}$ and $\vec{B}$) if these two vectors graphically add to equal $\vec{C}$, as illustrated in Figure 4.1.

Because forces are vectors, we can also replace a force $\vec{F}$ by two perpendicular forces $\vec{F}_x$ and $\vec{F}_y$ that graphically add to equal the original force. Suppose, for example, we place a small box on a very smooth surface. The box is stationary. Then we attach strings and spring scales to the box and pull in opposite directions, exerting a 5-N force on each string at angles of 37° relative to the plus or minus x-axis (**Figure 4.2a**). These two forces balance, and the box remains stationary.

Now, let's replace the string and spring scale pulling at 37° relative to the positive x-axis with two strings and two spring scales, one pulling along the x-axis and the other along the y-axis (Figure 4.2b). We find that if the x-axis scale pulls exerting a 4-N force and the y-axis scale pulls exerting a 3-N force, the box again remains stationary. Thus, these two forces have the same effect on the box as the single 5-N force exerted on the box at an angle of 37° above the positive x-direction.

This experimental result is consistent with graphical vector addition, where we find that a 4-N force in the positive x-direction plus a 3-N force in the positive y-direction add to produce a 5-N force at 37° relative to the positive x-axis (Figure 4.2c). The vectors form a "3-4-5" right triangle.

In summary, **we can always replace any force $\vec{F}$ with two perpendicular forces $\vec{F}_x$ and $\vec{F}_y$, as long as the perpendicular forces graphically add to $\vec{F}$** (Figure 4.2c). In this case, the perpendicular forces are along the perpendicular x- and y-axes and are called the x- and y-**vector components** of the original force $\vec{F}$. Since the vector components are also vectors, they are written with a vector symbol above them. Note that $\vec{F}_x + \vec{F}_y = \vec{F}$, just as displacement $\vec{C}$ was identical to the displacement $\vec{A} + \vec{B}$ in Figure 4.1.

Scalar components of a vector

When working with x- and y-axes that are perpendicular to each other, we do not need to use the vector components of $\vec{F}$ but can instead provide the same information about the vector by specifying the **scalar components** F_x and F_y of the force $\vec{F}$. The advantage of the scalar components is that they are numbers with signs, which can be added and subtracted more easily than vector quantities. We are familiar with scalar components from Chapters 2 and 3, but here we will learn how to find scalar components of a vector

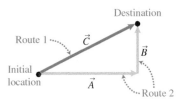

FIGURE 4.1 Two routes to get to the same destination.

FIGURE 4.2 Graphically adding force components.

(a)

Each string exerts a 5-N force on the box. The forces balance.

(b)

We replace one string and scale in (a) with two perpendicular strings and scales that have the same net effect on the box.

(c)

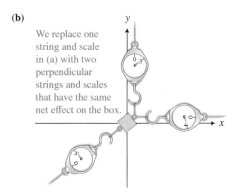

$\vec{F}$ (5 N)

$\vec{F}_y$ (3 N)

$\vec{F}_x$ (4 N)

The perpendicular 4-N and 3-N force component vectors graphically add to form the diagonal 5-N force.

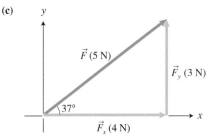

TIP When we replace a vector with the sum of its vector components, we sometimes say that we "resolve" the vector into components.

FIGURE 4.3 The vector and scalar components of a vector.

Horizontal vector component: the x-scalar component is $F_x = -4$ N.

that makes an angle with a coordinate axis. **Figure 4.3** shows how to find the scalar components of such a force vector. The force $\vec{F}$ is 5 N and points 37° above the negative x-axis. The x-vector component of force $\vec{F}$ is 4 N long and points in the negative x-direction. Thus, the x-scalar component of $\vec{F}$ is $F_x = -4$ N. Similarly, the y-vector component of $\vec{F}$ is 3 N in the positive y-direction—it pulls upward in the positive y-direction. Thus, we can specify the y-scalar component of $\vec{F}$ as $F_y = +3$ N. Thus, the x- and y-components $F_x = -4$ N and $F_y = +3$ N tell us everything we need to know about the force $\vec{F}$. Note that scalars are *not* written with a vector symbol above them.

Finding the scalar components of a force from its magnitude and direction

If we know the magnitude of a force F and the angle θ that the force makes above or below the positive or negative x-axis, we can determine its scalar components. Physics Tool Box 4.1 summarizes how to do this.

PHYSICS TOOL BOX 4.1 **Determining the scalar components of a vector**

1. Draw an arrow representing the vector $\vec{F}$.

2. Choose and draw perpendicular coordinate axes (e.g., a y-axis and an x-axis).

3. Represent the vector $\vec{F}$ as the sum of two perpendicular component vectors $\vec{F}_y$ and $\vec{F}_x$.

4. Finally, use the right triangle and trigonometry to calculate the values of the scalar components:

$$F_x = \pm F \cos \theta \qquad (4.1x)$$
$$F_y = \pm F \sin \theta \qquad (4.1y)$$

where F is the magnitude of the force vector and θ is the angle (90° or less) that $\vec{F}$ makes with respect to the $\pm$x-axis. F_x is positive if in the positive x-direction and negative if in the negative x-direction. F_y is positive if in the positive y-direction and negative if in the negative y-direction.

Notice that in the figure above, θ is the angle between the vector and the x-axis. If a different angle is given, use your knowledge of trigonometry to determine the components.

TIP Note that the sign ($+$ or $-$) of a scalar component depends on the orientation of the corresponding vector component relative to the axis. If the vector component points in the positive direction of the axis, the scalar component is positive. If the vector component points in the negative direction, the scalar component is negative.

QUANTITATIVE EXERCISE 4.1 **Components of forces exerted on a knot**

Three ropes pull on a knot. We draw the force diagram for this situation on a grid so that you can see the components of the three forces that the ropes exert on the knot. Use the diagram to visually determine the x- and y-scalar components of the force that rope 2 exerts on the knot. Then use the mathematical method described in

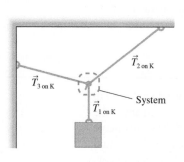

Physics Tool Box 4.1 to calculate the components of this force. The results of using the two methods should be consistent.

Represent mathematically We label the forces that the ropes exert on the knot using the letter T instead of F. This is a common way of noting the forces that ropes or strings exert on an object—tension forces. From the force diagram, we see that rope 2 pulls on the knot at a $37°$ angle above the positive x-axis, exerting a force of magnitude $T_{2 \text{ on } K} = 500$ N. The x-scalar component of this force is $+400$ N (4 grid units in the positive x-direction) and the y-scalar component of this force is $+300$ N (3 grid units in the positive y-direction). In symbols, $T_{2 \text{ on } K\,x} = +400$ N and $T_{2 \text{ on } K\,y} = +300$ N.

Solve and evaluate Now use the mathematical method to calculate the components of the force that rope 2 exerts on the knot:

$$T_{2 \text{ on } K\,x} = +(500 \text{ N})\cos 37° = +400 \text{ N}$$
$$T_{2 \text{ on } K\,y} = +(500 \text{ N})\sin 37° = +300 \text{ N}$$

The diagrammatical and the mathematical methods are consistent.

Try it yourself Use the same two methods to determine the components of the forces exerted by ropes 1 and 3 on the knot.

Answer

$T_{3 \text{ on } K\,x} = -400$ N and $T_{3 \text{ on } K\,y} = +100$ N, and $T_{1 \text{ on } K\,x} = 0$ and $T_{1 \text{ on } K\,y} = -400$ N.

TIP Be careful when applying Eqs. (4.1x) and (4.1y). The angles that appear in those equations must be measured with respect to the positive or negative x-axis.

REVIEW QUESTION 4.1 When does a vector have a positive scalar component? When does a vector have a negative scalar component?

4.2 Newton's second law in component form

In this section we will apply Newton's second law to more complex situations in which one or more of the forces exerted on the system do not point along one of the coordinate axes. The figure in Quantitative Exercise 4.1 is an example of such a situation. If the knot in that figure is not accelerating, the sum of the forces exerted on the knot is zero:

$$\Sigma \vec{F}_{\text{on } K} = \vec{T}_{1 \text{ on } K} + \vec{T}_{2 \text{ on } K} + \vec{T}_{3 \text{ on } K} = 0$$

It is difficult to use this vector equation to analyze the situation further—for example, to determine one unknown force if you know the other two forces. However, we can do such tasks if this situation is represented in scalar component form.

Notice in the force diagram in Quantitative Exercise 4.1 that the y-scalar components of the forces that the three ropes exert on the knot add to zero. Ropes 2 and 3 exert forces that have positive y-scalar components $+300$ N $+ 100$ N $= +400$ N, and rope 1 exerts a force with negative y-scalar component -400 N. Consequently:

$$a_{Ky} = \frac{T_{1 \text{ on } K\,y} + T_{2 \text{ on } K\,y} + T_{3 \text{ on } K\,y}}{m_K}$$
$$= \frac{-400 \text{ N} + 300 \text{ N} + 100 \text{ N}}{m_K} = 0$$

Similarly, the x-scalar components of the forces exerted on the knot also add to zero:

$$a_{Kx} = \frac{T_{1 \text{ on } K\,x} + T_{2 \text{ on } K\,x} + T_{3 \text{ on } K\,x}}{m_K}$$
$$= \frac{0 + 400 \text{ N} + (-400 \text{ N})}{m_K} = 0$$

Thus, we infer that when the x- and y-scalar components of the sum of the forces exerted on the system are zero, it does not accelerate:

$$a_x = 0 \quad \text{if} \quad \Sigma F_{\text{on } K\,x} = 0$$
$$a_y = 0 \quad \text{if} \quad \Sigma F_{\text{on } K\,y} = 0$$

Suppose that in general objects 1, 2, 3, and so forth exert forces $\vec{F}_{1\,\text{on S}}, \vec{F}_{2\,\text{on S}}, \vec{F}_{3\,\text{on S}}, \ldots$ on a system and that the forces point in arbitrary directions and do not add to zero. If the system has mass m_S, it has an acceleration $\vec{a}_S$ as a consequence of these forces (Newton's second law):

$$\vec{a}_S = \frac{\vec{F}_{1\,\text{on S}} + \vec{F}_{2\,\text{on S}} + \vec{F}_{3\,\text{on S}} + \cdots}{m_S}$$

Because this equation involves vectors, we can't work with it directly. However, we can split it into its x- and y-scalar component forms to use Newton's second law to determine the system's acceleration.

TIP In practice, we usually choose the axes so that the object accelerates along only one of these axes and has zero acceleration along the other axis.

Newton's second law rewritten in scalar component form becomes:

$$a_{Sx} = \frac{F_{1\,\text{on S}\,x} + F_{2\,\text{on S}\,x} + F_{3\,\text{on S}\,x} + \cdots}{m_S} \qquad (4.2x)$$

$$a_{Sy} = \frac{F_{1\,\text{on S}\,y} + F_{2\,\text{on S}\,y} + F_{3\,\text{on S}\,y} + \cdots}{m_S} \qquad (4.2y)$$

The process of analyzing a situation using Newton's second law in component form is illustrated in Physics Tool Box 4.2. (Note that for convenience we can drop the word *scalar* and use the simpler term *component form*.)

PHYSICS TOOL BOX 4.2   **Using Newton's second law in component form**

1. Draw a force diagram (the force arrows represent the forces that other objects exert on the system).

2. Visualize the x-components of the forces and apply the x-component form of Newton's second law to the force diagram:

$$a_x = \frac{\Sigma F_x}{m} = \frac{F_{1x} + F_{2x} + F_{3x} + F_{4x}}{m} = \frac{+F_1 \cos \theta + 0 + (-F_3) + 0}{m}$$

Notice that: $F_{1x} = +F_1 \cos \theta$
$F_{2x} = +F_2 \cos 90° = 0$
$F_{3x} = -F_3 \cos 0° = -F_3$
$F_{4x} = +F_4 \cos 90° = 0$

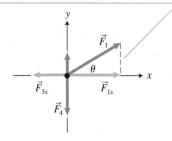

3. Visualize the y-components of the forces and apply the y-component form of Newton's second law to the force diagram:

$$a_y = \frac{\Sigma F_y}{m} = \frac{F_{1y} + F_{2y} + F_{3y} + F_{4y}}{m} = \frac{+F_1 \sin \theta + F_2 + 0 + (-F_4)}{m}$$

Notice that: $F_{1y} = +F_1 \sin \theta$
$F_{2y} = +F_2 \sin 90° = +F_2$
$F_{3y} = -F_3 \sin 0° = 0$
$F_{4y} = -F_4 \sin 90° = -F_4$

REVIEW QUESTION 4.2 The *x*- and *y*-components of Newton's second law for a specific scenario are:

$$x:\ 1.0\ \text{m/s}^2 = (60\ \text{N})\cos 30°/(50\ \text{kg})$$
$$y:\ 0\ \text{m/s}^2 = (50\ \text{kg}) \times (9.8\ \text{N/kg}) + (60\ \text{N})\sin 30° + (-520\ \text{N})$$

Draw a force diagram for the system and a sketch of a possible scenario.

4.3 Friction

Up to this point, we have assumed that objects move across absolutely smooth surfaces with no friction. In reality, the vast majority of situations involve some degree of friction. In this section we will examine the phenomenon of friction conceptually and construct mathematical models that allow us to take friction into account quantitatively.

Static friction

Consider a simple experiment as shown in Observational Experiment **Table 4.1**. A spring scale exerts an increasing force on the block. Observe carefully what happens to the block.

OBSERVATIONAL EXPERIMENT TABLE 4.1 Pulling a block with a spring scale

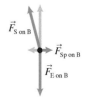

VIDEO
OET 4.1

Observational experiment	Analysis
A block is at rest on the horizontal surface of a desk.	The force diagram for the block is shown at right. $\vec{F}_{\text{S on B}}$ $\vec{F}_{\text{E on B}}$
A spring scale pulls lightly on the block; the block does not move.	The pulling force $\vec{F}_{\text{Sp on B}}$ is exerted parallel to the surface and $\vec{F}_{\text{E on B}}$ is perpendicular to the surface. Because the block is not accelerating, the sum of all forces exerted on it must be zero. We therefore conclude that the surface exerts a force $\vec{F}_{\text{S on B}}$ that, when added to the two forces above, gives zero. The vector component of this force perpendicular to the surface has the same magnitude as $\vec{F}_{\text{E on B}}$; its component parallel to the surface has the same magnitude as $\vec{F}_{\text{Sp on B}}$. $\vec{F}_{\text{S on B}}$ $\vec{F}_{\text{Sp on B}}$ $\vec{F}_{\text{E on B}}$

(CONTINUED)

Observational experiment	Analysis
The spring scale pulls harder on the block; the block still does not move.	The perpendicular component of $\vec{F}_{\text{S on B}}$ stays the same; the magnitude of the parallel vector component increases. Therefore, the direction of $\vec{F}_{\text{S on B}}$ changes, too.
The spring scale pulls even harder on the block; the block finally starts moving.	The force diagram for the moment just before the block begins moving is shown at right.

Patterns

- In each of the experiments, the surface exerted a force on the block that can be resolved into two vector components—one perpendicular to the surface and one parallel to it.
- The perpendicular component of $\vec{F}_{\text{S on B}}$ did not change with the increase of the pulling force, but the parallel component did. The parallel component seems to represent resistance to the motion of the block.
- The magnitude of the parallel component (the resistive force) can only increase up to a certain value. Thus the resistive force must be variable and have a maximum value.

FIGURE 4.4 The force that the surface exerts on an object.

The force that the surface exerts on the block is the sum of two vector components: the normal force $\vec{N}$ and the friction force $\vec{f}$.

TIP To simplify force notation in problem solving, we will use the symbol $\vec{N}_{\text{S on O}}$ for the normal force, $\vec{f}_{\text{s S on O}}$ for the static friction force, and $\vec{f}_{\text{k S on O}}$ for the kinetic friction force exerted by the surface on the object.

The patterns that we found in Table 4.1 allow us to make two important conclusions. First, we can always resolve the force that a surface exerts on an object into two vector components (see **Figure 4.4**):

1. the component perpendicular to the surface, which we will call the **normal force** (a *normal* line in geometry means a line perpendicular to a particular surface), and

2. the component parallel to the surface, which we will call the **friction force**.

Second, we can infer that if you place an object on a surface, and you pull on the object in an attempt to get it to slide across the surface, the friction component of the force that the surface exerts on the object is in the opposite direction to the pulling force so that the object does not begin to move. The friction force exerted on a stationary object is called the **static friction force**. It can vary in strength up to a maximum value. When the external force exerted on the object exceeds this maximum force, the object starts moving. This maximum resistive force that the surface can exert on the object is called the **maximum static friction force**. Once the object starts moving, we call the friction component of the force that the surface exerts on the object the **kinetic friction force**.

In some applications, such as car engines and bicycle chains, friction should be minimized. However, sometimes friction is a necessary phenomenon. For instance, walking on a flat horizontal sidewalk would not be possible if there were no static friction (**Figure 4.5a**). When one foot swings forward and contacts the sidewalk, static friction prevents it from continuing forward and slipping (the way it might on ice).

When your body is vertically above the foot, the friction force that the sidewalk exerts on your foot is momentarily zero (Figure 4.5b). As your body gets ahead of the back foot, the foot has a tendency to slip backward. However, the static friction force that the sidewalk exerts on your foot now points opposite that slipping direction, that is, forward, in the direction of motion (Figure 4.5c). As long as this foot does not slip, the surface continues exerting a forward static friction force on you, helping your body move forward.

FIGURE 4.5 The surface exerts a static friction force that helps you walk or run.

(a)

The surface pushes back on the front foot as it lands.

$\vec{f}_{s \text{ S on F}}$

(b)

When the body is vertically above the foot, the friction force that the surface exerts on the foot is momentarily zero.

$\vec{f}_{s} = 0$

(c)

The surface pushes forward on the back foot as it pushes off.

$\vec{f}_{s \text{ S on F}}$

What determines the magnitude of the maximum static friction force? To investigate this question, we use an experimental setup in Observational Experiment **Table 4.2** similar to that used in Observational Experiment Table 4.1, except that we will vary the characteristics of the block and the surfaces.

OBSERVATIONAL EXPERIMENT TABLE 4.2   What affects the maximum static friction force?

VIDEO
OET 4.2

Observational experiment	Analysis
Experiment 1. *Changes in the smoothness of the surfaces* We'll use the spring to pull a smooth plastic block that is resting on three different surfaces: (1) a glass tabletop, (2) a wood tabletop, and (3) a rubber exercise mat. The reading of the spring scale just before the block starts moving is largest for the rubber mat, next largest for the wood tabletop, and smallest for the glass tabletop.	$\vec{f}_{s \text{ R on B}} > \vec{f}_{s \text{ W on B}} > \vec{f}_{s \text{ G on B}}$
Experiment 2. *Changes in the surface area* We vary the size of contact area between the block and the rubber mat. The block is shaped like a brick and has faces of three different areas. We use the spring to pull the block while it is resting on each of these three different faces. The reading of the scale just before the block starts to move is the same for all three areas.	$\vec{f}_{s \text{ R on BA1}} = \vec{f}_{s \text{ R on BA2}} = \vec{f}_{s \text{ R on BA3}}$
Experiment 3. *Changes in the mass of the block* We take plastic blocks of 1.0 kg, 2.0 kg, and 3.0 kg and place them all on the same wood tabletop. We use a spring scale to pull each of them. The reading of the scale just before the blocks start moving is smallest for the 1.0-kg block, twice as large for the 2.0-kg block, and three times as large for the 3.0-kg block.	$\vec{f}_{s \text{ W on 3 kg B}} = 3\vec{f}_{s \text{ W on 1 kg B}}$ $\vec{f}_{s \text{ W on 2 kg B}} = 2\vec{f}_{s \text{ W on 1 kg B}}$ The maximum static friction force that the tabletop exerts on the block is directly proportional to the mass of the block.

Patterns

The maximum static friction force that the surface can exert on the block depends on the roughness of the contacting surfaces and the mass of the block, but does not depend on the area of contact between the surfaces.

The patterns in Table 4.2 make testable hypotheses. To test a hypothesis, we need to design an experiment in which we can vary one of the properties and make a prediction of the outcome based on the hypothesis being tested. We will test one hypothesis in Testing Experiment **Table 4.3** (on the next page), that the maximum static friction force is directly proportional to the mass of the object.

TESTING
EXPERIMENT TABLE 4.3 **Does the maximum static friction force depend on mass?**

Testing experiment	Prediction	Outcome
We use a string attached to a spring scale to pull a 1-kg block. The mass of the block does not change, but we push down on the block with a spring that exerts a series of downward forces on it. For each of these downward forces, we use the pulling string and spring scale to determine the maximum static friction force the surface exerts on the block. The downward pushing spring is on wheels to prevent it from exerting horizontal forces on the block.	If the friction force is proportional to the mass of the block, the friction force should remain constant during the experiment.	The friction force changes—the harder we press on the block, the higher the maximum static friction force.

The spring pushes down on the block.

The scale measures the extra downward push exerted by the spring on the block.

A spring scale measures the force required to slide the block.

Conclusion

The outcome of the experiment did not match the prediction; the hypothesis requires a revision.

Pulling vs. pushing Pulling a sled on snow is easier than pushing it. When we pull the sled, we lift it up slightly, decreasing the force that the sled exerts on the snow. Therefore, the force exerted by the snow on the sled decreases, too (both the perpendicular to the snow component and the parallel component).

The outcome of the testing experiment in Table 4.3 is not unexpected. Pushing down on the block perpendicularly to the surface increases the force that the block exerts on the surface, and according to Newton's third law, the surface therefore exerts a greater force on the block. Remember that the friction force is just the component parallel to the surface of the force that the surface exerts on the object; the other component of that same force is the normal force. Therefore, we expect these two components to be related. Table 4.4 shows the relationship between the normal force and maximum static friction force for our experiment in Table 4.3.

From the data in Table 4.4 we find a new pattern. The ratio of the maximum static friction force to the normal force

$$\frac{f_{s\ S\ on\ O\ max}}{N_{S\ on\ O}} \approx 0.31$$

and is about the same (within experimental uncertainty) for all the measurements (the last column in the table). It appears that **the maximum static friction force is directly proportional to the magnitude of the normal force**: $f_{s\ S\ on\ O\ max} \propto N_{S\ on\ O}$. This finding makes sense if you look back at Figure 4.4. If the normal force exerted by the surface on an object increases, the maximum static friction force the surface exerts on the object increases proportionally.

TABLE 4.4 Maximum static friction force when a block presses harder against a surface

Mass of the block	Extra downward force exerted on the 1-kg block	Normal force exerted by the surface on the block	Maximum static friction force	Ratio of maximum static friction force to normal force
1.0 kg	0.0 N	9.8 N	3.0 N	0.31
1.0 kg	5.0 N	14.8 N	4.5 N	0.30
1.0 kg	10.0 N	19.8 N	6.1 N	0.31
1.0 kg	20.0 N	29.8 N	9.1 N	0.31

If we repeat the Table 4.4 experiments using a different type of block on a different surface, we get similar results. The ratio of the maximum friction force to the normal force is the same for all the different values of the normal forces. However, the proportionality constant depends on the contacting surfaces: it is greater for rough surfaces and smaller for smoother surfaces.

This ratio

$$\mu_s = \frac{f_{s\,S\,on\,O\,max}}{N_{S\,on\,O}}$$

is called the **coefficient of static friction** μ_s for a particular pair of surfaces. The coefficient of static friction is a measure of the relative difficulty of sliding two surfaces across each other. The easier it is to slide one surface on the other, the smaller the value of μ_s. You experience different values of μ_s when you try to walk on ice and on rough snow. Where are you more likely to slide?

The coefficient of static friction μ_s has no units because it is the ratio of two forces. Although μ_s usually has values between 0 and 1, it can be greater than 1. Its value is about 0.8 for rubber car tires on a dry highway surface and is very small for bones in healthy body joints separated by cartilage and synovial fluid. Some values for different surfaces are listed in Table 4.5.

FIGURE 4.6 Force analyses when we choose different objects as the system.

(a)

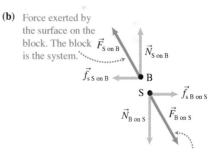

Static friction force When two objects are in contact and we try to move one across the other, they exert forces on each other. The components of these forces parallel to the surface are static friction forces, which oppose the tendency of the objects to move with respect to each other. Static friction force is exerted on both objects, but we are usually concerned with one of them, which we call the system (Figure 4.6). The static friction force changes magnitude to prevent motion of the system—up to a maximum value. This maximum static friction force depends on the roughness of the surfaces of the two objects (on the coefficient of static friction μ_s between the surfaces) and it is related to the magnitude of the normal force N exerted by one surface on the other. That is due to the fact that the friction force and normal force are two perpendicular vector components of the same force that the surface exerts on the system. The magnitude of the static friction force is always less than or equal to the product of these two quantities:

$$0 \le f_{s\,S\,on\,O} \le \mu_s N_{S\,on\,O} \qquad (4.3)$$

(b) Force exerted by the surface on the block. The block is the system.

Force exerted by the block on the surface. We are concerned with these forces when the surface is the system.

Keep in mind the assumptions we used when constructing this model of friction. We used relatively light objects resting on relatively firm surfaces. The objects never caused the surfaces to deform significantly (for example, it was not a car tire sinking into mud). Equation (4.3) is only reasonable in situations in which these conditions hold.

TABLE 4.5 **The coefficients of kinetic and static friction between two surfaces**

Contacting surfaces	Coefficient of static friction μ_s	Coefficient of kinetic friction μ_k
Rubber on concrete (dry)	1	0.6–0.85
Steel on steel	0.74–0.78	0.42–0.57
Aluminum on steel	0.61	0.47
Glass on glass	0.9–1	0.4
Wood on wood	0.25–0.5	0.20
Waxed skis on wet snow	0.14	0.1
Teflon on Teflon	0.04	0.04
Greased metals	0.1	0.06
Surfaces in a healthy human joint	0.01	0.003

Kinetic friction

If we repeat the previous friction experiments with a block that is already in motion, we find a similar relationship between the resistive friction force exerted by the surface on the block and the normal force exerted by the surface on the block. There are, however, two differences: (1) under the same conditions, the magnitude of the kinetic friction force is always lower than the magnitude of the maximum static friction force (see Table 4.5), and (2) the resistive force exerted by the surface on the moving object does not vary but has a constant value. As with the static friction force $\vec{f}_s$, the magnitude of the kinetic friction force f_k depends on the roughness of the contacting surfaces (indicated by a **coefficient of kinetic friction** μ_k) and on the magnitude N of the normal force exerted by one of the surfaces on the other, but not on the surface area of contact. The word *kinetic* indicates that the surfaces in contact are moving relative to each other.

> **Kinetic friction force** When an object slides along a surface, the components of the forces parallel to the surface that the surface exerts on the object are called kinetic friction forces. These forces oppose the motion of the object. The kinetic friction force depends on the surfaces themselves (on the coefficient of kinetic friction μ_k) and relates to the magnitude of the normal force component N of the total force that the surface exerts on the moving object as
>
> $$f_{k\,S\,on\,O} = \mu_k N_{S\,on\,O} \qquad (4.4)$$

As with any mathematical model, this expression for kinetic friction has its limitations. First, it is applicable only for sliding objects, not rolling objects. Second, it has the same assumption about the rigidity of the surfaces as the model for static friction. Third, predictions based on Eq. (4.4) fail for objects moving at high speed. Although this equation does not have general applicability, it is simple and useful for rigid surfaces and everyday speeds.

What causes friction?

FIGURE 4.7 A microscopic view of contacting surfaces.

Rough edges on the contacting surfaces cause friction.

Let's use the example of Velcro fastening material to help understand the phenomenon of friction. The hooks and loops on the two surfaces of Velcro connect with each other, making it almost impossible to slide one surface across the other. All objects are, in a less dramatic way, like Velcro. Even the slickest surfaces have tiny bumps that can hook onto the tiny bumps on another surface (**Figure 4.7**). Logically, smoother surfaces should have reduced friction. For example, a book with a glossy cover slides farther across a table than a book with a rough, unpolished cover. The friction force exerted by the table on the glossy book is smaller than that on the other book.

However, if the surfaces are too smooth (for example, two polished metal blocks), the friction increases again. Why? All substances are made of microscopic particles that are attracted to each other. This attraction keeps a solid object in whatever shape it has. But this attraction between particles works only at very short distances. If two surfaces are so smooth that the microscopic particles are close enough to attract each other strongly (almost close enough to form chemical bonds with each other), it becomes more difficult to slide one surface over the other (for example, two pieces of plastic wrap or two polished metal blocks).

CONCEPTUAL EXERCISE 4.2   Sitting on a slope

You decide to have a picnic on a grassy slope that makes an angle θ with the horizontal. You sit down and put your backpack on the grass, only to find that it starts to slide down the slope. Draw force diagrams for you and the backpack, and use them to explain why you do not start sliding but your backpack does.

Sketch and translate We draw a sketch of the situation and mark the angle of the slope with the horizontal on it. We choose two different systems, you and the backpack. We orient the coordinate axes so that the x-axis is parallel to the slope and the y-axis is perpendicular. This strategy is commonly used in inclined plane problems.

Simplify and diagram We assume that both you and the backpack can be considered point-like. There are two external objects interacting with both systems—Earth and the grassy slope. Because you are not accelerating, the forces exerted by these two objects on you ($\vec{F}_{\text{E on Y}}$ and $\vec{F}_{\text{S on Y}}$) must add to zero, while the sum of the forces exerted on the backpack ($\vec{F}_{\text{E on B}}$ and $\vec{F}_{\text{S on B}}$) must point down the slope, as shown in the left-hand figure at the top of the next column. The force that the slope exerts on both systems can be resolved into two components with respect to the slope's surface—the normal force and the static friction force. The force that Earth exerts on you can also be resolved into two vectors. Because neither object is accelerating along the y-axis, the normal force and the y-component of the force that Earth exerts on the system must be equal in magnitude. We also see that if the y-component of the force that Earth exerts on the system is greater than the maximum friction force, the system will accelerate down the slope. This is exactly what happens to the backpack.

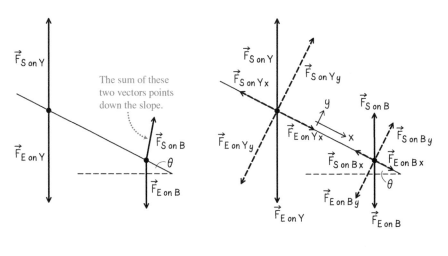

Try it yourself You press a book horizontally against a vertical wall so that the book is not sliding. Draw a force diagram for the book and indicate the magnitude and direction of the normal force and the static friction force components.

Answer

The magnitude of the static friction force is equal to the mass of the book times the gravitational coefficient. The magnitude of the normal force $N_{\text{W on B}}$ is thus $f_{\text{S W on B}} = \dfrac{\mu_s}{mg} = \dfrac{\mu_s}{mg}$.

Note that if the coefficient of friction is small, the force you have to exert on the book to keep it from sliding down the wall is much larger than the force needed to hold the book in your hand.

Other types of friction

There are other types of friction besides static and kinetic friction, such as *rolling friction*. Rolling friction is caused by the surfaces of rolling objects indenting slightly as they turn. This friction is decreased in tires that have been inflated to a higher pressure. In a later chapter (Chapter 14) we will learn about another type of friction, the friction that air or water exerts on a solid object moving through the air or water—a so-called *drag force*.

TIP From now on we will treat the normal component and the friction component of the force that the surface exerts on an object as two different forces.

REVIEW QUESTION 4.3 What is the force of friction that the floor exerts on your refrigerator? Assume that the mass of the refrigerator is 100 kg, the coefficient of kinetic friction is 0.30, and the coefficient of static friction is 0.35. What assumptions did you make in answering this question?

4.4 Skills for analyzing processes involving forces in two dimensions

Our analysis of processes in this and the previous chapter often involves the application of both Newton's second law and kinematics equations, thus relating the external forces exerted on an object and its changing motion. Below we describe a general method for analyzing processes that involve forces in two dimensions and illustrate its use for a simple example of pulling a sled.

PROBLEM-SOLVING
STRATEGY 4.1 > **Analyzing processes involving forces in two dimensions**

EXAMPLE 4.3 Pulling a sled

You pull a sled across a hard snowy surface. The sled and the two children sitting on it have a total mass of 60.0 kg. The rope you pull exerts a 100-N force on the sled and is oriented 37° above the horizontal. If the sled starts at rest, how fast is it moving after being pulled 10.0 m?

Sketch and translate

- Make a sketch of the process.
- Choose a system.
- Choose coordinate axes with one axis in the direction of acceleration and the other axis perpendicular to that direction.
- Indicate in the sketch everything you know about the process relative to these axes.
- Identify the unknown quantity of interest.

We first sketch the process. Choose the sled and children as the system. The process starts with the sled at rest $(v_{0x} = 0)$ and ends when it has traveled $x_1 - x_0 = 10.0$ m and has an unknown final speed v_{1x}. Thus the sled is accelerating.

Simplify and diagram

- Simplify the process. For example, can you model the system as a point-like object? Can you ignore friction?
- Represent the process diagrammatically with a motion diagram and a force diagram.
- Check for consistency of the diagrams—is the sum of the forces in the direction of the acceleration?

Consider the system as a point-like object. Since we have no information about friction, assume its effects on the sled are insignificant, and thus the snow only exerts a force perpendicular to its surface.

A motion diagram for the sled is shown at right. The sled moves at increasing speed toward the right.

Draw a force diagram for the system. Earth exerts a downward gravitational force on the system $\vec{F}_{\text{E on SL}}$, the rope pulls on the sled $\vec{T}_{\text{R on SL}}$ at a 37° angle above the horizontal, and the snow exerts a normal force on the sled perpendicular to the surface (in this case upward) $\vec{N}_{\text{S on SL}}$. Because the sled does not accelerate vertically, the y-components of the forces should add to zero; therefore, the length of the arrow for the force that the snow exerts on the sled is shorter than the arrow for the force that Earth exerts on the sled. The vertical component of the pulling force adds to the force exerted by the snow to balance the force exerted by Earth.

Are the diagrams consistent? The sum of the forces is toward the right in the direction of the acceleration. The vertical force components add to zero.

Represent mathematically

- Convert these qualitative representations into quantitative mathematical descriptions of the process using Newton's second law and kinematics equations.

The horizontal x-component form of Newton's second law is

$$ma_x = \Sigma F_x$$
$$m_{SL}a_x = T_{R \text{ on } SL\, x} + N_{S \text{ on } SL\, x} + F_{E \text{ on } SL\, x}$$

Substitute expressions for the x-components of these three forces:

$$m_{SL}a_x = T_{R \text{ on } SL} \cos 37° + N_{S \text{ on } SL} \cos 90° + F_{E \text{ on } SL} \cos 90°$$

Noting that $\cos 90° = 0$ and dividing both sides by the mass of the sled with the children, we find that

$$a_x = \frac{T_{R \text{ on } SL} \cos 37° + 0 + 0}{m_{SL}}$$

We could at this point use the y-component equation to find the magnitude of the normal force that the surface exerts on the sled, but we don't need to do that to answer the question.

   The above equation can be used to determine the sled's acceleration. We are then left with a kinematics problem to determine the speed v_{1x} of the sled after pulling it for $x_1 - x_0 = 10.0$ m. We can use kinematics Eq. (2.7):

$$v_{1x}^2 = v_{0x}^2 + 2a_x(x_1 - x_0)$$

Solve and evaluate

- Substitute the given values into the mathematical expressions and solve for the unknowns.
- Decide whether the assumptions that you made were reasonable.
- Finally, evaluate your work to see if it is reasonable (check units, limiting cases, and whether the answer has a reasonable magnitude).
- Make sure the answer is consistent with other representations.

The acceleration is

$$a_x = \frac{+T_{R \text{ on } SL} \cos 37°}{m_{SL}} = \frac{+(100 \text{ N})(\cos 37°)}{60.0 \text{ kg}} = +1.33 \text{ m/s}^2$$

The speed of the sled after being pulled for 10.0 m will be

$$v_{1x}^2 = v_{0x}^2 + 2a_x(x_1 - x_0) = 0^2 + 2(1.33 \text{ m/s}^2)(10.0 \text{ m} - 0)$$

or

$$v_{1x} = 5.2 \text{ m/s}$$

This is fast but not unreasonable. In real life, friction between the sled and the snow would probably cause the speed to be slower. The units are correct. If we examine a limiting case, in which the rope pulls vertically, the horizontal acceleration is zero, as it should be.

Try it yourself (a) Determine the y-component of the gravitational force that Earth exerts on the sled-children system. (b) Then determine the y-component of the force that the rope exerts on the system. (c) Finally, use the y-component form of Newton's second law to determine the y-component of the normal force that the snow exerts on the sled.

Answers

magnitude of the normal force less than $m_{SL}g$?

Looking at the force diagram, we see that these numbers make sense. Why is the

(c) $N_{S \text{ on } SL,\, y} = +530$ N.

(a) $F_{E \text{ on } SL,\, y} = -m_{SL}g \sin 90° = -590$ N; (b) $T_{R \text{ on } SL,\, y} = +T_{R \text{ on } SL,\, y} \sin 37° = +60$ N;

Processes involving friction

When solving problems involving friction, remember that the friction force is a vector component parallel to the surface of the force that the surface exerts on an object. The component of the same force perpendicular to the surface is the normal force.

EXAMPLE 4.4   Equation Jeopardy

The equations below are the horizontal x- and vertical y-component forms of Newton's second law applied to a physical process. Solve for the unknown quantities. Then work backward and construct a force diagram for the system of interest and invent a process and question for which the equations might provide an answer (there are many possibilities). Remember that the italicized N is the symbol for normal force, and the roman N is a symbol for the newton.

x-component equation:

$$(200 \text{ N})\cos 30° + 0 - 0.40N_{\text{S on O}} + 0 = (50 \text{ kg})a_x$$

y-component equation:

$$(200 \text{ N})\sin 30° + N_{\text{S on O}} + 0 - (50 \text{ kg})(9.8 \text{ N/kg}) = (50 \text{ kg})0$$

Solve Inserting the cosine and sine values, we get:

x-component equation:

$$(200 \text{ N})0.87 + 0 - 0.40N_{\text{S on O}} + 0 = (50 \text{ kg})a_x$$

y-component equation:

$$(200 \text{ N})0.50 + N_{\text{S on O}} + 0 - (50 \text{ kg})(9.8 \text{ N/kg}) = 0$$

We can solve the y-equation for the magnitude of the normal force: $N_{\text{S on O}} = 390 \text{ N}$. Inserting this value into the x-equation produces the following:

$$(200 \text{ N})0.87 + 0 - 0.40(390 \text{ N}) + 0 = (50 \text{ kg})a_x$$

This can now be solved for a_x:

$$a_x = (174 \text{ N} - 156 \text{ N})/(50 \text{ kg})$$
$$= +0.36 \text{ N/kg} = +0.36 \text{ m/s}^2$$

Simplify and diagram The equations provide the components for each of the four forces exerted on the system (since there are four terms on the left side of each equation). Consider the x- and y-scalar components of each force:

1. A 200-N force oriented 30° above the positive x-axis—maybe a rope is exerting a force on an object.
2. A 390-N normal force points along the y-axis, perpendicular to a surface.
3. It looks like a friction force of $0.40(390 \text{ N}) = 156 \text{ N}$ points in the negative x-direction with a coefficient of friction equal to 0.40.
4. A 490-N gravitational force points in the negative y-direction perpendicular to the surface.

We can now use these forces to construct a force diagram.

Sketch and translate The situation could involve a sled or wagon or crate that is being pulled along a horizontal surface, as shown below. Note that the rope is at an angle with respect to the horizontal, so the force that it exerts has a +100-N y-component. It combines with the normal force's +390-N y-component to balance the gravitational force's −490-N y-component.

One possible problem statement Determine the acceleration of a 50-kg sled being pulled across a horizontal surface by a rope oriented 30° above the horizontal and pulling with a force of 200 N. The coefficient of kinetic friction between the sled and the surface is 0.40.

EXAMPLE 4.5   Was the car speeding?

A car involved in a minor accident left 18.0-m skid marks on a horizontal road. After inspecting the car and the road surface, a police officer decided that the coefficient of kinetic friction was 0.80. The speed limit on that street is 15.6 m/s (35 mi/h). Was the car speeding?

Sketch and translate We first sketch the process (see the figure at top right). We choose the car as the system. Earth is the object of reference. The coordinate system consists of a horizontal x-axis pointing in the direction of the velocity and a vertical y-axis pointing upward.

Simplify and diagram Assume that the car can be modeled as a point-like object, that its acceleration while stopping was constant, and that the resistive force exerted by the air on the car is small compared with the other forces exerted on it. We sketch a motion diagram for the car at right.

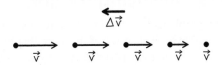

Our force diagram at right shows three forces exerted on the car. Earth exerts a downward gravitational force on the car $\vec{F}_{\text{E on C}}$, the road exerts an upward normal force on the car $\vec{N}_{\text{R on C}}$ (perpendicular to the road surface), and the road also exerts a backward kinetic friction force on the car $\vec{f}_{\text{k R on C}}$ (parallel to the road's surface and opposite the car's velocity). This friction force causes the car's speed to decrease.

Represent mathematically Use the force diagram as shown above to apply Newton's second law in component form. Use the expression for the kinetic friction force ($f_{\text{k R on C}} = \mu_k N_{\text{R on C}}$), the expression for the gravitational force ($F_{\text{E on C}} = m_C g$), and one of the kinematics equations. The car remains in contact with the road surface, so the y-component of its acceleration a_y is zero.

y-component equation:

$$0 = N_{\text{R on C} y} + F_{\text{E on C} y} + f_{\text{k R on C} y}$$

$$0 = N_{\text{R on C}} \sin 90° - m_C g \sin 90° + f_{\text{k R on C}} \sin 0°$$

Note that $\sin 90° = 1.0$ and $\sin 0° = 0$. Thus,

$$N_{\text{R on C}} = m_C g$$

The magnitude of the kinetic friction force is thus

$$f_{\text{k R on C}} = \mu_k N_{\text{R on C}} = \mu_k m_C g$$

x-component equation:

$$m_C a_x = N_{\text{R on C} x} + F_{\text{E on C} x} + f_{\text{k R on C} x}$$
$$= N_{\text{R on C}} \cos 90° + m_C g \cos 90° - f_{\text{k R on C}} \cos 0°$$

Substitute $\cos 90° = 0$ and $\cos 0° = 1.0$ into the above to get

$$m_C a_x = -f_{\text{k R on C}}$$

Combine the two equations for $f_{\text{k R on C}}$ to get

$$m_C a_x = -\mu_k m_C g$$

or

$$a_x = -\mu_k g$$

Now use kinematics to determine the car's velocity v_{0x} before the skid started:

$$v_x^2 - v_{0x}^2 = 2a_x(x - x_0)$$

Solve and evaluate The car's acceleration while stopping was

$$a_x = -\mu_k g = -(0.80)(9.8 \text{ m/s}^2) = -7.84 \text{ m/s}^2$$

We use the kinematics equation $0^2 - v_{0x}^2 = 2a_x(x - x_0)$ to determine the initial speed of the car before the skid started (recall that the final speed $v_x = 0$ and that the stopping distance was 18 m):

$$0^2 - v_{0x}^2 = 2a_x(x - x_0) = 2(18 \text{ m} - 0)(-7.84 \text{ m/s}^2)$$
$$= -282 \text{ m}^2/\text{s}^2 = -(16.8 \text{ m/s})^2$$

$$\Rightarrow v_{0x} = 16.8 \text{ m/s}$$
$$= 16.8 \text{ m/s}(3600 \text{ s/h})(1 \text{ mi}/1609 \text{ m})$$
$$= 38 \text{ mi/h}$$

This is slightly over the 15.6 m/s (35 mi/h) speed limit, but probably not enough for a speeding conviction. Note also that the answer had the correct units.

Try it yourself Describe a process that can be represented with the following equations:

$$0 = N_{\text{R on C}} - m_C g; a_x = -\mu_k g; 0^2 - v_{ix}^2 = 2(x - x_i)a_x$$

Answer

A car is skidding to a stop to avoid a collision.

This method of determining whether a car was speeding before an accident is used in real life. Police officers use the length of skid marks to estimate the speed of a vehicle at the time the driver applied the brakes. Police stations have charts listing the kinetic friction coefficients of various brands of car tires on different types of road surfaces.

Processes involving inclines

Processes involving inclines are common. For example, a skier skis down a slope or a wagon is pulled uphill. How do we use the component form of Newton's second law and kinematics to analyze such processes?

EXAMPLE 4.6 **Who wins a waterslide race?**

A child of mass m and an adult of mass $4m$ simultaneously start to slide from the top of a waterslide that has a downward slope of 30° relative to the horizontal. Which person reaches the bottom first?

Sketch and translate We can choose either the child or the adult as the system. Our sketch shows the person at one moment during

the downhill slide and includes the known information. The person who has the greater acceleration along the slide will win the race, so we will construct an expression for the

(CONTINUED)

person's acceleration and see how that acceleration depends on the person's mass. The object of reference is Earth. We choose the axes later when constructing the force diagram.

Simplify and diagram The person is modeled as a point-like object. Since a waterslide is very smooth and slippery, we assume that we can disregard the friction force exerted by the slide on the person.

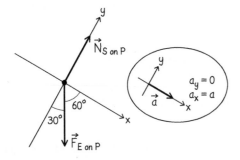

$\Delta \vec{v}$ and $\vec{a}$ point in the same direction.

Next, we build a motion diagram for the person, as shown to the right. The velocity arrows are parallel to the slide, and the person's velocity is increasing, so the velocity change arrow $\Delta \vec{v}$ (and the acceleration $\vec{a}$) points down the slide.

A force diagram shows that Earth exerts a downward gravitational force $\vec{F}_{\text{E on P}}$ on the person and the slide exerts a normal force $\vec{N}_{\text{S on P}}$ on the person perpendicular to the surface. We choose the x-axis parallel to the inclined surface and the y-axis perpendicular to the inclined surface. With this choice, the acceleration points in the positive x-direction. If we had chosen the standard horizontal and vertical directions for our coordinate axes, then the acceleration would have nonzero x- and y-components, making the situation more difficult to analyze.

Represent mathematically The motion of the person is along the x-axis. Use the force diagram to apply the x-component form of Newton's second law:

$$m_P a_x = F_{\text{E on P }x} + N_{\text{S on P }x}$$

The slide is inclined at 30° relative to the horizontal. This means that the gravitational force makes a 60° angle relative to the slide (and therefore relative to the x-axis, as shown in the diagram). We can calculate the x-component of the gravitational force in two ways: $F_{\text{E on P }x} = m_P g \cos 60°$ or $F_{\text{E on P }x} = m_P g \sin 30°$. The methods give the same result. The component equation becomes

$$m_P a_x = +(m_P g) \cos 60° + N_{\text{S on P}} \cos 90°$$

Solve and evaluate Substituting $\cos 90° = 0$, we get

$$m_P a_x = +(m_P g) \cos 60° + 0$$

Dividing by the m_P on each side of the equation, we get $a_x = g \cos 60°$. The person's acceleration is independent of her/his mass! Thus, the acceleration of the child and that of the adult are the same. If they start at the same time at the same position, both initially at rest, and have the same acceleration, they arrive at the bottom at the same time. The race is a tie (see the photo).

Try it yourself Suppose that the same child and adult slide down the same waterslide, but this time do not assume that the slide is frictionless. If they wear swimming trunks made of the same material, who will win the race? Explain your prediction and list assumptions.

Answer It is independent of the person's mass.

$$a_x = \frac{m g \sin \theta - \mu_k m g \cos \theta}{m} = g \sin \theta - \mu_k g \cos \theta$$

is acceleration of each person

Assuming that the coefficient of kinetic friction is the same for both people and that air resistance is negligible, there will be a tie again. The

---

The result of Example 4.6 may surprise you, but it is reasonable. The acceleration depends on both gravitational force and mass. The gravitational force that Earth exerts on the adult (only the x-component of this force determines the acceleration in this case) is four times that exerted by Earth on the child, but because the adult's mass is four times greater than the child's, the accelerations of the adult and the child are equal. This result is similar to what we learned in Chapters 1 and 2—objects of different masses have the same free-fall acceleration.

FIGURE 4.8 Complementary angles.

The angle $90° - \theta$ of the gravitational force relative to the x-axis is the complement to the angle θ of the axis relative to the horizontal.

TIP Notice in the last example that the inclined surface and the x-axis were at a 30° angle relative to the horizontal. The gravitational force exerted by Earth on the system then makes a 60° angle relative to the x-axis (60° is the complement of 30°). In general, the angle θ of the inclined surface relative to the horizontal is the complement of the angle that the gravitational force makes with respect to the inclined surface (see Figure 4.8).

Two objects linked together

We can apply Newton's second law to a process in which two objects are connected by a cable or rope, such as a van pulling a rope connected to a wagon that is pulling a rope connected to a second wagon. Perhaps the first scientific quantitative application of this type involved two blocks of different mass at the ends of a string that passed over a pulley (Figure 4.9). George Atwood invented this apparatus (now called the Atwood machine) in 1784. At that time, there were no motion sensors or precision stopwatches—nothing that would allow the accurate measurement of the motion of a rapidly accelerating object (9.8 m/s^2 would have been considered a rapid acceleration in those days). Atwood's machine allowed the determination of g despite these difficulties. We'll consider next a modified and simplified version of the Atwood machine, a machine that can serve the same purpose.

FIGURE 4.9 An Atwood machine.

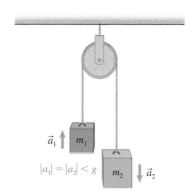

EXAMPLE 4.7 Slowing the fall

In a modified Atwood machine, one block rests on a smooth horizontal surface and the other hangs off the edge of the surface. A string attached to each block passes over a light, smooth pulley. Determine the force that the string exerts on the hanging block of mass m_2 and the magnitude of the acceleration of each block.

Sketch and translate We first sketch the modified Atwood machine, shown at right. For analysis we will use two systems—first block 1 and then block 2. The two blocks have masses m_1 and m_2. Block 2 starts moving down and pulls the string attached to block 1. Both blocks move with increasing speed— block 1 on the horizontal surface and block 2 downward. If the string does not stretch, then at every moment they move with the same speed. The sliding block 1 has the same magnitude acceleration as block 2, but pointing toward the right. Our goal is to find the force exerted by the string on block 2 and the magnitudes of the acceleration of the blocks.

Simplify and diagram Assume that the string is very light and does not stretch, that the pulley's mass is very small, and that the pulley is smooth and rotates free of friction. The pulley changes the direction of the low-mass string passing over it but does not change the magnitude of the force that the string exerts on the objects attached at either end. Assume also that there is no friction between the horizontal surface and block 1.

At right we show our force diagrams for each block and our choice of coordinate systems. For block 1, Earth exerts a downward gravitational force $\vec{F}_{E\,on\,1}$; the table surface exerts an equal-magnitude upward force on block 1 $\vec{N}_{T\,on\,1}$; and the string exerts an unknown horizontal force toward the right on the block $\vec{T}_{S\,on\,1}$. For block 2, Earth exerts a downward gravitational force $\vec{F}_{E\,on\,2}$ and the string exerts an unknown upward force $\vec{T}_{S\,on\,2}$. As noted earlier, the string exerts the same magnitude force on block 1 as it does on block 2: $T_{S\,on\,1} = T_{S\,on\,2} = T$.

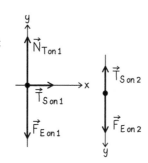

Represent mathematically Use the force diagrams to apply Newton's second law in component form for each block. For block 1, the forces in the vertical direction balance, since the vertical component of its acceleration is zero. We see by inspection of the force diagram that the gravitational force and the normal force have zero x-components and so we do not include them in the x-component equation. This leaves us with only one force in the horizontal direction—the force exerted by the string. The x-component form of Newton's second law becomes

$$m_1 a_{1x} = T_{S\,on\,1\,x} = +T \qquad (4.5)$$

For block 2, we only need to look at the vertical direction, as no forces are exerted in the horizontal direction. Notice that we've chosen the vertical y-axis pointing downward so that the y-component of block 2's acceleration will equal the x-component of block 1's acceleration. The y-component form of Newton's second law for block 2 is

$$m_2 a_{2y} = T_{S\,on\,2\,y} + F_{E\,on\,2\,y}$$

Noting that $T_{S\,on\,2\,y} = -T$ and $F_{E\,on\,2\,y} = m_2 g$, we get

$$m_2 a_{2y} = -T + m_2 g \qquad (4.6)$$

Because the two blocks are connected together by the string, they move with the same speed and have accelerations of the same magnitude, block 1 to the right and block 2 down.

Solve and evaluate After substituting a for the two accelerations in Eqs. (4.5) and (4.6), the two equations now have the same two unknown quantities T and a.

$$m_1 a = T$$
$$m_2 a = -T + m_2 g$$

We substitute the expression for T from the first equation into the second to get

$$m_2 a = -m_1 a + m_2 g$$

After moving the terms containing the acceleration to the left side, factoring a out, and dividing both sides by the sum of the masses, we have

$$a = \frac{m_2 g}{m_1 + m_2}$$

(CONTINUED)

Note that the acceleration is less than g! Also note that if the mass $m_1 \gg m_2$ (m_1 much greater than m_2), then the acceleration is almost zero and can be measured easily. We can now determine the force exerted by the string on the hanging block by inserting the above expression for the acceleration into $m_1 a = T$:

$$T = m_1 a = m_1 \frac{m_2 g}{m_1 + m_2} = \frac{m_1}{m_1 + m_2} m_2 g$$

Are the results reasonable? Consider the equation for acceleration. The only force pulling the hanging block down is the gravitational force $m_2 g$ that Earth exerts on the hanging block. But because the two blocks are connected, this force has to cause the sum of their masses to accelerate $[m_2 g = (m_1 + m_2)a]$, making the acceleration less than g.

Try it yourself Determine an expression for acceleration for a regular Atwood machine for which two objects of mass m_1 and m_2 move in the vertical direction ($m_1 > m_2$).

Answer

$$v = \frac{(m_1 - m_2)g}{m_1 + m_2}$$

There are two important ideas to take away from this example:

1. If we assume that the pulley is light, the magnitudes of the forces that the string on both sides of the pulley exerts on the two objects are the same.
2. If we assume that the string connecting the two objects does not stretch, the magnitudes of the accelerations of the two objects are the same.

If the above two assumptions are valid, the consequences apply to all problems involving pulleys.

We can see now how to use a modified or a regular Atwood machine to determine the acceleration g. If you measure the distance that one of the objects travels (Δy) during a particular time interval (Δt) after being released from rest, you can use the expression

$$\Delta y = \tfrac{1}{2} a (\Delta t)^2$$

to determine the acceleration of the objects. Using this value for acceleration and the values for their masses, you can use the expression

$$a = \frac{m_2 g}{m_1 + m_2}$$

for a modified machine or

$$a = \frac{(m_1 - m_2)g}{m_1 + m_2}$$

for a regular one to determine the value of g.

REVIEW QUESTION 4.4 For problems involving objects moving upward or downward along inclined surfaces, we choose the x-axis parallel to the surface and the y-axis perpendicular to the surface. Why not use horizontal and vertical axes?

4.5 Projectile motion

So far in this chapter, we have used the knowledge of Newton's laws and force components to explain the motion of objects moving along a straight line when the forces exerted on the objects have not all been in the direction of motion. Let's now look at some common objects that move along a curved path: *projectiles*. **Projectiles** are objects launched at an angle relative to a horizontal surface. We can use Newton's second law to analyze and explain their **projectile motion**. We will begin by constructing a qualitative explanation and then develop a quantitative description. Let's start by making some observations.

Qualitative analysis of projectile motion

You can analyze projectile motion with a basketball. First, throw the ball straight upward. It moves up and then down with respect to you and with respect to the floor and eventually returns to your hands (**Figure 4.10a**). Next, walk in a straight line at constant speed and throw the ball straight up again. Have a friend videotape you from the side. The ball goes up and returns to your hands as before, but this time you've moved along your walking path. Your friend sees a projectile—a ball traveling in an arc. Finally, put on Rollerblades and repeat the experiment several times, each time moving with greater speed. As long as you don't change your speed or direction while the ball is in flight, it lands back in your hands. The video shows that the ball moves in an arc with respect to the ground, and at every frame it is directly above your hands (Figure 4.10b). Why does this happen?

One possible explanation is that the ball moves vertically with respect to you independently of its horizontal motion with respect to the floor. Let us analyze projectile motion by considering the ball's vertical and horizontal motion independently. For you, the ball moves up and down like the ball in Figure 4.10a. An observer on the floor explains the vertical motion of the ball in the same way. Both observers are in inertial reference frames. Earth exerts a gravitational force on the ball, so its upward speed decreases until it stops at the highest point, and then its downward speed increases until it returns to your hands (the vertical component of the ball's acceleration is constant due to the gravitational force exerted on it by Earth).

In addition to this vertical motion, the ball also moves horizontally with respect to the floor, with the same constant speed as you. No object exerts a horizontal force on the ball. Thus, according to Newton's first law, the ball's horizontal velocity does not change once it is released. This velocity is the same as your velocity due to the way the ball was released. In every experiment the ball continues moving horizontally as if it were not thrown upward, and it moves up and down as if it does not move horizontally. It seems that the hypothesis that the horizontal and vertical motions of the ball are independent of each other can account for all our observations. Let's test this explanation in Testing Experiment **Table 4.6**.

FIGURE 4.10 A projectile launched by a moving person.

(a)

You throw the ball straight upward while standing at a fixed position.

(b) While moving, you throw the ball straight upward relative to yourself.

Notice that the ball is always above your hands.

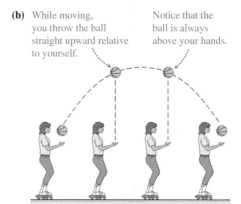

TESTING EXPERIMENT TABLE 4.6 Testing the independence of horizontal and vertical motions

 VIDEO TET 4.6

Testing experiment	Prediction	Outcome
At time zero, ball 1 is dropped. Simultaneously, ball 2 is shot horizontally when a compressed spring is released. Which ball hits the surface first?	Both balls start with zero initial vertical speed; thus their vertical motions are identical. Since we think that the vertical motion is independent of the horizontal motion, we predict that they will land at the same time.	When we try the experiment, the balls do land at the same time.

Conclusion

The outcome supports the hypothesis that horizontal and vertical motions are independent of each other.

The prediction may seem counterintuitive. However, the result matched the prediction. Ball 2 travels a longer path than ball 1, so why doesn't it land later? The vertical motions of both projectiles are identical; thus they land at the same time. The second ball moves forward at constant velocity while it is falling, but this horizontal motion does not affect its vertical fall.

Let us now analyze the experiment in Table 4.6 quantitatively. Ball 1 is dropped from a height of 1.0 m and ball 2 is simultaneously shot horizontally from the same height at a speed of 5.0 m/s (see **Figure 4.11**). We point the vertical axis down and the horizontal axis to the right. The origin for both axes is where the balls are released. Ball 1 falls at constant acceleration because the only object exerting a force on it is Earth; its equation of motion is $y_{1f} = y_{1i} + v_{1iy}t + \frac{1}{2}a_y t_1^2 = \frac{1}{2}g t_1^2$, where we used $y_{1i} = 0$; $v_{1iy} = 0$. The *time of flight* is therefore

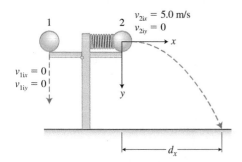

FIGURE 4.11 Finding the range of a projectile thrown horizontally.

$$t_1 = \sqrt{\frac{2y_{1f}}{g}} = \sqrt{\frac{2.0 \text{ m}}{9.8 \text{ m/s}^2}} \approx 0.5 \text{ s}$$

Ball 2 is also falling; its time of flight is the same as that for ball 1, $t_2 = t_1 = 0.5$ s. However, while falling, ball 2 is also moving horizontally. The displacement of the ball is $d_{2x} = v_{2ix}t_2 = (5.0 \text{ m/s})(0.5 \text{ s}) = 2.5$ m. This is a reasonable number because the ball does not land too far from the table. Notice that the time in which ball 2 moves horizontally is the same as the time in which it falls vertically. We can now write the expression for the horizontal distance that the ball travels during the flight as

$$d_{2x} = v_{2ix}\sqrt{\frac{2y_{1f}}{g}} \quad \text{where } y_{1f} = y_{2f}$$

This expression is true for a horizontal throw of any projectile from height *y*:

$$d_x = v_{ix}\sqrt{\frac{2y}{g}}$$

Distance *d* is called the *range*.

We now understand both conceptually and quantitatively the motion of a ball thrown horizontally. Let us next analyze the motion of the ball thrown at an angle to the horizontal in Conceptual Exercise 4.8.

CONCEPTUAL EXERCISE 4.8 ▸ **Throwing a ball**

You throw a tennis ball as a projectile. Draw an arrow or arrows representing its instantaneous velocity and acceleration and the force or forces exerted on the ball by other objects when at the three positions shown below.

Sketch and translate The ball is the system of interest.

Simplify and diagram To draw velocity arrows, we consider the velocity at each point as consisting of a constant horizontal vector component and a vertical vector component whose magnitude decreases as the ball moves up and increases as it moves down. The vertical component is zero at the highest point in the trajectory. If we do this very carefully, we find that the velocity arrows are

tangent to the ball's path at each position. If we ignore air resistance, only one object exerts a force on the ball—Earth. Thus in our sketch, the force arrows point down at all three positions as shown below.

The acceleration arrows point in the same direction as the sum of the forces—downward. Notice that at the top of the path, the velocity of the object is horizontal; however, the acceleration still points downward. The direction of velocity at each point and the direction of the sum of the forces exerted on the object at each point do not have to be the same.

Quantitative analysis of projectile motion

In Table 4.6, we launched a projectile horizontally. Now we will develop the mathematical description for the motion of a projectile launched at an arbitrary angle. To start, we need to choose the direction of our coordinate axes. Here, we show the analysis for the x-axis pointed to the right and the y-axis pointed up. However, we could have chosen the downward direction as positive for the y-axis. For every case, you need to decide how to point the axes for greatest convenience.

The x-component of a projectile's acceleration in the horizontal direction is zero $(a_x = 0)$. The y-component of the projectile's acceleration in the vertical direction is $a_y = -g$ (choosing the y-axis to point up). Consider the projectile shown in **Figure 4.12**. When the projectile is launched at speed v_0 at an angle θ relative to the horizontal, its initial x- and y-velocity components are $v_{0x} = v_0 \cos \theta$ and $v_{0y} = v_0 \sin \theta$. The x-component of velocity remains constant during the flight, and the y-component changes in the same way that the velocity of an object thrown straight upward changes. If the motions in the x- and y-directions are independent of each other, then the equations that describe the motion of a projectile become the following:

FIGURE 4.12 The initial velocity component vectors and their magnitudes.

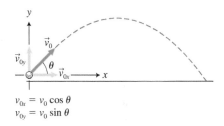

$v_{0x} = v_0 \cos \theta$
$v_{0y} = v_0 \sin \theta$

Projectile motion in the x-direction		Projectile motion in the y-direction	
$(a_x = 0)$		$(a_y = -g)$	
$v_x = v_{0x} = v_0 \cos \theta$	(4.7x)	$v_y = v_{0y} + a_y t = v_0 \sin \theta + (-g)t$	(4.7y)
$x = x_0 + v_{0x}t$		$y = y_0 + v_{0y} t + \frac{1}{2}a_y t^2$	
$= x_0 + (v_0 \cos \theta)t$	(4.8x)	$= y_0 + (v_0 \sin \theta)t - \frac{1}{2}gt^2$	(4.8y)

TIP Equations 4.7 and 4.8 include velocity components, whose signs depends on your choice of x- and y-axis directions. Be sure to check the signs carefully.

Equation (4.8y) can be used to determine the time interval for the projectile's flight, and Eq. (4.8x) can be used to determine how far the projectile travels in the horizontal direction during that time interval. If the projectile leaves ground level and returns to ground level (for example, when you hit a golf ball), then Eq. (4.8y) with $y_0 = 0$ and $y = 0$ can be solved for t, the time when the ball lands (assuming the ball is hit at time zero). Then Eq. (4.8x) can be used to determine the distance the projectile travels during that time interval.

EXAMPLE 4.9 **Best angle for longest distance**

You want to throw a rock the longest possible horizontal distance. You keep the initial speed of the rock constant and find that the horizontal distance it travels depends on the angle at which it leaves your hand. What is the angle at which you should throw the rock so that it travels the longest horizontal distance, assuming you throw it with the same initial speed?

Sketch and translate We sketch the rock's trajectory, including a coordinate system and the initial velocity of the rock when it leaves your hand. The origin is at the position of the rock at the moment it leaves your hand. We call that initial time $t_0 = 0$.

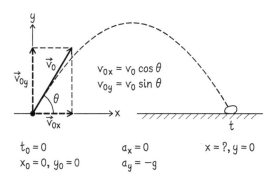

$v_{0x} = v_0 \cos \theta$
$v_{0y} = v_0 \sin \theta$

$t_0 = 0$ $a_x = 0$ $x = ?, y = 0$
$x_0 = 0, y_0 = 0$ $a_y = -g$

(CONTINUED)

Simplify and diagram Assume that the rock leaves at the same elevation as it lands. That's not actually what happens, since the rock leaves your hand several feet above the ground, but it is a reasonable assumption if the rock travels a long distance compared to its initial height above the ground. Assume also that the rock is a point-like object and that the drag force exerted by the air on the rock does not affect its motion.

Represent mathematically First, determine the time interval that the rock is in flight using Eq. (4.8y). Note that because of the coordinate system we chose, $y_0 = y = 0$ and $t_0 = 0$:

$$0 = 0 + (v_0 \sin \theta)t - \tfrac{1}{2}gt^2$$

Dividing both sides of the equation by t and then solving for t, we obtain

$$t = \frac{2v_0 \sin \theta}{g}$$

This is the time of flight—we see that it depends only on the initial velocity of the rock and the launch angle. To determine the distance that the rock travels in the horizontal direction, insert this expression for t into Eq. (4.8x), and note that $x_0 = 0$:

$$x = (v_0 \cos \theta)t = v_0 \cos \theta \left(\frac{2v_0 \sin \theta}{g} \right) = \frac{v_0^2 \, 2 \sin \theta \cos \theta}{g}$$

Solve and evaluate Recall from trigonometry that $2 \sin \theta \cos \theta = \sin(2\theta)$. This lets us rewrite the above equation as

$$x = \frac{v_0^2 \sin 2\theta}{g}$$

Our goal is to determine what launch angle will result in the rock traveling the farthest. Remember that the sine of an angle has a maximum value of 1 when the angle is 90°. Thus, $\sin 2\theta = 1$ when $2\theta = 90°$ or when $\theta = 45°$. The rock will travel the maximum horizontal distance when you launch it at an angle of 45° above the horizontal.

This answer seems reasonable. If you throw the rock at a smaller angle, the vertical component of the velocity is small and it spends less time in flight. If you launch it at an angle closer to 90°, it spends more time in flight, but its horizontal velocity component is small, so it does not travel very far horizontally. The 45° angle is a compromise between long time of flight and large horizontal velocity.

Try it yourself Will the rock travel farther if it is launched at a 60° angle or at a 30° angle?

Answer

It will travel the same distance in each case (see below).

A longer time in the air but a slower horizontal speed

A shorter time in the air but a faster horizontal speed

Human cannonballs

Calculating projectile motion is not always as simple as Eqs. (4.7) and (4.8) suggest. Human cannonball launches must use complicated formulas to account for the effect of air resistance. Results obtained from Eqs. (4.7) and (4.8) are correct only in the absence of air resistance or if air resistance is small. Even so, we will use the simplified equations in the next example.

EXAMPLE 4.10   **Shot from a cannon**

Stephanie Smith Havens (sister of David Smith, Jr.) is shot from a cannon at speed 20 m/s. The end of the cannon's barrel is 8.0 m above the ground, and the barrel is oriented at 53° above the horizontal. Stephanie enjoys her flight until she touches the net, positioned 2.0 m above ground level. The net stretches to reduce the force exerted on her while stopping her. (1) Determine the time Stephanie spends in flight. (2) Determine the horizontal distance from the end of the barrel where the net should be placed so that she lands safely.

Sketch and translate We make a sketch of Stephanie's flight, including a coordinate system, the known information, and the desired unknowns. We call the clock reading when she leaves the barrel $t_0 = 0$. The origin of the coordinate system is at ground level directly below the end of the barrel.

$t_0 = 0$ $a_x = 0$ $t = ?$
$x_0 = 0$ $a_y = -g$ $x = ?$
$y_0 = 8.0$ m $y = 2.0$ m
$v_0 = 20.0$ m/s
$\theta = 53°$

Simplify and diagram We assume that Stephanie can be treated as a point-like object and that air resistance can be ignored. (This is a reasonable assumption if her flight is not too long, but it might not be appropriate for the proposed flight. We will, however, ignore air resistance, as we do not yet have the tools to deal with it.) A force diagram for Stephanie as a system is shown at right.

$\vec{F}_{\text{E on S}}$

Represent mathematically (1) First determine the time that Stephanie is in flight using Eq. (4.8y):

$$y - y_0 = v_0 \sin 53° t + \tfrac{1}{2}(-9.8 \text{ m/s}^2)t^2$$

where $y = 2.0$ m, $y_0 = 8.0$ m, $v_0 = 20$ m/s, and $\sin 53° = 0.80$. (2) To determine the horizontal distance $x - x_0$ Stephanie travels to reach the net, use the time results from Part 1 and Eq. (4.8x):

$$x - x_0 = (20 \text{ m/s}) \cos 53° t \quad \text{where } \cos 53° = 0.60$$

Solve and evaluate (1) We solve the following quadratic equation to find the flight time:

$$2.0 \text{ m} - 8.0 \text{ m} = (20 \text{ m/s}) \sin 53° t + \tfrac{1}{2}(-9.8 \text{ m/s}^2)t^2$$

or

$$0 = (-4.9 \text{ m/s}^2)t^2 + (16 \text{ m/s})t + 6.0 \text{ m}$$

The above is a quadratic equation, $t_{1,2} = \dfrac{-b \pm \sqrt{b^2 - 4ac}}{2a}$, where $a = -4.9$ m/s^2, $b = 16$ m/s, and $c = 6.0$ m. The solutions are

$$t_{1,2} = \frac{-(16 \text{ m/s}) \pm \sqrt{(16 \text{ m/s})^2 - 4(-4.9 \text{ m/s}^2)(6.0 \text{ m})}}{2(-4.9 \text{ m/s}^2)}$$

$$\Rightarrow t_1 = -0.340 \text{ s}, t_2 = 3.61 \text{ s}$$

The negative solution is a hypothetical time before the actual process started. The time when she actually reaches the net is 3.61 s.

(2) The horizontal distance Stephanie travels during the 3.61 s is

$$x - x_0 = (20 \text{ m/s})(\cos 53°)t = 43 \text{ m}$$

This is a reasonable number and has the correct units. However, remember that we ignored the effects of air resistance in calculating this distance. In real life, air resistance needs to be taken into account.

Try it yourself Suppose that the following equations describe Stephanie's next cannonball performance, which has different orientation of the cannon, location of the end of the cannon barrel, and location of the net. Solve the equations for the unknowns and invent a problem that is consistent with the equations.

$$0 = 19.6 \text{ m} + 0 + \tfrac{1}{2}(-9.8 \text{ m/s}^2)t^2; \quad 40 \text{ m} - 0 = (v_0 \cos 0°)t$$

Answer Stephanie is shot horizontally from a horizontally oriented cannon barrel that is 19.6 m above the net. She leaves the barrel at speed 20 m/s. It takes 2.0 s for her to fall the 19.6 m to the level of the net.

REVIEW QUESTION 4.5 Why do we need to resolve the initial velocity of a projectile into components when we analyze situations involving projectiles?

4.6 Starting and stopping a car

Each second of our lives is affected by phenomena that can be described and explained using Newton's laws—the cycles of day and night, the four seasonal variations, walking, driving cars, riding bicycles, and throwing and catching balls. We have already explored walking as a function of Newton's laws. In this section, we will analyze another everyday phenomenon, starting and stopping a car.

When a car is moving, the wheels turn relative to the road surface. The part of the tire immediately behind the part in contact with the surface is lifting up off the road, and the part of the tire immediately in front of the part in contact is moving down to make new contact with the road. But the part of the tire in contact with the road is at rest with respect to the road—for just a short time interval (**Figure 4.13a**)—just as your foot is stationary relative to the ground as your body moves forward while walking or running.

Increasing or decreasing the car's speed involves static friction between the tire's region of contact and the pavement. How can this be? If you want to move faster, the tire turns faster and pushes back harder on the pavement (Figure 4.13b). The pavement in turn pulls forward more on the tire (Newton's third law) and helps accelerate the car forward. If you want to slow down, the tire turns slower and pulls forward on the pavement (Figure 4.13c). The pavement in turn pushes back on the tire (Newton's third law) and helps the car accelerate backwards. Thus, the static friction force exerted by the road on the car can cause the car to speed up or slow down depending on the direction of that force.

FIGURE 4.13 Static friction helps a car accelerate.

(a) Tire is moving to the right at constant speed.

Tire lifting off road · Tire making new contact · Tire not moving

(b) Tire is moving to the right and turning faster.

$\vec{f}_{\text{s Road on Tire}}$

If the tire turns faster, it pushes back harder on the road. The road in turn pushes forward on the tire, causing the car to accelerate to the right.

(c) Tire is moving to the right and turning slower.

$\vec{f}_{\text{s Road on Tire}}$

If the tire turns slower, it pulls forward on the road and the road pushes back on the tire, exerting a force that slows the car, causing it to accelerate backward.

Because the coefficient of static friction between the tire and the road is greater than the coefficient of kinetic friction, stopping is more efficient if your tires do not skid but instead roll to a stop. This is why vehicles are equipped with antilock brakes: to keep the tires rolling on the road instead of locking up and skidding.

Static friction not only helps a car speed up and slow down but also plays an important role in maintaining a car's constant speed. Without this additional force, the car would eventually slow down, because oppositely directed forces, such as air resistance and rolling resistance, point opposite the car's forward motion. A car's engine makes the axle and car tires rotate. The tires push back on the road and the road in turn pushes forward on the car in the opposite direction. The forward static friction force that the road exerts on the car keeps the car moving at constant speed. The presence of air resistance and rolling resistance explains why when you are driving you must continually give the engine gas; otherwise, you would not be able to maintain constant speed and your car would slow down.

> TIP Some people think that the car's engine exerts a force on the car that starts the car's motion and helps it maintain its constant speed despite the air resistance. However, the forces that the engine exerts on other parts of the car are internal forces. Only external forces exerted by objects in the environment can affect the car's acceleration. The engine does rotate the wheels, and the wheels push forward or back on the ground. However, it is the ground (an external object) that pushes backward or forward on the wheels causing the car to slow down or speed up. The force that is responsible for this backward or forward push is the static friction force that the road exerts on the car tires.

CONCEPTUAL EXERCISE 4.11 ⟩ **Car going down a hill**

Your car moves down a 6.0° incline at a constant speed of 16 m/s, and is not skidding. Describe everything you can about the friction force that the road exerts on your car.

Sketch and translate Sketch the situation as shown. Since the car is moving at constant velocity, the sum of the forces that other objects exert on the car must be zero. We use this information to help construct a force diagram for the car and then determine the direction and relative magnitude of the friction force that the road exerts on the car.

Simplify and diagram Model the car as a point-like object and assume that air resistance does not significantly affect the motion of the car. Make a force diagram for the car. The x-axis points in the direction of motion, and the y-axis points perpendicular to the road surface. Earth exerts a downward force on the car, and the surface exerts a force that we resolve into two vector components: the component perpendicular to the surface (the normal force) and the component parallel to the surface (the static friction force). This static friction force points in the direction opposite

to the car's motion, similarly to the situation in Figure 4.13c. To achieve such a direction of the static friction force, the driver must use the brakes to slow down the rotation of the wheels. According to the force diagram, the angle between the arrow for the gravitational force and the y-axis is 6°. If we had included air resistance, the static friction force would not have had to be as large in order to balance the x-component of the gravitational force.

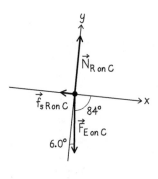

Try it yourself A car moves at a high, constant velocity on a horizontal, level road. Does the road exert a static friction force on the car, and if so, in what direction?

Answer

Air resistance becomes more important at high speeds and can exert a several hundred newton force on the car that opposes its motion. Thus, the road now has to exert a static friction force on the car tires that pushes the car in the forward direction—the tires push back on the road, and the road in turn pushes forward on the tires.

REVIEW QUESTION 4.6 You read in this section that it is the road and not the engine that is most directly responsible for a car speeding up. Why is this true?

Summary

Components of a vector quantity
A vector quantity such as force $\vec{F}$ can be broken into its scalar components F_x and F_y, which indicate the effect of the force in two perpendicular directions. (Section 4.1)

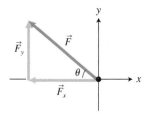

$$F_x = \pm F \cos \theta \qquad \text{Eq. (4.1x)}$$
$$F_y = \pm F \sin \theta \qquad \text{Eq. (4.1y)}$$

θ is the angle relative to the $\pm x$-axis. A scalar component is positive if it falls along a positive axis and negative if it falls along a negative axis.

Newton's second law in component form The acceleration of an object in the x-direction is the sum of the x-components of the forces exerted on it divided by its mass. The acceleration of the object in the y-direction is the sum of the y-components of the forces divided by its mass. (Section 4.2)

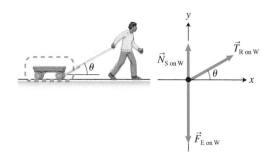

$$a_{Sx} = \frac{F_{1 \text{ on } Sx} + F_{2 \text{ on } Sx} + F_{3 \text{ on } Sx} + \cdots}{m_S} \qquad \text{Eq. (4.2x)}$$

$$a_{Sy} = \frac{F_{1 \text{ on } Sy} + F_{2 \text{ on } Sy} + F_{3 \text{ on } Sy} + \cdots}{m_S} \qquad \text{Eq. (4.2y)}$$

Static friction force When two objects are in contact and we try to move one across the other, they exert forces on each other. Components of these forces parallel to the surfaces are static friction forces; they oppose the tendency of the objects to move with respect to each other. Static friction forces are exerted on both interacting objects, but we are usually concerned with one of the objects, which we call the system. This static friction force changes magnitude as needed to prevent motion of the system, up to a certain maximum value. (Section 4.3)

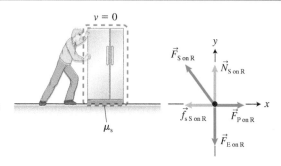

The maximum static friction force depends on the roughness of the two surfaces (on the coefficient of static friction μ_s between the surfaces), and it is related to the magnitude of the normal force $N_{S \text{ on } O}$ component exerted by one surface on the other. The magnitude of the static friction force is always less than or equal to the product of these two quantities:

$$0 \le f_{s \text{ S on O}} \le \mu_s N_{S \text{ on } O} \qquad \text{Eq. (4.3)}$$

Kinetic friction force When an object slides along a surface, the parallel to the surface components of the forces that the object and surface exert on each other are called kinetic friction forces. (Section 4.3)

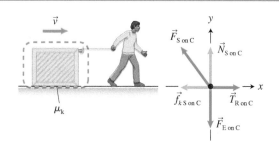

The kinetic friction force depends on the surfaces themselves (on the coefficient of kinetic friction μ_k) and is related to the magnitude of the normal force $N_{S \text{ on } O}$ component exerted by one surface on the other.

$$f_{k \text{ S on O}} = \mu_k N_{S \text{ on } O} \qquad \text{Eq. (4.4)}$$

Projectile motion Projectiles are objects launched either horizontally or at an angle with the horizontal. Horizontal and vertical motions of projectiles are independent of each other. If we ignore the resistive force that air exerts on the projectile, then its horizontal acceleration is zero ($a_x = 0$) and its vertical acceleration is the free-fall acceleration ($a_y = -g$). (Section 4.5)

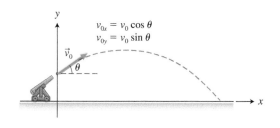

x-direction equations:
$$x = x_0 + v_{0x} t \qquad \text{Eq. (4.8x)}$$
$$v_x = v_0 \cos \theta = \text{constant} \qquad \text{Eq. (4.7x)}$$

y-direction equations (if positive direction up):
$$y = y_0 + v_{0y} t + \tfrac{1}{2} a_y t^2 \qquad \text{Eq. (4.8y)}$$
$$v_y = v_0 \sin \theta + (-g)t \qquad \text{Eq. (4.7y)}$$

Questions

Multiple Choice Questions

1. A car accelerates along a road. Identify the best combination of forces exerted on the car in the horizontal direction that explains the acceleration.
   (a) The force exerted by the engine in the direction of motion and the kinetic friction force exerted by the road in the opposite direction
   (b) The static friction force exerted by the road in the direction of motion and the kinetic friction force exerted by the road in the opposite direction
   (c) The static friction force exerted by the road in the direction of motion and the friction force exerted by the air in the opposite direction

2. A person pushes a 10-kg crate exerting a 200-N force on it, but the crate's acceleration is only 5 m/s². Explain.
   (a) The crate pushes back on the person.
   (b) Not enough information is given.
   (c) There are other forces exerted on the crate.

3. Compare the ease of pulling a lawn mower and pushing it.
   (a) They are the same.   (b) Pulling is easier.
   (c) Pushing is easier.   (d) Not enough information is given.

4. You simultaneously release two balls: one you throw horizontally, and the other you drop straight down. Which one will reach the ground first? Why?
   (a) The ball dropped straight down lands first, since it travels a shorter distance.
   (b) Neither. Their vertical motion is the same, so they will reach the ground at the same time.
   (c) It depends on the mass of the balls—the heavier ball falls faster so lands first.

5. You shoot an arrow with a bow. The following is Newton's second law applied for one of the instants in the arrow's trip: $m\vec{a} = m\vec{g}$. Choose the instant that best matches the equation description.
   (a) The arrow is accelerating while contacting the bowstring.
   (b) The arrow is flying up.
   (c) The arrow is slowing down while sinking into the target.
   (d) This equation does not describe any part of the arrow's trip.

6. In what reference frame does a projectile launched at speed v_0 at angle θ above the horizontal move only in the vertical direction?
   (a) There is no such reference frame.
   (b) In a reference frame that moves with the projectile
   (c) In a reference frame that moves horizontally at speed $v_0 \cos\theta$
   (d) In a reference frame that moves horizontally at speed v_0

7. In Table 4.6 we found that the two balls touched the ground at the same time. Which of the following quantities is/are also equal for both balls after they have been launched? Multiple answers could be correct.
   (a) Acceleration in the y-direction   (b) Acceleration in the x-direction
   (c) Speed at landing   (d) Average speed

8. While running at constant velocity, how should you throw a ball with respect to you so that you can catch it yourself, without needing to reach for the ball?
   (a) Slightly forward   (b) Slightly backward
   (c) Straight up   (d) It is impossible.

9. You hold a block on a horizontal, frictionless surface. It is connected by a string that passes over a pulley to a vertically hanging block (a modified Atwood machine). What is the magnitude of the acceleration of the hanging object after you release the block on the horizontal surface?
   (a) Less than g (b) More than g (c) Equal to g
   (d) Any of the answers a–c, depending on the masses of the blocks.

10. In the process described in the previous question, what is the magnitude of the force exerted by the string on the block on the horizontal plane after it is released?
   (a) Equal to the force that Earth exerts on the hanging block
   (b) Less than the force exerted by Earth on the hanging block
   (c) More than the force exerted by Earth on the hanging block

11. Suppose that two blocks are positioned on an Atwood machine so that the block on the right of mass m_1 hangs at a lower elevation than the block on the left of mass m_2. Both blocks are at rest. Based on this observation, what can you conclude?
   (a) $m_1 > m_2$ (b) $m_1 = m_2$ (c) $m_1 < m_2$
   (d) You cannot conclude anything with the given information.

12. A box containing some stones is resting on a rough inclined surface. Which of the following actions will definitely *not* make the box start moving down the incline? Multiple answers could be correct.
   (a) Push the box in the direction that is down the incline.
   (b) Remove the box, polish the inclined surface with sandpaper, and put the box back.
   (c) Increase the angle of the incline.
   (d) Take some stones out of the box.
   (e) Put more stones into the box.

13. For the following two questions, the answer options are the same. Select two correct answers for each question from the list below.
   I. A box containing some stones is resting on a rough inclined surface. What do you need to do to increase the force of static friction exerted on the box?
   II. A box containing some stones is moving with constant acceleration down a rough inclined surface. What do you need to do to increase the force of kinetic friction exerted on the box?
   (a) Increase the angle of the incline.
   (b) Decrease the angle of the incline.
   (c) Put more stones into the box.
   (d) Take some stones out of the box.

14. A block is resting on a rough inclined surface. You observe that by increasing the angle of the incline, you can make the block start moving down the incline. Which answer best explains why the block starts moving?
   (a) By increasing the angle of the incline, you increase the component of the force exerted by Earth on the block, which pulls the block down the incline.
   (b) By increasing the angle of the incline, you decrease the maximum static friction force exerted on the block by the surface.
   (c) By increasing the angle of the incline, you simultaneously increase the component of the force exerted by Earth on the block and decrease the maximum static friction force exerted on the block.

Conceptual Questions

15. A box with a heavy television set in it is placed against a box with a toaster oven in it. Both sit on the floor. Draw force diagrams for the TV box and the toaster oven box if you push the TV box to the right, exerting a force $\vec{F}_{\text{Y on TV}}$. Repeat for the situation in which you are exerting a force of exactly the same magnitude and direction, but pushing the toaster oven box, which in turn pushes the TV box.

16. How can an Atwood machine be used to determine the acceleration of freely falling objects if none of the objects used is in the state of free fall?

17. Your friend is on Rollerblades holding a pendulum. You gently push her forward and let go. You observe that the pendulum first swings in the opposite direction (backward) and then returns to the vertical orientation as she coasts forward. (a) Draw a force diagram to explain the behavior of the pendulum bob as your friend is being pushed. (b) How can you use it to determine the acceleration of your friend while you are pushing her? (c) Why does the pendulum return to its vertical orientation after you stop pushing your friend?

18. Explain why a car starts skidding when a driver abruptly applies the brakes.

19. Explain why old tires need to be replaced.

20. Explain how friction helps you to walk.

21. Explain why you might fall forward when you stumble.

22. Explain why you might fall backward when you slip.

23. Explain why the tires of your car can "spin out" when you are caught in the mud.

24. You throw two identical balls simultaneously at the same initial speed, one downward and the other horizontally. Describe and compare their motions in as much detail as you can.

25. Your friend says that the vertical force exerted on a projectile when at the top of its flight is zero. Why would he say this? Do you agree or disagree? If you disagree, how would you convince your friend that your opinion is correct?

26. Your friend says that a projectile launched at an angle relative to the horizontal moves forward because it retains the force of the launcher. Why would she say this? Do you agree or disagree? If you disagree, how would you convince your friend that your opinion is correct?

27. An object of mass m_1 placed on an inclined plane (angle θ relative to the horizontal) is connected by a string that passes over a pulley to a hanging object of mass m_2 ($m_2 \gg m_1$). Draw a force diagram for each object. Do not ignore friction.

28. An object of mass m_1 placed on an inclined plane (angle θ relative to the horizontal) is connected by a string that passes over a pulley to a hanging object of mass m_2 ($m_1 \gg m_2$). Draw a force diagram for each object. Do not assume that friction can be ignored.

29. A horse is pulling a sled. If the force exerted by the horse on the sled has the same magnitude as but the opposite direction to the force exerted by the sled on the horse, how are the horse and the sled able to start moving forward? Choose all the explanations that you consider to be correct.
    (a) If the system is the horse and the sled, then to explain the system's motion we need to consider the external forces exerted on it. The sum of these forces causes the forward acceleration of the system.

(b) If we consider the horse to be the system, and if the sum of the friction force and the force exerted by the sled points in the forward direction, the horse will accelerate in the same direction.
(c) For a brief instant right before the sled starts moving, the horse pulls slightly harder on the sled than the sled pulls back on it. This slight imbalance allows them both to start moving.

30. If you kick a block so that it starts moving up a rough incline, you observe that the acceleration of the block while moving up the incline is different from the acceleration when the block is sliding back down the incline. (a) Draw a force diagram for the block when it is moving up the incline and down the incline. (b) Explain why the accelerations are different and compare their magnitudes.

Problems

Below, **BIO** indicates a problem with a biological or medical focus. Problems labeled **EST** ask you to estimate the answer to a quantitative problem rather than derive a specific answer. Asterisks indicate the level of difficulty of the problem. Problems with no * are considered to be the least difficult. A single * marks moderately difficult problems. Two ** indicate more difficult problems.

4.1 Vectors in two dimensions and force components

1. Determine the x- and y-components of each force vector shown in **Figure P4.1**.

FIGURE P4.1   FIGURE P4.2

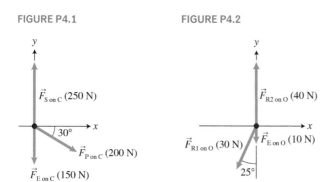

4.2 and 4.3 Newton's second law in component form and Friction

6. * Three ropes pull on a knot shown in **Figure P4.6a**. The knot is not accelerating. A partially completed force diagram for the knot is shown in Figure P4.6b. Use qualitative reasoning to determine the magnitudes of the forces that ropes 2 and 3 exert on the knot. Explain in words how you arrived at your answer.

FIGURE P4.6

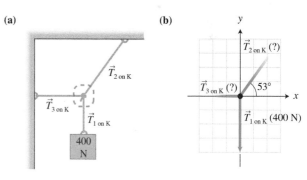

2. Determine the x- and y-components of each force vector shown in **Figure P4.2**.

3. * Determine the x- and y-components of each displacement shown in **Figure P4.3**.

FIGURE P4.3

4. * The x- and y-components of several unknown forces are listed below (F_x, F_y). For each force, draw on an x, y coordinate system the components of the force vectors. Determine the magnitude and direction of each force: (a) $(+100 \text{ N}, -100 \text{ N})$, (b) $(-300 \text{ N}, -400 \text{ N})$, and (c) $(-400 \text{ N}, +300 \text{ N})$.

5. * The x- and y-scalar components of several unknown forces are listed below (F_x, F_y). For each force, draw an x, y coordinate system and the vector components of the force vectors. Determine the magnitude and direction of each force: (a) $(-200 \text{ N}, +100 \text{ N})$, (b) $(+300 \text{ N}, +400 \text{ N})$, and (c) $(+400 \text{ N}, -300 \text{ N})$.

7. * **Figure P4.7** shows an unlabeled force diagram for a hockey puck. The length of the sides of the square grid corresponds to a force magnitude of 1 N. Draw a similar square grid on paper and then draw the vector for the force that should be exerted on the puck so that the puck (a) moves with constant speed, (b) moves with constant acceleration toward the north, (c) moves with constant acceleration toward the west, and (d) moves with constant acceleration toward the east.

FIGURE P4.7

8. For each of the following situations, draw the forces exerted on the moving object and identify the other object causing each force. (a) You pull a wagon along a level floor using a rope oriented 45° above the horizontal. (b) A bus moving on a horizontal road slows in order to stop. (c) You slide down an inclined waterslide. (d) You lift your overnight bag into the overhead compartment on an airplane. (e) A rope connects two boxes on a horizontal floor, and you pull horizontally on a second rope attached to the right side of the right box (consider each box separately).

9. * Write Newton's second law in component form for each of the situations described in Problem 4.8.

10. For the situations described here, construct a force diagram for the block, sled, and skydiver. (a) A cinder block sits on the ground. (b) A rope pulls at an angle of 30° relative to the horizontal on a sled moving on a horizontal surface. The sled moves at increasing speed toward the right (the surface is not smooth.) (c) A rope pulls on a sled parallel to an inclined slope (inclined at an arbitrary angle). The sled moves at increasing speed up the slope. (d) A skydiver falls downward at constant terminal velocity (air resistance is present).

11. * Write Newton's second law in component form for each of the situations described in Problem 4.10.

12. * Apply Newton's second law in component form for the force diagram shown in Figure P4.1.

13. * Apply Newton's second law in component form for the force diagram shown in Figure P4.2.

14. * **Equation Jeopardy 1** The three sets of equations below are the horizontal x- and vertical y-component forms of Newton's second law applied to three physical processes. For each case, solve for the unknowns. Then work backward and construct a force diagram for the object of interest and invent a problem for which the equations could be an answer (there are many possibilities). Note the difference between N (unit of force, newton) and N (the symbol for the normal force).

(a)
$$(5.0\,\text{kg})a_x = (50\,\text{N})\cos 30° + N\cos 90°$$
$$+ (5.0\,\text{kg})(9.8\,\text{N/kg})\cos 90°$$
$$(5.0\,\text{kg})0 = (-50\,\text{N})\sin 30° + N\sin 90°$$
$$- (5.0\,\text{kg})(9.8\,\text{N/kg})\sin 90°$$

(b) This process involves an object on an incline.
$$(5.0\,\text{kg})a_x = (30\,\text{N})\cos 30° + N\cos 90°$$
$$- (5.0\,\text{kg})(9.8\,\text{N/kg})\cos 60°$$
$$(5.0\,\text{kg})0 = (30\,\text{N})\sin 30° + N\sin 90°$$
$$- (5.0\,\text{kg})(9.8\,\text{N/kg})\sin 60°$$

(c)
$$(5.0\,\text{kg})a_x = (50\,\text{N})\cos 30° + N\cos 90°$$
$$+ (5.0\,\text{kg})(9.8\,\text{N/kg})\cos 90°$$
$$(5.0\,\text{kg})0 = (-50\,\text{N})\sin 30° + N\sin 90°$$
$$- (5.0\,\text{kg})(9.8\,\text{N/kg})\sin 90°$$
$$x - 0 = (2.0\,\text{m/s})(4.0\,\text{s}) + \tfrac{1}{2}a_x(4.0\,\text{s})^2$$

Provide all the information you know about your process.

15. * You exert a force of 100 N on a rope that pulls a sled across a very smooth surface. The rope is oriented 37° above the horizontal. The sled and its occupant have a total mass of 40 kg. The sled starts at rest and moves 10 m. List all the quantities you can determine using these givens and determine three of the quantities on the list.

16. * You exert a force of a known magnitude F on a grocery cart of total mass m. The force you exert on the cart points at an angle θ below the horizontal. If the cart starts at rest, determine an expression for the speed of the cart after it travels a distance d. Ignore friction.

17. * **Olympic 100-m dash start** At the start of his race, 86-kg Olympic 100-m champion Usain Bolt from Jamaica pushes against the starting block, exerting an average force of 1700 N. The force that the block exerts on his foot points 20° above the horizontal. Determine his horizontal speed after the force is exerted for 0.32 s. Indicate any assumptions you made.

18. * **Your own accelerometer** A train has an acceleration of magnitude 1.4 m/s² while stopping. A pendulum with a 0.50-kg bob is attached to a ceiling of one of the cars. Determine everything you can about the pendulum during the deceleration of the train.

19. * **EST** Finn and Hazel are using a battery-powered fan fixed on a low-friction cart. As the fan blades rotate, they exert a force on the air and therefore the air exerts an equal and opposite force on the blades, causing the cart to move at constant acceleration. They decide to investigate how

the acceleration of the cart depends on the tilt of the fan. They conduct two experiments: one with the fan oriented horizontally (see **Figure P4.19a**) and one by tilting the axis of the fan by a certain angle (Figure P4.19b). In both experiments, they hold the cart still, turn on the fan, then let the cart move for some time before stopping the cart. During the experiment they measure the acceleration of the cart using a motion detector. The graphs in Figure P4.19c show their measurements.

(a) Which graph (A or B) belongs to the cart with the horizontal fan and which to the cart with the tilted fan? Explain. (b) Based on the graph, estimate the angle between the axis of the fan and the horizontal in the second experiment. Indicate any assumptions that you made. Check how your result compares with the angle on the photo (Figure P4.19c) of the actual experimental setup (you will need a protractor). Indicate any assumptions that you made.

FIGURE P4.19

(a)

(b)

(c) a (m/s²)

20. A 91.0-kg refrigerator sits on the floor. The coefficient of static friction between the refrigerator and the floor is 0.60. What is the minimum force that one needs to exert on the refrigerator to start the refrigerator sliding?

21. A 60-kg student sitting on a hardwood floor does not slide until pulled by a 240-N horizontal force. Determine the coefficient of static friction between the student and floor.

22. **Car stopping distance and friction** A car traveling at 60 mi/h (97 km/h) can stop in 48 m on a level road. Determine the coefficient of friction between the tires and the road. Assume that the car starts skidding the moment the driver hits the brakes. Is this kinetic or static friction? Explain.

23. * A 50-kg box rests on the floor. The coefficients of static and kinetic friction between the bottom of the box and the floor are 0.70 and 0.50, respectively. (a) What is the minimum force a person needs to exert on the box to start it sliding? (b) After the box starts sliding, the person continues to push it, exerting the same force. What is the acceleration of the box?

24. * Marsha is pushing down and to the right on a 12-kg box at an angle of 30° below horizontal. The box slides at constant velocity across a carpet whose coefficient of kinetic friction with the box is 0.70. Determine three physical quantities using this information, one of which is a kinematics quantity.

25. * You want to determine the coefficient of kinetic friction between a 1 m × 1 m rug and a wooden floor. You obtain a piece of the same rug of size 10 cm × 10 cm (which has a surface area 100 times smaller than the entire rug), a scale for measuring mass, a force meter, and an object of mass m_O. Which of the following procedures would you choose to obtain the correct value of kinetic friction between the rug and the floor?

  (a) Put the object on the piece of rug. Measure the force $F_{H \text{ on RP}}$ exerted by the hand on the rug piece in the horizontal direction while moving it with constant speed along the wooden floor. Calculate the coefficient of kinetic friction as $\mu_k = \dfrac{F_{H \text{ on RP}}}{m_O g}$.

  (b) Use the same method described in (a) but calculate the coefficient of friction as $\mu_k = \dfrac{100 \cdot F_{H \text{ on RP}}}{m_O g}$.

  (c) Use the same method described in (a) but also measure the mass of the rug piece m_{RP} and calculate the coefficient of friction as $\mu_k = \dfrac{F_{H \text{ on RP}}}{(m_O + m_{RP})g}$.

  (d) Use the same method described in (c) but calculate the coefficient of friction as $\mu_k = \dfrac{100 \cdot F_{H \text{ on RP}}}{(m_O + m_{RP})g}$.

26. * A wagon is accelerating to the right. A book is placed against the interior back wall of the wagon and does not slide down (see **Figure P4.26**). Explain how this can be.

FIGURE P4.26

27. ** In Problem 4.26, the coefficient of static friction between the book and the vertical back of the wagon is μ_s. Determine an expression for the minimum acceleration of the wagon in terms of μ_s so that the book does not slide down. Does the mass of the book matter? Explain.

28. * A car has a mass of 1520 kg. While traveling at 20 m/s, the driver applies the brakes to stop the car on a wet surface with a 0.40 coefficient of friction. (a) How far does the car travel before stopping? (b) If a different car with a mass 1.5 times greater is on the road traveling at the same speed and the coefficient of friction between the road and the tires is the same, what will its stopping distance be? Explain your results.

29. ** A 20-kg wagon accelerates on a horizontal surface at 0.50 m/s² when pulled by a rope exerting a 120-N force on the wagon at an angle of 25° above the horizontal. Determine the magnitude of the effective friction force exerted on the wagon and the effective coefficient of friction associated with this force.

30. ** A crate of mass m sitting on a horizontal floor is attached to a rope that pulls at an angle θ above the horizontal. The coefficient of static friction between the crate and floor is μ_s. (a) Construct a force diagram for the crate when being pulled by the rope but not sliding. (b) Determine an expression for the smallest force that the rope needs to exert on the crate that will cause the crate to start sliding.

31. **EST** You absentmindedly leave your book bag on the top of your car. Estimate the safe acceleration of the car needed for the bag to stay on the roof. Describe the assumptions that you made.

32. * Block 1 is on a horizontal surface with a 0.29 coefficient of kinetic friction between it and the surface. A string attached to the front of block 1 passes over a light frictionless pulley and down to hanging block 2. Determine the mass of block 2 in terms of block 1 so that the blocks move at constant nonzero speed while sliding.

4.4 Skills for analyzing processes involving forces in two dimensions

33. * You want to use a rope to pull a 10-kg box of books up a plane inclined 30° above the horizontal. The coefficient of kinetic friction is 0.30. What force do you need to exert on the other end of the rope if you want to pull the box (a) at constant speed and (b) with a constant acceleration of 0.50 m/s² up the plane? The rope pulls parallel to the incline.

34. * A car with its wheels locked rests on a flatbed of a tow truck. The flatbed's angle with the horizontal is slowly increased. When the angle becomes 40°, the car starts to slide. Determine the coefficient of static friction between the flatbed and the car's tires.

35. * **Olympic skier** Olympic skier Tina Maze skis down a steep slope that descends at an angle of 30° below the horizontal. The coefficient of sliding friction between her skis and the snow is 0.10. Determine Maze's acceleration, and her speed 6.0 s after starting.

36. * **Another Olympic skier** Bode Miller, 80-kg downhill skier, descends a slope inclined at 20°. Determine his acceleration if the coefficient of friction is 0.10. How would this acceleration compare to that of a 160-kg skier going down the same hill? Justify your answer using sound physics reasoning.

37. * A book slides off a desk that is tilted 15° relative to the horizontal. What information about the book or the desk does this number provide?

38. * **Equation Jeopardy 2** The two sets of equations below are the horizontal x- and vertical y-component forms of Newton's second law applied to two physical processes. In both cases, solve for the unknowns. Then work backward and construct a force diagram for the object of interest and invent a problem for which the equations could be an answer (there are many possibilities). Note the difference between N (unit of force, newton) and N (symbol for a normal force).

$$(5.0 \text{ kg})a_x = (50 \text{ N})\cos 30° + N \cos 90° - 0.5N \cos 0°$$
$$+ (5.0 \text{ kg})(9.8 \text{ N/kg})\cos 90°$$
$$(5.0 \text{ kg})0 = (-50 \text{ N})\sin 30° + N \sin 90° + 0.5N \sin 0°$$
$$- (5.0 \text{ kg})(9.8 \text{ N/kg})\sin 90°$$

  (b) This process involves an object on an incline.

$$(5.0 \text{ kg})0 = +F \cos 0° + N \cos 90° - 0.50N \cos 0°$$
$$- (5.0 \text{ kg})(9.8 \text{ N/kg})\cos 60°$$
$$(5.0 \text{ kg})0 = +F \sin 0° + N \sin 90° - 0.50N \sin 0°$$
$$- (5.0 \text{ kg})(9.8 \text{ N/kg})\sin 60°$$

39. * Helge, Steve, and Heidi are sitting on a sled on a slope covered with a hard snow. The sled is stationary. The friends have different suggestions for how to make the sled start moving:

  Helge: If one of us gets off, the sled will start moving.

  Steve: We should invite another person to join us, and then the sled will start moving.

  Heidi: We should get off the sled, polish the bottom of the sled to make it smoother, and sit back down on it. The sled will then start moving.

  Comment on the students' suggestions and decide whose reasoning is correct. Explain using physics concepts, including appropriate diagrams, why those who were incorrect said what they did, and indicate what part of their reasoning was correct.

40. * When traveling on an airplane you get meals on a serving tray that has large coefficients of static and kinetic friction between the tray and dishes on it. For each case below, draw a force diagram for a cup on a serving tray as seen by a stationary observer on Earth and assuming the airplane is moving from left to right. Assume also that the tray is parallel to the velocity of the airplane.

  (a) The airplane is flying horizontally at constant speed; the cup is at rest on the tray.

  (b) The airplane is on a runway and slowing down; the cup is at rest on the tray.

  (c) The airplane is on a runway and slowing down; the cup is sliding. Next to the force diagram draw an arrow that shows the velocity of the cup relative to the airplane.

  (d) The airplane is on a runway and speeding up; the cup is at rest on the tray.

  (e) The airplane is descending at constant speed at an angle 30° with respect to horizontal; the cup is at rest on the tray.

  (f) The airplane is slowing down and descending at an angle 30° with respect to horizontal; the cup is at rest on the tray.

41. **Skier** A 52-kg skier starts at rest and slides 30 m down a hill inclined at 12° relative to the horizontal. List five quantities that describe the motion of the skier, and solve for three of them (at least one should be a kinematics quantity).

42. * **Ski rope tow** You agree to build a backyard rope tow to pull your siblings up a 20-m slope that is tilted at 15° relative to the horizontal. You must choose a motor that can pull your 40-kg sister up the hill. Determine the force that the rope should exert on your sister to pull her up the hill at constant velocity.

43. * **Soapbox racecar** A soapbox derby racecar starts at rest at the top of a 301-m-long track tilted at an average 4.8° relative to the horizontal. If the car's speed were not reduced by any structural effects or by friction, how long would it take to complete the race? What is the speed of the car at the end of the race?

44. * A person is pushing two carts that are connected with a metal bar so that the carts are moving at constant acceleration 0.2 m/s². The masses of the carts are 100 kg and 180 kg, and the mass of the connecting bar is 60 kg. Determine (a) the force that the bar exerts on the smaller cart and (b) the force that the bigger cart exerts on the bar. Assume that the friction forces are negligible.

45. * **BIO Whiplash experience** A car sitting at rest is hit from the rear by a semi-trailer truck moving at 13 m/s. The car lurches forward with an acceleration of about 300 m/s². **Figure P4.45** shows an arrow that represents the force that the neck muscle exerts on the head so that it accelerates forward with the body instead of flipping backward. If the head has a mass of 4.5 kg, what is the horizontal component of the force $\vec{F}$ required to cause this head acceleration? If $\vec{F}$ is directed 37° below the horizontal, what is the magnitude of $\vec{F}$?

FIGURE P4.45

46. **Iditarod race practice** The dogs of four-time Iditarod Trail Sled Dog Race champion Jeff King pull two 100-kg sleds that are connected by a rope. The sleds move on an icy surface. The dogs exert a 240-N force on the rope attached to the front sled. Find the acceleration of the sleds and the force the rope between the sleds exerts on each sled. The front rope pulls horizontally.

47. ** You pull a rope oriented at a 37° angle above the horizontal. The other end of the rope is attached to the front of the first of two wagons that have the same 30-kg mass. The rope exerts a force of magnitude T_1 on the first wagon. The wagons are connected by a second horizontal rope that exerts a force of magnitude T_2 on the second wagon. Determine the magnitudes of T_1 and T_2 if the acceleration of the wagons is 2.0 m/s².

48. * Rope 1 pulls horizontally, exerting a force of 45 N on an 18-kg wagon attached by a second horizontal rope to a second 12-kg wagon. Make a list of physical quantities you can determine using this information, and solve for three of them, including one kinematics quantity.

49. * Three sleds of masses m_1, m_2, m_3 are on a smooth horizontal surface (ice) and connected by ropes, so that if you pull the rope connected to sled 1, all the sleds start moving. Imagine that you exert a force of a known magnitude on the rope attached to the first sled. What will happen to all of the sleds? Provide information about their accelerations and all the forces exerted on them. What assumptions did you make?

50. ** Repeat Problem 4.49, only this time with the sleds on a slope inclined at θ.

51. * A skier is moving down a snowy hill with an acceleration of 0.4 m/s². The angle of the slope is 5° to the horizontal. What is the acceleration of the same skier when she is moving down a hill with a slope of 10°? Assume the coefficient of kinetic friction is the same in both cases.

52. ** A person holds a 200-g block that is connected to a 250-g block by a string going over a light pulley with no friction in the bearing (an Atwood machine). After the person releases the 200-g block, it starts moving upward and the heavier block descends. (a) What is the acceleration of each block? (b) What is the force that the string exerts on each block? (c) How long will it take each block to traverse 1.0 m?

53. ** Two blocks of masses m_1 and m_2 are connected to each other on an Atwood machine. A person holds one of the blocks with her hand. When the system is released, the heavier block moves down with an acceleration of 2.3 m/s² and the lighter object moves up with an acceleration of the same magnitude. What is one possible set of masses for the blocks?

54. ** The 20-kg block shown in **Figure P4.54** accelerates down and to the left, and the 10-kg block accelerates up. Find the magnitude of this acceleration and the force that the cable exerts on a block. There is no friction between the block and the inclined plane, and the pulley is frictionless and light.

FIGURE P4.54

4.5 Projectile motion

55. * A squirrel jumps off a roof in the horizontal direction. The origin of the coordinate system is at the point where the squirrel leaves the roof. Complete **Table P4.55** by drawing crosses in the cells that correctly connect the physical quantities in the first column that describe the motion of the squirrel and the descriptions of what is happening to these quantities while the squirrel is in flight. Consider the squirrel as a point-like object and assume that the resistive force exerted by the air is negligible.

56. * A frog jumps at an angle 30° above the horizontal. The origin of the coordinate system is at the point where the frog leaves the ground. Complete Table P4.55 by drawing check marks in the cells that correctly connect the quantities in the first column that describe the motion of the frog and the descriptions of what is happening to these quantities while the frog is moving. Consider the frog as a point-like object and assume that the resistive force exerted by the air is negligible.

TABLE P4.55

Physical quantity	Remains constant	Is changing	Increases only	Decreases only	Increases, then decreases	Decreases, then increases
x-coordinate magnitude						
y-coordinate magnitude						
Direction of velocity						
Magnitude of velocity						
Direction of acceleration						
Magnitude of acceleration						

57. A bowling ball rolls off a table. Draw a force diagram for the ball when on the table and when in the air at two different positions.

58. * A tennis ball is served from the back line of the court such that it leaves the racket 2.4 m above the ground in a horizontal direction at a speed of 22.3 m/s (50 mi/h). (a) Will the ball cross a 0.91-m-high net 11.9 m in front of the server? (b) Will the ball land in the service court, which is within 6.4 m of the net on the other side of the net?

59. * **Equation Jeopardy 3** The equations below describe a projectile's path. Solve for the unknowns and then invent a process that the equations might describe. There are many possibilities.

$$x = 0 + (20 \text{ m/s})(\cos 0°)t$$

$$0 = 8.0 \text{ m} + (20 \text{ m/s})(\sin 0°)t - \frac{1}{2}(9.8 \text{ m/s}^2)t^2$$

60. * **EST** An airplane is delivering food to a small island. It flies 100 m above the ground at a speed of 160 m/s. (a) Where should the parcel be released so it lands on the island? Neglect air resistance. (b) Estimate whether you should release the parcel earlier or later if there is air resistance. Explain.

61. A ball moves in an arc through the air (see Figure P4.61). Construct a force diagram for the ball when at positions (a), (b), and (c). Ignore the resistive force exerted by the air on the ball.

FIGURE P4.61

62. A marble is thrown as a projectile at an angle above the horizontal. Draw its path during the flight. Choose six different positions along the path so that one of them is at the highest point. For each position, indicate the direction of the marble's velocity, acceleration, and all of the forces exerted on it by other objects. Ignore the resistive force exerted by the air on the marble.

63. * Marbles are exiting a container through a horizontal nozzle positioned 1.3 m above sandy ground (see Figure P4.63). You notice that all the marbles land between 0.25 m and 0.45 m from the point directly below the end of the nozzle. How do you explain this observation? What can you determine based on the data given in the problem? Indicate any assumptions that you made.

FIGURE P4.63

64. * **Robbie Knievel ride** On May 20, 1999, Robbie Knievel easily cleared a narrow part of the Grand Canyon during a world-record–setting long-distance motorcycle jump—69.5 m. He left the jump ramp at a 10° angle above the horizontal. How fast was he traveling when he left the ramp? Indicate any assumptions you made.

65. * Daring Darless wishes to cross the Grand Canyon of the Snake River by being shot from a cannon. She wishes to be launched at 60° relative to the horizontal so she can spend more time in the air waving to the crowd. With what minimum speed must she be launched to cross the 520-m gap?

66. * A football punter wants to kick the ball so that it is in the air for 4.0 s and lands 50 m from where it was kicked. At what angle and with what initial speed should he kick the ball? Assume that the ball leaves 1.0 m above the ground.

67. * If you shoot a cannonball from the same cannon first at 30° and then at 60° relative to the horizontal, which orientation of the cannon will make the ball go farther? How do you know? Under what circumstances is your answer valid? Explain.

68. When you actually perform the experiment described in Problem 4.67, the ball shot at a 60° angle lands closer to the cannon than the ball shot at a 30° angle. Explain why this happens.

69. * You can shoot an arrow straight up so that it reaches the top of a 25-m-tall building. (a) How far will the arrow travel if you shoot it horizontally while pulling the bow in the same way? The arrow starts 1.45 m above the ground. (b) Where do you need to put a target that is 1.45 m above the ground in order to hit it if you aim 30° above the horizontal? (c) What assumptions did you make?

70. * Robin Hood wishes to split an arrow already in the bull's-eye of a target 40 m away. If he aims directly at the arrow, by how much will he miss? The arrow leaves the bow horizontally at 40 m/s.

4.6 Starting and stopping a car

71. * Three force diagrams for a car on a road are shown in Figure P4.71. Indicate as many situations as possible for the car in terms of its velocity and acceleration at that instant for each diagram.

FIGURE P4.71

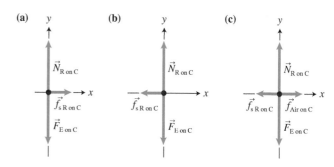

72. * A minivan of mass 1560 kg starts at rest and then accelerates at 2.0 m/s². (a) What is the object exerting the force on the minivan that causes it to accelerate? What type of force is it? (b) Air resistance and other opposing resistive forces are 300 N. Determine the magnitude of the force that causes the minivan to accelerate in the forward direction.

General Problems

73. ** A daredevil motorcycle rider hires you to plan the details for a stunt in which she will fly her motorcycle over six school buses. Assume the ride starts on flat ground. Provide as much information as you can to help the rider successfully complete the stunt.

74. * Emily pulls a 5-kg block across a rough horizontal surface, exerting a constant force on the block. The magnitude of the force is initially 5 N, and the block moves at a constant velocity of 2 m/s. While the block is moving, Emily instantly increases the force to 10 N. How will the block move now?
Andrew says: The block will move at a constant velocity of 4 m/s because it was initially moving with constant velocity and the final force is two times larger than the initial force.
Jasit says: The block will move at a constant acceleration of 2 m/s² because the sum of the forces exerted on the block is 10 N and the mass of the block is 5 kg.
Mary says: The block will move at a constant acceleration of 1 m/s² because the sum of the forces exerted on the block is 5 N and the mass of the block is 5 kg.
Explain which student is correct. Also comment on what is incorrect about the other statements.

75. * **EST** You abruptly push a 1.7-kg book along a table and let go. The book comes to a stop after a short distance. Figure P4.75 shows the acceleration-versus-time graph of the book as recorded by a motion detector. Estimate the following quantities: (a) the coefficient of kinetic friction between the book and the table and (b) the maximum force exerted by your hand on the book.

FIGURE P4.75

76. ** **EST** In the situation of Problem 4.75, estimate (a) the maximum speed of the book and (b) the clock reading when your hand stopped exerting the force on the book.

77. * **EST** Estimate the range of the horizontal force that a sidewalk exerts on you during every step while you are walking. Indicate clearly how you made the estimate.

78. * Two blocks of masses m_1 and m_2 hang at the ends of a string that passes over the very light pulley with low friction bearings shown in **Figure P4.78**. Determine an expression in terms of the masses and any other needed quantities for the magnitude of the acceleration of each block and the force that the string exerts on each block. Apply the equation for two cases: (a) the blocks have the same mass, but one is positioned lower than the other and (b) the blocks have different masses, but the heavier block is positioned higher than the light one. What assumptions did you make?

FIGURE P4.78

79. * A 3.5-kg object placed on an inclined plane (angle 30° above the horizontal) is connected by a string going over a pulley to a 1.0-kg hanging block. (a) Determine the acceleration of the system if there is no friction between the object and the surface of the inclined plane. (b) Determine the magnitude of the force that the string exerts on both objects.

80. ** A 3.5-kg object placed on an inclined plane (angle 30° above the horizontal) is connected by a string going over a pulley to a 1.0-kg object. Determine the acceleration of the system if the coefficient of static friction between object 1 and the surface of the inclined plane is 0.30 and equals the coefficient of kinetic friction.

81. ** An object of mass m_1 placed on an inclined plane (angle θ above the horizontal) is connected by a string going over a pulley to a hanging object of mass m_2. Determine the acceleration of the system if the coefficient of static friction between object 1 and the surface of the inclined plane is μ_s, and the coefficient of kinetic friction is μ_k. If the problem has multiple answers, explore all of them.

82. ** You are driving at a reasonable constant velocity in a van with a windshield tilted 120° relative to the horizontal (see **Figure P4.82**). As you pass under a utility worker fixing a power line, his wallet falls onto the windshield. Determine the acceleration needed by the van so that the wallet stays in place. When choosing your coordinate axes, remember that you want the wallet's acceleration to be horizontal rather than vertical. What assumptions and approximations did you make?

FIGURE P4.82   Wallet

83. * **EST** Figure P4.19c (from Problem 4.19 above) shows acceleration-versus-time graphs for two fan carts. For each cart, estimate (a) the final velocity of the cart and (b) the distance traveled by the cart while it was speeding up. In both experiments the cart started to move from rest. Indicate any assumptions that you made.

84. * In the situation of Problem 2.71 (Chapter 2) and using the data in the accompanying table, determine the coefficient of kinetic friction between Piglet and the wooden floor.

85. * You push a small box so that it starts moving with a speed of 1 m/s along a rough table. After 1 m, the box reaches the edge of the table and lands on the floor 20 cm from the table edge. The surface of the table is 0.8 m above the floor. Determine (a) the coefficient of friction between the box and the table and (b) the speed of the box at the moment it hits the floor.

86. ** Debbie wants to determine the initial speed at which a spring toy shoots a marble. She assumes the toy shoots the marble with the same speed every time. She shoots the marble horizontally from different heights, using different numbers of bricks with equal heights of 20 cm on a table as a support (see **Figure P4.86**). In every experiment she records the number of bricks and the distance D from the table the marble lands (see the table at right). When she finishes collecting the data, she realizes that she has forgotten to measure the distance between the surface of the table and the floor, h. Determine the initial speed of the marble and the distance h using Debbie's data. (Hint: This problem requires linearization. First, express distance D as a function of initial speed of the marble and the height from which the marble is shot. Then rearrange the equation to obtain linear dependence on the height, plot the best-fit line, and determine unknown quantities from the data that you obtained from the graph.)

# of bricks	D (m)
0	0.42
1	0.48
2	0.54
3	0.59
4	0.64

FIGURE P4.86

87. * Your friend has a pie on the roof of his van. You are standing on the ground and observe the van stopping abruptly for a red light. The pie does not slip off the roof. (a) Construct a motion diagram and a force diagram for the pie as the van approaches the red light, from your frame of reference and from the driver's frame of reference. (b) Repeat part (a) for the case when the light turns green. Be sure to specify the observer and identify the other object causing each force. (c) Are the motion diagrams consistent with the force diagrams for each case?

88. * A ledge on a building is 20 m above the ground. A taut rope attached to a 4.0-kg can of paint sitting on the ledge passes up over a pulley and straight down to a 3.0-kg can of nails on the ground. If the can of paint is accidentally knocked off the ledge, what time interval does a carpenter have to catch the can of paint before it smashes on the ground?

89. * You are hired to devise a method to determine the coefficient of friction between the ground and the soles of a shoe and of its competitors. Explain your experimental technique and provide a physics analysis that could be used by others using this method.

90. The mass of a spacecraft is about 480 kg. An engine designed to increase the speed of the spacecraft while in outer space provides 0.09-N thrust at maximum power. By how much does the engine cause the craft's speed to change in 1 week of running at maximum power? Describe any assumptions you made.

91. ** **EST** A 60-kg Rollerblader rolls 10 m down a 30° incline. When she reaches the level floor at the bottom, she applies the brakes. Use Newton's second law to estimate the distance she will move before stopping. Justify any assumptions you made.

92. Design, perform, and analyze the results of an experiment to determine the coefficient of static friction and the coefficient of kinetic friction between a penny and the cover of this textbook.

93. * **Tell all** A sled starts at the top of the hill shown in **Figure P4.93**. Add any information that you think is reasonable about the process that ensues when the sled goes down the hill and finally stops. Then tell everything you can about this process.

FIGURE P4.93

Reading Passage Problems

Professor tests airplane takeoff speed D. A. Wardle, a professor of physics from the University of Auckland, New Zealand, tested the takeoff speed of a commercial airliner. The pilot had insisted that the takeoff speed had to be 232 km/h.[1] To perform the testing experiment, Wardle used a pendulum attached to stiff cardboard (Figure 4.14). Prior to takeoff, when the plane was stationary, he marked the position of the pendulum bob on the cardboard to provide a vertical reference line (the dashed line in Figure 4.14). During the takeoff, he recorded the position of the bob at 5-s intervals. The results are shown in the table.

FIGURE 4.14 Wardle's device.

Using these data, Professor Wardle determined the acceleration at takeoff to be greater than $g/4$. Then he plotted an acceleration-versus-time graph and used it to find the takeoff speed. It turned out to be about 201 km/h. He was very satisfied—the day was windy, and the speed of the breeze was about 15–20 km/h. Thus the takeoff speed predicted by his simple pendulum was 215–220 km/h, very close to what the pilot said.

t (s)	θ (°)
0	9.9
5	14.8
10	13.8
15	13.0
20	12.0
25	11.4

94. Choose the best force diagram for the pendulum bob as the plane is accelerating down the runway (Figure P4.94).

FIGURE P4.94

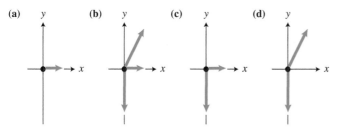

95. The professor used which of the following expressions for the pendulum bob acceleration (θ is the angle of the pendulum bob string relative to the vertical)?
    (a) $a = g \sin \theta$
    (b) $a = g \cos \theta$
    (c) $a = g \tan \theta$
    (d) None of the choices

96. Approximately when did the peak acceleration occur?
    (a) 25 s   (b) 20 s   (c) 10 s   (d) 5 s

97. Approximately when did the peak speed occur?
    (a) 25 s   (b) 20 s   (c) 10 s   (d) 5 s

98. Choose the best velocity-versus-time graph below for the airplane (Figure P4.98).

    FIGURE P4.98

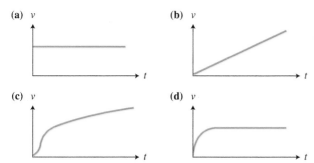

Ski jumping in Vancouver The 2010 Olympic ski jumping competition was held at Whistler Mountain near Vancouver. During a jump, a skier starts near the top of the in-run, the part down which the skier glides at increasing speed before the jump. The Whistler in-run is 116 m long and for the first part is tilted down at about 35° below the horizontal. There is then a curve that transitions into a take-off ramp, which is tilted 11° below the horizontal. The skier flies off this ramp at high speed, body tilted forward and skis separated (Figure 4.15). This position exposes a large surface area to the air, which creates lift, extends the time of the jump, and allows the jumper to travel farther. In addition, the skier pushes off the exit ramp of the in-run to get a vertical component of velocity when leaving the ramp. The skier lands 125 m or more from the end of the in-run. The landing surface has a complex shape and is tilted down at about 35° below the horizontal. The skier moves surprisingly close (2 to 6 m) above the snowy surface for most of the jump. The coefficient of kinetic friction between the skis and the snow on the in-run is about 0.05 ± 0.02, and skiers' masses are normally small—about 60 kg. We can make some rough estimates about an idealized ski jump with an average in-run inclination of $(35° + 11°)/2 = 23°$.

FIGURE 4.15 A ski jumper exposes a large surface area to the air in order to get more lift.

99. Which answer below is closest to the magnitude of the normal force that the idealized in-run exerts on the 60-kg skier?
    (a) 590 N   (b) 540 N   (c) 250 N   (d) 230 N

100. Which numbers below are closest to the magnitudes of the kinetic friction force and the component of the gravitational force parallel to the idealized inclined in-run?
    (a) 30 N, 540 N   (b) 27 N, 540 N   (c) 12 N, 540 N
    (d) 30 N, 230 N   (e) 27 N, 230 N   (f) 12 N, 230 N

101. Which answers below are closest to the magnitude of the skier's acceleration while moving down the idealized in-run and to the skier's speed when leaving its end?
    (a) 9.8 m/s², 48 m/s   (b) 4.3 m/s², 32 m/s   (c) 4.3 m/s², 28 m/s
    (d) 3.4 m/s², 32 m/s   (e) 3.4 m/s², 28 m/s

102. Assume that the skier left the ramp moving horizontally. Treat the skier as a point-like particle and assume the force exerted by air on him is minimal. If he landed 125 m diagonally from the end of the in-run, and the landing region beyond the in-run was inclined 35° below the horizontal for its entire length, which answer below is closest to the time interval that he was in the air?
    (a) 1.9 s   (b) 2.4 s   (c) 3.1 s
    (d) 3.8 s   (e) 4.3 s

103. Using the same assumptions as stated in Problem 4.102, which answer below is closest to the jumper's speed when leaving the in-run?
    (a) 37 m/s   (b) 31 m/s   (c) 27 m/s
    (d) 24 m/s   (e) 21 m/s

104. Which factors below would keep the skier in the air longer and contribute to a longer jump?
    1. The ramp at the end of the in-run is level instead of slightly tilted down.
    2. The skier extends his body forward and positions his skis in a V shape.
    3. The skier has wider and longer than usual skis.
    4. The skier pushes upward off the end of the ramp at the end of the in-run.
    5. The skier crouches in a streamline position when going down the in-run.

    (a) 1   (b) 5   (c) 1, 3, 5
    (d) 2, 3, 4   (e) 1, 2, 4, 5   (f) 1, 2, 3, 4, 5

---

[1] The information is taken from the article by D. A. Wardle "Measurement of aeroplane takeoff speed and cabin pressure" published in *The Physics Teacher*, 37, 410–411 (1999).

5

Circular Motion

- **Why do pilots sometimes black out when pulling out of a steep dive?**

- **Are astronauts really "weightless" while in orbit?**

- **Why do you tend to slide across the seat of your car when making a sharp turn?**

Kruti Patel, a civilian test pilot, wears a special flight suit to prevent dizziness, disorientation, and possibly passing out as she comes out of a steep dive. These symptoms characterize a blackout, which can occur when the head and brain do not receive a sufficient amount of blood. Why would pulling out of a dive cause a blackout, and how does a special suit prevent it from occurring? Our study of circular motion in this chapter will help us understand blackouts and other interesting phenomena.

BE SURE YOU KNOW HOW TO:

- Find the direction of acceleration using a motion diagram (Section 2.7).

- Draw a force diagram (Section 3.1).

- Use a force diagram to help apply Newton's second law in component form (Sections 4.1 and 4.2).

IN PREVIOUS CHAPTERS we studied motion in situations in which the sum of the forces exerted on a system had a constant magnitude and direction. More often, the forces exerted on an object continually change direction and magnitude as time passes. In this chapter we will focus on circular motion. It is the simplest example of motion in which the sum of the forces exerted on a system by other objects continually changes.

5.1 Qualitative dynamics of circular motion

Think of a person riding on a Ferris wheel or a merry-go-round, a car making a turn, or a swinging pendulum bob. All of these objects have one thing in common: they do not follow a straight line but move in a circle or a part of a circle (an arc). To study the motion of objects along a circular path, we will start with the simple case of an object moving around a circle such that the time to complete one full circle does not change—that is, the object moves with constant speed. An example of such an object is a car on a circular track whose speedometer reads 55 mph. We will also consider the object to be a point-like object. These assumptions allow us to model circular motion in the simplest way.

OBSERVATIONAL
EXPERIMENT TABLE 5.1   Forces exerted on an object moving in a circle at constant speed   ▶ VIDEO OET 5.1

Observational experiment	Analysis
Experiment 1. A bowling ball is rolled toward you over a smooth floor. You are asked to tap it with a mallet to make the ball roll in a circle with constant speed. You find that directing the mallet taps along the desired circular path doesn't work; the ball rolls wide. The only thing that works is to tap *directly toward the center of the desired circle*.	There are three objects interacting with the bowling ball: Earth, the floor, and the mallet. We assume that the force exerted by the floor is perpendicular to the floor's surface. The force diagrams for the ball at two locations are shown below. We see that at each location the sum of the forces points toward the center of the circle.

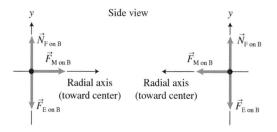

Experiment 2. You hold a bag by the handle and swing it in a horizontal circle at constant speed. You observe that your arms pulling the bag are angled down with respect to the horizontal.	Two forces are exerted on the bag as it moves around the circle: Earth pulls downward, and you pull at an angle with the vertical. The vertical component of the force that you exert on the bag must balance the force that Earth exerts on it because the bag does not accelerate in the vertical direction. Consequently, the sum of these two forces again points toward the center of the circle.

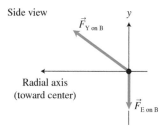

Pattern

In each case, as shown in the force diagrams, the sum of the forces exerted on the circling object points directly toward the center of the circle.

We can therefore propose a hypothesis: when an object is moving in a circle, the sum of all the forces exerted on it by other objects is directed toward the center of the circle. (We assume the speed of the object is constant.) Let's test this hypothesis with some more experiments.

TESTING
EXPERIMENT TABLE 5.2   Does the sum of the forces exerted on an object moving at constant speed in a circle point toward the center of the circle?

VIDEO
TET 5.2

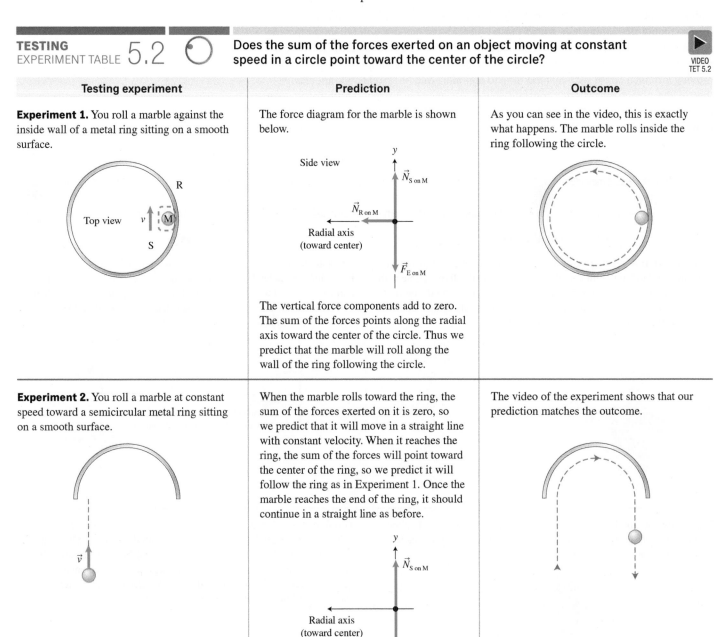

Testing experiment	Prediction	Outcome
Experiment 1. You roll a marble against the inside wall of a metal ring sitting on a smooth surface.	The force diagram for the marble is shown below.	As you can see in the video, this is exactly what happens. The marble rolls inside the ring following the circle.
	The vertical force components add to zero. The sum of the forces points along the radial axis toward the center of the circle. Thus we predict that the marble will roll along the wall of the ring following the circle.	
Experiment 2. You roll a marble at constant speed toward a semicircular metal ring sitting on a smooth surface.	When the marble rolls toward the ring, the sum of the forces exerted on it is zero, so we predict that it will move in a straight line with constant velocity. When it reaches the ring, the sum of the forces will point toward the center of the ring, so we predict it will follow the ring as in Experiment 1. Once the marble reaches the end of the ring, it should continue in a straight line as before.	The video of the experiment shows that our prediction matches the outcome.

Conclusion

The outcomes of both experiments are consistent with the pattern of the sum of the forces exerted on an object moving at constant speed in a circle pointing toward the center.

Now we need to explain why this pattern exists. Why would the sum of the forces exerted on an object moving in a circle at constant speed point toward the center of the circle? According to Newton's second law, the sum of the forces is in the same direction as the object's acceleration. Therefore, the object's acceleration should also point toward the center of the circle. We will learn how to determine the direction of acceleration during circular motion at constant speed in the next section.

REVIEW QUESTION 5.1 How do we know that the sum of the forces exerted on an object moving in a circle at constant speed points toward the center of the circle?

5.2 Analyzing velocity change for circular motion

Consider the motion of the car shown in **Figure 5.1** as it travels at constant speed around a circular track. The instantaneous velocity of the car is tangent to the circle at every point. Recall from Chapter 2 that an object has acceleration if its velocity changes—in magnitude, in direction, or in both. Even though the car is moving at constant speed, it is accelerating because the *direction* of its velocity changes from moment to moment.

To estimate the direction of acceleration of any object (including that of the car) while passing a particular point on its path (see Physics Tool Box 5.1), we use the **velocity change method**. Consider a short time interval $\Delta t = t_f - t_i$ during which the car passes that point. The velocity arrows $\vec{v}_i$ and $\vec{v}_f$ represent the initial and final velocities of the car, a little before and a little after the point (note that the velocity arrows are tangent to the curve in the direction of motion; this is Step 1 in the Tool Box). The velocity change vector is $\Delta \vec{v} = \vec{v}_f - \vec{v}_i$. You can think of $\Delta \vec{v}$ as the vector that you need to add to the initial velocity $\vec{v}_i$ in order to get the final velocity $\vec{v}_f$, that is, $\vec{v}_i + \Delta \vec{v} = \vec{v}_f$. Rearranging this, we get $\Delta \vec{v} = \vec{v}_f - \vec{v}_i$ (the change in velocity).

To estimate $\Delta \vec{v}$, place the $\vec{v}_i$ and $\vec{v}_f$ arrows tail to tail (Step 2 in the Tool Box) without changing their magnitudes or directions. $\Delta \vec{v}$ starts at the head of $\vec{v}_i$ and ends at the head of $\vec{v}_f$. The car's acceleration $\vec{a}$ is in the direction of the $\Delta \vec{v}$ arrow (Step 3 in the Tool Box) and is the ratio of the velocity change and the time interval needed for that change:

$$\vec{a} = \frac{\Delta \vec{v}}{\Delta t}$$

FIGURE 5.1 Since the direction of the velocity is changing, the car is accelerating.

The magnitude of the car's velocity is constant but its direction is changing.

PHYSICS TOOL BOX 5.1 ▶ **Estimating the direction of acceleration during two-dimensional motion**

1. Choose the point at which you want to determine the direction of acceleration and draw velocity vectors at equal distances before and after the point.

2. Place the $\vec{v}_i$ and $\vec{v}_f$ arrows tail to tail. Draw a $\Delta \vec{v}$ arrow from the head of $\vec{v}_i$ to the head of $\vec{v}_f$.

$$\vec{v}_i + \Delta \vec{v} = \vec{v}_f$$
or
$$\Delta \vec{v} = \vec{v}_f - \vec{v}_i$$

3. The acceleration arrow $\vec{a}$ is in the direction of $\Delta \vec{v}$.

$$\vec{a} = \frac{\Delta \vec{v}}{\Delta t} = \frac{\vec{v}_f - \vec{v}_i}{t_f - t_i}$$

TIP When using this diagrammatic method to estimate the direction of acceleration at a particular point during circular motion, make sure that you choose initial and final points at the same distance before and after the point at which you are estimating the acceleration direction. Draw long velocity arrows so that when you put them tail to tail, you can clearly see the direction of the velocity change arrow. Also, be sure that the velocity change arrow points from the head of the initial velocity to the head of the final velocity. Note that this method of estimating the direction of acceleration works for any motion, not just circular.

CONCEPTUAL EXERCISE 5.1 | **Direction of a racecar's acceleration**

Determine the direction of the racecar's acceleration at points A, B, and C in the figure below as the car travels at constant speed on the circular path.

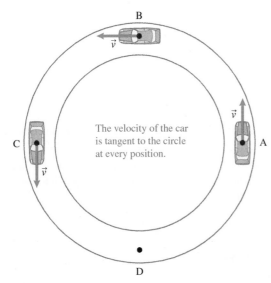

B

$\vec{v}$

C   A

The velocity of the car
is tangent to the circle
at every position.

$\vec{v}$ $\vec{v}$

D

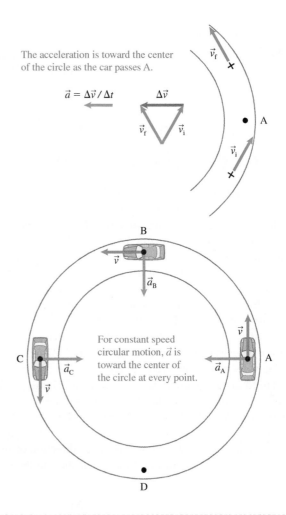

The acceleration is toward the center
of the circle as the car passes A.

$\vec{a} = \Delta\vec{v} / \Delta t$ $\Delta\vec{v}$

$\vec{v}_f$ $\vec{v}_i$

$\vec{v}_f$

A

$\vec{v}_i$

B

$\vec{v}$

$\vec{a}_B$

C $\vec{a}_C$ For constant speed
circular motion, $\vec{a}$ is
toward the center of
the circle at every point. $\vec{a}_A$ A

$\vec{v}$ $\vec{v}$

D

Sketch and translate A top view of the car's path is shown above. We are interested in the car's acceleration as it passes points A, B, and C.

Simplify and diagram To find the direction of the car's acceleration at each point, we use the velocity change method (shown for point A above, right). When done for all three points, notice that a pattern emerges: the acceleration at different points along the car's path has a different direction, but in every case it points toward the center of the circular path.

Try it yourself Find the direction of the car's acceleration at point D on the track.

Answer

The car's acceleration points toward the center of the circle here as well.

In Conceptual Exercise 5.1 we found an important pattern: when an object is moving in a circle at *constant speed,* its acceleration at any position points toward the center of the circle. This acceleration is called **radial acceleration** because it points along the radius of the circle.

The radial direction of acceleration is due to the fact that the direction of the velocity continuously changes. In Section 5.1 we found that the sum of the forces exerted on an object moving in a circle points toward the center of the circle at every point. This force finding makes sense now—for one object to continuously change its direction of motion, another object needs to exert a force on the first object. We can think of this force as a "hand" physically turning the velocity vector so that it is always tangent to the circle. Once this "hand" is removed, the object continues to move in a straight line, as we observed in Experiment 2 in Table 5.2. Our new knowledge concerning circular motion is consistent with Newton's second law.

REVIEW QUESTION 5.2 Why is it true that when an object is moving in a circle at constant speed, its acceleration at any point is perpendicular to its instantaneous velocity?

5.3 Radial acceleration and period

So far we have learned about the direction of the acceleration of an object moving in a circle at constant speed. Let's now determine the magnitude of the object's acceleration. We will begin by thinking about factors that might affect its acceleration.

Imagine a car following the circular curve of a highway. Our experience indicates that the faster a car moves along a highway curve, the greater the risk that the car will skid off the road. So the car's speed v matters. Also, the tighter the turn, the greater the risk that the car will skid. So the radius r of the curve also matters. But how exactly do speed and radius combine in the mathematical expression for radial acceleration?

Dependence of acceleration on speed

Let's begin by investigating how an object's acceleration depends on speed. In Observational Experiment Table 5.3, we use the diagrammatic velocity change method to investigate how the acceleration differs for objects moving at speeds v, $2v$, and $3v$ while traveling along the same circular path of radius r.

OBSERVATIONAL EXPERIMENT TABLE 5.3   How does acceleration depend on the object's speed?

Observational experiment	Analysis
Experiment 1. An object moves in a circle at constant speed.	The acceleration is toward the center of the circular path. $a_1 = \dfrac{\Delta v_1}{\Delta t_1}$
Experiment 2. An object moves in the same circle at a constant speed that is twice as fast as in Experiment 1.	When the object moves twice as fast between the same two points on the circle, the velocity change doubles. In addition, the velocity change occurs in one-half the time interval since it is moving twice as fast. Hence, the acceleration increases by a factor of 4. $a_2 = \dfrac{\Delta v_2}{\Delta t_2}$ $= \dfrac{2\Delta v_1}{\Delta t_1/2} = 4\dfrac{\Delta v_1}{\Delta t_1}$

(CONTINUED)

Observational experiment	Analysis
Experiment 3. An object moves in the same circle at a constant speed that is three times as fast as in Experiment 1.	Tripling the speed triples the velocity change and reduces to one-third the time interval needed to travel between the points. The acceleration increases by a factor of 9.

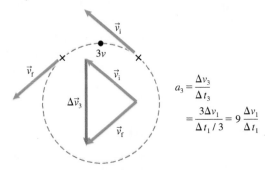

$$a_3 = \frac{\Delta v_3}{\Delta t_3}$$

$$= \frac{3\Delta v_1}{\Delta t_1 / 3} = 9\frac{\Delta v_1}{\Delta t_1}$$

Pattern

We find that doubling the speed of the object results in a fourfold increase of its radial acceleration; tripling the speed leads to a ninefold increase. Therefore, the radial acceleration of the object is proportional to its speed squared: $a_r \propto v^2$

From the pattern in Table 5.3 we conclude that the magnitude of radial acceleration is proportional to the speed squared. We can express this pattern mathematically:

$$a_r \propto v^2 \tag{5.1}$$

Dependence of acceleration on radius

To find how the magnitude of acceleration of an object moving in a circle at constant speed depends on the radius r of the circle, we use the velocity change method to consider two objects moving with the same speed but on circular paths of different radii (Observational Experiment Table 5.4). For simplicity, we make one circle twice the radius of the other. We arrange to have the same velocity change for the two objects by considering them while moving through the same angle θ (rather than through the same distance).

OBSERVATIONAL
EXPERIMENT TABLE 5.4 **How does acceleration depend on the radius of the curved path?**

Observational experiment	Analysis
Experiment 1. An object moves in a circle of radius r at speed v. Choose two points on the circle to examine the velocity change from the initial to the final location.	

$$a_1 = \frac{\Delta v}{\Delta t_1}$$

Observational experiment	**Analysis**
Experiment 2. An object moves in a circle of radius $2r$ at speed v. Choose two points so that the velocity change is the same as in Experiment 1. This occurs if the radii drawn to the location of the object at the initial position and the final position make the same angle as in Experiment 1.	To have the same velocity change as in Experiment 1, the object has to travel twice the distance because the radius is twice as long. Since the speed of the object is the same as in the first experiment, it takes the object twice as long to travel that distance. Hence, the magnitude of the acceleration is half that in Experiment 1.
	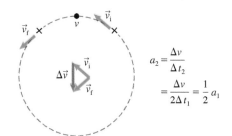

Pattern	
When an object moves in a circle at constant speed, its radial acceleration decreases by half when the radius of the circular path doubles—the bigger the circle, the smaller the acceleration. Therefore, the radial acceleration of the object is proportional to the inverse of the radius of its circular path:	$a_r \propto \dfrac{1}{r}$

If we carried out the same thought experiment with circular paths of other radii, we would get similar results. The magnitude of the radial acceleration of an object moving in a circular path is inversely proportional to the radius of the circle:

$$a_r \propto \frac{1}{r} \tag{5.2}$$

We now combine the two proportionalities in Eqs. (5.1) and (5.2):

$$a_r \propto \frac{v^2}{r}$$

A detailed mathematical derivation would show that the constant of proportionality is 1. Thus, we can turn this into an equation for the magnitude of the radial acceleration.

Radial acceleration For an object moving at constant speed v on a circular path of radius r, the magnitude of the radial acceleration is

$$a_r = \frac{v^2}{r} \tag{5.3}$$

The acceleration points toward the center of the circle. The units for v^2/r are the correct units for acceleration:

$$\left(\frac{\mathrm{m}^2/\mathrm{s}^2}{\mathrm{m}} \right) = \frac{\mathrm{m}}{\mathrm{s}^2}$$

This expression for radial acceleration agrees with our everyday experience. When a car is going around a highway curve at high speed, a large v^2 in the numerator leads to its large acceleration. When it is going around a sharp turn, the small radius r in the denominator also leads to a large acceleration.

Consider a limiting case: when the radius of the circular path is infinite—equivalent to the object moving in a straight line. The velocity of an object moving at constant speed in a straight line is constant, and its acceleration is zero. Notice that the acceleration in Eq. (5.3) is zero if the radius in the denominator is infinite.

TIP Sometimes the radial acceleration is called centripetal acceleration, and the product of radial acceleration and the mass of the object is called centripetal force. We will not use this terminology, as sometimes it makes students think that centripetal force is a new type of force exerted on an object.

TIP When you derive a new mathematical relation for physical quantities, it is always useful to examine how reasonable it is in limiting cases.

EXAMPLE 5.2   Pilot blackout

When a pilot pulls out from the bottom of a dive, his body moves at high speed along a segment of an upward-bending approximately circular path. However, while his body moves up, his blood tends to move straight ahead (tangent to the circle) and begins to fill the easily expandable veins in his legs. This can deprive his brain of blood and cause a blackout if the radial acceleration is $4g$ or more and lasts several seconds. Suppose during a dive an airplane moves at a modest speed of $v = 80$ m/s (180 mi/h) through a circular arc of radius $r = 150$ m. Is the pilot likely to black out?

Sketch and translate We sketch the situation. To determine if blackout occurs, we estimate the radial acceleration of the pilot as he passes along the lowest point of the circular path.

Simplify and diagram Assume that at the bottom of the plane's dive, the plane is moving in a circle at constant speed and the point of interest is the lowest point on the circle. Assume also that the magnitude of the plane's acceleration is constant for a quarter circle, that is, one-eighth of a circle on each side of the bottom point. The acceleration points up (see the velocity change method shown below).

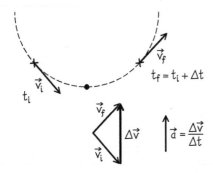

Represent mathematically We calculate the magnitude of the radial acceleration using Eq. (5.3):

$$a_r = \frac{v^2}{r}$$

To estimate the time interval that the pilot will experience this acceleration, we calculate the time interval for the plane to move through the part of the dive represented by a quarter circle (arc length $l = 2\pi r/4$):

$$\Delta t = \frac{l}{v} = \frac{2\pi r/4}{v}$$

Solve and evaluate Inserting the given quantities, we find:

$$a_r = \frac{(80 \text{ m/s})^2}{150 \text{ m}} = 43 \text{ m/s}^2$$

The acceleration is more than four times greater than g (more than 4×9.8 m/s^2). Thus, the pilot could potentially black out if the acceleration lasts too long. The time interval during which the pilot will experience this acceleration is about

$$\Delta t = \frac{[2\pi(150 \text{ m})]/4}{80 \text{ m/s}} = \frac{240 \text{ m}}{80 \text{ m/s}} \approx 3 \text{ s}$$

This is long enough that blackout is definitely a concern. Special flight suits, however, exert considerable pressure on the legs during such motion. This pressure stops blood from accumulating in the veins of the legs and thus prevents a pilot blackout.

- -

Try it yourself Imagine that you are a passenger on a roller coaster with a double loop-the-loop—two consecutive loops. You are traveling at speed 24 m/s as you move along the bottom part of a loop of radius 10 m. Determine (a) the magnitude of your radial acceleration while passing the lowest point of that loop and (b) the time interval needed to travel along the bottom quarter of that approximately circular loop. (c) Are you at risk of blacking out? Explain.

Answer

(a) 58 m/s^2, or almost six times the 9.8 m/s^2 free-fall acceleration; (b) about 0.7 s; (c) the acceleration is more than enough to cause blackout but does not last long enough, so you are safe from blacking out and will get a good thrill!

Period

TIP Do not confuse the symbol T for period with the symbol T for the tension force that a string exerts on an object.

When an object repeatedly moves in a circle, we can describe its motion with another useful physical quantity, its **period** T. The period equals the time interval that it takes an object to travel around the entire circular path one time and has the units of time. For example, suppose that a bicyclist racing on a circular track takes 24 s to complete a circle that is 400 m in circumference. The period T of the motion is 24 s.

For constant speed circular motion we can determine the speed of an object by dividing the distance traveled in one period (the circumference of the circular path, $2\pi r$) by the time interval T it took the object to travel that distance (its period), or $v = 2\pi r/T$. Thus,

$$T = \frac{2\pi r}{v} \tag{5.4}$$

We can express the radial acceleration of the object in terms of its period by inserting this expression for speed in terms of period $v = 2\pi r/T$ into Eq. (5.3):

$$a_r = \frac{v^2}{r} = \left(\frac{2\pi r}{T}\right)^2 \frac{1}{r} = \frac{4\pi^2 r^2}{T^2 r} = \frac{4\pi^2 r}{T^2} \qquad (5.5)$$

Let's use limiting case analysis to see if Eq. (5.5) is reasonable. If the speed of the object is extremely large, its period would be very short $(T = 2\pi r/v)$. Thus, according to Eq. (5.5), its radial acceleration would be very large. That makes sense—high speed and large radial acceleration. Similarly, if the speed of the object is small, its period will be very large and its radial acceleration will be small. That also makes sense.

We can use Eq. (5.5) to calculate our acceleration on the surface of Earth due to Earth's rotation around its axis. Imagine we are interested in calculating radial acceleration in Singapore, which is on the equator, with $r = 6400$ km and $T = 24$ h. Making the appropriate unit conversions, we find the magnitude of the acceleration:

$$a_r = \frac{4\pi^2 (6.4 \times 10^6 \text{m})}{(24 \text{ h} \times 3600 \text{ s/h})^2} = 0.034 \text{ m/s}^2$$

This is a very small number compared to the much greater free-fall acceleration of objects near Earth's surface—9.8 m/s². Because it is so small, the radial acceleration due to Earth's rotation around its axis can be ignored under most circumstances.

Remember that Newton's laws were formulated and are valid only for observers in inertial reference frames, that is, for observers who are not accelerating. But we have just discovered that observers on Earth's surface are accelerating due to Earth's rotation. Does this mean that Newton's laws do not apply? As we found above, the acceleration due to Earth's rotation is much smaller than the accelerations we experience from other types of motion. Thus, in most situations we can assume that Earth is not rotating and therefore *does* count as an inertial reference frame, and we can apply Newton's laws on Earth.

REVIEW QUESTION 5.3 Show that the two expressions for radial acceleration (v^2/r and $4\pi^2 r/T^2$) have the correct units of acceleration.

5.4 Skills for analyzing processes involving circular motion

The strategy for analyzing processes involving constant speed circular motion is similar to that used to analyze linear motion processes. However, when we analyze constant speed circular motion processes, we do *not* use traditional x- and y-axes but instead use a radial r-axis and sometimes a vertical y-axis. **The radial r-axis should always point *toward the center of the circle*.** The radial acceleration of magnitude $a_r = v^2/r$ is positive along this axis. The application of Newton's second law for circular motion using a radial axis is summarized below.

Circular motion component form of Newton's second law For the radial direction (the axis pointing toward the center of the circular path), the component form of Newton's second law is

$$a_r = \frac{\Sigma F_r}{m} \quad \text{or} \quad ma_r = \Sigma F_r \qquad (5.6)$$

where ΣF_r is the sum of the radial components of all forces exerted on the object moving in the circle (positive toward the center of the circle and negative away from the center), and $a_r = v^2/r$ is the magnitude of the radial acceleration of the object.

(CONTINUED)

For some situations, usually when friction forces are involved (for example, a car moving around a highway curve or a person standing on the platform of a merry-go-round), we also include in the analysis the force components along a perpendicular vertical y-axis:

$$ma_y = \Sigma F_y = 0 \tag{5.7}$$

When an object moves with uniform (constant speed) circular motion, both the y-component of its acceleration and the y-component of the sum of the forces exerted on it are zero.

PROBLEM-SOLVING STRATEGY 5.1

Processes involving constant speed circular motion

Sketch and translate

- Sketch the situation described in the problem statement. Label it with all relevant information.
- Choose the system and a specific position to analyze its motion.
- Identify the unknown that you need to find and label it with a question mark.

Simplify and diagram

- Decide if the system can be modeled as a point-like object.
- Determine if the constant speed circular motion approach is appropriate.
- Indicate with an arrow the direction of the object's acceleration as it passes the chosen position.
- Draw a force diagram for the system at the instant it passes that position.
- On the force diagram, draw an axis in the radial direction toward the center of the circle.

EXAMPLE 5.3 **Driving over a hump in the road**

Josh drives his car at a constant 12 m/s speed over a bridge whose shape is bowed in the vertical arc of a circle. Find the direction and the magnitude of the force exerted by the car seat on Josh as he passes the top of the 30-m-radius arc. His mass is 60 kg.

The sketch includes all of the relevant information: Josh's speed, his mass, and the radius of the arc along which the car moves. Josh is the system.

Consider Josh a point-like object and analyze him as he passes the highest part of the vertical circle along which he travels.

A velocity change diagram as he passes the top of the circular path indicates that he has a downward radial acceleration toward the center of the circle.

The forces exerted on him are shown in the force diagram. The sum of the forces must also point down. Thus, the upward normal force $\vec{N}_{S \text{ on } J}$ that the car seat exerts on Josh is smaller in magnitude than the downward gravitational force $\vec{F}_{E \text{ on } J}$ that Earth exerts on Josh. A radial r-axis points down toward the center of the circle.

Represent mathematically

- Convert the force diagram into the radial r-component form of Newton's second law.

- For objects moving in a horizontal circle (unlike this example), you may also need to apply a vertical y-component form of Newton's second law.

Apply the radial form of Newton's second law $ma_r = \Sigma F_r$.

The radial components of the two forces exerted on Josh are $F_{\text{E on J } r} = +F_{\text{E on J}} = +mg$ and $N_{\text{S on J } r} = -N_{\text{S on J}}$. The magnitude of the radial acceleration is $a_r = v^2/r$. Therefore, the radial form of Newton's second law is

$$m\frac{v^2}{r} = +mg + (-N_{\text{S on J}})$$

Solving for $N_{\text{S on J}}$, we have

$$N_{\text{S on J}} = mg - m\frac{v^2}{r}$$

Solve and evaluate

- Solve the equations formulated in the previous two steps.

- Evaluate the results to see if they are reasonable (the magnitude of the answer, its units, limiting cases).

Substituting the known information into the previous equation, we get

$$N_{\text{S on J}} = (60\text{ kg})(9.8\text{ m/s}^2) - (60\text{ kg})\frac{(12\text{ m/s})^2}{(30\text{ m})} = 588\text{ N} - 288\text{ N} = 300\text{ N}$$

The seat exerts a smaller upward force on Josh than Earth pulls down on him. You have probably noticed this effect when going over a smooth hump in a roller coaster or while in a car or on a bicycle when crossing a hump in the road—it almost feels like you are leaving the seat or starting to float briefly above it. This feeling is caused by the reduced upward normal force that the seat exerts on you.

Try it yourself Imagine that Josh drives on a road that has a dip in it. The speed of the car and the radius of the dip are the same as in Example 5.3. Find the direction and the magnitude of the force exerted by the car seat on Josh as he passes the bottom of the 30-m-radius dip.

Answer

The seat exerts an upward force of 880 N. This is almost 50% more than the 590-N force that Earth exerts on Josh—he sinks into the seat.

The next three examples involve processes in which some of the forces have nonzero radial r-components *and* some of the forces have nonzero vertical y-components. To solve these problems we will use the component form of Newton's second law in both the radial r- and vertical y-directions.

EXAMPLE 5.4   Toy airplane

You have probably seen toy airplanes flying in a circle at the end of a string. Once the plane reaches a constant speed, it does not move up or down, just around in a horizontal circle. Suppose you have an airplane attached to the end of a 46-cm string, which makes a 25° angle relative to the horizontal while the airplane is flying. A scale at the top of the string measures the force that the string exerts on the airplane. Find the period of the airplane's motion (the time interval for it to complete one circle).

Sketch and translate We sketch the situation, including the known information: the length of the string and the angle of the string relative to the horizontal. The airplane is the system.

$T = ?$

$L = 0.46$ m

Note: $r = L \cos 25°$

(CONTINUED)

Simplify and diagram Neglect the airplane's interaction with the air and assume that it is a point-like object moving at constant speed in a horizontal circle. To draw a force diagram, identify objects that interact with the airplane—in this case, Earth and the string. Include a vertical y-axis and a radial r-axis; note that the radial axis points horizontally toward the center of the circle—not along the string. Because the airplane moves at constant speed, its acceleration points toward the center of the circle—it has a radial component but does not have a vertical component. Thus, the vertical component of the tension force that the string exerts on the plane $\vec{T}_{\text{S on P }y}$ must balance the downward force that Earth exerts on the plane $\vec{F}_{\text{E on P}}$, which has magnitude $m_{\text{P}}g$ (see the force diagram broken into components at right). The radial component of the force that the string exerts on the plane is the only force with a nonzero radial component and causes the plane's radial acceleration.

Represent mathematically We use the force diagram to help apply Newton's second law in component form. First, apply the vertical y-component form of Newton's second law: $m_{\text{P}}a_y = \Sigma F_y$. The y-components of the forces exerted on the plane are $T_{\text{S on P }y} = +T_{\text{S on P}} \sin 25°$ and $F_{\text{E on P }y} = -m_{\text{P}}g$, and the y-component of the plane's acceleration is zero. We can now apply the y-component form of Newton's second law:

$$0 = +T_{\text{S on P}} \sin 25° + (-m_{\text{P}}g)$$

Next, apply the radial r-component version of Newton's second law: $m_{\text{P}}a_r = \Sigma F_r$. The force exerted by Earth on the plane does not have a radial component. The radial component of the force exerted by the string on the plane is $T_{\text{S on P }r} = +T_{\text{S on P}} \cos 25°$, and the magnitude of the radial acceleration is $a_r = 4\pi^2 r/T^2$, where T is the period of the plane's motion [Eq. (5.5)]. Thus:

$$m_{\text{P}}\frac{4\pi^2 r}{T^2} = +T_{\text{S on P}} \cos 25°$$

Note that $T_{\text{S on P}}$ represents the magnitude of the tension force that the string exerts on the airplane and T represents the period of the plane's motion—the time interval it takes the plane to travel once around its

circular path. Note that the radius of the circular path is not the length of the string but is instead

$$r = L \cos 25° = (0.46 \text{ m}) \cos 25° = 0.417 \text{ m}$$

where L is the length of the string (46 cm).

Solve and evaluate The goal is to find the period T of the plane's circular motion. Unfortunately, the above radial application of Newton's second law has three unknowns: $T_{\text{S on P}}$, T, and the mass of the plane. However, we can rearrange the vertical y-equation to get an expression for the magnitude of the tension force exerted by the string on the plane:

$$T_{\text{S on P}} = \frac{m_{\text{P}}g}{\sin 25°}$$

Insert this expression for $T_{\text{S on P}}$ into the radial component equation to get an expression that does *not* involve the force that the string exerts on the plane:

$$m_{\text{P}}\frac{4\pi^2 r}{T^2} = \frac{m_{\text{P}}g}{\sin 25°} \cos 25°$$

Divide both sides by m_{P}, multiply both sides by T^2, and do some rearranging to get

$$T^2 = \frac{4\pi^2 r \sin 25°}{g \cos 25°}$$

Taking the square root of each side, we find that

$$T = 2\pi \sqrt{\frac{r \sin 25°}{g \cos 25°}}$$

Substituting the known information, we calculate the period to be

$$T = 2\pi \sqrt{\frac{(0.417 \text{ m}) \sin 25°}{(9.8 \text{ m/s}^2) \cos 25°}} = 0.88 \text{ s}$$

- -

Try it yourself Imagine that another plane of mass 0.12 kg moves with a speed of 6.2 m/s. A spring scale measures the tension force that the string exerts on the plane as 5.0 N. Find the radius of the plane's orbit.

Answer   0.95 m.

Many carnivals and amusement parks have a "rotor ride" in which people stand up against the wall of a spinning circular room. The room (also called the "drum") spins faster until, at a certain speed, the floor drops out! Amazingly, the people remain up against the wall with their feet dangling. How is this possible?

EXAMPLE 5.5   Rotor ride

A 62-kg woman is a passenger in a rotor ride. A drum of radius 2.0 m rotates faster and faster about a vertical axis until it reaches a period of 1.7 s. When the drum reaches this turning rate, the floor drops away, but the woman does not slide down the wall of the drum. Which characteristics of the ride ensure that the woman remains stuck to the wall? Justify your answer quantitatively.

Sketch and translate Representing the situation in different ways should help us decide which characteristics of the drum are important. We start with a sketch with all the relevant information. The woman is the system. Her mass is $m_{\text{W}} = 62$ kg, the radius of her circular path is $r = 2.0$ m, and the period of her circular motion is $T = 1.7$ s.

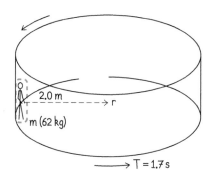

Simplify and diagram Model the woman as a point-like object and consider the situation once the drum has reached its maximum (constant) speed. Because she is moving in a circular path at constant speed, her acceleration points toward the center of the circle, and we can use our understanding of constant speed circular motion to analyze the situation.

Next, draw a force diagram for the woman as she passes one point along the circular path. She interacts with two objects—Earth and the drum. Earth exerts a downward gravitational force $\vec{F}_{E \text{ on } W}$ on her. The drum exerts a force that we can resolve into two vector components: a normal force $\vec{N}_{D \text{ on } W}$ perpendicular to the drum's surface and toward the center of the circle and an upward static friction force $\vec{f}_{s \, D \text{ on } W}$ parallel to its surface. Include a radial r-axis that points toward the center of the drum and a vertical y-axis pointing upward. Examining the force diagram, we see that if the maximum upward static friction force is less than the downward gravitational force, the woman will slip. Thus the question can be formulated as follows: how large must the coefficient of static friction be so that the static friction force exerted on the woman by the drum balances the force that Earth exerts on her?

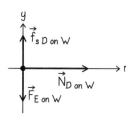

Represent mathematically Use the force diagram to help apply the component form of Newton's second law for the inward radial r-direction and for the vertical y-direction. The normal force is the only force that has a nonzero radial component.

Radial r-equation: $\qquad m_W a_r = \Sigma F_r$

$$\Rightarrow m_W \frac{v^2}{r} = N_{D \text{ on } W}$$

Now, consider the y-component form of Newton's second law. The static friction force $\vec{f}_{s \, D \text{ on } W}$ points in the positive y-direction, and the gravitational force $\vec{F}_{E \text{ on } W}$ points in the negative y-direction. Since these forces must balance for the woman not to slip, the y-component of the woman's acceleration should be zero.

Vertical y-equation: $\qquad m_W a_y = \Sigma F_y$

$$\Rightarrow 0 = f_{s \, D \text{ on } W} + (-F_{E \text{ on } W})$$

Assume that the static friction force is at its maximum possible value, meaning $f_{s \, D \text{ on } W} = \mu_s N_{D \text{ on } W}$. We wish to find the smallest (minimum) coefficient of static friction $\mu_{s \, \text{min}}$ that results in the woman remaining stationary. Substitute this minimum coefficient of static

friction and the earlier expression for the normal force exerted by the drum into this expression:

$$f_{s \, \text{max D on W}} = \mu_{s \, \text{min}} N_{D \text{ on } W} = \mu_{s \, \text{min}} m_W \frac{v^2}{r}$$

Substituting this expression for $f_{s \, \text{max D on W}}$ and $F_{E \text{ on } W} = m_W g$ into the vertical y-component application of Newton's second law, we get

$$0 = \mu_{s \, \text{min}} m_W \frac{v^2}{r} - m_W g$$

Notice that the woman's mass cancels out of this equation. This means that the rotor ride works equally well for any person, independent of the person's mass, and for many people at once as well!

To determine the minimum value of μ_s needed to prevent the woman from sliding, we divide both sides by $m_W v^2$, multiply by r, and rearrange to get

$$\mu_{s \, \text{min}} = \frac{gr}{v^2}$$

We find the woman's speed by using the information we have about the radius and period of her circular motion:

$$v = \frac{2\pi r}{T}$$

Plug this expression for the speed into the equation above for the minimum coefficient of friction:

$$\mu_{s \, \text{min}} = \frac{gr}{v^2} = \frac{grT^2}{4\pi^2 r^2} = \frac{gT^2}{4\pi^2 r}$$

Solve and evaluate We can now use the given information to find the minimum coefficient of friction:

$$\mu_{s \, \text{min}} = \frac{gT^2}{4\pi^2 r} = \frac{9.8 \text{ m/s}^2 \times (1.7 \text{ s})^2}{4 \times 9.87 \times 2.0 \text{ m}} = 0.36$$

The coefficient of friction is a number with no units (also called *dimensionless*), and our answer also has no units. The magnitude 0.36 is reasonable and easy to obtain with everyday materials (wood on wood ranges from 0.25 to 0.5). To be on the safe side, we probably want to make the surface rougher, with twice the minimum coefficient of static friction. The limiting case analysis of the final expression also supports the result. If the drum is stationary ($T = \infty$), the coefficient of static friction would be infinite—the person would have to be glued permanently to the vertical surface. Perhaps most important, the required coefficient of static friction does not depend on the mass of the rider. Earth exerts a greater gravitational force on a more massive person. But the drum also pushes toward the center with a greater normal force, which leads to an increased friction force. The effects balance, and the mass does not matter.

Try it yourself Merry-go-rounds have railings for people to hold when riding. What is the purpose of the railings? Answer this question by drawing a force diagram for an adult standing on a rotating merry-go-round.

Answer The railings exert a force on the person in the radial direction. This force helps provide the necessary radial acceleration.

Texas Motor Speedway

The Texas Motor Speedway is a 2.4-km (1.5-mile)-long oval track. One of its turns is about 200 m in radius and is banked at 24° above the horizontal. How fast would a car have to move so that no friction is needed to prevent it from sliding sideways off the raceway (into the infield or off the track)?

Sketch and translate We start by drawing top view and rear view sketches of the situation. The car is the system. For simplicity, we draw the speedway as a circular track.

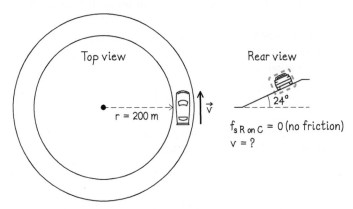

Simplify and diagram Model the car as a point-like object that moves along a circular path at constant speed. Earth exerts a downward force $\vec{F}_{\text{E on C}}$. The surface of the road R exerts on the car a normal force $\vec{N}_{\text{R on C}}$ perpendicular to the surface (see the force diagram at right). We assume the sideways friction force exerted by the road on the car is zero (see the problem statement). For constant speed circular motion, the sum of the forces exerted on the car should point toward the center of the circle. The horizontal radial component of the normal force $N_{\text{R on C }r}$ causes the radial acceleration toward the center. Since the vertical y-component of the car's acceleration is zero, the vertical y-component of the normal force $N_{\text{R on C }y}$ balances the downward gravitational force $\vec{F}_{\text{E on C}}$ that Earth exerts on the car.

Represent mathematically Use the force diagram to help apply the radial r- and vertical y-component forms of Newton's second law. Let's start with the vertical y-equation.

Vertical y-equation: $m_{\text{C}}a_y = \Sigma F_y = N_{\text{R on C }y} + F_{\text{E on C }y}$

Since $a_y = 0$, $N_{\text{R on C }y} = N_{\text{R on C}} \sin 66°$, and $F_{\text{E on C }y} = -m_{\text{C}}g$, we get

$$0 = +N_{\text{R on C}} \sin 66° - m_{\text{C}}g, \text{ or}$$

$$N_{\text{R on C}} \sin 66° = m_{\text{C}}g \qquad (5.8)$$

Now, apply the radial component equation.

Radial r-equation: $m_{\text{C}}a_r = \Sigma F_r = N_{\text{R on C }r} + F_{\text{E on C }r}$

Since $a_r = v^2/r$, $N_{\text{R on C }r} = N_{\text{R on C}} \cos 66°$, and $F_{\text{E on C }r} = 0$, we get:

$$m_{\text{C}}\frac{v^2}{r} = N_{\text{R on C}} \cos 66° + 0 \qquad (5.9)$$

Now combine Eqs. (5.8) and (5.9) to eliminate $N_{\text{R on C}}$ and determine an expression for the speed of the car. To do this, divide the first equation by the second with the $N_{\text{R on C}}$ sides of each equation on the left:

$$\frac{N_{\text{R on C}} \sin 66°}{N_{\text{R on C}} \cos 66°} = \frac{m_{\text{C}}g}{m_{\text{C}}\dfrac{v^2}{r}}$$

Canceling $N_{\text{R on C}}$ on the left side and the m_{C} on the right side and remembering that

$$\frac{\sin 66°}{\cos 66°} = \tan 66°$$

we get

$$\tan 66° = \frac{gr}{v^2}$$

Rearrange the above to find v^2:

$$v^2 = \frac{gr}{\tan 66°}$$

Solve and evaluate The road has a radius $r = 200$ m. Thus,

$$v = \sqrt{\frac{gr}{\tan 66°}} = \sqrt{\frac{(9.8 \text{ m/s}^2)(200 \text{ m})}{\tan 66°}}$$

$$= 29.5 \text{ m/s} \approx 30 \text{ m/s} = 66 \text{ mi/h}$$

This is clearly much slower than the speed of actual racecars. See the "Try it yourself" question.

- -

Try it yourself Construct a force diagram for a racecar that is traveling somewhat faster around the circular track and indicate what keeps the racecar moving in a circle at this higher speed.

Answer   to keep the car moving in a circle.

vides the much greater net radial force needed normal force and the static friction force provides the much greater net radial force needed. The combination of the increased radial components of the increased normal force to be bigger. The combination of the negative y-component and causes the normal force to be bigger. The combination of the track, as shown. The friction force has a negative y-component and causes the normal to the road surface and toward the center of the track, as shown. The friction force has a static friction has to push in on the car parallel to the road surface and toward the center of the static friction has to push in on the car parallel If the racecar is going faster than 30 m/s,

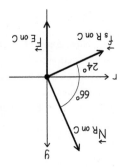

TIP Remember that there is no special force that causes the radial acceleration of an object moving at constant speed along a circular path. This acceleration is caused by all of the forces exerted on the system by other objects (Earth, the surface of a road, a rope. . .). Add the radial components of these regular forces. This sum is what causes the radial acceleration of the system.

Conceptual difficulties with circular motion

Although Newton's second law applies to circular motion just as it applies to linear motion, some everyday experiences seem to contradict that fact. Imagine that you sit on the left side of the back seat of a taxi moving at high speed on a straight road. After traveling straight for a while, the taxi starts to make a high speed *left* turn (Figure 5.2). The car seat is slippery and you aren't holding on to anything. As the car turns left, you start sliding across the seat *toward the right* with respect to the car until you hit the door on the other side of the car. This feeling of being thrown outward in a turning car seems inconsistent with the idea that sum of the forces (the net) points toward the center of the circle during circular motion. The feeling that there is a force pushing outward on you seems very real, but what object is exerting that force?

Remember that Newton's laws explain motion only when made by an observer in an inertial reference frame. Since the car is accelerating as it goes around the curve, it is not an inertial reference frame. To a roadside observer, you and the car were moving at constant velocity before the car started turning. The forces exerted on you by Earth and the car seat added to zero. The observer saw the car begin turning to the left and you continuing to travel with constant velocity because the sum of the forces that other objects exerted on you was zero (we assume that the seat is slippery and the friction force that the surface of the seat exerted on you was very small). When the door finally intercepted your forward-moving body, it started exerting a force on you toward the center of the circle. At that moment, you started moving with the car at constant speed but changing velocity along the circular path. Your velocity then changed because of the force exerted on you by the door.

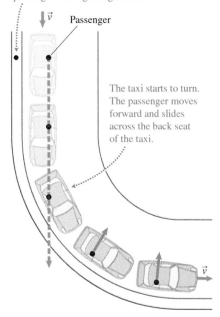

FIGURE 5.2 An observer's view of a passenger as a taxi makes a high-speed turn.

An observer sees the taxi and passenger moving straight ahead.

$\vec{v}$ Passenger

The taxi starts to turn. The passenger moves forward and slides across the back seat of the taxi.

$\vec{v}$

REVIEW QUESTION 5.4 Think back to Example 5.5 (the rotor ride). Use your understanding of Newton's laws and constant speed circular motion to explain why the woman in the ride seems to feel a strong force pushing her against the wall of the drum.

5.5 The law of universal gravitation

So far, we have investigated examples of circular motion on or near the surface of Earth. Now let's look at circular motion as it occurs for planets moving around the Sun and for satellites and our Moon moving around Earth.

Observations and explanations of planetary motion

By Newton's time, scientists already knew a great deal about the motion of the planets in the solar system. Planets were known to move around the Sun at approximately constant speed in almost circular orbits. (Actually, the orbits have a slightly elliptical shape.) However, no one had a scientific explanation for what was causing the Moon and the planets to travel in their nearly circular orbits.

Newton was among the first to hypothesize that the Moon moved in a circular orbit around Earth because Earth pulled on it, continuously changing the direction of the Moon's velocity. He wondered if the force exerted by Earth on the Moon was the same type of force as the force that Earth exerted on falling objects, such as an apple falling from a tree.

The dependence of gravitational force on distance Newton realized that directly measuring the gravitational force exerted by Earth on a distant object was impossible. However, his second law ($\vec{a} = \Sigma\vec{F}/m$) provided an indirect way to determine the gravitational force exerted by Earth: measure the acceleration of the object and use the second law to determine the gravitational force. He knew that the Moon was separated from Earth by about $r_M = 3.8 \times 10^8$ m, roughly 60 Earth radii $(60R_E)$. With this distance and the Moon's orbital period of 27.3 days, Newton was able to use Eq. (5.5) to determine the Moon's radial acceleration:

$$a_{r \text{ at } R=60R_E} = \frac{4\pi^2 R}{T^2} = \frac{4\pi^2(3.8 \times 10^8 \text{ m})}{\left[(27.3 \text{ days})(86,400 \text{ s/day}) \right]^2}$$
$$= \frac{4\pi^2(3.8 \times 10^8 \text{ m})}{(2.36 \times 10^6 \text{ s})^2} = 2.69 \times 10^{-3} \text{ m/s}^2$$

He then did a similar analysis for a second situation. He performed a thought experiment in which he imagined what would happen if the Moon were condensed to a small point-like object (while keeping its mass the same) located near Earth's surface. This would place the "Moon particle" at a distance of one Earth radius $(1\, R_E)$ from Earth's center. If the Moon only interacted with Earth, it would have the same free-fall acceleration as any object near Earth's surface (9.8 m/s^2), since this acceleration is independent of mass. Now Newton knew the Moon's acceleration at two different distances from Earth's center.

Newton then used this information to determine how the gravitational force that one object exerts on another depends on the separation of the objects, assuming that the force causing the acceleration of the Moon is the gravitational force exerted on it by Earth and that this force changes with the separation of the objects. He took the ratio of these two accelerations using his second law:

$$\frac{a_{r \text{ at } R=60R_E}}{a_{r \text{ at } R=1R_E}} = \frac{F_{\text{E on Moon at } R=60R_E}/m_{\text{Moon}}}{F_{\text{E on Moon at } R=1R_E}/m_{\text{Moon}}} = \frac{F_{\text{E on Moon at } R=60R_E}}{F_{\text{E on Moon at } R=1R_E}}$$

Notice that the Moon's mass cancels when taking this ratio. Substituting the two accelerations he had observed, Newton made the following comparison between the forces:

$$\frac{F_{\text{E on Moon at } R=60R_E}}{F_{\text{E on Moon at } R=1R_E}} = \frac{a_{r \text{ at } R=60R_E}}{a_{r \text{ at } R=1R_E}} = \frac{2.69 \times 10^{-3} \text{ m/s}^2}{9.8 \text{ m/s}^2} = \frac{1}{3600} = \frac{1}{60^2}$$

$$\Rightarrow F_{\text{E on Moon at } R=60R_E} = \frac{1}{60^2} F_{\text{E on Moon at } R=1R_E}$$

The force exerted on the Moon when the Moon is at a distance of 60 times Earth's radius is $1/60^2$ times the force exerted on the Moon when near Earth's surface (about 1 times Earth's radius). It appears that as the distance from Earth's center to the Moon *increases* by a factor of 60, the force decreases by a factor of $1/60^2 = 1/3600$. This suggests that the gravitational force that Earth exerts on the Moon depends on the inverse square of its distance from the center of Earth:

$$F_{\text{E on Moon at } r} \propto \frac{1}{r^2} \tag{5.10}$$

TIP You might be wondering why if Earth pulls on the Moon, the Moon does not come closer to Earth in the same way that an apple falls from a tree. The difference in these two cases is the speed of the objects. The apple is at rest with respect to Earth before it leaves the tree, and the Moon is moving tangentially. Think of what would happen to the Moon if Earth stopped pulling on it—it would fly away along a straight line.

The dependence of gravitational force on mass Newton next wanted to determine in what way the gravitational force depended on the masses of the interacting objects. Remember that the free-fall acceleration of an object near Earth's surface does not depend on the object's mass. All objects have the same free-fall acceleration of 9.8 m/s^2:

$$a_O = \frac{F_{\text{E on O}}}{m_O} = g$$

Mathematically, this only works if the magnitude of the gravitational force that Earth exerts on a falling object $F_{\text{E on O}}$ is proportional to the mass of that object: $F_{\text{E on O}} = m_O g$. Newton assumed that the gravitational force that Earth exerts on the Moon has this same feature—the gravitational force must be proportional to the Moon's mass:

$$F_{\text{E on M}} \propto m_{\text{Moon}} \tag{5.11}$$

The last question was to decide whether the gravitational force exerted by Earth on the Moon depends only on the Moon's mass or on the mass of Earth as well. Newton's own third law suggested an answer. If Earth exerts a gravitational force on the Moon, then according to Newton's third law the Moon must also exert a gravitational force on Earth that is equal in magnitude (see **Figure 5.3**). Because of this, Newton decided that the gravitational force exerted by Earth on the Moon should also be proportional to the mass of Earth:

$$F_{\text{E on M}} = F_{\text{M on E}} \propto m_{\text{Earth}} \tag{5.12}$$

Newton then combined Eqs. (5.10)–(5.12), obtaining a mathematical relationship for the dependence of the gravitational force on all these factors:

$$F_{\text{E on M}} = F_{\text{M on E}} \propto \frac{m_{\text{Earth}} m_{\text{Moon}}}{r^2} \tag{5.13}$$

Newton further proposed that this was not just a mathematical description of the gravitational interaction between Earth and the Moon, but that it was more general and described the gravitational interaction between any two objects; he called this relation the **law of universal gravitation**. Newton's extension of the applicability of Eq. (5.13) to celestial objects was based on his analysis of Kepler's laws, which you will read about in the next section.

When Newton devised Eq. (5.13), he did not know the masses of Earth and the Moon and could not measure the force that one exerted on the other. Thus, the best he could do was to construct the above proportionality relationship. It was much later that scientists determined the value of the constant of proportionality:

$$G = 6.67 \times 10^{-11} \, \text{N} \cdot \text{m}^2/\text{kg}^2$$

G, known as the **gravitational constant**, allows us to write a complete mathematical equation for the gravitational force that one mass exerts on another:

$$F_{g \, 1 \, \text{on} \, 2} = G \frac{m_1 m_2}{r^2} \tag{5.14}$$

Notice that the value of the gravitational constant is quite small. If you take two objects, each with a mass of 1.0 kg, and separate them by 1.0 m, the force that they exert on each other equals 6.67×10^{-11} N. This means that the gravitational force that objects like us (with masses of about 50–100 kg) exert on each other is extremely weak. However, because the masses of celestial objects are huge (Earth's mass is approximately 6.0×10^{24} kg and the Sun's mass is 2.0×10^{30} kg), the gravitational forces that they exert on each other are very large, despite the small value of the gravitational constant.

Newton's third law and the gravitational force

The mass of Earth is approximately 6.0×10^{24} kg. The mass of a tennis ball is approximately 50 g, or 5.0×10^{-2} kg. How does the gravitational force that Earth exerts downward on the ball compare to the upward force that the ball exerts on Earth? According to Newton's third law, interacting objects exert forces of equal magnitude and opposite direction on each other. The gravitational force exerted by Earth on the ball has the same magnitude as the force that the ball exerts on Earth. Equation (5.14) is consistent with this idea:

$$F_{\text{E on B}} = G \frac{M_E m_B}{r^2}$$

$$= (6.67 \times 10^{-11} \, \text{N} \cdot \text{m}^2/\text{kg}^2) \frac{(6.0 \times 10^{24} \, \text{kg})(5.0 \times 10^{-2} \, \text{kg})}{(6.4 \times 10^6 \, \text{m})^2} = 0.49 \, \text{N}$$

$$F_{\text{B on E}} = G \frac{m_B M_E}{r^2}$$

$$= (6.67 \times 10^{-11} \, \text{N} \cdot \text{m}^2/\text{kg}^2) \frac{(5.0 \times 10^{-2} \, \text{kg})(6.0 \times 10^{24} \, \text{kg})}{(6.4 \times 10^6 \, \text{m})^2} = 0.49 \, \text{N}$$

(In the denominator we use the distance between the center of Earth and the center of the ball.)

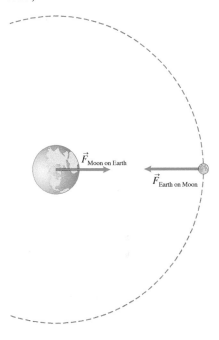

FIGURE 5.3 Earth and the Moon exert equal-magnitude and oppositely directed forces on each other. (Earth and Moon not drawn to scale.)

It might seem counterintuitive that the ball exerts a gravitational force on Earth. After all, Earth doesn't seem to react every time someone drops something. But according to Newton's second law, if the ball exerts a nonzero net force on Earth, then Earth should accelerate. What is the magnitude of this acceleration?

$$a_E = \frac{F_{B\,on\,E}}{M_E} = \frac{0.49\ N}{6.0 \times 10^{24}\ kg} = 8.2 \times 10^{-26} m/s^2$$

Earth's acceleration is so tiny that there is no known way to observe it. On the other hand, the acceleration of the ball is easily noticeable since the ball's mass is so much smaller.

$$a_B = \frac{F_{E\,on\,B}}{m_B} = \frac{0.49\ N}{(5.0 \times 10^{-2}\ kg)} = 9.8\ m/s^2$$

Notice that the ball has the familiar free-fall acceleration. In fact, the discussion above allows us to understand why free-fall acceleration on Earth equals $9.8\ m/s^2$ and not some other number. Use Newton's second law and the law of gravitation for an object falling freely near Earth's surface to determine its acceleration:

$$a_{object} = \frac{F_{Earth\,on\,object}}{m_{object}} = \frac{(GM_{Earth}m_{object})/R_{Earth}^2}{m_{object}} = \frac{GM_{Earth}}{R_{Earth}^2}$$

Inserting Earth's mass $M_{Earth} = 5.97 \times 10^{24}\ kg$ and its radius $R_{Earth} = 6.37 \times 10^6\ m$, we have

$$a_{object} = \frac{GM_{Earth}}{R_{Earth}^2}$$
$$= \frac{(6.67 \times 10^{-11}\ N \cdot m^2/kg^2)(5.97 \times 10^{24}\ kg)}{(6.37 \times 10^6\ m)^2} = 9.8\ N/kg = 9.8\ m/s^2$$

This is exactly the free-fall acceleration of objects near Earth's surface. The fact that the experimentally measured value of $9.8\ m/s^2$ agrees with the value calculated using the law of gravitation gives us more confidence in the correctness of the law. This consistency check serves the purpose of a testing experiment. The expression $F_{E\,on\,O} = m_O g$ is actually a special case of the law of gravitation used to determine the gravitational force that Earth exerts on objects when near Earth's surface. Using a similar technique we can determine the free-fall acceleration of an object near the surface of the Moon $(m_{Moon} = 7.35 \times 10^{22}\ kg$ and $R_{Moon} = 1.74 \times 10^6\ m)$ to be $1.6\ N/kg = 1.6\ m/s^2$.

Kepler's laws and the law of universal gravitation

About half a century before Newton's work, Johannes Kepler (1571–1630) studied astronomical data concerning the motion of the known planets and crafted three laws, called Kepler's laws of planetary motion. Newton knew about Kepler's work and applied the force equation developed for Earth-Moon interaction to the Sun and the known planets. By using the law of universal gravitation he could explain Kepler's laws. This fact contributed to scientists' confidence in the law of universal gravitation.

Kepler's first law of planetary motion The orbits of all planets are ellipses with the Sun located at one of the ellipse's foci (**Figure 5.4a**).[1]

Kepler's second law of planetary motion When a planet travels in an orbit, an imaginary line connecting the planet and the Sun continually sweeps out the same area during the same time interval, independent of the planet's distance from the Sun (Figure 5.4b).

FIGURE 5.4 Orbital models used by Kepler to develop his laws of planetary motion.

(a) The orbits of all planets are ellipses with the Sun located at one focus.

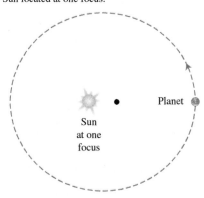

Sun
at one
focus

Planet

(b) An imaginary line connecting the planet and the Sun continually sweeps out the same area during the same time interval.

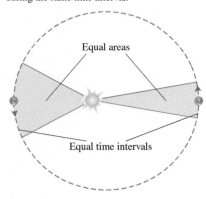

Equal areas

Equal time intervals

_____

[1] The shape of the planetary orbits is close to circular for most of the planets. In those cases, the foci of the ellipse are very close to the center of the circular orbit that most closely approximates the ellipse.

Kepler's third law of planetary motion The square of the period T of the planet's motion (the time interval to complete one orbit) divided by the cube of the semi-major axis of the orbit (which is half the maximum diameter of an elliptical orbit or the radius r of a circular one) equals the same constant for all the known planets:

$$\frac{T^2}{r^3} = K \qquad (5.15)$$

Newton was able to use his laws of motion and gravitation to derive all three of Kepler's laws of planetary motion. In addition, astronomers later used Newton's law of universal gravitation to predict the locations of the then-unknown planets Neptune and Pluto. They did it by observing subtle inconsistencies between the observed orbits of known planets and the predictions of the law of universal gravitation. At present, physicists and engineers use the law of universal gravitation to decide how to launch satellites into their desired orbits and how to successfully send astronauts to the Moon and beyond. Newton's law of universal gravitation has been tested countless times and is consistent with observations to a very high degree of accuracy.

Newton's law of universal gravitation The magnitude of the attractive gravitational force that an object with mass m_1 exerts on an object with mass m_2 separated by a center-to-center distance r is:

$$F_{g\,1\,\text{on}\,2} = G\frac{m_1 m_2}{r^2} \qquad (5.14)$$

where $G = 6.67 \times 10^{-11}\ \text{N} \cdot \text{m}^2/\text{kg}^2$ is known as the gravitational constant.

However, just because this relationship is called a "law" does not mean that the law is always valid. Every mathematical expression in physics is applicable only for certain circumstances, and the law of universal gravitation is no exception.

Limitations of the law of universal gravitation

The law of universal gravitation in the form of Eq. (5.14) only applies to objects with spherical symmetry or whose separation distance is much bigger than their size. An object that is spherically symmetric has the shape of a perfect sphere and a density that only varies with distance from its center. In these cases, we can model the objects as though all of their mass was located at a single point at their centers. However, if the objects are not spherically symmetric and are close enough to each other so that they cannot be modeled as point-like particles, then the law of universal gravitation does not apply directly. But using calculus, we can divide each object into a collection of very small point-like objects and use the law of universal gravitation on each pair to add by integration the forces of the small point-like particles on each other.

However, even with calculus, there are details of some motion for which the law cannot account. When astronomers made careful observations of the orbit of Mercury, for example, they noticed that its orbit exhibited some patterns that the law of universal gravitation could not explain. It wasn't until the early 20th century, when Einstein constructed a more advanced theory of gravity, that scientists could predict all of the details of the motion of Mercury.

Are astronauts weightless?

The Moon orbits Earth due to the gravitational force that provides the necessary radial acceleration. It is Earth's natural *satellite*, a celestial object that orbits a bigger celestial object (usually a planet). Artificial satellites are objects that are placed in orbit by humans. The first artificial satellite was launched in 1957, and since then thousands of satellites have been sent into orbit. Earth satellites make worldwide communication possible, allow us to monitor Earth's surface and the weather, help us find our way to unknown destinations with our global positioning systems (GPS), and provide access to hundreds of television stations. One such satellite is the International Space Station.

We are all familiar with videos of astronauts floating in the International Space Station. News reports commonly say that the astronauts are weightless or experience zero gravity. Is this true?

QUANTITATIVE EXERCISE 5.7   **Are astronauts in the International Space Station weightless?**

The International Space Station orbits about 0.40×10^6 m (250 miles) above Earth's surface, or $(6.37 + 0.40) \times 10^6$ m $= 6.77 \times 10^6$ m from Earth's center. Compare the force that Earth exerts on an astronaut in the station to the force when he is on Earth's surface.

Represent mathematically The gravitational force that Earth exerts on the astronaut (system) in the space station is

$$F_{\text{E on A in station}} = G\frac{m_E m_A}{\left(r_{\text{station-Earth}}\right)^2}$$

where $r_{\text{station-Earth}} = 6.77 \times 10^6$ m is the distance from Earth's center to the astronaut in the space station. The gravitational force exerted by Earth on the astronaut when on Earth's surface is

$$F_{\text{E on A on Earth's surface}} = G\frac{m_E m_A}{\left(r_{\text{surface-Earth}}\right)^2}$$

where $r_{\text{surface-Earth}} = 6.37 \times 10^6$ m is the distance from the center of Earth to its surface. To compare these two forces, take their ratio:

$$\frac{F_{\text{E on A in station}}}{F_{\text{E on A on Earth's surface}}} = \frac{G\dfrac{m_E m_A}{r_{\text{station-Earth}}^2}}{G\dfrac{m_E m_A}{r_{\text{surface-Earth}}^2}} = \left(\frac{r_{\text{surface-Earth}}}{r_{\text{station-Earth}}}\right)^2$$

Solve and evaluate Inserting the appropriate values gives

$$\frac{F_{\text{E on A in station}}}{F_{\text{E on A on Earth's surface}}} = \left(\frac{6.37 \times 10^6 \text{ m}}{6.77 \times 10^6 \text{ m}}\right)^2 = 0.89$$

The gravitational force that Earth exerts on the astronaut when in the space station is just 11% less than the force exerted on him when on Earth's surface. The astronaut is far from "weightless."

- -

Try it yourself At what distance from Earth's center should a person be so that her weight is half her weight when on Earth's surface?

Answer $1.4R_E$.

Quantitative Exercise 5.7 shows that astronauts are not actually weightless while in the International Space Station. So why do they float? Remember that Earth exerts a gravitational force on both the astronaut and the space station. This force causes them both to fall toward Earth at the same rate while they fly forward, thus staying on the same circular path. Both the astronaut and the space station are in free fall. As a result, if the astronaut stood on a scale placed on the floor of the orbiting space station, her weight, according to the scale, would be zero. The same thing happens on Earth when you stand on a scale inside an elevator that is in free fall, falling at the acceleration $\vec{g}$. Since both you and the elevator fall with the same acceleration, you do not press on the scale, and the scale does not press on you; it reads zero.

The confusion occurs because in physics weight is a shorthand way of referring to the gravitational force being exerted on an object, not the reading of a scale. The scale measures the normal force it exerts on any object with which it is in contact. So astronauts are not "weightless"; the word is actually being misused.

REVIEW QUESTION 5.5 A friend says he has heard that the Moon is falling toward Earth. What can you tell your friend to reassure him that all is well?

Summary

Velocity change method to find the direction of acceleration We can estimate the direction of an object's acceleration as it passes a point along its circular path by drawing an initial velocity vector v_i just before that point and a final velocity vector v_f just after that point. We place the vectors tail to tail and draw a velocity change vector from the tip of the initial to the tip of the final velocity. The acceleration points in the direction of the velocity change vector. (Section 5.2)

Top view

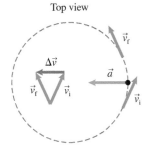

$$\vec{v}_i + \Delta\vec{v} = \vec{v}_f$$

or

$$\Delta\vec{v} = \vec{v}_f - \vec{v}_i$$

$$\vec{a} = \frac{\Delta\vec{v}}{\Delta t} = \frac{\vec{v}_f - \vec{v}_i}{t_f - t_i}$$

Radial acceleration for constant speed circular motion An object moving at constant speed along a circular path has acceleration that points toward the center of the circle and has a magnitude a_r that depends on its speed v and the radius r of the circle. (Section 5.3)

Top view

$$a_r = \frac{v^2}{r} \qquad \text{Eq. (5.3)}$$

Period and radial acceleration The radial acceleration can also be expressed using the period T of circular motion, the time interval needed for an object to complete one trip around the circle. (Section 5.3)

$$T = \frac{2\pi r}{v} \qquad \text{Eq. (5.4)}$$

$$a_r = \frac{4\pi^2 r}{T^2} \qquad \text{Eq. (5.5)}$$

Sum of the forces for constant speed circular motion The sum of the forces exerted on an object during constant speed circular motion points in the positive radial direction toward the center of the circle. The object's acceleration is the sum of the radial components of all forces exerted on an object divided by its mass—consistent with Newton's second law. In addition, for horizontal circular motion, you sometimes analyze the vertical y-components of forces exerted on an object. (Sections 5.1 and 5.4)

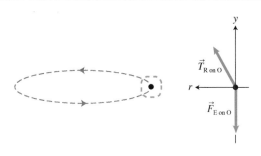

$$a_r = \frac{\Sigma F_r}{m} \quad \text{or} \quad ma_r = \Sigma F_r \qquad \text{Eq. (5.6)}$$

$$ma_y = \Sigma F_y = 0 \qquad \text{Eq. (5.7)}$$

Law of universal gravitation This force law is used primarily to determine the magnitude of the force that the Sun exerts on planets or that planets exert on satellites or on moons. The force depends on the masses of the objects and on their center-to-center separation r. (Section 5.5)

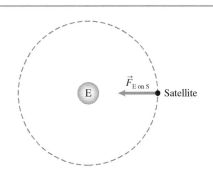

$$F_{g\,1\text{ on }2} = G\frac{m_1 m_2}{r^2} \qquad \text{Eq. (5.14)}$$

G is the gravitational constant,

$$G = 6.67 \times 10^{-11}\ \text{N}\cdot\text{m}^2/\text{kg}^2$$

Questions

Multiple Choice Questions

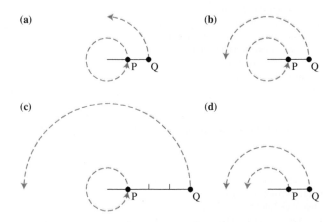

1. Which of the objects below is accelerating?
   (a) Object moving at constant speed along a straight line
   (b) Object moving at constant speed in a circle
   (c) Object slowing down while moving in a straight line
   (d) Both b and c describe accelerating objects.

2. The circle in **Figure Q5.2** represents the path followed by an object moving at constant speed. At four different positions on the circle, the object's motion is described using velocity and acceleration arrows. Choose the location where the descriptions are correct.
   (a) Point A   (b) Point B
   (c) Point C   (d) Point D

   FIGURE Q5.2

3. One of your classmates drew a force diagram for a pendulum bob at the bottom of its swing. He put a horizontal force arrow in the direction of the velocity. Evaluate his diagram by choosing from the statements below.
   (a) He is incorrect since the force should point partly forward and tilt partly down.
   (b) He is incorrect since there are no forces in the direction of the object's velocity.
   (c) He is correct since any moving object has a force in the direction of its velocity.
   (d) He is correct but he also needs to add an outward (down) force.

4. Why is it difficult for a high-speed car to negotiate an unbanked turn?
   (a) A huge force is pushing the car outward.
   (b) The magnitude of the static friction force might not be enough to provide the necessary radial acceleration.
   (c) The sliding friction force is too large.
   (d) The faster the car moves, the harder it is for the driver to turn the steering wheel.

5. How does a person standing on the ground explain why you, sitting on the left side of a slippery back car seat, slide to the right when the car makes a high-speed left turn?
   (a) You tend to move in a straight line and thus slide with respect to the seat that is moving to the left under you.
   (b) There is a net outward force being exerted on you.
   (c) The force of motion propels you forward.
   (d) The car seat pushes you forward.

6. A pilot performs a vertical loop-the-loop at constant speed. The pilot's head is always pointing toward the center of the circle. Where is a pilot blackout more likely to occur?
   (a) At the top of the loop
   (b) At the bottom of the loop
   (c) At both top and bottom
   (d) None of the answers is correct.

7. Why is the following an inaccurate statement about blackout? "As the g forces climb up toward 7 g's, ..."
   (a) There is no such thing as a g force.
   (b) 7 g's is not a big force.
   (c) g forces are constant and do not climb.
   (d) g forces are not responsible for blackout but can cause dizziness.

8. Two point-like objects P and Q are undergoing circular motion at constant speeds. **Figure Q5.8** shows four possible combinations of paths for the two objects during the same time interval. Choose which is correct for
   i. the periods of the objects to be equal,
   ii. the speeds of the objects to be equal, and
   iii. the radial accelerations of the objects to be equal.

9. Compare the magnitude of the normal force of a car seat on you with the magnitude of the force that Earth exerts on you when the car moves across the bottom of a dip in the road.
   (a) The normal force is more than Earth's gravitational force.
   (b) The normal force is less than Earth's gravitational force.
   (c) The two forces are equal.
   (d) It depends on whether the car is moving right or left.

10. If you put a penny on the center of a rotating turntable, it does not slip. However, if you place the penny near the edge, it is likely to slip off. Which answer below explains this observation?
    (a) The penny moves faster at the edge and hence needs a greater force to keep it moving.
    (b) The edge is more slippery since the grooves are farther apart.
    (c) The radial acceleration is greater at the edge and the friction force is not enough to keep the penny in place.
    (d) The outward force responsible for slipping is greater at the edge than at the middle.

11. Where on Earth's surface would you expect to experience the greatest radial acceleration as a result of Earth's rotation?
    (a) On the poles
    (b) On the equator
    (c) On the highest mountain
    (d) Since all points of Earth have the same period of rotation, the acceleration is the same everywhere.

12. What observational data might Newton have used to decide that the gravitational force is inversely proportional to the distance squared between the centers of objects?
    (a) The data on the acceleration of falling apples
    (b) The data describing the Moon's orbit (period and radius) and the motion of falling apples
    (c) The data on comets
    (d) The data on moonrise and moonset times

13. What observations combined with his second and third laws helped Newton decide that the gravitational force of one object on another object is directly proportional to the product of the masses of the interacting objects?
    (a) The data on the acceleration of falling apples
    (b) The data on Moon phases
    (c) The data on comets
    (d) The data on moonrise and moonset times

14. What would happen to the force exerted by the Sun on Earth if the Sun shrank and became half its present size while retaining the same mass?
    (a) The force would be half the present force.
    (b) The force would be one-fourth of the present force.
    (c) The force would double.
    (d) The force would stay the same.

15. James fixes a camera on a tripod and takes several photos of the full Moon. **Figure Q5.15** shows two photos taken during a time interval in which the Moon seems to moves one Moon diameter to the right. Which phenomenon is mainly responsible for this observed motion of the Moon?
    (a) The rotation of Earth around its axis
    (b) The revolution of Earth around the Sun
    (c) The revolution of the Moon around Earth
    (d) The motion of tectonic plates on Earth

FIGURE Q5.15

11:00:44

11:02:14

Conceptual Questions

16. Your friend says that an object weighs less on Jupiter than on Earth as Jupiter is far away from the center of Earth. Do you agree or disagree?

17. Your friend says that when an object is moving in a circle, there is a force pushing it out away from the center. Why would he say this? Do you agree or disagree? If you disagree, how would you convince your friend of your opinion?

18. Describe three everyday phenomena that are consistent with your knowledge of the dynamics of circular motion. Specify where the observer is. Then find an observer who will not be able to explain the same phenomena using the knowledge of Newton's laws and circular motion.

19. Two identical cars are moving with equal speeds on a wet road toward a turn, as shown in **Figure Q5.19**. Which car is more likely to slide off the road during the turn and why, assuming their speeds remain unchanged?

FIGURE Q5.19

    (a) The two cars are equally likely to slide off the road because their masses are equal.
    (b) The two cars are equally likely to slide off the road because their speeds are equal.
    (c) Car B because it is closer to the outer side of the turn
    (d) Car B because the distance it travels around the turn is larger
    (e) Car A because it moves through the turn with a smaller radius
    (f) Car A because the distance it travels around the turn is smaller

A   B

20. Astronauts on the space station orbiting Earth are said to be in "zero gravity." Do you agree or disagree? If you disagree, why?

21. In the movies you often see space stations with "artificial gravity." They look like big doughnuts rotating around an axis perpendicular to the plane of the doughnut (see **Figure Q5.21**). People walking on the outer rim inside the turning space station feel the same gravitational effects as if they were on Earth. How does such a station work to simulate artificial gravity?

FIGURE Q5.21

22. Give one example of a situation in which an object moves at constant speed and has zero acceleration and another of a situation in which the object moves at constant speed and has a nonzero acceleration.

23. Name a planet on which you would weigh less than on Earth. Name a planet on which you would weigh more than on Earth. Explain how you know.

24. A motorized cart is moving at a constant speed along a horizontal surface that at point P smoothly changes from being horizontal to rising upward in a circular arc (see **Figure Q5.24**). The cart passes point P at time t_P while maintaining the same constant speed v. Which graph correctly shows the time dependence of the sum of the forces exerted on the cart as it passes the point P, assuming the cart is a point-like object?

FIGURE Q5.24

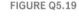

(a) $\Sigma F_{\text{on cart}}$

(b) $\Sigma F_{\text{on cart}}$

(c) $\Sigma F_{\text{on cart}}$

(d) $\Sigma F_{\text{on cart}}$

Problems

Below, **BIO** indicates a problem with a biological or medical focus. Problems labeled **EST** ask you to estimate the answer to a quantitative problem rather than derive a specific answer. Asterisks indicate the level of difficulty of the problem. Problems with no * are considered to be the least difficult. A single * marks moderately difficult problems. Two ** indicate more difficult problems.

5.1–5.3 Qualitative dynamics of circular motion, Analyzing velocity change for circular motion, and Radial acceleration and period

1. **Mountain biker** While mountain biking, you first move at constant speed along the bottom of a trail's circular dip and then at constant speed across the top of a circular hump. Assume that you and the bike are a system. Determine the direction of the acceleration at each position and construct a force diagram for each position (consistent with the direction of the acceleration). Compare at each position the magnitude of the force of the surface on the bike with the force Earth exerts on the system.

2. * You swing a rock tied to a string in a vertical circle. (a) Determine the direction of the acceleration of the rock as it passes the lowest point in its swing. Construct a consistent force diagram for the rock as it passes that point. How does the force that the string exerts on the rock compare to the force that Earth exerts on the rock? Explain. (b) Repeat this analysis as best you can for the rock as it passes the highest point in the swing. (c) If the string is tied around your finger, when do you feel a stronger pull—when the rock is at the bottom of the swing or at the top? Explain.

3. * **Loop-the-loop** You ride a roller coaster with a loop-the-loop. Compare *as best you can* the normal force that the seat exerts on you to the force that Earth exerts on you when you are passing the bottom of the loop and the top of the loop. Justify your answers by determining the direction of acceleration and constructing a force diagram for each position. Make your answers consistent with Newton's second law.

4. You start an old record player and notice a bug on the surface close to the edge of the record. The record has a diameter of 12 inches and completes 33 revolutions each minute. (a) What are the speed and the acceleration of the bug? (b) What would the bug's speed and acceleration be if it were halfway between the center and the edge of the record?

5. Determine the acceleration of Earth due to its motion around the Sun. What do you need to assume about Earth to make the calculation? How does this acceleration compare to the acceleration of free fall on Earth?

6. The Moon is an average distance of 3.8×10^8 m from Earth. It circles Earth once each 27.3 days. (a) What is its average speed? (b) What is its acceleration? (c) How does this acceleration compare to the acceleration of free fall on Earth?

7. **Aborted plane landing** You are on an airplane that is landing. The plane in front of your plane blows a tire. The pilot of your plane is advised to abort the landing, so he pulls up, moving in a semicircular upward-bending path. The path has a radius of 500 m with a radial acceleration of 17 m/s². What is the plane's speed?

8. **BIO** **Ultracentrifuge** You are working in a biology lab and learning to use a new ultracentrifuge for blood tests. The specifications for the centrifuge say that a red blood cell rotating in the ultracentrifuge moves at 470 m/s and has a radial acceleration of 150,000 g's (that is, 150,000 times 9.8 m/s²). The radius of the centrifuge is 0.15 m. You wonder if this claim is correct. Support your answer with a calculation.

9. * **EST** A tire-pressure monitoring system warns you with a dashboard alert when one of your car tires is significantly underinflated, possibly creating unsafe driving conditions. How does the car know when to turn on the alert? Many cars have sensors that measure the period of rotation of each wheel. Based on the period, the car's computer determines the relative size of the tires when the car is moving at a certain speed. (a) Explain how the period of the wheel rotation is related to the size of the tire. (b) Derive a mathematical equation that relates the period T of the wheel rotation to the tire radius r and the speed v of the car. (c) Figure P5.9 shows a properly inflated tire (left) and an underinflated tire (right) of the same car. Estimate the percent change of the period of the underinflated tire compared to the properly inflated tire. (d) Estimate the distance this car would have to travel for the difference between the numbers of turns of the two wheels to be equal to one turn. Which will undergo more turns, the underinflated or the properly inflated tire?

FIGURE P5.9

11. * Rolling is a combination of linear and circular motion. In order to study rolling motion, we perform the following experiment. We mark three points A, B, and C on a bicycle wheel and observe how these points move while the bicycle is pushed forward 2 cm, a distance much smaller than the radius of the wheel. Figure P5.11 shows a long-exposure photo of the experiment. The camera (i.e., the observer) is standing still on the ground. A is the point where the wheel is initially touching the ground, B is the point at the wheel's axis, and C is the point on the top of the wheel, directly above point A. (a) Determine the displacements of points A, B, and C as observed by the stationary observer. (Hint: Use similar triangles.) (b) Using your answers from part (a), determine the speed of the points A, B, and C with respect to the ground for the bicycle that is moving at constant speed v. How would the answers to part (b) change if the observer were moving in the same direction as the bicycle but (c) with the same speed as the bicycle or (d) twice as fast as the bicycle?

FIGURE P5.11

12. ** Figure P5.12 shows photos of car wheels that were obtained in four different ways:
    i. A person standing on the sidewalk took a photo of a car that was passing by.
    ii. A person in a moving car took a photo of another car. Both cars were moving with the same speed.
    iii. A person in a moving car took a photo of another car that was moving faster than the photographer's car.
    iv. A person in a moving car took a photo of another car that was moving slower than the photographer's car.
    Match each photo (a) to (d) with the corresponding photographing procedure. [Hint: If you have difficulties with this problem, solve the previous problem.]

FIGURE P5.12

13. * Three people are standing on a horizontally rotating platform in an amusement park. One person is almost at the edge, the second one is $(3/5)R$ from the center, and the third is $(1/2)R$ from the center. Compare their periods of rotation, their speeds, and their radial accelerations.

14. * Consider the scenario described in Problem 5.13. If the platform speeds up, who is more likely to have trouble staying on the platform? Support your answer with a force diagram and describe the assumptions that you made.

10. * Imagine that you are standing on a horizontal rotating platform in an amusement park (like the platform for a merry-go-round). The period of rotation and the radius of the platform are given, and you know your mass. Make a list of the physical quantities you could determine using this information, and describe how you would determine them.

15. * You want to determine the radial acceleration of a friend who is standing on the rotating platform of a merry-go-round. You have a stopwatch and a measuring tape. Which of the following procedures would best serve your task?

(a) Measure the distance d between the point where the person is standing and the axis of rotation, count the number of rotations N of the platform in 20 seconds, and calculate $d \cdot \left(\dfrac{2\pi N}{20 \text{ s}}\right)^2$.

(b) Measure the distance d between the point where the person is standing and the edge of the platform, count the number of rotations N of the platform in 20 seconds, and calculate $d \cdot \left(\dfrac{2\pi N}{20 \text{ s}}\right)^2$.

(c) Measure the distance d between the point where the person is standing and the axis of rotation, measure the time t for 20 revolutions of the platform, and calculate $d \cdot \left(\dfrac{20}{2\pi t}\right)^2$.

16. **Ferris wheel** You are sitting on a rotating Ferris wheel. Draw a force diagram for yourself when you are at the bottom of the circle and when you are at the top.

17. * **EST** Estimate the radial acceleration of the foot of a college football player in the middle of punting a football.

18. * **EST** Estimate the radial acceleration of the toe at the end of the horizontally extended leg of a ballerina doing a pirouette.

5.4 Skills for analyzing processes involving circular motion

19. * Is it safe to drive your 1600-kg car at speed 27 m/s around a level highway curve of radius 150 m if the effective coefficient of static friction between the car and the road is 0.40? Explain.

20. * You are fixing a broken rotary lawn mower. The blades on the mower turn 50 times per second. What is the magnitude of the force needed to hold the outer 2 cm of the blade to the inner portion of the blade? The outer part is 21 cm from the center of the blade, and the mass of the outer portion is 7.0 g.

21. * Your car speeds around the 80-m-radius curved exit ramp of a freeway. A 70-kg student holds the armrest of the car door, exerting a 220-N force on it in order to prevent himself from sliding across the vinyl-covered back seat of the car and slamming into his friend. How fast is the car moving in meters per second and miles per hour? What assumptions did you make?

22. How fast do you need to swing a 200-g ball at the end of a string in a horizontal circle of 0.5-m radius so that the string makes a 34° angle relative to the horizontal? What assumptions did you make?

23. ** A small ball is attached by a string to a smooth incline, as shown in Figure P5.23. The incline is initially at rest. (a) Draw a force diagram for the ball and label all the forces. (b) Describe what happens to the magnitude of each force (does it increase, decrease, or remain the same?) when the incline (with the ball and the string) rotates around the vertical axis at constant speed.

FIGURE P5.23

24. A coin rests on a record 0.15 m from its center. The record turns on a turntable that rotates at variable speed. The coefficient of static friction between the coin and the record is 0.30. What is the maximum coin speed at which it does not slip?

25. * **Roller coaster ride** A roller coaster car travels at speed 8.0 m/s over a 12-m-radius vertical circular hump. What is the magnitude of the upward force that the coaster seat exerts on a 48-kg woman passenger?

26. ** A person sitting in a chair (combined mass 80 kg) is attached to a 6.0-m-long cable. The person moves in a horizontal circle. The cable angle θ is 62° below the horizontal. What is the person's speed? Note: The radius of the circle is not 6.0 m.

27. * A car moves around a 50-m-radius highway curve. The road, banked at 10° relative to the horizontal, is wet and icy so that the coefficient of friction is approximately zero. At what speed should the car travel so that it makes the turn without slipping?

28. * A 20.0-g ball is attached to a 120-cm-long string and moves in a horizontal circle (see Figure P5.28). The string exerts a force on the ball that is equal to 0.200 N. What is the angle θ?

FIGURE P5.28

29. A 50-kg ice skater goes around a circle of radius 5.0 m at a constant speed of 3.0 m/s on a level ice rink. What are the magnitude and direction of the horizontal force that the ice exerts on the skates?

30. * A car traveling at 10 m/s passes over a hill on a road that has a circular cross section of radius 30 m. What is the force exerted by the seat of the car on a 60-kg passenger when the car is passing the top of the hill?

31. A 1000-kg car is moving at 30 m/s around a horizontal level curved road whose radius is 100 m. What is the magnitude of the frictional force required to keep the car from sliding?

32. * **Equation Jeopardy 1** Describe using words, a sketch, a velocity change diagram, and a force diagram two situations whose mathematical description is presented below.

$$700 \text{ N} - (30 \text{ kg})(9.8 \text{ m/s}^2) = \frac{(30 \text{ kg})v^2}{12 \text{ m}}$$

33. * **Equation Jeopardy 2** Describe using words, a sketch, a velocity change diagram, and a force diagram two situations whose mathematical description is presented below.

$$\frac{(2.0 \text{ kg})(4.0 \text{ m}^2/\text{s}^2)}{r} = 0.4 \times (2.0 \text{ kg}) \times (9.8 \text{ N/kg})$$

34. ** **Banked curve raceway design** You need to design a banked curve at the new circular Super 100 Raceway. The radius of the track is 800 m, and cars typically travel at speed 160 mi/h. What feature of the design ensures that all racecars can move around the track safely in any weather? (a) Provide a quantitative answer. (b) List your assumptions and describe whether the number you provided will increase or decrease if the assumption is not valid.

35. * A circular track is in a horizontal plane, has a radius of r meters, and is banked at an angle θ above the horizontal. (a) Develop an expression for the speed a person should Rollerblade on this track so that she needs zero friction to prevent her from sliding sideways off the track. (b) Should another person move faster or slower if her mass is 1.3 times the mass of the first person? Justify your answer.

36. ** Design a quantitative test for Newton's second law as applied to constant speed circular motion. Describe the experiment and provide the analysis needed to make a prediction using the law.

5.5 The law of universal gravitation

37. * Your friend says that the force that the Sun exerts on Earth is much larger than the force that Earth exerts on the Sun. (a) Do you agree or disagree with this opinion? (b) If you disagree, how would you convince your friend of your opinion?

38. Determine the gravitational force that (a) the Sun exerts on the Moon, (b) Earth exerts on the Moon, and (c) the Moon exerts on Earth. List at least two assumptions for each force that you made when you calculated the answers.

39. * (a) What is the ratio of the gravitational force that Earth exerts on the Sun in the winter and the force that it exerts in the summer? (b) How many correct answers can you give for part (a)? (Hint: Earth's orbit is an ellipse with the Sun located at one of the foci of the ellipse. Think about the location of the observer on Earth.)

40. ** **EST** Estimate (a) the average distance between Earth and the Moon and (b) the diameter of the Moon, using the following data: the period of the Moon's revolution around Earth ($T_M \approx 28$ days), the average radius of Earth ($R_E = 6400$ km), free-fall acceleration of objects near Earth's surface ($g = 9.8$ m/s^2), and the data in Figure Q5.15. The numbers on the photos are the times when the photos were taken (hh:mm:ss). Indicate any assumptions that you made. (Hint: Use Kepler's third law.)

41. * **EST** The average radius of Earth's orbit around the Sun is 1.5×10^8 km. The mass of Earth is 5.97×10^{24} kg, and it makes one orbit in approximately 365 days. (a) What is Earth's speed relative to the Sun? (b) Estimate the Sun's mass using Newton's law of universal gravitation and Newton's second law. What assumptions did you need to make?

42. * The Moon travels in a 3.8×10^5-km-radius orbit about Earth. Earth's mass is 5.97×10^{24} kg. Determine the period T for one Moon orbit about Earth using Newton's law of universal gravitation and Newton's second law. What assumptions did you make?

43. * Determine the ratio of Earth's gravitational force exerted on an 80-kg person when at Earth's surface and when 1000 km above Earth's surface. The radius of Earth is 6370 km.

44. * Determine the magnitude of the gravitational force Mars would exert on your body if you were on the surface of Mars.

45. * When you stand on a bathroom scale here on Earth, it reads 540 N. (a) What would your mass be on Mars, Venus, and Saturn? (b) What is the magnitude of the gravitational force each planet would exert on you if you stood on their surface? (c) What assumptions did you make?

46. The free-fall acceleration on the surface of Jupiter, the most massive planet, is 24.79 m/s^2. Jupiter's radius is 7.0×10^4 km. Use Newtonian ideas to determine Jupiter's mass.

47. * A satellite moves in a circular orbit a distance of 1.6×10^5 m above Earth's surface. Determine the speed of the satellite.

48. * Mars has a mass of 6.42×10^{23} kg and a radius of 3.40×10^6 m. Assume a person is standing on a bathroom scale on the surface of Mars. Over what time interval would Mars have to complete one rotation on its axis to make the bathroom scale have a zero reading?

49. * Determine the speed a projectile must reach in order to become an Earth satellite. What assumptions did you make?

50. ** Determine the distance above Earth's surface to a satellite that completes two orbits per day. What assumptions did you make?

51. *Determine the period of an Earth satellite that moves in a circular orbit just above Earth's surface. What assumptions did you need to make?

52. * A spaceship in outer space has a doughnut shape with 500-m outer radius. The inhabitants stand with their heads toward the center and their feet on an outside rim (see Figure Q5.21). Over what time interval would the spaceship have to complete one rotation on its axis to make a bathroom scale have the same reading for the person in space as when on Earth's surface? Indicate any assumptions that you made.

General Problems

53. * Using the velocity change method from Physics Tool Box 5.1, find the direction of a car's acceleration at the marked point in the middle of the circular turn shown in Figure P5.53 for the three cases (a) the car is moving at constant speed, (b) the car is speeding up, and (c) the car is slowing down.

FIGURE P5.53

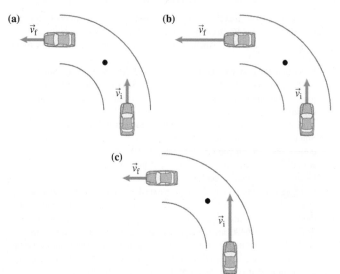

54. * **Loop-the-loop** You have to design a loop-the-loop for a new amusement park so that when each car passes the top of the loop inverted (upside-down), each seat exerts a force against a passenger's bottom that has a magnitude equal to 1.5 times the gravitational force that Earth exerts on the passenger. Choose some reasonable physical quantities so these conditions are met. Show that the loop-the-loop will work equally well for passengers of any mass.

55. ** **A Tarzan swing** Tarzan (mass 80 kg) swings at the end of an 8.0-m-long vine (Figure P5.55). When directly under the vine's support, he releases the vine and flies across a swamp. When he releases the vine, he is 5.0 m above the swamp and 10.0 m horizontally from the other side. Determine the force the vine exerts on him at the instant before he lets go (the vine is straight down when he lets go).

FIGURE P5.55

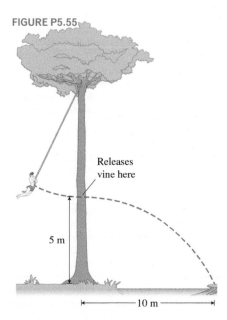

Releases vine here

5 m

10 m

56. * (a) If the masses of Earth and the Moon were both doubled, by how much would the radius of the Moon's orbit about Earth have to change if its speed did not change? (b) By how much would its speed have to change if its radius did not change? Justify each answer.

57. * **EST** Estimate the radial acceleration of the tread of a car tire. Indicate any assumptions that you made.

58. ** **EST** Estimate the force exerted by the tire on a 10-cm-long section of the tread of a tire as the car travels at speed 80 km/h. Justify any numbers used in your estimate.

59. ** **EST** Estimate the maximum radial force that a football player's leg needs to exert on his foot when swinging the leg to punt the ball. Justify any numbers that you use.

60. * **EST** Estimate the force exerted by the wheel of a car on the 20-g tire valve while the car is driving on a highway (see Figure P5.60). Justify any numbers used in your estimate.

FIGURE P5.60

Valve

61. Lucia's bathroom scale on the equator reads 110 lb when she stands on it. If her mass stays the same, will the scale read more or less at the North Pole? To answer this question, (a) draw a force diagram for Lucia. (b) With the assistance of this diagram, write an expression using Newton's second law that relates the forces exerted on Lucia and her acceleration along the radial direction. (c) Decide whether the reading of the scale would be different at the North Pole. List the assumptions that you made and describe how your answer might change if those assumptions were not valid.

62. **Design 1** Design and solve a circular motion problem for a roller coaster.

63. * **Design 2** Design and solve a circular motion problem for the amusement park ride shown in **Figure P5.63**.

FIGURE P5.63

64. ** **Demolition** An old building is being demolished by swinging a heavy metal ball from a crane. Suppose that the 100-kg ball swings from a 20-m-long wire at speed 16 m/s as the wire passes the vertical orientation. (a) What tension force must the wire be able to withstand in order not to break? (b) Assume the ball stops after sinking 1.5 m into the wall of the building. What was the average force that the ball exerted on the wall? Indicate any assumptions you made for each part of the problem.

65. **Designing a banked roadway** You need to design a banked curve for a highway in which cars make a 90° turn moving at 50 mi/h. Indicate any assumptions you make. Describe your design.

66. * **Evaluation question** You find the following in a report about blackouts: "The acceleration that causes blackouts in fighter pilots is called the maximum *g*-force. Fighter pilots experience this force when accelerating or decelerating quickly. At high *g*'s, the pilot's blood pressure changes and the flow of oxygen to the brain rapidly decreases. This happens because the pressure outside of the pilot's body is so much greater than the pressure a human is normally accustomed to." Indicate any incorrect physics (including the application of physics to biology) that you find.

67. * Suppose that Earth rotated much faster on its axis—so fast that people were almost weightless when at Earth's surface. How long would the length of a day be on this new Earth?

68. * On Earth, an average person's vertical jump is 0.40 m. What is it on the Moon? Explain.

69. * You read in a science magazine that on the Moon, the speed of a shell leaving the barrel of a modern tank is enough to put the shell in a circular orbit above the surface of the Moon (there is no atmosphere to slow the shell). What should be the speed for this to happen? Is this number reasonable?

70. * Determining the forces between powders and surfaces (*adhesive forces*) plays an important role in several industrial applications. One method to determine adhesive forces employs the following procedure. Powder is deposited on a flat surface that was previously covered with a layer of the material under study, for example, glue. The sample is then spun very quickly for some time. During the spinning, some particles fly off the surface and some remain on it. Examining the sample after the rotation allows scientists to determine the maximum force that the surface exerted on the powder particles. (a) Explain the physics principles that are employed in this method.

Figure P5.70a shows salt crystals (cubes of average mass 0.06 mg) stuck to a piece of cardboard by a layer of glue. Figure P5.70b shows the same piece of cardboard after it was spun at 80 rotations per second about the axis of rotation shown by the black dot in center of the photos. The yellow curve surrounds the area where the salt mostly remained on the cardboard. (b) Estimate the maximum force exerted by the glue on a salt crystal in this experiment. Indicate any assumptions that you made. Note: You will need a ruler to solve this problem!

FIGURE P5.70

71. ** Isabel notices that if she places a small plastic box on a horizontal wooden disk and slowly increases the rotational frequency of the disk, the box flies off the disk at a certain rotational frequency. She investigates this phenomenon by placing the box at different distances d from the edge of the disk and measuring the period of disk revolution T when the box flies off the disk. Her data are presented in the table at right. Based on these data Isabel is able to determine the radius of the disk and the coefficient of static friction between wood and plastic. Explain how she obtains these values and determine them using her data. (Hint: First, write an expression for T in the limiting case when the static friction force is the largest. Then rearrange your equation to obtain a linear dependence on d [this does not necessarily mean that quantity that depends linearly on d is to the first power], draw a graph, plot the best-fit line, and determine the unknown quantities from the information that you obtain from your graph.)

d (m)	T (s)
0.17	0.55
0.16	0.65
0.15	0.72
0.14	0.80
0.13	0.85

Reading Passage Problems

Texas Motor Speedway On October 28, 2000 Gil de Ferran set the single-lap average speed record for the then-named California Speedway—241.428 mi/h. The following year, on April 29, 2001, the Championship Auto Racing Teams (CART) organization canceled the inaugural Firestone Firehawk 600 race at the Texas Motor Speedway 2 hours before it was to start. During practice, drivers became dizzy when they reached speeds of more than 230 mph on the high-banked track. The Texas Motor Speedway has turns banked at 24° above the horizontal. By comparison, turns on the Indianapolis Motor Speedway are banked at 9°.

In qualifying runs for the race, 21 of 25 drivers complained of dizziness and disorientation. They experienced accelerations of 5.5*g* (5.5 times 9.8 m/s²) that lasted for several seconds. Such accelerations for such long time intervals have caused pilots to black out. Blood drains from their heads to their legs as they move in circles with their heads toward the center of the circle. CART felt that the track was unsafe for their drivers. The Texas Motor Speedway had tested the track with drivers before the race and thought it was safe. The Texas Motor Speedway sued CART.

72. Why did drivers get dizzy and disoriented while driving at the Texas Motor Speedway?
    (a) The cars were traveling at over 200 mi/h.
    (b) The track was tilted at an unusually steep angle.
    (c) On turns, the drivers' blood tended to drain from their brains into veins in their lower bodies.
    (d) The *g* force pushed blood into their heads.
    (e) The combination of a and b caused c.

73. What was the time for Gil de Ferran's car to complete one lap of 1.5 mi during his record-setting drive?
    (a) 16.8 s   (b) 18.4 s   (c) 22.4 s
    (d) 25.1 s   (e) 37.3 s

74. If the racecars had no help from friction, which expression below would describe the normal force of the track on the cars while traversing the 24° banked curves?
    (a) $mg \cos 66°$ (b) $mg \sin 66°$ (c) $mg/\cos 66°$
    (d) $mg/\sin 66°$ (e) None of these

75. For the racecars to stay on the road while traveling at high speed, how does a friction force need to be exerted?
    (a) Parallel to the roadway and outward
    (b) Parallel to the roadway and toward the infield
    (c) Horizontally toward the center of the track
    (d) Opposite the direction of motion
    (e) None of the above

76. The average speed reported in the reading passage has six significant digits, implying that the speed is known to within ±0.001 mi/h. If this is correct, which answer below is closest to the uncertainty in the time needed to travel around the 1.5-mi oval track? (Think about the percent uncertainties.)
    (a) ±0.0001 s   (b) ±0.001 s   (c) ±0.01 s
    (d) ±0.1 s   (e) ±1 s

77. What was the approximate radius of the part of the track where the drivers experienced the 5.5g acceleration?
    (a) 40 m   (b) 80 m   (c) 200 m
    (d) 400 m   (e) 1000 m

Halley's Comet Edmond Halley was the first to realize that the comets observed in 1531, 1607, and 1682 were really one comet (now called Halley's Comet) that moved around the Sun in an elongated elliptical orbit (see Figure 5.5). He predicted that the peanut-shaped comet would reappear in 1757. It appeared in March 1759 (attractions to Jupiter and Saturn delayed its trip by 618 days). More recent appearances of Halley's Comet were in 1835, 1910, and 1986. It is expected again in 2061.

FIGURE 5.5 The elongated orbit of Halley's Comet.

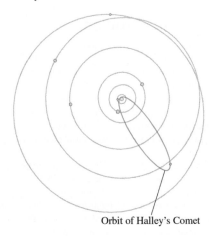

Orbit of Halley's Comet

The nucleus of Halley's Comet is relatively small (15 km long, 8 km wide, and 8 km thick). It has a low 2.2×10^{14}-kg mass with an average density of about 600 kg/m³. (The density of water is 1000 kg/m³.) The nucleus rotates once every 52 h. When Halley's Comet is closest to the Sun, temperatures on the comet can rise to about 77 °C and several tons of gas and dust are emitted each second, producing the long tail that we see each time it passes the Sun.

78. **EST** Use the velocity change method to estimate the comet's direction of acceleration when passing closest to the Sun (position I in Figure P5.78).
    (a) A   (b) B   (c) C   (d) D
    (e) The acceleration is zero.

FIGURE P5.78

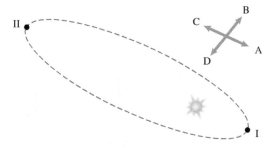

79. What object or objects exert forces on the comet as it passes position I (shown in Figure P5.78)?
    (a) The Sun's gravitational force toward the Sun
    (b) The force of motion tangent to the direction the comet is traveling
    (c) An outward force away from the Sun
    (d) a and b
    (e) a, b, and c

80. Suppose that instead of being peanut shaped, Halley's Comet was spherical with a radius of 5.0 km (about its present volume). Which answer below is closest to what your radial acceleration would be if you were standing on the "equator" of the rotating comet?
    (a) 10^{-5} m/s²   (b) 10^{-3} m/s²   (c) 0.1 m/s²
    (d) 10 m/s²   (e) 1000 m/s²

81. Approximately what gravitational force would the spherical-shaped 5-km radius comet exert on a 100-kg person on the surface of the comet?
    (a) 0.06 N   (b) 0.6 N   (c) 6 N
    (d) 60 N   (e) 600 N

82. At its closest point, the comet is 8.77×10^{10} m from the Sun (position I in Figure P5.78). Apply Newton's second law and the law of universal gravitation to determine which answer below is closest to the comet's speed when passing position I.
    (a) 1000 m/s   (b) 8000 m/s   (c) 20,000 m/s
    (d) 40,000 m/s   (e) 800,000 m/s

83. At its farthest point, the comet is 5.25×10^{12} m from the Sun (position II in Figure P5.78). Apply Newton's second law and the law of universal gravitation to determine which answer below is closest to the comet's speed when passing position II.
    (a) 800 m/s   (b) 5000 m/s   (c) 10,000 m/s
    (d) 50,000 m/s   (e) 80,000 m/s

Impulse and Linear Momentum

Car tires push against the ground and, in turn, the ground pushes the car, accelerating it. Ship propellers push water and, in turn, the water pushes the ship, accelerating it. But how does a rocket, far above Earth's atmosphere, change velocity with no object to push against?

- How does jet propulsion work?
- How can you measure the speed of a bullet?
- Why is radon gas in the basement dangerous?

WE CAN USE NEWTON'S second law $(\vec{a} = \Sigma\vec{F}/m)$ to relate the acceleration of a system to the forces being exerted on it. However, to use this law effectively we need quantitative information about the forces that objects exert on each other. Unfortunately, if two cars collide, we don't know the force that one car exerts on the other during the collision. When fireworks explode, we don't know the forces that are exerted on the pieces flying apart. In this chapter you will learn a new approach that helps us analyze and predict mechanical phenomena when the forces are not known.

BE SURE YOU KNOW HOW TO:

- Construct a force diagram for an object (Section 3.1).
- Use Newton's second law in component form (Section 4.2).
- Use kinematics to describe an object's motion (Section 2.9).

FIGURE 6.1 Experiments to explore the mass of a system.

(a)

The block balances the steel wool and flask.

Closed flask
Steel wool
Balancing block

(b)

The steel wool is burned in a closed flask. The block still balances.

(c)

When the steel wool is burned in the open flask, the mass in the flask increases.

6.1 Mass accounting

We begin our investigation by analyzing the physical quantity of mass. Earlier (in Chapter 3), we found that the acceleration of a system depended on its mass—the greater its mass, the less it accelerated due to the sum of the forces exerted on it. We ignored the possibility that a system's mass might change during some process. Is the mass in a system always a constant value?

You have probably observed countless physical processes in which mass seems to change. For example, the mass of a log in a campfire decreases as the log burns; the mass of a seedling increases as the plant grows. What happens to the "lost" mass from the log? Where does the seedling's increased mass come from?

A system perspective helps us understand what happens to the burning log. Remember that in physics we can choose what objects are in a system of interest. Objects outside the system interact with the system. If we choose only the log as the system, the mass of the system decreases as it burns. However, air is needed for burning. What happens to the mass if we choose the log and the surrounding air as the system?

Suppose that to explore this question we place steel wool in a closed flask on one side of a balance scale and a metal block of equal mass on the other side (**Figure 6.1a**). In one experiment, we burn the steel wool in the closed flask (the flask also contains air), forming an oxide of iron. We find that the total mass of the closed flask containing burned steel wool (iron oxide) is the same as the mass of the balancing metal block (Figure 6.1b). Next, we burn the steel wool in an open flask and observe that the mass of that flask increases (Figure 6.1c). The steel wool in the open flask absorbs some oxygen from the air as it burns.

Eighteenth-century French chemist Antoine Lavoisier actually performed such experiments. He realized that the choice of the system was very important. Lavoisier defined an **isolated system** as a group of objects that interact with each other but not with objects outside the system. The mass of an isolated system is the sum of the masses of all objects in the system. He then used the concept of an isolated system to summarize his (and our) experiments in the following way:

> **Law of constancy of mass in an isolated system** When a system of objects is isolated (a closed container system), its mass equals the sum of the masses of its components and does not change—it remains constant in time.

When the system is not isolated (an open container system), the mass might change. However, this change is not random—it is always equal to the amount of mass leaving or entering the system from the environment.

Thus, even when the mass of a system is not constant, we can keep track of the changes if we take into account how much is entering or leaving the system:

$$\begin{pmatrix} initial \text{ mass of} \\ \text{system at earlier clock} \\ \text{reading} \end{pmatrix} + \begin{pmatrix} new \text{ mass entering or} \\ \text{leaving system between} \\ \text{the two clock readings} \end{pmatrix} = \begin{pmatrix} final \text{ mass of} \\ \text{system at later clock} \\ \text{reading} \end{pmatrix}$$

The above equation describes the change of mass in any system. The mass is constant if there is no flow of mass in or out of the system; the mass changes in a predictable way if there is some flow of mass between the system and the environment. If we define our system to include the environment, then the mass of the system remains constant. Basically, mass cannot appear from nowhere and does not disappear without a trace. Imagine you have a system that has a total mass of $m_i = 3$ kg (say, a bag of oranges). In the environment (your refrigerator), you have another kilogram of oranges. You add these oranges to the bag ($\Delta m = 1$ kg). The final mass of the system equals exactly the sum of the initial mass and the added mass: $m_i + \Delta m = m_f$, or 3 kg $+ 1$ kg $= 4$ kg. Notice that the total mass of the oranges in the system and the environment did not change (**Figure 6.2a**). We can represent this process with a *bar chart* (Figure 6.2b). The bar on the left represents the initial mass of the system, the central bar represents the mass added or taken away, and the bar on the right represents the mass of the system in the final situation. As a result, the height of the left bar plus the height of the central bar equals the height of the right bar. The bar chart allows us to keep track of the changes in mass of a system even if the system is not isolated.

> **Conserved quantities** Mass is an example of a **conserved** quantity. A conserved quantity is constant in an isolated system. When the system is not isolated, the quantity may change. However, if it does, we can always find another system in which the quantity is constant.

Just as with every idea in physics, the law of constancy of mass in an isolated system does not apply in all cases. We will discover later in this book (Chapters 29 and 30) that in situations involving atomic particles, mass is not constant even in an isolated system; instead, what is constant is a new quantity that includes mass as a component.

REVIEW QUESTION 6.1 When you burn a log in a fire pit, the mass of wood clearly decreases. How can you define the system so that the mass of the objects in that system remains constant?

FIGURE 6.2 (a) The initial mass of the oranges plus the mass of the oranges that were added (or subtracted) equals the final mass of the oranges. (b) The mass change process is represented by a mass bar chart.

(a)

3 kg   +   1 kg   =   4 kg

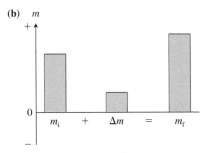

(b)

6.2 Linear momentum

We now know that mass is an example of a conserved quantity. Is there a quantity related to motion that is conserved? When you kick a stationary ball, there seems to be a transfer of motion from your foot to the ball. When you knock bowling pins down with a bowling ball, a similar transfer occurs. However, motion is not a physical quantity. What physical quantities describing motion are constant in an isolated system? Can we describe the changes in these quantities using a bar chart?

Let's conduct a few experiments to find out. In Observational Experiment **Table 6.1** we observe two carts of different masses that collide on a smooth track. For these experiments, the system will include both carts. A *collision* is a process that occurs when two (or more) objects come into direct contact with each other. The system is isolated since the forces that the carts exert on each other are internal, and external forces either add to zero (as do the vertical forces) or are negligible (the horizontal friction force).

OBSERVATIONAL EXPERIMENT TABLE 6.1 **Collisions in a system of two carts**

VIDEO
OET 6.1

Observational experiment	Analysis

Experiment 1. Cart A (0.20 kg) moving right at 1.0 m/s collides with cart B (0.20 kg), which is stationary. Cart A stops, and cart B moves right at 1.0 m/s.

$v_{Aix} = +1.0$ m/s $v_{Bix} = 0$ $v_{Afx} = 0$ $v_{Bfx} = +1.0$ m/s

A
0.20 kg

B
0.20 kg

A

B

The positive *x*-axis points right; motion to the right has a positive velocity *x*-component.

→ *x*

- *Speed:* The sum of the speeds of the objects is the same before and after the collision: 1.0 m/s $+ 0$ m/s $= 0$ m/s $+ 1.0$ m/s.
- *Mass · speed:* The sum of the products of mass and speed is the same before and after the collision: 0.20 kg$(1.0$ m/s$) + 0.20$ kg$(0$ m/s$) = 0.20$ kg$(0$ m/s$) + 0.20$ kg$(1.0$ m/s$)$.
- *Mass · velocity:* The sum of the products of mass and the *x*-component of velocity is the same before and after the collision: 0.20 kg$(+1.0$ m/s$) + 0.20$ kg$(0) = 0.20$ kg$(0) + 0.20$ kg$(+1.0$ m/s$)$.

(CONTINUED)

Observational experiment	Analysis
Experiment 2. Cart A (0.40 kg) moving right at 1.0 m/s collides with cart B (0.20 kg), which is stationary. After the collision, both carts move right, cart B at 1.2 m/s and cart A at 0.4 m/s.	• *Speed:* The sum of the speeds of the objects is not the same before and after the collision: 1.0 m/s + 0 m/s ≠ 0.4 m/s + 1.2 m/s. • *Mass · speed:* The sum of the products of mass and speed is the same before and after the collision: 0.40 kg(1.0 m/s) + 0.20 kg(0 m/s) = 0.40 kg(0.4 m/s) + 0.20 kg(1.2 m/s). • *Mass · velocity:* The sum of the products of mass and the x-component of velocity is the same before and after the collision: 0.40 kg(+1.0 m/s) + 0.20 kg(0) = 0.40 kg(+0.4 m/s) + 0.20 kg(+1.2 m/s).
Experiment 3. Cart A (0.20 kg) with a piece of clay attached to the front moves right at 1.0 m/s. Cart B (0.20 kg) moves left at 1.0 m/s. The carts collide, stick together, and stop.	• *Speed:* The sum of the speeds of the objects is not the same before and after the collision: 1.0 m/s + 1.0 m/s ≠ 0 m/s + 0 m/s. • *Mass · speed:* The sum of the products of mass and speed is not the same before and after the collision: 0.20 kg(1.0 m/s) + 0.20 kg(1.0 m/s) ≠ 0.20 kg(0 m/s) + 0.20 kg(0 m/s). • *Mass · velocity:* The sum of the products of mass and the x-component of velocity is the same before and after the collision: 0.20 kg(+1.0 m/s) + 0.20 kg(−1.0 m/s) = 0.20 kg(0 m/s) + 0.20 kg(0 m/s).

Pattern

One quantity remains the same before and after the collision in each experiment—the sum of the products of the mass and the x-velocity component of the objects.

In the three experiments in Table 6.1, only one quantity—the sum of the products of the mass and the x-component of velocity, $\Sigma m v_x$—remained the same before and after the carts collided. Note also that the sum of the products of the mass and the y-component of velocity, $\Sigma m v_y$, did not change—it remained zero. Perhaps $\Sigma m\vec{v}$ is the quantity characterizing some quantity of motion that is constant in an isolated system. But will this pattern persist in other situations? Let's test this idea by using it to predict the outcome of the experiment in Testing Experiment Table 6.2.

TESTING
EXPERIMENT TABLE 6.2 **Testing the idea that $\Sigma m\vec{v}$ in an isolated system remains constant**   VIDEO TET 6.2

Testing experiment	Prediction	Outcome
Cart A (0.40 kg) has a piece of modeling clay attached to its front and is moving right at 1.0 m/s. Cart B (0.20 kg) is moving left at 1.0 m/s. The carts collide and stick together. Predict the velocity of the carts after the collision.	The system consists of the two carts. The direction of velocity is noted with a plus or minus sign of the velocity component: $$(0.40\text{ kg})(+1.0\text{ m/s}) + (0.20\text{ kg})(-1.0\text{ m/s}) = (0.40\text{ kg} + 0.20\text{ kg})v_{fx}$$ or $$v_{fx} = (+0.20\text{ kg}\cdot\text{m/s})/(0.60\text{ kg}) = +0.33\text{ m/s}$$ After the collision, the two carts should move right at a speed of about 0.33 m/s.	After the collision, the carts move together toward the right at close to the predicted speed.

Conclusion

Our prediction matched the outcome. The quantity $m\vec{v}$ might be the quantity whose sum is constant in an isolated system.

This new quantity is called **linear momentum** $\vec{p}$.

Linear momentum The linear momentum $\vec{p}$ of a single object is the product of its mass m and velocity $\vec{v}$:

$$\vec{p} = m\vec{v} \tag{6.1}$$

Linear momentum is a vector quantity that points in the same direction as the object's velocity $\vec{v}$ (**Figure 6.3**). The SI unit of linear momentum is (kg $\cdot$ m/s). The total linear momentum of a system containing multiple objects is the vector sum of the momenta (plural of momentum) of the individual objects.

$$\vec{p}_{net} = m_1\vec{v}_1 + m_2\vec{v}_2 + \cdots + m_n\vec{v}_n = \Sigma m\vec{v}$$

FIGURE 6.3 Momentum is a vector quantity with components.

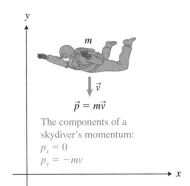

The components of a skydiver's momentum:
$p_x = 0$
$p_y = -mv$

Note the following three important points.

1. Unlike mass, which is a scalar quantity, $\vec{p} = m\vec{v}$ is a vector quantity. Therefore, it is important to consider the direction in which the colliding objects are moving before and after the collision. For example, because cart B in Table 6.2 was moving left along the x-axis, the x-component of its momentum was negative before the collision.

2. Because momentum depends on the velocity of the object, and the velocity depends on the choice of the reference frame, different observers will measure different momenta for the same object. As a passenger, the momentum of a car with respect to you is zero. However, it is not zero for an observer on the ground watching the car move away from him.

3. We chose an isolated system (the two carts) for our investigation. The sum of the products of mass and velocity $\Sigma m\vec{v}$ of all objects in the isolated system remained constant even though the carts collided with each other. However, if we had chosen the system to be just one of the carts, we would see that the linear momentum $\vec{p} = m\vec{v}$ of the cart before the collision differs from that after the collision. Thus, to establish that momentum $\vec{p}$ is a conserved quantity, we need to make sure that when the momentum of a chosen system changes, we can find a new system in which momentum is constant.

Momentum constancy of an isolated system The momentum of an isolated system is constant. For an isolated two-object system:

$$m_1\vec{v}_{1i} + m_2\vec{v}_{2i} = m_1\vec{v}_{1f} + m_2\vec{v}_{2f} \tag{6.2}$$

Because momentum is a vector quantity and Eq. (6.2) is a vector equation, we will work with its x- and y-component forms:

$$m_1v_{1ix} + m_2v_{2ix} = m_1v_{1fx} + m_2v_{2fx} \tag{6.3x}$$

$$m_1v_{1iy} + m_2v_{2iy} = m_1v_{1fy} + m_2v_{2fy} \tag{6.3y}$$

For a system with more than two objects, we simply include a term on each side of the equation for each object in the system. Let's test the idea that the momentum of an isolated system is constant in another situation.

EXAMPLE 6.1 Two Rollerbladers

Jen (50 kg) and David (75 kg), both on Rollerblades, push off each other abruptly. Each person coasts backward at approximately constant speed. During a certain time interval, Jen travels 3.0 m. How far does David travel during that same time interval?

Sketch and translate The process is sketched below. All motion is with respect to the floor and is along the x-axis. We choose the two Rollerbladers as the system. Initially, they are at rest. After pushing off, Jen (J) moves to the left and David (D) moves to the right. We can use momentum constancy to calculate David's velocity component and predict the distance he will travel during that same time interval.

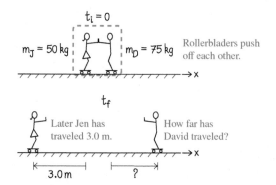

Simplify and diagram We model each person as a point-like object and assume that the friction force exerted on the Rollerblades does not affect their motion. Thus there are no horizontal external forces exerted on the system. In addition, the two vertical forces, an upward normal force $\vec{N}_{\text{F on P}}$ that the floor exerts on each person and an equal-magnitude downward gravitational force $\vec{F}_{\text{E on P}}$ that Earth exerts on each person, cancel, as we see in the force diagrams. Since the net external force exerted on the system is zero, the system is isolated. The forces that the Rollerbladers exert on each other are internal forces and should not affect the momentum of the system.

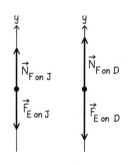

Represent mathematically The initial state (i) of the system is before they start pushing on each other, and the final state (f) is when Jen has traveled 3.0 m.

$$m_{\text{J}}v_{\text{Ji}x} + m_{\text{D}}v_{\text{Di}x} = m_{\text{J}}v_{\text{Jf}x} + m_{\text{D}}v_{\text{Df}x}$$

We choose the positive direction toward the right. Because the initial velocity of each person is zero, the above equation becomes

$$0 + 0 = m_{\text{J}}v_{\text{Jf}x} + m_{\text{D}}v_{\text{Df}x}$$

or

$$m_{\text{D}}v_{\text{Df}x} = -m_{\text{J}}v_{\text{Jf}x}$$

Solve and evaluate The x-component of Jen's velocity after the push is $v_{\text{Jf}x} = -(3.0 \text{ m})/\Delta t$, where Δt is the time interval needed for her to travel 3.0 m. We solve the above equation for David's final x-velocity component to determine how far David should travel during that same time interval:

$$v_{\text{Df}x} = -\frac{m_{\text{J}}v_{\text{Jf}x}}{m_{\text{D}}} = -\frac{m_{\text{J}}}{m_{\text{D}}}v_{\text{Jf}x} = -\frac{(50 \text{ kg})}{(75 \text{ kg})}\frac{(-3.0 \text{ m})}{\Delta t} = \frac{(2.0 \text{ m})}{\Delta t}$$

If momentum is constant in this isolated system, then David should travel 2.0 m in the positive direction during Δt. The measured value is very close to the predicted value.

- -

Try it yourself Estimate the magnitude of your momentum when walking and when jogging. Assume your mass is 60 kg.

Answer

When walking, you travel at a speed of about 1 to 2 m/s. So the magnitude of your momentum will be $p = mv \approx (60 \text{ kg})(1.5 \text{ m/s}) \approx 90 \text{ kg} \cdot \text{m/s}$. When jogging, your speed is about 2 to 5 m/s or a momentum of magnitude $p = mv \approx (60 \text{ kg})(3.5 \text{ m/s}) \approx 200 \text{ kg} \cdot \text{m/s}$.

Notice that in Example 6.1 we were able to determine David's velocity by using the principle of momentum constancy. We did not need any information about the forces involved. This is a very powerful result, since in all likelihood the forces they exerted on each other were not constant. The kinematics equations we have used up to this point have assumed constant acceleration of the system (and thus constant forces). Using the idea of momentum constancy has allowed us to analyze a situation involving nonconstant forces.

So far, we have investigated situations involving isolated systems. In the next section, we will investigate momentum in nonisolated systems.

REVIEW QUESTION 6.2 Ryan says, "Based on the experiments in Table 6.1, momentum of an isolated system during some collisions can disappear." What information in the table prompted Ryan to make this statement? Is Ryan correct? Explain your answer.

6.3 Impulse and momentum

So far, we have found that the linear momentum of a system is constant if that system is isolated (the net external force exerted on the system is zero). In this section, we will learn how to use Newton's laws to account for the changes in momentum when the system is not isolated.

Impulse due to a force exerted on a single object

When you push a bowling ball, you exert a force on it, causing the ball to accelerate. The average acceleration $\vec{a}$ is defined as the change in velocity $\vec{v}_f - \vec{v}_i$ divided by the time interval $\Delta t = t_f - t_i$ during which that change occurs:

$$\vec{a} = \frac{\vec{v}_f - \vec{v}_i}{t_f - t_i}$$

We can also use Newton's second law to determine an object's acceleration if we know its mass and the sum of the forces that other objects exert on it:

$$\vec{a} = \frac{\Sigma\vec{F}}{m}$$

We now have two expressions for an object's acceleration. Setting these two expressions for acceleration equal to each other, we get

$$\frac{\vec{v}_f - \vec{v}_i}{t_f - t_i} = \frac{\Sigma\vec{F}}{m}$$

Now multiply both sides by $m(t_f - t_i)$ and get the following:

$$m\vec{v}_f - m\vec{v}_i = \vec{p}_f - \vec{p}_i = \Sigma\vec{F}(t_f - t_i) \tag{6.4}$$

The left side of this equation is the change in momentum of the object. This change depends on the product of the sum of the external forces and the time interval during which the forces are exerted on the object (the right side of the equation). Note these two important points:

1. Equation (6.4) is just Newton's second law written in a different form—one that involves the physical quantity momentum. It assumes that the sum of the forces does not change during the time interval. Actually, Newton wrote the second law using momentum notation, not masses and acceleration, so historically Eq. (6.4) is Newton's second law.

2. The forces *and* time interval affect momentum—the longer the time interval, the greater the momentum change. A small force exerted for a long time interval can change the momentum of an object by the same amount as a large force exerted for a short time interval.

The product of any external force exerted on an object during a time interval and the time interval gives us a new quantity, the **impulse** of that force. When you kick a football or hit a baseball with a bat, your foot or the bat exerts an impulse on the ball. The forces in these situations are not constant but instead vary in time (see the example in **Figure 6.4**). The shaded area under the varying force curve represents the impulse of the force. We can estimate the impulse by drawing a horizontal line that is approximately the average force exerted during the time interval of the impulse. The area under the rectangular average force-impulse curve equals the product of the height of the rectangle (the average force) and the width of the rectangle (the time interval over which the average force is exerted). The product $F_{av}(t_f - t_i)$ equals the magnitude of the impulse.

FIGURE 6.4 The impulse of a force is the area under the F-versus-t graph line.

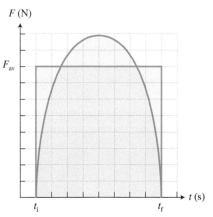

The area of the shaded rectangle is about the same as the area under the curved line and equals the impulse of the force.

Impulse The impulse $\vec{J}$ of a force is the product of the average force $\vec{F}_{av}$ exerted on an object during a time interval $(t_f - t_i)$ and that time interval:

$$\vec{J} = \vec{F}_{av}(t_f - t_i) \tag{6.5}$$

Impulse is a vector quantity that points in the direction of the force. The impulse has a plus or minus sign depending on the orientation of the force relative to a coordinate axis. The SI unit for impulse is $N \cdot s = (kg \cdot m/s^2) \cdot s = kg \cdot m/s$, the same unit as momentum.

It is often difficult to measure directly the impulse of the average of the sum of the forces during a time interval. However, we can determine the average of the sum of the forces on the right side of Eq. (6.4) indirectly by measuring or calculating the momentum change on the left side of the equation. For this reason, the combination of impulse and momentum change provides a powerful tool for analyzing interactions between objects. We can now write Eq. (6.4) as the **impulse-momentum equation** for a single object.

Impulse-momentum equation for a single object If several external objects exert forces on a single-object system during a time interval $(t_f - t_i)$, the sum of their impulses $\Sigma \vec{J}$ causes a change in momentum of the system:

$$\vec{p}_f - \vec{p}_i = \Sigma \vec{J} = \Sigma \vec{F}_{on\ System}(t_f - t_i) \tag{6.6}$$

The x- and y-scalar component forms of the impulse-momentum equation are

$$p_{fx} - p_{ix} = \Sigma F_{on\ System\ x}(t_f - t_i) \tag{6.7x}$$

$$p_{fy} - p_{iy} = \Sigma F_{on\ System\ y}(t_f - t_i) \tag{6.7y}$$

A few points are worth emphasizing. First, notice that Eq. (6.6) is a vector equation, as both the momentum and the impulse are vector quantities. Vector equations are not easy to manipulate mathematically. Therefore, we will use the scalar component forms of Eq. (6.6)—Eqs. (6.7x) and (6.7y).

Second, the time interval in the impulse-momentum equation is very important. When object 2 exerts a force on object 1, the momentum of object 1 changes by an amount equal to

$$\Delta \vec{p}_1 = \vec{p}_{1f} - \vec{p}_{1i} = \vec{F}_{2\ on\ 1}(t_f - t_i) = \vec{F}_{2\ on\ 1}\Delta t$$

The longer that object 2 exerts this force on object 1, the greater the momentum change of object 1. This explains why a fast-moving object might have a smaller effect on a stationary object during a collision than a slow-moving object interacting with the stationary object over a longer time interval. For example, a fast-moving bullet passing through a partially closed wooden door might not open the door (it will just make a hole in the door), whereas your little finger, moving much slower than the bullet, could open the door. Although the bullet moves at high speed and exerts a large force on the door, the time interval during which it interacts with the door is very small (milliseconds). Hence, it exerts a relatively small impulse on the door—too small to significantly change the door's momentum. A photo of a bullet shot through an apple illustrates the effect of a short impulse time (**Figure 6.5**). The impulse exerted by the bullet on the apple was too small to knock the apple off its support.

Third, if the magnitude of the force changes during the time interval considered in the process, we use the average force.

Finally, if the same amount of force is exerted for the same time interval on a large-mass object and on a small-mass object, the objects will have an equal change in momentum (the same impulse was exerted on them). However, the small-mass object will experience a greater change in velocity than the large-mass object.

FIGURE 6.5 When a bullet is shot through an apple balanced on a support, the bullet's time of interaction with the apple is very short, causing a small impulse that does not knock the apple over.

EXAMPLE 6.2   Abrupt stop in a car

A 60-kg person is traveling in a car that is moving at 16 m/s with respect to the ground when the car hits a barrier. The person is not wearing a seat belt, but is stopped by an air bag in a time interval of 0.20 s. Determine the *average* force that the air bag exerts on the person while stopping him.

Sketch and translate   First we draw an initial-final sketch of the process. We choose the person as the system since we are investigating a force being exerted on him.

$t_i = 0$
$v_{ix} = +16\ m/s$
$m_P = 60\ kg$

$F_{B\ on\ P\,x} = ?$
$t_f = 0.20\ s$
$v_{fx} = 0$

Simplify and diagram   The force diagram shows the average force $\vec{F}_{B\ on\ P}$ exerted in the negative direction by the bag on the person. The vertical forces exerted by Earth and the seat on the person add to zero.

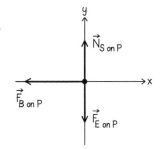

$\vec{N}_{S\ on\ P}$

$\vec{F}_{B\ on\ P}$

$\vec{F}_{E\ on\ P}$

Represent mathematically   The x-component form of the impulse-momentum equation is

$$m_P v_{Pix} + F_{B\ on\ P\,x}(t_f - t_i) = m_P v_{Pfx}$$

Solve and evaluate   Solve for the force exerted by the air bag on the person:

$$F_{B\ on\ P\,x} = \frac{m_P(v_{Pfx} - v_{Pix})}{(t_f - t_i)}$$

The person's initial x-component of velocity $v_{Pix} = +16$ m/s decreases to the final x-component of velocity $v_{Pfx} = 0$ in a time interval $(t_f - t_i) = 0.20$ s. Thus the average force exerted by the air bag on the person in the x-direction is

$$F_{B\ on\ P\,x} = \frac{(60\ kg)(0 - 16\ m/s)}{(0.20\ s - 0)} = -4800\ N$$

The negative sign in -4800 N indicates that the average force points in the negative x-direction. The magnitude of this force is about 1000 lb!

- -

Try it yourself   Suppose a 60-kg crash test dummy is in a car traveling at 16 m/s. The dummy is not wearing a seat belt and the car has no air bags. During a collision, the dummy flies forward and stops when it hits the dashboard. The stopping time interval for the dummy is 0.02 s. What is the average magnitude of the stopping force that the dashboard exerts on the dummy?

Answer

Why air bags save lives.
The average force that the hard surface exerts on the dummy is about 50,000 N, extremely unsafe for a human. Note that the momentum change of the person in Example 6.2 was the same. However, since the change for the dummy occurs during a shorter time interval (0.02 s instead of 0.20 s), the force exerted on the dummy is much greater. This is

Newton's laws and momentum constancy

Let's apply the impulse-momentum equation Eq. (6.6) to the scenario we described in Table 6.1 in order to explore momentum constancy in a two-object isolated system.

Two carts travel toward each other at different speeds, collide, and rebound (Figure 6.6). We first analyze each cart as a separate system and then analyze them together as a single system. Assume that the vertical forces exerted on the carts add to zero and that the friction force exerted by the surface on the carts does not significantly affect their motion.

Cart 1: In the initial state, before the collision, cart 1 with mass m_1 travels in the positive direction at velocity $\vec{v}_{1i}$. In the final state, after the collision, cart 1 moves with a different velocity $\vec{v}_{1f}$ in the opposite direction. To determine the effect of the impulse exerted by cart 2 on cart 1, we apply the impulse-momentum equation to cart 1 only:

$$m_1(\vec{v}_{1f} - \vec{v}_{1i}) = \vec{F}_{2\ on\ 1}(t_f - t_i)$$

Cart 2: We repeat this analysis with cart 2 as the system. Its velocity and momentum change because of the impulse exerted on it by cart 1:

$$m_2(\vec{v}_{2f} - \vec{v}_{2i}) = \vec{F}_{1\ on\ 2}(t_f - t_i)$$

Newton's third law provides a connection between our analyses of the two carts; interacting objects at each instant exert equal-magnitude but oppositely directed forces on each other:

$$\vec{F}_{1\ on\ 2} = -\vec{F}_{2\ on\ 1}$$

FIGURE 6.6 Analyzing the collision of two carts in order to develop the momentum constancy idea.

Initial

$\vec{v}_{1i}$ $\vec{v}_{2i}$

During collision

$\vec{F}_{2\ on\ 1}$ $\vec{F}_{1\ on\ 2}$

Final

$\vec{v}_{1f}$ $\vec{v}_{2f}$

Substituting the expressions for the forces from above and simplifying, we get

$$\frac{m_2(\vec{v}_{2f} - \vec{v}_{2i})}{t_f - t_i} = -\frac{m_1(\vec{v}_{1f} - \vec{v}_{1i})}{t_f - t_i}$$

Next, we multiply both sides by the time interval.

$$m_2(\vec{v}_{2f} - \vec{v}_{2i}) = -m_1(\vec{v}_{1f} - \vec{v}_{1i})$$

We now move the initial momentum for both objects to the left side and the final momentum for both objects to the right side, making sure we keep track of the signs:

$$m_1\vec{v}_{1i} + m_2\vec{v}_{2i} = m_1\vec{v}_{1f} + m_2\vec{v}_{2f}$$
$$\underbrace{\phantom{m_1\vec{v}_{1i} + m_2\vec{v}_{2i}}}_{\text{Initial momentum}} \qquad \underbrace{\phantom{m_1\vec{v}_{1f} + m_2\vec{v}_{2f}}}_{\text{Final momentum}}$$

This is Eq. (6.2), the equation we arrived at in Section 6.2, where we observed and analyzed collisions to understand the constant momentum of an isolated system. Here we have reached the same conclusions using only our knowledge of Newton's laws, momentum, and impulse. What is really important here is that while the momentum of each cart changes during the collision, if we choose the system to comprise both carts, the momentum of that system stays constant. We also see that the internal forces (in this case, the forces that carts exert on each other) cannot change the momentum of the system.

REVIEW QUESTION 6.3 An apple is falling from a tree. Why does its momentum change? Specify the external force responsible. Find a system in which the momentum is constant during this process.

6.4 The generalized impulse-momentum principle

TIP It is important to understand the difference between *conservation* and *constancy*. Mass and momentum are conserved quantities, but they are not necessarily constant in a particular system. When two carts of different masses collide, during the instant of collision, the momentum of each cart as a system changes by the amount of impulse exerted on it by the other cart (momentum is not constant). However, in the system that includes both carts, the momentum is constant. When the momentum of each cart changed, it did not disappear. It appeared in the addition to the momentum of the second cart (momentum is conserved).

Let's summarize what we have learned about momentum. The change in momentum of a system is equal to the net external impulse exerted on it. If the net impulse is zero, then the momentum of the system is constant. If the objects outside the system exert forces on it and change its momentum, we can always redefine the system by including all interacting objects to keep the momentum of this new system constant. Therefore, **momentum is a conserved quantity.** This idea, expressed mathematically as the **generalized impulse-momentum principle**, accounts for situations in which the system includes one or more objects and may or may not be isolated.

Generalized impulse-momentum principle For a system containing one or more objects, the initial momentum of the system plus the sum of the impulses that external objects exert on the system during the time interval $(t_f - t_i)$ equals the final momentum of the system:

$$\underbrace{(m_1\vec{v}_{1i} + m_2\vec{v}_{2i} + \cdots)}_{\substack{\text{Initial momentum of} \\ \text{the system}}} + \underbrace{\Sigma\vec{F}_{\text{on Sys}}(t_f - t_i)}_{\substack{\text{Net impulse exerted on} \\ \text{the system}}} = \underbrace{(m_1\vec{v}_{1f} + m_2\vec{v}_{2f} + \cdots)}_{\substack{\text{Final momentum of} \\ \text{the system}}} \qquad (6.8)$$

The x- and y-component forms of the generalized impulse-momentum principle are

$$(m_1 v_{1ix} + m_2 v_{2ix} + \cdots) + \Sigma F_{\text{on Sys}\,x}(t_f - t_i) = (m_1 v_{1fx} + m_2 v_{2fx} + \cdots) \quad (6.9x)$$

$$(m_1 v_{1iy} + m_2 v_{2iy} + \cdots) + \Sigma F_{\text{on Sys}\,y}(t_f - t_i) = (m_1 v_{1fy} + m_2 v_{2fy} + \cdots) \quad (6.9y)$$

Note: If the net impulse exerted in a particular direction is zero, then the component of the momentum of the system in that direction is constant.

Impulse-momentum bar charts

We can describe an impulse-momentum process mathematically using Eqs. (6.9x and y). These equations help us see that we can represent the changes of a system's momentum using a bar chart similar to the one used to represent the changes of a system's mass. Physics Tool Box 6.1 shows the steps for constructing an **impulse-momentum bar chart** for a simple system of two carts of equal mass traveling toward each other.

PHYSICS TOOL BOX 6.1 Constructing a qualitative impulse-momentum bar chart

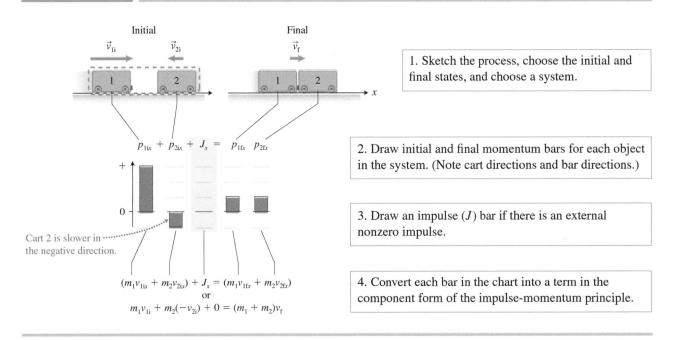

1. Sketch the process, choose the initial and final states, and choose a system.

2. Draw initial and final momentum bars for each object in the system. (Note cart directions and bar directions.)

3. Draw an impulse (J) bar if there is an external nonzero impulse.

4. Convert each bar in the chart into a term in the component form of the impulse-momentum principle.

$$p_{1ix} + p_{2ix} + J_x = p_{1fx}\; p_{2fx}$$

Cart 2 is slower in the negative direction.

$$(m_1v_{1ix} + m_2v_{2ix}) + J_x = (m_1v_{1fx} + m_2v_{2fx})$$
or
$$m_1v_{1i} + m_2(-v_{2i}) + 0 = (m_1 + m_2)v_f$$

Note that before constructing the bar chart, we represent the process in an initial-final sketch (Step 1 in the Tool Box). We then use the sketch to help construct the impulse-momentum bar chart. The lengths of the bars are *qualitative* indicators of the relative magnitudes of the momenta. In the final state in the example shown, the carts are stuck together and are moving in the positive direction. Since they have the same mass and velocity, they each have the same final momentum.

The middle shaded column in the bar chart represents the net external impulse exerted on the system during the time interval $(t_f - t_i)$—there is no impulse for the process shown. The shading reminds us that impulse does not reside in the system; it is the influence of the external objects on the momentum of the system. In our choice of the system, there is no bar in this column because the system is isolated. But if we choose only one cart as the system, then the influence of the second cart will be represented with a bar in the impulse column of the bar chart. Notice that the sum of the heights of the bars on the left plus the height of the shaded impulse bar should equal the sum of the heights of the bars on the right. This "conservation of bar heights" reflects the conservation of momentum.

We can use the bar chart to apply the generalized impulse-momentum principle (Step 4). Each nonzero bar corresponds to a nonzero term in the equation; the sign of the term depends on the orientation of the bar.

TIP When you draw a bar chart, always specify the reference frame (the object of reference and the coordinate system). The direction of the bars on the bar chart (up for positive and down for negative) should match the direction of the momentum or impulse based on the chosen coordinate system.

EXAMPLE 6.3 **Happy and sad balls**

You have two balls of identical mass and size that behave very differently. When you drop the so-called "sad" ball, it thuds on the floor and does not bounce at all. When you drop the so-called "happy" ball from the same height, it bounces back to almost the same height from which it was dropped. The difference in the bouncing ability of the happy ball is due to its internal structure; it is made of different material. You hang each ball from a string of identical length and place a wood board on its end so that the top of the board is next to the hanging ball, and if you pull the ball back and release it, it hits the top of the board. You pull each ball back to an equal height and release the balls one at a time. Which has the best chance of knocking the board over: the sad ball or the happy ball?

Sad ball   Happy ball

Sketch and translate Initial and final sketches of the process are shown below. The system is just the ball. In the initial state, the ball is just about to hit the board, moving horizontally toward the left (the balls are moving equally fast). The final state is just after the collision with the board. The happy ball (H) bounces back, whereas the sad ball (S) does not.

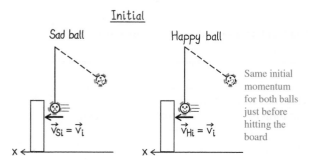

Same initial momentum for both balls just before hitting the board

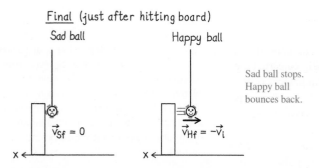

Sad ball stops. Happy ball bounces back.

Simplify and diagram Assume that the collision time interval Δt for each ball is about the same. We analyze only the horizontal x-component of the process, the component that is relevant to whether or not each of the boards is knocked over. Each board exerts an impulse on the ball that causes the momentum of the ball to change. Therefore, each ball, according to Newton's third law, exerts an impulse on the

board that it hits. A larger force exerted on the board means a larger impulse and a better chance to tip the board. A bar chart for each ball-board collision is shown below.

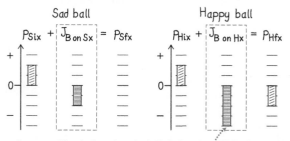

The impulse exerted by the board on the ball during the collision is equal in magnitude to the impulse that the ball exerts on the board. Thus, based on the bar charts the impulse exerted by the happy ball on the board is twice as large in magnitude as that of the sad ball and causes twice as large a momentum change.

Represent mathematically The x-component form of the impulse-momentum Eq. (6.9x) applied to each ball is as follows:

Sad ball: $mv_i + F_{B \text{ on } S \, x}\Delta t = m \cdot 0$

Happy ball: $mv_i + F_{B \text{ on } H \, x}\Delta t = m(-v_i)$

Note that the x-component of the final velocity of the sad ball is $v_{Sfx} = 0$ (it does not bounce) and that the x-component of the final velocity of the happy ball is $v_{Hfx} = -v_i$ (it bounces).

Solve and evaluate We can now get an expression for the force exerted by each board on each ball:

Sad ball: $F_{B \text{ on } S \, x} = \dfrac{m(0 - v_i)}{\Delta t} = -\dfrac{mv_i}{\Delta t}$

Happy ball: $F_{B \text{ on } H \, x} = \dfrac{m[(-v_i) - v_i]}{\Delta t} = -\dfrac{2mv_i}{\Delta t}$

Because we assumed that the time of collision is the same, the board exerts twice the force on the happy ball as on the sad ball, since the board causes the happy ball's momentum to change by an amount twice that of the sad ball. According to Newton's third law, this means that the happy ball exerts twice as large a force on the board as the sad ball. Thus, the happy ball has a greater chance of tipping the board.

▶ VIDEO 6.1

Try it yourself Is it less safe for a football player to bounce backward off a goal post or to hit the goal post and stop?

Answer Although any collision is dangerous, it is better to hit the goal post and stop. If the football player bounces back off the goal post, his momentum will have changed by a greater amount (like the happy ball in this example). This means that the goal post exerts a greater force on him, which means there is a greater chance for injury.

The pattern we found in the example above is true for all collisions—when an object bounces back after a collision, we know that a larger magnitude force is exerted on it than if the object had stopped and did not bounce after the collision. For that reason, bulletproof vests for law enforcement agents are designed so that the bullet embeds in the vest rather than bouncing off it.

REVIEW QUESTION 6.4 If in solving the problem in Example 6.3 we chose the system to be the ball and the board, how would the mathematical description for each ball-board collision change?

6.5 Skills for analyzing problems using impulse and momentum

Initial and final sketches and bar charts are useful tools to help analyze processes using the impulse-momentum principle. Let's investigate further how these tools work together. A general strategy for analyzing such processes is described on the left side of the table in Example 6.4 and illustrated on the right side for a specific process.

PROBLEM-SOLVING STRATEGY 6.1 ▷ **Applying the generalized impulse-momentum principle**

EXAMPLE 6.4 **Bullet hits wood block**

A 0.020-kg bullet traveling horizontally at 250 m/s embeds in a 1.0-kg block of wood resting on a table. Determine the speed of the bullet and wood block together immediately after the bullet embeds in the block.

Sketch and translate

- Sketch the initial and final states and include appropriate coordinate axes. Label the sketches with the known information and identify the unknowns. Decide on the object of reference.

- Choose a system based on the quantity you are interested in, for example, a multi-object isolated system to determine the velocity of an object, or a single-object nonisolated system to determine an impulse or force. Sometimes it is better to decide what is in your system and then identify initial and final states.

The left side of the sketch below shows the bullet traveling in the positive x-direction with respect to the ground; it then joins the wood. All motion is along the x-axis; the object of reference is Earth. The system includes the bullet and wood; it is an isolated system since the vertical forces balance. The initial state is immediately before the collision; the final state is immediately after.

Initial
$m_B = 0.020 \text{ kg}, m_W = 1.0 \text{ kg}$
$v_{Bix} = +250 \text{ m/s}, v_{Wix} = 0$
$\vec{v}_{Bi}$

Final
$v_{B\text{-}W fx} = ?$
$\vec{v}_{B\text{-}W f}$

(CONTINUED)

Simplify and diagram

- Determine if there are any external impulses exerted on the system. Drawing a force diagram could help determine the external forces and their directions.

- Draw an impulse-momentum bar chart for the system for the chosen direction(s) to help you understand the situation, formulate a mathematical representation of the process, and evaluate your results.

Represent mathematically

- Use the bar chart to apply the generalized impulse-momentum principle along the chosen axis. Each nonzero bar becomes a nonzero term in the equation. The orientation of the bar determines the sign in front of the corresponding term in the final equation.

- Remember that momentum and impulse are vector quantities, so include the plus or minus signs of the components based on the chosen coordinate system.

Solve and evaluate

- Insert the known information to determine the unknown quantity.

- Check if your answer is reasonable with respect to sign, unit, and magnitude. Also make sure it applies for limiting cases, such as objects of very small or very large mass.

Assume that the friction force exerted by the tabletop on the bottom of the wood does not change the momentum of the system during the very short collision time interval. The bar chart represents the process. The bar for the bullet is shorter than that for the block—their velocities are the same after the collision, but the mass of the bullet is much smaller. We do not draw a force diagram here, as the system is isolated.

$$m_B v_{Bix} + m_W \cdot 0 + (J_x) = m_B v_{B\text{-}Wfx} + m_W v_{B\text{-}Wfx}$$

Since $J_x = 0$,

$$v_{B\text{-}Wfx} = \frac{m_B v_{Bix}}{(m_B + m_W)}$$

$$v_{B\text{-}Wfx} = \frac{(0.020 \text{ kg})(250 \text{ m/s})}{(0.020 \text{ kg} + 1.0 \text{ kg})} = +4.9 \text{ m/s}$$

The magnitude of the answer seems reasonable given how fast the bullet was initially traveling. The plus sign indicates the direction, which makes sense, too. The units are also correct (m/s). We can test this using a limiting case: if the mass or speed of the bullet is zero, the block remains stationary after the collision.

Try it yourself A 0.020-kg bullet is fired horizontally into a 2.00-kg block of wood resting on a table. Immediately after the bullet joins the block, the block and bullet move in the positive x-direction at 4.0 m/s. What was the initial speed of the bullet?

Answer   400 m/s.

We could have worked Example 6.4 backward to determine the initial speed of the bullet before hitting the block (like the "Try it yourself" question). This exercise would be useful since the bullet travels so fast that it is difficult to measure its speed. Variations of this method are used, for example, to decide whether or not golf balls conform to the necessary rules. The balls are hit by the same mechanical launching impulse and the moving balls embed in another object. The balls' speeds are determined by measuring the speed of the object they embed in.

Determining the stopping time interval from the stopping distance

When a system collides with another object and stops—a car collides with a tree or a wall, a person jumps and lands on a solid surface, or a meteorite collides with Earth—the system travels what is called its **stopping distance**. By estimating the stopping distance of the system, we can estimate the stopping time interval.

Suppose that a car runs into a large tree and its front end crumples about 0.5 m. This 0.5 m, the distance that the center of the car traveled from the beginning of the impact to the end, is the car's stopping distance. Similarly, the depth of the hole left by a meteorite provides a rough estimate of its stopping distance when it collided with Earth. However, to use the impulse-momentum principle, we need the stopping time interval associated with the collision, not the stopping distance. Here's how we can use a known stopping distance to estimate the stopping time interval.

- Assume that the acceleration of the object while stopping is constant. In that case, the average velocity of the object while stopping is just the sum of the initial and final velocities divided by 2: $v_{\text{average }x} = (v_{fx} + v_{ix})/2$.
- Thus, the stopping displacement $(x_f - x_i)$ and the stopping time interval $(t_f - t_i)$ are related by the kinematics equation

$$x_f - x_i = v_{\text{average }x}\left(t_f - t_i\right) = \frac{\left(v_{fx} + v_{ix}\right)}{2}\left(t_f - t_i\right)$$

- Rearrange this equation to determine the stopping time interval:

$$t_f - t_i = \frac{2(x_f - x_i)}{v_{fx} + v_{ix}} \tag{6.10}$$

Equation (6.10) provides a method to convert stopping distance $x_f - x_i$ into stopping time interval $t_f - t_i$. Equation (6.10) can be applied to horizontal or vertical stopping.

EXAMPLE 6.5 **Stopping the fall of a stunt diver**

The record for the highest stunt fall without a parachute is 71 m (230 ft), held by 80-kg A. J. Bakunas. His fall was stopped by a large air cushion, into which he sank about 4.0 m. His speed was about 36 m/s (80 mi/h) when he reached the top of the air cushion. Estimate the average force that the cushion exerted on his body while stopping him.

Sketch and translate We focus only on the part of the fall when Bakunas is sinking into the cushion. The situation is sketched below. We choose Bakunas as the system and the y-axis pointing up. The initial state is just as he touches the cushion at position $y_i = +4.0$ m, and the final state is when the cushion has stopped him, at position $y_f = 0$. All motion is with respect to Earth. The other information about the process is given in the figure. Be sure to pay attention to the signs of the quantities (especially the initial velocity).

Simplify and diagram We draw a force diagram top right, modeling Bakunas as a point-like object. Since Bakunas's downward speed decreases, the cushion must be exerting an upward force on Bakunas of greater magnitude than the downward force that Earth exerts on him. Thus, the net force exerted on him points upward, in the positive y-direction. Using this information, we can draw a qualitative impulse-momentum bar chart for the process.

Each external force causes an impulse.

Bakunas has zero momentum in the final state.

Represent mathematically Since all motion and all of the forces are in the vertical direction, we use the bar chart to help construct the vertical y-component form of the impulse-momentum equation [Eq. (6.7y)] to determine the force that the cushion exerts on Bakunas as he sinks into it:

$$m_B v_{iy} + (N_{\text{C on B }y} + F_{\text{E on B }y})(t_f - t_i) = m_B v_{fy}$$

Using the force diagram, we see that the y-components of the forces are $N_{\text{C on B }y} = +N_{\text{C on B}}$ and $F_{\text{E on B }y} = -F_{\text{E on B}} = -m_B g$, where $N_{\text{C on B}}$ is the magnitude of the average normal force that the cushion exerts on Bakunas, the force we are trying to estimate. Noting that $v_{fy} = 0$ and substituting the force components into the above equation, we get

$$m_B v_{iy} + \left[(+N_{\text{C on B}}) + (-m_B g)\right](t_f - t_i) = m_B \cdot 0$$
$$\Rightarrow m_B v_{iy} + (N_{\text{C on B}} - m_B g)(t_f - t_i) = 0$$

We can find the time interval that the cushion takes to stop Bakunas using Eq. (6.10) and noting that $v_{fy} = 0$:

$$t_f - t_i = \frac{2(y_f - y_i)}{0 + v_{iy}}$$

(CONTINUED)

Solve and evaluate The stopping time interval while Bakunas sinks 4.0 m into the cushion is

$$t_f - t_i = \frac{2(0 - 4.0 \text{ m})}{0 + (-36 \text{ m/s})} = 0.22 \text{ s}$$

Solving for $N_{C \text{ on } B}$, we get

$$N_{C \text{ on } B} = \frac{-m_B v_{iy}}{(t_f - t_i)} + m_B g$$

$$= \frac{-(80 \text{ kg})(-36 \text{ m/s})}{(0.22 \text{ s})} + (80 \text{ kg})(9.8 \text{ N/kg})$$

$$= +13,000 \text{ N} + 780 \text{ N} = 14,000 \text{ N}$$

Wow, that is a huge force! To reduce the risk of injury, stunt divers practice landing so that the stopping force that a cushion exerts on them is distributed evenly over the entire body. The cushions must be deep enough so that they provide a long stopping time interval and thus a smaller stopping force. The same strategy is applied to developing air bags and collapsible frames for automobiles to make them safer for passengers during collisions.

Notice four important points. First, we've included only two significant digits since that is how many the data had. Second, it is very easy to make sign mistakes. A good way to avoid these is to draw a sketch that includes a coordinate system and labels showing the values of known physical quantities, including their signs. Third, the impulse due to Earth's gravitational force is small in magnitude compared to the impulse exerted by the air cushion. Lastly, the force exerted by the air cushion would be even greater if the stopping distance and consequently the stopping time interval were shorter.

Try it yourself Suppose that the cushion in this example stopped Bakunas in 1.0 m instead of 4.0 m. What would be the stopping time interval and the magnitude of the average force of the cushion on Bakunas?

Answer

The stopping time interval is 0.056 s, and the average stopping force is approximately 50,000 N.

Order-of-magnitude estimate—will bone break?

The strategy that we used in the previous example can be used to analyze skull fracture injuries that might lead to concussions. Laboratory experiments indicate that the human skull can fracture if the compressive force exerted on it per unit area is $1.7 \times 10^8 \text{ N/m}^2$. The surface area of the skull is much smaller than 1 m^2, so we will use square centimeters, a more reasonable unit of area for this discussion. Since $1 \text{ m}^2 = 1 \times 10^4 \text{ cm}^2$, we convert the compressive force per area to

$$(1.7 \times 10^8 \text{ N/m}^2)\left(\frac{1 \text{ m}^2}{1 \times 10^4 \text{ cm}^2}\right) = 1.7 \times 10^4 \text{ N/cm}^2$$

EXAMPLE 6.6   **Bone fracture estimation**[1]

A bicyclist is watching for traffic from the left while turning toward the right. A street sign hit by an earlier car accident is bent over the side of the road. The cyclist's head hits the pole holding the sign. Is there a significant chance that his skull will fracture?

Sketch and translate The process is sketched at the right. The initial state is at the instant that the head initially contacts the pole; the final state is when the head and body have stopped. The bicyclist is the system. We have been given little information, so we'll have to make some reasonable estimates of various quantities in order to make a decision about a possible skull fracture.

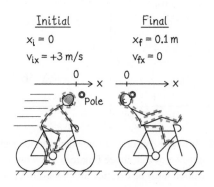

---

[1]This is a true story—it happened to one of the book's authors, Alan Van Heuvelen.

Simplify and diagram The bar chart illustrates the momentum change of the system and the impulse exerted by the pole that caused the change. The bicyclist was initially moving in the horizontal *x*-direction with respect to Earth, and not moving after the collision. The pole exerted an impulse in the negative *x*-direction on the cyclist. We'll need to estimate the following quantities: the

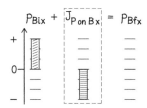

$$p_{Bix} + |J_{P\,on\,B\,x}| = p_{Bfx}$$

The pole's negative impulse on the bicyclist causes the momentum to decrease.

mass and speed of the cyclist in this situation, the stopping time interval, and the area of contact. Let's assume that this is a 70-kg cyclist moving at about 3 m/s. His body keeps moving forward for a short distance after the bone makes contact with the pole. The skin indents some during the collision. Because of these two factors, we assume a stopping distance of about 10 cm. Finally, we assume an area of contact of about 4 cm^2. All of these numbers have large uncertainties and we are not worrying about significant digits, because this is just an estimate.

Represent mathematically We now apply the generalized impulse-momentum principle:

$$m_B v_{Bi\,x} + (F_{P\,on\,B\,x})(t_f - t_i) = m_B v_{Bfx} = 0$$

$$\Rightarrow F_{P\,on\,B\,x} = -\frac{m_B v_{Bi\,x}}{t_f - t_i}$$

We can use the strategy from the last example to estimate the stopping time interval $t_f - t_i$ from the stopping distance $x_f - x_i$:

$$t_f - t_i = \frac{2(x_f - x_i)}{v_{fx} + v_{ix}}$$

where v_{ix} is the initial velocity of the cyclist and $v_{fx} = 0$ is his final velocity.

Solve and evaluate Substituting the estimated initial velocity and the stopping distance into the preceding equation, we get an estimate for the stopping time interval:

$$t_f - t_i = \frac{2(x_f - x_i)}{v_{fx} + v_{ix}} = \frac{2(0.1\text{ m} - 0)}{0 + 3\text{ m/s}} = 0.067\text{ s}$$

Since this stopping time interval is an intermediate calculated value, we don't need to worry about its number of significant digits. When we complete our estimate, though, we will keep just one significant digit.

We can now insert our estimated values of quantities into the expression for the force exerted by the pole on the bicyclist:

$$F_{P\,on\,B\,x} = -\frac{m_B v_{Bix}}{(t_f - t_i)}$$

$$= -\frac{(70\text{ kg})(3\text{ m/s})}{(0.067\text{ s})} = -3000\text{ N}$$

Our estimate of the force per area is

$$\frac{\text{Force}}{\text{Area}} = \frac{3000\text{ N}}{4\text{ cm}^2} \approx 800\text{ N/cm}^2$$

Is the bicyclist likely to fracture his skull? The force per area needed to break a bone is about $1.7 \times 10^4\text{ N/cm}^2 = 17,000\text{ N/cm}^2$. Our estimate could have been off by at least a factor of 10. The force per area is still too little for a fracture.

Try it yourself What would be the magnitude of the force exerted on the cyclist if he bounced back off the pole instead of stopping, assuming the collision time interval remains the same?

Answer   'N 0009

REVIEW QUESTION 6.5 As the bullet enters the block in Example 6.4, the block exerts a force on the bullet, causing the bullet's speed to decrease to almost zero. Why did we not include the impulse exerted by the block on the bullet in our analysis of this situation?

6.6 Jet propulsion

Cars change velocity because of an interaction between the tires and the road. Likewise, a ship's propellers push water backward; in turn, water pushes the ship forward. Once the ship or car is moving, the external force exerted by the water or the road has to balance the opposing friction force, or the vehicle's velocity will change.

What does a rocket push against in empty space to change its velocity? Rockets carry fuel that they ignite and then eject at high speed out of the exhaust nozzles (see **Figure 6.7**). Could this burning fuel ejected from the rocket provide the push to change its velocity? Choose the system to be the rocket and fuel together. If the rocket and fuel are at rest before the rocket fires its engines, then its momentum is zero. If there are no external impulses, then even after the rocket fires its engines, the momentum of the rocket-fuel system should still be zero. However, the burning fuel is ejected backward at high velocity from the exhaust nozzle and has a backward momentum. The rocket must now have a nonzero forward velocity. Let us analyze this idea quantitatively.

Imagine you are in a spacecraft traveling through space beside a rocket whose speed is the same as yours. All of a sudden you notice a burst of burning fuel that is ejected from the rocket. Let us predict what happens to the rocket's velocity. We choose

FIGURE 6.7 A rocket as it expels fuel.

Rocket moves left.   Expelled fuel moves right.

FIGURE 6.8 Analyzing a rocket ejecting fuel.

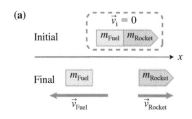

(a)

Initial $\vec{v}_i = 0$ — m_{Fuel} m_{Rocket}

Final m_{Fuel} $\vec{v}_{\text{Fuel}}$ — m_{Rocket} $\vec{v}_{\text{Rocket}}$

(b) Fuel and rocket system

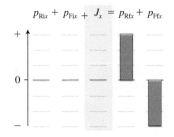

$p_{\text{Rix}} + p_{\text{Fix}} + J_x = p_{\text{Rfx}} + p_{\text{Pfx}}$

(c) Rocket alone as system

$p_{\text{Rix}} + J_{\text{F on R}\,x} = p_{\text{Rfx}}$

the rocket and its fuel as the system. Your spacecraft serves as the object of reference, and the $+x$-axis points in the direction of your motion. The rocket has zero velocity in the initial state with respect to you. Its final state is just after it expels fuel backward at high speed; the rocket in turn gains an equal magnitude of momentum in the forward direction (see **Figure 6.8a**). We can represent this process with an initial-final sketch and an impulse-momentum bar chart for the rocket-fuel system (Figure 6.8b) and use the bar chart to write momentum constancy mathematically:

$$m_{\text{Rocket}}v_{\text{Rocket}\,xi} + m_{\text{Fuel}}v_{\text{Fuel}\,xi} = m_{\text{Rocket}}v_{\text{Rocket}\,xf} + m_{\text{Fuel}}v_{\text{Fuel}\,xf}$$

The initial speeds of the rocket and fuel are zero with respect to you. Assuming that all the fuel is ejected at once at constant speed, we can find the velocity of the rocket:

$$0 = m_{\text{Rocket}}v_{\text{Rocket}\,x} + m_{\text{Fuel}}v_{\text{Fuel}\,x} \quad \Rightarrow \quad v_{\text{Rocket}\,x} = -\frac{m_{\text{Fuel}}v_{\text{Fuel}\,x}}{m_{\text{Rocket}}}$$

Here m_{Rocket} is the mass of the rocket without the fuel.

We can also choose the rocket alone as the system. The rocket pushes back on the fuel, expelling it backward at high speed ($v_{\text{Fuel}\,x} < 0$); the fuel in turn pushes forward against the rocket, exerting an impulse that causes the rocket's momentum and the velocity (assuming the mass of the rocket itself does not change) to increase ($v_{\text{Rocket}\,x} > 0$) (see Figure 6.8c).

The prediction matches the outcome of the experiment if all the fuel is ejected at once. If the fuel is ejected over some time interval, as happens in real rockets, the quantitative result is slightly different, but the essence will stay the same: you do not need anything to push against when trying to accelerate a rocket—expelling fuel is enough.

The main idea behind jet propulsion is that when an object ejects some of its mass in one direction, it accelerates in the opposite direction. This means that the same method that is used to speed up a rocket can also be used to slow it down. To do this, the fuel needs to be ejected in the same direction that the rocket is traveling.

The force exerted by the fuel on a rocket during jet propulsion is called **thrust.** Typical large rocket thrusts measure in mega-newtons (10^6 N), and exhaust speeds are more than 10 times the speed of sound. Thrust provides the necessary impulse to change a rocket's momentum.

You can observe the principles of jet propulsion using a balloon. Attach an inflated balloon to a cart and then open the valve—the cart will accelerate in the direction opposite to the ejection of air from the balloon. This example is not exactly how the rocket accelerates, as there is air in the room, but in this case the interaction of the balloon and cart with the air from the balloon is much stronger than the interactions with the outside air (see the video). You can become your own jet propulsion machine by standing on Rollerblades or a skateboard and throwing a medicine ball or a heavy book forward or backward.

▶ VIDEO 6.2

REVIEW QUESTION 6.6 The following equation is a solution for a problem. State a possible problem.

$$(2.0\,\text{kg})(-8.0\,\text{m/s}) + 0 = (2.0\,\text{kg} + 58\,\text{kg})\,v_x$$

6.7 Collisions in two dimensions

So far, the collisions we have investigated have occurred along one axis. However, many collisions in real life do not occur along a straight line. For instance, colliding billiard balls and crashing cars (as you will see in Example 6.7) rarely move along the same line. Some of the most important applications of collisions in two dimensions are in the field of physics studying the behavior of light, elementary particles (see Example 6.8), and their interactions. In all collisions, the generalized impulse-momentum principle helps determine velocities of moving objects either before or after the collision. In car accidents, momentum conservation can help determine whether the drivers were speeding before the collision; in the field of elementary particles, the same principle can help discover a new particle.

EXAMPLE 6.7 **Truck collides with car**

A 1600-kg pickup truck traveling east at 20 m/s collides with a 1300-kg car traveling north at 16 m/s. The vehicles remain tangled together after the collision. Determine the velocity (magnitude and direction) of the combined wreck immediately after the collision.

Sketch and translate We sketch the initial and final situations of the vehicles. We use a P subscript for the pickup and a C subscript for the car. The initial state is just before the collision; the final state is just after the vehicles collide and are moving together. We choose the two vehicles as the system. The object of reference is Earth; the positive x-axis points east and the positive y-axis points north.

Simplify and diagram Force diagrams represent the side view for each vehicle just before the collision. The net external force on the system is zero—it is an isolated system. We assume that the friction force exerted by the road is very small compared to the force that each vehicle exerts on the other. Thus, we ignore the impulse due to surface friction during the short collision time interval of about 0.1 s. We then apply momentum constancy in each direction. Impulse-momentum bar charts for the x-direction and for the y-direction are shown below.

The x-and y-components of momentum are independently constant.

Represent mathematically Now, convert each momentum bar in the x-component bar chart into a term in the x-component form of the impulse-momentum equation [Eq. (6.7x)] and each bar in the y-component bar chart into a term in the y-component form of the impulse-momentum equation [Eq. (6.7y)]. Notice that the x-component of the final velocity vector is $v_{\text{P-C }fx} = v_{\text{P-C }f} \cos \theta$ and the y-component is $v_{\text{P-C }fy} = v_{\text{P-C }f} \sin \theta$:

x-component equation:

$$m_\text{P} v_{\text{P}ix} + m_\text{C} v_{\text{C}ix} = (m_\text{P} + m_\text{C}) v_{\text{P-C }f} \cos \theta$$

y-component equation:

$$m_\text{P} v_{\text{P}iy} + m_\text{C} v_{\text{C}iy} = (m_\text{P} + m_\text{C}) v_{\text{P-C }f} \sin \theta$$

We have two equations and two unknowns ($v_{\text{P-C }f}$ and θ). We can solve for both unknowns.

Solve and evaluate

x-component equation:

$$(1600 \text{ kg})(20 \text{ m/s}) + (1300 \text{ kg})(0 \text{ m/s}) = (2900 \text{ kg}) v_{\text{P-C }f} \cos \theta$$

y-component equation:

$$(1600 \text{ kg})(0 \text{ m/s}) + (1300 \text{ kg})(16 \text{ m/s}) = (2900 \text{ kg}) v_{\text{P-C }f} \sin \theta$$

Divide the left side of the second equation by the left side of the first equation and the right side of the second equation by the right of the first, and cancel the 2900 kg and $v_{\text{P-C }f}$ on the top and bottom of the right side. We get

$$\frac{(1300 \text{ kg})(16 \text{ m/s})}{(1600 \text{ kg})(20 \text{ m/s})} = \frac{\sin \theta}{\cos \theta} = \tan \theta = 0.65$$

A 33° angle has a 0.65 tangent. Thus, the vehicles move off at 33° above the $+x$-axis (the north of east direction). We can now use this angle with either the x-component equation or the y-component equation above to determine the speed of the two vehicles immediately after the collision. Using the x-component equation, we get

$$v_{\text{P-C }f} = \frac{(1600 \text{ kg})(20 \text{ m/s})}{(2900 \text{ kg}) \cos 33°} = 13 \text{ m/s}$$

From the y-component equation, we have

$$v_{\text{P-C }f} = \frac{(1300 \text{ kg})(16 \text{ m/s})}{(2900 \text{ kg}) \sin 33°} = 13 \text{ m/s}$$

The two equations give the same result for the final speed, a good consistency check. For collisions in which vehicles lock together like this, police investigators commonly use the lengths of the skid marks along with the direction of the vehicles after the collision to determine their initial speeds. This allows them to determine whether either vehicle was exceeding the speed limit before the collision.

Try it yourself Use a limiting case analysis and the x- and y-component forms of the impulse-momentum equation to predict what would happen during the collision if the pickup had infinite mass. Is the answer reasonable?

Answer

If we place ∞ instead of the 1600-kg mass of the pickup in the following:

$$\frac{(1300 \text{ kg})(16 \text{ m/s})}{(1600 \text{ kg})(20 \text{ m/s})} = \frac{\sin \theta}{\cos \theta} = \tan \theta$$

the left side of the equation becomes zero. Then $\tan \theta = 0$. The pickup would move straight ahead when hitting the car. In other words, the collision with the car would not change the direction of travel of the pickup. The result seems reasonable if the mass of the pickup was large compared to the mass of the car.

An object breaks into parts (radioactive decay)

We will learn in the chapter on nuclear physics (Chapter 29) that the nuclei of some atoms are unstable and spontaneously break apart. In a process called *alpha decay*, the nucleus of the atom breaks into a daughter nucleus, which is slightly smaller and lighter than the original parent nucleus, and an even smaller alpha particle (given the symbol α, though it is actually a helium nucleus). For example, radon (the radioactive gas that is sometimes found in basements) decays into a polonium nucleus (the daughter) and an alpha particle.

Radon is produced by a series of decay reactions starting with heavier elements in the soil, such as uranium. Radon diffuses out of the soil and can enter a home through cracks in its foundation. If you are in your basement when a radon atom near you undergoes alpha decay, the fast-moving alpha particle it releases could cause a gene mutation in your body. In the next example, we will analyze alpha decay of radon using the idea of momentum conservation.

EXAMPLE 6.8 **Radioactive decay of radon**

The nucleus of a radioactive radon atom in the air near you decays to a polonium nucleus and an alpha particle. With what speed does the alpha particle move if the radon nucleus was moving at 4.00×10^5 m/s and the polonium nucleus produced by the decay moves at 4.99×10^5 m/s at an angle of 36.7° above the initial path of the radon nucleus? What is the angle between the velocity of the alpha particle and the velocity of the radon nucleus? Use the relative masses of the nuclei involved: radon has mass $222m$, polonium $218m$, and the alpha particle $4m$.

Sketch and translate An initial-final sketch of the situation is shown below. We choose the positive x-axis to point in the initial direction of the radon nucleus, and the system consists of the three nuclei. We do not know the exact value of the angle ϕ, but we hypothesize that the direction of the alpha particle is below the radon's path if the momentum is constant in this direction. We might need to revise our sketch after we determine the angle. The magnitude and direction of the velocity $\vec{v}_{\alpha f}$ of the alpha particle are the unknowns to be determined.

Simplify and diagram Assume that there are no external forces exerted on the system, meaning that the system is isolated and thus its momentum is constant. The impulse-momentum bar charts at top right represent the process separately for the x- and y-directions.

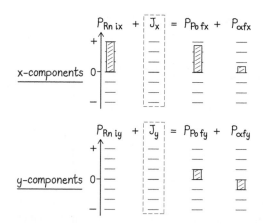

We don't know in advance which of the bars in the x-component bar chart is larger in the final state, but we know that their heights added together should be equal to the height of the initial momentum bar. For the y-component bar chart, we know that the bars in the final state should have the same height, but should be drawn in opposite directions. This means that the alpha particle does indeed move along the line below the x-axis.

Represent mathematically Use the bar charts to help apply the impulse-momentum principle for the process. The momentum constancy equation for the x-axis is

$$m_{\text{Rn}}v_{\text{Rn ix}} + 0 = m_{\text{Po}}v_{\text{Po f}} \cos 36.7° + m_\alpha v_{\alpha f} \cos \phi$$

The momentum constancy equation for the y-axis is

$$0 = m_{\text{Po}}v_{\text{Po f}} \sin 36.7° - m_\alpha v_{\alpha f} \sin \phi$$

Solve and evaluate Plugging values for the known quantities into the above, we have

$$222m(4.00 \times 10^5 \text{ m/s}) + 0 = 218m(4.99 \times 10^5 \text{ m/s})\cos 36.7° + 4mv_{\alpha f} \cos \phi$$

$$0 = 218m(4.99 \times 10^5 \text{ m/s})\sin 36.7° - 4mv_{\alpha f} \sin \phi$$

We now have two independent equations and two unknowns, $v_{\alpha f}$ and ϕ. Note that the mass m cancels from each equation. One way to solve for the two unknowns is to first insert the known information into each equation and get an expression involving $v_{\alpha f}$ and ϕ. From the x-equation we get

$$v_{\alpha f}\cos\phi = \frac{222m}{4m}(4.00 \times 10^5 \text{ m/s}) - \frac{218m}{4m}(4.99 \times 10^5 \text{ m/s})\cos 36.7°$$
$$= 3.95 \times 10^5 \text{ m/s}$$

From the y-equation we get

$$v_{\alpha f}\sin\phi = \frac{218m}{4m}(4.99 \times 10^5 \text{ m/s})\sin 36.7° = 162.5 \times 10^5 \text{ m/s}$$

Recall that $\sin^2\phi + \cos^2\phi = 1$. If we square each of the previous equations and add them together, we can factor out the $v_{\alpha f}^2$ from each term and cancel the $\sin^2\phi + \cos^2\phi$, which equals 1. We solve what is left for $v_{\alpha f}$:

$$v_{\alpha f}^2\cos^2\phi + v_{\alpha f}^2\sin^2\phi = (3.95 \times 10^5 \text{ m/s})^2 + (162.5 \times 10^5 \text{ m/s})^2$$
$$= 2.64 \times 10^{14} \text{ m}^2/\text{s}^2$$
$$v_{\alpha f}^2(\cos^2\phi + \sin^2\phi) = 2.64 \times 10^{14} \text{ m}^2/\text{s}^2$$
$$\Rightarrow v_{\alpha f} = \sqrt{2.64 \times 10^{14} \text{ m}^2/\text{s}^2}$$

Taking this square root, we find the speed of the alpha particle after the decay:

$$v_{\alpha f} = 162.5 \times 10^5 \text{ m/s}$$

We can then use the velocity components to find ϕ:

$$\frac{v_{\alpha f}\sin\phi}{v_{\alpha f}\cos\phi} = \tan\phi = \frac{162.5 \times 10^5 \text{ m/s}}{3.95 \times 10^5 \text{ m/s}} = 41.14$$

or $\phi = 88.6°$, that is, 88.6° below the x-axis. The alpha particle travels almost perpendicularly to the direction of the original radon nucleus—and pretty fast.

The speeding alpha particle can pass through tissue and collide with atoms and molecules in the body, dislodging electrons and creating damaging ions. Radon exposure causes approximately 15,000 cases of lung cancer each year.

Try it yourself Francium nuclei undergo radioactive decay by emitting either an alpha particle or a *beta particle* (an electron). When a particle is emitted from a nucleus, the nucleus starts moving, too (we say that it recoils, similar to a gun recoiling after shooting a bullet). An alpha particle is about 8000 times more massive than a beta particle. If the particles are emitted with the same speed, in which case is the recoil speed of the nucleus that is left after an alpha or a beta particle is emitted greater?

Answer

Since the mass of the alpha particle is much greater than the mass of the beta particle, and they are traveling with the same speed, the momentum of the alpha particle is much greater than the momentum of the beta particle. Therefore, the nucleus that is left will have a greater recoil speed during alpha decay.

REVIEW QUESTION 6.7 Object A, moving in the positive x-direction, collides with object B, moving in the negative y-direction. After colliding, the objects stick together. Is it possible for both objects to stop after the collision? Explain your answer.

Summary

Isolated system An isolated system is one in which the objects interact only with each other and not with the environment, or the sum of external forces exerted on it is zero. (Sections 6.1–6.2)

Isolated system

Nonisolated system

Mass is a conserved quantity. It is constant in an isolated system. If the system is not isolated and its mass changes, we can always find the system in which the mass is constant. (Section 6.1)

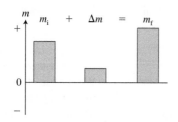

$$m_i + \Delta m = m_f$$

Linear momentum $\vec{p}$ is a vector quantity that is the product of an object's mass m and velocity $\vec{v}$. The total momentum of the system is the sum of the momenta of all objects in the system. (Section 6.2)

$$\vec{p} = m\vec{v} \qquad \text{Eq. (6.1)}$$
$$\vec{p}_{\text{system}} = \vec{p}_1 + \vec{p}_2 + \cdots$$

Impulse $\vec{J}$ is the product of the average external force $\vec{F}_{av}$ exerted on an object during a time interval Δt and that time interval. (Section 6.3)

$$\vec{J} = \vec{F}_{av}(t_f - t_i) \qquad \text{Eq. (6.5)}$$

Generalized impulse-momentum principle The change in momentum of a system is equal to the net external impulse exerted on it. If the net impulse is zero, then the momentum of the system is constant. If the objects outside the system exert forces on it and change its momentum, we can always redefine the system by including all interacting objects to keep the momentum of this new system constant. Therefore, momentum is a conserved quantity. (Section 6.4)

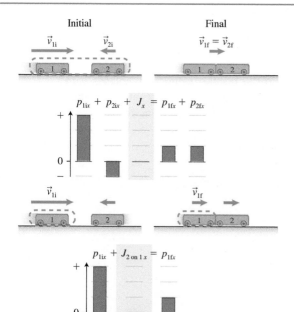

$$(m_1\vec{v}_{1i} + m_2\vec{v}_{2i} + \cdots) + \Sigma\vec{F}_{\text{on Sys}}\Delta t$$
$$= (m_1\vec{v}_{1f} + m_2\vec{v}_{2f} + \cdots) \qquad \text{Eq. (6.8)}$$

x- and y-component forms:

$$m_1v_{1ix} + m_2v_{2ix} + \Sigma F_{\text{on Sys }x}\Delta t$$
$$= m_1v_{1fx} + m_2v_{2fx} \qquad \text{Eq. (6.9}x\text{)}$$

$$m_1v_{1iy} + m_2v_{2iy} + \Sigma F_{\text{on Sys }y}\Delta t$$
$$= m_1v_{1fy} + m_2v_{2fy} \qquad \text{Eq. (6.9}y\text{)}$$

Questions

Multiple Choice Questions

1. The gravitational force that Earth exerts on an object causes an impulse of $+10\ \text{N} \cdot \text{s}$ in one experiment and $+1\ \text{N} \cdot \text{s}$ on the same object in another experiment. How can this be?
   (a) The mass of the object changed.
   (b) The time intervals during which the force was exerted are different.
   (c) The magnitudes of the force were different.

2. A bullet fired at a door makes a hole in the door but does not open it. Your finger does not make a hole in the door but does open it. Why?
   (a) The bullet is too small.
   (b) The force exerted by the bullet is not enough to open the door.
   (c) A finger exerts a smaller force but the time interval is much longer.
   (d) The bullet goes through the door and does not exert a force at all.

3. How would you convince somebody that the momentum of an isolated system is constant?
   (a) It is a law; thus, you do not need to convince anybody.
   (b) Use an example from a textbook to show that the sum of the initial and final velocities of the objects involved in a collision are the same.
   (c) Derive it from Newton's second and third laws.
   (d) Use it to make predictions about a new experiment, and then compare the outcome to the prediction.
   (e) Both (c) and (d) will work.

4. A wagon full of medicine balls is rolling along a street. Suddenly one medicine ball (3 kg) falls off the wagon. What happens to the speed of the wagon?
   (a) The wagon slows down.
   (b) The speed of the wagon does not change.
   (c) The wagon speeds up.
   (d) Additional information about the ball's motion is needed to answer.

5. When can you apply the idea that momentum is constant to solve a problem?
   (a) When the system is isolated
   (b) When the system is not isolated but the time interval when the external forces are exerted is very small
   (c) When the external forces are much smaller than the internal forces

6. Choose an example in which the momentum of a system is not constant.
   (a) A frog on a floating piece of wood jumping off the wood, with the frog and the wood as the system
   (b) A freely falling metal ball, with the ball as the system
   (c) A freely falling metal ball, with the ball and Earth as the system
   (d) It is not possible to give an example since the momentum of a system is always constant.

7. Figure Q6.7 shows the velocity-versus-time graphs of two carts on a horizontal track before and after a collision during which the carts stick together. The green and blue lines show the velocities of the carts before they collide, and the black lines show the velocity of the carts joined together after the collision. Which graph *cannot* represent the collision of the carts with equal masses?

   FIGURE Q6.7

8. Which is a safer car bumper in a collision: one that is flexible and retracts or one that is rigid? Why?
   (a) The retractable bumper, because softer things withstand collisions better
   (b) The retractable bumper, because it extends the time interval of the collision, thus reducing the force exerted on the car
   (c) The rigid bumper, because it does not change shape so easily

9. Why does an inflated balloon shoot across a room when air is released from it?
   (a) Because the outside air pushes on the balloon
   (b) Because the momentum of the balloon-air system is constant
   (c) Because the air inside the balloon pushes on the balloon, exerting the same force that the balloon exerts on the air
   (d) Both b and c are correct.

10. In which situation does the momentum of a tennis ball change more?
    (a) It hits the racket and stops.
    (b) It hits the racket and flies off in the opposite direction.
    (c) It misses the racket and continues moving.

11. A toy car with very low friction wheels and axles rests on a level track. In which situation will its speed increase more?
    (a) It is hit from the rear by a wad of clay that sticks to the car.
    (b) It is hit by a rubber ball with the same mass and velocity of the clay that rebounds in the opposite direction after hitting the car.

12. You hold a clay ball above a scale and then drop it. After hitting the scale, the ball sticks to it. In Figure Q6.12, which of the graphs (a) to (c) correctly shows the qualitative time dependence of the force exerted by the ball on the scale, and which of the graphs (d) to (f) correctly shows the qualitative time dependence of sum of the forces exerted on the ball? Assume the y-axis points vertically down.

FIGURE Q6.12

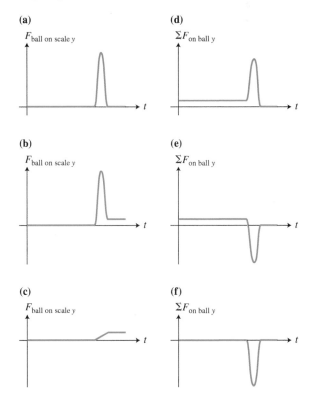

13. You hold a rubber ball above a scale and then drop it. After hitting the scale, the ball bounces up. In **Figure Q6.13**, which of the graphs (a) to (c) correctly shows the qualitative time dependence of the force exerted by the ball on the scale, and which of the graphs (d) to (f) correctly shows the qualitative time dependence of sum of the forces exerted on the ball? Assume the y-axis points vertically down.

FIGURE Q6.13

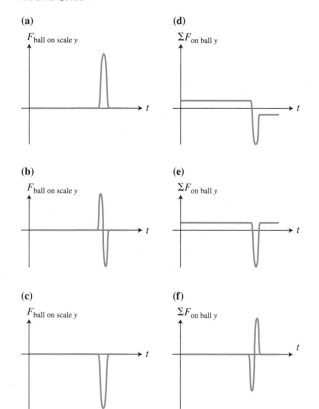

14. Two battery-powered fan carts are resting on a frictionless horizontal track. The mass of cart A is twice the mass of cart B. The carts are released so they start moving to the right. Which answer and explanation correctly compare the linear momentum of the carts after traveling the same distance? Assume that the force exerted by the air on the carts is the same for both carts and remains constant while the carts are moving.
    (a) $p_A = p_B$ because the sums of the forces exerted on the carts are equal.
    (b) $p_A = p_B$ because the distances traveled are equal.
    (c) $p_A > p_B$ because the air exerts a force on cart A for a longer time interval than on cart B.
    (d) $p_A > p_B$ because the heavier cart A pushes back on the air with a larger force than does cart B.
    (e) $p_A < p_B$ because the final velocity of cart B is larger than the final velocity of cart A.
    (f) $p_A < p_B$ because cart B has a larger acceleration than does cart A.

Conceptual Questions

15. According to a report on traumatic brain injury, woodpeckers smack their heads against trees at a force equivalent to 1200 g's without suffering brain damage. This statement contains one or more mistakes. Identify the mistakes in this statement.
16. Jim says that momentum is not a conserved quantity because objects can gain and lose momentum. Do you agree or disagree? If you disagree, what can you do to convince Jim of your opinion?
17. Say five important things about momentum (for example, momentum is a vector quantity). How does each statement apply to real life?
18. Three people are observing the same car. One person claims that the car's momentum is positive, another person claims that it is negative, and the third person says that it is zero. Can they all be right at the same time? Explain.
19. When would a ball hitting a wall have a greater change in momentum: when it hits the wall and bounces back at the same speed or when it hits and sticks to the wall? Explain your answer.
20. In the previous question, in which case does the wall exert a greater force on the ball? Explain.
21. Explain the difference between the concepts of constancy and conservation. Provide an example of a conserved quantity and a nonconserved quantity.
22. Why do you believe that momentum is a conserved quantity?
23. A heavy bar falls straight down onto the bed of a rolling truck. What happens to the momentum of the truck at the instant the bar lands on it? Explain by drawing sketches of possible situations and their corresponding impulse-momentum bar charts. How many correct answers do you think are possible? Make sure you think of what "falls straight down" means.
24. Construct impulse-momentum bar charts to represent a falling ball in (a) a system whose momentum is not constant and (b) a system whose momentum is constant. In the initial state, the ball is at rest; in the final state, the ball is moving.
25. A person moving on Rollerblades throws a medicine ball in the direction opposite to her motion. Construct an impulse-momentum bar chart for this process. The person is the system.
26. A small ball of mass m rolls at a constant speed v along a horizontal surface, then rolls down an incline and continues rolling along a lower horizontal surface with a constant speed $2v$ (see **Figure Q6.26**). Analyze the motion of the ball (the system) using a momentum approach. Identify segments of the ball's motion where the ball's momentum is changing (either in magnitude or in direction). Determine the direction of the impulse exerted on the ball during these segments and draw corresponding impulse vectors. Assume that the ball is in contact with the surface at all times and that it does not slip.

FIGURE Q6.26

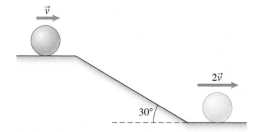

Problems

Below, **BIO** indicates a problem with a biological or medical focus. Problems labeled **EST** ask you to estimate the answer to a quantitative problem rather than derive a specific answer. Asterisks indicate the level of difficulty of the problem. Problems with no * are considered to be the least difficult. A single * marks moderately difficult problems. Two ** indicate more difficult problems.

6.2 Linear momentum

1. You and a friend are playing tennis. (a) What is the magnitude of the momentum of the 0.057-kg tennis ball when it travels at a speed of 30 m/s?

(b) At what speed must your 0.32-kg tennis racket move to have the same magnitude momentum as the ball? (c) If you run toward the ball at a speed of 5.0 m/s, and the ball is flying directly at you at a speed of 30 m/s, what is the magnitude of the total momentum of the system (you and the ball)? Assume your mass is 60 kg. In every case specify the object of reference.

2. You are hitting a tennis ball against a wall. The 0.057-kg tennis ball traveling at 25 m/s strikes the wall and rebounds at the same speed. (a) Determine the ball's original momentum (magnitude and direction). (b) Determine the ball's change in momentum (magnitude and direction). What is your object of reference?

3. A ball of mass m and speed v travels horizontally, hits a wall, and rebounds. Another ball of the same mass and traveling at the same speed hits the wall and sticks to it. Which ball has a greater change in momentum as a result of the collision? Explain your answer.

4. Figure P6.4 shows the velocity-versus-time graph before and after the collision of two carts on a horizontal track. After the collision the carts stick together. Green and blue lines show the velocities of cart 1 and cart 2, respectively, before the collision. The black line shows the velocity of the carts joined together after the collision. Determine the ratio of the carts' masses m_1/m_2.

FIGURE P6.4

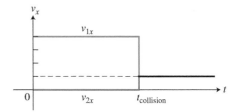

5. A 1300-kg car is traveling at a speed of 10 m/s with respect to the ground when the driver accelerates to make a green light. The momentum of the car increases by 12,800 kg·m/s. List all the quantities you can determine using this information and determine three of those quantities.

6. * The rules of tennis specify that the 0.057-kg ball must bounce to a height of between 53 and 58 inches when dropped from a height of 100 inches onto a concrete slab. What is the change in the momentum of the ball during the collision with the concrete? You will have to use some free-fall kinematics to help answer this question. Ignore air resistance.

7. * A cart of mass m moving right at speed v with respect to the track collides with a cart of mass $0.7m$ moving left. What is the initial speed of the second cart if after the collision the carts stick together and stop?

8. * A cart of mass m moving right collides with an identical cart moving right at half the speed. The carts stick together. What is their speed after the collision?

9. **EST** Estimate your momentum when you are walking at your normal pace.

6.3 Impulse and momentum

10. A 100-g apple is falling from a tree. What is the impulse that Earth exerts on it during the first 0.50 s of its fall? The next 0.50 s?

11. * The same 100-g apple is falling from the tree. What is the impulse that Earth exerts on it during the first 0.50 m of its fall? The next 0.50 m?

12. Why does Earth exert the same impulse during the two time intervals in Problem 6.10 but different impulses during the same distances traveled in Problem 6.11?

13. * **Van hits concrete support** In a crash test, a 2000-kg van collides with a concrete support. The stopping time interval for the collision is 0.10 s, and the impulse exerted by the support on the van is 7.5×10^3 N·s. (a) Determine everything you can about the collision using this information. (b) If the van is constructed to collapse more during the collision so that the time interval during which the impulse is exerted is tripled, what is the average force exerted by the concrete support on the van?

14. **BIO** **Force exerted by heart on blood** About 80 g of blood is pumped from a person's heart into the aorta during each heartbeat. The blood starts at rest with respect to the body and has a speed of about 1.0 m/s in the aorta. If the pumping takes 0.17 s, what is the magnitude of the average force exerted by the heart on the blood?

15. * The train tracks on which a train travels exert a 2.0×10^5-N friction force on the train, causing it to stop in 50 s. (a) Determine the average force needed to stop the train in 25 s. (b) Determine the stopping time interval if the tracks exert a 1.0×10^5-N friction force on the train.

16. ** **EST** Your friend is catching a falling basketball after it has passed through the basket. Her hands move straight down 0.20 m while catching the ball. Estimate (a) the time interval for the ball to stop as she catches it and (b) the average force that her hands exert on the ball while catching it. Indicate any assumptions or estimates you have to make in order to answer the questions.

17. **BIO** **Traumatic brain injury** According to a report on traumatic brain injury, the force that a professional boxer's fist exerts on his opponent's head is equivalent to being hit with a 5.9-kg bowling ball traveling at 8.9 m/s that stops in 0.018 s. Determine the average force that the fist exerts on the head.

18. * A 65-kg astronaut pushes against the inside back wall of a 2000-kg spaceship and moves toward the front. Her speed increases from 0 to 1.6 m/s. (a) If her push lasts 0.30 s, what is the average force that the astronaut exerts on the wall of the spaceship? (b) If the spaceship was initially at rest, with what speed does it recoil? (c) What was the object of reference that you used to answer parts (a) and (b)?

19. * You decide to use your garden hose to wash your garage door. Every second, 0.8 kg of water leaves the hose at a speed of 9.0 m/s with respect to the hose, hits the garage door perpendicularly, and spills down the door surface. Determine the magnitude of the force that the water jet from the hose exerts on the door. What assumptions did you make?

20. * An egg rolls off a kitchen counter and breaks as it hits the floor. How large is the impulse that the floor exerts on the egg, and how large is the force exerted on the egg by the floor when stopping it? The counter is 1.0 m high, the mass of the egg is about 50 g, and the time interval during the collision is about 0.010 s.

21. ** **Proportional reasoning** During a collision when an average stopping force F is exerted on a car by an obstacle, the car's bumper retracts a distance Δx. These two quantities are related by the equation $F\Delta x = 0.5mv_0^2$. Use proportional reasoning to determine (a) the necessary percent change in the retraction distance so that the average force required to stop a car is reduced by 20% and (b) the percent change in initial to final speed that would produce the same reduction in force.

22. (a) What force is required to stop a 1500-kg car in a distance of 0.20 m if it is initially moving at 2.2 m/s? (b) What if the car is moving at 4.5 m/s?

23. ** **EST** You drop a 78-g ball vertically onto a force sensor that is 0.4 m below the ball's initial position. The ball bounces off the force sensor after the collision. Figure P6.23 shows the time dependence of the force exerted by the ball on the force sensor. Estimate the expected peak magnitude of the force. Indicate any assumptions that you made, and discuss how they affect the estimated peak value. (Hint: The area under the force-versus-time curve has approximately the shape of a triangle.)

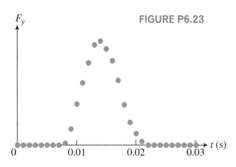

FIGURE P6.23

24. * **Air bag force on head** The graph in Figure P6.24 shows the time variation of the force that an automobile's air bag exerts on a person's head during a collision. The mass of the head is 8.0 kg. Determine (a) the total impulse of the force exerted by the air bag on the person's head and (b) the person's speed just before the collision occurred.

FIGURE P6.24

25. * **Equation Jeopardy 1** Invent a problem for which the solution is
$$(27 \text{ kg})(-3.0 \text{ m/s}) + (30 \text{ kg})(+4.0 \text{ m/s}) = (27 \text{ kg} + 30 \text{ kg})v.$$

26. * **Equation Jeopardy 2** Invent a problem for which the solution is
$$(0.020 \text{ kg})(300 \text{ m/s}) - (10 \text{ N})(0.40 \text{ s}) = (0.020 \text{ kg})(100 \text{ m/s}).$$

6.4 The generalized impulse-momentum principle

27. * Two carts (100 g and 150 g) on an air track are separated by a compressed spring. The spring is released. Represent the process with a momentum bar chart (a) with one cart as the system and (b) with both carts as the system. (c) Write expressions for all of the physical quantities you can from this information. Identify your object of reference.

28. * A tennis ball of mass m hits a wall at speed v and rebounds at about the same speed. Represent the process with an impulse-momentum bar chart for the ball as the system. Using the bar chart, develop an expression for the change in the ball's momentum. What is the object of reference?

29. * A tennis ball traveling at a speed of v stops after hitting a net. Represent the process with an impulse-momentum bar chart for the ball as the system. Develop an expression for the ball's change in momentum. What is the object of reference?

30. * You drop a happy ball and a sad ball of the same mass from height h (see Example 6.3). One ball hits the ground and rebounds almost to the original height. The other ball does not bounce. Represent each process with a bar chart, starting just before the balls hit the ground to just after the first ball rebounds and the other ball stops. Choose the ball as the system.

31. * You have two objects: a ball that is covered with the hook side of Velcro and an incline covered with the loop side of Velcro that is fixed onto a low-friction cart. The ball is dropped vertically onto the incline, which is at rest (see **Figure P6.31**). When the ball collides with the incline, it sticks to it without moving relative to it. What happens to the cart after the ball sticks to the incline? Will the cart move or remain stationary? Explain. Represent the process with impulse-momentum bar charts.

FIGURE P6.31

32. * You hold a beach ball above your head and then throw it upward. Represent the motion of the ball with an impulse-momentum bar chart for (a) the ball as the system and (b) the ball and Earth as the system.

33. * A basketball player drops a 0.60-kg basketball vertically so that it is traveling at 6.0 m/s when it reaches the floor. The ball rebounds upward at a speed of 4.2 m/s. (a) Determine the magnitude and direction of the ball's change in momentum. (b) Determine the average net force that the floor exerts on the ball if the collision lasts 0.12 s.

34. * **Bar chart Jeopardy** Invent a problem for each of the bar charts shown in Figure P6.34.

FIGURE P6.34

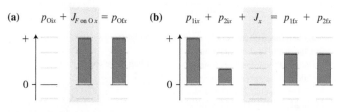

(a) $p_{Oix} + J_{F \text{ on } O\,x} = p_{Ofx}$ 　(b) $p_{1ix} + p_{2ix} + J_x = p_{1fx} + p_{2fx}$

6.5 Skills for analyzing problems using impulse and momentum

35. * A baseball bat contacts a 0.145-kg baseball for 1.3×10^{-3} s. The average force exerted by the bat on the ball is 8900 N. If the ball has an initial velocity of 36 m/s toward the bat and the force of the bat causes the ball's motion to reverse direction, what is the ball's speed as it leaves the bat?

36. A cannon mounted on the back of a ship fires a 50-kg cannonball in the horizontal direction at a speed of 150 m/s. If the cannon and ship have a combined mass of 40,000 kg and are initially at rest, what is the speed of the ship just after shooting the cannon? What assumptions did you make?

37. A 10-kg sled carrying a 30-kg child glides on a horizontal, frictionless surface at a speed of 6.0 m/s toward the east. The child jumps off the back of the sled, propelling it forward at 20 m/s. What was the child's velocity in the horizontal direction relative to the ground at the instant she left the sled?

38. A 10,000-kg coal car on the Great Northern Railroad coasts under a coal storage bin at a speed of 2.0 m/s. As it goes under the bin, 1000 kg of coal is dropped into the car. What is the final speed of the loaded car?

39. * **Avoiding chest injury** A person in a car during a sudden stop can experience potentially serious chest injuries if the combined force exerted by the seat belt and shoulder strap exceeds 16,000 N. Describe what it would take to avoid injury by estimating (a) the minimum stopping time interval and (b) the corresponding stopping distance, assuming an initial speed of 16 m/s. Indicate any other assumptions you made.

40. * **Bruising apples** An apple bruises if a force greater than 8.0 N is exerted on it. Would a 0.10-kg apple be likely to bruise if it falls 2.0 m and stops after sinking 0.060 m into the grass? Explain.

41. * **Fast tennis serve** The fastest server in women's tennis is Sabine Lisicki, who recorded a serve at 131 mph (211 km/h) in 2014. Suppose that the mass of her racket was 328 g and the mass of the ball was 57 g. If her racket was moving at 210 km/h when it hit the ball, approximately what was the racket's speed after hitting the ball? Indicate any assumptions you made.

42. * You are in an elevator whose cable has just broken. The elevator is falling at 20 m/s when it starts to hit a shock-absorbing device at the bottom of the elevator shaft. If you are to avoid injury, the upward force that the floor of the elevator exerts on your upright body while stopping should be no more than 8000 N. Determine the minimum stopping distance needed to avoid injury (do not forget to include your mass in the calculations). What assumptions did you make? Do these assumptions make the stopping distance smaller or larger than the real-world value?

43. ** You jump from the window of a burning hotel and land in a safety net that stops your fall in 1.0 m. Estimate the average force that the net exerts on you if you enter the net at a speed of 24 m/s. What assumptions did you make? If you did not make these assumptions, would the stopping distance be smaller or larger?

44. * After a 70-kg person steps from a boat onto the shore, the boat moves away with a speed of 0.7 m/s with respect to the shore. If the same person steps from the shore onto the boat, the person and the boat move away from the shore with the speed of 0.5 m/s. Assume that the initial speed of the person with respect to the shore is the same in both cases and that the person is moving in a horizontal direction. (a) Draw impulse-momentum bar charts for both situations and (b) determine the mass of the boat and the speed of the person.

45. * **BIO** **Leg injuries during car collisions** During a car collision, the knee, thighbone, and hip can sustain a force no greater than 4000 N. Forces that exceed this amount could cause dislocations or fractures. Assume that in a collision a knee stops when it hits the car's dashboard. Also assume that the mass of the body parts stopped by the knee is about 20% of the total body mass. (a) What minimum stopping time interval is needed to avoid injury to the knee if the person is initially traveling at 15 m/s (34 mi/h)? (b) What is the minimum stopping distance?

46. * **BIO** **Bone fracture** The zygomatic bone in the upper part of the cheek can be fractured by a 900-N force lasting 6.0 ms or longer. A hockey puck can easily exert such a force when hitting an unprotected face. (a) What change in velocity of a 0.17-kg hockey puck is needed to provide that impulsive force? What assumptions did you make? (b) A padded facemask doubles the stopping time. By how much does it change the force on the face? Explain.

47. ** You are investigating a newly discovered particle X that has an unknown mass M and moves with a constant but unknown speed u. In your experiments you observe collisions between particle X and different test particles with known masses m that are initially at rest. You let the particles collide head-on and stick together, and you measure the speed v of the combined-particle object. From other observations you know that in all experiments the objects move along the same straight line before and after the collision. Your data are shown in the table at right. Note that the masses are expressed in units of reference mass m_0 and the speed in units of reference speed v_0. (a) Derive an expression for v in terms of M, m, and u. Indicate any assumptions that you made. (b) Determine M and u using the data in the table. (Hint: Rearrange the expression that you derived in (a) to obtain a new equation with the linear dependence on m.)

m (m_0)	v (v_0)
2.0	0.55
4.0	0.45
6.0	0.35
8.0	0.30
10.0	0.25

48. * An impulse of 150 N · s stops your head during a car collision. (a) A crash test dummy's head stops in 0.020 s, when the cheekbone hits the steering wheel. What is the average force that the wheel exerts on the dummy's cheekbone? (b) Would this crash fracture a human cheekbone (see Problem 6.46)? (c) What is the shortest impact time that a person could sustain without breaking the bone?

49. A cart is moving on a horizontal track when a heavy bag falls vertically onto it. What happens to the speed of the cart? Represent the process with an impulse-momentum bar chart. Justify your choice of a system.

50. * A cart is moving on a horizontal track. A heavy bag falls off the cart and moves straight down relative to the cart. Describe what happens to the speed of the cart. Represent your answer with the impulse-momentum bar chart. Justify your choice of a system. (Hint: What reference frame will you use when you draw the bar chart?)

51. Your friend shoots an 80-g arrow through a 100-g apple balanced on William Tell's head. The arrow has a speed of 50 m/s before passing through the apple and 40 m/s after. Determine the final speed of the apple.

6.6 Jet propulsion

52. * **BIO** **EST** Cuttlefish use jet propulsion to move rapidly. To do this, a cuttlefish fills its muscular mantle cavity with water and then quickly expels the water through a tube-like opening. The mass of a cuttlefish is about the same as the mass of water of the same volume. A 15-cm-long and 5-cm-wide cuttlefish is moving at a speed of 3 m/s after expelling the water from its body. Estimate the speed of the expelled water relative to the sea bed. Indicate any assumptions that you made.

53. * **BIO** **Potassium decay in body tissue** Certain natural forms of potassium have nuclei that are radioactive. Each radioactive potassium nucleus decays to a slightly less massive daughter nucleus and a high-speed electron called a beta particle. If after the decay the daughter nucleus is moving at speed 200 m/s with respect to the decaying material, how fast is the electron (the beta particle) moving? Indicate any assumptions you made. The mass of the daughter nucleus is about 70,000 times greater than the mass of the beta particle.

54. **Drifting space mechanic** An astronaut with a mass of 90 kg (including spacesuit and equipment) is drifting away from his spaceship at a speed of 0.20 m/s with respect to the spaceship. The astronaut is equipped only with a 0.50-kg wrench to help him get back to the ship. With what speed and in what direction relative to the spaceship must he throw the wrench for his body to acquire a speed of 0.10 m/s and direct him back toward the spaceship? Explain.

55. * **Astronaut flings oxygen tank** While the astronaut in Problem 6.54 is trying to get back to the spaceship, his comrade, a 60-kg astronaut, is floating at rest a distance of 10 m from the spaceship when she runs out of oxygen and fuel to power her back to the spaceship. She removes her oxygen tank (3.0 kg) and flings it away from the ship at a speed of 15 m/s relative to the ship. (a) At what speed relative to the ship does she recoil toward the spaceship? (b) How long must she hold her breath before reaching the ship?

56. **Rocket stages** A 5000-kg rocket ejects a 10,000-kg package of fuel. Before ejection, the rocket and the fuel travel together at a speed of 200 m/s with respect to distant stars. If after the ejection, the fuel package travels at 50 m/s opposite the direction of its initial motion, what is the velocity of the rocket?

57. * A rocket has just ejected fuel. With the fuel and the rocket as the system, construct an impulse-momentum bar chart for (a) the rocket's increase in speed and (b) the process of a rocket slowing down due to fuel ejection. (c) Finally, draw bar charts for both situations using the rocket without the fuel as the system.

6.7 Collisions in two dimensions

58. * **Car collision** A 1180-kg car traveling south at 24 m/s with respect to the ground collides with and attaches to a 2470-kg delivery truck traveling east at 16 m/s. Determine the velocity (magnitude and direction) of the two vehicles when locked together just after the collision.

59. * **Ice skaters collide** While ice skating, you unintentionally crash into a person. Your mass is 60 kg, and you are traveling east at 8.0 m/s with respect to the ice. The mass of the other person is 80 kg, and he is traveling north at 9.0 m/s with respect to the ice. You hang on to each other after the collision. In what direction and at what speed are you traveling just after the collision?

60. ** **EST** **Meteorite impact with Earth** About 65 million years ago a 10-km-diameter 2×10^{15}-kg meteorite traveling at about 10 km/s crashed into what is now the Gulf of Mexico. The impact produced a cloud of debris that darkened Earth and led to the extinction of the dinosaurs. Estimate the change of the Earth's speed as a result of this impact and the average force that the meteorite exerted on Earth during the collision. Indicate any assumptions made in your calculations.

61. * Three friends play beach volleyball. The 280-g ball is flying east at speed 8.0 m/s with respect to the ground when one of the players bumps the ball north. The force exerted by the wrist on the ball has an average magnitude of 84 N and lasts for 0.010 s. Determine the ball's velocity (magnitude and direction) following the bump. Does your answer make sense?

62. ** Two forces exert impulses on a hockey puck, which can move with no friction on an icy surface. The graphs at left in **Figure P6.62** show the time dependence of the x- and y-components of the sum of the forces exerted on the puck. Which of the trajectories (a) to (d) of the puck's motion in the xy-plane shown at right in Figure P6.62 *cannot* be the result of these forces? The numbers correspond to successive clock readings that are marked on the force graphs.

FIGURE P6.62

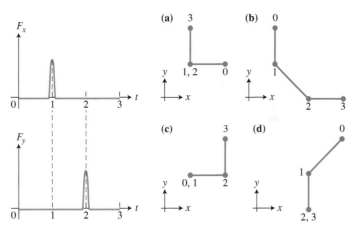

General Problems

63. ** **EST** Estimate the recoil speed of Earth if all of the inhabitants of Canada and the United States simultaneously jumped straight upward from Earth's surface (reaching heights from several centimeters to a meter or more). Indicate any assumptions that you made in your estimate.

64. * A cart of mass m traveling in the negative x-direction at speed v collides head-on with a cart that has triple the mass and is moving at 60% of the speed of the first cart. The carts stick together after the collision. In which direction and at what speed will they move?

65. ** Two cars of unequal mass moving at the same speed collide head-on. Explain why a passenger in the smaller mass car is more likely to be injured than one in the larger mass car. Justify your reasoning with the help of physics principles.

66. ** **Restraining force during collision** A 1340-kg car traveling east at 13.6 m/s (30 mi/h) has a head-on collision with a 1930-kg car traveling west at 20.5 m/s (46 mi/h). If the collision time is 0.10 s, what is the force needed to restrain a 68-kg person in the smaller car? In the larger car?

67. * **EST** A carpenter hammers a nail using a 0.80-kg hammerhead. He lifts the hammer and lets it fall down without exerting an additional force on the hammer. After the hammer hits the nail, they move together for 0.005 s until they stop. (a) Estimate the speed of the hammerhead before it hits the nail. (b) Estimate the average force that the hammerhead exerts on the nail. Indicate any assumptions that you made.

68. ** A 0.020-kg bullet traveling at a speed of 300 m/s embeds in a 1.0-kg wooden block resting on a horizontal surface. The block slides horizontally 4.0 m on a surface before stopping. Determine the coefficient of kinetic friction between the block and surface.

69. * Two identical lightweight arms are mounted so that they can revolve in a horizontal plane around a common axis. Two balls of masses m and $3m$ are fixed to the ends of the arms and connected with a light compressed spring (see the top view in **Figure P6.69**). When the spring is released, the arms are pushed apart, rotate in opposite directions, and finally collide. What is the angle through which each arm turns before they collide? Indicate any assumptions that you made.

FIGURE P6.69

70. * In a first experiment, a 30-g clay ball is shot at a speed of 1.3 m/s horizontally from the edge of a table. The ball lands on the floor 0.60 m from the table. In a second experiment, the same ball is shot at the same speed, but this time the ball hits a wooden block that is placed on the edge of the table. The ball sticks to the block, and the block lands on the floor 0.06 m from the table. (a) Represent the second experiment with impulse-momentum bar charts, treating the x- and y-components separately; draw two sets of bar charts, first taking the wooden block as a system and then taking the block and the clay ball as a system (initial state: just before the clay ball hits the block; final state: just before the block touches the floor). Determine (b) the mass of the block and (c) the height of the table. Indicate any assumptions that you made.

71. ** **EST** A record rainstorm produced 304.8 mm (approximately 1 ft) of rain in 42 min. Estimate the average force that the rain exerted on the roof of a house that measures 10 m × 16 m. Indicate any assumptions you made.

72. * While dangling a hairdryer by its cord, as shown in **Figure P6.72**, you observe that the cord is vertical when the hairdryer is off and that it makes an angle of 5° with the vertical when the hairdryer is turned on. Draw a force diagram for the hairdryer in both cases and explain the outcome of the experiment using momentum arguments.

FIGURE P6.72

73. ** While dangling a hairdryer by its cord, as shown in Figure P6.72, you observe that the cord is vertical when the hairdryer is off and that it makes an angle of 5° with the vertical when the hairdryer is turned on. In a different experiment, you determine that the same hairdryer is pushing 0.03 m³ of air through itself every second. The mass of the hairdryer is 420 g. Determine the speed of the air that is leaving the hairdryer, assuming that the mass of 1 m³ of air is 1.2 kg and that the hairdryer is blowing air perpendicular to the wire.

74. ** Martha learns that a helicopter is able to lift or hover by pushing air downward, resulting in air pushing up on the helicopter blades. She decides to investigate this mechanism using her remote-controlled toy helicopter. In her first experiment, she fixes the helicopter on a long stick, holds it directly above the plate of a scale (without touching the scale), and lets the blades rotate with maximum speed (see **Figure P6.74a**). The scale reading is zero before she switches the helicopter on and 54 g when the blades are rotating. In her second experiment, she fixes the helicopter legs to the scale plate using Scotch tape and again lets the blades rotate with maximum speed (see Figure P6.74b). The scale reading shows the mass of the helicopter as 123 g before the blades are rotating and 134 g when the blades are rotating. (a) Explain why the scale shows a nonzero reading in the first experiment even though the helicopter is not touching the scale. (b) Martha is surprised that, in the second experiment, the scale reading does not decrease when the blades start rotating—indeed, the reading *increases*. Why did Martha expect the reading to decrease? Explain the increase of the scale reading after the blades start rotating. (Note that the increase is much smaller than the increase in the first experiment.)

FIGURE P6.74

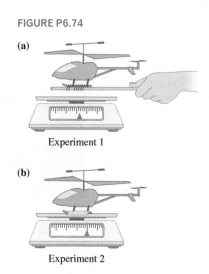

(a)

Experiment 1

(b)

Experiment 2

75. * A 2045-kg sports utility vehicle hits the rear end of a 1220-kg car at rest at a stop sign. The car and SUV remain locked together and skid 4.6 m before stopping. If the coefficient of friction between the vehicles and the road is 0.70, what was the SUV's initial velocity?

76. ** A car of mass m_1 traveling north at a speed of v_1 collides with a car of mass m_2 traveling east at a speed of v_2. They lock together after the collision. Develop expressions for the direction and the distance the cars will move until they stop if the coefficient of kinetic friction μ_k between the cars' tires and the road is about the same for both cars.

77. * **Write your own problem** Write and solve a problem that requires using the law of conservation of momentum in which it is important to know that momentum is a vector quantity.

78. ** You have two carts, a force probe connected to a computer, a motion detector, and an assortment of objects of different masses. Design three experiments to test whether momentum is a conserved quantity. Describe carefully what data you will collect and how you will analyze the data.

Reading Passage Problems

BIO **Heartbeat detector** A prisoner tries to escape from a Nashville, Tennessee prison by hiding in the laundry truck. The prisoner is surprised when the truck is stopped at the gate. A guard enters the truck and handcuffs him. "How did you know I was here?" the prisoner asks. "The heartbeat detector," says the guard.

A heartbeat detector senses the tiny vibrations caused by blood pumped from the heart. With each heartbeat, blood is pumped upward to the aorta, and the body recoils slightly, conserving the momentum of the blood-body system. The body's vibrations are transferred to the inside of the truck. Vibration sensors on the outside of the truck are linked to a geophone, or signal amplifier, attached to a computer. A wave analyzer program in the computer compares vibration signals from the truck to wavelets produced by heartbeat vibrations. The wave analyzer distinguishes a person's heartbeat from other vibrations in the truck or in the surrounding environment, allowing guards to detect the presence of a human in the truck.

79. What does the heartbeat detector sense?
    (a) Electric signals caused by electric dipole charges produced on the heart
    (b) Body vibrations caused by blood pumped from the heart
    (c) Sound caused by breathing
    (d) Slight uncontrollable reflexive motions of an enclosed person
    (e) All of the above

80. A heartbeat detector relies on a geophone placed against the exterior of a truck or car. What can cause the vibrations of the truck or car?
    (a) Wind
    (b) Ground vibrations due to other moving cars or trucks
    (c) Vibrations of passengers in the car or truck
    (d) All of the above

81. What can be used to analyze the motion of a person's body hidden inside a car or truck (choosing the body and blood as a system)?
    (a) The idea that mass of an isolated system is constant
    (b) The idea that momentum of an isolated system is constant
    (c) The impulse-momentum principle
    (d) a and b
    (e) b and c

82. During each heartbeat, about 0.080 kg of blood passes through the aorta in about 0.16 s. This blood's velocity changes from about 0.8 m/s upward toward the head to 0.8 m/s down toward the feet. What is the blood's acceleration?
    (a) zero   (b) 5 m/s^2 up   (c) 5 m/s^2 down
    (d) 10 m/s^2 up   (e) 10 m/s^2 down

83. Suppose 0.080 kg of blood moving upward in the aorta at 0.8 m/s reverses direction in 0.16 s when it reaches the aortic arch. If a prisoner is trying to escape from prison by hiding in a laundry truck, and the mass of his body is 70 kg, which is the closest to the speed his body is moving immediately after the blood changes direction passing through the aortic arch?
    (a) 0.0009 m/s   (b) 0.002 m/s   (c) 0.8 m/s
    (d) 0.08 m/s   (e) 0.01 m/s

Space Shuttle launch The mass of the Space Shuttle at launch was about 2.1×10^6 kg. Much of this mass was the fuel used to move the orbiter, which carried the astronauts and various items in the shuttle's payload. The Space Shuttle generally traveled from 3.2×10^5 m (200 mi) to 6.2×10^5 m (385 mi) above Earth's surface. The shuttle's two solid fuel boosters (the cylinders on the sides of the shuttle) provided 71.4% of the thrust during liftoff and the first stage of ascent before being released from the shuttle 132 s after launch at 48,000 m above sea level. The boosters continued moving up in free fall to an altitude of approximately 70,000 m and then fell toward the ocean to be recovered 230 km from the launch site. The shuttle's five engines together provided 3.46×10^7 N of thrust during liftoff.

84. Which number below is closest to the acceleration of the shuttle during liftoff? (Hint: Remember the gravitational force that Earth exerts on the shuttle.)
    (a) 3.3 m/s^2 (b) 6.6 m/s^2 (c) 9.8 m/s^2
    (d) 16 m/s^2 (e) 33 m/s^2

85. Which number below is closest to the average vertical acceleration of the shuttle during the first 132 s of its flight?
    (a) 3.3 m/s^2 (b) 5.5 m/s^2 (c) 9.8 m/s^2
    (d) 14 m/s^2 (e) 360 m/s^2

86. The boosters were released from the shuttle 132 s after launch. How did their vertical components of velocity compare to that of the shuttle at the instant of release?
    (a) The boosters' vertical component of velocity is zero.
    (b) The boosters' vertical component of velocity is about -9.8 m/s.
    (c) The vertical component of velocity of the boosters and that of the shuttle are the same.
    (d) There is too little information to decide.

87. What was the approximate impulse of the jet engine thrust exerted on the shuttle during the first 10 s of flight?
    (a) $980 \text{ N} \cdot \text{s}$ downward   (b) $980 \text{ N} \cdot \text{s}$ upward
    (c) $3.4 \times 10^7 \text{ N} \cdot \text{s}$ upward   (d) $3.4 \times 10^8 \text{ N} \cdot \text{s}$ upward
    (e) $3.4 \times 10^8 \text{ N} \cdot \text{s}$ downward

88. What was the approximate impulse of Earth's gravitational force exerted on the shuttle during the first 10 s of flight?
    (a) $980 \text{ N} \cdot \text{s}$ downward   (b) $980 \text{ N} \cdot \text{s}$ upward
    (c) $2.1 \times 10^7 \text{ N} \cdot \text{s}$ upward   (d) $2.1 \times 10^8 \text{ N} \cdot \text{s}$ upward
    (e) $2.1 \times 10^8 \text{ N} \cdot \text{s}$ downward

89. What was the momentum of the Space Shuttle 10 s after liftoff closest to?
    (a) $2.1 \times 10^6 \text{ kg} \cdot \text{m/s}$ down   (b) $2.1 \times 10^6 \text{ kg} \cdot \text{m/s}$ up
    (c) $2.1 \times 10^7 \text{ kg} \cdot \text{m/s}$ up   (d) $1.3 \times 10^8 \text{ kg} \cdot \text{m/s}$ up
    (e) $1.3 \times 10^8 \text{ kg} \cdot \text{m/s}$ down

90. What answer below is closest to the speed of the shuttle and boosters when they were released? Assume that the free-fall gravitational acceleration at this elevation is about 9.6 m/s^2 down.
    (a) 100 m/s   (b) 300 m/s   (c) 600 m/s
    (d) 1000 m/s   (e) 1400 m/s

Work and Energy

• How do pole vaulters reach a height of 5 or 6 m?

• Why does blood pressure increase when the walls of the aorta thicken?

• If our Sun were to become a black hole, how big would it be?

Yelena Isinbayeva from Russia has held the women's pole-vaulting record of 5.06 m since 2009. The men's record is 6.16 m. How can pole vaulters reach such great heights? If you watch the event, you will see three aspects of the vault that are essential: the vaulters run very fast before inserting the pole into the vaulting box; they jump at the moment of takeoff while bending the pole as they start their vertical ascent; and they crunch their muscles at the top of the flight to give a specific body shape as they pass over the bar. All these elements contribute to the height of the vault—in this chapter, we will learn how.

BE SURE YOU KNOW HOW TO:

• Choose a system and the initial and final states of a physical process (Sections 6.2–6.4).

• Use Newton's second law and your knowledge of momentum to analyze a physical process (Sections 4.4 and 6.4).

• Use kinematics to describe motion (Section 2.9).

SO FAR, WE HAVE learned how to use forces and momentum to explain and analyze different types of motion. However, not all questions concerning motion can be answered using these concepts. For example, imagine a roller coaster in an amusement park. It has no engine or battery, but it goes through multiple loops and many twists and turns. The force that the track exerts on it continuously changes direction and magnitude. How is it possible to design the track so the coaster will make it through the entire ride without falling or grinding to a halt? It turns out that there are several important physical quantities, and a new principle, that are important in designing a roller coaster. We will learn about these quantities and this principle in this chapter.

7.1 Work and energy

We will begin our new approach by conducting several experiments and looking for patterns to help explain what we observe. Specifically, in Observational Experiment Table 7.1 we investigate the changes that happen to a system when an external force is exerted on it over a certain distance. In the analysis we draw vectors indicating the external force causing the displacement and the resulting displacement of the system.

OBSERVATIONAL EXPERIMENT TABLE 7.1   External forces and system changes

VIDEO OET 7.1

Observational experiment	Analysis
Experiment 1. You hold a heavy block just above a piece of chalk (the initial state) and then release the block. The chalk does not break. Now you lift the block about 30 cm above the chalk (the final state). When you release the block from the higher elevation, the block falls and smashes the chalk.	The force you exerted and the block's displacement while being lifted were in the same direction and caused an increase in the block's elevation and in its ability to break the chalk.
Experiment 2. You push a cart initially at rest (the initial state) until it is moving fast about two-thirds of the way across a smooth track (the final state is where you stop pushing the cart). A piece of chalk is taped to the end of the track. The fast-moving cart (no longer being pushed) collides with the piece of chalk and breaks the chalk. When you repeat the experiment with a slow-moving cart, it does not break the chalk.	The force exerted on the cart and the cart's displacement were in the same direction and increased the cart's speed so it could break the chalk.
Experiment 3. A piece of chalk rests in the hanging sling of a slingshot (the initial state). You then pull it back until the slingshot is fully stretched (the final state). When released from the stretched sling, the chalk flies across the room, hits the wall, and smashes.	The force that you exerted on the sling and its displacement were in the same direction and made it possible for the stretched sling to cause the chalk to break.
Experiment 4. A heavy cooler sits on a shag carpet (the initial state). You pull the cooler across the carpet to a position several meters from where it started (the final state).	You exerted a force on the cooler in the direction of its displacement. After several meters of travel across the carpet, the bottom became warmer.

(CONTINUED)

Patterns

In each of these experiments, you exerted an external force $\vec{F}$ on an object in a system. The force $\vec{F}_{\text{You on Object}}$ and the object's displacement $\vec{d}_{\text{Object}}$ were in the *same direction* and caused the system to change so that it could do something that it could not do before, such as break the chalk or warm up the carpet. Specifically:

1. The *block at higher elevation* above Earth could break the chalk, but the block at low elevation could not.
2. The *fast-moving cart* could break the chalk, but the slow-moving cart could not.
3. The *stretched slingshot* could break the chalk, but the unstretched slingshot could not.
4. The *cooler and the carpet it was pulled across* became warmer than they were before the cooler was pulled across the carpet.

A photo of the carpet and cooler in Experiment 4 from Table 7.1 taken with a thermal camera. The temperature of the carpet and the cooler increases by about 2.5 °F. This shows the change in the thermal energy of both objects.

In the experiments in Table 7.1, the system could not break the chalk or warm the carpet in the initial state, but in the final state it could. What happened to each system to cause that change? In all cases, **an external force exerted on the system caused a displacement of an object in the system in the *same direction* as the force.** Using everyday words in a very specific way, physicists say that such a force does positive **work** on a system, and that this work done causes a change in what is called the system's **energy** (we will learn later that there can be negative work that also changes the system's energy). Energy is rather difficult to define at this point, but you can think of it loosely as a property of a system that changes when work is done on the system. Four types of energy changed in the systems in Table 7.1:

- Experiment 1: The energy of the object-Earth system associated with the elevation of the object above Earth is called **gravitational potential energy** U_g. The higher the object is above Earth, the greater the gravitational potential energy of that object-Earth system.
- Experiment 2: The energy due to an object's motion is called **kinetic energy** K. The faster the object is moving, the greater its kinetic energy.
- Experiment 3: The energy associated with an elastic object's degree of stretch is called **elastic potential energy** U_s. The greater the stretch (or compression), the greater the object's elastic potential energy.
- Experiment 4: The energy associated with both temperature and structure is called **internal energy** U_{int}. You will learn later that internal energy is the energy of motion and interaction of the microscopic particles making up the objects in the system. A component of internal energy is **thermal energy**, energy associated with the temperature of the objects in the system (see the photo). In this chapter, when we apply the concept of internal energy to people, we will also talk about muscles' **chemical energy**—this is another component of internal energy related to the interaction of particles in chemical compounds.

Negative and zero work

Is it possible to devise a process in which an external force causes the energy of a system to decrease or possibly causes no energy change at all? Let's try more experiments to investigate these questions. See Observational Experiment Table 7.2 on the next page.

OBSERVATIONAL EXPERIMENT TABLE 7.2 Negative and zero work

VIDEO
OET 7.2

Observational experiment	Analysis
Experiment 1. A friend holds a block high above a piece of chalk (the initial state) and then releases it. Your hand catches and stops the block (the final state). The block's potential to break the chalk is greater in the initial state than in the final state. 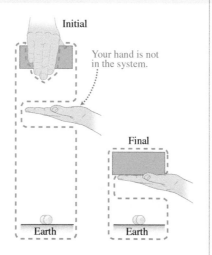	The direction of the force exerted by your hand on the block is opposite the block's displacement and reduced the system's potential to break the chalk.
Experiment 2. A cart is moving fast (the initial state) toward a piece of chalk taped on the wall. While it is moving, you push lightly on the moving cart opposite the direction of its motion, causing it to slow down and stop (final state). The cart's potential to break the chalk is greater in the initial state than in the final state.	The direction of the force exerted by your hand on the cart is opposite the cart's displacement and caused the moving cart to slow down and stop, thus reducing its potential to break the chalk.
Experiment 3. Your hand is slowly moving a block to the right, keeping the block just above the tabletop. The block passes above the piece of chalk on the left (initial state) and then over the piece of chalk on the right (final state). The block is too close to the chalk in both states to break it.	The direction of the force exerted by your hand on the block is perpendicular to the block's displacement and caused no change in the block's potential to break the chalk.

Patterns

- In Experiments 1 and 2, the direction of the external force exerted on the object in the system is opposite the object's displacement, and the system's ability to break the chalk decreases.
- In Experiment 3, the direction of the external force exerted on the object in the system is perpendicular to the object's displacement, and the system's ability to break the chalk is unchanged.

We can now summarize what we have learned about work:

- When the external force is in the direction of the object's displacement, the external force does *positive work,* causing the system to gain energy.
- When the external force points opposite the object's displacement, the external force does *negative work,* causing the system's energy to decrease.
- When the external force points perpendicular to the object's displacement, the external force does *zero work* on the system, causing no change to its energy.

Defining work as a physical quantity

We now want to find a mathematical relation to define the physical quantity of work. The definition needs to be consistent with all three of the features of work listed in the previous section. Intuitively, we know that a larger force should lead to larger work, as should a larger displacement. But how do we account for the angle dependence? It turns out that there is a mathematical function that will take care of this issue—the cosine of the angle between the direction of the force vector and displacement vector (this angle is θ in **Figure 7.1**).

Note that in the four experiments in Table 7.1, the force and displacement were in the same direction: $\theta = 0°$, and $\cos 0° = +1.0$. Positive work was done. In the first two experiments in Table 7.2, the force and displacement were in opposite directions: $\theta = 180°$, and $\cos 180° = -1.0$. Negative work was done. Finally, in Experiment 3 in Table 7.2, the force and displacement were perpendicular to each other: $\theta = 90°$, and $\cos 90° = 0$. Zero work was done.

FIGURE 7.1 The definition of work.

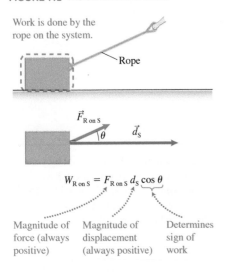

> **Work** The work done by a constant external force $\vec{F}$ exerted on an object in a system while that object undergoes a displacement $\vec{d}$ is
>
> $$W = Fd \cos \theta \tag{7.1}$$
>
> where F is the magnitude of the force in newtons (always positive), d is the magnitude of the displacement in meters (always positive), and θ is the angle between the direction of $\vec{F}$ and the direction of $\vec{d}$. The sign of $\cos \theta$ determines the sign of the work. Work is a scalar physical quantity. The unit of work is the joule (J); $1\text{ J} = 1\text{ N} \cdot \text{m}$.

The joule is named in honor of James Joule (1818–1889), one of many physicists who contributed to our understanding of work-energy relationships.

According to the definition of work, if the displacement of an object is zero, even if forces are exerted on it, no work is done. This idea might contradict your everyday experience. Imagine holding a heavy object (for example, a suitcase) in a steady position. The system is the suitcase and Earth; you are an external object. If the suitcase remains at rest, the work done by the force you exert on it is zero. But you still get tired. How is this possible? It turns out that even when you are holding something still, your muscles still contract and relax all the time, making your arm go up and down slightly (we cannot see this movement with the naked eye, but scientists use a special method called *sound myography* to detect such muscle contractions). Thus you are doing work on the suitcase by continuously lifting it a little and then dropping it, repeating the process many times. Let us now include you in the system. Careful observations show that while holding the suitcase, you breathe a little faster and small drops of perspiration appear on your face. These are the signs of increased metabolic activity—your chemical energy is converted into the kinetic and elastic potential energy of your muscles and also into thermal energy. Thus there is no contradiction between physics and everyday life.

QUANTITATIVE EXERCISE 7.1   Pushing a bicycle uphill

Two friends are cycling up a hill inclined at 8°—steep for bicycle riding. The stronger cyclist helps his friend up the hill by exerting a 50-N pushing force on his friend's bicycle and parallel to the hill while the friend moves a distance of 100 m up the hill. Determine the work done by the stronger cyclist on the weaker cyclist.

Represent mathematically Choose the system to be the weaker cyclist. The external force that the stronger cyclist (S) exerts on the weaker cyclist (W) $\vec{F}_{S \text{ on } W}$ is parallel to the hill, as is the 100-m displacement of the weaker cyclist. The work done by the stronger cyclist on the weaker cyclist is $W = Fd \cos \theta$. The hill is inclined at 8°

above the horizontal. Before reading on, decide what angle you would insert in this equation.

Solve and evaluate Note that the angle between $\vec{F}_{S \text{ on } W}$ and $\vec{d}$ is 0° and not 8° because the force exerted on the cyclist is parallel to the person's displacement. Thus:

$$W_{S \text{ on } W} = F_{S \text{ on } W} d \cos \theta = (50 \text{ N})(100 \text{ m})\cos 0°$$
$$= +5000 \text{ N} \cdot \text{m} = +5000 \text{ J}$$

- -

Try it yourself You pull a box 20 m up a 10° ramp. The rope is oriented 20° above the surface of the ramp. The force that the rope exerts on the box is 100 N. What is the work done by the rope on the box?

Answer the ramp itself.

$W_{R \text{ on } B} = (100 \text{ N})(20 \text{ m})\cos 20° = 1900 \text{ J}$. Note that you use the angle between the displacement (parallel to the ramp) and the force the rope exerts on the box (20° above the ramp). You can disregard the angle of

---

REVIEW QUESTION 7.1 Assuming that Earth's orbit around the Sun is circular, what information do you need to estimate the work done by the Sun on Earth while Earth travels 1000 km? Explain your answer.

7.2 Energy is a conserved quantity

We have found that the work done on an object in a system by an external force results in a change of one or more types of energy in the system: kinetic energy, gravitational potential energy, elastic potential energy, and internal energy. We can think of the *total energy E* of a system as the sum of all these energies in the system:

$$\text{Total energy} = E = K + U_g + U_s + U_{\text{int}} \qquad (7.2)$$

The energy of a system can be converted from one form to another. For example, the elastic potential energy of a stretched slingshot is converted into the kinetic energy of the chalk when the sling is released. Similarly, the gravitational potential energy of a separated block-Earth system is converted into kinetic energy of the block when the block falls. Here we assume that Earth does not move when the block moves toward it. What happens to the amount of energy when it is converted from one form to another? So far we have one mechanism through which the energy of the system changes—that mechanism is work. Thus it is reasonable to assume that if no work is done on the system, the energy of the system should not change; it should be *constant*. Let's test this hypothesis experimentally (see Testing Experiment **Table 7.3**, on the next page).

> **TIP** Remember that an *isolated* system is a system for which we can neglect interactions with the environment. Thus, when external objects do zero work on a system, the system can be considered isolated.

TESTING
EXPERIMENT TABLE **7.3**   **Is the energy of an isolated system constant?**

VIDEO
TET 7.3

Testing experiment	Prediction	Outcome
You have a toy car and a frictionless track. The bottom of the track sits near the edge of a table. You can tilt the track at different angles to vary its steepness. When the car reaches the end of the track, it flies horizontally off the table. Where should you release a car on the track so the car always lands the same distance from the table regardless of the slope of the track? 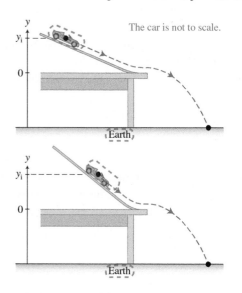	Consider car + Earth as the system. The initial state is just before we release the car and the final state is just as the car leaves the horizontal track. For the car to land on the floor at the same distance from the table's edge, the car needs to have the same horizontal velocity and hence the same kinetic energy when leaving the track. If energy is constant in the isolated system, to get the same kinetic energy at the bottom of the track, the car-Earth system must have the same initial gravitational energy when it starts. Since the gravitational potential energy depends on the separation between the car and Earth, the car should start at the same vertical elevation independent of the tilt angle of the track, $y_i = h$. Our prediction is based on the assumption that the gravitational potential energy of the system is only converted to the kinetic energy of the car and not to internal energy.	When released from the same vertical height with respect to the table, the car lands the same distance from the table.

Conclusion
The outcome of the experiment matches the prediction and thus supports our hypothesis that the energy of an isolated system is constant.

TIP It is important to note that in the above arguments we assumed that Earth is infinitely large and does not move when an object falls down or flies up.

So far we have experimental support for a qualitative hypothesis that the energy of an isolated system is constant and different processes inside the system convert energy from one form to another. We also reason that work is a mechanism through which the energy of a nonisolated system changes. This sounds a lot like our discussion of linear momentum (Chapter 6). In that chapter, we described a *conserved quantity* as one that is constant in an isolated system. We also said that when the system is not isolated and the conserved quantity changes, we can always redefine the system in such a way as to keep the quantity constant. Imagine that we did not include Earth in the system in the experiment above. We would find that the energy of the cart (our new system) changed during the experiment because it acquired kinetic energy moving down the track due to the work done by Earth on it. Thus the energy of the cart as a system was not constant. However, when we include Earth in the system, the energy remains constant.

Work-energy bar charts

We can represent work-energy processes with bar charts that are similar to the impulse-momentum bar charts we used to describe momentum. A work-energy bar chart indicates with vertical bars the relative amount of a system's different types of energy in the initial state of a process, the work done on the system by external forces during the process, and the relative amount of different types of energy in the system at the end of the process. The area for the work bar is shaded to emphasize that work is not a property of the system. The physical quantity of work characterizes the process through which energy is transferred in or out of the system due to the external forces that cause displacements of the objects inside the system. In terms of system change, work is similar to impulse in our study of momentum. In **Table 7.4** you see how bar charts can be used to represent the process when you push a clay ball off a frictionless table exerting a constant force (situation 1), the ball flies off the table (situation 2), and lands on the floor and stops (situation 3). We break the process into three parts and explain how we choose the system in each situation.

TABLE 7.4 Examples of system energy conversions

Process	Sketch of initial and final states and explanations of the system choice	Bar chart for the process
1. You push a clay ball along a frictionless table, exerting a constant force.	Initial state: The ball is at rest. Final state: The ball is moving and is just about to leave the table. System: The ball. We choose the ball as the system because there is no elevation change, thus we do not need to worry about gravitational potential energy. 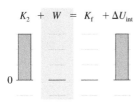	Your hand does positive work on the system. The ball gains kinetic energy. $$K_i + W = K_1$$
2. The clay ball falls down freely off the edge of the table.	Initial state: The ball is at the top of its flight. Final state: Just before the ball hits the floor (it is still moving). System: The ball and Earth. We included Earth in the system because there is elevation change and thus a change in gravitational potential energy. If we did not include Earth in the system, it would do work but there would be no gravitational potential energy in any state. 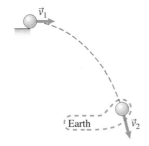	No work is done on the system. The system's initial gravitational potential energy is converted into kinetic energy. We assume that the gravitational potential energy of the system at floor level is zero, so there is no bar for U_{gf}. $$K_1 + U_{g1} + W = K_2 + U_{g2}$$
3. The clay ball lands on the floor and stops.	Initial state: Just before the moving ball hits the floor. Final state: Just after the ball stops (and its shape changes). System: The ball and the floor. The floor stopped the ball during landing, but it did not move; thus it would be difficult to account for the effects of the floor via work. Including it in the system removes the need to account for work done by the floor. Earth is not included in the system because the elevation did not change in this situation. 	No work is done on the system. The ball's kinetic energy is converted into internal energy of the ball and the floor. $$K_2 + W = K_f + \Delta U_{int}$$

From the examples in the table you can conclude that when choosing what objects to include in the system and what not to include, we can use the following approach: **If it is difficult to calculate the work done by that object on the system, we are usually better off including it in the system so that we describe the processes using different forms of energy.** As you proceed through the chapter you will see more examples and discussion of how to choose what to include and not to include in the system.

1. Choose and describe the initial and final states.

2. Choose a system.

3. Represent the energies in the initial state on the left side of the bar chart.

4. Represent the energies in the final state on the right side of the bar chart.

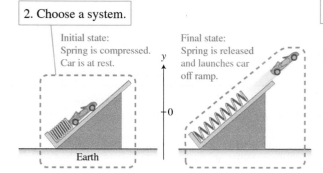

Initial state: Spring is compressed. Car is at rest.

Final state: Spring is released and launches car off ramp.

Earth

$$K_i + U_{gi} + U_{si} + W = K_f + U_{gf} + U_{sf} + \Delta U_{int}$$

5. Determine the work done by external forces. (No external work is done in this process.)

> TIP In Physics Tool Box 7.1, the system is isolated. No work is done on it. If we do not include Earth in the system, then Earth will do negative work on the cart. However, such a system will not have gravitational potential energy.

The generalized work-energy principle

We found experimentally that the energy of an isolated system is constant. We also defined work as a mechanism for energy transfer. Therefore, we can summarize what we have learned about work and energy with the **generalized work-energy principle**.

> **Generalized work-energy principle** The sum of the initial energies of a system plus the work done on the system by external forces equals the sum of the final energies of the system:
>
> $$E_i + W = E_f \tag{7.3}$$
>
> or
>
> $$(K_i + U_{gi} + U_{si}) + W = (K_f + U_{gf} + U_{sf} + \Delta U_{int}) \tag{7.3}$$
>
> Note that we have moved $U_{int\,i}$ to the right hand side ($\Delta U_{int} = U_{int\,f} - U_{int\,i}$) since values of internal energy are rarely known, while internal energy changes are. We will also call Eq. (7.3) the *work-energy equation*.

The generalized work-energy principle allows us to define the total energy of a system as the sum of the different types of energy in the system. Total energy is measured in the same units as work and changes when work is done on the system. In Eq. (7.3), kinetic energy and gravitational and elastic potential energy are forms of **mechanical energy**. Mechanical energy is associated with motion and interactions of macroscopic objects—objects that we can see with our eyes. Internal energy is associated with motion and interactions of microscopic objects—objects that we cannot see with our eyes. Unlike such physical quantities as position or force, energy is not directly observed, and there are no instruments that measure energy. We can only infer the amount of any type of energy if we can express it through measurable quantities.

The work in Eq. (7.3) can be positive or negative. Unlike velocity and acceleration, for which plus and minus signs signify direction, in the case of work the sign signifies another operation—addition or subtraction.

CONCEPTUAL EXERCISE 7.2   Pole vaulting, qualitatively

Qualitatively analyze a pole vault from an energy perspective by first drawing a sketch and then a work-energy bar chart.

Sketch and translate To analyze the process, we choose the final state to be when a pole vaulter is just clearing the bar. What should be the initial state? As we discussed at the beginning of the chapter, pole vaulters run very fast while holding an unbent pole, then insert the pole into the vaulting box. Next, using the bent pole, they lift themselves above the bar. When they have just cleared the bar, they are still holding the pole, but it is no longer bent. Thus it makes sense to choose the moment when the pole vaulter is running, just before she inserts the pole into the box, as the initial state. In both of these states the pole is not deformed, so we do not need to worry about changes in the pole's elastic energy or the pole doing work on the system. Therefore, it does not matter whether it is in the system or not. Let's choose the pole vaulter, Earth, and the pole as the system. The beauty of the energy approach is that we are only concerned with what is happening in the initial and final states, not with the processes in between. The sketch shows the pole-vaulting process with the initial and final states marked.

However, if we carefully watch videos of pole vaulters on the Internet, we see that right after inserting the pole into the box, they jump while bending the pole as they start their vertical ascent. They also crunch their muscles at the top of their flight to produce a specific body shape as they pass over the bar. During both of these actions, the vaulters convert some of their internal energy into the mechanical energy of the system. This means that the internal energy of the system is less in the final state than in the initial state. To represent this, we draw the bar for the internal energy *change* ΔU_{int} below the axis to indicate that the internal energy of the pole vaulter between the moment she inserts the pole in the vaulting box and when she clears the bar decreases. The pole vaulter does not cool down during the process, so where does this internal energy come from? It comes from the chemical energy stored in her body. From the bar chart, we reason that the greater the initial kinetic energy and the more internal energy the vaulter can "pump" into the vault, the greater the final gravitational potential energy, or the height of the vault. Note that we did not take into account the initial kinetic energy of the pole or the internal energy change due to its deformation (it could have gotten a little warmer).

$$K_i + U_{gi} + \boxed{W} = K_f + U_{gf} + \Delta U_{int}$$

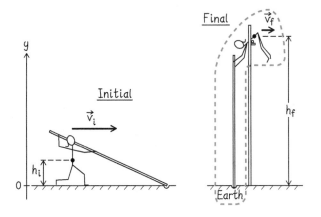

Simplify and diagram Assume that the pole vaulter can be modeled as a point-like object with her mass concentrated approximately in the middle of her body. Between the initial and final states there are no external forces exerted on the system; thus there is no work done. Using our understanding of energy, we can say that the gravitational potential energy U_{gi} of the system and the kinetic energy K_i of the vaulter in the initial state are converted to the final gravitational potential energy U_{gf} and a little bit of kinetic energy K_f (the vaulter needs to be moving horizontally a little in order not to fall straight down when no longer in contact with the pole) of the system in the final state. To draw the bars for gravitational potential energy, we assume that it is zero at ground level.

Try it yourself You throw a ball straight upward as shown at right. The system is the ball and Earth (but not your hand). Ignore interactions with the air. Draw a work-energy bar chart starting when the ball is at rest in your hand and ending when the ball is at the very top of its flight.

Hand does work on ball only when it is in contact with the ball.

Answer

See the diagram. Note that there is no kinetic energy represented in the bar chart because the ball was not moving in either the initial state or the final state. It does not matter that it was moving in the time interval between those two states.

$$K_i + U_{gi} + W_{H on b} = K_f + U_{gf}$$

In Table 7.4 we chose different systems to analyze different parts of a process. In general, how do we know what to include and what not to include in a system to make the analysis of a process easier? Generally, it is preferable to have a larger system so that the changes occurring can be included as energy changes within the system rather than as the work done by external forces. However, often it is best to exclude something like a motor from a system because its energy changes are complex. For instance, if you are

FIGURE 7.2 Energy changes in the motor pulling up on the elevator cable are very complicated.

We exclude the complicated motor from the system.

Earth

studying a moving elevator (**Figure 7.2**), you might choose to exclude from the system the motor that turns the cable while lifting the elevator. The motor's cable does positive work on the system by exerting a tension force on the elevator when the elevator is moving up and negative work when the elevator is moving down. You can take into account the effect of the motor by calculating the work done by the cable on the elevator.

REVIEW QUESTION 7.2 A system can possess energy but it cannot possess work. Why?

7.3 Quantifying gravitational potential and kinetic energies

In this section we will use what we know about work, forces, and kinematics to devise mathematical expressions for gravitational potential energy and kinetic energy.

Gravitational potential energy

FIGURE 7.3 Lifting a box at a negligible constant speed.

(a)

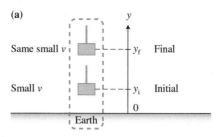

Same small v — y_f Final

Small v — y_i Initial

0

Earth

(b)

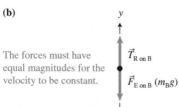

The forces must have equal magnitudes for the velocity to be constant.

$\vec{T}_{R \text{ on } B}$

$\vec{F}_{E \text{ on } B}$ $(m_B g)$

(c) $U_{gi} + W = U_{gf}$

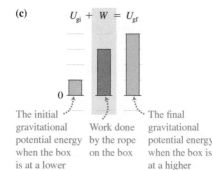

0

The initial gravitational potential energy when the box is at a lower position

Work done by the rope on the box

The final gravitational potential energy when the box is at a higher vertical position

Imagine that a rope lifts a heavy box upward at a constant negligible velocity (**Figure 7.3a**). The rope is attached to a motor above, which is not shown in the figure. First, we choose only the box as the system and apply Newton's second law to find the magnitude of the force that the rope exerts on the box. Since the box moves up at constant velocity, the upward tension force $\vec{T}_{R \text{ on } B}$ exerted by the rope on the box is equal in magnitude to the downward gravitational force $\vec{F}_{E \text{ on } B}$ exerted by Earth on the box (see the force diagram in Figure 7.3b), $m_B g$, we find that the magnitude of the tension force for this process is $T_{R \text{ on } B} = m_B g$.

To derive an expression for gravitational potential energy, we must change the boundaries of the system to include the box and Earth (if Earth is not included in the system, the system does not have gravitational potential energy). The origin of a vertical y-axis is the ground directly below the box with the positive direction upward. The initial state of the system is the box at position y_i moving upward at a negligible speed $v_i \approx 0$. The final state is the box at position y_f moving upward at the same negligible speed $v_f \approx 0$. According to work-energy Eq. (7.3),

$$E_i + W = E_f$$

The rope does work on the box, lifting the box from vertical position y_i to y_f:

$$W_{R \text{ on } B} = T_{R \text{ on } B} d \cos \theta = T_{R \text{ on } B} (y_f - y_i) \cos 0° = m_B g (y_f - y_i)$$

where we substituted $T_{R \text{ on } B} = m_B g$ and $\cos 0° = 1$. The kinetic energy did not change. Substituting this expression for work into the work-energy equation, we get

$$U_{gi} + m_B g (y_f - y_i) = U_{gf}$$

Figure 7.3c represents this information with an energy bar chart. We now have an expression for the change in the gravitational potential energy of the system: $U_{gf} - U_{gi} = m_B g y_f - m_B g y_i$. This suggests the following definition for the gravitational potential energy of a system.

> **Gravitational potential energy** The gravitational potential energy of an object-Earth system is
>
> $$U_g = mgy \qquad (7.4)$$
>
> where m is the mass of the object, $g = 9.8$ N/kg, and y is the position of the object with respect to the **zero level of gravitational potential energy** (the origin of the coordinate system is our choice). The units of gravitational potential energy are kg(N/kg)m = N · m = J (joule), the same unit used to measure work and the same unit for every type of energy.

TIP Three things are important here:

1. The value of U_g depends on our choice of zero for the y-coordinate. According to Eq. (7.4), this is where U_g is zero. However, the *change* in U_g does not depend on the choice of the zero level.
2. The derivation of Eq. (7.4) assumed that g is constant. We will look at nonconstant g later in the chapter.
3. If Earth is not included in the system, the system does not have any U_g, but Earth can do work on the system.

Kinetic energy

Next, we analyze a simple thought experiment to determine an expression for the kinetic energy of a system that consists of a single object. Imagine that your hand exerts a force $\vec{F}_{\text{H on C}}$ on a cart of mass m while pushing it toward the right a displacement $\vec{d}$ on a horizontal frictionless surface (**Figure 7.4a**). A bar chart for this process is shown in Figure 7.4b. There is no change in gravitational potential energy. The kinetic energy changes from the initial state to the final state because of the work done by the external force exerted by your hand on the cart:

$$K_i + W_{\text{H on C}} = K_f$$

or

$$W_{\text{H on C}} = K_f - K_i$$

We know that the work in this case equals $F_{\text{H on C}}d \cos 0° = F_{\text{H on C}}d$. Thus,

$$F_{\text{H on C}}d = K_f - K_i$$

This does not look like a promising result—the kinetic energy change on the right equals quantities on the left side that do not depend on the mass or speed of the cart. However, we can use dynamics and kinematics to get a result that does depend on these properties of the cart. The horizontal component form of Newton's second law is

$$m_C a_C = F_{\text{H on C}}$$

We can rearrange a kinematics equation ($v_f^2 = v_i^2 + 2ad$) to get an expression for the displacement of the cart in terms of its initial and final speeds and its acceleration:

$$d = \frac{v_f^2 - v_i^2}{2a_C}$$

Now, we insert these expressions for force and displacement into the left side of the equation $F_{\text{H on C}}d = K_f - K_i$:

$$F_{\text{H on C}}d = (m_C a_C)\left(\frac{v_f^2 - v_i^2}{2a_C}\right) = m_C\left(\frac{v_f^2}{2} - \frac{v_i^2}{2}\right) = \tfrac{1}{2}m_C v_f^2 - \tfrac{1}{2}m_C v_i^2$$

We can now insert this result into the equation $F_{\text{H on C}}d = K_f - K_i$:

$$\tfrac{1}{2}m_C v_f^2 - \tfrac{1}{2}m_C v_i^2 = K_f - K_i$$

It appears that $\tfrac{1}{2}m_C v^2$ is an expression for the kinetic energy of the cart.

FIGURE 7.4 The work done by the hand causes the cart's kinetic energy to increase.

(a)

The hand does positive work on the cart, leading to an increase in its kinetic energy.

(b)

| Initial kinetic energy | Work done by hand on cart | Final kinetic energy $K_f = K_i + W$ |

Kinetic energy The kinetic energy of an object is

$$K = \tfrac{1}{2}mv^2 \qquad (7.5)$$

where m is the object's mass and v is its speed relative to the chosen coordinate system. Notice that kinetic energy is always positive.

TIP Remember that speed is relative—it depends on the observer. A person on a Boeing 747 airplane would consider the kinetic energy of a 70-kg passenger next to her as zero, while a person on the ground would find the kinetic energy of the same person to be about 2.5 million joules.

FIGURE 7.5 Two fan carts of different masses with identical fans.

To check whether the unit of kinetic energy is the joule (J), we insert the units into Eq. (7.5): $\dfrac{\text{kg} \cdot \text{m}^2}{\text{s}^2} = \left(\dfrac{\text{kg} \cdot \text{m}}{\text{s}^2}\right)\text{m} = \text{N} \cdot \text{m} = \text{J}.$

It is important to understand the differences between force, acceleration, work, impulse, momentum, and kinetic energy. To illustrate this, we will perform the following thought experiment. Imagine two carts on a smooth surface that have identical fans attached but different masses. The fans rotate and push away the air around the carts; according to Newton's third law, the air pushes on the carts in the opposite direction, making them accelerate. Both carts start from rest, and we let them move the *same distance* on the floor (see **Figure 7.5**). **Table 7.5** compares the above physical quantities for the two carts. Notice that some quantities are the same for both carts, some are larger for cart 1, and some are larger for cart 2. Your task is to think about the explanations for the statements in the table.

TABLE 7.5 **Physical quantities describing the motion of two fan carts**

Quantity	Cart 1 (mass m_1)	Cart 2 (mass $m_2 = 2m_1$)
Force exerted by the air on the fan cart	$\vec{F}_{\text{A on 1}} = \vec{F}$	$\vec{F}_{\text{A on 2}} = \vec{F}_{\text{A on 1}} = \vec{F}$ Same as cart 1
Acceleration of the cart	$\vec{a}_1 = \dfrac{\vec{F}}{m_1}$	$\vec{a}_2 = \dfrac{\vec{F}}{m_2} = \dfrac{\vec{F}}{2m_1} = \tfrac{1}{2}\vec{a}_1$ Less than cart 1
Time interval to travel distance d	$\Delta t_1 = \sqrt{\dfrac{2d}{a_1}}$	$\Delta t_2 = \sqrt{\dfrac{2d}{a_2}} = \sqrt{\dfrac{4d}{a_1}} = \sqrt{2}\Delta t_1 > \Delta t_1$ Greater than cart 1
Work done by the fan on the cart	$W_{\text{A on 1}} = Fd\cos 0°$	$W_{\text{A on 2}} = Fd\cos 0° = W_{\text{A on 1}}$ Same as cart 1
Change of kinetic energy of the cart after moving distance d	$\Delta K_1 = \dfrac{m_1 v_{1f}^2}{2} = Fd$	$\Delta K_2 = \dfrac{m_2 v_{2f}^2}{2} = Fd = \Delta K_1$ Same as cart 1
Speed of the cart after moving distance d	$v_{1f} = \sqrt{\dfrac{2Fd}{m_1}}$	$v_{2f} = \dfrac{v_{1f}}{\sqrt{2}}$ Less than cart 1
Change of momentum of the cart after moving distance d	$m_1 v_{1f}$	$m_2 v_{2f} = 2m_1\dfrac{v_{1f}}{\sqrt{2}} = \sqrt{2}m_1 v_{1f}$ Greater than cart 1
Impulse exerted on the cart after moving distance d	$\vec{J} = \vec{F}\Delta t_1 = m_1\vec{v}_{1f}$	$\vec{J} = \vec{F}\Delta t_2 = m_2\vec{v}_{2f}$ Greater than cart 1

You might be surprised that the heavier, slower cart has a greater momentum gain than the lighter, faster cart but the same gain in kinetic energy. But if you think about the work done by the air on each cart (equal to force times distance because the cosine of the angle between the force and displacement is 1), it is the same; thus the change in kinetic energy must be the same.

EXAMPLE 7.3   **Mystic Timbers roller coaster**

The Mystic Timbers roller coaster in Ohio has a drop of 33.2 m. The maximum speed of the coaster at the bottom of this drop is 53 mph. Can we say that all work done by Earth on the roller coaster is converted to the coaster's kinetic energy during the ride?

Sketch and translate   We first draw a sketch of the roller coaster and mark the givens, assuming that the coaster starts from rest at the top of the track. We choose the initial state to be just before the roller coaster starts to drop and the final state to be at the very bottom of the drop. Because the problem asks about the work done by Earth, Earth should not be in the system—we include only the coaster and the track in the system.

Simplify and diagram   A work-energy bar chart representing the process is shown at right. Because Earth is not in the system, Earth does work on the coaster, but the system does not have gravitational potential energy in any state. In the initial state, the system has no kinetic energy. In the final state, it has kinetic energy and possibly a change in internal energy (the coaster wheels and the track could warm up), but we are not sure.

Because the force that Earth exerts on the coaster points down, it only does work on the coaster due to the coaster's *vertical* displacement. This is because the work done by Earth on the coaster during the horizontal displacement of the coaster is zero, and the total displacement of the coaster can be decomposed into vertical and horizontal displacement components, as shown below.

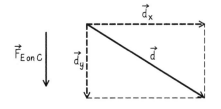

Represent mathematically   We use the bar chart to write the work-energy equation:

$$0 + W = K_f + \Delta U_{int}$$

Because we do not know the mass of the coaster, we cannot calculate the work done by Earth on the coaster or the coaster's final kinetic energy. However, we can check whether all work goes into the kinetic energy increase. If it does, the final speed of the coaster calculated using this assumption should be equal to the reported final speed of the coaster.

$$W_{E\,on\,C} = F_{E\,on\,C}d \cos \theta = F_{E\,on\,C}d \cos 0° = F_{E\,on\,C}d = m_C g d$$

$$\text{and } K_f = \tfrac{1}{2}m_C v_f^2$$

Inserting these into the work-energy equation with the assumption that $\Delta U_{int} = 0$ gives

$$0 + m_C g d = \tfrac{1}{2}m_C v_f^2 \implies v_f = \sqrt{2gd}$$

Solve and evaluate $v_f = \sqrt{2 \times (9.8 \text{ m/s}^2) \times (33.2 \text{ m})} = 25.5 \text{ m/s}.$

The reported final speed of the coaster is

$$53 \text{ mph} = \frac{(53 \text{ miles/h}) \times (1600 \text{ m/mile})}{3600 \text{ s/h}} = 23.6 \text{ m/s}$$

This reported final speed is about 7% $\left(\dfrac{25.5 - 23.6}{25.5} = 0.07\right)$ less than it would be if we assumed that all of the work done by the gravitational force went into the increase of the kinetic energy of the roller coaster. Thus some of this work went into an increase of the coaster and track's internal energy (and maybe some to the air around it).

Try it yourself   Change the system in this example to include Earth. Does the result change? How does the work-energy bar chart change?

Answer   The roller coaster speed calculated for the new system is the same. On the bar chart there is no work done, but the system instead has an initial gravitational potential energy bar of the same height as the previous work done bar. The kinetic energy and change in internal energy bars remain the same.

REVIEW QUESTION 7.3 When we use the work-energy equation, how do we incorporate the force that Earth exerts on an object?

7.4 Quantifying elastic potential energy

Our next goal is to construct a mathematical expression for the elastic potential energy stored by an elastic object when it has been stretched or compressed. We have a special problem in deriving this expression. Previously, we investigated work done by constant forces. However, when you stretch or compress an elastic spring-like object, you have to pull or push harder the more the object is stretched or compressed. The force is not constant. How does the force you exert to stretch a spring-like object change as the object stretches?

FIGURE 7.6 Measuring spring stretch x caused by a scale exerting force F on a spring.

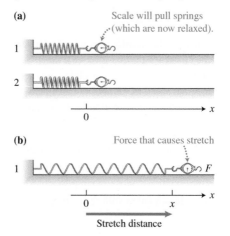

Hooke's law

To answer this question, we use two springs of the same length: a thinner and less stiff spring 1 and a thicker and stiffer spring 2 (**Figure 7.6a**). The springs are attached at the left end to a rigid object and placed on a smooth surface. We use a scale to pull on the right end of each spring, exerting a force whose magnitude F can be measured by the scale (Figure 7.6b). We record F and the distance x that each spring stretches from its unstretched position (see **Table 7.6**).

TABLE 7.6 Result of pulling on springs while exerting an increasing force

Force F exerted by the scale on the spring	Spring 1 stretch distance x	Spring 2 stretch distance x
0.00 N	0.000 m	0.000 m
1.00 N	0.050 m	0.030 m
2.00 N	0.100 m	0.060 m
3.00 N	0.150 m	0.090 m
4.00 N	0.200 m	0.120 m

FIGURE 7.7 A graph of the stretch of springs 1 and 2 when the same force is exerted on each spring.

We need to pull harder on spring 2 than on spring 1 to stretch it 0.1 m.

Figure 7.7 shows a graph of the Table 7.6 data. We use stretch distance x as an independent variable and the magnitude of the force $\vec{F}_{\text{Scale on Spring}}$ as a dependent variable. The magnitude of the force exerted by the scale on each spring is proportional to the distance that each spring stretches.

$$F_{\text{Scale on Spring}} = kx$$

The coefficient of proportionality k (the slope of the F-versus-x graph) is called the **spring constant**. The slope for the stiffer spring 2 is larger ($33\ \text{N/m}$) than the slope for spring 1 ($20\ \text{N/m}$). In other words, to stretch spring 1 by 1.0 m we would have to exert only a 20-N force, but we would need a 33-N force for spring 2.

Often we are interested not in the force that something exerts on the spring, but in the force that the spring exerts on something else, $\vec{F}_{\text{Spring on Scale}}$. Using Newton's third law we have $\vec{F}_{\text{Spring on Scale}} = -\vec{F}_{\text{Scale on Spring}}$. The x-component of this force is

$$F_{\text{Spring on Scale }x} = -kx$$

Note that if an object stretches the spring to the right in the positive direction, the spring pulls back on the object in the opposite negative direction (**Figure 7.8a**; in this case, the object is your finger). If the object compresses the spring to the left in the negative direction, the spring pushes back on the object in the opposite positive direction (Figure 7.8b). These observations are the basis for a relation first developed by Robert Hooke (1635–1703), called **Hooke's law**.

FIGURE 7.8 Hooke's law.

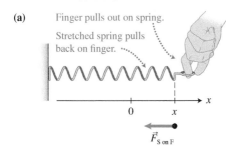

Elastic force (Hooke's law) If any object causes a spring to stretch or compress, the spring exerts an elastic force on that object. If the object stretches the spring along the x-direction, the x-component of the force the spring exerts on the object is

$$F_{\text{S on O}\,x} = -kx \qquad (7.6)$$

The spring constant k is measured in newtons per meter and is a measure of the stiffness of the spring (or any elastic object); x is the distance that the object has been stretched/compressed (not the total length of the object). The elastic force exerted by the spring on the object points in a direction opposite to the direction it was stretched (or compressed)—hence the negative sign in front of kx. The object in turn exerts a force on the spring:

$$F_{\text{O on S}\,x} = +kx$$

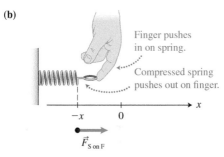

Elastic potential energy

Our goal now is to develop an expression for the elastic potential energy of an elastic stretched or compressed object (such as a stretched spring). Consider the constant slopes of the lines in the graph shown in Figure 7.7. While stretching the spring with your hand from zero stretch $(x = 0)$ to some arbitrary stretch distance x, the magnitude of the force your hand exerts on the spring changes in a linear fashion from zero when unstretched to kx when stretched.

To calculate the work done on the spring by such a variable force, we can replace this variable force with the average force $(F_{\text{H on S}})_{\text{average}}$:

$$(F_{\text{H on S}})_{\text{average}} = \frac{0 + kx}{2}$$

The force your hand exerts on the spring is in the same direction as the direction in which the spring stretches. Thus the work done by this force on the spring to stretch it a distance x is

$$W = (F_{\text{H on S}})_{\text{average}}\,x = \left(\tfrac{1}{2}kx\right)x = \tfrac{1}{2}kx^2$$

This work equals the change in the spring's elastic potential energy (see **Figure 7.9** for the bar chart). Assuming that the elastic potential energy of the unstretched spring is zero, the work we calculated above equals the final elastic potential energy of the stretched spring.

FIGURE 7.9 Bar chart showing that external work increases the elastic potential energy of the spring (the system).

The external force pulling the spring does work on it.

$$U_{\text{si}} + W = U_{\text{sf}}$$

The initial state is the unstretched spring ($U_{\text{si}} = 0$).

The final state is the stretched spring with a stretch distance x and an elastic potential energy of U_{sf}.

Elastic potential energy The elastic potential energy of a spring-like object with a spring constant k that has been stretched or compressed a distance x from its undisturbed position is

$$U_{\text{s}} = \tfrac{1}{2}kx^2 \qquad (7.7)$$

Just like any other type of energy, the unit of elastic potential energy is the joule (J).

EXAMPLE 7.4   **Shooting an arrow**

You load an arrow (mass = 0.090 kg) into a bow and pull the bowstring back 0.40 m. The bow has a spring constant $k = 900 \text{ N/m}$. Determine the arrow's speed as it leaves the bow.

Sketch and translate We sketch the process, as shown below. The system is the bow and arrow. In the initial state, the bowstring is pulled back 0.40 m. In the final state, the string has just relaxed and the arrow has left the string.

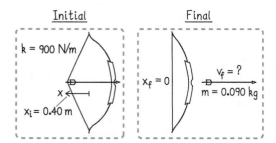

Simplify and diagram Since the arrow moves horizontally, we do not need to keep track of gravitational potential energy. The initial elastic potential energy of the bow is converted into the final kinetic energy of the arrow. We represent the process with the bar chart.

$$K_i + U_{gi} + U_{si} + W = K_f + U_{gf} + U_{sf} + \Delta U_{int}$$

Represent mathematically Use the bar chart to help apply the work-energy equation:

$$E_i + W = E_f$$
$$\Rightarrow K_i + U_{si} + 0 = K_f + U_{sf}$$
$$\Rightarrow 0 + \tfrac{1}{2}kx^2 = \tfrac{1}{2}mv^2 + 0$$

Multiply both sides of the above by 2, divide by m, and take the square root to get

$$v = \sqrt{\frac{k}{m}}x$$

Solve and evaluate

$$v = \sqrt{\frac{k}{m}}x = \sqrt{\frac{900 \text{ N/m}}{0.090 \text{ kg}}}(0.40 \text{ m}) = 40 \text{ m/s}$$

This is reasonable for the speed of an arrow fired from a bow. The units of $\sqrt{\dfrac{k}{m}}x$ are equivalent to $\dfrac{\text{m}}{\text{s}}$:

$$\sqrt{\frac{\text{N}}{\text{m} \cdot \text{kg}}}(\text{m}) = \sqrt{\frac{\text{kg} \cdot \text{m}}{\text{s}^2 \cdot \text{m} \cdot \text{kg}}}(\text{m}) = \frac{\text{m}}{\text{s}}$$

Try it yourself If the same arrow were shot vertically, how high would it go?

Answer   82 m.

REVIEW QUESTION 7.4 If the magnitude of the force exerted by a spring on an object is kx, why is it that the work done to stretch the spring a distance x is not equal to $kx \cdot x = kx^2$?

7.5 Friction and energy conversion

In nearly every mechanical process, objects exert friction forces on each other. Sometimes the effect of friction is negligible (for example, in an air hockey game), but most often friction is important (for example, when a driver applies the brakes to avoid a collision). Our next goal is to investigate how we can incorporate friction into work and energy concepts. Let's analyze a car skidding to avoid an accident.

Can a friction force do work?

Imagine that you pull a rope attached to a box (the system) so that the box moves at constant velocity on a rough carpet (**Figure 7.10a**). A force diagram for the box is shown in Figure 7.10b.

The bar chart (Figure 7.10c) shows that the force exerted by the rope does positive work on the box system; the friction force does negative work. These horizontal forces have exactly the same magnitude (the box is moving at constant velocity). Thus, the sum of the work done by those forces on the system is zero and the energy of the system should not change.

$$T_{\text{R on B}}d \cos 0° + f_{\text{k C on B}}d \cos 180° = T_{\text{R on B}}d + (-f_{\text{k C on B}}d) = 0$$

However, if you touch the bottom of the box at its final position, you find that it is warmer than before you started pulling and the box has scratches on its bottom. Again, the internal energy of the box increased. We get $0 = \Delta U_{\text{int}}$, where ΔU_{int} is greater than zero. The zero work done on the box does *not* equal the positive increase in internal energy. This is a contradiction of the work-energy principle. How can we resolve this?

The effect of friction as a change in internal energy

The key to resolving this problem is to change the system. The new system will include both surfaces that are in contact, for example, the box and the carpet. The friction force is then an internal force and therefore does no work on the system. But there is a change in the internal energy of the system caused by friction between the two surfaces.

When the rope is pulling the box across the rough carpet at constant velocity, we know that if the rope pulls horizontally on the box, the magnitude of the force that it exerts on the box $T_{\text{R on B}}$ must equal the magnitude of the friction force that the carpet exerts on the box $f_{\text{k C on B}}$. But now the box and carpet are both in the system—so the force exerted by the rope is the only external force. Thus the work done on the box-carpet system is

$$W = T_{\text{R on B}}d \cos 0°$$

Substituting $T_{\text{R on B}} = f_{\text{k C on B}}$ and $\cos 0° = 1.0$ into the above, we get

$$W = +f_{\text{k C on B}}d$$

The only system energy change is its internal energy ΔU_{int}. The work-energy equation for pulling the box across the surface is

$$W = \Delta U_{\text{int}}$$

After inserting the expression for the work done on the box and rearranging, we get

$$\Delta U_{\text{int}} = +f_{\text{k}}d$$

We have constructed an expression for the change in internal energy of a system caused by the friction force that the two contacting surfaces in the system exert on each other when one object moves a distance d across the other.

Increase in the system's internal energy due to friction

$$\Delta U_{\text{int}} = +f_{\text{k}}d \qquad (7.8)$$

where f_{k} is the magnitude of the average friction force exerted by the surface on the object moving relative to the surface, and d is the distance that the object moves across that surface. The increase in internal energy is shared between the moving object and the surface.

Including friction in the work-energy equation as an increase in the system's internal energy produces the same result as calculating the work done by friction force. In this new approach, there is an increase in internal energy $\Delta U_{\text{int}} = +f_{\text{k}}d$ in a system that includes the two surfaces rubbing against each other (this expression goes on the right side of the work-energy equation). When we include the two surfaces in the system, we can see why the moving box gets warmer and why there might be structural changes.

FIGURE 7.10 Pulling a box across a rough carpet.

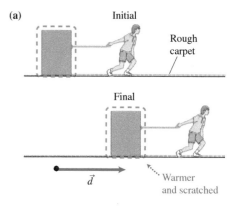

(a) Initial / Final

Rough carpet

$\vec{d}$ Warmer and scratched

(b) Because the velocity is constant, T and f_{k} have the same magnitudes.

$\vec{N}_{\text{C on B}}$

$\vec{f}_{\text{k C on B}}$ $\vec{T}_{\text{R on B}}$

$\vec{F}_{\text{E on B}}$

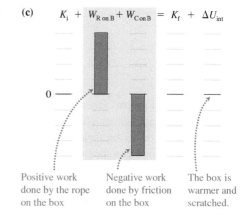

(c) $K_{\text{i}} + W_{\text{R on B}} + W_{\text{C on B}} = K_{\text{f}} + \Delta U_{\text{int}}$

0

Positive work done by the rope on the box | Negative work done by friction on the box | The box is warmer and scratched.

If we include only the box in the system, tracking the interactions between the system and the environment becomes very complicated because mass and thermal energy are transferred between the system and the environment. Therefore, in this book we prefer to include both surfaces in the system and consider the increase in internal energy caused by friction.

EXAMPLE 7.5   **Skidding to a stop**

You are driving your car when another car crosses the road at an intersection in front of you. To avoid a collision, you apply the brakes, leaving 24-m skid marks on the road while stopping. A police officer observes the near collision and gives you a speeding ticket, claiming that you were exceeding the 35 mi/h speed limit. She estimates your car's mass as 1390 kg and the coefficient of kinetic friction μ_k between your tires and this particular road as about 0.70. Do you deserve the speeding ticket?

Sketch and translate We sketch the process. We choose your car and the road surface as the system. We need to decide if your car was traveling faster than 35 mi/h at the instant you applied the brakes.

Simplify and diagram Assume that the process occurs on a horizontal level road and neglect interactions with the air. The initial state is just before the brakes are applied. The final state is just after your car has come to rest. We draw the energy bar chart to represent the process. In the initial state, the system has kinetic energy. In the final state, the system has no kinetic energy and has increased internal energy due to friction.

$$K_i + U_{gi} + U_{si} + \boxed{W} = K_f + U_{gf} + U_{sf} + \Delta U_{int}$$

Represent mathematically Convert the bar chart into an equation:

$$K_i + 0 = U_{int\,f}$$

$$\Rightarrow \tfrac{1}{2}m_C v_i^2 = f_{k\,R\,on\,C}\,d = (\mu_k N_{R\,on\,C})d$$

The magnitude of the upward normal force $N_{R\,on\,C}$ that the road exerts on the car equals the magnitude of the downward gravitational force $F_{E\,on\,C} = m_C g$ that Earth exerts on the car: $N_{R\,on\,C} = m_C g$. Thus the above becomes

$$\tfrac{1}{2}m_C v_i^2 = \mu_k m_C g d$$

Solve and evaluate Rearranging the above and canceling the car mass, we get

$$v_i = \sqrt{2\mu_k g d} = \sqrt{2(0.70)(9.8\text{ N/kg})(24\text{ m})}$$

$$= (18.1\text{ m/s})\left(\frac{1\text{ mi/h}}{0.447\text{ m/s}}\right) = 41\text{ mi/h}$$

It looks like you deserve the ticket. We ignored the resistive drag force that air exerts on the car, which could be significant for a car traveling at 41 mi/h. Thus, air resistance helps the car's speed decrease, which means you actually were traveling faster than 41 mi/h.

Try it yourself Imagine the same situation as above, only you are driving a 2000-kg minivan. Will the answer for the initial speed increase or decrease?

Answer   The speed does not change.

REVIEW QUESTION 7.5 Why, when friction cannot be neglected, is it useful to include both surfaces in the system when analyzing processes using the energy approach?

7.6 Skills for analyzing processes using the work-energy principle

The general strategy for solving problems with the work-energy principle is described on the left side of the table in Example 7.6 and illustrated on the right side for a specific process.

PROBLEM-SOLVING
STRATEGY 7.1   Applying the work-energy principle

EXAMPLE 7.6 An elevator slows to a stop

A 1000-kg elevator is moving downward. While moving down at 4.0 m/s, its speed decreases steadily until it stops in 6.0 m. Determine the magnitude of the tension force that the cable exerts on the elevator ($T_{C \text{ on } El}$) while it is stopping.

Sketch and translate

- Sketch the initial and final states of the process, labeling known and unknown information.
- Include the object of reference and the coordinate system.
- Choose the system of interest. Sometimes you might need to go back and redefine your system after you draw a bar chart if you find that the chosen system is not convenient.

- The elevator and Earth are in the system. We exclude the cable—its effect will be included as the work done by the cable on the elevator.

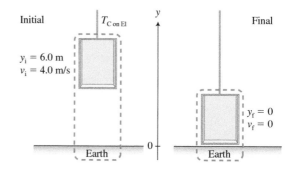

- The observer is on the ground; the coordinate system has a vertical axis pointing up with the zero at the bottom of the shaft.

Simplify and diagram

- What simplifications can you make to the objects, interactions, and processes?
- Decide which energy types are changing.
- Are external objects doing work?
- Use the initial-final sketch to help draw a work-energy bar chart. Include work bars (if needed) and initial and final energy bars for the types of energy that are changing. Specify the zero level of gravitational potential energy.

- We assume that the cable exerts a constant force on the elevator and that the elevator can be considered a point-like object since all of its points move the same way (no deformation).

- We will keep track of kinetic energy (since the elevator's speed changes) and gravitational potential energy (since the elevator's vertical position changes).

- The tension force exerted by the cable on the elevator does negative work (the tension force points up, and the displacement of the elevator points down).

- The zero gravitational potential energy is at the lowest position of the elevator. In its initial state the system has kinetic energy and gravitational potential energy. In its final state the system has no energy.

$$K_i + U_{gi} + W_{C \text{ on } El} = K_f + U_{gf}$$

Represent mathematically

- Convert the bar chart into a mathematical description of the process. Each bar in the chart will appear as a single term in the equation.

$$E_i + W = E_f$$

$$\tfrac{1}{2}mv_i^2 + mgy_i + T_{C \text{ on } El}(y_i - 0)\cos 180° = 0$$

$$\tfrac{1}{2}mv_i^2 + mgy_i - T_{C \text{ on } El}\, y_i = 0$$

Solve and evaluate

- Solve for the unknown and evaluate the result.

- Does it have the correct units? Is its magnitude reasonable? Do the limiting cases make sense?

$$T_{C\,on\,El} = mg + \frac{mv_i^2}{2y_i}$$

$$= (1000\text{ kg})(9.8\text{ N/kg}) + \frac{(1000\text{ kg})(4.0\text{ m/s})^2}{2(6.0\text{ m})}$$

$$= 11{,}000\text{ N}$$

- The result has the correct units. The force that the cable exerts is more than the 9800-N force that Earth exerts. This is reasonable since the elevator slows down while moving down; thus the sum of the forces exerted on it should point up.

- Limiting case: If the elevator slowed down over a much longer distance ($y_f =$ very large number instead of 6.0 m), then the force would be closer to 9800 N.

Try it yourself Solve the same problem using Newton's second law and kinematics.

Answer   11,000 N.

The "Try it yourself" question above demonstrates that we can obtain the same result for Example 7.6 using Newton's second law and kinematics. However, the energy approach is often easier and quicker.

We have used the work-energy approach to examine several real-world phenomena. Let's use it to examine how the body pumps blood.

BIO CONCEPTUAL EXERCISE 7.7   ▸ **Stretching the aorta**

Every time your heart beats, the left ventricle pumps about 80 cm³ of blood into the aorta, the largest artery in the human body. This pumping action occurs during a very short time interval, about 0.13 s. The elastic aorta walls stretch to accommodate the extra volume of blood. During the next 0.4 s or so, the walls of the aorta contract, applying pressure on the blood and moving it out of the aorta into the rest of the circulatory system. Represent this process with a qualitative work-energy bar chart.

Sketch and translate We sketch the process, as shown below. Choose the system to be the aorta and the 80 cm³ of blood that is being pumped. The left ventricle is not in the system. Choose the initial state to be just before the left ventricle contracts. Choose the final state to be when the blood is in the stretched aorta before moving out into the rest of the circulatory system.

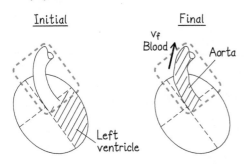

Simplify and diagram We keep track of the kinetic energy (the blood speed changes) and the elastic potential energy (the aorta wall stretches). We ignore the slight increase in the vertical elevation of the blood. The left ventricle is not in the system and does positive work pushing the blood upward in the direction of the blood's displacement. The work-energy bar chart below left represents this process for an aorta with flexible walls.

Flexible-walled aorta

$$W = K_f + U_{sf}$$

Try it yourself Modify the work-energy bar chart for a person with hardened and thickened artery walls.

Answer

$$W = K_f + U_{sf}$$

Stiff-walled aorta

Heart does more work.

If the person has stiff, thick arteries (a condition called atherosclerosis), more energy than normal is required to stretch the aorta walls. This process is represented by the bar chart to the right. In this case, the blood pressure is higher than it would be for a healthy cardiovascular system because the heart has to do more work to stretch the walls while pushing blood into the aorta.

In Conceptual Exercise 7.2 we discussed the physics of pole vaulting qualitatively; here we will solve a quantitative problem to figure out the contributions of different types of energy to the height of the jump. Make sure you review and continually consult Exercise 7.2, as both the sketch and the energy bar charts in that exercise are the same as in this problem.

EXAMPLE 7.8   **Pole vaulting, quantitatively**

65-kg, 1.74-m-tall Yelena Isinbayeva from Russia holds the women's pole-vaulting record of 5.06 m. Estimate the amount of internal energy that she converts into mechanical energy during the vault if before take-off she is running at a speed of 8.2 m/s, assuming that the change in the internal energy of the pole is negligible.

Sketch and translate See the sketch of the situation in Exercise 7.2. We choose the initial state as Yelena is running just before she inserts the pole into the vaulting box and the final state when she just clears the bar. The givens are $m = 65$ kg, $H_Y = 1.74$ m, $h_f = 5.06$ m, and $v_i = 8.2$ m/s. Yelena, Earth, and the pole are the system.

Simplify and diagram Again, we assume that Yelena can be modeled as a point-like object with her mass concentrated approximately in the middle of her body, which is moving at 8.2 m/s about $h_i = \frac{1}{2}H_Y \approx 0.9$ m above the ground. We also assume that she needs to be moving at a small but nonzero horizontal speed when she clears the bar, about 0.1 m/s ($v_f = 0.1$ m/s). In the final state, there should be some internal energy decrease because she converts this energy into mechanical energy when jumping as she starts her vertical ascent and at the top of the flight when she is crunching her muscles to give a specific body shape as she passes over the bar. We also assume that no other external forces are doing work on the system.

Represent mathematically We use the bar chart (see Exercise 7.2) to write the work-energy equation:

$$K_i + U_{gi} = K_f + U_{gf} + \Delta U_{int}$$
$$\Rightarrow \Delta U_{int} = K_i + U_{gi} - K_f - U_{gf} = (K_i - K_f) + (U_{gi} - U_{gf})$$
$$= \frac{m}{2}(v_i^2 - v_f^2) - mg(h_f - h_i)$$

Solve and evaluate We can now substitute the givens into the final equation:

$$\Delta U_{int} = \frac{(65.0\text{ kg})}{2}\left[(8.2\text{ m/s})^2 - (0.1\text{ m/s})^2\right]$$
$$- (65.0\text{ kg}) \times (9.8\text{ m/s}^2)(5.06\text{ m} - 0.90\text{ m})$$
$$= 2184.98\text{ kg} \cdot \text{m}^2/\text{s}^2 - 2649.92\text{ kg} \cdot \text{m}^2/\text{s}^2$$
$$= -464.95\text{ J} \approx -465\text{ J}$$

The negative sign shows that the internal energy *decreases*—in agreement with the bar chart. We could now adjust the length of the bars: The bar for ΔU_{int} should be about 20% of the height of the bar for K_i. The internal energy decrease is significant but reasonable—it is still a small part of the kinetic energy of the pole vaulter. This is the energy that Yelena converted to the mechanical energy to reach the desired height. However, she probably converted some more of her internal energy into other forms of energy because we ignored the increase in the internal energy of the pole as well as the air.

Try it yourself The men's record for pole vaulting is 6.16 m. Estimate how much internal energy the vaulter converts to mechanical energy assuming his mass is 69 kg, his height is 1.77 m, and he runs 100 m in 11.4 s (8.77 m/s). Specify all assumptions you made.

Answer

About 930 J, assuming that all the vaulter's mass is concentrated at a point at the middle of the vaulter's height, and that the vaulter is not moving at the top of the flight.

Pole vaulting may not be part of your daily life, but there are plenty of everyday processes that we can analyze using a work-energy approach, for example, walking up the stairs. Let's choose the system to be you and Earth. The initial state is you on the first floor and the final state is you on the 10th floor. The gravitational potential energy of the system definitely increases. In addition, on the 10th floor you are sweating and the soles of your shoes and the surface of the stairs warm up a little due to the repeated deformation of the materials; thus the thermal energy (part of the internal energy) also increases. We have two increases; where does the energy for these increases come from? You might think that the stairs do work on you. However, it is not clear what the displacement is when you are in contact with each stair—different parts of your body move differently, so it is not clear what work the staircase does on you. Therefore, it is best to redefine the system to include the entire staircase.

The increases in gravitational potential and thermal energy of the you-Earth-staircase system came from the chemical energy of your muscles. The bar chart in

Figure 7.11a shows the average energy conversion across several stairs. We show the bar for the initial U_{gi} and the taller bar for the final U_{gf}, and we split the bar for internal energy change ΔU_{int} into two bars: chemical energy ΔU_{ch} and thermal energy ΔU_{th}. The thermal energy increases, but the chemical energy decreases. This is why you need to eat lunch to get back the expended energy! Note that if we did not include Earth in the system, the system would not have any gravitational potential energy, but Earth would do negative work. Therefore, some energy will need to increase to compensate for this negative work (see Figure 7.11b). In this situation, we see that including or excluding Earth from the system does not make any difference to the analysis, but including the staircase in the system simplifies the analysis.

FIGURE 7.11 Work-energy bar charts describing the process of walking up the stairs for different systems. (a) You, Earth and the staircase are in the system. (b) Earth is not in the system.

(a) $K_i \ + \ U_{gi} \ + \ W \ = \ K_f \ + \ U_{gf} \ + \Delta U_{ch} + \Delta U_{th}$

(b) $K_i \ + \ U_{gi} \ + \ W \ = \ K_f \ + \ U_{gf} \ + \Delta U_{ch} + \Delta U_{th}$

FIGURE 7.12 Force diagram and impulse-momentum bar chart for you pushing upward off a stair.

(a)

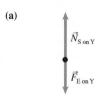

(b) $p_{iy} \ + \ J_y \ = \ p_{fy}$

The work-energy approach describes the process, but does not explain it. During every step, you start at rest and then accelerate upward. If you are the system, then what are the objects responsible for your acceleration? You interact with Earth and the stair. The force exerted by the stair on you ($\vec{N}_{S\ on\ Y}$) should be larger than the force exerted by Earth ($\vec{F}_{E\ on\ Y}$) for you to accelerate upward (see the force diagram in Figure 7.12a).

Although the sum of the forces exerted by the stair and Earth on you does zero work (as there is no displacement of your feet while they are in contact with the floor), the sum of these two forces exerts a nonzero impulse on you due to the nonzero time interval of contact. Figure 7.12b shows the impulse-momentum bar chart for the process of each step. The initial state is right before you start your step, and the final state is when you are moving upward, but before you reach the next stair.

CONCEPTUAL EXERCISE 7.9   **Dropping a steel ball**

You drop a steel ball on the floor and take a picture of the floor with a thermal camera. The camera shows an increase in the temperature of the floor of about 4 °F; the ball gets a little hotter, too. Choose an appropriate system to analyze the energy conversions for the process that starts when the ball is about to hit the floor and ends when the ball is at rest on the floor.

Sketch and translate The initial and final states are provided in the givens. We show on the sketch a ball that is moving downward at speed v_i and then at rest (speed $v_f = 0$) on the floor. We also show the increased temperatures of the floor and the ball $\Delta T_{floor} > 0$; $\Delta T_{ball} > 0$. Because the floor did not move during the collision, it could not do any work on the ball. Therefore, we will include it in the system. Earth is

irrelevant here because the distance between the ball and the floor does not change between the initial and final states.

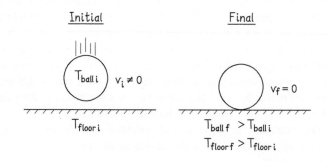

Simplify and diagram The bar chart for the process is shown below. The left side shows the initial kinetic energy; the right side shows the change in internal energy. We assume that all this change is due to the temperature change of the ball and the floor.

$$K_i + W = K_f + \Delta U_{int}$$

Try it yourself Imagine that the ball lands on a trampoline instead of the floor. Analyze the process using the ball and Earth as the system. The initial state is when the ball touches the trampoline, and the final state is when the trampoline is stretched downward and the ball stops momentarily.

Answer

After the ball first touches the trampoline, the trampoline stretches downward and exerts an increasing upward force on the ball in the direction opposite to the ball's downward motion. The trampoline therefore does negative work on the ball. This negative work decreases the kinetic energy of the ball. There is also a small decrease in the gravitational potential energy as the ball moves down with the trampoline.

REVIEW QUESTION 7.6 What would change in the solution to the problem in Example 7.8 if we did not include Earth in the system? Would the answer be different? Explain.

7.7 Collisions

A collision is a process that occurs when two (or more) objects are in direct contact with each other for a short time interval, such as when a baseball is hit by a bat (**Figure 7.13**). The ball compresses during the first half of the collision, then decompresses during the second half of the collision. We have already used impulse and momentum principles to analyze collisions (in Chapter 6). We learned that the forces that the two colliding objects exert on each other during the collision are complicated, nonconstant, and exerted for a very brief time interval—roughly 1 ms in the case of the baseball-bat collision. Can we learn anything new by analyzing collisions using the work-energy principle?

FIGURE 7.13 A baseball is compressed while being hit by a bat.

Analyzing collisions using momentum and energy principles

Observational Experiment **Table 7.7**, on the next page, shows three different observational experiments involving collisions. In each case a 1.0-kg object attached to the end of a string (the bob of a pendulum) swings down and hits a 4.0-kg wheeled cart at the lowest point of its swing (**Figure 7.14**). In each experiment, the pendulum bob and cart start at the same initial positions, but the compositions of the pendulum bob and cart are varied. After each collision, the cart moves at a nearly constant speed because of the smoothness of the surface on which it rolls.

FIGURE 7.14 A pendulum bob hits a cart, causing it to move forward.

OBSERVATIONAL
EXPERIMENT TABLE **7.7**   Analyzing momentum and energy during collisions

VIDEO
OET 7.7

Observational experiment	Analysis
Experiment 1. A 1.0-kg metal bob swings and hits a 4.0-kg metal cart. Their velocity components just before and just after the collision are shown below: $m_1 = 1.0$ kg; $v_{1ix} = 10$ m/s $v_{1fx} = -6.0$ m/s $m_2 = 4.0$ kg; $v_{2ix} = 0$ m/s $v_{2fx} = 4.0$ m/s	**Momentum:** Before collision: $(1.0 \text{ kg})(+10 \text{ m/s}) + (4.0 \text{ kg}) 0 = +10 \text{ kg} \cdot \text{m/s}$ After collision: $(1.0 \text{ kg})(-6.0 \text{ m/s}) + (4.0 \text{ kg})(+4.0 \text{ m/s}) = +10 \text{ kg} \cdot \text{m/s}$ **Kinetic energy:** Before collision: $(1/2)(1.0 \text{ kg})(+10 \text{ m/s})^2 + (1/2)(4.0 \text{ kg}) 0^2 = 50 \text{ J}$ After collision: $(1/2)(1.0 \text{ kg})(-6.0 \text{ m/s})^2 + (1/2)(4.0 \text{ kg})(+4.0 \text{ m/s})^2 = 50 \text{ J}$
Experiment 2. A 1.0-kg sand-filled balloon swings and hits a 4.0-kg flimsy cardboard cart. After the collision, the damaged cart moves across the table at constant speed. The side of the balloon that hit the cart is flattened in the collision. Their velocity components just before and just after the collision are shown below: $m_1 = 1.0$ kg; $v_{1ix} = 10$ m/s $v_{1fx} = -4.2$ m/s $m_2 = 4.0$ kg; $v_{2ix} = 0$ m/s $v_{2fx} = 3.55$ m/s	**Momentum:** Before collision: $(1.0 \text{ kg})(+10 \text{ m/s}) + (4.0 \text{ kg}) 0 = +10 \text{ kg} \cdot \text{m/s}$ After collision: $(1.0 \text{ kg})(-4.2 \text{ m/s}) + (4.0 \text{ kg})(+3.55 \text{ m/s}) = +10 \text{ kg} \cdot \text{m/s}$ **Kinetic energy:** Before collision: $(1/2)(1.0 \text{ kg})(+10 \text{ m/s})^2 + (1/2)(4.0 \text{ kg}) 0^2 = 50 \text{ J}$ After collision: $(1/2)(1.0 \text{ kg})(-4.2 \text{ m/s})^2 + (1/2)(4.0 \text{ kg})(+3.55 \text{ m/s})^2 = 34 \text{ J}$
Experiment 3. A 1.0-kg sand-filled balloon covered with Velcro swings down and sticks to a 4.0-kg Velcro-covered cardboard cart. The string holding the balloon is cut by a razor blade immediately after the balloon contacts the cart. The damaged cart and flattened balloon move off together across the table. Their velocity components just before and just after the collision are shown below: $m_1 = 1.0$ kg; $v_{1ix} = 10$ m/s $v_{1fx} = 2.0$ m/s $m_2 = 4.0$ kg; $v_{2ix} = 0$ m/s $v_{2fx} = 2.0$ m/s	**Momentum:** Before collision: $(1.0 \text{ kg})(+10 \text{ m/s}) + (4.0 \text{ kg}) 0 = +10 \text{ kg} \cdot \text{m/s}$ After collision: $(1.0 \text{ kg})(+2.0 \text{ m/s}) + (4.0 \text{ kg})(+2.0 \text{ m/s}) = +10 \text{ kg} \cdot \text{m/s}$ **Kinetic energy:** Before collision: $(1/2)(1.0 \text{ kg})(+10 \text{ m/s})^2 + (1/2)(4.0 \text{ kg}) 0^2 = 50 \text{ J}$ After collision: $(1/2)(1.0 \text{ kg})(+2.0 \text{ m/s})^2 + (1/2)(4.0 \text{ kg})(+2.0 \text{ m/s})^2 = 10 \text{ J}$

Patterns

Two important patterns emerge from the data collected from these different collisions.

● The momentum of the system is constant in all three experiments.
● The kinetic energy of the system is constant when no damage is done to the objects during the collision (Experiment 1 but not in Experiments 2 and 3).

We can understand the first pattern in Table 7.7 using our knowledge of impulse-momentum. The x-component of the net force exerted on the system in all three cases was zero; hence the x-component of momentum should be constant.

What about the second pattern? In Experiment 1, the objects in the system were very rigid, but in Experiments 2 and 3, they were more fragile and as a result were deformed during the collision. Using the data, we can determine the amount of kinetic energy that was converted to internal energy. But it is impossible to predict this amount ahead of time. Unfortunately, this means that in collisions where any deformation of the objects occurs, we cannot make predictions about the amount of kinetic energy converted to internal energy. However, we now know that even in collisions where the objects become damaged, the momentum of the system still remains constant. So even though the work-energy equation is less useful in these types of collisions, the impulse-momentum equation is still very useful.

Types of collisions

The experiments in Table 7.7 are examples of the three general collision categories: elastic collisions (Experiment 1), inelastic collisions (Experiment 2), and totally inelastic collisions (Experiment 3). **Table 7.8** summarizes these three types of collisions.

TABLE 7.8 Types of collisions

Elastic collisions	Inelastic collisions	Totally inelastic collisions
Both the momentum and kinetic energy of the system are constant. The internal energy of the system does not change. The colliding objects never stick together. Examples: There are no perfectly elastic collisions in nature, although collisions between very rigid objects (such as billiard balls) come close. Collisions between atoms or subatomic particles are almost exactly elastic.	The momentum of the system is constant but the kinetic energy is not. The colliding objects do not stick together. Internal energy increases during the collisions. Examples: A volleyball bouncing off your arms, or you jumping on a trampoline.	These are inelastic collisions in which the colliding objects stick together. Typically, a large fraction of the kinetic energy of the system is converted into internal energy in this type of collision. Examples: You catching a football, or a car collision where the cars stick together.

EXAMPLE 7.10   **Spiderman saves Mary Jane**

Spiderman is climbing up a building on one side of a 10-m-wide street when he spots Mary Jane 10 m below, standing on a platform on the other side of the street. The platform is about to explode. Spiderman swings down using his web and grabs Mary Jane, hoping to swing up with her to an open terrace, 5 m above the platform on the same building—where they both would be safe. Will they make it?

Sketch and translate Let's break the process into three parts: (I) Spiderman's swing from the first building down to Mary Jane (before they join together); (II) their brief impact, an inelastic collision (because they are traveling together after the impact); and (III) their swing together until they stop on the terrace. The givens are in the sketch below. We assume that the mass of Spiderman is the same as that of Mary Jane.

Simplify and diagram Consider Spiderman and Mary Jane to be point-like objects. For Part I, choose Spiderman, the web, and Earth as the system. During Spiderman's downward swing, the gravitational potential energy of the system is converted to his kinetic energy just before colliding with Mary Jane (see the work-energy bar chart, top right). We can use this to determine his speed just before impact. This way, we do not need to worry about the impulse exerted by the gravitational force on Spiderman to find his momentum at the bottom of the swing. The direction of the force exerted by Earth on Spiderman continually

changes relative to the direction of his motion as he swings down, making calculations of impulse very difficult.

$$K_i + U_{gi} + W = K_f + U_{gf}$$

For Part II, the inelastic collision of Spiderman with Mary Jane, there will probably be an increase in internal energy. We can't use an energy analysis to decide how much—we don't have a way to independently determine internal energy change. Thus, we use only momentum conservation and impulse-momentum bar charts to determine Mary Jane and Spiderman's speed immediately after the collision (see the impulse-momentum bar chart below—the system is Spiderman and Mary Jane, who we have assumed have equal masses).

$$p_{SMix} + p_{MJix} + J_x = p_{SMfx} + p_{MJfx}$$

For Part III, knowing their speed after the collision, we can now use energy conservation to determine the maximum height they reach after the collision (see the work-energy bar chart below). Again, momentum conservation is very difficult because it involves the impulse caused by the gravitational force, which continually changes direction relative to the changing direction of motion as they swing upward.

$$K_i + U_{gi} + W = K_f + U_{gf}$$

Represent mathematically Use the energy bar charts from Part I to find that

$$mgy_i = \tfrac{1}{2}mv_{SM}^2$$

where $y_i = 10$ m is Spiderman's original height, and v_{SM} is his unknown speed just before grabbing Mary Jane. Dividing both sides of the equation by Spiderman's (nonzero) mass and rearranging, we get

$$v_{SM} = \sqrt{2gy_i}$$

We now use the outcome of the Part II momentum conservation collision process to determine their joint speed $v_{SM\text{-}MJ}$ immediately after the collision:

$$mv_{SM} = (m + m)v_{SM\text{-}MJ} \quad \text{or} \quad v_{SM\text{-}MJ} = \tfrac{1}{2}v_{SM}$$

For Part III, the height to which they rise after the collision is determined using energy conservation:

$$\tfrac{1}{2}(m + m)v_{SM\text{-}MJ}^2 = (m + m)gy_f \quad \text{or} \quad y_f = \frac{v_{SM\text{-}MJ}^2}{2g}$$

Solve and evaluate We can work backward to find the maximum height they will reach:

$$y_f = \frac{v_{SM\text{-}MJ}^2}{2g} = \frac{[(1/2)v_{SM}]^2}{2g} = \frac{v_{SM}^2}{4 \times 2g} = \frac{y_i}{4}$$

Therefore, they will rise to only a quarter of Jane's original height, or $(10 \text{ m})/4 = 2.5$ m—too low to reach the terrace that is 5 m up on the other side of the street. The units are correct. Why is the answer so small? It turns out that half the initial gravitational potential energy is converted to internal energy during the collision.

Try it yourself Suppose Mary Jane's mass is half Spiderman's mass. Would there be more, less, or the same mechanical energy converted into internal energy during the collision?

Answer

Using the same process as above except with Mary Jane's mass $m/2$ instead of m, we find that the final height of Spiderman and Mary Jane is $4/9$ of the initial height, or about 4.4 m. Therefore, there is less mechanical energy converted into internal energy during this collision.

REVIEW QUESTION 7.7 Imagine that a collision occurs. You measure the masses of the two objects before the collision and measure the velocities of the objects both before and after the collision. Describe how you could use these data to determine which type of collision had occurred.

7.8 Power

FIGURE 7.15 Walking or running up stairs.

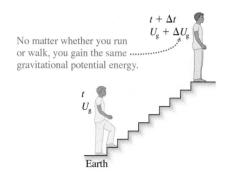

No matter whether you run or walk, you gain the same gravitational potential energy.

$t + \Delta t$
$U_g + \Delta U_g$

t
U_g

Earth

Why is it harder for a person to run up a flight of stairs than to walk up the stairs if the change in gravitational potential energy of the system person-Earth is the same? The *amount* of internal energy converted into gravitational energy is the same in both cases (as shown in **Figure 7.15**), but the *rate* of that conversion is not. When you run upstairs, you convert the energy at a faster rate. The rate at which the conversion occurs is called the **power**.

Power The power of a process is the amount of some type of energy converted into a different type divided by the time interval Δt in which the process occurred:

$$\text{Power} = P = \left| \frac{\Delta E}{\Delta t} \right| \tag{7.9}$$

If the process involves external forces doing work, then power can also be defined as the magnitude of the work W done on or by the system divided by the time interval Δt needed for that work to be done:

$$\text{Power} = P = \left| \frac{W}{\Delta t} \right| \tag{7.10}$$

The SI unit of power is the watt (W). 1 watt is 1 joule/second $(1 \text{ W} = 1 \text{ J/s})$.

A cyclist in good shape pedaling at moderate speed will convert about 400–500 J of internal chemical energy each second (400–500 W) into kinetic, gravitational potential, and thermal energies.

Power is sometimes expressed in *horsepower* (hp): 1 hp = 746 W. Horsepower is most often used to describe the power rating of engines or other machines. A 50-hp gasoline engine (typical in cars) converts the internal energy of the fuel into other forms of energy at a rate of 50 × 746 W = 37,300 W, or 37,300 J/s.

EXAMPLE 7.11   Car power

Miguel and Sierra are investigating a new toy car. The 350-g car is powered by a spring-based engine that delivers approximately constant power. They measure the time dependence of the speed of the car using a motion detector; their data are shown in the table at right. Estimate the power of the spring-based engine during the motion represented in the table. Indicate any assumptions that you made.

t (s)	v (m/s)
0.00	0
0.25	2.4
0.50	3.4
0.75	4.1
1.00	4.8

Sketch and translate We choose the car as the system. The initial state is at time $t = 0$ when the car starts moving from rest, and the final state is the moment at some later time t when the car is moving with the speed v.

Simplify and diagram Inside the car's engine, the elastic potential energy of the spring is converted into the kinetic energy of the car. There are two external forces that are exerted on the car, but neither of them does any work on the car. The force exerted by Earth on the car is perpendicular to the displacement of the car, and the point where the ground exerts a force on the car has zero displacement with respect to the car (the part of the car that is in contact with the ground is the tire—if it rolls without slipping, the point of contact is stationary). Ignoring air resistance and any friction in bearings, the work-energy bar chart for the process is shown at right.

$$K_i + U_{si} + W = K_f + U_{sf}$$

Represent mathematically We can express the power of the conversion of elastic potential energy into kinetic energy using the definition of power [see Eq. (7.9)]:

$$P = \left| \frac{\Delta E}{\Delta t} \right| = \frac{\frac{1}{2}mv^2 - 0}{t - 0} = \frac{mv^2}{2t}$$

Solve and evaluate We know the mass of the car (0.350 kg), and we have five data points for t and the corresponding values of v. We want to estimate the unknown power P. Could we just pick one pair of data (t, v) from the table and calculate P using the equation above? We could, but we would discard valuable information that is hidden in the other data points that we did not use. Note that each measurement in the table contains random uncertainties, inherent to all measurements. The more data we use, the more precise our result will be. How can we use all data in the table?

First, we rearrange the previous equation to express v as a function of t, as suggested in the table. We obtain

$$v = \sqrt{\frac{2Pt}{m}}$$

This equation suggests that the speed of the car powered by a constant-power engine is directly proportional to the square root of time. Since we are much more familiar with linear functions and have the tools to analyze them, we square both sides of this equation to obtain a linear dependence on time t:

$$v^2 = \frac{2P}{m}t$$

Next, we draw a graph by plotting the t values on the horizontal axis and the v^2 values on the vertical axis. The best-fit line through the points has a slope equal to 22.9 m²/s³.

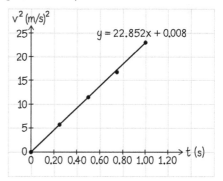

Since the slope of the line is given by $\frac{2P}{m}$, we can finally calculate the power:

$$P = \frac{(22.9 \text{ m}^2/\text{s}^3)(0.35 \text{ kg})}{2} = 4.0 \frac{\text{m} \cdot \text{kg} \cdot \text{m/s}^2}{\text{s}} = 4.0 \text{ W}$$

The units are correct. The qualitative dependence also makes sense: a heavier car with the same (t, v) data would need a more powerful engine. The actual power is probably larger than 4 W since we neglected air resistance and friction in the bearings.

Try it yourself A 1400-kg car is traveling on a level road at a constant speed of 27 m/s (60 mph). The drag force exerted by the air on the car and the rolling friction force exerted by the road on the car tires add to give a net force of 680 N pointing opposite the direction of motion of the car. Determine the rate of work done by the drag and rolling friction forces. Express your results in watts and in horsepower.

Answer $P = 1.8 \times 10^4 \text{ W} = 24 \text{ hp.}$

REVIEW QUESTION 7.8 Toyota says that the power of their Prius car is 121 hp. What does this number mean?

7.9 Improving our model of gravitational potential energy

So far, we have assumed that the gravitational force exerted by Earth on an object is constant $(F_{E\,on\,O} = mg)$. Using this equation we devised the expression for the gravitational potential energy of an object-Earth system as $U_g = mgy$ (with respect to a chosen zero level). This expression is only valid when an object is close to Earth's surface. We know from our study of gravitation (at the end of Chapter 5) that the gravitational force exerted by planetary objects on moons and satellites and by the Sun on the planets changes with the distance between the objects according to Newton's law of gravitation $(F_{1\,on\,2} = Gm_1m_2/r^2)$. How would the expression for the gravitational potential energy change when we take into account that the gravitational force varies with distance?

Gravitational potential energy for large separations between objects

Imagine that a "space elevator" has been built to transport supplies from the surface of Earth to the International Space Station (ISS). The elevator moves at constant velocity, except for the very brief acceleration and deceleration at the beginning and end of the trip. How much work must be done to lift the supplies from the surface to the ISS?

The initial state is the moment just as the supplies leave the surface. The final state is the moment just as they arrive at the ISS (Figure 7.16a). We choose Earth and the supplies as the system. The force that the elevator cable exerts on the supplies is an external force that does positive work on the system. We keep track of gravitational potential energy only; it is the only type of energy that changes between the initial and final states. Since the supplies are moving at constant velocity, the force exerted by the elevator cable on the supplies is equal in magnitude to the gravitational force exerted by Earth on the supplies (Figure 7.16b). Earth is the object of reference, and the origin of the coordinate system is at the center of Earth. The process can be described mathematically as follows:

$$U_{gi} + W = U_{gf} \implies W = U_{gf} - U_{gi}$$

At this point it would be tempting to say that the work done by the elevator cable on the system is $W = Fd\cos\theta$, where F is the constant force that the cable exerts on the supplies. However, as the supplies reach higher and higher altitudes, the force exerted by the elevator cable decreases—Earth exerts a weaker and weaker force on the supplies.

Determining the work done by a variable force requires a complex mathematical procedure. The outcome of this procedure is

$$W = \left(-G\frac{m_E m_S}{R_E + h_{ISS}}\right) - \left(-G\frac{m_E m_S}{R_E}\right)$$

where m_E is the mass of Earth, R_E is the radius of Earth, m_S is the mass of the supplies, h_{ISS} is the altitude of the ISS above Earth's surface, and $R_E + h_{ISS}$ is the distance of the ISS from the center of Earth. Before you move on, check whether this complicated equation makes sense (for example, check the units).

The work is written as the difference in two quantities; the latter describes the initial state and the former the final state. If we compare the above result with $W = U_{gf} - U_{gi}$, we see that each term is an expression for the gravitational potential energy of the Earth-object system for a particular object's distance from the center of Earth.

FIGURE 7.16 A cable lifts supplies to the International Space Station via a space elevator. (a) Determine the work required to lift the supplies. (b) The cable exerts the force in the direction of motion. In this force diagram, the supplies are the system.

> **Gravitational potential energy of a system consisting of Earth and any object**
>
> $$U_g = -G \frac{m_E m_O}{r_{E\text{-}O}} \qquad (7.11)$$
>
> where m_E is the mass of Earth (5.97×10^{24} kg), m_O is the mass of the object, $r_{E\text{-}O}$ is the distance from the center of Earth to the center of the object, and $G = 6.67 \times 10^{-11}$ N $\cdot$ m^2/kg^2 is Newton's gravitational constant.

We can use Eq. (7.11) to find the gravitational potential energy of the system of any two spherical or point-like objects if we know the masses of the objects and the distance between their centers.

Note that the cable did positive work on the system while pulling the supplies away from Earth. When the object (the supplies in this case) is infinitely far away, the gravitational potential energy is zero. The only way to add positive energy to a system and have it become zero is if it started with negative energy, for example, $-5 + 5 = 0$. Thus, for the case of zero energy at infinity, the gravitational potential energy is a negative number when the object is closer to Earth. We can represent the process of pulling an object from the surface of Earth to infinity using a work-energy bar chart (**Figure 7.17**). The initial state is when the object is near Earth, and the final state is when it is infinitely far away.

Now we can determine the amount of work needed to raise 1000 kg of supplies to the International Space Station, which is 3.50×10^5 m above Earth's surface.

$$W = \left(-G \frac{m_E m_S}{R_E + h_{ISS}}\right) - \left(-G \frac{m_E m_S}{R_E}\right)$$

$$= -G m_E m_S \left(\frac{1}{R_E + h_{ISS}} - \frac{1}{R_E}\right)$$

$$= -(6.67 \times 10^{-11}\, \text{N} \cdot \text{m}^2/\text{kg}^2)(5.97 \times 10^{24}\, \text{kg})(1000\, \text{kg})$$

$$\times \left(\frac{1}{6.37 \times 10^6\, \text{m} + 3.50 \times 10^5\, \text{m}} - \frac{1}{6.37 \times 10^6\, \text{m}}\right)$$

$$= 3.26 \times 10^9\, \text{J}$$

Let's compare this with what would have been calculated had we used our original expression for gravitational potential energy. Choose the zero level at the surface of Earth.

$$W = m_S g y_f - m_S g y_i = m_S g y_f - 0$$

$$= (1000\, \text{kg})(9.8\, \text{N/kg})(3.50 \times 10^5\, \text{m}) = 3.43 \times 10^9\, \text{J}$$

This differs by only about 5% from the more accurate result. Remember that $U_g = mgy$ is reasonable when the distance above the surface of Earth is a small fraction of the radius of Earth. The altitude of the ISS (350 km) is a small fraction of the radius of Earth (6371 km), so $U_g = mgy$ is still reasonably accurate.

Escape speed

The best Olympic high jumpers can leap over bars that are about 2.5 m (8 ft) above Earth's surface. Let's estimate a jumper's speed when leaving the ground in order to attain that height. We choose the jumper and Earth as the system and the zero level of gravitational potential energy at ground level. The kinetic energy of the jumper as he leaves the ground is converted into the gravitational potential energy of the system.

$$\tfrac{1}{2} m v^2 = mgy$$

$$\Rightarrow v = \sqrt{2gy} = \sqrt{2(9.8\, \text{N/kg})(2.5\, \text{m})} = 7.0\, \text{m/s}$$

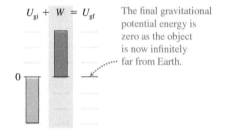

FIGURE 7.17 A bar chart representing the work needed to take an object from near Earth to infinitely far away. The system is the supplies and Earth.

$U_{gi} + W = U_{gf}$ The final gravitational potential energy is zero as the object is now infinitely far from Earth.

The gravitational coefficient of objects near the Moon's surface is $g_M = \dfrac{Gm_M}{R_M^2} = 1.6 \text{ N/kg}$. Using the equation above, we find the height of the jump to be 15.3 m. That's about 50 feet!

Is it possible to jump entirely off a celestial body—that is, jump up and never come down? What is the minimum speed you would need in order to do this? This minimum speed is called **escape speed**.

EXAMPLE 7.12   Escape speed

What vertical speed must a jumper have in order to leave the surface of a planet and never come back down?

Sketch and translate First, we draw a sketch of the process. The initial state is the instant after the jumper's feet leave the surface. The final state is when the jumper has traveled far enough away from the planet to no longer feel the effects of its gravity (at $r = \infty$). Choose the system to be the jumper and the planet.

Simplify and diagram We represent the process with the bar chart. In the initial state, the system has both kinetic and gravitational potential energy. In the final state, both the kinetic energy and the gravitational potential energy are zero.

Represent mathematically Using the generalized work-energy equation and the bar chart:

$$E_i + W = E_f$$
$$\Rightarrow K_i + U_{gi} + 0 = K_f + U_{gf}$$
$$\Rightarrow \tfrac{1}{2}m_J v^2 + \left(-G\dfrac{m_P m_J}{r_P}\right) + 0 = 0 + 0$$

where m_P is the mass of the planet, m_J is the mass of the jumper, r_P is the radius of the planet, and $G = 6.67 \times 10^{-11} \text{ N} \cdot \text{m}^2/\text{kg}^2$ is the gravitational constant.

Solve and evaluate Solving for the escape speed of the jumper,

$$v = \sqrt{\dfrac{2Gm_P}{r_P}} \qquad (7.12)$$

We can use the above equation to determine the escape speed for any celestial body. For example, the escape speed for the Moon is

$$v = \sqrt{\dfrac{2(6.67 \times 10^{-11} \text{ N} \cdot \text{m}^2/\text{kg}^2)(7.35 \times 10^{22} \text{ kg})}{1.74 \times 10^6 \text{ m}}}$$
$$= 2370 \text{ m/s}$$

The escape speed for Earth is

$$v = \sqrt{\dfrac{2Gm_E}{R_E}}$$
$$= \sqrt{\dfrac{2(6.67 \times 10^{-11} \text{ N} \cdot \text{m}^2/\text{kg}^2)(5.97 \times 10^{24} \text{ kg})}{6.37 \times 10^6 \text{ m}}}$$
$$= 11{,}200 \text{ m/s} = 11.2 \text{ km/s}$$

Try it yourself What is the escape speed of a particle near the surface of the Sun? The mass of the Sun is 2.0×10^{30} kg and its radius is 700,000 km.

Answer   620 km/s.

TIP Notice that the escape speed does not depend on the mass of the escaping object—a tiny speck of dust and a huge boulder would need the same initial speed to leave Earth. Why is that?

Black holes

Equation (7.12) for the escape speed suggests something amazing. If the mass of a star were large enough and/or its radius small enough, the escape speed could be made arbitrarily large. What if the star's escape speed were greater than light speed ($c = 3.00 \times 10^8 \text{ m/s}$)? What would this star look like? Light leaving the star's surface would not be moving fast enough to escape the star. The star would be completely dark.

Let's imagine what would happen if the Sun started shrinking so that its material were compressed into a smaller volume. How small would the Sun have to be for its escape speed to be greater than light speed? Use Eq. (7.12) with $v = c$ to answer this question.

$$v = c = \sqrt{\frac{2Gm_S}{r_S}} \tag{7.13}$$

or

$$r_S = \frac{2Gm_S}{c^2} = \frac{2(6.67 \times 10^{-11}\,\text{N}\cdot\text{m}^2/\text{kg}^2)(1.99 \times 10^{30}\,\text{kg})}{(3.00 \times 10^8\,\text{m/s})^2}$$

$$= 2.95 \times 10^3\,\text{m} \approx 3\,\text{km}$$

So, if the Sun collapsed to smaller than 3 km in radius, it would become invisible for the outside world. Could this happen? We will return to this question in later chapters.

Equation (7.13) was first constructed by the brilliant astronomer Pierre-Simon Laplace (1749–1827), who used classical mechanics to predict the presence of what he called "dark stars"—now more commonly called black holes.

We've been talking about objects whose escape speed is larger than light speed. Physicists once thought that since light has zero mass, the gravitational force exerted on it would always be zero. At the beginning of the 20th century, Albert Einstein's theory of general relativity improved greatly on Newton's law of universal gravitation. According to general relativity, light is affected by gravity. Amazingly, the theory predicts the same size for a black hole that is provided by the Newtonian theory.

REVIEW QUESTION 7.9 In this section you read that the gravitational potential energy of two large bodies (for example, the Sun and Earth) is negative. Why is this true?

Summary

Work (*W*) is a way to change the energy of a system. Work is done on a system when an external object exerts a force of magnitude *F* on an object in the system as it undergoes a displacement of magnitude *d*. The work depends on the angle θ between the directions of $\vec{F}$ and $\vec{d}$. It is a scalar quantity. (Section 7.1)

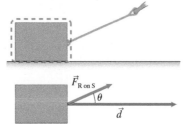

$$W = Fd\cos\theta \qquad \text{Eq. (7.1)}$$

Gravitational potential energy (*U*g) is the energy that a system has due to the relative separation of two objects with mass. It is a scalar quantity. *U*g depends on the gravitational interaction of the objects. A single object cannot have gravitational potential energy.

$$U_g = mgy \qquad \text{Eq. (7.4)}$$
(near Earth's surface; zero level is at the surface)
$$U_g = -G\frac{m_A m_B}{r_{AB}} \qquad \text{Eq. (7.11)}$$
(general expression; zero level is at infinity)

Kinetic energy (*K*) is the energy of an object of mass *m* moving at speed *v*. It is a scalar quantity.

$$K = \tfrac{1}{2}mv^2 \qquad \text{Eq. (7.5)}$$

Elastic potential energy (*U*s) is the energy of a stretched or compressed elastic object (e.g., coils of a spring or a stretched bow string).

$$U_s = \tfrac{1}{2}kx^2 \qquad \text{Eq. (7.7)}$$

Internal energy (*U*int) is the energy of motion and interaction of the microscopic particles making up the objects in the system. (Section 7.1)

$$\Delta U_{int} = +f_k d \qquad \text{Eq. (7.8)}$$
(conversion of mechanical energy to internal due to friction)

Total energy (*E*) is the sum of all the energies of the system. (Section 7.2)

$$E = K + U_g + U_s + U_{int} + \cdots \qquad \text{Eq. (7.2)}$$

Work-energy principle The energy of a system changes when external forces do work on it. Internal forces do not change the energy of the system. When there are no external forces doing work on the system, the system's energy is constant. (Section 7.2)

$$E_i + W = E_f \qquad \text{Eq. (7.3)}$$
$$(K_i + U_{gi} + U_{si}) + W = (K_f + U_{gf} + U_{sf} + \Delta U_{int})$$

$$K_i + U_{gi} + U_{si} + W = K_f + U_{gf} + U_{sf} + \Delta U_{int}$$

Collisions
- **Elastic:** momentum and kinetic energy of the system are constant—no changes in internal energy.
- **Inelastic:** momentum is constant but kinetic energy is not—internal energy increases and kinetic energy decreases.
- **Totally inelastic:** an inelastic collision in which the colliding objects stick together. (Section 7.7)

All collisions:
$$\Sigma\vec{p}_i = \Sigma\vec{p}_f$$
Elastic collisions only:
$$\Sigma K_i = \Sigma K_f$$
For other collisions,
$$\Delta U_{int} > 0$$

Power (*P*) is the rate of energy conversion, or rate of work done on a system during a process. (Section 7.8)

$$P = \left|\frac{\Delta E}{\Delta t}\right| \text{ or } P = \left|\frac{W}{\Delta t}\right| \quad \text{Eq. (7.9) or (7.10)}$$

Questions

Multiple Choice Questions

1. In which of the following is positive work done by a person on a suitcase?
   (a) The person holds a heavy suitcase.
   (b) The person lifts a heavy suitcase.
   (c) The person stands on a moving walkway carrying a heavy suitcase.
   (d) All of the above
   (e) None of the above

2. Which answer best represents the system's change in energy for the following process? The system includes Earth, two carts, and a compressed spring between the carts. The spring is released, and in the final state, one cart is moving up a frictionless ramp until it stops. The other cart is moving in the opposite direction on a horizontal frictionless track.
   (a) Kinetic energy to gravitational potential energy
   (b) Elastic potential energy to gravitational potential energy
   (c) Elastic potential energy to kinetic energy and gravitational potential energy
   (d) Elastic potential energy to kinetic energy

3. An Atwood machine is shown in Figure Q7.3. As the blocks are released and block 1 moves downward, the energy of the block 1-Earth system
   (a) increases.
   (b) decreases.
   (c) stays constant.
   (d) It's impossible to say without including block 2 in the system.

   FIGURE Q7.3

4. Below you see several statements analyzing the process described in the previous question. Match the energy analysis with the system choice for which the analysis is correct.
   I.   The total energy of the system decreases.
   II. The total energy of the system increases.
   III. The total energy of the system stays constant.
   Systems:
   (a) Block 2 and Earth   (b) Block 1 and Earth
   (c) Both blocks, the string, and Earth   (d) Both blocks and the string

5. Three processes are described below. Choose one process in which there is work done on the system. The spring, Earth, and the cart are part of the system.
   (a) A relaxed spring rests upright on a tabletop. You slowly compress the spring. You then release the spring and it flies up several meters to its highest point.
   (b) A cart at the top of a smooth inclined surface coasts at increasing speed to the bottom (ignore friction).
   (c) A cart at the top of a smooth inclined surface slides at increasing speed to the bottom where it runs into and compresses a spring (ignore friction).

6. Choose which statement describes a process in which an external force does negative work on the system. The person is not part of the system.
   (a) A person slowly lifts a box from the floor to a tabletop.
   (b) A person slowly lowers a box from a tabletop to the floor.
   (c) A person carries a bag of groceries horizontally from one location to another.
   (d) A person holds a heavy suitcase.

7. Which example(s) below involve(s) zero physics work? Choose all that apply.
   (a) A person holds a child.
   (b) A person pushes a car stuck in the snow but the car does not move.
   (c) A rope supports a swinging chandelier.
   (d) A person uses a self-propelled lawn mower on a level lawn.
   (e) A person pulls a sled uphill.

8. Estimate the change in gravitational potential energy when you rise from bed to a standing position.
   (a) No change (0 J)   (b) About 250 J
   (c) About 2500 J   (d) About 25 J

9. What does it mean if object 1 does +10 J of work on object 2?
   (a) Object 1 exerts a 10-N force on object 2 in the direction of its 1-m displacement.
   (b) Object 1 exerts a 1-N force on object 2 in the direction of its 10-m displacement.
   (c) Object 1 exerts a 10-N force on object 2 at a 60° angle relative to its 2-m displacement.
   (d) All of the above
   (e) None of the above

10. You pull on a spring, which obeys Hooke's law, in two steps. In step 1, you extend it by distance x_s. In step 2, you further extend it by the same distance x_s. Choose the answer that correctly compares the elastic potential energy of the spring after the first step (U_{s1}) with that after the second step (U_{s2}).
    (a) $U_{s2} = 2U_{s1}$ because the force exerted on the spring at the end of step 2 is two times larger than at the end of step 1.
    (b) $U_{s2} = 2U_{s1}$ because the extension of the spring at the end of step 2 is two times larger than the extension of the spring at the end of step 1.
    (c) $U_{s2} > 2U_{s1}$ because the more the spring is stretched, the larger the force that is needed to extend the spring for a unit distance.
    (d) $U_{s2} > 2U_{s1}$ because the average force exerted on the spring in step 2 is more than two times larger than the average force exerted on the spring in step 1.

11. The graph in Figure Q7.11 shows the time dependence of the vertical displacement of a lead ball with marked initial and final states. Choose all the work-energy bar charts (a) to (d) that can represent this process (multiple answers may be correct). Note that the y-axis can point either up or down.

    FIGURE Q7.11

    (a) $K_i + U_{gi} + W = K_f + U_{gf}$ (b) $K_i + U_{gi} + W = K_f + U_{gf}$

    (c) $K_i + U_{gi} + W = K_f + U_{gf}$ (d) $K_i + U_{gi} + W = K_f + U_{gf}$

    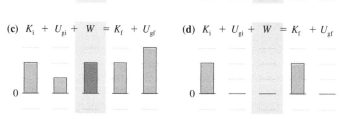

12. A 1400-kg car is traveling on a level road at a constant speed of 60 mi/h. The drag force exerted by the air on the car and the rolling friction force exerted by the road on the car tires add to give a net force of 680 N pointing opposite to the direction of the car's motion. What is the rate of work done by the air and the road on the car?
    (a) 820 kW   (b) 370 kW
    (c) 41 kW   (d) 18 kW

13. You place two toy cars on a horizontal table and connect them with a light compressed spring as shown in Figure Q7.13. The spring tries to push the cars apart, but they are tied together by a thread. When the thread is burned, the spring pushes the cars apart. You decide to investigate how the final speed of car 2 depends on the mass of car 1. You run several experiments changing m_1 and measuring v_2 while keeping the compression of the spring and the mass of car 2 constant. Which of the v_2-versus-m_1 graphs (a) to (c) do you expect to obtain? Evaluate the graphs by analyzing limiting cases.

FIGURE Q7.13

(a) v_2 (b) v_2 (c) v_2

14. Two clay balls are moving toward each other. The balls have equal kinetic energies ($K_1 = K_2$), but ball 1, which is coming from the left, is moving faster than ball 2 ($v_1 > v_2$). Which answer correctly explains how the balls will move after a totally inelastic collision?
    (a) The balls will not move because $K_1 = K_2$ before the collision.
    (b) The balls will move to the left because $p_2 > p_1$ before the collision.
    (c) The balls will move to the right because $v_1 > v_2$ before the collision.
    (d) We cannot predict how the balls will move after the collision.

Conceptual Questions

15. Is energy a physical phenomenon, a model, or a physical quantity? Explain your answer.
16. Your friend thinks that the escape speed should be greater for more massive objects than for less massive objects. Provide an argument for his opinion. Then provide a counterargument for why the escape speed is independent of the mass of the object.
17. Suggest how you can measure the following quantities: the power of a motor, the kinetic energy of a moving car, and the elastic potential energy of a stretched spring.
18. How can satellites stay in orbit without any jet propulsion system? Explain using work-energy ideas.
19. Why does the Moon have no atmosphere, but Earth does?
20. What will happen to Earth if our Sun becomes a black hole?
21. In the equation $U_g = mgy$, the gravitational potential energy is directly proportional to the distance of the object from a planet. In the equation
$$U_g = -G\frac{m_p m}{r},$$ it is inversely proportional. How can you reconcile those two equations?
22. You push a small cart by exerting a constant force F along a table. After you push the cart a distance s, the cart falls off the table and lands on the floor a distance D_1 away from the table edge. You repeat the same experiment (same F, same s) using a heavier cart and obtain a new distance D_2. Choose the answer that correctly compares the distances D_1 and D_2 and gives the best explanation of the outcome.
    (a) $D_1 = D_2$ because the same work was done on both carts before they flew off the table.
    (b) $D_1 = D_2$ because the carts had the same momentum before they flew off the table.
    (c) $D_1 > D_2$ because the carts had the same kinetic energy before they flew off the table, and consequently the speed of the lighter cart had been greater.
    (d) $D_1 > D_2$ because the kinetic energy of cart 1 had been greater than that of cart 2 before they flew off the table.

Problems

Below, **BIO** indicates a problem with a biological or medical focus. Problems labeled **EST** ask you to estimate the answer to a quantitative problem rather than derive a specific answer. Asterisks indicate the level of difficulty of the problem. Problems with no * are considered to be the least difficult. A single * marks moderately difficult problems. Two ** indicate more difficult problems.

7.1 Work and energy

1. Jay fills a wagon with sand (about 20 kg) and pulls it with a rope 30 m along the beach. He holds the rope 25° above the horizontal. The rope exerts a 20-N tension force on the wagon. How much work does the rope do on the wagon?
2. You have a 15-kg suitcase and (a) slowly lift it 0.80 m upward, (b) hold it at rest to test whether you will be able to move the suitcase without help in the airport, and then (c) lower it 0.80 m. What work did you do in each case? What assumptions did you make to solve this problem?
3. * You use a rope to slowly pull a sled and its passenger 50 m up a 20° incline, exerting a 150-N force on the rope. The system is the sled. (a) How much work will you do if you pull parallel to the incline? (b) How much work will you do if you exert the same magnitude force while slowly lowering the sled back down the incline and pulling parallel to the incline? (c) How much work did Earth do on the sled for the trip in part (b)?
4. A rope attached to a truck pulls a 180-kg motorcycle at 9.0 m/s. The rope exerts a 400-N force on the motorcycle at an angle of 15° above the horizontal. (a) What is the work that the rope does in pulling the motorcycle 300 m? (b) How will your answer change if the speed is 12 m/s? (c) How will your answer change if the truck accelerates?

5. You lift a 25-kg child 0.80 m, slowly carry him 10 m to the playroom, and finally set him back down 0.80 m onto the playroom floor. What work do you do on the child for each part of the trip and for the whole trip? List your assumptions.
6. A truck runs into a pile of sand, moving 0.80 m as it slows to a stop. The magnitude of the work that the sand does on the truck is 6.0×10^5 J. (a) Determine the magnitude of the average force that the sand exerts on the truck. (b) Did the sand do positive or negative work? (c) How does the average force change if the stopping distance is doubled? Indicate any assumptions you made.
7. A 0.50-kg block is placed in a straight gutter that allows the block to move only along the gutter. The friction between the gutter walls and the block is negligible. You are exerting a constant force of 3.0 N on the block at an angle of 60° with respect to the gutter (Figure P7.7 shows the top view). The block is initially at rest. Determine (a) the work done on the block if the distance it moved is 20 cm and (b) the final speed of the block.

FIGURE P7.7

$\vec{F}_{Y \text{ on } B}$

$v_i = 0$ 60°

20 cm

8. ** You are pushing a block of mass m at constant speed v over distance s up a smooth incline, which makes an angle θ with the horizontal. (a) Derive an expression for the work done by Earth on the block. (b) How would the answer to part (a) change if there is friction between the block and the incline? (c) How would the answer to part (a) change if you are pushing the block so that it is speeding up but there is no friction between the block and the incline?

9. ** It is a windy day. You are moving a 20-kg box at a constant speed over 10 m by pulling a rope that is attached to the box. The rope makes an angle of 25° with the horizontal. The coefficient of friction between the box and the ground is 0.4. The force exerted by the wind on the box is 15.0 N. Determine the work done by you on the box if (a) the wind is blowing toward you and (b) if the wind is blowing on your back.

7.2 and 7.3 Energy is a conserved quantity and Quantifying gravitational potential and kinetic energies

10. A 5.0-kg rabbit and a 12-kg Irish setter have the same kinetic energy. If the setter is running at speed 4.0 m/s, how fast is the rabbit running?

11. * **EST** Estimate your average kinetic energy when walking to physics class. What assumptions did you make?

12. * A pickup truck (2268 kg) and a compact car (1100 kg) have the same momentum. (a) What is the ratio of their kinetic energies? (b) If the same horizontal net force were exerted on both vehicles, pushing them from rest over the same distance, what is the ratio of their final kinetic energies?

13. * When does the kinetic energy of a car change more: when the car accelerates from 0 to 10 m/s or from 30 m/s to 40 m/s? Explain.

14. * When exiting the highway, a 1100-kg car is traveling at 22 m/s. The car's kinetic energy decreases by 1.4×10^5 J. The exit's speed limit is 35 mi/h. Did the driver reduce the car's speed enough? Explain.

15. * You are on vacation in San Francisco and decide to take a cable car to see the city. A 5200-kg cable car goes 360 m up a hill inclined 12° above the horizontal. The system is the car and Earth. Determine the change in the total energy of the system when the car moves from the bottom to the top. Indicate any assumptions that you made.

16. * **Flea jump** A 5.4×10^{-7}-kg flea pushes off a surface by extending its rear legs for a distance of slightly more than 2.0 mm, consequently jumping to a height of 40 cm. What physical quantities can you determine using this information? Make a list and determine three of them.

17. * **Roller coaster ride** A roller coaster car drops a maximum vertical distance of 35.4 m. (a) Determine the maximum speed of the car at the bottom of that drop. (b) Describe any assumptions you made. (c) Will a car with twice the mass have more or less speed at the bottom? Explain.

18. * **BIO** **EST** **Heart pumps blood** The heart does about 1 J of work while pumping blood into the aorta during each heartbeat. (a) Estimate the work done by the heart in pumping blood during a lifetime. (b) If all of that work was used to lift a person, to what height could an average person be lifted? Indicate any assumptions you used for each part of the problem.

19. * **Wind energy** Air circulates across Earth in regular patterns. A tropical air current called the Hadley cell carries about 2×10^{11} kg of air per second past a cross section of Earth's atmosphere while moving toward the equator. The average air speed is about 1.5 m/s. (a) What is the kinetic energy of the air that passes the cross section each second? (b) About 1×10^{20} J of energy was consumed in the United States in 2005. What is the ratio of the kinetic energy of the air that passes toward the equator each second and the energy consumed in the United States each second?

20. * **BIO** **Bone break** The tibia bone in the lower leg of an adult human will break if the compressive force on it exceeds about 4×10^5 N (we assume that the ankle is pushing up). Suppose that a student of mass 60 kg steps off a chair that is 0.40 m above the floor. If landing stiff-legged on the surface below, what minimum stopping distance does he need to avoid breaking his tibias? Indicate any assumptions you made in your answer to this question.

21. * **BIO** **EST** **Climbing Mt. Everest** In 1953 Sir Edmund Hillary and Tenzing Norgay made the first successful ascent of Mt. Everest. How many slices of bread did each climber have to eat to compensate for the increase of the gravitational potential energy of the system climbers-Earth? (One piece of bread releases about 1.0×10^6 J of energy in the body.) Indicate all of the assumptions used. Note: The body is an inefficient energy converter—see the reading passage at the end of this section.

7.4 Quantifying elastic potential energy

22. A door spring is difficult to stretch. (a) What maximum force do you need to exert on a relaxed spring with a 1.2×10^4-N/m spring constant to stretch it 6.0 cm from its equilibrium position? (b) How much does the elastic potential energy of the spring change? (c) Determine its change in elastic potential energy as it returns from the 6.0-cm stretch position to a 3.0-cm stretch position. (d) Determine its elastic potential energy change as it moves from the 3.0-cm stretch position back to its equilibrium position.

23. * A moving car has 40,000 J of kinetic energy while moving at a speed of 7.0 m/s. A spring-loaded automobile bumper compresses 0.30 m when the car hits a wall and stops. What can you learn about the bumper's spring using this information? Answer quantitatively and list the assumptions that you made.

24. * The force required to stretch a slingshot by different amounts is shown in the graph in Figure P7.24. (a) What is the spring constant of the sling? (b) How much work does a child need to do to stretch the sling 15 cm from equilibrium?

FIGURE P7.24

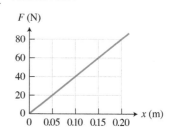

7.5 Friction and energy conversion

25. Jim is driving a 2268-kg pickup truck at 20 m/s and releases his foot from the accelerator pedal. The car eventually stops due to an effective friction force that the road, air, and other things exert on the car. The friction force has an average magnitude of 800 N. Determine the stopping distance of the truck.

26. * A car skids 18 m on a level road while trying to stop before hitting a stopped car in front of it. The two cars barely touch. The coefficient of kinetic friction between the first car and the road is 0.80. A policewoman gives the driver a ticket for exceeding the 35 mi/h speed limit. Can you defend the driver in court? Explain.

27. ** A waterslide of length l has a vertical drop of h. Abby's mass is m. An average friction force of magnitude f opposes her motion. She starts down the slide at initial speed v_i. Use work-energy ideas to develop an expression for her speed at the bottom of the slide. Then evaluate your result using unit analysis and limiting case analysis.

28. ** You are pulling a crate on a rug, exerting a constant force on the crate $\vec{F}_{\text{Y on C}}$ at an angle θ above the horizontal. The crate moves at constant speed. Represent this process using a motion diagram, a force diagram, a momentum bar chart, and an energy bar chart. Specify your choice of system for each representation. Make a list of physical quantities you can determine using this information.

29. * Your friend Devin has to solve the following problem: "You have a spring with spring constant k. You compress it by distance x and use it to shoot a steel ball of mass m into a sponge of mass M. After the collision, the ball and the sponge move a distance s along a rough surface and stop (see Figure P7.29). The coefficient of friction between the sponge and the surface is μ. Derive an expression that shows how the distance s depends on relevant physical quantities."

FIGURE P7.29

Devin derived the following equation:

$$s = \frac{kmx^2}{2(m + M)^2 g\mu}$$

Without deriving it, evaluate the equation that Devin came up with. Is it reasonable? How do you know?

7.6 Skills for analyzing processes using the work-energy principle

30. In a popular new hockey game, the players use small launchers with springs to move the 0.0030-kg puck. Each spring has a 120-N/m spring constant and can be compressed up to 0.020 m. Determine the maximum speed of the puck. First, represent the process with a work-energy bar chart and then solve the problem.

31. The top of a descending ski slope is 50 m higher than the bottom of the slope. A 60-kg skier starts from rest and skis straight to the bottom of the slope. (a) Represent the process with work-energy bar charts. Indicate the system, the initial state, and the final state. (b) Determine the speed of a skier at the bottom of the slope, assuming the friction is negligible. (c) Would a heavier skier have a larger, smaller, or the same speed at the bottom of the slope? Explain.

32. * If 20% of the gravitational potential energy change of the skier in the previous problem is converted into internal energy (due to friction and air drag), how fast is the 60-kg skier traveling at the bottom of the slope? Again, represent the process with work-energy bar charts indicating the system, the initial state, and the final state.

33. You exert a quick push on a 20-g ice cube horizontally along the top of a smooth incline so that it starts moving at a speed of 0.8 m/s, as shown in Figure P7.33. After you stop pushing it, the cube slides down the incline as shown in the figure. Calculate the speed of the ice cube when it reaches the bottom of the incline.

FIGURE P7.33

$v = 0.8$ m/s

20 cm

30 cm

30 cm

34. A driver loses control of a car, drives off an embankment, and lands in a canyon 6.0 m below. What was the car's speed just before touching the ground if it was traveling on the level surface at 12 m/s before the driver lost control?

35. * You are pulling a box so it moves at increasing speed. Compare the work you need to do to accelerate it from 0 m/s to speed v to the work needed to accelerate it from speed v to the speed of $2v$. Discuss whether your answer makes sense. How many different situations do you need to consider?

36. ** A cable lowers a 1200-kg elevator so that the elevator's speed increases from zero to 4.0 m/s in a vertical distance of 6.0 m. What is the force that the cable exerts on the elevator while lowering it? Specify the system, its initial and final states, and any assumptions you made. Then change the system and solve the problem again. Do the answers match?

37. ** **EST** **Hit by a hailstone** A 0.030-kg hailstone the size of a golf ball (4.3 cm in diameter) is falling at about 16 m/s when it reaches Earth's surface. Estimate the force that the hailstone exerts on your head—a head-on collision. Indicate any assumptions used in your estimate. Note that the cheekbone will break if something exerts a 900-N or larger force on the bone for more than 6 ms. Is this hailstone likely to break a bone?

38. * **BIO** **Froghopper jump** Froghoppers may be the insect jumping champs. These 6-mm-long bugs can spring 70 cm into the air, about the same distance as the flea. But the froghopper is 60 times more massive than a 12-mg flea. The froghopper pushes off for about 4 mm. What average force does it exert on the surface? Compare this to the gravitational force that Earth exerts on the bug.

39. * **Bar chart Jeopardy 1** Describe in words and with a sketch a process that is consistent with the qualitative work-energy bar chart shown in Figure P7.39. Then apply in symbols the generalized work-energy principle for that process.

FIGURE P7.39

$$K_i + U_{gi} + U_{si} + W = K_f + U_{gf} + U_{sf} + \Delta U_{int}$$

0

40. * **Bar chart Jeopardy 2** Describe in words and with a sketch a process that is consistent with the qualitative work-energy bar chart shown in Figure P7.40. Then apply in symbols the generalized work-energy principle for that process.

FIGURE P7.40

$$K_i + U_{gi} + U_{si} + W = K_f + U_{gf} + U_{sf} + \Delta U_{int}$$

0

41. * **Equation Jeopardy 1** Construct a qualitative work-energy bar chart for a process that is consistent with the equation below. Then describe in words and with a sketch a process that is consistent with both the equation and the bar chart.

$$(1/2)(400 \text{ N/m})(0.20 \text{ m})^2 = (1/2)(0.50 \text{ kg})v^2$$
$$+ (0.50 \text{ kg})(9.8 \text{ m/s}^2)(0.80 \text{ m})$$

42. * **Equation Jeopardy 2** Construct a qualitative work-energy bar chart for a process that is consistent with the equation below. Then describe in words and with a sketch a process that is consistent with both the equation and the bar chart.

$$(120 \text{ kg})(9.8 \text{ m/s}^2)(100 \text{ m})\sin 53°$$
$$= (1/2)(120 \text{ kg})(20 \text{ m/s})^2 + f_k(100 \text{ m})$$

43. * **Evaluation 1** Your friend provides a solution to the following problem. Evaluate his solution. Constructively identify any mistakes he made and correct the solution. Explain possible reasons for the mistakes.

The problem: A 400-kg motorcycle, including the driver, travels up a 10-m-long ramp inclined 30° above the paved horizontal surface holding the ramp. The cycle leaves the ramp at speed 20 m/s. Determine the cycle's speed just before it lands on the paved surface.

Your friend's solution:

$$(1/2)(400 \text{ kg})(20 \text{ m/s}) = (400 \text{ kg})(9.8 \text{ m/s}^2)(10 \text{ m})$$
$$+ (1/2)(400 \text{ kg})v^2$$
$$v = -13.2 \text{ m/s}$$

44. * **Evaluation 2** Your friend provides a solution to the following problem. Evaluate her solution. Constructively identify any mistakes she made and correct the solution. Explain possible reasons for the mistakes.

The problem: Jim (mass 50 kg) steps off a ledge that is 2.0 m above a platform that sits on top of a relaxed spring of force constant 8000 N/m. How far will the spring compress while stopping Jim?

Your friend's solution:

$$(50 \text{ kg})(9.8 \text{ m/s}^2)(2.0 \text{ m}) = (1/2)(8000 \text{ N/m})x$$
$$x = 0.25 \text{ m}$$

45. A crab climbs up a vertical rock with a constant speed. Students Iris and Alden represent this process using the work-energy bar charts in Figure P7.45. They choose the same initial and final states but different systems. On both bar charts ΔU_{ch} represents change in chemical energy. The students also assumed that the temperature of the crab did not change. Describe the system and the initial and final states that each student has chosen.

FIGURE P7.45

Iris:

$$K_i + U_{gi} + W = K_f + U_{gf} + \Delta U_{ch}$$

0

Alden:

$$K_i + W = K_f + \Delta U_{ch}$$

0

46. * Work-energy bar charts for a person going downstairs and upstairs are shown in Figure P7.46. The bar charts show the average energy conversion across several steps. The system is the person, Earth, and the stairs. (a) What are the initial and final states in each case? (b) Describe and explain the similarities in the bar charts. (c) Describe the differences in the bar charts. How do these differences explain why we are less tired going downstairs? (d) How do you explain that when we walk upstairs $|\Delta U_{\text{th}}| < |\Delta U_{\text{ch}}|$, but when we walk downstairs $|\Delta U_{\text{th}}| > |\Delta U_{\text{ch}}|$? (Hint: Think about what happens to our shoes and the stairs when we go downstairs.)

FIGURE P7.46

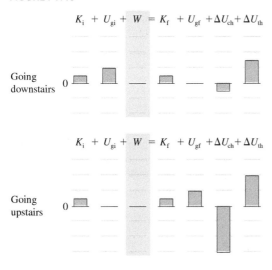

7.7 Collisions

47. A 4.0-kg block moving at 2.0 m/s west on a frictionless surface collides totally inelastically with a second 1.0-kg block traveling east at 3.0 m/s. (a) Determine the final velocity of the blocks. (b) Determine the kinetic energy of each block before and after the collision. (c) What happens to the difference in kinetic energies in this process?

48. * A 1060-kg car moving west at 16 m/s collides with and locks onto a 1830-kg stationary car. (a) Determine the velocity of the cars just after the collision. (b) After the collision, the road exerts a 1.2×10^4-N friction force on the car tires. How far do the cars skid before stopping? Specify the system and the initial and final states of the process.

49. * You fire an 80-g arrow so that it is moving at 80 m/s when it hits and embeds in a 10-kg block resting on ice. (a) What is the velocity of the block and arrow just after the collision? (b) How far will the block slide on the ice before stopping? A 7.2-N friction force opposes its motion. Specify the system and the initial and final states for (a) and (b).

50. * You fire a 50-g arrow that moves at an unknown speed. It hits and embeds in a 350-g block that slides on an air track. At the end, the block runs into and compresses a 4000-N/m spring 0.10 m. How fast was the arrow traveling? Indicate the assumptions that you made and discuss how they affect the result.

51. * To confirm the results of Problem 7.50, you try a new experiment. The 50-g arrow is launched in an identical manner so that it hits and embeds in a 3.50-kg block. The block hangs from strings. After the arrow joins the block, they swing up so that they are 0.50 m higher than the block's starting point. How fast was the arrow moving before it joined the block?

52. * Somebody tells you that Figure P7.52 shows a displacement-versus-time graph of two carts on a linear track before and after a collision, when friction and air drag are negligible. The mass of cart A is 2.0 kg and mass of cart B is 1.0 kg. (a) Is the graph in agreement with the laws of physics? Explain. (b) If yes, is the collision elastic, inelastic, or totally inelastic? Explain. You will need a ruler to solve this problem.

FIGURE P7.52

53. ** Some students are investigating an unknown device that is shooting objects in opposite directions, as shown in Figure P7.53. The objects can have different masses and are therefore ejected from the device with different speeds. The objects are always launched in pairs. The students propose two different explanations for the object-launching mechanism.

FIGURE P7.53

Mechanism 1: Two objects are touching opposite ends of a very light spring that is initially compressed always for the same amount (Figure P7.53a). When the spring is released, it pushes on the objects and makes them fly in opposite directions. The spring is not attached to anything.

Mechanism 2: Each object is touching its own light spring that is initially compressed always for the same amount. The springs are identical and are compressed by the same distance (Figure P7.53b). When the springs are released, they push on the objects and cause them to fly in opposite directions.

The students also measure the masses and speeds of the pairs of the objects that are launched together (see the table below). Can you reject one or both mechanisms using the data in the table? Are there any events in the table that are in agreement with both mechanisms? Explain.

Event #	m_1 (g)	v_1 (m/s)	m_2 (g)	v_2 (m/s)
1	1.0	3.0	2.0	1.5
2	1.0	2.0	1.0	2.0
3	2.0	1.5	3.0	1.0
4	2.0	2.0	1.0	4.0
5	3.0	1.0	2.0	1.5

7.8 Power

54. A roofing shingle elevator is lifting a 10.0-kg package to the top of a building at constant speed. The angle between the elevator track and the horizontal is 75°. The power of the elevator is 700 W. What is the vertical component of the speed with which the package is moving up? Indicate any assumptions that you made.

55. (a) What is the power involved in lifting a 1.0-kg object 1.0 m in 1.0 s? (b) While lifting a 10-kg object 1.0 m in 0.50 s? (c) While lifting the 10-kg object 2.0 m in 1.0 s? (d) While lifting a 20-kg object 1.0 m in 1.0 s?

56. * A fire engine must lift 30 kg of water a vertical distance of 20 m each second. What is the amount of power needed for the water pump for this fire hose?

57. * BIO **Internal energy change while biking** You set your stationary bike on a high 80-N friction-like resistive force and cycle for 30 min at a speed of 8.0 m/s. Your body is 10% efficient at converting chemical energy in your body into mechanical work. (a) What is your internal chemical energy change? (b) How long must you bike to convert 3.0×10^5 J of chemical potential while staying at this speed? (This amount of energy equals the energy released by the body after eating three slices of bread.)

58. * **Climbing Mt. Mitchell** An 82-kg hiker climbs to the summit of Mount Mitchell in western North Carolina. During one 2.0-h period, the climber's vertical elevation increases 540 m. Determine (a) the change in gravitational potential energy of the climber-Earth system and (b) the power of the process needed to increase the gravitational energy.

59. * BIO EST **Sears stair climb** The fastest time for the Sears Tower (now Willis Tower) stair climb (103 flights, or 2232 steps) is about 13 min. (a) Estimate the mechanical power in watts for a top climber. Indicate any assumptions you made. (b) If the body is 20% efficient at converting chemical energy into mechanical energy, approximately how many joules and kilocalories of chemical energy does the body expend during the stair climb? Note: 1 food calorie = 1 kilocalorie = 4186 J.

60. * BIO EST **Exercising so you can eat ice cream** You curl a 5.5-kg (12 lb) dumbbell that is hanging straight down in your hand up to your shoulder. (a) Estimate the work that your hand does in lifting the dumbbell. (b) Estimate the average mechanical power of the lifting process. Indicate any assumptions used in making the estimate. (c) Assuming the efficiency described at the end of Problem 7.59, how many times would you have to lift the dumbbell in order to burn enough calories to use up the energy absorbed by eating a 300-food-calorie dish of ice cream? (Problem 7.59 provides the joule equivalent of a food calorie.) List the assumptions that you made.

61. ** BIO **Salmon move upstream** In the past, salmon would swim more than 1130 km (700 mi) to spawn at the headwaters of the Salmon River in central Idaho. The trip took about 22 days, and the fish consumed energy at a rate of 2.0 W for each kilogram of body mass. (a) What is the total energy used by a 3.0-kg salmon while making this 22-day trip? (b) About 80% of this energy is released by burning fat and the other 20% by burning protein. How many grams of fat are burned? One gram of fat releases 3.8×10^4 J of energy. (c) If the salmon is about 15% fat at the beginning of the trip, how many grams of fat does it have at the end of the trip?

62. * EST Estimate the maximum horsepower of the process of raising your body mass as fast as possible up a flight of 20 stair steps. Justify any numbers used in your estimate. The only energy change you should consider is the change in gravitational potential energy of the system you-Earth.

63. * A 1600-kg car smashes into a shed and stops. The force that the shed exerts on the car as a function of a position at the car's center is shown in Figure P7.63. How fast was the car traveling just before hitting the shed?

FIGURE P7.63

7.9 Improving our model of gravitational potential energy

64. At what distance from Earth is the gravitational potential energy of a spaceship-Earth system reduced to half the energy of the system before the launch?

65. * **Possible escape of different air molecule types** (a) Determine the ratio of escape speeds from Earth for a hydrogen molecule (H_2) and for an oxygen molecule (O_2). The mass of the oxygen is approximately 16 times that of the hydrogen. (b) In the atmosphere, the average random kinetic energy of hydrogen molecules and oxygen molecules is the same. Determine the ratio of the average speeds of the hydrogen and the oxygen molecules. (c) Based on these two results, give one reason why our atmosphere lacks hydrogen but retains oxygen.

66. Determine the escape speed for a rocket to leave Earth's Moon.

67. Determine the escape speed for a rocket to leave the solar system.

68. If the Sun were to become a black hole, how much would it increase the gravitational potential energy of the Sun-Earth system?

69. * A satellite moves in elliptical orbit around Earth, which is one of the foci of the elliptical orbit. (a) The satellite is moving faster when it is closer to Earth. Explain why. (b) If the satellite moves faster when it is closer to Earth, is the energy of the satellite-Earth system constant? Explain.

70. * Determine the maximum radius Earth's Moon would have to have in order for it to be a black hole.

General Problems

71. You throw a clay ball vertically upward. The ball hits the ceiling and sticks to it. Let the initial state be just after the ball leaves your hand and the final state just after the ball sticks to the ceiling. Choose a system and represent the process with a work-energy bar chart. Indicate any assumptions that you made.

72. * A spherical street lamp accidentally explodes. Three equal pieces A, B, and C fly off the lamp holder with equal speeds but in different directions, as shown in Figure P7.72. (a) Compare the speeds with which each piece hits the ground. (b) Compare qualitatively the times needed for each piece to reach the ground. Indicate any assumptions that you made.

FIGURE P7.72

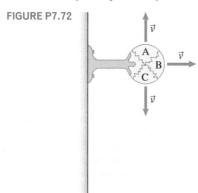

73. Assume that in the previous problem each piece of the lamp has mass m. Develop expressions for the magnitude of (a) the total kinetic energy and (b) the total momentum of the three pieces together just after the explosion.

74. * EST A "gravity force car" is powered by the force exerted by Earth on an object of mass M. The object is tied to a rope that is turning the axle of the rear wheels, as shown in Figure P7.74. The mass of the car (excluding the hanging object) is m. As the car accelerates, the object descends a distance h. (a) Ignoring air resistance and friction in the pulley and wheels, derive an expression for the final speed of the car. (b) Based on your answer to part (a), what would you suggest to the designers of such a car if the goal is to achieve the greatest final speed? (c) Estimate the maximum final speed of a tabletop version of a gravity force car (think of reasonable values for the variables in the equation that you derived).[1]

FIGURE P7.74

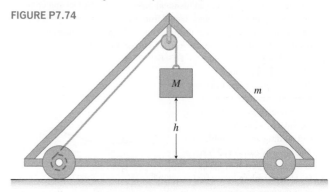

---

[1]This problem was inspired by Gordon Aubrecht, who sadly passed away in 2016.

75. * **Loop the loop** You are given a loop raceway for your Hot Wheels cars (as shown in Figure P7.75). While playing with the cars, you and your friends notice that you need to release the car from at least a height H of 1.30 diameters of the loop above the ground for the car not to fall off the track at the top of the circular loop. Two of your friends have different explanations for the observed pattern:

FIGURE P7.75

Jordan: If there were no friction forces exerted on the car, the minimum height H would be equal to the loop diameter d because the mechanical energy of the car-Earth system is constant in the process.

Leila: The minimum height H is larger than the loop diameter d even if the friction forces are negligible. The car's speed at the top of the loop should be large enough that the force exerted by Earth on the car alone keeps the car moving in a circle determined by the loop.

Analyze the problem and decide which of your friends is correct.

76. ** **Atwood machine** Two blocks of masses m_1 and m_2 hang at the ends of a string that passes over a very light pulley (see Figure Q7.3). Friction in the pulley can be ignored. The blocks are initially at rest at the same height (the initial state). After they are released, the heavier block m_2 moves through a distance d (the final state). Represent the process with work-energy bar charts choosing as a system (a) block 1 and Earth, (b) block 1, block 2, and Earth, and (c) block 1 and block 2. (d) Derive an expression for the final speed of the blocks in terms of m_1, m_2, and d, and evaluate the expression using unit analysis and extreme case analysis.

77. ** Two blocks of masses m_1 and m_2 are connected with a string that passes over a very light pulley (see Figure P7.77). Friction in the pulley can be ignored. Block 1 is resting on a rough table and block 2 is hanging over the edge. The coefficient of friction between the block 1 and the table is μ (assume static and kinetic friction have the same value). Block 1 is also connected to a spring with a constant k. In the initial state, the spring is re-laxed as a person is holding block 2, but the string is still taut. When block 2 is released, it moves down for a distance d until it stops (the final state). (a) Explain why block 2 eventually stops moving. (b) Choose a system and represent the process with a work-energy bar chart. (c) Derive the expression for distance d in terms of m_1, m_2, μ, and k, and evaluate the expression using extreme case analysis and unit analysis.

FIGURE P7.77

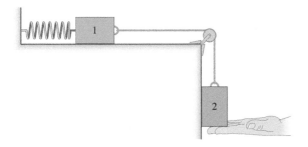

78. ** BIO EST **Impact extinction** 65 million years ago over 50% of all species became extinct, ending the reign of dinosaurs and opening the way for mammals to become the dominant land vertebrates. A theory for this extinction, with considerable supporting evidence, is that a 10-km-wide 1.8×10^{15}-kg asteroid traveling at speed 11 km/s crashed into Earth. Use this information and any other information or assumptions of your choosing to (a) estimate the change in velocity of Earth due to the impact; (b) esti-mate the average force that Earth exerted on the asteroid while stopping it; and (c) estimate the internal energy produced by the collision (a bar chart for the process might help). By comparison, the atomic bombs dropped on Japan during World War II were each equivalent to 15,000 tons of TNT (1 ton of TNT releases 4.2×10^9 J of energy).

79. ** Newton's cradle is a toy that consists of several metal balls touching each other and suspended on strings (Figure P7.79). When you pull one ball to the side and let it strike the next ball, only one ball swings out on the other side. When you use two balls to hit the others, two balls swing out. Can you account for this effect using your knowl-edge about elastic collisions?

FIGURE P7.79

80. * Pose a problem involving work-energy ideas with real numbers. Then solve the problem choosing two different systems and discuss whether the answers were different.

81. ** **Six Flags roller coaster** A loop-the-loop on the Six Flags Shockwave roller coaster has a 10-m radius (Figure P7.81). The car is moving at 24 m/s at the bottom of the loop. Determine the force exerted by the seat of the car on an 80-kg rider when passing inverted at the top of the loop.

FIGURE P7.81

82. ** **Designing a ride** You are asked to help design a new type of loop-the-loop ride. Instead of rolling down a long hill to generate the speed to go around the loop, the 300-kg cart starts at rest (with two passengers) on a track at the same level as the bottom of the 10-m radius loop. The cart is pressed against a compressed spring that, when released, launches the cart along the track around the loop. Choose a spring of the appropriate spring constant to launch the cart so that the downward force exerted by the track on the cart as it passes the top of the loop is 0.2 times the force that Earth exerts on the cart. The spring is initially compressed 6.0 m.

Reading Passage Problems

BIO **Metabolic rate** Energy for our activities is provided by the chemical energy of the foods we eat. The absolute value of the rate of conversion of this chemical energy into other forms of energy $(\Delta E / \Delta t)$ is called the metabolic rate. The metabolic rate depends on many factors—a person's weight, physical activity, the efficiency of bodily processes, and the fat-muscle ratio. Table 7.9 lists the metabolic rates of people under several different conditions and in several differ-ent units of measure: 1 kcal = 1000 calories = 4186 J. Dieticians call a kcal simply a Cal. A piece of bread provides about 70 kcal of metabolic energy.

In 1 hour of heavy exercise a 68-kg person metabolizes 600 kcal − 90 kcal = 510 kcal more energy than when at rest. Typically, reducing kilocalorie intake by 3500 kcal (either by burning it in exercise or not consuming it in the first place) results in a loss of 0.45 kg of body mass (the mass is lost through exhaling carbon dioxide—the product of metabolism).

TABLE 7.9 **Energy usage rate during various activities**

Type of activity	$\Delta E / \Delta t$ (watts)	$\Delta E / \Delta t$ (kcal/h)	$\Delta E / \Delta t$ (kcal/day)
45-kg person at rest	80	70	1600
68-kg person at rest	100	90	2100
90-kg person at rest	120	110	2600
68-kg person walking 3 mph	280	240	5800
68-kg person moderate exercise	470	400	10,000
68-kg person heavy exercise	700	600	14,000

83. Why is the metabolic rate different for different people?
    (a) They have different masses.
    (b) They have different body function efficiencies.
    (c) They have different levels of physical activity.
    (d) All of the above

84. A 50-kg mountain climber moves 30 m up a vertical slope. If the muscles in her body convert chemical energy into gravitational potential energy with an efficiency of no more than 5%, what is the chemical energy used to climb the slope?
    (a) 7 kcal   (b) 3000 J
    (c) 70 kcal   (d) 300,000 J

85. If 10% of a 50-kg rock climber's total energy expenditure goes into the gravitational energy change when climbing a 100-m vertical slope, what is the climber's average metabolic rate during the climb if it takes her 10 min to complete the climb?
    (a) 8 W   (b) 80 W   (c) 500 W
    (d) 700 W   (e) 800 W

86. **EST** A 68-kg person wishes to lose 4.5 kg in 2 months. Estimate the time that this person should spend in moderate exercise each day to achieve this goal (without altering her food consumption).
    (a) 0.4 h   (b) 0.9 h   (c) 1.4 h
    (d) 1.9 h   (e) 2.4 h

87. **EST** A 68-kg person walks at 5 km per hour for 1 hour a day for 1 year. Estimate the *extra* number of kilocalories of energy used because of the walking.
    (a) 40,000 kcal   (b) 47,000 kcal
    (c) 88,000 kcal   (d) 150,000 kcal

88. Suppose that a 90-kg person walks for 1 hour each day for a year, expending 50,000 extra kilocalories of metabolic energy (in addition to his normal resting metabolic energy use). What approximately is the person's mass at the end of the year, assuming his food consumption does not change?
    (a) 57 kg   (b) 61 kg   (c) 64 kg
    (d) 66 kg   (e) 67 kg

BIO Kangaroo hopping Hopping is an efficient method of locomotion for the kangaroo (see **Figure 7.18**). When the kangaroo is in the air, the Earth-kangaroo system has a combination of gravitational potential energy and kinetic energy. When the kangaroo lands, its Achilles tendons and the attached muscles stretch—a form of elastic potential energy. This elastic potential energy is used along with additional muscle tension to launch the kangaroo off the ground for the next hop. In the red kangaroo, more than 50% of the total energy used during

FIGURE 7.18 A kangaroo hopping.

Notice change in foot orientation when taking off and landing.

each hop is recovered elastic potential energy. This is so efficient that the kangaroo's metabolic rate actually decreases slightly as its hopping speed increases from 8 km/h to 25 km/h.

The horizontal and vertical force components exerted by a firm surface on a kangaroo's feet while it hops are shown in **Figure 7.19a**. The vertical force $N_{S \text{ on } K y}$ (Figure 7.19b) varies: when the kangaroo is not touching the surface S, the force is zero; when it is pushing off, the force is about three times the gravitational force that Earth exerts on the kangaroo. The surface exerts a backward horizontal force ($F_{S \text{ on } K x}$) on the kangaroo's foot while it lands and a forward horizontal force as it pushes off for the next hop (Figure 7.19c), similar to what happens to a human foot when landing in front of the body and when pushing off for another step when behind the body.

89. Why is hopping an energy-efficient mode of transportation for a kangaroo?
    (a) There is less resistance since there is less contact with the ground.
    (b) The elastic energy stored in muscles and tendons when landing is returned to help with the next hop.
    (c) The kangaroo has long feet that cushion the landing.
    (d) The kangaroo's long feet help launch the kangaroo.

90. Why does the horizontal force exerted by the ground on the kangaroo change direction as the kangaroo lands and then hops forward?
    (a) The backward force when it lands prevents it from slipping, and the forward force when taking off helps propel it forward.
    (b) One horizontal force is needed to help stop the kangaroo's fall and the other to help launch its upward vertical hop.
    (c) Both forces oppose the kangaroo's motion, but one looks like it is forward because the kangaroo is moving fast.
    (d) The kangaroo is not an inertial reference frame, and the forward force is not real.
    (e) All of the above

91. Which answer below is closest to the vertical impulse that the ground exerts on the kangaroo while it takes off?
    (a) 0   (b) +50 N·s   (c) +150 N·s
    (d) −50 N·s   (e) −150 N·s

92. Which answer below is closest to the vertical impulse due to the gravitational force exerted on the kangaroo by Earth during the short time interval while it takes off?
    (a) 0   (b) +50 N·s   (c) +150 N·s
    (d) −50 N·s   (e) −150 N·s

93. Suppose the net vertical impulse on the 50-kg kangaroo due to all external forces was +100 N·s. Which answer below is closest to its vertical component of velocity when it leaves the ground? Assume that the kangaroo is initially at rest.
    (a) +2.0 m/s   (b) +3.0 m/s   (c) +4.0 m/s
    (d) +8.0 m/s   (e) +10 m/s

94. Which answer below is closest to the vertical height above the ground that the kangaroo reaches if it leaves the ground traveling with a vertical component of velocity of 2.5 m/s?
    (a) 0.2 m   (b) 0.3 m   (c) 0.4 m
    (d) 0.6 m   (e) 0.8 m

FIGURE 7.19 The kangaroo stores elastic potential energy in its muscle and tendon when landing and uses some of that energy to make the next hop.

(a)

(b)
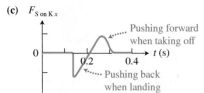

(c) $F_{S \text{ on } K x}$

Pushing forward when taking off

Pushing back when landing

Extended Bodies at Rest

Back pain is a major health problem. In 2010 in the United States, the medical costs related to back pain were around 80 billion dollars a year, about the same as the yearly cost of treating cancer. Back pain often results from incorrect lifting, which compresses the disks in the lower back, causing nerves to be pinched. Understanding the physics principles underlying lifting can help us develop techniques that minimize this compression and prevent injuries.

- Why is it best to lift heavy objects with your knees bent and the object near your body?

- Why are doorknobs located on the side of the door opposite the hinges?

- Why is the force that your biceps muscle exerts on your forearm when lifting a barbell about seven times stronger than the force that Earth exerts on it?

SO FAR IN THIS BOOK we have primarily been modeling moving objects as point-like with no internal structure. This method is appropriate when the shapes of objects do not affect the consequences of their interactions with each other—for example, an elevator moving up a shaft or an apple falling into a pile of leaves. However, objects in general and the human body in particular are extraordinarily complex, with many internal parts that rotate and move relative to one another. To study the body and other complex structures, we need to develop a new way of modeling objects and analyzing their interactions.

BE SURE YOU KNOW HOW TO:

- Define the point-like model for an object (Section 2.2).

- Draw a force diagram for a system (Section 3.1).

- Use the component form of Newton's second law (Section 4.2).

8.1 Extended and rigid bodies

In earlier chapters we focused on situations in which real objects that have nonzero dimensions could be reasonably modeled as point-like. Such modeling is possible when an object moves as a whole from one location to another, without turning. Such motion is called **translational motion**.

In this chapter we will be analyzing objects whose size and shape matter. We will start with the simplest case—objects at rest. (In Chapter 9 we will investigate their motion.) We'd like, for example, to understand how the contemporary dancers in **Figure 8.1** manage to maintain their unusual balance. Where should each force be exerted and how large should their magnitudes be in order for the dancers to remain stable? To answer these questions, we need a new model for extended objects and a new method for analyzing the forces that objects exert on each other. Our first task is to develop this new model for objects.

FIGURE 8.1 The point-like model of an object is not useful when we try to analyze the balance of these dancers.

Rigid bodies

Notice that at the instant shown in the photo the various parts of the dancers' bodies are not moving with respect to each other. They are acting as a single rigid object. This observation motivates a new model for an extended object, in which the size of the object is not zero (it is not a point-like object), but parts of the object do not move with respect to each other. In physics, this model is called a **rigid body**.

> **Rigid body** A rigid body is a model of a real extended object. When we model an extended object as a rigid body, we assume that the object has a nonzero size but the distances between all parts of the object remain the same (the size and shape of the object do not change).

Many bones in your body can be reasonably modeled as rigid bodies, as can many everyday objects—buildings, bridges, streetlights, and utility poles. However, many objects cannot be modeled as rigid bodies. Objects that cannot be modeled as rigid bodies include ones that contain liquid or moving interior parts, such as water balloons or boxes of loose candy. In this chapter we will investigate what conditions are necessary for a rigid body to remain at rest.

FIGURE 8.2 The cardboard is stable in (a) but not in (b). The place where the supporting force is exerted on the board matters.

(a)

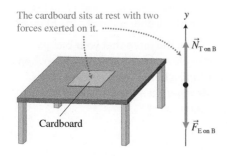

The cardboard sits at rest with two forces exerted on it. ....

y

$\vec{N}_{\text{T on B}}$

Cardboard

$\vec{F}_{\text{E on B}}$

(b)

Cardboard

Pencil

The cardboard tips if supported on the bottom off to the side.

Center of mass

Let's start with some simple experiments. Place a piece of thin, flat cardboard on a very smooth table. If we consider the cardboard to be a point-like object, on a force diagram the upward normal force that the table exerts on the cardboard will balance the downward gravitational force that Earth exerts on the cardboard (**Figure 8.2a**)—the cardboard will not accelerate. Now, place the cardboard on a very small surface—like the eraser of a pencil. The cardboard tilts and falls off (Figure 8.2b). Does this result mean that the eraser cannot exert the same upward force on the cardboard that the table did, or is there some other explanation? The model of the point-like object cannot explain the tilting, since point-like objects do not tilt. Perhaps we need to model the cardboard as a rigid body. Before we do this, let us learn a little more about rigid bodies in Observational Experiment **Table 8.1**.

Pushing a board so that it moves without turning
VIDEO
OET 8.1

Observational experiment	Analysis

Experiment 1. We push with a pencil eraser on the edge of a heart-shaped piece of flat cardboard at different locations such that for each push the cardboard moves on a smooth surface without turning. We also push against the cardboard at points where it turns as it moves.

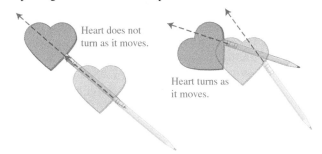

We draw lines across the top of the cardboard from the places and in the directions that the cardboard did not turn as it moved. The lines along which the forces are exerted all cross at one point.

Experiment 2. We now place a heavy object on the cardboard and repeat the experiment, each time pushing at different locations along the edge, trying to find the direction of pushing so that for each push the cardboard moves without turning.

The lines along which the forces are exerted all cross at one point. This point is located somewhere between the point in experiment 1 and the position of the added object in experiment 2.

Pattern

1. In both experiments, all of the lines along which we had to push to move the cardboard without it turning pass through a common point on the cardboard.

2. Pushing at the same locations in other directions causes the cardboard to turn as it moves.

The analysis we did in Table 8.1 indicates that a rigid body possesses a special point. If a force exerted on that object points directly toward or away from that point, the object will not turn. We call this point the object's **center of mass**. It is found experimentally that the center of mass is not necessarily located at the geometrical center of the object, but depends on the distribution of mass in the object.

Center of mass (qualitative definition) The center of mass of an object is a point where a force exerted on the object pointing directly toward or away from that point will not cause the object to turn. The location of this point depends on the mass distribution of the object.

TIP ▷ Although the location of the center of mass depends on the mass distribution of the object, the mass of the object is not necessarily evenly distributed around the center of mass. We will learn more about the properties of the center of mass; we just want to caution you against taking the name of this point literally.

Where is the gravitational force exerted on a rigid body?

At the beginning of the chapter we found that we could not balance cardboard on the eraser of a pencil. Why did the cardboard fall off? Imagine drawing the forces exerted on the cardboard. The cardboard interacts with two objects: the eraser and Earth. The eraser exerts an upward normal force on the cardboard at the point where it touches it. Earth exerts a downward gravitational force on every part of the cardboard. Is it possible to simplify the situation and to find one location at which we can assume that the entire gravitational force is exerted on the cardboard?

Let us go back to the heart-shaped cardboard from Experiment 1 in Table 8.1. If we place the eraser exactly under the cardboard's center of mass, the cardboard does not tip over and fall (**Figure 8.3**). If the cardboard does not tip, it means that all forces exerted on it, including the force exerted by Earth and the force exerted by the eraser, pass through the center of mass. The normal force exerted by the eraser passes through the place where it contacts the cardboard—below the center of mass (indicated on the figure with a checked circle).

Earth exerts a small force on each small part of the board, but we can assume that the total force is exerted exactly at the center of mass. That is why sometimes the object's center of mass is called the object's **center of gravity**. Note that the center of gravity is located at the same spot as the center of mass only when the object is so small that we can neglect the change of g throughout the object. For example, if you think of a large spaceship close to Earth, the value of g at the points closer to Earth is larger than at the points that are farther.

When we model something as a point-like object, we model it as if all of the object's mass is located at its center of mass. Likewise, we can apply what we know about translational motion for point-like objects to rigid bodies, as long as we apply the rules to their centers of mass. **When multiple forces are exerted on a rigid body, the center of mass of the rigid body accelerates translationally according to Newton's second law $\vec{a} = \Sigma\vec{F}/m$.**

FIGURE 8.3 Balancing cardboard.

The heart does not tip if supported under its center of mass.

Center of mass

We can assume that Earth exerts gravitational force entirely at the center of mass.

$\vec{N}_{\text{P on H}}$

$\vec{F}_{\text{E on H}}$

REVIEW QUESTION 8.1 You have an oval framed painting. How do you determine where you should insert a single nail into the frame so that the painting is correctly oriented in both the vertical and horizontal directions?

8.2 Torque: a new physical quantity

We learned in the previous section that the turning effect of an individual force depends on where and in which direction the force is exerted on an object. The translational acceleration of the object's center of mass is still determined by Newton's second law, independently of where the force is exerted. In this section we will learn about the turning ability of a force that an object exerts on a rigid body.

Axis of rotation

When objects turn around an axis, physicists say that they undergo **rotational motion**. The axis may be a fixed physical axis, such as the hinge of a door, or it may not, as in the case of a spinning top. In this chapter we will focus on the conditions under which objects that could potentially rotate do not do so.

Consider a door. When you push on the doorknob perpendicular to the door's surface ($\vec{F}_2$ in **Figure 8.4**), it rotates easily about the door hinges. We call the imaginary line passing through the hinges the **axis of rotation**. You know from experience that pushing a door at or near the axis of rotation ($\vec{F}_1$ in Figure 8.4) is not as effective as

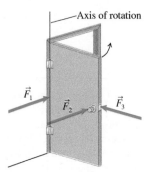

FIGURE 8.4 Different forces have different effects in turning a door about its axis of rotation.

Axis of rotation

$\vec{F}_1$ $\vec{F}_2$ $\vec{F}_3$

$\vec{F}_1$ and $\vec{F}_3$ do not rotate the door, whereas $\vec{F}_2$ moves it easily.

pushing the doorknob. You also know that the harder you push near the knob, the more rapidly the door starts moving. Lastly, pushing on the outside edge of the door toward the axis of rotation ($\vec{F}_3$ in Figure 8.4) does not move the door at all.

These observations suggest that three factors affect the turning ability of a force: (1) the place where the force is exerted, (2) the magnitude of the force, and (3) the direction in which the force is exerted. Next, let's construct a quantitative expression for this turning ability.

The role of position on the turning ability of a force

In order to quantify the turning ability of a force, we take a 0.10-kg meter stick and suspend it at its center of mass from spring scale 2 (**Figure 8.5**); $\vec{F}_2$ is the force exerted on the meter stick by scale 2 at the point of suspension. Spring scale 1 pulls perpendicularly on the stick, exerting a downward force $\vec{F}_1$ at different locations on the left side, and scale 3 exerts a downward perpendicular force $\vec{F}_3$ at different locations on the right side. Earth exerts a 1.0-N force $\vec{F}_{E\,on\,M}$ on the stick's center of mass, which is the point of suspension.

When either scale 1 or scale 3 pulls alone on the stick, the stick rotates (when scale 1 pulls on the stick, it rotates counterclockwise; when 3 pulls, it rotates clockwise). If we pull equally on scales 1 and 3 ($\vec{F}_1 = \vec{F}_3$) but the scales are located at different distances from the axis of rotation, the stick rotates (**Figure 8.6a**). We can use trial and error to find the combinations where the scales can be placed and pulled such that the stick does not rotate. For example, if scale 1 is twice as far from the axis of rotation and pulls half as hard as scale 3, the stick remains in equilibrium (Figure 8.6b). Similarly, if scale 1 is three times farther from the axis of rotation and exerts one-third of the force compared to scale 3, the stick remains in equilibrium (Figure 8.6c). When the stick does not rotate and does not move translationally, it is in a state of **static equilibrium**.

FIGURE 8.5 An experiment to determine a condition necessary for multiple forces to keep a meter stick in a horizontal position.

Scale 2 supports meter stick at center of mass.

Axis of rotation

We pull down on scales 1 and 3, exerting different forces at different places.

FIGURE 8.6 (a) The meter stick does not balance even though equal downward forces are exerted on each side. (b) and (c) A greater force on one side nearer the pivot point balances a smaller force on the other side farther from the pivot point.

(a)

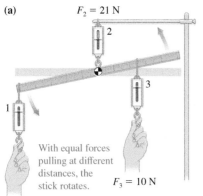

$F_2 = 21$ N

With equal forces pulling at different distances, the stick rotates.

$F_3 = 10$ N

$F_1 = 10$ N

(b)

$F_2 = 31$ N

l_1 $l_3 = l_1/2$

But the stick can be stable with unequal forces pulling at appropriate different distances.

Stick does not rotate.

$F_1 = 10$ N $F_3 = 20$ N $= 2F_1$

(c)

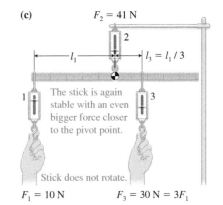

$F_2 = 41$ N

l_1 $l_3 = l_1/3$

The stick is again stable with an even bigger force closer to the pivot point.

Stick does not rotate.

$F_1 = 10$ N $F_3 = 30$ N $= 3F_1$

Static equilibrium An object is said to be in static equilibrium when it *remains at rest* (does not undergo either translational or rotational motion) with respect to a particular observer in an inertial reference frame.

If we explore more situations in which the stick is not rotating, we find that when we have two springs pulling perpendicular to the stick, the stick remains in static equilibrium when the product of the magnitude of the force exerted by the scale on the stick (F_1 or F_3 in our experiment) and the distance between where the force is exerted and the axis of rotation (l_1 or l_3 in our experiment) is the same for both forces:

$$F_1 l_1 = F_3 l_3$$

In other words, the turning ability of the force on the left cancels the turning ability of the force on the right, and the object is in static equilibrium.

The role of magnitude on the turning ability of a force

In addition, we notice that independently of whether the stick rotated or not, the reading of scale 2, F_2, supporting the stick always equals the sum of the readings of scales 1 and 3, $F_1 + F_3$, plus the magnitude of the force exerted by Earth on the stick $F_{E \text{ on } S}$. In other words, in all cases the sum of the forces exerted on the stick was zero ($\Sigma F_y = 0$). This finding is consistent with what we know from Newton's laws—an object does not accelerate translationally if the sum of the forces exerted on it is zero. If it is originally at rest and does not accelerate translationally, then it remains at rest.

However, as we have seen from the experiment with the cardboard on the eraser, this sum-of-forces-equals-zero rule does not guarantee the rotational stability of rigid bodies (see Figure 8.2). Even when the sum of the forces exerted on the cardboard was zero, it could still start turning.

Another simple experiment helps illustrate this idea. Take a book with a glossy cover, place it on a smooth table (to minimize friction), and push it, exerting the same magnitude, oppositely directed force on each of two corners (as shown in **Figure 8.7a**). The force diagram in Figure 8.7b shows that the net force exerted on the book is zero—there is no translational acceleration. However, the book starts turning. The forces exerted by Earth and the table on the book pass through the book's center of mass; thus they do not cause turning. But the forces that you exert on the corners of the book do cause it to turn. Notice that these forces are of the same magnitude and are exerted at the same distance from the center of mass. You can imagine that there is an invisible axis of rotation passing through the center of mass perpendicular to the desk's surface. You would think that the turning effect caused by each force around this imaginary axis is the same; thus the two turning effects should cancel, as they did for scales 1 and 3 in the experiment described earlier. However, this does not happen. The fact that the book turns tells us that not only are the magnitude and placement of the force on the object important to describe the turning ability of the force, but the direction in which this force causes turning (for example, clockwise or counterclockwise around an axis of rotation) is also important.

By convention, physicists call counterclockwise turning about an axis of rotation positive and clockwise turning negative. So far, this is what we know about the new quantity that characterizes the turning ability of a force:

(a) It is equal to the product of the magnitude of the force and the distance the force is exerted from the axis of rotation.

(b) It is positive when the force tends to turn the object counterclockwise and negative when the force tends to turn the object clockwise.

(c) When one force tends to rotate an object counterclockwise and the other force tends to rotate an object clockwise, their effects cancel if $(F_{\text{counterclockwise}} l_1) + (-F_{\text{clockwise}} l_2) = 0$. In this case the object does not rotate.

Let's apply what we have devised so far to check whether this new quantity is useful for explaining other situations. Here is another simple experiment that you can do at home. Place a full milk carton or something of similar mass into a grocery bag. The bag with the milk carton by itself is not too heavy and you can easily lift it with one hand. Now hang the grocery bag from the end of a broomstick (**Figure 8.8a**). Try to support it by holding only the handle end of the broomstick with your hands close together. It is very difficult. Why?

FIGURE 8.7 The turning effect of a force must depend on more than F and l.

(a)

The equal and opposite forces that you exert on the book at its corners cause the book to rotate.

(b)

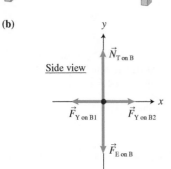

The broomstick, the object of interest, can turn around an axis through the hand that is closest to you. The bag exerts a force on the broomstick far from this axis of rotation. This means that l_{Bag}, a quantity that we must find in order to determine the turning ability of the force exerted by the bag on the broomstick, is very large (Figure 8.8b). Your other hand, which is very close to the axis of rotation (distance l_{Hand}), must balance the effect of the bag. But since l_{Hand} is so small, the force your hand exerts must be very large. The outcome of this experiment agrees with what we learned so far about the turning ability of a force.

Holding the broomstick perpendicular to your body is quite difficult. However, if you hold the broomstick at an angle above the horizontal (Figure 8.8c), you find that the bag becomes easier to support. Why? Perhaps this has to do with the direction the force is exerted relative to the broomstick (see Figure 8.8d).

The role of angle on the turning ability of a force

Our current mathematical model of the physical quantity that characterizes the turning ability of the force ($\pm Fl$) takes into account the direction in which an exerted force can potentially rotate an object (clockwise or counterclockwise) but does not take into account the actual direction of the force. However, our experiment with the broomstick indicates that the angle at which we exert a force relative to the object affects the turning ability of the force. We know from experience that pushing on a door on its outside edge directly toward the hinges does not cause it to rotate. The direction of the push must matter. How can we improve our model for the physical quantity to take the direction of the force into account?

To investigate this question we can change our experiment with the meter stick slightly by making scale 1 pull on the stick at an angle other than 90° (**Figure 8.9**). Scale 2, on the far right end of the meter stick, 0.50 m from the suspension point, will exert a constant force of 10.0 N downward at a 90° angle. Scale 1 on the far left end of the stick will pull at different angles θ so that the meter stick remains horizontal. The results are shown in **Table 8.2**. In all cases the stick is horizontal—therefore, the turning ability of the force on the right is balanced by the turning ability of the force on the left.

FIGURE 8.9 An experiment to determine the angle dependence of the turning ability caused by a force.

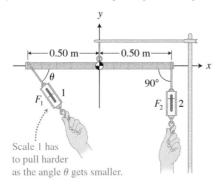

TABLE 8.2 Magnitude, location, and direction of force and its turning ability

Magnitude of $\vec{F}_1$	Distance to the axis of rotation	Angle θ between $\vec{F}_1$ and the stick	Turning ability produced by $\vec{F}_2$
10.0 N	0.50 m	90°	$-(10.0\,\text{N})(0.50\,\text{m}) = -5.0\,\text{N}\cdot\text{m}$
12.6 N	0.50 m	53°	$-(10.0\,\text{N})(0.50\,\text{m}) = -5.0\,\text{N}\cdot\text{m}$
14.2 N	0.50 m	45°	$-(10.0\,\text{N})(0.50\,\text{m}) = -5.0\,\text{N}\cdot\text{m}$
20.0 N	0.50 m	30°	$-(10.0\,\text{N})(0.50\,\text{m}) = -5.0\,\text{N}\cdot\text{m}$

FIGURE 8.8 Holding a bag at the end of a stick is more difficult when the stick is horizontal than when the stick is tilted up.

(a)

It is very difficult to support the broomstick.

(b)

(c)

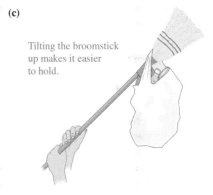

Tilting the broomstick up makes it easier to hold.

(d)

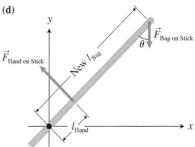

Using the data in the table, we see the effects of the magnitude and the angle of force $\vec{F}_1$ on its turning ability: the smaller the angle between the direction of the force and the stick, the larger the magnitude of the force that is necessary to produce the same turning ability. Thus we find that there are four factors that affect the turning ability of a force: (1) the direction (counterclockwise or clockwise) that the force can potentially rotate the object, (2) the magnitude of the force F, (3) the distance l of the point of application of the force from the axis of rotation, and (4) the angle θ that the force makes relative to a line from the axis of rotation to the point of application of the force. If we combine these four factors, the physical quantity characterizing the turning ability of a force takes a form such as

Counterclockwise (CCW) or clockwise (CW) ·········· $\pm Flf(\theta)$ ·········· Angle

Magnitude   Distance

where $f(\theta)$ is some function of the angle θ.

Consider the last row of Table 8.2. The force exerted by scale 1 is 0.50 m from the axis of rotation ($l = 0.50$ m), and the scale exerts a 20-N force F_1 on the stick at a 30° angle relative to the stick. This force produces the counterclockwise +5.0 N·m effect needed to balance the −5.0 N·m clockwise effect of scale 2. What value would $f(\theta)$ have to be to get this rotational effect?

$$+5.0 \text{ N} \cdot \text{m} = (20.0 \text{ N})(0.5 \text{ m}) \, f(30°)$$

or $f(30°)$ must be 0.50. Recall that $\sin 30° = 0.50$. Maybe the function $f(\theta)$ is the sine function, that is, $\tau = Fl \sin \theta$. Is this consistent with the other rows in Table 8.2?

$$+(10.0 \text{ N})(0.5 \text{ m})(\sin 90°) = +5.0 \text{ N} \cdot \text{m} \ (1.00) = +5.0 \text{ N} \cdot \text{m}$$
$$+(12.6 \text{ N})(0.5 \text{ m})(\sin 53°) = +6.3 \text{ N} \cdot \text{m} \ (0.80) = +5.0 \text{ N} \cdot \text{m}$$
$$+(14.2 \text{ N})(0.5 \text{ m})(\sin 45°) = +7.1 \text{ N} \cdot \text{m} \ (0.71) = +5.0 \text{ N} \cdot \text{m}$$

This expression ($\tau = Fl \sin \theta$) is the mathematical definition of the new physical quantity that characterizes the ability of a force to turn (rotate) a rigid body. This physical quantity is called a **torque**. The symbol for torque is τ, the Greek letter tau.

FIGURE 8.10 A method to determine the torque (turning ability) produced by a force.

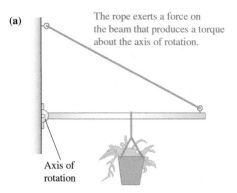

(a)

The rope exerts a force on the beam that produces a torque about the axis of rotation.

Axis of rotation

(b)

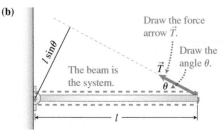

Draw the force arrow $\vec{T}$.

Draw the angle θ.

The beam is the system.

$l \sin\theta$

$\vec{T}$

θ

l

Write an expression for the distance l from the axis of rotation to the place the force is exerted.

> **Torque τ produced by a force** The torque produced by a force exerted on a rigid body about a chosen axis of rotation is
>
> $$\tau = \pm Fl \sin \theta \qquad (8.1)$$
>
> where F is the magnitude of the force, l is the magnitude of the distance between the point where the force is exerted on the object and the axis of rotation, and θ is the angle that the force makes relative to a line connecting the axis of rotation to the point where the force is exerted (see **Figure 8.10**).

Figure 8.10 illustrates the method for calculating the turning ability (torque) due to a particular force. In this case, we are calculating the torque due to the force that the slanted rope exerts on the end of a beam that supports a load hanging from the beam. The torque is positive if the force has a counterclockwise turning ability about the axis of rotation, and negative if the force has a clockwise turning ability. The SI unit for torque is newton · meter, N·m (the British system unit is lb·ft). Note that the expression $l \sin \theta$ gives us the shortest distance from the axis of rotation to the line along which the force producing the torque is exerted.

TIP Notice that the units of torque (N·m) are the same as the units of energy (N·m = J). Torque and energy are very different quantities. We will always refer to the unit of torque as newton · meter (N·m) and the unit of energy as joule (J).

QUANTITATIVE EXERCISE 8.1 **Rank the magnitudes of the torques**

Suppose that five strings pull one at a time on a horizontal beam that can pivot about a pin through its left end, which is the axis of rotation. The magnitudes of the tension forces exerted by the strings on the beam are either T or $T/2$. Rank the magnitudes of the torques that the strings exert on the beam, listing the largest magnitude torque first and the smallest magnitude torque last. Indicate if any torques have equal magnitudes. Try to answer the question before looking at the answer below.

Represent mathematically A mathematical expression for the torque produced by each force is shown below. To understand why each torque is positive, imagine in what direction each string would turn the beam about the axis of rotation if that were the only force exerted on it. You will see that each string tends to turn the beam counterclockwise (except string 5).

Torque due to string 1: $\tau_1 = +T(l/2)\sin 60° = +0.43\ Tl$

Torque due to string 2: $\tau_2 = +T(l/2)\sin 90° = +0.50\ Tl$

Torque due to string 3: $\tau_3 = +T(l/2)\sin 150° = +0.25\ Tl$

Torque due to string 4: $\tau_4 = +(T/2)l \sin 90° = +0.50\ Tl$

Torque due to string 5: $\tau_5 = +Tl \sin 0° = 0$

Solve and evaluate Notice that the angle used for the torque for the force exerted by rope 3 was 150° and not 60°—the force makes a 150° angle relative to a line from the pivot point to the place where the string exerts the force on the beam. String 5 exerts a force parallel to the line from the pivot point to the place where it is exerted on the beam; as a result, torque 5 is zero. The rank order of the torques is $\tau_2 = \tau_4 > \tau_1 > \tau_3 > \tau_5$.

Try it yourself Determine the torque caused by the cable pulling horizontally on the inclined draw-bridge shown below. The force that the cable exerts on the bridge is 5000 N, the bridge length is 8.0 m, and the bridge makes an angle of 50° relative to the vertical support for the cable system.

Answer

$\tau = +(5000\ \text{N})(8.0\ \text{m})\sin 40° = +26{,}000\ \text{N} \cdot \text{m}$. Note that we did not use 50° in our calculation. Why?

To decide the sign of the torque that a particular force exerts on a rigid body about a particular axis, pretend that a pencil is the rigid body. Hold it with two fingers at a place that represents the axis of rotation (**Figure 8.11**) and exert a force on the pencil representing the force whose torque sign you wish to determine. Does that force cause the pencil to turn counterclockwise (a + torque) or clockwise (a − torque) about the axis of rotation?

In this chapter we treat torque as a scalar quantity that can be positive or negative. The sign of the torque indicates the specific direction of rotation (compare this to work, the sign of which means adding or subtracting). The directional meaning of the sign hints that torque has a vector nature. You will learn more about the vector nature of torque in Chapter 9.

FIGURE 8.11 A method to determine the sign of a torque.

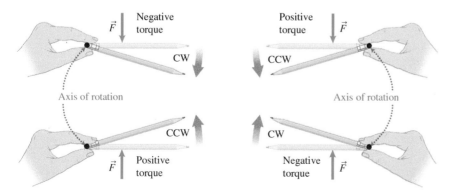

EXAMPLE 8.2   **A painter on a ladder**

A 75-kg painter stands on a 6.0-m-long 20-kg ladder tilted at 53° relative to the ground. He stands with his feet 2.4 m up the ladder. Determine the torque produced by the force exerted by the painter on the ladder for two choices of axis of rotation: (a) an axis parallel to the base of the ladder where it touches the ground and (b) an axis parallel to the top ends of the ladder where it touches the wall of the house.

Sketch and translate A sketch of the situation with the known information is shown below. The ladder is our system.

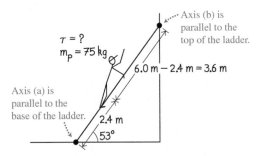

Simplify and diagram Four objects interact with the ladder: Earth, the painter's feet, the wall, and the ground. Our interest in this problem is only in the torque that the painter's feet exert on the ladder. Thus, we diagram for the ladder and the downward force $\vec{F}_{\text{P on L}}$ that the painter's feet exert on the ladder. The diagrams are for the two different axes of rotation.

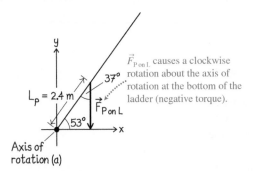

Represent mathematically We use Eq. (8.1) for the torque caused by the force that the painter's feet exert on the ladder.

Solve and evaluate (a) For the first calculation, we choose the axis of rotation at the place where the feet of the ladder touch the ground. The torque produced by the force exerted by the painter's feet on the ladder is

$$\tau = -F_{\text{P on L}}\, l\sin\theta = -(m_\text{p}g)l\sin\theta$$
$$= -(75\text{ kg})(9.8\text{ N/kg})(2.4\text{ m})\sin 37° = -1100\text{ N}\cdot\text{m}$$

Note that the force exerted by the painter's feet on the ladder has the same magnitude as the force mg that Earth exerts on the painter, but it is not the same force, as it is exerted on a different object and is a contact force and not a gravitational force. That force tends to rotate the ladder clockwise about the axis of rotation (a negative torque). The force makes a 37° angle relative to a line from the axis of rotation to the place where the force is exerted.

(b) We now choose the axis of rotation parallel to the wall at the top of the ladder where it touches the wall. The torque produced by the force exerted by the painter's feet on the ladder is

$$\tau = +F_{\text{P on L}}\, l\sin\theta = +(m_\text{p}g)l\sin\theta$$
$$= +(75\text{ kg})(9.8\text{ N/kg})(3.6\text{ m})\sin 143°$$
$$= +1600\text{ N}\cdot\text{m}$$

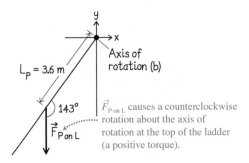

When we choose the axis of rotation at the top of the ladder, the downward force exerted by the painter's feet on the ladder tends to rotate the ladder counterclockwise about the axis at the top (a positive torque). This force makes a 143° angle with respect to a line from the axis of rotation to the place where the force is exerted.

Note that the torque depends on where we place the axis of rotation. You cannot do torque calculations without carefully defining the axis of rotation.

Try it yourself Write an expression for the torque produced by the upward normal force exerted by the floor on the ladder about the same two axes: (a) at the base of ladder and (b) at the top of the ladder.

Answer

(a) The torque will be 0 at the bottom, because the normal force passes through the axis. (b) About the axis at the top of the ladder, the torque is $\tau = -(6.0\text{ m})N_{\text{Floor on Ladder}}\sin 37°$. Note that the floor tends to push the ladder clockwise about an axis through the top of the ladder. The normal force makes a 37° angle relative to a line from the top axis to the place where the force is exerted.

REVIEW QUESTION 8.2 Give an example of a situation in which (a) a torque produced by a force is zero with respect to one choice of axis of rotation but not zero with respect to another; (b) a force is not exerted at the axis of rotation, but the torque produced by it is zero anyway; and (c) several forces produce nonzero torques on an object, but the object does not rotate.

8.3 Conditions of equilibrium

We can combine our previous knowledge of forces and our new knowledge of torque to determine under what conditions rigid bodies remain in static equilibrium, that is, at rest. Recall that we defined static equilibrium as a state in which an object *remains* at rest with respect to a particular observer in an inertial reference frame.

It is possible for an object to be at rest briefly. For example, a ball thrown upward stops for an instant at the top of its flight, but it does not remain at rest. Thus, the word *remains* is important in the expression "remains at rest"—the object has to stay where it is. The words "with respect to an observer in an inertial reference frame" are also an important part of the definition of static equilibrium. Recall from the chapter on Newtonian mechanics (Chapter 3) that if an observer is not in an inertial reference frame, an object can accelerate with respect to the observer even if the sum of the forces exerted on it is zero. In this chapter we will only consider observers who are at rest with respect to Earth, since that is the most common point of view for observing real-life situations involving static equilibrium.

We again suspend the same 0.1-kg meter stick from spring scale 2, as shown in **Figure 8.12**. However, the suspension point is no longer at the center of mass of the meter stick. You and your friend again pull on the stick at different positions with spring scales 1 and 3. When pulled as described in Observational Experiment **Table 8.3**, the stick does not rotate. Pulling at other positions while exerting the same forces, or pulling at the same positions while exerting different forces, causes the stick to rotate. We need to find a pattern in the combinations of forces and torques exerted on the meter stick that keep the stick in static equilibrium.

For most situations that we analyze in this chapter, we assume that the objects rotate in the *x-y* plane and that the axis of rotation goes through the origin of the coordinate system and is perpendicular to the *x-y* plane.

FIGURE 8.12 Multiple objects exert forces on a meter stick.

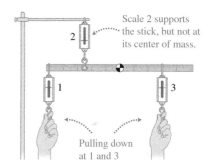

Scale 2 supports the stick, but not at its center of mass.

Pulling down at 1 and 3

OBSERVATIONAL EXPERIMENT TABLE 8.3   Meter stick in static equilibrium

VIDEO
OET 8.3

Observational experiment	Analysis

Observational experiment

Experiment 1. Three spring scales and Earth exert forces on a meter stick at locations shown below. Examine the forces and torques exerted on the stick. Choose the axis of rotation at the place where the string from scale 2 supports the stick. This choice determines the distances in the torque equations for each force.

F_2 (9 N)

Center of mass

$F_{\text{E on S}}$ (1 N)

F_1 (6 N) F_3 (2 N)

0.2 m

0.2 m   0.5 m

Analysis

$$\Sigma F_y = (-6.0\,\text{N}) + 9.0\,\text{N} + (-1.0\,\text{N}) + (-2.0\,\text{N}) = 0$$

Counterclockwise torques:

$$\tau_1 = (F_1)(l_1) = (6.0\,\text{N})(0.20\,\text{m}) = 1.2\,\text{N}\cdot\text{m}$$

Clockwise torques:

$$\tau_2 = (F_2)(l_2) = (9.0\,\text{N})(0) = 0$$

$$\tau_E = -(F_{\text{E on S}})(l_{\text{CM}}) = -(1.0\,\text{N})(0.2\,\text{m}) = -0.2\,\text{N}\cdot\text{m}$$

$$\tau_3 = -(F_3)(l_3) = -(2.0\,\text{N})(0.50\,\text{m}) = -1.0\,\text{N}\cdot\text{m}$$

$$\Sigma\tau = \tau_1 + \tau_2 + \tau_E + \tau_3$$

$$= +1.2\,\text{N}\cdot\text{m} + 0 - 0.2\,\text{N}\cdot\text{m} - 1.0\,\text{N}\cdot\text{m} = 0$$

(CONTINUED)

Observational experiment	Analysis
Experiment 2. Three spring scales and Earth exert forces on a meter stick at locations shown below. Examine the forces and torques exerted on the stick. Choose the axis of rotation at the place where the string from scale 2 supports the stick. 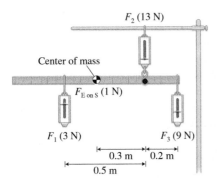	$\Sigma F_y = -3.0\,\text{N} + (-1.0\,\text{N}) + 13\,\text{N} + (-9.0\,\text{N}) = 0$ Counterclockwise torques: $\tau_1 = (F_1)(l_1) = (3.0\,\text{N})(0.50\,\text{m}) = 1.5\,\text{N}\cdot\text{m}$ $\tau_E = (F_{E\,\text{on}\,S})(l_{CM}) = (1.0\,\text{N})(0.3\,\text{m}) = 0.3\,\text{N}\cdot\text{m}$ Clockwise torques: $\tau_2 = (F_2)(l_2) = (13\,\text{N})(0) = 0$ $\tau_3 = -(F_3)(l_3) = -(9.0\,\text{N})(0.20\,\text{m}) = -1.8\,\text{N}\cdot\text{m}$ $\Sigma\tau = \tau_1 + \tau_E + \tau_2 + \tau_3$ $\quad = +1.5\,\text{N}\cdot\text{m} + 0.3\,\text{N}\cdot\text{m} + 0 - 1.8\,\text{N}\cdot\text{m} = 0$

Patterns

- In both cases the net force exerted on the meter stick in the vertical direction is zero: $\Sigma F_y = 0$.
- In both cases the sum of the torques exerted on the meter stick equals zero: $\Sigma\tau = 0$.

TIP ▸ Notice that you cannot determine the torque produced by a force without specifying the point at which the force is exerted on the object relative to the axis of rotation.

The first pattern in both experiments is familiar to us. It is simply Newton's second law applied to the vertical axis of the meter stick for the case of zero translational acceleration. Because the sum of the vertical forces exerted on the meter stick is zero, there is no vertical acceleration. We had no horizontal forces, so the meter stick could not accelerate horizontally. The second pattern shows that in both cases the net torque is zero and the meter stick does not start turning.

We can now summarize these conclusions as follows:

TIP ▸ Remember that all the gravitational forces exerted by Earth on the different parts of the rigid body can be combined into a single gravitational force being exerted on the center of mass of the rigid body.

Condition 1. Translational (force) condition of static equilibrium An object modeled as a rigid body is in translational static equilibrium with respect to a particular observer if it is at rest with respect to that observer and the components of the sum of the forces exerted on it in the perpendicular x- and y-directions are zero:

$$\Sigma F_{\text{on}\,O\,x} = F_{1\,\text{on}\,O\,x} + F_{2\,\text{on}\,O\,x} + \cdots + F_{n\,\text{on}\,O\,x} = 0 \tag{8.2x}$$

$$\Sigma F_{\text{on}\,O\,y} = F_{1\,\text{on}\,O\,y} + F_{2\,\text{on}\,O\,y} + \cdots + F_{n\,\text{on}\,O\,y} = 0 \tag{8.2y}$$

The subscript n indicates the number of forces exerted by external objects on the rigid body.

Condition 2. Rotational (torque) condition of static equilibrium A rigid body is in turning or rotational static equilibrium if it is at rest with respect to the observer and the sum of the torques $\Sigma\tau$ (positive counterclockwise torques and negative clockwise torques) about any axis of rotation produced by the forces exerted on the object is zero:

$$\Sigma\tau = \tau_1 + \tau_2 + \cdots + \tau_n = 0 \tag{8.3}$$

EXAMPLE 8.3 **Testing the conditions of static equilibrium**

Place the ends of a standard meter stick on two scales, as shown below. The scales each read 0.50 N. From this, we infer that the mass of the meter stick is about 0.10 kg (the gravitational force that Earth exerts on the meter stick would be $(0.10 \text{ kg})(9.8 \text{ N/kg}) = 1.0 \text{ N}$). Predict what each scale will read if you place a 5.0-kg brick 40 cm to the right of the left scale.

Scale 1: 0.5 N   Scale 2: 0.5 N

Sketch and translate A labeled sketch of the situation is shown below. We choose the stick as the system of interest and use a standard x-y coordinate system. We choose the axis of rotation at the place where the left scale touches the stick. By doing this, we are making the torque produced by the normal force exerted by the left scale on the stick zero—that force is exerted exactly at the axis of rotation. With this choice, we remove one of the unknown quantities from the torque condition of equilibrium and will be able to use that condition to find the force exerted by the right scale on the meter stick.

Simplify and diagram We model the meter stick as a rigid body with a uniform mass distribution (its center of mass is at the midpoint of the stick). We model the brick as a point-like object, and assume that the scales push up on the stick at the exact ends of the stick. We then draw a force diagram showing the forces exerted on the stick by Earth, the brick, and each of the scales. As noted, the left end of the stick has been chosen as the axis of rotation.

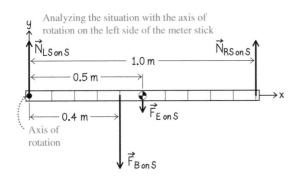

Represent mathematically According to the conditions of static equilibrium, the sum of the forces exerted on the meter stick should equal zero, as should the sum of the torques around the axis of rotation. The gravitational force exerted by Earth and the normal force exerted by the brick on the stick both have clockwise turning ability around the axis of rotation and produce negative torques. The force exerted by the right scale on the stick has counterclockwise turning ability and produces a positive torque. The force exerted by the left scale on the stick produces zero torque since it is exerted at the axis of rotation. The two conditions of equilibrium are then the following:

Translational (force) condition ($\Sigma F_y = 0$):

$$(-F_{\text{E on S}}) + (-F_{\text{B on S}}) + N_{\text{RS on S}} + N_{\text{LS on S}} = 0$$

Rotational (torque) condition ($\Sigma \tau = 0$):

$$-F_{\text{E on S}}(0.50 \text{ m}) - F_{\text{B on S}}(0.40 \text{ m}) + N_{\text{RS on S}}(1.00 \text{ m}) = 0$$

Since none of the forces have x-components, we didn't apply the x-component form of the force condition of equilibrium.

Earth exerts a downward gravitational force on the brick of magnitude:

$$F_{\text{E on B}} = m_{\text{B}}g = (5.0 \text{ kg})(9.8 \text{ N/kg}) \approx 50 \text{ N}$$

Thus, the stick must exert a balancing 50-N upward force on the brick. According to Newton's third law ($\vec{F}_{\text{B on S}} = -\vec{F}_{\text{S on B}}$), the brick must exert a downward 50-N force $\vec{F}_{\text{B on S}}$ on the stick.

Solve and evaluate We have two equations with two unknowns ($N_{\text{RS on S}}$ and $N_{\text{LS on S}}$). We first use the torque equilibrium condition to determine the magnitude of the force exerted by the right scale on the stick (we use 10 N/kg instead of 9.8 N/kg for simplicity):

$$-[(0.10 \text{ kg})(10 \text{ N/kg})](0.50 \text{ m})$$
$$- [(5.0 \text{ kg})(10 \text{ N/kg})](0.40 \text{ m}) + N_{\text{RS on S}}(1.00 \text{ m}) = 0$$

or $N_{\text{RS on S}} = 20.5 \text{ N}$

We can use this result along with the force equilibrium condition equation to determine the magnitude of the force exerted by the left scale on the stick.

$$-(0.10 \text{ kg})(10 \text{ N/kg}) - (5.0 \text{ kg})(10 \text{ N/kg})$$
$$+ 20.5 \text{ N} + N_{\text{LS on S}} = 0$$

or $N_{\text{LS on S}} = 30.5 \text{ N}$

These predictions make sense because the sum of these two upward forces equals the sum of the two downward forces that Earth and the brick exert on the meter stick. Also, the force on the left end is greater because the brick is positioned closer to it, which sounds very reasonable. Performing this experiment, we find that the outcome matches the predictions.

Using a different axis of rotation Remember that we had the freedom to choose whatever axis of rotation we wanted. Let's try it again with the axis of rotation at 40 cm from the left side, the location of the brick. See the force diagram on the next page. The force condition

(CONTINUED)

of equilibrium will not change since it does not depend on the choice of the axis of rotation:

$$(-F_{\text{E on S}}) + (-F_{\text{B on S}}) + N_{\text{RS on S}} + N_{\text{LS on S}} = 0$$

The axis of rotation is now at the 0.4-m position.

The torque condition will change:

$$\left[-F_{\text{E on S}}(0.10\ \text{m})\right] + \left[-N_{\text{LS on S}}(0.40\ \text{m})\right]$$
$$+ N_{\text{RS on S}}(0.60\ \text{m}) = 0$$

Now we have two unknowns in each of the two equations and have to solve them simultaneously to determine the unknowns. This will be somewhat harder than when we chose the axis of rotation at the left end of the stick. Let's solve the force condition equation for $N_{\text{RS on S}}$ and substitute the result into the torque condition equation:

$$N_{\text{RS on S}} = F_{\text{E on S}} + F_{\text{B on S}} - N_{\text{LS on S}}$$
$$\Rightarrow -F_{\text{E on S}}(0.10\ \text{m}) - N_{\text{LS on S}}(0.40\ \text{m})$$
$$+ (F_{\text{E on S}} + F_{\text{B on S}} - N_{\text{LS on S}})(0.60\ \text{m}) = 0$$

Combining the terms with $N_{\text{LS on S}}$ on one side, we get

$$N_{\text{LS on S}}(0.40\ \text{m} + 0.60\ \text{m}) = F_{\text{E on S}}(0.50\ \text{m}) + F_{\text{B on S}}(0.60\ \text{m})$$
$$= (1.0\ \text{N})(0.50\ \text{m}) + (50\ \text{N})(0.60\ \text{m}) = 30.5\ \text{N} \cdot \text{m}$$

or $N_{\text{LS on S}} = 30.5$ N. Substituting back into the force condition equation, we find that

$$N_{\text{RS on S}} = 1.0\ \text{N} + 50.0\ \text{N} - 30.5\ \text{N} = 20.5\ \text{N}$$

These are the same results we obtained from the original choice of the axis of rotation. The choice of the axis of rotation does not affect the results. This makes sense, in the same way that choosing a coordinate system does not affect the outcome of an experiment. The concepts of axes of rotation and coordinate systems are mental constructs and should not affect the outcome of actual experiments.

Try it yourself A uniform meter stick with a 50-g object on it is positioned as shown below. The stick extends 30 cm over the edge of the table. If you push the stick so that it extends slightly farther over the edge, it tips over. Use this result to determine the mass of the meter stick.

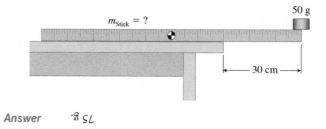

Answer   75 g.

REVIEW QUESTION 8.3 You read the following sentence in a book: "In problem solving, put the axis at the place on the rigid body where the force you know the least about is exerted." Explain why this hint is helpful.

8.4 Center of mass

FIGURE 8.13 Where is the jumper's center of mass?

Many extended bodies are not rigid—the human body is a good example. A high jumper crossing the bar is often bent into an inverted U shape (**Figure 8.13**). Why does this shape allow her to jump higher? At the moment shown in the photo, her legs, arms, and head are below the bar as the trunk of the body passes over the top. As each part of the body passes over the bar, the rest of the body is at a lower elevation so that her center of mass is always slightly below the bar. The high jumper does not have to jump as high because she is able to reorganize her body's shape so that her center of mass passes *under* or at least not significantly over the bar.

Without realizing it, we change the position of our center of mass with respect to other parts of the body quite often. Try this experiment. Sit on a chair with your back straight and your feet on the floor in front of the chair (see **Figure 8.14a**). Without using your hands, try to stand up; you cannot. No matter how hard you try, you cannot raise yourself to standing from the chair if your back is vertical.

Why can't you stand? The center of mass of an average person when sitting upright is near the front of the abdomen. Figure 8.14 explains how the location of the center of

FIGURE 8.14 Getting out of a chair without using your hands.

(a)

You sit on a chair with back straight and feet on the floor. Your center of mass is over the chair.

(b)

With your back straight as you lift yourself from the chair, the normal force exerted by the floor on your feet causes a counterclockwise torque about the center of mass. You fall backward.

(c)

Bending forward so that your center of mass is in front of the floor's normal force causes a clockwise torque so that you can stand.

mass with respect to your feet affects whether the torques exerted on you by the two forces are able to rotate you in the desired direction.

Calculating center of mass

How do we know that the center of mass of a sitting person is near the abdomen? In Section 8.1 we determined the location of an object's center of mass by investigating the directions along which one needs to push the object so it does not turn while being pushed on a flat smooth surface. When pushing in this way at different locations on the object, we found that lines drawn along the directions of these pushing forces all intersected at one point: the center of mass. This is a difficult and rather impractical way to find the center of mass of something like a human. Another method that we investigated consisted of balancing the object on a pointed support. This is also not very practical with respect to humans. Is there a way to predict where an object's center of mass is without pushing or balancing it? Our goal here is to develop a theoretical method that will allow us to determine the location of the center of mass of a complex object, such as the uniform meter stick with two apples shown in **Figure 8.15**. In the next example, we start with an object that consists of three other objects: two people of different masses and a uniform seesaw whose supporting fulcrum (point of support) can be moved. To determine the location of the center of mass of a system involving two people and a uniform beam, we find a place for the fulcrum to support the seesaw and the two people so that the system remains in static equilibrium.

FIGURE 8.15 Where is the center of mass of the meter stick with two apples?

Adjust the position of the fulcrum supporting the seesaw until the system is in static equilibrium.

EXAMPLE 8.4 **Supporting a seesaw with two people**

Find an expression for the position of the center of mass of a system that consists of a uniform seesaw of mass m_1 and two people of masses m_2 and m_3 sitting at the ends of the seesaw beam ($m_2 > m_3$).

Sketch and translate The figure at right shows a labeled sketch of the situation. The two people are represented as blocks. We choose the seesaw and two blocks as the system and construct a mathematical equation that lets us calculate the center of mass of that system. We place the x-axis along the seesaw with its origin at some arbitrary position on the left side of the seesaw. The center of mass of the seesaw beam is at x_1 and the two blocks rest at x_2 and x_3. At what *position x* should we place the fulcrum under the seesaw so that the system does not rotate—so that it remains in static equilibrium? At this position, the

sum of all torques exerted on the system is zero. This position is the center of mass of the three-object system.

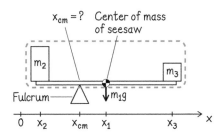

(CONTINUED)

Simplify and diagram We model the seesaw as a rigid body and model each of the people blocks as point-like objects. Assume that the fulcrum does not exert a friction force on the seesaw. As you can see in the force diagram, we know the locations of all forces except the normal force exerted at the fulcrum. We will calculate the unknown position x_{cm} of the fulcrum so the seesaw with two people on it balances—so that it satisfies the second condition of equilibrium.

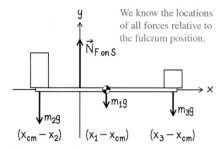

We know the locations of all forces relative to the fulcrum position.

Represent mathematically Apply the torque condition of equilibrium with the axis of rotation going through the unknown fulcrum position x_{cm}. Then determine the torques around this axis produced by the forces exerted on the system. The gravitational force exerted by Earth on the center of mass of the seesaw has magnitude m_1g, is exerted a distance $x_1 - x_{cm}$ from the axis of rotation, and has clockwise turning ability. The gravitational force exerted by Earth on block 2 has magnitude m_2g, is exerted a distance $x_{cm} - x_2$ from the axis of rotation, and has counterclockwise turning ability. The gravitational force exerted by Earth on block 3 has magnitude m_3g, is exerted a distance $x_3 - x_{cm}$ from the axis of rotation, and has clockwise turning ability. The torque condition of equilibrium for the system becomes

$$m_2g(x_{cm} - x_2) - m_1g(x_1 - x_{cm}) - m_3g(x_3 - x_{cm}) = 0$$

Solve and evaluate Divide all terms of the equation by the gravitational constant g and collect all terms involving x on one side of the equation to get

$$m_2x_{cm} + m_1x_{cm} + m_3x_{cm} = m_2x_2 + m_1x_1 + m_3x_3$$

or

$$x_{cm}(m_2 + m_1 + m_3) = m_2x_2 + m_1x_1 + m_3x_3$$

Divide both sides of the equation by $(m_2 + m_1 + m_3)$ to obtain an expression for the location of the center of mass of the three-object system:

$$x_{cm} = \frac{m_1x_1 + m_2x_2 + m_3x_3}{m_1 + m_2 + m_3}$$

Let's evaluate this result. The units of x are meters. Next check some limiting cases to see if the result makes sense. Imagine that there are no people sitting on the seesaw ($m_2 = m_3 = 0$). In this case, $x_{cm} = \dfrac{m_1x_1}{m_1} = x_1$. The center of mass of the seesaw-only system is at the center of mass of the seesaw x_1, as it should be since we assumed its mass was uniformly distributed. Finally, if we increase the mass of one of the people on the seesaw, the location of the center of mass moves closer to that person.

Try it yourself Where is the center of mass of a 3.0-kg, 2.0-m-long uniform beam with a 0.5-kg object on the right end and a 1.5-kg object on the left?

Answer

0.8 m from the left end of the beam.

In Example 8.4 we arrived at an expression for the location of the center of mass of a three-object system where all objects were located along one straight line. We can apply the same method to a system whose masses are distributed in a two-dimensional x-y plane. For such a two-dimensional system, we get the following:

> **Center of mass (quantitative definition)** If we consider an object as consisting of parts 1, 2, 3, ... n whose centers of masses are located at the coordinates $(x_1, y_1); (x_2, y_2); (x_3, y_3); \dots (x_n, y_n)$, then the center of mass of this whole object is at the following coordinates:
>
> $$x_{cm} = \frac{m_1x_1 + m_2x_2 + m_3x_3 + \cdots + m_nx_n}{m_1 + m_2 + m_3 + \cdots + m_n}$$
>
> $$y_{cm} = \frac{m_1y_1 + m_2y_2 + m_3y_3 + \cdots + m_ny_n}{m_1 + m_2 + m_3 + \cdots + m_n}$$
>
> (8.4)

You can think of the position of the center of mass of an object of mass M as a weighted average of the positions of parts 1, 2, 3, etc. ($m_1, m_2, m_3 \dots m_n$ etc.), or

$$x_{cm} = \frac{m_1}{M}x_1 + \frac{m_2}{M}x_2 + \frac{m_3}{M}x_3 + \cdots + \frac{m_n}{M}x_n$$

$$y_{cm} = \frac{m_1}{M}y_1 + \frac{m_2}{M}y_2 + \frac{m_3}{M}y_3 + \cdots + \frac{m_n}{M}y_n$$

Using Eq. (8.4) for an object with a continuous mass distribution is difficult and involves calculus. For example, in the first section we found the position of the center of mass of a cardboard heart empirically. Now we can calculate its position mathematically (**Figure 8.16**). We subdivide the cardboard into many tiny sections and insert the mass and position of each section into Eq. (8.4), and then add all the terms in the numerator and denominator together to determine the center of mass of the cardboard. For example, section 15 would contribute a term $m_{15}x_{15}$ in the numerator of the x_{cm} equation and a term m_{15} in the denominator. Usually, you will be given the location of the center of mass of such continuous mass distributions.

Knowledge of the center of mass helps you answer many questions: Why if you walk with a heavy backpack do you fall more easily? Why does a baseball bat have an elongated, uneven shape? Why do ships carry heavy loads in the bottom rather than near the top of the ship?

Mass distribution and center of mass

The term "center of mass" is deceiving. It might make you think that the center of mass of an object is located at a place where there is an equal amount of mass on each side. However, this is not necessarily the case. Consider again a uniform seesaw (see **Figure 8.17a**). Suppose that the mass of the seesaw is 20 kg, the length is 3.0 m, the mass of the person on the left end (m_1) is 60 kg, and the mass of the person on the right end (m_3) is 20 kg. Where is the center of mass of this two-person seesaw system, and how much mass is on the left side and the right side of the center of mass?

To find the center of mass, we need a coordinate system with an origin. The origin can be anywhere; we will put it at the location of the more massive person on the left side. The center of mass is then

$$x_{cm} = \frac{m_1x_1 + m_2x_2 + m_3x_3}{m_1 + m_2 + m_3}$$

$$= \frac{(60\text{ kg} \times 0\text{ m}) + (20\text{ kg} \times 1.5\text{ m}) + (20\text{ kg} \times 3.0\text{ m})}{60\text{ kg} + 20\text{ kg} + 20\text{ kg}} = 0.9\text{ m}$$

The seesaw will balance about a fulcrum located 0.9 m from m_1 and 2.1 m from m_2 (see Figure 8.17b). We see that the masses are not equal. The mass on the left side of the center of mass is much greater than on the right side—66 kg versus 34 kg. The larger mass on the left is a shorter distance from the center of mass than the smaller masses on the right, which on average are farther from the center of mass. However, the product of mass and distance on each side balances out, causing torques of equal magnitude. We could rename the center of mass as "the center of torque" to reflect the essence of the concept, but since this is not the term used in physics, we will continue to use the term center of mass.

REVIEW QUESTION 8.4 Jade says that the mass of an object is evenly distributed around the location of its center of mass. Do you agree with her? If you disagree, how would you convince her of your opinion?

8.5 Skills for analyzing situations using equilibrium conditions

We often use the equations of equilibrium to determine one or two unknown forces if all other forces exerted on an object of interest are known. Consider the muscles of your arm when you lift a heavy ball or push down on a desktop (**Figure 8.18**). When

FIGURE 8.16 Finding the center of mass of a continuous mass distribution.

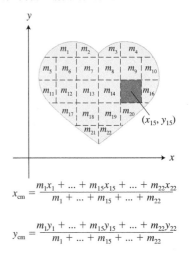

$$x_{cm} = \frac{m_1x_1 + ... + m_{15}x_{15} + ... + m_{22}x_{22}}{m_1 + ... + m_{15} + ... + m_{22}}$$

$$y_{cm} = \frac{m_1y_1 + ... + m_{15}y_{15} + ... + m_{22}y_{22}}{m_1 + ... + m_{15} + ... + m_{22}}$$

FIGURE 8.17 The masses on each side of a system's center of mass are unequal.

(a)

(b) Note that there is more mass on the left side of the center of mass than on the right side.

FIGURE 8.18 Muscles in the upper arm lift and push down on the forearm.

you hold a ball in your hand, your biceps muscle tenses and pulls up on your forearm in front of the elbow joint. When you push down with your hand on a desk, your triceps muscle tenses and pulls up on a protrusion of the forearm behind the elbow joint. The equations of equilibrium allow you to estimate these muscle tension forces—see the next example, which describes a general method for analyzing static equilibrium problems. The right side of the table applies the general strategies to the specific problem provided.

PROBLEM-SOLVING
STRATEGY 8.1   Applying static equilibrium conditions

EXAMPLE 8.5 Using the biceps muscle to lift

Imagine that you hold a 6.0-kg lead ball in your hand with your arm bent. The ball is 0.35 m from the elbow joint. The biceps muscle attaches to the forearm 0.050 m from the elbow joint and exerts a force on the forearm that allows it to support the ball. The center of mass of the 12-N forearm is 0.16 m from the elbow joint. Estimate the magnitude of (a) the force that the biceps muscle exerts on the forearm and (b) the force that the upper arm exerts on the forearm at the elbow.

Sketch and translate

- Construct a labeled sketch of the situation; mark knowns and unknowns. Choose an axis of rotation.
- Choose a system for analysis.

We choose the axis of rotation to be where the upper arm bone (the humerus) presses on the forearm at the elbow joint. This will eliminate from the torque equilibrium equation the unknown force that the upper arm exerts on the forearm.

We choose the system of interest to be the forearm and hand.

Simplify and diagram

- Decide whether you will model the system as a rigid body or as a point-like object.
- Construct a force diagram for the system. Include the chosen coordinate system and the axis of rotation (the origin of the coordinate system).

Model the system as a rigid body and draw a force diagram for the forearm and hand.

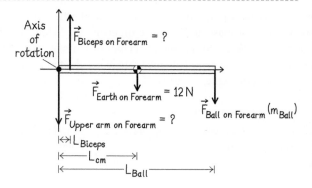

Represent mathematically

- Use the force diagram to apply the conditions of equilibrium.

$$\Sigma \tau = 0$$

$$(F_{\text{UA on FA}})(0) + (F_{\text{Biceps on FA}})(L_{\text{Biceps}} \sin 90°) + (-F_{\text{E on FA}}(L_{\text{cm}} \sin 90°))$$
$$+ (-F_{\text{Ball on FA}}(L_{\text{Ball}} \sin 90°)) = 0$$

$$\Sigma F_y = 0$$

$$(-F_{\text{UA on FA}}) + F_{\text{Biceps on FA}} + (-F_{\text{E on FA}}) + (-F_{\text{Ball on FA}}) = 0$$

Solve and evaluate

- Solve the equations for the quantities of interest.

- Evaluate the results. Check to see if their magnitudes are reasonable and if they have the correct signs and units. Also see if they have the expected values in limiting cases.

Substitute $\sin 90° = 1.0$ and rearrange the torque equation to find $F_{\text{Biceps on FA}}$.

$$F_{\text{Biceps on FA}} = \left[\,(F_{\text{E on FA}})(L_{\text{cm}}) + (F_{\text{Ball on FA}})(L_{\text{Ball}})\,\right]/L_{\text{Biceps}}$$
$$= \left[\,(12\text{ N})(0.16\text{ m}) + (59\text{ N})(0.35\text{ m})\,\right]/(0.050\text{ m}) = 450\text{ N}$$

Use the force equation to find $F_{\text{UA on FA}}$:

$$F_{\text{UA on FA}} = F_{\text{Biceps on FA}} - F_{\text{E on FA}} - F_{\text{Ball on FA}}$$
$$= 450\text{ N} - 12\text{ N} - 59\text{ N} = 380\text{ N}$$

The 450-N force exerted by the biceps on the forearm is much greater than the 59-N force exerted by the ball on the forearm. This difference occurs because the force exerted by the biceps is applied much closer to the axis of rotation than the force exerted by the lead ball.

If the center of mass of the forearm were farther from the elbow, the biceps would have to exert an even larger force.

Try it yourself How would the force exerted by the biceps on the forearm change if the biceps were attached to the forearm farther from the elbow?

Answer

A longer L_{Biceps} in the torque equilibrium equation would mean that the biceps muscle would need to exert a smaller force on the forearm when lifting something.

Standing on your toes

Most injuries to the Achilles tendon occur during abrupt movement, such as jumps and lunges. However, we will analyze what happens to your Achilles tendon in a less stressed situation.

EXAMPLE 8.6 **Standing with slightly elevated heel**

Suppose you stand on your toes with your heel slightly off the ground. In order to do this, the larger of the two lower leg bones (the tibia) exerts a force on the ankle joint where it contacts the foot. The Achilles tendon simultaneously exerts a force on the heel, pulling up on it in order for the foot to be in static equilibrium. What is the magnitude of the force that the tibia exerts on the ankle joint? What is the magnitude of the force that the Achilles tendon exerts on the heel?

Sketch and translate First we sketch the foot with the Achilles tendon and the tibia. We choose the foot as the system of interest. Three forces are exerted on the foot: the tibia is pushing down on the foot at the ankle joint; the floor is pushing up on the ball of the foot and the toes; and the Achilles tendon is pulling up on the heel. We choose the axis of rotation as the place where the tibia presses against the foot.

Tibia bone

$F_{\text{Bone on Foot}} = ?$

$T_{\text{Tendon on Foot}} = ?$

Achilles tendon

Axis of rotation

L_{Floor}

L_{Tendon}

Simplify and diagram Model the foot as a very light rigid body. The problem says that the foot is barely off the ground, so we will neglect the angle between the foot and the ground and consider the foot

horizontal. The gravitational force exerted on the foot by Earth is quite small compared with the other forces that are being exerted on it, so we will ignore it. A force diagram for the foot is shown at right. When you are standing on the ball and toes of both feet, the floor exerts an upward force on each foot equal to half the magnitude of the gravitational force that Earth exerts on your entire body: $F_{\text{Floor on Foot}} = \dfrac{m_{\text{Body}}g}{2}$. The Achilles tendon pulls up on the heel of the foot, exerting a force $T_{\text{Tendon on Foot}}$. The tibia bone in the lower leg pushes down on the ankle joint exerting a force $F_{\text{Bone on Foot}}$.

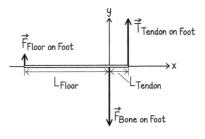

$\vec{F}_{\text{Floor on Foot}}$

$\vec{T}_{\text{Tendon on Foot}}$

L_{Floor}

L_{Tendon}

$\vec{F}_{\text{Bone on Foot}}$

Represent mathematically Let's apply the conditions of equilibrium to this system. Note that the distance from the toes to the joint L_{Floor} is somewhat longer than the distance from the joint to the Achilles tendon attachment point L_{Tendon}. The torque condition of equilibrium becomes

$$+\left[\,T_{\text{Tendon on Foot}}(L_{\text{Tendon}})\,\right] + F_{\text{Bone on Foot}}(0) - \frac{m_{\text{Body}}g}{2}(L_{\text{Floor}}) = 0$$

$$\Rightarrow T_{\text{Tendon on Foot}} = \frac{m_{\text{Body}}g}{2}\left(\frac{L_{\text{Floor}}}{L_{\text{Tendon}}}\right)$$

(CONTINUED)

Now, apply the y-scalar component of the force condition of equilibrium:

$$\Sigma F_y = T_{\text{Tendon on Foot}} + (-F_{\text{Bone on Foot}}) + \frac{m_{\text{Body}}g}{2} = 0$$

$$\Rightarrow F_{\text{Bone on Foot}} = T_{\text{Tendon on Foot}} + \frac{m_{\text{Body}}g}{2}$$

Solve and evaluate The distance from the place where the bone contacts the foot to where the floor contacts the foot is about 5 times longer than the distance from the bone to where the tendon contacts the foot. Consequently, the force that the Achilles tendon exerts on the foot is about

$$T_{\text{Tendon on Foot}} = \frac{m_{\text{Body}}g}{2}\left(\frac{L_{\text{Floor}}}{L_{\text{Tendon}}}\right) = \frac{m_{\text{Body}}g}{2}(5) = \frac{5}{2}m_{\text{Body}}g$$

or two and a half times the gravitational force that Earth exerts on the body. Using $g = 10$ N/kg for a 70-kg person, this force will be about 1750 N. That's a very large force for something as simple as standing

with your heel slightly elevated! The force exerted on the joint by the leg bone would be

$$F_{\text{Bone on Foot}} = T_{\text{Tendon on Foot}} + \left(\frac{m_{\text{Body}}g}{2}\right)$$

$$= 1750 \text{ N} + 350 \text{ N} = 2100 \text{ N}$$

This force is three times the weight of the person! The forces are much greater when moving. Thus, every time you lift your foot to walk, run, or jump, the tendon tension and joint compression are several times greater than the gravitational force that Earth exerts on your entire body.

Try it yourself Estimate the increase in the magnitude of the force exerted by the Achilles tendon on the foot of the person in this example if his mass were 90 kg instead of 70 kg.

Answer   An increase of at least 500 N (110 lb).

FIGURE 8.19 A bad way to lift.

Lifting from a bent position

Back problems often originate with improper lifting techniques—a person bends over at the waist and reaches to the ground to pick up a box or a barbell. In **Figure 8.19**, the barbell pulls down on the woman's arms far from the axis of rotation of her upper body about her hip area. This downward pull causes a large clockwise torque on her upper body. To prevent her from tipping over, her back muscles must exert a huge force on the backbone, thus producing an opposing counterclockwise torque. This force exerted by the back muscles compresses the disks that separate vertebrae and can lead to damage of the disks, especially in the lower back. We can use the equilibrium equations to estimate the forces and torques involved in such lifting.

EXAMPLE 8.7   **Lifting incorrectly from a bent position**

Estimate the magnitude of the force that the back muscle in the woman's back in Figure 8.19 exerts on her backbone and the force that her backbone exerts on the disks in her lower back when she lifts an 18-kg barbell. The woman's mass is 55 kg. Model the woman's upper body as a rigid body.

- The back muscle attaches two-thirds of the way from the bottom of her $l = 0.60$ m backbone and makes a 12° angle relative to the horizontal backbone.

- The mass of her upper body is $M = 33$ kg centered at the middle of the backbone and has uniform mass distribution. The axis of rotation is at the left end of the backbone and represents one of the disks in the lower back.

Sketch and translate

The figure at right is our mechanical model of a person lifting a barbell. We want to estimate the magnitudes of the force $T_{\text{M on B}}$ that the back muscle exerts on the backbone and the force $F_{\text{D on B}}$ that the disk in the lower back exerts on the backbone. The force that the disk exerts on the bone is equal in magnitude to the force exerted by the backbone on the disk. The upper body (including the backbone) is the system of interest, but we consider the back muscle to be external to the system since we

want to focus on the force it exerts on the backbone. The hinge where the upper body meets the lower body is the axis of rotation.

Simplify and diagram We next draw a force diagram for the upper body. The gravitational force that Earth exerts on the upper body $F_{\text{E on B}}$ at its center of mass is $Mg = (33 \text{ kg})(9.8 \text{ N/kg}) = 323$ N(73 lb). The barbell exerts a force on the upper body equal to $mg = (18 \text{ kg}) \times (9.8 \text{ N/kg}) = 176$ N(40 lb). Because of our choice of axis of rotation, the force exerted by the disk on the upper body $F_{\text{D on B}}$ will not produce a torque. The gravitational force exerted by Earth on the upper body and the force that the barbell exerts on the upper body have clockwise turning ability, while the tension force exerted by the back muscles on the upper body has counterclockwise turning ability.

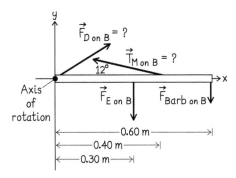

Represent mathematically The torque condition of equilibrium for the upper body is

$$\Sigma\tau = +(F_{\text{D on B}})(0) + [-(Mg)(l/2)\sin 90°]$$
$$+ (T_{\text{M on B}})(2l/3)\sin 12° + [-(F_{\text{Barb on B}})(l)\sin 90°]$$
$$= 0$$

The x- and y-component forms of the force condition of equilibrium for the backbone are

$$\Sigma F_x = F_{\text{D on B}x} + (-T_{\text{M on B}}\cos 12°) = 0$$
$$\Sigma F_y = F_{\text{D on B}y} + T_{\text{M on B}}\sin 12° + (-mg) + (-Mg) = 0$$

where $F_{\text{D on B}x}$ and $F_{\text{D on B}y}$ are the scalar components of the force that the disk exerts on the upper body.

Solve and evaluate We can solve the torque equation immediately to determine the magnitude of the force that the back muscle exerts on the backbone:

$$T_{\text{M on B}} = \frac{(Mg)(l/2)(\sin 90°) + (mg)(l)\sin 90°}{(2l/3)(\sin 12°)}$$

Note that the backbone length l in the numerator and denominator of all of the terms in this equation cancels out. Thus,

$$T_{\text{M on B}} = \frac{(Mg)(1/2)(\sin 90°) + (mg)(1)\sin 90°}{(2/3)(\sin 12°)}$$
$$= \frac{(323\text{ N})(0.50)(1.0) + (176\text{ N})(1)(1.0)}{(0.667)(0.208)}$$
$$= 2432\text{ N }(547\text{ lb})$$

We then find $F_{\text{D on B}x}$ from the x-component force equation:

$$F_{\text{D on B}x} = +T_{\text{M on B}}\cos 12°$$
$$= +(2432\text{ N})\cos 12° = +2380\text{ N}$$

and $F_{\text{D on B}y}$ from the y-component force equation:

$$F_{\text{D on B}y} = +Mg + mg - T_{\text{M on B}}\sin 12°$$
$$= +323\text{ N} + 176\text{ N} - (2432\text{ N})(\sin 12°) = -7\text{ N}$$

Thus, the magnitude of $F_{\text{D on B}}$ is

$$F_{\text{D on B}} = \sqrt{(2380\text{ N})^2 + (-7\text{ N})^2} = 2380\text{ N }(540\text{ lb})$$

The direction of $\vec{F}_{\text{D on B}}$ can be determined using trigonometry:

$$\tan\theta = \frac{F_{\text{D on B}y}}{F_{\text{D on B}x}} = \frac{-7\text{ N}}{2380\text{ N}} = -0.0029$$

or $\theta = 0.17°$ below the horizontal. We've found that the back muscles exert a force more than four times the gravitational force that Earth exerts on the person and that the disks of the lower back are compressed by a comparable force.

Try it yourself Suppose that a college football lineman stands on top of a 1-inch-diameter circular disk. How many 275-lb linemen, one on top of the other, would exert the same compression force on the disk as that exerted on the woman's disk when she lifts the 40-lb barbell?

Answer

The magnitude of the force exerted on the vertebral disk is equivalent to two linemen (2×275 lb $= 550$ lb) standing on the 1-inch-diameter disk.

To lift correctly, keep your back more vertical with the barbell close to your body, as in **Figure 8.20**. Bend your knees and lift with your legs. With this orientation, the back muscle exerts one-third of the force that is exerted when lifting incorrectly. The disks in the lower back undergo one-half the compression they would experience from lifting incorrectly.

FIGURE 8.20 A better way to lift things.

Backbone

Muscle

Barbell

REVIEW QUESTION 8.5 You are trying to hold a heavy dumbbell in one hand so that your arm is perpendicular to your body. Why is it easier to hold it with a bent arm than with a straight arm?

8.6 Stability of equilibrium

Often, objects can remain in equilibrium for a long time interval—you can sit comfortably for a long time on a living room couch without tipping. But sometimes equilibrium is achieved for only a short time interval—think of sitting on a chair and tilting it backward too far onto its rear legs.

FIGURE 8.21 Balancing on the subway.

(a) The train is at rest.

(b) The train is accelerating.

Equilibrium and tipping objects

You have probably observed that it is easier to balance and avoid falling while standing in a moving bus or subway train if you spread your feet apart in the direction of motion. By doing this you are increasing the **area of support**, the area of contact between an object and the surface it is supported by. To understand area of support, consider two people riding the subway (see **Figure 8.21**).

In Figure 8.21a, the man's feet are close together, whereas the woman's feet are farther apart. When the train accelerates toward the right, as shown in Figure 8.21b, the two people tilt to the left. The man falls over because the gravitational force exerted by Earth on his body is outside the area of support provided by his feet. The woman recovers because the gravitational force still points between her feet. These patterns lead us to a tentative rule about tipping:

> For an object in static equilibrium, if a vertical line passing through the object's center of mass is within the object's area of support, the object does not tip. If the line is not within the area of support, the object tips.

If this is a general rule, then we can use it to predict the angle at which an object with a known center of mass will tip over (see Testing Experiment **Table 8.4**).

TESTING EXPERIMENT TABLE 8.4   **Testing our tentative rule about tipping**

VIDEO
TET 8.4

Testing experiment	Prediction	Outcome
Experiment 1. Place a full box of crackers on a flat but rough surface. Its center of mass is at its geometric center. The box's height is 20 cm and the bottom surface is 13 cm × 6.5 cm.	1. The center of mass of a full box of crackers is at its geometric center.	
Tilt the box along the 13-cm side a little and release it.	2. When you release the slightly tilted box, it returns to the vertical position because of the torque due to the gravitational force exerted by Earth.	The box returns to the vertical position.
Tilt the box at larger and larger angles. Predict the angle at which the box will tip.	3. When you tilt the box more and more, eventually the line defined by the gravitational force passes over the support point for the box at its bottom edge. Tilting the box more than this critical angle causes it to tip over.	
	4. For a box with a side of 13 cm and a height of 20 cm, this angle will be $\theta_C = \tan^{-1}(13\ cm/20\ cm) = 33°$.	The outcome matches the prediction.

Testing experiment	Prediction	Outcome
Experiment 2. Repeat the experiment, but this time replace the crackers with a wooden block of the same mass. The block takes up about half the space that the crackers of the same mass did. Predict the angle at which the box will tip.	The center of mass should now be about one-fourth of the way from the bottom (if we neglect the mass of the box). The critical tilt angle should now be $\theta_C = \tan^{-1}(13\text{ cm}/10\text{ cm}) = 52°$.	The outcome matches the prediction within experimental uncertainty (about 1°).

Conclusion

The outcomes are consistent with the predictions. We've increased our confidence in the tipping rule: for an object to be in static equilibrium, the line defined by the gravitational force exerted by Earth must pass within the object's area of support. If it is not within the area of support, the object tips over.

We now know that an object will tip if it is tilted so that the gravitational force passes beyond its area of support. If the area of support is large or if the center of mass is closer to the ground, more tipping is possible without the object falling over—it is more stable. This idea is regularly used in building construction. Tall towers (like the Eiffel Tower) have a wide bottom and a narrower top. The Leaning Tower of Pisa does not tip because a vertical line through its center of mass passes within the area of support (**Figure 8.22**).

The equilibrium of a system is stable against tipping if the vertical line through its center of mass passes through the system's area of support.

Equilibrium and rotating objects

Objects that can rotate around a fixed axis can also have either stable or unstable equilibria. Consider a ruler with several holes in it. If you hang the ruler on a nail using a hole near one end, it hangs as shown in **Figure 8.23a**. If you pull the bottom of the ruler to the side and release it, the ruler swings back and forth with decreasing maximum displacement from the equilibrium position, but eventually hangs straight down. This equilibrium position is called *stable* because the ruler always tries to return to that position if free to rotate. However, if you turn the ruler 180° so that the axis of rotation is at the bottom of the ruler (see Figure 8.23b), it can stay in this position only if very carefully balanced. If disturbed, the ruler swings down and never returns. In this case the ruler is in *unstable* equilibrium.

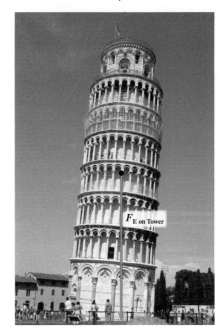

FIGURE 8.22 Because $F_{\text{E on Tower}}$ passes through the base of the tower, the Leaning Tower of Pisa does not tip over.

FIGURE 8.23 Stable, unstable, and neutral equilibrium.

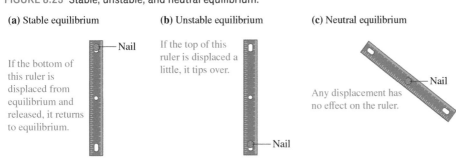

(a) Stable equilibrium

If the bottom of this ruler is displaced from equilibrium and released, it returns to equilibrium.

Nail

(b) Unstable equilibrium

If the top of this ruler is displaced a little, it tips over.

Nail

(c) Neutral equilibrium

Any displacement has no effect on the ruler.

Nail

The stability or lack of stability can be understood by considering the torque around the axis of rotation due to the force exerted by Earth on the object. In both positions a and b, the sum of the forces exerted on the ruler by Earth and by the nail is zero. The difference is that in the first case the center of mass is *below the axis of rotation*; in the second case it is *above the axis of rotation*. Torques produced by gravitational forces tend to lower the center of mass of objects. In other words, if it is possible for the object to rotate so that its center of mass becomes lower, it will tend to do so.

If we hang the ruler using the center hole (Figure 8.23c), it remains in whatever position we leave it. Both the force exerted by the nail and the gravitational force exerted by Earth produce zero torques. This is called *neutral* equilibrium.

The equilibrium of a system is stable against rotation if the center of mass of the rotating object is below the axis of rotation.

CONCEPTUAL EXERCISE 8.8   **Balancing a pencil**

Is it possible to balance with stable equilibrium the pointed tip of a pencil on your finger?

Sketch and translate The figure below shows a sketch of the situation.

Simplify and diagram The forces exerted on the pencil are shown at right. The tip of the pencil is the axis of rotation. When the pencil is tilted by only a small angle, the line defined by the gravitational force exerted by Earth on the pencil is not within the area of support of the pencil (which is just the pencil tip). This equilibrium is unstable. Having the center of mass *above* the axis of rotation leads to this instability. To make it stable, we need to lower the center of mass to below the axis of rotation. We can do it by attaching a pocketknife to the pencil, as shown below. Notice that the massive part of the knife is below the tip of the pencil and so is the center of mass of the system. Then when the pencil tilts, the torque due to the gravitational force brings it back to the equilibrium position.

FIGURE 8.24 Balancing a bicycle on a high wire may not be as dangerous as it looks.

Center of mass is below balance point. Bicycle will not tip.

TIP Always try to understand new situations in terms of ideas we have already discussed. The trick of balancing the pencil by adding a pocketknife can be understood with the rules we expressed above: (1) the equilibrium is most stable when the center of mass of the system is in the lowest possible position or, equivalently, (2) when the gravitational potential energy of the system has the smallest value.

The rules we have learned about equilibrium and stability have many applications, including circus tricks. Think about where the center of mass is located for the bicycle and the two people shown in **Figure 8.24**. Another application involves vending machines.

EXAMPLE 8.9   Tipping a vending machine

According to the U.S. Consumer Product Safety Commission, tipped vending machines caused 37 fatalities between 1978 and 1995 (2.2 deaths per year). Why is tipping vending machines so dangerous? A typical vending machine is 1.83 m high, 0.84 m deep, and 0.94 m wide and has a mass of 374 kg. It is supported by a leg on each of the four corners of its base. (a) Determine the horizontal pushing force you need to exert on its front surface 1.5 m above the floor in order to just lift its front feet off the surface (so that it will be supported completely by its back two feet). (b) At what critical angle would it fall forward?

Sketch and translate
(a) See the sketch of the situation at right. The axis of rotation is through the back support legs of the vending machine.

Find the angle θ at which the vending machine will tip.

Simplify and diagram
Model the vending machine as a rigid body. Three objects exert forces on the vending machine (shown in the side view force diagram at right; the tilt is not shown since the vending machine is barely off the floor): the person exerting force $\vec{F}_{\text{P on M}}$ on the machine, Earth exerting gravitational force $\vec{F}_{\text{E on M}}$, and the floor exerting a force on the machine that we represent with two vector components: normal force $\vec{N}_{\text{F on M}}$ and static friction force $\vec{f}_{\text{s F on M}}$ on the machine.

Represent mathematically
(a) We use the torque condition of equilibrium to analyze the force needed to tilt the vending

machine until the front legs are just barely off the floor. The force exerted by the person has a clockwise turning ability, while the gravitational force has counterclockwise turning ability. The force exerted by the floor on the back legs does not produce a torque, since it is exerted at the axis of rotation.

$$-F_{\text{P on M}}L_{\text{P}} \sin \theta_{\text{P}} + F_{\text{E on M}}L_{\text{E}} \sin \theta_{\text{E}} + N_{\text{F on M}}(0) + f_{\text{s F on M}}(0) = 0$$

(b) We can apply the analysis in Table 8.4 for this situation:

$$\theta_{\text{C}} = \tan^{-1}\left(\frac{\text{depth}}{\text{height}}\right)$$

Solve and evaluate (a) Using the torque equation, we can find the normal force that the person needs to exert on the vending machine to just barely lift its front off the floor:

$$-F_{\text{P on M}}L_{\text{P}}\left(\frac{1.5 \text{ m}}{L_{\text{P}}}\right) + (3700 \text{ N})L_{\text{E}}\left(\frac{0.42 \text{ m}}{L_{\text{E}}}\right)$$
$$+ N_{\text{F on M}}(0) + f_{\text{s F on M}}(0) = 0$$

or $F_{\text{P on M}} = 1000$ N or 220 lb.

(b) We find that the critical tipping angle is

$$\theta_{\text{tipping}} = \tan^{-1}\left(\frac{\text{depth}}{\text{height}}\right) = \tan^{-1}\left(\frac{0.84 \text{ m}}{1.83 \text{ m}}\right) \approx 24°$$

Both answers seem reasonable.

Try it yourself Determine how hard you need to push against the vending machine to keep it tilted at a 25° angle above the horizontal.

Answer   equilibrium is unstable and represents a dangerous situation. of rotation, so the net torque about that axis is zero. However, this a vertical line through the center of mass passes through the axis 0 N. A

The chance of being injured by a tipped vending machine is small since a large force must be exerted on it to tilt it up, and it must be tilted at a fairly large angle before it reaches an unstable equilibrium. A more common danger is falling bookcases in regions subject to earthquakes. The base of a typical 2.5-m-tall bookcase is less than 0.3 m deep. The shelves above the base that are filled with books are the same size as the base. If tilted by just

$$\theta_{\text{C}} = \tan^{-1}\left(\frac{\text{depth}}{\text{height}}\right) \approx \tan^{-1}\left(\frac{0.15 \text{ m}}{1.25 \text{ m}}\right) \approx 7°$$

the bookcase can tip over. In earthquake-prone regions, people often attach a bracket to the top back of the bookcase and then anchor it to the wall.

REVIEW QUESTION 8.6 Why is a ball hanging by a thread in stable equilibrium, while a pencil balanced on its tip is in unstable equilibrium?

Summary

Center of mass The gravitational force that Earth exerts on an object can be considered to be exerted entirely on the object's center of mass.

$$x_{cm} = \frac{m_1 x_1 + m_2 x_2 + m_3 x_3 + \cdots + m_n x_n}{m_1 + m_2 + m_3 + \cdots + m_n}$$

$$y_{cm} = \frac{m_1 y_1 + m_2 y_2 + m_3 y_3 + \cdots + m_n y_n}{m_1 + m_2 + m_3 + \cdots + m_n}$$

Eq. (8.4)

An external force pointing directly toward or away from the center of mass of a free object will not cause the object to turn or rotate. (Sections 8.1 and 8.4)

A **torque** τ around an *axis of rotation* is a physical quantity characterizing the turning ability of a force with respect to a particular axis of rotation. The torque is positive if the force tends to turn the object counterclockwise and negative if it tends to turn the object clockwise about the axis of rotation. (Section 8.2)

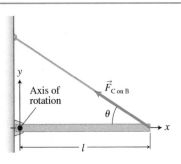

$$\tau = \pm Fl \sin \theta \qquad \text{Eq. (8.1)}$$

Static equilibrium is a state in which a rigid body is at rest and remains at rest both translationally and rotationally. (Section 8.3)

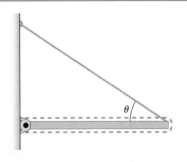

Translational (force) condition:

$$\Sigma F_{on\,O\,x} = F_{1\,on\,O\,x} + \cdots + F_{n\,on\,O\,x} = 0 \quad \text{Eq. (8.2x)}$$

$$\Sigma F_{on\,O\,y} = F_{1\,on\,O\,y} + \cdots + F_{n\,on\,O\,y} = 0 \quad \text{Eq. (8.2y)}$$

Rotational (torque) condition:

$$\Sigma \tau = \tau_1 + \tau_2 + \cdots + \tau_n = 0 \qquad \text{Eq. (8.3)}$$

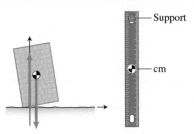

The **equilibrium** of a system is stable against tipping as long as the line through its center of mass passes through the system's area of support.

The **equilibrium** of a system is stable against rotating as long as the center of mass of the rotating object is below the axis of rotation. (Section 8.6)

Questions

Multiple Choice Questions

1. A falling leaf usually flutters while falling. However, we have learned that the force that Earth exerts on an object is exerted at its center of mass and thus should not cause rotational motion. How can you resolve this contradiction?
   (a) A leaf is not a rigid body and the rule does not apply.
   (b) There are other forces exerted on the leaf as it falls besides the force exerted by Earth.
   (c) Some forces were not taken into account when we defined the center of mass.
2. You have an irregularly shaped flat object. To find its center of mass you can do which of the following?
   (a) Find a point where you can put a fulcrum to balance it.
   (b) Push the object in different directions and find the point of intersection of the lines of action of the forces that do not rotate the object.
   (c) Separate the object into several regularly shaped objects whose center of masses you know, and use the mathematical definition of the center of mass to find it.
   (d) All of the above
   (e) a and b only
3. A hammock is tied with ropes between two trees. A person lies in it. Under what circumstances are its ropes more likely to break?
   (a) If stretched tightly between the trees
   (b) If stretched loosely between the trees
   (c) The ropes always have equal likelihood to break.
4. Where is the center of mass of a donut?
   (a) In the center of the hole
   (b) Uniformly distributed throughout the donut
   (c) Cannot be found
5. A physics textbook lies on top of a chemistry book, which rests on a table. Which force diagram below best describes the forces exerted by other objects on the chemistry book (Figure Q8.5)?

FIGURE Q8.5

6. What does it mean if the torque of a force is positive?
   (a) The object exerting the force is on the right side of the axis of rotation.
   (b) The object exerting the force is on the left side of the axis of rotation.
   (c) The force points up.
   (d) The force points down.
   (e) None of these choices is necessarily correct.

7. Constant force $\vec{F}_2$ is exerted on the lower arm of the object in Figure Q8.7. What is the magnitude of the force $\vec{F}_1$ such that the object is in equilibrium?

FIGURE Q8.7

   (a) $F_1 = F_2$
   (b) $F_1 = \sqrt{2}F_2$
   (c) $F_1 = \dfrac{\pi}{2}F_2$
   (d) $F_1 = 2F_2$
   (e) $F_1 = 0$
   (f) The object cannot be in equilibrium no matter how large the force F_1 is.
8. Why do you tilt your body forward when hiking with a heavy backpack?
   (a) The backpack pushes you down.
   (b) Bending forward makes the backpack press less on your back.
   (c) Bending forward moves the center of mass of you and the backpack above your feet.
9. What does it mean if the torque of a 10-N force is zero?
   (a) The force is exerted at the axis of rotation.
   (b) A line parallel to and passing through the place where the force is exerted passes through the axis of rotation.
   (c) Both a and b are correct.
10. What is the maximum angle to the horizontal you can tilt the candleholder in Figure Q8.10 before it tips over? The center of mass of the candleholder is marked on the figure.

FIGURE Q8.10

   (a) 26.6°
   (b) 18.4°
   (c) 14.0°
   (d) 9.4°
   (e) None of the above

11. Water starts dripping at a constant rate into an empty plastic can through a small opening at the top, as shown in **Figure Q8.11**. Which graph shows the correct qualitative dependence of the vertical position of the center of mass of the water-can system (y_{cm}) on the height of the water level in the can (y_{water})?

FIGURE Q8.11

(a)

(b)

(c)

(d)

(e)

Conceptual Questions

12. Is it possible for an object not to be in equilibrium when the net force exerted on it by other objects is zero? Give an example.

13. Explain the meaning of torque so that a friend not taking physics can understand.

14. Something is wrong with the orientation of the ropes shown in **Figure Q8.14**. Use the first condition of equilibrium for the hanging pulley to help explain this error, and then redraw the sketch as you would expect to see it.

FIGURE Q8.14

15. What are the two conditions of equilibrium? What happens if one or the other condition is not satisfied?

16. Give three examples of situations in which an object is starting to rotate even though the sum of the forces exerted on the object is zero.

17. The force that the body muscles exert on bones that are used to lift various objects is usually five to ten times greater than the gravitational force that Earth exerts on the object being lifted. Explain and give an example.

18. A ladder leans against a wall. Construct a force diagram showing the direction of all forces exerted on the ladder. Identify two interacting objects for each force.

19. Using a crowbar, a person can remove a nail by exerting little force, whereas pulling directly on the nail requires a large force to remove it (you probably can't). Why? Draw a sketch to support your answer.

20. Is it more difficult to do a sit-up with your hands stretched in front of you or with them behind your head? Explain.

21. Sit on a chair with your feet straight down at the front of the chair. Keeping your back perpendicular to the floor, try to stand up without leaning forward. Explain why it is impossible to do it.

22. Can you balance the tip of a wooden ruler vertically on a fingertip? Why is it so difficult? Design a method to balance the ruler on your fingertip. Describe any extra material(s) you will use.

23. Try to balance a sharp wooden pencil on your fingertip, point down. (Hint: A small pocketknife might help by lowering the center of mass of the system.)

24. Design a device that you can use to successfully walk on a tightrope.

25. Explain why it is easier to keep your balance while jumping on two feet than while hopping on one.

26. A carpenter's trick to keep nails from bending when they are pounded into a hard material is to grip the center of the nail with pliers. Why does this help?

Problems

Below, **BIO** indicates a problem with a biological or medical focus. Problems labeled **EST** ask you to estimate the answer to a quantitative problem rather than derive a specific answer. Asterisks indicate the level of difficulty of the problem. Problems with no * are considered to be the least difficult. A single * marks moderately difficult problems. Two ** indicate more difficult problems.

8.2 Torque: a new physical quantity

1. Determine the torques about the axis of rotation P produced by each of the four forces shown in **Figure P8.1**. All forces have magnitudes of 120 N and are exerted a distance of 2.0 m from P on some unshown object O.

FIGURE P8.1

2. Three 200-N forces are exerted on the beam shown in **Figure P8.2**. (a) Determine the torques about the axis of rotation on the left produced by forces $\vec{F}_{1 \text{ on B}}$ and $\vec{F}_{2 \text{ on B}}$. (b) At what distance from the axis of rotation must $\vec{F}_{3 \text{ on B}}$ be exerted to cause a torque that balances those produced by $\vec{F}_{1 \text{ on B}}$ and $\vec{F}_{2 \text{ on B}}$?

FIGURE P8.2

3. * A 2.0-m-long, 15-kg ladder is resting against a house wall, making a 30° angle with the vertical wall. The coefficient of static friction between the ladder feet and the ground is 0.40, and between the top of the ladder and the wall the coefficient is 0. Make a list of the physical quantities you can determine or estimate using this information and calculate them.

4. Figure P8.4 shows two different situations where three forces of equal magnitude are exerted on a square board hanging on a wall, supported by a nail. For each case, determine the sign of the total torque that the three forces exert on the board.

FIGURE P8.4

(a)

(b)

8.3 Conditions of equilibrium

5. Three friends tie three ropes in a knot and pull on the ropes in different directions. Adrienne (rope 1) exerts a 20-N force in the positive x-direction, and Jim (rope 2) exerts a 40-N force at an angle 53° above the negative x-axis. Luis (rope 3) exerts a force that balances the first two so that the knot does not move. (a) Construct a force diagram for the knot. (b) Use equilibrium conditions to write equations that can be used to determine $F_{L \, on \, K \, x}$ and $F_{L \, on \, K \, y}$. (c) Use equilibrium conditions to write equations that can be used to determine the magnitude and direction of $\vec{F}_{L \, on \, K}$.

6. Adrienne from Problem 8.5 now exerts a 100-N force $\vec{F}_{A \, on \, K}$ that points 30° below the positive x-axis, and Jim exerts a 150-N force in the negative y-direction. How hard and in what direction does Luis now have to pull the knot so that it remains in equilibrium?

7. * Kate joins Jim, Luis, and Adrienne in the rope-pulling exercise described in the previous two problems. This time, they tie four ropes to a ring. The three friends each pull on one rope, exerting the following forces: $\vec{T}_{1 \, on \, R}$ (50 N in the positive y-direction), $\vec{T}_{2 \, on \, R}$ (20 N, 25° above the negative x-axis), and $\vec{T}_{3 \, on \, R}$ (70 N, 70° below the negative x-axis). Kate pulls rope 4, exerting a force $\vec{T}_{4 \, on \, R}$ so that the ring remains in equilibrium. (a) Construct a force diagram for the ring. (b) Use the first condition of equilibrium to write two equations that can be used to determine $T_{4 \, on \, R \, x}$ and $T_{4 \, on \, R \, y}$. (c) Solve these equations and determine the magnitude and direction of $\vec{T}_{4 \, on \, R}$.

8. You hang a light in front of your house using an elaborate system to keep the 1.2-kg light in static equilibrium (see Figure P8.8). What are the magnitudes of the forces that the ropes must exert on the knot connecting the three ropes if $\theta_2 = 37°$ and $\theta_3 = 0°$? Rope 3 can be tied to the hook on the wall.

FIGURE P8.8

Rope 3   Rope 2
θ_3 θ_2
Rope 1
m

9. * Find the values of the forces the ropes exert on the knot if you replace the light in Problem 8.8 with a heavier 12-kg object and the ropes make angles of $\theta_2 = 63°$ and $\theta_3 = 45°$ (see Figure P8.8).

10. Redraw Figure P8.8 with $\theta_2 = 50°$ and $\theta_3 = 0°$. Rope 2 is found to exert a 100-N force on the knot. Determine m and the magnitudes of the forces that the other two ropes exert on the knot.

11. Determine the masses m_1 and m_2 of the two objects shown in Figure P8.11 if the force exerted by the horizontal cable on the knot is 64 N.

FIGURE P8.11

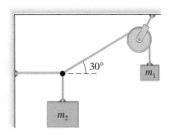

12. * **Lifting an engine** You work in a machine shop and need to move a huge 640-kg engine up and to the left in order to slide a cart under it. You use the system shown in Figure P8.12. How hard and in what direction do you need to pull on rope 2 if the angle between rope 1 and the horizontal is $\theta_1 = 60°$?

FIGURE P8.12

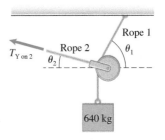

13. * **More lifting** You exert a 630-N force on rope 2 in the previous problem (Figure P8.12). Write the two equations (x and y) for the first condition of equilibrium using the pulley as the object of interest for a force diagram. Calculate θ_1 and θ_2. You may need to use the identity $(\sin\theta)^2 + (\cos\theta)^2 = 1$.

14. **Even more lifting** A pulley system shown in Figure P8.14 will allow you to lift heavy objects in the machine shop by exerting a relatively small force. (a) Construct a force diagram for each pulley. (b) Use the equations of equilibrium and the force diagrams to determine T_1, T_2, T_3, and T_4.

FIGURE P8.14

15. * **Tightrope walking** A tightrope walker wonders if her rope is safe. Her mass is 60 kg and the length of the rope is about 20 m. The rope will break if its tension exceeds 6700 N. What is the smallest angle at which the rope can bend up from the horizontal on either side of her to avoid breaking?

16. * **Lifting patients** An apparatus to lift hospital patients sitting at the sides of their beds is shown in Figure P8.16. At what angle above the horizontal does the rope going under the pulley bend while supporting the 78-kg person hanging from the pulley?

FIGURE P8.16

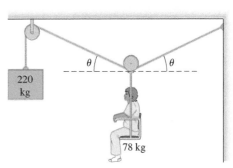

17. A father (80 kg), mother (56 kg), daughter (16 kg), and son (24 kg) try to occupy seats on the seesaw shown in Figure P8.17 so that the seesaw is in equilibrium. Can they succeed? Explain.

FIGURE P8.17

18. *You stand at the end of a uniform diving board a distance d from support 2 (similar to that shown in Figure P8.18). Your mass is m, and the distance between the two supports is a. What can you determine from this information? Make a list of physical quantities and show how you will determine them.

FIGURE P8.18

19. *You place a 3.0-m-long board symmetrically across a 0.5-m-wide chair to seat three physics students at a party at your house. If 70-kg Dan sits on the left end of the board and 50-kg Tahreen sits on the right end of the board, where should 54-kg Komila sit to keep the board stable? What assumptions did you make?

20. **Car jack** You've got a flat tire. To lift your car, you make a homemade lever (see Figure P8.20). A very light 1.6-m-long handle part is pushed down on the right side of the fulcrum and a 0.10-m-long part on the left side supports the back of the car. How hard must you push down on the handle so that the lever exerts an 8000-N force to lift the back of the car?

FIGURE P8.20

21. ***Mobile** You are building a toy mobile, copying the design shown in Figure P8.21. Object A has a 1.0-kg mass. What should be the mass of object B? The numbers in Figure P8.21 indicate the relative lengths of the rods on each side of their supporting cords.

FIGURE P8.21

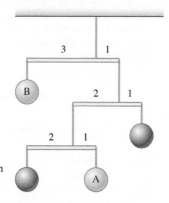

22. **Another mobile** You are building a toy mobile similar to that shown in Figure P8.21 but with different dimensions and replacing the objects with cups. The bottom rod is 20 cm long, the middle rod is 15 cm long, and the top rod is 8 cm long. You put one penny in the bottom left cup, three pennies in the bottom right cup, eleven pennies in the middle right cup, and five pennies in the top left cup. (a) Draw a force diagram for each rod. (b) Determine the cord attachment points and lengths on each side for each rod. (c) What assumptions did you make in order to solve the problem?

23. **EST** Compare the two different designs of nutcracker shown in Figure P8.23 and decide which one is more efficient in cracking a nut. Estimate the forces exerted by each cracker on the nut when a 30-N force is exerted on each handle. Indicate any assumptions that you made. You will need a ruler to solve this problem.

FIGURE P8.23

(a)   (b)

24. Ray decides to paint the outside of his uncle's house. He uses a 4.0-m-long board supported by vertical cables at each end to paint the second floor. The board has a mass of 21 kg. Ray (70 kg) stands 1.0 m from the left cable. What are the forces that each cable exerts on the board?

25. *A 2.0-m-long uniform beam of mass 8.0 kg supports a 12.0-kg bag of vegetables at one end and a 6.0-kg bag of fruit at the other end. At what distance from the vegetables should the beam rest on your shoulder to balance? What assumptions did you make?

26. *A uniform beam of length l and mass m supports a bag of mass m_1 at the left end, another bag of mass m_2 at the right end, and a third bag m_3 at a distance l_3 from the left end ($l_3 < 0.5l$). At what distance from the left end should you support the beam so that it balances?

8.4 Center of mass

27. A person whose height is 1.88 m is lying on a light board placed on two scales so that scale 1 is under the person's head and scale 2 is under the person's feet. Scale 1 reads 48.3 kg and scale 2 reads 39.3 kg. Where is the center of mass of the person?

28. *A seesaw has a mass of 30 kg, a length of 3.0 m, and a fulcrum beneath its midpoint. It is balanced when a 60-kg person sits on one end and a 75-kg person sits on the other end. Locate the center of mass of the seesaw. Where is the center of mass of a uniform seesaw that is 3.0 m long and has a mass of 30 kg if two people of masses 60 kg and 75 kg sit on its ends?

29. *You decide to cut an L-shaped object out of cardboard so that the object is in static equilibrium if hung as shown in Figure P8.29. Find the missing dimension of the object. Cut the object out of some cardboard and check if your result is correct.

FIGURE P8.29

30. *You have a 1-m-wide, 2-m-long 14-kg wooden board. If you place a 5.0-kg pot of soup in one corner of the board, where is the center of mass of the board-pot system?

31. *An 80-kg clown sits on a 20-kg bike on a tightrope attached between two trees. The center of mass of the clown is 1.6 m above the rope, and the center of mass of the bike is 0.7 m above the rope. A load of what mass should be fixed onto the bike and hang 1.5 m below the rope so that the center of mass of the clown-bike-load system is 0.5 m below the rope? What is the force that the rope exerts on each tree if the angle between the rope and the horizontal is 10°?

32. **Figure P8.32** shows a disk of radius R with a circular hole of radius r cut a distance a from the center of the disk. Where is the disk's center of mass? (Hint: You can think of cutting the hole as adding material of negative mass to the original object.)

FIGURE P8.32

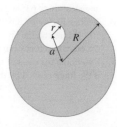

8.5 Skills for analyzing situations using equilibrium conditions

33. **Leg support** A person's broken leg is kept in place by the apparatus shown in Figure P8.33. If the rope pulling on the leg exerts a 120-N force on it, how massive should be the block hanging from the rope that passes over the pulley? First, derive a general expression for m in terms of relevant parameters and then determine the mass of the block.

FIGURE P8.33

20°
120 N
$m = ?$

34. * **Diving board** The diving board shown in Figure P8.18 has a mass of 28 kg and its center of mass is at the board's geometrical center. Determine the forces that support posts 1 and 2 (separated by 1.4 m) exert on the board when a 60-kg person stands on the end of the board 2.8 m from support post 2.

35. ** A uniform cubical box of mass m and side L sits on the floor with its bottom left edge pressing against a ridge. Derive the expression for the least force you need to exert horizontally at the top right edge of the box that will cause its bottom right edge to be slightly off the floor, as shown in Figure P8.35. (Note: With the right edge slightly off the floor, the ground and ridge exert their forces on the bottom left edge of the box.)

FIGURE P8.35

36. * If the force F shown in Figure P8.35 is 840 N and the bottom right edge of the box is slightly off the ground, what is the mass of the cubical box of side 1.2 m?

37. * You decide to hang a new 10-kg flowerpot using the arrangement shown in Figure P8.37. Can you use a slanted rope attached from the wall to the end of the beam if that rope breaks when the tension exceeds 170 N? The mass of the beam is not known, but it looks light.

FIGURE P8.37

30°
1.5 m
10 kg

38. * You decide to hang another plant from a 1.5-m-long 2.0-kg horizontal beam that is attached by a hinge to the wall on the left. A cable attached to the right end goes 37° above the beam to a connecting point above the hinge on the wall. You hang a 100-N pot from the beam 1.4 m away from the wall. What is the force that the cable exerts on the beam?

39. * Now you decide to change the way you hang the pot described in Problems 8.37 and 8.38. You orient the beam at a 37° angle above the horizontal and orient the cable horizontally from the wall to the end of the beam. The beam still holds the 2.0-kg pot and plant hanging 0.1 m from its end. Now determine the force that the cable exerts on the beam and the force that the wall hinge exerts on the beam (its x- and y-components and the magnitude and direction of that force).

40. * What mechanical work must you do to lift a log that is 3.0 m long and has a mass of 100 kg from the horizontal to a vertical position? (Hint: Use the work-energy principle.) You are lifting one end of the log while the other end is on the ground all the time.

8.6 Stability of equilibrium

41. * A 70-g meter stick has a 30-g piece of modeling clay attached to the end. Where should you drill a hole in the meter stick so that you can hang the stick horizontally in equilibrium on a nail in the wall? Draw a picture to help explain your decision.

42. * You are trying to tilt a very tall refrigerator (2.0 m high, 1.0 m deep, 1.4 m wide, and 100 kg) so that your friend can put a blanket underneath to slide it out of the kitchen. Determine the force that you need to exert on the front of the refrigerator at the start of its tipping. You push horizontally 1.4 m above the floor.

43. Can you put a 0.2-kg candle on a 0.6-kg candleholder, as shown in Figure P8.43, without tipping the candleholder over? Explain. The centers of mass of the candle and candleholder are marked on the figure.

FIGURE P8.43

35 cm
20 cm
10 cm   10 cm

44. * You have an Atwood machine (see Figure 4.9) with two blocks each of mass m attached to the ends of a string of length l. The string passes over a frictionless pulley down to the blocks hanging on each side. While pulling down on one block, you release it. Both blocks continue to move at constant speed, one up and the other down. Is the system still in equilibrium? Find the vertical component of the center of mass of the two-block system. Indicate all of your assumptions and the coordinate system used.

45. * **EST** You stand sideways in a moving train. Estimate how far apart you should keep your feet so that when the train accelerates at 2.0 m/s² you can still stand without holding anything. List all your assumptions.

General Problems

46. **EST** Your hand holds a liter of milk (mass about 1 kg) while your arm is bent at the elbow in a 90° angle. Estimate the torque caused by the milk on your arm about the elbow joint. Indicate all numbers used in your calculations. This is an estimate, and your answer may differ by 10 to 50% from the answers of others.

47. **EST** **Body torque** You hold a 4.0-kg computer. Estimate the torques exerted on your forearm about the elbow joint caused by the downward force exerted by the computer on the forearm and the upward 340-N force exerted by the biceps muscle on the forearm. Ignore the mass of the arm. Indicate any assumptions you make.

48. * **Using biceps to hold a child** A man is holding a 16-kg child using both hands with his elbows bent in a 90° angle. The biceps muscle provides the positive torque he needs to support the child. Determine the force that each of his biceps muscles must exert on the forearm in order to hold the child safely in this position. Ignore the triceps muscle and the mass of the arm.

49. **BIO** **Using triceps to push a table** A man pushes on a table exerting a 20-N downward force with his hand. Determine the force that his triceps muscle must exert on his forearm in order to balance the upward force that the table exerts on his hand. Ignore the biceps muscle and the mass of the arm. If you did not ignore the mass of the arm, would the force you determined be smaller or larger? Explain.

50. *BIO **Using biceps to hold a dumbbell** Find the force that the biceps muscle shown in Example 8.5 exerts on the forearm when you lift a 16-kg dumbbell with your hand. Also determine the force that the bone in the upper arm (the humerus) exerts on the bone in the forearm at the elbow joint. The mass of the forearm is about 5.0 kg and its center of mass is 16 cm from the elbow joint. Ignore the triceps muscle.

51. *BIO **Hamstring** You are exercising your hamstring muscle (the large muscle in the back of the thigh). You use an elastic cord attached to a hook on the wall while keeping your leg in a bent position (**Figure P8.51**). Determine the magnitude of the tension force $\vec{T}_{\text{H on L}}$ exerted by the hamstring muscles on the leg and the magnitude of compression force $\vec{F}_{\text{F on B}}$ at the knee joint that the femur exerts on the calf bone. The cord exerts a 20-lb force $\vec{F}_{\text{C on F}}$ on the foot.

FIGURE P8.51

52. *BIO **Lift with bent legs** You injure your back at work lifting a 420-N radiator. To understand how it happened, you model your back as a weightless beam (**Figure P8.52**), analogous to the backbone of a person in a bent position when lifting an object. (a) Determine the tension force that the horizontal cable exerts on the beam (which is analogous to the force the back muscle exerts on the backbone) and the force that the wall exerts on the beam at the hinge (which is analogous to the force that a disk in the lower back exerts on the backbone). (b) Why do doctors recommend lifting objects with the legs bent?

FIGURE P8.52

53. *BIO **Dumbbell lift I** A woman lifts a 3.6-kg dumbbell in each hand with her arm in a horizontal position at the side of her body and holds it there for 3 s (see **Figure P8.53**). What force does the deltoid muscle in her shoulder exert on the humerus bone while holding the dumbbell? The deltoid attaches 13 cm from the shoulder joint and makes a 13° angle with the humerus. The dumbbell in her hand is 0.55 m from the shoulder joint, and the center of mass of her 4.0-kg arm is 0.24 m from the joint.

FIGURE P8.53

Deltoid muscle

13°   Humerus

54. **BIO **Dumbbell lift II** Repeat the previous problem with a 7.2-kg dumbbell. Determine both the force that the deltoid exerts on the humerus and the force that the lifter's shoulder joint exerts on her humerus.

55. *BIO **Facemask penalty** The head of a football running back (see **Figure P8.55**) can be considered as a lever with the vertebra at the bottom of the skull as a fulcrum (the axis of rotation). The center of mass is about 0.025 m in front of the axis of rotation. The torque caused by the force that Earth exerts on the 8.0-kg head/helmet is balanced by the torque caused by the downward forces exerted by a complex muscle system in the neck. That muscle system includes the trapezius and levator scapulae muscles, among others (effectively 0.057 m from the axis of rotation). (a) Determine the magnitude of the force exerted by the neck muscle system pulling down to balance the torque caused by the force exerted by Earth on the head. (b) If an opposing player exerts a downward 180-N (40-lb) force on the facemask, what muscle force would these neck muscles now need to exert to keep the head in equilibrium?

FIGURE P8.55

56. * Eiichi has purchased an adjustable hand grip to use for strengthening hands, fingers, and wrists. The hand grip consists of two handles that can rotate around a common axis and a spring that connects the handles (see **Figure P8.56**). The force needed to squeeze the hand grip can be adjusted from small (Figure P8.56a) to large (Figure P8.56b) by changing the position at which the lower end of the spring is hooked to the left handle. Eiichi, Yuko, and Lars have different explanations for why it is harder to squeeze the hand grip in case (b) compared to (a).

Eiichi: It is harder to squeeze the hand grip because the distance between the axis of rotation and the point where the force is exerted on the left handle by the spring is larger in case (b) than in case (a).

Yuko: It is harder to squeeze the hand grip because the extension of the spring in case (b) is larger than in case (a).

Lars: It is harder to squeeze the hand grip because the angle θ in case (b) is smaller than in case (a).

First, comment on the students' explanations and decide whose ideas are correct and whose aren't. Then construct a complete correct explanation. Indicate any assumptions that you made.

FIGURE P8.56

(a)   (b)

θ

Axis of rotation

57. * **BIO** While browsing books on neurophysiology, you come across a book published in 1967 by Soviet neurophysiologist Nikolai Aleksandrovich Bernstein, *The Co-ordination and Regulation of Movements*, in which he describes a technique to determine the mass of a part of the human body if the position of the center of mass of that body part is known. His technique for determining the mass of the forearm is described as follows: The person lies on a support board, as shown in Figure P8.57. Two readings of the scale are taken: first with the forearm held in position 1 (m_1) and second with it in position 2 (m_2). Knowing the distance from the elbow to the center of mass of the forearm (d_c) and the distance between the knife edges supporting the board (D), the mass of the forearm and hand can be calculated from the following expression:

$$m_{fh} = \frac{D(m_2 - m_1)}{d_c}$$

(a) First, without deriving the expression, evaluate it to see if it is reasonable. Are the units correct? Is the sign of the expression positive? Are qualitative dependences reasonable? (b) Now derive the expression. (c) Why do you think it is important to place the support board on knife edges instead of rigid blocks?

FIGURE P8.57

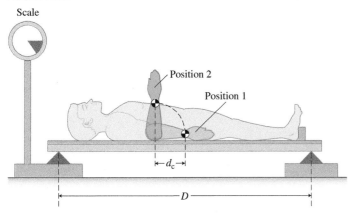

58. ** **Touch detector** You have two force sensors connected to a computer and a meter stick of known mass. The sensors are used to keep the stick horizontal; there are no other supports. You push on the stick with your finger in an arbitrary location. (a) Design an experimental setup that will allow you to determine the magnitude of your pushing force F and the location of your finger x based on the readings of the two force sensors. (b) Derive an expression that can be used as a computer algorithm to calculate x and F using the readings of the force sensors and the parameters of your setup. (c) Evaluate the expression, analyzing the limiting cases.

59. ** Design two experiments to determine the mass of a ruler, using different methods. Your available materials are the ruler, a spring, and a set of objects of standard mass: 50 g, 100 g, and 200 g. One of the methods should involve your knowledge of static equilibrium. After you design and perform the experiment, decide whether the two methods give you the same or different results.

60. * An 80-kg person stands at one end of a 130-kg boat. He then walks to the other end of the boat so that the boat moves 80 cm with respect to the bottom of the lake. (a) What is the length of the boat? (b) How much did the center of mass of the person-boat system move when the person walked from one end to the other? (Hint: Note that the total momentum of the person-boat system remains constant.)

61. **EST** Two people (50 kg and 75 kg) holding hands stand on Rollerblades 1.0 m apart. (a) Estimate the location of their center of mass. (b) The two people push each other and roll apart. Estimate the new location of the center of mass when they are 4.0 m apart. What assumptions did you make? (Hint: Note that the total momentum of the two people remains constant.)

62. ** Find the center of mass of an L-shaped object. The vertical leg has a mass of m_a of length a and the horizontal leg has a mass of m_b of length b. Both legs have the same width w, which is much smaller than a or b.

Reading Passage Problems

BIO **Muscles work in pairs** Skeletal muscles produce movements by pulling on tendons, which in turn pull on bones. Usually, a muscle is attached to two bones via a tendon on each end of the muscle. When the muscle contracts, it moves one bone toward the other. The other bone remains in nearly the original position. The point where a muscle tendon is attached to the stationary bone is called the *origin*. The point where the other muscle tendon is attached to the movable bone is called the *insertion*. The origin is like the part of a door spring that is attached to the doorframe. The insertion is similar to the part of the spring that is attached to the movable door.

During movement, bones act as levers and joints act as axes of rotation for these levers. Most movements require several skeletal muscles working in groups, because a muscle can only exert a pull and not a push. In addition, most skeletal muscles are arranged in opposing pairs at joints. Muscles that bring two limbs together are called flexor muscles (such as the biceps muscle in the upper arm in Figure 8.25). Those that cause the limb to extend outward are called extensor muscles (such as the triceps muscle in the upper arm). The flexor muscle is used when you hold a heavy object in your hand; the extensor muscle can be used, for example, to extend your arm when you throw a ball.

FIGURE 8.25 Muscles often come in flexor-extensor pairs.

63. You hold a 10-lb ball in your hand with your forearm horizontal, forming a 90° angle with the upper arm (Figure 8.25). Which type of muscle produces the torque that allows you to hold the ball?
    (a) Flexor muscle in the upper arm
    (b) Extensor muscle in the upper arm
    (c) Flexor muscle in the forearm
    (d) Extensor muscle in the forearm

64. In Figure 8.25, how far in centimeters from the axis of rotation are the forces that the ball exerts on the hand, that the biceps exerts on your forearm, and that the upper arm exerts on your forearm at the elbow joint?
    (a) 0, 5, 35   (b) 35, 5, 0   (c) 35, 5, 3
    (d) 35, 5, -3   (e) 30, 5, 0

65. Why is it easier to hold a heavy object using a bent arm than a straight arm?
    (a) More flexor muscles are involved.
    (b) The distance from the joint to the place where gravitational force is exerted by Earth on the object is smaller.
    (c) The distance from the joint to the place where force is exerted by the object on the hand is smaller.
    (d) There are two possible axes of rotation instead of one.

66. Why are muscles arranged in pairs at joints?
    (a) Two muscles can produce a bigger torque than one.
    (b) One can produce a positive torque and the other a negative torque.
    (c) One muscle can pull on the bone and the other can push.
    (d) Both a and b are true.

BIO **Improper lifting and the back** A careful study of human anatomy allows medical researchers to use the conditions of equilibrium to estimate the internal forces that body parts exert on each other while a person lifts in a bent position (see Figure 8.19). Suppose an 800-N (180-lb) person lifts a 220-N (50-lb) barbell in a bent position. The situation can be represented with a mechanical model (Figure 8.26a). The cable (the back muscle) exerts a tension force $\vec{T}_{\text{M on B}}$ on the backbone and the support at the bottom of the beam (the disk in the lower back) exerts a compression force $\vec{F}_{\text{D on B}}$ on the backbone. The backbone in turn exerts the same magnitude force on the 2.5-cm-diameter fluid-filled disks in the lower back-bone. Such disk compression can cause serious back problems. A force diagram of this situation is shown in Figure 8.26b. The magnitude of the gravitational force $\vec{F}_{\text{E on B}}$ that Earth exerts on the center of mass of the upper stomach-chest region is 300 N. Earth exerts a 380-N force on the head, arms, and 220-N barbell held in the hands. Using the conditions of equilibrium, we estimate that the back muscle exerts a 3400-N (760-lb) force $\vec{T}_{\text{M on B}}$ on the backbone and that the disk in the lower back exerts a 3700-N (830-lb) force $\vec{F}_{\text{D on B}}$ on the backbone. This is like supporting a grand piano on the 2.5-cm-diameter disk.

FIGURE 8.26 Analysis of a person's backbone when lifting from a bent position.

(a)   **(b)**

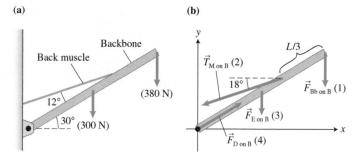

67. Rank in order the magnitudes of the torques caused by the four forces ex-erted on the backbone (see Figure 8.26b), with the largest torque listed first.
    (a) $1 > 2 > 3 > 4$ (b) $2 = 3 > 1 > 4$
    (c) $3 > 2 > 1 > 4$ (d) $2 > 1 > 3 > 4$
    (e) $1 = 2 = 3 = 4$

68. What are the signs of the torques caused by forces 1, 2, 3, and 4, respec-tively, about the origin of the coordinate system shown in Figure 8.26b?
    (a) $+, +, +, +$ (b) $-, +, -, 0$
    (c) $+, -, +, 0$ (d) $-, -, -, 0$
    (e) $+, -, +, -$

69. Which expression below best describes the torque caused by force $F_3 = F_{\text{E on B}}$, the force that Earth exerts on the upper body at its center of mass for the backbone of length L?
    (a) 0 (b) $F_3(2L/3)\sin 12°$
    (c) $F_3(L/2)\cos 30°$ (d) $-F_3(2L/3)\sin 12°$
    (e) $-F_3(L/2)\cos 30°$

70. Which expression below best describes the torque caused by force $F_2 = T_{\text{M on B}}$ exerted by the muscle on the backbone?
    (a) 0 (b) $F_2(2L/3)\sin 12°$
    (c) $F_2(L)\cos 30°$ (d) $-F_2(2L/3)\sin 12°$
    (e) $-F_2(L)\cos 30°$

Rotational Motion

In 1967, a group of astrophysicists from the University of Cambridge in England was looking for quasars using an enormous radio telescope. Jocelyn Bell, a physics graduate student, noticed a series of regular radio pulses in the midst of a lot of receiver noise. It looked like somebody was sending a radio message, turning the signal on and off every 1.33 seconds. This is an incredibly small time for astronomical objects. At first, the astrophysicists believed that they had found signals from extraterrestrial life. The group had, in fact, discovered a new class of astronomical objects, called pulsars, which emit radio signals every second or so. The study of rotational motion explains how pulsars can emit signals so rapidly.

- How can a star rotate 1000 times faster than a merry-go-round?
- Why is it more difficult to balance on a stopped bike than on a moving bike?
- How is the Moon slowing Earth's rate of rotation?

IN THE LAST CHAPTER, we only analyzed rigid bodies that remained at rest. In many cases, however, objects do not remain at rest when torques are exerted—they rotate. Think about a car tire that rotates around the axle as the car moves. In this chapter, we will learn how to describe, explain, and predict such motions.

BE SURE YOU KNOW HOW TO:
- Draw a force diagram for a system (Section 3.1).
- Determine the torque produced by a force (Section 8.2).
- Apply conditions of static equilibrium for a rigid body (Section 8.3).

FIGURE 9.1 Top views comparing the velocities of coins traveling on a rotating disk.

(a) The direction of the velocity $\vec{v}$ for each coin changes continually.

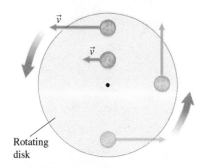

Rotating disk

(b) Coins at the edge travel farther during Δt than those near the center. The speed v will be greater for coins near the edge than for coins near the center.

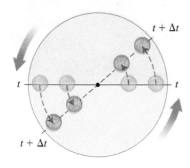

(c) All coins turn through the same angle in Δt, regardless of their position on the disk.

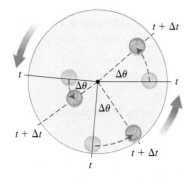

FIGURE 9.2 The rotational position of a point on a rotating disk.

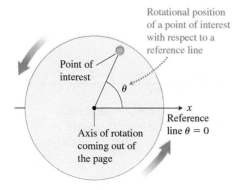

Rotational position of a point of interest with respect to a reference line

Point of interest

θ

x
Reference line $\theta = 0$

Axis of rotation coming out of the page

9.1 Rotational kinematics

In order to understand the motion of rotating rigid bodies, we will follow the same strategy that we used for linear motion. We start by investigating how to describe rotational motion and to explain how forces and torques cause objects to rotate in the way they do. This will help us understand why a bicycle is so stable when moving, how the gravitational pull of the Moon slows the rotation of Earth, and many other interesting phenomena.

One common example of a rotating rigid body is a spinning disk, such as a turning DVD. Suppose a horizontal disk is rotating on a lab bench in front of you and you are looking down on it. You wish to describe the counterclockwise motion of the disk quantitatively. This is trickier than it might seem at first. When we investigated the motion of point-like objects, we did not have to specify which part of the object we were describing, since the object was located at a single point. With a rigid body, there are infinitely many points to choose from. For example, imagine that you place small coins at different locations on the disk, as shown in **Figure 9.1a**. As the disk turns, you observe that the direction of the velocity of each coin changes continually (see the coins on the outer edge of the disk in Figure 9.1a). In addition, a coin that sits closer to the edge moves faster and covers a longer distance during a particular time interval than a coin closer to the center (Figure 9.1b). This means that different parts of the disk move not only in different directions, but also at different speeds relative to you.

On the other hand, there are similarities between the motions of different points on a rotating rigid body. In Figure 9.1c, we see that during a particular time interval, all coins at the different points on the rotating disk turn through the same angle. Perhaps we should describe the rotational position of a rigid body using an angle.

Rotational (angular) position θ

Consider again a disk that rotates on a lab bench about a fixed point. The axis of rotation passes through the center of and is perpendicular to the disk (**Figure 9.2**). A fixed line *perpendicular* to the axis of rotation (like the positive x-axis in Figure 9.2) is used as a reference line. We can draw another line on the disk from the axis of rotation to a point of interest, for example, to a coin sitting on the rotating disk. The angle θ in the counterclockwise direction between the reference line and the line to the point of interest is the **rotational position** (or **angular position**) of the point of interest. The observer is stationary beside the lab bench and looking down on the disk.

> **Rotational position θ** The rotational position θ of a point on a rotating object (sometimes called the angular position) is defined as an angle in the counterclockwise direction between a reference line (usually the positive x-axis) and a line drawn from the axis of rotation to that point. The units of rotational position can be either degrees or radians.

Units of rotational position

The degree (°) is the most familiar unit of rotational position. There are 360° in a circle. If a point on the turning object is at the top of the circle, its position is 90° from a horizontal, positive x-axis. When at the bottom of the circle, its position is 270° or, equivalently, −90°.

The unit for rotational position that is most useful in physics is the **radian**. It is defined in terms of the two lengths shown in **Figure 9.3**. The **arc length** s is the path length in the counterclockwise (CCW) direction along the circumference of the circle from the positive x-axis to the position of a point on the circumference of the rotating object. The other length is the radius r of the circle. The angle θ in units of radians (rad) is the ratio of s and r:

$$\theta \, (\text{in radians}) = \frac{s}{r} \tag{9.1}$$

Note that the radian unit has no dimensions; it is the ratio of two lengths. We can multiply by the radian unit or remove the radian unit from an equation with no consequence. If we put the unit **rad** in the equation, it is usually because it is a reminder that we are using radians for angles.

TIP From Eq. (9.1) we see that the arc length for a 1-rad angle equals the radius of the circle. For example, the 1-rad angle shown in **Figure 9.4** is the ratio of the 2-cm arc length and the 2-cm radius and is simply 1. If you use a calculator to work with radians, make sure it is in the radian mode.

One complete rotation around a circle corresponds to a change in arc length of $2\pi r$ (the circumference of the circle) and a change in rotational position of

$$\theta \, (\text{one complete rotation}) = \frac{s}{r} = \frac{2\pi r}{r} = 2\pi$$

Thus, there are 2π radians in one circle. We can now relate the two rotational position units:

$$360° = 2\pi \text{ rad}$$

We can use this equation to convert between degrees and radians.

We can use Eq. (9.1) to find the arc length s if the radius and rotational position θ are known:

$$s = r\theta \, (\text{for } \theta \text{ in radians only})$$

For example, if a car travels 2.0 rad around a highway curve of radius 100 m, the car travels a path length along the arc equal to

$$s = r\theta = (100 \text{ m})(2.0 \text{ rad}) = 200 \text{ m}$$

We dropped the radian unit in the answer because angles measured in radians are dimensionless.

Rotational (angular) velocity ω

When we were investigating the motion of a point-like object along a single axis, we defined the translational velocity of that object as the rate of change of its linear position. Thus, it seems natural to define the **rotational (angular) velocity** ω of a rigid body as the rate of change of each point's rotational position. Because all points on the rigid body rotate through the same angle in the same period of time (see Figure 9.1c), each point on the rigid body has the same rotational velocity. This means we can just refer to the rotational velocity of the rigid body itself, rather than to any specific point within it.

Rotational velocity ω The average rotational velocity (sometimes called angular velocity) of a turning rigid body is the ratio of its change in rotational position $\Delta\theta$ and the time interval Δt needed for that change (see **Figure 9.5**):

$$\omega = \frac{\Delta\theta}{\Delta t} \tag{9.2}$$

The sign of ω (omega) is positive for counterclockwise turning and negative for clockwise turning, as seen looking along the axis of rotation. *Rotational (angular) speed* is the magnitude of the rotational velocity. The most common units for rotational velocity and speed are radians per second (rad/s) and revolutions per minute (rpm).

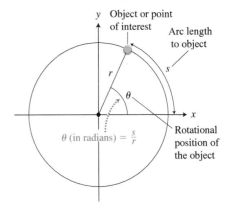

FIGURE 9.3 The rotational position θ in radians is the ratio of the arc length s and the radius r.

$$\theta \, (\text{in radians}) = \frac{s}{r}$$

FIGURE 9.4 The arc length for a 1-rad angle equals the radius of the circle.

A 1-rad rotational position has equal arc length s and radius r.

$s = 2$ cm
$\theta = 1$ rad
$r = 2$ cm

$$\theta = \frac{s}{r} = \frac{2 \text{ cm}}{2 \text{ cm}} = 1 \text{ rad}$$

TIP You cannot calculate arc length using $s = r\theta$ when θ is measured in degrees. You must first convert θ to radians.

FIGURE 9.5 Each point on a rigid body has the same rotational velocity ω.

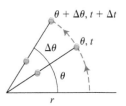

$\theta + \Delta\theta, t + \Delta t$
$\Delta\theta$
θ, t
θ
r

Rotational velocity ω:
$$\omega = \frac{(\theta + \Delta\theta) - \theta}{(t + \Delta t) - t} = \frac{\Delta\theta}{\Delta t}$$

TIP Rotational velocity is the same for all points of a rotating rigid body. It is independent of the radius—the distance of a chosen point on the rigid body from the axis of rotation.

To distinguish the rotational velocity from the familiar velocity that characterizes the linear motion of an object, the latter is called linear velocity. When an object rotates, each point of the object has *linear* velocity. If you examine Figure 9.1a you see that the linear velocity vectors are tangent to the circle. Thus the linear velocity of a point on a rotating object is sometimes called **tangential velocity**.

The **revolution** is a familiar unit from everyday life. One revolution (rev) corresponds to a complete rotation about a circle and equals 360°. The revolution is not a unit of rotational position. It is a unit of *change* in rotational position $\Delta\theta$. Revolutions are usually used to indicate change in rotational position per unit time. For example, a motor that makes 120 complete turns in 1 min is said to have a rotational speed of 120 revolutions per minute (120 rpm). Automobile engines rotate at about 2400 rpm.

> TIP The definition of average rotational velocity or rotational speed becomes the instantaneous values of these quantities if you consider a small time interval in Eq. (9.2) and the corresponding small change in the rotational position. See our earlier discussion of instantaneous velocity in Section 2.7.

Rotational (angular) acceleration α

When we investigated the linear motion of a point-like object along a single axis, we developed the physical quantity acceleration to describe the object's change in velocity. This was translational acceleration, as it described the changing velocity of the object while moving from one position to another. We could apply the same translational acceleration idea to the center of mass of a rigid body that is moving as a whole from one position to another. But usually we are interested in the rate of change of the rigid body's rotational velocity, that is, its **rotational acceleration**. In other words, when the rotation rate of a rigid body increases or decreases, it has a nonzero rotational acceleration.

> **Rotational acceleration α** The average rotational acceleration α (alpha) of a rotating rigid body (sometimes called angular acceleration) is its change in rotational velocity $\Delta\omega$ during a time interval Δt divided by that time interval:
>
> $$\alpha = \frac{\Delta\omega}{\Delta t} \tag{9.3}$$
>
> The unit of rotational acceleration is $(\text{rad}/\text{s})/\text{s} = \text{rad}/\text{s}^2$.

Figure 9.6 shows motion diagrams for three different types of rotational motion. Let's consider these rotational motion diagrams and try to develop a rule for how the sign of the rotational acceleration relates to the rotational velocity for a CCW-turning disk. Note that when the disk's rotational velocity is constant (the lengths of arcs between the dots are the same), its rotational acceleration is zero (Figure 9.6a). When its CCW rotational velocity (positive) is increasing (note that in Figure 9.6b the arcs between the dots increase in length), its rotational acceleration has the same sign (positive) as the rotational velocity. If the disk's CCW rotational velocity (positive) is decreasing (note that in Figure 9.6c the arcs between the dots are shrinking), its rotational acceleration has the opposite sign (negative). Similarly, when a disk is rotating CW (negative rotational velocity) faster and faster, its rotational acceleration has the same sign (negative). If the disk's clockwise rotational velocity (negative) is decreasing, its rotational acceleration has the opposite sign (positive).

What could we conclude about the signs of Earth's rotational velocity and acceleration if we were looking down on Earth from above the North Pole? The rotational velocity would have a positive sign (turning counterclockwise), and because the rotational velocity is constant, the rotational acceleration would be zero. (We will learn later that Earth's rotational velocity is not exactly constant.)

FIGURE 9.6 Three rotational motion diagrams and the corresponding signs of the rotational accelerations.

(a) $\Delta\theta$ is constant
ω is constant
$\alpha = 0$

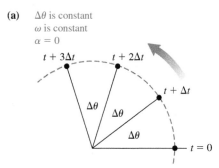

(b) $\Delta\theta$ is increasing
ω is positive (CCW) and increasing
$\alpha > 0$

(c) $\Delta\theta$ is decreasing
ω is positive (CCW) and decreasing
$\alpha < 0$

Relating translational and rotational quantities

Are there mathematical connections between physical quantities describing the rotational motion of a rigid body and the translational motion of different points on the body? Recall that the rotational position θ of a point on a turning object depends on the radial distance r of that point from the axis of rotation and the path length s measured along the arc connecting that point to the reference axis (see Figure 9.3):

$$s = r\theta \tag{9.1}$$

If the angle changes by $\Delta\theta$, the distance of the point of the object along the arc changes by Δs, so that

$$\Delta s = r\,\Delta\theta$$

Do similar relations exist for other quantities? Suppose, for example, that a point on the object changes rotational position by $\Delta\theta$ in a time interval Δt. Its rotational velocity is $\omega = \Delta\theta/\Delta t$. The change in arc length is Δs along its circular path, and its tangential speed (the speed of the object tangent to the circle, sometimes called linear speed) is $v_t = \Delta s/\Delta t$. Substituting for Δs, we get

$$v_t = \frac{\Delta s}{\Delta t} = \frac{r\,\Delta\theta}{\Delta t} = r\left(\frac{\Delta\theta}{\Delta t}\right) = r\omega \tag{9.4}$$

Notice that while the rotational speed of all points of the same rigid body is the same, the tangential (linear) speed of different points increases as their distance from the axis of rotation increases. A similar relationship can be derived that relates that point's acceleration a_t tangent to the circle and its rotational acceleration α:

$$a_t = \frac{\Delta v_t}{\Delta t} = \frac{r\,\Delta\omega}{\Delta t} = r\left(\frac{\Delta\omega}{\Delta t}\right) = r\alpha \tag{9.5}$$

The signs of the rotational position and velocity are positive for counterclockwise turning, and the signs of the translational position and velocity are also positive for counterclockwise motion.

To visualize this relationship, imagine five people (the point objects in **Figure 9.7**) holding on to a long stick that can rotate horizontally about a vertical pole to which it is attached on one end. These people hold on to the stick as it completes a full circle. The person closest to the pole moves the slowest, the next person moves a little faster, and the one at the free end has to almost run to keep the stick in his hands. At a particular time, all of them have the same rotational position θ and the same rotational velocity ω. However, the linear distances and speeds are larger for the people farther from the axis of rotation (larger values of r).

Black holes are an extreme case of rotational motion. Black holes form when some stars at the end of their lives collapse, forming small, very dense objects. If the star was spinning when it was young, it continues to spin when it becomes a black hole, only much faster. The matter near the outer edge of the black hole is usually extremely hot gas that orbits the black hole with a tangential speed near the speed of light. The figure below is an artist's rendition of what such a black hole might look like, if we could see it.

TIP ▸ You get the familiar translational quantities for motion along the circular path by multiplying the corresponding angular rotational quantities by the radius r of the circle.

FIGURE 9.7 A top-view diagram of five people (represented by dots) holding on to a stick that rotates about a fixed pole.

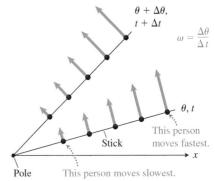

Top view

Five people (the dots) hold a stick that rotates about a fixed pole.

$\theta + \Delta\theta,$
$t + \Delta t$

$\omega = \dfrac{\Delta\theta}{\Delta t}$

θ, t

This person moves fastest.

Stick

Pole   This person moves slowest.

QUANTITATIVE EXERCISE 9.1   Orbiting a black hole

Black hole GRS 1915+105 in the constellation Aquila (the Eagle) is about 35,000 light-years from Earth. It was formed when the core of a star with about 14 solar masses (mass of 14 times the mass of our Sun) collapsed. The boundary of the black hole, called the event horizon, is a sphere with radius about 25 km. Surrounding the black hole is a stable gaseous cloud with an innermost 30-km-radius stable circular orbit. This cloud moves in a circle about the black hole about 970 times per second. Determine the tangential speed of matter in this innermost stable orbit.

Black hole

Gaseous cloud just outside black hole

(CONTINUED)

Represent mathematically We model the cloud as a rotating disk with radius $r = 30,000$ m; to find the rotational speed we convert the rotations in revolutions per second into radians per second: $\omega = (970 \text{ rev/s})(2\pi \text{ rad/rev}) = 6100 \text{ rad/s}$. We need to find the tangential speed v_t of the particles of gas in orbit around the black hole, which is related to the radius r of the circular orbit and the rotational speed ω of the matter:

$$v_t = r\omega$$

Solve and evaluate The speed of the matter in the innermost stable orbit is

$$v_t = r\omega = (30,000 \text{ m})(6100 \text{ rad/s}) = 1.8 \times 10^8 \text{ m/s}$$

This is slightly more than half the speed of light! Actually, the physics we have developed is only moderately applicable in this environment of extreme gravitational forces and high speeds. Our answer is about 20%

high compared to a more sophisticated analysis done using Einstein's theory of general relativity. Nevertheless, using ideas from uniform circular motion, we estimate that the radial acceleration of matter moving in that circular orbit is about $v^2/r \approx 10^{11}g$. It's not a good place to visit.

- -

Try it yourself You ride a carnival merry-go-round and are 4.0 m from its center. A motion detector held by your friend next to the merry-go-round indicates that you are traveling at a tangential speed of 5.0 m/s. What are the rotational speed and time interval needed to complete one revolution on the merry-go-round?

Answer $\omega = 1.3 \text{ rad/s}; T = 5.0 \text{ s}.$

Rotational motion at constant acceleration

Earlier, we developed equations that related the physical quantities t, x, v, and a, which we used to describe the translational motion of a point-like object along a single axis with constant acceleration. Similar equations relate the rotational kinematics quantities t, θ, ω, and α, assuming the rotational acceleration is constant. We're not going to develop them based on observations in the way we did for the equations of translational motion, since the process will be nearly the same. Instead, we will rely on the connections we have seen between the translational and rotational quantities. The analogous rotational motion equations are provided in **Table 9.1** along with the corresponding translational motion equations. Because the quantities that describe motion depend on the choice of the reference frame, always note the location of the observer in a particular situation.

TABLE 9.1 Equations of kinematics for translational motion with constant acceleration and the analogous equations for rotational motion with constant rotational acceleration

Translational motion	Rotational motion	
$v_x = v_{0x} + a_x t$	$\omega = \omega_0 + \alpha t$	(9.3)
$x = x_0 + v_{0x}t + \frac{1}{2}a_x t^2$	$\theta = \theta_0 + \omega_0 t + \frac{1}{2}\alpha t^2$	(9.6)
$2a_x(x - x_0) = v_x^2 - v_{0x}^2$	$2\alpha(\theta - \theta_0) = \omega^2 - \omega_0^2$	(9.7)

For rotational motion, θ_0 is an object's rotational position at time $t_0 = 0$; ω_0 is the object's rotational velocity at time $t_0 = 0$; θ and ω are the rotational position and rotational velocity at some later time t; and α is the object's constant rotational acceleration during the time interval from time zero to time t.

- The sign of the rotational position is positive for counterclockwise θ and negative for clockwise θ from the reference axis.
- The sign of the rotational velocity ω depends on whether the object is rotating counterclockwise $(+)$ or clockwise $(-)$.
- The sign of the rotational acceleration α depends on how the rotational velocity is changing; α has the same sign as ω if the magnitude of ω is increasing and the opposite sign of ω if ω's magnitude is decreasing.

REVIEW QUESTION 9.1 Visualize an ice skater rotating faster and faster in a clockwise direction. What are the signs of rotational velocity and rotational acceleration? As the skater starts slowing down, what are the signs of rotational velocity and acceleration?

9.2 Physical quantities affecting rotational acceleration

What causes a rigid body to have a particular rotational acceleration? When we investigated translational motion we learned that the acceleration of a point-like object was determined by its interactions with other objects, that is, forces that objects in the environment exerted on it. Perhaps there is an analogous way to think about what causes rotational acceleration. In the last chapter, we learned that the net torque produced by forces exerted on a system had to equal zero for the object to remain in static equilibrium, to not start rotating. What happens when the net torque isn't zero? Let's investigate this. In Observational Experiment **Table 9.2**, we perform experiments with a metal arm that can rotate freely around a vertical axis. We attach to the arm two fans that we can turn on and off and move along the arm (**Figure 9.8**).

FIGURE 9.8 The equipment for the experiments in Table 9.2.

OBSERVATIONAL
EXPERIMENT TABLE 9.2   **Turning effects of forces exerted on a rotating arm**

VIDEO
OET 9.2

Observational experiment	Analysis
Experiment 1. Two fans are fixed on the arm. One fan is switched on, and it pushes air along the arm. The arm does not rotate. (Note that all figures show the top view of the experimental setup.) 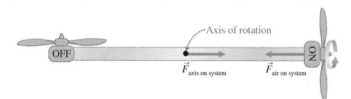	The arm with the fans is the system. The horizontal forces are shown in the figure at left. Because the arm does not accelerate translationally, $\vec{a}_1 = 0$. This is only possible when $\Sigma \vec{F} = 0$. The force exerted by the air on the system and the force exerted by the axis on the system therefore add to zero. The arm does not rotate ($\omega_1 = 0$) or accelerate rotationally ($\alpha_1 = 0$). This is only possible when $\Sigma \tau = 0$; in this case, the torques produced by both forces are equal to zero.
Experiment 2. The turned-on fan rotates so that it pushes air perpendicular to the arm. When the fan is on, the arm rotates faster and faster. We determine its rotational acceleration by measuring the change in rotational velocity and the time interval.	The air exerts a force on the fan and the arm and causes a positive torque, $\tau_2 > 0$. The arm has increasing positive rotational velocity and a positive rotational acceleration, $\alpha_2 > 0$. The arm does not accelerate translationally, $\vec{a}_2 = 0$; this is only possible when the sum of the forces exerted on it is zero, $\Sigma \vec{F} = 0$.
Experiment 3. Both fans are turned on so that air pushes on the fans and the arm in opposite directions. The arm rotates faster and faster, with the rotational acceleration twice as large as before.	The air pushing on the fans and the arm creates two torques of the same magnitude that are both positive: $$\Sigma \vec{F} = 0$$ $$\vec{a}_3 = 0$$ $$\tau_2 + \tau_3 = 2\tau_2$$ $$\alpha_3 = 2\alpha_2$$

(CONTINUED)

Observational experiment	Analysis
Experiment 4. While the arm is rotating, you simultaneously turn off the fans. The arm continues to rotate at a constant rotational velocity.	$\Sigma \vec{F} = 0$ $\vec{a}_4 = 0$ $\Sigma \tau = 0$ $\alpha_4 = 0$ $\omega \neq 0$

Patterns

- An external force that produces a zero torque on the arm does not change the arm's rotational velocity. If the arm is at rest, it remains at rest.
- When there are no external forces exerting torques on a rotating arm, its rotational velocity remains constant.
- External forces that produce a nonzero net torque on the arm cause rotational acceleration. Doubling the net torque doubles the rotational acceleration of the arm.

Using these patterns, we can hypothesize that the rotational acceleration of an object with fixed axis of rotation depends on the external net torque exerted on the object. But is it just the torque that determines the acceleration? In translational motion, the other quantity affecting translational acceleration was the mass of the object. What is the analogous quantity in rotational motion? Could it be the mass, too? Let's test the hypothesis that the rotational acceleration depends on the sum of the external torques and the mass of the object in Testing Experiment **Table 9.3**.

TESTING
EXPERIMENT TABLE 9.3 Testing the hypothesis that mass affects rotational acceleration
VIDEO
TET 9.3

Testing experiment	Prediction	Outcome
Experiment 1. Repeat Experiment 2 from Table 9.2, but this time move the turned-off fan closer to the axis of rotation. 	If the rotational acceleration depends on the external torques and the mass of the system and we have the same fans as in Experiment 2 from Table 9.2, then changing the location of the turned-off fan should not change the rotational acceleration of the system.	The rotational acceleration of the arm is greater than in Experiment 2 from Table 9.2.
Experiment 2. Remove the turned-off fan from the arm and repeat Experiment 2 from Table 9.2.	Because the mass of the system decreases, the rotational acceleration should increase.	The rotational acceleration of the arm is greater than in Experiment 2 from Table 9.2 and greater than in Experiment 1 above.

Conclusion

Although we found that the mass of the system affects its rotational acceleration, we rejected the hypothesis that the rotational acceleration depends *only* on the net exerted torque and the mass of the object. We found that the distribution of mass with respect to the axis of rotation is important, too.

We found that the distribution of mass with respect to the axis of rotation affects an object's rotational acceleration. A simple experiment that you can perform in your kitchen helps you "feel" the effect of mass distribution with respect to the axis of rotation. Lay a broom on a hard, smooth floor. First, try to increase the broom's rotational speed about a vertical axis by spinning it with one hand holding the broomstick near the end opposite the broom head. Then do it again while holding the broom in the middle, nearer the broom head. It is much easier to increase the rotational speed while holding it in the middle. You are turning the same mass, but the location of that mass with respect to the axis of rotation makes a difference—just as we found in Table 9.3.

We can now summarize two patterns that we have discovered so far:

Changes in rotational velocity Rotational acceleration depends on net torque. The greater the net torque, the greater the rotational acceleration. Rotational acceleration also depends on the total mass of the object and the mass distribution with respect to the axis of rotation.

Let us focus on the *distribution* of mass of the rotating object. The closer the mass of the object to the axis of rotation, the easier it is to change its rotational motion. We call the physical quantity characterizing the location of the mass relative to the axis of rotation the **rotational inertia** (also known as the **moment of inertia**) of the object. Rotational inertia I depends on both the total mass of the object and the distribution of that mass about its axis of rotation. For objects of the same mass, the more mass that is located near the axis of rotation, the smaller the object's rotational inertia will be. Likewise, the more mass that is located farther away from the axis of rotation, the greater the object's rotational inertia will be. For objects of different mass but the same mass distribution, the more massive object has more rotational inertia. The higher the rotational inertia of an object, the harder it is to change its rotational motion. In summary, this quantity is the rotational equivalent of mass.

We have now found two factors that affect the rotational acceleration of an object:

- The rotational inertia of the object
- The net torque produced by forces exerted on the object

Notice how this is similar to what we learned when studying translational motion. A nonzero net force (sum of the forces) needs to be exerted on an object to cause its velocity to change. The greater the net force, the greater the translational acceleration of the object. The mass of the object affects its acceleration, too—the greater the mass, the smaller the acceleration for the same net force.

Remember that in all of the experiments we have performed so far in this chapter, rotational motion was all that was possible. Each rigid body was rotating about a fixed axis through its center of mass. If the rigid body were not held fixed, then a change in both translational and rotational motion could occur. The translational acceleration of the center of mass of such an object is determined by Newton's second law $\vec{a} = \Sigma \vec{F}/m$. The rotational acceleration around its center of mass will be determined by the ideas we will investigate over the next several sections.

REVIEW QUESTION 9.2 A solid wooden ball and a smaller solid metal ball have equal mass (the metal ball is smaller because it is much denser than wood). Both can rotate on an axis going through their centers. You exert a force on each that produces the same torque about the axis of rotation. Which sphere's rotational motion will change the least? Explain.

9.3 Newton's second law for rotational motion

FIGURE 9.9 A top view of an experiment to relate torque and rotational acceleration.

Block moves at increasing speed.

m_B

r

$\vec{F}_{\text{F on B}}$

Axis of rotation

Your finger (not shown) pushes the block, causing its rotational acceleration.

To construct a quantitative relation between rotational acceleration, net torque, and rotational inertia we start with a simple example of rotational motion: a small block attached to a light stick that can move on a smooth surface in a circular path (Figure 9.9). The axis of rotation passes through a pin at the other end of the stick. After we analyze this case, we will generalize the result to the rotation of an extended rigid body. You push the block with your finger, exerting a small force $\vec{F}_{\text{F on B}}$ on the block tangent to the circular path. This push causes a torque, which in turn causes the block and stick's rotational velocity about the pin to increase.

The torque produced by the force $\vec{F}_{\text{F on B}}$ is

$$\tau = rF_{\text{F on B}} \sin \theta = rF_{\text{F on B}} \sin 90° = rF_{\text{F on B}}$$

Since the block is small, we can reasonably model it as a point-like object. This allows us to apply Newton's second law. Since the mass of the block is much larger than the mass of the stick, we assume that the stick has no mass. The finger exerts a force of constant magnitude pushing lightly in a direction tangent to the block's circular path. Thus, the tangential component of Newton's second law for the block is

$$a_t = \frac{F_{\text{F on B}}}{m_B}$$

There is a mathematical way to get the torque produced by the pushing force. Rearrange the above equation to get

$$m_B a_t = F_{\text{F on B}}$$

Then multiply both sides of the equation by r, the radius of the circular path:

$$m_B \, r \, a_t = rF_{\text{F on B}}$$

Recall from Eq. (9.5) that $a_t = r\alpha$. Thus,

$$m_B \, r \, (r\alpha) = rF_{\text{F on B}}$$

The right side of this equation equals the torque τ caused by $\vec{F}_{\text{F on B}}$:

$$(m_B r^2) \, \alpha = \tau$$

$$\Rightarrow \alpha = \frac{\tau}{m_B r^2}$$

Examine the above equation and compare it to Newton's second law for the same object—the block not only acquires translational acceleration a_t due to the force exerted on it by the finger, but also acquires rotational acceleration around the axis caused by the torque produced by that same force. This rotational acceleration is directly proportional to the torque produced by the force and inversely proportional to the mass of the block times the square of the distance between the block and the axis of rotation. The latter makes sense—we found experimentally that the farther the mass of the object is from the axis of rotation, the harder it is to change its rotational velocity. Thus the denominator in the equation above is an excellent candidate for the rotational inertia of the block about the pin (the axis of rotation in this situation).

In the above thought experiment, there was just a single force exerted on the object producing a torque about the axis of rotation. More generally, there could be several forces producing torques. It's reasonable that we should add the torques produced by all forces exerted on the object to determine its rotational acceleration:

$$\alpha = \frac{1}{m_B r^2} \Sigma \tau = \frac{1}{m_B r^2} (\tau_1 + \tau_2 + \dots) \qquad (9.8)$$

where $\tau_1, \tau_2, \dots$ are the torques produced by forces $\vec{F}_{1 \text{ on O}}, \vec{F}_{2 \text{ on O}}, \dots$ exerted on the object.

Analogy between translational motion and rotational motion

Notice how similar Eq. (9.8)

$$\alpha = \frac{1}{m_B r^2} \Sigma \tau$$

is to Newton's second law for translational motion

$$\vec{a} = \frac{1}{m} \Sigma \vec{F}$$

When forces are exerted on a point-like object, we can describe its motion using two acceleration-type quantities—translational and rotational acceleration when the axis around which the object rotates is located outside the object. The translational acceleration is determined by Newton's second law, and the rotational acceleration is determined by Eq. (9.8), called **Newton's second law for rotational motion**. There is a strong analogy between each of the three quantities in the two equations (see Table 9.4).

For translational motion, mass is a measure of an object's resistance to the changes in its translational motion. For the rotational motion of a point-like object, the object's mass times the square of its distance r from the axis of rotation (mr^2) is a measure of the object's resistance to the changes in its rotational motion. In summary, the quantity mr^2 is the rotational inertia I of a point-like object of mass m around the axis that is the distance r from the location of the object.

For translational motion, the net force $\Sigma \vec{F}$ exerted on an object of interest by other objects causes that object's velocity to change—it has a translational acceleration $(\vec{a} = \Delta \vec{v}/\Delta t)$. For rotational motion, the net torque $\Sigma \tau$ produced by forces exerted on the object causes its rotational velocity to change—it has a rotational acceleration $(\alpha = \Delta \omega / \Delta t)$.

TABLE 9.4 Analogy between translational and rotational quantities in Newton's second law

	Translational motion	Rotational motion
Inertia of a point-like object	m	mr^2
Cause of acceleration	$\Sigma \vec{F}$	$\Sigma \tau$
Acceleration	$\vec{a}$	α

EXAMPLE 9.2   Pushing a Rollerblader

A 60-kg Rollerblader holds a 4.0-m-long rope that is loosely tied around a metal pole. You push the Rollerblader, exerting a 40-N force on her, which causes her to move increasingly fast in a counterclockwise circle around the pole. The surface she skates on is smooth, and the wheels of her Rollerblades are well oiled. Determine the tangential and rotational acceleration of the Rollerblader.

Sketch and translate We sketch the situation as shown below. We choose the Rollerblader as the object of interest.

Simplify and diagram Since the size of the Rollerblader is small compared to the length of the rope, we can model her as a point-like object. The following figure shows a force diagram for the Rollerblader (viewed from above). Her tangential acceleration has no vertical component along a vertical axis (which would extend up and out of the page in the figure). The upward normal force $\vec{N}_{F \text{ on } R}$ that the floor exerts on her balances the downward gravitational force $\vec{F}_{E \text{ on } R}$ exerted by Earth on

her (these forces are not shown in the figure). The tension force exerted by the rope on the Rollerblader $\vec{T}_{\text{Rope on R}}$ points directly toward the axis of rotation, so that force produces no torque.

Represent mathematically From the force diagram we conclude that your push $\vec{F}_{Y \text{ on } R}$ on the Rollerblader is the only force that produces a nonzero torque $\tau = rF_{Y \text{ on } R} \sin 90° = rF_{Y \text{ on } R}$, where r is the radius of the Rollerblader's circular path.

Use Newton's second law in the tangential direction to determine the Rollerblader's tangential acceleration:

$$a_t = \frac{1}{m_R} \Sigma F_t = \frac{F_{Y \text{ on } R}}{m_R}$$

Use Newton's second law for rotational motion [Eq. (9.8)] to determine the Rollerblader's rotational acceleration:

$$\alpha = \frac{1}{m_R r^2} \Sigma \tau = \frac{1}{m_R r^2} \tau_{Y \text{ on } R} = \frac{1}{m_R r^2} (rF_{Y \text{ on } R}) = \frac{F_{Y \text{ on } R}}{m_R r}$$

Solve and evaluate For the tangential acceleration:

$$a_t = \frac{F_{Y \text{ on } R}}{m_R} = \frac{40 \text{ N}}{60 \text{ kg}} = 0.67 \text{ m/s}^2$$

(CONTINUED)

For the rotational acceleration:

$$\alpha = \frac{F_{Y \text{ on R}}}{m_R r} = \frac{40 \text{ N}}{(60 \text{ kg})(4.0 \text{ m})} = 0.17 \text{ rad/s}^2$$

Let's check the units; note that $\dfrac{\text{N}}{\text{kg m}} = \dfrac{\text{kg} \cdot \text{m}}{\text{s}^2 \cdot \text{kg m}} = \dfrac{1}{\text{s}^2}$. Remember that the radian is not an actual unit. It is dimensionless. It is just a reminder that this is the angle unit appropriate for these calculations. So the units are correct, and the magnitudes for both results are reasonable. The Rollerblader would have a rotational velocity of 0.17 rad/s after 1 s, 0.34 rad/s after 2 s, and so forth—the rotational velocity increases 0.17 rad/s each second.

Try it yourself Suppose you exerted the same force on your friend, but the friend is holding an 8.0-m-long rope instead of a 4.0-m-long rope. How will this affect the rotational acceleration?

Answer

The tangential acceleration will not change, but the rotational acceleration will be half as large. The torque will be doubled because the distance from the axis of rotation to the point where the force is applied will be doubled. But the rotational inertia (mr^2) in the denominator will be quadrupled because of the r^2. The combination will be a reduction of the rotational acceleration by half—0.08 rad/s^2.

Newton's second law for rotational motion applied to rigid bodies

FIGURE 9.10 Comparing the effect of rotational inertia on rotational acceleration.

(a) The rotational inertia I of the two-block system should be twice that of the one-block system.

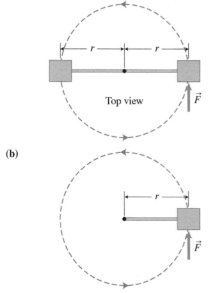

Top view

(b)

Thus, we predict that with equal torque and using $\alpha = \tau/I$, $\alpha_{\text{One block}} = 2\alpha_{\text{Two block}}$.

We know that the mass of an object composed of many small objects with masses m_1, m_2, m_3, etc. is the sum of the masses of its parts: $m = m_1 + m_2 + m_3 + \ldots$. Mass is a scalar quantity and therefore is always positive. The rotational inertia of a point-like object with respect to some axis of rotation is a scalar quantity. Thus, it is reasonable to think that the same rule applies to rigid bodies: the rotational inertia of a rigid body about some axis of rotation is the sum of the rotational inertias of the individual point-like objects that make up the rigid body.

To test this idea, we can use it to calculate the rotational inertia of a lightweight stick with a block attached to each end, as shown in **Figure 9.10a**. The axis of rotation is at the middle of the stick. If our reasoning is correct, the rotational inertia of this two-block rigid body should be twice the rotational inertia of a single block at the end of a stick that is half the length (Figure 9.10b): $I_{\text{Two block}} = 2I_{\text{One block}}$.

If this is correct, and we exert the same torque ($\tau = rF \sin 90° = rF$) on the two-block–stick system, the rotational acceleration of this system should be half the rotational acceleration of the one-block–stick system:

$$\alpha_{\text{Two block}} = \frac{\tau}{I_{\text{Two block}}} = \frac{\tau}{2I_{\text{One block}}} = \frac{1}{2}\left(\frac{\tau}{I_{\text{One block}}}\right) = \frac{1}{2}\alpha_{\text{One block}}$$

We check this prediction by performing a testing experiment. We exert the same force on the two-block system and the one-block system, thus producing the same torque, and measure the angular acceleration of each system. The outcome matches the above prediction. The rotational inertia of the two-block system is twice that of the one-block system.

It appears that the rotational inertia of a rigid body that consists of several point-like parts located at different distances from the axis of rotation is the sum of the mr^2 terms for each part:

$$I = m_1 r_1^2 + m_2 r_2^2 + \ldots$$

Let's apply this idea.

QUANTITATIVE EXERCISE 9.3   **Rotational inertia of a rigid body**

Use what you learned about rotational inertia to write an expression for the rotational inertia of the rigid body shown at right. Each of the five blocks has mass m. They are connected with lightweight sticks of equal length $L/4$.

$I = ?$

Axis of rotation

Represent mathematically Each block of mass m contributes differently to the rotational inertia of the system. The farther the block from the axis of rotation, the greater its contribution to the rotational inertia of the system of blocks. We can add the rotational inertia of each block of mass m about the axis of rotation:

$$I = m(0)^2 + m(L/4)^2 + m(2L/4)^2 + m(3L/4)^2 + m(4L/4)^2$$

Solve and evaluate When added together, the rotational inertia of the five-block system is $I = 1.88mL^2$. Each block (except the one located at the axis of rotation) contributes to the system's rotational inertia. However, blocks farther from the axis of rotation contribute much more than those near the axis. In fact, the block at the right side of the rod contributes more to the rod's rotational inertia (mL^2) than the other four blocks combined ($0.88mL^2$).

Try it yourself Calculate the rotational inertia of the same system altered so that the axis of rotation passes perpendicular through the central block.

Answer $0.625mL^2$.

Calculating rotational inertia

We can calculate the rotational inertia of a rigid body about a specific axis of rotation in about the same way we determined the rotational inertia of the five-block system in Quantitative Exercise 9.3, by adding the rotational inertias of each part of the entire system. However, for most rigid bodies, the parts are not separate objects but are instead parts in a continuous distribution of mass—like in a door or baseball bat. In such a case, we break the continuous distribution of mass into very small pieces and add the rotational inertias for all of the pieces. Consider the person's leg shown in **Figure 9.11**, which we model as a rigid body if none of the joints bend. For example, mass element 7 contributes an amount $m_7 r_7^2$ to the rotational inertia of the leg. The rotational inertia of the whole leg is then

$$I = m_1 r_1^2 + m_2 r_2^2 + \ldots + m_7 r_7^2 + \ldots + m_{18} r_{18}^2 \qquad (9.9)$$

All r's in the above are from the same axis of rotation. The rotational inertia would be different if we chose a different axis of rotation. There are other ways to do the summation process in Eq. (9.9); often it is done using integral calculus, and sometimes I is determined experimentally.

Table 9.5 gives the rotational inertias of some common *uniform* objects (a uniform object is made of the same material throughout) for specific axes of rotation. Notice the coefficients in front of the mR^2 and mL^2 expressions for the objects of different shapes. The value of the coefficient is determined by the mass distribution inside the object and the location of the axis of rotation.

FIGURE 9.11 Add the mr^2 of all the small parts to find the rotational inertia I of the leg.

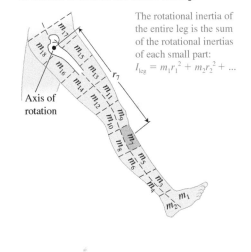

The rotational inertia of the entire leg is the sum of the rotational inertias of each small part:
$I_{leg} = m_1 r_1^2 + m_2 r_2^2 + \ldots$

Axis of rotation

TABLE 9.5 Expressions for the rotational inertia of standard shape objects

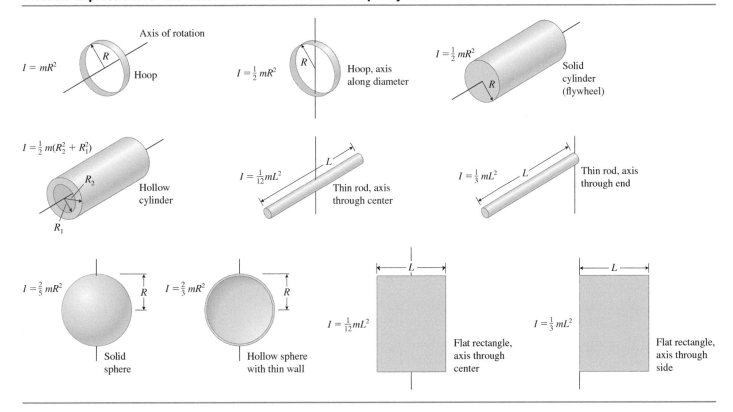

We can now rewrite the rotational form of Newton's second law in terms of the rotational inertia of the rigid body.

> **Rotational form of Newton's second law** One or more objects exert forces on a rigid body with rotational inertia I that can rotate about some axis. The sum of the torques $\Sigma\tau$ due to these forces about that axis causes the object to have a rotational acceleration α:
>
> $$\alpha = \frac{1}{I}\Sigma\tau \qquad (9.10)$$

TIP By writing Newton's second law in the form

$$\vec{a}_S = \frac{1}{m_S}\Sigma\vec{F}_{\text{on S}} = \frac{\vec{F}_{O_1\text{ on S}} + \vec{F}_{O_2\text{ on S}} + \ldots + \vec{F}_{O_n\text{ on S}}}{m_S}$$

we see the cause-effect relationship between the net force $\Sigma\vec{F}_{\text{on S}}$ exerted on the system and the system's resulting translational acceleration $\vec{a}_S$. The same idea is seen in Eq. (9.10), only applied to the rotational acceleration:

$$\alpha_S = \frac{1}{I_S}\Sigma\tau = \frac{\tau_1 + \tau_2 + \ldots + \tau_n}{I_S}$$

EXAMPLE 9.4   **Atwood machine**

In an Atwood machine, a block of mass m_1 and a less massive block of mass m_2 are connected by a string that passes over a pulley of mass M and radius R. What are the translational accelerations a_1 and a_2 of the two blocks and the rotational acceleration α of the pulley?

Sketch and translate A sketch of the situation is shown here. You might recall that we analyzed a similar situation previously (in Section 4.4). However, at that time we assumed that the pulley had negligible (zero) mass. We had no choice but to make that assumption because we had not yet developed the physics for rotating rigid bodies. Now that we have, we can analyze the situation in different ways depending on the choice of system. We analyze the situation using three separate systems: block 1, block 2, and the pulley, and then combine the analyses to answer the questions.

Simplify and diagram We model the blocks as point-like objects and the pulley as a rigid body. Force diagrams for all three objects are shown at top right. A string wrapped around the rim of a pulley (or any disk/cylinder) pulls purely tangentially, so the torque it produces is simply the product of the magnitude of the force and the radius of the pulley. Previously, we assumed that the force exerted by the string pulling down on each side of the massless and frictionless pulley was the same—the pulley just changed the direction of the string but not the tension it exerted on the blocks below. Now, with a pulley with nonzero mass, the forces that the string exerts on two sides are different. If they were not, the pulley would not have a rotational acceleration. We assume that the string does not stretch (the translational acceleration of the blocks will have the same magnitude $a_1 = a_2 = a$) and that the string does not slip on the pulley (a point on the edge of the pulley has the same

translational acceleration as the blocks). We also assume that the pulley's axle is oiled enough that the frictional torque can be ignored.

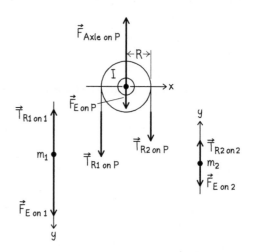

The translational acceleration of the hanging objects is due to the difference between the gravitational force that Earth exerts on them and the tension force that the string exerts on them. The rotational acceleration of the pulley is due to a nonzero net torque produced by the two tension forces exerted on the pulley.

We consider the pulley to be similar to a solid cylinder. Then according to Table 9.5, its rotational inertia around the axis that passes through its center is $I = \frac{1}{2}MR^2$, where R is the radius of the pulley and M is its mass.

Represent mathematically The force diagrams help us apply Newton's second law in component form for the two blocks and the rotational form for the pulley. The coordinate systems used in each case are shown. We choose the coordinate systems so the translational accelerations of both blocks are positive.

Block on the left: $m_1 a = +m_1 g + (-T_{R1 \text{ on } 1})$

Block on the right: $m_2 a = -m_2 g + T_{R2 \text{ on } 2}$

The pulley: $T_{R1 \text{ on } P} R + (-T_{R2 \text{ on } P} R) = I\alpha = \left(\dfrac{MR^2}{2}\right)\left(\dfrac{a}{R}\right)$

$$\Rightarrow (T_{R1 \text{ on } 1})R - (T_{R2 \text{ on } 2})R = \dfrac{MRa}{2}$$

$$\Rightarrow T_{R1 \text{ on } 1} - T_{R2 \text{ on } 2} = \dfrac{Ma}{2}$$

We now have three equations with three unknowns—the two tension forces exerted by the rope on the pulley and the magnitude of the acceleration a of the blocks. We can write expressions for $T_{R1 \text{ on } 1}$ and $T_{R2 \text{ on } 2}$ using the first two equations. We then have

$$T_{R1 \text{ on } 1} = m_1 g - m_1 a$$

$$T_{R2 \text{ on } 2} = m_2 g + m_2 a$$

After substituting these expressions for the rope forces into the pulley equation, we get

$$(m_1 g - m_1 a) - (m_2 g + m_2 a) = Ma/2$$

Solve and evaluate This equation can be rearranged to get an expression for the translational acceleration of the blocks:

$$a = \dfrac{m_1 - m_2}{\frac{1}{2}M + m_1 + m_2} g$$

Notice that if we neglect the mass of the pulley, the acceleration becomes

$$a = \dfrac{m_1 - m_2}{m_1 + m_2} g$$

a larger acceleration than with the pulley (and consistent with the result we got in Chapter 4). Thus, the massive pulley decreases the acceleration of the blocks, which makes sense. If the pulley mass M is much heavier than the masses of the hanging objects m_1 and m_2, the acceleration becomes very small—it is almost like the blocks are hanging from a fixed massive object and not moving at all. We can find the rotational acceleration of the pulley by dividing the translational acceleration by the radius of the pulley because the string does not slip on the pulley.

$$\alpha = \dfrac{a}{R} = \left(\dfrac{m_1 - m_2}{\frac{1}{2}M + m_1 + m_2}\right)\dfrac{g}{R}$$

- -

Try it yourself Determine the translational acceleration of the blocks and the rotational acceleration of the pulley for the following given information: $m_1 = 1.2$ kg, $m_2 = 0.8$ kg, $M = 1.0$ kg, and $R = 0.20$ m.

Answer $a = 1.6 \text{ m/s}^2 \text{ and } \alpha = 7.8 \text{ rad/s}^2.$

Notice the calculations of the translational acceleration of the blocks in Example 9.4. The acceleration is equal to the sum of the forces divided by an effective total mass of the system.

$$a = \dfrac{m_1 - m_2}{\frac{1}{2}M + m_1 + m_2} g$$

Here you see that the pulley only contributes half of its mass due to its mass distribution. We can say that the pulley's "effective mass" is $M/2$ because of how its mass is distributed.

EXAMPLE 9.5   **Throwing a bottle**

A woman tosses a 0.80-kg soft drink bottle vertically upward to a friend on a balcony above. At the beginning of the toss, her forearm rotates upward from the horizontal so that her hand exerts a 20-N upward force on the bottle. Determine the force that her biceps exerts on her forearm during this initial instant of the throw. The mass of her forearm is 1.5 kg, and its rotational inertia about the elbow joint is 0.061 kg·m². The attachment point of the biceps muscle is 5.0 cm from the elbow joint, the hand is 35 cm away from the elbow, and the center of mass of the forearm/hand is 16 cm from the elbow.

Sketch and translate A sketch of the situation is shown at right. There is no information given about the kinematics of the process (for example, no way to directly determine the rotational acceleration of her arm). How can we use the rotational form of Newton's second law to determine the unknown force that the biceps muscle exerts on the woman's forearm during the throw? We do know the force her hand exerts on the bottle and the bottle's mass. So, in the first part of the problem, we can first use the translational form of Newton's second law with the bottle as the system to find the bottle's vertical acceleration at the beginning of the throw. We can then use this acceleration to find the

angular acceleration of the arm and then finally use the rotational form of Newton's second law to find the force that the biceps muscle exerts on her arm during the throw. For this second part of the problem, we choose the lower arm and hand as the system of interest. The axis of rotation is at the elbow joint between the upper arm and the forearm.

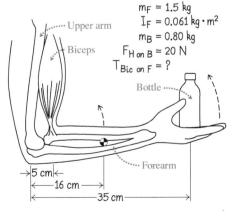

$m_F = 1.5$ kg
$I_F = 0.061$ kg·m²
$m_B = 0.80$ kg
$F_{H \text{ on } B} = 20$ N
$T_{Bic \text{ on } F} = ?$

Upper arm
Biceps
Bottle
Forearm
5 cm
16 cm
35 cm

(CONTINUED)

Simplify and diagram The figure at right is a force diagram for the bottle as a system. Earth exerts a downward 7.8-N gravitational force $\vec{F}_{E \text{ on } B}$ on the bottle and the woman's hand exerts an upward 20-N normal force $\vec{N}_{H \text{ on } B}$ on the bottle. Since these forces do not cancel, the bottle has an initial upward acceleration. Next, consider the forearm and hand as the system. Assume that the forearm and hand form a rigid body. The bottle exerts a downward 20-N force on her hand $\vec{N}_{B \text{ on } H}$. Earth exerts a downward gravitational force $\vec{F}_{E \text{ on } F}$ on the forearm at its center of mass. Her biceps muscle exerts an upward tension force $\vec{T}_{Bic \text{ on } F}$. The upper arm presses down on the forearm at the joint, exerting a force $\vec{F}_{UA \text{ on } F}$. If the upper arm did not push down, the forearm at the joint would fly upward when the biceps muscle pulled up on it.

Bottle is system

$\vec{N}_{H \text{ on } B}$ (20 N)

$\vec{F}_{E \text{ on } B}$ ($m_B g$)

Forearm is system

$\vec{T}_{Bic \text{ on } F}$

$\vec{F}_{E \text{ on } F}$

0.05 m

0.16 m

0.35 m

$\vec{F}_{UA \text{ on } F}$

$\vec{N}_{B \text{ on } H}$

$L_{\text{Joint to Bic}}$

$L_{\text{Joint to cm}}$

$L_{\text{Joint to H}}$

Represent mathematically We first analyze the bottle's motion to determine its translational acceleration; then determine the rotational acceleration of the forearm and hand system; and finally apply the rotational form of Newton's second law to find the force that the biceps needs to exert on the system to cause this rotational acceleration. Consider the initial instant of the bottle's upward trip. The y-component form of Newton's second law applied to the bottle can be used to determine the vertical acceleration a_{By} for the bottle:

$$a_{By} = \frac{N_{H \text{ on } B y} + F_{E \text{ on } B y}}{m_B} = \frac{N_{H \text{ on } B} + (-m_B g)}{m_B}$$

The rotational acceleration of the forearm/hand system at that instant is related to the vertical acceleration of the bottle:

$$\alpha_F = \frac{a_{By}}{r}$$

where r is the distance from the axis of rotation to the hand. The magnitude of the force that the biceps muscle exerts on the forearm $(T_{Bic \text{ on } F})$ can be determined using the rotational form of Newton's second law applied to the forearm/hand. Notice that here the system consists of different parts joined together.

$$T_{Bic \text{ on } F} L_{\text{Joint to Bic}} + (-F_{E \text{ on } F} L_{\text{Joint to cm}})$$
$$+ (-N_{B \text{ on } H} L_{\text{Joint to H}}) = I_F \alpha_F$$

Solve and evaluate We now use the known values of the quantities to solve the problem:

$$a_{By} = \frac{N_{H \text{ on } B} - m_B g}{m_B} = \frac{20 \text{ N} - (0.80 \text{ kg})(9.8 \text{ N/kg})}{(0.80 \text{ kg})}$$
$$= 15.2 \text{ m/s}^2$$

$$\alpha_F = \frac{a_{By}}{r} = \frac{(15.2 \text{ m/s}^2)}{(0.35 \text{ m})} = +43.4 \text{ rad/s}^2$$

$$T_{Bic \text{ on } F}(0.05 \text{ m}) - [(1.5 \text{ kg})(9.8 \text{ N/kg})](0.16 \text{ m})$$
$$- (20 \text{ N})(0.35 \text{ m}) = (0.061 \text{ kg} \cdot \text{m}^2)(43.4 \text{ rad/s}^2)$$

Solving the above equation, we find that $T_{Bic \text{ on } F} = 240 \text{ N} = 54$ lb, a reasonable magnitude for this force.

Try it yourself Determine the force that the woman's biceps exerts on her forearm during the initial instant of a vertical toss of a 100-g rubber ball if she is exerting a 10-N force on the ball.

Answer   430 N.

REVIEW QUESTION 9.3 How is Newton's second law for rotational motion similar to Newton's second law for translational motion? How is it different?

9.4 Rotational momentum

Earlier in this textbook (Chapters 6 and 7), we constructed powerful principles for momentum and energy that allowed us to analyze complex processes that involved translational motion. Is it possible to find analogous principles for the rotational (angular) momentum and rotational energy of extended bodies? Consider the experiments in Observational Experiment Table 9.6.

OBSERVATIONAL
EXPERIMENT TABLE 9.6 Observations concerning rotational motion

Observational experiment		Analysis

Experiment 1. A figure skater initially spins slowly with a leg and two arms extended. Then she pulls her leg and arms close to her body, and her spinning rate increases dramatically.

Initial situation: Large rotational inertia I and small rotational speed ω.

Final situation: Smaller rotational inertia I and larger rotational speed ω.

Experiment 2. A man sitting on a chair that can spin with little friction initially holds dumbbells far from his body and spins slowly. When he pulls the dumbbells close to his body, the spinning rate increases dramatically.

Initial situation: Large rotational inertia I and small rotational speed ω.

Final situation: Smaller rotational inertia I and larger rotational speed ω.

Pattern

- There are no external forces exerted on either person—no torques.
- As the mass distribution of the system moves closer to the axis of rotation, the system's rotational inertia I decreases and the system's rotational speed ω increases (even though the net torque on the system is zero).

For each experiment in Table 9.6, the rotational inertia I of the spinning person decreased (the mass moved closer to the axis of rotation). Simultaneously, the rotational speed ω of the person increased. When I increases and ω decreases (or vice versa), $I\omega$ remains constant. We tentatively propose that when the rotational inertia I of an extended body in an isolated system decreases, its rotational speed ω increases, and vice versa.

Below we will continue to explore this idea quantitatively and then test it experimentally.

Rotational momentum is constant for an isolated system

Note that I is the rotational analog of the mass m of a point-like object and ω is the rotational analog of the translational velocity $\vec{v}$. The linear momentum of an object is the product of its mass m and its velocity $\vec{v}$. Let's propose that a turning object's **rotational momentum** L (analogous to linear momentum $\vec{p} = m\vec{v}$) is defined as

$$L = I\omega \tag{9.11}$$

In the chapter on linear momentum (Chapter 6) we derived a relationship [Eq. (6.4)] between the net force exerted on an object and the change in its linear momentum:

$$\Sigma \vec{F}(t_f - t_i) = \vec{p}_f - \vec{p}_i \tag{6.4}$$

where $\vec{p} = m\vec{v}$. Torque τ is analogous to force $\vec{F}$. Thus, using the analogy between rotational and translational motion, we write

$$\Sigma\tau(t_f - t_i) = L_f - L_i$$

where $L = I\omega$. If a system with one rotating body is isolated, then the external torque exerted on the object is zero. In such a case, the rotational momentum of the object does not change $(0 = L_f - L_i)$, and the object's rotational momentum is constant $(L_f = L_i)$, or

$$I_i\omega_i = I_f\omega_f$$

Note that this is consistent with our tentative qualitative rule. If the final value of one quantity (I or ω) increases for an isolated system, then the other quantity must decrease.

The similarity between the reasoning concerning rotational momentum of an isolated system and the reasoning we used to study linear momentum leads us to believe that we can use the bar chart representation to analyze rotational situations. We can also hypothesize that the change of rotational momentum of a system that is not isolated should be equal to the rotational equivalent of impulse. When a sum of forces $\Sigma\vec{F}$ is exerted on an system over a time interval Δt, the momentum of the system changes by the amount of $\Sigma\vec{F} \cdot \Delta t$. We can use the analogy between translational and rotational motion to say that **when a sum of torques $\Sigma\tau$ is exerted on a system over a time interval Δt, the rotational momentum of the system changes by the rotational impulse $\Sigma\tau\,\Delta t$.**

> **Rotational momentum and rotational impulse** We now have a quantitative relation between rotational momentum $L = I\omega$ and rotational impulse.
>
> $$L_i + \Sigma\tau\Delta t = L_f \tag{9.12}$$
>
> The initial rotational momentum of a turning object plus the product of the net external torque exerted on the object and the time interval during which it is exerted equals the final rotational momentum of the object.
>
> If the net torque that external objects exert on the turning object is zero, or if the torques add to zero, then the rotational momentum L of the turning object remains constant:
>
> $$L_f = L_i \quad \text{or} \quad I_f\omega_f = I_i\omega_i \tag{9.13}$$

TIP Rotational momentum is sometimes called angular momentum.

To explain most of the applications of torque and rotational momentum in this book, we account for their directions using positive or negative signs:

- A torque is positive if it tends to rotate the object counterclockwise and negative if it tends to rotate the object clockwise about the axis of rotation.
- A body rotating counterclockwise has positive rotational momentum and one rotating clockwise has negative rotational momentum.

EXAMPLE 9.6   **Jumping on a merry-go-round**

A boy of mass m_B running at speed v_B steps tangentially onto the stationary circular platform of a merry-go-round that can rotate on a frictionless bearing about its central shaft. The radius of the platform is r and the rotational inertia of the merry-go-round is I_M. After stepping onto the platform, the boy stops moving with respect to the merry-go-round. Derive an expression for the rotational velocity of the merry-go-round ω_f after the boy steps on it.

Sketch and translate A sketch of the situation is shown at right. We choose the boy and the merry-go-round as the system and place the axis of rotation along the shaft of the merry-go-round. The initial state is just before the running boy steps onto the platform, and the final state is when the boy and the platform are rotating together.

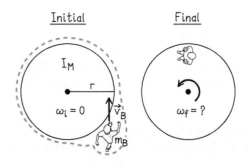

Simplify and diagram We model the boy as a point-like object. The process in the problem is similar to an inelastic collision of two objects.

The vertical forces exerted on the system by Earth and the ground cannot exert torques on the system because they are parallel to the axis of rotation. At the moment when the boy steps onto the platform, the ground exerts a horizontal static friction force on the axis (preventing the merry-go-round from moving translationally), but because this force is exerted on the axis of the merry-go-round, it exerts no torque on the system. Therefore, the sum of all torques exerted on the system is zero, and the rotational momentum of the boy and the merry-go-round should remain constant. We can represent the process with a rotational momentum bar chart as shown in the figure below. The external impulse is zero because the torque is zero; thus the initial momentum of the boy is equal to the final momentum of the boy and merry-go-round together. Because the boy does not move with respect to the merry-go-round after that, and there are no external torques, the rotational momentum of the system and therefore the speed of the system are constant. Here we have a process in which the linear momentum of the system is not constant but the rotational momentum is.

$$L_{Bi} + L_{Mi} + \Sigma_\tau \Delta t = L_{Bf} + L_{Mf}$$

Represent mathematically We use the bar chart to write the mathematical representation of the process: the initial and final rotational momenta of the system should be equal ($L_i = L_f$). Thus,

$$L_{Bi} + L_{Mi} = L_{B \text{ and } Mf}$$

From the bar chart: $L_{Mi} = 0$. The initial rotational momentum of the boy with respect to the axis of rotation (which is outside his body), just before he jumps onto the merry-go-round, can be expressed as

$$L_{Bi} = I_B \omega_B = m_B r^2 \cdot \frac{v_B}{r} = m_B v_B r$$

In the final state, the boy and the merry-go-round rotate as one rigid body. Therefore, the final rotational momentum of the system can be expressed as $L_{B \text{ and } Mf} = (m_B r^2 + I_M)\omega_f$.

Solve and evaluate Combining the equations above, we get

$$m_B v_B r = (m_B r^2 + I_M)\omega_f$$

and finally

$$\omega_f = \frac{m_B v_B r}{m_B r^2 + I_M}$$

To evaluate the expression we need to examine it for extreme cases. The expression predicts that the faster the boy runs, the larger the final rotational velocity of the merry-go-round—this makes sense. The larger the rotational inertia of the merry-go-round, the smaller the final rotational velocity—this prediction makes sense, too. Checking the units for ω_f in our expression, we get $\dfrac{\text{kg} \times \text{m/s} \times \text{m}}{\text{kg} \times \text{m}^2} = \dfrac{1}{\text{s}}$. The units for rotational velocity are rad/s, but as we noted earlier, radians have no dimension and so our unit analysis confirms the expression for ω_f.

Try it yourself In the example, we chose the boy and the merry-go-round as the system. How will the rotational momentum bar charts change if we choose (a) the boy as the system or (b) the merry-go-round as the system? Draw these two bar charts.

Answer

(a) $L_{Bi} + \Sigma_\tau \Delta t = L_{Bf}$ (b) $L_{Mi} + \Sigma_\tau \Delta t = L_{Mf}$

Rotational momentum of a shrinking object

At the beginning of the chapter, we described the discovery of pulsars, astronomical objects that rotate very quickly and emit repetitive radio signals with a very small time interval between them. The signals from the first discovered pulsar had a period of approximately 1.33 s. Astronomers could not explain at first how pulsars could rotate so rapidly. Most stars, including our Sun, rotate very much the way Earth does, usually taking several days to complete one rotation (about a month for our Sun).

However, as a star's core collapses and its mass moves closer to the axis of rotation, its rotational velocity increases because its rotational momentum is constant (assuming the star does not interact with any other objects). How much does a star need to shrink so that its period of rotation becomes seconds instead of days?

TIP Remember that the notation for period is T. Do not confuse this with T also used for the tension force.

EXAMPLE 9.7   **A pulsar**

Imagine that our Sun ran out of nuclear fuel and collapsed. What would its radius have to be in order for its period of rotation to be the same as the first discovered pulsar described above? The Sun's current period of rotation is 25 days.

Sketch and translate First, sketch the process. The Sun is the system. We can convert the present period of rotation of the Sun into seconds ($T_i = 25$ days $= 2.16 \times 10^6$ s). Its mass is $m = 2.0 \times 10^{30}$ kg and its radius is $R_i = 0.70 \times 10^9$ m. After the Sun collapses, its period of rotation will be $T_f = 1.33$ s. What will be its radius?

$$\omega_i = \frac{2\pi}{25 \text{ days}}$$

$m = 2.0 \times 10^{30}$ kg

$R_i = 0.70 \times 10^9$ m

$$\omega_f = \frac{2\pi}{1.33 \text{ s}}$$

$R_f = ?$

(CONTINUED)

Simplify and diagram Assume that the Sun is a sphere with its mass distributed uniformly. Assume also that it does not lose any mass as it collapses.

Represent mathematically Now, apply the principle of rotational momentum conservation (Eq. 9.13) to the Sun's collapse:

$$I_i\omega_i = I_f\omega_f$$

From Table 9.5 we find that the rotational inertia of a sphere rotating around an axis passing through its center is

$$I = \tfrac{2}{5}mR^2$$

The rotational velocity of an object is

$$\omega = \frac{\Delta\theta}{\Delta t} = \frac{2\pi}{T}$$

where T is the period for one rotation. Combining the above three equations, we get

$$\left(\tfrac{2}{5}mR_i^2\right)\frac{2\pi}{T_i} = \left(\tfrac{2}{5}mR_f^2\right)\frac{2\pi}{T_f}$$

Dividing by the $2/5$, 2π, and m on each side of the equation, we get

$$\frac{R_i^2}{T_i} = \frac{R_f^2}{T_f}$$

Solve and evaluate Multiply both sides of the above by T_f and take the square root:

$$
\begin{aligned}
R_f &= \sqrt{\frac{R_i^2 T_f}{T_i}} \\
&= \sqrt{\frac{(0.70 \times 10^9 \text{ m})^2(1.33 \text{ s})}{2.16 \times 10^6 \text{ s}}} \\
&= 5.5 \times 10^5 \text{ m} = 550 \text{ km}
\end{aligned}
$$

Although this is much smaller than the radius of Earth, models of stellar evolution actually do predict that the Sun's core will eventually shrink to this size and possibly smaller.

Try it yourself When massive stars explode, the collapse can shrink their radii to about 10 km. What would be the period of rotation of such a star if it originally had a mass twice the mass of the Sun, a radius that was 1.3 times the Sun's radius, and the same initial period of rotation as the Sun (25 days)?

Answer   2.6 × 10⁻⁴ s.

Vector nature of torque, rotational velocity, and rotational momentum

So far, we have treated torque, rotational velocity, and rotational momentum as scalar quantities, but at the same time we assigned signs to them reflecting CW or CCW direction of rotation. Thus the value of these quantities depends on the direction of rotation. In fact, all of those quantities are vector quantities, but for most applications in this chapter we do not need to treat them as such. However, there are a few applications for which the vector nature of torque, rotational velocity, and rotational momentum is important. The vector direction of both rotational velocity and rotational momentum can be determined using **a right-hand rule**.

Right-hand rule for determining the direction of torque, rotational velocity, and rotational momentum Curl the four fingers of your right hand in the direction of rotation of the turning object. Your thumb, held perpendicular to the fingers, then points in the direction of both the object's rotational velocity and rotational momentum (**Figure 9.12**). To determine the **vector direction** of the torque that a force produces on an object (as opposed to the clockwise/counterclockwise way of describing it) about an axis of rotation, first imagine that the object is at rest and that the torque you are interested in is the only torque exerted on the object. Next, curl the fingers of your right hand in the direction that the torque would make the object rotate. Your thumb, held perpendicular to the fingers, shows the direction of this torque.

Bicycling We can use the vector nature of torque and rotational momentum to understand why a bicycle is much more stable when moving fast—especially if the bicycle has massive tires. Consider an axis of rotation parallel to the ground that passes through the two contact points of the tires with the ground. This axis of rotation is below the center of mass of the stationary bike—an unstable equilibrium (**Figure 9.13**). When the bicycle is moving quickly, the rotating tires (and therefore the bicycle + rider system) have considerable rotational momentum, which will change only when an unbalanced torque

FIGURE 9.12 Using the right-hand rule to determine the direction of an object's rotational momentum $\vec{L}$.

Circle fingers in direction of rotation. Thumb points in the direction of rotational momentum.

FIGURE 9.13 A bicycle that is not moving is in unstable equilibrium.

Center of mass

Axis of rotation

Axis of rotation is below center of mass—unstable.

is exerted on the system. When a bicycle is moving on a smooth road, the rotational velocity and the rotational momentum vectors are perpendicular to the plane of rotation of the bike tires (Figure 9.14), and they are large due to the rapid rotation of the tires.

When the bike + rider system is balanced, the gravitational force exerted by Earth on the system produces no torque since that force points directly at the axis of rotation. If the rider's balance shifts a bit, or the wind blows, or the road is uneven, the system will start tilting. As a result, the gravitational force exerted on the system will produce a torque. However, since the rotational momentum of the system is large, this torque does not change its direction by much right away, but it takes only several tenths of a second for the torque to change the rotational momentum significantly. This is enough time for an experienced rider to make corrections to rebalance the system. The faster the person is riding the bike, the greater the rotational momentum of the system and the more easily the rider can keep the system balanced.

Gyroscopes Guidance systems for spaceships rely on the constancy of rotational momentum in isolated systems to help them maintain their chosen course. Once the ship is pointed in the desired direction, one or more heavy gyroscopes start rotating. The gyroscope is a wheel whose axis of rotation keeps the ship oriented in the chosen direction. The gyroscope is similar to the rotating bicycle tires that help keep a rider upright without tipping or changing direction. Gyroscopes are also used in cameras to prevent them from vibrating or moving while the camera lens is open.

REVIEW QUESTION 9.4 After a playground merry-go-round is set in motion, its rotational speed decreases noticeably if another person jumps on it. However, if a person riding the merry-go-round steps off, the rotational speed seems not to change at all. Explain.

9.5 Rotational kinetic energy

We are familiar with the kinetic energy $(1/2)mv^2$ of a single particle moving along a straight line or in a circle. It would be useful to calculate the kinetic energy of a rotating body—like Earth. Doing so would allow us to use the work-energy approach to solving problems involving rotation. Let's start by deriving an expression for the **rotational kinetic energy** of a single particle of mass m moving in a circle of radius r at speed v. According to the kinematics in Section 9.1, its linear speed v and rotational speed ω are related:

$$v = r\omega$$

Thus, the kinetic energy of this particle moving in a circle can be written as

$$K_{rot} = \tfrac{1}{2}mv^2 = \tfrac{1}{2}m(r\omega)^2 = \tfrac{1}{2}(mr^2)\omega^2 = \tfrac{1}{2}I\omega^2$$

where $I = mr^2$ is the rotational inertia of a particle moving a distance r from the center of its circular path. The expression for the translational kinetic energy $(1/2)mv^2$ of a particle is similar to the rotational version, which involves the product of a mass-like term I and the square of a speed-like term ω. Can we use the expression $\tfrac{1}{2}I\omega^2$ for the rotational kinetic energy of a rotating rigid body?

To test this idea, consider a solid sphere of known radius R and mass m that can rotate freely on an axis. We wrap a string around the sphere and pull the string with a force probe exerting a constant force of a known magnitude so that the sphere starting at rest completes 5.0 revolutions (Figure 9.15a). After we stop pulling, we measure the rotational speed ω of the sphere. But before measuring it, we predict its value using this expression for rotational kinetic energy. If we choose the sphere as the system, the string is the only external object that exerts a force that causes a nonzero torque on the sphere. This string force does work, which changes the sphere's kinetic energy from zero to some new value (Figure 9.15b). Thus, the initial rotational kinetic energy of

FIGURE 9.14 Rotating bicycle tires have rotational momentum that stabilizes the bicycle.

$\vec{L}$
$L = I\omega$ Axis of rotation

FIGURE 9.15 A string pulls a solid sphere (top view).

(a)

5.0 N
Axis of rotation
$R = 0.10$ m
$m = 10$ kg

(b)   Work done by the string causes the rotational kinetic energy of the sphere to increase.

$K_{rot\,i} + W = K_{rot\,f}$

0

the sphere (zero) plus the work done by the string on the sphere during these five turns equals the final rotational kinetic energy of the sphere:

$$K_{\text{rot i}} + W = K_{\text{rot f}}$$

The string pulls parallel to the displacement of the edge of the sphere during the entire time. We can use the expression for the rotational kinetic energy under test to predict the magnitude of the final rotational speed:

$$0 + F_{\text{String on Sphere}}(5 \cdot 2\pi R) \cos 0 = \tfrac{1}{2} I \omega^2$$

where $5 \cdot 2\pi R$ is the distance the string is pulled—five circumferences of the sphere. The above leads to a prediction of the final rotational speed:

$$\omega_{\text{f}} = \sqrt{\frac{F_{\text{String on Sphere}} \cdot 20\pi R}{I}}$$

From Table 9.5 we know that the rotational inertia of a solid sphere of radius R and mass m is $I = (2/5)mR^2$ (the axis passes through the center of the sphere); in our sphere, $m = 10$ kg and $R = 0.10$ m. Thus, the rotational inertia is

$$I = (2/5)mR^2 = (2/5)(10\text{ kg})(0.10\text{ m})^2 = 0.040\text{ kg} \cdot \text{m}^2$$

We pull the string so that it exerts a 5.0-N force on the edge of the sphere. Thus, the sphere's final speed should be

$$\omega_{\text{f}} = \sqrt{\frac{(5.0\text{ N})(20\,\pi)(0.10\text{ m})}{(0.040\text{ kg} \cdot \text{m}^2)}} = \left(28\,\frac{\text{rad}}{\text{s}}\right)\left(\frac{1\text{ rev}}{2\pi\text{ rad}}\right) = 4.5\,\frac{\text{rev}}{\text{s}}$$

When we measure the final angular velocity, it is about 4.5 rev/s.

We can apply our new understanding of rotational kinetic energy to predict the outcome of the experiment shown in **Figure 9.16a**. A solid cylinder and a hoop of the same radius and mass start rolling at the top of an inclined plane. Which object reaches the bottom of the plane first? What is the ratio of their speeds at the bottom?

Both Earth-object systems start with the same gravitational potential energy. As they roll, both acquire translational and rotational kinetic energies. We can represent the process in a work-energy bar chart (Figure 9.16b). The bar chart helps us construct a mathematical description:

$$mgy_{\text{i}} = \tfrac{1}{2}mv_{\text{f}}^2 + \tfrac{1}{2}I\omega_{\text{f}}^2$$

Since the objects are rolling without skidding, the rotational speed ω and translational speed v are related as $\omega = v/r$. We substitute this into the above energy equation and then rearrange:

$$mgy_{\text{i}} = \tfrac{1}{2}\left(m + \frac{I}{r^2}\right)v_{\text{f}}^2$$

$$\Rightarrow v_{\text{f}} = \sqrt{\frac{2mgy_{\text{i}}}{\left(m + \dfrac{I}{r^2}\right)}}$$

This expression for v_{f} suggests that at the bottom of the inclined plane, the object with greater rotational inertia I will have the smaller translational speed v_{f}. Using the information in Table 9.5, we have $I_{\text{cylinder}} = \tfrac{1}{2}mr^2$ and $I_{\text{hoop}} = mr^2$. Thus $I_{\text{hoop}} > I_{\text{cylinder}}$, giving the conclusion that the cylinder should reach the bottom first.

We can now find the ratio of their final speeds. We insert the expressions for the rotational inertia of the solid cylinder and of the hoop to find their final speeds. For the cylinder, $v_{\text{f}} = \sqrt{\tfrac{4}{3}gy_{\text{i}}}$, and for the hoop, $v_{\text{f}} = \sqrt{gy_{\text{i}}}$. The ratio of their final speeds is $\sqrt{\tfrac{4}{3}}$, with the cylinder moving faster. When we perform the experiment (see the video), we see that the cylinder reaches the bottom of the plane first.

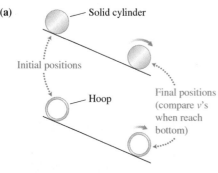

FIGURE 9.16 A race between a hoop and a cylinder of the same mass.

(a)   Solid cylinder

Initial positions

Hoop   Final positions (compare v's when reach bottom)

(b) $U_{\text{gi}} = K_{\text{tran f}} + K_{\text{rot f}}$

0

VIDEO 9.1

This is the second testing experiment involving rotational kinetic energy in which the outcome matched the prediction. Given this support for our prediction and the lack of counterevidence, we will use this mathematical expression for a rigid body's rotational kinetic energy.

> **Rotational kinetic energy** The rotational kinetic energy of an object with rotational inertia I turning with rotational speed ω is
>
> $$K_{rot} = \tfrac{1}{2}I\omega^2 \qquad (9.14)$$

TIP When you encounter a new physical quantity, always check whether its units make sense. In this particular case, the units for I are $kg \cdot m^2$ and the units for ω^2 are $1/s^2$. Thus, the unit for kinetic energy is $kg \cdot m^2/s^2 = (kg \cdot m/s^2)\, m = N \cdot m = J$, the correct unit for energy.

Flywheels for storing and providing energy

You stop your car at a stoplight. Before stopping, the car had considerable kinetic energy; after stopping, the kinetic energy is zero. It has been converted to internal energy due to friction in the brake pads. Unfortunately, this thermal energy cannot easily be converted back into a form that is useful. Is there a way to convert that translational kinetic energy into some other form of energy that would help the car regain translational kinetic energy when the light turns green?

Efforts are under way to use the rotational kinetic energy of flywheels (rotating disks) for this purpose. Instead of rubbing a brake pad against the wheel and slowing it down, the braking system would, through a system of gears or through an electric generator, convert the car's translational kinetic energy into the rotational kinetic energy of a flywheel. As the car's translational speed decreases, the flywheel's rotational speed increases. This rotational kinetic energy could then be used to help the car start moving, rather than relying entirely on the chemical potential energy of gasoline.

EXAMPLE 9.8 Flywheel rotational speed

A 1600-kg car traveling at 20 m/s approaches a stop sign. If it could transfer all of its translational kinetic energy to a 0.20-m-radius, 20-kg flywheel while stopping, what rotational speed would the flywheel acquire?

Sketch and translate A sketch of the situation is shown below. The system of interest is the car, including the flywheel.

Car: $v_i = +20$ m/s $v_f = 0$ $M = 1600$ kg
Flywheel: $\omega_i = 0$ $\omega_f = ?$ $m = 20$ kg
                                                 $r = 0.20$ m

Simplify and diagram The process is represented at right with a bar chart. The initial energy of the system is the car's translational kinetic energy; the final energy is the flywheel's rotational kinetic energy. Braking converts car's initial kinetic energy into the flywheel's rotational energy, which is saved for future use. Assume that the flywheel is a solid disk with rotational inertia of $(1/2)mr^2$ (see Table 9.5).

$K_{tran\,i} = K_{rot\,f}$

Represent mathematically Use the bar chart to help construct an energy conservation equation:

$$K_{tran\,i} = K_{rot\,f}$$

$$\tfrac{1}{2}Mv_i^2 = \tfrac{1}{2}I\omega_f^2$$

Multiplying both sides of the equation by 2 and dividing by I, we get

$$\omega_f^2 = \frac{Mv_i^2}{I}$$

The rotational inertia of the disk (a solid cylinder) is $I_{cylinder} = (1/2)mr^2$. Thus,

$$\omega_f^2 = \frac{Mv_i^2}{I} = \frac{Mv_i^2}{\tfrac{1}{2}mr^2} = \frac{2Mv_i^2}{mr^2}$$

Solve and evaluate To find the rotational speed, take the square root of both sides of the above equation:

$$\omega_f = \frac{v_i}{r}\sqrt{\frac{2M}{m}} = \frac{20\ \text{m/s}}{0.20\ \text{m}}\sqrt{\frac{2(1600\ \text{kg})}{(20\ \text{kg})}} = 1300\ \text{rad/s}$$

$$= 200\ \text{rev/s} = 12{,}000\ \text{rpm}$$

Try it yourself Could you store more energy in a rotating hoop or in a rotating solid cylinder, assuming they have the same mass, radius, and rotational speed?

Answer The hoop has a greater rotational inertia and therefore would have a greater rotational kinetic energy at the same rotational speed.

9.6 Tides and Earth's day

FIGURE 9.17 The ocean bulges on both sides of Earth along a line toward the Moon.

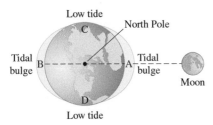

The level of the ocean rises and falls by an average of 1 m twice each day, a phenomenon known as the tides. Many scientists, including Galileo, tried to explain this phenomenon and suspected that the Moon was a part of the answer. Isaac Newton was the first to explain how the motion of the Moon actually creates tides. He noted that at any moment, different parts of Earth's surface are at different distances from the Moon and that the distance from a given location on Earth to the Moon varied as Earth rotated. As illustrated in **Figure 9.17**, point A is closer to the Moon than the center of Earth or point B are, and therefore the gravitational force exerted by the Moon on point A is greater than the gravitational force exerted on point B. Due to the difference in forces, Earth elongates along the line connecting its center to the Moon's. This makes water rise to a high tide at point A and surprisingly also at B. The water "sags" a little at points C and D, forming low tides at those locations. When the Sun is aligned with the Moon and Earth, the bulging is especially pronounced.

As the solid Earth rotates beneath the tidal bulges, it attempts to drag the bulges with it. A large amount of friction is produced, which converts the rotational kinetic energy of Earth into internal energy. The time interval needed for Earth to complete one turn on its axis increases by 0.0016 s every 100 years. In other words, the Earth day is slowly getting longer. In a very long time, Earth will stop turning relative to the Moon and an unmoving tidal bulge will face toward and away from the Moon. This "tidal locking" has already occurred on the Moon (although it is solid, the same principle applies), which is why on Earth we only see one side of the Moon. It rotates around its axis with the same period as it moves around Earth.

Let's use our new understanding of rotational dynamics to estimate the friction force the tides exert on Earth, causing Earth's rotation rate to decrease.

EXAMPLE 9.9 | **Tides slow Earth's rotation**

Estimate the effective tidal friction force exerted by ocean water on Earth that causes a 0.0016-s increase in Earth's rotation time every 100 years.

Sketch and translate The situation is already sketched in Figure 9.17. The solid Earth is the system; the water covering most of its surface is considered an external object for this estimate. Earth rotates counterclockwise, taking 24 hours for one revolution (one period), as seen looking down on the North Pole. Remember that when an object rotates counterclockwise, its angular velocity is considered to be positive. Since Earth's rotation is gradually slowing, the tidal friction force is producing a negative torque, opposite the positive sign of the rotational velocity. We need to find the magnitude of the force that would increase the time of one revolution by 0.0016 s in 100 years.

Simplify and diagram Assume that the solid Earth is a sphere covered uniformly with water on its surface. Assume also that the frictional force exerted by the water on Earth is constant in magnitude and exerted at the equator. The force exerted by the tidal bulge on Earth produces a torque that opposes Earth's rotation.

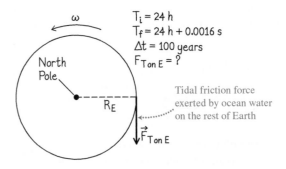

$T_i = 24$ h
$T_f = 24$ h $+ 0.0016$ s
$\Delta t = 100$ years
$F_{T \text{ on } E} = ?$

Tidal friction force exerted by ocean water on the rest of Earth

Represent mathematically To estimate the friction force exerted by the oceans on Earth, we need to determine the torque produced by that force. We can use the rotational form of Newton's second law if we can determine the rotational inertia of Earth and its rotational acceleration. The rotational acceleration of Earth is

$$\alpha = \frac{\Delta \omega}{\Delta t} = \frac{\omega_f - \omega_i}{\Delta t} = \frac{\left(+ \dfrac{2\pi}{T_f} \right) - \left(+ \dfrac{2\pi}{T_i} \right)}{\Delta t}$$

$$= \frac{2\pi}{\Delta t} \left(\frac{1}{T_f} - \frac{1}{T_i} \right)$$

where $T_i = 24 \text{ h} \left(\dfrac{3600 \text{ s}}{1 \text{ h}} \right) = 86{,}400 \text{ s}$, $T_f = T_i + \Delta T = 86{,}400 \text{ s} + 0.0016 \text{ s}$, and $\Delta t = 100 \text{ years} \left(\dfrac{365 \text{ days}}{1 \text{ year}} \right) \left(\dfrac{86{,}400 \text{ s}}{1 \text{ day}} \right) = 3.15 \times 10^9 \text{ s}$.

Because T_f and T_i are so close, your calculator will likely evaluate the rotational acceleration of Earth to be zero. To deal with this, we put the equation into another form.

$$\alpha = \frac{2\pi}{\Delta t} \left(\frac{1}{T_f} - \frac{1}{T_i} \right) = \frac{2\pi}{\Delta t} \left(\frac{T_i - T_f}{T_i T_f} \right)$$

$$= \frac{2\pi}{\Delta t} \left(\frac{-\Delta T}{T_i (T_i + \Delta T)} \right) = \frac{2\pi}{\Delta t} \left(\frac{-\Delta T}{T_i^2 + T_i \Delta T} \right)$$

Now look at the two terms in the denominator, T_i^2 and $T_i \Delta T$. Because ΔT is so small, the second term is much less than the first term. This means the second term can be dropped without affecting the result in any significant way. Thus,

$$\alpha = -\frac{2\pi \Delta T}{\Delta t T_i^2}$$

We can use the rotational form of Newton's second law to get an alternative expression for Earth's rotational acceleration:

$$\alpha = \frac{1}{I} \Sigma \tau = \frac{1}{\frac{2}{5} m R_E^2} (F_{T \text{ on } E} R_E \sin 90°)$$

The R_E in the numerator is the distance from the axis of rotation to the surface where the friction force is exerted, the radius of Earth. The R_E in the denominator is also the radius of Earth.

$$\alpha = \frac{1}{\frac{2}{5} m R_E^2} (F_{T \text{ on } E} R_E \sin 90°) = \frac{5 F_{T \text{ on } E}}{2 m R_E}$$

Setting the two expressions for the magnitude of the rotational acceleration equal to each other, we get

$$\frac{2\pi \Delta T}{\Delta t T_i^2} = \frac{5 F_{T \text{ on } E}}{2 m R_E}$$

Solve and evaluate Solve the previous equation for the force of the tides on Earth:

$$F_{T \text{ on } E} = \frac{4\pi m R_E \, \Delta T}{5 \, \Delta t T_i^2}$$

$$= \frac{4\pi (5.97 \times 10^{24} \text{ kg})(6.38 \times 10^6 \text{ m})(0.0016 \text{ s})}{5(3.15 \times 10^9 \text{ s})(86{,}400 \text{ s})^2}$$

$$= 6.5 \times 10^9 \text{ N}$$

The magnitude of this friction force seems big, and it is. But when exerted on an object of such large mass (5.97×10^{24} kg), the effect is extremely tiny. By comparison, the gravitational force that the Sun exerts on Earth is 3.5×10^{22} N.

Try it yourself Estimate the time interval in years that it will take for Earth's rotation to change from 24 hours to 27 days.

Answer

Using Eq. (9.3), we find $t = (\omega - \omega_0)/\alpha \approx -\omega_0/\alpha = (-7 \times 10^{-5} \text{ rad/s})/(-4 \times 10^{-22} \text{ rad/s}^2) = 2 \times 10^{17}$ s, almost 10 billion years.

REVIEW QUESTION 9.6 How can you explain the increasing length of a day on Earth?

Summary

Rotational kinematics The rotational motion of a rigid body can be described using quantities similar to those for translational motion—rotational position θ, rotational velocity ω, and rotational acceleration α. (Section 9.1)

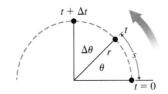

- Rotational position (in radians)
$$\theta = s/r \qquad \text{Eq. (9.1)}$$
- Rotational velocity (in rad/s)
$$\omega = \Delta\theta/\Delta t \qquad \text{Eq. (9.2)}$$
- Rotational acceleration (in rad/s^2)
$$\alpha = \Delta\omega/\Delta t \qquad \text{Eq. (9.3)}$$

Rotational inertia I is the physical quantity equal to the sum of the mr^2 terms for each part of an object and depends on the distribution of mass relative to an axis of rotation. (Sections 9.2 and 9.3)

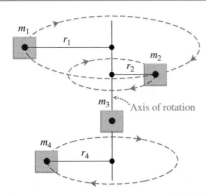

$$I = \Sigma mr^2 \qquad \text{Eq. (9.9)}$$

Rotational dynamics A rigid body's rotational acceleration equals the net torque produced by forces exerted on the body divided by its rotational inertia. (Section 9.3)

$$\alpha = \frac{1}{I}\Sigma\tau \qquad \text{Eq. (9.10)}$$

Rotational momentum L is the product of the rotational inertia I of an object and its rotational velocity ω, positive for counterclockwise rotation and negative for clockwise rotation. For an isolated system (zero net torque exerted on it), the rotational momentum of the system is constant. (Section 9.4)

$I_i\omega_i$ $\qquad$ $I_f\omega_f$

$$L_{Ai} + L_{Bi} + \Sigma\tau\Delta t = L_{Af} + L_{Bf}$$

$$L = I\omega \qquad \text{Eq. (9.11)}$$
Translational motion equivalent:
$$\vec{p} = m\vec{v} \qquad \text{Eq. (6.1)}$$
For isolated system,
$$I_i\omega_i = I_f\omega_f \qquad \text{Eq. (9.13)}$$
For nonisolated systems, rotational impulse $\Sigma\tau\,\Delta t$ is nonzero.

Rotational kinetic energy K_{rot} of a rigid body is energy due to the rotation of the object about a particular axis. This is another form of kinetic energy that is included in the work-energy principle. (Section 9.5)

$$U_{gi} = K_{tran\,f} + K_{rot\,f}$$

$$K_{rot} = \tfrac{1}{2}I\omega^2 \qquad \text{Eq. (9.14)}$$
Translational motion equivalent:
$$K_{tran} = \tfrac{1}{2}mv^2 \qquad \text{Eq. (7.5)}$$

Questions

Multiple Choice Questions

1. Is it easier to open a door that is made of a solid piece of wood or a door of the same mass made of light fiber with a steel frame?
   (a) Wooden door
   (b) Fiber door with a steel frame
   (c) The same difficulty
   (d) Not enough information to answer

2. You push a child on a swing. Why doesn't the child continue in a vertical loop over the top of the swing?
   (a) The torque of the force that Earth exerts on the child pulls him back.
   (b) The swing does not have enough kinetic energy when at the bottom.
   (c) The swing does not have enough rotational momentum.
   (d) All of the above are correct.

3. In terms of the torque needed to rotate your leg as you run, would it be better to have a long calf and short thigh, or vice versa?
   (a) Long calf and short thigh
   (b) Short calf and long thigh
   (c) Does not matter

4. Suppose that two bicycles have equal overall mass, but one has thin light-weight tires while the other has heavier tires made of the same material. Why is the bicycle with thin tires easier to accelerate?
   (a) Thin tires have less area of contact with the road.
   (b) With thin tires, less mass is distributed at the rims.
   (c) With thin tires, you don't have to raise the large mass of the tire at the bottom to the top.

5. When riding a 10-speed bicycle up a hill, a cyclist shifts the chain to a larger-diameter gear attached to the back wheel. What must be true when the cyclist shifts to a larger gear?
   (a) The torque exerted by the chain on the gear is larger.
   (b) The force exerted by the chain on the gear is larger.
   (c) The cyclist pedals more frequently to travel the same distance.
   (d) Both a and c are correct.

6. The objects in Figure Q9.6 are made of two identical paper cups glued together. Rank the rotational inertias I_1, I_2, I_3, and I_4 about the indicated axes using the signs $=$, $>$, and $<$. Explain.

FIGURE Q9.6

7. Select all the pairs below in which the two physical quantities have the same units.
   (a) Rotational velocity and translational velocity
   (b) Rotational kinetic energy and translational kinetic energy
   (c) Linear momentum and rotational momentum
   (d) Work and torque
   (e) Power and energy
   (f) Impulse (that changes linear momentum) and rotational impulse

8. If you turn on a coffee grinding machine sitting on a smooth tabletop, what do you expect it to do?
   (a) Start rotating in the same direction as the blades rotate
   (b) Start rotating in the direction opposite the blade rotation
   (c) Grind the coffee without any rotation of the machine

9. A bowling ball is rolling without skidding down an incline. While accelerating down the incline, the ratio of translational kinetic energy to rotational kinetic energy of the ball
   (a) remains constant and equal to about 0.4.
   (b) remains constant and equal to about 2.5.
   (c) remains constant; the value depends on the radius of the ball.
   (d) decreases.
   (e) increases.

10. The Mississippi River carries sediment from higher latitudes toward the equator. How does this affect the length of the day?
    (a) Increases the day
    (b) Decreases the day
    (c) Does not affect the day
    (d) There is no relation between the mass distribution and the length of the day.

11. Two disks are cut from the same uniform board. The radius of disk B is twice the radius of the disk A. The disks can rotate around axes with negligible friction. Two very light battery-powered fans are attached to the disks, as shown in Figure Q9.11. When switched on, the fans exert equal forces on the disks. Which of the following correctly compares and explains the rotational accelerations of the disks after the fans are switched on?
    (a) $\alpha_A = 4\alpha_B$ because $F_B = F_A$ and $m_B = 4m_A$.
    (b) $\alpha_A = \alpha_B$ because $\tau_B = 2\tau_A$ and $I_B = 2I_A$.
    (c) $\alpha_A = 2\alpha_B$ because $\tau_B = 2\tau_A$ and $I_B = 4I_A$.
    (d) $\alpha_A = 4\alpha_B$ because $\tau_B = 2\tau_A$ and $I_B = 8I_A$.
    (e) $\alpha_A = 8\alpha_B$ because $\tau_B = 2\tau_A$ and $I_B = 16I_A$.

FIGURE Q9.11

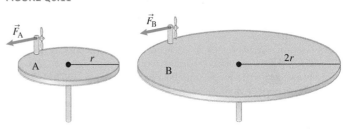

Conceptual Questions

12. Explain your choices for Questions 1–11 (your instructor will choose which ones).

13. A spinning raw egg, if stopped momentarily and then released by the fingers, will resume spinning. Explain. Will this happen with a hard-boiled egg? Explain.

14. Compare the magnitude of Earth's rotational momentum about its axis to that of the Moon about Earth. The tides exert a torque on Earth and the Moon so that eventually they will rotate with the same period. The object with the greater rotational momentum will experience the smaller percent change in the period of rotation. Will Earth's solar day increase more than the Moon's period of rotation decreases? Explain.

15. You lay a pencil on a smooth desk (ignore sliding friction). You push the pencil, exerting a constant force first directly at its center of mass and then close to the tip of the pencil. In both cases, the force is exerted perpendicular to the body of the pencil. If the forces that you exert on the pencil are exactly the same in magnitude and direction, in which case is the translational acceleration of the pencil greater in magnitude?

16. If you watch the dive of an Olympic diver, you note that she continues to rotate after leaving the board. However, her center of mass follows a parabolic curve. Explain why.

17. Explain why you do not tip over when riding a bicycle but do tip when stationary at a stoplight.

18. Sometimes a door is not attached properly and it will open by itself or close by itself. But it will never do both. Why?

19. Why do tightrope walkers carry long, heavy bars?

Problems

Below, **BIO** indicates a problem with a biological or medical focus. Problems labeled **EST** ask you to estimate the answer to a quantitative problem rather than derive a specific answer. Asterisks indicate the level of difficulty of the problem. Problems with no * are considered to be the least difficult. A single * marks moderately difficult problems. Two ** indicate more difficult problems.

9.1 Rotational kinematics

1. The sweeping second hand on your wall clock is 20 cm long. What is (a) the rotational speed of the second hand, (b) the translational speed of the tip of the second hand, and (c) the rotational acceleration of the second hand? Assume the second hand moves smoothly.

2. You find an old record player in your attic. The turntable has two readings: 33 rpm and 45 rpm. What do they mean? Express these quantities in different units.

3. * Consider again the turntable described in the last problem. Determine the magnitudes of the rotational acceleration in each of the following situations. Indicate the assumptions you made for each case. (a) When on and rotating at 33 rpm, it is turned off and slows and stops in 60 s. (b) When off and you push the play button, the turntable attains a speed of 33 rpm in 15 s. (c) You switch the turntable from 33 rpm to 45 rpm, and it takes about 2.0 s for the speed to change. (d) In the situation in part (c), what is the magnitude of the average tangential acceleration of a point on the turntable that is 15 cm from the axis of rotation?

4. You step on the gas pedal in your car, and the car engine's rotational speed changes from 1200 rpm to 3000 rpm in 3.0 s. What is the engine's average rotational acceleration?

5. You pull your car into your driveway and stop. The drive shaft of your car engine, initially rotating at 2400 rpm, slows with a constant rotational acceleration of magnitude 30 rad/s². How long does it take for the drive shaft to stop turning?

6. An old wheat-grinding wheel in a museum actually works. The sign on the wall says that the wheel has a rotational acceleration of 190 rad/s² as its spinning rotational speed increases from zero to 1800 rpm. How long does it take the wheel to attain this rotational speed?

7. **Centrifuge** A centrifuge at the same museum is used to separate seeds of different sizes. The average rotational acceleration of the centrifuge according to a sign is 30 rad/s². If starting at rest, what is the rotational velocity of the centrifuge after 10 s?

8. * **Potter's wheel** A fly sits on a potter's wheel 0.30 m from its axle. The wheel's rotational speed decreases from 4.0 rad/s to 2.0 rad/s in 5.0 s. Determine (a) the wheel's average rotational acceleration, (b) the angle through which the fly turns during the 5.0 s, and (c) the distance traveled by the fly during that time interval.

9. * During your tennis serve, your racket and arm move in an approximately rigid arc with the top of the racket 1.5 m from your shoulder joint. The top accelerates from rest to a speed of 20 m/s in a time interval of 0.10 s. Determine (a) the magnitude of the average tangential acceleration of the top of the racket and (b) the magnitude of the rotational acceleration of your arm and racket.

10. * An ant clings to the outside edge of the tire of an exercise bicycle. When you start pedaling, the ant's speed increases from zero to 10 m/s in 2.5 s. The wheel's rotational acceleration is 13 rad/s². Determine everything you can about the motion of the wheel and the ant.

11. * The speedometer on a bicycle indicates that you travel 60 m while your speed increases from 0 to 10 m/s. The radius of the wheel is 0.30 m. Determine three physical quantities relevant to this motion.

12. * You pedal your bicycle so that its wheel's rotational speed changes from 5.0 rad/s to 8.0 rad/s in 2.0 s. Determine (a) the wheel's average rotational acceleration, (b) the angle through which it turns during the 2.0 s, and (c) the distance that a point 0.60 m from the axle travels.

13. **Mileage gauge** The odometer on an automobile actually counts axle turns and converts the number of turns to miles based on knowledge that the diameter of the tires is 0.62 m. How many turns does the axle make when traveling 10 miles?

14. * **Speedometer** The speedometer on an automobile measures the rotational speed of the axle and converts that to a linear speed of the car, assuming the car has 0.62-m-diameter tires. What is the rotational speed of the axle when the car is traveling at 20 m/s (45 mph)?

15. * **Ferris wheel** A Ferris wheel starts at rest, acquires a rotational velocity of ω rad/s after completing one revolution and continues to accelerate. Write an expression for (a) the magnitude of the wheel's rotational acceleration (assumed constant), (b) the time interval needed for the first revolution, (c) the time interval required for the second revolution, and (d) the distance a person travels in two revolutions if he is seated a distance l from the axis of rotation.

16. * You push a disk-shaped platform tangentially on its edge 2.0 m from the axle. The platform starts at rest and has a rotational acceleration of 0.30 rad/s². Determine the distance you must run while pushing the platform to increase its speed at the edge to 7.0 m/s.

17. ** **EST** Estimate what Earth's rotational acceleration would be in rad/s² if the length of a day increased from 24 h to 48 h during the next 100 years.

9.3 Newton's second law for rotational motion

18. A 0.30-kg ball is attached at the end of a 0.90-m-long stick. The ball and stick rotate in a horizontal circle. Because of air resistance and to keep the ball moving at constant speed, a continual push must be exerted on the stick, causing a 0.036-N · m torque. Determine the magnitude of the resistive force that the air exerts on the ball opposing its motion. What assumptions did you make?

19. **Centrifuge** A centrifuge with a 0.40-kg · m² rotational inertia has a rotational acceleration of 100 rad/s² when the power is turned on. (a) Determine the minimum torque that the motor supplies. (b) What time interval is needed for the centrifuge's rotational velocity to increase from zero to 5000 rad/s?

20. **Airplane turbine** What is the average torque needed to accelerate the turbine of a jet engine from rest to a rotational velocity of 160 rad/s in 25 s? The turbine's rotating parts have a 32-kg · m² rotational inertia.

21. * A turntable turning at rotational speed 33 rpm stops in 50 s when turned off. The turntable's rotational inertia is 1.0×10^{-2} kg · m². How large is the resistive torque that slows the turntable?

22. * The solid pulley in Figure P9.22 consists of a two-part disk, which initially rotates counterclockwise. Two ropes pull on the pulley as shown. The inner part has a radius of 1.5a, and the outer part has a radius of 2.0a. (a) Construct a force diagram for the pulley with the origin of the coordinate system at the center of the pulley. (b) Determine the torque produced by each force (including the sign) and the resultant torque exerted on the pulley. (c) Based on the results of part (b), decide on the signs of the rotational velocity and the rotational acceleration.

FIGURE P9.22

23. * The pulley shown in Figure P9.22 is initially rotating clockwise. Compare the forces exerted by the ropes on the disk in order for the wheel's rotational velocity to (a) remain constant, (b) increase in magnitude, and (c) decrease in magnitude. The outer radius is 2.0a compared to 1.5a for the inner radius.

24. The pulley shown in Figure P9.22 is initially rotating in the clockwise direction. The force that the rope on the right exerts on it is 1.5F and the force that the rope on the left exerts on it is F. Determine the ratio of the maximum radius of the inner circle compared to that of the outer circle in order for the wheel's rotational speed to decrease.

25. A 2.0-kg metal cylinder on a table is placed inside a hoop that is fixed at the end of a meter stick. The stick can rotate around a vertical axis that is located 20 cm from the cylinder (see Figure P9.25). By exerting a force on the stick, you can make the cylinder rotate in a horizontal plane, with the bottom surface of the cylinder sliding on the rough table. The coefficient of kinetic friction between the cylinder and the table is 0.3. (a) Determine the magnitude of the force that you need to exert on the end of the stick (perpendicular to it) so that the cylinder moves with a constant speed. (b) Determine the work done by your hand on the stick while your hand makes one full turn around the axis. Then determine the work done by the stick on the cylinder during the same time. Compare the two values and comment on the result. Does the result surprise you?

FIGURE P9.25

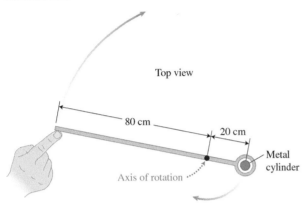

Top view

80 cm

20 cm

Metal cylinder

Axis of rotation

26. * **Equation Jeopardy 1** The equation below describes a rotational dynamics situation. Draw a sketch of a situation that is consistent with the equation and construct a word problem for which the equation might be a solution. There are many possibilities.

$$-(2.2 \text{ N})(0.12 \text{ m}) = [(1.0 \text{ kg})(0.12 \text{ m})^2]\alpha$$

27. * **Equation Jeopardy 2** The equation below describes a rotational dynamics situation. Draw a sketch of a situation that is consistent with the equation and construct a word problem for which the equation might be a solution. There are many possibilities.

$$-(2.0 \text{ N})(0.12 \text{ m}) + (6.0 \text{ N})(0.06 \text{ m}) = [(1.0 \text{ kg})(0.12 \text{ m})^2]\alpha$$

28. Derive an expression for the rotational inertia of the four balls shown in Figure P9.28 about an axis perpendicular to the paper and passing through point A. The mass of each ball is m. Ignore the mass of the rods to which the balls are attached.

FIGURE P9.28

29. * Repeat the previous problem for an axis perpendicular to the paper through point B.

30. Repeat the previous problem for axis BC, which passes through two of the balls.

31. * **Merry-go-round** A mechanic needs to replace the motor for a merry-go-round. What torque specifications must the new motor satisfy if the merry-go-round should accelerate from rest to 1.5 rad/s in 8.0 s? You can consider the merry-go-round to be a uniform disk of radius 5.0 m and mass 25,000 kg.

32. * A small 0.80-kg train propelled by a fan engine starts at rest and goes around a circular track with a 0.80-m radius. The air exerts a 2.0-N force on the train. Determine (a) the rotational acceleration of the train and (b) the time interval needed for it to acquire a speed of 3.0 m/s. Indicate any assumptions you made.

33. * **Motor** You wish to buy a motor that will be used to lift a 20-kg bundle of shingles from the ground to the roof of a house. The shingles are to have a 1.5-m/s² upward acceleration at the start of the lift. The very light pulley on the motor has a radius of 0.12 m. Determine the minimum torque that the motor must be able to provide.

34. ** A string wraps around a 6.0-kg wheel of radius 0.20 m. The wheel is mounted on a frictionless horizontal axle at the top of an inclined plane tilted 37° below the horizontal. The free end of the string is attached to a 2.0-kg block that slides down the incline without friction. The block's acceleration while sliding down the incline is 2.0 m/s². (a) Draw separate force diagrams for the wheel and for the block. (b) Apply Newton's second law (either the translational form or the rotational form) for the wheel and for the block. (c) Determine the rotational inertia for the wheel about its axis of rotation.

35. * Elena, a black belt in tae kwon do, is experienced in breaking boards with her fist. A high-speed video indicates that her forearm is moving with a rotational speed of 40 rad/s when it reaches the board. The board breaks in 0.0040 s and her arm is moving at 20 rad/s just after breaking the board. Her fist is 0.32 m from her elbow joint and the rotational inertia of her forearm is 0.050 kg·m². Determine the average force that the board exerts on her fist while breaking the board (equal in magnitude to the force that her fist exerts on the board). Ignore the gravitational force that Earth exerts on her arm and the force that her triceps muscle exerts on her arm during the break.

36. ** **Like a yo-yo** Sam wraps a string around the outside of a 0.040-m-radius 0.20-kg solid cylinder and uses it like a yo-yo (Figure P9.36). When released, the cylinder accelerates downward at $(2/3)g$. (a) Draw a force diagram for the cylinder and apply the translational form of Newton's second law to the cylinder in order to determine the force that the string exerts on the cylinder. (b) Determine the rotational inertia of the solid cylinder. (c) Apply the rotational form of Newton's second law and determine the cylinder's rotational acceleration. (d) Is your answer to part (c) consistent with the application of $a = r\alpha$, which relates the cylinder's linear acceleration and its rotational acceleration? Explain.

FIGURE P9.36

r = 0.040 m

37. * **Fire escape** A unique fire escape for a three-story house is shown in Figure P9.37. A 30-kg child grabs a rope wrapped around a heavy flywheel outside a bedroom window. The flywheel is a 0.40-m-radius uniform disk with a mass of 120 kg. (a) Make a force diagram for the child as he moves downward at increasing speed and another for the flywheel as it turns faster and faster. (b) Use Newton's second law for translational motion and the child force diagram to obtain an expression relating the force that the rope exerts on him and his acceleration. (c) Use Newton's second law for rotational motion and the flywheel force diagram to obtain an expression relating the force the rope exerts on the flywheel and the rotational acceleration of the flywheel. (d) The child's acceleration a and the flywheel's rotational acceleration α are related by the equation $a = r\alpha$, where r is the flywheel's radius. Combine this with your equations in parts (b) and (c) to determine the child's acceleration and the force that the rope exerts on the wheel and on the child.

FIGURE P9.37

38. ** An Atwood machine is shown in Example 9.4. Use $m_1 = 0.20$ kg, $m_2 = 0.16$ kg, $M = 0.50$ kg, and $R = 0.10$ m. (a) Construct separate force diagrams for block 1, for block 2, and for the solid cylindrical pulley. (b) Determine the rotational inertia of the pulley. (c) Use the force diagrams for blocks 1 and 2 and Newton's second law to write expressions relating the unknown accelerations of the blocks. (d) Use the pulley force diagram and the rotational form of Newton's second law to write an expression for the rotational acceleration of the pulley. (e) Noting that $a = R\alpha$ for the pulley, use the three equations from parts (c) and (d) to determine the magnitude of the acceleration of the hanging blocks.

39. ** A physics problem involves a massive pulley, a bucket filled with sand, a toy truck, and an incline (see **Figure P9.39**). You push lightly on the truck so it moves down the incline. When you stop pushing, it moves down the incline at constant speed and the bucket moves up at constant speed. (a) Construct separate force diagrams for the pulley, the bucket, and the truck. (b) Use the truck force diagram and the bucket force diagram to help write expressions in terms of quantities shown in the figure for the forces $F_{1 \text{ on Truck}}$ and $F_{2 \text{ on Bucket}}$ that the rope exerts on the truck and that the rope exerts on the bucket. (c) Use the rotational form of Newton's second law to determine if the tension force $F_{1 \text{ on Pulley}}$ that the rope on the right side exerts on the pulley is the same, greater than, or less than the force $F_{2 \text{ on Pulley}}$ that the rope exerts on the left side.

FIGURE P9.39

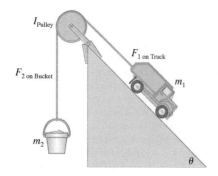

40. * A thin rod of length L and mass m rotates around an axis perpendicular to the rod, passing through the rod's left end. Treat the rod as an object made up of five rods of length $L/5$. Derive an expression for the rotational inertia of this five-piece object around the same axis, assuming each piece is a point-like object with mass $m/5$. Compare your result with the expression for the rotational inertia of the one-piece rod ($I = \frac{1}{3}mL^2$). Discuss the similarities and the differences.

9.4 Rotational momentum

41. * (a) Determine the rotational momentum of a 10-kg disk-shaped flywheel of radius 9.0 cm rotating with a rotational speed of 320 rad/s. (b) With what magnitude rotational speed must a 10-kg solid sphere of 9.0 cm radius rotate to have the same rotational momentum as the flywheel?

42. **Ballet** A ballet student with her arms and a leg extended spins with an initial rotational speed of 1.0 rev/s. As she draws her arms and leg in toward her body, her rotational inertia becomes 0.80 kg · m² and her rotational velocity is 4.0 rev/s. Determine her initial rotational inertia.

43. * A 0.20-kg block moves at the end of a 0.50-m string along a circular path on a frictionless air table. The block's initial rotational speed is 2.0 rad/s. As the block moves in the circle, the string is pulled down through a hole in the air table at the axis of rotation. Determine the rotational speed and tangential speed of the block when the string is 0.20 m from the axis.

44. * **Puck on a string** You attach a 100-g puck to a string and let the puck glide in a circle on a horizontal, frictionless air table. The other end of the string passes through a hole at the center of the table. You pull down on the string so that the puck moves along a circular path of radius 0.40 m. It completes one revolution in 4.0 s. If you pull harder on the string so the radius of the circle slowly decreases to 0.20 m, what is the new period of revolution?

45. * **Equation Jeopardy 3** The equation below describes a process. Draw a sketch representing the initial and final states of the process and construct a word problem for which the equation could be a solution.

$$\left(\frac{2}{5}mR^2\right)\left(\frac{2\pi}{30 \text{ days}}\right) = \left[\frac{2}{5}m\left(\frac{R}{100}\right)^2\right]\left(\frac{2\pi}{T_f}\right)$$

46. ** A student sits motionless on a stool that can turn friction-free about its vertical axis (total rotational inertia I). The student is handed a spinning bicycle wheel, with rotational inertia I_{wheel}, that is spinning about a vertical axis with a counterclockwise rotational velocity ω_0. The student then turns the bicycle wheel over (that is, through 180°). Estimate, in terms of ω_0, the final rotational velocity acquired by the student.

47. * **Neutron star** An extremely dense neutron star with mass equal to that of the Sun has a radius of about 10 km—about the size of Manhattan Island. These stars are thought to rotate once about their axis every 0.03 to 4 s, depending on their size and mass. Suppose that the neutron star described in the first sentence rotates once every 0.040 s. If its volume then expanded to occupy a uniform sphere of radius 1.4×10^8 m (most of the Sun's mass is in a sphere of this size) with no change in mass or rotational momentum, what time interval would be required for one rotation? By comparison, the Sun rotates once about its axis each month.

48. * A boy of mass m is standing on the edge of a merry-go-round platform, which is initially rotating with a constant rotational velocity ω_i (initial state). The rotational inertia of the merry-go-round is I, and the radius of the platform is r. After the boy steps down from the platform, tangentially to its edge, the merry-go-round continues rotating with a smaller rotational velocity ω_f, and the boy continues moving away from the platform with constant speed v (final state). (a) Draw three rotational momentum bar charts, choosing the boy, the merry-go-round, and both of them as a system. (b) Derive the expression for ω_f in terms of the relevant quantities.

49. * **Bar chart jeopardy** The rotational momentum bar chart in **Figure P9.49** describes a rotational dynamics situation. Draw a sketch of a situation that is consistent with the equation and write a word problem to which the bar chart could apply.

FIGURE P9.49

$$L_{1i} + L_{2i} + \Sigma \tau \Delta t = L_{1f} + L_{2f}$$

0

9.5 Rotational kinetic energy

50. A grinding wheel with rotational inertia I gains rotational kinetic energy K after starting from rest. Determine an expression for the wheel's final rotational speed.

51. * The rotational speed of a flywheel increases by 40%. By what percent does its rotational kinetic energy increase? Explain your answer.

52. ** Two uniform disks are made from the same material and have equal masses. The radius of disk B is three times larger than the radius of disk A. The disks touch each other and rotate at constant angular speed without skidding (**Figure P9.52**). Determine (a) the ratio of the disks' thicknesses d_B/d_A, (b) the ratio of the disks' angular velocities ω_B/ω_A, (c) the ratio of the disks' rotational inertias I_B/I_A, (d) the ratio of the disks' rotational momenta L_B/L_A, and (e) the ratio of the disks' rotational kinetic energies $K_{\text{rot B}}/K_{\text{rot A}}$.

FIGURE P9.52

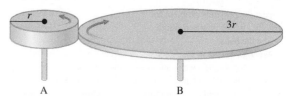

A   B

53. * **Flywheel energy for car** The U.S. Department of Energy had plans for a 1500-kg automobile to be powered completely by the rotational kinetic energy of a flywheel. (a) If the 300-kg flywheel (included in the 1500-kg mass of the automobile) had a 6.0-kg · m² rotational inertia and could turn at a maximum rotational speed of 3600 rad/s, determine the energy stored in the flywheel. (b) How many accelerations from a speed of zero to 15 m/s could the car make before the flywheel's energy was dissipated, assuming 100% energy transfer and no flywheel regeneration during braking? Ignore rotational kinetic energy of car wheels.

54. * **Flywheel energy** Engineers at the University of Texas at Austin are developing an Advanced Locomotive Propulsion System that uses a gas turbine and perhaps the largest high-speed flywheel in the world in terms of the energy it can store. The flywheel can store 4.8×10^8 J of energy when operating at its maximum rotational speed of 15,000 rpm. At that rate, the perimeter of the rotor moves at approximately 1,000 m/s. Determine the radius of the flywheel and its rotational inertia.

55. * **Equation Jeopardy 4** The equations below represent the initial and final states of a process (plus some ancillary information). Construct a sketch of a process that is consistent with the equations and write a word problem for which the equations could be a solution.

$$(80 \text{ kg})(9.8 \text{ N/kg})(16 \text{ m}) = \tfrac{1}{2}(80 \text{ kg})v_f^2 + \tfrac{1}{2}(240 \text{ kg} \cdot \text{m}^2)\omega_f^2$$
$$v_f = (0.40 \text{ m})\omega_f$$

56. ** **Rotating student** A student sitting on a chair on a circular platform of negligible mass rotates freely on an air table at initial rotational speed 2.0 rad/s. The student's arms are initially extended with 6.0-kg dumbbells in each hand. As the student pulls her arms in toward her body, the dumbbells move from a distance of 0.80 m to 0.10 m from the axis of rotation. The initial rotational inertia of the student's body (not including the dumbbells) with arms extended is $6.0 \text{ kg} \cdot \text{m}^2$, and her final rotational inertia is $5.0 \text{ kg} \cdot \text{m}^2$. (a) Determine the student's final rotational speed. (b) Determine the change of kinetic energy of the system consisting of the student together with the two dumbbells. (c) Determine the change in the kinetic energy of the system consisting of the two dumbbells alone without the student. (d) Determine the change of kinetic energy of the system consisting of the student alone without the dumbbells. (e) Compare the kinetic energy changes in parts (b) through (d).

57. * A turntable whose rotational inertia is $1.0 \times 10^{-3} \text{ kg} \cdot \text{m}^2$ rotates on a frictionless air cushion at a rotational speed of 2.0 rev/s. A 1.0-g beetle falls to the center of the turntable and then walks 0.15 m to its edge. (a) Determine the rotational speed of the turntable with the beetle at the edge. (b) Determine the kinetic energy change of the system consisting of the turntable and the beetle. (c) Account for this energy change.

58. ** Repeat the previous problem, only assume that the beetle initially falls on the edge of the turntable and stays there.

59. * A bug of a known mass m stands at a distance d cm from the axis of a spinning disk (mass m_d and radius r_d) that is rotating at f_i revolutions per second. After the bug walks out to the edge of the disk and stands there, the disk rotates at f_f revolutions per second. (a) Use the information above to write an expression for the rotational inertia of the disk. (b) Determine the change of kinetic energy in going from the initial to the final situation for the total bug-disk system.

60. * **Merry-go-round** A carnival merry-go-round has a large disk-shaped platform of mass 120 kg that can rotate about a center axle. A 60-kg student stands at rest at the edge of the platform 4.0 m from its center. The platform is also at rest. The student starts running clockwise around the edge of the platform and attains a speed of 2.0 m/s relative to the ground. (a) Determine the rotational velocity of the platform. (b) Determine the change of kinetic energy of the system consisting of the platform and the student.

61. * **EST** You hold an apple by its stem between your thumb and index finger and spin it so that the apple is rotating at approximately constant speed. Estimate the rotational kinetic energy and the rotational momentum of the apple. Indicate any assumptions that you made.

General Problems

62. * **Stopping Earth's rotation** Suppose that Superman wants to stop Earth so it does not rotate. He exerts a force on Earth $\vec{F}_{\text{S on E}}$ at Earth's equator tangent to its surface for a time interval of 1 year. What magnitude force must he exert to stop Earth's rotation? Indicate any assumptions you make when completing your estimate.

63. * **BIO** **EST** **Punting a football** Estimate the tangential acceleration of the foot and the rotational acceleration of the leg of a football punter during the time interval that the leg starts to swing forward in an arc until the instant just before the foot hits the ball. Indicate any assumptions that you make and be sure that your method is clear.

64. * **EST** Estimate the average rotational acceleration of a car tire as you leave an intersection after a light turns green. Discuss the choice of numbers used in your estimate.

65. * **BIO** **Triceps and darts** Your upper arm is horizontal and your forearm is vertical with a 0.010-kg dart in your hand (**Figure P9.65**). When your triceps muscle contracts, your forearm initially swings forward with a rotational acceleration of 35 rad/s². Determine the force that your triceps muscle exerts on your forearm during this initial part of the throw. The rotational inertia of your forearm is $0.12 \text{ kg} \cdot \text{m}^2$ and the dart is 0.38 m from your elbow joint. Your triceps muscle attaches 0.03 m from your elbow joint.

FIGURE P9.65

0.38 m

0.03 m

Triceps

66. * **BIO** **Bowling** At the start of your throw of a 2.7-kg bowling ball, your arm is straight behind you and horizontal (**Figure P9.66**). Determine the rotational acceleration of your arm if the muscle is relaxed. Your arm is 0.64 m long, has a rotational inertia of $0.48 \text{ kg} \cdot \text{m}^2$, and has a mass of 3.5 kg with its center of mass 0.28 m from your shoulder joint.

FIGURE P9.66

2.7 kg   cm

0.28 m

0.64 m

Axis of rotation

67. ** **BIO** **Leg lift** You are doing one-leg leg lifts (**Figure P9.67**) and decide to estimate the force that your iliopsoas muscle exerts on your upper leg bone (the femur) when being lifted (the lifting involves a variety of muscles). The mass of your entire leg is 15 kg, its center of mass is 0.45 m from the hip joint, and its rotational inertia is $4.0 \text{ kg} \cdot \text{m}^2$, and you estimate that the rotational acceleration of the leg being lifted is 35 rad/s². For calculation purposes assume that the iliopsoas attaches to the femur 0.10 m from the hip joint. Also assume that the femur is oriented 15° above the horizontal and that the muscle is horizontal. Estimate the force that the muscle exerts on the femur.

FIGURE P9.67

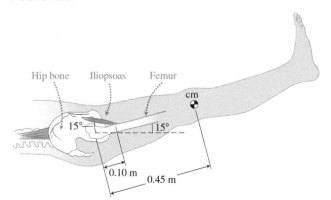

Hip bone   Iliopsoas   Femur

cm

15°   15°

0.10 m

0.45 m

68. * A horizontal, circular platform can rotate around a vertical axis at its center with negligible friction. You decide to use the rotating platform to design a procedure that will allow you to determine the unknown rotational inertias of different objects, for example, a bowl. You know the rotational inertia of the platform I_P. Which of the following procedures would best serve your task? (More than one answer could be correct.) Comment on the procedures that are not suitable and explain what is wrong with them.

(a) Exert a constant torque τ on the empty platform and measure its angular acceleration α_P. Then place the bowl on the platform at its center and repeat the previous experiment, obtaining α_{PB}. Calculate the rotational inertia of the bowl using the expression $I_B = \dfrac{\tau}{\alpha_{PB} - \alpha_P}$.

(b) Perform the same experiment as in (a) but use the expression
$$I_B = I_P\left(\frac{\alpha_P}{\alpha_{PB}} - 1\right).$$

(c) Spin the empty platform and measure its angular velocity ω_i. While the platform is rotating, place the bowl carefully on the platform at its center. Wait until the bowl rotates together with the platform. Measure the final angular velocity of the bowl-platform system ω_f. Calculate the rotational inertia of the bowl using the expression $I_B = I_P\left(\dfrac{\omega_i}{\omega_f} - 1\right)$.

(d) Perform the same experiment as in (c) but use the expression
$$I_B = I_P\left(\frac{\omega_i^2}{\omega_f^2} - 1\right).$$

(e) Measure the mass of the platform m_P, the mass of the bowl m_B, the radius of the platform r_P, and the radius of the bowl r_B. Calculate the rotational inertia of the bowl using the expression $I_B = I_P \cdot \dfrac{m_B r_B^2}{m_P r_P^2}$.

69. * You have an empty cylindrical metal can and two metal nuts. In your first experiment, you spin the can around its axis and place it on a rough table (Figure P9.69a). The can slows down due to friction, with the rotational acceleration $\alpha_1 = 30 \text{ rad/s}^2$ until it stops rotating. In your second experiment, you first fix two nuts near the center of the can (Figure P9.69b) and spin it again before placing it on the table. In this case, the can slows down with the rotational acceleration $\alpha_2 = 38 \text{ rad/s}^2$. In your third experiment, you spin the can with the nuts fixed at opposite sides of the can (Figure P9.69c). Now the can slows down with rotational acceleration $\alpha_3 = 26 \text{ rad/s}^2$. Explain (a) why $\alpha_2 > \alpha_1$, (b) why $\alpha_3 < \alpha_2$, and (c) why $\alpha_3 < \alpha_1$.

FIGURE P9.69

(a)

(b)

(c)

70. ** In the previous problem, each nut has a mass of 8.0 g. The distances between the centers of mass of the nuts and the axis of rotation are 0.8 cm in Experiment 2 and 3.5 cm in Experiment 3. Determine the rotational inertia of the can (without nuts). Indicate any assumptions that you made. (Hint: The torque due to the kinetic friction force in Experiments 2 and 3 is the same.)

71. * **Superball** If you give a superball backspin and throw it toward a horizontal floor, it is possible that the ball bounces backward, as shown in Figure P9.71. If the ball has a color pattern or stripes, you may also notice that during the collision with the ground, the direction of rotation of the ball changes, as indicated in the figure. Both changes (the change of the ball's translational velocity and the change of the ball's angular velocity) are the result of a force exerted on the ball by the ground during the collision. (a) In which approximate direction did the ground exert the force on the ball in the case shown in the figure? Choose the best answer from the options given in the figure and explain your answer. (b) The ball has initial and final velocity (with components v_{xi}, v_{yi} and v_{xf}, v_{yf}) and initial and final angular velocity (ω_i and ω_f). During the collision, the ground exerts a force

FIGURE P9.71

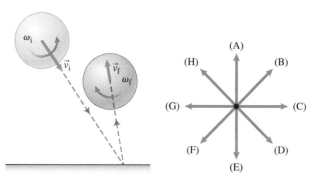

(determined by F_x and F_y) on the ball. Complete the table below by drawing crosses in the cells that correctly connect changes of the quantities in the first column and the components of the force during the collision. Explain your answers.

	F_x	F_y
$v_{xf} - v_{xi}$		
$v_{yf} - v_{yi}$		
$\omega_f - \omega_i$		

72. ** **Yo-yo trick** A yo-yo rests on a horizontal table. The yo-yo is free to roll but friction prevents it from sliding. When the string exerts one of the following tension forces on the yo-yo (shown in Figure P9.72), which way does the yo-yo roll? Try the problem for each force: (a) $\vec{T}_{A \text{ S on Y}}$; (b) $\vec{T}_{B \text{ S on Y}}$; and (c) $\vec{T}_{C \text{ S on Y}}$. (Hint: Think about torques about a pivot point where the yo-yo touches the table.)

FIGURE P9.72

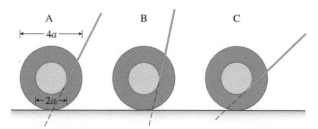

73. * **EST** **White dwarf** A star the size of our Sun runs out of nuclear fuel and, without losing mass, collapses to a white dwarf star the size of our Earth. If the star initially rotates at the same rate as our Sun, which is once every 25 days, determine the rotation rate of the white dwarf. Indicate any assumptions you make.

Reading Passage Problems

Rolling versus sliding Our knowledge of rotational kinetic energy helps explain a very simple but rather mysterious experiment that you can perform at home. For this experiment, you need two identical plastic water bottles (1 and 2). Fill bottle 1 with snow (if you do not have snow, you can use whipped cream). Fill bottle 2 with water so that the masses of filled bottles are the same. Place one bottle at the top of an inclined plane and let it roll down. Then repeat the experiment with the other bottle (Figure 9.18). You observe an interesting effect: when bottle 1 rolls down, it rotates and the solid snow inside rotates with the bottle. Bottle 2 rotates, too, but the water inside does not rotate much. Thus, in effect, the water slides down the incline and does not roll. When the snow-filled bottle rolls down, it rotates as a solid cylinder, acquiring rotational kinetic energy in addition to translational kinetic energy. In the case of the water-filled bottle, only the bottle rotates; the water inside just translates. Rolling is a combination of translation and rotation, whereas sliding involves only translation. The water bottle has almost no rotational kinetic energy and a larger translational kinetic energy at the end of the plane.

FIGURE 9.18

Bottle 1 containing snow   Bottle 2 containing water

74. Which bar chart in Figure P9.74 corresponds to the bottle filled with solid snow rolling from the top of the incline to the bottom?
   (a) Bar chart A
   (b) Bar chart B
   (c) Both charts could represent the trip.

FIGURE P9.74

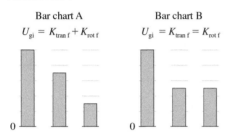

Bar chart A
$$U_{gi} = K_{tran\,f} + K_{rot\,f}$$

Bar chart B
$$U_{gi} = K_{tran\,f} = K_{rot\,f}$$

75. Which bar chart in Figure P9.74 corresponds to the bottle with the shortest time interval from the top of the incline to the bottom?
   (a) Bar chart A
   (b) Bar chart B
   (c) The charts represent the same time interval for the trip.
76. Based on the results from the two previous questions, which bottle should take the shortest time interval to complete the race?
   (a) Bottle 1
   (b) Bottle 2
   (c) Both bottles take the same time interval.
77. Which statement best explains your answer to the previous problem?
   (a) Most of the initial gravitational potential energy is converted into translational kinetic energy.
   (b) Some of the initial gravitational potential energy is converted to rotational kinetic energy.
   (c) All initial gravitational potential energy is converted to kinetic energy.

Tidal energy Tides are now used to generate electric power in two ways. In the first, huge dams can be built across the mouth of a river where it exits to the ocean. As the ocean tide moves in and out of this tidal basin or estuary, the water flows through tunnels in the dam (see Figure 9.19). This flowing water turns turbines in the tunnels that run electric generators. Unfortunately, this technique works best with large increases in tides—a 5-m difference between high and low tide. Such differences are found at only a small number of places. Currently, France is the only country that successfully uses this power source. A tidal basin plant in France, the Rance Tidal Power Station, makes 240 megawatts of power—enough energy to power 240,000 homes. Damming tidal basins can have negative environmental effects because of reduced tidal flow and silt buildup. Another disadvantage is that they can only generate electricity when the tide is flowing in or out, for about 10 hours each day.

A second method for collecting energy from the tidal flow (as well as all water flow) is to place turbines directly in the water—like windmills in moving water instead of in moving air. These water turbines have the advantages that they are much cheaper to build, they do not have the environmental problems of a tidal basin, and there are many more suitable sites for such water flow energy farms. Also, the energy density of flowing water is about 800 times the energy density of dry air flow. Verdant Power is developing turbine prototypes in the East River near New York City and in the Saint Lawrence Seaway in Canada, and they are looking at other sites in the Puget Sound and all over the world. The worldwide potential for hydroelectric power is about 25 terawatts = 25×10^{12} J/s—enough to supply the world's energy needs.

FIGURE 9.19 Dams built across tidal basins can generate electric power.

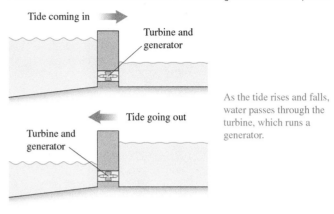

As the tide rises and falls, water passes through the turbine, which runs a generator.

78. If the Rance Tidal Power Station in France could produce power 24 hours a day, which answer below is closest to the daily amount of energy in joules that it could produce?
   (a) 240 J
   (b) 240×10^6 J
   (c) 6×10^9 J
   (d) 2.5×10^{10} J
   (e) 2×10^{13} J
79. Suppose a tidal basin is 5 m above the ocean at low tide and that the area of the basin is 4×10^7 m^2 (about 4 miles by 4 miles). Which answer below is closest to the gravitational potential energy change if the water is released from the tidal basin to the low-tide ocean level? The density of water is 1000 kg/m^3. (Hint: The level does not change by 5 m for all of the water.)
   (a) 5×10^8 J
   (b) 5×10^{11} J
   (c) 1×10^{12} J
   (d) 5×10^{12} J
   (e) 1×10^{13} J
80. The Rance tidal basin can only produce electricity when what is occurring?
   (a) Water is moving into the estuary from the ocean.
   (b) Water is moving into the ocean from the estuary.
   (c) Water is moving in either direction.
   (d) The Moon is full.
   (e) The Moon is full and directly overhead.
81. Why do water turbines seem more promising than tidal basins for producing electric energy?
   (a) Turbines are less expensive to build.
   (b) Turbines have less impact on the environment.
   (c) There are many more locations for turbines than for tidal basins.
   (d) Turbines can operate 24 hours/day versus only 10 hours/day for tidal basins.
   (e) All of the above
82. Why do water turbines have an advantage over air turbines (windmills)?
   (a) Air moves faster than water.
   (b) The energy density of moving water is much greater than that of moving air.
   (c) Water turbines can float from one place to another, whereas air turbines are fixed.
   (d) All of the above
   (e) None of the above
83. Which of the following is a correct statement about water turbines?
   (a) Water turbines can operate only in moving tidal water.
   (b) Water turbines can produce only a small amount of electricity.
   (c) Water turbines have not had a proof of concept.
   (d) Water turbines cause significant ocean warming.
   (e) None of the above are correct statements.

10

Vibrational Motion

- Why should soldiers avoid marching in step when they cross a bridge?
- Why do you need to "pump" your legs when you begin swinging on a park swing?
- How can you carry a full cup of coffee without splashing it?

On a cold day in January 1905, a Russian cavalry squadron of 50 soldiers was crossing a bridge over the Fontanka River in St. Petersburg when it collapsed. The stone bridge had stood for 79 years, with many more people simultaneously walking or riding horses across it than there were soldiers in that squadron. However, unlike ordinary crowds, the Russian soldiers were marching in step as they crossed the bridge. Why did their marching cause the bridge to collapse? We will find out in this chapter.

BE SURE YOU KNOW HOW TO:

- Apply Hooke's law to analyze forces exerted by springs (Section 7.4).
- Use radians to describe angles (Section 9.1).
- Draw an energy bar chart for a process and convert it into an equation (Sections 7.2 and 7.6).

WE HAVE STUDIED LINEAR MOTION—objects moving in straight lines at either constant velocity or constant acceleration. We have also studied objects moving at constant speed in a circle. In the case of circular motion, the direction of the velocity and acceleration of the objects changes, although the magnitude of both usually remains constant. In this chapter we will encounter a new type of motion during which both direction and speed change. You experience this motion every day—when you walk, when you talk, and when you listen to music.

10.1 Observations of vibrational motion

When you walk, your arms and legs swing back and forth. These motions repeat themselves; your arm or your leg passes all of the same points multiple times along its path relative to your body (**Figure 10.1**). The back-and-forth motion of an object that passes through the same positions, first moving in one direction and then in the opposite direction, is an important feature of *vibrational motion*, which we will investigate in this chapter.

Let's start with some observational experiments using a cart on a smooth surface attached to a spring (Observational Experiment **Table 10.1**). Recall that a spring exerts a force on an object that stretches or compresses the spring (Chapter 7). The force is variable in magnitude and points in the direction opposite to the direction in which the spring is stretched or compressed.

FIGURE 10.1 Vibrational motion. Your arms and legs undergo repetitive back-and-forth motion relative to the body while you walk.

OBSERVATIONAL EXPERIMENT TABLE 10.1 Some features of vibrational motion

 VIDEO OET 10.1

Observational experiment	Analysis
Experiment 1. A cart attached to a relaxed spring sits at rest on a frictionless horizontal surface at position 0.	In all experiments, we choose the cart as the system. The spring is relaxed, and the sum of the forces exerted on the cart by other objects is zero when at position 0. 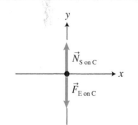
Experiment 2. (a) Pull the cart to the right and release it. The cart starts to move back toward position 0.	The spring exerts a force on the cart toward the left. The sum of the forces exerted on the cart now points to the left and causes the cart to start moving left toward position 0. 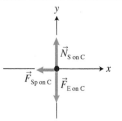
(b) The cart is moving fast when it reaches position 0 and overshoots that position.	As the cart passes position 0, the sum of the forces that other objects exert on the cart is again zero. But since it is moving, it continues moving. 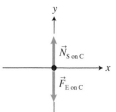
(c) The cart now slows down and eventually stops to the left of position 0, then starts moving back to the right toward position 0.	After passing position 0, the spring exerts a force to the right toward position 0; the sum of the forces points to the right, causing the cart to slow down, stop, and move back toward position 0. 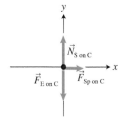
(d) The cart overshoots again and eventually stops where it started, on the right side of 0. The motion then repeats itself.	

Patterns

The system is at rest in a particular position when not vibrating. When the cart is at rest in this position, the sum of the forces that other objects exert on it is zero. When the cart is displaced from this position and released, the cart moves back and forth, passing through that position in two opposite directions. If displaced from the rest position, a force is exerted on the vibrating object that tends to return it to that position.

In Table 10.1, the cart attached to the spring vibrates back and forth through what is called its **equilibrium position**—the place where it resides when not disturbed. When the cart is displaced on either side of the equilibrium position, the spring exerts a so-called **restoring force** on the cart that tends to return it to that equilibrium position. However, the cart is moving when it reaches the equilibrium position and therefore over-shoots, moving to the other side before turning back toward the equilibrium position.

> **Equilibrium position** (or just **equilibrium**) The position at which a vibrating object resides when not disturbed. When resting at this position, the sum of the forces that other objects exert on it is zero. During **vibrational motion** the object passes back and forth through this position from two opposite directions.

> **Restoring force** When an object is displaced from equilibrium, some other object exerts a force with a component that always points opposite the direction of the vibrating object's displacement from equilibrium. This force tends to restore the vibrating object back toward equilibrium.

We have analyzed the vibrational motion of the cart-spring system using sketches and force diagrams. In Observational Experiment **Table 10.2**, we also use motion diagrams and energy bar charts to see if there are other common characteristics of vibrating systems.

OBSERVATIONAL EXPERIMENT TABLE 10.2 Multiple representation analysis of a cart on a spring
VIDEO OET 10.2

Observational experiment	Analysis
A cart attached to a spring is pulled to the right (position $+A$) and released. It moves past equilibrium (position 0) at high speed and briefly stops an equal distance to the left of equilibrium (position $-A$). The spring pushes the cart back toward the right. The process repeats over and over. We analyze the process using motion diagrams, force diagrams, and energy bar charts for one half of the period of vibration.	

Pattern

Restoring force The restoring force is zero as the vibrating object passes through the equilibrium position and has maximum magnitude when at the extreme positions on the left and right.

Potential and kinetic energy The energy of the vibrating system (the cart and spring) varies between maximum potential energy when at the extreme positions to maximum kinetic energy as the object passes equilibrium. In between, the energy is a combination of kinetic and potential energy.

In Table 10.2, the acceleration of the cart and the force that the spring exerted on the cart were greatest at its extreme positions—at $x = \pm A$. This distance from the equilibrium position to an extreme position of the motion is called the **amplitude** of the vibration (**Figure 10.2**).

FIGURE 10.2 The cart vibrates between $+A$ and $-A$. The distance from the equilibrium position to A is called the amplitude of the vibration.

> **Amplitude** The amplitude A of a vibration is the magnitude of the maximum displacement of the vibrating object from its equilibrium position.

The patterns described in the observational experiment tables are general features of vibrational motion. We summarize them below:

1. During vibrational motion, an object passes through the same positions, first moving in one direction and then in the opposite direction.

2. The object passes one position (the equilibrium position) at high speed, first in one direction, then the other. When it is not in the equilibrium position, a component of a restoring force exerted on it by some other object points back toward equilibrium.

3. A system composed of the vibrating object and the object exerting the restoring force has maximum potential energy when the vibrating object is at extreme positions ($x = \pm A$). The system has maximum kinetic energy when the vibrating object is passing through the equilibrium position ($x = 0$).

Many objects undergo vibrational motion: the strings of musical instruments, a drumhead, or keys on a key chain after you slam the door (see **Figure 10.3**). Vibrations are not always desirable. Tall buildings must be constructed so that they do not sway excessively during earthquakes or on very windy days, and cars must have shock absorbers to dampen the up-and-down vibrations that occur on a bumpy road. In other cases, vibrations are a desirable result: feel the surface of a speaker when music is playing.

FIGURE 10.3 Long-exposure photo of oscillating keys.

Period and frequency

Another important quantity is the time interval for one complete vibration. Consider the cart attached to a spring in Figure 10.2. The time interval for one complete vibration from $x = +A$ to $-A$ and back to $+A$ is called the **period** T of the vibration.

We are familiar with this quantity from our study of circular motion. For example, one day is the time interval (the period) for Earth to make one complete rotation on its axis—the time interval for the appearance of the Sun above the horizon on two consecutive mornings.

> **Period** The period T of a vibrating object is the time interval needed for the object to make one complete vibration—from the clock reading when it passes through a position while moving in a certain direction until the next clock reading when it passes through that *same* position moving in the *same* direction. The SI unit of period is the second.

The number of vibrations each second is called the **frequency** f of the vibration. If an object makes 4 vibrations each second, the frequency is 4 vib/s and the period is 0.25 s. The most commonly used frequency unit is the hertz (Hz), where 1 hertz equals 1 vibration per second: 1 Hz = 1 vib/s (or 1 cycle/s). The words vibration and cycle indicate what is being counted but are not themselves units. These words (cycle and vibration) can be removed from the units in equations involving frequency. Thus, an appropriate unit for frequency is $1/s$ or s^{-1}.

Frequency The frequency f of vibrational motion is the number of complete vibrations of the system during one second. Frequency is related to period:

$$f = \frac{1}{T} \tag{10.1}$$

The SI unit for frequency is the hertz (Hz), where $1 \text{ Hz} = 1 \text{ vib/s} = 1 \text{ s}^{-1}$.

QUANTITATIVE EXERCISE 10.1   **The vibration of a building**

The Empire State Building sways back and forth in high wind at a vibration frequency of 0.125 Hz. What is its period of vibration? Does your answer make sense?

Represent mathematically Since we know the vibration frequency, we can determine the period $T = 1/f$.

Solve and evaluate The period is

$$T = \frac{1}{f} = \frac{1}{0.125 \text{ Hz}} = \frac{1}{0.125 \text{ vib/s}} = 8.0 \text{ s/vib} = 8.0 \text{ s}$$

We find that the period of vibration is rather long. Observations of the tower show that the amplitude of vibration at the top of the building in a

high wind is less than 4 cm. The long period of vibration and relatively small amplitude mean that this vibration is unlikely to damage the building or affect its occupants.

- - - - - - - - - - - - - - - - - - - -

Try it yourself You sit on a rocking chair. It takes 12 s to complete six rocks—all the way back and all the way forward again. Determine the period and the frequency of the rocking.

Answer   2.0 s period and 0.5 rocks/s = 0.5 s^{-1} = 0.5 Hz frequency.

REVIEW QUESTION 10.1 Can we say that the period of vibration depends on the frequency or that the frequency depends on the period? Explain your answer.

FIGURE 10.4 Kinematics graphs for vibrational motion.

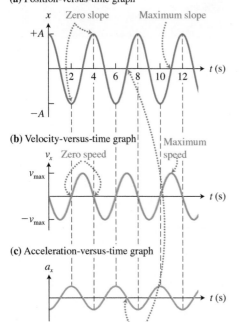

(a) Position-versus-time graph

(b) Velocity-versus-time graph

(c) Acceleration-versus-time graph

The acceleration-versus-time is proportional to the negative of the position-versus-time.

10.2 Kinematics of vibrational motion

To help describe vibrational motion quantitatively, let's look at the data from an experiment in which a motion detector collects position-versus-time, velocity-versus-time, and acceleration-versus-time data for a cart vibrating on a spring. Graphs of the data are shown in **Figure 10.4**.

The position of the object changes periodically, as does the slope of the graph. Note that the period for all three graphs is 4.0 s. Note also the relationship between the position and acceleration graphs. The acceleration has its maximum negative value when the position has its maximum positive value and vice versa. This makes sense. When the cart is at the rightmost positive x-position, the spring is exerting its maximum leftward negative x-force on the cart, causing maximum leftward negative acceleration.

Consistency of motion diagram and graphs

Is a motion diagram for vibrational motion consistent with the graphs in Figure 10.4? We can rotate the motion diagram in Table 10.2 so that it is oriented vertically beside a position-versus-time graph (**Figure 10.5a**) with the first five points numbered 0–4. We can move the $\vec{v}$ arrows from the motion diagram down to the velocity-versus-time graph (Figure 10.5b) and the $\vec{a}$ arrows from the diagram down to the acceleration-versus-time graph (Figure 10.5c). The graphs and the motion diagram are consistent with each other.

Notice also that the three graphs in Figure 10.4 are consistent with each other. Recall that velocity is the rate of change of position $v_x = (\Delta x / \Delta t)$—the slope of the position-versus-time graph. The velocity has its maximum positive value when the position-versus-time graph has its maximum positive slope. The slope of the position-versus-time graph is zero when $x = A$, as is the velocity. Note that at other clock readings, the slope of the position-versus-time graph has the same sign and relative magnitude as the reading of the velocity-versus-time graph.

Recall that acceleration is the rate of change of velocity $a_x = (\Delta v_x / \Delta t)$. The acceleration-versus-time graph is also consistent with the velocity-versus-time graph. For example, the acceleration is maximum when the velocity-versus-time graph has its maximum slope at $t = 2$ s and 6 s. The acceleration is zero when the slope of the velocity-versus-time graph is zero at $t = 3$ s.

Mathematical description of position as a function of time

Our goal now is to devise a periodic mathematical function whose graphical representation is similar to the graphs shown in Figure 10.4. What kind of function would work? The function should be periodic and vary between the amplitude of the vibrations and the amplitude with the negative sign. It should have zero value twice during one period.

Sinusoidal functions (in which the value of the function is proportional to the sin or cos of an angle) are the simplest periodic functions. The meanings of sine and cosine can be best understood using a unit circle (**Figure 10.6**) whose **radius vector** has a length of 1 unit and makes an angle θ with the x-axis (Figure 10.6a). We can then resolve the radius vector into its x- and y-components (Figure 10.6b):

$$r_x = r \cos \theta = (1) \cos \theta = \cos \theta$$
$$r_y = r \sin \theta = (1) \sin \theta = \sin \theta$$

Imagine now that the radius vector rotates counterclockwise around the origin of the coordinate system. The angle θ changes continuously from 0 to 360°, or from 0 to 2π rad. Consequently, the x-component of the radius vector changes from +1 to 0 to −1 to 0 and then back to +1; the y-component changes from 0 to +1 to 0 to −1 and then back to 0 (Figure 10.6c). If the radius vector continues rotating, the functions $\cos \theta$ and $\sin \theta$ become periodic functions with period 2π rad, or 360°.

If instead of a unit circle, you have a circle of radius A, the rotating radius vector will now have components:

$$R_x = A \cos \theta$$
$$R_y = A \sin \theta$$

as in Figure 10.6d. In these functions A is the maximum value or amplitude of the function, and θ is an angle that changes by 2π radians (2π rad $= 360°$) during each period of the function. We can think of R_x and R_y as the functions $x = A \cos \theta$ and

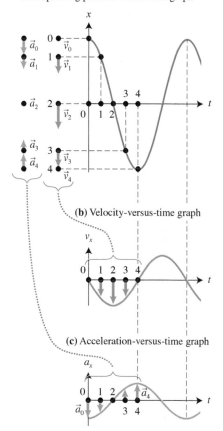

FIGURE 10.5 The motion diagram and kinematics graphs are consistent with each other.

(a) Comparing a motion diagram to its corresponding position-versus-time graph

(b) Velocity-versus-time graph

(c) Acceleration-versus-time graph

FIGURE 10.6 (a–c) A unit circle used to define the cos and sin functions and (d) a circle of radius A used to define sinusoidal functions.

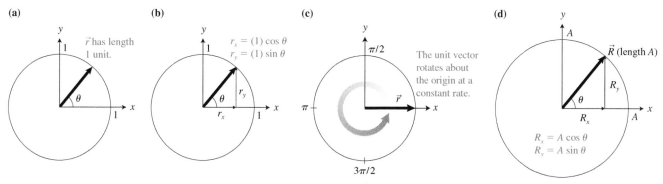

FIGURE 10.7 A graph of an $A \cos \theta$ function.

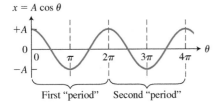

This graph has the same shape as the position-versus-time graph in Figure 10.4a.

$y = A \sin \theta$. A graph of $x = A \cos \theta$, shown in **Figure 10.7**, looks very similar to the position-versus-time graph produced by the motion detector shown in Figure 10.4a. However, in Figure 10.7 the independent variable is the angle θ, not the time t. We need a sinusoidal function of time to describe the motion of a vibrating object.

In **Table 10.3**, we compare the changes in position of the vibrating object as a function of clock reading shown in Figure 10.4a and the changes in position as a function of angle of the sinusoidal function shown in Figure 10.6d.

TABLE 10.3 **Position of a vibrating object as a function of time**

Clock reading t of the vibrating object shown in Figure 10.4a	Position x of the vibrating object shown in Figure 10.4a	Angle of the radius vector θ (radians) for the function $x = A \cos \theta$	Value of the function $x = A \cos \theta$
$0 \, (0 \text{ s})$	A	0	A
$\frac{1}{4}T \, (1 \text{ s})$	0	$\pi/2$	0
$\frac{1}{2}T \, (2 \text{ s})$	$-A$	π	$-A$
$\frac{3}{4}T \, (3 \text{ s})$	0	$3\pi/2$	0
$T \, (4 \text{ s})$	A	2π	A
$2T \, (8 \text{ s})$	A	4π	A
$3T \, (12 \text{ s})$	A	6π	A

The function $x = A \cos \theta$ matches the actual vibrational motion well. If we compare more times, the match remains strong. Now, we just need to replace θ with an appropriate function of time. Let's take the ratio of the angle and clock reading at any of the clock readings listed in the table. For example,

$$\frac{\theta}{t} = \frac{2\pi}{T}$$

or

$$\theta = \left(\frac{2\pi}{T}t\right)$$

We can write a periodic function $x(t)$ to represent the position-versus-time graph:

$$x = A \cos\left(\frac{2\pi}{T}t\right) \tag{10.2}$$

Notice that the above function has $x = +A$ at $t = 0$. If an object is at $x = 0$ at $t = 0$, you can either adjust the cos function by adding $-(\pi/2)$ or use the sine function to describe the motion:

$$x = A \sin\left(\frac{2\pi}{T}t\right)$$

since $\sin \theta = \cos\left(\theta - \frac{\pi}{2}\right)$.

Model of simple harmonic motion Motion described by Eq. (10.2) is called **simple harmonic motion** (SHM) since the cosine and sine functions are known as harmonic functions. SHM is a *mathematical model* of motion. Based on position-versus-time graphs, we see that a cart attached to a spring undergoes motion that is similar to SHM.

QUANTITATIVE EXERCISE 10.2 Describing the motion of two objects

The graphical descriptions of the vibrational motions of two objects are shown in the figure. Write mathematical descriptions (equations) for these motions.

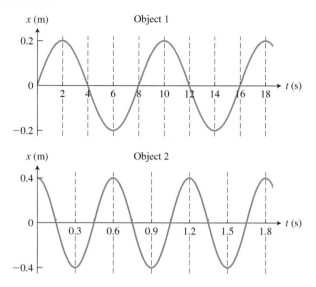

Represent mathematically (a) Vibrating object 1 has its maximum displacement of $A_1 = 0.2$ m and a period $T_1 = 8$ s. At time 0 the object is at zero and then its position increases. Thus, the function that describes its motion is a sine function:

$$x_1 = A_1 \sin\left(\frac{2\pi}{T_1}t\right)$$

Alternatively, we could write it as

$$x_1 = A_1 \cos\left(\frac{2\pi}{T_1}t - \frac{\pi}{2}\right)$$

(b) Vibrating object 2 has a maximum displacement $A_2 = 0.4$ m at time zero and has period $T_2 = 0.6$ s. Thus, the function

$$x_2 = A_2 \cos\left(\frac{2\pi}{T_2}t\right)$$

describes its motion. Alternatively, we could write it as

$$x_2 = A_2 \sin\left(\frac{2\pi}{T_2}t + \frac{\pi}{2}\right)$$

Solve and evaluate The mathematical description of vibrating object 1 is

$$x_1 = (0.2 \text{ m}) \sin\left(\frac{2\pi}{8 \text{ s}}t\right)$$

The mathematical description of vibrating object 2 is

$$x_2 = (0.4 \text{ m}) \cos\left(\frac{2\pi}{0.6 \text{ s}}t\right)$$

- -

Try it yourself Write a mathematical description for the displacement of an object that has amplitude 0.10 m and frequency of 5.0 Hz, and starts vibrating from position $x = -A$.

Answer

$x = -(0.10 \text{ m}) \cos\left(\frac{2\pi}{0.20 \text{ s}}t\right)$. Note that the period is $T = \frac{1}{f} = \frac{1}{5.0 \text{ s}^{-1}} = 0.20$ s.

Mathematical description of velocity and acceleration as a function of time

We see in Figure 10.4 that the position-, velocity-, and acceleration-versus-time graphs are all sinusoidal. They have the same period but have different amplitudes and reach their respective maxima and minima at different clock readings. We discussed the qualitative relationships among the three graphs earlier in this section. What are the mathematical descriptions of the three functions? When you have a function of time, you can find its rate of change by taking the derivative of the function. Those operations are studied in calculus; we will not show them in this book but will provide you with the final results. If the position function is given by Eq. (10.2)

$$x = A \cos\left(\frac{2\pi}{T}t\right)$$

then the velocity and acceleration functions are

$$v_x = -\left(\frac{2\pi}{T}\right)A \sin\left(\frac{2\pi}{T}t\right) \tag{10.3}$$

$$a_x = -\left(\frac{2\pi}{T}\right)^2 A \cos\left(\frac{2\pi}{T}t\right) \tag{10.4}$$

TIP Instead of the cosine in Eq. (10.2), you can represent the position of the object as a sine function (see Quantitative Exercise 10.2). This replacement will change the argument (or the independent variable) of the function but will not change the relationship between position, velocity, and acceleration. The velocity function will be a cosine and the acceleration will be a sine function again, but with a negative sign.

where A is the amplitude of the vibration and T is the period of the vibration. Notice that the velocity function has amplitude $(2\pi/T)A$ while the acceleration function has amplitude $(2\pi/T)^2 A$. Note also that the function for acceleration looks exactly the same as the function for position multiplied by $-(2\pi/T)^2$. We can say that the acceleration of a vibrating object is directly proportional to its displacement from equilibrium. Specifically, $a_x = -(2\pi/T)^2 x$. Finally, we see that the clock readings when the velocity and acceleration reach their respective maxima and minima match those in Figure 10.4.

QUANTITATIVE EXERCISE 10.3   **Equation Jeopardy**

The position of a vibrating cart attached to a spring is described by the equation

$$x = (0.050 \text{ m}) \cos\left[(12 \text{ s}^{-1})t\right]$$

What can you determine about the motion?

Represent mathematically Compare the above equation with the position-versus-time equation for simple harmonic motion:
$x = A \cos\left(\dfrac{2\pi}{T}t\right)$. We see that the amplitude A of the vibration is

0.050 m and $\dfrac{2\pi}{T} = 12 \text{ s}^{-1}$. The period is then $T = (2\pi/12) \text{ s} = (\pi/6) \text{ s}$.
The frequency of vibration is $f = 1/T = (6/\pi) \text{ s}^{-1} = (6/\pi) \text{ Hz}$.
The object has its maximum positive displacement at time zero.

Solve and evaluate The velocity of the vibrating cart is

$$v_x = -\frac{2\pi}{(\pi/6) \text{ s}}(0.050 \text{ m}) \sin\left[\frac{2\pi}{(\pi/6) \text{ s}}t\right]$$

$$= -(0.60 \text{ m/s}) \sin\left[(12 \text{ s}^{-1})t\right]$$

The amplitude of the velocity (the cart's maximum speed) is 0.60 m/s. The cart's acceleration is

$$a_x = -\left[\frac{2\pi}{(\pi/6) \text{ s}}\right]^2 (0.050 \text{ m}) \cos\left[\frac{2\pi}{(\pi/6) \text{ s}}t\right]$$

$$= -(7.2 \text{ m/s}^2) \cos\left[(12 \text{ s}^{-1})t\right]$$

The amplitude of the acceleration (the cart's maximum acceleration) is 7.2 m/s².

- -

Try it yourself Suppose the amplitude of vibration in this exercise remained the same, but the period was reduced by half. By what factors would this affect the cart's maximum speed and maximum acceleration? Try to answer without plugging numbers into equations.

Answer

Since the maximum speed is proportional to $1/T$, the maximum speed would double if the period were halved. Since the maximum acceleration is proportional to $1/T^2$, the maximum acceleration would quadruple if the period were halved.

REVIEW QUESTION 10.2 The velocity of an object attached to a spring changes as

$$v_x = -(0.84 \text{ m/s}) \sin\left[(4.19 \text{ s}^{-1})t\right]$$

Say all you can about its motion.

FIGURE 10.8 Dynamics of simple harmonic motion.

(a)

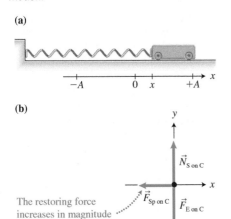

(b)

The restoring force increases in magnitude as the object is farther from equilibrium.

10.3 Dynamics of simple harmonic motion

So far, we have described simple harmonic motion (SHM) using sketches, motion diagrams, kinematics graphs, and kinematics equations. Now we will investigate the mechanism behind SHM.

Forces and acceleration

Consider again a cart on a frictionless surface attached to a spring (**Figure 10.8a**). We use a light (massless) spring that when stretched exerts a force on the cart. According to Hooke's law, that force is $F_{\text{Sp on C} x} = -kx$. Position x is equal to the displacement of the end of the spring from its unstretched position to where it is attached to the cart. The other end of the spring is fixed. When the cart is at position x relative to the unstretched equilibrium position, the spring exerts a restoring force on the cart (the object) that pulls it toward the equilibrium position (Figure 10.8b). The force component along the x-direction is

$$F_{\text{Sp on C }x} = -kx \qquad (10.5)$$

The spring is the only object that exerts a force with a nonzero x-component on the cart. Thus, the x-component form of Newton's second law $(a_x = \Sigma F_x/m)$ becomes

$$a_x = \frac{-kx}{m} = -\frac{k}{m}x \qquad (10.6)$$

where m is the mass of the cart. The above expression shows that the cart's acceleration a_x is proportional to the negative of its displacement x from the equilibrium position. This is exactly how the displacement and acceleration of an object undergoing SHM are related [Eq. (10.4)]. If the cart is on the positive side of the equilibrium position, the spring pulls it in the negative direction so that its acceleration component is negative. If the cart is on the negative side of the equilibrium position, the spring pushes it in the positive direction so its acceleration component is positive. Note that the farther the cart is from equilibrium, the greater the magnitude of the force that the spring exerts on the cart and the greater the magnitude of the cart's resulting acceleration.

We now understand *why* the acceleration-versus-time graph in Figure 10.4c is the same shape as the position-versus-time graph in Figure 10.4a with the sign of a being opposite the sign of x at all times; the restoring force exerted on the vibrating object is proportional to the displacement of the object from equilibrium but opposite in direction. Whenever this relation holds for a system, that system's motion can be described mathematically as simple harmonic motion.

Stop for a moment and think how unusual Eq. (10.6) is. When we studied linear motion at constant acceleration (Chapter 2), the acceleration of a moving object was the same at any location and did not depend on the position. When we studied constant speed circular motion (Chapter 5), the direction of acceleration changed with changing position along the circle, but the magnitude remained the same. In SHM the acceleration of an object changes in time in both magnitude and direction but is synchronized with the displacement of the object. Thus the sum of the forces exerted on an object moving in a circle at constant speed is constant in magnitude and only changes direction (it always points toward the center of the circle). The sum of the forces exerted on an object undergoing SHM not only changes direction during the motion but also changes magnitude.

Period of vibration of a cart attached to a spring

We now understand that Eq. (10.6) explains the relationship between the position and the acceleration of the cart. Can it also explain the frequency or period of vibration of a spring-cart system? What properties of a vibrating system affect the period of the vibration? Intuitively, you probably think that the period depends on the amplitude—the larger the amplitude, the longer it should take the cart to go through a full cycle. This is easy to check by simply starting the vibration with various amplitudes. We find that the period is essentially constant even when the amplitude is doubled or tripled (provided the amplitude is not too large). This seems surprising, but on reflection is reasonable. The farther we pull the cart from its equilibrium position, the greater the elastic force exerted on it by the spring, the greater the acceleration of the cart, and the faster it moves toward the equilibrium position. Thus it moves faster but over a longer distance—the two effects compensate. The period is independent of the amplitude.

Two other factors could affect the period of the vibration. The stiffness of the spring should make a difference, because a stiffer spring pulls harder on the cart, increasing the acceleration of the cart. The mass of the cart should also affect the period, as a more massive cart is harder to accelerate. We can use Newton's second law [Eq. (10.6)] and the kinematics equations that describe position and acceleration to derive an expression for the period of vibration of a cart with mass m at the end of a spring of spring constant k. We start with Eq. (10.6):

$$a_x = -\frac{k}{m}x$$

Substitute the expressions for x and a_x [Eqs. (10.2) and (10.4)] in the above to get

$$-\left(\frac{2\pi}{T}\right)^2 A \cos\left(\frac{2\pi}{T}t\right) = -\frac{k}{m}A \cos\left(\frac{2\pi}{T}t\right)$$

Divide both sides of the equation by $-A \cos\left(\frac{2\pi}{T}t\right)$ and solve for T:

$$\left(\frac{2\pi}{T}\right)^2 = \frac{k}{m} \quad \Rightarrow \quad 4\pi^2 \frac{m}{k} = T^2$$

Thus, the period is

$$T = 2\pi \sqrt{\frac{m}{k}} \qquad (10.7)$$

Notice that in this expression for period, there is no dependency on the amplitude. We can also check the units: $\sqrt{\dfrac{\text{kg}}{\text{N/m}}} = \sqrt{\dfrac{(\text{kg} \times \text{m})\text{s}^2}{\text{kg} \times \text{m}}} = \text{s}$, the correct unit for period. The frequency f of vibration of an object attached to a spring is $\dfrac{1}{T}$, or

$$f = \frac{1}{2\pi} \sqrt{\frac{k}{m}} \qquad (10.8)$$

In our derivation we assumed that the spring obeys Hooke's law, that the spring has zero mass, and that the cart is a point-like object. We also neglected friction. In real life, most springs obey Hooke's law if you do not stretch them too far, but it is impossible to find a massless spring or to eliminate frictional effects entirely. However, even with these assumptions, the above equation agrees closely with experiments in which friction is minimal and the mass of the spring is small compared with the mass of the attached object (see the video).

VIDEO
10.1

QUANTITATIVE EXERCISE 10.4 ▶ **Vibrating chair**

You want to design a bouncy seat for your nephew using a bungee cord (see figure at right). The chair should allow a 12-kg baby (including the mass of the seat) to bounce up and down naturally at a frequency of about 0.40 Hz. Estimate the spring constant the cord should have.

$k = ?$

$m = 12$ kg

$f = 0.40$ Hz

Represent mathematically Model the cord as a linear spring and the child + seat as a point-like object. We know the requested vibration frequency and the proposed mass of the child and seat. We can use Eq. (10.8) to find k:

$$f = \frac{1}{2\pi}\sqrt{\frac{k}{m}}$$

Solve and evaluate Square both sides of the above equation to get $f^2 = (1/4\pi^2)(k/m)$. Then, multiplying both sides by $4\pi^2 m$, we have

$$k = 4\pi^2 m f^2 = 4\pi^2(12 \text{ kg})(0.4 \text{ s}^{-1})^2$$
$$\approx 80 \text{ kg/s}^2 = 80 \text{ N/m}$$

- -

Try it yourself If the total mass of the child + seat were increased to 24 kg without changing the spring constant, what would be the frequency of vibration? Describe any additional difficulties that would result from this change.

Answer

$f = 0.28 \text{ s}^{-1}$ or $T = 3.5$ s. The cord would stretch twice as far—it may not fit in a doorway.

REVIEW QUESTION 10.3 What will happen to the period of a cart on a spring if you attach a second spring next to the first one in parallel, making two of them pull on the cart together?

10.4 Energy of vibrational systems

So far we have used the force approach to analyze the motion of vibrational systems quantitatively. Let us investigate how the energy approach can help us understand the process better. Consider again the vibration of a cart at the end of a spring (our system). The cart starts one vibration cycle at time $t = 0$ at position $x = +A$.

As the cart-spring system vibrates, the energy of the system continuously changes from all elastic when at $x = +A$ to all kinetic when passing through the equilibrium position at $x = 0$ to all elastic again when stopping for an instant at the far left of the vibration at $x = -A$ (see the far right column of Table 10.4). The cycle repeats on the way back to its starting position. Thus, the energy cycles between elastic and kinetic twice during each period. Assuming that no external forces do work on the system, the system's total energy is constant. Its value equals the maximum elastic potential energy when at $x = \pm A$, the maximum kinetic energy when at $x = 0$, and some combination of the two at other positions.

$$E = \tfrac{1}{2}kA^2 = \tfrac{1}{2}mv_{\text{max}}^2 = \tfrac{1}{2}mv^2 + \tfrac{1}{2}kx^2 \tag{10.9}$$

Notice that Table 10.4 shows only the states in which the system possesses either only elastic potential energy or only kinetic energy. These states are grouped together. In all other states (not shown in the table due to their infinite number), the system possesses some combination of both energies.

TABLE 10.4 Variation of energy during one vibration

Clock reading t	Displacement	Elastic potential energy U_s	Kinetic energy K	Total energy E_{tot}
$\tfrac{1}{2}T$	$-A$	$\tfrac{1}{2}kA^2$	0	$E_{\text{tot}} = \tfrac{1}{2}kA^2$
$\tfrac{1}{4}T$	0	0	$\tfrac{1}{2}mv_{\text{max}}^2$	$E_{\text{tot}} = \tfrac{1}{2}mv_{\text{max}}^2$
$\tfrac{3}{4}T$	0	0	$\tfrac{1}{2}mv_{\text{max}}^2$	
0	A	$\tfrac{1}{2}kA^2$	0	$E_{\text{tot}} = \tfrac{1}{2}kA^2$
T	A	$\tfrac{1}{2}kA^2$	0	

The second equality in Eq. (10.9) can be rearranged to give a relationship between the amplitude of the vibration and the cart's maximum speed:

$$v_{\text{max}} = \sqrt{\frac{k}{m}}\,A \tag{10.10}$$

This is exactly what follows from the kinematics equations describing simple harmonic motion [Eqs. (10.2) and (10.3)]. It also makes sense conceptually: when the mass of the cart is large, it should move slowly through the equilibrium position—having the mass in the denominator makes sense. If the spring is stiff, the cart will move faster—having the spring constant in the numerator also makes sense.

TIP In this discussion we neglected the interactions of the system with the surface of the track and with the air. If we take these interactions into account and include the surface and air in the system, the mechanical energy of the system will be converted into internal energy, eventually bringing the vibrating system to rest.

FIGURE 10.9 An astronaut Body Mass Measurement Device.

Astronauts in space can lose significant body mass because their muscles do not get sufficient use and begin to atrophy. Since traditional scales do not function in orbit (both the scale and the astronaut are in free fall), NASA developed the Body Mass Measurement Device (BMMD) to monitor the astronauts' mass. The BMMD consists of a chair on which the astronaut sits (**Figure 10.9**). A spring holds the chair in place. When the chair is pulled back and released, it vibrates at a frequency that depends on the mass of the chair plus astronaut.

EXAMPLE 10.5   **Astronaut mass**

The BMMD chair (mass 32 kg) has a vibrational period of 1.2 s when empty. When an astronaut sits on the chair, the period changes to 2.1 s. Determine (a) the effective spring constant of the chair's spring, (b) the mass of the astronaut, and (c) the maximum vibrational speed of the astronaut if the amplitude of vibration is 0.10 m.

Sketch and translate We choose the chair and the spring as the system for (a) and the astronaut, chair, and spring for (b) and (c).

$M = ?$
$v_{max} = ?$
$k = ?$
$T_{Chair} = 1.2$ s
$T_{Ast+Chair} = 2.1$ s
$m = 32$ kg

Simplify and diagram Assume that the chair's spring obeys Hooke's law and that the spring's mass is much smaller than the mass of the chair plus astronaut. We determine the effective spring constant as though there were one spring pushing and pulling the chair. For (c) we represent the process using a bar chart starting from the state when the chair is pulled the farthest from equilibrium and ending as the chair and person pass equilibrium.

$K_i \quad U_{si} \quad K_f \quad U_{sf}$

Represent mathematically (a) Use $T_{Chair} = 2\pi\sqrt{\dfrac{m}{k}}$ to determine the spring constant with the chair empty.

(b) Use $T_{Ast+Chair} = 2\pi\sqrt{\dfrac{m + M}{k}}$ to determine the astronaut mass.

(c) Determine the energy in the spring-astronaut-chair system when at the extreme position $x = +A$: $E = \frac{1}{2}kA^2$. Then use this energy to determine the maximum speed as the cart passes through equilibrium using $E = \frac{1}{2}kA^2 = \frac{1}{2}(m + M)v_{max}^2$.

Solve and evaluate (a) Square $T_{Chair} = 2\pi\sqrt{\dfrac{m}{k}}$ or $T_{Chair}^2 = 4\pi^2\dfrac{m}{k}$. Rearrange to get

$$k = 4\pi^2\frac{m}{T_{Chair}^2} = 4\pi^2\frac{32 \text{ kg}}{(1.2 \text{ s})^2} = 877 \text{ N/m}$$

(b) Square $T_{Ast+Chair} = 2\pi\sqrt{\dfrac{m + M}{k}}$ or $T_{Ast+Chair}^2 = 4\pi^2\dfrac{m + M}{k}$. Rearrange to get

$$m + M = \frac{T_{Ast+Chair}^2 k}{4\pi^2} \quad \text{or} \quad M = \frac{(2.1 \text{ s})^2(877 \text{ N/m})}{4\pi^2} - 32 \text{ kg} = 66 \text{ kg}$$

(c) From $E = \frac{1}{2}kA^2 = \frac{1}{2}(m + M)v_{max}^2$, $kA^2 = (m + M)v_{max}^2$.

Thus $v_{max} = A\sqrt{\dfrac{k}{m + M}} = (0.10 \text{ m})\sqrt{\dfrac{(877 \text{ N/m})}{32 \text{ kg} + 66 \text{ kg}}} = 0.30 \text{ m/s}$.

Check the units: $\text{m}\sqrt{\dfrac{\text{N/m}}{\text{kg}}} = \sqrt{\dfrac{\text{m}^2(\text{N/m})}{\text{kg}}} = \sqrt{\dfrac{\text{m(kg} \times \text{m)}}{\text{kg} \times \text{s}^2}} = \text{m/s}$.

The units are correct. The magnitudes are reasonable—a person can have a mass of 66 kg, and the speed of 0.3 m/s is a realistic speed for such a process.

- -

Try it yourself What would be the period of vibration of the machine if used by an 80-kg astronaut?

Answer   About 2.25 s.

REVIEW QUESTION 10.4 The period of vibration of a cart on a spring is 2.0 s. The elastic potential energy fluctuates in a repetitive pattern. What is the period of the elastic potential energy variation of the system? What is the period of the total energy of the system?

10.5 The simple pendulum

Springs are common in many objects that we use every day; they are an important component of mattresses, doors, shock absorbers, and office chairs. However, we seldom observe the springs themselves—most of them are hidden. One vibrating system in which the motion is very apparent is a pendulum. Grandfather clocks and mechanical metronomes use pendulums, of course. Another version of a pendulum is the oscillating movement of our arms and legs when we walk. In fact, much of animal locomotion involves complicated pendulum-like motions. Imagine a walking giraffe—its long legs swing as it walks. For now, let's consider a simplified model of a pendulum system that has a compact object (called a *bob*) at the end of a comparatively long and massless string and undergoes small amplitude vibrations. This idealized system is called a **simple pendulum**.

When the bob shown in **Figure 10.10** hangs straight down at rest, the sum of the forces exerted on it is zero. This is the equilibrium position for the bob. If you pull the bob to the right side (Figure 10.10a) and release it, it swings to the left (Figure 10.10b), overshooting the equilibrium position (Figure 10.10c), and then stops briefly at about the same distance to the left as it started on the right. The bob then swings back toward the equilibrium position, overshoots, stops about where it started and the motion repeats again. The pendulum undergoes circular motion with changing speed.

Two objects interact with the bob of the pendulum. The string S exerts a force in the radial direction that is always perpendicular to the path of the bob. Earth exerts a downward gravitational force that we can resolve into a radial component r along the string and a tangential component t perpendicular to the string. The tangential component of the force exerted by Earth on the bob points toward the equilibrium position—toward the left when the bob is on the right side of equilibrium (Figure 10.10b) and toward the right when on the left side (Figure 10.10d). This component of the gravitational force is what is always trying to restore the pendulum bob to its equilibrium position.

However, the tangential component of the force exerted by Earth on the bob is not equal to the sum of the forces exerted on the bob. Figure 10.10d shows an intermediate position of the bob. In this position you can see clearly that the sum of the forces does not point toward the equilibrium position, but the tangential component of the force exerted by Earth does. Hence, the tangential component of the gravitational force exerted by Earth on the bob is the restoring force. Thus, the ideas of equilibrium position and restoring force also apply to the vibrational motion of a pendulum. The bar charts in the figure also show the conversion of the bob-Earth system's energy from potential to kinetic, similar to the energy conversions for a cart attached to a spring.

FIGURE 10.10 Analyzing the forces exerted on a pendulum bob during pendulum motion.

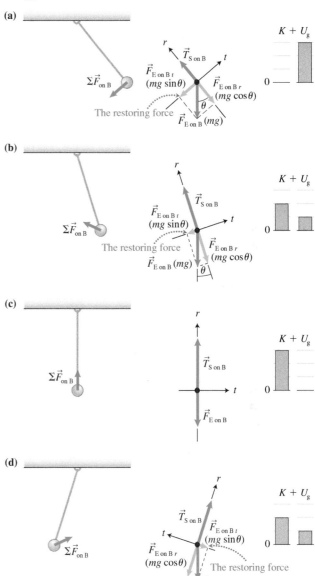

Experimental investigation of the period T of a simple pendulum

To investigate how the properties of a simple pendulum affect the period, we record its period for different vibration amplitudes A, bob masses m, and string lengths L (changing them one at a time). The results appear in Table 10.5. Notice that for the pendulum the amplitude of the angular displacement is measured in terms of the angle θ that the string makes with the vertical direction. Quantities that have been changed are boldfaced.

It appears that the period depends only on the string length. It does not depend on the mass of the bob or the amplitude of vibration. Let's make a formal derivation of the period and see if it agrees with the above.

TABLE 10.5 Effect of bob mass, string length, and amplitude on pendulum period

Bob mass m (kg)	String length L (m)	Amplitude θ (°)	Period T (s)
1.0	1.0	10.0	2.0
2.0	1.0	10.0	2.0
3.0	1.0	10.0	2.0
1.0	**2.0**	10.0	2.8
1.0	**3.0**	10.0	3.5
1.0	**4.0**	10.0	4.0
1.0	1.0	**15.0**	2.0
1.0	1.0	**20.0**	2.0
1.0	1.0	**25.0**	2.0

Deriving the period T of a simple pendulum

Suppose that the pendulum string has an angular displacement θ from the vertical—positive to the right and negative to the left (Figure 10.10). We already know that the string exerts a force on the bob that only has a radial component (see Figures 10.10b and c). The force exerted by Earth can be resolved into radial and tangential components. The radial component of magnitude $F_{\text{E on B } r} = -mg \cos \theta$ is along the radial axis and points outward, away from the center of the circle. The two radial components provide the bob with the radial acceleration necessary to keep the bob following a circular path. The tangential component of magnitude $F_{\text{E on B } t} = -mg \sin \theta$ serves as a restoring force (it is responsible for the change in the bob's speed along the arc). The negative sign indicates that the restoring force points opposite the direction of the angular displacement and back toward the equilibrium position.

Note that for small angles, the value of $\sin \theta$ is approximately equal to the value of θ (you can check this using your calculator). For example, when $\theta = 0.3491$ radians (equal to 20°), $\sin \theta = 0.3420$. θ and $\sin \theta$ differ by only 2%, even for this moderately large angle. Because $\theta \approx \sin \theta$ for small angles, the restoring force can be written as $F_{\text{E on B } t} = -mg\, \theta$, where θ must be in radians.

The angular displacement θ is related to the linear position x along the arc of the swinging bob (the solid line on the arc in **Figure 10.11**):

$$x = L\theta$$

where L is the length of the pendulum string. Thus, we find that the tangential component of the restoring force is

$$F_{\text{E on B } t} = -mg\, \theta = -mg \left(\frac{x}{L} \right) = -\left(\frac{mg}{L} \right) x \qquad (10.11)$$

Equation (10.11) for the restoring force exerted on a vibrating pendulum bob is very similar to the restoring force exerted on a mass at the end of a spring ($F_x = -kx$); the only difference is that the spring constant k for a spring has been replaced by mg/L. Since the restoring force is directly proportional to the bob's displacement from equilibrium, the bob is in simple harmonic motion. Consequently, the period of a pendulum is given by Eq. (10.7) with mg/L replacing k:

$$T = 2\pi \sqrt{\frac{m}{k}} = 2\pi \sqrt{\frac{m}{mg/L}} = 2\pi \sqrt{\frac{L}{g}}$$

Thus, the period of a pendulum is

$$T = 2\pi \sqrt{\frac{L}{g}} \qquad (10.12)$$

Notice that the period of a simple pendulum depends only on its length and on the magnitude of the gravitational constant. It does not depend on the mass of the object

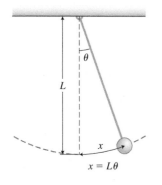

FIGURE 10.11 The linear displacement x of the bob is related to the string's angular displacement θ and length L.

$x = L\theta$

hanging at its end or on the amplitude of vibration. This agrees with the patterns we discovered in Table 10.5. The restoring gravitational force is proportional to the bob's mass, while the bob's acceleration $a_t = \Sigma F_t/m = -mg\theta/m$ is inversely proportional to its mass. Since mass cancels in the equation, the pendulum's motion does not depend on the bob's mass.

EXAMPLE 10.6 **Leg swinging period**

Estimate the number of steps that is "natural" for a leg to take per second while walking: in other words, the natural swinging frequency of the leg. We will treat the leg as a rigid body and model it as a simple pendulum.[1] The calculations in this example are approximate.

Sketch and translate Imagine a leg that is about 1 m long swinging like a pendulum from the hip. To estimate the frequency, the only thing we need to know is the length of the pendulum.

Simplify and diagram Although the leg swings as a pendulum, it does not resemble a simple pendulum since its mass is concentrated not at the foot but can be thought to be concentrated at the center of mass of the leg. We'll assume that the center of mass of the leg is at the knee and that the knee is at the midpoint of the leg.

Represent mathematically If a leg is 1 m long, then $L = 0.5$ m and its natural swinging frequency is calculated using Eq. (10.12).

$$f = \frac{1}{T} = \frac{1}{2\pi}\sqrt{\frac{g}{L}}$$

[1]In actuality, the leg is not a simple pendulum but is an example of what is called a physical pendulum—an extended body that swings back and forth about a pivot point. The swinging period of a physical pendulum is $f = (1/2\pi)\sqrt{mgL/I}$, where L is the distance from the pivot point to its center of mass, m is its mass, and I is the rotational inertia of the pendulum about its pivot point.

Solve and evaluate Inserting the estimated length in the above, we get

$$f \approx \frac{1}{2\pi}\sqrt{\frac{9.8 \text{ m/s}^2}{0.5 \text{ m}}} = 0.7 \text{ s}^{-1}$$

The period of one step is the inverse of f, or 1.4 s. According to our analysis, it is natural to take one step with your right leg (or with your left leg) approximately each 1.4 s. To check if this estimate is close to the real value, you can stand on the toes of one foot and swing your other leg freely back and forth at what seems like its natural unforced frequency (see figure). Count the number of swings in 60 s. When you divide 60 s by this number, you have the period for one swing.

Longer legs have longer periods of vibration. Giraffes take relatively few steps per second because of their long legs (a low walking frequency). A small dog like a Chihuahua has short legs and a high walking frequency. Each leg of the dog in the photo at right takes about two steps for each step taken by the man.

Try it yourself What should happen to your walking frequency when you walk wearing ankle weights? Why?

Answer The natural swinging frequency should decrease as the center of mass of the leg is lower and the pendulum becomes longer.

REVIEW QUESTION 10.5 Your grandfather's pendulum clock is slow. How should you adjust the position of the bob to correct for this?

10.6 Skills for analyzing processes involving vibrational motion

We will now look at how to apply our general problem-solving strategy to solve problems involving vibrational motion. In addition to following the details of the steps, pay attention to the nature of the problem. It allows for two different answers depending on initial conditions.

PROBLEM-SOLVING
STRATEGY 10.1

> ## Analyzing vibrational motion

EXAMPLE 10.7 **Vibrations of a vertical spring: a multiple possibility problem**

A long, light spring is hooked to the ceiling and hangs vertically. When we hang a 0.40-kg clay ball (ball 1) at the bottom end, the spring extends by 4.0 cm. A 0.20-kg clay ball (ball 2) rests on a table 30 cm below the hanging clay ball. You kick ball 1 down from its equilibrium position (you give the ball initial velocity). Determine the frequency of oscillations of the vibrating system.

Sketch and translate

- Sketch the process described in the problem statement. Label known and unknown physical quantities.
- Visualize the processes described in the problem statement.
- Choose an object (or objects) as the system of interest. Depending on the process, you might need to analyze several systems.

We sketch the process and label the givens. From the sketch, we see that the problem has two possible outcomes depending on the initial velocity of ball 1. If the velocity is such that the amplitude of vibration of ball 1 is less than 30 cm, the period of vibrations will only depend on the mass of ball 1 and spring constant k of the spring. If the amplitude of vibration is greater than 30 cm, the vibrating ball 1 will hit the resting ball 2, they will stick together, and the frequency of the system will be lower. We need to consider both possibilities.

Because we do not know k for the spring, we first choose ball 1 as the system and do a force analysis to find k. Then we choose Earth, the spring, and ball 1 as the second system to do the amplitude analysis of the vibrating system. To determine the frequency, we need to take into account the mass of the system, which in turn depends on whether the amplitude of vibration is large enough for the balls to stick together. We will either proceed with the second system (if the amplitude is less than 30 cm) or include ball 2 in a new system (if the amplitude is larger than 30 cm). Therefore, we have two conditional answers for the frequency.

Simplify and diagram

- Identify and evaluate assumptions and approximations.
- Represent the process with force diagrams and/or bar charts if needed.

Assume that the spring obeys Hooke's law and that the spring's mass is much smaller than the masses of the balls. We also assume that the size of the balls is negligible. To determine the spring constant, we draw a force diagram for ball 1 when it is just hanging, before it is pulled down.

$\vec{F}_{S \text{ on } 1}$

$\vec{F}_{E \text{ on } 1}$

To calculate the initial amplitude of ball 1, we need to do an energy analysis of the process using a bar chart. We point the coordinate axis down and choose the zero level for the position and the gravitational potential energy to be at the equilibrium position. The initial state of the ball 1-Earth-spring system is when the spring is stretched and ball 1 is moving with initial velocity v_i through the equilibrium position. The final state is when ball 1 is in its lowest position and not moving (the position of ball 1 from equilibrium is A). The system has elastic potential energy and negative gravitational potential energy. The stretch of the spring is now $\Delta y_i + A$.

$$K_i + U_{gi} + U_{si} + W = K_f + U_{gf} + U_{sf}$$

Represent mathematically

- If necessary, use kinematics equations to describe the changing motion of the object.

- If necessary, use the force diagrams to apply the component form of Newton's second law to the problem or use the bar charts to apply work-energy principles.

- If necessary, use the expressions for the period of an object attached to a spring or to a pendulum.

We use the force diagram for the hanging ball 1 to write Newton's second law and to find the spring constant of the spring:

$$m_1\vec{a} = \vec{F}_{\text{E on 1}} + \vec{F}_{\text{S on 1}}$$

Using the direction of the y-axis, the equilibrium condition (the ball is not accelerating), and the expressions for the two forces, we can rewrite Newton's second law as:

$$0 = m_1 g + (-k\Delta y_i)$$

From this we can determine k:

$$k = \frac{m_1 g}{\Delta y_i}$$

We use the energy bar chart to write the mathematical representation of the process:

$$\tfrac{1}{2}k\Delta y_i^2 + \tfrac{1}{2}m_1 v_i^2 = \tfrac{1}{2}k(A + \Delta y_i)^2 + (-m_1 g A)$$

$$\Rightarrow \tfrac{1}{2}ky_i^2 + \tfrac{1}{2}m_1 v_i^2 = \tfrac{1}{2}kA^2 + kAy_i + \tfrac{1}{2}ky_i^2 - m_1 g A$$

$$\Rightarrow \tfrac{1}{2}m_1 v_i^2 = \tfrac{1}{2}kA^2 + kA\Delta y_i - m_1 g A$$

Using the expression for k we found above,

$$kA\Delta y_i = \frac{m_1 g}{\Delta y_i}A\Delta y_i = m_1 g A$$

and thus the second term and the third term on the right side of the last energy equation add to zero. We now have $\tfrac{1}{2}m_1 v_i^2 = \tfrac{1}{2}kA^2$ and therefore the amplitude of the vibrations is equal to

$$A = v_i\sqrt{\frac{m_1}{k}}$$

This equation helps us find the value of the "critical" velocity that determines which of the two possibilities will occur. If $A < 30$ cm, only the mass of ball 1 will affect the frequency; if $A \geq 30$ cm, the frequency of the oscillations will be determined by the sum of the masses of ball 1 and ball 2. The amplitude will depend on the initial velocity of ball 1.

Solve and evaluate

- Solve for the unknowns.

- Evaluate the solution—is it reasonable? Consider the magnitude of the answer, its units, limiting cases, etc.

$$k = \frac{m_1 g}{\Delta y_i} = \frac{0.40 \text{ kg} \times 9.8 \text{ N/kg}}{0.040 \text{ m}} = 98.0 \text{ N/m}$$

$$A_{\text{critical}} = h = 0.30 \text{ m}$$

$$v_{i \text{ critical}} = h\sqrt{\frac{k}{m_1}} = 0.30 \text{ m}\sqrt{\frac{98.0 \text{ N/m}}{0.40 \text{ kg}}} = 4.70 \text{ m/s}$$

Therefore, if the initial velocity of ball 1 is less than 4.70 m/s, the frequency of the system is

$$f_1 = \frac{1}{2\pi}\sqrt{\frac{k}{m_1}} = 2.49 \text{ Hz}$$

If the initial velocity of the ball is equal to or greater than 4.70 m/s, the frequency of the system is

$$f_1 = \frac{1}{2\pi}\sqrt{\frac{k}{m_1 + m_2}} = 2.03 \text{ Hz}$$

We expected the frequency to be smaller when the mass is larger. The magnitudes are reasonable—a system with a relatively stiff spring and relatively heavy object attached to it can reasonably make two oscillations every second.

Try it yourself What happens to the equilibrium position if ball 2 sticks to ball 1 and they start vibrating together?

Answer   The new equilibrium position is 2.0 cm below the initial equilibrium position.

Note that the relation between velocity and amplitude, $A = v_i\sqrt{\dfrac{m_1}{k}}$, is exactly the same as that for the horizontal spring oscillator [see Eq. (10.10)]. This means that changing the orientation of a spring oscillator from horizontal to vertical does not affect its frequency or amplitude. However, it does affect the equilibrium position.

EXAMPLE 10.8 **Skiing and vibrations**

Imagine that you ski down a slope wearing a Velcro ski vest and then continue skiing on a horizontal surface at the bottom of the hill. There you run into a padded, Velcro-covered cart, which is also on skis (see the figure below). A 1280-N/m spring is attached to the other end of the cart and also to a wall. The spring compresses after your 60-kg body hits and sticks to the 20-kg cart. Your speed is 16 m/s just before you hit the cart. (a) What is your maximum speed after joining with the cart? (b) What is the maximum compression of the spring? (c) What is the period of the vibrational motion? (d) What is your maximum acceleration and where does it occur?

Sketch and translate Draw a labeled sketch.

(a) We represent the collision first (see the sketch below). The collision with the cart is a completely inelastic collision because you stick to the cart. If you and the cart are the system, then during the moment of the collision, the momentum of the system is constant (the same before and after the collision). We can use impulse-momentum ideas to determine your speed immediately after the collision.

(b) Now choose a different system: you, the cart, and the spring. The kinetic energy of you and the cart is converted into the spring's elastic potential energy as the spring compresses. (c) This is now a spring-object system that vibrates—you can determine the period of the vibration and the frequency. (d) The maximum acceleration is when the spring is compressed or stretched the most—when $x = \pm A$. We can determine the magnitude of the maximum acceleration using Newton's second law.

Simplify and diagram Assume that the snowy surface is perfectly smooth (zero friction), and that the spring has zero mass and obeys Hooke's law. (a) A momentum bar chart represents your collision with the cart (see the figure at right). There are no horizontal forces exerted on the you + cart system

along the x-direction during the instant of collision. Thus the x-component of momentum is constant. (b) Immediately after the collision, you and the cart, now joined, compress the spring. An energy bar chart represents this process (see the figure at right).

Represent mathematically (a) Use momentum constancy for the collision with the cart to find the speed of you + cart after the collision:

$$m_{You}v_{You\,i\,x} + m_{Cart}v_{Cart\,i\,x} = m_{You+Cart}v_{fx}$$

(b) Use energy constancy to determine the maximum compression distance:

$$\tfrac{1}{2}m_{You+Cart}v_{fx}^2 = \tfrac{1}{2}kA^2$$

(c) Determine the period using

$$T = 2\pi\sqrt{\dfrac{m_{You+Cart}}{k}}$$

(d) Use Newton's second law to determine the maximum acceleration during the vibrations after the collision:

$$a_x = \dfrac{-kx}{m_{You+Cart}}$$

a is maximum when the spring has its maximum compression (when $x = -A$).

Solve and evaluate (a) The velocity of you and the cart immediately after the collision is

$$v_{fx} = \dfrac{m_{You}v_{You\,i\,x} + m_{Cart}v_{Cart\,i\,x}}{m_{You+Cart}} = \dfrac{(60\text{ kg})(16\text{ m/s}) + 0}{(60\text{ kg} + 20\text{ kg})}$$

$$= 12\text{ m/s}$$

The speed seems reasonable, and the sign matches the original direction of motion.

(b) Using the equation $\tfrac{1}{2}m_{You+Cart}\,v_f^2 = \tfrac{1}{2}k\,A^2$, we can cancel the $(1/2)$, divide both sides of the equation by k, and take the square root. We find that the maximum displacement of the cart from its equilibrium position is

$$A = \sqrt{\dfrac{m_{You+Cart}}{k}}v_f = \sqrt{\dfrac{80\text{ kg}}{1280\text{ N/m}}}(12\text{ m/s}) = 3.0\text{ m}$$

The compression seems a little high, but the skier was moving with high speed; the units are fine.

(c) The period for the you + cart vibration is

$$T = 2\pi\sqrt{\dfrac{m_{You+Cart}}{k}} = 2\pi\sqrt{\dfrac{80\text{ kg}}{1280\text{ N/m}}} = 1.6\text{ s}$$

(d) The cart started at its equilibrium position ($x = 0$) and was moving at 12 m/s after you collided with it. This is its maximum speed.

The amplitude of the vibration was 3.0 m. Thus, the maximum acceleration of the you + cart system when at the $x = -A$ position is

$$a_{max} = \frac{-k(-A)}{(m_{You+Cart})} = \frac{(1280\ \text{N/m})(3.0\ \text{m})}{(60\ \text{kg} + 20\ \text{kg})} = 48\ \text{m/s}^2$$

The maximum acceleration of you and the cart is five times free-fall acceleration. The units for the result are the units of acceleration:

$$\frac{(\text{N/m}) \times (\text{m})}{(\text{kg})} = \frac{\text{N}}{\text{kg}} = \frac{\text{kg} \times \text{m/s}^2}{\text{kg}} = \text{m/s}^2$$

Try it yourself (a) How would the period of vibration change if you hit the cushion at half the velocity in the example? (b) What would be the period of vibration for a person whose mass is half of yours and whose speed before the collision with the cart is 0.8 times your speed?

Answer

(b) The period would be 1.24 s.

(a) The period is independent of the amplitude.

Radiation absorption by molecules

We can use what we know about vibrational motion to understand some aspects of global climate change. To maintain a consistent climate, Earth and its atmosphere as a system rely on a balanced exchange of energy in and out of the system. The energy absorbed by the Earth-atmosphere system, which enters the system in the form of visible light and infrared radiation from the Sun, must be balanced by the energy transferred out of that system. The energy is transferred out when vibrating molecules on Earth emit lower frequency infrared radiation that travels away from Earth. The radiation from the Sun is a much higher temperature than that emitted by Earth.

Some of the radiation emitted by molecular vibration on Earth is absorbed by CO_2 molecules in the atmosphere—a form of resonant energy transfer. This absorption causes the vibrational energy of these atmospheric molecules to increase. The excited CO_2 molecules re-emit the infrared radiation, much of which is reabsorbed by molecules on Earth. Thus, less energy is transferred out of the system than is transferred in—the planet warms. This **greenhouse effect** is similar to what happens to a person sitting in a car on a hot summer day. The energy from the Sun in the form of sunlight enters the car and is converted into additional thermal energy of the air in the car. The hot air in the car emits lower frequency infrared radiation, which is absorbed and reflected by the car's windows. Thus, less energy is radiated out of the car—the temperature inside the car rises.

Let's look briefly at the CO_2 vibrations that are involved in this energy absorption.

EXAMPLE 10.9 **CO_2 vibration**

Within molecules, atoms interact with each other via attractive and repulsive electric forces. These forces maintain a fairly consistent distance between the atoms. When the atoms get close to each other they repel; when they are pulled away they attract. Thus we can model the bond between the atoms as a spring. One of the vibrations in a CO_2 molecule involves an asymmetric stretching and compression of the CO bonds between the carbon and oxygen atoms (see the figure below). This bond has an effective spring constant of 1400 N/m. The vibration of the molecule has a natural frequency of 7.0×10^{13} Hz. (a) Determine the effective mass of the CO_2 molecule when vibrating in this way. (b) When a CO_2 molecule in the atmosphere absorbs infrared energy emitted by Earth, the energy of the vibration is 4.7×10^{-20} J. Estimate the amplitude of the vibration.

k = 1400 N/m
f = 7.0 × 10¹³ Hz
U = 4.7 × 10⁻²⁰ J
A = ?

One CO bond is compressed while the other stretches, and the process then reverses at high frequency.

Sketch and translate The vibration is depicted in the figure at the bottom of the previous column. We will use the expression for the vibration frequency of a cart with mass m and spring with spring constant k to estimate an effective mass of the vibrating molecule. We can use $\frac{1}{2}kA^2$ for the energy of vibration to estimate the amplitude.

Simplify and diagram Assume that the molecule can be modeled as a system undergoing SHM with a mass m at the end of a spring of spring constant k. We are also assuming that the physics ideas we are using apply to atomic-scale objects (which turns out not to be a very good assumption).

Represent mathematically The frequency of vibration of a mass at the end of a spring is

$$f = \frac{1}{2\pi}\sqrt{\frac{k}{m}}$$

The energy of vibration is

$$U_{total} = \frac{1}{2}kA^2$$

(CONTINUED)

Solve and evaluate The effective mass of the vibrating system is

$$m = \frac{k}{4\pi^2 f^2} = \frac{(1400 \text{ N/m})}{4\pi^2 (7.0 \times 10^{13} \text{ s}^{-1})^2} = 7.2 \times 10^{-27} \text{ kg}$$

The amplitude of vibration is

$$A = \sqrt{\frac{2U_{\text{total}}}{k}} = \sqrt{\frac{2(4.7 \times 10^{-20} \text{ J})}{1400 \text{ N/m}}} = 8.2 \times 10^{-12} \text{ m}$$

Are these numbers reasonable? The mass 7.2×10^{-27} kg is approximately the mass of an oxygen atom. The amplitude is a few percent of the typical distance between atoms in a molecule. Both answers thus seem reasonable.

Try it yourself Check the expressions used for each answer to see if they have the correct units.

Answer The correct unit

for amplitude.

$$\sqrt{\frac{f}{\text{N/m}}} = \sqrt{\frac{1}{\text{N} \cdot \text{m}} \cdot \frac{\text{m}}{\text{m}}} = (\text{m}^2)^{1/2} = \text{m, the correct unit}$$

the correct unit for mass, $A = \sqrt{\frac{2U_{\text{total}}}{k}}$ has units

$$m = \frac{k}{4\pi^2 f^2} \text{ has units } \frac{(\text{N/m})}{(\text{s}^{-1})^2} = \left(\frac{\text{kg} \cdot \text{m}}{\text{s}^2}\right)\left(\frac{1}{\text{m}}\right)\left(\frac{\text{m}}{1}\right)\left(\frac{\text{s}^2}{1}\right) = \text{kg},$$

REVIEW QUESTION 10.6 Why was it important to assume that the springs obey Hooke's law in Examples 10.7 and 10.8?

10.7 Including friction in vibrational motion

In the preceding sections, we have mostly ignored the effect of friction on vibrating objects. Without it, though, a car would continue vibrating on its suspension system for miles after crossing a bump on a road, and tall buildings would continue to sway even after the wind had died down.

You can observe the effects of friction on a simple system. Imagine that you have three identical containers, one filled with water, the second with honey, and the third with oil. You also have four identical springs and four identical blocks to hang on those springs. Hang the first block in air, the second in the water, the third in the honey, and the fourth one in the oil (**Figure 10.12**).

Set all four blocks in motion by pulling each one down and releasing it. The position-versus-time graphs in the figure show the motion of the blocks. The block in the air continues to vibrate with almost constant period and with very slowly reducing amplitude. The block in the water makes several vibrations with significantly decreasing amplitude before stopping. (Its frequency also slightly decreases, but this is difficult to observe.) The block in the honey does not even complete one full vibration; it returns to the equilibrium position after a very long time. The block in the oil quickly returns to the equilibrium position, but does not go past it. This phenomenon of decreasing vibration amplitude and increasing period is called **damping**. In practice, all vibrational systems are subject to some damping.

When we discuss vibrations with damping, vibrational systems are called *oscillators*. If the oscillator oscillates with decreasing amplitude, it is referred to as *underdamped* (the block in water and, to a lesser degree, the block in air). The damping of underdamped oscillators can vary to some degree (smaller or larger damping) as long as the system passes the equilibrium positions several times. If the displacement of the oscillator is smoothly (asymptotically) approaching the equilibrium position (without oscillations), it is referred to as *overdamped* (the block in honey and some door mechanisms). For a particular degree of damping, the oscillator reaches the equilibrium position in the shortest time and remains there. This case is referred to as *critical damping* (the block in oil). Critical damping has several important applications in industry and engineering (for example, in robots' arms that have to be designed to reach the target objects in the shortest time).

Damping is a useful aspect in the design of vehicles and bridges. For example, a mountain bike has shock absorbers to decrease the force exerted by the surface on the

FIGURE 10.12 Differently damped oscillators. (a) An underdamped oscillator (small damping); (b) an underdamped oscillator (larger damping); (c) an overdamped oscillator; (d) a critically damped oscillator.

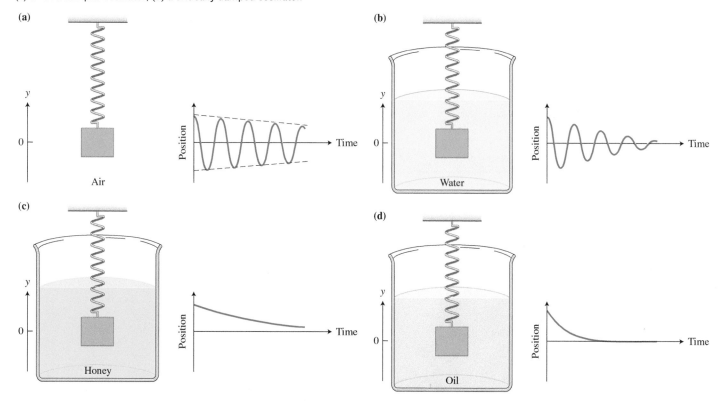

front tire when the tire hits a bump or a hole in the trail. The shock absorbers consist of springs and oil-filled dampers in the front fork of the bicycle and underneath the seat. The spring is wrapped around a piston filled with oil (**Figure 10.13**). During a collision, the spring compresses and increases the stopping distance, thus smoothing the effect of the collision on the rider. When the spring compresses, the piston pushes oil through a small hole. This process converts the spring's elastic potential energy into internal energy in the oil. Instead of continuing to bounce up and down, the spring returns quickly to its noncompressed position.

Damping is not desirable in all cases. Spiders that spin webs depend on vibrations to locate and restrain prey. An insect striking the web will produce vibrations that allow the spider to locate the insect. The farther these vibrations travel before damping out, the more likely it is that the spider will locate its prey (**Figure 10.14**).

FIGURE 10.13 A shock absorber for a mountain bike.

FIGURE 10.14 Vibrations in the spiderweb help the spider locate the captured prey.

REVIEW QUESTION 10.7 What features of damped vibrational motion make it different from SHM?

10.8 Vibrational motion with an external driving force

All real vibrations are damped and eventually stop unless energy is added to the system to compensate. In Observational Experiment **Table 10.6** (on the next page), we investigate how external interactions of the environment with a vibrating system could lead to continuous vibrations.

A forced vibration

VIDEO
OET 10.6

Observational experiment	Analysis
Experiment 1. A young child sits on a swing. You exert a steady force on the child and swing until the swing cable is at such an angle that the swing stops (this angle depends on the construction of the swing). There are no vibrations.	The swing, child, and Earth are the system (but not you). You exerted a constant force that did positive work on the swing. But no vibrations occurred during this time.
Experiment 2. A young child swings with a period $T = 2\pi\sqrt{L/g}$. You push her gently and briefly every time just after she reaches the peak of her vibration. Each brief push is in the same direction as her displacement. The pushes have the same period as her swinging.	You exerted a variable force that did positive work on the swing. The energy of the system increased. The amplitude of the swing vibrations increases. $U_{\text{system i}} + W = U_{\text{system f}}$
Experiment 3. The child still swings with period $T = 2\pi\sqrt{L/g}$. This time, you push her gently and briefly as she swings back toward you, just before she reaches the peak of her vibration. You push opposite the direction of her displacement.	You did negative work on the swing. The energy of the system decreased. The amplitude of the swing vibrations decreases. $U_{\text{system i}} + W = U_{\text{system f}}$
Experiment 4. The child still swings with period $T = 2\pi\sqrt{L/g}$. This time, you push her gently with a different period. Sometimes you push as she is moving back and at other times as she is moving forward. Her amplitude becomes small and stays small.	You do positive work sometimes and negative work at other times (the net work over time is zero). Friction and air resistance cause the swinging amplitude to decrease.

Patterns

For an external force exerted on the swinging system to cause the amplitude to increase:
1. The external force has to be a *variable force*.   2. The external force must do *positive work* on the vibrating system.
3. The *period (and therefore frequency)* of the external force *must match* that of the vibrating system.

To test these patterns, we use them to predict the results of the following experiment. A string is tied between two ring stands—the string sags a little (**Figure 10.15a**). Six pendulums of different lengths are attached to the string (80 cm, 10 cm, 50 cm, 80 cm, 100 cm, and 120 cm). What happens if we pull the 80-cm-long pendulum on the left in the direction perpendicular to the orientation of the string and let it go?

The swinging 80-cm-long pendulum causes the horizontal string to move back and forth at the **natural frequency** of the 80-cm-long pendulum:

$$f_0 = \frac{1}{2\pi}\sqrt{\frac{9.8 \text{ m/s}^2}{0.80 \text{ m}}} = 0.56 \text{ s}^{-1} = 0.56 \text{ Hz}$$

This 0.56-Hz motion of the vertical string causes it to pull the horizontal string back and forth at that frequency, which in turn tugs on the other vertical strings and thus does work on the other five pendulums. All pendulums start swinging at the 0.56-Hz frequency. However, since only the other 80-cm pendulum has the same natural frequency, we predict that only that pendulum will eventually swing with large amplitude. This is what we observe, a phenomenon called **resonance**—the increase in the amplitude of vibration of one of the pendulums while all other pendulums vibrate with smaller amplitudes. Resonance occurs in vibrating systems when some object in the environment exerts a force on it that varies in time and does net positive work over time. This work leads to an increase in the total energy of the system and therefore to the increase of the amplitude of vibration. This increase in energy occurs when the frequency of the external force is either the same as the natural frequency of the system without damping or close to the natural frequency for the systems with damping. The child on the swing exhibited another example of resonance, when the frequency of the variable force exerted on her was close to the natural frequency of the swing.

As we continue to observe the pendulums, an even more unusual thing happens. As time passes, the amplitude of the original 80-cm pendulum decreases to zero as the amplitude of the second 80-cm pendulum reaches a maximum. Then the first pendulum begins vibrating again as the second pendulum's amplitude decreases. Energy is being gradually transferred from one 80-cm-long pendulum to the other. This is an example of **resonant energy transfer**.

Finally, it is important to notice that the amplitudes of vibration of the other pendulums are not the same. The closer the length of the pendulum is to 80 cm, the larger the amplitude of vibration of that pendulum (see Figure 10.15b). This pattern can be generalized for any oscillator driven by a variable force. The closer the frequency of the external force to the natural frequency of the oscillator, the greater the amplitude.

We can see an example of energy transfer through resonance in our chapter-opening story about the Russian soldiers marching across a bridge. When the frequency of their steps approached the natural vibration frequency of the bridge, the marching initiated large-amplitude vibrations in the bridge, which led to its collapse. Any small vibrations that the bridge had were amplified by the positive work done by the soldiers on the bridge. As the amplitude of the vibrations increased, parts of the bridge twisted and pulled apart, and eventually the bridge collapsed.

Resonance caused the collapse of the Tacoma Narrows Bridge in Washington only four months after its completion. Unbeknownst to the engineers who designed it, vibrations near one of the bridge's resonant frequencies would increase in amplitude rather than damp out. On November 7, 1940, a strong wind combined with the bridge's natural twisting vibration caused an amplitude vibration of over 3 m at the sides of the bridge (**Figure 10.16a**). It collapsed after about an hour (Figure 10.16b).

Vocalists have been known to break wine glasses by producing sounds of particular frequencies. To do so, they must match the natural frequency of the wine glass (about 560 Hz). You are probably familiar with another example of energy transfer through resonance: spilling a beverage while carrying it across a room.

FIGURE 10.15 Connected 80-cm-long pendulum oscillators.

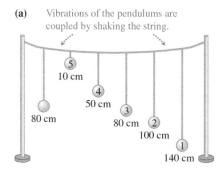
(a) Vibrations of the pendulums are coupled by shaking the string.

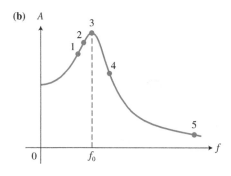
(b)

FIGURE 10.16 The collapse of the Tacoma Narrows bridge.

(a)

(b)

CONCEPTUAL EXERCISE 10.10   **Carrying a glass of juice**

Snehal is crossing a room while carrying a glass of cranberry juice. Estimate the stepping frequency that might cause the juice to splash onto the carpet. The natural frequency of vibrations of the juice sloshing in the nearly full glass is about 3 vibrations/s.

Sketch and translate Snehal's arm is a vibrating object that interacts with the glass of juice as well as with his legs.

Simplify and diagram Assume that Snehal's hand and arm oscillate at the same frequency at which his feet hit the ground. To avoid spilling the cranberry juice, he should try to walk so that his feet do not hit the ground 3 times per second (or 1.5 steps/s for each foot). That is a rapid pace, so it is unlikely that Snehal will cause resonant vibrations in the glass.

Try it yourself Do a follow-up investigation. First, determine the natural sloshing frequency in some container of liquid. Then walk with each of your legs swinging at the same frequency to see if the liquid spills. Think about what other factors affect the possibility of a coupled resonance between your stepping and the sloshing vibrations of liquid in a container. List the possible factors and then do experiments to determine the effect of changing those factors.

REVIEW QUESTION 10.8 Describe the phenomenon of energy transfer by resonance and explain why it happens.

Summary

Vibrational motion is the repetitive movement of an object back and forth about an **equilibrium position**. This vibration is due to the **restoring force** exerted by another object that tends to return the first object to its equilibrium position. An object's maximum displacement from equilibrium is the **amplitude** A of the vibration. **Period** T is the time interval for one complete vibration, and **frequency** f is the number of complete vibrations per second (in hertz). The frequency is the inverse of the period. (Section 10.1)

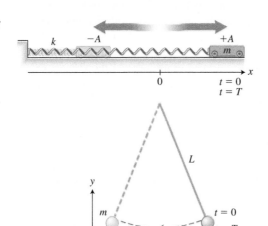

Object at end of spring obeying Hooke's law:

$$F_{\text{Restoring } x} = -kx \qquad \text{Eq. (10.5)}$$

$$T = 2\pi\sqrt{\frac{m}{k}} = \frac{1}{f} \qquad \text{Eq. (10.7)}$$

Simple pendulum:

$$F_{\text{Restoring } x} = -\left(\frac{mg}{L}\right)x \qquad \text{Eq. (10.11)}$$

$$T = 2\pi\sqrt{\frac{L}{g}} = \frac{1}{f} \qquad \text{Eq. (10.12)}$$

Simple harmonic motion is a mathematical model of vibrational motion when position x, velocity v, and acceleration a of the vibrating object change as sine or cosine functions with time. (Section 10.2)

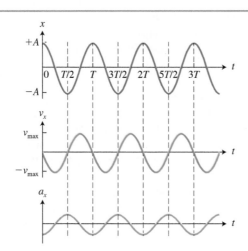

$$x = A\cos\left(\frac{2\pi}{T}t\right) \qquad \text{Eq. (10.2)}$$

$$v_x = -\left(\frac{2\pi}{T}\right)A\sin\left(\frac{2\pi}{T}t\right) \qquad \text{Eq. (10.3)}$$

$$a_x = -\left(\frac{2\pi}{T}\right)^2 A\cos\left(\frac{2\pi}{T}t\right) \qquad \text{Eq. (10.4)}$$

The **energy of a spring-object system** vibrating horizontally converts continuously from elastic potential energy when at the extreme positions to maximum kinetic energy when passing through the equilibrium position to a combination of energy types at other positions. (Section 10.3)

Object at end of spring:

$x = \pm A \quad x = 0 \quad \text{Other } x$

$U_s \;=\; K \;=\; K + U_s$

$$E = \tfrac{1}{2}kx^2 + \tfrac{1}{2}mv^2$$

$$E = \tfrac{1}{2}kA^2$$

$$= \tfrac{1}{2}mv_{\text{max}}^2 \qquad \text{Eq. (10.9)}$$

The **energy of a pendulum-Earth system** converts continuously from gravitational potential energy when it is at the maximum height of a swing to kinetic energy when it is passing through the lowest point in the swing to a combination of energy types at other positions. (Section 10.5)

Simple pendulum:

$\pm A \quad 0 \quad \text{At other places}$

$U_g \;=\; K \;=\; K + U_g$

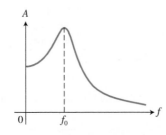

$$E = mgy + \tfrac{1}{2}mv^2$$

$$E = mgy_{\text{max}}$$

$$= \tfrac{1}{2}mv_{\text{max}}^2$$

Resonant energy transfer occurs when the frequency of the variable external force driving the oscillations is close to the **natural frequency** f_0 of the vibrating system. (Section 10.8)

Questions

Multiple Choice Questions

1. What are the features that make vibrational motion *different* from circular motion? Choose all that apply.
   (a) Vibrational motion is periodic.
   (b) Vibrational motion repeats itself.
   (c) During vibrational motion, there is a periodic change in the form of system energy.
   (d) Vibrational motion has a specific equilibrium point through which the system passes from different directions.

2. What does it mean if the amplitude of an object's vibration is constant and equal to 0.10 m?
   (a) The object's maximum distance from the starting point is 0.10 m.
   (b) The object's displacement from the equilibrium position is ±0.10 m.
   (c) The magnitude of the object's maximum displacement from the equilibrium position is 0.10 m.

3. What does it mean if the period of an object's vibration equals 0.60 s?
   (a) The motion lasts for 0.60 s.
   (b) The time interval needed for the object to move between the two farthest points from the equilibrium position is 0.60 s.
   (c) The time interval needed for the object to move through the same point in the same direction is 0.60 s.

4. What is the period of the kinetic or the potential energy change if the period of position change of an object attached to a spring is 2.0 s?
   (a) The energy does not change.   (b) 4.0 s
   (c) 2.0 s   (d) 1.0 s

5. A cart undergoing simple harmonic motion has a 2.0-Hz frequency and 3.0-cm amplitude. At time zero, the cart is at $x = +A$. Which of the three position-versus-time graphs shown in **Figure Q10.5** best represents the motion?

FIGURE Q10.5

6. The period of the object attached to a spring is T. How much time does the object need to move from the equilibrium position to half the amplitude?
   (a) $T/2$ (b) $T/6$
   (c) $T/8$ (d) $T/12$
   (e) There are not enough data to answer.

7. You have a simple harmonic oscillator. Where is its acceleration zero?
   (a) At the equilibrium position
   (b) At the maximum distances from equilibrium
   (c) At $t = 0$
   (d) Because the oscillator is constantly moving, there is no place where its acceleration is zero.

8. You have a simple harmonic oscillator. Where is the magnitude of its acceleration at its maximum?
   (a) At the equilibrium position
   (b) At the maximum distances from equilibrium
   (c) At $t = 0$
   (d) The acceleration is constant.

9. Which of the following arguments can be used to support the statement that an object attached to a light spring undergoes simple harmonic motion after it is displaced from the equilibrium position?
   (a) Because the position-versus-time graph is a sinusoidal-type function
   (b) Because the acceleration of the object is proportional to its displacement with a negative sign
   (c) Because the motion is periodic and has a constant period
   (d) Because the speed of the object is largest when it passes the equilibrium position.

Conceptual Questions

10. (a) Give three common examples of vibrational motion with damping. Explain the reasons for the damping. (b) Give a common example of a forced vibration. (c) Give a common example of resonance. Do not use any examples already used in this chapter.

11. An object of known mass hangs at the end of a spring, causing it to stretch a distance ΔL. How can you determine the frequency of vibration of this object if you know only the mass and the stretching distance?

12. Describe two different ways to estimate the spring constant of a rubber band using only a 100-g object, a stopwatch, and a meter stick. Describe the assumptions that you need to make in each estimate.

13. You have a small metal ball attached to a 1.0-m string. Explain two different methods that you can use with this apparatus to measure free-fall acceleration at this location.

14. A pendulum clock is running too fast. Explain how you can fix the problem.

15. What simplifications were used to derive the formula for the period of vibration of an object at the end of a spring and of a simple pendulum?

16. A pendulum clock is moved from the Mississippi Delta region to the Rocky Mountains. Will the clock run faster or slower in the Rockies? Why? How can it be adjusted to compensate for this change? Describe your assumptions.

17. Oil is often found in a geological structure called a salt dome. The rock in the salt dome is less dense than surrounding rock and is pushed up so that it "floats" on surrounding rock. Explain how a sensitive pendulum can detect a salt dome.

18. By experiment and by calculation, estimate the natural period and frequency of vibration of your arm. Carefully explain the values of the numbers used in each estimate.

19. A pendulum and a block hanging at the end of a spring are both carefully adjusted to make one vibration each second when on Earth's surface. Will the period of vibration of the pendulum or of the block on the spring be affected if they are placed on the Moon? Explain.

20. Will the frequency of vibration of a swing when you stand on the swing be greater or less than when you sit? Explain.

21. Estimate the time interval required for a Chihuahua to take a single step. Estimate the time interval required for a giraffe to take a single step.

22. The amplitude of vibration of a swing slowly decreases to zero if you do not pump your legs while swinging. Explain what happens to the energy of the system.

23. If you walk with your arms hanging down, they often oscillate forward and back at the same frequency as your stepping frequency. However, if your arms are bent at the elbow so that the forearms are horizontal, they often do not oscillate. Why?

24. You have a pendulum with a 1-m string. What is the period of vibration of the pendulum when it is placed on a spaceship orbiting Earth? Explain.

Problems

Below, **BIO** indicates a problem with a biological or medical focus. Problems labeled **EST** ask you to estimate the answer to a quantitative problem rather than derive a specific answer. Asterisks indicate the level of difficulty of the problem. Problems with no * are considered to be the least difficult. A single * marks moderately difficult problems. Two ** indicate more difficult problems.

10.1 Observations of vibrational motion

1. A low-friction cart is placed between two identical springs that are attached to rigid walls, as shown in **Figure P10.1**. If you push the cart against one spring and release it, it continues moving back and forth between the springs. Is this simple harmonic motion? Explain why or why not. Compare this case with the case where the cart is attached to the spring (such as the cart in Figure 10.8). Describe similarities and differences.

FIGURE P10.1

2. * You have a ball bearing and a bowl. You let the ball roll down from the top of the bowl; it moves up and down the bowl's wall for some time. (a) Is this vibrational motion? Explain why or why not. (b) Draw a force diagram for the ball for four different positions and indicate what force or force component is responsible for the acceleration toward the equilibrium position. (c) Draw energy bar charts for these four ball-Earth system positions.

3. Draw a sketch of a pendulum. Indicate the equilibrium position. Show this object vibrating and indicate the amplitude of vibration on the sketch.

4. * (a) Draw a sketch of an object attached to a vertical spring. Indicate the equilibrium position. (b) Show this object vibrating up and down and indicate the amplitude of vibration on the sketch. (c) Describe an experiment you would perform to determine the period of vibration of the object. (d) Perform the experiment. Include experimental uncertainty in your result.

10.2 Kinematics of vibrational motion

5. Draw a graph showing the position-versus-time curve for a simple harmonic oscillator (a) with twice the frequency of that shown in **Figure P10.5** and (b) with the same frequency but twice the amplitude as shown in the figure.

FIGURE P10.5

6. Suppose that at time zero the cart attached to the spring such as shown in Figure 10.8 is released from rest at position $x = +A$ and that its period of vibration is 8.0 s. Draw nine sketches showing the cart's approximate position each second starting at 0 s and ending at 8.0 s. Draw and label arrows indicating the relative velocity and acceleration of the cart at each position.

7. * (a) Sketch a motion diagram and a position-versus-time graph for the motion of a cart attached to a spring during one period. It passes at high speed through the equilibrium position at time zero. (b) Sketch a velocity-versus-time graph for the cart. (c) Sketch an acceleration-versus-time graph for the cart. (d) Draw another representation of the process; explain how this representation allows you to learn more about the process than the kinematics graphs do.

8. * Devise a position-versus-time function that describes the simple harmonic motion of an object of your choice. Choose the amplitude and the period of vibration, and draw corresponding velocity-versus-time and acceleration-versus time graphs. Draw a motion diagram for one period of its motion.

9. * The position of a vibrating object changes as a function of time as $x = (0.2\ \text{m})\cos[(\pi\ \text{s}^{-1})t]$. Say everything you can about this motion. Write expressions for the velocity and acceleration as functions of time. Draw graphs for each function.

10. * The velocity of a vibrating object changes as a function of time as $v = -(0.6\ \text{m/s})\cos[(2\pi\ \text{s}^{-1})t]$. (a) Say everything you can about the motion. Draw a sketch of the situation when the motion starts. (b) Represent the motion with a position-versus-time graph and a work-energy bar chart.

11. * A cart at the end of a spring undergoes simple harmonic motion of amplitude 10 cm and frequency 5.0 Hz. (a) Determine the period of vibration. (b) Write an expression for the cart's position at different times, assuming it is at $x = -A$ when $t = 0$. Determine its position (c) at 0.050 s and (d) at 0.100 s.

12. ** Refer to the situation in Problem 10.1. (a) Draw position-versus-time and velocity-versus-time graphs for two cycles of the motion of the cart. (b) Derive the expression for the period of the motion of the cart as a function of relevant parameters.

10.3 Dynamics of simple harmonic motion

13. You exert a 100-N pull on the end of a spring. When you increase the force by 20% to 120 N, the spring's length increases 5.0 cm beyond its original stretched position. What is the spring constant of the spring and its original displacement?

14. **Metronome** You want to make a metronome for music practice. You use a 30-g object attached to a spring to serve as the time standard. What is the desired spring constant of the spring if the object needs to make 1.00 vibrations each second? What are the assumptions that you made?

15. Determine the frequency of vibration of the cart shown in **Figure P10.15**.

FIGURE P10.15

16. * A spring with a cart at its end vibrates at frequency 6.0 Hz. (a) Determine the period of vibration. (b) Determine the frequency if the cart's mass is doubled while the spring constant remains unchanged and (c) the frequency if the spring constant doubles while the cart's mass remains the same.

17. A cart with mass m vibrating at the end of a spring has an extra block added to it when its displacement is $x = +A$. What should the block's mass be in order to reduce the frequency to half its initial value?

18. * A 300-g apple is placed on a horizontal platform that can vibrate in the vertical direction with a fixed amplitude of 1.0 cm. What is the maximum frequency of the platform's vibration for the apple to remain in contact with the platform? (Hint: Think about how the platform needs to move to lose contact with the apple.)

19. ** A 2.0-kg cart vibrates at the end of an 18-N/m spring with the amplitude of 3.0 cm. (a) Make a list of physical quantities you can determine about the vibrations and determine two of them. (b) If a second 18-N/m spring is attached beside the first one, what will be the period of the vibration?

20. * What were the main ideas that we used to derive the expression for the period of an object vibrating at the end of a spring? Explain the assumptions that we made. When is the expression for the period not valid for a spring of known k?

10.4 Energy of vibrational systems

21. * A spring with a spring constant of 1200 N/m has a 55-g ball at its end. (a) If the energy of the system is 6.0 J, what is the amplitude of vibration? (b) What is the maximum speed of the ball? (c) What is the speed when the ball is at a position $x = +A/2$? What assumptions did you make to solve the problem? How do you know if your answer makes sense?

22. * A person exerts a 15-N force on a cart attached to a spring and holds the cart steady. The cart is displaced 0.060 m from its equilibrium position. When the person stops holding the cart, the system cart+spring undergoes simple harmonic motion. (a) Determine the spring constant of the spring. (b) Determine the energy of the system. (c) Write an expression $x(t)$ for the motion of the cart. (d) Draw as many graphical representations of the motion as you can.

23. A spring with spring constant 2.5×10^4 N/m has a 1.4-kg cart at its end. (a) If its amplitude of vibration is 0.030 m, what is the total energy of the cart+spring system? (b) What is the maximum speed of the cart? (c) If the energy is tripled, what is the new amplitude? (d) What is the new maximum speed of the cart? (e) What assumptions did you make to solve the problem? If the assumptions were not reasonable, how would the answers change?

24. * **Proportional reasoning** By what factor must we increase the amplitude of vibration of an object at the end of a spring in order to double its maximum speed during a vibration? Explain.

25. **Proportional reasoning** By what factor must we increase the amplitude of vibration of an object at the end of a spring in order to double the total energy of the system? Explain. What will happen to the speed of the object? What assumptions did you make?

26. **Monkey trick at zoo** A monkey has a cart with a horizontal spring attached to it that she uses for different tricks. In one trick, the monkey sits on the vibrating cart. When the cart reaches its maximum displacement from equilibrium, the monkey picks up a 0.30-kg cantaloupe from a trainer. The mass of the monkey and the cart together is 3.0 kg. The spring constant is 660 N/m. The amplitude of horizontal vibrations is 0.24 m. Determine the ratio of the maximum speed of the monkey before and after she picks up the cantaloupe.

27. * A frictionless cart attached to a spring vibrates with amplitude A. (a) What fraction of the total energy of the cart-spring system is elastic potential energy and what fraction is kinetic energy when the cart is at position $x = A/2$? (b) At what position is the cart when its kinetic energy equals its elastic potential energy?

28. A 2.0-kg cart attached to a spring undergoes simple harmonic motion so that its displacement is described by $x = (0.20 \text{ m})\sin[(2\pi/2.0 \text{ s})t]$. Construct energy bar charts for the cart-spring system at times $t = 0$, $t = T/4$, $t = T/2$, $t = 3T/4$, and $t = T$.

29. * The motion of a cart attached to a horizontal spring is described as $x = (0.17 \text{ m}) \sin(\pi s^{-1})t$. Say everything you can about this motion and represent it using graphs, motion diagrams, and energy bar charts.

10.5 The simple pendulum

30. **Pendulum clock** Shawn wants to build a clock whose pendulum makes one swing back and forth each second. (a) What is the desired length of the rod (assumed to have negligible mass) holding the metal ball at its end? (b) Will the rod need to be shorter or longer if he includes the mass of the rod in the calculations? Explain.

31. Show that the expression for the frequency of a pendulum as a function of its length is dimensionally correct.

32. A pendulum swings with amplitude 0.020 m and period of 2.0 s. What is its maximum speed?

33. * **Proportional reasoning** You are designing a pendulum clock whose period can be adjusted by 10% by changing the length of the pendulum. By what percent must you be able to change the length to provide this flexibility in the period? Explain.

34. * **Building demolition** A 500-kg ball at the end of a 30-m cable suspended from a crane is used to demolish an old building. If the ball has an initial angular displacement of 15° from the vertical, determine its speed at the bottom of the arc.

35. * You have a pendulum with a long string whose length you can vary but cannot measure, a small ball, and a stopwatch. Describe two experiments that you can design to determine the height of a ladder. Indicate the assumptions that you use in each method.

36. * **Variations in g** The frequency of a person's pendulum is 0.3204 Hz when at a location where g is known to be exactly 9.800 m/s^2. Where might the same pendulum be when its frequency is 0.3196 Hz? What is g at that location?

37. **EST** A graph of position versus time for an object undergoing simple harmonic motion is shown in Figure P10.37. Estimate from the graph the amplitude and period of the motion and determine the object's frequency. If the object is a pendulum, what is its length?

FIGURE P10.37

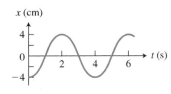

38. Determine the period of a 1.3-m-long pendulum on the Moon where the free-fall acceleration is 1.624 m/s^2.

39. * You have a simple pendulum that consists of a small metal ball attached to a long string. You push the ball so that at time $t = 0$ it moves with a speed of 2 m/s through the equilibrium position in the positive x-direction and continues oscillating with the amplitude of 20 cm. (a) Draw labeled position-versus-time and velocity-versus-time graphs for the first two periods of oscillation. (b) Determine the length of the string.

10.6 Skills for analyzing processes involving vibrational motion

40. * **Equation Jeopardy** The following expression describes a situation involving vibrational motion. Sketch a process and devise a problem for which the expression might be an answer.

$$\tfrac{1}{2}(20{,}000 \text{ N/m})(0.20 \text{ m})^2 = \tfrac{1}{2}(100 \text{ kg})v^2$$
$$+ \tfrac{1}{2}(20{,}000 \text{ N/m})(0.10 \text{ m})^2$$

41. * **Trampoline vibration** When a 60-kg boy sits at rest on a trampoline, it sags 0.10 m at the center. (a) What is the effective spring constant for the trampoline? (b) The trampoline is pulled downward an extra 0.050 m by a strap sewed under the center of the trampoline. When the strap is released, what are the energy and frequency of the boy-trampoline system? What assumptions did you make?

42. * **Proportional reasoning** If you double the amplitude of vibration of an object at the end of a spring, how does this affect the values of k, T, U, and v_{max}?

43. * **Pendulum on Mars** The frequency of a pendulum is 39% less when on Mars than when on Earth's surface. Use this fact to determine Mars's gravitational constant.

44. * **BIO** **EST** **Annoying sound** Low-frequency vibrations (less than 5 Hz) produced by a certain buzzer are annoying for humans near the buzzer if the product of the buzzer's amplitude and frequency squared (Af^2) exceeds the value 0.5×10^{-2} m·s^{-2}. Estimate the smallest frequency and amplitude of a buzzer that produces these annoying vibrations if the device vibrates a total mass of 0.12 kg and has a vibrational energy of 5.0×10^{-4} J.

45. ** A 1.2-kg block sliding at 6.0 m/s on a frictionless surface runs into and sticks to a spring. The spring is compressed 0.10 m before stopping the block and starting its motion back in the opposite direction. What can you determine about the vibrations that start after the collision? Make a list of physical quantities and determine four of them.

46. ** **EST** **Willis Tower vibration** The mass of the Willis (formerly Sears) Tower in Chicago is about 2×10^8 kg. The tower sways back and forth at a frequency of about 0.10 Hz. (a) Estimate the effective spring constant for this swaying motion. Explain why you included a particular number of significant figures. (b) A gust of wind hitting the building exerts a force of about 4×10^6 N. By approximately how much is the top of the building displaced by the wind? Explain why you included a particular number of significant figures. State the assumptions that you made. Does your answer make sense?

47. ** You shoot a 0.050-kg arrow into a 0.50-kg wooden cart that sits on a horizontal, frictionless surface at the end of a spring that is attached to the wall at the other end. The arrow hits the cart and sticks into it. The cart and arrow together compress the spring and start the system vibrating at a frequency of 2.0 Hz with a 0.20-m amplitude. How fast was the arrow moving? State any assumptions you made to solve the problem.

10.7 Including friction in vibrational motion

48. * You have a pendulum whose length is 1.3 m and bob mass is 0.20 kg. The amplitude of vibration of the pendulum is 0.07 m. (a) Determine the maximum energy of the bob-Earth system. Explain why you included a particular number of significant figures. (b) How much mechanical energy is converted to internal energy before the pendulum stops?

49. * You hang a 0.10-kg block from a spring, causing the spring to stretch 0.070 m. When the block is hanging at rest at this equilibrium position, you pull it down with a spring scale and let go of the block when the scale reading is 3.0 N. (a) Where was the block when you let go? (b) Determine the work you did stretching the spring. (c) What was the energy of the spring-block-Earth system when you let go? (d) How far will the block rise after you release it? (e) The vibrations last for 50 cycles. Qualitatively represent the beginning of cycle #1 and the beginning of cycle #25 with an energy bar chart.

50. * Imagine that you have a cart on a spring that moves on a horizontal surface with no friction. (a) Represent the cart's motion with a motion diagram for one period. (b) Draw force diagrams for each quarter of a period. (c) Draw energy bar charts for each quarter of a period. (d) Draw position-, velocity-, and acceleration-versus-time graphs for each period. (e) On these graphs, use a dashed line to sketch graphs for the same cart on the same spring, assuming there is a friction force between the cart and the surface.

51. Describe one situation from everyday life in which large damping is desirable and one in which it is not desirable.

10.8 Vibrational motion with an external driving force

52. **EST** **Twins on a swing** How frequently do you need to push a swing with twin brothers on it compared to when pushing the swing with one on it? What assumptions did you make?

53. (a) Determine the maximum speed of a girl on a 3.0-m-long swing when the amplitude of vibration is 1.2 m. (b) What assumptions did you make? Is the child's vibrational motion damped? Explain. (c) Under what conditions is the motion a forced vibration? Explain.

54. * **Sloshing water** You carry a bucket with water that has a natural sloshing period of 1.7 s. At what walking speed will water splash if your step is 0.90 m long?

55. * **Feeling road vibrations in a car** If the average distance between bumps on a road is about 10 m and the natural frequency of the suspension system in the car is about 0.90 Hz, at what speed will you feel the bumps the most?

56. * **EST** A vertical spring oscillator consists of a 0.3-m-long spring and the 30-g metal cube attached to its lower end. The upper end of the spring is attached to a mechanism that vibrates with the adjustable amplitude A_0 and period T. Some students measure how the amplitude of vibration of the lower end of the spring A_S depends on different settings of the mechanism. Their data are shown in the table at right. (a) Estimate the natural frequency of the spring oscillator. (b) Estimate the spring constant of the spring. (c) When the period of vibration of the mechanism is 0.040 s, is the period of vibration of the metal cube less than 0.04 s, larger than 0.04 s, or about 0.04 s?

T (s)	A_0	A_S
0.100	1	4.3
0.100	2	8.7
0.100	3	13.0
0.100	4	17.3
0.200	1	3.0
0.067	1	6.7
0.050	1	2.8
0.040	1	1.4

General Problems

57. A spring oscillator and a simple pendulum have equal natural frequencies on Earth. Determine the ratio of their periods $T_{Spr\ Osc}/T_{Sim\ Pen}$ on the Moon.

58. * You attach a block (mass m) to a spring (spring constant k) that oscillates in a vertical direction. Determine an expression for the period of its vibrations. Sketch a position-versus-time graph for the vibrations, assuming that the motion starts when the block is at its lowest maximum displacement from the equilibrium position. What assumptions did you make?

59. * You attach a 1.6-kg object to a spring, pull it down 0.12 m from the equilibrium position, and let it vibrate. You find that it takes 5.0 s for the object to complete 10 vibrations. Make a list of physical quantities that you can determine about the motion of the object and determine five of them.

60. * **Traveling through Earth** A hole is drilled through the center of Earth. The gravitational force exerted by Earth on an object of mass m as it goes through the hole is $mg(r/R)$, where r is the distance of the object from Earth's center and R is the radius of Earth (6.4×10^6 m). Will an object dropped into the hole execute simple harmonic motion? If yes, find the period of the motion. Would this be a comfortable ride? How does one-half of this time compare with the time needed to fly in an airplane halfway around Earth? What assumptions did you make to solve the problem?

61. * **EST** Estimate the effective spring constant of the suspension system of a car. Describe your technique carefully. How can you test your answer to determine if it makes sense?

62. * **Galileo's pendulum** The length L of a pendulum is shortened when the string encounters a peg placed a distance b below the support (see Figure P10.62). (a) What is the height to which the bob swings on the other side compared to the height from which it was released? Explain. (b) Derive the expression for the period of the pendulum with the inserted peg. (c) Draw a qualitative position-versus-time graph for three periods of the pendulum's motion. (d) What happens to the force exerted by the string on the bob when the bob passes the equilibrium position? Explain.

FIGURE P10.62

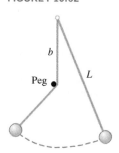

63. * A 0.5-kg low-friction cart is moving at constant speed of 1.7 m/s to the left along a horizontal table toward a relaxed spring with $k = 120$ N/m that is fixed at the left end to a wall. At $t = 0$, the front of the cart is 20 cm from the free end of the spring. The cart hits the spring, and the spring compresses and expands, making the cart move to the right with the same speed as the initial speed. (a) Sketch the situation. Determine the clock reading at the moment when the front (left end) of the cart passes the position where it was at $t = 0$ again. (b) Choose the cart and the spring as a system and represent the process using energy and momentum bar charts. (Hint: Think about which part of the cart's motion can be treated as vibrational motion.)

64. * A circular metal frame is fixed in a vertical position as shown in Figure P10.64a. If a small ice cube is placed inside the frame not too far from the lowest point, it exhibits simple harmonic motion. (a) Draw force diagrams for five positions of the ice cube and compare them with the force diagrams for a simple pendulum in Figure 10.10. (b) Derive the expression for the period of oscillations of the ice cube. (c) How would the melting of the ice cube affect the period of the oscillation, assuming friction can be neglected? Explain.

FIGURE P10.64

65. ** **"Reduced gravity pendulum"** The circular metal frame from Problem 10.64 is now fixed on a smooth incline. If a small ice cube is placed inside the frame as shown in Figure P10.65 and released, it exhibits simple harmonic motion similar to the vertical situation. (a) Draw force diagrams for three positions of the ice cube and compare them with the force diagrams that you obtained in Problem 10.64. (b) Show that the period of oscillation of the ice cube is given by the expression $T = 2\pi\sqrt{\dfrac{R}{g \sin \theta}}$, where R is the radius of the circular frame and θ is the angle of the incline. Why do you think this device is called a "reduced gravity pendulum"?

FIGURE P10.65

66. ** You drop a 57-g tennis ball from 1.0 m above a hard concrete floor. For small deformations the ball approximately follows Hooke's law and behaves like a spring with a constant 17×10^3 N/m. Determine (a) by how much the ball compresses during the collision with the floor and (b) the duration of the ball-floor impact. Indicate any assumptions that you made.

67. * A 5.0-g bullet traveling horizontally at an unknown speed hits and embeds itself in a 0.195-kg block resting on a frictionless table. The block slides into and compresses a 180-N/m spring a distance of 0.10 m before stopping the block and bullet. Determine the initial speed of the bullet. State any assumptions that you made to solve the problem.

68. ** You have a pendulum of mass m and length L that is displaced an angle θ at the start of the swinging. (a) Determine an expression for the energy of the bob-Earth system. (b) Determine an expression for the maximum vertical height of the bob with respect to the equilibrium position. (c) Determine an expression for the maximum speed of the bob. (d) What assumptions did you make? (e) Discuss how the answer to each of the questions changes if relevant assumptions are not valid.

69. ** A pendulum clock works for many days before the swinging stops. Describe several mechanisms that allow an actual pendulum clock to continue working without decreasing amplitude. Write a report explaining how it works.

70. ** **Foucault's pendulum** In 1851, the French physicist Jean Foucault hung a large iron ball on a wire about 67 m (220 ft) long to show that Earth rotates. The pendulum appears to continuously change the plane in which it swings as time elapses. Determine the swinging frequency of this pendulum. Explain why the behavior of Foucault's pendulum provides supporting evidence for the hypothesis that Earth is a noninertial reference frame and helps reject the model of a geocentric universe (Earth at the center).

71. ** A low-friction cart of mass m rests on a horizontal table. The cart is attached to a relaxed light spring with spring constant k. At distance d from the first cart rests a second identical cart (see **Figure P10.71**). Both carts are covered with Velcro so they stick together if they collide or touch. The first cart is pushed to the left with initial speed v_0. Determine the final frequency and amplitude of a vibrating system. (Note: This is a multiple possibility problem. Analyze and present all possible solutions.)

FIGURE P10.71

Reading Passage Problems

Transportation to the other side of Earth It has been suggested that an efficient way to travel from the United States to Australia would be to dig one or more holes through Earth and fall through one of the holes in a capsule. The gravitational force that Earth exerts on you and on the capsule when the capsule is at a distance r ($r < R$ of Earth) equals that of the mass M inside a sphere of radius r with its center at Earth's center. At Earth's center, the force is zero. At some other distance r from the center of Earth, the force would be

$$F_{\text{E on Y}} = \frac{GM_{\text{inside } r}m_{\text{Y}}}{r^2}$$

where $M_{\text{inside } r}$ equals the volume of a sphere of radius r times the density of Earth, assumed to be constant. Substituting for $M_{\text{inside } r}$, you find that the force that Earth exerts on you and on the capsule is proportional to r and provides a restoring force. You start at rest at Earth's surface and accelerate as you get closer to the center. At the center you are moving at maximum speed and Earth exerts no force on you. After you pass the center, Earth starts pulling back on you, causing your speed to decrease. With no friction or air resistance, you should stop on the other side of Earth. If you can find an expression for the restoring force that Earth exerts on you, you can substitute that in place of k in the expression for period $T = 2\pi(m/k)^{1/2}$. The travel time will be half the period.

72. Which expression below represents the mass m of Earth inside a sphere of radius r smaller than the radius R of Earth? Note that ρ is the density of Earth, assumed uniform.
    (a) $(4/3)\pi r^3\rho$ (b) $G(4/3)\pi r^3\rho$
    (c) GM_{Earth}/r^2 (d) $G(4/3)\pi r\rho$
    (e) mg

73. Which expression below represents the magnitude of the restoring force that Earth exerts on an object of mass m when a distance r from the center of Earth? Note that ρ is the density of Earth, assumed uniform.
    (a) $m(4/3)\pi r^3\rho$ (b) $mG(4/3)\pi r^3\rho$
    (c) $GM_{\text{E}}m/r^2$ (d) $mG(4/3)\pi r\rho$
    (e) mgr

74. Which expression below represents the period T for one oscillation from Earth's surface through the center, to the other side, and then back again?
    (a) $2\pi(3/G4\pi\rho)^{1/2}$ (b) $2\pi(G4\pi\rho/3)^{1/2}$
    (c) $2\pi(3m/G4\pi\rho)^{1/2}$ (d) $2\pi(mG4\pi\rho/3)^{1/2}$
    (e) $2\pi(m/k)^{1/2}$

75. During the trip, when will your acceleration be greatest?
    (a) At the beginning and end of the trip
    (b) At the end of the trip
    (c) When passing through the center of Earth
    (d) The same for entire the trip

76. During the trip, when will your speed be greatest?
    (a) At the beginning and end of the trip
    (b) At the end of the trip
    (c) When passing through the center of Earth
    (d) The same for entire the trip

77. If the radius of Earth is 6.4×10^6 m, its mass is 6.0×10^{24} kg, and $G = 6.67 \times 10^{-11}$ N·m^2/kg^2, what is the time interval for one trip through Earth?
    (a) 0.7 h   (b) 1.4 h
    (c) 3.2 h   (d) 4.8 h
    (e) 6.3 h

BIO **Resonance vibration transfer and the ear** When you push a person on a swing, a series of small pushes timed to match the swinger's swinging frequency makes the person swing with larger amplitude. If timed differently, the pushing is ineffective. The board shown in **Figure 10.17** (from the Exploratorium in San Francisco) is made of rods of different length with identical balls on the ends of each rod. Each rod vibrates at a different natural frequency, the long rod on the left at lower frequency and the short rod on the right at higher frequency. If you shake the board at the high frequency at which the short rod vibrates, the short rod swings with large amplitude while the others swing a little. If you shake the board at the middle frequency at which the two center rods vibrate, the center rods undergo large-amplitude vibrations and the rods on each end do not vibrate. Imagine now that you have a fancy board with 15,000 rods, each of slightly different length, the shortest on the left and the longest on the right. Shaking the board at a particular frequency causes the rods in one small region of the board to vibrate at this frequency and has little effect on the others.

FIGURE 10.17 Coupled oscillations of the shaking board and rods.

The short rod with a ball on the end responds to higher frequency shaking of the board, whereas the middle length rods respond to intermediate frequency shaking.

The inner ear (the cochlea) is a little like this fancy board. Sound reaching the tympanic membrane, or eardrum, is greatly amplified by three tiny bones in the middle ear—the hammer, anvil, and stirrup (Figure 10.18). These bones vibrate, pushing on the fluid in the inner ear and causing vibrations along its entire length. A basilar membrane with about 15,000 hair cells passes along the center of the inner ear. The basilar membrane is narrow and stiff near the entrance to the inner ear and wide and more flexible near the end. When a single-frequency vibration causes the fluid to vibrate, the membrane and the hair cells respond best at a single place—high frequencies near the oval window and low frequencies near the end of the basilar membrane. The bending of these hairs causes those nerve cells to fire. Thus, we detect the frequency of the sound by the part of the membrane from which the nerve signal comes.

FIGURE 10.18 Anatomy of the human ear.

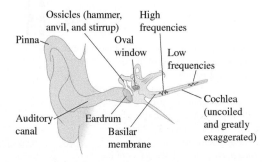

78. What causes vibrations in the fluid in the cochlea?
    (a) The eardrum pushes against the cochlea.
    (b) Sound waves push against the cochlea.
    (c) A bone in the middle ear pushes against the cochlea.
    (d) The fluid vibrates when something outside the ear vibrates.
    (e) The fluid does not vibrate—the hair cells vibrate.

79. How does the basilar membrane distinguish different frequency vibrations?
    (a) The hair cells have different lengths.
    (b) The dimensions and stiffness of the cochlea vary along its length.
    (c) The basilar membrane vibrates where the cochlea dimension and stiffness match the vibration frequency.
    (d) The cochlea fluid resonates in only one part of the cochlea.
    (e) All the hair cells bend back and forth at only the frequency of the vibration.

80. If you shake the board shown in Figure 10.17 at a frequency higher than the natural frequency of the rod on the right, then what happens?
    (a) None of the rods vibrate.
    (b) All of the rods vibrate.
    (c) The shortest rod vibrates.
    (d) The longest rod vibrates.
    (e) The middle rods vibrate.

81. If you were to shake the special board (the one that has 15,000 rods of varying length) at one particular frequency, then what would happen?
    (a) None of the rods would vibrate.
    (b) All of the rods would vibrate.
    (c) A small number of rods at one location would vibrate.
    (d) A disturbance would travel back and forth along the board.

82. You hang four pendulum bobs from strings connected to a wooden dowel. The strings are different lengths. How can you get the second longest pendulum bob to vibrate while the other three do not—without touching the pendulums?
    (a) Shake the dowel back and forth.
    (b) Shake the dowel back and forth at the resonant frequency of that pendulum.
    (c) Move the dowel sideways at any frequency.
    (d) Blow air on that bob.

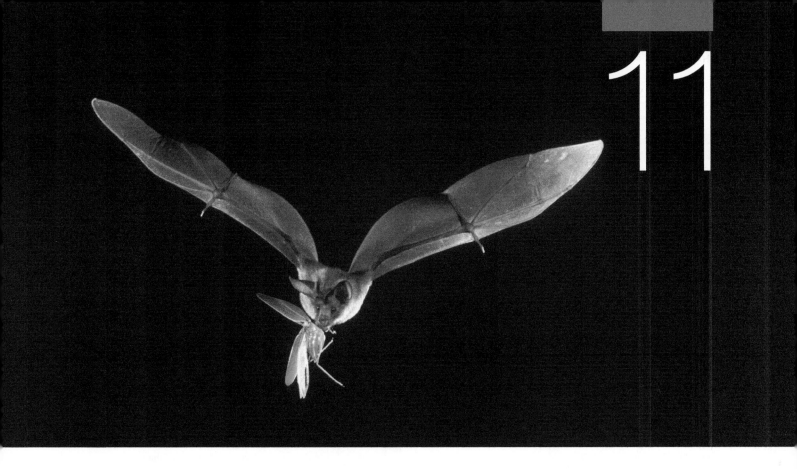

Mechanical Waves

You have probably heard the expression "blind as a bat." It turns out that bats are not blind; many species have very sensitive eyes that can see even when there is very little light. However, the organs that bats use to avoid obstacles at night and find their prey are their ears. While flying, they emit sound that travels to the prey, reflects, and travels back to the bat. The bat's brain processes the delay time and automatically calculates the distance to the prey. Why don't we ever hear the sounds produced by bats?

- How do bats "see" in the dark?
- How can you transmit energy without transmitting matter?
- Why does the pitch of a train whistle change as the train passes you?

PREVIOUSLY (IN CHAPTER 10), we studied the vibrational motion of one object such as a pendulum or a cart attached to a spring. We have not considered what happens to the environment that is in contact with the vibrating object. In this chapter we will focus on the effect that the vibrating object has on the medium that surrounds it. An example of such an effect is the propagation of sound.

BE SURE YOU KNOW HOW TO:
- Describe simple harmonic motion verbally, graphically, and algebraically (Section 10.3).
- Determine the period and amplitude of vibrations using a graph (Section 10.2).
- Determine the power of a process (Section 7.8).

FIGURE 11.1 A vibration disturbs the surrounding medium.

FIGURE 11.1 A vibration disturbs the surrounding medium.

Cylinder vibrates ...

... causing waves that travel across the water.

11.1 Observations: pulses and wave motion

Our everyday experience indicates that when an object vibrates, it also disturbs the medium surrounding it. We call the object that causes the vibration the **source**. For example, place a heavy metal cylinder attached to a spring so that the cylinder is partially submerged in a tub with water (**Figure 11.1**). Push down on the cylinder a little and release it. The vibrating cylinder (the source) sends ripples (waves) across the tub. In Observational Experiment Table 11.1 we observe the simplest case of wave motion—when the source of waves vibrates only once, sending a short disturbance, or a **pulse**, into the surrounding medium.

OBSERVATIONAL EXPERIMENT TABLE 11.1 Disturbances in different media

VIDEO OET 11.1

Observational experiment	Analysis
Experiment 1. Tie a rope to a doorknob and then shake the end that you are holding one time, producing a quick disturbance—a pulse. The pulse travels along the rope horizontally. The part of the rope marked with a ribbon moves up and down when the pulse reaches it.	The source of the pulse is your hand pulling the rope; the medium in which the pulse propagates is the rope itself. The pulse moves along the rope, but the rope fibers (see the little ribbon on the rope) return to their original position after the disturbance passes—they do not travel along the rope. It takes time for the pulse to travel to the other end of the rope.
Experiment 2. Place a beach ball and polystyrene foam pieces on the surface of a swimming pool. Push the ball down once. A pulse in the shape of a circular disturbance spreads outward. The foam pieces bob up and down but do not travel across the pool.	The source of the pulse is the ball; the medium is the surface layer of water in the pool. The circular disturbance produced by the ball's downward motion spreads; the water itself does not move across the pool, as the foam pieces on the water surface only move up and down. It takes time for the pulse to reach the far sides of the pool.

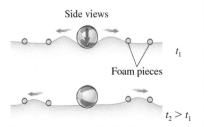

Side views

Foam pieces

t_1

$t_2 > t_1$

Patterns

In the two experiments, we saw that
- a moving disturbance was created by a moving source—your hand or the ball.
- the disturbance moved through a material at an observable speed.
- the particles of the medium vibrate close to their original position, but do not travel far.

Waves and wave fronts

In the beach ball experiment, when the vibrating ball (the source) moves down, the water (the medium) under the ball's bottom surface is pushed out to the side of the ball, forming a hump in the water at the sides of the ball. This hump falls back toward its equilibrium level and overshoots, pushing water slightly farther away upward into a hump. This disturbance propagates outward, but the water at a particular location mostly vibrates up and down, and not outward in the direction the wave travels. The vibrating source and the interactions of the neighboring sections of the medium result in a coordinated vibration—a wave. The interaction between the parts of the medium is crucial—without it, the wave would not propagate. If you cut the rope just after shaking its end, the disturbance will not propagate. We can now define wave motion:

> **Wave motion** involves a disturbance produced by a vibrating object (a source). The disturbance moves, or propagates, through a medium and causes points in the medium to vibrate. When the disturbed medium is physical material (solid, liquid, or gas), the wave is called a **mechanical wave**.

The moving circular disturbance itself created by the beach ball on the water is called a **wave front** (see the top view in Figure 11.2a). Every point on the wave front has the same displacement from equilibrium at the same clock reading. If instead of pushing the ball one time, you push it down and up in a regular pattern, you will see a continuous group of humps and dips (waves consisting of many wave fronts) of circular shape moving outward on the surface of the water (Figure 11.2b).

Two kinds of waves

The stretchy toy called a Slinky provides another good example of wave motion. If we (the source) push or pull the end coil of a stretched Slinky (the medium) several times, it will pull the coil attached to it. That coil pulls the next coil, which in turn pulls the next coil, and so on. We see a compressing and stretching pattern moving along the Slinky toy in the direction of wave propagation. The pattern is illustrated in Figure 11.3 at a single moment in time. Such a wave is called a **longitudinal wave**.

We can disturb a Slinky in a different way. If we vibrate the coil up and down perpendicular to the length of the Slinky, the coils move up and down perpendicular to the direction that the disturbance travels (Figure 11.4). This type of wave is called a **transverse wave**.

> **Longitudinal and transverse waves** In a longitudinal wave the vibrational motion of the particles or layers of the medium is parallel to the direction of propagation of the disturbance. In a transverse wave the vibrational motion of the particles or layers of the medium is perpendicular to the direction of propagation of the disturbance.

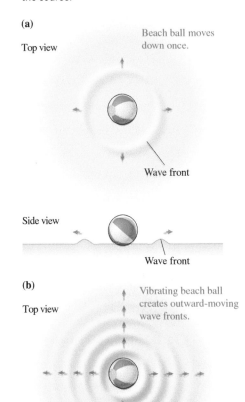

FIGURE 11.2 Wave fronts move outward from the source.

(a)

Top view   Beach ball moves down once.

Wave front

Side view

Wave front

(b)

Top view   Vibrating beach ball creates outward-moving wave fronts.

Wave fronts

FIGURE 11.3 A longitudinal wave on a Slinky.

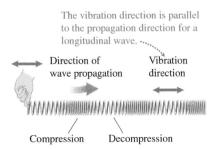

The vibration direction is parallel to the propagation direction for a longitudinal wave.

Direction of wave propagation   Vibration direction

Compression   Decompression

FIGURE 11.4 A transverse wave on a Slinky.

y

Vibration direction

x

Direction of wave propagation

The vibration direction and propagation direction are perpendicular for a transverse wave.

Reflection of waves

We observe another important property of waves when watching waves on the surface of the water, on a Slinky, or on a rope. When a wave reaches the wall of the container or the end of the Slinky or rope, it reflects off the end and moves in the opposite direction. In fact, when a wave encounters any boundary between different media, some of the wave is reflected back. Figure 11.5a, on the next page, shows this reflection for one pulse on a rope when the rope is attached to a wall. However, even if there is no wall, the pulse reflects off the end of the rope. Think of a rope that ends with a ring that can slide without friction along a vertical pole (Figure 11.5b). The reflected pulse is now upright and not inverted as it was in Figure 11.5a. If you continue shaking the rope, continuously producing pulses, you will create a wave. The wave traveling to the right and the reflected wave traveling to the left exist in the rope simultaneously, complicating the analysis. (Watch the video of a similar experiment using a Slinky). Thus, if we want to study simple wave motion and not worry about reflected waves, we need to have a very long medium.

VIDEO
11.1

FIGURE 11.5 A pulse on a rope at different times.

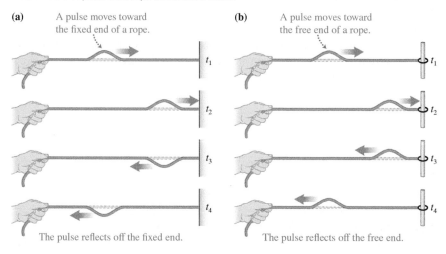

(a) A pulse moves toward the fixed end of a rope.

(b) A pulse moves toward the free end of a rope.

The pulse reflects off the fixed end.

The pulse reflects off the free end.

Sound waves

You have probably seen a tuning fork, a device used to tune musical instruments. If you strike a tuning fork, you hear sound. You can feel the vibrations of the prongs if you touch them. In fact, the moment you touch them, the sound stops, since you've stopped their vibrations. If you place the vibrating prongs in water, you see ripples going out (**Figure 11.6**). The tuning fork is a source of waves. What happens to cause our ears to hear a sound? The vibration of the tuning fork compresses and decompresses air around it. We will learn in the next chapter that in compressions, air molecules come closer together, and in decompressions, air molecules spread farther apart. These longitudinal compressions and decompressions (similar to what we observed on the Slinky) move through the air and reach our ears, and we perceive them as sound. Sound propagates in water (it's how dolphins communicate) and even in solid objects (seismic waves in Earth). If a tuning fork vibrates in a vacuum, there will be no medium for the tuning fork prongs to vibrate against and consequently no way for air compressions and decompressions to occur. Thus sound cannot propagate in a vacuum. We will study sound waves in more detail later in the chapter.

REVIEW QUESTION 11.1 How do you produce a longitudinal wave on a Slinky? How do you produce a transverse wave?

FIGURE 11.6 A vibrating tuning fork produces sound along with waves on the surface of the water.

The vibrating prongs of a tuning fork disturb the water and cause ripples.

11.2 Mathematical descriptions of a wave

To understand waves and predict their behavior, we need to describe them using physical quantities and find relationships between those physical quantities.

We will use the waves traveling along a very long rope (so long that we can ignore reflected waves) and a wave source that vibrates with simple harmonic motion (SHM; discussed in Chapter 10). This motion can be created by a motor that vibrates the end of a rope up and down perpendicular to the rope. The motor (the source) produces a transverse wave in the rope (the medium). At time zero, the source starts at $y = +A$. The source vibrates sinusoidally; thus the source displacement is described by a sinusoidal function of time (**Figure 11.7a**):

$$y = A \cos\left(\frac{2\pi}{T}t\right)$$ (11.1)

where y is the displacement of that end of the rope from its equilibrium position, A is the amplitude of the vibration, and T is the period of vibration. The left end of the

FIGURE 11.7 A wave produced on a rope by a vibrating source.

(a) The displacement-versus-time of one position on the rope (the source position)

(b) A snapshot of a wave at one instant in time

rope moves up and down, pulling the next part of the rope, which pulls its neighboring part, which pulls its neighboring part, and so on. The wave propagates along the rope at a particular speed, so that when the disturbance reaches a point at a distance x from the source, that point starts vibrating at the same period T as the source. Figure 11.7b shows a snapshot of the resulting wave at some arbitrary time t. We now have four quantities that describe the wave started by this vibrating source. The first three are familiar to us from Chapter 10: they describe the motion of a vibrating object. The fourth quantity (speed), as it applies to waves, is new.

Period T in seconds is the time interval for one complete vibration of a point in the medium anywhere along the wave's path.

Frequency f in Hz (s^{-1}) is the number of vibrations per second of a point in the medium as the wave passes.

Amplitude A is the maximum distance of a point of the medium from its equilibrium position as the wave passes.

Speed v in m/s is the distance a disturbance travels during a time interval, divided by that time interval.

TIP Note that the definition of the amplitude given here only works for waves on rope or a Slinky. For sound waves, we will learn a new definition.

Mathematical description of a one-dimensional traveling sinusoidal wave

Our goal is to construct a description of a traveling sinusoidal wave at different places along its path and at different times (clock readings). For simplicity, we only consider one-dimensional waves, such as waves on ropes or Slinkys, and we will assume that the mechanical energy of the source is not converted to the internal energy of the medium. In other words, no damping occurs—the amplitude of the vibration is the same at all points. We also assume that the speed of the wave propagation is constant throughout the medium. We choose the positive x-direction as the direction in which the disturbance travels at speed v. The left end of the rope at $x = 0$ is attached to the vibrating source. According to Eq. (11.1), the source oscillates up and down with a vertical displacement

$$y(0, t) = A \cos\left(\frac{2\pi}{T}t\right)$$

The zero in the parentheses indicates that the function $y(0, t)$ describes the motion of the medium at the position $x = 0$ and clock reading t. The shape of the rope at five different times is shown in **Figure 11.8**. If you look at any other position along the x-axis, the rope vibrates in the same way as it does at $x = 0$ but at a different instant in the vibration cycle. For example, at the first tick mark to the right of $x = 0$, you see that at time zero, the rope displacement is $-A$; at time $t = T/4$, its y-displacement is zero, then $+A$, then 0, and finally back to $-A$.

We can mathematically describe the disturbance $y(x, t)$ of a point of the rope at some positive position x to the right of $x = 0$ at some arbitrary time t. Every point to the right of $x = 0$ vibrates with the same frequency, period, and amplitude as the rope at $x = 0$. But there is a time delay $\Delta t = x/v$ for the disturbance at $x = 0$ to reach the location x. The vibrational disturbance at position x at time t is the same as the disturbance at $x = 0$ was at the earlier time $t - x/v$. Thus, the vibration of the rope at position x can be described as

$$y(x, t) = A \cos\left[\frac{2\pi}{T}\left(t - \frac{x}{v}\right)\right] \tag{11.2}$$

FIGURE 11.8 The shape of a transverse wave at five consecutive times.

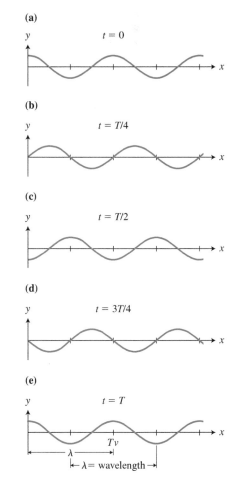

The snapshot graph of the disturbance (Figure 11.8e) shows that the disturbance pattern repeats at a distance Tv (period multiplied by the wave speed). This distance separates neighboring locations on the rope that have the same displacement y and the same slope. This distance is called the **wavelength** of the wave and is represented by the Greek letter λ (lambda):

$$\lambda = Tv = \frac{v}{f}$$

where v is the speed of travel of the wave, T is the period of vibration of each part of the rope, and f is the frequency of vibration $(1/T)$.

> TIP You can think of a wavelength as the distance that the wave propagates during one period. It is helpful to remember the equation $\Delta x = v \Delta t$.

> **Wavelength** λ equals the distance between two nearest points on a wave that at any clock reading have exactly the same displacement and shape (slope). It is also the distance between two consecutive wave fronts:
>
> $$\lambda = Tv = \frac{v}{f} \tag{11.3}$$

We can substitute this expression for wavelength into Eq. (11.2) and rearrange a little to get a function of the two variables t and x that allows us to predict the displacement y of any point x at any time t in a medium through which the wave propagates:

$$y(x, t) = A \cos\left[\frac{2\pi}{T}\left(t - \frac{x}{v}\right)\right] = A \cos\left[2\pi\left(\frac{t}{T} - \frac{x}{Tv}\right)\right] = A \cos\left[2\pi\left(\frac{t}{T} - \frac{x}{\lambda}\right)\right]$$

> **Mathematical description of a traveling sinusoidal wave** The displacement from equilibrium y of a point at location x at time t when a wave of period T travels at speed v in the positive x-direction through a medium is described by the function
>
> $$y = A \cos\left[2\pi\left(\frac{t}{T} - \frac{x}{\lambda}\right)\right] \tag{11.4}$$
>
> The wavelength λ of this wave equals $\lambda = T \cdot v$.

> TIP Notice that Eq. (11.4) is mathematically symmetric with respect to the period of the wave and its wavelength because there are two repetitive processes in a wave. If the location x is fixed, that point in the medium vibrates in time with period T. If the time t is fixed (as in a photograph), the space has a wavelike appearance with wavelength λ (see Figure 11.8). We replaced the notation for the function $y(x, t)$ with just y for simplicity.

In Eq. (11.4), at any fixed time t, the term $2\pi x/\lambda$ describes the difference, in units of radians, between the displacement of a vibrating point at position x and the displacement at the origin, and is called a **phase**. This term has the effect of shifting the cosine function by some number of radians. When x is equal to an integer multiple n of the wavelength, the term equates to an integer multiple of 2π. Shifting the cosine function by a multiple of 2π has no effect, so a point vibrating at position $x = n\lambda$ is said to be *in phase* with the origin. Therefore, two points separated by a distance equal to an integer multiple of wavelengths (one wavelength, two wavelengths, etc.) are always in phase (**Figure 11.9**).

FIGURE 11.9 Two snapshots of the same wave. Points a and b have the same phase at any clock reading.

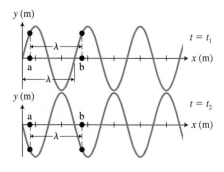

CONCEPTUAL EXERCISE 11.1   **Making sense of a mathematical description of a wave**

A transverse wave on an infinitely long Slinky is described mathematically as

$$y = (0.10 \text{ m}) \cos\left[2\pi\left(\frac{t}{0.70 \text{ s}} - \frac{x}{1.8 \text{ m}}\right)\right]$$

Say everything you can about this wave and represent it graphically.

Sketch and translate Compare the given equation with the wave equation [Eq. (11.4)]: $y = A \cos\left[2\pi\left(\frac{t}{T} - \frac{x}{\lambda}\right)\right]$. Therefore, $A = 0.10$ m, $T = 0.70$ s, and $\lambda = 1.8$ m. We can also say that at $t = 0$, the starting point with the coordinate $x = 0$ had a displacement of

$$y(0, 0) = (0.10 \text{ m}) \cos\left[2\pi\left(\frac{0}{0.70 \text{ s}} - \frac{0}{1.8 \text{ m}}\right)\right] = 0.10 \text{ m. We can}$$

write a mathematical equation for the motion of the wave source at location $x = 0$: $y_{\text{source}} = (0.10 \text{ m}) \cos\left[2\pi\left(\frac{t}{0.70 \text{ s}}\right)\right]$. Notice that at $x = 0$, y is a function of time only, as we are looking at the disturbance at only one location. Rearranging the equation $\lambda = Tv$, we get $v = \lambda/T = (1.8 \text{ m})/(0.70 \text{ s}) = 2.6$ m/s, a reasonable number.

Simplify and diagram Knowing the amplitude and the period, we can draw a displacement-versus-time (clock reading) graph for one particular location in the medium (we do it for $x = 0$ in the first figure at right). Knowing the amplitude and the wavelength, we can draw a

graph that represents an instantaneous picture of the whole Slinky at a particular time (clock reading) (shown for $t = 0$ in the second figure).

(a) y (m)   The displacement at different times at one location

(b) y (m)   The displacement at different locations at one time

Try it yourself A Slinky vibrates with amplitude 15 cm and period 0.50 s. The speed of the wave on the Slinky is 4.0 m/s. Write the wave equation for the Slinky.

Answer

$$y = (0.15 \text{ m}) \cos\left[2\pi\left(\frac{t}{0.50 \text{ s}} - \frac{x}{2.0 \text{ m}}\right)\right].$$

TIP   Because Eq. (11.4) is a function of two variables, when we represent waves graphically we need two graphs, such as those in the figures in Exercise 11.1—one showing how one point of the medium changes its position with time (y versus t; $x = $ constant) and the other one showing the displacements of multiple points of the medium at the same clock reading—a snapshot of the wave (y versus x; $t = $ constant).

REVIEW QUESTION 11.2 Compare and contrast the speed of a vibrating object and the speed of a wave.

11.3 Dynamics of wave motion: speed and the medium

We have defined the speed of a wave as the distance that a disturbance travels in the medium during a time interval divided by that time interval. This operational definition of speed tells us how to determine the speed but does not explain what determines the wave speed. To answer this question, we will investigate transverse waves on a rope or a Slinky via the following experiment.

Figure 11.10 shows two people holding the ends of a Slinky pulling on both ends so that the Slinky is taut. The person on the left shakes the end of the Slinky abruptly and sends several repeated pulses along the Slinky; the person on the right measures the force pulling on the Slinky and the time interval Δt needed for a pulse to travel the distance L. They vary the amplitude, the frequency, and the magnitude of the pulling force

FIGURE 11.10 Investigating the speed of pulse propagation on a Slinky.

one at a time. The data that they collect are presented below in Table 11.2. Notice that when they increase the force, the distance that the pulse travels also increases, because they have to stretch the Slinky more.

TABLE 11.2 How do different physical quantities affect the speed of pulse propagation?

Experiment	A (m)	f (Hz)	F (N)	L (m)	Δt (s)	v (m/s)
1: Changing A	0.3	4.0	5.0	1.3	0.36	3.6
2: Changing A	0.2	4.0	5.0	1.3	0.36	3.6
3: Changing A	0.1	4.0	5.0	1.3	0.36	3.6
4: Changing f	0.1	3.0	5.0	1.3	0.36	3.6
5: Changing f	0.1	2.0	5.0	1.3	0.36	3.6
6: Changing F	0.1	2.0	6.0	1.6	0.36	4.3
7: Changing F	0.1	2.0	7.0	1.9	0.36	5.0

There are three important patterns that we can infer from the table:

● Experiments 1–3: the speed of the wave does not depend on the amplitude of the pulses.
● Experiments 3–5: the speed of the wave does not depend on the frequency of the pulses.
● Experiments 5–7: the speed of the wave changes when both the length and the force exerted on the Slinky change.

These patterns are surprising—the speed of the pulse does not depend on the amplitude of the pulse or on its frequency. The amplitude and the frequency only depend on the source of the waves. Therefore, we can conclude that the source of the waves does not affect the speed of the wave on the Slinky!

However, pulling harder on the end of the Slinky does increase the wave speed. In addition, as we stretch the Slinky, the coils spread farther apart, reducing the Slinky's *linear density*—its mass per unit length. The force pulling on the Slinky and its linear density seem to affect the speed of a pulse or wave on the Slinky. Knowing that both of these quantities affect the speed, we should be careful not to conclude from Table 11.2 that the speed is linearly proportional to the force. As you pull harder on the spring, the length increases proportionally (with no change in mass), so the linear density decreases proportionally to the force.

If we do similar experiments with stiff springs that stretch less than a Slinky, we find that the wave speed is proportional to the square root of the force pulling on the end of the spring. This applies to other objects as well, such as strings:

$$v \propto \sqrt{F_{\text{M on M}}}$$

The subscripts M and M indicate the pulling force that one part of the medium (M) exerts on a neighboring part of the medium (M) when two people stretch the spring. This result makes sense—the more you pull on the spring, the more strongly adjacent parts interact with each other and therefore the faster they transfer the disturbance.

Other experiments indicate that speed is also affected by the linear density of the medium. The speed of a wave is inversely proportional to the square root of the linear density of the medium:

$$v \propto \frac{1}{\sqrt{\dfrac{m}{L}}} = \frac{1}{\sqrt{\mu}}$$

Here the Greek letter μ is used for the linear density (m/L) (note that in other contexts this symbol is used to denote the coefficient of friction). The dependence on mass

TIP Do not confuse the force you exert to stretch the Slinky to create some tension with the force you exert on its end when you are creating a pulse. The speed of the wave is affected by the former, not the latter.

makes sense in terms of Newton's second law. The greater the mass of a part of the medium, the smaller the acceleration of that part and the longer the time needed to move the next part. We can combine these two factors into one equation:

$$v = \sqrt{\frac{F_{M \text{ on } M}}{\mu}}$$

Let us check whether the units are correct:

$$v = \sqrt{\frac{N}{kg/m}} = \sqrt{\left(\frac{kg \cdot m}{s^2}\right)\left(\frac{m}{kg}\right)} = \frac{m}{s}$$

The units are correct. To make sure that the speed does in fact depend on the above properties, we perform the following experiment. We take a rope of known mass and length, attach one end to a wall, and pull the other end with a spring scale, exerting a known force. Then we pluck the rope once, sending a pulse. We determine the speed of the pulse using the time and distance data and compare to the speed that we calculate using the expression above. The match of the results shows that the coefficient of proportionality in the equation is 1.

Wave speed The speed v of a wave on a string or in some other one-dimensional medium depends on how hard the string or medium is pulled $F_{M \text{ on } M}$ (one part of the medium pulling on an adjacent part) and on the mass per unit length $\mu = m/L$ of the medium:

$$v = \sqrt{\frac{F_{M \text{ on } M}}{\mu}} \tag{11.5}$$

TIP Note two mathematical expressions for the speed of the pulse:

$$v = d/\Delta t \quad \text{and} \quad v = \sqrt{\frac{F_{M \text{ on } M}}{\mu}}$$

The first expression is an operational definition for speed, while the second is a cause-effect relationship.

We have just learned that the speed of waves in a medium in which no mechanical energy is converted into internal energy depends only on the properties of the medium. The period and frequency of a wave depend on the vibration source. Since the medium vibrates at the same frequency as the source, the wavelength of the wave depends on both the frequency of the source and the speed of the wave through the medium. If the vibration frequency is low and the wave speed high, the crest of the wave will travel a greater distance before the next crest leaves the source. Thus the wavelength is large:

$$\lambda = \frac{v}{f}$$

However, if the wave speed is low and/or the frequency high, the wavelength is small:

$$\lambda = \frac{v}{f}$$

The wave speed also depends on the type of wave (or pulse). In the same medium, transverse waves (and pulses) travel more slowly than longitudinal waves (and pulses). The reason for the difference is in the types of deformation of the medium that occur when the waves propagate. The transverse waves that can only propagate in solids require shear deformation—the layers of the medium slide across each other—while the longitudinal waves require compressions and decompressions. The respective forces that parts of the medium exert on each other for the shear deformations are weaker than for the compressions/decompressions.

TIP The equation $\lambda = Tv = \dfrac{v}{f}$ is a cause-effect relationship for the wavelength.

QUANTITATIVE EXERCISE 11.2   **Effects of tension force exerted on a violin G string**

A violin G string is 0.33 m long and has a mass of 2.2 g. Determine the tension force exerted on the ends of the string in order for a wave to travel at the speed of 128 m/s.

Represent mathematically We know that

$$v = \sqrt{\frac{F_{\text{S on S}}}{\mu}} = \sqrt{\frac{F_{\text{S on S}}}{m/L}}$$

Rearrange this to determine the magnitude of the tension force $F_{\text{S on S}}$ that adjacent parts of the string exert on each other and that the peg at the end of the violin exerts on the string.

Solve and evaluate Rearranging the above equation, we get

$$F_{\text{S on S}} = \frac{m}{L}v^2 = \frac{(2.2 \times 10^{-3}\,\text{kg})}{(0.33\,\text{m})}(128\,\text{m/s})^2$$

$$= 110\,\text{kg}\cdot\text{m/s}^2 = 110\,\text{N}$$

This is a reasonable tension (about 20 lb), and the unit is correct.

Try it yourself Would the force exerted on the end of a 1.4-g violin A string be higher or lower if the waves traveled at a speed of 288 m/s?

Answer   Higher—about 350 N.

REVIEW QUESTION 11.3 How do we define wavelength operationally? What affects the value of the wavelength of a wave? How do you write a cause-effect relationship for the wavelength?

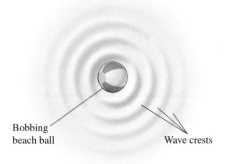

FIGURE 11.11 Wave crests are shown at one instant in time on the surface of water.

Top view

Bobbing beach ball

Wave crests

Side view

The farther the wave from the source, the smaller the crest amplitude.

FIGURE 11.12 The wave crest amplitude decreases on a two-dimensional surface with distance from the source.

Snapshot of wave crests at one instant in time

Bobbing beach ball

R

$2R$

The same energy/time passes two rings that have different circumferences.

11.4 Energy, power, and intensity of waves

So far we have investigated quantitatively waves that propagate in a one-dimensional medium such as a rope. If there is no damping of the vibrations as the wave propagates, then the amplitude of the vibration at any point in the one-dimensional medium is the same as that of the source. What happens if the wave propagates without damping through a two-dimensional medium (like the surface of a small lake) or a three-dimensional medium (like the air surrounding a balloon that pops and produces sound)? Do the amplitudes of these waves remain constant as they propagate outward from the source?

Wave amplitude and energy in a two-dimensional medium

Consider **Figure 11.11**. A beach ball bobs up and down in water in simple harmonic motion, producing circular waves that travel outward across the water surface in all directions. The shaded circles represent the peaks of the waves. We observe that the amplitudes of the crests decrease as the waves move farther from the source. Why?

Suppose that the wave source uses 10 J of energy E every second to produce a pulse, and that all of that energy is transferred to the pulse (one pulse per second). As the pulse propagates outward in all directions, more particles vibrate, since the circumference of the circular pulse increases as it gets farther from the source. However, the total energy of all vibrating particles in the pulse cannot exceed 10 J. Since there are more vibrating particles as the pulse moves outward, the vibrational energy per particle has to decrease.

To do a quantitative analysis, we surround the source with two imaginary rings (the dashed lines in **Figure 11.12**), one at a distance R from the source and the other at distance $2R$. If the energy output of the source per unit time is 10 J/s, then this energy first moves through the circumference $C_1 = 2\pi R$ and then through the second larger circumference $C_2 = 2 \cdot 2\pi R = 2C_1$. The circumference of the second ring is two times greater than the first, but the same energy per unit time moves through it. Consequently, the energy per unit circumference length (energy/C = energy/$2\pi r$)

passing the second ring is one-half that passing the first ring. We see that the energy per unit time and per unit circumference length is inversely proportional to the distance from the source (assuming that there is no damping).

> **Two-dimensional waves produced by a point source** The energy E per unit circumference length and per unit time crossing a line perpendicular to the direction that the wave travels decreases as $1/r$, where r is the distance from the point source of the wave.

In Chapter 10 we learned that the amplitude of vibrations and the energy of the vibrating system are related to each other. Thus, if the energy per circumference length and per unit time of a two-dimensional wave decreases as the wave propagates, so should the amplitude. Remember that another reason for amplitude decrease is damping.

Wave amplitude and energy in a three-dimensional medium

Consider a three-dimensional medium. You hear a cicada (an insect that makes a loud sound in the summer) sitting on top of a bush that is about the same height as you. It is a source of sound that is propagating in all directions. Consider two imaginary spheres that surround the source (**Figure 11.13**), one at distance R from the source and the other at distance $2R$. Assume that the energy output of the source per unit time is 10 J/s. This energy first moves through the surface area $A_1 = 4\pi R^2$ and then through the second larger area $A_2 = 4\pi(2R)^2 = 4A_1$. The area of the second sphere is four times the first, but the same energy per unit time moves through it. Consequently, the energy per unit area through the second sphere is one-fourth that through the first sphere. The energy per unit time per unit area is inversely proportional to the distance squared from the source. This conclusion is consistent (qualitatively) with our experience: we know that the farther we are from the cicada, the quieter the sound becomes.

FIGURE 11.13 The sound decreases in amplitude and loudness at a greater distance from the source.

The sound travels outward, crossing two imaginary spheres.

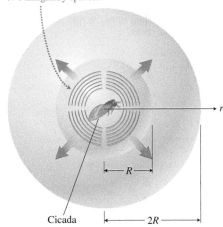

> **Three-dimensional waves produced by a point source** The energy E per unit area per unit time passing across a surface perpendicular to the direction that the wave travels decreases as $1/r^2$, where r is the distance from the point source of the wave.

TIP The symbol for the area and the amplitude of vibrations is the same—A. Make sure you interpret the meaning of A based on the context of the situation.

QUANTITATIVE EXERCISE 11.3 **Spreading waves from a cicada**

A cicada sits on a bush that is about your height. How far away from the bush should you stand so that the energy of the cicada's sound that you hear is 10 times less than when you are standing 5.0 m away from the bush?

Represent mathematically The energy of the sound waves emitted by the cicada moves through larger and larger surface areas. We know that the energy per area per unit time passing across a surface perpendicular to the direction that the wave travels decreases as $1/r^2$, where r is the distance from the point source of the wave. Thus we can represent this decrease of energy passing through a unit area mathematically as

$$\frac{E_{r2}}{E_{r1}} = \frac{(1/r_2^2)}{(1/r_1^2)} = \frac{r_1^2}{r_2^2}$$

Taking the area of the opening of your ear as a unit area in this case, and the ratio of the energies to be $\dfrac{E_{r2}}{E_{r1}} = \dfrac{1}{10}$, we can use this equation to find the distance.

Solve and evaluate Inserting the known r_1:

$$\frac{E_{r2}}{E_{r1}} = \frac{(r_1^2)}{(r_2^2)} = \frac{(5.0 \text{ m})^2}{r_2^2} = \frac{1}{10} \quad \Rightarrow \quad r_2 = \sqrt{25.0 \text{ m} \times 10} = 15.8 \text{ m}$$

We find that you only need to be about 16 m away for the energy reaching your ear to be 10 times less. We roughly triple the distance from the source to get a reduction in energy of 10 times. Given $3^2 = 9$, our answer makes sense.

Wave power and wave intensity

Earlier (in Chapter 7), we defined the power P of a process as energy change per unit time interval ($P = \Delta E / \Delta t$). Since this power is spread over the area the waves travel through, a more useful quantity for describing the wave is **intensity**. The intensity of a wave is defined as the energy per unit area per unit time interval that crosses perpendicular to an area in the medium through which it travels.

$$\text{Intensity} = \frac{\text{Energy}}{\text{time} \cdot \text{area}} = I = \frac{\Delta E}{\Delta t \cdot A} = \frac{P}{A} \tag{11.6}$$

The unit of intensity I is equivalent to joules per second per square meter $(\text{J/s} \cdot \text{m}^2)$. Since J/s is a watt, the unit of intensity is then watts per square meter (W/m^2). The intensity of one-dimensional waves propagating without damping does not depend on the distance from the source.

REVIEW QUESTION 11.4 Is the following statement true? The energy of three-dimensional waves decreases by a factor of 4 and the amplitude decreases by a factor of 2 when the distance from the point source doubles. Explain your answer.

FIGURE 11.14 Snapshots of the reflection and transmission of a pulse at an interface (the wave is traveling from the thin rope to the thick rope).

FIGURE 11.15 Snapshots of the reflection and transmission of a pulse at an interface (the wave is traveling from the thick rope to the thin rope).

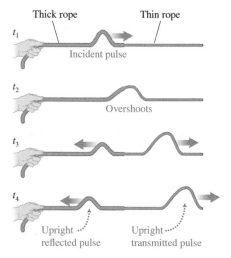

11.5 Reflection and impedance

Until now, we have considered pulses and waves moving through a homogeneous medium, a medium with the same properties everywhere. What happens to a wave when there is an abrupt change from one medium to another? Figure 11.5a showed one such case—a pulse on a rope reflected off the end attached to a wall. The reflected pulse pointed down, opposite to the incident pulse. We say that the pulse was *inverted*. What other changes to the pulse can occur?

Imagine holding a thin rope (small mass per unit length) that is attached to a thicker rope (larger mass per unit length) on the right. You send an upright transverse pulse to the right in the thin rope (Figure 11.14). A partially reflected inverted pulse (oriented like the pulse reflected off the fixed end) returns to your hand, and a partially transmitted upright pulse of smaller amplitude travels in the thicker rope.

With the ropes connected in the opposite order (the thick rope on the left and thin rope on the right), you send an upright transverse pulse along the thick rope toward the right (Figure 11.15). Now, a partially reflected upright pulse returns to your hand, and a partially transmitted upright pulse of larger amplitude travels in the thin rope. Why?

In the first experiment (Figure 11.14), because of its larger mass per unit length, the right-hand thick rope is harder to accelerate upward than a thin rope would be. A small-amplitude upward pulse is initiated in the thick rope. Because the thick rope does not "give" as easily, it exerts a larger downward force than a thin rope would exert, starting an inverted reflected pulse back toward the left in the thin rope. The same happened in Figure 11.5a—we can think of a wall to which the rope is fixed (or your hand holding the rope) as an infinitely thick rope.

In the second experiment (Figure 11.15), when the upright pulse in the left-hand thick rope reaches the interface of the two ropes, the thin rope restrains the thick rope less than the thick rope restrains itself. The thick rope initiates a large upright transmitted pulse in the thin rope (which "gives" more easily because of its smaller mass per unit length). The thin rope overshoots its position relative to the thick rope and starts a small upright reflected pulse back in the thick rope. The extreme case of this happens when the pulse on a uniform rope reaches an end of the rope that is free to move (Figure 11.5b). When the pulse reaches the end of the rope, the rope overshoots and starts an upright reflected pulse.

Understanding the patterns that occur when waves travel from one medium to another allows us to better transmit information and protect ourselves from unwanted signals.

In the first experiment, a significant amount of the pulse energy traveled back from the thicker medium—thus less energy was transmitted forward. Similarly, your hard skull reflects most of the energy from sound waves traveling through the air. However, the structure of the ear is such that some of this energy can reach the inner ear and be detected.

The pattern of traveling waves in varying media also allows geologists to "see" under Earth's surface. Radio waves reflected from the bottom of the Antarctic ice sheet are characteristic of an ice-rock interface rather than an ice-water interface. Thus, geologists have determined that the ice sheet rests on rock rather than water. Geologists have also used the pattern of motion of seismic waves to determine that there is a sharp, distinctive boundary between the Earth's mantle and its outer core.

Impedance

A physical quantity called **impedance** characterizes the degree to which waves are reflected and transmitted at the boundary between different media. When a wave travels through a medium, parts of the medium move with respect to each other, or the medium gets distorted. The impedance Z of a medium is a measure of the difficulty a wave has in distorting the medium and depends on both the medium's elastic and inertial properties. The elastic property is related to the interactions between the particles in the medium (the stronger the interactions, the greater the impedance). The inertial property characterizes the density of the medium. Impedance is proportional to the product of the elastic and inertial properties of the medium.

If two media connected together at an interface have different elastic and inertial properties (for example, two different linear density ropes woven together), a wave traveling in one medium is partially reflected and partially transmitted at the boundary between the media. If the product of the elastic and inertial properties of the two media is the same, a wave moves from one medium to the other as if there is no change. If the impedances of two media are very different, most of the wave energy is reflected back into the first medium and does not travel into the second medium. One application of this effect is the use of **ultrasound** (a compression wave with frequencies higher than 20,000 Hz), which takes advantage of the differing densities of internal structures to "see" inside the body, for example, to determine the shape of a fetus inside the mother's uterus.

Ultrasound is poorly transmitted from air outside the body to tissue inside the body. The impedance of tissue is much greater than that of air. To overcome this problem, the area of the body to be scanned is covered with a gel that helps "match" the impedance between the emitter and the body surface. The emitter is then held directly against the gel-covered area. This matching allows the ultrasound wave to propagate into the body instead of being reflected off the body's surface.

Many animals can detect seismic signals. Elephants have stiff cartilage and dense fat in their feet, which matches the impedance of Earth's surface. Elephants at times lean forward on their front feet, seemingly to detect vibrations in the ground—for example, prior to approaching a water hole. Reports describe Asian elephants as vigorously responding to earthquakes, even trumpeting at the approach of an earthquake before humans feel the quake.

REVIEW QUESTION 11.5 Why is it impossible to create a traveling wave on a Slinky that propagates in only one direction?

11.6 Superposition principle and skills for analyzing wave processes

We have studied the behavior of a single wave traveling in a medium and initiated by a simple harmonic oscillator. Now we investigate what happens when two or more waves simultaneously pass through a medium.

FIGURE 11.16 The meeting of two pulses oriented (a) in opposite directions and (b) in the same direction.

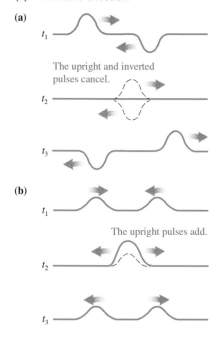

(a)

The upright and inverted pulses cancel.

(b)

The upright pulses add.

FIGURE 11.17 The superposition at point C of two waves that add.

(a)

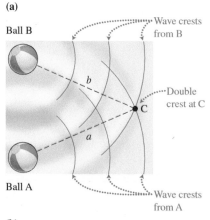

Ball B

Wave crests from B

Double crest at C

Ball A

Wave crests from A

(b)

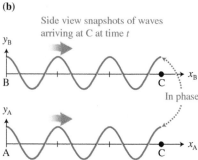

Side view snapshots of waves arriving at C at time t

In phase

Figure 11.16a shows a string with an incident pulse traveling right and the reflected pulse (inverted) traveling left. When these pulses meet, the string looks undisturbed. Figure 11.16b shows the meeting of two upright pulses. The string bulges more than for each of them individually. In both cases, after meeting, both pulses continue traveling as if they never met. How can we explain these observations? In Figure 11.16a, the part of the string where the pulses met was pulled up by the pulse coming from the left and down by the pulse coming from the right, causing a zero net disturbance. In Figure 11.16b, the part of the string where the pulses met is pulled up by both the right- and left-moving pulses, resulting in a double-amplitude disturbance. We can hypothesize that when pulses meet, the resultant disturbance is the sum of the two (pulse orientation is important), as if each pulse pulls the rope separately.

The final displacement of the rope at a particular time and location is just a combination of the two pulses at that time and location—adding them together:

$$y_{net} = y_1 + y_2 + \cdots \qquad (11.7)$$

This looks similar to the other kind of superposition we've encountered before—the superposition of forces. We can test this superposition idea by using it to predict the outcomes of new experiments.

Imagine we have two vibrating sources in water—for example, you push beach balls A and B up and down in exactly the same way. (**Figure 11.17a**). The sources A and B vibrate with the same amplitude and period, each sending a wave toward point C. The source A wave will cause C to vibrate as

$$y_A = A_C \cos\left[\frac{2\pi}{T}\left(t - \frac{a}{v}\right)\right]$$

The source B wave will cause C to vibrate as

$$y_B = A_C \cos\left[\frac{2\pi}{T}\left(t - \frac{b}{v}\right)\right]$$

where A_C is the amplitude at location C (smaller than the initial).

If distances a and b are equal, then the two vibrations arrive in the same phase at point C:

$$2\pi\frac{a}{Tv} = 2\pi\frac{b}{Tv}$$

If one wave causes C to move up, the other one will also cause it to move up (the displacements from the A and B waves at C are shown in Figure 11.17b). *If* the idea of superposition is correct *and* point C is at the same distance from both sources, *then* the amplitude of vibration at C will be twice the amplitude caused by one wave source at the same location. There are many points at the same distance from both sources, and the vibrations at these points will have twice the amplitude caused by one wave.

More generally, the superposition principle predicts a similar effect for any point where the difference in distance from source A and from source B is an integer multiple of the wavelength (if $a - b = n\lambda$ for $n = 0, \pm1, \pm2, \ldots$). At any of these points, the resultant vibration amplitude due to both waves will be the sum of the amplitudes of the individual waves.

Suppose now that we choose a point D located so that the wave from A arrives there with a maximum positive displacement and the wave from B arrives with a maximum negative displacement (see **Figure 11.18a**). Then the A wave pushes point D up while the B wave pushes it down the same distance. The net displacement is almost zero. This situation happens only if one wave has to travel a distance that is one-half, or three-halves, or five-halves, etc. times the wavelength λ longer than the other and we can neglect the difference in the amplitudes of the two waves due to a small difference in the distances that they travel to location D (Figure 11.18b). We conclude that at locations where $a - b = (n + 1/2)\lambda$ for $n = 0, \pm1, \pm2, \ldots$, there will be little vibration because the two waves will nearly cancel each other.

FIGURE 11.18 The superposition at point D of two waves that cancel.

(a)

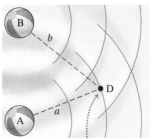

Crest and trough overlap—no disturbance

(b)

At a particular time, the two waves at D are out of phase and cancel.

Let us see whether this actually happens. Take two speakers at different locations and predict places between them where our ears will hear little sound. Then put your head in the predicted locations and check! The example below shows you how to make these predictions.

PROBLEM-SOLVING
STRATEGY 11.1 ⟩ Analyzing wave processes

EXAMPLE 11.4 ⟩ Where is there no sound?

Two sound speakers separated by 100 m face each other and vibrate in unison at a frequency of 85 Hz. Determine three places on a line between the speakers where you cannot hear any sound.

Sketch and translate

- Sketch the situation. Label all known quantities.
- Label the unknown quantities.

If x is your distance from the left speaker, we can write the distances the sound travels from the left and right speakers to you, a and b, respectively, in terms of x.

Simplify and diagram

- Decide what simplifying assumptions you should make.
- Draw displacement-versus-time or displacement-versus-position graphs to represent the waves if necessary.

Assume that the amplitudes of the waves are equal when they pass through the point of interest. Assume that both of your ears are located at the same point. We have two waves of the same frequency passing through the same locations. We can use the superposition principle to find the locations of destructive interference. You should hear no sound at places where the distance from the left speaker is an odd multiple of a half wavelength farther than the distance from the right speaker. The displacement-versus-position graph for this situation is shown in Figure 11.18b.

Represent mathematically

- Represent the problem using mathematical relationships between physical quantities.

You should hear no sound if

$$a - b = \left(n + \frac{1}{2}\right)\lambda \text{ for } n = 0, 1, 2, \ldots$$

Here a is x and b is 100 m $- x$: $x - (100 \text{ m} - x) = \dfrac{\lambda}{2}, \dfrac{3\lambda}{2}, \dfrac{5\lambda}{2} \ldots,$

where $\lambda = \dfrac{v_{sound}}{f} = \dfrac{340 \text{ m/s}}{85 \text{ Hz}} = \dfrac{340 \text{ m/s}}{85 \text{ s}^{-1}} = 4.0 \text{ m.}$

(CONTINUED)

Solve and evaluate

- Solve the equations for the unknown quantity.

- Evaluate whether the results are reasonable (the magnitude of the answer, its units, how the solution behaves in limiting cases, and so forth).

Solve the equation in the last step for places where we hear no sound:

$$x - (100 \text{ m} - x) = \frac{\lambda}{2}, \frac{3\lambda}{2}, \frac{5\lambda}{2} \cdots$$

$$\Rightarrow 2x - 100 \text{ m} = \frac{\lambda}{2}, \frac{3\lambda}{2}, \frac{5\lambda}{2} \cdots$$

$$\Rightarrow 2x = 100 \text{ m} + \frac{\lambda}{2}, \frac{3\lambda}{2}, \frac{5\lambda}{2} \cdots$$

$$\Rightarrow x = \left[50 \text{ m} + \frac{\lambda}{4} \right], \left[50 \text{ m} + \frac{3\lambda}{4} \right], \left[50 \text{ m} + \frac{5\lambda}{4} \right] \cdots$$

Using $\lambda = 4.0$ m, we expect no sound at

$$x = 51 \text{ m}, 53 \text{ m}, 55 \text{ m}, \ldots$$

Note that the first place is 51 m from the left speaker and 49 m from the right (one half wavelength different). The next place is 53 m − 47 m = 6 m or 1.5 wavelengths. The answers seem reasonable. The distance between the points with no sound is much larger than the size of your head, so the assumption of your two ears being at the same location is justified. When we conduct the experiment we find that at the predicted locations we hear almost no sound.

- -

Try it yourself For the situation described in this example, where should you stand to hear sound with twice the amplitude?

Answer

The condition for these places is that the distances from the ear to each source should differ by an integer multiple (including zero) of the wavelength of the waves: $x - (100 \text{ m} - x) = 0, \lambda, 2\lambda, \ldots$. This condition is satisfied at positions 50 m, 52 m, 54 m, ...

We can now assert the superposition principle for waves with greater confidence.

> **Superposition principle for waves** When multiple waves pass through the same medium at the same time, the net displacement at a particular time and location in the medium is the sum of the displacements that would be caused by each wave if it were alone in the medium at that time. Mathematically, this statement can be written as
>
> $$y_{\text{net}} = y_1 + y_2 + \cdots \tag{11.7}$$

The phenomenon that occurs when two or more waves are present at the same location at the same time is called **interference**. If the waves have the same frequency, the resulting pattern is stable in time. The points where the amplitude of the resultant wave is large are points of **constructive interference** (point C in Figure 11.17); the points where the waves add to produce a smaller amplitude are called points of **destructive interference** (point D in Figure 11.18).

What happens to the energy of the waves when they meet? When the interference pattern is stable in time, the energy gets redistributed in space—more energy goes into the points of constructive interference and none to the points of destructive interference. However, these are extreme cases. At most points, the interference is neither completely destructive nor completely constructive.

Huygens' principle

When a beach ball bobs up and down in water, circular waves move outward. We can explain the formation of the waves using the superposition principle, first explained this way by Christiaan Huygens (1629–1695). It begins with a wave front represented by curved line AB moving at speed v away from a point source S (**Figure 11.19**). To determine the location of the wave front a short time interval Δt later, draw a large number of circular arcs, each centered on a different point of wave front AB. Each arc represents a **wavelet** that moves away from a point on the original wave front. The addition of these wavelets produces a new wave front whose shape can be determined by adding the disturbances due to all wavelets. The new wave front is determined by drawing a tangent to the front edges of the wavelets—the curved line CD in Figure 11.19. To test Huygens' idea, consider two new experiments in Testing Experiment **Table 11.3**.

FIGURE 11.19 Using Huygens' principle to create a new wave front.

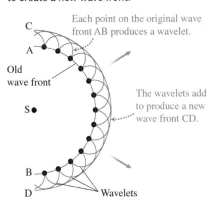

Each point on the original wave front AB produces a wavelet.

The wavelets add to produce a new wave front CD.

Old wave front

Wavelets

TESTING EXPERIMENT TABLE 11.3   Testing Huygens' principle

VIDEO TET 11.3

Testing experiment	Prediction	Outcome
Take a strip of cardboard with many closely spaced pushpins pushed through it from the bottom in a line. Place the cardboard in a tray filled with water just above the water so that when it vibrates, the pins move up and down, each creating a wavelet in the water. Then shake the cardboard to produce vibrations of the pins in the water.	When the resulting circular wavelets are added, they should produce a straight wave front. Top view. We predict a straight wave front.	We observe the predicted straight wave front.
Remove the pins from the cardboard and shake it. When the cardboard vibrates in the tray of water, it creates a wave with wave fronts that look like lines (plane waves). Place a barrier with a narrow slit opening in the path of the plane waves.	The vibrations of water beyond the narrow slit should be circular wavelets. We predict circular wavelets.	We observe circular wavelets spreading beyond the narrow slit.

Conclusion

Predictions based on Huygens' principle match the outcomes of the testing experiments. These experiments do not disprove the principle but rather give us more confidence in it.

REVIEW QUESTION 11.6 Your friend says that it is impossible for two waves to arrive at the same point and still have no vibrational motion at this point. How can you convince him that this is possible?

11.7 Sound

We hear sound when compressions and decompressions of air at the eardrum cause it to vibrate (**Figure 11.20**). Amplification mechanisms in the outer ear and middle ear increase the vibrations up to a factor of 180 and transform them into vibrations in the fluid inside the inner ear. Nerve cells in the inner ear can distinguish sounds differing by just 3 Hz over a range of approximately 20 Hz to 20,000 Hz. Compression waves in this frequency range are called **sound**. Higher and lower frequency waves are called **ultrasound** and **infrasound**, respectively. Many animals, such as bats, dolphins, whales, dogs, and some fish, can sense ultrasonic or infrasonic frequencies.

FIGURE 11.20 A simplified sketch of the human ear.

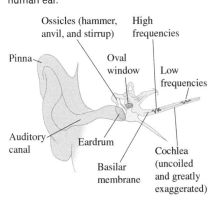

Ossicles (hammer, anvil, and stirrup)   High frequencies

Pinna

Oval window   Low frequencies

Auditory canal

Eardrum

Basilar membrane

Cochlea (uncoiled and greatly exaggerated)

Pressure waves

When physicists describe sound waves mathematically, they use a new physical quantity, *pressure* (P), that is equal to the magnitude of the perpendicular component of the force $F_\perp$ that a gas, liquid, or solid exerts on a surface divided by the contact area A over which that force is exerted:

$$P = \frac{F_\perp}{A}$$

TIP Although they have the same symbol P, do not confuse the ideas of pressure and power: use the context to decide which concept is being used. Similarly, A is used for both area and amplitude.

The units of pressure are N/m^2. The air surrounding us in normal conditions exerts about 10^5 N per every square meter of surface; thus normal atmospheric pressure is $P_{Atm} = 10^5 \, N/m^2$. We will learn more about pressure in Chapter 12.

The compressions and decompressions of air that we hear as sound are described by the variations of air pressure relative to the normal pressure when no sound is present (**Figure 11.21**). A young person with healthy hearing can hear sound that corresponds to only $10^{-5} \, N/m^2$ variation of pressure, which is less than one-billionth of atmospheric pressure. This pressure variation is equivalent to the force that Earth exerts on one ant distributed over the area of one square meter. We are amazingly sensitive creatures!

FIGURE 11.21 A pressure-versus-time graph of sound from one piano note.

Humans perceive the amplitude of pressure variations in sound waves as the *loudness* of sound. However, it is not only the amplitude of the waves that affects how loud a sound "feels" for us. Our sensitivity to different frequencies of sound waves also affects how loud we perceive a sound to be. Humans are most sensitive to frequencies around 4000 Hz and least sensitive to sounds of 20 Hz and 20,000 Hz. This means that humans do not perceive equal-amplitude sounds of different frequency as equally loud. As a result, loudness is not a physical quantity. It is a person's subjective impression of the intensity of sound. Instruments that inform us about the loudness of a sound do not measure the sound's pressure amplitude but instead measure its intensity.

QUANTITATIVE EXERCISE 11.5   **Sound energy at a construction site**

The intensity of sound at a construction site is 0.10 W/m^2. If the area of the eardrum is 0.20 cm^2, how much sound energy is absorbed by one ear in an 8-h workday?

Represent mathematically The intensity I of the sound, the area A of the eardrum, and the time interval Δt of the exposure have been given. We wish to determine the sound energy ΔE absorbed by the ear. Equation (11.6) relates all of these quantities:

$$\text{Intensity} = I = \frac{\text{Energy}}{\text{time} \cdot \text{area}} = \frac{\Delta E}{\Delta t \cdot A}$$

To find the energy, multiply both sides by the time interval and the area:

$$\Delta E = I \cdot \Delta t \cdot A$$

Solve and evaluate In using the above, we need to convert the time interval into seconds and the area into m^2.

$$\Delta E = (0.10 \, W/m^2)\left[(8.0 \, h)\left(\frac{3600 \, s}{1 \, h}\right)\right]$$

$$\times \left[(0.20 \, cm^2)\left(\frac{10^{-4} \, m^2}{1 \, cm^2}\right)\right] = 0.058 \, J$$

This looks like a small number. However, our perception of sound energy depends on the power per unit area, not the total energy. Our environment includes sounds that differ in intensity from about $10^{-12} \, W/m^2$ (barely audible) to nearly 1 W/m^2 (a painful sound). The sound intensity in a typical classroom is about $10^{-7} \, W/m^2$.

- -

Try it yourself How would the answer differ for a child with an eardrum that is half the area?

Answer   The sound energy would be half as big.

Intensity level

Because of the wide variation in the range of sound intensities, a quantity called **intensity level** is commonly used to compare the intensity of one sound to the intensity of a reference sound. Intensity level is defined on a base 10 logarithmic[1] scale as follows:

$$\beta = 10 \log_{10} \frac{I}{I_0} \tag{11.8}$$

where I_0 is a reference intensity to which other intensities I are compared (for sound, $I_0 = 10^{-12} \, W/m^2$, the smallest sound intensity typically detectable by humans).

---

[1]If $x = b^y$, then $y = \log_b(x)$. For common logarithms $b = 10$. Thus for $y = \log(x)$, $x = 10^y$.

I and I_0 in Eq. (11.8) have the same units, and the intensity level is a dimensionless quantity. Intensity level does have a dimensionless unit associated with it, called the decibel (dB). An increase of 10 dB is equivalent to a 10-fold increase in intensity. The unit serves as a reminder of what we are quantifying, much like the radian is a dimensionless reminder of one way of quantifying angles. The intensity levels β of several familiar sounds are listed in Table 11.4.

TABLE 11.4 Intensities and intensity levels of common sounds

Source of sound	Intensity (W/m^2)	Intensity level (dB)	Description
Large rocket engine (nearby)	10^6	180	
Jet takeoff (nearby)	10^3	150	
Rock concert with amplifiers (2 m); jet takeoff (60 m)	1	120	Pain threshold
Heavy truck (15 m)	10^{-3}	90	Constant exposure endangers hearing
Niagara Falls (nearby)	10^{-3}	90	
Inside an average factory	10^{-4}	80	
Busy traffic	10^{-5}	70	
Normal conversation (1 m)	10^{-6}	60	
Quiet office	10^{-7}	50	Quiet
Rustling leaves	10^{-10}	20	Barely audible
Normal breathing	10^{-11}	10	
	10^{-12}	0	Hearing threshold

QUANTITATIVE EXERCISE 11.6   Busy classroom

The sound in a classroom has an intensity of 10^{-7} W/m^2. (a) Determine the intensity level of that sound. (b) If the sound intensity doubles, what is the new sound intensity level?

Represent mathematically To calculate the intensity level β, we need to compare the given sound intensity I with the reference intensity for sound $I_0 = 10^{-12}$ W/m^2 and then take 10 times the log of that ratio (see Appendix A for more information on logarithms):

$$\beta = 10 \log\left(\frac{I}{I_0}\right)$$

Solve and evaluate (a) The ratio of the sound intensity I and the reference intensity I_0 is

$$\frac{I}{I_0} = \frac{10^{-7} \text{ W/m}^2}{10^{-12} \text{ W/m}^2} = 10^5$$

Since $\log 10^5$ is 5, the sound intensity level is

$$\beta = 10 \log 10^5 = 10(5) = 50 \text{ dB}$$

The classroom is pretty quiet. Maybe the students are taking a test? (b) If I is doubled, the ratio of I and I_0 becomes

$$\frac{I}{I_0} = \frac{2 \times 10^{-7} \text{ W/m}^2}{10^{-12} \text{ W/m}^2} = 2 \times 10^5$$

The sound intensity level is then

$$\beta = 10 \log(2 \times 10^5) = 10(5.3) = 53 \text{ dB}$$

A doubling of intensity corresponds to an intensity level increase of just 3 dB.

Try it yourself The sound intensity at a concert is 0.10 J/s · m^2. What is the intensity level?

Answer   110 dB.

Pitch, frequency, and complex sounds

Another perceived physiological characteristic of sound is *pitch*. It has no units, but we all know what the terms high and low pitch mean. Pitch relates to our perception of the frequency of sound—a higher pitch means a sound of higher frequency.

FIGURE 11.22 Pressure-versus-time graphs produced by one or two tuning forks.

(a) Wave 1 produced by the first tuning fork

(b) Wave 2 produced by the second tuning fork

(c) Combined wave produced by both tuning forks

FIGURE 11.23 A frequency spectrum produced by two tuning forks.

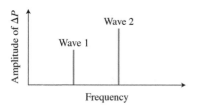

FIGURE 11.24 Waveforms and frequency spectra for a piano and a violin playing concert A.

However, there is more to sound than the sensations of its loudness and pitch. For example, an oboe and a violin playing the note concert A equally loudly at 440 Hz sound very different. Sound must have some other property that causes these different sensations in our ears. Let's use a microphone and a computer to look at a pressure-versus-time graph of a sound wave produced by playing one note on a piano. Notice that the graph in Figure 11.21 has negative values; thus it must be representing the pressure difference from atmospheric pressure. There is a characteristic frequency, but the wave is not sinusoidal. What causes this different shape?

To answer this question, we use two tuning forks with the second fork vibrating at twice the frequency of the first. Each fork separately produces a sinusoidal wave (**Figures 11.22a** and b). When we hit the two forks simultaneously, the wave looks different (Figure 11.22c)—it is more like that recorded when playing the piano. It must be a superposition of the waves from the two tuning forks. This complex pattern is called a **waveform**—a combination of two or more waves of different frequencies and possibly different amplitudes.

It is possible, using a microphone and a device known as a spectrum analyzer, to decompose a complex waveform into its component frequencies. The amplitude of each component can be plotted versus the frequency of that component, a graph called the **frequency spectrum** of the waveform. A frequency spectrum of the waveform in Figure 11.22c is shown in **Figure 11.23**. The height of the line at each frequency is proportional to the amplitude of the wave at that frequency. In this case, the graph shows that the amplitude and frequency of wave 2 are greater than those of wave 1.

Most sounds we hear are complex waves consisting of sinusoidal waves of many different frequencies. "White noise"—the sound you hear if you tune a radio to a frequency that doesn't match a nearby radio station—includes waves of a broad range of frequencies. We seldom associate a pitch with white noise. However, if the complex sound is the result of the addition of a relatively large-amplitude lowest frequency component (called the **fundamental**) with several higher frequency components that are whole-number multiples of the fundamental (called **harmonics**), then we usually identify the pitch of the sound as being that of the fundamental frequency. The waveforms and frequency spectra of complex waves from a piano and a violin while playing concert A at 440 Hz are shown in **Figure 11.24**. Notice that the violin A has many frequency components, which are all multiples of 440 Hz (880 Hz, 1320 Hz, 1760 Hz, 2200 Hz, and even higher frequencies). Different musical instruments have different frequency spectrums that contribute in part to the quality of the sound we associate with that instrument.

Beat and beat frequencies

What happens when two sound waves of close but not identical frequencies are present at the same time at the same location? Suppose that two sound sources are equidistant from a microphone that records the air pressure variations due to the two sound sources as a function of time (**Figure 11.25**). Source 1 vibrates at 10 Hz and source 2 vibrates at 12 Hz. Figures 11.25a and b show the pressure variations due to wave 1 and wave 2 during a 1.0-s interval. Note that in each case only one wave source is playing at a time. Consider the pressure-versus-time wave pattern we would get if both sound sources played at the same time.

We add the two waves shown in Figures 11.25a and b to see the pattern produced when both sounds play simultaneously (Figure 11.25c). Note in Figure 11.25 that at time 0.00 s the positive pressure caused by wave 1 cancels the negative pressure caused by wave 2. The net disturbance at time 0.00 s is zero and there is no sound. At 0.25 s, both waves have a maximum negative pressure variation, so the net disturbance is twice that of a single wave—the sound is loud. At 0.50 s, the two wave pressure variations again cancel each other, causing a zero pressure disturbance and no sound.

The pattern resulting from adding two waves of close frequencies shown in Figure 11.25c from 0.00 s to 0.50 s is called a **beat**. The vibration has a frequency of 11 Hz (the average of the two wave frequencies), but its amplitude is modulated from zero to approximately twice the normal amplitude and back to zero again at 0.50 s. A second beat is produced from 0.50 s to 1.00 s. The **beat frequency** is the number of beats per second—in this case, 2 beats/s—the sound goes on and off twice each second. The beat frequency is just the difference in the two wave frequencies: 12 Hz − 10 Hz = 2 Hz. This modulated vibration frequency is easily heard if the beat frequency is a few hertz or less.

> **Beats and beat frequency** A beat is a wave that results from the superposition of two waves of about the same frequency. The beat (the net wave) has a frequency equal to the average of the two frequencies and has variable amplitude. The frequency with which the amplitude of the net wave changes is called the beat frequency f_{beat}; it equals the difference in the frequencies of the two waves:
>
> $$f_{beat} = |f_1 - f_2| \qquad (11.9)$$

Beats are useful in precise frequency measurements. For example, a piano tuner can easily set the frequency of the middle C string on the piano to 262 Hz by listening to the beat frequency produced when the piano string and a 262-Hz tuning fork are sounded simultaneously. If the tuner hears a beat frequency of 3 Hz, then the piano string must be vibrating at either 259 Hz or 265 Hz—a difference of 3 Hz from the frequency of the tuning fork. The string is then tightened to increase the string frequency or loosened to lower the frequency until the beat frequency is reduced to zero (you do not hear beats anymore); then the string is vibrating at exactly 262 Hz.

REVIEW QUESTION 11.7 Is the following sentence true? When two sinusoidal sounds of the same amplitude but different frequencies are added, the result is not a sinusoidal wave. Explain your answer.

11.8 Standing waves on strings

The superposition of waves is the basis for understanding how musical instruments produce sound. Let's consider stringed instruments first. Start by observing the vibration of a rope with one end attached to a fixed support (a wall) while you hold the other end your hand. You shake that end up and down. Depending on the frequency with which you shake your hand, the rope has either very small amplitude irregular vibrations or large-amplitude sine-shaped vibrations. The latter appear not to be traveling waves but are instead coordinated vibrations, with parts of the rope vibrating up and down with large amplitude and other parts not vibrating at all. Such waves are called **standing waves**. Displacement-versus-position graphs for several of those waves for one instant are shown in **Figure 11.26**. Because stringed musical instruments have both ends of their strings fixed, the motion of the strings forms standing waves. How is the first standing wave vibration in Figure 11.26 produced? We will explore this question in Observational Experiment **Table 11.5**, on the next page.

FIGURE 11.25 Beats.

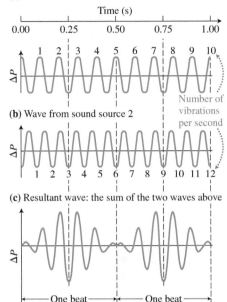

(a) Wave from sound source 1

(b) Wave from sound source 2

(c) Resultant wave: the sum of the two waves above

FIGURE 11.26 Standing waves on a string.

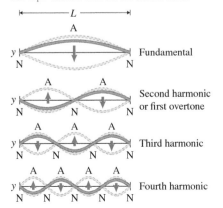

The rope vibrates between the dashed lines.

Observational experiment	Analysis
You shake a rope that is fixed at one end at a relatively low frequency, and at one particular frequency you observe the large-amplitude vibration.	A pulse starts moving to the right (b–d). When it reaches the fixed end, the upward pulse is reflected and inverted to a downward pulse traveling back toward your hand (e–g). When it reaches your hand, it once again reflects and inverts to an upward pulse (h). To make this pulse bigger, you begin another upward shake, timing the shake so that it adds to the previous upward pulse that has just been reflected (h).

Pattern

- You must shake the rope upward each time a previous pulse returns. The amplitude of the disturbance traveling along the rope grows.
- You need one up and down shake during the time interval $(2L/v)$ that it takes for the pulse to travel down the string and back. Thus the period of the one vibration is $T_1 = \dfrac{2L}{v}$, and consequently the wave frequency is

$$f_1 = \frac{1 \text{ vibration}}{\dfrac{2L}{v}} = \frac{v}{2L}$$

The subscript 1 in the last equation in Table 11.5 means the lowest standing wave frequency. The numerator in that equation is the number of vibrations, but *vibration* is not a unit (similar to the radian and the decibel). Thus, this lowest frequency vibration of the rope is one up and down shake per time interval $\tau = 2L/v$, or

$$f_1 = \frac{1}{\tau} = \frac{1}{\dfrac{2L}{v}} = \frac{v}{2L} \tag{11.10}$$

This frequency is called the **fundamental frequency**. If the shaking frequency is just a little greater or a little less than this, each new pulse will interfere destructively with previous pulses, and there will no longer be a time-independent pattern; the standing wave will disappear and multiple traveling waves will bump here and there, like splashing water. At frequency $f_1 = v/(2L)$, each new pulse constructively interferes with previous pulses and creates a large-amplitude vibration. The wavelength of this fundamental frequency vibration is

$$\lambda_1 = \frac{v}{f_1} = \frac{v}{\dfrac{v}{2L}} = 2L$$

The length of the rope equals one-half of the wavelength of the wave that has this fundamental frequency. We will test these ideas in the following experiment.

A banjo's D string has a mass of 0.252 g and length 0.690 m. The wooden peg pulls on its end, exerting a 60.0-N force. Can we predict the frequency of its fundamental vibration? We can find the speed of the wave on the string using Eq. (11.5):

$$v = \left[\frac{F_{\text{Peg on String}}}{(m_{\text{String}}/L_{\text{String}})}\right]^{1/2} = \left[\frac{(60.0 \text{ N})}{(0.252 \times 10^{-3} \text{ kg})/(0.690 \text{ m})}\right]^{1/2} = 405 \text{ m/s}$$

Then we can use this speed to predict the fundamental frequency:

$$f_1 = \frac{v}{2L} = \frac{405 \text{ m/s}}{2(0.690 \text{ m})} = 294 \text{ s}^{-1} = 294 \text{ Hz}$$

When we pluck a banjo D string and compare it to the frequency of different tuning forks, we find that the matching fork has the frequency $f = 294$ Hz. The agreement is excellent!

Now consider how the second harmonic can be produced (Figure 11.26). You vibrate the rope so that your hand has two up and down shakes during the time interval that is needed for a pulse to make a round trip down the rope and back (**Figure 11.27**). The new pulse adds to a pulse started earlier that has completed a round trip back to your hand. Thus, the frequency f_2 of the second harmonic:

$$f_2 = \frac{2 \text{ vibrations}}{\dfrac{2L}{v}} = 2\frac{v}{2L}$$

We can see a pattern now.

Standing wave frequencies on a string The frequencies f_n at which a string with fixed ends can vibrate are

$$f_n = n\left(\frac{v}{2L}\right) \text{ for } n = 1, 2, 3, \ldots \qquad (11.11)$$

where v is the speed of the wave on the string and L is the string length.

Since $\lambda = v/f$, we can write an expression for the wavelengths of these allowed standing wave vibrations:

$$\lambda_n = \frac{v}{f_n} = \frac{v}{n\left(\dfrac{v}{2L}\right)} = \frac{2L}{n} \text{ for } n = 1, 2, 3, \ldots \qquad (11.12)$$

Note that a rope or string will vibrate with large amplitude only at those frequencies for which a pulse produced by the wave source (for example, the vibrating hand) is reinforced by a new pulse after completing a round trip up the string and back again. Thus standing waves are caused by the superposition of waves moving in opposite directions.

How are standing waves different from traveling waves? When a traveling wave moves along the string, each point on the string vibrates between the maximum and minimum displacements (between $y = \pm A$). However, different points reach the maximum displacements at different times (the phase of the wave is different at different

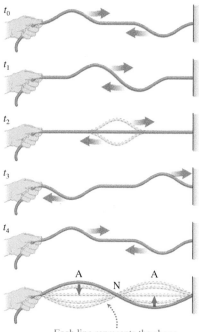

FIGURE 11.27 Producing the second harmonic on a string or rope.

t_0

t_1

t_2

t_3

t_4

Each line represents the shape of the rope at a different time.

locations). In a standing wave, only very specific points on the string, called **antinodes** (labeled A in Figure 11.27), reach the maximum displacement. Other points, called **nodes** (labeled N in Figure 11.27), do not vibrate at all. However, each point reaches its respective maximum displacement at the same time—the points vibrate in phase. When you play a musical instrument, you excite multiple standing wave vibrations. For example, when you bow a violin string, you simultaneously excite 10 or 20 standing wave vibrations.

EXAMPLE 11.7   **Playing a banjo**

A banjo D string is 0.69 m long and has a fundamental frequency of 294 Hz. Shortening the string (by holding the string down over a fret) causes it to vibrate at a higher fundamental frequency. Where should you press to play the note F-sharp, which has a fundamental frequency of 370 Hz?

Sketch and translate We know that when the length of the string is $L = 0.69$ m, the fundamental frequency is $f_1 = 294$ Hz. At what length of the string L' is the fundamental frequency $f_1' = 370$ Hz?

Simplify and diagram Assume that the speed of the wave on the string is independent of where the finger pushes down on the string.

Represent mathematically The fundamental frequency f_1 relates to the string length L by the equation $f_1 = v/(2L)$, or $L = v/(2f_1)$. We cannot solve immediately for the unknown length when the string is vibrating at 370 Hz since we do not know the speed v of the wave on the string. But we can use this idea to write an expression for the speed of the wave for both cases (the speed is the same for both cases):

$$v = f_1(2L) = f_1'(2L')$$

Solve and evaluate Dividing both sides of the expression for the speed by f_1' and by 2, we get an expression for L':

$$L' = \frac{f_1 L}{f_1'} = \left(\frac{f_1}{f_1'}\right)L = \left(\frac{294\ \text{Hz}}{370\ \text{Hz}}\right)(0.69\ \text{m}) = 0.55\ \text{m}$$

The answer is reasonable, as it is less than 0.69 m. To find where to press on the string, we need to find the difference between the original length L and the new length L':

$$L - L' = 0.69\ \text{m} - 0.55\ \text{m} = 0.14\ \text{m}$$

You should press 0.14 m from the end of the string.

- -

Try it yourself If tuned correctly, the fundamental frequency of the A string of a violin is 440 Hz. List the frequencies of several other standing wave vibrations of the A string.

Answer

The other standing wave frequencies are whole-number multiples of the fundamental: 880 Hz, 1320 Hz, 1760 Hz, and so forth.

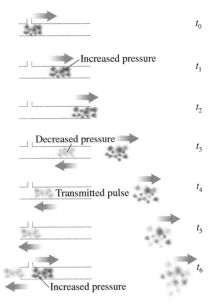

FIGURE 11.28 Producing a standing wave vibration in an open-open pipe.

t_0

Increased pressure t_1

t_2

Decreased pressure t_3

Transmitted pulse t_4

t_5

t_6

Increased pressure

REVIEW QUESTION 11.8 One end of a horizontal string of length L passes over a pulley with a block hanging at the end. A motor attached to the other end of the string exerts a constant horizontal force on the string. When the motor is turned on, it vibrates the end of the string up and down at frequencies that increase slowly from zero to a high frequency. What do you observe?

11.9 Standing waves in air columns

Take a bottle, partially fill it with water, and blow across the top of the bottle. You hear sound with a pitch that depends on the amount of water in the bottle. The same phenomenon underlies how sound is made in musical instruments made of pipes or tubes, such as organs, clarinets, and flutes. How can we explain this phenomenon?

Standing waves in open-open pipes

A flute is a wind instrument that is an example of an **open-open** pipe, so called because it is open on both ends. Blowing into the open-open pipe initiates pressure pulses at that end of it. Alternating increased and decreased pressure pulses move down the air column, compressing the air in the instrument in a repetitive fashion (**Figure 11.28**, $t_0 \rightarrow t_2$). Near the opposite end, the flute is also open at the nearest open valve. The high-pressure pulse expands into open space outside of the valve, leaving behind a pulse of decreased pressure that reflects back into the tube.

This change from high to low pressure is similar to the phase change of a pulse on a string reflected off a fixed end. This phase change does not occur when a pressure pulse reaches a closed end.

The pulse of decreased pressure moves back toward the other end of the pipe where it started ($t_3 \rightarrow t_5$) and is reflected at the open end on the left (t_6). The higher pressure outside the pipe causes the decreased pressure pulse to compress as air rushes in, becoming a high-pressure pulse. If a new high-pressure pulse is initiated at this time, it interferes constructively with the reflected pulse, and its amplitude increases. If this is done repeatedly, a large-amplitude standing wave vibration builds up in the open-open pipe. If, however, the new air pressure pulse is too early or too late, it interferes destructively with the reflected pulse, and the amplitude decreases.

The frequency of standing wave vibrations in pipes depends on the time interval needed for a pulse to travel down the pipe and back again. If the sound pulse travels at a speed v along a pipe of length L, then the time interval T needed to travel a distance $2L$ is $T = 2L/v$. To reinforce the vibration of air in the pipe, and to create a standing wave of fundamental frequency, pressure pulses should occur once each time period T, or at a frequency $f = 1/T = v/(2L)$.

Just as we can make a string vibrate at multiples of the fundamental frequency, the pipe can also vibrate with large amplitude if excited by air pressure pulses at whole-number multiples of the fundamental frequency.

Standing wave vibration frequencies in open-open pipes The resonant frequencies f_n of air in an open-open pipe of length L in which sound travels at speed v are

$$f_n = n\left(\frac{v}{2L}\right) \text{ for } n = 1, 2, 3, \ldots \tag{11.13}$$

where v is the speed of sound in the medium that fills the pipe.

Standing waves in one-end-closed pipes—an open-closed pipe

Next, consider a pipe that is closed at one end and open at the other end, an **open-closed** pipe, such as a clarinet or the human throat (the closed end being the reed and vocal cords, respectively). Blowing into the reed initiates a high-pressure pulse near the closed end of the pipe. After reflection off the open end, a low-pressure pulse returns to the closed end of the pipe and reflects. However, reflection off the closed end causes it to remain a low-pressure pulse. The pulse has to make another complete round trip in the pipe before a high-pressure pulse returns to the reed. Thus, the pulse has to make two round trips, or travel a distance $4L$, before a new high-pressure pulse from the reed produces a fundamental frequency vibration. Thus, the fundamental frequency of an open-closed pipe is $f_1 = v/(4L)$ instead of $f_1 = v/(2L)$ for an open-open pipe. The fundamental frequency of an open-closed pipe clarinet is half the fundamental frequency of an open-open pipe flute of about the same length. Another difference is that the standing wave frequencies of open-closed pipes are not integer multiples of the fundamental but instead are odd whole-number multiples, as the even whole numbers would cause waves in the open-closed pipe that interfere destructively.

We can test our understanding of standing waves in pipes. If sound in pipe-like instruments is due to standing waves inside of them, and we change the medium (which changes the wave speed) inside a pipe, then we should observe the change of the fundamental frequency produced by the pipe even though the length of the pipe has not changed.

What should happen to a person's voice after inhaling helium and singing a note?

Sketch and translate The speed v of sound in helium gas (930 m/s) is much greater than in air (340 m/s), and the fundamental frequency f_1 is proportional to the speed of sound in the medium.

Simplify and diagram Model the vocal chamber as an open-closed pipe with the vocal cords at the closed end and the mouth at the open end. The length of the chamber does not change during the experiment. Since the speed of sound in helium is almost three times greater than in air, the fundamental frequency of the sound should increase by a factor of 3 and we should hear a higher pitch. Inhaling helium before singing

or talking creates a very high pitched sound—as predicted. However, without measuring the frequency we cannot say how much greater the frequency is.

Try it yourself In our conceptual exercise with helium, we used gas with a greater speed of sound. What would happen if we could inhale a gas in which the speed of sound is lower than it is in air, such as CO_2?

Answer

The pitch of one's voice would be lower. However, this gas is dangerous to inhale; nobody should perform this experiment. Thus this exercise is purely theoretical.

We can now formulate the conditions for standing waves in pipes with more confidence.

Standing wave vibration frequencies in open-closed pipes The resonant frequencies f_n of air in an open-closed pipe of length L in which sound travels at speed v are

$$f_n = n\left(\frac{v}{4L}\right) \text{ for } n = 1, 3, 5, \ldots \tag{11.14}$$

Exciting the sound and changing the pitch of a wind instrument

We can apply our understanding of standing waves in pipes to explain how wind instruments produce sounds of a particular frequency. In clarinets and saxophones, the sources of vibrations are the reeds. In trumpets, trombones, and French horns, the sources of vibrations are the vibrating lips and mouthpiece. The vibrations produced by the reeds and mouthpieces are not pure tones but are instead pressure pulses at a variety of frequencies. The pipes attached to the reeds and mouthpieces reinforce only those input frequency vibrations that are at resonant frequencies of the instruments.

A resonant frequency for a particular instrument can be changed by opening one or more valves or keys. Opening the valve makes a hole that serves as the open end of the pipe. Thus, the air column length in the pipe is changed, as is its fundamental frequency. Because usually more than one resonant frequency is excited at a time, sound from the instrument consists of multiple resonant frequencies. The combination of resonant frequencies contributes in part to the quality of the sound from the instrument.

REVIEW QUESTION 11.9 When we studied traveling waves, we decided that the wavelength of a wave depends on the frequency of vibration and the speed of the wave in the medium. Is this statement true for standing waves in pipes?

11.10 The Doppler effect

When you hear the sound from the horn of a passing car, its pitch is noticeably higher than its pitch at rest as it approaches and is noticeably lower as it moves away. This phenomenon is an example of the **Doppler effect**. The Doppler effect occurs when a source of sound and an observer move with respect to each other and/or with respect to the medium in which the sound travels.

Is there a pattern in these observed frequency shifts?

In Observational Experiment Table 11.6, we will investigate the Doppler effect, using a battery-powered whistle (the source) that generates a source frequency f_S and a detector (the observer) that records the observed sound and determines its frequency f_O using a special device called a spectrum analyzer. We will place each on a cart that can move toward or away from the other. In all experiments the air in which the sound propagates is at rest with respect to Earth. When both carts are at rest, the spectrum analyzer records frequency f_S.

OBSERVATIONAL EXPERIMENT TABLE 11.6   Effects of motion on observed frequency

Observational experiment	Analysis
Place a battery-powered whistle (the source) on one cart and an observer with a spectrum analyzer on another cart. The carts can move toward or away from each other. **(a) Source** stationary with respect to the ground **Observer** moving toward source	The graph shows the frequency spectrum for the signal heard by the observer (f_O). The frequency of the source f_S is shown as a comparison. This is not the frequency that the observer records. $f_S < f_O$
(b) Source moving toward observer **Observer** stationary with respect to the ground	The frequency spectrum: $f_S < f_O$
(c) Source stationary with respect to the ground **Observer** moving away from source	The frequency spectrum: $f_S > f_O$
(d) Source moving away from observer **Observer** stationary	The frequency spectrum: $f_S > f_O$
(e) Source moving with respect to the ground **Observer** moving at the same speed in the same direction	The frequency spectrum: $f_S = f_O$

Patterns

- (a) and (b) Source and/or observer are moving *toward* each other: the observed frequency is *greater than* the emitted source frequency.
- (c) and (d) Source and/or observer are moving *away from* each other: the observed frequency is *less than* the source frequency.
- (e) There is no relative motion: the observed and emitted frequencies are the same.

Let's develop a quantitative explanation for the patterns observed in Table 11.6.

Doppler effect for the source moving relative to the medium

We can use the waves created by a water beetle bobbing up and down at a constant frequency on the water to explain the Doppler effect. If the beetle bobs up and down in the same place, the pattern of wave crests appears as shown in **Figure 11.29a**. The crests move away from the source at a constant speed. The distance between adjacent crests is one wavelength and is the same toward observer A as toward observer B. The frequency of waves reaching both observers is also the same.

Now suppose the bobbing beetle moves to the right at a speed slower than the wave speed. Each new wave produced by the beetle originates from a point farther to the right (Figure 11.29b). As a result, B receives four wave crests in a shorter time interval than the source took to produce them, $f_O > f_S$. A receives four wave crests in a longer time than the source took to produce them, $f_O < f_S$.

FIGURE 11.29 (a) A bobbing beetle creates wave crests. (b) The wave crests created as the beetle hops to the right.

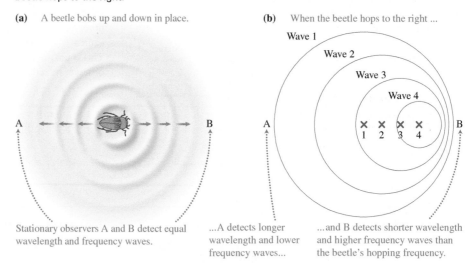

(a)   A beetle bobs up and down in place.

Stationary observers A and B detect equal wavelength and frequency waves.

(b)   When the beetle hops to the right ...

Wave 1
Wave 2
Wave 3
Wave 4

...A detects longer wavelength and lower frequency waves...

...and B detects shorter wavelength and higher frequency waves than the beetle's hopping frequency.

FIGURE 11.30 The wavelength is reduced in front of the hopping beetle.

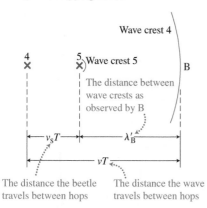

Wave crest 4

Wave crest 5

B

The distance between wave crests as observed by B

v_sT λ'_B

vT

The distance the beetle travels between hops

The distance the wave travels between hops

To derive an equation for calculating the observed frequency, we first calculate the separation of crests in front of the beetle. Consider the separation of wave crest 4 and the wave initiated by the beetle's fifth step (wave crest 5, shown in **Figure 11.30**). If the beetle bobs up and down at frequency f_S, the time between the start of wave 4 and the start of wave 5 is $T = 1/f_S$. During this time interval, the beetle moves a distance v_sT to the right, where v_s is the speed of the beetle (the wave source). Wave 4 has moved a distance vT during that same time interval, where v is the speed of the wave. The crests of waves 4 and 5 arriving at observer B are separated by a distance λ'_B (observed wavelength), given by

$$\lambda'_B = vT - v_sT$$

Similarly, the beetle is moving away from A (see Figure 11.29), and the wavelength of the wave reaching A is

$$\lambda'_A = vT + v_sT$$

The waves pass observer B at a frequency

$$f_{OB} = \frac{\text{wave speed}}{\text{wavelength}} = \frac{v}{\lambda'_B} = \frac{v}{T(v - v_S)} = \frac{1}{T}\frac{v}{(v - v_S)} = f_S\frac{v}{v - v_S}$$

where $f_S = 1/T$ is the source frequency. The waves are incident on observer A at frequency

$$f_{OA} = \frac{\text{wave speed}}{\text{wavelength}} = \frac{v}{\lambda'_A} = \frac{v}{T(v + v_S)} = \frac{1}{T}\frac{v}{(v + v_S)} = f_S\frac{v}{v + v_S}$$

Combining these two equations, we get

$$f_O = f_S \frac{v}{v \mp v_S} \tag{11.15}$$

We use the minus sign when the source is moving toward the observer and the plus sign when the source is moving away from the observer. This equation is consistent with the patterns discovered in Table 11.6.

Doppler effect for the observer moving relative to the medium

If the observer moves toward the source at speed v_O relative to the medium, the wave fronts appear to move past the observer at speed $v + v_O$ and the observer encounters the fronts more frequently than if she had remained stationary. If the observer moves away from the source, the wave fronts appear to move past the observer at speed $v - v_O$ and she encounters the fronts less frequently.

Mathematically,

$$f_O = \frac{v \pm v_O}{\lambda} = \frac{v \pm v_O}{v/f_S} = f_S \frac{v \pm v_O}{v} \tag{11.16}$$

where λ is the wavelength that the observer who is not moving with respect to the source will record (λ in the medium). This equation can be used to calculate the frequency f_O detected by an observer moving at speed v_O toward (plus sign) or away from (minus sign) a stationary source emitting waves at frequency f_S.

General equation for the Doppler effect for sound

We arrive at a general equation for the Doppler effect by combining Eqs. (11.15) and (11.16).

Doppler effect for sound When a sound source and sound observer move, the observed sound frequency f_O differs from the source sound frequency f_S:

$$f_O = f_S \left(\frac{v \pm v_O}{v \mp v_S} \right) \tag{11.17}$$

where v is the speed of waves through the medium, v_O is the speed of the observer relative to the medium (use the plus sign if the observer moves toward the source and a minus sign if the observer moves away from the source), and v_S is the speed of the source relative to the medium (use the minus sign if the source moves toward the observer and a plus sign if the source moves away from the observer).

TIP Once you have chosen which signs to use in Eq. (11.17), check for consistency. When the source and observer are getting closer, $f_O > f_S$, and so on.

Measuring the speed of blood flow in the body

The Doppler effect is used to measure the speed of blood flow in the body. Ultrasound waves ($f > 20,000$ Hz) are directed into an artery (**Figure 11.31**). The waves are reflected by red blood cells back to a receiver. The frequency detected at the receiver f_R relative to that emitted by the source f_S indicates the blood cell's speed and the speed of the blood flow. This process has two parts. (1) The waves leave the source at frequency f_S and strike the cell at frequency f'. (2) The waves reflect from the cell at frequency f' and return to the receiver where they are detected at frequency f_R.

FIGURE 11.31 Using the Doppler effect to measure blood flow speed.

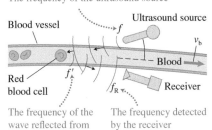

The frequency of the ultrasound source

The frequency of the ultrasound f' striking a red blood cell is related to the frequency f_S emitted from the source by the equation

$$f' = f_S\left(\frac{v + 0}{v - v_b}\right) \tag{11.18}$$

where v is the speed of sound in blood and v_b is the speed of the blood. The red cells reflect this sound, acting like a source with frequency f'. Note that in this case the source is not moving with respect to the ground, but the medium is moving with respect to the source. Thus we have $-v_b$ in the equation. The frequency detected at the receiver f_R is related to f' by the equation

$$f_R = f'\left(\frac{v + v_b}{v - 0}\right)$$

Substituting for f' from Eq. (11.18), we find the frequency detected at the receiver f_R relative to the frequency emitted by the sound source f_S:

$$f_R = f_S\left(\frac{v + 0}{v - v_b}\right)\left(\frac{v + v_b}{v - 0}\right) = f_S\left(\frac{v + v_b}{v - v_b}\right) \tag{11.19}$$

This result allows us to measure the speed of blood using the Doppler effect.

EXAMPLE 11.9 **Buzzer moves in a circle**

A friend with a ball attached to a string stands on the floor and swings it in a horizontal circle. The ball has a 400-Hz buzzer in it. When the ball moves toward you on one side, you measure a frequency of 412 Hz. When the ball moves away on the other side of the circle, you measure a frequency of 389 Hz. Determine the speed of the ball.

Sketch and translate Draw a sketch of the situation. You detect a higher frequency when the ball moves toward you and lower frequency when it moves away, consistent with our understanding of the Doppler effect.

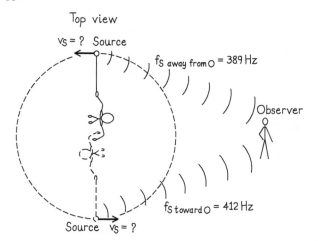

Simplify and diagram The speed of sound in air is constant and equals 340 m/s. The air is stationary with respect to the ground.

Represent mathematically We use Eq. (11.17) to find an expression for the ratio of the frequency when the source (the buzzer) is moving toward you (the observer) and when it is moving away. This ratio can then be rearranged to get an expression for the speed of the source (the buzzer):

$$\frac{f_{S \text{ moving toward O}}}{f_{S \text{ moving away from O}}} = \frac{f_S\dfrac{v + 0}{v - v_S}}{f_S\dfrac{v + 0}{v + v_S}} = \frac{v + v_S}{v - v_S}$$

where v is the speed of sound in air and v_S is the speed of the source.

Solve and evaluate Solving for v_S, we get

$$v_S = \left(\frac{f_{S \text{ moving toward O}} - f_{S \text{ moving away from O}}}{f_{S \text{ moving toward O}} + f_{S \text{ moving away from O}}}\right)v$$

$$= \left(\frac{412 \text{ Hz} - 389 \text{ Hz}}{412 \text{ Hz} + 389 \text{ Hz}}\right)(340 \text{ m/s}) = 9.8 \text{ m/s}$$

The ball was moving at 9.8 m/s—the correct unit and a reasonable magnitude.

- -

Try it yourself What frequency do you hear when the ball is moving perpendicular to your line of sight directly in front of you?

Answer 400 Hz. At that instant, the ball is not moving toward or away from you—it moves in the perpendicular direction—there is no Doppler shift and you should hear the buzzer's frequency of 400 Hz.

REVIEW QUESTION 11.10 An ambulance siren blares continuously as the ambulance approaches you and then moves away. How does the sound from the siren change and why?

Summary

Wave speed v depends on the properties of the medium (such as a force $F_{M \text{ on } M}$ that one part of the medium exerts on another part and a mass/length quantity $\mu = m/L$). The *wavelength* λ is equal to the shortest distance between two points that vibrate in phase and depends on the frequency f and the wave speed v. (Sections 11.2 and 11.3)

$$\lambda = Tv = \frac{v}{f}$$ Eq. (11.3)

$$v = \sqrt{\frac{F_{M \text{ on } M}}{\mu}}$$ Eq. (11.5)

Traveling sinusoidal wave The traveling wave expression allows us to determine the displacement from equilibrium y of a point in the medium at location x at time t when a wave of wavelength λ travels at speed v. In a traveling wave all points vibrate with the same amplitude but have different phases. (Section 11.2)

$$y = A \cos\left[2\pi\left(\frac{t}{T} - \frac{x}{\lambda}\right)\right]$$ Eq. (11.4)

Intensity I of a wave is the energy per unit area per time interval that crosses an area perpendicular to the direction of travel of the wave.
Intensity level β is a logarithmic measure of the intensity I relative to reference intensity I_0. (Sections 11.4 and 11.7)

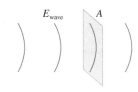

$$I = \frac{\Delta E}{\Delta t \cdot A} = \frac{P}{A}$$ Eq. (11.6)

$$\beta = 10 \log_{10}\left(\frac{I}{I_0}\right)$$ Eq. (11.8)

where for sound $I_0 = 10^{-12} \text{ J/m}^2 \cdot \text{s}$

Superposition principle When two or more waves pass through the same medium at the same time, the net displacement of any point in the medium is the sum of the displacements that would be caused by each wave if alone in the medium. (Section 11.6)

$$y_{\text{net}} = y_1 + y_2 + y_3 + \cdots$$ Eq. (11.7)

Huygens' principle Every vibrating point of a medium produces a wavelet surrounded by a circular wave front. The new wave front is the superposition of all wavelets due to the vibrations of all points on the previous wave front. (Section 11.6)

Standing waves are the result of the superposition of reflected waves of frequencies f_n traveling at speed v in both directions in a medium of length L. The vibrating points in a standing wave have different amplitudes, but all vibrate in phase. (Sections 11.8 and 11.9)

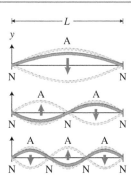

Strings: $f_n = n\left(\dfrac{v}{2L}\right)$ for $n = 1, 2, 3, \ldots$ Eq. (11.11)

Open-open pipes: $f_n = n\left(\dfrac{v}{2L}\right)$ for $n = 1, 2, 3, \ldots$ Eq. (11.13)

Open-closed pipes: $f_n = n\left(\dfrac{v}{4L}\right)$ for $n = 1, 3, 5, \ldots$ Eq. (11.14)

Doppler effect for sound When a sound source and sound observer move relative to each other, the frequency f_O of the observed sound differs from the frequency f_S of the source. v_O and v_S are the observer and source speeds (magnitudes of the velocities) and v is the speed of the wave through the medium. (Section 11.10)

$$f_O = f_S\left(\frac{v \pm v_O}{v \mp v_S}\right)$$ Eq. (11.17)

Use $+$ in front of v_O if observer moves toward source and $-$ if away; use $-$ in front of v_S if source moves toward observer and $+$ if away. All speeds in this equation are relative to the medium.

Questions

Multiple Choice Questions

1. What does it mean if the speed of a wave is 300 m/s?
   (a) Each particle moves at a speed of 300 m/s.
   (b) The wave disturbance travels 300 m during each second.
   (c) The wave disturbance travels 300 m for each wavelength.
   (d) The wave disturbance travels 300 m during each period.

2. What does it mean if the wavelength of a wave is 4.0 m?
   (a) The wave continues to travel for 4.0 m before stopping.
   (b) The period of the wave is 4.0 m.
   (c) The distance between the nearest particles that have identical displacements is 4.0 m.

3. If you wish to represent one period of a wave on a string, which two variables should you have on the axes?
   (a) v and T (b) y and x
   (c) y and t (d) T and λ

4. If you wish to graph the disturbance pattern of a wave on a long string at one clock reading, which two variables should you have on the axes?
   (a) y and λ (b) y and x
   (c) y and t (d) T and λ

5. Which mathematical expression represents a cause-effect relationship for a traveling wave?
   (a) $f = v/\lambda$ (b) $v = \lambda \cdot f$ (c) $\lambda = v/f$

6. Two small spheres vibrate up and down on the surface of a large tank of water. We observe places in the water where the surface does not vibrate. What are these statements?
   (a) A hypothesis
   (b) A description of a physical phenomenon
   (c) A physics law   (d) A physical quantity

7. What is Huygens' principle?
   (a) A law describing the behavior of waves
   (b) A hypothesized mechanism of wave propagation
   (c) A physical phenomenon involving waves
   (d) A physical quantity describing waves

8. **Figure Q11.8** shows the displacement-versus-time graph of the left end of a 12-m-long rope. The wave speed on the rope is 2 m/s. Which graph below correctly shows a snapshot of the rope at a clock reading of $t = 5$ s?

   FIGURE Q11.8

9. A small battery-operated buzzer emits sound at a constant frequency of $f_b = 3000$ Hz. If you drop the buzzer from the top of a tall building, what will you observe?
   (a) Sound at constant frequency $f_o = f_b$
   (b) Sound at constant frequency $f_o > f_b$
   (c) Sound at constant frequency $f_o < f_b$
   (d) Sound at increasing frequency $f_o > f_b$
   (e) Sound at decreasing frequency $f_o < f_b$
   (f) Sound at beat frequency $|f_o - f_b|$

Conceptual Questions

10. **Figure Q11.10** shows a snapshot of two pulses that are propagating with speed $v = 1.0$ m/s along a rope fixed to a wall at the right end. Which of the following shows the correct snapshot of the rope 1.0 s later?

    FIGURE Q11.10

    (d) None of the above.

11. **Figure Q11.11** shows a snapshot of two pulses that are propagating with speed $v = 1.0$ m/s along a rope. The right end of the rope is fixed to a very light ring that can move along the vertical pole with no friction. Which of the following shows the correct snapshot of the rope 1.0 s later?

    FIGURE Q11.11

    (d) None of the above.

12. Can a wave have a period of 2.0 s, a speed of 20 m/s, and a wavelength of 10 m?

13. What physics ideas were necessary to construct Eq. (11.4)? What math ideas did you need?

14. How do you know that the wavelength of a wave can be calculated by multiplying the speed of the wave in a medium by the period of the vibrating source?

15. What conditions are necessary to create a sinusoidal wave in a certain medium? What is needed to create a wave of a different frequency? Different speed? Different wavelength?

16. Invent and describe an experiment to estimate the speed of sound in air. You can use only everyday items.

17. How can you use the time delay between a lightning flash and the sound of thunder to estimate the distance to the place where the flash occurred? Justify any assumptions you make in an example calculation.

18. Describe two useful types of information a geologist can obtain by detecting sound reflected from different materials under Earth's surface.

19. Two speakers hang from racks placed in an open field. When sound of the same frequency comes from both speakers, no sound is heard in some locations. The sound is loud in other places at about the same distance from the speakers. Explain.

20. Two identical sound waves are sent down a long hall. Does the resultant wave necessarily have amplitude twice that of the constituent waves? Explain. Under what conditions would we hear no sound?

21. Sound waves of all frequencies in the audio frequency range (20–20,000 Hz) travel at the same speed in air. Explain how some common experiences support this statement.

22. How can you show that an object producing sound can be used to detect sound?

23. Describe the common features and differences between traveling waves and standing waves.

24. Why do different guitar strings sound different when you pluck them, even though they are the same length?

Problems

Below, **BIO** indicates a problem with a biological or medical focus. Problems labeled **EST** ask you to estimate the answer to a quantitative problem rather than derive a specific answer. Asterisks indicate the level of difficulty of the problem. Problems with no * are considered to be the least difficult. A single * marks moderately difficult problems. Two ** indicate more difficult problems.

11.1 and 11.2 Observations: pulses and wave motion and Mathematical descriptions of a wave

Assume that the speed of sound in air is 340 m/s for all of these problems unless stated otherwise.

1. Imagine that you and your friend pull tightly on the ends of a rope lying on a smooth floor. You shake your end of the rope, causing a transverse pulse. (a) Draw a displacement-versus-time graph representing the motion of your end of the rope. (b) Draw two displacement-versus-position graphs representing all of the points on the rope at two different clock readings.

2. Explain the meaning of each symbol in the equation

$$y = A \cos\left[\frac{2\pi}{T}\left(t - \frac{x}{v}\right)\right]$$

and summarize how we devised this equation.

3. * **Tell all** You have a sinusoidal wave that is described by the function

$$y = (0.2 \text{ m}) \sin\left[\frac{\pi}{2}\left(t + \frac{x}{20 \text{ m/s}}\right)\right]$$

(a) Say everything you can about this wave. Pay attention to the positive sign in the equation. (b) Draw a y-versus-t graph and a y-versus-x graph for the wave.

4. * A transverse wave of amplitude 3.0 cm, frequency 2.0 Hz, and speed 3.0 m/s travels on an infinitely long Slinky. How far apart are the two nearest points on the Slinky that at one particular time both have the maximum displacements from their equilibrium positions? Explain your reasoning.

5. A boat is moving up and down in the ocean with a period of 1.7 s caused by a wave traveling at a speed of 4.0 m/s. What other physical quantities relevant to this wave can you determine using this information? Determine them.

6. * A large goose lands in a lake and bobs up and down for a short time. A fisherman notices that the first wave created by the goose reaches the shore in 8.0 s. The distance between two wave crests is 80 cm, and in 2.0 s he sees four waves hit the shore. How can the fisherman use these observations to determine how far from the shore the goose landed?

7. * **Equation Jeopardy** The equation below describes the variation of pressure at different positions and times (relative to atmospheric pressure) caused by a sound wave:

$$\Delta P = (2.0 \text{ N/m}^2) \cos\left[2\pi\left(\frac{t}{0.010 \text{ s}} - \frac{x}{3.4 \text{ m}}\right)\right]$$

Say everything you can about this wave.

8. **BIO** **Hearing** People can hear sounds ranging in frequency from about 20 Hz to 20,000 Hz. Determine the wavelengths of the sounds at these two extremes. What assumptions are you making? If these assumptions are not correct, how will your answer change?

9. **BIO** **Dolphin sonar** A dolphin has a sonar system that emits sounds with a frequency of 2.0×10^5 Hz. Determine everything you can about this sound wave. Remember that the sounds emitted by the dolphin travel in water.

10. **Antarctic ice** Radio waves travel at a speed of 1.7×10^8 m/s through ice. A radio wave pulse sent into the Antarctic ice reflects off the rock at the bottom and returns to the surface in 32.9×10^{-6} s. How deep is the ice? What assumptions did you make to solve the problem?

11. **Lightning and thunder** You see a flash of lightning, and 2.4 s later you hear thunder coming from the same location. (a) Why is there this long delay? (b) How far away did the lightning flash occur? Justify any assumptions you make in your calculations.

12. * See Question 11.8. Draw a v_y-versus-x graph for the points on the rope at time $t = 5$ s if you know that $A = 0.10$ m.

13. Three waves are described by the following equations: $y_1 = 0.1 \text{ m} \times \sin[(2 \text{ s}^{-1})t + (3 \text{ m}^{-1})x]$, $y_2 = -0.2 \text{ m} \times \sin[(3 \text{ s}^{-1})t - (2 \text{ m}^{-1})x]$, $y_3 = 0.3 \text{ m} \times \sin[(4 \text{ s}^{-1})t - (4 \text{ m}^{-1})x]$. (a) Rank the speeds of the waves from largest to smallest; (b) rank the wavelengths of the waves from largest to smallest; (c) rank the displacement magnitudes of the waves at $t = 0.5$ s, $x = 0.5$ m from largest to smallest.

11.3 Dynamics of wave motion: speed and the medium

14. * **Ratio reasoning** Two ropes have equal length and are stretched the same way. The speed of a pulse on rope 1 is 1.4 times the speed on rope 2. Determine the ratio of the masses of the two ropes (m_1/m_2).

15. **Telephone line** A telephone lineman is told to stretch the wire between two poles so the poles exert an 800-N force on the wire. As the lineman does not have a scale to measure forces, he decides to measure the speed of a pulse created in the wire when he hits it with a wrench. The pulse travels 60 m from one pole to the other and back again in 2.6 s. The 60-m wire has a mass of 15 kg. Should the wire be tightened or loosened? Explain. What assumptions did you make?

16. * A pulse travels at speed v on a stretched rope. By what factor must you increase the force you exert on the rope to cause the speed to increase by a factor of 1.30? Indicate any assumptions that you made.

17. A 0.62-kg Slinky has 185 coils. When you and your friend stretch the Slinky by exerting a 2.0-N force on its ends, the separation between the adjacent coils is 7 cm. Determine the speed of the transverse wave on the Slinky.

18. * **Design** Describe two experiments to determine the speed of propagation of a transverse wave on a rope. You have the following tools to use: a stopwatch, a meter stick, a mass-measuring scale, and a force-measuring device. Use whatever other items you need for your experiments.

19. ** A 3-m rope is made of three 1-m-long sections. The first and last sections are made from rope with $\mu_1 = 100$ g/m, and the middle section is made from rope with $\mu_2 = 200$ g/m. When a pulse is produced at one end of the rope, it reaches the other end of the rope after 0.3 s. (a) Determine the speed of the transverse waves in both types of rope. (b) Determine the force exerted on the ends of the rope.

11.4 Energy, power, and intensity of waves

20. Show using a sketch and mathematics that the intensity of a two-dimensional wave is inversely proportional to the distance from a source. Give three examples that will explain the meaning of "inversely proportional to the distance" in this case. Assume there is no conversion of mechanical energy into thermal energy.

21. Show using a sketch and mathematics that the intensity of a three-dimensional wave is inversely proportional to the distance squared from the source. Give three examples that will explain what it means to be inversely proportional to the distance squared in this case. Assume there is no conversion of mechanical energy into thermal energy.

22. * You are standing at position A and your friend at position B on a lake shore (see Figure P11.22). Someone throws a ball so that it lands on the lake surface at point S. The ball produces a circular pulse that propagates along the water surface with a speed of 0.3 m/s. (a) Determine the time that elapses from the moment the pulse reaches you to the moment it reaches your friend. (b) Determine the ratio of the pulse amplitudes observed by you and your friend, assuming there is no damping.

FIGURE P11.22

23. *A cicada is at the top of a 15.0-m-high tree, producing sound that is propagating in all directions. How far away from the tree should you stand so that the intensity of the sound that you hear is 10 times less than when you were standing at the bottom of the tree? Assume your height is 1.7 m.

11.5 and 11.6 Reflection and impedance and Superposition principle and skills for analyzing wave processes

24. Explain why the transverse pulse traveling on a rope held by two people reflects in the opposite orientation each time it reaches a person.

25. **Design** Use your knowledge of waves to explain why echoes occur. Use your explanation to devise a system to measure distances to objects that cannot be reached directly.

26. **BIO** **Bat echo** A bat receives a reflected sound wave from a fly. If the reflected wave returns to the bat 0.042 s after it is sent, how far is the fly from the bat? Would you expect the reflected wave pulse to be in or out of phase with the incident wave pulse? Explain.

27. **Sound wave in Earth** A sound wave created by an explosion at Earth's surface is reflected by a discontinuity of some type under Earth's surface (**Figure P11.27**). Determine the distance from the surface to the discontinuity. Does the discontinuity have a greater or lesser impedance to sound than the surface above it? Explain. Sound travels at about 3000 m/s through the top layer of Earth.

FIGURE P11.27

Time (s)

28. A 5.0-kg rope that is 20 m long is woven to an 8.0-kg rope that is 16 m long. The ropes are pulled taut and a pulse initiated in one is reflected at their interface. Draw a snapshot of the rope just after the pulse reaches the interface between the ropes.

29. The pulses shown in **Figure P11.29** (shown at time zero) are moving toward each other at speeds of 10 m/s. Draw $y(x)$ graphs showing the resultant pulse at 0.10 s, 0.20 s, and 0.30 s.

FIGURE P11.29

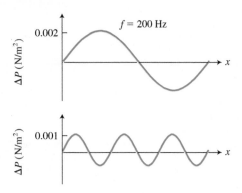

30. Repeat the previous problem for the case where the pulse on the right is upright rather than inverted.

31. * Find a resultant wave for the two waves shown in **Figure P11.31** at the instant they are traveling through the same medium.

FIGURE P11.31

32. Two waves shown in **Figure P11.32** at zero clock reading move toward each other at a speed of 10 m/s. Draw graphs of the resultant wave at times of 0.10 s, 0.20 s, and 0.30 s.

FIGURE P11.32

33. * Use Huygens' principle to find the shape of the wave fronts of a wave generated by the long edge of a flat piece of plastic floating horizontally and vibrating up and down in a swimming pool.

34. * Use Huygens' principle and a wave front representation of waves to show that if you place a screen with a small circular hole in the path of a wave with flat wave fronts, the wave fronts beyond the screen will be circular.

35. * Use Huygens' principle and a wave front representation of waves to find the locations in the swimming pool where no vibrations occur if two identical point-like objects synchronously vibrate in the pool. Assume the pool is infinitely large. In this case, neglect the decrease of amplitude with distance.

36. * You have two synchronously vibrating objects in an infinitely large pool. The distance between them is 6.0 m. Their frequency of vibration is 2.0 Hz and the wave speed is 4.0 m/s. The vibrations are sinusoidal. Find a location between them where the water does not vibrate and another location between them where the water vibrates with the largest amplitude. In this case, neglect the decrease of amplitude with distance.

11.7 Sound

37. **Design** Describe an experiment to convince a friend that sound is a wave.

38. The energy of a sound wave is proportional to its amplitude squared. Why is it difficult to hear poolside sounds when swimming underwater?

39. * **Warming with sound** The sound intensity at a gasoline station next to a freeway averages 10^{-3} W/m^2. The owner decides to collect this energy and convert it to thermal energy for heating his building. Assuming that this conversion is 100% efficient, what is the length of one side of a square sound collector that is needed to provide thermal energy at a rate of 500 W? Is this a practical idea for the owner? Explain.

40. * **Supersonic jet** The sound intensity 5 km from the place where a supersonic jet takes off is 0.60 W/m^2. Determine the area of a sound collector you would need to run a 40-W lightbulb from the energy collected. What might require you to create a larger collector?

41. * You are in an open field investigating how sound propagates. You place a rubber balloon 40.0 m away from you and pop the balloon with an air gun while while recording the sound produced by the gun and the popping balloon. **Figure P11.41** shows your recording represented as a time dependence of the sound amplitude. (a) Explain what determines the amplitude of each pulse, what determines the time interval between $t = 0$ (when you triggered the gun) and the first pulse, and what determines the time interval between the pulses. (b) Can you estimate the speed of the bullet in this experiment based on the given data? Explain your answer. (c) Draw a new graph that you would expect to get if you repeated the experiment with the balloon 80.0 m away from you. Try to make your prediction as quantitative as possible. Indicate any assumptions that you made.

FIGURE P11.41

42. One loudspeaker is producing a tone of frequency 236.30 Hz and a second loudspeaker a tone of frequency 236.50 Hz. What is the shortest time interval that you have to listen to the sound from the two loudspeakers to hear three beats?

43. * Tovi is playing a flute and Dawn is playing a violin. At the beginning of the performance, they tuned their instruments by ear, so the difference between the fundamental frequencies of the A notes played on each instrument was less than 1 Hz. At the end of the performance, they measured the A note fundamental frequencies of their instruments using a device that measures sound frequency with uncertainty 0.05 Hz. Dawn's A was 442.6 Hz and Tovi's was 439.3 Hz. (a) Can you say for sure that the A note frequency of at least one musical instrument changed during the performance? (b) Can you say for sure that the A note frequency of both musical instruments changed during the performance? (c) What do Dawn and Tovi hear when they simultaneously play the A note at the end of the performance? Give a quantitative description.

44. **Music** In music a very soft sound called "pianississimo" (ppp) might have an intensity of about 10^{-8} W/m^2. A very loud sound called fortississimo (fff) might have an intensity of about 10^{-2} W/m^2. Convert these intensities to intensity levels (units of dB).

45. Two sounds differ by 1 dB. What is the difference in their intensities?

46. Calculate the change in intensity level when a sound intensity is increased by a factor of 8, by a factor of 80, and by a factor of 800.

11.8 and 11.9 Standing waves on strings and Standing waves in air columns

47. **Banjo string** A banjo D string is 0.69 m long and has a fundamental frequency of 294 Hz. (a) Determine the speed of a wave or pulse on the string. (b) Identify three other frequencies at which the string can vibrate.

48. **Banjo fret** How far from the end of the banjo string discussed in the previous problem must a fret be so that the string's fundamental frequency is 330 Hz when you hold it down at the fret?

49. * **Violin string** A 0.33-m-long violin string has a mass of 0.89 g. The peg exerts a 45 N force on it. What can you determine about the sound produced by that string?

50. A person secures a 5.0-m-long rope of mass 0.40 kg at one end and pulls on the rope, exerting a 120-N force. The rope vibrates in three segments with nodes separating each segment. List the physical quantities you can determine using this information and determine three of them.

51. * Laura and Elana are discussing how to solve the following problem: "A canary sits 10 m from the end of a 30-m-long clothesline, and a grackle sits 5 m from the other end. The rope is pulled by two poles that each exert a 200-N force on it. The mass per unit length is 0.10 kg/m. At what frequency must you vibrate the line in order to dislodge the grackle while allowing the canary to sit undisturbed?"
    Laura: "First we have to determine the wave speed on the rope. Then we have to find the frequency that will produce a standing wave with a node located at 10 m from the end."
    Elana: "I think there is no way that the canary would be undisturbed. You cannot make a standing wave without first sending a traveling wave along the rope."
    Comment on the students' suggestions. Explain how each student reached her conclusion and decide which parts of their statements are correct or incorrect.

52. * **Ratio reasoning** Two wires on a piano are the same length and are pulled by pegs that exert the same force on the wires. Wire 1 vibrates at a fundamental frequency 1.5 times that of wire 2. Determine the ratio of their masses (m_1/m_2).

53. * **Ratio reasoning** By what percent does the frequency of a piano string change if the force that the peg exerts on it increases by 10%?

54. (a) Determine the first three standing wave frequencies of a 40-cm-long open-open pipe. (b) Do the same for a 40-cm-long open-closed pipe.

55. * **Brooklyn-Battery Tunnel** The 2779-m Brooklyn-Battery Tunnel, connecting Brooklyn and Manhattan, is one of the world's longest underwater vehicular tunnels. (a) Determine its fundamental frequency of vibration. (b) What harmonic must be excited so that it resonates in the audio region at 20 Hz or greater?

56. * **Flute** A wooden flute, open at both ends, is 0.48 m long. (a) Determine its fundamental vibration frequency. (b) How far from one end should a finger hole be placed to produce a sound whose frequency is four-thirds that calculated in part (a)? Be sure to justify how you arrive at your answer.

57. **Organ pipe** The lowest three standing wave vibration frequencies of an organ pipe are 120 Hz, 360 Hz, and 600 Hz. (a) Is the pipe open or closed, and what is its length? (b) Determine the frequencies of the first two harmonic vibrations on a pipe of the same length but of the other type than that described in part (a).

58. The speed of sound can be measured using the apparatus shown in **Figure P11.58.** A 440-Hz tuning fork vibrating above a tube partially filled with water initiates sound waves in the tube. The standing sound waves are formed inside the tube when the water in the tube is lowered 0.20 m and 0.60 m from the top of the tube. Use this information to determine the speed of sound in air.

FIGURE P11.58

Tuning fork

Movable reservoir

Resonant tube

Water

59. * Four strings of equal lengths are pulled with equal forces. All four strings are made to vibrate with standing wave frequencies. The table shows the vibration frequencies of the strings and the corresponding number of antinodes. Rank the masses of the strings from smallest to largest.

String #	Number of antinodes	f (Hz)
1	1	220
2	2	670
3	1	440
4	3	360

60. * A rope of length L is attached to a vibrating mechanism that vibrates with a constant frequency f. If a force F_1 is exerted on the ends of the rope, the rope is observed to vibrate in standing waves with three antinodes. Determine the force F_2 that has to be exerted on the end of the rope to make the rope vibrate in standing waves with two antinodes.

61. * A 3.0-m-long rope with a mass of 100 g is tied to a vibrating mechanism at one end and to a hook in a wall at the other end. The observed fundamental frequency of the rope is $f_1 = 7.0$ Hz. You modify the rope by attaching 2.0-g lead fishing weights every 10.0 cm along the entire rope. Determine the fundamental frequency of the modified rope. (Note: Since the wavelength is much larger than the separation between the nearest weights, we can assume that the mass is uniformly distributed along the modified rope.)

62. * A 1.2-m-long open-closed pipe is producing sound by vibrating with its third harmonic. (a) Determine the frequency of the sound. (b) How long should a pipe of the same type be to produce sound of the same frequency while at the fundamental frequency? (c) How would the answers to parts (a) and (b) change if the pipe was filled with helium, in which the speed of sound is about three times larger than in air? Explain.

63. * **Figure P11.63** shows the spectrum of sound that is produced by two pipes simultaneously vibrating with their fundamental frequencies. Each pipe can be either an open-open or an open-closed pipe. Determine all possible lengths of the pipes.

FIGURE P11.63

64. * See Problem 11.63. (a) Determine the length of the pipes if you know that the pipes have equal lengths. (b) Which frequency belongs to the open-closed pipe?

65. * See the spectrum in Figure P11.63. (a) Can this be a spectrum of the sound of an open-open pipe vibrating in its fundamental and next higher harmonic? Explain. (b) Can this be a spectrum of the sound of an open-closed pipe vibrating in its fundamental and first harmonic? Explain.

11.10 The Doppler effect

66. **Car horn** A car horn vibrates at a frequency of 250 Hz. Determine the frequency a stationary observer hears as the car (a) approaches at a speed of 20 m/s and (b) departs at 20 m/s. If the car is stationary, what frequency is heard (c) by an observer approaching the car at 20 m/s and (d) by an observer departing from the car at 20 m/s?

67. **Train whistle** A car drives at a speed of 25 m/s along a road parallel to a railroad track. A train traveling at 15 m/s sounds a horn that vibrates at 300 Hz. (a) If the train and car are moving toward each other, what frequency of sound is heard by a person in the car? (b) If the train and car are moving away from each other, what frequency of sound is heard in the car?

68. * **BIO** **Speed of blood** A source of ultrasound emits waves at a frequency of 2.00×10^6 Hz. The waves are reflected by red blood cells moving toward the source at a speed of 0.30 m/s. Determine the frequency of sound detected at a receiver next to the source. The speed of sound in the blood is 1500 m/s.

69. * **Circular motion sound source** A whistle with frequency 400 Hz moves at speed 20 m/s in a horizontal circle at the end of a rotating stick. Determine the highest and lowest frequencies heard by a person riding a bicycle at speed 10 m/s toward the whistle.

70. **BIO** **Bat echo** A bat emits short pulses of sound at a frequency of 1.60×10^5 Hz. As the bat swoops toward a flat wall at speed 30 m/s, this sound is reflected from the wall back to the bat. (a) What is the frequency of sound incident on the wall? (b) Consider the wall as a sound source at the frequency calculated in part (a). What frequency of sound does the bat hear coming from the wall?

71. ** **Detecting speed of baseball** A Doppler speed meter operating at exactly 1.02×10^5 Hz emits sound waves and detects the same waves after they are reflected from a baseball thrown by the pitcher. The receiver "mixes" the reflected wave with a small amount of the emitted wave and measures a beat frequency of 0.30×10^5 Hz. How fast is the ball moving?

General Problems

72. * **Violin strings** The speed of a wave on a violin A string is 288 m/s and on the G string is 128 m/s. Use this information to determine the ratio of mass per unit length of the strings. What assumptions did you make?

73. * Use Huygens' principle and a wave front representation of waves to show that a plane wave incident on a barrier traveling at an angle θ relative to a line perpendicular to the barrier (the so-called normal line) reflects, forming a wave front that travels at angle θ on the other side of the normal line.

74. * Use Huygens' principle and a wave front representation of waves to show that when a plane wave traveling in one medium hits a border with another different-speed medium that is not perpendicular to the direction of wave propagation, the wave changes its direction of propagation.

75. ** **Design** Describe an experiment that you can perform to measure the speed of sound in air using a graduated cylinder and a tuning fork that produces sound of a known frequency. Draw a picture. Carefully outline your experimental and mathematical procedure.

76. ** **BIO** **EST** **Reaction time** A basketball player's teammate shouts at her to catch a ball. Estimate the time required for the sound to travel a distance of 10 m, be detected by the ear, travel as a nerve impulse to the brain, be processed, then travel as a nerve signal for muscle action back to an arm, and finally cause a muscle contraction. Nerve impulses travel at a speed of about 120 m/s in humans. You will have to make reasonable assumptions about quantities not stated in the problem.

77. ** **EST** While camping, you record a thunderclap whose intensity is 10^{-2} W/m^2. The clap reaches you 3.0 s after a flash of lightning. Estimate the total acoustical power generated by the bolt of lightning. Clearly state all assumptions that you made.

78. ** **BIO** **Blood speed** A red blood cell travels at speed 0.40 m/s in a large artery. A sound of frequency 2.00×10^6 Hz enters the blood opposite the direction of flow. (a) Determine the frequency of sound reflected from the cell and detected by a receiver. (b) If the emitted and received sounds are combined in the receiver, what beat frequency is measured?

79. ** One end of a horizontal string of mass m_{st} and length $L = 0.8$ m passes over a pulley with a metal bolt of mass M hanging at the end. A motor attached to the other end of the string vibrates with a constant frequency $f = 18.0$ Hz. Some students add metal disks on the bolt and observe the vibration of the string. For certain masses of the added disks m_{disk}, they observe standing waves. When this happens, the students count the number of antinodes on the string n (**Figure P11.79** shows the setup for the $n = 2$ case). The students' data are presented in the table.

n	m_{disk} (kg)
4	0.100
3	0.250
2	0.700
1	3.100

(a) Derive an equation that relates m_{st}, L, f, n, m_{disk}, and M in cases when standing waves are observed. (b) Using the given data, estimate the mass of the string m_{st} and the mass of the bolt M. (Hint: Rearrange the equation that you derived in (a) to obtain the equation for m_{disk}. Then choose a new variable for which the equation represents linear dependence, plot the graph, and determine unknown quantities from the data that you obtained from the best-fit line.)

FIGURE P11.79

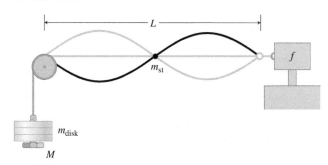

Reading Passage Problems

BIO **Bats and echolocation** If you shout "hello" near a vertical canyon wall, an echo "hello" returns after a short delay. If it returns in 2.0 s and sound travels through air at 340 m/s, then the total distance the signal traveled is 680 m. Therefore, the canyon wall must be half that distance from you (340 m). Bats use a similar process to hunt their prey. Bats travel and locate food at night by emitting ultrasound waves that bounce off prey and return to the bat. The time interval between signal emission and its return to the bat tells the bats which insects are closer than others. The amplitude of the reflected wave indicates the size of the prey— a large animal causes a large-amplitude reflected wave. The bat can tell the left-right location of the object by the relative time delays of the return sound between the left and right ears. Rows of horizontal folded sound sensors in the bat's ears allow them to determine the up-down location of the prey. Finally, the bat can determine if the prey is moving toward it, is moving away from it, or is stationary by the Doppler shift of the returning signal. The relationship between the source frequency of the bat and the frequency of the reflected sound is given by Eq. (11.19):

$$f_R = f_S\left(\frac{v + 0}{v - v_{P\,to\,B}}\right)\left(\frac{v + v_{P\,to\,B}}{v - 0}\right) = f_S\left(\frac{v + v_{P\,to\,B}}{v - v_{P\,to\,B}}\right)$$

80. You have a bat-like echolocation system on your car, which emits a 20,000-Hz frequency sound that returns to the car in 0.18 s after being reflected by another stationary car. Which answer below is closest to the distance from your car to the other car?
(a) 10 m   (b) 20 m   (c) 30 m
(d) 40 m   (e) 60 m

81. If the car from Problem 11.80 is moving at 20 m/s toward the stationary car, which answer below is closest to the frequency of the reflected wave detected by the moving car from the stationary car?
(a) 18,000 Hz   (b) 19,000 Hz   (c) 20,000 Hz
(d) 21,000 Hz   (e) 23,000 Hz

82. Which answer below is closest to the distance that the moving car in Problem 11.81 travels during the 0.18 s?
(a) 0.18 m   (b) 0.36 m   (c) 1.8 m
(d) 3.6 m   (e) 5.4 m

83. Compare your answers to Problems 11.80 and 11.82. If you want a more accurate measure of the object's distance from you, you must consider the distance you travel during the time delay for the sound to get from you to the object and back to you. By approximately what percent does your change in position affect the distance calculated in Problem 11.80?
   (a) 6%   (b) 12%   (c) 18%
   (d) 24%   (e) 30%

84. While your car from Problem 11.80 is stationary, you emit a 20,000-Hz signal and get a 22,000-Hz signal back from a reflecting object. What can you say about the object?
   (a) It is moving away from you.
   (b) It is stationary.
   (c) It is moving toward you.

85. Your echolocation system has one transmitter in the middle of the front of the car and two detectors of reflected waves, one by each headlight (like the bat's ears). You detect a reflected signal in the left detector slightly before you detect the signal from the right detector. Where must the reflecting object be?
   (a) To the left of center
   (b) Straight in front of you
   (c) To the right of center

BIO Human ear In the human ear (shown in Figure 11.32), the exterior ear flap, called the pinna, gathers and guides sound waves into the auditory canal. The sound waves create a pressure variation at the eardrum, a variation that is about two times as great as it would be without the pinna and auditory canal. This pressure variation causes the eardrum to vibrate. A large fraction of the vibrational energy is transmitted through the three small bones in the middle ear called the hammer, anvil, and stirrup. These three bones, collectively known as the ossicles, constitute a lever system that increases the pressure exerted on a small membrane of the inner ear known as the oval window. The pressure increase is possible for two reasons.

FIGURE 11.32

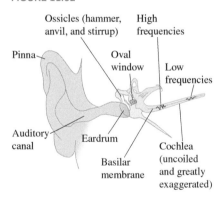

(1) The hammer feels a small pressure variation over the large area of its contact with the eardrum, whereas the stirrup exerts a large pressure variation over the small area of its contact with the oval window. This difference in areas increases the pressure by a factor of 15 to 30. (2) The three bones act as a lever that increases the force that the stirrup exerts on the oval window, although the displacement resulting from the force is less than if the bones were not present. Because of these effects, pressure fluctuation of the stirrup against the oval window can be 180 times greater than the pressure fluctuation of a sound wave in air before it reaches the ear.

The increased pressure fluctuation against the oval window causes a fluctuating pressure in the fluid inside the cochlea of the inner ear. The fluctuating pressure is sensed by nerve cells along the basilar membrane. Nerves nearest the oval window respond to high-frequency sounds, whereas nerves farther from the window respond to low-frequency sounds. Thus, our ability to distinguish the frequency of a sound depends on the variation in sensitivity of different cells to different frequencies along the basilar membrane. The basilar membrane is only about 3 cm long, yet a normal ear can distinguish sounds that differ in frequency by about 0.3%. This ability to distinguish sounds of slightly different frequency spans the human hearing range from about 20 Hz to 20,000 Hz.

86. The human ear can detect sound waves whose intensity is about 10^{-12} W/m². Where should you be with respect to a stereo speaker that produces sound of intensity 10^{-5} W/m² when 1 m from the speaker if you do not want to listen to the music?
   (a) About 3 m away   (b) About 30 m away
   (c) About 3 km away
   (d) It depends on the size of the speaker.

87. What amplifies the air pressure in the ear?
   (a) Pinna (2 times) and ossicles (30 times)
   (b) Pinna (2 times), auditory canal (3 times), and ossicles (30 times)
   (c) Pinna and auditory canal (2 times), the area change from the eardrum to the oval window (30 times), and the lever action of the ossicles (3 times)

88. Where is the mechanism that allows the ear to distinguish between low-frequency and high-frequency sounds?
   (a) The fluid inside the cochlea of the inner ear
   (b) In the oval window
   (c) In the basilar membrane   (d) Both b and c

89. What frequency difference can the human ear distinguish near 1000 Hz?
   (a) 100 Hz   (b) 10 Hz   (c) 3 Hz
   (d) 1 Hz

90. The threshold for pressure variation of a barely audible sound is about 2×10^{-5} N/m². Which answer below is closest to the pressure variation of a barely audible sound in the cochlea of the inner ear?
   (a) 1×10^{-7} N/m²   (b) 1×10^{-6} N/m²
   (c) 2×10^{-5} N/m²   (d) 4×10^{-3} N/m²
   (e) 4×10^{-2} N/m²

Gases

- Why does a plastic bottle left in a car overnight look crushed on a chilly morning?
- How hard is air pushing on your body?
- How long can the Sun shine?

When you inflate the tires of your bicycle in a warm basement in winter, they tend to look a bit flat when you take the bike outside. The same thing happens to a basketball—you need to pump it up before playing outside on a cold day. An empty plastic bottle left in a car looks crushed on a chilly morning. What do all those phenomena have in common, and how do we explain them?

BE SURE YOU KNOW HOW TO:
- Draw force diagrams (Section 3.1).
- Use Newton's second and third laws to analyze interactions of objects (Section 3.7 and 3.8).
- Use the impulse-momentum principle (Section 6.3).

IN CHAPTER 11, we learned that sound propagates due to the compression and decompression of air. But what exactly is being compressed? To answer this question and the ones above, we need to investigate what makes up a gas and how certain properties of gases can change.

12.1 Structure of matter

When we look at objects that surround us, we do not see their internal structure—water in a cup looks homogeneous, and the cup itself looks like one piece of material. However, they must be made up of *something*. What are the building blocks of matter? To begin to answer this question, consider a simple observational experiment.

Imagine that you dip a cotton ball in rubbing alcohol and wipe it across a piece of paper (**Figure 12.1**). The wet alcohol strip disappears gradually, with the edges of the strip disappearing first. You observe the same behavior when you wipe water or acetone on the paper, except that the water strip disappears more slowly and the acetone strip disappears more quickly. This phenomenon is the same one we observe with wet clothes and puddles as they dry.

Since the alcohol disappeared gradually, it is reasonable to suggest that it is made of "pieces" too small to be seen. If the alcohol were composed of one piece, it would be gone all at once. However, the model of small pieces does not explain *how* the disappearance occurs. Let's try to construct some possible *mechanisms* that explain how the alcohol disappears (these mechanisms would also be applicable to the drying of water or acetone). Three of the many possible mechanisms are described below:

Mechanism 1. The little pieces of liquid move to the *inside* of the paper and are still there, even though the paper looks dry.

Mechanism 2. The air surrounding the paper somehow pulls the liquid pieces out of the paper.

Mechanism 3. The pieces of liquid are moving—they bump into each other and slowly bump each other out of the paper one by one.

All of these mechanisms seem viable. Testing Experiment **Table 12.1** will help us decide if we can rule out any of them. To make predictions based on these three mechanisms, we will assume that the pieces comprising the alcohol have mass.

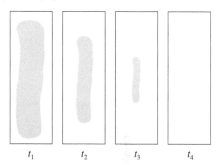

FIGURE 12.1 A disappearing moist strip on a piece of paper.

t_1 t_2 t_3 t_4

TESTING EXPERIMENT TABLE 12.1   Testing various mechanisms for the drying of wet objects   ▶ VIDEO TET 12.1

Testing experiment 1	Prediction	Outcome
Weigh a dry strip of paper on a sensitive scale. Record its mass. Then moisten the paper with alcohol and record the mass of the wet paper. What happens to the mass of the strip when it dries? Moist paper strip 0.0160 g	**Mechanism 1.** If the alcohol pieces go "inside" the paper, then the mass of the paper after it dries should be greater than it was before it was made wet and exactly the same as it was just after it was made wet. **Mechanism 2.** If the air is somehow responsible for removing the alcohol pieces, then after the paper dries, it should have the same mass as it had before it was made wet. **Mechanism 3.** If the moving alcohol pieces bump each other out of the paper, then after the paper dries it should have the same mass as it had before it was made wet.	After the paper dries, its mass is the same as before it was made wet.

Conclusion 1

The alcohol pieces left the paper. Mechanism 1 is disproved by this experiment, but Mechanisms 2 and 3 are consistent with it. This leaves us with only two possible mechanisms.

(CONTINUED)

Testing experiment 2	Prediction	Outcome
Moisten two strips of paper. Place one inside a sealed glass jar attached to a vacuum pump. Place the other strip outside the jar. Pump the air from the jar.	**Mechanism 2.** If the air is somehow responsible for removing the alcohol pieces out of the strip, then the strip inside the vacuum jar with little air should dry more slowly than the strip outside the jar. **Mechanism 3.** If the moving alcohol pieces bump each other out of the paper, then the strip inside the vacuum jar should not dry more slowly than the strip outside the jar.	The paper inside the evacuated jar dries faster.

Conclusion 2

Mechanism 2 is disproved by this experiment. Mechanism 3 is not disproved by it.[1] Now we have only one mechanism that has not been disproved.

Testing experiment 3	Prediction	Outcome
Add a droplet of colored alcohol to a glass of clear alcohol. What happens to the colored droplet?	**Mechanism 3.** If the small pieces of clear alcohol are moving and bumping each other, then they should bump the colored pieces and cause them to spread.	The color slowly spreads throughout the clear alcohol.

Conclusion 3

Mechanism 3 is supported by this experiment.

[1]Note that while Mechanism 3 was not disproved in Experiment 2, the mechanism does not explain why the strip in the vacuum jar actually dried faster.

FIGURE 12.2 Random motion of water particles and pollen granules.

(a) The shaded water particle moves randomly due to collisions with other water particles.

(b) Water molecules collide with the pollen granule and cause it to move in random directions.

Based on these experiments, it is reasonable to assume that alcohol and other liquids are composed of smaller objects, called **particles**, which move randomly in all directions. These particles need empty space between them so that particles of other materials can move between them, as happened in Experiment 3 in Table 12.1. This model of the internal structure of alcohol can be used to explain many other phenomena that we encounter—the way some liquids mix or smells spread. In fact, experiments such as those described in Table 12.1 could have led the Greek philosopher Democritus (460–370 B.C.) to the *atomistic* model. *Atomos* in Greek means *indivisible*. According to Democritus, matter was composed of small, indivisible pieces with different shapes and properties. Democritus also suggested that the pieces were separated by tiny regions of completely empty space, which is a crucial idea as we understand now. Democritus proclaimed, "There is nothing in the world but atoms and empty space."

In 1827, Scottish botanist Robert Brown was observing pollen in a droplet of water through a microscope. To his surprise, the pollen granules were moving randomly, stopping, and then continuing in a different direction after each stop. To explain his observations, Brown used Mechanism 3—that water itself is composed of particles smaller than the granules and these particles move randomly between frequent collisions (Figure 12.2a). The water particles randomly hit the pollen granules from all directions and caused random changes in the position of a granule (Figure 12.2b). This experiment supported the model of water composed of invisible particles that were in continual random motion.

In 1905, Albert Einstein constructed a quantitative model to describe the phenomenon observed by Brown, called *Brownian motion*. He predicted the average distance that a granule of a certain size would move in a given time interval under specific conditions. Later, Jean Perrin conducted experiments whose outcomes matched Einstein's predictions. Physicists consider Perrin's experiments strong support for the *particle structure of matter*.

We now know that **atoms** are the smallest objects that still retain the chemical properties of a particular element (hydrogen, oxygen, carbon, iron, gold, etc.). Atoms of these various elements combine to build solids, liquids, and gases. A **molecule** is a certain combination of atoms that bond together in a particular arrangement. Molecules may consist of two atoms (such as oxygen, O_2, and nitrogen, N_2), three atoms (such as water—two hydrogen atoms and one oxygen atom, H_2O), or many atoms. Protein molecules can consist of thousands of atoms. Atoms are composed of fundamental particles—protons, neutrons, and electrons. We will discuss these fundamental particles later in the book (Chapters 28 and 29). For now, we will use the word *particle* to indicate an object approximately the size of a molecule or smaller. Our investigations and the history of physics bring us to a *particle model* of the internal structure of matter. In this model the matter is made of small particles that move randomly. The space between the particles is completely empty. We will learn later what conditions affect particle motion. We will also learn how these particles interact with each other.

The particle model explains how we can smell things even when we are not near them. Suppose we open a bottle of perfume while standing in the middle of a room. Several minutes later, people all over the room can smell the perfume. According to the particle model, the little particles of perfume leave the bottle and gradually disperse, eventually arriving at our nostrils. Since everyone in the room eventually smells the perfume, the perfume particles must move in all directions. However, it takes time to smell the perfume if you are far away from the bottle. Why is that? Perfume particles leaving the bottle move quickly but collide with air particles along the way and reach your nose only after many collisions.

If air is composed of tiny particles, and those particles have mass, then air must have mass. We can test this hypothesis with another simple experiment. Take two identical rigid metal cylinders—one that has had the air evacuated with a vacuum pump, and the other filled with air. Then weigh them. If the hypothesis that air has mass is correct, the evacuated cylinder should weigh less than the unevacuated cylinder. The cylinder filled with air is indeed heavier than the evacuated cylinder (**Figure 12.3**). However, the difference in masses is small, about 7 g for cylinders of about $6 \times 10^{-3} \, m^3$ (1.5 gallons). Air particle composition can also help us explain the propagation of sound. The propagating disturbances are the regions of air particles closer together or farther apart.

Gases, liquids, and solids

We know from experience that gases are easy to compress, while liquids and solids are almost incompressible (**Figure 12.4**). The particle model helps us explain this difference: we assume that matter in all states is composed of small particles, but the amount of empty space between the particles is different in solids, liquids, and gases. In solids and liquids, the particles are closely packed (almost no empty space between them), while in gases the particles are packed more loosely (lots of empty space).

Gases tend to occupy whatever volume is available. If you take the air that fills a small cylinder and move it to a much larger cylinder, the air also fills the large cylinder (**Figure 12.5a**). In contrast, if we move the liquid filling a small container to a much larger container, the liquid volume remains the same independent of the container's shape (Figure 12.5b). Solids maintain not only their volume but also their shape. To distinguish them from solids, gases and liquids are sometimes called *fluids*. We will learn more about fluid behavior in Chapters 13 and 14.

In order for particles to form a substance, the particles must somehow be attracted to each other. We will learn later (in Chapters 17 and 28) that although the nature of this attractive force between particles is different from their gravitational attraction for each other, there are some similarities between the two types of attraction. For instance,

FIGURE 12.3 Air has mass.

The cylinder with air has more mass than the evacuated cylinder.

FIGURE 12.4 A gas can be compressed, but a liquid cannot.

Gas

Gas is compressed.

Liquid is not compressible.

FIGURE 12.5 Although gases and liquids are both fluids, they do not share all the same properties.

(a) The same gas completely fills a different volume.

(b) The liquid volume remains the same regardless of the container.

in both cases the particles are attracted to each other at a distance, without any direct contact. The behavior of gases indicates that particles in gases must be so far from each other that such attractive forces are very small or even negligible. In liquids, the particles are closer together, and these new attractive forces between them are strong. In solids they are even stronger.

Ideal gas model

From the observed behavior of gases and our explanations based on the particle nature of matter, we can conclude that gases will be the easiest to study, as we can neglect the interactions of the particles at a distance. We start by constructing a simplified model of a gas as a system.

In this model, we assume that the average distance between the particles (molecules or atoms) is much larger than the size of the particles. We assume that the gas particles are point-like objects that only interact with each other and with any surfaces in their containers during collisions (**Figure 12.6**), similar to the impulsive interactions of a billiard ball with a side of a pool table. In our model the particles do not attract each other at a distance the way they do in solids and liquids. We also assume that the motions of these particles obey Newton's laws. Thus, between collisions the particles move in straight lines at constant velocity and only change their velocities during collisions. Altogether, these simplifying assumptions make up the **ideal gas model**.

FIGURE 12.6 Ideal gas model.

> **Ideal gas model** A model of a system in which gas particles are considered point-like and only interact with each other and the walls of their container through collisions. This model also assumes that the particles and their interactions are accurately described using Newton's laws.

Ideal in this context does not mean perfect; it means *simplified*. This is a simplified model with certain assumptions. Whether or not this model can be used to represent a real gas remains to be seen. Only testing experiments can resolve that issue.

How useful is this new model of a gas? As with our previous models of objects (point-like objects, rigid bodies), we will use it to describe and explain known phenomena and then to predict new phenomena. However, so far our model is only qualitative. We need to devise physical quantities to represent the features and behavior of the model. This process will then allow us to construct a mathematical description of an ideal gas and predict its behavior.

REVIEW QUESTION 12.1 Use the particle model to explain how moist objects dry out.

12.2 Pressure, density, and the mass of particles

In this section we will identify several new physical quantities that are useful for describing the properties of the model of an ideal gas. These quantities are also applicable to the properties of real gases, liquids, and solids.

Gas pressure

Imagine you are holding an air-filled balloon. Try to crush it a little bit. You feel the balloon resisting the crushing, as if something inside it pushes back on your fingers. You might think it is the rubber resisting your squeeze, but what makes the rubber bulge out in the same way in all directions? The balloon is filled with air. As we will learn later in this section, the particles of matter (atoms and molecules) are rather small. As air particles move randomly in space, they eventually collide with the solid surfaces of any objects in that space. In each of these collisions, the particle exerts an impulsive force on the object—like a tennis ball hitting a practice wall (see Figure 12.7a). However, when a huge number of particles bombard a solid surface at a constant rate, these collisions collectively exert an approximately constant force on the object (Figure 12.7b). This impulsive force must be what we feel when we are trying to squeeze the balloon. Notice that we have now constructed a *model of a process* that explains how the motion of the particles of gas inside the balloon accounts for the observational evidence—the apparent resistance of the balloon to squeezing. This ideal process that we imagined serves as a mechanism for what we observe.

We can test this process model with a simple experiment. When we blow up a balloon, filling it with air, we should not forget that there is also air outside the balloon. It is this outside air that will help us do the testing experiment described in Testing Experiment Table 12.2.

FIGURE 12.7 Impulsive forces during collisions cause an approximately constant force against a wall.

(a)

Each ball or particle exerts an impulsive force on the wall during a collision.

(b)

Many particles hitting the wall cause a near-constant force.

  12.2   Testing the model of moving gas particles pushing on the surface   VIDEO TET 12.2

Testing experiment	Prediction	Outcome
Place a partially inflated balloon inside a vacuum jar. Seal the jar. What happens to the balloon's shape when you start pumping air out of the jar?	As we remove air particles from outside the balloon, the collisions of particles on the outside of the balloon are less frequent and exert less force on each part of the balloon's outer surface. The collision rate of the particles inside the balloon does not decrease. Therefore, the balloon should expand.	As air outside of the balloon is removed, the balloon expands.

Conclusion

The model of air consisting of moving particles colliding with objects exposed to the air has not been disproved.

In normal situations, an extremely large number of gas particles collide each second with the surface that is in contact with the gas. For example, about 10^{23} particles of air collide each second with each square centimeter of your skin. Making a force diagram for the skin by including these individual particle collision forces is not practical. A different approach is needed.

As we have discussed, although each particle collision is impulsive, the forces are so small and so frequent that the force exerted by the gas on the walls of the container can be modeled as a single constant force. The force also depends on the area of that surface—the bigger the area, the more particles push on it during collisions. Thus, instead of using force to describe gas processes, we use the physical quantity of **pressure**, which we learned about in Chapter 11.

FIGURE 12.8 The quantity pressure.

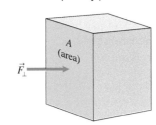

Pressure P Pressure is a physical quantity equal to the magnitude of the perpendicular component of the force, $F_\perp$, that a gas, liquid, or solid exerts on a surface divided by the contact area A over which that force is exerted (see Figure 12.8):

$$P = \frac{F_\perp}{A}$$

(12.1)

$1 \text{ atm} = 1.01 \times 10^5 \text{ N/m}^2 = 14.7 \text{ lb/in.}^2$

$= 760 \text{ mm Hg}$

$1 \text{ Pa} = 1 \text{ N/m}^2 = 1.45 \times 10^{-4} \text{ psi}$

The SI unit of pressure is the pascal (Pa), where $1 \text{ Pa} = 1 \text{ N/m}^2$. In British units, pressure is measured in pounds per square inch (psi). Table 12.3 gives common pressure unit conversions.

Pressure is easy to visualize when you think about two solid surfaces that contact each other. Compare the magnitude of force you exert on soft snow when wearing street shoes versus the magnitude of force that you exert on the snow when wearing snowshoes. The magnitude of force that you exert on the snow is the same, but the corresponding pressure is not. The snowshoes decrease the pressure you exert on the snow, since the force is spread over a much larger area. Similarly, when you decrease the area over which the force is spread, the pressure increases. Scissors and knives increase pressure on a surface because they decrease the area over which they exert a force.

Does a gas exert pressure inside itself, or does it only press on the walls of its container? Let us think about the air in a room. The air particles hitting one wall in the room collectively exert an average force on the wall, and therefore a pressure. However, the air particles in the center of the room are not interacting with the wall at all, at least not for the moment. Yet the air there has a pressure as well. If a table were placed in the center of the room, the air particles would exert an average force, and therefore a pressure, on the top, bottom, and sides of the table. Thus, air has a pressure whether or not a solid object is present.

Measuring pressure

FIGURE 12.9 An aneroid barometer.

Pointer

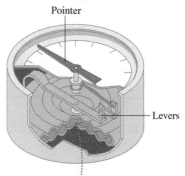

Levers

An evacuated aneroid cell gets thicker or thinner depending on the outside air pressure.

Many instruments are used to measure pressure. An aneroid barometer is used to measure gas pressure directly. The barometer contains a small aneroid cell (Figure 12.8). Inside, the cell has almost no air. A lever is attached to the cell's moveable wall. As the outside air pressure on this wall changes, the cell thickness changes, and the lever causes a pointer needle on the aneroid barometer to turn, indicating the outside air pressure. Measurements show that the pressure of the atmospheric air at sea level is on average 10^5 N/m^2, or 10^5 Pa. This atmospheric pressure defines yet another unit of pressure, called an atmosphere: $1.0 \text{ atm} = 1.0 \times 10^5 \text{ Pa} = 1.0 \times 10^5 \text{ N/m}^2$.

Another way to measure gas pressure is to compare the pressure of a gas in contact with a gauge to the atmospheric pressure, as we did in Chapter 11 when studying sound waves. For example, when you use a tire gauge to measure the air pressure in a car tire, you are comparing the air pressure inside the tire to that of the atmosphere outside the tire. The pressure in the tire is called **gauge pressure**. If the pressure in a container is 1.0 atm, its gauge pressure is zero, because there is no difference between the pressure inside the container and outside of it.

> **Gauge pressure** P_{gauge} Gauge pressure is the difference between the pressure in some container and the atmospheric pressure outside the container:
>
> $$P_{\text{gauge}} = P - P_{\text{atm}} \qquad (12.2)$$
>
> where $P_{\text{atm}} = 1.0 \text{ atm} = 1.0 \times 10^5 \text{ N/m}^2$.

We are continually immersed in a fluid—the gaseous atmosphere. What is the magnitude of the force that the air exerts on one side of your body?

QUANTITATIVE EXERCISE 12.1   **The force that air exerts on your body**

Estimate the total force that air exerts on the front side of your body, assuming that the pressure of the atmosphere is constant.

Air particles colliding with the front of the body cause a net force on it.

Represent mathematically To calculate the total force exerted by air on the front of your body, multiply your front surface area by 10^5 N/m^2, the atmospheric air pressure. To estimate the surface area of the front of your body, model your body as a rectangular box—your height times your width. Assume a height of 1.8 m and a width of 0.3 m. Use Eq. (12.1) to estimate the force:

$$F_{\text{A on F}} = P \cdot A$$

Solve and evaluate $F_{A \text{ on } F} = P \cdot A = (10^5 \text{ N/m}^2)[(1.8 \text{ m})(0.3 \text{ m})]$
$\approx 5 \times 10^4 \text{ N}$. That's $5 \times 10^4 \text{ N} \times 0.22 \text{ lb/N} = 1.1 \times 10^4 \text{ lb}$,
or 10,000 lb. So why aren't you immediately thrown backward?

(Hint: Think about the force that the air behind you is exerting on you as well.)

Try it yourself What is the minimum force that you must exert to lift a 10 cm × 20 cm 0.10-kg rubber sheet that is stuck flat to a tabletop? Assume that there is no air below the sheet.

Answer

$F_{y \text{ on } s} = (P \cdot A) + mg = (10^5 \text{ N/m}^2)[(0.10 \text{ m})(0.20 \text{ m})] + (0.10 \text{ kg})(9.8 \text{ N/kg}) = 2000 \text{ N}$, or about 450 lb. Since it is not really this difficult (although it is not easy!) to lift a rubber sheet off the table, under real circumstances there must be a small amount of air trapped between the rubber sheet and the table, exerting an additional upward force on the sheet.

Density

The quantity *mass* helps describe solid objects that have discrete real boundaries—a person, a car, or a ball. However, air is all around us. We can't see it, and it doesn't have well-defined boundaries. If we used the quantity mass to describe air, would it be the mass of one molecule of air, the mass of the air in a room, the mass of air over the street to a certain height, or something else? It's difficult to visualize air as a macroscopic object. For gases, a much more useful physical quantity is the mass of one unit of volume—volumetric mass density, or simply **density**.

Density measures the mass of one cubic meter of a substance. For example, at sea level and 0 °C, the mass of 1.0 m³ of air is 1.3 kg. We say that the density of air is 1.3 kg/m³. If we had 2.0 m³ of air at sea level, its mass would be 2.6 kg. Its density is still 1.3 kg/m³, since

$$\frac{2.6 \text{ kg}}{2.0 \text{ m}^3} = 1.3 \text{ kg/m}^3$$

Density ρ The density ρ (lowercase Greek letter "rho") of a substance or of an object equals the ratio of the mass m of a volume V of the substance (for example, air or water) divided by that volume V:

$$\rho = \frac{m}{V} \qquad (12.3)$$

The unit of density is kg/m³.

TIP Density is different from mass. Air in a room has a particular mass and density. If you divide the room into two equal parts using a screen, the mass of air in each part will be half the total mass, but the density in each part will remain the same.

QUANTITATIVE EXERCISE 12.2 **The density of a person**

Estimate the average density of a person.

Represent mathematically Assume the following about the person: mass is 80 kg; dimensions are 1.8 m tall, 0.3 m wide, and 0.1 m thick; and volume is $V \approx 1.8 \text{ m} \times 0.3 \text{ m} \times 0.1 \text{ m} = 0.054 \text{ m}^3$.

The person's average density is

$$\rho = \frac{m}{V}$$

Solve and evaluate Substitute the person's mass and volume into the above to get

$$\rho = \frac{80 \text{ kg}}{0.054 \text{ m}^3} = 1500 \text{ kg/m}^3$$

This is the correct unit for density. In Chapter 13, you will learn how to decide whether this is a reasonable estimate.

Try it yourself An iron ball with radius 5.0 cm has a mass of 2.0 kg. Determine the ball's average density. Explain your answer.

Answer

The density of the ball is 3800 kg/m³, much less than the 7860-kg/m³ density of iron. The ball must be hollow.

Mass and size of particles

In 1811, an Italian scientist named Amedeo Avogadro proposed that equal volumes of different types of gas, when at the same temperature and pressure, contain the same number of gas particles. Using Avogadro's hypothesis, scientists could determine the relative masses of different types of particles by comparing the masses of equal volumes of the gases. Presently, scientists use Avogadro's number N_A to indicate the number of atoms or molecules present in 22.4 L $(22.4 \times 10^3 \text{ cm}^3)$ of any gas at $0°$ C and standard atmospheric pressure.

The mass in grams of any substance that has exactly Avogadro's number of particles is equal to the **atomic mass** listed in the periodic table of chemical elements. For example, the atomic mass of molecular hydrogen is 2, that of molecular oxygen is 32, and that of lead is 207; therefore, 2 g of molecular hydrogen, 32 g of molecular oxygen, and 207 g of lead all have the same number of particles—exactly $N_A = 6.02 \times 10^{23}$. This number of particles is called a **mole** (**Figure 12.10**).

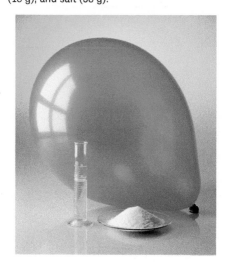

FIGURE 12.10 A mole $(6 \times 10^{23}$ particles) of helium (4 g of helium in the balloon), water (18 g), and salt (58 g).

> **Avogadro's number and the mole** Avogadro's number $N_A = 6.02 \times 10^{23}$ particles is called a mole. The number of particles in a mole is the same for all substances and is the number of particles whose total mass in grams is equal in magnitude to the molecular mass as measured in atomic mass units.

The mass of one mole of particles of any substance is called **molar mass**. One mole of hydrogen (H_2) has a mass of 2.0 g, one mole of oxygen (O_2) has a mass of 32 g, and one mole of lead has a mass of 207 g.

We can now easily determine the mass of a single gas particle of any substance by dividing the molar mass of the substance by 6.02×10^{23}, the number of particles in one mole of the substance:

$$m_{\text{particle}} = \frac{m_{\text{mole}}}{N_A}$$

We find that air (typically 70% N_2 with $m_{N_2} = 28$ g, 29% O_2 with $m_{O_2} = 32$ g, and small percentages of other gases) has a molar mass of about 29 g. Thus, the mass per air particle is approximately

$$m_{\text{air particle}} = \frac{29 \times 10^{-3} \text{ kg}}{6.02 \times 10^{23} \text{ air particles}} = 4.8 \times 10^{-26} \text{ kg/air particle}$$

In addition to a particle's mass, its size is also important. The size of particles was estimated much later than the mass—in the 1860s by Josef Loschmidt. Loschmidt found the linear size of the particles that made up gases, liquids, and solids to be about $d \approx 10^{-9} \text{ m} = 10^{-7} \text{ cm} = 1 \text{ nm}$. Contemporary methods indicate that the size of nitrogen and oxygen particles is about 0.3 nm.

EXAMPLE 12.3   **The average distance between air particles**

What is the average separation between nearby gas particles in the air, and how does it compare to the size of the particles themselves?

Sketch and translate Sketch the situation as a mole of air divided into equal-sized cubes as shown at the right, with one particle located at the center of each cube. The diameter of an individual particle is d; the average distance between them is D.

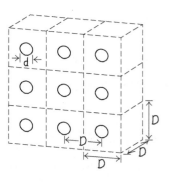

Simplify and diagram Assume standard conditions, with one mole of gas particles occupying a volume of 22.4×10^3 cm³. Now, imagine this volume divided equally between the particles (the cubes in the figure) and each particle being contained in a cube whose volume is D^3.

Represent mathematically We determine the volume of the cube corresponding to each particle and then take the cube root of that volume to estimate the average distance between them.

$$V_{\text{per particle}} = \frac{V_{\text{one mole}}}{N_{\text{particles in a mole}}}$$

$$D = \sqrt[3]{V_{\text{per particle}}} = \sqrt[3]{\frac{V_{\text{one mole}}}{N_{\text{particles in a mole}}}}$$

Solve and evaluate

$$D = \sqrt[3]{\frac{V_{\text{one mole}}}{N_{\text{particles in a mole}}}} = \sqrt[3]{\frac{22.4 \times 10^3 \text{ cm}^3}{6.02 \times 10^{23} \text{ particles}}}$$

$$= \sqrt[3]{3.72 \times 10^{-20} \text{ cm}^3/\text{particle}} = \sqrt[3]{37.2 \times 10^{-21} \text{ cm}^3/\text{particle}}$$

$$= 3.34 \times 10^{-7} \text{ cm}$$

Recall that the average diameter of a single particle of air is 3×10^{-8} cm. Thus, the approximate distance between the particles of air is on average

$$\frac{3.34 \times 10^{-7} \text{ cm}}{3 \times 10^{-8} \text{ cm}} \approx 10 \text{ times the size of the particles } (D/d = 10).$$

This is a lot of empty space. A macroscopic analogy would be two apples separated by about four feet. The empty space between the apples compared to the apple size is similar to the empty space between the particles of air compared to their size.

Try it yourself Estimate the distance between water molecules in liquid water and compare this distance to their dimensions. The density of water is 1.0×10^3 kg/m³. The molar mass of water is 18×10^{-3} kg/mole.

Answer

The volume occupied by one mole of water is about

$$V_{\text{one mole}} \approx \frac{(18 \text{ g})(10^{-3} \text{ kg}/\text{g})}{1000 \text{ kg}/\text{m}^3} = 18 \times 10^{-6} \text{ m}^3$$

The volume occupied by one molecule is

$$V_{\text{one molecule}} \approx \frac{18 \times 10^{-6} \text{ m}^3}{6 \times 10^{23}} = 3 \times 10^{-29} \text{ m}^3$$

The distance between particles is $D \approx 3 \times 10^{-8}$ cm. This is just a little larger than the size of the molecules, which means there is very little space between the water molecules.

REVIEW QUESTION 12.2 The distance between air particles is very small—about 3×10^{-7} cm. How can we say that there is considerable empty space in air?

12.3 Quantitative analysis of an ideal gas

We can use the quantities pressure, density, and the mole to construct a mathematical description of an ideal gas that will allow us to make predictions about new phenomena.

To start, we make a few more simplifying assumptions. First, in addition to modeling the particles (atoms or molecules) as point-like objects whose motion is governed by Newton's laws, assume that the particles do not collide with each other—they only collide with the walls of the container, exerting pressure on the walls (in other words, they move like the model depicted in Figure 12.6 but with no particle collisions). This is a reasonable assumption for a gas of low density. Second, assume that the collisions of particles with the walls are elastic. This makes sense, as the pressure of the gas in a closed container remains constant, which would not happen if the particles' kinetic energy decreased during inelastic collisions.

Now, let's construct a mathematical description of an ideal gas. Imagine the gas inside a cubic container with sides of length L (see **Figure 12.11**). A particle moves at velocity $\vec{v}$ with respect to a vertical wall. When it hits the wall, the wall exerts a force on the particle that causes it to reverse direction. Since the kinetic energy of the particle is the same before and after the collision, the same is true for its speed. The wall exerts a force on the particle, and the particle in turn exerts an equal-magnitude and oppositely directed force on the wall (Newton's third law).

Let's use impulse-momentum ideas to analyze the particle–wall collision. During the collision with the right wall, it exerts a normal force on the particle in the negative x-direction. Before the collision, the particle has a positive x-component of velocity v_{xi}. After the collision, the particle has a negative x-component of velocity v_{xf}. The impulse-momentum equation gives

$$mv_{xi} + F_{\text{W on P } x} \Delta t_{\text{c}} = mv_{xf}$$

FIGURE 12.11 A gas particle bouncing back and forth between the walls of a container.

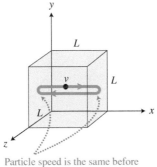

Particle speed is the same before and after these elastic collisions.

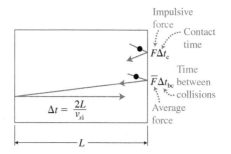

FIGURE 12.12 A method to find the average force of particles colliding with the container wall.

The impulsive force exerted by a particle on the wall during the short contact time interval equals the average force exerted by the particle against the wall during the long time interval between collisions: $F \Delta t_c = \overline{F} \Delta t_{bc}$.

Because the particle's speed is the same before and after the collision, $v_{xf} = -v_{xi}$. In addition, because of Newton's third law, $F_{\text{W on P } x} = -F_{\text{P on W } x}$. Thus,

$$mv_{xi} + (-F_{\text{P on W } x} \Delta t) = -mv_{xi}$$
$$2mv_{xi} = F_{\text{P on W } x} \Delta t_c$$

In the above equation, $F_{\text{P on W } x}$ is the impulsive force that the particle exerts on the wall during the very short time interval Δt that the particle is actually touching and colliding with the wall. How can we determine an average effect of these impulsive collisions in order to determine the pressure of the gas on the wall? We rewrite the right side of the equation as the product of the *average force* exerted by the particle on the wall from one collision to the next, multiplied by the time that *passes between collisions*. Note that this average force is much smaller than the impulsive force, since most of the time the particle is flying through the container and is not in contact with the wall (**Figure 12.12**). However, the time interval between collisions is longer than the impulsive time interval. The product of the big impulsive force and short time interval equals the product of the small average force and the long time interval between collisions.

Looking at the x-component of the motion of the particle, we see that the time interval between collisions with the wall is $\Delta t_{bc} = (2L/v_{xi})$ since the particle must travel a distance $2L$ in the x-direction before colliding with the same wall again. So our equation becomes

$$2mv_{xi} = F_{\text{P on W } x} \Delta t_c = \overline{F}_{\text{P on W } x} \Delta t_{bc} = \overline{F}_{\text{P on W } x} \frac{2L}{v_{xi}}$$

The bar above the force in the last two expressions indicates that it is the *average force* exerted over the time interval between collisions. Multiplying by v_{xi} and dividing by 2 gives

$$mv_{xi}^2 = \overline{F}_{\text{P on W } x} L$$

To relate this microscopic relationship to macroscopic quantities, such as the pressure of the gas and its volume, multiply both sides of the equation by N, the number of particles of gas in the container:

$$Nmv_{xi}^2 = N\overline{F}_{\text{P on W } x} L$$

Because the particles do not all move with the same speed, we also replace the quantity v_{xi}^2 (the square of the x-component of the velocity of an individual particle) with its average value for all the particles:

$$Nm\overline{v}_x^2 = N\overline{F}_{\text{P on W } x} L$$

and then multiply both sides of the equation by L^2/L^2:

$$Nm\overline{v}_x^2 = \left(N\frac{\overline{F}_{\text{P on W } x}}{L^2} \right) L^3$$

The term in parentheses is the pressure exerted by the gas on the wall (the force exerted by all N particles divided by the wall area L^2.) The L^3 outside the bracket is the volume occupied by the gas. Thus, the above equation becomes

$$Nm\overline{v}_x^2 = PV$$

Note that this equation refers only to the motion of the particles along the x-direction. The particles also move in the y- and z-directions. Those equations are

$$Nm\overline{v}_y^2 = PV$$
$$Nm\overline{v}_z^2 = PV$$

To make an equation that simultaneously takes into account the motion of the particles in all three directions, we just add these three equations:

$$Nm(\overline{v}_x^2 + \overline{v}_y^2 + \overline{v}_z^2) = 3PV$$

How does the sum $\overline{v}_x^2 + \overline{v}_y^2 + \overline{v}_z^2$ relate to $\overline{v}^2$, the average of the squared speed of a particle? We can think of the square of the speed of any particle at any time as the sum of components of velocity along the three axes: $v^2 = v_x^2 + v_y^2 + v_z^2$ (using the geometrical argument in **Figure 12.13**). These components are independent of each other; this means that the average of the right side of the above v^2 equation over time is simply the sum of the averages: $\overline{v}^2 = \overline{v}_x^2 + \overline{v}_y^2 + \overline{v}_z^2$. Because all directions are equivalent, the average of the square of a particle's velocity in the x-direction is the same as in any other direction: $\overline{v}_x^2 = \overline{v}_y^2 = \overline{v}_z^2$. Therefore, $\overline{v}_x^2 = \frac{1}{3}\overline{v}^2$. Using this new average of the squared speed, we can write

$$Nm\overline{v}^2 = 3PV$$

The left side of the equation is the average kinetic energy of the particles if we multiply by $\frac{1}{2}$. After multiplying by $\frac{1}{2}$ and rearranging the equation, we get

$$PV = \frac{2}{3}N\left(\frac{1}{2}m\overline{v}^2\right) = \frac{2}{3}N\overline{K}$$

Dividing by V, we get

$$P = \frac{2}{3}\left(\frac{N}{V}\right)\left(\frac{1}{2}m\overline{v}^2\right) \tag{12.4}$$

When the particles have high speed, they (1) hit the walls of the container more frequently and (2) exert a greater force during the collisions. Both factors lead to a greater pressure. Thus, it is the speed squared and not just the speed of the particles that affects the pressure.

Suppose the number of particles N and the average of the squared speed of a particle $\overline{v}^2$ remain the same, but the volume of the box decreases. Then the pressure must increase. This seems reasonable. The smaller the container, the more frequently particles collide with the walls, and the greater the pressure. A unit analysis with a little manipulation indicates that both sides of Eq. (12.4) have the units $kg/(m \cdot s^2)$, so the equation checks out from the point of view of dimensional analysis.

FIGURE 12.13 Using the Pythagorean theorem, we can relate the square of the speed of an individual particle at an arbitrary time to the squares of the x-, y-, and z-components of velocity.

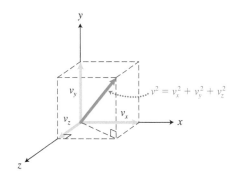

TIP Every time you derive a new equation, ask yourself whether it makes sense. In Eq. (12.4), the pressure is proportional to the average squared speed of the particles. This means that doubling the speed of all the particles quadruples the pressure. Try to explain this dependence before you read on.

QUANTITATIVE EXERCISE 12.4   **How fast do they move?**

Estimate the average speed of air particles at standard conditions, when the air is at atmospheric pressure $(1.0 \times 10^5 \text{ N/m}^2)$, and one mole of the air particles $(6.02 \times 10^{23}$ molecules) occupies 22.4 L or 22.4×10^{-3} m^3. Although air is composed of many types of particles, we will assume that the air particles have an average mass $m_{air} = 4.8 \times 10^{-26}$ kg/particle.

Represent mathematically We can estimate the average speed using Eq. (12.4):

$$P = \frac{2}{3}\left(\frac{N}{V}\right)\left(\frac{1}{2}m_p\overline{v}^2\right)$$

Solve and evaluate Multiply both sides by $3V$, divide by Nm_p, and take the square root to get

$$\overline{v} = \sqrt{\frac{3PV}{Nm_p}}$$

Inserting the appropriate values gives

$$\overline{v} = \sqrt{\frac{3PV}{Nm_p}} = \sqrt{\frac{3(1.0 \times 10^5 \text{ N/m}^2)(22.4 \times 10^{-3} \text{ m}^3)}{(6.02 \times 10^{23})(4.8 \times 10^{-26} \text{ kg})}}$$

$$= 480 \text{ m/s}$$

This number seems very high. We know that it takes several minutes for the smell of perfume to propagate across a room. How could it take so long if the perfume molecules move at hundreds of meters per second (about 1000 mph)? We will find out in the next section.

Try it yourself The pressure in a diver's full oxygen tank is about 4×10^7 N/m^2. What happens to the average speed of the particles when some of the oxygen is used up and the pressure in the tank drops to half this value? What assumptions did you make?

Answer If we assume that the only change is in the number of particles in the tank, then their average speed should stay the same. However, when a gas expands (as this one does when the valve to the tank is opened), the gas inside the tank has to push outward against the environment in order to leave the tank. This requires energy, which lowers the average kinetic energy (and therefore the speed) of the particles inside.

We will learn later that the speed we calculated in Exercise 12.4 has a special name—it is called the *root mean square* speed, or *rms* speed.

Time interval between collisions

We estimated that the average speed of air particles is $v = 480$ m/s, or about 1000 mph. Then why does it take 5 to 10 minutes for the smell of perfume to travel across a room? Remember that we estimated that the average distance between particles in a gas at normal conditions is about $D = 3.3 \times 10^{-7}$ cm. While deriving our mathematical descriptions of the ideal gas model, we assumed that the particles do not collide with each other. Perhaps the gas particles actually do collide with neighboring particles and change direction due to each collision, thus making little progress in crossing the room. More detailed estimates show that particles collide about 10^9 times a second under typical atmospheric conditions. They change direction at each collision, and even though they are moving very fast, their migration from one place to another is very slow.

REVIEW QUESTION 12.3 In the expression $PV = \frac{1}{3}N(m_p \bar{v}^2)$, pressure is proportional to the particle mass times the average squared speed of the particles. Thus, doubling the average speed of particles leads to quadrupling the pressure. Explain why this makes sense.

12.4 Temperature

So far, we have said nothing about the temperature of the gas. Could there be a connection between the temperature of a gas and the average kinetic energy of its particles? We now consider this idea.

As with other physical quantities, temperature can be measured. A common way to measure temperature is with a liquid thermometer. A liquid thermometer consists of a narrow tube connected to a bulb at the bottom (**Figure 12.14**). The bulb and part of the tube are filled with a liquid that expands predictably when heated and shrinks when cooled. To calibrate a thermometer, one marks the height of the liquid at the freezing and boiling conditions of water and then divides this interval by a set number of degrees. On the Celsius scale, 100 °C separates the boiling point and freezing point of water. On the Fahrenheit scale, 180 °F separates these same points (212 °F for boiling and 32 °F for freezing). We'll discuss the Kelvin scale shortly.

FIGURE 12.14 Thermometers calibrated for (a) Celsius scale, (b) Fahrenheit scale, and (c) absolute (Kelvin) scale.

What does temperature really quantify?

We measure the temperature of an object indirectly by measuring the changing volume of a liquid that contacts the object. But what is actually different about an object at higher temperature compared to an object at lower temperature? Consider the following experiments.

Imagine you have kept your inflated beach ball in a cold garage in the winter. When you pick it up, it feels soft and looks a little wrinkly. You then bring it into your warm house, and it suddenly becomes firmer and bouncier. If you take it back to the garage, it becomes soft and wrinkly again. Assuming that the air pressure inside the house and garage is the same, the particles of air inside the ball must exert a higher pressure when the gas inside the ball gets warmer and a lower pressure when the gas inside the ball gets colder. Changing the temperature of the gas therefore seems to change its pressure.

How can we explain that a warmer gas seems to exert a greater pressure on its container? We can hypothesize that the particles in the warm gas exert a greater impulsive force on the container because they are moving faster, and therefore in addition to exerting a greater force during every collision, they collide more frequently with the container walls. Conversely, the particles in a cooler gas exert a smaller force on their container's walls because they are moving more slowly and colliding less frequently. Based on this reasoning, we can hypothesize that the temperature of a gas is related to the speed of the random motion of its particles. Is the temperature of the gas related to any other properties of the gas (the pressure or volume of the gas, or perhaps how many particles comprise the gas and how massive they are)?

Let's do more formal observational experiments. We place three different gases in three containers of different but known volumes. A pressure gauge measures the pressure inside each container. The number N of particles (atoms or molecules) in each container is determined by measuring the mass of the gas m_{gas} in each container and then calculating

$$N = \frac{m_{gas}}{m_{molar\ mass}} N_A$$

where N_A is Avogadro's number. Each container, with known V, P, and N for each gas, is placed first in an ice water bath and then in boiling water, as depicted in **Figure 12.15**. Notice that the volume, pressure, and the number of particles in each container are different, but the temperature of the matter in the three containers is the same. Collected data show the following pattern: independently of the type of gas in a container, the ratio PV/N is identical for all of the gases in the containers when they are at the same temperature:

$$\frac{P_N V_N}{N_N} = \frac{P_O V_O}{N_O} = \frac{P_{He} V_{He}}{N_{He}}$$

The ratio for the gases in the containers at $T_1 = 0\ °C$ is smaller than that for the gases in the containers at $T_2 = 100\ °C$.

From these experiments we can conclude that if you have any amount of a particular type of gas and know its pressure, volume, and the number of particles, then the ratio PV/N only depends on the temperature of the gas. Maybe it equals the temperature? Consider the units of this quantity:

$$\frac{(N/m^2)m^3}{particle} = \frac{N \cdot m}{particle} = J/particle$$

The joule is a unit of energy, not temperature! Perhaps gas particles at the same temperature have the same average energy per particle. Remember that in the ideal gas model, the particles do not have any potential energy between them. This suggests that temperature is related to the average kinetic energy per particle of the gas.

Can we mathematically relate the energy per particle of the gas molecules to the temperature of the gas? The simplest relationship is a direct proportionality to the temperature:

$$\frac{PV}{N} = k_B T \tag{12.5}$$

where k_B is a proportionality constant whose value we need to determine (you'll see shortly why we are using the subscript B). Notice that this relationship immediately leads to a difficulty. The kinetic energy per particle is always a positive number. But in the Celsius and Fahrenheit scales, temperatures can have negative values. So the particle energy cannot be proportional to temperature if measured using either of those scales.

Absolute (Kelvin) temperature scale and the ideal gas law

We need a scale in which the zero point is the lowest possible temperature. That way, all temperatures will be positive. This lowest possible temperature can be found by applying Eq. (12.5) to measurements with a container of gas at two different reference temperatures—the freezing and boiling temperatures of water. The data in **Table 12.4** on the next page were collected when a constant volume metal container with 1 mol

FIGURE 12.15 The ratio PV/N seems to depend only on the gas temperature.

Ice water 0 °C

$N_2(N_N, V_N, P_N)$ $He(N_{He}, V_{He}, P_{He})$

$O_2(N_O, V_O, P_O)$

The ratio of $\frac{PV}{N}$ is the same for different gases if in the same temperature bath.

Boiling water 100 °C

of nitrogen ($N = N_A = 6.02 \times 10^{23}$ particles) was placed in baths at two different temperatures. If we assume that the ratio PV/N is proportional to the absolute temperature of the gas, we can find the coefficient of proportionality.

TABLE 12.4 PV/N **for one mole of gas in a 22.4-L container at two different temperatures**

Conditions in the bath	Pressure	Volume	$\dfrac{PV}{N} = k_B T$
Ice water (T)	$1.013 \times 10^5 \text{ N/m}^2$	$22.42 \times 10^{-3} \text{ m}^3$	$3.773 \times 10^{-21} \text{ J}$
Boiling water ($T + 100$)	$1.384 \times 10^5 \text{ N/m}^2$	$22.42 \times 10^{-3} \text{ m}^3$	$5.154 \times 10^{-21} \text{ J}$

We now have two equations with two unknowns: the constant k_B and the water temperature T on the new scale.

$$3.773 \times 10^{-21} \text{ J} = k_B T$$
$$5.154 \times 10^{-21} \text{ J} = k_B(T + 100)$$

We subtract the first equation from the second to get $k_B = 1.38 \times 10^{-23}$ J/degree. The freezing temperature T of the water is then

$$T = \frac{3.773 \times 10^{-21} \text{ J}}{k_B} = \frac{3.773 \times 10^{-21} \text{ J}}{1.381 \times 10^{-23} \text{ J/degree}} = 273.2 \text{ degrees}$$

On this new scale, water freezes at $T = 273.2$ degrees above the lowest possible temperature. The lowest possible temperature on the new scale is 0 and on the Celsius scale should be $-273.2\,°C$. Considerable modern research has refined this value to $-273.15\,°C$. This temperature scale is called the absolute temperature scale or the Kelvin scale (because it was invented by William Thomson (Lord Kelvin) in 1848). Temperatures are described in kelvin (see the right scale in Figure 12.14). Temperature intervals on the Kelvin scale are the same as on the Celsius scale: a change in temperature of 1 C° is equivalent to a change in temperature of 1 K. Celsius temperatures are related to kelvin as follows:

$$T_K = T_C + 273.15 \tag{12.6}$$

In other words, a temperature of 273.15 K is equivalent to 0 °C.

The constant k_B that we determined using the data in Table 12.4 is called **Boltzmann's constant** after the German physicist Ludwig Boltzmann (1844–1906):

$$k_B = 1.38 \times 10^{-23} \text{ J/K}$$

Temperature conversions

$T_F = (9/5)T_C + 32°$

$T_C = (5/9)(T_F - 32°)$

$T_K = T_C + 273.15°$

TIP Note that Eq. (12.7) implies that when the absolute temperature of the ideal gas is zero, its pressure must be zero.

We can now rewrite Eq. (12.5) using the absolute temperature T and the value of the constant k_B:

$$PV = Nk_B T \tag{12.7}$$

Equation (12.7) is called **the ideal gas law**. It is also commonly used in a slightly different form. Rather than referring to the number of particles that comprise the gas (typically an extremely large number), we refer to the number n of moles of the gas. Since one mole has Avogadro's number of particles, $N = nN_A$. Substituting this into Eq. (12.7), we get

$$PV = nN_A k_B T$$

The product of the two constants $N_A k_B$ is another constant called the **universal gas constant** R. Equation (12.7) then becomes

$$PV = nRT \tag{12.8}$$

where $R = N_A k_B = \left(6.02 \times 10^{23} \dfrac{\text{particles}}{\text{mole}}\right)\left(1.38 \times 10^{-23} \dfrac{\text{J}}{\text{K}}\right) = 8.3 \dfrac{\text{J}}{\text{K} \cdot \text{mole}}$

This is the more common form of the ideal gas law.

Ideal gas law For an ideal gas, the quantities pressure P, volume V, number of particles N, temperature T (in kelvins), and Boltzmann's constant $k_B = 1.38 \times 10^{-23}$ J/K are related in the following way:

$$PV = Nk_B T \qquad (12.7)$$

The law can also be written in terms of the number of moles n, and the universal gas constant $R = 8.3 \dfrac{J}{K \cdot mole}$:

$$PV = nRT \qquad (12.8)$$

Temperature and particle motion

Let us look back at what we have done so far. First, we found the relation between the pressure and volume of an ideal gas and the average kinetic energy of the particles that comprise it, $PV = \frac{2}{3}N\overline{K}$ [Eq. (12.4)]. This was a reasonable finding: the faster the particles move inside the gas, the more often and the harder they hit the walls. The particle mass and the number of them per unit volume N/V also affect the pressure.

Next, we found that the product of the pressure and volume of a gas is related to the temperature of the gas, $PV = Nk_B T$ [Eq. (12.7)]. We can now connect the average kinetic energy of the gas particles to the absolute temperature of the gas. Rearrange Eqs. (12.4) and (12.5) so they each have PV/N on the left side. Insert the average kinetic energy $\overline{K} = \frac{1}{2}m\overline{v}^2$ in the right side of Eq. (12.4) and then set the right sides of the two equations equal to each other to get

$$\overline{K} = \frac{3}{2}k_B T \qquad (12.9)$$

The temperature of a gas is an indication of the average random translational kinetic energy of the particles in the ideal gas. Note that temperature is an indication of not only the particle's speed but also its kinetic energy—the mass of the particle also matters. One implication of this discovery is that when you have a mixture of particles of different gases in one container (for example, in air there are nitrogen molecules, oxygen molecules, carbon dioxide molecules, etc.), the lighter molecules move faster than the heavier ones, though each species of particles has the same average kinetic energy (since each species will have the same temperature once the gas has mixed together thoroughly).

Before we move to an example, let's think more about temperature. Imagine that you have two metal containers with identical gases that have been sitting in the same room for a long time. One container is large and the other one is small. Which one has the higher temperature? Since the average kinetic energy per particle is the same in each container, the temperatures of the two gases are the same. However, the total kinetic energy of the particles in the large container is larger because it contains more particles.

Imagine another scenario: you have two containers with the same type of gas. In one container the gas is hot, and in the other it is cold. What will happen if you mix those two gases together? The faster moving particles of the hot gas will collide with the slower moving particles of the cold gas. If we use the laws of momentum and kinetic energy conservation (assuming that collisions are elastic), we find that following a collision, the faster moving particle on average is moving slower than before the collision, and the slower particle is on average moving faster than before. Eventually, the particles of the two gases have the same average kinetic energy and therefore the same temperature. Physicists say that the gases are in *thermal equilibrium*.

TIP Only when temperature is measured in kelvins can Eq. (12.9) be used to calculate the average kinetic energy of the particles.

Using Eq. (12.9) we can determine the square root of the average speed squared of gas particles $\sqrt{\overline{v^2}}$, called the root-mean-square speed, or **rms speed**:

$$\overline{K} = \tfrac{1}{2}m\overline{v}^2 = \tfrac{3}{2}k_B T$$

$$\Rightarrow \sqrt{\overline{v^2}} = \sqrt{\frac{3k_B T}{m}} \qquad (12.10)$$

Calculating rms speeds of particles allows us to estimate quantities involved in gas processes.

QUANTITATIVE EXERCISE 12.5   **Comparing speeds for the same temperature**

Two different gases (gas 1 has particles of mass m_1, and gas 2 has less massive particles, of mass m_2) are at the same temperature. How do the rms speeds of their particles compare?

Represent mathematically Because the gases are at the same temperature, their particles have the same average kinetic energy (Eq. 12.9):

$$\overline{K} = \tfrac{3}{2}k_B T$$

Therefore, their rms speeds are $\sqrt{\overline{v^2}} = \sqrt{\dfrac{3k_B T}{m}}$.

Solve and evaluate For gas 1, $\sqrt{\overline{v_1^2}} = \sqrt{\dfrac{3k_B T}{m_1}}$; for gas 2, $\sqrt{\overline{v_2^2}} = \sqrt{\dfrac{3k_B T}{m_2}}$. The ratio of the rms speeds is

$$\frac{\sqrt{\overline{v_1^2}}}{\sqrt{\overline{v_2^2}}} = \frac{\sqrt{m_2}}{\sqrt{m_1}}$$

The ratio of the rms speeds is the inverse ratio of the square roots of their masses. Therefore, the molecules of smaller mass move faster at the same temperature.

REVIEW QUESTION 12.4 Ken says that the temperature of a gas measured in kelvins is the average kinetic energy of the gas particles. Do you agree with him? Support your answer with an explanation.

12.5 Testing the ideal gas law

In order to determine if the ideal gas law describes the behavior of real gases, we will use the law to predict the outcomes of some testing experiments. If the predictions match the outcomes, we gain confidence in the ideal gas law. In the experiments below we will keep one of the variables (T, V, or P) constant and predict the relation between the two other variables. Processes in which T, V, or P are constant are called **isoprocesses**. The three types of isoprocesses we will investigate are **isothermal** (T = constant), **isochoric** (V = constant), and **isobaric** (P = constant).

TESTING EXPERIMENT TABLE 12.5 **Does the ideal gas law apply to real gases?**

Testing experiment	Prediction	Outcome
Experiment 1: Isothermal process. n moles of gas are in a variable volume V container that is held in an ice bath at constant 0 °C (273 K) temperature T. How does the pressure of the gas change as we change the volume of the container? We push the piston slowly so that the temperature of the gas is always the same as the ice bath.	According to the ideal gas law $PV = nRT$, during a constant temperature process, the product of PV should remain constant. We predict that as the volume decreases, the pressure will increase so that the product remains constant.	Data collected: $V(\mathrm{m}^3)$ $P(\mathrm{N/m}^2)$; 3.0×10^{-4} 2.0×10^5 ; 6.0×10^{-4} 1.0×10^5 ; 9.0×10^{-4} 0.67×10^5 . The product of volume and pressure remains constant in all experiments.

Testing experiment	Prediction	Outcome
Experiment 2: Isochoric process. n moles of gas and the gas volume V are kept constant. The container is placed in different-temperature baths. How does the gas pressure change as the temperature changes?	According to the ideal gas law $PV = nRT$, during a constant volume process, the ratio $\dfrac{P}{T} = \dfrac{nR}{V}$ should remain constant. We predict that the pressure should increase in proportion to the temperature.	Data collected: $T\,(\text{K})$ $\quad$ $P\,(\text{N/m}^2)$ 300 $\quad$ 1.00×10^5 350 $\quad$ 1.17×10^5 400 $\quad$ 1.33×10^5 The ratio of pressure and temperature is constant in all experiments.
Experiment 3: Isobaric process. n moles of gas and the gas pressure P are held constant, as a frictionless piston in the gas container can move freely up and down keeping the pressure constant. The pressure inside the container is the sum of the constant atmospheric pressure and the pressure exerted by the object on top of the piston. How does the gas volume change as the temperature changes?	According to the ideal gas law $PV = nRT$, during a constant pressure process, the ratio $\dfrac{V}{T} = \dfrac{nR}{P}$ should remain constant. We predict that the volume should increase in proportion to the temperature.	Data collected: $T\,(\text{K})$ $\quad$ $V\,(\text{m}^3)$ 300 $\quad$ 3.0×10^{-4} 350 $\quad$ 3.5×10^{-4} 400 $\quad$ 4.0×10^{-4} The ratio of volume and temperature remains constant in all experiments.

Conclusion

The outcomes of all three experiments are consistent with the predictions.

- In the first experiment, the product of pressure and volume remains constant, as predicted.
- In the second experiment, the pressure increases in direct proportion to the temperature, as predicted.
- In the third experiment, the volume increases in direct proportion to the temperature, as predicted.

The outcomes of the experiments in Testing Experiment Table 12.5 were consistent with predictions based on the ideal gas law, giving us increased confidence that the law applies to real gases in the range of temperatures and pressures used in the experiments (however, we cannot say that we proved it). A summary of gas processes (some of which are not isoprocesses) is provided in Table 12.6, on the next page.

Reflection on the process of construction of knowledge

Let's pause here and reflect on the process through which we arrived at the mathematical version of the ideal gas law. The first step was to construct a simplified model of a system that could represent a real gas—the ideal gas model. This involved making assumptions about the internal structure of gases. This model was based on some observations and also on the knowledge of particle motion and interactions developed earlier—Newton's laws of motion. We used this model to devise a mathematical description of the behavior of gases, the ideal gas law. We then tested its applicability to real gases by using it to predict how macroscopic quantities describing the gas (temperature, pressure, volume, and the amount of gas) would change during specific processes (isothermal, isobaric, and isochoric) and used the ideal gas law to construct equations that described those processes. These predictions were consistent with the outcomes of the new testing experiments.

TABLE 12.6 A summary of ideal gas law processes

Name	Constant quantities	Changing quantities	Equation	Graphical representations
Isothermal	N or n, T	P, V	$PV = \text{constant}$ $P_1V_1 = P_2V_2$	
Isochoric	N or n, V	P, T	$\dfrac{P}{T} = \text{constant}$ $\dfrac{P_1}{T_1} = \dfrac{P_2}{T_2}$	
Isobaric	N or n, P	V, T	$\dfrac{V}{T} = \text{constant}$ $\dfrac{V_1}{T_1} = \dfrac{V_2}{T_2}$	
[No name]	N or n	P, V, T	$\dfrac{PV}{T} = \text{constant}$ $\dfrac{P_1V_1}{T_1} = \dfrac{P_2V_2}{T_2}$	
[No name]		P, V, T, N or n	$\dfrac{PV}{NT} = k_\text{B}$ $\dfrac{PV}{nT} = R$	

The process of constructing the knowledge described above looks relatively smooth and straightforward—observe, simplify, explain, test. However, in real physics, knowledge construction is not that simple and straightforward. For example, the isoprocesses mentioned above were known in physics long before the ideal gas model was constructed. They were discovered in the 17th and 18th centuries and carried the names of the people who discovered them through patterns found in observational experiments. The relation $PV = \text{constant}$ for constant temperature processes is called **Boyle's law** and was discovered experimentally by Robert Boyle in 1662. The relation $V/T = \text{constant}$ for constant pressure processes is called **Charles's law** and was discovered by Jacques Charles in 1787, though the work was published by Joseph Gay-Lussac only in 1802. Gay-Lussac also discovered the relation $P/T = \text{constant}$, now called **Gay-Lussac's law**. When these relations were discovered empirically, there was no explanation for why gases behaved in these ways. The explanations arrived much later via the ideal gas model. The real process of knowledge construction is often more complicated and nonlinear than how it is presented in a textbook. The skills you are learning by constructing knowledge through experimentation will prepare you for those more complicated situations.

Applications of the ideal gas law

The ideal gas law has numerous everyday applications. Try to explain the following phenomena using the ideal gas law. Clearly state your assumptions.

- A sealed, half-full plastic bottle of water shrinks when placed in the refrigerator.
- A refrigerator's door is difficult to open again right after you opened and closed it the first time.
- A bubble of gas expands as it rises from the bottom of a lake.
- Air rushes into your lungs when your diaphragm, a dome-shaped membrane, contracts.

Below we consider in greater detail two of these phenomena.

Breathing During inhalation, our lungs absorb oxygen from the air, and during exhalation, they release carbon dioxide, a metabolic waste product. Yet the lungs have no muscle to push air in or out. The muscle that makes inhaling and exhaling possible, the diaphragm, is not part of the lungs. The diaphragm is a large, dome-shaped muscle that separates the rib cage from the abdominal cavity (see **Figure 12.16**).

The diaphragm works like a bellows. As the diaphragm contracts and moves down from the base of the ribs, the volume of the chest cavity and lungs increases. If we assume that for a brief instant both the temperature and the number of particles are constant (an isothermal process),

$$P = \frac{nRT}{V}$$

then the pressure in the cavity and in the lungs will decrease. The pressure of the outside air is greater than that of the inside. Because of this, the outside air at normal atmospheric pressure enters the mouth or nostrils and fills the lungs with fresh new air (an increase $\Delta n > 0$ of the amount of air). During exhalation, the opposite occurs. The diaphragm relaxes and rises, decreasing the volume of the chest cavity and the lungs, thus increasing the pressure inside (again assuming that for a brief instant both the temperature and the number of particles are constant),

$$P = \frac{nRT}{V}$$

Because the inside pressure is higher than the outside pressure, air is then forced out of the lungs.

Your water bottle on an airplane The behavior of an empty plastic water bottle on an airplane is another example of an isothermal process.

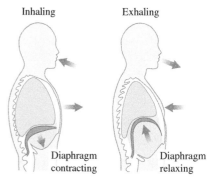

FIGURE 12.16 The diaphragm is a large, dome-shaped muscle that separates the rib cage from the abdominal cavity. Relaxing or contracting the diaphragm changes the volume of the lungs and chest cavity.

Inhaling   Exhaling

Diaphragm contracting   Diaphragm relaxing

Diaphragm contraction and relaxation cause air to enter and leave the lungs.

CONCEPTUAL EXERCISE 12.6   **A shrunk bottle**

On a flight from New York to Los Angeles you drink all of a bottle of water, close the cap, and then store it in your seat pocket. You take it out right after landing. The bottle has changed shape; it looks like someone crushed it. How can we explain this shape change?

Sketch and translate The bottle's volume decreased, though the temperature in the cabin didn't change much during the flight.

Simplify and diagram If we assume the bottle was perfectly sealed, and the temperature of the gas inside the bottle remained constant, then we can model the process as an isothermal process, represented graphically in the figure below. Although the cabin is pressurized, at higher elevations air pressure and air density inside the cabin are less than at lower elevations. You closed the bottle when it was filled with low-density air at high elevation. As the plane descended, the air density and pressure inside the cabin increased. More air particles

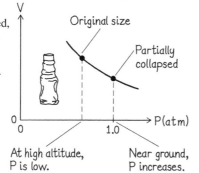

At high altitude, P is low.   Near ground, P increases.

were hitting the outside walls of the empty bottle than were hitting the inside. The higher pressure from outside partially crushed the bottle. If you open the cap after landing, the bottle will pop back to its original shape.

- -

Try it yourself A different process is represented on a graph at right. Describe the process in words. Assume the mass to be constant.

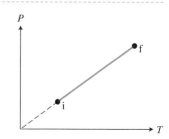

Answer

The graph represents an isochoric process. The graph line, if extended, passes through the origin; thus the pressure is directly proportional to temperature. The gas was in a sealed container. The gas container was first placed in a water bath at low temperature and then transferred to a water bath at high temperature. The volume of the gas is constant. The pressure increases as the particles move faster and collide with the walls of the container more often.

Comparing different gases

How does the type of gas that is in a container affect the pressure of the gas? Consider the next Conceptual Exercise.

CONCEPTUAL EXERCISE 12.7   Analyzing two types of gas

You have two containers, both with pistons of equal mass that can move up and down depending on the pressure of the gas below. Each container has a nozzle that allows you to add gases to the containers. Container 1 holds nitrogen (molar mass M_1); we label its volume V_1. Container 2 (V_2) holds helium (molar mass $M_2 < M_1$). The volumes are the same ($V_1 = V_2$), and the containers sit in the same room. What physical quantities describing the gases inside the two containers can we compare?

Sketch and translate We sketch the situation in the figure below. Quantities that we can try to compare between the two containers are the following:

(a) the pressure inside the containers
(b) the average kinetic energy of a particle of each type of gas
(c) the mass of individual gas particles
(d) the rms speed of the particles in each container
(e) the number of particles in each container
(f) the mass of the gas in each container
(g) the density of the gas in the containers

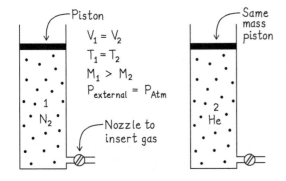

Simplify and diagram Assume that the ideal gas model accurately describes the gases.

(a) Consider the pressures inside the containers by analyzing the identical movable pistons above the gases. Force diagrams for both pistons are shown at above right. Earth exerts a downward force $\vec{F}_{\text{E on P}}$ on the piston, the atmospheric gas above a piston pushes down on the piston $\vec{F}_{\text{Atm on P}}$, and the gases inside the containers push up on their pistons $\vec{F}_{\text{N}_2\text{ on P}}$ or $\vec{F}_{\text{He on P}}$. As the pistons are not accelerating, the net force exerted on each is zero. Because the downward forces exerted by Earth and by the outside atmospheric air on each piston are the same, the upward forces exerted by the gases in each container on the piston

must also be the same. Since the surface areas of the two pistons are the same, the pressure of the gas inside each container must be the same, $P_{\text{N}_2} = P_{\text{He}}$.

The downward forces on each piston are the same, so the upward force is also the same.

(b) Since the temperature of a gas depends only on the average kinetic energy of the particles, and the temperature of the two gases is the same, the average kinetic energy of each particle is the same:

$$\frac{3}{2}k_\text{B}T = \frac{1}{2}m_{\text{N}_2\text{ in 1}}\overline{v}^2_{\text{N}_2\text{ in 1}} = \frac{1}{2}m_{\text{He in 2}}\overline{v}^2_{\text{He in 2}}$$

(c) and (d) Since the N_2 particles in gas 1 have a higher molar mass than the He particles in gas 2, the N_2 molecular mass $m = M_1/N_\text{A}$ is also the higher mass ($m_{\text{N}_2\text{ in 1}} > m_{\text{He in 2}}$). Using this result and that from (b), we find that $\overline{v}^2_{\text{N}_2\text{ in 1}} < \overline{v}^2_{\text{He in 2}}$. The more massive particles move slower.

(e) The pressure, volume, and temperature are the same for both gases. Thus, according to the ideal gas law, both gases have the same number of particles and the same number of moles of gas in their containers ($PV/k_\text{B}T = N$; $PV/RT = n$).

(f) and (g) Since the mass of a nitrogen molecule is greater than the mass of a helium atom, and there are equal numbers of particles in each container, the total mass of the gas in the nitrogen container must be greater than the total mass of the gas in the helium container ($M_{\text{N}_2\text{ in 1}} > M_{\text{He in 2}}$). Since the volumes of the containers are equal, the density $\rho = M/V$ of the gas in the nitrogen container must be greater than that in the helium container ($\rho_{\text{N}_2\text{ in 1}} > \rho_{\text{He in 2}}$).

- -

Try it yourself Why are the gases at the same temperature?

Answer

Both containers sit in the same room temperature environment.

REVIEW QUESTION 12.5 What is the difference between the following two equations: $PV = \frac{1}{3}Nm\overline{v}^2$ and $PV = nRT$?

12.6 Speed distribution of particles

In our previous analysis, we defined the rms speed of gas atoms or molecules:

$$v_{rms} = \sqrt{\overline{v^2}} = \sqrt{\frac{3k_B T}{m}} \tag{12.10}$$

We found that for air molecules at room temperature, the rms speed was about 500 m/s. When we derived relationships such as these, we assumed for simplicity that gas particles do not collide with each other, just with the walls of a container. However, we know that if gas particles did not collide, the smells of food and perfume would spread at hundreds of meters per second—almost instantly. The smells spread slowly, so the particles must be colliding. What happens if we no longer ignore collisions of particles with each other?

Maxwell speed distribution

In 1860 James Clerk Maxwell included the collisions of the particles in his calculations involving an ideal gas. This inclusion led to the following prediction: at a particular temperature, the collisions of gas particles with each other cause a very specific distribution of speeds. When we were deriving Eq. (12.4) we assumed that the speeds of the particles were different, but we did not have any idea of why they were different. Maxwell's work explained this variability of speeds by the collisions of the particles with each other. Consider the red H_2 line (hydrogen atoms) in **Figure 12.17**. On the vertical axis, we plot the percentage of particles that have a particular speed. According to Maxwell, a certain percentage of the particles should have speeds around 1000 m/s, but the most around 1500 m/s, and fewer at higher speeds. Very few particles have extremely low or extremely high speeds. Most should have intermediate speeds. The most probable speed is at the highest point on the curve in Figure 12.17. Surprisingly, the rms speed of the particles at a particular temperature is the same as that predicted by the model with no collisions—Eq. (12.10).

Maxwell's distribution of molecular speeds allows us to explain some very interesting observational evidence. You know that our atmosphere contains molecules of oxygen, nitrogen, carbon dioxide, water, and so forth, but it has almost no hydrogen molecules. Why is there so little free hydrogen in the Earth's atmosphere? Recall from Chapter 7 that every planet has a specific value of escape speed—the speed that an object needs to have to escape the gravitational pull of that planet. For Earth, this speed is about 11 km/s = 11,000 m/s. In Quantitative Exercise 12.4 we estimated the rms speed of an air molecule to be about 480 m/s. The most probable molecular speed according to Maxwell's distribution is a little smaller. We also know that the lighter the molecules, the faster they move at the same temperature. Assuming that a hydrogen molecule is about 15 times less massive than an average air molecule, we can say that its rms speed at the same temperature should be about 4 times higher (see the reasoning in Quantitative Exercise 12.5), but it is still much smaller than the escape speed. Based on this estimate, we should have an atmosphere full of hydrogen molecules. What is wrong with our reasoning?

We did not take into account the *shape* of Maxwell's distribution of molecular speeds (Figure 12.17). The tail of the distribution shows that there are molecules with a speed that is much faster than the most probable speed. These molecules might escape. As soon as the fastest molecules leave, more molecules, colliding with other molecules, might acquire the speeds in the tail of the distribution and leave, too. This process will repeat until almost all light molecules are gone. Is this reasoning correct? Surely there are some oxygen or nitrogen molecules that have very high speeds—why don't they eventually escape? It turns out that the answer lies in the word *probability*. The likelihood (or probability) of the escape of a hydrogen molecule at 10 °C (50 °F) is 300 orders of magnitude higher than that for the molecules of oxygen or nitrogen. It is the interplay between escape speed (gravitational interaction) and the Maxwell distribution of molecular speeds (gas processes) during the 4.6-billion-year history of our solar system that determines what kind of atmosphere a particular planet has.

FIGURE 12.17 The Maxwell particle speed distributions at a particular temperature for two gases.

Limitations of the ideal gas law

Isoprocesses were discovered empirically with experiments long before the model of an ideal gas was constructed. However, the fact that the ideal gas law predicts those empirical results is evidence that the law itself and the ideal gas model on which the law is based can be applied to real gases in certain conditions (although real gases are made of molecules that have internal structure and the ideal gas model assumes point-like particles). Nevertheless, the ideal gas law has limitations.

For real gases such as air, the measurements of pressure and volume at the conditions of normal pressure (1.0×10^5 N/m^2, as elsewhere) and temperature (room temperature) are consistent with predictions of the ideal gas law. But at very high pressures or very low temperatures, real measurements differ from those predictions (for example, the ideal gas law does not predict that one can turn a gas into a liquid). The ideal gas law describes gases accurately only over certain temperature and pressure ranges.

REVIEW QUESTION 12.6 Why do we need the term "probability" when we explain the absence of free hydrogen in Earth's atmosphere?

12.7 Skills for analyzing processes using the ideal gas law

In this section, we adapt our problem-solving strategy to analyze gas processes. A general strategy for analyzing such processes is described on the left side of the table in Example 12.8 and illustrated on the right side for a specific process.

PROBLEM-SOLVING
STRATEGY 12.1   ⟩ Applying the ideal gas law

EXAMPLE 12.8   Scuba diver returns to the surface

The pressure of the air in a scuba diver's lungs when she is 15 m under the water surface is 2.5×10^5 N/m^2 (equal to the pressure of the water at that depth), and the air occupies a volume of 4.8 L. Determine the volume of the air in the diver's lungs when she reaches the surface, where the pressure is 1.0×10^5 N/m^2.

- -

The gas inside the diver's lungs is the system. The initial state of the system is when the diver is underwater; the final state is when she is at the surface. As the diver swims upward, the pressure of the system decreases and its volume increases.

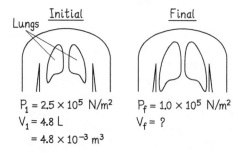

Sketch and translate

- Sketch the process. Choose a system and describe the initial and final states of the system.
- Label the knowns and the unknowns on the sketch.
- Describe the process (a word description of the changes in the system between the initial and final states) in terms of macroscopic quantities (pressure, volume, temperature, moles of gas).

Simplify and diagram

- Decide if the system can be modeled as an ideal gas.
- Decide which macroscopic quantities remain constant and which do not.
- If helpful, draw P vs. V, P vs. T, and/or V vs. T graphs to represent the process.

Represent mathematically

Use your sketch or the graphs to help construct a mathematical description—some version of the ideal gas law.

Solve and evaluate

- Solve for the unknowns.
- Evaluate the answer: is it reasonable? (For example, evaluate the magnitude of the answer, its units, and how the solution behaves in limiting cases.)

Assume that the gas inside the lungs can be modeled as an ideal gas. Assume that the diver does not exhale, which means that the moles of gas remain constant. Assume the temperature of the gas is constant at body temperature. A pressure-versus-volume graph for the process is shown at right.

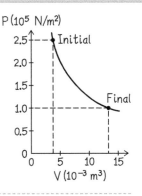

$$P_i V_i = P_f V_f \implies V_f = \frac{P_i V_i}{P_f}$$

$$V_f = \frac{P_i V_i}{P_f} = \frac{(2.5 \times 10^5 \, \text{N/m}^2)(4.8 \times 10^{-3} \, \text{m}^3)}{(1.0 \times 10^5 \, \text{N/m}^2)} = 12 \times 10^{-3} \, \text{m}^3$$

We found the lung volume to be 12 L, much larger than seems possible. Is the answer realistic? Remember that we assumed that the mass of the gas inside remains constant—the diver does not exhale. As we see from the solution, it is important for divers to exhale as they are ascending and the gas expands. If not, they can suffer severe internal damage.

Try it yourself How many moles of gas are in the diver's lungs and how many should she exhale so the final volume is only 6 L instead of 12? Assume that the gas temperature is 37 °C.

Answer   0.47 moles; 0.23 moles.

TIP Notice that the graph lines for isobaric processes (constant pressure) pass through the origin in V-versus-T graphs and those for isochoric (constant volume) processes pass through the origin in P-versus-T graphs.

EXAMPLE 12.9   **Further investigation of an airplane bottle**

In Conceptual Exercise 12.6 you qualitatively investigated the behavior of a plastic bottle on a long flight. On your next long flight, you decide to investigate the shrinking bottle phenomenon quantitatively. In the middle of the flight you open a bottle (Figure a) and then close it again. When you land, you observe that the bottle shrinks (Figure b). You take the closed bottle home, turn it upside down into a bowl of water (Figure c), and open it. The bottle expands to its normal size, letting 150 mL of water into the bottle (Figure d). You measure the total volume of the bottle to be 550 mL. Use this information to estimate the pressure in the airplane cabin.

(a)

(b)

(c)

(d)

(CONTINUED)

Sketch and translate The air inside the bottle is the system. The initial state is when you open the bottle on the cruising airplane. The air fills the whole bottle ($V_1 = 550$ mL) at the pressure inside the airplane (P_1). When the plane lands, the volume decreases to V_L (the bottle is crushed) and the pressure inside the bottle increases to P_L. The L state is an intermediate state in the process. The final state is when you open the bottle at home and water is let in. The volume of air in the bottle is now $V_2 = 550$ mL $-$ 150 mL $= 400$ mL. The pressure inside is P_2. We need to determine the pressure P_1. We already know why the bottle shrinks when the plane lands. But why does the water enter the bottle? Even though the atmospheric air crushes the bottle, the pressure inside is probably still less than atmospheric due to the semiflexible walls of the bottle that tend to return the bottle to the original shape. When the bottle is opened under the water, the pressure difference pushes water into the bottle. In addition, the bottle expands, returning to its original shape, which helps let water into the bottle. This process stops when the pressure of the air in the bottle is equal to the atmospheric pressure outside the bottle. Thus $P_2 = P_{atm}$.

Simplify and diagram We assume that the temperature in the airplane was the same as at your home ($T_1 = T_2$). Thus the process is isothermal. We can represent the process that the gas in the bottle went through with the P-versus-V graph at right.

Represent mathematically We can now express the graphical representation with an equation for an isothermal process: $P_1V_1 = P_2V_2$. Dividing both sides by V_1 we get $P_1 = P_2 \dfrac{V_2}{V_1}$. Now we see that we do not need to convert ml of volume into m^3 because the units for volume cancel.

Solve and evaluate

$$P_1 = P_2 \frac{V_2}{V_1} = (1.0 \times 10^5 \text{ N/m}^2)\frac{400 \text{ mL}}{550 \text{ mL}} = 0.73 \times 10^5 \text{ N/m}^2$$

The result seems reasonable: the pressure during flight is lower than the atmospheric pressure at the ground.

Try it yourself Your sister happens to measure the temperature while you are performing the experiments on the airplane and at home. Her data are $T_{flight} = 80\ °F$ and $T_{home} = 67\ °F$. Use her data to improve your previous estimation of the pressure during the flight.

Answer   0.75 × 10⁵ N/m².

REVIEW QUESTION 12.7 Why is it helpful to know whether the mass of the gas is constant during a particular process?

12.8 Thermal energy, the Sun, and diffusion

We've seen that the ideal gas model can be applied to many real-world situations as long the gas particles do not interact with each other (they are far apart and/or the pressure and temperature are not extreme). In this section we will look at two more everyday phenomena: the Sun shining and gas diffusion. In order to get started, we need to quantify something for gases that we've only discussed qualitatively in the textbook so far: thermal energy.

We learned in Chapter 7 that thermal energy is the component of the internal energy of a system that is associated with temperature. We know from Eq. (12.9) that the temperature of an ideal gas is determined by the average kinetic energy $\bar{K}$ of its randomly moving particles. We can now therefore define the gas's **internal thermal energy** or just **thermal energy**, U_{th}, as the sum of all the individual particles' kinetic energies. This is equal to $N\bar{K}$, where N is the number of particles. Using Eq. (12.9), this gives

$$U_{th} = N\left(\tfrac{3}{2}k_B T\right)$$

How long can the Sun shine?

The thermal energy of gases can be used to help analyze many kinds of systems, such as automobile engines, Earth's atmosphere, and the Sun. The mass of the Sun is about 2×10^{30} kg, its radius is about 7×10^8 m, and the temperature of the Sun's surface is about 6×10^3 K. Its core temperature is about 10^7 K. Every second, the Sun radiates about 4×10^{26} J as visible light and other forms of radiation. How long can our Sun shine if its particles possess *only* thermal energy?

We first need to determine how much thermal energy the Sun possesses. Then we can divide this amount of energy by the amount it loses every second from the emission of visible light and other forms of radiation to find the number of seconds the Sun can shine if its thermal energy is the only source of this radiative energy. Although in reality there are some complications with modeling the material of the Sun as an ideal gas, we will do so in this case for simplicity.

To estimate the thermal energy of the Sun, we need to know the number of particles and the average temperature of these particles. The Sun consists mostly of hydrogen atoms. To find the number of hydrogen atoms in the Sun, we estimate the number of moles of hydrogen gas and then multiply by the number of particles in one mole (note that 1 mole of hydrogen has a mass of 1 g = 10^{-3} kg):

$$N = \frac{m_{Sun}}{m_{hydrogen\ atom}} = \frac{m_{Sun}}{M_{molar\ mass\ hydrogen}/N_A}$$

$$= \frac{(2 \times 10^{30}\,kg)}{(10^{-3}\,kg/mole)/(6 \times 10^{23}\,particles/mole)}$$

$$= \frac{(2 \times 10^{30}\,kg)}{(1.7 \times 10^{-27}\,kg/particle)} = 12 \times 10^{56}\,particles$$

At the very high temperatures within the Sun, each hydrogen atom separates into two smaller particles called an electron and a proton—subjects of later study. This separation of hydrogen atoms doubles the number of particles to 24×10^{56}. We can show that the average distance between these particles is much greater than their dimensions, and thus we can apply the ideal gas model to the Sun's material.

These particles do not spread out into space or collapse in toward the center of the Sun because two competing forces remain in balance. All parts of the Sun exert a gravitational force on all other parts. If we select a small volume inside the Sun as the system of interest and add the forces that all other particles of the Sun exert on the system (see **Figure 12.18**), the net gravitational force points toward the center of the Sun. The second force exerted on the system is the pressure force exerted by other particles on the system. This force points outward and balances the gravitational force. As long as these forces balance each other, the Sun is in equilibrium and does not expand or collapse.

Now consider the total thermal energy of the Sun's particles for two extreme cases. A lower bound for the Sun's thermal energy assumes its temperature throughout equals its surface temperature. An upper bound for the Sun's thermal energy assumes its temperature throughout equals its core temperature.

$$U_{th\ min} = \tfrac{3}{2}Nk_B T_{min}$$

$$= \tfrac{3}{2}(24 \times 10^{56}\,particles)(1.38 \times 10^{-23}\,J/K)(6 \times 10^3\,K) = 3 \times 10^{38}\,J$$

$$U_{th\ max} = \tfrac{3}{2}Nk_B T_{max}$$

$$= \tfrac{3}{2}(24 \times 10^{56}\,particles)(1.38 \times 10^{-23}\,J/K)(10^7\,K) = 5 \times 10^{41}\,J$$

Both numbers are so large that they imply that the Sun will shine for a long time. The life expectancy of the Sun in seconds (based on the thermal energy alone) equals the total thermal energy divided by the energy radiated per second, $L = 4 \times 10^{26}$ J/s. We can convert our result to years, noting that there are about 3×10^7 s in 1 year.

FIGURE 12.18 Gas in the Sun is in equilibrium due to two forces.

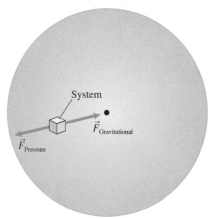

This small volume of gas (the system) is kept in equilibrium by the gravitational force and pressure force exerted on it by other particles in the Sun.

System

$\vec{F}_{Gravitational}$

$\vec{F}_{Pressure}$

$$\Delta t_{min} = \frac{E_{min}}{L} = \frac{3 \times 10^{38} \text{ J}}{4 \times 10^{26} \text{ J/s}} = 7.5 \times 10^{11} \text{ s} = \frac{7.5 \times 10^{11} \text{ s}}{3 \times 10^7 \text{ s/year}} = 2 \times 10^4 \text{ years}$$

$$\Delta t_{max} = \frac{E_{max}}{L} = \frac{5 \times 10^{41} \text{ J}}{4 \times 10^{26} \text{ J/s}} = 1.2 \times 10^{15} \text{ s} = \frac{1.2 \times 10^{15} \text{ s}}{3 \times 10^7 \text{ s/year}} = 4 \times 10^7 \text{ years}$$

The maximum possible lifetime for the Sun if it simply converts its thermal energy into light and other forms of radiation is 4×10^7 years, or 40 million years. Yet according to astronomical knowledge of stars and our solar system, the Sun has been shining in the same way for over 4 billion years. Thus this simple estimate suggests that either the Sun's material cannot be modeled as an ideal gas or some source of energy other than the thermal energy of the Sun's particles has been supporting its existence for a time interval equal to the age of Earth. As the ideal gas model explains many phenomena occurring on the Sun, it means that there must be another energy source within the Sun that far exceeds the thermal energy present. (We will learn about it in Chapter 29.)

Diffusion

Imagine that you fill a balloon with air and leave it for one hour. The balloon looks the same; it did not change. You fill another balloon with carbon dioxide (CO_2); after one hour the balloon is significantly smaller. Why do the balloons look different?

One possible explanation is that the balloon rubber interacts differently with different molecules. If we hypothesize that CO_2 molecules can easily pass through it, but other molecules in air cannot, then we can create a simple model of the process. To indicate that CO_2 molecules interact differently with the balloon than other air molecules, we will represent CO_2 molecules with red dots and other air molecules (mostly nitrogen and oxygen) with blue dots (**Figure 12.19**). Blue dots cannot pass through the wall. They bounce off the wall either immediately or after some time. This happens to air molecules inside the first balloon (Experiment 1) and outside both balloons. Red dots inside the balloon in Experiment 2 move randomly inside the balloon and slowly migrate through the wall to the outer surface, where their concentration is smaller. They then leave the surface, traveling randomly in all directions.

In our model, CO_2 molecules outside the balloon can as easily get into the balloon as get out of it, so why does CO_2 leak out? Remember that the concentration of CO_2 molecules inside the balloon is much higher than that outside. Many more CO_2 molecules leave the balloon than come back in (Experiment 2), while the air molecules from outside cannot get into the balloon at all. This imbalance in the behavior of different molecules explains why the balloon deflates.

We can test this explanation with another simple experiment. Place an air-filled balloon into a container filled with CO_2. If our explanation is correct, then the CO_2 molecules from outside the balloon should pass through the rubber into the balloon. Air molecules cannot get out of the balloon (Experiment 3). The balloon should therefore expand and eventually burst. This is exactly what happens!

The process of molecules moving, due to their random motion, from a region of higher concentration to a region of lower concentration is called **diffusion**. The explanation of diffusion follows from the ideal gas model. Diffusion plays an important role in biological processes. Oxygen is carried by hemoglobin in the blood from the heart to the tiny capillary vessels spread throughout the body. Oxygen diffuses from the oxygen-rich blood inside the capillaries to the oxygen-poor cells near the capillaries. Some of these cells may be muscle fibers. Since a muscle fiber needs oxygen to twitch, the action of muscles may be limited by the rate of oxygen diffusion into the fiber. Thus, diffusion limits the rate of some processes in our bodies.

FIGURE 12.19 Testing the model of air molecule movement through semipermeable balloons.

Experiment 1: Air balloon in air

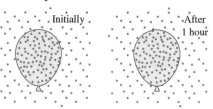
Initially · · · · · After 1 hour

Experiment 2: CO_2 balloon in air

Experiment 3: Air balloon in CO_2

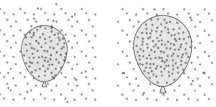

REVIEW QUESTION 12.8 How do we know that the Sun's thermal energy is not the main source of the light energy it produces?

Summary

Pressure P The perpendicular component of the force F that another solid object or a gas or liquid exerts perpendicular to a surface of area A divided by that area. (Section 12.2)

$$P = \frac{F_\perp}{A}$$ Eq. (12.1)

Density ρ The mass m of a substance divided by the volume V that the substance occupies. (Section 12.2)

$m' = 8m$
$V' = 8V$

$\rho = \dfrac{m}{V}$ $\rho' = \dfrac{8m}{8V} = \rho$

$$\rho = \frac{m}{V}$$ Eq. (12.3)

Moles n and **Avogadro's number** N_A A mole of any type of particle equals Avogadro's number of that type of particle. Molar mass is the mass of one mole. (Section 12.2)

≈ 56 g iron
(1 mole)

$$N_A = 6.02 \times 10^{23} \text{ particles/mole}$$

Temperature T and **temperature scales** Temperature measures how hot or cool a substance is. When measured in kelvins, the temperature is directly proportional to the average random kinetic energy of a particle in that substance. (Section 12.4)

The average kinetic energy per atom or molecule in a gas is

$$\overline{K} = \frac{3}{2}k_B T$$

$T_F = (9/5)T_C + 32°$
$T_C = (5/9)(T_F - 32°)$
$T_K = T_C + 273.15°$ Eq. (12.6)

Thermal energy of an ideal gas U_{th} The sum of average kinetic energies of the particles of the ideal gas. (Section 12.8)

N particles in the system

$$U_{th} = N\left(\frac{3}{2}k_B T\right)$$

Ideal gas model and ideal gas law The ideal gas model is a simplified model of gas in which atoms/molecules are considered to be point-like objects that obey Newton's laws. They only interact with each other and with the walls of the container during collisions exerting pressure. The ideal gas law relates the macroscopic quantities of such a gas. (Sections 12.1, 12.3, 12.4, and 12.5)

$$P = \frac{2}{3}\left(\frac{N}{V}\right)\left(\frac{1}{2}m\overline{v}^2\right)$$ Eq. (12.4)

$$PV = Nk_B T$$ Eq. (12.7)

$$PV = nRT$$ Eq. (12.8)

Assume that N, n, and P are constant. Then

$$\frac{V_1}{T_1} = \frac{V_2}{T_2}$$ (Isobaric)

Assume that N, n, and V are constant. Then

$$\frac{P_1}{T_1} = \frac{P_2}{T_2}$$ (Isochoric)

Assume that N, n, and T are constant. Then

$$P_1 V_1 = P_2 V_2$$ (Isothermal)

Questions

Multiple Choice Questions

1. What experimental evidence rejects the explanation that wet clothes become dry because the air absorbs the water?
   (a) The clothes dry faster if you blow air across them.
   (b) They do not dry if you put the wet clothes in a plastic bag.
   (c) The clothes dry faster under a vacuum jar with the air pumped out.

2. What is the difference between the words particle, molecule, and atom?
   (a) A particle is bigger than a molecule or an atom.
   (b) Particles can be microscopic and macroscopic, while atoms and molecules are only microscopic.
   (c) Molecules are made of atoms; both can be called particles.
   (d) All are correct.
   (e) Both b and c are correct.

3. You have a basketball filled with gas. Which method below changes its volume because of a mass change of the gas inside?
   (a) Put it into a refrigerator.   (b) Squeeze it.
   (c) Pump more gas into it.   (d) Hold it under water.
   (e) Leave it in the sunshine.

4. Choose the quantities describing the air inside a bike tire that do *not* change when you pump the tire.
   (a) Mass   (b) Volume
   (c) Density   (d) Pressure
   (e) Particle mass   (f) Particle concentration

5. Which answer below does *not* explain the decrease in size of a basketball after you take it outside on a cold day?
   (a) The pressure inside the ball decreases.
   (b) The temperature of the gas inside the ball decreases.
   (c) The volume of the ball decreases.
   (d) The number of gas particles inside the ball decreases.

6. What causes balloons filled with helium to deflate as time passes?
   (a) The elasticity of the rubber decreases.
   (b) The temperature inside decreases.
   (c) The pressure outside the balloon increases.
   (d) The gas from inside diffuses into the atmosphere.

7. From the list below, choose the assumption that we did not use in deriving the ideal gas law.
   (a) Gas particles can be treated as objects with zero size.
   (b) The particles do not collide with each other inside the container.
   (c) The particles collide partially inelastically with the walls of the container.
   (d) The particles obey Newton's laws.

8. You have a mole of oranges. The mass of each orange is m. Imagine that you split each orange in half. What will be the molar mass of this pile of half oranges in kilograms?
   (a) $m/2$ (b) $m/(6.02 \times 10^{23})$
   (c) $2m/(6.02 \times 10^{23})$ (d) $(m/2)/(6.02 \times 10^{23})$

9. How might physicists have come to know that at a constant temperature and constant mass, the pressure of an ideal gas is inversely proportional to its volume?
   (a) They could have conducted an experiment maintaining the gas as described above and made a pressure-versus-volume graph.
   (b) They could have derived this relationship using the equations describing the ideal gas model and the relationship between the speed of the particles and the gas temperature.
   (c) Both a and b are correct.

10. A cylindrical container is filled with a gas. On the top of the container sits a heavy piston that is free to move (see the experimental setup in Table 12.5, Experiment 3). When the gas is heated, the piston moves up, causing the volume of the gas to increase. Which equation below best describes this process?
    (a) $V_{initial}/T_{initial} = V_{final}/T_{final}$
    (b) $P_{initial}V_{initial} = P_{final}V_{final}$
    (c) $P_{initial}V_{initial} = nRT_{final}$
    (d) $P_{initial}V_{initial} = Nk_BT_{final}$

11. A cylindrical container is filled with a gas. On the top of the container sits a heavy piston that is free to move. In a certain process, you observe that with constant external pressure pushing down on the piston, the piston moves up, causing the volume of the gas to increase. Which graph below best represents this process (**Figure Q12.11**)?

FIGURE Q12.11

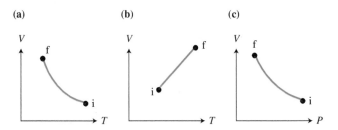

12. A completely closed rigid container of gas is taken from the oven and placed in ice water. Which graph below does not represent this process (**Figure Q12.12**)?

FIGURE Q12.12

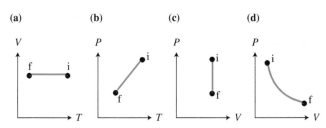

   (e) None of them do.

13. Which graph below does not represent a process described by the equation $V_{initial}/T_{initial} = V_{final}/T_{final}$ (**Figure Q12.13**)?

FIGURE Q12.13

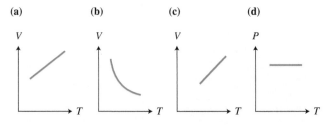

   (e) a and b

14. What does contracting and relaxing the diaphragm allow?
    (a) Greater O_2 absorption by humans and mammals compared to other species
    (b) Greater pressure difference between air in the lungs and outside air
    (c) The only way to facilitate air intake into and out of the lungs
    (d) a and b
    (e) a, b, and c

15. When is air inhaled into the lungs?
    (a) When lung muscles cause them to expand
    (b) When the diaphragm contracts
    (c) When the chest expands
    (d) b and c
    (e) a, b, and c

16. Which of the following conditions are crucial for performing an isothermal compression of a gas (see the experimental setup in Table 12.5, Experiment 1)?
    (a) The piston should be moved very slowly.
    (b) There should be no friction between the piston and the gas container.
    (c) The gas container should be kept in a vertical position.
    (d) The volume of the gas container should be much smaller than the volume of the ice bath.

Conceptual Questions

17. (a) How do we know if a real gas can be described as an ideal gas? (b) How can you decide if air in the physics classroom can be described using the ideal gas model?
18. Why does it hurt to walk barefoot on gravel?
19. In the magic trick in which a person lies on a bed of nails, why doesn't the person get hurt by the nails?

20. What does it mean if the density of a gas is 1.29 kg/m³?
21. How many oranges would you have if you had two moles of oranges?
22. Imagine that you have an unknown gas. What experiments do you need to do and what real equipment do you need to determine the mass of one molecule of this gas?
23. Describe how temperature and one degree are defined on the Celsius scale.
24. Describe how temperature and one degree are defined on the Kelvin scale.
25. Why does sugar dissolve faster in hot tea than in cold water?
26. (a) Describe experiments that were used to test the predictions of the molecular kinetic theory. (b) What experiments revealed its limitations?
27. Give three examples of diffusion that are important for human life.
28. Why do very light gases such as hydrogen not exist in Earth's atmosphere but do exist in the atmospheres of giant planets such as Uranus and Saturn?
29. Why does the Moon have no atmosphere? Explain.
30. Explain why Earth has almost no free hydrogen in its atmosphere.

Problems

Below, **BIO** indicates a problem with a biological or medical focus. Problems labeled **EST** ask you to estimate the answer to a quantitative problem rather than derive a specific answer. Asterisks indicate the level of difficulty of the problem. Problems with no * are considered to be the least difficult. A single * marks moderately difficult problems. Two ** indicate more difficult problems.

12.2 Pressure, density, and the mass of particles

1. What are the molar masses of molecular and atomic hydrogen, helium, oxygen, and nitrogen? What are their molecular masses?
2. **EST** Estimate the number of hydrogen atoms in the Sun. The mass of the Sun is 2×10^{30} kg. About 70% of it by mass is hydrogen, 30% is helium, and there is a negligible amount of other elements.
3. The average particle density in the Milky Way galaxy is about one particle per cubic centimeter. Express this number in SI units (kg/m³). Indicate any assumptions you made.
4. * (a) What is the concentration (number per cubic meter) of the molecules in air at normal conditions? (b) What is the average distance between molecules compared to the dimensions of the molecules? (c) Can you consider air to be an ideal gas? Explain your answer.
5. * What is the mass of a water molecule in kilograms? What is the mass of an average air molecule in kilograms?
6. You find that the average gauge pressure in your car tires is about 35 psi. (a) How many newtons per square meter is it? What is gauge pressure? (b) Determine the absolute pressure in the tires.
7. **BIO** **Forced vital capacity** Physicians use a machine called a spirometer to measure the maximum amount of air a person can exhale (called the forced vital capacity). Suppose you can exhale 4.8 L. How many kilograms of air do you exhale? What assumptions did you make to answer the question? How do these assumptions affect the result?
8. A container is at rest with respect to a desk. Inside the container a particle is moving horizontally at a speed v with respect to the desk. It collides with a vertical wall of the container elastically and rebounds. Compare the direction and the speed of the particle if the wall is (a) at rest with respect to the desk; (b) moving in the same direction as the particle at a speed smaller than the particle's; and (c) moving in the direction opposite to the motion of the particle at a smaller speed.

12.3 Quantitative analysis of ideal gas

9. **Oxygen tank for mountains** Consider an oxygen tank for a mountain climbing trip. The mass of one molecule of oxygen is 5.3×10^{-26} kg. What is the pressure that oxygen exerts on the inside walls of the tank if its concentration is 10^{25} particles/m³ and its rms speed is 600 m/s? What assumptions did you make?
10. You have five molecules with the following speeds: 300 m/s, 400 m/s, 500 m/s, 450 m/s, and 550 m/s. (a) What is their average speed? (b) What is their rms speed? Compare it with the average speed.

11. Two gases in different containers have the same concentration and same rms speed. The mass of a molecule of the first gas is twice the mass of a molecule of the second gas. What can you say about their pressures? Explain.
12. Four molecules are moving with the following velocities: 300 m/s south, 300 m/s north, 400 m/s east, and 400 m/s west. (a) What is their average velocity? (b) What is their rms speed?
13. * **Hitting tennis balls against a wall** A 0.058-kg tennis ball, traveling at 25 m/s, hits a wall, rebounds with the same speed in the opposite direction, and is hit again by another player, causing the ball to return to the wall at the same speed. The collision of the ball with the wall lasts 1 ms. The ball returns to the wall once every 0.60 s. (a) Determine the average force exerted on the ball during a single collision and the force that the ball exerts on the wall averaged over the time between collisions. State the assumptions that you made. (b) If 10 people are practicing against a wall with an area of 30 m², what is the average pressure of the 10 tennis balls against the wall?
14. * Friends throw snowballs at the wall of a 3.0 m × 6.0 m barn. The snowballs have mass 0.10 kg and hit the wall moving at an average speed of 6.0 m/s. They do not rebound. Determine the average pressure exerted by the snowballs on the wall if 40 snowballs hit the wall each second. Which problem, this or the previous problem, resembles the actions of the molecules of an ideal gas hitting the walls of their container?
15. * A ball moving at a speed of 3.0 m/s with respect to the ground hits a stationary wall at a 30° angle with respect to the surface of the wall. Determine the direction and the magnitude of the velocity of the ball after it rebounds. Explain carefully what physics principles you used to find the answer. What assumptions did you make? How will the answer change if one or more of them are not valid?
16. * **Oxygen tank for mountain climbing** An oxygen container used by mountain climbers has a 90-min oxygen supply at a rate of 6 L/min (with the volume measured at atmospheric pressure). Determine everything you can about the gas in the container. Make reasonable assumptions.

12.4 Temperature

17. Your temperature, when taken orally, is 98.6 °F. When taken under your arm, it's 36.6 °C. Are these results consistent? Explain.
18. Air consists of many different molecules, for example, N_2, O_2, H_2O, and CO_2. Which molecules are the fastest on average? The slowest on average? Explain.
19. What is the average kinetic energy of a particle of air at standard conditions?
20. Air is a mixture of molecules of different types. Compare the rms speeds of the molecules of N_2, O_2, and CO_2 at standard conditions. What assumptions did you make?
21. * How many moles of air are in a regular 1-L water bottle when you finish drinking the water? What assumptions did you make? How do these assumptions affect your result?

22. * At approximately what temperature does the average translational kinetic energy of a N_2 molecule in an ideal gas equal the macroscopic translational kinetic energy of a copper atom in a penny that is dropped from the height of 1.0 m?

23. ** A molecule moving at speed v_1 collides head-on with a molecule of the same mass moving at speed v_2. Compute the speeds of the molecules after the collision. What assumptions did you make? How does the answer to this problem explain why the mixing of hot and cold gases causes the cold gas to become warmer and the hot gas to become cooler?

24. * **Balloon flight** For a balloon ride, the balloon must be inflated with helium to a volume of 1500 m^3 at sea level. The balloon will rise to an altitude of about 12 km, where the temperature is about $-52\,°C$ and the pressure is about 20 kPa. How many kg of helium should be put into the balloon? What assumptions did you make?

25. **BIO** **Ears pop** The middle ear has a volume of about 6.0 cm^3 when at a pressure of $1.0 \times 10^5\ N/m^2$. Determine the volume of that same air when the air pressure is $0.83 \times 10^5\ N/m^2$, as it is at an elevation of 1500 m above sea level (assume the air temperature remains constant). If the volume of the middle ear remains constant, some air will have to leave as the elevation increases. That is why ears "pop."

26. * Even the best vacuum pumps cannot lower the pressure in a container below 10^{-15} atm. How many molecules of air are left in each cubic centimeter in such a "vacuum"? Assume that the temperature is 273 K.

27. **Pressure in interstellar space** The concentration of particles (assume neutral hydrogen atoms) in interstellar gas is 1 particle/cm^3, and the average temperature is about 3 K. What is the pressure of the interstellar gas? How does it compare to the best vacuum that can be achieved on Earth (see the previous problem)?

12.5, 12.6, and 12.7 Testing the ideal gas law, Speed distribution of particles, and Skills for analyzing processes using the ideal gas law

28. * Describe experiments to determine if each of the three gas isoprocess laws works. The experiments should be ones that you could actually carry out.

29. * The following data were collected for the temperature and volume of a gas. Can this gas be described by the ideal gas model? Explain how you know.

Temperature (°C)	Volume (mL)
11	95.0
25	100.0
47	107.5
73	116.0
159	145.1
233	170.0
258	177.9

30. * Explain the microscopic mechanisms for the relation of macroscopic variables for an isothermal process, an isobaric process, and an isochoric process occurring in an ideal gas.

31. ** **Scuba diving** The pressure of the air in a diver's lungs when he is 20 m under the water surface is $3.0 \times 10^5\ N/m^2$, and the air occupies a volume of 4.8 L. How many moles of air should he exhale while moving to the surface, where the pressure is $1.0 \times 10^5\ N/m^2$?

32. * When surrounded by air at a pressure of $1.0 \times 10^5\ N/m^2$, a basketball has a radius of 0.12 m. Compare its volume at this condition with the volume that it would have if you took it 15 m below the water surface, where the pressure is $2.5 \times 10^5\ N/m^2$. What assumptions did you make?

33. * Some students are given the following problem: "A 5000-cm^3 cylinder is filled with nitrogen gas at 1.0×10^5 Pa and 300 K and closed with a movable piston. The gas is slowly compressed at constant temperature to a final volume of 5 cm^3. Determine the final pressure of the gas." (a) Explain why the ideal gas law cannot be applied to solve this problem. Present quantitative arguments. (b) Modify the problem so that it *can* be solved using the ideal gas law and give your solution.

34. ** You have gas in a container with a movable piston. The walls of the container are thin enough so that its temperature stays the same as the temperature of the surrounding medium. You have baths of water of different temperatures, different objects that you can place on top of the piston, etc. (a) Describe how you could make the gas undergo an isothermal process so that the pressure inside increases by 10%, then undergo an isobaric process so that the new volume decreases by 20%, and finally undergo an isochoric process so that the temperature increases by 15%. (b) Represent this three-step process in P-versus-T, V-versus-T, and P-versus-V graphs. (c) Express the new pressure, volume, and temperature in terms of their initial values.

35. * You want to determine the temperature in a freezer using the following equipment: an empty plastic bottle with a cap, a graduated measuring cylinder, and a bowl of water. You also know your room temperature T_R. (a) Which of the procedures described below will give data from which you can determine the temperature in the freezer? (b) For the procedure that you have chosen in part (a), derive the expression for the temperature in the freezer as a function of relevant physical quantities. Indicate any assumptions that you made.

    I.   Put the open plastic bottle into the freezer for half an hour. Take the bottle out, turn it upside down into the water, and measure the volume of the water sucked into the bottle.
    II. Put the open plastic bottle in the freezer for half an hour. Take the bottle out and close it tightly with a cap. Wait until the bottle's shape stops changing. Turn the bottle upside down into the water, open the cap, and measure the volume of the water that entered the bottle.
    III. Put the closed plastic bottle in the freezer for half an hour. Take the bottle out, turn it upside down into the water, and then open the cap. Measure the volume of the water sucked into the bottle.
    IV. Put the closed plastic bottle in the freezer for half an hour. Take the bottle out, open the cap, and turn the bottle upside down into the water. Measure the volume of the water sucked into the bottle.

36. * **Bubbles** While snorkeling, you see air bubbles leaving a crevice at the bottom of a reef. One of the bubbles has a radius of 0.060 m. As the bubble rises, the pressure inside it decreases by 50%. Now what is the bubble's radius? What assumptions did you make to solve the problem?

37. ** **Diving bell** A cylindrical diving bell, open at the bottom and closed at the top, is 4 m tall. Scientists fill the bell with air at the pressure of $1.0 \times 10^5\ N/m^2$. The pressure increases naturally by $1.0 \times 10^5\ N/m^2$ for each 10 m that the bell is lowered below the surface of the water. If the bell is lowered 30 m below the ocean surface, how many meters of air space are left inside the bell? Why doesn't water enter the entire bell as it goes under water? Draw several sketches for this problem. Assume the temperature remains constant.

38. * **Mount Everest** (a) Determine the number of molecules per unit volume in the atmosphere at the top of Mount Everest. The pressure is $0.31 \times 10^5\ N/m^2$, and the temperature is $-30\,°C$. (b) Determine the number of molecules per unit volume at sea level, where the pressure is $1.0 \times 10^5\ N/m^2$ and the temperature is $20\,°C$.

39. * **EST** **Breathing on Mount Everest** Using the information from Problem 12.38, estimate how frequently you need to breathe on top of Mount Everest to inhale the same amount of oxygen as you do at sea level. The pressure is about one-third the pressure at sea level.

40. * **Capping beer** You would like to make homemade beer, but you are concerned about storing it. Your beer is capped into a bottle at a temperature of $27\,°C$ and a pressure of $1.2 \times 10^5\ N/m^2$. The cap will pop off if the pressure inside the bottle exceeds $1.5 \times 10^5\ N/m^2$. At what maximum temperature can you store the beer so the gas inside the bottle does not pop the cap? List the assumptions that you made.

41. **Car tire** With a tire gauge, you measure the pressure in a car tire as $2.1 \times 10^5\ N/m^2$. How can this be if you know that absolute pressure in the tire is three times higher than atmospheric? The tire looks okay. What's the deal?

42. * **Car tire dilemma** Imagine a car tire that contains 5.1 moles of air when at a gauge pressure of $2.1 \times 10^5\ N/m^2$ (the pressure above atmospheric pressure) and a temperature of $27\,°C$. The temperature increases to $37\,°C$, the volume decreases to 0.8 times the original volume, and the gauge pressure decreases to $1.6 \times 10^5\ N/m^2$. Can these measurements be correct if the tire did not leak? If it did leak, then how many moles of air are left in the tire?

43. * There is a limit to how much gas can pass through a pipeline, because the pipes can only tolerate so much pressure on the walls. To increase the amount of gas going through the pipeline, engineers decide to cool the gas (to reduce its pressure). Suggest how much they should lower the temperature of the gas if they want to increase the mass per unit time by 1.5 times.

44. Explain how you know that the volume of one mole of gas at standard conditions is 22.4 L.

45. * At what pressure is the density of $-50\,°C$ nitrogen gas (N_2) equal to 0.01 times the density of water?

46. * In the morning, the gauge pressure in your car tires is 35 psi. During the day, the air temperature increases from $20\,°C$ to $30\,°C$ and the pressure increases to 36.5 psi. By how much did the volume of one of the tires increase? What assumptions did you make?

47. **Equation Jeopardy 1** The equation below describes a process. Construct a word problem for a process that is consistent with the equations (there are many possibilities). Provide as much detailed information as possible about your proposed process.

$$\frac{1.2 \times 10^5 \ \text{N/m}^2}{293 \ \text{K}} = \frac{2.0 \times 10^5 \ \text{N/m}^2}{T}$$

48. * **Equation Jeopardy 2** The equation below describes a process. Construct a word problem for a process that is consistent with the equations (there are many possibilities). Provide as much detailed information as possible about your proposed process.

$$\Delta n = \frac{(0.67 \times 10^5 \ \text{N/m}^2)(0.60 \times 10^{-6} \ \text{m}^3)}{(8.3 \ \text{J/mole} \cdot \text{K})(303 \ \text{K})}$$
$$- \frac{(1.00 \times 10^5 \ \text{N/m}^2)(0.60 \times 10^{-6} \ \text{m}^3)}{(8.3 \ \text{J/mole} \cdot \text{K})(310 \ \text{K})}$$

49. ** The *P*-versus-*T* graph in **Figure P12.49** describes a cyclic process comprising four hypothetical parts. (a) What happens to the pressure of the gas in each part? (b) What happens to the temperature of the gas in each part? (c) What happens to the volume of the gas in each part? (d) Explain each part microscopically. (e) Use the information from (a)–(c) to represent the same parts in *P*-versus-*V* and *V*-versus-*T* graphs. (Hint: It helps to align the *P*-versus-*V* graph *beside* the *P*-versus-*T* graph using the same *P* values on the ordinate (vertical axis) and to place the *V*-versus-*T* graph *below* the *P*-versus-*T* graph using the same *T* values on the abscissa (horizontal axis). This helps keep the same scale for the variables.)

FIGURE P12.49

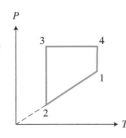

50. ** The *V*-versus-*T* graph in **Figure P12.50** describes a cyclic process comprising four hypothetical parts. (a) What happens to the pressure of the gas in each part? (b) What happens to the temperature of the gas in each part? (c) What happens to the volume of the gas in each part? (d) Explain each part microscopically. (e) Use the information from (a)–(c) to represent the same process in a *P*-versus-*T* graph (below the *V*-versus-*T* graph) and a *P*-versus-*V* graph (beside the *P*-versus-*T* graph). See the hint in the previous problem about the graph alignments.

FIGURE P12.50

51. * **BIO Breathing** You are breathing heavily while hiking up the mountain. To inhale, you expand your diaphragm and lungs. Explain, using your knowledge of gas pressure, why this mechanical movement leads to the air flowing into your nose or mouth. Support your reasoning with diagrams if necessary.

12.8 Thermal energy, the Sun, and diffusion

52. * The temperature of the Sun's atmosphere near the surface is about 6000 K, and the concentration of atoms is about 10^{15} particles/m³. What are the average pressure and density of its atmosphere? What assumptions did you make to solve the problem?

53. * **BIO EST Breathing and metabolism** We need about 0.7 L of oxygen per minute to maintain our resting metabolism and about 2 L when standing and walking. Estimate the number of breaths per minute for a person to satisfy this need when resting and when standing and walking. What assumptions did you make? Remember that oxygen is about 21% of the air.

54. * **EST Sun's life expectancy** (a) Estimate the average kinetic energy of the particles in the Sun. Assume that it is made of atomic hydrogen and that its average temperature is 100,000 K. The mass of the Sun is 2×10^{30} kg. (b) For how long would the Sun shine using this energy if it radiates 4×10^{26} W/s? Is your answer reasonable? Explain.

55. ** A gas that can be described by the ideal gas model is contained in a cylinder of volume *V*. The temperature of the gas is *T*. The mass of the gas is *m*, and the molar mass is *M*. Write an expression for the total thermal energy of the gas. Now, imagine that the exact same gas has been placed in a container of volume 2*V*. What happens to its pressure? What happens to its temperature? What happens to its thermal energy?

56. * **Equation Jeopardy 3** The three equations below describe a physical situation. Construct a word problem for a situation that is consistent with the equations (there are many possibilities). Provide as much detailed information as possible about the situation.

$$m = (1.3 \ \text{kg/m}^3)(3.0 \ \text{m} \times 5.0 \ \text{m} \times 2.0 \ \text{m})$$
$$N = m/[(29 \times 10^{-3} \ \text{kg})/(6.0 \times 10^{23} \ \text{particles})]$$
$$U_{\text{th}} = N(3/2)(1.38 \times 10^{-23} \ \text{J/K})(273 \ \text{K})$$

General Problems

FIGURE P12.57

57. * **EST** A simple experimental setup for performing isobaric processes can be made with a thin glass tube filled with a gas and closed with a drop of water (**Figure P12.57**). If the pressure of the gas changes, the drop moves until the sum of the forces exerted on the drop is zero (there is no friction between the drop and the tube). The inner diameter of the glass tube is 3.0 mm and the length of the tube is 200.0 mm. At normal room temperature, a 5.0-mm-long water drop is 30.0 mm from the open end of the tube. Estimate the temperature range in which this setup can be used to show isobaric processes. (Hint: What happens to water at low temperature?)

VIDEO 12.1

58. * See the previous problem. Explain how the force exerted by Earth on the water drop and the orientation of the tube affects the performance of the setup.

59. * Jeff and Natalie notice that a rubber balloon, which is first in a warm room, shrinks when they take it into the garden on a cold winter day. They propose two different explanations for the observed phenomenon: (a) the balloon is slowly leaking; (b) the balloon shrinks due to decreased temperature while the pressure in the balloon remains constant (isobaric compression). In order to test their proposed explanations, Jeff and Natalie perform three consecutive experiments: they measure the volume of the balloon and the temperature of the air near the balloon (1) in the room, (2) in the garden, and (3) again in the room. Their measurements, including uncertainties, are presented in the table below.

Exp. #	Location	Temperature	Volume of the balloon
1	Room	$26.2\,°C \pm 0.1\,°C$	$7500 \ \text{cm}^3 \pm 400 \ \text{cm}^3$
2	Garden	$-15.3\,°C \pm 0.1\,°C$	$6400 \ \text{cm}^3 \pm 400 \ \text{cm}^3$
3	Room	$26.2\,°C \pm 0.1\,°C$	$7300 \ \text{cm}^3 \pm 400 \ \text{cm}^3$

Based on the data, can Jeff and Natalie reject any of their hypotheses? Explain. Make sure you include uncertainties in your answer.

60. * **Different planet compositions** Explain why planets closer to the Sun have low concentrations of light elements, but light elements are relatively abundant in giant planets such as Jupiter, Uranus, and Saturn, which are far from the Sun.

61. ** **EST** **Density of our galaxy** Estimate the average density of particles in our galaxy, assuming that the most abundant element is atomic hydrogen. There are about 10^{11} stars in the galaxy and the size of the galaxy is about 10^5 light-years. A light-year is the distance light travels in 1 year moving at a speed of 3×10^8 m/s. What do you need to assume about the stars in order to answer this question?

62. * **BIO** **Breathing** During each breath you probably inhale about 0.50 L of air. How many oxygen molecules do you inhale if you are at sea level?

63. * **Car engine** During a compression stroke of a cylinder in a diesel engine, the air pressure in the cylinder increases from 1.0×10^5 N/m^2 to 50×10^5 N/m^2, and the temperature increases from 26 °C to 517 °C. A typical compression ratio (the ratio of the largest and the smallest volume of a cylinder) for a diesel engine is 20:1. Using this information, how would you convince your friends that knowledge about ideal gases can help explain how hot gases burned in the car engine affect the motion of the car?

64. * **EST** How can the pressure of air in your house stay constant during the day if the temperature rises? Estimate the volume of your house and the number of moles of air that leave the house during the daytime. Assume that nighttime temperature and daytime temperature differ by about 10 °C. List all other assumptions that are necessary to answer the question.

65. * **Tell-all problem** Tell everything you can about a process that was performed with 2×10^{-3} moles of a gas and that is described by the pressure-versus-volume graph shown in Figure P12.65.

FIGURE P12.65

66. ** Two massless, frictionless pistons are inside a horizontal tube opened at both ends. A 10-cm-long thread connects the pistons. The cross-sectional area of the tube is 20 cm^2. The pressure and temperature of gas between the pistons and the outside air are the same and are equal to $P = 1.0 \times 10^5$ N/m^2 and $T = 24$ °C. At what temperature of the gas will the thread break if it breaks when the tension reaches 30 N?

67. * A closed cylindrical container is divided into two parts by a light, movable, frictionless piston. The container's total length is 100 cm. Where is the piston located when one side is filled with nitrogen (N_2) and the other side with the same mass of hydrogen (H_2) at the same temperature?

68. How many molecules are there in 1 g of air at normal conditions? If these molecules were distributed uniformly on Earth's surface, *estimate* the number that would be under your feet right now. The radius of Earth is about 6400 km.

69. ** The speed of sound in an ideal gas is given by the relationship

$$v = \sqrt{\frac{\gamma RT}{M}}$$

where $R =$ the universal gas constant $= 8.314$ J/mol · K; $T =$ the absolute temperature; $M =$ the molar mass of the gas in kg/mol; and γ is a characteristic of the specific gas. For air, $\gamma = 1.4$ and the average molar mass for dry air is 28.95 g/mol. (a) Show that the equation gives you correct units. (b) Give reasons why the temperature of the gas is in the numerator and the molar mass of the gas is in the denominator.

70. * Using the information from problem 12.69, calculate the speed of sound in the air. What assumptions are you making?

71. * **Speed of sound in summer and winter** Using the information in problem 12.69, estimate how much faster sound travels in summer than in winter. Explain how you arrived at your answer and the assumptions that you made.

72. ** A chamber of unknown volume is filled with an unknown amount of ideal gas. A cylinder with an open bottom and calibrated volume is connected to the chamber and is closed with a movable piston. A gas pressure sensor is also mounted on the chamber (Figure P12.72). The whole experimental setup is kept at a constant temperature of 300 K. Doris moves the piston from mark 0 to mark 200 mL in steps of 50 mL and records the corresponding pressure in the chamber. She moves the piston very slowly and waits several minutes before recording the pressure values. Use Doris's data in the table to determine the volume of the chamber and the number of moles of gas. (Note: This is a problem that requires linearization of data.)

V_{Cyl} (mL)	P (10^5 N/m^2)
0	4.35
50	3.85
100	3.50
150	3.15
200	2.90

FIGURE P12.72

Reading Passage Problems

BIO **Vascular wall tension and aortic blowout** The walls of blood vessels contain varying amounts of elastic fibers that allow the vessels to expand and contract as the pressure and amount of fluid inside vary (these fibers are more prevalent in the aorta and large arteries than in the small arterioles and capillaries). These fibers in the cylindrical walls produce a wall tension T, defined as

$$T = \frac{F}{L}$$

where L is the length of an imaginary cut parallel to the axis of the vessel and F is the magnitude of the force that each side of the cut must exert on the other side to hold the two sides together.

Three forces are exerted on a short section of wall fiber—the system. (1) The fluid inside pushes outward, due to fluid pressure from inside $P_{\text{inside fluid pushing out}} = P_{out}$; (2) fluid outside the vessel pushes inward, due to fluid pressure from outside $P_{\text{outside fluid pushing in}} = P_{in}$; and (3) the wall next to the system exerts wall tension T on each side of that wall. The pressure difference across the wall $\Delta P = P_{out} - P_{in}$, the wall tension T, and the radius R of the cylindrical vessel are related by Laplace's law:

$$\Delta P = \frac{T}{R} \quad \text{or} \quad T = \Delta P \cdot R$$

The inward gauge pressure P_{in} of tissue surrounding the vessels is approximately zero. Thus, the pressure difference $\Delta P = P_{out} - P_{in} = P_{vessel} - 0 = P_{vessel}$ is the gauge pressure in the blood vessel. We can now estimate the wall tension for different types of vessels. The tension in the aorta is approximately

$$T = \Delta P \cdot R \approx (100 \text{ mm Hg})\left(\frac{133 \text{ N/m}^2}{1 \text{ mm Hg}}\right)(1.3 \times 10^{-2} \text{ m}) = 170 \text{ N/m}$$

Using similar reasoning, the wall tension in the low-pressure, very small radius capillaries is about 0.016 N/m—about 0.0003 times the tension needed to tear a facial tissue. Because such little tension is needed to hold a capillary together, its wall can be very thin, allowing easy diffusion of various molecules across the wall.

The walls of a healthy aorta can easily provide the tension needed to support the increased blood pressure when it fills with blood from the heart during each heartbeat. However, aging and various medical conditions may weaken the aortic wall in a short section, and increased blood pressure can cause it to stretch. The weakened wall bulges outward; this is called an aortic aneurism. The increased radius causes increased tension, which can increase bulging. This cycle can result in a rupture to the aorta: an aortic blowout.

73. Why is the wall tension in capillaries so small?
   (a) There are so many capillaries.
   (b) Their radii are so small.
   (c) The outward pressure of the blood inside is so small.
   (d) b and c
   (e) a, b, and c

74. Which answer below is closest to the wall tension in a typical arteriole of radius 0.15 mm and 60 mm Hg blood pressure?
   (a) 0.001 N/m   (b) 0.01 N/m
   (c) 0.1 N/m   (d) 1 N/m
   (e) 10 N/m

75. According to Laplace's law, elevated blood pressure in an artery should cause the wall tension in the artery to do what?
   (a) Increase   (b) Remain unchanged
   (c) Decrease   (d) Impossible to decide

76. As a person ages, the fibers in arteries become less elastic and the wall tension increases. According to Laplace's law, this will cause the blood pressure to do what?
   (a) Increase   (b) Remain unchanged
   (c) Decrease   (d) Impossible to decide

77. Aortic blowout occurs when part of the wall of the aorta becomes weakened. What does this cause?
   (a) A bulge and increased radius of the aorta when the blood pressure inside increases
   (b) An increased radius of the aorta, which causes increased tension in the wall
   (c) An increased tension in the aorta, which causes the radius to increase
   (d) a and b
   (e) a, b, and c

BIO **Portable hyperbaric chamber** In 1997, a hiking expedition was stranded for 38 days on the Tibetan side of Mount Everest at altitudes between 5200 m and 8000 m. While at those altitudes, 10 climbers suffered acute mountain sickness. A 37-year-old climber with acute pulmonary edema (buildup of fluid in the lungs) was treated with a portable Gamow bag (see Figure 12.20), named after its inventor, Igor Gamow. The Gamow bag is a windowed cylindrical portable hyperbaric chamber constructed of nonpermeable nylon that requires constant pressurization with a foot pump attached to the bag. The climber enters the bag, and a person outside pumps air into the bag so that the air pressure inside is somewhat higher than the outside pressure.

FIGURE 12.20 A Gamow bag, used to help climbers with acute altitude sickness.

The bag and pump have a 6.76-kg mass. The volume of the inflated bag is 0.476 m^3. The maximal bag pressure is 0.15×10^5 N/m^2 above the air pressure at the site where it is used. In the 1997 climb, with the temperature at $-20°$C, the bag was filled in about 2 min with 10–20 pumps per minute. This raised the pressure in the bag to 0.58×10^5 N/m^2 (equivalent to an elevation of 4400 m) instead of the actual outside pressure of 0.43×10^5 N/m^2 at the 6450-m elevation at which the climber was treated. The treatment lasted for 2 h, with the climber inhaling about 15 times/min at about 0.5 L/inhalation, and was successful—the pulmonary edema disappeared.

78. What is closest to the volume of the Gamow bag?
   (a) 50 L   (b) 100 L
   (c) 200 L   (d) 500 L
   (e) 1000 L

79. What is closest to the temperature at the 6450-m elevation on the day described in the problem?
   (a) 37 K   (b) 253 K
   (c) 20 K   (d) -20 K
   (e) 273 K

80. What is closest to the number n of moles of air in the filled bag when at 4400 m?
   (a) 3 moles   (b) 10 moles
   (c) 13 moles   (d) 110 moles
   (e) 170 moles

81. What is closest to the number n of moles of air in the bag if its pressure is at the 6450-m level?
   (a) 3 moles   (b) 10 moles
   (c) 13 moles   (d) 110 moles
   (e) 170 moles

82. Estimate the fraction of the air in the Gamow bag that an occupant would inhale in 1 h, assuming no replacement of the air.
   (a) 0.01   (b) 0.1
   (c) 0.3   (d) 0.5
   (e) 1.0

13

Static Fluids

- **How can a hot air balloon fly for hours?**
- **Why is it dangerous for scuba divers to ascend from a deep-sea dive too quickly?**
- **Why, in water, does a 15-g nail sink but a cargo ship float?**

The first hot air balloon with a person on board was launched by the Montgolfier brothers in 1783. Hot air is less dense than cold air. By carefully balancing the density of the balloon and its contents with the density of air, the pilot can control the force that the outside cold air exerts on the balloon. What do pressure, volume, mass, and temperature have to do with this force?

BE SURE YOU KNOW HOW TO:

- Draw a force diagram for a system of interest (Section 3.1).
- Apply Newton's second law in component form (Section 4.2).
- Define pressure (Section 12.2).
- Apply the ideal gas law (Sections 12.3 and 12.4).

IN THE PREVIOUS CHAPTER, we constructed the ideal gas model and used it to explain the behavior of gases. The temperature of the gas and the pressure it exerted on surfaces played an important role in the phenomena we analyzed. We ignored the effect of the gravitational force exerted by Earth on the gas particles. This simplification was reasonable, since in most of the processes we analyzed the gases had little mass and occupied a relatively small region of space, like in a piston. In this chapter, our interest expands to include phenomena in which the force exerted by Earth plays an important role. We will confine the discussion to static fluids—fluids that are not moving.

13.1 Density

We are familiar with the concept of density (from Chapter 12). To find the density of an object or a substance, determine its mass m and volume V and then calculate the ratio of the mass and volume:

$$\rho = \frac{m}{V} \tag{13.1}$$

Archimedes (Greek, 287–212 B.C.) discovered how to determine the density of an object of irregular shape. First determine its mass using a scale. Then determine its volume by submerging it in a graduated cylinder with water (Figure 13.1). Finally, divide the mass in kilograms by the volume in cubic meters to find the density in kilograms per cubic meter. Using this method we find that the density of an iron nail is 7860 kg/m³—relatively large. A gold coin has an even larger density—19,300 kg/m³. The universe, though, contains much denser objects. For example, the rapidly spinning neutron star known as a pulsar (discussed in Chapter 9) has a density of approximately 10^{18} kg/m³. Table 13.1 lists the densities of various solids, liquids, and gases.

FIGURE 13.1 Measuring the density of an irregularly shaped object.

1. Measure mass of object.

2. Place the object in water in a graduated cylinder.

3. Measure the volume change of the water. Volume change of water = volume of object.

4. Density = $\rho = m/V$

TABLE 13.1 **Densities of various solids, liquids, and gases**

Solids		Liquids[1]		Gases[2]	
Substance	Density (kg/m³)	Substance	Density (kg/m³)	Substance	Density (kg/m³)
Aluminum	2700	Acetone	791	Dry air 0 °C	1.29
Copper	8920	Ethyl alcohol	789	10 °C	1.25
Gold	19,300	Methyl alcohol	791	20 °C	1.21
Iron	7860	Gasoline	726	30 °C	1.16
Lead	11,300	Olive oil	800–920	Helium	0.178
Platinum	21,450	Mercury	13,600	Hydrogen	0.090
Silver	10,500	Milk	1028–1035	Oxygen	1.43
Bone	1700–2000	Seawater	1025	Carbon dioxide	1.98
Brick	1400–2200	Water 0 °C	999.8		
Cement	2700–3000	3.98 °C	1000.00		
Clay	1800–2600	20 °C	998.2		
Glass	2400–2800	Blood plasma	1030		
Ice	917	Blood whole	1050		
Styrofoam	25–100				
Balsa wood	120				
Oak	600–900				
Pine	500				
Planet Earth	5515				
Moon	3340				
Sun	1410				
Universe (average)	10^{-26}				
Pulsar	10^{11}–10^{18}				

[1]Densities of liquids are at 0 °C unless otherwise noted.
[2]Densities of gases are at 0 °C and 1 atm unless otherwise noted.

QUANTITATIVE EXERCISE 13.1   **Ping-pong balls with different densities**

Saturn has the lowest density of all the planets in the solar system ($M_{\text{Saturn}} = 5.7 \times 10^{26}$ kg and $V_{\text{Saturn}} = 8.3 \times 10^{23}$ m^3). The average density of a neutron star is 10^{18} kg/m^3. Compare the mass of a ping-pong ball filled with material from Saturn with that of the same ball filled with material from a neutron star. An empty ping-pong ball has a 0.037-m diameter (0.020-m radius) and a 2.7-g mass.

Represent mathematically To find the mass of a ping-pong ball filled with a particular material, we add the mass of the ball alone and the calculated mass of the material inside:

$$m_{\text{filled ball}} = m_{\text{ball}} + m_{\text{material}}$$

where $m_{\text{material}} = \rho_{\text{material}} V_{\text{ball}}$.

The density of the neutron star is given, and the density of Saturn can be found using the operational definition $\rho_{\text{Saturn}} = \dfrac{m_{\text{Saturn}}}{V_{\text{Saturn}}}$. The interior of the ping-pong ball is a sphere of volume $V_{\text{sphere}} = \frac{4}{3}\pi R^3$. Assume that the plastic shell of the ball has negligible volume. The mass of either filled ball is

$$m_{\text{filled ball}} = m_{\text{ball}} + m_{\text{material}} = m_{\text{ball}} + \rho_{\text{material}} V_{\text{ball}}$$
$$= m_{\text{ball}} + \rho_{\text{material}} \tfrac{4}{3}\pi R_{\text{ball}}^3$$

Solve and evaluate For the neutron-star-filled ball:

$$m_{\text{neutron star ball}} = (0.003 \text{ kg}) + (10^{18} \text{ kg/m}^3)\tfrac{4}{3}\pi(0.020 \text{ m})^3$$
$$= 3.4 \times 10^{13} \text{ kg}$$

For the Saturn-filled ball:

$$m_{\text{Saturn ball}} = m_{\text{ball}} + \left(\frac{m_{\text{Saturn}}}{V_{\text{Saturn}}}\right)\left(\tfrac{4}{3}\pi R_{\text{ball}}^3\right)$$
$$= 0.003 \text{ kg} + \left(\frac{5.7 \times 10^{26} \text{ kg}}{8.3 \times 10^{23} \text{ m}^3}\right)\left(\tfrac{4}{3}\pi(0.020 \text{ m})^3\right)$$
$$= 0.003 \text{ kg} + 0.023 \text{ kg} = 0.026 \text{ kg}$$

The material from Saturn has less mass than an equal volume of water. The ball filled with the material from a neutron star has a mass of more than a billion tons!

- -

Try it yourself The mass of the ping-pong ball filled with soil from Earth's surface is 0.050 kg. What is the density of the soil?

Answer   1400 kg/m^3.

Density and floating

FIGURE 13.2 Less dense matter floats on denser matter.

(a)

Oil

Water

(b)

Water

Corn syrup

Understanding density allows us to pose questions about phenomena that we observe almost every day. For example, why does oil form a film on water? If you pour oil into water or water into oil, they form layers (see **Figure 13.2a**) independently of which fluid is poured first—the layer of oil is always on top of the water. The density of oil is less than the density of water. If you pour corn syrup and water into a container, the corn syrup forms a layer below the water (Figure 13.2b); the density of corn syrup is 1200 kg/m^3, greater than the density of water. Why, when mixed together, is the lower density substance always on top of the higher density substance?

Similar phenomena occur with gases. Helium-filled balloons accelerate upward in air, while air-filled balloons accelerate (slowly) downward. The mass of helium atoms is much smaller than the mass of any other molecules in the air. (Recall that at the same pressure and temperature, atoms and molecules of gas have the same concentration; because helium atoms have much lower mass, their density is lower.) The air-filled balloon must be denser than air. The rubber with which the skin of any balloon is made is denser than air. We can disregard the slight compression of the gas by the balloon, because even though it increases the density of the gas, the effect is the same for both the air and helium in the balloons.

Ice (solid water) floats in liquid water

The solid form of a particular substance is almost always denser than the liquid form of the substance, with one very significant exception: liquid water and solid ice. Since ice floats on liquid water, we can assume that the density of ice is less than that of water. This is in fact true: the density of water changes slightly with temperature and is the highest at 4 °C: 1000 kg/m^3. The density of ice is 917 kg/m^3. Ice has a lower density because in forming the crystal structure of ice, water molecules spread apart. The fact that water expands when it forms ice is important for life on Earth (see the second Reading Passage at the end of this chapter). Fish and plants living in lakes survive cold winters in liquid water under a shield of ice and snow at the surface. Water absorbed in the cracks of rocks freezes and expands in the winter, cracking the rock. Over the years, this process of liquid water absorption, freezing, and cracking eventually converts the rock into soil.

Why do denser forms of matter sink in less dense forms of matter? We learned (in Chapter 12) that the quantity *pressure* describes the forces that fluids exert on each other and on the solid objects they contact. Let us investigate whether pressure explains, for example, why a nail sinks in water or why a hot air balloon rises.

REVIEW QUESTION 13.1 How would you determine the density of an irregularly shaped object?

13.2 Pressure inside a fluid

We know that as gas particles collide with the walls of the container in which they reside, they exert pressure. In fact, if you place any object inside a gas, the gas particles exert the same pressure on the object as the gas exerts on the walls of the container. Do liquids behave in a similar way? In the last chapter we learned that the particles in a liquid are in continual random motion, somewhat similar to particles in gases.

Let's conduct a simple observational experiment. Take a plastic water bottle and poke several small holes at the same height along its perimeter. Close the holes with tacks, fill the bottle with water, open the cap, remove the tacks, and observe what happens (**Figure 13.3**). Identically shaped parabolic streams of water shoot out of the holes. The behavior of the water when the tacks are removed is analogous to a person leaning on a door that is suddenly opened from the other side—the person falls through the door. Evidently, the water inside must push out perpendicular to the wall of the bottle, just as gas pushes out perpendicular to the wall of a balloon. Due to their similar behaviors, liquids and gases are often studied together and are collectively referred to as **fluids**. In addition, since the four streams are identically shaped, the pressure at all points at the same depth in the fluid is the same.

Pascal's first law

Many practical applications involve situations in which an external object (for example, a piston) exerts a force on a particular part of a fluid. What happens to the pressure at other places inside the fluid? To investigate how the pressures at different points in a fluid are related, we use a special instrument that consists of a round glass bulb with holes in it connected to a glass tube with a piston on the other side (**Figure 13.4a**). When we fill the apparatus with water and push the piston, water comes out of all of the holes, not only those that align with the piston. When we fill the apparatus with smoke and push the piston, we get the same result (Figure 13.4b). The liquid and the gas behave similarly.

How can we explain this observation? Pushing the piston in one direction caused a greater pressure in the fluid close to the piston. It seems that almost immediately the pressure throughout the fluid increased as well, as the fluid was pushed out of *all* of the holes in the bulb in the same way. This phenomenon was first discovered by French scientist Blaise Pascal in 1653 and is called Pascal's first law.

> **Pascal's first law** An increase in the pressure of a static, enclosed fluid at one place in the fluid causes a uniform increase in pressure throughout the fluid.

The above experiment describes Pascal's first law macroscopically. We can also explain Pascal's first law at a microscopic level using gases as an example. Gas particles inside a container move randomly in all directions. When we push harder on one of the surfaces of the container, the gas compresses near that surface. The molecules near

FIGURE 13.3 Arcs of water leaving holes at the same level in a bottle.

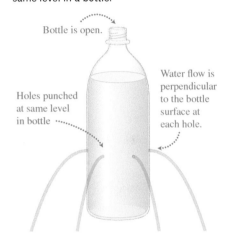

Bottle is open.

Holes punched at same level in bottle

Water flow is perpendicular to the bottle surface at each hole.

FIGURE 13.4 Pascal's first law: increasing the pressure of a fluid at one location causes a uniform pressure increase throughout the fluid.

(a)

Pushing the piston in...

...causes water to come out of the holes.

(b)

Similarly, pushing the piston...

...causes smoke to come out of the holes.

FIGURE 13.5 Glaucoma is an increase in intraocular pressure, caused by blockage of the ducts that normally drain aqueous humor from the eye.

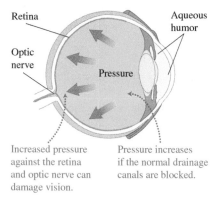

Retina

Optic nerve

Aqueous humor

Pressure

Increased pressure against the retina and optic nerve can damage vision.

Pressure increases if the normal drainage canals are blocked.

that surface collide more frequently with their neighbors farther from the surface. They in turn collide more frequently with their neighbors. The extra pressure exerted at the one surface quickly spreads, and soon there is increased pressure throughout the gas.

Glaucoma

Pascal's first law can help us understand a common eye problem—*glaucoma*. A clear fluid called *aqueous humor* fills two chambers in the front of the eye (**Figure 13.5**). In a healthy eye, new fluid is continually secreted into these chambers while old fluid drains from the chambers through sinus canals. A person with glaucoma has closed drainage canals. The buildup of fluid causes increased pressure throughout the eye, including at the retina and optic nerve, which can lead to blindness. Ophthalmologists diagnose glaucoma by measuring the pressure at the front of the eye. The eye pressure of a person with healthy eyes is about $1.2 \times 10^5 \text{ N/m}^2$; a person with glaucoma has an elevated pressure of $1.3 \times 10^5 \text{ N/m}^2$.

Hydraulic lift

One of the technical applications of Pascal's first law is a *hydraulic lift*, a form of simple machine that converts small forces into larger forces, or vice versa. Automobile mechanics use hydraulic lifts to lift cars, and dentists and barbers use them to raise and lower their clients' chairs. The hydraulic brakes of an automobile are also a form of hydraulic lift. Most of these devices work on the simple principle illustrated in **Figure 13.6**, although the actual devices are usually more complicated in construction.

In Figure 13.6, a downward force $\vec{F}_{1 \text{ on L}}$ is exerted by piston 1 (with small area A_1) on the liquid. This piston compresses a liquid (usually oil) in the lift. The pressure in the liquid just under piston 1 is

$$P_1 = \frac{F_{1 \text{ on L}}}{A_1}$$

Because the pressure changes uniformly throughout the liquid, the pressure under piston 2 is also $P = F_{1 \text{ on L}}/A_1$, with a key assumption that the pistons are at the same elevation. Since piston 2 has a greater area A_2 than piston 1, the liquid exerts a greater upward force on piston 2 than the downward force on piston 1:

$$F_{\text{L on 2}} = PA_2 = \left(\frac{F_{1 \text{ on L}}}{A_1}\right)A_2 = \left(\frac{A_2}{A_1}\right)F_{1 \text{ on L}} \qquad (13.2)$$

Since A_2 is greater than A_1, the lift provides a significantly greater upward force $F_{\text{L on 2}}$ on piston 2 than the downward push of the smaller piston 1 on the liquid $F_{1 \text{ on L}}$.

FIGURE 13.6 Schematic of a hydraulic lift.

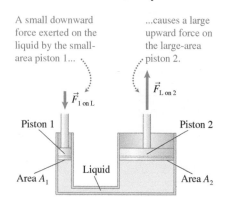

A small downward force exerted on the liquid by the small-area piston 1...

...causes a large upward force on the large-area piston 2.

$\vec{F}_{\text{L on 2}}$

$\vec{F}_{1 \text{ on L}}$

Piston 1

Piston 2

Liquid

Area A_1

Area A_2

EXAMPLE 13.2   **Lifting a car with one hand**

A hydraulic lift similar to that described above has a small piston with surface area 0.0020 m² and a larger piston with surface area 0.20 m². Piston 2 and the car placed on piston 2 have a combined mass of 1800 kg. What is the minimal force that piston 1 needs to exert on the fluid to slowly lift the car?

Sketch and translate The situation is similar to that shown in Figure 13.6. We need to find $F_{1 \text{ on L}}$ so the fluid exerts a force great enough to support the mass of the car and piston 2. The hydraulic lift Eq. (13.2) should then allow us to determine $F_{1 \text{ on L}}$.

Simplify and diagram Assume that the levels of the two pistons are the same and that the car is being lifted at constant velocity. Use the force diagram for the car and piston 2 (see diagram at right) and Newton's second law to determine $F_{\text{L on 2}}$. Note that the force that the liquid exerts on the large piston 2 $F_{\text{L on 2}}$ is equal in magnitude to the

force that piston 2 and the car exert on the liquid $F_{2 \text{ on L}}$, which equals the downward gravitational force that Earth exerts on the car and piston:

$F_{2 \text{ on L}} = m_{\text{Car+Piston}}g$.

$\vec{F}_{1 \text{ on L}}$

$\vec{F}_{\text{L on Car + Piston 2}}$

$\vec{F}_{\text{E on Car + Piston 2}}$

Piston 1

Piston 2

Liquid

$A_1 = 0.0020 \text{ m}^2$ $A_2 = 0.20 \text{ m}^2$

$F_{1 \text{ on L}} = ?$ $m_{\text{Car + Piston 2}} = 1800 \text{ kg}$

Represent mathematically We rewrite the hydraulic lift Eq. (13.2) to determine the unknown force:

$$F_{1 \text{ on L}} = \left(\frac{A_1}{A_2}\right) F_{2 \text{ on L}} = \left(\frac{A_1}{A_2}\right) m_{\text{Car+Piston}} g$$

Solve and evaluate

$$F_{1 \text{ on L}} = \frac{(0.0020 \text{ m}^2)}{(0.20 \text{ m}^2)}\left[(1800 \text{ kg})(9.8 \text{N/kg})\right] = 180 \text{ N}$$

That is the force equal to lifting an object of mass 18 kg, which is entirely possible for a person. The units are also consistent.

Try it yourself If you needed to lift the car about 0.10 m above the ground, what distance would you have to push down on the small piston, assuming the model of the hydraulic lift applies to industrial lifts? Is the answer realistic, and what does it tell you about the operation of real hydraulic lifts?

Answer

10 m. This answer is too large to be realistic. There must be some other technical details involved in the operation of industrial hydraulic lifts.

REVIEW QUESTION 13.2 If you poke many small holes in a closed toothpaste tube and squeeze it, the paste comes out equally from all holes. Why?

13.3 Pressure variation with depth

Pascal's first law states that an increase in the pressure in one part of an enclosed fluid results in an increase at all other parts of the fluid. Does that mean that the pressure is the same throughout a fluid—for example, in a vertical column of fluid? To test this hypothesis, consider another experiment with a water bottle. This time, poke vertical holes along one side of the bottle. Place tacks in the holes, and fill the bottle with water. Leave the cap off. If the pressure is the same throughout the fluid, when the tacks are removed the water should come out of each hole making an arc of the same shape, similar to projectiles thrown horizontally at the same speed, as shown in **Figure 13.7a**.

FIGURE 13.7 Water seems to be pushed harder from holes deeper in the water.

However, when the tacks are removed, we observe that the shapes of the arcs are different. The arcs produced by the water coming out from the lower holes resemble the trajectories of projectiles thrown at higher speeds (Figure 13.7b). Should we abandon Pascal's first law now because the prediction based on it did not match the outcome of the experiment? In such cases, scientists do not immediately throw out the principle but first examine the additional assumptions that were used to make the prediction. In our first experiment with the water bottle, we did not consider the impact of poking holes at different heights. Maybe this was an important factor in the experiment. Let's investigate this in Observational Experiment **Table 13.2**, on the next page.

How does the location of the holes affect the streams leaving the holes?

VIDEO
OET 13.2

Observational experiment	Analysis
Experiment 1. Place two tacks on each side of a plastic bottle, one hole above the other, and fill the bottle with water above the top tack. Remove the tacks. Water comes out on the left and right, and the stream from the lower holes resembles a projectile thrown at a higher speed.	There must be greater pressure inside than outside. The pressure must be greater at the bottom holes than at the top holes.
Experiment 2. Repeat Experiment 1 but this time fill the bottle with water to the same distance above the bottom tack as it was filled above the top tack in Experiment 1. Remove the tacks. The stream comes out the bottom holes with the same arc as it came out of the top holes in Experiment 1.	The total water depth seems not to matter, just the height of the water above the hole.
Experiment 3. Repeat Experiment 1 using a thinner bottle with the water level initially the same distance above the top tack as it was in Experiment 1. Remove the tacks. The water streams are identical to those in Experiment 1.	Because the water comes out in exactly the same arc in a bigger bottle and in a smaller bottle when the water level above the top tack is at the same height, we can conclude that the mass of the water in the bottle does not affect the pressure.

Patterns

The stream shape at a particular level:
- Depends on the depth of the water above the hole.
- Is the same in different directions at the same level.
- Does not depend on the amount of water (volume or mass) above the hole.
- Does not depend on the amount (mass or volume) or depth of the water below the hole.

FIGURE 13.8 Pressure increases with depth.

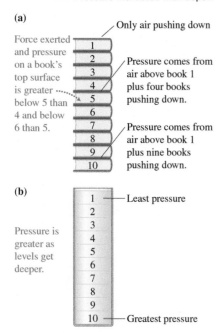

(a)
Force exerted and pressure on a book's top surface is greater below 5 than 4 and below 6 than 5.

Only air pushing down

Pressure comes from air above book 1 plus four books pushing down.

Pressure comes from air above book 1 plus nine books pushing down.

(b)
Pressure is greater as levels get deeper.

Least pressure

Greatest pressure

From the patterns above we reason that the pressure of the fluid at the hole depends only on the *depth* of the fluid above the hole, and not on the *mass* of the fluid above. We also see that the pressure at a given depth is the same in all directions. This is consistent with the experience you have when you dive below the surface of the water in a swimming pool, lake, or ocean. The pressure on your ears depends only on how deep you are below the surface. Pascal's first law fails to explain this pressure variation at different depths below the surface. Now we need to understand why the pressure varies with depth and to devise a rule to describe this variation quantitatively.

Why does pressure vary at different levels?

To explain the variation of pressure with depth we can use an analogy of stacking ten books on a table (**Figure 13.8a**). Imagine that each book is a layer of water in a cylindrical tube (see Figure 13.8b). Consider the pressure (force per unit area) from above on the top surface of each book. The only force exerted on the top surface of the top book is due to air pushing down from above. However, there are in effect two forces exerted on the top surface of the second book: the force that the top book exerts on it (equal in magnitude to the weight of the book) plus the force exerted by the air on the top book. The top surface of the bottom book in the stack must balance the force exerted by the nine books above it (equal in magnitude to the weight of nine books) plus the pressure force exerted by the air on the top book. So the pressure increases on the top surface of each book in the stack as we go lower in the stack.

Similar reasoning applies for the fluid-filled tube divided into a number of imaginary thin layers in Figure 13.8b. Air pushes down on the top layer. The second layer balances the weight of the top layer plus the force exerted by the air pushing down on the top layer, and so on. The pressure is lowest at the top of the fluid and greatest at the bottom.

Note that, at each layer, the pressure in a fluid is the same in all directions. If we could take a pressure sensor and place it inside the container of water, the readings of the sensor would be the same independent of the orientation of the sensor as long as its depth remains the same.

How can we quantify pressure change with depth?

We know that pressure increases with depth, and according to our model of pressure using layers, we can hypothesize that it increases with depth linearly (assuming that the density of the fluid remains constant). But what is the slope of the pressure-versus-depth graph, and what is the intercept? Consider the shaded cylinder C of water shown in Figure 13.9a as our system of interest. The walls on opposite sides of the cylinder push inward, exerting equal-magnitude and oppositely directed forces—the forces exerted by the sides cancel. What about the forces exerted by the water above and below? If the pressure at elevation y_2 is P_2 and the cross-sectional area of the cylinder is A, then the fluid above pushes down, exerting a force of magnitude $F_{\text{fluid above on C}} = P_2 A$ (Figure 13.9b). Similarly, fluid from below the shaded section of fluid at elevation y_1 exerts on the cylinder an upward force of magnitude $F_{\text{fluid below on C}} = P_1 A$. Earth exerts a third force on the shaded cylinder $\vec{F}_{\text{E on C}}$ equal in magnitude to $m_C g$, where m_C is the mass of the fluid in the cylinder. Since the fluid is not accelerating, these three forces add to zero. Choosing the y-axis pointing up, we have

$$\Sigma F_y = (-F_{\text{fluid above on C}}) + F_{\text{fluid below on C}} + (-m_C g) = 0$$

Substituting the earlier expressions for the forces, we have

$$(-P_2 A) + P_1 A + (-m_C g) = 0$$

The mass of the fluid in the shaded cylinder is the product of the fluid's density and the volume of the cylinder (assuming the density of the fluid is the same throughout the cylinder):

$$m_C = \rho_{\text{fluid}} V = \rho_{\text{fluid}}[A(y_2 - y_1)]$$

Substituting this expression for the mass in the above expression for the forces, we get

$$-P_2 A + P_1 A - \rho_{\text{fluid}}[A(y_2 - y_1)]g = 0$$

Divide by the common A in all of the terms and rearrange to get

$$P_1 = P_2 + \rho_{\text{fluid}}(y_2 - y_1)g \tag{13.3}$$

This is Pascal's second law. As we see, pressure varies linearly with depth.

Pascal's second law—variation of pressure with depth The pressure P_1 in a static fluid at position y_1 can be determined in terms of the pressure P_2 at position y_2 as follows:

$$P_1 = P_2 + \rho_{\text{fluid}}(y_2 - y_1)g \tag{13.3}$$

where ρ_{fluid} is the fluid density, assumed constant throughout the fluid, and $g = 9.8 \text{ N/kg}$. The positive y-direction is up. If y_2 is chosen to be the top of the fluid, then Eq. (13.3) can be simplified as

$$P_1 = P_{\text{atm}} + \rho_{\text{fluid}} g d \tag{13.4}$$

where P_{atm} is the atmospheric pressure and d is the depth from the top of the fluid to the level at which we want to determine the pressure.

FIGURE 13.9 Using Newton's second law to determine how fluid pressure changes with the depth in the fluid.

(a)

(b)

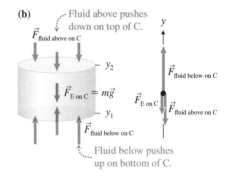

TIP ▸ When using Pascal's second law [Eq. (13.3)], picture the situation and be sure to include a vertical y-axis that points upward and has a defined origin, or zero point. Then choose the two points of interest and identify their vertical y-positions relative to the axis. This lets you relate the pressures at those two points.

Pascal burst a barrel by filling a long, narrow tube with water.

P_{atm}

Very high P

FIGURE 13.10 Pascal tests his second law.

Some history of physics books say that Blaise Pascal conducted the following testing experiment for his second law. Pascal filled a barrel with water and inserted a long, narrow vertical tube into the water from above. He then sealed the barrel (see Figure 13.10). He predicted that when he filled the tube with water, the barrel would burst even though the mass of water in the thin tube was small, because the pressure of the water in the barrel would depend on the depth, not the mass, of the water column. The barrel burst, matching Pascal's prediction, thus supporting the idea that the depth of the fluid above, not its mass determined the pressure.

We derived Eqs. (13.3) and (13.4) using liquid as an example. The density of liquids does not change with depth because they are incompressible. Gases are compressible, and thus their density changes with depth. However, in all our examples, the changes in depth will be small enough so that we can neglect the density change in gases. Applying Eq. (13.3) to atmospheric air, we can explain why the pressure at the top of a mountain is less than at the bottom.

CONCEPTUAL EXERCISE 13.3 **Closed water bottle with tacks**

You have a closed plastic water bottle with two holes in the side closed with tacks, one tack near the bottom of the bottle and the other in the middle. You remove the bottom tack, and a few drops of water come out of the hole but then the leaking stops. Then you remove the top tack (with the bottom hole still open), and an air bubble enters the bottle through the top hole. Draw depth-versus-pressure graphs and force diagrams to explain this phenomenon.

Sketch and translate We sketch the bottle with two tacks in the holes. Let's choose the cap of the bottle to be the origin of the y-axis that points down; the horizontal axis will represent the pressure (Figure a).

Simplify and diagram We consider the pressure above the water level to be atmospheric at the beginning of the experiment (Figure a). Because only a few drops of water leave the bottle, we will neglect the changes in water level on the sketches and graphs. Figures b–d show the removal of the tacks and the changes in pressure on the graphs. We draw force diagrams for a tiny portion of water at the hole as our system of interest. There are two horizontal forces exerted on that system element of water: the force exerted by the outside air pushing inward $\vec{F}_{A\ on\ S}$ and the force of the inside water pushing outward $\vec{F}_{W\ on\ S}$. If one of the forces is smaller, then the tiny system of water will either accelerate outward and we will see the bottle leaking, or it will accelerate inward and we will see bubbles of air coming in. When a little bit of water leaks out, the volume of air above the water in the bottle increases, and its pressure decreases. This decrease leads to a decrease of the pressure everywhere inside the bottle.

(a) Closed bottle
$P_1 = P_{atm}$ before bottom tack removed

Tacks — Water inside the bottle pushes out more than air outside pushes in ($\Delta P = P_{inside} - P_{outside} > 0$)

$-\Delta P > 0$

Pressure outside bottle — Pressure inside bottle

(b) $P_2 < P_{atm}$ after some water leaks out

When the bottom tack is removed, water starts leaking out. When some water leaks out, volume of air at the top increases and P decreases.

$-\Delta P > 0$

$\vec{F}_{A\ on\ S}$ $\vec{F}_{W\ on\ S}$

(c) $P_3 < P_2 < P_{atm}$

When pressure inside the bottle at the hole becomes equal to air pressure outside ($\Delta P = 0$), water stops leaking.

$\Delta P = 0$

$\vec{F}_{A\ on\ S}$ $\vec{F}_{W\ on\ S}$

(d) Second tack removed

$\vec{F}_{A\ on\ S}$ $\vec{F}_{W\ on\ S}$

$\Delta P < 0$

...When we remove the top tack, the water pressure inside is less than the outside pressure ($\Delta P < 0$). Air pushes into the bottle from outside. Bubbles start to enter the bottle.

Try it yourself Predict what will happen after an air bubble enters the bottle through the top hole.

Answer

As new air enters the bottle, the air pressure above the water surface increases and so does the pressure elsewhere in the bottle. This results in more water leaking through the lower hole, which again decreases the pressure and causes more bubbles to enter the bottle, and so on. The process repeats until the water level drops to the upper hole. From then on, the water leaks out continuously until it reaches the level of the bottom hole.

Pop your ears

If your ears did not pop, then what would be the net force exerted by the inside and outside air on your eardrum at the top of a 1000-m-high mountain? You start your hike from sea level. The area of your eardrum is 0.50 cm². The density of air at sea level at standard conditions is 1.3 kg/m³. Assume the air density remains constant during the hike. The situation at the start at $y_1 = 0$ and at the end of the hike at $y_2 = 1000$ m is sketched below.

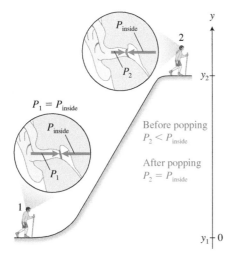

Represent mathematically We use an upward-pointing vertical y-axis with the origin at sea level. Assume that the air pressure inside the eardrum remains constant at its sea level value $P_{inside} = P_1$. The air pressure difference between the top of the mountain and sea level is

$$P_2 - P_1 = \rho(y_1 - y_2)g$$

When at 1000 m above sea level, the outside air exerts a lower pressure P_2 on the eardrum. So the net force exerted by the air on the drum is the pressure difference of air pushing out $P_{inside} = P_1$ and air pushing in P_2 times the area of the eardrum: $F_{net\ air\ on\ drum} = (P_1 - P_2)A$. This pressure difference can be determined using Pascal's second law.

Solve and evaluate

$$P_1 - P_2 = \rho(y_2 - y_1)g$$
$$= (1.3 \text{ kg/m}^3)(1000 \text{ m} - 0)(9.8 \text{ N/kg})$$
$$= +0.13 \times 10^5 \text{ N/m}^2$$

This is 0.13 atm!

$$F_{net\ air\ on\ drum} = (P_1 - P_2)A$$
$$= (0.13 \times 10^5 \text{ N/m}^2)(0.50 \text{ cm}^2) \times (1 \text{ m}/100 \text{ cm})^2$$
$$= +0.60 \text{ N}$$

The net force is exerted outward and is about half the gravitational force that Earth exerts on an apple. No wonder it can hurt until you get some air out of that middle ear!

Try it yourself Determine the difference in water pressure on your ear when you are 1.0 m underwater compared to when you are at the surface. The density of water is 1000 kg/m³.

Answer

or 0.1 atm.
+9800 N/m² greater pressure when under the water.

REVIEW QUESTION 13.3 Pascal's first law says that an increase in pressure in one part of an enclosed liquid results in an increase in pressure throughout all parts of that fluid. Why then does the pressure differ at different heights?

13.4 Measuring atmospheric pressure

We can now use Pascal's second law to develop a method for measuring atmospheric air pressure.

Torricelli's experiments

In the 1600s, suction pumps were used to lift drinking water from wells and to remove water from flooded mines. The suction pump was like a long syringe. The pump consisted of a piston in a long cylinder that pulled up water (**Figure 13.11a**). Such pumps could lift water a maximum of 10.3 m. Why 10.3 m?

Evangelista Torricelli (1608–1647), one of Galileo's students, hypothesized that the pressure of the air in the atmosphere could explain the limit to how far water could be lifted. Torricelli did not know of Pascal's second law, which was published in the year that Torricelli died. However, it is possible that Torricelli's work influenced Pascal. Let's analyze the situation shown in Figure 13.11b.

FIGURE 13.11 A piston pulled up a cylinder causes water to rise to a maximum height of 10.3 m.

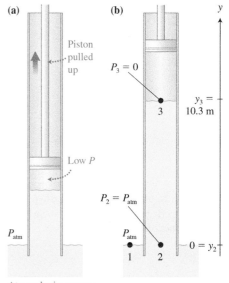

Consider the pressure at three places: point 1, at the water surface in the pool outside the cylinder; point 2, at the same elevation but inside the cylinder; and point 3, in the cylinder 10.3 m above the pool water level. The pressure at point 1 is atmospheric pressure. The pressure at point 2, according to Pascal's second law, is also atmospheric pressure, since it is at the same level as point 1. To get the water to the 10.3-m maximum height, the region above the water surface inside the cylinder and under the piston must be at the least possible pressure—essentially a vacuum. Thus, we assume that the pressure at point 3 is zero.

Now we will use Eq. (13.3) with $y_3 - y_2 = 10.3$ m to predict the pressure P_2:

$$P_2 = P_3 + \rho_{water}(y_3 - y_2)g = 0 + (1.0 \times 10^3 \, \text{kg/m}^3)(10.3 \, \text{m} - 0)(9.8 \, \text{N/kg})$$
$$= 1.01 \times 10^5 \, \text{N/m}^2$$

This number is exactly the value of the atmospheric pressure that we encountered in our discussion of gases (in Chapter 12). The atmospheric pressure pushing down on the water outside the tube can push water up the tube a maximum of 10.3 m if there is a vacuum (absence of any matter) above the water in the tube.

At Torricelli's time, the value of normal atmospheric pressure was unknown, so the huge number that came out of this analysis surprised Torricelli. Not believing the result, he tested it using a different liquid—mercury. Mercury is 14 times denser than water $(\rho_{Hg} = 13,600 \, \text{kg/m}^3)$; hence, the column of mercury should rise only $1/14$ times as high in an evacuated tube. However, instead of using a piston to lift mercury, Torricelli devised a method that guaranteed that the pressure at the top of the column was about zero. Consider Testing Experiment **Table 13.3**.

TESTING
EXPERIMENT TABLE 13.3   **Testing Torricelli's hypothesis using mercury**

Testing experiment	Prediction	Outcome
• Torricelli filled a long glass tube closed at one end with mercury. • He put his finger over the open end and placed it upside down in a dish filled with mercury. He then removed his finger. Predict what he observed based on the hypothesis that atmospheric pressure limits the height of the liquid in a suction pump.	Mercury should start leaking from the tube into the dish. When it leaks, it leaves an empty evacuated space at the top of the tube. It will leak until the height of the mercury column left in the tube produces the same pressure as the atmosphere at the bottom of the column at position 1. The height of the mercury in the tube should be $$y_2 - y_1 = \frac{P_1 - P_2}{\rho_{mercury}g}$$ $$= \frac{(1.01 \times 10^5 \, \text{N/m}^2 - 0)}{(13.6 \times 10^3 \, \text{kg/m}^3)(9.8 \, \text{N/kg})}$$ $$= 0.76 \, \text{m}$$	Torricelli observed some mercury leaking from the tube and then the process stopped. He measured the height of the remaining mercury to be $0.76 \, \text{m} = 760 \, \text{mm}$, in agreement with the prediction.

Conclusion

The outcome of the experiment was consistent with the prediction based on Torricelli's hypothesis that atmospheric pressure limits the height of the liquid being lifted in a suction pump. Thus the hypothesis is supported by evidence.

Torricelli also used his understanding of pressure and fluids to predict that in the mountains, where the atmospheric pressure is lower, the height of the mercury column should be lower. Experiments have shown that the mercury level indeed decreases at higher elevation. These experiments supported the explanation that atmospheric air pushes the liquids upward into the tubes.

Torricelli's apparatus with the mercury tube became a useful device for measuring atmospheric pressure—called a barometer. However, since mercury is toxic, the Torricelli device has since been replaced by the aneroid barometer (described in Chapter 12).

Eugenia, one of the book's authors, stands under a 17th-century diving bell displayed in the Vasa Museum in Stockholm, Sweden.

TIP We now understand why pressure is often measured and reported in mm Hg and why atmospheric pressure is 760 mm Hg. The atmospheric pressure $(101{,}000 \text{ N/m}^2)$ can push mercury (density $13{,}600 \text{ kg/m}^3$) 760 mm up a column.

Diving bell

Our understanding of atmospheric pressure allows us to explain many simple experiments that lead to important practical applications. For example, have you ever submerged a transparent container upside down under water? If you have, you have seen that at first little water enters the container; then as the inverted container is pushed deeper into the water, more water enters the container. One practical application of this phenomenon is a diving bell—a large, bottomless chamber lowered under water with people and equipment inside. Divers use the diving bell to take a break and refill on oxygen. In the past, diving bells were used for underwater construction, such as building bridge foundations.

EXAMPLE 13.5 ▶ **Diving bell**

The bottom of a 4.0-m-tall cylindrical diving bell is at an unknown depth underwater. The pressure of the air inside the bell is $2.0 \times 10^5 \text{ N/m}^2$ (having been about $1.0 \times 10^5 \text{ N/m}^2$ before the bell entered the water). The density of ocean water is slightly higher than fresh water: $\rho_{\text{ocean water}} = 1027 \text{ kg/m}^3$. How high is the water inside the bell, and how deep is the bottom of the bell under the water?

Sketch and translate A labeled sketch shows the situation. We want to find the height h of the water in the bell and the depth d that the bottom of the bell is under the water.

Simplify and diagram Assume that the temperature is constant so we can apply our knowledge of an isothermal process (PV is constant for a constant temperature gas) to the air inside the bell and use it to relate the state of the air inside the bell before it is submerged to its state

after it is submerged. This will let us find the ratio of the air volumes before and after, and from that we can determine h. Once we have that result, we can use Pascal's second law to determine d.

Represent mathematically Apply the mathematical expression for an isothermal process to the process of submerging the bell, where the initial state is just before the bell starts to be submerged in the water and the final state is when submerged to some unknown depth. We have

$$P_3 V_i = P_2 V_f \quad \Rightarrow \quad V_f = \frac{P_3}{P_2} V_i$$

Next, use Pascal's second law to determine d. The bottom of the bell is at $y_1 = 0$; the water surface is at $y_3 = d$. Compare the pressure at the ocean surface P_3 to the pressure at the water surface inside the bell P_2:

$$P_2 = P_3 + \rho(y_3 - y_2)g$$

Solve and evaluate $V_f = \dfrac{P_3}{P_2} V_i = \dfrac{1.0 \times 10^5 \text{ N/m}^2}{2.0 \times 10^5 \text{ N/m}^2} V_i = \dfrac{1}{2} V_i$

Thus, the air volume inside the diving bell is half of what it was before entering the water. The bell is therefore half full of water, which means the height of the water level inside is

$$h = \frac{1}{2}(4.0 \text{ m}) = 2.0 \text{ m} = y_2$$

(CONTINUED)

Rearrange Pascal's second law to solve for the position y_2 of the water level inside the bell:

$$y_2 = 2.0 \text{ m} = y_3 - \frac{P_2 - P_3}{\rho g} = d - \frac{(2.0 \times 10^5 \text{ Pa} - 1.0 \times 10^5 \text{ Pa})}{(1027 \text{ kg/m}^3)(9.8 \text{ N/kg})}$$

$$= d - 10 \text{ m}$$

Thus, $d = 2.0 \text{ m} + 10 \text{ m} = 12 \text{ m}$. The position of the bottom of the bell is 12 m below the ocean's surface.

Try it yourself Suppose the air pressure in the bell is 3.0 atm. In this case, how high is the water in the bell, and how deep is the bottom of the bell?

Answer   The water is 2.7 m high in the bell (there is 1.3 m of air at the top), and the bottom of the bell is 23 m below the ocean's surface.

REVIEW QUESTION 13.4 What does it mean if atmospheric pressure is 760 mm Hg?

13.5 Buoyant force

Pascal's first law tells us that pressure changes in one part of a fluid result in pressure changes in other parts. Pascal's second law describes how the pressure in a fluid varies depending on the depth in the fluid. Do these laws explain why some objects float and others sink? Consider Observational Experiment **Table 13.4**.

OBSERVATIONAL EXPERIMENT TABLE 13.4 **Effect of depth of submersion on a steel block suspended in water**   VIDEO OET 13.4

Observational experiment	Analysis
Experiment 1. Hang a 1.0-kg block from a spring scale. The force that the scale exerts on the block balances the downward force that Earth exerts on the block ($mg = (1.0 \text{ kg})(9.8 \text{ N/kg}) = 9.8 \text{ N}$). 9.8 N	Force diagram for the block B: y $\vec{F}_{\text{S on B}}$ $\vec{F}_{\text{E on B}}$
Experiment 2. Lower the block into a container of water, so it is partially submerged. The water level rises. The reading of the scale decreases. 9.0 N	We explain the decreased reading on the scale by the water pushing upward a little on the block. y $\vec{F}_{\text{S on B}}$ $\vec{F}_{\text{W on B}}$ $\vec{F}_{\text{E on B}}$
Experiment 3. Lower the same block into the container of water to the point where the block is completely submerged. As the water level rises, the reading of the scale decreases. 8.0 N	The upward force exerted by the water increases. y $\vec{F}_{\text{S on B}}$ $\vec{F}_{\text{W on B}}$ $\vec{F}_{\text{E on B}}$

Observational experiment		Analysis
Experiment 4. Lower the block into the container of water so that the block is completely submerged near the bottom. The water level and the reading of the scale do not change.		The upward force exerted by the water does not change once it is completely submerged.

Patterns

We notice two effects:
1. The level of the water in the container rises as more of the block is submerged in the water.
2. The scale reading decreases as more of the block is submerged. The water exerts an upward force on the block. The magnitude of this force depends on how much of the block is submerged. After it is totally submerged, the force does not change, even though the depth of submersion changes.

CONCEPTUAL EXERCISE 13.6   Qualitative force and energy analysis for an object in a fluid

You hang a solid metal cube (with side a) by a string over a rectangular container filled with water up to height a. The cube barely fits inside the container. Starting when the cube is just above the water in the container, you very slowly lower the cube into the water until it is totally submerged. Represent this process by drawing force diagrams for the cube when it is above the water and when it is totally submerged. Also draw an energy bar chart for the cube, Earth, and the water as the system for those states.

Sketch and translate We draw a sketch of the situation; the system and the initial and final states are specified in the problem statement. The submerged cube displaces water that is now above the cube. The volumes of the cube and water are the same, but the water has less mass due to its smaller density.

Simplify and diagram Because you are moving the cube very slowly, we can neglect the kinetic energy of the cube and the water in this process. We also assume that the cube has no acceleration during this motion and all friction effects are negligible. On the force diagrams

we show the three forces exerted by the three objects with which the cube interacts: Earth, the water, and the string. Because there is no acceleration, the sum of the forces is zero.

The string does negative work on the cube because the force it exerts on the cube points opposite to the cube's displacement. Given that both water and the cube interact with Earth, we consider the gravitational potential energies of these interactions separately. We take zero gravitational potential energy to be at the bottom of the container. In the initial state, the system has initial gravitational potential energy due to the location of the cube and water. The gravitational potential energy of cube-Earth interaction decreases as the cube moves down; the gravitational potential energy of water-Earth interaction increases as some water moves up (being replaced by the cube). However, the total gravitational potential energy of the system must decrease due to the negative work done by the string.

$$U_{giC} + U_{giW} + W = U_{gfC} + U_{gfW}$$

Try it yourself Draw the bar charts for the two states, choosing the cube and Earth as a system.

Answer

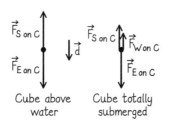

In Table 13.4, after the block is completely submerged, the scale reads 8.0 N instead of 9.8 N. Evidently, the water exerts a 1.8-N upward force on the block. What is the mechanism responsible for this force?

FIGURE 13.12 A fluid exerts an upward buoyant force on the block.

(a) The force exerted on the block by the fluid below is greater than the force exerted by the fluid above.

(b) The upward force of the fluid on the bottom surface is greater than the downward force of the fluid on the top surface.

The magnitude of the force the fluid exerts on a submerged object

Consider *only* the fluid forces exerted on the block shown in **Figure 13.12a**. The fluid pushes inward on the block from all sides, including the top and the bottom. The forces exerted by the fluid on the vertical sides of the block cancel, since the pressure at a specific depth is the same magnitude in all directions.

What about the fluid pushing down on the top and up on the bottom of the block? The pressure is greater at elevation y_1 at the bottom of the block than at elevation y_2 at the top surface of the block. Consequently, the force exerted by the fluid pushing up on the bottom of the block is greater than the force exerted by the fluid pushing down on the top of the block. Arrows in Figure 13.12b represent the forces that the fluid exerts on the top and bottom of the block. The vector sum of these two fluid forces always points up and is called a buoyant force $\vec{F}_{\text{F on B}}$ (fluid on block).

To calculate the magnitude of the upward buoyant force $F_{\text{F on B}}$ exerted by the fluid on the block, we use Eq. (13.3) to determine the pressure P_1 of the fluid on the bottom surface of the block compared to the pressure P_2 of the fluid on the top surface (see Figure 13.12):

$$P_1 = P_2 + \rho_{\text{fluid}}(y_2 - y_1)g$$

The magnitudes of the forces exerted by the fluid on the top and on the bottom of the block are the products of the pressure P and the area A of the top and bottom surfaces of the block:

$$P_1 A = P_2 A + \rho_{\text{fluid}}(y_2 - y_1)Ag$$

or

$$F_1 = F_2 + \rho_{\text{fluid}}(y_2 - y_1)Ag$$

The volume of the block is

$$V_{\text{B}} = A(y_2 - y_1)$$

where A is the cross-sectional area of the block and $(y_2 - y_1)$ is its height. Substitute this volume into the above force equation and rearrange it to get an expression for the magnitude of the total upward **buoyant force** $F_{\text{F on B}}$ that the fluid exerts on the block (the object of interest):

$$F_{\text{F on B}} = F_1 - F_2 = \rho_{\text{fluid}}gV$$

Note that for a totally submerged block, V is the volume of the block. However, when it is partially submerged, V is the volume of the submerged part.

We can now understand the results of the experiments in Table 13.4. When submerging the block further into the water, the scale reading decreased because the buoyant force was increasing. The scale reading stopped changing after the block was completely under the water. Once completely underwater, the submerged volume did not change; the upward buoyant force exerted by the fluid on the block remained constant. Fluids of different densities exert different upward forces on the same object submerged to the same depth (**Figure 13.13**).

FIGURE 13.13 Buoyant force dependence on the depth of the object and the density of the fluid.

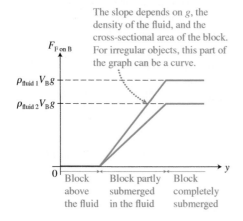

The slope depends on g, the density of the fluid, and the cross-sectional area of the block. For irregular objects, this part of the graph can be a curve.

TIP The derivation of Eq. (13.5) was for a solid cube, but the result applies to objects of any shape, though calculus is needed to establish that.

> **Archimedes' principle—the buoyant force** A stationary fluid exerts an upward buoyant force on an object that is totally or partially submerged in the fluid. The magnitude of the force is the product of the fluid density ρ_{fluid}, the volume $V_{\text{displaced}}$ of the fluid that is displaced by the object, and the gravitational constant g:
>
> $$F_{\text{F on O}} = \rho_{\text{fluid}}V_{\text{displaced}}g = \rho_{\text{fluid}}V_{\text{subm}}g \qquad (13.5)$$

For simplicity, we will always use the volume of the submerged part of the object in our calculations and label it V_{subm}. Because $\rho_{fluid}V_{displaced} = m_{displaced\ fluid}$, we see that **the buoyant force is equal in magnitude to the force that Earth exerts on the amount of fluid displaced by the object.** However, the natures of these two forces are different. Earth exerts a gravitational force at a distance; the buoyant force is a contact force.

REVIEW QUESTION 13.5 Why does a fluid exert an upward force on an object submerged in it?

13.6 Skills for analyzing static fluid problems

In this section we adapt our problem-solving strategy to analyze processes involving static fluids.

**PROBLEM-SOLVING
STRATEGY 13.1** > Analyzing situations involving static fluids

---

EXAMPLE 13.7 **Buoyant force exerted by air on a human**

Suppose your mass is 70.0 kg and your density is 970 kg/m³. If you could stand on a scale in a vacuum chamber on Earth's surface, the reading of the scale would be $mg = (70.0\ \text{kg})(9.80\ \text{N/kg}) = 686\ \text{N}$. What will the scale read when you are completely submerged in air of density 1.29 kg/m³?

Sketch and translate

- Make a labeled sketch of the situation and choose the system of interest. If applicable, decide on the initial and final states.
- Include all known information in the sketch and indicate the unknown(s) you wish to determine.

You are the system.

The scale reads 686 N when in a vacuum. Your density is 970 kg/m³ and the density of air is 1.29 kg/m³. What does it read when you are submerged in air?

Simplify and diagram

- Indicate any assumptions you are making.
- Identify objects outside the system that interact with it.
- Construct a force diagram for the system, including a vertical coordinate axis. The buoyant force is just one of the forces included in the diagram.
- Construct a bar chart or any other graphical representation that might help solve the problem.

Assume that the air density is uniform.

Three objects exert forces on you. Earth exerts a downward gravitational force $F_{E\ on\ Y} = mg = 686\ \text{N}$. The air exerts an upward buoyant force $F_{A\ on\ Y} = \rho_{air}gV_Y$. The scale exerts an unknown upward normal force of magnitude $N_{S\ on\ Y}$.

Represent mathematically

- Use the force diagram to help apply Newton's second law in component form.
- Use the energy bar chart to calculate work and energy if needed.
- Use the expression for the buoyant force and the definitions of pressure and density if needed; sometimes you might need the ideal gas law.

Solve and evaluate

- Insert the known information and solve for the desired unknown.
- Evaluate the final result in terms of units, reasonable magnitude, and whether the answer makes sense in limiting cases.

The y-component form of Newton's second law for your body with zero acceleration is (assuming the upward direction as positive)

$$0 = N_{\text{S on Y}} + F_{\text{A on Y}} + (-F_{\text{E on Y}})$$

or

$$N_{\text{S on Y}} = +F_{\text{E on Y}} - F_{\text{A on Y}}$$

The buoyant force that the air exerts on your body has magnitude $F_{\text{A on Y}} = \rho_{\text{air}} V_{\text{Y}} g$. The volume of your body is $V_{\text{Y}} = (m/\rho_{\text{body}})$. The magnitude of the buoyant force that the air exerts on you is

$$F_{\text{A on Y}} = \rho_{\text{air}} \left(\frac{m}{\rho_{\text{body}}} \right) g = mg \left(\frac{\rho_{\text{air}}}{\rho_{\text{body}}} \right)$$

Thus the reading of the scale should be:

$$N_{\text{S on Y}} = mg - mg \left(\frac{\rho_{\text{air}}}{\rho_{\text{body}}} \right)$$

$$N_{\text{S on Y}} = +686 \text{ N} - (686 \text{ N}) \frac{1.29 \text{ kg/m}^3}{970 \text{ kg/m}^3} = 685 \text{ N}$$

According to Newton's third law, the force that you exert on the scale $N_{\text{Y on S}}$ is equal in magnitude to the force the scale exerts on you.

The reading of the scale is actually 0.1% less when you step on the scale in air—not a big deal. We can usually neglect air's buoyant force. Notice that overall the atmospheric air pushes up on objects, not down.

Try it yourself What will the scale read if you weigh yourself in a swimming pool with your body completely submerged?

Answer

0 N. Because you are less dense than water, the buoyant force exerted by the water on you will completely support you, and the scale will not push upward on you at all.

We will use these strategies to analyze several more situations. In the next example, it might not be obvious how to arrive at the answer to the question with the information provided. Following the suggested problem-solving routine will help you arrive at a solution.

EXAMPLE 13.8   **Is the crown made of gold?**

You need to determine if a crown is made from pure gold or some less valuable metal. From Table 13.1 you know that the density of gold is 19,300 kg/m³. You find that the force that a string attached to a spring scale exerts on the crown is 25.0 N when the crown hangs in air and 22.6 N when the crown hangs completely submerged in water.

Sketch and translate We draw a sketch of the situation and label the givens. If you could measure the mass m_{C} of the crown and its volume V_{C}, you could calculate the density $\rho_{\text{C}} = m_{\text{C}}/V_{\text{C}}$ of the crown—it should be 19,300 kg/m³.

You can determine the mass of the crown easily from the measurement of the scale when the crown hangs in air. But how can you determine the volume from the given information? Crowns have irregular shapes, and it would be difficult to determine its volume by simple measurements and calculations.

Simplify and diagram Let's follow the recommended strategy and see what happens. First, we draw force diagrams for the crown hanging in air and again when hanging in water. When the crown is in air, the upward force exerted by the string attached to the spring scale balances the downward force exerted by Earth. We ignore the buoyant force that air exerts on the crown when hanging in air, since it will be very small in magnitude compared with the other forces exerted on the crown.

When the crown is in water, the upward force exerted by the string (the force measured by the scale) and the upward buoyant force that the water exerts on the crown combine to balance the downward gravitational force that Earth exerts on the crown.

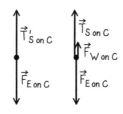

Represent mathematically Since the crown is in equilibrium, the forces exerted on it must add to zero in both cases. When the crown is hanging in air, the vertical component form of Newton's second law is

$$0 = \Sigma F_y = T'_{\text{S on C}} + (-F_{\text{E on C}})$$

where $T'_{\text{S on C}}$ is the 25.0-N string tension force exerted on the crown when it is suspended in air. Thus,

$$F_{\text{E on C}} = m_{\text{C}}g = 25.0 \text{ N}$$

or

$$m_{\text{C}} = \frac{25.0 \text{ N}}{9.8 \text{ N/kg}} = 2.55 \text{ kg}$$

The vertical component form of Newton's second law when the crown hangs in water becomes (the upward direction is positive):

$$\Sigma F_y = T_{\text{S on C}} + F_{\text{W on C}} + (-F_{\text{E on C}}) = 0$$

where $T_{\text{S on C}} = 22.6 \text{ N}$ is the magnitude of the string tension force, and the buoyant force that the water exerts on the crown is $F_{\text{W on C}} = \rho_{\text{W}} V_{\text{C}} g$. Substituting in the above, we get

$$T_{\text{S on C}} + \rho_{\text{W}} V_{\text{C}} g - m_{\text{C}} g = 0$$

Solve and evaluate We see now that the last equation can be used to determine the volume of the crown:

$$V_{\text{C}} = \frac{m_{\text{C}}g - T_{\text{S on C}}}{\rho_{\text{W}}g}$$

$$= \frac{25.0 \text{ N} - 22.6 \text{ N}}{(1000 \text{ kg/m}^3)(9.8 \text{ N/kg})} = 0.000245 \text{ m}^3$$

We now know the crown mass and volume and can calculate its density:

$$\rho = \frac{m}{V} = \frac{2.55 \text{ kg}}{0.000245 \text{ m}^3} = 10{,}400 \text{ kg/m}^3$$

Oops! Since $10{,}400 \text{ kg/m}^3$ is much less than the $19{,}300 \text{ kg/m}^3$ density of gold, the crown is not made of pure gold. The goldsmith must have combined the gold with some less expensive metal.

- -

Try it yourself What is the density of the crown if the scale reads 0 when submerged in water?

Answer 1000 kg/m^3.

REVIEW QUESTION 13.6 Two objects have the same volume, but one is heavier than the other. When they are completely submerged in oil, on which one does the oil exert a greater buoyant force?

13.7 Ships, balloons, climbing, and diving

As we learned in Section 13.1, whether an object floats or sinks depends on its density relative to the density of the fluid. The reason for this lies in the interactions of the object with the fluid and Earth. Specifically, the magnitude of the buoyant force is $F_{\text{F on O}} = \rho_{\text{fluid}} V_{\text{subm}} g$, and the magnitude of the force exerted by Earth is $F_{\text{E on O}} = \rho_{\text{object}} V_{\text{object}} g$. These forces are exerted in the opposite directions. The relative densities of the fluid and the object and consequently the relative magnitudes of the forces determine what happens to the object when placed in the fluid.

- If the object's density is less than that of the fluid $\rho_{\text{object}} < \rho_{\text{fluid}}$, then $\rho_{\text{object}} V_{\text{object}} g < \rho_{\text{fluid}} V_{\text{object}} g$; the object floats *partially* submerged since the buoyant force can balance the gravitational force with less than the entire object below the surface of the fluid.

- If the densities are the same $\rho_{\text{object}} = \rho_{\text{fluid}}$, then $\rho_{\text{object}} V_{\text{object}} g = \rho_{\text{fluid}} V_{\text{object}} g$; the sum of the forces exerted on the object is zero and it remains wherever it is placed totally submerged at any depth in the fluid.

- If the object is denser than the fluid $\rho_{\text{object}} > \rho_{\text{fluid}}$, then $\rho_{\text{object}} V_{\text{object}} g > \rho_{\text{fluid}} V_{\text{object}} g$; the magnitude of the gravitational force is always greater than the magnitude of the buoyant force. The object sinks until it reaches the bottom of the container.

FIGURE 13.14 Making a bottle float with stable equilibrium.

(a)   Bottle partially filled with sand floats upright if the center of mass is below the geometrical center of bottle.

Geometrical center of portion of bottle that is under water

Center of mass of bottle

(b)   If the bottle is tipped, torques due to forces exerted on the bottle return it to the upright position.

(c)   If this bottle tips slightly, torques due to forces exerted on the bottle cause it to overturn.

Clay stuck in upper part of bottle

Air

These cases show that by changing the average density of an object relative to the density of the fluid, the object can be made to float or sink in the same fluid. In this section we investigate this phenomenon and its many practical applications.

Building a stable ship

For years ships were made of wood. In the middle of the 17th century, people decided to try building metal ships. Many thought that this idea was absurd: iron is denser than water and an iron boat would certainly sink. In 1787 British engineer John Wilkinson succeeded in building the first iron ship that did not sink. Since the middle of the 19th century, large ships have been made primarily of steel, which is less dense than iron but much denser than water. These ships can float because part of the volume of a ship is filled with air, which reduces the average density of the ship to a density lower than that of water.

Making a ship or a raft float is only part of the challenge of building watercraft. Another problem is to maintain stable equilibrium for the ship, allowing it to right itself if it tilts to one side due to wind or rough seas. Refresh your knowledge of stable equilibrium (Section 8.6) before you read on.

Consider a floating bottle partially filled with sand. Earth exerts a gravitational force at the center of mass of the bottle (**Figure 13.14a**). The buoyant force exerted by the water on the bottle is effectively exerted at the geometrical center of the part of the bottle that is underwater, which equals the center of mass of the displaced water. If this point is above the center of mass of the bottle, then any slight tipping causes these forces to produce a torque that attempts to return the bottle to an upright position (see Figure 13.14b). However, if the geometrical center of the part of the bottle that is underwater is below the center of mass of the bottle, slight tipping causes the gravitational force to produce a torque that enhances the tipping—unstable equilibrium (Figure 13.14c).

Although ships are more complicated than water bottles filled with sand, it is important to load a ship in such a way that when it begins to heel and one side of the hull begins to rise from the water, the center of mass of the displaced water is above the center of mass of the ship. This is why ships have their cargo stored at the bottom.

EXAMPLE 13.9   Should we take this trip?

The top of an empty life raft of cross-sectional area 2.0 m × 3.0 m is 0.36 m above the waterline. How many 75-kg passengers can the raft hold before water starts to flow over its top? The raft is in seawater of density 1025 kg/m³.

Sketch and translate We make a sketch of the unloaded raft. As people get on the raft, it sinks deeper into the water, and the upward buoyant force increases until the raft reaches a maximum submerged volume, when the maximum number of people are on board. The maximum submerged volume is $V_{subm} = 2.0 \text{ m} \times 3.0 \text{ m} \times 0.36 \text{ m} = 2.16 \text{ m}^3$. We need to determine the maximum buoyant force the seawater can exert on the raft and then decide how to convert this into the number of passengers the raft can hold. The raft and the passengers are our system of interest.

Empty raft

3.0 m   2.0 m wide

0.36 m

$\rho_{Water} = 1025 \text{ kg/m}^3$

Simplify and diagram We have no information about the mass of the raft; thus we will assume it is negligible. We draw a sketch of the filled raft and a force diagram for the raft with passengers (the system). The vertical axis points up. There are two forces exerted on the system: the upward force exerted by the water $\vec{F}_{W \text{ on } S}$ of magnitude $\rho_{water} g V_{subm}$ and the downward force exerted by Earth $\vec{F}_{E \text{ on } S}$. The magnitude of the force Earth exerts on N people is $N_{people} m_{person} g$. As the system is in equilibrium, the net force exerted on it is zero.

Raft sinks deeper in water as more people enter it.

System

N = ?

$\vec{F}_{W \text{ on } S}$

$\vec{F}_{E \text{ on } S}$

The maximum submerged volume is (2.0 m)(3.0 m)(0.36 m)

Represent mathematically Using the upward direction as positive, apply the vertical component form of Newton's second law:

$$\Sigma F_y = F_{W \text{ on S}} + (-F_{E \text{ on S}}) = 0$$

$$\rho_{water} g V_{subm} - N_{people} m_{person} g = 0$$

Assuming that all people have the same mass, we find the number of people:

$$N_{people} = \frac{\rho_{water} V_{subm} g}{m_{person} g} = \frac{\rho_{water} V_{subm}}{m_{person}}$$

Solve and evaluate

$$N_{people} = \frac{\rho_{water} V_{subm}}{m_{person}}$$

$$= \frac{(1025 \text{ kg/m}^3)(0.36 \text{ m} \times 2.0 \text{ m} \times 3.0 \text{ m})}{75 \text{ kg}} = 29.5$$

The raft can precariously hold 29 passengers, which is a reasonable number. The number is inversely proportional to the mass of a person. This makes sense—the heavier the people, the fewer of them the raft should hold. The units, dimensionless, also make sense. We assumed that the raft has negligible mass. If we take the mass into account, the number of people will be smaller.

Try it yourself Suppose that 10 people of average mass 80 kg entered the raft. Now how far would the water line be from the top of the raft?

Answer

The raft would sink 0.13 m into the water, and the water line would be 0.23 m below the top of the raft.

Ballooning

Balloons used for transportation are filled with hot air. Why hot air? The density of 100 °C air is 0.73 times the density of 0 °C air. Thus, balloonists can adjust the average density of the balloon (the balloon's material, people, equipment, etc.) to match the density of air so that the balloon can float at any location in the atmosphere (up to certain limits). A burner under the opening of the balloon regulates the temperature of the air inside the balloon and hence its volume and density. This allows control over the buoyant force that the outside cold air exerts on the balloon. The same approach that we used in Example 13.9 allows us to predict that a balloon with radius of 5.0 m and a mass of 20.0 kg filled with air at the temperature of 100 °C can carry about 160 kg (three slim 53-kg people or two medium-mass 80-kg people). This is not a heavy load.

At one time, hydrogen was used in closed balloons instead of air. The density of hydrogen is 1/14 times the density of air. Unfortunately, hydrogen can burn explosively in the presence of oxygen. The hydrogen-filled Hindenburg, a German airship, caught fire and exploded in 1937, killing 36 people (see **Figure 13.15**). Balloonists then turned to helium—an inert gas that does not interact readily with other types of atoms.

FIGURE 13.15 The Hindenburg explosion.

Effects of altitude on humans

Table 13.5 on the next page shows that atmospheric pressure decreases with altitude. Therefore, climbers and balloonists have to guard against altitude sickness, caused by the low pressure and lack of oxygen. Below 3000 m, altitude has little effect on performance. Between 3000 m and 4600 m, climbers experience compensated hypoxia—increased heart and breathing rates. Between 4600 m and 6000 m, manifest hypoxia sets in. Heart and breathing rates increase dramatically, and cognitive and sensory function and muscle control decline. Climbers may feel lethargy and euphoria and even experience hallucinations. Between 6000 m and 8000 m, climbers undergo critical hypoxia, characterized by rapid loss of muscular control, loss of consciousness, and possibly death.

These symptoms were exhibited clearly on April 15, 1875 by three French balloon pioneers attempting to set an altitude record. They carried bags of oxygen with them, but as their elevation increased, slowly lost the mental awareness needed to use the bags. Instruments indicate that the balloon reached a maximum elevation of 8600 m twice. During the second time, two of the balloonists died. The third lost consciousness but survived.

TABLE 13.5 **Pressure of air and pressure due to oxygen in the air (called partial pressure) at different elevations**

Location	Elevation (m)	P_{air} (atm)	P_{oxygen} (atm)
Sea level	0	1.0	0.21
Mount Washington	1917	0.93	0.18
Pikes Peak	4301	0.59	0.12
Mount McKinley	6190	0.47	0.10
Mount Everest	8848	0.34	0.07
Jet travel	12,000	0.23	0.05

Scuba diving

The sport of scuba diving depends on an understanding of fluid pressure and buoyant force to avoid cases of excess internal pressure, oxygen overload, and decompression sickness.

Assuming that the surface area of your body is about 2 m², the air exerts a 200,000-N force (about 20 tons) on the surface of your body. Fortunately, fluids inside the body push outward and balance the force exerted by the outside air. For example, the pressure inside your lungs is approximately atmospheric pressure. What would happen if the fluid pressure on the inside remained constant while the pressure on the outside doubled or tripled? Would you be crushed, the way a can or barrel is crushed by outside air pressure when the air pressure inside the can is much lower than the pressure outside (**Figure 13.16**)? Scuba divers face this problem.

We know that atmospheric pressure (1 atm) is equivalent to the pressure of a 10-m column of water. Therefore, at depth $d = 10$ m, the water pressure is 2 atm. At 40 m below the water surface, the pressure is about 5 atm. This would surely be a problem for a scuba diver if the internal pressure were only 1 atm!

To avoid this problem, divers breathe compressed air. While moving slowly downward, a diver adjusts the pressure outlet from the compressed air tank in order to accumulate gas from the cylinder into her lungs and subsequently into other body parts, increasing the internal pressure to balance the increasing external pressure. If a diver returns to the surface too quickly, the great gas pressure in the lungs can force bubbles of gas into the bloodstream. These bubbles can behave like blood clots, blocking blood flow to the brain and possibly causing death. Blood vessels can rupture if the pressure difference between the inside and outside of the vessel is too great. Thus, a diver rises to the surface slowly so that pressure changes gradually and bubbles of gas do not form. This gradual process is called *decompression*.

When humans travel to dangerous environments (mountaintops, the deep sea, or outer space), physics intersects with human physiology. A careful understanding of gases, liquids, and the effects of changing pressures on the human body is needed to allow humans to survive in these places. As we explore the universe, we will need to learn to adapt to ever more challenging environments.

FIGURE 13.16 External air pressure collapses an evacuated can.

When air is pumped out of a can, outside air pressure causes it to collapse.

REVIEW QUESTION 13.7 A ship's waterline marks the maximum safe depth of the ship in the water when it has a full cargo. An empty ship is at the dock with its waterline somewhat above the water level. How could you estimate its maximum cargo?

Summary

Density ρ The ratio of the mass m of a substance divided by the volume of that substance. (Section 13.1)

$$\rho = \frac{m}{V}$$ Eq. (13.1)

Pascal's first law—hydraulic lift An increase in the pressure in one part of an enclosed fluid increases the pressure throughout the fluid. In a hydraulic lift, a small force F_1 exerted on a small piston of area A_1 can cause a large force F_2 to be exerted on a large piston of area A_2. (Section 13.2)

For a hydraulic lift:

$$F_{\text{F on 2}} = PA_2 = (A_2/A_1)\, F_{\text{1 on F}}$$ Eq. (13.2)

Pascal's second law—variation of pressure with depth On a vertical upward-pointing y-axis, the pressure of a fluid P_1 at position y_1 depends on the pressure P_2 at position y_2 and on the density of the fluid. (Section 13.3)

$$P_1 = P_2 + \rho_{\text{fluid}}(y_2 - y_1)g$$ Eq. (13.3)

Buoyant force A fluid exerts an upward-pointing buoyant force on an object totally or partially immersed in the fluid. The force depends on the density of the fluid and on the volume of the fluid displaced. Once the object is totally submerged, the buoyant force does not change. (Section 13.5)

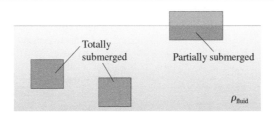

$$F_{\text{F on O}} = \rho_{\text{fluid}}V_{\text{displaced}}g$$ Eq. (13.5)

Newton's second law Use the standard problem-solving strategies with this law (sketches, force diagrams, math descriptions) to find some unknown quantity. The problems often involve the buoyant force, pressure, and density. (Section 13.6)

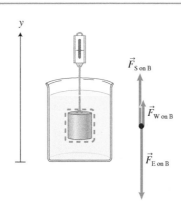

$$a_y = \frac{\Sigma F_y}{m}$$ Eq. (3.7y)

Questions

Multiple Choice Questions

1. Rank in increasing order the pressure that the *italicized objects* exert on the surface.
   I.   A person standing with *bare feet* on the floor
   II. A person in *skis* standing on snow
   III. A person in *Rollerblades* standing on a road
   IV. A person in *ice skates* standing on ice
   (a) I, II, III, IV   (b) IV, III, I, II   (c) IV, III, II, I
   (d) III, II, IV, I   (e) II, I, III, IV

2. Choose a device that reduces the pressure caused by a force.
   (a) Scissors   (b) Knife   (c) Snowshoes
   (d) Nail   (e) Syringe

3. What does it mean if the density of a material equals 2000 kg/m³?
   (a) The mass of the material is 2000 kg.
   (b) The volume of the material is 1 m³.
   (c) The ratio of the mass of any amount of this material to the volume is equal to 2000 kg/m³.

4. An upside-down mug with some air trapped in it is fixed under water, as shown in **Figure Q13.4**. Which qualitative pressure-versus-position graph correctly shows how the gauge pressure changes along the dashed line through the mug from A to D?

FIGURE Q13.4

(a) *P*

(b) *P*

(c) *P*

(d) *P*

5. If you hold a cylinder vertically, what is the net force exerted by the atmospheric pressure on it?
   (a) Downward   (b) Upward   (c) Zero

6. How do we know that a fluid exerts an upward force on an object submerged in the fluid?
   (a) Fluid pushes on the object in all directions.
   (b) The reading of a scale supporting the object when submerged in the fluid is less than when not in the fluid.
   (c) The fluid pressure on the bottom of the object is greater than the pressure on the top.
   (d) Both b and c are correct.

7. When you suspend an object from a spring scale, it reads 15 N. Then you place the same object and scale under a vacuum jar and pump out the air. What happens to the reading of the scale?
   (a) It increases slightly.
   (b) It decreases slightly.
   (c) It says the same.
   (d) Don't have enough information to answer

8. Why can't a suction pump lift water higher than 10.3 m?
   (a) Because it does not have the strength to pull up higher
   (b) Because the atmospheric pressure is equal to the pressure created by a 10.3-m-high column of water
   (c) Because suction pumps are outdated lifting devices
   (d) Because most suction cups have an opening to the bulb that is too narrow

9. If Torricelli had a wider tube in his mercury barometer, what would the height of the mercury column in the tube do?
   (a) Decrease   (b) Increase   (c) Stay the same

10. A wooden cube is floating in a fish tank that is filled with water. Imagine that you take this setup to a space station on the Moon. Air pressure and temperature inside the station are similar to conditions on Earth. After bringing the setup from Earth to the space station, you will observe that (choose all correct statements)
    (a) the amount of water displaced by the cube increases.
    (b) the amount of water displaced by the cube decreases.
    (c) the amount of water displaced by the cube stays the same.
    (d) the buoyant force exerted on the cube increases.
    (e) the buoyant force exerted on the cube decreases.
    (f) the buoyant force exerted on the cube stays the same.
    (g) the pressure at the bottom of the fish tank is about 1/6 of the value on Earth.
    (h) the pressure at the bottom of the fish tank is significantly less than 1/6 of the value on Earth.
    (i) the pressure at the bottom of the fish tank is slightly less than the value on Earth.
    (j) the pressure at the bottom of the fish tank is the same as on Earth.

11. Two identical beakers with the same amount of water sit on the arms of an equal arm balance. A wooden block floats in one of them. What does the scale indicate?
    (a) The beaker with the block is heavier.
    (b) The beaker without the block is heavier.
    (c) The beakers weigh the same.

12. A piece of steel and a bag of feathers are suspended from two spring scales in a vacuum. Each scale reads 100 N. What happens when you repeat the experiment outside under normal conditions?
    (a) The scale with feathers reads more than the scale with steel.
    (b) The scale with feathers reads less that the scale with steel.
    (c) The scales have the same reading, but the reading is less than the reading in a vacuum.

13. A metal boat floats in a pool. What happens to the level of the water in the pool if the boat sinks?
    (a) It rises.
    (b) It falls.
    (c) It stays the same.

14. When a boat sails from seawater to fresh water, the buoyant force exerted on the boat
    (a) decreases.   (b) increases.   (c) stays the same.

15. Three blocks are floating in oil as shown in **Figure Q13.15**. Which block has the highest density?
    (a) A   (b) B   (c) C
    (d) All blocks have the same density.

16. Three blocks are floating in oil as shown in Figure Q13.15. On which block does the oil exert the greatest buoyant force?
    (a) A   (b) B   (c) C
    (d) The oil exerts the same force on all of them.

FIGURE Q13.15

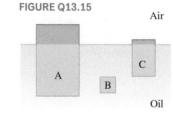

Conceptual Questions

17. Describe a method to measure the density of a liquid.
18. How can you determine the density of air?
19. Design an experiment to determine whether air has mass.
20. Does air exert a net upward force or a net downward force on an object submerged in the air? How can you test your answer experimentally?
21. What causes the pressure that air exerts on a surface that is in the air?
22. Why, when you fill a teapot with water, is the water always at the same level in the teapot and in the spout?
23. What experimental evidence supports Pascal's first law?

24. Fill a plastic cup to the very top with water. Put a piece of paper on top of the cup so that the paper covers the cup at the edges and is not much bigger than the surface of the cup. Turn the cup and paper upside down (practice over the sink first) and hold the bottom of the cup (now on the top). Why doesn't the water fall out of the cup?

25. Why does a fluid exert a net upward force on an object submerged in the fluid?

26. Describe how you could predict whether an object will float or sink in a particular liquid without putting it into the liquid.

27. Why can you lift objects while in water that are too heavy to lift when in the air?

28. When placed in a lake, a solid object either floats on the surface or sinks. It does not float at some intermediate location between the surface and the bottom of the lake. However, a weather balloon floats at some intermediate distance between Earth's surface and the top of its atmosphere. Explain.

29. A flat piece of aluminum foil sinks when placed under water. Take the same piece and shape it so that it floats in the water. Explain why the method worked.

30. Ice floats in water in a beaker. Will the level of the water in the beaker change when the ice melts? Explain.

31. The density of ice at 0 °C is less than the density of water at 0 °C. How is this related to the existence of life on Earth?

32. How would you determine the density of an irregular-shaped unknown object if (a) it sinks in water and (b) it floats in water? List all the steps and explain the reasoning behind them.

33. Why do people sink in fresh water and in most seawater (if they do not make an effort to stay afloat) but do not sink in the Dead Sea?

34. A bucket filled to the top with water has a piece of ice floating in it. Will the pressure on the bottom of the bucket change when the ice melts? Explain.

35. Marjory thinks that the mass of a fluid above a certain level should affect the pressure at this level. Describe how you will test her idea.

36. You are holding a brick that is completely submerged in water. Draw a force diagram for the brick. Why does it feel lighter in water than when you hold it in the air?

37. A bucket filled with water has a piece of ice floating in it. Will the level of water rise when the ice melts? Justify your answer.

38. Explain qualitatively and quantitatively how we drink through a straw. Make sure you can account for the water going up the length of the straw.

39. Three test tubes are inverted in a Petri dish as shown in Figure Q13.39. The first is completely filled with water, the second has an air bubble at the top, and the third has oil at the same level as the air bubble. (a) Rank the pressures at points A, B, and C from largest to smallest. (b) Rank the pressures at points A′, B′, and C′ from largest to smallest.

FIGURE Q13.39

Problems

Below, **BIO** indicates a problem with a biological or medical focus. Problems labeled **EST** ask you to estimate the answer to a quantitative problem rather than derive a specific answer. Asterisks indicate the level of difficulty of the problem. Problems with no * are considered to be the least difficult. A single * marks moderately difficult problems. Two ** indicate more difficult problems.

13.1 Density

1. Determine the average density of Earth. What data did you use? What assumptions did you make?

2. * **EST** **Height of atmosphere** Use data for the normal pressure and the density of air near Earth's surface to estimate the height of the atmosphere, assuming it has *uniform* density. Indicate any additional assumptions you made. Are you on the low or high side of the real number?

3. **EST** A single-level home has a floor area of 200 m^2 with ceilings that are 2.6 m high. Estimate the mass of the air in the house.

4. * **BIO** A diet decreases a person's mass by 5%. Exercise creates muscle and reduces fat, thus increasing the person's density by 2%. Determine the percent change in the person's volume.

5. **Pulsar density** A pulsar, an extremely dense rotating star made of neutrons, has a density of 10^{18} kg/m^3. Determine the mass of a pulsar contained in a volume the size of your fist (about 200 cm^3).

6. Use the graph lines in Figure P13.6 to determine the densities of the three liquids, A, B, and C, in SI units. If you place them in one container, how will they position themselves? How does the density of each liquid change as its volume increases? As its mass decreases? Compare the masses of the three liquids when they occupy the same volume. Compare the volumes of the three liquids when they have the same mass.

FIGURE P13.6

7. Imagine that you have gelatin cut into three cubes: the side of cube A is *a* cm long, the side of cube B is double the side of A, and the side of cube C is three times the side of A. Compare the following properties of the cubes: (a) density, (b) volume, (c) surface area, (d) cross-sectional area, and (e) mass.

8. An object made of material A has a mass of 90 kg and a volume of 0.45 m^3. If you cut the object in half, what would be the density of each half? If you cut the object into three pieces, what would be the density of each piece? What assumptions did you make?

9. You have a steel ball that has a mass of 6.0 kg and a volume of 3.0×10^{-3} m^3. How can this be?

10. * A material is made of molecules of mass 2.0×10^{-26} kg. There are 2.3×10^{29} of these molecules in a 2.0-m^3 volume. What is the density of the material?

11. You compress all the molecules described in Problem 13.10 into 1.0 m^3. Now what is the density of the material? What type of material could possibly behave this way?

12. * Bowling balls are heavy. However, some bowling balls float in water. Use available resources to find the dimensions of a bowling ball and explain why some balls float while others do not.

13. * **EST** Estimate the average density of a glass full of water and then the glass when the water is poured out (do not forget the air that now fills the glass instead of water).

13.2 Pressure inside a fluid

14. * Anita holds her physics textbook and complains that it is too heavy. Andrew says that her hand should exert no force on the book because the atmosphere pushes up on it and balances the downward pull of Earth on the book (the book's weight). Jim disagrees. He says that the atmosphere presses down on things and that is why they feel heavy. Who is correct? Approximately how large is the force that the atmosphere exerts on the bottom of the book? Why does this force not balance the force exerted by Earth on the book?

15. * The air pressure in the tires of a 980-kg car is 3.0×10^5 N/m². Determine the average area of contact of each tire with the road. Indicate any assumptions that you made.

16. * **EST** Estimate the pressure that you exert on the floor while wearing hiking boots. Now estimate the pressure under each heel if you change into high-heeled shoes. Indicate any assumptions you made.

17. **Hydraulic car lift** You are designing a hydraulic lift for a machine shop. The average mass of a car it needs to lift is about 1500 kg. What should be the specifications on the dimension of the pistons if you wish to exert a force on a smaller piston of not more than 500 N? How far down will you need to push the piston in order to lift the car 30 cm?

18. **EST** **Force of air on forehead** Estimate the force that air exerts on your forehead. Describe the assumptions you made.

19. * **EST** A 30-cm-diameter cylindrical iron plunger is held against the ceiling, and the air is pumped from inside it. A 72-kg person hangs by a rope from the plunger (**Figure P13.19**). List the quantities that you can estimate about the situation and estimate them. Make assumptions if necessary.

FIGURE P13.19

Flat surface of hollow cylinder

20. You have a rubber pad with a handle attached to it (**Figure P13.20**). If you press the pad firmly on a smooth table, it is impossible to lift it off the table. Why? What force would you need to exert on the handle to lift it? The surface area of the pad is 0.023 m².

FIGURE P13.20

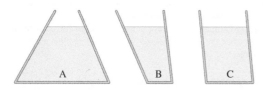

21. * **EST** **Toy bow and arrow** A child's toy arrow has a suction cup on one end. When the arrow hits the wall, it sticks. Draw a force diagram for the arrow stuck on the wall and estimate the magnitudes of the forces exerted on it when it is in equilibrium. The mass of the arrow is about 10 g. Why are the words "suction cup" not appropriate?

13.3 Pressure variation with depth

22. * **Pressure on the Titanic** The Titanic rests 4 km (2.5 miles) below the surface of the ocean. What physical quantities can you determine using this information?

23. You have three reservoirs (**Figure P13.23**). Rank the pressures at the bottom of each and explain your rankings. Then rank the net force that the water exerts on the bottom of each reservoir. Explain your rankings.

FIGURE P13.23

24. **Water reservoir and faucet** The pressure at the top of the water in a city's gravity-fed reservoir is 1.0×10^5 N/m². Determine the pressure at the faucet of a home 42 m below the reservoir.

25. **Dutch boy saves Holland** An old story tells of a Dutch boy who used his fist to plug a 2.0-cm-diameter hole in a dike that was 3.0 m below sea level, thus preventing the flooding of part of Holland. What physical quantities can you determine from this information? Determine them.

26. **BIO** **EST** **Blood pressure** Estimate the pressure of the blood in your brain and in your feet when standing, relative to the average pressure of the blood in your heart of 1.3×10^4 N/m² above atmospheric pressure.

27. * **BIO** **Intravenous feeding** A glucose solution of density 1050 kg/m³ is transferred from a collapsible bag through a tube and syringe into the vein of a person's arm. The blood pressure in the arm exceeds the atmospheric pressure by 1400 N/m². How high above the arm must the top of the liquid in the bottle be so that the pressure in the glucose solution at the needle exceeds the pressure of the blood in the arm?

28. * **Mountain climbing** Determine the change in air pressure as you climb from elevation of 1650 m at the timberline of Mount Rainier to its 4392-m summit, assuming an average air density of 0.82 kg/m³. Will the real change be more or less than the one you calculated? Explain.

29. **BIO** **EST** **Giraffe raises head** Estimate the pressure change of the blood in the brain of a giraffe when it lifts its head from the grass to eat a leaf on an overhead tree. Without special valves in its circulatory system, the giraffe could easily faint when lifting its head.

30. * A truck transporting chemicals has crashed, and some dangerous liquid has spilled onto the ground and possibly entered a water well. An inspector fixes a pressure sensor to the end of a long string and lets the sensor slowly descend from the top of the well to the bottom. Using this device, he obtains the graph in **Figure P13.30** that shows how the pressure P in the well changes with distance d measured from the top of the well. (a) Explain what features of the graph support the idea that there is another liquid in the well in addition to water. (b) Determine the density of the unknown liquid. Is the liquid above the water or below the water? (c) Determine the depth of the water, the depth of the unknown liquid, and the depth of the well.

FIGURE P13.30

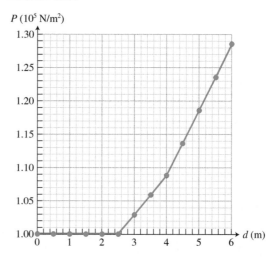

31. **Drinking through a straw** You are drinking water through a straw in an open glass. Select a small volume of water in the straw as a system and draw a force diagram for the water inside this volume that explains why the water goes up the straw.

32. * **More straw drinking** While you are drinking through the straw, the pressure in your mouth is 30 mm Hg below atmospheric pressure. What is the maximum length of a straw in an open glass that you can use to drink a fruit drink of density 1200 kg/m³?

33. * Your office has a 0.020 m³ cylindrical container of drinking water. The radius of the container is about 14 cm. When the container is full, what is the gauge pressure that the water exerts on the sides of the container halfway down from the top? All the way down?

34. * **BIO** **EST** **Eardrum** Estimate the net force on your 0.5-cm² eardrum that air exerts on the inside and the outside after you drive from Denver, Colorado (elevation 1609 m) to the top of Pikes Peak (elevation 4301 m). Assume that the air pressure inside and out are balanced when you leave Denver and that the average density of the air is 0.80 kg/m³. What other assumptions did you make?

35. **BIO** **Eardrum again** You now go snorkeling. What is the net force exerted on your eardrum when you are 2.4 m under the water, assuming the pressure was equalized before the dive?

36. Water and oil are poured into opposite sides of an open U-shaped tube. The oil and water meet at the exact center of the U at the bottom of the tube. If the column of oil of density 900 kg/m³ is 16 cm high on one side, how high is the water on the other side?

37. * Examine the vertical cross section of the Hoover Dam shown in Figure P13.37. Explain why the dam is thicker at the bottom than at the top.

FIGURE P13.37

38. * A test tube of length L and cross-sectional area A is submerged in water with the open end down so that the edge of the tube is a distance h below the surface. The water goes up into the tube so its height inside the tube is l. Describe how you can use this information to decide whether the air that was initially in the tube obeys the mathematical description for an isothermal process (Boyle's law). List your assumptions.

39. Half of a 20-cm-tall beaker is filled with glycerol (density 1259 kg/m³) and the other half with olive oil (density 800 kg/m³). (a) Draw a graph that shows how the density of the liquids in the beaker changes with the distance from the surface to the bottom of beaker. (b) Draw a graph that shows how the pressure in the liquids changes with the distance from the surface to the bottom of the beaker.

40. Blaise Pascal found a seemingly paradoxical situation when he poured water into the apparatus shown in Figure P13.40. The water level was the same in all four parts of the apparatus despite differences in the shapes of the parts and the masses of water in each part. Explain qualitatively the outcome of Pascal's experiment.

FIGURE P13.40

41. Four containers are filled with different volumes of water as shown in Figure P13.41. Rank the containers in order of decreasing pressure that the water exerts on the bottom of the containers.

FIGURE P13.41

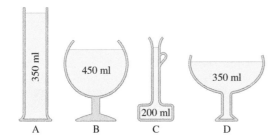

42. **Venus pressure and underwater pressure** Atmospheric pressure on Venus is 9.0 × 10⁶ N/m². How deep underwater on Earth would you have to go to feel the same pressure?

13.4 Measuring atmospheric pressure

43. The reading of a barometer in your room is 780 mm Hg. What does this mean? What is the pressure in pascals?

44. How long would Torricelli's barometer have had to be if he had used oil of density 950 kg/m³ instead of mercury?

45. Sometimes gas pressure is measured with a device called a *liquid manometer* (Figure P13.45). Explain how this instrument can be used to measure the pressure of gas in a bulb attached to one of the tubes.

FIGURE P13.45   FIGURE P13.46

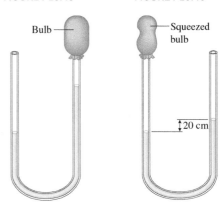

46. You use a liquid manometer with water to measure the pressure inside a rubber bulb. Before you squeeze the bulb, the water is at the same level in both legs of the tube. After you squeeze the bulb, the water in the opposite leg rises 20 cm with respect to the leg connected to the bulb (Figure P13.46). What is the pressure in the bulb? What assumptions did you make? How will the answer change if the assumptions are not valid?

47. * In a mercury-filled manometer (Figure P13.47), the open end is inserted into a container of gas and the closed end of the tube is evacuated. The difference in the height of the mercury is 80 mm. The radius of the connecting tube is 0.50 cm. (a) Determine the pressure inside the container in newtons per square meter. (b) An identical manometer has a connecting tube that is twice as wide. If the difference in the height of the mercury is the same, then what is the pressure in the container?

FIGURE P13.47

48. Examine the reading of the manometer that you use to measure the pressure inside car tires. What are the units? Does the manometer measure the absolute pressure of the air inside the tires or gauge pressure? How do you know?

13.5 Buoyant force

49. Draw a force diagram for an object that is floating at the surface of a liquid. Is the force exerted by air on the object included in your diagram? Explain.

50. Draw a cubic object that is completely submerged in a fluid but not resting on the bottom of the container. Then draw arrows to represent the forces exerted by the fluid on the top, sides, and bottom of the object. Make the arrows the correct relative lengths. What is the direction of the total force exerted by the fluid on the object?

51. Three people are holding three identical sealed metal containers, ready to release them. Container A is filled with air at atmospheric pressure and room temperature, container B is filled with helium at atmospheric pressure and room temperature, and container C is evacuated (all air has been pumped out). Draw force diagrams for the containers after they are released.

52. * Four cubes of the same volume are made of different materials: lead (density 11,300 kg/m³), aluminum (density 2700 kg/m³), wood (density 800 kg/m³), and Styrofoam (density 50 kg/m³). You place the cubes in a large container filled with water. Rank the buoyant forces that the water exerts on the cubes from largest to smallest.

53. * You place four identical cubes made of oak (density 900 kg/m³) in water, olive oil (density 880 kg/m³), alcohol (density 790 kg/m³), and mercury (density 13,600 kg/m³). Rank the buoyant forces that the liquids exert on the cubes from largest to smallest.

54. ** You fill a 20-cm-tall container with glycerol so that the glycerol (density 1260 kg/m^3) reaches the 10-cm mark. You place an oak cube (density 900 kg/m^3) into the container. Each side of the cube is 10 cm. (a) What is the distance x from the upper face of the cube to the glycerol surface assuming the cube is in an upright position? (b) Make a prediction using qualitative reasoning (without mathematics) about what will happen to the cube in (a) if you now add olive oil (density 800 kg/m^3) to the container until it is completely full. Will the distance x decrease, increase, or stay the same? (c) Determine the new distance x using mathematics (physics laws) and compare your result with the prediction in (b).

55. * A 30-g ball with volume 37.5 cm^3 is attached to the bottom of a glass beaker with a light string. When the beaker is filled with water, the ball floats fully submerged under the water surface. Draw a force diagram for the ball and determine the force exerted on the ball by the string.

56. ** You have a ball (volume V, average density ρ_B) and a glass beaker. The ball is attached to the bottom of the beaker with a light spring (coefficient k). When you fill the beaker with a liquid of density ρ_L, the ball floats fully submerged under the surface with the spring extended by x. Draw a force diagram for the ball and derive the expression for x in terms V, ρ_B, ρ_L, and k. Evaluate the expression using unit analysis and limiting case analysis.

57. * This textbook says that the upward force that a fluid exerts on a submerged object is equal in magnitude to the product of the density of the fluid, the gravitational constant g, and the volume of the submerged part of the object. Where did this equation come from?

58. * **Design** This textbook says that the upward force that a fluid exerts on a submerged object is equal in magnitude to the product of the density of the fluid, the gravitational constant g, and the volume of the submerged part of the object. Design an experiment to test this expression, including a prediction about the outcome of the experiment.

59. * You have four objects at rest, each of the same volume. Object A is partially submerged, and objects B, C, and D are totally submerged in the same container of liquid, as shown in **Figure P13.59**. Draw a force diagram for each object. Rank the densities of the objects from least to greatest and indicate whether any objects have the same density.

FIGURE P13.59

60. * **Does air affect what a scale reads?** A 60-kg woman with a density of 980 kg/m^3 stands on a bathroom scale. Determine the reduction of the scale reading due to air.

61. * When analyzing a sample of ore, a geologist finds that it weighs 2.00 N in air and 1.13 N when immersed in water. What is the density of the ore? What assumptions did you make to answer the question? If the assumptions are not correct, how would the answer be different?

62. * A pin through a hole in the middle supports a meter stick. Two identical blocks hang from strings at an equal distance from the center so the stick is balanced. What happens to the stick if one block is submerged in water of density 1000 kg/m^3 and the other block in kerosene of density 850 kg/m^3?

63. * A meter stick is supported by a pin through a hole in the middle. (a) Two blocks made of the same material but different sizes hang from strings at different positions in such a way that the stick balances. What happens when the blocks hang entirely submerged in beakers of water? (b) Next you hang two blocks of different masses but the same volume at different positions so the stick balances. What happens when these blocks hang completely submerged in beakers of water? Support your answer for each part using force diagrams with arrows drawn with the correct relative lengths.

13.6 Skills for analyzing static fluid problems

64. **Goose on a lake** A 3.6-kg goose floats on a lake with 40% of its body below the 1000-kg/m^3 water level. Determine the average density of the goose.

65. ** **Floating in seawater** A person of average density ρ_1 floats in seawater of density ρ_2. What fraction of the person's body is submerged? Explain.

66. * **Floating in seawater** A person of average density 980 kg/m^3 floats in seawater of density 1025 kg/m^3. What can you determine using this information? Determine it.

67. ** (a) Determine the force that a vertical string exerts on a 0.80-kg rock of density 3300 kg/m^3 when it is fully submerged in water of density 1000 kg/m^3. (b) If the force exerted by the string supporting the rock increases by 12% when the rock is submerged in a different fluid, what is that fluid's density? (c) If the density of another rock of the same volume is 12% greater, what happens to the buoyant force the water exerts on it?

68. * **Snorkeling** A 60-kg snorkeler (including snorkel, mask, and other gear) displaces 0.058 m^3 of water when 1.2 m under the surface. Determine the magnitude of the buoyant force exerted by the 1025-kg/m^3 seawater on the person. Will the person sink or drift upward?

69. * A helium balloon of volume 0.12 m^3 has a total mass (the helium plus the balloon) of 0.12 kg. Determine the buoyant force exerted on the balloon by the air if the air has density 1.13 kg/m^3. Determine the initial acceleration of the balloon when released.

70. * **BIO** **Protein sinks in water** A protein molecule of mass 1.1×10^{-22} kg and density 1.3×10^3 kg/m^3 is placed in a vertical tube of water of density 1000 kg/m^3. (a) Draw a motion diagram and a force diagram at the moment immediately after the molecule is released. (b) Determine the initial acceleration of the protein.

71. * How can you determine if a steel ball of known radius is hollow? List the equipment that you will need for the experiment, and describe the procedure and calculations. Can you determine how big the hollow part is if present in the ball?

72. ** **Crown composition** A crown is made of gold and silver. The scale reads its mass as 3.0 kg when in air and 2.75 kg when in water. Determine the masses of the gold and the silver in the crown. The density of gold is 19,300 kg/m^3 and that of silver is 10,500 kg/m^3.

73. * You place an open bottle filled with olive oil (density 880 kg/m^3) in a container filled with water so that the surfaces of both liquids are at the same level. The bottle has a hole in it 10 cm below the surface that is initially closed with adhesive tape (**Figure P13.73**). (a) Using Pascal's laws, predict what will happen when you remove the tape. Indicate any assumptions that you made. (b) Draw pressure-versus-depth graphs (similar to the graphs in Conceptual Exercise 13.3) for the states before and after you remove the tape. (c) If you predict any changes, determine their numerical values.

FIGURE P13.73

74. * You hold a ping-pong ball fully underwater (the initial state). When you release the ball, it jumps out of the water to a certain height above the surface. Let the final state be when the ball is just above the water surface, moving upward. Represent the process with energy bar charts (a) choosing the water, the ball, and Earth as the system, and (b) choosing the ball and Earth as the system. Indicate any assumptions that you made.

75. * You hang a steel ball on a string above a beaker that is filled to the top with water (the initial state). The beaker is sitting on a large empty tray. You slowly lower the ball until it reaches the bottom of the beaker. Some water spills over the rim of the beaker to the tray (the final state). (a) Represent the process with an energy bar chart choosing the water, the ball, and Earth as the system. (b) Repeat the analysis for a similar process with the same ball and the same amount of water, but a beaker tall enough so that no water spills over into the tray. Indicate any assumptions that you made.

76. * One end of a light spring is attached to a ping-pong ball, the other end to a heavy metal block that is fixed to a thin-wire holder (see Figure P13.76). You hold this setup so that the metal block is above the surface of water in a beaker (the initial state). You then slowly lower the setup until the metal block touches the bottom of the beaker (the final state). Represent the process with an energy bar chart, choosing the water, the ball, the spring, the block, and Earth as the system. Assume that the mass of the ball and the mass of the spring are much smaller than the mass of the block and so can be ignored. Indicate any other assumptions that you made.

FIGURE P13.76

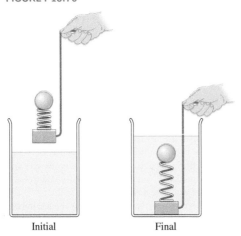

Initial   Final

13.7 Ships, balloons, climbing, and diving

77. * **Wood raft** Logs of density 600 kg/m³ are used to build a raft. What is the maximum mass of the load that can be supported by a raft built from 300 kg of logs?

78. * **EST** **Standing on a log** A floating log is L long and d in diameter. What is the mass of a person who can stand on the log without getting her feet wet?

79. * **Ferryboat** A ferryboat is 12 m long and 8 m wide. Two cars, each of mass 1600 kg, ride on the boat for transport across the lake. How much farther does the boat sink into the water?

80. **EST** **Iceberg** Icebergs are large pieces of freshwater ice. Estimate the percentage of the volume of an iceberg that is underwater. Indicate any assumptions that you made.

81. * **Life preserver** A life preserver is manufactured to support a 70-kg person with 20% of his volume out of the water. If the density of the life preserver is 100 kg/m³ and it is completely submerged, what must its volume be? List your assumptions.

82. ** To increase the effect of the buoyant force on a submarine, the crew replaces seawater in the ballast tanks with air. To push the seawater out of the tanks, they use compressed air from air flasks. The submarine is located 10 m below the surface in seawater with density 1030 kg/m³ and temperature 6 °C. How many kilograms of air should the crew let out from the flasks to increase the difference between the gravitational force and the buoyant force exerted on the submarine by 5×10^6 N? Assume the temperature of air in the flasks is the same as the seawater temperature.

General Problems

83. * Compare the density of water at 0 °C to the density of ice at 0 °C. Suggest possible explanations in terms of the molecular arrangements inside the liquid and solid forms of water that would account for the difference. If necessary, use extra resources to help answer the question.

84. * **Collapsing star** The radius of a collapsing star destined to become a pulsar decreases by 10% while at the same time 12% of its mass escapes. Determine the percent change in its density.

85. * **Deep dive** The *Trieste* research submarine traveled 10.9 km below the ocean surface while exploring the Mariana Trench in the South Pacific, the deepest place in the ocean. Determine the force needed to prevent a 0.10-m-diameter window on the side of the submarine from imploding. Assume that the pressure inside the submarine is 1 atm and the density of the water is 1025 kg/m³.

86. * **EST** **Bursting a wine barrel** Pascal placed a long 0.20-cm-radius tube in a wine barrel of radius 0.24 m. He sealed the barrel where the tube entered it. When he added wine of density 1050 kg/m³ to the tube so the column of wine was 8.0 m high, the cover of the barrel burst off the top of the barrel. Estimate the net force that caused the cover to come off.

87. * **BIO** **Lowest pressure in lungs** Experimentally determine the maximum distance you can suck water up a straw. Use this number to determine the pressure in your lungs above or below atmospheric pressure while you are sucking. Be sure to indicate any assumptions you made and show clearly how you reached your conclusion.

88. ** **Measuring ocean depth with a soda bottle** You can use an empty soda bottle with a cap to measure the depth to which you dive in the ocean. Dive to a certain depth, turn the closed bottle upside down (with the cap toward the sea bottom), and open the cap. A certain volume of seawater will enter the bottle, compressing the air that was trapped in it. Keep the bottle in an upside-down position and carefully close the bottle. Then swim up to the surface and measure the volume V_{water} of water in the bottle and the total volume V_0 of the bottle. (a) Derive the expression for the depth to which you dived in terms of V_{water}, V_0, and other relevant parameters. Indicate any assumptions that you made. Evaluate the expression using unit analysis and limiting case analysis.

89. * Justin is observing pearl-like strings of bubbles that move upward in his father's glass of champagne. He notices that as the bubbles rise, their volume increases (see Figure P13.89). He proposes the following explanation: "The pressure inside the bubble is equal to the pressure in the liquid surrounding the bubble. Because the pressure in the liquid decreases toward the surface of the liquid, the pressure in the ascending bubble decreases and therefore the volume of the bubble increases." Justin estimates that the size of a bubble near the surface is approximately twice that of a bubble 10 cm below the surface. Can Justin reject his explanation based on these data? Explain. If you answered yes, propose a different explanation that is consistent with the data and describe how you could test it. (Hint: Champagne contains dissolved carbon dioxide.)

FIGURE P13.89

90. ** You have an empty water bottle. Predict how much mass you need to add to it to make it float half-submerged. Then add the calculated mass and explain any discrepancy that you found. How did you make your prediction?

91. ** **BIO** **Flexible bladder helps fish sink or rise** A 1.0-kg fish of density 1025 kg/m³ is in water of the same density. The fish's bladder contains 10 cm³ of air. The bladder compresses to 4 cm³. Now what is the density of the fish? Will it sink or rise? Explain.

92. * **Plane lands on Nimitz aircraft carrier** When a 27,000-kg fighter airplane lands on the deck of the aircraft carrier Nimitz, the carrier sinks 0.25 cm deeper into the water. Determine the cross-sectional area of the carrier.

93. Derive an equation for determining the unknown density of a liquid by measuring the magnitude of a force $T_{S \text{ on } O}$ that a string needs to exert on a hanging object of unknown mass m and density ρ to support it when the object is submerged in the liquid.

Reading Passage Problems

BIO **Free diving** So-called "no-limits" free divers slide to deep water on a weighted sled that moves from a boat down a vinyl-coated steel cable to the bottom of a dive site. The diver reaches depths where a soda can would implode. After reaching the target depth, the diver releases the sled and an air bag opens and brings the diver quickly back to the surface. The divers have no external oxygen supply—just lungs full of air at the start of the dive. In August 2002, Tanya Streeter of the Cayman Islands held the women's no-limits free dive record at 160 m. In 2005 Patrick Musimu set the men's record with a 209.6-m free dive in the Red Sea just off the Egyptian coast (the record was later broken by Herbert Nitsch of Austria).

Musimu's 2005 dive took 3 minutes 28 seconds. He began the dive with his 9-L lungs full of air. By the time he passed the 200-m mark, Musimu's lungs had contracted to the size of a tennis ball. His body transferred blood from his limbs to essential organs such as the heart, lungs, and brain. This "blood shift" occurs when mammals submerge in water. Blood plasma fills the chest cavity, especially the lungs. Without this adaptation, the lungs would shrink and press against the chest walls, causing permanent damage. When he reached his target, Musimu released the weighted segment of the specialized sled that had taken him down and opened an airbag, which began his return to the surface at an average speed of 3–4 m/s.

94. Assuming Musimu weighs 670 N (150 lb) and is 1.6 m tall, 0.30 m wide, and 0.15 m thick, which answer below is closest to the magnitude of the force that the deep water exerted on one side of his body?
    (a) 0
    (b) 670 N (130 lb)
    (c) 15,000 N (3000 lb)
    (d) 10^5 N (20,000 lb)
    (e) 10^6 N (200,000 lb)

95. Musimu's training allows him to hold up to 9 L = 9000 cm³ of air when in a 1 atm environment. Which answer below is closest to the volume of that air if at pressure 22 atm?
    (a) 100 cm³
    (b) 200 cm³
    (c) 400 cm³
    (d) 9000 cm³
    (e) 2×10^5 cm³

96. As Musimu descends, the buoyant force that the water exerts on him
    (a) remains approximately constant.
    (b) increases a lot because the pressure is so much greater.
    (c) decreases significantly because his body is being compressed and made much smaller.
    (d) is zero for the entire dive.
    (e) There is not enough information to answer the question.

97. Why don't his lungs, heart, and chest completely collapse?
    (a) The return balloon helps counteract the external pressure.
    (b) There is no external force pushing directly on the organs.
    (c) The sled that helps him descend protects the front of his body.
    (d) Blood plasma moves from his extremities to his chest and the organs in it.
    (e) The air originally in the lungs is transferred to the vital organs.

Lakes freeze from top down We all know that ice cubes float in a glass of water. Why? Virtually every substance contracts when it solidifies—the solid is denser than the liquid. If this happened to water, ice cubes would sink to the bottom of a glass, and ice sheets would sink to the bottom of a lake. Fortunately, this doesn't happen. Liquid water expands by 9% when it freezes into solid ice at 0 °C, from a liquid density of a little less than 1000 kg/m³ to a solid density of 917 kg/m³.

But this is not the only special thing about water. While the density of most substances increases when they are cooled, water density shows a very peculiar temperature dependence (see **Figure 13.17**). As the temperature decreases, water density increases, but only until 4 °C, when water density reaches its maximum value, 1000.0 kg/m³. As the temperature decreases further, water density decreases until at 0 °C water freezes and density abruptly decreases to 917 kg/m³.

FIGURE 13.17

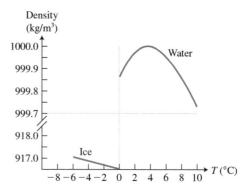

It is this peculiar water-density-temperature dependence that plays a vital role in the survival of animals and plants that live in water. In the winter when the water in a lake freezes, the solid ice stays at the top, forming an ice sheet. Water just below the ice sheet cools, but when it reaches 4 °C it becomes the most dense and sinks to the bottom of the lake. Since water colder than 4 °C is less dense, it stays above, keeping the bottom of the lake at a constant 4 °C. Note that if water were most dense at the freezing point, then in the winter the very cold water at the surface of lakes would sink. In this case the lake would freeze from the bottom up, and all life in it would be destroyed.

The expansion of water when it freezes has another important environmental benefit: the so-called freeze-thaw effect on sedimentary rocks. Water is absorbed into cracks in these rocks and then freezes in cold weather. The solid ice expands and cracks the rock, like a wood-cutter splitting logs. This continual process of liquid water absorption, freezing, and cracking releases mineral and nitrogen deposits into the soil and can eventually break the rock down into soil.

98. When is water densest?
    (a) When liquid at 0 °C
    (b) When solid ice at 0 °C
    (c) When liquid at 4 °C
    (d) Water density is always 1000.0 kg/m³.

99. Why does water freeze from the top down?
    (a) The denser water at 0 °C sinks to the bottom of the lake.
    (b) The less dense ice at 0 °C rises above the liquid water at 0 °C.
    (c) The denser water at 4 °C sinks to the bottom of the lake.
    (d) Because of both a and b
    (e) Because of both b and c

100. Using Newton's second law, expressions for buoyant force and other forces, and the densities of liquid and solid water at 0 °C, find the fraction of an iceberg or an ice cube that is under liquid water.
    (a) 0.84
    (b) 0.88
    (c) 0.92
    (d) 0.96
    (e) 1.00

101. A swimming pool at 0 °C has a very large chunk of ice floating in it—like an iceberg in the ocean. When the ice melts, what happens to the level of the water at the edge of the pool?
    (a) It rises.
    (b) It stays the same.
    (c) It drops.
    (d) It depends on the size of the chunk.

102. Which of the following is/are benefits of the temperature dependence of the density of water?
    (a) Fish and plants can survive winters without being frozen.
    (b) Over time, soil is formed from sedimentary rocks.
    (c) Water pipes when frozen in the winter do not burst.
    (d) Two of the above three
    (e) All of the first three

Fluids in Motion

Plaque (fatty deposits) accumulates on the walls of arteries as cholesterol-laden blood flows by. As plaque grows, blood flows past at higher speed. If the blood is moving fast enough, it can dislodge deposits, which may then become lodged downstream and stop blood flow. If this stoppage occurs in the heart, it can cause a heart attack. In this chapter, we will learn why blood flows faster through an artery clogged with plaque and why the fast-moving stream of blood tends to pull the plaque off the artery wall.

- How does blood flow dislodge plaque from an artery?
- Why can a strong wind cause the roof to blow off a house?
- Why do people snore?

IN THE PREVIOUS CHAPTER, we investigated the behavior of static fluids. What happens when a gas or liquid moves across a surface—for example, when air moves across the roof of a house or when blood moves through a blood vessel? In this chapter, we will investigate and explain phenomena involving moving fluids—*fluid dynamics*.

BE SURE YOU KNOW HOW TO:
- Draw work-energy bar charts (Section 7.2).
- Apply the concept of pressure to explain the behavior of liquids (Section 13.2).
- Draw force diagrams and apply Newton's second law (Section 3.5).

14.1 Fluids moving across surfaces— qualitative analysis

You learned in Chapters 12 and 13 that a key property of a static fluid is its pressure. What happens to that pressure when a fluid moves? To investigate this, we will analyze the simple experiment described in Observational Experiment Table 14.1.

OBSERVATIONAL
EXPERIMENT TABLE 14.1 Pressure inside a moving fluid

VIDEO
OET 14.1

Observational experiment	Analysis
Hold two pieces of paper separated by plastic blocks and blow down directly between them. You see the pieces of paper coming close together, as if they are pushed toward the moving air. Left paper Right paper	We consider the piece of paper on the right to be our system and examine the moment when the paper starts moving to the left. We draw forces exerted on the paper in the horizontal direction. For the system to start moving, the sum of the horizontal forces must point to the left—from the region of stationary air to the region of moving air. 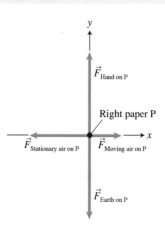

Pattern

Based on the analysis of the experiment and using the force diagram, we can infer that air moving along the surface of an object exerts a smaller force on the object. Given that force divided by the area over which it is exerted is pressure, we conclude that the air moving along a surface exerts less pressure on the object than stationary air.

Bernoulli's effect

Extrapolating from the pattern in Table 14.1, could it be that for any fluid, the speed with which fluid is moving along the object and the pressure exerted by this fluid on the object are related: the greater the speed, the smaller the pressure? Let's test this hypothesis experimentally.

It is important to note here that in all of the experiments, the movement of the fluid was parallel to the surface of interest. When the fluid is moving perpendicular to the surface of interest, it exerts a force in the direction of motion (you can think of the momentum of fluid particles changing as the particles hit the surface). You can clearly observe this effect if you repeat the experiment in Testing Experiment Table 14.2 with the hairdryer blowing directly downward into the tube—the water level inside the tube lowers.

TESTING
EXPERIMENT TABLE 14.2 How are the speed of a fluid and its pressure related?

VIDEO
TET 14.2

Testing experiment	Prediction	Outcome
We cut the top and bottom off a plastic bottle to make a plastic tube and put it into a container of water. The level of the water in the tube is the same as in the container. Then using a hair dryer set to cold, we blow cold air parallel to the surface of the water right above the tube, first using a low-speed setting and then a high-speed setting. 	If a faster-moving fluid exerts less pressure than a stationary fluid, and we blow air parallel to the opening of the tube, then the pressure at the top of the tube (but also inside the tube) should be lower than atmospheric pressure. When the speed of the moving air increases, the pressure should decrease. The pressure outside the tube, above the surface of the water in the container is atmospheric; therefore, the sum of the forces on an element of water inside the tube should point upward and the water in the tube should rise until the sum of the forces exerted on the raised water inside the tube is zero. The height of the water inside the tube should increase when the hair dryer is switched to high.	The outcomes are shown in the figures. They match the predictions.

Conclusion

The hypothesis that the pressure exerted by a faster-moving fluid along an object is less than the pressure exerted by a slower-moving fluid was not disproved.

It is important to note here that in all of the experiments, the movement of the fluid was parallel to the surface of interest. When the fluid is moving perpendicular to the surface of interest, it exerts a force in the direction of motion (you can think of the momentum of fluid particles changing as the particles hit the surface). You can clearly observe this effect if you repeat the experiment in Table 14.2 with the hairdryer blowing directly downward into the tube—the water level inside the tube lowers.

At this point, without evidence to the contrary, we can say that as a fluid's speed parallel to a surface increases, the pressure that the moving fluid exerts on the surface decreases. This statement is a qualitative version of a principle formulated in 1738 by Daniel Bernoulli and named in his honor.

Bernoulli's effect The pressure that a fluid exerts on a surface *decreases* as the speed with which the fluid moves parallel to the surface *increases*.

FIGURE 14.1 Snoring occurs when the soft palate opens and closes due to the starting and stopping of airflow across it.

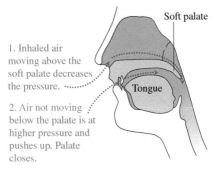

Soft palate

1. Inhaled air moving above the soft palate decreases the pressure.

Tongue

2. Air not moving below the palate is at higher pressure and pushes up. Palate closes.

3. When airflow stops, the pressures equalize and the soft palate reopens. The moving air causes the process to repeat.

4. The vibrating palate and airflow cause the snoring sound.

Bernoulli's effect has important fluid-flow implications in biological systems—for example, in the flow of blood through blood vessels. The blood pressure against the wall of a vessel depends on how fast the blood is moving—pressure is lower when the blood is moving faster. Let's look at another biological application of Bernoulli's effect.

Snoring

A snoring sound occurs when air moving through the narrow opening above the soft palate at the back of the roof of the mouth has lower pressure than nonmoving air below the palate (**Figure 14.1**). The normal air pressure below the soft palate, where the air is not moving, pushes the palate closed. When airflow stops, the pressures equalize and the passage reopens. The rhythmic opening and closing of the soft palate against the throat leads to the snoring sound.

REVIEW QUESTION 14.1 What is the empirical evidence for Bernoulli's effect?

14.2 Flow rate and fluid speed

We have found qualitatively that the pressure of a fluid along the direction of flow depends on the speed of the moving fluid—the greater the speed, the lower the pressure. In this section, we will learn new physical quantities that we will need to describe the effect quantitatively. The quantitative analysis is much easier if we limit our investigations to fluids in confined regions—pipes or tubes.

Perhaps you have taken a shower in which little water flowed from the showerhead. In physics, we would say the water flow rate was low. The **flow rate Q** is an important consideration in designing showerheads. A smaller flow rate will save water, but a larger flow rate will help you rinse off faster. Flow rate is defined as the volume V of fluid that moves through a cross section of a pipe divided by the time interval Δt during which it moved (see **Figure 14.2**):

$$Q = \frac{V}{\Delta t} \tag{14.1}$$

The SI unit of flow rate is m^3/s, but you may also see it as ft^3/s, ft^3/min, gallons/min, L/min, or any unit of volume divided by any unit of time interval. Notice that flow rate in m^3/s is different from the speed of the fluid v in m/s.

FIGURE 14.2 Flow rate is the volume of fluid that passes a cross section of a vessel in a given time interval.

(a) $t = 0$

A volume $V = lA$ of fluid flows past cross section A in time interval Δt.

A

l

(b) $t = 0 + \Delta t$

V

l

TIP The symbols V, t, and Q are also used in other aspects of physics. For example, a lowercase v denotes speed, the capital letter T is used for temperature, and in future chapters we will use Q for two other unrelated quantities. Because these symbols are often used to indicate different quantities, it is important when working with equations to try to visualize their meaning with concrete images (for example, the volume of water flowing out of a faucet during 1 s).

How does the flow rate relate to the speed of the moving fluid? To explore the relationship, consider Figure 14.2a. Over a certain time interval Δt the shaded volume of fluid passes a cross section of area A at some position along the pipe. Thus, after a time Δt, the back part of this fluid volume has in effect moved forward to the position shown in Figure 14.2b. The volume V of fluid in the shaded portion of the cylinder is the product of its length l and the cross-sectional area A of the pipe:

$$V = lA$$

Thus, the fluid flow rate is

$$Q = \frac{V}{\Delta t} = \frac{lA}{\Delta t} = \left(\frac{l}{\Delta t}\right)A$$

However, l is also the distance the fluid moves in a time interval Δt. Thus, $l/\Delta t$ is the average fluid speed v. Substituting $v = l/\Delta t$ into the above equation, we find that

$$Q = vA \qquad (14.2)$$

The flow rate is equivalent to the average fluid speed multiplied by the cross-sectional area of the pipe.

When an incompressible fluid (such as water) flows through a pipe with variable cross-sectional area (**Figure 14.3**), the amount of fluid entering at cross section A_1 must equal the amount of fluid leaving at cross section A_2. Thus the flow rate should remain constant. What must change is the speed of the fluid as it travels through the narrower part of the pipe. In the narrow section of the pipe (A_2), the speed will be greater in order to keep the flow rate constant. Thus, the flow rate past cross section A_1 equals that past cross section A_2:

$$Q_1 = v_1 A_1 = v_2 A_2 = Q_2 \qquad (14.3)$$

where v_1 is the average speed of the fluid passing cross section A_1, and v_2 is the average speed of the fluid passing cross section A_2. Equation (14.3) is called the **continuity equation** and is used to relate the cross-sectional area and average speed of fluid flow in different parts of a rigid pipe carrying an incompressible fluid.

TIP The definition of flow rate is an operational definition, not a cause-effect relationship, because it is the speed of the fluid that depends on the flow rate and the cross-sectional area. The flow rate is determined by the source of the fluid (for example, how much you open the faucet).

FIGURE 14.3 The flow speed $v_2 > v_1$ depends on the cross-sectional area of pipe carrying the fluid.

Fluid speed is lower through a wide opening ...

... and greater through a narrow opening.

$$Q = v_1 A_1 = v_2 A_2$$

QUANTITATIVE EXERCISE 14.1   Speed of blood flow in the aorta

The heart pumps blood at an average flow rate of 80 cm³/s into the aorta, which has a diameter of 1.5 cm. Determine the average speed of blood flow in the aorta.

Represent mathematically The flow rate can be determined by rearranging Eq. (14.2):

$$v = \frac{Q}{A}$$

where the cross-sectional area of the aorta is

$$A = \pi r^2 = \pi \left(\frac{d}{2}\right)^2$$

Solve and evaluate Combining the above two equations, we find that the average speed of blood flow in the aorta is

$$v = \frac{Q}{A} = \frac{Q}{\pi (d/2)^2} = \frac{(80 \text{ cm}^3/\text{s})}{\pi (1.5 \text{ cm}/2)^2} = 45 \text{ cm/s}$$

The unit is correct. The magnitude is reasonable—about half a meter each second.

- -

Try it yourself Determine the average speed of blood flow if the diameter is reduced from 1.5 cm to 1.0 cm but the flow rate is the same.

Answer   100 cm/s.

Notice that blood speed more than doubles when the aorta diameter decreases by 33%. Vessel diameter has a very significant effect on the flow rate of fluid through a vessel, including those in biological systems. The narrower the blood vessel, the faster the blood flows, increasing the risk of dislodging plaque. Likewise, the narrower the airway from the nose to the mouth, the faster the air moves and the more likely you are to snore. These effects depend on the speed of the fluid, like blood or air in different parts of a vessel or pipe in which the diameter changes from one section to another.

FIGURE 14.4 Why is $v_2 > v_1$?

Outlet

REVIEW QUESTION 14.2 Why does water in a river flow more slowly just before a dam than it does while passing through the outlet of the dam (**Figure 14.4**)?

FIGURE 14.5 Laminar flow.

FIGURE 14.6 Turbulent flow.

Eddies

FIGURE 14.7 Streamlining a truck.

FIGURE 14.8 Measuring blood pressure

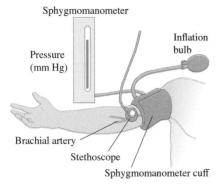

Sphygmomanometer

Pressure
(mm Hg)

Inflation
bulb

Brachial artery

Stethoscope

Sphygmomanometer cuff

14.3 Types of fluid flow

Before we can proceed to describing Bernoulli's effect quantitatively, we need to learn more about fluid flow.

What kinds of flow can occur? There is the smooth flow that we see in a wide river and the whitewater flow that we see as water rushes and swirls through a narrow channel. Studies of fluid flow in wind tunnels indicate that there are two primary kinds of flow: laminar flow and turbulent flow. In **laminar flow**, every particle of fluid that passes a particular point follows the same path as the particles that preceded it. These paths are called **streamlines**. This is the smooth flow that we see in a wide river (see the fluid flowing smoothly in the wide tube in **Figure 14.5**). **Turbulent flow**, on the other hand, is characterized by agitated, disorderly motion. Instead of following a given path, the fluid forms whirlpool patterns called eddies, which come and go randomly, or sometimes become semi-stable (see **Figure 14.6**). Turbulent flow occurs when a fluid moves around objects and through pipes at high speed.

The force exerted by the fluid on objects is called the **drag force**. This is the force you "feel" when running against the wind. Due to the drag force, more kinetic energy is converted into thermal energy in turbulent flow than in laminar flow. Designing a car so that air moves over it with laminar flow reduces the drag force that the air exerts on the car and improves gasoline mileage. Placing a curved dome above the cab of a truck deflects air up and over the trailer, reducing turbulent flow and increasing gas mileage by more than 10% (see **Figure 14.7**).

Measuring blood pressure

To measure a person's blood pressure, a nurse uses a device called a *sphygmomanometer* (see **Figure 14.8**). The nurse places a cuff around the upper arm of a patient at about the level of the heart and places a stethoscope on the inside of the elbow above the brachial artery in the arm. The nurse then increases the gauge pressure in the cuff to about 180 mm Hg by pumping air into the cuff. The expanded cuff pushes on the brachial artery and stops blood flow in the arm. Then the nurse slowly releases the air from the cuff, decreasing the pressure of the air in it. When the pressure in the cuff is equal to the *systolic* pressure (120 mm Hg if the systolic blood pressure is normal), blood starts to squeeze through the artery past the cuff. The flow is intermittent and turbulent and causes a sound heard with the stethoscope. This turbulent sound continues until the cuff pressure decreases below the *diastolic* pressure (80 mm Hg for normal diastolic blood pressure). At that point the artery is continually open and blood flow is laminar and makes no sound. The systolic and diastolic pressure numbers together make up the blood pressure measurement.

REVIEW QUESTION 14.3 Is it easier for the heart to pump blood if the flow of the blood through the blood vessels is laminar or if it is turbulent? Explain.

14.4 Bernoulli's equation

Earlier in this chapter we investigated Bernoulli's effect qualitatively. In this section we will learn how to describe it quantitatively.

To achieve this goal, we use the case of a fluid flowing through a pipe, as shown in **Figure 14.9a**. We assume that (1) the fluid is incompressible, (2) there are no resistive forces exerted on the flowing fluid, and (3) the flow is laminar. We can apply the work-energy principle to describe the behavior of the fluid as it moves a short distance along the pipe. Consider the system to be composed of the shaded volume of fluid pictured in Figures 14.9a and b and Earth.

FIGURE 14.9 Applying the work-energy principle to fluid flow.

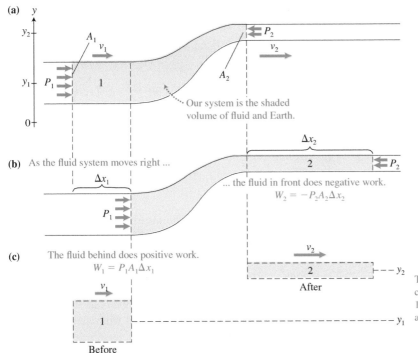

(a)

Our system is the shaded volume of fluid and Earth.

(b) As the fluid system moves right ...

... the fluid in front does negative work.
$$W_2 = -P_2 A_2 \Delta x_2$$

(c) The fluid behind does positive work.
$$W_1 = P_1 A_1 \Delta x_1$$

Before

After

The work done on the system has effectively caused the volume of fluid to move from position 1 to position 2. The gravitational potential energy and the kinetic energy have increased:
$$(U_{g1} + K_1) + W_1 + W_2 = (U_{g2} + K_2)$$

Figure 14.9a shows us the initial state of the system. As the volume of fluid flows to the right, the fluid behind it at position 1 exerts a force of magnitude $F_1 = P_1 A_1$ to the right, where P_1 is the fluid pressure against the left side of the volume and A_1 is the cross-sectional area of the left side of the volume. Simultaneously, the fluid ahead of the system at position 2 exerts a force in the opposite direction of magnitude $F_2 = P_2 A_2$, where P_2 is the fluid pressure against the right side of the volume and A_2 is the cross-sectional area of the right side of the volume.

In Figure 14.9b, the shaded volume of fluid has moved to the right. Because the pipe is narrower at 2, the right side of the fluid at position 2 moves a greater distance than the left side of the same volume of fluid in the wider part of the pipe at position 1. The net effect of the movement of the fluid a short distance to the right is summarized in Figure 14.9c. The volume of fluid initially at position 1 moving at speed v_1 has now been transferred, in effect, to position 2 where it moves at speed v_2. The volume of fluid stays constant, since we assume the fluid is incompressible. The fluid is now moving faster through the narrow tube at position 2 than it was earlier when moving through the wider tube at position 1, thus we have an increase in kinetic energy. The fluid at position 2 is at a higher elevation than when at position 1, thus the gravitational potential energy of the system increases. The energies changed as a result of the work done by the forces exerted by the fluid behind and ahead of the shaded volume. We can represent this quantitatively using the generalized work-energy equation [Eq. (7.3)].

$$(K_1 + U_{g1}) + W = (K_2 + U_{g2})$$

If we move the terms with the subscript 1 to the right side of the equation, we have

$$W = (K_2 - K_1) + (U_{g2} - U_{g1})$$

or

$$W = \Delta K + \Delta U_g \tag{14.4}$$

Let us now write expressions for each of the terms in the above equation.

Work done Two forces are doing work on the system. The fluid behind the system exerts a force F_1 to the right on the left side of the system over a distance Δx_1. The fluid ahead of the system exerts a force F_2 to the left on the right side of the system over a distance Δx_2. (Figures 14.9a and b show fluid pressures; the corresponding forces have magnitudes $F_1 = P_1A_1$ and $F_2 = P_2A_2$.) The force F_1 does positive work since it points in the direction of the motion of the system. The force F_2 does negative work since it points in the direction opposite the motion of the system. The total work done on the system is

$$W = F_1 \Delta x_1 \cos 0° + F_2 \Delta x_2 \cos 180°$$
$$= P_1A_1 \Delta x_1 - P_2A_2 \Delta x_2$$

The volume of fluid V that has moved from the left to the right is $V = A_1 \Delta x_1 = A_2 \Delta x_2$ since the fluid is incompressible. The preceding expression for work becomes

$$W = P_1V - P_2V = (P_1 - P_2)V$$

Change in kinetic energy The mass m of the system (the moving volume of fluid) is related to its density ρ and volume V:

$$m = \rho V$$

As the system moves from the initial to the final state, the speed of the element of fluid that effectively moved changes from v_1 to v_2. Thus, the kinetic energy change of the mass m of fluid shown in Figure 14.9c is

$$\Delta K = \tfrac{1}{2}mv_2^2 - \tfrac{1}{2}mv_1^2 = \tfrac{1}{2}\rho Vv_2^2 - \tfrac{1}{2}\rho Vv_1^2$$

Change in gravitational potential energy The gravitational potential energy of the system has also changed because part of the system has moved from elevation y_1 to elevation y_2. The change in gravitational potential energy is then

$$\Delta U_g = mg(y_2 - y_1) = \rho Vg(y_2 - y_1)$$

We can now substitute the above three expressions into Eq. (14.4) to get

$$(P_1 - P_2)V = (\tfrac{1}{2}\rho Vv_2^2 - \tfrac{1}{2}\rho Vv_1^2) + \rho Vg(y_2 - y_1)$$

If the common V is canceled from each term, we find that

$$P_1 - P_2 = \tfrac{1}{2}\rho(v_2^2 - v_1^2) + \rho g(y_2 - y_1)$$

By dividing by V in that last step we have changed the units of each term in the equation from energy (measured in joules) to **energy density** (measured in joules per cubic meter). Energy density is the same as energy per unit volume of fluid and appears on the right side of the above equation. The left-hand side represents the amount of work done on the fluid per unit volume of fluid.

Bernoulli's equation relates the pressures, speeds, and elevations at two points on the same streamline in laminar flow in a fluid:

$$P_1 - P_2 = \tfrac{1}{2}\rho (v_2^2 - v_1^2) + \rho g(y_2 - y_1) \tag{14.5}$$

The equation can be rearranged into an alternate form:

$$\tfrac{1}{2}\rho v_1^2 + \rho gy_1 + P_1 = P_2 + \tfrac{1}{2}\rho v_2^2 + \rho gy_2 \tag{14.6}$$

The sum of the kinetic and gravitational potential energy densities and the pressure at position 1 equals the sum of the same three quantities at position 2.

It is important to remember the assumptions that we used to derive Bernoulli's equation. It describes quantitatively the flow of a frictionless, nonturbulent, incompressible fluid.

Using Bernoulli bar charts to understand fluid flow

Bernoulli's equation looks fairly complex and might be difficult to use for visualizing fluid dynamics processes. However, since Bernoulli's equation is based on the work-energy principle, we can represent such processes using energy bar charts similar to the ones used in Chapters 7 and 13 (here the bars represent pressures and energy densities; we use a ~ over the energy symbols to emphasize that they are energy densities, not energies). Physics Tool Box 14.1 describes how to construct a fluid dynamics bar chart for the following process. A fire truck pumps water through a big hose up to a smaller hose on the ledge of a building. Water sprays out of the smaller hose onto a fire. Compare the pressure in the hose at the lowest point of the larger hose to the pressure at the exit of the smaller hose.

PHYSICS TOOL BOX 14.1   Constructing a bar chart for a moving fluid

$$\tilde{K}_1 + \tilde{U}_{g1} + P_1 = P_2 + \tilde{K}_2 + \tilde{U}_{g2}$$

$$0.5\rho v_1^2 + 0 + P_1 = P_2 + 0.5\rho v_2^2 + \rho g y_2$$

| 1. Sketch the situation. Include an upward-pointing y-coordinate axis. | 2. Choose points 1 and 2 at positions in the fluid that will help you achieve the goal of your analysis. | 3. Construct a Bernoulli bar chart. | 4. Use the bar chart and the sketch to help apply Bernoulli's equation. |

To start, first draw a sketch of the process. Choose positions 1 and 2 at appropriate locations in order to help answer the question. One of the positions might be a place where you want to determine the pressure in the fluid and the other position a place where the pressure is known. For the water pump-hose process, it is useful to choose position 1 (the location of the unknown pressure) and position 2 at the exit of the water from the small hose (at known atmospheric pressure).

Represent this process by placing bars of appropriate relative lengths on the chart (the absolute lengths are not known). It is often easiest to start by analyzing the gravitational potential energy density. Use a vertical y-axis with a well-defined origin to keep track of the gravitational potential energy densities. For the fire hose process, choose position 1 as the origin of the vertical coordinate system. The gravitational potential energy density at position 1 is then zero. The exit of the water from the small hose is at higher elevation; thus, there is a positive gravitational potential energy density bar for position 2 (in the bar chart we arbitrarily assign it one positive unit of energy density).

Next, consider the kinetic energy. The water flows from a wider hose at position 1 to a narrower hose at position 2. Thus the kinetic energy density at 2 is greater than at 1—thus the longer bar at position 2. We arbitrarily assume that the kinetic energy density bar for position 1 is one unit and that for position 2 is three units.

Notice now that the total length of the bars on the right side of the chart is much higher than on the left side (the difference is three units). To account for the difference we need to consider the change in pressure. Since the fluid pressures at 1 and 2 are analogous to the work done on the system in an ordinary work-energy bar chart, P_1 and P_2 appear in the shaded box in the center of the bar chart, where work is represented. The difference in the pressure heights should account for the total difference in the energy densities. Thus, we draw the bar for P_1 three units higher than for P_2. The bar chart is now complete. We can use it to write a mathematical description for the process and solve for any unknown quantity.

REVIEW QUESTION 14.4 Compare and contrast work-energy bar charts and Bernoulli bar charts.

14.5 Skills for analyzing processes using Bernoulli's equation

In this section, we will adapt our problem-solving strategy to analyze processes involving moving fluids. In this case, we describe and illustrate a strategy for finding the speed of water as it leaves a bottle. The general strategy is on the left side of the table and the specific process is on the right.

PROBLEM-SOLVING STRATEGY 14.1   Applying Bernoulli's equation

---

EXAMPLE 14.2   Removing a tack from a water bottle

What is the speed with which water flows from a hole punched in the side of an open plastic bottle? The hole is 10 cm below the water surface.

Sketch and translate

- Sketch the situation. Include an upward-pointing y-coordinate axis. Choose an origin and positive direction for the coordinate axis.
- Choose points 1 and 2 at positions in the fluid where you know the pressure/speed/position or that involve the quantity you are trying to determine.
- Choose a system.

- Choose the origin of the vertical y-axis to be the location of the hole.
- Choose position 1 to be the place where the water leaves the hole and position 2 to be a place where the pressure, elevation, and water speed are known—at the water surface $y_2 = 0.10$ m. The pressure in Bernoulli's equation at both positions 1 and 2 is atmospheric pressure, since both positions are exposed to the atmosphere ($P_1 = P_2 = P_{atm}$).
- Choose Earth and the water as the system.

Simplify and diagram

- Identify any assumptions you are making. For example, can we assume that there are no resistive forces exerted on the flowing fluid?
- Construct a Bernoulli bar chart.

- Assume that no resistive forces are exerted on the flowing fluid.
- Assume that y_2 and y_1 stay constant during the process, since the elevation of the surface decreases slowly compared to the speed of the water as it leaves the tiny hole.
- Because the water at the surface is moving very slowly relative to the hole, assume that $v_2 = 0$.
- Draw a bar chart that represents the process.

Represent mathematically

- Use the sketch and bar chart to help apply Bernoulli's equation.
- You may need to combine Bernoulli's equation with other equations, such as the equation of continuity $Q = v_1 A_1 = v_2 A_2$ and the definition of pressure $P = F/A$.

- We see from the sketch and the bar chart that the speed of the fluid at position 2 is zero (zero kinetic energy density) and that the elevation is zero at position 1 (zero gravitational potential energy density). Also, the pressure is atmospheric at both 1 and 2. Thus

$$\tfrac{1}{2}\rho(0)^2 + \rho g y_2 + P_{atm} = P_{atm} + \tfrac{1}{2}\rho v_1^2 + \rho g(0)$$

$$\Rightarrow \rho g y_2 = \tfrac{1}{2}\rho v_1^2$$

Solve and evaluate

- Solve the equations for an unknown quantity.
- Evaluate the results to see if they are reasonable (the magnitude of the answer, its unit, how the answer changes in limiting cases, and so forth).

- Solve for v_1:

$$v_1 = \sqrt{2gy_2}$$

Substituting for g and y_2, we find that

$$v_1 = \sqrt{2(9.8 \text{ m/s}^2)(0.10 \text{ m})} = 1.4 \text{ m/s}$$

- The unit m/s is the correct unit for speed. The magnitude seems reasonable for water streaming from a bottle (if we obtained 120 m/s it would be unreasonably high).

Try it yourself In the above situation the water streams out of the bottle onto the floor a certain horizontal distance away from the bottle. The floor is 1.0 m below the hole. Predict this horizontal distance using your knowledge of projectile motion. (Hint: Use Eqs. (4.7) and (4.8).)

Answer The equations yield a result of 0.63 m. However, if we were to actually perform this experiment with a tack-sized hole, the water would land short of our prediction because of resistive forces exerted by the hole on the water. In order to make the water land closer to 0.63 m from the bottle, we must increase the diameter of the hole. We discuss the effect of resistive forces on fluid flow later in the chapter.

Blowing the roof off a house

You've no doubt seen images of roofs being blown from houses during tornadoes or hurricanes. How does that happen? On a windy day, the air inside the house is not moving, whereas the air outside is moving very rapidly. The air pressure inside the house is therefore greater than the air pressure outside, creating a net pressure that pushes outward against the roof and windows. If the net pressure becomes great enough, the roof and/or the windows will blow outward off of the house. In the following example, we do a quantitative estimate of the net force exerted by the inside and outside air on a roof.

EXAMPLE 14.3 **Effect of high-speed air moving across the roof of a house**

During a storm, air moves at speed 45 m/s (100 mi/h) across the top of the 200-m^2 flat roof of a house. Estimate the net force exerted by the air pushing up on the inside of the roof and the outside air pushing down on the outside of the roof. Indicate any assumptions made in your estimate.

Sketch and translate The situation is shown at right. We need to determine the pressure just above and below the roof.

Simplify and diagram A force diagram for the roof is shown here. The air above the house exerts a downward force on the roof $F_{1 \text{ on R}} = P_1 A$, where P_1 is the air pressure above the house and A is the area of the roof. The air inside the house pushes up, exerting a force on the roof $F_{2 \text{ on R}} = P_{\text{atm}} A$, where P_{atm} is the assumed atmospheric pressure of the stationary air inside the house. We assume that the air is incompressible and flows

without friction or turbulence and that the roof is fairly thin so that the air has approximately the same gravitational potential energy density at points 1 and 2.

Represent mathematically With the y-axis oriented upward, the net force exerted by the air on the roof is

$$F_{\text{net Air}} = F_{2 \text{ on R}} - F_{1 \text{ on R}} = P_{\text{atm}} A - P_1 A$$
$$= (P_{\text{atm}} - P_1)A$$

We use Bernoulli's equation to find this pressure difference.

$$P_2 + \tfrac{1}{2}\rho v_2^2 + \rho g y_2 = \tfrac{1}{2}\rho v_1^2 + \rho g y_1 + P_1$$
$$\Rightarrow P_{\text{atm}} + 0 + \rho g y_2 = \tfrac{1}{2}\rho v_1^2 + \rho g y_1 + P_1$$
$$\Rightarrow P_{\text{atm}} - P_1 = \tfrac{1}{2}\rho v_1^2 + (\rho g y_1 - \rho g y_2) = \tfrac{1}{2}\rho v_1^2 + 0$$

We can now determine the net force exerted by the air on the roof.

$$F_{\text{net Air}} = (P_{\text{atm}} - P_1)A = \tfrac{1}{2}\rho v_1^2 A$$

$v_1 = 45$ m/s

$A = 200$ m^2

(CONTINUED)

Solve and evaluate

$$F_{\text{net Air}} = \tfrac{1}{2}\rho v_1^2 A$$

$$= \tfrac{1}{2}(1.3 \text{ kg/m}^3)(45 \text{ m/s})^2(200 \text{ m}^2) = 2.6 \times 10^5 \text{ N}$$

The result is an upward net force that is enough to lift more than ten cars of combined mass 27,000 kg.

We were a little lax in applying Bernoulli's equation in this example. The equation relates the properties of a fluid at two points along the same streamline. A streamline does not flow between just below the roof and just above the roof. We could, with some more complex reasoning, use the equation correctly by considering two streamlines

that start far from the house at the same pressure. One ends up in the house under its roof with the air barely moving. The other passes just above the roof with the air moving fast. We would get the same result in a somewhat more cumbersome manner.

Try it yourself A 2.0 m × 2.0 m canvas covers a trailer. The trailer moves at 29 m/s (65 mi/h). Determine the net force exerted on the canvas by the air above and below it.

Answer ˙pɹɐʍʇno suooʃʃɐq ʎɐʍɥᵷᴉɥ ɐ uo ᵷuᴉʌoɯ ɹǝʃᴉɐɹʇ ʞɔnɹʇ ɐ ᵷuᴉɹǝʌoɔ
sɐʌuɐɔ ǝɥʇ ɹǝpuoʍ oN ˙(qʃ 00ϛ ʇnoqɐ) N 00ᘔᘔ ɟo ǝɔɹoɟ pɹɐʍdn u∀

Dislodging plaque

The physical principles of a roof being lifted from a house also explain how plaque can become dislodged from the inner wall of an artery. Plaque may block a considerable portion of the area where blood normally flows. Suppose the radius of the vessel opening is one-third its normal value because of the plaque. Then the area available for blood flow, proportional to r^2, is about one-ninth the normal value. The speed of flow in the narrowed portion of the artery will be about nine times greater than in the unblocked part of the vessel. The kinetic energy density term in Bernoulli's equation is proportional to v^2 and therefore is 81 times greater in the constricted area than in the open part of the vessel.

Notice that in Bernoulli's equation, the sum of the gravitational potential energy density, the kinetic energy density, and the pressure at one location should equal the sum of the same three terms at some other location along a streamline in the blood. As blood speeds by the plaque, its kinetic energy density is 81 times greater, and consequently its pressure is much less than the pressure in the open vessel just before and just after the plaque. This pressure differential could cause the plaque to be pushed off the wall and tumble downstream, causing a blood clot (a process called thrombosis). Let's estimate the net force that the blood exerts on the plaque.

EXAMPLE 14.4   **A clogged artery**

Blood flows through the unobstructed part of a blood vessel at a speed of 0.5 m/s. The blood then flows past a plaque that constricts the cross-sectional area to one-ninth the normal value. The surface area of the plaque parallel to the direction of blood flow is about 0.60 cm^2 = 6.0 × 10^{-5} m^2. Estimate the net force that the fluid exerts on the plaque.

Sketch and translate A simplified two-dimensional sketch of three-dimensional the situation is shown above right. The third dimension is perpendicular to the page of the book. The plaque is as large in this dimension as in the x-dimension, but the area where it is attached to the top wall is very small. Point 1 is above the plaque in stationary blood pooled in a channel where the plaque attaches to the artery wall. Point 2 is in the bloodstream below the plaque, where blood flows rapidly past it. The net force depends on the differences in pressure at points 1 and 2. Thus, we need to first find the pressure difference. Since points 1 and 2 are not on the same streamline, we cannot automatically use Bernoulli's equation. But since the streamlines that do go through points 1 and 2 were side by side before they reached the plaque, each streamline will have the same $P + (1/2)\rho v^2 + \rho gy$ value. This means we can equate the $P + (1/2)\rho v^2 + \rho gy$ values at points 1 and 2.

Simplify and diagram Assume for simplicity that the blood is nonviscous and flows with laminar flow without turbulence. Assume also that the vertical distance between points 1 and 2 is small ($y_1 - y_2 \approx 0$) and that the area of the stationary blood above the plaque is the same as the area where the blood moves below the plaque. A bar chart represents the process. The blood pressure at position 2 is less than at position 1 because the blood flows at high speed through the constricted artery, whereas it sits at rest in the channels at position 1 ($v_1 = 0$).

Represent mathematically Compare the two points using Bernoulli's equation:

$$P_1 - P_2 = \tfrac{1}{2}\rho(v_2^2 - v_1^2) + \rho g(y_2 - y_1)$$
$$= \tfrac{1}{2}\rho(v_2^2 - 0) + 0 = \tfrac{1}{2}\rho v_2^2$$

Since the cross-sectional area of the vessel at the location of the plaque is one-ninth its normal area (the radius is one-third the normal value), the blood must be flowing at nine times its normal speed. The net force exerted by the blood on the plaque downward and perpendicular to the direction the blood flows will be

$$F_{\text{net blood on P } y} = F_{\text{blood 1 on P}} + (-F_{\text{blood 2 on P}})$$
$$= P_1 A - P_2 A = (P_1 - P_2)A$$
$$\Rightarrow F_{\text{net blood on P } y} = \tfrac{1}{2}\rho v_2^2 A$$

Solve and evaluate

$$F_{\text{net blood on P } y} = \tfrac{1}{2}(1050 \text{ kg/m}^3)(9 \times 0.5 \text{ m/s})^2(6.0 \times 10^{-5} \text{ m}^2)$$
$$= 0.64 \text{ N} \approx 0.6 \text{ N}$$

This is about the weight of one-half of an apple pulling on this tiny plaque. In addition, an "impact" force caused by blood hitting the plaque's upstream side contributes to the risk of breaking the plaque off the side wall of the vessel. The loose plaque can then tumble downstream and block blood flow in a smaller vessel in the heart (causing a heart attack) or in the brain (causing a stroke).

- -

Try it yourself Air of density 1.3 kg/m³ moves at speed 10 m/s across the top surface of a clarinet reed that has an area of 3 cm². The air below the reed is not moving and is at atmospheric pressure. Determine the net force exerted on the reed by the air above and below it.

Answer ·ǝɔǝıdɥʇnoɯ ǝɥʇ ɟo ǝpısuı ǝɥʇ pɹɐʍoʇ 'pɹɐʍdn N 20·0

Using Bernoulli's equation to explain how airplanes fly

You may have heard that airplanes can fly because of the special shape of their wings, which causes the air above the wing to move faster than the air below the wing ($\vec{v}_{\text{above}} > \vec{v}_{\text{below}}$, shown in **Figure 14.10a**). Let us apply Bernoulli's equation to an airplane wing to evaluate this claim. A Boeing 747-8's maximum takeoff mass is 442 metric tons, its takeoff speed is 330 km/h, and its wing surface area is 554 m². We assume that the speed of air above the wing is equal to the takeoff speed and the speed of air below the wing is zero (this is a rather unrealistic assumption, but it allows us to estimate the largest possible force due to Bernoulli's effect). In this case we find

$$\Delta p = \tfrac{1}{2}\rho_{\text{air}} v_{\text{takeoff}}^2 = 0.5 \times (1.3 \text{ kg/m}^3) \times (92 \text{ m/s})^2 = 5501 \text{ N/m}^2$$

and

$$F_{\text{air on airplane}} = \Delta p \cdot A = 3.1 \times 10^6 \text{ N}$$

The force exerted by Earth on the airplane is

$$F_{\text{E on airplane}} = m_{\text{airplane}} g = 4.3 \times 10^6 \text{ N}$$

This force is much larger than the largest possible lifting force exerted on the plane due to Bernoulli's effect.

In other words, Bernoulli's effect is at work, but it does not provide a complete explanation of how airplanes fly. The important missing force comes from the change of direction of the air's motion. Due to the shape and the tilt of the wing, the air passing the wing changes its direction of motion from horizontal ($\vec{v}_i$) to slightly downward ($\vec{v}_f$) (Figure 14.10b). For this to happen, the wing must exert a downward force on the air; therefore (remember Newton's third law), the air exerts an equal and opposite force upward on the wing. It is this force that provides the main lift and (together with the force due to Bernoulli's effect) makes the airplane fly. Note that the origin of this force is the change of the direction of motion of air and has nothing to do with Bernoulli's effect.

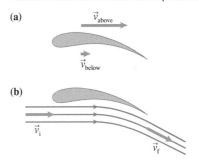

FIGURE 14.10 Airflow over an airplane wing.
(a)
(b)

REVIEW QUESTION 14.5 In Example 14.2 we said that the pressure was the same at two levels when we drew the bar chart. Doesn't the pressure in a fluid increase with depth?

14.6 Viscous fluid flow

In our previous discussions and examples in this chapter, we assumed that there were no resistive forces exerted on the moving fluid or that fluids flow without friction. That is, we assumed no interaction either between the fluid and the walls of the pipes it flows in, or between the layers of the fluid. However, in Example 14.2 we found that this assumption was not reasonable. In fact, for many processes, such as the transport of blood in the small vessels in our bodies, fluid friction is very important. When we cannot neglect this friction inside the fluid, we call the fluid **viscous**.

Consider the following situation. You have an object that can slide on a frictionless horizontal surface, say, a puck on smooth ice. You push the puck abruptly and then let go. What happens to the puck? Once in motion, the puck will continue to slide at constant speed with respect to the ice even if nothing else pushes it. However, if there is friction between the contacting surfaces (there is a little sand in the ice), then the puck starts slowing down; for it to continue moving at constant speed, someone or something has to push it forward to balance the opposing friction force.

By analogy, if a fluid flows through a horizontal tube without friction (nonviscous fluid), we would expect it to continue to flow at a constant rate with no additional forward pressure. But if friction is present (the fluid is viscous), there must be greater pressure at the back of the fluid than at the front of the fluid to maintain a constant flow rate. If this is the case, the force exerted on any volume of the fluid due to the forward pressure is greater than the force exerted on the same volume of the fluid due to the pressure in the opposite direction.

Factors that affect fluid flow rate

What factors affect the flow rate in the vessel with friction? What is the functional dependence of those factors? Let's think about the physical properties of the fluid and the vessel that can affect the flow rate. The following quantities might be important.

Pressure difference The flow rate should depend on how hard the fluid is pushed forward, that is, on the difference between the fluid pressure pushing forward from behind and the fluid pressure pushing back from in front of the fluid, or $(P_1 - P_2)$.

Radius of the tube The radius r of the tube carrying the fluid should affect the flow rate. From everyday experience we know that it is more difficult to push (a greater pressure difference) fluid through a tube of tiny radius than through a tube with a large radius.

Length of the tube The length l of the tube might also affect the ease of fluid flow. A long tube offers more resistance to flow than a shorter tube.

Fluid type Water flows much more easily than molasses does. Thus some property of a fluid that characterizes its "thickness" or "stickiness" should affect the flow.

Let's design an experiment to investigate exactly how the first three of these four factors (P, r, and l) affect the fluid flow rate Q. As shown in **Figure 14.11**, a pump that produces an adjustable pressure P_1 causes fluid to flow through tubes of different radii r and lengths l. We collect the fluid exiting the tube and measure the flow rate Q, which is the volume V of fluid leaving the tube in a certain time interval Δt divided by that time interval. The results of the experiments are reported in **Table 14.3**. How is the flow rate affected by each of the three factors?

Pressure difference Looking at the first three rows of the table, we notice that the flow rate is proportional to the pressure difference ($Q \propto P_1 - P_2$).

FIGURE 14.11 How do $P_1 - P_2$, r, and l affect the flow rate Q?

We can change $P_1 - P_2$, r, and l.

Pump

We collect water to measure the flow rate $Q = V/\Delta t$.

TABLE 14.3 **Different quantities affect the flow rate Q of fluid through a tube (data are reported in relative units)**

$P_1 - P_2$ (Pressure difference)	r (Radius)	l (Length)	Q (Flow rate)
1	1	1	1
2	1	1	2
3	1	1	3
1	2	1	16
1	3	1	81
1	1	2	0.5
1	1	3	0.33

Radius of the tube Looking at rows 1, 4, and 5, we notice that the flow rate increases rapidly as the radius increases. Doubling the radius causes the flow rate to increase by a factor of 16 (2^4). Tripling the radius causes the flow rate to increase by a factor of 81 (3^4). It seems that the flow rate is proportional to the fourth power of the radius of the tube ($Q \propto r^4$).

Length of the tube Looking at rows 1, 6, and 7, we notice that the flow rate decreases as the length of the tube increases. It seems that the flow rate is proportional to the inverse of the length ($Q \propto 1/l$).

These three relationships can be combined in a single equation:

$$Q \propto \frac{r^4(P_1 - P_2)}{l}$$

The equation has been confirmed by numerous experiments.

Viscosity and Poiseuille's law

In this experiment we did not investigate the fourth factor: the type of fluid. Under the same conditions, water flows faster than oil, which flows faster than molasses. If we use the same pressure difference to push different fluids through the same tube, we find that the fluids have different flow rates. The quantity by which we measure this effect on flow rate is called the **viscosity** η of the fluid. The flow rate is inversely proportional to viscosity:

$$Q \propto \frac{1}{\eta}$$

In 1840, using an experiment similar to that described above, French physician and physiologist Jean Louis Marie Poiseuille established a relationship between these physical quantities. However, instead of writing the flow rate in terms of the other four quantities, he wrote an expression for the pressure difference needed to cause a particular flow rate.

Poiseuille's law The forward-backward pressure difference $P_1 - P_2$ needed to cause a fluid of viscosity η to flow at a rate Q through a vessel of radius r and length l is

$$P_1 - P_2 = \left(\frac{8}{\pi}\right)\frac{\eta l}{r^4}Q \qquad (14.7)$$

The pressure difference term $P_1 - P_2$ on the left side of Poiseuille's law determines the net force pushing the fluid. The flow rate Q on the far right side is a consequence of this net push on the fluid. The term before Q on the right side (the $\left(\frac{8}{\pi}\right)\frac{\eta l}{r^4}$ term) can be thought of as the resistance of the fluid to flow: the same pressure difference will produce a smaller flow rate if the resistance is greater. The resistance is greater if the fluid has greater viscosity η, is greater for a longer tube (greater l), and is far more resistive if the vessel through which the fluid flows has a smaller radius ($1/r^4$ is much greater

TIP Notice that the pressure difference needed to cause a particular flow rate is proportional to the inverse of the fourth power of the radius of the vessel. If the radius of a vessel carrying fluid is reduced by a factor of 0.5, the pressure difference needed to cause the same flow rate must increase by $(1/0.5)^4 = 16$. We need 16 times the pressure difference to cause the same flow rate.

TABLE 14.4 Viscosities of some liquids and gases

Substance	Viscosity η (N·s/m²)
Air (30 °C)	1.9×10^{-5}
Water vapor (30 °C)	1.25×10^{-5}
Water (0 °C)	1.8×10^{-3}
Water (20 °C)	1.0×10^{-3}
Water (40 °C)	0.66×10^{-3}
Water (80 °C)	0.36×10^{-3}
Blood, whole (37 °C)	4×10^{-3}
Oil, SAE No. 10	0.20

for small r). This idea has many applications relative to the circulatory system—see the example a little later in this section.

From Poiseuille's law we can determine the unit for viscosity. To do this, we express the viscosity using other quantities in Eq. (14.7):

$$P_1 - P_2 = \left(\frac{8}{\pi}\right)\frac{\eta l}{r^4}Q \Rightarrow \eta = \frac{(P_1 - P_2)r^4\pi}{8Ql}$$

We use the latter equation to find the units for the viscosity. Remember that the units of pressure are

$$\text{Pa} = \frac{\text{N}}{\text{m}^2} = \frac{\text{kg}\cdot\text{m}}{\text{s}^2\cdot\text{m}^2} = \frac{\text{kg}}{\text{s}^2\cdot\text{m}}$$

and the units for flow rate are m³/s. Using these units, we get

$$\eta = \frac{(\text{kg})(\text{m}^4)(\text{s})}{(\text{s}^2\cdot\text{m})(\text{m}^3)(\text{m})} = \frac{\text{kg}}{\text{s}\cdot\text{m}}$$

We can also rewrite the last combination of units as N·s/m². Remember that N/m² = Pa; thus the unit for viscosity can be written Pa·s. A list of viscosities of several fluids appears in **Table 14.4** using N·s/m² for the units of η.

QUANTITATIVE EXERCISE 14.5   **Blood flow through a narrow artery**

Because of plaque buildup, the radius of an artery in a person's heart decreases by 40%. Determine the ratio of the present flow rate to the original flow rate if the pressure across the artery, its length, and the viscosity of blood are unchanged.

Represent mathematically In this exercise, we are interested in the change in the flow rate and not in the change in pressure. Consequently, we rearrange Poiseuille's law for the flow rate in terms of the other quantities:

$$Q = \left(\frac{\pi}{8}\right)\left(\frac{\Delta P}{\eta l}\right)r^4$$

If the radius decreases by 40%, the new radius is 100% − 40% = 60% of the original. Thus the radius r of the vessel at the present time is related to the radius r_0 years earlier by the equation $r = 0.60r_0$.

Solve and evaluate The ratio of the flow rates is

$$\frac{Q}{Q_0} = \frac{\left(\frac{\pi}{8}\right)\left(\frac{\Delta P}{\eta l}\right)r^4}{\left(\frac{\pi}{8}\right)\left(\frac{\Delta P}{\eta l}\right)r_0^4} = \frac{r^4}{r_0^4} = \left(\frac{r}{r_0}\right)^4 = (0.60)^4 = 0.13$$

The flow rate is only 13% of the original flow rate! To compensate for such a dramatically reduced flow rate, the person's blood pressure will increase.

- -

Try it yourself Determine the reduction in flow rate, assuming a constant pressure difference, if the radius of the vessel is reduced 90% (to 0.10 times its original value). This is not an unusual reduction for people with high blood pressure.

Answer $Q/Q_0 = 0.0001$, or 0.01 % of its original value!

Limitations of Poiseuille's law: Reynolds number

Poiseuille's law describes the flow of a fluid accurately only when the flow is laminar. Experiments indicate that to determine when the flow is laminar or turbulent, we need to calculate what is called the pipe **Reynolds number** R_e:

$$R_e = \frac{2\bar{v}r\rho}{\eta} \tag{14.8}$$

where $\bar{v}$ is the average speed of the fluid, ρ is its density, η is the viscosity, and r is the radius of the pipe that carries the fluid. Experiments show that if the Reynolds number is less than 2000, the fluid flow is laminar; if it is more than 3000, the flow is turbulent; and between 2000 and 3000 the flow is unstable and can be either laminar or turbulent.

The Reynolds number, used as a criterion to decide whether the flow inside a pipe is laminar or turbulent, came from experiments conducted in the 1880s by Osborne Reynolds. Using a glass pipe with flowing water inside, he adjusted a valve to control the speed of the water flow. He then added colored water to the stream and observed that when the speed of the water was low, the colored layer of water could be clearly seen inside the pipe. When the speed of water flow increased beyond a certain limit, the colored part would break apart into vortices and mix with the rest of the water. Reynolds expressed the criterion for the type of flow with a unitless number (hence the name the Reynolds number), which he derived by taking the ratio of the net force pushing the fluid and the resistive force opposing it.

The transition from laminar to turbulent flow is also observed when an object and fluid are moving relative to each other. Reynolds derived a similar criterion (expressed as a unitless number) for this phenomenon. In general, the flow in pipes remains laminar up to much larger R_e numbers than does the flow of a fluid around an object.

REVIEW QUESTION 14.6 Describe some of the physics-related effects on the cardiovascular system of medication that lowers the viscosity of blood.

14.7 Drag force

So far in this chapter all of our analyses have focused on a moving fluid. In Section 14.6 we were concerned with the resistive forces as fluids move through a tube. Now we focus on solid objects moving through a fluid—for example, a swimmer moving through water, a skydiver falling through the air, or a car traveling through air. As you know from experience, the fluid in these and in other cases exerts a resistive **drag force** on the object moving through the fluid. So far we have been neglecting this force in our mechanics problems. Now we will not only learn how to calculate this force but also learn whether our assumptions were reasonable: for example, is the resistive drag force indeed insignificant when people and objects fall from small and large heights?

Laminar drag force Imagine that an object moves relatively slowly through a fluid (for example, a marble sinking in oil). In this case the fluid flows around the solid object in streamline laminar flow, with no turbulence. However, the fluid does exert a drag force on the object. For a spherical object O of radius r moving at speed v through a liquid of viscosity η, the magnitude of this nonturbulent drag force $F_{\text{D F on O}}$ exerted by fluid on the object is given by the equation

$$F_{\text{D F on O}} = 6\pi\eta r v \qquad (14.9)$$

This equation is called **Stokes's law**. Notice that the drag force is proportional to the speed of the object relative to the fluid and to the radius of the sphere.

Turbulent drag force A rock sinks in oil much faster than a small marble. In this case the motion of oil past the sinking rock is turbulent, and Eq. (14.9) does not apply. A different Reynolds number, called the object Reynolds number, can be used to determine whether the flow of fluid past an object is laminar or turbulent:

$$R_e = \frac{v l \rho}{\eta} \qquad (14.10)$$

where v is the object's speed with respect to the fluid, l is the characteristic dimension of the object, in most cases the diameter, and ρ and η are the density and viscosity of the fluid. When the Reynolds number is calculated using this equation, the threshold value for laminar flow is 1. If the Reynolds number is much more than 1, the flow is turbulent and we cannot use Stokes's law. In this case, a new equation for drag force applies:

$$F_{\text{D F on O}} \approx \tfrac{1}{2}C_D\rho A v^2 \tag{14.11}$$

where ρ is the density of fluid, A is the cross-sectional area of the object as seen along its line of motion, and C_D is a dimensionless number called the *drag coefficient*. The drag coefficient depends on the shape of the object (the lower the number, the smaller the drag force and the more laminar the flow past the object). For example, the drag coefficient for a sphere is 0.5 and for a dolphin it is 0.005.

Drag force exerted on a moving vehicle

Does Stokes's law apply to moving cars? At 60 mi/h (about 27 m/s), for a car about 2 m wide in air with density 1.3 kg/m³ and viscosity 2×10^{-5} N·s/m², the estimated Reynolds number is

$$R_e = \frac{vl\rho}{\eta} = \frac{(27 \text{ m/s})(2 \text{ m})(1.3 \text{ kg/m}^3)}{(2 \times 10^{-5} \text{ N·s/m}^2)} \approx 4 \times 10^6$$

This is much more than 1. We need to use Eq. (14.11) for the drag force.

We can estimate the magnitude of the drag force that air exerts on a compact car traveling at 27 m/s (60 mi/h) assuming that the drag coefficient C_D is approximately 0.3 for a well-designed car, the cross-sectional area of a car is 2 m², and the air density is 1.3 kg/m³.

Because the flow of air past the car is turbulent, we use Eq. (14.11) to estimate the drag force that the air exerts on the car:

$$F_{\text{D Air on Car}} = \tfrac{1}{2}C_D\rho A v^2 = \tfrac{1}{2}(0.3)(1.3 \text{ kg/m}^3)(2 \text{ m}^2)(27 \text{ m/s})^2 = 280 \text{ N}$$

or a force of about 60 lb. Designing cars to minimize the drag force improves fuel efficiency.

> **TIP** Note that if the speed of a car doubles, the drag force exerted on it quadruples. Thus, because of air drag, when you increase your driving speed, you reduce your gas mileage.

Terminal speed

As a skydiver falls through the air, her speed increases, as does the drag force that the air exerts on her. Eventually, the diver's speed becomes so great that the magnitude of the upward resistive drag force that the air exerts on the diver equals the magnitude of the downward gravitational force that Earth exerts on the diver. The sum of the forces exerted on the diver is zero, so the diver moves downward at a constant speed, known as **terminal speed**. Let's estimate the terminal speed for a skydiver.

EXAMPLE 14.6   **Terminal speed of skydiver**

Estimate the terminal speed of a 60-kg skydiver falling through air of density 1.3 kg/m³, assuming a drag coefficient $C_D = 0.6$.

Sketch and translate The situation is sketched here. When the diver is moving at terminal speed, the forces that the air exerts on the diver and that Earth exerts on the diver balance—the net force is zero. We choose the diver as the system of interest with vertical y-axis pointing upward.

$\rho = 1.3$ kg/m³ $C_D = 0.6$
$m = 60$ kg
$v_{\text{terminal}} = ?$

Simplify and diagram A force diagram for the diver is shown to the right. Assume that the buoyant force that the air exerts on the diver is negligible in comparison to the other forces exerted on her and that the drag force involves turbulent airflow past the diver.

Represent mathematically Use the force diagram to help apply Newton's second law for the diver:

$$ma_y = \Sigma F_y$$
$$\Rightarrow 0 = F_{\text{D Air on D}} + (-F_{\text{E on D}})$$
$$= \tfrac{1}{2}C_D\rho A v_{\text{terminal}}^2 - mg$$

Solve and evaluate Solving for the diver's terminal speed gives

$$v_{\text{terminal}} = \sqrt{\frac{2mg}{C_D\rho A}}$$

All of the quantities in the above expression are known except the cross-sectional area of the diver along her line of motion. If we assume that she is 1.5 m tall and 0.3 m wide, her cross-sectional area is about 0.5 m². We find that her terminal speed is

$$v_{\text{terminal}} = \sqrt{\frac{2(60\ \text{kg})(9.8\ \text{N/kg})}{(0.6)(1.3\ \text{kg/m}^3)(0.5\ \text{m}^2)}} = 55\ \text{m/s}$$

The unit is correct. The magnitude seems reasonable—about 120 mi/h.

Note that we assumed a turbulent drag force (Eq. (14.11). Was this assumption appropriate? To assess this, we will estimate the maximum speed of the diver for the drag force to be laminar. According to Eq. (14.10),

$$R_e = \frac{v l \rho}{\eta} \implies v = \frac{R_e \eta}{l\rho}$$

Estimating the characteristic dimension for the person to be about 1 m (somewhere between her 1.5-m height and 0.3-m width) and the critical value of the Reynolds number to be 1 for a laminar drag force, we obtain the following estimate for the greatest speed of an object for which the drag force exerted by the fluid is laminar:

$$v = \frac{R_e \eta}{l\rho} \approx \frac{1 \times (1.9 \times 10^{-5}\ \text{N·s/m}^2)}{(1\ \text{m}) \times (1.3\ \text{kg/m}^3)} \approx 1.5 \times 10^{-5}\ \text{m/s}$$

This is a tiny speed—much too slow for a skydiver. Therefore, we could safely use the equation for turbulent drag force to estimate her speed.

- -

Try it yourself Suppose the diver pulled her legs to her chest so she was more in the shape of a ball. How qualitatively would that affect her terminal speed? Explain.

Answer Her cross-sectional area would be smaller, and according to the above equation, her terminal speed would be greater.

Below is an example of a problem whose solution depends on our assumptions. Being able to consider the effects of assumptions is a very useful skill.

EXAMPLE 14.7   **Multiple possibility problem**

A small metal ball is launched vertically downward, with initial velocity $\vec{v}_{\text{Bi}}$, from just below the surface of oil that fills a very deep container. Draw a qualitative velocity-versus-time and an acceleration-versus-time graph for the motion of the ball.

Sketch and translate We draw a sketch of the situation; let's choose the y-axis to point down. In order to draw velocity- and acceleration-versus-time graphs, we need to know the sum of the forces exerted on the ball.

Simplify and diagram The term "simplify" relates to the assumptions that we need to make to solve the problem. Given that one of the forces exerted on the ball is variable (the drag force), the assumptions we make about the initial velocity will determine the outcome. Therefore, in this case it is more useful to first identify the forces exerted on the ball and then consider different assumptions.

First, we need to identify the forces that are exerted on the ball and draw a force diagram (at right). Earth exerts a force $F_{\text{E on B}} = m_B g$ that points down. The oil exerts a buoyant force $F_{\text{B O on B}} = \rho_{\text{oil}} g V_B$ that points up. Since the ball is made of metal we know that $F_{\text{E on B}} > F_{\text{B O on B}}$. The oil also exerts a drag force $F_{\text{D O on B}}$ that is proportional to v_B and points opposite to $\vec{v}_B$ (up in our case). Given that we are interested in qualitative graphs, it is enough to remember that the drag force exerted by the oil on the ball is proportional to the speed of the ball with respect to the oil. This is true for a laminar as well as for a turbulent drag force.

Since $a_y = \dfrac{\Sigma F_{\text{O on B }y}}{m_B}$, and $F_{\text{E on B }y}$ and $F_{\text{B O on B }y}$ are constant, the sign and magnitude of $\vec{a}_y$ will depend on the magnitude of $F_{\text{D O on B }y}$ or on the speed of the ball. Let us consider the following assumptions.

1. The ball is initially moving such that the magnitude of the drag force is exactly equal to the difference between the gravitational force and the buoyant force:

$$F_{\text{D O on B }y} = F_{\text{E on B }y} - F_{\text{B O on B }y} \Rightarrow \Sigma F_{\text{O on B }y} = 0;\ a_y = 0$$

In this special case, the ball will continue to move with the same speed with which it was launched. This is the terminal speed, which we will denote v_{BT}. The graphs for this motion are shown below.

2. The ball is initially moving such that the magnitude of the drag force is *smaller* than the difference between the gravitational force and the buoyant force:

$$F_{\text{D O on B }y} < F_{\text{E on B }y} - F_{\text{B O on B }y} \Rightarrow \Sigma F_{\text{O on B }y} > 0;\ a_y > 0$$

Since $F_{\text{D O on B}}$ is proportional to v_B, this case will occur when $v_{\text{Bi}} < v_{\text{BT}}$. In this case, the ball will initially accelerate downward,

(CONTINUED)

but as the speed of the ball increases, $F_{D\,O\,on\,B}$ will increase until the sum of the forces is zero. As explained in case 1, at this point the ball reaches terminal velocity and continues moving with constant speed (zero acceleration). The graphs for this motion are shown below.

3. The ball is initially moving such that the magnitude of the drag force is *larger* than the difference between the gravitational force and the buoyant force:

$$F_{D\,O\,on\,B\,y} > F_{E\,on\,B\,y} - F_{B\,O\,on\,B\,y} \Rightarrow \Sigma F_{O\,on\,B\,y} < 0\,;\,a_y < 0$$

Since $F_{D\,O\,on\,B}$ is proportional to v_B, this case will occur when $v_{Bi} > v_{BT}$. In this case, the acceleration of the ball will initially point up (the ball will slow down). As the speed of the ball decreases, $F_{D\,O\,on\,B}$ will decrease until the sum of the forces is zero.

Therefore, the ball will slow down until it reaches the terminal velocity and then continues moving with constant speed (zero acceleration). The graphs for this motion are shown below.

Try it yourself Describe what will happen if we shoot the ball upward from the bottom of the oil-filled container. How does the scenario depend on the initial speed of the ball?

Answer No matter what the initial speed of the ball, the ball will move upward and slow down because all the forces exerted on it point in the direction opposite to its motion. Then it will start moving downward in the same way as described in case 2 in this example.

REVIEW QUESTION 14.7 When a skydiver falls at constant terminal speed, shouldn't the magnitude of the resistive drag force that the air exerts on the skydiver be a little less than the magnitude of the downward gravitational force that Earth exerts on the diver? If they are equal, shouldn't the diver stop falling? Explain.

Summary

Flow rate The flow rate Q of a fluid is the volume V of fluid that passes a cross section in a tube divided by the time interval Δt needed for that volume to pass. The flow rate also equals the product of the average speed v of the fluid and the cross-sectional area A of the vessel. (Section 14.2)

$$Q = vA$$ Eq. (14.2)

Continuity equation If fluid does not accumulate, the flow rate into a region (position 1) must equal the flow rate out of the region (position 2). At position 1 the fluid has speed v_1 and the tube has cross-sectional area A_1. At position 2 the fluid has speed v_2 and the tube has cross-sectional area A_2. (Section 14.2)

$$Q = v_1 A_1 = v_2 A_2$$ Eq. (14.3)

Bernoulli's equation For a fluid flowing without resistive forces or turbulence, the sum of the kinetic energy density $(1/2)\rho v^2$, the gravitational potential energy density ρgy, and pressure P of the fluid is a constant. (Section 14.4)

$$\frac{1}{2}\rho v_1^2 + \rho g y_1 + P_1$$
$$= P_2 + \frac{1}{2}\rho v_2^2 + \rho g y_2$$ Eq. (14.6)

Poiseuille's law For viscous fluid flow, the pressure drop $(P_1 - P_2)$ across a fluid of viscosity η flowing in a tube depends on the length l of the tube, its radius r, and the fluid flow rate Q. (Section 14.6)

$$P_1 - P_2 = \left(\frac{8}{\pi}\right)\frac{\eta l}{r^4}Q$$ Eq. (14.7)

Laminar drag force When a spherical object (like a marble sinking in oil) moves slowly through a fluid, the fluid exerts a resistive drag force on the object that is proportional to the object's speed v. **Stokes's law** describes the force. (Section 14.7)

$$F_D = 6\pi\eta r v$$ Eq. (14.9)

Turbulent drag force For an object moving at faster speed through a fluid (like a parachutist with an open parachute), turbulence occurs and the resistive drag force is proportional to the square of the speed. (Section 14.7)

$$F_D = \frac{1}{2}C_D\rho A v^2$$ Eq. (14.11)

Questions

Multiple Choice Questions

1. A roof is blown off a house during a tornado. Why does this happen?
   (a) The air pressure in the house is lower than that outside.
   (b) The air pressure in the house is higher than that outside.
   (c) The wind is so strong that it blows the roof off.
2. A river flows downstream and widens, and the flow speed slows. As a result, the pressure of the water against a dock downstream compared to upstream will be
   (a) higher.   (b) lower.   (c) the same.

3. Why does the closed top of a convertible bulge when the car is riding along a highway?
   (a) The volume of air inside the car increases.
   (b) The air pressure is greater outside the car than inside.
   (c) The air pressure inside the car is greater than the pressure outside.
   (d) The air blows into the front part of the roof, lifting the back part.

4. How does Bernoulli's principle help explain air going up the chimney of a house?
   (a) Air blowing across the top of the chimney reduces the pressure above the chimney.
   (b) The air above the chimney attracts the ashes.
   (c) The hot ashes seek the cooler outside air.
   (d) The gravitational potential energy is lower above the chimney.

5. As a river approaches a dam, the width of the river increases and the speed of the flowing water decreases. What can explain this effect?
   (a) Bernoulli's equation
   (b) The continuity equation
   (c) Poiseuille's law

6. What is an incompressible fluid?
   (a) A law of physics
   (b) A physical quantity
   (c) A model of an object

7. What is viscous flow?
   (a) A physical phenomenon
   (b) A law of physics
   (c) A physical quantity

8. The heart does about 1 J of work pumping blood during one heartbeat. What is the immediate first and main type of energy that increases due to the heart's work?
   (a) Kinetic energy
   (b) Thermal energy
   (c) Elastic potential energy

9. Several air bubbles are present in water flowing through a pipe of variable cross-sectional area (Figure Q14.9). What happens to the volume of an air bubble when it arrives at the narrow part of the pipe, where the cross-sectional area is half that of the wider part? Assume the temperature of the gas in the bubble stays constant.

FIGURE Q14.9

   (a) The volume of the bubble remains unchanged because water is incompressible.
   (b) The volume of the bubble remains unchanged because the temperature of the gas in the bubble is constant.
   (c) The volume of the bubble decreases because the pressure in water increases.
   (d) The volume of the bubble increases because the pressure in water decreases.

10. A small metal ball is released from just below the surface of oil that fills a very deep container. The y-axis points down. Which of the acceleration-versus-time graphs in Figure Q14.10 represents the motion of the ball after it is released?

FIGURE Q14.10

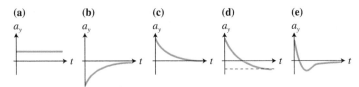

11. A small metal ball is launched downward from just below the surface of oil that fills a very deep container. The initial speed of the ball is larger than the terminal speed of the ball in the oil. The y-axis points down. Which of the graphs in Figure Q14.10 represents the motion of the ball after it is launched?

Conceptual Questions

12. You have two identical large jugs with small holes on the side near the bottom. One jug is filled with water and the other with liquid mercury. The liquid in each jug, sitting on a table, squirts out the side hole into a container on the floor. Which container, the one catching the water or the one catching the mercury, must be closer to the table in order to catch the fluid? Or should they be placed at the same distance? Which jug will empty first, or do they empty at the same time? Explain. Indicate any assumptions that you made.

13. Why does much of the pressure drop in the circulatory system occur across the arterioles (small vessels carrying blood to the capillaries) and capillaries as opposed to across the much larger diameter arteries?

14. If you partly close the end of a hose with your thumb, the water squirts out farther. Give at least one explanation for why this phenomenon occurs.

15. Compare and contrast work-energy bar charts, which you learned about in Chapter 7, with Bernoulli bar charts.

16. Consider Bernoulli's equation, Poiseuille's law, and Stokes's law. Which of these are applicable to viscous fluids? Explain.

17. You need a liquid that will exhibit turbulent flow in a tube even at lower speeds. Which properties of liquids will you evaluate when choosing a liquid? Explain.

Problems

Below, **BIO** indicates a problem with a biological or medical focus. Problems labeled **EST** ask you to estimate the answer to a quantitative problem rather than derive a specific answer. Asterisks indicate the level of difficulty of the problem. Problems with no * are considered to be the least difficult. A single * marks moderately difficult problems. Two ** indicate more difficult problems. Unless stated otherwise, assume in these problems that atmospheric pressure is $1.01 \times 10^5 \text{ N/m}^2$ and that the densities of water and air are 1000 kg/m^3 and 1.3 kg/m^3, respectively.

14.1 and 14.2 Fluids moving across surfaces—qualitative analysis and Flow rate and fluid speed

1. **Watering plants** You water flowers outside your house. (a) Determine the flow rate of water moving at an average speed of 32 cm/s through a garden hose of radius 1.2 cm. (b) Determine the speed of the water in a second hose of radius 1.0 cm that is connected to the first hose.

2. **Irrigation canal** You live near an irrigation canal that is filled to the top with water. (a) It has a rectangular cross section of 5.0-m width and 1.2-m depth. If water flows at a speed of 0.80 m/s, what is its flow rate?

(b) If the width of the stream is reduced to 3.0 m and the depth to 1.0 m as the water passes a flow-control gate, what is the speed of the water past the gate?

3. **Fire hose** During a fire, a firefighter holds a hose through which 0.070 m³ of water flows each second. The water leaves the nozzle at an average speed of 25 m/s. What information about the hose can you determine using these data?

4. The main waterline for a neighborhood delivers water at a maximum flow rate of 0.010 m³/s. If the speed of this water is 0.30 m/s, what is the pipe's radius?

5. * **BIO** **Blood flow in capillaries** The average flow rate of blood in the aorta is 80 cm³/s. Beyond the aorta, this blood eventually travels through about 6×10^9 capillaries, each of radius 8.0×10^{-4} cm. What is the average speed of the blood in the capillaries?

6. * **Irrigating a field** It takes a farmer 2.0 h to irrigate a field using a 4.0-cm-diameter pipe that comes from an irrigation canal. How long would the job take if he used a 6.0-cm pipe? What assumption did you make? If this assumption is not correct, how will your answer change?

14.4 Bernoulli's equation

7. Represent the process sketched in Figure P14.7 using a qualitative Bernoulli bar chart and an equation (include only terms that are not zero).

FIGURE P14.7   **FIGURE P14.8**

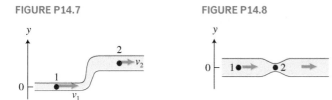

8. * Represent the process sketched in Figure P14.8 using a qualitative Bernoulli bar chart and an equation (include only terms that are not zero).
9. **Fluid flow problem** Write a symbolic equation (include only terms that are not zero) and draw a sketch of a situation that could be represented by the qualitative Bernoulli bar chart shown in Figure P14.9 (there are many possibilities).

FIGURE P14.9   **FIGURE P14.10**

$$\tilde{K}_1 + \tilde{U}_{g1} + (P_1 - P_2) = \tilde{K}_2 + \tilde{U}_{g2} \qquad \tilde{K}_1 + \tilde{U}_{g1} + (P_1 - P_2) = \tilde{K}_2 + \tilde{U}_{g2}$$

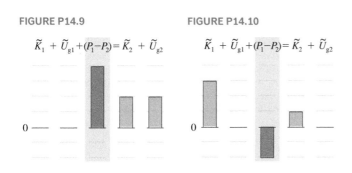

10. Repeat Problem 14.9 using the bar chart in Figure P14.10.
11. * Repeat Problem 14.9 using the bar chart in Figure P14.11.

FIGURE P14.11   **FIGURE P14.12**

$$\tilde{K}_1 + \tilde{U}_{g1} + (P_1 - P_2) = \tilde{K}_2 + \tilde{U}_{g2} \qquad \tilde{K}_1 + \tilde{U}_{g1} + (P_1 - P_2) = \tilde{K}_2 + \tilde{U}_{g2}$$

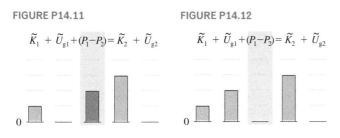

12. Repeat Problem 14.9 using the bar chart in Figure P14.12.
13. An application of Bernoulli's equation is shown below. Construct a qualitative Bernoulli bar chart that is consistent with the equation and draw a sketch of a situation that could be represented by the equation (there are many possibilities).
    $$\rho g y_2 = 0.5\rho v_1^2$$
14. Repeat Problem 14.13 using the equation $0.5\rho v_1^2 + (P_1 - P_2) = 0.5\rho v_2^2$ and $P_1 < P_2$.
15. * Repeat Problem 14.13 using the equation below.
    $$0.5\rho v_1^2 + (P_1 - P_2) =$$
    $$0.5\rho v_2^2 + \rho g y_2 \text{ and } P_1 < P_2.$$
16. * **Wine flow from barrel** While visiting a winery, you observe wine shooting out of a hole in the bottom of a barrel. The top of the barrel is open. The hole is 0.80 m below the top surface of the wine. Represent this process in multiple ways (a sketch, a bar chart, and an equation) and apply Bernoulli's equation to a point at the top surface of the wine and another point at the hole in the barrel.

17. **Water flow in city water system** Water is pumped at high speed from a reservoir into a large-diameter pipe. This pipe connects to a smaller diameter pipe. There is no change in elevation. Represent the water flow from the large pipe to the smaller pipe in multiple ways—a sketch, a bar chart, and an equation.

14.5 Skills for analyzing processes using Bernoulli's equation

18. * The pressure of water flowing through a 0.060-m-radius pipe at a speed of 1.8 m/s is $2.2 \times 10^5 \text{ N/m}^2$. What is (a) the flow rate of the water and (b) the pressure in the water after it goes up a 5.0-m-high hill and flows in a 0.050-m-radius pipe?
19. * **Siphoning water** You want to siphon rainwater and melted snow from the cover of an above-ground swimming pool. The cover is 1.4 m above the ground. You have a plastic hose of 1.0-cm radius with one end in the water on the pool cover and the other end on the ground. (a) At what speed does water exit the hose? (b) If you want to empty the pool cover in half the time, what new hose radius should you use? (c) How much faster does the water flow through this wider pipe?
20. * **Cleaning skylights** You are going to wash the skylights in your kitchen. The skylights are 8.0 m above the ground. You connect two garden hoses together—a 0.80-cm-radius hose to a 1.0-cm-radius hose. The smaller hose is held on the roof of the house and the wider hose is attached to the faucet on the ground. The pressure at the opening of the smaller hose is 1 atm, and you want the water to have the speed of 6.0 m/s. What should be the pressure at ground level in the large hose? What should be the speed?
21. * **BIO** **Blood flow in artery** Blood flows at an average speed of 0.40 m/s in a horizontal artery of radius 1.0 cm. The average pressure is $1.4 \times 10^4 \text{ N/m}^2$ above atmospheric pressure (the gauge pressure). (a) What is the average speed of the blood past a constriction where the radius of the opening is 0.30 cm? (b) What is the gauge pressure of the blood as it moves past the constriction? What assumption did you make to answer these questions?
22. * **Straw aspirator** A straw extends out of a glass of water by a height h. How fast must air blow across the top of the straw to draw water to the top of the straw?
23. * **Gate for irrigation system** You observe water at rest behind an irrigation dam. The water is 1.2 m above the bottom of a gate that, when lifted, allows water to flow under the gate. Determine the height h from the bottom of the dam that the gate should be lifted to allow a water flow rate of $1.0 \times 10^{-2} \text{ m}^3/\text{s}$. The gate is 0.50 m wide.
24. * **BIO** **Flutter in blood vessel** A person has a 5200-N/m² gauge pressure of blood flowing at 0.50 m/s inside a 1.0-cm-radius main artery. The gauge pressure outside the artery is 3200 N/m². When using his stethoscope, a physician hears a fluttering sound farther along the artery. The sound is a sign that the artery is vibrating open and closed, which indicates that there must be a constriction in the artery that has reduced its radius and subsequently reduced the internal blood pressure to less than the external 3200-N/m² pressure. What is the maximum artery radius at this constriction?
25. * **BIO** **Effect of smoking on arteriole radius** The average radius of a smoker's arterioles, the small vessels carrying blood to the capillaries, is 5% smaller than those of a nonsmoker. (a) Determine the percent change in flow rate if the pressure across the arterioles remains constant. (b) Determine the percent change in pressure if the flow rate remains constant.
26. * **Roof of house in wind** The mass of the roof of a house is 2.1×10^4 kg and the area of the roof is 160 m². At what speed must air move across the roof of the house so that the roof is lifted off the walls? Indicate any assumptions you made.
27. * You have a U-shaped tube open at both ends. You pour water into the tube so that it is partially filled. You have a fan that blows air at a speed of 10 m/s. (a) How can you use the fan to make water rise on one side of the tube? Explain your strategy in detail. (b) To what maximum height can you get the water to rise? Note: You cannot touch the water yourself.

28. * Engineers use a venturi meter to measure the speed of a fluid traveling through a pipe (see Figure P14.28). Positions 1 and 2 are in pipes with surface areas A_1 and A_2, with A_1 greater than A_2, and are at the same vertical height. How can you determine the relative speeds at positions 1 and 2 and the pressure difference between positions 1 and 2?

FIGURE P14.28

14.6 Viscous fluid flow

29. * A 5.0-cm-radius horizontal water pipe is 500 m long. Water at 20 °C flows at a rate of 1.0×10^{-2} m³/s. (a) Determine the pressure drop due to viscous friction from the beginning to the end of the pipe. (b) What radius pipe must you use if you want to keep the pressure difference constant and double the flow rate?

30. **Fire hose** A volunteer firefighter uses a 5.0-cm-diameter fire hose that is 60 m long. The water moves through the hose at 12 m/s. The temperature outside is 20 °C. What is the pressure drop due to viscous friction across the hose?

31. **Another fire hose** The pump for a fire hose can develop a maximum pressure of 6.0×10^5 N/m². A horizontal hose that is 50 m long is to carry water of viscosity 1.0×10^{-3} N·s/m² at a flow rate of 1.0 m³/s. What is the minimum radius for the hose?

32. * **Solar collector water system** Water flows in a solar collector through a copper tube of radius R and length l. The average temperature of the water is T °C and the flow rate is Q cm³/s. Explain how you would determine the viscous pressure drop along the tube, assuming the water does not change elevation.

33. * **BIO** **Blood flow through capillaries** Your heart pumps blood at a flow rate of about 80 cm³/s. The blood flows through approximately 9×10^9 capillaries, each of radius 4×10^{-4} cm and 0.1 cm long. Determine the viscous friction pressure drop across a capillary, assuming a blood viscosity of 4×10^{-3} N·s/m².

34. * Determine the ratio of the flow rate through capillary tubes A and B (that is, Q_A/Q_B). The length of A is twice that of B, and the radius of A is one-half that of B. The pressure across both tubes is the same.

35. * A piston pushes 20 °C water through a horizontal tube of 0.20-cm radius and 3.0-m length. One end of the tube is open and at atmospheric pressure. (a) Determine the force needed to push the piston so that the flow rate is 100 cm³/s. (b) Repeat the problem using SAE 10 oil instead of water.

36. * How can you use the venturi meter system (see Problem 14.28) to determine whether viscous fluid needs an additional pressure difference to flow at the same speed as a nonviscous fluid?

37. * A syringe is filled with water and fixed at the edge of a table at height $h = 1.0$ m above the floor (Figure P14.37). The diameter of the piston is $d_1 = 20.0$ mm, the needle has length $L = 50.00$ mm, and the inner diameter of the needle is $d_2 = 1.0$ mm. You press on the piston so that it moves at constant speed of 10.00 mm/s. Determine (a) the distance D at which the water jet hits the floor and (b) the pressure difference between the ends of the needle. Assume the viscosity of water is 1.0×10^{-3} N·s/m².

FIGURE P14.37

14.7 Drag force

38. **Car drag** A 2300-kg car has a drag coefficient of 0.60 and an effective frontal area of 2.8 m². Determine the air drag force on the car when traveling at (a) 24 m/s (55 mi/h) and (b) 31 m/s (70 mi/h).

39. * **EST** **Air drag when biking** Estimate the drag force opposing your motion when you ride a bicycle at 8 m/s.

40. **BIO** **Drag on red blood cell** Determine the drag force on an object the size of a red blood cell with a radius of 1.0×10^{-5} m that is moving through 20 °C water at speed 1.0×10^{-5} m/s. (Assume laminar flow.)

41. * **BIO** **EST** **Protein terminal speed** A protein of radius 3.0×10^{-9} m falls through a tube of water with viscosity $\eta = 1.0 \times 10^{-3}$ N·s/m². Earth exerts a constant downward 3.0×10^{-22}-N force on the protein. (a) Use Stokes's law and the information provided to estimate the terminal speed of the protein. Assume no buoyant force is exerted on the protein. (b) How many hours would be required for the protein to fall 0.10 m?

42. * **EST** Earth exerts a constant downward force of 7.5×10^{-13} N on a clay particle falling in water. The particle settles 0.10 m in 820 min. Estimate the radius of a clay particle. Assume no buoyant force is exerted on the clay particle. The viscosity of water is 1.0×10^{-3} N·s/m².

43. * A sphere falls through a fluid. Earth exerts a constant downward 0.50-N force on the sphere. The fluid exerts an opposing drag force on the fluid given by $F_D = 2v$, where F_D is in newtons if v is in meters per second. Determine the terminal speed of the sphere.

44. * **Terminal speed of balloon** A balloon of mass m drifts down through the air. The air exerts a resistive drag force on the balloon described by the equation $F_D = 0.03v^2$ where F_D is in newtons if v is in meters per second. What is the terminal speed of the balloon?

45. You observe four different liquids (listed with their viscosities and densities in the table below) as they flow with the same speed through identical tubes. You gradually increase the speed of the liquids until their flows become turbulent. Rank the liquids in the order in which you see their flow become turbulent.

Liquid	Viscosity (N·s/m²)	Density (kg/m³)
Blood	4×10^{-3}	1.05×10^3
Water	1×10^{-3}	1.00×10^3
Ethanol	1×10^{-3}	0.79×10^3
Acetone	3×10^{-4}	0.79×10^3

General Problems

46. ** Describe how you will design an experimental procedure that will help you decide whether the drag force exerted on a coffee filter falling at terminal velocity depends on the filter's speed v or on its speed squared v^2.

47. * If the speed of air that is blowing across the upper end of a vertical tube is great enough, a thin plate that is placed at the lower end of the tube will remain touching the end of the tube, without falling down. Derive an expression for the minimum speed of air that needs to blow across a tube of diameter d to keep the plate of mass m from falling down.

48. **Mariotte's bottle** Figure P14.48 shows a device called Mariotte's bottle that can deliver a constant flow rate. It consists of a tube that releases air bubbles into a sealed bottle or tank at a height b above the point where the liquid exits. (a) Under what conditions does Mariotte's bottle deliver a constant flow rate and thus a constant speed of liquid? Explain. (b) How does this constant speed of liquid leaving the bottle depend on b, assuming that there is no friction at the opening where liquid leaves the bottle? (Hint: Note that the pressure at the point where the bubbles are entering the bottle is equal to atmospheric pressure.)

FIGURE P14.48

49. ** BIO **Pressure needed for intravenous needle** A glucose solution of viscosity 2.2×10^{-3} N·s/m² and density 1030 kg/m³ flows from an elevated open bag into a vein. The needle into the vein has a radius of 0.20 mm and is 3.0 cm long. All other tubes leading to the needle have much larger radii, and viscous forces in them can be ignored. The pressure in the vein is 1000 N/m² above atmospheric pressure. (a) Determine the pressure relative to atmospheric pressure needed at the entrance of the needle to maintain a flow rate of 0.10 cm³/s. (b) To what elevation should the bag containing the glucose be raised to maintain this pressure at the needle?

50. ** **Viscous friction with Bernoulli** We can include the effect of viscous friction in Bernoulli's equation by adding a term for the thermal energy generated by the viscous retarding force exerted on the fluid. Show that the term to be added to Eq. (14.5) for flow in a vessel of uniform cross-sectional area A is

$$\frac{\Delta U_{Th}}{V} = \frac{4\pi \eta l v}{A}$$

where v is the average speed of the fluid of viscosity η along the center of a pipe whose length is l.

51. ** (a) Show that the work W done per unit time Δt by viscous friction in a fluid with a flow rate Q across which there is a pressure drop ΔP is

$$\frac{W}{\Delta t} = \Delta P Q = Q^2 R = \frac{\Delta P^2}{R}$$

where $R = 8\eta l/\pi r^4$ is called the *flow resistance* of the fluid moving through a vessel of radius r. (b) By what percentage must the work per unit time increase if the radius of a vessel decreases by 10% and all other quantities including the flow rate remain constant (the pressure does not remain constant)?

52. ** BIO EST **Thermal energy in body due to viscous friction** Estimate the thermal energy generated per second in a normal body due to the viscous friction force in blood as it moves through the circulatory system.

53. ** BIO **Essential hypertension** Suppose your uncle has hypertension that causes the radii of his 40,000,000 arterioles to decrease by 20%. Each arteriole initially was 0.010 mm in radius and 1.0 cm long. By what factor does the resistance $R = 8\eta l/\pi r^4$ to blood flow through an arteriole change because of these decreased radii? The pressure drop across all of the arterioles is about 60 mm Hg. If the flow rate remains the same, what now is the pressure drop change across the arteriole part of the circulatory system?

54. * **Parachutist** A parachutist weighing 80 kg, including the parachute, falls with the parachute open at a constant 8.5-m/s speed toward Earth. The drag coefficient $C_D = 0.50$. What is the area of the parachute?

55. A 0.20-m-radius balloon falls at terminal speed 40 m/s. If the drag coefficient is 0.50, what is the mass of the balloon?

56. ** **Terminal speed of skier** A skier going down a slope of angle θ below the horizontal is opposed by a turbulent drag force that the air exerts on the skier and by a kinetic friction force that the snow exerts on the skier. Show that the terminal speed is

$$v_T = \left[\frac{2mg(\sin\theta - \mu\cos\theta)}{C_D \rho A} \right]^{1/2}$$

where μ is the coefficient of kinetic friction between the skis and the snow, ρ is the density of air, A is the skier's frontal area, and C_D is the drag coefficient.

57. ** A grain of sand of radius 0.15 mm and density 2300 kg/m³ is placed in a 20 °C lake. Determine the terminal speed of the sand as it sinks into the lake. Do not forget to include the buoyant force that the water exerts on the grain.

58. ** EST **Comet crash** On June 30, 1908, a monstrous comet fragment of mass greater than 10^9 kg is thought to have devastated a 2000-km² area of remote Siberia (this impact was called the Tunguska event). Estimate the terminal speed of such a comet in air of density 0.70 kg/m³. State all of your assumptions.

Reading Passage Problems

BIO **Intravenous (IV) feeding** A patient in the hospital needs fluid from a glucose nutrient bag. The glucose solution travels from the bag down a tube and then through a needle inserted into a vein in the patient's arm

(Figure 14.12a). Your study of fluid dynamics makes you think that the bag seems a little low above the arm and the narrow needle seems long. You wonder if the glucose is actually making it into the patient's arm. What height should the bag (open at the top) be above the arm so that the glucose solution (density 1000 kg/m³ and viscosity 1.0×10^{-3} N·s/m²) drains from the open bag down the 0.6-m-long, 2.0×10^{-3}-m radius tube and then through the 0.020-m-long, 4.0×10^{-4}-m radius needle and into the vein? The gauge pressure in the vein in the arm is +930 N/m² (or 7 mm Hg)? The nurse says the flow rate should be 0.20×10^{-6} m³/s (0.2 cm³/s).

FIGURE 14.12 (a) A glucose solution flowing from an open container into a vein. (b) The analysis of the needle in this system.

(a)

(b)

59. Which answer below is closest to the speed with which the glucose should flow out of the end of the needle at position C in Figure 14.12b?
    (a) 0.0004 m/s   (b) 0.004 m/s   (c) 0.04 m/s
    (d) 0.4 m/s   (e) 4 m/s

60. Which answer below is closest to the speed with which the glucose should flow through the end of the tube just to the right of position B in Figure 14.12b?
    (a) 0.0002 m/s   (b) 0.002 m/s   (c) 0.02 m/s
    (d) 0.2 m/s   (e) 2 m/s

61. Assume that there is no resistive friction pressure drop across the needle (as could be determined using Poiseuille's law). Use the Bernoulli equation and the results from Problems 14.59 and 14.60 to determine which answer below is closest to the change in pressure between positions B and C $(P_B - P_C)$ in Figure 14.12b.
    (a) 8 N/m²   (b) 80 N/m²   (c) 800 N/m²
    (d) 8000 N/m²   (e) 80,000 N/m²

62. Now, in addition to the Bernoulli pressure change from position B to position C calculated in Problem 14.61, there may be a Poiseuille resistive friction pressure change across the needle from position B to position C. Which answer below is closest to that pressure change?
    (a) 0.4 N/m²   (b) 4 N/m²   (c) 40 N/m²
    (d) 400 N/m²   (e) 4000 N/m²

63. The blood pressure in the vein at position C in Figure 14.12b at the exit of the needle into the blood is 930 N/m². Use this value and the results of Problems 14.61 and 14.62 to determine which answer below is closest to the gauge pressure at position B in the tube carrying the glucose to the needle.
    (a) 1010 N/m²   (b) 1410 N/m²   (c) 1980 N/m²
    (d) 2800 N/m²   (e) 4620 N/m²

64. Suppose that there is no Poiseuille resistive friction pressure decrease from the top of the glucose solution in the open bag (position A in Figure 14.12a) through the tube and down to position C near the entrance to the needle. Which answer below is closest to the minimum height of the top of the bag in order for the glucose to flow down from the tube and through the needle into the blood? Remember that the pressure at position A is atmospheric pressure, which is zero gauge pressure.
    (a) 0.04 m   (b) 0.08 m   (c) 0.14 m
    (d) 0.27 m   (e) 0.60 m

65. Suppose there is a Poiseuille resistive friction pressure decrease from the top of the glucose solution (position A in Figure 14.12a) through the tube and down to position C near the entrance to the needle. How will this affect the placement of the bag relative to the arm?
    (a) The bag will need to be higher.
    (b) The bag can remain the same height above the arm.
    (c) The bag can be placed lower relative to the arm.
    (d) Too little information is provided to answer the question.

BIO **The human circulatory system** In the human circulatory system, depicted in Figure 14.13, the heart's left ventricle pumps about 80 cm³ of blood into the aorta every second. The blood then moves into a larger and larger number of smaller radius vessels (aorta, arteries, arterioles, and capillaries). After the capillaries, which deliver nutrients to the body cells and absorb waste products, the vessels begin to combine into a smaller number of larger radius vessels (venules, small veins, large veins, and finally the vena cava). The vena cava returns blood to the heart (see Table 14.5).

FIGURE 14.13 A schematic representation of the circulatory system including the pressure variation across different types of vessels.

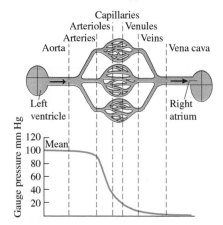

TABLE 14.5 The different types of vessels in the circulatory system

Vessel type	Number of vessels	Approximate radius (mm)
Aorta	1	5
Large arteries	40	2
Smaller arteries	2400	0.4
Arterioles	40,000,000	0.01
Capillaries	1,200,000,000	0.004
Venules	80,000,000	0.02
Small veins	2400	1
Large veins	40	3
Vena cava	1	6

The flow rate Q of blood through the arteries equals the flow rate through the arterioles, which equals the flow rate through the capillaries, and so forth. The average blood gauge pressure in the aorta is about 100 mm Hg. The pressure drops as blood passes through the different groups of vessels and is approximately 0 mm Hg when it returns to the heart at the vena cava.

A working definition of the resistance R to flow by a group of vessels is the ratio of the gauge pressure drop ΔP across those vessels divided by the flow rate Q through the vessels:

$$R = \frac{\Delta P}{Q} \qquad (14.12)$$

The gauge pressure drop across the whole system is (100 mm Hg – 0), and the total resistance is

$$R_{\text{total}} = \frac{\Delta P_{\text{total}}}{Q} = \frac{100 \text{ mm Hg}}{80 \text{ cm}^3/\text{s}} = 1.25 \frac{\text{mm Hg}}{\text{cm}^3/\text{s}}$$

The gauge pressure drop across the whole system is the sum of the drops across each type of vessel:

$$\Delta P_{\text{total}} = \Delta P_{\text{aorta}} + \Delta P_{\text{arteries}} + \Delta P_{\text{arterioles}} + \Delta P_{\text{capillaries}} + \cdots + \Delta P_{\text{vena cava}}$$

Now rearrange and insert Eq. (14.12) into the above for the pressure drop across each part:

$$QR_{\text{total}} = QR_{\text{aorta}} + QR_{\text{arteries}} + QR_{\text{arterioles}} + QR_{\text{capillaries}} + \cdots + QR_{\text{vena cava}}$$

Canceling the common flow rate through each group of vessels, we have an expression for the total resistance of the circulatory system:

$$R_{\text{total}} = R_{\text{aorta}} + R_{\text{arteries}} + R_{\text{arterioles}} + R_{\text{capillaries}} + \cdots + R_{\text{vena cava}}$$

The measured gauge pressure drop across the arterioles is about 50 mm Hg, and the arteriole resistance is

$$R_{\text{arterioles}} = \frac{\Delta P_{\text{arterioles}}}{Q} = \frac{50 \text{ mm Hg}}{80 \text{ cm}^3/\text{s}} = 0.62 \frac{\text{mm Hg}}{\text{cm}^3/\text{s}}$$

or about 50% of the total resistance. The next most resistive group of vessels is the capillaries, at about 25% of the total resistance. These percentages vary significantly from person to person. A person with essential hypertension has arterioles and capillaries that are reduced in radius. The resistance to blood flow increases dramatically (the resistance has a $1/r^4$ dependence). The blood pressure has to be greater (for example, double the normal value) in order to produce reasonable flow to the body cells. Even with the increased pressure, the flow rate may still be lower than normal.

66. The capillaries typically produce about 25% of the resistance to blood flow. Which pressure drop below is closest to the pressure drop across the group of capillaries?
    (a) 5 mm Hg   (b) 15 mm Hg   (c) 25 mm Hg
    (d) 35 mm Hg   (e) 45 mm Hg

67. We found that the arteriole resistance to fluid flow was about 0.62 mm Hg/(cm³/s). By what factor would you expect the resistance of all the arterioles to change if the radius of each arteriole decreased to 80% of the original value?
    (a) 1.3   (b) 1.6   (c) 2.4
    (d) 0.4   (e) 0.6

68. Why is the resistance to fluid flow through unobstructed arteries relatively small compared to resistance to fluid flow through the arterioles and capillaries?
    (a) The arteries are nearer the heart.
    (b) There is a relatively small number of arteries.
    (c) The artery radii are relatively large.
    (d) b and c
    (e) a, b, and c

69. The huge number of capillaries and venules is needed to
    (a) provide nutrients (such as O_2) and remove waste products from all of the body cells.
    (b) distribute water uniformly throughout the body.
    (c) reduce the resistance of the circulatory system.
    (d) b and c
    (e) a, b, and c

70. Which number below best represents the ratio of the resistance of a single capillary to the resistance of a single arteriole, assuming they are equally long?
    (a) 40   (b) 6   (c) 2.5
    (d) 0.4   (e) 0.026

First Law of Thermodynamics

The photos above show the changes that occurred to the Muir Glacier in Alaska between 1941 and 2004. Earth's average temperature is increasing every year. The rise in temperature correlates with the concentration of carbon dioxide (CO_2) in Earth's atmosphere. Before the Industrial Revolution of the 1800s, the concentration of atmospheric CO_2 ranged from 200 ppm (parts per million) during relatively cold periods to 280 ppm during relatively warm periods. The CO_2 concentration is now over 400 ppm and is increasing at a rate of about 2 ppm/year. How do we explain the connection between the concentration of CO_2 and the changes in Earth's average global temperature?

- **How is the CO_2 concentration in the atmosphere related to the melting of glaciers?**

- **How does sweating protect our bodies from overheating?**

- **Why does the thermal energy of a spoon increase when you place it in a cup of hot coffee even though no external objects do work on it?**

WHEN WE STUDIED WORK AND ENERGY (Chapter 7), we learned that a system's energy could change when external objects do work on it. However, in real life we know many examples of processes that can change the internal energy of a system without external objects exerting forces and doing work. Imagine, for example, that you put a room-temperature spoon into a cup of hot coffee. Our understanding of work-energy processes does not explain why the spoon gets hot. The branch of physics called **thermodynamics** helps us explain this phenomenon by extending the concepts of work and energy. Thermodynamics is the subject of this chapter.

BE SURE YOU KNOW HOW TO:

- Identify a system and decide on the initial and final states of a process (Section 3.1).

- Draw an energy bar chart and use it to help apply the work-energy principle (Section 7.2).

- Apply your understanding of molecular motion to explain gas processes (Section 12.6).

15.1 Internal energy and work in gas processes

The ideal gas model describes a gas consisting of atoms and molecules that are assumed to be identical point objects—particles that do not interact with each other at a distance. We determined (in Chapter 12) an expression for the energy of the particles in a container of such a gas.

Thermal energy of ideal gas

Imagine a bottle filled with air. The air is our system. What energy does this system possess? It does not have any gravitational potential energy (Earth is not in the system), and it has no organized kinetic energy (when all parts of the system are moving in the same direction). However, the gas consists of individual particles that are moving randomly.

Earlier (in Chapter 12) we reasoned that each gas particle has some average kinetic energy

$$\overline{K}_{\text{particle}} = \tfrac{3}{2}k_{\text{B}}T$$

due to its random motion. T is the absolute (kelvin) temperature of the gas. Consequently, the N molecules in the container have a total random kinetic energy (called *thermal energy*):

$$U_{\text{th}} = N\left(\tfrac{3}{2}k_{\text{B}}T\right)$$

The number N of particles can be written as the number of moles n of the gas times Avogadro's number of particles N_{A} in one mole $(N = nN_{\text{A}})$. Also, recall that $N_{\text{A}}k_{\text{B}} = R$, the universal gas constant. Thus we can rewrite the equation as

$$U_{\text{th}} = N\left(\tfrac{3}{2}k_{\text{B}}T\right) = nN_{\text{A}}\left(\tfrac{3}{2}k_{\text{B}}T\right) = n\left(\tfrac{3}{2}N_{\text{A}}k_{\text{B}}T\right) = \tfrac{3}{2}nRT \qquad (15.1)$$

Note that the thermal energy of this gas depends only on its absolute temperature and on the number of moles of gas. It does not depend on the volume it occupies. The thermal energy of a gas is called a **state function**. The term state refers to a set of measurements that describe a system. The term state function means that the value of the quantity depends only on the state of the system and not on how the system is brought to this state. For example, the thermal energy of the ideal gas depends only on the temperature and not on the way that the gas was brought to this temperature. We will see later that not every quantity has this property.

Do gas molecules possess any potential energy? We assumed that ideal gas particles do not interact at a distance; thus, the system has no potential energy due to particle interactions. **Therefore, in an ideal gas the total internal energy of the gas particles equals its thermal energy.** This means that in Eq. (15.1), we can use either the subscript th or int.

TIP ╟ We distinguish between the thermal energy of all N particles in a container of gas $U_{\text{th}} = N\left(\tfrac{3}{2}k_{\text{B}}T\right)$ and the average random kinetic energy of a single particle $\overline{K} = \tfrac{3}{2}k_{\text{B}}T$. Imagine a small container filled with gas. You separate half of the gas by placing a wall inside the container. The gas in the whole container has more thermal energy than the half even though the average kinetic energy of individual molecules in the separated part is the same as that in the whole container.

EXAMPLE 15.1   **Thermal energy of the gas in a room**

Construct an expression that could be used to estimate the thermal energy of air at pressure P in a room of volume V. Use the expression to estimate the thermal energy of the air filling a small bedroom.

Sketch and translate We have an expression for the thermal energy of a container of gas in terms of its absolute temperature T and the number n of moles of particles (Eq. 15.1). We need to connect these quantities to the gas pressure P and the room volume V. The ideal gas law seems promising for relating these quantities.

Simplify and diagram Assume that the air in the room is an ideal gas, that the air pressure is $P_{\text{atm}} = 1.0 \times 10^5 \text{ N/m}^2$, and that the dimensions of a small bedroom are $4 \text{ m} \times 3 \text{ m} \times 3 \text{ m} = 36 \text{ m}^3$.

Represent mathematically The thermal energy of the gas is given by Eq. (15.1): $U_{\text{th}} = (3/2)nRT$. Using the ideal gas law, $PV = nRT$, we find that

$$U_{\text{th}} = \tfrac{3}{2}nRT = \tfrac{3}{2}PV$$

We can use this expression to estimate the thermal energy of the gas.

Solve and evaluate The thermal energy of the gas in a small bedroom is approximately

$$U_{th} = \tfrac{3}{2}PV = \tfrac{3}{2}(1.0 \times 10^5 \, \text{N/m}^2)(36 \, \text{m}^3)$$

$$= 54 \times 10^5 \, \text{N} \cdot \text{m} \approx 5 \times 10^6 \, \text{J}$$

The result is equivalent to the energy given off by a 100-J/s (100-watt) lightbulb during a 50,000-s (14-hour) time interval—a significant amount of energy! Notice that when estimating, we keep only one significant figure.

Try it yourself What is the thermal energy of the air in an empty 1-L soft drink bottle? State your assumptions.

Answer

150 J. We assumed that the gas pressure inside the bottle equaled atmospheric pressure and that the air inside the bottle could be modeled as an ideal gas.

Work done on a gas

Imagine that you have a gas (the system) at high pressure in a cylinder with a movable piston (**Figure 15.1**). You hold the piston and allow it to move slowly outward (to the left in Figure 15.1). The gas pushes to the left on the piston and the piston in turn pushes toward the right on the gas, exerting a force on the gas $\vec{F}_{\text{P on G}}$. The gas inside the piston expands slowly. What work is done by the force that the piston exerts on the gas? Assume that the piston moves outward a distance d, allowing the gas to expand from its initial volume V_i to its final larger volume V_f.

Recall that work is the product of the magnitude of the force exerted on an object, the object's displacement, and the cosine of the angle between the directions of the force and the displacement ($W = Fd \cos \theta$). This equation only works for a *constant* magnitude force. In our case, as the high-pressure gas expands, there is a decrease in the frequency of collisions of particles with the piston, and the pressure decreases. Thus, the piston exerts a *changing* force on the expanding gas.

It is also possible to determine an expression for the work done on the changing volume of gas by breaking the big volume change into many small increments and adding the work done during each small change. Let's start with a process in which the piston moves outward in small steps each of distance Δx starting at position x_i and ending at position x_f (**Figure 15.2a**). For each step the gas volume change is small, the gas pressure changes little, and the force that the piston exerts on the gas remains approximately constant. The gas pushes outward on the piston whether the piston is causing an expansion or contraction of the gas. The piston in turn exerts a force on the gas in the opposite direction (toward the right in Figure 15.2a). If the gas is expanding, then the piston in Figure 15.2a is moving left, and the work done by the piston on the gas for the Δx expansion is

$$W_{\text{Piston on Gas}} = F_{\text{P on G}} \Delta x \cos(180°) = -F_{\text{P on G}} \Delta x$$

The force that the piston exerts on the gas and the piston's displacement point in opposite directions—hence the minus sign.

The change in the volume of the gas when the piston moves out a distance Δx is $\Delta V = A \Delta x$, where A is the cross-sectional area of the piston. Rearranging the above, we find that $\Delta x = (\Delta V)/A$. Recall also the definition of pressure: $P = (F/A)$ or $F = P \cdot A$. Insert these expressions for F and Δx into the above expression for work to get

$$W_{\text{Piston on Gas}} = -F_{\text{P on G}} \Delta x = -(P \cdot A)\left(\frac{\Delta V}{A}\right) = -P \Delta V \qquad (15.2)$$

As the gas has expanded and has a larger volume, ΔV is positive. Since the piston is pushing opposite the expansion, it does negative work on the gas and the energy of the gas decreases. If instead the piston exerted a force that caused the gas to compress, the force would do positive work on the system (ΔV would be negative and it would cancel the negative sign in Eq. (15.2); the gas would have increased energy).

FIGURE 15.1 The piston does work on the gas, changing its thermal energy.

The gas in the cylinder expands as the piston moves to the left.

FIGURE 15.2 Determine the work done on a changing volume of gas.

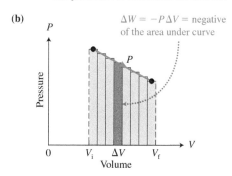

Remember also that this is the work done *on the gas* by the external force that the piston exerts on the gas during a small change in the gas volume. This work done by the piston on the gas during a small expansion of the gas is the negative of the area under the narrow dark-shaded column of the P-versus-V graph line in Figure 15.2b. This area is the height P of the narrow column times its width ΔV.

We split the large change in gas volume for the whole process into n tiny volume changes. For each small volume change, assume that the pressure is approximately constant (Figure 15.2b). Then add the work done during each of these volume changes to get the total work done by the piston on the gas:

$$W_{\text{Piston on Gas}} = (-P_1 \Delta V_1) + (-P_2 \Delta V_2) + \cdots + (-P_n \Delta V_n)$$

Each term is the negative of the area of a narrow column in Figure 15.2b. Thus, the work done during the expansion from volume V_i to volume V_f is the negative area under the entire graph line between those initial and final volumes.

FIGURE 15.3 Work depends on the way in which the system changes from the initial to the final state.

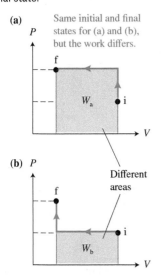

FIGURE 15.3 Work depends on the way in which the system changes from the initial to the final state.

(a) Same initial and final states for (a) and (b), but the work differs.

(b)

Different areas

TIP Notice that the thermal energy of the gas in a cylinder depends only on the temperature of the gas when in that state (and on the number of particles). Temperature and internal thermal energy are state functions.

Work depends on the process between the initial and final states

Many different processes can take a gas from an initial pressure P_i and volume V_i to a final pressure P_f and volume V_f (different curves or lines on a P-versus-V graph). The amount of work done on the gas can differ in each case (see the examples in **Figure 15.3a and b**). In Figure 15.3a, to go from the initial to the final state, the gas first has a constant volume and increasing pressure and then a constant pressure and decreasing volume. On the other hand, in (b), the gas first has a constant pressure volume decrease and then a constant volume pressure increase to get to the same final state. Note that the work done for each process, represented by the shaded area under the graph lines, is different—it is greater for the process in (a) than for the process in (b), even though the initial and final states are the same. The work done on or by a gas depends on how the pressure and volume change in going from the initial to the final state. The work depends on the process that occurs in changing state and is *not* a state function. It does not describe a single state or the difference between the two states of the system.

Work done *by* a gas *on* its environment

The gas mixture in the cylinders of your automobile engine is part of a motor. In this case we are not interested in the work done by the environment (the piston) on the gas (the system) but are instead interested in the work done by that gas on the environment. When the gas in the cylinder expands, it does positive work on its environment. The environment in turn does negative work of the same magnitude on the gas. The work that the piston does on the gas is equal in magnitude and opposite in sign to the work that the gas does on the piston—an example of Newton's third law. Thus, we can define work involving gas changes as follows:

Work done on or by a gas The magnitude of the work done by the environment on a gas $W_{\text{Environment on Gas}}$ when a gas changes from volume V_i to V_f is equal to the area under the P-versus-V graph for that process. The sign of the work is negative if the gas volume increases and positive if its volume decreases. The work done by the gas on the environment $W_{\text{Gas on Environment}}$ has the same magnitude but the opposite sign. If the pressure is constant, the work done by the environment on the gas is

$$W_{\text{Environment on Gas}} \, (\text{constant pressure process}) = -P(V_f - V_i) = -P\Delta V \quad (15.2)$$

and the work done by the gas on the environment is

$$W_{\text{Gas on Environment}} \, (\text{constant pressure process}) = +P(V_f - V_i) = +P\Delta V \quad (15.3)$$

When the pressure is not constant, the work done on the gas by the environment can be found by using the graphical method shown in Figure 15.2, or mathematically by breaking the process into small steps for which the pressure is constant, finding the works done over these steps using Eq. (15.2) for each step, and then adding them together, or by finding the average pressure and using Eq. (15.2) where the pressure is equal to the average pressure (the latter method gives an estimate).

REVIEW QUESTION 15.1 Imagine that a balloon expands when brought from a cold garage into a warm room. Both the room and the garage are at atmospheric pressure. The change in the volume of the balloon is ΔV. What is the work that the air in the room does on the balloon during the process?

15.2 Two ways to change the energy of a system

Throughout this text, we have stressed the importance of defining a system—a carefully identified object or group of objects on which we focus our attention. Everything outside the system is the environment. A system is characterized by physical quantities, such as the system's internal energy, mass, volume, temperature, and number of particles. The interaction of the environment with the system is defined in terms of the work done by the environment on the system. We devised the generalized work-energy equation [Eq. (7.3)] that summarizes these ideas:

$$E_i + W_{\text{Environment on System}} = E_f$$

Using this equation, we find that the total initial energy E_i of the system plus the work done on the system by the environment $W_{\text{Environment on System}}$ equals the total final energy E_f of the system. The energy of the system can be in many different forms, but the only way the total energy of the system can change is if the environment does work on the system.

$$(K_i + U_{gi} + U_{si} + \cdots) + W_{\text{E on S}} = (K_f + U_{gf} + U_{sf} + \cdots + \Delta U_{int})$$

Are there situations where these ideas are inadequate? Consider Observational Experiment Table 15.1. To analyze these experiments we will use a rewritten version of the generalized work-energy equation:

$$W_{\text{Environment on System}} = E_f - E_i$$

OBSERVATIONAL EXPERIMENT TABLE 15.1 Using the work-energy principle to analyze two processes

Observational experiment	Analysis
Experiment 1. Place a fixed amount of gas in a fixed-volume container. A flame below causes the gas to warm from temperature T_i to T_f. [Constant V]	• The temperature of the gas increased; hence its internal thermal energy increased: $\Delta U_{int} > 0$. • The work done on the gas was zero because the gas volume did not change: $W = 0$. • Thus, $0 = \Delta U_{int}$. *According to the generalized work-energy principle, this process cannot occur—the left side is zero and the right side is a positive number.*

(CONTINUED)

Observational experiment	Analysis
Experiment 2. Place the same gas (same initial conditions) in a cylindrical enclosure with a heavy piston that can move freely up or down in the cylinder, thus keeping the pressure exerted on the gas constant. A flame below causes the gas to expand and warm from temperature T_i to T_f. It takes longer to reach T_f than it did in Experiment 1. T_i and T_f are the same in Experiment 1 and Experiment 2. Piston moves up as gas warms. Constant P	• The temperature of the gas increased; hence its thermal energy increased: $\Delta U_{int} > 0$. • The piston pushed down on the gas opposite the direction of its expansion; thus negative work was done on the gas ($W_{\text{Piston on Gas}} < 0$). • Thus, $-|W| = +|U_{int}|$. *According to the generalized work-energy principle, this process cannot occur—the left side is negative and the right side is positive.*

Pattern
Analyzing both experiments using the work-energy principle, we find that according to this principle the processes cannot possibly occur. However, since we observe these processes in real life, they must be possible. We conclude that either the principle is incorrect or it needs modification.

The generalized work-energy principle is unable to account for the energy changes in the two experiments in Table 15.1. What modification in the principle would help us explain what happened in those two experiments?

Particle motion explains the change in thermal energy of the gas

To reconcile the work-energy principle with the observational experiments described in Table 15.1, we need to assume that energy can be transferred to the gas by means other than work. This new way of transferring energy from an object at one temperature (the hot flame) to an object at a different temperature (the cool gas) is called **heating**. We will discuss the mechanism through which the process of heating occurs later in the chapter.

Heating Q is a physical quantity that characterizes a process of transferring energy from the environment to a system, which is at a different temperature. The environment does no work on the system.

TIP Note that we use the term *heating* rather than *heat*. We do this deliberately. Heating does not mean the actual increase of the temperature of the system; it means the process by which energy is transferred from a hot object to a cooler object (without any work being done).

TIP As the unit of calorie is very small, the unit of kilocalorie, or kcal, is often used. It is written as Cal. 1 Cal = 1000 cal. You see the units of Calories on food packaging.

The SI unit of heating is the joule. However, another common unit of heating is the **calorie**. One calorie (1 cal) is the amount of energy that must be transferred to 1 g of water to increase its temperature by 1 degree C. If you have a known mass of water that changes its temperature by a known number of degrees, you can determine how many calories of energy were transferred to this water through the process of heating. To do this you must multiply one calorie by the mass and the temperature difference; the answer will be in units of calories:

$$Q_{\text{Environment to Water}} = \left(1 \frac{\text{cal}}{\text{g} \cdot {}^\circ\text{C}}\right)(m_{\text{water}})(\Delta T)$$

Testing the equivalence of heating and work as a means of energy transfer

Recall that in the second experiment in Table 15.1 negative work was done on the system, but the system's thermal energy increased. How do we explain this effect using our new process of transferring energy by heating? German physician and physicist Julius Robert von Mayer (1814–1878) and English brewer and physicist James Joule (1818–1889) were the first scientists to speculate that heating and mechanical work were similar processes in terms of energy transfer. The difference between the two processes, they hypothesized, is that for the energy to be transferred to a system through heating, *there must be some temperature difference between the system and the environment.* This is not the case with mechanical work. Both men performed experiments to test this idea.

We describe James Joule's testing experiment in Testing Experiment Table 15.2 (see the modern version of this experiment in the video).

TESTING **EXPERIMENT TABLE** 15.2 Testing the equivalence of work and heating as a means for energy transfer

VIDEO TET 15.2

Testing experiment	Prediction	Outcome
Joule placed a paddle wheel in water. The wheel was connected with a string passing over a pulley tied to a heavy block. When the block moved down, the paddle wheel rotated in the water. 	Our system is the block, paddle wheel, and water. The initial state is just before the block starts to move downward. As it moves, Earth does positive work on the block, and the paddle wheel rotates and warms up the water. The work done by Earth on the system causes a change in the thermal energy of the water. If we assume that all the work done leads to a thermal energy change, we can predict the temperature change of the water. $$m_{block}gh = \left(1\frac{cal}{g\cdot °C}\right)(m_{water})(\Delta T)$$ or $$\Delta T = \frac{m_{block}gh}{\left(1\frac{cal}{g\cdot °C}\right)(m_{water})}$$ To validate the assumption and minimize the transfer of energy to the container, pulleys, strings, and surrounding air, Joule covered the container in blankets, used light and smooth pulleys, and made other adjustments.	Joule lifted the block multiple times to get a bigger temperature increase and repeated the experiment many times. In all experiments the observed change in temperature of the water agreed with the predicted change.

Conclusion

The amount that the system's energy changes when external objects do mechanical work on it is the same as when external objects transferred the same amount of energy by heating.

Based on Table 15.2 we can conclude that the thermal energy of a system can be changed by two means: mechanical work and heating. Mechanical work and heating are both ways of transferring energy between the environment and the system. Joule found that 1.0 J of work done by the person lifting the block and transferred to the kinetic energy of the paddle wheel inside the liquid caused the same effect as the transfer of 0.24 cal through the process of heating to the same liquid. In independent experiments others found the same result:

$$1.0\ J = 0.24\ cal$$
$$1.0\ cal = 4.2\ J$$

Consequently, the amount of energy that we need to transfer to 1 kg of water to change its temperature by 1 °C is 1000 cal, or 4200 J.

REVIEW QUESTION 15.2 Suppose that heating with an electric burner transfers 400 cal of energy to a pan of water. How many joules of mechanical energy would produce the same change in the water's thermal energy?

15.3 First law of thermodynamics

Now that we have established the equivalence of heating and work as mechanisms through which energy can be transferred between an environment and a system, we can incorporate the concept of heating into the generalized work-energy principle. Let's try now to explain the outcomes of the experiments in Table 15.1 by including a heating term in the equation:

$$W_{\text{Environment on System}} + Q_{\text{Environment to System}} = \Delta E_{\text{System}} \qquad (15.4a)$$

Here ΔE_{System} stands for the total energy change of the system. In thermodynamics, this change is usually equal to ΔU_{int} because the mechanical energy of the system does not change. The internal energy can change due to a change in the *motion* of particles (the kinetic energy of the particles changes, causing a change ΔU_{th}) or in the *interaction* of particles (the potential energy of particle interactions changes). An example of potential energy of particle interactions is the familiar chemical energy. In real gases, liquids, and solids, both energies (kinetic energy of particle motion and potential energy of their interaction) can change, but in an ideal gas, the internal energy has only one component—the kinetic energy of gas particles. Therefore, for ideal gases, $\Delta E = \Delta U_{\text{th}}$. The thermal energy in liquids is due to the vibrational and translational motion of particles; in solids, because particles do not move translationally, the thermal energy is due to the vibrational motion only.

Let us see if this new equation helps us explain the experiments in Table 15.1. In Experiment 1 we placed a fixed amount of gas in a fixed-volume container. A flame below caused the gas to warm from its initial to its final temperature. How can we use Eq. (15.4a) to explain the process?

The temperature of the gas increased; hence its internal thermal energy increased: $\Delta E_{\text{System}} = \Delta U_{\text{int 1}} > 0$. The work done on the gas was zero, because the gas volume did not change: $W_1 = 0$. The hot flame transferred energy to the gas by heating: $Q_1 > 0$. Thus, $0 + Q_1 = \Delta U_{\text{int 1}}$. *Equation (15.4a) is consistent with the experimental outcome.*

In Experiment 2 we placed a fixed amount of gas in a cylindrical enclosure with a piston that kept the pressure exerted on the gas constant. A flame below caused the gas to expand and warm from the same initial temperature to the same final temperature as in Experiment 1. We found that it took longer to reach the final temperature than in Experiment 1. How do we explain this? The temperature of the gas increased; hence the thermal energy increased by the same amount as in Experiment 1: $\Delta U_{\text{int 2}} = \Delta U_{\text{int 1}} > 0$. The piston pushed down on the gas opposite the direction of its expansion; thus negative work was done on the gas: $(W_2 < 0)$. Because $-|W_2| + Q_2 = +\Delta U_{\text{int 2}}$, to achieve the same change in thermal energy the hot flame had to transfer more energy to the gas by heating than in Experiment 1—that is why the gas was exposed to the flame for a longer period of time: $Q_{\text{Flame to Gas 2}} > Q_{\text{Flame to Gas 1}} > 0$. *Equation (15.4a) is consistent with the experimental outcome.*

Equation (15.4a) succeeds in explaining many phenomena that the old work-energy principle could not—such as the change of the temperature of water (the system) placed on a hot electric stove (the environment) and the change in the temperature of a hot hard-boiled egg (the system) placed in a bowl of cold water (the environment). In each of these cases, the energy of the system changes with no work done on it. But in all cases, the system contacts a part of the environment that is *at a different temperature* than the system.

We can use work-energy bar charts with the added concept of heating as another process of energy transfer. **Figure 15.4** shows how to use bar charts to represent the second process in Table 15.1. In thermodynamics, the gravitational potential energy and kinetic energy of the whole system often do not change. We therefore do not show bars for them on the bar chart (unless necessary). The only type of energy that changes is the internal energy (either thermal energy or internal potential energy (including chemical) or both). The bars for the mechanisms through which the energy changes—work and heating—are on the left side of the bar chart, and the internal energy change bar (if it's nonzero) is on the right. The bar chart shows that the change in the thermal energy is equal to the sum of the energy transferred through work (which is negative) and through heating (which is positive).

FIGURE 15.4 A work-heating-energy bar chart for Experiment 2 in Table 15.1.

A note about temperature, thermal energy, and heating

We've learned that energy can only be transferred spontaneously through heating when the temperatures of the system and the environment are different. This is different than saying that their thermal energies must be different. Imagine a small spoon at room temperature placed in a bathtub filled with water at room temperature. Although the water in the bathtub has far more thermal energy than the spoon does, there is no energy transfer from the water to the spoon.

The spoon and bathtub example highlights the difference between three physical quantities: temperature, thermal energy, and heating. Temperature is the physical quantity that quantifies the *average* random kinetic energy of the individual particles that comprise the object. Temperature does not depend on the number of particles in the object. Thermal energy is a physical quantity that quantifies the *total* random kinetic energy of all the particles. Adding more particles at the same temperature to the system increases the system's thermal energy. Heating is the physical quantity that quantifies the *process* through which some amount of thermal energy is transferred between the system and the environment when they are at different temperatures. If two systems have the same temperatures but different thermal energies, there is no transfer of energy between them through heating.

Equation (15.4a) is formally known as the **first law of thermodynamics**, a fundamental principle of physics.

TIP ▸ When the temperature of a system is lower than that of the environment, "positive heating" (cooling) occurs—the system gains energy from the environment; when the temperature of the system is higher than the environment's, "negative heating" (cooling) occurs—the system loses energy to the environment.

First law of thermodynamics Consider a process involving a system of interest. The system's internal energy change $(U_f - U_i) = \Delta U_{int}$ is equal to the amount of work W done on the system plus the amount of energy transferred to the system through the process of heating Q:

$$W_{\text{Environment on System}} + Q_{\text{Environment to System}} = \Delta U_{int} \qquad (15.4b)$$

If we rewrite the first law by including all types of energy in a system, we get the familiar, more general equation, Eq. (15.4a):

$$(U_{gi} + K_i + U_{si}) + W + Q = (U_{gf} + K_f + U_{sf} + \Delta U_{int}) \qquad (15.4a)$$

TIP ▸ Heating, like work, is a physical quantity characterizing a process during which energy is transferred between a system and its environment. Both heating and work lead to a change in the energy of a system. Remember, heating and work are not things that reside within the system.

The difference between Eq. (15.4b) and Eq. (15.4a) is that the former considers the mechanical energy of the system to be unchanged during the process, whereas the latter takes changes in mechanical energy into account. Thus Eq. (15.4a) is more general.

Think of how you can warm your right hand (the system) after coming from a trip outside on a cool windy day. You can press your other hand against it and rub them together—transferring energy to your right hand by doing work on it (notice that both of your hands are at the same temperature). You can also submerge your right hand in warmer water—transferring energy to it through heating.

EXAMPLE 15.2 **Quantitative analysis of work and heating**

The pressure-versus-volume graph at right represents the process undergone by one mole of ideal gas. Determine the work done on the gas during this process, the energy transferred to the gas through heating, and the change in the internal energy of the gas.

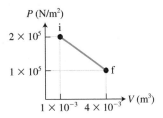

Sketch and translate First we need to visualize the process. The volume is increasing, so the gas is expanding. However, it is not an expansion at constant pressure. It is also not an expansion at constant temperature because the volume increased by a factor of 4 and the pressure decreased by only a factor of 2. Since the product of volume and pressure is larger in the final state, the final temperature of the gas is greater than the initial temperature (recall the ideal gas law equation). This means that in this process some energy needs to be added to the gas through heating. Therefore, to make this process happen, something (a flame, for example) should be providing energy by heating, and some mechanical device (a hand pushing on the top of the container, for example) should be doing work. Now we can draw a sketch and label the knowns and the unknowns.

Simplify and diagram Because the gas is ideal, the change in internal energy is the same as the change in the thermal energy. The mass of the gas is constant during the process; thus we can use the ideal gas law. We know the gas is expanding; thus the work done on the gas is negative. Because the final temperature of the gas is greater than the initial, the change in the thermal energy is positive. We represent these conclusions on the bar chart above, and immediately see that the total heating should be positive.

Represent mathematically We can determine the work from the pressure-versus-volume graph and the change in the thermal energy using the data from the graph and the givens. Once we know ΔU_{int} and W, we can determine Q using the first law of thermodynamics.

Work: Since the gas is expanding, the work done on the gas is equal to the negative of the area under the graph connecting the initial and final state:

$$W = -P\Delta V = -\frac{(P_i + P_f)}{2}(V_f - V_i)$$

Thermal energy change: $\Delta U_{int} = \frac{3}{2}nR(T_f - T_i)$; and $Q + W = \Delta U_{int}$; thus $Q = \Delta U_{int} - W$.

Solve and evaluate

$$W = -P\Delta V = -\frac{(P_i + P_f)}{2}(V_f - V_i)$$
$$= -(1.5 \times 10^5 \text{ N/m}^2)(3 \times 10^{-3} \text{ m}^3)$$
$$= -4.5 \times 10^2 \text{ J}$$

Using the ideal gas law $PV = nRT$, we can write the expressions for the initial and final temperatures of the gas, using the fact that the number of moles (n) is 1:

$$T_f = \frac{P_f V_f}{nR}, T_i = \frac{P_i V_i}{nR}$$

Therefore,

$$\Delta U_{int} = \frac{3}{2}nR\left(\frac{P_f V_f}{nR} - \frac{P_i V_i}{nR}\right) = \frac{3}{2}(P_f V_f - P_i V_i)$$
$$= \frac{3}{2}[(1 \times 10^5 \text{ N/m}^2)(4 \times 10^{-3} \text{ m}^3)$$
$$- (2 \times 10^5 \text{ N/m}^2)(1 \times 10^{-3} \text{ m}^3)]$$
$$= 3 \times 10^2 \text{ J}$$

This gives $Q = \Delta U_{int} - W = 7.5 \times 10^2$ J. The total heating is positive, as we had deduced using the bar chart. The gas has done work on the environment and received energy from heating. The temperature of the gas increased. Note that during this process none of the variables P, T, or V were constant.

Try it yourself What would be the final volume of the gas in the same type of process if the final temperature of the gas were the same as the initial? Assume all other parameters are the same as in the example above. Can we call such a process an isothermal process?

Answer $V_f = 2 \times 10^{-3}$ m^3. The process cannot be called isothermal. The temperature of the gas first increased and then decreased so that the final temperature was equal to the initial temperature.

REVIEW QUESTION 15.3 How are the first law of thermodynamics and the work-energy principle that we learned in Chapter 7 similar? How are they different?

15.4 Applying the first law of thermodynamics to gas processes

In Chapter 12 we learned about gas processes and described them using macroscopic quantities. We can now add an energy-based description of the processes and a microscopic explanation of the mechanism of heating as the transfer of energy to a gas.

Mechanism of energy transfer to a gas through heating

We consider a gas inside a container that is being heated by a flame (see Figure 15.5). The initial temperature of the gas is lower than the temperature of the flame. As you read the following description of the heating process, remember that when we talk about the velocities and kinetic energies of molecules, we always have average values in mind.

Transfer of thermal energy from the flame's hot gas to the container wall The container wall is made of atoms that cannot move around (solids do not flow) but can vibrate. As we heat the container (in this case with a flame), hot gas molecules hit the outer surface of the container wall. If we assume that these collisions are elastic, as we did when we developed the microscopic model of an ideal gas in Chapter 12, then each molecule rebounds with the same speed without transferring any energy to the wall. To explain how the molecules of the hot gas in the flame transfer energy to the container walls, we need to model the collisions differently. In this new model, when a molecule hits the surface of the wall, instead of rebounding immediately, it sticks to the wall for a very short time and then leaves the surface because it is "knocked" by the vibrating wall atoms (see the magnified part of Figure 15.5). Hot gas molecules have a large velocity (and thus kinetic energy). When they stick to the surface, their kinetic energy is transferred to the wall atoms. As a result, the wall atoms start to vibrate with larger amplitudes. When the gas molecules detach from the surface and fly away, their velocity (and therefore kinetic energy) is smaller than it was before they stuck to the surface.

Transfer of thermal energy through the container wall Atoms at the outer surface of the container wall vibrate with large amplitudes. Since the atoms in the container wall are bound to their nearest neighbors as if they were connected with little springs, the vibrations of outer surface atoms are transferred to the atoms in deeper layers in the wall (Figure 15.5). The amplitude of vibration decreases as we move away from the outer surface (and into the wall), but everywhere in the wall the amplitude of vibration is larger than before we started heating the container with the flame. The atoms on the inner surface of the container and molecules of the gas inside the container undergo a similar process to that at the outer surface of the container. The only difference is that now slow-moving gas molecules stick to the inner surface of the wall, where they receive energy from the vibrating atoms of the wall. As a result, they leave the inner surface with a larger kinetic energy than they had before they stuck to the surface. We can describe this process using the language of energy: the high-temperature flame is transferring energy through heating to the container wall. Thermal energy is transferred through the wall to the gas inside the container. As a result, the temperature of the gas increases.

Isothermal process (constant temperature)

During an *isothermal* process, the gas pressure and volume change while the temperature remains constant. We will limit our analysis to volume expansion only. To demonstrate this process, we use a cylinder with a movable piston. The cylinder is immersed in an ice water bath to keep all parts at the same temperature. The cylinder walls are made of a thermally conductive material (a material that allows thermal energy transfer—aluminum is a good example, whereas Styrofoam has low thermal conductivity).

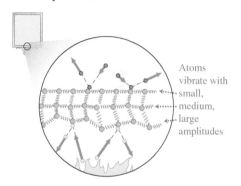

FIGURE 15.5 Microscopic view of a gas being heated by a flame.

Atoms vibrate with small, medium, large amplitudes

FIGURE 15.6 Analysis of an isothermal process.

FIGURE 15.6 Analysis of an isothermal process.

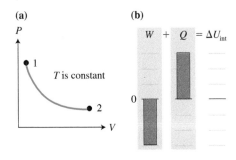

We move the piston very slowly so that the volume of the gas inside the cylinder is increasing (alternatively, we can say that we are letting the gas push the piston by expanding). A graph of this process is shown in **Figure 15.6a**. The gas pushes on the piston to the right, and the piston pushes on the gas to the left. Let the gas inside the cylinder be our system. In this case, the piston does negative work on the system. Therefore, the thermal energy of the gas should decrease and the temperature should decrease, but because the gas is in a bath of constant temperature, the ice water transfers energy to the gas through the walls of the container, keeping the temperature constant. The energy bar chart for the process is shown in Figure 15.6b. Because the temperature of the gas stays constant, $\Delta U_{int} = 0$. The external work done on the gas is negative ($W < 0$). Transfer of energy through heating ($Q > 0$) is needed to make the thermal energy change zero. The first law of thermodynamics can be written as $-|W| + Q = 0$.

Now let us explain the same process microscopically. We will first look at the interactions of the gas particles with the moving piston. As described before, the gas molecules stick to the inner walls (and the piston) and leave a short time later. Molecules that leave from the *moving piston* (expansion) have on average slightly smaller velocities than before they stuck to the piston. (This is analogous to when somebody throws a ball toward you from a car that is moving away from you. The speed of the ball is smaller than it would be if the person had thrown the ball from a stationary car.) The kinetic energy (and therefore the temperature) of the molecules of the gas that leave the piston is slightly smaller than the kinetic energy of the molecules before we started to move the piston. Such a "cold molecule" eventually hits the inner surface of the wall of the cylinder, which is kept at the constant temperature. When a molecule sticks to the wall, it gains energy from the vibrating atoms and leaves the wall with a velocity that is larger than the velocity it had before it hit the wall. Because the volume of the gas is increasing, the density of the gas is decreasing and so is the frequency with which the molecules are hitting the wall (note that the speed of the molecules remains constant).

Isochoric process (constant volume)

FIGURE 15.7 Analysis of an isochoric process.

During an *isochoric* process, the volume of the gas remains constant. To demonstrate this process, we now fix the piston so that it cannot move and place the container in a higher temperature bath. The temperature of the gas increases. Because the volume is constant, there is no work done on the gas ($W = 0$). Energy is transferred to the gas through heating ($Q > 0$). The transfer of energy through heating leads to an increase in the internal thermal energy of the gas ($\Delta U_{int} > 0$). The first law of thermodynamics can be written as $0 + Q = +\Delta U_{int}$. **Figure 15.7** shows the bar chart for this process.

Microscopically speaking, the container wall transfers thermal energy to the gas through the mechanism explained above. The faster moving molecules of the gas collide with the walls of the container more often and more violently. The gas pressure increases.

Isobaric process (constant pressure)

During an *isobaric* process, the pressure of the gas remains constant (**Figure 15.8a**). To demonstrate this process, we place the gas in a container that has a heavy piston on top that can move freely up (or down), keeping the gas at constant pressure. The container is placed in a bath at a higher temperature, causing the gas to expand at constant pressure.

During this process, the environment does negative work on the gas ($W < 0$). Energy is transferred through heating ($Q > 0$) and is greater than the negative work. The internal thermal energy of the gas increases ($\Delta U_{int} > 0$). The first law of thermodynamics can be written as $-|W| + Q = +\Delta U_{int}$. Figure 15.8b shows the bar chart for the process.

To explain the process microscopically, we again need to consider how the walls of the container exchange energy with the warm bath. The walls transfer energy to the gas molecules through collisions, but the gas warms less because the piston moves outward, slowing the particles in a way similar to the mechanism for the isothermal process. The greater velocity of the molecules is compensated for by less frequent collisions between the molecules and the walls so that the pressure remains constant.

FIGURE 15.8 Analysis of an isobaric process.

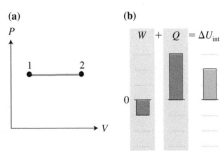

Adiabatic process

In addition to the above three isoprocesses, another gas process, called *adiabatic*, is important. This is a process in which the gas cannot exchange thermal energy with the environment. This can be achieved either by placing good thermal insulation around the system or by compressing or expanding the gas so quickly so that it does not have time to exchange energy through heating with the environment. Therefore, in an adiabatic process not only the volume and the pressure of the gas change, but the temperature changes, too (**Figure 15.9a**). As you can see in the video, the quick compression of air causes the cotton to catch fire.

Using the energy approach, we describe an adiabatic compression as follows. The environment does positive work on the gas ($W > 0$). Since the process happens very quickly, or the gas is in an insulated container, we assume that there is no transfer of energy through heating ($Q = 0$). The internal thermal energy of the gas increases ($\Delta U_{int} > 0$). The first law of thermodynamics can be written as $W + 0 = +\Delta U_{int}$. Figure 15.9b shows the bar chart for the process.

The microscopic explanation uses everything we learned above. Because the gas is being compressed, the molecules collide with the piston moving inward; the speed of the molecules that bounce off the incoming piston is greater than before, and the gas's temperature increases.

FIGURE 15.9 Analysis of an adiabatic process.

Skills for solving gas problems using the first law of thermodynamics

Everyday life offers many examples of isoprocesses and adiabatic processes, where one quantity T, P, V, or Q is constant (in an adiabatic process $Q = $ const $= 0$). A useful strategy for analyzing such processes using the first law of thermodynamics is described on the left side of the table in Example 15.3 and illustrated for the problem in that example.

PROBLEM-SOLVING
STRATEGY 15.1   **Using thermodynamics for gas law processes**

EXAMPLE 15.3   **Hot air balloon**

A burner heats 1.0 m³ of air inside a small hot air balloon. Initially, the air is at 37 °C and atmospheric pressure. Determine the amount of energy that needs to be transferred to the air through heating (in joules) to make it expand from 1.0 m³ to 1.2 m³.

Sketch and translate

- Make a labeled sketch of the process.
- Choose a system of interest and the initial and final states of the process.

The system is the gas inside the balloon. The initial state is before the flame is turned on; the final state is after the balloon has partially expanded.

(CONTINUED)

Simplify and diagram

- Decide whether you can model the system as an ideal gas.
- Decide whether the gas undergoes one of the isoprocesses.
- Determine whether you can ignore any interactions of the environment with the system.
- Draw the following representations if appropriate: a work-heating-energy bar chart and/or a P-versus-V graph.

- We can model the air as an ideal gas since conditions are close to normal atmospheric conditions.
- Assume that the gas expands slowly so that it remains at constant atmospheric pressure—an isobaric process. We also assume that the skin of the balloon does not exert a significant additional inward pressure on the gas. Assume that the number of moles of air in the balloon remains constant.

- For this process, the initial and final gravitational potential and (nonrandom) kinetic energies of the gas are equal, and there are no springs in the system. Thus the first law equation becomes $W + Q = \Delta U_{int}$. We need to determine the amount of energy transferred by heating: $Q = \Delta U_{int} - W$. The gas expands; thus the environment pushes in the direction opposite to the motion of the balloon walls and does negative work (area under P-versus-V graph line with a negative sign): $W = -P\Delta V$.

Represent mathematically

- Use the bar chart to help apply the first law of thermodynamics:

$$U_{gi} + K_i + U_{si} + W + Q$$
$$= U_{gf} + K_f + U_{sf} + \Delta U_{int}$$

- Decide if the mechanical energy of the system is constant (kinetic, gravitational potential, or elastic).
- If the system can be modeled as an ideal gas, use, if needed, gas law equations such as $PV = nRT = Nk_BT$.

- To find the change in internal energy, we need to use the expression for the internal energy of the ideal gas $U_{int} = \frac{3}{2}nRT$. To find the change, we need to determine the number of moles of gas and then the change in temperature. We can use the ideal gas law to determine the number of moles of air initially in the 1.0 m³ of air: $n = P_iV_i/RT_i$. Then use the ideal gas law to determine the final temperature of the air when it expands to the final volume: $T_f = P_fV_f/nR$. From this, determine the change in internal energy of the air: $\Delta U_{int} = \frac{3}{2}nR(T_f - T_i)$.

- We can now put everything together into the equation $Q = \Delta U_{int} - W$.

Solve and evaluate

- Solve the equations for the unknown quantity.
- Evaluate the results to see if they are reasonable (the magnitude of the answer, its units, and how the solution changes in limiting cases).

The work done on the gas is

$$W = -P\Delta V = -(1.0 \times 10^5 \text{ N/m}^2)(1.2 \text{ m}^3 - 1.0 \text{ m}^3)$$
$$= -0.2 \times 10^5 \text{ N} \cdot \text{m} = -2.0 \times 10^4 \text{ J}$$

- The number of moles of air is

$$n = P_iV_i/RT_i = (1.0 \times 10^5 \text{ N/m}^2)(1.0 \text{ m}^3)/(8.3 \text{ J/mole} \cdot \text{K})(310 \text{ K}) = 39 \text{ moles}$$

- The final temperature of the air is

$$T_f = P_fV_f/nR$$
$$= (1.0 \times 10^5 \text{ N/m}^2)(1.2 \text{ m}^3)/(39 \text{ mole})(8.3 \text{ J/mole} \cdot \text{K}) = 370 \text{ K}$$

- The internal thermal energy change of the gas is

$$\Delta U_{int} = \frac{3}{2}nR(T_f - T_i)$$
$$= (3/2)(39 \text{ mole})(8.3 \text{ J/mole} \cdot \text{K})[(370 - 310)\text{K}] = +3.0 \times 10^4 \text{ J}$$

- The energy transferred to the gas through heating is

$$Q = \Delta U_{int} - W = +3.0 \times 10^4 \text{ J} - (-2.0 \times 10^4 \text{ J}) = +5.0 \times 10^4 \text{ J}$$

The final temperature is reasonable, as is the number of moles of gas. Notice that almost half of the energy transferred by heating is needed to compensate for negative work done on the system by the environment as the gas system expands.

Try it yourself Air rushing out of the open valve of a pumped bicycle tire feels cool. Why?

Answer

This is an approximately adiabatic process ($Q = 0$) since it happens so rapidly. The gas is expanding, so the environment does negative work on the gas ($W < 0$). Thus, the internal energy change should also be negative ($W + 0 = \Delta U_{\text{int}}$). The gas cools as it expands adiabatically.

REVIEW QUESTION 15.4 Describe two situations in which the same amount of energy is supplied through heating to the same mass of gas at the same initial conditions, but the gas ends up at a different final temperature in each situation.

15.5 Specific heat

We need to transfer 1000 cal or 4200 J to 1 kg of water to change its temperature by 1 °C. Do all materials need the same amount of energy to change their temperature by 1 °C per unit mass? Consider the experiments in Observational Experiment Table 15.3.

OBSERVATIONAL EXPERIMENT TABLE 15.3 Energy transferred to different materials to achieve the temperature change of 1 °C

Observational experiment	Analysis
Experiment 1. A closed Styrofoam container holds 1.0 kg of water at 20 °C and a thermometer. A large pan of boiling water (at 100 °C) holds a 1.0-kg iron block. You transfer the block quickly to the container with 20 °C water. You observe the temperature rising and then stabilizing at around 28 °C.	Although the masses of the water and iron block were the same, the water temperature increased by 8 °C, whereas the iron block's temperature decreased by 100 °C − 28 °C = 72 °C. We assume that the water received 4200 J of energy for each 1 °C temperature change, or a total of $8(4200 \text{ J}) = 3.4 \times 10^4$ J. The iron must have lost the same amount of energy. Thus the energy loss per 1 °C for the 1.0-kg iron block was $$\frac{3.4 \times 10^4 \text{ J}}{(1.0 \text{ kg})(72 \text{ °C})} = \frac{470 \text{ J}}{\text{kg} \cdot \text{°C}}$$
Experiment 2. You repeat Experiment 1, replacing the iron block with a 1.0-kg aluminum block. The final temperature is 34 °C.	The masses of the water and aluminum block were the same (1.0 kg), but the water temperature increased by 14 °C, whereas the aluminum block's temperature decreased by 100 °C − 34 °C = 66 °C. We assume that the water received 4200 J of energy for each 1 °C temperature change, or a total of $14(4200 \text{ J}) = 5.88 \times 10^4$ J. The aluminum must have lost the same amount of energy. Thus the energy loss per 1 °C for the 1.0-kg aluminum block was $$\frac{5.88 \times 10^4 \text{ J}}{(1.0 \text{ kg})(66 \text{ °C})} = \frac{890 \text{ J}}{\text{kg} \cdot \text{°C}}$$

Pattern

The same amount of energy added to or removed from objects of the same mass by heating or cooling leads to different changes of temperature of these objects. Alternatively, different amounts of energy are gained or lost for each 1 °C temperature change, depending on the type of material.

TIP Every time you encounter a new physical quantity, think about its meaning. For example, what does it mean that the specific heat of ethanol is 2430 J/kg · °C? It means that we would have to transfer 2430 joules of energy to 1 kg of ethanol to change its temperature by 1 °C.

To explain the patterns in Table 15.3 we hypothesize that there is a physical quantity that characterizes how much additional energy 1 kg of a substance needs to change its temperature by 1 °C, and this physical quantity is different for different substances. As we found, the amount of energy is different for water, iron, and aluminum and is about

$$4200 \ \frac{J}{kg \cdot °C}, \ 470 \ \frac{J}{kg \cdot °C}, \ \text{and} \ 890 \ \frac{J}{kg \cdot °C},$$

respectively. These values are the values of a physical quantity, called the **specific heat** of each substance.

Specific heat c is the physical quantity equal to the amount of energy that needs to be added to 1 kg of a substance to increase its temperature by 1 °C. The symbol for specific heat is c and the units are $\frac{J}{kg \cdot °C}$. This energy is added through heating or work or both.

TABLE 15.4 Specific heats (c) of various solid and liquid substances

Solid substance	c (J/kg · °C)
Lead	130
Copper	390
Iron, steel	450
Brick	~800
Sand	~800
Silica glass	840
Sodium chloride	860
Aluminum	900
Concrete	~900
Wood	~1700
Ice	2090
Human body	~3500

Liquid substance	c (J/kg · °C)
Benzene	1730
Ethylene glycol	2420
Ethanol	2430
Methanol	2510
Water	4180

Table 15.4 lists specific heat capacities for several solid and liquid materials. Because 1 °C is equal to 1 K, the change in the temperature of an object in °C is the same as the change in the temperature of the object in K. The units $\frac{J}{kg \cdot °C}$ and $\frac{J}{kg \cdot K}$ are identical.

Water has the largest specific heat in Table 15.4. Notice that the specific heats of sand, bricks, and concrete are about one-fifth that of water. These materials have much greater temperature changes than water when equal masses of these materials absorb the same energy. This is one reason why the sand on a beach or the concrete beside a swimming pool feels so much hotter than the adjacent water on a sunny day. The sand and concrete have bigger temperature changes than the water when exposed to the same sunlight. Note that the values for specific heat of iron and aluminum that we obtained in the experiments in Table 15.3 are slightly different from those in Table 15.4. There are many reasons for the discrepancy. For example, the values in Table 15.4 are the averages of many experiments. Also, in our experiments, for simplicity we did not take instrumental uncertainty into account.

The physical quantity specific heat allows us to construct an equation that can be used to determine the total amount of energy ΔU that needs to be transferred to a mass m of some type of material to cause its temperature to change by ΔT. If the amount of energy equal to the specific heat c in joules (J) needs to be transferred to 1 kg of material to change its temperature by 1 °C, then to change the energy of m kg of the same material we need to transfer $c \cdot m$ J. If we now want to change the temperature by ΔT degrees instead of 1 degree, we need to transfer $c \cdot m \Delta T$ joules. Note that the unit of the product of these three quantities is the unit of energy:

$$\frac{J}{kg \cdot °C} \times kg \times °C = J$$

Therefore, we conclude that an amount of energy ΔU must be added to a substance of mass m and specific heat c to cause its temperature to change by ΔT:

$$\Delta U = cm \Delta T \tag{15.5}$$

The change in the energy of the system can occur both due to the process of heating and due to external objects doing work on the system. This leads to an interesting consequence. Imagine that you are heating a gas in a rigid container. No work is done on the gas since it is not expanding, so the change in thermal energy of the gas is exactly equal to the heating. However, if the container is very flexible and there is atmospheric air outside, that atmospheric air does negative work on the system, reducing the change in thermal energy. To achieve the same change in temperature as in the previous process and thus the same change in thermal energy of the gas, one needs to transfer more energy through heating. However, in most applications, we deal with solids and liquids whose volume changes very little when they warm up or cool down.

In this case the thermal energy changes only due to the process of heating, and thus $Q + W = Q + 0 = \Delta U = cm\,\Delta T$.

Specific heat is a useful quantity. For example, we can use it to determine how much thermal energy a car engine would need to gain by heating in order to warm it to the boiling temperature of the water in its cooling system. Normally, a radiator cooling system prevents the engine from getting too hot, but what if the radiator isn't functioning?

QUANTITATIVE EXERCISE 15.4 **Heating a car engine**

A car has a 164-kg mostly aluminum engine that when operating at highway speeds converts chemical potential energy into thermal energy at a rate of $P = 1.8 \times 10^5$ J/s (240 horsepower). Suppose the cooling system shuts down and all of the thermal energy continually generated by the engine remains in the engine. Determine the time interval for the engine's temperature to change from $T_i = 20\,°C$ to $T_f = 100\,°C$, the boiling temperature of water.

Represent mathematically Assume that the energy is transferred from the burning fuel to the engine through the mechanism of heating, which produces the temperature change in the engine:

$$Q_{\text{Fuel to Engine}} = c_{\text{Engine}} m_{\text{Engine}} (T_{\text{Engine f}} - T_{\text{Engine i}})$$

The burning fuel converts chemical energy into thermal energy at a rate of

$$P = \frac{\Delta U_{\text{th}}}{\Delta t} = \frac{\Delta Q}{\Delta t} = \frac{Q_{\text{Fuel to Engine}}}{\Delta t}$$

Thus, the time interval Δt needed for the burning fuel to warm the engine is

$$\Delta t = \frac{Q_{\text{Fuel to Engine}}}{P}$$

Solve and evaluate The engine is made mostly of aluminum with specific heat $c_{\text{aluminum}} = 900$ J/kg·°C. Thus,

$$Q_{\text{Fuel to Engine}} = \left(900\,\frac{\text{J}}{\text{kg}\cdot°C}\right)(164\text{ kg})(100\,°C - 20\,°C)$$
$$= 1.2 \times 10^7\text{ J}$$

The engine will reach boiling temperature in

$$\Delta t = \frac{Q_{\text{Fuel to Engine}}}{P} = \frac{(1.2 \times 10^7\text{ J})}{(1.8 \times 10^5\text{ J/s})}$$
$$= 67\text{ s}$$

The unit is correct. The engine warms quickly. An operating cooling system is important!

Try it yourself You have two identical insulated cups filled with the same amount of hot water at the same temperature. You put two identically shaped room-temperature spoons in the cups. One spoon is made of steel and the other is aluminum. In which cup will the water reach a lower temperature?

Answer

The cup with the steel spoon will be cooler. Aluminum has a much lower density than steel (about three times lower); thus the steel spoon is about three times more massive. Although steel has a lower specific heat capacity, it is about half of that for aluminum, not enough to compensate for a larger mass.

Transferring energy through the process of heating when objects are in contact

If you have a fever, bathing in cool water lowers your temperature and raises the water temperature. After a swim in a cold lake, to warm up you can bathe in hot water—the hot water warms your body and your cool body cools the water. These changes in the thermal energies of two objects can be explained using the idea of energy transfer: the hot object loses thermal energy to the cold object through the process of heating, and the cold object gains energy through the process of heating from the hot object. We can summarize the above discussion as follows:

$$Q_{\text{Hot to Cold}} + Q_{\text{Cold to Hot}} = 0$$

assuming that no thermal energy leaves the system of these two objects; in other words, the system is isolated. It is important to note that when a cold object and a hot object are in contact, the energy transfers from the hot object to the cold object until their temperatures become the same. At that point the average random kinetic energies of their particles are the same, and there is no more energy transfer. The objects are said to be in **thermal equilibrium**. Remember that it is the temperatures and not the internal thermal energies that become equal when thermal equilibrium is achieved.

TIP When object 1 at high temperature is in contact with object 2 at lower temperature, the process of energy transfer from 1 to 2 is still called heating, despite the fact that the temperature of object 1 goes down.

EXAMPLE 15.5 **Warming the body of a hypothermic person**

The kayak of a 70-kg man tips and he falls into a cold stream. When he is rescued from the cold water, his body temperature is 33 °C (91.4 °F). He is placed in 50 kg of warm bath water at temperature 41 °C (105.8 °F). What is the final temperature of the man and the water?

Sketch and translate Draw a sketch of the situation. Since the initial bath water temperature is higher than the man's initial temperature, the decrease in the thermal energy of the water equals the increase in the thermal energy of the man. The specific heats of water and of humans are listed in Table 15.4.

$M_{Man} = 70 \text{ kg}$ $M_{Water} = 50 \text{ kg}$
$T_{Man\,i} = 33\,°C$ $T_{Water\,i} = 41\,°C$
$T_f = ?$

Simplify and diagram Assume that no energy is transferred from the water and man to the surroundings. In addition, assume that the amount of chemical energy converted into thermal energy in the man's body is small and can be ignored during the short time interval in the water.

Represent mathematically The thermal energy change of the man equals the energy transferred from the water by heating to the man: $Q_{Water\ to\ Man} = m_{Man}c_{Man}(T_{Man\,f} - T_{Man\,i})$. The thermal energy change of the water equals the energy transferred by heating from the man to the water: $Q_{Man\ to\ Water} = m_{Water}c_{Water}(T_{Water\,f} - T_{Water\,i})$. We also know that $T_{Water\,f} = T_{Man\,f} = T_f$ and that $Q_{Man\ to\ Water} + Q_{Water\ to\ Man} = 0$. Putting this all together, we get

$$m_{Water}c_{Water}(T_f - T_{Water\,i}) + m_{Man}c_{Man}(T_f - T_{Man\,i}) = 0$$

Notice that the first expression on the left side of the above equation will be negative, as the water's final temperature is less than its initial temperature. The above can now be solved for the final temperature (try to do it yourself before you look at the expression shown below).

Solve and evaluate Solving for the final temperature of the system, we get

$$T_f = \frac{m_{Water}c_{Water}T_{Water\,i} + m_{Man}c_{Man}T_{Man\,i}}{m_{Water}c_{Water} + m_{Man}c_{Man}}$$

$$= \frac{(50 \text{ kg})(4180 \text{ J/kg} \cdot °C)(41\,°C) + (70 \text{ kg})(3500 \text{ J/kg} \cdot °C)(33\,°C)}{(50 \text{ kg})(4180 \text{ J/kg} \cdot °C) + (70 \text{ kg})(3500 \text{ J/kg} \cdot °C)}$$

$$= 37.0\,°C$$

The answer is reasonable: the final temperature is 4.0 °C above the initial temperature of the man and 3.0 °C below the water's initial temperature.

Try it yourself Fill a Styrofoam cup with 100 g of water at 90 °C and then place a 30-g aluminum spoon at 25 °C into it. The water cools to 86 °C and the spoon warms to 86 °C, the same temperature as the water. Use these results to estimate the specific heat of aluminum. What assumptions did you make?

Answer

900 J/kg · °C. We assumed that the spoon is completely submerged in the water and no energy is transferred through heating to the cup or any other part of the environment.

REVIEW QUESTION 15.5 What experimental evidence indicates that the specific heat of aluminum or iron is much smaller that the specific heat of water?

15.6 Phase change

FIGURE 15.10 An experiment to study phase changes and temperature changes due to heating.

We have found that providing energy to a system through heating or by doing work causes its temperature to change. Are there any processes in which the thermal energy of a system changes but the temperature of the system does not? Suppose you warm an ice cube in a Styrofoam cup so that some of it melts and becomes liquid water. Is the melted water warmer than the solid ice cube that remains? The transformation of ice into water is a **phase change**—a process during which a substance changes from one phase to another. The term phase refers to the bulk properties of solids, liquids, gases, plasmas, etc. Plasma is what fills neon lights and is the main phase of matter in stars. We will learn about plasma in later chapters of the book. Let's consider an experiment involving the heating of water that starts as very cold solid ice.

Assume that you have 1.0 kg of water, which you place in a foam container with the top initially open (Figure 15.10). You insert a thermometer and an electric heating coil into the water. You place all of this in a freezer that is set at −40 °C. After the water freezes and reaches −40 °C, you place a foam cover over the container with the thermometer sticking out. You then remove the container from the freezer and turn on the electric coil. You observe the temperature change with time. The graph in **Figure 15.11** shows the relationship between the temperature and energy transferred to the container from the electric coil. Note that although the temperature depends on the energy transferred to the system, we represent temperature on the horizontal, independent variable axis in this case. Therefore, the graph does not represent a cause-effect relationship.

Warming from − 40 °C to 0 °C

Between −40 °C and 0 °C, the contents of the container remain frozen (A in Figure 15.11). The energy transferred to the ice from the electric coil causes the ice to warm and eventually reach its melting temperature at 0 °C. Because you know the power output of the coil, you learn that it takes 2090 J to warm 1.0 kg of ice by 1.0 °C. In other words, it takes 8.4×10^4 J to warm 1.0 kg of ice from −40 °C to 0 °C. This means that the specific heat of ice is 2090 J/kg·°C. It takes less energy to warm ice by 1.0 °C than it does to warm the same amount of liquid water. So far the transfer of energy to the ice led to a change of its temperature—no surprises here.

Melting and freezing

At 0 °C the graph changes suddenly, becoming vertical (B in Figure 15.11). We observe that when the ice reaches 0 °C, it starts to melt, but the thermometer reading does not change. According to the graph, it takes 3.35×10^5 J of energy to completely melt 1.0 kg of ice. While this energy is added, the temperature of the system does not change. Why?

In a solid, neighboring molecules are close to each other and are attracted and bonded to each other—that is why solids keep their shape. Their potential energy of interaction is negative, meaning they are bound to one another much like the Earth-Moon system is bound by negative gravitational potential energy. The molecules vibrate and therefore have kinetic energy of random motion (though they do not fly around freely as they would in a gas). As the temperature of the solid rises toward its melting temperature, the vibrations of the neighboring water molecules become larger in amplitude and less ordered. When the solid reaches its melting temperature, the vibrations of neighboring molecules become so disruptive that the molecules begin to separate from one another and are no longer bound (**Figure 15.12**). There is still considerable interaction between neighboring molecules, but the interaction is less than in the rigid solid. Therefore, when melting or freezing occurs, it is the potential energy of particle interactions that changes. When we transfer energy to the solid material at the melting temperature, all of this energy goes into changing the potential energy of particle interactions, not the kinetic energy—thus the temperature does not change.

The energy needed to melt 1.0 kg of ice into water is called the **heat of fusion** of water and is given the symbol L_f ($L_{f\,water} = 3.35 \times 10^5$ J/kg). (The heat of fusion is sometimes called the *latent heat of fusion* to underscore that this energy does not lead to temperature change; the change is in a hidden or latent form.) When a liquid freezes, it releases the same amount of energy (in other words, the reaction occurs in the opposite direction). The heat of fusion differs for different materials; their melting-freezing temperatures also differ (see **Table 15.5** on the next page). The unit of heat of fusion is joules per kilogram.

FIGURE 15.11 Effect of heating on ice, liquid water, and steam.

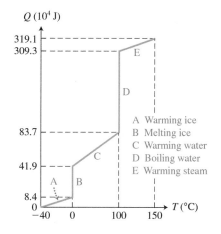

FIGURE 15.12 Melting. Depiction of structural change as ice in (a) melts to form liquid water in (b).

Energy to melt or freeze The energy in joules needed to melt a mass m of a solid at its melting temperature, or the energy released when a mass m of the liquid freezes (solidifies) at that same temperature, is

$$\Delta U_{int} = \pm m L_f \qquad (15.6)$$

L_f is the heat of fusion of the substance (see Table 15.5). The plus sign is used when the substance melts and the minus sign when it freezes, or solidifies.

TIP The term heat of fusion is confusing, as the word *heat* is often associated with warming, that is, temperature change. In the case of melting or freezing, there is no temperature change. However, there is still change in internal energy—the potential energy of interactions of the particles in the system.

TABLE 15.5 Heats of fusion and vaporization

Substance	Melting temperature (°C)	L_f, heat of fusion at the melting temperature (J/kg)	Boiling temperature (°C)	L_v, heat of vaporization at the boiling temperature (J/kg)
Water	0	3.35×10^5	100	2.256×10^6
Ethanol	−114	1.01×10^5	78	0.837×10^6
Hydrogen (H_2)	−259	0.586×10^5	−253	0.446×10^6
Oxygen (O_2)	−218	0.138×10^5	−183	0.213×10^6
Nitrogen (N_2)	−210	0.257×10^5	−196	0.199×10^6
Aluminum	660	3.999×10^5	2520	10.9×10^6
Copper	1085	2.035×10^5	2562	4.73×10^6
Iron	1538	2.473×10^5	2861	6.09×10^6

Warming liquid water from 0 °C to 100 °C

Once the ice has completely melted (C in Figure 15.11), further transfer of energy by heating causes the liquid water to warm. The graph in this region has a slope of 4180 J/°C and indicates that the specific heat of liquid water is 4180 J/kg·°C. With continued heating, the water temperature continues to rise.

Boiling and condensation

At 100 °C, the graph again changes abruptly (D in Figure 15.11). We observe that the water is boiling and steam is forming. The temperature of the water remains at the boiling temperature until all of the liquid water boils into vapor. From the graph we see that when the water reaches 100 °C, we must add 2.256×10^6 J of energy to completely convert 1.0 kg of liquid water to gaseous water vapor (steam). Once again, energy has been added to the water, but its temperature does not change. Where does the energy go? For a molecule to leave the surface of a liquid, it must have enough kinetic energy to break away from the neighboring molecules, which are exerting attractive forces on it (**Figure 15.13a**). A fast-moving water molecule can escape, while a slow-moving one cannot (Figure 15.13b). Escape of the fast-moving molecules leaves the remaining slow-moving "cooler" molecules behind. A continuous supply of energy through heating when the substance is at the boiling temperature causes a continual number of water molecules to gain the energy needed to leave the water surface but does not change the water temperature until all molecules have become water vapor. The energy transferred to the liquid leads to a change in the potential energy component of the internal energy.

FIGURE 15.13 Boiling. (a) Water molecules are attracted toward other water molecules. (b) Fast moving molecules can escape.

(a)

A water molecule at the surface of the water is prevented from leaving the water by its attraction to neighboring molecules.

(b)

A fast-moving water molecule can escape from a liquid when it hits the surface. A slow-moving one cannot escape.

Liquids can evaporate at any temperature, not just at the boiling temperature. At any temperature some molecules have enough kinetic energy to escape the liquid. However, as they leave, the temperature of the liquid left behind decreases, since the average kinetic energy of the molecules left behind decreases. This is how we cool ourselves by evaporation on hot days.

The **heat of vaporization** of a substance L_v (sometimes called the *latent heat of vaporization*) is the energy needed to transform 1 kg of a liquid at its boiling temperature into its gaseous state at that same temperature. For water, $L_{v\,water} = 2.256 \times 10^6$ J/kg. When 1.0 kg of the gas turns into liquid (condenses) at its boiling temperature, it releases the same amount of energy. The heat of vaporization differs for different substances; their boiling temperatures also differ (see Table 15.5). The unit of heat of vaporization is joules per kilogram.

Energy to boil or condense The energy in joules needed to vaporize a mass m of a liquid at its boiling temperature, or the energy released when a mass m of a gas condenses at that same temperature, is

$$\Delta U_{int} = \pm mL_v \qquad (15.7)$$

L_v is the heat of vaporization of the substance (see Table 15.5). The plus sign is used when the substance vaporizes and the minus sign when it condenses.

The heat of fusion, melting temperature, heat of vaporization, and boiling temperature depend on the type of substance, as you can see in Table 15.5.

Notice that the values of the heats of fusion and vaporization are much larger than the specific heat of these substances. This means that much more energy is needed to change the *phase* of a substance than to change the *temperature* of the substance by one degree. Also notice that the heats of vaporization are significantly larger than the heats of fusion; more energy is required to boil the same mass of the same substance than to melt it. Finally, notice that the values are provided for boiling temperatures. Liquids evaporate at temperatures below the boiling temperature; however, the energy needed to keep them at the same temperature during this evaporation is greater. For example, the heat of vaporization at the boiling temperature for water is 2.256×10^6 J/kg, whereas at body temperature (when sweat evaporates) the heat of vaporization is $L_{v\,water} \approx 2.4 \times 10^6$ J/kg.

Warming gaseous water vapor above 100 °C

We cannot perform the next part of the experiment (E in Figure 15.11) with the apparatus shown in Figure 15.10 since the water vapor would escape. But the experiment could be performed if the container were airtight. In that case, if we continued to add energy to 1.0 kg of water vapor, we would find that its temperature increased 1 °C for each approximately 2000 J of additional energy added to it. This tells us that the specific heat of water vapor is approximately $c_{water\,vapor} = 2000$ J/kg·°C.

Reversing the process

We now put the 1.0 kg of hot water vapor back in our −40 °C freezer and slowly remove thermal energy to reverse the process shown graphically in Figure 15.11. We find that the water first cools to its boiling temperature (100 °C for water). It then releases considerable energy as it condenses to form liquid water. Next, it cools to 0 °C as more energy is removed. The liquid water again releases considerable energy as it freezes. The ice finally cools to a final temperature of −40 °C. The same amount of energy that was supplied through heating to make vapor out of ice is released (or removed from the system) when vapor turns into ice (assuming the same initial and final temperatures).

Eugenia carries a 10-kg backpack on a hike down into the Grand Canyon and then back up. Sweat evaporates off her skin at a rate of 0.28 g/s. At what rate is thermal energy transferred from inside her body to the surface of her skin in order to provide the energy needed to vaporize the sweat on her skin at this rate? At body temperature, the heat of vaporization for sweat is about $L_{v\,\text{Sweat}} \approx 2.4 \times 10^6$ J/kg.

Represent mathematically The sweat on the skin is the system. The energy needed to evaporate a small mass Δm of the sweat that is added to the sweat from inside Eugenia's body is $Q_{\text{Eugenia to Sweat}} = +\Delta m \cdot L_v$. Dividing each side by a small time interval Δt, we get

$$\frac{Q_{\text{Eugenia to Sweat}}}{\Delta t} = \frac{\Delta m_{\text{Sweat}} L_{v\,\text{Sweat}}}{\Delta t} = \left(\frac{\Delta m_{\text{Sweat}}}{\Delta t}\right) L_{v\,\text{Sweat}}$$

where the rate of sweating is $\dfrac{\Delta m_{\text{Sweat}}}{\Delta t} = 0.28$ g/s $= 0.28 \times 10^{-3}$ kg/s.

Solve and evaluate We find that the rate of energy transfer through evaporative heating is

$$\frac{Q_{\text{Eugenia to Sweat}}}{\Delta t} = \left(\frac{\Delta m_{\text{Sweat}}}{\Delta t}\right) L_{v\,\text{Sweat}}$$
$$= (0.28 \times 10^{-3} \text{ kg/s})(2.4 \times 10^6 \text{ J/kg})$$
$$= 670 \text{ J/s} = 670 \text{ W}$$

This is the rate that energy is being transferred from inside her body to the moisture on her skin. Another way to think about this is that Eugenia's body is being cooled at a rate of 670 W!

The evaporation of sweat is a natural cooling mechanism that protects us from overheating. Increasing the rate of evaporation from the skin increases the body's cooling rate. A sick child's fever is often reduced by rubbing the skin with alcohol, since alcohol evaporates much faster than water.

Try it yourself After swimming, you exit the pool with 300 g of water absorbed in your swimsuit. How much energy will this water absorb from the environment to evaporate? How can you evaluate your answer?

Answer

7.2×10^5 J. If we assume that the swimsuit dries using only the energy generated by your body at a rate of 100 J/s, it should take about 7200 s to dry, which is about 2 hours—this is a reasonable time. When the Sun is out, a swimsuit dries somewhat faster, so the absorption of energy from the Sun's radiation must be significant.

We can often solve problems in multiple ways. We could solve the problem in the following example by treating the two objects (hot coffee and an ice cube) as two different systems. However, this time we will treat them as a single system. The hot object loses thermal energy as its temperature decreases, and the cooler object gains an equal amount of internal energy (potential and thermal) as it melts and warms.

EXAMPLE 15.7   Cooling hot coffee

You add 10 g of ice at temperature 0 °C to 200 g of coffee at 90 °C. Once the ice and coffee reach equilibrium, what is their temperature? Indicate any assumptions you made.

Sketch and translate A sketch of the process is shown below. We choose the coffee and ice as the system. The ice gains thermal energy and melts at 0 °C. Then the melted ice, now liquid water, warms to the final equilibrium temperature. Simultaneously, the coffee loses an equal amount of thermal energy as it cools until the mixture is at a single final equilibrium temperature.

Simplify and diagram

The coffee and ice are the system. Assume that the coffee-ice system is isolated; there is no energy transferred between the system and the environment, including the cup. Assume also that the coffee has the same specific heat as water, which is reasonable since coffee is mostly water. The process is represented qualitatively (the straight lines do not represent linear relations) in a temperature-versus-time (T-versus-t) graph using two lines for the ice and one line for the coffee. Notice that after some time, they reach the same thermal equilibrium temperature.

$m_{\text{Coffee}} = 200$ g
$T_{\text{Coffee i}} = 90$ °C
$m_{\text{Ice}} = 10$ g
$T_{\text{Ice i}} = 0$ °C

Represent mathematically The energy of the system is constant, so the internal energy change of the ice plus the thermal energy change of the coffee must add to zero.

$$\Delta U_{int} = \Delta U_{int\ Ice} + \Delta U_{th\ Coffee} = 0$$

The ice first melts, then warms. The coffee cools.

$$(m_{Ice}L_{Ice} + m_{Ice}c_{Water}(T_f - T_{Ice\ i})) + m_{Coffee}c_{Coffee}(T_f - T_{Coffee\ i}) = 0$$

Note that the system reaches equilibrium, so $T_{Ice\ f} = T_{Coffee\ f} = T_f$. Also, the initial temperature of the melted ice is $T_{Ice\ i} = 0\ °C$.

Solve and evaluate Solve for the final temperature T_f:

$$T_f = \frac{m_{Coffee}c_{Coffee}T_{Coffee\ i} + m_{Ice}c_{Water}(0\ °C) - m_{Ice}L_{Ice}}{m_{Ice}c_{Water} + m_{Coffee}c_{Coffee}}$$

$$= \frac{(0.2\ kg)(4180\ J/kg \cdot °C)(90\ °C) + 0 - (0.01\ kg)(3.35 \times 10^5\ J/kg)}{(0.01\ kg)(4180\ J/kg \cdot °C) + (0.2\ kg)(4180\ J/kg \cdot °C)}$$

$$= 82\ °C$$

This is a reasonable temperature considering the small amount of ice used.

Try it yourself You pour 200 g water at 5.0 °C into a Styrofoam container and add 500 g of ice at −40 °C. What is the result? Describe the steps in solving the problem and determine the numerical answer.

Answer

Four possible outcomes are (a) all ice melts and the resulting water is above 0 °C; (b) all water freezes and the resulting ice is below 0 °C; (c) some but not all of the water freezes and the resulting mixture is at 0 °C; (d) some but not all ice melts and the mixture is at 0 °C. To decide which outcome is correct, first determine how much energy the water will release to cool to 0 °C and how much energy the ice will need to warm to 0 °C. Calculations show that the water will release 4.2 × 10³ J and the ice will require 42 × 10³ J. Thus cases (b) and (c) are possible. Final answer: the mixture will be at 0 °C and the amount of water that freezes is 110 g.

REVIEW QUESTION 15.6 Why are the units for specific heat and heat of fusion/vaporization different?

15.7 Heating mechanisms

So far we have discussed the mechanism of heating gas through walls of a container or heating the walls of the container itself through contact with a substance at a different temperature. This is one of the mechanisms of energy transfer through heating. There are several other mechanisms, an understanding of which is important for our well-being. They help us reduce energy losses from our homes during the winter and confront climate change.

We define heating rate H as the energy transfer by heating Q from one object to another divided by the time interval Δt needed for that heating energy transfer:

$$\text{Heating rate} = H = \frac{Q}{\Delta t}$$

This transfer can occur through several mechanisms.

Conduction

You are boiling water for morning coffee at your campsite. You have two cups: one is Styrofoam and the other aluminum. Which cup would you prefer to hold just after filling it with boiling water? Experience tells us that the Styrofoam cup will feel more comfortable, while the aluminum cup may burn our fingers.

Later, you cook scrambled eggs in a frying pan on the camp stove. You start to remove the pan from the stove, but the handle is very hot, even though it hasn't touched the stove's hot burner. How can you explain these observations?

The mechanism behind them is exactly the same as the one we learned about in Section 15.3. The high-energy, high-speed hot water molecules hit the inside wall of the aluminum or Styrofoam cup, making the particles on the inside wall of the cup vibrate faster. These more energetic particles in the wall then bump into neighboring

Why is the handle hot, even though it has not touched the flame?

TABLE 15.6 Thermal conductivity of materials (at 25 °C)

Material	k_{th} (W/m · °C)
Metals	
Aluminum	205
Brass	109
Copper	400
Stainless steel	16
Other solids	
Attic insulation	0.04
Bone	0.4
Brick	1.31
Concrete, stone	1.7
Fat	0.2
Glass, window	0.9
Ice (at 0 °C)	2.18
Muscle	0.4
Soil	0.6–4
Styrofoam	0.03
Wood	0.04–0.14
Other materials	
Air	0.024
Water	0.58

particles in the wall, which in turn bump into their neighbors, which are closer to the outside wall. Thus, the thermal energy transfers from particle to particle from the inside wall to the outside wall. Your fingers touch the outside wall. When its surface is hot, the high-energy particles on the surface bump the particles on the surface of your skin, causing them to move faster. The high-energy vibrations of the particles on your skin can produce a burning sensation. The process by which thermal energy is transferred through physical contact is called **conductive heating** or **cooling**.

Why does the aluminum cup of hot water feel hotter than the Styrofoam cup filled with equally hot water? Aluminum transfers thermal energy from particle to particle quickly. Styrofoam transfers thermal energy more slowly. We say that aluminum is a good thermal energy conductor and Styrofoam is a poor thermal energy conductor. A quantity called **thermal conductivity** k_{th} characterizes the rate at which a particular material transfers thermal energy (see Table 15.6). The larger the thermal conductivity, the faster thermal energy transfers from molecule to molecule in that material. You want to use a material with low thermal conductivity to make a cup for holding hot beverages. Likewise, you want to use a material with low thermal conductivity to insulate your home—to prevent thermal energy from leaving the house during the winter as well as to prevent it from entering during the summer. We say that materials with low thermal conductivity are good thermal insulators.

Conductive heating/cooling The rate of conductive heating $H_{Conduction}$ from the hot side of a material to the cool side is proportional to the temperature difference between the two sides $(T_{Hot} - T_{Cool})$, the thermal conductivity k_{th} of the material separating the sides, and the cross-sectional area A across which the thermal energy flows. The rate is inversely proportional to the distance l that the thermal energy travels.

$$H_{Conduction} = \frac{Q_{Hot\ to\ Cool}}{\Delta t} = \frac{k_{th}A(T_{Hot} - T_{Cool})}{l} \qquad (15.8)$$

CONCEPTUAL EXERCISE 15.8   Snow on the roof

In the morning on a cloudy winter day your house with an attached garage is covered with light snow. What might you expect to see 24 hours later, assuming that the temperature stays low and there is no new snow?

Sketch and translate The melting on different parts of the roof depends on the conductive thermal energy transfer through the roof from inside the house to the outside.

Simplify and diagram Assume that the energy delivered to the roof comes from inside the house and not from the Sun. Assume also that the insulation and roofing material is the same in all parts of the house (k_{th} is the same for the roof over all parts of the house). Now divide the house into two parts: the living quarters and the garage. Thermal energy transfer occurs from the warmer regions inside the house through the roof to cooler regions outside of the house. The conductive heating transfer rate depends on the temperature difference $(T_{Hot\ inside} - T_{Cold\ outside})$. The temperature difference is greater from inside the living area to the outside than it is from the inside of the garage to the outside. Thus, the thermal energy transfer rate should be greater through the roof above the living quarters than it is through the roof over the garage. Because of this, there will be more snow melting above the living area than above the garage.

The photo confirms this reasoning. The faster melting over the living area tells us that energy is being transferred out of the house more quickly than out of the garage.

Can we reduce the amount of energy we are releasing to the environment (and therefore keep more of the energy—and warmth—in the house)? To reduce energy transfer, we could slow the heating transfer rate by using lower thermal conductivity k_{th} insulation or thicker insulation. When we use thicker insulation [thereby increasing length l in Eq. (15.8)], the thermal energy has to travel a longer distance across the interface.

Try it yourself Your skin temperature is about 35 °C. Imagine that you touch the wooden top of a desk in a room where the temperature is 20 °C. The desk feels comfortable. Then you touch the steel leg of the desk, and it feels cold. Does it mean that the legs of the desk are at a lower temperature than the wooden top?

Answer

The legs and the desktop are at the same temperature—room temperature. When you touch each of them with your hand, which is somewhat warmer than these objects, thermal energy transfers from your hand to the objects. The larger thermal conductivity of steel makes this transfer occur at a higher rate, which is why the steel feels cooler to your hand.

FIGURE 15.14 How a thermos bottle works. A thermos consists of an inner chamber surrounded by a vacuum.

— Near vacuum

The concept of thermal conductivity explains why a thermos keeps drinks hot or cold. The thermos depicted in **Figure 15.14** consists of two metal cups separated by a layer of low-density air (as close to a vacuum as we can get). Because the particles of air in the nearly evacuated space between the metal cups are very far from each other, they slowly transfer energy by conduction, resulting in a very small thermal conductivity. (In a perfect vacuum k_{th} would be zero because there would be no particles to interact with each other to transfer thermal energy.) Thus, if a liquid in the inner cup is cold, it stays cold; if it is warm, it stays warm.

Another application of the concept of thermal conductivity is fur, which allows animals to survive cold winters. The fur consists of hair strands separated by air. The air trapped between strands is a poor thermal conductor; therefore, the strands work almost like a thermos in preventing the conductive energy transfer from a warm body to a cold environment.

Convection

Conduction as the heating mechanism works well in materials in which particles interact with each other. In gases, for example, the particles are so far apart that these interactions are almost nonexistent. How does thermal energy transfer in those materials? Consider the experiments in Observational Experiment **Table 15.7**.

OBSERVATIONAL EXPERIMENT TABLE 15.7   **Convective heating/cooling**

Observational experiment	Analysis
Experiment 1. You light a candle and hold your hand near one side of it. Your hand does not feel much warmer than when it is far away from the candle. However, if you place your hand above the candle, you can only hold it there for a short time interval before you risk burning yourself.	The air near the flame quickly warms and expands. Since its density is now less than the density of the surrounding air, it rises and this warmed air contacts the skin of your hand directly above the candle.

Ouch!

Hot air moves up.

(CONTINUED)

Observational experiment	Analysis
Experiment 2. You place the apparatus shown below at left over a flame. You then put some potassium permanganate powder into the water ($KMnO_4$ is a dark purple powder that dissolves in water, turning it pink). Colored water first spreads upward, then moves through the horizontal pipe, down the other vertical tube, and back to the area above the flame.	Thermal conduction warms the water near the flame. This warm water expands and rises. The warm water that moved up starts a *flow* of warm liquid that moves around the apparatus. As the pink water moves horizontally, it cools and sinks, replacing the water heated by the flame. The pink water is then heated again, and the process repeats.

Pattern
In both processes the warmed fluid becomes less dense and rises. As the warm fluid rises, it is replaced by cooler fluid.

The processes described in Table 15.7 are called **natural convective heating**.

> **Natural convective heating/cooling** Natural convective heating/cooling is the process of energy transfer by a fluid moving in a vertical direction away from/toward the center of Earth. Fluids such as air or water rise because their density decreases as their temperature increases and sink because their density increases as their temperature decreases.

FIGURE 15.15 A car's engine is cooled mostly by forced convection.

1. Cool water travels from radiator to engine (forced convection).

2. Water passes through hot engine, cooling the engine and warming water (conduction).

Radiator

Air

Engine

4. Air passes through radiator, causing hot water to cool (convection).

3. Hot water travels from engine to radiator (forced convection).

While convection can occur as a natural process, it can also be induced. For example, in a forced-air heating system, air in the furnace is warmed by a gas or electric heater. The warmed air is forced by a fan through ducts to different rooms in the house. Other ducts return cooler air back to the furnace to be warmed. In cars, burning gasoline makes the engine very hot. To cool it, a pump moves fluid (water or a coolant) from the radiator through hoses into and through the engine (**Figure 15.15**). Thermal energy is transmitted by conduction to the coolant from the engine. The energy is then carried away by the fluid (forced convective thermal energy transfer) through a radiator hose and back to the radiator, where outside air moves past the fins in the radiator and cools the fluid again. This process is called forced convective heating (or cooling). The forced-air heating system in a house and the cooling system in cars are examples of **forced convection**.

> **Forced convective heating/cooling** In forced convective heating, a hot fluid is forced to flow past a cooler object, transferring thermal energy to the object. The fluid moves away with less thermal energy. In forced convective cooling, a cool fluid moves past a warmer object (for example, a cold winter wind moving across your warm skin). The object transfers thermal energy to the cooler fluid. The now-warmed fluid carries the thermal energy away from the object.

Calculating the rate of convective energy transfer is a complex subject and depends on many factors. Often, the calculations are estimates and involve the use of tables of data assembled by energy transfer specialists. There are many practical examples of natural and forced convective heating and cooling.

A *convection oven* has advantages over a traditional oven. In a traditional oven, a thin layer of air surrounds the food. This air is at a temperature between that of the food and the hotter air in the rest of the oven. Thus, cooking occurs at a lower temperature than the temperature of air in the oven. In a convection oven, a fan at the back of the oven moves heated air around the cooking compartment. This causes fresh hot air to continually move past the food. Cooking time is reduced by about 20% and cooking temperature can be 10–20 °C lower than in a traditional oven. You get the same result with less energy use.

Radiation

Consider another phenomenon, a campfire on a cool evening. The air above the flames warms and rises with the smoke from the fire, an example of natural convective heating. You sit on a log at the side of the fire and are also warmed by it. The warming is not the result of the convection from the fire—the warm air is moving upward and does not move across your skin. You notice that the part of your body facing the flame is warmed, while the part of your body facing away from the flame remains cool. Why is that?

Consider a similar process. Outside Earth's atmosphere is an approximate vacuum with low concentrations of atomic particles. The particles are very far from each other and therefore essentially do not transfer energy by conduction. The Earth has remained warm enough for habitation because the Sun irradiates Earth in much the same way that a campfire warms the part of your body facing it. The Earth is warmer on the side facing the Sun and cooler on the side facing away from it. Both the Sun and the campfire emit infrared radiation and some light that is the source of this thermal energy. We will study what is "traveling" in this process later, in Chapters 25 and 27.

Radiative heating/cooling All objects, hot and cool, emit different forms of radiation that can travel through a vacuum. The hotter the object, the greater the rate of radiation emission. Objects in the path of this radiation may absorb it, gaining energy.

FIGURE 15.16 A passive solar heating system.

Passive solar heating is an important application of radiative heating (and other heating mechanisms). Passive solar buildings are constructed so that they have walls with large heat capacities for storing radiative energy (sunlight) that passes through windows during the daytime (Figure 15.16). The light energy absorbed by the wall transforms its energy into the thermal energy of the surface atoms and molecules in the wall. These warmer atoms and molecules transfer their energy by conduction to other atoms and molecules deeper in the wall. A well-designed wall has large mass m with a high specific heat c. As a result, the wall's temperature change

$$\Delta T = \frac{Q}{m \cdot c}$$

will be small during the day, because both m and c are large. The large amount of energy transferred to the wall by heating during the day is released slowly at night when it is needed. Energy is transferred to the building at night by natural convective heating as cool air moves up past the warm wall and into the building as well as from radiative heating from the wall into the building.

Radiation is also used for diagnostic methods in what is called *thermography*. One application is breast thermography, which detects differences in infrared radiation from warmer and cooler surfaces. Thermography detects "hot spots" caused by the increased metabolic activity in cancer cells and combines advanced digital technology with ultrasensitive infrared cameras for safe and early screening for breast cancer.

Thermography was also employed to take the picture in Figure 15.17 of a house at night. The parts of the house that are radiating energy at the fastest rate are yellow. The next fastest radiating parts are orange, then pink, purple, violet, and black. None of these regions are sources of visible light; all lights in the house are off. This radiation is being emitted simply because the house is not at zero absolute temperature.

FIGURE 15.17 An infrared photo of a house at night (no lights are on).

FIGURE 15.18 Evaporation rates for different conditions.

Evaporation

Why are the hot, dry summers in Arizona less uncomfortable than hot, humid summers elsewhere in the United States? The answer is that we more effectively transfer energy by evaporation in a dry climate. On a humid day, water not only evaporates from the skin but also condenses on the skin from the gaseous water vapor in the air. When gaseous water vapor converts to liquid water (condenses), energy is released and returned to your body, raising your temperature. To stay cool, you want the rate of evaporation to be somewhat greater than the rate of condensation. In a "dry" climate with low humidity, not much water vapor condenses on your skin. You evaporate more water than condenses on you, and there is a net cooling effect. **Figure 15.18** indicates some typical evaporation rates under different circumstances.

Evaporative heating/cooling Evaporative heating occurs when a gas condenses on a surface. Evaporative cooling occurs when a liquid evaporates from a surface. The evaporative energy transfer rate $H_{Evaporation}$ is proportional to the rate at which liquid mass evaporates or condenses $\Delta m / \Delta t$ on the surface and to the heat of vaporization L_v of the liquid at the temperature of the liquid:

$$H_{Evaporation} = \pm \left(\frac{\Delta m}{\Delta t} \right) L_v \qquad (15.9)$$

VIDEO
15.2

We use the positive sign for condensation (energy transferred to the system) and the negative sign for evaporation (energy transferred out of the system). Watch the video to see the cooling due to evaporation of two paper towels, one wet with water and the other wet with acetone.

The greenhouse effect and climate change

Figure 15.19 shows a strong correlation between the CO_2 concentration in Earth's atmosphere and Earth's average temperature during the last 400,000 years. During the coldest periods, when much of Earth was covered with glaciers, the CO_2 levels were lowest. During warm periods, some parts of Earth became much dryer while other parts experienced flooding. During these warm periods, the CO_2 levels were highest. Glaciers melted, leading to a rising ocean and coastal flooding. More frequent and severe weather events occurred (such as hurricanes, typhoons, droughts, and blizzards). Scientific evidence indicates that the concentration of CO_2 in our atmosphere plays a big part in these dramatic climate changes.

FIGURE 15.19 Earth's average temperature and CO_2 concentration for the past 400,000 years.

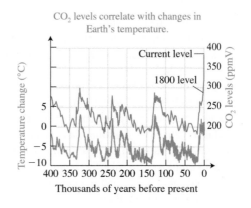

The cause of the increase in atmospheric CO_2 is a complex subject. Part of the increase is due to human activities, such as the increasing use of fossil fuels and the removal of forests and plant life, which absorb CO_2.

We can qualitatively investigate the role of CO_2 concentration in Earth's changing temperature by applying the first law of thermodynamics to the system pictured in **Figure 15.20**. The system inside the dashed lines consists of Earth and its atmosphere. The Earth/atmosphere system gains energy through radiative heating from the Sun ($Q_{\text{Sun to Earth}} > 0$) and loses energy through radiation that it emits into space ($Q_{\text{Space to Earth}} < 0$):

$$Q_{\text{Sun to Earth}} + Q_{\text{Space to Earth}} = \Delta U_{\text{int System}}$$

If we divide both sides of this equation by an arbitrary time interval Δt during which heating occurs, we get an equation that involves the heating rates and the rate of the system's energy change:

$$\frac{Q_{\text{Sun to Earth}}}{\Delta t} + \frac{Q_{\text{Space to Earth}}}{\Delta t} = \frac{\Delta U_{\text{int System}}}{\Delta t}$$

or

$$H_{\text{Sun to Earth}} + H_{\text{Space to Earth}} = \frac{\Delta U_{\text{int System}}}{\Delta t}$$

The terms on the left are the radiative heating rates from the Sun to Earth ($H_{\text{Sun to Earth}} > 0$) and from Earth to space ($H_{\text{Space to Earth}} < 0$). If $H_{\text{Sun to Earth}} + H_{\text{Space to Earth}} = 0$, then the internal energy of the system (Earth and its atmosphere) remains constant. If $H_{\text{Sun to Earth}} + H_{\text{Space to Earth}} > 0$, the system's energy increases. If $H_{\text{Sun to Earth}} + H_{\text{Space to Earth}} < 0$, its energy decreases.

Which type of radiation is emitted by an object depends on the temperature of the object. Hot objects such as the Sun emit more energy as short-wavelength infrared radiation and visible light. Cooler objects such as Earth emit longer wavelength infrared radiation and very little short-wavelength infrared radiation and visible light. Carbon dioxide is a strong absorber of long-wavelength infrared radiation, but not a strong absorber of short-wavelength infrared radiation and visible light. This means that carbon dioxide does not absorb much of the radiation Earth receives from the Sun, but it does absorb considerable radiation emitted by Earth. Since the magnitude of $H_{\text{Space to Earth}}$ decreases as CO_2 concentration increases, $H_{\text{Sun to Earth}} + H_{\text{Space to Earth}} > 0$ and the thermal energy of the Earth/atmosphere system increases, which manifests as a global temperature increase. Our simple model based on the first law of thermodynamics and the mechanisms of energy transfer predicts global warming if CO_2 concentration is increasing.

The concentration of CO_2 in Earth's atmosphere is now at the highest level it has been in the last 400,000 years (Figure 15.19). The hottest years since Earth temperature data were first recorded in 1861 have all occurred since 1990, with each decade getting slightly warmer than the previous decades.

FIGURE 15.20 Earth's temperature (including the atmosphere) depends on the balance of incoming and outgoing radiation.

Earth emits longer wavelength radiation into space. Earth absorbs shorter wavelength radiation from the Sun.

Earth's atmosphere

$H_{\text{Sun to Earth}}$

$H_{\text{Space to Earth}}$

System is Earth and its atmosphere.

REVIEW QUESTION 15.7 How does the concentration of CO_2 in the atmosphere contribute to the rise of the atmosphere's temperature?

Summary

Thermal energy U_{th} The random kinetic energy of the atoms and molecules in an ideal gas. (Section 15.1)

$$U_{th} = \tfrac{3}{2}nRT = \tfrac{3}{2}Nk_BT \qquad \text{Eq. (15.1)}$$

(Ideal gas, T in kelvins)

Work W The process of energy transfer from the environment to the system due to mechanical motion. When a gas expands, the work done on the gas is negative. (Section 15.1)

For constant P

$$W_{\text{Environment on System}} = -P\Delta V \quad \text{Eq. (15.2)}$$

For nonconstant P, the work is the negative of the area under the P-versus-V graph for the process.

Heating Q The process of energy transfer from the environment to the system due to differences in their temperatures. $Q > 0$ if energy is transferred into the system; $Q < 0$ if energy is transferred out of the system. (Section 15.2)

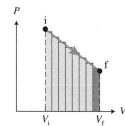

Consequences of heating (no work done on the system)
- **Temperature change** The system's temperature changes when there is no phase change occurring.
- **Phase changes** System melts ($\Delta U > 0$) or freezes ($\Delta U < 0$) at constant temperature. System boils ($\Delta U > 0$) or condenses ($\Delta U < 0$) at constant temperature. (Sections 15.5 and 15.6)

$$\Delta U = cm\,\Delta T \qquad \text{Eq. (15.5)}$$
$$\Delta U = \pm mL_f \qquad \text{Eq. (15.6)}$$
$$\Delta U = \pm mL_v \qquad \text{Eq. (15.7)}$$

First law of thermodynamics A system's energy can change $\Delta U_{int} = (U_f - U_i)$ during a process due to the work W done on the system by the environment and/or due to the heating Q of the system by the environment. (Section 15.3)

$$W_{\text{Environment on System}}$$
$$+\, Q_{\text{Environment to System}}$$
$$= \Delta U_{int} \qquad \text{Eq. (15.4b)}$$

Thermodynamic energy transfer mechanisms
- **Conduction $H_{\text{Conduction}}$** Energy transfer from particle to particle via contact.
- **Convection $H_{\text{Convection}}$** Energy transfer by particles moving from one place to another.
- **Evaporation $H_{\text{Evaporation}}$** Energy transfer due to evaporation / condensation.
- **Radiation $H_{\text{Radiation}}$** Energy transfer via absorption or emission of radiation. (Section 15.7)

$$H = Q/\Delta t$$

Conduction:
$$H_{\text{Conduction}} = k_{th}A(T_{\text{Hot}} - T_{\text{Cold}})/l \quad \text{Eq. (15.8)}$$

Convection: *No equation*

Evaporation:
$$H_{\text{Evaporation}} = \pm(\Delta m/\Delta t)L_v \qquad \text{Eq. (15.9)}$$

Radiation: *No equation*

Questions

Multiple Choice Questions

1. An ideal gas in a container is separated with a divider into two identical half-size containers, each with the same amount of gas. Which of the following statements are correct? (More than one statement may be correct.)
   - (a) The temperature of the gas in each container is half the previous temperature.
   - (b) The mass of gas in each container is half the original mass.
   - (c) The density of the gas in each container is half the original density.
   - (d) The thermal energy of gas in each container is half the original thermal energy.

2. A container of gas has a movable piston, which you hold with your hand. Imagine that you remove your hand and simultaneously place the container in a vacuum. Which of the following statements are correct? (More than one statement may be correct.)
   - (a) The gas volume increases.
   - (b) The internal energy of the gas does not change.
   - (c) The temperature of the gas decreases.
   - (d) The internal energy of the gas decreases.

3. Which gas has the greatest specific heat?
   - (a) A gas in a constant volume container
   - (b) The same gas in a constant pressure container with a movable piston
   - (c) The specific heat is the same for both containers.
   - (d) There is not enough information to answer the question.

4. The specific heat of water is 4200 J/kg · °C. Which of the following statements is true?
   - (a) One kilogram of water stores 4200 J of energy when its temperature is 1 °C.
   - (b) If you add 4200 J to water, its temperature will increase by 1 °C.
   - (c) If you add 42,000 J to 1 kg of water, its temperature will increase by 10 °C.

5. How much heat is stored in 10 kg of water at 95 °C?
   - (a) 42,000 J
   - (b) 399,000 J
   - (c) The question makes no sense; heat is not energy, but is a process for transferring thermal energy.

6. We define the specific heat of a material as the energy that must be transferred to 1.0 kg of that material in order to cause it to warm 1.0 °C. What happens to the specific heat if we transfer twice that much energy?
   - (a) The specific heat doubles.
   - (b) The specific heat halves.
   - (c) The specific heat does not change.

7. Dublin, Ireland and Edmonton, Alberta, Canada are at the same latitude (53.5°). The January and July average temperatures in Dublin are 5 °C and 16 °C, respectively; in Edmonton they are −12 °C and 21 °C, respectively. Why is the temperature variation so much greater in Edmonton than in Dublin?
   - (a) Dublin and Edmonton are in different hemispheres.
   - (b) Dublin is a bigger city.
   - (c) Dublin is on the ocean.

8. **Figure Q15.8** shows a *P*-versus-*V* graph for two processes, A and B, that are performed with the same amount of ideal gas. Process A is an isothermal process. Which of the following statements are correct? (More than one statement may be correct.)

   FIGURE Q15.8

   - (a) The final and the initial temperatures in process B can be different than in process A.
   - (b) The final and the initial temperatures in process B are always the same as in process A.
   - (c) The work done by the gas on the environment is the same in both processes.
   - (d) The work done by the gas on the environment in process B is greater than that in process A.
   - (e) The work done by the gas on the environment in process B is smaller than that in process A.
   - (f) The temperature of the gas in process B is first increasing and then decreasing.
   - (g) The temperature of the gas in process B is first decreasing and then increasing.
   - (h) The temperature of the gas in process B is constant.

9. An electric heater is keeping the inside of a hut a constant temperature 68 °F, while the temperature outside is 10 °F. Which statement is correct?
   - (a) The thermal energy released by the heater in 1 hour is equal to the thermal energy that leaves the hut in the same time period.
   - (b) The thermal energy released by the heater in 1 hour is larger than the thermal energy that leaves the hut in the same time period.
   - (c) The thermal energy released by the heater in 1 hour is smaller than the thermal energy that leaves the hut in the same time period.
   - (d) All thermal energy released by the heater remains as thermal energy of objects in the hut.

Conceptual Questions

10. Match each heating mechanism (left column) with a corresponding phenomenon (right column). There will be more than one match for some mechanisms.

Mechanism	Phenomenon
a. Conduction	1. You feel cold when wearing wet clothes.
b. Convection	2. A mother rubs a sick child with alcohol.
c. Evaporation	3. A farmer tills the soil around a fruit tree in preparation for a cold winter.
d. Radiation	4. You wear white clothes in the summer.
	5. Ice chests are made of Styrofoam.
	6. An electric heating coil is inserted at the bottom of a water container, not at the top.

11. Your friend says, "Heat rises." Do you agree or disagree? If you disagree, what can you tell your friend to convince him of your opinion?

12. Suggest practical ways for determining the specific heats of different liquids and solids.

13. Suggest practical ways to measure heats of melting and evaporation for ice and water.

14. A farmer's fruit storage cellar is unheated. To prevent the fruit from freezing, the farmer places a barrel of water in the cellar. Explain why this helps prevent the fruit from freezing.

15. Why does an egg take the same time interval to cook in water that is just barely boiling as in water that boils vigorously?

16. Why does food cook faster in a pressure cooker than in an open kettle?

17. A potato into which several nails have been pushed bakes faster than a similar potato with no nails. Explain. List the physics ideas and principles that you used to answer this question.

18. Explain why double-paned windows help reduce winter energy losses in a home.

19. The water in a paper cup can be boiled by placing the cup directly over the flame from a candle or from a Bunsen burner. Explain why this is possible without burning a hole in the cup.

20. Provide two reasons why blowing across hot soup or coffee helps lower its temperature. How can you test your explanations?

21. Placing a moistened finger in the wind can help identify the wind's direction. Explain why.

22. Why does covering a keg of beer with wet towels on a warm day help keep the beer cool?

23. Explain why dogs can cool themselves by panting.

24. Some houses are heated by hot oil or water circulating through baseboards. Such a heating system is typically positioned close to the floor. Why?

25. If on a hot summer day you place one bare foot on a hot concrete swimming pool deck and the other bare foot on an adjacent rug at the same temperature as the concrete, the concrete feels hotter. Why?

26. A woman has a cup of hot coffee and a small container of room-temperature milk, which she plans to add to the coffee. If the woman must wait 10 minutes before drinking the coffee, and she wants it to be as hot as possible at that time, should she add the milk and wait 10 minutes or wait 10 minutes and add the milk? Explain.

27. Look carefully at your surroundings for one or more days. Make four recommendations of ways to reduce loss of energy by heat transfer. What physical ideas and principles did you use to make these recommendations?

Problems

Below, **BIO** indicates a problem with a biological or medical focus. Problems labeled **EST** ask you to estimate the answer to a quantitative problem rather than derive a specific answer. Asterisks indicate the level of difficulty of the problem. Problems with no * are considered to be the least difficult. A single * marks moderately difficult problems. Two ** indicate more difficult problems.

15.1 Internal energy and work in gas processes

1. * **EST** Estimate the thermal energy of the air in your bedroom. List all of the assumptions you make.

2. A balloon of volume 0.010 m³ is filled with 1.0 mol of helium. The temperature in the room and in the balloon is 20 °C. What are the average speed and the average kinetic energy of a particle of helium inside the balloon? What is the thermal energy of the helium?

3. * Imagine that the helium balloon from the previous problem was placed in an evacuated container of volume 0.020 m³ and that the balloon popped when it touched a sharp edge on the inside of the container. (a) How much work is done on the helium gas? (b) What happens to the temperature of the helium, its density, the gas pressure, the average kinetic energy of each particle, and the thermal energy of the helium gas? Provide quantitative answers.

4. * You accidentally release a helium-filled balloon that rises in the atmosphere. As it rises, the temperature of the helium inside decreases from 20 °C to 10 °C. What happens to the average speed of helium atoms in the balloon and the thermal energy of the helium inside the balloon? Describe the assumptions you made.

5. * Air in a cylinder with a piston and initially at 20 °C expands at constant atmospheric pressure. (a) What is the work that the piston does on the gas if the air expands from 0.030 m³ to 0.043 m³? (b) How many moles of gas are in the container? (c) Suppose that the work leads to a corresponding change in thermal energy (there is no heating). What is the final temperature of the gas?

15.3–15.5 First law of thermodynamics, Applying the first law of thermodynamics to gas processes, and Specific heat

In the problems in this section, clearly describe your system (S) and indicate important objects in its environment (E). Label the work as $W_{E \text{ on } S}$, heating as $Q_{E \text{ to } S}$, and the change in internal energy as ΔU_{int}.

6. ** Gas in a container with a movable piston initially at volume V_1, pressure P_1, and a very high temperature T_1 expands at constant pressure until its temperature and volume became T_2 and V_2. (a) Describe the process using the concepts of work, heating, and internal energy. (b) Draw a bar chart representing the process. (c) Calculate the work that the environment did on the gas. (d) Explain the process from a microscopic point of view. (e) Represent the process using P-versus-V, P-versus-T, and V-versus-T graphs. (f) Repeat steps (a)–(e) for a situation in which the gas started with the same initial state but expanded at constant temperature instead of constant pressure.

7. ** Gas in a closed container undergoes a cyclic process from state 1 to state 2 and then back to state 1 (**Figure P15.7**). Describe the processes 1-2 and 2-1 qualitatively using the concepts of work, heating, and internal energy. (a) What happened to the thermal energy of the gas as it went from 1 to 2 and then from 2 to 1? What is the net change in the internal energy after the gas returned to state 1? (b) On the P-versus-V graph, show the magnitude of the work that was done on the gas by the environment during process 1-2 and during process 2-1. (c) Was the total work done on the gas positive, negative, or zero during the entire process 1-2-1? (d) Discuss the heating of the gas during process 1-2 and then 2-1. Was the total heating of the gas positive, negative, or zero during the whole process 1-2-1?

FIGURE P15.7

8. * **Jeopardy problem** A gas process is described mathematically as follows: 100 J + (−P)(0.001 m³) = 0. Pose a problem for which this description could be the answer. Describe the process macroscopically and microscopically.

9. * **Jeopardy problem** A gas process is described mathematically as follows: Q + 120 J = 50 J. Pose a problem for which this description could be the answer. Describe the process macroscopically and microscopically.

10. Use the first law of thermodynamics to devise a mathematical description of a process in which gas is being heated (positive heating) but its temperature does not change. Represent the process with a bar chart.

11. * Use the first law of thermodynamics to devise a mathematical description of a process in which gas is being cooled (negative heating) but its temperature increases. Represent the process with a bar chart.

12. ** One of the experiments that Joule used to test the idea of energy conservation was measurement of the temperature of a gas during the process of gas expansion into a vessel from which all air was evacuated. (a) Why did he choose this experiment? (b) What outcome would he predict based on the first law of thermodynamics? (c) Draw a picture of the experiment and provide macroscopic and microscopic reasoning for your answer.

13. ** Derive an expression for the amount of energy that must be provided through heating for one mole of gas of molar mass M to have a temperature increase of 1 K. Consider different processes and decide whether the amount of heating is independent of the process.

14. * You are making a table for specific heats of gases. Compared to the specific heats of solid and liquid substances, what additional information do you need to provide when you are listing the values of specific heat for each gas (oxygen, hydrogen, etc.)? Explain your answer.

15. **EST** **Temperature change in Carrier Dome** On March 5, 2006, a new college basketball attendance record of 33,633 was set in Syracuse University's Carrier Dome in the last regular-season game against Villanova. The volume of air in the dome is about 1.5×10^6 m³. Estimate the temperature change of the air in 2 h, if all the seats in the dome are filled and each person transfers his or her metabolic thermal energy to the air in the dome at a rate of 100 W (100 J/s). Assume that no thermal energy leaves the air through the walls, floor, or ceiling of the dome.

16. * **BIO** **EST** **Body temperature change** A drop in temperature of the human body core from 37 °C to about 31 °C can be fatal. Estimate the thermal energy that must be removed from a human body to cause this temperature change.

17. * **BIO** **Temperature change of a person** A 50-kg person consumes about 2000 kcal of food in one day. If 10% of this food energy could be converted to thermal energy that does not leave the body, what is the person's temperature change? What assumptions did you make?

18. Determine the amount of thermal energy provided by heating to raise the temperature of (a) 0.50 kg of water by 10 °C, (b) 0.50 kg of ethanol by 10 °C, and (c) 0.50 kg of iron by 10 °C.

19. **EST** Estimate the time interval required for a 600-kg cast iron car engine to warm from 30 °C to 1500 °C (approximately the melting temperature of iron) if burning fuel in the engine as it idles produces thermal energy at a rate of 8000 J/s and none of the energy escapes the car engine.

20. * A lead bullet of mass m traveling at v_i penetrates a wooden block and stops. (a) Represent the process with a bar chart. What system did you choose? (b) Assuming that 50% of the initial kinetic energy of the bullet is converted into thermal energy in the bullet, write an expression that would allow you to determine the bullet's temperature increase. (c) List all of the physics ideas that you used to solve this problem.

21. * **BIO** **Exercising warms body** A 50-kg woman repeatedly lifts a 20-kg barbell 0.80 m from her chest to an extended position above her head. (a) If her body retains 10 J of thermal energy for each joule of work done while lifting, how many times must she lift the barbell to warm her body 0.50 °C? (b) State any assumptions you used. (c) List all of the physics ideas that you used to solve this problem.

22. * You add 25 g of milk at 10 °C to 200 g of coffee (essentially water) at 70 °C. The coffee is in a Styrofoam cup. If the specific heat of milk is 3800 J/kg·°C, by how much will the coffee temperature decrease when the milk is added? Indicate any assumptions you made.

23. * You add 20 °C water to 0.20 kg of 40 °C soup. After a little mixing, the water and soup mixture is at 34 °C. The specific heat of the soup is 3800 J/kg·°C. Determine everything you can using this information.

24. **BIO** **Cooling a hot child** A 30-kg child has a temperature of 39.0 °C (102.2 °F). How much thermal energy must be removed from the child's body by some heating process to lower his temperature to the normal 37.0 °C (98.6 °F) body temperature?

25. * **Impact of extinction-causing meteorite** Scientists have proposed that 65 million years ago in what is now the Yucatan peninsula of Mexico, a 1.2×10^{16}-kg meteorite moving at speed 11 km/s collided with Earth, and the resulting harsh conditions led to the extinction of many species, including the dinosaurs. (a) Calculate the kinetic energy of the meteorite before the collision. (b) If 20% of this energy was converted to thermal energy in the meteorite, which had a specific heat of 900 J/kg·°C, by how much did its temperature increase?

26. * You pour 250 g of tea into a Styrofoam cup, initially at 80 °C, and stir in a little sugar using a 100-g aluminum 20 °C spoon and leave the spoon in the cup. What is the highest possible temperature of the spoon when you finally take it out of the cup? What is the temperature of the tea at that time? What assumptions did you make to answer the questions?

27. * A 500-g aluminum container holds 300 g of water. The water and aluminum are initially at 40 °C. A 200-g iron block at 0 °C is added to the water. What can you determine using this information? State any assumptions you used. List all physics ideas that you used to solve this problem.

28. * A 150-g insulated aluminum container holds 250 g of water initially at 20 °C. A 200-g metal block at 60 °C is added to the water, resulting in a final temperature of 22.8 °C. What type of metal is the block? What assumptions did you make to answer the question?

15.6 Phase change

29. Determine the energy needed to change a 0.50-kg block of ice at 0 °C into water at 20 °C.

30. * When 1.4×10^5 J of energy is removed from 0.60 kg of water initially at 20 °C, will all the water freeze? If not, how much remains unfrozen?

31. * An electric heater warms ice at a rate of H. (a) What do you need to do to determine the mass of ice that melts in Δt min? (b) What assumptions did you use? How will your answer change if you use different assumptions?

32. ** Determine the number of grams of ice at 0 °C that must be added to a cup with 250 g of tea at 40 °C to cool the tea to 35 °C.

33. An ice-making machine removes thermal energy from ice-cold water at a rate of H. Determine the time interval needed to form m kg of ice at ice melting temperature.

34. * **Preventing freezing in canning cellar** A tub containing 50 kg of water is placed in a farmer's canning cellar, initially at 10 °C. On a cold evening the cellar loses thermal energy through the walls at a rate of 1200 J/s. Without the tub of water, the fruit would freeze in 4 h (the fruit freezes at −1 °C because the sugar in the fruit lowers the freezing temperature). By what time interval does the presence of the water delay the freezing of the fruit? Indicate any assumptions that you made.

35. * **Passive solar energy storage material** A Dow Chemical product called TESC-81 (primarily a salt, calcium chloride hexahydrate) is used as an energy-storage material for solar applications. Energy from the Sun raises the temperature of the solid material, causing it to melt at 27 °C (81 °F). At night the energy is released as the salt cools and returns to the solid state. (a) Determine the energy required to raise the temperature of 1.0 kg of solid TESC-81 from 20 °C to the liquid state at 27 °C. (b) How warm would 1.0 kg of water become if it started at 20 °C and absorbed the same energy? (c) Discuss the desirability of TESC-81 as a thermal energy-storage material compared to water. For TESC-81, c (solid) = 1900 J/kg·°C and $L_f = 1.7 \times 10^5$ J/kg.

36. How much energy is required to convert (a) 0.10 kg of water at 100 °C to steam at 100 °C and (b) 0.10 kg of liquid ethanol at 78 °C to ethanol vapor at 78 °C?

37. **Cooling with alcohol rub** During a back rub, 80 g of ethanol (rubbing alcohol) is converted from a liquid to a gas. Determine the thermal energy removed from a person's body by this conversion. Indicate any assumptions you made.

38. **Energy in a lightning flash** A lightning flash releases about 10^{10} J of electrical energy. If all this energy is added to 50 kg of water (the amount of water in a 165-lb person) at 37 °C, what are the final state and temperature of the water?

39. A kettle containing 0.75 kg of boiling water absorbs thermal energy from a gas stove at a rate of 600 J/s. What time interval is required for the water to boil away, leaving a charred kettle?

40. **Cooling nuclear power plant** A nuclear power plant generates waste thermal energy at a rate of 1000 MW = 1000×10^6 W. If this energy is transferred by hot water passing through tubes in the water in an evaporative cooling tower, how much water must evaporate to cool the plant (a) per second and (b) per day?

15.7 Heating mechanisms

In the problems in this section, clearly identify the system (S) and objects in the environment (E) that interact with the system. Label the work as $W_{E \text{ on } S}$, heating as $Q_{E \text{ to } S}$, and the change in internal energy as ΔU_{int}.

41. * **EST** **Energy changes when it rains** Estimate the energy that is released or absorbed as water condenses and falls to Earth. Use the following information. Clouds are formed when moisture in the gaseous state in the air condenses. A rainstorm follows, dropping 2 cm of rain over an area 2 km × 2 km. Note that the mass of 1 m³ of water is 1000 kg.

42. * **Insulating a house** You insulate your house using insulation rated as R-12, which will conduct 1/12 Btu/h of thermal energy through each square foot of surface if there is a 1 °F temperature difference across the material (R-12 insulation is said to have a thermal resistance R of 12 h·ft²·°F/Btu). The conduction rate through a material of area A, across which there is a temperature difference $T_2 - T_1$, is

$$H_{Conduction} = (1/R)A (T_2 - T_1)$$

Use this information to determine the conductive energy flow rate across (a) an 8.0 ft × 16.0 ft R-15 wall; (b) a 3.0 ft × 7.0 ft R-4 door; and (c) a 3.0 ft × 4.0 ft R-1.5 window. Assume that the inside temperature is 68 °F and the outside temperature is 20 °F. (d) Convert each answer from Btu/h to J/s = W.

43. ** **Igloo thermal energy conduction** A typical snow igloo (thermal conductivity about 1/10 of ice) is shaped like a hemisphere of inner radius 1.5 m with 0.36-m thick walls. What is the conductive heating rate through the walls if the inside temperature is 10 °C and the outside temperature is −10 °C?

44. To cool hot soup, you blow across the top of a bowl of soup. Assuming that the soup is the system, what are the signs (+ or −) of the different heating mechanisms and the effect of this energy transfer on the soup?

45. While blowing across the bowl of soup in the previous problem, you wonder how efficiently the soup can cool by itself through evaporation. You notice that the bowl of hot soup loses 0.40 g of water by evaporation in 1 min. What is the average evaporative heating rate of the soup during that minute? Assume that soup is primarily water.

46. ** **EST** **Solar collector** You wish to install a solar panel that will run at least five LED bulbs, an LCD TV, and a microwave. How large should the panel be if the average sunlight incident on a photoelectric solar collector on a roof for the 8-h time interval is 700 W/m² and the radiant energy is converted to electricity with an efficiency of 20%?

47. **BIO** **Marathon** You are training for a marathon. While training, you lose energy by evaporation at a rate of 380 W. How much water mass do you lose while running for 3.5 h?

48. * **Cooling beer keg** A keg of beer is covered with a wet towel. Imagine that the keg gains energy from its surroundings at a rate of 20 W. (a) At what rate in grams per second must water evaporate from a towel placed over the keg to cool the keg at the same rate that energy is being absorbed? (b) How much water in grams is evaporated in 2.0 h?

49. * A canteen is covered with wet canvas. If 15 g of water evaporates from the canvas and if 50% of the thermal energy used to evaporate the water is supplied by the 400 g of water in the canteen, what is the temperature change of the water in the canteen?

50. * **EST** **Evaporative cooling** Each year a layer of water of average depth 0.8 m evaporates from each square meter of Earth's surface. Estimate the average energy transfer rate in watts needed to continue this process.

51. * The rate of water evaporation from a fish bowl is 0.050 g/s and the natural thermal energy transfer rate to the bowl by conduction, convection, and radiation is +36 W. What power electric heater must you buy to keep the temperature in the fish bowl constant?

52. **BIO** **Tree leaf** A tree leaf of mass of 0.80 g and specific heat of 3700 J/kg · °C absorbs energy from sunlight at a rate of 2.8 J/s. If this energy is not removed from the leaf, how much does the temperature of the leaf change in 1 min? Note: Do not be surprised if your answer is large. A leaf clearly needs other heat transfer mechanisms to control its temperature.

53. **Warming a spaceship** Your friend says that natural convection would not work on a spaceship orbiting the Earth. Do you agree or disagree with her statement? Explain.

54. * **Passive solar energy storage** Solar energy entering the windows of your house is absorbed and stored by a concrete wall of mass m. The wall's temperature increases by 10 °C during the sunlight hours. What mass of water, in terms of m, would have the same temperature increase if it absorbed an equal amount of energy? What assumptions did you make?

55. Which is less dense: dry or wet air? Explain your answer.

56. * **BIO** **Losing liquid while running** While running, you need to transfer 320 J/s of thermal energy from your body to the moisture on your skin in order to remain at the same temperature. What mass of perspiration must you evaporate each second? Indicate any assumptions you made.

57. * **BIO** **Running a marathon** When you run a marathon, the opposing force that air exerts on you does −150 J of work each second. You convert 1000 J of internal chemical energy to thermal energy each second. (a) How much thermal energy must be removed from your body per second to keep your temperature constant? (b) If 50% of this energy loss is caused by evaporation, how much water do you lose per second? (c) How much water do you lose in 3 h?

58. ** **EST** **Global climate change** Assume that because of increasing CO_2 concentration in the atmosphere, the net radiation energy transfer rate for Earth and its atmosphere is +0.002 × 1350 W/m², corresponding to a 0.2% decrease in radiation leaving Earth. (a) Estimate the extra thermal energy added to Earth and its atmosphere in 10 years. Note: The radiation falls on an area approximately equal to πr_{Earth}^2 where $r_{Earth} = 6.4 \times 10^6$ m. (b) If 30% of this energy is used to melt the polar ice caps, how many kilograms of ice will melt in 10 years? (c) How many cubic meters of ice will melt? (d) By how much will the level of the oceans rise in 10 years?

59. * **Standard house 1** On an average winter day (3 °C or 38 °F) in a typical house, energy already in the house is transferred at the following rates: (i) 2.1 kW is transferred through partially insulated walls and the roof by conduction; (ii) 0.3 kW is transferred through the floor by conduction; and (iii) 1.9 kW is transferred by conduction through the windows. Additional heating is also needed at the following rates: (iv) 2.3 kW to heat the air infiltrating the house through cracks, flues, and other openings and (v) 1.1 kW to humidify the incoming air (because warm air must contain more water vapor than cold air for people to be comfortable). What is the total rate at which energy is transferred from this house?

60. * **Standard house 2** On the same day in the same house described in the previous problem, some thermal energy is supplied by heating in the following amounts: (i) sunlight through windows, 0.5 kW; (ii) thermal energy given off by the inhabitants, 0.2 kW; and (iii) thermal energy from appliances, 1.2 kW. How many kilowatts must be supplied to this standard house by the heating system to keep its temperature constant?

61. * **Standard house 3** Suppose that the following design changes are made to the house described in the previous two problems: (i) additional insulation of walls, roof, and floors, cutting thermal losses by 60%; (ii) tightly fitting double-glazed windows with selective coatings to reduce the passage of infrared light, cutting conduction losses by 70%; (iii) elimination of cracks, closing of flues, and so on, cutting infiltration losses by 70%; and (iv) improving the air humidifier, cutting humidifying losses by 70%. What is the total rate at which energy is lost from this house?

62. * **Standard house 4** After further improvements (shifting windows from the north to the south sides and replacing outmoded appliances), thermal energy is supplied to the house described above at the following rates: (i) sunlight through windows, 1.0 kW; (ii) people's warmth, 0.2 kW; and (iii) appliance warmth, 0.8 kW. How many kilowatts must be supplied by the heating system of this house to keep it at constant temperature?

General Problems

63. ** An ideal gas is placed in a cylinder with a movable heavy piston. The motion of the piston is limited by two stoppers, as shown in Figure P15.63. The data describing the initial state of the gas are $P_i = 1.1 \times 10^5$ N/m², $T_i = 300$ K, and $V_i = 1000$ ml. The initial volume of the gas is half of the maximum volume to which the gas can expand. A total of 500 J of energy is transferred to the gas through heating. (a) Determine the final temperature and the final pressure of the gas. (b) Represent the process by drawing P-versus-V, P-versus-T, and V-versus-T graphs.

FIGURE P15.63

64. ** **BIO** **EST** **Metabolism warms bedroom** Because of its metabolic processes, your body continually emits thermal energy. Suppose that the air in your bedroom absorbs all of this thermal energy during the time you sleep at night. Estimate the temperature change you expect in this air. Indicate any assumptions you make.

65. * **EST** You have an 850-W electric kettle. Estimate the least amount of time you have to boil water before 10 guests arrive for a tea break. State clearly all numbers that you use in your estimate.

66. * **EST** **House ventilation** For purposes of ventilation, the inside air in a home should be replaced with outside air once every 2 hours. This air infiltration occurs naturally by leakage through tiny cracks around doors and windows, even in well-caulked and weather-stripped homes. (a) Estimate the mass of air lost every 2 h and each second. (b) Estimate the energy per second needed to warm outside air leaking into the house during a winter night. State your assumptions.

67. **BIO** **Frostbite** When exposed to very cold temperatures, the human body maintains core body temperature by reducing blood circulation to the skin and extremities, and skin and extremity temperatures drop. This can eventually lead to frostbite. Explain why this helps conserve thermal energy.

68. ** **EST** **Heating an event center with metabolic energy** Estimate the temperature change in some enclosure on your campus during an athletic event. Assume that there is no thermal energy transfer into or out of the building during the event. Indicate any other assumptions you make.

69. ** Hector read that the traditional drink in Tibet is tea with added butter. Inspired by this idea, he decides to investigate how a thin layer of fat on the top of a hot liquid affects the cooling of the liquid. He performs two different experiments.

Experiment 1: Hector pours 100 ml of boiling water into each of two Styrofoam cups and adds a thin (2-mm) layer of hot oil (at about 100 °C) to one of the cups. He measures the initial temperature in the cups, waits for 10 minutes, and then measures the temperature again. His measurements are shown at right.

	Cup A (no oil)	Cup B (with layer of oil)
T_i	98 °C	98 °C
T_f	84 °C	95 °C

Experiment 2: Hector puts two empty Styrofoam cups on separate precision scales and zeroes the scales. Then he pours 100 ml of boiling water into each cup and adds a 2-mm layer of hot oil to one of the cups. He records the initial readings of the scales, waits for 10 minutes, and then records the readings of the scales again. His data are shown at right.

	Cup A (no oil)	Cup B (with layer of oil)
m_i	99.00 g	104.00 g
m_f	97.00 g	104.00 g

The surface area of the water in the cups is 3.0×10^{-3} m². (a) Explain why hot water with a layer of hot oil cools more slowly than water without oil. Show that your explanation is consistent with the data from both experiments. Present quantitative arguments. (b) Why does Hector use oil at 100 °C instead of room-temperature oil in his experiments? Describe qualitatively how Hector's measurements would change if he used room-temperature oil. (c) What other quantities relevant to the investigation can you determine or estimate from Hector's data? List as many as you can and determine two of them.

Reading Passage Problems

Cloud formation Air consists mostly of nitrogen (N_2), with a molecular mass of 28, and oxygen (O_2), with a molecular mass of 32. A water molecule (H_2O) has molecular mass 18. According to the ideal gas law ($N/V = P/k_BT$), dry air at a particular pressure and temperature has the same particle density (number of particles per unit volume) as humid air at the same pressure and temperature. Consequently, humid air, whose low-mass water molecules replace more massive nitrogen and oxygen molecules, is less dense than dry air—the humid air rises. Atmospheric pressure decreases with elevation since there is less air above that is pushing down. At about 5000 m above Earth's surface, the pressure is about 0.5 atm. Assuming an ideal gas, the gas volume V increases as pressure P decreases.

What happens to the air temperature as humid air rises? Air is a poor thermal energy conductor, and there is little heating from neighboring air ($Q \approx 0$, an adiabatic expansion process). As the rising gas expands, the neighboring environmental gas pushes in the opposite direction of the increasing volume of this rising gas. The environment does negative work on the rising air system ($W_{\text{by Environment on System}} < 0$). According to the first law of thermodynamics ($W + Q = \Delta U_{\text{System}}$): $-|W| + 0 = \Delta U_{\text{System}}$. The system's internal thermal energy decreases with a corresponding temperature decrease—about $-10\,°C$ for each 1000-m increase in altitude.

When the humid air reaches its dew point temperature, it starts to condense into water droplets (cloud formation). When condensation occurs, energy is released. There is a competition between decreasing thermal energy as the air expands and increasing thermal energy as the water vapor condenses. The air now cools at a lower rate of about $-5\,°C$ for each 1000-m increase in elevation.

70. If no condensation occurred, how high would $40\,°C$ humid air have to rise before its temperature decreased to $10\,°C$?
    (a) 1000 m (b) 2000 m (c) 3000 m
    (d) 6000 m (e) 10,000 m

71. When rising humid air starts to condense, why does its temperature change less rapidly with increasing elevation?
    (a) Its density increases, making it more difficult to change temperature.
    (b) Thermal energy released during condensation causes less thermal energy change.
    (c) The gas expands less, causing less negative work.
    (d) The temperature of the surrounding air is changing less.

72. After crossing a mountain top on a warm sunny day, what should cool dry air do?
    (a) Sink, because it is denser than warmer air below
    (b) Warm, because the surrounding gas does positive work in causing it to contract
    (c) Not change, as the surrounding air transfers little thermal energy by heating ($Q \approx 0$)
    (d) a and b are correct.

73. Why does humid air rise in dry air?
    (a) Water is attracted to the clouds above.
    (b) A water molecule has lower mass than other air molecules.
    (c) $1\,m^3$ of humid air has fewer molecules than $1\,m^3$ of dry air at the same T and P.
    (d) b and c are correct.
    (e) None of the above is correct.

74. If $1\,m^3$ of dry air rises 5000 m and has no temperature change, what would its volume be?
    (a) $0.3\,m^3$ (b) $0.5\,m^3$
    (c) $1.5\,m^3$ (d) $2.0\,m^3$
    (f) Too little information to answer

75. The magnitude of the thermal energy released from the water molecules to the air if 1.0 g of water vapor condensed to 1.0 g of liquid water is closest to which of the following?
    (a) 300 J (b) 500 J
    (c) 1000 J (d) 2000 J
    (e) $2 \times 10^6\,J$

Controlling body temperature A healthy human body maintains a central core of about $37.0\,°C$ ($98.6\,°F$). To do so, the body must shed excess thermal energy that is converted from chemical energy due to metabolic processes. This is especially challenging during vigorous exercise, when the body's core

temperature can increase by as much as $6\,°C$ in only 15 min, which can lead to convulsions or even brain damage.

Energy transfer is the key to maintaining a consistent temperature. We can analyze how the human body uses heating mechanisms of energy transfer by applying the first law of thermodynamics ($W_{\text{E on S}} + Q_{\text{E to S}} = \Delta U_{\text{int}}$) to the body (the system) during a short time interval Δt. Divide the quantities in the first law by the time interval during which the changes occurred:

$$\frac{W_{\text{E on S}}}{\Delta t} + \frac{Q_{\text{E to S}}}{\Delta t} = \frac{\Delta U_{\text{int}}}{\Delta t}$$

The second term in the above equation is the heating energy transfer rate ($Q_{\text{E to S}}/\Delta t$) from the environment to the system. We can separate Q into four separate terms, which represent the rates of conduction, convection, radiation, and evaporation (represented by H symbols with subscripts):

$$\frac{W_{\text{E on S}}}{\Delta t} + (H_{\text{Conduction}} + H_{\text{Convection}} + H_{\text{Radiation}} + H_{\text{Evaporation}}) = \frac{\Delta U_{\text{th}}}{\Delta t} + \frac{\Delta U_{\text{ch}}}{\Delta t}$$

We also substituted the two types of internal energy change that can occur in a person's body: the rate of thermal energy change and the rate of chemical energy change. This latter term is the metabolic rate, the rate of energy released by the chemical reactions occurring in the body.

We apply the above equation to a real process: a person running at moderate speed on an indoor air-conditioned track. The table below shows quantities that are representative of a real process.

Type of energy change	Work and heating rates (watts = J/s)
Work done by drag force exerted by air	-100
Conduction	≈ 0
Thermal energy transfer due to convection	-300
Transfer of energy by radiative cooling	-75
Cooling due to evaporation of sweat	-325

76. What should be the metabolic rate to keep the runner's body temperature constant?
    (a) $100\,J/s$ (b) $300\,J/s$
    (c) $400\,J/s$ (d) $800\,J/s$

77. If all of the conditions are the same except that the person runs in a plastic suit, what should be the metabolic rate so that the person does not overheat? Note that plastic (even if transparent for visible light) absorbs most of the infrared radiation.
    (a) $100\,J/s$ (b) $300\,J/s$
    (c) $400\,J/s$ (d) $800\,J/s$

78. Imagine that the runner consumes a candy bar of 200 kcal. How much time should the person run to transfer this energy to the environment through the processes described in the table above?
    (a) 0.25 s (b) 17.5 min
    (c) 4.2 min (d) 2.3 h

79. If the person runs outside on a windy day, which rates in the table above might change, and why?
    (a) The work done by the drag force might increase because wind can exert a greater force than stationary air.
    (b) Cooling by evaporation might increase because the wind will speed up the evaporation process.
    (c) Cooling by convection might increase in the same way as with forced convection.
    (d) Cooling by radiation might increase because the temperature of the skin surface will decrease.
    (e) All of the above
    (f) (a)–(c) are correct, but not (d).

80. When running a marathon, how much water should the person drink to replace the water lost by evaporation only? An average runner covers 26 miles in 4 h.
    (a) 0.5 L (b) 2 L
    (c) 4 L (d) 12 L

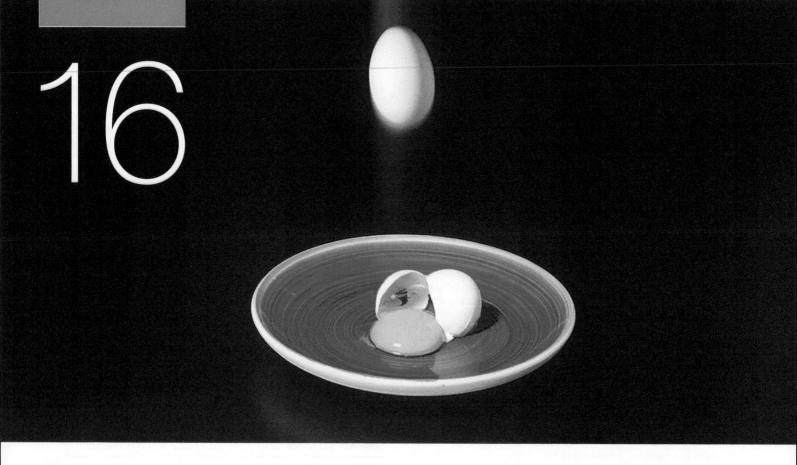

16

Second Law of Thermodynamics

- Why is the statement "we need to conserve energy" incorrect in terms of physics?

- How does the inside of a refrigerator stay cold?

- You add a droplet of blue food coloring to a glass of water, and after a while the water gains a bluish tint. Why doesn't the food coloring spontaneously form a new droplet?

A rubber ball compresses when the ball hits the ground and then uncompresses when it bounces back up. But a falling egg becomes a mess of broken shell and yolk when it hits a plate—it can never reassemble to form a whole egg and fly back up into the air. Why is the ball's bounce reversible, but the egg's fate irreversible? Similarly, a roller coaster converts gravitational potential energy into kinetic energy as it falls, and converts kinetic energy back into gravitational potential energy as it climbs. But why do we say that the thermal energy exhausted by our car's engine is "lost"? Why can't it be converted back into useful kinetic or potential energy?

BE SURE YOU KNOW HOW TO:

- Identify a system and the initial and final states of a well-defined process (Sections 7.1 and 7.2).

- Determine the energy transferred through heating to, the work done on, and the internal energy change of a system (Sections 15.1 and 15.3).

- Define thermal energy and apply the first law of thermodynamics to a process (Sections 12.8 and 15.4).

IN THIS CHAPTER, we will learn why some processes are irreversible (they can only occur in one direction). So far, we have treated all forms of energy as equal—one form can be converted to another, but the total never changes. We will learn in this chapter, however, that that is not the full story.

16.1 Irreversible processes

Let's consider another example of an irreversible process. We will describe it and use a bar chart to analyze the energy conversions. **Figure 16.1a** shows a waterfall; the system is the water, Earth, and the rocks. The initial state is when the water is at the top of the waterfall, and the final state is when the water stops after hitting the rocks. The gravitational potential energy of the system in the initial state converts to internal thermal energy in the final state (as shown in the bar chart in Figure 16.1b). Watch the video at right for the reverse process. The warmer water and the rocks at the bottom cool, and the released internal thermal energy is converted into the gravitational potential energy of the system (see Figure 16.1c). While this bar chart is consistent with the energy constancy of the system, physically it would mean that the water cools down and flows back up in the reverse direction. While you are watching the video, you intuitively know that the reversed process is unrealistic—it does not happen in real life. A waterfall is an example of an irreversible process.

VIDEO
16.1

FIGURE 16.1 Energy conversions and analysis for a waterfall.

(a)

(b) Energy analysis of the water
hitting the rocks

$$U_{gi} + W = U_{gf} + \Delta U_{int}$$

(c) Energy analysis of the
reversed process

$$U_{gi} + W = U_{gf} + \Delta U_{int}$$

Think about other examples of irreversible processes: a dropped ball that bounces lower and lower after every bounce and eventually stops; a droplet of food coloring in water that spreads with time, coloring all the water; two cars colliding and stopping.

Are there general characteristics of irreversible processes? Yes. The original state involves some form of organized energy—kinetic energy or gravitational potential energy. All of the particles in the object of interest are either moving together (kinetic energy) or at rest in the same location (gravitational potential energy). In the final state, much of that original coordinated energy is converted to thermal energy, which is associated with the random movement of the system's particles. It appears that this random energy cannot be spontaneously converted back to the coordinated energy needed to reverse the process.

CONCEPTUAL EXERCISE 16.1   **Irreversible processes?**

Use the ideas of "organized" and "random" energy to decide which of the following processes are reversible. Then compare the results with your everyday experience. (a) A car skids to a stop; (b) a rock is thrown upward and reaches a height h above the ground.

Sketch and translate The initial and final states of the processes and the choices of systems are sketched at right.

Simplify and diagram A process in which a system goes from more organized energy to less organized random energy is not reversible. In (a), the kinetic energy of the car (more organized energy) converts into the thermal energy (random energy) of the tires and road as the car is stopping and into internal potential energy as the rubber is rubbed off the tires while skidding to a stop. Thermal and

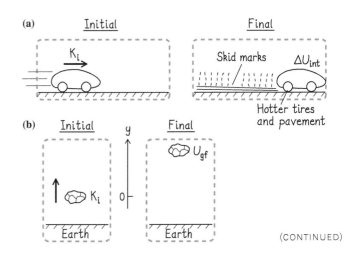

(CONTINUED)

internal potential energy are less organized forms of energy; thus the reverse process (the hot road and tires and the skid marks coming back together to get the car moving) should not occur, and definitely does not. (b) The thrown rock had organized kinetic energy that goes into the organized gravitational potential energy of the rock-Earth system; thus the reverse process (the rock falling back down) should be able to occur. We are assuming that air resistance is small, since the increase in thermal energy of the air would be a conversion from more organized to less organized energy.

Try it yourself You place a droplet of food coloring in a glass of water. Describe the process and explain why it is irreversible.

Answer in an organized motion in a specific direction.
the container; to reform a droplet, the molecules need to participate
its molecules participate in a nonorganized chaotic motion throughout
The process is irreversible because when the food coloring spreads,
it spontaneously reform into its original organized droplet.
The food coloring spreads throughout the water. We never see

> **Reversible and irreversible processes** Many processes in nature occur in only one direction—they are called *irreversible processes*. In the allowed direction, energy converts from more organized forms to less organized forms. In the reverse, unallowed process, energy would have to convert from a less organized form to a more organized form. However, the unallowed reverse process does not contradict energy conservation.

So far, in our examples of reversible and irreversible processes the interacting objects initially had the same temperature. We have not considered how interactions of objects of different temperature relate to the direction of energy conversion. Observational Experiment Table 16.1 provides examples from which we will infer a pattern for the direction.

OBSERVATIONAL EXPERIMENT TABLE 16.1 Direction of energy transfer

Observational experiment	Analysis
Experiment 1. You place a room-temperature spoon in a cup of hot tea. The spoon gets warmer and the tea gets cooler until they reach the same temperature.	Energy is transferred from the hot tea to the cool spoon until they are at the same temperature.
Experiment 2. The air-conditioning in a building fails on a very hot day. The temperature in the building increases until it reaches a temperature similar to the outside temperature.	Energy is transferred through the walls from the hot outside of the building to the cool inside until the inside and outside are at the same temperature.

Pattern

In both experiments, energy is transferred from a high-temperature region to a low-temperature region.

In the experiments in Table 16.1, you do not observe the reverse of the processes, in which a cool object becomes cooler and a hot object hotter. Although the first law of thermodynamics allows this, such processes do not happen. This leads us to another formulation concerning the direction of energy conversion in isolated systems.

> **Direction of thermal energy transfer** In an isolated system, energy always transfers from a warmer region to a cooler region.

Thermodynamic (or heat) engines

Why are we interested in organized versus nonorganized forms of energy in the first place? Historically, once people understood that hot compressed gas can push pistons and thus do work (maximizing this work being the main goal), they started dreaming of making machines that use this property. These machines are usually called *heat engines*, but we will use the term *thermodynamic engine* to underscore that the laws of thermodynamics explain their operation. The main difficulty of creating such engines is that the gas cannot expand and push something indefinitely. In order for an engine to work continuously, the piston needs to be brought back to its original state, or some new portion of gas needs to be acquired so it can now expand and push something again.

This is exactly what all thermodynamic engines have in common: they repeat the same process over and over again. We will call such a process a *cyclic* process, and we will call the one "round-trip" during which the device returns to the original state a *cycle*. In any cyclic process, the gas returns to the original state, meaning that the total energy transferred to or from the engine during one complete cycle is zero. If it were not zero, the engine would have had more or less energy at the end of the cycle, meaning that it would not have been in the same state as at the beginning. Therefore, in a cyclic process *all* energy that is transferred to the engine should also leave the engine. But remember— the goal is to maximize the portion of energy that leaves the engine as work. What is the difference between organized and nonorganized forms of energy for this purpose?

It is easier to understand the answer if we use an example. Imagine we want to construct an engine that will lift sugar cubes from a conveyor belt at one level to a higher level. Let's compare two different designs. The first engine uses mostly organized forms of energy, and the second engine uses mostly less organized thermal energy.

The first engine uses a spring to do work. A compressed spring (the system) stores elastic potential energy, which is also an example of organized energy. All parts of a deformed spring exert forces in the same direction on the neighboring parts. The initial state is a compressed spring with a sugar cube on it; the spring decompresses and pushes the cube up. We then push down on the spring and it returns into its initial state. Figure 16.2 shows one cycle of operation. Each cycle consists of a spring compression and spring expansion during which the spring does work on the cube (considered to be in the environment in this case).

The second engine uses an ideal gas to do work. The gas is inside a closed plastic bottle, and the sugar cube sits on the side of the bottle. The initial state is when the gas is cold (T_C) and the bottle is squished; we then use a hair dryer (HD) to warm up the gas. The positive heating of the gas ($Q_{HD \text{ to Gas}}$) causes the gas to warm up to temperature T_H and to expand. During the expansion the gas does work on the sugar cube. To return the bottle to the initial squished state, we need to cool it to temperature T_C, for example, by placing it in cold water (CW), which means that that the energy provided to the gas through the process of heating at this stage is negative ($Q_{CW \text{ to Gas}}$), as the gas transfers positive thermal energy to cold water. Figure 16.3 and the video at right show one cycle of operation.

Note that this engine (as with the first engine) ends the cycle in the same state as at the beginning of the cycle. Let's analyze energy conversions during one cycle for both engines. In the first engine, the compressed spring that stored elastic potential energy does work on the sugar cube. The amount of work done is almost equal to the stored energy.

In the second engine, the hot gas that stored internal thermal energy transferred to it by the hair dryer ($Q_{HD \text{ to Gas}}$) did work on the cube; however, the amount of work done is less than the amount of energy transferred to the gas through heating. Since the thermal energy of the gas is less organized, we can use only part of the initial thermal energy to do work. Because molecules in the gas are moving randomly, only a small fraction of molecules at a time exert forces that contribute to the work done on the environment. The rest of the energy transferred to the gas ($Q_{HD \text{ to Gas}}$) remains in the form of thermal energy, which has to be "dumped" (transferred) from the engine by cooling ($Q_{CW \text{ to Gas}} = -Q_{Gas \text{ to CW}}$). We cannot avoid this "dumping" of thermal energy from the engine because, in order to complete the cycle, we need to bring the engine back to the initial low-temperature state. Given that, at the end of the cycle, the engine is in the same state as at the beginning, this means that the total work done by the gas on the environment is equal to $W_{Gas \text{ on Cube}} = Q_{HD \text{ to Gas}} + Q_{CW \text{ to Gas}} = Q_{HD \text{ to Gas}} - Q_{Gas \text{ to CW}}$, where $Q_{CW \text{ to Gas}}$ is a negative number.

From the above analysis, we see that we can use all of the organized energy in the system to do work on the environment in a cyclic process, but not all of the nonorganized energy can be used for the same purpose. Unfortunately, the processes in fuel-based

FIGURE 16.2 One cycle of an engine that involves only organized energy.

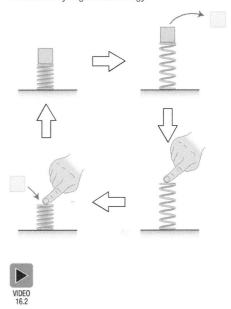

VIDEO
16.2

FIGURE 16.3 One cycle of an engine that involves nonorganized energy.

engines involve some form of nonorganized energy. Therefore, although less efficient, cyclic processes using nonorganized energy sources are more common in real life.

Carnot's principle

The person who first understood this idea, that is, the link between how organized the energy is in a system and how much work it can do, was the French engineer Sadi Carnot (1796–1832). Carnot was interested in the maximum work that a steam engine with a certain amount of fuel could do. Using reasoning similar to ours above, he proposed two things:

1. It is impossible to use the engine's fuel to directly do work. The fuel needs to first warm a gas. This gas then expands and pushes a piston that moves in an environment that is cooler than the gas. If the piston is considered the system of interest, then the gas does work on the piston.

2. When the fuel transfers a certain amount of energy to the gas, only a fraction of this energy can do work on the piston. There is a theoretical limit on the work that can be done. The remaining energy is transferred to the environment, where it manifests as thermal energy. Additionally, all real-world engines fail to reach the theoretical limit when they operate.

Carnot devised the following rule:

> **Carnot's principle** It is impossible to build an engine that uses all of its thermal energy to do mechanical work on its environment.

Direction of processes: the second law of thermodynamics

Let's review several important ideas we have learned so far. We have determined that as the energy of a system becomes less organized, the amount of work the system can do on other systems decreases. We found that in an isolated system, energy always transfers from a warmer region to a cooler region. Finally, we learned Carnot's principle, which tells us that it is impossible to build an engine that uses all of its thermal energy to do mechanical work on its environment. All of these statements are different versions of **the second law of thermodynamics**. The second law describes what processes can and cannot occur in an isolated system. While the first law of thermodynamics is a principle about energy conservation, the second law is a principle of *energy transfer* between two regions at different temperatures and *energy conversion* from one form of energy to another. In order to make a more quantitative version of the second law, we must construct a new physical quantity that describes the degree of organization of a system's energy.

REVIEW QUESTION 16.1 When your car burns gasoline, it can gain organized kinetic energy as it moves faster and faster. Does this contradict the second law of thermodynamics?

16.2 Entropy: the microscopic approach

One way to understand why organized energy in the system is "better" if we want the system to do work on some other system is to do a statistical analysis of the probability that a system is in one state versus another. Consider a very simple thought experiment with a box holding identical atoms that are labeled so we can tell them apart. We drop a thin divider down a slot in the middle of the box, thus dividing it in half. Let's start with four atoms in the box. At regular time intervals we insert the divider, count the number of atoms on each side, and then remove the divider. The five possible distributions are shown in Figure 16.4. Each distribution is characterized using the symbol i, which has integer values 0 through 4 indicating the number of atoms in the left half. For example, the $i = 0$ distribution has zero atoms on the left and four on the right; the $i = 1$ distribution has one on the left and three on the right; and so forth. Each of these five distributions is called a **macrostate** of the system of four atoms. Each macrostate has a particular number of atoms n_l on the left and n_r on the right. In thermodynamics we call the variable quantities n_l and n_r *state variables*; they distinguish one particular macrostate of a system from another.

Earlier, we said that each atom is labeled so we can tell them apart. Real atoms don't have labels, but this is a thought experiment so we can construct the situation however we like. Our goal now is to determine the different unique ways for each macrostate to exist. For example, there are four different ways to get one atom on the left and three on the right—each of the four atoms could be alone on the left with the other three on the right. We call this number of ways of getting a particular macrostate the count Ω_i of that state. Macrostate 0 with zero atoms on the left and four on the right can occur in only one way—if all four atoms are on the right. For macrostate 0, $\Omega_0 = 1$. As we said above, macrostate 1 can occur in four different ways, $\Omega_1 = 4$. Column 3 in Figure 16.4 lists the number of ways of getting the different macrostates. Each of these ways is called a **microstate** of that particular macrostate. All microstates are equally probable. The probability of a macrostate occurring depends on the number of microstates it has. For example, macrostate 2 has six unique ways to arrange four atoms to get two on each side.

In general, for any number of atoms we can determine the number Ω_i of microstates of the ith macrostate by using the equation

$$\Omega_i = \frac{n!}{(n_l!)(n_r!)} \tag{16.1}$$

where n is the total number of atoms and n_l and n_r are the number of atoms on the left and on the right, respectively.

We can check that Eq. (16.1) works for the four-atom example, whose results are shown in Figure 16.4. Consider the number of microstates for macrostate 2:

$$\Omega_2 = \frac{4 \times 3 \times 2 \times 1}{(2 \times 1)(2 \times 1)} = 6$$

The equation works. You can try the equation for the other macrostates.

If we add all the numbers in the third column of Figure 16.4, we find that there are 16 microstates, the number of unique ways in which the four atoms can be arranged in the two halves of the box. The *probability* P_i of a particular macrostate occurring is the number of microstates for that macrostate (the count) divided by the total number of microstates. For example, the probability of finding zero atoms in the left box is 1 in 16; the probability of finding one atom in the left half is 4 in 16; and the probability of observing two atoms in each half is 6 in 16.

For the box with four atoms, there is a small chance (1 in 16) that you will observe all atoms on one side of the box. With a larger number of atoms, the situation is quite different. With 10 atoms, for instance, the state with 10 on the left and 0 on the right is 252 times less likely to occur than the macrostate with 5 on each side. That is,

$$\frac{\Omega_0}{\Omega_5} = \frac{1}{252}$$

With 100 atoms, the state with 100 on the left and zero on the right is $1/10^{29}$ as likely to occur as the state with 50 on each side. If we increase the number of atoms to one million, states with equal numbers of atoms on each side are overwhelmingly more probable than states with an unequal number on each side. For example, for one million atoms there is almost a 10^{87} times better chance of observing an equal number on each

FIGURE 16.4 The possible states of four atoms found on each side of a divider after its insertion in the box.

TIP ▸ The exclamation marks (!) in Eq. (16.1) indicate the factorial function. It means that we multiply the number before the exclamation mark by all the positive integers equal to and smaller than the number. Thus, $8! = 8 \times 7 \times 6 \times 5 \times 4 \times 3 \times 2 \times 1$, and $2! = 2 \times 1$. The notation $8!$ is called eight factorial, and $n!$ is called n factorial. Zero factorial is defined as one ($0! = 1$).

side than of observing 51% of the atoms on the left and 49% on the right. This 50-50 distribution is the state with the highest probability—the state that is overwhelmingly most likely to occur. This most probable state is the state with the greatest number of microstates and has the greatest randomness, especially compared to the state with all particles on the same side.

QUANTITATIVE EXERCISE 16.2   **Five atoms**

Suppose a box with a divider contains five atoms. (a) Identify the six macrostates that can occur. (b) Determine the count of the 0 on left, 5 on right macrostate and that of the 3 on left, 2 on right macrostate.

Represent mathematically Use Eq. (16.1) with $n = 5$ to determine the count for the different macrostates:

$$\Omega_i = \frac{5 \times 4 \times 3 \times 2 \times 1}{(n_l!)(n_r!)}$$

Solve and evaluate (a) The macrostates (n_l, n_r) are $(0, 5)$, $(1, 4)$, $(2, 3)$, $(3, 2)$, $(4, 1)$, and $(5, 0)$.

(b) The count for the $i = 0$ macrostate [the $(0, 5)$ state] is

$$\Omega_0 = \frac{5 \times 4 \times 3 \times 2 \times 1}{(0!)(5 \times 4 \times 3 \times 2 \times 1)} = 1$$

The count for the $i = 3$ macrostate [the $(3, 2)$ state] is

$$\Omega_3 = \frac{5 \times 4 \times 3 \times 2 \times 1}{(3 \times 2 \times 1)(2 \times 1)} = 10$$

Try it yourself Determine the count for a container with seven atoms with 2 on the left side and 5 on the right side.

Answer   21.

Connecting atoms in a box to physics

In general, an isolated system with a large numbers of particles will evolve toward states that have a higher probability of occurring. The highest probability state is the one in which the particles are evenly distributed—in other words, the state with the highest level of randomness. This state of maximum probability is often called the *equilibrium state* of the system. An isolated system in the most probable state is said to have reached equilibrium.

The same rules apply in the real world. The state of a system with particles moving randomly (random thermal energy) is much more probable than the state with the particles all moving together in the same direction. To characterize this randomness or disorder in the system, physicists use a physical quantity called **entropy**, a measure of the probability of a particular macrostate.

Entropy is a physical quantity that quantifies the probability of a state occurring and the degree of disorder or disorganization of a system when in that state. In statistical thermodynamics, the entropy S_i of the ith macrostate is defined in terms of the count Ω_i of that macrostate:

$$\text{Entropy of } i\text{th macrostate} = S_i = k_B \ln \Omega_i \qquad (16.2)$$

in which $\ln \Omega_i$ is the natural logarithm of the count Ω_i and k_B is a proportionality constant (Boltzmann's constant) with a value $k_B = 1.38 \times 10^{-23}$ J/K (joules/kelvin).[1]

The entropy of the five macrostates for the distribution of four atoms on the left or right side of a box is $S_0 = 0$, $S_1 = 1.4k_B$, $S_2 = 1.8k_B$, $S_3 = 1.4k_B$, and $S_4 = 0$. Try to use Eq. (16.2) to confirm these results. (The use of logarithms is reviewed briefly in Appendix A.)

---

[1] A J/K is sometimes called an entropy unit (eu). Thus, 1 J/K = 1 eu.

At the beginning of this section we set out to construct a physical quantity that described the degree of organization of a system's energy. We learned that isolated systems spontaneously evolve toward states whose energy is more disorganized. We now have a way to quantify this statement: isolated systems evolve toward states with higher entropy.

Statistical approach to entropy of an expanding gas

Ultimately, we would like a quantitative way to understand the concept of entropy in terms of macroscopic quantities such as temperature T and heating Q rather than in terms of counts of microstates. In working toward this, let's first examine from a statistical point of view a familiar process: a gas expanding at constant temperature.

EXAMPLE 16.3 **Entropy change of an expanding gas**

Suppose you have a gas composed of just six atoms, each moving randomly and with the same amount of energy. Initially, the atoms are grouped together in one tiny box (situation 1). The box then doubles in size (situation 2), triples in size (situation 3), and finally reaches six times the original size (situation 4). In all these expansions the temperature of the gas remains constant. The gas has expanded its volume to six times its original value. Determine the count and the entropy of the maximum probability state for each of the four situations.

Sketch and translate The situations are pictured below. For a gas with six atoms and a box that is twice as big, the maximum probability state has three atoms in each half of the box. With a box that is three times bigger, the maximum probability state has two atoms in each third of the box. The maximum probability state with a box that is six times bigger has one atom in each sixth of the box.

Volume of box	Most probable distribution	Count	Entropy
1		1	0
2	As the box gets bigger, the count of the most probable distribution increases dramatically.	20	$3.0k_B$
3		90	$4.5k_B$
6		720	$6.6k_B$

Simplify and diagram We assume that the atoms are labeled. Then for situation 2, the distribution 1, 2, 3 on side 1 and 4, 5, 6 on side 2 differs from 1, 2, 4 on side 1 and 3, 5, 6 on side 2—these are two distinct microstates.

Represent mathematically The count of the number of microstates with n atoms in N distinct parts of a container is

$$\Omega_N = \frac{n!}{n_1!n_2!\cdots n_N!}$$

Solve and evaluate The count for each maximum probability macrostate, as calculated using the above equation, is reported in the figure. For example, for situation 3 the count is

$$\Omega_3 = \frac{6!}{2!2!2!} = 90$$

You can check the numbers for the other maximum probability macrostates. The entropy of each state is determined using

$$S_n = k_B \ln \Omega_n$$

and is reported in the figure. Note that as the number of possible locations for the atoms increases, the maximum probability count and entropy increase dramatically. As the gas expands, its entropy increases. From an energy point of view, we would say that the energy of the gas evolved from being localized in a smaller region and potentially being more useful for doing work to being distributed evenly over a larger region and being much less useful.

Try it yourself Four atoms start in a small box. They then expand into a box that is twice as large and finally into a box that is four times as large. Determine the count of the maximum probability macrostate for each situation. Label the first box as a single location, the second box as having two places the atoms can reside, and the largest box as having four places the atoms can reside.

Answer 1: 6: 24.

This last example illustrates quantitatively the second law of thermodynamics expressed using entropy.

Second law of thermodynamics Spontaneous irreversible processes in an isolated system proceed in the direction of increasing entropy.

Remember that entropy is a measure of the probability of a particular macrostate. High-entropy states are more disorganized than lower entropy states. High-probability disorganized states with higher entropy are less able to do work on their environment than less probable organized states with lower entropy. Notice that as we progress through the chapter, we will meet new formulations of the second law of thermodynamics. This is a unique law in this sense—it can be expressed in multiple ways, and all of them relate to the direction of processes that occur spontaneously.

REVIEW QUESTION 16.2 Explain why turning on an extra lightbulb in your apartment does not prevent energy from being conserved but still affects the entropy of the apartment.

16.3 Entropy: the macroscopic approach

So far, we have related entropy to the number of microstates (the count) for a particular macrostate. The greater the number of microstates, the greater the disorder of the system when in that macrostate and the greater the entropy of the system. In systems with large entropy, the energy is very disorganized (for example, there is more random thermal energy) and the system can do less work on its environment.

These ideas are all based on our observations of real-world phenomena and on the statistical analysis of simple microscopic systems. We can actually calculate the entropy of these simple systems (involving 10 or even 100 particles). But real macroscopic systems involve vastly larger numbers of particles (10^{23} or more, for example). It would be difficult to count the number of microstates in a macrostate of such a large system. We need another method to determine the entropy in real-world macroscopic systems with large numbers of particles. Can we express entropy using macroscopic physical quantities such as temperature?

Macroscopic definition of entropy change

Consider a macroscopic process during which a gas (modeled as an ideal gas) is heated by a Bunsen burner flame and expands at constant temperature. When the gas in a system expands, it pushes outward on the environment. Therefore, the environment (in this case, a piston) exerts a force on the system that does negative work (**Figure 16.5a**). The bar chart in Figure 16.5b depicts this process. The system's internal energy does not change since this process is isothermal. Since the environment is doing negative work on the system, the energy provided to the gas by the environment through heating must transfer energy so that the system's thermal energy remains constant.

The volume of the gas increases, so the gas particles now have more places to reside than before—the entropy of the system should increase even though the internal energy does not change. We can understand this connection through a limiting case analysis. Imagine that at the beginning, the volume of the gas is almost zero—all particles are crammed together and the number of microstates of that macrostate of the system is 1. If the gas expands, the particles have more room and can occupy this room in many different configurations. Thus, the count of microstates and therefore the entropy of the system increase.

We can use the bar chart in Figure 16.5b to apply the first law of thermodynamics to this isothermal process:

$$Q_{\text{Env to Sys}} + W_{\text{Env on Sys}} = \Delta U_{\text{int}} = 0$$

or

$$Q_{\text{Env to Sys}} = -W_{\text{Env on Sys}}$$

FIGURE 16.5 Isothermal expansion. (a) The gas expands isothermally. (b) The negative work done by the piston on the gas is balanced by the positive heating.

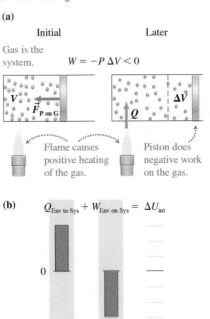

(a)

Initial   Later

Gas is the system. $W = -P\,\Delta V < 0$

Flame causes positive heating of the gas.

Piston does negative work on the gas.

(b) $Q_{\text{Env to Sys}} + W_{\text{Env on Sys}} = \Delta U_{\text{int}}$

0

Because the pressure is changing during the isothermal process, we cannot determine the work done by the environment on the gas as $W_{\text{Env on Sys}} = -P\Delta V$. We will therefore replace P in this equation with P_{eq} (where the subscript "eq" stands for "equivalent"). P_{eq} is some intermediate pressure in the process such that the product of P_{eq} and the volume change with the negative sign is equal to the work done by the environment on the gas. The volume of the gas corresponding to this pressure in the ideal gas law will be V_{eq}. Thus,

$$Q_{\text{Env to Sys}} = +P_{\text{eq}}\Delta V$$

As the gas expands at constant temperature, its volume increases and its pressure decreases. We use the ideal gas law $P_{\text{eq}} = nRT/V_{\text{eq}}$ to substitute for P_{eq} in the above:

$$Q_{\text{Env to Sys}} = +P_{\text{eq}}\Delta V = \frac{nRT}{V_{\text{eq}}}\Delta V$$

Solving for $\Delta V/V_{\text{eq}}$, we get

$$\frac{\Delta V}{V_{\text{eq}}} = \left(\frac{1}{nR}\right)\frac{Q_{\text{Env to Sys}}}{T}$$

Remember that this is an isothermal process in which positive energy transfer through heating is required to balance the negative work done by the environment as the gas expands. Because of the increasing volume, more space is available for the same number of gas particles at the same temperature. We can say that the entropy increases because

$$\frac{\Delta V}{V_{\text{eq}}} > 0$$

It appears that either $\Delta V/V_{\text{eq}}$ or Q/T could be used as macroscopic measures of the increase in disorder of the system—measures of the system's entropy change. The units of Q/T are J/K, the same as the units of entropy from the microscopic entropy definition using Eq. (16.2): $S_i = k_B \ln \Omega_i$. Note that the natural logarithm of any quantity is unitless, and k_B (Boltzmann's constant) has units J/K. Thus, the unit of statistical entropy is the same as the unit of Q/T. It seems reasonable to define the entropy change of a macroscopic system as Q/T.

Often, when energy is transferred between an environment and a system through heating, the temperature of the system changes. What temperature should be used to determine the system's entropy change? To deal with this, let's define the entropy change ΔS as a small amount of heating Q when the system is at a particular temperature T. Since this is a small amount of energy, the temperature of the system will not change significantly during the process. Alternatively, we can use the average temperature of the process to find the change in entropy. Here's the summary.

Macroscopic definition of entropy change The entropy change ΔS_{Sys} of a system during a process equals the ratio of the energy transferred to the system from the environment through heating Q divided by the system's temperature T_{Sys} while the process is occurring:

$$\Delta S_{\text{Sys}} = \frac{Q_{\text{Env to Sys}}}{T_{\text{Sys}}} \tag{16.3}$$

T_{Sys} is either the constant temperature of the system or the average temperature if the temperature changes, as long as the temperature changes smoothly.

TIP ▸ When you use Eq. (16.3), the temperature must have units of kelvin.

Consider a very simple example. You leave a 1-L water bottle outside on a sunny day. Let's estimate the change in the entropy of the water in the bottle. Assume that the mass of the water in the bottle is about 1 kg, and the temperature inside changes from

20 °C to 30 °C. To change the temperature of water, the environment must have transferred energy to it through heating:

$$\Delta U_{\text{Sys}} = Q_{\text{Env to Sys}} = mc\Delta T = \left(4200 \, \frac{\text{J}}{\text{kg} \cdot {}^{\circ}\text{C}}\right)(1 \text{ kg})(10 \, {}^{\circ}\text{C}) = 42{,}000 \text{ J}$$

The average temperature is 25 °C. Thus the entropy change is

$$\Delta S_{\text{Sys}} = \frac{Q_{\text{Env to Sys}}}{T_{\text{Sys}}} = \frac{42{,}000 \text{ J}}{(25 + 273) \text{ K}} = 140 \, \frac{\text{J}}{\text{K}}$$

Let's apply this definition to a process in which objects at different temperatures reach thermal equilibrium.

Entropy change of a system and its environment

Consider a process in which a warm object in the environment at temperature T_{Env} transfers a small amount of energy through heating $Q > 0$ to a system that is at a lower temperature T_{Sys}. What is the net entropy change for the system plus the environment, known as the entropy change of the universe: $\Delta S_{\text{Universe}} = \Delta S_{\text{Sys}} + \Delta S_{\text{Env}}$?

$$\Delta S_{\text{Sys}} = \frac{Q_{\text{Env to Sys}}}{T_{\text{Sys}}} = \frac{Q}{T_{\text{Sys}}}$$

and

$$\Delta S_{\text{Env}} = \frac{Q_{\text{Sys to Env}}}{T_{\text{Env}}} = \frac{-Q}{T_{\text{Env}}}$$

or

$$\Delta S_{\text{Universe}} = \frac{Q}{T_{\text{Sys}}} + \frac{(-Q)}{T_{\text{Env}}} = Q\left(\frac{1}{T_{\text{Sys}}} - \frac{1}{T_{\text{Env}}}\right) = Q\left(\frac{T_{\text{Env}} - T_{\text{Sys}}}{T_{\text{Sys}}T_{\text{Env}}}\right) > 0$$

The heating occurred because the environment was at a higher temperature than the system $T_{\text{Env}} > T_{\text{Sys}}$. But this also means that the magnitude of the system's entropy change was greater than the magnitude of the environment's entropy change. Thus, the entropy change of the system and its environment is positive; the entropy of the system and environment increased. This process occurred when objects of different temperature came in contact and achieved thermal equilibrium. This leads to yet another version of the second law of thermodynamics.

> **Second law of thermodynamics** During any process that involves the transfer of energy through heating, the net change in entropy of the system and its environment is always greater than zero.

EXAMPLE 16.4 **Mixing hot and cool water**

We add 0.50 kg of water at 70 °C to 0.50 kg of water at 30 °C. Estimate the net entropy change of all the water once the system has come to equilibrium.

Sketch and translate The process is sketched in the figure below. The warm water cools and the cool water warms.

Equal amounts of warm and cool water combine to form intermediate-temperature water.

Simplify and diagram Consider this as a two-system process; the warm water (the warm system) transfers energy to the cool water (the cool system) through heating. Assume that the combined system is isolated from its environment. Notice that in this process, the temperature of both systems continually changes. Therefore, we use the average temperature of each system as they undergo the cooling or warming process. Since there are equal amounts of water, the equilibrium temperature is 50 °C. The average temperature of the warm system is then

$$\frac{70 \, {}^{\circ}\text{C} + 50 \, {}^{\circ}\text{C}}{2} = 60 \, {}^{\circ}\text{C}$$

while the average temperature of the cool system is

$$\frac{30 \, {}^{\circ}\text{C} + 50 \, {}^{\circ}\text{C}}{2} = 40 \, {}^{\circ}\text{C}$$

Represent mathematically The thermal energy transferred from the warm water to the cool water is determined using Eq. (15.5):

$$Q_{\text{Warm to Cool}} = mc_{\text{water}}(50\,°C - 30\,°C)$$

Similarly, the negative heating of the warm water is

$$Q_{\text{Cool to Warm}} = mc_{\text{water}}(50\,°C - 70\,°C)$$

We now use Eq. (16.3) to add the entropy changes of the warm water and of the cool water:

$$\Delta S_{\text{Total}} = \Delta S_{\text{Cool}} + \Delta S_{\text{Warm}} \approx \frac{Q_{\text{Warm to Cool}}}{T_{\text{Cool}}} + \frac{Q_{\text{Cool to Warm}}}{T_{\text{Warm}}}$$

Solve and evaluate The specific heat of water is $c = 4180\ \text{J/kg} \cdot °C$. The magnitude of the energy transferred through heating is

$$Q_{\text{Warm to Cool}} = mc_{\text{water}}(50\,°C - 30\,°C)$$

$$= (0.50\ \text{kg})\left(\frac{4180\ \text{J}}{\text{kg} \cdot °C}\right)(20\,°C) = 41{,}800\ \text{J}$$

The cool to warm water heating is $-41{,}800$ J. We can now make an estimate of the total entropy change:

$$\Delta S_{\text{Total}} = \Delta S_{\text{Cool}} + \Delta S_{\text{Warm}}$$

$$\approx \frac{+41{,}800\ \text{J}}{(273 + 40)\ \text{K}} + \frac{-41{,}800\ \text{J}}{(273 + 60)\ \text{K}}$$

$$= +134\ \text{J/K} - 126\ \text{J/K} = +8\ \text{J/K}$$

Since the net entropy of the combined system increases, this process can occur spontaneously (the second law of thermodynamics.)

Suppose you started with all the water at 50 °C and waited for half the water to cool to 30 °C while the other half warmed to 70 °C (the reverse of the above process). This process is consistent with energy conservation (the first law of thermodynamics). The entropy change for such a process would be

$$\Delta S_{\text{Total}} = \Delta S_{\text{Cool}} + \Delta S_{\text{Warm}}$$

$$\approx \frac{-41{,}800\ \text{J}}{(273 + 40)\ \text{K}} + \frac{+41{,}800\ \text{J}}{(273 + 60)\ \text{K}}$$

$$= -134\ \text{J/K} + 126\ \text{J/K} = -8\ \text{J/K}$$

The entropy of the system would decrease. This process will not occur spontaneously (the second law of thermodynamics.)

Try it yourself Determine the entropy change of a 0.10-kg chunk of ice at 0 °C that melts and becomes a 0.10-kg puddle of liquid water at 0 °C. The heat of fusion of water is $3.35 \times 10^5\ \text{J/kg}$.

Answer   +120 J/K.

Entropy and complex organic species

Living beings develop from less complex forms to more complex forms; more complex means less entropy, because there is an order in the functioning of the parts of the system. Does this contradict what we have learned about entropy and the second law of thermodynamics? What about when an organism grows? Clearly, its organization and complexity increase.

The growth of an organism does not violate the second law of thermodynamics. The key reason is that the second law applies only to an isolated system. A growing organism is definitely not an isolated system; it interacts with the environment. If we consider the environment as part of the system, then the analysis is quite different.

Plant tissues, for example, absorb energy from the Sun and carbon in the air and water from the soil (**Figure 16.6**), initiating complicated processes that lead to the eventual formation of carbohydrates, amino acids, fatty acids, and thousands of other compounds by a myriad of other reactions. The formation of these complex useful compounds seems like a decrease in entropy and a violation of the second law of thermodynamics. In each reaction in the complex set of reactions, two or more atoms or molecules combine to form a more complex, lower entropy molecule. However, when these reactions occur, they release energy through heating into the environment. The corresponding entropy increase of the environment is over two times greater than the corresponding decrease due to the formation of the larger molecule. Photosynthesis produces a net increase in the entropy of the Earth system.

What about humans? A person grows and develops into a complex, highly ordered adult by consuming and converting complex molecules with low entropy, such as carbohydrates, fats, and proteins, into simpler molecules with higher entropy. Each year a healthy adult consumes about 500 kg, or a half-ton, of this low-entropy food. During consumption, over the course of a year a human transfers roughly two billion joules of energy to the environment, increasing its thermal energy, a form of energy that has little ability to do work. The entropy of the human may decrease, becoming more complex

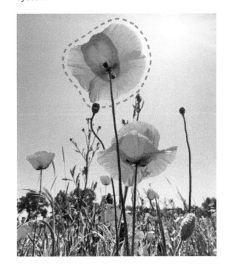

FIGURE 16.6 The flower is not an isolated system.

and organized, but the entropy of the environment increases significantly, more than enough to compensate.

In industrialized countries that burn low-entropy fuels such as petroleum and coal to meet their energy needs, the entropy increase of the environment is almost 100 times greater per person than in nonindustrialized countries. The scientific basis of the "energy crisis" rests with the second law of thermodynamics and the increase in entropy of the universe as a result of the conversion of useful energy to much less useful forms.

REVIEW QUESTION 16.3 Why doesn't the development of more and more complex life forms on Earth with lower entropy than less complex forms contradict the second law of thermodynamics?

16.4 Quantitative analysis of thermodynamic engines and pumps

We have learned that over time an isolated system's energy converts into less useful forms—for example, mechanical energy converts into thermal energy. Is it possible, though, to use the thermal energy of one system to do work on another system?

Suppose that a system has two "reservoirs" at different temperatures. A reservoir could be a pond of cool water, a flame that converts water into steam, or the burning fuel in the cylinder of a car engine. If the hot and cold reservoirs are in contact, thermal energy flows spontaneously from the warmer reservoir to the cooler one. We can harness this energy to do work on another system. Such a device is already familiar to us: the **thermodynamic engine** from Section 16.1. Even more surprising, we can create a device that makes a cool object even colder, such as a refrigerator.

We will begin by first examining in more depth the operation of a thermodynamic engine. We will then take a look at how a refrigerator works.

Quantitative analysis of a thermodynamic engine

A thermodynamic engine (**Figure 16.7a**) has three main parts: (1) a reservoir with hot material at temperature T_H, (2) some kind of working substance that can expand and contract and do useful mechanical work, and (3) a reservoir with cool material at a lower temperature T_C. The working substance undergoes thermodynamic processes that cause some desired effect—for example, doing useful work such as turning an electric generator. At the end of each cycle, the working substance returns to its starting state. It then repeats the sequence.

The hot reservoir is maintained at temperature T_H usually by burning some kind of fuel (for example, coal, natural gas, or nuclear fuel in a power plant, or the gasoline in a car engine). The working substance is usually a gas that can expand significantly to do work. The cold reservoir may be a pool of cold water that continually flows to transfer energy away from the working substance.

In the first step of the cycle, the hot reservoir transfers thermal energy to the working substance by heating $Q_{H\,to\,Sub}$. In the second step, the working substance, which is now hot and has expanded, pushes a piston or turns the turbine of a generator. Thus, the working substance does positive work on the environment $W_{Sub\,on\,Env}$. This means that the environment does work of the same magnitude with the opposite sign on the substance: $W_{Sub\,on\,Env} = -W_{Env\,on\,Sub}$. The same applies to heating. According to the second law of thermodynamics, it is impossible to use 100% of the energy transferred by heating from the hot reservoir into work done by the working substance. What happens to the rest of the energy?

FIGURE 16.7 Thermodynamic engine. (a) A schematic illustration. (b) A bar chart for a cycle. (c) An electric power plant is a thermodynamic engine.

(a)

1. The hot reservoir heats the working substance $Q_{H\,to\,Sub}$.

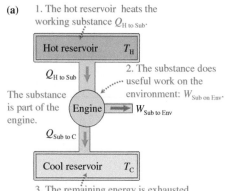

2. The substance does useful work on the environment: $W_{Sub\,on\,Env}$.

The substance is part of the engine.

3. The remaining energy is exhausted by heating the cold reservoir $Q_{Sub\,to\,C}$.

(b)

$$Q_{H\,to\,Sub} + W_{Env\,on\,Sub} + Q_{C\,to\,Sub} = \Delta U_{int\,Sub}$$

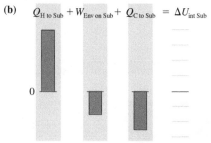

(c)

2. Turbine generator does work on the environment.

Working substance

Steam line

W

Pump

Cool reservoir

1. Fuel heats the working substance $Q_{H\,to\,Sub}$.

3. The working substance heats the cool water $Q_{Sub\,to\,C}$, causing the working substance to condense.

In the third step, the energy is transferred from the substance to the cold reservoir $Q_{\text{Sub to C}}$. Since the working substance is the system, this is negative heating of the substance by the cold reservoir $Q_{\text{C to Sub}}$. The working substance, now cooler, contracts. A compressor also helps in this contraction, doing some work on the substance but somewhat less work than the substance did on the environment in step 2. In other words, during step 3 the environment does positive work on the working substance; however, the net work done by the environment on the working substance during the whole process is negative.

In summary, we have positive heating of the substance by the hot reservoir $(Q_{\text{H to Sub}})$, the positive net work done by the working substance on the environment $(W_{\text{Sub on Env}})$, and the positive heating of the cold reservoir by the substance $(Q_{\text{Sub to C}})$. The bar chart in Figure 16.7b represents the process. Since the working substance returns to the same state at the end of each loop around the engine, the net internal energy change of the working substance is zero. Thus, we can put this into an equation form using the first law of thermodynamics as applied to the working substance:

$$Q_{\text{H to Sub}} + W_{\text{Env on Sub}} + Q_{\text{C to Sub}} = 0$$

The second and third terms in this equation are negative (see the bar chart). We can switch their subscripts and the signs of those two terms:

$$Q_{\text{H to Sub}} - W_{\text{Sub on Env}} - Q_{\text{Sub to C}} = 0$$

Thus, the net positive work that the substance does on the environment is

$$W_{\text{Sub on Env}} = Q_{\text{H to Sub}} - Q_{\text{Sub to C}} \qquad (16.4)$$

All of the quantities in the above equation have positive values. A schematic model of the thermodynamic engine in an electric power plant is shown in Figure 16.7c.

CONCEPTUAL EXERCISE 16.5   **Air-filled bottle as a thermodynamic engine**

In Section 16.1 we met a simple version of a thermodynamic engine. A closed, squished air-filled bottle is heated with a hair dryer. After being heated for some time, the bottle suddenly expands and lifts up a sugar cube. Then the bottle is placed in a cold-water bath. It sits there for a few seconds and then collapses, returning to its original state. Describe the processes that occur during one cycle using physical quantities and draw a P-versus-V graph for the cycle. Show on the graph for each step when the energy was transferred between the system and the environment.

Sketch and translate A sketch of the process was given in Figure 16.3. The system is the gas, which is the working substance in this thermodynamic engine; we therefore use subscript "Sub" for the gas. We can break the process into several steps. First, the gas is heated at constant volume. After that the volume of the bottle increases, and the pressure drops. The gas then cools at constant volume, and finally the volume decreases and pressure increases.

Simplify and diagram We assume that changes between the steps occur instantly. We also neglect the changes in the gravitational potential energy of the gas in the bottle. The P-versus-V graph is shown on the right, with the following steps:

1-2: Heating the gas at constant volume. Heating from the environment to the gas is positive.

2-3: Sudden expansion of the bottle. The gas pressure drops, and its volume increases. The work done on the gas is negative.

3-4: Cooling the gas at constant volume. Heating from the environment to the gas is negative.

4-1: Sudden compression of the bottle. The volume of the gas decreases, and the gas pressure increases. The environment does positive work on the gas.

Try it yourself Represent the process in the exercise with bar charts. Draw two bar charts—one for steps 1-2-3 (the expansion process) and one for steps 3-4-1 (the compression process).

Answer

Steps 1-2-3   Steps 3-4-1

$$Q_{\text{H to Sub}} + W_{\text{expansion}} = \Delta U_{\text{int Sub}}$$

$$Q_{\text{C to Sub}} + W_{\text{compression}} = \Delta U_{\text{int Sub}}$$

whole cycle is zero.

change of the internal energy of the system (the engine) through the

work done is positive and the heating is negative. Note that the total

is positive and the work done during the compression, the

See the bar charts in the figure below. During the expansion, the heating

We can also analyze the total work done *by the substance on the environment* in the above exercise quantitatively: $W_{\text{Sub on Env}} = W_{\text{expansion}} + W_{\text{compression}}$, where $W_{\text{compression}} < 0$. Recall from Chapter 15 that we can determine the work done by the environment on the gas (the substance in this case) as the area under the pressure-versus-time graph with the negative sign; therefore, the work that the substance does on the environment is equal to the area under the pressure-versus-time graph. In the case of the cyclic process, we can determine the total work by adding the area under the expansion part of the graph (**Figure 16.8a**) to the area under the compression part of the graph with a negative sign (Figure 16.8b). The result $W_{\text{Sub on Env}}$ **is equal to the area inside the *P*-versus-*V* cycle** (Figure 16.8c).

The **efficiency** of a thermodynamic engine is the work done by the working substance on the environment divided by the energy transferred from the hot reservoir to the working substance:

$$e = \frac{W_{\text{Sub on Env}}}{Q_{\text{H to Sub}}} \qquad (16.5)$$

We are dividing the useful work done by the engine by the energy supplied to it, in other words, the energy "spent" to run the engine.

FIGURE 16.8 Graphical representation of work done in one cycle of a thermodynamic engine.

EXAMPLE 16.6 ▶ **Efficiency of a thermodynamic engine**

In a car engine, gasoline is ignited by a spark plug to generate an explosion in a cylinder. Each explosion transfers 700 J of energy from the burning fuel to the gas in the cylinder. During the expansion (and later contraction) of the gas, the gas does a net 200 J of mechanical work on the environment (on a piston that pushes outward and helps propel the car). (a) How much energy is transferred through heating to the car's cooling system during this cycle? (b) What is the efficiency of the engine?

Sketch and translate The working substance is the gas (mostly air) inside the cylinder. The hot reservoir is the burning fuel in the cylinder that is ignited by the spark plug. The cooling system is the cold reservoir. One cycle of the process can be represented as shown in the figure at right. $Q_{\text{H to Sub}} = +700$ J (the energy transferred from the burning fuel to the gas inside the cylinder) and $W_{\text{Sub on Env}} = +200$ J (the hot air in the cylinder expands and contracts, doing positive work against the piston in the cylinder, which ultimately causes the wheels to turn.) We need to find the energy transferred through heating to the cooling system of the car and the efficiency of the engine.

Simplify and diagram We model the car's engine as a thermodynamic engine and represent the process with a bar chart in the figure below.

$$Q_{\text{H to Sub}} + W_{\text{Env on Sub}} + Q_{\text{C to Sub}} = \Delta U_{\text{int Sub}}$$

Represent mathematically Rearrange Eq. (16.4) to determine how much energy is transferred through heating to the car's cooling system:

$$W_{\text{Sub on Env}} = Q_{\text{H to Sub}} - Q_{\text{Sub to C}}$$

Simplifying the subscripts:

$$\Rightarrow W_{\text{Sub}} = Q_{\text{H}} - Q_{\text{C}}$$
$$\Rightarrow Q_{\text{C}} = Q_{\text{H}} - W_{\text{Sub}}$$

Then use Eq. (16.5) to determine the efficiency:

$$e = \frac{W_{\text{Sub}}}{Q_{\text{H}}}$$

Solve and evaluate Using the above equations, and noting that the substance does positive work on the environment (the environment does negative work on the system), we find that

$$Q_C = Q_H - W_{Sub} = +700 \text{ J} - 200 \text{ J} = +500 \text{ J}$$

Over 70% of the energy that the substance receives from the hot reservoir is moved to the cold reservoir. The engine's efficiency is

$$e = \frac{W_{Sub}}{Q_H} = \frac{200 \text{ J}}{700 \text{ J}} = 0.29$$

The result agrees with what you know from your everyday experience when you stand near a parked car when the engine is running. The car's exhaust system pumps hot gas into the atmosphere, carrying away energy that has served no useful purpose. The hood and radiator grill of the car are also warm, releasing even more energy to the environment.

Try it yourself Another car engine has the same efficiency as the car in this example. If the engine does work on its environment at a rate of 40 hp (1 hp = 746 J/s), how much work does the engine do on its environment each second, and how much energy through heating must be transferred from the burning fuel to the gas in the cylinders each second?

Answer

During this second, there might be as many as 100 explosions in the cylinders of the engine.

$$Q_{H \text{ to Sub}} = \frac{W_{Sub \text{ on Env}}}{e} = \frac{30{,}000 \text{ J}}{0.29} = 1.0 \times 10^5 \text{ J}$$

and

$$W_{Sub \text{ on Env}} = (40 \text{ hp})(1.0 \text{ s})\left(\frac{746 \text{ J/s}}{1 \text{ hp}}\right) = 30{,}000 \text{ J}$$

Maximum efficiency of a thermodynamic engine

The 29% efficiency of the automobile engine in the last example is a little depressing—most of the chemical potential energy stored in the gasoline ends up being transferred to the environment rather than being used to do work. Is it just that car engines are poorly designed, or is there a more fundamental reason why they are so inefficient? Recall that the efficiency of a thermodynamic engine is

$$e = \frac{W_{Sub \text{ on Env}}}{Q_{H \text{ to Sub}}}$$

where $W_{Sub \text{ on Env}}$ is the useful work done by the working substance in the engine on the environment and $Q_{H \text{ to Sub}}$ is the energy transferred to the engine by the hot reservoir.

According to the first law of thermodynamics, the net work done by the engine on its environment is the energy transferred from the hot reservoir to the working substance through heating minus the energy transferred from the working substance to the cold reservoir:

$$W_{Sub \text{ on Env}} = Q_{H \text{ to Sub}} - Q_{Sub \text{ to C}}$$

Thus, the efficiency of a thermodynamic engine can be written as

$$e = \frac{W_{Sub \text{ on Env}}}{Q_{H \text{ to Sub}}} = \frac{Q_{H \text{ to Sub}} - Q_{Sub \text{ to C}}}{Q_{H \text{ to Sub}}}$$

In England, when thermodynamic engines were being developed to pump water out of mines, scientists and engineers wondered if they could improve the efficiency of the engines. Sadi Carnot, using ideas of entropy and the second law of thermodynamics, showed that the maximum efficiency that a thermodynamic engine could possibly have was

$$e_{max} = \frac{T_H - T_C}{T_H} \tag{16.6}$$

where T_H is the temperature of the hot reservoir and T_C is the temperature of the cold reservoir (all temperatures are in units of kelvin). Reducing internal friction between the engine parts or attempting any other sort of optimization may improve efficiency somewhat but never above the value given by Eq. (16.6). The maximum efficiency is determined entirely by the temperatures of the hot and cold reservoirs.

Diesel engines are more efficient than standard gasoline engines.

The temperature of the exploding fuel in the cylinders of cars is about 200 °C, while the temperature of the exhaust system of the car is about 60 °C. Thus, the maximum efficiency of such an engine is

$$e_{max} = \frac{T_H - T_C}{T_H} = \frac{(200 + 273)\, K - (60 + 273)\, K}{(200 + 273)\, K} = \frac{473\, K - 333\, K}{473\, K} \approx 0.30$$

We know that gas mileage in cars can vary significantly in efficiency. The above number is for a relatively low-efficiency car—the number can vary from below 0.3 to above 0.4.

The equation on the previous page indicates ways we might increase the maximum efficiency of an engine.

1. Increase the temperature of the hot reservoir. For example, a pebble bed nuclear power plant reactor under development can operate at a temperature up to 1600 °C with efficiency up to 50% (supposedly more safely than traditional nuclear reactors). The "pebble bed" name comes from the shape of the fuel elements, which are tennis ball-sized pebbles. Diesel fuel is burned at a higher temperature than gasoline, making diesel car engines more efficient than gasoline engines.

2. Reduce the temperature of the cold reservoir. Doing so requires a more advanced cooling system. For example, turbocharged and supercharged car engines have a device known as an intercooler, a special radiator through which the compressed air passes and is cooled before it enters the cylinder.

Thermodynamic pumps (refrigerators or air conditioners)

What happens if we reverse the thermodynamic engine process so that the work done by the environment on the gas is positive? Then we obtain an engine in which positive work done on the working substance transfers thermal energy from the cool reservoir to the hot reservoir (**Figure 16.9a**). If we use an energy bar chart to analyze this new type of thermodynamic engine, we conclude that the cold reservoir gets cooler and the hot reservoir gets warmer (Figure 16.9b).

By reversing the thermodynamic engine process, we have invented a machine called a **thermodynamic pump**, sometimes known as a *heat pump*. A refrigerator is a type of thermodynamic pump. So is an air conditioner. As you can see from the schematic drawing in **Figure 16.10**, in a thermodynamic pump a working substance called a refrigerant takes thermal energy from a "cold reservoir" (position 1 in Figure 16.10). The substance moves through the pipes in the direction of the small blue and red arrows and delivers some of the thermal energy to a hot reservoir outside the pump (position 3 in the figure). The net effect is that thermal energy is transferred from the cool reservoir to the substance $(Q_{C\,to\,Sub})$, which transfers some of the thermal energy to the warmer air outside the pump $(Q_{Sub\,to\,H})$.

The substance can take energy from the cold reservoir that is at lower temperature and deliver it to the hot reservoir at higher temperature because it is not an isolated system. In addition to the two reservoirs at low and high temperatures, the substance passes through a motor (a compressor) that does positive work on it $(W_{Comp\,on\,Sub})$.

Let's consider in detail how a refrigerator works. The refrigerant is typically a fluid with a boiling point below 0 °C, such as a hydrofluorocarbon (HFC). The boiling temperature of HFCs is −29.9 °C; below this temperature most of the substance is in liquid form, while above this temperature it is a gas. The food compartment is the "cold reservoir," which transfers thermal energy to the substance, which runs through coils in the appliance. As the substance's temperature increases, it partially evaporates. As the substance flows through the inside of the refrigerator, more of it evaporates. Energy transfers from the food to the substance, leading to a decrease in the food temperature. As the substance leaves the interior of the refrigerator, it is warmer and much more of it is in the gaseous state than when it entered.

FIGURE 16.9 To transfer energy from a cooler reservoir to a warmer reservoir, work must be done.

(a)

(b) $W_{Comp\,on\,Sub} + Q_{C\,to\,Sub} + Q_{H\,to\,Sub} = \Delta U_{int\,Sub}$

FIGURE 16.10 A schematic drawing of a refrigerator.

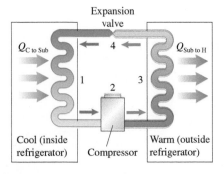

The refrigerant then passes through a compressor (position 2) that does positive work on the substance ($W_{\text{Comp on Sub}}$), adiabatically compressing it to higher internal energy and temperature ($Q = 0$, $\Delta V < 0$, and $P|\Delta V| + 0 = \Delta U_{\text{Sub}}$). The substance is now hotter than the environment outside the refrigerator (considered the "hot reservoir"). The substance passes through a condenser coil (position 3) on the outside of the refrigerator, where it transfers thermal energy to the outside environment ($Q_{\text{Sub to H}}$), which is the hot reservoir. At this point, the substance temperature decreases, and it condenses in part back from a gas to a liquid.

The condensed substance passes through a pressure-lowering device, the expansion valve (position 4). There, it expands adiabatically ($Q = 0$, $\Delta V > 0$, and $-P\Delta V + 0 = \Delta U_{\text{Sub}}$), causing its internal energy to decrease. At this point, even more of the substance condenses into a cool liquid, and this liquid reenters the refrigerator for another trip around the cycle.

The flow diagram and bar chart in Figure 16.9 represent the refrigerator (the thermodynamic pump). At the end of one cycle, the internal energy of the substance is unchanged—no change in the temperature or thermal energy. The process can be summarized mathematically as follows:

$$Q_{\text{C to Sub}} + W_{\text{Comp on Sub}} + Q_{\text{H to Sub}} = 0$$

The first two terms are positive and the latter is negative. We can reverse the subscripts on the latter term and change its sign to get an equation with all positive terms:

$$Q_{\text{C to Sub}} + W_{\text{Comp on Sub}} - Q_{\text{Sub to H}} = 0 \qquad (16.7)$$

Air conditioners operate in precisely the same way. The efficiencies of refrigerators and air conditioners are rated in terms of a **performance coefficient** K:

$$K_{\text{Ref}} = \frac{Q_{\text{C to Sub}}}{W_{\text{Comp on Sub}}} = \frac{Q_{\text{C to Sub}}}{Q_{\text{Sub to H}} - Q_{\text{C to Sub}}} \qquad (16.8)$$

The second law of thermodynamics limits this performance coefficient:

$$K_{\text{Ref max}} \leq \frac{T_{\text{C}}}{T_{\text{H}} - T_{\text{C}}} \qquad (16.9)$$

The temperatures must be in kelvin (K).

Using a thermodynamic pump to warm a house

A thermodynamic pump can also be used to warm a home in the winter. The only difference is that the outside of the house (cold in the winter) plays the role of the inside of the refrigerator. Energy is transferred through heating from the cooler air outside (the cold reservoir at temperature T_{C}) to the warmer air inside (the warm reservoir at temperature T_{H}). A compressor plugged into a wall outlet does work on the working substance (a refrigerant) to make the transfer possible. A sketch of this operation is shown in Figure 16.11a. The process is represented by a flow diagram in Figure 16.11b and by a bar chart in Figure 16.11c. The system is the working substance (like the refrigerant). The goal of this device is to warm something (the house) rather than cool something. The compressor does the work that makes possible the transfer of energy through heating to the house. Thus, the performance coefficient $K_{\text{Home pump}}$ is the ratio of the energy provided through heating to the house and the work that must be done for this energy transfer to occur:

$$K_{\text{Home pump}} = \frac{Q_{\text{Sub to H}}}{W_{\text{Comp on Sub}}} \leq \frac{T_{\text{H}}}{T_{\text{H}} - T_{\text{C}}} \qquad (16.10)$$

Again, the temperatures must be in kelvin (K). From Eq. (16.10), you see that the performance coefficient is always larger than 1 (typical values range between 3 and 5).

FIGURE 16.11 Thermodynamic pump.

(a)

(b)

(c) $W_{\text{Comp on Sub}} + Q_{\text{C to Sub}} + Q_{\text{H to Sub}} = \Delta U_{\text{int Sub}}$

The term "thermodynamic pump" does not mean that any sort of physical material is being moved around. It is energy that is being "pumped" from a cooler region to a warmer region. This transfer is possible because of the work being done on the working substance by the environment—by the compressor.

QUANTITATIVE EXERCISE 16.7 **Warming a house in the winter**

A thermodynamic pump transfers 14,000 J of energy through heating into a house. The air temperature outside is $-10\,°C$ and inside is $20\,°C$. Determine the maximum performance coefficient of the thermodynamic pump and the minimum work that the pump's compressor must do on the working substance to cause this energy transfer.

Represent mathematically Energy $Q_{\text{Sub to H}} = +14,000\,\text{J}$ is transferred from the working substance through heating to the house. Our goal is to determine the maximum performance coefficient of the pump and the work $W_{\text{Comp on Sub}}$ that the pump's compressor must do if the pump is operating at maximum efficiency. The maximum performance coefficient is given by Eq. (16.10):

$$K_{\text{Pump max}} = \frac{T_{\text{H}}}{T_{\text{H}} - T_{\text{C}}}$$

We can now determine the minimum work that must be done using the definition of the performance coefficient.

$$K_{\text{Pump}} = \frac{Q_{\text{Sub to H}}}{W_{\text{Comp on Sub}}} \Rightarrow W_{\text{Comp on Sub}} = \frac{Q_{\text{Sub to H}}}{K_{\text{Pump max}}}$$

Solve and evaluate

$$K_{\text{Pump max}} = \frac{T_{\text{H}}}{T_{\text{H}} - T_{\text{C}}} = \frac{(273 + 20)\,\text{K}}{(273 + 20)\,\text{K} - (273 - 10)\,\text{K}}$$
$$= 9.8$$

$$W_{\text{Comp on Sub}} = \frac{Q_{\text{Sub to H}}}{K_{\text{Pump max}}} = \frac{14,000\,\text{J}}{9.8} = 1430\,\text{J}$$

The units are correct for the respective quantities, and the numbers are reasonable. Unfortunately, thermodynamic pumps do not usually work at maximum performance, but they are still an efficient way to warm up a house in the winter and cool it in the summer (when behaving as an air conditioner).

Try it yourself A refrigerator has a performance coefficient of 4. How much work must the compressor do to transfer 4000 J of energy from inside the refrigerator to the outside air?

Answer   1000 J.

REVIEW QUESTION 16.4 Why is there no contradiction between the second law of thermodynamics and the operation of refrigerators and air conditioners that transfer thermal energy from cooler objects to warmer objects?

Summary

Statistical definition of entropy The entropy S of an energy state is a measure of the probability of that state occurring. The greater the entropy, the less useful the energy in the system is for doing work. The entropy depends on how many ways we can construct the given state, the count Ω_i of a state. (Section 16.2)

Microstate arrangements	Count Ω_i	Entropy S_i
① ②③④	4	$1.4k_B$
② ①③④		
③ ①②④		
④ ①②③		

$$\Omega_i = \frac{n!}{n_1! n_r!} \quad \text{Eq. (16.1)}$$

$$S_i = k_B \ln \Omega_i \quad \text{Eq. (16.2)}$$

where

$$k_B = 1.38 \times 10^{-23}\,\text{J/K}$$

Thermodynamic definition of entropy A system's entropy changes by ΔS_Sys when energy is transferred through heating $Q_\text{Env to Sys}$ from the environment while the system is at a temperature T_Sys. (Section 16.3)

$$\Delta S_\text{Sys} = \frac{Q_\text{Env to Sys}}{T_\text{Sys}} \quad \text{Eq. (16.3)}$$

where T_Sys is in kelvin

Second law of thermodynamics The entropy of any system plus its environment increases during any irreversible process. The increase of entropy is an indicator that energy was converted from a more organized form to a less organized form. The entropy in reversible processes remains constant. (Sections 16.2 and 16.3)

Initial — $m = 0.5$ kg, 70 °C; $m = 0.5$ kg, 30 °C. Final — $m = 1.0$ kg, 50 °C; $S_f > S_i$

During any irreversible process

$$\Delta S_\text{Sys} + \Delta S_\text{Env} > 0$$

Thermodynamic engines and pumps Energy transferred through heating $Q_\text{H to Sub}$ from a hot reservoir at temperature T_H to a working substance allows the substance to do work $W_\text{Sub on Env}$ on the environment. Energy is then transferred through heating from the working substance to a cold reservoir $Q_\text{Sub to C}$ at temperature T_C. During one cycle of operation, the working substance's internal energy does not change, $\Delta U_\text{int Sub} = 0$. The maximum efficiency of this process e_max is determined only by T_H and T_C. A thermodynamic pump is the reverse of this process. (Section 16.6)

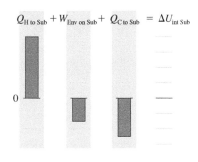

$$Q_\text{H to Sub} + W_\text{Env on Sub} + Q_\text{C to Sub} = \Delta U_\text{int Sub}$$

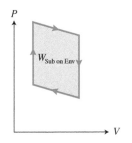

Thermodynamic engine:
$$W_\text{Sub on Env} = Q_\text{H to Sub} - Q_\text{Sub to C} \quad \text{Eq. (16.4)}$$

All quantities are positive.

$$e = \frac{W_\text{Sub on Env}}{Q_\text{H to Sub}} \quad \text{Eq. (16.5)}$$

$$e_\text{max} = \frac{T_H - T_C}{T_H} \quad \text{Eq. (16.6)}$$

where T is in kelvin

Refrigeration:
$$Q_\text{C to Sub} + W_\text{Comp on Sub} - Q_\text{Sub to H} = 0 \quad \text{Eq. (16.7)}$$

$$K = \frac{Q_\text{C to Sub}}{W_\text{Comp on Sub}} \leq \frac{T_C}{T_H - T_C} \quad \text{Eqs. (16.8) and (16.9)}$$

Home heating:
$$K = \frac{Q_\text{Sub to H}}{W_\text{Comp on Sub}} \leq \frac{T_H}{T_H - T_C} \quad \text{Eq. (16.10)}$$

Questions

Multiple Choice Questions

1. Which of the following processes is reversible?
   (a) The room temperature collision of two atoms
   (b) The collision of two cars
   (c) The collision of a pool cue with a pool ball
   (d) The collision of two billiard balls

2. In physics the collision of billiard balls is usually considered to be elastic. What *observable* reason could you have for not choosing the collision of billiard balls as an example of a reversible process?
   (a) You hear sound produced by the collision.
   (b) There is change in the internal energy of the billiard balls.
   (c) All macroscopic processes are irreversible.

3. Why can't all of the thermal energy in a system be converted into mechanical work done on an environment?
   (a) The environment would do equal negative work on the system.
   (b) Energy would not be conserved.
   (c) The disordered energy in the system cannot be completely converted into ordered energy that can do work.

4. When driving a car (the system), what object does work on the car to make it move at constant speed in the presence of air resistance and rolling friction?
   (a) The engine
   (b) The surface of the road
   (c) The fuel in the car's tank

5. The law of energy conservation says that energy is always conserved; it can only be transferred from one system to another or converted from one form to another. Why then do environmentalists argue for energy conservation?
   (a) Because they are talking about isolated systems
   (b) Because the environmentalists mean the energy that is useful for doing work
   (c) Because when a car burns fuel, its energy is used up

6. If you leave a refrigerator door open for a long time, the temperature in the room does what?
   (a) Increases   (b) Decreases   (c) Stays the same

7. Entropy can be calculated using which of the following expressions?
   (a) Q/T (b) $(T_H - T_C)/T_H$ (c) $\Delta Q/T$
   (d) None of these

8. In what way does human life contradict the second law of thermodynamics?
   (a) The law does not apply to living organisms.
   (b) The human body is an isolated system.
   (c) When a human body interacts with the environment, the entropy of the body decreases.
   (d) There is no contradiction.

9. When a drop of ink enters a glass of water and spreads out, the entropy of the ink-water system does what?
   (a) Increases   (b) Decreases   (c) Stays the same

10. Choose the best reason why the following statement is incorrect: When you clean your room and decrease its disorder, you violate the second law of thermodynamics.
    (a) The room is an isolated system. Entropy does not change.
    (b) Rooms are macroscopic objects; the second law applies only to molecules.
    (c) The room is not an isolated system. Your work leads to considerable entropy increase.

11. In which ideal gas process does the entropy of the gas remain constant?
    (a) An isothermal process
    (b) An isobaric process
    (c) An adiabatic process
    (d) An isochoric process
    (e) Entropy changes in all these processes.

12. Which of the following changes will always increase the efficiency of a thermodynamic engine? Choose all correct statements.
    (a) Decrease the hot reservoir temperature T_H while keeping the cool reservoir temperature T_C constant.
    (b) Increase T_H while keeping T_C constant.
    (c) Decrease T_C while keeping T_H constant.
    (d) Increase T_C while keeping T_H constant.
    (e) Increase the difference $T_H - T_C$.
    (f) Decrease the ratio T_C/T_H.

13. Entropy change is easier to determine for processes in which the temperature of the system is constant. Which of the following processes are of this type? Choose all correct statements.
    (a) Some ice is melting in a cup in a warm room. The system is the ice.
    (b) Some ice is melting in a cup in a warm room. The system is the ice and the cup.
    (c) A box is slowing down due to friction with a rough table. The system is the box and the table.
    (d) Some alcohol is boiling in a beaker over a burner. The system is the alcohol in the beaker.
    (e) An ideal gas in a syringe is very slowly compressed while the syringe is held in an ice bath. The system is the gas.
    (f) An ideal gas is in a closed, rigid container. Energy is transferred to the gas at a constant rate. The system is the gas.

Conceptual Questions

14. Describe five everyday examples of processes that involve increases in entropy. Be sure to state all parts of the system and environment involved in these entropy increases.

15. A cup of hot coffee sits on your desk. Assume that some of the thermal energy in the coffee can be converted to gravitational potential energy of the coffee-Earth system—enough so that the coffee rises up above the rim of the cup. Estimate the drop in temperature needed to accomplish this energy conversion. Why does this conversion not occur?

16. In terms of the statistical definition of entropy, why is there a good chance that the entropy due to the distribution of five atoms in a box can decrease, whereas the chance of an entropy decrease for redistribution of 10^6 atoms in a box is negligible?

17. The entropy of the molecules that form leaves on a tree decreases in the spring of each year. Is this a contradiction of the second law of thermodynamics? Explain.

18. Give three examples of processes, other than those described in this book, in which potentially more organized forms of energy are converted to less organized forms of energy.

Problems

Below, **BIO** indicates a problem with a biological or medical focus. Problems labeled **EST** ask you to estimate the answer to a quantitative problem rather than derive a specific answer. Asterisks indicate the level of difficulty of the problem. Problems with no * are considered to be the least difficult. A single * marks moderately difficult problems. Two ** indicate more difficult problems.

16.1 Irreversible processes

1. * **BIO** **Types of energy and reversibility of a process** Describe the types of energy that change, the work done on the system, and the energy transferred through heating during the following processes. Indicate whether a reverse process can occur. (a) Water at the top of Niagara Falls cascades onto the blades of an electric generator near the bottom of the falls, rotating the blades and generating an electric current that causes a lightbulb to glow. The water, generator, lightbulb, and Earth are the system. (b) Each second, your body converts 100 J of metabolic energy (converting complex molecules from food) to thermal energy transferred to the air surrounding your body. The system is your body and the surrounding air. (c) The hot gas in a cylinder pushes a piston, which causes the blades of an electric generator to turn, which in turn causes a lightbulb to glow briefly. The system is the original hot gas (which cools while pushing the piston), the generator, and the lightbulb (which first glows and then stops glowing and cools down).

2. * For the following processes, choose the initial and final states and describe the process using the physical quantities internal energy, work, and heating. Explain why the process is irreversible. (a) A large foam ball is moving vertically up at speed v and reaches a maximum height h' somewhat less than $\sqrt{v^2/2g}$. The ball, Earth, and air are the system. (b) Two cups of water, one cold and the other hot, are mixed in an insulated bowl. The mixture reaches an intermediate temperature. The water in the cups is the system.

3. **Hourglass** An hourglass starts with all of the sand in the top bulb. During the next hour, the sand slowly leaks into the bottom bulb. Describe the energy changes in a system that includes the glass, sand, and Earth. Is this a reversible or irreversible process? Explain.

4. * **Car hits tree** Your car slides on ice and runs into a tree, causing the front of the car to become slightly hotter and crumpled. The car and ice are the system. Indicate what object does the work on the system. Indicate whether heating occurs. Identify the types of energy that change. Are these quantities positive or negative? Explain.

5. **BIO** **EST** **Human metabolism** A 60-kg person consumes about 2000 kcal of food in one day. If 10% of this food energy is converted to thermal energy and cannot leave the body, estimate the temperature change of the person. Note: 1 kcal = 4180 J. Is this a reversible or irreversible process?

16.2 Entropy: the microscopic approach

6. (a) Identify all of the macrostate distributions for five atoms located in a box with two halves. (b) Determine the number of microstates for each macrostate. (c) Determine the entropy of each state.

7. * Repeat the previous problem for a system with six atoms.

8. * Determine the ratio of the number of microstates (count) of a system of eight atoms when in a macrostate with four atoms on the left side of a container and four on the right side and when in a macrostate with seven atoms on the left and one on the right. Which state has the greater entropy? Explain.

9. * **Person lost on island** The probability that a lost person wandering about an island will be on the north part is one-half and on the south part is also one-half. (a) Determine the probability that three lost people wandering about independently will all be on the south half. (b) Repeat part (a) for the probability of six lost people all being on the south half of the island.

10. * **Parachutists landing on island** Parachutists have an equal chance of landing on the south half of a small island or the north half. If eight parachutists jump at one time, what is the ratio of the probability that six land on the north half and two on the south half to the probability that four land on each half?

11. * Determine the ratio of the counts of a system of 20 atoms when in a macrostate with 10 atoms on the left half of a box and 10 on the right half and when in a macrostate with 18 atoms on the left and 2 on the right. (b) Do the same for a system with 10 coins for the states with 5 coins on the left and 5 on the right compared to 9 coins on the left and 1 on the right. (c) When you compare your answers to parts (a) and (b), what do you infer about a similar ratio for a system with 10,000 atoms?

12. * Nine numbered balls are dropped randomly into three boxes. The numbers of balls falling into each box are labeled n_1, n_2, and n_3. (a) Identify five of the many possible arrangements or macrostates of the balls. (b) Determine the ratio of the count for the equal distribution ($n_1 = 3$, $n_2 = 3$, and $n_3 = 3$) and for the 0, 0, 9 distribution. (c) Determine the ratio of the count for the equal distribution and for the 2, 3, 4 distribution. Note: The count is given by

$$\Omega = \frac{n!}{n_1! n_2! n_3!}$$

where n is the total number of balls.

13. * **Rolling dice** Two dice are rolled. Macrostates of these dice are distinguished by the total number for each roll (that is, 2, 3, 4, . . . , 12). (a) Determine the number of microstates for each macrostate. For example, there are three microstates for macrostate 4: (2, 2), (3, 1), and (1, 3). (b) What is the macrostate with greatest entropy? (c) What is the macrostate with least entropy? Explain.

14. * (a) Apply your knowledge of probability to explain why a drop of food coloring in a glass of clear water spreads out so that all of the water has an even color after some time. (b) Discuss whether after the food coloring spreads evenly in a glass of clear water it could condense back to an original droplet.

15. Explain using your knowledge of probability why a gas always occupies the entire volume of its container.

16.3 Entropy: the macroscopic approach

16. * **EST** Estimate the total change in entropy of two containers of water. One container holds 0.1 kg of water at 70 °C and is warmed to 90 °C by heating from contact with the other container. The other container, also holding 0.1 kg of water, cools from 30 °C to 10 °C. Is this energy transfer process allowed by the first law of thermodynamics? By the second?

17. * **EST** (a) You add 0.1 kg of water at 0 °C to 0.3 kg of iced tea at 70 °C. Determine the final temperature of the mixture after it reaches equilibrium. The specific heat of iced tea is the same as water. (b) Estimate the entropy change of this system during this process. Is it allowed by the second law of thermodynamics? Explain.

18. * **Entropy change of a house** A house at 20 °C transfers 1.0×10^5 J of thermal energy to the outside air, which has a temperature of −15 °C. Determine the entropy change of the house-outside air system. Is this process allowed by the second law of thermodynamics?

19. ** **Barrel of water in cellar in winter** A barrel containing 200 kg of water sits in a cellar during the winter. On a cold day, the water freezes, releasing thermal energy to the room. This energy passes from the cellar to the outside air, which has a temperature of −20 °C. Determine the entropy change for this process if the cellar remains at 0 °C.

20. * **EST** (a) Determine the final temperature when 0.100 kg of water at 10 °C is added to 0.300 kg of soup at 50 °C. What assumptions did you make? (b) Estimate the entropy change of this water-soup system during the process. Does the second law of thermodynamics allow this process?

21. * A 5.0-kg block slides on a level surface and stops because of friction. Its initial speed is 10 m/s and the temperature of the surface is 20 °C. Determine the entropy change of the block, which is the system in this process.

22. ** A 5.0-kg block slides from an initial speed of 8.0 m/s to a final speed of zero. It travels 12 m down a plane inclined at 15° with the horizontal. Determine the entropy change of the block-inclined plane-Earth system for this process if originally the block and the inclined plane were at 27 °C. Why did the block stop?

16.4 Quantitative analysis of thermodynamic engines and pumps

23. **Maximum efficiencies** Determine the maximum efficiencies of the thermodynamic engines described below. (a) Burning coal heats the gas in the turbine of an electric power plant to 700 K. After turning the blades of the generator, the gas is cooled in cooling towers to 350 K. (b) An inventor claims to have a thermodynamic engine that attaches to a car's exhaust system. The temperature of the exhaust gas is 90 °C and the temperature of the output of this proposed heat engine is 20 °C. (c) Near Bermuda, ocean water is about 24 °C at the surface and about 10 °C at a depth of 800 m.

24. * **BIO** **Efficiency of woman walking** A 60-kg woman walking on level ground at 1 m/s metabolizes energy at a rate of 230 W. When she walks up a 5° incline at the same speed, her metabolic rate increases to 370 W. Determine her efficiency at converting chemical energy into gravitational potential energy.

25. * **Nuclear power plant** A nuclear power plant operates between a high-temperature heat reservoir at 560 °C and a low-temperature stream at 20 °C. (a) Determine the maximum possible efficiency of this thermodynamic engine. (b) Determine the heating rate (J/s) from the high-temperature heat reservoir to the power plant so that it produces 1000 MW of power (work/time).

26. ** A cyclic process involving 1 mole of ideal gas is shown in Figure P16.26. (a) Determine the work done on the gas by the environment during each step of the cycle. (b) Determine the net work done on the gas $W_{\text{Env on Gas}}$. (c) Find the work that the gas does on the environment; it is equal to $W_{\text{Gas on Env}} = -W_{\text{Env on Gas}}$. (d) Use the ideal gas law to determine the temperature of the gas at each corner of the process (1, 2, 3, and 4). (e) Use the temperatures found in (d) to determine the thermal energy of the gas at each corner of the process and the change in internal thermal energy during each step of the process. (f) Use the results recorded in the previous parts and the first law of thermodynamics to determine the heating during each step of the process. (g) Determine the efficiency of the process.

FIGURE P16.26

27. ** A cyclic process involving 1 mole of ideal gas is shown in Figure P16.27 (a) Determine the work done on the gas by the environment during each step of the cycle. (b) Determine the net work done on the gas $W_{\text{Env on Gas}}$. (c) Find the work that the gas does on the environment equal to $W_{\text{Gas on Env}} = -W_{\text{Env on Gas}}$. (d) Use the ideal gas law to determine the temperature of the gas at each corner of the process (1, 2, and 3). (e) Use the temperatures found in (d) to determine the thermal energy of the gas at each corner of the process and the change in internal thermal energy during each step of the process. (f) Use the results recorded in the previous parts and the first law of thermodynamics to determine the heating during each step of the process. (g) Determine the efficiency of the process.

FIGURE P16.27

28. **Home thermodynamic pump** A heat pump collects thermal energy from outside air at 5 °C and delivers it into a house at 40 °C. (a) Determine the maximum coefficient of performance (the maximum coefficient is determined in the same way as for a thermodynamic engine). (b) If the motor of the heating pump uses 1000 J of electrical energy to do work during a certain time interval, how much thermal energy is delivered into the house through heating, assuming the heating pump works at the maximum coefficient of performance? (c) Repeat (b) for a coefficient of performance of 2.0.

29. ** **Ice-making machine** An ice-making machine needs to convert 0.20 kg of water at 0 °C to 0.20 kg of ice at 0 °C. The room temperature surrounding the ice machine is 20 °C. (a) How much thermal energy must be removed from the water? (b) Determine the minimum work needed to extract this energy by the ice-making machine. (c) How much energy is deposited in the room?

30. * **Automobile engine** An automobile engine has a power output for doing work of 150 kW (about 200 hp). The efficiency of the engine is 0.32. Determine the heating input per second to the engine by burning gasoline and the heating rate of the engine to the environment.

31. **Diesel car engine** A diesel engine in a car does 1000 J of work due to 2800 J of heating caused by the combustion of diesel fuel in its cylinders. Determine the efficiency of the engine and the thermal energy emitted by the engine to the environment.

32. Rank the engines that operate with the following conditions in order of decreasing efficiency: engine 1: $T_H = 800$ K, $T_C = 300$ K; engine 2: $T_H = 700$ °C, $T_C = 30$ °C; engine 3: $T_H - T_C = 350$ °C, $T_C = -5$ °C; engine 4: $T_H = 5T_C$.

33. **Nuclear power plant** A nuclear power plant generates electric energy at a rate of 500 MW. The energy transfer rate to the electric generator from the high-temperature nuclear fuel (the hot reservoir) is 1200 MW. Determine the efficiency of the power plant and the rate at which the working substance in the plant transfers energy through heating to the cold water (the cool reservoir).

34. * **Nuclear power plant** A nuclear power plant warms water to 500 °C and emits it at 100 °C. You want to get work done by the plant at a rate of 1.0×10^9 J/s. Determine the rate at which the nuclear fuel must provide energy to the working substance and the rate of thermal energy exhausted from the plant to the environment.

General Problems

35. * **BIO** **EST** Humans and warm-blooded animals maintain a constant body temperature, which is typically higher than their environment. (a) Estimate the maximum efficiency of a thermodynamic engine that is powered by this temperature difference in moderate climate conditions (the living organism is the hot reservoir and the environment is the cold reservoir). (b) Determine the output power of such an engine if the rate at which the energy is transferred from the living organism to the engine is 100 W.

36. ** The following equations represent the four parts (A, B, C, and D) of a cyclic process with a gas. In this case we consider the work done by the system on the environment and consequently $Q - W = \Delta U_{\text{int}}$.

$$W = (3.0 \times 10^5 \text{ N/m}^2)(0.020 \text{ m}^3 - 0.010 \text{ m}^3) + 0$$
$$+ (1.0 \times 10^5 \text{ N/m}^2)(0.010 \text{ m}^3 - 0.020 \text{ m}^3) + 0$$

$$\Delta U_{\text{int}} = (3/2)(1.0 \text{ mole})(8.3 \text{ J/mole} \cdot \text{K})[(700 \text{ K} - 360 \text{ K})$$
$$+ (480 \text{ K} - 700 \text{ K}) + (240 \text{ K} - 480 \text{ K})$$
$$+ (360 \text{ K} - 240 \text{ K})]$$

$$Q = Q_A + Q_B + Q_C + Q_D$$

(a) Draw a P-versus-V graph for the process with labeled axes (including a scale). (b) Determine the net change in the internal energy during the entire cycle. (c) Determine the heating of the system for each of the four parts of the process.

37. * A thermodynamic engine operates between two temperatures that differ from each other by 200 °C. Assuming that the engine operates at a maximum possible efficiency of 0.15, determine the temperatures of its hot and cold reservoirs.

38. * A refrigerator transfers 700 J of thermal energy every second to a room while 580 J of thermal energy is transferred every second from its cool reservoir. The inside of the refrigerator is 4 °C and the temperature in the room is 23 °C. Determine (a) the performance coefficient of the refrigerator and (b) the power of the refrigerator engine.

Reading Passage Problems

Fuel used to counter air resistance The resistive drag force that the air exerts on an automobile is

$$F_{\text{Air on Car}} = (1/2)CA\rho v^2$$

where C is a drag coefficient that depends on how streamlined the car is, ρ is the density of the air, A is the cross-sectional area of the car along its line of motion, and v is the car's speed relative to the air. Consider the effect of air resistance on fuel consumption in a typical car during a 160-km (100-mi) trip while traveling at 22 m/s (50 mi/h). The resistive force exerted by the air on the car is about 200 N. Thermal energy produced due to this resistive force during the trip is

$$\Delta U_{\text{thermal}} = F_{\text{Air on Car}}d = (200\ \text{N})(1.6 \times 10^5\ \text{m})$$
$$= 3.2 \times 10^7\ \text{J}$$

One liter of gas releases about 3.25×10^7 J of energy. So the gasoline energy that is equivalent to the resistive thermal energy produced during this trip is

$$(3.2 \times 10^7\ \text{J})/(3.25 \times 10^7\ \text{J/L}) = 0.98\ \text{L} = 0.26\ \text{gallons}$$

In summary, you use about one-quarter gallon of gasoline to counter air resistance during the 160-km trip. Unfortunately, the car is only about 13% efficient, where

$$\text{Efficiency} = \frac{\text{useful energy output}}{\text{energy input}} = 0.13$$

Consequently, your useful output of 0.98 liters requires an energy input of $(0.98\ \text{L})/0.13 = 7.5\ \text{L}$ of gas to overcome this air resistance. This is about half of the gasoline a typical car consumes when driving at a steady speed for 160 km. The other half is needed to overcome rolling resistance.

39. Assuming a 200-N drag force when traveling at 22 m/s through air of density 1.3 kg/m³, what is the closest value to the product CA in the drag force equation for the vehicle?
 (a) 0.50 m² (b) 0.62 m² (c) 0.86 m²
 (d) 1.1 m² (e) 1.5 m²

40. At 22 m/s the magnitude of the resistive force that air exerts on the car is about 200 N. Which answer below is closest to the magnitude of the drag force when the car is traveling at 31 m/s?
 (a) 100 N (b) 140 N (c) 280 N
 (d) 400 N (e) 520 N

41. The amount of fuel used to counter air resistance should do what?
 (a) Increase in proportion to the speed squared
 (b) Increase in proportion to the speed
 (c) Be the same independent of the speed
 (d) Decrease in proportion to the inverse of the speed
 (e) Decrease in proportion to the inverse of the speed squared

42. The 200-N resistive force of the air in this problem is closest to which answer below?
 (a) 800 lb (b) 90 lb (c) 60 lb
 (d) 45 lb (e) 30 lb

43. Why is the resistive force of the air on a Hummer H3 traveling at 22 m/s about 600 N instead of 200 N?
 (a) The Hummer has a bulky, less streamlined shape.
 (b) The Hummer has a greater cross-sectional area along the line of motion.
 (c) The Hummer has greater mass.
 (d) a and b
 (e) a, b, and c

44. The value of CA for a Ford Escape Hybrid is 1.08 m². Which answer below is closest to the drag force on this car when traveling at 22 m/s?
 (a) 130 N (b) 180 N (c) 270 N
 (d) 350 N (e) 440 N

17

Electric Charge, Force, and Energy

- How does a photocopier work?
- Why is your hair attracted to a plastic comb after combing?
- How is a metal soda can different from a plastic bottle on a microscopic level?

When you make a photocopy, the toner inside the copier reproduces your original image on a new sheet of paper. How does the toner stick to the paper, and how does the toner "know" where to stick? To attract small particles of toner to its drum, a photocopier uses a fundamental mechanism that is similar to the one that causes a rubbed balloon to stick to your sweater. This mechanism involves a new type of force that we will learn about in this chapter.

BE SURE YOU KNOW HOW TO:

- Identify a system and construct a force diagram for it (Section 3.1).
- Identify a system and construct an energy bar chart for it (Section 7.2).
- Convert a force diagram and an energy bar chart into a mathematical statement (Sections 3.3 and Section 7.6).

IN CHAPTERS 12–16, we learned that all objects are made of tiny particles. What holds these particles together? For that matter, what holds the particles that make up the paper in this textbook together? The interaction responsible for holding particles together is the same interaction that makes toner stick to a copier drum and a balloon stick to your sweater, as well as many other phenomena that we observe in our everyday world.

17.1 Electrostatic interactions

If you rub a balloon with a wool sweater (**Figure 17.1a**) and then bring the sweater near the balloon, the sweater attracts the balloon (Figure 17.1b). If you bring a second balloon rubbed the same way near the first balloon, they repel each other (Figure 17.1c). Let's investigate this phenomenon in detail, in Observational Experiment **Table 17.1**.

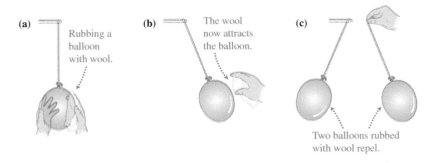

(a) Rubbing a balloon with wool.

(b) The wool now attracts the balloon.

(c) Two balloons rubbed with wool repel.

FIGURE 17.1 Rubbing different materials causes an interaction between them.

OBSERVATIONAL EXPERIMENT TABLE 17.1   Experimenting with rubbed objects

VIDEO
OET 17.1

Observational experiment	Analysis

Experiment 1. Hang two small balloons from two thin strings. Rub each balloon with felt. The balloons repel each other and settle into the position shown at right. Bringing the balloons closer to each other increases the angle between the strings.

The force diagram shows that the balloons are in the equilibrium position. To settle in the position shown at left, they must exert a repulsive force on each other. To explain the increase in the angle between the strings, we need to assume that as the balloons are brought closer together, the forces that they exert on each other increase in magnitude.

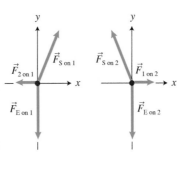

Experiment 2. Bring the felt used to rub the balloons close to one of the rubbed balloons. The balloon is attracted to the felt. The attraction strengthens as the felt is held closer.

Felt used to rub balloon

Rubbed balloon

The felt exerts an unknown attractive force on the rubbed balloon. The force strengthens as the felt gets closer to the balloon.

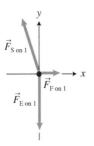

Experiment 3. Repeat Experiment 1, only this time rub the two small balloons with plastic wrap. The balloons repel. Bringing the balloons closer to each other increases the angle between the strings.

Each balloon exerts an unknown repulsive force on the other balloon. This force increases in magnitude as the balloons get closer to each other.

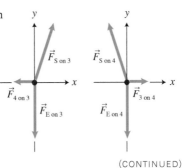

(CONTINUED)

Observational experiment		Analysis
Experiment 4. Bring the plastic wrap close to one of the balloons that was rubbed with plastic wrap in Experiment 3. The balloon is attracted to the plastic. The attraction gets stronger as the plastic wrap is held closer.	 Plastic wrap used to rub balloon	The plastic wrap exerts an unknown attractive force on the rubbed balloon. The force is stronger the closer the plastic wrap is to the balloon.
Experiment 5. Bring a balloon rubbed with felt near a balloon rubbed with plastic wrap. The balloons are attracted to each other. The attraction strengthens as the balloons are held closer.	 Balloon rubbed with felt Balloon rubbed with plastic wrap	An unknown force pulls each balloon toward the other. This force increases in magnitude as the balloons get closer to each other. 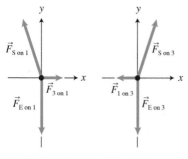

Patterns

- Balloons rubbed with the same material repel each other.
- A balloon rubbed with a particular material is attracted to that material.
- A balloon rubbed with felt attracts a balloon rubbed with plastic wrap.
- All of these effects are stronger the closer the objects are to each other.

Many other objects exhibit similar effects. For example, you can use plastic rods instead of balloons and rub them with felt (like Experiment 1) or glass rods and rub them with silk. In all of these experiments, we find that more vigorous rubbing increases the magnitude of the forces that the objects exert on each other.

Ancient people observed similar effects with different materials. The Greeks noticed that amber, a fossilized resin-like material, attracted small, light objects after it was rubbed with wool. The word for amber in Greek is "electron"; thus this attraction was called electrical. Over time, the new property acquired by the materials due to rubbing came to be called **electric charge**. Physicists explained the observations by saying that the objects acquired **positive electric charge** or **negative electric charge**. Objects that interact with each other because they are charged are said to exert **electric forces** on each other. When the charged objects are at rest, the force is called an **electrostatic force**; the word *static* emphasizes that the objects are not moving with respect to the observer.

By convention, the charge that appears on a balloon rubbed with plastic wrap or on a glass rod rubbed with silk is called positive charge; the charge that appears on a balloon rubbed with wool or felt or on a plastic rod rubbed with felt is called negative charge. We can now reinterpret the patterns in Table 17.1:

TIP ▶ In figures, we will draw red pluses to note positive charges and blue minuses to note negative charges.

1. Materials rubbed against each other acquire electric charge.
2. Two objects with the same type of charge repel each other.
3. Two objects with opposite types of charge attract each other.
4. Two objects made of different materials rubbed against each other acquire opposite charges.
5. The magnitude of the force that the charged objects exert on each other increases when the distance between the objects decreases.
6. Often more vigorous rubbing leads to a greater force exerted by the rubbed objects on each other.

Remember, we have *not* observed electric charge in any of these experiments. We *have* observed attraction and repulsion and used the concept of electric charge to explain these observations. The model of two electric charges is sufficient to explain all of the experiments that we have done so far.

CONCEPTUAL EXERCISE 17.1 > **Mysterious Scotch tape**

Take two 9-inch-long pieces of Scotch tape and place them sticky side down on a plastic, glass, or wooden tabletop. Now pull on one end of each piece to remove them from the table. Bring the pieces of tape near each other. They repel, as shown at right. Can we explain the repulsion of the pieces of tape as an electric interaction?

The pieces of tape pulled from the table repel.

Glass rod rubbed with silk ······The same tape is attracted to this rod.

Plastic rod rubbed with felt ······The tape pulled from the table is repelled by the rod.

Sketch and translate The pieces of tape repel each other, exhibiting the behavior of like-charged objects. Thus, it is possible that the pieces of tape have electric charge. To test this hypothesis, we need to design an experiment whose outcome we can predict using the idea that the pieces are electrically charged. *If* they are, and we bring them one at a time next to a glass rod rubbed with silk (positively charged) and a plastic rod rubbed with felt (negatively charged), *then* they should be attracted to one and repelled by the other.

Simplify and diagram When we perform the experiment, we see that a piece of tape is repelled by the plastic rod rubbed with felt and attracted to the glass rod rubbed with silk. Does it mean that the pieces are electrically charged? We cannot say for sure, but our experiment does not rule out the hypothesis that the pieces of tape are electrically charged.

Try it yourself Take two pieces of tape and fold the end of each piece over about half a centimeter to make a "handle." Place one piece of tape on top of a table and the second piece on top of the first. Then pull them together off the table and separate them by pulling the "handles" away from each other. The pieces of tape attract each other. Design an experiment to determine the sign of each of their charges.

Answer

According to our charge sign convention, a plastic rod rubbed with felt is negatively charged and a glass rod rubbed with silk is positively charged. Hold both rods near the two pieces of tape to observe repulsion or attraction—this will tell you the signs of the charges of the tape pieces.

Charged objects attract uncharged objects

So far, we have learned that objects rubbed with different materials can have opposite charges and attract each other electrically. However, when we bring a plastic comb that has been rubbed with felt near small bits of paper, the bits of paper are attracted to the comb (**Figure 17.2**). Small pieces of aluminum foil are attracted to the comb even more readily than paper. Why are uncharged objects attracted to charged objects? We explore this question in Observational Experiment **Table 17.2**.

FIGURE 17.2 Uncharged paper pieces are attracted to a charged comb.

OBSERVATIONAL **17.2** Interactions of charged and uncharged objects
EXPERIMENT TABLE

VIDEO
OET 17.2

Observational experiment	Analysis
Experiment 1. Hang a balloon from a string next to a wall (neither the balloon nor the wall are charged). The balloon hangs straight down. Next, hang a negatively charged balloon near the same wall. The negatively charged balloon is attracted to the wall. Uncharged wall	The uncharged wall exerts an attractive force on the negatively charged balloon. $\vec{F}_{\text{S on B}}$ $\vec{F}_{\text{W on B}}$ $\vec{F}_{\text{E on B}}$

(CONTINUED)

Observational experiment		Analysis
Experiment 2. Hang a positively charged balloon near the uncharged wall. The balloon is attracted to the wall.	Uncharged wall	The uncharged wall exerts an attractive force on the positively charged balloon. $\vec{F}_{\text{S on B}}$ $\vec{F}_{\text{W on B}}$ $\vec{F}_{\text{E on B}}$
Experiment 3. Repeat Experiment 1, only this time bring the negatively charged balloon near an uncharged metal plate. You may observe that the balloon has a stronger attraction to the plate than to the wall.	Uncharged metal plate	The magnitude of the force that the uncharged metal plate exerts on the negatively charged balloon is larger than the magnitude of the force that the wall in Experiment 1 exerts on the balloon. $\vec{F}_{\text{S on B}}$ $\vec{F}_{\text{M on B}}$ $\vec{F}_{\text{E on B}}$
Experiment 4. Repeat Experiment 2, only this time bring the positively charged balloon near an uncharged metal plate. You may observe that the balloon has a stronger attraction to the plate than to the wall.	Uncharged metal plate	The magnitude of the force that the uncharged metal plate exerts on the positively charged balloon is larger than the magnitude of the force that the wall in Experiment 2 exerts on the balloon. $\vec{F}_{\text{S on B}}$ $\vec{F}_{\text{M on B}}$ $\vec{F}_{\text{E on B}}$

Patterns

- The positively and negatively charged balloons were attracted to uncharged materials (the wall and the metal).
- The intensity of interaction was stronger with the metal plate than with the nonmetal wall.

Using the patterns in Table 17.2, we add a new statement to our conceptual model of the electric interaction: uncharged objects are attracted to both positively and negatively charged objects; uncharged metal objects are usually attracted more strongly than nonmetal objects.

In the next section, we will develop conceptual explanations for these and other observations.

REVIEW QUESTION 17.1 To decide whether an object is electrically charged, we need to observe its repulsion from some other object, not its attraction. Why is attraction insufficient?

17.2 Explanations for electrostatic interactions

Electrically charged objects interact with each other: they repel or attract. They also attract uncharged objects. What is the mechanism behind this interaction?

Is the electric interaction a magnetic interaction?

Do you recall observing any other interaction that involved repulsion? If you've ever played with magnets, you probably know that magnets have poles (north and south). Like poles repel, while unlike poles attract. Both poles attract objects that are not magnets but contain iron—nails, paper clips, etc. Electrically charged objects also attract and repel each other and attract uncharged objects. Perhaps the electric interaction is actually the magnetic interaction, just described using different terminology. We test this hypothesis in Testing Experiment Table 17.3.

TESTING EXPERIMENT TABLE 17.3 Testing the electric = magnetic interaction hypothesis

VIDEO
TET 17.3

Testing experiment	Prediction	Outcome
Experiment 1. Bring a negatively charged plastic rod near the north pole of a magnet that is free to rotate about a pivot.	If the electric interaction is the same as the magnetic interaction, then the negatively charged rod should either attract or repel the north pole of the magnet.	The negatively charged rod attracts the north pole.
Experiment 2. Bring the negatively charged plastic rod near the south pole of a magnet that is free to rotate about a pivot.	If the electric interaction is the same as the magnetic interaction, then the negatively charged rod should repel the south pole of the magnet.	The negatively charged rod also attracts the south pole.

Conclusion

The outcome of the second experiment is inconsistent with the prediction. Thus we can reject the hypothesis that electric interaction is the same as magnetic. However, both experiments can be explained using our knowledge of electric interactions if we assume that a magnet *also* behaves like a piece of metal and thus attracts any charged object, the way all metals do.

The experiments in Table 17.3 show that the same charged rod attracts both the north and the south poles of a magnet, which is very different from the behavior of a magnet. This disproves our hypothesis that electric interactions are the same as magnetic interactions. We need a different mechanism to explain electrostatic interactions.

Fluid models of electric charge

Benjamin Franklin (1706–1790) proposed that a weightless electric fluid is present in all objects. According to Franklin, too much electric fluid in an object makes it positively charged and too little makes it negatively charged. If you rub two different types of material against each other, the electric fluid may move from one material to the other. The material that loses some electric fluid becomes negatively charged, and the material that gains some electric fluid becomes positively charged. The fluid can also move inside the uncharged object when another charged object is present, making its sides positive or negative. This movement of the fluid inside explains why an uncharged object is attracted to a charged object.

According to another model, there are two types of weightless electric fluids—positive and negative. In an object that is not rubbed, the two fluids are in balance.

After rubbing, one object loses some positive fluid and becomes negative, while the other one gains some positive fluid and becomes positive. Objects with an excess of the same fluid repel each other, and objects with an excess of different fluids attract. The fluids can move inside objects, making their sides charged even when the object overall is uncharged. Both of these fluid models account for all of our observations so far.

Particle model of electric charge

In 1897, J.J. Thomson, explaining the results of his experiments with cathode rays (you will learn more about cathode rays in Chapter 27), proposed that electric charge was not associated with a weightless fluid but was carried by the particles that comprise matter. In 1909, Robert Millikan and his graduate student Harvey Fletcher tested this hypothesis. They placed tiny droplets of oil between two oppositely charged horizontal metal plates. As the tiny droplets of oil moved between the plates, they picked up ions or electrons from the air. The drops drifted up or down, depending on the charge on the plates. Millikan and Fletcher adjusted the electric charge on the plates so that the drop velocity was constant because of the balance of three forces: viscous drag, the electric force, and Earth's gravitational force. If the magnitude of the charge of the droplets changed, the researchers would adjust the charge of the plates to maintain the steady speed of the droplets (**Figure 17.3**).

Thousands of measurements demonstrated that the electric charge of each oil droplet changed only by multiples of some discrete unit of electric charge. This pattern indicated to Millikan and Fletcher that electric charge was carried by microscopic particles. Removing or adding these particles to an object changed the charge of the object. Each charged particle had a charge equal to some smallest indivisible amount of electric charge. These particles were called **electrons**. The electron also had a very small mass, several thousand times less than the mass of an average atom.

Contemporary model

Today we know that the explanation of charging by rubbing is actually very complicated. Atoms, the basic units of matter, consist of positively charged nuclei and negatively charged electrons. In normal conditions, the atoms are **neutral**, meaning that the total charge of each atom is zero because the positive charge of the nucleus is equal in magnitude to the negative charge of the electrons. Inside nuclei reside positively charged protons and neutral neutrons. You will learn about the structure of atoms and nuclei in Chapters 28 and 29, respectively (**Figure 17.4a**). It is possible for an electron to leave one object and move to a different object. Such a transfer occurs because the atoms of the material that gains electrons hold them more tightly than the atoms of the material that loses them. An atom that has lost one electron has a net charge of $+e$ (it has one more proton than it has electrons) and is called a **positive ion** (a positively charged sodium ion is shown in Figure 17.4b). An atom that gains an extra electron has a net charge of $-e$ and is called a **negative ion**.

We now understand why rubbed objects acquire opposite charges. Two objects start as neutral—the total electric charge of each is zero. During rubbing, one object gains electrons and becomes negatively charged. The other loses an equal number of electrons and with this deficiency of electrons becomes positively charged. Sometimes when you rub two objects against each other, no transfer of electrons occurs. When the electrons in both materials are bound equally strongly to their respective atoms, no transfer occurs during rubbing. Note that atomic nuclei or individual protons are never transferred during charging.

We can represent this process using integers. A neutral object can be represented with a zero. A zero can be made up of a sum of positive and negative numbers—for example, $+10 + (-10) = 0$. When we rub object A with a second object B that pulls an electron from object A, the negative number of object A becomes smaller while its positive number stays the same: $+10 + (-10) - (-1) = +10 + (-9) = +1$.

FIGURE 17.3 A simplified model of Millikan and Fletcher's oil drop experiment.

Three forces are exerted on the drop D: gravitational force (by Earth, E), electric force (by the plates, P), and drag force (by air, A). When the sum of the forces is zero, the drop moves at constant velocity.

Positively charged plate

$\vec{F}_{\text{A on D}}$ $\vec{F}_{\text{P on D}}$
Oil drop $\vec{v}$
$\vec{F}_{\text{E on D}}$

Negatively charged plate

FIGURE 17.4 A simplified model of (a) a carbon atom and (b) a sodium ion.

(a)

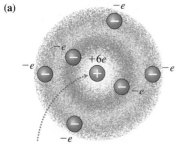

The charge of the carbon nucleus is $+6e$.

(b) The sodium ion Na$^+$ has 10 negative electrons and a nucleus with charge $+11e$.

Free electron

Object A is positively charged. The model of charging by rubbing described above allows us to explain why balloons rubbed with different materials acquire different electric charges. A balloon rubbed with plastic wrap becomes positively charged because its electrons transfer to the wrap (which becomes negatively charged). The opposite effect occurs when the balloon is rubbed with felt. The plastic wrap must attract electrons more strongly than the balloon, and the felt must attract them less strongly.

CONCEPTUAL EXERCISE 17.2   **Attraction of shirt and sweater**

You pull your sweater and shirt off together and then pull them apart. You notice that they attract each other—a phenomenon called "static" in everyday life. Explain the mechanism behind this attraction and suggest an experiment to test your explanation.

The garments attract after they are pulled apart.

Sketch and translate Electric attraction between the sweater and shirt is a possible explanation. We need to explain how pulling the sweater and shirt apart leads to their opposite charges.

Simplify and diagram According to the particle model of electric charge, some of the electrons from one garment transfer to the other garment when they rub against each other. Before rubbing, both objects were neutral. If some negative electrons transfer from one object to the other, the former object now has a net positive charge while the latter has a net negative charge. If this hypothesis is correct, then a Scotch tape strip pulled off a table should be repelled by one of the garments and attracted to the other. When we perform the experiment, shown at top right, the outcome matches this prediction.

Negatively charged tape is attracted to the sweater . . .   . . . and repelled by the shirt.

Try it yourself You suspect that some objects after rubbing acquire a third type of charge. How can you test this hypothesis?

Answer

Imagine that rubbing a balloon with some new type of material (spandex, for example) gives the balloon a third type of charge. Then according to our pattern, two balloons rubbed with spandex should repel—and they do. But they should also attract both the balloon rubbed with felt and the balloon rubbed with plastic (if we assume that all differently charged objects attract each other). However, this never happens. If the balloon rubbed with spandex attracts the balloon rubbed with felt, it always repels the balloon rubbed with plastic, and vice versa. No experiment ever performed has required the existence of a third type of electric charge in order to explain its outcome.

REVIEW QUESTION 17.2 The model of charging by rubbing involved the transfer of negatively charged electrons. How then can a neutral object become positively charged?

17.3 Conductors and insulators (dielectrics)

We found in Table 17.2 that an electrically charged balloon was attracted to an uncharged wall and was slightly more strongly attracted to an uncharged sheet of metal. The former is a so-called insulator (dielectric) and the latter a conductor. The meaning of these terms will be apparent by the end of this section.

Conductors

Suppose that in metals, some electrons are not bound to their respective atoms but can move freely throughout the metal. When we bring a positively charged rod next to a metal bar (without touching it), these **free electrons** in the metal bar should move closer to the positively charged rod, leaving the other side of the metal bar with a deficiency of electrons and therefore positively charged (**Figure 17.5a**). The negatively charged side of the bar is closer to the positively charged rod and thus should be attracted to it more strongly than the more distant positively charged side of the bar is repelled by it. When a negatively charged object is brought near an uncharged metal bar, the free electrons are repelled, resulting in similar charge separation (but with opposite orientation) and attraction (Figure 17.5b). We can use the same model of the internal structure of a metal to explain the experiments in Table 17.2. The magnet, which is made of metal, was attracted to both positively and negatively charged objects. We now see that the attraction is electric in nature and has nothing to do with magnetism.

So far, we have explained the experiments that we already performed. Can we use the model of free electrons in metals to explain new experiments?

Consider the following experiment. An uncharged ball of aluminum foil hangs from a thin insulating string. When a negatively charged plastic rod is brought near the metal foil, the ball first moves toward the rod, touches it, and swings away. **Figure 17.6** helps explain this experiment using our model of free electrons in metals.

FIGURE 17.5 A metal bar is attracted to charged objects.

(a)

(b)

FIGURE 17.6 A neutral foil ball interacts with a negatively charged plastic rod.

(a)
Negatively charged plastic rod

Small ball of aluminum foil

(b)
The positive side of the foil ball is closer to the rod than the negative side. There is a net force toward the left on the foil ball.

(c)
The metal foil ball then touches the rod and acquires some of the negative charge from it.

(d)
The negatively charged foil ball is now repelled from the negatively charged rod.

It looks as if our model of the freely moving charged particles in metals can explain a variety of experiments. It turns out that this model can be generalized to other materials. Metals are examples of electric **conductors**, materials in which some of the charged particles can move. In metals the moving charges are negatively charged free electrons. Some nonmetallic materials, such as graphite and silicon, can be conductors under certain conditions. In liquid conductors, such as nondistilled water or human blood, the moving charged particles can be positively or negatively charged ions. In hot gases, the moving charged particles can be both free electrons and ions.

The electroscope

FIGURE 17.7 An electroscope.

Electroscopes, such as that shown in **Figure 17.7**, are useful tools for studying electrostatic interactions. They rely on the movement of free electrons in metal conductors. An electroscope consists of a metal ball attached to a metal rod that passes from the outside through an insulating support into a glass-fronted metal enclosure. A very lightweight needle-like metal rod is connected on a pivot near the bottom of the larger rod.

Imagine that we bring a negatively charged object near the top of the electroscope ball without touching it. Then some of the free electrons in the ball and rod move away from the negatively charged object, leaving the ball with a positive charge. Electrons move to the lower part of the rod and to the needle, and as a result both become negatively charged. They repel each other, and the needle deflects away from the rod (Figure 17.8a). If we then take the negatively charged object away from the top of the electroscope, the needle returns to its original vertical position.

FIGURE 17.8 Charging an electroscope.

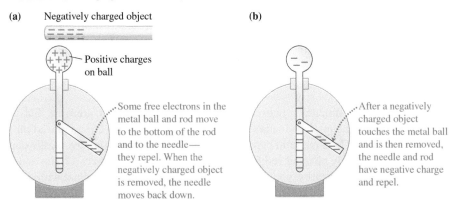

Now, we repeat the experiment, only this time, we touch the top of the electroscope with the charged rod. We observe the same deflection; only this time when the rod is removed, the needle stays deflected (Figure 17.8b). Touching a negatively charged object to the ball on the top of the electroscope transfers electrons to the electroscope. If we then remove the negatively charged object from the electroscope, the electroscope ball, rod, and needle are now negatively charged throughout, and again the rod and needle repel each other. When we repeat the experiments with a positively charged object instead, we observe the same outcomes.

If instead of just touching the top of the electroscope with the charged rod, we carefully rub the charged rod against the top of the electroscope as if we are scrubbing the charges from the rod, we see that the angle of the deflection increases—the angle of deflection is related to the magnitude of the electric charge of the electroscope.

Dielectrics

The presence of electrically charged particles that can freely move inside conductors explains how a charged object attracts a neutral metal object. How can we explain the attraction of a neutral nonmetal object and a charged object, such as observed in Table 17.2?

Plastic, glass, and other nonmetal materials do not have free electrons or any other charged particles that are free to move inside. Such materials are called **insulators** or **dielectrics**. All the electrons are tightly bound to their atoms/molecules (Figure 17.9a). If a charged object (let's make it positive) is brought nearby, it exerts forces on the negatively charged electrons and on the positively charged nuclei inside the atoms of the dielectric material. Negatively charged electrons are pulled slightly closer to the charged bar, and the positively charged nuclei are pushed slightly away. This causes a slight atomic-scale separation of the negative and positive charges in the material, a process called **polarization**. When the atoms are in this polarized state, they are called **electric dipoles**. An electric dipole is any object that is overall electrically neutral but has its negative and positive charges separated. Polarization leads to

FIGURE 17.9 A simplified model of how a charged rod polarizes atoms in an insulating material.

(a)

Neutral atoms with charge uniformly distributed

(b) When a positively charged object is brought near the dielectric material, its atoms and molecules become polarized.

There is now a weak attraction between the charged object and the dielectric material.

(c) Polarization also occurs when we bring a negatively charged object near a nonconducting material.

FIGURE 17.10 The effect of a charged rod on metal and plastic cups.

(a)

The rod polarizes the charge on the cups so that there are excess electrons on the right cup and a deficiency of electrons on the left cup.

(b)

Atoms become polarized but there is no net charge separation.

(c)

When separated, the metal cups are oppositely charged.

(d)

When separated, the plastic cups are uncharged.

FIGURE 17.11 Free electrons in the body move toward a positively charged cup.

(a)

If your hand touches the cup, free electrons on the hand transfer to the cup until the net charge on the cup is zero.

(b)

a small accumulation of charge on the surface of the object (depicted in Figure 17.9b). A similar polarization occurs if you bring a negatively charged object near an insulating material (Figure 17.9c).

Due to this polarization, electrically neutral pieces of paper are attracted to a charged comb that has been run through your hair, and an electrically neutral wall attracts a charged balloon.

Let's summarize what we have learned about the electric properties of materials.

Conductors and insulators Materials can be divided into two groups: conductors and insulators (dielectrics). Both materials are composed of oppositely charged particles with a total charge of zero. In conductors, some of the charged particles can move freely. In dielectrics, charges can only be redistributed slightly a process called polarization.

We can test the difference between conductors and dielectrics. Imagine that you have two aluminum cups (conductors) and two plastic cups (dielectrics) placed on an insulating surface. The aluminum cups are touching each other, as are the plastic cups. You bring a negatively charged rod next to each pair of cups (**Figures 17.10a** and b). While holding the rod nearby, you separate the cups by pushing them apart with a plastic ruler. If our understanding of conductors and insulators is correct, the metal cups will be oppositely charged; the cup closer to the rod will be positively charged and the cup farther from the rod will be negatively charged (Figure 17.10c). The plastic cups should remain uncharged (Figure 17.10d). We can check this prediction using a strip of Scotch tape pulled off a table. The strip of tape is attracted to one of the metal cups and repelled from the other. However, the charged tape is attracted slightly to both plastic cups (each plastic cup becomes slightly polarized and, thus, attracts the negatively charged tape). Since the outcome of the experiment matched the prediction, we gain confidence in our model.

Experiments with metal cups give us another example of charging. We charged both cups without touching them with a charged object. We did not transfer any extra charge to both cups; we merely separated the existing charges. Such a process is called charging *by induction*. We are already familiar with charging *by rubbing* from the balloon experiments in Section 17.1, and with charging *by touching* from the electroscope experiments earlier in this section.

> TIP Electric charge is completely defined by the effects it produces. You cannot see electric charge.

Is the human body a conductor or a dielectric?

In the previous experiment you used a plastic ruler to separate the metal cups. If you had touched the metal cups with your hand instead of the ruler, the charged tape would have been attracted to both. This is because the human body is a conductor. As your hand approaches a positively charged cup, the free electrons on the surface of your hand and arm move toward it (**Figure 17.11a**). When you touch the positively charged metal cup, electrons from the surface of your hand travel to the cup until the cup has a net charge of approximately zero (Figure 17.11b). A similar process happens when you touch a negatively charged cup, only this time the electrons transfer from the cup to your skin. Sometimes when you touch a charged object, you feel a shock and might even see a spark. We will explain these phenomena later in the book.

Grounding

Instead of touching the charged metal cup with your hand, you can safely discharge it (make its net charge effectively zero) by connecting a wire to both the cup and a metal pipe that is driven into the ground (**Figure 17.12**). This process is called **grounding**. To briefly explain why a grounded metal cup discharges, let's assume that Earth is a large conductor. If the cup is charged positively, negative electrons in the ground are attracted toward the cup and travel from the ground to the cup, causing it to become neutral. For a negatively charged cup, electrons in the cup travel through the grounding wire into the ground. Grounding electric appliances is extremely important in order to avoid electric shocks. A more detailed explanation of grounding will be given in Chapter 18.

Discharging by moisture in the air

If you leave a negatively or positively charged object alone for a long time interval, the object can become neutral without being grounded. This can happen for two reasons. First, the air, which is made of neutral molecules and should ideally be a dielectric, in fact has a few electrically charged particles, which are attracted to any object that has an opposite charge. The second reason is the presence of water vapor in the air. The electric charge in water molecules is distributed in such a way that the molecule is a natural electric dipole even without the polarizing influence of an external charged object (**Figure 17.13a**). The hydrogen side of the water molecule is slightly positive, while the oxygen side is slightly negative, thus producing an electric dipole (Figure 17.13b). When water molecules are near a charged object (in this case negative), they will be attracted to it. On a humid day, a very thin film of water forms on all objects. This water layer works just like a wire in grounding. This is why it is a good idea to dry equipment for electrostatic experiments with a hair dryer before use.

Properties of electric charge

Think back to our explanation of charging by rubbing. When a plastic rod is rubbed with felt, the rod becomes negatively charged. The felt becomes positively charged. However, rubbing did not create the charges—it just transferred some electrons from one object to the other. If we assume that each transferred electron has the same electric charge independent of the presence of other electrons, then the total electric charge of an isolated system should be constant. The charge of a nonisolated system changes only if charged objects are transferred in or out of the system. Thus, similar to mass, momentum, and energy, *electric charge is a conserved quantity*; if it changes in one system, we can always find another system in which it is constant.

Another property of electric charge can be deduced from the Millikan-Fletcher experiments. Electric charge cannot be divided infinitely—the smallest amount of charge that an object can have is the charge of one electron. This property is called charge **quantization**.

> **Electric charge** Electric charge (symbol q or Q) is a property of objects that participate in electrostatic interactions. Electric charge is quantized—you can only change an object's charge by increments, not continuously. Electric charge is conserved. The unit for electric charge is the coulomb (C). The smallest increment of charge is that of one electron $-e = -1.6 \times 10^{-19}$ C. The magnitude of the charge of one electron is sometimes called the *elementary charge*.

Our next goal is to develop a quantitative description of the forces that charged objects exert on each other.

REVIEW QUESTION 17.3 One cannot charge a held metal object by rubbing. Why?

FIGURE 17.12 A wire that is "grounded" on one end to a pipe in the ground can discharge the positively charged cup.

The wire provides a path for electrons from Earth toward the positively charged cup.

Cup

Pipe into ground

FIGURE 17.13 A schematic sketch of a water molecule.

(a)

H $^+$

$-$

O   Water molecule (H_2O)

$-$

H

$+$

(b)

The net effect is an electric dipole.

TIP Notice how small the charge of the electron is. Even if we place a billion electrons on an object, it will still have only a -1.6×10^{-10} C charge. Typical charges of objects through rubbing are about 10 times larger, but still are very small. 1 C is a *huge* charge!

17.4 Coulomb's force law

We have observed in the experiments so far that the nearer rubbed objects are to each other, the greater the forces they exert on each other. The forces also increase when the magnitude of the charges of the interacting objects increases. So the force that a charged object exerts on another must depend on the distance r between them and on their electric charges q_1 and q_2.

Charles Coulomb determined the relationship between these quantities in 1785. The experimental apparatus he used is called a torsion balance (**Figure 17.14**). Coulomb hung a light glass rod from a thin wire. At the ends of the rod he attached identical small metal spheres. He then charged a third metal sphere identical in size to the first two and touched one of the spheres attached to the rod. Because the two touching spheres were identical, after touching they had the same electric charge, half of the original, and they repelled. The wire twisted until the torque exerted by the wire on the glass rod balanced the torque exerted by one charged sphere on the other. Coulomb measured the angle of this twist and used it to determine the electric force exerted by one sphere on the other. He measured the distance between the repelling spheres and varied it to see how the force of repulsion depended on the distance between them.

In Coulomb's time, scientists did not know how to directly measure electric charge, and the unit for electric charge did not exist. That is why Coulomb used relative charges rather than absolute charges. Imagine two identical metal spheres—one charged and one uncharged. If you bring them into contact, what would be the charge on each? Coulomb reasoned that since the spheres were identical, the charge would distribute equally between them. Using the same method described above to get the charge of one sphere equally distributed between two identical spheres and thus obtaining half of the original charge on both, he could achieve fractions of charge by touching a charged sphere to identical uncharged spheres many times. This way he could have charges of $q, \frac{1}{2}q, \frac{1}{4}q$, and so forth on the original sphere. He needed to have spheres of different charges to find out how the force exerted by one on the other depended on the magnitude of their charges. **Table 17.4** shows a simplified version of data that Coulomb might have collected. It indicates how the repulsive force that one metal sphere exerted on another depended on their separation and on their charges. (In reality, Coulomb's data did not follow such clear patterns and included experimental uncertainty.)

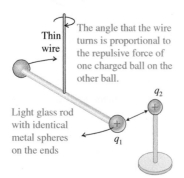

FIGURE 17.14 Coulomb's torsion balance, an apparatus for measuring the electric force between charged objects.

Thin wire

The angle that the wire turns is proportional to the repulsive force of one charged ball on the other ball.

q_2

Light glass rod with identical metal spheres on the ends

q_1

TABLE 17.4 Coulomb's data

Experiment	Charge q_1	Charge q_2	Distance r	Force $F_{q_1 \text{ on } q_2}$
1	1 (unit)	1 (unit)	1 (unit)	1 (unit)
2	$\frac{1}{2}$	1	1	$\frac{1}{2}$
3	$\frac{1}{4}$	1	1	$\frac{1}{4}$
4	1	$\frac{1}{2}$	1	$\frac{1}{2}$
5	1	$\frac{1}{4}$	1	$\frac{1}{4}$
6	$\frac{1}{2}$	$\frac{1}{2}$	1	$\frac{1}{4}$
7	$\frac{1}{4}$	$\frac{1}{4}$	1	$\frac{1}{16}$
8	1	1	2	$\frac{1}{4}$
9	1	1	3	$\frac{1}{9}$
10	1	1	4	$\frac{1}{16}$

To find patterns in the data presented in Table 17.4, let's focus on one variable at a time. In Experiments 1 and 2, the only quantity that changes is the magnitude of the charge q_1. When the charge is halved, the force exerted by q_1 on q_2 is also halved. Looking at Experiment 3, we see that dividing the charge by four reduced the force to one-fourth. The force seemed to be directly proportional to the magnitude of the charge q_1. Experiments 4 and 5 show the same pattern for the magnitude of the charge q_2. If a quantity is directly proportional to two independent quantities, it has to be proportional to their product. Thus, we infer from the table that $F_{q_1 \text{ on } q_2} \propto q_1 q_2$. All 10 experiments are consistent with this conclusion.

From the data we also see that doubling the distance between the objects decreased the force by a factor of 4, and tripling the distance decreased the force by a factor of 9. It appears that the force is inversely proportional to the square of the separation between the objects:

$$F_{q_1 \text{ on } q_2} \propto \frac{1}{r^2}$$

We can now combine these patterns into one mathematical expression, called **Coulomb's law**.

> **Coulomb's law** The magnitude of the electric force that point-like object 1 with electric charge q_1 exerts on point-like object 2 with electric charge q_2 when they are separated by a distance r is given by the expression
>
> $$F_{q_1 \text{ on } q_2} = k_C \frac{|q_1||q_2|}{r^2} \qquad (17.1)$$
>
> where k_C is a proportionality constant called **Coulomb's constant** whose value using SI units is
>
> $$k_C = 9.0 \times 10^9 \ \frac{\text{N} \cdot \text{m}^2}{\text{C}^2}$$
>
> When using the above to determine the magnitude of the electric force, we do not include the signs of the electric charges q_1 and q_2.
>
> The force that object 1 exerts on object 2 points away from object 1 if they have same sign charges and toward object 1 if they have opposite sign charges.

Notice that, mathematically, the expression for the electric force that two charged objects exert on each other,

$$F_{q_1 \text{ on } q_2} = k_C \frac{|q_1||q_2|}{r^2}$$

is analogous to the expression for the gravitational force that any two objects with mass exert on each other,

$$F_{m_1 \text{ on } m_2} = G \frac{m_1 m_2}{r^2}$$

The gravitational force depends on the masses of the objects, and the electric force depends on the charges of the objects. Also, the gravitational force is always an attractive force, whereas the electric force can be attractive or repulsive. The proportionality constants have very different values.

TIP If the interacting objects are large conductors and near each other, the electrons will redistribute on their surfaces, making the effective distance smaller than the distance between the centers of the objects (and difficult to estimate). Thus, we only use Coulomb's law for point-like objects. These objects can be either conductors or dielectrics.

Comparing the magnitude of the electric force to the gravitational force

Let's compare the electric force to the gravitational force exerted by the nucleus, which consists of just one proton, on the electron in a hydrogen atom. A proton has a charge of $+1.6 \times 10^{-19}$ C and mass of 1.7×10^{-27} kg. An electron has a charge of -1.6×10^{-19} C and mass of 9.1×10^{-31} kg. They are separated in the atom by about 10^{-10} m. We call the proton charge 1 and the electron charge 2.

Electric force:

$$F_{q_1 \text{ on } q_2} = k_C \frac{q_1 q_2}{r^2} = (9.0 \times 10^9 \, \text{N} \cdot \text{m}^2/\text{C}^2) \frac{(1.6 \times 10^{-19} \, \text{C})^2}{(10^{-10} \, \text{m})^2}$$

$$= 2.3 \times 10^{-8} \, \text{N}$$

Gravitational force:

$$F_{m_1 \text{ on } m_2} = G \frac{m_1 m_2}{r^2} = (6.67 \times 10^{-11} \, \text{N} \cdot \text{m}^2/\text{kg}^2) \frac{(1.7 \times 10^{-27} \, \text{kg})(9.1 \times 10^{-31} \, \text{kg})}{(10^{-10} \, \text{m})^2}$$

$$= 1.0 \times 10^{-47} \, \text{N}$$

If we divide the electric force by the gravitational force, the result is about 2×10^{39}. The electric force that the proton exerts on the electron is about 2×10^{39} times greater than the gravitational force that the proton exerts on the electron! What about the gravitational force exerted by Earth on the electron?

$$F_{E \text{ on } m_2} = G \frac{m_E m_2}{r^2} = (6.67 \times 10^{-11} \, \text{N} \cdot \text{m}^2/\text{kg}^2) \frac{(6.0 \times 10^{24} \, \text{kg})(9.1 \times 10^{-31} \, \text{kg})}{(6.4 \times 10^6 \, \text{m})^2}$$

$$= 8.9 \times 10^{-30} \, \text{N}$$

TIP   Since Coulomb's law contains the product of the two charges, the electric force that object 1 exerts on object 2 is exactly the same in magnitude as the electric force that object 2 exerts on object 1. Coulomb's law is consistent with Newton's third law.

The gravitational force exerted by Earth on the electron is 18 orders of magnitude larger than the gravitational force exerted by the proton on the electron, but still about 22 orders of magnitude weaker than the electric force exerted by the proton on the electron. That is why physicists can confidently ignore gravitational forces when dealing with atomic size particles.

We have learned that the electric force that electrons and nuclei in atoms exert on each other is what holds atoms together. Although the atoms themselves are electrically neutral and should not interact with each other via electric forces, they do in fact interact. This interaction occurs because the electrons of one or more atoms can interact with the nuclei and electrons of other atoms. These forces are responsible for the existence of groups of bound atoms—molecules. In addition, the electric forces that the charged parts of electrically neutral molecules exert on each other are responsible for molecules forming liquids and solids at sufficiently low temperatures. Many forces that we encounter in everyday life—tension forces, friction forces, normal forces, buoyant forces, etc.—can be explained using Coulomb's law. You can use Coulomb's law to explain why a liquid exerts an upward buoyant force on a submerged object, or why carpet exerts a friction force on your feet.

CONCEPTUAL EXERCISE 17.3   **Interactions of charged objects**

Two equal-mass small aluminum foil balls A and B have electric charges $+q$ (on A) and $+3q$ (on B). (a) Compare the magnitude of the electric force that the foil ball with the smaller charge exerts on the ball with the larger charge $(F_{A \text{ on } B})$ to the force that the larger charged ball exerts on the smaller charged ball $(F_{B \text{ on } A})$. Justify your answer. (b) If the balls are suspended by equal-length strings from the same point, will one string hang at a greater angle from the vertical than the other? Justify your answer.

Sketch and translate A labeled sketch of the situation is shown at right. We choose each ball as a separate system of interest. Ball A

interacts with Earth, the string, and ball B. Ball B interacts with Earth, the string, and ball A.

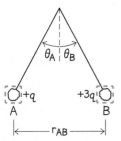

Simplify and diagram We model the balls as point-like objects (see the force diagrams below).

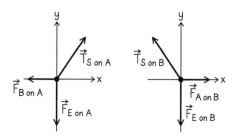

(a) The electric forces that the balls exert on each other have the same magnitude, according to Newton's third law and Coulomb's

law ($F_{A \text{ on } B} = k_C q_A q_B / r_{AB}^2 = k_C q_B q_A / r_{AB}^2 = F_{B \text{ on } A}$). (b) Since the balls have the same mass, Earth exerts the same force on each. Thus the forces exerted by the strings must also have the same magnitude since the net force exerted on each ball equals zero. Therefore, the angles of the strings relative to the vertical should also be the same.

Try it yourself Assume that the balls have twice the mass of the balls in the above exercise. By what factor should we increase the charges on the balls so that the angle made by the strings does not change?

Answer   √2.

QUANTITATIVE EXERCISE 17.4   **Coulomb's law and proportional reasoning**

Three identical metal spheres A, B, and C are on separate insulating stands (Figure a below). You charge sphere A with charge $+q$. Then you touch sphere B to sphere A and separate them by distance d (Figure b). (a) Write an expression for the electric force that the spheres exert on each other. What assumptions did you make? (b) You now take

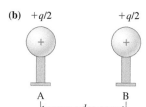

(a) $+q$

Charge A with $+q$.

A   B   C

(b) $+q/2$ $+q/2$

Touch B to A and separate them by distance d.

A ├──── d ────┤ B

(c) $+q/4$ $+q/4$ $+q/2$

Touch C to A and remove C.

C

A   B

(d) $+q/4$ $+q/2$

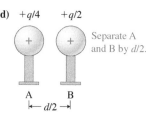

Separate A and B by $d/2$.

A   B
├─ $d/2$ ─┤

sphere C, touch it to sphere A (Figure c), and remove sphere C. You then separate spheres A and B by a distance $d/2$ (Figure d). Write an expression for the magnitude of the force that spheres A and B exert on each other.

Represent mathematically Assume that we can model the spheres as point-like objects. (a) Assume that after sphere A touches B, each of them has a charge of $+q/2$. (b) After C touches A, the charge on A halves again, so that A now has charge $+q/4$. The distance between A and B is $d/2$. We can use Coulomb's law to write an expression for the force that each charge exerts on the other for each part of the problem.

Solve and evaluate (a) The magnitude of the electric force that A and B exert on each other is

$$F_{A \text{ on } B} = F_{B \text{ on } A} = k_C \frac{(q/2)(q/2)}{d^2} = k_C \frac{q^2/4}{d^2} = k_C \frac{q^2}{4d^2}$$

(b) Now, the force that A and B exert on each other is

$$F'_{A \text{ on } B} = F'_{B \text{ on } A} = k_C \frac{(q/4)(q/2)}{(d/2)^2} = k_C \frac{q^2/8}{d^2/4} = k_C \frac{q^2}{2d^2}$$

Although one of the charges got smaller, the decrease in the distance between A and B more than compensated for this, and the force became twice as great as in the first case.

Try it yourself Repeat this exercise, starting with charge $+q$ on sphere A and no charge on spheres B and C. Touch B to A. Then touch C to B. Separate spheres B and C by a distance $d/2$. Write an expression for the force that B exerts on C.

Answer

$$F_{B \text{ on } C} = F_{C \text{ on } B} = k_C \frac{(q/4)(q/2)}{(d/2)^2} = k_C \frac{q^2}{4d^2}.$$

EXAMPLE 17.5 Net electric force

The metal spheres on the insulating stands from the previous exercise have the following electric charges: $q_A = +2.0 \times 10^{-9}$ C, $q_B = +2.0 \times 10^{-9}$ C, and $q_C = -4.0 \times 10^{-9}$ C. The spheres are placed at the corners of an equilateral triangle whose sides have length $d = 1.0$ m with C at the top of the triangle. What is the magnitude of the total (net) electric force that spheres A and B exert on C?

Sketch and translate A labeled sketch of the top view of the situation is shown below. All three angles of the triangle are 60°.

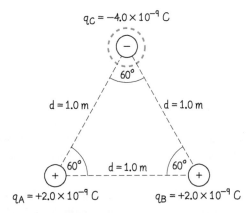

Simplify and diagram Assume that the distance between the spheres is much bigger than their radii so that they can be modeled as point-like objects. Choose sphere C as the system of interest and draw a force diagram for it. We are only interested in the magnitude of the net electric force exerted on C. Thus, we include in the diagram only the electric forces exerted on C by A and by B. We see that the vector sum of $\vec{F}_{A \text{ on } C}$ and $\vec{F}_{B \text{ on } C}$ points straight down (in the negative y-direction).

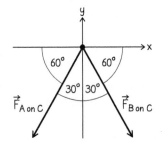

Represent mathematically Each of the spheres A and B exerts a force on C of magnitude

$$F_{A \text{ on } C} = F_{B \text{ on } C} = k_C \frac{|q_A||q_C|}{d^2}$$

where d is the length of a side of the triangle. Each force vector points along the line connecting the respective spheres. The x-components of the two force vectors cancel each other, and the resultant force has only a y-component. Using trigonometry, we find that the net y-component force that A and B exert on C is

$$F_{\text{Net on C } y} = (-F_{A \text{ on } C} \sin 60°) + (-F_{B \text{ on } C} \sin 60°)$$

$$= -2F_{A \text{ on } C} \sin 60° = -2\left(k_C \frac{|q_A||q_C|}{d^2}\right) \sin 60°$$

Solve and evaluate Inserting the appropriate values gives

$$F_{\text{Net on C } y} = -2\Big((9.0 \times 10^9 \text{ N} \cdot \text{m}^2/\text{C}^2)$$

$$\times \frac{(2.0 \times 10^{-9} \text{ C})(4.0 \times 10^{-9} \text{ C})}{1.0 \text{ m}^2} \Big) \sin 60°$$

$$= -1.2 \times 10^{-7} \text{ N}$$

The force has magnitude 1.2×10^{-7} N and points straight down. This is a very small force. If the spheres had relatively small masses (a few hundred grams) and were mounted on stands on which the surface exerted a regular friction force, we would probably not observe any effects of these electric forces.

Try it yourself What charges should you place at the corners of an equilateral triangle with a horizontal base so that the net force exerted on the top charge points horizontally to the left?

Answer The bottom left could be $-q$, the bottom right $+q$, and the top center sphere could have any positive charge.

REVIEW QUESTION 17.4 Two charged objects (1 and 2) with charges $q_1 = q$ and $q_2 = 28q$ are placed r meters away from each other. What is the ratio of the electric force that object 1 exerts on 2 to the force that 2 exerts on 1? Explain your answer.

17.5 Electric potential energy

In the previous section we learned to describe the interactions of electric charges in terms of forces. In this section we will learn to describe these interactions in terms of energy.

When you have a system of two objects exerting gravitational forces on each other (an apple-Earth system, for example), the system possesses gravitational potential energy. Since Coulomb's law is mathematically very similar to the law of universal gravitation, it is reasonable to suggest that a system of electrically charged objects also possesses some sort of electric energy. Let's investigate this idea.

Electric potential energy: a qualitative analysis

Let's begin by looking at the electric potential energy of two like-charged objects that are part of the hypothetical "electric cannon" shown in **Figure 17.15a**. A positively charged cannonball is held near another fixed positively charged object in the barrel of the cannon. This situation is similar to that of an object pressed against a compressed spring. The cannonball, the other charged object, and Earth are the system. Since the cannonball is repelled from the fixed-charge object, the cannonball when released accelerates up the barrel and out of its end (Figure 17.15b). During this acceleration, the system's kinetic (K) and gravitational potential (U_g) energies both increase. If we assume that the system is isolated, then some other type of energy has to decrease so that these two can increase. This other type of energy must be the electric energy suggested above. We will call it **electric potential energy** U_q.

FIGURE 17.15 The electric potential energy of two like-charged objects.

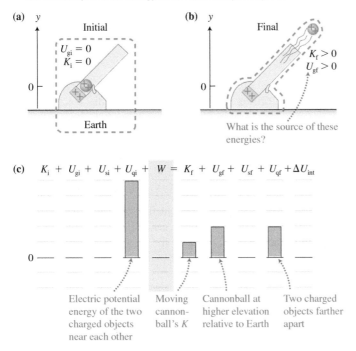

The system comprising the two charged objects has electric potential energy. It seems that the electric potential energy of like-charged objects decreases as they get farther apart from each other. As the cannonball moves farther from the fixed-charge object, the electric potential energy of the system decreases as it is converted into kinetic and gravitational potential energy.

We can represent this process using a bar chart (Figure 17.15c). The initial state is the moment at which the cannonball is near the other fixed-charge object; the final state is when the cannonball is moving at the end of the barrel. For this process we assign the zero level for gravitational potential energy to be at the initial position of the cannonball. We must also make a zero-level assignment for electric potential energy. We say that when two charged objects are infinitely far apart, so that they essentially do not interact, they have zero electric potential energy.

Let us consider a situation in which two oppositely charged objects interact: a hypothetical "electrostatic nutcracker" (**Figure 17.16a**, on the next page). The system consists of two oppositely charged blocks, one of which can slide without friction. When the negatively charged block is released and moves nearer the nut, the kinetic energy of the system increases (Figure 17.16b). Thus electric potential energy must decrease. Assuming the zero level is when the two oppositely charged blocks are infinitely far apart, we conclude that the electric potential energy of a pair of unlike charges is less than zero.

FIGURE 17.16 The electric potential energy of two oppositely charged objects.

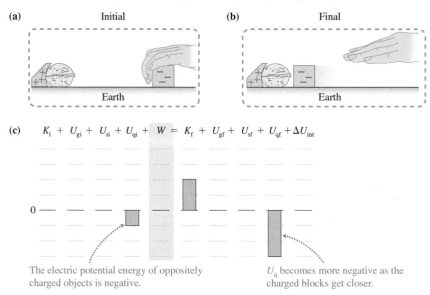

(c)
$$K_i + U_{gi} + U_{si} + U_{qi} + W = K_f + U_{gf} + U_{sf} + U_{qf} + \Delta U_{int}$$

0 —

The electric potential energy of oppositely charged objects is negative.

U_q becomes more negative as the charged blocks get closer.

The bar chart in Figure 17.16c represents this process. The initial electric potential energy of the system is negative. As the objects come closer together, the kinetic energy of the system increases and its electric potential energy decreases, becoming even more negative. Our next step is to devise a mathematical expression for electric potential energy.

Electric potential energy: a quantitative analysis

To derive an expression for the change in electric potential energy of two electric charges whose separation changes, we use the generalized work-energy principle from Chapter 7 (we assume that all objects are at the same temperature, thus there is no energy transferred by heating):

$$W = \Delta E_{system} \tag{7.3}$$

where W is the work done on the system by objects in the environment and ΔE_{system} is the resulting change in the system's energy. To derive an expression for ΔU_q, we choose a hypothetical system that has the following feature: when work is done on this system, only its electric potential energy changes. The system shown in **Figure 17.17** consists only of the point-like charged objects 1 and 2 with like charges q_1 and q_2, initially separated by a distance r_i. The like charges exert a repulsive force on each other of magnitude

$$F_{q_1 \text{ on } q_2} = k_C \frac{|q_1||q_2|}{r^2} \tag{17.1}$$

Imagine that we wish to prevent object 2 from flying away from 1 by exerting a force on 2 toward 1 of the same magnitude that the charged object 1 exerts on 2 (Figure 17.17a). If we push object 2 just a tiny bit harder, object 2 can be displaced a small distance Δr toward object 1 (Figure 17.17b). We use Eq. (7.1) to calculate the work ΔW_1 (here the symbol delta Δ indicates a small amount of work and not the change in work) done by the pushing force during this small displacement (the force and displacement are in the same direction; thus the cosine of the angle between them is 1):

$$\Delta W_1 = F_{P \text{ on } 2} \Delta r = \frac{k_C q_1 q_2}{r_i^2} \Delta r$$

Notice that there are no magnitude symbols for the charges because their signs are now important for the sign of work. Another important point is that this calculation is only

FIGURE 17.17 The work done in pushing two like-charged objects closer together.

(a)

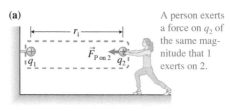

A person exerts a force on q_2 of the same magnitude that 1 exerts on 2.

(b)

Pushing slightly harder moves q_2 to the left.

(c)

To move q_2 even closer to q_1 requires a much harder push (more work).

(d)

The total work is the sum of the work done in each tiny displacement from the initial to the final position.

approximate because the force needed to push object 2 increases as its separation from object 1 decreases. If Δr is small, the equation is a good approximation for the work done during that small displacement of object 2. If we wish to push object 2 through the next small displacement Δr (Figure 17.17c), we must exert a larger force since object 2 is now closer to object 1. For each step closer, the amount of work needed to move object 2 closer to object 1 increases.

The total work done in moving the charge from an initial separation r_i to a final separation r_f (Figure 17.17d) equals the sum of the work for each small step. This type of infinitesimal addition is done easily using calculus. The result of such a calculation is

$$W = \Delta W_1 + \Delta W_2 + \Delta W_3 + \cdots = \frac{k_C q_1 q_2}{r_f} - \frac{k_C q_1 q_2}{r_i} = k_C q_1 q_2 \left(\frac{1}{r_f} - \frac{1}{r_i} \right)$$

At each step of the process represented in Figure 17.17, there is no acceleration and therefore no change in kinetic energy or any other kind of energy. Thus we can reason that the only energy change of the system due to this work is the electric potential energy change. By substituting the above expression for W into Eq. (7.3) and zeros for all energy changes except the electric potential energy change, we find

$$k_C q_1 q_2 \left(\frac{1}{r_f} - \frac{1}{r_i} \right) = 0 + 0 + 0 + \Delta U_q + 0 + \cdots$$

Therefore,

$$\Delta U_q = k_C q_1 q_2 \left(\frac{1}{r_f} - \frac{1}{r_i} \right)$$

$$\Rightarrow U_{qf} - U_{qi} = \frac{k_C q_1 q_2}{r_f} - \frac{k_C q_1 q_2}{r_i}$$

Evidently, the electric potential energy of a system with two charged point-like objects separated by distance r is

$$U_q = \frac{k_C q_1 q_2}{r}$$

where q_1 and q_2 are the actual electric charges, not their magnitudes. This equation is consistent with the convention that we established earlier: the electric potential energy of interaction of two charged objects is zero when the distance between them is infinite. It is also consistent with our qualitative understanding of electric potential energy from the previous section. The energy of interaction of two like charges is positive, and that of two unlike charges is negative.

Electric potential energy The change in electric potential energy ΔU_q of a system of two charged objects q_1 and q_2 when they are moved from an initial separation r_i to a final separation r_f is

$$\Delta U_q = U_{qf} - U_{qi} = k_C q_1 q_2 \left(\frac{1}{r_f} - \frac{1}{r_i} \right) \tag{17.2}$$

Electric potential energy is measured in units of joules. Equation (17.2) is valid for both positively and negatively charged objects, provided the signs of the charges are included.

Two points are worth noting: (a) electric potential energy is proportional to $1/r$ and not to $1/r^2$ (as is the force in Coulomb's law), and (b) mathematically the expression for the electric potential energy is similar to the expression for the gravitational potential energy of a system with two objects with mass, $U_g = -G \dfrac{m_1 m_2}{r}$.

Changes in electric potential energy

Two oppositely charged objects (with positive charge $+q$ and negative charge $-q$) are separated by distance r_i. Will the electric potential energy of the system decrease or increase if you pull the objects farther apart? Explain.

Sketch and translate The initial state is the two oppositely charged objects (the system) a distance r_i apart. You pull the $-q$ charged object so that it is farther from the positively charged object—the final state—as shown below.

Because the direction of the force that you exert on the $-q$ charged object is the same as the direction of its displacement, positive work is being done on the system, increasing the electric potential energy of the system. If you pulled $-q$ infinitely far from $+q$, the final electric potential energy of the system would be zero. Thus, the initial electric potential energy must be negative. To understand this, imagine that you add a positive number to an unknown number and get zero: $x + 5 = 0$. What is the unknown number? $x = -5$ because $(-5) + 5 = 0$. The initial energy was negative.

Simplify and diagram We model the two objects as point-like particles, and we assume that only the electric potential energy changes. A bar chart representing the process is shown at top right.

Represent mathematically We now use the bar chart and the generalized work-energy principle to represent the process mathematically:

$$E_i + W = E_f$$
$$\Rightarrow \frac{k_C(+q)(-q)}{r_i} + W = \frac{k_C(+q)(-q)}{r_f}$$

Solve and evaluate Notice that if r_i is less than r_f, the electric potential energy term on the left side is more negative than that on the right side—consistent with the bar chart. This means that the electric potential energy of the system is negative and changes from a larger negative value to a smaller negative value as the charges get farther apart—an increase in energy because positive work was done on the system. This outcome is consistent with our conceptual understanding of the process.

Try it yourself In a hydrogen atom, a proton and an electron have charges of the same magnitude, 1.6×10^{-19} C, but opposite sign. The distance between them is 0.53×10^{-10} m. What is the change in electric potential energy if the electron is moved far away from the nucleus?

Answer 4.3×10^{-18} J.

FIGURE 17.18 The electric potential energy-versus-distance graphs for two charged objects.

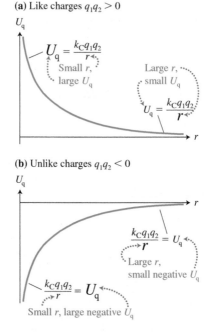

(a) Like charges $q_1q_2 > 0$

(b) Unlike charges $q_1q_2 < 0$

Graphing the electric potential energy versus distance

So far, we have represented electric potential energy with bar charts and with mathematics. Using a graph is another way to conceptualize this abstract physical quantity. Consider first how the electric potential energy changes as the separation r between two like-charged objects is varied. Because of the $1/r$ dependence, the electric potential energy approaches positive infinity when the separation approaches zero (see **Figure 17.18a**). The electric potential energy becomes less positive and approaches zero as the like charges are moved far apart. For a system with two unlike-charged objects, the electric potential energy approaches negative infinity when their separation approaches zero and becomes less negative (increases) and approaches zero as they are moved far apart (Figure 17.18b).

Electric potential energy of multiple charge systems

When a system has several interacting electrically charged objects, to determine the electric potential energy of such a system, we consider the electric potential energy of each pair of objects and then sum up all of the energies to find the total. Suppose that the system has three charged objects labeled 1, 2, and 3. Then the total electric potential energy is

$$U_q = U_{q12} + U_{q13} + U_{q23}$$

$$= \frac{k_C q_1 q_2}{r_{12}} + \frac{k_C q_1 q_3}{r_{13}} + \frac{k_C q_2 q_3}{r_{23}} \qquad (17.3)$$

where r_{12} is the distance between object 1 and 2, and similarly for the other terms. Remember that the signs of all charges must be included.

REVIEW QUESTION 17.5 How can we reduce the magnitude of the electric potential energy of a system of two electrically charged objects?

17.6 Skills for analyzing processes involving electric charges

While solving problems with electrically charged objects, you can use a problem-solving strategy similar to that used in the dynamics and energy chapters. The process is described on the left side of Problem-Solving Strategy 17.1 and illustrated on the right side for Example 17.7.

PROBLEM-SOLVING STRATEGY 17.1 > **Analyzing processes involving electric charges**

EXAMPLE 17.7 **Where to put the charge?**

Two electrically charged objects with charges $q_1 = +1.0 \times 10^{-9}$ C and $q_2 = +2.0 \times 10^{-9}$ C are separated by 1.0 m. Where, with respect to charge 1, should you place a third electrically charged object so that the net electric force exerted on it by the first two objects is zero?

Sketch and translate

- Sketch the process described in the problem statement. Label the known physical quantities.
- Identify the unknowns.
- Choose an appropriate system.

We do not yet know where to place the third object. It could be in any of the three regions shown, but it has to be on the line connecting objects 1 and 2 so that the forces that 1 and 2 exert on this third object are parallel and pointing in opposite directions. We choose the object of unknown electric charge q as the system. The system interacts with objects 1 and 2. We need to determine the location r of the third charge.

Simplify and diagram

- Decide whether you can consider the charged objects to be point-like.
- Decide what other interactions you will consider and what interactions you can ignore.
- Construct a force diagram for the system. Choose appropriate coordinate axes.
- If you are using the work-energy principle, construct an energy bar chart. Decide where the zeros for potential energies are.

Consider all objects to be smaller than the distances between them and, thus, point-like.

Draw force diagrams for the charged object q in the three possible regions. The net electric force exerted on q is the vector sum of the two forces exerted by q_1 and q_2.

(I) If q is positioned to the left of both charged objects, the forces due to q_1 and q_2 both point left and cannot add to zero.

(III) If q is positioned to the right of the two charges, both forces point right and cannot add to zero.

(II) If q is between the charges, the forces point in opposite directions and can add to zero. If q is closer to the smaller charge, the reduced distance will compensate for the smaller q_1.

Represent mathematically

- Use the force diagram to apply the component form of Newton's second law to the process (or use the energy bar chart to apply the generalized work-energy equation).
- If necessary, use kinematics equations to describe the motion of the object.

Both forces are horizontal (their y-components are zero). Thus, we use only the horizontal x-component form of Newton's second law. With the positive x-direction toward the right and taking the signs of the components into account, we get

$$F_{q_1 \text{ on } q} + (-F_{q_2 \text{ on } q}) = 0$$

Using Coulomb's law for electric force, the fact that the total distance between the charges is 1.0 m, and that $q_2 = 2q_1$ and labeling the distance between q_1 and q as r, the above equation becomes

$$\frac{k_C q_1 q}{r^2} = \frac{k_C 2q_1 q}{(1.0 \text{ m} - r)^2}$$

Divide both sides of the equation by $k_C q q_1$ to get

$$\frac{1}{r^2} = \frac{2}{(1.0 \text{ m} - r)^2}$$

Take the square root of both sides of the equation:

$$\frac{1}{r} = \frac{\sqrt{2}}{1.0 \text{ m} - r}$$

Solve and evaluate

- Rearrange the equation and solve for the unknown quantity.
- Verify that your answer is reasonable with respect to sign, unit, and magnitude.
- Also make sure the equation applies for limiting cases, such as objects having very small or very large charge.

Rearranging the last equation, we get

$$\sqrt{2}r = 1.0 \text{ m} - r \quad \text{or} \quad r = 0.41 \text{ m}$$

The location where a net electric force of zero will be exerted on q is a distance 0.41 m from charged object 1 in the direction of object 2. The result looks reasonable, as the unknown charge q should be closer to the smaller magnitude charge q_1 than to the larger charge q_2.

The distance that we found does not depend on the magnitude or sign of the charge q. The net force exerted on q by the two other objects will be zero regardless of the sign of q.

One limiting case is if one of the charges q_1 or q_2 is zero. If $q_1 = 0$, then only q_2 could exert a force on q and the net force could *not* be zero. The force equation becomes $0 + (-F_{q_2 \text{ on } q}) = 0$, which has no solution because the right and left sides are never equal to each other. So this limiting case is consistent with our mathematics.

Try it yourself Suppose we placed q slightly closer to q_1 than the position calculated in this example. What happens to q?

Answer

The magnitude of the force that q_1 exerts on q increases slightly, and the magnitude of the force that q_2 exerts on q decreases slightly. The net force exerted on q is no longer zero—if released, q will accelerate away from the desired position if it is negative and toward the desired position if it is positive.

In the next example we apply the steps of the problem-solving strategy to a process whose analysis requires an energy approach.

EXAMPLE 17.8   Radon decay in lungs

Suppose that a radon atom in the air in a home is inhaled into the lungs. While in the lungs, the nucleus of the radon atom undergoes radioactive decay, emitting an alpha (α) particle, which is composed of two protons and two neutrons. (These high-energy alpha particles can damage lung tissue—a reason for keeping radon concentration low in homes.) During this process, the radon nucleus turns into a polonium nucleus with charge $+84e$ and mass 3.6×10^{-25} kg. The alpha particle has charge $+2e$ and mass 6.6×10^{-27} kg. Suppose the two particles are initially separated by 1.0×10^{-15} m (the size of the nucleus) and are at rest. How fast is the alpha particle moving when it is very far from the polonium nucleus? (Note: 1 mm would be very far for such a process since even 1 mm is much larger than the size of a nucleus.)

Sketch and translate We draw a sketch showing the initial and final states of the system, choosing the product polonium nucleus and the

escaping alpha particle as the system. We need to determine the speed of the alpha particle when it is infinitely far from the polonium nucleus.

Simplify and diagram Model the nuclei and alpha particle as point-like objects and neglect the gravitational attraction between them. Assume that they are at rest at the start of the process and that we can ignore the final kinetic energy of the massive polonium nucleus. A work-energy bar chart for the process is shown at the right. In the initial state, the system has electric potential energy; in the final state, the alpha particle has kinetic energy, and since the particles are comparatively far apart we assume the system has zero electric potential energy.

Represent mathematically Each nonzero bar in the bar chart turns into a nonzero term in the generalized work-energy equation.

$$\frac{k_C(+84e)(+2e)}{r_i} = \frac{1}{2}m_\alpha v_f^2$$

Rearranging the above, we get an expression for the final speed of the alpha particle:

$$v_f = \sqrt{\frac{2k_C(84e)(2e)}{m_\alpha r_i}}$$

Initial

$m_{Po} = 3.6 \times 10^{-25}$ kg
$m_\alpha = 6.6 \times 10^{-27}$ kg

Alpha particle at rest near polonium nucleus

Final

Alpha particle moving fast after being repelled by the polonium nucleus

$v_f = ?$

$\approx \infty$

(CONTINUED)

Solve and evaluate Substituting the appropriate values:

$$v_f = \sqrt{\frac{2k_C(84e)(2e)}{m_\alpha r_i}}$$

$$= \sqrt{\frac{2(9.0\times10^9\,\text{N}\cdot\text{m}^2/\text{C}^2)(84.0\times1.6\times10^{-19}\,\text{C})(2.0\times1.6\times10^{-19}\,\text{C})}{(6.6\times10^{-27}\,\text{kg})(1.0\times10^{-15}\,\text{m})}}$$

$$= 1.1 \times 10^8 \text{ m/s}$$

We estimate that the final alpha particle speed is about 10^8 m/s. (Note that when we study relativity in Chapter 26, we will find that we need a new expression for kinetic energy at these speeds, which reduces our answer by a factor of three [which is still a very high speed].)

Try it yourself In this example we assumed that the massive polonium nucleus remained at rest. Use impulse-momentum ideas to determine how fast the polonium nucleus was actually moving following the decay process. Compare the kinetic energies of the nucleus and the alpha particle.

Answer

in this example.

were justified in ignoring the final kinetic energy of the polonium

the speed squared, are related by $K_{Po} = 0.0183K_\alpha$. We

on depend which energies, kinetic The $v_{Po} = 0.0183v_\alpha$.

FIGURE 17.19 A charged Van de Graaff generator.

REVIEW QUESTION 17.6 How would our reasoning in Example 17.8 change if we chose the alpha particle alone as the system instead of both the nucleus and the escaping alpha particle?

17.7 Charge separation and photocopying

We can separate charge by rubbing together two objects that are made of different materials. However, the magnitude of the resulting charge is quite low. In humid weather the separated charges on these objects are quickly neutralized, making charge-related investigations difficult. To study the behavior of charged objects more easily, we need charges of large magnitudes. Specialized equipment such as Van de Graaff generators (Figure 17.19) and Wimshurst machines is commonly used for such studies in physics courses.

Van de Graaff generator

A Van de Graaff generator is made of a plastic cylinder with a motorized moving belt inside and a hollow metal dome at the top. When the belt is moving, you can hear a cracking sound and see sparks around the dome. Some generators have an attachment—a metal ball on an insulating stand. The sparks between the dome and this ball or any nearby metal objects can be large, implying large charge separation. How does this device function?

A Van de Graaff generator belt moves on two rollers (one on the top inside of the metal dome and the other on the bottom—see Figure 17.20). The lower plastic roller becomes negatively charged due to a repeated contact and then separation from the moving belt. The inner surface of the belt acquires an equal amount of positive charge. Because the positive charge on the belt is spread over a larger area than the negative charge on the plastic roller, the electrons in the lower metal comb are repelled more by the negative charge on the roller than they are attracted by the positive charge on the belt. Thus the electrons are pushed from the lower metal comb to the ground. Now, the positively charged metal comb attracts and removes electrons from the outer surface of the belt. As the positively charged outer surface of the belt reaches the dome, it attracts electrons from the upper metal comb, which is connected to the dome—the dome therefore becomes positively charged.

FIGURE 17.20 How a Van de Graaff generator works.

CONCEPTUAL EXERCISE 17.9   Making your hair stand on end

The woman in the photo has her hand on the dome of a Van de Graaff generator and her feet on an electrically insulated footstool. Why is her hair standing on end?

Sketch and translate Hair can stand up when the individual strands are repelled from each other and from the body. For this to happen, the hair and body need to have the same sign of electric charge.

Simplify and diagram If we assume that the human body is a fairly good conductor of electric charge, the woman's body is acting like an extension of the dome. Electrons flow from her to the dome, leaving her and each strand of her hair positively charged. This makes them repel from her body and from the other strands—they stand on end.

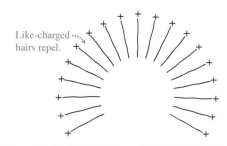

Like-charged hairs repel.

Try it yourself You place a stack of aluminum pie plates on top of an uncharged Van de Graaff generator and then turn on the generator. Predict what happens to the plates and explain your prediction.

Answer

If we assume that electrons transfer from the plates to the positively charged dome, the now positively charged plates repel one another and the dome. They fly up off the dome.

Now imagine that you are holding a grounded metal ball about 10 cm from the dome of a charged Van de Graaff generator. What is actually happening in the air that results in a flash and a sharp cracking sound? The explanation involves what are known as **cosmic rays**, high-energy particles that continually rain down on Earth's atmosphere from space. These cosmic rays ionize atoms, producing free electrons and positively charged ions. High-energy particles produced during naturally occurring radioactive decay can do the same. The positive charge on the dome of a Van de Graaff generator attracts the free electrons already present in the air, causing them to accelerate toward the dome, colliding with other atoms and causing some of them to ionize. When these free electrons recombine with atoms, light is produced and we see a spark.

EXAMPLE 17.10   What causes the sparking from a Van de Graaff generator?

The energy needed to remove an electron from a hydrogen atom is about 2×10^{-18} J (about the same for other atoms, too). The average distance a free electron in air will travel between collisions, called the mean free path, is about 10^{-6} m. The dome has a 0.15-m radius and a $+10^{-5}$-C charge. Could a free electron in the air gain enough kinetic energy to ionize an atom and cause a spark as it travels that short distance toward the charged dome?

Sketch and translate The situation is sketched below (not to scale). We need to compare the kinetic energy that the electron gains between collisions to the energy of ionization of the atom. Given that the mean free path is much smaller than the size of the dome, what values should we put for the initial and final location of the electron?

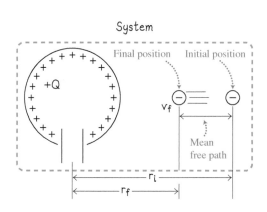

System

Simplify and diagram Assume that the electron is near the surface of the dome. The system is the electron and the dome. We model both as point-like objects. This is reasonable for the electron and reasonable for the dome as well, but only because it is spherical in shape. (We will discuss that in more detail when we discuss the electric field in the next chapter.) An energy bar chart representing the process is shown at right. The zero level of electric potential energy is at infinity. The electron-dome system starts with negative electric potential energy and zero kinetic energy. As the electron accelerates toward the dome, its kinetic energy increases and electric potential energy decreases.

$K_i + U_{qi} + \boxed{W} = K_f + U_{qf}$

Represent mathematically Use the bar chart to help apply the generalized work-energy equation to the process:

$$\frac{k_C(Q)(-e)}{r_i} = \tfrac{1}{2}mv_f^2 + \frac{k_C(Q)(-e)}{r_f}$$

Solve and evaluate Our question concerns the kinetic energy $(1/2)mv_f^2$ of the electron—is the energy as large as or does it exceed the energy needed to ionize an atom $\Delta U \approx 2 \times 10^{-18}$ J?

$$\tfrac{1}{2}mv_f^2 = \frac{k_C(Q)(-e)}{r_i} - \frac{k_C(Q)(-e)}{r_f} = -k_C Qe\left(\frac{1}{r_i} - \frac{1}{r_f}\right)$$

$$= k_C Qe\left(\frac{r_i - r_f}{r_i r_f}\right)$$

(CONTINUED)

The value for $r_i - r_f$ is 10^{-6} m—the distance that the electron travels between the collisions. But what values should we choose for r_i and r_f to calculate their product? Given that the difference between the initial and final positions is tiny compared to the radius of the dome (0.15 m), we can assume that r_i and r_f are both equal to the radius of the dome, giving $r_i r_f = (0.15 \text{ m})^2$. Therefore,

$$\tfrac{1}{2}mv_f^2 = (9 \times 10^9 \text{ N} \cdot \text{m}^2/\text{C}^2)(10^{-5} \text{ C})(1.6 \times 10^{-19} \text{ C})\left(\frac{10^{-6} \text{ m}}{(0.15 \text{ m})^2}\right)$$

$$\approx (10^{10})(10^{-5})(10^{-19})\left(\frac{10^{-6}}{10^{-2}}\right) \text{ N} \cdot \text{m}$$

$$\approx 10^{-18} \text{ N} \cdot \text{m} = 10^{-18} \text{ J}$$

This is comparable to the energy needed to ionize an atom. Note that some electrons will travel farther than the average distance and therefore will have even more kinetic energy.

Try it yourself Why are we ignoring the positive ions in the air and considering only the acceleration of the free electrons due to the positive charge on the dome?

Answer

The positive ions are much more massive than the electrons and do not accelerate to sufficient speeds to have the kinetic energy needed to ionize other atoms.

Wimshurst machine

FIGURE 17.21 A Wimshurst machine.

Electrodes

The Wimshurst machine (**Figure 17.21**) is another device that can produce large charge separations. The device, invented in the 1880s, consists of two plastic or glass disks that rotate in opposite directions. Charge separation in the Wimshurst machine occurs by a complex process that we will not explain here. As a result of this process, one of the electrodes becomes positively charged and the other becomes negatively charged. When the two electrodes are brought near one another, we see sparks as long as 5–10 cm. The Wimshurst machine not only allows us to create large charges on the electrodes but also allows us to study the creation of the spark and its dependence on the distance between the electrodes—one goal of the next chapter.

A photocopier

The photocopying process depends on the electrical attraction of oppositely charged particles—like how a balloon rubbed on a wool sweater attracts tiny bits of paper or sugar off a tabletop. In a copy machine, the charged balloon is a drum that becomes electrically charged with an image of the page being copied, and the paper bits are dark toner particles that stick to the places where the drum is charged. The toner then transfers to paper in the exact location of the letters or pictures on the copied page.

Let's analyze this process in more detail. The drum is covered with a special photoconductive material. When we turn on the photocopier, the drum becomes positively charged beneath the photoconductive material and negatively charged on the outside of the photoconductive material (**Figure 17.22a**).

FIGURE 17.22 A schematic representation of the process of making a photocopy.

(a)

Photoconductor

Cylindrical drum

The photoconductor layer on the drum is charged negatively on the outside and positively on the inside.

(b)

Light

Light reflects from white parts of the page being copied and neutralizes the photoconductive layer. The remaining negative charge is an electrical image of the copied page.

(c)

Toner

Positive toner particles stick only to the negatively charged part of the photoconductive layer.

(d)

Paper

A negatively charged blank white paper attracts the dark toner particles, which are baked onto the page before its ejection from the machine.

During copying, a strong source of light moves across the page from under the glass cover. This light reflects off white regions of the page but does not reflect from dark regions. The reflected light reaches the photoconductive material where electrons on the surface of the photoconductive material absorb the light and move inside the photoconductive material to neutralize the positive charge. The drum surface is now neutral in places where light was reflected from the original page being copied. The drum remains negatively charged in areas below the dark text of the page that was being copied. Thus, the negative electrons on the top of the drum form an electrical image of the text being copied (Figure 17.22b).

The next step is covering the drum with the toner, which is positively charged. The toner sticks to the negatively charged "electrical image" on the drum (Figure 17.22c). Then a blank negatively charged paper wraps around the drum and pulls off the toner (Figure 17.22d). The drum and the paper are heated and pressed together to make this transfer more effective and to bake the toner on the surface of the paper. Finally, the new copy is ejected and a rubber material wipes the drum clean of remaining toner and illuminates it with light to remove all remaining charge.

REVIEW QUESTION 17.7 In a Van de Graaff generator, where does the energy emitted as light in a spark come from?

Summary

Electric charge is a physical quantity that characterizes how charged objects participate in electrostatic interactions. The unit of charge in the SI system is the coulomb (C).

- Charge comes in two types—positive and negative.
- Like charged objects repel and unlike charged objects attract.
- The smallest charge is the charge of an electron ($-e = -1.6 \times 10^{-19}$ C).
- Electric charge is constant in an isolated system. (Sections 17.1 and 17.2)

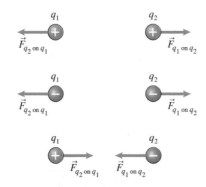

Conductors are materials in which electrically charged particles can move freely. (Section 17.3)

Free electrons move from one side of the conductor to the other.

Excess free electrons

Deficiency of free electrons

Insulators (dielectrics) are materials in which electrically charged particles cannot move freely. However, the electric charges in the atoms and molecules in the material can rearrange slightly (a process called polarization), allowing them to participate in electric interactions. (Section 17.3)

The charge in neutral atoms is slightly rearranged to create electric dipoles.

Coulomb's law is used to determine the magnitude of the electric force that point-like objects with charges q_1 and q_2 exert on each other when separated by a distance r. (Section 17.4)

$$|F_{q_1 \text{ on } q_2}| = |F_{q_2 \text{ on } q_1}| = \frac{k_C |q_1||q_2|}{r^2} \quad \text{Eq. (17.1)}$$

$$k_C = 9.0 \times 10^9 \frac{\text{N} \cdot \text{m}^2}{\text{C}^2}$$

The **electric potential energy** U_q of a system of point-like objects with charges q_1 and q_2 separated by a distance r is positive for like charges and negative for unlike charges (include the signs of the charges when using this equation.) The zero level of electric potential energy is usually chosen to be when they are infinitely far apart. (Section 17.5)

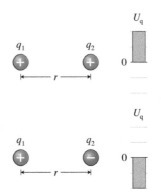

$$\Delta U_q = U_{qf} - U_{qi} = k_C q_1 q_2 \left(\frac{1}{r_f} - \frac{1}{r_i} \right) \quad \text{Eq. (17.2)}$$

Questions

Multiple Choice Questions

1. Which of the following occurs when two objects are rubbed against each other?
   (a) Both acquire the same sign charge.
   (b) They acquire opposite sign charge.
   (c) No new charges are created.
   (d) Both b and c can be correct.

2. With which statements do you disagree?
   (a) If two objects repel, they have like charges.
   (b) If two objects attract, they have opposite charges.
   (c) If two objects attract, one of them can be charged and the other one can be neutral.
   (d) None of them

3. Which explanation agrees with the contemporary model of electric charge?
   (a) If an object is charged negatively by rubbing, it has lost positively charged particles.
   (b) If an object is charged positively by rubbing, it has lost negatively charged particles.
   (c) If an object is charged positively by rubbing, it has acquired positively charged particles.
   (d) If an object is charged negatively by rubbing, it has acquired negatively charged particles.
   (e) a and c
   (f) b and d

4. When an object gets charged by rubbing, where does the electric charge originate?
   (a) It comes from the air surrounding the object.
   (b) It is created by the rubbing energy.
   (c) The process of rubbing leads to a redistribution of charge, not the creation of charge.

5. Choose all of the quantities that are constant in an isolated system.
   (a) Mass
   (b) Entropy
   (c) Electric charge
   (d) Momentum
   (e) Mechanical energy
   (f) Total energy
   (g) Temperature

6. Identically charged point-like objects A and B are separated by a distance d. You measure the force F that A exerts on B. If you now separate the objects by distance $1.4d$, what will be the new force that A exerts on B?
   (a) $1.4F$
   (b) $2.0F$
   (c) $\frac{1}{2}F$
   (d) $\frac{1}{1.4}F$
   (e) F

7. When separated by distance d, identically charged point-like objects A and B exert a force of magnitude F on each other. If you reduce the charge of A to one-half its original value, and the charge of B to one-fourth, and reduce the distance between the objects by half, what will be the new force that they exert on each other?
   (a) $2F$
   (b) $\frac{1}{2}F$
   (c) $4F$
   (d) $\frac{1}{4}F$
   (e) F

8. Balloon A has charge q, and identical mass balloon B has charge $10q$. You hang them from threads near each other. Choose all of the statements with which you agree.
   (a) The force that A exerts on B is $1/10$ the force that B exerts on A.
   (b) The force that A exerts on B is 10 times the force that B exerts on A.
   (c) A and B exert the same magnitude forces on each other.
   (d) The angle between the thread supporting A and the vertical is less than the angle between the thread supporting B and the vertical.

9. Imagine that two charged objects are the system of interest. When the objects are infinitely far from each other, the electric potential energy of the system is zero. When the objects are close to each other, the electric potential energy is positive. Which of the following statements is(are) incorrect?
   (a) Both objects are positively charged.
   (b) Both objects are negatively charged.
   (c) One object is negatively charged and the other one is positively charged.

10. Two objects with charges $+q$ and $-2q$ are separated by a distance r. You slowly move one of the objects closer to the other so that the distance between them decreases by half. Considering the two objects as the system, which statements are *incorrect*?
    (a) The electric potential energy of the system doubles.
    (b) The electric potential energy of the system decreases by half.
    (c) The magnitude of the electric potential energy doubles.
    (d) The magnitude of the electric potential energy decreases by half.

11. Charged point-like objects A and B are separated by a distance d. Object A is fixed in place, while object B can be moved. Choose object A alone as the system. You slowly push object B closer to A, decreasing the distance between them to $0.3d$. Based on the given information, choose all of the correct statements.
    (a) The electric potential energy of the system increases.
    (b) The electric potential energy of the system decreases.
    (c) The electric potential energy of the system does not change.
    (d) You do positive work on the system.
    (e) You do negative work on the system.
    (f) You do zero work on the system.

12. If you move a negatively charged balloon toward a neutral dielectric wall, the electric potential energy of the balloon-wall system
    (a) increases.
    (b) decreases.
    (c) is constant but nonzero.
    (d) is constant and zero.

Conceptual Questions

13. Describe the differences between the electric force and the gravitational force and some experiments that can be explained through these differences.

14. Describe the difference between charged objects and magnets and some experiments that can be explained through these differences.

15. At one time it was thought that electric charge was a weightless fluid. An excess of this fluid resulted in a positive charge; a deficiency resulted in a negative charge. Describe an experiment for which this hypothesis provides a satisfactory explanation and another experiment for which it does not.

16. What experiments can you do to show that there are only two kinds of charge?

17. An object becomes positively charged due to rubbing. Does its mass increase, decrease, or remain the same? Explain. How is its mass affected if it becomes negatively charged? Explain.

18. List everything that you know about electric charges.

19. What experimental evidence supports the idea that conducting materials have freely moving electrically charged particles inside them?

20. You have an aluminum pie pan with pieces of aluminum foil attached to it. Predict what will happen if you touch the pan with a plastic rod rubbed with wool.

21. You have a charged metal ball. How can you reduce its charge by half?

22. You have a foam rod rubbed with felt and a small aluminum foil ball attached to a thread. Describe what happens when you slowly approach the ball with the rod and then touch the ball. Explain why this happens.

23. A positively charged metal ball A is placed near metal ball B. Measurements demonstrated that the force between them is zero. Is ball B charged? Explain.

24. Show that if the charge on B in the previous question is positive but of small magnitude, the balls will be attracted to each other.

25. Two metal balls of the same radius are placed a small distance apart. Will they interact with the same magnitude force when they have like charges as when they have opposite charges? Explain your answer. Include a sketch showing the charge distribution.

26. Describe the experiments that were first used to determine a quantitative expression for the magnitude of the force that one electrically charged object exerts on another electrically charged object. What are the limitations of this expression?

27. The electrical force that one electric charge exerts on another is proportional to the product of their charges—that is, $F_{1\text{ on }2} \propto q_1 q_2$. How would Coulomb's observations be different if the force were proportional to the sum of their charges $(q_1 + q_2)$?

28. Why isn't Coulomb's law valid for large conducting or dielectric objects, even if they are spherically symmetrical?

29. How is electric potential energy similar to spring potential energy? How is it different?

Problems

Below, **BIO** indicates a problem with a biological or medical focus. Problems labeled **EST** ask you to estimate the answer to a quantitative problem rather than derive a specific answer. Asterisks indicate the level of difficulty of the problem. Problems with no * are considered to be the least difficult. A single * marks moderately difficult problems. Two ** indicate more difficult problems.

In some of these problems you may need to know that the mass of an electron is 9.1×10^{-31} kg and the charge is -1.6×10^{-19} C.

17.1–17.3 Electrostatic interactions, Explanations for electrostatic interactions, and Conductors and insulators (dielectrics)

1. **BIO** **Ventricular defibrillation** During ventricular fibrillation, the muscle fibers of the heart's ventricles undergo uncoordinated rapid contractions, resulting in little or no blood circulation. To restore the heart's normal rhythm, a defibrillator sends an abrupt jolt of about −0.20 C of electric charge through the chest into the heart. How many electrons pass through the body during this defibrillation?

2. * You rub two 2.0-g balloons with a wool sweater. The balloons hang from 0.50-m-long very light strings. When you attach the strings together at the top, the balloons hang away from each other each string making an angle of 37° with the vertical. (a) Represent the situation with the force diagram for each balloon and determine the magnitudes of the forces on the diagram. (b) What can you say about the magnitudes of the forces that the balloons exert on each other? Explain. (c) Will the ratio of the forces that the balloons exert on each other change if the charge on one balloon is two times larger than on the other? How do you know?

3. * Two balloons of different mass hang from strings near each other. You charge them about the same amount by rubbing each balloon with wool. Draw a force diagram for each of the balloons. Compare the angles of the threads with the vertical. How do your answers depend on whether the balloons have the same or different magnitude charge?

4. * **Lightning** A cloud has a large positive charge. Assume that this is the only cloud in the sky over Earth and that Earth is a good electrical conductor. Draw a sketch showing electric charge distribution on Earth due to the cloud's electric charge. Explain why a person's hair might stand on end before a lightning strike.

5. Sodium chloride (table salt) consists of sodium ions of charge $+e$ arranged in a crystal lattice with an equal number of chlorine ions of charge $-e$. The mass of each sodium ion is 3.82×10^{-26} kg and that of each chlorine ion 5.89×10^{-26} kg. Suppose that the sodium ions could be separated into one pile and the chlorine ions into another. What mass of salt would be needed to get 1.00 C of charge into the sodium ion pile and −1.00 C into the chlorine ion pile?

17.4 Coulomb's force law

6. * **EST** (a) Earth has an excess of 6×10^5 electrons on each square centimeter of surface. Determine the electric charge of Earth in coulombs. (b) If, as you walk across a rug, about 10^{-22} kg of electrons transfer to your body, estimate the number of excess electrons and the total charge in coulombs on your body.

7. Determine the electrical force that two protons in the nucleus of a helium atom exert on each other when separated by 2.0×10^{-15} m.

8. * Determine the number of electrons that must be transferred from Earth to the Moon so that the electrical attraction between them is equal in magnitude to their present gravitational attraction. What is the mass of this number of electrons?

9. **BIO** **Ions on cell walls** The membrane of a body cell has a positive ion of charge $+e$ on the outside wall and a negative ion of charge $-e$ on the inside wall. Determine the magnitude of the electrical force between these ions if the membrane thickness is 0.80×10^{-9} m. Ignore the effect of the material in which the ions are located.

10. * **Hydrogen atom** In a simplified model of a hydrogen atom, the electron moves around the proton nucleus in a circular orbit of radius 0.53×10^{-10} m. Determine at least four physical quantities related to this information.

11. * Three 100 nC charged objects are equally spaced on a straight line. The separation of each object from its neighbor is 0.3 m. Find the force exerted on the center object if (a) all charges are positive, (b) all charges are negative, and (c) the rightmost charge is negative and the other two are positive.

12. ** Two objects with charges q and $4q$ are separated by 1.0 m. (a) Determine the charge (both sign and magnitude) and position of a third object that causes all three objects to remain in equilibrium (all three objects are on the same line). (b) Is the equilibrium stable or unstable? How do you know? (Hint: An object is in a stable equilibrium if a small displacement of the object from the equilibrium results in a force exerted on the object that points opposite to the direction of displacement.)

13. * **Salt crystal** Four ions (Na^+, Cl^-, Na^+, and Cl^-) in a row are each separated from their nearest neighbor by 3.0×10^{-10} m. The charge of a sodium ion is $+e$ and that of a chlorine ion is $-e$. Determine the electric force exerted on the chlorine ion at the right end of the row due to the other three ions.

14. * A $+10^{-6}$-C charged object and a $+2 \times 10^{-6}$-C charged object are separated by 100 m. Where should a -1.0×10^{-6}-C charged object be located on a line between the positively charged objects so that the net electrical force exerted on the negatively charged object is zero?

15. ** **BIO** **Bee pollination** Bees acquire an electric charge in flight from friction with the air, which causes pollen to cling to them. The pollen is then attracted to the stigma of the next flower (Figure P17.15). Suppose the bee's body has a charge of -10^{-9} C and is about 3×10^{-3} m from the front edge of a spherical granule of pollen of diameter 5×10^{-5} m. Charged particles in the pollen become polarized with $+10^{-11}$ C on the front edge and -10^{-11} C on the backside of the pollen ($3 \times 10^{-3} + 5 \times 10^{-5}$) m from the bee. What useful physical quantities can you determine?

FIGURE P17.15

16. * A triangle with equal sides of length 10 cm has −2.0-nC charged objects at each corner. Determine the electrical force (magnitude and direction) exerted on the object at the top corner due to the two objects at the base of the triangle.

17. You have a small metal sphere fixed on an insulating stand and another sphere attached to a plastic handle. You charge one of the spheres and then touch it with the second sphere. Draw a force-versus-time graph for the force that one sphere exerts on the other as you move the sphere on the handle slowly and steadily away from the sphere on the stand. What information do you need in order to determine the magnitudes of the quantities on the graph?

18. * After the experiment in Problem 17.17, you touch the sphere on the stand with your hand. Then you touch that sphere with the sphere on the handle. Draw a force-versus-time graph for the force that one sphere exerts on the other as you move the sphere on the handle slowly and steadily away from the sphere on the stand. How is this graph different from the graph in the previous problem?

19. ** Coulomb's law is formulated for point-like charged objects. Imagine that you have two point-like charged objects q_1 and q_2 of unspecified sign separated by a distance d and you also have two large spheres of radius R with the same charges q_1 and q_2. The distance between the centers of the spheres is d. Compare the electric force that the point-like objects exert on each other to the force that the two spheres exert on each other. Consider all possibilities.

17.5 Electric potential energy

20. (a) Determine the change in electric potential energy of a system of two charged objects when a −15.0-nC charged object and a −40.0-nC charged object move from an initial separation of 50.0 cm to a final separation of 10.0 cm. (b) What other quantities can you calculate using this information?

21. You have a system of two positively charged objects separated by some arbitrary finite distance. (a) What is the sign of their potential energy? (Remember that charges that are infinitely far from each other have zero potential energy.) (b) What can you do to decrease this energy? (c) Draw an energy bar chart for this process of decreasing the energy.

22. You have a system of two negatively charged objects separated by some arbitrary finite distance. (a) What is the sign of their potential energy? (Remember that charges that are infinitely far from each other have zero potential energy.) (b) What can you do to decrease this energy? (c) Draw an energy bar chart for this process of decreasing the energy.

23. Repeat (a)–(c) of Problem 17.22 for a system with a negatively charged object and a positively charged object separated by some arbitrary finite distance.

24. The metal sphere on the top of a Van de Graaff generator has a relatively large positive charge. In which direction must a positively charged ion in the air move relative to the sphere in order for the electrical energy of the ion-generator system to decrease? Justify your answer.

25. * **EST** An electron is 0.10 cm from an object with electric charge of $+3.0 \times 10^{-6}$ C. (a) Determine the magnitude of the electrical force $F_{\text{O on }e}$ that the object exerts on the electron. (b) The electron is pulled so that it moves to a distance of 0.11 cm from the charged object. Determine the magnitude of the electrical force $F'_{\text{O on }e}$ exerted by the charged object on the electron when at this distance. (c) Estimate the work done by the average force pulling the electron

$$\left(\frac{F_{\text{O on }e} + F'_{\text{O on }e}}{2} \right) \Delta x$$

(d) Compare this number to the change in electric potential energy of the electron–charged-object system as the electron moves away from the object. Why should the numbers be approximately equal?

26. * (a) An object with charge $q_4 = +3.0 \times 10^{-9}$ C is moved to position C from infinity (Figure P17.26). $q_1 = q_2 = q_3 = +10.0 \times 10^{-8}$ C. Determine as many work-energy quantities as you can that characterize this process. Make sure you specify the system. (b) Repeat your calculations, but for $q_1 = q_3 = -10.0 \times 10^{-9}$ C and $q_2 = +10.0 \times 10^{-9}$ C.

FIGURE P17.26

27. * An object with charge $+2.0 \times 10^{-9}$ C is moved from position C to position D in Figure P17.26. $q_1 = q_3 = +10.0 \times 10^{-9}$ C and $q_2 = -20.0 \times 10^{-9}$ C. All four charged objects are the system. What work-energy-related quantities can you determine for the process?

28. ** Two small spheres with charges $q_1 = +8$ nC and $q_2 = -4$ nC are placed at marks $x_1 = -2$ cm and $x_2 = 13$ cm. Where along the x-axis should you place a third sphere with charge q_3 so that (a) the sum of the forces exerted by the first two spheres on the third one is zero, and (b) the total electric potential energy of the three spheres does not depend on q_3?

29. * Two small objects with charges $+Q$ and $-Q$ are fixed a distance a apart. (a) Show that if a third small object with an arbitrary charge q is placed halfway between the first two objects, the total electric potential energy does not change even though we have added a new object to the system. (b) While solving this problem, your friend came to the following conclusion: "If adding the third charged object does not change the total electric potential energy of the system (the system comprises all three objects), then there must be a way to remove the third object from the system without doing any work." Do you agree or disagree with your friend? If you disagree, explain why. If you agree, describe a path along which you can remove the third object from the system so that the total work done on the third object is zero.

30. * A stationary block has a charge of $+6.0 \times 10^{-8}$ C. A 0.80-kg cart with a charge of $+4.0 \times 10^{-8}$ C is initially at rest and separated from the block by 0.4 m. The cart is released and moves along a frictionless surface to a distance of 1.0 m from the block. Determine as many values of the physical quantities describing the motion of the cart as you can.

17.6 Skills for analyzing processes involving electric charges

31. Figure P17.31 shows four different configurations with equal-magnitude charges that are placed at the corners of equal-size squares. Rank the configurations A to D in order of increasing electric potential energy.

FIGURE P17.31

32. * **Equation Jeopardy 1** The solution to a problem is represented by the following equation:

$(9.0 \times 10^9$ N $\cdot$ m^2/C$^2)(+20.0 \times 10^{-9}$ C)$(+50.0 \times 10^{-9}$ C)/(0.1 m)^2
$+ (9.0 \times 10^9$ N $\cdot$ m^2/C$^2)(+10.0 \times 10^{-9}$ C)$(+50.0 \times 10^{-9}$ C)/(0.5 m)$^2 = (10$ kg$)a_x$

Sketch a situation that the equation might represent and formulate a problem for which it is a solution (there are multiple possibilities).

33. * **Equation Jeopardy 2** The solution to a problem is represented by the following equation:

$(9.0 \times 10^9$ N $\cdot$ m^2/C$^2)(+20.0 \times 10^{-9}$ C)$(+50.0 \times 10^{-9}$ C)/(5 cm) $=$
$\frac{1}{2}(10$ g$)v_x^2 + (9.0 \times 10^9$ N $\cdot$ m^2/C$^2)(+20.0 \times 10^{-9}$ C)$(50.0 \times 10^{-9}$ C)/(20 cm)

Sketch a situation that the equation might represent and formulate a problem for which it is a solution (there are multiple possibilities).

34. * **Equation Jeopardy 3** The solution to a problem is represented by the following equation:

$(9.0 \times 10^9$ N $\cdot$ m^2/C$^2)(+20.0 \times 10^{-9}$ C)$(+50.0 \times 10^{-9}$ C)/(10 cm)^2
$- (9.0 \times 10^9$ N $\cdot$ m^2/C$^2)(+10.0 \times 10^{-9}$ C)$(+50.0 \times 10^{-9}$ C)/(5 cm)$^2 = (10$ g$)a_x$

Sketch a situation that the equation might represent and formulate a problem for which it is a solution (there are multiple possibilities).

35. * **Equation Jeopardy 4** The solution to a problem is represented by the following equation:

$(9.0 \times 10^9$ N $\cdot$ m^2/C$^2)(-20.0 \times 10^{-9}$ C)$(+50.0 \times 10^{-9}$ C)/(20 cm)
$= \frac{1}{2}(10$ g$)v_x^2 + (9.0 \times 10^9$ N $\cdot$ m^2/C$^2)(-20.0 \times 10^{-9}$ C)$(+50.0 \times 10^{-9}$ C)/(5 cm)

Sketch a situation that the equation might represent and formulate a problem for which it is a solution (there are multiple possibilities).

36. * **Evaluate the solution** Metal sphere 1 has charge q and is attached to an insulating stand that is fixed on a table. Metal sphere 2 has charge $3q$ and is attached to an insulating stand that can be moved along the table. The distance between the centers of the spheres is initially $a = 35$ cm. If we slowly move sphere 1 toward sphere 2 until the distance between their centers is equal to $b = 25$ cm, we do 10^{-3} J work on the system consisting of the two spheres. (a) Determine the magnitudes and the signs of the charges on the spheres. (b) Your friend provides the following solution for (a): The signs of the charges are either both positive or both negative because work is done on the system as they are moved closer together. The magnitudes of the charges can be determined using the generalized work-energy principle:

$$k_C \frac{q \cdot 3q}{a^2} + W = k_C \frac{q \cdot 3q}{b^2}$$

$$\Rightarrow q = \sqrt{\frac{W}{3k_C}\left(\frac{1}{b^2} - \frac{1}{a^2}\right)^{-1}} = \sqrt{\frac{10^{-3} \text{ J} \cdot \text{C}^2}{3(9 \times 10^9 \text{ N} \cdot \text{m}^2)}(16.0 \text{ m}^{-2} - 8.2 \text{ m}^{-2})^{-1}}$$

$$= 6.9 \times 10^{-8} \text{ C}$$

Evaluate the solution and correct any errors you find.

37. * Construct separate force diagrams for each charged object shown in Figure P17.37. Use two-letter subscripts identifying each force.

FIGURE P17.37

$q_1 = -2q \qquad q_2 = +q \qquad q_3 = +q$

38. ** The six objects shown in **Figure P17.38** have equal-magnitude electric charge. Adjacent objects are separated by distance a. Write an expression in terms of q and a for the force that the five objects on the right exert on the positive charge on the left.

FIGURE P17.38

39. * A small metal ball with positive charge $+q$ and mass m is attached to a very light string, as shown in **Figure P17.39**. A larger metal ball with negative charge $-Q$ is securely held on a plastic rod to the ceiling. Write an expression for the magnitude of the force T that the string exerts on the ball. Define any other quantities used in your expression.

FIGURE P17.39

40. * Four objects each with charge $+1.0 \times 10^{-7}$ C are located at the corners of a square whose sides are 2.0 m long. Determine the values of two physical quantities describing the situation but not provided in the givens.

41. * Two 5.0-g aluminum foil balls hang from 1.0-m-long threads that are suspended from the same point at the top. The charge on each ball is $+5.0 \times 10^{-9}$ C. Make a list of the physical quantities that you can determine using this information. Determine the values of two of those physical quantities.

42. * A 6.0-g ball with charge $+3.0 \times 10^{-8}$ C hangs from a 0.50-m-long string at an angle of 87° below the horizontal. The string is attached at the top on the wall above a second charged object, as shown in **Figure P17.42**. (a) Determine the charge on the second object. (b) Make a list of other physical quantities that you can determine using the information in the problem. Describe how you can determine two of those quantities.

FIGURE P17.42

43. * A 0.40-kg cart with charge $+4.0 \times 10^{-8}$ C starts at rest on a horizontal frictionless surface 0.50 m from a fixed object with charge $+2.0 \times 10^{-8}$ C. When the cart is released, it moves away from the fixed object. (a) How fast is the cart moving when very far (infinity) from the fixed charge? (b) When 2.0 m from the fixed object?

44. A dust particle has an excess charge of 4×10^6 electrons and a mass of 5×10^{-8} g. It is initially held at $d = 10^{-3}$ m directly above a fixed object of the same size that is charged with the same number of electrons. Both objects are in an evacuated container. (a) Determine the maximum height of the dust particle after its release. (b) Show how you can determine two other physical quantities relevant to the process.

45. **Electric accelerator** A micro-transporter moves from one side of an evacuated chamber to another. It is powered by equal-magnitude, opposite-sign charges Q and $-Q$ on top of two station points (see **Figure P17.45**). If a 7×10^{-3}-N friction force opposes the transporter's motion, what must be the magnitude of Q to make the transporter move with an initial acceleration of 1.0 m/s²? The transporter's mass is 1.0 g and it is carrying a positive charge of 8×10^{-10} C.

FIGURE P17.45

46. * You are holding at rest a small sphere A with electric charge $+q$ and mass m at a distance d from another small sphere B with charge Q fixed on top of an insulating support. What are the largest and smallest magnitudes of the force that your hand exerts on sphere A? Make sure you analyze different possibilities.

17.7 Charge separation and photocopying

47. ** A Van de Graaff generator is placed in rarefied air at 0.04 times the density of air at atmospheric pressure. The average distance that free electrons move between collisions (the mean free path) in that air is about 6×10^{-6} m. Determine the positive charge needed on the generator dome so that a free electron located 0.20 m from the center of the dome will gain at the end of the mean free path length the 2.0×10^{-18} J of kinetic energy needed to ionize a hydrogen atom during a collision.

48. * Two protons each of mass 1.67×10^{-27} kg and charge $+e$ are initially at rest and separated by 1.0×10^{-14} m (approximately the radius of the largest nucleus). When released, the protons fly apart. (a) Determine the change in their electric potential energy when they are 1.0×10^{-10} m apart (approximately the radius of an atom). (b) If the electric potential energy change is converted entirely into the kinetic energy of the protons (shared equally), what is the speed of one proton when 1.0×10^{-10} m from the other proton?

49. * Two protons, initially separated by a very large distance (r_i is infinity), move directly toward each other with the same initial speed v_i. (a) Determine their initial speeds if the distance of closest approach when their speeds are zero is 4.0×10^{-14} m. (b) Determine some other physical quantity relevant to the process.

50. * An alpha particle consists of two protons and two neutrons together in one nucleus with a mass of 6.64×10^{-27} kg and charge $+2e$. The alpha particle flies at 3.0×10^7 m/s from a large distance toward the nucleus of a stationary gold atom (charge $+79e$). (a) Make a list of physical quantities that you can determine using this information and explain how you will calculate one of them. (b) Determine the distance of the alpha particle from the gold nucleus when it stops.

51. * Determine the speed that the proton shown in **Figure P17.51** must be moving in order to get within 1.0×10^{-15} m of the helium-3 nucleus that has two protons and one neutron. Assume that the helium nucleus is a point-like object and does not move.

FIGURE P17.51

Proton	Helium-3 nucleus

Initial

v

$q = +e$
m

$q = +2e$
$3m$

$r_0 = \infty$

$v = 0$

Final (the proton is at rest)

r

General Problems

52. ** Suppose that Earth and the Moon initially have zero charge. Then 1000 kg of electrons are transferred from Earth to the Moon. Determine the radius of a stable Moon orbit when both the electrical and gravitational forces of attraction are exerted on the Moon and it completes one rotation about Earth in 27.5 days.

53. * **BIO** **Calcium ion synapse transfer** Children have about 10^{16} synapses that can transfer signals between neurons in the brain and between neurons and muscle cells. Suppose that these synapses simultaneously transmit a signal, sending 1000 calcium ions (Ca^{2+}) across the membrane at each synaptic ending. Determine the total electric charge transfer in coulombs during that short time interval. By comparison, a lightning flash involves about 5 C of charge transfer. (Note: This is a fictional scenario. All human neurons do not simultaneously produce signals.)

54. A small ball D has a charge of -10 nC and cannot move. You move another small ball E that has a charge of -5 nC from point 1 to point 4 as shown in Figure P17.54. Determine the work done by you on ball E when you move it (a) from 1 to 4 along the path shown via 2 and 3 (determine the work for each step separately), and (b) along a straight line from 1 to 4. (c) Compare your two answers for parts (a) and (b) and explain the comparison.

FIGURE P17.54

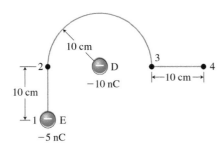

55. * Two small balls A and B with equal charges $+q$ are fixed at distances a and $2a$ from the origin of a coordinate system as shown in Figure P17.55. You move a third equally charged ball C along the z-axis from negative infinity to positive infinity. Draw a qualitative graph that shows how the electric potential energy of the three-ball system depends on the position of ball C along the z-axis.

FIGURE P17.55

56. ** **EST** **Shocking your friend**
(a) You shuffle across the rug and then place your finger near your friend's nose, causing a small spark that transfers about 10^{-9} C of charge from you to your friend. Determine the number of electrons transferred. (b) Estimate the fraction of electrons in your body that were transferred to the friend. Note that the electron mass is about $1/20{,}000$ the mass of the atom—the nuclei are much more massive—and the mass of an average atom in the body is about 2×10^{-26} kg. (Note: Don't get hung up on minutia—this is a rough estimate.)

Reading Passage Problems

Static cling You pull your clothes from the dryer and find that they stick together. You take off your coat and find that your pants stick to your legs. Static cling like this can occur for two reasons. First, different types of atoms have greater or lesser affinity for additional electrons. When two different materials are rubbed together, the atoms with a greater electron affinity attract electrons from the material with a lesser electron affinity. Because of this, the substance that gains electrons becomes negatively charged and the substance that loses electrons becomes positively charged. The differently charged materials attract and stick together.

Second, static cling can occur between charged and uncharged objects. For instance, you may notice that a sock removed from the dryer is attracted to an uncharged sweater you are wearing. Or sometimes your skirt sticks to your legs. This happens because the molecules in a charged piece of clothing cause the electric charge inside the molecules of the nearby uncharged objects to slightly redistribute (to become polarized) so that the unlike charge of the molecule moves closer to the charged object and is attracted more than the same molecular charge of the same sign, which is slightly farther away (see Figures 17.9b and c).

Some people use fabric softener to prevent static cling. This product coats cloth fibers with a thin layer of electrically conductive molecules, thus preventing buildup

of static electricity. You can also use a metal clothes hanger to remove electric charge from already clean charged clothes. Brush the hanger on the inside of the garment from top to bottom.

Shoes scuffing on different surfaces can also cause electric charge transfer. For that reason, hospital personnel wear special shoes in hospital operating rooms to avoid sparking that might ignite flammable gases in the room.

57. You rub a balloon against your wool sweater and then place the balloon on the wall—it sticks. Why?
(a) The balloon and wall have opposite electric charges.
(b) The molecules on the wall redistribute their charge so that the charge opposite that on the balloon is nearest the balloon.
(c) Electric charge in Earth is pulled to the part of the wall nearest the balloon.
(d) a and c are correct.
(e) a, b, and c are correct.

58. As you unload the clothes dryer, you find a sock clinging to a shirt. As you pull the sock off the shirt, you do positive work on the sock. What is the main form of energy increase?
(a) Gravitational   (b) Elastic
(c) Electric potential   (d) Thermal
(e) c and d

59. Table salt, Na^+Cl^-, is made of ionized sodium and chlorine atoms. Which atom is most attractive for an excess electron?
(a) Na   (b) Cl
(c) They are equally attractive.

60. You add fabric softener to your next load of wash. Your clothes do not cling after they emerge from the dryer. What do the water softener molecules do?
(a) Carry away all the excess electrons.
(b) Cause sparks in the dryer that discharge the excess charge on the clothes.
(c) Form a protective coating that prevents the charge from joining the clothes.
(d) Make the cloth fluffy so the charge comes off naturally.
(e) Join the cloth molecules and make a conductive layer that prevents excess charge from accumulating on the clothes.

61. You put strips of aluminum foil in the dryer along with your clothes. Which answer below best represents the condition of the clothes after leaving the dryer?
(a) The clothes are uncharged because the excess charge is on the aluminum strips.
(b) The clothes are uncharged because they transfer excess charge to the strips, which transfer it to the metal dryer walls.
(c) The clothes are charged but the strips are not because they are conductive.
(d) The clothes are charged because the strips are not connected to anything.
(e) None of these answers is reasonable.

62. You remove electric charge from your clean pants by rubbing a metal clothes hanger down the inside of the pants. Which answer below represents the best explanation for why this works?
(a) The charge travels from the cloth to the metal to your hand through your body to the ground.
(b) The metal hanger absorbs all the charge.
(c) The metal causes tiny sparks that send the charge into the air.
(d) The metal provides charge to the cloth that neutralizes its charge.
(e) None of these answers is reasonable.

Electrostatic exploration Geologists sometimes analyze the distribution of materials under Earth's surface, materials such as iron, water, oil, or dry soil. The process they use is called electrostatic exploration. Electrodes are placed in the ground about 800 m apart. An electric generator is connected to the electrodes and causes them to become oppositely charged. The opposite-sign charges on the electrodes cause electrically charged ions in the matter below the surface to move. The moving ions are detected by equipment on the surface. This helps the geologist decide what type of matter is below the surface. What causes this motion? To help answer this question, determine the net electric force exerted on ions at different places below the surface.

63. Which arrow in **Figure P17.63** best represents the force exerted on a positive ion at position A?
    (a) I   (b) II   (c) III
    (d) IV   (e) V   (f) VI

FIGURE P17.63

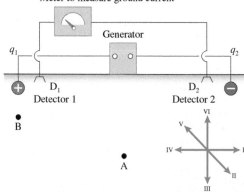

Meter to measure ground current

Generator

q_1 q_2

D_1 D_2
Detector 1   Detector 2

B

A

64. Which arrow in Figure P17.63 best represents the force exerted on a negative ion at position A?
    (a) I   (b) II   (c) III
    (d) IV   (e) V   (f) VI

65. Which arrow in Figure P17.63 best represents the force exerted on a positive ion at position B?
    (a) I   (b) II   (c) III
    (d) IV   (e) V   (f) VI

66. Which arrow in Figure P17.63 best represents the force exerted on a negative ion at position B?
    (a) I   (b) II   (c) III
    (d) IV   (e) V   (f) VI

67. Based on the analysis in the first four questions, which charge detector in Figure P17.63 will be collecting negative ions?
    (a) D_1 (b) D_2
    (c) Both D_1 and D_2 (d) Neither D_1 nor D_2

68. The resistance to the motion of negative electrons through different types of materials is, in decreasing order, dry soil, moist soil, underground water, and iron ore. How can this knowledge and the measurement of the charge reaching the detector per unit time help identify what type of material is under Earth's surface?
    (a) More electrons will reach the detector from iron than from dry soil.
    (b) The electrons reaching the detector will carry water with them if water is under the surface.
    (c) The electrons reaching the detector are affected very little by what's under the surface.
    (d) The detector will get more electric charge if the resistance to flow is less, and resistance is related to the type of material.
    (e) a and d

The Electric Field

Imagine that you are in a car during a thunderstorm. Lightning strikes the road just in front of you. Are you in danger? Many people think that cars are safe because the rubber tires isolate them from the ground. We will learn in this chapter why it is not the rubber tires but the metal body of the car that protects passengers from the effects of the lightning.

* **Why is it safe to sit in a car during a lightning storm?**
* **Why do we need Earth or a large conductor for grounding?**
* **How does electrocardiography work?**

PREVIOUSLY (IN CHAPTER 17) we learned how to describe electrostatic interactions in two ways: (1) with a force exerted by one charged object on another and (2) with the electric potential energy possessed by systems of charged objects. We also discovered that charged objects can exert forces on each other without being in direct contact. This is only the second interaction we have encountered with this property, the gravitational interaction being the first. How does one charged object "know" about the presence of another when they are not in direct contact? This chapter helps us answer this question and develop a new approach for analyzing electric processes.

BE SURE YOU KNOW HOW TO:

* Find the force that one charged object exerts on another charged object (Section 17.4).
* Determine the electric potential energy of a system (Section 17.5).
* Explain the differences in the internal structure of electric conductors and dielectrics (Section 17.3).

18.1 A model of the mechanism for electrostatic interactions

When we pull two pieces of Scotch tape off a table and place them near each other, they repel. How does an electrically charged object, such as a piece of tape, affect the other piece of tape without touching it? You could ask the same question about gravitational interactions—how does the Sun pull on Earth even though they are millions of kilometers apart?

Historically, there have been two answers to this question. The first model was called *interaction at a distance*. Isaac Newton supported this model for the gravitational interaction, and Charles Coulomb supported it for the electrostatic interaction. In this model, both interactions just happen. For example, if you move one of the two interacting charges, the other charge instantaneously "senses" that movement and responds accordingly.

The second model for electric interactions, suggested by Michael Faraday, involved an agent that acts as an intermediary between the charges. It was some sort of *electric disturbance* in the region surrounding them. Physicists call this electric disturbance an **electric field**. According to this **electric field model**, a charged object q_1 electrically disturbs the region surrounding itself (**Figure 18.1a**). If you place a second charged object q_2 in this region, the electric field due to object 1 exerts a force on object 2 (Figure 18.1b). The same is true for object 2—it creates its own field that exerts a force on object 1.

We can apply the field model to the gravitational interaction. For example, how does the Sun pull on Earth, located about 150 million km away? According to the field model, the Sun gravitationally disturbs the region around it in an invisible way. This disturbance is called a **gravitational field**. The Sun's gravitational field exerts a force on Earth that causes it to travel in an almost circular orbit around the Sun.

An analogy may help you visualize the idea of a field. Imagine a very large horizontal elastic sheet that has been pulled tight. The sheet represents space with no charges (no field). Now imagine a billiard ball resting on the sheet. The ball causes the sheet to bend, more so closer to the ball than farther away from it (see **Figure 18.2a**). The presence of the ball causes a disturbance of space (the sheet), and the bending of the sheet represents the field due to the billiard ball. Now suppose a much smaller object such as a ball bearing is placed on the sheet, near the billiard ball. The ball bearing is so small that it does not disturb the sheet field. But the billiard ball's field disturbs the ball bearing—less when it is farther away and more when it is closer to the billiard ball. If the ball bearing is released, it will move toward the billiard ball as if they attract each other. An analogous model for visualizing repulsion can be obtained by stretching the elastic sheet over a pole and again placing the ball bearing on the bent surface (see Figure 18.2b). The billiard ball's or the pole's distortion of the elastic sheet acts as an intermediary, allowing the ball bearing and the billiard ball or pole to interact with each other, even though they are not in contact. Similarly, the electric field acts as an intermediary between charged objects, allowing the objects to interact without direct contact with each other.

Making analogies serves two purposes: (1) to understand a new phenomenon using a phenomenon that we already understand and (2) to predict new phenomena. The elastic sheet analogy allows us to make a prediction based on the electric field model: the interaction between the objects changes at a finite, nonzero speed. When you move one of the two interacting charges (object 1), the other one does not "sense" that movement instantly—it takes time for the change in the field to reach object 2. According to the model of interaction at a distance, the prediction is different: any change in the position of object 1 leads to the *immediate* change in the force exerted on object 2. Which prediction matches the outcome of an experiment testing these models? We will learn the answer to this question when we study electromagnetic waves in Chapter 25.

It is important to remember that every analogy we make has limitations. The two most important limitations of the elastic sheet analogy are the following: (1) In the analogy, there would be no visible interaction between the billiard ball and the ball bearing if Earth did not exert gravitational forces on both balls, while the interaction

FIGURE 18.1 The electric field model.

(a)

q_1 creates a field.

The field is stronger near q_1 than farther from it.

(b)

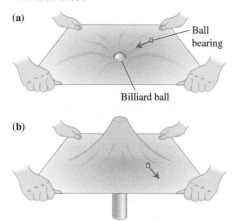

The field exerts an electric force on q_2.

$\vec{F}_{\text{Field on } q_2}$

FIGURE 18.2 Visualizing an electric field using an elastic sheet.

(a)

Ball bearing

Billiard ball

(b)

between charged objects does not need a "third agent." (2) In the analogy, the ball bearing can move only on the two-dimensional surface of the sheet, while charged objects can move freely in three-dimensional space.

Gravitational field due to a single object with mass

Let's develop a gravitational field approach for the gravitational force that one object with mass exerts on another. In the region near Earth, Earth's contribution to the gravitational field is the dominant one. Imagine Earth and one of three small objects A, B, or C. We call these objects *test objects*, because we use them as detectors to probe the field (**Figure 18.3a**). We place the test objects one at a time at the same location near Earth. The masses of these test objects are $m_A = m$, $m_B = 2m$, and $m_C = 3m$. Consider the gravitational force that Earth exerts on each test object (Figure 18.3b).

$$F_{\text{E on A}} = G\frac{m_E}{r^2}m_A = m_A\left(G\frac{m_E}{r^2}\right)$$

$$F_{\text{E on B}} = G\frac{m_E}{r^2}m_B = m_B\left(G\frac{m_E}{r^2}\right) = 2F_{\text{E on A}}$$

$$F_{\text{E on C}} = G\frac{m_E}{r^2}m_C = m_C\left(G\frac{m_E}{r^2}\right) = 3F_{\text{E on A}}$$

The directions of these forces are all toward the center of Earth. However, the magnitudes of the forces differ: they are proportional to the masses of the test objects. Despite the differences in magnitude, however, the ratio of the magnitude of the force exerted on each object and the mass of that object is identical for all three objects:

$$\frac{F_{\text{E on A}}}{m_A} = \frac{F_{\text{E on B}}}{m_B} = \frac{F_{\text{E on C}}}{m_C} = G\frac{m_E}{r^2}$$

Consider the objects to be near Earth's surface. When we substitute the values of G, m_E, and Earth's radius r_E, we find that

$$G\frac{m_E}{r_E^2} = \left(6.67 \times 10^{-11}\,\frac{\text{N}\cdot\text{m}^2}{\text{kg}^2}\right)\frac{(5.97 \times 10^{24}\,\text{kg})}{(6.37 \times 10^6\,\text{m})^2} = 9.8\,\text{N/kg}$$

Since this value does not depend on the mass of any test object, we speculate that this value might be a mathematical description of the "strength" of Earth's gravitational field at a particular location. Since the gravitational force has direction, we say that the gravitational field close to Earth's surface at a particular location has a magnitude of 9.8 N/kg and points directly toward the center of Earth. Until now, we have called this quantity free-fall acceleration. Now we characterize the gravitational field using the quantity $\vec{g}$ field. We define the $\vec{g}$ **field** at any location as the gravitational force exerted by the field on a test object at that location, divided by the mass of that object:

$$\vec{g} = \frac{\vec{F}_{\text{Field on Object}}}{m_{\text{Object}}} \tag{18.1}$$

The magnitude of the field at any location does not depend on the test object; it depends on the mass of the object(s) creating the field (in this case, Earth) and the location where the field is measured: $g = G(m_E/r_E^2)$.

Electric field due to a single point-like charged object

Let's use a similar approach to construct a physical quantity for the "strength" of the electric field. Imagine an object with positive electric charge Q (we use capital Q to denote the object whose field we are investigating), and one of three test objects K,

FIGURE 18.3 Earth exerts a gravitational force on each of the three test objects placed at the same location.

(a)

We place test objects A, B, or C one at a time at the same location relative to Earth.

(b)

$\vec{F}_{\text{E on test object}}$ ◄———● The gravitational force that Earth exerts on a test object

TIP Equation (18.1) is an operational definition for the $\vec{g}$ field; the expression $g = G(m_E/r^2)$ is the cause-effect relationship for g.

FIGURE 18.4 An object with charge Q exerts an electric force on one of three charged objects placed at the same location.

(a)

Q

+q

We place test objects K, L, or M one at a time at the same location relative to Q.

(b)

$\vec{F}_{Q \text{ on } O}$

Q exerts an electrical force on the test object.

L, or M placed at a distance r from the center of object Q (**Figure 18.4a**). Objects K, L, and M have small positive charges $q_K = q$, $q_L = 2q$, and $q_M = 3q$ (here we use lowercase q to denote that these are test objects). Use Coulomb's law to compare the magnitudes of the electric forces that Q exerts on K, L, or M (Figure 18.4b):

$$F_{Q \text{ on } K} = k_C \frac{Q q_K}{r^2} = q_K \left(k_C \frac{Q}{r^2} \right)$$

$$F_{Q \text{ on } L} = k_C \frac{Q q_L}{r^2} = q_L \left(k_C \frac{Q}{r^2} \right) = 2 F_{Q \text{ on } K}$$

$$F_{Q \text{ on } M} = k_C \frac{Q q_M}{r^2} = q_M \left(k_C \frac{Q}{r^2} \right) = 3 F_{Q \text{ on } K}$$

Notice that Q exerts electric forces proportional to the electric charge of the test objects K, L, or M. However, the ratio of the magnitude of the electric force exerted on objects K, L, or M and the electric charge of each object equals the same value:

$$\frac{F_{Q \text{ on } K}}{q_K} = \frac{F_{Q \text{ on } L}}{q_L} = \frac{F_{Q \text{ on } M}}{q_M} = k_C \frac{Q}{r^2}$$

Since these ratios all have the same value, the ratio $(F_{Q \text{ on } K}/q_K)$ could be a mathematical description of the strength of the electric field produced by the charge Q at that location.

We can now think of the objects K, L, and M as probes of the electric field—called **test charges**. We use these test charges to examine the electric field produced by some object Q—the **source charge**. If we agree that a test charge is always positive and relatively small in magnitude so that it has little effect on the source charge, we can define the physical quantity that characterizes the electric field at a location as a vector with magnitude $(F_{Q \text{ on } q_{\text{test}}}/q_{\text{test}})$ that points in the direction of the electric force that is exerted on the positive test charge placed at that location. This physical quantity is called the $\vec{E}$ **field**.

The $\vec{E}$ **field** is a physical quantity that characterizes properties of space around charged objects. To determine the $\vec{E}$ field at a specific location, place an object with a small positive test charge q_{test} at that location and measure the electric force exerted on that object. The $\vec{E}$ field at that location equals the ratio

$$\vec{E} = \frac{\vec{F}_{Q \text{ on } q_{\text{test}}}}{q_{\text{test}}} \tag{18.2}$$

and points in the direction of the electric force exerted on the positive test charge. The $\vec{E}$ field is independent of the test charge used to determine the field. The unit of the electric field is newtons per coulomb (N/C).

TIP Equation (18.2) is an operational definition for the $\vec{E}$ field; the expression $E = k_C Q / r^2$ is the cause-effect relationship.

Using the above operational definition of the $\vec{E}$ field at a point, we found that the magnitude of the $\vec{E}$ field produced by a point-like source object with charge Q was

$$E = \frac{F_{Q \text{ on } q_{\text{test}}}}{q_{\text{test}}} = k_C \frac{|Q| q_{\text{test}}}{r^2 q_{\text{test}}} = k_C \frac{|Q|}{r^2}$$

We can interpret this field as follows:

$$E = k_C \frac{|Q_{\text{Source}}|}{r^2} \tag{18.3}$$

Effect   Cause

The object with electric charge Q is the cause, and the $\vec{E}$ field produced by it is the effect. Note also that the $\vec{E}$ field vector at any location points *away* from the object creating the field if Q is positive and *toward* it if Q is negative.

$\vec{E}$ field due to multiple charged objects

To determine the magnitude and direction of the $\vec{E}$ field at a particular point caused by more than one charged object, we start by investigating the field created by two charged objects in Observational Experiment Table 18.1.

OBSERVATIONAL
EXPERIMENT TABLE **18.1** $\vec{E}$ field due to two charged objects

VIDEO
OET 18.1

Observational experiment	Analysis
Experiment 1. Two identical metal spheres are on insulating stands with equal positive charge. You charge a small aluminum foil ball positively and hang it from an insulated string between the two spheres. When hanging closest to the left sphere, the foil ball is repelled away, toward the right sphere. When hanging nearer the right sphere, it is repelled away, toward the left sphere. At the exact middle between the spheres, the foil ball hangs straight down. (Note that the figure shows the outcomes of three independent experiments.)	Regard the foil ball as a test charge to determine the direction of the $\vec{E}$ field at different places between the spheres. The $\vec{E}$ field points right when nearest the left sphere, is zero in the exact middle between the spheres, and points left when nearest the right sphere.

Experiment 2. Repeat Experiment 1, only this time with the left sphere charged negatively and the right sphere charged positively. When the foil ball hangs at any place between the spheres, the ball is attracted toward the left negative sphere and repelled from the right sphere. (Note that the figure shows the outcomes of three independent experiments.)	The foil ball indicates that the $\vec{E}$ field points left at all points between the spheres. The different tilts of the strings indicate slightly different magnitudes of the $\vec{E}$ field vector.

Patterns

- In Experiment 1, the zero field at the exact middle must have been due to the vector addition of equal-magnitude but oppositely directed $\vec{E}$ fields caused by the two spheres.
- In Experiment 1, when on either side of the middle, the nearer sphere caused a greater repulsion than the farther sphere. The opposing $\vec{E}$ fields only partially canceled each other.
- In Experiment 2, both spheres caused an $\vec{E}$ field to the left, which added to form a stronger $\vec{E}$ field to the left.

It appears that when multiple charged objects are present, each object creates its own contribution to the $\vec{E}$ field. At a particular location, the field is the *vector* sum of the $\vec{E}$ field contributions due to all of these charges. This summation effect is an example of the superposition principle. The fact that we add the $\vec{E}$ field contributions means that they do not affect each other, but they all affect a test charge for $\vec{E}$ fields placed at a particular location.

Superposition principle for $\vec{E}$ fields The $\vec{E}$ field at a point of interest is the *vector* sum of the individual contributions to the $\vec{E}$ field of each charged object, and it is called $\vec{E}_{net}$:

$$\vec{E}_{net} = \vec{E}_1 + \vec{E}_2 + \vec{E}_3 + \cdots \qquad (18.4)$$

Physics Tool Box 18.1 illustrates step by step how to use the superposition principle and graphical vector addition to estimate the direction of the $\vec{E}$ field at a point of interest.

PHYSICS TOOL BOX 18.1 ▸ Estimating the $\vec{E}$ field at a position of interest

1. Place an imaginary positive test charge at the point of interest, A. Use this test charge to estimate the direction of the $\vec{E}$ field contribution of each charged object creating the field at A, which is the same as the direction of the force that each object exerts on the test charge.

2. Draw an arrow in the direction of the $\vec{E}$ field contribution of each charged object q_1, q_2, and q_3.

3. Estimate the approximate relative lengths of the $\vec{E}$ fields due to each charged object.

$\vec{E}_1$ (due to q_1)

A $\vec{E}_3$ (due to q_3)

$\vec{E}_2$ (due to q_2)

$q_3 = +q$ $q_2 = -2q$ $q_1 = +q$

4. Graphically add $\vec{E}_1$, $\vec{E}_2$, and $\vec{E}_3$ to estimate the net $\vec{E}$ field at point A.

$\vec{E}_2$ $\vec{E}_1$

A

$\vec{E}_{net}$

$\vec{E}_3$

CONCEPTUAL EXERCISE 18.1 ▸ The $\vec{E}$ field in body tissue due to the heart dipole charge

The muscles of the heart continually contract and relax, making the heart an electric dipole with equal-magnitude positive and negative electric charges, such as shown in the figure at right. Estimate the direction of the $\vec{E}$ field at position A in the body tissue to the left side of the midpoint between the dipole charges.

A•– – – – – – –   –Q
                    +Q

Sketch and translate Imagine placing a positive test charge at position A to estimate the direction of the forces exerted on it by each dipole charge. We use the directions of those forces to find the direction of the corresponding $\vec{E}$ fields.

Simplify and diagram We draw arrows representing the $\vec{E}$ field at position A due to each dipole charge (see figure at right). Then we graphically add the $\vec{E}$ fields to find the direction of the net $\vec{E}$ field.

$\vec{E}_{net}$ is the vector sum of the $-Q$ and $+Q$ $\vec{E}$ fields.

$\vec{E}_{net}$

⊖ –Q

$\vec{E}_{+Q}$ A $\vec{E}_{-Q}$

⊕ +Q

$\vec{E}_{+Q}$ is the field at A due to $+Q$. $\vec{E}_{-Q}$ is the field at A due to $-Q$.

Try it yourself Repeat this exercise for a position that is in the body tissue several centimeters to the right side of the center of the heart dipole.

Answer

The net $\vec{E}$ field on the right side also points up in the same direction as the $\vec{E}$ field on the left side of the heart dipole.

$\vec{E}$ field lines

We have just learned how to estimate the $\vec{E}$ field at a single location in the vicinity of several charged objects. To represent the overall shape of the $\vec{E}$ field, we use electric field lines, or $\vec{E}$ *field lines*. Imagine two source charges ($+Q$ and $-Q$) arranged as shown in **Figure 18.5a**. We start by using the superposition principle and graphical vector addition to find the net electric field $\vec{E}_{net}$ at position A due to both charges (see Figure 18.5a). Next, we use the same method to draw another $\vec{E}_{net}$ vector for a position near the tip of our first $\vec{E}_{net}$ vector (Figure 18.5b). The field at this location points toward the right and downward a little, since the negative charge $-Q$ is slightly closer and more attractive to a positive test charge at that point than the slightly more distant positive charge $+Q$ is repulsive to it.

Next, we draw a third $\vec{E}$ field vector for a position near the tip of the second $\vec{E}$ field vector. We continue drawing the $\vec{E}_{net}$ vectors for adjacent positions along the direction of the previous $\vec{E}_{net}$ vector, eventually getting a series of $\vec{E}_{net}$ vectors that seem to follow one after the other from the positive charge $+Q$ to the negative charge $-Q$, as shown in Figure 18.5c. The $\vec{E}_{net}$ vectors from the positive charge to A have a similar shape and are also shown in Figure 18.5c. Now, draw a single line that passes through each position tangent to these vectors. The line looks as shown in Figure 18.5d. If you repeat the process for a series of positions farther and closer to a line between the two charges, you get a series of lines such as shown in Figure 18.5e—the $\vec{E}$ field lines for the source charges shown in Figure 18.5a.

TIP In this book all figures are two-dimensional, but in real life the situations are three-dimensional. The $\vec{E}$ field vectors and $\vec{E}$ field lines are really in a three-dimensional space.

Using a similar process, you could construct the $\vec{E}$ field lines for the single positively charged object shown in **Figure 18.6a**, for the single negatively charged object shown in Figure 18.6b, and for the two equal-magnitude positively charged objects shown in Figure 18.6c. Note that at locations closer to the charged objects, the lines are closer together than when the locations are farther away. This means that the number of lines per unit area (the density of the lines) is larger where the $\vec{E}$ field is stronger. Similarly, if you have two source charges, then at the same relative location with respect to the charges, the magnitude of the $\vec{E}$ field is larger next to a bigger charge. Thus the number of lines emanating from the charge represents the magnitude of that charge. If you have two charged objects next to each other with the magnitude of one charge twice the magnitude of the other, and you draw six lines coming out of the first object, then the second object should have 12 lines coming out of it.

Summary—$\vec{E}$ field lines

- $\vec{E}$ field lines are drawn so that the $\vec{E}$ field vectors at positions on those lines are tangent to the lines.
- $\vec{E}$ field lines begin on positively charged objects and end on negatively charged objects (these charged objects do not necessarily have to be present on the sketches).
- The number of $\vec{E}$ field lines beginning or ending on a charged object is proportional to the magnitude of that object's charge. Therefore, on an electric field diagram you will see twice as many $\vec{E}$ field lines leaving a charge of $+2q$ as leaving a charge of $+1q$.
- The density of the $\vec{E}$ field lines in a region is proportional to the magnitude of the $\vec{E}$ field in that region.
- The lines do not cross; if they did, the force exerted on a test charge placed in that location would be undetermined.

FIGURE 18.5 Constructing $\vec{E}$ field lines.

(a) The net field at A due to the source charges

(b) We draw the $\vec{E}_{net}$ vector for a position to the right of A. We repeat the process.

(c)

A series of $\vec{E}_{net}$ vectors

(d)

We draw the $\vec{E}$ field line tangent to $\vec{E}_{net}$ vectors at each position on the line

(e) A less dense and weaker field

A denser and stronger field

Lines point from $+$ to $-$ charge.

FIGURE 18.6 $\vec{E}$ field lines for (a) a single positive charge, (b) a single negative charge, and (c) two positive charges.

(a)

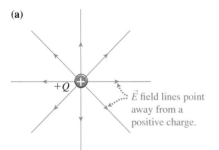

$\vec{E}$ field lines point away from a positive charge.

(b)

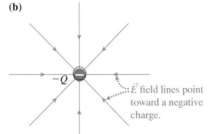

$\vec{E}$ field lines point toward a negative charge.

(c)

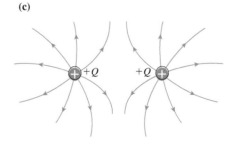

> TIP   The $\vec{E}$ field lines give information about the force that the electric field exerts on a test charge placed at any location, and the resulting acceleration of the test charge, but not the direction of motion of the test charge.

CONCEPTUAL EXERCISE 18.2   Uniform $\vec{E}$ field

Draw $\vec{E}$ field lines for a large, uniformly charged plate of glass.

Sketch and translate   A uniformly charged plate of glass is shown in the figure at right. "Uniformly charged" means that the amount of electric charge located on each unit area of the surface is constant throughout the plate.

Simplify and diagram   Assume that the plate of glass is infinitely large in both perpendicular directions in the plane of its surface. Choose a position of interest to the right of the plate. Think of the $\vec{E}$ field at this position as caused by each small segment of positive source charge on the plate. Notice that the y-component of the $\vec{E}$ field from one of the charge segments on the plate (for example, the position 1 segment in the figure below) is canceled by the y-component of the $\vec{E}$ field from a charge segment at some other location on the plate (position 2). This occurs for every charge segment. Therefore, the y-component of the $\vec{E}$ field is zero at every position to the right of the plate.

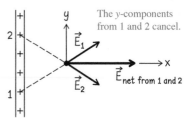

Consequently, the $\vec{E}$ field is perpendicular to the plate (see the first figure in the next column) at every point to the right of the plate (close or far away). The $\vec{E}$ field lines are parallel at every position. Therefore,

the $\vec{E}$ field lines must point away from the plate and be perpendicular to it (see the second figure below).

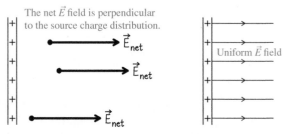

The plate is infinitely large, so the $\vec{E}$ lines should look the same at any point on the right side of the plate. If the $\vec{E}$ field decreased in magnitude as you moved farther from the plate, the lines would have to spread farther apart. But they don't. The field lines must have a uniform density—the same separation between the $\vec{E}$ field lines everywhere. This means the $\vec{E}$ field has the same magnitude everywhere in this region. The uniform charge on the plate produces a uniform $\vec{E}$ field.

--- --- --- --- ---

Try it yourself   Imagine that an infinitely large, uniformly positively charged plate oriented in the plane perpendicular to the page produces an $\vec{E}$ field of magnitude E. You add another plate, negatively charged, at a distance d from the first one and parallel to it. What is the magnitude of the $\vec{E}$ field to the left of the plates, between them, and to the right? Note that the fields created by the two plates are independent of each other.

Answer   0, 2E, 0.

The analysis of the uniform $\vec{E}$ field in the last Conceptual Exercise will be very useful later in the chapter, especially when we study capacitors.

REVIEW QUESTION 18.1   How do you estimate the direction of the $\vec{E}$ field at a point located near two point-like charged objects?

18.2 Skills for analyzing processes involving $\vec{E}$ fields

You will encounter two main types of problems involving $\vec{E}$ fields. In the first type, you will know the charge distribution and must determine the $\vec{E}$ field at one or more points caused by these source charges. In the second type, you will need to analyze a process involving a charged object that is stationary or moving in a given $\vec{E}$ field. Let's start by describing a strategy for solving the first type of problem.

Determining the $\vec{E}$ field produced by given source charges

When you have several objects (with charges Q_1, Q_2, etc.) contributing to the electric field at a particular location, to determine the direction and magnitude of the $\vec{E}_{net}$ field vector at that location, you first need to follow the steps in Physics Tool Box 18.1 to qualitatively determine the direction of $\vec{E}_{net}$. You can then start the calculations. Choose coordinate axes and write expressions for the magnitudes of the $\vec{E}$ field vector contributions of each charged object:

$$E_1 = k_C \frac{|Q_1|}{r_1^2}; \; E_2 = k_C \frac{|Q_2|}{r_2^2}; \text{ etc.}$$

Then find the x- and y-components of each contribution and add the respective components algebraically to find $E_{x \, net}$ and $E_{y \, net}$, which are the components of $\vec{E}_{net}$. Knowing these components, you can use the Pythagorean theorem to determine the magnitude of $\vec{E}_{net}$:

$$E_{net} = \sqrt{E_{x \, net}^2 + E_{y \, net}^2}$$

and next use trigonometry to determine as needed the angles that $\vec{E}_{net}$ makes with the coordinate axes.

This approach is implemented in Example 18.3. Follow the example closely and focus on each step in turn.

EXAMPLE 18.3 $\vec{E}$ **field due to multiple charges**

Two small metal spheres attached to insulating stands reside on a table a distance d apart. The left sphere has positive charge $+q$ and the right has negative charge $-q$. Determine the magnitude and direction for the net $\vec{E}$ field at a distance d above the center of the line connecting the spheres.

Sketch and translate The sketch shows the source charges $+q$ and $-q$ that are producing the $\vec{E}$ field and the place (the point of interest) where we want to determine the net $\vec{E}$ field—both its magnitude and direction. To estimate the direction of the vectors, imagine placing a positively charged test object q_T at that point. (Remember, the fields exist without the test charge.)

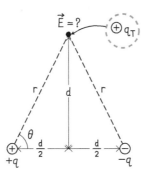

Simplify and diagram Next, construct an $\vec{E}$ field diagram for the point of interest, showing an $\vec{E}$ field arrow for each source charge that contributes to the $\vec{E}$ field at that point (see figure at right). Include coordinate axes in the diagram. Notice that the $\vec{E}$ field vectors due to each source charge point in the direction of the electric force that each of the source charges would exert on the imaginary positive charge placed at the location of interest.

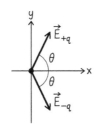

Represent mathematically Use Eq. (18.3) to determine the magnitude of the $\vec{E}$ field contributions from both $+q$ and $-q$: $E_{+q} = E_{-q} = k_C(q/r^2)$.

Use the Pythagorean theorem to determine the distance r from $+q$ and $-q$ to the point of interest:

$$r^2 = \left(\frac{d}{2}\right)^2 + d^2 = \frac{5d^2}{4}$$

Thus, the magnitude of the $\vec{E}$ field contributions from both $+q$ and $-q$ is

$$E_{+q} = E_{-q} = \frac{4k_C q}{5d^2}$$

The angle θ that each $\vec{E}$ field vector makes relative to the positive x-axis can be determined using the triangle in the first figure:

$$\tan \theta = \frac{\text{opposite side}}{\text{adjacent side}} = \frac{d}{d/2} = 2.0 \quad \text{or} \quad \theta = 63.4°$$

Now determine the x- and y-components of the $\vec{E}$ field.

$$E_{x \, net} = E_{+qx} + E_{-qx} = E_{+q} \cos \theta + E_{-q} \cos \theta$$
$$= 2E_{+q} \cos \theta = 2\left(\frac{4k_C q}{5d^2}\right) \cos 63.4° = 0.716 \frac{k_C q}{d^2}$$

$$E_{y \, net} = E_{+qy} + E_{-qy}$$
$$= +E_{+q} \sin 63.4° + (-E_{-q}) \sin 63.4° = 0$$

The y-components of the two $\vec{E}$ field contributions add to zero.

Solve and evaluate The magnitude of the net $\vec{E}$ field is

$$E_{net} = \sqrt{E_{x \, net}^2 + E_{y \, net}^2} = \sqrt{\left(\frac{0.716 \, k_C q}{d^2}\right)^2 + 0^2} = 0.72 \frac{k_C q}{d^2}$$

and it points in the positive x-direction.

(CONTINUED)

Try it yourself Determine the $\vec{E}$ field at point I in the figure at right. (a) Determine the magnitudes of the $\vec{E}$ field contributions at point I from charges $+q$ and $-q$. (b) Determine the x- and y-components of those $\vec{E}$ field contributions. (c) Determine the net x- and y-components of the $\vec{E}$ field at point I. (d) Finally, determine the magnitude and the direction of the $\vec{E}$ field at point I.

Answer

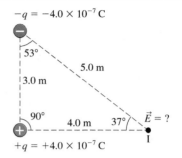

(a) 225 N/C, 144 N/C; (b) (225 N/C, 0) and (−115 N/C, +87 N/C); (c) (+110 N/C, +87 N/C); (d) 140 N/C, 38° above +x-axis (toward the right).

Analyzing a process involving a known $\vec{E}$ field

If the $\vec{E}$ field at some location is known, we can use Eq. (18.2) to determine the electric force that the field exerts on an object with charge q if at that location:

$$\vec{F}_{\vec{E} \text{ on } q} = q\vec{E} \tag{18.5}$$

We can now combine our understanding of electric field and force with our knowledge of Newton's laws to solve problems.

PROBLEM-SOLVING
STRATEGY 18.1   Electric field and Newton's second law

EXAMPLE 18.4 **A charged object in a known $\vec{E}$ field**

A spring made of a dielectric material with spring constant 5.0 N/m hangs from a large, uniformly charged horizontal plate. The uniform $\vec{E}$ field produced by the plate has magnitude 1.0×10^6 N/C and points down. A 0.020-kg ball with charge $+4.0 \times 10^{-8}$ C hangs at the end of the spring. Determine the distance the spring is stretched from its equilibrium length. Assume that $g = 10$ N/kg.

Sketch and translate

- Draw a labeled sketch of the situation. Include the symbols for the known and unknown quantities that you plan to use.
- Choose the system.

The situation is sketched at the right. The spring has stretched distance Δy from its equilibrium length. The charged ball (B) is the system.

Simplify and diagram

- Determine the $\vec{E}$ field produced by the environment. Is it produced by point-like charges (making it nonuniform) or by large charged plates (making it uniform)?
- Construct a force diagram for the system if necessary. Do not forget the axes.
- State any assumptions you have made.

The field is uniform. The upward direction is positive.
Assume that the spring has no mass.
A force diagram for the ball is shown at the right. S stands for the spring and P for the plate.

Represent mathematically

- Use the force diagram to help apply Newton's second law in component form. Use component addition to determine the net force along each coordinate axis. Determine the acceleration of the system if needed.

- If necessary, use kinematics equations to describe the motion of the system.

Solve and evaluate

- Combine the above equations and complete the solution.

- Check the direction, magnitude, and units and decide whether the result makes sense in limiting cases.

The ball is not accelerating, so the forces exerted on the ball add to zero. None of the forces have x-components:

$$F_{S \text{ on B } y} + F_{E \text{ on B } y} + F_{P \text{ on B } y} = 0$$
$$\Rightarrow k\,\Delta y + (-mg) + (-qE) = 0$$
$$\Rightarrow \Delta y = \frac{mg + qE}{k}$$

$$\Delta y = \frac{(0.02\ \text{kg})(10\ \text{N/kg}) + (4.0 \times 10^{-8}\ \text{C})(1.0 \times 10^{6}\ \text{N/C})}{5.0\ \text{N/m}}$$

$$= 0.048\ \text{m} = 48\ \text{mm}$$

The units for the stretch distance are the units of length, as they should be. Check using limiting cases. As both the gravitational force and the electric force point in the same direction, eliminating either of them should reduce the distance that the spring stretches. Note in the equation above that Δy decreases if $m = 0$ or if $q = 0$. Also, a stiff spring (larger k) should stretch less if we have a ball of the same mass and the same electric charge—also consistent with our result.

Try it yourself If the plate were negatively charged, producing an $\vec{E}$ field with the same magnitude but now pointing up, what would the spring stretch or compression be?

Answer   ·(pǝɥɔʇǝɹʇs sı ƃuıɹds ǝɥʇ) ɯɯ zε+

EXAMPLE 18.5   Electric field deflects a tiny ink ball

Inside an inkjet printer, a tiny ball of black ink of mass 1.1×10^{-11} kg with charge -6.7×10^{-12} C moves horizontally at 40 m/s. The ink ball enters an upward-pointing uniform $\vec{E}$ field of magnitude 1.0×10^{4} N/C produced by a negatively charged plate above and a positively charged plate below. The plates are used to deflect the ink ball so that it lands at a particular spot on a piece of paper. Determine the deflection of the ink ball after it travels 0.010 m in the $\vec{E}$ field. Assume that $g = 10$ N/kg.

Sketch and translate After sketching the situation (see figure below), we choose the charged ink ball as the system. We break the problem into two parts: first we determine the acceleration of the ball due to the forces being exerted on it, and then we use kinematics to determine its vertical displacement as it passes through the region with the electric field.

$m_B = 1.1 \times 10^{-11}$ kg
$q_B = -6.7 \times 10^{-12}$ C
$x_0 = 0$
$y_0 = 0$
$v_{0x} = 40$ m/s
$E = 1.0 \times 10^{4}$ N/C
$y = ?$

0.01 m

A negatively charged ink ball is deflected downward by the electric field.

Simplify and diagram Assume that the ink ball is a point-like object and that the plates produce a uniform $\vec{E}$ field. A force diagram for the ball is shown at right. Earth and the electric field created by the plates (P) exert downward forces on the negatively charged ball (B).

Represent mathematically The forces exerted on the ink ball do not have x-components, so we use only the y-component form of Newton's second law to determine the y-component of the acceleration (we choose the positive y-direction pointing down):

$$a_y = \frac{1}{m_B}\sum F_y = \frac{1}{m_B}\left(F_{P \text{ on B } y} + F_{E \text{ on B } y}\right)$$

$$= \frac{1}{m_B}\left(q_B E_y + m_B g_y\right) = \frac{q_B E_y}{m_B} + g_y$$

The ink ball does not accelerate in the x-direction, so the x-component of its velocity remains constant at 40 m/s during the time interval it takes to traverse the region with nonzero $\vec{E}$ field. During this time interval, $\Delta t = \Delta x / v_{0x}$, the ball's displacement in the y-direction is

$$y - y_0 = v_{0y}\,\Delta t + \frac{1}{2}a_y\,\Delta t^2$$

(CONTINUED)

Solve and evaluate Substituting for a_y, Δt, and $v_{0y} = 0$ gives

$$y - y_0 = 0 + \frac{1}{2}\left(\frac{q_B E_y}{m_B} + g_y\right)\left(\frac{\Delta x}{v_{0x}}\right)^2$$

We can now insert the known information ($E_y = -1.0 \times 10^4\,\text{N/C}$, $q_B = -6.7 \times 10^{-12}\,\text{C}$, $m_B = 1.1 \times 10^{-11}\,\text{kg}$, and $g_y = +10\,\text{N/kg}$) into the above to determine the ink ball deflection:

$$y - y_0 = \frac{1}{2}\left(\frac{(-6.7 \times 10^{-12}\,\text{C})(-1.0 \times 10^4\,\text{N/C})}{1.1 \times 10^{-11}\,\text{kg}} + 10\,\text{N/kg}\right)$$

$$\times \left(\frac{0.010\,\text{m}}{40\,\text{m/s}}\right)^2 = 1.9 \times 10^{-4}\,\text{m} = 0.19\,\text{mm}$$

The distance unit is correct, and the value seems reasonable given the size of a letter on a printed page. To get a bigger deflection, we can use a larger electric field.

- -

Try it yourself What should the mass of the ink ball be so that if the direction of the $\vec{E}$ field is reversed, the ball will have zero acceleration?

Answer $6.7 \times 10^{-9}\,\text{kg}.$

REVIEW QUESTION 18.2 You have a point-like object with charge $+Q$ on the left and a second object with charge $-2Q$ at a distance d on the right. Sketch the $\vec{E}$ field vector at the same distance d directly above the midpoint between the charged objects.

18.3 The V field: electric potential

So far, we have been studying electric fields using a force approach. However, in many problems in mechanics and electrostatics, an energy approach allows us to focus on the initial and final stages of a process without knowing what happened in between. Can we describe properties of space around charged objects using the concepts of work and energy? In order to do this, we need to describe space around charged objects not as the force-related $\vec{E}$ field, but instead as an energy-related field.

Electric potential due to a single charged object

FIGURE 18.7 Consider the electric potential energy of a system with source charge Q and test charge $q = q_K$, q_L, or q_M.

To construct this new way of describing space around charged objects, we will think of the electric interaction in terms of the electric potential energy. The source charge Q creates an electric field around it, and charged objects K, L, or M with different small electric charges q_K, q_L, or q_M (test charges) are placed one at a time at a distance r from Q (Figure 18.7). Recall (from Chapter 17) that if we consider the electric potential energy of two charged objects to be zero when they are infinitely far from each other, then the electric potential energy of two point-like charged objects 1 and 2 separated by a distance r is

$$U_{q_1 q_2} = \frac{k_C q_1 q_2}{r}$$

Thus, the electric potential energy of charges Q and q_K, q_L, or q_M is

$$U_{Q q_K} = \frac{k_C Q q_K}{r} = q_K\left(\frac{k_C Q}{r}\right)$$

$$U_{Q q_L} = \frac{k_C Q q_L}{r} = q_L\left(\frac{k_C Q}{r}\right)$$

$$U_{Q q_M} = \frac{k_C Q q_M}{r} = q_M\left(\frac{k_C Q}{r}\right)$$

If we divide the electric potential energy of the Q-q_K, Q-q_L, or Q-q_M charge pair by the charges q_K, q_L, or q_M (that is, U_{qQ}/q), we find that all three ratios equal the same quantity:

$$\frac{U_{Q q_K}}{q_K} = \frac{U_{Q q_L}}{q_L} = \frac{U_{Q q_M}}{q_M} = \frac{k_C Q}{r}$$

Since these ratios all have the same value (kQ/r), this ratio could be a mathematical representation of another physical quantity characterizing space around charge Q at a distance r from Q. We call this physical quantity the **V field** or **electric potential** at a particular location that is a distance r from the charge Q creating the field.

V field (or electric potential) due to a single charge To determine the *V* field due to a single source charge at a specific location, place a test charge q_{test} at that location and determine the electric potential energy $U_{Qq_{test}}$ of a system consisting of the test charge and the source charge that creates the field. The *V* field at that location equals the ratio

$$V = \frac{U_{Qq_{test}}}{q_{test}} = \frac{k_C Q}{r} \qquad (18.6a)$$

The unit of electric potential is joule/coulomb (J/C) and is called the **volt** (V).

We can also think of the *V* field at a specific location (let's say location A) as the ratio of the work that an external force needs to do on a positive test charge q_{test} to bring it to the location A from infinity and that charge:

$$V_A = \frac{W_{from \infty to A}}{q_{test}} \qquad (18.6b)$$

TIP

- Both the *V* field and the $\vec{E}$ field at a specific location are independent of the test charge and characterize the properties of space at that location.
- Unlike the $\vec{E}$ field, which is a vector quantity, the *V* field is a scalar quantity with a sign that depends on the sign of the charge that creates it.

The superposition principle and the *V* field due to multiple charges

Now that we can determine mathematically the *V* field produced by a single point-like charged object [Eq. (18.6)], we can use the superposition principle to determine the *V* field at a specific location produced by several charged objects. Using the same superposition principle idea we used for the $\vec{E}$ field, we have

$$V = V_1 + V_2 + V_3 + \cdots = \frac{k_C Q_1}{r_1} + \frac{k_C Q_2}{r_2} + \frac{k_C Q_3}{r_3} + \cdots \qquad (18.7)$$

where $Q_1, Q_2, Q_3, \ldots$ are the source charges (including their signs) creating the field and $r_1, r_2, r_3, \ldots$ are the distances between the source charges and the location where we are determining the *V* field. Because the *V* field is a scalar field rather than a vector field (like the $\vec{E}$ field is), it is much easier to apply the superposition principle.

QUANTITATIVE EXERCISE 18.6 **Electric potential due to heart dipole**

Suppose that the heart dipole charges $-Q$ and $+Q$ at a particular instant (the heart does change during the cardiac cycle) are separated by distance d as in the figure. Write an expression for the *V* field due to both charges at point A, a distance d to the right of the $+Q$ charge.

$V_A = ?$

Represent mathematically Use Eq. (18.7) to write an expression for the *V* field:

$$V = V_{+Q} + V_{-Q} = \frac{k_C(+Q)}{r_{+Q}} + \frac{k_C(-Q)}{r_{-Q}}$$

Solve and evaluate Inserting the distances, we get

$$V = \frac{k_C(+Q)}{d} + \frac{k_C(-Q)}{2d} = +\frac{k_C Q}{2d}$$

Try it yourself Write an expression for the *V* field at a distance d to the left of $-Q$.

Answer

$$V = \frac{k_C(-Q)}{d} + \frac{k_C(+Q)}{2d} = -\frac{k_C Q}{2d}.$$

Finding the electric potential energy when the V field is known

If we know the electric potential at a specific location, we can rearrange the definition of the V field [Eq. (18.6)] to determine the electric potential energy of a system that includes the source charges and another charge q that is in the vicinity of the source charges:

$$U_{\text{Source } Q \text{ and } q} = qV_{\text{Due to source } Q} \qquad (18.8)$$

For example, we could use this equation to determine the electric potential energy of the heart dipole in Quantitative Exercise 18.6 and a sodium ion in the tissue near the dipole.

Since the value of the electric potential depends on the choice of zero level (as does electric potential energy), we often use the difference in electric potential between two points, called the **potential difference**, to analyze processes using electric potential.

> **Potential difference** ΔV between two points A and B is equal to the difference in the values of electric potential at those points: $\Delta V = V_B - V_A$. Potential difference is also called **voltage**. This term for potential difference is often used in electric circuits, the subject of the next chapter.

Potential difference is a very useful quantity—if we know the potential difference between two points A and B, we can predict the direction of acceleration of a charged object. A positively charged object accelerates from regions of higher electric potential toward regions of lower potential (like an object falling to lower elevation in Earth's gravitational field). A negatively charged particle tends to do the opposite, accelerating from regions of lower potential toward regions of higher potential. Consider the next example.

EXAMPLE 18.7   X-ray machine

Inside an X-ray machine is a wire (called a filament) that, when hot, ejects electrons. Imagine one of those electrons, now located outside the wire. It starts at rest and accelerates through a region where the V field increases by 40,000 V. The electron stops abruptly when it hits a piece of tungsten at the other side of the region, producing X-rays (we will learn more about that in Chapter 27). How fast is the electron moving just before it reaches the tungsten?

Sketch and translate The situation is sketched in the figure. The system consists of the electron and the electric field of the tube. We choose the zero level of the V field to be at the initial position of the electron, just outside the filament.

Simplify and diagram Neglect the gravitational force that Earth exerts on the electron given that the time of flight is very short. A work-energy bar chart for the process is shown at right. In the initial state, the system has neither electric potential energy (the electric

potential is zero there) nor kinetic energy (the electron is not moving). In the final state, the system has positive kinetic energy and negative electric potential energy. The electric potential energy is determined using $U_q = qV$ with $q < 0$ and $V_f > 0$.

Represent mathematically Each nonzero bar in the bar chart turns into a term in the generalized work-energy equation:

$$0 = \tfrac{1}{2} m_e v_f^2 + (-e)V_f$$

Solve and evaluate

$$v_f = \sqrt{-\frac{2(-e)V_f}{m_e}}$$

$$= \sqrt{\frac{-2(-1.6 \times 10^{-19}\,\text{C})(4.0 \times 10^4\,\text{V})}{9.11 \times 10^{-31}\,\text{kg}}} = 1.2 \times 10^8\,\text{m/s}$$

Limiting case check: If the electric charge of the electron were zero, it would not participate in the electric interaction at all and its final speed should be zero; it is. If the electric potential at the final location were the same as at the initial location, the final speed of the electron should be zero; and it is. Additionally, the speed of the electron is very high, similar to Example 17.8; the result obtained with the corrections for higher speeds that we will learn in Chapter 26 gives us an estimate that is three times smaller but still very high.

Try it yourself A 0.10-kg cart with a charge of $+6.0 \times 10^{-5}$ C rests on a 37° incline. Through what potential difference must the cart move along the incline so that it travels 2.0 m up the incline before it stops? The cart starts at rest. Assume that $g = 10$ N/kg.

Answer The potential difference needed is $-20{,}000$ V.

Equipotential surfaces—representing the *V* field

$\vec{E}$ field lines help us visualize the $\vec{E}$ field. We will now develop a new way to visualize the *V* field at different points surrounding one or more charged objects. We start by considering a positively charged point-like charged object $+Q$ (shown in **Figure 18.8**). The values of the *V* field at points A, B, and C are all the same. The values of the *V* field at D, E, and F are all the same but different than the values at A, B, and C (in this case, they are smaller than the values for A, B, and C). If we imagine the surface to which all the points of equal *V* field value belong (for example, points A, B, and C), that surface will be a sphere surrounding the source charge $+Q$. These surfaces of constant electric potential *V* are called **equipotential surfaces**. The surfaces are spheres (they look like circles on a two-dimensional page) because the *V* field produced by a point-like charged object has the same value at all points equidistant from the charged object. The electric potential energy of a system with the source charge $+Q$ and a test charge $+q$ is constant if the test charge remains on an equipotential surface. No work is needed to move the test charge $+q$ to some other place on that equipotential surface.

The $\vec{E}$ field lines for a single positively charged source charge point outward along radial lines. The spherical equipotential surfaces are perpendicular to the $\vec{E}$ field lines at every point (**Figure 18.9a**). This is a general feature of every charge distribution; the $\vec{E}$ field lines and the equipotential surfaces are perpendicular to each other at every point in the region surrounding the charges (see the $\vec{E}$ field lines and the *V* field surfaces for the dipole and two equal-magnitude positive charge distributions in Figures 18.9b and c). Note also that the values of the *V* field of the surfaces decrease in the direction of the $\vec{E}$ field lines.

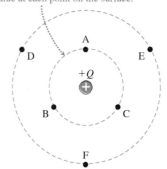

FIGURE 18.8 Equipotential surfaces produced by an object with charge $+Q$.

An equipotential surface has the same *V* field value at each point on the surface.

FIGURE 18.9 The equipotential surfaces and $\vec{E}$ field lines produced by (a) a positively charged point object, (b) an electric dipole, and (c) two equal-magnitude positively charged objects.

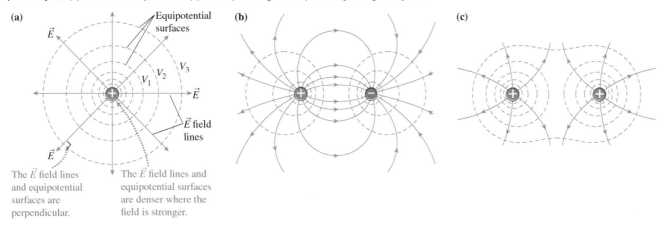

(a)
Equipotential surfaces

V_1 V_2 V_3

$\vec{E}$ field lines

The $\vec{E}$ field lines and equipotential surfaces are perpendicular.

The $\vec{E}$ field lines and equipotential surfaces are denser where the field is stronger.

The relative distances between neighboring equipotential surfaces that vary by the same amount (for example, 10 V) are smaller where the $\vec{E}$ field is greater—the *V* field changes faster where the $\vec{E}$ field is stronger. Consider the equipotential surfaces and the

FIGURE 18.10 Changes in the V field created by point-like charge Q shown (a) on a V-versus-r graph and (b) using equipotential surfaces.

(a) V (V)

The equipotential surfaces are close when the electric potential changes rapidly.

(b)

FIGURE 18.11 Visualizing equipotential surfaces using an elastic sheet.

FIGURE 18.12 An experiment to derive a relationship between the $\vec{E}$ field and the V field.

(a) A stick pushes on a positively charged object in a uniform $\vec{E}$ field.

(b) The charged object is not accelerating, so the forces add to zero.

(c) The stick (not in the system) does negative work on the charged object.

(d) ΔV is constant in this case because E is constant.

graph of V field-versus-distance from $+Q$ (V-versus-r graph) shown in **Figure 18.10**. The charge creating the field is $+1.1 \times 10^{-11}$ C. The closer to $+Q$, the more rapidly the V field changes with changing distance r. The adjacent equipotential surfaces vary by the same amount of potential from one surface to the next and are closer together when nearer $+Q$ and farther apart when farther away from $+Q$. The density of the $\vec{E}$ field lines is greatest near the $+Q$ charge and less dense when far from it, as shown in Figures 18.9b and c.

The elastic sheet analogy that we used in Section 18.1 to visualize an electric field can also help us to visualize V fields and equipotential surfaces.

The disturbance of the elastic sheet due to the billiard ball can also be described by assigning to each point on the sheet a number that tells us the energy of the billiard ball–ball bearing system. In our case, the analog of the electric potential energy is gravitational potential energy. We choose this energy to be zero when the ball bearing is very far away from the billiard ball, where the sheet is undisturbed (horizontal). As the ball bearing approaches the billiard ball, the sheet is deformed and the energy of the system is no longer zero. If we divide this energy by the mass of the ball bearing, the value will depend only on the location, not on the mass, of the ball bearing. In this way, each point on the sheet can be assigned a number (a scalar). These numbers represent the analogous V field on the elastic sheet. In our case, points that have the same analogous V value form concentric circles around the billiard ball (see **Figure 18.11**). When the ball bearing is moved around one of these circles, the energy of the system remains the same regardless of where on the circle the ball bearing is. A different circle indicates a different value of analogous V. Thus these circles are analogous to equipotential surfaces in the electric field.

REVIEW QUESTION 18.3 Compare the work needed to move an object with charge q from point A to point B to the work needed to move the same object from point C to point D in the electric potential region depicted in Figure 18.10a. Explain.

18.4 Relating the $\vec{E}$ field and the V field

We know qualitatively that the V field varies most rapidly with position where the $\vec{E}$ field is strongest. We can also examine this idea quantitatively.

Deriving a relation between the $\vec{E}$ field and ΔV

Consider the uniform $\vec{E}$ field produced by an electrically charged infinitely large glass plate. We attach a small object with charge $+q$ to the end of a wooden stick and place the charged object and stick in the electric field produced by the plate (**Figure 18.12a**). The electric field exerts a force on the charged object in the positive x-direction $F_{\vec{E} \text{ on O } x} = +qE_x$. The stick exerts a force on the charged object with an x-component $F_{\text{S on O } x} = -F_{\text{S on O}}$ pointing toward the plate (Figure 18.12b) in the negative x-direction. If the charged object does not move or moves slowly with zero acceleration, the x-component form of Newton's second law applied to the charged object is $\Sigma F_x = F_{\vec{E} \text{ on O } x} + F_{\text{S on O } x} = +qE_x - F_{\text{S on O}} = 0$, or

$$F_{\text{S on O}} = qE_x$$

Now, let's do a work-energy analysis for a process in which the charged object (still attached to the stick) is moved slowly a small distance Δx farther from the plate (Figure 18.12c). For our system we choose the charged object and the electric field, but not the stick, which is part of the environment and does work on the system. The stick exerts a force on the charged object opposite its displacement and does negative work on the system:

$$W = F_{\text{S on O}} \Delta x \cos(180°) = -F_{\text{S on O}} \Delta x = -qE_x \Delta x$$

The only energy change is the system's electric potential energy because the positively charged object moves farther away from the positively charged plate. Applying the generalized work-energy principle, we get $U_{qi} + W = U_{qf}$, or

$$W = \Delta U_q = q\,\Delta V$$

Setting these two expressions for work equal to each other and canceling the common q, we get

$$\Delta V = -E_x\,\Delta x \qquad (18.9)$$

Equivalently, the component of the $\vec{E}$ field along the line connecting two points on the x-axis is the negative change of the V field divided by the distance between those two points:

$$E_x = -\frac{\Delta V}{\Delta x} \qquad (18.10)$$

TIP Equation (18.10) suggests that another unit for the $\vec{E}$ field is V/m.

The magnitude of the $\vec{E}$ field component in a particular direction indicates how fast the V field (electric potential) changes in that direction (Figure 18.12d). The $\vec{E}$ field vector points in the direction in which the V field decreases fastest with position, hence the minus sign in Eq. (18.10). Similar equations apply for other directions if the situation is in two or three dimensions. Although we derived Eqs. (18.9) and (18.10) using the example of a uniform $\vec{E}$ field, the equations represent a general result that relates the component of the $\vec{E}$ field in the chosen direction to the rate of change of the V field in that direction. The relation between the $\vec{E}$ field and V field tells us two things: (1) in a region where the V field is constant, the $\vec{E}$ field is zero, and (2) if you have two points at different potentials, the closer those points are, the stronger the $\vec{E}$ field between them will be.

QUANTITATIVE EXERCISE 18.8   **Connecting electric potential to force and energy**

The figure below shows a region of space with an electric field, represented using vertical lines for equipotential surfaces with the corresponding values of the potential. (a) What are the forces exerted by the electric field on a particle with charge -20 nC placed at locations A and B? (b) The particle is released from rest at point A. What is its kinetic energy at point B? The mass of the particle is very small, so you can ignore any changes in gravitational potential energy.

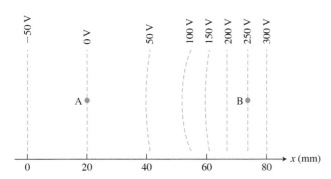

Represent mathematically (a) Because the problem situation is one-dimensional, we can use Eq. (18.10) to determine the component of the $\vec{E}$ field in the x-direction and then use this information to determine the forces exerted on the particle at two locations.

$$E_x = -\frac{\Delta V}{\Delta x}$$

To determine the forces, we use Eq. (18.2) in component form:

$$E_x = \frac{F}{q_{\text{test}}} \;\Rightarrow\; F = E_x q_{\text{test}}$$

In our case, this means $F_x = qE_x$.

(b) Using the work-energy principle, we can write $U_{qi} + K_i = U_{qf} + K_f$ for the system consisting of the charged particle and the electric field. Given that the initial kinetic energy is zero, the final kinetic energy is equal to the change in electric potential energy:

$$K_f = U_{qi} - U_{qf} = q(V_i - V_f)$$

Solve and evaluate (a) From the figure we see that in the region near point A, the V field changes by 50 V over the distance of 20 mm. From these data, we get

$$E_{Ax} = -\frac{\Delta V}{\Delta x} = -2500\text{ V/m}$$

Near B, the V field changes by 50 V over the distance of 20/3 mm; therefore, we get

$$E_{Bx} = -7500\text{ V/m}$$

The minus signs indicate that the $\vec{E}$ field points in the negative x-direction.

(CONTINUED)

Next, we determine the forces exerted on the particle at two locations:

$$F_{Ax} = qE_{Ax} = (-20 \times 10^{-9}\,\text{C})(-2500\,\text{N/C}) = 5 \times 10^{-5}\,\text{N}$$
$$F_{Bx} = qE_{Bx} = (-20 \times 10^{-9}\,\text{C})(-7500\,\text{N/C}) = 15 \times 10^{-5}\,\text{N}$$

Both force components are positive; this means that the forces point in the positive x-direction. Note that when calculating the force it is more convenient to use units N/C than V/m for the $\vec{E}$ field.

(b) $K_f = q(V_i - V_f) = (-20 \times 10^{-9}\,\text{C})(0\,\text{V} - 250\,\text{V}) = 5 \times 10^{-6}\,\text{J}$.

The components of the forces exerted on the particle are positive: does this make sense? Since the V field is decreasing in the negative x-direction, the $\vec{E}$ field points in this direction. Therefore, the force exerted on the negative charge at A and B must point in the positive x-direction. Thus our answer about forces makes sense. Given that the negatively charged particle accelerates in the direction of decreasing potential, the increase in kinetic energy also makes sense.

Try it yourself What is the maximum kinetic energy of an electron launched from point B toward A such that the electron does not pass the 50-V equipotential line?

Answer $K < 2.4 \times 10^{-17}\,\text{J}$.

REVIEW QUESTION 18.4 Imagine that you have an electric field in some large region of space. The electric potential at location P is 10 V and at another faraway location N is 3 V. Your friend says that the $\vec{E}$ field magnitude at point P is greater than that at N. Do you agree with your friend? Justify your opinion.

18.5 Conductors in electric fields

We can use the concepts of the $\vec{E}$ field and the V field to understand more precisely the behavior of electrically conducting materials and applications such as shielding and grounding.

Electric field of a charged conductor

Imagine a metal sphere of radius R that we touch with a negatively charged plastic rod (**Figure 18.13a**). Some of the excess electrons on the rod move to the metal sphere (Figure 18.13b). The electrons transferred to the sphere create an electric field in the metal that causes free electrons that are already in the metal to accelerate away from that spot. As a result of repulsion, the free electrons in the conductor quickly redistribute until equilibrium is reached, at which point the $\vec{E}$ field inside the conductor becomes zero (Figure 18.13c). If it were not zero, the electrons would continue to move. Given that the electrons stop moving, the component of the $\vec{E}$ field at the sphere's surface that is parallel to the surface is zero as well. Hence, outside the sphere, the $\vec{E}$ field vector is

FIGURE 18.13 Charging a metal sphere.

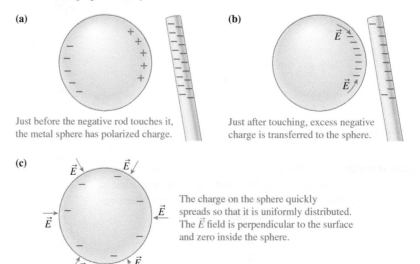

(a) Just before the negative rod touches it, the metal sphere has polarized charge.

(b) Just after touching, excess negative charge is transferred to the sphere.

(c) The charge on the sphere quickly spreads so that it is uniformly distributed. The $\vec{E}$ field is perpendicular to the surface and zero inside the sphere.

perpendicular to the surface. Since the $\vec{E}$ field is zero within and parallel to the surface of the sphere, this also means that the V field has the same value at all points on the sphere. If there were an electric field within or parallel to the surface of the sphere, it would exert a force on free electrons inside that would cause them to move.

Outside the charged conductor the field is not zero. **Figures 18.14a** and **b** illustrate the $\vec{E}$ field lines and equipotential surfaces for a point-like charged object with charge $+q$ and for a metal sphere of radius R with the same charge $+q$ on its surface. We won't do the calculation here since it requires calculus, but the magnitude of the $\vec{E}$ field and the value of the V field outside the sphere are the same as that produced by a point-like charged object:

$$E = k_C \frac{|q|}{r^2} \text{ and } V = k_C \frac{q}{r}$$

where r is the distance from the center of the sphere $(r \geq R)$. **Figures 18.15a** and **b** show graphs of $\vec{E}$ field and V field magnitudes as functions of the distance from the center of the sphere with charge q. Inside the sphere, $E = 0$ and $V = k_C (q/R)$, the same value as on the surface.

How do we know the potential is constant throughout the sphere and on the surface? When $\vec{E} = 0$ inside the sphere, no work is needed to bring a test charge to the surface. Therefore, the potential inside the sphere should be equal to the potential on the surface. What we have found here for one case is true in general:

1. When any conductor is charged, all electric charges reside on the surface.
2. The $\vec{E}$ field inside the conductor is zero.
3. The interior and the surface of the conductor are at the same potential; the surface is always an equipotential surface.

The above summary helps explain an important application—*grounding*.

Grounding

Grounding discharges an object made of conducting material by connecting it to Earth. Here we will explain how grounding works by applying our new understanding of conductors and electric potential.

Imagine that we have two conducting metal spheres of radii R_1 and R_2 with charges q_1 and q_2, respectively (**Figure 18.16a**). If we connect them with a long metal wire, electrons will move between and within the spheres until the V field on the surfaces

FIGURE 18.16 (a) When you connect the different size spheres by a metal wire, they are at the same electric potential. (b) The charge/area and the $\vec{E}$ field are greater on the small sphere.

(a) The conducting spheres are at the same electric potential V.

(b)

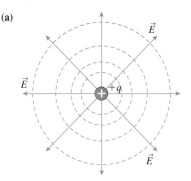

FIGURE 18.14 The $\vec{E}$ field lines and equipotential surfaces for (a) a point charge and (b) a charged metal sphere.

(a)

(b)

$\vec{E}$ field lines and equipotential surfaces outside the metal sphere are the same as for the point-like charge.

FIGURE 18.15 Graphs showing how E and V change with the distance r from the center of a charged metal sphere of radius R.

(a)

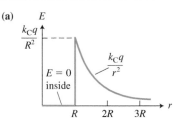

(b)

of and within both spheres achieves the same value. (As long as there is potential difference between the spheres, there will be a net force exerted on an electron and it will continue to accelerate; the process stops when the electrons are in equilibrium.) Because

$$V_1 = \frac{k_C q_1}{R_1} \text{ and } V_2 = \frac{k_C q_2}{R_2}$$

and $V_1 = V_2$, the ratio of the charges on the surface of each sphere is

$$\frac{q_1}{q_2} = \frac{R_1}{R_2}$$

This result tells us that when two spheres are connected, the bigger sphere has a greater electric charge than the smaller sphere. For example, suppose that sphere 1 has a radius that is 10 times greater than sphere 2 ($R_1 = 10R_2$). Large sphere 1 is originally uncharged and small sphere 2 has charge q. If we connect the two spheres with a wire, we find that $10/11$ of small sphere 2's original charge q will be transferred to large sphere 1 $[Q_1 = (10/11)q]$ and $1/11$ will remain on small sphere 2 $[Q_2 = (1/11)q]$. This result follows from two conditions:

$$\frac{Q_1}{Q_2} = \frac{R_1}{R_2} \text{ and } Q_1 + Q_2 = q$$

When we ground a conducting object, we are effectively connecting it with a wire to a sphere of radius $R = 6400$ km $= 6,400,000$ m, the radius of Earth. The object, the wire, and Earth become a single large equipotential surface, and the electric charge on the conducting object approaches zero. Therefore, you can safely touch the grounded conducting object (which might be a toaster or a high-definition television) while standing on the ground without experiencing an electric shock.

Another important consequence of connecting different size conductors is that the $\vec{E}$ field lines will be denser on the surface of the small sphere and the $\vec{E}$ field will be stronger. In the above example with spheres of radii $R_1 = 10R_2$, the large sphere had a greater charge $Q_1 = (10/11)q$ and the small sphere had charge $Q_2 = (1/11)q$. The *charge per unit surface area* (or surface charge density) for the large sphere is

$$\frac{Q_1}{A_1} = \frac{10q}{11 \cdot 4\pi(10R_2)^2} = \frac{q}{11 \cdot 4\pi \cdot 10R_2^2}$$

while for the smaller sphere, the charge per unit surface area is

$$\frac{Q_2}{A_2} = \frac{q}{11 \cdot 4\pi R_2^2}$$

The surface charge density is 10 times greater on the small sphere (Figure 18.16b). The $\vec{E}$ field lines are 10 times denser on the surface of the small sphere, and the electric field is 10 times stronger. As a result, on a pointed surface, such as a lightning rod, the electric field is strongest at the tip.

Uncharged conductor in an electric field—shielding

In our chapter-opening story we asked if you would be safe in a car during a lightning storm. Now we can answer that question. Imagine that we take a noncharged hollow conducting object and place it in a region with a uniform $\vec{E}$ field whose value is $\vec{E}_0$ (Figure 18.17a). The free electrons inside the object redistribute due to electric forces, producing their own contribution $\vec{E}_1$ to the $\vec{E}$ field (Figure 18.17b). This continues until the net $\vec{E}$ field within the conducting object is reduced to zero and the field outside the conductor has field lines perpendicular to the surface of the conductor (Figure 18.17c):

$$\vec{E}_{net} = \vec{E}_0 + \vec{E}_1 = 0$$

FIGURE 18.17 A hollow conducting object in an external $\vec{E}_0$ field.

The $\vec{E}$ field produced by the environment $\vec{E}_0$ is cancelled by the contribution produced by the conductor. The net $\vec{E}$ field outside the conducting object is now a superposition of two $\vec{E}$ fields. Because the conductor effectively "destroys" the external $\vec{E}$ field inside itself by creating the oppositely directed $\vec{E}_1$ field, the interior is protected from the external field. This effect is called **shielding**.

The hollow conducting object we just described might well be a car. Note that because the $\vec{E}$ field inside the conductor is zero, people inside the car during a lightning storm are completely unaffected by the outside electric processes as long as they do not touch the surface of the car (Figure 18.17d).

REVIEW QUESTION 18.5 In this section you read that the $\vec{E}$ field inside the conductor is always zero. Why is this true?

18.6 Dielectric materials in an electric field

In this section we will investigate the behavior of dielectric (insulating) materials. Recall from Chapter 17 that dielectric materials are attracted to both positively and negatively charged objects. How does our concept of the electric field contribute to the explanation of this phenomenon?

All atoms are composed of positively charged nuclei surrounded by negatively charged electrons (**Figure 18.18a**). If an atom in a dielectric material resides in a region with an external electric field $\vec{E}_0$, the field exerts a force on the atom's positive nucleus in the direction of the field $\vec{E}_0$ and a force on the atom's negatively charged electrons in the opposite direction. The nucleus and the electrons are displaced slightly in opposite directions until the force that the field exerts on each of them is balanced by the force they exert on each other due to their attraction (Figure 18.18b). Now the centers of the positive and negative charges are spatially separated. Such a system is familiar to us as an electric dipole (Figure 18.18c). We can represent an electric dipole using a vector $\vec{p}$ that points from the negative charge center of the system to its positive charge center.

However, some molecules, such as water, are natural electric dipoles even when the external $\vec{E}$ field is zero (**Figure 18.19a**, on the next page). Without an external electric

FIGURE 18.18 The electric field polarizes an atom.

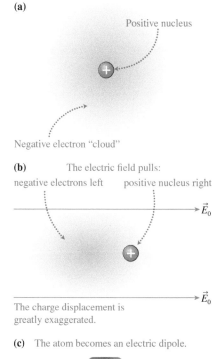

(a)

Positive nucleus

Negative electron "cloud"

(b)   The electric field pulls:
negative electrons left   positive nucleus right

$\vec{E}_0$

$\vec{E}_0$

The charge displacement is greatly exaggerated.

(c)   The atom becomes an electric dipole.

$\vec{p}$

FIGURE 18.19 Polar water molecules in an external electric field.

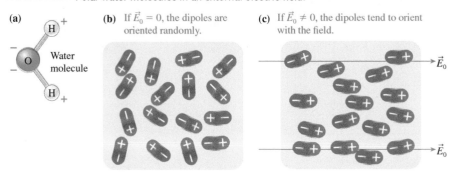

(a) Water molecule

(b) If $\vec{E}_0 = 0$, the dipoles are oriented randomly.

(c) If $\vec{E}_0 \neq 0$, the dipoles tend to orient with the field.

FIGURE 18.20 The field $\vec{E}_0$ causes polarization and an internal field $\vec{E}_1$.

(a) The field $\vec{E}_0$ causes the polarization of molecules.

(b) The net effect of the polarization is a net negative charge on the left, a positive charge on the right, and an internal field $\vec{E}_1$.

TABLE 18.2 Dielectric constants for different materials[1]

Material	Dielectric constant (κ)
Vacuum	1.0000
Dry air	1.0006
Wax	2.25
Glass	4–7
Paper	3–6
Axon membrane	8
Body tissue	8
Ethanol	26
Water	81

[1] At 20° C.

field, water molecules bump into each other and the dipoles are oriented randomly (Figure 18.19b). When the external $\vec{E}$ field is nonzero, the electric field exerts forces in opposite directions on the ends of the molecules, producing a torque that tends to rotate the molecules so that their electric dipoles align with the $\vec{E}$ field (Figure 18.19c).

When a dielectric material is placed in a nonzero $\vec{E}$ field, the atomic and molecular electric dipoles that result form a layer of positive electric charge on one surface of the material and a negative layer on the other surface (**Figure 18.20a**). These layers produce an additional contribution to the $\vec{E}$ field, an $\vec{E}_1$ field inside the dielectric material that points in the opposite direction to the external field $\vec{E}_0$ (Figure 18.20b). The net electric field inside is $\vec{E} = \vec{E}_0 + \vec{E}_1$. Since $\vec{E}_1$ points opposite $\vec{E}_0$, $E < E_0$. Thus, dielectric materials reduce the magnitude of the $\vec{E}$ field inside the material, similar to what happens in conductors. However, unlike conductors, the $\vec{E}$ field within does not decrease to zero. Thus, a dielectric material cannot completely shield its interior from an external electric field.

The ability of a dielectric material to decrease the $\vec{E}$ field varies from material to material. Physicists use a physical quantity to characterize this ability—the **dielectric constant** κ (Greek letter kappa). The dielectric constant of a material determines how much the material reduces the external $\vec{E}$ field (the field produced by the environment) within itself. The larger the value of κ, the more the $\vec{E}_0$ external field is reduced in the material.

The dielectric constant of a material is defined as follows:

$$\kappa = \frac{E_0}{E} \tag{18.11}$$

where E_0 is the magnitude of the electric field produced by the environment and $E = E_0 - E_1$ is the magnitude of the electric field within the dielectric material. You can see from Eq. (18.11) that the dielectric constant is dimensionless. **Table 18.2** lists sample dielectric constants for various materials.

Using Eq. (18.11), we can calculate the magnitude of the electric force that one charged object (object 1) exerts on another charged object (object 2) when both are inside a material with a dielectric constant κ:

$$F_{1\text{ on }2\text{ in dielectric}} = q_2 E_{1\text{ in dielectric}} = q_2 \frac{E_{1\text{ in vacuum}}}{\kappa} = \frac{F_{1\text{ on }2\text{ in vacuum}}}{\kappa}$$

The force that object 1 exerts on object 2 is reduced by κ compared with the force it would exert in a vacuum. Inside the dielectric material, Coulomb's law is now written as

$$F_{1\text{ on }2} = k_C \frac{|q_1||q_2|}{\kappa r^2} \tag{18.12}$$

Note in **Figure 18.21** that the positive ends of the dipole water molecules group around a negative ion and in effect reduce its net charge. The same thing happens to a positive ion in water. There is a reduction in the force that charged objects exert on each other when in that dielectric material. We can interpret Eq. (18.12) in this way:

$$F_{1 \text{ on } 2} = k_C \frac{\left(\dfrac{q_1 q_2}{\kappa}\right)}{r^2}$$

The charge product $q_1 q_2$ in Coulomb's law has been reduced by a factor of $1/\kappa$.

Salt dissolves in blood but not in air

The presence of dielectric materials that have polar molecules (like water) has dramatic effects on processes occurring inside the human body. One important function of water is to dissolve compounds, making their elements available for physiological use. For instance, the nervous system uses sodium ions in the transmission of information. Solid salt (NaCl) dissolves into sodium (Na^+) and chlorine (Cl^-) ions in the blood, which is made up primarily of water.

Table salt comes in the form of an *ionic crystal*. Because a chlorine atom is more attractive to electrons than a sodium atom, an electron is transferred from a sodium atom (leaving it with electric charge $+1.6 \times 10^{-19}$ C) to the chlorine atom (giving it electric charge -1.6×10^{-19} C). The electric force that the oppositely charged sodium and chlorine ions exert on each other holds the salt crystal together. The distance between the two ions is about the same as the characteristic size of a molecule (about 10^{-10} m). The magnitude of the electrical force that the ions exert on each other is then

$$
\begin{aligned}
F_{\text{Na on Cl}} = F_{\text{Cl on Na}} &= k_C \frac{|q_{\text{Na}}||q_{\text{Cl}}|}{r^2} \\
&= (9 \times 10^9 \text{ N} \cdot \text{m}^2/\text{C}^2)\frac{(1.6 \times 10^{-19} \text{ C})^2}{(10^{-10} \text{ m})^2} \approx 2 \times 10^{-8} \text{ N}
\end{aligned}
$$

The electric potential energy of a sodium-chlorine ion pair is about

$$
\begin{aligned}
U_q &= k_C \frac{q_{\text{Na}} q_{\text{Cl}}}{r} \\
&= (9 \times 10^9 \text{ N} \cdot \text{m}^2/\text{C}^2)\frac{(+1.6 \times 10^{-19} \text{ C})(-1.6 \times 10^{-19} \text{ C})}{(10^{-10} \text{ m})} \approx -2 \times 10^{-18} \text{ J}
\end{aligned}
$$

The energy is negative because of the opposite charges of the ions. To separate the ions, approximately $+2 \times 10^{-18}$ J must be added to the system.

Atoms, molecules, and ions also have positive kinetic energy due to their random thermal motion. A rough estimate of this energy is $K = (3/2)k_B T$, an expression we learned (in Chapter 12) for the kinetic energy of the particles in an ideal gas, where k_B is Boltzmann's constant (do not confuse it with the k_C used in Coulomb's law). The average positive kinetic energy at room temperature (300 K) is approximately

$$K = \frac{3}{2}k_B T = \frac{3}{2}(1.4 \times 10^{-23} \text{ J/K})(300 \text{ K}) \approx 6 \times 10^{-21} \text{ J}$$

This positive kinetic energy of the individual ions is much smaller than the negative electric potential energy holding them together. Thus, the total energy of the system is negative and the ions remain bound together when salt is in air.

When the salt is placed in water or blood, however, two changes occur. First, in water, there are many more collisions between molecules than there are in air. Although most of the collisions will not break an ion free from the crystal, some might. Remember, $K = (3/2)k_B T$ is the *average* kinetic energy of the molecules. A few molecules will have enough random kinetic energy to knock an ion free from the crystal during a collision.

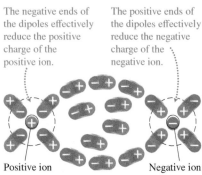

FIGURE 18.21 The force that positive and negative ions exert on each other is reduced by the polar water molecules.

The negative ends of the dipoles effectively reduce the positive charge of the positive ion.

The positive ends of the dipoles effectively reduce the negative charge of the negative ion.

Positive ion

Negative ion

Polar water molecule

Second, any ions that do become separated from the crystal by collisions do not exert nearly as strong an attractive force on each other because of the dielectric effect of the blood. If we assume that blood is mostly water, the attractive force that the two ions exert on each other decreases by a factor of 81, the dielectric constant of water:

$$F_{\text{Na on Cl in water}} = k_C \frac{e^2}{\kappa r^2} = (9 \times 10^9\,\text{N}\cdot\text{m}^2/\text{C}^2)\frac{(1.6 \times 10^{-19}\,\text{C})^2}{81(10^{-10}\,\text{m})^2} \approx 3 \times 10^{-10}\,\text{N}$$

The electric potential energy is also smaller by a factor of 81:

$$U_q = k_C \frac{-e^2}{\kappa r} = (9 \times 10^9\,\text{N}\cdot\text{m}^2/\text{C}^2)\frac{-(1.6 \times 10^{-19}\,\text{C})^2}{81(10^{-10}\,\text{m})} \approx -3 \times 10^{-20}\,\text{J}$$

TIP You can think of a conductor as a medium with a dielectric constant of infinity.

This energy is almost the same as the average kinetic energy of the random motion of the water molecules. This means that the random kinetic energy of the water molecules is sufficient to keep the sodium and chlorine ions from recombining. This allows the nervous system to use the freed sodium ions to transmit information.

Dielectric breakdown

Imagine that you shuffle across a rug and then reach for a metal doorknob. Just before you touch the doorknob, a spark jumps between you and the knob. You estimate that the distance the spark jumps is about 1 cm. What happened?

Air (a dielectric under normal conditions) can turn into a conductor if free electrons in the air are accelerated to such speeds that they can knock an electron out of an air molecule on the next collision. The two electrons then move on and knock additional electrons out of other air molecules, creating a cascade called **dielectric breakdown**. When these electrons recombine with molecules that have lost an electron, light is produced—a spark. Experiments indicate that dielectric breakdown occurs in dry air when the magnitude of the $\vec{E}$ field exceeds about $3 \times 10^6\,\text{N/C} = 3 \times 10^6\,\text{V/m}$.

Dielectric breakdown explains why you see the spark when your hand gets relatively close to the knob, not long before. When walking on a rug, you acquire some electric charge (let's say it's negative). When you approach a metal doorknob, you polarize the metal, and an excess of positive charge appears on the side of the knob closest to you. However, the $\vec{E}$ field in the region between you and the knob is not strong enough to cause the dielectric breakdown. As you bring your hand closer to the knob, the magnitude of the field increases until it reaches the breakdown value—and the spark occurs! The value for the dielectric breakdown of air allows us to roughly estimate the potential difference between your body and the doorknob:

$$\Delta V = E\,\Delta x = (3 \times 10^6\,\text{V/m})(0.01\,\text{m}) = 3 \times 10^4\,\text{V} = 30{,}000\,\text{V}$$

This is a huge potential difference. Fortunately, the potential difference is not dangerous—it is the amount of electric charge that flows that is dangerous, and in this case it is very small.

REVIEW QUESTION 18.6 What are the differences between conducting and dielectric materials when they are placed in a region with a nonzero $\vec{E}$ field? Explain these differences.

18.7 Capacitors

We have learned about practical applications of conductors in electric fields, such as grounding and shielding. Another important application involving electric fields and conductors is storing energy in the form of electric potential energy. When positively and negatively charged objects are separated, the system's electric potential energy increases. How can this charge separation be maintained so that the electric potential energy can be stored for useful purposes? This is accomplished with a device known as a *capacitor*.

A **capacitor** consists of two conducting surfaces separated by an insulating material. Although a variety of configurations are possible, the simplest are parallel plate capacitors made of two metal plates separated by air, rubber, paper, or some other dielectric material (Figure 18.22a). The role of a capacitor is to store electric potential energy.

To charge a capacitor, one usually connects its plates to the terminals of a battery, a device that creates potential difference between two terminals. The positive battery terminal is connected by a conducting wire to one plate, for example, the left capacitor plate. Then an $\vec{E}$ field produced by the battery causes electrons to flow from the left capacitor plate toward the positive battery terminal (Figure 18.22b). Electrons likewise flow from the negative battery terminal to the right capacitor plate. This charge transfer continues until the capacitor plate, conducting wire, and battery terminal on each side are at the same potential (Figure 18.22c). Effectively, the battery removes electrons from one plate and deposits them on the other. The plate with a deficiency of electrons is positively charged with charge $+q$, and the plate with excess electrons is negatively charged with charge $-q$. The total charge of the capacitor is zero.

If we consider the capacitor plates to be large, flat conductors, the charges should distribute evenly on the plates. Each charged plate produces a uniform $\vec{E}$ field (Figure 18.23a). When the plates are close to each other, their two fields overlap between the plates and outside the plates. Outside the plates, the $\vec{E}$ field from the positively charged plate points opposite the $\vec{E}$ field from the negatively charged plate (Figure 18.23b) and they cancel each other. Between the plates, the $\vec{E}$ fields are in the same direction and add. Thus, the net $\vec{E}$ field is strong between the plates and zero outside (Figure 18.23c). The equipotential surfaces are parallel to the plates and equidistant as the $\vec{E}$ field is uniform, and thus the V field increases linearly with the distance from the 0 V plate to the $+V$ plate.

According to Eq. (18.10), the magnitude of the $\vec{E}$ field between the plates relates to the potential difference ΔV from one plate to the other and the distance d separating them:

$$E = \left| -\frac{\Delta V}{\Delta x} \right| = \left| \frac{\Delta V}{d} \right|$$

Thus, if the potential difference across the plates doubles (for example, you connect the capacitor to a 12-V battery instead of a 6-V battery), the magnitude of the $\vec{E}$ field between the plates also doubles. Recall that $\vec{E}$ field lines are created by charges on the plates. Thus to double the $\vec{E}$ field, the charge on the plates has to double. We conclude that the magnitude of the charge q on the plates ($+q$ on one plate and $-q$ on the other) should be proportional to the potential difference V across the plates: $q \propto \Delta V$. Or, if we use a proportionality constant,

$$q = C|\Delta V| \tag{18.13}$$

The proportionality constant C in this equation is called the **capacitance** of the capacitor. In the above, q is the magnitude of the charge on each plate. The unit of capacitance is 1 coulomb/volt = 1 farad (1 C/V = 1 F) in honor of Michael Faraday, whose experiments helped establish the atomic nature of electric charge.

Capacitance of a capacitor

What properties of capacitors determine their capacitance? It might seem that Eq. (18.13) answers this question. However, the capacitance of a capacitor does not change when the charge on the plates or the potential difference across them is changed. These two quantities are proportional to each other ($q \propto \Delta V$); to double the charge on the plates we need to double the potential difference across the same capacitor. But the ratio $C = q/|\Delta V|$ remains the same. This ratio is an operational definition of capacitance.

So then what does affect the capacitance of a particular capacitor? Imagine two capacitors whose plates have different surface areas and are connected to the same potential difference source. The capacitor with the larger surface area A plates should

FIGURE 18.22 Charging a capacitor.

(a) A capacitor

Metal plates / Dielectric material

(b) In the process of charging

$+V$ ⊕ ⊖ 0
Battery

(c) Capacitor is now fully charged.

$+V$ 0
$+q$ $-q$

$+V$ ⊕ ⊖ 0

FIGURE 18.23 Electric field inside a charged capacitor.

(a) The $\vec{E}$ fields produced on each plate

$\vec{E}$

(b) The $\vec{E}$ fields for both plates when together

Fields from the $+$ and $-$ plates cancel.

(c) The net $\vec{E}$ field and equipotential surfaces

$V \quad \frac{2V}{3} \quad \frac{V}{3} \quad 0$

$\vec{E}_{net}$

FIGURE 18.24 Capacitance depends on A, d, and κ.

(a)

(b)

(c)

be able to maintain more charge separation because there is more room for the charge to spread out (**Figure 18.24a**).

Now imagine two capacitors with the same surface area and the same potential difference across the plates but with different distances d between the plates (Figure 18.24b). Since $|\Delta V| = Ed$ (based on Equation 18.9) and ΔV is the same for both capacitors because they are connected to the same potential difference source, a larger distance d between the plates leads to a smaller magnitude $\vec{E}$ field between the plates. But since the magnitude of this $\vec{E}$ field is proportional to the amount of electric charge on the plates, a larger plate separation leads to a smaller magnitude of electric charge on the plates ($+q$ on one and $-q$ on the other). This means the capacitance of the capacitor decreases with increasing d.

In addition to surface area and distance between the plates, capacitance is affected by the dielectric constant κ of the material between the plates. The presence of a dielectric reduces the $\vec{E}$ field between the plates (Figure 18.24c). Thus the potential difference between the plates for the same amount of charge decreases. Therefore, the battery transfers additional charge to the capacitor until the potential difference across the capacitor becomes the same as across the battery again. The capacitance increases in proportion to the dielectric constant κ of the material.

Thus we conclude that the capacitance of a particular capacitor should increase if the surface area A of the plates increases, decrease if the distance d between them is increased, and increase if the dielectric constant κ of the material between them increases. A careful derivation (which we will not go through) provides the following result for a parallel plate capacitor:

$$C_{\text{Parallel plate capacitor}} = \frac{\kappa A}{4\pi k_C d} \tag{18.14}$$

where $k_C = 9.0 \times 10^9 \, \text{N} \cdot \text{m}^2/\text{C}^2$. Let's check the units for capacitance:

$$\frac{(\text{m}^2)}{\left(\dfrac{\text{N} \cdot \text{m}^2}{\text{C}^2}\right)\text{m}} = \frac{\text{C}^2}{\text{N} \cdot \text{m}} = \frac{\text{C}^2}{\text{J}} = \frac{\text{C}}{\text{J}}\text{C} = \frac{\text{C}}{\text{V}} = \text{F}$$

The units check.

QUANTITATIVE EXERCISE 18.9 **Capacitance of a textbook**

Estimate the capacitance of your physics textbook, assuming that the front and back covers (area $A = 0.050 \, \text{m}^2$, separation $d = 0.040 \, \text{m}$) are made of a conducting material. The dielectric constant of paper is approximately 6.0. Second, determine what the potential difference must be across the covers in order for the textbook to have a charge separation of $10^{-6} \, \text{C}$ (one plate has charge $+10^{-6} \, \text{C}$ and the other has charge $-10^{-6} \, \text{C}$).

Represent mathematically The capacitance of a parallel plate capacitor is

$$C = \frac{\kappa A}{4\pi k_C d}$$

The potential difference ΔV needed to produce a charge separation of $q = 10^{-6} \, \text{C}$ is

$$\Delta V = \frac{q}{C}$$

Solve and evaluate

$$C = \frac{\kappa A}{4\pi k_C d} = \frac{(6.0)(0.050 \, \text{m}^2)}{4\pi(9.0 \times 10^9 \, \text{N} \cdot \text{m}^2/\text{C}^2)(0.040 \, \text{m})}$$

$$= 7.0 \times 10^{-11} \, \text{F} = 70 \, \text{pF}$$

(Here, p stands for "pico," the metric prefix for 10^{-12}.)

$$\Delta V = \frac{q}{C} = \frac{10^{-6} \, \text{C}}{7.0 \times 10^{-11} \, \text{F}} = 1.4 \times 10^4 \, \text{V} = 14 \, \text{kV}$$

This is a rather small capacitance, but reasonable given the large plate separation and the low dielectric value. The farad is a very large unit, and capacitors with picofarad or nanofarad capacitances are quite common.

- -

Try it yourself Approximately how long must the book cover be to have a 1-F capacitance?

Answer

If we assume the same plate separation, plate width, and dielectric constant of the paper, then the covers should be about 3×10^6 m long, or about 3 million km long! One farad is a large capacitance.

The previous exercise shows that making a capacitor with a large capacitance requires a large surface area. It is possible to make capacitors of much larger capacitance by making multiple very thin layers of specially designed materials. These are called supercapacitors, and they can reach thousands of farads in capacitance.

Body cells as capacitors

Capacitors are used in all devices that require some instant action—such as camera flashes or computer keyboards. They are also present in the tuning circuits of radios, music amplifiers, etc. Biological capacitors are found inside our bodies. Cells, including nerve cells, have capacitor-like properties (see Example 18.10). The conducting "plates" are the fluids on either side of a moderately nonconducting cell membrane. In this membrane, chemical processes cause ions to be "pumped" across the membrane. As a result, the membrane's inner surface becomes slightly negatively charged while the outer surface becomes slightly positively charged.

EXAMPLE 18.10   Capacitance of and charge on body cells

Estimate (a) the capacitance C of a single cell and (b) the charge separation q of all of the membranes of the human body's 10^{13} cells. Assume that each cell has a surface area of $A = 1.8 \times 10^{-9}\,m^2$, a membrane thickness of $d = 8.0 \times 10^{-9}\,m$, a $\Delta V = 0.070$-V potential difference across the membrane wall, and a membrane dielectric constant $\kappa = 8.0$. There is conducting fluid inside and outside the cell.

Sketch and translate A sketch of a cell is given below. Because the dielectric constant of the membrane is provided, it must be made of a dielectric material.

Membrane
$d = 8.0 \times 10^{-9}\,m$
$A = 1.8 \times 10^{-9}\,m^2$
$\Delta V = 0.070\,V$

Simplify and diagram The thickness of the membrane $d = 8.0 \times 10^{-9}\,m$ is much smaller than the dimensions of the cell (roughly the square root of the surface area $\sqrt{A} = 4.2 \times 10^{-5}\,m$). Up close, the cell membrane appears almost flat, just as Earth appears flat to people on its surface. Thus, we can use the expression for the capacitance of a parallel plate capacitor to make our estimate.

Represent mathematically The capacitance of one cell is

$$C = \frac{\kappa A}{4\pi k_C d}$$

The total charge separation q on the 10^{13} cells (biological capacitors) in the body is then

$$q_{total} = (10^{13})C|\Delta V|$$

Solve and evaluate The capacitance of one cell is

$$C = \frac{(8.0)(1.8 \times 10^{-9}\,m^2)}{4\pi(9.0 \times 10^9\,N \cdot m^2/C^2)(8.0 \times 10^{-9}\,m)} = 1.6 \times 10^{-11}\,F$$

The total charge separated by the membranes of all of the cells is approximately

$$q_{total} = 10^{13}C|\Delta V| = 10^{13}(1.6 \times 10^{-11}\,F)(0.070\,V) \approx 10\,C$$

Although these calculations are approximate, it is clear that the separation of electric charge (-10 C total on the inside walls of cell membranes and $+10$ C on the outside walls) is an important part of our metabolic processes. This 10-C electric charge separation is huge—about the same as the charge transferred during a lightning flash. Fortunately, our bodies' cells do not all discharge simultaneously in one surge.

Try it yourself Suppose you doubled the membrane thickness of all body cells. How would the potential difference across them have to change in order for the cells to maintain the same charge separation?

Answer Doubling the membrane thicknesses would reduce the capacitance by half. Thus, you would need to double the potential difference across the membranes to maintain the same charge separation.

Energy of a charged capacitor

A capacitor is essentially a system of positively and negatively charged objects that have been separated from each other so that the system as a whole is electrically neutral yet has electric potential energy. To determine electric potential energy in a charged capacitor, we start with an uncharged capacitor and then calculate the amount of work that must be done on the capacitor to move electrons from one plate to the other.

FIGURE 18.25 The energy stored while charging a capacitor.

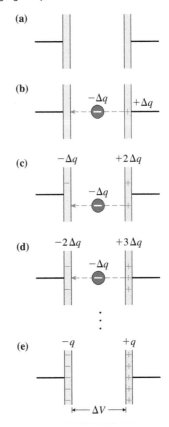

This process is sketched in **Figure 18.25** in the form of a thought experiment. It's not possible for us to manually grab electrons and move them from one capacitor plate to the other. Usually, this work is performed by an external source such as a battery. However, in this thought experiment, we will move increments of charge $-\Delta q$ from one plate to the other. Note that the charge is negative because these charge increments are composed of electrons. After the first charge increment has been moved, the plate from which it was taken has a charge of $+\Delta q$ and the plate to which it is taken has a charge of $-\Delta q$.

Only a very small amount of work must be done to move this first $-\Delta q$ since the left plate is uncharged while moving it (Figure 18.25b). The next $-\Delta q$ is more difficult to move (Figure 18.25c) since the left plate now has a charge of $-\Delta q$ that repels the $-\Delta q$ we are trying to move there. Additionally, the $+2\Delta q$ charge on the right plate pulls back on it. The more charge increments we move from one plate to the other, the more difficult it becomes to move the next one. Eventually, with increasing effort, we will have transferred a total (negative) charge of $-q = (-\Delta q) + (-\Delta q) + \dots$ to the left plate, leaving the right plate with a charge of $+q$. We can represent the whole process of charging the capacitor with a work-energy bar chart (**Figure 18.26**). The external object (us moving the charge increments) does work on the system (the capacitor), and the electric potential energy of the system increases.

The electric potential energy increase can be calculated using Eq. (18.8):

$$\Delta U_q = q\,\Delta V_{average}$$

where $\Delta V_{average}$ is the average potential difference from one plate to the other during the process of charging. Since the initial potential difference is zero and the final is ΔV, the average potential difference between the plates is

$$\Delta V_{average} = \frac{0 + \Delta V}{2} = \frac{\Delta V}{2}$$

Substituting this expression for $\Delta V_{average}$ into Eq. (18.8), we get

$$U_q = q\,\Delta V_{average} = q\,\frac{\Delta V}{2}$$

(We dropped the Δ, as $U_{qi} = 0$.) The above expression can be written in three different ways using Eq. (18.13), $q = C|\Delta V|$:

$$U_q = \tfrac{1}{2}q|\Delta V| = \tfrac{1}{2}C|\Delta V|^2 = \frac{q^2}{2C} \qquad (18.15)$$

Note that the process of charging a capacitor is similar to stretching a spring: at the beginning a smaller force is needed to stretch the spring by a certain amount compared to the much greater force needed when the spring is already stretched.

FIGURE 18.26 A bar chart representing the work done to charge a capacitor.

Energy needed to charge human body cells

In Example 18.10, we estimated that the total charge separated across all of the cell membranes in a human body was about 11 C. Recall that the potential difference across the cell membranes is 0.070 V. Estimate the work that must be done to separate the charges across the membranes of these approximately 10^{13} cells.

Represent mathematically We can use the first expression in Eq. (18.15) to answer this question:

$$U_q = \tfrac{1}{2}q\,\Delta V$$

where $q = 11$ C and $\Delta V = 0.070$ V.

Solve and evaluate

$$U_q = \tfrac{1}{2}q\,\Delta V = \tfrac{1}{2}(11\ \text{C})(0.070\ \text{V}) = 0.40\ \text{J}$$

These cells are continually being charged and discharged as part of the body's metabolic processes.

- -

Try it yourself Estimate the energy needed to charge body cells each day assuming that each cell discharges once per second.

Answer   result is a rough estimate.

times per second, though some might discharge fairly infrequently. The
bread. This is not much. However, some cells might discharge up to 400
35,000 J = 8 kcal, or about one-tenth the energy provided by a piece of

Energy density of electric field

We have found that the energy stored in a charged capacitor is

$$U_q = \tfrac{1}{2}C|\Delta V|^2$$

Where is this energy stored? The difference between a charged and an uncharged capacitor is the presence of the electric field in the region between the plates. We hypothesize that it is the electric field that possesses this energy.

What physical quantities affect the amount of this stored energy? It takes more energy to charge a larger capacitor (for example, a capacitor with larger plates) that has the same electric field between the plates. To have a measure of energy independent of the capacitor volume, we will use the physical quantity of **energy density**. Energy density quantifies the electric potential energy stored in the electric field per cubic meter of volume. The $\vec{E}$ field energy density u_E is

$$u_E = \frac{U_q}{V}$$

where U_q is the electric potential energy stored in the electric field in that region, and V is the volume of the region.

Let us rewrite the above equation for the electric field energy density in terms of the $\vec{E}$ field, using our knowledge of the electric energy of a parallel plate capacitor:

$$u_E = \frac{U_q}{V} = \frac{\tfrac{1}{2}C|\Delta V|^2}{V} = \frac{\tfrac{1}{2}\left(\dfrac{\kappa A}{4\pi k_C d}\right)(Ed)^2}{V}$$

We used Eqs. (18.14) and (18.9) to substitute for C and ΔV. A is the plate area, d is the plate separation, E is the magnitude of the $\vec{E}$ field between the plates, and κ is the dielectric constant of the material between the plates. Simplifying this, we have

$$u_E = \frac{\kappa A(Ed)^2}{8\pi k_C dV} = \frac{\kappa}{8\pi k_C}E^2\frac{Ad}{V}$$

Since the volume of the region between the plates is $V = Ad$, this equation simplifies to

$$u_E = \frac{\kappa}{8\pi k_C}E^2 \qquad (18.16)$$

The energy density depends only on the magnitude of the $\vec{E}$ field, the properties of the dielectric, and a few mathematical and physical constants. We'll use this idea in a later chapter on electromagnetic waves.

TIP Notice that here the letter V stands for volume, not for the electric potential. Sometimes similar or identical symbols are used for different physical quantities. When working with mathematical expressions, always ask yourself: "What does each of these symbols represent?"

TIP Although we derived the expression for energy density for a parallel plate capacitor, it can be applied for any electric field.

REVIEW QUESTION 18.7 A parallel plate capacitor has two oppositely charged plates, each producing its own electric field. Why is there no electric field outside the capacitor?

18.8 Electrocardiography

The human heart has four chambers. During a 1-s heartbeat, each chamber pumps in a specific sequence. For example, the lower right chamber (the right ventricle) pumps blood through the lungs, where carbon dioxide is exchanged for oxygen, and the lower left chamber (the left ventricle) pumps freshly oxygenated blood into the aorta for a new trip around the circulatory system to provide oxygen for the body cells.

An electric charge separation occurs when muscle cells in the heart contract during the pumping process. For example, when the left ventricle pumps blood into the aorta, many left ventricle muscle cells contract. As each muscle cell contracts, positive and negative charges separate (this process is represented schematically in **Figure 18.27**). The simultaneous contraction of the large number of cells in the heart muscle of the left ventricle results in a relatively large charge separation—an electric dipole. On the other

FIGURE 18.27 An electric dipole is produced by each heart muscle cell contraction.

The right atrium receives blood from the circulatory system and sends it to the right ventricle.

The lungs return blood to the left atrium, which sends blood to the left ventricle.

The right ventricle pumps blood to the lungs.

The left ventricle pumps blood to the circulatory system.

Contracting muscle cells produce an electric dipole.

hand, the number of simultaneously contracting muscle cells is much smaller when the left atrium pumps blood into the left ventricle. Thus, during the 1-s cycle of a heartbeat, there is an electric dipole with continually changing magnitude and orientation that depends at each instant on the number and the orientation of the muscle cells that are contracting at that instant.

This electric dipole produces an electric field that extends outside the heart to the body tissue. A device called an electrocardiogram (ECG) detects potential difference between selected points in that field. Data from an ECG can identify abnormalities, such as enlarged left or right ventricles. How does an ECG work? Electrodes, in the form of pads, are placed on the skin. These pads transmit data to the ECG about the potential difference between the various pairs of pads. The ECG uses these data to reconstruct the changing electric dipole on the heart.

A physician looks at the varying potential difference between different pairs of electrode pads and from this can learn about the heart's changing electric dipole. This dipole in turn indicates the number of muscle cells contracting at different times. A person with left ventricular hypertrophy (an enlarged left ventricle) will have a larger than normal dipole and potential difference during contraction of the left ventricle. Figure 18.28 shows a normal ECG.

FIGURE 18.28 Electrocardiogram signals.

CONCEPTUAL EXERCISE 18.12   **Heart's electric dipole and ECG**

The figure below shows a simplified electric dipole charge distribution on a heart at one instant during a heartbeat and two ECG pads, I and II, on opposite shoulders of the person's body. (a) What do we expect to be the sign of the potential difference between the pads at that particular instant? (b) Determine the direction of the forces exerted by the electric field on a positive sodium ion (Na^+) and on a negative chlorine ion (Cl^-) in the body tissue located exactly halfway between the ECG pads. (c) What do you expect to happen to the potential difference between the pads you found in part (a) if you assume that the ions move in the direction of these forces?

Sketch and translate The figure above gives a sketch of the situation.

Simplify and diagram We assume no other electric charges and fields are present.

(a) In the figure at right, we draw equipotential lines for the heart's dipole across the chest. Comparing the positions of the two electrodes with respect to the two outer lines of equal potential, we see that $V_{II} > V_I \Rightarrow V_{II} - V_I > 0$. We can also characterize the electric field created by the heart's dipole with the $\vec{E}_{heart}$ that points from the region of V_{II} to the region of V_I.

(b) In an electric field, the force exerted on a positive ion points in the direction of lower electric potential, and the force exerted on the negative ion points toward the region of higher electric potential. Adding the sodium and chlorine ions to our sketch (see figure below), and given that electrode II is at higher electric potential than electrode I, we see that the force on Na^+ points left and the force on Cl^- points right.

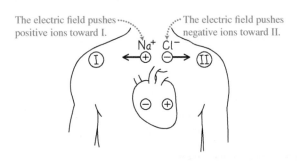

The electric field pushes positive ions toward I.   The electric field pushes negative ions toward II.

(c) The forces on the ions will cause them to move. As the ions redistribute, they create their own electric field, with the direction of $\vec{E}_{ions}$ opposite to the $\vec{E}_{heart}$; thus the potential V_I will increase and the potential V_{II} will decrease. Therefore, the measured difference $V_{II} - V_I$ will become smaller.

Try it yourself What will the potential difference between the electrodes be if the polarization of the heart is such that the dipole is vertical with the positive charge above the negative one?

Answer   The potential difference between the electrodes will be zero.

REVIEW QUESTION 18.8 Why do heart contractions produce electric dipoles?

Summary

$\vec{E}$ field The $\vec{E}$ field is a force-like quantity characterizing the electric field at a specific location. A source charge creates an electric field that exerts a force on a test charge placed in that field at that location.

The $\vec{E}$ field at the chosen location is equal to the electric force exerted by a source charge on a positively charged test object divided by the test object's charge (Section 18.1).

$$E = \frac{\vec{F}_{Q \text{ on } q_{test}}}{q_{test}}$$ Eq. (18.2)

$$\vec{F}_{Q \text{ on } q_{test}} = q_{test}\vec{E}$$ Eq. (18.5)

Superposition principle The $\vec{E}$ field at a location is the *vector* sum of the contributions of n source charges to the $\vec{E}$ field at that location (Section 18.1).

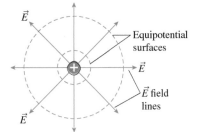

$$\vec{E}_{net} = \vec{E}_1 + \vec{E}_2 + \vec{E}_3 + \cdots$$ Eq. (18.4)

V field (electric potential) is an energy-like quantity characterizing the electric field at a point. The V field (electric potential) at a position of interest is equal to the electric potential energy U_q of a small positively charged test charge q_{test} and the source charge divided by the test object's charge. Include the signs of charges when calculating U_q (Section 18.3).

$$V = \frac{U_{Qq_{test}}}{q_{test}} = \frac{k_C Q}{r}$$ Eq. (18.6a)

For one source charge:

$$V = \frac{k_C Q_1}{r_1}$$

$$U_{\text{Source } Q \text{ and } q} = qV_{\text{Due to source } Q}$$ Eq. (18.8)

Representing the $\vec{E}$ field and the V field $\vec{E}$ field lines and equipotential surfaces (where V is constant) represent the field. $\vec{E}$ field lines are perpendicular to equipotential surfaces (Section 18.4).

Equipotential surfaces

$\vec{E}$

$\vec{E}$ field lines

or

$$\Delta V = -E_x \Delta x$$ Eq. (18.9)

$$E_x = -\frac{\Delta V}{\Delta x}$$ Eq. (18.10)

Dielectric constant κ A dielectric material partially cancels the $\vec{E}$ field within it produced by the environment. This reduces the magnitude of the force exerted by charged objects on each other within the material (Section 18.6).

$\vec{E}_0$

$\vec{E} = \vec{E}_0 - \vec{E}_1$ $\vec{E}_1$

$$\kappa = \frac{E_0}{E}$$ Eq. (18.11)

$$F_{1 \text{ on } 2} = k_C \frac{|q_1||q_2|}{\kappa r^2}$$ Eq. (18.12)

Capacitors A capacitor is a device that (when charged) stores electric potential energy. It has two conducting surfaces separated by an insulating material. To charge a capacitor, one connects its plates to the points at different electric potential (like the terminals of a battery). The charge separation stores electric potential energy in the capacitor (Section 18.7).

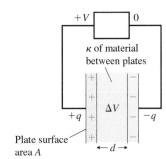

$+V$ 0

κ of material between plates

ΔV

$+q$ $-q$

Plate surface area A

$\leftarrow d \rightarrow$

$$q = C|\Delta V|$$ Eq. (18.13)

For parallel plate capacitors:

$$C = \frac{\kappa A}{4\pi k_C d}$$ Eq. (18.14)

$$U_q = \tfrac{1}{2}q|\Delta V|$$

$$= \tfrac{1}{2}C|\Delta V|^2 = \frac{q^2}{2C}$$ Eq. (18.15)

Questions

Multiple Choice Questions

1. What does the $\vec{E}$ field at point A, which is a distance d from the source charge, depend on? (More than one statement may be correct.)
   (a) the magnitude of the test charge
   (b) the sign of the test charge
   (c) the magnitude and the sign of the source charge
   (d) the magnitude and sign of the test charge

2. Why can you shield an object from an external electric field using a conductor?
   (a) There are freely moving electric charges inside conductors.
   (b) There are positive and negative charges inside conductors.
   (c) Both a and b are essential.

3. If you place a block made of a conducting material in an external uniform electric field $\vec{E}_0$ so that two opposite sides of the block are perpendicular to $\vec{E}_0$, what will be the magnitude and direction of the net $\vec{E}$ field inside the block?
   (a) magnitude: E_0; direction: same as $\vec{E}_0$
   (b) magnitude: E_0; direction: opposite to $\vec{E}_0$
   (c) magnitude: less than E_0 but more than zero; direction: same as $\vec{E}_0$
   (d) magnitude: more than E_0; direction: same as $\vec{E}_0$
   (e) magnitude: zero

4. If you place a block made of a dielectric material in an external uniform electric field $\vec{E}_0$ so that two opposite sides of the block are perpendicular to $\vec{E}_0$, what will be the magnitude and direction of the net $\vec{E}$ field inside the material?
   (a) magnitude: E_0; direction: same as $\vec{E}_0$
   (b) magnitude: E_0; direction: opposite to $\vec{E}_0$
   (c) magnitude: less than E_0 but more than zero; direction: same as $\vec{E}_0$
   (d) magnitude: less than E_0 but more than zero; direction: opposite to $\vec{E}_0$
   (e) magnitude: zero

5. Two identical positive charges are located at a distance d from each other. Where are both the $\vec{E}$ field and the electric potential zero?
   (a) exactly between the charges
   (b) at a distance d from both charges (all three charges making an equilateral triangle)
   (c) Both a and b are correct.
   (d) None of these choices is correct.

6. An electric dipole is placed between the oppositely charged plates as shown in Figure Q18.6. At which of the marked points could the net $\vec{E}$ field be zero? Choose all possible answers.

FIGURE Q18.6

7. A positive charge is fixed at some distance d from the origin of the z-axis, as shown in Figure Q18.7. Which graph shows how E_z, the z-component of $\vec{E}$ field, depends on the position along the z-axis?

FIGURE Q18.7

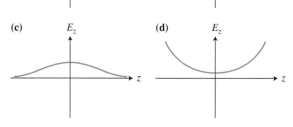

8. Figure Q18.8 shows $\vec{E}$ field lines in a region of space. Select two correct statements about the $\vec{E}$ field and V field at the points P, R, and S.
   (a) $E_P = E_R = E_S$
   (b) $E_P > E_S > E_R$
   (c) $E_R > E_S > E_P$
   (d) $V_P > V_R > V_S$
   (e) $V_P > V_S > V_R$
   (f) $V_R > V_S > V_P$

FIGURE Q18.8

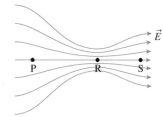

Conceptual Questions

9. How do we use the model of the electric field to explain the interaction between charges?

10. Describe a procedure to determine the $\vec{E}$ field at some point.

11. What does it mean if the $\vec{E}$ field at a certain point is 5 N/C and points north?

12. A very small positive charge is placed at one point in space. There is no electric force exerted on it. (a) What is the value of the $\vec{E}$ field at that point? (b) Does this mean there are no electric charges nearby? Explain. (c) Suggest one charge distribution (two or more charges) that would produce a zero $\vec{E}$ field at a point.

13. How do we create an $\vec{E}$ field with parallel lines and uniform density (equally spaced)?

14. Draw a sketch of the $\vec{E}$ field lines caused by (a) two positive charges of the same magnitude, (b) two negative charges of the same magnitude, and (c) two positive charges of different magnitudes.

15. Draw a sketch of the $\vec{E}$ field lines caused by (a) a positive and a negative charge of the same magnitude, (b) a positive and a negative charge of different magnitudes, and (c) a positive charge between two negative charges.

16. Jim thinks that $\vec{E}$ field lines are the paths that test charges would follow if placed in the $\vec{E}$ field. Do you agree or disagree with his statement? If you disagree, give a counter example.

17. Can $\vec{E}$ field lines cross? Explain why or why not.

18. An electron moving horizontally from left to right across the page enters a uniform $\vec{E}$ field that points toward the top of the page. Draw vectors indicating the direction of the electric force exerted by the field on the electron and its acceleration. Draw a sketch indicating the path of the electron in this field. What motion in the gravitational field is similar to the motion of the electron?

19. (a) What does it mean if the electric potential at a certain point in space is 10 V? (b) What does it mean if the potential difference between two points is 10 V?

20. Explain how grounding works.

21. Explain how shielding works.

22. Explain the difference between the microscopic structures of polar and of nonpolar dielectric materials. Give an example of each.

23. Explain why, for charged objects submerged in a dielectric material, the electric force they exert on each other needs to be divided by the dielectric constant.

24. What does it mean if the dielectric constant κ of water is 81?

25. What is the dielectric constant of a metal?

26. Describe the relation between the quantities $\vec{E}$ field and V field.

27. If the V field in a region is constant, what is the $\vec{E}$ field in this region? If the $\vec{E}$ field is constant in the region, what does the V field look like? Explain your answers.

28. Why are uncharged pieces of a dielectric material attracted to charged objects?

29. Draw equipotential surfaces and label them in order of decreasing potential for (a) one positive charge, (b) one negative charge, (c) two identical positive point charges at a distance d from each other, and (d) a negatively charged infinitely large metal plate.

30. Show a charge arrangement and a point in space where the potential produced by the charges is zero but the $\vec{E}$ field is not zero. Then repeat for the case where the potential produced by the charges is not zero but the $\vec{E}$ field is zero.

31. Explain what happens when you place a conductor in an external electric field. How can you test your explanation?

32. (a) Explain what happens when you place a dielectric material in an external electric field. How can you test your explanation?

33. Explain why the excess charge on an electrical conducting material is located on its surface.

34. Draw a microscopic representation of the charge distribution in two identical conducting plates that are touching each other and placed in an external uniform electric field as shown in Figure Q18.34. What happens if you move the plates apart (while keeping them in the external field) and then move them out of the field? How can you test your answer? Repeat the steps described above for two dielectric plates.

FIGURE Q18.34

Problems

Below, **BIO** indicates a problem with a biological or medical focus. Problems labeled **EST** ask you to estimate the answer to a quantitative problem rather than derive a specific answer. Asterisks indicate the level of difficulty of the problem. Problems with no * are considered to be the least difficult. A single * marks moderately difficult problems. Two ** indicate more difficult problems.

In some of these problems you may need to know that the mass of an electron is 9.11×10^{-31} kg, the mass of a proton is 1.67×10^{-27} kg, the electric charge of the electron is -1.6×10^{-19} C, and the charge of a proton is 1.6×10^{-19} C.

18.1 A model of the mechanism for electrostatic interactions

1. * (a) Construct a graph of the magnitude of the $\vec{E}$ field-versus-position for the $\vec{E}$ field created by a point-like object with charge $+Q$, representing the $\vec{E}$ field magnitude as a function of the distance from the object. (b) Using the same set of axes, draw a graph for the field produced by an object of charge $+2Q$. (c) Using the same set of axes, draw a graph for the field produced by an object of charge $-2Q$.

2. * A uranium nucleus has 92 protons. (a) Determine the magnitude of the $\vec{E}$ field at a distance of 0.58×10^{-12} m from the nucleus (about the radius of the innermost electron orbit around the nucleus). (b) What is the magnitude of the force exerted on an electron by this $\vec{E}$ field? (c) What assumptions did you make? If the assumptions are not valid, will the magnitude in part (b) be an overestimate or underestimate of the actual value?

3. The electron and the proton in a hydrogen atom are about 10^{-10} m from each other. What quantities related to the $\vec{E}$ field can you determine using this information?

4. * Use the superposition principle to draw $\vec{E}$ field lines for the two objects whose charges are given. Consider all objects to be point-like and choose the distance you want between them: (a) $+q$ and $+q$; (b) $+q$ and $+3q$; (c) $+q$ and $-q$.

5. * Use the superposition principle to draw $\vec{E}$ field lines for the following objects whose charges are given. Consider all objects to be point-like and choose the distance you want between them: (a) $+q$ and $-3q$; (b) $+q$, $+q$, and $-3q$.

6. * $\vec{E}$ field lines for a field created by an arrangement of charged objects are shown in Figure P18.6. (a) Where are these objects located, and what are the signs of their electric charge? (b) What else can you determine using the information? Give two examples.

FIGURE P18.6

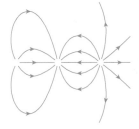

18.2 Skills for analyzing processes involving $\vec{E}$ fields

7. * Two objects with charges $+4.0 \times 10^{-9}$ C and $+3.0 \times 10^{-9}$ C are 50 cm from each other. Find a location where the $\vec{E}$ field due to these two charged objects is zero.

8. * A $+4.0 \times 10^{-9}$-C charged object is 6.0 cm along a horizontal line toward the right of a -3.0×10^{-9}-C charged object. Determine the $\vec{E}$ field at a point 4.0 cm to the left of the negative charge.

9. ** A $+4.0 \times 10^{-9}$-C charged object is 4.0 cm along a horizontal line toward the right of a -3.0×10^{-9}-C charged object. Determine the $\vec{E}$ field at a point 3.0 cm directly above the negative charge.

10. ** A distance d separates two objects, each with charge $+q$. Determine an expression for the $\vec{E}$ field at a point that is a distance d from both charges.

11. * A point-like charged object with a charge $+q$ is placed in an external horizontal uniform electric field $\vec{E}_0$ that points to the right. Determine an expression for the net $\vec{E}$ field at a distance d from the charge (a) to the right of it and along a horizontal line going through the charge $+q$ and (b) to the left of it along that same line.

12. * A 3.0-g aluminum foil ball with a charge of $+4.0 \times 10^{-9}$ C is suspended on a string in a uniform horizontal $\vec{E}$ field. The string makes an angle of 30° with the vertical. What information about the $\vec{E}$ field can you determine for this situation?

13. ** (a) If the string in the previous problem is cut, how long will it take the ball to fall to the floor 1.5 m below the level of the ball? (b) How far will the ball travel in the horizontal direction while falling? Indicate all the assumptions that you made.

14. * **EST** **Using Earth's $\vec{E}$ field for flight** Earth has an electric charge on its surface that produces a 150-N/C $\vec{E}$ field, which points down toward the center of Earth. Estimate the electric charge that a person would need so that the electric force that the field exerts on the person will support the person in the air above Earth. Indicate all of your assumptions. Is this a reasonable idea? Explain.

15. * An electron moving with a speed v_0 enters a region where an $\vec{E}$ field points in the same direction as the electron's velocity. What will happen to the electron in this field? Answer the question qualitatively and quantitatively (using symbols). Ignore the force that Earth exerts on the electron.

16. ** A 1.0-g aluminum foil ball with a charge of 1.0×10^{-9} C hangs freely from a 1.0-m-long thread. What happens to the ball when a horizontal 5000-N/C $\vec{E}$ field is turned on? Answer the question assuming the stabilized situation.

17. A 0.50-g oil droplet with charge $+5.0 \times 10^{-9}$ C is in a vertical $\vec{E}$ field. In what direction should the $\vec{E}$ field point and what magnitude should it have so that the droplet moves at constant speed? Ignore the forces that air exerts on the droplet.

18. * An electron is ejected into a horizontal uniform $\vec{E}$ field at a parallel horizontal velocity of 1000 m/s. Describe everything you can about the motion of the electron. Ignore the force that Earth exerts on the electron.

19. * **Equation Jeopardy 1** The equations below describe one or more physical processes. Solve the equations for the unknowns and write a problem statement for which the equations are a satisfactory solution.

$$-(0.001 \text{ kg})(9.8 \text{ N/kg}) + T \sin 88° = 0$$
$$(-5.0 \times 10^{-8} \text{ C})E_x + T \cos 88° = 0$$

20. * **Equation Jeopardy 2** The equations below describe one or more physical processes. Solve the equations for the unknowns and write a problem statement for which the equations are a satisfactory solution.

$$(+1.0 \times 10^{-8} \text{ C})(-4.0 \times 10^4 \text{ N/C}) = (0.20 \text{ kg})a_x$$
$$0 = (8.0 \text{ m/s}) + a_x t$$

18.3 and 18.4 The V field and Relating the $\vec{E}$ field and the V field

21. During a lightning flash, -15 C of charge moves through a potential difference of 8.0×10^7 V. Determine the change in electric potential energy of the field-charge system.

22. * (a) Construct a graph of the V field created by a point-like object with charge $+Q$, representing the V field as a function of the distance from the object. (b) Using the same set of axes, draw a graph for an object with charge $+2Q$ and for an object with charge $-2Q$.

23. * A horizontal distance d separates two objects each with charge $+q$. Determine the value of the electric potential at a point that is located at a distance d from each charge.

24. * Two objects with charges $-q$ and $+q$ are separated by a distance d. Determine an expression for the V field at a point that is located at a distance d from each charge.

25. * Four objects with the same charge $-q$ are placed at the corners of a square of side d. Determine the values of the $\vec{E}$ field and V field in the center of the square.

26. **Spark jumps to nose** An electric spark jumps from a person's finger to your nose. While passing through the air, the spark travels across a potential difference of 2.0×10^4 V and releases 3.0×10^{-7} J of electric potential energy. What is the charge in coulombs, and how many electrons flow?

27. * Two -3.0×10^{-9}-C charged point-like objects are separated by 0.20 m. Determine the potential (assuming zero volts at infinity) at a point (a) halfway between the objects and (b) 0.20 m to the side of one of the objects (and 0.40 m from the other) along a line joining them.

28. **BIO** **Electric field in body cell** The electric potential difference across the membrane of a body cell is $+0.070$ V (higher on the outside than on the inside). The cell membrane is 8.0×10^{-9} m thick. Determine the magnitude and direction of the $\vec{E}$ field through the cell membrane. Describe any assumptions you made.

29. * **BIO** **EST** **Energy used to charge nerve cells** A nerve cell is shaped like a cylinder. The membrane wall of the cylinder has a $+0.07$-V potential difference from the inside to the outside of the wall. To help maintain this potential difference, sodium ions are pumped from inside the cell to the outside. For a typical cell, 10^9 ions are pumped each second. (a) Determine the change in chemical energy each second required to produce this increase in electric potential energy. (b) If there are roughly 7×10^{11} of these cells in the body, how much chemical energy is used in pumping sodium ions each second? (c) Estimate the fraction of a person's metabolic rate used to pump these ions.

30. * **Equation Jeopardy 3** The equation below describes one or more physical processes. Solve the equation for the unknown and write a problem statement for which the equation is a satisfactory solution.

$$0 = (1/2)(1.0 \times 10^{-4} \text{ kg})(6.0 \text{ m/s})^2 + (2.0 \times 10^{-7})\Delta V$$

31. * **Equation Jeopardy 4** The equation below describes one or more physical situations. Solve the equation for the unknown and write a problem statement for which the equation is a satisfactory solution.

$$(9.0 \times 10^9 \text{ N} \cdot \text{m}^2/\text{C}^2)(+2.0 \times 10^{-8} \text{ C})/(2.0 \text{ m})$$
$$+ (9.0 \times 10^9 \text{ N} \cdot \text{m}^2/\text{C}^2)(-2.0 \times 10^{-8} \text{ C})/(1.0 \text{ m}) = \Delta V$$

32. * While a sphere with positive charge q_1 remains fixed, a second sphere with positive charge q_2 is successively placed at the points A, B, C and D (see **Figure P18.32**). Rank (a) the magnitudes of the $\vec{E}$ field at point P and (b) the V field values at P for the four positions of the second sphere.

FIGURE P18.32

33. * **Figure P18.33** shows $\vec{E}$ field lines in a region of space. Note that the distances between A and B, between C and D, and between F and G are all equal. (a) Rank the magnitudes of the $\vec{E}$ field at points A to G from largest to smallest. (b) Rank the electric potentials at points A to G from largest to smallest. (c) A small negatively charged object is moved (I) from A to B, (II) from C to D, (III) from A to D, and (IV) from F to G. Rank the work done on the object in these four cases from largest to smallest.

FIGURE P18.33

18.5 and 18.6 Conductors and dielectrics in electric fields

34. * A metal sphere has no charge on it. A positively charged object is brought near, but does not touch the sphere. Show that this object can exert a force on the sphere even though the sphere has no net charge. How can you test your answer experimentally?

35. ** **EST** A Van de Graaff generator of radius 0.10 m has a charge of about 1.0×10^{-7} C on it. The Van de Graaff generator is then turned off and grounded. How many excess electrons remain on its dome? Indicate any assumptions that you made.

36. ** A metal ball of radius R_1 has a charge Q. Later it is connected to a neutral metal ball of radius R_2. What is the fraction of the charge Q that remains on the first ball?

37. * Positively charged metal sphere A is placed near metal sphere B. B has a very small positive charge. Explain why the spheres could attract each other. Draw sketches of the spheres and their charge distribution to support your answer.

38. * Two small metal spheres A and B have different electric potentials (A has a higher potential). Describe in words and mathematically what happens if you connect them with a wire. What assumptions did you make?

39. * An electric dipole such as a water molecule is in a uniform $\vec{E}$ field. (a) Will the force exerted by the field cause the dipole to have a linear acceleration along a line in the direction of $\vec{E}$? Explain. (b) Will the field exert a torque on the dipole? Explain.

40. ** **BIO** **Electric field of a fish** An African fish called the aba has a charge $q = +1.0 \times 10^{-7}$ C at its head and an equal magnitude negative charge $-q$ at its tail (see Figure P18.40). Determine the magnitude and direction of the electric field at position A and the force exerted on a hydroxide ion (charge $-e$) at that point. The fish and ion are in water. Indicate any assumptions that you made.

FIGURE P18.40

41. **BIO** **Body cell membrane electric field** (a) Determine the average magnitude of the $\vec{E}$ field across a body cell membrane. A 0.07-V potential difference exists from one side to the other and the membrane is 7.5×10^{-9} m thick. (b) Determine the magnitude of the electrical force on a sodium ion (charge $+e$) in the membrane. Assume that the dielectric constant is 1.0 (it is actually somewhat larger).

42. ** **Earth's electric field** Earth has an electric charge of approximately -5.7×10^5 C distributed relatively uniformly on its surface. Determine as many quantities as possible about the electrical properties of the space around Earth. Use any additional information that you need. Indicate any assumptions that you made.

18.7 Capacitors

43. You have a parallel plate capacitor. (a) Determine the average $\vec{E}$ field between the plates if a 120-V potential difference exists across the plates. Their separation is 0.50 cm. (b) A spark will jump if the magnitude of the $\vec{E}$ field exceeds 3.0×10^6 V/m when air separates the plates. What is the closest the plates can be placed to each other without sparking?

44. * A capacitor of capacitance C with a vacuum between the plates is connected to a source of potential difference ΔV. (a) Write an expression for the charge on each of the plates and for the total energy stored by the capacitor. (b) You then fill the capacitor with a dielectric material of dielectric constant κ while the capacitor remains connected to the same potential difference. What are the new charge on the plates and the new energy stored by the capacitor?

45. * A capacitor of capacitance C with a vacuum between the plates is connected to a source of potential difference ΔV. (a) Write an expression for the charge on each of the plates and for the total energy stored by the capacitor. (b) The capacitor is then disconnected from the potential difference source and you fill it with a dielectric that has a dielectric constant κ. What are the new charge on the plates and the new energy?

46. How does the capacitance of a parallel plate capacitor change if you double the magnitude of the charge on its plates? If you triple the potential difference across the plates? What assumptions did you make?

47. ** **BIO** **EST** **Axon capacitance** The long thin cylindrical axon of a nerve carries nerve impulses. The axon can be as long as 1 m. (a) Estimate the capacitance of a 1.0-m-long axon of radius 4.0×10^{-6} m with a membrane thickness of 8.0×10^{-9} m. The dielectric constant of the membrane material is about 6.0. (b) Determine the magnitude of the charge on the inside (negative) and outside (positive) of the membrane wall if there is a 0.070-V potential difference across the wall. (c) Determine the energy stored in this axon capacitor when charged.

48. ** **Sphere capacitance** A metal sphere of radius R has an electric charge $+q$ on it. Determine an expression for the electric potential V on the sphere's surface. Use the definition of capacitance to show that the capacitance of this isolated sphere is R/k, where k is the constant used in Coulomb's law. (Hint: Assume the other plate is infinitely far away.)

49. ** **EST** Two identical metal plates that are touching each other are placed in a uniform electric field as shown in Figure Q18.34. The electric field is created by a parallel plate capacitor with capacitance 150 pF and potential difference 3600 V between the plates. The surface area of the capacitor plates is 10 times larger than the surface area of the metal plates. You move the metal plates apart (while still in the electric field) and then move them out of the field. Estimate the final charge on each metal plate.

50. * **BIO** **EST** **Capacitance of red blood cell** Assume that a red blood cell is spherical with a radius of 4×10^{-6} m and with wall thickness of 9×10^{-8} m. The dielectric constant of the membrane is about 5. Assuming the cell is a parallel plate capacitor, estimate the capacitance of the cell and determine the positive charge on the outside and the equal-magnitude negative charge inside when the potential difference across the membrane is 0.080 V.

51. **BIO** **Defibrillator** During ventricular fibrillation the heart muscles contract randomly, preventing the coordinated pumping of blood. A defibrillator can often restore normal blood pumping by discharging the charge on a capacitor through the heart. Paddles are held against the patient's chest, and a 6×10^{-6}-F charged capacitor is discharged in several milliseconds. If the capacitor energy is 250 J, what potential difference was used to charge the capacitor?

18.8 Electrocardiography

52. * **EST** The dielectric strength of air is 3×10^6 V/m. As you walk across a synthetic rug, your body accumulates electric charge, causing a potential difference of 6000 V between your body and a doorknob. What can you can estimate using this information?

53. * **Charged cloud causes electric field on Earth** The electric charge of clouds is a complex subject. Consider the simplified model shown in Figure P18.53. A positive charge is near the top of the cloud and a negative charge is near the bottom. Determine the direction of the $\vec{E}$ field on Earth at point P below the cloud and explain it so that a classmate can understand why there is positive charge on the ground directly below the cloud.

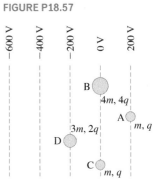
FIGURE P18.53

54. * **BIO** **Heart's dipole charge** The heart has a dipole charge distribution with a charge of $+1.0 \times 10^{-7}$ C that is 6.0 cm above a charge of -1.0×10^{-7} C. Determine the $\vec{E}$ field (magnitude and direction) caused by the heart's dipole at a distance of 8.0 cm directly above the heart's positive charge. All charges are located in body tissue of dielectric constant 7.0. What is the force exerted on a sodium ion (charge $+1.6 \times 10^{-19}$ C) at that point?

General Problems

55. * In a hot water heater, water warms when electric potential energy is converted into thermal energy. (a) Determine the energy needed to warm 180 kg of water by 10 °C. (b) If -10.0 C of electric charge passes through a $+120$-V potential difference in the heating coils each second, determine the time needed to warm the water by 10 °C.

56. ** **EST** **Lightning warms water** A lightning flash occurs when -40 C of charge moves from a cloud to Earth through a potential difference of 4.0×10^8 V. Estimate how much water can boil as a result of energy released during the process. Describe the assumptions you made.

57. * Four charged particles A, B, C, and D are placed in an electric field as shown in Figure P18.57. Vertical lines indicate equipotential surfaces of the V field. All particles are initially at rest. When they are released, they all start moving to the left. (a) Determine the signs of the charges on the particles. Rank (b) the kinetic energies of the particles and (c) the speeds of the particles as they cross the -600-V equipotential line. Relative masses and charges of the particles are indicated in the figure. Assume the particles are far from each other so they do not interact, and ignore forces exerted by Earth on the particles.

FIGURE P18.57

58. * **EST** Figure P18.58 shows a region of space with an electric field. Vertical lines indicate equipotential surfaces. A particle with charge -3.0 nC is initially at the location of the -20-V equipotential line. At time $t = 0$ the particle is released from rest. (a) Estimate the force exerted by the electric field on the particle when it passes the 0-V equipotential line and when it passes the 80-V equipotential line. (b) Draw a qualitative velocity-versus-time graph for the motion of the particle until it reaches the potential 120 V. (c) Determine the kinetic energy of the particle when it passes the 60-V equipotential line. (d) Do the data in the problem allow you to estimate the time needed for the particle to move from the -20-V line to the 120-V line? If you think the answer is no, list the additional data that are needed. Ignore the force exerted by Earth on the particle.

FIGURE P18.58

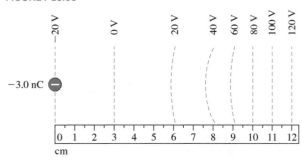

59. ** A small object of unknown mass and charge is tied to an insulating string that is connected to a sensitive force sensor. The object is placed between two large conducting plates, and the string is kept taut by charging the plates as shown in Figure P18.59. The mass of the string is 0.030 g, and the distance between the plates is 15.0 cm. As the potential difference between the plates increases, the force exerted by the force sensor on the string changes as shown in the table. Determine the mass and charge of the particle. (Note: This is a problem that requires linearization of data.)

ΔV (V)	$F_{\text{FS on S}}$ (10^{-3} N)
5000	52
5500	67
6000	82
6500	97
7000	110

FIGURE P18.59

60. * **BIO** **Can a shark detect an axon $\vec{E}$ field?** A nerve signal is transmitted along the long, thin axon of a neuron in a small fish. The transmission occurs as sodium ions (Na^+) transfer like tipping dominos across the axon membrane from outside to inside. Each short section of axon gets an excess of about 6×10^8 sodium ions/mm. Determine the $\vec{E}$ field 4.0 cm from the axon produced by the excess sodium ions on the inside of the axon and an equal number of negative ions on the outside of a 1-mm length of axon. The ions are separated by the 8×10^{-9}-m-thick axon membrane. Will a shark that is able to detect fields as small as 10^{-6} N/C be able to detect that axon field? Explain.

61. ** **BIO** **Electrophoresis** Electrophoresis is used to separate biological molecules of different dimensions and electric charge (the molecules can have different charge depending on the pH of the solution). A particular molecule of radius R with charge q is in a viscous solution that has an $\vec{E}$ field across it. The field exerts an electric force on the molecule and the viscous solution exerts an opposing drag force $F_{\text{drag}} = DRv$, where D is a constant drag coefficient that depends on the shape and other features of the molecule and the solution, and v is the molecule's speed in the solution. When the molecule gets up to speed, the electric force exerted on it by the field is equal in magnitude and opposite in direction to the drag force. Experiments show that once the molecule reaches terminal velocity, the distance Δx traveled by the molecule during time interval Δt is determined by the expression $\Delta x = \dfrac{qE\Delta t}{DR}$. Describe the assumptions that were made when deriving this equation, and show the derivation.

62. ** **BIO** **Energy stored in axon electric field** An axon has a surface area of 5×10^{-6} m² and the membrane is 8×10^{-9} m thick. The dielectric constant of the membrane is 6. (a) Determine the capacitance of the axon considered as a parallel plate capacitor. (b) If the potential difference across the membrane wall is 0.080 V, determine the magnitude of the charge on each wall. (c) Determine the energy needed to charge that axon capacitor. (d) Determine the magnitude of the $\vec{E}$ field across the membrane due to the opposite sign charges across the membrane walls. (e) Calculate the energy density of that field. (f) Multiply the volume of space occupied by that field by the volume of the membrane to get the total energy stored in the field. How do the answers to (c) and (f) compare?

Reading Passage Problems

BIO **Electric discharge by eels** In several aquatic animals, such as the South American electric eel, electric organs produce 600-V potential difference pulses to ward off predators as well as to stun prey. Figure 18.29 illustrates the key component that produces this electric shock—an **electrocyte**. The interior of an inactive electrocyte (Figure 18.29a) has an excess of negatively charged ions. The exterior has an excess of positively charged sodium ions (Na^+) on the left side and positively charged potassium ions (K^+) on the right. There is a -90-mV electric potential difference from outside the cell membrane to inside the electrocyte cell membrane, but zero potential from one exterior side to the other exterior side. How then does an eel produce the 600-V potential difference necessary to stun an intruder?

FIGURE 18.29 An electric eel's shocking system.

The eel's long trunk and tail contain many electrocytes placed one after the other in columns (Figures 18.29b and c). Each electrocyte contains several types of ion channels, which when activated by a nerve impulse allow sodium ions to pass through channels *on the left flat side* of each electrocyte from outside the cell to the inside. This causes the electric potential across that cell membrane to change from -90 mV to $+50$ mV. The electric potential from the left external side to the right external side of an electrocyte is now about 100 mV (Figure 18.29c). With about 6000 electrocytes placed one after the other (in series), the eel is able to produce an electric impulse of over 600 V (6000 electrocytes in series times 0.10 V per electrocyte). The discharge lasts about 2–3 ms.

63. Suppose you have a 1.0-F capacitor (very large capacitance) with a 0.10-V potential difference across the capacitor. What is the magnitude of the electric charge on each plate of the capacitor?
    (a) 0.05 C   (b) 0.10 C   (c) 0.20 C
    (d) 1.0 C   (e) 10 C

64. Suppose you place two of these 1.0-F capacitors with the charge calculated in the previous question as shown in Figure P18.64a. What is the net potential difference across the two capacitors (from one dot to the other)?
    (a) 0.05 V   (b) 0.10 V
    (c) 0.20 V   (d) none of these
    (e) not enough information

FIGURE P18.64

65. If both capacitors are discharged simultaneously, how much electric charge goes through the wire shown in pink in Figure P18.64a?
    (a) 0.05 C
    (b) 0.10 C
    (c) 0.20 C
    (d) 1.0 C
    (e) 10 C

66. Suppose you place the two capacitors with the 0.10-V potential difference across each capacitor as shown in Figure P18.64b. What is the net potential difference across the two capacitors?
    (a) 0.05 V
    (b) 0.10 V
    (c) 0.20 V
    (d) none of these
    (e) not enough information

67. Look at the electrocyte shown in Figure 18.29c. What causes the 0.10-V potential difference from the outer left to the outer right side of the cell?
    (a) The membrane is thicker on the left than on the right.
    (b) The ion distribution across the left membrane is different than across the right membrane.
    (c) The left and right membranes have different capacitances.
    (d) b and c
    (e) a, b, and c

68. Suppose that one cell of the electrocyte is regarded as a small capacitor with a 0.10-V potential difference across it. How should we arrange 10 cells to get a 1.0-V potential difference across them?
    (a) in series, as in Figure P18.64a
    (b) in parallel, as in Figure P18.64b
    (c) randomly so that they do not cancel each other
    (d) not enough information

Electrostatic precipitator (ESP) Electrostatic precipitators are a common form of air-cleaning device. ESPs are used to remove particle emissions from smoke moving up smokestacks in coal and oil-fired electricity-generating plants and pollutants from the boilers in oil refineries. You can buy portable ESPs or whole-house ESPs that connect to the cold-air return on the furnace. These devices remove about 95% of dirt and 85% of microscopic particles from the air.

A basic electrostatic precipitator contains a negatively charged horizontal metal grid (made of thin wires) and a stack of large, flat, vertically oriented metal collecting plates, with the plates typically spaced about 1 cm apart (only two plates are shown in Figure 18.30). Air flows across the charged grid of wires and then passes between the plates. A large negative potential difference (tens of thousands of volts) is applied between the wires and the plates, creating a strong electric field that ionizes particles in the air around the thin wires. Negatively charged smoke particles flow upward between the plates. The charged particles are attracted to and stick to the oppositely charged plates and are thus removed from the moving gas.

FIGURE 18.30 An electrostatic precipitator.

Waste gases without smoke particles

Smoke particles are attracted to the positively charged collecting plates.

Collecting plates are bumped so that the smoke particles fall into the trays placed below the plates.

Smoke particles pick up a negative charge as they pass the negatively charged metal grid.

Tray

Waste gases containing smoke particles

69. An $\vec{E}$ field of approximately 3×10^6 V/m is needed to ionize dry air molecules. If the potential difference between the thin wires and metal plates is 60,000 V, what minimum distance must the wires be from the plates in order to cause dry air to ionize?
    (a) 1.0 cm
    (b) 2.0 cm
    (c) 4.0 cm
    (d) 50 cm
    (e) 50 m

70. Why are the smoke particles attracted to the closely spaced plates?
    (a) The particles are negatively charged.
    (b) The particles are positively charged.
    (c) The plates are positively charged.
    (d) The plates are negatively charged.
    (e) a and c
    (f) b and c

71. Suppose a 2.0×10^{-6}-kg dust particle with charge -4.0×10^{-9} C is moving vertically up a chimney at speed 6.0 m/s when it enters the $+2000$-N/C $\vec{E}$ field near the metal collecting plate, pointing perpendicularly to the plate. If the particle is 2.0 cm from the plate at that instant, which answer below is closest to the magnitude of its horizontal acceleration toward the plate?
    (a) 2.0×10^{-5} m/s^2
    (b) 1.2×10^{-4} m/s^2
    (c) 4.0×10^{-4} m/s^2
    (d) 4.0 m/s^2
    (e) 24 m/s^2

72. Suppose everything is the same as in the previous problem. Which answer below is closest to the time needed for the particle to hit the plate?
    (a) 0.01 s
    (b) 0.10 s
    (c) 0.001 s
    (d) 1×10^{-6} s
    (e) 1.0 s

73. What is the purpose of giving a negative charge to the particles collected by the precipitator?
    (a) The particles are then attracted to the positively charged plates while moving upward.
    (b) The particles are then repelled from the positively charged plates while moving upward.
    (c) The extra mass slows the upward movement of the particles.
    (d) The negative charge cancels the positive charge the particles have when entering the precipitator.
    (e) a and c

DC Circuits

- Why are modern buildings equipped with electrical circuit breakers?
- Why are LEDs replacing incandescent lightbulbs?
- Why is it dangerous to use a hair dryer while taking a bath?

On a cold Sunday morning you turn on a space heater and the toaster oven. Then you start the washing machine and the dishwasher. Finally, you turn on the coffee grinder, and suddenly all the appliances turn off. What happened? A device called a circuit breaker disconnected that part of the house from the external supply of electric energy to prevent a house fire. How does a circuit breaker "know" when to turn off the energy supply, and why might keeping all of those appliances on lead to a fire? We will find out in this chapter.

BE SURE YOU KNOW HOW TO:

- Apply the concept of the electric field to explain electric interactions (Section 18.1).
- Define the V field (electric potential) and electric potential difference ΔV (Section 18.3).
- Explain the differences in internal structure between conducting materials and insulating materials (Section 17.3).

IN PREVIOUS CHAPTERS (Chapters 17 and 18) we learned to explain processes that involved electrostatic phenomena. In electric devices such as cell phones, computers, and lightbulbs, charged particles are continually moving. Similar movements occur inside the human body's nervous system. In this chapter we will learn how to explain phenomena that involve these moving, microscopic, charged particles.

19.1 Electric current

In the previous chapter we studied processes such as grounding that occur when a charged conducting object is connected with a wire to an uncharged conductor. We found that the excess electric charge on the charged object redistributes itself until the electric potential V at both conductors becomes equal. Let us look at similar experiments using a Wimshurst machine (Figure 17.21 shows one example of a Wimshurst machine). Recall that cranking the machine's handle generates opposite charges in the two metal spheres. The charge separation leads to a potential difference between the spheres (**Figure 19.1**). In the new Wimshurst machine experiments, we'll try to identify common features that might relate to transferring electric charge for useful purposes. Consider the experiments in Observational Experiment **Table 19.1**.

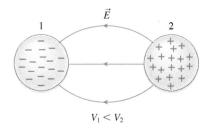

FIGURE 19.1 The charged Wimshurst spheres produce an $\vec{E}$ field.

$V_1 < V_2$

OBSERVATIONAL EXPERIMENT TABLE 19.1 Electric potential difference and charge transfer

VIDEO
OET 19.1

Observational experiment	Analysis
Experiment 1. Crank the handle of a Wimshurst machine and then bring the oppositely charged spheres of the machine close to each other (about 5 cm apart). You see a big spark. After the spark, when you again bring the spheres close, no further sparking occurs.	There is charge separation and nonzero potential difference ΔV between the two spheres. The spark means that the air between the spheres becomes a conductor, leading to the rapid discharge and the production of light. The original electric potential energy is converted to light and sound energy.
Experiment 2. Crank the handle of the Wimshurst machine and hang a light aluminum foil ball from an insulating thread between the oppositely charged spheres. The ball swings back and forth from one sphere to the other for a few minutes and then stops. When you remove the foil ball and bring the spheres near each other, no spark occurs.	The foil ball must acquire a small amount of negative charge from the negative sphere each time it touches it. The ball carries the negative charge to the positive sphere, deposits it there, and then returns to the negative sphere to repeat the process. This continues until the spheres are discharged and the potential difference between them is zero. The original electric potential energy is converted to mechanical energy and internal energy.
Experiment 3. Crank the handle of the Wimshurst machine and connect the leads of a neon bulb between the charged spheres. There is a flash of light from the bulb. When you remove the bulb and bring the spheres close, no spark occurs.	Before the bulb touches them, the spheres are charged and at different potentials. The bulb and its leads provide a conduction path to discharge the spheres. The discharge causes a flash of light from the bulb. The original electric potential energy is converted to light energy.

Patterns

- In all three experiments, the Wimshurst machine started with negative charge on one sphere and positive charge on the other sphere. There was a potential difference ΔV between the spheres.
- This charge separation and potential difference led to a flow of charge from one sphere to the other.
- The charge flow involved different observable consequences: a spark of light, the swinging ball of foil, and the flash of the bulb.
- After the charge flow, the Wimshurst spheres were discharged and the potential difference between them was zero. No more sparking or movement could occur.

The observable events in Table 19.1 were only able to occur because of the following factors:

1. An initial charge separation and resulting potential difference ΔV between the oppositely charged spheres. After the discharges occurred, resulting in equipotential spheres, nothing further could happen.

2. The presence of a charge conduction pathway (the foil ball, the leads of the neon bulb, and the conducting neon inside). No flash or movement occurred if the charged spheres were far apart and the electric field was too weak between the spheres to ionize the air.

3. A process that converted the electric potential energy of the spheres into some other form of energy: the flash of light from the air, the mechanical energy of the foil ball, and the flash of light from the neon bulb.

Each of these three processes ended very quickly. If we want to convert electrical energy to light or mechanical energy for a substantial time interval, we need to learn how to keep the observable consequences described in Table 19.1 happening continuously.

Fluid flow and charge flow

A fluid flow analogy may help us better understand the electric potential difference and conduction pathways of these electrical processes. You have two containers with water—A and B. A is almost full, and B is almost empty. You connect a hose between the two containers (**Figure 19.2a**). Water starts flowing from A to B until the levels are the same (Figure 19.2b), at which point the water flow stops. The volume of water in container A is analogous to the excess positive charge on Wimshurst sphere 2, and the difference in water pressure in the hose is analogous to the potential difference between the spheres. The hose provides a pathway for water to flow until the pressure at both ends of the hose is the same; the electric charge flows until the electric potential at each sphere is the same.

Notice that it is the pressure difference and not the difference in the volume or mass of water in each container that makes the water flow. Imagine a large container A full of water with the water level the same as in small container B—the water will not flow (**Figure 19.3a**). Similarly, it is not the total charge difference between the spheres of the Wimshurst machine but the potential difference that makes the charge flow. Imagine a large sphere A and a small sphere B. Suppose the charge is large on A and small on B; but if the V fields (electric potentials) on the surfaces of these two spheres are the same (recall the discussion of grounding in Chapter 18),

$$V = k_C \frac{q}{R}$$

there will be no charge flow through the wire connecting them (Figure 19.3b).

FIGURE 19.2 A water flow analogy for electric charge flow.

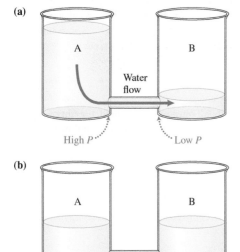

(a)

A   B

Water flow

High P Low P

(b)

A   B

Water flow stops when the two ends of the hose are at equal pressure.

FIGURE 19.3 There is no (a) water flow if $P_A = P_B$ or (b) charge flow if $V_A = V_B$.

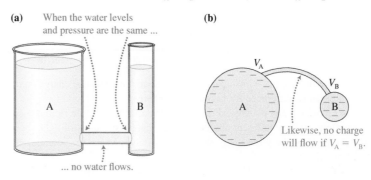

(a) When the water levels and pressure are the same ...

A   B

... no water flows.

(b)

V_A

V_B

A   B

Likewise, no charge will flow if $V_A = V_B$.

Making the process continuous

We can make fluid flow continuously by connecting the containers with another hose attached to a pump that moves water from B back to A (**Figure 19.4**). Pumping water from B back to A maintains a pressure difference between the ends of the hose and results in a continuous flow from A to B.

To achieve a steady flow of electric charge, we need a device that can maintain a steady potential difference between, for example, the leads of the neon bulb, effectively moving the negative charges back to the negative terminal. We attach wires from the positive and negative terminals of 10 batteries connected to the leads of a neon bulb. (We use 10 because one is not enough to make a neon bulb glow) (see **Figure 19.5**). The bulb does in fact steadily glow. Thus, it appears that the battery produces a steady potential difference across its terminals, which in turn causes a steady flow of electric charge through the neon bulb. The battery is equivalent to the pump in the water analogy.

Electric current

The flow of charged particles moving through a wire between two locations that are at different electric potentials is a physical phenomenon called **electric current**. When the charged particles always move in the same direction, the current is called **direct current**. A system of devices such as a battery, wires, and a bulb that allows for the continuous flow of charge is called an **electric circuit**. An electric circuit that has a direct current in it is called a **DC circuit**. In most electric circuits, the moving charged particles are free electrons in the wires and circuit elements. However, as we will learn later in the chapter, in some circuits those moving particles can be positively or negatively charged ions.

According to a simplified model of the internal structure of a wire, the wire consists of a **crystal lattice** composed of positive metal ions and **free electrons**. Free electrons (lost by the atoms that became ions) move within the lattice structure (**Figure 19.6a**). The ions constitute the mass of the material and are relatively stationary except for minor vibrations due to their thermal motion. In this model, the electrons move chaotically inside the wire like a swarm of gnats on a day when the air is still—each gnat moves, but the swarm remains at rest above your head. When the wire is connected to the terminals of the battery, there is an electric field inside the wire, and the electrons accelerate opposite the direction of the $\vec{E}$ field. They slow down when they "collide" with the ions and then accelerate again in the same direction (Figure 19.6b). This motion, called **drift**, occurs in the same direction for all of the electrons—just as a cloud of mosquitoes collectively moves in a single direction on a windy day. The collisions of electrons with ions make the ions vibrate faster, warming the wire. You can test this phenomenon by connecting two terminals of a battery with a wire for a brief moment—both the battery and the wire get hot!

FIGURE 19.6 (a) Free electrons drift randomly, moving fast between collisions but with no preferred direction. (b) When an electric field is present, the electrons continue this random motion but now drift toward the area of higher potential, opposite the direction of the $\vec{E}$ field.

(a) Positive ions form a crystal lattice structure.
In the absence of an electric field, the electrons move randomly within the wire.

(b)
In the presence of an electric field, the electrons drift toward the higher V region.

Free electrons in a wire drift toward the direction of higher V field (toward the positive terminal of a battery)—see **Figure 19.7a** on the next page. However, traditionally, the direction of electric current in a circuit is defined as opposite the direction of the electrons'

FIGURE 19.4 A pump maintains flow.

The pump returns water to container A. Water flows through the hose from A back to B.

FIGURE 19.5 Ten batteries create a constant potential difference across a neon bulb and charge flows through the bulb.

FIGURE 19.7 The direction of electric current is defined as opposite the direction that electrons move.

(a) Electrons travel around the circuit toward the positive terminal.

Inside the battery, the electrons move from + to −.

Battery

(b) The electric current I is defined in terms of the direction in which positive charges would move.

$I = \dfrac{|q|}{\Delta t}$

Battery

drifting motion. Imagine that positively charged particles are flowing from the higher electric potential terminal (labeled "+" on the battery) through the wires and electrical devices in the circuit to the terminal at lower electric potential (labeled "−"), then back through the battery to the positive terminal, where the cycle repeats (Figure 19.7b).

This definition of the direction of the electric current as a flow of positively charged particles is a historical quirk. When scientists first studied electric processes, they did not know about electrons and instead thought that an invisible positively charged electric fluid flowed around the circuit. For most situations involving electric circuits, it does not matter if we model the process in terms of electrons flowing one way or positive charges flowing the other way, so this historical convention is not a problem. Physicists use the term electric current to describe the drifting of electrically charged particles as well as the physical quantity that characterizes how much charge is moved through a conductor per unit time.

Electric current The magnitude of the physical quantity of electric current I in a wire equals the magnitude of the electric charge q that passes through a cross section of the wire divided by the time interval Δt needed for that charge to pass:

$$I = \frac{|q|}{\Delta t} \tag{19.1}$$

The unit of current is the **ampere** A, equivalent to 1 coulomb per second (C/s). A current of 1 A (one ampere, or amp) means that 1 C of charge passes through a cross section of the wire every second. The direction of the current is from higher potential to lower potential; therefore, it is in the direction that positive charges would move.

In SI units, the unit for electric current is a fundamental unit. The unit of electric charge (1 coulomb) is defined in terms of the ampere as $1\ C = (1\ A)(1\ s)$. In the next chapter, we will study the phenomenon that can be used to define the ampere.

QUANTITATIVE EXERCISE 19.1 **Electric current calculation**

Each second, 1.0×10^{17} electrons pass from right to left past a cross section of a wire connecting the two terminals of a battery. Determine the magnitude and direction of the electric current in the wire.

Represent mathematically The electric charge of an electron is $-1.6 \times 10^{-19}\ C$. Since we know how many electrons (N) pass a cross section of the wire in 1 s, we can find the total electric charge passing that cross section each second and therefore the current:

$$I = \frac{|q|}{\Delta t} = \frac{|-eN|}{\Delta t}$$

Solve and evaluate

$$I = \frac{|-eN|}{\Delta t} = \frac{|-(1.6 \times 10^{-19}\ C)(1.0 \times 10^{17})|}{1\ s}$$

$$= 1.6 \times 10^{-2}\ A = 16\ mA$$

The direction of the current is from left to right, opposite the direction of electron flow.

- -

Try it yourself If there is a 2.0-A current through a wire for 20 min, what is the total charge that moved through the wire during this time interval?

Answer   2400 C.

REVIEW QUESTION 19.1 What condition(s) is/are needed for electric charge to continuously travel from one place to another?

19.2 Batteries and emf

We have found that a battery creates a potential difference that can cause a neon bulb to glow continuously, analogous to how a water pump can cause water to flow continuously. If the current is a moving positive charge, the battery does work on the charge to move it within the battery from the lower electric potential negative terminal to the higher electric potential positive terminal. Once there, the charge can travel around the external circuit (the bulb and connecting wires) and cause the bulb to glow continuously. Batteries maintain a potential difference by means of a nonelectrostatic chemical process. This chemical process inside the battery separates neutral atoms at the negative terminal into electrons and positive ions, and moves the positive ions to the positive terminal.

The work done by a battery per unit charge moved inside the battery from one terminal to the other is the battery's **emf $\mathcal{E}$**. When first conceived, the term for this work was electromotive force, but now it is just called emf. (See the Tip in the margin.)

If you connect a neon bulb to the battery terminals, the bulb will glow and get a little warmer as work done by the battery results in the production of light and an increase in the thermal energy of the external circuit (bulb and connecting wires). If instead of a neon bulb, you were to connect a toaster to the battery, the work would be converted into thermal energy and light energy—the toaster's coils would glow red and get hot.

Emf $\mathcal{E}$ The emf $\mathcal{E}$ equals the work W done by a battery per coulomb of electric charge q that needs to be moved through the battery from one terminal to the other in order to maintain a potential difference at the battery terminals:

$$\mathcal{E} = \frac{W}{q} \tag{19.2}$$

TIP Electromotive force (emf), despite its name, is not a force. Emf is work done per coulomb of charge. The term *electromotive force* was coined by Alessandro Volta (1745–1827), who invented the battery. At that time the terms "force," "energy," and "power" were used somewhat interchangeably. The linguistic distinction between force and energy had not yet been made clear. For example, kinetic energy was called "live force" and potential energy was called "dead force."

The unit of emf is $J/C = V$, the same as the unit of electric potential difference. The emf of standard AAA, AA, C, and D batteries is 1.5 V. The small rectangular batteries used in smoke detectors have an emf of 9.0 V. The physical size of the battery is not related to the emf but to its storage capacity—the total charge it can move before it must be replaced or recharged. The bigger the size, the larger the product of $q = I\Delta t$, and the longer the time interval during which the battery works (assuming constant current). The storage capacity of an AA battery is about 2000 mAh (which is 2 Ah, or the charge transferred by the current of 2 amperes during 3600 s, which is about 7200 C).

CONCEPTUAL EXERCISE 19.2   Graphing electric potential in a circuit

You connect a 9.0-V battery to a small motor. Describe the changes in electric potential in the circuit with a graph.

Sketch and translate A sketch of a motor connected to the battery is shown at right. The battery has two terminals, + and −. The emf of the battery is 9.0 V.

Simplify and diagram To represent the changes in electric potential in the circuit, use a V-versus-location graph for an imaginary positively charged particle moving through the circuit (see figure at right). Assume that the electric potential of the − terminal at A is zero and that of the + terminal at B is 9.0 V. The battery does work on the charged particle-circuit system, increasing its electric potential energy. This work is similar to a person lifting a ball up a hill. When the ball reaches the top of the hill and starts rolling down, the potential energy of the system begins to decrease. Similarly, after leaving the battery at B, the imaginary positively charged particle travels along a wire to the

motor at C. We don't know yet the decrease in potential in the connecting wire, but will learn later that it is very small. There is also a small decrease in potential as the charge travels along the wire from D back to the battery at A. Thus, most of the potential decrease should occur across the motor. The potential decreases back to zero at A.

From A to B, the electric potential inside the battery increases.

Try it yourself How would the sketch and graph differ if there were two identical small motors attached one after the other to the battery?

(CONTINUED)

Answer

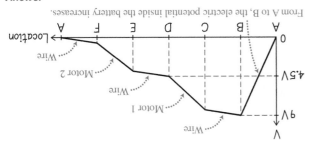

From A to B, the electric potential inside the battery increases.

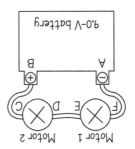

It's the same battery, so there is still a 9.0-V potential increase from its negative to its positive terminal. While passing through the two-motor circuit, about half of the decrease in potential (4.5 V) is across the first motor and the other 4.5-V decrease is across the second motor (see figure).

REVIEW QUESTION 19.2 Describe the changes in electric potential through a circuit and relate them to the motion of charged particles.

19.3 Making and representing simple circuits

An electric circuit is a system of devices such as a battery, wires, and a bulb (or some other elements, such as a motor) that allows for the continuous flow of charge. However, just having those elements next to each other does not produce a circuit—the electrical connections among them are important. In this section we investigate how to make a circuit. However, first we need to learn more about lightbulbs—common indicators of current in a circuit.

Neon and incandescent bulbs

All lightbulbs have a mechanism that converts electrical energy into light and thermal energy when electric charge flows through them. Here we will consider two types of bulbs—neon and incandescent—that use different mechanisms for this energy conversion.

A neon bulb consists of a glass bulb filled with low-pressure neon gas. Two thin wires (called **electrodes**) extend from the bottom of the bulb (**Figure 19.8a**). The leads outside the bulbs are the extensions of the electrodes. If the potential difference between the electrodes is high enough, the gas inside undergoes dielectric breakdown and becomes a conductor connecting the two electrodes. As a result, ions and free electrons are now present in the space between the electrodes. As the electric field associated with the potential difference causes them to travel in opposite directions, some electrons and ions will collide with atoms and change the energies of those atoms. Atoms release this extra energy by emitting light (you will learn more about this process in Chapter 28). Even if the potential difference across the terminals of the neon bulb only lasts for a very short time, we still see a flash of light.

An incandescent bulb has a metal filament inside of it. When there is current through the filament, the interactions of free electrons with the lattice of atoms make the filament, and thus the bulb, extremely hot (Figure 19.8b). When the filament of an incandescent bulb is hot enough to glow, we see light. The bulb is filled with inert gas; this gas prevents the filament from oxidizing and burning.

Unlike incandescent bulbs, a neon bulb is relatively cool to the touch. Also unlike the neon bulb, an incandescent bulb usually has no outside leads. The filament of an incandescent bulb is connected on one side to the metal screw-like base of the bulb (one terminal) and on the other side to a separate metal contact (the second terminal) at the very bottom of the base. We will learn the importance of this feature in Observational Experiment **Table 19.2**.

FIGURE 19.8 Comparing a neon bulb and an incandescent bulb.

(a) Neon bulb

The electric field between the electrodes causes free electrons to ionize neon atoms. Light is produced when the electrons rejoin the ionized neon atoms.

Neon gas

Bulb stays relatively cool

Electrodes

(b) Incandescent bulb

Current through the metal filament energizes the electrons in the filament, increasing thermal energy and generating light.

Filament glows when hot

Filament

Electrodes

Terminals

OBSERVATIONAL
EXPERIMENT TABLE 19.2 **Making a bulb glow**

Observational experiment	Analysis
Connection 1—No light is produced.	One bulb terminal is connected to one battery terminal with a wire.
Connection 2—no light	Both bulb terminals are connected to the same battery terminal with a wire.
Connection 3—no light	One bulb terminal is connected to both battery terminals with wires.
Connection 4—light	Both bulb terminals are connected to both battery terminals with wires.
Connection 5—light	The same as connection 4, only the bulb terminals are connected to the opposite battery terminals.
Connection 6—light	The same as connection 5, only one bulb terminal is directly touching the battery.

Pattern

The incandescent bulb glows if one battery terminal is connected to one lightbulb terminal and the other battery terminal is connected to the other bulb terminal, either with a metal wire or by direct contact, making a *complete closed* loop. This arrangement is an example of a **complete circuit**.

In Table 19.2 we identified conditions necessary to have a circuit with an electric current in it. To check whether any circuit you build is complete, trace a path of an imaginary positive charge moving from the positive terminal of the battery to the negative terminal (either in a real circuit or in the circuit you drew on paper). The path has to pass along conducting material at every location, making a complete closed loop. Check the circuits in which the bulb did not light in the table above using this rule.

Symbols for the elements in electric circuits

It is inconvenient to represent electric circuits by drawing realistic likenesses of actual batteries and bulbs. Instead, scientists and engineers use simplified symbols for the elements in an electric circuit. Diagrams with these symbols are called **circuit diagrams**. The symbols for circuit elements used in this chapter are shown in **Figure 19.9** (some elements are already familiar to you; others we will learn about later). Notice that the positive terminal of a battery is represented with a longer line than the negative terminal. Straight lines represent connecting wires. Bulbs are represented by circles with a loop in the middle (representing the filament). A sawtooth line represents any *resistive* device, an element in a circuit that converts electric energy into some other type of energy. We will discuss resistive devices—often called **resistors**—later in the chapter. Lightbulbs and toasters are all examples of resistors. Switches—devices that allow us to insert a dielectric (such as air) into a conductive path of the circuit to stop the current—have an open and closed position. Devices that measure current through a circuit element or the potential difference across it are called ammeters and voltmeters, respectively. Their symbols are shown in Figure 19.9.

FIGURE 19.9 The symbols used in circuit diagrams.

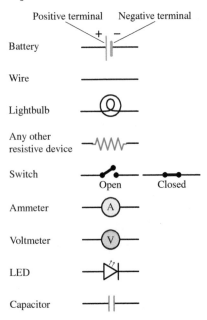

Ammeters

Recall our water analogy. An **ammeter** acts like a water flow meter. If you wish to measure how much water flows through a cross section of a pipe each second (the *flow rate*), you insert a flow meter into the pipe—just before or after the cross section that interests you. The water must pass through the flow meter. Similarly, to measure electric current through a circuit element, you place the ammeter next to that element. This means creating a break in the circuit next to that element, inserting the ammeter into the break, and reconnecting the wires to the terminals of the ammeter (**Figure 19.10a**). The circuit diagram in Figure 19.10b shows the arrangement. Ammeters have positive and negative terminals, which determine the sign of the ammeter reading. It is positive when the current is in the positive to negative terminal direction and negative when the current is in the negative to positive terminal direction.

FIGURE 19.10 Connecting an ammeter to measure electric current.

Voltmeters

A voltmeter measures the electric potential difference between two points in a circuit. Using a voltmeter to measure potential difference is analogous to using a pressure meter to measure the water pressure difference between two points in a water system. To measure the pressure difference you would submerge one terminal of the pressure meter into the water at the first point and the other terminal into the water at the second point. To measure the electric potential difference between two points in a circuit, you place a voltmeter so that its terminals are connected to those two points (**Figure 19.11a**). A circuit diagram for that connection is shown in Figure 19.11b. The voltmeter has a positive reading when the terminal labeled with a + sign is connected to the point with a higher electric potential than the terminal labeled with the − sign.

FIGURE 19.11 Connecting a voltmeter to measure potential difference.

A multimeter is a device that combines the functions of an ammeter with a voltmeter. When you use a multimeter be sure you check the settings—accidentally measuring potential difference when the multimeter is on the ammeter setting may harm the device.

REVIEW QUESTION 19.3 Explain the meaning of the term "a complete circuit."

19.4 Ohm's law

Is there a relationship between the potential difference across a circuit element and the electric current through it? To determine such a relationship, we conduct an experiment (Observational Experiment Table 19.3). We use a wire made of constantan (an alloy of nickel and copper), the resistive device in this case, and connect the ends of the wire (wire 1) to a variable emf indicated by the symbol ↗ (Figure 19.12). A variable emf can change both in magnitude and in direction. We vary the potential difference across wire 1 and record the resulting current through it. We then repeat the experiment with a longer constantan wire (wire 2). We focus only on the current through and potential difference across the two constantan wires, not the connecting wires.

FIGURE 19.12 Measuring the current through and the potential difference across a resistive constantan wire.

OBSERVATIONAL EXPERIMENT TABLE 19.3 Relating current through and potential difference across a resistive element

Observational experiment	Analysis			
	Potential difference across the wire (volts)	**Wire 1** I **(amps)**	**Wire 2** I **(amps)**	
The electric circuit is shown in Figure 19.12. The second constantan wire, wire 2, is twice as long as wire 1. By changing the setting of the variable emf, we vary the potential difference ΔV across the ends of the first resistive constantan wire and measure the current I through it. Then we repeat the experiment using constantan wire 2. Finally, we graph the results.	0	0.00	0.00	The graphs I versus ΔV for both wires are straight lines that pass through the origin.
	1.0	0.37	0.19	
	1.5	0.56	0.28	
	2.0	0.74	0.37	
	2.5	0.94	0.47	
	−1.0	−0.37	−0.19	
	−1.5	−0.56	−0.28	
	−2.0	−0.74	−0.37	
	−2.5	−0.94	−0.47	

Pattern

The graphs show that the current through both resistive wires is directly proportional to the potential difference across them. When the potential difference reverses direction, so does the current. The current through the shorter wire (wire 1) is greater than the current through the longer wire (wire 2) when the potential difference across them is the same.

The relationship we observed in Table 19.3 is reasonable given our conceptual understanding. The larger the potential difference across the wire, the larger the $\vec{E}$ field inside. This means that a larger electric force is exerted on the electrons, resulting in larger accelerations of the electrons between the collisions with the ions in the lattice and thus a larger drift velocity and a larger current through the wire. Since the pattern is the same in both directions, it seems that the electric properties of the metal are independent of the direction of current.

We notice that for the same potential difference across each wire, there is always twice as much current through wire 1 as through 2. Thus the ratio of current through and potential difference across a wire, which is called G, seems to quantify how easily the charged particles in the circuit flow (the less they interact with the ions, the less their kinetic energy is converted into the thermal energy of the lattice). The quantity G is called **conductance**. Circuit elements with higher conductance require a smaller potential difference across them to have the same current through them.

To characterize how a circuit element resists a current, we use another quantity, called **resistance R**:

$$R = \frac{1}{G}$$

We can write the relationship between current and potential difference using resistance R as

$$I = \frac{\Delta V}{R} \tag{19.3}$$

where R—resistance—is the physical quantity that characterizes the degree to which an object resists a current. The unit of R is called the **ohm** Ω (Greek omega), where 1 ohm = 1 volt/ampere (1 Ω = 1 V/A). A 1.0-ohm resistor has a one-ampere (1-A) electric current passing through it when a one-volt (1-V) potential difference is placed across it. Note that R is always positive. When we write Eq. (19.3), we usually use the absolute values of current and potential difference.

TIP Note that ΔV is the potential difference between one side of the device and the other (hence the Δ), whereas current I is the flow rate of charge through the device.

> **Ohm's law** The current I through a circuit element (other than a battery) can be determined by dividing the potential difference ΔV across the circuit element by its resistance R:
>
> $$I = \frac{\Delta V}{R} \tag{19.3}$$

Ohmic and non-ohmic devices

Equation (19.3) is called Ohm's law in honor of 19th-century physicist Georg Ohm, who discovered it experimentally. It is a cause-effect relationship that predicts the value of the current through a resistive device when the potential difference across it and its resistance are known.

TIP The equation $R = \Delta V/I$, which allows us to determine the resistance of a circuit element, is an operational definition of resistance. It is not a cause-effect relationship. Resistance is the ratio of the potential difference across the element and the current through it, *not* the slope of the I-versus-ΔV graph.

Ohm's law provides us with a method to determine the resistances of different circuit elements. We connect each circuit element into a circuit, close the circuit, and measure the current through and potential difference across the element to determine its resistance $R = \Delta V/I$. Using the above method we can determine the resistance of the two constantan wires used in the experiments in Table 19.3: $R_1 = 2.7 \, \Omega$ and $R_2 = 5.3 \, \Omega$. Physically, the second wire is twice as long as the first; the length of the wire must affect its resistance.

The resistance of each wire remains constant even though the potential difference across and the current through it change. Is this true for other circuit elements, such as commercial resistors that you see in many circuits inside analog and digital electronic equipment and lightbulbs? **Figure 19.13** shows graphs representing collected data for three elements: two commercial resistors and a lightbulb. Notice that the slopes of the graphs for the commercial resistors are constant. They behave in a similar fashion to constantan wires with resistances $R_1 = 25.0 \, \Omega$ and $R_2 = 50.0 \, \Omega$, independent of the current through them. If the resistance of a circuit element does not depend on the potential difference across it, the element is called **ohmic**. Circuit elements that cannot be modeled as ohmic devices are called **non-ohmic**.

Unlike for the commercial resistors, however, the slope of the graph for the lightbulb decreases as the potential difference increases. It takes more potential difference

FIGURE 19.13 I-vs-ΔV graphs for ohmic and non-ohmic circuit elements.

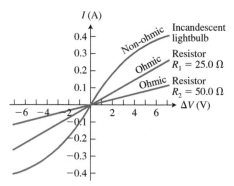

to cause the same increase in current. In other words, the resistance of the lightbulb increases with increasing potential difference and current. This relationship must be taken into account when building a circuit because the resistance of such a circuit element depends on the current through it. Note that for all of these elements, including the lightbulb, reversing the polarity of the battery does not change the shape of the curve. An incandescent lightbulb is an example of a non-ohmic device.

QUANTITATIVE EXERCISE 19.3   **Comparing an incandescent bulb and a resistor**

The graph shows how current through an incandescent lightbulb and a resistor changes with the changes in potential difference across them. The data for each were collected separately, although we present them on the same graph. (a) What are the resistances of the resistor and the lightbulb when the potential difference across each element is first 1.0 V and then 1.5 V? (b) What are the resistances of the resistor and the lightbulb when the current through each element is 50 mA?

Represent mathematically We rearrange Eq. (19.3) and use data from the graphs to answer the questions:

$$I = \frac{\Delta V}{R} \quad \Rightarrow \quad R = \frac{\Delta V}{I}$$

Solve and evaluate (a) When $\Delta V = 1.0$ V, $I_{Res} = 100$ mA, thus $R_{Res} = 1.0$ V$/0.100$ A $= 10\ \Omega$; for the bulb, $I_{Bulb} = 0.125$ A, thus $R_{Bulb} = 1.0$ V$/0.125$ A $= 8.0\ \Omega$. When $\Delta V = 1.5$ V, $I_{Res} = 150$ mA; thus, as before, $R_{Res} = 1.5$ V$/0.150$ A $= 10\ \Omega$. For the bulb, the current and potential difference are the same as for the resistor, thus its resistance at this moment is also $R_{Bulb} = 1.5$ V$/0.150$ A $= 10\ \Omega$.

(b) When the current through each element is 50 mA, the potential difference across the resistor is 0.5 V; therefore, its resistance is $R_{Res} = 0.5$ V$/0.05$ A $= 10.0\ \Omega$. The potential difference across the bulb is 0.25 V; therefore, its resistance is $R_{Bulb} = 0.25$ V$/0.05$ A $= 5.0\ \Omega$. We see that the resistance of the resistor remains constant, while the resistance of the lightbulb increases with the current through it.

Try it yourself In which interval of potential difference is the resistance of the bulb greater than that of the resistor?

Answer   Between 1.5 V and 2.0 V.

Light-emitting diodes (LEDs)

Another non-ohmic device that is present in every house is an LED–a **light-emitting diode**. LEDs are the basis of the lightbulbs that are replacing incandescent lightbulbs. How are LEDs different from incandescent bulbs? Examining an LED shows that it has two leads, one short and one long (**Figure 19.14a**). LEDs also come in different colors; we choose a green LED for our investigations in this chapter. We now connect the LED to a variable power supply, as we did with the resistor in Table 19.3, and collect data in the same way using an ammeter and voltmeter (see Figure 19.14b).

We observe several interesting patterns. First, when we connect the long lead of the LED to the *negative* terminal of the power supply, the LED does not light, no matter how much we increase the potential difference across it (as long as the potential difference across it does not exceed 10 V; above that, LED breakdown occurs, which leads to a large current and damage of the LED). Second, when we connect the long lead of the LED to the *positive* terminal of the power supply, at first the LED does not light. When the potential difference across it reaches a certain value (about +2.1 V for a green LED), it suddenly lights, and when we increase the potential difference by a small amount, its brightness increases dramatically. Can it be that the LED only allows current to pass through it in one direction, and that it glows when there is a current through it and the potential difference exceeds +2.1 V?

When we analyze the *I*-versus-ΔV graph from our experiment in Figure 19.14c, we find support for both aspects of our hypothesis: (1) there is no current through the LED when its long lead is connected to the negative terminal of the power supply (we will call this the "wrong" direction), and (2) the current is still zero even if the long lead of the LED is connected to the positive terminal (the "right" direction) for all potential differences below about +2.1 V.

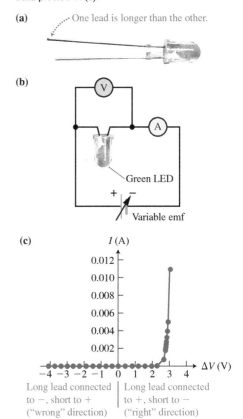

FIGURE 19.14 A green LED. The electric circuit in (b) is used to collect the *I*-versus-ΔV data plotted in (c).

We also notice that the current "shoots up" when we increase the potential difference above +2.1 V. Note that at a certain voltage above 3 V, the LED stops glowing, even though it is connected the right way, and never lights again, even if we decrease the potential difference back to +2.1–2.3 V. The current through it becomes so large that it burns and destroys the LED.

Based on the *I*-versus-ΔV graph, we can say that the LED is a non-ohmic element and has asymmetric resistive properties. The resistance of the LED is infinitely large when its long lead is at a lower potential than the short lead, and it is variable (it depends on the voltage across the LED) when its long lead is at a higher potential than the short lead. For example, the resistance of a green LED is very large (infinite) below +2.1 V and becomes very small when the potential difference across it is above +2.1 V. The 2.1-V potential difference is called the **opening voltage** of the LED. Note that the opening voltage does not have a precisely defined value. When the potential difference across the LED is approaching this value, the current through the LED starts to increase, and eventually the LED starts glowing.

LEDs come in different colors. The color of an LED's light does not come from the plastic cover, but from the physical materials that comprise it. LEDs of different colors have similar properties, but different opening voltages. For example, the opening voltage for a red LED is about 1.6 V and that for a blue LED is about 2.3 V.

To represent an LED in a circuit diagram, we use the symbol shown in **Figure 19.15**. The triangle resembles an arrow and points in the direction in which the LED could conduct current. The long lead of the LED must be at higher potential than the short lead for the LED to glow (assuming that the potential difference across it is above its opening voltage).

FIGURE 19.15 The LED circuit symbol.

Resistances of connecting wires and switches

So far, we have investigated the resistances of resistors, incandescent lightbulbs, and LEDs. However, almost every electric circuit contains connecting wires and switches. What can we learn about their resistance? First, we will investigate a slightly more complicated circuit (**Figure 19.16**). A battery and switch are connected to two resistors, $R_1 = 25\ \Omega$ and $R_2 = 50\ \Omega$. There are four ammeters in the circuit and five voltmeters. When the switch is closed, the meters have the readings as shown in the figure. The current through all the elements is the same. The same current enters the 25-Ω resistor then leaves it, and that same current enters the 50-Ω resistor and then leaves it. Current is not used up!

FIGURE 19.16 Examine the ammeter and voltmeter readings.

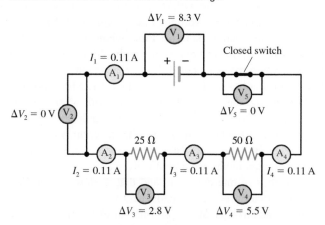

Voltmeter 2 across a connecting wire and voltmeter 5 across the closed switch both have zero readings. This is only possible if the resistance of the connecting wire and the closed switch is very small, or almost zero: $\Delta V = IR \approx I \cdot 0 = 0$. Now let's open the switch and use Ohm's law to make predictions about the meter readings.

TESTING EXPERIMENT TABLE 19.4

Applying Ohm's law to an open circuit

VIDEO
TET 19.4

Testing experiment	Prediction	Outcome			
		Ammeter readings		**Voltmeter readings**	
Use the circuit in Figure 19.16 but with the switch open.	We predict that the current in the circuit is zero, the ammeters will measure zero current $I = 0$, and according to Ohm's law, voltmeters 2–5 will measure zero potential difference $\Delta V = IR = 0R = 0$.	Ammeter 1	0 A	Voltmeter 1	8.5 V
		Ammeter 2	0 A	Voltmeter 2	0 V
		Ammeter 3	0 A	Voltmeter 3	0 V
		Ammeter 4	0 A	Voltmeter 4	0 V
				Voltmeter 5	8.5 V

Conclusion

We predicted 0 for voltmeter 5, but it measured 8.5 V. The outcome of the experiment does not match our prediction. We need to either revise Ohm's law or examine how we apply it.

When an outcome does not match a prediction, we need to evaluate the reasoning that led to the prediction. When the switch is open, there is no current in the circuit, so on first glance Eq. (19.3) written as $\Delta V = IR$ predicts a zero potential difference. Let us look at the battery first. Even when there is no current in the circuit, the battery has potential difference across the terminals equal to its emf. Thus, the nonzero reading of voltmeter 1 makes sense. But why would there be potential difference across the switch when there is no current in the circuit? When the switch is open, it stops being a conductor; its resistance becomes infinite. We cannot apply relation $\Delta V = IR$, as we do not know the result of multiplying zero by infinity. Therefore, when we made a prediction that did not match the outcome, it was not Ohm's law that failed us, but how we applied it. How do we explain the existence of the potential difference across the switch? When the switch is open, the whole part of the circuit connected to the positive terminal of the battery is at a potential of 8.5 V. The part of the circuit connected to the negative battery terminal is at a potential of 0.0 V (**Figure 19.17**).

We can find a similarity in the behavior of a switch in a circuit and an LED. Remember that for the "wrong" direction of an LED, the current through the LED is zero. The same holds also for an LED that is connected in the "right" direction but has a potential difference across it that is less than its opening voltage. Thus we can equate an LED connected in the "wrong" direction and an LED connected in the "right" direction for voltages below opening voltage to an open switch. If this is correct and we measure the potential difference across the LED for these conditions, we should measure the same voltage as that of the battery. This is exactly what happens (see **Figure 19.18**)! However, once there is electric current through a switch, the potential difference across it is zero for any current through it, whereas the potential difference across an LED changes very little while the current might change by many factors of 10. This is due to the rapidly decreasing resistance of an LED in the range of glowing potential differences. Therefore, it is useful to remember that when an LED is glowing, the potential difference across it is always approximately the opening voltage.

TIP Any burned out lightbulb causes the current in the line with the bulb to be zero. It is like an open switch. If the light switch on the wall is on, there may be a 120-V potential difference across the contact points in the bulb canister. It is not safe to touch the contact points!

FIGURE 19.17 Electric potential in a circuit with an open switch.

Minus signs indicate that the electric potential of this part of the circuit is 8.5 V less than the potential of the positive terminal of the battery.

FIGURE 19.18 An LED is similar to an open switch when (a) the LED is connected in the "wrong" direction, and (b) the LED is connected in the "right" direction but the potential difference across it is less than the opening voltage.

(a)

(b)

REVIEW QUESTION 19.4 Why does it make sense that all ammeters in Figure 19.16 have the same reading?

19.5 Qualitative analysis of circuits

So far, most of our analyses have applied to circuits with one element connected to a battery. Real-world circuits contain many elements. In this section we will learn about common arrangements of circuit elements and analyze the effect of these arrangements on the electric current in the circuits.

FIGURE 19.19 The bulbs are in series.

Circuits in series

In Observational Experiment **Table 19.5** we use small incandescent lightbulbs as circuit elements. We start by analyzing bulbs arranged in **series**, one after the other, as shown in **Figure 19.19**.

OBSERVATIONAL
EXPERIMENT TABLE 19.5   Multiple bulbs arranged in series

VIDEO
OET 19.5

Observational experiment		Analysis
Experiment 1. One lightbulb connected to a battery is bright.		The bulb is lit when the circuit is complete.
Experiment 2. Two identical bulbs connected one after the other (in series) to the battery from Experiment 1 are each dimmer than the bulb in 1 but are equally bright.		• There must be less current through the bulbs than in Experiment 1. • Each bulb must have the same lower current through it (since they are equally bright).
Experiment 3. Three identical bulbs connected in series to the same battery are each dimmer than the bulbs in Experiments 1 and 2 but are equally bright.		• There must be less current through the bulbs than in Experiments 1 and 2. • Each bulb must have the same even lower current through it (since they are equally bright.)

Patterns

• The brightness of all identical bulbs arranged in series is the same.
• Adding more bulbs arranged in series decreases the brightness of all bulbs.

The first pattern in Table 19.5 makes sense if we assume that the magnitude of the current through a bulb affects how bright the bulb is. Electric current is the flow of electrons. Since none of those electrons can escape the circuit or accumulate somewhere in the circuit, the current must be the same in the first bulb as in the last bulb in the series circuit—it is not used up. This reasoning is consistent with the results of the experiment in Figure 19.16 that showed us that in a similar circuit with resistors instead of the bulbs, the current was the same through all of the resistors.

The second pattern we observed in Table 19.5 indicated that adding more bulbs in series reduced the brightness of each bulb. Evidently, there is less current through series bulbs when there are more of them. Less current means that the amount of charge passing through a cross-sectional area of any element of a circuit per unit time is smaller. Adding more bulbs must increase the resistance of the circuit, reducing the electric current *everywhere* in the circuit (similar to how putting more filters in a hose slows the flow of water through the whole water system). **A battery is not a source of constant current**.

Circuits in parallel

We can also arrange bulbs in **parallel**: side by side with the terminals on each side connected together. We investigate simple parallel circuits in Observational Experiment Table 19.6.

OBSERVATIONAL
EXPERIMENT TABLE 19.6 **Multiple bulbs arranged in parallel**

VIDEO
OET 19.6

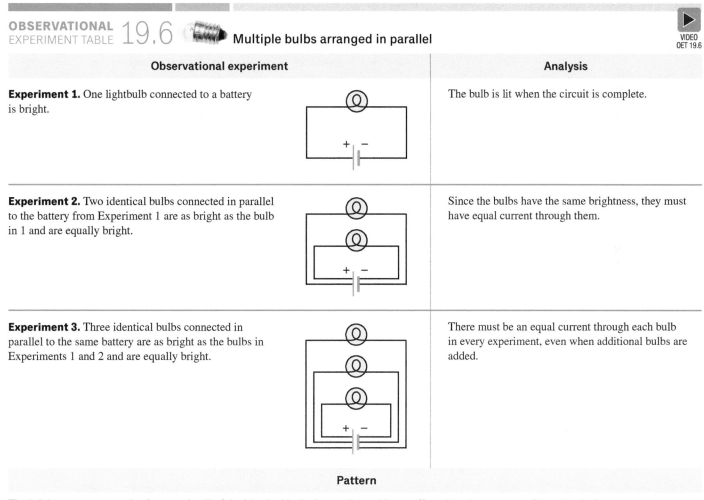

Observational experiment	Analysis
Experiment 1. One lightbulb connected to a battery is bright.	The bulb is lit when the circuit is complete.
Experiment 2. Two identical bulbs connected in parallel to the battery from Experiment 1 are as bright as the bulb in 1 and are equally bright.	Since the bulbs have the same brightness, they must have equal current through them.
Experiment 3. Three identical bulbs connected in parallel to the same battery are as bright as the bulbs in Experiments 1 and 2 and are equally bright.	There must be an equal current through each bulb in every experiment, even when additional bulbs are added.

Pattern

The brightness appears to be the same in all of the identical bulbs in parallel and is not affected by the presence of the other bulbs.

How can we explain the pattern in Table 19.6? In order for the identical bulbs to be equally bright and have the same brightness, the current must be the same through each of them. In other words, when bulbs are arranged in parallel, they behave as though they were connected to the battery all by themselves and are not affected by the presence of the other bulbs. Since the bulbs are identical, they have identical resistance (same resistance R for the same current I); the connecting wires have almost zero resistance, independent of their length. Therefore, the potential difference across all bulbs should be the same ($I = \Delta V/R$) and equal to the potential difference across the battery. The battery must drive a current through the bulbs simultaneously—the total current driven by the battery must double for two parallel bulbs and triple for three parallel bulbs. We see again that the battery is *not* a constant current source—it sends a smaller current through bulbs arranged in series and a larger total current through bulbs in parallel. This also means that multiple bulbs connected in parallel must have less collective resistance than a single bulb because as external resistance decreases, current through the battery increases.

Conventionally, circuits such as those in Table 19.6 are represented in two different ways; see **Figure 19.20**. The two circuits in the figure are equivalent. The points that have the same potential are connected by a wire, as in the circuit on the left, or connected directly to each other, as in the circuit on the right.

FIGURE 19.20 The bulbs connected in parallel to the battery have the same potential difference across them as a single bulb.

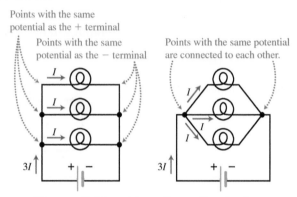

We can now summarize our qualitative investigations of electric circuits:

- When circuit elements are connected in series, the current through each element is the same; when circuit elements are connected in parallel, the battery is connected across each parallel element and the potential difference across each element is the same.
- Adding more circuit elements in series *decreases* the total current through all the elements and increases the effective total resistance of the elements.
- Adding more elements in parallel *increases* the total current through the parallel elements and reduces the effective total resistance of the elements.
- The battery is *not* a source of constant current: it drives a smaller current through circuit elements arranged in series and a larger total current through circuit elements in parallel.
- Changing a part of the circuit (by adding or removing an element) may change the currents through and potential differences across all other elements.

CONCEPTUAL EXERCISE 19.4 **Current versus charge storage**

You have two identical resistors with resistance R and a battery with a *storage capacity* Q (the total charge it can move before it becomes "dead"). Compare the current through the battery and the total time before the battery dies if (a) one resistor is connected to the battery, (b) both resistors are connected to the battery in series, and (c) both resistors are connected to the battery in parallel.

Sketch and translate The sketches of the circuits at right include the givens and the quantities to compare.

Compare I_R, I_{Ser}, I_{Par}
Compare t_R, t_{Ser}, t_{Par}

Simplify and diagram We drew the circuit diagrams in the previous step, so we use this step to reason about the process. Given that the resistors are identical, we know that the current through the resistors and the battery in series, I_{Ser}, is smaller than the current through a single resistor connected to the battery, I_R. That is, $I_{Ser} < I_R$. When the resistors are in parallel, the current through each resistor is equal to I_R. The battery needs to send these currents simultaneously; therefore, the current through the battery must be equal to $I_{Par} = 2I_R$. To summarize: $I_{Ser} < I_R < I_{Par}$.

The storage capacity of the battery is equal to the product of the current and time interval during which the charges are moving through the battery: $Q = I \Delta t$, or $\Delta t = Q/I$. Therefore, for a given storage capacity Q, the time interval before the battery is dead is inversely proportional to the current through the battery. Combining the relationship

between the currents and the last equation, we get $t_{Par} < t_R < t_{Ser}$. This makes sense: the smaller the current through the battery, the longer the time before the battery dies.

Try it yourself You have three identical resistors with resistance R and a battery with a storage capacity Q. Compare the current through the battery and the total time before the battery dies if one resistor is connected to the battery compared to (a) when three resistors are connected to the battery in series, and (b) when three resistors are connected to the battery in parallel.

Answer

(a) The current through the battery is three times smaller, and the total time before the battery dies is three times longer. (b) The current through the battery is three times larger, and the battery time is three times shorter.

How is your house wired?

Although the current in the electric circuit in our homes is different in many ways from the current we study in this chapter, there are many similarities that allow us to apply what we are learning in this book to the electrical wiring in a home.

You turn on a lightbulb, a computer, and a washing machine. Are they connected in series or in parallel? There are three possible arrangements (**Figure 19.21**): (a) all the devices are in series, (b) all are in parallel, or (c) some combination of the two. If they are all in series, when you turn one of the devices off, the circuit will open. There will be no current through the circuit, and all of the devices will be off. From your everyday experience, you know that does not happen. The devices must not be in series. If they are all in parallel, as in (b), then turning any one device on or off will not affect the others. This is what we observe, so parallel wiring is a possibility. In (c) the devices are arranged neither all in series nor all in parallel. Still, turning off one device (the computer in our case) also turns off others. This is not what happens in your house, so this sort of arrangement is not possible. The only choice that is consistent with the way devices in your house behave is (b), parallel.

REVIEW QUESTION 19.5 What experimental evidence supports the hypothesis that in your house the appliances are connected in parallel and not in series?

19.6 Joule's law

We used identical incandescent lightbulbs in our experiments in the previous section. Identical bulbs in series in a closed circuit have the same brightness. However, if we connect the two bulbs shown in **Figure 19.22a** in series to a battery, we observe that bulb 1 is brighter than 2. If we reverse the order of the bulbs, bulb 1 is still the brighter one (Figure 19.22b). We know that in a series circuit, there is the same current through each bulb. Thus, it is not just the current through a bulb that determines its brightness. In this section we will learn what other quantities affect a bulb's brightness.

Electric power

An incandescent lightbulb filament is hot and glows when there is current through it (the free electrons drift through the lattice and interact with ions, making the ions vibrate faster). Consider the bulb and connecting wires (electrons and electric field) as the system and the battery as an external object. The battery does work, increasing

FIGURE 19.21 The possible ways to wire household appliances.

(a) All in series

(b) All in parallel

(c) A combination

FIGURE 19.22 The bulb's brightness depends on something other than the current through the bulb.

Bulb 1 is brighter than bulb 2 in both circuits.

the electric potential energy of the system. When there is current through a circuit element, the electric potential energy is continuously converted into other forms of energy (thermal energy, light energy, mechanical energy, etc.). Earlier (in Chapter 7) we learned that power is the physical quantity that describes the rate of energy conversion. In this case it is the rate at which the device converts electric energy into thermal energy (or other forms of energy):

$$P = \left| \frac{\Delta U_{th}}{\Delta t} \right|$$

This conversion leads to the decrease of the electric potential energy of the system U_q. Therefore,

$$P = \left| \frac{-\Delta U_q}{\Delta t} \right|$$

As electrons (total electric charge $-Q$) travel through a resistive device, they travel from the side that is at low electric potential to the side that is at high electric potential. This movement of electrons results in a change in the electric potential energy of the system $\Delta U_q = q \, \Delta V = -Q \, \Delta V$. Therefore,

$$P = \left| \frac{Q \, \Delta V}{\Delta t} \right| = \frac{Q}{\Delta t} |\Delta V| = I |\Delta V|$$

The above has the correct power units (watts; $1 \, W = 1 \, J/s$).

$$A \cdot V = \frac{C}{s} \cdot \frac{J}{C} = \frac{J}{s} = W$$

James Joule determined this expression in experiments using a current-carrying wire to warm water. Later, Heinrich Lenz and Alexandre-Edmond Becquerel repeated the experiments using alcohol instead of water and achieved similar results. Now the expression for the rate of electric potential energy conversion into thermal energy is called **Joule's law**.

> **Joule's law** The rate P at which electric potential energy is converted into other forms of energy ΔU_{th} in a resistive device equals the magnitude of the potential difference ΔV across the device multiplied by the current I through the device:
>
> $$P = \left| \frac{\Delta U_{th}}{\Delta t} \right| = I |\Delta V| \qquad (19.4)$$

Because potential difference and current are related through Ohm's law, $I = \Delta V/R$, Joule's law can be written in two alternate forms:

$$P = I |\Delta V| = I(IR) = I^2 R$$

$$P = I |\Delta V| = \left(\frac{\Delta V}{R} \right) |\Delta V| = \frac{(\Delta V)^2}{R} \qquad (19.5)$$

Examine the different forms of Eq. (19.5). The top version says that the power is proportional to the resistance of the circuit element; the bottom version says that it is inversely proportional. Do these contradict each other? As the current through and potential difference across the element are not independent of each other, there is no contradiction. It is usually most convenient to use the $I^2 R$ expression when comparing power in elements connected in series and $(\Delta V)^2/R$ when elements are connected in parallel.

If we assume that the brightness of a bulb is directly related to its power, we can use Joule's law to explain the experiment at the beginning of this section. The lightbulbs in Figure 19.22 had different brightnesses even though the current through them was the

---

TIP The word "electricity" is used in many ways. People say that electricity flows, that electricity is used to heat buildings, and that electricity comes out of a plug in the wall. When people say that electricity flows, they are referring to the flow of charged particles in wires (electric current); when they talk about heating with electricity, they mean electric potential energy being transformed into thermal energy that heats the environment; and when they talk about wall sockets, they usually refer to a 120-V effective potential difference across any appliance plugged into the socket. When you use the word electricity, make sure you understand which of these ideas you are referring to.

same. Based on Eq. (19.4), $P = I|\Delta V|$, the potential difference across bulb 1 had to be greater than across bulb 2. This could only happen if bulb 1 had a greater resistance than bulb 2, $R_1 > R_2$, as $I_1 = I_2$; thus $P_1 = I^2 R_1 > P_2 = I^2 R_2$. More resistance means a higher rate of conversion of electric energy to thermal energy!

If our understanding of electric power is correct, then the same two resistors also will have different relative powers when connected differently. Consider two resistors: A is a 25-Ω resistor and B is a 50-Ω resistor. They are connected to the same power supply that has potential difference of 9 V across its terminals first in series and then in parallel. When the resistors are in series, the current through them is the same; therefore, the power of resistor B is greater than that of A because

$$P = I^2 R; \quad I_A = I_B; \quad R_A < R_B \implies P_A < P_B$$

When they are in parallel, the potential difference across them is the same; therefore, the power of resistor A is greater than that of B:

$$P = \frac{\Delta V^2}{R}; \quad \Delta V_A = \Delta V_B; \quad R_A < R_B \implies P_B < P_A$$

Thus the same resistor (or any electric appliance) can be more or less powerful depending on the circuit it is in; power is not a fixed characteristic. This is important to know when you buy electric appliances.

Power of incandescent lightbulbs and LEDs

In the last few years, LED lightbulbs have almost completely replaced incandescent lightbulbs in our homes. Why? A simple observation can give us a hint. When we touch a bright incandescent lightbulb, it is very hot; however, LED-based bulbs are not hot to touch. Therefore, we can hypothesize that traditional bulbs convert a significant fraction of the electric energy to thermal energy rather than to light, while LEDs convert most of the electric energy delivered to them to light energy. Both light energy and thermal energy leave the bulbs and LEDs through the process of radiation.

We can conduct an experiment to investigate this difference quantitatively using a variable power supply, a small incandescent lightbulb, and a white LED. Both light sources are enclosed in white ping-pong balls to ensure that the light they emit is spread evenly, making it easier to compare their brightness (**Figure 19.23a**). In order to compare the light sources more objectively, we will use a special device (a light sensor) that measures the brightness of both sources; its sensitivity to light is similar to that of our eyes. We will change the potential difference across the power supply until the light sensor has the same reading for both sources when placed at the surface of the ping-pong ball. (In Figure 19.23a, the LED appears to have a somewhat different color than the bulb, but despite this, their brightness as measured by the sensor is the same). The data collected in the experiment are shown in Figure 19.23b. A pattern is clearly visible from the data: the white LED needs about 10 times less electric power to emit light of the same brightness as an incandescent lightbulb. Thus the old-time incandescent lightbulb must be converting the rest (90%) of the energy into thermal energy or some other form of light that cannot be detected by our eyes. Figure 19.23c shows thermal images for both light sources: the LED is at room temperature and the bulb is much warmer. We will learn in Chapter 28 the mechanism behind the high efficiency of LEDs.

Paying for electric energy

Electric power companies charge their customers according to the amount of electric potential energy that is converted into other energy forms in the devices that the customer uses. Utility companies do not use the joule. Instead, they use an energy unit called the **kilowatt-hour**. A kilowatt-hour (kW·h) is the electric potential energy that a 1000-W device transforms to other energy forms in a time interval of 1 h (1 kW·h $= 3.6 \times 10^6$ J).

FIGURE 19.23 Comparing energy conversions for an LED versus an incandescent lightbulb.

(a) Experimental setup

(b) Table of measurements taken

	LED	Incandescent lightbulb
ΔV (V)	3.1	3.6
I (A)	0.020	0.190
P (W)	0.062	0.680
B (lux)*	450	450

*Brightness of light measured at the surface of the ping-pong ball

(c) Thermal images of the experiment

19.7 Kirchhoff's rules

Most electric circuits consist of combinations of several resistive elements, switches, one or more power supplies that may have a variable potential difference output, and a variety of other circuit elements. In this section we will develop techniques to determine the electric current through each circuit element in a complicated circuit.

Kirchhoff's loop rule

FIGURE 19.24 The changing electric potential around an electric circuit.

(a)

(b)

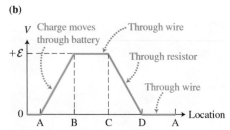

Let us first examine the changes in electric potential in a simple circuit. The current in the circuit shown in **Figure 19.24a** is counterclockwise. We trace the change in electric potential as we move with these charges around the circuit (Figure 19.24b). The potential when we return to the starting position must be the same as when we left that position (the potential cannot have multiple values at one location). We see from the graph that

Changes in electric potential

$$
\underset{\substack{\text{Across} \\ \text{battery} \\ \text{A to B} \\ +\mathcal{E}}}{} + \underset{\substack{\text{Across} \\ \text{connecting} \\ \text{wire B to C} \\ 0}}{} + \underset{\substack{\text{Across} \\ \text{resistor} \\ \text{C to D} \\ (-IR)}}{} + \underset{\substack{\text{Across} \\ \text{connecting} \\ \text{wire D to A} \\ 0}}{} = 0
$$

A pattern (that we shall call the loop pattern) accounts for the data:

> The algebraic sum of the changes in electric potential around any closed circuit loop is zero.

Let us test this pattern.

EXAMPLE 19.5   **Testing the loop pattern**

Use the loop pattern to predict the potential difference between point A and point B in the electric circuit shown in the figure.

Sketch and translate To predict the potential difference between points A and B, we first use the loop pattern to determine the current in the circuit shown in the figure.

Simplify and diagram Assume that the wires and batteries have negligible (zero) resistance compared to that of the resistors. To determine the direction of the current, note that the 10-V battery has a larger emf than the 4-V battery. Therefore, the current in the circuit will be clockwise.

Represent mathematically We apply the loop pattern going clockwise from A:

$$\Delta V_{10\,V} + \Delta V_{20\,\Omega} + \Delta V_{4\,V} + \Delta V_{60\,\Omega} + \Delta V_{40\,\Omega} = 0$$
$$(+10\,\text{V}) + [-I(20\,\Omega)] + (-4\,\text{V}) + [-I(60\,\Omega)] + [-I(40\,\Omega)] = 0$$

Notice that the potential difference across the 4.0-V battery was negative since we moved from the positive to the negative terminal of that battery. Also, the potential differences across the resistors were negative because we moved across these resistors in the clockwise downstream direction of the electric current.

Solve and evaluate We can now determine the current:

$$I = \frac{10.0\,\text{V} - 4.0\,\text{V}}{20\,\Omega + 60\,\Omega + 40\,\Omega} = 0.050\,\text{A}$$

To determine the potential difference from A to B, we follow the circuit along any continuous path from A to B and add the potential differences across each element in the circuit along that path. There are two possible paths. We first move clockwise from A through the 10.0-V battery and 20-Ω resistor:

$$\Delta V_{AB} = \Delta V_{10\,V} + \Delta V_{20\,\Omega} = +10\,V - (0.050\,A)(20\,\Omega)$$
$$= +9.0\,V$$

The V field value at B is 9.0 V higher than at A.

We can also calculate the potential change from A to B by moving counterclockwise upstream through the 40-Ω and 60-Ω resistors (opposite the current) and through the 4.0-V battery from its negative to positive terminal.

$$\Delta V_{AB} = \Delta V_{40\,\Omega} + \Delta V_{60\,\Omega} + \Delta V_{4\,V}$$
$$= +(0.050\,A)(40\,\Omega) + (0.050\,A)(60\,\Omega) + 4.0\,V$$
$$= +9.0\,V$$

The potential difference from A to B is the same no matter which path we take. When we take a voltmeter and connect its terminals to A and B (its negatively marked terminal to A and its positively marked terminal to B), we can see if the outcome and prediction are consistent. The voltmeter reads about 9 V, supporting the loop pattern.

Try it yourself Suppose that in this circuit, there was an open switch inserted in the vertical wire on the left side before position A. Apply the loop pattern to determine the potential difference across the switch.

Answer

Using Table 19.4, finding $\Delta V_{switch} = -6$ V. The top of the switch closer to the 10-V battery will be at a lower potential. The presence of potential difference across the switch when it is open is consistent with our experimental finding in Table 19.4.

T|P To indicate the signs of the electric potential changes across the resistors in the previous example, we chose and indicated in the circuit drawing the assumed direction of the current. The potential change across a resistor if moving in the direction of the current is $\Delta V = -IR$; if moving across the resistor opposite the current it is $\Delta V = +IR$. We *must* indicate a current direction in the circuit diagram and keep the signs of I and ΔV consistent while making a trip around the circuit.

This "loop pattern" was first developed in 1845–1846 by German physicist Gustav Robert Kirchhoff (1824–1887) and is known as **Kirchhoff's loop rule**. The rule is general: it applies to any loop in any circuit, even when a circuit has multiple loops.

> **Kirchhoff's loop rule** The sum of the electric potential differences ΔV across the circuit elements that make up a closed path (called a loop) in a circuit is zero.
>
> $$\sum_{\text{Loop}} \Delta V = 0 \qquad (19.6)$$

EXAMPLE 19.6   LEDs and the loop rule

The opening voltage for a green LED is about 2.1 V, and the current that will make it glow with full brightness is about 20 mA. What resistor should we connect in series with the LED so we can safely power the LED using a 6.0-V battery?

Sketch and translate We draw a circuit diagram and label knowns and unknowns. The green LED would burn out if we connected it directly to the 6.0-V battery because the I-versus-ΔV graph for an LED is very steep once the potential difference across it is larger than the opening voltage, so the current would be too large. If we connect a suitable resistor in series with the LED, then we can limit the current to the allowed value of 20 mA.

Simplify and diagram We assume the resistance of the connecting wires to be zero and the potential difference across the 6.0-V battery to be 6.0 V when there is current in the circuit.

Represent mathematically Based on the loop rule,

$$\Delta V_{LED} + \Delta V_R = 6.0\,V$$

In addition, we know that the voltage across the normally glowing LED is about the same as the opening voltage of the LED:
$\Delta V_{LED} = \Delta V_{Open} = 2.1\,V$.

We can determine the potential difference across the resistor using Ohm's law:

$$I = \frac{\Delta V_R}{R} \implies \Delta V_R = IR$$

In our case, the current is given: $I = 0.020\,A$.

Solve and evaluate We combine the two equations,
$\Delta V_{LED} + \Delta V_R = 6.0\,V$ and $\Delta V_R = (0.020\,A)R$, to get

$$2.1\,V + (0.020\,A) \cdot R = 6.0\,V \implies R = 195\,\Omega$$

(CONTINUED)

We can evaluate our result using extreme-case analysis. We know that the current of a circuit with no LED should be larger than a circuit with the LED. This is exactly what we obtain: $\frac{6 \text{ V}}{195 \text{ }\Omega} = 0.03$ A, larger than with the LED. If there were no resistor, the LED would burn out because the current would exceed 0.02 A. Resistors are therefore essential elements in LED circuits.

Try it yourself What is the voltage across the LED in this circuit if we switch the terminals of the battery?

Answer   ᴧ 0˙9

FIGURE 19.25 Investigating a 9.0-V battery.

(a)

$\Delta V = 9.0$ V

(b)

I

$\Delta V = 7.5$ V

$I = 1.5$ A

$R = 5.0$ Ω

(c)

Battery

r

I

$I = 1.5$ A

$R = 5.0$ Ω

Internal resistance of a battery

In all our previous examples, the emf and potential difference across the battery terminals were assumed to be the same. Let us investigate whether this is truly the case in real life. Imagine that you take a 9.0-V battery and use a voltmeter to measure potential difference across its terminals. The voltmeter reads 9.0 V (**Figure 19.25a**). Then you build the circuit shown in Figure 19.25b. An ammeter reads a 1.5-A current through the 5.0-Ω resistor. A voltmeter reads 7.5 V across the battery and −7.5 V across the resistor. Notice that we found two unexpected results: the potential difference is smaller than the emf of the battery, and the current is less than expected $\left(I_{exp} = \frac{\mathcal{E}}{R} = \frac{9.0 \text{ V}}{5.0 \text{ }\Omega} = 1.8 \text{ A} \right)$. What are possible reasons for these anomalous data?

The current would be smaller than predicted if there were some other resistor connected to the battery. Perhaps the battery has its own **internal resistance** r (see Figure 19.25c). Using this hypothetical resistance, we can write the loop rule as

$$+\mathcal{E} - Ir - IR = 0$$

and use it to find the internal resistance of the battery,

$$r = \frac{\mathcal{E} - IR}{I}$$

In our case, the internal resistance of the battery is

$$r = \frac{\mathcal{E} - IR}{I} = \frac{9.0 \text{ V} - (1.5 \text{ A})(5.0 \text{ }\Omega)}{1.5 \text{ A}} = 1.0 \text{ }\Omega$$

This explains why the measured current is less than 1.8 A:

$$I_1 = \frac{\mathcal{E}}{R + 1.0 \text{ }\Omega} = \frac{9.0 \text{ V}}{5.0 \text{ }\Omega + 1.0 \text{ }\Omega} = 1.5 \text{ A}$$

The idea of internal resistance is supported by the observation that batteries feel warm to the touch after they have been operating for a while. We can also explain why the voltmeter reading across the battery is smaller in a closed circuit than in an open circuit. When there is current in the circuit, the loop rule is

$$\mathcal{E} - Ir = IR \tag{19.7}$$

IR is the potential difference across the resistor, and at the same time it is the potential difference measured across the battery, because the resistor is connected directly to the battery and we assume that the resistance of the wires is negligible.

$$\Delta V_{batt} = IR = (1.5 \text{ A})(5.0 \text{ }\Omega) = 7.5 \text{ V}$$

We can write that the potential difference across the battery is equal to $\Delta V_{batt} = \mathcal{E} - Ir$. Notice that whenever there is a current through the battery, the potential difference across it is less than its emf; how much less depends on the current through the battery and on its internal resistance. If there is no circuit attached to the battery (no current), then the potential difference across it equals its emf. If $r = R$, then $\Delta V_{batt} \approx \mathcal{E}/2$. If $r \ll R$, then $\mathcal{E} \approx \Delta V_{batt}$.

TIP When using the loop rule for a closed circuit, follow these sign conventions:

- The potential difference across a resistor is $-IR$ when the loop traverses the resistor in the direction of the current through it and $+IR$ when the loop traverses the resistor in the direction opposite the current through it.
- The potential difference across a battery (assuming zero internal resistance) is $+\mathcal{E}$ when the loop traverses the battery from its negative terminal to its positive terminal and $-\mathcal{E}$ when the loop traverses the battery from its positive terminal to its negative terminal. The potential difference instead is $\pm(\mathcal{E} - Ir)$ if the battery's internal resistance cannot be neglected.
- Assuming the resistance of a connecting wire is zero, all points along that wire are at the same potential.

Kirchhoff's junction rule

We developed the loop rule for a circuit with only one loop. However, many electric circuits have several loops. In multiloop circuits, the charged particles moving along a wire can arrive at a junction where several wires meet. The charged particles comprising the current cannot vanish or be created out of nothing (electric charge is a conserved quantity). Thus, the sum of the currents into a junction must equal the sum of the currents out of a junction.

Consider the multiloop circuit shown in **Figure 19.26**. The battery at the top causes a counterclockwise current I in the circuit. The current entering junction A equals the sum of the currents leaving A. Expressed mathematically:

$$I = I_1 + I_2$$

When the currents I_1 (through R_1) and I_2 (through R_2) meet at junction B, the two currents recombine to form a single larger current I that leaves junction B:

$$I_1 + I_2 = I$$

For the circuit in Figure 19.26, the current I moving upward through the wire on the right side of the circuit equals the downward current I through the wire on the left side of the circuit; the current leaving the positive terminal of the battery equals the current entering the negative terminal. The current is not used up in a circuit. We can summarize this splitting and joining of currents at junctions as **Kirchhoff's junction rule**.

FIGURE 19.26 Kirchhoff's junction rule for electric currents.

> **Kirchhoff's junction rule** The total rate at which electric charge enters a junction equals the total rate at which electric charge leaves the junction:
>
> Sum of currents into junction = Sum of currents out
>
> In symbols:
>
> $$\sum_{\text{In}} I = \sum_{\text{Out}} I \tag{19.8}$$

Short circuit

Imagine that you accidentally connect the terminals of a battery with a connecting wire of approximately zero resistance $R = 0$ (**Figure 19.27**). The wire and the battery become very hot. Notice that when the external resistance is $R = 0$, the current through the battery is

$$I = \frac{\mathcal{E}}{R + r} = \frac{\mathcal{E}}{0 + r} = \frac{\mathcal{E}}{r}$$

FIGURE 19.27 A short circuit across a battery can cause a very large current through it.

Because the internal resistance of the battery is usually small, the current becomes large; it can burn the connecting wire and destroy the battery. This situation is called a **short circuit**. Every time you build an electric circuit, before you turn it on, check that you did not connect a single wire directly to the two battery terminals without any other resistive devices in series with it.

REVIEW QUESTION 19.7 Where is the electric potential higher: where the current "enters" a resistor or where it "leaves" it? Where is the electric potential higher: where the current "enters" a conducting wire or where it "leaves" it?

19.8 Resistor and capacitor circuits

In the previous sections we learned about resistors in series and parallel. Here we will learn how to determine mathematically the equivalent resistor with which we can replace a combination of several resistors.

Resistors in series

FIGURE 19.28 A single equivalent resistor replaces the three series resistors.

$$R_{\text{eq, series}} = R_1 + R_2 + R_3$$

Suppose that you arrange three resistors (R_1, R_2, and R_3) in series with a battery of emf $\mathscr{E}$. Can we replace these three resistors with a single resistor without changing the current through the battery (**Figure 19.28**)? Use Kirchhoff's loop rule and imagine traveling counterclockwise around the circuit starting at the negative terminal of the battery:

$$\Delta V - IR_1 - IR_2 - IR_3 = 0$$

where ΔV is the potential difference across the battery when there is current in the circuit.

$$\Rightarrow I = \frac{\Delta V}{R_1 + R_2 + R_3}$$

The current through each element in this circuit is the same since there are no junctions. It appears that a single resistor $R = R_1 + R_2 + R_3$ would have the same effect on the current in the circuit as the three separate resistors R_1, R_2, and R_3.

Series resistance When resistors are connected in series,

- The current through each resistor is the same: $I_1 = I_2 = I_3 = \cdots$
- The potential difference across each resistor is $\Delta V_1 = IR_1$, $\Delta V_2 = IR_2$, $\cdots$
- The potential difference across the entire series arrangement of resistors is $\Delta V = \Delta V_1 + \Delta V_2 + \Delta V_3 + \cdots$
- The equivalent resistance of the resistors arranged in series is the sum of the resistances of the individual resistors:

$$R_{\text{eq, series}} = R_1 + R_2 + R_3 + \cdots \qquad (19.9)$$

Resistors in parallel

When resistors are arranged in parallel, one side of each resistor is connected to one common junction and the other side of each resistor is connected to another common junction (**Figure 19.29**). We know from Kirchhoff's junction rule that the current I into the left junction equals the sum of the currents through each resistor:

$$I = I_1 + I_2 + I_3$$

The potential difference between the junctions A and B at the sides of this group of parallel resistors is the same regardless of the path taken between those two points:

$$\Delta V_1 = \Delta V_2 = \Delta V_3 = \Delta V$$

We can use Ohm's law to determine the current through each resistor:

$$I_1 = \frac{\Delta V}{R_1}, I_2 = \frac{\Delta V}{R_2}, I_3 = \frac{\Delta V}{R_3}$$

Inserting the above into Kirchhoff's junction rule, we get

$$I = I_1 + I_2 + I_3 = \frac{\Delta V}{R_1} + \frac{\Delta V}{R_2} + \frac{\Delta V}{R_3}$$

Now, suppose that we replace these three resistors by a single equivalent parallel resistor $R_{\text{eq, parallel}}$ that causes the same current I in the circuit. The current will be

$$I = \frac{\Delta V}{R_{\text{eq, parallel}}}$$

Combining these two results, we get

$$\frac{\Delta V}{R_{\text{eq, parallel}}} = \frac{\Delta V}{R_1} + \frac{\Delta V}{R_2} + \frac{\Delta V}{R_3}$$

$$\Rightarrow \frac{1}{R_{\text{eq, parallel}}} = \frac{1}{R_1} + \frac{1}{R_2} + \frac{1}{R_3}$$

$$\Rightarrow R_{\text{eq, parallel}} = \left(\frac{1}{R_1} + \frac{1}{R_2} + \frac{1}{R_3} \right)^{-1}$$

FIGURE 19.29 A single equivalent resistor replaces the three parallel resistors.

$$R_{\text{eq, parallel}} = \left(\frac{1}{R_1} + \frac{1}{R_2} + \frac{1}{R_3} \right)^{-1}$$

$$I = I_1 + I_2 + I_3$$

Parallel resistance When resistors are connected in parallel,

- The potential difference across each individual resistor is the same:

$$\Delta V_1 = \Delta V_2 = \Delta V_3 = \Delta V$$

- The sum of the currents through them equals the total current through the parallel arrangement:

$$I = I_1 + I_2 + I_3 = \frac{\Delta V}{R_1} + \frac{\Delta V}{R_2} + \frac{\Delta V}{R_3}$$

- The equivalent resistance of resistors in parallel is

$$R_{\text{eq, parallel}} = \left(\frac{1}{R_1} + \frac{1}{R_2} + \frac{1}{R_3} + \cdots \right)^{-1} \qquad (19.10)$$

TIP Do not forget the (-1) exponent when you calculate the equivalent resistance of resistors in parallel.

Note that when you connect resistors in series, the equivalent resistance is larger than the largest resistance; when you connect resistors in parallel, the equivalent resistance is smaller than the smallest resistance.

Quantitative reasoning about electric circuits

In Section 19.5 we learned to reason qualitatively about currents in electric circuits. Here we analyze similar circuits quantitatively.

EXAMPLE 19.7 **Power in series and parallel circuits**

Assume that we have a battery and two identical resistors each of resistance R. We connect the resistors to the battery in three different configurations: (a) one resistor, (b) two resistors in series, and (c) two resistors in parallel. Determine the equivalent resistance of a single resistor that will produce the same current through the battery as produced in arrangements (b) and (c). Next, compare the total power output of the resistors in arrangements (b) and (c) with the output in (a).

Sketch and translate We draw the three arrangements (see figures). We first determine the equivalent resistance for (b) and (c) and then use Joule's law to determine the power in each case.

(a)   (b)   (c)

Simplify and diagram Assume that the resistance of the connecting wires and the internal resistance of the battery do not affect the current.

Represent mathematically The two identical resistors in series have an equivalent series resistance:

$$R_{eq,\ series} = R + R = 2R$$

The two identical resistors in parallel have an equivalent parallel resistance:

$$R_{eq,\ parallel} = \left(\frac{1}{R} + \frac{1}{R}\right)^{-1} = \left(\frac{2}{R}\right)^{-1} = \frac{R}{2}$$

The current in each simplified circuit (with the multiple resistors replaced by a single equivalent resistor) equals the battery emf divided by the circuit equivalent resistance:

$$I = \frac{\mathcal{E}}{R_{eq}}$$

The power output of the equivalent resistor (the rate at which electric potential energy is transformed into thermal energy) is

$$P = I^2 R_{eq}$$

Solve and evaluate (a) Circuit (a) has resistance R. The current in this circuit is $I_{One} = \mathcal{E}/R$. The power output of this circuit is

$$P_{One} = I_{One}^2 R = \left(\frac{\mathcal{E}}{R}\right)^2 R = \frac{\mathcal{E}^2}{R}$$

(b) The equivalent resistance of circuit (b) is $R_{Series} = 2R$. The current in this circuit is

$$I_{Series} = \frac{\mathcal{E}}{R_{eq,\ series}} = \frac{\mathcal{E}}{2R} = \frac{1}{2}I_{One}$$

The power output of this circuit is

$$P_{Series} = I_{Series}^2 R_{eq,\ series} = \left(\frac{\mathcal{E}}{R_{eq,\ series}}\right)^2 R_{eq,\ series} = \left(\frac{\mathcal{E}}{2R}\right)^2 2R$$

$$= \frac{1}{2}\frac{\mathcal{E}^2}{R} = \frac{1}{2}P_{One}$$

(c) The equivalent resistance of circuit (c) is $R_{eq,\ parallel} = R/2$. The current in this circuit is

$$I_{Parallel} = \frac{\mathcal{E}}{R_{eq,\ parallel}} = \frac{\mathcal{E}}{R/2} = 2I_{One}$$

The power output of this circuit is

$$P_{Parallel} = I_{Parallel}^2 R_{eq,\ parallel} = \left(\frac{\mathcal{E}}{R_{eq,\ parallel}}\right)^2 R_{eq,\ parallel}$$

$$= \left(\frac{\mathcal{E}}{R/2}\right)^2 R/2 = 2\frac{\mathcal{E}^2}{R} = 2P_{One}$$

In general, the total power of two devices connected in parallel to the battery is four times larger than when they are connected in series.

Try it yourself Two equally bright identical incandescent bulbs in series are shown in the first figure. What happens to the brightness of the bulbs and the power of the circuit if you place a wire (called a shorting wire) in parallel with bulb 2 (see the second figure)? Explain.

(a)   (b)

Answer Bulb 1 becomes much brighter and bulb 2 goes out. Assuming that the shorting wire has zero resistance, all current is through the shorting wire, not through 2. The resistance of the new circuit is half the original resistance. Thus, the current in the circuit doubles and the brightness of the remaining bulb quadruples; the power of the circuit doubles.

FIGURE 19.30 Investigating capacitor circuits.

(a)   (b)

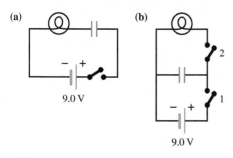

Capacitor circuits

In Chapter 18 we learned about capacitors, devices you can charge to store electric potential energy. What if we place a capacitor in an electric circuit? To investigate what happens, we make the circuit shown in **Figure 19.30a**, with a 9-V battery, a switch, a lightbulb, and a capacitor in series. When we close the switch, the bulb glows for a very short time interval and then goes out. Using a voltmeter, we measure the potential difference across each element after the bulb goes out. It measures 9 V across the battery, 9 V across the capacitor, and 0 V across the bulb and the switch. How can we explain this observation?

When we close the switch, the capacitor starts charging, which leads to a current through the bulb, so the bulb glows. Once the capacitor is charged, the current stops because the charged capacitor acts like a battery connected in the opposite direction to the original battery. We can test this reasoning in the following experiment: we make the new circuit shown in Figure 19.30b. When switch 1 is closed and switch 2 is open, the capacitor is connected to the battery; when 1 is open and 2 is closed, the capacitor is connected to the bulb. We first close switch 1 and wait for a few seconds. Then we open switch 1 and close switch 2. If our explanation of the first experiment is correct, then we should observe the bulb glowing for a short time once switch 2 is closed. This is exactly what happens (see the video).

We can investigate the phenomenon quantitatively by building the circuit shown in Figure 19.31a. We will focus on the process of discharging. Notice that we are using the same circuit as in Figure 19.30b, but we have replaced the bulb with a resistor to make sure that the resistance of the circuit remains constant. Such a circuit with a resistor and a capacitor is called an **RC circuit**. We use a voltmeter to measure potential difference across the capacitor. The potential difference-versus-time graph is shown in Figure 19.31b.

It takes about half a second for the capacitor to discharge. Changing capacitors to have different capacitance and changing resistors to have different resistance shows that the time for the potential difference to go to zero varies. Larger capacitance and larger resistance lead to a longer discharge time (when the capacitor is charged using the same battery). This makes sense: larger capacitance means that the capacitor can accumulate more charges on the plates (higher Q); larger resistance means smaller current and therefore smaller charge per unit time decrease of the original charge (smaller I).

Since the time dependence of this process is repeatable and predictable with high precision, an RC circuit is often used as a timer in simple electronic circuits to periodically switch on and off the current through other devices. Examples of such timers are blinking LEDs in bicycle lights and the switching on and off of windshield wipers (depending on the intensity of the rain, the period of the wipers changes with the help of different resistor-capacitor combinations). Capacitors in DC circuits are also used when we need a large current in a short time, for example, in a camera flash. A battery cannot provide a large current due to its internal resistance, but it can slowly move a lot of charge from one capacitor plate to another. The capacitor is first slowly charged from the battery, and then it is quickly discharged through the flash lamp.

Circuit breakers and fuses

Our study of parallel circuits allows us to understand how fuses protect our houses from fires caused by electric current. Although the current through appliances connected to wall sockets (called **alternating current**, or AC) differs in some ways from the current we are studying in this chapter (**direct current**, or DC), many of the same physics ideas apply to both. As noted earlier, household appliances are connected in parallel. When we connect more appliances in parallel, the total resistance of the circuit decreases. As a result, the total current in the main line (the connection to the outside power grid) of the circuit increases. The main line provides a constant effective potential difference of 120 V. When the current in the main line becomes too large, the line can become hot enough (due to its own resistance) that the insulation around the wire melts; that hot material can set adjacent objects on fire. To prevent this from occurring, the electrical systems in buildings are installed with circuit interrupters: either **fuses** or **circuit breakers**.

A fuse is a piece of wire made of an alloy of lead and tin. This alloy melts at a relatively low temperature. Thus, when the current through the fuse gets too large to be safe, the increase in thermal energy of the fuse melts the wire. The melted fuse is a gap in the circuit, which causes the current to quickly drop to zero. In order to reconnect the circuit, the fuse must be replaced.

Modern electrical wiring uses circuit breakers rather than fuses (Figure 19.32). There are different kinds of circuit breakers, but all of them are based on the same principle—when the current in a circuit increases to a critical value, the circuit breaker opens a switch, cutting off the current through the circuit. Circuit breakers are far more convenient than fuses because they are not self-destructive. A circuit breaker that has been tripped can quickly be reset instead of having to be replaced.

VIDEO
19.1

FIGURE 19.31 Discharging a capacitor in an RC circuit.

(a)

9.0 V

(b) ΔV_c (V)

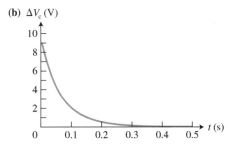

FIGURE 19.32 A circuit breaker panel box and a 30-A breaker switch.

Branch circuit breaker

Main power disconnect (at top)

FIGURE 19.33 Review Question 19.8.

REVIEW QUESTION 19.8 Rank the four identical bulbs shown in **Figure 19.33** from brightest to dimmest. Explain your reasoning.

19.9 Skills for solving circuit problems

The left side of the table below describes a general strategy for quantitatively solving circuit problems. The right side illustrates its application to the following problem.

PROBLEM-SOLVING STRATEGY 19.1	Applying Kirchhoff's rules

> **EXAMPLE 19.8**   **Solving circuit problems**
>
> Determine the currents in each branch of the circuit shown in the figure. The 9.0-V battery has a 2.0-Ω internal resistance and the 3.0-V battery has negligible internal resistance.

Sketch and translate

- Draw the electric circuit described in the problem statement and label all the known quantities.
- Decide which resistors are in series with each other and which are in parallel.

- We first draw the circuit with known information.
- We cannot simplify the circuit using the series or parallel equivalent resistance equations because there are batteries in two of the branches and the parallel equivalent resistance rule applies for branches with resistors only.
- Since there are three unknown currents, I_1, I_2, and I_3, we need three independent equations to solve for the currents. We use the loop rule twice and the junction rule once.

Battery with internal resistance

Simplify and diagram

- Decide whether you can neglect the internal resistance of the battery and/or the resistance of the connecting wires.
- Draw an arrow representing the direction of the electric current in each branch of the circuit.

- The text of the problem does not provide information about the resistances of the wires, so we will assume their resistance is approximately zero.
- Add an arrow and a label for the current in each branch of the circuit. Each arrow points in the direction that we think is the direction of the current in that branch.

Represent mathematically

- If possible, replace combinations of resistors with equivalent resistors.
- Apply the loop rule for the potential changes as you move around one or more different loops of the circuit. Each additional loop you choose must include at least one branch of the circuit that you have not yet included.
- Once you have included branches, apply the junction rule for one or more junctions. In total, you will need the same number of independent equations as the number of unknown currents.
- If necessary, use expressions for electric power.

Apply the loop rule for two different loops, First, start at B and move clockwise around the bottom loop:

$$\Delta V_{3\,\text{V}} + \Delta V_{18\,\Omega} = 0$$
$$+3.0\,\text{V} - I_2(18\,\Omega) = 0$$

Next, start at B and move counterclockwise around the outside loop:

$$\Delta V_{2\,\Omega} + \Delta V_{9\,\text{V}} + \Delta V_{12\,\Omega} + \Delta V_{3\,\text{V}} = 0$$
$$-I_1(2\,\Omega) + 9.0\,\text{V} - I_1(12\,\Omega) - 3.0\,\text{V} = 0$$

Notice that we cross the 3.0-V battery from the positive to the negative side, a decrease in potential.

Apply the junction rule to junction A: $I_1 + I_3 = I_2$

Solve and evaluate

- Solve the equations for the unknown quantities. Check whether the directions of the current and the magnitude of the quantities make sense.

Using the first loop: $I_2 = \dfrac{3.0\ \text{V}}{18\ \Omega} = 0.17\ \text{A}$

Using the second loop: $I_1 = \dfrac{9.0\ \text{V} - 3.0\ \text{V}}{2\ \Omega + 12\ \Omega} = 0.43\ \text{A}$

Using the junction A: $I_3 = 0.17\ \text{A} - 0.43\ \text{A} = -0.26\ \text{A}$

Since I_3 turned out negative, it means our initial choice for its direction was incorrect. The magnitude is still correct.

Try it yourself Determine the potential difference from A to B in two ways: (1) through the 18-Ω resistor and (2) through the 12-Ω resistor, the 2-Ω resistor, and the 9-V battery.

Answer   Both paths give us −3.0 V.

In the next example we will use the idea of equivalent resistance to simplify a circuit and determine the current through the battery.

EXAMPLE 19.9   **Using equivalent resistors to solve a circuit problem**

Determine the total current I through the battery in the circuit in part (a) of the figure below.

Sketch and translate To find the current I, first simplify the circuit by combining the resistors into a single equivalent resistor.

(a)

(b) Replace the parallel resistors with a single equivalent resistor.

(c) Replace the three series resistors with a single equivalent resistor.

Simplify and diagram Assume that the battery and the connecting wires have negligible resistance. The circuit is a combination of series and parallel parts, so simplify the circuit in steps. First, replace the parallel 3.0-Ω and 6.0-Ω resistors in (a) by the 2.0-Ω equivalent resistor in (b). Then replace the three resistors in series in (b) with the one equivalent resistor in (c). The new circuit is shown in (c). Once we have a circuit with a single resistor and a single battery, we can determine the current through the battery.

Represent mathematically First we combined the two resistors in parallel:

$$R_{\text{eq, parallel}} = \left(\frac{1}{3.0\ \Omega} + \frac{1}{6.0\ \Omega} \right)^{-1}$$

We then added this combined resistor to the other resistors in series to determine the equivalent resistance of the entire circuit:

$$R_{\text{eq}} = 5.0\ \Omega + \left(\frac{1}{3.0\ \Omega} + \frac{1}{6.0\ \Omega} \right)^{-1} + 8.0\ \Omega$$

Our simplified circuit has the single equivalent resistor attached to the 12-V battery. The current I is

$$I = \frac{\mathcal{E}}{R_{\text{eq}}}$$

Solve and evaluate Completing these calculations gives

$$R_{\text{eq, parallel}} = \left(\frac{1}{3.0\ \Omega} + \frac{1}{6.0\ \Omega} \right)^{-1} = \left(\frac{2}{6.0\ \Omega} + \frac{1}{6.0\ \Omega} \right)^{-1}$$

$$= \left(\frac{3}{6.0\ \Omega} \right)^{-1} = 2.0\ \Omega$$

$$R_{\text{eq}} = 5.0\ \Omega + 2.0\ \Omega + 8.0\ \Omega = 15.0\ \Omega$$

$$I = \frac{12.0\ \text{V}}{15.0\ \Omega} = 0.8\ \text{A}$$

This value for the current is reasonable given the resistances and battery emf.

Try it yourself Find the equivalent resistance of the combination of resistors shown in the figure at right.

12.0 V

2.0 Ω   4.0 Ω

3.0 Ω

Answer   2 Ω.

QUANTITATIVE EXERCISE 19.10 **Current caused by parallel appliances**

A 1380-W electric heater, a 180-W computer, a 20-W LED bulb, and a 1200-W microwave are all plugged into wall sockets in the same part of a house (see figure). When the current through the main line exceeds 20 A, the circuit breaker opens the circuit. Will the circuit breaker open the circuit if all these appliances are in operation at the same time?

I_4
1200-W Microwave

I_3
20-W LED bulb

I_2
180-W Computer

I_1
1380-W Heater

I

Main line

120 V

Represent mathematically The connecting wires and the circuit breaker have negligible resistance. Consequently, there is no electric potential difference across them, and as a result the potential difference across each appliance is 120 V. We can use one form of Joule's law to determine the current in each appliance:

$$I = \frac{P}{|\Delta V|} = \frac{P}{120 \text{ V}}$$

Kirchhoff's junction rule can be used to determine the total current through the circuit (which equals the current through the main line):

$$I = I_1 + I_2 + I_3 + I_4$$

Solve and evaluate The current through each appliance is

Heater: $I_1 = \dfrac{P_1}{|\Delta V_1|} = \dfrac{1380 \text{ W}}{120 \text{ V}} = 11.5 \text{ A}$

Computer: $I_2 = \dfrac{P_2}{|\Delta V_2|} = \dfrac{180 \text{ W}}{120 \text{ V}} = 1.5 \text{ A}$

LED bulb: $I_3 = \dfrac{P_3}{|\Delta V_3|} = \dfrac{20 \text{ W}}{120 \text{ V}} = 0.17 \text{ A}$

Microwave: $I_4 = \dfrac{P_4}{|\Delta V_4|} = \dfrac{1200 \text{ W}}{120 \text{ V}} = 10.0 \text{ A}$

The total current through the main line I is

$$I = I_1 + I_2 + I_3 + I_4$$
$$= 11.5 \text{ A} + 1.5 \text{ A} + 0.17 \text{ A} + 10.0 \text{ A} = 23.2 \text{ A}$$

Since the current I exceeds 20 A, the circuit breaker will open and cause the current to stop.

Try it yourself Determine the resistance of each of the above appliances.

Answer

Heater, 10 Ω; computer, 80 Ω; LED bulb, 710 Ω; microwave, 12 Ω.

REVIEW QUESTION 19.9 What does it mean when you get a negative value for the current in the circuit when using Kirchhoff's rules?

19.10 Properties of resistors

What properties of a resistor affect the value of its resistance? Consider again our water flow analogy, in which a hose connects two containers of water. A wider hose should allow more water to pass through a cross section of the hose each second, and a

Go

longer hose should reduce the amount of water passing through per unit time (because of friction exerted by the walls of the hose on the water). By analogy, we suspect that the length and the cross-sectional area of a wire will affect its electrical resistance. Also, if a resistor is made of a dielectric material, then the current through it will be nearly zero, meaning it has a huge resistance. So it seems that resistance must also depend on the internal properties of the material of which it is made. The experimental setup in Figure 19.34a will help us decide how the resistance of a circuit element depends on its geometric properties (its length L and cross-sectional area $A = \pi r^2$; see Figure 19.34b.). We make measurements on wires of different lengths and radii. The measurements yield the data shown graphically in Figure 19.35. It appears that the resistance increases in proportion to the wire's length (Figure 19.35a) and inversely in proportion to its radius squared, that is, to the cross-sectional area of the wire (Figure 19.35b).

$$R \propto L, \text{ and } R \propto \frac{1}{r^2} \propto \frac{1}{A}$$

FIGURE 19.34 What does resistance depend on? (a) A circuit used to measure resistance. (b) Geometrical properties affect resistance.

FIGURE 19.35 Wire resistance depends on (a) its length and (b) its cross-sectional area.

To investigate how the type of material affects the resistance, we measure the resistance of wires with the same length and cross-sectional area but made from different materials. We find that copper and silver wires have small resistance, and constantan and tungsten wires have relatively large resistance.

The dependence of the resistance R on the type of material of which the resistor is made is characterized in terms of a physical quantity called the **resistivity** ρ of that material. Microscopically, a material's resistivity depends on the number of free electrons per atom in the material, the degree of "difficulty" the electrons have moving through the material due to their interactions with it, and other factors. Copper has a low resistivity because of its large concentration of free electrons (two per atom) and the relative lack of difficulty they experience while moving through the copper. On the other hand, the resistivity of glass (a dielectric) is about 10^{20} times greater than that of copper because it contains almost no free electrons.

Alternatively, materials are sometimes characterized in terms of another physical quantity called **conductivity** σ. The conductivity of a material is the reciprocal of its resistivity ($\sigma = 1/\rho$).

We can assemble an equation for the resistance of a resistive circuit element by combining these three dependencies.

TIP The Greek letter ρ is used to designate resistivity as well as the density of a substance. Do not confuse the two!

TABLE 19.7 The resistivity ρ of different materials

Material	ρ ($\Omega \cdot$ m, at 20 °C)
Metals	
Silver	1.6×10^{-8}
Copper	1.7×10^{-8}
Aluminum	2.8×10^{-8}
Tungsten	5.2×10^{-8}
Constantan	50×10^{-8}
Nichrome	130×10^{-8}
Dielectrics	
Ordinary glass	9×10^{11}
Dry wood	$10^{14} - 10^{16}$
Hard rubber	1×10^{16}
Human tissue	
Blood	1.5
Trunk	5
Lung tissue	20
Fat	25
Skin	5000–50,000
Geological materials	
Ground water	≈ 10
Sedimentary rocks	$1 - 10^5$
Igneous rocks	$10^2 - 10^7$

Electrical resistance The electrical resistance R of a resistive circuit element depends on its geometric structure (its length L and cross-sectional area A) and on the resistivity ρ of the material of which it is made:

$$R = \rho \frac{L}{A} \qquad (19.11)$$

Eq. (19.11) helps us understand why when you connect resistors in series, the total resistance of the circuit increases, and when you connect the resistors in parallel, the total resistance decreases. Connecting the resistors in series is similar to increasing the length of the resistor ($R \propto L$), and connecting the resistors in parallel is similar to increasing the cross-sectional area of the resistor $R \propto 1/A$.

TIP **Table 19.7** lists resistivity for materials at a particular temperature. As we know, the resistance of a lightbulb is greater when it is hot than when cold; its resistivity changes with temperature. The resistivity of some other materials, such as constantan, does not change much with temperature.

In the previous sections we disregarded the resistance of connecting wires in circuits. Since wires are long and thin, according to Eq. (19.11) they could have quite high resistance. Let's investigate the resistance of such wires.

QUANTITATIVE EXERCISE 19.11 **Resistance of connecting wires**

The connecting wires that we use in electric circuit experiments are usually made of copper. What is the resistance of a 10-cm-long piece of copper connecting wire that has a diameter of 2.0 mm = 2.0×10^{-3} m?

Represent mathematically The resistance of a wire is $R = \rho \frac{L}{A}$. The cross-sectional area of a round wire is $A = \pi r^2$. Therefore, the resistance is

$$R = \rho \frac{L}{\pi r^2} = \rho \frac{L}{\pi (d/2)^2}$$

Solve and evaluate Insert the value for the resistivity of copper from Table 19.7 and the wire dimensions into the above:

$$R = \rho \frac{L}{\pi (d/2)^2} = (1.7 \times 10^{-8} \, \Omega \cdot \text{m}) \frac{0.1 \text{ m}}{\pi (1.0 \times 10^{-3} \text{ m})^2}$$

$$= 5.4 \times 10^{-4} \, \Omega$$

This is a tiny amount of resistance. If the current through this wire was 1 A, the potential difference across it would be very small:

$$\Delta V = IR = (1.0 \text{ A})(5.4 \times 10^{-4} \, \Omega) = 5.4 \times 10^{-4} \text{ V}$$

The filaments of old household incandescent lightbulbs have resistances of 100–300 Ω and will therefore have a potential difference of 100–300 V across them when the current through them is 1 A. Connecting wires are deliberately made of material with low resistivity so that the potential difference across them is negligible.

Try it yourself A power line from the power plant to your home has the same cross-sectional area as in this exercise but is 50 km long. What is the resistance of the power line and the potential difference across it if the current through it is 1 A?

Answer 270 Ω, 270 V. However, power lines are actually designed with much greater cross-sectional areas so that their resistance is kept low.

Microscopic model of resistivity

Recall the microscopic structure of a metal: a crystal structure of metal atoms whose outer electrons are free to roam throughout the bulk of the metal. The free electrons moving chaotically inside the metal are similar to ideal gas particles. This model suggests a microscopic explanation for resistance. As the electrons drift through the wire, they collide with the metal ions. These collisions are also the reason why the current in a wire that is connected to a battery does not increase but rather has a constant value. Although the electric field inside the metal accelerates the free electrons, they are continuously slowed down by the collisions with ions. These collisions increase the internal energy of the metal ions, causing them to vibrate more vigorously. These vibrating ions become even more of an obstacle for the electrons. That explains why a hotter metal has a higher resistivity than a cooler one and thus explains why the resistance of the lightbulb filament increases with temperature. Paul Drude first suggested this model in 1900, now known as the Drude model. The Drude model does not correctly predict the details of many properties of metals (quantum mechanical models do much better). However, the Drude model is still a concrete and convenient way to understand qualitatively the reasons for electrical resistance and the mechanism of transforming electrical energy to thermal energy.

Superconductivity

In the early 1900s Dutch physicist Kamerlingh Onnes was attempting to liquefy helium, the only gas that had not yet been liquefied at that time. In 1908 he achieved the desired result—the helium condensed from a gas to a liquid at 4.2 K. Onnes continued experimenting with helium and focused his attention on measuring the electric resistivity of materials at this low temperature. Many physicists, including Lord Kelvin himself, thought that at very low temperatures, electrons in metals should completely stop moving (as the average kinetic energy of the particles is proportional to temperature), thus making the resistivity infinite. Others, including Onnes, hypothesized that the resistivity decreased with temperature and would eventually become zero at zero K (Figure 19.36a).

In 1911 Onnes made an electric circuit by connecting the terminals of a battery to a sample of mercury. He measured the current through it and the potential difference across it. When Onnes submerged the mercury sample in liquid helium at 4.2 K and repeated his measurement, he found that there was still current through the sample, but the potential difference across it abruptly dropped to zero—this meant that the resistance of the sample was zero. Basically, he observed that the resistivity of mercury became zero not at 0 K, but at 4.2 K (Figure 19.36b). Onnes had discovered what became known as **superconductivity**. The temperature at which the resistivity of a material abruptly decreases to zero is called the critical temperature, T_c. Superconductivity was a surprising result that was not predicted and wasn't successfully explained until about 50 years later, using quantum mechanics. At very low temperatures, the electrons as a whole no longer transfer energy to the metal lattice ions, allowing the electrons to move through the lattice with exactly zero resistance.

In the meantime, Onnes and other physicists continued experimenting with different metals and found that many of them, including aluminum and tin, become superconductors at very low temperatures, but that others, including gold and silver, do not.

Unfortunately, the temperatures at which superconductivity was achieved were very low, never having a critical temperature above 20 K. This made the practical applications of superconductivity almost impossible. In addition, the existing theory predicted that superconductivity could not occur above about 30 K.

In 1986 a seemingly improbable event occurred. J. Georg Bednorz and K. Alex Müller observed superconductivity in a material at 35 K. The material they used was not even a metal; it was a complex copper-based ceramic compound. With this as guidance, physicists manufactured similar compounds and raised the high-temperature superconductivity record to 95 K. In 1993, a complex ceramic compound was found to have a critical temperature of 138 K. In 2008 a new family of iron-based high-temperature superconductors was discovered. While there is a long way to go, the goal of discovering a room-temperature superconductor no longer seems completely impossible.

The applications of superconductors are numerous. One is transportation by magnetic levitation, the unique property of superconductors to float suspended above magnets. Superconductors can also be used to make super-strong electromagnets. Using superconductors to transmit information will bypass limitations on data rates caused by electrical resistance. Finally, superconductors can be used as a basis for supersensitive thermometers.

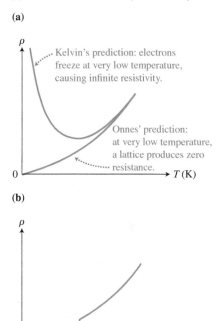

FIGURE 19.36 Superconductivity.
(a) Predictions about low temperature resistivity.
(b) The measured resistivity for mercury.

(a)

Kelvin's prediction: electrons freeze at very low temperature, causing infinite resistivity.

Onnes' prediction: at very low temperature, a lattice produces zero resistance.

(b)

Zero resistance (superconducting) below the critical temperature (T_c)

Semiconductors

So far, we have learned that the resistivity of conductors made of metal (including the wires in incandescent lightbulbs) increases with temperature. Do LEDs behave in a similar way? To investigate this question, we conduct the following experiment. We make two circuits: one consisting of a battery and an incandescent lightbulb with removed glass cover, and the other made of a green LED and a battery. We measure the potential difference across and current through the bulb and the LED at room temperature. We make sure that the current through the lightbulb is so small that it does not glow and does not warm up (we know that an LED is cold even when glowing, so we don't need to worry about warming due to its own current). Then without opening the circuits, we warm up both the bulb and the LED with a hair dryer and measure the same quantities (**Figure 19.37**). Our collected data are given in **Table 19.8**.

FIGURE 19.37 Experimental setups for comparing temperature dependence of resistance of (a) an incandescent lightbulb and (b) an LED.

(a)

(b)

TABLE 19.8 Data collected from the experiment in Figure 19.37

	Physical quantity	Incandescent lightbulb	Green LED
Initial conditions	$T\,(°\text{C})$	18	18
	$\Delta V\,(\text{V})$	0.020	3.00
	$I\,(10^{-3}\text{A})$	16.6	11.0
	$R\,(\Omega)$	**1.2**	**270**
After warming up with the hair dryer	$T\,(°\text{C})$	60	60
	$\Delta V\,(\text{V})$	0.020	3.00
	$I\,(10^{-3}\text{A})$	14.3	12.8
	$R\,(\Omega)$	**1.4** ($\approx 17\%$ increase)	**230** ($\approx 15\%$ decrease)

From the table, we see that the resistance of the filament of the incandescent bulb increases with temperature (as we expected). However, the resistance of the LED decreases with temperature, rather than increases, as we are used to with metals. Why? The material inside the LED is not metal. The glowing part of an LED is made of two different materials that belong to a class called **semiconductors**. Although we will not go into the details behind the mechanism through which these two semiconductors emit light in this chapter (we will do that in Chapter 27), it is important for you to learn about their basic properties. Silicon is the best example of a semiconductor; LEDs are usually made of gallium alloys.

When cooled to near 0 K, silicon has practically infinite resistivity. As its temperature increases, its resistivity decreases. Unlike in metals, the outer electrons of silicon atoms are not free—they are tightly bound to the atoms. When the temperature of a silicon crystal increases, some of the electrons do become free, leaving behind a positively charged ion in the crystal lattice. The location of this missing electron is called a **hole** (**Figure 19.38**). If charged objects outside the silicon crystal are used to produce a nonzero $\vec{E}$ field within the silicon, these newly freed conduction electrons accelerate in a direction opposite the $\vec{E}$ field, similar to the way they do in metals. In addition, the electrons of atoms next to the holes can move into the hole. The hole then moves to the atom where the electron came from. In effect, these holes behave similarly to positively charged particles, accelerating in the direction of the $\vec{E}$ field. Because the number of conduction electrons and holes increases as the temperature of the semiconductor increases, the resistivity of a semiconductor decreases with increasing temperature. Just like in metals, in semiconductors charge carriers collide

FIGURE 19.38 Migration of holes in a semiconductor in the presence of an electric field. Parts (a) to (d) show successive steps as a hole and bound electrons swap places. Part (e) shows the trajectory of the hole during the entire time interval.

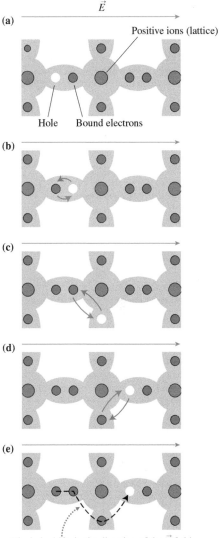

(a)

$\vec{E}$

Positive ions (lattice)

Hole   Bound electrons

(b)

(c)

(d)

(e)

The hole drifts in the direction of the $\vec{E}$ field, while the electrons move in the opposite direction.

with ions. The fact that resistivity decreases with temperature tells us that production of free electrons at higher temperature wins over the increased resistance due to collisions with vibrating ions.

Semiconductors are used not only in LEDs but in nearly all modern technology, including consumer electronics, medical electronic products, and solar power. We will learn later in the book how LEDs convert electric energy into light energy, and how solar cells convert light energy into electric energy.

Electrolytes

Electrolytes are substances containing free ions that make the substance electrically conductive. Several electrolytes are critical in human physiology. For example, free sodium (Na^+) and chlorine (Cl^-) ions in the body's tissues move in opposite directions, forming currents $(I^+$ and $I^-)$ in response to electric fields produced by electric dipoles in the heart. Both types of ions contribute to the current $(I_{total} = I^+ + I^-)$. In nerve cells, Na^+, Cl^-, and other ions move across nerve membranes, transmitting electrical signals from the brain to other parts of the body—to activate muscles, for example.

REVIEW QUESTION 19.10 Why does the resistance of a lightbulb increase as the current through it increases?

Summary

Electric current Electric current I equals the electric charge q that passes through a cross section of a circuit element divided by the time interval Δt needed for that amount of charge to pass. (Section 19.1)

$$I = \frac{|q|}{\Delta t}$$ Eq. (19.1)

Unit: 1 A (ampere) = 1 C/s

Emf is the work done per unit charge by a power source (such as a battery) to move charge from one terminal to other. (Sections 19.2 and 19.7)

$$\mathcal{E} = \frac{W}{q}$$ Eq. (19.2)

Unit: 1 V (volt) = 1 J/C

$$\mathcal{E} - Ir = IR$$ Eq. (19.7)

Electrical resistance The electrical resistance R of a wire depends on its geometrical structure: its length L, cross-sectional area A, and the resistivity of the material ρ. (Section 19.10)

$$R = \rho \frac{L}{A}$$ Eq. (19.11)

$$I_1 = I_2 = I_3 = \cdots$$

$$R_{\text{eq, series}} = R_1 + R_2 + R_3 + \cdots$$ Eq. (19.9)

$$\Delta V_1 = \Delta V_2 = \Delta V_3 = \cdots$$

Series and parallel resistance Multiple resistors can be connected in series or in parallel in a circuit. The current through series resistors is the same; the potential difference across parallel resistors is the same. Resistor combinations can be replaced with a single resistor with equivalent resistance. (Section 19.8)

$$R_{\text{eq, parallel}} = \left(\frac{1}{R_1} + \frac{1}{R_2} + \frac{1}{R_3} + \cdots \right)^{-1}$$ Eq. (19.10)

Ohm's law The current I through a circuit element equals the potential difference ΔV across it divided by its resistance R. For ohmic devices, the resistance is independent of the current through the resistor. Incandescent lightbulbs and LEDs are examples of non-ohmic devices. (Section 19.4)

Joule's law determines the rate of conversion of electric potential energy into thermal energy. (Section 19.6)

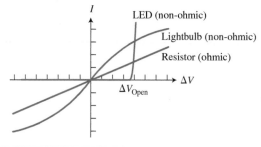

$$I = \frac{\Delta V}{R}$$ Eq. (19.3)

$$P = \left| \frac{\Delta U_{\text{th}}}{\Delta t} \right| = I|\Delta V|$$ Eq. (19.4)

$$= I^2 R = (\Delta V)^2 / R$$ Eq. (19.5)

Unit: 1 W (watt) = 1 J/s

Kirchhoff's junction rule The algebraic sum of all currents into a junction in a circuit equals the algebraic sum of the currents out of the junction. (Section 19.7)

$$\sum_{\text{in}} I = \sum_{\text{out}} I$$ Eq. (19.8)

Kirchhoff's loop rule The algebraic sum of the potential differences across circuit elements around any closed circuit loop is zero. For resistors, the potential difference is $-IR$ if moving across the resistor in the direction of the current and $+IR$ if moving across the resistor opposite the direction of current. (Section 19.7)

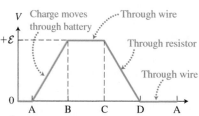

$$\sum_{\text{Loop}} \Delta V = 0$$ Eq. (19.6)

Battery: $\Delta V = \pm \mathcal{E}$, assuming no internal resistance
$\Delta V = \pm (\mathcal{E} - Ir)$ (including the internal resistance of the battery r)

Resistor: $\Delta V = \mp IR$

Questions

Multiple Choice Questions

1. Two identical bulbs are connected in parallel across the battery shown in **Figure Q19.1**. There is an open switch next to bulb 2. Initially, bulb 1 is bright (strong current) and bulb 2 is dark (no current). How do the electric currents change when the switch is closed? Multiple answers could be correct.

   FIGURE Q19.1

   (a) The current through the battery is the same but now splits between 1 and 2.
   (b) The current in 1 is still bigger than that in 2.
   (c) The current in 1 does not change and is the same as in 2.
   (d) The current through the battery doubles.

2. Compare the potential difference across bulbs 1 and 2 in Figure Q19.1 after the switch is closed. Which statement is correct?
   (a) The potential difference is greater across bulb 1.
   (b) The potential difference is greater across bulb 2.
   (c) The potential difference is the same across bulbs 1 and 2.
   (d) The potential difference across the battery doubles.
   (e) None of the above.

3. Two identical bulbs are in series as shown in **Figure Q19.3**. Which statement is correct?

   FIGURE Q19.3

   (a) The current through the battery is twice as great as when one bulb was present.
   (b) The current through the battery is half as great as when one bulb was present.
   (c) The current is greater in bulb 1 than in bulb 2.
   (d) The current is greater in bulb 2 than in bulb 1.
   (e) The currents in bulbs 1 and 2 are identical.
   (f) b and e are correct.
   (g) a and c are correct.

4. Which statement below about the potential difference across the battery and across the identical bulbs 1 and 2 shown in Figure Q19.3 is correct? Multiple answers could be correct.
   (a) The potential difference is greater across bulb 1 than across bulb 2.
   (b) The potential difference is greater across bulb 2 than across bulb 1.
   (c) The potential difference is the same across bulbs 1 and 2.
   (d) The potential difference across the battery is twice what it was with only one bulb present.

5. Three circuits with identical bulbs and emf sources are shown in **Figure Q19.5**. Rank the circuits in terms of the ammeter readings, with the largest ammeter reading listed first.
   (a) A > B > C (b) A > C > B (c) B > A > C
   (d) B > C > A (e) C > A > B

   FIGURE Q19.5

6. Rank in order the potential differences across each of the five bulbs in the three circuits in Figure Q19.5, listing the largest potential difference first. Indicate equal potential differences with equal signs. Assume that there is zero potential difference across the ammeters.
   (a) $\Delta V_1 > \Delta V_2 = \Delta V_3 > \Delta V_4 = \Delta V_5$
   (b) $\Delta V_2 > \Delta V_3 > \Delta V_1 > \Delta V_4 > \Delta V_5$
   (c) $\Delta V_2 = \Delta V_3 > \Delta V_1 > \Delta V_4 = \Delta V_5$
   (d) $\Delta V_1 = \Delta V_2 = \Delta V_3 > \Delta V_4 = \Delta V_5$
   (e) None of these

7. Rank in order the five identical bulbs in the circuit shown in **Figure Q19.7**, listing the brightest bulb first. Indicate equally bright bulbs with an equal sign.
   (a) $B_1 > B_5 > B_2 > B_3 = B_4$
   (b) $B_1 = B_5 = B_2 > B_3 = B_4$
   (c) $B_1 > B_5 > B_2 = B_3 = B_4$
   (d) $B_1 = B_2 = B_3 = B_4 = B_5$
   (e) None of these is correct.

   FIGURE Q19.7

8. Four identical bulbs are shown in the circuit in **Figure Q19.8**. With the switch open, bulbs 1, 2, and 3 are equally bright and bulb 4 is dark. How does the brightness of bulbs 1, 2, and 3 change when the switch is closed?
   (a) 1 and 2 stay the same and 3 gets brighter.
   (b) 1, 2, and 3 stay the same.
   (c) 1 and 3 get brighter and 2 gets dimmer.
   (d) 1, 2, and 3 all get brighter.
   (e) None of these is correct.

   FIGURE Q19.8

9. Four identical bulbs are shown in the circuit in **Figure Q19.9** with the switch open. How does the brightness of bulbs 1, 3, and 4 change when the switch is closed?
   (a) 1 gets dimmer, 4 gets brighter, and 3 stays the same.
   (b) 1 stays the same, 4 gets brighter, and 3 gets dimmer.
   (c) 1 stays the same, 4 gets brighter, and 3 stays the same.
   (d) 1, 3, and 4 all stay the same.
   (e) None of these is correct.

   FIGURE Q19.9

10. Consider the circuit in **Figure Q19.10**. The switch is open. What is the potential difference across the switch $\Delta V_{A \text{ to } B}$?
    (a) Zero (b) 9.0 V
    (c) 4.5 V (d) −4.5 V

    FIGURE Q19.10

    100 Ω   100 Ω
    B A
    $\mathcal{E} = 9.0\ V$

11. **Figure Q19.11** shows I-versus-ΔV graphs for an incandescent lightbulb and a resistor. Choose all correct statements.
    (a) For $\Delta V = 3$ V, the resistance of the lightbulb and the resistor are equal ($R_B = R_R$).
    (b) For $\Delta V = 3$ V, $R_B > R_R$.
    (c) For $\Delta V = 3$ V, $R_B < R_R$.
    (d) R_B increases with increasing current through it.
    (e) R_B decreases with increasing current through it.
    (f) For any potential difference $0 < \Delta V < 3$ V, $R_B > R_R$.
    (g) For any potential difference $0 < \Delta V < 3$ V, $R_B < R_R$.
    (h) For potential difference $0 < \Delta V < 3$ V, R_B can be larger than, smaller than, or equal to R_R.

    FIGURE Q19.11

12. If an electric current were due to electrons moving only on the surface of a wire rather than through the bulk of the wire, which expression would describe the resistance of the wire? r is the wire's radius, L is its length, and C is a constant with suitable units.
    (a) $R = C_a \dfrac{L}{r^2}$ (b) $R = C_b L r$ (c) $R = C_c L$
    (d) $R = C_d \dfrac{L}{r}$ (e) None of the above

13. Three light sources (a lightbulb, a blue LED $[\Delta V_{\text{Open B}} = 2.3\text{ V}]$, and a red LED $[\Delta V_{\text{Open R}} = 1.6\text{ V}]$) are connected with two 50-Ω resistors and a 3.0-V battery, as shown in Figure Q19.13. Which light source will definitely not glow?
    (a) The blue LED
    (b) The red LED
    (c) The lightbulb
    (d) Both the blue LED and the red LED
    (e) All light sources will glow.

FIGURE Q19.13

14. Figure Q19.14 shows a top view of a flat resistor that is cut from a thin metal foil of uniform thickness. Which circuit can be used as a model for this resistor? You may use a ruler to solve this problem.

FIGURE Q19.14

(a) $\overset{R\quad 2R\quad 3R\quad 4R}{\bullet\!-\!\text{www}\!-\!\text{www}\!-\!\text{www}\!-\!\text{www}\!-\!\bullet}$

(b) $\overset{4R\quad 3R\quad 2R\quad R}{\bullet\!-\!\text{www}\!-\!\text{www}\!-\!\text{www}\!-\!\text{www}\!-\!\bullet}$

(c) $\overset{R\quad R/2\quad R/3\quad R/4}{\bullet\!-\!\text{www}\!-\!\text{www}\!-\!\text{www}\!-\!\text{www}\!-\!\bullet}$

(d) $\overset{R/4\quad R/3\quad R/2\quad R}{\bullet\!-\!\text{www}\!-\!\text{www}\!-\!\text{www}\!-\!\text{www}\!-\!\bullet}$

Conceptual Questions

15. What is the role of a battery in an electric circuit? Describe a mechanical analogy.
16. Compare and contrast the physical quantities emf and potential difference.
17. **Birds on high power lines** Why can birds perch on a 100,000-V power line with no adverse effects?
18. **Preventing electric shock** When a person is repairing electrical equipment and can encounter high potential differences, he or she should keep one hand in a back pocket and work only with the other hand. Why?
19. (a) Using a voltmeter, how can you determine the electric current through a resistor? (b) Using an ammeter, how can you determine the potential difference across a resistor? (c) Using a voltmeter and an ammeter, how can you determine the resistance of a resistor?

20. (a) What does it mean if the current through a resistor is 3 A? (b) What does it mean if the potential difference across the resistor is 3 V? (c) What does it mean if the resistance of the resistor is 3 Ω?
21. Resistors become warm when there is an electric current through them. Why doesn't this phenomenon depend on the direction of the current? Give a microscopic explanation.
22. At one time aluminum rather than copper wires were used to carry electric current through homes. Which wire must have the larger radius if they are the same length and need to have the same electrical resistance? Explain.
23. How do you connect an ammeter in a circuit to measure the current? How do you connect a voltmeter in a circuit to measure the potential difference between two points in the circuit? What implications do these connections have for the resistances of these measuring instruments?
24. Why do we connect electric devices in a home in parallel rather than in series?
25. Dylan thinks that conductance is equal to the slope of the I-versus-ΔV graph at a particular point; Finn thinks resistance is equal to the ratio between the ΔV and I values at that point. Explain for which types of circuit element both Dylan and Finn are correct, and for which type only Finn is correct.
26. Construct an electric circuit that is analogous to your circulatory system. Indicate the corresponding parts of the two systems.
27. Most Christmas tree lights with incandescent bulbs are connected in parallel. (a) Describe one advantage of lights connected in parallel. (b) Suppose you connect these lights in series. Will they be brighter or darker? Explain. What modification would make the lights equally bright? Explain.
28. Two students are arguing. Student A says that Ohm's law is only valid for ohmic resistors. Student B says that Ohm's law is valid for any circuit element other than a battery. Which student do you agree with? Explain your choice.
29. Use the laws of energy and charge conservation to explain Kirchhoff's rules.
30. When you close the switch in the circuit in Figure Q19.30, the capacitor charges. When you open the switch again, the capacitor discharges. (a) Explain qualitatively what happens to the current in each process. (b) Which process takes a longer time, charging or discharging? Explain.

FIGURE Q19.30

Problems

Below, **BIO** indicates a problem with a biological or medical focus. Problems labeled **EST** ask you to estimate the answer to a quantitative problem rather than derive a specific answer. Asterisks indicate the level of difficulty of the problem. Problems with no * are considered to be the least difficult. A single * marks moderately difficult problems. Two ** indicate more difficult problems. **If not stated otherwise, the internal resistance of batteries and the resistance of wires are negligible.**

19.1–19.3 Electric current, Batteries and emf, and Making and representing simple circuits

1. A bulb in a table lamp has a current of 0.50 A through it. Determine two physical quantities related to the electrons passing through the wires leading to the bulb.
2. A long wire is connected to the terminals of a battery. In 8.0 s, 9.6×10^{20} electrons pass a cross section along the wire. The electrons flow from left to right. What physical quantities can you determine using this information?
3. A typical flashlight battery will produce a 0.50-A current for about 3 h before losing its charge. Determine the total number of electrons that have moved past a cross section of wire connecting the battery and lightbulb.
4. * Four friends each have a battery, a bulb, and one wire. They use the materials to try to light their bulb. Two succeed and two do not. What possible circuit arrangements could they have made?

5. Draw a circuit that has a battery, a lightbulb, and connecting wires. Draw a schematic of a water flow system that allows a continual circulation of water and compare the function of each element of the circuit to each element of the water flow system.
6. Add another battery to the circuit described in Problem 19.5. In how many different ways can you do this? Draw analogous changes in the water flow system.
7. Add another lightbulb to the circuit with one battery described in Problem 19.5. In how many different ways can you connect the second bulb? What do you need to add to the water flow system to make it similar to the circuit?
8. A 9.0-V battery is connected to a resistor so that there is a 0.50-A current through the resistor. For how long should the battery be connected in order to do 200 J of work?
9. ** **Design an experiment I** When sugar is poured from a metal container, sugar particles and the container become charged with opposite charges due to rubbing. Design an experiment that will allow you to estimate the magnitude and the sign of the charge transferred by 1 kg of sugar, using only the following measuring devices: a stopwatch, a scale, and an ammeter that can measure very small current.

19.4 and 19.5 Ohm's law and Qualitative analysis of circuits

10. * A graph of the electric potential versus location in a series circuit with 1.0 A of current is shown in **Figure P19.10**. Draw a circuit in which such changes could occur.

FIGURE P19.10

11. Sketch a potential-versus-location graph for the circuit shown in **Figure P19.11**. Start at A and move clockwise around the circuit.

FIGURE P19.11

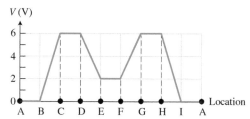

12. **BIO** **Electric currents in the body** A person accidentally touches a 120-V electric line with one hand while touching a ground wire with the other hand. Determine the current through the body when the hands are dry (100,000-Ω resistance) and when wet (5000-Ω resistance). If the current exceeds about 10 mA, muscular contractions may prevent the person from releasing the wires—a dangerous situation. Is the person in danger with dry hands? With wet hands? Explain.

13. An automobile lightbulb has a 1.0-A current when it is connected to a 12-V battery. What can you determine about the lightbulb using this information?

14. * If a long wire is connected to the terminals of a 12-V battery, 6.4×10^{19} electrons pass a cross section of the wire each second. Make a list of the physical quantities whose values you can determine using this information and determine three of them.

15. Determine the current through a 2.5-Ω resistor when connected across two 1.5-V batteries in series (a potential difference of 3.0 V).

16. * You have a circuit with a 50-Ω, a 100-Ω, and a 150-Ω resistor, and a battery connected in series. (a) Rank the current through them from highest to lowest. (b) Rank the potential difference across them from highest to lowest. Explain your rankings.

17. You have a circuit with a 50-Ω, a 100-Ω, and a 150-Ω resistor connected in parallel to the same battery. (a) Rank the current through them from highest to lowest. (b) Rank the potential difference across them from highest to lowest.

18. * A toy has two red LEDs ($\Delta V_{\text{Open R}} = 1.6$ V), two green LEDs ($\Delta V_{\text{Open G}} = 2.1$ V), and two blue LEDs ($\Delta V_{\text{Open B}} = 2.3$ V) connected in series. (a) What is the smallest number of 1.2-V rechargeable batteries needed to make all the LEDs glow? (b) How do you need to connect the batteries (in series or in parallel)? (c) For how many hours will the toy work with that number of batteries if the current through the LEDs is 20 mA and the total storage capacity of the batteries is 1200 mAh?

19. * You want to power a green LED ($\Delta V_{\text{Open G}} = 2.1$ V) and a red LED ($\Delta V_{\text{Open R}} = 1.6$ V) using a 9.0-V battery. You decide to connect the LEDs in series. In order to limit the current to the allowed value of 20 mA, you need to add a resistor in series with the LEDs. (a) Draw a circuit diagram, and (b) determine the resistance of the resistor that you need to connect to limit the current.

20. ** A circuit consists of a green LED ($\Delta V_{\text{Open G}} = 2.1$ V) and a 100-Ω resistor connected in series with a variable power supply. Draw a labeled I-versus-ΔV graph for this circuit for the interval -4.0 V $< \Delta V < +4.0$ V. (Hint: Remember that once the voltage across an LED reaches the opening voltage, an LED behaves like a closed switch, except that the voltage across it is almost constant and equal to ΔV_{Open}.)

19.6 Joule's law

21. * You connect a 50-Ω resistor to a 9-V battery whose internal resistance is 1 Ω. (a) Determine the electric power dissipated by the resistor. (b) Will a lightbulb with a 10-Ω resistance dissipate more or less power if connected to the same battery? Explain.

22. * **EST** **Making tea** You use an electric teapot to make tea. It takes about 2 min to boil 0.5 L of water. (a) Estimate the power of the heater. What are your assumptions? (b) Estimate the current through the heater. State your assumptions.

23. * If a long wire is connected to the terminals of a 12-V battery, 6.4×10^{19} electrons pass a cross section of the wire each second. What is the rate of work being done by the battery?

24. ** Three friends are arguing with each other. Adam says that a battery will send the same current to any circuit you connect it to. Alice says that the battery will produce the same potential difference at its terminals independent of the circuit connected to it. Paul says that the battery will always produce the same electric power independent of the circuit connected to it. Describe experiments that you can perform to convince all of them that their opinions are wrong in at least one situation.

25. * You have a 40-W lightbulb and a 100-W bulb. What do these readings mean? Can we say that the 100-W lightbulb is always brighter? Explain and give examples.

26. * Does a 60-W lightbulb have more or less resistance than a 100-W bulb? Explain.

19.7 and 19.8 Kirchhoff's rules and Resistor and capacitor circuits

27. * (a) Write two loop rule equations and one junction rule equation for the circuit in **Figure P19.27**. (b) Use these equations to determine the current in each branch of the circuit for the case in which $R_1 = 50$ Ω, $R_2 = 30$ Ω, $R_3 = 15$ Ω, and $\mathcal{E} = 120$ V.

FIGURE P19.27

28. * (a) Write Kirchhoff's loop rule for the circuit shown in **Figure P19.28** for the case in which $\mathcal{E}_1 = 20$ V, $\mathcal{E}_2 = 8$ V, $R_1 = 30$ Ω, $R_2 = 20$ Ω, and $R_3 = 10$ Ω. (b) Determine the current in the circuit. (c) Using this value of current, start at position A and move clockwise around the circuit, calculating the electric potential change across each element in the circuit (be sure to indicate the sign of each change). (d) Add these potential changes around the whole circuit. Explain your result.

FIGURE P19.28

29. * Repeat parts (a) and (b) of the previous problem for the case in which $\mathcal{E}_1 = 12$ V, $\mathcal{E}_2 = 3$ V, $R_1 = R_2 = 1$ Ω, and $R_3 = 16$ Ω. (c) Determine the potential difference from A to B. (d) Draw a potential-versus-position graph starting at any location in the circuit and returning to the same point.

30. * (a) Determine the value of $\mathcal{E}_1$ so that there is a clockwise current of 1.0 A in the circuit shown in Figure P19.28, where $\mathcal{E}_2 = 12$ V, $R_1 = 2$ Ω, $R_2 = 1$ Ω, and $R_3 = 12$ Ω. (b) Determine the potential difference from B to A. (c) Draw a graph representing the potential at different points of the circuit.

31. ** The current through resistor R_1 in Figure P19.27 is 2.0 A. Determine the currents through resistors R_2 and R_3 and the emf of the battery. $R_1 = 4$ Ω, $R_2 = 10$ Ω, and $R_3 = 40$ Ω.

32. ** Use Kirchhoff's rules to show in general that the currents I_2 and I_3 through resistors R_2 and R_3 shown in Figure P19.27 satisfy the relation $I_2/I_3 = R_3/R_2$.

33. * (a) Write the loop rule for two different loops in the circuit shown in Figure P19.33 and the junction rule for point A. Solve the equations to find the current in each loop when $\mathcal{E}_1 = 3$ V, $\mathcal{E}_2 = 6$ V, $R_1 = 10$ Ω, $R_2 = 20$ Ω, and $R_3 = 30$ Ω. (b) Determine the potential difference from A to B. Check your answer by taking a different path from A to B. (c) Sketch a potential-versus-position graph for each of the three loops, assuming the potential is zero at B.

FIGURE P19.33

34. ** Determine the value of R_2, shown in Figure P19.33, so that the current through R_3 equals twice that through R_2. The values of other circuit elements are $\mathcal{E}_1 = 12$ V, $\mathcal{E}_2 = 15$ V, $R_1 = 15$ Ω , and $R_3 = 30$ Ω.

35. * Determine (a) the equivalent resistance of resistors R_1, R_2, and R_3 in Figure P19.35 for $R_1 = 28$ Ω, $R_2 = 30$ Ω, and $R_3 = 20$ Ω and (b) the current through the battery if $\mathcal{E} = 10$ V.

FIGURE P19.35

36. (a) Determine the equivalent resistance of resistors R_1, R_2, and R_3 in Figure P19.35 for $R_1 = 50$ Ω, $R_2 = 30$ Ω, and $R_3 = 15$ Ω.
    (b) Determine the current through R_1 if $\mathcal{E} = 120$ V.
    (c) Use Kirchhoff's loop rule and your result from part (b) to determine the current through R_2 and through R_3.

37. * Determine the equivalent resistance of the resistors shown in Figure P19.37 if $R_1 = 60$ Ω, $R_2 = 30$ Ω, $R_3 = 20$ Ω, $R_4 = 20$ Ω, $R_5 = 60$ Ω, $R_6 = 20$ Ω, and $R_7 = 10$ Ω.

FIGURE P19.37

FIGURE P19.38

38. * Determine (a) the equivalent resistance of the resistors in the circuit in Figure P19.38 and (b) the current through the battery.

39. ** Write a problem for which the following mathematical statement can be a solution:

$$1\text{ A} = \frac{\mathcal{E} - (1\text{ A})(3\text{ }\Omega)}{R}$$

40. You close the switch in the circuit in Figure P19.40 and wait until the currents stop changing. (a) Determine the charge on the capacitor. (b) What happens to that charge when you open the switch again? Explain qualitatively.

FIGURE P19.40

FIGURE P19.41

41. * You close the switch in the circuit in Figure P19.41 and wait until the currents stop changing. Determine the charge on each capacitor.

19.9 Skills for solving circuit problems

42. * **Home wiring** A simplified electrical circuit for a home is shown in Figure P19.42. (a) Determine the currents through the circuit breaker, the lightbulb, the microwave oven, and the toaster. (b) Determine the electric power used by each appliance.

FIGURE P19.42

43. ** (a) Write Kirchhoff's rules for two loops and one junction in the circuit shown in Figure P19.43. (b) Solve the equations for the current in each branch of the circuit when $\mathcal{E}_1 = 10$ V, $\mathcal{E}_2 = 2$ V, $R_1 = 50$ Ω, $R_2 = 200$ Ω, and $R_3 = 20$ Ω. (c) Determine the potential difference across resistor R_3 from point B to point A.

FIGURE P19.43

44. ** **BIO** **Electric eel** The South American eel can generate electric current that can stun and even kill nearby fish. The eel has 140 parallel rows of electric cells (0.15 V per cell). Each row has 5000 such cells for a total emf per row of 5000(0.15 V) = 750 V. Each row of cells also has about 1250 Ω of internal resistance. Because each row has the same emf and the rows are connected together on each side, the eel's circuit can be represented as shown in Figure P19.44—a 750-V emf source in series with 140 1250-Ω parallel internal resistances all connected across an external resistance (the seawater from the front of the eel to its back) of about 800 Ω. Can the eel produce enough current to be dangerous to a person? Explain your answer. Identify all assumptions that you made.

FIGURE P19.44

45. **Home wiring** A 120-V electrical line in a home is connected to a 60-W lightbulb, a 180-W television set, a 300-W desktop computer, a 1050-W toaster, and a 240-W refrigerator. How much current is flowing in the line?

46. * **Tree lights** Nine tree lights are connected in parallel across a 120-V potential difference. The cord to the wall socket carries a current of 0.36 A. (a) Determine the resistance of one of the bulbs. (b) What would the current be if the bulbs were connected in series?

47. * Two lightbulbs use 30 W and 60 W, respectively, when connected in parallel to a 120-V source. How much power does each bulb use when connected in series across the 120-V source, assuming that their resistances remain the same?

48. * Three identical resistors, when connected in series, transform electrical energy into thermal energy at a rate of 15 W (5 W per resistor). Determine the power consumed by the resistors when connected in parallel to the same potential difference.

49. ** **Impedance matching** A battery has an emf of 12 V and an internal resistance of 3 Ω. (a) Determine the power delivered to a resistor R connected to the battery terminals for values of R equal to 1, 2, 3, 4, 5, and 6 Ω. (b) Plot on a graph the calculated values of P versus the different values of R. Connect the points by a smooth curve. Confirm that the maximum power is delivered when R has the same resistance as the internal resistance of the power source (3 Ω in this example).

50. * Determine the equivalent resistance of the circuit in Figure P19.50 when (a) both switches are opened, (b) switch 1 is closed and switch 2 is opened, and (c) both switches are closed.

FIGURE P19.50

51. ** The resistors in the circuit shown in Figure P19.51 all have the same resistance. The current through the battery is I_0. You make the changes to the original circuit that are described in the table below, and each time record the current through the battery. Rank the currents I_0 to I_4 from largest to smallest. Assume all wires have negligible resistance and the battery has negligible internal resistance.

FIGURE P19.51

Connect the following points with a wire	Current through the battery
A and B	I_1
C and D	I_2
B and E	I_3
C and E	I_4

52. **Figure P19.52** shows a real circuit that consists of four identical lightbulbs that are mounted in yellow holders (each lightbulb holder has two terminals) and a battery. (a) Draw a diagram for this circuit. (b) Predict which lightbulb(s) will glow if you connect the free red alligator clip to the free terminal of the battery. (c) Now move the far left red alligator clip from the upper to the lower terminal of the bulb P (as explained on the figure). Draw a new circuit diagram and predict which lightbulb(s) now glows.

FIGURE P19.52

Connect this red clip to the battery for (b).

Move this red clip to the lower P terminal for (c).

19.10 Properties of resistors

53. * A 100-m-long copper wire of radius 0.12 mm and mass 40 g is connected across a 1.5-V battery. Make a list of the physical quantities that you can determine using this information and determine the values of three of them. The resistivity of copper is $1.7 \times 10^{-8} \ \Omega \cdot m$.

54. * **BMT subway rail resistance** The BMT subway line in New York City stretches roughly 30 km from the Bronx to Brooklyn. The electrified rail on which it runs has a cross section of about 40 cm^2 and is made of steel with a resistivity of $10 \times 10^{-8} \ \Omega \cdot m$. Make a list of all the physical quantities describing different properties of the rail that you can determine using this information. Then calculate the values of one quantity related to the electrical properties and one quantity not related to the electrical properties.

55. * **Thermometer** A platinum resistance thermometer consists of a 0.10-mm-diameter platinum wire wrapped in a coil. Determine the length of wire needed so that the coil's resistance at 20 °C is 25 Ω. The resistivity of platinum at this temperature is $1.0 \times 10^{-7} \ \Omega \cdot m$.

56. As the potential difference in volts across a thin platinum wire increases, the current in amperes changes as follows: $(\Delta V, I) =$ (0, 0), (1.0, 0.112), (3.0, 0.337), and (6.1, 0.675). Plot a graph of potential difference as a function of current and indicate whether the platinum wire is an ohmic element. Explain how you made your decision.

57. * **BIO** **Respiration detector** A respiration detector monitors a person's breathing. One type consists of a flexible hose filled with conductive salt water (resistivity of 5.0 $\Omega \cdot m$). Electrodes at the ends of the tube measure the resistance of the fluid in the tube. The tube is wrapped around a person's chest. When the person inhales and exhales, the tube stretches and contracts and its resistance changes. Determine the factor by which the resistance of the fluid changes when the hose is stretched so that its length increases by a factor of 1.1. The water volume remains constant.

58. * A wire whose resistance is R is stretched so that its length is tripled while its volume remains unchanged. Determine the resistance of the stretched wire.

59. * **Ratio reasoning** Determine the ratio of the resistances of two wires that are identical except that (a) wire A is twice as long as wire B, (b) wire A has twice the radius of wire B, and (c) wire A is made of copper and wire B is made of aluminum. Be sure to show clearly how you arrive at each answer.

60. ** **Electronics detective** You need to determine the mass and length of the wire inside a particular electronic device. You cannot take the wire out (to measure its length), but you can reach its free ends and connect devices to them. Devise a method to do this by using a battery, ammeter, voltmeter, and micrometer (a device that measures small distances.)

61. * A battery produces a 2.0-A current when connected to an unknown resistor of resistance R. When a 10-Ω resistor is connected in series with R, the current drops to 1.2 A. (a) Determine the emf of the battery and the resistance R. (b) What assumptions did you make? Do your assumptions make each of the values you determined greater than or less than their actual values? Explain.

62. * **BIO** **Resistance of human nerve cell** Some human nerve cells have a long, thin cylindrical cable (the axon) from their inputs to their outputs. Consider an axon of radius 5×10^{-6} m and length 0.6 m. The resistivity of the fluid inside the axon is 0.5 $\Omega \cdot m$. Determine the resistance of the fluid in this axon.

63. * **Conductive textiles** Metal strands can be woven into textiles to make them conductive. You have some 30-mm-wide nickel textile tape with a resistance of 3.0 Ω per meter length. You cut two rectangular strips from this tape. The first strip is 30 mm by 100 mm, and the second strip is 15 mm by 120 mm. You connect the strips in series to a 1.5-V battery as shown in **Figure P19.63**. Determine the total power output of the two strips. Assume the resistance of the wires and the internal resistance of the battery are negligible.

FIGURE P19.63

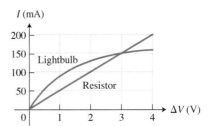

General Problems

64. * **EST** **Figure P19.64** shows an I-versus-ΔV graph for an incandescent lightbulb and a resistor (data are measured separately but presented in the same graph). In one case, the lightbulb and the resistor are connected in parallel to a battery. The current through the battery is 300 mA. Estimate the emf of the battery, ignoring the internal resistance of the battery.

FIGURE P19.64

65. * **EST** Figure P19.64 shows an I-versus-ΔV graph for an incandescent lightbulb and a resistor (data are measured separately but presented in the same graph). In one case, the lightbulb and the resistor are connected in series and the current through them is 100 mA. Estimate (a) the emf of the battery and (b) the power output of the lightbulb and of the resistor. Ignore the internal resistance of the battery.

66. * **EST** Figure P19.64 shows an I-versus-ΔV graph for an incandescent lightbulb and a resistor (data are measured separately but presented in the same graph). In one case, the lightbulb and the resistor are connected in series to a 1.5-V battery. Estimate (a) the voltage across the lightbulb and the resistor and (b) the current through the battery. Ignore the internal resistance of the battery. (Hint: The sum of the voltages across the elements is 1.5 V and the current through both elements should be the same.)

67. * **Wiring high-fidelity speakers** Your high-fidelity amplifier has one output for a speaker of resistance 8 Ω. How can you arrange two 8-Ω speakers, one 4-Ω speaker, and one 12-Ω speaker so that the amplifier powers all speakers and their equivalent resistance when connected together in this way is 8 Ω? Compare the power output of your arrangement with the power output of a single 8-Ω speaker. Express your answer as a ratio $P_{Arrange}/P_{8\Omega}$.

68. * **BIO** **EST** **Lifting forearm by electric current** You are asked to evaluate an idea for making it possible for a stroke patient to contract the biceps muscle in order to lift the forearm. The muscle will contract if a current of about 20 mA flows through the upper arm. Estimate the potential difference across electrodes, one inserted near the top of the arm and the other near the elbow, that will produce this current. Indicate any assumptions you made.

69. * **EST** **Switches** You have a power supply, a 10-W bulb, and a 15-W bulb. You also have two types of switches: a single pole single throw (SPST) switch and a single pole double throw (SPDT) switch (shown in **Figure P19.69**). Build a circuit satisfying the following conditions: when the switch is in one position, the 10-W bulb is on; when the switch is in the other position, the 10-W bulb no longer glows, but the 15-W bulb does.

FIGURE P19.69

SPST switch

SPDT switch

70. ** **Wiring a staircase** Devise an electric circuit that will allow you to turn a stairway lightbulb on and off from the bottom and from the top of the stairway. You can use switches such as those shown in Figure 19.69.

71. ** **Gas or electric heating?** What information should you collect to decide whether to use electricity or gas to heat your house in the winter? Search relevant information and write a report that will convince your roommates that one way is better than the other. Make sure you include quantitative arguments.

72. ** **EST** **Electric water heater** An electric hot water heater has a 7000-W resistive heating element. (a) Determine the energy needed to warm the 120 kg of water in the heater from 20 °C to 60 °C. (b) What time interval is needed to warm the water? (c) Estimate the electric energy cost of warming the water at recent rates.

73. ** **BIO** **EST** **The hands and arms as a conductor** While doing laundry you reach to turn on the light above the washing machine. Unfortunately, you touch an exposed 120-V wire in the broken switch box. Your other hand is supporting you on the top of the grounded washing machine. Assume that the resistance is 100,000 Ω across the dry skin of each hand and that the resistivity of tissue inside your arms and body is 5 Ω · m. (a) Estimate the total resistance of the tissue inside your arms from one side to the other (ignore the resistance across your chest). (b) Estimate the total resistance across the skin of one hand through the tissue in your arms and then across the skin of your other hand. (c) Are you in danger?

74. * **Design an experiment II** You have a 1.5-V battery, a 1.0-F capacitor, a blue LED (with opening voltage 2.3 V) and connecting wires. Using only this equipment, design an experiment that will make the blue LED glow. Describe your design using words and circuit diagrams.

75. * A nickel wire of length L and a voltmeter are connected to a battery as shown in **Figure P19.75**. One terminal of the voltmeter is connected to the left end of the nickel wire and the other terminal can slide along the wire. The battery has negligible internal resistance and an emf $\mathcal{E}$. Assume the resistance of the connecting wires is negligible. Derive an expression that describes how the reading of the voltmeter ΔV depends on the position x of its right terminal, and evaluate the solution.

FIGURE P19.75

76. ** Solve the previous problem if the internal resistance of the battery is r and the resistance of the nickel wire is R. Evaluate the solution.

77. * **EST** **Figure P19.77** shows an I-versus-ΔV graph of a circuit that consists of a red LED ($\Delta V_{\text{Open R}} = 1.6$ V), a blue LED ($\Delta V_{\text{Open B}} = 2.3$ V), and a resistor. (a) Choose the circuit in the lower part that will produce the I-versus-ΔV graph of the figure and explain your choice. (b) Which terminal of the circuit (left or right) is at the higher potential and which is at the lower potential when $\Delta V = +3.0$ V? (c) Estimate the resistance of the resistor.

FIGURE P19.77 I (A)

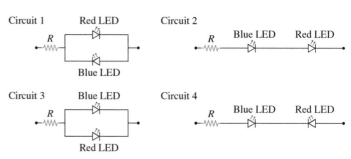

78. ** Jesse needs to determine the resistance of a resistor. Which method should he use? If you think that more than one of the described methods is useful, compare your choices and explain which would give the most accurate result. Describe any assumptions that you made. The final step in all methods is to calculate the resistance of the resistor as $R = \dfrac{\Delta V}{I}$.

(a) Connect a voltmeter to a battery's terminals. Record the voltage reading ΔV. Then (in a different experiment) connect the battery, the resistor, and an ammeter in series. Record the current reading I.

(b) Connect a battery, the resistor, and an ammeter in series and connect a voltmeter in parallel to the resistor. Record the current reading I and the voltage reading ΔV.

(c) Connect a battery, the resistor, an ammeter, and a voltmeter in series. Record the current reading I and the voltage reading ΔV.

(d) Connect a battery, the resistor, and an ammeter in series and connect a voltmeter in parallel to the ammeter. Record the current reading I and the voltage reading ΔV.

(e) Connect a battery, the resistor, and an ammeter in series. Record the current reading I. Then (in a different experiment) connect the resistor to the battery and a voltmeter in parallel to the resistor. Record the voltage reading ΔV.

79. ** **People current** Suppose that all the people on the Earth moved at the same speed around a circular track. Approximately how many times per second would each person pass the starting line if that "people current" equaled the number of electrons passing a cross section in a wire when there is a 1.0-A current through it?

80. * A 5.0-A current caused by moving electrons flows through a wire. (a) Determine the number of electrons and the number of moles of electrons that flow past a cross section each second. (b) The same number of water molecules moves along a stream. Determine the volume of water that moves each second under a bridge that passes over the stream.

81. ** **BIO** **Current across membrane wall of axon** An axon cable that connects the input and output ends of a human nerve cell is 5×10^{-6} m in radius and 0.5 m long. The thickness of the membrane surrounding the fluid in the axon is 6×10^{-9} m, its resistivity is 1.6×10^{7} Ω · m, and there is a 0.070-V potential difference across the membrane. (a) Determine the current through the membrane. (b) If the current is due to sodium ions, how many of these ions leak across the membrane wall per second? (c) How does the current change if the membrane is twice as thick?

Reading Passage Problems

BIO **Signals in nerve cells stimulate muscles** The input end of a human nerve cell is connected to an output end by a long, thin, cylindrical axon. A signal at the input end is caused by a stretch sensor, a temperature sensor, contact with another cell or nerve, or some other stimulus. At the output end, the nerve signal can stimulate a muscle cell to perform a function (to contract, provide information to the brain, etc.).

The axon of a so-called unmyelinated human nerve cell has a radius of 5×10^{-6} m and a membrane that is 6×10^{-9} m thick. The membrane has a resistivity of about $1.6 \times 10^7 \, \Omega \cdot m$. The fluid inside the axon has resistivity of about $0.5 \, \Omega \cdot m$. The membrane wall has proteins that pump three sodium ions (Na^+) out of the axon for each two potassium ions (K^+) pumped into the axon. In the resting axon, the concentration of these ions results in a net positive charge on the outside of the membrane compared to negative charge on the inside. Because of the unequal charge distribution, there is a -70 mV potential inside compared to outside the axon.

When an external source stimulates the input end of the nerve cell so the potential inside reaches about -50 mV, gates or channels in the membrane walls near that input open and sodium ions rush into the axon. This stimulates neighboring gates to swing open and sodium ions rush into the axon farther along. This disturbance quickly travels along the axon—a nerve impulse. The potential across the inside of the membrane changes in 0.5 ms from -70 mV to $+30$ mV relative to the outside. Immediately after this depolarization, potassium ion gates open and positively charged potassium ions rush out of the axon, repolarizing the axon. Sodium and potassium ion pumps then return the axon and its membrane to their original configuration.

82. Which answer below is closest to the resistance of the fluid inside a 0.5-m-long unmyelinated axon?
    (a) $10^9 \, \Omega$ (b) $10^6 \, \Omega$ (c) $10^3 \, \Omega$ (d) $10^{-3} \, \Omega$ (e) $10^{-6} \, \Omega$

83. An electric signal transmitted in a metal wire involves electrons moving parallel to the wire. What does an electric signal in an axon involve?
    (a) Sodium ions moving across the axon membrane perpendicular to its length
    (b) Potassium ions moving across the axon membrane perpendicular to its length
    (c) Sodium and potassium ions moving inside the axon parallel to its length
    (d) Only sodium ions moving inside the axon parallel to its length
    (e) a and b

84. Which answer below is closest to the magnitude of the $\vec{E}$ field in the resting axon membrane?
    (a) $10 \, V/m$ (b) $10^3 \, V/m$ (c) $10^5 \, V/m$ (d) $10^7 \, V/m$ (e) $10^9 \, V/m$

85. The charge density on the axon membrane walls (positive outside and negative inside) is $1.0 \times 10^{-4} \, C/m^2$. Suppose the membrane of a 1.0-m-long axon discharged completely in 0.02 s. What answer below is closest to the electric current across the membrane wall of one such hypothetical axon discharge?
    (a) 0.2 A (b) 0.02 A (c) $2 \times 10^{-4} \, A$ (d) $2 \times 10^{-5} \, A$ (e) $2 \times 10^{-7} \, A$

86. The horizontal 4-Ω resistors in the two circuits in **Figure P19.86** represent the resistance of a small horizontal length of fluid inside an axon. The 15-Ω resistors represent the resistance across the axon membrane for a tiny length of axon. The 10-Ω resistor to the right side of each "ladder" is the effective resistance in front of the axon sections under consideration. Determine the resistance between B and G of the small length of axon shown in Figure P19.86a and between A and G of the slightly longer length axon shown in Figure P19.86b.
    (a) $4.2 \, \Omega, 8.4 \, \Omega$ (b) $10 \, \Omega, 10 \, \Omega$ (c) $29 \, \Omega, 48 \, \Omega$
    (d) $10 \, \Omega, 12 \, \Omega$ (e) $4.2 \, \Omega, 12 \, \Omega$

FIGURE P19.86

(a)

(b)

87. Suppose nerve impulses travel at 100 m/s in the axons of nerve cells from your fingers to your brain and then back again to your fingers in order to stimulate muscles that lift your fingers off the hot burner of a stove. Which answer below is closest to the time interval needed for the nerve signal transmission along the axons?
    (a) 0.01 s (b) 0.001 s (c) 0.2 s (d) 0.02 s (e) 0.002 s

BIO **Effect of electric current on human body** Nerve impulses are initiated at the input end of a nerve cell, travel along a relatively long axon (cable), and then cause an effect at the output end of the cell—for example, the initiation of a muscle contraction in a muscle cell. The nerve impulse is initiated by a stimulus that lowers the potential difference from outside the cell to inside from its normal -70 mV to about -50 mV.

A potential difference across two parts of the body (for example, the 120-V potential difference from a wall socket from one hand to the other or from the

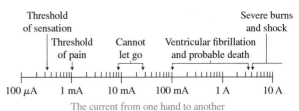

FIGURE 19.39 The effects of electric current on the human body.

The current from one hand to another

hands to the feet) can initiate an electric current in the body that stimulates nerve endings and triggers nerve signals that cause muscular contraction. Even worse, the current in the body can upset the rhythmic electrical operation of the heart. The heart muscles might be stimulated randomly in what is called ventricular fibrillation—a random contraction of the ventricles, which can be deadly. A rough guide to the effects of electric current on the body at different current levels is provided in **Figure 19.39**. Under dry conditions, human skin has high electrical resistance. Wet skin dramatically lowers the body's resistance and makes electrocution more likely to occur.

88. The electrical resistance across dry skin is about 100,000 Ω. Suppose a person with dry skin puts one hand on a 120-V power cord from a home wall socket while the other hand is touching a metal object at 0 V (at what is called ground). Which condition described below is most likely to occur?
    (a) No sensation (b) Threshold of pain
    (c) Cannot let go (d) Ventricular fibrillation
    (e) Severe burns and shock

89. The electrical resistance across wet skin is about 1000 Ω. Suppose a person with wet skin puts one hand on a 120-V power cord from a home wall socket while the other hand is touching a metal object at 0 V (at what is called ground). Which condition described below is most likely to occur?
    (a) No sensation (b) Threshold of pain
    (c) Cannot let go (d) Ventricular fibrillation
    (e) Severe burns and shock

90. Suppose the electrical resistance across your wet skin is about 1000 Ω. Which answer below is closest to the least potential difference from one hand that will cause slight pain?
    (a) 0.1 V (b) 1 V (c) 10 V (d) 100 V (e) 1000 V

91. Suppose the electrical resistance across your wet skin is about 1000 Ω. Which answer below is closest to the least potential difference from one hand to the other that will cause ventricular fibrillation?
    (a) 0.1 V (b) 1 V (c) 10 V (d) 100 V (e) 1000 V

92. When muscular contraction caused by electrical stimulation prevents a person from releasing contact with the potential difference sources, lower current can be extremely dangerous. For example, a 100-mA current for 3 s causes about the same effect as a 900-mA current for 0.03 s. In which of these situations is the least electric charge transferred through the body?
    (a) 100 mA for 3 s
    (b) 900 mA for 0.03 s
    (c) Too little information to decide

93. Why is it dangerous to place a hair dryer, radio, or other electric appliance that is plugged into a wall socket near a bathtub?
    (a) The water provides a conductive path for current, which heats the metal cover on the appliance and can cause burns.
    (b) If the appliance is accidentally knocked into the tub while a person is bathing, large currents could pass through the person's low-resistance body because of the 120-V potential difference that powers the appliance.
    (c) There is no potential for danger, because electric appliances are grounded.
    (d) a and b

94. Occasionally, the electric circuit that produces a coordinated pumping of blood from the four chambers of the heart becomes disturbed. Ventricular fibrillation can occur—random muscle contractions that produce little or no blood pumping. To stop the fibrillation, two defibrillator pads are placed on the chest and a large current (about 14 amps) is sent through the heart, restarting its normal rhythmic pattern. The current lasts 10 ms and transfers 140 J of electric energy to the body. Which answer below is closest to the potential difference between the defibrillator pads?
    (a) 100 V (b) 400 V (c) 1000 V (d) 5000 V (e) 10,000 V

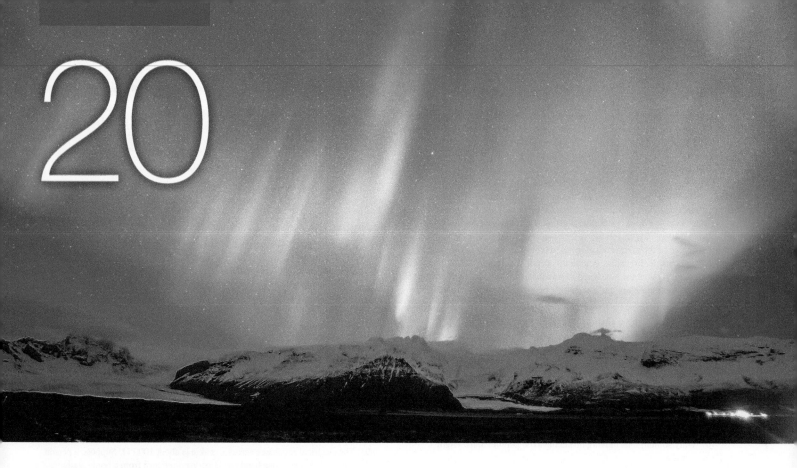

Magnetism

- What causes the northern lights?
- How does Earth protect us from solar wind and cosmic rays?
- Are we really walking northward when we follow a compass needle?

The aurora borealis—the northern lights—illuminates the autumn Arctic sky with a flickering greenish glow that is streaked with red and sometimes blue or violet. A comparable phenomenon—the aurora australis—appears in the autumn sky above southern Australia, southern South America, and the Antarctic. What causes auroras, and why do we see auroras near Earth's poles and only rarely anywhere else?

BE SURE YOU KNOW HOW TO:

- Use the electric field concept to explain how electrically charged objects exert forces on each other (18.1).
- Find the direction of the electric current in a circuit (19.1).
- Apply Newton's second law to a particle moving in a circle (5.3).

IN PREVIOUS CHAPTERS we learned that electrically charged objects do not exhibit magnetic properties. Until 1820, physicists believed that electric and magnetic phenomena were completely different. Since then, scientists have made many discoveries about the relationships between electricity and magnetism. These discoveries have made possible electromagnets, electric motors, magnetic resonance imaging (MRI), and many other devices. We begin this chapter by learning about the connections between electricity and magnetism and how scientists discovered them.

20.1 Magnetic interactions

Magnets become familiar through toys, refrigerator magnets, and countless household applications. Very early we discover that, when brought near each other, magnets can both attract and repel. The two sides of a bar magnet are called **poles**—the **north pole** and the **south pole**. If you bring the like poles of two magnets (north to north and south to south) near each other, they repel. If you bring opposite poles near each other, they attract (**Figures 20.1a** and **b**). Additionally, both sides of a magnet attract objects made from steel or iron, even if those objects aren't magnets themselves.

Magnets always have two poles. If you break a magnet into two pieces, each piece still has two poles—a north pole and a south pole (**Figure 20.2**). If you break one of those pieces again, each smaller piece has two poles, and so on. Unlike electrically charged objects that can be either negatively or positively charged, there are no magnets with a single pole (a so-called monopole).

The labels "north" and "south" originated in 11th-century China. People noticed that if they put a tiny magnet on a low-friction pivot (or attached it to a piece of wood floating in water), one end always pointed toward geographic north. This property of a magnet resulted in the names for the two ends: the north pole points toward geographical north; the south pole points toward geographical south. The device became known as a **compass**.

Since the north pole of a compass is attracted to the south pole of another magnet, and since compasses that are not near other magnets always point approximately toward Earth's geographic north pole, we can infer that Earth itself acts as a giant magnet with its magnetic south pole close to its geographic north pole and its magnetic north pole close to its geographic south pole (note the separation between the magnetic and geographical poles in **Figure 20.3**).

The magnetic interaction depends on separation

We can use a compass to investigate not only the direction of the magnetic influence on it but also the strength. **Figure 20.4** shows a sequence of four simple experiments all performed on a wooden table. In each experiment we place a compass at a specific location with respect to a magnet and observe the compass's behavior. When we put the compass at location A, it vibrates slowly and then stops in the orientation shown. Placed at location B, the compass vibrates a little faster (shorter period T) and comes to the same orientation as in A. In C, its vibrations have even a smaller period, and finally, at D, very close to the magnet, the compass vibrates very quickly until its tip points toward the magnet. From these observations, we can infer that when closer to the magnet, the interaction is stronger (in Chapter 10 we learned that the period of vibrations is proportional to the square root of the restoring force, Eq. 10.12).

TIP   In Chapter 17 we learned that although electrically charged objects attract and repel in a similar way to magnets, the interactions involved are different. Make sure you do not call magnetic poles charges, and do not confuse north and south poles with positive and negative electric charges. Review Table 17.3 before moving forward.

REVIEW QUESTION 20.1 What aspects of compass behavior near a magnet give us information about the direction of magnetic influence, and what aspects give us information about the strength?

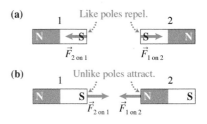

FIGURE 20.1 Like magnetic poles repel and unlike poles attract.

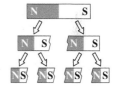

FIGURE 20.2 Breaking a magnet into pieces produces more magnets, each with a north and south pole.

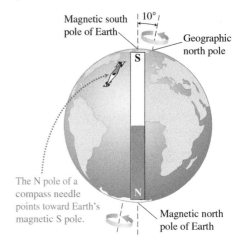

FIGURE 20.3 The magnetic and geographic poles of Earth.

FIGURE 20.4 Investigating the strength of magnetic interactions using vibrations of a compass.

$$T_A > T_B > T_C > T_D$$

FIGURE 20.5 Compasses indicate the $\vec{B}$ field directions at various locations near (a) a bar magnet and (b) a horseshoe magnet.

(a)

(b)

FIGURE 20.6 Compasses indicate $\vec{B}$ field lines.

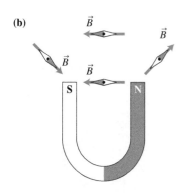

FIGURE 20.7 Iron filings indicate $\vec{B}$ field lines.

Iron filings lie on a transparent sheet lying on a bar magnet.

2 mm

Above the magnet, the filings align with the S-N direction of the magnet.

FIGURE 20.8 $\vec{B}$ field lines.

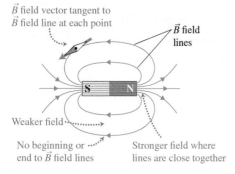

$\vec{B}$ field vector tangent to $\vec{B}$ field line at each point

$\vec{B}$ field lines

Weaker field

No beginning or end to $\vec{B}$ field lines

Stronger field where lines are close together

20.2 Magnetic field

The electric field explains how electrically charged objects interact without contact. Since magnets also interact without contact, it is reasonable to suggest the existence of some kind of field, in this case a **magnetic field**, as the mechanism behind magnetic interactions. Thus we can assume that a magnet produces a magnetic field—a magnetic disturbance with which other objects with magnetic properties (another magnet, anything made of iron, etc.) interact. In this chapter we will learn how to describe this field qualitatively and quantitatively and how to predict its effects on objects that possess their own magnetic properties.

Direction and strength of the magnetic field

To describe the magnetic field, we use a vector physical quantity called the $\vec{B}$ **field**. The $\vec{B}$ field has a magnitude and direction. We can use a compass to detect the direction of the $\vec{B}$ field at a particular location, just as we can use a test charge for the $\vec{E}$ field. The direction of the $\vec{B}$ field is defined as the direction that the north pole of the compass needle points. We can also determine the direction of the $\vec{B}$ field by drawing a vector from the south pole to the north pole of the compass inside the compass. **Figure 20.5a** shows the direction of the $\vec{B}$ field at several points near a bar magnet. Figure 20.5b shows the $\vec{B}$ field vectors near a horseshoe magnet. We can determine the relative strength of the field at a particular location by observing the vibrations of a compass needle placed there. The shorter the period of vibrations, the stronger the field.

Representing the magnetic field: field lines

To represent the magnetic field in a region, we draw $\vec{B}$ field lines (similar to $\vec{E}$ field lines). We can visualize the lines by placing multiple compasses around a bar magnet, as shown in **Figure 20.6**. Following the direction of the south-north vector for each compass as we move from compass to compass, we find that the compasses are tangent to lines surrounding the magnet—the blue lines in Figure 20.6. Instead of compasses, we can use hundreds of tiny iron sticks or needles called iron filings (which act like tiny compasses) sprinkled on a thin piece of transparent plastic placed on top of the magnet. The filings form a pattern that looks similar to the lines formed by the compasses (**Figure 20.7**; the inset shows a photo of two filings magnified under a microscope). These lines are so-called $\vec{B}$ **field lines**. They are used to represent the $\vec{B}$ field produced by the magnet in the whole region, not just at a point, as one $\vec{B}$ field vector indicates. Similar to $\vec{E}$ field lines, $\vec{B}$ field lines represent not only the direction of the $\vec{B}$ field, but also its magnitude. The lines are closer together in regions where the $\vec{B}$ field magnitude is greater (where the compass needle vibrated with shorter periods)—see the pattern shown in **Figure 20.8**.

The magnetic $\vec{B}$ field and its representation by $\vec{B}$ field lines The direction of the magnetic $\vec{B}$ field at a particular location is defined as the direction that the north pole of a compass needle points when at that location. The $\vec{B}$ field vector at a location is tangent to the direction of the $\vec{B}$ field line at that location. Both point in the same direction. The density of lines in a region represents the magnitude of the $\vec{B}$ field in that region—where the $\vec{B}$ field is stronger, the lines are closer together. Unlike $\vec{E}$ field lines, which begin on positively charged objects and end on negatively charged objects, $\vec{B}$ field lines do not have beginnings or ends. Figure 20.8 shows that they are closed loops that pass through the source of the magnetic field.

What are other ways to produce a magnetic field?

We can now explain the interaction between two magnets (let's call them magnet A and magnet B) in the following way. Magnet A creates a magnetic field around itself, which exerts a force (and possibly a torque) on magnet B. Magnet B creates its own magnetic field that exerts a force (and possibly a torque) on magnet A.

Do any other objects besides magnets create magnetic fields? We found earlier that stationary electrically charged objects do not. Could electric currents (moving electric charges) produce magnetic fields and be affected by magnetic fields? Observational Experiment Table 20.1 describes simplified versions of two experiments performed in 1820 by Hans Oersted that investigated this idea.

OBSERVATIONAL EXPERIMENT TABLE 20.1 Do electric currents produce magnetic fields?

VIDEO OET 20.1

Observational experiment	Analysis
Experiment 1. Connect a battery, a switch, some wires, and a lightbulb in a circuit as shown at left below. The bulb indicates the presence of an electric current in the circuit. Note the north-south orientation of the right-hand wire. With the switch open (the bulb is off), place compasses beneath, on top of, and to the sides (not shown) of one of the wires. Notice the direction the compasses point.	Without current in the circuit, the needles point toward geographic north (magnetic south). Thus, the wire does not produce a magnetic field.
Close the switch (the bulb is on) as shown at right below. Notice the directions that the compasses above and beneath the wire now point.	The current in the wire affects the orientation of the compasses placed nearby. Therefore, the current-carrying wire might produce a magnetic field.
Experiment 2. Reverse the direction of the current. The compasses on top of and beneath the wire point as shown.	The change in the orientation of the compass indicates that the wire might produce a magnetic field and the direction of the current affects the field.

Pattern

Electric current affects the orientation of a compass. If we placed the compasses on the sides of the wire, we would have seen a complete pattern: with the current in the wire in Experiment 1, the compasses orient in the direction of the circles around the wire shown at right: toward the right when under the wire, up on the right side of the wire, toward the left on top of the wire, and down on the left side of the wire. As the $\vec{B}$ field lines are closed loops, we can use the circles to represent the effect of the current on the compasses.

In Experiment 2, the compass orientations are reversed. Circles in the opposite direction would represent the effect of this current on the compasses.

FIGURE 20.9 The right-hand rule for the direction of the $\vec{B}$ field produced by an electric current.

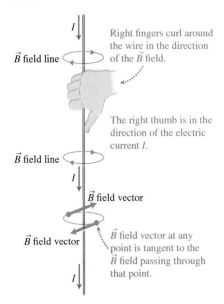

We conclude that an electric current does produce a magnetic field. The direction of the $\vec{B}$ field depends on the direction of the current. The $\vec{B}$ field lines form closed circles around the current. Their direction changes when the direction of the current changes.

Since the current is made of electrically charged particles that are in collective motion with respect to the compass, this means that charged objects in motion produce a magnetic field, and stationary charged objects do not. Through the experiments in Table 20.1 we have deduced a pattern in the direction of the $\vec{B}$ field vectors and $\vec{B}$ field lines around a current-carrying wire. Imagine grasping the wire with your right hand, with your thumb pointing in the direction of the current. When you do so, your fingers wrap around the current in the direction of the $\vec{B}$ field lines (**Figure 20.9**). This method for determining the shape of the $\vec{B}$ field produced by the electric current in a wire is called the **right-hand rule** for the $\vec{B}$ field.

Right-hand rule for the $\vec{B}$ field   To determine the direction of the field lines produced by a current-carrying straight wire, grasp the wire with your right hand, with your thumb pointing in the direction of the current. Your four fingers wrap around the current in the direction of the $\vec{B}$ field lines. At each point on a $\vec{B}$ field line, the $\vec{B}$ field vector is tangent to the line and points in the direction of the line.

CONCEPTUAL EXERCISE 20.1   **Magnetic field of a solenoid**

Draw the magnetic field lines of a solenoid connected to a battery. (A **solenoid** is a wire with a large number of loops in a cylindrical shape. For simplicity in this exercise we will use a solenoid with eight loops.)

Sketch and translate   To draw the $\vec{B}$ field lines produced by a current in the solenoid, we use the right-hand rule for the $\vec{B}$ field lines produced by each loop as shown in the figure. We then combine them for a complete picture of the $\vec{B}$ field produced by the solenoid.

Simplify and diagram   Consider three adjacent loops near the center of the solenoid. The $\vec{B}$ field lines inside each loop point in the same direction (as shown in the figure below). In the adjacent loops the lines point in the same direction. We assume that the superposition principle applies to the $\vec{B}$ field as it does for the $\vec{E}$ field (see Chapter 18). This means that the total

$\vec{B}$ field at any location can be found as the sum of the $\vec{B}$ field vectors due to individual loops. The final result is shown in the figure at right. To check the answer, we place a compass close to the solenoid at several locations. At each location, the compass needle points in the direction of the $\vec{B}$ field.

When you add the $\vec{B}$ fields for each turn, you get a uniform field inside.

Try it yourself   Use the right-hand rule for the $\vec{B}$ field and the superposition principle to predict the direction of the magnetic field exactly in the middle between two straight wires. The two wires are oriented horizontally in the plane of the page, with currents as shown below.

Answer

The $\vec{B}$ field contribution of each wire at the location of interest points into the page. Therefore, the $\vec{B}$ field at that location points into the page.

Each turn produces a $\vec{B}$ field.

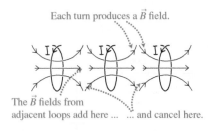

The $\vec{B}$ fields from adjacent loops add here ...   ... and cancel here.

Notice that the $\vec{B}$ field produced by the current in a loop or a coil (a wire that is bent into multiple loops) (Figure 20.10a) and that produced by a bar magnet (Figure 20.10b) are very similar. The $\vec{B}$ field in the region to the right of the plane of the coil (or loop) looks just like the $\vec{B}$ field around the north pole region of the bar magnet. Likewise, the region to the left of the plane of the coil looks just like the $\vec{B}$ field in the south pole region of the bar magnet. Wire coils with current act magnetically in a very similar way to bar magnets and are known as **electromagnets**. We will discuss electromagnets further in Section 20.7.

FIGURE 20.10 A current loop or coil produces a $\vec{B}$ field that is similar to that of a bar magnet.

(a) $\vec{B}$ field produced by a loop

REVIEW QUESTION 20.2 What is the direction of the $\vec{B}$ field at a point halfway between two parallel wires with currents in the same direction?

(b) $\vec{B}$ field produced by a bar magnet

20.3 Magnetic force on a current-carrying wire

If a current-carrying wire is similar in some ways to a magnet, then a magnetic field should exert a magnetic force on a current-carrying wire similar to the force it exerts on another magnet. When we place a bar magnet near a circuit with long connecting wires, we find that the magnet sometimes pulls on the wire and sometimes does not—the effect depends on the relative directions of the $\vec{B}$ field and the current in the wire.

The magnetic field exerts a force on a current-carrying wire

To investigate how the relative directions of the $\vec{B}$ field and current-carrying wire affect their interactions, we conduct the experiments described in Observational Experiment Table 20.2. We use a horseshoe magnet because between the poles of such a magnet the $\vec{B}$ field lines are almost parallel to each other, representing an example of a uniform magnetic field (see Figure 20.11).

FIGURE 20.11 A horseshoe magnet.

Between the poles the $\vec{B}$ field lines are almost parallel to each other.

OBSERVATIONAL EXPERIMENT TABLE 20.2   Direction of magnetic force

VIDEO OET 20.2

Observational experiment	Analysis
Experiment 1. Hang a horizontal straight wire between the poles of a horseshoe magnet using conducting support wires connected to a battery. Orient the wire parallel to the $\vec{B}$ field lines. Attach a battery and open switch in series to the wire. Turn the current on by closing the switch. The wire does not move. If you reverse the current, the wire still does not move.	The $\vec{B}$ field lines for both experiments are parallel to the direction of the current. The magnet does not exert a force on the current-carrying wire.

(CONTINUED)

Observational experiment		Analysis
Experiment 2. Orient the wire perpendicular to the $\vec{B}$ field lines and turn the current on by closing the switch. The wire bends down.		The field lines are perpendicular to the direction of the current in the wire. The force exerted on the wire by the magnetic field is downward and perpendicular to both the current and the $\vec{B}$ field.
Experiment 3. Repeat Experiment 2, but reverse the orientation of the battery, reversing the direction of the current. The wire bends up.		The field lines are perpendicular to the direction of the current in the wire. The force exerted on the wire by the magnetic field is upward and perpendicular to both the current and the $\vec{B}$ field.

Pattern

- The magnetic field does not exert a force on the current-carrying wire if the $\vec{B}$ field is parallel to the wire.
- When the $\vec{B}$ field is perpendicular to the direction of the current, the magnetic field exerts a force on the current-carrying wire that is perpendicular to both the direction of the current and the $\vec{B}$ field. The direction of this force depends on the direction of the current-carrying wire and the direction of the $\vec{B}$ field.

FIGURE 20.12 A method to determine the direction of the force that a magnetic field exerts on a current-carrying wire.

(a) Magnetic force

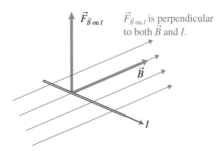

$F_{\vec{B} \text{ on } I}$ is perpendicular to both $\vec{B}$ and I.

(b) Right-hand rule for magnetic force

1. Your fingers point in the direction of the $\vec{B}$ field.

Palm

3. Your palm points in the direction of the force.

2. Your thumb points in the direction of the current I.

In Table 20.2 we found that the magnetic field does exert a force on a current-carrying wire when the wire is oriented perpendicular to the $\vec{B}$ field and does not exert a force when the wire is parallel. Other experiments indicate that when the current-carrying wire and the $\vec{B}$ field form any nonzero angle, the magnetic field exerts a force on the current-carrying wire that is perpendicular to both the direction of the current and the direction of the $\vec{B}$ field. The larger the angle between the current-carrying wire and the $\vec{B}$ field, the larger the magnetic force exerted on it (for angles up to 90°).

The direction of the magnetic force that a magnetic field exerts on a current-carrying wire is illustrated in **Figure 20.12a**. The right-hand rule for determining the force direction is described below and in Figure 20.12b.

Right-hand rule for the magnetic force Hold your right hand flat with your thumb extended from the four fingers. Orient your hand so that your fingers point in the direction of the $\vec{B}$ field and your right thumb points along the direction of the current. The direction of the magnetic force exerted by the magnetic field on the current is the direction your palm faces—perpendicular to both the direction of the current and the direction of the $\vec{B}$ field.

TIP Do not confuse the two right-hand rules. The right-hand rule for the field describes the $\vec{B}$ field produced by a known current. The right-hand rule for the force describes the direction of the magnetic force exerted by a known magnetic field on a current-carrying wire.

Forces that current-carrying wires exert on each other

If a current-carrying straight wire produces a magnetic field, the field should exert a force on a second current-carrying straight wire placed nearby. Similarly, the magnetic field produced by the second wire's current should exert a force on the first wire's current. According to Newton's third law, the forces that these wires exert on each other should point in opposite directions and have the same magnitudes. Let us test this reasoning with the experiments described in Testing Experiment Table 20.3. In the table, the crosses X represent the $\vec{B}$ field pointing into the page, and the circled dots ⊙ represent the $\vec{B}$ field coming out of the page. This convention, which we will use throughout the book, applies to any field.

TESTING EXPERIMENT TABLE 20.3   Do two current-carrying wires interact as magnets?   VIDEO TET 20.3

Testing experiment	Prediction	Outcome
Two vertical strips of aluminum foil (A and D) are next to each other, each connected by wires at their ends to the terminals of their own batteries. We use aluminum strips to detect this rather weak interaction instead of wires because the strips are much lighter and more flexible. Predict what happens when the currents in the strips are in the same direction.	Choose strip D as the system. Using the right-hand rule for the $\vec{B}$ field, we find that the $\vec{B}_A$ field produced by strip A at D points into the page. Using the right-hand rule for force, we find that this field exerts a force on strip D $\vec{F}_{A\,on\,D}$ that points to the left. Repeat the procedure for strip A as the system. The $\vec{B}$ field $\vec{B}_D$ produced by strip D at A points out of the page. This field exerts a force $\vec{F}_{D\,on\,A}$ that points to the right. The strips should bend toward each other.	When we do the experiment, the strips bend toward each other.
Predict what happens when the currents in the strips are in the opposite directions.	We repeat the same process to make our prediction. When the current in D is in the downward direction instead of up, the same analysis shows that the strips should repel.	When the currents are in the opposite directions, the strips repel.

Conclusion

We found that two current-carrying wires exert forces on each other whose direction can be predicted using both right-hand rules. The results are consistent with Newton's third law—the forces that the strips exert on each other point in opposite directions.

TIP In Table 20.3 the wires attract and repel each other via magnetic forces. The wires do not interact via electric forces, because the net electric charge of each wire is zero.

FIGURE 20.13 Ampere's experiment for the unit of electric current.

The currents are each defined as 1.0 A if the magnetic force that each meter of a long wire exerts on the other is 2.0×10^{-7} N.

$$F_{1 \text{ on } 2} = F_{2 \text{ on } 1} = 2.0 \times 10^{-7} \text{ N}$$

André-Marie Ampère in 1820 conducted numerous such experiments and found that the magnitude of the force that one current-carrying wire exerts on another parallel current was directly proportional to the magnitude of each current and inversely proportional to the distance between the wires. This relationship became the basis for the unit of the electric current in the SI system—the **ampere** (**Figure 20.13**).

Ampere (A) When two long parallel wires are separated by 1.0 m (as in Figure 20.13) and there are equal-magnitude electric currents through the wires so that each meter of one wire exerts a force of 2.0×10^{-7} N on 1.0 m of the neighboring wire, the current I in each wire is defined to have a magnitude of 1.0 A.

From this definition it is apparent that the magnetic forces that current-carrying wires exert on each other are rather weak. Unless the currents are very large, the magnetic forces they exert on each other will be small and difficult to detect.

Expression for the magnetic force that a magnetic field exerts on a current-carrying wire

To determine a mathematical expression for the magnetic force that a magnetic field exerts on a current-carrying wire, we hang a horizontal wire at its ends from conducting springs. The tops of the springs are connected to a battery so that the current through the wire is in the direction shown in **Figure 20.14a**. We place the wire between the poles of an electromagnet (not shown), which produces an approximately uniform horizontal variable $\vec{B}$ field in the region surrounding the wire. The right-hand rule for magnetic force indicates that the magnetic field exerts a downward magnetic force $\vec{F}_{\vec{B} \text{ on } W}$ on the current in the wire. Earth exerts a downward gravitational force on the wire $\vec{F}_{E \text{ on } W}$. These two forces are balanced by the net upward force of the two springs on the wire $\vec{F}_{S \text{ on } W}$, as shown in the force diagram in Figure 20.14b. Knowing the spring constant of the springs and the mass of the wire, we can use the stretch of the springs to deduce the magnitude of the magnetic force exerted on different length wires when different currents are in them. The collected data are shown in **Table 20.4**.

FIGURE 20.14 By varying the $\vec{B}$ field, we can develop an expression for the force that the magnetic field exerts on the current-carrying wire.

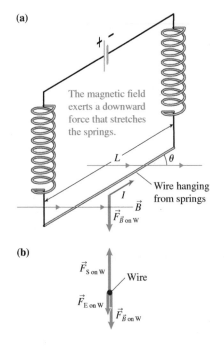

TABLE 20.4 Magnetic force data for the experiment in Figure 20.14

Current I in the wire (A)	Length L of wire (m)	Orientation angle θ between the current and the $\vec{B}$ field	Magnitude of magnetic force F exerted on the wire (N)
I	L	90°	F_1
$2I$	L	90°	$2F_1$
$3I$	L	90°	$3F_1$
I	L	90°	F_1
I	$2L$	90°	$2F_1$
I	$3L$	90°	$3F_1$
I	L	0°	0
I	L	30°	$0.5F_1$
I	L	60°	$0.87F_1$
I	L	90°	F_1

Notice that the magnitude of the magnetic force exerted on the wire is proportional to the current I through the wire (the first three rows), to the length L of the wire (the second three rows), and to the sine of the angle θ between the direction of the current and the direction of the $\vec{B}$ field (the last four rows). We can represent this mathematically:

$$F_{\vec{B} \text{ on W}} \propto IL \sin \theta \qquad (20.1)$$

When one quantity is proportional to another quantity, the ratio of the two is constant: $\dfrac{F_{\vec{B} \text{ on W}}}{IL \sin \theta} = \text{constant}$. If you place the same current-carrying wire in the magnetic field of an electromagnet in which you can increase the magnetic field by increasing the electric current in the magnet's coil windings, you find that the proportionality constant increases as the $\vec{B}$ field increases. We can hypothesize that the proportionality constant is the magnitude of the $\vec{B}$ field.

We can use the above mathematical relation to define the magnitude of the $\vec{B}$ field in a particular region as the ratio of the magnitude of the maximum force that the field exerts on a current-carrying wire to the product of the length L of the wire and the current I in the wire. This maximum force is exerted when the wire is perpendicular to the direction of the $\vec{B}$ field.

$$B = \frac{F_{\vec{B} \text{ on W max}}}{IL} \qquad (20.2)$$

Equation (20.2) allows us to define a unit for the $\vec{B}$ field, known as the **tesla, T** (named in honor of the Serbian inventor Nikola Tesla [1856–1943]). In a particular region (assuming a uniform field), the $\vec{B}$ field of magnitude 1 T exerts a force of 1 N on a 1-m-long wire with a 1-A current through it when the wire is oriented perpendicular to the $\vec{B}$ field: $1 \text{ T} = 1 \text{ N/A} \cdot \text{m}$. A 1-T $\vec{B}$ field is very strong. By comparison, the average value of the $\vec{B}$ field produced by Earth at the surface is 5×10^{-5} T. The strongest magnets, neodymium-iron-boron (NIB) magnets, can produce a $\vec{B}$ field near their poles of about 0.5 T in magnitude.

We can now use the definition of the magnitude of the $\vec{B}$ field to rewrite the expression for the magnetic force exerted by the field on a current.

Magnetic force exerted on a current The magnitude of the magnetic force $F_{\vec{B} \text{ on W}}$ that a uniform magnetic field $\vec{B}$ exerts on a straight current-carrying wire of length L with current I is

$$F_{\vec{B} \text{ on W}} = ILB \sin \theta \qquad (20.3)$$

where θ is the angle between the directions of the $\vec{B}$ field and the current I. The direction of this magnetic force is given by the right-hand rule for the magnetic force.

EXAMPLE 20.2 **How strong is a strong magnet?**

Neodymium magnets can produce magnetic fields of about 0.5 T near their poles. They are cylindrical, about 1 cm in diameter and 1 cm high. You use a wire to short circuit a battery and achieve a 3-A current through the wire. If you want to use a neodymium magnet to lift the wire, how should you orient the magnet? Can the magnet lift a 1-g wire?

Sketch and translate The maximum force is exerted when the wire is oriented perpendicular to the $\vec{B}$ field vector, which is perpendicular to the cylinder's axis. Given that we want to lift the wire, the force exerted on it must be upward. Thus, using the right-hand rule for the magnetic force, we conclude that the poles of the magnet (i.e., the cylinder) need to be oriented along a horizontal line. The necessary orientation of the

cylinder and wire and the direction of the current are shown on the sketch below, as well as the knowns and unknowns.

(CONTINUED)

Simplify and diagram We assume that the magnetic field at the end of the cylinder is uniform and the length of the wire in this field is 1 cm. We neglect the field on the sides of the magnet that can potentially affect the wire. From the force diagram for the wire, we can infer that to lift the wire, the magnetic force must be larger than the gravitational force.

Represent mathematically The magnitude of the force exerted by the magnet on the wire is given by Eq. (20.3). We know all the values needed to calculate the force. The angle is 90°. The magnitude of the force exerted by Earth is equal to $m_W g$.

$\vec{F}_{\vec{B} \text{ on W}}$

$\vec{F}_{\vec{E} \text{ on W}}$

Solve and evaluate

$$F_{\vec{B} \text{ on W}} = ILB \approx (3 \text{ A})(10^{-2} \text{ m})(0.5 \text{ T}) = 1.5 \times 10^{-2} \text{ N}$$

The magnitude of the force exerted by Earth is $m_W g \approx (1 \times 10^{-3} \text{ kg}) \times (10 \text{ N/kg}) = 10^{-2} \text{ N}$. Therefore, the magnet can potentially lift the wire.

- -

Try it yourself A 2.0-m-long wire has a 10-A current through it. The wire is oriented south to north and located near the equator. Earth's $\vec{B}$ field has a 5.0×10^{-5}-T magnitude in the vicinity of the wire. What is the magnetic force exerted by the field on the wire?

Answer The wire is parallel to the $\vec{B}$ field. Thus, $\sin \theta = \sin(0°) = 0$ and the magnetic force exerted on the wire is zero.

Summary of the differences between gravitational, electric, and magnetic forces

- The gravitational and electric forces exerted on objects do not depend on the direction or the speed of motion of those objects, whereas the magnetic force does. If the direction of the electric current is parallel or antiparallel (pointing exactly in the opposite direction) to the $\vec{B}$ field, no magnetic force is exerted on it.
- While the forces exerted by the gravitational and the electric fields are always in the direction of the $\vec{g}$ or $\vec{E}$ field (or opposite that direction in the case of a negatively charged object), the force exerted by the magnetic field on a current-carrying wire is perpendicular to both the $\vec{B}$ field and the direction of the electric current (**Figure 20.15**).

FIGURE 20.15 The directions of different forces relative to their fields.

The gravitational force is parallel to the $\vec{g}$ field.

The electric force is parallel to the $\vec{E}$ field.

The magnetic force is perpendicular to the magnetic field and to the electric current.

The direct current electric motor

A **motor** is a device that converts electric energy into mechanical energy, specifically rotational or translational kinetic energy. In order to understand the principle behind the operation of an electric motor, we will first study the motion of a current-carrying coil in a uniform magnetic field (**Figure 20.16a**). The coil rotates around an insulating axle. How does this device convert electric energy into the rotational energy of the coil?

Let us start with the rectangular coil oriented so that the plane of the coil is parallel to the $\vec{B}$ field. The currents through sides 1 and 3 of the coil are perpendicular to the $\vec{B}$ field, so the field exerts a force on them. The currents through sides 2 and 4 are parallel to the $\vec{B}$ field: the force exerted on them is zero.

According to the right-hand rule for magnetic force, the magnetic field exerts an upward force on wire 1 of the coil and a downward force on wire 3 of the coil

(Figure 20.16b). These forces each produce a torque around the axle that causes the coil to start rotating clockwise.

As the coil turns, the orientations of the currents relative to the $\vec{B}$ field change, as do the magnetic forces exerted by the field on these sides. As the coil reaches an orientation with its surface perpendicular to the $\vec{B}$ field, the magnetic field exerts forces on each side of the coil that tend to stretch it but are not able to turn it (Figure 20.16c). Because the coil is in motion, it continues moving and turns past this orientation, reaching the one shown in Figure 20.16d. Using the right-hand rule for magnetic forces again, we find that the torques produced by the magnetic forces exerted on sides 1 and 3 cause the coil to accelerate in the counterclockwise direction, slowing down and reversing the direction of its rotation. Given that, for a motor, we need a continuous rotation of the coil, we need to reverse the direction of the current when the plane of the coil passes the perpendicular to the $\vec{B}$ field orientation (shown in Figure 20.16c) so that the net magnetic torque remains in the clockwise direction. Consequently, for the net torque produced by the magnetic force exerted on the coil to always remain clockwise, the current through the coil must change direction each time the coil passes the vertical orientation.

This reversal of the current is made possible using a device known as a **commutator** (**Figure 20.17**). A commutator consists of two semicircular rings that are attached to the rotating coil. Sliding contacts (metal brushes) connect the battery to the rotating commutator rings. Every time the rings rotate 180° (a half-turn), the brush touching the ring changes, and so does the direction of the current in the coil. Thus the direction of the torque exerted on the coil remains constant.

The motor described above, known as a **direct current electric motor**, is a much more complex version of a "motor" first invented by Michael Faraday in 1821. Faraday also invented the electric generator and was the first to describe the magnetic field. He had no formal education and did not use complex mathematics to describe his ideas.

Torque exerted on a current-carrying coil

We see that the torque produced by the magnetic forces exerted on a coil depends on the orientation of the coil relative to the $\vec{B}$ field. The magnitude of this torque depends on how far from the coil's rotation axis these magnetic forces are exerted. To derive a mathematical expression for the torque that a magnetic field exerts on a current-carrying coil, we will start with a single loop of wire. The torque that the magnetic force exerts on sides 1 and 3 of the loop in Figure 20.16b is directly proportional to the distance $D/2$ from the axis of rotation to where the force is exerted—shown in Figure 20.16a. As there are two equal-magnitude torques exerted on the loop with turning ability in the same direction, the total torque is proportional to D. The magnitude of the magnetic force exerted on wire 1 and on wire 3 depends on the length L of that side of the loop (Figure 20.16a). Therefore, the total torque should be proportional to the product of D and L, which is the area A of the loop. If we have a coil with N loops, the torque will be N times the torque exerted on each loop.

We arrive at an expression for the magnitude of the torque that magnetic forces exert on a current-carrying coil:

$$\left|\tau_{\vec{B}\text{ on Coil}}\right| = NBAI\sin\theta \tag{20.4}$$

where N is the number of turns in the coil, B is the magnitude of the $\vec{B}$ field in the region of the coil (assumed uniform), A is the area of the coil, I is the electric current through the coil, and θ is the angle between a vector perpendicular to the coil's surface (called the **normal vector**) and the direction of the $\vec{B}$ field.

Magnetic dipole moment

The product of the current I, the area A, and the number of turns N in a coil in Eq. (20.4) is called the magnitude of the **magnetic dipole moment** $\vec{p}_m$ of the coil ($|\vec{p}_m| = NIA$). The direction of the dipole moment vector is perpendicular to the surface of the

FIGURE 20.16 A magnetic field exerts a torque on a current-carrying coil.

(a)

(b) The magnetic forces exerted on sides 1 and 3 cause a clockwise torque on the coil.

(c) The magnetic forces due to the magnetic field stretch the coil but cause zero torque.

(d) The magnetic field now exerts forces on sides 1 and 3 that cause a counterclockwise torque.

FIGURE 20.17 A commutator ring.

The commutator ring reverses the current in the coil each half turn of the coil. The torque is then always in the same direction.

FIGURE 20.18 Magnetic dipole moment.

(a) Magnetic dipole moment vector

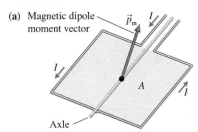

Axle

(b) The normal vector is perpendicular to the surface of the loop or coil.

A torque rotates the loop in the clockwise direction.

FIGURE 20.20 An electron beam in a cathode ray tube is deflected by a magnetic field.

(a) Cathode ray tube

(b) $\vec{B}$ into page

(c) $\vec{B}$ out of page

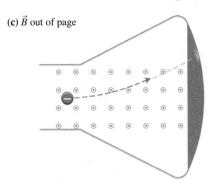

coil and in the direction of the $\vec{B}$ field produced by current at the center of the coil (Figure 20.18a). The bigger the dipole moment of a coil, the greater the torque that an external magnetic field exerts on it:

$$\left|\tau_{\vec{B}\text{ on Coil}}\right| = NBIA \sin\theta = Bp_{m}\sin\theta \qquad (20.4)$$

where θ is the angle between the normal vector perpendicular to the surface of the current loop (same direction as the dipole moment vector) and the external $\vec{B}$ field vector (see Figure 20.18b).

The direction of the magnetic dipole moment of a bar magnet is the direction from the south to the north pole inside the magnet (Figure 20.19). We can use the same equation for the torque exerted by an external magnetic field on a magnet if we know its magnetic dipole moment.

FIGURE 20.19 A bar magnet's dipole moment.

REVIEW QUESTION 20.3 Equation (20.2) defines the magnitude of the $\vec{B}$ field at a point as $B = F_{\vec{B}\text{ on W max}}/IL$. Explain the meaning of every symbol in this equation. Do any quantities on the right side of the equation cause the $\vec{B}$ field to have a bigger or smaller magnitude? Explain.

20.4 Magnetic force exerted on a single moving charged particle

We have learned that the magnetic field exerts a force on a current-carrying wire. The current through a wire is the result of the collective motion of a huge number of electrically charged particles—electrons. Thus, it is reasonable to conclude that the magnetic field also exerts a force on each individual electron.

Direction of the force that the magnetic field exerts on a moving charged particle

A device called a **cathode ray tube** (Figure 20.20a) demonstrates how a magnetic field affects moving electrons. A cathode ray tube consists of two electrodes—a **cathode** and a hollow **anode**—in an evacuated glass enclosure. The cathode is connected to the negative terminal of a power source and the anode to the positive terminal. A current caused by a separate power source keeps the cathode hot. Due to its high temperature, the cathode emits electrons (more on this in Chapter 27). The potential difference between the cathode and the anode causes the electrons to accelerate toward the anode, through the hole in the anode, and onto a screen. The screen is treated with a material that glows green when hit by electrons. We can use the location of the green dot on the screen to infer the path of the electrons within the cathode ray tube (see the video).

If a magnetic field due to a hand-held bar magnet (not shown in the figure) exerts a force on individual electrons similar to the force it exerts on an electric current in a wire, then the right-hand rule for the magnetic force should predict the direction that the electrons will be deflected. The rule was formulated in terms of electric current, which by convention is the direction that positively charged particles move. The magnetic force exerted on negative electrons moving toward the screen should be in the opposite direction to the one given by the right-hand rule for the magnetic force exerted on an imaginary current-carrying wire with the current toward the screen. To conduct the testing experiment, we orient the magnet so that its $\vec{B}$ field points perpendicular to the electron path shown in Figure 20.20a (into the page, as in Figure 20.20b).

The right-hand rule for the magnetic force predicts that positively charged particles should be deflected upward, while the negatively charged electrons inside the tube should deflect downward. If we reverse the direction of the magnetic field, the electrons should be deflected upward, and the green spot does not deflect if the $\vec{B}$ field points parallel to the velocity of the electrons (Figure 20.20c). When the experiment is performed, the outcome is consistent with the prediction. This result supports the idea that the magnetic force exerted by the magnetic field on individual charged objects is similar to the one it exerts on currents. Let's construct a quantitative relation for the magnitude of the magnetic force exerted by the magnetic field on an individual charged particle.

Magnitude of the force that a magnetic field exerts on a moving charged particle

We know that a magnetic field exerts a force on a current-carrying wire of magnitude

$$F_{\vec{B} \text{ on W}} = ILB \sin\theta$$

We use this to develop an expression for the magnitude of the force that the magnetic field exerts on a single charged object with charge q moving at speed v (the current I in the wire consists of a large number of moving charged particles—Figure 20.21). Although we know that electrons move in the wire when current is present, for simplicity we use positively charged particles in this calculation.

Imagine that between the two dashed lines there are N moving charged particles. In a time interval Δt, all of them pass through the dashed line on the right. Thus, the electric current through this wire is

$$I = \frac{Nq}{\Delta t}$$

The speed of the charged particles is $v = L/\Delta t$ since a particle at the left dashed line takes a time interval Δt to reach the right dashed line. Rearrange this for $\Delta t = L/v$ and substitute in the above:

$$I = \frac{Nqv}{L}$$

Inserting this into $F_{\vec{B} \text{ on W}} = ILB \sin\theta$, we get

$$F_{\vec{B} \text{ on W}} = \left(\frac{Nqv}{L}\right)LB \sin\theta = N(qvB \sin\theta)$$

This is the magnitude of the force exerted by the magnetic field on all N moving charged particles. The magnitude of the force exerted by the field on a single charged particle is then given in Eq. (20.5) below.

FIGURE 20.21 Imaginary positively charged particles moving in a wire.

N positive charges

All charged particles between the dashed lines pass the right dashed line in a time interval Δt.

Magnetic force exerted by the magnetic field on an individual charged particle The magnitude of the magnetic force that the magnetic field of magnitude B exerts on a particle with electric charge q moving at speed v is

$$F_{\vec{B} \text{ on } q} = |q|vB \sin\theta \qquad (20.5)$$

where θ is the angle between the direction of the velocity of the particle and the direction of the $\vec{B}$ field. The direction of this force is determined by the right-hand rule for the magnetic force. If the particle is negatively charged, the force points opposite the direction given by the right-hand rule.

TIP Note that if the velocity and the $\vec{B}$ field are parallel, $\sin 0° = 0$ and the force is zero. The force is maximum when the object's velocity and the $\vec{B}$ field are perpendicular.

The direction of the magnetic force that a magnetic field exerts on a moving positively charged particle is illustrated in **Figure 20.22a**. The right-hand rule for determining the direction of the force is the same as we learned before for the current-carrying wire. It is described in Figure 20.22b.

FIGURE 20.22 The right-hand rule for the magnetic force exerted by a magnetic field on a positively charged particle.

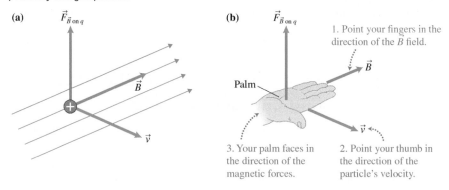

(a) $\vec{F}_{\vec{B} \text{ on } q}$ $\vec{B}$ $\vec{v}$

(b) $\vec{F}_{\vec{B} \text{ on } q}$ 1. Point your fingers in the direction of the B field. $\vec{B}$ Palm $\vec{v}$ 3. Your palm faces in the direction of the magnetic forces.   2. Point your thumb in the direction of the particle's velocity.

QUANTITATIVE EXERCISE 20.3   **Particles in a magnetic field**

Each of the lettered dots shown in the figure represents a small object with an electric charge $+2.0 \times 10^{-8}$ C moving at speed 200 m/s in the directions shown. Determine the magnetic force (magnitude and direction) that a 0.10-T $\vec{B}$ field exerts on each object. The $\vec{B}$ field points in the positive y-direction.

Represent mathematically First, use the right-hand rule for the magnetic force to determine the directions of the magnetic force exerted on each object. (a) For object A, point the fingers of your right hand toward the top of the page in the direction of $\vec{B}$. Then orient your hand so that your thumb points to the left in the direction of $\vec{v}$. Your palm points into the page. This is the direction of the magnetic force exerted on object A. (b) Object B moves in a direction opposite to $\vec{B}$ ($\theta = 180°$); thus, the magnetic force is zero. (c) For C, point your fingers toward the top of the page and the thumb out of the page. Your palm then faces left, so the magnetic force exerted on object C points in the negative x-direction. (d) For object D, point your fingers toward the top of the

page and your thumb parallel to the page pointing 37° above rightward. Your palm faces out of the page, the direction of the force exerted on D.

Solve and evaluate Use Eq. (20.5) to determine the magnitude of each force:

$$F_{\vec{B} \text{ on A}} = (+2.0 \times 10^{-8}\,\text{C})(200 \text{ m/s})(0.10 \text{ T}) \sin(90°)$$
$$= 4.0 \times 10^{-7}\,\text{N}$$
$$F_{\vec{B} \text{ on B}} = (+2.0 \times 10^{-8}\,\text{C})(200 \text{ m/s})(0.10 \text{ T}) \sin(180°)$$
$$= 0$$
$$F_{\vec{B} \text{ on C}} = (+2.0 \times 10^{-8}\,\text{C})(200 \text{ m/s})(0.10 \text{ T}) \sin(90°)$$
$$= 4.0 \times 10^{-7}\,\text{N}$$
$$F_{\vec{B} \text{ on D}} = (+2.0 \times 10^{-8}\,\text{C})(200 \text{ m/s})(0.10 \text{ T}) \sin(53°)$$
$$= 3.2 \times 10^{-7}\,\text{N}$$

Try it yourself The equation below represents the solution to a problem. Devise a possible problem that is consistent with the equation:

$$(1.6 \times 10^{-19}\,\text{C})v(0.50 \times 10^{-5} \text{ T}) \sin(30°) = 1.0 \times 10^{-18}\,\text{N}$$

Answer   ¿pəəds s,uoʇoɹd əɥʇ sᴉ ʇɐɥM ·ʎɹʇuə ɟo ʇuᴉod əɥʇ ʇɐ uoʇoɹd proton at the point of entry. What is the proton's speed? əɥʇ uo əɔɹoɟ N 8₁₋0I × 0·I ɐ sʇɹəxə pləᴉɟ əɥ⊥ ·pləᴉɟ $\vec{B}$ əɥʇ ɟo uoᴉʇɔəɹᴉp direction of the $\vec{B}$ field. The field exerts a 1.0 × 10⁻¹⁸ N force on the əɥʇ ɥʇᴉʍ əlƃuɐ °0Ɛ ɐ səʞɐɯ ʎʇᴉɔolə𝑣 s,uoʇoɹd əɥ⊥ ·əɔɐɟɹns əɥʇ ə𝑣oqɐ above the surface. The proton's velocity makes a 30° angle with the ɹɐɟ pləᴉɟ ɔᴉʇəuƃɐɯ s,ɥʇɹɐƎ sɹəʇuə uoʇoɹd ∀ :ɯəlqoɹd əlqᴉssod əuO One possible problem: A proton enters Earth's magnetic field far

---

Circular motion in a magnetic field

We know that the force that a magnetic field exerts on a moving charged particle is always perpendicular to the particle's velocity. This is a characteristic of circular motion (discussed in Chapter 5). Consider an example of such motion where a positively charged particle moves to the left across the top of your open textbook (position **a** in

Figure 20.23) when a uniform magnetic field pointing into the page turns on. Using the right-hand rule for the magnetic force, we find that the field exerts a magnetic force on the particle that points downward. The particle continues to move forward, but the direction of its velocity changes and now points slightly downward. The force exerted by the magnetic field always points perpendicular to the particle's velocity. This deflects the particle further. Once it has made a quarter turn and reaches position **b**, the particle is moving toward the bottom of the page, and the magnetic field exerts a force toward the right. This pattern persists—the magnetic force exerted on the particle always points toward the center of the particle's path. Thus, in a uniform $\vec{B}$ field a charged particle that initially moves perpendicular to the $\vec{B}$ field will move along a circular path in a plane perpendicular to the field. Since the force that the magnetic field exerts on a particle is perpendicular to its velocity at every instant, it is also perpendicular to its displacement during each short time interval; therefore, the work done by this force on the particle is zero. As the work done by the magnetic field on the particle is zero, its kinetic energy should not change. This is exactly what we observe: the particle continues to move in a circle at constant speed.

FIGURE 20.23 A charged particle moving perpendicular to a magnetic field moves in a circle.

The magnetic field exerts a force perpendicular to the velocity, leading to circular motion.

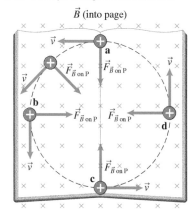

Cosmic rays

The deflection of charged particles in a magnetic field has an important everyday application. Charged particles zoom past and *through* us every minute, bombarding Earth and its inhabitants. These particles, called **cosmic rays**, are usually electrons, protons, and other elementary particles produced by various astrophysical processes including those occurring in the Sun and sources outside the solar system. Every minute about 20 of these fast-moving charged particles pass through a person's head. These particles can cause genetic mutations, which can lead to cancer and other unpleasant effects. Fortunately, our bodies have multiple mechanisms to repair most of the damage.

Thousands more particles would pass through our heads each minute without the protection provided by Earth itself. Earth's magnetic field serves as a shield against harmful cosmic rays, causing them to deflect from their original trajectory toward Earth. How does this shield work?

EXAMPLE 20.4   **Motion of protons in Earth's magnetic field**

Determine the path of a cosmic ray proton flying into Earth's atmosphere above the equator at a speed of about 10^7 m/s and perpendicular to Earth's magnetic field. The average magnitude of Earth's $\vec{B}$ field in this region is approximately 5×10^{-5} T. The mass m of a proton is approximately 10^{-27} kg.

Sketch and translate The situation is sketched below.

Simplify and diagram Consider a short distance that the proton travels high above Earth and assume that in this region the $\vec{B}$ field vectors are parallel to Earth's surface and have a constant magnitude of 5×10^{-5} T. Our calculations will be much easier if the magnetic force is the

$\vec{F}_{\vec{B}\,on\,P}$ is into page

only force exerted on the proton. Can we neglect the gravitational force exerted by Earth? We do a quick estimate:

$$mg = (10^{-27}\,\text{kg})(10\,\text{N/kg}) \approx 10^{-26}\,\text{N}$$

$$qvB = (1.6 \times 10^{-19}\,\text{C})(10^7\,\text{m/s})(5 \times 10^{-5}\,\text{T}) = 8 \times 10^{-17}\,\text{N}$$

The magnetic force is more than a billion times larger than the gravitational force, so we can neglect the gravitational force.

Represent mathematically When the velocity of a charged particle is perpendicular to the $\vec{B}$ field, it moves in a circular path at constant speed. We can use the radial r component form of Newton's second law to relate the magnetic force exerted on the proton to its resulting motion. The force exerted by the magnetic field points toward the center of the proton's circular path.

$$a_r = \frac{v^2}{r} = \frac{1}{m}\sum F_r = \frac{1}{m}(F_{\vec{B}\,on\,P\,r}) = \frac{1}{m}(|q|vB\sin\theta)$$

$$= \frac{|q|vB\sin(90°)}{m} = \frac{|q|vB}{m}$$

(CONTINUED)

This equation can be used to determine the radius r of the proton's circular path and to determine the period T of its motion, noting that

$$v = \frac{2\pi r}{T}$$

Solve and evaluate We used Newton's second law to develop an expression describing the proton's circular motion:

$$\frac{v^2}{r} = \frac{|q|vB}{m}$$

Multiply both sides by the product of r and m and then rearrange to get an expression for r:

$$\frac{v^2 mr}{r} = \frac{|q|vBmr}{m} \Rightarrow mv^2 = |q|vBr$$

$$\Rightarrow r = \frac{mv}{|q|B}$$

$$\approx \frac{(10^{-27}\,\text{kg})(10^7\,\text{m/s})}{|1.6 \times 10^{-19}\,\text{C}|(5 \times 10^{-5}\,\text{T})} \approx 10^3\,\text{m}$$

The period T of the proton's motion is then

$$T = \frac{2\pi r}{v} \approx \frac{2\pi(10^3\,\text{m})}{10^7\,\text{m/s}} \approx 10^{-3}\,\text{s}$$

Something interesting happens if we combine the equations for r and T:

$$T = \frac{2\pi r}{v} = \frac{2\pi}{v}\left(\frac{mv}{|q|B}\right) = \frac{2\pi m}{|q|B}$$

We find that the period of the proton's motion depends only on the ratio of the mass and charge of the proton and the magnitude of the $\vec{B}$ field. It does not depend on the speed of the proton. (Do not confuse the period T with the unit for the magnetic field, the tesla T.)

Try it yourself What happens to the motion of a proton that enters a uniform magnetic field parallel to the $\vec{B}$ field?

Answer

The motion of the proton will not be affected by the magnetic field.

Earth's magnetic field acts as a shield, protecting life on the surface from cosmic ray particles. A cosmic ray proton entering the magnetic field of Earth at one-third light speed, or 1×10^8 m/s, will move in a helical path with a radius of 1 or more km. Earth's magnetic field extends several tens of thousands of kilometers above the surface, so life on Earth is well protected. The Earth's atmosphere itself serves as an additional shield, absorbing those protons that enter it. However, astronauts traveling outside of Earth's atmosphere have neither the atmospheric shield nor the magnetic shield. They are exposed to higher levels of cosmic radiation than are present on Earth's surface. This exposure is an important consideration in planning future long missions to Mars and other planets.

The auroras

Charged particles moving in Earth's magnetic field actually follow helical paths around the $\vec{B}$ field lines and enter the Earth's atmosphere near the poles (**Figure 20.24**). These particles collide with molecules in the atmosphere, ionizing the molecules. When the electrons recombine with the ionized molecules, the excess energy is radiated as light. We see this light in the upper atmosphere in the region of the magnetic poles—the auroras mentioned at the beginning of the chapter.

Often these charged particles come from solar flares caused by the interactions of the Sun's hot ionized gas with its magnetic field. Therefore, when magnetic activity on the Sun is high, the auroras become more intense. On occasion, the auroras are visible far from the magnetic pole regions, sometimes even quite close to the equator.

FIGURE 20.24 Paths of charged particles that can cause auroras.

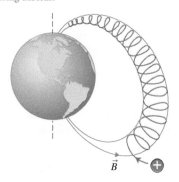

Particles entering slightly less than perpendicular to Earth's magnetic field follow helical paths toward the poles, causing auroras.

$\vec{B}$

REVIEW QUESTION 20.4 If the magnetic force is always perpendicular to the velocity of a charged particle, can it do any work on the particle? Explain your answer.

20.5 Magnetic fields produced by electric currents

So far, we have qualitatively investigated the $\vec{B}$ fields produced by current-carrying straight wires, loops, and solenoids. To analyze problems such as the danger of magnetic fields due to high-power transmission lines, we need to know how to quantitatively determine the magnitude and direction of a $\vec{B}$ field produced by a particular current configuration (the source)—the subject of this section.

The $\vec{B}$ field produced by an electric current in a long straight wire

An electric current in a long straight wire produces a magnetic field whose $\vec{B}$ field lines circle around the wire (Figure 20.25). What is the magnitude of that field at various locations near the wire? To answer this question we need to place some kind of a detector (test object) of magnetic field at different locations and investigate the effects of the field on this object. In Section 20.3 we learned that a magnetic field will produce a maximum torque $\left| \vec{\tau}_{\vec{B} \text{ on C max}} \right| = NBAI$ on a small current-carrying coil used to probe a magnetic field at different locations. We can use such a coil as a detector to determine the relative strength of the magnetic field produced by a long straight current-carrying wire.

We place the coils at different distances from the field-producing current-carrying wire and measure the maximum torque that the magnetic field created by the wire exerts on the coil. We use this torque, the magnitude of the current through the coil, its area, and the number of turns to determine the $\vec{B}$ field surrounding a long straight wire with different currents I and at different distances r from the wire. The data are in Table 20.5.

From the data, we can identify a pattern: the magnitude of the $\vec{B}$ field created by a long straight current-carrying wire is directly proportional to the magnitude of the current I and inversely proportional to the distance r between the wire and the location where the field is measured.

We can express this pattern mathematically as follows:

$$B_{\text{straight wire}} \propto \frac{I_{\text{wire}}}{r}$$

Since we can experimentally measure all three of the quantities appearing in this relationship, it's possible to determine the constant of proportionality that will turn it into an equation. Traditionally, this constant is written as $\mu_0/2\pi$, where $\mu_0 = 4\pi \times 10^{-7}$ T·m/A is called the **magnetic permeability** of a vacuum. We now have an expression for the magnitude of the $\vec{B}$ field at a perpendicular distance r from a long straight current-carrying wire:

$$B_{\text{straight wire}} = \frac{\mu_0}{2\pi} \frac{I_{\text{wire}}}{r} \qquad (20.6)$$

Note that the farther you move from the current-carrying wire (larger r), the smaller the magnitude of the $\vec{B}$ field; the greater the current (larger I_{wire}), the larger the magnitude of the $\vec{B}$ field.

Figure 20.26 shows the information about magnetic fields created by different configurations of current-carrying wires. It presents the equations for a loop/coil and solenoid without the derivation. We will use the magnitude of the $\vec{B}$ field at the center of a single current-carrying loop next to analyze the magnetic field produced by the motion of an electron in a hydrogen atom.

$\vec{B}$ field due to electron motion in an atom

In an early 20th-century model of the hydrogen atom, the electron was thought to move rapidly in a tiny circular path around the nucleus of the atom (similar to the figure in Example 20.5). Although the model itself is outdated, it allows us to make reasonable predictions about magnetic properties of individual atoms and objects made from them. In this model the electron motion is like a circular electric current I that produces a $\vec{B}$ field. The $\vec{B}$ field at the center of a current-carrying loop of radius r is

$$B = \frac{\mu_0 I}{2r}$$

The electron's circular motion also produces a magnetic dipole moment of magnitude $p_m = IA$ equal to the product of the electron current I and the area A of the loop. If these dipole moments are significant in magnitude, they could potentially help

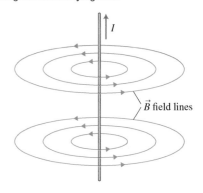

FIGURE 20.25 The magnetic field caused by a long current-carrying wire.

TABLE 20.5 Magnitude of the $\vec{B}$ field at distance r from a straight wire carrying current I

Current I	Distance	B
I_0	r_0	B_0
$2I_0$	r_0	$2B_0$
$3I_0$	r_0	$3B_0$
I_0	$2r_0$	$B_0/2$
I_0	$3r_0$	$B_0/3$
I_0	$r_0/2$	$2B_0$

FIGURE 20.26 Expressions for the $\vec{B}$ field produced by (a) a straight wire, (b) a loop or coil, and (c) a solenoid.

(a) Straight wire

$$B_{\text{straight wire}} = \frac{\mu_0}{2\pi} \frac{I_{\text{wire}}}{r}$$

(b) Loop or coil

$B = \dfrac{\mu_0 I}{2r}$ at center of loop

$B = \dfrac{N\mu_0 I}{2r}$ at center of coil with N turns

(c) Solenoid

$B = \dfrac{N\mu_0 I}{l}$ inside solenoid

explain magnetic properties of materials. In the next example we will make a rough estimate of the hydrogen atom's electric current I and the $\vec{B}$ field and dipole moment produced by this current.

EXAMPLE 20.5 **Magnetic field produced by the electron in a hydrogen atom**

In the above-mentioned early 20th-century model of the hydrogen atom, the electron was thought to move in a circle of radius 0.53×10^{-10} m, orbiting once around the nucleus every 1.5×10^{-16} s. Determine the magnitude of the $\vec{B}$ field produced by the electron at the center of its circular orbit and its dipole moment.

Sketch and translate A sketch of the moving electron is shown along with the known information.

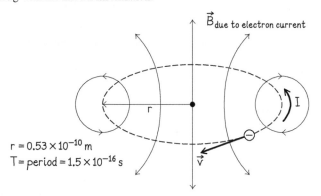

$r = 0.53 \times 10^{-10}$ m
$T = $ period $= 1.5 \times 10^{-16}$ s

Simplify and diagram The electron's motion corresponds to a counterclockwise current (opposite the direction of travel of the negatively charged electron). This is similar to a single current loop. The direction of the $\vec{B}$ field at the center can be determined by the right-hand rule for the $\vec{B}$ field. This is also the direction of the electron's dipole moment.

Represent mathematically The magnitude of the $\vec{B}$ field at the center of a circular current I is

$$B = \frac{\mu_0 I}{2r}$$

To determine B, we have to determine the electric current due to the electron's motion. Electric current is $I = q/\Delta t$, where q is the magnitude of the total electric charge that passes a cross section of a wire in time Δt. The cross section in this case is a single point along the electron's orbit. The single electron passes that point once every 1.5×10^{-16} s.

Solve and evaluate Using these ideas, we can calculate the magnitude of the current:

$$I = \frac{q}{\Delta t} = \frac{1.6 \times 10^{-19}\,\text{C}}{1.5 \times 10^{-16}\,\text{s}} = 1.1 \times 10^{-3}\,\text{A}$$

The magnitude of the $\vec{B}$ field at the center of the loop is then

$$B = \frac{\mu_0 I}{2r} = \frac{(4\pi \times 10^{-7}\,\text{T}\cdot\text{m/A})(1.1 \times 10^{-3}\,\text{A})}{2(0.53 \times 10^{-10}\,\text{m})} = 13\,\text{T}$$

This is a huge $\vec{B}$ field, especially if we remember that the strongest permanent magnets can produce fields only up to 1 T. The magnitude of the dipole moment of an electron is

$$p_m = IA = (1.1 \times 10^{-3}\,\text{A})[\pi(0.53 \times 10^{-10}\,\text{m})^2]$$
$$= 9.7 \times 10^{-24}\,\text{A}\cdot\text{m}^2$$

This is a tiny quantity compared to the magnetic moment of a macroscopic current-carrying loop. However, one object can easily contain a trillion trillion electrons.

- -

Try it yourself We found in this example that a moving electron in an atom produces a very strong magnetic field. Suggest an explanation for why all materials are not strong magnets.

Answer

One possible reason is that the electron orbits of the individual atoms are oriented randomly and therefore so are their dipole moments. If this is correct, their dipole moments would cancel each other. Another reason could be that most atoms have more than one electron, and magnetic moments produced by each of these electrons add to zero so that the $\vec{B}$ field produced by each individual atom is zero.

REVIEW QUESTION 20.5 The definition of a 1-A current states that two parallel 1.0-m-long current-carrying wires separated by 1.0 m with a current of 1 A in each exert an attractive force on each other of magnitude 2.0×10^{-7} N. Where does this value of 2.0×10^{-7} N originate?

20.6 Skills for analyzing magnetic processes

Problems involving magnetic interactions are often of two main types: (1) determine the magnetic force exerted on a current or on an individual moving charged object by the magnetic field or (2) determine the $\vec{B}$ field produced by a known source such as an electric current. In this section we will develop skills needed to solve such problems.

PROBLEM-SOLVING STRATEGY 20.1 | Analyzing processes involving magnetic fields and forces

---

EXAMPLE 20.6 | **Magnetic force problem**

A horizontal metal wire of mass 5.0 g and length 0.20 m is supported at its ends by two very light conducting threads. The wire hangs in a 49-mT magnetic field, which points perpendicular to the wire and out of the page. The maximum force each thread can exert on the wire before breaking is 39 mN. What is the minimum current through the wire that will cause the threads to break?

Sketch and translate

- Sketch the process described in the problem. Label the known and unknown quantities.
- Choose the system of interest.
- Show the direction of the $\vec{B}$ field and the direction of the electric current (or the velocity of a charged particle) if known.
- Decide whether the problem asks to find a $\vec{B}$ field produced by an electric current or to find a magnetic force exerted by an external field on a moving charged particle or wire with electric current.

- A sketch of the situation is shown along with the known information and the unknown quantity. We do not yet know the direction of the current.

- The horizontal wire is the system. What downward force exerted by the magnetic field, when added to the gravitational force exerted by Earth, will be enough to break the threads? Then what current will produce this downward magnetic force?
- This is a problem about finding the force exerted by the external field on an object.

Simplify and diagram

- Decide whether the $\vec{B}$ field can be considered uniform in the region of interest.
- Draw a force diagram for the system (the object in the field region) if necessary. Use the right-hand rule for the magnetic force to find an unknown force, current, velocity, or field direction if needed.
- Use the right-hand rule for the $\vec{B}$ field if the problem is about the field of a known source.

- Nothing specific is mentioned about the $\vec{B}$ field, so we consider it uniform in the vicinity of the wire.
- Construct a force diagram for the wire. The wire interacts with the threads ($\vec{F}_{\text{T on W}}$), Earth ($\vec{F}_{\text{E on W}}$), and the magnetic field ($\vec{F}_{\vec{B}\text{ on W}}$). With the $\vec{B}$ field pointing out of the page, the magnetic force points down in the desired direction if the wire current is toward the right. We choose the y-axis as pointing down.

Represent mathematically

- Describe the situation mathematically using the expressions for magnetic force exerted on a current or charged particle and the expressions for the $\vec{B}$ field produced by currents.
- If necessary, use Newton's second law in component form and/or kinematics.

- The wire is in equilibrium ($\Sigma F_y = 0$). In component form,

$$F_{\text{E on W }y} + 2F_{\text{T on W }y} + F_{\vec{B}\text{ on W }y} = 0$$
$$\Rightarrow mg + 2(-F_{\text{T on W}}) + ILB = 0$$

Move the terms that do not contain I to the right side of the equation: $ILB = 2F_{\text{T on W}} - mg$. Then divide both sides by LB:

$$I = \frac{2F_{\text{T on W}} - mg}{LB}$$

Solve and evaluate

● Use the mathematical representation of the process to determine the unknown quantity.

● Evaluate the results—units, magnitude, and limiting cases—to make sure they are reasonable.

● Inserting the appropriate values:

$$I = \frac{2(39 \times 10^{-3}\,\text{N}) - (5.0 \times 10^{-3}\,\text{kg})(9.8\,\text{N/kg})}{(0.20\,\text{m})(49 \times 10^{-3}\,\text{T})} = 3.0\,\text{A}$$

This is a large current for a thin wire but is not completely unreasonable.

● Looking at a limiting case: if the wire mass is such that $mg = 2F_{T\,\text{on W}}$, then the required current is zero—the gravitational force that Earth exerts would be enough to break the wire.

- -

Try it yourself If you want the magnetic force to lift the wire, what changes to the experiment do you need to make? Determine the minimum current.

Answer

We need to change either the direction of the current or the $\vec{B}$ field, but not both. The minimum current is 5.0 A assuming that the magnetic field remains unchanged.

EXAMPLE 20.7   Determine the $\vec{B}$ field

Determine the $\vec{B}$ field 5.0 cm from a long straight wire that is connected in series with a 5.0-Ω resister and a 9.0-V battery.

Sketch and translate Make a sketch of the situation including the electric circuit connected to the wire. The current in the circuit is clockwise. The problem is about finding the field produced by a known source.

Simplify and diagram Assume that the only contribution to the $\vec{B}$ field at the point of interest comes from the long wire. The other wires in the circuit contribute as well, but since they are somewhat farther away by comparison, we will neglect their contributions. Assume also that the other connecting wires and the battery have zero resistance. Using the right-hand rule for the $\vec{B}$ field, we find that below the wire the field points into the paper, and above the wire it points out of the paper.

Represent mathematically The magnitude of the $\vec{B}$ field produced by a long straight current-carrying wire is given by Eq. (20.6):

$$B_{\text{straight wire}} = \frac{\mu_0}{2\pi}\frac{I}{r}$$

Using Ohm's law and the loop rule (traversing the loop clockwise), we can determine the current through the wire:

$$\Delta V_{\text{batt}} + \Delta V_{\text{R}} = 0$$
$$\Rightarrow \mathcal{E} + (-IR) = 0$$

Solve and evaluate Combining the above equations gives

$$B_{\text{straight wire}} = \frac{\mu_0}{2\pi}\frac{I}{r} = \frac{\mu_0}{2\pi}\frac{(\mathcal{E}/R)}{r} = \frac{\mu_0}{2\pi}\frac{\mathcal{E}}{rR}$$

$$= \frac{4\pi \times 10^{-7}\,\text{T}\cdot\text{m/A}}{2\pi}\frac{9.0\,\text{V}}{(0.050\,\text{m})(5.0\,\Omega)}$$

$$= 7.2 \times 10^{-6}\,\text{T}$$

We can do a limiting case analysis for the case of a discharged battery. If $\mathcal{E} = 0$, then there should be no current and the $\vec{B}$ field should be zero, consistent with the equation.

- -

Try it yourself Estimate the magnitude of the $\vec{B}$ field produced by an electron beam in a cathode ray tube at a point 1.0 m to the side of the beam. Assume that 10^{10} electrons hit the screen every second and that they move at a speed of $10^7\,\text{m/s}$.

Answer $3.2 \times 10^{-16}\,\text{T}$.

Mass spectrometer

In Section 20.4 we learned that when a charged particle's velocity is perpendicular to the direction of the $\vec{B}$ field, the particle moves in a circular path. The radius r of the circle depends on the particle's mass:

$$r = \frac{mv}{|q|B}$$

This relationship is the foundation for the **mass spectrometer**. The mass spectrometer helps determine the mass of ions, molecules, and even elementary particles such as protons and electrons. It also can determine the relative concentration of atoms of the same chemical element that have slightly different masses (known as isotopes, which we will learn about in Chapter 29).

A mass spectrometer produces ions through heating, collisions, or some other mechanism. An electric field accelerates the ions into a device called a *velocity selector*. A velocity selector consists of a region with uniform magnetic and electric fields whose vectors $\vec{B}$ and $\vec{E}$ are perpendicular to each other (**Figure 20.27a**; the $\vec{E}$ field vector points upward on the page and the $\vec{B}$ field vector points out of the page). When a positively charged particle (an ion) enters this region in the horizontal direction pointed to the right, both fields will exert forces on the particle. The electric force will accelerate the particle upward, and the magnetic force will accelerate it downward (Figure 20.27b).

If the forces have the same magnitude, the particle will pass through the region unaffected. However, the magnetic force exerted on the particle depends on its velocity; thus only particles traveling at a specific velocity will be unaffected. All others will be deflected either upward or downward on the page. To determine the speed of the unaffected particles, we will use the force diagram in Figure 20.27b. Using Newton's second law and assuming that the gravitational force exerted on the particle is negligible, we can write:

$$\vec{F}_{B \text{ on } q\,y} + \vec{F}_{E \text{ on } q\,y} = 0 \Rightarrow (-F_{B \text{ on } q}) + (+F_{E \text{ on } q}) = 0$$
$$\Rightarrow |q|vB - |q|E = 0$$
$$\Rightarrow vB - E = 0$$

The speed of unaffected particles is therefore $v = E/B$. The velocity selector will only allow particles that move with a speed equal to the ratio E/B to pass through. Note that the value of the speed does not depend on the mass of the particle or the magnitude of the charge as long as the particle is charged.

After passing through the velocity selector, the ions then enter a region with a different uniform $\vec{B}_1$ field, traveling perpendicular to the field lines. As a result, the ions move in circular paths. The radius of the circle is measured by observing the place where the ions strike a detector after moving halfway around a circle (**Figure 20.28**). The ion mass is then determined using the method in the exercise below.

Mass spectrometry has many scientific uses. For example, researchers can use the mass spectrometer to measure the concentration of two isotopes of oxygen in glacial ice: oxygen-16 and oxygen-18. The oxygen-16/18 isotope ratio changes over geologic time due to global climate conditions, and therefore can be used as a way to track climate change and to determine the age of plant and animal remains found in the ice.

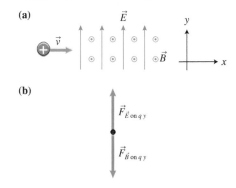
FIGURE 20.27 How a velocity selector works.

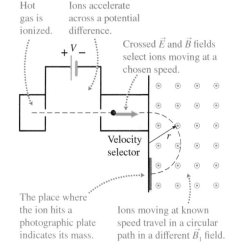
FIGURE 20.28 A mass spectrometer.

QUANTITATIVE EXERCISE 20.8 **Mass spectrometer**

An atom or molecule with a single electron removed (an ion) is traveling at 1.0×10^6 m/s when it enters a mass spectrometer's 0.50-T uniform $\vec{B}$ field region. Its electric charge is $+1.6 \times 10^{-19}$ C. It moves in a circle of radius 0.20 m until it hits the detector (see figure). Determine (a) the magnitude of the magnetic force that the magnetic field exerts on the ion and (b) the mass of the ion.

Represent mathematically The magnitude of the force that the magnetic field exerts on the moving ion is determined using Eq. (20.5):

$$F_{B \text{ on } q} = |q|vB \sin\theta$$

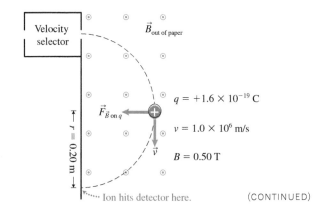

(CONTINUED)

This force perpendicular to the ion's velocity causes its radial acceleration:

$$ma_r = |q|vB \implies m\frac{v^2}{r} = |q|vB$$

The mass of the ion is

$$m = \frac{|q|rB}{v}$$

Solve and evaluate Now insert the appropriate values to determine the two unknowns:

$$F_{\vec{B} \text{ on } q} = |q|vB \sin\theta$$
$$= (+1.6 \times 10^{-19}\text{C})(1.0 \times 10^6\text{ m/s})(0.50\text{ T})\sin(90°)$$
$$= 8.0 \times 10^{-14}\text{ N}$$

and

$$m = \frac{|q|Br}{v} = \frac{(+1.6 \times 10^{-19}\text{C})(0.50\text{ T})(0.20\text{ m})}{1.0 \times 10^6\text{ m/s}}$$
$$= 1.6 \times 10^{-26}\text{ kg}$$

The magnitude of the force exerted on the ion is small but reasonable since it is the force exerted on a single ion. The mass of the ion is also reasonable (the mass of a single proton is 1.67×10^{-27} kg).

Try it yourself What would be the radius of the circular path of a mass 1.7×10^{-26}-kg particle?

Answer   0.21 m.

FIGURE 20.29 An IMRT machine.

The machine produces a stream of electrons ...

... that are accelerated to high speed ...

... and then bent by a magnetic field.

The electrons collide with a tungsten block, producing high-energy X-rays.

Cancerous tissue

Intensity modulated radiation therapy

Intensity modulated radiation therapy (IMRT) is a powerful cancer-fighting technology in which a high dose of X-rays is directed at a tumor, leaving most of the surrounding tissue untouched. The IMRT machine accelerates electrons to the desired kinetic energy, then uses a magnetic field to bend them 90° into a tungsten alloy target, resulting in the production of X-rays. Movable metal leaves then shape the X-ray beam to match the shape of the tumor. The accelerator and radiation device rotate around the patient with the leaves continually changing to match the 3-D shape of the tumor (**Figure 20.29**). IMRT works well for prostate cancer as well as for tumors of the head, neck, and other delicate areas. In the next example, we consider the magnetic field that bends the electron beam inside the IMRT.

EXAMPLE 20.9 **Magnetic field that bends the electron beam in an IMRT device**

Estimate the magnitude of the $\vec{B}$ field needed for the IMRT machine. For the estimate, assume that the electrons are moving at 2×10^8 m/s, the mass of the electrons is 9×10^{-31} kg, and the radius of the turn is 5 cm = 0.05 m.

Sketch and translate The bending process is sketched in the figure below.

B (into paper) = ?

$m \approx 9 \times 10^{-31}$ kg
$q = -e = -1.6 \times 10^{-19}$ C
$v \approx 2 \times 10^8$ m/s

⊢ r = 0.05 m ⊣

Simplify and diagram This is an estimate, and we assume that the physics principles we will use apply without modification to these electrons traveling at a significant fraction of light speed. A force diagram for the electron part of the way around the 90° curved arc is shown in the figure at right. We neglect the gravitational force that Earth exerts on the

electron. The magnetic force points in the radial direction perpendicular to the electron's velocity at each point along its path.

Represent mathematically Apply Newton's second law to the radial direction. The magnitude of the acceleration is determined by the magnitude of the sum of the forces exerted on the electron in the radial direction and by the mass of the electron:

$$m_{\text{El}}a_r = m_{\text{El}}\frac{v^2}{r} = \Sigma F_r$$

The only force with a nonzero radial component is the magnetic force. The magnitude of that force is $F_{\vec{B} \text{ on El}} = evB$, where e is the magnitude of the electron's electric charge. We get

$$m_{\text{El}}\frac{v^2}{r} = evB$$

Solve and evaluate Divide each side of the equation by ev to determine the magnitude of the magnetic field B:

$$B = \frac{m_{\text{El}}v}{er} = \frac{(9 \times 10^{-31}\text{ kg})(2 \times 10^8\text{ m/s})}{(1.6 \times 10^{-19}\text{ C})(0.05\text{ m})} = 0.023\text{ T} \approx 0.02\text{ T}$$

This is an easily attained magnetic field. Thus, there should be no difficulty bending the electron beam by 90°.

Try it yourself Suppose that this is a beam of protons instead of electrons. How would this affect the required magnetic field?

Answer   Protons are about 2000 times more massive than electrons. Thus, the magnetic field would need to be about 2000 times greater.

REVIEW QUESTION 20.6 What is the difference between the right-hand rule for the magnetic force and the right-hand rule for the $\vec{B}$ field?

20.7 Magnetic properties of materials

Materials, even metals, have widely varying magnetic properties. For example, magnets strongly attract objects made from iron, such as paper clips, but do not exert an observable magnetic force on objects made from aluminum, such as soda cans. Iron has the ability to greatly amplify the $\vec{B}$ field surrounding it. How can we explain this? Let us examine the following experiments (be sure to watch them in the video).

VIDEO
20.2

Experiment 1. You place a piece of pyrolytic carbon (PC; a synthetic material similar to graphite) on a thick piece of Styrofoam (to minimize any direct interaction between the magnet and the scale), which is then placed on a high-precision scale. You hold a strong magnet right above the PC, but not touching it, and observe that the scale reading *increases* (Figure 20.30a). When you repeat the experiment with the magnet flipped around (reversing its poles), the outcome is the same.

To explain the increase in the reading of the scale, we consider the pyrolytic carbon piece as the system and draw a force diagram for the situation when the magnet is above it (Figure 20.30b). Given that the force that the scale exerts on the PC increases in the presence of the magnet, the magnet must exert a weak repulsive (downward) force on the PC. It seems that the PC has become a magnet with its poles oppositely aligned to the magnet above it (its north pole facing the north pole of the magnet).

FIGURE 20.30 Experiment 1. Interaction of a piece of pyrolytic carbon with a magnet.

Experiment 2. You place a piece of copper sulfate (CS; a blue crystal with chemical formula $CuSO_4 \cdot 5H_2O$) on the thick piece of Styrofoam, and then place the Styrofoam on the high-precision scale (Figure 20.31a). When you hold a strong magnet right above the CS, without touching it, the scale reading *decreases*. When the experiment is repeated with the magnet flipped around (reversing its poles), the outcome is the same.

To explain the outcome of the experiment, we consider the copper sulfate crystal as the system and draw a force diagram for the situation when the magnet is above it (Figure 20.31b). The magnet must exert a weak attractive force on the CS. It seems that the CS has become a magnet with its poles in the same orientation as the magnet above it (its south pole facing the north pole of the magnet).

FIGURE 20.31 Experiment 2. Interaction of a copper sulfate crystal with a magnet.

Experiment 3. You place an iron nail on the thick piece of Styrofoam, which is then placed on the high-precision scale (Figure 20.32a). When you hold a strong magnet above the nail, the scale reading *decreases* to zero—the nail is so strongly attracted to the magnet, it flies off the scale. When the experiment is repeated with the magnet flipped around (reversing its poles), the outcome is the same.

To explain the outcome, we consider the nail as the system and draw a force diagram for the situation when the magnet is above it (Figure 20.32b). Because the nail flies off the scale, the scale does not exert a force on it anymore, but the magnet seems to exert a large attractive force on the iron nail. It seems that the iron has become a strong magnet with its poles in the same orientation as the magnet above it (its south pole facing the north pole of the magnet).

These experiments show that all of the materials have magnetic properties. Some become "magnetized" opposite to the direction of the strong external magnet and some in the same direction as the magnet. The amount of this "magnetization" varies significantly in strength.

FIGURE 20.32 Experiment 3. Interaction of an iron nail with a magnet.

It turns out that *all* materials belong to one of the three types of materials encountered in the experiments: the materials that are repelled by magnets are called **diamagnetic** (like pyrolytic carbon or water); those that are weakly attracted are called **paramagnetic** (like copper sulfate or aluminum); and those that are strongly attracted are called **ferromagnetic** (like iron). Because copper sulfate is paramagnetic, we did not observe a significant interaction between the CS and the magnet. Ferromagnetic materials also retain their magnetization to a certain degree even after the external magnetic field that magnetized them is removed. Let's look at the mechanisms behind diamagnetism, paramagnetism, and ferromagnetism. We will learn about the quantitative analysis of magnetic materials in the next chapter.

Magnetic properties of atoms

To explain these three magnetic properties of materials, we have to understand the magnetic behavior of individual atoms. The model of an atom that we use has a point-like positively charged nucleus at its center and point-like negatively charged electrons moving at high speeds in circular orbits around the nucleus. Due to this motion, each electron has a magnetic dipole moment (**Figure 20.33**) and acts like a tiny bar magnet. In addition to this electron orbital magnetic moment, the electron itself is like a tiny magnet, which also contributes to the total magnetic moment of the atom.

In atoms with more than one electron, the magnetic moments produced by the individual electrons often cancel each other. This happens because the electrons tend to pair up and orient themselves in opposite directions.

FIGURE 20.33 Electron motion in an atom effectively produces a tiny "bar magnet."

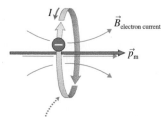

The electron motion produces a circular electric current.

Diamagnetic materials

In diamagnetic materials such as water (and therefore all living organisms), graphite, bismuth, and pyrolytic carbon, the magnetic moments of individual electrons in the atoms cancel each other, making the total field produced by the atom zero. When diamagnetic materials are placed in a region with a nonzero external $\vec{B}_{ex}$ field, the motion of the electrons in the individual atoms changes slightly. This change in electron motion causes the atoms in the diamagnetic material to become magnetic dipoles, each with their own magnetic field pointing in the direction opposite to $\vec{B}_{ex}$. The change in the net magnetic field of the atoms from zero without an external field to nonzero with an external field can only be explained using quantum physics that is beyond the scope of this book. However, we can use the knowledge of such a change to qualitatively explain diamagnetism as follows.

When we place a diamagnetic sample in a uniform external magnetic field $\vec{B}_{ex}$, the field exerts a torque on each atom's magnetic moment. However, the sum of the forces exerted on each atom in the sample is zero because the external magnetic field exerts forces that are equal in magnitude but opposite in direction on the two ends of the dipole. Thus a *uniform* magnetic field has no effect on diamagnetic materials. So why did we observe the repulsion of pyrolytic carbon by a bar magnet? The reason is that the magnetic field of a bar magnet is not uniform. When we place a diamagnetic sample in a *nonuniform* $\vec{B}_{ex}$, the net force exerted by this nonuniform field on each atom in the sample points in the direction of decreasing $\vec{B}_{ex}$ magnitude, independent of the $\vec{B}_{ex}$ direction. For a bar magnet, this net force points away from the magnet, causing the observed repulsion.

Diamagnetic materials held inside an external magnetic field slightly reduce the external magnetic field in the space occupied by the material because their magnetic dipoles create their own magnetic field in the opposite direction. In other words, they try to "expel" the external magnetic field from the material. The reduction of the field is very small, except with superconductive materials. Superconductors expel all external magnetic fields—they are like an ideal diamagnetic material. This is another special property of superconductors, in addition to their zero electrical resistance, which we discussed in Chapter 19.

Paramagnetic materials

In most atoms, the orbital magnetic moments of the electrons and the intrinsic moments of the electrons themselves usually cancel. If they don't cancel exactly, the atom will have a magnetic moment similar to that of a small bar magnet (**Figure 20.34a**). These paramagnetic materials include aluminum, sodium, and oxygen.

When a paramagnetic material is placed in a $\vec{B}_{ex}$ field, the atoms behave like tiny (but weak) compasses and tend to align with that $\vec{B}_{ex}$ field. The magnetic moments of individual atoms no longer add to zero but instead contribute a small field that adds to the $\vec{B}_{ex}$ field (Figure 20.34b). In most materials, this paramagnetic effect produces only a slight enhancement of the external $\vec{B}_{ex}$ field (about 10^{-5} times greater). The random motion of the particles causes the magnetic moments of the atoms to remain mostly randomly oriented. Therefore, as the temperature decreases, the magnetic moments become more aligned with the external field, producing a larger enhancement.

Because paramagnetic atoms in an external magnetic field possess a net magnetic moment that points in the direction of $\vec{B}_{ex}$, when we place a paramagnetic sample in a uniform external magnetic field, the field exerts a torque on the atoms. However, the sum of the forces on each atom in the sample is zero (as for diamagnetic materials). But when we place a paramagnetic sample in a nonuniform $\vec{B}_{ex}$, the net force exerted on the atoms in the sample points in the direction of *increasing* $\vec{B}_{ex}$ magnitude, independently of the $\vec{B}_{ex}$ direction. This interaction explains why both poles of a magnet attract paramagnetic materials, although very weakly.

It is important to note that diamagnetic properties are present in all materials, but in paramagnetic materials the diamagnetic effects are masked by the paramagnetic properties.

Ferromagnetic materials

Ferromagnetic materials, such as iron, nickel, and cobalt, have individual atoms with magnetic moments, just like in paramagnetic materials. However, the "magnetization" effect in an external magnetic field is thousands of times stronger in ferromagnetic materials.

Even when not in an external magnetic field, neighboring atoms in ferromagnetic materials tend to line up in small, localized regions called **domains**. Each domain may include 10^{15} to 10^{16} atoms and occupy a space less than a millimeter on a side. In a piece of iron that is unmagnetized, like a nail, the domains are oriented randomly, and the magnetic moments of the domains add to zero (**Figure 20.35a**). If an unmagnetized piece of iron is placed in a region with a nonzero $\vec{B}_{ex}$ field, the domains with magnetic moments oriented in the direction of the $\vec{B}_{ex}$ field increase in size, while those oriented in other directions decrease in size (Figure 20.35b). When the external magnetic field is removed, the magnetic moments of the domains remain aligned, and the iron now produces its own strong $\vec{B}$ field (Figure 20.35c). This is how steel nails become magnets after being placed in external magnetic fields and also how permanent magnets are created. This alignment of domains also explains why each piece of a broken magnet is still a complete magnet. Each piece has its domains aligned before splitting. Thus, each piece has its own north and south pole.

Understanding ferromagnetism helps explain how a device called an **electromagnet** works. Picture a current through a solenoid (a wire that has been shaped into a series of coils) that produces a $\vec{B}$ field that resembles the field of a bar magnet. Now, insert an iron bar into the solenoid. The $\vec{B}$ field produced by the wire causes the magnetic domains within the iron to line up. The now-magnetized iron produces its own contribution to the $\vec{B}$ field, a contribution that is up to several thousands of times stronger than the $\vec{B}$ field produced by the current. This is an electromagnet.

The permanent magnetization of ferromagnetic materials has many practical applications. Hard disk drives use them to store data. Airport metal detectors, transformers, electric motors, loudspeakers, electric generators, and permanent magnets all depend on the magnetization of ferromagnetic materials.

REVIEW QUESTION 20.7 Why is there a difference in the behavior of paramagnetic and diamagnetic materials when they are placed in a region with nonuniform $\vec{B}_{ex}$ field?

FIGURE 20.34 Paramagnetism.

(a) $\vec{B}$ field of randomly oriented atomic "magnets"

(b) In an external $\vec{B}$ field, the atoms make a slight contribution to the $\vec{B}$ field.

FIGURE 20.35 Ferromagnetism.

(a)

$\vec{B}_{ex} = 0$
Unmagnetized

(b)

Magnetized

(c)

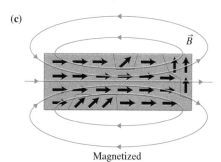

Magnetized

Summary

Magnets always have a north pole and a south pole. A magnetic field is produced by magnets or electric charges moving relative to the observer. (Section 20.1)

A magnetic field is represented by $\vec{B}$ field vectors and $\vec{B}$ field lines. The $\vec{B}$ field vector at a point is tangent to the direction of the $\vec{B}$ field line at that point. The separation of lines in a region represents the magnitude of the $\vec{B}$ field in that region—the closer the lines, the stronger the $\vec{B}$ field. (Sections 20.1–20.3, 20.5)

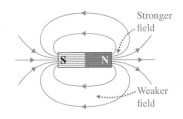

Stronger field

Weaker field

The magnitude of a magnetic field $\vec{B}$ can be found as

$$B = \frac{F_{\vec{B}\text{ on }I\text{ max}}}{IL} \qquad \text{Eq. (20.2)}$$

The fields produced by a long wire and at the center of a coil or solenoid are:

$$B_{\text{straight wire}} = \frac{\mu_0 I}{2\pi r} \qquad \text{Eq. (20.6)}$$

$$B_{\text{center of coil}} = \frac{N\mu_0 I}{2r}$$

$$B_{\text{inside solenoid}} = \frac{N\mu_0 I}{l}$$

Right-hand rule for the $\vec{B}$ field: To find the orientation of the $\vec{B}$ field produced by a current, grasp the wire with your right hand so that your thumb points in the direction of the current. Your four fingers will wrap around the wire in the direction of the $\vec{B}$ field lines. (Section 20.2)

I

$\vec{B}$ Thumb

Right hand

Right-hand rule for the magnetic force: Point the fingers of your open right hand in the direction of the magnetic field. Orient your hand so that your thumb points along the direction of motion of the electric current or charged particle. If the particle is positively charged (or is a current) the magnetic force exerted on it is in the direction your palm is facing. If the particle is negatively charged, it points in the opposite direction. (Section 20.3)

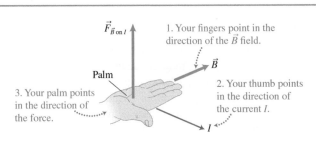

$\vec{F}_{\vec{B}\text{ on }I}$

1. Your fingers point in the direction of the $\vec{B}$ field.

Palm

$\vec{B}$

3. Your palm points in the direction of the force.

2. Your thumb points in the direction of the current I.

I

Magnitude of magnetic force depends on the object on which the magnetic field exerts a force (either a current-carrying wire or an individual moving charged particle), the magnitude of the $\vec{B}$ field vector, the relative orientation of the $\vec{B}$ field vector, and the direction of motion of charged objects. (Sections 20.3–20.5)

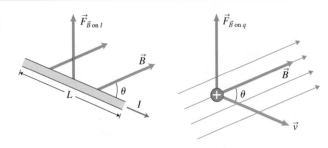

$\vec{F}_{\vec{B}\text{ on }I}$

$\vec{B}$

L

θ

I

$\vec{F}_{\vec{B}\text{ on }q}$

$\vec{B}$

θ

$\vec{v}$

$$F_{\vec{B}\text{ on }I} = ILB\sin\theta \qquad \text{Eq. (20.3)}$$

where θ is the angle between I and $\vec{B}$.

$$F_{\vec{B}\text{ on }q} = |q|vB\sin\theta \qquad \text{Eq. (20.5)}$$

where θ is the angle between $\vec{v}$ and $\vec{B}$.

Magnetic torque exerted on current-carrying coil: A magnetic field exerts forces on the current passing through the wires of a coil, resulting in a torque on the coil. (Section 20.3)

I

$\vec{p}_m$

θ

$\vec{B}$

Area A

I

N turns

$$|\tau_{\vec{B}\text{ on }C}| = NBAI\sin\theta \qquad \text{Eq. (20.4)}$$

$$|\vec{p}_m| = NIA$$

Questions

Multiple Choice Questions

1. You place a metal bar magnet on a swivel and bring a negatively charged plastic rod near the north pole and then near the south pole. What do you observe?
   (a) The north pole turns toward the rod.
   (b) The south pole turns toward the rod.
   (c) The poles do not interact with the rod.
   (d) Both poles are attracted to the rod.

2. An electron moves at constant speed from left to right in the plane of the page when it enters a magnetic field going into the page. The acceleration of the electron is
   (a) upward.   (b) downward.
   (c) in the direction of motion. (d) opposite to the direction of motion.
   (e) into the page.   (f) out of the page.

3. What is one tesla?
   (a) $1\,N/(1\,m \cdot 1\,A)$ (b) $(1\,N \cdot m)/(1\,A \cdot 1\,m^2)$
   (c) $(1\,N)/(1\,C \cdot 1\,m/s)$ (d) All of the above

4. Choose all that apply. Objects that produce magnetic fields include which of the following?
   (a) Current-carrying wires
   (b) Permanent magnets
   (c) A compass
   (d) A glass rod rubbed with silk sitting on the table observed by a person standing on the floor
   (e) A glass rod rubbed with silk when placed on a moving truck and observed by a person standing on the ground
   (f) Current in a solenoid

5. What is one difference between magnetic and electric field lines?
   (a) Magnetic lines start on the poles and electric field lines start on electric charges.
   (b) Magnetic lines do not start or end anywhere, whereas electric field lines do have a beginning and end.
   (c) Magnetic field lines are shorter than electric field lines.

6. Two parallel straight current-carrying wires are lying on a table, 12 cm apart. The total magnetic field produced by the currents is zero at a distance of 3 cm from the left wire. Which of the following statements are correct? Select all that apply.
   (a) The current in the left wire is larger than the current in the right wire.
   (b) The current in the left wire is smaller than the current in the right wire.
   (c) The currents are in the same direction.
   (d) The currents are in opposite directions.
   (e) The currents can be either in the same or opposite directions.

7. Choose all of the units that are fundamental, rather than derived, SI units.
   (a) Coulomb   (b) Volt
   (c) Ampere   (d) Newton
   (e) Meter   (f) Second
   (g) Gram   (h) Kilogram

8. Particles of various masses, charges, and speeds are injected into a region in which a uniform $\vec{E}$ field and a uniform $\vec{B}$ field are perpendicular to each other. All the particles are initially moving in the same direction. Which two conditions must be simultaneously fulfilled for the particles to continue moving in a straight line after entering the region?
   (a) The particles must have the same mass.
   (b) The particles must have the same charge.
   (c) The particles must have the same speed.
   (d) The particles must have the same kinetic energy.
   (e) The particles must have the same momentum.
   (f) The particles' velocities must be parallel to the $\vec{E}$ field.
   (g) The particles' velocities must be parallel to the $\vec{B}$ field.
   (h) The particles' velocities must be perpendicular to the $\vec{E}$ field and the $\vec{B}$ field.
   (i) The particles' velocities must be perpendicular to the $\vec{B}$ field.
   (j) The particles' velocities must be perpendicular to the $\vec{E}$ field.

9. When a diamagnetic material is placed in an external $\vec{B}$ field, the force exerted on the material by the magnetic field always points
   (a) in the $\vec{B}$ field direction.
   (b) in the direction opposite to the $\vec{B}$ field direction.
   (c) toward the region where the magnitude of the $\vec{B}$ field is greater.
   (d) toward the region where the magnitude of the $\vec{B}$ field is smaller.

Conceptual Questions

10. If you triple the speed of a particle entering a magnetic field, what happens to the radius of the helix that it makes? Select the correct answer, then describe the possible reasoning that could lead a student to choose each of the other answers. You do not need to agree with that reasoning.
    (a) It triples.
    (b) It becomes one-third as big.
    (c) It increases by 9 times.
    (d) It will not change if the magnetic field does not change.

11. In 1911 physicists measured a magnetic field around a beam of electrons. Draw $\vec{B}$ field lines for this field.

12. Describe two experiments that will allow you to show that electric currents create magnetic fields.

13. How can you determine if there is a magnetic field in a certain region?

14. You have a magnet on which the poles are not marked. How can you determine which pole is north and which is south if in addition to the magnet you have (a) another magnet only, (b) a current-carrying coil only, or (c) neither a magnet nor a coil, but you are familiar with the geographical data of your location? Describe what you will do so a reader can repeat the experiment and get the same results.

15. List as many ways as you can to detect a magnetic field in a particular region. Explain how they work and how you can use them to determine the magnitude and the direction of the $\vec{B}$ field.

16. List as many ways as you can to produce a magnetic field.

17. A current-carrying wire is placed in a magnetic field, as shown in Figure Q20.17. In which direction does the magnetic field exert a force on the wire?

FIGURE Q20.17

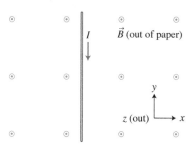

18. An electron flies through the magnetic field shown in Figure Q20.18. In which direction does the magnetic field exert a force on it? What effect does the field have on the magnitude of the velocity of the electron? Describe the path of the electron. Describe what would be different if a proton flew in the same direction through the same field.

FIGURE Q20.18   FIGURE Q20.19

19. Find the direction of the magnetic field whose effect on a charged object is as shown in Figure Q20.19.

20. A beam of electrons is not deflected as it moves through a region of space in which a magnetic field exists. Give two explanations why the electrons might move along a straight path.

21. A beam of electrons moving toward the east is deflected upward by a magnetic field. Determine the direction in which the magnetic field $\vec{B}$ points. Repeat for the case with an $\vec{E}$ field instead of a $\vec{B}$ field. Describe the effect of each field on the speed of the electron.

22. Why are residents of northern Canada less shielded from cosmic rays than are residents of Mexico?

23. A U-shaped wire with a current in it hangs with the bottom of the U between the poles of an electromagnet (see Figure Q20.23). When the magnitude of the $\vec{B}$ field in the magnet is increased, does the U swing to the right or to the left, or does it remain at rest? Explain.

FIGURE Q20.23

24. An electron enters a solenoid at a small angle relative to the magnetic field inside. Describe the electron's motion.

25. Two parallel wires carry electric current in the same direction. Does the moving charge in one wire cause a magnetic force to be exerted on the moving charge in the other wire? If so, in what direction is the force relative to the wires? Explain. Repeat for currents moving in opposite directions.

26. Why is the magnetic field so much greater at the end of a solenoid or long coil that has an iron core than at the end of a similar solenoid or long coil with the same current through it but with an air core?

27. Describe a situation in which an electron will be affected by an external electric field but will not be affected by an external magnetic field. Is it possible that an electron is affected by an external magnetic field but not by an external electric field? Explain.

Problems

Below, **BIO** indicates a problem with a biological or medical focus. Problems labeled **EST** ask you to estimate the answer to a quantitative problem rather than derive a specific answer. Asterisks indicate the level of difficulty of the problem. Problems with no * are considered to be the least difficult. A single * marks moderately difficult problems. Two ** indicate more difficult problems.

20.1 and 20.2 Magnetic interactions and Magnetic field

1. When a switch is closed, a compass needle deflects from the initial to final direction, as shown in Figure P20.1. Say everything you can about this circuit.

FIGURE P20.1

Compass under wire

2. You have a lightbulb connected to a battery. (a) What happens if a compass is placed under the constant current cable, made of two twisty wires that connect the bulb with the battery? (b) What happens if you separate the wires and place the compass under one of the separated wires? Explain your answers for both parts.

3. The current through a circuit is shown in Figure P20.3. The deflection of a compass needle is shown in the figure. Is the picture correct? If not, what is wrong?

FIGURE P20.3

Compass above wire

4. Draw $\vec{B}$ field lines for the magnetic field produced by the objects shown in Figure P20.4.

FIGURE P20.4

(a)   (b)   (c)

5. * You need to determine the direction of the $\vec{B}$ field at two points in space and compare the magnitudes of the $\vec{B}$ field at those two points. Describe the experiments that will allow you to accomplish this task.

6. * Two compass needles are fixed at the ends of a straw (the north pole of one needle is placed above the south pole of the other needle) and are hung at the end of a light string as shown in Figure P20.6. The needles are brought close to a long horizontal wire connected to a circuit, as shown in the figure. (a) How will the needles point after the switch is closed? Explain. (b) Your friend claims that Earth's magnetic field does not affect the two-needle device. Do you agree or disagree? Explain your reasoning.

FIGURE P20.6

20.3 Magnetic force exerted by the magnetic field on a current-carrying wire

7. * In Houston, Earth's $\vec{B}$ field has a magnitude of 5.2×10^{-5} T and points in a direction 57° below a horizontal line pointing north. Determine the magnitude and direction of the magnetic force exerted by the magnetic field on a 10-m-long vertical wire carrying a 12-A current straight upward.

8. * A 15-g 10-cm-long wire is suspended horizontally between the poles of a horseshoe magnet. When the 0.50-A current in the wire is turned on, the wire jumps up and out of the magnet. What can you learn about the magnet using this information?

9. ** A metal rod of length l and mass m is suspended from two light wires that are connected to a battery with emf $\mathcal{E}$. When a vertical magnetic field $\vec{B}$ pointing up is turned on, the supporting wires make an angle θ with the vertical. What can you learn about the circuit using this information? Derive a mathematical expression that relates the magnitude of the magnetic field $\vec{B}$ to other relevant quantities.

10. * A metal rod is connected to a battery through two stiff metal wires that hold the rod horizontally. The rod is between the poles of a horseshoe magnet that is sitting on a mass-measuring platform scale, which reads 100 g. Draw the magnetic poles of the magnet and the battery connected to the metal rod so that when you turn the current in the circuit on (a) the reading of the scale supporting the magnet increases; (b) the reading decreases.

11. * After you turned on the current in the circuit described in the previous problem, the scale supporting the magnet read 106 g. The part of the rod that is exposed to the magnetic field is 7.0 cm long and the current through it is 1.0 A. What can you learn about the magnet using this information?

12. * **Equation Jeopardy** Describe a problem for which the following equation is a solution:

$$0.70 \text{ A} = \frac{3.0 \text{ N}}{(0.20 \text{ m})B}$$

13. ** A square coil with 30 turns has sides that are 16 cm long. When it is placed in a 0.30-T magnetic field, a maximum torque of 0.60 N·m is exerted on the coil. What can you learn about the coil from this information?

14. A 5.0-A current runs through a 0.12 m²-area loop. The loop is in a 0.15-T $\vec{B}$ field. Determine the torque exerted on the loop (the magnitude and the direction it tends to turn the coil if initially at rest) for each orientation shown in **Figure P20.14**.

FIGURE P20.14

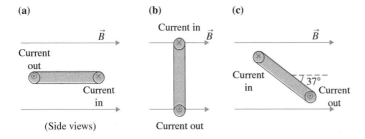

(a)   (b)   (c)

(Side views)   Current out

15. * (a) Determine the magnetic force (magnitude and direction) that a 0.15-T magnetic field exerts on segments a, b, and c of the wire shown in **Figure P20.15**. Segments a and c are 2.0 cm long, and segment b is 10.0 cm long. A 3.0-A current flows through the wire. (b) Determine the net torque caused by forces $\vec{F}_{\vec{B} \text{ on a}}$ and $\vec{F}_{\vec{B} \text{ on c}}$ and the torque caused by force $\vec{F}_{\vec{B} \text{ on b}}$.

FIGURE P20.15

16. * A 500-turn square coil of wire is hinged to the top of a table, as shown in **Figure P20.16**. Each side of the coil has a length of 0.50 m. (a) In which direction should a magnetic field point to help lift the free end of the coil off the table? (b) Determine the torque caused by a 0.70-T field pointing in the direction described in part (a) when there is a 0.80-A current through the wire and when the coil is parallel to the table.

FIGURE P20.16

17. * **Electric motor 1** An electric motor has a square armature with 500 turns. Each side of the coil is 12 cm long and carries a current of 4.0 A. The $\vec{B}$ field inside the motor is uniform with a magnitude of 0.60 T. Choose two orientations of the coil for which you can calculate force-related and torque-related quantities. Draw pictures to show the relative orientations of the coil and the magnetic field vectors.

18. ** You make a seesaw by placing a 50-g magnet (whose poles' faces are 2-cm-by-2-cm squares) at one end of a 50-cm-long ruler and a small 50-g metal object at the other end and balancing the whole setup by placing a pencil directly under the middle of the ruler. You know that the magnet is horizontally magnetized, but you don't know which face is north and which face is south. When you place a horizontal wire close to the pole of the magnet and send a 14-A current through the wire in the direction shown in **Figure P20.18**, the end of the seesaw with the magnet goes up and the other end goes down. You can get the seesaw to go back to equilibrium (while keeping the current in the wire) by placing another 10-g metal object 6.0 cm from the axis of rotation, also shown in the figure. (a) Determine the polarity of the magnet and (b) the magnitude of the $\vec{B}$ field at the location of the wire. Indicate any assumptions that you made.

FIGURE P20.18

Magnet

19. * **Electric motor 2** An electric motor has a circular armature with 250 turns of radius 5.0 cm through which an 8.0-A current flows. The maximum torque exerted on the armature as it turns in a magnetic field is 1.9 N·m. Show the orientation of the armature coil in the magnetic field when this torque is exerted on it and determine the magnitude of the field.

20.4 Magnetic force exerted on a single moving charged particle

20. Each of the lettered dots a–d shown in **Figure P20.20** represents a $+1.0 \times 10^{-18}$-C charged particle moving at speed 2.0×10^7 m/s. A uniform 0.50-T magnetic field points in the positive z-direction. Determine the magnitude of the magnetic force that the field exerts on each particle and indicate carefully, in a drawing, the direction of the force.

FIGURE P20.20

21. **Duck gets a lift** A duck accumulates a positive charge of 3.0×10^{-8} C while flying north at speed 18 m/s. Earth's magnetic field at the duck's location has a magnitude of 5.3×10^{-5} T and points in a direction 62° below a horizontal line pointing north. Determine the magnitude and direction of the force exerted by Earth's magnetic field on the duck.

22. An electron of mass 9.1×10^{-31} kg moves horizontally toward the north at 3.0×10^7 m/s. Determine the magnitude and direction of a $\vec{B}$ field that will exert a magnetic force that balances the gravitational force that Earth exerts on the electron.

23. A 1000-kg car moves west along the equator. At this location Earth's magnetic field is 3.5×10^{-5} T and points north parallel to Earth's surface. If the car carries a charge of -2.0×10^{-3} C, how fast must it move so that the magnetic force balances 0.010% of Earth's gravitational force exerted on the car?

24. * **BIO** **Magnetic force exerted by Earth on ions in the body** A hydroxide ion (OH⁻) in a glass of water has an average speed of about 600 m/s. (a) Determine the magnitude of the electrical force between the hydroxide ion (charge −e) and a positive ion (charge +e) that is 1.0×10^{-8} m away (about the separation of 30 atoms). (b) Determine the magnitude of the maximum magnetic force that Earth's 3.0×10^{-5}-T magnetic field can exert on the ion. (c) On the basis of these two calculations, does it seem likely that Earth's magnetic field has much effect on the biochemistry of the body?

25. ** **EST** **Design a magnetic shield** An alien from a planet in the galaxy M31 (Andromeda) has a ray gun that shoots protons at a speed of 2.0×10^5 m/s. Design a magnetic shield that will deflect the protons away from your body. Using rough estimates, show that your magnetic shield will, in fact, protect you. Indicate the orientation of your protective device's magnetic field relative to the direction of the proton beam and the direction in which the ions are deflected.

26. ** **EST** An electron beam moves toward a cathode ray tube screen, which is 30 cm away from the negative electrode. The electrons are accelerated by a potential difference of 14 kV. Estimate the maximum displacement of the electron beam caused by Earth's magnetic field.

27. * An electron and a proton, moving side by side at the same speed, enter a 0.020-T magnetic field. The electron moves in a circular path of radius 7.0 mm. Describe the motion of the proton qualitatively and quantitatively.

20.5 Magnetic fields produced by electric currents

28. An east-west electric power line carries a 500-A current toward the east. The line is 10 m above Earth's surface. Determine the magnitude and direction of the magnetic field at Earth's surface directly under the wire. What assumptions did you make?

29. * **Pigeons** A solenoid of radius 1.0 m with 750 turns and a length of 5.0 m surrounds a pigeon cage. What current must be in the solenoid so that the solenoid field just cancels Earth's 4.2×10^{-5}-T magnetic field (used occasionally by the pigeons to determine the direction they travel)?

30. * A horizontal current-carrying wire that is perpendicular to the plane of the paper is placed in a uniform magnetic field of magnitude 2 mT that is parallel to the plane of the paper and points down. At the distance 1.0 cm to the right of the wire, the total magnetic field is zero. (a) What physical quantity can you determine using this information? (b) How will this quantity change if the zero field point is 0.4 cm to the left of the wire?

31. * A coil of radius r is made of N circular loops. It is connected to a battery of known emf and internal resistance. The coil produces a magnetic field whose magnitude at the center of the coil is $\vec{B}$. Make a list of physical quantities you can determine using this information and show how to determine two of them.

32. * **Magnetic field sensor on your mobile phone** You place a long wire next to your mobile phone on a table as shown in Figure P20.32. You run an application that allows you to measure the time dependence of the component of the $\vec{B}$ field that is perpendicular to the screen of the phone (let's call it B_z). While recording the magnetic field, you repeat the following steps: connect the wire to an AA battery, disconnect the battery, flip the battery (to swap + and − terminals), again connect the wire to the battery … and so on. Using an ammeter, you also determine the current through the wire. Average values of your measurements are summarized in the table below.

FIGURE P20.32

Wire

Connection	I (A)	B_z (T)
Battery $(+, -)$	4.9	122×10^{-6}
No battery	0	52×10^{-6}
Battery $(-, +)$	4.9	-19×10^{-6}

(a) Estimate the distance between the magnetic field sensor in your phone and the wire. (b) If the experiment was performed in New Jersey, determine the direction of the current in the wire (up or down) when the magnetic field reading is -19×10^{-6} T. Explain your reasoning. Indicate any assumptions that you made.

20.6 Skills for analyzing magnetic processes

33. * **Equation Jeopardy 1** The equation below describes a process involving magnetism. Solve for the unknown quantity and draw a sketch that represents a possible process described by the equation.

$$(1.6 \times 10^{-19}\,C)(2.0 \times 10^7\,m/s)(3.0 \times 10^{-5}\,T)$$
$$= (1.7 \times 10^{-27}\,kg)(2.0 \times 10^7\,m/s)^2/r$$

34. * **Equation Jeopardy 2** The equation below describes a process involving magnetism. Solve for the unknown quantity and draw a sketch that represents a possible process described by the equation. Note: T is a symbol that denotes a force (for example, force exerted by a rope, tension), and T is the unit (Tesla).

$$2T - (0.020\,kg)(9.8\,N/kg) + (10\,A)(0.10\,m)(0.10\,T) = 0$$

35. * **Equation Jeopardy 3** The equation below describes a process involving magnetism. Solve for the unknown quantity and draw a sketch that represents a possible process described by the equation.

$$100(2.0\,A)(4.0 \times 10^{-2}\,m^2)(0.20\,T) - m(9.8\,N/kg)(0.10\,m) = 0$$

36. ** **EST** The magnitude of the $\vec{B}$ field inside a long solenoid is given by the equation $B = \mu_0 I\,(N/l)$, where N is the number of turns and l is the length of the solenoid. (a) Describe an experiment that can help you test this relation. (b) Explain whether this equation is an operational definition of the magnitude of the $\vec{B}$ field or a cause-effect relationship. (c) Powerful industrial solenoids produce $\vec{B}$ field magnitudes of about 30 T. Estimate the relevant physical quantities for such solenoids.

37. * **Electron current and magnetic field in H atom** In a simplified model of the hydrogen atom, an electron moves with a speed of 1.09×10^6 m/s in a circular orbit with a radius of 2.12×10^{-10} m. (a) Determine the time interval for one trip around the circle. (b) Determine the current corresponding to the electron's motion. (c) Determine the magnetic field at the center of the circular orbit and the magnetic moment of the atom.

38. * Two long, parallel wires are separated by 2.0 m. Each wire has a 30-A current, but the currents are in opposite directions. (a) Determine the magnitude of the net magnetic field midway between the wires. (b) Determine the net magnetic field at a point 1.0 m to the side of one wire and 3.0 m from the other wire.

39. * **Minesweepers** During World War II, explosive mines were dropped by the Nazis in the harbors of England. The mines, which lay at the bottom of the harbors, were activated by the changing magnetic field that occurred when a large metal ship passed above them. Small English boats called minesweepers would tow long, current-carrying coils of wire around the harbors. The field created by the coils activated the mines, causing them to explode under the coils rather than under ships. (a) Determine the current in one long, straight wire to create a 0.0050-T magnetic field at a depth of 20 m under the water. The magnetic permeability of water is about the same as that of air. (b) How might the field be created using a smaller current?

40. An electron moves at the speed of 8.0×10^6 m/s toward the velocity selector shown in Figure 20.27a. A 0.12-T $\vec{B}$ field points into the paper. (a) Determine the magnitude and direction of the magnetic force that the magnetic field exerts on the electron. (b) What $\vec{E}$ field magnitude and direction are required so that the electric force exerted on the electron is equal in magnitude and opposite in direction to the magnetic force?

41. * **Mass spectrometer** A mass spectrometer has a velocity selector that allows ions traveling at only one speed to pass with no deflection through slits at the ends. While moving through the velocity selector, the ions pass through a 60,000-N/C $\vec{E}$ field and a 0.0500-T $\vec{B}$ field. The quantities $\vec{v}$, $\vec{E}$, and $\vec{B}$ are mutually perpendicular. (a) Determine the speed of the ions that are not deflected. (b) After leaving the velocity selector, the ions continue to move in the 0.0500-T magnetic field. Determine the radius of curvature of a singly charged lithium ion, whose mass is 1.16×10^{-26} kg.

42. ** **Mass spectrometer 2** One type of mass spectrometer accelerates ions of charge q, mass m, and initial speed zero through a potential difference ΔV. The ions then enter a magnetic field where they move in a circular path of radius r. How is the mass of the ions related to these other quantities?

43. * An ion with charge 1.6×10^{-19} C moves at speed 1.0×10^6 m/s into and perpendicular to the 0.30-T magnetic field of a mass spectrometer. After entering the field, the ion moves in a circular path of radius 0.31 m. What physical quantities describing the field and the ion can you determine using this information? Determine them.

General Problems

44. * A box has either an electric field or a magnetic field inside. Describe experiments that you might perform to determine which field is present and its orientation. You are allowed to make small holes in the box.

45. ** A piece of wire, shown in Figure P20.45, moves downward perpendicular to a magnetic field. (a) In what direction will the electrons in the wire move? (b) After the electrons are forced in one direction, leaving positive charges behind, an $\vec{E}$ field and potential difference ΔV develop from one end of the wire to the other. The electrons no longer move, since the electric and magnetic forces exerted on them balance. Show that this happens when the potential difference from one end of the wire to the other is $\Delta V = vLB$, where v is the speed of the wire, L is its length, and B is the magnitude of the magnetic field. Describe two real-world situations where this phenomenon might occur.

FIGURE P20.45

46. ** [EST] Particles in cosmic rays are mostly protons, which have energies up to about 10^{20} eV, where 1 eV is the change in the kinetic energy of an electron moving through a potential difference of 1 V. Use this information to estimate the path of these particles in Earth's atmosphere. State the assumptions that you are making in your estimate.

Reading Passage Problems

[BIO] **Magnetic resonance imaging** In magnetic resonance imaging (MRI), a patient lies in a strong uniform constant magnetic field $\vec{B}$ oriented parallel to the body. Typical magnitudes of this field are between 1.0 T and 3.0 T. This field is produced by a large superconducting solenoid. The MRI measurements depend on the magnetic dipole moment $\vec{\mu}$ of a proton, the nucleus of a hydrogen atom. The proton magnetic dipoles can have only two orientations: either with the field or against the field. The energy needed to reverse this orientation ("flip" the protons) from with the field to against the field is exactly $\Delta U = 2\mu B$ (like the energy needed to turn a compass needle from north to south).

The pulse of an alternating magnetic field of frequency several tens of MHz irradiates the patient's body in the region to be imaged. When this alternating field is tuned correctly so that its energy equals the $\Delta U = 2\mu B$ needed to reverse the orientation of the protons from with the external $\vec{B}$ field to against it, a reasonable number of protons will flip. When the protons return to their initial orientation and a lower energy state, they produce an alternating magnetic field with almost exactly the same frequency as that of the original radiation. This alternating magnetic field produced by the protons is detected and provides information about the spatial distribution of the concentration of protons as well as information about the structure of the material surrounding the protons in the region irradiated by the alternating magnetic field. The proton concentration as well as the protons' characteristic interactions with their surroundings differ in fat, muscle, and bone tissue, and in healthy and diseased tissue.

The MRI image of an internal body part is made by adjusting an auxiliary magnetic field, which varies the external B field over the region being examined so that the probe field energy equals the flipping energy $\Delta U = 2\mu B$ in only a thin slice of the body. A measurement is made at this slice. The external magnetic field is then adjusted to flip protons in a neighboring slice. Continual shifts in the magnetic field and detection of proton concentrations at different slices produce a map of proton concentration in the body. The MRI image of the lower back in Figure 20.36 indicates an L45 disc that has partially collapsed— it has lost water, and because it contains fewer protons, it produces a darker MRI image.

FIGURE 20.36 An MRI image of a patient's lower back.

47. The uniform magnetic field in an MRI apparatus is greater than Earth's 5.0×10^{-5}-T magnetic field by about how many times?
(a) 40
(b) 400
(c) 4000
(d) 40,000
(e) 400,000

48. The energy of the probe field causes protons to flip when in a 1.50000-T magnetic field. A ± 0.001-T variation in the magnetic field causes a mismatch with the radio frequency flipping field and no flipping at a distance of 0.002 m from where the magnetic field is matched to the flipping field. Which answer below is closest to the change in the B field per unit distance?
(a) 0.08 T/m
(b) 0.8 T/m
(c) 8 T/m
(d) 0.003 T·m
(e) 0.000003 T/m

49. Using your answer to Problem 20.48, determine the quantity closest to the amount that the auxiliary magnetic field causes the magnetic field to vary over a 0.20-m region of the body being mapped by MRI.
(a) 4 T
(b) 2 T
(c) 0.002 T
(d) 0.02 T
(e) 0.2 T

50. The MRI apparatus is able to look at proton concentration (and hence hydrogen concentration) at one tiny part of the body by doing what?
(a) Aiming the probe field at the whole body
(b) Varying the probe field frequency so that only protons in one place are flipped
(c) Varying the B field over the body so $\Delta U = 2\mu B$ matches the probe field energy at one small location
(d) Placing a small hole in a body shield so the probe field reaches only one part of body
(e) All of the above

51. Which answer below is closest to the energy of the probe field needed to flip protons? The magnetic dipole moment of a proton is 1.41×10^{-26} J/T.
(a) 4 J
(b) 4×10^{-10} J
(c) 4×10^{-25} J
(d) 4×10^{-26} J
(e) 4×10^{-27} J

52. Why might a herniated disc projecting slightly out from between two vertebrae look different in an MRI image than a nonherniated disc?
(a) The vertebrae adjacent to a herniated disc are closer than vertebrae beside a nonherniated disc.
(b) There is a different concentration of hydrogen atoms in bone and in discs.
(c) Protons in the herniation produce an image that can be seen.
(d) b and c
(e) a, b, and c

[BIO] **Power lines—do their magnetic fields pose a risk?** Power lines produce both electric and magnetic fields. The interior of the human body is an electrical conductor, and as a result, electric fields are greatly reduced in magnitude within the body. The electric fields inside the body from power lines are much smaller than electric fields normally existing in the body.

However, magnetic fields are not reduced in the body. Earth's magnetic field, approximately 5×10^{-5} T, is very small and not regarded as a health threat. Thus, it is interesting to compare Earth's magnetic field to fields produced by high power lines. The magnetic field B produced at a distance r from a straight wire with an electric current I is

$$B = (2 \times 10^{-7} \text{ T·m/A})I/r$$

The magnetic field from a high-voltage power line located 40 m above the ground carrying a 100-A current is much smaller than Earth's $\vec{B}$ field.

Wires that provide electric power for household appliances also produce electric and magnetic fields. The current in the wire for a 500-W space heater is about 5 A. With the wire located several meters from your body, the magnetic field of such an appliance is somewhat smaller than Earth's magnetic field. By comparison, laboratory mice lived for several generations in 0.0010-T magnetic fields (20 times Earth's magnetic field) without any adverse effects.

During the last three decades, electric power use has increased the magnitudes of the $\vec{B}$ field created by power lines to which Americans are exposed by roughly a factor of 20. Yet during that same time interval, leukemia rates have slowly declined. It seems unlikely that magnetic fields produced by home appliances or high-voltage power lines are a hazard for causing leukemia.

53. Which answer below is closest to the ratio of the power line $\vec{B}$ field on the ground 50 m below a 100-A current and Earth's $\vec{B}$ field at the same location?
    (a) 0.001
    (b) 0.01
    (c) 0.1
    (d) 10
    (e) 100

54. A 550-W toaster oven is connected to a 110-V wall outlet. Which answer below is closest to the electric current in one of the wires from the wall outlet to the oven?
    (a) 0.2 A
    (b) 50,000 A
    (c) 1 A
    (d) 5 A
    (e) 20 A

55. Which answer below is closest to the ratio of the magnetic field produced 0.4 m from the cable for the toaster oven in the last problem and Earth's magnetic field? (This is assuming the current flows in only one direction—which it does not.)
    (a) 0.001
    (b) 0.003
    (c) 0.05
    (d) 0.5
    (e) 5

56. Leukemia rates have declined in recent years, whereas the magnitudes of the $\vec{B}$ fields created by power lines in our environment have increased significantly. Why does this not necessarily rule out power line magnetic fields as a contributing cause of leukemia?
    (a) Correlation studies have no cause-effect relationship.
    (b) There is no known mechanism for power line magnetic field-induced deaths.
    (c) The power line magnetic fields cannot penetrate clothing and skin.
    (d) Perhaps the power line magnetic field-induced cancers have increased from 0.001 to 0.025 of the cases and other causes have decreased.
    (e) a, b, and d
    (f) All of the above

57. Why would scientists be more concerned about the relationship between magnetic fields and human health than about the relationship between electric fields and human health?
    (a) The human body is a conductor.
    (b) Many molecules are dipoles.
    (c) There are moving electrically charged particles inside the body.
    (d) Magnetic fields, if dangerous, are more likely to penetrate into the body.
    (e) Electric fields are reflected from clothing but magnetic fields are not.

Electromagnetic Induction

Because human cells, including nerve cells, contain ions that can move, they are electrically conductive. Transcranial magnetic stimulation (TMS) is a noninvasive technology that allows doctors to electrically stimulate the brain, which helps treat certain diseases such as mood disorders, Parkinson's disease, and Huntington's disease. Given that the skull is a fairly good electrical insulator, how is it possible that changing current moving through a coil placed on the patient's scalp can produce a current in the brain without any direct electrical connection?

- How might electrical stimulation of the brain help treat certain diseases?
- How does a microphone work?
- We put metal cans and glass jars into the same recycling bin. How are they separated at the dump?

IN THE LAST CHAPTER, we learned that an electric current through a wire produces a magnetic field. Could the reverse happen? Could a magnetic field produce a current? It took scientists many years to answer this question. In this chapter, we will investigate the conditions under which this can happen.

BE SURE YOU KNOW HOW TO:
- Find the direction of the $\vec{B}$ field produced by an electric current (Section 20.2).
- Find the direction of the magnetic force exerted on moving electric charges (Sections 20.3 and 20.4).
- Explain how an electric field produces a current in a wire and how that current relates to the resistance of the wire (Sections 19.1 and 19.4).

21.1 Inducing an electric current

In the chapter on circuits (Chapter 19), we learned that an electric current results when a battery or some other device produces an electric field in a wire. The field in turn exerts an electric force on the free electrons in the wire connected to the battery. As a result, the electrons move in a coordinated manner around the circuit—an electric current.

In this section we will learn how to produce a current in a circuit without a battery—a process called *inducing* a current. We start by analyzing some simple experiments in Observational Experiment Table 21.1. See if you can find any patterns in the outcomes.

OBSERVATIONAL
EXPERIMENT TABLE 21.1   **Inducing an electric current using a magnet**

VIDEO
OET 21.1

Observational experiment	Analysis
Experiment 1. You hold a bar magnet motionless in front of a coil. The coil is not connected to a battery but is connected to a galvanometer that detects both the presence of a current and its direction. A galvanometer works like an ammeter, but it is usually not calibrated in specific units.	The galvanometer reads zero. There is no current through the coil.
Experiment 2. You move the magnet toward the coil (or you could move the coil toward the magnet).	While the magnet is moving, the galvanometer needle moves to the right, indicating a current through the coil.
Experiment 3. You move the magnet away from the coil.	While the magnet is moving, the galvanometer needle moves to the left, indicating a current through the coil but opposite the direction in Experiment 2.
Experiment 4. You turn the magnet 90° so that the poles are now perpendicular to their previous position.	While the magnet is turning, the galvanometer registers a current.
Experiment 5. You push the sides of the coil together so its opening becomes very small. You pull open the sides of the coil so the opening becomes large again.	In both cases, while the coil's area is changing, the galvanometer registers a current, but the direction is different in each case.

Patterns
Although no battery was used, an electric current was induced in the coil when the magnet moved toward or away from the coil. Current was also induced when the coil's orientation relative to the magnet or the area of the coil changed.

In Table 21.1 there was no battery, yet the galvanometer registered an electric current through a coil. For the current to exist, there must be some source of emf. What produced the emf in these experiments?

Recall from our study of magnetism (in Chapter 20) that a magnetic field can exert a force on moving electrically charged particles. The force exists only if the magnetic field or a component of the field is perpendicular to the velocity of the electrically charged particles. Let's consider again Experiment 2 in Table 21.1 in which the magnet moves toward the coil; for simplicity, we use a square loop made of conducting wire (**Figure 21.1**). Inside the wire there are positively charged ions that make up the lattice of the metal (and cannot leave their locations) and negatively charged free electrons. The $\vec{B}$ field produced by the bar magnet points away from the magnet and spreads out as shown in the figure.

Notice that at the top and bottom sections of the loop, a component of the $\vec{B}$ field is perpendicular to the velocity of the loop as it moves toward the right. As a result, the magnetic field exerts a force on each electron in the wire. This force causes the electrons in the wire to accelerate clockwise around the loop as viewed from the magnet (use the right-hand rule for the magnetic force on the negatively charged electrons, as discussed in Section 20.4). Note also that electrons in the vertical section of the loop closest to us accelerate upward, while the electrons in the vertical section farthest from us accelerate downward. The overall result is that due to the relative motion between the loop and the bar magnet, the electrons start moving around the loop in a coordinated fashion—we have an induced electric current.

Thus, it seems that the currents produced in Table 21.1 can be explained using the concept of the magnetic force we have previously developed (in Chapter 20). However, the magnetic force approach does not explain how transcranial magnetic simulation (TMS) works. The magnetic force-based explanation requires motion of the loop relative to the magnetic field, but in the TMS procedure, the coil is not moving relative to the brain.

Perhaps there is another explanation. Let's examine Table 21.1 from a different perspective, one that focuses on the $\vec{B}$ field itself rather than on any sort of motion. When the magnet moved or rotated with respect to the magnetic field, the number of $\vec{B}$ field lines going through the area of the coil increased or decreased (**Figure 21.2a**). The number of $\vec{B}$ field lines through the coil also changed when the area of the coil changed (Figure 21.2b). Thus, an alternative explanation for the pattern we observed in the experiments is that when the number of $\vec{B}$ field lines through the coil's area changes, there is a corresponding emf produced in the coil, which leads to the induced current.

FIGURE 21.1 Magnetic field exerts a force on the electrons in the moving loop.

The component of the magnetic field that is perpendicular to the velocity exerts a magnetic force on the electrons.

The free electrons in the loop move clockwise as seen from the magnet.

FIGURE 21.2 Changing $\vec{B}$ or A causes an induced current.

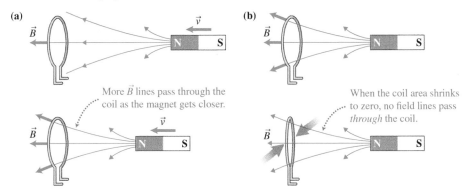

(a)

More $\vec{B}$ lines pass through the coil as the magnet gets closer.

(b)

When the coil area shrinks to zero, no field lines pass *through* the coil.

FIGURE 21.3 A current-carrying loop and a bar magnet create magnetic fields that have the same distribution of $\vec{B}$ field lines.

(a)

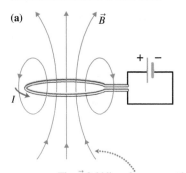

The $\vec{B}$ field lines for a current loop are about the same as those for a bar magnet.

(b)

We can summarize these two explanations for the induced current as follows:

Explanation 1: The induced current is caused by the magnetic force exerted on moving electrically charged objects (for example, the free electrons in conducting wires that are moving relative to the magnet).

Explanation 2: Any process that changes the number of $\vec{B}$ field lines through a coil's area induces a current in the coil. The mechanism explaining how it happens is unclear at this point.

Let's test these explanations with an experiment involving a change in the number of $\vec{B}$ field lines through a coil's area, but with no relative motion. Recall that we can create a magnetic field using either a wire that carries current or a permanent magnet (Chapter 20). The current-carrying coil in **Figure 21.3a** has a $\vec{B}$ field that resembles the $\vec{B}$ field of the bar magnet in Figure 21.3b.

Testing Experiment **Table 21.2** uses two coils, such as those shown in **Figure 21.4**. Coil 1 on the bottom is connected to a battery and has a switch to turn the current on and off. When the switch is open, there is no current in coil 1. When the switch is closed, the current in coil 1 produces a magnetic field whose $\vec{B}$ field lines pass through coil 2's area. For each of the experiments we will use the two explanations to predict whether or not there should be an induced electric current in coil 2.

FIGURE 21.4 An arrangement of coils for the testing experiment.

TESTING EXPERIMENT TABLE 21.2 Testing the explanations for induced current ▶ VIDEO TET 21.2

| Testing experiment | Will a current be induced in coil 2? | | Outcome |
	Prediction based on Explanation 1	Prediction based on Explanation 2	
Experiment 1. The switch in the circuit for coil 1 is open, so there is no current in coil 1. Is there any current in coil 2?	There is no current in coil 1, thus there is no magnetic field at coil 2. Neither coil is moving. No current will be induced in coil 2.	There is no current in coil 1; therefore, there is no change in the number of $\vec{B}$ field lines through coil 2's area. No current will be induced in coil 2.	Coil 2 — Coil 1 — The galvanometer registers no current in coil 2.
Experiment 2. You close the switch in the circuit for coil 1. While the switch is being closed, the current in coil 1 increases rapidly from zero to a steady final value. Is there any current in coil 2 while the switch is being closed?	Neither coil is moving, thus no current will be induced in coil 2.	The increasing current in coil 1 produces an increasing $\vec{B}$ field. This changes the number of $\vec{B}$ field lines through coil 2's area. A brief current should be induced in coil 2.	$\vec{B}_1$ increasing I_{in} Coil 2 — Coil 1 — I_1 increasing Just as the switch closes, the galvanometer needle briefly moves to the left and then returns to vertical, indicating a brief induced current in coil 2.

| Testing experiment | Will a current be induced in coil 2? | | Outcome |
	Prediction based on Explanation 1	Prediction based on Explanation 2	
Experiment 3. You keep the switch in the circuit for coil 1 closed. The current in coil 1 has a steady value. Is there current in coil 2?	Neither coil is moving, thus no current will be induced in coil 2.	There is a steady current in coil 1, which results in a steady $\vec{B}$ field. Thus, the number of $\vec{B}$ field lines through coil 2's area is not changing. Therefore, no current will be induced in coil 2.	The galvanometer registers no current in coil 2.
Experiment 4. You open the switch again. Is there any current in coil 2 while the switch is being opened?	Neither coil is moving, thus no current will be induced in coil 2.	The decreasing current in coil 1 produces a decreasing $\vec{B}$ field. This changes the number of $\vec{B}$ field lines through coil 2's area, which should induce a brief current in coil 2.	Just as the switch opens, the galvanometer needle briefly moves to the right (opposite the direction in Experiment 2) and then returns to vertical, indicating a brief induced current in coil 2.

Conclusion

The predictions based on Explanation 2 matched the outcomes in all four experiments. The predictions based on Explanation 1 did not match the outcomes in two of the four experiments.

Motion is not necessary to have an induced current. In contrast, when the number of $\vec{B}$ field lines through a coil's area changes, there is an induced current in that coil. Explanation 1 has been found not to be valid in all cases, but Explanation 2 has been found to be valid for all examined cases.

We have learned that it is possible to have a current in a closed loop of wire without using a battery. This phenomenon of inducing a current using a changing $\vec{B}$ field is called **electromagnetic induction**. The discovery of electromagnetic induction, in 1831, is credited to two scientists: a British physicist, Michael Faraday, and an American physicist, Joseph Henry. Faraday spent 10 years trying to induce electric currents using magnetic fields, and it was only the invention of wire insulation by Henry that enabled Faraday to make multi-loop coils, leading to his success.

Electromagnetic induction explains how transcranial magnetic stimulation (TMS) works. When the physician closes the switch in the circuit with the coil that rests on the outside of the skull, the increasing current in the coil produces a changing $\vec{B}$ field within the brain. A current is briefly and noninvasively induced in a small region of the brain's electrically conductive tissue.

CONCEPTUAL EXERCISE 21.1 **Moving loops in a steady magnetic field**

The figure shows four wire loops 1–4 moving at constant velocity $\vec{v}$ relative to a region with a constant $\vec{B}$ field (within the dashed lines). In which of these loops will electric currents be induced?

Sketch and translate An electric current is induced in a loop whenever the number of $\vec{B}$ field lines through the loop's area changes.

Simplify and diagram For each loop, the current should be as follows:

1. Loop 1 is moving into the field region, so the number of $\vec{B}$ field lines through its area is increasing. A current will be induced in the loop.

(CONTINUED)

2. A current will be induced in loop 2 for the same reasons as (1).

3. Loop 3 is completely within the field region, thus the number of $\vec{B}$ field lines is not changing. As a result, no current is induced.

4. No current is induced in loop 4 for the same reason as in (3).

Try it yourself What will happen to loops 3 and 4 as they leave the magnetic field region?

Answer   The number of $\vec{B}$ field lines through each loop will change and a brief current will be induced as the loops leave the $\vec{B}$ field region.

Dynamic microphones and seismometers

FIGURE 21.5 A schematic of a dynamic microphone.

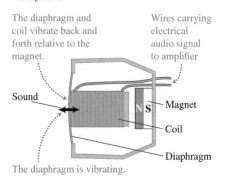

The diaphragm and coil vibrate back and forth relative to the magnet.

Wires carrying electrical audio signal to amplifier

Sound

N S — Magnet

— Coil

— Diaphragm

The diaphragm is vibrating.

A dynamic microphone that converts sound vibrations into electrical oscillations employs the principle of electromagnetic induction (**Figure 21.5**). When a sound wave, such as that produced by a singer's voice, strikes the diaphragm inside the microphone, the diaphragm oscillates. This oscillation moves a coil of wire attached to the diaphragm alternately closer to and farther from a magnet in the microphone, corresponding to locations with stronger and weaker magnetic field. This changing magnetic field through the coil induces a changing current in the coil. The changes in the current mirror the sound waves that led to the production of this current; thus this current can be used to store the details of the sound electronically.

A seismometer operates by the same principle. A seismometer is a sensor that detects seismic waves during an earthquake. It has a massive base with a magnet that vibrates as seismic waves pass. At the top, a coil attached to a block hangs at the end of a spring (**Figure 21.6**). The spring acts as a shock absorber that reduces the vibrations of the hanging block and coil relative to the base. The motion of the base relative to the coil induces a current through the coil, which produces a signal that is recorded on a seismograph.

FIGURE 21.6 A simple seismometer.

— Spring

— Block

— Coil

N

— Magnet

S

— Base

REVIEW QUESTION 21.1 Your friend thinks that relative motion of a coil and a magnet is absolutely necessary to induce current in a coil that is not connected to a battery. Support your friend's point of view with a physics argument. Then provide a counterargument and describe an experiment you could perform to disprove your friend's idea.

21.2 Magnetic flux

In the last section we found that an electric current is induced in a coil when the number of $\vec{B}$ field lines through the coil's area changes. This occurred when

- the strength of the $\vec{B}$ field in the vicinity of the coil changed, or
- the area A of the coil changed, or
- the orientation of the $\vec{B}$ field relative to the coil changed.

In this section we will construct a physical quantity describing the number of $\vec{B}$ field lines through a coil's area. Based on the analysis in the last section, changes in that quantity should cause an induced electric current in the coil. Physicists call this physical quantity **magnetic flux** Φ (Φ is the greek letter "phi").

The greater the magnitude of the $\vec{B}$ field passing through a particular two-dimensional area, the greater the number of field lines through the area. Additionally, if the area itself is larger, the number of field lines through the area is greater. If we double one or the other, the number of lines through the area should double. This suggests that the magnetic flux is proportional to the magnitude of the $\vec{B}$ field passing through the area and to the size of the area itself. Mathematically,

$$\Phi \propto BA$$

How do we include the dependence on the orientation of the loop relative to the $\vec{B}$ field lines? Imagine a rigid loop of wire in a region with a uniform $\vec{B}$ field. If the plane

TIP Often we discuss magnetic flux through a particular area, such as the area inside a wire loop. But the loop itself is not needed for the flux to exist.

of the loop is perpendicular to the $\vec{B}$ field lines, then a maximum number of field lines pass through the area (**Figure 21.7a**). If the plane of the loop is parallel to the $\vec{B}$ field, zero field lines pass through the area (Figure 21.7b). In between these two extremes, the magnetic flux takes on intermediary values (Figure 21.7c).

To describe this relative orientation we use a line perpendicular to the plane of the loop—the black normal vector shown in Figure 21.7. The angle θ between this vector and the $\vec{B}$ field lines quantifies this orientation. Since the cosine of an angle is at a maximum when the angle is zero and a minimum when the angle is 90°, the magnetic flux through the area is also proportional to $\cos \theta$. This leads to a precise definition for the magnetic flux through an area.

> **Magnetic flux Φ** The magnetic flux Φ through a region of area A is
>
> $$\Phi = BA \cos \theta \qquad (21.1)$$
>
> where B is the magnitude of the uniform magnetic field throughout the area and θ is the angle between the direction of the $\vec{B}$ field and a normal vector perpendicular to the area. The SI unit of magnetic flux is the unit of the magnetic field (the tesla T) times the unit of area (m^2), or $T \cdot m^2$. This unit is also known as the weber (Wb).

Equation (21.1) assumes that the magnetic field throughout the area is uniform and that the area is flat. For situations in which this is not the case, first split the area into small subareas within which the $\vec{B}$ field is approximately uniform; then add together the magnetic fluxes through each. This book will not address such cases.

In Section 21.1, we proposed that when the number of magnetic field lines through a wire loop changes, a current is induced in the loop. We can now refine that idea and say that *current is induced when there is a change in the total magnetic flux through the loop's area*. In other words, if the magnetic flux throughout the loop's area is steady, no current will be induced.

TIP In everyday life, flux often refers to something that is changing. "Our plans for the weekend are in flux." However, in physics, the word flux means the value of a particular physical quantity. When the $\vec{B}$ field, the orientation of the loop, and the area of the loop are constant, the flux through the area of the loop is constant.

FIGURE 21.7 Flux depends on the angle between the normal vector to the loop surface and the direction of the $\vec{B}$ field.

(a) Maximum flux: $\cos 0° = 1.0$

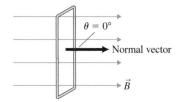

(b) Zero flux: $\cos 90° = 0$

$\vec{B}$ lines pass across the loop but *not* through it.

(c) Intermediate flux: $0 < \cos \theta < 1.0$

QUANTITATIVE EXERCISE 21.2   **Flux through a book's cover**

A book is positioned in a uniform 0.20-T $\vec{B}$ field that points from left to right parallel to the plane of the page, shown in the figures on the right. (For simplicity, we depict the book as a rectangular loop.) Each side of the book's cover measures 0.10 m. Determine the magnetic flux through the cover when (a) the cover is in the plane of the page (figure a), (b) the cover is perpendicular to the plane of the page and the normal vector makes a 60° angle with the $\vec{B}$ field (figure b), and (c) the cover is perpendicular to the plane of the page and the normal vector points toward the top of the page (figure c).

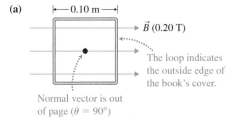

The loop indicates the outside edge of the book's cover.

Normal vector is out of page ($\theta = 90°$)

Represent mathematically The magnetic flux through an area is determined by Eq. (21.1) $\Phi = BA \cos\theta$. The angle in each of the three situations is (a) $\theta = 90°$, (b) $\theta = 60°$, and (c) $\theta = 90°$.

Solve and evaluate The magnetic flux through the cover in each case is

(a) $\Phi = (0.20\,\text{T})(0.10\,\text{m})^2 \cos(90°) = 0$

(b) $\Phi = (0.20\,\text{T})(0.10\,\text{m})^2 \cos(60°) = 1.0 \times 10^{-3}\,\text{T} \cdot \text{m}^2$

(c) $\Phi = (0.20\,\text{T})(0.10\,\text{m})^2 \cos(90°) = 0$

We can evaluate these results by comparing the calculated fluxes to the number of $\vec{B}$ field lines through the cover's area. Note that for the orientation of the book in (a) and (c), the $\vec{B}$ field lines are parallel to the book's area and therefore do not go through it. Those positions are consistent with our mathematical result. The orientation for the book in (b) is such that some $\vec{B}$ field lines do pass through the book, which is also consistent with the nonzero mathematical result.

(CONTINUED)

Try it yourself A circular ring of radius 0.60 m is placed in a 0.20-T uniform $\vec{B}$ field that points toward the top of the page. Determine the magnetic flux through the ring's area when (a) the plane of the ring is perpendicular to the surface of the page and

its normal vector points to the right and (b) the plane of the ring is perpendicular to the surface of the page and its normal vector points toward the top of the page.

Answer   (a) 0; (b) 0.23 T·m².

REVIEW QUESTION 21.2 You have a bar magnet and a gold ring. How should you position the ring relative to the magnet so that the magnetic flux through the circular area inside the ring is zero?

21.3 Direction of the induced current

Recall from Section 21.1 that the galvanometer registered current in one direction for some of the experiments and in the opposite direction for others. In this section we will focus on the pattern in the direction of the induced current.

Figure 21.8 shows the results of two experiments in which the number of $\vec{B}$ field lines through a wire coil's area is changing. As the bar magnet moves closer to the coil in (a), the number of $\vec{B}$ field lines through the coil's area increases (the magnetic flux through the coil's area increases). As expected, there is a corresponding induced current. An arrow along the coil indicates the direction of this induced current as measured by a galvanometer.

FIGURE 21.8 How is the changing $\vec{B}_{ex}$ related to the direction of I_{in} and $\vec{B}_{in}$? The number of the $\vec{B}_{ex}$ field lines passing though the coil increases in (a) and decreases in (b).

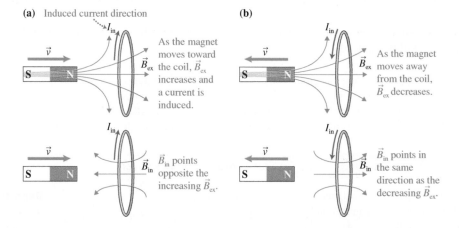

(a) Induced current direction

As the magnet moves toward the coil, $\vec{B}_{ex}$ increases and a current is induced.

$\vec{B}_{in}$ points opposite the increasing $\vec{B}_{ex}$.

(b)

As the magnet moves away from the coil, $\vec{B}_{ex}$ decreases.

$\vec{B}_{in}$ points in the same direction as the decreasing $\vec{B}_{ex}$.

Because electric currents produce a magnetic field, the induced current in the coil must also produce a magnetic field and a corresponding magnetic flux through the coil. We call this second magnetic field $\vec{B}_{induced}$ or $\vec{B}_{in}$. The direction of $\vec{B}_{in}$ can be determined using the right-hand rule for the $\vec{B}$ field. Notice that in the case shown in Figure 21.8a, the flux through the coil due to the magnet (called $\vec{B}_{external}$ or $\vec{B}_{ex}$) is increasing and the magnetic field due to the induced current $\vec{B}_{in}$ points in the opposite direction of $\vec{B}_{ex}$.

In Figure 21.8b, the bar magnet is moving away from the coil. As a result, the number of external field lines through the coil's area (and therefore the magnetic flux through it) is decreasing. Again, there is a corresponding induced current (see Figure 21.8b). In this case, however, $\vec{B}_{in}$ (produced by the induced current) points in the same direction as $\vec{B}_{ex}$ (produced by the magnet). Can we find a pattern in these data?

Notice that in both cases $\vec{B}_{in}$ points in the direction that diminishes the change in the external flux through the coil. In the first experiment the flux through the coil was increasing. In that situation, $\vec{B}_{in}$ pointed in the opposite direction to $\vec{B}_{ex}$, as if to resist the increase. In the second experiment, the external flux through the coil was decreasing, and $\vec{B}_{in}$ pointed in the same direction as $\vec{B}_{ex}$, as if to resist the decrease. In both situations, the $\vec{B}_{in}$ due to the induced current resisted the *change* in the external flux through the coil. Put another way, $\vec{B}_{in}$ points in whatever direction is necessary to try to keep the magnetic flux through the coil's area constant.

Consider what would happen if the reverse occurred. Suppose an increasing external magnetic flux through the loop led to an induced magnetic field in the same direction as the external field. In that case, the magnetic flux due to the induced field would augment rather than reduce the total flux through the loop. This would cause an even greater induced current, which would cause yet a greater induced magnetic field and a steeper increase in magnetic flux. In other words, just by lightly pushing a bar magnet toward a loop of wire, you would cause a runaway induced current that would continually increase until the wire melted. Of course, this would violate the conservation of energy. If such a scenario were possible, we could heat water by simply moving a bar magnet over a coil in a large tank of water.

This pattern concerning the direction of the induced current was first developed in 1833 by the Russian-German physicist Heinrich Lenz.

Lenz's law The direction of the induced current in a coil is such that its $\vec{B}_{in}$ field opposes the change in the magnetic flux through the coil's area produced by other objects. If the magnetic flux through the coil is increasing, the direction of the induced current's $\vec{B}_{in}$ field leads to a decrease in the flux. If the magnetic flux through the coil is decreasing, the direction of the induced current's $\vec{B}_{in}$ field leads to an increase in the flux.

Lenz's law lets us determine the direction of the induced current's magnetic field $\vec{B}_{in}$. From there we can use the right-hand rule for the $\vec{B}$ field to determine the direction of the induced current itself. This process is summarized in Physics Tool Box 21.1, which shows how to determine the direction of the induced current in a loop of wire when the magnetic flux though the loop changes. In this Tool Box, the loop of wire is in a decreasing external field $\vec{B}_{ex}$ that is perpendicular to the plane of the loop.

Note that the magnetic flux through a loop or coil can change because of a change in the external magnetic field, a change in the area of a loop or coil, or a change in its orientation.

PHYSICS TOOL BOX 21.1   Determining the direction of an induced current

1. Determine the initial external magnetic flux $\Phi_{ex\,i}$ through the coil (represented below by the number and direction of the external magnetic field lines).	2. Determine the final external magnetic flux $\Phi_{ex\,f}$ through the coil (represented below by the smaller number of external magnetic field lines).	3. The induced flux opposes the *change* in the external flux (a decrease in upward external flux in this case; $\vec{B}_{in}$ points up so that the net upward flux decreases less than it would in the absence of the induced flux).

4. Use the right-hand rule for the magnetic field to determine the direction of the induced current that will produce the $\vec{B}_{in}$ and the induced flux.

CONCEPTUAL EXERCISE 21.3   **Practice using Lenz's law**

A loop in the plane of the page is being pulled to the right at constant velocity out of a region of a uniform magnetic field $\vec{B}_{ex}$. The field is perpendicular to the loop and points out of the paper in the region inside the rectangular dashed line (see the first figure below). Determine the direction of the induced electric current in the loop when it is halfway out of the field, as shown in the second figure below.

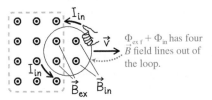

Sketch and translate The number of $\vec{B}_{ex}$ lines passing through the coil's area decreases by half during the time interval from when the coil

is completely in the field to when it is halfway out. Thus, the external magnetic flux through the loop's area has decreased.

Simplify and diagram According to Lenz's law, the induced magnetic flux and induced magnetic field $\vec{B}_{in}$ should point out of the paper, as shown below, thus keeping the flux through the loop closer to the initial flux. The direction of the induced current that causes this induced magnetic field is determined using the right-hand rule for the $\vec{B}$ field; the induced electric current through the loop is counterclockwise.

Try it yourself Notice that the induced current in the left side of the loop is still in the external magnetic field $\vec{B}_{ex}$ (as shown in the figure above). Determine the direction of the force that the external magnetic field $\vec{B}_{ex}$ exerts on the induced current in the left side of the loop.

Answer   To the left, opposite the direction of the loop's velocity.

Eddy currents: an application of Lenz's law

Conceptual Exercise 21.3 shows an example of a phenomenon called an **eddy current**. An eddy current usually occurs when a piece of metal moves through a magnetic field. If you were to hold a sheet of aluminum or copper between the poles of a strong horseshoe magnet, you would find that neither of the poles attracts the sheet. However, when you move the sheet out from between the poles of the magnet, especially if you pull it quickly, you encounter resistance. This force is similar to the force exerted on the loop leaving the magnetic field region in the "Try it yourself" part of the exercise.

Let's examine this phenomenon. **Figure 21.9a** shows a metal sheet between the poles of an electromagnet. Pulling the sheet to the right decreases the external magnetic flux through area 1. This is similar to the situation in Conceptual Exercise 21.3, although in that exercise the external $\vec{B}$ field was out of the page instead of down.

FIGURE 21.9 Inducing an eddy current by pulling a metal sheet through a magnetic field. (a) Area 1 is leaving the field and area 2 is entering the field. (b) Changing flux through both areas induces eddy currents. The external magnetic field exerts a force on the eddy currents opposite the direction of motion of the sheet.

This decrease in flux induces an eddy current in the metal sheet around area 1, which according to Lenz's law circles or curls clockwise in this region (see Figure 21.9b) to produce an induced field that points in the same direction as the external $\vec{B}$ field.

At the same time, area 2 of the sheet is entering the magnetic field region, so the magnetic flux through that area is increasing. This change in magnetic flux induces a counterclockwise eddy current.

How do we explain the force that points opposite the direction of the sheet's motion? The left side of the eddy current in area 1, shown in Figure 21.9b, is still in the magnetic field region. Using the right-hand rule for the magnetic force, we find that the force exerted by the magnet on the left side of the eddy current in area 1 points toward the left when the sheet is pulled to the right, in agreement with what was observed. Similarly, the magnet also exerts a force on the part of the eddy current in area 2 that is in the magnetic field region. The right-hand rule for the magnetic force determines that this force points to the left as well. Both of these forces point opposite the direction the sheet is moving, acting as a sort of "braking" force.

What would happen if you were to push the sheet to the left instead of pulling it to the right? Using Lenz's law, we find that the eddy currents reverse direction, and the magnetic forces exerted on them also reverse direction, again resulting in a magnetic braking effect.

Many technological applications rely on the phenomenon of resistance to the motion of a nonmagnetic metal material through a magnetic field. For instance, the braking system used in some cars, trains, and amusement park rides consists of strong electromagnets with poles on either side of the turning metal wheels of the vehicle (Figure 21.10). When the electromagnet is turned on, it exerts magnetic forces on the eddy currents in the wheels that oppose their motion. As the turning rate of the wheels decreases, the eddy currents decrease, and therefore the braking forces decrease.

Another very important application of eddy currents is the eddy current waste separator. When general waste is delivered to a dump, it is placed on a conveyer belt. First, strong magnets lift objects containing ferromagnetic materials (iron nails, for example), separating them from the rest of the waste, which continues along the belt. Under the end of the conveyer belt are new magnets (usually electromagnets that produce variable magnetic field), which induce eddy currents in all non-ferromagnetic moving metal objects (aluminum cans, for example) and therefore exert repulsive upward forces on them. These forces lift the metal objects above the belt (see Figure 21.11), causing them to fly off the belt horizontally while the nonmetal objects continue moving with the belt and fall straight down off its end. The metal objects land on a new conveyer belt placed below the original belt and slightly forward. Each type of waste lands in a different place, and metal objects are separated from nonmetal objects.

REVIEW QUESTION 21.3 What difficulty would occur if the $\vec{B}$ field produced by the induced current enhanced the change in the external field rather than opposed the change? Give a specific example.

21.4 Faraday's law of electromagnetic induction

In the first two sections of this chapter, we found that when the magnetic flux through a coil's area changes, there is an induced electric current in the coil. The flux depends on the magnitude of the $\vec{B}$ field, the area of the coil, and the orientation of the coil relative to the $\vec{B}$ field. Our next goal is to construct a quantitative version of this idea that will allow us to predict the magnitude of the induced current through a particular coil. We begin with Observational Experiment Table 21.3 (on the next page), in which we will determine what factors affect the magnitude of the induced current produced by the flux change.

FIGURE 21.10 Magnetic braking is used to stop the Giant Drop ride at Dreamworld in Queensland, Australia.

FIGURE 21.11 An eddy current waste separator. Metal objects fly forward while non-metals fall off the end of the conveyer belt.

Factors affecting the magnitude of induced current

Observational experiment	Analysis
Experiment 1. Rapidly move a bar magnet toward a coil and observe the galvanometer needle. Repeat the process, this time moving the magnet slowly.	The galvanometer registers a larger induced current when the magnet moves rapidly toward the coil compared with when it moves slowly.
Experiment 2. Rapidly rotate the magnet by 90 degrees in front of the stationary coil. Repeat the process, this time rotating the magnet slowly.	The galvanometer registers a larger induced current when the magnet rotates rapidly compared with when it rotates slowly.
Experiment 3. Use two coils, each with a different number of turns. Move the magnet toward the coil with more turns. Then move the magnet at the same speed toward the coil with fewer turns.	The galvanometer registers a larger induced current in the coil with more turns.

Patterns

- The speed at which the magnet moved or rotated affected the induced current. The shorter the time interval for the change of the flux through the coil, the greater the induced current.
- The induced current in a coil with a larger number of turns is greater than in a coil with a smaller number of turns.

TIP Notice that even though we were investigating the effects of two quantities on the magnitude of the induced current, we changed only one of them at a time so that we could investigate those effects separately.

In the experiments in Table 21.3, we found that the induced current was greater if the same change in magnetic flux through a coil occurred in a shorter time interval Δt. Additionally, the induced current through the coil was greater in a coil with a larger number of turns N.

Quantitatively, what is the relationship between the magnitude of the induced current in a coil and the change in magnetic flux through that coil's area? Let's connect a single circular loop with zero resistance in series with a 100-Ω resistor (**Figure 21.12a**). The loop is placed between the poles of an electromagnet with its surface perpendicular to the magnetic field. An ammeter (not shown) measures the current in the loop and resistor. The upward magnetic field (up is defined as the positive direction) decreases steadily for 2.0 s (we decrease the magnitude of the $\vec{B}$ field by decreasing the current in the coils of the electromagnet; note that this is not the induced current), after which the field and flux remain constant at a smaller positive value. The magnetic flux-versus-time and the induced current-versus-time graphs are shown in Figure 21.12b.

When the flux is *changing* at a constant rate, the current through the loop and resistor has a constant value. For example, for the first 2.0 s the *slope* of the magnetic flux-versus-time graph has a constant negative value of $-1.0\ \mathrm{T \cdot m^2/s}$, and the induced current has a constant positive value of $+0.010$ A. If we replace the 100-Ω resistor with a 50-Ω resistor, the current-versus-time graph has the same shape, but its magnitude during the first 2.0 s doubles to $+0.020$ A. Thus, the same flux change produces different size currents in the same loop depending on the resistance of the circuit. However, the product of the current and resistance for both situations has the same value:

$$(0.010 \text{ A})(100 \ \Omega) = (0.020 \text{ A})(50 \ \Omega) = 1.0 \text{ A} \cdot \Omega = 1.0 \text{ V}$$

It almost seems that there must be a 1.0-V battery in series with the resistor, but there isn't.

The changing flux through the loop caused the induced current. However, the unit of the *slope* of the flux-versus-time graph is not the ampere, but the volt:

$$\frac{\Delta \Phi}{\Delta t} = \frac{-2.0 \text{ T} \cdot \text{m}^2}{2.0 \text{ s}} = -1.0 \left(\frac{\text{N}}{\text{C} \cdot (\text{m/s})} \right) \frac{\text{m}^2}{\text{s}} = -1.0 \frac{\text{N} \cdot \text{m}}{\text{C}} = -1.0 \frac{\text{J}}{\text{C}} = -1.0 \text{ V}$$

(We used the expression for the magnetic force $F_{\text{magnetic}} = qvB$ or $B = F_{\text{magnetic}}/qv$ to convert the magnetic field in tesla (T) to other units.) Thus the induced current is a consequence of the emf induced in the coil. The changing flux acts as a "battery" that produces a 1.0-V emf that causes the electric current. You could apply Kirchhoff's loop rule to the circuit shown in Figure 21.12a. There is a +1.0-V potential change across the loop and a −1.0-V potential change across the resistor.

The magnitude of the emf in the loop depends only on the rate of change of flux through the loop (the slope of the flux-versus-time graph $\mathcal{E}_{\text{in}} = |\Delta\Phi/\Delta t|$). When we repeat this experiment for a coil with N loops, we find that the magnitude of the induced emf increases in proportion to the number N of loops in the coil. Thus,

$$\mathcal{E}_{\text{in}} = N \left| \frac{\Delta \Phi}{\Delta t} \right| \tag{21.2}$$

This expression is known as **Faraday's law of electromagnetic induction**. However, the mathematical expression was actually developed by James Clerk Maxwell.

> **Faraday's law of electromagnetic induction** The average magnitude of the induced emf $\mathcal{E}_{\text{in}}$ in a coil with N loops is the magnitude of the ratio of the magnetic flux change through the loop $\Delta\Phi$ to the time interval Δt during which that flux change occurred multiplied by the number N of loops:
>
> $$\mathcal{E}_{\text{in}} = N \left| \frac{\Delta \Phi}{\Delta t} \right| = N \left| \frac{\Delta (BA \cos \theta)}{\Delta t} \right| \tag{21.3}$$

The direction of the current induced by this emf is determined using Lenz's law. A minus sign is often placed in front of the N to indicate that the induced emf opposes the change in magnetic flux. However, we will not use this notation.

Faraday realized that the phenomenon of electromagnetic induction has the same effect on devices attached to a coil as a battery. This idea enabled him to build the first primitive electric generator that produced an induced current in a coil.

FIGURE 21.12 An experiment to quantitatively relate induced emf ($\mathcal{E}_{\text{in}}$) and induced current (I_{in}) to flux change.

(a)

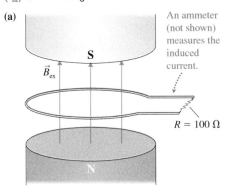

An ammeter (not shown) measures the induced current.

$R = 100 \ \Omega$

(b) Resistance $R = 100 \ \Omega$

$\vec{B}_{\text{ex}}$ decreases for 2.0 s and is then constant.

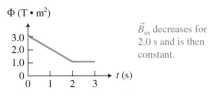

The induced current in the coil is a constant positive value for 2.0 s and is then zero.

> **TIP** Note that the value of the emf does not depend on the properties of the material of the coil. It depends on the properties of the magnetic field and the geometry of the coil.

QUANTITATIVE EXERCISE 21.4   **Changing flux produces emf**

The figure below shows the magnetic flux-versus-time graph for the magnetic flux through a 100-turn coil that is moved with a constant speed through an external magnetic field. Draw the corresponding induced emf-versus-time graph.

Represent mathematically The magnitude of the induced emf in the coil is

$$\mathcal{E}_{\text{in}} = N \left| \frac{\Delta \Phi}{\Delta t} \right| = 100 \left| \frac{\Delta \Phi}{\Delta t} \right|$$

where $\left| \dfrac{\Delta \Phi}{\Delta t} \right|$ is equal to the absolute value of the slope of the magnetic flux-versus-time graph. The graph consists of three regions (I, II, and III); we therefore need to determine three different magnitudes. To

(CONTINUED)

determine the slope within each region, we find the change in magnetic flux for the region and divide it by the time interval during which the change occurred.

Solve and evaluate

Region I: $\Delta\Phi_I = (10 - 0) \times 10^{-7}\ \text{T}\cdot\text{m}^2;\ \Delta t_I = (0.2 - 0)\ \text{s}$

$$\Rightarrow\ \mathcal{E}_{\text{in I}} = 100\left|\frac{\Delta\Phi}{\Delta t}\right| = \frac{100 \times 10 \times 10^{-7}\ \text{T}\cdot\text{m}^2}{0.2\ \text{s}} = 0.5 \times 10^{-3}\ \text{V}$$

Region II: $\Delta\Phi_{II} = (-20 - 10) \times 10^{-7}\ \text{T}\cdot\text{m}^2;\ \Delta t_{II} = (0.3 - 0.2)\ \text{s}$

$$\Rightarrow\ \mathcal{E}_{\text{in II}} = 100\left|\frac{\Delta\Phi}{\Delta t}\right| = \frac{100 \times 30 \times 10^{-7}\ \text{T}\cdot\text{m}^2}{0.1\ \text{s}} = 3.0 \times 10^{-3}\ \text{V}$$

Region III: $\Delta\Phi_{III} = -20 - (-20) \times 10^{-7}\ \text{T}\cdot\text{m}^2;\ \Delta t_{III} = (0.4 - 0.3)\ \text{s}$

$$\Rightarrow\ \mathcal{E}_{\text{in III}} = 100\left|\frac{\Delta\Phi}{\Delta t}\right| = \frac{100 \times 0}{0.1\ \text{s}} = 0\ \text{V}$$

The induced emf is negative when the magnetic flux through the loop is increasing and positive when it is decreasing (by Lenz's law). Therefore, the induced emf in region I is negative and in region II is positive. The graph of $\mathcal{E}_{\text{in}}$-versus-t for the three regions is shown below.

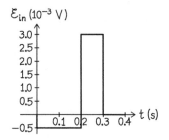

Try it yourself The previous experiment is repeated, only this time the coil is moved twice as fast. Determine the maximum induced emf in the coil.

Answer $\mathcal{E}_{\text{in max}} = 6.0 \times 10^{-3}\ \text{V}.$

REVIEW QUESTION 21.4 Why do we write the law of electromagnetic induction in terms of emf rather than in terms of induced current?

21.5 Skills for analyzing processes involving electromagnetic induction

Faraday's law enables us to design and understand practical applications of electromagnetic induction. For example, to design an automobile ignition system that uses spark plugs, engineers must estimate how quickly the magnetic field through a coil must be reduced to zero to produce a large enough emf to ignite a spark plug. An engineer designing an electric generator will be interested in the rate at which the generator coil must turn relative to the $\vec{B}$ field to produce the desired induced emf. The general strategy for analyzing questions like these is described and illustrated in Example 21.5.

**PROBLEM-SOLVING
STRATEGY 21.1**   Problems involving electromagnetic induction

EXAMPLE 21.5   Determine the $\vec{B}$ field produced by an electromagnet

To determine the $\vec{B}$ field produced by an electromagnet, you use a 30-turn circular coil of radius 0.10 m (30-Ω resistance) that rests between the poles of the magnet and is connected to an ammeter. When the electromagnet is switched off, the $\vec{B}$ field decreases to zero in 1.5 s. During this 1.5 s, the ammeter measures a constant current of 18 mA. How can you use this information to determine the initial $\vec{B}$ field produced by the electromagnet?

Sketch and translate

- Create a labeled sketch of the process described in the problem. Show the initial and final situations to indicate the change in magnetic flux. Label knowns and unknowns.
- Determine which physical quantity is changing ($\vec{B}$, A, or θ), thus causing the magnetic flux to change.

The changing quantity (from time 0.0 s to 1.5 s) is the magnitude of the $\vec{B}$ field produced by the electromagnet. Due to this change, the flux through the coil's area changes; thus there is an induced current in the coil. We can use the law of electromagnetic induction to find the magnitude of the initial $\vec{B}$ field produced by the electromagnet.

Simplify and diagram

- Decide what assumptions you are making. Does the flux change at a constant rate? Is the magnetic field uniform?
- If useful, draw a graph of the flux and the corresponding induced emf-versus-clock reading.
- If needed, use Lenz's law to determine the direction of the induced current.

Assume the following:

- The current in the electromagnet changes at a constant rate; thus the flux through the coil does also.
- The $\vec{B}$ field in the vicinity of the coil is uniform.
- The $\vec{B}$ field is perpendicular to the coil's surface.

Represent mathematically

- Apply Faraday's law and indicate the quantity ($\vec{B}$, A, or $\cos\theta$) that causes a changing magnetic flux.
- If needed, use Ohm's law and Kirchhoff's loop rule to determine the induced current.

$$\mathcal{E}_{in} = N\left|\frac{\Phi_f - \Phi_i}{t_f - t_i}\right| = N\left|\frac{0 - B_i A \cos\theta}{\Delta t}\right|$$

The number of turns N and the angle θ remain constant. The magnitude of the magnetic field changes.
Substitute $\mathcal{E}_{in} = IR$ and $A = \pi r^2$ and solve for B_i:

$$B_i = \frac{IR\Delta t}{N\pi r^2 \cos\theta}$$

Solve and evaluate

- Use the mathematical representation to solve for the unknown quantity.
- Evaluate the results—units, magnitude, and limiting cases.

The plane of the coil is perpendicular to the magnetic field lines, so $\theta = 0$ and $\cos 0° = 1$. Inserting the appropriate quantities:

$$B_i = \frac{IR\Delta t}{N\pi r^2} = \frac{0.018 \text{ A} \times 30 \text{ } \Omega \times 1.5 \text{ s}}{30 \times \pi \times (0.10 \text{ m})^2} = 0.86 \text{ T}$$

The answer is reasonable. Let's check the units:

$$\frac{A \cdot \Omega \cdot s}{m^2} = \frac{V \cdot s}{m^2} = \frac{J \cdot s}{C \cdot m^2} = \frac{N \cdot m \cdot s}{C \cdot m^2} = \frac{N}{C \cdot (m/s)} = T$$

The units match. As a limiting case, a coil with fewer turns would require a larger $\vec{B}$ field to induce the same current. Also, if the resistance of the circuit is larger, the same $\vec{B}$ field change induces a smaller current.

Try it yourself Determine the current in the loop if the plane of the loop is parallel to the magnetic field. Everything else is the same.

Answer   Zero.

In the rest of this section, we will analyze practical applications of electromagnetic induction that involve a change in (1) the magnitude of the $\vec{B}$ field, (2) the area of the loop or coil, or (3) the orientation of the coil relative to the $\vec{B}$ field. All of these processes involve the same basic idea: **a changing magnetic flux through the area of a coil or single loop is accompanied by an induced emf around the coil or loop.** In turn, this emf induces an electric current in the coil or loop.

EXAMPLE 21.6 Changing the magnitude of the $\vec{B}$ field

The magnitude of the $\vec{B}$ field from a TMS coil increases from 0 T to 0.2 T in 0.002 s. The $\vec{B}$ field lines pass through the scalp into a small region of the brain, inducing a small circular current in the conductive brain tissue in the plane perpendicular to the field lines. Assume that the radius of the circular current in the brain is 0.0030 m and that the tissue in this circular region has an equivalent resistance of 0.010 Ω. What are the direction and magnitude of the induced electric current around this circular region of brain tissue?

Sketch and translate We sketch the situation and mark knowns and unknowns. The change in magnetic flux through the circular region of brain tissue is caused by the increasing $\vec{B}$ field produced by the TMS coil (called $\vec{B}_{ex}$). This change in flux causes an induced emf, which produces an induced current in the brain tissue. The direction of the current can be determined using Lenz's law.

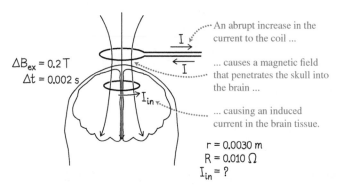

An abrupt increase in the current to the coil ...

... causes a magnetic field that penetrates the skull into the brain ...

... causing an induced current in the brain tissue.

$\Delta B_{ex} = 0.2$ T
$\Delta t = 0.002$ s

$r = 0.0030$ m
$R = 0.010$ Ω
$I_{in} = ?$

Simplify and diagram Assume that $\vec{B}_{ex}$ throughout the disk-like region of brain tissue is uniform and increases at a constant rate. Model the disk as a single-turn coil. Viewed from above, $\vec{B}_{ex}$ points into the page (shown as X's in the figure below). Since the number of $\vec{B}$ field lines is increasing into the page, the $\vec{B}_{in}$ field produced by the induced current points out of the page (shown as dots). Using the right-hand rule for the $\vec{B}_{in}$ field, we find that the direction of the induced current is counterclockwise.

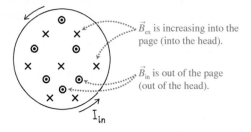

$\vec{B}_{ex}$ is increasing into the page (into the head).

$\vec{B}_{in}$ is out of the page (out of the head).

I_{in}

Represent mathematically To find the magnitude of the induced emf, use Faraday's law:

$$\mathcal{E}_{in} = N \left| \frac{\Phi_f - \Phi_i}{t_f - t_i} \right|$$

where the magnetic flux through the loop at a specific clock reading is $\Phi = B_{ex}A \cos\theta$. The area A of the loop and the orientation angle θ between the loop's normal vector and the $\vec{B}_{ex}$ field are constant, so

$$\mathcal{E}_{in} = N \left| \frac{B_{ex\,f}A \cos\theta - B_{ex\,i}A \cos\theta}{t_f - t_i} \right|$$

$$= NA \cos\theta \left| \frac{B_{ex\,f} - B_{ex\,i}}{t_f - t_i} \right|$$

Using our understanding of electric circuits, we relate this induced emf to the resulting induced current:

$$I_{in} = \frac{\mathcal{E}_{in}}{R}$$

Solve and evaluate Combine these two equations and solve for the induced current:

$$I_{in} = \frac{1}{R}\mathcal{E}_{in} = \frac{1}{R}\left(NA \cos\theta \left| \frac{B_{ex\,f} - B_{ex\,i}}{t_f - t_i} \right|\right)$$

$$= \frac{N\pi r^2 \cos\theta}{R}\left| \frac{B_{ex\,f} - B_{ex\,i}}{t_f - t_i} \right|$$

$$= \frac{(1)\pi(0.0030\text{ m})^2(1)}{0.010\text{ Ω}}\left| \frac{0.2\text{ T} - 0}{0.002\text{ s} - 0} \right| = 0.28\text{ A}$$

Note that the normal vector to the loop's area is parallel to the magnetic field; therefore, $\cos\theta = \cos(0°) = 1$. This is a significant current and could affect brain function in that region of the brain.

Try it yourself Derive the expression for the electric charge that has been moved around the circular region in this example.

Answer

$$\mathcal{Q} = N\frac{\Phi_f - \Phi_i}{R}.$$

EXAMPLE 21.7 Changing the area of the loop

A 10-Ω lightbulb is connected between the ends of two parallel conducting rails that are separated by 1.2 m, as shown in the figure below. A metal rod is pulled along the rails so that it moves to the right at a constant speed of 6.0 m/s. The two rails, the lightbulb, its connecting wires, and the rod make a complete rectangular loop circuit. A uniform 0.20-T magnitude B_{ex} field points downward, perpendicular to the loop's area. Determine the direction of the induced current in the loop, the magnitude of the induced emf, the magnitude of the current in the lightbulb, and the power output of the lightbulb.

Area increases as the rod moves right.

10-Ω bulb

$L = 1.2$ m

$\vec{v}$ (6.0 m/s)

x

0

$\vec{B}_{ex}$ (0.20 T)

Sketch and translate We sketch the situation as shown below. Choose the normal vector for the loop's area to point upward. The loop's area increases as the rod moves away from the bulb. Because of this, the magnitude of the magnetic flux through the loop's area is increasing as the rod moves to the right.

Top view $\vec{B}_{ex}$ into page

$\vec{B}_{in}$ out of page

The $\vec{B}_{ex}$ flux increases because the area with $\vec{B}_{ex}$ is increasing.

Simplify and diagram Assume that the rails, rod, and connecting wires have zero resistance. The induced field $\vec{B}_{in}$ due to the loop's induced current should point upward, resisting the change in the downward increasing magnetic flux through the loop's area. Using the right-hand rule for the $\vec{B}_{in}$ field, we find that the direction of the induced current is counterclockwise.

Represent mathematically To find the magnitude of the induced emf, use Faraday's law:

$$\mathcal{E}_{in} = N \left| \frac{\Phi_f - \Phi_i}{t_f - t_i} \right|$$

The angle between the loop's normal line and the B_{ex} field is 180°. The magnitude of the B_{ex} field is constant. Therefore,

$$\mathcal{E}_{in} = (1) \left| \frac{B_{ex} A_f \cos(180°) - B_{ex} A_i \cos(180°)}{t_f - t_i} \right|$$

$$= B_{ex} \left| \frac{A_f - A_i}{t_f - t_i} \right|$$

The area A of the loop at a particular clock reading equals the length L of the sliding rod times the x-coordinate of the sliding rod (the origin of the x-axis is placed at the bulb.) Therefore,

$$\mathcal{E}_{in} = B_{ex} \left| \frac{Lx_f - Lx_i}{t_f - t_i} \right| = B_{ex} L \left| \frac{x_f - x_i}{t_f - t_i} \right| = B_{ex} L v$$

The quantity inside the absolute value is the x-component of the rod's velocity, the absolute value of which is the rod's speed v. Using Ohm's law, the induced current depends on the induced emf and the bulb resistance:

$$I_{in} = \frac{\mathcal{E}_{in}}{R}$$

The power output of the lightbulb is

$$P = I_{in}^2 R$$

Solve and evaluate The magnitude of the induced emf around the loop is

$$\mathcal{E}_{in} = B_{ex} L v = (0.20 \text{ T})(1.2 \text{ m})(6.0 \text{ m/s}) = 1.44 \text{ V}$$

The current in the bulb is

$$I_{in} = \frac{1.44 \text{ V}}{10 \ \Omega} = 0.14 \text{ A}$$

The lightbulb should glow: its power output is

$$P = I_{in}^2 R = (0.14 \text{ A})^2 (10 \ \Omega) = 0.21 \text{ W}$$

Try it yourself Suppose the rod in this example moves at the same speed but in the opposite direction, so that the loop's area decreases. The rails make a 45° angle with the direction of the $\vec{B}$ field. Determine the magnitude of the induced emf, the magnitude of current in the bulb, and the direction of the current.

Answer   1.0 V; 0.1 A; clockwise.

Limitless electric energy?

We have just found that the induced emf depended on the speed with which a metal rod is pulled along the rails. By accelerating the rod to a speed of our choosing, we could induce whatever emf we desired. Would this method of obtaining the emf by moving a conductor in the external magnetic field maintain the current indefinitely?

FIGURE 21.13 An alternative way to analyze the motion of the rod in terms of electric and magnetic forces.

(a) $\vec{B}_{ex}$ into page

The magnetic field exerts a magnetic force on free electrons in the moving rod.

(b)

The magnetic force causes excess electrons to accumulate at I.

(c)

$F_{\vec{E} \text{ on } q} = qE$

$F_{\vec{B} \text{ on } q} = qvB$

The opposite charges at I and II produce an electric force that balances the magnetic force.

FIGURE 21.14 A simple version of an electric generator.

The steam turns the turbine, which turns the wire loop with respect to the magnet.

Steam

Turbine

Burner

The moving rod resulted in an emf and an induced current in the rails, bulb, and rod. However, if we use the right-hand rule for magnetic force, we find that the magnetic field exerts a force on the induced current in the rod toward the left, causing it to slow down. From an energy perspective, the kinetic energy of the rod is being transformed into light and thermal energy in the bulb. In order to keep the rod moving at constant speed, some other object must exert a force on it to the right, doing positive work on the rod.

Motional emf

The emf produced in Example 21.7 is sometimes called **motional emf**; it is caused by the motion of an object through the region of a magnetic field. We explained this emf using the idea of electromagnetic induction. Is it possible to understand it just in terms of magnetic forces? When an electrically charged object with charge q moves within a region with nonzero $\vec{B}$ field, the field exerts a magnetic force on it ($F_{\vec{B} \text{ on } q} = |q|vB \sin\theta$). Consider a metal rod, shown in **Figure 21.13a**, moving at velocity $\vec{v}$ in the horizontal direction. The external magnetic field $\vec{B}_{ex}$ points into the page. Inside the rod are fixed positively charged ions and negatively charged free electrons. When the rod slides to the right, all of its charged particles move with it. According to the right-hand rule for the magnetic force, the external magnetic field exerts a magnetic force on the electrons toward end I. The positive charges cannot move inside the rod, but the free electrons can. The electrons accumulate at end I, leaving end II with a deficiency of electrons (a net positive charge). The ends of the rod become charged, as shown in Figure 21.13b.

These separated charges create an electric field $\vec{E}$ in the rod that exerts a force of magnitude $F_{\vec{E} \text{ on } q} = qE$ on other electrons in the rod; the electric field exerts a force on negative electrons toward II (Figure 21.13c) opposite the direction of the magnetic force. When the magnitude of the electric force equals the magnitude of the magnetic force, the accumulation of opposite electric charge at the ends of the rod ceases. Then

$$qvB = qE \quad \text{or} \quad E = vB$$

Magnetic force   Electric force

An electric potential difference is produced between points I and II that depends on the magnitude of the electric field $\vec{E}$ in the rod and the distance L between ends I and II:

$$\mathcal{E}_{\text{motional emf}} = |\Delta V_{\text{I–II}}| = EL = vBL \qquad (21.4)$$

This expression for motional emf is the same expression we derived in Example 21.7 using Faraday's law. Thus, for problems involving conducting objects moving in a magnetic field, we can use either Faraday's law or the motional emf expression to determine the emf produced—either method will provide the same result.

The electric generator: changing the loop's orientation

Worldwide, we convert an average of 1000 J per person of electric potential energy into other less useful energy forms every second. Electric generators make this electric potential energy available by converting mechanical energy (such as water rushing through a hydroelectric dam) into electric potential energy.

To understand how an electric generator works, consider a very simple device that consists of a loop of wire attached to a turbine (a propeller-like object that can rotate). The loop is positioned between the poles of an electromagnet that produces a steady uniform $\vec{B}$ field. A Bunsen burner next to the turbine heats a flask of water (**Figure 21.14**). The water is converted to steam, which strikes the blades of the turbine, causing the turbine to rotate. The loop of wire attached to the turbine rotates in the $\vec{B}$ field region. When the loop's surface is perpendicular to the $\vec{B}$ field, the magnetic flux through the loop's area is at a maximum. One quarter turn later, the $\vec{B}$ field lines are parallel to the loop's area, and the flux through it is zero. After another quarter turn, the flux is again at its maximum magnitude, but negative in value since the orientation of the loop's area is opposite what it was originally. This changing magnetic flux through the loop's area causes a corresponding induced emf, which produces a current that changes direction each time the loop rotates one half turn. Current that periodically changes direction in this way is known as *alternating current* (AC).

A coal-fired power plant is based on this process. Coal is burned to heat water, converting it to steam. The high-pressure steam pushes against turbine blades, causing the turbine and an attached wire coil to rotate in a strong $\vec{B}$ field. The resulting emf drives the electric power grid. Nuclear, natural gas, and geothermal power plants use the same procedure except they use different fuel to heat water.

Emf of a generator

To determine an expression for the emf produced by an electric generator, consider the changing magnetic flux through a loop's area as it rotates with constant rotational speed ω in a constant uniform $\vec{B}_{ex}$ field (**Figure 21.15**). If there is an angle θ between the loop's normal vector and the $\vec{B}_{ex}$ field, then the flux through the loop's area is

$$\Phi = B_{ex}A \cos\theta$$

Since the loop is rotating, θ is continuously changing. The loop is rotating with zero rotational acceleration ($\alpha = 0$); we can describe the motion with rotational kinematics (see Chapter 9):

$$\theta = \theta_0 + \omega t + \tfrac{1}{2}\alpha t^2 = \theta_0 + \omega t + \tfrac{1}{2}(0)t^2 = \theta_0 + \omega t$$

If we define the initial orientation θ_0 to be zero, then

$$\theta = \omega t$$

where ω is the constant rotational speed of the loop. This means that the magnetic flux Φ through the loop's area as a function of time t is

$$\Phi = B_{ex} A \cos(\omega t)$$

For a side view of the rotating loop, see **Figure 21.16a**. Figure 21.16b shows a graph of the magnetic flux through the loop's area as a function of time.

According to Faraday's law Eq. (21.2), the induced emf around a coil (a multi-turn loop) is

$$\mathcal{E}_{in} = N\left|\frac{\Delta\Phi}{\Delta t}\right| = N\left|\frac{\Phi_f - \Phi_i}{t_f - t_i}\right|$$

Since Φ is continually changing, we should use calculus to write the above equation. However, in this text, we will simply show the result:

$$\mathcal{E}_{in} = NB_{ex} A\omega \sin(\omega t) \tag{21.5}$$

where N is the number of turns in a coil rotating between the poles of the magnet.

Figure 21.16c shows a graph of the induced emf as a function of clock reading. Comparing Figure 21.16b and c, you will see a pattern. The value of $\mathcal{E}_{in}$ at a particular clock reading equals the negative value of the slope of the Φ-versus-t graph at that same clock reading. This makes sense, since the induced emf is related to the rate of change of the magnetic flux through the loop's area. Slopes represent exactly that, rates of change.

Electric power plants in the United States produce an emf with frequency f equal to 60 Hz (Hz is a unit of frequency; 60 Hz means the emf undergoes 60 full cycles in 1 second). This corresponds to a generator coil with a rotational speed

$$\omega = 2\pi f = 2\pi(60 \text{ Hz}) = 120\pi \text{ rad/s}$$

FIGURE 21.15 The magnetic flux through the loop changes as the angle between the normal vector to the loop and the $\vec{B}$ field lines changes.

As the loop rotates in the magnetic field, the flux through the loop continually changes.

FIGURE 21.16 Three representations of the loop in Figure 21.15.

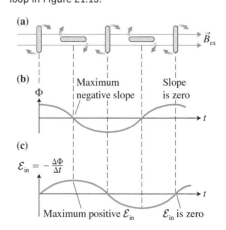

These power plants can produce a peak (maximum) emf as high as 20 kV. The peak emf produced by a generator occurs when $\sin(\omega t) = 1$ and when $\sin(\omega t) = -1$. At those times,

$$\mathcal{E}_{\text{in max}} = NB_{\text{ex}}A\omega.$$ (21.6)

REVIEW QUESTION 21.5 How does the law of electromagnetic induction explain why there is an induced emf in a rotating generator coil?

21.6 AC circuits

In the previous sections we learned that electromagnetic induction allows us to produce variable emf, and thus **alternating current**, or AC (we also use the term AC when we talk about variable voltage). As we discussed above, U.S. electric power plants produce sinusoidally changing emf with a frequency of 60 Hz. Driven by the variable emf, the electric current in household wires also changes magnitude and direction sinusoidally 60 times per second. Direct current (DC) remains constant in direction.

Today, all home appliances operate using AC, but the path to this outcome was not straightforward. Why is AC better for general (massive) use than DC? The battle between AC and DC emerged at the end of the 19th century, when the understanding of electricity and magnetism and the level of related technology were sufficient to enable practical applications. There are two important figures in this story, American physicists Nikola Tesla (Serbian born) and Thomas Edison. Edison was the proponent of DC, and Tesla of AC. The battle was finally won by AC because it proved to be much more efficient in transferring electric power over large distances. We will learn how this is done in the next section, but first we need to study some basic phenomena that are characteristic of AC.

Before we begin, we will agree to call a regular battery a DC power supply and a device that produces alternating voltage an AC power supply.

Resistors in AC circuits

We start with experimental investigations comparing the properties of resistors in DC and AC circuits.

Experiment 1. We connect an incandescent lightbulb to a DC power supply and measure the potential difference across it to be ΔV_{DC} (**Figure 21.17a**). Then we connect an identical lightbulb to an AC power supply (with a frequency of 60 Hz) and change the amplitude of the variable voltage until the bulb appears equally bright (Figure 21.17b). We use a special kind of voltmeter called an oscilloscope, which shows on the screen how voltage changes with time. We measure the amplitude of the variable voltage across the AC power supply ΔV_{AC} and find that when the bulbs appear equally bright, $\Delta V_{\text{AC}} \approx 1.4\Delta V_{\text{DC}}$.

Does this result make sense? Qualitatively, it does. The voltage across the DC power supply is constant all the time, while the voltage across the AC power supply is changing 60 times every second. Therefore, most of the time the voltage across the bulb is less than the amplitude. The higher amplitude AC voltage should thus produce the same warming effect as the lower DC voltage.

Can we explain the factor of 1.4? The energy transferred from the DC power supply to the lightbulb during time interval Δt is equal to the product of power and time interval:

$$P_{\text{DC}}\Delta t = \frac{\Delta V_{\text{DC}}^2}{R}\Delta t$$

FIGURE 21.17 An incandescent lightbulb in a DC and in an AC circuit.

(a)

(b)

An AC power supply allows the voltage amplitude to be changed.

Symbol for AC power supply

We can represent this energy as the area under the plotted line on the P-versus-t graph (**Figure 21.18a**). We use the same method to determine the energy transferred to the bulb by the AC power supply:

$$P_{AC} \Delta t = \frac{\Delta V_{AC}^2 \sin^2(\omega t)}{R} \Delta t$$

The graph of AC power as a function of time is shown in Figure 21.18b. Note that the power is never negative because it follows the square of the sinusoidal function. The output power of the bulb is continuously changing from zero to the maximal value, 120 times a second. We do not detect these changes in brightness because our eyes can only see variations that happen about 24 or fewer times per second. For us, the bulb appears to glow with constant brightness. Therefore, it makes sense to replace this time-varying power with an equivalent average constant power (remember, we did this before in Chapter 12, when we derived the relation between the speed of gas molecules and the pressure on the container wall).

As you can see in Figure 21.18b, the area under the sinusoidal curve is equal to the area of the shaded rectangle whose height is equal to the half of the maximal power. This means that the average output power of a resistive element that is connected to an AC source is equal to half of the maximal power output of that element.

Now we can write

$$\bar{P}_{AC} \Delta t = \frac{\Delta V_{AC}^2}{2R} \Delta t$$

Comparing this equation with the equation

$$P_{DC} \Delta t = \frac{\Delta V_{DC}^2}{R} \Delta t$$

we see that when both power supplies transfer the same energy to the bulb,

$$\bar{P}_{AC} \Delta t = P_{DC} \Delta t, \frac{\Delta V_{AC}^2}{2R} \Delta t = \frac{\Delta V_{DC}^2}{R} \Delta t \implies \Delta V_{AC} = \sqrt{2}\, \Delta V_{DC} \approx 1.4\, \Delta V_{DC}$$

This is in agreement with the earlier result we found experimentally. The DC power supply has to have a smaller voltage across it to produce the same average power as the AC power supply. This finding has important practical value because it reduces the analysis of AC circuits with resistive elements to the analysis of the equivalent DC circuits, which we have already learned. For this reason, we define **root-mean-square voltage** ΔV_{rms} as

$$\Delta V_{rms} = \frac{1}{\sqrt{2}} \Delta V_{AC} \tag{21.7}$$

We could do a similar analysis for the current through the bulb using the relation $P = I^2 R$ and define **root-mean-square current** I_{rms} as

$$I_{rms} = \frac{1}{\sqrt{2}} I_{AC} \tag{21.8}$$

Using these new quantities, we can write the average power as

$$P_{AC} = I_{rms} \Delta V_{rms}$$

The above equation has the same form as our DC definition of power; therefore, we can use rms values to analyze AC circuits the same way we did with DC circuits.

What is happening inside the wires and other resistive elements in AC circuits? In Chapter 19, we learned that free electrons in metals accelerate in the presence of an external electric field. When a constant potential difference is applied across a metal conductor and there is direct current through it, the free electrons, drifting in one direction (due to the DC), interact with ions of the crystal lattice, and the whole wire warms up. What happens when there is an alternating current through the wire? The changing electric field in the wire exerts a variable force on the free electrons,

FIGURE 21.18 Power-versus-time graphs for an incandescent lightbulb in a DC and in an AC circuit.

(a) P_{DC}

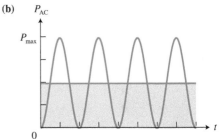

(b) P_{AC}

which in turn vibrate. While vibrating, the free electrons interact with the ions of the lattice and transfer some energy to them. Note that in direct current, the free electrons move along the wire from one terminal to another terminal of the power supply, while in alternating current, all free electrons just vibrate. It is due to these vibrations that the wires warm up in AC circuits.

How are current through and voltage across the resistor related? Imagine that you apply variable voltage across a metal resistor. Free electrons inside the metal immediately respond to the changes in the electric field. Therefore, current through and voltage across the resistor vibrate in phase (they have the same period T, as shown in **Figure 21.19**, where the two quantities are presented on the same time axis).

The instantaneous values of the current through and voltage across the resistor are related by Ohm's law:

$$I_R(t) = \frac{\Delta V_R(t)}{R}$$

Since the rms values are directly proportional to the peak values, we can write

$$I_{R\,rms} = \frac{\Delta V_{R\,rms}}{R}$$

FIGURE 21.19 Current through and voltage across a metal resistor in an AC circuit.

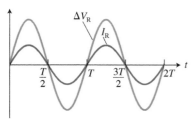

The current and the voltage are in phase.

Capacitors in an AC circuit

VIDEO
21.1

To investigate the behavior of capacitors in AC circuits, we will conduct the following experiments.

Experiment 1. Connect an incandescent lightbulb and a capacitor in series to a DC power supply. The lightbulb does not glow (**Figure 21.20a**). This makes sense—the capacitor behaves like an open switch. If we connect a wire across the capacitor, the lightbulb glows.

Experiment 2. Connect the (same) lightbulb and the capacitor to an AC power supply (see the video; the experiment in the video starts with a frequency of 9 MHz). The lightbulb glows, but not very brightly (Figure 21.20b).

Experiment 3. Increase the frequency of the AC, keeping the amplitude the same as in Experiment 2. The lightbulb becomes brighter (Figure 21.20c).

Experiment 4. Set the AC frequency to the initial value and place a dielectric between the capacitor plates (as you know, this will increase its capacitance). The bulb glows even more brightly than in Experiment 2 (Figure 21.20d).

FIGURE 21.20 An incandescent lightbulb and a capacitor in series in DC and AC circuits.

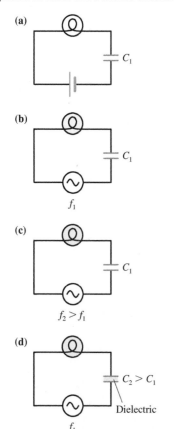

Based on the above experiments, we can conclude that **a capacitor behaves like a resistor in an AC circuit.** The larger the value of C, the smaller the resistance of the capacitor; the larger the frequency, the smaller the resistance of the capacitor. This resistive property of a capacitor is called **capacitive reactance** and is denoted with the symbol X_C.

How can we explain why capacitors conduct AC but not DC? As we discussed earlier, the electrons do not travel but just vibrate when a wire is connected across an AC power supply. As the electrons in the conducting wires vibrate, they charge and discharge the capacitor. The fact that the electrons cannot move across the capacitor (from one plate to the other) does not prevent electrons from vibrating in other parts of the wire, including the electrons in the bulb. Hence, the bulb can glow.

Why does the bulb glow more brightly when the AC frequency increases? With all other quantities unchanged, the larger AC frequency means that the same electric charge is transferred to and from the capacitor in a shorter time; therefore, there is a larger peak current and consequently a brighter bulb.

Why does the bulb glow more brightly when we use a capacitor with a larger capacitance? With all other quantities unchanged, the larger capacitance means that more charge is transferred to and from the capacitor plates (remember that $q = C\Delta V$), and that is why we observe a larger peak current and thus a brighter bulb.

How are current through and voltage across the capacitor related? Measurements with an oscilloscope connected to a capacitor connected to an AC power supply show time dependence of both variables: the current and voltage across the capacitor change sinusoidally with the same period T (these are both shown on the same time axis in **Figure 21.21**). However, the maximal values of current and voltage are achieved at different times.

Explanation. The current to the capacitor plates is equal to the electric charge that is transferred to the plates divided by the time interval for that amount of charge to pass. In other words, the current is equal to the rate of change of charge on the capacitor plates. Because the charge on the capacitor plates is directly proportional to the voltage across the capacitor, we can say that the current through the capacitor is proportional to the rate of change of the voltage across the capacitor. The current is maximal when the rate of change of the voltage is maximal. Since the rate of change of sinusoidally changing voltage is largest every time it passes through zero (the steepest part of the sine function), the current through and the voltage across the capacitor are not in phase. They should be shifted by one quarter of a period $(T/4)$. This is exactly what is observed: Figure 21.21 shows that the peak values of the current are to the left of the peak values of the voltage.

Quantitative analysis of capacitive reactance The derivation of the cause-effect expression for capacitive reactance is beyond the scope of this book. We therefore show it without the derivation but do discuss how we can evaluate the expression. Capacitive reactance is given by

$$X_C = \frac{1}{2\pi f C} \tag{21.9}$$

where $f\ (=1/T)$ is the frequency of the AC power supply and C is the capacitance of the capacitor. To make sense of this expression, we first look at the units. Remember that the unit for frequency is Hz (1 Hz is equal to $1/s$); the unit of capacitance is F (1 F is equal to 1 C/1 V, and 1 C is equal to 1 A $\times$ 1 s). Looking at the right side of Eq. (21.9), we get the units

$$\frac{1}{\text{Hz} \cdot \text{F}} = \frac{\text{s} \cdot \text{V}}{\text{C}} = \frac{\text{V}}{\text{A}} = \Omega$$

The units are therefore correct. The expression is also consistent with our previous observations when the capacitance and frequency were increased independently: both lead to larger current in the circuit. We can next check the extreme cases. When the frequency is close to zero, the reactance should be infinite because the capacitor becomes an open switch; this is exactly what the expression predicts. When the capacitance is close to zero, the reactance should be infinite because no charge will accumulate on the plates.

We can relate the reactance X_C to $I_{C\,rms}$ through and $\Delta V_{C\,rms}$ across the capacitor with an expression that is equivalent to Ohm's law:

$$I_{C\,rms} = \frac{\Delta V_{C\,rms}}{X_C} \tag{21.10}$$

We can use Eq. (21.10) to write an operational definition for capacitive reactance:

$$X_C = \frac{\Delta V_{C\,rms}}{I_{C\,rms}} \tag{21.11}$$

Note that the *instantaneous* voltage across and current through the capacitor are not related to the reactance as described above since they are out of phase.

FIGURE 21.21 Current through and voltage across a capacitor in an AC circuit.

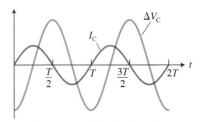

The current peaks before the voltage.

VIDEO
21.2

FIGURE 21.22 An incandescent lightbulb and wires of different shapes in series in DC and AC circuits.

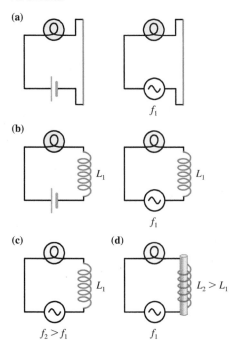

(a)

f_1

(b)

L_1 L_1

f_1

(c)   (d)

L_1 $L_2 > L_1$

$f_2 > f_1$ f_1

Solenoids in AC circuits

Another important element of AC circuits is a solenoid. We will investigate its behavior in the following experiments.

Experiment 1. We connect a lightbulb to a battery using a long straight wire. The lightbulb glows. We then prepare another setup using an identical lightbulb and wire, but this time we connect them to an AC voltage source. We adjust the AC voltage amplitude so the lightbulbs in both circuits are equally bright (**Figure 21.22a**). In the experiment in the video, the frequency is about 1 MHz. Now we change the shape of the long wire by winding it as a solenoid without changing its length (the solenoid is often called an *inductor*). The brightness of the bulb connected to the DC power supply does not change, while the bulb connected to the AC power supply glows more dimly (Figure 21.22b).

Experiment 2. In the same circuit, we increase the AC frequency while keeping the amplitude constant. The lightbulb connected to the AC source glows even more dimly (Figure 21.22c). Now we return the AC frequency to its initial value and insert an iron core into the solenoid. The bulb glows more dimly than with the inductor without the iron core in the last step in Experiment 1 (compare Figure 21.22d and 21.22b).

From these experiments, we can conclude that the inductor behaves like a resistor in an AC circuit. The inductor must possess resistive properties that depend on the properties of the inductor and the frequency of the AC, similar to a capacitor. This resistive property of an inductor is called **inductive reactance** and is denoted with the symbol X_L.

In order to understand how the properties of the inductor affect its inductive reactance, we need to go back to our studies of magnetic fields. In Chapter 20 we learned that the magnetic field inside a solenoid (assumed to be very long) is described by the equation

$$B_{\text{solenoid}} = \mu_0 \frac{N}{l} I$$

where N is the number of turns, l is the length of the solenoid, and μ_0 is the magnetic permeability of vacuum $(\mu_0 = 4\pi \times 10^{-7} (\text{T} \cdot \text{m})/\text{A})$. Knowing the magnetic field inside the solenoid, we can write an expression for the magnetic flux produced by the solenoid when there is current I through it:

$$\Phi = NBA = N\left(\mu_0 \frac{N}{l} I\right) A = \frac{\mu_0 N^2 A}{l} I \tag{21.12}$$

We see that the magnetic flux depends on the current through the solenoid and a combination of physical quantities dependent only on the geometry of the solenoid and the medium inside it. We can rewrite the above equation as

$$\Phi = LI$$

where L is a quantity called **inductance** that contains only properties of the solenoid:

$$L = \frac{\mu_0 N^2 A}{l} = \mu_0 \left(\frac{N}{l}\right)^2 V \tag{21.13}$$

Here, V is the volume of the solenoid $(V = Al)$. The symbol L for inductance is in honor of Heinrich Lenz, and the unit for inductance, the henry (H), is in honor of Joseph Henry.

As we noted above, μ_0 in Eq. (21.13) is the magnetic permeability of vacuum. If we place an iron core inside the solenoid, the magnetic flux in the solenoid increases by a factor of about 5000. This is also the constant by which we need to multiply μ_0 to find the inductance of a solenoid with an iron core. The constant is called the **relative magnetic permeability** μ_r. The relative magnetic permeability of iron is around 5000.

When we place a material of relative permeability μ_r inside the solenoid, Eq. (21.13) becomes

$$L = \frac{\mu_0 \mu_r N^2 A}{l} = \mu_0 \mu_r \left(\frac{N}{l}\right)^2 V$$

Equation (21.13) is the cause-effect relationship for inductance because every physical quantity in it can be varied separately. The inductive properties of a solenoid can be increased by increasing the number of turns per unit length of the solenoid, by increasing the volume of the solenoid, or by filling the inside with a material with a higher relative magnetic permeability.

Now that we have learned the new physical quantity of inductance, we can develop a microscopic explanation of the observed behavior of an inductor in an AC circuit. When an inductor is connected to an AC power supply, the electrons in the wire of the solenoid vibrate. They produce a changing magnetic field, which in turn creates an electric field (induced emf) that opposes the changes in the motion of the same electrons (by Lenz's law). Hence, the bulb glows more dimly.

To explain why the bulb glows more dimly at a larger AC frequency, we need to consider what larger frequency means for the above process. If everything else is unchanged, the larger AC frequency means faster changes of the current and a larger induced emf, which opposes the voltage produced by the AC source and therefore results in a dimmer bulb.

To explain why the bulb glows more dimly when we use an inductor with a larger L, we need to consider how the inductance affects the above process. If everything else is unchanged, the larger L of the solenoid means a larger B inside the solenoid and a larger magnetic flux produced by the solenoid for a given current through it ($\Phi = LI$). Larger magnetic flux means a larger induced emf that opposes the changes in AC source voltage, and this leads to a dimmer bulb.

How are current through and voltage across the inductor related? Using the same equipment that we used for the capacitor, we can record with an oscilloscope the time dependence of both variables (these are both presented in **Figure 21.23**, sharing the same time axis).

Explanation. We assume that the resistance of the inductor is negligible (in real life, this is often not true, which is why real inductors are more complicated to describe than real capacitors). In this case, the voltage across the inductor is equal to the induced emf due to the changing magnetic flux through the inductor. We know that the induced emf is proportional to the rate of change of the magnetic flux through the inductor (for example, a solenoid). Since the magnetic flux is directly proportional to the current through the inductor, we can say that the induced emf across the inductor is proportional to the rate of change of the current through the inductor. Since the rate of change of a sinusoidally vibrating current is largest every time it passes through zero, the voltage across and the current through the inductor are not in phase. They should be shifted by one quarter of a period ($T/4$). This is exactly what we observe: Figure 21.23 shows that the peak values of the current are to the right of the peak values of the voltage.

Quantitative analysis of inductive reactance The derivation of the cause-effect expression for inductive reactance is beyond the scope of this book. We therefore show it without the derivation, but do discuss how we can evaluate the expression. The expression for the inductive reactance is

$$X_L = 2\pi f L \qquad (21.14)$$

You can check whether the units come out as ohms and whether the limiting cases predict reasonable results in the same way we did for capacitive reactance. Note that, unlike capacitive reactance, inductive reactance increases with both the frequency of current through the inductor and the inductance.

FIGURE 21.23 Current through and voltage across an inductor in an AC circuit.

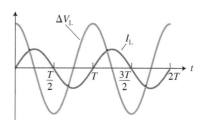

The current peaks after the voltage.

TIP In observational experiments with capacitors and inductors, we were using high-frequency AC (frequency in the range of MHz) to make the phenomena easily observable. However, all the expressions given also hold for lower frequencies, such as 60 Hz.

The inductive reactance X_L is related to the $I_{L\,rms}$ through and $\Delta V_{L\,rms}$ across the inductor with an expression that is equivalent to Ohm's law:

$$I_{L\,rms} = \frac{\Delta V_{L\,rms}}{X_L} \qquad (21.15)$$

Note that the *instantaneous* voltage across and current through the inductor are not related to the inductive reactance as described above since they are out of phase.

One important consequence of the phase shift between the current through and voltage across an ideal inductor or a capacitor is that the power output of these elements (the product of voltage and current) is changing sinusoidally. Therefore, the average power output of the capacitor or inductor connected to AC source is zero. A mechanical analogy would be a spring that is first compressed and then stretched by an external object, the average work (and thus the power) done on the spring is zero.

EXAMPLE 21.8 **A capacitor in an AC circuit**

You have a set of different capacitors. (a) What should be the capacitance of a capacitor that allows a 20-mA rms current to pass through when it is connected to a household electric outlet? (b) If the maximum of the current appears at $t_0 = 0$, at what later clock reading will the first maximum of voltage across the capacitor occur? (c) What is the value of the maximum voltage?

Sketch and translate We sketch the circuit at right. The knowns include $I_{C\,rms} = 20$ mA and the house AC. We need to determine the capacitance, t for the maximum voltage, and its value.

$$\Delta V_{C\,rms} = 120\ V$$
$$C = ?$$
$$I_{C\,rms} = 20\ mA$$
$$f = 60\ Hz$$

Simplify and diagram Properties of household AC voltage depend on the country. In the United States, household power outlets provide AC voltage vibrating with a frequency of 60 Hz and an rms voltage of 120 V. In Europe, most of Africa, most of Asia, most of South America, and Australia, AC voltage vibrates at 50 Hz and rms voltage of 230 V. We will use the data for the United States.

Represent mathematically

(a) We have two equations for the capacitive reactance, an operational definition and a cause-effect relationship:

$$X_C = \frac{\Delta V_{C\,rms}}{I_{C\,rms}} \quad \text{and} \quad X_C = \frac{1}{2\pi f C}$$

We determine the reactive capacitance using the first equation and use the second to determine the capacitance.

(b) The maximum voltage across the capacitor appears one quarter of a period later than the maximum current. Remember that $T = \dfrac{1}{f}$.

(c) To find the maximum voltage, we can use the rms value $\Delta V_{C\,max} = \sqrt{2}\,\Delta V_{C\,rms}$.

Solve and evaluate

(a) $X_C = \dfrac{\Delta V_{C\,rms}}{I_{C\,rms}} = \dfrac{120\ V}{0.020\ A} = 6000\ \Omega$

$\Rightarrow C = (2\pi \times 60\ s^{-1} \times 6000\ \Omega)^{-1} = 4.4 \times 10^{-7}\ F$

(b) $T = \dfrac{1}{f} = 0.0167\ s;\ t = \dfrac{1}{4}T = 0.0042\ s.$

(c) The maximum voltage across the capacitor is $\Delta V_{C\,max} = \sqrt{2}\,\Delta V_{C\,rms} = 170\ V.$

- -

Try it yourself Repeat the problem above but this time for an inductor (replace the words "capacitor" with "inductor" and "capacitance" with "inductance"). Assume that the resistance of the inductor is negligible.

Answer $L = 15.9\ H,\ t = 0.0125\ s,\ \text{and}\ \Delta V_{L,\,max} = 170\ V.$

REVIEW QUESTION 21.6 A capacitor in an electric circuit works as an open switch because its plates are separated by a dielectric material. How can there be current in a circuit with an open switch?

21.7 Transformers

Knowledge of AC helps us understand the operation of an important device, the **transformer**, which allows us to increase or decrease the maximum value of an alternating emf.

A transformer consists of two coils, each wrapped around an iron core (ferromagnetic) (**Figure 21.24**). An alternating emf across the **primary coil** is converted into a larger or smaller alternating emf across the **secondary coil**, depending on the number of loops in each coil.

Transformers are used in many electronic devices. They are also essential for transmitting electric energy from a power plant to your house. The rate of this electric energy transmission is proportional to the product of the emf across the power lines and the electric current in the lines. If the emf is low, considerable electric current is needed to transmit a considerable amount of energy. However, due to the electrical resistance of the power lines, much of the electric energy is converted into thermal energy. The rate of this conversion is $P = I^2R$. To reduce the I^2R losses, the transmission of electric energy is done at high peak emf (about 200,000 V) and low current. Transformers then reduce this peak emf to about 170 V for use in your home. How does a transformer change the peak emf?

Suppose there is an alternating current in the primary coil, the coil connected to an external power supply. The secondary coil is connected to an electrical device, but this device requires an emf that is different from what the external power supply produces. The alternating current in the primary coil produces a $\vec{B}$ field within the transformer core. Since the current is continuously changing, the magnetic flux through the primary coil's area is also continuously changing. Thus, an emf is induced in the primary coil:

$$\mathcal{E}_p = N_p \left| \frac{\Delta \Phi_p}{\Delta t} \right|$$

where the p subscript refers to the primary coil.

The primary coil is connected to an AC power supply. Assuming that the resistance of the coil is zero, the emf of the power supply and the induced emf in the coil should add to zero according to Kirchhoff's loop rule. Therefore, the emf across the primary coil is the same in magnitude as the emf of the AC power supply.

In an efficient transformer, nearly all of the $\vec{B}$ field lines passing through the primary coil's area also pass through the secondary coil's area. This is due to the presence of the ferromagnetic core around which both coils are wound. The core serves as a "guide" for the magnetic field. As a result, there is a changing magnetic flux through the secondary coil's area as well as a corresponding emf produced in it:

$$\mathcal{E}_s = N_s \left| \frac{\Delta \Phi_s}{\Delta t} \right|$$

If the transformer is perfectly efficient, then the rates of change of the magnetic flux through one turn of each coil are the same:

$$\left| \frac{\Delta \Phi_p}{\Delta t} \right| = \left| \frac{\Delta \Phi_s}{\Delta t} \right|$$

Combining the three equations above,

$$\frac{\mathcal{E}_p}{N_p} = \frac{\mathcal{E}_s}{N_s}$$

$$\Rightarrow \mathcal{E}_s = \frac{N_s}{N_p} \mathcal{E}_p \qquad (21.16)$$

We see that the emf in the secondary coil can be substantially larger or smaller than the emf in the primary coil depending on the number of turns in each coil. Engineers use this result to design transformers for specific purposes. For example, a **step-down transformer** can convert the 120-V alternating emf from a wall socket to a 9-V alternating emf, which is then converted to DC to power a laptop computer.

FIGURE 21.24 A transformer changes the input/output emf depending on the ratio of the turns.

Alternating power output

Secondary coil with N_2 turns

Iron core

Primary coil with N_1 turns

$\mathcal{E}_s$

$\mathcal{E}_p$

Transformer for laptop

Your laptop requires a 24-V emf to function. What should be the ratio of primary coil turns to secondary coil turns if this transformer is to be plugged into a standard house AC outlet (effectively a 120-V emf)?

Represent mathematically   The ratio we are looking for is related to the coil emfs by Eq. (21.16):

$$\mathcal{E}_s = \frac{N_s}{N_p} \mathcal{E}_p$$

Solve and evaluate   Solving for the ratio and inserting the appropriate values:

$$\frac{N_p}{N_s} = \frac{\mathcal{E}_p}{\mathcal{E}_s} = \frac{120 \text{ V}}{24 \text{ V}} = 5$$

The primary coil needs to have five times the number of turns as the secondary coil. This is a step-down transformer, since the resulting secondary coil peak emf is lower than the primary coil peak emf.

Try it yourself   If the primary coil had 200 turns, how many turns should the secondary coil have to reduce the peak emf from 120 V to 6 V?

Answer   10 turns.

Some transformers are designed to increase rather than decrease emf. In a car that uses spark plugs for ignition, a transformer converts the 12-V potential difference of the car battery to the 20,000-V potential difference needed to produce a spark in the engine's cylinder (**Figure 21.25**). The battery supplies a steady current in the transformer. An electronic switching system in the circuit can open the circuit, stopping the current in the primary coil in a fraction of a millisecond. This causes an abrupt change in the magnetic flux through the primary coil of the transformer, which leads to an induced emf in the secondary coil. The secondary coil is attached to a spark plug that has a gap between two conducting electrodes. When the potential difference across the gap becomes sufficiently high, the gas between the electrodes ionizes. When the ionized atoms recombine with electrons, the energy is released in the form of light—a spark that ignites the gasoline.

FIGURE 21.25   How a transformer converts the emf from a 12-V car battery to a 20,000-V spark.

1. Current in the primary circuit produces a magnetic field in the magnetic core.

2. Opening the switch causes an abrupt decrease in the primary current and in the core magnetic field ...

Electronic switch

I

12-V battery   Primary coil   Secondary coil

20,000-V spark

3. ... which causes a large induced potential difference in the secondary coil and dielectric breakdown (spark) across the spark plug gap.

The induced emf in the secondary coil is much greater than the 12 V in the primary coil for three reasons. First, the secondary coil has many more turns than the primary coil ($N_s \gg N_p$). Second, the magnetic flux through the primary coil's area decreases very quickly (the Δt in the denominator of Faraday's law is very small), resulting in a large induced emf $\mathcal{E}_p$ to which $\mathcal{E}_s$ is proportional [see Eq. (21.2)]. Third, the ferromagnetic core (usually iron) passing through the two coils significantly increases the $\vec{B}$ field within it (Section 20.8). For these three reasons, it is possible for $\mathcal{E}_s \gg \mathcal{E}_p$; the 12-V car battery can produce a 20,000-V potential difference across the electrodes of a spark plug.

REVIEW QUESTION 21.7   How does a transformer achieve different induced peak emfs across its primary and secondary coils?

21.8 Mechanisms explaining electromagnetic induction

Faraday's law *describes* how a changing magnetic flux through a wire loop is related to an induced emf, but it does not *explain* how the emf comes about. In this section we will explain the origin of the induced emf.

A changing $\vec{B}_{ex}$ field has a corresponding $\vec{E}$ field

We know that a changing magnetic flux induces an electric current in a stationary loop (Figure 21.26). Because the loop is not moving, there is no net magnetic force exerted on the free electrons in the wire. Thus, an electric field must be present. The electric field that drives the current exists throughout the wire. This electric field is not produced by charge separation, but by a changing magnetic field.

If we were to represent it with $\vec{E}$ field lines, those lines would have no beginning or end—they would form closed loops. This electric field essentially "pushes" the free electrons along the loop. We can describe it quantitatively with the emf. But this emf is very different from the emf produced by a battery. For the battery, the emf is the result of charge separation across its terminals. For the induced emf, the electric field that drives the current does not originate on the charges but is distributed throughout the entire loop; thus the emf is also actually distributed throughout the entire loop. You might visualize it as an electric field "gear" with its teeth hooked into the electrons in the wire loop, pushing the free electrons along the wire at every point.

What do we now know about electricity and magnetism?

We have learned a great deal about electric and magnetic phenomena. We learned about electrically charged objects that interact via electrostatic (Coulomb) forces. Stationary electrically charged objects produce electric fields, and electric field lines start on positive charges and end on negative charges (Figure 21.27a). In our study of magnetism (Chapter 20), we learned that moving electrically charged objects and permanent magnets interact via magnetic forces and produce magnetic fields. Magnetic field lines do not have beginnings or ends (Figure 21.27b); they are continuous loops. So far, individual magnetic charges, magnetic monopoles, have not been found.

When we studied electric circuits (Chapter 19) we learned that electric fields cause electrically charged particles inside conductors to move in a coordinated way— electric currents. Later we learned that electric currents produce magnetic fields (Figure 21.27c). In this chapter, we learned about the phenomenon of electromagnetic induction and its explanation: a changing magnetic field is always accompanied by a corresponding electric field (Figure 21.27d). However, this new electric field is not produced by electric charges, and its field lines do not have beginnings or ends.

Except for the lack of magnetic charges, there is symmetry between electric and magnetic fields. This symmetry leads us to pose the following question: if in a region where the magnetic field is changing there is a corresponding electric field, is it possible that in a region where the electric field is changing there could be a corresponding magnetic field (Figure 21.27e)? This hypothesis, suggested in 1862 by James Clerk Maxwell, led to a unified theory of electricity and magnetism, a subject we will investigate in our chapter on electromagnetic waves (Chapter 25).

REVIEW QUESTION 21.8 Explain how (a) an electric current is produced when only a part of a single wire loop moves through a magnetic field and how (b) an electric current is produced when an external magnetic flux changes through a closed loop of wire.

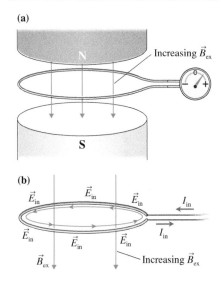

FIGURE 21.26 A changing $\vec{B}_{ex}$ creates an electric field $\vec{E}$ that induces an electric current.

(a)

(b)

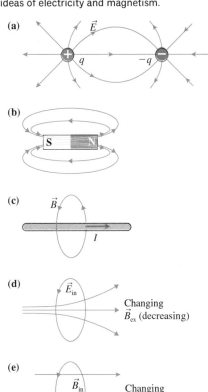

FIGURE 21.27 A summary of some of the main ideas of electricity and magnetism.

(a)

(b)

(c)

(d)

Changing $\vec{B}_{ex}$ (decreasing)

(e)

Changing $\vec{E}_{ex}$ (decreasing)

Summary

The **magnetic flux** through a loop's area depends on the size of the area, the $\vec{B}$ field magnitude, and the orientation of the loop relative to the $\vec{B}$ field. (Sections 21.2 and 21.4)

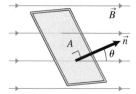

$$\Phi = BA \cos\theta \qquad \text{Eq. (21.1)}$$

when B is uniform through the loop area

Electromagnetic induction In a region with a changing magnetic field, there is a corresponding induced electric field. When a wire loop or coil is placed in this region, the magnetic flux through that loop changes, and an electric current is induced in the loop. (Section 21.4)

$\vec{B}_{ex}$ increasing at location of coil

$\vec{B}_{ex}$ decreasing when switch is opened

$$\mathcal{E}_{in} = N\left|\frac{\Delta\Phi}{\Delta t}\right| = N\left|\frac{\Phi_f - \Phi_i}{t_f - t_i}\right|$$

Eq. (21.2)

Lenz's law When current is induced, its direction is such that the $\vec{B}$ field it produces opposes the change in the magnetic flux through the loop. (Section 21.3)

An **electric generator** produces an emf by rotating a coil within a region of strong $\vec{B}$ field, an important application of electromagnetic induction. (Section 21.5)

$$\mathcal{E}_{in} = NB_{ex}A\omega \sin(\omega t) \quad \text{Eq. (21.5)}$$

Alternating current (AC) circuits Alternating current is a current that varies in magnitude and direction, in contrast to direct current (DC), which remains constant in direction. AC is produced by special power supplies whose emf varies with frequency f. Resistors, capacitors, and inductors (solenoids) behave differently in DC and AC circuits. (Section 21.6)

Resistor   **Capacitor**   **Inductor**

$$X_C = \frac{1}{2\pi f C} \qquad \text{Eq. (21.9)} \qquad X_L = 2\pi f L \qquad \text{Eq. (21.14)}$$

$$I_{C\,rms} = \frac{\Delta V_{C\,rms}}{X_C} \quad \text{Eq. (21.10)} \qquad I_{L\,rms} = \frac{\Delta V_{L\,rms}}{X_L} \quad \text{Eq. (21.15)}$$

$$\Delta V_{rms} = \frac{1}{\sqrt{2}}\Delta V_{AC} \qquad \text{Eq. (21.7)}$$

$$I_{rms} = \frac{1}{\sqrt{2}}I_{AC} \qquad \text{Eq. (21.8)}$$

Transformers are electrical devices used to increase or decrease the peak value of an alternating emf. (Section 21.7)

$$\mathcal{E}_s = \frac{N_s}{N_p}\mathcal{E}_p \qquad \text{Eq. (21.16)}$$

Questions

Multiple Choice Questions

1. In which of the experiments with a loop and a bar magnet shown in Figure Q21.1 is an electric current induced in the loop at the instant shown in the figures? In experiments (d) to (f), the magnet rotates around the given axis; in (g) to (i), the loop collapses as indicated by the arrows.

FIGURE Q21.1

2. If you move the coil in Figure Q21.2 toward the N pole of the large electromagnet, which produces a $\vec{B}$ field that is constant (in time) and uniform (in space), would an electric current be induced?
   (a) Yes
   (b) No
   (c) It depends on the current in the electromagnet.

FIGURE Q21.2

3. The magnetic flux through a 10.0-cm^2 loop is 8.0 T·cm^2, and the magnetic field at the location of the loop is uniform. Choose all correct answers.
   (a) The magnitude of the $\vec{B}$ field must satisfy the condition $B \leq 0.8$ T.
   (b) The direction of the $\vec{B}$ field cannot be perpendicular to the surface of the loop.
   (c) If the direction of $\vec{B}$ field is parallel to the normal to the loop's surface, then $B = 0.8$ T.
   (d) If $B = 1.0$ T, then the angle between the $\vec{B}$ field and the normal to the loop's surface is greater than 0°.

4. Your friend says that the emf induced in a coil supports the changing flux through the coil rather than opposes it. According to your friend, what happens when the magnetic flux increases slightly?
   (a) The induced current will increase continuously.
   (b) The coil will get hot and eventually melt.
   (c) The induced emf will get larger.
   (d) None of the above.
   (e) (a), (b), and (c) will occur.

5. A metal ring lies on a table. The S pole of a bar magnet moves down toward the ring from above and perpendicular to its surface. Which answer and explanation correctly predict the direction of the induced current as seen from above?
   (a) Clockwise because the $\vec{B}$ field is down and increasing
   (b) Clockwise because the $\vec{B}$ field is up and increasing
   (c) Counterclockwise because the $\vec{B}$ field is down and increasing
   (d) Counterclockwise because the $\vec{B}$ field is up and increasing.
   (e) There is no current; it only changes when the N pole approaches.

6. One coil is placed on top of another. The bottom coil is connected in series to a battery and a switch. With the switch closed, there is a clockwise current in the bottom coil. When the switch is opened, the current in the bottom coil decreases abruptly to zero. What is the direction of the induced current in the top coil, as seen from above while the current in the bottom coil decreases?
   (a) Clockwise   (b) Counterclockwise
   (c) Zero—the current is induced only when the coils move relative to each other
   (d) There is not enough information to answer this question.

7. Two coils are placed next to each other flat on the table. The coil on the right is connected in series to a battery and a switch. With the switch closed, there is a clockwise current in the right coil as seen from above. When the switch is opened, the current in the right coil decreases abruptly to zero. What is the direction of the induced current in the coil on the left as seen from above while the current in the right coil decreases?
   (a) Clockwise   (b) Counterclockwise
   (c) Zero because the current is only present when the coils move relative to each other
   (d) Zero because there is no magnetic field through the coil on the left

8. Two identical bar magnets are dropped vertically from the same height. One magnet passes through an open metal ring, and the other magnet passes through a closed metal ring. Which magnet will reach the ground first?
   (a) The magnet passing through the closed ring
   (b) The magnet passing through the open ring
   (c) The magnets arrive at the ground at the same time.
   (d) There is too little information to answer this question.

9. A window's metal frame is essentially a metal loop through which a magnetic field can change when the window swings shut abruptly. The metal frame is 1.0 m × 0.50 m and Earth's 5.0×10^{-5}-T magnetic field makes an angle of 53° relative to the horizontal. Which answer below is closest to the average induced emf when the window swings shut 90° in 0.20 s from initially parallel to the field? Assume that it's a vertical window that faces north when closed.
   (a) 9.9×10^{-5} V   (b) 7.5×10^{-5} V
   (c) 9.9×10^{-2} V   (d) 12.5×10^{-4} V

10. Four identical loops move at the same velocity toward the right in a uniform magnetic field inside the dashed lines, as shown in Figure Q21.10. Which choice below best represents the ranking of the magnitudes (largest to smallest) of the induced currents in the loops at the snapshot shown in the figure?

    FIGURE Q21.10

    (a) $I_1 = I_2 > I_3 = I_4$
    (b) $I_2 = I_1 = I_4 > I_3$
    (c) $I_3 > I_4 > I_2 > I_1$
    (d) $I_1 > I_2 > I_3 = I_4$
    (e) $I_1 = I_2 = I_3 = I_4$

11. A 12-V automobile battery provides the thousands of volts needed to produce a spark across the gap of a spark plug. Which of the following mechanisms is involved?
    (a) The secondary loop connected to the spark plug has many more turns than the primary loop attached to the battery.
    (b) The current in the primary coil is reduced quickly.
    (c) An iron core increases the magnetic flux through the primary and secondary coils.
    (d) All three mechanisms are involved.
    (e) Only mechanisms (a) and (b) are involved.

12. A respiration detector consists of a coil placed on a person's chest and another placed on the person's back. There is a constant current in one coil. What causes an induced current to be produced in the other coil?
    (a) The person's heart beats.
    (b) The person's breathing causes the coil separation to change.
    (c) The person moves.
    (d) All three of the above occur.

13. A parallel plate capacitor and a lightbulb are connected in series to an AC power supply. Which of the following will lead to an increase in the I_{rms} through the capacitor?
    (a) Decrease the distance between the capacitor plates.
    (b) Decrease the frequency of the AC.
    (c) Insert a material with a smaller dielectric constant between the capacitor plates.
    (d) Decrease the surface area of the capacitor plates.
14. A solenoid and a lightbulb are connected in series to an AC power supply. Which of the following will lead to a decrease in the I_{rms} through the solenoid (assuming the resistance of the solenoid wires is negligible)? Multiple answers are possible.
    (a) Decrease the length of the solenoid while keeping the number of turns and the radius of the solenoid constant.
    (b) Increase the frequency of the AC.
    (c) Increase the $\Delta V_{AC\ rms}$.
    (d) Insert a paramagnetic material into the solenoid.
    (e) Increase the number of turns in the solenoid per unit length by a factor of 2 and decrease its radius by a factor of 2.

Conceptual Questions

15. A bar magnet falling with the north pole facing down passes through a coil held vertically. Sketch flux-versus-time and emf-versus-time graphs for the magnet approaching the coil and passing through it. What assumptions did you make?
16. An induction cooktop has a smooth surface. When on high, the surface does not feel warm, yet it can quickly cook soup in a metal bowl. However, it cannot cook soup in a ceramic or glass bowl. Explain how the cooktop works.

17. Describe three common applications of electromagnetic induction.
18. Two rectangular loops A and B are near each other. Loop A has a battery and a switch. Loop B has no battery. Imagine that a current starts increasing in loop A. Will there be a current in loop B? Samir argues that there will be current. Ariana argues that there will be no current. Describe experiments that support the claims of both students.
19. A simple metal detector has a coil with an alternating current in it. The current produces an alternating magnetic field. If a piece of metal is near the coil, eddy currents are induced in the metal. These induced eddy currents produce induced magnetic fields that are detected by a magnetic field detection device. Draw a series of sketches representing this process, including the appropriate directions of the magnetic fields at one instant of time, and indicate two applications for this device.
20. Construct flux-versus-time and emf-versus-time graphs that explain how an electric generator works.
21. How is it possible to get a 2000-V emf from a 12-volt battery?
22. You connect a capacitor and a lightbulb in series with an AC power supply. The lightbulb does not glow. Describe what changes you can make to the capacitor or to the AC power supply that will cause the lightbulb to glow. Explain your reasoning.
23. You connect a solenoid and a lightbulb in series with an AC power supply. The lightbulb glows. How will the brightness of the lightbulb change if you squeeze the solenoid hard between two large books so that the solenoid flattens? Explain your reasoning.

Problems

Below, **BIO** indicates a problem with a biological or medical focus. Problems labeled **EST** ask you to estimate the answer to a quantitative problem rather than derive a specific answer. Asterisks indicate the level of difficulty of the problem. Problems with no * are considered to be the least difficult. A single * marks moderately difficult problems. Two ** indicate more difficult problems.

21.1 Inducing an electric current

1. * You and your friend are performing experiments in a physics lab. Your friend claims that in general, something has to move in order to induce a current in a coil that has no battery. What experiments can you perform to support her idea? What experiments can you perform to reject it?
2. You decide to use a metal ring as an indicator of induced current. If there is a current, the ring will feel warm in your hand. You place the ring around a solenoid as shown in **Figure P21.2**, position I. Will the ring feel warm if the solenoid is connected to (a) a DC power supply or (b) an AC power supply? (c) Answer questions (a) and (b) for the case when the ring is parallel to the solenoid (position II). Explain your answers.
3. * To check whether a lightbulb permanently attached to a coil is still good, you place the coil next to another coil that is attached to a battery, as shown in **Figure P21.3**. Explain how or whether each of the following actions can help you determine if the lightbulb works.
    (a) Close the switch in circuit A.
    (b) Keep the switch in circuit A closed.
    (c) Open the switch in circuit A. Indicate any assumptions that you made.

4. * **Flashlight without batteries** A flashlight that operates without batteries is lying on your desk. The light illuminates only when you repeatedly squeeze the flashlight's handle. You notice that paper clips tend to stick to the outside of the flashlight. What physical mechanism might control the operation of the flashlight?
5. You need to invent a practical application for a coil of wire that detects the vibrations or movements of a nearby magnet. Describe your invention. (The application should not repeat any described in this book.)
6. * **Detect burglars entering windows** Describe how you will design a device that uses electromagnetic induction to detect a burglar opening a window in your ground floor apartment. Include drawings and a word description.
7. * A coil connected to an ammeter can detect alternating currents in other circuits. Explain how this system might work. Could you use it to eavesdrop on a telephone conversation being transmitted through a wire?

21.2 Magnetic flux

8. * The $\vec{B}$ field in a region has a magnitude of 0.40 T and points in the positive z-direction, as shown in **Figure P21.8**. Determine the magnetic flux through (a) surface abcd, (b) surface bcef, and (c) surface adef.
9. **EST** How do you position a bicycle tire so that the magnetic flux through it due to Earth's magnetic field is as large as possible? Estimate this maximum flux. What assumptions did you make?
10. * **EST** Estimate the magnetic flux through your head when the $\vec{B}$ field of a 1.4-T MRI machine passes through your head.
11. * **EST** Estimate the magnetic flux through the south- and west-facing windows of a house in British Columbia, where Earth's $\vec{B}$ field has a magnitude of 5.8×10^{-5} T and points roughly north with a downward inclination of 72°. Explain how you made the estimates.

FIGURE P21.2

FIGURE P21.3

FIGURE P21.8

21.3 Direction of the induced current

12. You perform experiments using an apparatus that has two insulated wires wrapped around a cardboard tube (Figure P21.3). Determine the direction of the current in the bulb when (a) the switch is closing and the current in coil A is increasing, (b) the switch has just closed and there is a steady current in coil A, and (c) the switch has just opened and the current in coil A is decreasing.

13. You have the apparatus shown in **Figure P21.13**. A circular metal plate swings past the north pole of a permanent magnet. The metal consists of a series of rings of increasing radius. Indicate the direction of the current in one ring (a) as the metal swings down from the left into the magnetic field and (b) as the metal swings up toward the right out of the magnetic field. Use Lenz's law to justify your answers.

FIGURE P21.13

Magnet

Metal plate

14. * You suggest that eddy currents can stop the motion of a steel disk that vibrates while hanging from a spring. Explain how you can do this without touching the disk.

15. * Your friend thinks that an induced magnetic field is *always* opposite the changing external field that induces an electric current. Provide a counterexample to your friend's claim. Discuss why your friend's idea would violate conservation of energy.

21.4 and 21.5 Faraday's law of electromagnetic induction and Skills for analyzing processes involving electromagnetic induction

16. The magnetic flux through three different coils is changing as shown in **Figure P21.16**. For each situation, draw a corresponding graph showing qualitatively how the induced emf changes with time.

FIGURE P21.16

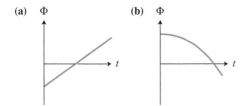

17. The magnetic flux through three different coils is changing as shown in **Figure P21.17**. For each situation, draw a corresponding graph showing quantitatively how the induced emf changes with time.

FIGURE P21.17

18. A magnetic field passing through two coils of the same diameter and length decreases from a magnitude of B_{ex} to zero in the time interval Δt. The first coil has twice the number of turns as the second. (a) Compare the emfs induced in the coils. (b) How can you change the experiment so that the emfs produced in them are the same?

19. **BIO Stimulating the brain** In transcranial magnetic stimulation (TMS) an abrupt decrease in the electric current in a small coil placed on the scalp produces an abrupt decrease in the magnetic field inside the brain. Suppose the magnitude of the $\vec{B}$ field changes from 0.80 T to 0 T in 0.080 s. Determine the induced emf around a small circle of brain tissue of radius 1.2×10^{-3} m. The $\vec{B}$ field is perpendicular to the surface area of the circle of brain tissue.

20. * To measure a magnetic field produced by an electromagnet, you use a circular coil of radius 0.30 m with 25 loops (resistance of 25 Ω) that rests between the poles of the magnet and is connected to an ammeter. While the current in the electromagnet is reduced to zero in 1.5 s, the ammeter in the coil shows a steady reading of 180 mA. Draw a picture of the experimental setup and determine everything you can about the electromagnet.

21. You want to use the idea of electromagnetic induction to make the bulb from your small flashlight glow; it glows when the potential difference across it is 1.5 V. You have a small bar magnet and a coil with 100 turns, each with area 3.0×10^{-4} m^2. The magnitude of the $\vec{B}$ field at the front of the bar magnet's north pole is 0.040 T and reaches 0 T when it is about 4 cm away from the pole. Can you make the bulb light? Explain.

22. * **BIO Breathing monitor** An apnea monitor for adults consists of a flexible coil that wraps around the chest (**Figure P21.22**). When the patient inhales, the chest expands, as does the coil. Earth's $\vec{B}$ field of 5.0×10^{-5} T passes through the coil at a 53° angle relative to a line perpendicular to the coil. Determine the average induced emf in such a coil during one inhalation if the 300-turn coil area increases by 42 cm^2 during 2.0 s.

FIGURE P21.22

23. * A bar magnet induces a current in an N-turn coil as the magnet moves closer to it (**Figure P21.23**). The coil's radius is R cm, and the average induced emf across the bulb during the time interval is $\mathcal{E}$ V. (a) Make a list of the physical quantities that you can determine using this information; (b) Is the direction of the induced current from lead a to b, or from b to a? Explain.

FIGURE P21.23

24. * You have a coil of wire with 10 turns each of 1.5-cm radius. You place the plane of the coil perpendicular to a 0.40-T $\vec{B}$ field produced by the poles of an electromagnet (Figure Q21.2). (a) Find the magnitude of the average induced emf in the coil when the magnet is turned off and the field decreases to 0 T in 2.4 s. (b) Is the direction of the induced current in the coil from lead a to b, or from b to a? Explain.

25. * An experimental apparatus has two parallel horizontal metal rails separated by 1.0 m. A 2.0-Ω resistor is connected from the left end of one rail to the left end of the other. A metal axle with metal wheels is pulled toward the right along the rails at a speed of 20 m/s. Earth's uniform 5.0×10^{-5}-T $\vec{B}$ field points down at an angle of 53° below the horizontal. Make a list of the physical quantities you can determine using this information and determine two of them.

26. A Boeing 747 with a 65-m wingspan is cruising northward at 250 m/s toward Alaska. The $\vec{B}$ field at this location is 5.0×10^{-5} T and points 60° below its direction of travel. Determine the potential difference between the tips of its wings.

27. * A circular loop of radius 9.0 cm is placed perpendicular to a uniform 0.35-T $\vec{B}$ field. You collapse the loop into a long, thin shape in 0.10 s. What is the average induced emf while the loop is being reshaped? What assumptions did you make?

28. ** **BIO EST Magnetic field and brain cells** Suppose a power line produces a 6.0×10^{-4}-T peak magnetic field 60 times each second at the location of a neuron brain cell of radius 6.0×10^{-6} m. Estimate the maximum magnitude of the induced emf around the perimeter of this cell during one-half cycle of magnetic field change.

29. * You need to test Faraday's law. You have a 12-turn rectangular coil that measures 0.20 m × 0.40 m and an electromagnet that produces a 0.25-T magnetic field in a well-defined region that is larger than the area of the coil. You also have a stopwatch, an ammeter, a voltmeter, and a motion detector. (a) Describe an experiment you will design to test Faraday's law. (b) How will you calculate the measurable outcome of this experiment using the materials available? (c) Describe how you can test Lenz's law with this equipment.

30. * You build a coil of radius r (m) and place it in a uniform $\vec{B}$ field oriented perpendicular to the coil's surface. What is the total electric charge that passes through the coil's wire loops if the $\vec{B}$ field decreases at a constant rate to zero? The resistance of the coil's wire is R (Ω).

31. * **Equation Jeopardy 1** Invent a problem for which the following equation might be a solution.

$$0.01 \text{ V} = (100)\frac{(A)\cos 0°(0.12 \text{ T} - 0)}{(1.2 \text{ s} - 0)}$$

32. * **Equation Jeopardy 2** Invent a problem for which the following equation might be a solution.

$$0.01\text{V} = 100\frac{\pi (0.10 \text{ m})^2(0.12 \text{ T})(\cos 0° - \cos 90°)}{(t - 0)}$$

33. * **Equation Jeopardy 3** Invent a problem for which the following equation might be a solution.

$$\mathcal{E} = -(35)\frac{(0.12 \text{ T})(\cos 0°)\left[(0)^2 - \pi (0.10 \text{ m})^2\right]}{(3.0 \text{ s} - 0)}$$

34. * **EST** **Generator for space station** Astronauts on a space station decide to use Earth's magnetic field to generate electric current. Earth's $\vec{B}$ field in this region has the magnitude of 3.0×10^{-7} T. They have a coil that rotates $90°$ in 1.2 s. The area inside the coil measures 5000 m². Estimate the number of loops needed in the coil so that during that $90°$ turn it produces an average induced emf of about 120 V. Indicate any assumptions you made. Is this a feasible way to produce electric energy?

35. * A toy electric generator has a 20-turn circular coil with each turn of radius 1.8 cm. The coil resides in a 1.0-T magnitude $\vec{B}$ field. It also has a lightbulb that lights if the potential difference across it is about 1 V. You start rotating the coil, which is initially perpendicular to the $\vec{B}$ field. (a) Determine the time interval needed for a $90°$ rotation that will produce an average induced emf of 1.0 V. (b) Use a proportion technique to show that the same emf can be produced if the time interval for one rotation is reduced to one-fourth while the radius of the coil is reduced by one-half.

36. * A generator has a 450-turn coil that is 10 cm long and 12 cm wide. The coil rotates at 8.0 rotations per second in a 0.10-T magnitude $\vec{B}$ field. Determine the generator's peak voltage.

37. * You need to make a generator for your bicycle light that will provide an alternating emf whose peak value is 4.2 V. The generator coil has 55 turns and rotates in a 0.040-T magnitude $\vec{B}$ field. If the coil rotates at 400 revolutions per second, what must the area of the coil be to develop this emf? Describe any problems with this design (if there are any).

38. * **Evaluating a claim** A British bicycle light company advertises flashing bicycle lights that require no batteries and produce no resistance to riding. A magnet attached to a spoke on the bicycle tire moves past a generator coil on the bicycle frame, inducing an emf that causes a light to flash. The magnet and coil never touch. Does this lighting system really produce no resistance to riding? Justify your answer.

39. * A generator has a 100-turn coil that rotates in a 0.30-T magnitude $\vec{B}$ field at a frequency of 80 Hz (80 rotations per second) causing a peak emf of 38 V. (a) Determine the area of each loop of the coil. (b) Write an expression for the emf as a function of time (assuming the emf is zero at time zero). (c) Determine the emf at 0.0140 s.

40. * A 10-Hz generator produces a peak emf of 40 V. (a) Write an expression for the emf as a function of time. Indicate your assumptions. (b) Determine the emf at the following times: 0.025 s, 0.050 s, 0.075 s, and 0.100 s. (c) Plot these emf-versus-time data on a graph and connect the points with a smooth curve.

41. * A rectangular wire loop is moving with constant speed through a region of uniform $\vec{B}$ field that points into the page (Figure P21.41). The longer side of the loop is twice the width of the magnetic field region. Draw a qualitative graph that shows how the induced current in the loop depends on time during the period in which the loop moves from the position shown in the figure to the position where the left side of the loop exits the magnetic field. Assume that a counterclockwise direction of the current is positive.

FIGURE P21.41

42. ** A wire loop consists of two equal parts that are connected so that they form a "rectangular figure eight." The loop is moving with constant speed through a region of uniform $\vec{B}$ field that points into the page (Figure P21.42). The longer side of the loop is twice the width of the magnetic field region. Draw a qualitative graph that shows how the induced current in the loop depends on time during the period in which the loop moves from the position shown in the figure to the position where the left side of the loop exits the magnetic field. Assume that a counterclockwise direction of the current in the right half of the loop is positive.

FIGURE P21.42

21.6 AC circuits

43. The voltage across an AC power supply is given by $\Delta V_{AC} = (100 \text{ V}) \sin (2\pi 50t)$. Determine (a) ΔV_{rms}, (b) the period with which the voltage vibrates, (c) the amplitude of the current through the 200-Ω resistor that is connected to this power supply, and (d) the average power transferred by the power supply to the resistor.

44. * The alternating current through a 2.0-μF capacitor is given by $I_C = (0.04 \text{ A}) \sin (2\pi 400t)$. Determine (a) the rms voltage across the capacitor and (b) the shortest time interval between the peak of the voltage across the capacitor and the peak of the current through the capacitor. (c) How will the answers to questions (a) and (b) change if the capacitance is doubled? Explain.

45. * The alternating current through a solenoid is given by $I_L = (3.2 \text{ A}) \sin (2\pi 180t)$, and the rms voltage across the solenoid is 190 V. Determine (a) the inductance of the solenoid and (b) the shortest time interval between the peak of the voltage across the solenoid and the peak of the current through the solenoid, assuming that the resistance of the solenoid can be ignored. (c) A nickel cylinder with a relative magnetic permeability of 300 is inserted into the solenoid. Determine the new amplitude of the current through the solenoid, assuming that the solenoid remains connected to the same power supply.

46. * The rms voltage of household AC in Europe is 230 V. Determine (a) the amplitude of the alternating voltage across and (b) the amplitude of the alternating current through a 100-W European incandescent lightbulb. Determine (c) the resistance of this lightbulb and (d) the power output of the lightbulb when you use it in the United States, where the rms voltage is 120 V. Indicate any assumptions that you made.

47. * You build a parallel plate capacitor by placing a 0.1-mm-thick sheet of paper between two 100-cm² pieces of aluminum foil. The dielectric constant of paper is 2.3. Determine (a) the capacitance of the capacitor and (b) the frequency of the alternating current at which the capacitive reactance of the capacitor is equal to 5 Ω.

48. * You have a 100-m copper wire with cross-sectional area 0.20 mm² and resistivity $1.7 \times 10^{-8} \Omega \cdot$ m. You make a 780-turn solenoid by winding the wire on a plastic tube (diameter 4.0 cm, length 10.0 cm). Determine (a) the resistance of the solenoid, (b) the inductance of the solenoid, and (c) the frequency of the alternating current at which the inductive reactance of the solenoid is equal to its resistance.

21.7 Transformers

49. You need to build a transformer that can step the emf up from 120 V to 12,000 V to operate a neon sign for a restaurant. What will be the ratio of the secondary to primary turns of this transformer?

50. Your home's electric doorbell operates on 10 V. Should you use a step-up or step-down transformer in order to convert the home's 120 V to 10 V? Determine the ratio of the secondary to primary turns needed for the bell's transformer.

51. A 9.0-V battery and switch are connected in series across the primary coil of a transformer. The secondary coil is connected to a lightbulb that operates on 120 V. Draw the circuit. Describe in detail how you can get the bulb to light—not necessarily continuously.

52. * You are fixing a transformer for a toy truck that uses an 8.0-V rms voltage to run it. The primary coil of the transformer is broken; the secondary coil has 30 turns. The primary coil is connected to a 120-V wall outlet. (a) How many turns should you have in the primary coil? (b) If you then connect this primary coil to a 240-V rms voltage, what will be the amplitude of the alternating voltage across the secondary coil?

21.8 Mechanisms explaining electromagnetic induction

53. * A wire loop has a radius of 10 cm. A changing external magnetic field causes an average 0.60-N/C electric field in the wire. (a) Determine the work that the electric field does in pushing 1.0 C of electric charge around the loop. (b) Determine the induced emf caused by the changing magnetic field. (c) You measure a 0.10-A electric current. What is the electrical resistance of the loop?

General Problems

54. * **Ice skater's flashing belt** You are hired to advise the coach of the Olympic ice-skating team concerning an idea for a costume for one of the skaters. They want to put a flat coil of wire on the front of the skater's torso and connect the ends of the coil to lightbulbs on the skater's belt. They hope that the bulbs will light when the skater spins in Earth's magnetic field. Do you think that the system will work? If so, could you provide specifications for the device and justification for your advice?

55. **BIO** **Hammerhead shark** A hammerhead shark (**Figure P21.55**) has a 0.90-m-wide head. The shark swims north at 1.8 m/s. Earth's $\vec{B}$ field at this location is 5.0×10^{-5} T and points 30° below the direction of the shark's travel. Determine the potential difference between the two sides of the shark's head.

FIGURE P21.55

56. * **Car braking system** You are an inventor and want to develop a braking system that not only stops the car but also converts the original kinetic energy to some other useful energy. One of your ideas is to connect a rotor coil (the rotating coil of the generator) to the turning axle of the car. When you press on the brake pedal, a switch turns on a steady electric current to a stationary coil (an electromagnet called the stator) that produces a steady magnetic field in which the rotor turns. You now have a generator that produces an alternating current and an induced emf—electric power. Make a simple drawing of the rotor and stator at one instant and determine the direction of the magnetic force exerted on the rotor. Does this force help stop the car? Explain.

57. ** Your professor asks you to help design an electromagnetic induction sparker (a device that produces sparks). Include drawings and word descriptions for how it might work, details of its construction, and a description of possible problems.

58. ** In a new lab experiment, two parallel vertical metal rods are separated by 1.0 m. A 2.0-Ω resistor is connected from the top of one rod to the top of the other. A 0.20-kg horizontal metal bar falls between the rods and makes contact at its ends with the rods. A $\vec{B}$ field of 0.50×10^{-4} T points horizontally between the rods. The bar should eventually reach a terminal falling velocity (constant speed) when the magnetic force of the magnetic field on the induced current in the bar balances the downward force due to the gravitational pull of the Earth. (a) Develop in symbols an expression for the current through the bar as it falls. (b) Determine in symbols an expression for the magnetic force exerted on the falling bar (and determine the direction of that force). Remember that an electric current passes through it, and the bar is falling in the magnetic field. (c) Determine the final constant speed of the falling bar (d) Is this process realistic? Explain.

59. ** You have a 12-V battery, some wire, a switch, and a separate coil of wire. (a) Design a circuit that will produce an emf around the coil even though it is not connected to the battery. (b) Show, using appropriate equations, why your system will work. (c) Describe one application for your circuit.

60. * You want to build a generator for a multi-day canoe trip. You have a fairly large permanent magnet, some wire, and a lightbulb. Design a generator and provide detailed specifications for it. (Ideas for the design could include cranking a handle or placing a paddle wheel that turns a coil in a nearby stream.)

61. * **EST** A sparker used to ignite lighter fluid in a barbeque grill is shown in **Figure P21.61**. You compress a knob at the end of the sparker. This compresses a spring, which when released moves a magnet at the end of the knob quickly into a 200-turn coil. The change in flux through the coil induces an emf that causes a spark across the 0.10-mm gap at the end of the sparker. (a) Estimate the time interval needed for the change in flux in order to produce this spark. Indicate any assumptions you made. (b) Is this a realistic process? Explain.

FIGURE P21.61 Note: The size of the gap is not to scale.

62. * **EST** **Design a magnetometer** Your friend needs to design an experiment that will allow her to measure the magnitude of the $\vec{B}$ field near a large permanent magnet. She has a small coil (50 turns, diameter 1.0 cm), a small motor that can spin at 300 revolutions per second, and an AC voltmeter that can measure voltage with 0.001-V precision. (a) Describe in detail the experimental setup (using this equipment) that will allow your friend to estimate the magnitude of the $\vec{B}$ field. (b) Estimate the smallest value of the $\vec{B}$ field magnitude that she will be able to measure. Indicate any assumptions that you made.

63. * **EST** **Can a distant strike of lightning damage your computer?** Lightning is a sudden electrical discharge that occurs during a storm. In a typical cloud-to-ground lightning flash, a discharge current changes at a rate of about 10^{11} A/s. Estimate the maximum induced emf around the metal box of a typical computer during a lightning flash, assuming the lightning strikes the ground 200 m from the computer. Describe any assumptions that you made. (Hint: Model the lightning as a long straight wire carrying current that changes in time.)

64. **BIO** **EST** **MRI** Jose needs an MRI (magnetic resonance imaging) scan. During the exam, Jose lies in a region of a very strong 1.5-T horizontal magnetic field pointing from Jose's waist toward his head. A medical assistant moves Jose from the scanner, reducing the magnetic field from 1.5 T to 0.1 T in 1.0 s. Consequently, the $\vec{B}$ field through Jose's 0.3-m by 0.4-m chest decreases. The conductive tissue inside his body around the outer part of his chest is a loop, with the chest as the area inside this loop. (a) Estimate the induced emf around this conducting loop as the $\vec{B}$ field decreases. (b) If the resistance of his body tissue around this loop is 5 Ω, what is the induced current passing around his body? (c) What is the direction of the current?

65. * **Magstripe reader** A magstripe reader used to read a credit card number or a card key for a hotel room has a tiny coil that detects a changing magnetic field as tiny bar magnets pass by the coil. Calculate the magnitude of the induced emf in a magstripe card reader coil. Assume that the magstripe magnetic field changes at a constant rate of 500 mT/ms as the region between two tiny magnets on the stripe passes the coil. The reader coil is 2.0 mm in diameter and has 5000 turns.

66. Show that when a metal rod L meters long moves at speed v perpendicular to $\vec{B}$ field lines, the magnetic force exerted by the field on the electrically charged particles in the rod produces a potential difference between the ends of the rod equal to the product BLv.

67. ** **EST** **The Tower of Terror ride** Magnetic braking brings the car of the Tower of Terror ride to a stop from a speed of 161 km/h. (a) Is its 161-km/h speed what you would expect of an object after a 115-m fall? Explain. (b) Estimate the time interval for the free-fall part of its trip. (c) Estimate the average acceleration of the car while stopping due to its magnetic braking.

Reading Passage Problems

BIO **Magnetic induction tomography (MIT)** Magnetic induction tomography is an imaging method used in mineral, natural gas, oil, and groundwater exploration; as an archaeological tool; and for medical imaging. MIT has also been used to measure topsoil depth in agricultural soils. Topsoil depth is information that many farmers need: for instance, corn yield is much higher in soil that has a deep topsoil layer above the underlying, impermeable claypan. Using a trailer attached to a tractor, a farmer can map an 80,000-m² (about 20-acre) field for topsoil depth in about 1 hour.

Figure 21.28 shows how MIT works. A time-varying electric current in a source coil (Figure 21.28a) induces a changing magnetic field that passes into the region to be imaged—in this case, the soil (Figure 21.28b). This changing magnetic field induces a weak induced electric current in topsoil and a stronger induced current in the more conductive claypan soil at the same depth. (Figure 21.28c; the current direction here is drawn as though the source current and source fields are increasing). This changing induced electric current in turn produces its own induced magnetic field (Figure 21.28d). The induced magnetic field passes out of the region being mapped to a detector coil (Figure 21.28e) near the source coil. The nature of the signal at the detector (its magnitude and phase) provides information about the region being mapped. A strong signal returned to the detector coil indicates a claypan layer near the surface; a weak signal returns if the clay layer is deeper below the surface.

FIGURE 21.28 Magnetic induction tomography.

(a) A changing current in the source coil above the soil ...

(b) ... causes a changing external magnetic field in the soil ...

(c) ... which induces a current in the soil.

(d) The induced current causes an induced magnetic field out of the soil ...

(e) ... which induces a current in a detector coil above the soil.

68. Why is the detected signal from an MIT apparatus greater if a moist conductive layer is near the surface?
    (a) The signal is reflected better from the top of a nearby conductive layer.
    (b) The induced current is greater if the soil is moist and conductive.
    (c) The induced magnetic field from the induced current is bigger if its source is near the detection coil.
    (d) All three of the above reasons
    (e) b and c

69. All other conditions being equal, why is the induced current greater in claypan soil than in topsoil?
    (a) Claypan soil has a higher concentration of magnetic ions compared to topsoil.
    (b) Claypan soil is partly metallic in composition.
    (c) Claypan soil has greater density than loose topsoil.
    (d) The clay is closely packed, moist, and a better electrical conductor than loose, dry topsoil.

70. Why is MIT used to search for mineral deposits (iron, copper, zinc)?
    (a) The minerals are good conductors of electricity and produce strong induced currents and strong returning magnetic fields.
    (b) The minerals absorb the incident magnetic field, indicating their presence by a lack of returning signal.
    (c) The minerals produce their own returning magnetic fields.
    (d) The minerals attract the incoming magnetic field and reflect it directly above the minerals.

71. Which of the statements below about magnetic induction tomography (MIT) and transcranial magnetic stimulation (TMS), studied in Section 21.5, are true?
    (a) Both MIT and TMS have source currents in coils, source magnetic fields, and induced currents.
    (b) MIT detects the induced magnetic field produced by the induced current, and TMS does not.
    (c) MIT provides information directly about the imaged area, whereas TMS disrupts some brain activity, and the disruption is measured in some other way.
    (d) a and c only   (e) a, b, and c

72. Describe all the changes that would occur if the source current were in the direction shown in Figure 21.28 but decreasing instead of increasing.
    (a) The induced current would be in the opposite direction.
    (b) The induced magnetic field would be in the opposite direction.
    (c) The detected current would be in the opposite direction.
    (d) a and b   (e) a, b, and c

BIO **Measuring the motion of flying insects** Studying the motion of flying animals, particularly small insects, is difficult. One method researchers use involves attaching a tiny coil with miniature electronics to the neck of an insect and another coil to its thorax (Figure 21.29). They place the insect in a strong magnetic field and observe the changing orientations and induced emfs of the two coils in the field as the insect flies. Suppose that a 50-turn coil of radius 2.0×10^{-3} m is attached to a tsetse fly that is flying in a 4.0×10^{-3}-T magnetic field. The tsetse fly makes a 90° turn in 0.020 s. Consider the average magnitude of the induced emf that occurs due to the turn of the tsetse fly and its coil.

FIGURE 21.29 The coil changes its orientation with respect to the external magnetic field as the fly makes a turn.

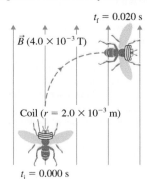

73. Which of the quantities B_{ex}, A, or θ is changing as the fly turns?
    (a) B_{ex} (b) A (c) θ
    (d) All of them   (e) None of them

74. Which answer is closest to the magnitude of the flux change?
    (a) 1×10^{-4} T·m²   (b) 2×10^{-6} T·m²
    (c) 3×10^{-7} T·m²   (d) 5×10^{-8} T·m²

75. Which answer is closest to the induced emf on the tsetse fly coil during the 90° turn?
    (a) 6×10^{-2} V   (b) 1×10^{-4} V
    (c) 4×10^{-6} V   (d) 2×10^{-7} V

76. Which of the following could double the emf produced when the fly turns 90°?
    (a) Double the number of turns in the coil.
    (b) Double the coil's area.
    (c) Double the magnitude of the external magnetic field.
    (d) Get the tsetse fly to take twice as long to turn.
    (e) a, b, and c

Reflection and Refraction

A magician breaks a glass vial with a hammer and drops the pieces of broken glass into a beaker full of oil. Then she reaches into the beaker—and takes out an intact vial! Is this really magic? No. The photo above shows the same phenomenon: the glass pipette is visible in the right-hand bottle of water, but invisible in the left-hand bottle of oil. In this chapter you will learn how to explain this trick.

- How can you make something become invisible?
- What causes a mirage?
- What is fiber optics?

IN THIS CHAPTER we will begin our investigation of light. Is light different from everything we have studied, or can we use the principles and tools we have already developed to understand light and its useful applications? Many of the remaining chapters in this textbook involve using and improving the model of light we will develop in this chapter as well as investigating many physical processes involving light.

BE SURE YOU KNOW HOW TO:

- Define rotational velocity (Section 9.1).
- Apply ideas of impulse-momentum to explain collisions (Section 6.4).
- Draw a wave front and apply Huygens' principle (Sections 11.1 and 11.6).

22.1 Light sources, light propagation, and shadows

Some ancient people thought that humans emitted special invisible rays from their eyes. These rays reached out toward objects and wrapped around them to collect information about the objects. The rays then returned to the person's eyes with this information. If this ancient model were accurate, humans should be able to see in total darkness. However, a simple experiment disproves this model. If you sit for a while in a room with absolutely no light sources, no matter how long you wait, you will still see nothing. There must be some other explanation for how we see things. The experiments in Observational Experiment **Table 22.1** use a laser pointer as a source of light to explore the phenomena of light and seeing.

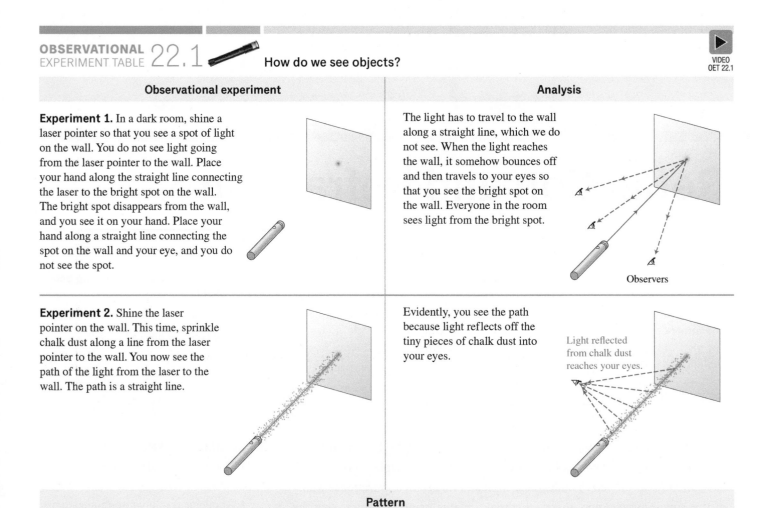

OBSERVATIONAL
EXPERIMENT TABLE 22.1 **How do we see objects?**
VIDEO
OET 22.1

Observational experiment	Analysis
Experiment 1. In a dark room, shine a laser pointer so that you see a spot of light on the wall. You do not see light going from the laser pointer to the wall. Place your hand along the straight line connecting the laser to the bright spot on the wall. The bright spot disappears from the wall, and you see it on your hand. Place your hand along a straight line connecting the spot on the wall and your eye, and you do not see the spot.	The light has to travel to the wall along a straight line, which we do not see. When the light reaches the wall, it somehow bounces off and then travels to your eyes so that you see the bright spot on the wall. Everyone in the room sees light from the bright spot.
Experiment 2. Shine the laser pointer on the wall. This time, sprinkle chalk dust along a line from the laser pointer to the wall. You now see the path of the light from the laser to the wall. The path is a straight line.	Evidently, you see the path because light reflects off the tiny pieces of chalk dust into your eyes.

Pattern

We can see objects (even tiny ones such as dust) illuminated by light. The path of light is a straight line from the source of light to the object and then (assuming that the behavior of light does not change) another straight line of reflected light from the object to our eyes.

Table 22.1 indicates that to see something, we need a source of light and an object off which the light bounces (reflects) and then reaches the eyes of the observer. Light travels in a straight-line path between the source of the light and the object reflecting the light, then in another straight line between that object and our eyes.

How do we represent the light sent by a light source? How do objects illuminated by a light source reflect light into our eyes? We will start by investigating the first question.

Representing light emitted by different sources

A laser pointer is useful for studying light propagation because the emitted light emerges as one narrow beam. However, most light sources, such as lightbulbs and candles, do not emit light as a single beam. These **extended light sources** consist of multiple points, each of which emits light. When we turn on a lightbulb in a dark room, the walls, floor, and ceiling of the room are illuminated (**Figure 22.1**). Obviously, the bulb sends light in all directions. But does one point of the shining bulb send light in one direction, or does each point send light in multiple directions? Both of these ideas can explain why the walls, floor, and ceiling are illuminated. Let's investigate those two possible models of how extended sources emit light. To do this we will represent the travel of light from one location to another with a **light ray**, drawn as a straight line and an arrow. Diagrams that include light rays are called **ray diagrams**. A ray is not a real thing; it is a model that allows us to show the direction of light.

One-ray model: Each point of an extended light source emits light that can be represented by one outward-pointing ray (**Figure 22.2a**). Different points send rays in different directions.

Multiple-ray model: Each point on an extended light source emits light in multiple directions represented by multiple rays (Figure 22.2b).

To help us determine which model better explains real phenomena, we will use the models to make predictions about the outcome of three experiments (Testing Experiment Table 22.2). All of the experiments are conducted in an otherwise dark room.

FIGURE 22.1 A lightbulb illuminates the ceiling of a room.

FIGURE 22.2 Two models of light emission from a bulb.

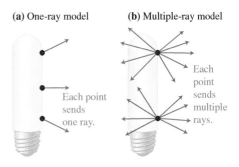

(a) One-ray model   (b) Multiple-ray model

Each point sends one ray.

Each point sends multiple rays.

TESTING
EXPERIMENT TABLE 22.2   How many rays does each point on a light source emit?

VIDEO
TET 22.2

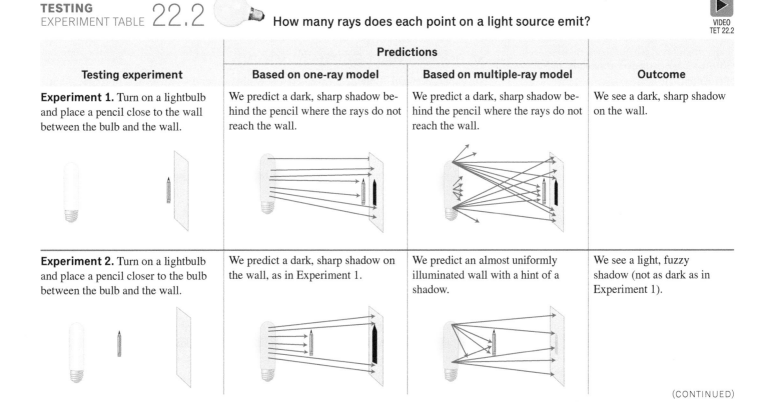

	Predictions		
Testing experiment	**Based on one-ray model**	**Based on multiple-ray model**	**Outcome**
Experiment 1. Turn on a lightbulb and place a pencil close to the wall between the bulb and the wall.	We predict a dark, sharp shadow behind the pencil where the rays do not reach the wall.	We predict a dark, sharp shadow behind the pencil where the rays do not reach the wall.	We see a dark, sharp shadow on the wall.
Experiment 2. Turn on a lightbulb and place a pencil closer to the bulb between the bulb and the wall.	We predict a dark, sharp shadow on the wall, as in Experiment 1.	We predict an almost uniformly illuminated wall with a hint of a shadow.	We see a light, fuzzy shadow (not as dark as in Experiment 1).

(CONTINUED)

	Predictions		
Testing experiment	**Based on one-ray model**	**Based on multiple-ray model**	**Outcome**
Experiment 3. Cover the bulb with aluminum foil and poke a hole in the foil in the middle of the bulb facing the wall. Turn the bulb on. Foil with hole in it.	We predict that we will see only a spot on the wall directly in front of the hole.	We predict that the whole wall will be dimly lit. 	We observe that the wall is dimly lit. If we cover the first hole and poke a hole in a different place, the result remains the same.

Conclusions

Experiment 1. Both models gave predictions that matched the outcome of this experiment. Neither is disproved.

Experiments 2 and 3. The predictions based on the one-ray model did not match the outcomes of the experiments; thus, that model is disproved. The predictions based on the multiple-ray model matched the outcomes of both experiments; thus, the model is supported (not proved).

The testing experiments above disprove the one-ray model and support the idea that each point on an extended light source emits light in many different directions. This light can be represented by multiple rays diverging from that point.

TIP ▸ In Experiment 3 in Table 22.2 we made a small hole in the aluminum foil covering the bulb. Light coming through the hole traveled in all directions. Thus, light from a point-like light source must be represented using multiple rays—with the exception of laser light, which can be represented using one ray.

Shadows and semi-shadows

The experiments in Table 22.2 revealed a new phenomenon—shadows. In Experiment 1 we found a sharp **shadow**, or **umbra**, behind the pencil on the wall. A shadow is a region behind an object where no light reaches. In Experiment 2 we found that there was no extremely dark shadow anywhere on the wall. What we observed is called a **semi-shadow**, also called a **penumbra**. A semi-shadow is a region where some light reaches and some does not. It appears as a fuzzy shadow.

The shadows of our bodies on a sunny day are sharp—there are almost no semi-shadows. Why? The Sun is so *far* from Earth that the only light from the Sun that reaches a small region on Earth is the light inside a narrow cylinder that extends from the Sun to Earth (**Figure 22.3a**). We can model sunlight that reaches Earth as a collection of parallel rays (Figure 22.3b). Similarly, we can represent the laser's very narrow beam of light with just one ray (Figure 22.3c).

Pinhole camera

If you hold a candle flame about 1 m from a blank wall, you do not see a projection of the flame on the wall. Since each point on the flame is a point source of light and each point source emits light in all directions, the wall is illuminated by light coming from all of the points on the candle flame.

However, we can use a piece of cardboard with a very small hole in it to make a sharp projection of the flame on a wall in a dark room. Conceptual Exercise 22.1 explains how this projection is formed.

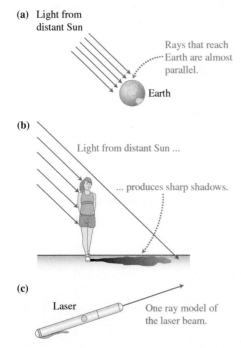

FIGURE 22.3 Ray models of sunlight and laser light.

(a) Light from distant Sun

Rays that reach Earth are almost parallel.

Earth

(b)

Light from distant Sun ...

... produces sharp shadows.

(c)

Laser

One ray model of the laser beam.

CONCEPTUAL EXERCISE 22.1 Simulating a pinhole camera

Use the ray model of light propagation to predict what you will see in the following experiment. You place a lit candle several meters from the wall in an otherwise dark room. Between the candle and the wall (close to the candle), you place a piece of stiff paper (or cardboard) with a small hole in it (see figure). What do you see on the wall?

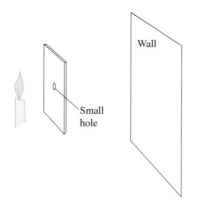

Sketch and translate

We already have a sketch of the situation, shown above. The candle flame is an extended light source. We need to predict what happens when light from the candle flame reaches the wall.

Simplify and diagram Assume that each point of the candle flame sends light in all directions. Represent this light with multiple light rays emitted from only the top and the bottom of the flame. Most of these rays are blocked by the paper and do not reach the wall, as shown in the figure at top right. However, ray 1 from the bottom of the flame passes through the hole and reaches the wall. Ray 2 from the top of the flame reaches the wall but below where ray 1 did. Because of where these

rays reach the wall, we predict that we will see an upside-down projection of the flame on the wall. Of course, many more rays reach the wall between rays 1 and 2, but these two rays help us find the top and bottom of the flame. That is what you actually see when you perform the experiment: the candle flame upside down on the wall.

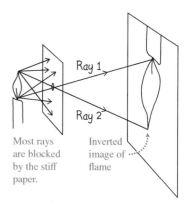

- -

Try it yourself Use the ray diagram to predict what you will see if you (a) move the candle closer to the hole; (b) move it farther from the hole; and (c) keep the candle and the stiff paper fixed relative to each other but move them together away from the wall.

Answer

(a) The upside-down image of the flame on the wall gets bigger; (b) it gets smaller; (c) it gets bigger.

The cardboard with a small hole in it is the foundation of the pinhole camera, also called a **camera obscura.** Such a camera consists of a lightproof box with a very small hole in one wall and a photographic plate or film inside the box on the opposite wall. Before the invention of modern cameras that use lenses, pinhole cameras were used to make photographs. To photograph a person, you would shine intense light on the person for a long time. A small amount of light that reflected off the person would pass through the hole and form an inverted projection of the person on the film. In **Figure 22.4**, the outside world is projected through a pinhole onto the wall of a cardboard box. We used a GoPro camera to take a picture of the image on the wall. In the video you can see an upside-down view of a harbor.

FIGURE 22.4 A pinhole view of the outside world. Note that the woman is upside down.

REVIEW QUESTION 22.1 A light source placed in front of a screen with a small hole in it will be projected upside down on the wall behind the screen. Why?

22.2 Reflection of light

In the last section we learned that people see objects because light is reflected off of objects into our eyes. In this section we will investigate the phenomenon of reflection. We will start with a simple case of a narrow beam of light that we can represent with a single ray and a very good, smooth reflecting surface—a mirror.

TIP Light propagates along straight lines, which we represent as *rays*. We use the term "a ray of light" as a model of a very narrow beam of light that describes and helps to predict light phenomena. We use purple lines to represent light rays and red lines to represent light beams.

FIGURE 22.5 Light reflection from a mirror.

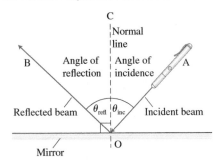

FIGURE 22.5 Light reflection from a mirror.

In Figure 22.5, light from a laser pointer at A shines on a mirror supported vertically on a horizontal surface (a top view is shown). Light striking the mirror is called **incident** light. After reflection, the laser light beam is *reflected* as shown. Notice line OC in the figure—it is perpendicular to the surface of the mirror at the point where the incident light hits the mirror. This line is called a **normal line** (recall that in mathematics and physics, "normal" means perpendicular). Table 22.3 lists some angles related to the incident and reflected beams as measured by a protractor lying on the table.

TABLE 22.3 Angles indicating directions of incident and reflected light beams from the mirror in Figure 22.5

Angle AOB between the incident and reflected beams	Angle between the incident beam AO and the normal line CO	Angle between the reflected beam BO and the normal line CO
0°	0°	0°
40°	20°	20°
60°	30°	30°
90°	45°	45°
120°	60°	60°
160°	80°	80°

The angle between the incident beam AO and the normal line CO is the **angle of incidence**. The angle between the reflected beam BO and the normal line CO is the **angle of reflection**. The data in the table show that the angles in columns two and three (the incident angle and the reflected angle, respectively) are always equal to each other. We can formulate this pattern as a mathematical rule or law for reflection:

TIP The angle of incidence and the angle of reflection are always the angles that the light beams form with the normal line and not the angles that they form with the surface of the mirror.

Law of reflection When a narrow beam of light, represented by one ray, shines on a smooth surface such as a mirror, the angle between the incident ray and the normal line perpendicular to the surface equals the angle between the reflected ray and the normal line (the angle of reflection equals the angle of incidence). The incident beam, reflected beam, and the normal line are always in the same plane.

$$\theta_{\text{reflection}} = \theta_{\text{incidence}} \qquad (22.1)$$

Specular and diffuse reflection, and absorption

When we shine a laser beam on a wall in a dark room, everyone in the room can see the bright spot. According to the law of reflection, the light beam should reflect at a particular angle determined by the angle of incidence. Does the fact that everyone sees the bright spot disprove the law of reflection?

Before we reject the law of reflection, we must carefully examine the assumptions we made when developing it: the reflecting surface is very smooth (like a mirror), and the light beam shining on it can be represented by one light ray. What if the reflecting surface is bumpy, and the laser beam is wider than the bumps on the surface (Figure 22.6a)? In that case, we cannot represent the reflected laser light as a single ray. Various parts of the surface are oriented at many different angles, so the light is reflected in many different directions. This explains why everyone in the room can see some of the light reflected from the wall. If we repeated the experiment using a perfectly clean mirror instead of the wall, only one person would be able to see the light reflected from the spot on the mirror illuminated by the laser.

FIGURE 22.6 Reflection from a bumpy surface.

The phenomenon in which a beam of light is reflected by a smooth surface (like a mirror) is called **specular reflection**. During specular reflection, a narrow incident beam represented by parallel rays is reflected as a narrow beam of light represented by parallel rays according to the law of reflection (**Figure 22.7a**).

The phenomenon in which light is reflected by a rough surface is called **diffuse reflection**. During diffuse reflection, each of the parallel incident rays of a light beam strikes a part of the surface oriented at a different angle with respect to the incident light. Thus the reflected light spreads in multiple directions (Figure 22.7b).

FIGURE 22.7 (a) Specular light reflection from a smooth surface. (b) Diffuse light reflection from a bumpy surface.

We can now explain how we saw the path of the laser beam in the experiments in Table 22.1. Each tiny speck of dust reflects incident light in multiple directions. As there are myriads of dust particles, reflected rays reach our eyes and we can see the "path" of light. Understanding diffuse reflection helps us explain the phenomenon captured in the photo in **Figure 22.8**. Sunlight coming through the church windows reflects off the dust in the air and reaches different observers below.

In addition to reflection, another process occurs when light encounters the surface of a medium. The surface warms up a little bit as some of the light energy is converted into thermal energy of the medium. This process is called **absorption**. When light interacts with a mirror's surface, very little absorption occurs, but for other media, such as black velvet, almost all of the light may be absorbed.

In the next exercise, we use diffuse and specular reflection and absorption to understand an everyday phenomenon.

FIGURE 22.8 A ray diagram represents how observers see dust particles in a church.

CONCEPTUAL EXERCISE 22.2 ▶ **Dark window on a sunny day**

On a sunny day, a house's uncovered window looks almost black, but its outside walls do not. How can we explain this difference?

Sketch and translate If the window looks black, it must reflect very little sunlight to your eyes. Our goal is to explain why.

Simplify and diagram When light shines on the rough walls of the house, it reflects back diffusely at different angles, and some of the light reaches your eyes—you see the walls easily. When light reaches the transparent surface of the window, most of it passes through the window and into the room and then reflects diffusely many times. During each reflection, some of the light is absorbed; because of this absorption, little comes back out. A small amount reflects off the smooth glass window as specular reflection (see the front view of the house and the top view inside the room in the figures on the right). If you are not standing in the right spot, you see very little light coming from the window—it appears dark. But if you are standing in that one location, you see a bright reflection of the Sun in the window.

Try it yourself Why is the pupil of your eye dark?

Answer

The pupil is a hole in the eye—similar to a window in a house. Incident light enters the pupil, and very little is reflected back—the pupil looks dark.

FIGURE 22.9 The red-eye effect.

Red-eye effect

You are familiar with the so-called "red-eye" effect that occurs occasionally when using flash photography (see **Figure 22.9**). The red-eye effect is especially common at night and in dim lighting, when the pupil is wide open. When the flash illuminates the open iris, light reflects from the red blood vessels in the retina on the back of the eye. Some of this reflected light passes back out of the pupil and makes the pupil appear red.

REVIEW QUESTION 22.2 How can we test the law of reflection?

22.3 Refraction of light

At the shore of a lake, you see sunlight reflecting off the water's surface. But you also see rocks and plants under the surface. Because you see them, light must have entered the water, reflected off the rocks and plants, returned to the water surface, and then traveled from the surface to your eyes. If you have ever tried to use a stick to touch a rock under the surface of a pond or lake, you know that it is not easy. Although you carefully point the stick, you miss—you tend to extend the stick farther away from you than is needed to touch the rock. Why is that?

To answer this question, we will observe what happens when we shine a laser beam into water at different angles (see Observational Experiment **Table 22.4**).

OBSERVATIONAL EXPERIMENT TABLE 22.4 — **The path of light changes when moving from air to water**   VIDEO OET 22.4

Observational experiment	Analysis
Experiment 1. Shine a laser beam straight down into a glass container filled with water with a few drops of milk added to make the beam visible. The container sits on a supporting ring so that light can leave through the bottom. We see red spots on the ceiling and floor of the room and on the bottom of the container.	Draw light rays to describe the appearance of the red spots. For simplicity we do not draw rays from the spots to our eyes.
Experiment 2. Shine the laser at an angle, and all four spots change their locations.	The path of the ray changes as it enters the water—it bends toward the normal. The path changes again when the ray emerges into the air—it bends away from the normal.
Experiment 3. Increase the angle, and the spots move farther from their original locations.	Although the incident angles change, the main features of the pattern from Experiment 2 remain.

Patterns

When light reaches the air-water boundary at the top surface, the incident light beam represented by ray 1

- partially reflects (ray 2) at the same angle as the angle of incidence and
- partially passes into the second medium (ray 3), bending at the interface *toward* the normal line.

Similar things happen to the light beam represented by ray 3. However, there are some differences. When ray 3 reaches the bottom water-air interface, it

- partially reflects (ray 4) at the same angle as the angle of incidence and
- partially passes from the water into the air below the container (ray 5), bending at the interface *away* from the normal line.

Note in Table 22.4 that when the incident light represented by rays 1 and 3 is perpendicular to the boundary of the surfaces (Experiment 1), light reflects along the same line (rays 2 and 4) and passes into the second medium without bending (ray 5). However, if the light is not perpendicular to the surface, it bends and travels in a different direction than in the previous medium. If we were to replace the water with a thick piece of transparent glass, we would observe a similar phenomenon. The change in the path of light when light travels from one medium to another is called **refraction**.

To develop a mathematical relationship between the angle of incidence and the **angle of refraction** (the angle the refracted beam makes with the normal line), we can do an experiment similar to the one we did when studying reflection. Such an experiment is shown in **Figure 22.10**. We shine a laser beam at horizontal air-water and air-glass interfaces and record the angles of incidence and refraction. The setup in Figure 22.10 allows us to do this easily because the light, after bending at the air-glass interface, always travels along the radius of the hemispheric container. Because this direction is perpendicular to the glass-air interface for the light leaving the container, light does not bend anymore. The results are shown below in **Table 22.5**.

FIGURE 22.10 An experiment to measure the incident and refracted angles of light at an air-water (or air-glass) interface.

TABLE 22.5 Angles of incidence and refraction between laser beams and the normal line

Incident angle (ray in air) θ_{air}	Refraction angle (air into water) θ_{water}	Refraction angle (air into glass) θ_{glass}
0°	0°	0°
10°	8°	7°
20°	15°	13°
30°	22°	19°
40°	29°·	25°
50°	35°	30°
60°	40°	35°

Notice that as the angle of incidence increases, the angle of refraction also increases, but not in an obvious pattern. The pattern is also not the same for all materials.

In 1621 the Dutch scientist Willebrord Snell (1580–1626) found a pattern. The ratio of the sine of the incident angle θ_1 to the sine of the refraction angle θ_2 remains constant for light traveling from air to water and for light traveling from air to glass. However, the constants are different for each pair of media (see columns four and five of Table 22.6, on the next page).

Notice that the ratio of the sines for air/water is about 1.3 and for air/glass is about 1.5. From these observations, Snell formulated a mathematical model for refraction phenomena:

$$\frac{\sin \theta_1}{\sin \theta_2} = n_{1 \text{ to } 2} \qquad (22.2)$$

where $n_{1 \text{ to } 2}$ is a number that depends on the two materials the light is traveling through. If we split $n_{1 \text{ to } 2}$ into a ratio of two numbers, one number that depends on the material

TABLE 22.6 Pattern found by Snell for the ratio of the sines of the incident and refraction angles

Air $\sin \theta_1$	Water $\sin \theta_2$	Glass $\sin \theta_2$	Air/water $(\sin \theta_1)/(\sin \theta_2)$	Air/glass $(\sin \theta_1)/(\sin \theta_2)$
0.000	0.000	0.000		
0.174	0.131	0.114	1.33	1.53
0.342	0.259	0.225	1.32	1.52
0.500	0.374	0.326	1.34	1.53
0.643	0.485	0.423	1.33	1.52
0.766	0.573	0.500	1.34	1.53
0.866	0.649	0.569	1.33	1.52

through which the incident ray travels (n_1) and the other that depends on the material through which the refracted ray travels (n_2), we get

$$\frac{\sin \theta_1}{\sin \theta_2} = \frac{n_2}{n_1}$$

We call n_1 and n_2 the **refractive indexes** (or the indexes of refraction) of medium 1 and medium 2, respectively. Rearranging the equation, we get what has become known as **Snell's law** (or the law of refraction).

TABLE 22.7 Refractive indexes of yellow light in various substances

Material	Refractive index n
Vacuum	1.00000
Air	1.00029
Carbon dioxide	1.00045
Water (20 °C)	1.333
Ethyl alcohol	1.362
Glucose solution (25%)	1.372
Glucose solution (50%)	1.420
Glucose solution (75%)	1.477
Glass	1.517–1.647
Diamond	2.417

Snell's law Snell's law (the law of refraction) relates the refraction angle θ_2 to the incident angle θ_1 and the indexes of refraction of the incident medium n_1 and the refracted medium n_2:

$$n_2 \sin \theta_2 = n_1 \sin \theta_1 \qquad (22.3)$$

The angles of incidence and refraction are measured with respect to the normal line where the ray hits the interface between the two media. If the refracted ray is closer to the normal than the incident ray, then medium 2 is more optically dense (with a higher refractive index) than medium 1. The incident ray, refracted ray, and normal line are in the same plane.

TIP Light going from a material with a lower to that with a higher index of refraction will bend toward the normal, but light going from a material with a higher to that with a lower index of refraction will bend away from the normal.

If we define the index of refraction of air as 1.00, then the indexes of refraction of water and of the glass used in Tables 22.5 and 22.6 are 1.33 and 1.53, respectively. Notice that glass refracted the light ray more toward the normal line than did water; the glass is more optically dense than water. The refractive indexes of several materials are given in Table 22.7, using yellow light. We will learn later in the chapter that light of different colors has slightly different indexes of refraction. Thus specifying the color of light in the table is important.

At the beginning of this section we discussed how it is difficult to touch an object under water with a stick. Figure 22.11 and our knowledge of refraction help us understand why. Because the light coming from the rock bends at point A, our brain perceives that the light is coming from point B′. If we aim the stick along AB′, instead of AB, we miss the rock.

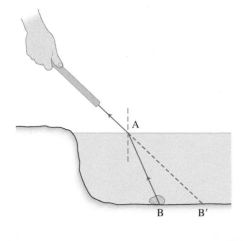

FIGURE 22.11 The bending of light makes it difficult to touch the rock.

EXAMPLE 22.3   Concentration of glucose in blood

The refractive index of blood increases as the blood's glucose concentration increases (for normal glucose content it ranges from 1.34 to 1.36; it is higher than the 1.33 for water because of all the additional organic components in the blood). Therefore, measuring the index of refraction of blood can help determine its glucose concentration.

In a hypothetical process, a hemispheric container holds a small sample of blood. A narrow laser beam enters from the bottom of the container perpendicular to the curved surface and into the sample. The light reaches the blood-air interface at a 40.0° angle relative to the normal line. The light leaves the blood and passes through the air toward a row of light detectors indicating that the light beam left the blood at a 61.7° angle. Determine the refractive index of the blood.

Sketch and translate First, we sketch the situation (see the figure). We use Snell's law to determine the index of refraction of the blood

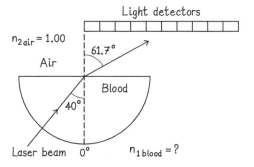

$n_{1\,\text{blood}}$. The incident angle is $\theta_{1\,\text{blood}} = 40.0°$. The refraction angle is $\theta_{2\,\text{air}} = 61.7°$. The index of refraction of the air is $n_{2\,\text{air}} = 1.00$.

Simplify and diagram The diagram in the figure can be used as a ray diagram if we assume that the laser beam is narrow.

Represent mathematically Snell's law for this situation is

$$n_{1\,\text{blood}} \sin \theta_{1\,\text{blood}} = n_{2\,\text{air}} \sin \theta_{2\,\text{air}}$$

Solve and evaluate Dividing both sides of Snell's law equation by $\sin \theta_{1\,\text{blood}}$, we have

$$n_{1\,\text{blood}} = \frac{n_{2\,\text{air}} \sin \theta_{2\,\text{air}}}{\sin \theta_{1\,\text{blood}}} = \frac{1.00 \sin 61.7°}{\sin 40.0°} = 1.37$$

This result is higher than the normal index of refraction for blood.

Try it yourself Pure blood plasma has a higher index of refraction than blood. Suppose the refractive index of the patient's plasma is 1.43 instead of 1.37. What then would the refraction angle be if the incident angle were still 40.0°?

Answer   66.8°.

QUANTITATIVE EXERCISE 22.4   Laser beams in the pool

A waterproof laser device produces three laser beams in the same vertical plane, one pointing directly upward and two at 60.0° above the horizontal. You put the device on the bottom of your swimming pool. The directions of the beams emerging from the water change as shown in the figure. At what angles relative to the horizontal do the laser beams emerge from the pool? Does the depth of the pool affect the answer?

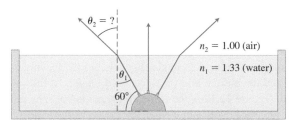

Represent mathematically The vertical beam does not change direction because it is perpendicular to the water's surface. The side beams are symmetrical, thus we can analyze one of them and apply the result to the other one.

$$n_1 \sin \theta_1 = n_2 \sin \theta_2 \implies \sin \theta_2 = \frac{n_1 \sin \theta_2}{n_2}$$

The incident angle $\theta_1 = 90° - 60° = 30°$. The angle with the horizontal is $90° - \theta_2$.

Solve and evaluate

$$\sin \theta_2 = \frac{n_1 \sin \theta_2}{n_2} = \frac{1.33 \sin 30.0°}{1.00} = 0.665$$

Since $\sin \theta_2 = 0.665$, we use a calculator to determine the angle: $\theta_2 = 41.7°$. The side beams emerge from water at angles $90° - 41.7° = 48.3°$ relative to horizontal. Because air has a smaller index of refraction than water, the rays traveling from water to air should bend away from the normal. Therefore, they should make a smaller angle with the horizontal than when in the water. Our answer therefore makes sense. The answer does not depend on the depth of water in the pool because we are only concerned with the angles, not with the trajectories of the beams. If the pool were deeper, the beams would emerge in different locations, but their directions would not change.

Try it yourself The index of refraction of water increases by 0.007 when you add 35 g of salt per 1 kg of water. How much will the angles of the beams in this exercise change as a result of such an addition of salt?

Answer   The change for the vertical beam angle will be zero. The side beams will emerge from the water at angles 48.0° relative to horizontal; therefore, their angles will decrease by 0.3°.

TIP When you draw ray diagrams, remember the following:

a. Most objects do not emit light; they reflect light shining on them. For simplicity, we draw them as light-emitting objects, though really they are just sources of reflected light.

b. Each point of an object emits rays in all directions. We only draw the ones that are the most convenient for describing the situation.

c. No rays come from our eyes. We see objects because light from them reaches our eyes. When drawing a ray diagram, think of which rays will reach the eyes of the observer.

Restoring a broken glass

Imagine that you place a piece of glass in water. Will you be able to see it? As long as light reflects off the glass and reaches your eye, you will. Since the index of refraction of glass is different from that of water, there is an **optical boundary** between them. When light traveling in water hits the glass, it refracts at the water-glass boundary and also reflects some light back to the eyes of an observer, thus making the glass visible. Now, imagine that you have a liquid whose refractive index is exactly equal to that of the glass. A good example is vegetable oil. Light traveling through this oil will not reflect off the piece of glass submerged in it or refract at the surface of the glass because there is effectively no optical boundary between the two materials. Light will just continue traveling in straight lines as if the glass were not there. Thus a piece of glass becomes invisible in vegetable oil! This explains the magic trick described in the chapter-opening story. (Note: If you decide to try this experiment, remember to fill the unbroken vial with oil before you put it in the oil bath; otherwise, it will be full of air and the air will make it visible!)

In summary, in order for us to see things, they must radiate light (like lightbulbs or fire), reflect light (like planets, trees, and snowflakes), or refract light (like glass in water). Reflection and refraction occur off a transparent object only if the optical density of the reflecting object is different from that of the material that surrounds it.

REVIEW QUESTION 22.3 Why is the expression "light travels in a straight line" not entirely accurate?

22.4 Total internal reflection

In two examples in the last section, light traveled from a more optically dense medium to a less optically dense medium, for example, from water to air. In such cases the light bent away from the normal line when entering the less dense medium. This behavior has some very important applications in the transmission of light by optical fibers.

Consider the situation represented by the ray diagram in **Figure 22.12a**. In a series of experiments, an incident ray under water hits a water-air interface at an increasingly larger angle relative to the normal line. As the incident angle (θ_1) gets bigger, the angle in the air between the refracted ray and the normal line (θ_2) gets even bigger. At the so-called **critical angle of incidence** θ_c (Figure 22.12c), the refraction angle is 90°. The refracted ray travels along the water-air interface. Notice that when this happens, the incident angle is still less than 90°. At incident angles larger than the critical angle, $\theta_1 > \theta_c$, the light is totally reflected back into the water—none escapes into the air (Figure 22.12d).

This behavior occurs whenever light attempts to travel from a more optically dense medium 1 of refractive index n_1 to a less optically dense medium 2 of refractive index n_2 ($n_1 > n_2$). We can use Snell's law to determine the critical angle in terms of the two indexes of refraction: $n_1 \sin \theta_1 = n_2 \sin \theta_2$, where $\theta_1 = \theta_c$ and $\theta_2 = 90°$. Remember that $\sin 90° = 1$. Inserting these values and solving for $\sin \theta_c$, we have

$$n_1 \sin \theta_c = n_2 \sin 90°$$

$$\sin \theta_c = \frac{n_2 \sin 90°}{n_1} = \frac{n_2(1)}{n_1} = \frac{n_2}{n_1}$$

FIGURE 22.12 Total internal reflection.

(a)

(b)

(c)

(d)

If the incident angle θ_1 is greater than θ_c, there is no solution to Snell's law, as $\sin \theta_2$ would be greater than 1.00 (the maximum value of sine is 1.00). For $\theta_1 > \theta_c$, all of the light is reflected back into the water. There is no refracted ray. Remember that this applies only for situations when $n_1 > n_2$. This phenomenon is called **total internal reflection**.

> **Total internal reflection** When light travels from a more optically dense medium 1 of refractive index n_1 into a less optically dense medium 2 of refractive index n_2 ($n_1 > n_2$), the refracted light beam in medium 2 bends away from the normal line. If the incident angle θ_1 is greater than a critical angle θ_c, *all* of the light is reflected back into the denser medium. The critical angle is determined by
>
> $$\sin \theta_c = \frac{n_2}{n_1} \qquad (22.4)$$

Figure 22.13 shows light traveling from a less dense medium (air) into a more dense medium (Plexiglas) at an incident angle of a little less than 90°. Due to this incident angle, the light refracts into the more dense medium at an angle that is a little less than the critical angle. Given that the piece of plastic is rectangular, the incident angle at the upper edge, going from plastic to air this time, is a little *larger* than the critical angle (it is equal to 90° minus the earlier angle of refraction). We therefore observe total internal reflection at this point.

The refractive index is a fundamental physical property of a substance and is often used to identify an unknown substance, confirm a substance's purity, or measure its concentration. Refractometers, instruments that measure refractive index, have medical and industrial applications. Veterinarians use portable refractometers to measure the total serum protein in blood and urine and also to detect drug tampering in racehorses.

FIGURE 22.13 Total internal reflection in Plexiglas. Note the initial incident angle.

EXAMPLE 22.5 **Another method to measure glucose concentration in blood**

Let's look at another hypothetical method for determining the glucose concentration in blood by measuring its refractive index. Light travels in a narrow beam through a hemispheric block of high-refractive-index glass, as shown in the figure at right. A thin layer of blood is placed between it and another hemispheric block on top. For this example we assume that the refractive index of the blood is 1.360 and the refractive index of the glass is 1.600. What pattern of light reaching the light detectors on the top curved surface of the hemispheric block will you observe as you move the light source clockwise around the edge of the curved surface of the bottom block?

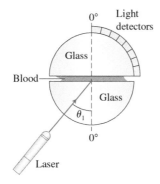

toward the same point on the glass-blood interface. We also assume that the blood layer is thin. Next, we draw a ray diagram for four rays (as shown in the figure below).

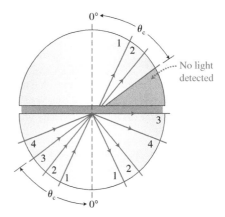

Sketch and translate We already have a sketch of the apparatus. Our goal is to predict what happens to the refracted beam if we vary the angle of incidence.

Simplify and diagram We assume that the light source is always oriented perpendicular to the curved surface on the bottom and points

Ray 1 bends away from the normal line at the first glass-blood interface. It is partially reflected and partially refracted into the blood. On the top surface of the blood, it bends back toward the normal line as

(CONTINUED)

it moves into the hemispheric glass block above (part of it is reflected at the second interface—not shown). The net angular deflection of the ray is zero—ray 1 in the lower block is parallel to ray 1 in the top block. Ray 2 has a greater incident angle with the first glass-blood interface and also bends away from the normal line as it enters the blood. It is partially transmitted into the blood and partially reflected. On the top surface of the blood it bends back toward the normal line as it moves into the hemispheric glass block above—to the right of ray 1. Ray 3 has a greater incident angle and is refracted at 90° in the blood. Hence, the incident angle is the critical angle. Ray 4 has an even greater incident angle than the critical angle and is totally reflected back into the lower hemispheric block.

The detectors on the top surface stop detecting light when the incident angle is larger than the critical angle. Thus, you can detect the critical angle by the place where the light stops arriving at the top block.

Represent mathematically We can use Eq. (22.4) to determine the critical angle:

$$\sin \theta_c = \frac{n_2}{n_1}$$

Solve and evaluate Using the above with the glass index of refraction $n_1 = 1.600$ and the blood index of refraction $n_2 = 1.360$, we find that

$$\sin \theta_c = \frac{n_2}{n_1} = \frac{1.360}{1.600} = 0.850$$

A 58.2° angle has a sine equal to 0.850. When incident light reaches this angle, the detectors on the top hemisphere will stop detecting light. As long as the incident angle is smaller than 58.2°, the detectors on the top of the apparatus will detect light. Thus the apparatus allows us to measure the critical angle and use it to determine the index of refraction and glucose concentration of the blood.

Try it yourself For a different sample of blood, you do not detect light at angles of 59.2° and larger. What is the refractive index of this new blood sample?

Answer   1.374.

REVIEW QUESTION 22.4 Why did we study total internal reflection after our investigation of refraction and not after reflection?

22.5 Skills for analyzing reflective and refractive processes

TIP When drawing ray diagrams, be sure to use a ruler.

When you solve problems involving light, use ray diagrams to help in your reasoning and quantitative work. The diagrams will also help you evaluate the final answer. In this section we will work through three exercises: a quantitative exercise, an example that requires the use of all the problem-solving steps, and, finally, an Equation Jeopardy example that shows how to interpret the mathematical description of light phenomena using ray diagrams.

QUANTITATIVE EXERCISE 22.6 ⟩ **Rotating mirror**

A laser beam is incident on a mirror that rotates with rotational velocity of 1.0 rad/s around an axis that is parallel to the plane of the mirror, as shown in the figure. Determine the rotational velocity of the laser beam that reflects from the mirror.

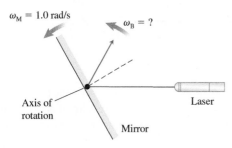

Represent mathematically According to the law of reflection, $\theta_{refl} = \theta_{inc}$. Suppose the initial angle of incidence of the laser beam is $\theta_{inc\ 1}$. The initial angle between the incident and reflected beam is therefore $2\theta_{inc}$. When the mirror rotates by an angle θ_M, the normal to the mirror rotates by the same angle, and because the laser does not move, the new angle of incidence becomes $\theta_{inc\ 2} = \theta_{inc\ 1} + \theta_M$. According to the law of reflection, the new reflected angle is $\theta_{refl\ 2} = \theta_{inc\ 2} = \theta_{inc\ 1} + \theta_M$. Therefore, the angle between the incident beam and reflected beam is $2\theta_{inc} + 2\theta_M$, and the change in this angle from the initial value is $\Delta\theta_B = 2\theta_M$.

Solve and evaluate Rotational velocity is given by $\omega = \frac{\Delta\theta}{\Delta t}$ (Eq. 9.2). Thus if the rotational velocity of the mirror is

$\omega_M = \dfrac{\Delta\theta_M}{\Delta t} = 1$ rad/s, the rotational velocity of the laser beam reflected off the mirror will be $\omega_B = \dfrac{\Delta\theta_B}{\Delta t} = \dfrac{2\Delta\theta_M}{\Delta t} = 2.0$ rad/s.

Answer

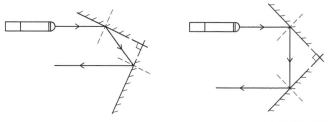

Try it yourself You send a laser beam horizontally from right to left. You have two mirrors. What are the possible arrangements of the mirrors that will allow you to send the reflected beam horizontally from left to right above the incident beam? What is the general pattern that any arrangement that achieves this goal should follow?

The figure below shows two possible arrangements of the mirrors. Any arrangement in which the mirrors are at a 90° angle to each other will work.

PROBLEM-SOLVING STRATEGY 22.1

Analyzing processes involving reflection and refraction

EXAMPLE 22.7   Hiding mosquito fish

A mosquito fish hides from a kingfisher bird at the bottom of a shallow lake, 0.40 m below the surface. A leaf floats above the mosquito fish. How big would the leaf need to be so that the kingfisher could not see its prey from any location above the water?

Sketch and translate

- Sketch the described situation or process.
- Indicate all the known and unknown quantities.
- Outline a solution for the problem as best you can.

- The situation is sketched below.

$n_2 = 1.00$

$n_1 = 1.33$

$h = 0.40$ m $R = ?$

- We want to know what size leaf is needed so that any light reflected from the fish and reaching the water surface undergoes total internal reflection and does not leave the water. Light incident at a smaller angle relative to the normal hits the leaf.

Simplify and diagram

- Indicate any assumptions you are making.
- Draw a ray diagram directly on the sketch showing all relevant paths the light travels. Consider the light rays originating from the object and eventually reaching the observer. In the diagram you can represent the observer by an eye.

- Model the fish as a shining point particle and the leaf as circular.

- Draw a ray diagram. θ_c is the critical angle. For incident angles greater than θ_c, there is no refracted ray—all light is totally reflected. Note that here we show the ray diagram in a separate figure, but you only need to draw one combined sketch and ray diagram.

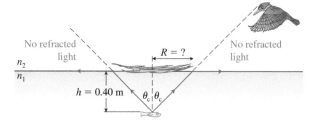

No refracted light $R = ?$ No refracted light

n_2
n_1

$h = 0.40$ m $\theta_c \, | \, \theta_c$

(CONTINUED)

Represent mathematically

- Use the sketch and ray diagram to help construct a mathematical description using the law of reflection, Snell's law for refraction, or the application of Snell's law for total internal reflection.

- Use Eq. (22.4), $\sin \theta_c = n_2/n_1$, where $n_1 = 1.33$ (the refractive index of water) and $n_2 = 1.00$ (the refractive index of air).

- Knowing the angle θ_c, we can determine the radius of the leaf: $R = h \tan \theta_c$.

Solve and evaluate

- Solve the equations for the unknown quantities.

- Evaluate the results to see if they are reasonable (the magnitude of the answer, its units, how the answer changes in limiting cases, and so forth).

- We find that $\sin \theta_c = 1.00/1.33 = 0.752$. The angle whose sin is 0.752 equals 48.8°.

- Thus $\tan \theta_c = \tan 48.8° = 1.14$.

- We can now determine the radius of the leaf: $R = h \tan \theta_c = (0.40 \text{ m})(1.14) = 0.46 \text{ m}$.

- This is an unrealistically big leaf. The fish is not safe. Note that the ray diagram shows the large radius of the leaf compared to the depth of the fish. In addition, even if we could find such a leaf, we assumed that the fish is a point-like object with no size. For a real fish, we would need an even a bigger leaf.

Try it yourself If the leaf radius is 0.20 m, how far below the leaf would the mosquito fish have to be so that it could not be seen by the kingfisher bird?

Answer   0.18 m or less.

Equation Jeopardy

In the following example, you are asked to interpret the mathematical description of a process and construct a ray diagram that represents that process. Then you are asked to invent a word problem that the equation could be used to solve. (The problem-solving procedure is reversed in an Equation Jeopardy problem.)

EXAMPLE 22.8   Equation Jeopardy

The two equations below describe a physical process. Invent a problem for which the equations would provide a solution.

$$1.00 \sin 30° = 1.60 \sin \theta_2$$
$$1.60 \sin \theta_2 = 1.33 \sin \theta_3$$

Solve and evaluate We solve for the unknown quantities:

$$\sin \theta_2 = \frac{1.00 \sin 30°}{1.60} = 0.313, \text{ or } \theta_2 = 18.2°$$

$$\sin \theta_3 = \frac{1.60 \sin \theta_2}{1.33} = 0.377, \text{ or } \theta_3 = 22.1°$$

Represent mathematically The equations are the applications of Snell's law.

Simplify and diagram Based on the indexes of refraction, the first medium in which light travels is air, then glass, then water. The ray diagram for the process is shown at top right.

Sketch and translate One possible problem is as follows. A narrow beam of light moves through air and hits the glass bottom of an aquarium at a 30° angle relative to the normal line. After passing through the glass wall, the beam hits the glass-water interface and refracts

into the water, as shown in the sketch. Determine the angle relative to the normal at which the light beam propagates in water. The index of refraction of the glass is 1.60 and that of the water is 1.33.

Try it yourself Show that the angle relative to the normal at which the light beam propagates in water in the previous problem does not depend on the index of refraction of glass.

Answer

$$n_1 \sin\theta_1 = n_2 \sin\theta_2 = n_3 \sin\theta_3;$$
$$\Rightarrow n_1 \sin\theta_1 = n_3 \sin\theta_3.$$

REVIEW QUESTION 22.5 What is the critical angle for total internal reflection for light going from water into glass of refractive index 1.56?

22.6 Fiber optics, prisms, mirages, and the color of the sky

In this section we will consider several applications of reflection, refraction, and total internal reflection: fiber optics, prisms, mirages, and the color of the sky.

Fiber optics

Imagine light traveling inside a long, thin, flexible, glass cylinder (like a wire made of flexible glass): such a cylinder is called an **optical fiber**. Fiber optic filaments are used in telecommunications to transmit high-speed light-based data and in medicine to see inside the human body during surgery. The following example will help you understand the physics behind fiber optics.

EXAMPLE 22.9 **Trapping light inside glass**

Imagine that you have a long glass block of refractive index 1.56 surrounded by air. Light traveling inside the block hits the top horizontal surface at a 41° angle. What happens next?

Sketch and translate The situation is sketched below. As light travels from glass into air, we need to compare the incident angle to the critical angle for total internal reflection.

Air $n_2 = 1.00$

Glass $n_1 = 1.56$ $\theta_1 = 41°$

Simplify and diagram Assume that the top and bottom surfaces of the block are parallel to each other. The light is hitting the top of the glass at a 41° angle. There are three possibilities: the light leaves the block, the light refracts so it moves along the surface, or the light is totally reflected back into the block. The last possibility is shown in the sketch.

Represent mathematically The critical angle for the glass-air interface according to Eq. (22.4) is

$$\sin\theta_c = \frac{n_2}{n_1} = \frac{1.00}{1.56} = 0.64, \text{ or } \theta_c = 40°$$

If the incident angle of light in the glass is greater than this critical angle, all light is reflected at the glass-air interface. When it reaches the opposite side of the block, it hits the bottom surface at the same angle because the sides are parallel and thus reflects back again.

Solve and evaluate We found above that $\theta_c = 40°$. The given 41° angle is slightly greater than the critical angle for total internal reflection. Thus the light is totally internally reflected during the first incidence on the upper surface. From there it moves down and to the right and hits the bottom surface at a 41° angle, where it is totally internally reflected again. The process of total internal reflection continues as the light travels the length of the block. However, if the material of the block is flexible and bends, then the incident angle changes. The incident angle might become so small that total internal reflection does not occur and some light refracts out of the block.

Try it yourself What happens if the light hits the top surface at 45°? At 38°?

Answer   the intensity of light within the block to diminish.
some light refracts out of the top and bottom at each incidence, causing
incident at 40° or more). For 38° incidence (since it is less than 40°),
For 45° incidence, total internal reflection occurs (it occurs for all light

FIGURE 22.14 (a) A fiber optic view inside a knee during surgery. (b) The fiber optic endoscope used to look inside the body.

(a)

(b)

Outer fibers conduct light inside body

Inner fibers conduct image to observer

Object

Within optical fibers, light is totally internally reflected, even when the fiber is bent (unless you bend it too much), allowing rapid transmission of light and data. In surgery, fiber optics can guide the surgeon's hands (**Figure 22.14a**). A thin bundle of tiny glass fibers transmits light into the area being illuminated and then transmits the reflected light out for viewing (Figure 22.14b). A tiny tool can be inserted along with the fibers to, for example, clean cartilage or repair tendons. Only a small incision is needed to insert the fibers and the tool, reducing the amount of trauma to the joint and surrounding tissue.

Prisms

Centuries ago, Isaac Newton observed a thin beam of light coming into a room through a tiny hole in one of the shutters. By chance, this beam illuminated a prism sitting on a desk. To Newton's surprise, he saw a rainbow band of colored light on the opposite wall (**Figure 22.15a**). The band was much wider than the original beam, with violet light on the bottom and red on top. Was the prism creating these colors? To test this idea, he put an identical prism after the first one, but upside down (Figure 22.15b), and the colored band disappeared—the spot on the wall was white and small. He concluded that the prism did not create the colors; it somehow could separate the different colors out of white light and then recombine them back into white light. Newton further reasoned that the refractive index was greater for violet light and smaller for red, explaining the separation of colors (see **Table 22.8**).

FIGURE 22.15 (a) A prism separates white light into a light spectrum. (b) A second prism combines the spectrum back into white light.

(a)

(b)

TABLE 22.8 **Refractive indexes of glass for different colors**

Color of light	n_{glass}
Red	1.613
Yellow	1.621
Green	1.628
Blue	1.636
Violet	1.661

FIGURE 22.16 Reflecting prisms.

(a)

(b)

(c)

Since the critical angle for a glass-air interface is less than 45° (see Example 22.9), glass prisms with 45-45-90° right angles are used in many optical instruments such as telescopes and binoculars to achieve total reflection of light through 90° or 180° angles. Examples of the reflecting ability of prisms are shown in **Figure 22.16**. Prisms are better for reflection than mirrors for several reasons. First, prisms reflect almost 100% of the light incident on the prism via total internal reflection, whereas mirrors reflect somewhat less than 100%. Second, most mirrors are made of glass over a thin sheet of metal, which tarnishes and loses its reflective ability with age. Prisms retain their reflective ability. Finally, prisms reverse the order of rays, as seen in Figure 22.16.

Mirages

An interesting consequence of the refraction of light is the formation of a **mirage**, such as the distant shimmering on a roadway that looks like water (**Figure 22.17**). How can we explain this observation?

On a hot day, hot air may hover just above the pavement. The air above is cooler. The hot air is less dense and has a lower index of refraction than the cooler air above it. When light from the sky passes through air with a gradually changing index of refraction, its path gradually bends. This bending of light leads us to perceive that the source of light is at a different location than it actually is (see **Figure 22.18**).

For simplicity we will consider only one ray coming from point A in the sky (Figure 22.18a). Instead of a continuous variation in the refractive index as the light slants downward, we will assume that it passes through several layers of different refractive index (represented as parallel layers). Its path changes according to Snell's law. At points 1 and 2, the light moving into layers of lower refractive index bends away from the normal line, and its angle with the pavement decreases. At some point (point 3), the incident angle becomes so large that the ray undergoes total internal reflection and starts going up. After passing through several layers, it enters the eye of the observer, whose brain perceives the ray as traveling along the straight line that originated at point B on the ground.

FIGURE 22.17 A mirage.

FIGURE 22.17 A mirage.

FIGURE 22.18 The formation of a mirage. An observer perceives that light is coming from the pavement, when in fact the source of light is the sky.

(a)

Ray coming from point A bends at the boundary of two layers of air. The observer perceives light coming from point B.

$n_1 > n_2 > n_3 > n_4$

(b)

Blue region is perceived as water, but is really formed by light from the sky.

If we now consider more rays coming from a section of the sky (Figure 22.18b), the observer would see them originating in the vicinity of B. This location will look blue and will shimmer due to convection in the air above the road surface. The result looks like water on the road (though it is actually light from the sky).

Color of the sky

We learned in this chapter that we could represent light reaching Earth from the Sun as a beam of parallel rays. Sunlight contains all visible colors. Why then does the entire sky look blue?

Molecules, dust particles, and water droplets in the atmosphere that are along the path of light from the Sun to Earth reflect sunlight in all directions, similar to the chalk dust reflecting laser light in the first experiments in this chapter. If the atmosphere reflected all colors of light similar to the way chalk dust does, the atmosphere would be the same color as the Sun.

Instead, we see a sky that is primarily blue. Due to their sizes, atmospheric particles reflect blue light more efficiently than other colors (**Figure 22.19**). Thus, all colors other than blue pass through the atmosphere without changing direction as much. Because the blue light is reflected in all directions, even when we look at the sky away from the Sun, reflected blue light from the atmosphere still reaches our eyes. This explanation is supported by probes sent to other planets, where the atmospheres have different chemical and physical compositions than ours. The skies of Venus and Mars are different colors even though illuminated by the same sunlight as Earth.

If our atmosphere reflected all colors the same way, we would see the sky as white—this is in fact what we see when the sky is covered with clouds. The water droplets in clouds reflect all colors equally.

FIGURE 22.19 Why the sky is blue. Blue light scatters off atmospheric particles and reaches our eyes from all directions. Light of other colors is not scattered as much.

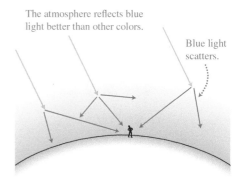

The atmosphere reflects blue light better than other colors.

Blue light scatters.

REVIEW QUESTION 22.6 Why is the sky blue? Why are most clouds white?

FIGURE 22.20 Newton's particle-bullet model of light.

(a) "Bullets" of light

"Bullet" semi-shadow or no shadow

(b)

"Bullet" shadow

Obstacle

(c)

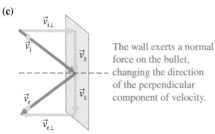

The wall exerts a normal force on the bullet, changing the direction of the perpendicular component of velocity.

FIGURE 22.21 Explanation of light refraction using the particle-bullet model.

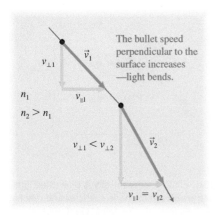

The bullet speed perpendicular to the surface increases —light bends.

FIGURE 22.22 Wavelets produce a new wave front.

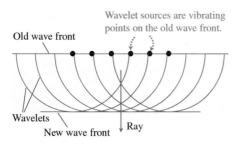

Wavelet sources are vibrating points on the old wave front.

Old wave front

Wavelets

New wave front   Ray

22.7 Explanation of light phenomena: two models of light

We have developed a ray model of light that describes the way light behaves. What other models can we use to explain the way light behaves?

Particle-bullet model

Isaac Newton modeled light as a stream of very small, low-mass particles moving at very high speeds, like bullets. According to this model, light bullets are affected by Earth's gravitational pull and move like projectiles, but since they move very quickly, their deflection from straight lines is not noticeable. Let us use this model to explain light phenomena familiar to us. We will start with shadows.

Imagine an extended light source such as a bulb sending light bullets in all directions from each point. If we place an obstacle close to the light source, bullets of light will still reach each part of the screen—the result will be a semi-shadow or no shadow (**Figure 22.20a**). If the obstacle is farther away, there will be a place on the screen where no bullets reach the screen—a shadow will form (Figure 22.20b). We can explain the reflection of light if we imagine that the bullets bounce elastically off surfaces. For the bullet model of light, the normal force exerted by a surface on the light bullets can only change the component of velocity perpendicular to the wall (because the acceleration of an object is in the same direction as the force exerted on it); the component parallel to the wall stays the same (Figure 22.20c). This is consistent with the law of reflection. Newton's particle-bullet model of light does successfully explain light propagation and reflection.

Can we explain refraction using this model of light? In order for the model to be consistent with Snell's experiments on refraction, the light bullets will have to speed up when they refract into a more optically dense medium (see **Figure 22.21**). The particle-bullet model suggested that the denser medium exerts an attractive force on the light particles, causing their speeds to increase. Accurate measurements of the speed of light turned out to be very challenging, and for many years the speed of light could only be determined in air or in a vacuum (it was found to be 299,792,458 m/s—about 3×10^8 m/s). In addition, the particle-bullet model requires an additional interaction between the surface and the light particles that causes them to speed up when entering the denser medium. Scientists prefer explanations that are as simple as possible. Another model of light that did not require this additional interaction was proposed.

Wave model of light

Simultaneously with Newton's development of the particle-bullet model of light, Christiaan Huygens was constructing a wave model of light. The motivation for the model could have come from the observations that light reflects off objects similar to the way sound reflects. Recall (from Chapter 11) that Huygens' wave propagation ideas involved disturbances of a medium caused by each point on a wave front producing a circular wavelet. Imagine that we have a wave with a wave front moving downward parallel to the page (see **Figure 22.22**). We choose six dots on this wave front, and from each dot we draw a wavelet originating from it (as shown in Figure 22.22). Each dot represents the source of a circular wavelet produced by the wave disturbance passing that point. According to Huygens' principle, each small wavelet disturbance produces its own circular disturbance that moves down the page in the direction the wave is traveling. The distance between two consecutive wavelets is equal to the wavelength. Now, note places where the wavelets add together to form bigger waves. These places are where the wave front will be once it has moved a short distance down the page. We also draw an arrow (a ray) indicating the direction the wave is traveling perpendicular to the wave fronts and consistent with the ray model of light propagation. In this model, the ray is the line perpendicular to the wave front.

When the waves emitted by the same source travel in different media, their frequency remains the source frequency, but their speed changes depending on the medium. For example, water waves travel more slowly in shallow water than in deeper water. Sound travels more slowly in cold air than in warm air. What does the wave model predict will happen when waves propagate through media with different wave speeds and with zero incidence angles? Imagine that the speed of the wave shown in **Figure 22.23** is greater in medium A than in medium B. When the wavelets from medium A reach the boundary with medium B, their radii reduce because the wave has a smaller speed in medium B. The new wave fronts are therefore closer together.

Wave model and refraction

Imagine a light wave moving in a less optically dense medium A and reaching an interface with a denser medium B (where it travels at a slower speed) at a nonzero angle of incidence (**Figure 22.24**). Using the wave model of light, we can now draw wave fronts for this wave as it travels from medium A into medium B. During a certain time interval, the wavelet that departed from the right edge of the wave front is still traveling in medium A. The wavelet that departed from the lower left edge of the wave front at the same time travels a shorter distance in medium B since it is moving slower. Once the point on a wave front reaches the boundary, the radius of the wavelet in medium B becomes smaller than in medium A, while the wavelets for the points on the same wave front that is still in medium A have larger radii. The wave front breaks at the boundary, as shown in Figure 22.24a.

What if a light wave travels from a slower to a faster medium, as shown in Figure 22.24b? Using similar reasoning, we find that the light bends away from the normal. If the wave model is going to describe refraction properly, we must conclude that light travels more slowly in water than in air.

According to the wave model of light, the speed of light should be slower in water than in the air—exactly opposite the prediction of the particle-bullet model. Which model is more accurate? An obvious way to answer this question is to measure the speed of light in water. But this experiment is difficult. Thus, we are left with two models of light that both explain the reflection and refraction of light but lead to different predictions about its speed in water. This dilemma existed in physics for a long time due to the difficulty in measuring the speed of light in different media. Ultimately, the resolution came not from measurements of the speed of light but from overwhelming experimental support for the wave model. We will discuss the wave model in more detail in a later chapter (Chapter 24).

Unanswered questions

Although we have learned a great deal about light, there are still questions that we have not answered. We do not know why different media bend light differently. We still do not know whether light propagates faster or slower in water and other media compared with air. We do not know how objects radiate light. Why do some stars look white but others look red? We also have not decided which model of light is better—the particle-bullet model or the wave model. We will investigate these questions in the coming chapters.

REVIEW QUESTION 22.7 What is the difference between the two models of light and the predictions they make for the speed of light in water?

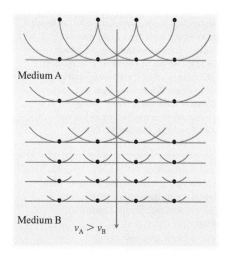

FIGURE 22.23 Wavelets crossing perpendicularly to the boundary between two regions with different wave speeds.

Medium A

Medium B $v_A > v_B$

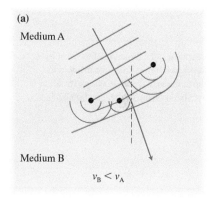

FIGURE 22.24 Huygens' principle explains why the wave changes directions when it reaches the interface between two media.

(a)

Medium A

Medium B

$v_B < v_A$

The wave travels farther on the right side in the less optically dense medium A than on the left side in denser medium B.

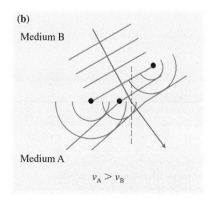

(b)

Medium B

Medium A

$v_A > v_B$

Medium B is less optically dense than medium A, causing the wave to bend out rather than in.

Summary

Light sources and light rays Lightbulbs, candles, and the Sun are examples of **extended sources** that emit light. Light from such sources illuminates other objects, which reflect the light. We see an object because incident light reflects off of it and reaches our eyes. We represent the travel of light with **rays** (drawn as lines and arrows). Each point of a shining object or a reflecting object sends rays in all directions. (Section 22.1)

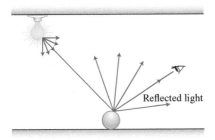

Reflected light

Law of reflection When a ray strikes a smooth surface such as a mirror, the **angle of incidence** equals the **angle of reflection**. This phenomenon is called **specular reflection**. (Section 22.2)

Smooth surface

θ_{incident}

$\theta_{\text{reflected}}$

$\theta_{\text{reflection}} = \theta_{\text{incidence}}$ Eq. (22.1)

Diffuse reflection If light is incident on an irregular surface, the incident light is reflected in many different directions. This phenomenon is called **diffuse reflection**. (Section 22.2)

Refraction If the direction of travel of light changes as it moves from one medium to another, the light is said to refract (bend) as it moves between the media. (Section 22.3)

Air
$n = 1.00$

Glass
$n = 1.60$

Water
$n = 1.33$

Snell's law Light going from a lower to a higher index of refraction will bend toward the normal, but when going from a higher to a lower index of refraction it will bend away from the normal. (Section 22.3)

θ_1

n_1

n_2

θ_2

$n_2 \sin \theta_2 = n_1 \sin \theta_1$ Eq. (22.3)

Total internal reflection If light tries to move from a more optically dense medium 1 of refractive index n_1 into a less optically dense medium 2 of refractive index n_2 ($n_1 > n_2$), and the incident angle in medium 1 θ_1 is greater than a critical angle θ_c, *all* of the light is reflected back into the denser medium. (Section 22.4)

$90°$

n_2

n_1

θ_c

$n_1 > n_2$

$\sin \theta_c = \dfrac{n_2}{n_1}$ Eq. (22.4)

Questions

Multiple Choice Questions

1. How can you convince your friend that a beam of light from a laser pointer travels to a wall along a straight line?
   (a) Tell him that light rays travel in straight lines.
   (b) Sprinkle chalk dust along the laser beam.
   (c) Try to move along the laser beam blocking its light and note the light path.
   (d) Both b and c will work.
   (e) All answers a–c will work.

2. Each point of a light-emitting object
   (a) sends one ray.
   (b) sends two rays.
   (c) sends an infinite number of rays.

3. To test the model mentioned in Question 22.2, what can one use? Multiple answers are possible.
   (a) A pinhole camera
   (b) A small source of light and a large obstacle near a distant screen
   (c) A large source of light and a small obstacle close to the source of light

4. What is a light ray?
   (a) A thin beam of light
   (b) A model invented by physicists to represent the direction of travel of light
   (c) A physical law

5. What is a semi-shadow?
   (a) A shadow that is half the size of the original shadow
   (b) A place where some light rays from an object reach and rays from other parts of the object do not.
   (c) A scientific term for a shadow

6. You fix a point-like light source 3.0 m away from a large screen and hold a basketball 1.0 m away from the screen so that the line connecting the center of the light source and the center of the basketball is perpendicular to the screen. You observe a shadow of the basketball on the screen. Select two correct statements.
   (a) Moving the light source away from the screen will produce a larger shadow.
   (b) Moving the basketball closer to the screen will produce a smaller shadow.
   (c) Moving the basketball and the light source away from the screen (while keeping the distance between the light source and the basketball fixed) will not change the size of the shadow.
   (d) Moving the light source up will result in moving the shadow down.
   (e) Moving the basketball up will result in moving the shadow down.

7. What is a normal line?
   (a) A line parallel to the boundary between two media
   (b) A vertical line separating two media
   (c) A line perpendicular to the boundary between two media

8. A light ray travels through air and then passes through a rectangular glass block. It exits
   (a) parallel to the original direction.
   (b) bent toward the normal line.
   (c) along the identical path that it entered the block.
   (d) bent away from the normal.

9. A right triangular prism sits on a base. A narrow light beam from a laser travels through air and then passes through the slanted side of the prism and out the vertical back side. It now travels
   (a) parallel to the original direction.
   (b) bent downward toward the base of the prism.
   (c) along the identical path that it entered the block.
   (d) bent upward away from the base.

10. A laser beam travels through oil in a horizontal direction. Which of the following objects could you put in the oil in the path of the laser beam in order to tilt the direction of the laser beam upward? Ignore partial reflections at the boundaries between media. Multiple answers are possible.
    (a) A rectangular glass plate
    (b) A mirror
    (c) A triangular plastic prism with index of refraction less than the oil's index of refraction
    (d) A triangular plastic prism with index of refraction equal to the oil's index of refraction

11. We can observe total internal reflection when light travels
    (a) from air to glass.
    (b) from water to glass.
    (c) from glass to water.
    (d) from air to water to glass.

12. You point a laser beam perpendicularly to the surface of water in a glass aquarium that is tilted at 30° (see Figure Q22.12). Where will you observe the laser beam hitting the table? Ignore any effects of the glass on the beam path. (Point P lies directly under the laser, on the extension of the incident beam in air.) (a) At point P (b) To the right of point P (c) To the left of point P (d) The laser beam will not hit the table because it will undergo total internal reflection.

FIGURE Q22.12
Laser
Water
30°
P

Conceptual Questions

13. What effects of light radiation and reflection are responsible for the fact that we can see objects in a room?

14. Why does the inside of a well look black?

15. What do you need to do to create different size shadows of the same stick?

16. Explain how a sundial works (a sundial is just a vertical stick in the ground).

17. You are trying to see what is happening in a room through a keyhole in a door. Where should you place your eye to see the most: closer or farther away from the hole? Explain.

18. If you stand near a light pole at night, the shadow of your foot (if slightly off the ground) is rather sharp while the shadow of your head looks fuzzy. Why?

19. In what cases can you see only a semi-shadow of an object and not a shadow?

20. Why when moving away from a lightbulb, toward a screen, does your shadow on the screen become sharper and sharper?

21. Why can't you see a shadow of a pencil on the wall if you place it close to a lightbulb?

22. The visible diameters of the Moon and the Sun are almost the same. What should you see if the Moon were in the same line of sight as the Sun? Explain.

23. The shadow of the Moon on Earth is 200 km wide. Describe what you would see if you were in that area during the moment described in the previous question.

24. You stand at the side of a road. Why is light from a passing car's headlights easier to see in foggy weather than in clear weather?

25. During the day, you can see the trees in your garden through the window, but you can hardly see yourself in the window. At night, you can hardly see the trees through the same window, but you can clearly see yourself in the window. Explain.

26. You look at a fish underwater. Draw a ray diagram of the light that enters your eye from the fish.

27. Take a pencil and try to touch a penny on the bottom of a large pan of water while looking down into the pan at an angle. Explain why it is difficult to do this.

28. Will a beam of light experience total internal reflection going from glass into water or from water into glass? Explain (include drawings).

29. Explain how a prism turns a beam of white light into a rainbow.

30. On a hot sunny day, a highway sometimes looks wet. Why?

31. Why can't you see stars during the day?

32. What light phenomena can be explained using the particle-bullet model of light?

33. What phenomena can be explained using a wave model of light?

34. How is it possible that two different models can explain the same phenomenon? Give an example of another phenomenon that can be successfully explained by two different models (think of areas other than optics).

35. Oliver has finished building a wall in a house. He inspects how flat it is by placing a small lightbulb next to the wall. Explain how this method can help Oliver see how flat the wall is.

Problems

Below, **BIO** indicates a problem with a biological or medical focus. Problems labeled **EST** ask you to estimate the answer to a quantitative problem rather than derive a specific answer. Asterisks indicate the level of difficulty of the problem. Problems with no * are considered to be the least difficult. A single * marks moderately difficult problems. Two ** indicate more difficult problems.

22.1 Light sources, light propagation, and shadows

1. **Tree height** You are standing under a tree. The tree's shadow is 34 m long and your shadow is about twice your height. How tall is the tree?
2. * **Lighting for surgeon** A shadow from a surgeon's hand obstructs her view while operating. Make suggestions for an alternative light source that avoids this difficulty. Include one or more sketches for your proposed plan.
3. **Lunar eclipse** A lunar eclipse happens when the Moon, Earth, and Sun are aligned in that order (the Moon is in the shadow of Earth). Aristotle used this phenomenon to determine the shape of Earth. He proposed that Earth has a round shape. Draw a picture to describe his reasoning.
4. * **Shadows during romantic dinner** You and a friend are having a romantic candlelight dinner. You notice that the light shadows of your hands on the wall look fuzzy. However, the shadow of a glass is very sharp and crisp. Where are you and your friend sitting with respect to the candle and the wall? Where is the glass? To answer these questions, draw ray diagrams assuming that the candle is an extended light source.
5. * **Pinhole camera (camera obscura)** You want to make a pinhole camera with a blank wall as the screen and you (or a friend) as the object of interest. Draw a sketch showing the wall, the pinhole, the place you or your friend will sit, the best location for the Sun or some other light source, and the location of people viewing the wall. Will the image appear upright or inverted? Explain.
6. * **Solar eclipse** Only observers in a very narrow region on Earth (about 200 km diameter) can see a total solar eclipse. In the region of such an eclipse, there is no sunlight and a person can see stars during daytime. Draw the arrangement of the Sun, the Moon, and Earth during a total solar eclipse.
7. * **Tree height 2** Your summer ecology research job involves documenting the growth of trees at an experimental site. One day you forget your tree-height-measuring instrument. How can you determine the height of trees without it? Provide a sketch for your method.
8. An extended light source can be modeled as a group of several point-like light sources. **Figure P22.8** shows a linear light source (such as a fluorescent lamp) modeled by three point-light sources, a sheet of black paper, and a screen. Copy the sketch in your notebook. Determine the regions of shadow and semi-shadow on the screen by drawing rays from the three point-like light sources. Qualitatively compare the relative brightnesses (or darknesses) of the regions.

FIGURE P22.8

Light source

Side view

—— Paper

———— Screen

22.2 Reflection of light

9. * You have a small mirror. While holding the mirror, you see a light spot on a wall at the same height as the mirror. At what angle are you holding the mirror if the Sun is 50° above the horizontal? Draw a ray diagram to answer the question. What assumptions did you make?
10. * **Design** Design an experiment that you can perform to test the law of reflection. Describe the instruments that you are going to use and their experimental uncertainty (half of the smallest division). How certain will you be of your results?

FIGURE P22.11

Laser

60°

11. * You place one mirror on a table and a second mirror at 60° above it, as shown in **Figure P22.11**. You direct a laser beam horizontally toward the second mirror as shown in the figure.

(a) Draw the trajectory of the laser beam after reflection from both mirrors. (b) Determine the angle between the incoming and outgoing beams. (c) What should be the angle between the mirrors so that the outgoing beam trajectory overlaps with the incoming trajectory?
12. Design a mirror arrangement so that light from a laser pointer will travel in exactly the opposite direction after it reflects off the mirror(s), even if you change the direction the laser pointer is pointing.
13. Two mirrors are oriented at right angles. A narrow light beam strikes the horizontal mirror at an incident angle of 65°, reflects from it, and then hits the vertical mirror. Determine the angle of incidence at the vertical mirror and the direction of the light after leaving the vertical mirror. Include a sketch with your explanation.
14. * You are driving along a street on a sunny day. Only a few apartment building windows appear bright; the rest are pitch black. Explain this difference and include a ray diagram to help with your explanation.
15. A flat mirror is rotated 17° about an axis in the plane of the mirror. What is the angle change of a reflected light beam if the direction of the incident beam does not change?

22.3 Refraction of light

16. (a) A laser beam passes from air into a 25% glucose solution at an incident angle of 35°. In what direction does light travel in the glucose solution? (b) The beam travels from ethyl alcohol to air at an incident angle of 12°. Determine the angle of the refracted beam in the air. (c) Draw pictures for (a) and (b) showing the interface between the media, the normal line, the incident rays, the reflected rays, the refracted rays, and the angles of these rays relative to the normal line.
17. A beam of light passes from glass with refractive index 1.58 into water with a refractive index 1.33. The angle of the refracted ray in water is 58.0°. Draw a sketch of the situation showing the interface between the media, the normal line, the incident ray, the reflected ray, the refracted ray, and the angles of these rays relative to the normal line.
18. A beam of light passes from air into a transparent petroleum product, cyclohexane, at an incident angle of 48°. The angle of refraction is 31°. What is the index of refraction of the cyclohexane?
19. * **Moving laser beam** An aquarium open at the top has 30-cm-deep water in it. You shine a laser pointer into the top opening so it is incident on the air-water interface at a 45° angle relative to the vertical. You see a bright spot where the beam hits the bottom of the aquarium. How much water (in terms of height) should you add to the tank so the bright spot on the bottom moves 5.0 cm?
20. ** **Lifting light** You have a V-shaped transparent empty container such as shown in **Figure P22.20**. When you shine a laser pointer horizontally through the empty container, the beam goes straight through and makes a spot on the wall. (a) What happens to this spot if you fill the container with water just a little above the level at which the laser beam passes through the container? (b) What happens if you fill the container to the very top? Indicate any assumptions used and draw a ray diagram for each situation. Note: This is a multiple-possibility problem.

FIGURE P22.20

21. * A light beam hits the interface between air and an unknown material at an angle of 43° relative to the normal. The reflected ray and the refracted ray make an angle of 108° with respect to each other. What is the index of refraction of the material?
22. * A light ray passes from air through a glass plate with refractive index 1.60 into water. The angle of the refracted ray in the water is 42.0°. Determine the angle of the incident ray at the air-glass interface.
23. * **BIO** **Vitreous humor** Behind the lens of the eye is the vitreous humor, a jellylike substance that occupies most of the eyeball. The refractive index of the vitreous humor is 1.35 and that of the lens is 1.44. A narrow beam of light traveling in the lens comes to the interface with the vitreous humor at a 23° angle. What is its direction relative to the interface when in the vitreous humor?

24. * You watch a crab in an aquarium. Light traveling in air enters a sheet of glass at the side of the aquarium and then passes into the water. If the angle of incidence at the air-glass interface is 22°, what are the angles at which the light wave travels in the glass and in the water? Indicate any assumptions you made.

25. * Light moving up and toward the right in air enters the side of a cube of gelatin of refractive index 1.30 at an incident angle of 80°. Determine the angle at which the light leaves the top surface of the cube. How does the angle change if the refractive index of the gelatin is slightly greater? Explain.

26. * A laser beam is incident at 30° with respect to the normal on a stack of five transparent plates made of materials with the following indexes of refraction (from the top of the stack to the bottom): $n_1 = 1.20$, $n_2 = 1.30$, $n_3 = 1.40$, $n_4 = 1.50$, $n_5 = 1.60$. The plates touch each other. Determine the angle between the outgoing beam and the normal to the last plate. Assume the first and the last plate are in contact with air. (Hint: Before you start calculating, write the expression for the final angle and see if you can simplify it.)

22.4 Total internal reflection

27. * **Can your light be seen?** You swim under water at night and shine a laser pointer so that it hits the water-air interface at an incident angle of 52°. Will a friend see the light above the water? Explain.

28. * Light is incident on the boundary between two media at an angle of 30°. If the refracted light makes an angle of 42°, what is the critical angle for light incident on the same boundary?

29. **Diamond total reflection** Determine the critical angle for light inside a diamond incident on an interface with air.

30. Determine the refractive index of a glucose solution for which the critical angle for light traveling in the solution incident on an interface with air is 42.5°. How would the critical angle change if the glucose concentration were slightly greater? Explain.

31. * You wish to use a prism to change the direction of a beam of light 90° with respect to its original direction. Describe the shape of the prism and its orientation with respect to the original beam to achieve this goal.

32. * You aim a laser beam (in air) at 80.0° with respect to the normal of the surface of a transparent plastic cube. You observe that the refracted beam inside the plastic makes an angle of 42.8° with the normal. Determine the critical angle for light inside the plastic cube (a) incident on an interface with air and (b) incident on an interface with water.

33. * **Prism total reflection** What must be the minimum value of the refractive index of the prism shown in Figure P22.33 in order that light is totally reflected where indicated? Will some of the light make it out of the top surface? Explain.

FIGURE P22.33

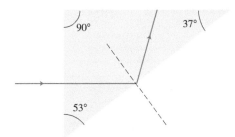

34. **Gems and critical angles** In gemology, two of the most useful pieces of information concerning an unknown gem are the refractive index of the stone and its mass density. The refractive index is often determined using a critical angle measurement. Determine the refractive index of the gemstone shown in Figure P22.34. The critical angle for the total reflection of light at the gem-air interface is 37.28°.

FIGURE P22.34

35. (a) The refractive index for the gem aquamarine is 1.57. Determine the critical angle for light traveling inside aquamarine when reaching an air interface. (b) You have a tourmaline gem and find that a laser beam in air incident on the air-tourmaline interface at a 50° angle has a refracted angle in the tourmaline of 28.2°. Determine the refractive index and critical angle of tourmaline.

36. * You have three transparent media with indexes of refraction $n_1 = 1.33$, $n_2 = 1.46$, and $n_3 = 1.66$. Rank the pairs of media according to decreasing critical angle for a light beam incident on the border between the two media.

37. (a) Rays of light are incident on a glass-air interface. Determine the critical angle for total internal reflection ($n_{glass} = 1.58$). (b) If there is a thin, horizontal layer of water ($n_{water} = 1.33$) on the glass, will a ray incident on the glass-water interface at the critical angle determined in part (a) be able to leave the water? Justify your answer.

22.5 Skills for analyzing reflective and refractive processes

38. * **Invisible in pool?** A swimming pool is 1.4 m deep and 12 m long. Is it possible for you to dive to the very bottom of the pool so people standing on the deck at the end of the pool do not see you? Explain.

39. **Equation Jeopardy 1** Tell all that you can about a process described by the following equation: $1.33 \sin 30° = 1.00 \sin \theta_2$.

40. **Equation Jeopardy 2** Tell all that you can about a process described by the following equation: $1.33 \sin \theta_c = 1.00 \sin 90°$.

41. * **Equation Jeopardy 3** Tell all that you can about a process described by the following equation: $1.00 \sin 53° = 1.60 \sin \theta_2 = 1.33 \sin \theta_3$.

42. ** When reaching a boundary between two media, an incident ray is partially reflected and partially refracted. At what angle of incidence is the reflected ray perpendicular to the refracted ray? The indexes of refraction for the two media are known.

43. * A laser beam travels from air ($n = 1.00$) into glass ($n = 1.52$) and then into gelatin. The incident ray makes a 58.0° angle with the normal as it enters the glass and a 36.4° angle with the normal in the gelatin. (a) Determine the angle of the refracted ray in the glass. (b) Determine the index of refraction of the gelatin.

44. ** The height of the Sun above the horizon is 25°. You sit on a raft and want to orient a mirror so that sunlight reflects off the mirror and travels at an angle of 45° in the water of refractive index 1.33. How should you orient the mirror?

45. ** **Rain sensor** Many cars today are equipped with a rain sensor that automatically switches on the windshield wipers when rain starts to fall. The most common modern rain sensors are based on the principle of total internal reflection. Light from an LED is beamed at a 45° angle onto the windshield glass using a semicircular piece of glass that is glued on the windshield from inside the car. When there is no rain, the light beam undergoes total internal reflection and illuminates the detector (see Figure P22.45a). If the glass is wet, the condition for total internal reflection is no longer fulfilled, and some light escapes to the outside (see Figure P22.45b). This results in a decrease in the light intensity at the detector, which is used as a signal to the car electronics to switch on the wipers. Determine the range of values of the index of refraction of the windshield glass (i.e., the minimum and the maximum values) that will allow the described operation of the sensor shown in the figure.

FIGURE P22.45

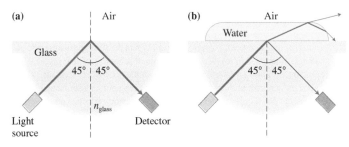

22.6 Fiber optics, prisms, mirages, and the color of the sky

46. * The prism shown in Figure P22.33 is immersed in water of refractive index 1.33. Determine the minimum value of the refractive index for the prism so that the light is totally internally reflected where shown. Will any light leave the top surface of the prism? Explain.

47. * **Light pipe** Rays of light enter the end of a light pipe from air at an angle θ_1 (Figure P22.47). The refractive index of the pipe is 1.64. Determine the greatest angle θ_1 for which the ray is totally reflected at the top surface of the glass-air interface inside the pipe.

FIGURE P22.47

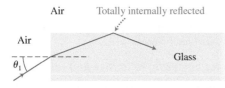

48. * A light ray is incident on a flat piece of glass. The angle of incidence is θ_1 and the thickness of the glass is d. Determine the distance that the ray travels in the glass. At what angle will the ray emerge from the glass on the other side? Draw a ray diagram to help explain your solution.

49. * **Prism** You have a triangular prism made of glass of refractive index 1.60, with angles of 30°-90°-60°. The short side is oriented vertically. A horizontal ray hits the middle of the slanted side of the prism. Draw the path of a ray as it passes into and through the prism. Determine all angles for its trip through the prism.

General Problems

50. * You have a candle and a large piece of paper with a triangular hole slightly larger than the candle flame cut in it. You place the paper between the candle and the wall. Draw ray diagrams to show what you see on the wall when the paper is placed (a) near the candle, (b) halfway between the candle and wall, and (c) near the wall.

51. ** **Design** You wish to investigate the properties of different transparent materials and decide to design a refractometer, a device used to measure the refractive index of an unknown liquid. A light beam in a liquid is refracted into a material whose refractive index is known and is less than that of the unknown liquid. The incident beam is adjusted for total internal reflection, and the equation for the angle of total internal reflection is used to determine the unknown index. Provide suggestions for a design for such a device and perform a sample calculation to show how it might work.

52. * You place a point-like source of light at the bottom of a container filled with vegetable oil of refractive index 1.60. At what height from the light source do you need to place a circular cover with a diameter of 0.30 cm so no light emerges from the liquid?

53. ** There is a light pole on one bank of a small pond. You are standing up while fishing on the other bank. After reflection from the surface of the water, part of the light from the bulb at the top of the pole reaches your eyes. Use a ray diagram to help find a point on the surface of the water from where the reflected ray reaches your eyes. Determine an expression for the distance from this point on the water to the bottom of the light pole if the height of the pole is H, your height is h, and the distance between you and the light pole is l.

54. ** **Coated optic fiber** An optic fiber of refractive index 1.72 is coated with a protective covering of glass of refractive index 1.50. (a) Determine the critical angle for the fiber-glass interface. (b) Determine the critical angle for the glass-air interface. (c) Determine the critical angle for a fiber-air interface (no glass covering). (d) Suppose a ray hits the fiber-glass interface at the angle calculated in (c). What is the angle of refraction of that ray when it reaches the glass-air interface? Will it leave the optic fiber? Explain.

55. ** You put a mirror at the bottom of a 1.4-m-deep pool. A laser beam enters the water at 30° relative to the normal, hits the mirror, reflects, and comes back out of the water. How far from the water entry point will the beam come out of the water? Draw a ray diagram.

56. ** A scuba diver stands at the bottom of a lake that is 12 m deep. What is the distance to the closest points at the bottom of the lake that the diver can see due to light from these points being totally reflected by the water surface? The height of the diver is 1.8 m and $n_{water} = 1.33$.

Reading Passage Problems

Rainbows How is a rainbow formed? Recall that the index of refraction of a medium is slightly different for different colors. When white light from the Sun enters a spherical raindrop, as shown in **Figure 22.25**, the light is refracted, or bent. After reflecting off the back surface of the drop, the light is refracted again as it leaves the front surface.

Each drop separates the colors of light. An observer on the ground with her back to the Sun sees at most one color of light coming from a particular drop (see **Figure 22.26**). If the observer sees red light from a drop (for example, the top drop in Figure 22.26), the violet light for that same drop is deflected above her head. However, if she sees violet light coming from a drop lower in the sky, the red light from that drop is deflected below her eyes onto the ground. She sees red light when her line of view makes an angle of 42° with the beam of sunlight and violet light when the angle is 40°. Other colors of light are seen at intermediate angles.

FIGURE 22.25 Different colors of light are refracted and reflected by different amounts by a spherical raindrop. The color violet is refracted more than red.

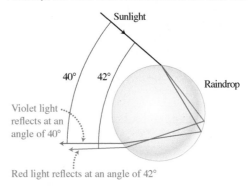

FIGURE 22.26 An observer on the ground sees only one color of light from each raindrop. The drops that send the color red are above the drops that send violet.

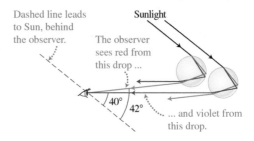

57. Why do you see rainbows only when you are between the Sun and the location of the rainbow?
    (a) The sunlight must be refracted and reflected from raindrops and needs to move back and downward from its original direction.
    (b) You need to intercept the light coming from the Sun.
    (c) If you are looking toward the Sun, rainbows would block the sunlight.
    (d) The sunlight is refracted by the raindrops and changes direction.

58. Suppose that light entered one side of a square water droplet of refractive index 1.33 at an incident angle of 20°. Some of the light reflects off the back surface and then goes back out on the same side as it entered. What is the angle of refraction of light leaving the square droplet?
    (a) 15°   (b) 18°   (c) 20°   (d) 24°   (e) 35°

59. Raindrops reflect different colors of light at different angles. Why do we see the parts of the rainbow as different colors of light rather than all colors coming from each?
    (a) The net deflection of light seen from one drop depends on its refractive index.
    (b) All of the raindrops reflecting a particular color have the same angular deflection relative to the direction of the Sun.
    (c) Your eye sees only one color coming from a particular raindrop.
    (d) The different colors are reflected and refracted at different angles.
    (e) All of the above

60. What is the minimum angle of incidence of red light against the back wall of the water drop in Figure 22.25 so that it is totally reflected at that wall?
    (a) 16°   (b) 21°   (c) 41°   (d) 45°   (e) 49°

61. Repeat the previous problem, only for the violet light.
    (a) 15°   (b) 20°   (c) 41°   (d) 45°   (e) 49°

62. Why is violet light refracted more than red light?
    (a) The violet light travels a shorter distance in the drop than the red light.
    (b) The red light travels more slowly than the violet light.
    (c) The refractive index of water for violet light is greater than that for red light.
    (d) All of the above

Earth energy balance Gases in Earth's atmosphere, such as carbon dioxide and water vapor, act like a blanket that reduces the amount of energy that Earth radiates into space. This phenomenon is called the greenhouse effect. Without the greenhouse effect, most of Earth would have a climate comparable to that of the polar or subpolar regions. What would Earth's mean surface temperature be, in the absence of the gases causing the greenhouse effect?

The Sun continually irradiates our upper atmosphere with an intensity of about 1360 W/m². About 30% of this energy is reflected back into space, leaving an average of 950 W/m² to be absorbed by the surface area of Earth that is exposed to the Sun. This exposed area is πR_{Earth}^2, the Earth's circular cross section that intercepts the sunlight. All objects, including Earth, emit radiation at a rate proportional to the fourth power of the surface temperature (T^4) in kelvins (K). Thus, the radiation is emitted from Earth's surface, which as a sphere has the area $4\pi R_{Earth}^2$.

To maintain a constant temperature, Earth's radiation rate must equal its energy absorption rate from the Sun. A fairly simple calculation indicates that the two rates are equal when the average surface temperature of Earth is 255 K, or about 0 °F. However, Earth's mean surface temperature is much warmer than that—about 288 K—because the calculation neglected the effect of greenhouse gases in the atmosphere. These gases absorb infrared radiation emitted by Earth and reflect some of it back to Earth. Thus, Earth emits less energy into space than it would without greenhouse gases. The energy absorbed from the Sun and the reduced energy emitted into space have caused the Earth to warm to its mean surface temperature of 288 K.

Over the past two centuries the concentration of carbon dioxide in our atmosphere has increased from a pre-industrial level of about 270 parts per million to 380 parts per million. This increase in carbon dioxide and other greenhouse gases has been caused by the burning of fossil fuels and the removal of forests, which absorb carbon dioxide. The carbon dioxide concentration in the atmosphere is expected to reach 600–700 parts per million by 2100. If that occurs, it will be warmer in 2100 than at any time in the last half million years.

63. Why is there essentially zero greenhouse effect on the Moon?
    (a) There is no photosynthesis on the Moon.
    (b) There is no carbon dioxide on the Moon.
    (c) There is no gaseous atmosphere on the Moon.
    (d) Answers b and c are correct.
    (e) Answers a, b, and c are correct.

64. The average Earth surface temperature without its atmosphere would be 255 K. Why?
    (a) At that temperature, the emission rate of radiation from Earth would just balance the absorption rate of radiation from the Sun.
    (b) The emission of Earth radiation is from a sphere of area $4\pi R_{Earth}^2$, whereas the absorption of radiation from the Sun is from an area πR_{Earth}^2.
    (c) Earth's cross section would be significantly reduced because of the lack of the atmosphere.
    (d) Answers a and b are correct.
    (e) Answers a, b, and c contribute about equally to this temperature calculation.

65. The Sun irradiates Earth's outer atmosphere at what rate?
    (a) 1×10^{16} J/s   (b) 4×10^{16} J/s
    (c) 2×10^{17} J/s   (d) 7×10^{17} J/s
    (e) 1400 J/s

66. Because of the greenhouse effect, Earth's average surface temperature is 288 K instead of 255 K. Because of this higher temperature, Earth's surface emits radiation at a rate that is higher by a factor of approximately which of the following?
    (a) 1.13   (b) 1.28   (c) 1.63   (d) 1.87   (e) 2.21

67. Because of the increased temperature of Earth's surface (288 K compared to 255 K) due to the greenhouse effect, its energy emission rate has increased significantly. Why hasn't Earth cooled down as a result?
    (a) It is cooling down.
    (b) The Sun absorbs much of Earth's extra radiation, and the Sun is warming and emitting more radiation.
    (c) There is a long delay between the change in temperature and the increased rate of radiation by Earth.
    (d) Earth's atmosphere absorbs some of the outgoing radiation and emits it back to Earth's surface.
    (e) All of the above are correct.

68. What will the increasing concentration of greenhouse gases do?
    (a) Increase the reflection rate of sunlight.
    (b) Increase the reflection of Earth's radiation back to the surface.
    (c) Increase the cross-sectional area of the Earth, causing more sunlight to be absorbed.
    (d) Shield Earth from dangerous ultraviolet radiation.
    (e) Provide extra nourishment for plants on Earth.

23

Mirrors and Lenses

- How do eyeglasses correct your vision?

- When you look in a mirror, where is the face you see?

- How is a cell phone camera similar to a human eye?

Your friend wears glasses when she is driving, while your physics professor puts her glasses on when she is using her computer. When they buy glasses, your friend gets the glasses labeled − 2.0 and your professor gets the ones labeled + 1.5. What vision problems do they have, and what do those numbers mean? In this chapter you will learn how to answer these questions.

BE SURE YOU KNOW HOW TO:

- Apply the properties of similar triangles (Appendix A).

- Draw ray diagrams and normal lines (Sections 22.1, 22.2, and 22.3).

- Use the laws of reflection and refraction (Sections 22.2 and 22.3).

IN THIS CHAPTER we will apply what we have learned about reflection and refraction to mirrors and lenses. In the process we will learn how cameras, telescopes, microscopes, human vision, and corrective lenses work.

23.1 Plane mirrors

The simplest mirror is a **plane mirror**—a flat, reflective surface, often a metal film covered in glass. When you stand in front of a plane mirror, you see a reflection of yourself. How does the second "you" appear?

Recall the model that we created to describe how extended objects emit light. Each point on a luminous object sends light rays in all directions. Some of these rays enter our eyes and we see the object at the place where those rays originate (**Figure 23.1**). In other words, to locate an object, at least two rays should come into our eyes; the object is at the intersection of those two rays. Now, suppose that a small shining object is in front of a mirror. What happens to the rays emitted by the object when they reach the mirror? Consider Observational Experiment **Table 23.1**.

FIGURE 23.1 Seeing light from an object.

Each point on the surface of the bulb emits light in all directions.

We perceive each point on the light source as the origination point for the light reaching our eyes.

OBSERVATIONAL EXPERIMENT TABLE 23.1   **Seeing a point object in a plane mirror**

 VIDEO OET 23.1

Observational experiment	Analysis
We place a small lightbulb on a table about 20 cm in front of a plane mirror that is held perpendicular to the table, with one side resting on the table. Observers A, B, and C place rulers on the table and point them in the direction of the bulb that they see in the mirror. 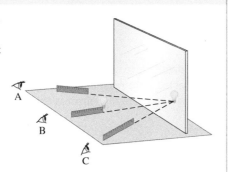 Note that the dashed extensions of the rulers all intersect at one point behind the mirror.	For observer A to see an image of the lightbulb when looking at the mirror, one or more rays of light reflected from the mirror must reach his eyes. Rays between 1 and 2 do reach observer A's eyes. 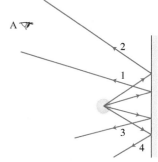

Patterns

- The rulers all point to the same spot behind the mirror. Rays between 1 and 2 reaching observer A seem to originate from that spot. This is the location of the perceived image of the bulb produced by the mirror.
- The ray diagram shows that the perceived image of the bulb is produced at the same distance behind the mirror as the bulb is in front of it.
- To locate the image, at least two rays need to reach the observer's eyes because the perceived image is at the intersection of those rays.

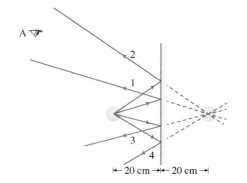

In Table 23.1 we found that the image of the real object seen in the mirror is located where light reflected from the mirror to the eye of the observer *seems* to originate. This perceived image is behind the mirror and not on the surface of the mirror. We also found, using the ray diagram, that the image is exactly the same distance behind the plane mirror as the object is in front of it. Let us test these findings in Testing Experiment **Table 23.2**, on the next page.

Testing the image location of a plane mirror
VIDEO
TET 23.2

Testing experiment	Prediction	Outcome
We repeat the Table 23.1 experiment for observer A but cover the part of the mirror directly in front of the bulb.	If the image is due to the reflected rays, then even if we cover part of the mirror, some of the reflected rays will still reach the observer's eyes. The location of the image should not change. 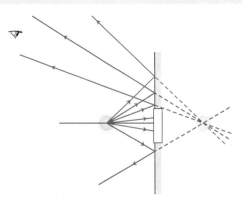	The observer sees the image of the bulb exactly where it was before.

Conclusions

The position of the image formation is consistent with our previous experiment. It also disproves the common idea that the image forms on the surface of the mirror.

The image we see in a plane mirror is not real; there is no real light coming from behind the mirror. The reflected light reaching your eyes appears to originate from a point behind the mirror. What we see in the mirror is called a **virtual image**.

> **Plane mirror virtual image** A plane mirror produces a virtual image that is the same distance behind the mirror as the object is in front of it. The reflected light seems to diverge from the image behind the mirror. But no light actually leaves that image—you see light reflected from the mirror.

So far, we have learned how to find images in plane mirrors of objects that emit light. But when you stand in front of a mirror, you can see your image even though you do not seem to be a source of light. To explain this phenomenon, think back to Chapter 22. All objects reflect light diffusely. Thus, if the room is illuminated by sunlight or you have a lightbulb on, light from those sources reaches you and reflects off you, effectively making you a new (secondary) source of light. To test this hypothesis, go into an absolutely dark room and stand in front of a mirror—you will not see anything. Once you switch on a flashlight, however, you are able to see your virtual image in the mirror as well as an image of the flashlight. All objects in lit rooms become secondary sources of light due to their diffuse reflection of incident light.

CONCEPTUAL EXERCISE 23.1 **Where is the lamp?**

You place a lamp in front of a mirror and tilt it so that the top and the bottom of the lamp are at different distances from the mirror, at the position shown in the figure at right. Where do you see the image of the lamp produced by the mirror?

Mirror

Sketch and translate We use the figure in the problem statement to visualize the situation. We need to find the virtual image of the lamp produced by the mirror. The virtual image of each point of an object is

the place from which the reflected light, represented by rays, appears to diverge.

Simplify and diagram Assume that all points of the lamp (including the base) send out rays in all directions. For simplicity, we will consider the top and the bottom of the lamp and locate the images of each of these two end points. Assume that the images of all other points on the lamp are formed between the end point images. As you can see from the following ray diagram, the image of each point is behind the mirror, and each point is at the same distance behind the mirror as the point on the lamp is in front of the mirror.

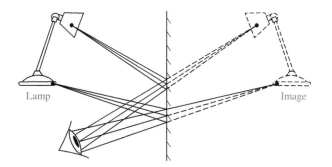

Lamp   Image

Try it yourself What happens to the distance between an observer and her image in a plane mirror if the observer moves backward, doubling her distance from the mirror?

Answer   The distance between the observer and the image doubles.

CONCEPTUAL EXERCISE 23.2   **Buying a mirror**

Your only requirement when buying a mirror is that you can see yourself from head to toe. What should be the minimum length of the mirror you buy?

Sketch and translate Draw a sketch of the situation, as shown below. Let h be your height. Express the mirror height H as a fraction or a multiple of your height. Let s be your distance from the mirror. Since the goal is to see your entire body in the mirror, rays from the top of your head and from the bottom of your feet must reflect off the mirror into your eyes.

Simplify and diagram Assume that the mirror is mounted on a vertical wall and that all parts of your body are the same distance from the mirror. Also assume that you are a shining object of a particular height h. Finally, let h_1 be the vertical distance between your feet and eyes and h_2 be the vertical distance between the top of your head and your eyes. This means $h_1 + h_2 = h$.

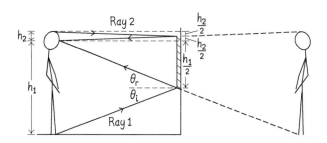

You will see your toes if ray 1 from your toes reaches your eyes after reflection. Ray 1 hits the mirror at a height $h_1/2$ above your toes. You will see the top of your head if ray 2 from the top of your head reaches your eyes after reflection. Ray 2 hits the mirror at a distance $h_2/2$ below a horizontal line from the top of your head to the mirror. From the diagram we see that you can cut off both the bottom $h_1/2$ and the top $h_2/2$ of the mirror and still see your whole body. Thus the length of the smallest mirror is $h_1/2 + h_2/2 = h/2$. Note that the size of the mirror does not depend on the distance s between you and the mirror—you can see your entire body in the mirror no matter how far you stand from it.

Try it yourself You are 1.6 m tall and stand 2.0 m from a plane mirror. How tall is your image? How far from the mirror on the other side is it? What is your image height if you double your distance from the mirror?

Answer   1.6 m; 2.0 m; the same size.

REVIEW QUESTION 23.1 A mirror is hanging on a vertical wall. Will you see more or less of your body if you step closer to the mirror?

23.2 Qualitative analysis of curved mirrors

Curved mirrors are cut from a spherically shaped piece of glass backed by a metal film. **Concave mirrors** are often used for magnification, in telescopes and cosmetic mirrors. The curve of a concave mirror bulges away from the light it reflects (see **Figure 23.2a** on the next page). In a **convex mirror**, the curve bulges toward the light it

FIGURE 23.2 Features of (a) concave and (b) convex mirrors.

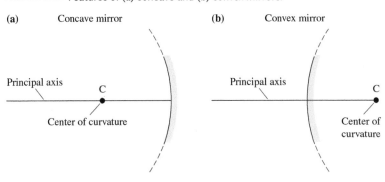

(a) Concave mirror

Principal axis

C

Center of curvature

(b) Convex mirror

Principal axis

C

Center of curvature

FIGURE 23.3 A convex mirror provides visibility at blind spots and around corners.

reflects (see Figure 23.2b). Convex mirrors are used as passenger-side rearview mirrors and to provide visibility at blind spots, such as hallway corners and driveway exits (**Figure 23.3**).

In both kinds of curved mirrors, the center of the sphere of radius R from which the mirror is cut is called the **center of curvature** C of the mirror. The imaginary line connecting the center of curvature with the center of the mirror's surface is called the **principal axis.**

FIGURE 23.4 A concave mirror reflects a laser beam.

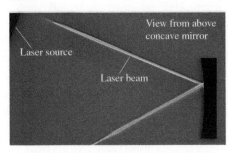

Concave mirrors

In Observational Experiment Table 23.3, we cut a narrow band from a concave mirror and lay it on its edge on a piece of paper (**Figure 23.4**). The paper allows us to observe the paths of laser light incident and reflected from the mirror.

OBSERVATIONAL EXPERIMENT TABLE 23.3 Reflection of light from a concave mirror

VIDEO OET 23.3

Observational experiment	Analysis
Experiment 1. We shine a laser beam parallel to the plane of the page toward a concave mirror and trace the path of the incident and reflected light on the page. 	We use the law of reflection to draw a ray diagram. The dashed line from the place where the ray hits the mirror to the center of curvature is a normal line perpendicular to the surface of the mirror. The law of reflection accounts for the path of the reflected ray.
Experiment 2. We send several beams parallel to the principal axis. Reflected beams all pass near the same point on the principal axis. 	We draw a ray diagram and use the law of reflection to analyze the paths of the light. The normal lines here are radii of the mirror. We find that all reflected rays pass through the same point on the principal axis (called the *focal point* F of the mirror), matching the outcome of the experiment. 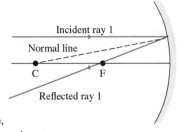

Observational experiment	Analysis
Experiment 3. This time we send several beams parallel to each other but not parallel to the principal axis. One ray passes though the center of curvature C. All reflected rays pass through the same point.	We use the law of reflection to draw a ray diagram to analyze the paths of the light. The ray passing through the center of curvature is perpendicular to the mirror and thus reflects back on itself. All other rays after reflection pass through the 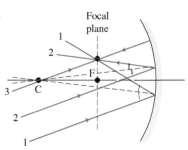

We use the law of reflection to draw a ray diagram to analyze the paths of the light. The ray passing through the center of curvature is perpendicular to the mirror and thus reflects back on itself. All other rays after reflection pass through the same point on a line perpendicular to the principal axis and through the focal point. This point is in the *focal plane* of the mirror. The ray diagram matches the outcome of the experiment.

Patterns

- The mirror reflects light according to the law of reflection. The normal line at any point on the mirror goes through the center of curvature. The normal lines are in the radial directions.
- After reflection, all incident rays traveling parallel to the principal axis pass through the same point on the principal axis—the focal point of the mirror.
- After reflection, all incident rays traveling parallel to each other but not parallel to the principal axis intersect at a common point. This point is on the line perpendicular to the principal axis through the focal point.

Table 23.3 shows us that a curved mirror reflects light in agreement with the law of specular reflection. But unlike a flat mirror, it causes parallel incident rays to pass through a single point after reflection. If the incident rays are also parallel to the principal axis, this point is called the **focal point** F of the mirror (see Experiment 2 in Table 23.3). If they are not parallel to the principal axis (but are still parallel to each other), they all pass through another point on the **focal plane** of the mirror (see Experiment 3 in Table 23.3). The focal plane is the plane that is perpendicular to the principal axis and passes though the focal point. The distance from the focal point to the surface of the mirror where it intersects the principal axis is called the **focal length** f (**Figure 23.5**). A concave mirror is sometimes called a *converging mirror* because incident parallel rays converge after reflection from it. Since we can consider the rays of the Sun to be almost parallel, we can easily determine the location of the focal point and the focal plane of any converging mirror if we use the Sun as the light source (be careful if you perform this experiment—you might start a fire!).

If a mirror has a large radius of curvature compared to its size and the incident rays are close to the principal axis, the focal point is approximately halfway between the center of the mirror and the center of curvature ($f = R/2$). See the geometric proof in **Figure 23.6**.

FIGURE 23.5 The effect of a concave mirror on parallel rays.

Rays parallel to the principal axis reflect and pass through the focal point.

Principal axis

Rays parallel to each other but not to the principal axis reflect and pass through a point in the focal plane.

FIGURE 23.6 The focal length f of a curved mirror is half the radius of curvature R.

The normal line goes through the center of curvature C.

Ray 1

Equal incident and reflected angles

For small angles,
$\sin \theta/2 \approx \theta/2 \approx h/R$ and $\sin \theta \approx \theta \approx h/f$.
Thus, $2(\theta/2) = 2(h/R) = h/f$, or $f = R/2$.

TIP The verb *converge* does not mean that the rays reach the focal point and stop there. They continue traveling after passing through this point.

Using a ray diagram to locate the image formed by a concave mirror

We can use a ray diagram to locate the image of an object that is placed in front of a concave mirror. The method is illustrated in Physics Tool Box 23.1, on the next page. The object in this case is a lightbulb that is placed beyond the center of curvature of a concave mirror. We will represent the object with an arrow that originates on the principal axis. For the time being we only locate the image of the top of the arrow and assume that the image of the base of the object (the bottom of the arrow) is also on the axis. We will validate this assumption later.

Constructing a ray diagram to locate the image of an object produced by a concave mirror

1. Place a vertical arrow on the principal axis to represent the object. Each point on the object emits light in all directions. We find the image location of only the tip of the object arrow.

2. Choose two or three rays from the tip of the arrow. You know the directions of these rays after reflection from the mirror.

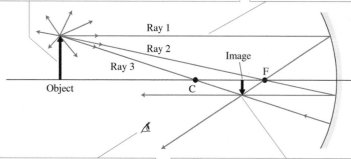

Ray 1 travels from the tip of the object toward the mirror parallel to the principal axis. After reflection, it passes through the focal point F.
Ray 2 travels from the tip of the object through the focal point F. After reflection, it travels parallel to the principal axis.
Ray 3 passes through the center of curvature C. It hits the mirror perpendicular to its surface and reflects back in the same direction.

4. The observer sees the image at the place from which the reflected rays seem to diverge.

3. A *real image* is formed where these two or three rays intersect.

TIP Remember that each point on an object emits an infinite number of rays that reflect off the mirror according to the law of reflection. Rays 1, 2, and 3 in the Physics Tool Box were chosen because we know how they reflect without needing to draw the normal line to the point of incidence and measure the angles of incidence and reflection.

In Physics Tool Box 23.1 we used three special rays to locate the image of an object produced by a concave mirror. We will use these three rays throughout our study of mirrors.

The image formed in the Tool Box is a **real inverted image**: it is **inverted**, meaning that the image points opposite the direction of the object; it is also a **real image**, meaning that the reflected rays actually pass through that image location and then reach the eyes of the observer. If you placed a small screen (a piece of paper, your hand, etc.) at that image location, you would see a sharp (focused), inverted (upside-down) image projected onto it. If you place the screen closer to the mirror or farther away than the image location, reflected light will reach it, but the image will be fuzzy because the rays pass through different points on the screen. Note that we only considered the case when the object is located behind the focal point of the mirror. We will consider the case when the object is between the mirror and the focal point below.

Not all reflected rays result in a real image. Consider the situation in **Figure 23.7a**, in which a pencil is closer to a concave mirror than the focal point F. We will use two rays to construct a ray diagram. In Figure 23.7b we see that reflected rays 1 and 2 diverge. For a person holding the pencil (the eye in the diagram), the light seems to originate from a location behind the mirror. This location is the location of the **virtual image** produced by the mirror. The image is *virtual* (similar to the images formed by plane mirrors) because the light does not actually come from that point—it just appears to. The image is magnified (bigger than the object) and upright. You see such an image when you use a concave mirror to apply makeup or shave.

FIGURE 23.7 Using a ray diagram to locate the image of an object located near a concave mirror.

(a)

(b)

Locating the image of the base of an object produced by a concave mirror

So far, our ray diagrams have only included the image of the top of the object. We have assumed that the image of the base of the object is always on the principal axis directly below the image of the top of the object. To validate this assumption we will use a ray from the base of the object traveling along the principal axis as ray 1. It reflects back on itself. Since ray 1 stays on the principal axis, any other ray that we use to locate the image of the base will also have to intersect the principal axis. To find where on the axis the image of the base is located, first draw an arbitrary imaginary ray through the center of curvature. This is ray 3 from Physics Tool Box 23.1. That ray reflects back through C (**Figure 23.8**). Now draw another ray, which we will call ray 4, from the base of the object and oriented parallel to ray 3. Ray 4 will pass through the same point on the focal plane as ray 3. Ray 4 intersects the principal axis at the location of the image of the base of the object. This locates the image of the base. It does in fact lie directly above the image of the top of the arrow (not shown in the figure). The method that we used to locate the base of the object works for locating the image of any point of an object that resides on the principal axis.

FIGURE 23.8 Locating the base of the image produced by a curved mirror.

A line parallel to ray 4 that passes through C also passes through the focal plane at the same place that the reflected ray passes through that plane.

Convex mirrors

Let's investigate convex mirrors. Consider Observational Experiment **Table 23.4**.

OBSERVATIONAL EXPERIMENT TABLE 23.4   **Where is the focal point for a convex mirror?**   VIDEO OET 23.4

Observational experiment	Analysis
We send two laser beams parallel to the principal axis of a convex mirror. We observe that the reflected beams diverge. 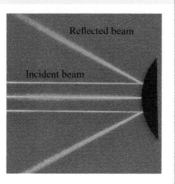	The ray diagram and the law of reflection help us analyze the situation. Using the normal lines passing though the center of curvature, we find that the reflected rays diverge. If we extend the reflected rays behind the mirror, their extensions intersect at one point, which we will call the virtual focal point F. 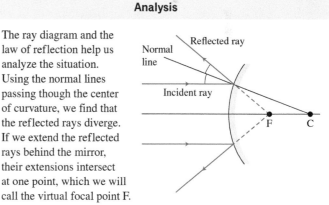

Pattern

- Light rays moving parallel to the principal axis of a convex mirror reflect and diverge away from each other.
- Lines drawn backward along the direction of the reflected light cross the axis at a focal point F behind the mirror.

Recall that for a concave mirror, rays parallel to the principal axis reflect through a focal point in *front* of the mirror. However, for a convex mirror, rays parallel to the principal axis diverge after reflection. They diverge from what is called a **virtual focal point** F behind the mirror. As with a concave mirror, the focal length f equals half the radius of curvature R of the mirror ($f = R/2$), as shown in **Figure 23.9**. Let's use a new set of ray diagrams to investigate images formed by convex mirrors.

FIGURE 23.9 A convex mirror reflects rays parallel to the principal axis.

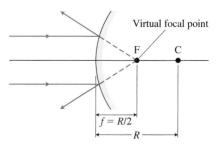

1. Place a vertical arrow on the principal axis to represent the object. Each point on the object emits light in all directions. We find the image location of only the tip of the object arrow.

2. Choose two rays from the tip of the arrow. You know the directions of these rays after reflection from the mirror. The rays diverge after reflection.

Ray 1 travels from the tip of the object toward the mirror parallel to the principal axis. After reflection, it diverges as if it were coming from the focal point F behind the mirror. Ray 2 passes toward the center of curvature C. It hits the mirror perpendicular to its surface and reflects back in the opposite direction.

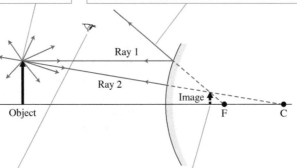

4. The observer sees the image at the place from which the reflected rays seem to diverge.

3. Extend the rays behind the mirror (the dashed lines). A *virtual image* is formed at the point behind the mirror from which these two rays seem to diverge.

In Physics Tool Box 23.2 the object was about $1.6R$ from the mirror. An upright virtual image was formed about $0.4R$ behind the mirror. What happens to the image when the object is moved closer to the mirror?

You hold a convex mirror $0.7R$ behind a pencil. Approximately where is the image of the pencil, and what are the properties of the image?

Sketch and translate To solve the problem, we draw a ray diagram on which we represent the pencil with an arrow. The given distance in the problem statement is the distance between the object and the mirror along the principal axis. We use the ray diagram to locate the image of the top of the pencil (the tip of the arrow).

Simplify and diagram Draw a ray diagram. We use rays 1 and 2 described in Physics Tool Box 23.2 to locate the image. The image is virtual and upright about $0.3R$ behind the mirror and is smaller than the object. This is what you would see if you looked at your reflection on the back of a shiny tablespoon. The images formed by convex mirrors are always upright, virtual (behind the mirror), and reduced in size compared to the objects.

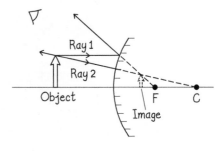

Try it yourself In order to obtain the following types of images of an object, should you use a concave or a convex mirror? Where should you place the object relative to the mirror? (a) Real, bigger than the object; (b) real, smaller than the object; (c) virtual, bigger than the object; (d) virtual, smaller than the object; (e) real and the same size as the object.

Answer

(a) Concave: the object should be between R and f from the mirror; (b) concave: the object should be farther than R from the mirror; (c) concave: the object should be closer than a focal length from the mirror; (d) convex: the object can be at any location; (e) concave: the object should be at $2f$ from the mirror—right at the center of curvature.

23.3 The mirror equation

We have learned that the distance to the image from the mirror depends on where the object is located and on the mirror's type and focal length. We will use the ray diagram in **Figure 23.10** to help derive a mathematical relationship between these three quantities:

FIGURE 23.10 A ray diagram to help develop the mirror equation.

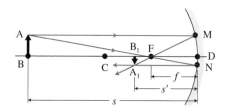

- the distance of the object from the mirror s,
- the distance of the image from the mirror s', and
- the focal length of the mirror f.

Note the following notations in Figure 23.10: AB is the object, and A_1B_1 is the image. M and N are points on the mirror where light rays strike it. We assume that the mirror isn't curved very much ($MD \approx AB$) and the rays are incident close to the principal axis. Now, using this notation, we make the following steps to complete the derivation.

- $AM \approx BD = s$
- $A_1N \approx B_1D = s'$
- $MD \approx AB$ and $ND \approx A_1B_1$.
- ABF and NDF are similar triangles. Thus,

$$\frac{ND}{DF} = \frac{A_1B_1}{DF} = \frac{AB}{BF} \quad \text{or} \quad \frac{AB}{A_1B_1} = \frac{BF}{DF} = \frac{s-f}{f}$$

- A_1B_1F and MDF are similar triangles. Thus,

$$\frac{A_1B_1}{B_1F} = \frac{MD}{DF} = \frac{AB}{DF} \quad \text{or} \quad \frac{AB}{A_1B_1} = \frac{DF}{B_1F} = \frac{f}{s'-f}$$

- Setting the previous two AB/A_1B_1 ratios equal to each other, we find that

$$\frac{s-f}{f} = \frac{f}{s'-f} \tag{23.1}$$

After some algebra (see the Tip), we get a relationship called the **mirror equation**:

$$\frac{1}{s} + \frac{1}{s'} = \frac{1}{f} \tag{23.2}$$

The mirror equation [Eq. (23.2)] allows us to predict the distance s' of the image from the mirror given the distance s of the object from the mirror and the mirror's focal length f. Before we test the equation experimentally, let us check to see if it is consistent with an extreme case. We know that a concave mirror causes rays parallel to the principal axis to pass through the focal point after reflection. If an object is extremely far away along the principal axis, we can assume that rays from the object reaching the mirror are parallel to the principal axis. *If* Eq. (23.2) is correct, and we use infinity for the object distance, *then* we should find that the image distance for that object equals the focal length distance. Using Eq. (23.2), we get

$$\frac{1}{\infty} + \frac{1}{s'} = 0 + \frac{1}{s'} = \frac{1}{f}$$

Therefore, the image is at a distance equal to the focal length away from the mirror. This result is consistent with the prediction based on the extreme case analysis.

Where should the screen be?

You place a candle 0.80 m from a concave mirror with a radius of curvature of 0.60 m. Where should you place a paper screen to see a sharp image of the candle?

Sketch and translate Draw a sketch of the situation, as shown at right. Assemble all parts of the experiment on a meter stick so the distances can be easily measured. The known quantities are $s = 0.80$ m and $f = R/2 = 0.30$ m. To find where to place the screen, we need to determine s'.

Simplify and diagram Model the candle as an arrow. We use a ray diagram to locate the image of the tip of the arrow, as shown at right. For this ray diagram, we use rays 1 and 3 as described in Physics Tool Box 23.1. You can estimate the image distance from the diagram—it looks like the image distance is a little less than two-thirds of the object distance.

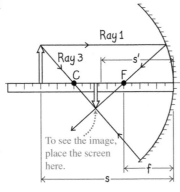

Represent mathematically We see from the ray diagram that the image is real, inverted, and closer to the mirror than the candle. Rearrange Eq. (23.2) to find s':

$$\frac{1}{s'} = \frac{1}{f} - \frac{1}{s}$$

Solve and evaluate

$$\frac{1}{s'} = \frac{1}{f} - \frac{1}{s} = \frac{1}{0.30 \text{ m}} - \frac{1}{0.80 \text{ m}}$$

$$= \frac{0.80 \text{ m} - 0.30 \text{ m}}{(0.30 \text{ m})(0.80 \text{ m})} = 2.08 \text{ m}^{-1}$$

Thus, $s' = 1/(2.08 \text{ m}^{-1}) = 0.48$ m. The image is closer to the mirror than the object, as predicted by our ray diagram. Thus, the predictions using mathematics and the ray diagram are consistent. When we place a screen where we predicted, we see an inverted real image of the flame. We will make the screen small and place it below the principal axis to avoid blocking light from the candle.

- -

Try it yourself Determine the distance of the image when the candle is 0.20 m from the mirror. Does the answer make sense to you?

Answer   the negative sign means that the image is virtual.
−0.60 m. The image is virtual and behind the mirror. Evidently,

---

In the "Try it yourself" part of Example 23.4, the mirror equation led to a negative value for the image distance. The image was virtual. We need to agree on some sign conventions when using Eq. (23.2).

- We put a negative sign in front of the distance s' for a virtual image that appears to be behind a mirror.
- We put a negative sign in front of the focal length f for a convex mirror.

Let us see if these sign conventions work when we use the mirror equation for a convex mirror.

Testing the mirror equation for a convex mirror

A friend's face is 0.60 m from a convex 0.50-m-radius mirror. Where does the image of her face appear to you when you look at the mirror?

Sketch and translate The situation is sketched below. The known information is $s = 0.60$ m and $f = -R/2 = -(0.50 \text{ m})/2 = -0.25$ m. Note that for a convex mirror, the focal length is a negative number.

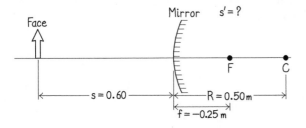

Simplify and diagram Draw a ray diagram representing your friend's face as an arrow. The image is upright, virtual, behind the mirror, and closer to the mirror than the object. Let us see if the mathematics matches this prediction.

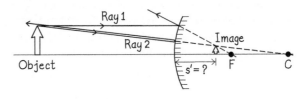

Represent mathematically Rearrange the mirror equation [Eq. (23.2)] to locate the image of your friend's face.

$$\frac{1}{s'} = \frac{1}{f} - \frac{1}{s}$$

Solve and evaluate Insert the known information:

$$\frac{1}{s'} = \frac{1}{-0.25 \text{ m}} - \frac{1}{0.60 \text{ m}} = -5.67 \text{ m}^{-1}$$

The image distance is $s' = 1/(-5.67 \text{ m}^{-1}) = -0.18$ m. We see that the image distance is negative and its magnitude is less than the magnitude of the object distance—consistent with the ray diagram.

Try it yourself The image of your friend's face in a convex mirror is upright, virtual, and 0.30 m behind the mirror when the mirror is 1.0 m from her face. Determine the radius of the sphere from which the mirror was cut.

Answer

The object distance is $s = 1.0$ m, the image distance is $s' = -0.30$ m, and the focal length is $f = -0.43$ m. Consequently, $R = -0.86$ m. The minus sign indicates a convex mirror.

It appears that the mirror equation works equally well for concave and convex mirrors provided that appropriate sign conventions are used.

> **Mirror equation** The distance s of an object from the surface of a mirror, the distance s' of the image from the surface of the mirror, and the focal length f of the mirror are related by the mirror equation:
>
> $$\frac{1}{s} + \frac{1}{s'} = \frac{1}{f} \tag{23.2}$$
>
> The focal length is half the radius of curvature: $f = R/2$. The following sign conventions apply for mirrors:
>
> - The focal length f is positive for concave mirrors and negative for convex mirrors.
> - The image distance s' is positive for a real image and negative for a virtual image.

Magnification

You have probably noticed that the size of an image produced by a curved mirror is sometimes bigger and sometimes smaller than the size of the object. The change in the size of the image compared to the size of the object is a quantity called **linear magnification** m. Linear magnification is defined in terms of the image height h' and the object height h, where the heights are the perpendicular sizes of the image and object relative to the principal axis:

$$\text{linear magnification} = \frac{\text{image height}}{\text{object height}} = m = \frac{h'}{h} \tag{23.3}$$

A height is positive if the image or object is upright and negative if inverted.

It is often not possible to calculate magnification as defined above since the object and image heights are sometimes unknown. However, we can determine the magnification using the mirror equation and the object distance and image distance. Consider **Figure 23.11**, which will be used to derive a relation between h' and h and s' and s. In Figure 23.11a we draw the image of the object using rays 1 and 2. Then in Figure 23.11b we draw a new ray that travels from the top of the object to the center of the mirror and then to the top of the inverted image. The reflected ray makes the same angle θ with the principal axis as does the incident ray. Thus,

$$\tan \theta = \frac{h}{s} = \left| \frac{h'}{s'} \right|$$

Therefore, the absolute value of the magnification is $|m| = |h'/h| = |s'/s|$. If the image is real, it will be inverted. Remember, the height of the inverted image is

FIGURE 23.11 Determine linear magnification.

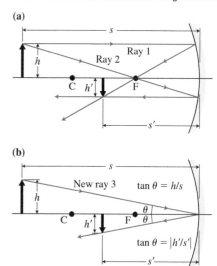

considered negative. Hence h'/h will be negative, whereas s' will be positive—the ratios have the opposite signs. To account for this, we add a negative sign. Thus, $m = (h'/h) = -(s'/s)$.

Linear magnification m is the ratio of the image height h' and object height h and can be determined using the negative ratio of the image distance s' and object distance s:

$$m = \frac{h'}{h} = -\frac{s'}{s} \qquad (23.4)$$

Remember that a height is positive for an upright image or object and negative for an inverted image or object. Also note that m is a unitless physical quantity.

EXAMPLE 23.6   Your face in a mirror

You use a concave mirror with a radius of curvature of 0.32 m for putting on makeup or shaving. When your face is 0.08 m from the mirror, what are the image size and magnification of a 0.0030-m-diameter birthmark on your face?

Sketch and translate Sketch a ray diagram, such as the one shown below. The givens are $h = 0.30$ cm, $s = 0.08$ m, and $f = (R/2) = +0.16$ m. We need to determine the linear magnification m and then the diameter h' of the birthmark image.

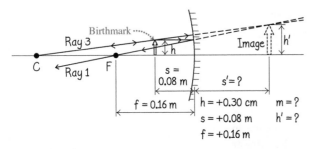

Simplify and diagram Represent the birthmark as an arrow. The reflected rays diverge. Extend the reflected rays back through the mirror to find where they intersect. This is the place where reflected light seems to come from and is the location of the image of the arrow's tip. The image is enlarged, upright, and virtual. We do not use ray 2 from Physics Tool Box 23.1 in this diagram because it does not hit the mirror.

Represent mathematically First use Eq. (23.2) to determine the image distance s'. Then use Eq. (23.4) to find the magnification m and the image height h'.

Solve and evaluate Rearranging Eq. (23.2), we find

$$\frac{1}{s'} = \frac{1}{f} - \frac{1}{s} = \frac{1}{(0.16 \text{ m})} - \frac{1}{(0.08 \text{ m})} = -6.25 \text{ m}^{-1}$$

or $s' = 1/(-6.25 \text{ m}^{-1}) = -0.16$ m. Use Eq. (23.4) to determine the linear magnification:

$$m = -\frac{s'}{s} = -\frac{(-0.16 \text{ m})}{(0.08 \text{ m})} = +2.0$$

The magnification is also $m = (h'/h)$, so

$$h' = mh = (+2.0)(0.30 \text{ cm}) = +0.60 \text{ cm}$$

The birthmark image is upright and two times bigger than the object.

Try it yourself You hold a 1.0-cm-tall coin 0.20 m from a concave mirror with focal length +0.60 m. Is the image of the coin upright or inverted, and what are its magnification and height?

Answer

The image is upright; $m = 1.5$ and $h' = 1.5$ cm.

TIP   Notice that the physical size of the mirror does not enter into the mirror equation and does not affect the magnification. However, its focal length (and therefore its radius of curvature) does. If you cut the same mirror in half, the size of the image will not change.

REVIEW QUESTION 23.3 You place a concave mirror on a stand so it is vertical and slowly move a candle closer to the mirror from a relatively large distance away. What does your friend need to do to make an image appear on a paper screen as you move the candle closer to the mirror?

23.4 Qualitative analysis of lenses

Flat and curved mirrors allow us to create magnified images of objects using the reflection of light. We can also create images using a lens. A **lens** is a piece of glass or other transparent material with two curved surfaces that produces images of objects by changing the direction of light through refraction. Lenses, like curved mirrors, can be concave or convex. A convex lens (**Figure 23.12a**) is thicker in the middle than at its edges; it consists of two sections of spheres faced with the convex parts pointing out. A concave lens (Figure 23.12b) is thinner in the middle than at its edges; it consists of two sections of spheres faced with the convex parts pointing inward. In this book we will only investigate lenses in which the two surfaces are identical sections of spheres. However, there are practical uses for lenses with one curved surface and one flat, or with two differently curved surfaces. We will start with convex lenses, as they are easier to analyze.

FIGURE 23.12 Lenses are made from spherical surfaces.

Convex lenses

In Observational Experiment Table 23.5, we show the paths of several parallel laser beams passing through a convex lens made of transparent plastic. The index of refraction of plastic is higher than that of air. So when a ray of light moves from air to plastic, it bends toward the normal line; when the ray leaves the lens, it bends away from the normal line. The overall result is that the ray bends toward the principal axis.

OBSERVATIONAL EXPERIMENT TABLE 23.5 Laser beams passing through a convex lens

VIDEO OET 23.5

Observational experiment	Analysis
Experiment 1. We shine three laser beams from the left parallel to the principal axis of a convex plastic lens. 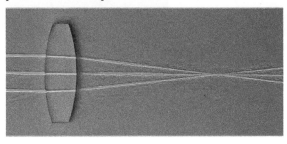	After passing through the lens, the rays pass through the same point on the principal axis. We use the law of refraction to analyze the paths of the rays. Ray 2 is perpendicular to the boundary of the two media and thus does not bend. Rays 1 and 3 bend toward the normal at the air-plastic interface and away from the normal at the plastic-air interface. 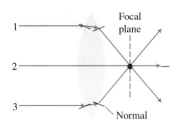
Experiment 2. We shine three laser beams from the left parallel to each other at an arbitrary angle relative to the principal axis of the lens.	The rays pass through a point on the plane perpendicular to the principal axis that passes through the focal point found above. We use the law of refraction to analyze the paths of the rays. Although ray 2 is not perpendicular to the boundary, it passes straight through the lens—a small bending toward the normal on the air-plastic interface is cancelled by the small bending away from the normal at the plastic-air interface 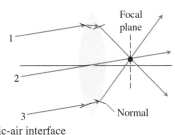

Patterns

1. In all experiments, the rays passing through the center of the lens do not bend.
2. The rays parallel to the principal axis pass through the same point after the lens. We call this point the focal point, similar to the focal point of a curved mirror.
3. The rays parallel to each other pass through a point on the plane perpendicular to the principal axis through the focal point. We call this plane the focal plane.

FIGURE 23.13 More and less curved convex lenses.

(a)

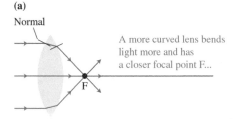

A more curved lens bends light more and has a closer focal point F...

(b)

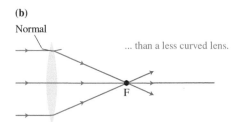

... than a less curved lens.

We have found that a convex lens made of plastic is similar to a concave mirror where incident rays parallel to the principal axis intersect at a **focal point** after passing through the lens. Note in **Figure 23.13** that the rays converge more steeply when passing through a lens with highly curved surfaces than when passing through a lens with less curvature. The focal point is closer to a lens with more curvature than for a lens with less curvature.

Therefore, a lens with two symmetrical curved surfaces has two focal points at equal distances on each side of the lens. In all of our experiments, we shine light from the left side of the lens. As a result, the focal point is on the right side of the lens. If we repeat the experiments, shining the light from the right, the rays will converge on the left side of the lens.

So far, we have used only glass and plastic lenses surrounded by air. However, other materials can act as lenses, such as a round plastic bottle filled with water. Its shape is slightly different from that of a regular lens, but if our understanding of the processes that involve lenses is correct, it should work as one. We fill the bottle with water and shine laser beams on it as shown in **Figure 23.14a**. We observe that after passing through the bottle, the beams converge at one point and then diverge again. Figure 23.14b shows why the rays converge (notice the direction of the normal at each interface). The bottle seems to work as a convex lens with a focal length of about 5 cm.

FIGURE 23.14 Experiments using a glass bottle as a lens.

What happens if we place an empty bottle in a container filled with water (Figure 23.14c)? The laser beams diverge. This experiment shows us that it is not only the shape of the lens but also the refractive index of the material from which it is made and the index of the surrounding material that determine whether the lens is converging or diverging. In addition, the focal length of a lens depends not only on the radius of curvature of the surfaces but also on the refractive indexes of the material between the surfaces and the outside matter. The closer the refractive indexes are to each other, the less the lens bends light rays, and therefore the longer its focal length.

Concave lenses

FIGURE 23.15 The focal point of a concave lens.

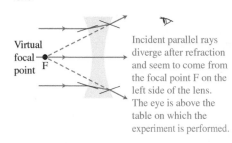

Virtual focal point F

Incident parallel rays diverge after refraction and seem to come from the focal point F on the left side of the lens. The eye is above the table on which the experiment is performed.

We can perform an experiment comparable to that in Table 23.5, sending parallel laser beams through a concave glass lens placed flat on a table. We use the rays to represent the laser beams in **Figure 23.15**. The rays parallel to the principal axis diverge after passing through the concave lens. The dashed lines extended back reveal that these rays seem to diverge from a single point on the axis. This point is called the **virtual focal point**.

Lenses are used in magnifying glasses, cameras, eyeglasses, microscopes, telescopes, and many other devices. Some of these devices produce an image that is smaller than the object (camera), and some produce an image that is larger than the object (magnifying glass). Lens ray diagrams help us understand how these devices work. The method for constructing ray diagrams for lenses is summarized in Physics Tool Box 23.3.

PHYSICS TOOL BOX 23.3 ▶ **Constructing ray diagrams for single-lens situations**

Convex lens

1. Draw the principal axis for the lens and a vertical line perpendicular to the axis representing the location of the lens. Label the focal points on each side of the lens.

2. Place an arrow at the object position with its base on the axis a distance s from the center of the lens.

3. *Convex lenses* Draw two or three rays.
Ray 1 moves parallel to the principal axis and after refraction through the lens passes through the focal point on the right side of the lens.
Ray 2 passes directly through the middle of the lens and its direction does not change.
Ray 3 passes through the focal point on the left and refracts through the lens so that it moves parallel to the axis on the right.

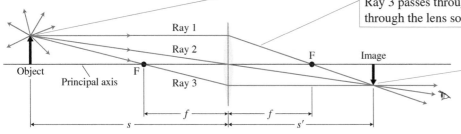

4. The place where the rays intersect on the right side at a distance s' from the lens is the location of a real image of the object.

Concave lens

Steps 1 and 2 are the same as above.
3. *Concave lenses* Draw two or three rays.
Ray 1 moves parallel to the principal axis and is refracted away from the principal axis. Ray 1 appears to come from the focal point on the left side of the lens.
Ray 2 passes directly through the middle of the lens and for thin lenses its direction does not change.
Ray 3 is refracted before it reaches the focal point on the right. The ray moves parallel to the axis on the right.

4. If the rays diverge after passing through the lens, the place from which they seem to diverge is on the left side at distance s' from the lens at the location of a virtual image.

Table 23.6 (on the next page) summarizes possible combinations of the locations of objects and images produced by different types of lenses. For the convex lenses, the points that are located at twice the focal distance from the center of the lens are labeled 2F. This notation is used to emphasize the importance of these points for predicting where the image of the object will be. Specifically, when the object is behind the 2F point, the image is real, inverted, and reduced. When the object is between F and 2F, the image is real, inverted, and enlarged. When the object is between the lens and the focal point, the image is virtual, upright, and enlarged. These points are marked on both sides of the lens because the object can be on either side. Note that the 2F point is not marked for concave lenses because for an object located at any distance from the lens, the result is the same—the image is virtual, upright, and reduced.

The right-hand column of the table lists common applications for each lens type. For example, in a camera we need an image that is reduced and real. In a digital projector we need an image that is magnified and real. In a magnifying glass we want an upright enlarged virtual image.

TABLE 23.6 Ray diagrams for various lenses

Situation	Ray diagram and description of the image	Application
		Camera, human eye
		Digital projector
		Magnifying glass
		Glasses for seeing distant objects (for correcting myopia)
		No known applications

Image location of the base of the object

So far, in all of our examples we have found the image location of the top of the object and assumed that the image of the base of the object was directly beneath it on the principal axis. We will now check this assumption, using the rays that were described in Experiment 2 in Table 23.5, parallel to each other but not to the principal axis. After refraction, these rays pass through a point on the focal plane of the lens—a plane perpendicular to the principal axis and passing through the focal point (**Figure 23.16**).

Now let's draw a ray diagram for the base of an object that lies on the principal axis (**Figure 23.17a**). An infinite number of rays leave the base of the object. Ray 1, parallel to the principal axis, passes through the center of the lens and does not bend. This means that the image of the base point should be on the principal axis—but where? We can draw an imaginary ray 3 (Figure 23.17b) parallel to ray 2 passing through the center of the lens (undeflected). After leaving the lens, *parallel* rays 2 and 3 pass through the same point in the focal plane.

The image of the object appears where bent ray 2 intersects ray 1 on the principal axis (Figure 23.17b). Now we can draw the image of the arrow using the top and the bottom points (Figure 23.17c).

FIGURE 23.16 Parallel rays pass though the same point in the focal plane.

The path of each ray can be explained by refraction relative to the normal lines to each air-glass interface. The rays converge at the same distance from the lens as rays that are parallel to the principal axis.

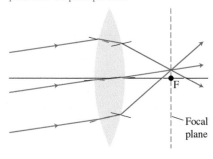

FIGURE 23.17 Locating the image of the base of an object.

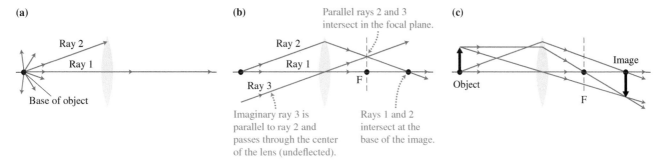

CONCEPTUAL EXERCISE 23.7 **Lens Jeopardy**

The image of a shining point-like object S is produced by a lens (see figure below). The line is the principal axis of the lens. Where is the lens, what kind of lens is it, and where are its focal points?

Image

Sketch and translate Let's first decide the location of the lens that produced the image S′. We know that the shining point S sends light in all directions. Rays that reach the lens bend and pass through the image point S′. We need to find rays with paths that will help us to find that lens and its focal points.

Simplify and diagram A ray that passes through the center of the lens does not bend. This ray also passes through the image point. Thus,

we draw a straight line from the object to the image. The point where the line crosses the principal axis will be the location of the center of the lens, as shown at bottom left. The lens must be convex since the image is on the opposite side of the lens from the object.

To locate a focal point, draw a ray from the object and parallel to the principal axis. This ray will refract through the lens and pass through the focal point on the right side of the lens and then through the image, as shown below. The other focal point of the lens must be the same distance to the left of the lens.

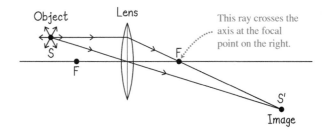

Try it yourself A ray from the object that passes through the focal point on the left side of the lens should bend at the plane of the lens (it actually passes below the lens) and move parallel to the axis on the right side of the lens. Does this ray also pass through the image point?

Answer It should. Sketch it to see if it does.

CONCEPTUAL EXERCISE 23.8 **A partially covered lens**

Imagine that you have an object, a lens, and a screen. You place an object at a position $s > 2f$ from the lens and cover the top half of the lens. Half of the object is above the principal axis, half below. Describe what you will see on the screen.

Sketch and translate To answer the question, we need to draw a ray diagram and see what happens to the image with the top part of the lens covered.

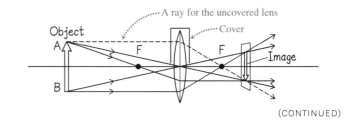

(CONTINUED)

Simplify and diagram We represent the object with an arrow and find the image by examining rays radiated from points A and B (the top and the bottom of the object). With the top of the lens covered, the rays moving toward the top of the lens do not pass through, but rays moving toward the bottom half do. Remember that *all* rays from a point on the object that pass through the lens pass through the corresponding image point. Covering the top half of the lens still allows some light from all points on the object to reach the image location, so the entire image will form. However, the image brightness will decrease as less light passes through the lens.

Try it yourself Imagine that you cover the central part of the lens. You have a shining object far away from the lens. How will the location of the image and its size change compared to the situation when not covered?

Answer

The size and location will not change, but the brightness will decrease.

REVIEW QUESTION 23.4 How do we know how many rays an object sends onto a lens?

23.5 Thin lens equation and quantitative analysis of lenses

We wish to derive a relation between s, s', and f that can be used to make a quantitative prediction for the location and properties of the image formed by a lens. To do this, we use the same technique that was used for curved mirrors. We use a convex lens in the derivation, but the results apply to concave lenses as well. As we did with mirrors, we will assume that the radii of curvature of the spheres from which the sides of the lens were cut are much larger than the size of the lens. Such a lens is called a **thin lens**. Both convex and concave lenses can be thin. Let's look at the geometry of an object at a known distance from a convex lens and its image. AB is the object and A_1B_1 is the image (**Figure 23.18**).

FIGURE 23.18 We can use a ray diagram to develop a relation between the location of an object, the image, and the focal distance of a lens.

- We assume that the lens is very thin and that all of the rays are close to the principal axis. Triangles BAO and B_1A_1O are similar triangles. Thus $(AB/s) = (A_1B_1/s')$. After rearranging, we get

$$\frac{A_1B_1}{AB} = \frac{s'}{s}$$

- Triangles NOF and A_1B_1F are similar. Thus $(NO/f) = [A_1B_1/(s' - f)]$. Also, $NO = AB$. Thus

$$\frac{AB}{f} = \frac{A_1B_1}{s' - f}$$

- Rearranging this last equation, we get

$$\frac{A_1B_1}{AB} = \frac{s' - f}{f}$$

- Setting the above two equations with (A_1B_1/AB) on the left equal to each other, we get

$$\frac{s'}{s} = \frac{s' - f}{f} \tag{23.5}$$

Using algebra similar to what we used to derive the mirror equation, we get a relationship that is called the **thin lens equation**:

$$\frac{1}{s} + \frac{1}{s'} = \frac{1}{f} \tag{23.6}$$

Similar to the mirror equation, the thin lens equation will help us predict the location of the image when we know the location of the object and the focal length of the lens.

Let's see if this equation works for extreme cases. Imagine an object that is located at the focal point. Rays from this object will refract through the lens so that they are parallel. For an object at a focal point $(s = f)$:

$$\frac{1}{s} + \frac{1}{s'} = \frac{1}{f} \quad \text{or} \quad \frac{1}{f} + \frac{1}{s'} = \frac{1}{f}$$

Therefore, $(1/s') = 0$ or $s' = \infty$: the image will form at infinity. The formation of an image by a lens is summarized below, including sign conventions for lenses.

Thin lenses The distance s of an object from a lens, the distance s' of the image from the lens, and the focal length f of the lens are related by the thin lens equation:

$$\frac{1}{s} + \frac{1}{s'} = \frac{1}{f} \tag{23.6}$$

Several sign conventions are important when using the thin lens equation:

1. The focal length f is positive for convex lenses and negative for concave lenses.
2. The image distance s' is positive for real images and negative for virtual images.

FIGURE 23.19 The lens produces a sharp image only at s'.

(a)

(b)

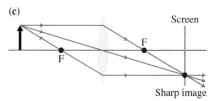

(c)

TIP Remember that all of the preceding reasoning applies to convex and concave lenses made of glass or transparent plastic when they are used in air (or made of any material of higher index of refraction than the surrounding medium).

How can we explain why a sharp, clear image of an object appears in only one place for a particular lens for a specific object-lens distance? If we place a screen closer to the lens than distance s' (**Figure 23.19a**), then the rays leaving the tip of the arrow have not yet converged to a single point. As a result, the light from the tip of the arrow is spread out on the screen and the image will be blurry. If we place the screen at a distance beyond s' (Figure 23.19b), where rays from the tip are diverging after having converged at s', we again get a blurry image. Only at distance s' from the lens do we see a sharp, clear image (Figure 23.19c). Being able to locate the position where a sharp image will appear is important for all applications involving optical instruments.

Digital projectors

A digital projector creates an enlarged real image of a small object inside the projector, an image that can be seen on an external screen. How do these devices work? Here we will focus on their optical components.

EXAMPLE 23.9 **Digital projector**

An object (a small display of the desired image inside the projector) is 0.200 m from a +0.190-m focal length lens. Where should we place the external screen in order to get a sharp image?

Sketch and translate See the labeled sketch of the situation, shown at right. The goal is to determine the correct distance s' of the screen from the projector lens.

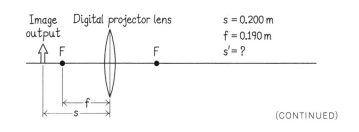

$s = 0.200$ m
$f = 0.190$ m
$s' = ?$

(CONTINUED)

Simplify and diagram Draw a ray diagram for the situation, such as the one shown below. Notice that the image is enlarged, real, and inverted relative to the object. We must invert what is shown on the internal display to obtain an upright image.

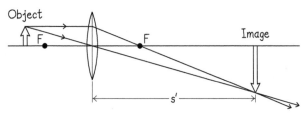

Object

F   F   Image

s'

Represent mathematically We use the thin lens equation [Eq. (23.6)] to determine the unknown image distance s':

$$\frac{1}{s} + \frac{1}{s'} = \frac{1}{f}$$

Solve and evaluate Solve for s' and substitute the known quantities to get

$$\frac{1}{s'} = \frac{1}{f} - \frac{1}{s} = \frac{1}{(0.190 \text{ m})} - \frac{1}{(0.200 \text{ m})}$$

$$= \frac{0.200 \text{ m} - 0.190 \text{ m}}{(0.190 \text{ m})(0.200 \text{ m})} = 0.263 \text{ m}^{-1}$$

Inverting, we get

$$s' = \frac{1}{(0.263 \text{ m}^{-1})} = 3.80 \text{ m}$$

The screen should be placed 3.80 m from the lens—a reasonable distance for a projector in a small conference room.

Try it yourself Suppose the screen needs to be 4.8 m from the lens. What distance should the small display inside the projector be from the same lens to produce a sharp image on the more distant screen?

Answer

$s = 0.198$ m (the lens and internal display need to be moved closer to each other).

TIP Forgetting to invert when solving for s' is a common mistake.

Linear magnification in lenses

Just as with mirrors, lenses can produce images that are bigger or smaller in size than the original objects. To characterize the relative sizes of the object and the image produced by a lens, we use the same physical quantity for magnification that we used for curved mirrors called **linear magnification**:

$$\text{linear magnification} = \frac{\text{image height}}{\text{object height}} = m = \frac{h'}{h} \tag{23.3}$$

where the heights h' and h are positive if the image or object is upright and negative if inverted. As with mirrors, we can relate the linear magnification to the image distance s' and the object distance s. Using the similar triangles in Figure 23.18, we can write

$$\text{linear magnification} = \frac{\text{image height}}{\text{object height}} = m = \frac{h'}{h} = -\frac{s'}{s} \tag{23.4}$$

EXAMPLE 23.10 Looking at an insect through a magnifying glass

You use a convex lens of focal length +10.0 cm to look at a tiny insect on a book page. The lens is 5.0 cm from the paper. Where is the image of the insect? If the insect is 1.0 cm in size, how large is the image?

Sketch and translate Draw a labeled sketch of the situation, as shown at right.

$s = 0.05$ m
$f = 0.10$ m
$h = 0.01$ m
$s' = ?$

Simplify and diagram Assume that the lens is parallel to the paper looking down at the insect. For convenience we draw the ray diagram horizontally, as shown here. From the ray diagram, we conclude that the image is virtual and is on the same side of the lens as the real insect. It is farther away, upright, and enlarged. This means that the image is *beneath* the book where the insect lies. This is where the light reaching your eye appears to originate.

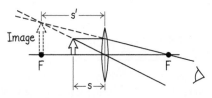

Represent mathematically Use the thin lens equation [Eq. (23.6)] to find the image distance s':

$$\frac{1}{s} + \frac{1}{s'} = \frac{1}{f}$$

Then use the linear magnification equation [Eq. (23.4)] to determine the magnification and the height of the insect's image:

$$m = \frac{h'}{h} = -\frac{s'}{s}$$

Solve and evaluate Insert $f = 0.10$ m and $s = 0.05$ m into the lens equation to get

$$\frac{1}{s'} = \frac{1}{f} - \frac{1}{s} = \frac{1}{0.10 \text{ m}} - \frac{1}{0.05 \text{ m}} = -10 \text{ m}^{-1}$$

Therefore, $s' = 1/(-10 \text{ m}^{-1}) = -0.10$ m. The minus sign indicates that the image is virtual, consistent with the ray diagram. The linear magnification is

$$m = -\frac{s'}{s} = -\frac{(-0.10 \text{ m})}{(+0.05 \text{ m})} = +2.0$$

The image is upright (the plus sign) and twice the object size, or $2.0(1.0 \text{ cm}) = 2.0$ cm, consistent with the ray diagram.

Try it yourself An image seen through a 10-cm focal length convex lens is exactly the same size as the object but inverted and on the opposite side of the lens. Where must the object and image be located?

Answer

The object must be $s = 2f = 20$ cm from the lens on one side, while the image must be $s' = 2f = 20$ cm from the lens on the other side.

EXAMPLE 23.11   Find the image

You place an object 20 cm to the left of a concave lens whose focal length is -10 cm (the negative sign indicates a concave lens). Where is the image located? Is it real or virtual?

Sketch and translate A labeled sketch of the situation is shown below.

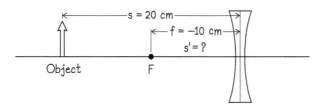

Simplify and diagram Draw a ray diagram for the situation (see below). The image is smaller than the object, virtual, and closer to the lens than the object. So, in your quantitative results, the image distance should be negative and the absolute value should be less than 10 cm.

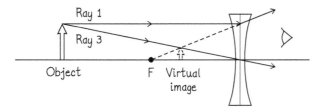

Represent mathematically We can use the thin lens equation [Eq. (23.6)] to determine the image location:

$$\frac{1}{s} + \frac{1}{s'} = \frac{1}{f}$$

Solve and evaluate Insert $f = -0.10$ m and $s = 0.20$ m into the above to get

$$\frac{1}{s'} = \frac{1}{f} - \frac{1}{s} = \frac{1}{-0.10 \text{ m}} - \frac{1}{0.20 \text{ m}} = -15 \text{ m}^{-1}$$

or $s' = 1/(-15 \text{ m}^{-1}) = -0.067$ m or -6.7 cm, consistent with the ray diagram.

Try it yourself Where will the image be if you place an object exactly at the focal point of a concave lens? How big will the height of the image be compared to the height of the object?

Answer

The image will be at a distance $f/2$, on the same side of the lens as the object. The height of the image will be half the height of the object.

The answer to the "Try it yourself" part gives us an idea of how to estimate the focal distance of a concave lens: when the image appears to be half the size of the object, the object-lens distance is equal to the focal distance.

REVIEW QUESTION 23.5 Where should you place an object with respect to a convex lens to have an image at exactly the same distance from the lens? What size will the image be?

23.6 Skills for analyzing processes involving mirrors and lenses

Example 23.12 illustrates the strategies for solving problems involving mirrors and lenses.

PROBLEM-SOLVING
STRATEGY 23.1

Processes involving mirrors and lenses

EXAMPLE 23.12 **A camera**

A camera made in the 1880s consisted of a single lens and light-sensitive film placed at the image location (**Figure 23.20**). To focus light on the image, the photographer would change the image distance—the distance from the lens to the film. Imagine that the image distance is 20 cm and that the film is a 16 cm × 16 cm square. A 1.9-m-tall person stands 8.0 m from the camera. What should be the focal length of the camera's lens for the image to be sharp? Would you be able to see the entire body of the person in the picture?

FIGURE 23.20 A single-lens camera made in the late 1800s.

Sketch and translate

• Sketch the situation in the problem statement.

• Include the known information and the desired unknown(s) in the sketch.

• The situation is sketched below.

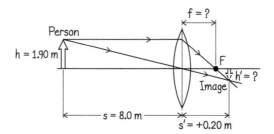

Simplify and diagram

• Assume the lens/mirror is only slightly curved and the rays are incident near the principal axis.

• Draw a ray diagram representing the situation in the problem.

• A ray diagram with the person represented as a shining arrow is included in the sketch.

• The object distance is $s > 2f$, so the image is real, inverted, and smaller than the object.

Represent mathematically

• Use the picture and ray diagram to help construct a mathematical description of the situation.

• Use the lens equation to find the focal length:

$$\frac{1}{f} = \frac{1}{s} + \frac{1}{s'}$$

• Use the linear magnification equation [Eq. (23.4)] to find the image height:

$$h' = mh = -\left(\frac{s'}{s}\right)h$$

Solve and evaluate

- Solve the equations for an unknown quantity.

- Evaluate the results to see if they are reasonable (the magnitude of the answer, its units, how the answer changes in limiting cases, and so forth).

$$\frac{1}{f} = \frac{1}{s} + \frac{1}{s'} = \frac{1}{8.0 \text{ m}} + \frac{1}{0.20 \text{ m}}$$

$$= 0.125 \text{ m}^{-1} + 5.0 \text{ m}^{-1} = 5.125 \text{ m}^{-1}$$

$$f = \frac{1}{5.125 \text{ m}^{-1}} = 0.20 \text{ m}$$

The image size is

$$h' = -\left(\frac{s'}{s}\right)h = -\left(\frac{0.20 \text{ m}}{8.0 \text{ m}}\right)(1.9 \text{ m}) = -0.048 \text{ m}$$

The inverted image size is about 5 cm and will easily fit on the film.

Try it yourself Suppose the 1.9-m-tall person stands 4.0 m from the +0.20-m focal length lens. Now where is the image formed relative to the lens, and what is the image height? Will you get a good picture? Explain.

Answer

that is 0.20 m away from the lens.
The image distance is larger than 0.20 m; thus, the image will not be sharp on the screen
$s' = 0.21$ m from the lens, $m = -0.053$ (the image is inverted), and $h' = -0.10$ m.

REVIEW QUESTION 23.6 If we have a mathematical equation for lenses and mirrors, what is the purpose of drawing ray diagrams when solving lens and mirror problems?

23.7 Single-lens optical systems

We have already discussed several applications for mirrors and lenses, from the magnifying glass to the digital projector. In this section we will investigate three single-lens optical systems: cameras, the human eye, and two kinds of corrective lenses.

Photography and cameras

A simple camera (**Figure 23.21**) has a lens of fixed focal length. Light from an object enters the camera through the lens, which focuses the light on a surface that has light-sensitive properties. In analog cameras, this surface is film; in digital cameras, the film has been replaced by an image sensor that consists of a two-dimensional array of light-sensitive elements—pixels. There are two types of image sensors: CCD (charge-coupled devices) and CMOS (complementary metal oxide semiconductors). To produce sharp images of objects located at different distances from the lens, the lens is moved relative to the film or the image sensor. When the picture includes multiple objects at different distances, you have to choose the object to focus on when you take the photo. But what if you could take the picture and decide what object you wanted in focus later?

A new camera provides that capability using an idea called **light field photography**. In light field photography, the image sensor records *all* the light entering the camera, not just what would produce a sharp image on the focal plane. In fact, a light field camera

FIGURE 23.21 The physics of a simple camera.

FIGURE 23.22 A light field camera allows you to choose which object to focus on after you have taken the picture.

FIGURE 23.23 The human eye.

FIGURE 23.24 As the image distance is fixed in the eye, the lens changes shape to change the focal length, so that the image of any object is always on the retina.

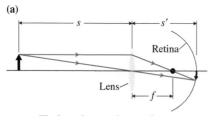

The lens changes shape to focus on objects at different distances.

doesn't have a conventional lens at all (**Figure 23.22**). Instead, a two-dimensional array of tiny micro-lenses is placed just in front of the image sensor. Behind each micro-lens is its own personal portion of the image sensor, its own grid of pixels. Since each micro-lens has its own grid of pixels, the camera records not only how much light of each color reaches each micro-lens but also the direction of each ray reaching that micro-lens. Using this camera, a photographer can choose an object to focus on after the picture has been taken since the camera is effectively focusing on all objects at once.

Optics of the human eye

In many ways, the human eye resembles an expensive digital video camera. It is equipped with a built-in cleaning and lubricating system, an exposure meter, an automatic field finder, and about 130 million photosensitive elements that continuously send electric signals to the brain (the equivalent of a 130-megapixel digital camera CCD) in real time. Light from an object enters the **cornea** (**Figure 23.23**) and passes through a transparent lens. An **iris** in front of the lens widens or narrows, like the aperture on a camera that regulates the amount of light entering the device. The **retina** plays the role of the detector.

A normal human eye can produce sharp images of objects located anywhere from about 10–25 cm to hundreds of kilometers away (effectively at infinity). Unlike a camera, which moves a fixed focal length lens to accommodate different object distances, the eye has a fixed image distance of about 2.1 cm (the distance from the lens to the retina) and a variable focal length lens system.

The changing focal length of the eye's lens is illustrated in **Figure 23.24**. When the eye looks at distant objects, muscles around the lens of the eye relax, and the lens becomes less curved (Figure 23.24a), increasing the focal length and allowing an image to form on the retina. As the object moves closer, the corresponding image moves behind the retina, appearing blurry. To compensate, the eye muscles contract, increasing the curvature of the lens, reducing the focal length (Figure 23.24b) and moving the image back onto the retina. The contraction of these muscles causes your eyes to become tired after reading for many hours.

The **far point** is the most distant point at which the eye can form a sharp image of an object on the retina with the eye muscles relaxed. The **near point** of the eye is the nearest an object can be to the eye and still have a sharp image produced on the retina with the eye muscles tensed. For the normal human eye, the far point is effectively at infinity (we can focus on the Moon and on distant stars) and the near point is at approximately 25 cm from the eye.

When you swim with your eyes open, objects look blurry. If you wear goggles, however, you can see objects clearly. Why can't the eye produce sharp images on the retina underwater? What is the role of the goggles? The answer is that the lens is not the only optical element in the eye. The moist, curved surface of the eye (the cornea) acts as a sort of lens as well. Under normal conditions, the cornea works with the rest of the eye to form a sharp image on the retina. Without the cornea, your eye cannot bend the light passing through it to make a sharp image on your retina.

In water, the cornea stops acting as a lens because its refractive index is about the same as the refractive index of water. Therefore, the eye cannot produce sharp images on the retina, and everything looks blurry. When you wear goggles, you put air between the eye and the water, and your cornea can operate as a lens again.

Corrective lenses

Corrective lenses compensate for the inability of the eye to produce a sharp image on the retina. The two most common vision abnormalities corrected with lenses are **myopia** (nearsightedness) and **hyperopia** (farsightedness).

Myopia A person with myopia, or nearsightedness, can see close-up objects with clarity but not those that are distant. The far point of a nearsighted person may be only a few meters away rather than at infinity.

Myopia is caused by a larger-than-normal eyeball (greater than 2.5 cm in diameter) or a lens that is too curved. In such cases the image of a distant object is formed in front of the retina (**Figure 23.25a**), even when the eye muscles are relaxed. Placing a concave lens in front of the eye corrects nearsighted vision by causing rays from an object to diverge more than without the lens, so that when they pass through the eye lens, an image is formed farther back in the eye (Figure 23.25b). When an object is very distant (that is, $s = \infty$), the focal length of the concave lens is chosen so that its virtual image is formed at the far point of the eye (Figure 23.25c). Light passing through the eyeglass lens appears to come from the image at the far point, not from the more distant object, which allows the eye to produce an image on the retina. To calculate the desired focal length of the eyeglass lens, we set $s = \infty$ and $s' = -($far point distance$)$. The negative sign accounts for the fact that the image produced by the eyeglass is virtual. We can then calculate the value of the desired focal length f of the concave lens using the thin lens equation.

FIGURE 23.25 Correcting myopia (nearsightedness): (a) uncorrected; (b) corrected; (c) glasses help "move" the distant object to the eye's far point.

(a) Relaxed eye

Distant object   Far point

In a nearsighted eye, the image of a distant object forms in front of the retina.

(b) Final image on retina

Far point

With a concave lens, the image forms farther back, on the retina.

(c)

Image formed by eyeglass lens at eye's far point

The person is actually looking at the eyeglass image at the far point of the eye.

EXAMPLE 23.13   **Glasses for a nearsighted person**

Alex is nearsighted, and his far point is 2.0 m from his eyes. What should be the focal length of his eyeglass lens so that he can focus on a very distant object $(s = \infty)$?

Sketch and translate A sketch of the eyeglass lens is shown in Figure 23.25b. The image of an object at infinity is to be formed at the far point 2.0 m in front of the lens.

Simplify and diagram Assume that the distance between the eyeglass lens and the eye lens is small compared to 2.0 m. The ray diagram in Figure 23.25c represents the optics of the eyeglass lens. Without glasses, the image of a distant object is formed before the light reaches Alex's retina and appears blurred. When Alex wears glasses, the object at infinity $(s = \infty)$ has an image 2.0 m to the left of the eyeglass lens $(s' = -2.0 \text{ m})$.

Represent mathematically We can use the given values of s and s' in the lens equation [Eq. (23.6)] to determine the focal length of the eyeglass lenses:

$$\frac{1}{s} + \frac{1}{s'} = \frac{1}{f}$$

Solve and evaluate Insert the object and image distances into the above equation:

$$\frac{1}{f} = \frac{1}{\infty} + \frac{1}{(-2.0 \text{ m})}$$

Therefore, $f = -2.0$ m. The number is negative. This means that the lens should be concave. The focal length of the lens for a nearsighted person equals the negative of the distance to the person's far point.

Try it yourself If Alex is able to see very distant objects with glasses of focal length –1.5 m, what is his far point distance?

Answer   1.5 m.

Hyperopia Unlike nearsighted people, hyperopic (farsighted) people are able to produce sharp images of distant objects on the retina but cannot do it for nearby objects, such as a book or a cell phone screen. Whereas the normal eye has a near point at about 25 cm, a farsighted person may have a near point several meters from the eye. If the

FIGURE 23.26 Correcting hyperopia (farsightedness): (a) uncorrected; (b) corrected; (c) glasses help "move" the closely positioned object to the eye's near point.

(a)

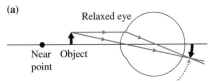

In a farsighted eye, the image of a near object forms behind the retina.

(b)

With a convex lens, the image forms farther forward, on the retina.

(c)

Image formed by eyeglass lens at eye's near point

The person is actually looking at an eyeglass virtual image at the near point of the eye.

object is closer than the person's near point, the image is formed behind the retina (**Figure 23.26a**) and appears blurred. Farsightedness may occur if the diameter of a person's eyeball is smaller than normal or if the lens is unable to curve sufficiently when the surrounding muscles contract.

Convex lenses are used to correct farsighted vision. The eyeglass lens slightly converges light from a nearby object so that the image produced by the eye lens moves forward onto the retina (Figure 23.26b). The lens produces a virtual image of the object at or beyond the person's near point (Figure 23.26c).

To determine the focal length of the eyeglass lens, we need to solve the thin lens equation for f. A farsighted person's eyeglass lens needs to produce an image at their near point, $s' = -(\text{near point distance})$. The negative sign means that the image is virtual.

> **Optical power** An optometrist prescribes glasses using a different physical quantity—the optical power P, measured in **diopters**:
>
> $$P = \frac{1}{f} \qquad (23.7)$$
>
> where f is measured in meters.

For example, the power of a concave lens of focal length $f = -0.50$ m is $P = 1/(-0.50 \text{ m}) = -2.0$ diopters. A lens of large positive power has a small positive focal length and causes rays to converge rapidly after passing through the lens. A lens of large negative power is concave. It has a small negative focal length and causes rays to diverge rapidly.

TIP   Optical power P is not related in any way to energy conversion rate P, which we learned about in earlier chapters.

EXAMPLE 23.14   **Glasses for a farsighted person**

Eugenia is farsighted with a near point of 1.5 m. If you were an optometrist, what glasses would you prescribe for her so she can read a book held 25 cm from her eyes?

Sketch and translate Eugenia holds a book 0.25 m away (see Figure 23.26b) and wants her glasses to make her eyes think it is 1.5 m away at her near point. Thus $s = 0.25$ m and $s' = -1.5$ m for her glasses (on the same side of the lens as the book). We need to find their focal length f and power P.

Simplify and diagram A ray diagram for the optics of the eyeglass system is shown in Figure 23.26c.

Represent mathematically Use Eq. (23.6) to find the focal length of the lenses:

$$\frac{1}{s} + \frac{1}{s'} = \frac{1}{f}$$

Solve and evaluate Insert the known object and image distances into the previous equation to determine the focal length of the desired glasses:

$$\frac{1}{f} = \frac{1}{0.25 \text{ m}} + \frac{1}{-1.50 \text{ m}}$$
$$= 4.0 \text{ m}^{-1} - 0.67 \text{ m}^{-1} = 3.33 \text{ m}^{-1}$$

Thus,

$$f = \frac{1}{3.33 \text{ m}^{-1}} = 0.30 \text{ m}$$

and the power of the glasses is $P = (1/f) = 1/(0.30 \text{ m}) = 3.3$ diopters.

- -

Try it yourself Eugenia went to a drugstore to buy glasses, but her only choice was 4.0-diopter glasses. Where will Eugenia need to hold a book when wearing her new glasses so its image is at her 1.5-m near point?

Answer   21 cm.

REVIEW QUESTION 23.7 What is the main difference between how a camera produces sharp images of objects at different distances and how the human eye does it?

23.8 Angular magnification and magnifying glasses

The linear magnification of an optical system compares only the heights of the image and the object, but the apparent size of an object as judged by the eye depends not only on its height but also on its distance from the eye. For example, a pencil held 25 cm from your eye appears longer than one held 100 cm away (Figure 23.27). In fact, the pencil may appear longer than a 100-story building if the building is several miles away, even though the pencil is actually much shorter.

The impression of an object's size is quantified by its **angular size** θ, shown in Figure 23.27. The angular size depends on an object's height h and its distance r from the eye:

$$\theta = \frac{h}{r} \tag{23.8}$$

Remember that an angle θ, in radians, is the ratio of an arc length ($\approx h$ in Figure 23.27) and the radius of a circle (r in Figure 23.27: $\theta = (h/r)$). We can use h for the length of the straight line between the corners of the triangle only if h is very close to the actual arc length. This is only true for small angles. Thus Eq. (23.8) is an approximation for small angle θ.

If a person looks at an object through one or more lenses, he or she sees light that appears to come from the final image of the system of lenses. If the person's eye is close to the last lens (as in a typical telescope or microscope), then the angular size θ' of the image is

$$\theta' = \frac{h'}{s'} \tag{23.9}$$

where h' is the height of the final image and s' is the distance of the final image from the last lens in the system. The **angular magnification** M of an optical system is defined as the ratio of θ' and θ:

$$M = \frac{\theta'}{\theta} = \frac{\text{Angular size of the final image of optical system}}{\text{Angular size of object as seen by the unaided eye}} \tag{23.10}$$

A magnifying glass consists of a single convex lens. To use the lens as a magnifying glass, we position the object between its focal point and the lens, as shown in Figure 23.28a. A magnified virtual image is formed behind the object. The light appears to be coming from this enlarged image, which is farther away from the lens than the object. In other words, when you use a magnifying glass to look at small print on a page on a desk, the image you are viewing is located under the desk!

Note that in Figure 23.28a both the object and the image make the same angle θ' with the principal axis ($\theta' = (h'/s') = (h/s)$). The object has the same angular size when we look directly at it when it is a distance s from our eye as when we look through the magnifying glass at the image a distance s' from our eye. Using the magnifying glass lets you place the object closer than your near point (and hence have a larger angular size) and use the magnifying glass to focus it clearly on your retina (shown in Figure 23.28b).

To calculate the angular magnification of the magnifying glass, compare the angular size θ' of the image seen through the magnifying glass and the angular size θ of the object seen with the unaided eye when the object is as close as possible and still appears clear (at your near point). The angular size of the image viewed through the magnifying glass is $\theta' = (h'/s')$. By considering ray 2 in Figure 23.28a, we find that $(h'/s') = (h/s)$. Thus,

$$\theta' = \frac{h}{s}$$

where h is the actual height of the object and s is its distance from the magnifying glass.

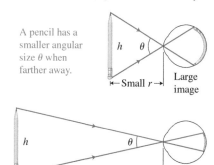

FIGURE 23.27 The angular size θ of an object.

A pencil has a smaller angular size θ when farther away.

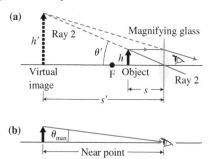

FIGURE 23.28 A magnifying glass produces an enlarged virtual image of the closely positioned object.

FIGURE 23.29 Maximum angular size θ_{max} of an object whose image is on the retina. If we bring the object closer, the image is behind the retina and what we see is blurred.

(a)

(b)

(c)

(d)

FIGURE 23.30 Analyzing a two-lens optical system. The image produced by lens 1 becomes the object for lens 2.

(a) Object of first lens

(b) Image of first lens

(c) Object of second lens

(d) Image of second lens

Next consider the angular size of the object as seen with the unaided eye. The angular size θ of an object of height h that we hold a distance r from the eye is $\theta = (h/r)$ (**Figure 23.29a**). As we bring the object closer to the eye (Figure 23.29b and c), its angular size increases. If we bring the object closer than the near point, the image on the retina is blurred (Figure 23.29d). The *maximum* angular size θ_{max} of an object of height h, when viewed by the unaided eye and when focused on the retina, is

$$\theta_{max} = \frac{h}{\text{Near point distance}} \tag{23.11}$$

The angular magnification of the image is the ratio of the angular size of the image θ' and the maximum angular size θ_{max} as seen with the unaided eye. Thus,

$$M = \frac{\theta'}{\theta_{max}} = \frac{h/s}{h/(\text{Near point distance})} = \frac{\text{Near point distance}}{s} \tag{23.12}$$

REVIEW QUESTION 23.8 If a person with normal vision and a farsighted person use the same magnifying glass, who will get the greater magnification? Why?

23.9 Telescopes and microscopes

Telescopes and microscopes use multiple lenses to produce images that are magnified more than is possible with a single lens. The technique for locating the final image and calculating its magnification requires careful attention to several details—provided below for a system with two lenses. This process applies to both telescopes and microscopes.

To find the final image of a two-lens system:

Step 1: Construct a sketch of the system of lenses (**Figure 23.30a**). The lenses are separated by a distance d and have focal lengths f_1 and f_2. The object is a distance s_1 from lens 1.

Step 2: Use Eq. (23.6) to determine the location s_1' of the image formed by the first lens (Figure 23.30b). If s_1' is positive, the image is to the right of lens 1. If s_1' is negative, the image is to the left of lens 1. In our example, s_1' is positive.

Step 3: Next, note that the image of lens 1 is now the object for lens 2. The object distance s_2 (Figure 23.30c) is $s_2 = d - s_1'$, where d, the separation of the lenses, is a positive number. It is possible for s_2 to be negative if the image formed by lens 1 is to the right of lens 2.

Step 4: Use the thin lens equation to determine the image distance s_2' of the image formed by lens 2 (Figure 23.30d):

$$\frac{1}{s_2'} = \frac{1}{f_2} - \frac{1}{s_2}$$

This is the location of the final image relative to the second lens. If s_2' is positive, the final image is real and located to the right of lens 2. If s_2' is negative, the final image is virtual and to the left of lens 2. In our example, the final image is real and to the right of lens 2.

Step 5: The total linear magnification m of the two-lens system equals the product of the linear magnifications of each lens: $m = m_1 m_2$. If m is positive, the final image has same orientation as the original object. If m is negative, the final image is inverted relative to the original object. Techniques for calculating the total angular magnification M are illustrated in the discussions and examples that follow.

Telescopes

Galileo is believed to have been the first to study the night sky and the Sun with a telescope. He discovered mountains and craters on the Moon, the moons of Jupiter, stars inside the Milky Way, and sunspots. His telescope consisted of two lenses—one convex and the other concave. Today, a more common version has two convex lenses separated by a distance slightly less than the sum of their focal lengths (**Figure 23.31a**). When you observe a distant object, the first lens produces a real image just beyond its own focal length (Figure 23.31b). The second lens is located so that the image from the first lens is just inside the focal point of the second lens. The image from the first lens becomes the object for the second lens. The second lens produces a magnified inverted virtual image (Figure 23.31c), which we observe by looking through the telescope.

FIGURE 23.31 A telescope.

(a)

(b)

(c)

EXAMPLE 23.15 **Looking at a lion with a telescope**

A 1.2-m-tall lion stands 50 m from the first lens of a telescope (Figure 23.31a). Locate the final image of the lion. Determine the linear magnification, height, and angular magnification of the final image.

Sketch and translate Figure 23.21a shows that the givens are $f_1 = +20.0$ cm, $f_2 = +5.0$ cm, $d = 24.5$ cm, $s_1 = 50$ m $= 5000$ cm, and $h = 1.2$ m. We need to find s_2', m, and M.

Simplify and diagram The ray diagrams in Figure 23.31b and c help us locate the first and second images.

Represent mathematically Use the procedure described on the previous page to determine the locations of the first image, the second object, and the second image s_1', s_2, and s_2'. Then use the values of s_1, s_1', s_2 and s_2' to determine the linear magnification $m = m_1 m_2$ of the two-lens system, where $m_1 = -(s_1'/s_1)$ and $m_2 = -(s_2'/s_2)$. Next, determine the height of the final image of the lion $h' = mh$. To find the angular magnification, compare the angular size of the lion as seen through the optical system $\theta' = (h_2'/s_2')$ and its angular size as seen with the unaided eye $\theta = (h/r_1)$, where $r_1 = s_1 + d$. Then take the ratio of the angular sizes in order to determine the angular magnification: $M = (\theta/\theta')$.

Solve and evaluate This procedure gives us results of $s_1' = 20.08$ cm, $s_2 = 4.42$ cm, and $s_2' = -38.1$ cm. Notice that s_2' is negative, indicating a virtual image, in agreement with the ray diagram. To determine the linear magnification:

$$m = m_1 m_2 = \left(-\frac{s_1'}{s_1}\right)\left(-\frac{s_2'}{s_2}\right)$$

$$= \frac{(20.08 \text{ cm})(-38.1 \text{ cm})}{(5000 \text{ cm})(4.42 \text{ cm})} = -0.035$$

Notice that the linear magnification is less than 1; the final image is smaller than the original object. The image height is $h_2' = mh = (-0.035)(1.2 \text{ m}) = -0.042 \text{ m} = -4.2 \text{ cm}$.

Finally, determine the angular magnification M:

$$\theta' = \frac{h_2'}{s_2'} = \frac{(-4.2 \text{ cm})}{(-38.1 \text{ cm})} = 0.11 \text{ rad}$$

$$\theta = \frac{h}{s_1} = \frac{(1.2 \text{ m})}{(50 \text{ m})} = 0.024 \text{ rad}$$

$$M = \frac{\theta'}{\theta} = \frac{0.11 \text{ rad}}{0.024 \text{ rad}} = 4.6$$

Although the image of the lion is smaller than the actual lion, its angular size is almost five times bigger when viewed through the telescope than when seeing it without a telescope from 50 m away. This telescope succeeds in making the lion appear bigger than it really is.

Try it yourself How can the lion appear larger when its final image is smaller?

Answer

The final image was much closer than the lion (131 times closer) and thus looked bigger through the telescope, even though it was smaller (29 times smaller).

The compound microscope

A compound microscope, like a telescope, has two convex lenses and passes light reflected from the object through two lenses to form the image. A microscope magnifies tiny nearby objects instead of large distant ones.

Both lenses have relatively short focal lengths and are separated by 10–20 cm (Figure 23.32a). The object is placed just outside and very close to the focal point of the first lens (called the **objective** lens). The real, enlarged, inverted image produced by the first lens is just inside the focal point of the second lens (called the **eyepiece** lens), as shown in Figure 23.32b. When an observer looks through the eyepiece, she sees an inverted, enlarged, virtual image of the object (Figure 23.32c).

FIGURE 23.32 Compound microscope.

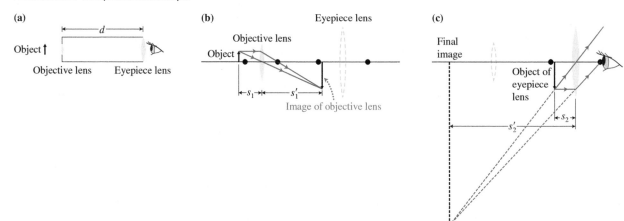

Consider the magnification of such a system. The image produced by the objective lens has linear magnification $m_1 = -(s_1'/s_1)$. The image produced by the objective lens becomes the object for the eyepiece lens. Since this object is located just inside the focal point of the eyepiece lens, the eyepiece lens acts as a magnifying glass used to view the real image produced by the objective lens (Figure 23.32c). The angular magnification of this magnifying glass is, according to Eq. (23.12), $M_2 = (\text{Near point distance}/s_2)$. The total angular magnification of the microscope is

$$M = m_1 M_2 = -\frac{s_1'}{s_1}\frac{\text{Near point distance}}{s_2}$$

FIGURE 23.33 Developing an alternative expression for angular magnification.

This expression for angular magnification can be rewritten in terms of the focal lengths of the lenses and their separation d. Consider Figure 23.33. Notice that $s_1 \approx f_1$; $s_1' \approx d - f_2$; and $s_2 \approx f_2$. By substituting these values in the previous equation, we have

$$M = \left(-\frac{d - f_2}{f_1}\right)\left(\frac{\text{Near point distance}}{f_2}\right) \tag{23.13}$$

This expression is easy to use because the magnification depends only on the focal lengths of the lenses, their separation, and the near point distance of the eye. The very best optical microscopes provide angular magnifications of about 1000.

EXAMPLE 23.16   **Seeing a cell with a microscope**

A compound microscope has an objective lens of focal length 0.80 cm and an eyepiece of focal length 1.25 cm. The lenses are separated by 18.0 cm. If a red blood cell is located 0.84 cm in front of the objective lens, where is the final image of the cell, and what is its angular magnification? The viewer's near point is 25 cm.

Sketch and translate We have the following information: $f_1 = +0.80$ cm, $f_2 = +1.25$ cm, $d = 18.0$ cm, and $s_1 = +0.84$ cm. We need to find s_2' and the angular magnification M of the microscope. We can use Figure 23.32a as a sketch of the situation.

Simplify and diagram Assume that the red blood cell is brightly illuminated and can be modeled as a shining arrow. We use the ray diagrams in Figure 23.32b and c.

Represent mathematically First find the image produced by the first lens:

$$\frac{1}{s_1'} = \frac{1}{f_1} - \frac{1}{s_1}$$

This image becomes the object for the second lens. The object distance for the second lens is $s_2 = d - s_1'$. Use this distance to find the location of the image produced by the second lens (the final image):

$$\frac{1}{s_2'} = \frac{1}{f_2} - \frac{1}{s_2}$$

To find the angular magnification M we can use

$$M = m_1 M_2 = -\frac{s_1'}{s_1} \frac{\text{Near point distance}}{s_2}$$

Solve and evaluate

$$\frac{1}{s_1'} = \frac{1}{f_1} - \frac{1}{s_1} = \frac{1}{0.80 \text{ cm}} - \frac{1}{0.84 \text{ cm}} = 0.0595 \text{ cm}^{-1}$$

Thus, $s_1' = 1/(0.0595 \text{ cm}^{-1}) = +16.8$ cm. The object distance for the second lens is $s_2 = d - s_1' = 18.0 \text{ cm} - 16.8 \text{ cm} = 1.2$ cm. The final image distance is

$$\frac{1}{s_2'} = \frac{1}{f_2} - \frac{1}{s_2} = \frac{1}{1.25 \text{ cm}} - \frac{1}{1.20 \text{ cm}} = -0.033 \text{ cm}^{-1}$$

or $s_2' = 1/(0.033 \text{ cm}^{-1}) = -30$ cm. The total angular magnification is

$$M = m_1 M_2 = -\frac{s_1'}{s_1} \frac{\text{Near point distance}}{s_2}$$

$$= -\frac{16.8 \text{ cm}}{0.84 \text{ cm}} \frac{25 \text{ cm}}{1.2 \text{ cm}} = -(20)(21) = -420$$

The negative sign means that the final image is inverted.

Try it yourself How does the result using the magnifications for each lens compare to the approximate magnification equation [Eq. (23.13)] for a microscope?

Answer   You get the same result, −420.

REVIEW QUESTION 23.9 Why is saying that a telescope magnifies simultaneously a correct and an incorrect statement?

Summary

Plane mirror: A plane mirror produces a virtual image at the same distance behind the mirror as that of the object in front. A **virtual image** is at the position where the paths of the reflected rays seem to originate from behind the mirror. (Section 23.1)

$$s = s'$$

Concave and convex mirrors: The distance s of an object from the surface of a concave or convex mirror, the distance s' of the image from the surface of the mirror, and the focal length f of the mirror are related by the mirror equation. The **focal point** F is the location through which all incident rays parallel to the principal axis pass after reflection (concave mirrors) or appear to have come from (convex mirrors). (Sections 23.2 and 23.3)

$$\frac{1}{s} + \frac{1}{s'} = \frac{1}{f} \qquad \text{Eq. (23.2)}$$

The *sign conventions* for these quantities are:

- The **focal length** f is positive for concave mirrors and negative for convex mirrors.
- The **image distance** s' is positive for real images and negative for virtual images.

Convex and concave lenses: The distance s of an object from a lens, the distance s' of the image from the lens, and the focal length f of the lens are related by the thin lens equation. (Sections 23.4 and 23.5)

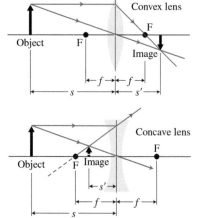

$$\frac{1}{s} + \frac{1}{s'} = \frac{1}{f} \qquad \text{Eq. (23.6)}$$

The *sign conventions* are:

- The **focal length** f is positive for convex lenses and negative for concave lenses.
- The **image distance** s' is positive for real images and negative for virtual images.

Linear magnification m is the ratio of the height h' of the image and the height h of the object. The heights are positive if upright and negative if inverted. (Sections 23.3, 23.5, 23.7, and 23.8)

$$m = \frac{h'}{h} = -\frac{s'}{s} \qquad \text{Eq. (23.4)}$$

Angular magnification M is the ratio of the angular size θ' of an image as seen through a lens or lenses and the maximum angular size θ_{max} of the object as seen with the unaided eye. (Section 23.8)

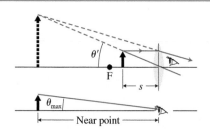

$$M = \frac{\theta'}{\theta_{max}}$$

$$= \frac{\text{Near point distance}}{s} \qquad \text{Eq. (23.12)}$$

Questions

Multiple Choice Questions

1. Where does the image of an object in a plane mirror appear?
   (a) In front of the mirror
   (b) On the surface of the mirror
   (c) Behind the mirror
2. Where does the image of an object that is s meters in front of a plane mirror appear?
   (a) s meters away from the object
   (b) $2s$ meters away from the object
   (c) There is no image formed by the plane mirror as the reflected rays do not converge.
3. A plane mirror produces an image of an object that is which of the following?
   (a) Real and enlarged
   (b) Virtual and the same size
   (c) Virtual and enlarged
   (d) Virtual and smaller
4. A concave mirror can produce an image that is which of the following?
   (a) Real, enlarged, and upright
   (b) Virtual, smaller, and inverted
   (c) Virtual, enlarged, and inverted
   (d) None of the answers is correct
5. A convex mirror can produce an image that is which of the following?
   (a) Real, enlarged, and inverted
   (b) Real, smaller, and inverted
   (c) Real, smaller, and upright
   (d) Virtual, smaller, and upright
   (e) Virtual, enlarged, and upright
6. A virtual image is the image produced
   (a) on a screen by reflected rays.
   (b) on a screen by refracted rays.
   (c) at a point where diverging rays appear to originate.
7. To see an image of an object that is enlarged, real, and inverted, you need to place the object in front of a convex lens in which region?
   (a) $s > 2f$
   (b) $2f > s > f$
   (c) $f > s$
   (d) None of these
8. To see an image of an object that is enlarged, real, and inverted, you need to place the object in front of a concave lens in which region?
   (a) $s > 2f$
   (b) $2f > s > f$
   (c) $f > s$
   (d) None of these
9. When we derived the thin lens equation [Eq. (23.6)], what did we use?
   (a) A ray diagram
   (b) Similar triangles
   (c) An assumption that the radius of curvature of the lens is much less than the distances f, s, and s'
   (d) a, b, and c
   (e) a and b only
10. When drawing images of objects produced by curved mirrors and lenses, which of the following ideas is required?
    (a) Each point of an object sends two rays.
    (b) Each point of an object sends three rays.
    (c) Each point of an object sends an infinite number of rays.
11. The focal length of a glass lens is 10 cm. When the lens is submerged in water, what is its new focal length?
    (a) 10 cm
    (b) less than 10 cm
    (c) more than 10 cm
    (d) Not enough information to answer

12. A microbiologist uses a microscope to look at body cells because the microscope will do which of the following?
    (a) Magnify the linear size of the cells
    (b) Magnify the angular size of the cells
    (c) Bring the cells closer
    (d) Make the cells brighter
    (e) Both (a) and (b) are correct.
13. The human eye works in a similar way to which of the following?
    (a) A microscope
    (b) An overhead projector
    (c) A telescope
    (d) A camera
14. Which of the following changes will result in a decrease of the focal length of a concave mirror? Select all correct answers.
    (a) Increase the radius of curvature of the mirror.
    (b) Decrease the radius of curvature of the mirror.
    (c) Immerse the mirror in water.
    (d) Reduce the diameter of the mirror without changing the curvature.

Conceptual Questions

15. When we draw a ray passing through the center of a lens, it does not bend independent of its direction. Explain why and state the implicit assumptions that we use to draw such rays.
16. You run toward a building with walls of a metallic, reflective material. Your speed relative to the ground is 1.5 m/s. How fast does your image appear to move toward you? Explain.
17. A tiny plane mirror can produce an image similar to a pinhole camera. Explain how.
18. Explain how we derived the mirror equation.
19. Explain how we derived the thin lens equation.
20. Explain the difference between a real and a virtual image.
21. You stand in front of a fun house mirror. You see that your head is enlarged and your legs and feet are shrunk. How can this be? Provide a ray diagram to support your answer.
22. A bubble of air is suspended underwater. Draw a ray diagram showing the approximate path followed by a horizontal light ray passing from the water (a) through the upper half of the bubble, (b) through its center, and (c) through its lower half. Does the bubble act like a horizontal converging or diverging lens? Explain.
23. A bubble of oil is suspended in water. Draw ray diagrams as in Question 23.22 and explain the difference between the diagrams in Questions 23.22 and 23.23.
24. A typical person underwater cannot focus clearly on another object under the water. Yet if the person wears goggles, he or she can see underwater objects clearly. Explain.
25. In a video projector, the picture that appears on a small liquid-crystal display is projected onto a screen. The lens of the video projector is adjusted so that a clear image on the display is formed on a screen. If the screen is moved farther away, must the lens be moved toward or away from the display? Explain.
26. The retina has a blind spot at the place where the optic nerve leaves the retina. Design and describe a simple experiment that allows you to perceive the presence of this blind spot. Why doesn't the blind spot affect your normal vision?

Problems

Below, **BIO** indicates a problem with a biological or medical focus. Problems labeled **EST** ask you to estimate the answer to a quantitative problem rather than derive a specific answer. Asterisks indicate the level of difficulty of the problem. Problems with no * are considered to be the least difficult. A single * marks moderately difficult problems. Two ** indicate more difficult problems.

23.1 Plane mirrors

1. You need to teach your friend how to draw rays to locate the images of objects produced by a plane mirror. Outline the steps that she needs to take.
2. Place a pencil in front of a plane mirror so that it is not parallel to the mirror. Draw an image that the mirror forms of the pencil and show using rays how the image is formed.
3. * Use geometry to prove that the virtual image of an object in a plane mirror is at exactly the same distance behind the mirror as the object is in front.
4. * You are 1.8 m tall. Where should you place the top of a mirror on the wall so you can see the top of your head? Where should you stand with respect to the wall?
5. * Two people are standing in front of a rectangular plane mirror. Each of them claims that she sees her own image but not the image of the other person. Draw a ray diagram to find out if this is possible.
6. * **Test an idea** Describe an experiment that you can conduct to test that the image produced by a plane mirror is virtual.

23.2 and 23.3 Qualitative analysis of curved mirrors and The mirror equation

7. * Describe in detail an experiment to find the image of a candle produced in a curved mirror. Draw pictures of the experimental setup and a ray diagram. Show on the ray diagram where your eye is.
8. * Explain with a ray diagram how (a) a concave mirror and (b) a convex mirror produce images of objects. Make sure that you explain the choice of rays and the location of your eye.
9. * **Test an idea** Describe an experiment to test the idea that a convex mirror never produces a real image of an object.
10. * **Test an idea** Describe an experiment to test the mirror equation. What are you going to measure? What are you going to calculate?
11. * **Tablespoon mirror** You look at yourself in the back (convex shape) of a shiny steel tablespoon. Describe and explain what you see as you bring the spoon closer to your face. Then turn the spoon around and repeat the steps.
12. * Use ray diagrams and the mirror equation to locate the position, orientation, and type of image of an object placed in front of a concave mirror of focal length 20 cm. The object distance is (a) 200 cm, (b) 40 cm, and (c) 10 cm.
13. Repeat Problem 23.12 for a convex mirror of focal length −20 cm.
14. Use ray diagrams and the mirror equation to locate the images of the following objects: (a) an object that is 10 cm from a concave mirror of focal length 7 cm and (b) an object that is 10 cm from a convex mirror of focal length −7 cm.
15. * **Sinking ships** A legend says that Archimedes once saved his native town of Syracuse by burning the enemy's fleet with mirrors. Describe quantitatively the type of mirrors that Archimedes could have used to burn ships that were 150 m away. Justify your answer.
16. * **EST** **Fortune-teller** A fortune-teller looks into a silver-surfaced crystal ball with a radius of 10 cm and focal length of −5 cm. (a) If her eye is 30 cm from the ball's surface, where is the image of her eye? (b) Estimate the size of that image.
17. * You view yourself in a large convex mirror of −1.2-m focal length from a distance of 3.0 m. (a) Locate your image. (b) If you are 1.7 m tall, what is your image height?
18. * **Seeing the Moon in a mirror** The Moon's diameter is 3.5×10^3 km, and its distance from Earth is 3.8×10^5 km. Determine the position and size of the image formed by the Hale Telescope reflecting mirror, which has a focal length of +16.9 m.
19. * You view your face in a +20-cm focal length concave mirror. Where should your face be in order to form an image that is magnified by a factor of 1.5?

20. * **Buying a dental mirror** A dentist wants to purchase a small mirror that will produce an upright image of magnification +4.0 when placed 1.6 cm from a tooth. What mirror should she order? Say everything you can about the mirror.
21. * **Using a dental mirror** A dentist examines a tooth that is 1.0 cm in front of the dental mirror. An image is formed 2.0 cm behind the mirror. Say everything you can about the mirror and the image of the tooth.
22. * **BIO** **EST** **Spooky fish** The glasshead barreleye (*Rhynchohyalus natalensis*) is a fish that lives in the oceans around the world at depths of 200 to 600 m. This fish and the brownsnout spookfish are the only two vertebrates known to use mirror-like reflective surfaces to form the images in their eyes. These fish have eyes with reflective surfaces to observe regions below them and eyes with lenses to observe regions above and beside them. Figure P23.22 is a sketch of the cross section of an eye that uses reflection and shows how the eye focuses parallel beams of light that enter the eye at two extreme angles. Use the dimensions in the figure to estimate the range of values for the radius of curvature of the reflective surface. You will need a ruler to solve this problem.

FIGURE P23.22

Retina   Reflective surface   10 mm

23.4 Qualitative analysis of lenses

23. * If you place a point-like light source on the axis of a convex lens, you obtain on the screen the pattern shown in Figure P23.23a. If you repeat the experiment with a concave lens, you obtain the pattern shown in Figure P23.24b. Explain qualitatively how the patterns are formed, using ray diagrams. Note that there are no frames around the lenses.

FIGURE P23.23

24. * You have a convex lens and a candle. Describe in detail an experiment that you will perform to find the image of the candle that this lens produces. Draw pictures of the experimental setup and a ray diagram. Show on the ray diagram where your eye is.
25. * Explain how to draw ray diagrams to locate images produced by objects in front of convex and concave lenses. Focus on the choice of rays and how you know where and what type of image is produced.
26. * Draw ray diagrams to show how a convex lens can produce (a) a real image that is smaller than the object, (b) a real image larger than the object, and (c) a virtual image.
27. * Use a ruler to draw ray diagrams to locate the images of the following objects: (a) an object that is 30 cm from a convex lens of +10-cm focal length, (b) an object that is 14 cm from the same lens, and (c) an object that is 5 cm from the same lens. (Choose a scale so that your drawing fills a significant portion of the width of a paper.) Measure the image locations on your drawings and indicate if they are real or virtual, upright or inverted.

28. * Repeat the procedure described in Problem 23.27 for the following lenses and objects: (a) an object that is 30 cm from a concave lens of −10-cm focal length, (b) an object that is 14 cm from the same lens, and (c) an object that is 5 cm from the same lens.

29. * Repeat the procedure described in Problem 23.27 for the following lenses and objects: (a) an object that is 7 cm from a convex lens of +10-cm focal length and (b) an object that is 7 cm from a concave lens of −10-cm focal length.

30. * Repeat the procedure in Problem 23.27 for the following lenses and objects: (a) an object that is 20 cm from a convex lens of +10-cm focal length, (b) an object that is 5 cm from the same lens, (c) an object that is 20 cm from a concave lens of −10-cm focal length, and (d) an object that is 5 cm from the lens in part (c).

31. * **Partially covering lens** Your friend thinks that if she covers one half of a convex lens she will only be able to see half of the object. Do you agree with her opinion? Why would she have such an opinion? Provide physics arguments. Design an experiment to test her idea.

23.5 and 23.6 Thin lens equation and quantitative analysis of lenses and Skills for analyzing processes involving mirrors and lenses

32. * Use ray diagrams to locate the images of the following objects: (a) an object that is 10 cm from a convex lens of +15-cm focal length and (b) an object that is 10 cm from a concave lens of −15-cm focal length. (c) Calculate the image locations for parts (a) and (b) using the thin lens equation. Check for consistency.

33. * Use ray diagrams to locate the images of the following objects: (a) an object that is 6.0 cm from a convex lens of +4.0-cm focal length and (b) an object that is 8.0 cm from a diverging lens of −4.0-cm focal length. (c) Calculate the image locations for parts (a) and (b) using the thin lens equation. Check for consistency.

34. Light passes through a narrow slit, and then through a lens and onto a screen. The slit is 20 cm from the lens. The screen, when adjusted for a sharp image of the slit, is 15 cm from the lens. What is the focal length of the lens?

35. * Describe two experiments that you can perform to determine the focal length of a glass convex lens. Is it a converging or a diverging lens? How do you know?

36. * **Shaving/makeup mirror** You wish to order a mirror for shaving or makeup. The mirror should produce an image that is upright and magnified by a factor of 2.0 when held 15 cm from your face. What type and focal length mirror should you order?

37. **Dentist lamps** Dentists use special lamps that consist of a concave mirror and a small, bright light source that is fixed on the principal axis of the mirror. When the light source is placed 5.0 cm from the mirror, the reflected light is focused in a bright spot at distance 70.0 cm from the mirror. Determine the radius of curvature of the mirror.

38. * A large concave mirror of focal length 3.0 m stands 20 m in front of you. Describe the changing appearance of your image as you move from 20 m to 1.0 m from the mirror. Indicate distances from the mirror where the change in appearance is dramatic.

39. * **EST** Two convex mirrors on the side of a van are shown in Figure P23.39. Estimate the ratio f_{upper}/f_{lower} of the focal lengths of the mirrors. Explain the reasoning behind your estimate, stating what assumptions you had to make.

FIGURE P23.39

40. Your friend gives you the following description of a method that apparently allows you to estimate the focal length of a concave lens: "Place a ruler on a table. Hold the lens above the ruler and parallel to the table so that you can simultaneously observe the image of the ruler (through the lens) and the part of the ruler that extends beyond the lens. Move the lens up and down until the width of the image of the ruler is half the width of the ruler that extends beyond the lens. The distance between the lens and the ruler is equal to the focal distance of the lens." Do you agree that this method gives you the focal length of the concave lens? Explain using a ray diagram.

23.7 Single-lens optical systems

41. **Camera** You are using a camera with a lens of focal length 6.0 cm to take a picture of a painting located 3.0 m from the camera lens. Where should the image sensor be positioned in relation to the lens in order to capture the image?

42. * **Camera** A camera with an 8.0-cm focal length lens is used to photograph a person who is 2.0 m tall. The height of the image on the image sensor must be no greater than 3.5 cm. (a) Calculate the closest distance the person can stand to the lens. (b) For this object distance, how far should the image sensor be located from the lens?

43. **Video projector** An LCD video projector (LCD stands for liquid-crystal display) produces real, inverted, enlarged images of a small LCD picture on a screen (the display picture is upside down in the projector). If the display is located 12.6 cm from the 12.0-cm focal length lens of the projector, what are (a) the distance between the screen and the lens and (b) the height of the image of a person on the screen who is 2.0 cm tall on the display?

44. **Photo of carpenter ant** You take a picture of a carpenter ant with an old fashioned camera with a lens 18 cm from the film. How far from the 4.0-cm focal length convex camera lens should the ant be located so you see a sharp image of the ant on the film?

45. * **Photo of secret document** A secret agent uses a camera with a 5.0-cm focal length lens to photograph a document whose height is 10 cm. At what distance from the lens should the agent hold the document so that an image 2.5 cm high is produced on the image sensor? [Note: The real image is inverted.]

46. * **Photo of landscape** To photograph a landscape 2.0 km wide from a height of 5.0 km, Joe uses an aerial camera with a lens of 0.40-m focal length. What is the width of the image on the detector surface?

47. * Make a rough graph of image distance versus object distance for a convex lens of a known focal length as the object distance varies from infinity to zero.

48. * Make a rough graph of linear magnification versus object distance for a convex lens of 20-cm focal length as the object distance varies from infinity to zero. Indicate in which regions the image is real and in which regions it is virtual.

49. * Repeat Problem 23.48 for a concave lens of −20-cm focal length.

50. **BIO Eye** The image distance for the lens of a person's eye is 2.10 cm. Determine the focal length of the eye's lens system for an object (a) at infinity, (b) 500 cm from the eye, and (c) 25 cm from the eye.

51. **BIO Lens-retina distance** Fish and amphibians accommodate their eyes to see objects at different distances by altering the distance from the lens system to the retina (to learn more about this, see the Reading Passage at the end of this chapter). If the lens system has a focal length of 2.10 cm, what is the lens-retina distance needed to view objects at (a) infinity, (b) 300 cm, and (c) 25 cm?

52. **BIO Nearsighted and farsighted** (a) A woman can produce sharp images on her retina only of objects that are from 150 cm to 25 cm from her eyes. Indicate the type of vision problem she has and determine the focal length of eyeglass lenses that will correct her problem. (b) Repeat part (a) for a man who can produce sharp images on his retina only of objects that are 3.0 m or more from his eyes. He would like to be able to read a book held 30 cm from his eyes.

53. * **BIO Prescribe glasses** A man who can produce sharp images on his retina only of objects that lie from 80 cm to 240 cm from his eyes needs bifocal lenses. (a) Determine the desired focal length of the upper half of the glasses used to see distant objects. (b) Determine the focal length of the lower half used to read a paper held 25 cm from his eyes.

54. * **BIO Correcting vision** A woman who produces sharp images on her retina only of objects that lie from 100 to 300 cm from her eyes needs bifocal lenses. (a) Determine the desired power of the upper half of the glasses used to see distant objects. (b) Determine the power of the lower half used to read a book held 30 cm from her eyes.

55. * **BIO Where are the far and near points?** (a) A woman wears glasses of 50-cm focal length while reading. What eye defect is being corrected and what approximately are the near and far points of her unaided eye? (b) Repeat part (a) for a man whose glasses have a −350-cm focal length. He wears the glasses while driving a car.

56. * **BIO** **Age-related vision changes** A 35-year-old patent clerk needs glasses of 50-cm focal length to read patent applications that he holds 25 cm from his eyes. Five years later, he notices that while wearing the same glasses, he has to hold the patent applications 40 cm from his eyes to see them clearly. What should be the focal length of new glasses so that he can read again at 25 cm?

23.8 Angular magnification and magnifying glasses

57. **Looking at an aphid** You examine an aphid on a plant leaf with a magnifying glass of +6.0-cm focal length. You hold the glass so that the final virtual image is 40 cm from the lens. If you assume that your near point is at 30 cm, then what is the angular magnification? How will the magnification change if you are farsighted? Nearsighted?

58. * **Reading with a magnifying glass** You examine the fine print in a legal contract with a magnifying glass of focal length 5.0 cm. (a) How far from the lens should you hold the print to see a final virtual image 30 cm from the lens (at the eye's near point)? (b) Determine the angular magnification of the magnifying glass.

59. * **Seeing an image with a magnifying glass** A person has a near point of 150 cm. (a) What is the nearest distance at which she needs to hold a magnifying glass of 5.0-cm focal length from print on a page and still have an image formed beyond her near point? (b) Determine the angular magnification for an image at her near point.

60. * **Stamp collector** A stamp collector is viewing a stamp through a magnifying glass of 5.0-cm focal length. Determine the object distance for virtual images formed at (a) negative infinity, (b) −200 cm, and (c) −25 cm. (d) Determine the angular magnification in each case.

23.9 Telescopes and microscopes

61. * You place a +20-cm focal length convex lens at a distance of 30 cm in front of another convex lens of focal length +4.0 cm. Then you place a candle 100 cm in front of the first lens. Find (a) the location of the final image of the candle, (b) its orientation, and (c) whether it is real or virtual.

62. * You place a +25-cm focal length convex lens at a distance of 50 cm in front of a concave lens with a focal length of −40 cm. Then you place a small lightbulb (2 cm tall) 30 cm in front of the convex lens. Determine (a) the location of the final image, (b) its orientation, and (c) whether it is real or virtual.

63. * **EST** You place a candle 10 cm in front of a convex lens of focal length +4.0 cm. Then you place a second convex lens, also of focal length +4.0 cm, at a distance of 12 cm from the first lens. (a) Use a ray diagram to locate the final image (keep the scale). (b) Using measurements on your ray diagram, estimate the linear magnification of the object. Be sure to show your rays and/or estimation technique for each step. Do not use equations!

64. * **EST** Repeat Problem 23.63 for an object located 6.0 cm from a convex lens of focal length 3.0 cm separated by 11 cm from another convex lens of focal length 1.0 cm.

65. ** You measure the focal length of a concave lens by first forming a real image of a light source using a convex lens. The image is formed on a screen 20 cm from the lens. You then place the concave lens halfway between the convex lens and the screen. To obtain a sharp image, you need to move the screen 15 cm farther away from the lenses. How does this experiment help you determine the focal length of the concave lens? What is the focal length?

66. ** **Telescope** A telescope consists of a +4.0-cm focal length objective lens and a +0.80-cm focal length eyepiece that are separated by 4.78 cm. Determine (a) the location and (b) the height of the final image for an object that is 1.0 m tall and is 100 m from the objective lens.

67. ** **Yerkes telescope** The world's largest telescope made only from lenses (with no mirrors) is located at the Yerkes Observatory near Chicago. Its objective lens is 1.0 m in diameter and has a focal length of +18.9 m. The eyepiece has a focal length of +7.5 cm. The objective lens and eyepiece are separated by 18.970 m. (a) What is the location of the final image of a Moon crater 3.8×10^5 km from Earth? (b) If the crater has a diameter of 2.0 km, what is the size of its final image? (c) Determine the angular magnification of the telescope by comparing the angular size of the image as seen through the telescope and the object as seen by the unaided eye.

68. * **Telescope** A telescope consisting of a +3.0-cm objective lens and a +0.60-cm eyepiece is used to view an object that is 20 m from the objective lens. (a) What must be the distance between the objective lens and eyepiece to produce a final virtual image 100 cm to the left of the eyepiece? (b) What is the total angular magnification?

69. ** **Design a telescope** You are marooned on a tropical island. Design a telescope from a cardboard map tube and the lenses of your eyeglasses. One lens has a +1.0-m focal length and the other has a +0.30-m focal length. The telescope should allow you to view an animal 100 m from the objective with the final image being formed 1.0 m from the eyepiece. Indicate the location of the lenses and the expected angular magnification.

70. * **Microscope** A microscope has a +0.50-cm objective lens and a +3.0-cm eyepiece that is 20 cm from the objective lens. (a) Where should the object be located to form a final virtual image 100 cm to the left of the eyepiece? (b) What is the total angular magnification of the microscope, assuming a near point of 25 cm?

71. ** **BIO** **Dissecting microscope** A dissecting microscope is designed with a larger than normal distance between the object and the objective lens. The microscope has an objective lens of +5.0-cm focal length and an eyepiece of +2.0-cm focal length. The lenses are separated by 15 cm. The final virtual image is located 100 cm to the left of the eyepiece. (a) Determine the distance of the object from the objective lens. (b) Determine the total angular magnification.

72. ** **Microscope** A microscope has an objective lens of focal length +0.80 cm and an eyepiece of focal length +2.0 cm. An object is placed 0.90 cm in front of the objective lens. The final virtual image is 100 cm from the eyepiece at the position of minimum eyestrain. (a) Determine the separation of the lenses. (b) Determine the total angular magnification.

73. ** **Microscope** Determine the lens separation and object location for a microscope made from an objective lens of focal length +1.0 cm and an eyepiece of focal length +4.0 cm. Arrange the lenses so that a final virtual image is formed 100 cm to the left of the eyepiece and so that the angular magnification is −260 for a person with a near point of 25 cm.

General Problems

74. Paul needs to determine the diameter of the wire used to make the filament of an incandescent lightbulb. He decides to project a magnified image of the filament onto a screen using a lens. (a) Describe Paul's experimental setup in words and with a sketch. (b) What additional data (other than those that can be obtained from the image on the screen) does Paul need to determine the diameter of the wire?

75. * **Figure P23.75** shows three cases of the primary axis of a lens (the lens is not shown) and the location of a shining object and its image. In each case, find the location and the type of the lens (convex or concave) that could produce the image and find the focal points of the lens.

FIGURE P23.75

(a)

Object
•

‒ ‒

•
Image

(b)   Image
•
Object
•

‒ ‒

(c)   Object
•
•
Image

‒ ‒

76. * **Jeopardy problem** The equations below describe a process that involves one lens. Determine the unknown quantities and write a word description of an optics situation that is consistent with the equations.

$$\frac{1}{4.0\text{ m}} + \frac{1}{s'} = \frac{1}{0.10\text{ m}}$$

$$h' = -\left(\frac{s'}{4.0\text{ m}}\right)(1.6\text{ m})$$

77. * **Jeopardy problem** The equations below describe a process involving more than one lens. Determine the unknown quantities and write a word description of an optics situation that is consistent with the equations.

$$\frac{1}{100\text{ m}} + \frac{1}{s_1'} = \frac{1}{0.10\text{ m}}$$

$$s_2 = (0.14\text{ m}) - s_1'$$

$$\frac{1}{s_2} + \frac{1}{s_2'} = \frac{1}{0.042\text{ m}}$$

78. **Image Jeopardy** Figure P23.78 shows the primary axis of a lens; the path of ray A as it approaches the lens, gets refracted, and continues to travel; and the path of ray B before it hits the lens. (a) Determine the type of lens (concave or convex) and its location, (b) determine the location of both focal points of the lens, and (c) draw the path of ray B after it passes through the lens.

FIGURE P23.78

79. ** **Two-lens camera** A two-lens camera (see Figure P23.79) has one lens with focal length +15.0 cm located 12.0 cm from the film and a second lens of focal length +13.0 cm a variable distance d of 5.0 to 10.0 cm from the film. Determine the range of distances at which you can photograph objects and achieve sharp images on the film.

FIGURE P23.79

80. ** You have a small spherically shaped bottle made of plastic that is so thin that we can disregard its effects for optical processes. Show that the bottle works as a converging lens when filled with water, but when the same bottle is empty, closed, and submerged in water, it becomes a diverging lens. If you have access to the appropriate materials, test your ray diagram experimentally.

81. **BIO** **Find a farsighted person.** Design an experiment to find out what lenses will help this person read a book holding it 25 cm away. Borrow this person's glasses and invent a simple experiment to measure the approximate focal lengths of the lenses of his or her glasses. Perform the experiment. Are the real glasses close to your "prescription?"

82. **BIO** **Find a nearsighted person.** Design an experiment to find out what lenses will help this person see objects that are far away. Borrow this person's glasses and invent a simple experiment to measure the approximate focal lengths of the lenses of his or her glasses. Perform the experiment. Are the real glasses close to your "prescription?"

Reading Passage Problems

BIO **Laser surgery for the eye** LASIK (laser-assisted *in situ* keratomileusis) is a surgical procedure intended to reduce a person's dependency on glasses or contact lenses. Laser eye surgery corrects common vision problems, such as myopia (nearsightedness), hyperopia (farsightedness), astigmatism (blurred vision resulting from corneal irregularities), or some combination of these. In myopia, the cornea, the clear covering at the front of the eye, is often too highly curved, causing rays from distant objects to form sharp images in front of the retina (Figure 23.34). LASIK refractive surgery can flatten the cornea so that images of distant objects form on the retina. A knife cuts a flap in the cornea with a hinge left at one end of this flap. The flap is folded back, exposing the middle section of the cornea. Pulses from a computer-controlled laser vaporize a portion of the tissue and the flap is replaced.

FIGURE 23.34 LASIK surgery to correct nearsighted vision.

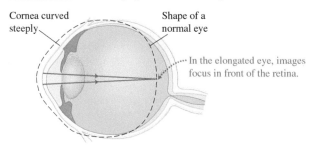

In farsighted people, an object held at the normal near point of the eye (about 25–50 cm from the eye) forms an image behind the retina (Figure 23.35). Increasing the curvature of the cornea causes rays from near objects to produce an image on the retina. Conductive keratoplasty (CK) can increase the curvature of the cornea to correct farsightedness. In CK, a tiny probe emits radio waves in a circular pattern on the outside of the cornea. Exposure to radio waves causes shrinking in the ring around the cornea, which acts like a constrictive band, increasing the overall curvature of the cornea.

FIGURE 23.35 Conductive keratoplasty (CK) surgery to correct farsighted vision.

83. Your eyeball is 2.10 cm from the cornea to the retina. You look at a sign on the freeway that is effectively an infinite distance away. The image for the sign is 1.96 cm from the cornea within the eye. What is the focal length of the cornea-lens system?
    (a) +1.96 cm   (b) +2.10 cm   (c) +2.26 cm
    (d) +2.42 cm   (e) +27.90 cm

84. In Problem 23.83, what should the cornea-lens focal length be in order to form a sharp image on the retina of the sign?
    (a) +1.96 cm   (b) +2.10 cm   (c) +2.26 cm
    (d) +2.42 cm   (e) +27.90 cm

85. What is one surgical way to correct nearsighted vision?
    (a) Reduce the curvature of the cornea, thus reducing the focal length of the eye's optics system.
    (b) Increase the curvature of the cornea, thus increasing the focal length of the eye's optics system.
    (c) Increase the iris opening to allow more light to enter.
    (d) Use the ciliary muscle to increase the thickness of the lens.

86. Your eyeball is 2.10 cm from the cornea to the retina. A book held 30 cm from your eyes produces an image 2.26 cm from the entrance to the image location. What is the focal length of the cornea-lens system?
    (a) +1.96 cm   (b) +2.10 cm   (c) +2.26 cm
    (d) +2.42 cm   (e) +27.90 cm

87. In Problem 23.86, what should the cornea-lens focal length be in order to form a sharp image of the book on the retina?
    (a) +1.96 cm   (b) +2.10 cm   (c) +2.26 cm
    (d) +2.42 cm   (e) +27.90 cm

88. What is one surgical way to correct the farsighted vision of the person in the last two problems?
    (a) Reduce the curvature of the cornea, thus increasing the focal length of the eyes optics system.
    (b) Increase the curvature of the cornea, thus reducing the focal length of the eyes optics system.
    (c) Open the iris wider to allow more light to enter.
    (d) Cause the ciliary muscle to stretch the lens so it becomes thinner.

BIO **Bulging fish eyes** Fish eyes (see Figure 23.36) are evolutionary precursors to human eyes. Fish eyes have a dome-shaped cornea covering an iris that has a fixed opening and a spherical lens, which protrudes from the iris. Behind the lens is a retina with many rod and cone cells that detect light. Muscles attached to the lens change the lens location relative to the retina. Because water has a refractive index similar to that of the fluid inside fish eyes, there is little bending of light entering the eye at the water-cornea interface. The lens has a different refractive index than the fluid surrounding the lens and bends and focuses light so that images are formed on the retina. The fish eye lens is spherical in shape and collects light from a 180° field of view (Figure 23.37).

FIGURE 23.36 A fish eye.

FIGURE 23.37 A fisheye lens collects light from a 180° field of view.

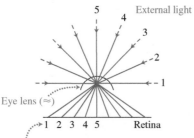

Camera manufacturers now make fisheye lenses that produce an image from a 180° spread in front of the lens. Fisheye lens applications include IMAX projection, hemispherical photography, and peepholes that provide 180° views outside doors.

89. Why is the fish eye in Figure 23.36 black?
    (a) The lens is black.
    (b) No light reflects out from behind the lens.
    (c) The fluid behind the lens is black.
    (d) The retina is black.

90. Which of the following is the same for the fish eye and the human eye?
    (a) The muscles attached to the lenses change their shape.
    (b) The corneas help focus the light.
    (c) The lenses have a spherical shape.
    (d) The irises help focus the light.

91. Which ray in Figure P23.91 best represents the path of the incident light through the fisheye lens?
    (a) I   (b) II   (c) III   (d) IV
    (e) V

FIGURE P23.91

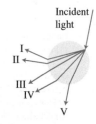

92. Which of the following statements regarding the fish eye and the human eye is not true?
    (a) They both have a cornea.
    (b) They both have an iris.
    (c) They both have a lens.
    (d) They both rely on the cornea and the lens for image formation.

93. How can the peephole lens in a hotel door have an approximately 180° view?
    (a) The light moves straight through the lens into the eye.
    (b) Light from about 90° from the straight-ahead direction is refracted into the observer's eye if very close to the lens.
    (c) Light from about 90° from the straight-ahead direction is refracted into the observer's eye if looking at the lens from inside the room.
    (d) The outside hall is illuminated so reflected light from the objects inside the 180° angle goes into the lens.
    (e) Both b and d

Wave Optics

Soap bubbles display a remarkable array of colors. Watch a bubble hanging from a bubble wand, and you will see horizontal bands appear on its surface and repeatedly change colors. The bands of color on the top of the bubble will widen until a dark band takes their place—and then the bubble pops. In this chapter you will learn why soap bubbles display such brilliant colors and why those colors disappear just before the bubble pops.

IN OUR CHAPTER on reflection and refraction (Chapter 22) we developed two models of light: a particle-bullet model and a wave model. The particle model predicted that the speed of light in water should be greater than in air; the wave model predicted the opposite. Eventually, in 1850, Hippolyte Fizeau and Léon Foucault established that light travels more slowly in water than in air, finally disproving the particle model. However, even before Fizeau and Foucault, the wave model was gaining wide acceptance because of experiments performed by Thomas Young in the early 1800s.

- Why do we see colors in soap bubbles?
- Which best explains the behavior of light: the particle-bullet model or the wave model?
- What limits our ability to perceive as separate two closely positioned objects?

BE SURE YOU KNOW HOW TO:

- Draw wave fronts for circular and plane waves (Section 11.1).
- Relate wave properties such as frequency, speed, and wavelength (Section 11.2).
- Apply Huygens' principle and the superposition principle (Section 11.6).

24.1 Young's double-slit experiment

In 1801, Englishman Thomas Young performed a testing experiment for the particle model of light. Testing Experiment Table 24.1 describes a contemporary version of the experiment and the prediction of its outcome based on the particle model.

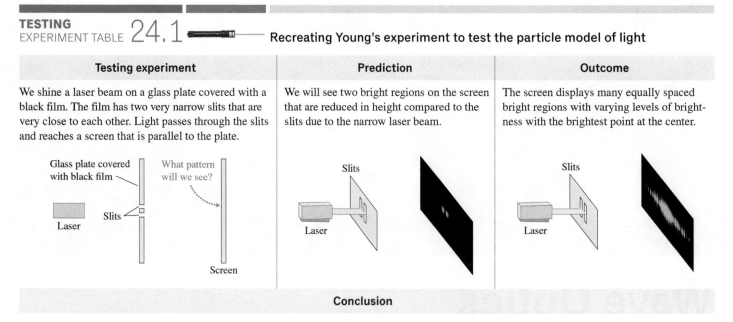

TESTING
EXPERIMENT TABLE 24.1 ⸻ **Recreating Young's experiment to test the particle model of light**

Testing experiment	Prediction	Outcome
We shine a laser beam on a glass plate covered with a black film. The film has two very narrow slits that are very close to each other. Light passes through the slits and reaches a screen that is parallel to the plate.	We will see two bright regions on the screen that are reduced in height compared to the slits due to the narrow laser beam.	The screen displays many equally spaced bright regions with varying levels of brightness with the brightest point at the center.

Conclusion

The prediction based on the particle model of light does not match the outcome, disproving the particle model.

In Table 24.1, light from the two slits produced a pattern of alternating bright regions on the screen (called **bands** or **fringes**). This outcome is inconsistent with the model of light that represents light as a stream of fast-moving bullets. In addition to alternating bright and dark regions, we also saw variation in the levels of brightness on a larger scale. We will focus on the equally spaced bright spots first, and discuss the variation in their brightness later in the chapter.

Thomas Young explained the bright fringes using a wave model and Huygens' principle (see Chapters 11 and 22). According to this principle, every point on a wave front is the source of a circular-shaped wavelet that moves in the forward direction at the same speed as the wave. A new wave front results from adding these wavelets together—the superposition (*interference*) of these wavelets.

We can use the wave model of light to explain the outcome of the testing experiment. According to this model, the laser emits some kind of wave. The brighter the light, the larger the amplitude of the wave. At the location where there is no light, we assume the amplitude of the light wave to be zero.

Because the laser beam does not spread much, we can assume that a laser produces a wave with plane wave fronts. The crests and troughs of this wave reach the two closely spaced slits and produce synchronized (in-phase) disturbances within each slit. According to Huygens' principle, the slits become sources of in-phase wavelets (**Figure 24.1a**). The new wavelets generated from each slit spread outward and eventually irradiate the screen. There, the disturbances due to both wavelets combine according to the superposition principle. At a particular time, the wavelet disturbances produced by each slit appear as shown in Figure 24.1b. (Note that the angles between the directions to the bright spots are greatly exaggerated in order to make the wave front representation easier to understand. You can infer the real angles from the photo in Figure 24.1a.) At some points, the crests from the wavelets from each slit overlap (the positions of solid dots), resulting in double-sized crests. The troughs between the crests also overlap, resulting in double-sized troughs. This *constructive interference* results in a double-amplitude wave. These alternating double crests and double troughs arrive at the screen shown in Figure 24.1b, causing a large-amplitude light wave—a bright band (or fringe). This occurs along line 0 and along the two lines labeled 1. It also occurs along several other lines not shown in the figure.

Along lines about halfway between these bright bands, you find that the crest of one wavelet overlaps with the trough of the other (the open circles in Figure 24.1c). The wavelets are out of phase along those lines and therefore cancel each other. This *destructive interference* means no light is present—a dark band (or fringe) is present.

Between the centers of the bright and dark bands, the wavelets are neither exactly in nor exactly out of phase with each other, and the wave amplitudes are between double and zero; the light has intermediate brightness. Since the pattern on the screen can be explained using the idea of interference of the wavelets from the two slits, it is called an **interference pattern**. The wave model of light explains the fringes observed by Thomas Young; the particle-bullet model does not.

> **Qualitative wave-based explanation of Young's experiment** The interference pattern of dark and bright bands produced by light passing through two closely spaced narrow slits can be explained by the addition or superposition of circular wavelets originating from the two slits. The length of the path that each wave travels to a particular spot on the screen determines the phase of the wave at that location. Locations where two waves arrive in phase are the brightest; locations where two waves arrive with different phases are darker; and locations where the waves arrive completely out of phase are dark.

What is a light wave—what is being disturbed?

Imagine that we want to draw a graphical representation of a light wave—a disturbance-versus-position graph (similar to the graphs we drew in Chapter 11 for mechanical waves). We know, for example, that a sound wave involves a moving disturbance of air pressure and density and that a wave on a Slinky involves a moving disturbance of the closeness of the Slinky coils. What is being disturbed when a light wave propagates?

In the early 1800s, physicists thought that a medium was needed for the propagation of any wave. They proposed the existence of a massless invisible medium for light called **ether**, which supposedly filled the entire universe and vibrated when light traveled through it. This vibrating ether model was accepted until the late 1800s. Later in this book (in Chapters 25 and 26), we will learn about a contemporary theory of light and the experiments that disproved the existence of ether. For now, we will assume that light is a wave-like disturbance and that the nature of that disturbance is unknown. We will also assume that the disturbances from different waves obey the superposition principle.

Quantitative analysis of the double-slit experiment

A mathematical description of the phenomenon observed by Young involves calculating the location of bright and dark bands on a screen for a particular light wavelength λ, slit separation d, and distance L of the screen from the slits. As discussed earlier, we consider the slits to be sources of synchronized circular wavelets. At all locations where the waves arrive in phase, there will be constructive interference; where they arrive completely out of phase, there will be destructive interference. For simplicity, we will represent the waves traveling to the screen from the slits as wiggles. You can think of these wiggles as cross-sections of the three-dimensional waves. The waves are in phase at a particular location if they have the same shape at this location.

FIGURE 24.1 (a) Double slits are the source of wavelets that can interfere. (b) When wave crests or troughs overlap, a large-amplitude wave occurs. (c) When a crest overlaps a trough, they cancel and there is no wave disturbance.

(a)

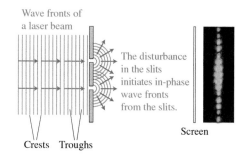

Wave fronts of a laser beam

The disturbance in the slits initiates in-phase wave fronts from the slits.

Screen

Crests   Troughs

(b)   Overlap of two crests produces a double crest—constructive interference.

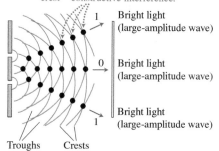

Bright light (large-amplitude wave)

Bright light (large-amplitude wave)

Bright light (large-amplitude wave)

Troughs   Crests

(c)

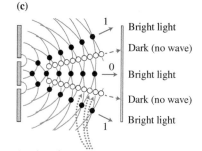

Bright light

Dark (no wave)

Bright light

Dark (no wave)

Bright light

Overlap of crest and trough produces zero disturbance—destructive interference.

Now consider the brightest locations on the screen, as shown in **Figure 24.2a**. The wavelets are in phase at the location of the slits, and the waves travel the same distance from the slits to point 0. Thus the waves arrive at this point in phase no matter how far the screen is from the slits. Figure 24.2b represents the wave disturbances from the two slits at one instant of time. This bright band or fringe in the center is called the **0th order maximum** (or zeroth maximum).

FIGURE 24.2 Double-slit wave interference.

(a)

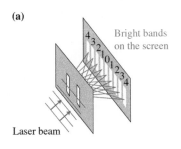

Bright bands on the screen

Laser beam

(b)

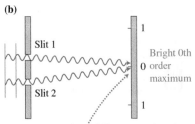

Slit 1

Slit 2

Bright 0th order maximum

The waves travel equal distances and are in phase when they reach the screen.

(c)

Dark

Dark

$\lambda/2$

The lower wave travels $\lambda/2$ farther than the upper wave. They are out of phase when they reach the screen.

(d)

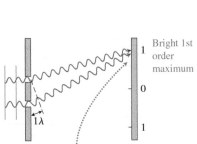

Bright 1st order maximum

1λ

The wave from the lower slit travels one wavelength farther than the wave from the upper slit. At the screen, they are in phase and interfere constructively.

(e)

Bright 2nd order maximum

Waves are in phase.

2λ

The wave from the lower slit travels two wavelengths farther.

(f)

y

Intensity

The peaks correspond to the maximum brightness on the screen, and the troughs correspond to dark regions.

Just to either side of this location, the waves arrive slightly out of phase and the screen is dimmer. The brightness of the screen changes slowly, from a maximum at the center, where the waves arrive exactly in phase, to zero where waves arrive completely out of phase. At the two locations marked "Dark" in Figure 24.2c, the difference in the path length traveled by the two waves is one-half wavelength.

Moving away from the dark band on each side of the center, the screen becomes brighter again as the path length difference between the waves becomes larger. When the path length difference reaches one whole wavelength at location 1, there is another point of maximum brightness (Figure 24.2d). The wave from the lower slit travels exactly one wavelength farther than the distance traveled by the wave from the upper slit. This bright band is called the **1st order maximum**. A similar maximum appears at the symmetrical location 1 below location 0.

We can use similar reasoning to find the 2nd, 3rd, and 4th order maxima. The path length differences for the centers of these bright bands are 2λ, 3λ, and 4λ, respectively (see the second bright band in Figure 24.2e). We can use this thinking to predict the locations of these bands using geometry. Figure 24.2f shows the change in the intensity of light incident on the screen (measured by a photosensitive device) as a function of position for the 0th order maximum and the 1st and 2nd order maxima on both sides.

Mathematical location of the mth bright band

Consider the general case in **Figure 24.3**. The small shaded triangle next to the slits has a hypotenuse d equal to the slit separation. The two red lines are the distances that the two waves travel to get to the center of the mth bright band on the screen (position P_m in Figure 24.3). The path length difference between the waves equals an integer number of whole wavelengths—m wavelengths in this case. The side of the small shaded triangle opposite angle θ is the path difference for the two waves reaching location P_m on the screen. If there is an interference maximum at that location on the screen, then the path length difference Δ that the waves originating at the two slits travel to P_m should equal an integer number of wavelengths:

$$\Delta = m\lambda, \text{ where } m = 0, \pm 1, \pm 2, \pm 3, \text{ etc.}$$

From now on, we will use a negative sign to distinguish the maxima below the 0th band from the maxima above it. Using trigonometry: $\sin \theta = (\Delta/d) = (m\lambda/d)$. The angle θ in the small triangle equals the angle θ_m between the horizontal and the mth bright band. Thus,

$$\sin \theta_m = \frac{m\lambda}{d} \tag{24.1}$$

This derivation has led us to a quantitative relation between the slit separation d, the angle θ_m at which the mth maximum occurs, and the wavelength λ of the light. If we know the distance L between the slits and the screen, we can find the distance y_m between the 0th maximum on the screen and the mth maximum in Figure 24.3:

$$\tan \theta_m = \frac{y_m}{L} \tag{24.2}$$

If the angle θ_m is very small, the tangent of the angle equals the sine of the angle and also equals the angle itself if measured in radians: $\tan \theta_m \approx \theta_m = (y_m/L)$ and $\sin \theta_m \approx \theta_m = (m\lambda/d)$. Combining these two relationships, we find that for small angles relative to the horizontal, the bright bands on the screen are located at

$$y_m = \frac{m\lambda L}{d} \tag{24.3}$$

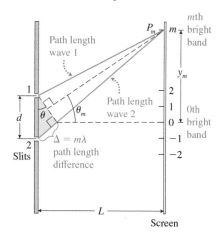

FIGURE 24.3 The geometry for finding the position of the mth bright band on a screen.

Testing expressions for bright band locations on a screen

Since it is not easy to independently measure the wavelength of light, we will instead test the inverse relationship between the location of the fringes y_m and the slit separation d (Testing Experiment Table 24.2).

TESTING EXPERIMENT TABLE 24.2 — **Slit separation and locations of the bright spots**

VIDEO TET 24.2

Testing experiment	Prediction	Outcome
• We shine red laser light through slits with separation $d = 0.050$ mm onto a screen at $L = 2.0$ m from the slits. The first bright fringe on the screen from the center maximum is at $y_1 = 2.75$ cm. • Using the same laser, we predict the location of the first bright band on the screen if using slits with a different separation $d' = 0.030$ mm.	We solve Eq. (24.3) for λ as $L \gg y$: $$\lambda = \frac{y_m d}{mL}$$ m is 1 since we are interested in the first bright fringe. Since the same laser is used for both slit separations, we can equate the right-hand sides of this equation for each case. $$\frac{y_1 d}{(1)L} = \frac{y_1' d'}{(1)L}$$ Multiplying by L and solving for y_1', we have our prediction: $$y_1' = \frac{y_1 d}{d'} = \frac{(2.75 \text{ cm})(0.050 \text{ mm})}{0.030 \text{ mm}} = 4.6 \text{ cm}$$	We measure a distance of 4.6 ± 0.1 cm between the center $m = 0$ maximum and the $m = 1$ maxima.

Conclusion

• The outcome of the experiment matches the prediction from Eq. (24.3).
• The result supports the wave model of light.

Because the outcome of the experiment agreed with the prediction that followed from the equations we derived using the wave model of light, we gain confidence in that model. Many other experiments have also agreed with predictions based on the wave model of light.

Now that we have some trust in Eq. (24.3), we can use it to determine the wavelength of the red laser light used in Table 24.2.

$$\lambda = \frac{y_m d}{mL} = \frac{(2.75 \text{ cm})(0.050 \text{ mm})}{(1)(2.0 \text{ m})} = 6.9 \times 10^{-7} \text{ m}$$

To simplify the expression of such small numbers, physicists record wavelengths of light in nanometers; 1 nm = 10^{-9} m. The wavelength for the red light described in Table 24.2 is about 690 nm. We can now summarize our findings about double-slit interference.

Double-slit interference When light of wavelength λ passes through two narrow slits separated by distance d, it forms a series of bright and dark bands on a screen beyond the slits. The angular deflection θ_m of the center of the mth bright band to the side of the central $m = 0$ bright band is determined using the equation

$$d \sin \theta_m = m\lambda \tag{24.1}$$

where $m = 0, \pm 1, \pm 2, \pm 3$, etc. The distance y_m on a screen from the center of the central bright band to the center of the mth bright band depends on the angular deflection θ_m and on the distance L of the screen from the slits:

$$L \tan \theta_m = y_m \tag{24.2}$$

For small angles:

$$y_m = \frac{m\lambda L}{d} \tag{24.3}$$

We see from Eq. (24.1) that for light of wavelength λ, the greater the slit separation d, the smaller the angular deflection of the bright spots θ_m. Thus, to observe distinct bands, we need slits that are very close to each other. Also, the longer the wavelength, the larger the angular deflection for the same slit separation. Finally, using Eq. (24.2), we see that the greater the distance L of the screen from the slits, the greater the distance y_m of the mth bright band from the central bright band on the screen.

TIP We could have done the derivation for the minima (the locations where two waves arrive out of phase and cancel each other) instead of for the maxima and obtained a relation for the angles at which the dark bands occur. This happens when the path length difference from the two slits to the dark band equals an odd number of half wavelengths: $\Delta = (2m + 1)(\lambda/2)$, where $m = 0, \pm 1, \pm 2, \pm 3$, etc., and consequently

$$\sin \theta = \frac{(2m + 1)\lambda/2}{d}$$

FIGURE 24.4 Pattern of white light produced by two slits on a screen.

Thomas Young used sunlight in his original experiments since lasers had not yet been invented. He found that the bands of light on the screen were colored like a rainbow, except for the central band, which was white. Each band that he observed had blue light on the inside edge (closest to the central maximum) and red light on the outside edge (**Figure 24.4**). We found that the angular deflection to the center of the mth bright band should be $\sin \theta_m = (m\lambda/d)$. If the wavelength λ is longer, the angular deflection θ_m should be greater. Evidently, red light must have a longer wavelength than blue light, with green in between. We can check this idea.

Wavelength of green light

You repeat the experiment described in Table 24.2 using slits separated by $d = 0.050$ mm and a screen at $L = 2.0$ m from the slits. This time, you use a green laser instead of a red laser. The second maximum ($m = 2$) appears on the screen at $y_2 = 4.4$ cm from the center of the 0th order maximum. Determine the wavelength of this light.

Represent mathematically Since $L \gg y$, we can use Eq. (24.3) $y_m = (m\lambda L/d)$ to determine the wavelength.

Solve and evaluate To determine λ, multiply both sides by d and divide by mL to get $(y_m d/mL) = \lambda$. Now substitute the known values:

$$\lambda = \frac{y_m d}{mL} = \frac{(4.4 \times 10^{-2}\,\text{m})(5.0 \times 10^{-5}\,\text{m})}{2 \times 2.0\,\text{m}}$$

$$= 5.5 \times 10^{-7}\,\text{m} = 550\,\text{nm}$$

The order of magnitude for the wavelength is the same as that for red light, but the value is smaller, consistent with Young's observations.

Try it yourself What is the angle between the direction of light going toward the central maximum and the direction to the first *dark* fringe?

Answer 5.5 × 10⁻³ rad.

REVIEW QUESTION 24.1 Explain why we observe multiple bright bands of light on a screen after light of a particular wavelength passes through two narrow, closely spaced slits.

24.2 Refractive index, light speed, and wave coherence

In Chapter 22, we learned that the wave model of light explains refraction of light and predicts that light should travel more slowly in a medium with a higher refractive index compared to a medium with a lower index. In this section, we will learn how to use the wave model of light to explain refraction quantitatively.

Relating the refractive index and the speed of light in a substance

In the chapter on reflection and refraction (Chapter 22), we learned that when light travels from medium 1 to medium 2, we can determine the change in the direction of light propagation using Snell's law: $n_2 \sin \theta_2 = n_1 \sin \theta_1$ (**Figure 24.5a**). Because $n_2 > n_1$, $\sin \theta_2 < \sin \theta_1$ and $\theta_2 < \theta_1$. If we are using the wave model to explain this change, we need to consider that the wave fronts that are perpendicular to the propagation of the wave also bend at the interface. For a wave to bend toward the normal, its speed must be smaller in medium 2 than in medium 1. We know that the frequency of a wave depends only on the source of the wave and does not change as it passes from medium 1 to medium 2. Therefore, the faster moving light in medium 1 has a longer wavelength than the slower moving light in medium 2.

Consider the right triangles DAB and ADC in Figure 24.5b. We know that the angle of incidence θ_1 is equal to the angle DAB because the corresponding half lines of θ_1 and of DAB are perpendicular to each other. The angle of refraction θ_2 is equal to the angle ADC for the same reason. The side opposite angle DAB (that is, DB) is one wavelength λ_1 of the light in medium 1. Similarly, the side opposite angle ADC (that is, AC) is one wavelength λ_2 of the light in medium 2. Thus, the sines of the angles are

$$\sin \text{DAB} = \sin \theta_1 = \frac{\lambda_1}{\text{DA}} \quad \text{and} \quad \sin \text{ADC} = \sin \theta_2 = \frac{\lambda_2}{\text{DA}}$$

FIGURE 24.5 (a) The wave slows as it moves from medium 1 to medium 2, causing the wavelength to decrease. (b) The angles of incidence and refraction are related to the wavelengths.

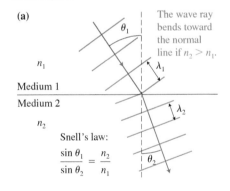

(a)

The wave ray bends toward the normal line if $n_2 > n_1$.

Medium 1 n_1

Medium 2 n_2

Snell's law:

$$\frac{\sin \theta_1}{\sin \theta_2} = \frac{n_2}{n_1}$$

(b)

Medium 1 n_1

Medium 2 n_2

$$\lambda_1 = \frac{v_1}{f}$$

$$\frac{v_2}{f} = \lambda_2$$

The angle of incidence θ_1 and angle of refraction θ_2 also depend on the wave speeds.

Note that the wave frequency does not change: $f_1 = f_2 = f$.

Geometry: $\dfrac{\sin \theta_1}{\sin \theta_2} = \left(\dfrac{v_1/f}{v_2/f}\right)\left(\dfrac{\text{AD}}{\text{AD}}\right) = \dfrac{v_1}{v_2}$

Now, use the above to take the ratio of the sines of the angles θ_1 and θ_2:

$$\frac{\sin \theta_1}{\sin \theta_2} = \frac{\lambda_1/DA}{\lambda_2/DA} = \frac{\lambda_1}{\lambda_2}$$

We can use the relation between wavelength λ, frequency f, and speed v of a light wave to rewrite the above in terms of the speeds of the light in the two media:

$$\lambda_1 = \frac{v_1}{f} \quad \text{and} \quad \lambda_2 = \frac{v_2}{f}$$

After substitution, we get

$$\frac{\sin \theta_1}{\sin \theta_2} = \frac{\lambda_1}{\lambda_2} = \frac{v_1/f}{v_2/f} = \frac{v_1}{v_2}$$

From Snell's law,

$$\frac{\sin \theta_1}{\sin \theta_2} = \frac{n_2}{n_1}$$

Thus, the ratio of the speeds of light in two media 1 and 2 should be equal to the inverse ratio of their indexes of refraction:

$$\frac{v_1}{v_2} = \frac{n_2}{n_1}$$

We know that the index of refraction of water is 1.33 and that of air is 1.00. Thus the speed of light in water should equal the ratio of its speed in air c (3.0×10^8 m/s) divided by the $n = 1.33$ index of refraction of water:

$$v = \frac{c}{1.33} = \frac{(3.0 \times 10^8 \text{ m/s})}{1.33} = 2.26 \times 10^8 \text{ m/s}$$

This is exactly the speed that French physicists Hippolyte Fizeau and Leon Foucault obtained in their 1850 experiments to determine the speed of light in water.

We found that the wave model of light not only explains why light bends at the boundary of two media but also explains Snell's law by connecting the medium's index of refraction to the speed of light in that medium.

> **Refractive index** The refractive index n of a medium equals the ratio of the speed of light c in vacuum (or in air) to the speed of light v in the medium:
>
> $$n = \frac{c}{v} \tag{24.4}$$

Refractive index and the color of light

We already know that when white light passes through a prism, it comes out as a set of colored bands (see **Figure 24.6a**). In other words, different colors of light are refracted differently by the prism. How does the wave model explain this? Snell's law of refraction [Eq. (22.3)] as applied to Figure 24.6b is

$$n_{glass} = \frac{n_{air} \sin \theta_{air}}{\sin \theta_{glass}}$$

Because the refracted angle in glass for blue light θ_{glass} is smaller than that for red light, the refractive index n_{glass} for blue light must be greater than that for red light. The refractive index is related to the speed of light in the medium by Eq. (24.4). The different refractive indexes for different colors mean that light of different colors or light waves of different frequencies travel at different speeds in the same medium. Only in a vacuum does light of all frequencies travel at the same speed.

FIGURE 24.6 The refractive index n of a medium is different for different wavelengths λ of light.

(a)

(b)

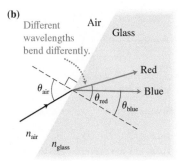

QUANTITATIVE EXERCISE 24.2 **Speed of light in a medium**

A laser pointer emits green light that has a 550-nm wavelength when in air. (a) What is the light's frequency? (b) What are the frequency, speed, and wavelength of the laser light as it travels through glass of refractive index 1.60?

Represent mathematically The frequency of a wave can be determined using its speed and wavelength $f = (c/\lambda)$. The speed of light in glass can be found using $n = (c/v)$.

Solve and evaluate (a) We are given the wavelength of the light when in air: $\lambda = 550 \times 10^{-9}$ m. The speed of the light when in air is $c = 3.0 \times 10^{8}$ m/s. Thus, the frequency of the light is

$$f = \frac{c}{\lambda} = \frac{3.0 \times 10^{8}\,\text{m/s}}{550 \times 10^{-9}\,\text{m}} = 5.5 \times 10^{14}\,\text{Hz}$$

(b) This light has the same 5.5×10^{14}-Hz frequency when in glass because the frequency of a traveling wave is determined by its source, not by the medium. However, its speed and wavelength change:

$$v_{\text{in glass}} = \frac{c}{n_{\text{glass}}} = \frac{3.0 \times 10^{8}\,\text{m/s}}{1.60} = 1.9 \times 10^{8}\,\text{m/s}$$

$$\lambda_{\text{in glass}} = \frac{v_{\text{in glass}}}{f} = \frac{1.9 \times 10^{8}\,\text{m/s}}{5.5 \times 10^{14}\,\text{Hz}} = \frac{1.9 \times 10^{8}\,\text{m/s}}{5.5 \times 10^{14}\,\text{s}^{-1}}$$

$$= 3.5 \times 10^{-7}\,\text{m} = 350\,\text{nm}$$

Try it yourself What is the wavelength of green light traveling through water of refractive index 1.33?

Answer   410 nm.

TIP Note that when light travels into a medium with a different index of refraction, its frequency does not change, but its wavelength does.

Chromatic aberration in lenses—a practical problem in optical instruments

High-contrast photos sometimes have purple edges—an effect called *purple fringing*. Purple fringing occurs because the refractive index is wavelength dependent, resulting in a lens having a slightly smaller focal length for purple light than for red light ($f_{\text{purple}} < f_{\text{red}}$, where f stands for the focal length, not frequency). Thus, the location of an image for the purple part of an object is slightly different from the image location for the red part of that object, an effect called **chromatic aberration**. Since the image locations are slightly different, their magnifications, which depend on image location, will also be slightly different. When the magnification of the purple image is slightly greater than the other colors, purple fringing occurs (**Figure 24.7**).

FIGURE 24.7 (a) Chromatic aberration is evident in this magnification of newsprint. (b) The purple and red lights form images at different places due to differences in their refractive indexes.

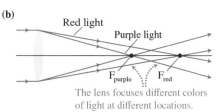

The lens focuses different colors of light at different locations.

Monochromatic and coherent waves

Our observations of double-slit interference (Table 24.1) involved laser light passing through two very narrow slits, each of which acted as a light source. If we replace the laser light with two small lightbulbs, we do not observe an interference pattern. Instead, we see a uniformly illuminated screen (**Figure 24.8**). Why?

FIGURE 24.8 Two lightbulbs do not produce an interference pattern.

(a) With the laser light, we observed bright and dark bands.

(b) With two incandescent lightbulbs, we observe a uniformly lit screen.

FIGURE 24.9 The vibrating beach balls are monochromatic and coherent and produce a stable interference pattern.

Two balls vibrating in phase produce a stable interference pattern.

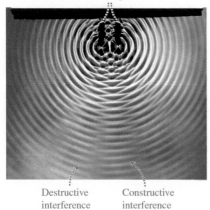

Destructive   Constructive
interference   interference

FIGURE 24.10 A laser produces coherent and monochromatic disturbances at the slits.

Laser   Monochromatic beam

Laser wave fronts produce
coherent disturbances at the slits.

FIGURE 24.11 The bright interference bands sharpen with an increasing number of slits.

(a) Distance between the slits in all cases is 0.2 mm.

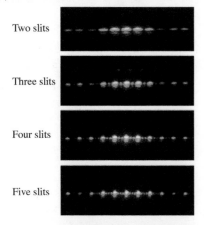

Two slits

Three slits

Four slits

Five slits

(b) Distance between the slits is 0.1 mm.

Grating

In order for two waves to interfere consistently over time, two conditions must be met. First, the waves must vibrate at the same constant frequency, producing what are called **monochromatic waves.** Second, the phases of the waves must be synchronized, meaning that as time passes they must arrive at a particular location consistently in phase, out of phase, or somewhere in between. The phase difference between the two waves must remain constant in time. When this occurs, the waves are called **coherent waves.** The synchronized vibrating beach balls shown in **Figure 24.9** produce coherent monochromatic waves and interference. If one of the balls changes phase with respect to the other but still vibrates with the same frequency, the waves produced by the balls will be monochromatic but not coherent. A laser emits plane waves that oscillate with essentially the same frequency. When such waves are incident on two narrow slits, the same wave front triggers the circular wavelets. Therefore, the circular waves behind the slits are coherent and monochromatic (**Figure 24.10**).

When two small light sources such as lightbulbs are placed at the same separation as the two slits in Figure 24.8a, the waves emitted by each source are not monochromatic (they consist of a mixture of frequencies), and they are not coherent because they produce light waves independently of each other. Thus the waves add randomly and produce no interference pattern on the wall (Figure 24.8b).

You might wonder how Young got coherent sources in his double-slit experiment when he was using sunlight, which is not monochromatic. Because the slits were very narrow and very close, they became two sources on the same wave front. Thus they produced waves that were in phase for every wavelength of sunlight. The resulting pattern on the screen had bands of different color at different locations.

Now we can summarize.

> **Coherent monochromatic waves** Only waves of constant frequency (*monochromatic*) and having constant phase difference (*coherent*) can add to produce an interference pattern. In Young's double-slit experiment, the light from the slits is coherent because the disturbance at each slit is caused by the same passing wave front.

REVIEW QUESTION 24.2 The refractive index of Crown glass is about 1.51 for a yellow color. What does this number tell us about the speed of light in glass? What does it tell us about the frequency of the light? Its wavelength?

24.3 Gratings: an application of interference

Thomas Young's double-slit experiments not only helped establish the wave model of light but also helped explain how a device invented earlier, called a **grating**, works. A grating is a plate of glass, plastic, or metal that has many very narrow parallel grooves or slits through which light can travel. Gratings allow us to analyze the wavelengths of light emitted by biological materials, particles in Earth's atmosphere, and even stars and galaxies.

How does a grating work? When light of wavelength λ passes through two narrow closely spaced slits, we observe the interference maxima at angular deflections determined by the equation $\sin \theta_m = (m\lambda/d)$, where $m = 0, \pm 1, \pm 2, \pm 3, \ldots$. In **Figure 24.11a** (two slits), notice the central five bright spots. Inside each spot there is a gradual change of brightness from the maxima at the locations determined by the equation above to zero at the locations where $\sin \theta = [(2m + 1)\lambda/2]/d$.

If we use three slits of the same width instead of two, we again see five bright spots on the screen at the same locations. But the photo looks different (Figure 24.11a, three slits).

We see a small spot between the bright bands, and the bands are narrower and brighter. The latter is due to the fact that the total amount of light reaching the screen has increased by a factor of 1.5. If we increase to four and then to five slits, the bright bands still do not change positions (Figure 24.11a), but they become even brighter and narrower. There are more minima between the bright bands. If we increase the number of slits above 10, we will see almost no light on the screen between the bright bands. This pattern is due to the complicated interference of the coherent wavelets coming from many slits. Having such narrow bright bands allows us to distinguish more clearly between different frequencies of light passing through the slits.

A typical grating has hundreds of slits per millimeter, and the bright bands are very intense and narrow with almost complete darkness between them. Figure 24.11b shows an image of bands on a screen for a grating with 80 slits per millimeter. The locations of the bright bands are determined using Eq. (24.3). In gratings, the distance between the slits can be very small, for example, $d = 0.001$ mm. Consequently, the bright bands are much farther apart.

Thus, for a grating with spacing between adjacent slits d, the bright bands (interference maxima) occur at the angles determined from the expression

$$\sin \theta_m = \frac{m\lambda}{d}, \text{ where } m = 0, \pm 1, \pm 2, \pm 3, \ldots \quad (24.5)$$

QUANTITATIVE EXERCISE 24.3 Double-slit and grating interference

You have a slide with two narrow slits separated by 0.10 mm and a grating with 100 slits/mm. Compare the angular deflections for the $m = 1$ maximum (see figure below) using the double slit and the grating for red laser light of wavelength 670 nm. Describe the differences in the interference pattern that you observe.

Double slit ($d = 0.10$ mm)

Light wave fronts

Grating (100 slits/mm)

$\lambda = 670$ mm

Intensity

$$\text{Double slit: } \sin \theta_1 = \frac{\lambda}{d_{\text{slits}}} = \frac{670 \times 10^{-9} \text{ m}}{10^{-4} \text{ m}}$$

$$= 6.7 \times 10^{-3}$$

$$\text{Grating: } \sin \theta_1' = \frac{\lambda}{d_{\text{grating}}} = \frac{670 \times 10^{-9} \text{ m}}{10^{-5} \text{ m}}$$

$$= 6.7 \times 10^{-2}$$

The figure shows the intensity-versus-location graph. The peaks correspond to the bright regions, and the places with zero intensity correspond to completely dark regions. As expected, the bright regions are much farther apart for the grating than for the double slit. The bright regions from the grating look like sharp dots, and those from the double slit look like fuzzy bands.

- - - - - - - - - - - - - - -

Try it yourself What is the angle of deflection of the 1st order maximum for a slide with three very narrow slits illuminated by the same light source as above, each separated from each other by 0.10 mm?

Represent mathematically We need to find the angle θ_1 for both cases. Use the equation $\sin \theta_m = (m\lambda/d)$ with $m = 1$.

Solve and evaluate To find the slit separation for the grating, divide 1 by the number of slits per millimeter:

$$d_{\text{grating}} = \frac{1}{100 \text{ slits/mm}} = 10^{-2} \text{ mm} = 10^{-5} \text{ m}$$

For the double slit, $d_{\text{slits}} = 10^{-4}$ m.

Answer

Adding or removing slits without changing their separation does not change the location of the maxima. With more slits, the spacing between bright bands is the same as for the two slits in this exercise; the maxima are brighter and narrower.

White light incident on grating

What happens when white light shines on a grating? After passing through the grating, light of all wavelengths produces a bright band at the center (the $m = 0$ band); thus on a screen beyond the grating, you see a central maximum of white light. However, according to $\sin\theta_m = (m\lambda/d)$, different wavelengths of the white light are deflected at different angles for each m (other than $m = 0$). The greater the wavelength, the larger the angle from the center of the $m = 0$ white band to the mth bright band. Consequently, each $m \neq 0$ band is a rainbow of colors on the screen, with the blue color always closer to the central maximum than the red, because light of smaller wavelength has a smaller deflection angle—$\sin\theta_m = (m\lambda/d)$ (**Figure 24.12a**). If the slit separation d is smaller, the colors are separated more. The colored band, called a **spectrum**, looks similar to the spectrum produced when white light passes through a prism, but the order of colors is reversed. The prism bends blue light the most (Figure 24.12b). However, the mechanisms by which these spectra are produced are different. A spectrum produced by a prism is a result of the different speeds with which light waves of different frequencies travel in a medium. A spectrum produced by a grating is a result of the light of different wavelengths interfering constructively at different locations.

FIGURE 24.12 A white light diffraction pattern caused by a grating.

FIGURE 24.13 White light reflection from a CD.

CDs and DVDs—reflection gratings

The reflective surfaces of CDs and DVDs (**Figure 24.13**) consist of spirals of closely spaced grooves. This makes a CD a type of grating that reflects light instead of transmitting it. Since the distance between the grooves is constant, the grooves play the role of the slits: the reflected white light forms interference maxima for different colors at different angles. Looking at the reflected light, your eye intercepts only one color of the light reflected from each part of the CD. The grooves on DVDs are closer together (that is why they store more information than CDs), and that is why their color pattern is more spread out and their colors are less pronounced.

FIGURE 24.14 A spectrometer.

Spectrometer

Analyzing the wavelengths of light from different sources is an important tool in science. For example, astronomers determine which chemical elements are present in distant stars by identifying the wavelengths of light coming from those stars. Chemists measure the wavelengths of different chemical compounds to determine properties of the compounds. The tool they use is a **spectrometer** (**Figure 24.14**). Light from a source passes through a narrow slit S_1 and then through a lens L_1 (Figure 24.14a), which focuses the light into a parallel beam. This beam then passes through a grating that separates the light into its various wavelengths. The light bent at a particular angle is collected and focused by lens L_2 and slit S_2 onto a recording surface such as a film or some other light-recording device (Figure 24.14b). The detector lens and slit mechanism rotate to produce on the film the spectrum of wavelengths of light coming from the source (Figure 24.14c).

A spectrometer can be used to analyze light from any source. The Sun and incandescent lightbulbs emit a broad range of light frequencies—a continuous spectrum of light. However, hot, low-density gases do not. Instead, they emit light of specific wavelengths that indicates that specific types of atoms and molecules are present in the gas. If the light entering the spectrometer consists only of specific wavelengths, the spectrum will contain bright colored lines against a black background. For example, mercury vapor, whose spectrum is shown in Figure 24.14c, emits a pair of purple lines near 408 nm, a blue line at 436 nm, a greenish-yellow line at 546 nm, a yellowish pair of lines near 579 nm, and two dim red lines near 620 nm. We will learn more about why atoms emit specific wavelengths of light in later chapters.

Spectra also provide information about the expansion of the universe. In the spectra of light from distant galaxies, the characteristic spectral lines are shifted to longer wavelengths (lower frequencies). This so-called red shift is an example of the Doppler effect for light and indicates that distant galaxies are moving away from us—supporting the idea that the universe is expanding.

EXAMPLE 24.4   Hydrogen spectrum

You wish to determine the frequencies of light emitted by atoms in a hydrogen gas-discharge lamp. You use a spectrometer with a 4000 slits/cm grating. The light-sensitive detector of the spectrometer is 0.500 m from the grating, as shown. The image produced by the detector looks as shown in the upper part of the figure below. Determine the wavelengths of the hydrogen lines in the red and blue-green parts of the visible spectrum.

Sketch and translate The blue-green and red bands shown should be the $m = 1$ 1st order maxima for these colors. The y_1s in the figure are the distances between the central maximum and the 1st order $m = 1$ maxima. The slit separation is

$$d = \frac{1}{4000 \text{ lines/cm}} = 2.50 \times 10^{-4} \text{ cm} = 2.5 \times 10^{-6} \text{ m}$$

Simplify and diagram Assume that the slits of the grating are coherent sources of light. The path of the bright bands of light from the grating to the detector is depicted in the lower part of the figure below.

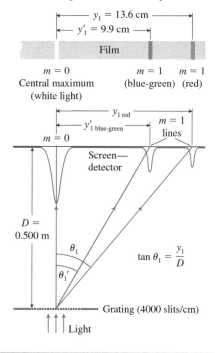

Represent mathematically Using the diagram and geometry, we get for each color:

Red: $\tan \theta_1 = \dfrac{y_1}{D}$

Blue-green: $\tan \theta_1' = \dfrac{y_1'}{D}$

Use these relations to find the corresponding angles. Then use the value for d and Eq. (24.3) $\lambda = d \sin \theta_1$ to find the wavelength of each line.

Solve and evaluate

Red: $\tan \theta_1 = \dfrac{y_1}{D} = \dfrac{0.136 \text{ m}}{0.500 \text{ m}} = 0.272$ or $\theta_1 = 15.2°$

$\lambda = d \sin \theta_1 = (2.50 \times 10^{-6} \text{ m})\sin 15.2°$

$= 6.55 \times 10^{-7} \text{ m} = 655 \text{ nm}$

Blue-green: $\tan \theta_1' = \dfrac{y_1'}{D} = \dfrac{0.099 \text{ m}}{0.500 \text{ m}}$

$= 0.198$ or $\theta_1 = 11.2°$

$\lambda' = d \sin \theta_1' = (2.50 \times 10^{-6} \text{ m})\sin 11.2°$

$= 4.86 \times 10^{-7} \text{ m} = 486 \text{ nm}$

The wavelength for red light is greater than that for blue-green, as expected.

Try it yourself Suppose a different atom emits yellow light of 580 nm. Where would it fall on the above pattern?

Answer   About 12 cm from the $m = 0$ central maximum.

REVIEW QUESTION 24.3 How do the locations of the maxima change when we add more slits of the same width to an interference device (assuming that we make them the same distance from each other)? How does the width of each maximum change?

24.4 Thin-film interference

FIGURE 24.15 The colorful peacock tail feathers are caused by thin-film interference.

The beautiful, swirling colors that occur on the surfaces of soap bubbles also appear in the thin oil films that float on water, the shimmering patterns of some butterflies, the brilliant feathers of many hummingbirds, and the peacock's elegant tail (**Figure 24.15**). This array of color is not caused by pigment but by the interference of light waves reflected off of the boundaries of a very thin surface—a film.

Bright and dark bands due to reflected monochromatic light

Figure 24.16a shows a photo of a thin soap film with white light shining on it. Let's start by analyzing the pattern produced when we irradiate the soap film with red light only (Figure 24.16b). We see a regular pattern of bright and dark bands produced by the reflected red light, a little like a double-slit interference pattern. What causes this pattern?

FIGURE 24.16 Soap bubble patterns caused by (a) white light and (b) red light. (c) Rays that interfere to cause the colors and fringes.

(a) Pattern caused by white light

(b) Pattern caused by red light

(c)

Note that we neglect refraction because the film is very thin.

We have learned that light is partially transmitted and partially reflected at interfaces between media with different refractive indexes. For a soap film and other thin films, light reflection occurs at both the front surface and the back surface (Figure 24.16c). Two factors affect the outcome of the superposition of these two reflected waves:

1. *Phase change upon reflection.* We know from our study of mechanical waves (Chapter 11) that a reflected wave changes phase by $180° = \pi$ rad if the second medium is denser than the first. For light, this would happen if the second medium had a greater refractive index than the first. However, the phase of the reflected wave remains the same if the second medium has a smaller refractive index than the first. Thus, when light traveling in air reflects off the front surface of the soap bubble film of refractive index 1.3, there is a phase change of $180° = \pi$ rad (**Figure 24.17a**). When light traveling inside the soap bubble film reflects off the back surface, there is no phase change (Figure 24.17b) since the refractive index of air is less than 1.3. So only the light reflecting off of the front of the bubble undergoes a phase change.

2. *Phase difference due to path length difference.* In Figure 24.17c we see that wave 2, which reflects from the back surface of the thin film, travels a longer distance than wave 1, which reflects from the front surface. Because of this path length difference, the two waves have different phases when they recombine after reflecting from the thin film. For example, if the film is one half wavelength thick, wave 2 travels a total of one wavelength farther than wave 1. Then the two waves should be in phase when they recombine (assuming no phase changes occur due to reflection). If the film is one-fourth of a wavelength thick, the path length difference is one-half of the wavelength. The waves are $180° = \pi$ rad out of phase when they recombine. Often those waves are said to be one half wavelength out of phase to underscore that when they add, the resulting wave has zero amplitude.

There is an additional important feature of this path length difference. The speed v of light in a medium of refractive index n is $v = (c/n)$, where c is the speed of light in air. If the refractive index n of the thin film is greater than 1.0, then the wavelength of the light in the film is

$$\lambda_{medium} = \frac{v_{medium}}{f} = \frac{c/n_{medium}}{f} = \frac{c/f}{n_{medium}} = \frac{\lambda_{air}}{n_{medium}}$$

To determine the phase difference between the waves, we have to use the light's wavelength in the film (λ_{medium}). To determine the total phase difference, we need to account for both factors (1) and (2). Let's apply this reasoning to soap bubbles.

Soap bubble The film of the soap bubble has a greater refractive index than air. Thus, light reflected from outside the air-soap bubble interface has a $180° = \pi$ rad phase change (equivalent to the one half wavelength path difference). Part of the transmitted light reflects from inside the soap bubble-air interface and has no phase change. Hence, the relative phase change from reflection between the two waves is $180° = \pi$ rad (equivalent to one half wavelength in terms of path length difference). For a *bright band* to appear on the surface of the bubble (constructive interference), the net phase change of light coming from that direction should be an integer multiple of $360° = 2\pi$ rad. Thus, the path length difference due to the light traveling back and forth through the film a distance $2t$ must also be one half wavelength, or three half wavelengths, or five half wavelengths, etc. so that when added together, those path length differences produce an integer number of whole wavelengths. Thus, $2t = (1, 3, 5, \ldots)(\lambda/n)/2$, where λ/n is the wavelength of light inside the bubble film. We've assumed through all this that the incident light is almost perpendicular to the film surface.

For a *dark band* to appear on the surface of the bubble, destructive interference must occur between the two reflected waves. The net phase change needs to be an odd multiple of $180° = \pi$ rad, or an odd number of half wavelengths in terms of path lengths. Thus, the path length difference $2t$ must be an integer number of wavelengths: 0, 1, 2, ... wavelengths. In short, for a dark band, $2t = (0, 1, 2, \ldots)(\lambda/n)$.

Thus for a specific wavelength of light, the thickness of the bubble film determines if light will be reflected from the film. Since the film thickness varies over the bubble, some locations look shiny (constructive interference) and other locations look dark (destructive interference). We can test this reasoning experimentally. Examine the chapter-opening photo showing a soap film across a frame just before it breaks. Notice the black area at the top of the film. The black color means that no light comes from this part of the film into our eyes. Could we have predicted this phenomenon using our knowledge of thin films?

As soon as the frame is lifted out of the soap solution, the soap film across the frame starts thinning due to evaporation and dripping from the bottom. Because of gravitational effects, the film is always thinnest at the top. Just before it breaks, the thickness at the top is nearly zero (t is almost zero). When we look at the film at an almost zero degree angle with the normal line, the waves of all wavelengths (colors) reflected off the front and back of the film are out of phase because of reflection (the phase changes

FIGURE 24.17 Light incident on a film at almost zero angle with the normal. (a) and (b) Phase changes due to reflection. (c) Phase difference due to path length difference.

(a) Reflection off surface with increasing n (like air-soap interface)

(b) Reflection off surface with decreasing n (like soap-air interface)

(c)

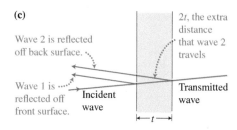

TIP When light reflected off a soap bubble interferes constructively, incident light is reflected. Soap bubbles are transparent, which means that most of the incident light passes through, even when constructive interference occurs.

by 180° during the reflection off the front of the film and does not change during the reflection off the back). When two waves exactly out of phase meet, they cancel each other out. We therefore predict that the top of the film should appear dark just before it breaks. And this is exactly what we observe in the photo!

Now that we are more confident in our reasoning about thin films, let's use it to develop a method for reducing the glare from glass surfaces, such as windows or camera lenses.

Thin film on glass surface In an optical instrument with several lenses, light reflection at each air-glass interface reduces the amount of light getting to the detector, whether that be the light detector in a digital camera or your eye looking through a microscope. To reduce this effect and increase the amount of light reaching the detector, the glass surfaces are often covered with a thin film. Waves reflecting from the film interfere destructively, minimizing reflected light. All incident light passes through.

Consider light incident almost perpendicularly on a thin film with refractive index $n_{film} = 1.4$ on the front surface of a glass lens whose refractive index is $n_{glass} = 1.6$ (**Figure 24.18**). Wave 1 undergoes a $180° = \pi$ rad phase change upon reflection, since $n_{film} > n_{air}$. Wave 2 also undergoes a $180° = \pi$ rad phase change, since $n_{glass} > n_{film}$. The net effect is that the two waves remain in phase. To minimize the amount of reflected light, we need to create destructive interference between these two waves. Thus, the path length difference must equal an odd integer multiple of $180° = \pi$ rad (an odd integer number of half wavelengths of the light in terms of path length difference). Choosing the proper thickness of the film t, so that $2t = (1, 3, 5, \ldots)[(\lambda/n)/2]$, where n is the refractive index of the film, achieves this goal. If for some reason we wanted to maximize the reflected light, then the two waves must constructively interfere. This will happen when $2t = (0, 2, 4, \ldots)[(\lambda/n)/2]$.

Table 24.3 summarizes the ways in which the refractive index of a thin film and its thickness combine to produce the bright and dark bands of reflected light.

FIGURE 24.18 Light reflecting from a thin film on glass surface. Light passing through the glass is not shown.

$\lambda/2$ phase change t

Wave 2

Wave 1

$\lambda/2$ phase change

Air Film Glass
$n_1 = 1.0 < n_2 < n_3$

TABLE 24.3 Examples of thin-film interference for monochromatic incident light

Type of thin film	Changes in path length difference	Total path length difference and outcome
Soap bubble in air or oil film on water 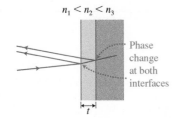 $n_1 < n_2 > n_3$ No phase change · Phase change t	**Due to reflection** $\Delta_1 = (\lambda/2)$ for reflection on front surface of film and $\Delta_2 = 0$ for reflection on back surface of film. Net reflection path length difference is $$\Delta_{ref} = \frac{\lambda}{2}$$ **Due to thickness** $$\Delta_t = 2t$$	$\Delta = \Delta_{ref} + \Delta_t = (\lambda/2) + 2t$ • *Bright band* if twice the thickness is an odd multiple of half wavelengths: $$2t = m\frac{\lambda/n}{2} \text{ for } m = 1, 3, 5 \ldots$$ • *Dark band* if twice the thickness is an even multiple of half wavelengths: $$2t = m\frac{\lambda/n}{2} \text{ for } m = 0, 2, 4 \ldots$$
Thin-film coating on lens, window, windshield, or computer screen $n_1 < n_2 < n_3$ Phase change at both interfaces t	**Due to reflection** $\Delta_1 = (\lambda/2)$ for reflection on front surface of film; $\Delta_2 = (\lambda/2)$ for reflection on back surface of film. Net reflection path length difference is $\Delta_{ref} = \lambda$. **Due to thickness** $$\Delta_t = 2t$$	$\Delta = \Delta_{ref} + \Delta_t = \lambda + 2t$ • *Bright band* if twice the thickness is an even multiple of half wavelengths: $$2t = m\frac{\lambda/n}{2} \text{ for } m = 0, 2, 4, \ldots$$ • *Dark band* if twice the thickness is an odd multiple of half wavelengths: $$2t = m\frac{\lambda/n}{2} \text{ for } m = 1, 3, 5 \ldots$$

Reflection patterns on a soap bubble in white light

So far, we have assumed that the surface of a bubble or a camera lens film is illuminated by monochromatic light (light of one wavelength). However, usually it is white light that is incident on the thin film. Consider the soap film in **Figure 24.19**. The white light incident on the film includes light wavelengths from about 400 nm to 700 nm. When the light of different wavelengths reflects and refracts, for one wavelength the net path difference might be equal to an odd number of half wavelengths. For another wavelength it could be exactly an even number of half wavelengths. For most wavelengths, however, the interference is neither completely constructive nor completely destructive. Also, when we look at a different spot on the film, the waves arriving at our eyes are incident on the film at different angles, and the film has different thicknesses at different locations.

Because of all these factors, light of only a small wavelength range is destructively reduced in intensity at a particular location on the bubble when viewed in a particular direction. The color that is reduced differs from place to place. At other locations, other colors are destructively reduced in intensity. As a result, each observer will see a different distribution of colors reflected from each location on the film. The colors that are left when light of a small wavelength range is subtracted from the white light are called **complementary colors**. Complementary colors are different from the spectrum of rainbow colors produced by a grating or a prism. These devices separate in space the primary colors that are combined spatially inside a beam of white light. The video shows the change of colors of a soap bubble as its thickness changes with time. Watch the top of the bubble become black just before the bubble bursts.

For example, if blue light is subtracted from white light, we see the remaining light as its complement, yellow. When red is subtracted from white light, we see cyan. Long-wavelength light (red) needs a thicker bubble wall for its reflections to destructively interfere than short-wavelength light (blue). This means that as the bubble's thickness decreases, first red is canceled out, leaving blue-green; then green is canceled, leaving magenta; and finally blue is canceled, leaving yellow. Eventually, the bubble becomes so thin that all wavelengths are canceled and the bubble appears black, just as the film in the testing experiment became black at the top just before it broke.

FIGURE 24.19 A soap bubble in white light.

Colors are the result of the absence of one or more colors from white light.

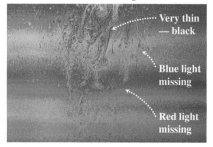

- Very thin — black
- Blue light missing
- Red light missing

VIDEO 24.1

TIP Remember that it is the frequency of light that determines its color. The frequency does not change when light travels from one medium to another. The wavelength does change. Thus when you hear "the wavelength of green light is 550 nm," check whether this statement assumes that the light is propagating in a vacuum.

Lens coatings

As we have learned, we can reduce the reflected light of a particular wavelength if we have a film of a particular thickness. The thickness of the coating on glass lenses for cameras, microscopes, and eyeglasses is usually chosen to reduce reflection of light at a wavelength of 550 nm, the center of the visible spectrum. The film reduces the reflected light from about 4% of the intensity of the incident light (without the coating) to less than 1%. The coating is less effective at reducing the reflection at extreme visible wavelengths (red at 700 nm and violet at 400 nm). A lens with a thin-film coating has a purple hue because it reflects red and violet light more than other colors; when combined, these colors appear purple.

QUANTITATIVE EXERCISE 24.5 How thin is the film?

In the optical industry, a thin film of magnesium fluoride (MgF_2) with a refractive index of 1.38 is used to coat a glass lens of refractive index 1.50. What should be the thickness of the coating to eliminate the reflection of 550-nm light (the wavelength of green light in a vacuum)?

Represent mathematically The purpose of the coating (a thin film) is to produce destructive interference for the reflected light. The refractive index increases at the air-film interface and at the film-glass interface. Thus there is a net zero phase change due to reflection. For destructive interference, the path difference (twice the thickness t)

should be $2t = (\lambda_{air}/2n)$, $3(\lambda_{air}/2n)$, $5(\lambda_{air}/2n)$, and so forth, where n is the refractive index of the film.

Solve and evaluate The thinnest possible film would be when $m = 1$: $2t = (\lambda_{air}/2n)$. Thus $t = (1/4)(\lambda_{air}/n) = (1/4)(550 \text{ nm}/1.38) = 100 \text{ nm}$. This is very thin. Making the film slightly thicker might still achieve the purpose of canceling the reflected light and provide greater durability for the film. Possible thicknesses are 300 nm and 500 nm—any positive odd integer times the minimal thickness.

(CONTINUED)

Try it yourself How thick should the film in this exercise be to increase the amount of reflected light?

Answer

We could use a 200-nm thin film of the material used in this exercise. Alternatively, we could cover a glass surface with a thin film of greater refractive index than the glass, achieving greater light reflection with the same 100-nm film. Increasing reflection is useful when building devices that allow us to look at bright light sources, such as sun visors for astronauts.

Limitations of the thin-film interference model

Window glass does not display color patterns. Why is that? Window glass is several millimeters thick, several thousand times the wavelength of light, like a *very* thick walled bubble. Natural sources of light such as the Sun do not emit light waves continuously. Instead, they emit light in short wave bursts over a very short time interval. Two consecutive bursts have a random phase difference between them. Imagine that light from one burst travels through the window glass and reflects back. While it was traveling inside the glass, a new burst reached the surface of the glass. As a result, the phase difference between the two waves is determined not only by reflections and path length differences but also by the time between bursts. The bursts are therefore not coherent and do not interfere in a way that remains constant in time. However, this limitation does not exist for laser light, which is much more coherent than natural light. You can observe interference on window glass using a laser.

Bird and butterfly colors

FIGURE 24.20 A Morpho butterfly.

Many colors in the natural world, such as those of flower petals and leaves, are caused by organic pigments that absorb certain colors and reflect others. Chlorophyll in leaves absorbs most colors but reflects green, which makes the leaves look green. However, some natural color is the result of the thin-film interference of light. Some feathers and insect bodies consist of microscopic translucent structures that act like thin films to produce destructive and constructive interference of light. We can see the colors that result from these structures in the tail feathers of peacocks, the head feathers of hummingbirds, and the wings of Morpho butterflies (**Figure 24.20**).

> TIP Remember that interference patterns for light that are stable in time and thus observable appear when two coherent waves arrive at the same location at the same time. These coherent waves can be created in two ways:
> 1. by combining waves from two sources that produce coherent, monochromatic waves (as in double-slit interference), and
> 2. by dividing a monochromatic wave into two waves and combining them (as in thin-film interference).

REVIEW QUESTION 24.4 If we look through a grating at a source of white light and then look at a thin film also illuminated by white light, we see different colors. What is the same and what is different about how these colors appear?

FIGURE 24.21 Laser light reaching a screen after passing through two slits.

Intensity modulated

Double-slit interference bright bands

24.5 Diffraction of light

Earlier in the chapter we analyzed interference phenomena involving light that passed through two narrow slits. We modeled those slits as infinitely narrow, and when light shined on them we considered them to be point-like sources of light. This model explained the observed interference pattern on the screen (**Figure 24.21**). However,

FIGURE 24.22 The single-slit diffraction pattern widens as the slit width narrows.

if you look carefully at the pattern produced by light passing through two slits on the screen, you will notice that in addition to the alternating bright and dark bands, there is also an overall periodic modulation of the brightness in the pattern. The wider the slits (not their separation, but their individual widths), the more pronounced this modulation is. In addition to the interference of light coming from the two slits, something else is going on that is related to the individual slits themselves. Let us investigate this phenomenon. We start with a series of experiments in which we shine laser light through one slit. The width of the slit is adjustable so that we can change it during the experiments. **Figure 24.22a** shows the setup, and Figures 24.22b–d show the outcomes of the experiments.

We observe in all three experiments that light reaching the screen from a relatively narrow slit makes a wide interference pattern on the screen, including dark and bright bands at the sides of the central wide band. The width of the central maximum increases as the width of the slit decreases. When the slit is wider, the bright band in the center is almost equal to the width of the slit (Figure 24.22b). When the slit is narrower, the band is much wider (Figures 24.22c and d). This spreading of light combined with the additional bright and dark regions is called **diffraction**. It becomes noticeable when the slit's width is roughly 1000 times the wavelength of the light or less. It is possible that this effect might be caused by the interference of wavelets produced by different mini-slit regions within the slit—an idea we will now evaluate quantitatively.

Quantitative analysis of single-slit diffraction

When we analyzed the interference pattern in a double-slit experiment, we modeled each slit as a point-like source of secondary wavelets on the same wave front using Huygens' principle. In the single-slit situation, the slit is not infinitely narrow. It is somewhat wider and produces an interference pattern like that shown in **Figure 24.23a**. If we see an interference pattern, then several waves might be adding to produce it. Where do those waves come from if we only have one slit? As the slit has a nonzero width, we can model it as consisting of multiple imaginary tiny mini-slit regions that become sources of the secondary waves on the same wave front when light shines on the slit. Because they are on the same wave front, light waves emitted by all the mini-slits must all be in phase at the slit location. If this is the case, light emitted by those different mini-slits can interfere when reaching the screen because they travel different distances to reach the same spot on the screen.

Imagine that the screen is very far away. In this case, the waves emitted by different mini-slits are traveling nearly parallel to each other when they reach a specific location

FIGURE 24.23 Single-slit diffraction. (a) Alternating bright and dark bands observed coming from light passing through a single slit. (b) Light from different parts of the slit cancels, causing the first dark band at y_1.

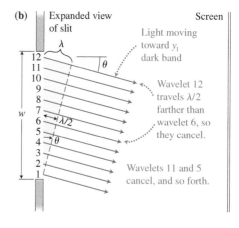

on the screen. Thus, to explain the existence of a bright or a dark spot on the screen, we only need to consider the phase difference of the parallel waves traveling from the slit to the location on the screen. Consider two wavelets coming from mini-slits to the first dark band below the central bright band (position y_1 in Figure 24.23a). One wavelet comes from the top mini-slit of the top half of the slit (mini-slit 12 in Figure 24.23b), and the second wavelet comes from the top mini-slit of the bottom half of the slit (mini-slit 6 in Figure 24.23b). The distance between these mini-slits is $(w/2)$, where w is the slit width.

If the path length difference from these two sources to the screen equals one half wavelength, then these two wavelets cancel each other, resulting in zero light at that location on the screen. If we repeat the procedure for the next pair of wavelets (11 and 5), which travel from the second mini-slit region in the top half of the slit and the second mini-slit region in the bottom half of the slit, we get the same result. The same will happen for all other pairs of waves chosen in this manner. We get total cancellation at that point on the screen—a dark band.

We see from the shaded triangle in Figure 24.23b that this first interference minimum occurs when $(w/2) \sin \theta = (\lambda/2)$, or when $w \sin \theta = \lambda$. Note that the same reasoning applies for the first dark band above the central maximum.

We designate the dark bands above the central maximum with positive signs and the dark bands below with negative signs. Thus, the first dark bands on both sides of the central bright spot occur at

$$w \sin \theta_{\pm 1} = \pm \lambda$$

We can next divide the slit into four mini-slit segments and draw lines from them to the second dark band on either side of the central maximum. We find that the angular deflection to this second dark band occurs when $(w/4)\sin \theta_2 = (\lambda/2)$, or when

$$w \sin \theta_{\pm 2} = \pm 2\lambda$$

We can repeat this process by dividing the slit into six mini-slit segments with lines drawn from them to find the third dark band on either side of the central maximum. These dark bands occur when $w \sin \theta_{\pm 3} = \pm 3\lambda$.

We can summarize this analysis as follows. The angular positions of the dark bands relative to the central maximum are determined by

$$w \sin \theta_m = m\lambda, \text{ where } m = \pm 1, \pm 2, \pm 3 \ldots \qquad (24.6)$$

The first dark bands on both sides of the central maximum appear at the angles $\theta_{\pm 1}$ when $m = \pm 1$ is substituted into the above equation. All the space between these angles belongs to the central bright fringe. Therefore, we do not use $m = 0$ in the above equation.

Equation (24.6), $w \sin \theta_m = m\lambda$, was derived using a wave model of light and can be tested by performing an experiment using a shorter wavelength green laser with the same slit. The angular deflection to the first dark band on each side of the central maximum should be less, since the green laser produces light of shorter wavelength than the red laser. When you perform the experiment, you observe that the first dark bands are closer to the central bright band—the central bright band is narrower.

We can now describe the variation in the brightness of the double-slit interference pattern. Real slits have a finite width, which we did not account for in our original analysis. Thus, in addition to the two-slit interference pattern with the closely spaced alternating bright and dark bands, these bands are modulated in intensity by the additional single-slit diffraction due to the interference of light coming from different mini-slits within each slit. This single-slit pattern combines with the double-slit pattern, resulting in a variation in brightness of the double-slit maxima.

TIP ▶ Notice that $w \sin \theta_m = m\lambda$ is similar to Eq. (24.1) for double-slit interference. However, for a double slit, Eq. (24.1) describes the angles at which the maxima (bright bands) are observed, and $m = 0$ is allowed. For a single slit, Eq. (24.6), $w \sin \theta_m = m\lambda$, describes angles at which the minima (dark bands) are observed, and $m = 0$ is not allowed since there is no dark band at $\theta = 0°$.

Single-slit diffraction When monochromatic light is incident on a slit whose width is approximately 1000 wavelengths of light or less, we observe a series of bright and dark bands of light on a screen beyond the slit. The bands are caused by the interference of light from different mini-slit regions within the slit. The angle between lines drawn from the slit to the minima, the dark bands, and a line drawn from the slit to the central maximum is determined using the equation

$$w \sin \theta_m = m\lambda \qquad (24.6)$$

where w is the slit width, λ is the wavelength of the light, and $m = \pm 1, \pm 2, \pm 3, \ldots$ (not zero). The dark bands on the screen are located at positions $y_m = L \tan \theta_m$ relative to the center of the pattern. L is the slit-screen distance.

The Poisson spot—testing the wave model of light

The year 1818 was a decisive one for the acceptance of the wave model of light. Simeon-Denis Poisson, a talented mathematician and proponent of the particle model, set out to disprove the wave model of light. He suggested shining a narrow beam of light at a small round obstacle. Though Poisson did not believe in the wave model explanation, he still used it to predict the outcome of this testing experiment. His reasoning was as follows: according to the wave model, when the obstacle is just smaller than the beam, the light should illuminate the edges of it. The edges could then be considered the sources of in-phase secondary wavelets. If so, a bright spot should also appear in the middle of the obstacle's shadow because the center of the shadow is equidistant from each secondary wavelet source. This phenomenon is similar to why a bright fringe appears at the center of a double-slit interference pattern (it is equidistant from both slits). This prediction is in direct contradiction to the particle model, which does not predict such a spot. Common sense also tells us that a bright spot should not appear at the center of the dark shadow. The French Academy of Sciences conducted the experiment, and much to their (and Poisson's) surprise, they saw a bright spot right in the middle of the shadow, consistent with the wave model.

We repeated Poisson's experiment using a small ball bearing as the obstacle (**Figure 24.24a**). We expanded the laser light to make sure that the beam was wider than the bearing (Figure 24.24b). You see the resulting pattern on the screen in Figure 24.24c. Notice the tiny bright spot at the center and fringes on the outside.

Diffraction and everyday experience

Since the wavelengths of visible light are so small, we seldom observe light diffraction in everyday life. However, diffraction is a phenomenon that all waves exhibit, including sound waves. In Eq. (24.6), $w \sin \theta_m = m\lambda$, if w is nearly equal to the wavelength of the waves, the angular position of the $m = \pm 1$ minima could reach almost 90°, in which case the waves would spread into the entire region beyond the slit with no regions of destructive interference. This full-screen diffraction seldom occurs with light because slits are seldom that small. But the wavelengths of sound waves are much larger. The wavelength of concert A (frequency 440 Hz) is $\lambda = v_{sound}/f = (340 \text{ m/s})/440 \text{ Hz} = 0.77 \text{ m}$, about the width of a doorway. Thus, a 90° central maximum will fill the entire region beyond the door. Diffraction explains why we can hear a person talking around a corner in another room when the door is open.

FIGURE 24.24 Poisson's spot tested the wave theory of light.

(a)

(b) Side view of Poisson's experimental design (the distance between the wave fronts is greatly exaggerated)

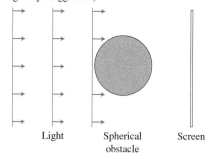

Light   Spherical   Screen
                obstacle

(c)

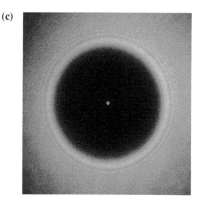

QUANTITATIVE EXERCISE 24.6 **Light and sound passing through a small slit in a door**

Sound of wavelength 0.85 m (approximately the wavelength of 400-Hz sound) and light of wavelength 400 nm (blue color) are incident on a 1.0-cm slit in a door. Will light and sound diffract significantly after passing through the slit?

Represent mathematically To answer the question we need to determine the angular size and the width of the central maximum for each type of wave. The angular deflection $\theta_{\pm 1}$ to the first dark band on either side of the central maximum is determined using the equation $w \sin \theta_{\pm 1} = \pm \lambda$.

Solve and evaluate Rearranging this equation and applying it to light, we get

$$\sin \theta = \frac{\lambda}{w} = \frac{400 \times 10^{-9} \text{ m}}{1.0 \times 10^{-2} \text{ m}} = 4.0 \times 10^{-5}$$

Taking the inverse sine of both sides,

$$\theta = \sin^{-1}(4.0 \times 10^{-5}) = 0.0023°$$

The first minima appear at 0.0023° from either side of the central bright band. The diffraction is so small that it will not be noticed. You will see a sharp image of the slit opening and no interference.

For sound, $\sin \theta = \dfrac{\lambda}{w} = \dfrac{0.85 \text{ m}}{1.0 \times 10^{-2} \text{ m}} = 85$. This impossible result ($\sin \theta \leq 1$) means that the central "bright" region for sound is spread over all angles on the other side of the opening and there will be no regions of destructive interference beyond the door. There will be almost no variation in the amplitude of sound waves behind the door due to diffraction. This phenomenon occurs whenever the wavelength is greater than the width of the opening, when $\lambda / w > 1$.

Try it yourself White light is incident on a slit that is 0.10 mm wide. A screen is 3.0 m from the slit. What is the width of the red bright band adjacent to the central white band compared to the width of the corresponding blue band? The width on the screen of the first colored bright band is the distance between the $m = 1$ and $m = 2$ dark bands. Consider the wavelength of blue light to be 480 nm and the wavelength of red light to be 680 nm.

Answer The angular positions for the first and second blue dark bands on one side of the central bright band are $\theta_1 = 0.28°$ and $\theta_2 = 0.55°$, and the distance on the screen between them is 14 mm. For red light $\theta_1 = 0.39°$ and $\theta_2 = 0.78°$, and they are separated by 20 mm.

REVIEW QUESTION 24.5 Equation (24.6), $w \sin \theta_m = m\lambda$, where $m = \pm 1, \pm 2, \pm 3 \ldots$ describes the angles at which one can see dark fringes produced by light passing through a single slit. Write a similar equation describing the angles at which one can see dark fringes produced by monochromatic light passing through two narrow slits.

24.6 Resolving power

The wave-like behavior of light limits our ability to see two distant closely spaced objects as separate objects or to discern the details of an individual distant object. Can we see the fine structure of a cell with a microscope? Can a reconnaissance aircraft take photographs of a missile site with sufficient detail? This section will help us answer such questions.

Consider light passing through the small circular hole shown in **Figure 24.25**. Light leaving the hole produces a diffraction pattern on a screen beyond the hole. The pattern resembles that formed by light passing through a narrow slit, except that alternating rings of bright light and darkness are formed rather than parallel bands of light and darkness. We can calculate the angle θ between a line drawn from the hole to the center of the pattern and a line drawn toward the first dark ring using the following equation (which we will not derive):

$$\sin \theta = \frac{1.22\lambda}{D} \tag{24.7}$$

FIGURE 24.25 The diffraction pattern due to light passing through a small hole.

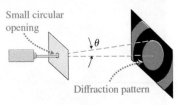

Small circular opening

Diffraction pattern

where λ is the wavelength of the wave passing through the hole and D is the diameter of the hole. The shiny disk at the center of the pattern is called the Airy disk, named after George Airy (1801–1892), who first derived Eq. (24.7) in 1831.

The equation is similar to Eq. (24.6), used to calculate the angular position of the dark bands in a single-slit diffraction pattern. The factor 1.22 that appears in Eq. (24.7) and not in Eq. (24.6) is a result of the circular geometry of the opening. The intensities of the bright secondary rings around the central disk are much lower than that of the central bright spot.

Resolving ability of a lens

How does diffraction affect the ability of a lens to produce a sharp image? Imagine that light from a very distant object enters a lens of diameter D as plane waves and diffracts as it passes through the opening (**Figure 24.26**). The diffraction produces an angular spread given by Eq. (24.7): $\sin \theta = 1.22\lambda/D$. Thus, instead of passing through a focal *point* a distance f from the lens, parallel light rays form a central *disk* of radius $R = f \tan \theta$ at the focal point. The radius of that central diffraction pattern depends on the diameter of the lens, the focal distance of the lens, and the wavelength of the light. The wave-like properties of light make it *impossible* to form a perfectly sharp image.

This result is important for surveillance, astronomical studies, microscopy, and human vision. When looking at two closely spaced objects, we see central disks whose radii depend on the diameter of the aperture the light passes through (the pupil of an eye, the objective lens of a telescope, etc.). If the central disks of the diffraction patterns from the two objects overlap, we cannot perceive them as two distinct objects. Physicists say that we cannot **resolve** them. In **Figure 24.27a**, we see a photograph of stars as seen through a telescope with a small-diameter objective lens. How many stars are in the field of vision? It is difficult to say by looking at Figure 24.27a. Figures 24.27b and c show the same stars photographed with telescopes with larger lenses that make it easier to resolve the individual stars.

What is the smallest separation of two images that can still be perceived as distinct? Consider **Figure 24.28a**, which shows the diffraction patterns of distant point-like objects produced on a film by light of wavelength λ passing through a lens of diameter D. In (b), the objects are far enough apart (large enough angular separation) that diffraction does not prevent their images from being resolved. In (c), the images can just barely be resolved. In (d), the images cannot be resolved.

FIGURE 24.26 A lens focusing light from a distant point source produces an Airy disk.

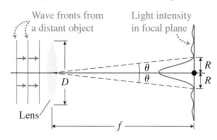

FIGURE 24.27 Telescope lenses of increasing diameter resolve two stars.

(a) Not resolved

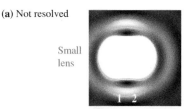

Small lens

(b) Barely resolved

Medium lens

(c) Well resolved

Large lens

FIGURE 24.28 The angular separation of two sources affects the ability to resolve them.

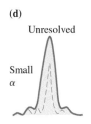

According to Lord Rayleigh (1842–1919), two objects can just barely be resolved when their angular separation α_{res} (called the **limit of resolution**) satisfies the following criterion:

Rayleigh criterion The minimal angular separation of two objects that can be resolved (perceived as separate) is the limit of resolution α_{res} of the instrument:

$$\alpha_{res} = \frac{1.22\lambda}{D} \qquad (24.8)$$

where D is the diameter of the opening through which light enters and λ is the wavelength of the light. In this equation, α_{res} is measured in radians.

TIP ▷ When the limit of resolution of the instrument increases, it means that its resolving power decreases. The larger α_{res}, the more difficult it is to resolve closely separated objects.

Equation (24.8) helps us understand the role of the opening diameter D of any instrument. Increasing the size of the opening of any instrument (such as a telescope) allows you to resolve more closely positioned objects or details on a single object.

The Rayleigh criterion is a factor in building various optical devices. Telescopes are built with objective lenses that are as large as possible. Large lenses are difficult to make; they break easily and also sag under their own weight. Thus, high-resolution telescopes use mirrors instead. Another limitation in producing sharp images of stars is the movement of the atmosphere, which results in images that continuously distort and shimmer. This is why observatories are built on mountains, where they are above a significant amount of the atmosphere.

Resolving the fine details of really small objects, such as parts of biological cells, is also challenging. We can improve the resolving power by decreasing the wavelength of the light. For example, we can use UV light instead of visible light. Much larger resolving power can be achieved with an electron microscope. Electron microscopes, discussed in Chapter 28, resolve images using the wavelengths of electrons, which are much smaller than those of visible light.

REVIEW QUESTION 24.6 Stars are so far away that they can be considered as point sources. Why then do the images of stars produced by a telescope and recorded on a film look like bright disks, not dots?

24.7 Skills for applying the wave model of light

In this section we will practice problem-solving skills while investigating some new phenomena. A problem-solving strategy is outlined on the left side of the table below and illustrated in Example 24.7.

PROBLEM-SOLVING STRATEGY 24.1   Analyzing processes using the wave model of light

EXAMPLE 24.7 Purchase a grating

You are to purchase a grating that will cause the deflection of the $m = 2$ 2nd order bright band for 650-nm wavelength red light to be 42°. How many lines per centimeter should your grating have? How many bright bands in total will you see on the screen?

Sketch and translate

- Visualize the situation and then sketch it.
- Identify given physical quantities and unknowns.

Simplify and diagram

- Decide if the sources in the problem produce coherent waves.
- Decide if the small-angle approximation is valid.
- Decide if the slit widths for multiple slits are wide enough that you have to consider single-slit diffraction as well as multiple-slit interference.
- If useful, represent the situation with a wave front diagram showing the overlapping crests and troughs of the light waves from different sources.

- Grating slits can be considered coherent light sources.
- With gratings, the angular deflection is often 10° or more, so we cannot use the small-angle approximation.
- Because the slits in gratings are very narrow compared to the distance between them, we can consider multiple-slit interference only and ignore single-slit diffraction.
- The maximum number of bright bands is determined by the maximum path length difference that the waves from two adjacent slits can have, which is equal to the distance d between the slits (see the figure at right). Thus n is equal to the integer part of d/λ.

Represent mathematically

- Describe the situation mathematically.
- Use geometry if needed.

- Apply Eq. (24.5) to the second bright band to determine the slit separation: $\sin \theta_m = m(\lambda/d)$.
- Find the number of slits/cm: $\# = \dfrac{1\ \text{cm}}{d}$
- Find the number n of bright bands on each side by taking the integer part of $\dfrac{d}{\lambda}$. We then need to add one more for the central maximum; thus the total number of bands is $2n + 1$.

Solve and evaluate

- Use the mathematical description of the process to solve for the desired unknown quantity.
- Evaluate the result. Does it have the correct units? Is its magnitude reasonable? Do the limiting cases make sense?

- From Eq. (24.5) with $m = 2$, we get

$$d = m\,\frac{\lambda}{\sin\theta_m} = 2\,\frac{650 \times 10^{-9}\ \text{m}}{\sin 42°} = 1.9 \times 10^{-6}\ \text{m} = 1.9 \times 10^{-4}\ \text{cm}$$

- The number of slits in one centimeter is

$$\# = \frac{1\ \text{cm}}{d} = \frac{1\ \text{cm}}{1.9 \times 10^{-4}\ \text{cm}} = 5100$$

The units match. This is a reasonable number of slits. A limiting case analysis is not needed, as we did not derive any new expressions.

- The number of bright bands on each side is

$$n = \frac{d}{\lambda} = \frac{1.9 \times 10^{-6}\ \text{m}}{650 \times 10^{-9}\ \text{m}} = 2.93$$

- We cannot round up to three because the 3rd band may not be fully visible, so we conclude that there are two bands on each side. Adding the central maximum, the total number of bands seen is five. We can check this number by using another method: the maximum angle of deflection is 90°, thus $d \sin 90° = n\lambda \Rightarrow d = n\lambda$. We arrived at the same equation. In addition, five bands is a reasonable number; if we got 50, it would be a warning that we had done something wrong in the calculation.

Try it yourself Suppose you used a spectrometer using this grating to analyze light coming from a star and found an $m = 2$ band of light at 34°. What is the wavelength of this light?

Answer   530 × 10⁻⁹ m = 530 nm.

EXAMPLE 24.8   **LED light passing through a slit**

Three LEDs (red, green, and blue) are mounted on a vertical plate, one above another. You observe the LEDs through a narrow vertical slit that is 2.50 m away from the LEDs. The photo below shows a magnified image of what you see (in reality, the pattern is small but clearly visible). Knowing that the red LED mostly emits light with a wavelength of about 630 nm, estimate the wavelength of light emitted by the green and the blue LEDs.

Sketch and translate We draw a sketch of the situation below and label the knowns and unknowns.

Simplify and diagram Because the LEDs emit light of clearly distinguishable colors and are far away from the slit, we can assume the light is coherent. Since the pattern is very small, we can also assume that the angular deflections of the minima are small. In this case, the widths of the bright central bands on the photo are directly proportional to the deflection angles to the first minima for each LED.

Represent mathematically Using our last assumption, we can write

$$y_{red} = l\theta_{red} = l\frac{\lambda_{red}}{w}$$

$$y_{green} = l\theta_{green} = l\frac{\lambda_{green}}{w}$$

$$y_{blue} = l\theta_{blue} = l\frac{\lambda_{blue}}{w}$$

where l is distance between the observer and the slit. Dividing the last two equations by the first one (notice that l does not enter the final equation), we get

$$\frac{y_{green}}{y_{red}} = \frac{\lambda_{green}}{\lambda_{red}}; \quad \frac{y_{blue}}{y_{red}} = \frac{\lambda_{blue}}{\lambda_{red}}$$

Solve and evaluate Using the white grid overlaid on the photo, we estimate

$$\frac{y_{green}}{y_{red}} = \frac{6}{7} \text{ and } \frac{y_{blue}}{y_{red}} = \frac{5}{7}$$

and find $\lambda_{green} = \frac{6}{7}\lambda_{red} = \frac{6}{7}(630 \text{ nm}) = 540 \text{ nm}$ and $\lambda_{blue} = \frac{5}{7}\lambda_{red} = \frac{5}{7}(630 \text{ nm}) = 450 \text{ nm}$. These answers seem reasonable—both wavelengths are in the regions of the spectrum where these colors belong.

A final note about light

In this chapter we applied the wave model of light to explain situations in which light passes through small openings or around small objects. The particle model of light cannot explain them. We found that the effects of light's wave-like behavior are most dramatic when the sizes of the holes and obstacles are comparable to the wavelength of the light. The wave model of light also explains the interaction of light with large objects (reflection and refraction). Therefore, it is tempting to accept the wave model as the "correct" model. However, we have not yet established a mechanism for this model. What is actually vibrating when light waves propagate? That question will be the topic of the next chapter.

REVIEW QUESTION 24.7 Why is it especially important to keep track of units when solving problems in wave optics?

Summary

Speed of light and refractive index The ratio of the speed of light in air c and its speed v in a medium equals the index of refraction n of the medium. (Section 24.2)

$$\frac{c}{v} = n$$

Eq. (24.4)

Monochromatic and coherent light sources Only waves of the same constant frequency and wavelength (*monochromatic*) and of constant phase difference (*coherent*) will add to produce a constant interference pattern. (Section 24.2)

Double-slit interference In the wave model of light, double-slit interference is explained by the superposition of light waves passing through two closely spaced slits. The superposition of those waves creates a pattern of dark and bright bands on a screen beyond the slits (rather than images of the slits). (Section 24.1)

Bright bands at
$$d \sin \theta_m = m\lambda$$
$$m = 0, \pm 1, \pm 2, \ldots$$
Eq. (24.1)

$$L \tan \theta_m = y_m$$
Eq. (24.2)

For small angles:
$$\tan \theta_m \approx \sin \theta_m$$

Grating A grating has many closely spaced slits, each separated by a distance d from its neighbors. The maxima produced by gratings are much narrower and brighter than those for two slits of the same slit separation. (Section 24.3)

Bright bands at
$$d \sin \theta_m = m\lambda$$
$$m = 0, \pm 1, \pm 2, \ldots$$
$$d = 1/\#\,\text{lines/m}$$
Eq. (24.1)

$$L \tan \theta_m = y_m$$
Eq. (24.2)

Thin-film interference Thin films such as soap bubbles and lens coatings produce interference effects by reflecting light from the front and back surfaces of the film. (Section 24.4)

$$n_1 < n_2 > n_3$$

Air Film Air

See Table 24.3.

Single-slit diffraction When monochromatic light of wavelength λ is incident on a slit of width w, a series of bright and dark bands of light is formed on a screen a distance L from the slit. The bands are caused by the interference of light from different regions within the slit. (Section 24.5)

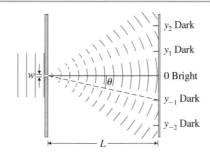

Dark bands at
$$w \sin \theta_m = m\lambda$$
$$m = \pm 1, \pm 2, \pm 3 \ldots$$
Eq. (24.6)

$$y_m = L \tan \theta_m$$
Eq. (24.2)

Rayleigh criterion for resolving objects with optical instruments The minimal angular separation for resolving two objects with an instrument of aperture diameter D is α_{res}. If the objects are closer together than this, they appear as a single object (cannot be resolved). (Section 24.6)

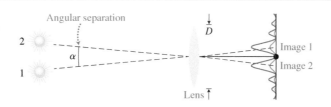

$$\alpha_{res} = \frac{1.22\lambda}{D}$$
Eq. (24.8)

Questions

Multiple Choice Questions

1. You shine a flashlight on two wide slits cut in cardboard. What do you observe on a screen beyond the slits?
   (a) A sharp image of the two slits
   (b) An interference pattern consisting of bright and dark bands
   (c) A fuzzy image of the slits
   (d) Depending on the distance from the slits to the screen, the answer could be (a) or (c).

2. When you shine a very narrow beam of white light on a grating in a dark room, what do you see on the screen behind the slits?
   (a) A sharp image of the grating's slits
   (b) A pattern of white and dark spots of approximately equal width
   (c) A colored stripe that contains rainbow colors
   (d) A set of rainbow-colored stripes separated by dark regions
   (e) A white spot in the center and a set of rainbow-colored stripes on each side of it, separated by dark regions

3. When green light travels from air to glass, what quantities change?
   (a) Speed only   (b) Frequency only
   (c) Wavelength only   (d) Speed and frequency
   (e) Speed and wavelength

4. If you add a third slit to two slits, with the same slit separation, and shine the same laser beam on the three slits, how will the three-slit pattern compare to the double-slit pattern?
   (a) It will be the same, only brighter.
   (b) It will be the same, only the spots will be farther apart.
   (c) The centers of its bright fringes will be the same distance from each other, but their width will be narrower; additional spots will appear between them.
   (d) The centers of its bright fringes will be the same distance from each other, and their width will be narrower; no additional spots will appear.

5. Why don't two flashlights about 1 m apart shining on a wall produce an interference pattern?
   (a) They are not bright enough.
   (b) The light waves are not coherent.
   (c) The fringes are too narrow to see.

6. You shine a laser beam through a grating with a known number of slits per millimeter and observe a pattern on a screen. Then you cover half of the grating with nontransparent material. What will you see?
   (a) There will be half as many bright fringes on the screen.
   (b) The bright fringes will be twice as close.
   (c) The bright fringes will be two times farther apart.
   (d) The location of the fringes will not change, but they will be a little bit wider and less bright.

7. What does the resolution limit of an optical system depend on? Choose all answers that are correct.
   (a) The wavelength of light   (b) The diameter of the aperture
   (c) The distance to the object being viewed
   (d) The distance from the aperture to the light detectors

8. You shine light of different colors on the same lens. The focal length of the lens for blue light
   (a) is shorter than that for red light.
   (b) is longer than that for red light.
   (c) is the same as that for red light because focal length is a property of the lens, not of the light shining on it.
   (d) depends on the location of the source of light.

9. You shine a green laser beam ($\lambda = 530$ nm) on a grating and observe seven bright beams behind the grating. Which of the following changes to the experiment (keeping the other parameters constant) could increase the total number of bright beams? Select all that are correct.
   (a) Using a grating with more slits per millimeter
   (b) Using a grating with fewer slits per millimeter
   (c) Using a red laser ($\lambda = 650$ nm)
   (d) Using a blue laser ($\lambda = 450$ nm)
   (e) Increasing the distance between the laser and the grating
   (f) Decreasing the distance between the laser and the grating
   (g) Immersing all of the equipment in water

Conceptual Questions

10. Describe a double-slit interference experiment for sound waves using one or more speakers, driven by an electronic oscillator, as your sources of sound waves.

11. You are investigating a pattern produced on a screen by laser light passing through two narrow slits. What would happen to the pattern if you covered one of the slits?

12. Give examples of phenomena that can be explained by a particle-bullet model of light. How can a wave model explain the same phenomena?

13. Give examples of phenomena that *cannot* be explained by a particle-bullet model of light. How can a wave model explain these phenomena?

14. How would you explain Huygens' principle to a friend who is not taking physics?

15. Draw a point-like source of light. What is the shape of wave fronts leaving that point, assuming a uniform medium? Does the distance between them change as you move away from the source? What is the direction of the rays? How are the wave fronts for laser light different from those produced by the point source of light? How are rays oriented with respect to the wave fronts?

16. Draw two coherent light sources next to each other. What will you see if you place light-sensitive detectors at several different locations around the sources? How do you know? Indicate any assumptions that you made.

17. Use the wave front representation to explain what happens to the interference pattern produced by two coherent light sources if their separation increases. Then evaluate the result using the ray representation.

18. Use the wave front representation to explain what happens to the interference pattern produced by two coherent light sources if their wavelength increases.

19. Compare the interference pattern produced by two coherent light sources in air with the case when they are both immersed in water.

20. Draw 10 coherent point-like sources of light placed along a straight line. Construct a wave front for a wave that is a superposition of the waves produced by the sources.

21. If you see green light of 520-nm wavelength when looking at the part of a soap bubble that is closest to you of uniform thickness, why will you see longer wavelength light when looking at the sides of this green region?

22. Imagine that you have a very thin uniform oil film on the surface of water. The thickness of oil is much smaller than the wavelength of blue light. White light is shining on the film. Will the film appear bright or dark if you look at it from above? Explain.

23. (a) Draw a picture of what you will see on a screen if you shine a red laser beam through a very narrow slit. (b) Redraw the picture for a wider slit. (c) Redraw the picture for a green laser for each case. Keep the scale the same.

24. Describe three situations that you can analyze using wave fronts, rays, and disturbance-versus-position graphs for a particular time. In what situation is one of the representations more helpful than the others?

25. Why can you hear a person who is around a corner talking but not see her?

26. Astronomers often called the resolution limit described by Eq. (24.8) a "theoretical resolution limit." Why do they add the adjective "theoretical" to the description? What in real life can affect the resolution limit of a telescope?

Problems

Below, **BIO** indicates a problem with a biological or medical focus. Problems labeled **EST** ask you to estimate the answer to a quantitative problem rather than derive a specific answer. Asterisks indicate the level of difficulty of the problem. Problems with no * are considered to be the least difficult. A single * marks moderately difficult problems. Two ** indicate more difficult problems.

24.1 and 24.2 Young's double-slit experiment and Index of refraction, light speed, and wave coherence

1. * **Sound interference** Two sources of sound waves are 2.0 m apart and vibrate in phase, producing sinusoidal sound waves of wavelength 1.0 m. (a) Use the wave front representation to explain what happens to the amplitude of sound along a line equidistant from each source and perpendicular to the line connecting the sources. (b) Use a graphical representation (pressure-versus-position graph) to explain what happens along that line. (c) Which representation is more helpful? Explain.

2. * Green light of wavelength 540 nm is incident on two slits that are separated by 0.50 mm. (a) Make a list of physical quantities you can determine using this information and determine three of them. (b) Describe three changes in the experiment that will each result in doubling the distance between the 0th and the first bright spot on the screen. Explain.

3. * Blue light of wavelength 440 nm is incident on two slits separated by 0.30 mm. Determine (a) the angular deflection to the center of the 3rd order bright band and (b) its distance from the 0th order band when the light is projected on a screen located 3.0 m from the slits. (c) Draw a sketch (not to scale) that schematically represents this situation and label all known distances and angles.

4. * Red light of wavelength 630 nm passes through two slits and then onto a screen that is 1.2 m from the slits. The center of the 3rd order bright band on the screen is separated from the central maximum by 0.80 m. What can you determine using this information?

5. **Sound from speakers** Sound of frequency 680 Hz is synchronized as it leaves two speakers that are separated by 0.80 m on an open field. Draw a sketch of this arrangement and draw a line from between the speakers to a location where the sound is intense and equidistant from the two speakers (the 0th order maximum). Determine the angular deflection of a line from between the speakers to the 1st order intensity maximum to the side of this 0th order maximum. The speed of sound is 340 m/s.

6. * **EST** **Sketch a moving wave front** Draw a sketch of two narrow slits separated by 4.0 cm. On the sketch, show the crests of six waves of wavelength 1.0 cm that have left each slit. Draw lines from a point halfway between the slits in the directions of the center of the 0th order and 1st order intensity maxima. Draw a screen 7.0 cm from the slits and use your sketch to estimate the distance between the center of the central maximum on the screen and a 1st order bright spot to the side. Check your results using equations from the text. Were your sketch-based estimated results within 20% of the mathematical results? Indicate any assumptions that you made.

7. * An LED bulb emits light that looks red. After passing through a narrow single slit, the light strikes two very narrow slits separated by distance d and located a distance D from the single slit. After passing through the pair of slits, the light strikes a screen a distance L away. (a) Make a sketch of the pattern that you would expect to see on the screen if light behaves like a wave. (b) Below the interference pattern, sketch the pattern that you would expect to see if light behaves like a stream of very light particles.

8. **Characteristics of laser light when in glass** A laser light in air has a wavelength of 670 nm. What is the frequency of the light? What is the frequency of this light when it travels in glass? In water? What is the wavelength of light in these media? (Use Table 22.7 if needed.)

9. * **Prism converts white light** Use the wave model of light to explain why white light striking a side of a triangular prism emerges as a spectrum.

24.3 Gratings: an application of interference

10. Light of wavelength 520 nm passes through a grating with 4000 lines/cm and falls on a screen located 1.6 m from the grating. (a) Draw a picture (not to scale) that schematically represents the process and label all known distances and angles. (b) Determine the angular deflection of the second bright band. (c) Determine the separation of the 2nd order bright spot from the central maximum. (d) Determine the total number of constructively interfering beams that are formed after the light passes the grating.

11. **Hydrogen light grating deflection** Light of wavelength 656 nm and 410 nm emitted from a hot gas of hydrogen atoms strikes a grating with 5300 lines per centimeter. Determine the angular deflection of both wavelengths in the 1st and 2nd orders.

12. * **Purchase a grating** How many lines per centimeter should a grating have to cause a 38° deflection of the 2nd order bright band of 680-nm red light?

13. **Only half a grating** You cover the left side of a grating with black tape so that half of the slits are covered. How will it affect the location of the 2nd order bright band?

14. * **EST** **Design** Design a quick way to estimate which one of two gratings has more lines per centimeter.

15. **Laser light on grating 1** The 630-nm light from a helium-neon laser irradiates a grating. The light then falls on a screen where the first bright spot is separated from the central maximum by 0.51 m. Light of another wavelength of light produces its first bright spot 0.43 m from its central maximum. Determine the second wavelength.

16. **Laser light on grating 2** Light of wavelength 630 nm passes through a grating and then onto a screen located several meters from the grating. The 1st order bright band is located 0.28 m from the central maximum. Light from a second source produces a band 0.20 m from the central maximum. Determine the wavelength of the second source. Show your calculations. Hint: If the angular deflection is small, $\tan \theta = \sin \theta$.

17. * **EST** **Memory capacity of CD** The reflective surface of a CD consists of spirals of equally spaced grooves. If you shine a laser pointer on a CD, each groove reflects circular waves that look exactly like the circular waves transmitted by the slits in a grating. You shine a green laser pointer ($\lambda = 532$ nm) perpendicularly to the surface of a CD and observe a diffraction pattern on a screen that is 3.0 m away from the CD. You observe that the 1st order maximum ($m = 1$) appears 1.1 m away from the central maximum ($m = 0$). (a) Determine the distance between the adjacent grooves on a CD. (b) Estimate the total number of bits (units of information) on a CD, assuming that each bit occupies a square with sides that are equal to the distance between the adjacent grooves. Compare the result with the typical data storage of a CD.

18. * **EST** You shine a green laser beam ($\lambda = 530$ nm) on a grating and observe 11 bright beams (due to constructive interference) behind the grating. What can you estimate about the grating?

24.4 Thin-film interference

19. You point a green laser beam perpendicularly toward a soap film, as shown in Figure P24.19. Above the laser you place a curved white screen. You slowly rotate the laser through 90° while pointing the beam toward the center of the soap film, as shown in the figure. While you perform the experiment, your friend takes a long-exposure photo of the screen. Describe what you will see in the photo. Explain your answer.

FIGURE P24.19

20. * **Representing thin-film interference** (a) Draw a ray diagram for a laser beam incident from air on a thin film that has air on the other side. Make sure to take into account the processes occurring on each surface of the film. (b) Discuss in words what happens to the wave phase at each boundary. (c) Under what conditions will a person observe no reflected light?

21. * **Oil film on water** A thin film of vegetable oil ($n = 1.45$) is floating on top of water ($n = 1.33$). Describe in words the processes occurring to a laser beam at the top and bottom surfaces of the oil.

22. * **Soap bubble 1** You look at a soap bubble film perpendicular to its surface. Describe the changes in colors of the film that you observe as the film thins and eventually breaks. Support your explanation with a ray diagram.

23. * **Soap bubble 2** A soap bubble of refractive index 1.40 appears blue-green when viewed perpendicular to its surface (blue-green appears when red light is missing from the continuous spectrum, where λ_{red} is about 670 nm in a vacuum). Does the light change phase when reflected from (a) the outside surface of the bubble and (b) the inside surface? (c) Determine the wavelength of red light when passing through the bubble. (d) Determine the thickness of the thinnest bubble for which the 670-nm red light reflected from the outside surface of the bubble interferes destructively with light reflected from the inside surface.

24. * **Thin-film coated lens** A lens coated with a thin layer of material having a refractive index 1.25 reflects the least amount of light at wavelength 590 nm. Determine the minimum thickness of the coating.

25. * **Thin-film coated glass plate** A film of transparent material 120 nm thick and having refractive index 1.25 is placed on a glass sheet having refractive index 1.50. Determine (a) the longest wavelength of light that interferes destructively when reflected from the film and (b) the longest wavelength that interferes constructively.

26. Two flat glass surfaces are separated by a 150-nm gap of air. (a) Explain why 600-nm-wavelength light illuminating the air gap is reflected brightly. (b) What wavelength of radiation is not reflected from the air gap?

27. * Jeff wants to solve the following problem: a light beam with $\lambda = 500$ nm is incident perpendicularly on a layer of oil ($n_{oil} = 1.45$, thickness $d = \frac{\lambda}{2} = 250$ nm). The oil layer is floating on top of water ($n_{water} = 1.33$). Compare (a) the total intensity I_R of the light that *reflects* from the oil and (b) the intensity I_T of the light that goes *through* the oil into the water to the intensity of the incident light I_0. Explain your answer. Ignore any absorption of light in the oil. Jeff's solution is as follows: the total path length difference between a ray that reflects from the air-oil boundary and a ray that reflects from the oil-water boundary is equal to λ. Therefore, the reflected rays interfere constructively, giving (a) $I_R = I_0$ and (b) $I_T = 0$. Evaluate Jeff's solution, identify any errors, and, if there are any errors, provide a corrected solution.

28. ** Dawn wants to solve the following problem: a light beam with $\lambda = 500$ nm is incident perpendicularly on a layer of oil ($n_{oil} = 1.45$, thickness $d = \frac{\lambda}{2} = 250$ nm). The oil layer is on top of glass ($n_{glass} = 1.55$). Compare (a) the total intensity I_R of the light that *reflects* from the oil and (b) the intensity I_T of the light that goes *through* the oil into the glass to the intensity of the incident light I_0. Explain your answer. Ignore any absorption of light in the oil. Dawn's solution is as follows: the total path length difference between a ray that reflects from the air-oil boundary and a ray that reflects from the oil-glass boundary is equal to 2λ. Therefore, the reflected rays interfere constructively, giving (a) $I_R = I_0$ and (b) $I_T = 0$. Evaluate Dawn's solution, identify any errors, and, if there are any errors, provide a corrected solution.

24.5 Diffraction of light

29. * **Explain diffraction** Draw a ray diagram and show path length differences to explain how wavelets originating in different parts of a slit produce the third dark fringe on a distant screen.

30. * **How did we derive it?** Explain how we derived the equation for the first dark fringe for single-slit diffraction. Draw a ray diagram or a disturbance-versus-position graph to help in your explanation. Show the path length difference and the phase differences.

31. * **Explain a white light diffraction pattern** White light passing through a single slit produces a white bright band at the center of the pattern on a screen and colored bands at the sides. Explain.

32. Light of wavelength 630 nm is incident on a long, narrow slit. Determine the angular deflection of the first diffraction minimum if the slit width is (a) 0.020 mm, (b) 0.20 mm, and (c) 2.0 mm.

33. * Light of wavelength of 430 nm is incident on a long, narrow slit of width 0.050 mm. Determine the angular deflection of the 5th order diffraction minimum.

34. * **Sound diffraction through doorway** Sound of frequency 440 Hz passes through a doorway opening that is 1.2 m wide. Determine the angular deflection to the first and second diffraction minima ($v_{sound} = 340$ m/s).

35. * Light of wavelength 624 nm passes through a single slit and then strikes a screen that is 1.2 m from the slit. The thin first dark band is 0.60 cm from the middle of the central bright band. Determine the slit width.

24.6 Resolving power

36. ** **Explain resolution** Explain the term "resolution limit." Illustrate your explanation with pictures and ray diagrams, and explain what characteristics of an optical device and the light passing through it affect the resolution limit.

37. **Resolution of telescope** A large telescope has a 3.00-m-radius mirror. What is the resolution limit of the telescope? What assumptions did you make?

38. * Laser light of wavelength 630 nm passes through a tiny hole. The angular deflection of the first dark band is 26°. What can you learn about the hole using this information?

39. * **Size of small bead** Infrared radiation of wavelength 1020 nm passes a dark, round glass bead and produces a circular diffraction pattern 0.80 m beyond it. You cannot see this light with the naked eye, but you can observe it with a cell phone camera. The diameter of the first bright circular ring is 6.4 cm. What can you learn about the bead using this information?

40. * **Resolution of telescope** How will the resolution limit of a telescope change if you take pictures of stars using a blue filter as opposed to using a red filter? Explain.

41. * **Detecting visual binary stars** Struve 2725 is a double-star system with a visual separation of 5.7 arcseconds in the constellation Delphinus. Determine the minimum diameter of the objective lens of a telescope that will allow you to resolve 400-nm violet light from the two stars.

42. **Hubble Telescope resolving power** The objective mirror of the Hubble Telescope is 2.4 m in diameter. Could it resolve the binary stars Lambda Cas in the constellation Cassiopeia? The stars have an angular separation of 0.5 arcseconds.

43. * Draw a graphical representation of Rayleigh's criterion. Explain how this criterion relates to the concept of the resolution limit.

24.7 Skills for applying the wave model of light

44. * Red light from a helium-neon gas laser has a wavelength of 630 nm and passes through two slits. (a) Draw a ray diagram to explain why you see a pattern of bright and dark bands on the screen. Show the path length difference. (b) Determine the angular deflection of the light to the first three bright bands when incident on narrow slits separated by 0.40 mm. (c) Determine the distance between the centers of the 0th and 2nd order bright bands when projected on a screen located 5.0 m from the slits. (d) List all of the assumptions that you made in your calculations.

45. * **EST** Figure P24.45 shows the diffraction pattern on a screen obtained when a narrow beam of white light is incident on a grating with 100 lines/mm. If you carefully examine the color pattern, you will notice narrow magenta bands (indicated by the arrows) that appear where the blue region from the 2nd order maximum overlaps with the red region from the 3rd order maximum (a mixture of blue and red light is perceived as magenta). Note that there are more magenta bands in higher orders, but there is no magenta between the 1st and 2nd order maxima. (a) Is there a magenta band between the 2nd and 3rd order maxima for any grating, or does the presence of the magenta color depend on the distance between the slits in the grating? Explain. (b) The 1st order maxima are about 20 cm from the central white band. Estimate how far from the screen the grating in this experiment is. Explain.

FIGURE P24.45

46. * Red light of wavelength 630 nm is incident on a pair of slits. The interference pattern is projected on a wall 6.0 m from the slits. The fourth bright band is separated from the central maximum by 2.8 cm. (a) Draw a ray diagram to represent the situation; show the path length difference. (b) What can you learn about the slit pair using this information? (c) What can you learn about the pattern on the screen using this information?

47. * Monochromatic light passes through a pair of slits separated by 0.025 mm. On a screen 2.0 m from the slits, the 3rd order bright fringe is separated from the central maximum by 15 cm. (a) Draw a ray diagram to represent the situation. Show the path length difference. (b) What can you learn about the light source using this information?

48. * **Ratio reasoning** Two different wavelengths of light shine on the same grating. The 3rd order line of wavelength A (λ_A) has the same angular deflection as the 2nd order line of wavelength B (λ_B). Determine the ratio (λ_A/λ_B). Be sure to show the reasoning leading to your solution.

49. * **Design** Design an experiment to use a grating to determine the wavelengths of light of different colors. Draw a picture of the apparatus. List the quantities that you will measure. Describe the mathematical procedure that you will use to calculate the wavelengths.

50. * **Fence acts as a grating for sound** A fence consists of alternating slats and openings, the openings being separated from each other by 0.40 m. Parallel wave fronts of a single-frequency sound wave irradiate the fence from one side. A person 20 m from the fence walks parallel to it. She hears intense sound directly in front of the fence and in another region 15 m farther along a line parallel to the fence. Determine everything you can about the sound used in this problem (the speed of sound is 340 m/s).

51. * **BIO Morpho butterfly reflection grating wings** A reflection grating reflects light from adjacent lines in the grating instead of allowing the light to pass through slits, as in a transmission grating. If we assume perpendicular incidence, then we can determine the angular deflection of bright bands the same way we did for a transmission grating. White light is incident on the wing of a Morpho butterfly (whose wings act as reflection gratings). Red light of wavelength 660 nm is deflected in the 1st order at an angle of 1.2°. (a) Determine the angular deflection in the 1st order of blue light (460 nm). (b) Determine the angular deflection in the 3rd order of yellow light (560 nm).

52. * **Ratio reasoning** Laser monochromatic light is used to illuminate two different gratings. The angular deflection of the 2nd order band of light leaving grating A equals the angular deflection of the 3rd order band from grating B. Determine the ratio of the number of lines per centimeter for grating A and for grating B.

53. * **Soap bubble interference** Light of 690-nm wavelength interferes constructively when reflected from a soap bubble having refractive index 1.33. Determine two possible thicknesses of the soap bubble.

54. * **Oil film on water** A film of oil with refractive index 1.50 is spread on water whose refractive index is 1.33. Determine the smallest thickness of the film for which reflected green light of wavelength 520 nm interferes destructively.

55. **Babinet's principle** Babinet's principle states that the diffraction pattern of complementary objects is the same. For example, a rectangular slit produces the same diffraction pattern on a screen as a rectangular obstacle of the same size as the slit, provided that the incident beam is wider than the slit or the obstacle. Determine the width of a hair that, when irradiated with laser light of wavelength 630 nm, produces a diffraction pattern on a screen with the first minimum 2.5 cm on the side of the central maximum. The screen is 2.0 m from the hair.

56. * **Diffraction from a loudspeaker** The opening of a loudspeaker can be modeled as a circular hole through which sound waves propagate into space. Determine the maximum diameter of the speaker such that the first diffraction minimum of sound is at least 45° from the direction in which the speaker points. Perform the calculations for sound waves of frequency (a) 200 Hz, (b) 1000 Hz, and (c) 10,000 Hz. The speed of sound is 340 m/s.

57. * The angular deflection of the 1st order bright band of light passing through double slits separated by 0.20 mm is 0.15°. Determine the angular deflection of the 3rd order diffraction minimum when the same light passes through a single slit of width 0.30 mm.

58. * **EST Diffraction of sound from the mouth** (a) Estimate the diameter of your mouth when open wide. (b) Determine the angular deflection of 200-Hz sound and of 15,000-Hz sound as it leaves your mouth. If, during your calculations, you find that sin θ > 1, explain the meaning of this result. The speed of sound is 340 m/s.

59. **Determine body cell size** Light of 630 nm wavelength from a helium-neon laser passes two different-size body cells. The angular deflection of the light as it passes the cells is (a) 0.060 radians and (b) 0.085 radians. Determine the size of each cell. (See the description of the Babinet principle in Problem P24.55.)

60. * **EST** A sound of frequency 1000 Hz passes a basketball. Estimate the angular deflection from the basketball to the first ring around the central maximum in the diffraction pattern. The speed of sound is 340 m/s. (See the description of the Babinet principle in Problem P24.55.)

61. * **BIO Ability of a bat to detect small objects** Bats emit ultrasound in order to detect prey. Ultrasound has a much smaller wavelength than sound, which improves the resolution of small objects. What is the diameter of the smallest object that forms a diffraction pattern when irradiated by the 8.0×10^4-Hz ultrasound from a bat? The first diffraction minimum from the smallest object that still produces diffraction forms at 90° with respect to the direction of the incident waves.

General Problems

62. As Jess and Matt are taking an evening walk in late December, they stop by a house that is decorated with dancing red and green stars (see **Figure P24.62**). They agree that the device that projects the stars is probably using red and green lasers, but they have different explanations for how the stars are formed. Jess thinks that the stars are produced by some kind of a grating that rotates in front of the lasers. Matt thinks that the device uses rotating prisms such as those sometimes used in kaleidoscope glasses. Who do you think is correct? Support your answer with data that you can estimate from the photo.

FIGURE P24.62

63. * Monochromatic light passes through two slits and then strikes a screen. The distance separating the central maximum and the first bright fringe at the side is 2.0 cm. Determine the fringe separation when the following quantities change simultaneously: the slit separation is doubled, the wavelength of light is increased 30%, and the screen distance is halved.

64. **Sound from speakers** Two stereo speakers separated by a distance of 0.8 m play the same musical note at frequency 1000 Hz. A listener starts from position 0 (**Figure P24.64**) and walks along a line parallel to the speakers. (a) Can the listener easily hear the sound at position 0? Explain. (b) Calculate the distance from position 0 to positions 1 and 2 where intense sound is also heard. The speed of sound is 340 m/s.

FIGURE P24.64

Speaker

$d = 0.8$ m

Speaker

8.0 m

65. * **Astronomer's spectrograph** An astronomer has a grating spectrograph with 5000 lines/cm. A film 0.500 m from the grating records bands of light passing through the grating. (a) Determine the wavelength and frequency of the H_α line of hydrogen gas in a laboratory discharge tube that produces a 1st order band separated on the film by 17.4 cm from the central maximum. (b) Determine the wavelength and frequency of the H_α line coming from a galaxy in the cluster Hydra A. The 1st order band of the H_α line of this light is 18.4 cm from the central maximum. (c) Suggest a possible reason for the differences in the frequency of the H_α line from a lab source and the H_α line from the galaxy.

66.   Diffraction of water waves entering a harbor   The wavelength of water waves entering a harbor is 14 m. The angular deflection of the 1st order diffraction minimum of the waves in the water beyond the harbor is 38°. Determine the width of the opening into the harbor.

67. * **EST** You observe a green LED   **FIGURE P24.67**
through a rectangular opening of width $w = 0.08$ mm and unknown height h. You see the pattern shown in Figure P24.67 (the actual pattern is very small, but you can see it clearly with the naked eye). The distance between the LED and the opening is 3.0 m. Estimate the height of the opening h. You will need a ruler to solve this problem. (Hint: A rectangular slit can be modeled as two single slits placed perpendicularly to each other.)

68. ** **Variable thickness wedge**   A wedge of glass of refractive index 1.64 has a silver coating on the bottom, as shown in Figure P24.68. Determine the smallest distance x to a position where 500-nm light reflected from the top surface of the glass interferes constructively with light reflected from the silver coating on the bottom. The light changes phase when reflected at the silver coating.

FIGURE P24.68

Silver coating   Glass 1.0×10^{-5} m

$x = ?$

0.20 m

69. * **EST** **Resolving car headlights**   Estimate the farthest away a car can be at night so that your eyes can resolve the two headlights. Indicate any assumptions you made.

70.   Looking at Moon rocks   You have a home telescope with a 3.0-cm objective lens. Determine the closest distance between two large boulders on the Moon that you can distinguish as separate objects. Indicate any assumptions you made.

71. * **BIO** **EST** **Diffraction-limited resolving power of the eye**   You look at closely spaced lines on a wall 5.0 m from your eyes. Estimate the closest the lines can be to each other and still be resolved by your eyes as separate lines. Indicate any assumptions you made in making your estimate.

72. * **Resolving sunspots**   You are looking at sunspots. They usually appear in pairs on the surface of the Sun. The Sun is about 1.5×10^{11} m from Earth. How close can two sunspots be so you can distinguish them when you observe them though an amateur telescope whose aperture (objective lens) is about 20 cm? Describe all of the assumptions that you made.

73. ** **The Moon's Mare Imbrium**   The outermost ring of mountains surrounding the Mare Imbrium on the Moon has a diameter of 1300 km. What diameter objective lens telescope would allow an astronomer to see the ring of mountains as a distinct feature of the Moon's landscape? What assumptions did you make? The average center-to-center distance from the Earth to the Moon is 384,403 km, which is about 30 times the diameter of the Earth. The Moon has a diameter of 3474 km.

74. * **Can you see atoms with a light-based microscope?**   Explain how you can use your knowledge of the wave model of light to explain why you cannot use an optical microscope to see atoms.

75. * **Detecting insects by diffraction of sound**   A biologist builds a device to detect and measure the size of insects. The device emits sound waves. If an insect passes through the beam of sound waves, it produces a diffraction pattern on an array of sound detectors behind the insect. What is the lowest-frequency sound that can be used to detect a fly that is about 3 mm in diameter? The speed of sound is 340 m/s.

Reading Passage Problems

BIO **What is 20/20 vision?**   Vision is often measured using the Snellen eye chart, devised by Dutch ophthalmologist Herman Snellen in 1862 (see Figure 24.29). With normal vision (20/20 vision), you can distinguish a letter that is 8.8 mm high from other letters of similar height at a distance of 6.1 m (20 ft) (the Snellen chart in the figure is smaller than normal size). If your vision is 20/40, the letters must be twice as high to be distinguishable. Alternatively, a person with 20/40 vision could distinguish letters from 20 ft that a person with 20/20 vision can distinguish at 40 ft. Someone with 20/60 vision could distinguish letters at 20 ft that someone with 20/20 vision could distinguish at 60 ft.

Does the Rayleigh criterion limit visual resolution? Assume that the eye's pupil is 5.0 mm in diameter for 500-nm light. The Rayleigh criterion angular deflection for such light entering the eye's pupil is

FIGURE 24.29 A Snellen eye chart used to detect visual acuity.

$$\alpha = 1.22 \frac{500 \times 10^{-9} \text{ m}}{5.0 \times 10^{-3} \text{ m}} = 1.2 \times 10^{-4} \text{ rad}$$

If the Rayleigh criterion limited visual resolution, then from a distance of 6.1 m, you should be able to distinguish details in shapes of size

$$y = L \tan \alpha = (6.1 \text{ m})\left[\tan(1.2 \times 10^{-4} \text{ rad}) \right]$$
$$= 0.7 \times 10^{-3} \text{ m} = 0.7 \text{ mm}$$

or about one-tenth the size of the 8.8-mm-tall letter in row 8 (20/20 vision) of the full-size Snellen eye chart. Thus, according to the Rayleigh criterion, we should easily be able to resolve different 8.8-mm-tall letters. Other factors, such as chromatic aberration, irregularities in the cornea-lens-retina shapes, the density of rods and cones in the retina, and air currents, limit visual resolution more than the Rayleigh criterion diffraction limit.

76. Which answer below is closest to the height of the letters that a person with 20/80 vision can distinguish when 20 ft from the wall chart?
(a) 2.2 mm   (b) 4.4 mm
(c) 8.8 mm   (d) 18 mm
(e) 34 mm

77. Suppose that a person with 20/20 vision stands 30 ft from a Snellen eye chart. Which answer below is closest to the minimum height of the letters the person can distinguish?
(a) 4.4 mm   (b) 6.6 mm
(c) 8.8 mm   (d) 13 mm
(e) 18 mm

78. What is the visual acuity of a person with 20/20 vision mainly limited by?
(a) The Rayleigh criterion
(b) Chromatic aberration
(c) Focal length of the eye lens
(d) Diameter of the eye's pupil
(e) The density of cones and rods
(f) A combination of these factors

79. A hawk's vision is said to be 20/5. If so, the hawk can distinguish 8.8-mm-tall letters from about what distance?
(a) 5 ft   (b) 10 ft
(c) 40 ft   (d) 80 ft
(e) 120 ft

80. If the vision of a hawk is 20/5, then what is the angular resolution of the hawk? Assume that it can distinguish objects about 8.8 mm from a distance of 60 m.
(a) 0.2×10^{-4} rad   (b) 0.5×10^{-4} rad
(c) 1×10^{-4} rad   (d) 2×10^{-4} rad

Thin-film window coatings for energy conservation and comfort Thin-film coatings are applied to eyeglasses, computer screens, and automobile instrument panels to reduce glare and to binocular and camera lenses to increase the amount of light transmitted. Thin-film coatings are also used to add color to architectural glass, to reduce surface friction, and to improve the energy efficiency of windows. Consider this latter application.

FIGURE 24.30 Over 50% of solar radiation reaching Earth's surface is infrared radiation.

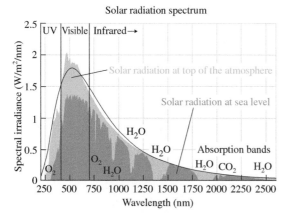

Figure 24.30 shows that over 50% of the radiation reaching Earth from the Sun is long-wavelength infrared radiation, which we perceive as warmth, not light. Thin-film coatings on windows serve two purposes. (1) On warm days, the coatings reflect almost all of the infrared radiation from outside while allowing almost all of the visible light through the window (the glass absorbs most of the ultraviolet). The rooms inside stay cooler and require less air conditioning. (2) On cold days, the same thin-film coating allows visible light to enter the room and reflects the thermal infrared radiation from the interior back into the room, keeping it warm. The reflectance of the thin-film window coating as a function of wavelength is shown in Figure 24.31.

FIGURE 24.31 A thin-film window coating causes reflectance of infrared radiation.

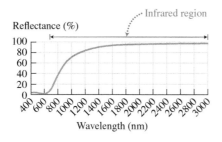

81. Which of the following are benefits of thin-film coatings on windows?
    (a) They keep ultraviolet light out of the house in summer and winter.
    (b) They keep infrared radiation out of the house in the summer.
    (c) They keep infrared radiation in the house in the winter.
    (d) b and c   (e) a, b, and c

82. A 1.25 refractive index thin film on a 1.50 refractive index glass window is made to reflect infrared radiation of wavelength 1000 nm. What is the net reflective phase change of infrared radiation reflected off the front surface of the thin film relative to the radiation reflected off the back surface of the film and returning to the front?
    (a) Zero   (b) 1/4 wavelength
    (c) 1/2 wavelength   (d) None of these

83. A 1.25 refractive index thin film on a 1.50 refractive index glass window is made to reflect infrared radiation of wavelength 1000 nm. What is the wavelength of the infrared radiation while in the thin film?
    (a) 670 nm   (b) 800 nm
    (c) 1000 nm   (d) 1250 nm
    (e) 1500 nm

84. A 1.25 refractive index thin film on a 1.50 refractive index glass window is made to reflect infrared radiation of wavelength 1000 nm. What is the desired thickness of the thin film?
    (a) 200 nm   (b) 250 nm
    (c) 400 nm   (d) 500 nm
    (e) 1000 nm

85. The actual thin film built for windows reflects about what percent of the incident infrared radiation at 1000 nm?
    (a) 10   (b) 30
    (c) 50   (d) 70
    (e) 90

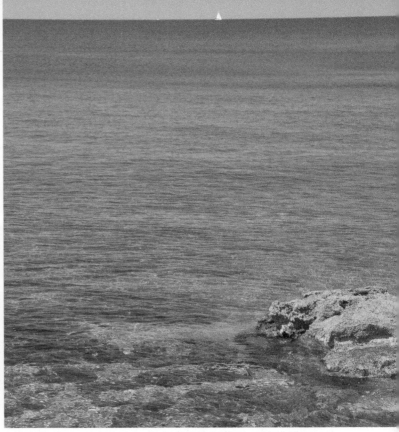

25

Electromagnetic Waves

- **How do polarized sunglasses work?**
- **How can a low-flying plane avoid radar detection?**
- **How is the screen of your calculator similar to the sky?**

The left-hand image above is a regular photograph of the sea and the sky. Next to it is a photograph of the same scene taken with polarized sunglasses oriented in a particular direction in front of the camera lens. Comparing the two photos, you can see differences in the brightness of the sky and in the brightness of the water. But there are other important differences as well. The sailboat that is almost invisible on the left is clearly visible on the right. The same is true for the underwater rocks: in the left photo, the glare of the reflected light masks them almost completely, while on the right, there is no glare. How do the polarized sunglasses cause the right-hand image to be so different?

BE SURE YOU KNOW HOW TO:

- Explain how a capacitor and a transformer work (Sections 18.7 and 21.7).
- Explain how an electric current can be generated without a battery (Section 21.1).
- Describe wave motion using the quantities frequency, speed, wavelength, and intensity (Sections 11.2 and 11.4).

IN THE PREVIOUS THREE CHAPTERS we constructed different models of light to explain its behavior and found that a wave model explained reflection, refraction, and interference. But a mystery remains: what is vibrating in a light wave? We continue to investigate that question in this chapter. We'll resolve the question when we learn about special relativity (in Chapter 26).

25.1 Polarization of waves

In our previous study of mechanical waves, we neglected a phenomenon that is crucial for answering the question posed in the chapter opening. Consider the experiments in Observational Experiment **Table 25.1**.

OBSERVATIONAL
EXPERIMENT TABLE **25.1** Rope waves and Slinky waves

VIDEO
OET 25.1

Observational experiment	Analysis
Experiment 1. A rope passes through the open ends of a narrow rectangular box. Shake the rope in a vertical plane, producing a transverse wave. The long sides of the box are parallel to the shaking direction. The rope wave is unaffected by the box.	Vectors represent the displacement of each section of rope at one instant. They are parallel to the long sides of the box.

Experiment 2. Rotate the slotted box 90° with respect to the original orientation. Shake the rope the same way as in Experiment 1. The long sides of the box are perpendicular to the shaking direction. The wave does not pass through the box.	The displacement vectors of the rope sections are perpendicular to the long sides of the box. The box exerts forces on the rope in the direction opposite to the motion of the parts of the rope inside the box, therefore decreasing the displacement of the parts of the rope.

Experiment 3. Now use two slotted boxes. Rotate the first box 30° from the vertical plane. Orient box 2 perpendicular to box 1. Shake the rope as in Experiment 1. Part of the wave travels through box 1 but is eliminated by box 2.	The component of the rope wave parallel to the first box travels through the first box. Once the rope wave reaches the second box, its displacement vectors are perpendicular to the box.

Experiment 4. We achieve the same results as in Experiments 1–3 if we replace the rope with a Slinky and produce a transverse wave. However, if we compress and decompress the Slinky horizontally, producing a longitudinal wave, the wave passes through the boxes in all of the experiments.	The longitudinal displacement vectors of the Slinky coil always pass through the boxes. They point along the direction of the Slinky.

Patterns

- If the displacement vectors of a transverse wave are parallel to the slit (the box), the wave passes through undisturbed.
- If the displacement vectors are in the plane perpendicular to the slit, the wave does not pass through.
- When the displacement vectors make an angle with the slit, only the component of the vectors parallel to the slit passes through.
- The displacement vectors of a longitudinal wave always pass through the slit. Longitudinal waves are not sensitive to the orientation of the slit.

We learned in Table 25.1 that slotted boxes with the proper orientation can block transverse mechanical waves but cannot block longitudinal waves. This phenomenon is referred to as **polarization of waves**—polarization is a property of waves that describes the orientation of the oscillations of the wave. As we found in Table 25.1, only transverse waves have this property. Thus, if we observe polarization of light waves, we know that the waves are transverse.

In a **linearly polarized** mechanical wave, the individual particles of the vibrating medium vibrate along only one axis that is perpendicular to the direction the wave travels (there are other types of polarization, but we will confine our discussion to linear polarization). In our experiments, the vibrating medium was the rope. The slotted box is called a **polarizer**. A polarizer is a device that allows only a single component of transverse waves to pass through it. The component that can pass through defines the **axis** of the polarizer.

The waves in Experiments 1–3 in Table 25.1 were linearly polarized. An **unpolarized** wave is one in which the particles of the medium vibrate in all possible directions in the plane perpendicular to the direction the wave is traveling (**Figure 25.1a**), often caused by a collection of differently polarized waves. If an unpolarized wave is incident on a polarizer, the wave that emerges from the other side is linearly polarized (Figure 25.1b). A second polarizer whose axis is perpendicular to the axis of the first polarizer blocks the wave completely. Thus, a linearly polarized wave can be reduced to zero by a polarizer whose axis is perpendicular to the polarization direction of the wave. An unpolarized wave is completely blocked by two successive perpendicularly oriented polarizers.

Are light waves transverse or longitudinal? According to our observations in Table 25.1, if we can polarize light waves, they must be transverse.

We can test this hypothesis using a semiprecious crystal called tourmaline, which affects light in the same way that our slotted box affected transverse rope waves. We can compare light from a bulb (**Figure 25.2a**) to the same light after it passes through the tourmaline crystal; the light looks dimmer. Rotating the crystal does not change the intensity of the light. However, when we place a second crystal at a 45° angle to the first, we observe that the intensity of light decreases (Figure 25.2b). When the second crystal is placed perpendicular to the first, the intensity of the light reduces to zero (Figure 25.2c). To explain this experiment, we hypothesize that light is a transverse wave and that a lightbulb emits unpolarized light waves. We also hypothesize that the tourmaline crystal allows only one component of the wave to pass through the first crystal. That component is completely blocked by the second crystal.

FIGURE 25.1 The effect of polarizers on a polarized wave.

(a) The particles in an unpolarized wave vibrate in all directions in the plane perpendicular to the direction the wave travels.

(b)

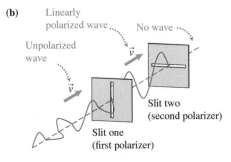

Linearly polarized wave

No wave

Unpolarized wave

Slit two (second polarizer)

Slit one (first polarizer)

FIGURE 25.2 The polarizing effects on light of tourmaline crystal.

(a) Bright light

(b) One polarizer allows only linearly polarized light to pass through.

The second polarizer reduces light passing through.

(c) No light passes through two perpendicular polarizers.

Most commonly used polarizers are made of a material that contains elongated molecules that serve as slits for passing light. When these molecules are oriented parallel to each other, their combined effect leads to polarization of passing light. The lenses of polarized sunglasses, for instance, are covered with a film containing such a material. Physics teaching labs often have "polarizers," sheets of plastic that are translucent and thus dark like sunglasses. Light seen through these polarizers behaves similarly to Figure 25.2. Based on these observations we can say that light's behavior is so similar to the behavior of transverse mechanical waves passing through mechanical polarizers that it supports the hypothesis that light waves are transverse waves.

Quantitative description of the effect of polarizers

Observational Experiment Table 25.2 investigates the amount of dimming caused by one or two polarizers oriented at different angles relative to each other.

OBSERVATIONAL EXPERIMENT TABLE 25.2 ◄▶■ **Effect of polarizers on light intensity**

VIDEO
OET 25.2

Observational experiment	Analysis

Observational experiment:

We place a light meter a fixed distance from a lightbulb (a source of unpolarized light).

Light sensor

Analysis:

Light intensity $= I_0$

We place a polarizer between the bulb and the detector.

Light sensor

Light intensity $= I_0/2$

We place two polarizers between the bulb and the detector, with varying angles between their axes.

Light sensor

Angle	Light intensity
0°	$I_0/2$
30°	$0.75\, I_0/2$
45°	$0.50\, I_0/2$
60°	$0.25\, I_0/2$
90°	0

Patterns

- One polarizer reduces the light intensity by half: $I = (I_0/2)$.
- The second polarizer seems to further reduce the intensity depending on the angle between the polarizers. We need to determine the relation between this reduction and the angle.

The patterns in Table 25.2 indicate that the intensity of light passing through the second polarizer decreases as some function of the angle between the axes of the first and second polarizer. Looking at the experiments with waves on a rope (Table 25.1), we realize that the box polarizer allows the component of the wave amplitude that vibrates in the direction of the long side of the box to pass through the box. If the amplitude of the incident wave is A, the component that will pass through the box is $A \cos\theta$, where θ is the angle between the direction of vibration of incident waves and the direction of the long side of the box.

Can we apply this to light? Looking at the patterns in Table 25.2, we see that the dependence follows a $\cos^2 \theta$ function rather than a $\cos \theta$ function ($\cos 30° = 0.86$, but $\cos^2 30° = (0.86)^2 = 0.75$). How can we explain this? We will see later in this chapter that the intensity of light is proportional to the amplitude squared of the quantity that vibrates. What we observe and measure is the light intensity and not the amplitude. Thus we accept $\cos^2 \theta$ as a function of the angle that we are looking for, where θ is the angle between the axes of the two polarizers. The intensity of light that passes through two polarizers is therefore $I = I' \cos^2 \theta$, where I' is the intensity of light that passes through both polarizers when their axes are parallel. An *ideal* polarizer reduces the unpolarized incident light intensity by one-half; therefore, $I' = I_0/2$, where I_0 is the intensity of unpolarized light. Real polarizers always absorb some light independent of polarization direction; therefore, in practice $I' < I_0/2$.

If a beam of linearly polarized light of intensity $I_{\text{polarized}}$ is incident on an ideal polarizer, then the intensity of light passing through the polarizer is $I = I_{\text{polarized}} \cos^2 \theta$, where θ is the angle between the direction of the light's initial polarization and the axis of the polarizer. This statement is called *Malus's law*.

Hypotheses concerning light and polarization

In Table 25.1 we found that polarization effects can only be observed with transverse waves and not longitudinal waves. We just found from the light polarization experiments in Table 25.2 that light seems to behave like a transverse wave with amplitude A and intensity $I \propto A^2$. However, since light can travel through a vacuum (for example, when it travels from the Sun to Earth), what is actually vibrating in a light wave?

To answer this question, we need to look carefully at how a transverse wave travels through a medium, such as a Slinky. When one coil is displaced, it pulls the next coil in a direction perpendicular to the direction of propagation of the wave. That coil then does the same to the coil next to it, and so on. The individual vibrating coils (analogous to the particles of the medium) exert elastic forces on each other. These elastic forces point perpendicular to the propagation direction of the wave and accelerate the displaced coils back toward equilibrium.

However, light can travel through a vacuum—a medium with no particles at all. How then can it be a transverse wave? Here are two hypotheses.

1. Light is actually a mechanical vibration that travels through an elastic medium. This medium is completely transparent and has exactly zero mass (just like the vacuum). This medium will be called *ether* (not related to the chemical compound of the same name).
2. A light wave is some new type of vibration that does not involve physical particles vibrating around equilibrium positions due to restoring forces being exerted on them.

The next section will show how studies of electricity and magnetism helped test the second hypothesis.

REVIEW QUESTION 25.1 What is the difference between a linearly polarized mechanical wave and an unpolarized one?

25.2 Discovery of electromagnetic waves

Before the second half of the 19th century, the investigations of light and electromagnetic phenomena proceeded independently. However, around 1860 the work of a British physicist, James Clerk Maxwell (1831–1879), led to the unification of those phenomena and helped finally answer the question of how a transverse wave can propagate in a vacuum. We know from our study of electromagnetic induction (Chapter 21) that Michael Faraday introduced the concept of a field and the relationship between electric and magnetic fields. According to Faraday, *a changing magnetic field can produce an electric field.* Subsequently, in 1865 Maxwell suggested a new field relationship: *a changing electric field can produce a magnetic field.* This idea was motivated by a thought experiment devised by Maxwell in which he imagined what would happen in the space between the plates of a charging or discharging capacitor. He suggested that the changing electric field between the capacitor plates could be viewed as a special nonphysical current, but one that would still produce a magnetic field (**Figure 25.3**). This magnetic field was first detected in 1929 but not measured precisely until 1985 due to its extremely tiny magnitude. Maxwell summarized this new idea and other electric and magnetic field ideas mathematically in a set of four equations, now known as Maxwell's equations. The equations are written using calculus, but we can summarize them conceptually:

1. Stationary electric charges produce a constant electric field. The $\vec{E}$ field lines representing this electric field start on positive changes and end on negative charges.

2. There are no individual magnetic charges (magnetic monopoles).

3. A magnetic field is produced either by electric currents or by a changing electric field (Figures 25.3 and **25.4a**). The $\vec{B}$ field lines that represent the magnetic field form closed loops and have no beginnings or ends.

4. A changing magnetic field produces an electric field. The $\vec{E}$ field lines representing this electric field are closed loops (Figure 25.4b).

Producing an electromagnetic wave

Maxwell's equations had important consequences. First, the equations led to an understanding that a changing electric field can produce a changing magnetic field, which in turn can produce a changing electric field, and on and on in a sort of feedback loop (**Figure 25.5**). This feedback loop does not require the presence of any electric charges or currents. Maxwell investigated this idea mathematically using the four equations, and to his surprise he found they led to a wave equation similar to Eq. (11.4) in which the electric and magnetic fields themselves were vibrating. The speed of propagation of these waves in a vacuum turned out to be a combination of two familiar constants: $v = \left(1/\sqrt{\epsilon_0\mu_0}\right)$, where the constants $\epsilon_0 = 8.85 \times 10^{-12}\ \mathrm{C^2/N \cdot m^2}$ and $\mu_0 = 4\pi \times 10^{-7}\ \mathrm{N/A^2}$ relate to the electric and magnetic interactions of electrically charged particles in a vacuum. The constant ϵ_0 is the **vacuum permittivity** and is related to Coulomb's constant k_C through the relationship

$$\epsilon_0 = \frac{1}{4\pi k_C}$$

(See Chapter 17 for more on vacuum permittivity.) The constant μ_0 is the **vacuum permeability**. We discussed vacuum permeability in the chapter on magnetism (Chapter 20). When Maxwell inserted the values of the constants into the expression for the speed of electromagnetic waves, he obtained

$$v = \frac{1}{\sqrt{\epsilon_0\mu_0}} = \frac{1}{\sqrt{(8.85 \times 10^{-12}\ \mathrm{C^2/N \cdot m^2})(4\pi \times 10^{-7}\ \mathrm{N/A^2})}}$$

$$= \sqrt{9.00 \times 10^{16}\ \frac{\mathrm{N \cdot m^2 \cdot A^2}}{\mathrm{C^2 \cdot N}}} = \sqrt{9.00 \times 10^{16}\ \frac{\mathrm{m^2 \cdot A^2}}{\mathrm{A^2 \cdot s^2}}} = 3.00 \times 10^8\ \mathrm{m/s}$$

At the time Maxwell did this calculation, the speed of light in air had already been measured and was consistent with this value. Could it be that light is an electromagnetic wave? This was the second testable consequence of Maxwell's model. *If* the relationship between the changing electric and magnetic fields suggested by the model is correct, *then* a change in either of these fields could generate an electromagnetic wave

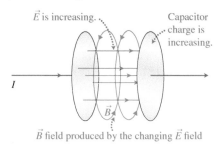

FIGURE 25.3 A changing electric field produces a magnetic field.

$\vec{E}$ is increasing.   Capacitor charge is increasing.

I

$\vec{B}$ field produced by the changing $\vec{E}$ field

FIGURE 25.4 Changing $\vec{E}$ and $\vec{B}$ fields produce $\vec{B}$ and $\vec{E}$ fields.

(a)

A changing $\vec{E}$ field produces a $\vec{B}$ field.

$\vec{E}$

$\vec{B}$

$\dfrac{\Delta E}{\Delta t} > 0$

(b)

A changing $\vec{B}$ field produces an $\vec{E}$ field.

$\vec{B}$

$\vec{E}$

$\dfrac{\Delta B}{\Delta t} > 0$

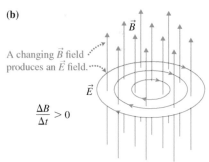

FIGURE 25.5 The changing $\vec{B}$ and $\vec{E}$ fields can spread without any charges or currents.

A changing $\vec{B}$ field ...   ... produces a changing $\vec{E}$ field ...   ... produces a changing $\vec{B}$ field.

$\vec{E}''$ $\vec{B}''$ $\vec{B}$ $\vec{B}'$ $\vec{E}'$

$\vec{E}$

The $\vec{B}$ and $\vec{E}$ fields propagate themselves.

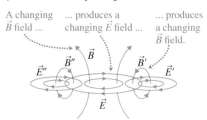

that would propagate at the speed of light. This prediction, if consistent with the experiment, would support the hypothesis that light is an electromagnetic wave composed of vibrating electric and magnetic fields!

Testing the hypothesis that light can be modeled as an electromagnetic wave

The German physicist Heinrich Hertz (1857–1894) was the first to test the hypothesis. In 1888 Hertz built a device called a **spark gap transmitter** and used it in his experiments. His work is summarized in Testing Experiment **Table 25.3**.

TESTING
EXPERIMENT TABLE **25.3**   Hertz's experiments

Testing experiment	Prediction	Outcome
Hertz used a switch to connect a transmitter with a charged capacitor to the primary coil of a transformer. He connected the secondary coil of the transformer to two metal spheres. When he discharged the capacitor, the changing current through the primary coil induced a huge emf across the secondary coil, which in turn charged the spheres, causing them to spark. A receiver that consisted of two small metal spheres separated by a small gap and connected to a loop of wire was placed some distance from the metal transmitter spheres.	A spark between the transmitter spheres would indicate a large changing electric field between the spheres of the transmitter. If Maxwell's hypothesis was correct, then this changing electric field would produce an electromagnetic wave. When the wave reached the receiver loop, it would induce a current in the loop and cause a weak spark between the spheres of the receiver.	After hours in a dark room watching for a spark, Hertz eventually saw a spark between the receiver spheres.

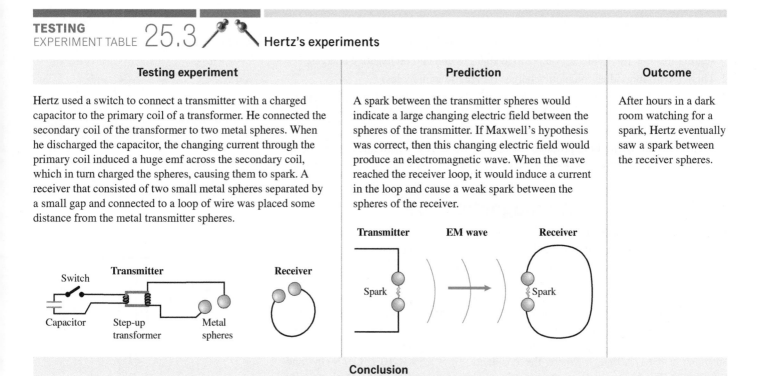

Conclusion

That Hertz could see a spark between the spheres of the receiver in response to the spark between the spheres of the transmitter meant that something had traveled from the transmitter to the receiver.

Hertz's experiment was the first supporting evidence for the idea that electromagnetic waves existed. He performed many additional experiments to determine if the electromagnetic waves generated in his experiment had the same properties as light waves. The only difference he found was in their frequency—Hertz's waves had a much smaller frequency than visible light waves. Hertz used metal sheets of different shapes to observe reflection. He let the waves pass through different media and observed refraction. He performed an analog of a double-slit experiment and observed interference. He even observed polarization of the waves using a metal fence. Finally, he performed experiments to measure the speed of the propagation of the waves, which turned out to be 3.00×10^8 m/s. These experiments supported the idea that light could be modeled as a transverse wave of vibrating electric and magnetic fields.

Maxwell's model of light as an electromagnetic wave seems to resolve the problem of what is vibrating in the light wave: the $\vec{E}$ field and the $\vec{B}$ field. Each field is vibrating perpendicular to the direction of travel of the wave. Thus we have a new model of light as a *transverse electromagnetic wave* that travels in a vacuum and in other media. In a vacuum, the speed of light is 3.00×10^8 m/s.

The waves predicted by Maxwell and experimentally discovered by Hertz soon found a practical application. In 1892 Nikola Tesla used an improved version of Hertz's transmitter to conduct the first transmission of information via radio waves, one form of electromagnetic waves. By 1899 there was successful radio wave-based communication across the English Channel.

You might be curious about what happened to the idea of ether. Although it was no longer needed to explain the nature of light, it still remained in physics until 1905. We will return to the fate of ether in Chapter 26.

Antennas are used to start electromagnetic waves

The most common device used to produce electromagnetic waves is an antenna. Cell phones, two-way radios, and over-the-air radio and television stations all use them. A simple type of antenna, called a **half-wave electric dipole antenna**, can be made from a pair of electrical conductors, one connected to each terminal of a power supply that is producing an alternating emf (**Figure 25.6**). The alternating emf leads to the continuous charging and discharging of the two ends of the antenna (somewhat similar to the action of a capacitor). How does this antenna produce an electromagnetic wave?

The source of alternating emf produces alternating current. Let the period of this oscillation be T. Assume that just after the power supply is turned on, the current is downward. This produces a $\vec{B}$ field ($t = 0$). After a very short time interval (nanoseconds for radio waves), this current causes the bottom of the antenna to become increasingly positively charged and the top increasingly negatively charged. This produces an $\vec{E}$ field ($t = T/8$). Both fields are shown in **Figure 25.7a**. Since the antenna is connected to a source of alternating emf, the current reverses direction periodically.

FIGURE 25.6 A half-wave dipole antenna.

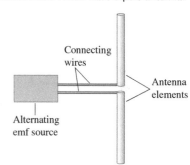

FIGURE 25.7 The electric and magnetic fields produced by a dipole antenna.

(a) $t = T/8$

Charge separation produces $\vec{E}$ field. Current produces $\vec{B}$ field.

(b) $t = T/4$

Current zero and $\vec{B}$ field zero. $\vec{E}$ field is maximum.

(c) $t = 3T/8$

Current and $\vec{B}$ field have reversed. $\vec{E}$ field is decreasing.

(d) $t = T/2$

No charge separation and $\vec{E}$ field is zero. $\vec{B}$ field is maximum.

(e)

$\vec{B}$ out of page $B = 0$ $\vec{B}$ into page B is maximal $B = 0$

$\vec{E}$ and $\vec{B}$ fields at a point near the front of the antenna

$t = 0$ $T/8$ $T/4$ $3T/8$ $T/2$ $5T/8$ $3T/4$ $7T/8$ T

When this happens, the current momentarily vanishes along with its corresponding $\vec{B}$ field (Figure 25.7b, $t = T/4$). When the current starts upward, the charge separation begins to decrease along with the $\vec{E}$ field produced by it (Figure 25.7c, $t = 3T/8$). Next, the charge separation vanishes along with its $\vec{E}$ field (Figure 25.7d, $t = T/2$). The process then repeats in the opposite direction and continues to repeat billions of times each second. Figure 25.7e shows what happens at one point in space during one period T of oscillation. The $\vec{E}$ and $\vec{B}$ fields at that point are shown in sequence as time passes. Note that the second part in (e) corresponds to the fields shown in (a). The third part in (e) corresponds to the fields in (b), and so forth.

Near the antenna, the oscillating electric charge and current in the antenna produce the $\vec{E}$ and $\vec{B}$ fields as described above (near fields). Farther from the antenna, the fields co-create each other in the way Maxwell's equations describe (far fields). There is a noticeable difference between these two situations. In Maxwell's electromagnetic (EM) waves, the $\vec{E}$ field and $\vec{B}$ field oscillate in phase. That is, when the $\vec{E}$ field at a point is at a maximum, so is the $\vec{B}$ field. Close to the antenna, the $\vec{E}$ field and $\vec{B}$ field are out of phase, the $\vec{E}$ field being at a maximum while the $\vec{B}$ field is zero, and vice versa.

Both near and far $\vec{E}$ and $\vec{B}$ fields contribute to the net $\vec{E}$ and $\vec{B}$ fields at a particular distance from the antenna: (1) the fields produced by the oscillating charges and current in the antenna and (2) the electromagnetic waves produced by the oscillating $\vec{E}$ field and $\vec{B}$ field in the region surrounding the antenna. The amplitudes of the $\vec{E}$ field and $\vec{B}$ field produced by contribution 1 get weaker with distance $(1/r^2)$ much faster than the amplitude of the EM waves produced by contribution 2 $(1/r)$. In other words, the farther from the antenna, the less significant the out-of-phase contribution compared to the in-phase contribution. Far from the antenna, only Maxwell's in-phase EM waves are significant.

The size and shape of antennas are carefully selected so that they can efficiently send and receive EM waves with a particular range of wavelengths. In the case of the dipole antenna shown in Figure 25.6, the total length of the antenna is related to the desired wavelength of EM waves that it will produce. It is called a *half-wave* dipole antenna because its length is half the wavelength of the oscillating electric field within the antenna that drives the current. However, since this electric field exists in a different material (the metal antenna), its wavelength isn't the same as the wavelength of the EM wave produced by the antenna (those fields are in air), but they are related. Recall from our study of refraction (Chapter 22) that different materials have different indexes of refraction, which affects the wavelength of the EM waves passing through them. Knowing the index of refraction of the antenna material allows us to choose an antenna length that will very efficiently produce EM waves of the desired wavelengths. For example, most cell phones operate at frequencies between 850 and 1900 MHz (called a *frequency band*), corresponding to wavelengths of about 10–30 cm. Within the antenna the wavelength is about one-third of its value in air. As a result, cell phone antennas need to be roughly 3–10 cm in length, consistent with the size of modern cell phones.

"Reflecting" on models of light

Let us review how our understanding of light has evolved over the past four chapters. First we learned about shadows and the phenomena of reflection and refraction. To explain these, we used a particle-bullet model, which assumed that light consisted of tiny bullets that travel in straight lines through a medium. The model did not specify the nature of these bullets. Then we found new phenomena that the bullet model could not explain, such as what happens when light passes through narrow openings. So we progressed to a wave model, but it was not clear what was vibrating in a light wave. In this chapter, observations of polarization of light led us to the idea that a light wave is a transverse wave. Maxwell's and Hertz's work led to the electromagnetic wave model of light, which proposes that light can be modeled as a transverse wave in which the electric and magnetic fields themselves are vibrating. The next section will give examples of practical applications of the new model of light.

REVIEW QUESTION 25.2 What needs to happen to produce an electromagnetic wave?

25.3 Applications of electromagnetic waves

We use electromagnetic waves every day: they are present when a police officer uses radar to catch speeding motorists, when we use GPS to find a new destination, and when we use a microwave to cook food.

Radar

The EM waves produced by Hertz in his experiments are called **radio waves**, based on their frequency $(10^4 – 10^6 \text{ Hz})$. Today, these and even higher frequency waves are used to transmit speech, music, and video over long distances. Another practical application of radio waves is radar (an acronym of radio detection and ranging). **Radar** is a way of determining the distance to a faraway object by reflecting radio wave pulses off the object. Radar devices help locate airplanes, determine distances to planets and other objects in the solar system, find schools of fish in the ocean, and perform many other useful functions.

In radar, an antenna (called the **emitter**) emits a short radio wave pulse, which reflects off the surface of the object and is detected by a **receiver**. This receiver is just an antenna that is connected to a sophisticated type of ammeter that can measure the current produced when the reflected electromagnetic wave arrives. Most often, the emitting antenna and the receiving antenna are actually the same antenna.

Imagine that a radar system is being used to detect aircraft. The time interval for the round trip of the pulse indicates the distance to the aircraft. The longer it takes the pulse to return, the farther away the aircraft is. By emitting a sequence of pulses and determining how the distance to and direction of the aircraft changes from one pulse to the next, the system can determine the velocity of the aircraft.

Radar has some limitations. It is less effective in tracking objects that are very close, very far away, or flying very low to the ground. The difficulty of tracking close objects is related to the time interval τ that it takes a single pulse to be emitted (called the *pulse width*) and the time interval T that passes between the start of each successive pulse (called the *period* of the pulse; **Figure 25.8**). If the object is so close that the end of the pulse has not been emitted before the beginning of the pulse returns, then the distance to the object is difficult to determine accurately. This limitation has been alleviated somewhat by advanced electronic switching systems that produce very short pulse widths.

The source of the second limitation, measuring distant objects, is the time interval T. If the object is so far away that the reflected pulse returns after the next pulse is sent, then it becomes unclear which reflected pulse corresponds to which emitted pulse. In modern radar the period is between $T = 10^{-3}$ s and $T = 10^{-6}$ s. The overall effect of these two difficulties is that there is both a minimum and a maximum range within which an object can be accurately detected, determined by the pulse width τ and the period T.

The third limitation, tracking objects that are flying close to the ground, is a result of Earth's curved surface. Because of the curvature, the distance to the horizon is about 5 km as seen by a person standing on the ground. Thus, a target that is farther away than 5 km will not be visible to radar because the ground will block the line of sight. In order to obtain a range of 20 km, the radar system must be mounted on a tall tower in order to increase the horizon distance.

FIGURE 25.8 The radar pulse width τ and the time interval T between pulses determine the near and far distances at which a radar system can detect an airplane.

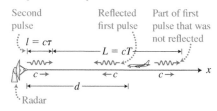

Snapshot of radar pulses at one instant

The reflected first pulse returns after the new second pulse has left.

EXAMPLE 25.1 ▸ **Radar**

Determine the range of distances to a target that can be reliably detected by a radar system with pulse width $\tau = 10.8\ \mu s$ ($1\ \mu s = 10^{-6}$ s) and a pulse emission frequency of 7.5 kHz, that is, 7500 radio wave pulses each second.

Sketch and translate The pulse width of 10.8 μs should be less than the time interval needed for the EM wave to travel to the target object and back. That way, the back of the 10.8-μs pulse will have left the transmitter before the front of the reflected pulse returns. The pulse width determines the minimum distance of the target from the radar.

  The pulse repetition period is $T = 1/f = 1/(7500\ \text{Hz}) = 1.33 \times 10^{-4}$ s. This period should be more than the time interval needed for the pulse to travel to the target and back. That way, the reflected pulse returns before a new pulse leaves. The period determines the maximum distance of the target from the radar.

Simplify and diagram Assume that pulses do not reflect off other objects while traveling to or from a target. Assume also that the radio wave pulses travel at the speed of light in vacuum (they actually travel just slightly slower in air).

Represent mathematically If the target is a distance d away, the time interval for a pulse to travel to the target and back is $\Delta t = (2d/c)$. This time interval needs to be longer than the pulse width and shorter than the pulse period in order for the radar to accurately determine the distance to the target. Mathematically,

$$\tau < \frac{2d}{c} < T$$

Solve and evaluate The pulse width time interval $\tau = 10.8\ \mu s = 10.8 \times 10^{-6}$ s and the period $T = 1.33 \times 10^{-4}$ s. Solving the above inequality for d, we get

$$\frac{c\tau}{2} < d < \frac{cT}{2}$$

The range of distances is

$$\frac{(3.00 \times 10^8\ \text{m/s})(10.8 \times 10^{-6}\ \text{s})}{2}$$
$$< d < \frac{(3.00 \times 10^8\ \text{m/s})(1.33 \times 10^{-4}\ \text{s})}{2}$$

or

$$1.62 \times 10^3\ \text{m} < d < 20.0 \times 10^3\ \text{m}$$

A target outside this 1.6-km to 20-km range will not be detectable by this radar system. However, military radars have significantly shorter pulse widths and can detect objects much closer than 1.6 km.

Try it yourself What is the maximum range of a 15-kHz pulse emission rate radar system mounted on a tall tower?

Answer   10 km.

Global Positioning System

Global positioning systems enable us to determine our location using electromagnetic waves and satellites. In the United States, the Global Positioning System (GPS) is operated by the U.S. Air Force, and its network of satellites is called Navstar. GPS has three main components:

1. A minimum of 24 satellites that orbit 20,200 km above Earth's surface and transmit signals using **microwaves**: electromagnetic waves at frequencies 1575 MHz and 1227 MHz ($1\ \text{MHz} = 10^6\ \text{Hz}$) (**Figure 25.9**).
2. Receivers that detect signals from multiple satellites.
3. A control system that maintains accurate information about the locations of the satellites.

The receiver detects signals from at least three satellites in order to determine your position on the ground. With signals from four or more satellites, the receiver can also determine your altitude. The clocks on the satellites and a clock in your unit are synchronized to within a nanosecond (10^{-9} s). When you turn on your GPS, it starts a measurement at the same time each satellite sends a microwave signal. Your unit measures the time of arrival of the signals from the satellites. Using the known positions of the satellites, your GPS unit is able to calculate your position by a process called **trilateration**.

  Trilateration is easier to understand in a two-dimensional space than in a three-dimensional space. Suppose a visitor from overseas is traveling in the United States and is unsure of his location. He comes across a sign that says Atlanta 430 miles, Minneapolis 510 miles, and Philadelphia 580 miles. The visitor has a map of the

FIGURE 25.9 Twenty-four satellites circle Earth as part of the Global Positioning System. Three or four GPS satellites trilaterate to determine your location.

United States and draws a circle centered on each city such that the radius of each circle is the distance to that city (**Figure 25.10**). There is only one place on the map where the three circles intersect—Indianapolis.

Your GPS system locates your current position in much the same way. The unit determines the distance between you and each of at least three satellites. The unit then constructs spheres (rather than circles) around each of them with corresponding radii equal to those distances. The spheres intersect at one unique location on Earth's surface (which is in effect a fourth sphere), your location. Most handheld GPS units can determine your position to within 3 to 10 m. More advanced receivers use a method called Differential GPS (DGPS) to obtain an accuracy of 1 m or better.

Microwave cooking

Percy Spencer accidentally discovered microwave cooking in 1945 while constructing magnetron tubes at Raytheon Corporation. While standing in front of one of the tubes, Spencer felt a candy bar in his pocket start to melt. Intrigued, he placed kernels of popcorn in front of the tube, and they popped. He placed a raw egg in front of the tube, and it exploded. How could this be explained?

Water, fat, and other substances in food absorb energy from microwaves in a process called **dielectric heating**. To understand what's happening, we have to think microscopically. Water is a polar molecule (**Figure 25.11**) and is a permanent electric dipole. When a microwave passes by a water molecule, the electric field of the microwave exerts oscillating electric forces on the positively and negatively charged regions of the molecule. This causes the water molecule to flip over billions of times each second (the frequency of microwaves). This molecular movement spreads to neighboring molecules through collisions (even to molecules that are not electric dipoles). The electric and magnetic energy in the microwaves is transformed into the internal energy of the food, which manifests as an increase in the food's temperature. Microwave heating is most efficient in liquid water and much less so in fats and sugars (which have smaller electric dipoles) and ice (where the water molecules are not as free to rotate). Substances that are composed of nonpolar molecules do not warm up (or warm up much less) in microwave ovens. Mineral oil (which you can find in a drugstore), for instance, barely warms up at all in a microwave.

FIGURE 25.10 Surface trilateration is analogous to that used in GPS.

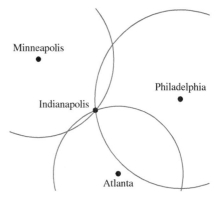

Minneapolis

Philadelphia

Indianapolis

Atlanta

Only one place (Indianapolis) is the known distance from the other three cities.

FIGURE 25.11 A water molecule is an electric dipole with more positive charge on one end and negative charge on the other.

Protons

H   H

O

Electron
pair

An equivalent model
of the water dipole

REVIEW QUESTION 25.3 How are GPS and radar similar, and how are they different?

25.4 Frequency, wavelength, and the electromagnetic spectrum

Recall that waves are described by frequency f, speed v, wavelength λ, and a relationship between these three quantities [Eq. (11.3)]:

$$\lambda = \frac{v}{f}$$

All electromagnetic waves travel at the speed of light c in a vacuum. However, in media other than a vacuum, the speed of electromagnetic waves changes to $v = c/n$, where n is the refractive index of the medium. In addition, light waves of different frequencies have slightly different refractive indexes and hence travel at different speeds in non-vacuum media such as glass or water. Therefore, in the equation $v = c/n$, n is not a constant for a particular medium, but has a small dependence on the frequency of the electromagnetic wave.

According to Maxwell's equations, electromagnetic waves can have essentially any frequency. AM radio stations emit relatively low-frequency EM waves (about 10^6 Hz = 1 MHz). Supernova remnants and extragalactic sources emit the highest frequency cosmic ray EM waves ever detected (up to about 10^{26} Hz).

Wavelengths of FM radio stations

The FM radio broadcast band is from 88 MHz to 108 MHz. What is the corresponding wavelength range?

Represent mathematically Assume that these radio waves are traveling at the speed of light in a vacuum. This is a reasonable assumption since the refractive index of air is very close to the refractive index of a vacuum (both are ≈ 1). Apply $\lambda = c/f$ to find the range of wavelengths for FM radio waves.

Solve and evaluate

$$\lambda_1 = \frac{3.0 \times 10^8 \text{ m/s}}{88 \times 10^6 \text{ s}^{-1}} = 3.4 \text{ m}$$

$$\lambda_2 = \frac{3.0 \times 10^8 \text{ m/s}}{108 \times 10^6 \text{ s}^{-1}} = 2.8 \text{ m}$$

The electromagnetic waves emitted by FM radio stations have wavelengths ranging from 2.8 m to 3.4 m.

Try it yourself AM radio stations broadcast in the wavelength range from 556 m to 186 m. Determine the frequency range of AM radio station broadcasting.

Answer   540 kHz to 1610 kHz.

"Hearing" FM radio waves

Human ears are not sensitive to electromagnetic waves of any frequency. Thus, for us to hear music on the radio, the electromagnetic waves need to be converted into sound waves. This is the role of a speaker or headphones. However, even if we convert an electromagnetic signal into a sound wave, our ears can only hear sounds with frequencies ranging from about 20 Hz to 20,000 Hz, much lower than the frequency of FM or satellite radio stations. How can we hear the music from those stations? The high-frequency EM waves used by FM radio stations are known as *carrier waves*. FM stands for "frequency modulation"; the information that is converted into the sounds we hear is encoded as tiny variations in the frequency of the carrier wave. The receiver decodes the variations and converts them into an electric signal that a speaker can then convert into sound. A similar process occurs in AM (amplitude modulation) radios, only in that case the amplitude of the carrier wave is varied rather than its frequency.

The electromagnetic spectrum

The range of frequencies and wavelengths of electromagnetic waves is called the **electromagnetic spectrum** (Figure 25.12). As you see, visible light occupies a very narrow range of the spectrum (wavelengths of 400 nm to 700 nm). In this chapter we've already encountered radio waves and microwaves. We encountered another type of radiation briefly (in Chapter 15) when we considered heating by radiation. Historically, scientists thought that visible light and thermal radiation (now called infrared [IR] radiation) were completely different. In other words, they thought that the light emitted by a fire in a fireplace was a different phenomenon than the warmth that the fire gave off. We now understand that they are both EM waves, just with different frequency ranges.

FIGURE 25.12 The electromagnetic (EM) spectrum.

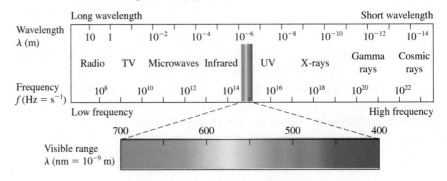

Table 25.4 lists the broad categories of EM radiation and some of the ways in which it is produced and detected. We will discuss EM radiation production in more detail in later chapters.

TABLE 25.4 **The electromagnetic spectrum**

Type of EM radiation and frequency/ wavelength range	Produced by	Detected by	Used in
Radio waves and microwaves 10^4–10^{10} Hz 10^3–10^{-3} m	Accelerated electrically charged particles in wires with oscillating current	Metal wires (antennas)	Radio and cell phone communication Microwave ovens
Infrared waves 10^{11}–10^{14} Hz 10^{-4}–10^{-5} m	Objects with a temperature between 10 K and 1000 K	Detectors whose properties (for example, electric conductivity) change with temperature	Toasters Conventional ovens
Visible light 10^{14}–10^{15} Hz 10^{-6} m	Low-density ionized gas Objects whose temperatures are between 1000 K and 10,000 K, such as typical stars	Human eyes Charge-coupled devices (CCDs) and complementary metal-oxide semi-conductor (CMOS) image sensors	Visual communication
UV light 10^{16}–10^{17} Hz 10^{-7}–10^{-8} m	Stars Black lights	Charge-coupled devices (CCDs) that can detect UV radiation	Forensics Decontamination
X-rays 10^{18}–10^{19} Hz 10^{-9}–10^{-10} m	Accelerating particles in X-ray tubes Very hot objects Energetic stellar events such as supernovas	Photographic plates Geiger counters	Medical imaging Studies of material properties Cosmology
Gamma rays 10^{20} Hz 10^{-12} m	Radioactive materials (such as uranium and its compounds) Nuclear reactions Stellar explosions	Solid-state detectors that transform gamma rays into optical or electronic signals	Detection of nuclear explosions Cosmology

Don't interpret Table 25.4 too literally. For example, most objects emit many of these types of EM waves, but in different proportions. For example, cold gas clouds in our galaxy primarily emit radio waves but also emit a small amount of infrared waves and visible light. Stars similar to our Sun emit mostly infrared, visible, and UV, but also produce some radio waves and X-rays. Even your body emits all types of EM waves: primarily infrared radiation and only very small amounts of UV, X-ray, and gamma radiation.

REVIEW QUESTION 25.4 If the frequency of one electromagnetic wave is twice that of another, how do the speeds and wavelengths of the two waves differ?

25.5 Mathematical description of EM waves and EM wave energy

When analyzing electric dipole antennas, we found that they produce electric and magnetic fields whose $\vec{E}$ and $\vec{B}$ vectors vibrate perpendicular to each other and perpendicular to the direction the wave travels (Figure 25.13). Maxwell's equations also predict that the amplitudes of the changing $\vec{E}$ field and $\vec{B}$ field vectors are related:

$$E_{max} = cB_{max} \qquad (25.1)$$

FIGURE 25.13 Electromagnetic wave $\vec{E}$ and $\vec{B}$ fields (a) at one time and (b) at one location.

(a)

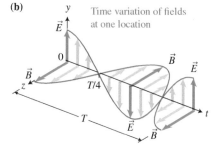

(b)

where $c = 3.00 \times 10^8$ m/s is the speed of electromagnetic waves in vacuum. In other words, not only do the electric and magnetic fields oscillate in phase, but also their magnitudes are related through the speed of light.

We can check the above equation with unit analysis. The units for electric field are newtons/coulomb, or, equivalently, volts/meter. The first combination comes from the definition of the $\vec{E}$ field; the second comes from the mathematical relationship between the $\vec{E}$ field and the V field (see Chapter 18). The units for magnetic field are T (tesla) = (newton $\cdot$ second)/(coulomb $\cdot$ meter). This unit can be deduced from the expression for the magnetic force exerted on a changed particle moving in a magnetic field ($F = qvB \sin\theta$ or $B = F/(qv\sin\theta)$, as discussed in Chapter 20). We can now insert the units into the right-hand side of Eq. (25.1):

$$\left(\frac{m}{s}\right)\left(\frac{N \cdot s}{C \cdot m}\right) = \frac{N}{C}$$

which are the units of the left-hand electric field side of the equation. Thus, the equation survives a unit analysis.

Mathematical description of EM waves

We now have enough background to make a quantitative description of EM waves that involves their speed c, wavelength λ, and frequency f and the relationship between the amplitudes of the vibrating $\vec{E}$ and $\vec{B}$ fields, $E_{max} = cB_{max}$. To do this, we use the mathematical descriptions of waves (developed in Chapter 11) as a guide.

Imagine the simplest EM wave—a sinusoidal wave of a single frequency. It travels along the x-axis in the positive direction and does not diminish in amplitude as it travels. Now, imagine that you place electric and magnetic field detectors at a point on the positive x-axis. As the wave passes, the detectors register the magnitude of the oscillating $\vec{E}$ field and $\vec{B}$ field as

$$E = E_{max}\cos\left(\frac{2\pi}{T}t\right) = E_{max}\cos(2\pi f t)$$

$$B = B_{max}\cos\left(\frac{2\pi}{T}t\right) = B_{max}\cos(2\pi f t)$$

We chose $t = 0$ at a moment when the fields have their maximum values E_{max} and B_{max}. The frequency f of the wave is the inverse of the period T of the wave. A convenient right-hand rule relates the direction of the $\vec{E}$ field, the direction of the $\vec{B}$ field, and the direction of propagation of the wave. Hold your right hand flat with your thumb pointing out at a 90° angle relative to your fingers. If you point your fingers in the direction of the $\vec{E}$ field, your palm will face in the direction of the $\vec{B}$ field. The direction of your thumb will determine the direction of the velocity vector of the electromagnetic wave. For example, if at a particular moment the $\vec{E}$ field points in the +y-direction and the $\vec{B}$ field points in the +z-direction, then the wave is traveling in the +x-direction at speed c (**Figure 25.14**).

The above two equations describe the wave vibrations at one specific point. However, the $\vec{E}$ field vectors and $\vec{B}$ field vectors are oscillating at every point along the wave's path, which means that the equation for the wave must be a function not only of time t but also of position x. The wave equation that we developed in Chapter 11 [Eq. (11.4)] takes this into account. Applied to the $\vec{E}$ field and $\vec{B}$ field, the wave equation becomes

$$E_y = E_{max}\cos\left[2\pi\left(\frac{t}{T} \pm \frac{x}{\lambda}\right)\right]$$

$$B_z = B_{max}\cos\left[2\pi\left(\frac{t}{T} \pm \frac{x}{\lambda}\right)\right] \tag{25.2}$$

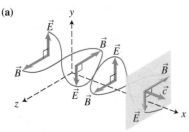

FIGURE 25.14 (a) The $\vec{E}$, $\vec{B}$, and $\vec{v} = \vec{c}$ vectors of an electromagnetic wave. (b) The right-hand rule for the $\vec{E}$, $\vec{B}$, and $\vec{v}$ EM wave directions.

The $\vec{E}$ and $\vec{B}$ fields vibrate perpendicular to each other and in a plane perpendicular to the wave's velocity $\vec{v} = \vec{c}$.

TIP The relation between E, B, and c [Eq. (25.1)] is applicable only for electromagnetic waves and not for fields in general and is correct only if the three quantities are measured using SI units.

The minus signs are used to describe a wave traveling in the positive x-direction, and the plus signs are used to describe a wave traveling in the negative x-direction. We can represent these equations graphically (shown for a wave moving in the positive x-direction at $t = 0$ in (a) and at $x = 0$ in (b) in Figure 25.13).

Suppose we wish to describe light from a laser pointer that emits light of frequency 4.5×10^{14} Hz with an oscillating $\vec{B}$ field of amplitude 6.0×10^{-4} T. What else can you determine about this electromagnetic wave from this information?

(a) The period of the wave is

$$T = \frac{1}{f} = \frac{1}{4.5 \times 10^{14} \text{ Hz}} = 2.2 \times 10^{-15} \text{ s}$$

(b) The wavelength of the wave is

$$\lambda = \frac{c}{f} = \frac{3.00 \times 10^8 \text{ m/s}}{4.5 \times 10^{14} \text{ Hz}} = 6.7 \times 10^{-7} \text{ m} = 670 \text{ nm}$$

(c) A wavelength of 670 nm corresponds to red light (see Figure 25.12).

(d) The amplitude of electric field oscillation [Eq. (25.1)] is

$$E_{max} = cB_{max} = (3.00 \times 10^8 \text{ m/s})(6.0 \times 10^{-4} \text{ T}) = 1.8 \times 10^5 \text{ N/C}$$

(e) Assuming the laser light travels in the positive x-direction, we can write wave equations for it:

$$E_y = (1.8 \times 10^5 \text{ N/C}) \cos\left[2\pi\left(\frac{t}{2.2 \times 10^{-15} \text{ s}} - \frac{x}{6.7 \times 10^{-7} \text{ m}}\right)\right]$$

$$B_z = (6.0 \times 10^{-4} \text{ T}) \cos\left[2\pi\left(\frac{t}{2.2 \times 10^{-15} \text{ s}} - \frac{x}{6.7 \times 10^{-7} \text{ m}}\right)\right]$$

Energy of electromagnetic waves

In previous chapters (Chapters 17 and 18) we learned that a system of electrically charged objects possesses electric potential energy, and that same idea can be expressed through the concept of electric field energy density. This energy density u_E quantifies the energy stored in the electric field per cubic meter of volume:

$$u_E = \frac{\kappa}{8\pi k_C} E^2 \tag{18.16}$$

Rewriting $k_C = \dfrac{1}{4\pi\epsilon_0}$ and assuming that the region with electric field is a vacuum ($\kappa = 1$), the above becomes

$$u_E = \frac{1}{8\pi k_C} E^2 = \frac{1}{8\pi\left(\dfrac{1}{4\pi\epsilon_0}\right)} E^2 = \frac{1}{2}\epsilon_0 E^2$$

What about the magnetic field? Does it have energy as well? To answer this question, think of what happens to the electric current when you connect a solenoid to a DC power supply. The current through the solenoid increases from the initial value of zero to the final value I. This changing current produces a changing magnetic flux through the solenoid, and this, in turn, produces an induced emf, which (according to Lentz's law) opposes the change in the current through the circuit. Let the power supply and the circuit, including the connecting wires and the solenoid, be the system. While the current created by the DC power supply through the solenoid is increasing, the DC power supply needs to add more energy to the circuit to counteract the induced current created in the solenoid, which is in the opposite direction to the current created by the battery. This extra energy coming from the power supply results in an increase of the magnetic field energy of the system—the energy that is stored in the magnetic field created inside and around the solenoid.

We can find this energy by calculating the energy that the power supply needs to add to the circuit to increase the current from zero to the final value. Assuming that the current increases linearly with time, the total charge transferred through the solenoid can be expressed as $q = I_{\text{average}}\Delta t = \dfrac{I}{2}\Delta t$. Assuming also that the initial energy of the magnetic field is zero, the final energy of the magnetic field is

$$u_B = \mathcal{E}_{\text{in}}q = \mathcal{E}_{\text{in}}\frac{I}{2}\Delta t = \frac{\Delta\Phi}{\Delta t}\frac{I}{2}\Delta t = \frac{LI^2}{2} \tag{25.3}$$

where we used Faraday's law of electromagnetic induction and expressed the magnetic flux with the inductance of the solenoid ($\Phi = LI$, Chapter 21). Using the expression for the inductance of a long solenoid ($L = \mu_0\dfrac{N^2}{l^2}V$, where N is the number of turns, l is the length of the solenoid, and V is the volume of the solenoid) and the expression for the magnetic field in the center of the solenoid $\left(B = \mu_0\dfrac{NI}{l}\right)$, the above expression for the energy of the magnetic field of the solenoid can be expressed in the following form:

$$u_B = \frac{1}{2\mu_0}\left(\frac{\mu_0 NI}{l}\right)^2 V = \frac{B^2}{2\mu_0}V$$

Because the magnetic field of a long solenoid is approximately uniform inside the solenoid and very small outside the solenoid, we can divide this expression by the volume of the solenoid and obtain the expression for the magnetic field energy density:

$$u_B = \frac{B^2}{2\mu_0} \tag{25.4}$$

Although we derived this expression for the case of a long solenoid, it turns out that it is valid for magnetic fields in general.

We can now compare the magnetic field energy density with the electric field energy density. Because $E = cB$ and the electric field energy density is $u_E = \frac{1}{2}\epsilon_0 E^2$, $u_E = \frac{1}{2}\epsilon_0 c^2 B^2$. But we know that the speed of light is $c = \dfrac{1}{\sqrt{\epsilon_0\mu_0}}$ or $c^2 = \dfrac{1}{\epsilon_0\mu_0}$; thus,

$$u_E = \frac{1}{2\mu_0}B^2 = u_B!$$ We find that the electric and magnetic fields provide equal contributions to the total energy density of an electromagnetic wave.

The total energy density of the electric and magnetic fields in an EM wave is the sum of the energy densities of each:

$$u = u_E + u_B$$

$$\Rightarrow u = \frac{1}{2}\epsilon_0 E^2 + \frac{1}{2\mu_0}B^2$$

However, since these two terms are equal, we can write the energy density of the EM wave as twice the first term or twice the second term:

$$u = 2\left(\frac{1}{2}\epsilon_0 E^2\right) = \epsilon_0 E^2 \quad \text{or} \quad u = 2\left(\frac{1}{2\mu_0}B^2\right) = \frac{1}{\mu_0}B^2 \tag{25.5}$$

Average values of the energy densities

Remember that in the equations above, E and B are the instantaneous values of the magnitudes of the $\vec{E}$ and $\vec{B}$ fields. In an EM wave of a specific wavelength, both of those fields oscillate sinusoidally (**Figure 25.15a**). This means that the energy density at each point in the wave fluctuates as a sine-squared function since the field appears squared in the equations. During one vibration of the field, the energy density starts with a maximum value; then after one-quarter of a period the energy density is zero; and after half a period the energy is again maximal (Figure 25.15b).

FIGURE 25.15 Variation of the $\vec{E}$ and $\vec{B}$ fields at one point and (b) the energy density variation with time of an EM wave at one point.

(a)

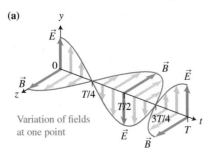

Variation of fields at one point

(b)

Variation of energy density at one point

Since the energy density at each point in the EM wave fluctuates with extremely high frequency, it's usually more useful to think in terms of the time-averaged energy density at each point in the EM wave. In order to do this, we need to calculate the average value of the sine-squared function shown in Figure 25.15b. We already know this value from Chapter 21: the average value of the square of a sinusoidal function (either sine or cosine) is one-half the maximum value. Therefore, the average value (indicated by the bar above u) of the energy density along the direction of the EM wave is

$$\bar{u} = \tfrac{1}{2}u_{max} = \tfrac{1}{2}(\epsilon_0 E_{max}^2) = \tfrac{1}{2}\epsilon_0 E_{max}^2 \qquad (25.6)$$

Intensity of an EM wave

Now that we understand the average energy density of the EM wave, we can understand quantitatively what is meant by its *intensity*. When you look at a source of light, it might appear dim to you, like a distant star. Or it might appear bright to you, like a car with its high beams on. The physical quantity that characterizes this brightness is called *intensity*. (We encountered this quantity in Chapter 11.)

Remember that intensity is the amount of energy (not energy density) that passes through a unit area during a unit time interval [see Eq. (11.6)].

$$\text{Intensity} = \frac{\text{Energy}}{\text{time interval} \times \text{area}} = \frac{\Delta E}{\Delta t \times A} = \frac{P}{A}$$

where P is the power of the wave. Technically, since the energy density of the field fluctuates, the intensity of the wave fluctuates as well, but this fluctuation is so rapid that it makes much more sense to think about the average intensity of the EM wave. The average intensity depends on the wave's average energy density $\bar{u} = (1/2)\epsilon_0 E_{max}^2$ and the speed c of the wave through the area (the faster the wave travels, the more energy passes through the area, though for EM waves this speed is always c). It is important to note that the intensity of all waves (sound waves, waves on rope, etc.), not just electromagnetic waves, is proportional to the square of the amplitude of vibrations.

One way to think of intensity and energy density is to imagine wind blowing into a sail. The wind speed is similar to the speed of the wave, and the kinetic energy per cubic meter of air is similar to the energy density of the wave. During a time interval Δt, the energy within the volume $V = A(c\,\Delta t)$ passes through the area A (**Figure 25.16**). The intensity equals this energy divided by the time interval and by the area:

$$I = \frac{\text{Energy}}{\text{time interval} \times \text{area}} = \frac{\bar{u}V}{\Delta t A} = \frac{\bar{u}Ac\,\Delta t}{\Delta t A} = \bar{u}c \qquad (25.7)$$

FIGURE 25.16 Intensity involves the rate of energy crossing a surface.

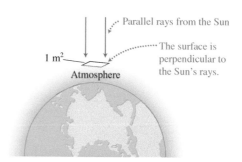

"Box" of electromagnetic energy

u_{total} A

$c\,\Delta t$ = the distance the energy moves to the right during a time interval Δt.

EXAMPLE 25.3   Solar constant

The *solar constant* is a fundamental constant in environmental science and astronomy that describes the intensity of the Sun's EM radiation when it arrives at Earth. It is expressed in terms of the total electric and magnetic field energy incident each second on a 1.0-m² surface located above the atmosphere directly facing the Sun:

$$\text{Solar constant} = 1.37 \text{ kW/m}^2$$

What is the amplitude of the $\vec{E}$ field oscillation just above the atmosphere due to the Sun's illumination of Earth?

Sketch and translate   Imagine the Sun's radiation striking a 1.0-m² surface above Earth's atmosphere at a zero angle of incidence. Remember that we define the angle of incidence as the angle between the ray and the normal to the surface. To determine the amplitude E_{max} of the $\vec{E}$ field, we need to relate the solar constant to the average energy density of the EM radiation, which in turn depends on the $\vec{E}$ field amplitude E_{max}.

Parallel rays from the Sun

The surface is perpendicular to the Sun's rays.

1 m²

Atmosphere

Simplify and diagram   Assume that light from the Sun travels to Earth at speed c. Assume also that all of the EM radiation is of a single wavelength. This is a poor assumption since the different wavelength EM waves arriving at Earth each have different electric field amplitudes. Therefore, what we are determining is the electric field amplitude averaged over all wavelengths.

(CONTINUED)

Represent mathematically We know from Eq. (25.5) that $I = \bar{u}c$ or $\bar{u} = (I/c)$. Using Eq. (25.4), we have $\bar{u} = (1/2)\epsilon_0 E_{max}^2$. Thus $(I/c) = (1/2)\epsilon_0 E_{max}^2$.

Solve and evaluate We can now determine E_{max} by multiplying both sides of the previous equation by 2, dividing by ϵ_0, and then taking the square root:

$$E_{max} = \sqrt{\frac{2I}{c\epsilon_0}}$$

$$= \sqrt{\frac{2(1.37 \times 10^3 \text{ W/m}^2)}{(3.00 \times 10^8 \text{ m/s})(8.85 \times 10^{-12} \text{ C}^2/(\text{N} \cdot \text{m}^2))}}$$

$$= 1010 \text{ N/C}$$

Is the number reasonable? We could estimate the $\vec{E}$ field amplitude a certain distance from a 60-W lightbulb in a room and compare it to this result. We expect the $\vec{E}$ field amplitude produced by the bulb to be significantly less since the Sun appears to be significantly brighter than a 60-W lightbulb. See the following "Try it yourself" question.

Try it yourself Estimate the amplitude of the $\vec{E}$ field produced by a 60-W electric bulb if you are 2.0 m away from it. Assume that the energy emitted by the bulb each second is completely in the form of EM waves.

Answer

30 N/C. As expected, this is significantly less than the $\vec{E}$ field amplitude produced by the Sun above Earth's atmosphere.

REVIEW QUESTION 25.5 Electromagnetic waves are sinusoidal vibrations of electric and magnetic fields. The average value of a sine function is zero. How can electromagnetic waves carry any energy?

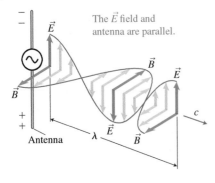

FIGURE 25.17 An antenna produces $\vec{E}$ field oscillations that far from the antenna are parallel to the antenna.

The $\vec{E}$ field and antenna are parallel.

Antenna

FIGURE 25.18 Light from a bulb is unpolarized.

(a)   Side view

Waves produced by independent atomic oscillators have random polarization directions.

(b)   Front view

Unpolarized light

25.6 Polarization and light reflection

We now have a more powerful model of light as an electromagnetic wave. Let's use this model to understand how light is produced, how polarized sunglasses work, how to reduce glare from reflected sunlight, and the role of polarization in liquid crystal displays (LCDs) on computer and calculator screens.

Producing unpolarized light

Many sources of light produce unpolarized light. In this section we discuss the mechanism behind this phenomenon. To understand unpolarized light, we first review the production of polarized light by an antenna.

Electrons vibrating up and down in an antenna produce radio waves. Far from the antenna, these waves take the form of plane waves with the $\vec{E}$ field vector polarized in the direction parallel to the antenna (**Figure 25.17**). A polarizer can reduce the amplitude of this radio wave to zero if the polarizer is oriented in a specific direction relative to the antenna.

We can think of light coming from an incandescent lightbulb as being produced by the motion of charged particles within the filament, each acting as a tiny "antenna." The light emitted by the bulb consists of many waves originating at random times with random polarizations (**Figure 25.18a**). If we could observe the many separate EM waves as a beam of unpolarized light moving directly toward our eyes, the oscillations of both the $\vec{E}$ and $\vec{B}$ fields would look something like a porcupine (the head-on appearance of the $\vec{E}$ field vectors is shown in Figure 25.18b). These oscillations explain why light produced by an incandescent bulb is unpolarized.

Light polarizers

How do polarizers interact with light? A polarizer directly affects the electric field of the passing electromagnetic wave and does not affect the magnetic field. The reason is that the magnetic field exerts a force only on a *moving* charged object. On the other hand, the electric field exerts a force that is independent of the motion of the object. This means that the electric field has a much more significant effect on matter since, most of the time, the protons and electrons that make up matter don't have any sort of collective motion.

A polarizer absorbs one component of the $\vec{E}$ field of the EM wave passing through it, allowing the perpendicular component to pass. To understand which component is absorbed, consider unpolarized microwaves passing through a grill of narrow metal bars (**Figure 25.19**). The component of the oscillating $\vec{E}$ field of the incident microwaves that is parallel to the grill produces an oscillating electric current in the grill. The resistive heating in the metal grill rods due to this current causes this component of the $\vec{E}$ field to diminish. The $\vec{E}$ field perpendicular to the metal grill rods does not cause a current and is not diminished. So the axis of the polarizing grill is actually perpendicular to the grill rods. The microwaves leaving the metal grill polarizer are linearly polarized in the direction perpendicular to the grill rods, and that is what defines the axis of the polarizer.

The situation is similar for light, only the dimensions are much smaller. A polarizing film such as that used in polarized sunglasses is made of long polymer molecules infused with iodine. The polymer chains are stretched in one direction to form an array of aligned linear molecules. Electrons can move along the lengths of the polymer chains, but there is resistance—just like the metal wires in the microwave polarizing grill. As a result, the $\vec{E}$ field component of unpolarized light parallel to the polymer chains is diminished. The $\vec{E}$ field perpendicular to the chains is not diminished. Thus, the axis of the light polarizer is perpendicular to the direction of the stretched polymer chains.

Polarization by reflection and Brewster's angle

The left-hand chapter-opening photo shows light reflected off a body of water. If you look at this light through a polarizer, as in the right-hand photo, the intensity of reflected light varies depending on the orientation of the polarizer as we rotate it around the direction of the reflected beam (see **Figure 25.20**).

This indicates that the reflected light is partially polarized even though the incident light was completely unpolarized. When David Brewster, a Scottish physicist and mathematician who lived in the middle of the 19th century, investigated this phenomenon more deeply, he found that the intensity of reflected light that passes through the polarizer is medium bright when the axis of the polarizer is parallel to the reflective surface (water or glass, for example), but the reflected light becomes dimmer when the axis of the polarizer is rotated by 90°. We will call the orientation of the polarizing axis parallel to the reflective surface the x-direction and the direction perpendicular to the surface the y-direction (marked in Figure 25.20). Brewster found that when light was incident on the surface at a particular angle, called the **polarizing angle** θ_p, the reflected light was completely blocked by the polarizer whose axis is oriented in the y-direction. Given that the reflected light is blocked by a polarizer oriented in the y-direction, the reflected light must be completely polarized in the x-direction. Brewster also found that when the angle of incidence is θ_p, the reflected and the refracted light beam directions always make a 90° angle relative to each other.

To help you understand what happens, we set up a glass table on a balcony on a sunny day. In **Figure 25.21a** you can see metal silverware, a wooden plate filled with water, and two dark lines that are the reflections of the shadow cast on the table by the closed balcony door. The black arrow shows the orientation of the polarizer placed in front of the camera. The axis is parallel to the reflective surface of the glass table and to the water surface in the plate (it is in the x-direction marked in Figure 25.20). Notice how bright the reflected light is. The table is shiny, and you cannot see that there is something in the water. The situation changes when the polarizer is rotated 90°, that is, its axis is in the y-direction marked by the arrow in Figure 25.21b. The glare from the table surface and water is less noticeable, and you can see a slice of carrot in the water that was previously masked by the reflected light. Because the polarizer with its axis in the y-direction blocks some light, we conclude that the blocked light (the reflected light) is polarized in the x-direction.

Now examine the bottom part of the photo in Figure 25.21b. You can see that the glare from the table and the reflection of the balcony door shadows are barely noticeable.

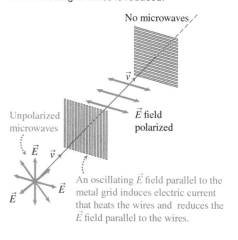

FIGURE 25.19 The $\vec{E}$ field component parallel to the metal grill wires is reduced.

An oscillating $\vec{E}$ field parallel to the metal grid induces electric current that heats the wires and reduces the $\vec{E}$ field parallel to the wires.

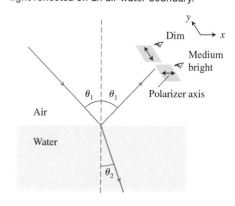

FIGURE 25.20 Investigating polarization of light reflected off an air-water boundary.

FIGURE 25.21 Photos taken on a sunny day (a) through a polarizer with its axis in the x-direction and (b) through the polarizer with its axis in the y-direction.

FIGURE 25.22 Reflection and refraction of light incident on a surface at the polarizing angle: the $\vec{E}$ field direction in the incident and reflected beams.

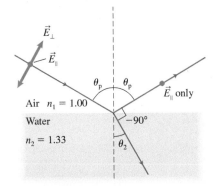

FIGURE 25.22 Reflection and refraction of light incident on a surface at the polarizing angle: the $\vec{E}$ field direction in the incident and reflected beams.

This is because light reflected into the polarizer was incident on that part of the surface at the polarizing angle θ_p. **Figure 25.22** shows the ray diagram for the situation.

There is one more important observation based on those two photos. Note that the reflections off the metal silverware did not change with the orientation of the polarizer. Light becomes partially or completely polarized only when it reflects off nonmetallic surfaces.

How can we explain these observations? We know that a polarizer blocks light in which the $\vec{E}$ field oscillates perpendicular to the axis of the polarizer. When light reflected at the angle θ_p is blocked when we hold the polarizer with its axis in the y-direction, this means that the reflected light is linearly polarized with the $\vec{E}$ field vibrating in the direction parallel to the reflective surface (Figure 25.22).

We found that when the angle of incidence is equal to the so-called polarizing angle θ_p, the reflected ray is linearly polarized in the plane parallel to the reflecting surface and the reflected and the refracted rays are perpendicular to each other.

From Figure 25.22 we see that

$$\theta_p + \theta_2 + 90° = 180°$$
$$\Rightarrow \theta_2 = 90° - \theta_p$$

We can use Snell's law to relate the angle of incidence θ_p to the angle of refraction θ_2:

$$n_1 \sin \theta_p = n_2 \sin \theta_2 = n_2 \sin(90° - \theta_p)$$

From trigonometry we know that $\sin(90° - \theta_p) = \cos \theta_p$. Substituting in the above, we get

$$n_1 \sin \theta_p = n_2 \cos \theta_p$$

or

$$\frac{\sin \theta_p}{\cos \theta_p} = \tan \theta_p = \frac{n_2}{n_1} \qquad (25.8)$$

This relationship is called **Brewster's law**.

Brewster's law Light is traveling in medium 1 when it reflects off medium 2. The reflected light is totally polarized along an axis in the plane parallel to the surface when the tangent of the incident polarizing angle θ_p equals the ratio of the indexes of refraction of the two media:

$$\tan \theta_p = \frac{n_2}{n_1} \qquad (25.8)$$

Brewster's law allows us to explain how polarizing sunglasses help reduce glare. When you are driving your car toward the Sun, it is often difficult to see because of the glare of reflected light from the car's hood and dashboard, from the pavement, and from other cars. This reflected light is partially polarized parallel to those surfaces. Polarizing sunglasses reduce the glare by absorbing light polarized in that direction.

EXAMPLE 25.4   Reducing glare

You are facing the Sun and looking at the light reflected off the ocean. At what angle above the horizon should the Sun be so that you get the most benefit from your polarizing sunglasses?

Sketch and translate The situation is illustrated on the next page. The water surface is horizontal and perpendicular to the page.

The sunlight incident on the water surface is unpolarized, represented by the dots pointing into and out of the page and the $\vec{E}$ field lines up and down perpendicular to the ray. We can answer the question by finding the polarization angle θ_p at which light reflected off the water is completely polarized parallel to the water surface—represented by the dots on the reflected ray.

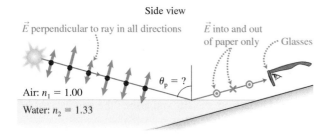

Side view

$\vec{E}$ perpendicular to ray in all directions

$\vec{E}$ into and out of paper only

Glasses

$\theta_p = ?$

Air: $n_1 = 1.00$

Water: $n_2 = 1.33$

Simplify and diagram The polarizing axis of your glasses and the polarization of the reflected light are shown below. Assume that the water reflects rays from the Sun as if it were a mirror. The reflected light $\vec{E}$ field oscillates parallel to the water surface—in and out of the page. The reflected light will not pass though the glasses if the polarizing axis of the glasses is perpendicular to the oscillating $\vec{E}$ field.

Side view

Polarizing axis of glasses

$\vec{E}$ in and out of paper is absorbed by glasses.

Polymer molecules parallel to $\vec{E}$ field

Represent mathematically We can use Brewster's law to find the angle at which parallel rays from the Sun that hit the water surface should produce reflected light that is completely polarized parallel to the air-water surface:

$$\tan \theta_p = \frac{n_2}{n_1}$$

Solve and evaluate Inserting $n_2 = 1.33$ for water and $n_1 = 1.00$ for air, we get

$$\tan \theta_p = 1.33$$

or an angle $\theta_p = 53.1°$. This is the angle between the vertical normal line and the direction of the sunlight. Thus, the direction of the sunlight above the horizontal is

$$90° - 53.1° = 36.9°$$

For all other incident angles, the reflected light will be partially polarized and the glare will not be completely eliminated by the sunglasses.

- -

Try it yourself A light beam travels through glass ($n = 1.60$) into the air. At what angle of incidence is the light that reflects back into the glass completely polarized?

Answer 32°.

Polarization by scattering

We learned in a previous chapter (Chapter 22) that the sky appears blue from Earth's surface because the molecules in the atmosphere scatter blue light more than red into your eyes. If you look through polarized sunglasses at the clear sky in an arbitrary direction and rotate the glasses, the intensity of the light passing through the glasses changes. For example, observer A in **Figure 25.23** sees a change in the light intensity when turning polarized sunglasses while looking at the sky. This means that light scattered by the atmosphere is partially polarized. In fact, if you look in a direction that makes a 90° angle with a line toward the Sun (observer B in Figure 25.23), you can orient the glasses so the sky appears almost dark. Evidently, the sunlight scattered toward you by molecules in that direction is almost completely polarized.

To explain this observation, we need to assume that molecules in the atmosphere are like the tiny dipole antennas described at the beginning of this section. The Sun emits unpolarized light in which all $\vec{E}$ field orientations perpendicular to the direction the light travels are present in equal amounts. When this light strikes a molecule consisting of particles carrying opposite charges (electrons and nuclei), these particles vibrate as the EM wave passes. In **Figure 25.24a** (on the next page), the electric field component E_x produces molecular vibrations along the x-direction. Such a vibration is like that produced by a charge vibration in an antenna and produces an EM wave straight downward with an E_x component (it also produces an upward wave, but we are not interested in the upward wave here). The sunlight also has an electric field component E_y in the y-direction (Figure 25.24b). This component causes atoms and molecules in the atmosphere to vibrate up and down in the y-direction. However, an antenna with vibrating electric charge (the vibrating atom or molecule) cannot emit an $\vec{E}$ field along the axis that the charge vibrates since EM waves are transverse. The result is that the downward 90° wave caused by the vibrating atoms and molecules is linearly polarized in the x-direction. This explains why sunlight is almost completely polarized when you look at a point in the sky in a direction perpendicular to a line from the Sun to that point.

FIGURE 25.23 Scattered light is partially or almost completely polarized.

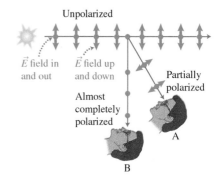

Unpolarized

$\vec{E}$ field in and out

$\vec{E}$ field up and down

Partially polarized

Almost completely polarized

A

B

FIGURE 25.24 Light scattered from a clear sky at 90° is linearly polarized.

(a)

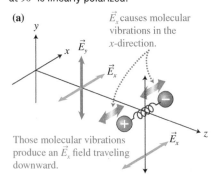

$\vec{E}_y$ causes molecular vibrations in the x-direction.

Those molecular vibrations produce an $\vec{E}_x$ field traveling downward.

(b)

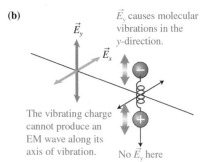

$\vec{E}_y$ causes molecular vibrations in the y-direction.

The vibrating charge cannot produce an EM wave along its axis of vibration.

No $\vec{E}_y$ here

If you look at clouds through polarized sunglasses, the light passing through the glasses is dimmer, but varies very little in intensity as you rotate the glasses. Evidently, light scattered from clouds is unpolarized. Clouds consist of many tiny water droplets. Light entering a cloud is randomly scattered many times by water droplets before it leaves. This multiple scattering keeps the light unpolarized. Photographers take advantage of this effect to take pictures of clouds using polarized "filters." The polarizers remove more of the partially polarized light from the sky and less of the unpolarized light from the clouds, making the clouds stand out (see **Figure 25.25**).

FIGURE 25.25 A polarizer affects light from the clear sky but not from the clouds.

Without polarizer   With polarizer

Polarized LCDs

Nearly all computer, TV, calculator, and cell phone screens are LCDs—liquid crystal displays. Liquid crystals are made of substances that have properties of both a solid and a liquid. They can flow like a liquid, but their elongated molecules are aligned or oriented in an orderly crystal-like manner. Polarization plays an important role in the operation of these screens.

For simplicity, imagine an LCD showing an image that consists of different shades of gray. Some parts of the display screen are white, and other parts are gray or black. The display consists of a grid with many thousands of tiny pixels. Each pixel consists of five parallel layers: two polarizing filters (polarizers), two transparent electrodes, and a liquid crystal between them (**Figure 25.26a**). The axes of the polarizers are perpendicular to each other. If there is no liquid crystal between the polarizers, light from behind the display cannot pass through the polarizers. But with the liquid crystal between the polarizers, light passes through even though the two polarizers are oriented perpendicular to each other! How is this possible?

In the absence of an electric field between the electrodes, the liquid crystal (LC) molecules that are touching the electrodes tend to align with the molecules on the surface of the electrodes. Using a special procedure, these molecules are aligned so that their direction on one electrode is perpendicular to their direction on the other electrode. As a result, the LC molecules change orientation from one electrode to the other electrode, making something similar to a spiral. When polarized light passes through such a spiral, the direction of the $\vec{E}_{light}$ field of the polarized light rotates by 90° so that most of it can pass through the second polarizer even though it is perpendicular to the first (Figure 25.26b). The screen looks bright at that pixel.

FIGURE 25.26 A pixel of a liquid crystal (LC) display.

(a)

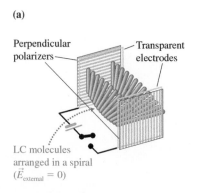

Perpendicular polarizers

Transparent electrodes

LC molecules arranged in a spiral ($\vec{E}_{external} = 0$)

(b)

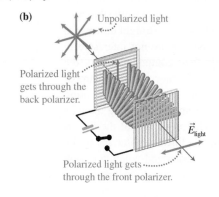

Unpolarized light

Polarized light gets through the back polarizer.

$\vec{E}_{light}$

Polarized light gets through the front polarizer.

(c)

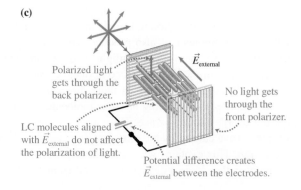

$\vec{E}_{external}$

Polarized light gets through the back polarizer.

No light gets through the front polarizer.

LC molecules aligned with $\vec{E}_{external}$ do not affect the polarization of light.

Potential difference creates $\vec{E}_{external}$ between the electrodes.

When a strong external electric field $\vec{E}_{external}$ is created between the transparent electrodes, the LC molecules align in the direction of this field, and thus the direction of the light's polarization (the direction of vibration of the $\vec{E}_{light}$ field vector) does not change as it passes through the liquid crystal (Figure 25.26c). Because the direction of $\vec{E}_{light}$ does not change, light does not pass through the second polarizer, making the screen look dark at that pixel.

Finally, if the external electric field does not have large enough magnitude $E_{external}$, then only some of the LC molecules align along the external $\vec{E}_{external}$ field and the rest make spirals, allowing some of the light to pass through. In this case, the pixel appears gray. When you press a key on your laptop, it turns on an electric field $\vec{E}_{external}$ that is applied to the liquid crystals of different pixels, making some parts of the screen appear black and others bright or gray. To produce colors, individual pixels are covered with red, green, or blue filters, but the intensity of light that passes through each of them is controlled in exactly the same way as described above.

Now we can answer the question at the beginning of the chapter concerning the similarities between the sky and the calculator screen: both are sources of polarized light.

3-D movies

Polarization also plays a role in one way of making a 3-D movie. Two projectors that send light of two perpendicular polarizations produce two distinct images on the screen. Each image consists of polarized light, and the polarizing axes of the two images are rotated by 90° relative to each other. The two images reflect from the metallized screen to theater viewers, who wear polarized glasses. The polarization axis of the left lens of the glasses is oriented at 90° relative to the axis of the right lens (**Figure 25.27**). Thus, each eye sees a distinct image. The two images are combined in the brain to convey the 3-D effect in much the same way that you see the real world without the help of 3-D glasses.

REVIEW QUESTION 25.6 Explain why polarizing glasses reduce the glare of sunlight reflected from the dashboard of your car or from the metal surface of another car.

FIGURE 25.27 3-D glasses to view a 3-D movie.

A different image goes through each of the perpendicularly polarized glass lenses.

Summary

Electromagnetic waves EM radiation consists of electric and magnetic fields whose vector $\vec{E}$ field and $\vec{B}$ field vibrate perpendicular to each other and to the direction the wave travels. Accelerating charged particles, often oscillating electric dipoles, produce the waves. Once formed, the waves are self-propagating at speed $c = 3.00 \times 10^8$ m/s in a vacuum or slightly less in air. (Sections 25.2 and 25.5)

$$E_{max} = cB_{max} \qquad \text{Eq. (25.1)}$$

$$E_y = E_{max} \cos\left[2\pi\left(\frac{t}{T} \pm \frac{x}{\lambda}\right)\right]$$

$$B_z = B_{max} \cos\left[2\pi\left(\frac{t}{T} \pm \frac{x}{\lambda}\right)\right] \quad \text{Eq. (25.2)}$$

Intensity of light The intensity I of light is the electromagnetic energy that crosses a perpendicular cross-sectional area A in a time interval Δt divided by that area and time interval. (Section 25.5)

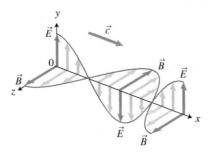

$$u_B = \frac{B^2}{2\mu_0} \qquad \text{Eq. (25.4)}$$

$$\bar{u} = \tfrac{1}{2}u_{max} = \tfrac{1}{2}\left(\epsilon_0 E_{max}^2\right)$$

$$= \tfrac{1}{2}\epsilon_0 E_{max}^2 \qquad \text{Eq. (25.6)}$$

$$I = \frac{\text{Energy}}{\text{time interval} \times \text{area}}$$

$$= \frac{\bar{u}V}{\Delta t A} = \frac{\bar{u}Ac\,\Delta t}{\Delta t A} = \bar{u}c \qquad \text{Eq. (25.7)}$$

Polarization Randomly vibrating wave sources produce *unpolarized* light. When light travels through a polarizer, it becomes *linearly polarized*—only the component of the $\vec{E}$ field perpendicular to the chains of molecules of the polarizer material passes through. The intensity of linearly polarized light is reduced when it travels through a second polarizer oriented at an angle θ relative to the first. (Sections 25.1 and 25.6)

$$I' = \frac{I_0}{2}$$

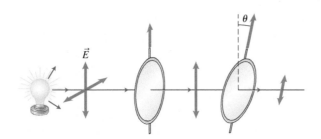

$$I = I' \cos^2 \theta$$

Light polarization by reflection off a surface Light reflected off a surface is partially polarized parallel to the surface. When light is incident at the Brewster (polarization) angle θ_p, the light is totally polarized parallel to the surface. (Section 25.6)

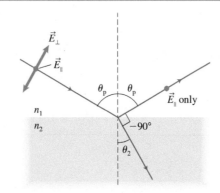

$$\frac{\sin\theta_p}{\cos\theta_p} = \tan\theta_p = \frac{n_2}{n_1} \qquad \text{Eq. (25.8)}$$

where n_1 and n_2 are the refractive indexes of the incident and reflecting media.

Questions

Multiple Choice Questions

1. The fact that light can be polarized means which of the following? (a) Light behaves like a transverse wave. (b) Light behaves like a longitudinal wave. (c) Light does not behave like a wave because it propagates in a vacuum.
2. What does a beam of unpolarized light lose after passing through one polarizer? (a) One-half of its intensity (b) One-quarter of its intensity (c) No intensity, as it was not polarized before
3. What does Faraday's law of electromagnetic induction describe? (a) A steady electric current producing a magnetic field (b) A changing magnetic field producing an electric field (c) A changing electric field producing a magnetic field
4. Maxwell's hypothesis contributed which of the following ideas? (a) A steady electric current produces a magnetic field. (b) A changing magnetic field produces an electric field. (c) A changing electric field produces a magnetic field.
5. What does a simple antenna with a changing electric current radiate? (a) An electric field whose $\vec{E}$ field vector is parallel to the antenna's orientation (b) A magnetic field whose $\vec{B}$ field vector is parallel to the antenna's orientation (c) An electric field whose $\vec{E}$ field vector is perpendicular to the antenna's orientation (d) It is impossible to say, as we do not have information about the magnitude of the current through the antenna.
6. An electrically charged particle radiates electromagnetic waves when (a) it is stationary. (b) it moves at a constant velocity. (c) it moves at changing velocity. (d) it is inside a wire with a DC current in it.
7. An electromagnetic wave propagates in a vacuum in the x-direction. Where does the $\vec{E}$ field oscillate? (a) In the x-y plane (b) In the x-z plane (c) In the y-z plane (d) Not enough information to answer
8. If the amplitude of an $\vec{E}$ field in a linearly polarized wave doubles, what will happen to the energy density of the wave? (a) It will double. (b) It will quadruple. (c) Not enough information to answer
9. You notice that unpolarized light reflected off a lake is completely blocked when observed through a polarizer. Which of the following statements are correct? Select all that apply
   (a) The $\vec{E}$ field in the reflected light is oscillating in a horizontal direction.
   (b) The $\vec{B}$ field in the reflected light is oscillating in a horizontal direction.
   (c) The $\vec{E}$ field and the $\vec{B}$ field in the reflected light are perpendicular to each other.
   (d) The axis of your polarizer is parallel to the water surface.
   (e) The axis of your polarizer is in the same plane as the plane defined by the incident and reflected beam.
   (f) The axis of your polarizer is parallel to the direction of the reflected beam.
10. You have two green lasers, A and B. A's light has twice the intensity of B's. If c is the speed of light in air and E_{max} is the amplitude of the electric field in light, then which of the following statements are correct? Select all that apply.

(a) $c_A = 2c_B$ because the intensity of light is proportional to c.
(b) $c_A = c_B$ because the speed of light in air is constant.
(c) $E_{max\ A} = 2E_{max\ B}$ because the intensity of light is proportional to E_{max}.
(d) $E_{max\ A} = \sqrt{2}E_{max\ B}$ because the intensity of light is proportional to E_{max}^2.
(e) $E_{max\ A} = 4E_{max\ B}$ because the intensity of light is proportional to E_{max}^2.

Conceptual Questions

11. Can light phenomena be better explained by a transverse wave model or by a longitudinal wave model? Explain how you know.
12. What are two models that explain how light can propagate through air? Describe their features in detail.
13. Summarize Maxwell's equations conceptually and describe a way to experimentally illustrate each of them.
14. What testable predictions followed from Maxwell's equations?
15. Describe the conditions that are necessary for the emission of electromagnetic waves.
16. Explain how radar works to determine distances to objects.
17. What determines the nearest and farthest distances of objects that can be distinguished by pulsed radar?
18. How was the hypothesis that light can be modeled as an electromagnetic wave tested experimentally?
19. What is the difference between the following two statements: Light is a wave. Light behaves like a wave. Which one do you think is more accurate and why?
20. How do polarized glasses work?
21. You bought a pair of glasses that are marketed as having polarizing filters. How can you test this claim if (a) you have another pair of identical glasses, and (b) you have only one pair of glasses and no other polarizer (and you are not allowed to break the glasses)?
22. Why, when we use polarized glasses, is the glare from reflective surfaces reduced? What parameters does the phenomenon depend on?
23. How does a polarizer for mechanical waves work, and how does a polarizer for light waves work?
24. What is an LCD and how does it work?
25. Jim does not understand the wave model of light. It does not make sense to him that light can propagate in a vacuum where there is no medium to transport vibrations from one place to another. How can you help Jim reconcile his understanding of waves with the wave model of light?
26. Make a list of phenomena that can be explained by the particle model of light. Then make a list of phenomena that can be explained using the wave model of light. Now that we have a wave model of light, does it mean that we should stop using the particle model? Explain your opinion.
27. Describe an experiment that you can perform to determine whether the light from a particular source is unpolarized or linearly polarized. If the latter, then how can you determine the polarization direction?

Problems

Below, **BIO** indicates a problem with a biological or medical focus. Problems labeled **EST** ask you to estimate the answer to a quantitative problem rather than derive a specific answer. Asterisks indicate the level of difficulty of the problem. Problems with no * are considered to be the least difficult. A single * marks moderately difficult problems. Two ** indicate more difficult problems.

25.1 and 25.2 Polarization of waves and Discovery of electromagnetic waves

1. The coils of a horizontal Slinky vibrate vertically up and down; the wave propagates in the horizontal direction. The amplitude of vibrations is 20 cm. You thread the Slinky through a board with a 50-cm slot at an angle of 60° relative to the vertical. What is the amplitude of the wave that moves past the slot? In what direction do the coils now vibrate? Indicate any assumptions that you made.
2. A 40-W lightbulb is 2.0 m from a screen. What is the intensity of light incident on the screen? What assumptions did you make?

3. * Assume that the bulb in Problem 25.2 radiates unpolarized light. You place a tourmaline crystal in front of the screen. (a) What is the intensity of light hitting the screen after passing through the crystal? (b) You place a second tourmaline crystal between the first crystal and the screen so that it is oriented at an angle of 50° relative to the axis of the first. What is the intensity of light hitting the screen? What assumptions did you make?
4. * (a) A uniform $\vec{B}$ field whose lines are oriented in the S-N direction increases steadily. Draw the $\vec{E}$ field lines of the induced electric field. (b) A uniform magnetic field whose lines are oriented E-W decreases steadily. Draw the $\vec{E}$ field lines of the induced electric field. (c) Repeat part (b) for a case when the $\vec{B}$ field changes at twice the rate.
5. * Investigate in detail how Hertz's apparatus worked and describe how it was used to produce and detect electromagnetic waves.
6. One dipole antenna points in the x-direction and the other identical antenna points in the y-direction. The currents in the antennas oscillate in phase and with the same frequency. Determine the angle between the x-axis and the direction of oscillation of the $\vec{E}$ field and $\vec{B}$ field at a point on the z-axis, far from the antennas.

7. * The current in an antenna is oscillating with a period 2.0×10^{-15} s. Using the same coordinate system, draw (a) two graphs that show how the $\vec{E}$ field and $\vec{B}$ field magnitudes at a point near the antenna change with time (label only the horizontal axis of the graph) and (b) two graphs that show how the $\vec{E}$ field and $\vec{B}$ field magnitudes at a point very far from the antenna change with time (again label only the horizontal axis of the graph).

8. **Milky Way** The Sun is a star in the Milky Way galaxy. When viewed from the side, the galaxy looks like a disk that is approximately 100,000 light-years in diameter (a light-year is the distance light travels in one year) and about 1000 light-years thick (Figure P25.8). What is the diameter and thickness of the Milky Way in meters? In kilometers? In miles?

FIGURE P25.8

Sun and solar system's location in the Milky Way

9. We observe an increase in brightness of a star that is 5.96×10^{19} m away. When did the actual increase in brightness take place? Is the star in the Milky Way galaxy? What assumptions did you make?

10. * **Milky Way neighbors** Two neighboring galaxies to the Milky Way, the Large and Small Magellanic Clouds, are about 180,000 light-years away. How far away are they in meters? In miles?

25.3 Applications of electromagnetic waves

11. **EST** **Radar distance** Estimate the range of distances at which you can detect an object using radar with a pulse width of 12 μs and a pulse repetition of 15 kHz.

12. * **EST** **Radar antenna height** Why do you need to raise a radar emission antenna above the ground to be able to detect objects near the surface of Earth at the farthest range of the radar? Support your answer with a sketch. Estimate how high you need to raise the radar with a pulse width of 12 μs and a pulse repetition of 15 kHz.

13. * **More radar detection** Radar uses radio waves of a wavelength of 2.0 m. The time interval for one radiation pulse is 100 times larger than the time of one oscillation; the time between pulses is 10 times larger than the time of one pulse. What is the shortest distance to an object that this radar can detect?

14. * **Communicating with Mars** Imagine that you have a vehicle traveling on Mars. The shortest distance between Earth and Mars is 56×10^6 km; the longest is 400×10^6 km. What is the delay time for the signal that you send to Mars from Earth? Can you use radio signals to give commands to the vehicle?

15. * **TV tower transmission distance** The height of a TV tower is 500 m and the radius of Earth is 6371 km. What is the maximum distance along the ground at which signals from the tower can be received? Indicate any assumptions that you made.

16. ** **Radar parameters** Radar is characterized by a pulse width and pulse repetition period. (a) Explain what these words mean and how you can use them to estimate the range of distances that radar can measure. (b) If you wish to be able to detect objects at a close distance, what parameter of the radar should you change and how? If you wish to be able to detect objects at a long distance, what parameter of the radar should you change and how?

25.4 and 25.5 Frequency, wavelength, and the electromagnetic spectrum and Mathematical description of EM waves and EM wave energy

17. * (a) The amplitude of the $\vec{E}$ field oscillations in an electromagnetic wave traveling in air is 20.00 N/C. What is the amplitude of the $\vec{B}$ field oscillations? (b) The amplitude of the $\vec{B}$ field oscillations in an electromagnetic wave traveling in a vacuum is 3.00×10^{-3} T. What is the amplitude of the oscillations of the $\vec{E}$ field?

18. * **Limits of human vision** The wavelength limits of human vision are 400 nm to 700 nm. What range of frequencies of light can we see? How do the wavelength and frequency ranges change when we are underwater? The speed of light in water is 1.33 times less than in the air.

19. **AM radio** An AM radio station has a carrier frequency of 600 kHz. What is the wavelength of the broadcast?

20. * The electric field in a sinusoidal wave changes as

$$E = (25 \text{ N/C}) \cos\left[(1.2 \times 10^{11} \text{ rad/s})t + (4.2 \times 10^2 \text{ rad/m})x \right]$$

(a) In what direction does the wave propagate? (b) What is the amplitude of the electric field oscillations? (c) What is the amplitude of the magnetic field oscillations? (d) What is the frequency of the wave? (e) What is the wavelength? (f) What is the speed? (g) What other information can you infer from the equation?

21. ** A sinusoidal electromagnetic wave propagates in a vacuum in the positive x-direction. The $\vec{B}$ field oscillates in the z-direction. The wavelength of the wave is 30 nm and the amplitude of the $\vec{B}$ field oscillations is 1.0×10^{-2} T. (a) What is the amplitude of the $\vec{E}$ field oscillations? (b) Write an equation that describes the $\vec{B}$ field of the wave as a function of time and location. (c) Write an equation that describes the $\vec{E}$ field in the wave as a function of time and location. (d) What type of electromagnetic radiation is this?

22. ** For the previous problem determine (a) the frequency with which the electric energy in the wave oscillates; (b) the frequency at which magnetic field energy oscillates; (c) the maximum energy density; (d) the minimal energy density; (e) the average energy density; and (f) the intensity of the wave.

23. **BIO** **Ultraviolet A and B** UV-A rays are important for the skin's production of vitamin D; however, they tan and damage the skin. UV-B rays can cause skin cancer. The wavelength range of UV-A rays is 320 nm to 400 nm and that of UV-B rays is 280 to 320 nm. What is the range of frequencies corresponding to the two types of rays?

24. **BIO** **X-rays used in medicine** The wavelengths of X-rays used in medicine range from about 8.3×10^{-11} m for mammography to shorter than 6.2×10^{-14} m for radiation therapy. What are the frequencies of the corresponding waves? What assumption did you make in your answer?

25. ** **Power of sunlight on Earth** The Sun emits about 3.9×10^{26} J of electromagnetic radiation each second. (a) Estimate the power that each square meter of the Sun's surface radiates. (b) Estimate the power that 1 m^2 of Earth's surface receives. (c) What assumptions did you make in part (b)? The distance from Earth to the Sun is about 1.5×10^{11} m and the diameter of the Sun is about 1.4×10^9 m.

26. ** **Light from an incandescent bulb** Only about 10% of the electromagnetic energy from an incandescent lightbulb is visible light. The bulb radiates most of its energy in the infrared part of the electromagnetic spectrum. If you place a 100-W lightbulb 2.0 m away from you, (a) what is the intensity of the infrared radiation at your location? (b) What is the infrared energy density? (c) What are the approximate magnitudes of the infrared electric and magnetic fields?

27. * Explain how the information about energy radiated by a lightbulb in Problem 25.26 can be used to compare the magnitudes of $\vec{E}$ fields oscillating at different frequencies. Pose a problem that requires the use of this information.

28. * **EST** Estimate the amplitude of oscillations of the $\vec{E}$ and $\vec{B}$ fields for sunlight near the surface of the Sun. List all of the assumptions that you made.

29. ** **BIO** **New laser to treat cancer** The HERCULES pulsed laser has the potential to help in the treatment of cancer, as it focuses its power on a tiny area and essentially burns individual cancer cells. The HERCULES can be focused on a surface that is 0.8×10^{-6} m across. This pulsed laser provides 1.2 J of energy during a 27×10^{-15}-s time interval at a wavelength of approximately 800 nm. (a) Determine the power provided during the pulse. (b) Determine the magnitude of the maximum $\vec{E}$ field produced by the pulse.

30. * **Equation Jeopardy 1** Tell everything you can about the wave described by the equations below.

$$E_y = (2.4 \times 10^5 \,\text{N/C}) \cos\left[2\pi\left(\frac{t}{1.5 \times 10^{-15}\,\text{s}} - \frac{x}{4.5 \times 10^{-7}\,\text{m}}\right)\right]$$

$$B_z = (8.0 \times 10^{-4} \,\text{T}) \cos\left[2\pi\left(\frac{t}{1.5 \times 10^{-15}\,\text{s}} - \frac{x}{4.5 \times 10^{-7}\,\text{m}}\right)\right]$$

31. * **Equation Jeopardy 2** Tell everything you can about the wave described by the equations below.

$$E_y = (9.0 \times 10^5 \,\text{N/C}) \cos\left[2\pi\left(\frac{t}{5.3 \times 10^{-15}\,\text{s}} - \frac{x}{1.6 \times 10^{-6}\,\text{m}}\right)\right]$$

$$B_z = B_{max} \cos\left[2\pi\left(\frac{t}{5.3 \times 10^{-15}\,\text{s}} - \frac{x}{1.6 \times 10^{-6}\,\text{m}}\right)\right]$$

32. * **Equation Jeopardy 3** Tell everything you can about the situation described below.

$$E_{max} = \sqrt{\frac{2 \cdot 640 \,\text{kW/m}^2}{(3.0 \times 10^8 \,\text{m/s})(8.85 \times 10^{-12} \,\text{C}^2/\text{N} \cdot \text{m}^2)}}$$

25.6 Polarization and light reflection

33. * An unpolarized beam of light passes through two polarizing sheets that are initially aligned so that the transmitted beam is maximal. By what angle should the second polarized sheet be rotated relative to the first to reduce the transmitted intensity to (a) one-half and (b) one-tenth the intensity that was transmitted through both polarizing sheets when aligned?

34. * **BIO** **Spider polarized light navigation** The gnaphosid spider *Drassodes cupreus* has evolved a pair of lensless eyes for detecting polarized light. Each eye is sensitive to polarized light in perpendicular directions. Near sunset, the spider leaves its nest in search of prey. Light from overhead is partly polarized and indicates the direction the spider is moving. After the hunt, the spider uses the polarized light to return to its nest. Suppose that the spider orients its head so that one of these two eyes detects light of intensity 800 W/m² and the other eye detects zero light intensity. What intensities do the two eyes detect if the spider now rotates its head 20° from the previous orientation?

35. * Two polarizing sheets are oriented at an angle of 60° relative to each other. (a) Determine the factor by which the intensity of an unpolarized light beam is reduced after passing through both sheets. (b) Determine the factor by which the intensity of a linearly polarized beam with the $\vec{E}$ field oscillating at 30° relative to each polarizing sheet is reduced after passing through both sheets. Indicate any assumptions that you made.

36. * **Light reflected from a pond** At what angle of incidence (and reflection) does light reflected from a smooth pond become completely polarized parallel to the pond's surface? How do you know? In which direction does the $\vec{E}$ field vector oscillate in this reflected light wave?

37. * **Light reflected from water in a cake pan** At what angle is reflected light from a water-glass interface at the bottom of a cake pan holding water completely polarized parallel to the surface of the glass whose refractive index is 1.65?

38. * Unpolarized light passes through three polarizers. The second makes an angle of 25° relative to the first, and the third makes an angle of 45° relative to the first. The intensity of light measured after the third polarizer is 40 W/m². Determine the intensity of the unpolarized light incident on the first polarizer. Indicate any assumptions that you made.

39. ** You have two pairs of polarized glasses. Make a list of experimental questions you can answer using one of them or both. Describe the experiments you will design to answer them and discuss how you will ensure that the solutions (answers) you find make sense.

40. * A beam of unpolarized light with intensity 100 W/m² is incident on a pair of ideal crossed (perpendicular) polarizers. You insert a third ideal polarizer between the two polarizers with its polarizing axis at 45° to the others. Determine the intensity of the emerging light before and after you insert the third polarizer.

General Problems

41. ** **EST** **Density of Milky Way** What is the average density of the Milky Way assuming that it contains about 200 billion stars with masses comparable to the mass of the Sun? What additional information do you need to know to answer the question? What assumptions should you make?

42. ** **EST** **Supernova in neighboring galaxy** On February 23, 1987 astronomers noticed that a relatively faint star in the Tarantula Nebula in the Large Magellanic Cloud suddenly became so bright that it could be seen with the naked eye. Astronomers suspected that the star exploded as a supernova. Late-stage stars eject material that forms a ring before they explode as supernovas. The ring is at first invisible. However, following the explosion the light from the supernova reaches the ring. Astronomers observed the ring exactly one year after the phenomenon itself occurred (Figure P25.42). (a) Use geometry and the speed of light to estimate the distance to the supernova. The angular radius of the ring is 0.81 arcseconds. (b) How do you know whether your answer makes sense? (c) Use the width of the ring to estimate the uncertainty in the value of the distance.

FIGURE P25.42

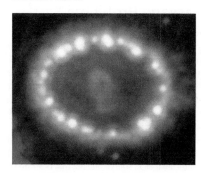

43. ** **BIO** **EST** **Human vision power sensitivity** A rod in the eye's retina can detect light of energy 4×10^{-19} J. Estimate the power of this light that the rod can detect. Indicate any assumptions you made. You will need more information than what is provided.

44. ** **Effect of weather on radio transmission** Weather affects the transmission of AM stations but does not affect FM stations. FM stations do not broadcast in remote areas. Suggest possible reasons for these phenomena. (Hint: Find information about the wavelengths of the waves and decide how they might propagate in Earth's atmosphere.)

45. ** **Measuring speed of light** In 1849 Hippolyte Fizeau conducted an experiment to determine the speed of light in a laboratory. He used an apparatus described in Figure P25.45. Light from the source S went through an interrupter K and after reflecting from the mirror M returned to the rotating wheel again. If light passed between the teeth of the wheel, Fizeau could see it; if light on the way back hit the tooth of the wheel, Fizeau could not see it. He could measure the speed of light by relating the time interval between teeth crossing the beam and the distance light traveled during those time intervals. The information that Fizeau had about the system was L = the distance between the wheel and the mirror; T = the period of rotation of the wheel; and N = the number of teeth in the wheel (the width of one tooth was equal to the width of the gap between the teeth). Using these parameters Fizeau derived a formula for calculating the speed of light: $c = (4LN/T)$. Explain how he arrived at this equation.

FIGURE P25.45

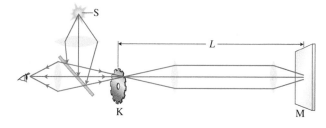

46. ** **Speed of light** In Fizeau's experiment (described in Problem 25.45) the distance between the wheel and the mirror was 3.733 km, the wheel had 720 teeth, and the wheel made 29 rotations per second. What was the speed of light determined by Fizeau? What were the uncertainties in this value?

47. * A sinusoidal electromagnetic wave in air has a 20-N/C $\vec{E}$ field amplitude. Determine everything you can about this wave.

48. * **EST** A microwave oven produces electromagnetic waves with a frequency of 2.45 GHz and emits energy at a rate of 700 W. The microwaves enter the cooking compartment through a rectangular waveguide of dimensions 5 cm × 10 cm. (a) Determine the wavelength of the microwaves, (b) estimate the amplitude of the $\vec{E}$ field at the waveguide opening, and (c) estimate the amplitude of the $\vec{E}$ field when the waves spread over the 30 cm × 20 cm cross section of the oven.

49. ** A beam of unpolarized light of intensity I_0 is incident on an ideal pair of crossed (perpendicular) polarizers. You insert a third ideal polarizer between the two polarizers, with its polarizing axis at an angle θ with respect to the axis of the first polarizer. Derive the expression for the intensity of the emerging light as function of θ.

Reading Passage Problems

BIO **Amazing honeybees** The survival of a bee colony depends on the ability of bee scouts to locate food and to convey that information to the hive. After finding a promising food source, a honeybee scout returns to the hive and uses a waggle dance to tell its worker sisters the direction and distance to the food.

Recall that light coming directly from the Sun is unpolarized. Bees use the direction of the Sun as a reference for their travel. In the hive, the scout bee's waggle dance is in the shape of a flat figure eight (Figure 25.28). The upward direction in the vertical hive represents the direction toward the Sun. The middle line of the scout's figure eight points in the direction of the food relative to the direction of the Sun. Thus a 50° middle waggle dance to the right of the vertical in the hive would indicate a food source outside that is 50° to the right of the direction toward the Sun. The distance to the food depends on the length of time the scout takes while wiggling her tail and wings through this middle of the figure eight. The other bees know the direction to the food even if the Sun is behind a cloud—they can detect the degree of polarization in other open parts of the sky and deduce the angle to the Sun from that position, and they learned from the scout the angle from the Sun's direction to the food.

FIGURE 25.28 A scout bee's waggle dance.

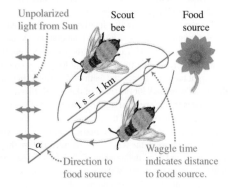

50. Bees know the direction to the Sun because
    (a) direct sunlight is linearly polarized.
    (b) direct sunlight is unpolarized.
    (c) they detect infrared radiation from the sun.
    (d) a and c.   (e) b and c.

51. Suppose that the scout bee in a waggle dance indicates that a good food source is at a 90° angle from the direction to the Sun. Bees who have never been to the food source when leaving the hive head toward a region of the sky that
    (a) has a flowery odor.
    (b) has completely unpolarized light coming from it.
    (c) has completely linearly polarized light coming from it.
    (d) has partially unpolarized light coming from it.
    (e) None of these answers

52. If the direction of the middle line of the scout's waggle dance figure eight points 53° to the left side of the upward direction in the hive, in what direction to the left should bees leaving the nest for the food source head relative to light coming from the Sun?
    (a) Where the light is 36% linearly polarized to the left of the Sun's direction
    (b) Where the light is 60% linearly polarized to the left of the Sun's direction
    (c) Where the light is 64% linearly polarized to the left of the Sun's direction
    (d) Where the light is 80% linearly polarized to the left of the Sun's direction
    (e) Where the light is 100% linearly polarized to the left of the Sun's direction

53. Polar molecules are caused to vibrate in all directions perpendicular to the direction of travel of sunlight. When the Sun is rising in the east and you look at the molecules directly overhead, the light from these molecules is
    (a) unpolarized.
    (b) linearly polarized along an axis between you and the molecules overhead.
    (c) linearly polarized parallel to a plane with you, the Sun, and the molecules.
    (d) linearly polarized perpendicular to a plane with you, the Sun, and the molecules.
    (e) All of these are correct.

54. In what direction(s) is the light from the sky almost completely unpolarized?
    (a) Looking directly at the Sun
    (b) Looking directly away from the Sun
    (c) Looking at 90° relative to the Sun
    (d) a and b   (e) b and c

Incandescent lightbulbs—soon to disappear Australia, Canada, New Zealand, and the European Union started phasing out incandescent lightbulbs in 2009. The United States phased them out in 2014. These bulbs have provided light for the world for more than 90 years. What's the problem?

Incandescent lightbulbs produce light when electrons (electric current) move through the filament and collide with tungsten atoms and ions in the filament, causing them to vibrate violently. Accelerating charged particles emit electromagnetic radiation. This is true for vibrating and accelerating particles in the filament, in your skin, and on the Sun. The 35 °C skin of a person emits most of its radiation in the low-frequency, long-wavelength infrared—sometimes called thermal radiation. At 2500 °C, the filament of an incandescent lightbulb emits most of its radiation as higher frequency infrared radiation and just 10% as visible light (very inefficient if the goal is to produce light). At 6000 °C, the Sun emits about 45% in the infrared range and 40% in the visible light range.

Banning incandescent bulbs will reduce energy usage. According to the Department of Energy, about 10^{18} J of electric energy is used each year in the United States for household lighting. Compact fluorescent lightbulbs (CFLs) and light-emitting diode (LED) bulbs use about one-fourth the energy of incandescent bulbs. Switching to CFLs and LEDs would also reduce greenhouse gas emissions by reducing the need for power produced by coal-burning electric power plants.

55. Why are the United States and other countries banning the use of incandescent lightbulbs?
    (a) The bulbs get too hot.
    (b) The bulbs have a tungsten filaments.
    (c) Ninety percent of the electric energy used is converted to thermal energy.
    (d) The bulbs are only 10% efficient in converting electric energy to light.
    (e) c and d

56. What is the rate of visible light emission from a 100-W incandescent lightbulb?
    (a) 10 W   (b) 20 W   (c) 50 W
    (d) 100 W   (e) Not enough information to make a determination

57. What does the surface of the body at about 35 °C primarily emit?
    (a) No electromagnetic radiation
    (b) Long-wavelength infrared radiation
    (c) Short-wavelength infrared radiation
    (d) Light   (e) Invisible ultraviolet light

58. Suppose you changed all incandescent lightbulbs to more energy-efficient bulbs that used one-fourth the amount of energy to produce the same light. About how many 3.3×10^9-kW·h/year electric power plants could be removed from the power grid?
    (a) 3   (b) 10   (c) 20   (d) 50   (e) 100

59. How much money will you save on your electric bill each year if you replace five 100-W incandescent bulbs with five CFL or LED 25-W bulbs that produce the same amount of light? Assume that the bulbs are on 3.0 h/day and that electric energy costs $0.12/kW·h.
    (a) $5   (b) $10   (c) $20   (d) $50   (e) $100

Special Relativity

At the end of the 19th century, Albert Michelson measured the speed of light quite accurately, but thought it needed a medium (called ether) through which to travel, just as sound needs air to propagate. Michelson and Edward Morley built a device that would demonstrate the existence of this invisible substance. Despite their efforts, they failed to show that ether exists. This negative result became the experimental foundation for a new theory—the subject of this chapter, and one that most of us use on a daily basis with our GPS devices.

- Does ether, the proposed medium for light waves, exist?
- If you were in a spaceship moving at near light speed away from a light source, at what speed would light pass you?
- How can atomic clocks on satellites help determine your location on Earth?

SO FAR IN THIS BOOK, we have assumed that time passes in the same way for all observers. When a second passes for me, the same second passes for you. Similarly, we have assumed that length measurements of objects are the same for all observers. These assumptions sound so reasonable that it is tempting to think of them as "absolute truths." In this chapter we will find that we need to reconsider our ideas—not just of time and space, but also of momentum and energy.

BE SURE YOU KNOW HOW TO:

- Distinguish between inertial and noninertial reference frames (Section 3.4).
- Explain why observers in noninertial reference frames cannot use Newton's laws to explain mechanics phenomena (Section 3.4).
- Calculate the change in kinetic energy of a charged particle that travels across a potential difference ΔV (Section 18.3).

26.1 Ether or no ether?

Recall (from Chapter 24) that in 1801, Thomas Young performed the double-slit experiment that strongly supported the wave model of light. All waves known at that time required a medium to travel through: ocean waves through water, sound waves through air or other materials, and so on. The role of the medium is to transfer the disturbance produced by a source at one location to other locations. The particles of the medium interact with each other, allowing this to occur. Light, then, it was reasoned, must also require a medium in which to travel. In the 19th century, physicists thought that this medium was *ether*.

Ether was considered to be invisible but able to undergo shear deformations so that transverse light waves could travel through it. We know (from Chapter 25) that the work of Maxwell and Hertz led to the conclusion that light propagation could be explained by changing $\vec{E}$ and $\vec{B}$ fields that do not require any medium to travel. However, that work was done in the late 1880s. Before that, physicists were searching for ether. This search, like many investigations in physics, produced an unexpected outcome that eventually changed the way we think about space and time.

An analogy: a boat race

To understand how one of those ether-testing experiments was done, we will use a mechanics analogy. Consider a process involving two identical boats in a race on a wide river. Each boat is to travel a distance of 1.6 km relative to the shore, then 1.6 km back (**Figure 26.1**). Boat 1 travels from a starting dock to another dock 1.6 km upstream from the first, then back again to where it started. Boat 2 starts at the same dock but instead travels 1.6 km across the river to a dock on the opposite side, then travels back to where it started. Each boat travels at speed 10 km/h relative to the water, and the water travels at speed 6 km/h relative to the shore. Which boat returns to the starting dock first?

You may think that the boats will arrive back at the dock at the same time. Since all the speeds are the same and are constant, won't everything just balance out? As you will see shortly, the reality is something different.

Travel time for boat 1 going upstream and back The displacement of boat 1 with respect to the shore as it moves upstream $\vec{d}_{\text{boat 1 to shore}}$ is the result of two displacements—the displacement of the boat with respect to the water $\vec{d}_{\text{boat 1 to water}}$ and the displacement of the water with respect to the shore $\vec{d}_{\text{water to shore}}$. Mathematically:

$$\vec{d}_{\text{boat 1 to shore}} = \vec{d}_{\text{boat 1 to water}} + \vec{d}_{\text{water to shore}}$$

The time intervals during which these three displacements occur are all the same:

$$\Delta t_{\text{boat to shore}} = \Delta t_{\text{boat to water}} = \Delta t_{\text{water to shore}} = \Delta t$$

If we divide all three parts of the displacement equation by the time interval Δt during which the displacements occurred, we obtain

$$\frac{\vec{d}_{\text{boat 1 to shore}}}{\Delta t} = \frac{\vec{d}_{\text{boat 1 to water}}}{\Delta t} + \frac{\vec{d}_{\text{water to shore}}}{\Delta t}$$

or

$$\vec{v}_{\text{boat 1 to shore}} = \vec{v}_{\text{boat 1 to water}} + \vec{v}_{\text{water to shore}}$$

Note that these are velocities, not speeds, which is why they are labeled with vector symbols. In **Figure 26.2** we summarize the situation using vector components. The upstream velocity of the boat relative to the shore is then (with the upstream direction positive)

$$v_{\text{boat 1 to shore }y} = +10 \text{ km/h} + (-6 \text{ km/h}) = +4 \text{ km/h}$$

FIGURE 26.1 Two identical length boat trips on a river.

1.6 km

$\vec{v}_{\text{water}}$

Boat 1 travels up the river and back.

1.6 km

Boat 2 travels across the river and back.

FIGURE 26.2 The velocity of boat 1.

(a) y Boat 1 traveling upstream

$v_{\text{boat to water }y} = +10 \text{ km/h}$

$v_{\text{water to shore }y} = -6 \text{ km/h}$

$v_{\text{boat to shore }y} = +4 \text{ km/h}$

The net speed of boat 1 relative to the shore when traveling upstream

(b) y Boat 1 traveling downstream

The net speed of boat 1 when traveling downstream

$v_{\text{boat to water }y} = -10 \text{ km/h}$

$v_{\text{boat to shore }y} = -16 \text{ km/h}$

$v_{\text{water to shore }y} = -6 \text{ km/h}$

You can see that the water moving against the boat's direction of travel slows the boat down relative to the shore. The time interval it takes the boat to reach the upstream dock is

$$\Delta t_{\text{upstream}} = 1.6 \text{ km}/(4 \text{ km/h}) = 0.4 \text{ h}$$

The downstream velocity relative to the shore is

$$v_{\text{boat 1 to shore } y} = -10 \text{ km/h} + (-6 \text{ km/h}) = -16 \text{ km/h}$$

The time interval it takes the boat to return to the downstream dock is

$$\Delta t_{\text{downstream}} = -1.6 \text{ km}/(-16 \text{ km/h}) = 0.1 \text{ h}$$

As you would expect, the return trip takes less time because the velocity of the water relative to the shore is in the same direction as that of the boat. The total time interval for boat 1's round trip is therefore

$$\Delta t = \Delta t_{\text{upstream}} + \Delta t_{\text{downstream}} = 0.4 \text{ h} + 0.1 \text{ h} = 0.5 \text{ h}$$

Travel time for boat 2 going across the stream and back The velocity of boat 2 relative to the shore is perpendicular to the velocity of the water with respect to the shore (Figure 26.3). In order for the boat to travel directly across the river, the 10-km/h boat velocity relative to the water has to point at an angle so that it has a 6-km/h upstream component to cancel the downstream velocity of the water relative to the shore. The magnitudes of the three velocities shown are related through the Pythagorean theorem:

$$v_{\text{boat 2 to water}}^2 = v_{\text{boat 2 to shore}}^2 + v_{\text{water to shore}}^2$$

or

$$v_{\text{boat 2 to shore}}^2 = v_{\text{boat 2 to water}}^2 - v_{\text{water to shore}}^2 = (10 \text{ km/h})^2 - (6 \text{ km/h})^2$$
$$= 64 \; (\text{km/h})^2$$
$$\Rightarrow v_{\text{boat 2 to shore}} = 8 \text{ km/h}$$

The crossing time interval is then

$$\Delta t_{\text{crossing}} = 1.6 \text{ km}/(8 \text{ km/h}) = 0.2 \text{ h}$$

This crossing velocity is the same in both directions. The time interval for the complete round trip, therefore, is just twice this:

$$\Delta t = 2\Delta t_{\text{crossing}} = 2(0.2 \text{ h}) = 0.4 \text{ h}$$

We have shown that boat 1 takes longer to make its round trip than boat 2, even though both boats travel a distance of 3.2 km relative to the shore. The moving water affects the time intervals differently depending on the direction of the boat's travel.

Testing the existence of ether

An experiment analogous to the boat experiment can be used to test for the existence of ether. Imagine that ether fills the solar system and is stationary with respect to the Sun. Because Earth moves around the Sun at a speed of about $3.0 \times 10^4 \text{ m/s}$, ether should be moving past Earth at this speed. Shining light waves parallel and perpendicular to the ether's motion relative to Earth is similar to sending boats parallel and perpendicular to a flowing river. Therefore, the travel time for light waves (1) up and back along the direction of the ether's motion and (2) across and back perpendicular to the ether's motion should differ if they travel the same distance. If one could measure this time interval difference experimentally, it would serve as strong support for the existence of ether as a medium for light wave propagation.

In 1887, Albert Michelson and Edward Morley set up such an experiment (although they themselves firmly believed in ether and were trying to measure Earth's speed relative to it). They used a device called an **interferometer** to detect

FIGURE 26.3 The velocity of boat 2.

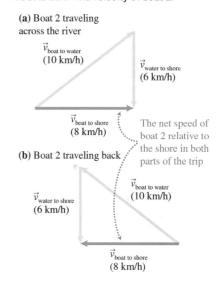

(a) Boat 2 traveling across the river

$\vec{v}_{\text{boat to water}}$ (10 km/h)

$\vec{v}_{\text{water to shore}}$ (6 km/h)

$\vec{v}_{\text{boat to shore}}$ (8 km/h)

The net speed of boat 2 relative to the shore in both parts of the trip

(b) Boat 2 traveling back

$\vec{v}_{\text{water to shore}}$ (6 km/h)

$\vec{v}_{\text{boat to water}}$ (10 km/h)

$\vec{v}_{\text{boat to shore}}$ (8 km/h)

ether (see Testing Experiment **Table 26.1**). A light beam is sent to a beam splitter (a half-silvered mirror) that causes the light beam to split into two beams. One beam continues forward and the other beam reflects perpendicularly, each beam moving along a separate arm of the device. Mirrors at the end of the arms reflect the two beams back to the half-silvered mirror, where they recombine and are detected.

Because the two beams are formed by splitting a single light beam, the two beams are coherent (have a constant phase difference with respect to each other). When the beams recombine after reflection, an interference pattern results. The details of this interference pattern depend on the difference in the travel time intervals between the two beams along each arm of the interferometer. *If* the ether hypothesis is correct and the interferometer is carefully rotated (causing the direction of each beam of light to change relative to the ether), *then* the time interval it takes each beam to travel along each arm and back should change. As a result, the interference pattern produced by the recombined light should change as the device rotates.

TESTING
EXPERIMENT TABLE 26.1   **Testing the existence of ether**

Testing experiment	Prediction	Outcome
Assume that the ether is moving to the left past the interferometer. We shine a beam of light from the left onto the beam splitter. One beam moves parallel to the ether's velocity and the other beam moves perpendicular to its velocity.	It should take the light longer to travel against and with the ether than to travel sideways across the moving ether. Rotating the interferometer in the plane of the page should lead to a change in the interference pattern formed by the two light beams at the detector.	No matter how we change the interferometer orientation, the interference pattern does not change.

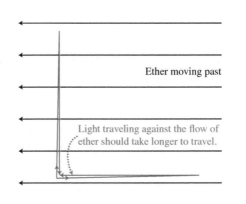

Possible conclusions
1. There is no ether through which light travels.
2. There is ether, but it is "stuck" to Earth's surface and does not move relative to the interferometer.

Michelson and Morley repeated the experiment several times, but the phase difference between the two beams never changed. It seemed that light traveled at the same speed independent of its direction. This appeared to clearly indicate that ether as a medium for light to propagate did not exist.

Physicists, including Michelson, were reluctant to accept this no-ether result. They tried to revise their model to fit the experimental data. Perhaps the ether was dragged along through space by Earth. This would mean that the ether was at rest relative to the interferometer, explaining the null result of their experiment. However, if the ether is attached to Earth in some way, then as Earth rotates around its axis and orbits the Sun, it should cause the ether in the solar system to become twisted. This would cause light coming from stars to be slightly deflected on its way to Earth (like a water wave would be deflected by a whirlpool in the water). However, no one observed such an effect. Another explanation was that the length of the side of the interferometer parallel to

Earth's motion changed due to the motion of Earth relative to ether, thus causing the change in the interference pattern. No such change was measured. It seemed that light did not require a medium to travel through.

REVIEW QUESTION 26.1 Why is the historical role of the Michelson-Morley experiment that of a testing experiment, not an observational experiment?

26.2 Postulates of special relativity

By the end of the 19th century, the laws of Newtonian physics had been well developed and accepted. Physicists knew that using Newton's laws to analyze a process would yield consistent results, regardless of the inertial (nonaccelerating) reference frames used. This feature is known as **invariance**.

Invariance is a result of acceleration (rather than velocity) being proportional to the net force exerted on an object. For example, when you are on an airplane moving smoothly at constant velocity, you can not detect the airplane's motion; all the events around you occur as if the airplane is at rest on the ground. If you drop a book, it lands at your feet; if you push a suitcase, it accelerates in the same way as if the airplane were parked on the runway. However, for observers in noninertial reference frames (accelerating frames), such as airplane passengers when the plane is taking off, Newton's laws do not explain observed phenomena, such as a loose food cart starting to roll without being pushed.

Invariance did not apply to electromagnetism and in particular to the observation of light in different reference frames. The speed of light in different reference frames was addressed by a teenager named Albert Einstein, who became a master at designing *thought experiments*. In 1895, at age 16, he tried to visualize himself traveling beside a beam of light. If he traveled through air at speed 3×10^8 m/s, it would be possible to stay beside one of the crests of the light wave, which would appear as a static electromagnetic field oscillating in space but not in time. Such a field would contradict the mechanism that explains EM waves. Changing fields create each other; these are not changing. We can visualize an even more disturbing difficulty. If Einstein then turned on a lamp held in his hand, the light from this lamp should move away from him at speed 3×10^8 m/s. Thus, to another observer at rest with respect to the source of the original light wave, there would be two beams of light: the one beside Einstein traveling at 3×10^8 m/s and the one from Einstein's lamp moving at Einstein's speed plus the speed of the light leaving his lamp, that is, at 3×10^8 m/s $+ 3 \times 10^8$ m/s $= 6 \times 10^8$ m/s (**Figure 26.4**). These thought experiments meant that Maxwell's equations (which we discussed at the end of Chapter 21) had to be written differently for these different observers, with a different speed of light in each case.

FIGURE 26.4 A thought experiment. Imagine a person traveling at light speed beside the crest of a light wave. His lamp is shining forward in the direction of travel. Would an observer detect one light wave traveling at c and light from the moving light source traveling at 2c?

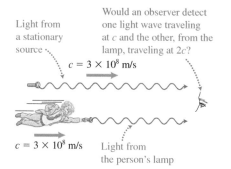

Einstein's two postulates

Einstein believed that the invariance principle was fundamental to any physics idea. He also believed that Maxwell's equations were correct. To resolve this contradiction, in 1905 Einstein proposed a new theory, called **the special theory of relativity**. He could have based it on the results of the Michelson-Morley experiment that suggested that the speed of light in air or a vacuum was independent of the motion of the observer, but he did not. In fact, it is not certain whether Einstein was aware of Michelson and Morley's results at all. Although there is some historical controversy about this, the general consensus is that Einstein based his ideas on his thought experiments.

Einstein's theory started with the following two *postulates*. A postulate is a statement that is assumed to be true (it is not derived from anything). It is usually used as the starting point for a logical argument.

1. *The laws of physics are the same in all inertial reference frames.* This was not a new idea in mechanics. Newton's second law,

$$\vec{a}_O = \frac{\sum \vec{F}_{\text{on O}}}{m_O}$$

remains the same regardless of the inertial reference frame in which you choose to apply it. But it was a new idea for Maxwell's equations of electromagnetism.

2. *The speed of light in a vacuum is measured to be the same in all inertial reference frames.* The speed of light in a vacuum measured by observers in different inertial reference frames is the same regardless of the relative motion of those observers.

FIGURE 26.5 All inertial reference frame observers measure the same speed for light *c*.

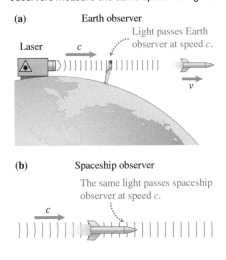

TIP As strange as postulate 2 may seem, remember that our everyday ideas about how to combine velocities have come from observing objects traveling very slowly compared with light speed. We should always be ready for our ideas to require improvements when applied in new situations.

Although postulate 2 is supported by the Michelson-Morley experiment and is consistent with Maxwell's equations, it is still counterintuitive. Think back to the boats in Section 26.1. The speed with respect to the shore of the boat traveling parallel to the shore depended on its velocity with respect to the water and the velocity of the water with respect to the shore. Now, suppose that an intense pulse of laser light is shot from Earth toward a spaceship moving away. A stationary observer on Earth measures the light's speed *c* as 3.0×10^8 m/s (**Figure 26.5a**). If the spaceship is moving away from Earth at a speed *v* that is just a little less than light speed, then our intuition tells us that the laser light will have a difficult time reaching the ship. After all, isn't the laser light's speed just $c - v$ relative to the ship (similar to the boat going upstream)? However, according to Einstein's second postulate, an observer on the spaceship should also measure the laser light speed to be 3.0×10^8 m/s (Figure 26.5b). Although it sounds impossible, this is exactly the outcome of the Michelson-Morley experiment!

FIGURE 26.6 Bob and Alice sending pulses.

REVIEW QUESTION 26.2 Alice is standing on the ground; Bob is above her in a high-speed rocket. They each shoot laser pulses to the right and left when Bob is directly above Alice (see **Figure 26.6**). Draw a sketch that shows these pulses a moment later (draw the relative locations of Alice, Bob, and their four laser pulses). Explain your reasoning.

26.3 Simultaneity

In our boat analogy, we assumed that the time interval during which each boat traveled during its trip was the same independent of the reference frame (shore or water) ($\Delta t_{\text{boat to shore}} = \Delta t_{\text{boat to water}} = \Delta t$). However, this seemingly reasonable way of thinking about time turns out to contradict the second postulate proposed by Einstein (and therefore it contradicts the results of the Michelson-Morley experiment).

The second postulate of the special theory of relativity made physicists completely rethink their ideas about time. Consider the following thought experiment.

Simultaneity of events in two different inertial reference frames

Suppose Alice is standing on the ground, and Bob is in a rocket (**Figure 26.7a**). They each have two clocks that they want to synchronize by sending light pulses to them. The clocks start when a pulse reaches them. If they start at the same time, the clocks are considered synchronized. All the clocks are *attached to the ground* and are equidistant from Alice (Bob's clocks are raised on poles so that his pulses reach them as they travel in straight lines to the right and left). When Bob is directly above Alice, they both send their signals. Who ends up with a synchronized pair of clocks, Alice or Bob? Let's consider the two perspectives.

Because the clocks are equidistant from Alice and the clocks are at rest with respect to her (that is, both Alice and the clocks are in the same inertial reference frame), she notes that the light reaches her front clock (on the right) and her back clock (on the left) at the same time (Figure 26.7b). However, Bob sees his front clock moving toward him while the light pulse he sent is moving toward it. He also sees his back clock moving away from him while the light pulse is moving toward it (Figure 26.7c). Therefore, the light pulse headed to Bob's (approaching) front clock travels a shorter distance before reaching the clock than the light pulse headed to Bob's (receding) back clock. Because the speed of light is finite and is the same in all directions for all observers, according to Bob, the light pulse reaches his front clock before it reaches his back clock. We conclude that the pulses arrive at the clocks at the same time for Alice, but do not arrive at the clocks at the same time for Bob. Alice's clocks are synchronized; Bob's are not.

Implications of the difference in observations

Think about what this means. We are focusing on two events for each person: (1) their light pulse reaching their front clock and (2) their light pulse reaching their back clock. In Alice's reference frame (attached to the ground), these two events are simultaneous. In Bob's reference frame (the flying rocket), these two events are not simultaneous.

This really occurs! Events that happen at the same time in one reference frame do not necessarily occur at the same time in another. At regular rocket speeds, however, the times of the events won't vary by more than 10^{-12} to 10^{-13} s between observers. This difference gets larger the faster the rocket travels relative to the ground, but the effect only becomes significant if the rocket travels at a substantial fraction of light speed.

Phenomena that only become significant in high-speed circumstances like this are called **relativistic effects**. This so-called **breakdown of simultaneity** is just the first of many relativistic effects that we will investigate in this chapter.

> **Breakdown of simultaneity** Events that occur simultaneously for an observer in one inertial reference frame do not necessarily occur simultaneously for observers in other inertial reference frames.

REVIEW QUESTION 26.3 You hear in your physics class the following statement: "Time is relative." How would you explain the meaning of this statement to a friend who is not taking physics?

26.4 Time dilation

In this section we will examine how the time interval between events differs in different inertial reference frames. An event is anything that happens: a person's departure for a trip, the person's return from that trip, light pulses reaching clocks, as in the example from the previous section, and so on. To specify an event quantitatively, we need to indicate where it occurs (a position) and when it occurs (a clock reading).

Our everyday experience supports the idea that the time interval between two events is something that all observers will agree on. For example, if your physics class lasts 60 min according to your clock, won't astronauts on the International Space Station watching the webcast find that 60 min passes on their clocks as well? It shouldn't matter that the astronauts are moving at 7700 m/s (17,000 mi/h) relative to you, right? The thought experiment in the previous section, however, suggests that it does matter.

FIGURE 26.7 Laser pulses used to synchronize clocks. (a) Alice, Bob, and the clocks. (b) Pulses arrive at the clocks at the same time for Alice. (c) Pulses do not arrive at the clocks at the same time for Bob.

(a)

(b)

(c)

FIGURE 26.8 An experiment for two observers (a) to detect the time interval between two events. The time interval between events as seen (b) in the proper reference frame, where they occur at the same place and (c) in another reference frame, where the events occur at different places.

(a)

Laser light travels upward, reflects off of a mirror, and is detected below.

Motion of train

Mirror

Laser Detector   You
on train

Observer on platform

(b)   You on train

Mirror

L_0

The light flash and detection as seen from the train

Flash Δt_0

Detector

(c)   Observer on platform

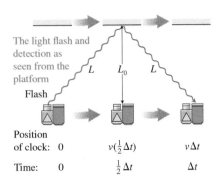

The light flash and detection as seen from the platform

Flash L L_0 L

Position of clock: 0 $v(\frac{1}{2}\Delta t)$ $v\Delta t$

Time: 0 $\frac{1}{2}\Delta t$ Δt

We can use another thought experiment to further investigate this. Suppose you are standing on a moving train. Your friend stands on the platform at the station. On the train is a laser pointing upward, a mirror located directly above the laser, and a light detector next to the laser (**Figure 26.8a**). The train is moving to the right at speed v, a significant fraction of light speed relative to the platform. A flash of light is emitted from the laser (event 1), reflects off the mirror, and is then detected (event 2).

In your reference frame on the train, you see the laser light travel vertically upward, then downward (Figure 26.8b). In your friend's reference frame on the platform, however, the laser light must move upward at an angle in order to reflect off the mirror, then downward at an angle to hit the detector—because the train is moving (Figure 26.8c).

What is the time interval between the flash and its detection (the two events), as measured by each of the two observers? For you, the light travels at speed c a distance $2L_0$, where L_0 is the distance from the light source to the mirror. The time interval between the flash and its detection is then

$$\Delta t_0 = \frac{2L_0}{c}$$

According to your friend on the platform, the light travels a longer distance, $2L$. Using the Pythagorean theorem, the distance L is the square root of the sum of the lengths squared of the two sides of the triangle with L as its hypotenuse:

$$L = \sqrt{L_0^2 + \left(\tfrac{1}{2}v\,\Delta t\right)^2}$$

The base of the triangle $(1/2)v\,\Delta t$ is one half the distance the train travels at speed v during the time interval Δt during which the light travels from its source up to the mirror and back down to the receiver. The time interval between the flash and its detection is

$$\Delta t = \frac{2L}{c} = \frac{2\sqrt{L_0^2 + \left(\tfrac{1}{2}v\,\Delta t\right)^2}}{c}$$

Rearrange $\Delta t_0 = \dfrac{2L_0}{c}$ to $L_0 = \dfrac{c\Delta t_0}{2}$ and substitute this into the above equation to get

$$\Delta t = \frac{2\sqrt{\dfrac{c^2\Delta t_0^2}{4} + \dfrac{v^2\Delta t^2}{4}}}{c}$$

Solving for Δt, we get

$$\Delta t = \frac{\Delta t_0}{\sqrt{1 - \dfrac{v^2}{c^2}}}$$

The term

$$\sqrt{1 - \frac{v^2}{c^2}}$$

is always less than or equal to 1. (It is equal to 1 only when the train is at rest with respect to the platform.) Therefore, the time interval Δt between the flash and its detection measured by your friend on the platform is greater than the time interval Δt_0 measured by you on the train.

Notice that we used $c = 3.00 \times 10^8$ m/s for the speed of light in both reference frames (according to Einstein's second postulate). This analysis results in a striking conclusion—the time interval during which the laser light was in flight differs depending on the reference frame in which it is measured. This relativistic effect is called **time dilation**.

Notice that your reference frame on the train is the one reference frame where the two events—the flash and the detection—occur at the same position. The time interval Δt_0 between the events measured in this reference frame is called the **proper time interval**, and this reference frame is known as the **proper reference frame**. The time interval Δt between these same two events measured in any other reference frame will always be longer. That is, $\Delta t > \Delta t_0$. The term "proper" comes from a mistranslation of "proprietary," which implies that only the person holding the clock would experience that much time.

We certainly don't observe this occurring in everyday life—if a particular boat trip takes 2 h according to the captain on the boat, won't it also take 2 h according to people on shore? Not if the idea of time dilation is correct. Actually, time dilation is happening all the time, though the effect only becomes substantial when the two reference frames are moving relative to each other at significant fractions of light speed. For example, for Δt and Δt_0 to be one-millionth of a percent different, the two reference frames would have to be moving relative to each other at over 4×10^4 m/s (over 90,000 mi/h).

Time dilation can be tested using what is known as a muon decay experiment, described in Testing Experiment **Table 26.2**. A muon is an elementary particle that occurs naturally in the streams of particles coming to us from the Sun and distant stars (called cosmic rays); it also can be produced during reactions in high-energy particle accelerators. Once produced in an accelerator, muons "live" for a very short time. If 1000 muons are produced, then 1.5×10^{-6} s later, approximately 500 muons remain. This time is called the **half-life** because half of the particles remain after this time.

Consider an experiment in which 1000 muons are produced and move across the lab at $0.95c$. Event 1 will be the production of the muons. Event 2 will be the decay of the 500th muon.

TIP Relativistic effects are important when objects are observed to move at speeds that are significant fractions of light speed and are not important at "everyday" speeds. A good rule of thumb is that if an object is observed to move at a speed $0.1c$ or greater, then relativistic effects should be included.

TESTING
EXPERIMENT TABLE 26.2   **Testing the idea of time dilation**

Testing experiment	Prediction	Outcome
A beam of 1000 muons is produced by a source (event 1) and emitted at a speed of $0.95c$. How far from the source will the muons travel before their number is reduced to 500 (event 2)? Muon just after it is produced — $\vec{v} = 0.95c$ — Muon decays to electron and energy. Electron, Energy, $d = ?$	*If time passes the same way for all observers (no time dilation),* then in the laboratory reference frame, the muons should travel a distance equal to the product of their speed with respect to the lab reference frame and their half-life: $$d = (0.95c)(1.5 \times 10^{-6} \text{ s})$$ $$= (0.95)(3.0 \times 10^8 \text{ m/s})(1.5 \times 10^{-6} \text{ s}) = 430 \text{ m}$$ *If time passes differently for observers in different reference frames and obeys the time dilation equation,* then in the laboratory reference frame, one half of the muons should travel a distance equal to the product of their speed with respect to the lab reference frame and their half-life *in the lab reference frame*, which will be different from their half-life in the proper reference frame. Their half-life in the lab reference frame is $$\Delta t = \frac{1.5 \times 10^{-6} \text{ s}}{\sqrt{1 - \frac{(0.95c)^2}{c^2}}} = 4.8 \times 10^{-6} \text{ s}$$ Because they move at speed $0.95c$ in the lab frame, one half of the muons travel $$d = (0.95c)(4.8 \times 10^{-6} \text{ s})$$ $$= (0.95)(3.0 \times 10^8 \text{ m/s})(4.8 \times 10^{-6} \text{ s}) = 1400 \text{ m}$$ Thus if time dilation occurs, one half of the muons should travel a significantly greater distance through the lab than they would if there were no time dilation.	The muons actually traveled about 1400 m before their number was reduced to one-half.

Conclusion

The outcome matches the prediction based on the time dilation equation. Many experiments confirm that short-lived particles travel much farther than would be possible given their proper half-lives. The idea of time dilation is not rejected. We have gained more confidence in it.

Muons are produced in nature as well. They are created when cosmic rays (high-energy protons) hit molecules in the atmosphere approximately 60,000 m above Earth's surface. If there were no time dilation, these muons would have a 1 in 10^{40} chance of reaching Earth from that altitude. However, because of time dilation and the fact that they are moving at nearly the speed of light, about 1 out of every 10 muons reaches Earth's surface.

Time dilation The reference frame in which two events occur at the same position is called the *proper reference frame* for these events. The time interval between these events measured by an observer in the proper reference frame is the *proper time interval,* Δt_0. The time interval Δt between the same two events as measured by an observer moving at constant velocity $\vec{v}$ relative to the proper reference frame is

$$\Delta t = \frac{\Delta t_0}{\sqrt{1 - \dfrac{v^2}{c^2}}} \tag{26.1}$$

where v is the speed of the observer relative to the proper reference frame.

The words "moving at constant velocity $\vec{v}$ relative to the proper reference frame" are especially important. Consider as an example a spaceship on which a light flashes at regular time intervals. The time interval between flashes as measured by a passenger on the ship is the proper time interval because the flashes occur at the same place relative to the passenger. For an observer on Earth, the flashes occur at different places as the spaceship moves past Earth. Therefore, Earth is not the proper reference frame. We say that Earth moves at speed v relative to the proper reference frame of the spaceship.

REVIEW QUESTION 26.4 You are on a train eating an apple. When you start eating, the train is passing a station, and a clock there reads 5 pm. When you finish, the train is passing the next station, where the clock reads 5:05 pm. Assuming that these two clocks are synchronized, did it take you 5 min to eat the apple? More than 5 min? Less than 5 min? Explain.

FIGURE 26.9 A moving arrow (a) starts and (b) stops a clock.

(a)

The arrowhead starts the clock.

(b)

The arrow's feathers stop the clock.

26.5 Length contraction

The measured time interval between two events depends on the reference frame from which the measurement is made. In addition, other physical quantities that are normally thought to be independent of reference frame, such as the length of an object, in fact do depend on the frame of reference.

Consider an arrow flying across a lab that moves past a clock at rest with respect to the lab (**Figure 26.9**). As the arrow's head passes the clock (event 1) it triggers the clock to start (Figure 26.9a). As the arrow's feathers pass the clock (event 2) the clock stops (Figure 26.9b). The two events occurred at the same place in the lab reference frame, so that is the proper reference frame, and the time interval Δt_0 measured by the clock is the proper time interval between the two events. In contrast, if the arrow had a tiny clock attached to it, the time interval Δt between the events would be

$$\Delta t = \frac{\Delta t_0}{\sqrt{1 - \dfrac{v^2}{c^2}}}$$

where v is the speed at which the lab moves past the arrow.

Could the length of the arrow depend on the reference frame from which it is measured? We'll call the arrow's **proper length,** L_0, its length in a reference frame in which the arrow is stationary. In this case, it's the reference frame defined by the arrow itself. Imagine that a fly has landed on the arrow and is traveling with it. Because the arrow is at rest relative to the fly, the fly would measure the length of the arrow to be the arrow's proper length L_0 (**Figure 26.10a**).

To an observer in the clock's (and lab's) reference frame, however, the length L of the moving arrow may be different (Figure 26.10b). We can relate L to L_0 by noting that the arrow's speed relative to the lab is the same as the lab's speed relative to the arrow (although their velocities are in opposite directions). We can calculate this speed in two different ways. In the arrow's reference frame, its speed is the proper length L_0 divided by the time interval Δt:

$$v = \frac{L_0}{\Delta t}$$

In the lab reference frame, the speed is the length L divided by the proper time interval Δt_0:

$$v = \frac{L}{\Delta t_0}$$

Because these two speeds are the same, we can set them equal to each other:

$$\frac{L}{\Delta t_0} = \frac{L_0}{\Delta t}$$

Using the time dilation equation, we can substitute for Δt:

$$\frac{L}{\Delta t_0} = \frac{L_0 \sqrt{1 - \dfrac{v^2}{c^2}}}{\Delta t_0}$$

Multiplying both sides of the above by Δt_0, we get a relationship between the proper length of the arrow, L_0, and its length L when measured in another reference frame:

$$L = L_0 \sqrt{1 - \frac{v^2}{c^2}} \tag{26.2}$$

The square root that appears in the above equation will always have a value less than 1. As a result, the length of the arrow measured in a reference frame where the arrow is moving will always be less than the arrow's proper length. In other words, the arrow will appear shorter along the direction that it is moving. This is known as **length contraction**.

Length contraction An object has a length L_0 when measured in a reference frame at rest relative to the object. This length is called the object's *proper length*. In a reference frame moving at speed v relative to the proper reference frame, the object's length L is

$$L = L_0 \sqrt{1 - \frac{v^2}{c^2}} \tag{26.2}$$

FIGURE 26.10 Measuring the arrow length (a) in the proper reference frame, where the arrow is at rest, and (b) in the clock's reference frame, where the arrow is moving.

(a)

A fly on the arrow measures the proper length, L_0.

(b)

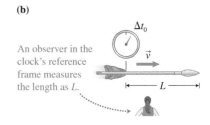

An observer in the clock's reference frame measures the length as L.

TIP Notice that length contraction only occurs parallel to the direction the object is moving. In the perpendicular direction, this effect does not occur.

Muons travel far!

In Table 26.2 we found that muons survive longer as a result of time dilation. However, that explanation does not work in the frame of the muon. How can a muon created in the upper atmosphere make it all the way to Earth if it should travel only about 450 m, given that it lives only 1.5 μs? The upper atmosphere is approximately 60 km deep. Cosmic rays are very energetic and create muons that move at $0.995c$.

Represent mathematically In the frame of the muon, its average lifetime is 1.5 μs and Earth is moving toward it at $0.995c$. But some variable has to change for it to reach Earth, with the long distance involved. Specifically, according to the muon, the distance to Earth is shorter according to length contraction. Using Eq. (26.2), we can write

$$L = L_0\sqrt{1 - \frac{v^2}{c^2}}$$

Solve and evaluate

$$L = L_0\sqrt{1 - \frac{v^2}{c^2}} = (60 \text{ km})\sqrt{1 - \frac{(0.995c)^2}{c^2}} = 6.0 \text{ km}$$

This distance is much closer to 450 m than 60 km and, given the statistics, an average muon now has a high probability of making it to Earth.

Try it yourself Suppose you live in a strange universe in which the speed of light is constant, but is only 20 m/s. Your car is 3.0 m long when parked by the curb in front of your house. How long will it look to your friend who is driving by your car, traveling at 10 m/s relative to your car?

Answer   2.6 m.

REVIEW QUESTION 26.5 You hold a 1-m-long stick. Describe those conditions under which an observer will measure the stick to be shorter than 1 m and those conditions under which an observer will measure the stick to be longer than 1 m.

26.6 Spacetime diagrams

FIGURE 26.11 World lines for two objects and two light beams drawn on a spacetime diagram.

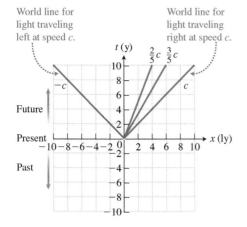

In relativity, as in other areas, a graph is useful for keeping track of information. This is especially true when there are multiple objects. There is a tool from Chapter 2 that, with only slight modification, comes in handy here. Recall that a position-versus-time graph shows where an object is at any given time. To represent the motion of a stationary object on a position-versus-time graph, we draw a horizontal line. If a position-versus-time graph is a straight line with a positive slope, it means that the object is moving in a positive direction with constant speed. On these graphs, time is represented on the horizontal axis and position on the vertical axis. When physicists are accounting for relativity, the axes are traditionally switched, so that time is represented on the vertical axis and position on the horizontal axis. This new representation is often referred to as a **spacetime diagram** to emphasize that space and time are intertwined. The line showing the time-versus-position graph of an object is called a **world line** for that object (**Figure 26.11**). In Figure 26.11 you see world lines for two objects and two light beams sent in opposite directions. Notice that the two objects move at speeds less than the speed of light and that their world lines are more steeply angled. In general, the slower the object, the steeper its world line. Thus the line for the object moving at $\frac{2}{5}c$ is steeper than that for the object moving at $\frac{3}{5}c$.

Because relativistic effects do not typically appear until close to the speed of light, we will be choosing units that highlight this aspect. And because the ultimate speed limit is the speed of light c, the units are also selected with this in mind. The motion of light is represented by a line with a slope of $\pm 45°$ depending on which direction the light is traveling. There are several systems of units that would give these results. For example, if the unit on the time axis is years (y), then the unit on the horizontal axis will be a **light-year** (ly), defined as the distance light travels in a year. Alternatively, we could have the unit of time as a second and the unit of distance the distance that light travels in 1 s—one *light-second*.

One result of switching the axes is that the slope of the line is no longer velocity; rather, it is the *inverse* of velocity. However, because we chose years (or seconds) for our time units and light-years (or light-seconds) for distance units, a line with the slope of ± 1 disregarding the units ($\pm 45°$) corresponds to the motion of an object traveling at the speed of light. Thus, an object that is moving at $\frac{1}{2}c$ would have a world line with slope 2. This leads to an important tip for building spacetime diagrams: **the slope of the world line for an object moving at speed v is c/v.**

CONCEPTUAL EXERCISE 26.2 **Drawing spacetime diagrams**

Draw a spacetime diagram for the situation in Review Question 26.2, with Bob traveling at 86.6% the speed of light. Include all objects (including light pulses) and draw the diagram with Alice as the observer.

Sketch and translate We have a sketch in Figure 26.6. It is important to remember that we are considering Alice's reference frame.

Simplify and diagram The problem statement does not say where Alice is standing, so we assume for simplicity that she is standing at the origin. Alice's world line is therefore vertical (her position does not change as time elapses). Bob is traveling to the right at 0.866c. The slope of his world line is therefore $\frac{c}{(0.866c)} = 1.15$. In other words, his world line makes an angle to the horizontal of slightly more than 45°.

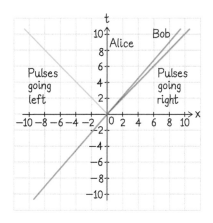

The four light pulses are created when Bob is directly above Alice. The pulses keep up with one another because of postulate 2. Because the graph only represents one dimension, each line represents two pulses. They each have a slope of ± 1. (Notice we have not specified the units yet. This diagram would work no matter which units we chose.) The spacetime diagram from Alice's perspective is shown in the left column.

Try it yourself Draw the spacetime diagram with Bob as the observer. Explain your reasoning.

Answer

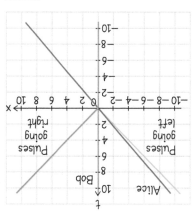

To get Bob's perspective, we first have to realize that Bob does not consider himself to be moving, so he would draw his world line as a vertical line, say, at the origin. Bob would say Alice is moving to the left. And Bob would still say that the light pulses travel away from him at the speed of light. This gives us the diagram at right.

Spacetime diagrams deserve one more note. When switching frames of reference, there are two effects that we have not yet incorporated: time dilation and length contraction. In principle, you could calculate either of these to adjust the amount of time between two events or distance between two objects. It turns out that by adjusting the distance between objects with length contraction, the amount of time that elapses between events always works itself out. This will also account for any mismatches in simultaneity experienced by the two observers. The following example shows how the process works.

EXAMPLE 26.3 **Using spacetime diagrams to analyze simultaneity**

Explain what you need to do to add world lines for the two clocks equidistant from Alice in Figure 26.7 to the spacetime diagrams in the previous exercise (including that in the "Try it yourself" section) and do the required calculations.

Sketch and translate The situation in the problem is sketched in Figures 26.7a and b. We need to represent the same process with spacetime diagrams.

Simplify and diagram The problem does not specify the distance between Alice and the clocks, so we will choose a distance consistent with our previous diagrams; let's say 6 units. It is helpful to pick a unit at this point, so let's agree that 1 unit of distance is 300 m. It takes light 1 $\mu s = 10^{-6}$ s to go that far, so our choice of 1 unit of time is 10^{-6} s. This puts each set of clocks at (6 units) $\times$ (300 m)/unit = 1800 m away from Alice. The clocks are not moving, so they should be

(CONTINUED)

represented by vertical lines on the diagram corresponding to Alice's perspective, as shown below.

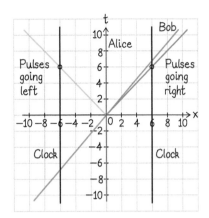

On the above diagram, the clocks are represented as black lines, and the times of the intersection of these lines with the lines of the light pulses indicate when those clocks register the reading (that is, when the pulses arrive at the clocks). You can see that the clocks register the arrival of the pulses simultaneously at a time of 6.0 units.

The spacetime diagram for Bob will be different. Once again, we need to add the lines for the clocks to the diagram, but this time the clocks are moving at the same speed as Alice and are equidistant from Alice. This would make the clocks' lines parallel to Alice's world line, which is tilted to the left. However, we do not know the distance between Alice and the clocks as seen by Bob; thus we cannot draw the diagram before we do the calculation.

Represent mathematically and solve Given that the clocks and Alice are moving with respect to Bob, he registers that there are not

6 units of length between them, but a smaller length because of length contraction. This length is calculated using the expression for length contraction:

$$L = L_0 \sqrt{1 - \frac{v^2}{c^2}} = (6 \text{ units}) \sqrt{1 - \frac{(0.866c)^2}{c^2}} = \pm (6 \text{ units}) \times 0.5$$
$$= \pm 3 \text{ units}$$

Evaluate The distance to the clocks is smaller for Bob than for Alice. We can now draw the spacetime diagrams for the clocks as seen by Bob. Notice that the clock that is approaching Bob (on the right) starts where the dot is—this is where the line for the light pulse crosses the line for the clock (just under 2 time units)—while the other light pulse is still chasing the receding clock. It doesn't catch it until sometime after the graph ends (where the yellow line for light pulses going left crosses the left-hand black line).

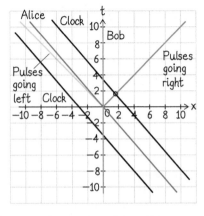

- -

Try it yourself At what time (in units of time) will the pair of right clocks fly by Bob? Show on the spacetime diagram and confirm with the calculation.

Answer   3.5 units of time.

A few final comments about spacetime diagrams are in order. First, each observer has a simple explanation for what happened. Alice says the clocks are synchronized because they were equidistant and not moving, whereas Bob declares that they are not synchronized because they were moving. Einstein would say that both Alice and Bob are correct! Each person just needs to explain what happened in their own reference frame. When we switch between reference frames, we encounter length contraction (we also experience time dilation).

Below is a summary of our knowledge of spacetime diagrams:

1. The slope of the world line for an object moving at speed v is c/v.
2. The world line for light is tilted at $\pm 45°$ (this means that the units on the axes need to be chosen for this to be true).
3. When you switch reference frames, length contraction needs to be accounted for. (Time dilation needs to be considered as well, but getting length contraction correct takes care of that.) If an object is not moving in one frame, it will be moving in the new frame, and its length will be contracted. Similarly, if two objects are stationary, the distance between them will be contracted.

REVIEW QUESTION 26.6 We encounter vertical and tilted world lines on spacetime diagrams, but never horizontal lines. Why?

26.7 Velocity transformations

In this section we will learn how to reconcile the idea that light speed is the same for all observers even if they are moving relative to each other at nearly the speed of light. Imagine you are on a skateboard moving to the left at velocity $\vec{v}_{\text{skateboarder to ground}}$ relative to the ground. You throw a ball you are carrying in the same direction at velocity $\vec{v}_{\text{ball to skateboarder}}$. Your friend (the ball catcher) standing on the ground detects the ball moving at a different velocity: $\vec{v}_{\text{ball to catcher}} = \vec{v}_{\text{ball to skateboarder}} + \vec{v}_{\text{skateboarder to catcher}}$ (Figure 26.12). This result (similar to the result for the boat traveling upstream or downstream) is based on the assumption that the time interval of the ball in flight is the same for the skateboarder and for the ball catcher on the ground (the same assumption that we made in the boat example in Section 26.1). We can generalize this equation as

$$\vec{v}_{\text{OS}} = \vec{v}_{\text{OS}'} + \vec{v}_{\text{S}'\text{S}} \tag{26.3}$$

Here $\vec{v}_{\text{OS}}$ is the velocity of the object (the ball) with respect to reference frame S (the ball catcher); $\vec{v}_{\text{OS}'}$ is the velocity of the object with respect to a second reference frame S' (the skateboarder); and $\vec{v}_{\text{S}'\text{S}}$ is the velocity of the reference frame S' relative to the reference frame S. Equation (26.3) is called the **classical** (or **Galilean**) **velocity transformation equation**. It is consistent with our everyday experience. The derivation of the equation is based on the assumption that the time interval during which the motion occurs is the same for all observers.

Difficulties with the classical velocity transformation equation

According to the ideas of special relativity and the results of the Michelson-Morley experiment, the speed of light is the same in all inertial reference frames. This idea contradicts the classical velocity transformation equation. Suppose you are in reference frame S' moving toward the right at speed $v_{\text{S}'\text{S}} = 0.99c$ relative to another reference frame S. You shine a laser beam straight ahead. The light moves at speed $v_{\text{LS}'} = c$ forward in your S' reference frame. According to the classical velocity transformation equation, the speed v_{LS} of the light in reference frame S should be

$$v_{\text{LS}} = v_{\text{LS}'} + v_{\text{S}'\text{S}} = c + 0.99c = 1.99c$$

This result contradicts Einstein's second postulate. We need an improved velocity transformation equation that respects the postulates of relativity.

Relativistic velocity transformation equation

Deriving this equation requires techniques beyond the scope of this book, so we'll just provide the result for one-dimensional processes.

Relativistic velocity transformation Suppose that inertial reference frame S' is moving relative to inertial reference frame S at velocity $\vec{v}_{\text{S}'\text{S}}$ (positive to the right and negative to the left). An object O moves at velocity $\vec{v}_{\text{OS}'}$ relative to reference frame S' (also positive to the right and negative to the left). The velocity $\vec{v}_{\text{OS}}$ of O in reference frame S is then

$$v_{\text{OS}} = \frac{v_{\text{OS}'} + v_{\text{S}'\text{S}}}{1 + \dfrac{v_{\text{OS}'} v_{\text{S}'\text{S}}}{c^2}} \tag{26.4}$$

Notice that in Eq. (26.4) the numerator is the same as in the classical equation, but the denominator is different. Does the above equation satisfy postulate 2 of special relativity and agree with the classical equation for low speeds? Let's check some limiting cases.

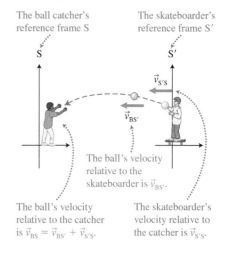

FIGURE 26.12 The classical velocity transformation. Velocities are simply additive in everyday situations.

The ball catcher's reference frame S

The skateboarder's reference frame S'

$\vec{v}_{\text{S}'\text{S}}$

$\vec{v}_{\text{BS}'}$

The ball's velocity relative to the skateboarder is $\vec{v}_{\text{BS}'}$.

The ball's velocity relative to the catcher is $\vec{v}_{\text{BS}} = \vec{v}_{\text{BS}'} + \vec{v}_{\text{S}'\text{S}}$.

The skateboarder's velocity relative to the catcher is $\vec{v}_{\text{S}'\text{S}}$.

TIP Because Eq. (26.3) is a vector equation, we must break it into components before using it. We have to choose a coordinate system that determines the signs of the individual vector components.

Limiting case analysis of Eq. (26.4)

Case 1: Suppose that an object O moves slowly in reference frame S' $(v_{OS'} \ll c)$, or that reference frame S' moves slowly relative to reference frame S $(v_{S'S} \ll c)$. In either case, the denominator of Eq. (26.4) is approximately 1 and the equation becomes the classical velocity transformation Eq. (26.3).

Case 2: Suppose that a rocket (reference frame S') passing Earth emits a laser pulse in the forward direction. The light travels at velocity $v_{LS'} = c$ relative to the rocket. Suppose the rocket (reference frame S') travels at velocity $v_{S'S} = 0.99c$ relative to the Earth (reference frame S). What is the velocity v_{LS} of the laser pulse relative to Earth?

$$v_{LS} = \frac{v_{LS'} + v_{S'S}}{1 + \dfrac{v_{LS'}v_{S'S}}{c^2}} = \frac{c + 0.99c}{1 + \dfrac{(c)(0.99c)}{c^2}} = \frac{(1 + 0.99)c}{(1 + 0.99)} = c$$

The laser pulse is observed to travel at speed c in both inertial reference frames, consistent with the postulates of special relativity.

REVIEW QUESTION 26.7 What is the meaning of the term "velocity transformation," and why is the classical velocity transformation different from the relativistic transformation?

26.8 Relativistic momentum

We know (from Chapter 6) that if the net impulse exerted on a system is zero, the momentum of the system is constant. This principle is very useful for analyzing collisions of various kinds. Our past use of the principle involved objects moving at nonrelativistic speeds $(v \ll c)$. Can we use the classical expression for the momentum of an object

$$\vec{p} = m\vec{v} = m\frac{\Delta \vec{x}}{\Delta t}$$

and the impulse-momentum principle to analyze situations where objects are moving at relativistic velocities?

When we use the above classical definition of momentum to analyze collisions of elementary particles moving at high speed, we find that even for an isolated system, the momentum of the system is constant in some reference frames but not in others. Thus, we have to either give up momentum as a conserved quantity or find a new expression that restores consistency across all reference frames. We'll take the second approach and find a new relativistic expression for momentum.

To redefine momentum, consider the classical expression $(\vec{p} = m\vec{v} = m(\Delta \vec{x}/\Delta t))$. In that expression, m is the mass of the object and $\Delta \vec{x}$ is the displacement of the object during the time interval Δt, both measured in some inertial reference frame. To get an improved relativistic expression for momentum, we try replacing Δt with the proper time interval Δt_0, where the relevant events are the object at the beginning and end of its displacement:

$$\vec{p} = m\vec{v} = m\frac{\Delta \vec{x}}{\Delta t_0}$$

If we substitute the expression for Δt_0 shown in Eq. (26.1) into the above definition, we get

$$\vec{p} = m\frac{\Delta \vec{x}}{\Delta t \sqrt{1 - (v/c)^2}} = \frac{m}{\sqrt{1 - (v/c)^2}}\frac{\Delta \vec{x}}{\Delta t} = \frac{m\vec{v}}{\sqrt{1 - (v/c)^2}}$$

Note that the $\vec{v}$ in the numerator is the velocity of the object, while the v in the denominator is its speed. This new expression reduces to the classical expression if the speed of the object is much less than the speed of light. Using this new expression for momentum, it is possible to show (though we will not do it here) that when the net impulse exerted on a system is zero, the system's momentum is constant regardless of inertial reference frame.

Relativistic momentum The relativistic momentum of an object of mass m in a reference frame where the object is moving at velocity $\vec{v}$ is

$$\vec{p} = \frac{m\vec{v}}{\sqrt{1 - \left(\dfrac{v}{c}\right)^2}} \qquad (26.5)$$

TIP Equation (26.5) is an improved expression for the momentum of objects traveling at all speeds rather than just speeds that are small compared with light speed. However, if the speed of the object is not a significant fraction of light speed, the simpler nonrelativistic expression can be used.

In Eq. (26.5) the denominator gets smaller and smaller as the object's speed gets closer and closer to the speed of light. This means that as speed increases toward light speed, the momentum of the object approaches infinity. This result is consistent with the idea that no object can travel faster than light speed in a vacuum. No such restriction was present in the nonrelativistic expression for the momentum of an object ($\vec{p} = m\vec{v}$).

EXAMPLE 26.4   Cosmic ray hits nitrogen

Some cosmic rays are high-energy protons produced during the explosive collapse of stars near the ends of their lives. Suppose a cosmic ray proton traveling at $0.90c$ enters our atmosphere and collides with a nitrogen atom. After the collision, the proton recoils at speed $0.70c$ in the opposite direction. The mass of the nitrogen atom is 14 times greater than the proton's mass. How fast is the nitrogen atom moving after the collision?

Sketch and translate We first sketch the process (see below). The system of interest is the proton and the nitrogen atom. Choose an inertial reference frame that is at rest relative to the nitrogen atom before it is hit by the proton and choose the positive x-axis to be the direction the proton is initially traveling. In that case, the initial and final velocity components of the proton and nitrogen atom along that axis are $v_{\text{pix}} = +0.90c$, $v_{\text{pfx}} = -0.70c$, and $v_{\text{Nix}} = 0$, and v_{Nfx} is the unknown final velocity component of the nitrogen atom. Let $m_{\text{p}} = m$; then $m_{\text{N}} = 14m$.

Initial situation

Final situation

Simplify and diagram Assume that there are no interactions between the system and the environment so that the momentum of the system is

constant. Because of the high speeds of the proton before and after the collision, we need to use the relativistic expression for the proton momentum. But we'll have to wait and see whether we need to treat the nitrogen atom relativistically; it depends on its speed relative to c. A momentum bar chart represents the process.

Represent mathematically The bar chart helps us apply the conservation of momentum principle along the x-axis:

$$p_{\text{pix}} + p_{\text{Nix}} + J_x = p_{\text{pfx}} + p_{\text{Nfx}}$$

or

$$\frac{m_{\text{p}}v_{\text{pix}}}{\sqrt{1 - (v_{\text{pi}}/c)^2}} + 0 + 0 = \frac{m_{\text{p}}v_{\text{pfx}}}{\sqrt{1 - (v_{\text{pf}}/c)^2}} + \frac{m_{\text{N}}v_{\text{Nfx}}}{\sqrt{1 - (v_{\text{Nf}}/c)^2}}$$

Solve and evaluate Substitute the known information and solve for v_{Nf}:

$$\frac{m(0.90c)}{\sqrt{1 - (0.90c/c)^2}} = \frac{m(-0.70c)}{\sqrt{1 - (0.70c/c)^2}} + \frac{14mv_{\text{Nf}}}{\sqrt{1 - (v_{\text{Nf}}/c)^2}}$$

Cancel factors of m and c and simplify to get

$$\frac{0.90}{\sqrt{1 - (0.90)^2}}c = \frac{-0.70}{\sqrt{1 - (0.70)^2}}c + \frac{14v_{\text{Nf}}}{\sqrt{1 - (v_{\text{Nf}}/c)^2}}$$

$$\Rightarrow \frac{14v_{\text{Nf}}}{\sqrt{1 - (v_{\text{Nf}}/c)^2}} = \left(\frac{0.90}{\sqrt{1 - (0.90)^2}} + \frac{0.70}{\sqrt{1 - (0.70)^2}}\right)c$$

$$= (2.06 + 0.98)c$$

(CONTINUED)

Finally, we get

$$\frac{v_{\text{Nf}}}{\sqrt{1 - (v_{\text{Nf}}/c)^2}} = 0.22c$$

To solve for v_{Nf}, multiply both sides by $\sqrt{1 - (v_{\text{Nf}}/c)^2}$ and square both sides:

$$v_{\text{Nf}}^2 = (0.22c)^2(1 - (v_{\text{Nf}}/c)^2)$$

Now collect all terms with v_{Nf} on the left side:

$$v_{\text{Nf}}^2 + (0.22v_{\text{Nf}})^2 = (0.22c)^2$$

Now factor out v_{Nf}^2 and solve for it:

$$v_{\text{Nf}}^2 = \frac{(0.22c)^2}{1 + 0.22^2}$$

v_{Nf} is then

$$v_{\text{Nf}} = \frac{0.22}{\sqrt{1 + 0.22^2}}c = 0.21c = 6.4 \times 10^7 \text{ m/s}$$

This is a significant fraction of light speed, but not at all unreasonable for a cosmic ray collision.

- -

Try it yourself Calculate the speed of the nitrogen atom after the collision using the nonrelativistic expression for the momentum of the proton and nitrogen. How does it compare to the relativistic value?

Answer   Relativistic effects are clearly relevant in this situation.
3.4×10^7 m/s, compared to the relativistic value of 6.4×10^7 m/s.

REVIEW QUESTION 26.8 Why must the classical expression of momentum be modified in special relativity?

26.9 Relativistic energy

Our ideas about space, time, velocity, and now momentum have been modified to include relativistic effects. A simple thought experiment shows that ideas concerning energy must also be modified. Imagine that we accelerate an electron through a potential difference of 300,000 V (large, but still achievable). The kinetic energy of this electron will then be $(1/2)mv^2 = e\,\Delta V$. This means the speed of the electron once it has finished accelerating will be

$$v = \sqrt{\frac{2e\,\Delta V}{m}} = \sqrt{\frac{2(1.6 \times 10^{-19} \text{ C})(300,000 \text{ V})}{9.11 \times 10^{-31} \text{ kg}}} = 3.2 \times 10^8 \text{ m/s}$$

This speed is faster than the speed of light. Something is wrong. Before we begin resolving this, let's introduce a little notation that will save time in our future analysis.

Special notations used in relativity

You've probably noticed that certain expressions commonly show up in relativistic equations. For example, the speed v of an object relative to the speed of light c often occurs as the ratio v/c. It is common to define this ratio as beta β:

$$\beta = \frac{v}{c} \tag{26.6}$$

Relativistic effects start becoming significant when $\beta > 0.1$, that is, at speeds $v > 0.1c$.

The quantity $\sqrt{1 - (v/c)^2}$ appears frequently in relativistic expressions as well, most often in the denominator. To make relativistic equations more compact and easier to work with, physicists define the symbol gamma γ as

TIP   Notice that γ is always >1 and β is always <1.

$$\gamma = \frac{1}{\sqrt{1 - (v/c)^2}} = \frac{1}{\sqrt{1 - \beta^2}} \tag{26.7}$$

Using these abbreviations, the time dilation, length contraction, and momentum equations become $\Delta t = \gamma \Delta t_0$, $L = L_0/\gamma$, and $p = \gamma mv$, respectively.

Relativistic kinetic energy

Deriving the relativistically correct expression for kinetic energy from first principles is rather complicated. Instead, we will provide the accepted expression and evaluate its low-speed limiting case to check for consistency with $(1/2)mv^2$.

The system of interest is a point-like object of mass m moving at constant speed v and not interacting with any other objects. The system therefore has no potential energy and no internal energy but does have kinetic energy. The correct relativistic expression for the energy of the system is

$$\gamma mc^2 = \frac{mc^2}{\sqrt{1 - \beta^2}} = \frac{mc^2}{\sqrt{1 - (v/c)^2}}$$

Note that mc^2 (mass times velocity squared) has energy units. Gamma $\gamma = 1/\sqrt{1 - (v/c)^2}$ and beta $\beta = v/c$ are dimensionless numbers.

To see if this expression is consistent with $K = (1/2)mv^2$ at low speeds ($\beta \ll 1$), we use the binomial expansion to rewrite the part of the expression that involves the square root. For situations where the quantity x is small,

$$\frac{1}{\sqrt{1 - x}} = (1 - x)^{-1/2} \approx 1 + \tfrac{1}{2}x$$

In our case $x = \beta^2$. Therefore, for small $\beta = v/c$, the expression becomes

$$\gamma mc^2 = \frac{mc^2}{\sqrt{1 - \beta^2}} = mc^2(1 - \beta^2)^{-1/2} \approx mc^2(1 + \tfrac{1}{2}\beta^2)$$

$$= mc^2\left(1 + \frac{v^2}{2c^2}\right) = mc^2 + \tfrac{1}{2}mv^2$$

Notice that the last term on the right is just the classical kinetic energy of an object of mass m moving at speed v. Based on the above, it appears that

$$K = \gamma mc^2 - mc^2$$

This is the relativistically correct expression for kinetic energy K that a precise derivation produces.

What is the meaning of the mc^2 term? Apparently, it is a kind of energy that depends only on the mass of the object. This is a new and very profound idea—an object has energy just because of its mass. The term mc^2 is called the **rest energy** of the object. The expression γmc^2 represents the total energy (rest plus kinetic) of a point-like object. The kinetic energy of the object equals its total energy minus its rest energy. These three ideas are summarized below.

Relativistic energy A point-like object of mass m has so-called rest energy because it has mass:

$$\text{Rest energy } E_0 = mc^2 \tag{26.8}$$

The total energy of the object is

$$E = \gamma mc^2 = \frac{mc^2}{\sqrt{1 - (v/c)^2}} \tag{26.9}$$

The object's kinetic energy is the object's total energy minus its rest energy:

$$\text{Kinetic energy } K = \gamma mc^2 - mc^2 = (\gamma - 1)mc^2 \tag{26.10}$$

You can see from Eq. (26.9) that when the speed of an object approaches the speed of light in a vacuum, its kinetic energy approaches infinity. Thus the relativistic expression for kinetic energy explains why no object can move faster than the speed of light in a vacuum—the work that needs to be done to accelerate the object to this speed is infinite.

FIGURE 26.13 One electron volt (1 eV) is the kinetic energy gained by an electron that accelerates across a 1-V potential difference.

The kinetic energy of the electron is 1 eV.

Relativistic energy effects are most commonly relevant in the context of atomic and nuclear processes. The joule is an inconvenient energy unit to use when analyzing these processes because the energies involved are so small. Instead, we use another energy unit—the electron volt. One electron volt is the kinetic energy of an electron that accelerates from rest across a potential difference of 1 V (from lower potential at position 1 to higher potential at position 2 in **Figure 26.13**). Using the work-energy principle:

$$K_1 + U_{q1} = K_2 + U_{q2}$$

$$\Rightarrow K_2 = U_{q1} - U_{q2} = q(V_1 - V_2) = (-1.6 \times 10^{-19}\,\text{C})(0.0\,\text{V} - 1.0\,\text{V})$$

$$= 1.6 \times 10^{-19}\,\text{J}$$

Electron volt An electron volt (1 eV) is the increase in kinetic energy of an electron when it moves across a 1.0-V potential difference:

$$1\,\text{eV} = 1.6 \times 10^{-19}\,\text{J}$$

In nuclear and elementary particle physics, energies much higher than this are common (but still small compared with everyday-life energies). In those cases the mega-electron volt $(1\,\text{MeV} = 1.6 \times 10^{-13}\,\text{J})$ and giga-electron volt $(1\,\text{GeV} = 1.6 \times 10^{-10}\,\text{J})$ are commonly used.

Rest energy of particles

Any object with mass has rest energy, from elementary particles such as electrons to massive objects such as the Sun. Rest energy can be converted into other forms of energy. In fact, the rest energy of the Sun is being slowly converted via nuclear fusion reactions into internal energy, which accounts for the Sun's high temperature and brightness.

QUANTITATIVE EXERCISE 26.5 ▸ **Mass equivalent of energy to warm and cool house**

On average, each year about 2×10^{10} J of electric and chemical potential energy are converted to cool and warm your home. If rest energy could be converted for this purpose, how much mass equivalent of rest energy would be needed?

Represent mathematically According to Eq. (26.8):

$$m = \frac{E_0}{c^2}$$

Solve and evaluate Inserting the appropriate values:

$$m = \frac{2 \times 10^{10}\,\text{J}}{(3.00 \times 10^8\,\text{m/s})^2} = 2 \times 10^{-7}\,\text{kg}$$

This is approximately one-tenth the mass of one of the hairs on your head.

Try it yourself Estimate the mass equivalent of the chemical potential energy that is transformed into other energy types when you drive your car for 1 year. Note that 3.8 L (1 gallon) of gasoline produces about 1.2×10^8 J.

Answer

Assume that you drive 13,000 km (8000 mi) in one year and get 15 km/L (35 mi/gallon). You would use 870 L (230 gallons) of gasoline, or about 3×10^{10} J; this is about 0.3 μg of rest energy (far less than the mass of the gasoline itself).

Relationship between mass and energy

Earlier (in Chapter 6) we learned that mass is a conserved quantity. Now we understand that that statement is not exactly correct. Rest energy (mass) can be converted into other energy forms, so only the *total* energy of the system, $E_{\text{total}} = \Sigma E_{\text{particles}} = \Sigma \gamma mc^2$, is a conserved quantity. However, when objects are not moving at relativistic speeds, mass and energy are conserved separately almost exactly.

EXAMPLE 26.6 Electron particle accelerator

An electron in a particle accelerator accelerates through a potential difference of 10^6 V. What is its final speed?

Sketch and translate We sketch the process to help visualize the motion of the electron, marking knowns and unknowns. We know the electric charge and the mass of the electron and need to find its final speed.

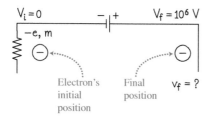

Simplify and diagram Assume that the electron starts at rest. The system is the electron and the electric field. The initial energy of the system is the system's rest energy; the final energy is the sum of the electron's positive kinetic energy, the system's negative electric potential energy, and the system's rest energy. Because no work is done on the system by the environment, the total energy of the system is constant. The energy bar chart below represents the process.

$$mc^2 + K_i + U_{qi} + \boxed{W} = mc^2 + K_f + U_{qf}$$

Represent mathematically We can now convert the energy bar chart into an equation:

$$E_{0i} + 0 + 0 + 0 = E_{0f} + K_f + U_{qf}$$

Because the rest energy of the system doesn't change during this process, we can subtract it from both sides of the equation. Inserting the appropriate expressions for each type of energy (including Eq. (26.10) for relativistic kinetic energy), we get

$$0 = (\gamma m_e c^2 - m_e c^2) + (-eV_f)$$

where m_e is the mass of the electron. We can simplify this equation by dividing both sides by $m_e c^2$ and solving for γ:

$$\gamma = \frac{eV_f}{m_e c^2} + 1$$

Remember:

$$\gamma = \frac{1}{\sqrt{1 - (v_f/c)^2}}$$

Solve and evaluate Now we need to solve for the final speed of the electron. We usually derive the expression for the needed quantity first and then substitute in the numbers. In this more complicated case, however, it is easier to first determine a numerical value for γ and then use it to solve for v_f. Insert the appropriate values in the first expression for γ above:

$$\gamma = \frac{eV_f}{m_e c^2} + 1$$

$$= \frac{(1.6 \times 10^{-19}\,\text{C})(10^6\,\text{V})}{(9.11 \times 10^{-31}\,\text{kg})(3.00 \times 10^8\,\text{m/s})^2} + 1 = 3.0$$

Now we can find the electron's final speed:

$$\frac{1}{\sqrt{1 - (v_f/c)^2}} = 3.0$$

Invert both sides of the equation:

$$\sqrt{1 - (v_f/c)^2} = \frac{1}{3.0}$$

Solve for v_f/c:

$$1 - (v_f/c)^2 = \frac{1}{9.0} \quad \Rightarrow \quad (v_f/c)^2 = 1 - \frac{1}{9.0}$$

$$\Rightarrow v_f/c = \sqrt{1 - \frac{1}{9.0}} = 0.94$$

or

$$v_f = 0.94c$$

Our result is reasonable—although the speed is very high, it is less than the speed of light.

Try it yourself Protons are accelerated from rest to a speed of $0.993c$ in the "small" 150-m-diameter booster accelerator that is part of the Fermi National Accelerator Laboratory in Batavia, Illinois. (a) Determine the proton's total energy when it leaves the booster accelerator. (b) Determine its kinetic energy as it leaves the accelerator.

Answer   (a) $E_f = 1.27 \times 10^{-9}$ J; (b) $K_f = 1.12 \times 10^{-9}$ J.

REVIEW QUESTION 26.9 If we did not use the relativistic expression for kinetic energy to calculate the speeds of elementary particles in accelerators, would the calculated nonrelativistic speeds be greater or less than the calculated relativistic speeds? (Hint: Estimate the values of any physical quantities that you need.)

26.10 Doppler effect for EM waves

The observed frequency of a mechanical wave (for example, a sound wave) is different from the frequency of the wave source if the observer and/or the source are moving relative to each other. A similar phenomenon exists for light (EM) waves; however, due to relativistic effects, the Doppler effect for light is somewhat different.

In the Doppler effect for sound [Eq. (11.17)], if the source or observer moves through the medium at a speed equal to or faster than the speed of sound, or if their motion relative to each other is equal to or faster than this speed, then the sound waves might not reach the observer at all. But the speed of EM waves, $c = 3.00 \times 10^8$ m/s, is the same for all observers no matter how they are moving relative to the source. As a result, the equation that describes the Doppler effect for EM waves differs from the equation for the Doppler effect for sound.

We won't derive the equation for the Doppler effect for EM waves. Instead, we just provide the equation and use it to analyze several processes.

Doppler effect for EM waves

$$f_O = f_S \sqrt{\frac{1 + v_{rel}/c}{1 - v_{rel}/c}} \qquad (26.11)$$

where f_O is the EM wave frequency detected by an observer, f_S is the frequency emitted by the source, and v_{rel} is the component of the relative velocity between the observer and the source along the line connecting them (the component is positive when the observer and source are approaching each other and negative when moving apart).

The major difference between this and the Doppler effect equation for sound is the square root. The need for this alteration comes about due to time dilation: the source and observer are in different inertial reference frames. The square root can be simplified for the Doppler effect for EM waves if v_{rel} is small compared to the speed of light. Using the binomial expansion, the equation becomes the following.

Doppler effect for EM waves (slow observer-source relative motion):

$$f_O = f_S \left(1 + \frac{v_{rel}}{c} \right) \qquad (26.12)$$

The sign conventions for Eq. (26.11) apply here.

When the observer and source are approaching, the observed frequency is higher than the source frequency. When they are moving apart, the observed frequency is lower.

Suppose, for example, that the source emits light in the blue part of the spectrum and that the source and observer are moving apart. The light detected by the observer could be in the yellow or red part of the spectrum. In astronomy this is known as a **red shift**, and it was important in discovering that the universe is expanding. Conversely, if the source and observer are moving toward each other, the light detected by the observer would be shifted toward the blue end of the spectrum.

Modern police radar and sports radar measure speeds by knowledge of the source frequency and the ability to measure the frequency of a reflected wave. Let's look at such practical examples of the Doppler effect for EM waves.

Doppler radar and speeding tickets

Police radar uses microwaves, one type of electromagnetic wave. Consider the next example.

EXAMPLE 26.7   **The speeding ticket**

Physicist Dr. R. Wood ran a red light while driving his car and was pulled over by a police officer. Dr. Wood explained that because he was driving toward the red light, he observed it as green due to the Doppler effect. Dr. Wood was then given a very expensive speeding ticket. Should Dr. Wood go to court to fight the ticket?

Sketch and translate The traffic light is the source, and Dr. Wood is the observer. The situation is sketched below. We are interested in determining the speed of Dr. Wood's car relative to the traffic light. Dr. Wood was traveling fast enough so that the red light emitted by the traffic signal was observed by him to be green. Because he is moving toward the source, the relative velocity v_{rel} will be positive and equal to Dr. Wood's car's speed relative to the traffic light.

Frequency f_S of red light traveling toward Dr. Wood's car.

What relative velocity would make the light appear green?

$\vec{v}_{car} = \vec{v}_{rel} = ?$

Simplify and diagram Assume that the traffic light emits red monochromatic light at a frequency of about $f_S = 4.5 \times 10^{14}$ Hz and that the frequency of green light is about $f_O = 5.4 \times 10^{14}$ Hz. Assume also that the car moves with constant velocity.

Represent mathematically Because we are talking about a low-speed car, we can use the low-speed Doppler effect $f_O = f_S(1 + (v_{rel}/c))$ to determine the speed of the car. Multiply both sides of the equation by c and then solve for v_{rel}:

$$cf_O = cf_S\left(1 + \frac{v_{rel}}{c}\right) = f_S(c + v_{rel}) = f_Sc + f_Sv_{rel}$$

$$\Rightarrow f_Sv_{rel} = cf_O - f_Sc$$

$$\Rightarrow v_{rel} = \frac{c(f_O - f_S)}{f_S} = c\left(\frac{f_O}{f_S} - 1\right)$$

Solve and evaluate Inserting the appropriate values:

$$v_{rel} = c\left(\frac{f_O}{f_S} - 1\right) = (3.0 \times 10^8 \text{ m/s})\left(\frac{5.4 \times 10^{14} \text{ Hz}}{4.5 \times 10^{14} \text{ Hz}} - 1\right)$$

$$= 6.0 \times 10^7 \text{ m/s}$$

This speed is approaching the speed of light. Dr. Wood was clearly just trying to get out of a ticket. His explanation that the red light appeared green to him isn't reasonable. Note also that the speed we found contradicts our assumption for using the low-speed Doppler effect. However, even if we used the equation appropriate for such high speeds, we would still obtain a speed close to the speed of light.

- -

Try it yourself Can you detect the change in frequency of a red light if you are driving at a speed of 34 m/s (76 mi/h)?

Answer The change is 5×10^7 Hz, which is

$$\left(\frac{5 \times 10^7 \text{ Hz}}{4.5 \times 10^{14} \text{ Hz}}\right) \times 100\% = 1 \times 10^{-5}\%$$

which is not detectable by the human eye (although detectors do exist that can register this small change).

Measuring car and ball speeds using the Doppler effect

A modern radar gun used by police emits microwaves (EM waves with frequencies much smaller than visible light) in the frequency range 33.4–36.0 GHz. Sports radar guns use a frequency of 10.525 GHz. The speed detection process involves three distinct steps outlined below for a baseball in flight.

1. The radar gun emits source microwaves at frequency $f_S = 10.525$ GHz. These microwaves are "detected" at frequency f_D by the ball (**Figure 26.14a**).
2. The waves reflect from the ball, making the ball act as a source of microwaves at frequency f_D. Some of these waves are reflected back toward the radar gun, which observes or detects them at frequency f_O (Figure 26.14b).
3. The observed frequency f_O and source frequency f_S are often combined, resulting in a beat frequency $f_{beat} = |f_O - f_S|$, which is then used to determine the speed of the baseball.

FIGURE 26.14 How a radar detector determines ball speed.

(a)

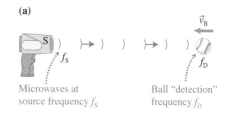

Microwaves at source frequency f_S

Ball "detection" frequency f_D

(b)

Ball reflects waves at frequency f_D

Observer at source detects reflected waves at frequency f_O

Hubble's discovery of the expansion of the universe

FIGURE 26.15 The expanding universe is analogous to an expanding loaf of raisin bread.

Bread loaf expands while baking in an oven.

All raisins move farther apart as the dough expands.

At the beginning of the 20th century it was believed that the universe was static. Planets moved around the Sun, but the distances between galaxies remained constant. This static universe model seemed consistent with observations. However, the gravitational interaction is always attractive, so why didn't the universe collapse? Even Einstein's ideas about gravity (which you will learn about in the next section) had trouble accommodating a static universe.

During this same time, astronomers Edwin Hubble and Milton Humason worked for years studying the light emitted by stars in other galaxies. They found that these stars emitted light wavelengths that were nearly the same as the light emitted from similar nearby stars—but with one major difference. With few exceptions, the light emitted by the distant galaxies was shifted toward longer wavelengths (smaller frequencies). Hubble concluded that the shift was not because the stars in other galaxies were different, but because significant relative motion existed between those stars and Earth, and nearly all those stars and the galaxies they were in were moving away from us. The universe was expanding.

The idea of an expanding universe can be visualized as a loaf of raisin bread while it's rising. The galaxies are the raisins and the space between the galaxies is the dough (Figure 26.15). As the bread expands, the raisins embedded in the dough all move farther and farther apart from each other.

Hubble's law

Hubble used the EM wave Doppler effect equations to determine the speeds of the other galaxies. They turned out to range from 0 km/s to 1500 km/s, and all but the closest galaxies were moving away from us. Additionally, Hubble found that the farther away the galaxy, the faster it was moving away. Notice in the expanding bread loaf in Figure 26.15 that the nearest two raisins on the top right changed positions from about 1 cm apart to 2 cm apart (a change in separation of 1 cm). However, the raisin farthest away from the top right raisin changed from about 2 cm apart to 4 cm apart (a change in separation of 2 cm). Hubble found a similar pattern for galaxies and described it mathematically as

$$v = Hd \qquad (26.13)$$

where v is the galaxy's recessional speed away from Earth, d is the distance between the galaxy and Earth, and H is a constant, now known as the **Hubble constant**. It has units of kilometers per second per unit of distance between the galaxies. In astronomy, distances to other galaxies are usually measured in megaparsecs (Mpc), with 1 parsec equal to 3.09×10^{16} m, or about 3.3 light-years. The current accepted value for the Hubble constant, coming from multiple experiments, is about 70 ± 6 (km/s)/Mpc.

TIP   If we measure distance in meters and speed in meters per second, then the value for the Hubble constant is 2.3×10^{-17} Hz—a number much harder to remember than 70.

It might seem that Hubble's result implies that our galaxy is the center of the universe and all other galaxies are moving away from that center. We can see it another way, however. Going back to the analogy of the raisin bread, if Earth were one of the raisins, we would observe all other raisins moving away from us. But that's exactly what each of the other raisins would observe as well! There is no center of the universe.

Figure 26.16 helps us visualize the expansion of the universe. It shows spacetime diagrams (one-dimensional representations) of several galaxies as observed by two different observers for a time interval between 0 and 1 time units, assuming that $H = 1$ with the units of 1/unit of time. The galaxies of the two observers are marked in red and blue. Notice that the observer in the red galaxy (Figure 26.16a) sees other galaxies moving away, as does the observer in the blue galaxy (Figure 26.16b). Although a particular galaxy (any black dot) moves at a different speed for these two observers, both observers describe the relation between the galaxy's speed and distance to them using the same Hubble constant, $H = 1$. Two galaxies that are separated by a certain distance will move away from each other at the same relative speed no matter where in the universe they are or where the observer is (follow three pairs of dots separated by half a unit of distance at $t = 0$ in Figures 26.16a and b).

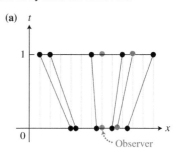

FIGURE 26.16 Expansion of the universe described by different observers.

> **Hubble's law** The average recessional speed v of two distant galaxies a distance d apart is determined by
>
> $$v = Hd \qquad (26.13)$$
>
> where $H = 70 \pm 6 \, (\text{km/s})/\text{Mpc}$ is Hubble's constant. One megaparsec (Mpc) equals 3.3×10^6 light-years (ly). Nearby galaxies do not necessarily obey this rule.

Age of the universe

If galaxies are moving away from each other at measurable speeds, and we can measure the distances to them, then we can estimate how long it took the galaxies to move apart to their present positions. This would give an estimate for the age of the universe. Choose two distant galaxies a distance d apart. Assuming they have moved away from each other at a constant speed $v = Hd$, then the time interval Δt since they were together is

$$\Delta t = \frac{d}{v} = \frac{d}{Hd} = \frac{1}{H}$$

The approximate age of the universe is the inverse of the Hubble constant. Substituting $70 \pm 6 \, (\text{km/s})/\text{Mpc}$ for H and converting to years from seconds, we obtain the following:

$$\Delta t = \frac{1}{H} = \cfrac{1}{\left(70 \, \dfrac{\text{km/s}}{\text{Mpc}}\right)\left(\dfrac{1000 \text{ m}}{1 \text{ km}}\right)\left(\dfrac{1 \text{ Mpc}}{10^6 \text{ pc}}\right)\left(\dfrac{1 \text{ pc}}{3.09 \times 10^{16} \text{ m}}\right)\left(\dfrac{3.16 \times 10^7 \text{ s}}{1 \text{ year}}\right)}$$

$$= 14.0 \text{ billion years}$$

This result is uncertain by $(\pm 6/70) \times 100\% = \pm 8.6\%$. This result is compatible with geological studies that estimate the age of Earth at approximately 4.5×10^9 years old (4.5 billion years); that is, geological studies do not conclude that Earth is older than the universe. Rather, the findings imply that Earth formed approximately 9.5 billion years after the universe came into being.

Returning briefly to our galaxies in Figure 26.16, the age of the universe in this model is $\frac{1}{H} = 1$ because $H = 1$. Therefore, our universe was born at $t = -1$. You can verify this by extending backward the world lines in Figure 26.16a that show expansion—they all meet at the same point, at time $t = -1$ (**Figure 26.17**). If you repeat the same procedure for Figure 26.16b you will again see that both observers determine the same age of the universe.

The assumption about the universe expanding at a constant rate is questionable. In fact we have learned in the last 15 years that the universe is accelerating slightly in its expansion. The 2011 Nobel Prize in physics was awarded to Saul Perlmutter, Brian P. Schmidt, and Adam G. Riess for this stunning and unexpected discovery.

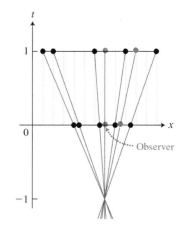

FIGURE 26.17 How to determine the age of the universe.

REVIEW QUESTION 26.10 Why can't we use the equations for the Doppler effect for sound to describe the Doppler effect for electromagnetic waves?

26.11 General relativity

The special theory of relativity allows us to compare measurements made by observers in two reference frames who move at constant velocity relative to each other. In these inertial reference frames, the laws of physics have the same form (that is, they are said to be **invariant**). Einstein was able to generalize this invariance to all reference frames, including noninertial (accelerated) reference frames. The result is the **general theory of relativity**, which is also an improved theory of the gravitational interaction. The cornerstone of general relativity is the **principle of equivalence**.

> **Principle of equivalence** No experiment can be performed that can distinguish between an accelerating reference frame and the presence of a uniform gravitational field.[1]

Consider a spaceship at rest or drifting at constant velocity far from any stars or planets. A passenger inside the ship would float freely (**Figure 26.18a**). Suppose now that the spaceship's rockets fired, causing the ship to accelerate at 9.8 m/s². Initially, the floating passenger "falls" to the spaceship's floor (Figure 26.18b) with acceleration 9.8 m/s². From there, the passenger would be able to stand, jump "up and down," and throw a ball "upward" (away from the floor) only to have it come back "down." The passenger would have exactly the same sensations as if he or she were standing on Earth's surface.

If the spaceship had no windows, and the engines provided a perfectly smooth ride, the passenger would have no way of knowing whether the ship was accelerating at 9.8 m/s² or was resting on Earth's surface in the presence of a nearly uniform gravitational field. The principle of equivalence says that there is no experiment that the passenger can perform to distinguish between these two situations.

From the principle of equivalence, Einstein reasoned that gravitation should be understood by a different mechanism entirely. He suggested that objects with mass cause the space and time around them to become curved and that objects moving in this curved space will not travel along straight paths at constant speed. These objects do not actually have a gravitational force exerted on them; rather, they are simply moving along natural paths through a curved spacetime. The Moon orbits Earth not because Earth exerts a force on it, but because Earth curves space and the Moon then naturally moves in a curved path. This path in the new spacetime is the shortest line connecting two points (called a *geodesic line*, similar to a straight line in Euclidean geometry). This also very neatly explains why all objects, regardless of size, mass, or composition, move identically in a vacuum. It also explains why astronauts do not notice any forces being exerted on them as they orbit Earth: because there aren't any!

An early testing experiment

An object of large mass such as the Sun causes space to curve more than an object of smaller mass such as Earth does (see **Figure 26.19**). In 1915 Einstein used general relativity to predict that light from distant stars that passes close to the Sun's surface would be deflected by an angle of $\theta = 4.86 \times 10^{-4}$ degrees. During a total solar eclipse, the observed position of those stars will be different because light bends as it passes close to the Sun (in **Figure 26.20**, the dashed line is the star's light path when the Sun is not present).

In 1919, astronomers under the leadership of Sir Arthur Eddington tried to measure such a deflection. They announced that the observations were in total agreement with

FIGURE 26.18 Acceleration produces the same effect as gravity.

(a)

When moving at constant velocity very far from massive objects, the person and the ball float in the spaceship. $\vec{v}$ = constant

(b)

When the ship is accelerating toward the right at $\vec{a} = \vec{g}$, the left floor of the ship presses against the person, causing the same sensation as if standing on Earth's surface.

$\vec{a} = \vec{g}$

FIGURE 26.19 The Sun's mass causes curvature of space.

The distortion of space caused by the Sun

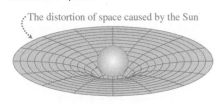

FIGURE 26.20 Starlight deflection caused by Sun's space curvature.

Light from star   Light path if no Sun

Sun θ (not to scale)

The light path bends due to space curvature caused by the Sun.

---

[1]Einstein's statement of the principle of equivalence is more general than this, but this restricted version is sufficient for our discussion.

the prediction of general relativity. Einstein and his theory of general relativity became famous almost overnight. However, despite the announcement, Eddington's results were not as conclusive as he had wanted them to be. It took many years and multiple experiments with solar eclipses until finally, by the 1960s, the measurements were convincingly in agreement with the predictions of general relativity.

Another testing experiment

General relativity solved another problem that had plagued astronomers since the early 1800s. The elliptical path of Mercury around the Sun was known to slowly rotate (**Figure 26.21**), a phenomenon called **precession**. This effect is so small that it takes about 3 million years for the elliptical orbit to complete one full precession. Newton's theory of gravitation predicts a precession from slight gravitational tugs due to the other planets (mostly Jupiter). However, Mercury's precession rate had been observed to be slightly larger than that predicted by Newton's theory. But if general relativity is used to predict the precession rate, the result is consistent with the observations. The additional precession results from the motion of Mercury through the curved space produced by the Sun.

FIGURE 26.21 Mercury's orbit about the Sun.

Precession of Mercury's elliptical orbit about the Sun

Gravitational time dilation and red shift

Another prediction of general relativity is **gravitational time dilation**. Objects do not just curve space around them, but also alter the rates at which time passes around them as well. Time passes at different rates at different points near a massive object. The closer a point is to the massive object, the slower time passes there. For example, a person working in an office on the first floor of a tall building would age more slowly relative to an observer on the top floor. However, this effect would be only about 10^{-5} s over an entire human lifetime.

One consequence of gravitational time dilation is an effect known as **gravitational redshift**. If EM waves are emitted from a region closer to a massive object and observed in a region farther away, the observed frequency will be lower than the emitted frequency. In 1960, R. V. Pound and G. A. Rebka Jr. at Harvard University tested this prediction. They observed a small reduction in the vibration frequency of gamma rays emitted from radioactive nuclei on the bottom floor of a laboratory building at Harvard compared to the frequency of gamma rays emitted from the top floor—yet another testing experiment supporting general relativity.

Gravitational waves and black holes

General relativity predicts that both space and time are curved by the presence of objects with mass. Experiments support this idea. These two ideas are unified by general relativity into the idea that space and time form a single entity known as **spacetime**. Newton imagined that space and time were a rigid background for the processes of the universe. Einstein showed that spacetime is curved by the presence of mass and changes shape when that mass moves. These changes in the curvature of spacetime can propagate as gravitational waves that ripple through the vacuum at the speed of light. In 1915, Einstein predicted the existence of such waves and what the properties of these waves should be, and after 100 years of searching, scientists found the first direct evidence of their existence in 2015.

General relativity also predicts the existence of black holes (objects we first encountered in Chapter 7). In Newtonian mechanics, black holes are hypothetical objects whose escape speed is equal to or greater than light speed. In general relativity, black holes are regions of spacetime that are so extremely curved that all matter and even light within that region cannot escape (see **Figure 26.22**). The most common black holes form when large stars run out of nuclear fuel and collapse under their own gravity to extreme densities. Much larger black holes have been detected at the centers

FIGURE 26.22 Extreme space curvature near a black hole does not allow passing objects (including light) to escape.

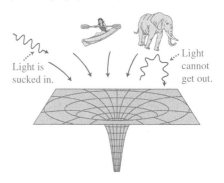

Light is sucked in.   Light cannot get out.

of most galaxies. Astronomers cannot detect black holes directly but can find them by the extreme influence they have on their immediate environment.

It turns out that black holes are the objects that helped scientists detect gravitational waves. Imagine that two massive black holes orbit each other. According to general relativity, they will continuously emit gravitational waves until they collide and join together, making one rotating super-massive black hole. When they collide, even more of their energy is converted into gravitational waves. The gravitational waves from such a collision were detected on September 14, 2015 by the two interferometers of the Laser Interferometer Gravitational-Wave Observatory (LIGO) in Louisiana and Washington. Based on the detected signals, scientists estimated that the two black holes were about 29 and 36 times more massive than the Sun and that the collision occurred around 1.3 billion years ago. The energy of the gravitational waves emitted during the event was equal to the rest energy of an object with the mass of three Suns, and it was emitted during a time interval of milliseconds. This was the first direct observation of the disturbances in space and time that gravitational waves passing through Earth make.

The discovery of gravitational waves by the LIGO experiment was the result of the joint effort of thousands of scientists from almost a hundred universities all over the world. Such significant collaboration was needed because to find gravitational waves, scientists needed to be able to design an experiment that would detect an incredibly tiny change, 10^{-19} m, in the length of the arm of an interferometer (similar to that of Michelson and Morley, but much larger, at 2.5 miles long) while eliminating all other factors that might contribute to much bigger changes in the length. To understand this difficulty in design, imagine that you need to measure a change in the distance from our solar system to the nearest star of the width of one hair—this is the relative distance change caused by gravitational waves that scientists in the LIGO experiment detected.

REVIEW QUESTION 26.11 What is the general relativity explanation for why Earth orbits the Sun?

26.12 Global Positioning System (GPS)

You will likely never travel at speeds close to the speed of light, never travel to a black hole, and never worry about your biological clock becoming out of sync by a thousandth of a second over the course of your life because you live on the top floor of a building. In later chapters, however, we will explore many examples of the impact of relativistic effects on our daily lives. Meanwhile, here's an everyday example of both special and general relativity that affects your life: determining your location using the Global Positioning System (GPS).

Each GPS satellite is about 20,000 km from the surface of Earth and has an orbital speed of about 3900 m/s. An atomic clock on each satellite "ticks" every nanosecond $(1 \text{ ns} = 10^{-9} \text{ s})$, which allows it to transmit precisely timed signals. A GPS receiver in a car or on an airplane determines its position on Earth to within 5 to 10 m using these signals.

To achieve this level of precision, the clock ticks from the GPS satellites must be measured to an accuracy of 20 to 30 ns. Because the satellites move with respect to us, their clocks tick slightly slower than ours (special relativity's time dilation). Using the time dilation equation [Eq. (26.1)], we can estimate by how much the clocks on the satellites should fall behind during 1 day.

The reference frame of the satellite is the proper reference frame, and a 1-ns tick is the proper time interval. The satellite, as stated, moves at a speed of 3900 m/s. This time interval as measured from Earth is

$$\Delta t = \frac{\Delta t_0}{\sqrt{1 - \dfrac{v^2}{c^2}}} = \frac{1 \text{ ns}}{\sqrt{1 - \dfrac{(3.90 \times 10^3 \text{ m/s})^2}{(3.00 \times 10^8 \text{ m/s})^2}}} = 1.0000000001 \text{ ns}$$

The difference between the proper time interval and the measured time interval is

$$\Delta t - \Delta t_0 = 1.0000000001 \text{ ns} - 1.0000000000 \text{ ns} = 1.0 \times 10^{-10} \text{ ns}$$

This means that with each 1-ns tick of the satellite's atomic clock, that clock falls behind clocks on Earth by 1.0×10^{-10} ns. The number N of nanosecond ticks in each day is

$$N = \left(\frac{3600 \text{ s}}{\text{h}} \right) \left(\frac{24 \text{ h}}{\text{day}} \right) \left(\frac{10^9 \text{ ticks}}{\text{s}} \right) = 8.64 \times 10^{13} \text{ ticks}$$

Therefore, in 1 day, time dilation will cause the satellite clock to fall behind Earth clocks by a total of

$$(\Delta t - \Delta t_0)N = (1.0 \times 10^{-10} \text{ ns})(8.64 \times 10^{13} \text{ ticks}) \approx 8.6 \times 10^3 \text{ ns} = 8.6 \text{ } \mu\text{s}$$

This result is just an estimate because the GPS satellites are not in exactly inertial reference frames. But because their accelerations are low, they are very close to inertial.

In addition, the satellites are in orbits high above Earth. There, the curvature of spacetime is smaller than on Earth's surface, and thus the clocks will tick faster than clocks on Earth. General relativity predicts that the space clocks get ahead by about 45 μs per day.

Together, the two effects result in the GPS satellite clocks running at about $(+45 - 8.6)$ μs per day, or about 36 to 37 μs per day faster than Earth clocks, or about 0.014 s/year. Thus, in 1 year, the satellite positions (moving at about 3900 m/s) would be off by about (0.014 s)(3900 m/s) = 50 m. The position the receiver on Earth would report would be off by about the same amount.

The precise locations of the satellites are determined by tracking stations around the world. Those data are collected, processed, and regularly transmitted up to the satellites, which then update their internal orbit models. Along with the orbit updates, transmissions are also sent to synchronize the atomic clocks on the satellites. This is where relativistic effects are taken into account. GPS receivers use the time and position data from multiple satellites to trilaterate their location. An understanding of both special and general relativity is essential for the GPS system to work.

REVIEW QUESTION 26.12 What are two relativistic effects that must be accounted for so that the GPS system can function accurately?

Summary

Postulates of special relativity
1. The laws of physics are the same in all inertial reference frames.
2. The speed of light in a vacuum is the same in all inertial reference frames. (Section 26.2)

Time dilation In the **proper reference frame**, events 1 and 2 occur at the same place. The time interval between events 1 and 2 measured by an observer in the proper reference frame is the **proper time interval** Δt_0. The time interval between events 1 and 2 as measured by an observer moving at constant speed v relative to the proper reference frame is Δt. (Section 26.4)

$$\Delta t = \frac{\Delta t_0}{\sqrt{1 - \dfrac{v^2}{c^2}}} \quad \text{Eq. (26.1)}$$

Length contraction An object's **proper length** L_0 is its length measured in an inertial reference frame in which the object is at rest. The length of the object when measured in a reference frame moving at speed v relative to the proper reference frame is L. (Section 26.5)

$$L = L_0\sqrt{1 - \frac{v^2}{c^2}} \quad \text{Eq. (26.2)}$$

A **spacetime diagram** is a representation of relativistic motion on a time-versus-position graph. The motion of an object is shown by its **world line**. The world line for a light beam is tilted at 45° in the first and second quadrants, representing two light beams traveling in opposite directions. All other objects have steeper world lines.

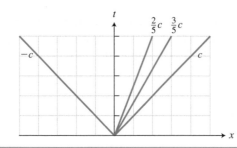

Relativistic momentum The relativistic momentum $\vec{p}$ of an object of mass m must be used when an object is moving at speed $v \geq 0.1c$. (Section 26.8)

$$\vec{p} = \frac{m\vec{v}}{\sqrt{1 - (v/c)^2}} \quad \text{Eq. (26.5)}$$

Relativistic energy
- **Rest energy** An object of mass m has rest energy E_0 that depends only on the mass m of the object and the speed of light c.
- **Kinetic energy** The relativistic kinetic energy K of an object is its total energy minus its rest energy.
- **Total energy** is the total relativistic energy E of an object moving at speed v. (Section 26.9)

$$E_0 = mc^2 \quad \text{Eq. (26.8)}$$

$$K = \frac{mc^2}{\sqrt{1 - (v/c)^2}} - mc^2 \quad \text{Eq. (26.10)}$$

$$E = \frac{mc^2}{\sqrt{1 - (v/c)^2}} \quad \text{Eq. (26.9)}$$

Doppler effect for electromagnetic radiation If a source emits EM waves at frequency f_S and the source and observer are moving at speed v_{rel} relative to each other (+ if toward each other and − if away from each other), then the observer detects EM waves at frequency f_O. (Section 26.10)

$$f_O = f_S\sqrt{\frac{1 + v_{rel}/c}{1 - v_{rel}/c}} \text{ for } v_{rel} \geq 0.1c \quad \text{Eq. (26.11)}$$

$$f_O = f_S\left(1 + \frac{v_{rel}}{c}\right) \text{ for } v_{rel} < 0.1c \quad \text{Eq. (26.12)}$$

Questions

Multiple Choice Questions

1. The Michelson-Morley experiment had as its goal to
   (a) measure the speed of light in different reference frames.
   (b) find out whether the Earth moves with respect to ether.
   (c) find whether ether exists.
   (d) All of the above
   (e) a and b only

2. On what did Michelson and Morley base their conclusions?
   (a) The measurement of the time intervals light takes to travel the same distance in different directions with respect to the moving ether
   (b) The analysis of an interference pattern for two beams of light
   (c) Comparison of the intensities of two perpendicular light beams

3. Physicists explained the results of the Michelson-Morley experiment by saying that
   (a) the motion of Earth through ether is undetectable.
   (b) there is no ether.
   (c) the length of experimental apparatus parallel to Earth's motion changes.
   (d) All of the above
   (e) None of the above

4. What is a proper time interval?
   (a) A correctly measured time interval
   (b) The shorter time interval of the two measured times
   (c) The time interval measured by observers with respect to whom the events occurred in the same location

5. You stand on a train platform and watch a person on a passing train through a train window. The person is eating a sandwich. The time interval it takes her to eat the sandwich as measured by you
   (a) is longer than measured by her.
   (b) is shorter than measured by her.
   (c) is the same.
   (d) depends on the orientation of the sandwich.

6. In Question 26.5, the length of the sandwich as measured by you
   (a) is longer than measured by the passenger.
   (b) is shorter than measured by the passenger.
   (c) is the same.
   (d) depends on the orientation of the sandwich.

7. A supernova explodes in a distant galaxy. The galaxy is moving away from us. The speed of light from the explosion with respect to Earth is
   (a) smaller than the speed of light with respect to the galaxy.
   (b) the same as light speed with respect to the galaxy.
   (c) larger than light speed with respect to the galaxy.

8. We used the muon experiment to test which of the following hypotheses? Multiple answers may be correct.
   (a) The existence of ether
   (b) Time dilation
   (c) Length contraction

9. The measurement of the coordinates of a star during a solar eclipse can be used as a testing experiment for
   (a) the existence of ether.
   (b) time dilation.
   (c) length contraction.
   (d) space-time curvature.

10. Suppose an early GPS system did not take the laws of special relativity into account. A 1-ns time interval measured by clocks on the GPS satellite would be
    (a) smaller than intervals measured on Earth.
    (b) larger than intervals measured on Earth.
    (c) exactly the same for both.

11. What are cosmic rays?
    (a) Light rays sent by cosmic objects
    (b) Electromagnetic radiation produced during supernova explosions
    (c) Signals sent by extraterrestrial civilizations
    (d) Subatomic particles such as protons and neutrons

12. Which of the blue world lines in the spacetime diagrams in Figure Q26.12 represent motion of an object that is impossible? The straight black line on each diagram represents a light pulse traveling in the $+x$-direction.

FIGURE Q26.12

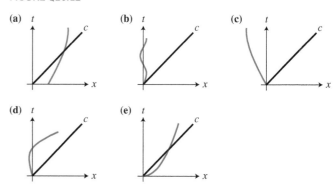

Conceptual Questions

13. What is an inertial reference frame? How can you convince someone that Newton's laws work the same way for observers in two different inertial reference frames?

14. Give an example of a phenomenon that an observer in a noninertial reference frame cannot explain using Newton's laws.

15. Explain the difference between a proper reference frame and other inertial reference frames in which the time interval between two events is measured.

16. What are the two postulates of the special theory of relativity? How would you explain them to a friend using your own words?

17. What does it mean to say that the speed of something is the same in different reference frames? Give an example.

18. You move toward a star at a speed of $0.99c$. At what speed does light from the star pass you? What if you are moving away from the star?

19. You pass Earth in a spaceship that moves at $0.99c$ relative to Earth. Do you notice a change in your heartbeat rate? Does an observer on Earth think your heart is beating at the normal rate? Do you think the heartbeat of Earth's observer is normal? Explain all three of your answers.

20. It takes light approximately 10^{10} years to reach Earth from the edge of the observable universe. Would it be possible for a person to travel this distance during a lifetime? Explain.

21. A person holds a meter stick in a spaceship traveling at $0.95c$ past Earth. The person rotates the meter stick so that it is first parallel and then perpendicular to the ship's velocity. Describe its changing length to
    (a) an observer at rest on the Earth and
    (b) the person in the spaceship.

22. Name several ways in which your life would be different if the speed of light were 20 m/s rather than its actual value.

23. If the speed of light were infinite, how would time dilation, length contraction, and the ways we calculate momentum and energy be affected? Justify your answer carefully.

24. The classical equation for calculating kinetic energy [Eq. (7.5)] and the relativistic equation for calculating kinetic energy [Eq. (26.10)] appear quite different. Under what conditions is each equation appropriate? Invent a simple example to show that they produce the same result at speeds less than $0.001c$.

25. How did the Doppler effect for light help scientists estimate the age of the universe?

26. What is the principle of equivalence? How would you explain this principle to a friend?

27. Can any of the blue world lines in Figure Q26.12 represent the motion of an object that is slowing down at constant acceleration? Explain.

Problems

Below, **BIO** indicates a problem with a biological or medical focus. Problems labeled **EST** ask you to estimate the answer to a quantitative problem rather than derive a specific answer. Asterisks indicate the level of difficulty of the problem. Problems with no * are considered to be the least difficult. A single * marks moderately difficult problems. Two ** indicate more difficult problems.

You may need to know the following masses to work some of these problems: $m_{electron} = 9.11 \times 10^{-31}$ kg and $m_{proton} = 1.67 \times 10^{-27}$ kg.

26.1 and 26.2 Ether or no ether? and Postulates of special relativity

1. **Relative motion on an airport walkway** A person is walking on a moving walkway in the airport. Her speed with respect to the walkway is 2 m/s. The speed of the walkway is 1 m/s with respect to the floor. What is her speed with respect to a person walking on the floor in the opposite direction at the speed of 2 m/s?

2. **Running on a treadmill** Explain how you can run on a treadmill at 3 m/s and remain at the same location.

3. * Describe the important parts of the Michelson-Morley experimental setup and explain how this setup could help them determine whether Earth moves with respect to ether. Explain which reference frames are involved and how the setup relates to the previous problem.

4. * Describe what Michelson and Morley would have observed when they rotated their interferometer if Earth were moving through ether compared to what they would have observed if Earth were not moving through ether.

5. **Person on a bus** A person is sitting on a bus that stops suddenly, causing her head to tilt forward. (a) Explain the acceleration of her head from the point of view of an observer on the ground. (b) Explain the acceleration of her head from the point of view of another person on the bus. (c) Which observer is in an inertial reference frame?

6. * **Turning on a rotating turntable** A matchbox is placed on a rotating turntable. The turntable starts turning faster and faster. At some instant the matchbox flies off the turning table. (a) Draw a force diagram for the box when still on the rotating turntable. (b) Draw a force diagram for the box just before it flies off. (c) Explain why the box flies off only when the turntable reaches a certain speed. In what reference frame are you when providing this explanation? (d) How would a bug sitting on the turntable explain the same situation?

26.3–26.6 Simultaneity, Time dilation, Length contraction, and Spacetime diagrams

7. * A particle called Σ^+ lives for 0.80×10^{-10} s in its proper reference frame before transforming into two other particles. How long does the Σ^+ seem to live according to a laboratory observer when the particle moves past the observer at a speed of 2.4×10^8 m/s?

8. A Σ^+ particle discussed in the previous problem appears to a laboratory observer to live for 1.0×10^{-10} s. How fast is it moving relative to the observer?

9. A person on Earth observes 10 flashes of the light on a passing spaceship in 22 s, whereas the same 10 flashes seem to take 12 s to an observer on the ship. What can you determine using this information?

10. A spaceship moves away from Earth at a speed of $0.990c$. The pilot looks back with a telescope and measures the time interval for one rotation of Earth on its axis. What time interval does the pilot measure? What assumptions did you make?

11. **Extending life?** A free neutron lives about 1000 s before transforming into an electron, a proton, and an anti-neutrino. If a neutron leaves the Sun at a speed of $0.999c$, (a) how long does it live according to an Earth observer? (b) Will such a neutron reach Pluto (5.9×10^{12} m from the Sun) before transforming? Explain your answers.

12. * A Σ^+ particle lives 0.80×10^{-10} s in its proper reference frame. If it is traveling at $0.90c$ through a bubble chamber, how far will it move before it disintegrates?

13. * **Extending the life of a muon** A muon that lives 2.2×10^{-6} s in its proper reference frame is created 10,000 m above Earth's surface. At what speed must it move to reach Earth's surface at the instant it disintegrates?

14. * **Effect of light speed on the time interval for a track race** Suppose the speed of light were 15 m/s. You run a 100-m dash in 10 s according to the timer's clock. How long did the race last according to your watch?

15. * Explain why an object moving past you would seem shorter in the direction of motion than when at rest with respect to you. Draw a sketch to illustrate your reasoning.

16. ** Explain why the length of an object that is oriented perpendicular to the direction of motion would be the same for all observers whose velocity is parallel to the object's velocity. Draw a sketch to illustrate your reasoning.

17. * You sit in a spaceship moving past the Earth at $0.97c$. Your arm, held straight out in front of you, measures 50 cm. How long is it when measured by an observer on Earth?

18. **Length of a javelin** A javelin hurled by Wonder Woman moves past an Earth observer at $0.90c$. Its proper length is 2.7 m. What is its length according to the Earth observer?

19. At what speed must a meter stick move past an observer so that it appears to be 0.25 m long?

20. **Changing the shape of a billboard** A billboard is 10 m high and 15 m long according to a person standing in front of it. At what speed must a person in a fast car drive by parallel to the billboard's surface so that the billboard appears to be square?

21. * A classmate says that time dilation and length contraction can be remembered in a simple way if you think of a person eating a foot-long "sub" sandwich on a train (the sandwich is oriented parallel to the train's motion). The person on the train finishes the sandwich in 20 min. You, standing on the platform, observe the person eating a shorter sandwich but for a longer time interval. Do you agree with this example? Explain your answer.

22. * **A pole and barn "paradox"** A barn of width 4.0 m is at rest in reference frame S. Let us take a pole of length 8.0 m and accelerate it to a velocity such that its length in frame S becomes equal to 4.0 m. Then at a certain moment the pole, flying through the barn (let's say from left to right), fits entirely within it. However, for an observer in reference frame S′ attached to the pole, the barn is moving to the left, and so it is the barn width that becomes reduced by half, and consequently the pole (8.0 m) does not fit in the barn (2.0 m). (a) Determine the speed of the pole relative to the barn. (b) Determine the time interval between the following two events, as observed by the observer in reference frame S and the observer in reference frame S′: *event 1*: the right end of the pole meets the right side of the barn; *event 2*: the left end of the pole meets the left side of the barn. (c) Why is there no paradox in this situation? Explain. (b) Represent the situation with spacetime diagrams for the observer in the reference frame S and the observer in the reference frame S′.

23. * **Space travel** An explorer travels at speed 2.90×10^8 m/s from Earth to a planet of Alpha Centauri, a distance of 4.3 light-years as measured by an Earth observer. (a) How long does the trip last according to the Earth observer? (b) How long does the trip last for the explorer on the spaceship? (c) Represent the situation with spacetime diagrams for the person on Earth and the explorer in the spaceship.

26.7 Velocity transformations

24. * Give examples of cases in which two observers record the motion of the same object to have different speeds. Sketch each example and explain how each observer arrives at the value of the measured speed.

25. * Now repeat Problem 26.24, only this time instead of a moving object, use a light flash. Describe what speeds of light different observers should measure according to the second postulate of special relativity.

26. * **Life in a slow-light-speed universe** Imagine that you live in a universe where the speed of light is 50 m/s. You sit on a train moving west at speed 20 m/s relative to the track. Your friend moves on a train in the opposite direction at speed 15 m/s. What is the speed of his train with respect to yours?

27. * **More slow-light-speed universe** In the scenario described in Problem 26.26, you and your friend listen to music on the same radio station. What is the speed of the radio waves that your antenna is registering compared to the speed of the waves that your friend's antenna registers if the station is 100 miles to the west of you?

28. * You are on a spaceship traveling at $0.80c$ with respect to a nearby star sending a laser beam to a spaceship following you, which is moving at $0.50c$ in the same direction. (a) What is the speed of the laser beam registered by the second ship's personnel according to the classical addition of the velocities? (b) What is the speed of the laser beam registered by the second ship's personnel according to the relativistic addition of the velocities? (c) What is the speed of the second ship with respect to yours according to the classical addition of the velocities? (d) What is the speed of the second ship with respect to yours according to the relativistic addition of the velocities?

29. ** Your friend says that it is easy to travel faster than the speed of light; you just need to find the right observer. Give physics-based reasons for why your friend would have such an idea. Then explain whether you agree or disagree with him.

26.8 Relativistic momentum

30. An electron is moving at a speed of $0.90c$. Compare its momentum as calculated using a nonrelativistic equation and using a relativistic equation.

31. * Explain why a relativistic expression is needed for fast-moving particles. Why can't we use a classical expression?

32. * If a proton has a momentum of 3.00×10^{-19} kg $\cdot$ m/s, what is its speed?

26.9 Relativistic energy

33. Determine the ratio of an electron's total energy to rest energy when moving at the following speeds: (a) 300 m/s, (b) 3.0×10^6 m/s, (c) 3.0×10^7 m/s, (d) 1.0×10^8 m/s, (e) 2.0×10^8 m/s, and (f) 2.9×10^8 m/s.

34. **Solar wind** To escape the gravitational pull of the Sun, a proton in the solar wind must have a speed of at least 6.2×10^5 m/s. Determine the rest energy, the kinetic energy, and the total energy of the proton.

35. * At what speed must an object move so that its total energy is 1.0% greater than its rest energy? 10% greater? Twice its rest energy?

36. * A person's total energy is twice his rest energy when he moves at a certain speed. By what factor must his speed now increase to cause another doubling of his total energy?

37. * A proton's energy after passing through the accelerator at Fermilab is 500 times its rest energy. Determine the proton's speed.

38. * A rocket of mass m starts at rest and accelerates to a speed of $0.90c$. Determine the change in energy needed for this change in speed.

39. * Determine the total energy, the rest energy, and the kinetic energy of a person with 60-kg mass moving at speed $0.95c$.

40. * An electron is accelerated from rest across 50,000 V in a machine used to produce X-rays. Determine the electron's speed after crossing that potential difference.

41. * A particle originally moving at a speed $0.90c$ experiences a 5.0% increase in speed. By what percent does its kinetic energy increase?

42. * An electron is accelerated from rest across a potential difference of 9.0×10^9 V. Determine the electron's speed (a) using the nonrelativistic kinetic energy equation and (b) using the relativistic kinetic energy equation. Which is the correct answer?

43. ** A particle of mass m initially moves at speed $0.40c$. (a) If the particle's speed is doubled, determine the ratio of its final kinetic energy to its initial kinetic energy. (b) If the particle's kinetic energy increases by a factor of 100, by what factor does its speed increase?

44. * Determine the mass of an object whose rest energy equals the total yearly energy consumption of the world (5×10^{20} J).

45. * **Mass equivalent of energy to separate a molecule** Separating a carbon monoxide molecule CO into a carbon and an oxygen atom requires 1.76×10^{-18} J of energy. (a) Determine the mass equivalent of this energy. (b) Determine the fraction of the original mass of a CO molecule 4.67×10^{-26} kg that was converted to energy.

46. * **Hydrogen fuel cell** A hydrogen-oxygen fuel cell combines 2 kg of hydrogen with 16 kg of oxygen to form 18 kg of water, thus releasing 2.5×10^8 J of energy. What fraction of the mass has been converted to energy?

47. ** **Mass to provide human energy needs** Determine the mass that must be converted to energy during a 70-year lifetime to continually provide electric power for a person at a rate of 1000 W. The production of the electric power from mass is only about 33% efficient.

48. * **EST** An electric utility company charges a customer about 6–7 cents for 10^6 J of electrical energy. At this rate, estimate the cost of 1 g of mass if converted entirely to energy.

49. * **Mass to produce electric energy in a nuclear power plant** A nuclear power plant produces 10^9 W of electric power and 2×10^9 W of waste heating. (a) At what rate must mass be converted to energy in the reactor? (b) What is the total mass converted to energy each year?

50. * **BIO** **EST** **Metabolic energy** Estimate the total metabolic energy you use during a day. (You can find more on metabolic rate in the first Reading Passage in Chapter 7.) Determine the mass equivalent of this energy.

51. ** **Energy from the Sun** (a) Determine the energy radiated by the Sun each second by its conversion of 4×10^9 kg of mass to energy. (b) Determine the fraction of this energy intercepted by Earth, which is 1.50×10^{11} m from the Sun and has a radius of 6.38×10^6 m.

26.10 Doppler effect for EM waves

52. **Why no color change?** Why don't the colors of buildings and tree leaves change when we look at them from a flying plane? Shouldn't the trees ahead look more bluish when you are approaching and reddish when you are receding?

53. * **Change red light to green** In a parallel universe the speed of light in a vacuum is 70,000 m/s. How fast should a driver's car move so that a red light looks green?

54. * **Effect of the Hubble constant on age and radius of the universe** How would the estimated age of the universe change if the new accepted value of the Hubble constant became 100 (km/s)/Mpc? How would the visible radius of the universe change?

55. * **Expanding faster** New observations suggest that our universe does not expand at a constant rate but instead is expanding at an increasing rate. How does this finding affect the estimation of the age of the universe using Hubble's law?

56. **Baseball Doppler shift** In September of 2010 Aroldis Chapman threw what may be the fastest baseball pitch ever recorded at 105 mi/h (47 m/s). What would the observed frequency of microwaves reflected from the ball be if the source frequency were 10.525 GHz? What would be the beat frequency between the source frequency and the observed frequency?

57. ** **Were you speeding?** A police officer stops you in a 29 m/s (65 mi/h) speed zone and says you were speeding. The officer's radar has source frequency 33.4 GHz and observed a 3900-Hz beat frequency between the source frequency and waves reflected back to the radar from your car. Were you speeding? Explain.

General Problems

58. * **Boat trip** A boat's speed is 10 m/s. It makes a round trip between stations A and B and then another between stations A and C. Stations A and B are on the same side of the river 0.5 km apart. Stations A and C are on the opposite sides of the river across from each other and also 0.5 km apart. The river flows at 1.5 m/s. What time interval is the round trip between stations A and B and then between A and C?

59. * **Space travel** An explorer travels at speed 2.90×10^8 m/s from Earth to a planet of Alpha Centauri, a distance of 4.3 light-years as measured by an Earth observer. (a) How long does the trip last according to an Earth observer? (b) How long does the trip last for the person on the ship?

60. ** **EST** **Extending life** Suppose that the speed of light is 8.0 m/s. You walk slowly to all of your classes during one semester while a classmate runs at a speed of 7.5 m/s during the time you are walking. Estimate your classmate's change in age, as judged by you, and your change in age according to you during that walking time. Indicate how you chose any numbers used in your estimate.

61. ** **Racecar when c is 100 m/s** Suppose that the speed of light is 100 m/s and that you are driving a racecar at speed 90 m/s. What time interval is required for you to travel 900 m along a track's straightaway (a) according to a timer on the track and (b) according to your own clock? (c) How long does the straightaway appear to you? (d) Notice that the speed at which the track moves past is your answer to part (c) divided by your answer to part (b). Does this speed agree with the speed as measured by the stationary timer?

62. ** **EST** Cherenkov radiation is electromagnetic radiation emitted when a fast-moving particle such as a proton passes through an insulator at a speed faster than the speed of light in that insulator. The Cherenkov radiation looks like a blue glow in the shape of a cone behind the particle. The radiation is named after Soviet physicist Pavel Cherenkov, who received a Nobel Prize in 1958 for describing the radiation. Estimate the smallest speed of a proton moving in oil that will produce Cherenkov radiation behind it.

63. ** A pilot and his spaceship of rest mass 1000 kg wish to travel from Earth to planet Scot ML, 30 light-years from Earth. However, the pilot wishes to be only 10 physiological years older when he reaches the planet. (a) At what constant speed must he travel? (b) What is the total energy of his spaceship and the rest energy, according to an Earth observer, while making the trip?

64. * Alice's friends Bob and Charlie are having a race to a distant star 10 light-years away. Alice is the race official who stays on Earth, and her friend Darien is stationed on the star where the race ends. Bob is in a rocket that can travel at 0.7c, whereas Charlie's rocket can reach a speed of 0.866c. Bob and Charlie start to move at the same time. Draw a spacetime diagram from each person's perspective (four diagrams altogether).

65. * Determine how long it takes each person in the previous problem to finish the race according to each observer. Compare all answers with the spacetime diagrams. Assume that both rockets can reach their speed instantly. (Hint: Don't forget that you can always use the definition of velocity, $v = \Delta x/\Delta t$.)

66. ** **Space travel** A pilot and her spaceship have a mass of 400 kg. The pilot expects to live 50 more Earth years and wishes to travel to a star that requires 100 years to reach even if she were to travel at the speed of light. (a) Determine the average speed she must travel to reach the star during the next 50 Earth years. (b) To attain this speed, a certain mass m of matter is consumed and converted to the spaceship's kinetic energy. How much mass is needed? (Ignore the energy needed to accelerate the fuel that has not yet been consumed.) Indicate any assumptions that you made.

67. ** (a) A container holding 4 kg of water is heated from 0 °C to 60 °C. Determine the increase in its energy and compare this to the rest energy when at 0 °C. (b) If the water, initially at 0 °C, is converted to ice at 0 °C, determine the ratio of its energy change to its original rest energy.

68. ** Your friend argues that Einstein's special theory of relativity says that nothing can move faster than the speed of light. (a) Give physics-based reasons for why your friend would have such an idea. (b) What examples of physical phenomena do you know of that contradict this statement? (c) Restate his idea so it is accurate in terms of physics. (Hint: See Problem 26.62 on Cherenkov radiation.)

Reading Passage Problems

Venus Williams's record tennis serve Venus Williams broke the women's tennis ball serving speed record at the European Indoor Championships at Zurich, Switzerland, on October 16, 1998 with a 57 m/s (127 mi/h) serve. She has since broken her own record more than once. The Doppler radar gun used in 1998 to measure the speed had a source microwave frequency of 10.525 GHz. Answer the following questions about the radar gun used on that serve.

69. Which principle can we use to determine the frequency f_D "detected" by the ball as it moved toward the source waves from the radar?
    (a) The beat frequency equation
    (b) The high-speed Doppler effect equation
    (c) The low-speed Doppler effect equation
    (d) The time dilation equation
    (e) The relationship between wave speed, frequency, and wavelength

70. Which frequency is closest to the frequency f_D detected by the ball as it moved toward the radar source waves?
    (a) 10.525 GHz (b) 10.525 GHz + 2.0 × 10⁻⁶ GHz
    (c) 10.525 GHz − 2.0 × 10⁻⁶ GHz (d) 3 × 10⁻⁷ Hz

71. Which principle can we use to determine the frequency f_O detected by the radar from waves reflected from the ball?
    (a) The beat frequency equation
    (b) The high-speed Doppler effect equation
    (c) The low-speed Doppler effect equation
    (d) The time dilation equation
    (e) The relationship between wave speed, frequency, and wavelength

72. Which answer is closest to the frequency f_O detected by the radar from the waves reflected from the ball?
    (a) Exactly 10.525 GHz (b) 10.525 GHz + 4.0 × 10⁻⁶ GHz
    (c) 10.525 GHz − 4.0 × 10⁻⁶ GHz (d) 3 × 10⁻⁷ Hz

73. Which principle is used to determine the frequency that the radar measures of the combined source and observed waves?
    (a) The beat frequency equation
    (b) The high-speed Doppler effect equation
    (c) The low-speed Doppler effect equation
    (d) The time dilation equation
    (e) The relationship between wave speed, frequency, and wavelength

74. Which answer is closest to the frequency that the radar measures of the combined source and observed waves?
    (a) 2.0×10^3 Hz (b) 4.0×10^3 Hz
    (c) 10.525 GHz (d) 3×10^{-7} Hz
    (e) 6×10^{-7} Hz

Quasars In 1963, Maarten Schmidt of the California Institute of Technology found the most distant object that had ever been seen in the universe so far. Called 3C 273, it emitted electromagnetic radiation with a power of 2×10^{13} Suns, or 100 times that of the entire Milky Way galaxy! Schmidt called 3C 273 a "quasi-stellar radio source," a name that was soon shortened to "quasar." Since then, astronomers have found many quasars, some substantially more powerful and distant than 3C 273.

Quasar 3C 273 is moving away from Earth at about speed 0.16c. The Sun is 1.5×10^{11} m from Earth and emits energy at a rate of 3.8×10^{26} J/s caused by the conversion of 4.3×10^9 kg/s of mass into light and other forms of electromagnetic energy. The Hubble constant is

$$H = 70.8 \ \frac{\text{km/s}}{\text{Mpc}} = 70.8 \ \frac{10^3 \ \text{m/s}}{3.09 \times 10^{22} \ \text{m}} = 22.9 \times 10^{-19} \ \text{s}^{-1}$$

75. What principle would you use to estimate the distance of 3C 273 from Earth?
    (a) The high-speed Doppler effect equation
    (b) The low-speed Doppler effect equation
    (c) Hubble's law
    (d) The time dilation equation
    (e) The relationship between wave speed, frequency, and wavelength

76. Which answer below is closest to the distance of 3C 273 from the Earth in terms of the distance of the Sun from Earth?
    (a) $\approx 10^3$ Sun distances (b) $\approx 10^6$ Sun distances
    (c) $\approx 10^9$ Sun distances (d) $\approx 10^{12}$ Sun distances
    (e) $\approx 10^{14}$ Sun distances

77. Which answer below is closest to the power of light and other forms of radiation emitted by 3C 273?
    (a) $\approx 10^8$ J/s (b) $\approx 10^{18}$ J/s
    (c) $\approx 10^{25}$ J/s (d) $\approx 10^{32}$ J/s
    (e) $\approx 10^{40}$ J/s

78. Which answer below is closest to the mass of 3C 273 that is converted to light and other forms of radiation each second? By comparison, the mass of Earth is 6×10^{24} kg.
    (a) $\approx 10^{11}$ kg/s (b) $\approx 10^{15}$ kg/s
    (c) $\approx 10^{19}$ kg/s (d) $\approx 10^{23}$ kg/s
    (e) $\approx 10^{29}$ kg/s

79. What is the speed of light emitted by 3C 273 as detected by an observer on 3C 273?
    (a) $1.15c$ (b) c
    (c) $0.85c$ (d) None of these is correct.

80. If 3C 273 is moving away from Earth at 0.16c, what speed below is closest to the light speed we on Earth detect coming from 3C 273?
    (a) $1.15c$ (b) c
    (c) $0.85c$ (d) None of these is correct.

Quantum Optics

Max Planck was a gifted young musician and composer. However, he chose physics as a field of study when he entered the University of Munich in 1874 at age 16. One of his professors discouraged him from studying physics because "almost everything is already discovered, and all that remains is to fill a few holes." Planck disregarded the advice; by 1879 he had defended his Ph.D. thesis, and in 1900 he made a discovery that totally changed how physicists understood the world.

- **What investigations started the transition from classical physics to quantum physics?**

- **Why do some stars look red and some look blue?**

- **How does a solar cell work?**

WE HAVE STUDIED the behavior of light and investigated various models of light, ultimately arriving at an electromagnetic wave model. That model explained the propagation of light in different media and its interactions with other objects, but it did not explain the mechanisms of light emission and absorption. To explain those, a new model of light was needed, and Planck was among those who helped develop it. This new model initiated the revolutionary *quantum physics* that is the subject of this and subsequent chapters.

BE SURE YOU KNOW HOW TO:

- Connect an ammeter and a voltmeter in a circuit to measure current and potential difference (Section 19.3).

- Relate the change of kinetic energy of an electrically charged particle to the potential difference across which it travels (Section 18.3).

- Relate the wavelength of a wave to its frequency and speed (Section 11.2).

27.1 Black body radiation

In the last quarter of the 19th century, physicists became interested in how hot objects such as stars and the coals of a fire emit thermal radiation (what we now call **infrared radiation**). Scientists studied the absorption and emission of infrared radiation and visible light from a hot object by modeling that object as a so-called **black body**. A black body absorbs all incident (incoming) light and converts the light's energy into thermal energy (no light is reflected). The black body then radiates electromagnetic (EM) waves based solely on its temperature.

What characterizes a black body?

FIGURE 27.1 An infrared photo taken at night with the house lights off.

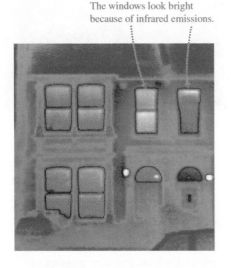

The windows look bright because of infrared emissions.

FIGURE 27.2 A black body spectrum.

(a)

Thermometer

The device records the radiation emitted from the black body hole. The power spectrum is shown in **(b)**.

Black body

(b)

To understand how the black body model works, imagine a small window in a house whose lights are off, or a small opening in a box, or the pupil of the human eye. Small openings like these look black to an outside observer because light entering the openings is not reflected—it is trapped inside. A small window, for example, admits sunlight. That light is absorbed inside the room, making the room warmer. As a result, more and more infrared radiation also leaves the room through the window. Eventually, the rate at which sunlight energy enters the room through the window equals the rate at which infrared radiation energy leaves the room. At this point the window becomes "a black body." In **Figure 27.1** an infrared camera has detected the infrared radiation being emitted through the windows of an unlit house.

We will model a black body as the surface of a small hole in one side of a closed container (**Figure 27.2a**). Imagine that this container also has a thermometer inside. The hole looks black, and the box is at temperature $T_1 = 310$ K. If you measure the power output per unit radiating surface area of the EM radiation coming from the hole at different wavelengths, you find that the hole emits a continuous EM spectrum (the lower curve in Figure 27.2b).

The graph in Figure 27.2b is called a **spectral curve**. The quantity on the vertical axis is the power output per unit radiating surface area of the black body per small wavelength interval in units of $W/m^2/m$. On the horizontal axis is the radiation's wavelength. The *total* power output per unit radiating surface area is the area under the black body spectral curve; this quantity is also known as the **intensity** (I) of the EM radiation. (Notice that at a particular wavelength we see the greatest intensity of EM radiation coming from the hole.)

Now imagine that you place the box on a hot stove—the box's temperature rises and the thermometer in the box indicates a higher temperature $T_2 = 373$ K. If we again measure the spectrum of the EM radiation being emitted from the hole, we find that

- the total power output (intensity multiplied by the area of the hole) from the hole is now greater,
- the spectral curve rises (the upper curve in Figure 27.2b) at all wavelengths, and
- the peak of the spectral curve shifts to a shorter wavelength.

As the temperature of the box increases further, these patterns persist. The higher the temperature, the shorter the wavelength at which the maximum of the black body spectral curve occurs, and the greater the power radiated at all wavelengths.

A black body is not a real object; it is a model. It turns out, however, that many real objects emit EM radiation similar to that of a black body. For example, the metal filament of an incandescent lightbulb gets very warm when electric current is passed through it. The bulb radiates some visible light, a lot of invisible infrared radiation, and a little bit of UV. The spectral curve of a lightbulb is very close to the spectral curve of a black body of the same temperature. Thus, studying black body radiation might allow us to understand how objects emit and absorb light and how the radiation they emit might relate to the objects' temperatures.

Studies of black body radiation

Several physicists in the late 1800s used the black body model to determine characteristics of EM radiation. Among their findings were the total power output of a black body and the wavelength at which the peak of the emitted radiation power occurs.

Total power output In 1879 a Slovenian physicist, Joseph Stefan (1835–1893), using his own experiments and data collected by John Tyndall, investigated the power of light emission from a black body—the total power output per unit of surface area—which is equal to the area under the black body spectral curve (see **Figure 27.3**). As the graph lines for different temperatures show, the total radiation output (power per unit area of the emitting object) increases dramatically as the temperature of the black body increases. Stefan found that the total power output per unit of surface area in Figure 27.3 is proportional to the fourth power of the temperature. For example, if you double the temperature of the object (in kelvins), the total radiation power increases by 16 times. The area under the 6000-K curve in Figure 27.3 is 16 times the area under the 3000-K curve.

The total energy emitted by a 1-m^2 black body every second can be written as $I \propto T^4$ or $I = \sigma T^4$, where σ (Greek letter sigma) is the coefficient of proportionality. To find the power P radiated by the whole object, we multiply the intensity of its emitted radiation at its surface by the radiating surface area A of the object. In 1884 Ludwig Boltzmann (1844–1906) derived theoretically the empirical relationship found by Stefan by modeling a black body as a thermodynamic engine with light instead of gas.

The work of Stefan and Boltzmann resulted in the following relationship between the total emitted power P in watts (W), the temperature of the black body T in kelvins, and its surface area A in square meters (**Stefan-Boltzmann's law**):

$$P = IA = \sigma T^4 A \tag{27.1}$$

where sigma (σ) stands for the coefficient of proportionality (Stefan-Boltzmann's constant): $\sigma = 5.67 \times 10^{-8} \, \text{W/m}^2 \cdot \text{K}^4$.

Wavelength at which maximum intensity occurs In 1893, the German physicist Wilhelm Wien (1864–1928) quantified a second aspect of black body radiation using data similar to those in the graphs in Figures 27.2 and 27.3. The wavelength λ_{max} at which a black body emits radiation of maximum power per wavelength depends on the temperature of the black body:

$$\lambda_{\text{max}} = \frac{2.90 \times 10^{-3} \, \text{m} \cdot \text{K}}{T} \tag{27.2}$$

This became known as **Wien's law.** Note that the wavelength λ_{max} at which the black body emits maximum power per wavelength interval becomes shorter as the black body temperature T becomes greater.

FIGURE 27.3 Black body radiation at different temperatures. The total power is proportional to the fourth power of the temperature in kelvins.

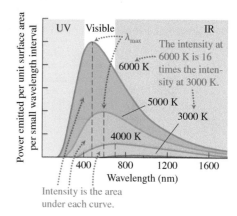

TIP ⟩ A black body (the system) emits electromagnetic waves with power given by Stefan-Boltzmann's law even if it is at the same temperature as the environment—but it also absorbs energy at the same rate. If the body is hotter than the environment, it radiates more energy than it absorbs. If it is cooler, it absorbs more than it radiates. The black body continues to emit and radiate different amounts of energy until it reaches thermal equilibrium with the environment. After that, emission and absorption continue, but their rates become the same.

QUANTITATIVE EXERCISE 27.1 ⟩ **The surface temperature of the Sun**

The maximum power per wavelength interval of light from the Sun is at a wavelength of about 510 nm, which corresponds to yellow light. What is the surface temperature of the Sun?

Represent mathematically Rearrange Wien's law solving for T in terms of λ_{max}:

$$T = \frac{2.90 \times 10^{-3} \, \text{m} \cdot \text{K}}{\lambda_{\text{max}}}$$

Solve and evaluate The temperature of a black body with its maximum power per wavelength interval at 510 nm = 510 × 10^{-9} m is

$$T = \frac{2.90 \times 10^{-3} \, \text{m} \cdot \text{K}}{510 \times 10^{-9} \, \text{m}} = 5.69 \times 10^3 \, \text{K}$$

This is about the temperature of the object in Figure 27.3 with the highest intensity (the uppermost curve).

At this temperature, matter is in a gaseous state. But the huge mass (2.0 × 10^{30} kg) of the Sun produces a strong gravitational field that prevents the gaseous atoms and molecules from easily leaving.

Try it yourself The surface temperature of Earth is about 288 K. What is the wavelength at which Earth radiates EM waves with maximum power per wavelength interval?

Answer

Infrared radiation with a wavelength of 10,000 nm.

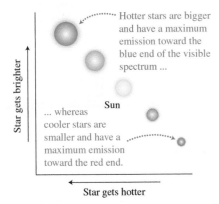

FIGURE 27.4 The color of stars. The color of a star is an indication of how hot it is. Hotter stars are bigger and appear bluer than cooler stars.

Hotter stars are bigger and have a maximum emission toward the blue end of the visible spectrum ...

... whereas cooler stars are smaller and have a maximum emission toward the red end.

Sun

Star gets brighter

Star gets hotter

As we saw in the "Try it yourself" question of Quantitative Exercise 27.1, the Earth radiates EM waves primarily in the infrared part of the electromagnetic spectrum, invisible to our eyes. Thus, when we see Earth and most objects on its surface, we are seeing sunlight reflected from them. The same mechanism (reflection of sunlight) makes the Moon visible to us. The Sun and other stars, however, do not need an external source of EM waves to illuminate them.

A star that has a lower surface temperature than the Sun's radiates with a maximum intensity wavelength that is longer than 510 nm, and the star looks red instead of yellow. A star that is hotter than the Sun radiates with maximum intensity at shorter wavelengths, a combination of blue and other visible wavelengths that cause it to look whiter. Very hot stars (around 10,000 K) look blue. Stars even hotter than these are generally invisible to our eyes, but they can be detected with instruments sensitive to ultraviolet radiation.

Figure 27.4 shows a pattern characteristic of the majority of the stars. You can see that hotter stars are bigger and cooler stars are smaller. This relationship is true for stars of ages similar to that of the Sun. Older stars can have lower temperatures and large sizes (red giants) or very high temperatures and smaller sizes (white dwarfs). The fascinating topic of stellar evolution is unfortunately beyond the scope of this book.

EXAMPLE 27.2 | **What is the power of surface radiation from the Sun?**

The radius of the Sun is about 7×10^8 m. Estimate the total amount of electromagnetic energy emitted by the Sun every second if its average surface temperature is about 6×10^3 K.

Sketch and translate We can visualize the Sun as a shining sphere of a given radius. If we know the total surface area of the Sun, we can use Stefan-Boltzmann's law to calculate the total energy radiated per unit time (power).

Simplify and diagram To apply Stefan-Boltzmann's law, assume that the Sun radiates as a black body. To find the surface area, assume that the Sun is a perfect sphere whose every square meter emits electromagnetic waves equally.

Represent mathematically According to Stefan-Boltzmann's law, the power radiated by a black body with a surface area A is

$$P = \sigma A T^4$$

The temperature is given, and the surface area of a sphere is $A = 4\pi R^2$. Thus the total power radiation by the Sun's surface is

$$P = \sigma(4\pi R^2) T^4$$

Solve and evaluate Now insert the appropriate values. Because we are doing an estimation, we will only calculate to one significant digit:

$$P \approx (6 \times 10^{-8}\,\text{W/m}^2\cdot\text{K}^4)4\pi(7 \times 10^8\,\text{m})^2(6 \times 10^3\,\text{K})^4$$
$$= 5 \times 10^{26}\,\text{W}$$

The actual number is close to 4×10^{26} W. Either way, this is a huge power output. Power plants on Earth typically have an output of $10^{12}-10^{13}$ W. The Sun has maintained (approximately) this power output for billions of years. You'll learn more about the mechanism of the Sun's power output in a later chapter (Chapter 29).

- -

Try it yourself Estimate how much solar energy reaches Earth every second. The effective surface area of Earth facing the Sun is a disk of radius 6.4×10^6 m (the radius of Earth). The average Sun-Earth distance is about 1.5×10^{11} m.

Answer 2×10^{17} W $= 2 \times 10^{17}$ J/s.

The ultraviolet catastrophe

By the end of the 19th century, physicists had devised a mechanism to explain how objects emitted electromagnetic radiation, namely, that the atoms and molecules inside objects are made of charged particles, and when these charged particles vibrate, they radiate electromagnetic waves. The higher the vibration frequency, the higher the frequency of the electromagnetic waves that are emitted. The atoms and molecules in a black body at temperature T vibrate at many different frequencies centered on the frequency of the peak radiation on the black body spectrum curve.

Maxwell's electromagnetic wave theory (Chapter 25), together with some assumptions about the structure of matter, predicts that the intensity of the radiation emitted by a black body should increase with the frequency of the radiation—greater intensity for ultraviolet light than for visible light; greater intensity for X-rays than for ultraviolet light. However, actual spectral curves show that at high frequencies (short wavelengths) the emitted intensity actually drops off (**Figure 27.5**). No models built on the physics of the late 19th century made predictions that were consistent with this drop. This problem became known as the **ultraviolet catastrophe.**

FIGURE 27.5 Classical black body radiation theory failed to match the experimental data at short wavelengths.

Planck's hypothesis

After working on the problem for several years, Max Planck (1858–1947) devised an ingenious model that correctly predicted the spectral curve for a black body. In Planck's model, the charged particles inside radiating objects were still responsible for producing electromagnetic radiation. However, they could radiate energy only in discrete portions called **quanta** (one portion is one quantum). Each quantum of emitted energy was equal to a multiple of some fundamental portion of energy: E_0, $2E_0$, $3E_0$, $4E_0$, etc. This idea is known as **Planck's hypothesis**.

Although the idea of this so-called **energy quantization** was completely revolutionary, the idea of quantization itself was not new to physics. Think of mechanical waves. When waves travel back and forth on a string of finite length with fixed ends, the string can vibrate only at discrete standing wave frequencies:

$$f_n = n\,\frac{v}{2L}$$

where $n = 1, 2, 3, \ldots$. The frequencies depend on the speed v of the waves on the string and the string length L. The standing wave frequencies are therefore said to be **quantized**. The energy of these standing waves can take on any positive value, however, because the energy of the wave depends on the *amplitude* of the wave and not on its frequency.

Planck's hypothesis is different. He suggested that it was *the energy of the oscillator* that was quantized, rather than its frequency.

Using his energy quantization model, Planck was able to derive a mathematical function describing the black body spectral curve. The derivation is complicated, so we won't show it. When he matched the calculated values from the function to experimental results, he was able to determine the constant of proportionality between the oscillating frequency of a charged particle f and its smallest fundamental energy E_0. The relationship is

$$E_0 = hf \tag{27.3}$$

The proportionality constant h became known as **Planck's constant**. The units of h in the SI system are $J/Hz = J/(1/s) = J \cdot s$. Its value was determined experimentally to be $h = 6.63 \times 10^{-34}\ J \cdot s$.

Planck viewed his energy quantization model as a mathematical trick of some sort rather than an explanation of actual physical phenomena. How could something oscillate smoothly at a particular frequency yet emit waves of that frequency only in discrete energy portions, as though it were emitting some sort of energy "particle"? He was almost embarrassed to have suggested such a model, and for much of his life he harbored doubts that real molecules or atoms radiated energy in this way. However, other scientists, including Albert Einstein, successfully applied this idea to explain other phenomena.

QUANTITATIVE EXERCISE 27.3   **Energy of light quanta**

Estimate the energy of a quantum of radio waves (frequency of 10^6 Hz), infrared radiation (3×10^{13} Hz), visible light (5×10^{14} Hz), and UV radiation (10^{15} Hz).

Represent mathematically According to Planck's hypothesis, the energy of a quantum of EM radiation is $E_0 = hf$.

Solve and evaluate The estimated energies are

$$E_{0\,\text{radio}} = hf_{\text{radio}} = (6.63 \times 10^{-34}\,\text{J} \cdot \text{s})(1 \times 10^6\,\text{Hz})$$
$$= 7 \times 10^{-28}\,\text{J}$$

$$E_{0\,\text{infrared}} = hf_{\text{infrared}} = (6.63 \times 10^{-34}\,\text{J} \cdot \text{s})(3 \times 10^{13}\,\text{Hz})$$
$$= 2 \times 10^{-20}\,\text{J}$$

$$E_{0\,\text{visible}} = hf_{\text{visible}} = (6.63 \times 10^{-34}\,\text{J} \cdot \text{s})(5 \times 10^{14}\,\text{Hz})$$
$$= 3 \times 10^{-19}\,\text{J}$$

$$E_{0\,\text{ultraviolet}} = hf_{\text{ultraviolet}} = (6.63 \times 10^{-34}\,\text{J} \cdot \text{s})(1 \times 10^{15}\,\text{Hz})$$
$$= 7 \times 10^{-19}\,\text{J}$$

These are very small energies. For perspective, we can estimate how many quanta a 60-watt incandescent lightbulb emits every second. Its filament (at about $T = 2500$ K) produces a spectrum very close to a black body spectrum. Using Wien's law we can estimate the wavelength of the maximum intensity radiation emitted by the filament and the corresponding frequency of that radiation:

$$\lambda_{\max} = \frac{c}{f} = \frac{2.90 \times 10^{-3}\,\text{m} \cdot \text{K}}{T}$$

or

$$f = \frac{c}{\lambda_{\max}} = \frac{cT}{2.90 \times 10^{-3}\,\text{m} \cdot \text{K}}$$
$$= \frac{(3.0 \times 10^8\,\text{m/s})(2500\,\text{K})}{2.90 \times 10^{-3}\,\text{m} \cdot \text{K}} \approx 2.6 \times 10^{14}\,\text{Hz}$$

This frequency corresponds to infrared radiation. The energy of one quantum of infrared radiation at this frequency is approximately 2×10^{-19} J. If we assume that all of the filament's radiated power is radiated at this frequency, we can estimate the number n of quanta emitted every second as the total energy per second divided by the energy of one quantum:

$$n = \frac{60\,\text{J/s}}{2 \times 10^{-19}\,\text{J/quantum}} \approx 3 \times 10^{20}\,\text{quanta/s}$$

Try it yourself Estimate how many quanta of electromagnetic radiation from the Sun reach Earth each second (use the data from the "Try it yourself" question in Example 27.2) and assume that all of the Sun's energy is radiated as visible light.

Answer

About 10^{36} quanta each second—equivalent to the output of about 3×10^{15} 60-watt bulbs.

Planck's quantum idea added a new twist to the study of light, and this divergence made him uncomfortable even though he knew the mathematics was correct. Table 27.1 presents a summary of that history. Note that as of the early 20th century, there was no explanation for the absorption of light.

TABLE 27.1 **Evolution of ideas concerning the nature of light**

Approximate time period	Model	Experimental evidence that the model explains
1600s	*Particle (bullet) model* Light can be modeled as a stream of bullet-like particles.	Reflection, shadows, and possibly refraction
Early 1800s	*Wave model* Particle model insufficient. Light modeled as wave. Wave mechanism unknown.	Double-slit interference, diffraction, light colors, reflection, and refraction
Middle 1800s	*Electromagnetic wave (ether model)* Light modeled as transverse vibrations of ether medium.	Polarization and electromagnetic wave transmission and all of the above evidence
Late 1800s	*Electromagnetic wave (no-ether model)* Light modeled as transverse vibrations of electric and magnetic fields. No medium required.	Michelson-Morley experiment
Early 1900s	*Quantum model* Electromagnetic wave model insufficient. Light modeled as stream of discrete energy quanta.	Emission of light by black bodies

Today, Planck's quantum model is understood as the birth of quantum physics. To understand how this transformation from classical to quantum physics occurred and what model of light is currently accepted, we now turn the discussion to a new phenomenon.

REVIEW QUESTION 27.1 Planck's hypothesis of light emission, like the classical explanation, stated that objects emit light through the accelerated motion of electrically charged particles. What in Planck's hypothesis brought the theory in line with observed values of light emission?

27.2 Photoelectric effect

Planck's quantum hypothesis contradicted the wave model of light and remained untested for several years. However, in the interim the hypothesis proved instrumental in understanding an interaction between metals and electromagnetic radiation. In order to understand this interaction, we will first review what we know about the structure of metals.

The structure of metals

In an earlier chapter (Chapter 19) you learned that metals are made of atoms whose outermost one or two electrons do not remain bound to the nucleus but are loosely bound to the whole crystal (**Figure 27.6a**). The positively charged ions (the metal atoms minus their outermost electrons) form a crystal lattice. The negatively charged electrons that have left the atoms move freely through the lattice (Figure 27.6b). These free electrons have positive kinetic energy $K = +(1/2)mv^2$. Why don't they leave the metal?

An electron and a positive ion have opposite electric charges ($-e$ for the electron and $+e$ for the positive ion). If we simplify the situation and consider a single electron and a single ion as the system of interest, the electric potential energy U_q of this system is negative (Section 17.5):

$$U_q = k\frac{q_e q_{ion}}{r} = k\frac{(-e)(+e)}{r} = -k\frac{e^2}{r} < 0$$

In addition to the electric potential energy of the electron-lattice system, the electrons possess kinetic energy. This energy must be smaller in magnitude than the electric energy because the electrons do not spontaneously leave the metal; thus the total energy of the system must be negative.

Because the energy of the system is negative, energy must be added to the system to remove the free electron from the metal (Figure 27.6c). The minimum energy needed to remove a free electron from a metal is called the **work function** ϕ (Greek letter phi) of the metal (Figure 27.6d).

The work function ϕ has a specific, known value for many metals. Although the work function has units of energy, it is not measured in joules because typical work function values are very small. Instead, we use the **electron volt** (eV), where $1\text{ eV} = 1.6 \times 10^{-19}$ J. As you may recall (from Chapter 26), one electron volt is the magnitude of the change in kinetic energy of an electron of charge $e = -1.6 \times 10^{-19}$ C when it moves across a potential difference of 1 V.

For most metals (silver, tin, aluminum, etc.), the value of ϕ ranges from about 4.30 eV to 4.60 eV. However, the ϕ values of some metals, such as cesium and lithium, are in the range of 2.10 eV to 2.90 eV. Thus it is significantly easier to remove electrons from these metals. The greater the work function of a particular metal, the more strongly the free electrons are bound to the metal as a whole, and the more energy that must be added to the electron-metal system to separate them.

Detection of the photoelectric effect

Now that we have an understanding of how metals hold on to their free electrons, let's discuss some experiments that eventually helped support Planck's hypothesis. We learned from our investigation of black body radiation that EM radiation is emitted as quanta. Is EM radiation absorbed as quanta as well? We investigate this idea in

FIGURE 27.6 The structure of metals. (a and b) Positive ions and free electrons. (c) Energy bar chart for a lattice-free electron system. (d) Adding the energy equal to the work function makes the total energy of the system zero.

(a) Neutral atom   Positive ion

Free electron

In a metal, a neutral atom becomes a positive ion and a free electron.

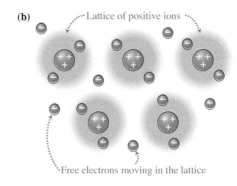

(b) Lattice of positive ions

Free electrons moving in the lattice

(c) $K_e + U_q < 0$

Because the electron-lattice energy is less than zero, the electron cannot escape.

(d) $K_e + U_q + \phi = 0$

The work function ϕ is the least energy needed to raise the electron-lattice energy to zero so that the electron can escape.

TIP Even though ϕ is called the work "function," it's not a function in the mathematical sense. It is a constant that differs from material to material. It is also not work because it is a property of a particular metal. Finally, although some electrons in metals are called "free" electrons, they are not free to leave the metal; they are free to move around within the metal, however. From this analysis of the vernacular, you can see how important it is to question the terms we use in order not to be confused.

Observational Experiment **Table 27.2**. These experiments were performed originally by Wilhelm Hallwachs (1859–1922) in Germany in 1888. He used a gold leaf electroscope, in which two thin "leaves" of gold foil repel each other if they carry a like charge and hang straight down if they are uncharged. The first experiment is similar to one we performed earlier (in Chapter 17).

  Discharging an electroscope

VIDEO
OET 27.2

Observational experiment	Analysis
Experiment 1. Take a plastic rod, rub it with felt, and touch it to the metal (zinc) sphere of an electroscope. Repeat the experiment using a glass rod rubbed with silk and another electroscope. Observe that the leaves of both electroscopes are deflected and remain deflected.	The negatively charged plastic rod rubbed with felt transfers electrons to the electroscope, making it negatively charged. The positively charged glass rod rubbed with silk takes free electrons away from the electroscope, leaving it positively charged.
Experiment 2. Shine visible light from an incandescent bulb on both the positively charged and the negatively charged electroscope. Nothing happens to them.	Visible light does not affect the charge of the electroscope.
Experiment 3. Shine UV radiation on the positively charged electroscope. The leaves of the electroscope stay deflected. Shine UV radiation on the negatively charged electroscope. The leaves quickly return to their original uncharged position.	The UV light discharges a negatively charged electroscope but does not discharge a positively charged electroscope.

Patterns

- UV radiation makes the negatively charged electroscope discharge but has no effect on the positively charged electroscope.
- Lower frequency visible light does not discharge either electroscope.

The sphere on the top of the electroscope in Table 27.2 was made of zinc. It turns out that if the same sphere had been covered with a thin layer of cesium-antimony, green light would have had an effect on it similar to the effect the UV light had in the experiments above. It is the combination of the frequency of light and the material on which it shines that allows the negatively charged electroscope to discharge. A positively charged electroscope does not discharge in the presence of visible or UV light.

The interaction described in Table 27.2 is an example of the **photoelectric effect**. Let's try to invent an explanation for the patterns that we've observed.

Explanation 1: The light's vibrating electric field interacts with the electric charges in the molecules of air surrounding the electroscope. It shakes them back and forth until the molecules ionize into electrons and positively charged ions. The positive ions are attracted to the negative electroscope and attach to it, making it neutral overall.

This sounds reasonable. However, if it is correct, then both the positively and negatively charged electroscopes should discharge, because the newly freed electrons in the air should neutralize the positively charged electroscope. But this was not observed. We can rule out this explanation for the photoelectric effect.

Explanation 2: The light's vibrating electric field interacts with the electrons in the electroscope and can cause them to be ejected from it, but it cannot cause positive ions to be ejected. This hypothesis makes sense when we recall that free electrons in metals are negatively charged particles, and when we charge an object by touching, we either add or remove these electrons. The positive ions in the metal make up its crystal lattice and probably cannot be removed easily.

Explanation 2 seems reasonable, but it does not explain why the frequency of light is important: in Table 27.2 light from the incandescent lightbulb did not discharge the electroscope, whereas UV radiation did. We need more observational experiments.

More experiments

Additional observational experiments were conducted by a German physicist, Philipp von Lenard (1862–1947), in 1902. He used an apparatus similar to that shown in **Figure 27.7**. The apparatus, called a *phototube*, consists of an evacuated glass container with a window that allows both visible light and UV radiation to shine on a metal plate. This plate is normally connected to the negative terminal of a battery and is at an electric potential of zero; it is known as a **cathode**. A corresponding plate, termed an **anode**, is usually connected to the positive terminal of the battery at potential V. However, the connection can be reversed by connecting the plates to the opposite terminals of the battery. When reversed, the potential of the former anode potential is negative relative to the zero cathode potential. The ammeter shown in Figure 27.7 can detect changes in the current in the circuit.

We can change the frequency of light passing through the window by inserting color filters in front of the window. We can also change the material of the cathode and the anode. In all further experiments, we will assume that the anode is made of the same material as the cathode. When they are made of different materials, the difference in the work functions of the two metals makes the analysis more complicated and beyond the scope of this book.

When there is no light shining on the cathode, the ammeter detects no current. When light is shining and we vary the filters, allowing light of different frequencies to illuminate the cathode, the ammeter detects current in the circuit for some filters but not for others. For example, when the cathode is made of zinc, only UV light leads to current in the circuit (visible light does not), whereas for a cesium-antimony cathode, green light leads to current (but red light does not).

Observational Experiment **Table 27.3** allows us to analyze the data that von Lenard collected.

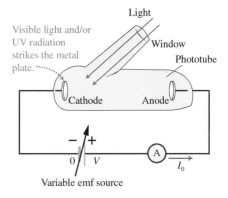

FIGURE 27.7 A phototube connected to a variable power supply and an ammeter.

Quantitative observations of the photoelectric effect

Observational experiment	Analysis
Experiment 1. Connect the cathode on which light from a light source with a variable aperture shines to a negative battery terminal. Find the frequency of light (using filters) for which the ammeter detects current. Keep the potential of the anode at a constant positive value relative to the zero potential at the cathode ($V_{anode} > V_{cathode} = 0$). Without changing the frequency, steadily increase the intensity of radiation shining on the cathode by opening the aperture. The ammeter measures current.	Electric current from light exposure (the **photocurrent**) registered by the ammeter is directly proportional to the intensity of the light radiation. When no light is shining on the cathode, no current is measured by the ammeter. 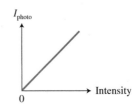
Experiment 2. Use the frequency of EM radiation for which the photoelectric effect occurs. Using the same cathode and the same light filter, detect current in the circuit. Keeping the intensity and the frequency of the light or UV radiation constant, vary $V_{anode} - V_{cathode}$. Notice that a negative value means that the electric potential at the anode is lower than at the cathode. The ammeter measures current.	Assume that the potential at the cathode is 0. The photocurrent increases as the positive potential at the anode increases. After the anode potential reaches a particular positive value, the current stops increasing. At a particular negative anode potential $-V_s$ (the **stopping potential**), the current stops. Note that V_s is the magnitude of the stopping potential. 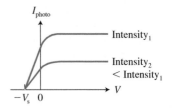
Experiment 3. Use the frequency of EM radiation for which the photoelectric effect occurs. Repeat Experiment 2 for an increased intensity of light or UV radiation by opening the aperture on the light source.	When the intensity of UV radiation increases, the maximum value of the photocurrent increases, but the negative stopping potential $-V_s$ does not change. 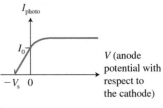
Experiment 4. Repeat Experiment 2 with radiation in both the visible light range and UV range, using different frequencies, and each time recording the stopping potential.	The stopping potential (now shown on the y-axis) is plotted against frequency (the x-axis). When the frequency of electromagnetic radiation increases, the negative stopping potential $-V_s$ becomes more negative. For higher frequency light, it takes a greater opposing potential difference to stop the photocurrent. Also, there is no photocurrent when the frequencies are less than a value termed the **cutoff frequency**.

Patterns

The photocurrent in the circuit depends on three factors: the potential difference across the electrodes (we assume that the potential of a cathode is zero), the intensity of the light, and the frequency of the light. If for a particular frequency of light, photocurrent is present in the circuit, then

1. For constant positive anode potential V, the photocurrent is directly proportional to the intensity of the radiation.
2. When the V_{anode} is zero, a nonzero photocurrent I_0 is still present in the circuit.
3. For the same intensity and frequency of radiation, the photocurrent increases with increasing positive anode potential V until it reaches some maximum value.
4. A well-defined negative stopping potential $-V_s$ stops the photocurrent.
5. Increasing the intensity of radiation increases the photocurrent I_0 and the maximum value of the photocurrent, but does not affect the negative anode stopping potential $-V_s$.
6. $-V_s$ is affected only by the frequency of the EM radiation incident on the cathode. The relationship is linear. When the frequency is below a certain cutoff value f_{cutoff}, the photocurrent is zero for any potential difference between the cathode and anode.

Von Lenard noted another interesting outcome. He saw that the ammeter recorded a current in the circuit immediately after the light was turned on.

Attempting to explain these patterns with the EM wave model

Let's try to use the EM wave model of light to explain the patterns reported for the frequencies of light for which the photoelectric effect occurred in Table 27.3. Remember that in all experiments, the cathode and the anode are made of the same material.

Pattern 1 For constant $V_{anode} - V_{cathode} > 0$, the photocurrent is directly proportional to the intensity of the radiation.

EM wave model explanation: Evidently, electrons in the cathode absorb enough energy from EM radiation that they are ejected from the cathode. The greater the intensity of the light, the more energy is transferred to the cathode. More electrons are freed from the cathode, causing a greater photocurrent. The wave model explains pattern 1.

Pattern 2 When $V_{anode} - V_{cathode} = 0$, there is still a nonzero photocurrent I_0 in the circuit.

EM wave model explanation: Evidently, the electrons in the metal gain enough energy from the EM radiation that they can escape the metal and have energy left over to reach the anode. The wave model explains pattern 2.

Pattern 3 For the same intensity of EM radiation, the photocurrent increases with increasing positive $V_{anode} - V_{cathode}$ until the current reaches a maximum value.

EM wave model explanation: It seems that with constant intensity EM radiation, the same number of electrons is ejected from the metal independent of $V_{anode} - V_{cathode}$. The electric force pushing the electrons to the anode increases as $V_{anode} - V_{cathode}$ increases. However, if the wave model is correct, it seems that a minimum $\vec{E}$ field is needed to draw all of the electrons to the anode. Increasing the $\vec{E}$ field beyond that would have no effect. The wave model explains pattern 3.

Pattern 4 A well-defined negative stopping potential $-V_s$ stops the photocurrent.

EM wave model explanation: With a negative $-V_s$, the anode is at a lower electric potential than the cathode. The electric field now opposes the motion of electrons across the tube to the anode **(Figure 27.8a)**. Suppose that $V_{anode} - V_{cathode} = (-V_s) - 0$ is the smallest negative value it can be while still preventing electrons from reaching the anode. This process can be quantified using the generalized work-energy principle—see the bar chart in Figure 27.8b. Mathematically:

$$K_i = U_{qf}$$

Inserting the expressions for kinetic energy and electric potential energy gives

$$\tfrac{1}{2}m_e v^2_{e\,i\,max} = (-e)(-V_s)$$

where $-e$ and m_e are the electron charge and mass, respectively, and $v_{e\,i\,max}$ is the maximum speed of the electron emitted from the metal. The magnitude of the stopping potential is then

$$V_s = \frac{m_e v^2_{e\,i\,max}}{2e}$$

The electrons have *one* particular $v_{e\,i\,max}$ and one stopping potential difference for low and high EM radiation intensity. It seems that with high-intensity EM radiation, electrons absorb more energy and escape at a higher speed. Thus the wave model explains pattern 4 qualitatively but does not explain why there is no intensity dependence in the expression for the stopping potential.

Pattern 5 Increasing the intensity of EM radiation increases the photocurrent I_0 and the maximum value of the photocurrent but does not affect the value of $-V_s$.

FIGURE 27.8 (a) The negative potential stops the electron. (b) A work-energy bar chart represents the stopping process.

(a)

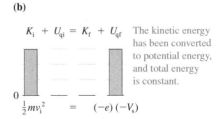

$U_{qi} = 0$ An electron is $U_{qf} > 0$
$K_i > 0$ knocked out of $K_f = 0$
the metal and has
kinetic energy.

v_i

As the electron gets closer
to the negative electrode,
it loses kinetic energy.

$0 \quad -V_s$

(b)

$K_i + U_{qi} = K_f + U_{qf}$ The kinetic energy
has been converted
to potential energy,
and total energy
is constant.

0
$\tfrac{1}{2}mv_i^2 \quad = \quad (-e)(-V_s)$

EM wave model explanation: The greater light intensity results in more electrons knocked out of the cathode and therefore a greater photocurrent. According to the EM wave model of light, a higher intensity electromagnetic wave should also transfer more energy to electrons, thus making it more difficult to stop them from reaching the anode. This is not what happens. The light intensity has no effect on the stopping potential. Thus the wave model does not explain pattern 5.

Pattern 6 The value of $-V_s$ is only affected by the frequency f of the EM radiation incident on the anode. The relationship is linear. When the frequency is below a certain cutoff value f_{cutoff}, the photocurrent does not appear at all.

EM wave model explanation: According to the wave model, the energy of an EM wave increases with the intensity of the wave, so it should be possible to use very intense but low-frequency light to knock electrons out of the metal. This does not happen. The EM wave model cannot explain pattern 6 at all and predicts something that contradicts the experiments.

In addition, for frequencies above the cutoff frequency, no delay occurs between the electromagnetic waves hitting the cathode and the release of the electrons, even for a very low intensity of EM radiation (for the frequencies above the cutoff frequency for a particular cathode). This is somewhat puzzling. Shouldn't it take a while for sufficient energy to be transferred to electrons in the cathode before they start being released, especially when the EM radiation intensity is low?

To summarize, the EM wave model of light can explain some aspects of the photoelectric effect, but it can't explain the following: (a) the lack of dependence of the stopping potential $-V_s$ on the intensity of the light, (b) the presence of the cutoff frequency f_{cutoff} below which there is no photocurrent (despite high intensity of light), and (c) the appearance of a photocurrent immediately after the EM radiation first strikes the cathode, even for low EM radiation intensity.

The photoelectric effect became the second phenomenon that made scientists question whether the EM wave model of light was sufficient to explain all EM wave behavior. In the next section you'll learn how to explain the photoelectric effect in its entirety. First, work through the following example to become comfortable with the notion of stopping potential and an electron's kinetic energy in the photoelectric effect.

EXAMPLE 27.4 **Kinetic energy of the electrons and the stopping potential**

In one of von Lenard's experiments, the value of the stopping potential was -0.20 V. What is the maximum kinetic energy of one of the electrons leaving the cathode?

Sketch and translate Figure 27.7 shows a schematic of von Lenard's experiment with electrons being released from the cathode by incident UV. The situation with negative potential across the electrodes is shown in Figure 27.8a. The left electrode (the cathode) is at zero potential and the right electrode (the anode) is at $-V_s$. The electric force exerted on the electrons is toward the left, causing them to slow down and stop. The electron has zero kinetic energy when it reaches the anode.

Simplify and diagram Assume for simplicity that the electrons travel in a straight path across the tube from the cathode to the anode. Assume also that the cathode from which an electron is ejected is at zero electric potential, and the anode that the electron flies toward is at -0.20 V. The work-energy bar chart in Figure 27.8b represents the process.

The electron and electric field produced by the electrodes are the system of interest. The initial state is when the electron has maximum kinetic energy just after being knocked out of the cathode and the system has zero electric potential energy; the final state is when the electron reaches the anode with zero kinetic energy. Thus, the final electric potential energy of the system is equal to the initial kinetic energy (Figure 27.8b).

Represent mathematically Using the generalized work-energy principle and the bar chart, we have

$$K_i = (-e)(-V_s)$$

Solve and evaluate The above equation can be used to determine the kinetic energy of the electron when it leaves the cathode; the energy of the electron is given in coulombs times volts (C · V):

$$K_i = (-e)(-V_s) = (-1.6 \times 10^{-19} \text{C})(-0.20 \text{ V})$$
$$= 3.2 \times 10^{-20} \text{C} \cdot \text{V}$$

Note that $1 \text{ V} = 1 \text{ J/C}$. Therefore, the electron's initial kinetic energy is

$$K_i = 3.2 \times 10^{-20} \text{ C} \cdot \left(\frac{\text{J}}{\text{C}}\right) = 3.2 \times 10^{-20} \text{ J}$$

This is a very tiny amount of energy, but is reasonable for a single electron. Converting this value to electron volts, we have

$$K_i = (3.2 \times 10^{-20} \text{ J})\left(\frac{1 \text{ eV}}{1.6 \times 10^{-19} \text{ J}}\right) = 0.20 \text{ eV}$$

Try it yourself What is the stopping potential for the electrons starting with the same kinetic energy if the distance between the cathode and anode is halved?

Answer

The stopping potential does not depend on the distance between the electrodes; it depends only on the change in the kinetic energy of the individual electrons. The potential will be the same: −0.20 V (with respect to the zero potential of the cathode).

REVIEW QUESTION 27.2 What features of the photoelectric effect could not be explained by the electromagnetic wave model of light?

27.3 Quantum model explanation of the photoelectric effect

Philipp von Lenard thought that the absorption of UV radiation (and in some cases visible light) by the electron-metal system (the cathode) increased the energy of the system in accordance with the generalized work-energy principle. If the energy is large enough to raise the negative total energy of the system to zero $(-\phi + E_{light} = 0)$, then an electron could just barely be ejected from the metal (see the bar chart representing the process in **Figure 27.9a**). If the system gained more energy from the EM radiation than it needed to eject the electron, then the electron would have a nonzero kinetic energy after leaving the cathode, and it might reach the anode even if the electric field produced by the electrodes pushed it in the opposite direction (Figure 27.9b). Mathematically:

$$-\phi + E_{light} = K_{ef} \qquad (27.4)$$

where K_{ef} is the kinetic energy of the electron immediately after leaving the metal. However, von Lenard could not explain why the stopping potential did not depend on the intensity of the incident light. In these experiments, light could mean either visible light or UV—whichever is causing the photoelectric effect. According to his understanding, increased light intensity should lead to greater energy of the light and cause the electron to leave with more kinetic energy, thus requiring a greater stopping potential to prevent it from reaching the anode.

In 1905, Albert Einstein (1879–1955) made a breakthrough similar to Planck's hypothesis. At that time working as a clerk in a Swiss patent office, Einstein suggested that light can be viewed as a stream of quanta of energy not only when it is *emitted* by objects (as proposed by Planck), but also when it is *absorbed* by them. According to Einstein's hypothesis, an electron-lattice system (a metal) can only absorb light energy one whole quantum at a time. If the value of the light energy of a particular quantum is more than the magnitude of the metal's work function, an electron could leave the lattice. If the quantum of light energy is less than that, the electron would stay even if the number of incident quanta is very high (corresponding to high-intensity light.)

According to Planck, the energy of a quantum of EM radiation equals $E_0 = hf$. Using this idea, Einstein rewrote Eq. (27.4) as

$$-\phi + hf = K_{ef} \qquad (27.5)$$

where K_{ef} is the kinetic energy of a single electron just after leaving the metal. This equation explains why for most metals UV radiation causes the photoelectric effect, whereas visible light does not—each quantum of higher frequency UV radiation has

FIGURE 27.9 The electron-lattice system gains energy from light and UV. (a) The energy gained from light offsets the negative electron-lattice energy, barely allowing the electron to escape. (b) If the energy gained from the light (or UV) is more than the amount needed to escape, then the electron has kinetic energy when it escapes.

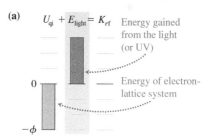

(a) $U_{qi} + E_{light} = K_{ef}$ Energy gained from the light (or UV)

Energy of electron-lattice system

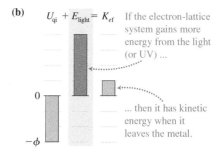

(b) $U_{qi} + E_{light} = K_{ef}$ If the electron-lattice system gains more energy from the light (or UV) ...

... then it has kinetic energy when it leaves the metal.

sufficient energy to remove an electron from the metal, whereas a lower frequency visible light quantum does not.

To check the feasibility of Einstein's hypothesis, let's calculate the energy of a quantum of UV radiation and compare it to typical work functions of metals. As mentioned in Section 27.2, the work function for most metals (silver, tin, aluminum, etc.) falls within the range $\phi = 4.30$ eV–4.60 eV. The energy of one UV radiation quantum is about

$$E_{0\,\text{UV}} = hf_{\text{UV}} = (6.63 \times 10^{-34}\,\text{J}\cdot\text{s})(1 \times 10^{15}\,\text{Hz}) \approx 7 \times 10^{-19}\,\text{J} \approx 4.4\,\text{eV}$$

This is encouraging! The frequency lies within the right range. Similarly, the energy of one green light quantum is about

$$E_{0\,\text{visible}} = hf_{\text{visible}} = (6.63 \times 10^{-34}\,\text{J}\cdot\text{s})(6 \times 10^{14}\,\text{Hz}) = 4 \times 10^{-19}\,\text{J} \approx 2.5\,\text{eV}$$

which is close to the magnitude of the work function for a few metals (cesium, lithium, sodium). This supports Einstein's hypothesis. We will test it more carefully in Testing Experiment **Table 27.4**.

TESTING
EXPERIMENT TABLE 27.4 **Testing Einstein's hypothesis using the photoelectric effect**

Testing experiment	Prediction	Outcome
We repeat von Lenard's experiments, but this time we use a sodium cathode and anode with a work function of 2.3 eV and use red and blue visible light.	If Einstein's hypothesis is correct, then • 1.9 eV red quanta of light (650-nm wavelength) will not have enough energy to eject electrons, and thus no photocurrent will be detected in the circuit. • 2.7 eV blue photons (450-nm wavelength) will be able to eject electrons from the sodium and thus will produce a photocurrent in the circuit.	• When red light shines on the cathode, the ammeter registers no photocurrent. • When blue light shines on the cathode, the ammeter registers a photocurrent.

Conclusion

The outcome of the experiment matches the prediction. Thus we have not disproved Einstein's hypothesis. Instead, we have more confidence in it.

The outcome of the testing experiment supports Einstein's hypothesis. Let's now consider other features of the photoelectric effect that the wave model could not explain, as described on pages 857 and 858: (a) the lack of dependence of the stopping potential $-V_s$ on the intensity of the light, (b) the presence of the cutoff frequency f_{cutoff} of light, and (c) the immediate appearance of the photocurrent once light is incident on the cathode.

(a) *Stopping potential is independent of intensity:* The negative stopping potential $(-V_s)$ of the anode with respect to the zero cathode potential depends on the kinetic energy of the electrons leaving the metal:

$$(-e)\big[(-V_s) - 0\big] = \tfrac{1}{2} m_e v_{ei}^2$$

According to Eq. (27.5), the kinetic energy of an electron knocked out of the metal is

$$\tfrac{1}{2} m_e v_{ei}^2 = -\phi + hf$$

Thus, the kinetic energy of the electron depends only on the energy hf of a single quantum and not on the intensity of EM radiation incident on the cathode. Since the stopping potential depends on the initial kinetic energy of a freed electron, the stopping potential also depends only on the frequency of the radiation and not on the intensity of the radiation. Einstein's hypothesis explains this pattern.

(b) *No photocurrent below cutoff frequency:* The cutoff frequency occurs if the energy hf of a light quantum equals the work function ϕ, that is, if $hf = \phi$. We can then express the cutoff frequency in terms of the work function of the metal and Planck's constant:

$$f_{\text{cutoff}} = \frac{\phi}{h} \qquad (27.6)$$

Under these circumstances the electron leaves the metal with zero kinetic energy. If the frequency of light is such that $hf < \phi$, then the energy of one quantum is less than the work function of the metal, and no photocurrent is produced. Einstein's hypothesis explains this pattern, too.

(c) *Photoelectric current produced immediately:* Absorbing one quantum of light of sufficient energy could instantly eject an electron out of the metal. Because the photoelectric effect is the result of an interaction between a single quantum of EM radiation and a single electron, energy does not need to accumulate in the metal before we see an effect. Here again, Einstein's hypothesis accounts for the experimental evidence.

Einstein's hypothesis explains the features of the photoelectric effect that the EM wave model could not. Let's see if it accounts for features that the wave model did explain.

CONCEPTUAL EXERCISE 27.5   **Applying Einstein's hypothesis**

Use Einstein's hypothesis to explain why the photocurrent in von Lenard's experiments is directly proportional to the intensity of the EM radiation incident on the cathode.

Sketch and translate According to Einstein's hypothesis we can visualize EM radiation as a stream of quanta, each with a specific amount of energy.

Simplify and diagram
We represent the quanta as little "bubbles" in the figure on the right. The greater the intensity of the light, the greater the density (number per unit volume) of bubbles. Each bubble can potentially liberate one electron

Quanta of light and UV   Cathode

Each free electron can absorb one quantum and leave the cathode.

from the metal; thus more bubbles lead to more liberated electrons and therefore a larger photocurrent. Einstein's hypothesis does indeed predict that the photocurrent is directly proportional to the intensity of EM radiation.

Try it yourself Explain why there is a photocurrent even when the potential difference between the electrodes in the tube is zero.

Answer The kinetic energy of a liberated electron is $K_{e'} = hf - \phi$. If the energy of one quantum is more than the magnitude of the work function, then the electron will have nonzero kinetic energy after being separated from the metal. There is a chance that some of these electrons will arrive at the anode and produce the photocurrent even in the absence of a potential difference between the electrodes.

Einstein's hypothesis was successful in explaining the features of the photoelectric effect that the EM wave model for light could not explain. Einstein's hypothesis also leads to another testable prediction. Equation (27.6) $f_{\text{cutoff}} = \phi/h$ implies that for every metal, the ratio of its work function to its cutoff frequency should equal Planck's constant:

$$h = \frac{\phi}{f_{\text{cutoff}}}$$

Robert Millikan (1868–1953) used this equation to determine the value of Planck's constant experimentally. Millikan's goal was to disprove Einstein's hypothesis and show that the absorption of light was not explained by a quantum hypothesis.

Millikan's reason for opposing Einstein's hypothesis was very similar to Planck's reason for not believing in his own quantum idea. The model of light as an electromagnetic wave could explain so many different phenomena that returning to a particle-like model seemed inconceivable. As Millikan wrote himself in a paper published in 1916: "Einstein's photoelectric equation . . . cannot in my judgment be looked upon at present as resting upon any sort of a satisfactory theoretical foundation," even though "it actually represents very accurately the behavior" of photoelectricity.[1]

Later, when accepting the Nobel Prize in 1923 for this work and for his measurement of the charge of the electron, Millikan said: "this work resulted, contrary to my own expectation, in the first direct experimental proof . . . of the Einstein equation, and the first direct photoelectric determination of Planck's h."

The quanta of energy first invented by Planck to explain the emission of light and later applied by Einstein to explain the absorption of light later became known as **photons**.

> **Photon** A photon is a discrete portion of electromagnetic radiation that has energy
>
> $$E_{photon} = hf \qquad (27.7)$$
>
> where f is the frequency of the electromagnetic radiation and $h = 6.63 \times 10^{-34}$ J $\cdot$ s is Planck's constant.

Using the language of photons, we can now summarize Einstein's photoelectric effect equation.

> **Einstein's equation for the photoelectric effect** During the photoelectric effect, the kinetic energy K_e of the emitted electron equals the difference in the energy of the photon $E_{photon} = hf$ absorbed by the metal and the metal's work function ϕ:
>
> $$-\phi + hf = K_e \qquad (27.5)$$

It is worth noting that Einstein received the Nobel Prize for the explanation of the photoelectric effect in 1921, two years before Millikan did. In the next example, we use Einstein's explanation of the photoelectric effect to understand some additional experimental results.

---

[1]Millikan, R. A. (1916). A direct photoelectric determination of Planck's "h." *Phys. Rev. 7*, 355. http://focus.aps.org/story/v3/st23

EXAMPLE 27.6   **Ejecting electrons from different metals**

Light and UV shine on three different metals, each metal being a cathode in a photoelectric tube. The anode in each experiment is made of the same material as the cathode. Graph lines representing the stopping potential $-V_s$ versus the incident light or UV frequency are shown in the figure at right. The work functions of the metals are sodium (Na), 2.3 eV; iron (Fe), 4.7 eV; and platinum (Pt), 6.4 eV. Use Einstein's hypothesis of light absorption to explain the results.

Sketch and translate The graph lines all have the same slope but have different intercepts on the horizontal frequency axis. Does Einstein's explanation of the photoelectric effect produce a mathematical equation that relates the observed stopping potential $-V_s$ and the frequency f of the incident light and UV? The experimental setup is shown below.

Note that the cathode electric potential is at zero and the anode potential V is negative (necessary to stop the photoelectrons). The system of interest is the electron and the metal lattice—the cathode. The energy of each absorbed photon of incident light may cause an electron to be ejected from the metal. If the energy of the photon is greater than the work function of the metal, then the ejected electron has some kinetic energy after leaving the metal.

Simplify and diagram We represent the process with a bar chart, as shown in the first figure below. After ejection, the electron travels to the anode. We adjust the potential difference between the electrodes so that the electron stops just as it reaches the anode. The anode is at potential $-V_s$, less than the cathode (which is at zero). The travel of the ejected electron to the anode is represented by the second bar chart below.

Represent mathematically Use the first bar chart above to represent the electron ejection process mathematically as

$$-\phi + hf = K_{ef}$$

The kinetic energy K_{ef} of the electrons after ejection equals their kinetic energy K_{ei} at the beginning of the next process (the movement of the electron from the cathode to the anode). Use the bar chart to represent that process mathematically:

$$K_{ei} = U_{qf} = qV_f = (-e)(-V_s)$$

where $-V_s$ is the negative anode potential. We combine the two equations to get

$$-\phi + hf = (-e)(-V_s)$$

Here $-V_s$ is the specific value of the stopping potential.

Solve and evaluate If we want to put the previous equation in a slope-intercept form so it can be compared with the graph, we need to replace the specific value of the stopping potential $(-V_s)$ with the variable V_s that we are plotting on the y-axis:

$$V_s = \left(-\frac{h}{e}\right)f + \left(\frac{\phi}{e}\right) \tag{27.8}$$

According to this equation, regardless of the metal used to make the cathode and anode, the slope will be $-h/e$. This explains why all the graph lines have the same slope, as seen in the first figure. Millikan measured these slopes to determine Planck's constant h. Because the photon energy hf is greater in magnitude than the work function ϕ, the stopping potential, $-V_s$, must be negative.

Try it yourself If blue light of frequency 6.7×10^{14} Hz is incident on a sodium target, what is the value of the stopping potential?

Answer   −0.5 V.

In the above example, the x-intercept of each graph line is different. According to Eq. (27.8), the intercept when $V_s = 0$ depends on the work function of each metal. Put $V_s = 0$ into Eq. (27.8) to get

$$0 = \left(-\frac{h}{e}\right)f + \left(\frac{\phi}{e}\right)$$

Solving for f, we get

$$f_{cutoff} = \frac{\phi}{h}$$

where f_{cutoff} is the cutoff frequency for a particular metal, the lowest frequency photon needed to eject an electron with no extra kinetic energy. The electron in that case is stopped with zero stopping potential $V_s = 0$. Measuring this cutoff frequency allows us to determine the work function of the metal. Or, if we know the work function, we can determine the cutoff frequency. We do this in Table 27.5 (on the next page) for the three metals shown in Example 27.6.

TABLE 27.5 **Cutoff frequencies for selected metals**

Metal (ϕ)	Cutoff frequency	Part of EM spectrum
Sodium (2.3 eV)	$f_{Na} = \dfrac{\phi_{Na}}{h} = \dfrac{(2.3\text{ eV})\left(\dfrac{1.6 \times 10^{-19}\text{ J}}{1\text{ eV}}\right)}{6.63 \times 10^{-34}\text{ J} \cdot \text{s}} = 5.6 \times 10^{14}\text{ Hz}$	Visible
Iron (4.7 eV)	$f_{Fe} = \dfrac{\phi_{Fe}}{h} = \dfrac{(4.7\text{ eV})\left(\dfrac{1.6 \times 10^{-19}\text{ J}}{1\text{ eV}}\right)}{6.63 \times 10^{-34}\text{ J} \cdot \text{s}} = 1.1 \times 10^{15}\text{ Hz}$	Ultraviolet
Platinum (6.4 eV)	$f_{Pt} = \dfrac{\phi_{Pt}}{h} = \dfrac{(6.4\text{ eV})\left(\dfrac{1.6 \times 10^{-19}\text{ J}}{1\text{ eV}}\right)}{6.63 \times 10^{-34}\text{ J} \cdot \text{s}} = 1.5 \times 10^{15}\text{ Hz}$	Ultraviolet

The results in Table 27.5 make sense because the x-intercept has a greater cutoff frequency for a metal with a larger work function.

REVIEW QUESTION 27.3 How does Einstein's hypothesis explain the cutoff frequency observed for a particular metal cathode in a photoelectric experiment?

27.4 Photons

In the first three decades of the 20th century, scientists faced a serious dilemma. The EM wave model of light explained a wide variety of phenomena, but not black body radiation and not the photoelectric effect. Explaining those required the construction of a new quantum model of light in which light was emitted and absorbed in discrete portions. Physicists started to think of light as composed of particle-like photons (the quanta of light). However, to explain interference and diffraction phenomena, photons had to have wave-like properties as well. The dual particle-wave properties of photons are summarized in Table 27.6.

TABLE 27.6 **Wave-like and particle-like properties of photons**

Experimental evidence	Wave-like properties f, λ	Particle-like properties $E = hf$
Double-slit interference	Superposition (including destructive interference)	
Diffraction	Single-slit interference and bending around obstacles	
Doppler effect	Change in frequency and wavelength	
Black body radiation		Photon (quanta) emission
Photoelectric effect		Photon (quanta) absorption

If we accept this model, photons must have both wave-like and particle-like behaviors. Experiments performed in 1932 tested the dual nature of photons in a dramatic way.

Low-intensity experiments support the hypothesis of the dual nature of photons

Experiments by Soviet physicists Sergei Vavilov and Eugeny Brumberg supported the hypothesis that photons have both particle and wave properties. Vavilov and Brumberg used a double-slit setup with light of extremely low intensity. Their goal was to reduce the number of photons that pass through the slits each second to make them easier to detect as individual photons. For detection, they used the extreme sensitivity of human eyes to low intensities of light. A person sitting in a dark room for extended periods of time develops increasing visual sensitivity to the extent that he or she can eventually see the individual photons hitting a screen that has been covered with a special material.

In Testing Experiment Table 27.7 we describe the experimental setup, including the predictions for the outcome made by the particle model of light, by the wave model, and by the dual wave-particle model.

TESTING EXPERIMENT TABLE 27.7 Do photons have both wave-like and particle-like properties?

Testing experiment	Prediction	Outcome
Use a light source with variable intensity. Shine the light on two narrow slits. Place a fluorescent screen beyond the slits that indicates where light hits the screen. 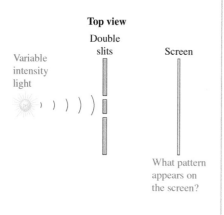	**Particle model:** Only two bright bands should appear—images of the slits themselves. As intensity decreases, we expect to see individual flashes caused by single photons at the slit image locations. High intensity Low intensity **Wave model:** Many alternating bright and dark bands should appear. As the intensity decreases, we expect all of the bright bands to become uniformly dimmer until they disappear. High intensity Low intensity **Dual wave-particle model:** Many alternating bright and dark bands should appear. As the intensity decreases, we expect to see individual flashes due to single photons at bright band locations only. At the locations of the dark bands no flashes should be seen. High intensity Low intensity	The experimental outcome is identical to the dual wave-particle model's prediction.

Conclusion

- Photons hit the screen, indicated by the flashes at low intensity.
- The photons only reach the screen at places where constructive wave interference occurs.
- Photons are simultaneously exhibiting both wave-like and particle-like behaviors.

Vavilov and Brumberg's experimental results were astounding—individual photons somehow "know" that they should only reach screen locations where coherent waves from the two slits would interfere constructively. It is as if each individual photon passes through both slits, interfering with itself and thereby producing a flash only at a location on the screen where constructive interference occurs.

Earlier in this chapter we constructed the idea of a photon as a single quantum of EM radiation. If it is a single quantum, then it seems impossible that it could pass through both slits simultaneously. However, Vavilov and Brumberg's experiment supports this seemingly strange idea.

Photon momentum

We have seen that photons participate in collisions with electrons inside metals (the photoelectric effect). This suggests that photons must have momentum. Otherwise, how could they knock an electron out of the cathode of a phototube during a collision? Because photons travel at light speed, they must be treated relativistically. The relativistic expression for the magnitude of the momentum of a particle of mass m moving at speed v is

$$p = \frac{mv}{\sqrt{1 - (v/c)^2}}$$

This expression has difficulties when applied to a photon. For a photon $v = c$, and the denominator becomes zero. In addition, if photons are quanta of electromagnetic radiation, then they would be composed of nothing but electric and magnetic fields. Their mass should be zero, and the numerator is also zero. So we get $p = 0/0$. That clearly doesn't work. Is there anything we can do to get a reasonable expression for the momentum of a photon?

We need an expression that does not have that square root factor in the denominator. Einstein resolved this by looking at the relativistic equations for the total energy and momentum of an object with nonzero mass:

$$E = \frac{mc^2}{\sqrt{1 - (v/c)^2}}$$

$$p = \frac{mv}{\sqrt{1 - (v/c)^2}}$$

Einstein eliminated v from these equations by solving for $(v/c)^2$ in the second equation and then substituting the result into the first equation. This takes a fair amount of algebra. The final result is

$$E = \sqrt{(pc)^2 + (mc^2)^2} \tag{27.9}$$

What we have here is a new equation for the total energy of an object. It is equivalent to

$$E = \frac{mc^2}{\sqrt{1 - (v/c)^2}}$$

but has an important new feature: the mass of the object can be set to zero without the equation misbehaving:

$$E = \sqrt{(pc)^2 + (0 \cdot c^2)^2} = \sqrt{(pc)^2} = pc$$

In other words, according to Eq. (27.9), the total energy of an object with zero mass (like a photon) is $E = pc$. But we already know from Einstein's explanation of the photoelectric effect that the energy of a photon is $E = hf = (hc/\lambda)$. Therefore,

$$\frac{hc}{\lambda} = pc$$

TIP Notice what happens when Eq. (27.9) is used to determine the total energy of an object at rest. Then $p = 0$ and $E = mc^2$. The total energy of an object at rest just equals its rest energy, $E_0 = mc^2$.

Solving for p, we get an expression for the momentum of a photon:

$$p = \frac{h}{\lambda} \qquad\qquad (27.10)$$

Following is an application of this expression.

QUANTITATIVE EXERCISE 27.7 **Revisiting $E_0 = mc^2$ with the Sun's changing mass**

Each second the Sun emits 3.8×10^{26} J of energy, mostly in the form of visible, infrared, and UV photons. Estimate the change in mass of the Sun as a result of the energy these photons carry away each second.

Represent mathematically Use the idea of energy conservation for the system that includes the Sun and the emitted photons:

$$E_{0i} = E_{0f} + E_{photons}$$

The rest energy of an object is $E_0 = mc^2$. Therefore,

$$m_i c^2 = m_f c^2 + E_{photons}$$

Solve and evaluate Solving for $\Delta m = m_f - m_i$:

$$m_i c^2 - m_f c^2 = E_{photons}$$
$$\Rightarrow m_f c^2 - m_i c^2 = -E_{photons}$$
$$\Rightarrow (m_f - m_i)c^2 = -E_{photons}$$
$$\Rightarrow \Delta m = -\frac{E_{photons}}{c^2}$$

Inserting the appropriate values gives

$$\Delta m = -\frac{3.8 \times 10^{26} \text{ J}}{(3.00 \times 10^8 \text{ m/s})^2} = -4.2 \times 10^9 \text{ kg}$$

The minus sign indicates that the Sun loses more than 4 billion kg each second! Won't the Sun "disappear" quickly while losing billions of kilograms each second?

The mass of the Sun is $m_{Sun} = 2.0 \times 10^{30}$ kg. How long will it take for the Sun to "disappear"?

$$\frac{2.0 \times 10^{30} \text{ kg}}{4.2 \times 10^9 \text{ kg/s}} = 4.8 \times 10^{20} \text{ s}$$

There are about 3.2×10^7 s in 1 year, which means the Sun will last

$$\frac{4.8 \times 10^{20} \text{ s}}{3.2 \times 10^7 \text{ s/year}} = 1.5 \times 10^{13} \text{ years}$$

The age of the universe (as we know from Chapter 26) is approximately 13.8×10^9 years, which means the Sun will last more than 1000 times the current age of the universe if it continues to radiate energy at this rate (assuming that the mechanism behind it can continue for that long).

Try it yourself Estimate the order of magnitude of the number of photons the Sun emits every second.

Answer Over 10^{45} photons/s are emitted from the Sun. Assumption: All photons have a frequency of 10^{14} Hz. If we make a different assumption, the estimated number will be different.

Our understanding of the momentum of photons can explain an important aspect of the photoelectric effect that we have so far overlooked. This aspect is described by the **quantum efficiency** of the photoelectric effect. Quantum efficiency is the number of electrons ejected per single incident photon. The efficiency depends on many factors. The most important is the work function of the metal—if the energy of the photon is less than the magnitude of the work function, the efficiency is zero. However, even if the energy of the photon is larger than the work function, it might still fail to eject an electron if the photon is reflected by the metal or absorbed by the lattice. Finally, due to the initial momentum of the incident photons being directed into the lattice, electrons that absorb the photons mostly move within the lattice, and only a few leave the metal.

REVIEW QUESTION 27.4 Explain how the outcome of the Vavilov-Brumberg experiment supports the idea that a photon has both wave-like and particle-like behaviors.

27.5 X-rays

The experiments and theory described in the previous section supported the idea that photons have momenta of magnitude $p = h/\lambda$. We will test this expression using X-rays in the next section. First, we need to learn about X-rays.

Cathode ray tubes

The story of X-rays is another example of how persistence and attention to detail often lead to groundbreaking discoveries. At the beginning of Section 27.2 we assumed that free electrons reside inside metals, and we used this idea to help explain the photo-electric effect. In this section we'll find how the electron was discovered and how its discovery led to the accidental discovery of X-rays.

In the middle of the 19th century, physicists did not know that metals consisted of a crystal lattice and free electrons; furthermore, they did not know that electric current in metals was caused by the movement of those free electrons.

At that time, some physicists started experimenting with what are called cathode ray (Crookes) tubes—evacuated glass tubes with two electrodes embedded inside (**Figure 27.10**). The electrodes were connected to a battery that produced a potential difference across the electrodes. The tubes were similar in some ways to the tube that von Lenard used to study the photoelectric effect. The difference was that the cathode (connected to the negative terminal of the battery) could be heated to high temperatures, rather than being illuminated by light.

Physicists discovered that although a physical gap was present between the cathode and the anode, when the cathode was heated, a current appeared in the circuit and the tube would glow. Because the tube's interior was a vacuum, they thought that the cathode must emit some kind of rays that traveled from the cathode to the anode. These rays were called *cathode rays*.

J. J. Thomson (1856–1940), at the Cavendish Laboratory at Cambridge University, was one of the scientists studying cathode rays. He observed that when the rays hit a metal target, the target became negatively charged. Through numerous experiments, Thomson found that the rays could be deflected by electric and magnetic fields as though they consisted of negatively charged particles. He determined their charge-to-mass ratio (q/m) and found it to be over 1000 times greater than that of a hydrogen ion (the modern value being $q/m = -1.76 \times 10^{11}$ C/kg), suggesting that the cathode rays either were very highly charged or had a very small mass. Later experiments by other scientists supported the latter hypothesis. The cathode rays' charge-to-mass ratio was also independent of the choice of cathode material. This led Thomson to conclude that there was only one type of cathode ray rather than many different types.

Let's model a cathode ray as a stream of charged particles and see if we can explain the behavior of the cathode ray tubes. Remember, the cathode in the tube is hot. This means that the particles inside the cathode have a large amount of kinetic energy of random motion (thermal energy) and could knock each other out of the cathode. Once outside the cathode, the electric field between the cathode and anode would cause them to accelerate toward the anode.

At about the same time, chemists were trying to find out what comprised electric currents. What was moving inside metals? They concluded that the charge carriers were tiny particles, and in 1894 George Johnstone Stoney called them "electrons." Experiments in electrochemistry helped determine the value of the smallest charge carried by these particles to be -1.6×10^{-19} C.

We can use the value of the charge of the electron found by Stoney and the charge-to-mass ratio found by Thomson to determine the mass of the electron:

$$m_e = \frac{1}{q_e/m_e} q_e = \frac{1}{-1.76 \times 10^{11}\,\text{C/kg}} (-1.6 \times 10^{-19}\,\text{C}) = 9.1 \times 10^{-31}\,\text{kg}$$

The mass of the electron is about 1/2000th the mass of a hydrogen atom.

The accidental discovery of X-rays

Experiments with cathode ray tubes led to the accidental discovery of X-rays. In 1895 Wilhelm Roentgen (1845–1923), like von Lenard, was experimenting with the rays outside the vacuum tubes by letting them pass through special windows

FIGURE 27.10 A cathode ray tube. (a) An actual cathode ray tube. (b) The electrical schematic of a cathode ray tube.

(a)

(b)

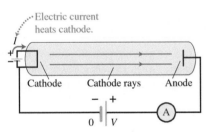

and observing them when they hit a fluorescent screen. The screen glowed when the cathode rays hit it. One night when Roentgen was ready to leave the lab, he covered the cathode ray tube with cardboard and turned off the lights. He noticed a faint glow on the fluorescent screen that was lying nearby on the lab table (**Figure 27.11a**). This was strange; the room was dark, and the cathode ray tube was covered with cardboard. Roentgen turned the lights back on and checked all of the equipment. He turned the tube off and turned off the lights again. The fluorescent screen was dark.

At this point most people would have breathed a sigh of relief and gone home. But Roentgen went back to the tube with the lights still off and turned it on. The glow on the screen appeared again even though the tube was completely covered with cardboard! When Roentgen placed his hand near the screen, he could see a strange shadow made by his hand. The shadow looked as though the flesh of his hand was completely transparent, and only his bones blocked whatever was making the screen glow.

Roentgen stayed in the lab all night trying to figure out what rays could possibly come out of the completely covered tube and make the fluorescent screen glow. He called these mysterious rays X-rays since at the time he couldn't explain what they were. Roentgen took the first X-ray image of a human on December 22, 1895; he photographed his wife's hand (Figure 27.11b). Although we now understand what X-rays are, they are still called X-rays.

Explanation of X-rays

Roentgen and others continued their experiments and found the following about X-rays:

1. X-rays are not deflected by either electric or magnetic fields.
2. X-rays do not cause the viewing screen to become electrically charged.
3. X-rays cause photographic paper to darken.
4. X-rays produce a diffraction pattern of dark and bright bands on the screen after passing through a single narrow slit.
5. X-rays can be polarized.
6. X-rays can go through many materials that are opaque to visible light.
7. X-rays can ionize gases.

Analysis of the single-slit diffraction patterns of the rays showed that the wavelengths of X-rays were about 10^{-10} to 10^{-11} m, a much smaller wavelength than any other EM radiation known at the time—much smaller than visible light. Given all of these characteristics, it is reasonable to hypothesize that X-rays are streams of very high energy photons. But how are these photons produced?

Producing X-ray photons

Consider **Figure 27.12**. The hot metal cathode is a source of negatively charged electrons. Due to their high random motion (the metal is very hot), the kinetic energy of some of the free electrons in the metal exceeds the metal's work function. These electrons escape the cathode and accelerate toward the anode because of the electric field within the evacuated cathode ray tube. (The electric field is produced by the negatively charged cathode and positively charged anode.) When the electrons reach the anode, they collide with it and stop abruptly. Recall that when a charged object accelerates (such as an electron stopping when it collides with the anode), it emits electromagnetic radiation (Chapter 25). The radiation emitted in this case might be in the form of X-ray photons. Let's estimate the energy of the photons produced by this mechanism and compare it to the energy of photons of visible light. We want to determine whether the energy of the electrons stopping in the cathode ray tube is enough to produce an X-ray photon.

FIGURE 27.11 Roentgen discovers X-rays. (a) Although the cathode ray tube was covered, the fluorescent screen still glowed in the dark. (b) Roentgen took an X-ray photo of his wife's hand.

(a)   High voltage on

Completely covered cathode ray tube

Light flashes

Fluorescent screen

(b)

FIGURE 27.12 Diagram of an X-ray tube. The collision of high-energy electrons with the anode produces X-ray photons.

Cathode becomes hot.

90,000 V

X-ray tube

Electron is ejected.

X-rays are emitted when electrons strike the anode.

EXAMPLE 27.8   X-ray photons

In a cathode ray tube used to produce X-rays for imaging in a hospital, the potential difference between the cathode and anode is 90.0 kV (90,000 V). If the X-rays are generated by electrons in the tube that stop abruptly when they collide with the anode and emit photons, what are the energy and wavelength of each photon?

Sketch and translate The process is sketched in three steps (states) at top right.

Simplify and diagram We consider the system to consist of a single electron, the electric field produced by the electrodes, the electrodes themselves, and a single X-ray photon. We represent each part of the process using a work-energy bar chart. The initial state shown below is just as the electron leaves the cathode. We assume that the electron has zero kinetic energy and the system has zero electric potential energy—that is, the electric potential at the cathode is zero.

The intermediate state is just before the fast-moving electron smashes into the anode. The final state is when the electron has stopped inside the anode and just after the X-ray photon has been emitted. We assume that none of the kinetic energy of the electron is converted into the internal energy of the cathode.

$$K_1 + U_{q1} + hf = K_2 + U_{q2} + hf = K_3 + U_{q3} + hf$$

Represent mathematically Use the first and second state bar charts to apply the generalized work-energy principle:

$$0 + 0 + 0 = \tfrac{1}{2}mv_2^2 + (-e)V$$

Use the second and third state bar charts to write

$$\tfrac{1}{2}mv_2^2 + (-e)V = (-e)V + hf$$

where $h = 6.63 \times 10^{-34}$ J·s is Planck's constant. Combining these two results, we get

$$0 = (-e)V + hf$$

Solve and evaluate The photon's energy is

$$hf = eV = (1.6 \times 10^{-19}\,\mathrm{C})(90.0 \times 10^3\,\mathrm{V})$$
$$= 90.0 \times 10^3\,\mathrm{eV}$$

This is much greater than the energy characteristic of visible light photons. The photon's frequency is

$$f = \frac{eV}{h} = \frac{(1.6 \times 10^{-19}\,\mathrm{C})(90.0 \times 10^3\,\mathrm{V})}{6.63 \times 10^{-34}\,\mathrm{J \cdot s}}$$
$$= 2.2 \times 10^{19}\,\mathrm{Hz}$$

which is much higher than visible light frequencies. The photon's wavelength is

$$\lambda = \frac{c}{f} = \frac{3.00 \times 10^8\,\mathrm{m/s}}{2.2 \times 10^{19}\,\mathrm{Hz}} = 1.4 \times 10^{-11}\,\mathrm{m}$$

Try it yourself What potential difference between the electrodes in a cathode ray tube would produce 10^{-10}-m wavelength X-rays?

Answer   12,000 V.

Based on the result of our calculation we can say that in a cathode ray tube, an electron can acquire enough kinetic energy to produce an X-ray photon when it stops.

Are X-rays dangerous?

Short-wavelength UV radiation, X-rays, and shorter wavelength (higher frequency) gamma ray photons are called ionizing radiation because each photon has enough energy to knock electrons out of atoms or to break bonds that hold atoms together in molecules. Ionizing radiation can damage genetic material (DNA), increasing the risk of mutations that lead to cancer. Fortunately, the body has potent DNA repair mechanisms, reducing the chance of serious harm from UV and X-ray exposure. However, people who work around ionizing radiation (dental hygienists and radiology technicians, for example) must take precautions to avoid overexposure.

QUANTITATIVE EXERCISE 27.9   Number of photons absorbed in a medical X-ray

Determine the number of photons absorbed during a single chest X-ray if the body absorbs a total of 10^{-3} J of 0.025-nm wavelength X-ray photons.

Represent mathematically The energy of a single X-ray photon is

$$E_1 = hf = \frac{hc}{\lambda}$$

The number of photons is then

$$N = \frac{E_{total}}{E_1}$$

Solve and evaluate Each photon has energy

$$E_1 = \frac{(6.63 \times 10^{-34}\ \text{J} \cdot \text{s})(3.00 \times 10^8\ \text{m/s})}{0.025 \times 10^{-9}\ \text{m}} = 8 \times 10^{-15}\ \text{J}$$

Thus, the number of photons absorbed in one chest X-ray is approximately

$$N = \frac{10^{-3}\ \text{J}}{8 \times 10^{-15}\ \text{J}} \approx 10^{11}\ \text{photons}$$

Try it yourself Suppose the wavelength of the X-rays was three times as long. Then approximately how many photons would pass through your body in a 10^{-3}-J X-ray exam?

Answer   About 3×10^{11} photons.

QUANTITATIVE EXERCISE 27.10   High-energy photons absorbed from environment

Your body is continuously exposed to natural radiation from sources such as radioactive materials in the environment and in the food you eat, and from cosmic rays coming from supernovae and other objects in the universe. Each year on average you absorb about 10^{-3} J of this radiation per kilogram of body mass. Estimate the number of these ionizing photons absorbed each second.

Represent mathematically Assume that your body mass $m = 70.0$ kg. The rate at which energy from this radiation is absorbed is $R = 10^{-3}$ (J/kg)/year. Assume for example that the radiation consists of $\lambda = 6 \times 10^{-13}$ m gamma ray photons each with energy $E_1 = 3 \times 10^{-13}$ J. The number N of photons absorbed in 1 year is then

$$N = \frac{Rm}{E_1}$$

We can then determine the number absorbed each second by dividing by the number of seconds in 1 year:

$$1\ \text{year} \left(\frac{365\ \text{days}}{1\ \text{year}}\right)\left(\frac{24\ \text{h}}{1\ \text{day}}\right)\left(\frac{3600\ \text{s}}{1\ \text{h}}\right) = 3.2 \times 10^7\ \text{s}$$

Solve and evaluate Inserting the appropriate values gives

$$N = \frac{(10^{-3}\ (\text{J/kg})/\text{year})(70.0\ \text{kg})}{3 \times 10^{-13}\ \text{J}}$$
$$= 2.3 \times 10^{11}\ \text{photons/year}$$

The number of photons absorbed each second is then

$$\frac{2.3 \times 10^{11}\ \text{photons/year}}{3.2 \times 10^7\ \text{s/year}} \approx 7 \times 10^3\ \text{photons/s}$$

As mentioned earlier, the body has ample repair mechanisms to take care of such radiation exposure.

Try it yourself Exposure to a 1.0-s burst of 0.010-nm gamma rays depositing 1.0 J of energy per kilogram into a human body has a 10% chance of being fatal within 30 days. This level of radiation is 10 billion times greater than the ambient natural radiation our bodies are continuously exposed to. How many gamma ray photons are absorbed by the body during this exposure? (Some workers present at the 1986 Chernobyl nuclear power plant disaster experienced as much as 16 times this exposure.)

Answer 3–4×10^{15} photons.

X-ray interference

We have mostly focused on the particle-like behavior of X-ray photons. However, X-rays can exhibit wave-like behavior if they can be made to interact with objects whose size is comparable to the wavelength of the X-rays. The spacing between atoms and ions in a crystal is the same order of magnitude as the wavelength of X-ray photons. When X-rays shine on a crystal, the X-rays reflected off the crystal lattice form a pattern very similar to that of visible light passing through an interference grating.

FIGURE 27.13 A photocell.

Photocell

Visible or UV photons knock electrons out of the emitter.

Electrons go from the emitter to the collector, causing a current.

Collector (anode)

Emitter (cathode)

FIGURE 27.14 A photoelectric smoke detector.

No smoke

Light ··· No smoke in tube

Light source

No light reaches the photocell— no alarm sounds.

Smoke

··· Smoke scatters light

Scattered light reaches the photocell— alarm sounds.

Alarm

FIGURE 27.15 Light shining light on silicon produces free electrons and "holes."

Effect of light on silicon

Photon knocks valence electron out of a silicon atom, leaving a hole ...

No light

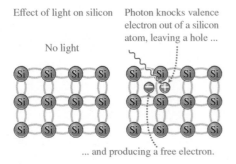

... and producing a free electron.

The details of the interference pattern allow scientists to determine the internal structure of the crystal. X-ray scattering can also be used to determine the structure of proteins and DNA. X-ray diffraction images produced by British biophysicist Rosalind Franklin played a key role in determining the helical structure of DNA in the early 1950s.

REVIEW QUESTION 27.5 What physical mechanism produces X-rays in a cathode ray tube?

27.6 Photocells, solar cells, and LEDs

The photoelectric effect has numerous applications. One of them is the **photocell**, which functions similarly to the tube that von Lenard used for his experiments. The cathode of the photocell (also called the *emitter*) is connected to the negative terminal of a battery and therefore is at a lower electric potential than the anode (called the *collector*). The cathode emitter is shaped like a cylindrical concave mirror to focus emitted electrons toward the anode, which has the shape of a thin wire that collects the electrons (**Figure 27.13**). When light of a frequency greater than the cutoff frequency shines on the emitter, electrons are ejected and absorbed by the collector, causing an electric current in the circuit. The magnitude of the current is a measure of the visible light or UV intensity being directed at the photocell. Products that use photocells include light meters (used in photography to adjust the size of a camera lens aperture), motion detectors, and photoelectric smoke detectors (**Figure 27.14**). In such a smoke detector, light from the source shoots across the top of the T-shaped detector. A cathode sits at the bottom of the T. If the air is clear, no light is scattered to the cathode and hence no photocurrent will be produced. However, smoke particles scatter light out of the beam, causing it to reach the cathode of the photocell and producing a photocurrent. When this photocurrent reaches a certain level, the alarm is triggered.

Solar cells

Solar cells, such as the rooftop panels used to generate electric current, are another application of the photoelectric effect. Solar cells are based on the semiconductor technology we discussed in an earlier chapter (Chapter 19). Semiconductors are materials that act as insulators under some conditions and conductors under others.

Silicon is a commonly used semiconductor. A silicon atom has four valence electrons that form covalent chemical bonds with neighboring atoms (each electron is shared between them). As a result, a silicon atom has no free electrons; therefore, silicon is an electric insulator. However, when silicon is heated or when light of sufficient frequency shines on it, some of the electrons gain enough energy to become free to roam around the crystal. The spaces vacated by these electrons lack negative charge and become positively charged "holes." These holes can be filled by other bound electrons that are not energetic enough to become free but can nevertheless move among the still-bound valence electrons. Both free electrons and holes can potentially contribute to a current if the silicon is placed in an external electric field. **Figure 27.15** shows this process— light photons deliver energy to the silicon, increasing its energy and allowing some of its valence electrons to become free. Whether electrons will be freed depends on the frequency of the light.

Adding impurities to silicon—a process called doping—is another way to turn it into a conductor. There are two types of impurities: electron donors and electron acceptors. Phosphorus, an example of an electron donor, contains five valence electrons, one more than silicon's four. The result is one "extra" electron that does not form a strong bond with a neighboring silicon atom and is essentially free to move about the bulk of

FIGURE 27.16 Doping of silicon. (a) Phosphorus is an n-type donor of free electrons and (b) boron is a p-type acceptor of bound electrons—a creator of "holes." (c) The contact of p- and n-type semiconductors causes charge transfer across the junction. (d) Charge transfer and recombination across the junction produces an internal electric field $\vec{E}_{int}$.

(a) n-type silicon when doped with phosphorus

Phosphorus has five valence electrons, four bonding to silicon atoms, leaving one free electron.

After the free electron drifts away, the phosphorus atom becomes a positive ion.

(b) p-type silicon when doped with boron

Boron has three valence electrons, three bonding to silicon atoms, leaving a hole at the unfilled bond.

After the hole drifts away, the boron atom becomes a negative ion.

We use square symbols to distinguish ions (that come from doped atoms) from Si atoms.

(c) p-n junction right after p and n are brought together

Boron ions   Phosphorus ions

Free electrons and holes move randomly. When they meet, they recombine. Those near the junction recombine first.

(d) p-n junction after some time, when equilibrium is reached

In the region with no electrons and holes, the remaining ions create the internal electric field that eventually prevents electrons and holes from entering the region.

the crystal (**Figure 27.16a**). The opposite is true for an electron acceptor. Boron, for example, has only three valence electrons. It can capture nearby electrons to form a strong fourth bond with the adjacent silicon atoms, creating extra holes (Figure 27.16b). These holes act as free positively charged particles (they lack the presence of a negatively charged electron). Silicon doped with phosphorus or any other electron donor is called an **n-type semiconductor**; the boron-doped version is called a **p-type semiconductor**. Note that these crystals are still electrically neutral, since they are still completely composed of neutral atoms.

When a p-type semiconductor and an n-type semiconductor are brought into contact, they form a **p-n junction** (Figure 27.16c). Free electrons in the n-type silicon migrate toward the p-type silicon, while the holes in the p-type silicon migrate toward the n-type silicon. When the free electrons and holes meet, they recombine. Therefore, in the n-type we have positively charged ions (the result of electrons leaving), and in the p-type we have negatively charged ions (the result of the disappearance of positively charged holes). In the region near the boundary, the n-type silicon becomes positively charged and the p-type silicon becomes negatively charged. This charge separation produces an electric field. As a result, an internal electric field appears in the boundary region (Figure 27.16d).

If we shine light of sufficient frequency on the p-n junction, both halves of the junction will absorb photons whose energy is transferred to valence electrons that then become free. These electrons leave behind holes—virtual positively charged particles (**Figure 27.17**). The electric field created by the p-n junction then separates the freed electrons and holes by accelerating them in opposite directions. The p-n junction in the presence of light results in an electric current—it behaves like a battery! If we now connect a lightbulb to the two ends of the p-n junction, we will observe a current in the circuit, and the bulb will light. Thus the energy of light photons is converted into electric energy.

FIGURE 27.17 The free electrons and the holes produced by absorbing the Sun's photons are separated by the electric field of the p-n junction. The p-n junction in the presence of light acts like a battery.

A silicon atom absorbs a photon, producing a hole and a free electron.

The electric field at the p-n junction causes the hole to move left and the electron right.

FIGURE 27.18 What is inside an LED?

VIDEO 27.1

"Sandwich"

FIGURE 27.19 A microscopic explanation of how an LED emits light.

Light-emitting diodes

The process described above is reversed in light-emitting diodes (LEDs), which you studied in Chapter 19. They also consist of two doped semiconductor materials that are joined together and form a p-n junction. However, the semiconductor material that is doped is not silicon; it is usually gallium arsenide (a semiconductor alloy). In contrast to solar cells, which transform light into electric energy, LEDs transform electric energy into light. Which part of the LED emits light, and how does the process work?

To answer the first question, we need to examine the inside of an LED. If we look at an LED with the naked eye, we do not see anything inside—whatever is there is too small. If we place an LED under a microscope, will we then see its interior? **Figure 27.18a** shows that a microscope does not help—the plastic dome refracts light and makes the interior hard to see. To get rid of refraction, we can submerge the LED in a liquid with the same index of refraction as the plastic (in this case, glycerin). Figure 27.18b shows the result. We can see an object inside a dome. If we increase the magnification, we see the details more clearly. Figure 27.18c shows the interior of the LED when it is off. We see two different materials in contact with each other that look like a sandwich: a thicker layer of one material covered with a thinner layer of another material. Figure 27.18d shows the same LED when the potential difference between its leads is just enough to make the LED glow. We see that light comes from the side of the "sandwich," where the materials join together. The top layer is the p-doped material and the bottom layer is n-doped. The video shows this process step by step.

As you know from the solar cells, when a p-n junction is formed, the free electrons and holes near the junction recombine, and the region near the junction becomes free of charge carriers (this region is called the *depletion region*). As a result, positive and negative ions (that came from the doped atoms) are the only charged particles in the depletion region—positive on the n side and negative on the p side. These ions create an $\vec{E}$ field that points from the n-type to the p-type. The more electrons and holes that recombine, the thicker the depletion region and the larger the magnitude of this $\vec{E}$ field. Because the forces due to this electric field push electrons back toward the n-type and holes toward the p-type of the semiconductor, the growth of the region eventually stops and the $\vec{E}$ field reaches its final value $\vec{E}_{int}$.

When an LED is connected to a battery as shown in **Figure 27.19a**, the external $\vec{E}_{ex}$ field points in the same direction as the $\vec{E}_{int}$ field of the junction and prevents electrons and holes from crossing the junction. There is no current through the p-n junction (note that in this figure the positive and negative ions are not shown, although the internal field is produced by the ions). However, when you connect the LED as shown in Figure 27.19b, the external $\vec{E}_{ex}$ field points in the opposite direction, and if its magnitude is large enough, it cancels the $\vec{E}_{int}$ field and accelerates free electrons toward the p-type semiconductor and holes toward the n-type semiconductor. When the free electrons and holes meet, they recombine, reducing the energy as they become bound electrons, not free. This excess energy is emitted as photons of light. For this process to happen, the magnitude of $\vec{E}_{ex}$ must be larger than the magnitude of $\vec{E}_{int}$; thus the potential difference across the LED created by the power supply must be larger than a certain value (the opening voltage) to make the LED glow. This explains the opening voltage of LEDs. Interestingly, if you do not connect an LED to a battery but instead shine light on it, it behaves like a solar cell and produces some voltage across its terminals.

Our knowledge of the internal structure of LEDs explains not only their electric properties but also the optical properties that we used in Chapters 2 and 3 when we attached a blinking LED to a moving object to keep track of the object's motion. The rapid switching on and off of an LED is possible because the processes inside LEDs (the movement and recombination of electric charges) happen almost instantly, unlike the warming up and cooling down of an incandescent lightbulb filament, which takes time. This is why incandescent bulbs are not used in processes where changes in brightness need to occur quickly.

REVIEW QUESTION 27.6 What characteristics would a material need in order to be used to make cathodes in photocells?

Summary

Stefan-Boltzmann's law for total radiating power The total black body radiation power P (all wavelengths) from an object with surface area A and kelvin temperature T is proportional to the fourth power of the temperature. (Section 27.1)

$$P = \sigma T^4 A \qquad \text{Eq. (27.1)}$$
$$\text{where } \sigma = 5.67 \times 10^{-8} \, \text{W/m}^2 \cdot \text{K}^4$$

Wien's law The wavelength λ_{max} at which a black body emits the maximum intensity radiation in a small wavelength interval depends on the temperature T in kelvins of the black body. (Section 27.1)

$$\lambda_{max} = \frac{2.90 \times 10^{-3} \, \text{m} \cdot \text{K}}{T} \qquad \text{Eq. (27.2)}$$

A **photon** is a discrete quantum of electromagnetic radiation that has both wave-like properties and particle-like properties. The energy of a photon depends on its frequency f and Planck's constant h. (Sections 27.3 and 27.4)

$$E_{photon} = hf \qquad \text{Eq. (27.7)}$$
$$\text{where } h = 6.63 \times 10^{-34} \, \text{J} \cdot \text{s}$$

Photoelectric effect A photon of energy hf hits a metal cathode of work function ϕ (the minimum energy needed to remove a free electron from the metal). The electron-lattice system absorbs the photon to free the electron from the metal. If the photon energy is greater than ϕ, the electron has nonzero kinetic energy when it leaves the metal. (Sections 27.2 and 27.3)

$$-\phi + hf = K_e \qquad \text{Eq. (27.5)}$$

Stopping potential $-V_s$ is the minimum negative electric potential that stops the maximum speed electron from reaching the anode. (This assumes that the potential at the cathode is zero.) (Section 27.2)

$$\tfrac{1}{2} m_e v_{e\,\text{i max}}^2 = (-e)(-V_s)$$

where V_s is the positive magnitude of the stopping potential

---

Cutoff frequency f_{cutoff} is the minimum frequency that a photon must have to remove an electron from the metal. (Sections 27.2 and 27.3)

$$f_{cutoff} = \frac{\phi}{h} \qquad \text{Eq. (27.6)}$$

Photon momentum The magnitude of a photon's momentum p_{photon} depends on its frequency f (or wavelength λ), Planck's constant h, and the speed of light c. (Section 27.4)

$$p_{photon} = \frac{h}{\lambda} = \frac{hf}{c} \qquad \text{Eq. (27.10)}$$

Questions

Multiple Choice Questions

1. What is a black body? (a) An object painted with black, non-reflective paint (b) An object that does not radiate any electromagnetic waves (c) A model of an object that emits electromagnetic waves with a characteristic frequency distribution

2. If you triple the temperature of a black body (in kelvins), the total amount of energy that it radiates increases how many times? (a) 3 times (b) 9 times (c) 27 times (d) 81 times

3. If you triple the temperature of a black body, the wavelength at which it emits the maximum amount of energy (a) increases 3 times. (b) increases 9 times. (c) increases 27 times. (d) increases 81 times. (e) decreases 3 times.

4. Among the items listed below, choose all of the photoelectric effect observations that could not be explained by the wave model of light. (a) Photocurrent appears instantly after the light is turned on. (b) The magnitude of the photocurrent is directly proportional to the intensity of light. (c) For each intensity of light there is a maximum value of the photocurrent. (d) The stopping potential difference does not depend on the intensity of light. (e) The cutoff frequency does not depend on the intensity of light.

5. What is the work function equal to? (a) The positive work one needs to do to remove an electron from a metal (b) The positive potential energy of interaction of an electron and a metal (c) The negative potential energy of interaction of an electron and a metal (d) a and b (e) a and c

6. What is a photon? (a) A physical quantity (b) A phenomenon (c) A model (d) A law of physics

7. What did the Vavilov-Brumberg experiments show? (a) Photons behave like waves. (b) Photons behave like particles. (c) Photons simultaneously behave like waves and like particles.

8. Which statement describes most accurately our present understanding of light? (a) Light is like a stream of particles. (b) Light is like an electromagnetic wave. (c) Light is a complex phenomenon that cannot be modeled by only one of the above models.

9. Which graph in **Figure Q27.9** correctly shows the dependence of photocurrent on the frequency of EM radiation that is incident on the cathode of a photocell, assuming that the intensity of the EM radiation and the anode potential are kept constant? Assume that quantum efficiency is independent of the frequency of light.

FIGURE Q27.9

10. In which of the following cases does an electron emit EM radiation? (a) A fast-moving electron hits metal and slows down. (b) An electron speeds up from rest in a uniform electric field. (c) An electron enters a uniform magnetic field; the initial velocity of the electron is perpendicular to the field. (d) An electron enters a uniform magnetic field; the initial velocity of the electron is parallel to the field.

11. Which of the following options correctly describes how an increase in the frequency of light above the cutoff frequency affects the number of electrons ejected per unit time from a cathode when the intensity of light remains constant? Assume that each photon ejects one electron from the cathode independent of photon frequency as long as its frequency is larger than the cutoff frequency. (a) The number of electrons remains the same because one photon can only eject one electron. (b) The number of electrons increases because each photon can be absorbed by two neighboring electrons. (c) The number of electrons increases because a photon can collide with an electron, transfer a part of its energy to the electron as it scatters, and then collide with other surrounding electrons. As long as the photon has enough energy, it will eject an electron at each collision. (d) The number of electrons decreases because we need fewer photons with larger frequency to maintain the same intensity of the light. (e) The number of electrons decreases because photons with a higher frequency have a lower probability of colliding with electrons than low-frequency photons.

12. A light source is shining light on a cathode in a phototube. The potential of the cathode is 0 V, and the potential of the anode is 10 V. The current through the phototube is zero. Which of the following changes could produce a current through the phototube due to photoelectrons? (a) Use more light sources aimed at the same position on the cathode. (b) Increase the anode potential. (c) Decrease the anode potential. (d) Replace the metal of the cathode and the anode with a metal with a smaller work function. (e) Replace the metal of the cathode and the anode with a metal with a larger work function. (f) Use light with a smaller wavelength. (g) Use light with a larger wavelength. (h) Decrease the distance between the cathode and the anode. (i) Increase the distance between the cathode and the anode.

Conceptual Questions

13. How do we know that photons possess wave-like properties?

14. How do we know that photons possess particle-like properties?

15. What is the difference between the photon model of light and the wave model of light?

16. What is the difference between the photon model of light and the particle model of light?

17. The Sun and other celestial objects emit X-rays. Why are we, on the Earth's surface, not concerned about them?

18. When photographic film is being developed, technicians work in a darkroom in which a special red light is used for illumination, not a regular white light. Explain why.

19. Why is the photon model of light supported by the fact that light below a certain frequency, when striking a metal plate, does not cause electrons to be freed from the metal?

20. Explain how we know that cathode rays are low-mass negatively charged particles. Draw pictures and field representations to illustrate your explanation.

Problems

Below, **BIO** indicates a problem with a biological or medical focus. Problems labeled **EST** ask you to estimate the answer to a quantitative problem rather than derive a specific answer. Asterisks indicate the level of difficulty of the problem. Problems with no * are considered to be the least difficult. A single * marks moderately difficult problems. Two ** indicate more difficult problems.

27.1 Black body radiation

1. **Wavelength of radiation from a person** If a person could be modeled as a black body, at what wavelength would his or her surface emit the maximum energy?

2. * (a) A surface at 27 °C emits radiation at a rate of 100 W. At what rate does an identical surface at 54 °C emit radiation? (b) Determine the wavelength of the maximum amount of radiation emitted by each surface.

3. **Maximum radiation wavelength from star, Sun, and Earth** Determine the wavelengths for the following black body radiation sources where they emit the most energy: (a) A blue-white star at 40,000 K; (b) the Sun at 6000 K; and (c) Earth at about 300 K.

4. * **EST** **Star colors and radiation frequency** The colors of the stars in the sky range from red to blue. Assuming that the color indicates the frequency at which the star radiates the maximum amount of electromagnetic energy, estimate the surface temperature of red, yellow, white, and blue stars. What assumptions do you need to make about white stars to estimate the surface temperature?

5. * **EST** Estimate the surface area of a 60-watt lightbulb filament. Assume that the surface temperature of the filament when it is plugged into an outlet of 120 V is about 3000 K and the power rating of the bulb is the electric energy it consumes (not what it radiates). Incandescent lightbulbs usually radiate in visible light about 10% of the electric energy that they consume.

6. * **EST** **Photon emission rate from human skin** Estimate the number of photons emitted per second from 1.0 cm^2 of a person's skin if a typical emitted photon has a wavelength of 10,000 nm.

7. * **Balancing Earth radiation absorption and emission** Compare the average power that the surface of Earth facing the Sun receives from it to the energy that Earth emits over its entire surface due to it being a warm object. Assume that the average temperature of Earth's surface is about 15 °C. The distance between Earth and the Sun is about 1.5×10^{11} m.

27.2 and 27.3 Photoelectric effect and Quantum model explanation of the photoelectric effect

8. (a) Explain how you convert energy in joules into energy in electron volts. (b) The kinetic energy of an electron is 2.30 eV. What is its kinetic energy in joules?

9. Determine the average kinetic energy of an atom in an ideal gas at room temperature (20 °C) and convert the result into electron volts. Compare the result with the energy of a photon of visible light. Indicate any assumptions that you made.

10. * The stopping potential for an ejected photoelectron is −0.50 V. What is the maximum kinetic energy of the electron ejected by the light?

11. Light shines on a cathode and ejects electrons. Draw an energy bar chart describing this process. Explain why the frequency of incident light determines whether the electrons will be ejected or not.

12. * What is the cutoff frequency of light if the cathode in a photoelectric tube is made of iron?

13. The work function of cesium is 2.1 eV. (a) Determine the lowest frequency photon that can eject an electron from cesium. (b) Determine the maximum possible kinetic energy in electron volts of a photoelectron ejected from the cesium that absorbs a 400-nm photon.

14. * Visible light shines on the metal surface of a phototube having a work function of 1.30 eV. The maximum kinetic energy of the electrons leaving the surface is 0.92 eV. Determine the light's wavelength.

15. * **Equation Jeopardy 1** Solve for the unknown quantity in the equation below and write a problem for which the equation could be a solution.

$$-3.9 \text{ eV} + f(6.63 \times 10^{-34} \text{ J} \cdot \text{s})\left(\frac{1 \text{ eV}}{1.6 \times 10^{-19} \text{ J}}\right) = (-e)(-1.0 \text{ V})$$

16. **Camera film exposure** In an old-fashioned camera, the film becomes exposed when light striking it initiates a complex chemical reaction. A particular type of film does not become exposed if struck by light of wavelength longer than 670 nm. Determine the minimum energy in electron volts needed to initiate the chemical reaction.

27.4 Photons

17. **Breaking a molecular bond** Suppose the bond in a molecule is broken by photons of energy 5.0 eV. Determine the frequency and wavelength of these photons and the region of the electromagnetic spectrum in which they are located.

18. * A 1.0-eV photon's wavelength is 1240 nm. Use a ratio technique to determine the wavelength of a 5.0-eV photon.

19. **BIO** **Tanning bed** In a tanning bed, exposure to photons of wavelength 300 nm or less can do considerable damage. Determine the lowest energy in electron volts of such photons.

20. * Determine the number of 650-nm photons that together have energy equal to the rest energy of an electron. (Hint: See Section 26.9.)

21. * **EST** You create a light pulse by switching on a green laser ($\lambda = 532$ nm) for 1 μs. The pulse contains 10^{10} photons. (a) Determine the power output of the laser. (b) Estimate the distance between two neighboring photons, assuming the photons travel as localized particles and the light pulse has uniform intensity.

22. ** **BIO** **Laser surgery** Scientists studying the use of lasers in various surgeries have found that very short 10^{-12}-s laser pulses of power 10^{12} W with 65 pulses every 200×10^{-6} s produced much cleaner welds and ablations (removals of body tissues) than longer laser pulses. Determine the number of 10.6-μm photons in one pulse and the average power during the 65 pulses delivered in 200×10^{-6} s.

23. * A laser beam of power P in watts consists of photons of wavelength λ in nanometers. Determine in terms of these quantities the number of photons passing a cross section along the beam's path each second.

24. What is the total momentum of the photons passing the cross section every second in the previous problem?

25. * **BIO** **Pulsed laser replaces dental drills** A laser used for many applications of hard surface dental work emits 2780-nm wavelength pulses of variable energy (0–300 mJ) about 20 times per second. Determine the number of photons in one 100-mJ pulse and the average power of these photons during 1 s.

26. ** **Light hitting Earth** The intensity of light reaching Earth is about 1400 W/m^2. Determine the number of photons reaching a 1.0-m^2 area each second. What assumptions did you make?

27. * **EST** **Lightbulb** Roughly 10% of the power of a 100-watt incandescent lightbulb is emitted as visible light, the rest being emitted as infrared radiation. Estimate the number of photons of light coming from a bulb each second. What assumptions did you make? How will the answer change if the assumptions are not valid?

28. * **BIO** **Human vision sensitivity** To see an object with the unaided eye, the light intensity coming to the eye must be about 5×10^{-12} J/m$^2 \cdot$ s or greater. Determine the minimum number of photons that must enter the eye's pupil each second in order for an object to be seen. Assume that the pupil's radius is 0.20 cm and the wavelength of the light is 550 nm.

29. * A laser produces a short pulse of light whose energy equals 0.20 J. The wavelength of the light is 694 nm. (a) How many photons are produced? (b) Determine the total momentum of the emitted light pulse.

30. * **Levitation with light** Light from a relatively powerful laser can lift and support glass spheres that are 20.0×10^{-6} m in diameter (about the size of a body cell). Explain how that is possible.

27.5 X-rays

31. * Explain how a cathode ray tube works. Draw a picture and an electric circuit. Label the important elements and explain how they work together to produce cathode rays.

32. * An X-ray tube emits photons of frequency 1.33×10^{19} Hz or less. (a) Explain how the tube creates the X-ray photons. (b) Determine the potential difference across the X-ray tube.

33. * Electrons are accelerated across a 40,000-V potential difference. (a) Explain why X-rays are created when the electrons crash into the anode of the X-ray tube. (b) Determine the frequency and wavelength of the maximum-energy photons created.

34. * A small 1.0×10^{-5}-g piece of dust falls in Earth's gravitational field. Determine the distance it must fall so that the change in gravitational potential energy of the dust-Earth system equals the energy of a 0.10-nm X-ray photon.

35. **BIO** **X-ray exam** While being X-rayed, a person absorbs 3.2×10^{-3} J of energy. Determine the number of 40,000-eV X-ray photons absorbed during the exam.

36. * **BIO** **Body cell X-ray** (a) A body cell of 1.0×10^{-5}-m radius absorbs 4.2×10^{-14} J of X-ray radiation. If the energy needed to produce one positively charged ion is 100 eV, how many positive ions are produced in the cell? (b) How many ions are formed in the 3.0×10^{-6}-m-radius nucleus of that cell (the place where the genetic information is stored)? Indicate any assumptions that you made.

27.6 Photocells, solar cells, and LEDs

37. * **BIO** **EST** **Light detection by human eye** The dark-adapted eye can supposedly detect one photon of light of wavelength 500 nm. Suppose that 100 such photons enter the eye each second. Estimate the intensity of light in W/m^2. Assume that the diameter of the eye's pupil is 0.50 cm.

38. * Light of wavelength 430 nm strikes a metal surface, releasing electrons with kinetic energy equal to 0.58 eV or less. Determine the metal's work function.

39. * **Solar cell** A 0.20-m × 0.20-m photovoltaic solar cell is irradiated with 800 W/m^2 sunlight of wavelength 500 nm. (a) Determine the number of photons hitting the cell each second. (b) Determine the maximum possible electric current that could be produced. (c) Explain how a solar cell converts the energy of sunlight into electric energy.

40. * Mark and Kat are discussing doped semiconductors. Mark says that p-doped semiconductors are positively charged and n-doped semiconductors are negatively charged. Kat disagrees; she says that both types are neutral. Discuss the reasoning that could lead each of them to their opinions. Who do you agree with? Explain.

41. * Figure Q27.41 shows the sketch of the processes in a solar cell. Evaluate the sketch and make suggestions for improvements.

FIGURE P27.41

General Problems

42. * **EST** Estimate the temperature of the Sun's surface knowing the solar constant (1.37 kW/m^2). Decide what other data you need to make the estimate.

43. * If you shine UV radiation on the positively charged zinc sphere on top of an electroscope, the leaves of the electroscope stay deflected and do not move. Lisa and Shumaila are trying to explain the outcome of this experiment.
Lisa: "The positive charges increase the work function of the zinc. As a result, the photons of UV light do not have enough energy to knock electrons from the zinc sphere."
Shumaila: "UV light knocks electrons from the sphere, but as soon as they leave the positively charged sphere they slow down and return to the sphere." Who do you think is correct? Propose a testing experiment whose outcome might reject one of the explanations. Predict the outcome of your testing experiment based on each student's hypothesis.

44. A light source consists of a green (530 nm) and a blue (430 nm) LED. Both LEDs emit light of the same intensity. Of 10^3 photons emitted by the light source, (a) how many are photons of green light and (b) how many are photons of blue light?

45. * A light source consists of a green (530 nm) and a blue (430 nm) LED. The intensity of light emitted by the green LED is two times larger than the intensity of the light emitted by the blue LED. The light from this source is incident on a cathode with work function 2.2 eV. (a) Determine the stopping potential between the anode and the cathode that will stop the current through the phototube. (b) When the potential of the anode is set to +10.0 V with respect to the cathode and the cathode is illuminated with the light source, you measure a photocurrent of 5.0 μA. Determine the photocurrent if the green LED in the light source is turned off. Assume that the number of photons required to eject an electron from the cathode does not depend on the frequency of light as long as the frequency is larger than the cutoff frequency, and that the cathode and the anode are made of the same metal.

46. * **BIO** **EST** **Fireflies** Fireflies emit light of wavelengths from 510 nm to 670 nm. They are about 90% efficient at converting chemical energy into light (compared to about 10% for an incandescent lightbulb). Most living organisms, including fireflies, use adenosine triphosphate (ATP) as an energy molecule. Estimate the number of ATP molecules a firefly would use at 0.5 eV per molecule to produce one photon of 590-nm wavelength if all the energy came from ATP.

47. * **BIO** **EST** **Owl night vision** Owls can detect light of intensity 5×10^{-13} W/m^2. Estimate the minimum number of photons an owl can detect. Indicate any assumptions you used in making the estimate.

48. ** **BIO** **Photosynthesis efficiency** During photosynthesis in a certain plant, eight photons of 670-nm wavelength can cause the following reaction: $6CO_2 + 6H_2O \rightarrow C_6H_{12}O_6 + 6O_2$. During respiration, when the plant metabolizes sugar, the reverse reaction releases 4.9 eV of energy per CO_2 molecule. Determine the ratio of the energy released (respiration) to the energy absorbed (photosynthesis), a measure of photosynthetic efficiency.

49. * Suppose that light of intensity 1.0×10^{-2} W/m^2 is made of waves rather than photons and that the waves strike a sodium surface with a work function of 2.2 eV. (a) Determine the power in watts incident on the area of a single sodium atom at the metal's surface (the radius of a sodium atom is approximately 1.7×10^{-10} m). (b) How long will it take for an electron in the sodium to accumulate enough energy to escape the surface, assuming it collects all light incident on the atom?

50. * **Force of light on mirror** A beam of light of wavelength 560 nm is reflected perpendicularly from a mirror. Determine the average force that the light exerts on the mirror when 10^{20} photons hit the mirror each second. (Hint: Refer to the impulse-momentum equation (Chapter 6). You may assume that the magnitude of the photons' momenta is unchanged by the collision, but their directions are reversed.)

51. * **EST** **Sirius radiation power** Sirius, a star in the constellation of Canis Major, is the second brightest star of the northern sky (the brightest is the Sun). Its surface temperature is 9880 K and its radius is 1.75 times greater than the radius of the Sun. Estimate the energy that Sirius emits every second from its surface and compare this energy to the energy that the Sun emits. The radius of the Sun is about 7.0×10^8 m and the energy emitted per second is about 3.9×10^{26} W.

52. ** **Force of sunlight on Earth** We wish to determine the net force on Earth caused by the absorption of light from the Sun. (a) Determine the net area of the surface of Earth exposed to sunlight (Earth's radius is 6.38×10^6 m). (b) The solar radiation intensity is 1400 $J/s \cdot m^2$. Determine the momentum of photons hitting Earth's surface each second. (c) Use the impulse-momentum equation to determine the average force of this radiation on Earth.

53. ** **EST** **Levitating a person** Suppose that we wish to support a 70-kg person by levitating the person on a beam of light. (a) If all of the photons striking the person's bottom surface are absorbed, what must be the power of the light beam, which is made of 500-nm-wavelength photons? (b) Estimate the person's temperature change in 1 s.

54. ** You use an incandescent lightbulb to shine light onto a cesium cathode in a phototube. How will increasing the light intensity affect the kinetic energy of photoelectrons? Explain. Assume the quantum efficiency is independent of the frequency of light.

Reading Passage Problems

BIO **Capturing energy from sunlight—photosynthesis** Green plants capture and store the energy of photons from the Sun to help build complex molecules such as glucose. The process starts with the photoelectric effect. Chloroplasts in plant cells contain many photosynthetic units, each of which has about 300 pigment "antenna" molecules that absorb sunlight (Figure 27.20).

Suppose that one of the antenna molecules absorbs a photon. The electrons in the molecule are now in an excited state, which means that they are temporarily storing the extra energy gained from the photon. Typically, such a molecule reemits a photon and returns to its original energy state in about 10^{-8} s. If the energy of the photon is simply released, the plant has lost the Sun's energy.

In the photosynthetic units, the antenna molecules are linked closely to each other. When a photon is absorbed by one antenna molecule, its excitation energy is transferred to a neighbor antenna molecule. This excited neighbor in turn excites

FIGURE 27.20 Photosynthetic unit inside plant chloroplasts. "Antenna" molecules absorb photons and become excited. The energy of this excitation is passed from one antenna to another until it arrives at an "acceptor" molecule. The acceptor passes the energy to an electron transport chain, where it is captured and stored in biomolecules like ATP.

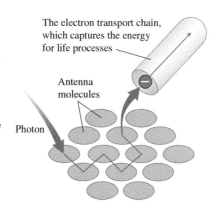

The electron transport chain, which captures the energy for life processes

Antenna molecules

Photon

another of its neighbors, in a more or less random fashion. Eventually, an electron in an excited antenna molecule is transferred to a primary acceptor, carrying the extra energy of the photon along with it. This energetic electron now passes through a complex electron transport chain. At several places along this pathway, the energy is given up to help form stable chemical bonds (Figure 27.21).

FIGURE 27.21 The electron transport chain. Electrons are passed between intermediate molecules, and energy is captured at certain steps for the ultimate production of ATP, the cellular energy molecule.

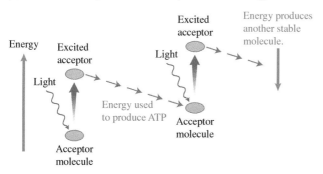

Energy

Excited acceptor

Light

Acceptor molecule

Energy used to produce ATP

Excited acceptor

Light

Acceptor molecule

Energy produces another stable molecule.

Life on Earth depends on the absorption of the photon, the "random walk" of the excitation energy from antenna molecule to antenna molecule, and the capture of the excited electron energy, all in a fraction of a second.

55. What is the number of antenna molecules that can absorb light in a photosynthetic unit?
(a) 1   (b) About 10   (c) Over 100
(d) Over 10,000   (e) About 10^8

56. Suppose an antenna molecule absorbs a 430-nm photon and that this energy is transferred directly to the acceptor molecule. Which answer below is closest to the energy that the photoelectron brings to the electron transport chain?
(a) 1 eV   (b) 2 eV   (c) 3 eV   (d) 4 eV   (e) 5 eV

57. Suppose that the excited energy of one antenna molecule is transferred 100 times between neighboring antenna molecules. Which answer below is closest to the maximum time interval for the transfer between neighboring antenna molecules?
(a) 10^{-3} s   (b) 10^{-6} s   (c) 10^{-8} s   (d) 10^{-10} s   (e) 10^{-14} s

58. Suppose that neighboring antenna molecules are separated by about 10^{-10} m and that the excitation energy is transferred 100 times between neighboring antenna molecules before the photoelectric transfer of an electron to the electron transport chain. Which answer below is closest to the minimum speed of the excitation energy through the photosynthetic unit?
(a) 1 m/s   (b) 10^2 m/s   (c) 10^4 m/s
(d) 10^{-2} m/s   (e) 10^{-4} m/s

59. The high-energy electron that transfers into an electron transport chain from a photosynthetic unit
(a) comes from the antenna molecule that absorbed the photon.
(b) comes from the acceptor molecule, which absorbed the photon.
(c) comes from the acceptor molecule that is excited by a nearby antenna molecule.
(d) is produced by the oxidation of a water molecule.
(e) is produced in the electron transport chain as other molecules react.

BIO **Radiation from our bodies** A person's body can be modeled as a black body that radiates electromagnetic radiation. Wien's law gives the peak wavelength of this radiation: $\lambda_{max} = (2.90 \times 10^{-3} \text{ m} \cdot \text{K}/T)$, which is mostly infrared radiation.

The net radiated power is the difference between the power emitted and the power absorbed ($P_{net} = P_{emit} - P_{absorb}$) and is given by Stefan-Boltzmann's law: $P_{net} = A\sigma\epsilon(T^4 - T_0^4)$, where $\sigma = 5.67 \times 10^{-8}$ W/m$^2 \cdot$ K^4, T is the absolute temperature of the skin, and T_0 is the absolute temperature of the surroundings. The surface area A of a human body is about 2 m^2, and the emissivity ϵ of the skin is about 1 (it is an almost perfect emitter and perfect absorber of infrared radiation). The skin temperature for a nude person is about 306 K and the temperature in a room is about 20 °C (293 K). Thus, for a nude person, $P_{net} \approx 160$ W. For a clothed person, the effective skin temperature is cooler, about 28 °C (301 K). The net radiated power is then $P_{net} \approx 100$ W.

60. The wavelength of maximum light emission from the body is closest to
(a) 500 nm.   (b) 700 nm.   (c) 1200 nm.
(d) 4500 nm.   (e) 9500 nm.

61. During one day, the total radiative energy loss by a clothed person having a 2 m^2 surface area in a 20 °C room in kcal (1 kcal = 4180 J) is closest to
(a) 0.5 kcal.   (b) 100 kcal.   (c) 400 kcal.
(d) 2000 kcal.   (e) 3000 kcal.

62. Photographs taken with a regular camera and with an infrared camera are shown in Figure Q27.62. The man's arm is covered with a black plastic bag. Why is his arm visible in the infrared picture?
(a) Light does not pass through black plastic, but infrared radiation does.
(b) His arm is very warm under the black plastic bag and emits much more infrared radiation.
(c) The bag temperature is similar to his arm temperature.
(d) The black bag absorbs light and becomes warm and is a good thermal emitter.
(e) None of the above

FIGURE P27.62

Visible light photo

Infrared photo

63. The man's glasses appear clear with the regular camera photo and black with the infrared camera photo in Figure P27.62. Why?
(a) Light does not pass through glass, but infrared radiation does.
(b) Light passes through glass, but infrared radiation does not.
(c) The lenses of the glasses are cool compared to the man's face, and thus they emit little infrared radiation.
(d) a and c   (e) b and c

64. What is the ratio of the emitted radiative power from a 310 K surface and the same surface at 300 K?
(a) 0.86   (b) 0.97   (c) 1.03   (d) 1.07   (e) 1.14

Atomic Physics

- How do scientists determine the chemical composition of stars?

- How do lasers work?

- Why is it impossible to know exactly where an electron is?

The photo above shows a spectrum of the Sun obtained using a spectrometer. It looks like a continuous spectrum with lots of dark lines. The dark lines mean that little light of those wavelengths reaches Earth. What causes these dark lines in the spectrum, and how do scientists use them to identify chemical elements present in the Sun's atmosphere as well as to study the Sun's magnetic field?

BE SURE YOU KNOW HOW TO:

- Write an expression for the rotational momentum of a point-like object of mass m moving at constant speed v in a circular orbit of radius r (Section 9.4).

- Apply Coulomb's law for the interaction between charged point-like objects (Section 17.4).

- Explain how a spectrometer works (Section 24.3).

PREVIOUSLY, you learned about cathode ray experiments that led to the discovery of the electron, a very low mass, negatively charged particle. However, most forms of matter we encounter on Earth are electrically neutral, which implies that there must be a positively charged component as well. The relationship between these positively charged components and electrons is the subject of this chapter.

28.1 Early atomic models

The ancient Greek philosopher Democritus proposed that "atoms" were the smallest indivisible components of matter. Indeed, much of what we now know about matter supports some of Democritus's claims. From experiments involving evaporation, we have learned that substances are made of individual point-like objects that move randomly and are surrounded by empty space. We also have learned that both dielectrics and conductors contain electrically charged particles. These particles move freely inside conductors and only a little inside dielectrics. How did these ideas motivate early models of the atomic nature of matter? By the end of the 19th century scientists had several hypotheses.

Dalton model: atom as billiard ball

In 1803 John Dalton proposed a billiard ball model of the atom as a small solid sphere. According to Dalton, a sample of a pure element was composed of a large number of atoms of a single kind. Compounds were substances that contained more than one kind of atom in specific integer ratios. In this model, each atom, regardless of type, contains an equal amount of positively and negatively charged subcomponents so that each atom is electrically neutral.

Thomson model: atom as plum pudding

In 1897 J. J. Thomson's experiments supported the model that cathode rays were electrons. Because cathodes made of different materials emitted the same electrons, Thomson hypothesized that these electrons were the negatively charged components of atoms.

According to his model, an atom is similar to a spherically shaped plum pudding. The plums represent electrons embedded in a massive sphere of diffuse positive charge represented by the pudding (**Figure 28.1**). This model explained the electrical neutrality of an atom and provided a qualitative explanation of how atoms could emit light by assuming that electrons could absorb and emit light through oscillations. At the time, it looked like a perfect model.

FIGURE 28.1 The "plum pudding" model of the atom.

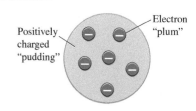

Rutherford model: atom as planetary system

Alpha particles are emitted by uranium and other radioactive elements. These particles have a mass four times greater than a hydrogen atom and a charge twice that of an electron, but positive, not negative. According to the Thomson model, alpha particles beamed at a thin layer of metal foil should pass nearly straight through the foil because the atoms are soft and their mass and positive charge are evenly distributed. In 1909, Ernest Rutherford, a New Zealand-born British physicist, noticed that such a beam came out of the foil much wider than it had been when it entered. Thomson's model could not explain this phenomenon.

The atomic nucleus Rutherford and his postdoctoral colleague Hans Geiger hypothesized that the atom's mass and positive charge were not spread uniformly throughout the atom but were instead concentrated in a small region. If this were true, then the beam should spread as observed. This new model also predicted that the alpha particles that come very close to the small dense region inside the atom should recoil backward at very large angles, possibly even in the reverse direction. However, Rutherford and Geiger did not believe that they would find this recoil. After all, how could an atom cause a very fast-moving alpha particle to almost instantaneously reverse direction? The idea seemed absurd. Geiger and Ernest Marsden set out to perform a testing experiment to disprove this hypothesis. In Testing Experiment Table 28.1 (on the next page) we look at their experiment.

TESTING
EXPERIMENT TABLE 28.1   Structure of the atom

Testing experiment	Prediction	Outcome
The experimental apparatus is shown below. 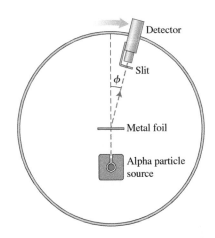	**Thomson model:** If the atom's mass and positive charge are dispersed evenly, the alpha particles will only undergo small deflections. 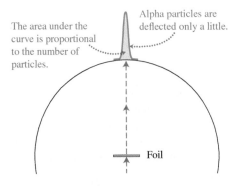 **Rutherford model:** If the atom's mass and positive charge are concentrated in a small region, then most of the alpha particles will pass straight through. But those that pass near the region of concentrated mass will recoil at larger angles, resulting in a wider beam. Some will recoil at very large angles. 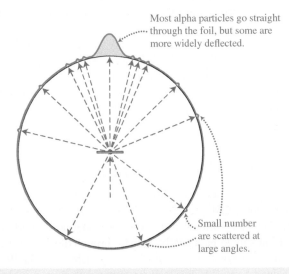	Geiger and Marsden recorded a significant widening of the beam and a small number of alpha particles recoiling at large angles.

Conclusion

The outcome is not consistent with the prediction based on the Thomson model. The outcome is consistent with the prediction based on the Rutherford model. It seems that the mass and positive charge of the atom are concentrated in a small region, with the rest of the atom mostly empty space.

Rutherford and his colleagues were surprised to find that some alpha particles recoiled backward at such large angles. Reflecting on these experiments, Rutherford once said, "It was quite the most incredible event that happened to me in my life. It was almost as incredible as if you fired a 15-inch shell at a piece of tissue paper and it came back and hit you."[1] Following these experiments, Rutherford developed a new model

---

[1]Rutherford, E. *Background to Modern Science*. New York: Macmillan, 1938.

of the atom in which a tiny **nucleus** contained nearly all of the mass of the atom and all of its positive charge. To be consistent with experiments, the diameter of the nucleus would have to be $1/100,000$ times the diameter of the atom, which was known at that time to be about 10^{-10} m. This made the nucleus extremely tiny—about 10^{-15} m.

Rutherford used this quantitative model to make predictions about the number of alpha particles that would be scattered at a particular angle for a specific thickness of foil, charge of the nucleus, and velocity of alpha particles. Geiger and Marsden performed experiments to test Rutherford's model and found the outcomes consistent with the model's predictions.

The location of the electrons Now that the nuclear model of the atom had emerged, questions remained about the electrons. Where in the atom were the electrons? Rutherford understood that electrons could not just sit stationary at the edge of the atoms, far from the nucleus. The positive nucleus would attract the negative electrons, causing the atom to collapse. Rutherford avoided this problem by proposing what became known as the **planetary model** of the atom. In this model, the electrons orbited the nucleus just as the planets orbit the Sun (**Figure 28.2**). The speed of the electrons could then be calculated using Newton's second law and the mathematical description of circular motion.

In the hydrogen atom, the nucleus exerts an electric force on the atom's lone electron. We can determine the magnitude of this force using Coulomb's law:

$$F_{\text{N on } e} = k_{\text{C}} \frac{q_{\text{N}} q_e}{r^2}$$

where r is the distance between the electron and the nucleus, which in this case is equal to the radius of the atom. The notations q_{N} and q_e are the magnitudes of the charges of the nucleus and the electron, respectively. In uniform circular motion, the radial component of Newton's second law is

$$a_r = \frac{v^2}{r} = \frac{1}{m_e} \sum F_r$$

where m_e is the mass of the electron, v is the magnitude of the tangential component of the electron's velocity (in this case, it equals the electron's speed, v) and a_r is its radial acceleration. As the only force exerted on the electron in the radial direction is the electric force, we can write

$$\frac{v^2}{r} = \frac{1}{m_e} \left(k_{\text{C}} \frac{q_{\text{N}} q_e}{r^2} \right)$$

Here q_{N} is the magnitude of the electric charge of the nucleus, q_e is the magnitude of the electron's charge, r is the radius of the electron's orbit, and k_{C} is the Coulomb's law constant. Multiplying both sides of the equation by r and taking the square root, we solve for v:

$$v = \sqrt{\frac{k_{\text{C}} q_{\text{N}} q_e}{m_e r}}$$

The above expression gives the constant speed of an electron moving in a circle at a distance r from the nucleus. Thus if the electron is moving fast enough, it can avoid falling onto the nucleus.

A difficulty with the planetary model The total energy of the nucleus-electron system, E, is the sum of its electric potential energy (U_{q}) and the kinetic energy of the electron (K_e):

$$E = U_{\text{q}} + K_e$$

Assuming the electron is not moving relativistically:

$$E = k_{\text{C}} \frac{q_{\text{N}} q_e}{r} + \tfrac{1}{2} m_e v^2 = -k_{\text{C}} \frac{e^2}{r} + \tfrac{1}{2} m_e v^2$$

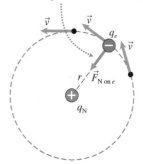

FIGURE 28.2 Rutherford's orbital model of the atom. The sizes of the electron and the nucleus are not to scale.

The inward force that the electric charge of the nucleus exerts on the electron keeps it in a circular orbit.

FIGURE 28.3 A difficulty with the planetary model. An accelerating electron should emit EM waves and spiral inward, causing the atom to collapse.

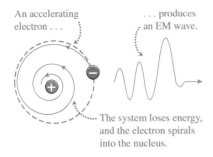

An accelerating electron . . .

. . . produces an EM wave.

The system loses energy, and the electron spirals into the nucleus.

FIGURE 28.4 A continuous spectrum and a line spectrum.

Continuous spectrum

Screen

Grating

Aperture

Lens

Lightbulb with heated filament

Line spectrum

Screen

Grating

Aperture

Lens

Lamp with heated gas

FIGURE 28.5 A hydrogen atom line spectrum.

For a hydrogen atom $q_N = +e$ and $q_e = -e$. Because the atom is a bound system, its total energy should be negative (as we learned in Chapter 7, the zero level of potential energy is at infinity; if the total energy were zero or positive, the system would fly apart).

As discussed earlier (in Chapter 25), an accelerating electron emits electromagnetic (EM) radiation, and this radiation carries energy. Because the electron orbiting the nucleus is continually accelerating, the electron-nucleus system should continuously lose energy. As the electron emits electromagnetic radiation, the electron should get closer to the nucleus (**Figure 28.3**). EM theory predicts that the electron would spiral into the nucleus in about 10^{-12} s. This definitely contradicts our experience. The Rutherford model could not explain the stability of atoms. It also could not explain another phenomenon that was well known to physicists at that time—the light emitted by low-density gases.

Spectra of low-density gases and the need for a new model

By the end of the 19th century, physicists knew that many objects emit a continuous spectrum of light whose properties depend on the temperature of the object—that is, the random motion of their particles. Room-temperature objects primarily emit infrared photons. Objects at several thousand degrees kelvin (such as an incandescent lightbulb filament) emit a larger number of visible light photons. We learned about this type of spectrum in Chapter 27 when studying black body radiation.

If atoms are in a gaseous form and the gas's temperature is increased so that it emits visible light, we observe a **line spectrum** (**Figure 28.4**), in which only specific wavelengths of light ("lines") are present. For example, if you throw a dash of table salt into a flame and observe the emitted light with a spectrometer, you see a line spectrum—two closely positioned yellow lines.

Another way to observe a gas spectrum is to put a strong electric field across a tube filled with gas. Neon signs use this method; the spectrum of neon gas is linear.

Observations show that different gases produce different sets of spectral lines. These line combinations do not depend on how the gases are made to glow—whether heated or placed in a strong electric field. When gases are mixed together, the spectral lines characteristic of each individual gas are present.

Scientists could not explain where the lines came from, but they could determine which chemical elements emitted which lines. The most carefully studied element was hydrogen because of the simplicity of its line spectra: just four distinct lines in the visible light region of the spectrum (**Figure 28.5**).

364.6 nm   410.2 nm   434.1 nm   486.1 nm   656.3 nm

In the ultraviolet spectrum

In the visible region of the spectrum

In 1885 Johann Balmer found a pattern in the wavelengths of the visible spectral lines produced by hydrogen and represented it with the following equation:

$$\frac{1}{\lambda} = R\left(\frac{1}{2^2} - \frac{1}{n^2}\right) \tag{28.1}$$

where λ is the wavelength of the spectral line, n has integer values of 3, 4, 5, and 6, and R is a constant experimentally determined to equal $R = 1.097 \times 10^7 \text{ m}^{-1}$ when the wavelength is measured in meters. The constant R is called the **Rydberg constant**. The Balmer equation produces the four wavelengths shown in Figure 28.5. However, Balmer could not explain why this pattern existed.

The Thomson model could explain the emission of light by assuming that electrons can vibrate inside the atom, but it did not provide a quantitative account for the specific frequencies. The Rutherford model predicted that individual atoms should radiate a continuous spectrum until the electron spiraled into the nucleus. Physicists needed a new atomic model that was stable and that could explain the line spectra; we will discuss this model in Section 28.2.

QUANTITATIVE EXERCISE 28.1   **Calculate photon energies**

Use Balmer's formula [Eq. (28.1)] to determine the energies in joules and in electron volts of the possible visible photons emitted by hydrogen atoms.

Represent mathematically Balmer's formula [Eq. (28.1)] provides an expression for the inverse of the emitted photon wavelength:

$$\frac{1}{\lambda} = R\left(\frac{1}{2^2} - \frac{1}{n^2}\right)$$

The four visible spectral lines correspond to $n = 3, 4, 5,$ and 6. The energy of a photon is

$$E = hf = \frac{hc}{\lambda} = hc\frac{1}{\lambda}$$

Solve and evaluate Combine these two equations to get

$$E = hcR\left(\frac{1}{2^2} - \frac{1}{n^2}\right)$$

Inserting the appropriate values for Planck's constant h, the speed of light c, and Rydberg's constant R, we get for the combined constant hcR

$$hcR = (6.63 \times 10^{-34}\,\text{J}\cdot\text{s})(3.00 \times 10^8\,\text{m/s})(1.097 \times 10^7\,\text{m}^{-1})$$
$$= 2.18 \times 10^{-18}\,\text{J}$$

This combined constant shows up frequently when we deal with line spectra, and it is worth remembering for convenience.

Inserting the four different n values and converting to electron volts ($1\,\text{eV} = 1.6 \times 10^{-19}\,\text{J}$), we get the following:

$$n = 3:\quad E_3 = (2.18 \times 10^{-18}\,\text{J})\left(\frac{1}{2^2} - \frac{1}{3^2}\right) = 3.03 \times 10^{-19}\,\text{J}$$
$$= (3.03 \times 10^{-19}\,\text{J})\left(\frac{1\,\text{eV}}{1.6 \times 10^{-19}\,\text{J}}\right) = 1.89\,\text{eV}$$

$$n = 4:\quad E_4 = (2.18 \times 10^{-18}\,\text{J})\left(\frac{1}{2^2} - \frac{1}{4^2}\right) = 4.09 \times 10^{-19}\,\text{J}$$
$$= (4.09 \times 10^{-19}\,\text{J})\left(\frac{1\,\text{eV}}{1.6 \times 10^{-19}\,\text{J}}\right) = 2.55\,\text{eV}$$

$$n = 5:\quad E_5 = (2.18 \times 10^{-18}\,\text{J})\left(\frac{1}{2^2} - \frac{1}{5^2}\right) = 4.58 \times 10^{-19}\,\text{J}$$
$$= (4.58 \times 10^{-19}\,\text{J})\left(\frac{1\,\text{eV}}{1.6 \times 10^{-19}\,\text{J}}\right) = 2.86\,\text{eV}$$

$$n = 6:\quad E_6 = (2.18 \times 10^{-18}\,\text{J})\left(\frac{1}{2^2} - \frac{1}{6^2}\right) = 4.84 \times 10^{-19}\,\text{J}$$
$$= (4.84 \times 10^{-19}\,\text{J})\left(\frac{1\,\text{eV}}{1.6 \times 10^{-19}\,\text{J}}\right) = 3.03\,\text{eV}$$

To determine whether these results are reasonable, complete the "Try it yourself" question below and see if the wavelengths correspond to the observed visible light wavelengths for hydrogen.

Try it yourself Determine the wavelengths in nanometers (nm) of the above spectral lines, and identify the colors.

Answer

Using the Balmer equation, we get 656 nm for $n = 3$ (orange), 486 nm for $n = 4$ (greenish blue), 434 nm for $n = 5$ (blue), and 410 nm for $n = 6$ (violet). All of these wavelengths fall within the visible part of the EM spectrum. Compare them to the values in Figure 28.5.

REVIEW QUESTION 28.1 Rutherford made the following statement: "I was perfectly aware when I put forward the theory of the nuclear atom that . . . the electron ought to fall onto the nucleus." Why must the electrons in Rutherford's model fall into the nucleus?

28.2 Bohr's model of the atom: quantized orbits

In 1913, Danish physicist Niels Bohr succeeded in creating a new model that explained the line spectra and the stability of the atom. He kept the structure of Rutherford's model in terms of the small nucleus and electrons moving around it, but imposed a restriction on the electron orbits.

TIP Before you read the description of Bohr's ideas, review the concept of rotational (angular) momentum (Chapter 9).

Bohr's model applies only to hydrogen and to one-electron ions. Here are the fundamental ideas of Bohr's model, known as **Bohr's postulates**.

1. The atom is made up of a small nucleus and an orbiting electron. The electron can occupy only certain orbits, called **stable orbits**, which are labeled by the positive integer n. When in these orbits, the electron moves around the nucleus but *does not* radiate electromagnetic waves. All other orbits are prohibited. Each of these stable orbits results in a specific value of the total energy (kinetic plus electric potential) of the atom, designated E_n.

2. When an electron transitions from one stable orbit to another, the atom's energy changes. When the energy of the atom decreases, the atom emits a photon whose energy equals the decrease in the atom's energy ($hf = E_i - E_f$, where "i" is the energy state number for the initial state and "f" is the energy state number for the final state). For the atom's energy to increase, the atom must absorb some energy, often by absorbing a photon whose energy equals the increase in the atom's energy. Because the stable orbits are discrete, the atom can radiate or absorb only certain specific amounts of energy.

3. The stable electron orbits are the orbits where the magnitude of the electron's rotational (angular) momentum L is given by

$$L = mvr = n\frac{h}{2\pi}, \quad n = 1, 2, 3, \ldots \tag{28.2}$$

In this equation m is the mass of the electron, v is its speed, r is the radius of its orbit, h is Planck's constant, and n is any positive integer. In this equation, rotational momentum L is **quantized**, meaning that it can have only specific discrete values (multiples of $h/2\pi$ in this case).

While Bohr's model explained the line spectra and stability of the hydrogen atom, it does not explain why electrons do not emit EM waves when accelerating in a circular orbit. It also does not explain why the only stable orbits are those with the above rotational momenta [Eq. (28.2)]. Let's see if Bohr's model correctly predicts the size of the hydrogen atom and the specific wavelengths present in its line spectrum.

Size of the hydrogen atom

To calculate the size of the atom using Bohr's postulates, assume that the electron and nucleus obey Newton's laws and interact only via the electrostatic (Coulomb) force, that gravitational forces exerted on the atom are negligible, and that the nucleus has charge $q_N = +e = 1.6 \times 10^{-19}$ C. A sketch of the atom and a force diagram for the electron are shown in **Figure 28.6**. The nucleus exerts an electric force on the electron, $F_{N \text{ on } e}$, of this magnitude:

$$F_{N \text{ on } e} = k_C \frac{|q_N q_e|}{r_n^2} = k_C \frac{|(+e)(-e)|}{r_n^2}$$

$$\Rightarrow F_{N \text{ on } e} = k_C \frac{e^2}{r_n^2} \tag{28.3}$$

This force points toward the nucleus, resulting in the electron's uniform circular motion. Recall that an object's radial acceleration a_r for uniform circular motion is given by the expression $a_r = (v^2/r)$. The radial r component of Newton's second law is

$$a_r = \frac{F_{N \text{ on } e}}{m_e} = \frac{1}{m_e}\left(k_C \frac{e^2}{r_n^2}\right) \Rightarrow \frac{v_n^2}{r_n} = \frac{k_C e^2}{m_e r_n^2}$$

Here m_e is the mass of the electron and a_r is the radial acceleration. We use the symbol e for the magnitudes of the charge of both the electron and the nucleus. The radius of

FIGURE 28.6 Force-acceleration analysis of the Bohr atomic model.

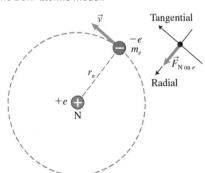

The electrical force that the nucleus exerts on the electron is much greater than any gravitational force that the nucleus exerts on the electron.

the electron's orbit is r_n, and v_n is the corresponding speed; k_C is the Coulomb's law constant. Multiplying both sides by r_n, we get

$$v_n^2 = \frac{k_C e^2}{m_e r_n} \tag{28.4}$$

Equation (28.4) relates two unknown quantities, v_n and r_n. Bohr's third postulate involving the rotational momentum L_n of the electron in the nth orbit provides a second relationship between these same two quantities:

$$L_n = m_e v_n r_n = n\frac{h}{2\pi}, \quad \text{with } n = 1, 2, 3, \ldots$$

Solving for v_n, we find

$$v_n = \frac{nh}{2\pi m_e r_n}, \quad n = 1, 2, 3, \ldots$$

Insert this expression for v_n into Eq. (28.4) to get

$$\left(\frac{nh}{2\pi m_e r_n}\right)^2 = \frac{k_C e^2}{m_e r_n}$$

Solving the above for r_n, we get

$$r_n = \frac{h^2}{4\pi^2 k_C e^2 m_e} n^2, \quad \text{with } n = 1, 2, 3, \ldots \tag{28.5}$$

Let's check the units of this result. The units for h are $\text{J} \cdot \text{s}$ and for k_C are $(\text{N} \cdot \text{m}^2/\text{C}^2)$. Thus, the unit of the right side of the equation is

$$\frac{(\text{J}^2 \, \text{s}^2)}{\left(\dfrac{\text{N} \cdot \text{m}^2}{\text{C}^2}\right) \text{C}^2 \, \text{kg}} = \frac{(\text{N} \cdot \text{m})^2 \, \text{s}^2}{\text{N} \cdot \text{m}^2 \, \text{kg}} = \frac{\text{N} \cdot \text{s}^2}{\text{kg}} = \frac{\left(\dfrac{\text{kg} \cdot \text{m}}{\text{s}^2}\right) \cdot \text{s}^2}{\text{kg}} = \text{m}$$

which is the correct unit for a radius. Substituting the values of the known constants into Eq. (28.5), we get

$$r_n = \frac{(6.63 \times 10^{-34} \, \text{J} \cdot \text{s})^2}{4\pi^2\left(9.00 \times 10^9 \, \dfrac{\text{N} \cdot \text{m}^2}{\text{C}^2}\right)(1.6 \times 10^{-19} \, \text{C})^2(9.11 \times 10^{-31} \, \text{kg})} n^2$$

or

$$r_n = (0.53 \times 10^{-10} \, \text{m})n^2, \quad \text{for } n = 1, 2, 3, \ldots \tag{28.6}$$

The smallest possible electron orbit radius corresponds to the $n = 1$ energy state:

$$r_1 = 0.53 \times 10^{-10} \, \text{m}$$

This result is in agreement with the measurements of atomic sizes, about 10^{-10} m.

This smallest radius orbit is called the **Bohr radius** r_1. The other stable orbit radii can be calculated using Eq. (28.6). Notice that the radii depend on the square of n:

$$r_2 = 4r_1; \, r_3 = 9r_1; \, r_4 = 16r_1; \, \text{therefore, } r_n = n^2 r_1$$

As a result, the radius increases dramatically as n increases. The term n is known as a **quantum number** and must be a positive integer. Because of this, only certain radii represent stable electron orbits (for example, $4r_1$ and $9r_1$, which correspond to $n = 2$ and $n = 3$, are stable orbits, but $5r_1$ is not); a stable radius is said to be **quantized**. The radii of allowed electron orbits in the Bohr model of the hydrogen atom are represented in **Figure 28.7**. (Note that the way the nucleus is drawn in the figure dramatically exaggerates its size in comparison to the electron orbits.)

FIGURE 28.7 Radii of stable Bohr electron orbits.

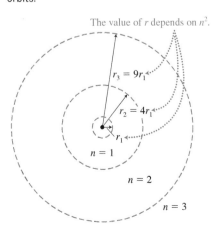

The value of r depends on n^2.

$r_3 = 9r_1$

$r_2 = 4r_1$

r_1

$n = 1$

$n = 2$

$n = 3$

Energy states of the Bohr model

Now that we have gained some confidence in Bohr's model, we can use it to determine the energy states of the hydrogen atom (or, put another way, the total energy of the atom for each energy state n). We'll make the reasonable assumption that the nucleus remains at rest as the electron orbits because of its much larger mass. We'll also assume that relativistic effects are negligible. The total energy of the electron-nucleus system is the sum of its electric potential energy plus the electron's kinetic energy (we neglect the motion of the nucleus due to the interactions with the electron). The electric potential energy U_{qn} is

$$U_{qn} = k_C \frac{q_N q_e}{r_n} = k_C \frac{(+e)(-e)}{r_n} = -k_C \frac{e^2}{r_n}$$

The total energy of the system in the nth energy state is then

$$E_n = U_{qn} + K_{en} = -k_C \frac{e^2}{r_n} + \frac{1}{2} m_e v_n^2 \tag{28.7}$$

From Eq. (28.4), note that $m_e v_n^2 = k_C \left(e^2 / r_n \right)$. Therefore, Eq. (28.7) becomes

$$E_n = -k_C \frac{e^2}{r_n} + \frac{1}{2} \left(k_C \frac{e^2}{r_n} \right) = -k_C \frac{e^2}{2r_n}$$

The Bohr model has predicted that the total energy of the atom is negative. This is good news because the total energy of a bound system should be negative, as explained earlier. Now use Eq. (28.5) for the radius of the atom to write an equation for its total energy:

$$E_n = -k_C \frac{e^2}{2r_n} = -k_C \frac{e^2}{2 \left(\dfrac{h^2}{4\pi^2 k_C e^2 m_e} n^2 \right)} = -\left(\frac{2\pi^2 e^4 k_C^2 m_e}{h^2} \right) \frac{1}{n^2} \tag{28.8}$$

Evidently, the energy of the atom is inversely proportional to the square of the quantum number n. Substituting the values of the constants into the above and converting from joules to electron volts, we get

$$E_n = \frac{-13.6 \text{ eV}}{n^2} \quad \text{for } n = 1, 2, 3, \ldots \tag{28.9}$$

To help visualize the energy states corresponding to these possible total energy values, we use a new representation—an energy state diagram (**Figure 28.8**). On this diagram the vertical axis represents the total energy of the hydrogen atom. Each line represents a specific energy state of the atom labeled by the quantum number n. The -13.6 eV horizontal line at the very bottom corresponds to the $n = 1$ lowest energy state, also called the **ground state**. The next horizontal line $(-13.6 \text{ eV})/2^2 = -3.4$ eV corresponds to the $n = 2$ energy state, also called the first **excited state**. This pattern repeats for the different integral values of n. This representation allows us to see that the higher the quantum number, the less negative the energy of the system. Equation 28.7 shows that the less negative the energy of the system, the larger the radius of the electron's orbit. The $n = \infty$ state has an infinite radius and zero energy, indicating the electron is unbound, an ionized hydrogen atom $\left(H^+ + e^- \right)$.

The energy state diagram in Figure 28.8 also allows us to represent the process of emission of light by the atom. If the atom is currently in the $n = 2$ energy state and makes a transition to the $n = 1$ energy state, the energy of the atom decreases. According to Bohr's second postulate, the energy is carried away by a newly created photon. This **emission process** is represented in **Figure 28.9** by a downward arrow pointing from the initial (higher energy) state of the atom to the final (lower energy) state. This energy, equal to the energy decrease of the atom, is indicated with the symbol E_γ. The subscript γ (Greek letter gamma) is often used to indicate a photon.

The atom's energy can increase if it absorbs a photon. This **absorption process** is represented on the energy diagram with an upward arrow pointing from the initial

FIGURE 28.8 Energy state diagram. The atom's energy is inversely proportional to the square of n.

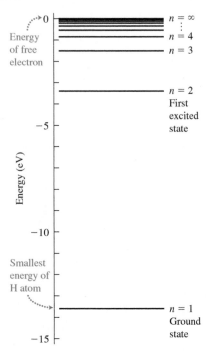

FIGURE 28.9 An example of photon emission. An atom transitions from the first excited state to the ground state, and a photon is emitted.

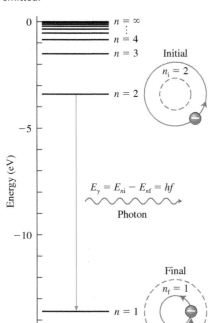

FIGURE 28.10 An example of photon absorption. An atom transitions from the ground state to the first excited state by absorbing the energy of a photon.

state of the atom to its final state (**Figure 28.10**). An atom can also move from one energy state to another (increasing or decreasing its energy) when it collides with another atom.

We have not yet tested whether Bohr's model correctly predicts the visible light photons emitted by hydrogen; this test is the goal of Example 28.2.

TIP Remember that according to Bohr's first postulate, the hydrogen atom can exist in states whose energies can be calculated by Eq. (28.8). States with energies between those values are not allowed.

EXAMPLE 28.2 **Photons emitted by hydrogen atoms**

Use the Bohr model to predict the energies and wavelengths of photons that a group of hydrogen atoms would emit if the atoms were all initially in the $n = 3$ state. In which parts of the electromagnetic spectrum are each of these photons?

Sketch and translate According to Bohr's model, an atom emits photons when the energy of the atom decreases. This emission occurs when the electron in the atom transitions from an orbit with a higher quantum number (n_i) to an orbit with a lower number (n_f). Three possible transitions could occur: $3 \rightarrow 1$, $3 \rightarrow 2$, and $2 \rightarrow 1$ (after the atom has already made the $3 \rightarrow 2$ transition).

Simplify and diagram The three processes are represented in the energy state diagram at right. Downward vertical arrows represent the energy transitions.

Represent mathematically Using the principle of energy conservation, the initial energy E_{n_i} of the atom must equal the final energy E_{n_f} plus the photon's energy E_γ:

$$E_{n_i} = E_{n_f} + E_\gamma$$

or

$$E_\gamma = E_{n_i} - E_{n_f}$$

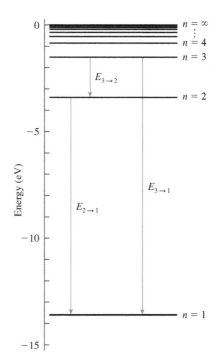

(CONTINUED)

Inserting the expressions for the total energy of the atom, we have

$$E_\gamma = \frac{-13.6 \text{ eV}}{n_i^2} - \frac{-13.6 \text{ eV}}{n_f^2} = -13.6 \text{ eV}\left(\frac{1}{n_i^2} - \frac{1}{n_f^2}\right)$$

Once we determine the energies E_γ of the emitted photons, we can determine their wavelengths λ using

$$E_\gamma = hf = \frac{hc}{\lambda} \text{ or } \lambda = \frac{hc}{E_\gamma}$$

Solve and evaluate We can now determine the energies in electron volts and joules of the photons emitted in each of the three processes. Remember that $1 \text{ eV} = 1.6 \times 10^{-19}$ J.

$$n_i = 3 \text{ to } n_f = 1: \quad E_\gamma = -13.6 \text{ eV}\left(\frac{1}{3^2} - \frac{1}{1^2}\right)$$
$$= 12.1 \text{ eV} = 1.94 \times 10^{-18} \text{ J}$$

$$n_i = 3 \text{ to } n_f = 2: \quad E_\gamma = -13.6 \text{ eV}\left(\frac{1}{3^2} - \frac{1}{2^2}\right)$$
$$= 1.89 \text{ eV} = 3.02 \times 10^{-19} \text{ J}$$

$$n_i = 2 \text{ to } n_f = 1: \quad E_\gamma = -13.6 \text{ eV}\left(\frac{1}{2^2} - \frac{1}{1^2}\right)$$
$$= 10.2 \text{ eV} = 1.63 \times 10^{-18} \text{ J}$$

We can determine the photon wavelengths from their energies:

$n_i = 3$ to $n_f = 1$:

$$\lambda = \frac{hc}{E_\gamma} = \frac{(6.63 \times 10^{-34} \text{ J} \cdot \text{s})(3.00 \times 10^8 \text{ m/s})}{1.94 \times 10^{-18} \text{ J}}$$
$$= 1.03 \times 10^{-7} \text{ m} = 103 \text{ nm}$$

$n_i = 3$ to $n_f = 2$:

$$\lambda = \frac{hc}{E_\gamma} = \frac{(6.63 \times 10^{-34} \text{ J} \cdot \text{s})(3.00 \times 10^8 \text{ m/s})}{3.02 \times 10^{-19} \text{ J}}$$
$$= 6.59 \times 10^{-7} \text{ m} = 659 \text{ nm}$$

$n_i = 2$ to $n_f = 1$:

$$\lambda = \frac{hc}{E_\gamma} = \frac{(6.63 \times 10^{-34} \text{ J} \cdot \text{s})(3.00 \times 10^8 \text{ m/s})}{1.63 \times 10^{-18} \text{ J}}$$
$$= 1.22 \times 10^{-7} \text{ m} = 122 \text{ nm}$$

The 3 to 2 transition is in the visible (red) part of the electromagnetic spectrum, and the 2 to 1 and 3 to 1 transitions are in the ultraviolet part of the spectrum.

Try it yourself Show that the wavelengths of emission lines calculated using Balmer's empirical formula [Eq. (28.1)] are consistent with Bohr's model. The transitions described by Balmer's equation are indicated in **Figure 28.11**. Series of emission lines corresponding to other possible transitions are also shown.

Answer

Balmer's equation is

$$\frac{1}{\lambda} = R\left(\frac{1}{2^2} - \frac{1}{n^2}\right)$$

which produces the same wavelength transitions as calculated above.

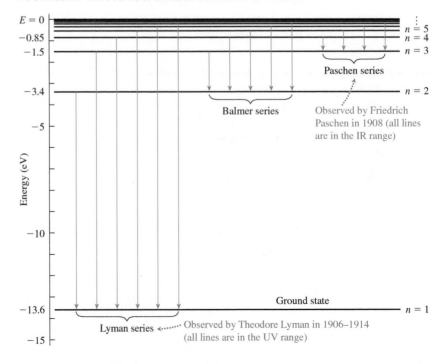

FIGURE 28.11 Three series of emission transitions for H atoms.

Consistency of Balmer's equation with Bohr's model

Let's find whether Balmer's result is consistent with the Bohr model prediction for emission transitions.

$$E_\gamma = -13.6 \text{ eV}\left(\frac{1}{n_i^2} - \frac{1}{n_f^2}\right)$$

If we multiply both sides of the Balmer equation by hc, we get

$$\frac{hc}{\lambda} = hcR\left(\frac{1}{2^2} - \frac{1}{n^2}\right)$$

The left-hand side is the energy E_γ of the emitted photon. The constant hcR must then have units of energy. This constant in electron volts is

$$hcR = (6.63 \times 10^{-34} \text{ J} \cdot \text{s})(3.00 \times 10^8 \text{ m/s})(1.097 \times 10^7 \text{ m}^{-1})\left(\frac{1 \text{ eV}}{1.6 \times 10^{-19} \text{ J}}\right)$$

$$= 13.6 \text{ eV}$$

Balmer's equation becomes

$$E_\gamma = 13.6 \text{ eV}\left(\frac{1}{2^2} - \frac{1}{n^2}\right) = -13.6 \text{ eV}\left(\frac{1}{n^2} - \frac{1}{2^2}\right)$$

This is exactly the result from the Bohr model with $n_i = n$ and $n_f = 2$.

Because there are infinitely many energy states (n can be any positive integer), there are infinitely many possible transitions. However, only four of these transitions emit a visible wavelength photon. The 3 to 2 transition is one. The others are 4 to 2, 5 to 2, and 6 to 2. All transitions from states 7 and higher to 2 result in ultraviolet photons. Any transition to the 1 state also results in an ultraviolet photon.

Notice that the photon emitted by the 3 to 2 transition closely matches the 656-nm line shown in the hydrogen line spectrum in Figure 28.5. The 4 to 2, 5 to 2, and 6 to 2 transitions produce the other visible light spectral lines in Figure 28.5.

Limitations of the Bohr model

We saw in Example 28.2 that the wavelengths of the hydrogen spectral lines predicted by Bohr's model are in good agreement with the observational evidence. Unfortunately, the model does not provide predictions that account for the spectral lines emitted by other atoms. It does, however, work well if the atoms are ionized and have only one electron (such as singly ionized helium, He^+, or doubly ionized lithium, Li^{2+}).

Also, note that Bohr's model arbitrarily imposes restrictions on the motion of the electron (its quantized rotational angular momentum), which results in a restriction on the allowed energy states of the atom. However, Bohr's model provides no explanations for these restrictions. Therefore, physicists continued searching for a new model of the atom that could provide them.

REVIEW QUESTION 28.2 Why is an atom's total energy negative? Why does this make sense?

28.3 Spectral analysis

Bohr's model, as you have seen, helps us understand the **emission spectra** of gases (see the examples of spectra in **Figure 28.12**). These spectra allow scientists to analyze the chemical composition of different materials according to the light they emit. So far we have discussed only the emission spectrum of hydrogen. We can understand the spectra of other atoms and molecules in a qualitatively similar way.

FIGURE 28.12 Emission spectra of elements in a gaseous state. Each type of atom has a unique spectrum, allowing its identification.

When an atom transitions from a higher energy state to a lower energy state, a photon of a specific wavelength is emitted. Thus, to observe an emission spectrum, some mechanism is needed to put the atom in an excited state. We observed the spectra of heated gases and gases placed in electric fields. What are the mechanisms that explain their glow in these conditions?

Thermal excitation

Heating can significantly increase the random kinetic energy of atoms (**Figure 28.13a**). If the kinetic energies of the atoms are high enough, a collision can cause one or both of the atoms to transition to an excited state (Figure 28.13b). When the atom returns to a lower energy state, it emits a photon of a characteristic wavelength (Figure 28.13c). The bright yellow, red, and blue colors of fireworks come from photon emission by excited sodium, strontium, and copper compounds, respectively.

FIGURE 28.13 Thermal excitation and production of a photon.

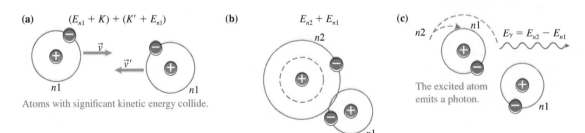

Discharge tube excitation

Another process that causes atoms to transition to higher energy states involves filling a discharge tube similar to a cathode ray tube with gas. There are always some free electrons (or other charged particles) present in our atmosphere due to cosmic rays. These free electrons are accelerated across an electric potential difference from the cathode to the anode (Figure 28.14a). The electrons then collide with atoms of the gas in the tube, causing the atoms to reach excited states (Figure 28.14b). When the excited atoms return to the ground state, they emit photons whose energy equals the difference in the energy of the atom between its initial and final states (Figure 28.14c). Looking at a discharge tube, you see a glow that appears to be of a single color. But if you look at the light through a grating, you see a line spectrum.

FIGURE 28.14 A discharge tube produces light.

(a) Cathode Electrons accelerate. Anode

(b) An electron excites an atom during a collision.

(c) An excited atom emits a photon. Photon

Figure 28.15 shows what you see if you observe a mercury discharge tube through a grating that has 500 lines per millimeter. The blue central maximum shows the color of the tube that you see with the naked eye. Different color lines on both sides are the 1st order maxima of the brightest lines that together produce the color of the central maximum.

FIGURE 28.15 Mercury spectrum as seen through a grating.

Mercury lamp

1st order maxima   1st order maxima

EXAMPLE 28.3 Discharge tube excitation of hydrogen

Use the results of Example 28.2 to estimate the potential difference needed between the cathode and anode in a discharge tube filled with hydrogen to produce hydrogen's visible light spectral lines.

Sketch and translate An illustration of the gas in the discharge tube is shown in Figure 28.14a–c. For the atom to emit a visible photon, it must transition from the $n = 3$, 4, 5, or 6 state to the $n = 2$ state. Therefore, the colliding electrons must have enough kinetic energy to induce the transition of the electrons in the hydrogen atoms from the $n = 1$ ground state to the $n = 3$, 4, 5, or 6 state. Because the transition from $n = 1$ to $n = 6$ requires the most energy, the potential difference between cathode and anode must be high enough to induce that transition. The system is the colliding electron, the electrodes and the electric field they produce, the hydrogen atoms, and the emitted photons.

The overall process has three parts.

I. The electric potential energy of the system is converted into the kinetic energy of the electron, as shown in part (a) of the figure on the top of the right column.

II. The electron collides with and excites the atom [part (b)].

III. The excited atom returns to the ground state, emitting a photon [part (c)].

(a) Initial   Final

$$K_i + U_{qi} = K_f + U_{qf}$$

The electron gains kinetic energy while crossing a potential difference.

(b) Initial   Final

$$K_{ei} + E_{\text{Atom i}} = K_{ef} + E_{\text{Atom f}}$$

The electron excites an atom when it collides with it.

(c)

$$E_{\text{Atom i}} + E_\gamma = E_{\text{Atom f}} + E_\gamma$$

The excited atom emits a photon.   (CONTINUED)

Simplify and diagram Assume first that the electron's initial kinetic energy is zero and that it increases only as a result of the decrease in electric potential energy of the system. Assume also that the electric potential at the initial position of the electron is 0 V. Before the colliding electron encounters the hydrogen atom, assume the atom is in the ground state and that the Bohr model is a reasonable way to model the atom. Lastly, assume the electron is not moving relativistically. Energy bar charts shown earlier represent the three stages of this process.

Represent mathematically We can use the bar charts to help us represent these three processes mathematically.

Part I: In the initial state, the electron is at rest at zero electric potential. In the final state, the electron has positive kinetic energy and the system's electric potential energy has decreased.

$$K_i + U_{qi} = K_f + U_{qf}$$

Using the expressions for nonrelativistic kinetic energy and electric potential energy:

$$0 + q_e V_i = \tfrac{1}{2} m_e v^2 + q_e V_f$$

Subtracting $q_e V_i$ from both sides, we have

$$0 = \tfrac{1}{2} m_e v^2 + q_e V_f - q_e V_i$$
$$= \tfrac{1}{2} m_e v^2 + q_e \, \Delta V = \tfrac{1}{2} m_e v^2 - e \, \Delta V$$

Part II: The electron collides with and excites an atom from the $n = 1$ to the $n = 6$ state. In the initial state, the colliding electron has considerable kinetic energy, and the atom is in the $n = 1$ ground state. In the final state, the colliding electron stops, and the atom is in the $n = 6$ excited state.

$$K_i + E_{\text{Atom i}} = K_f + E_{\text{Atom f}}$$

Inserting the expressions for kinetic energy and the total energy of the hydrogen atom:

$$\tfrac{1}{2} m_e v^2 + \left(\frac{-13.6 \text{ eV}}{n_i^2} \right) = 0 + \left(\frac{-13.6 \text{ eV}}{n_f^2} \right)$$

Part III: The excited atom emits a photon and transitions to the $n = 2$ state. Initially, the atom is in the $n = 6$ state. In the final state, the atom is in the $n = 2$ state and a photon has been emitted.

$$E_{\text{Atom i}} = E_{\text{Atom f}} + E_\gamma$$

Insert the expressions for the total energy of the atom and the energy of the photon:

$$\left(\frac{-13.6 \text{ eV}}{n_i^2} \right) = \left(\frac{-13.6 \text{ eV}}{n_f^2} \right) + \frac{hc}{\lambda}$$

Solve and evaluate We are interested in finding ΔV. Using our mathematical representation of Part I and solving for ΔV:

$$\Delta V = \frac{1}{e} \left(\tfrac{1}{2} m_e v^2 \right)$$

We can determine the kinetic energy from Part II. Solving for $(1/2) m_e v^2$ gives

$$\tfrac{1}{2} m_e v^2 = \left(\frac{-13.6 \text{ eV}}{n_f^2} \right) - \left(\frac{-13.6 \text{ eV}}{n_i^2} \right)$$
$$= (-13.6 \text{ eV}) \left(\frac{1}{n_f^2} - \frac{1}{n_i^2} \right)$$

Substituting this into the expression for ΔV:

$$\Delta V = \frac{1}{e} (-13.6 \text{ eV}) \left(\frac{1}{n_f^2} - \frac{1}{n_i^2} \right)$$

Inserting the appropriate values and converting to joules:

$$\Delta V = \frac{1}{1.6 \times 10^{-19} \text{ C}} (-13.6 \text{ eV}) \left(\frac{1}{6^2} - \frac{1}{1^2} \right) \left(1.6 \times 10^{-19} \, \frac{\text{J}}{\text{eV}} \right)$$
$$= 13.2 \text{ V}$$

This value is the smallest potential difference that the electrons need to cross in order to acquire enough kinetic energy to excite hydrogen atoms to the maximum $n = 6$ state so that they can emit photons in the visible spectrum. The potential difference we found is not the potential difference across the discharge tube, but rather the potential difference across the average distance that the colliding electrons travel between two consecutive collisions. The required potential difference across the cathode and anode of the tube is much greater than this.

Try it yourself Estimate the minimum potential difference required to allow the hydrogen in the tube to emit photons from the $n = 4$ to the $n = 2$ transition.

Answer   12.8 V.

Another way to produce visible light is by the collisions of hot gas atoms with each other. In order for visible light to be produced, the collisions must cause one of the atoms to become excited to the $n = 3$ state or higher. How hot does the gas need to be in order for this process to occur?

QUANTITATIVE EXERCISE 28.4   **Temperature at which an H atom emits visible photons**

Estimate the temperature of hydrogen gas at which you might observe visible line spectra.

Represent mathematically A collision between two hydrogen atoms must provide enough energy to excite one of the atoms to the $n = 3$ state or higher. From there it can make transitions to the $n = 2$

state and emit visible photons. Recall that the average kinetic energy (thermal energy) of a gas particle at absolute temperature T is given by this expression (from Section 12.4):

$$\overline{K} = \tfrac{3}{2} k_B T$$

Here $k_B = 1.38 \times 10^{-23}$ J/K, Boltzmann's constant. A hydrogen atom in the ground state $(n = 1)$ with energy E_1 needs additional energy ΔE for it to be excited to the state with energy E_3 or higher. Choosing the system of interest to be one of the colliding hydrogen atoms, we can use the idea of energy conservation to represent this process mathematically:

$$E_1 + \Delta E = E_3$$

Assume that ΔE equals the average kinetic energy of a single hydrogen atom, meaning one of the colliding particles converts all of its kinetic energy to excite the other. E_1 and E_3 are the energies of the atom in the indicated states:

$$E_n = \frac{-13.6 \text{ eV}}{n^2}$$

Solve and evaluate Putting the above equations together, we have

$$\frac{-13.6 \text{ eV}}{1^2} + \tfrac{3}{2} k_B T = \frac{-13.6 \text{ eV}}{3^2}$$

Solving for T, inserting the appropriate values, and converting to joules, we get

$$T = \frac{2(-13.6 \text{ eV})}{3(1.38 \times 10^{-23} \text{ J/K})}\left(\frac{1.6 \times 10^{-19} \text{ J}}{1 \text{ eV}}\right)\left(\frac{1}{3^2} - \frac{1}{1^2}\right)$$

$$= 93{,}000 \text{ K}$$

At this temperature, collisions will excite almost all of the atoms into the $n = 3$ state. However, even at lower temperatures, some will be excited to the $n = 3$ state (remember the Maxwell distribution). It is clearly easier to excite atoms using an electric field to accelerate free electrons in a cathode ray tube (which then collide with the atoms) than it is to raise the temperature of all of them.

Try it yourself Estimate the temperature at which an average hydrogen atom is ionized.

Answer At a temperature of 10^5 K or less, the collisions between atoms are violent enough to easily transfer the 13.6 eV of energy needed to ionize an atom.

The "Try it yourself" question of the last exercise indicates that the temperature at which hydrogen ionizes as a result of collisions with other atoms is almost the same as the temperature needed for hydrogen atoms to become excited to higher energy states so they can emit visible light. Ionized hydrogen gas no longer emits spectral lines, instead emitting a continuous spectrum. This is because the energy states of free electrons can have any positive value of energy, so when they combine with protons to form neutral hydrogen again, the emitted photons do not have discrete wavelengths. A gas with most of its atoms ionized due to collisions is called a *high-temperature plasma*. The cores of stars are at millions of kelvins and consist mostly of high-temperature plasma. Gas can also be ionized if placed in a strong electric field, producing a *low-temperature plasma* such as exists in neon lights or discharge tubes.

Continuous spectra from stars

The Sun consists almost entirely of hydrogen and helium atoms, so we would expect its light to consist primarily of hydrogen and helium emission lines. Instead, light from the Sun and many other high-temperature objects is close to a continuous black body spectrum that depends only on the object's surface temperature (Figure 28.16).

The continuous spectrum from these objects must depend on the distribution of the random kinetic energy (thermal energy) and temperature of the surface particles and not on their atomic compositions. On the hot Sun (5700 K on the surface), the material is dense and the atoms are continually colliding with each other. Evidently, these collisions distort the shapes of atoms during the collisions and excite them in unpredictable ways so that the light emitted when they de-excite is characteristic of the temperature of the gas and not of the types of atoms in the gas. These continuous (black body) spectra have the same shape for all objects at a particular temperature, but they differ significantly for objects at different temperatures (Figure 28.16).

Other stars emit similar black body radiation with the peak intensity at a wavelength that depends on their temperatures. However, if you look closely at radiation from the Sun and other stars, you discover important details. Observational Experiment Table 28.2 (on the next page) helps us analyze them.

FIGURE 28.16 As the temperature increases, the black body radiation increases in intensity and the peak moves toward shorter wavelength.

OBSERVATIONAL
EXPERIMENT TABLE 28.2   Stellar spectra

Observational experiment	Analysis
We observed a spectrum of the Sun using a handheld spectrometer and then recorded the same spectrum using a computer-controlled light detector.	The figure shows a photo of the spectrum from the first experiment and the spectral curve produced by the computer in the second experiment. In the visible region, we see a continuous spectrum but with several dark lines. The lines look dark in the photo, but the matching spectral curve (intensity versus wavelength) shows they are relative reductions in brightness: less light is emitted at particular wavelengths compared to the continuous spectrum.
	If we look at a narrow part of the visible spectrum of the Sun, we see dips in the intensity at particular wavelengths of the intensity-versus-wavelength graph. The wavelengths of the dips in the solar spectrum match the wavelengths of the line spectra emitted by a sodium vapor lamp in a lab. Intensity is shown in relative units. You can see that light from the lamp is much brighter than light from the Sun.

Pattern

If we observe a narrow region (516.6–517.4 nm) of the solar spectrum, we see dips that match the wavelengths of the line spectra of specific elements (iron, magnesium, etc.). The spectra of other stars are similar—continuous emission spectra with specific dark lines.

Why would photons be missing at particular wavelengths? The first explanation that might come to mind is that sodium is not present on the surface of the Sun. However, this explanation does not work, because we find dark lines matching the emission lines for almost all elements known on Earth.

Another explanation is that the dark lines are not the result of missing elements, but of the absorption of photons traveling from the Sun to Earth. Imagine that some sodium atoms in the ground state exist in the region between the Sun and the observer on Earth. Sodium atoms can absorb those photons coming from the Sun's surface whose energies match possible energy transitions in the atoms (see Figure 28.10). Soon after absorption, the atoms reemit these photons but in random directions. Thus the original photons that were absorbed would be "missing" from the continuous spectrum arriving at Earth. If we take a picture of the arriving spectrum, it should have dark lines at the locations of the missing photons. This is known as an **absorption spectrum**—a continuous spectrum with missing photons of specific wavelengths. This explanation is currently accepted as the correct one for the absorption spectrum of the Sun and other stars.

Satellite spectrometers above Earth's atmosphere show the same dark lines in the Sun's spectrum that we observe on Earth. Thus, the absorption must occur just outside the Sun's hot surface, indicating the presence of different types of atoms in the gas surrounding the Sun.

Spectral analysis is used in chemistry, engineering, transportation, medicine, and other technological fields. The main principles for these applications are the same: capture and analyze the wavelengths of the photons, and compare their wavelengths to the wavelengths of the known elements emitted or absorbed by a particular object to determine its composition.

Why are plants green?

Light is needed for the process of photosynthesis in plants. The process starts when pigments absorb energy delivered by light photons. Chlorophyll extracted from soft leaves absorbs light most strongly in the blue portion of the electromagnetic spectrum, followed by the red portion. It is a poor absorber of green and near-green portions of the spectrum, which it reflects, producing the green color of chlorophyll-containing tissues. **Figure 28.17** shows the percentage of light energy absorbed by chlorophyll molecules at the wavelengths of red and blue light. The peaks are essentially inverted dips on an absorption spectrum that you are familiar with. Notice two plots, marked chlorophyll a and chlorophyll b. These represent two chemical variants of chlorophyll that are universal constituents of wild vascular plants. Both chlorophylls show absorption maxima at wavelengths corresponding to blue and red.

FIGURE 28.17 Absorption of light by chlorophyll.

REVIEW QUESTION 28.3 What materials and measuring instruments do you need in order to observe the absorption spectrum of a gas?

28.4 Lasers

When excited gas atoms emit light, they produce photons at different times and in different directions; thus the corresponding waves are incoherent (have different phases) and spread out in many directions. This process is called **spontaneous emission**. Lasers, on the contrary, emit very narrow, almost coherent beams of light. How is this accomplished?

In a laser, many excited atoms return from the same excited state to the same ground state simultaneously. The result is emission of in-phase photons traveling in the same direction. This process, called **stimulated emission**, was first suggested by Einstein. The word *laser* is an acronym for "light amplification by stimulated emission of radiation."

How lasers work

Imagine that you have an atom in an excited state. You also have a photon whose energy equals exactly the energy difference between the atom's excited state and the ground state (**Figure 28.18a**). This photon can cause the excited atom to "de-excite" and emit a photon of exactly the same frequency (Figure 28.18b)—the stimulated emission mentioned above. The passing photon does not lose any energy; it only triggers the process. The result is two photons of the same frequency traveling in the same direction.

FIGURE 28.18 Stimulated emission and production of laser light. (a) In the initial state, a photon whose energy exactly equals the difference in energy between the excited and ground states passes through, stimulating the excited atom. (b) In the final state, the excited atom's electron drops to the lower energy state, emitting a photon. Now two synchronized photons have the same energy.

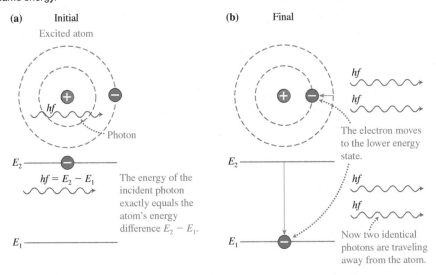

FIGURE 28.19 Steps leading to the production of a laser beam.

(a) Excitation

(b) Transition to metastable state

E_3

E_2

E_1

(c) An inverted population

E_3

E_2

Transition from state 2 to 1 is unlikely.

E_1

(d) An unlikely spontaneous transition from 2 to 1 starts the laser beam.

E_3

E_2

E_1

(e) A stimulated transition adds another photon to the beam.

E_3

E_2

E_1

To make stimulated emission effective, many more atoms must be in the same excited state than in the ground state, a situation known as **population inversion**. It is generally difficult to create population inversion because the average time that an atom spends in an excited state is very short (about 10^{-8} s). However, some atoms enter excited states called **metastable states**, in which they can remain for a significantly longer time (about 10^{-3} s). These atoms are good candidates for population inversion.

Suppose that the atoms in a material make the transition from ground state 1 to excited state 3 (**Figure 28.19a**) by absorbing photons. Instead of quickly returning to state 1, they make a more probable spontaneous transition to a metastable state 2 (Figure 28.19b). These atoms will remain in that state for a relatively long time since transitions from state 2 to state 1 are comparatively unlikely to occur. Thus, more atoms will be in state 2 than in the ground state 1 (Figure 28.19c). This is an example of population inversion.

One of these excited atoms eventually transitions to ground state 1 and emits a photon of energy $E_2 - E_1$ (Figure 28.19d). That photon passes through the material, many of whose atoms are in excited metastable energy state 2. This photon causes another atom in the metastable state 2 to de-excite via stimulated emission, resulting in two synchronized photons moving through the material (Figure 28.19e). Each of these photons can then stimulate another transition, resulting in a total of four photons. This process continues and creates an avalanche of new photons by stimulated emission—an intense laser beam (**Figure 28.20**).

The *lasing* material is placed between two mirrors that reflect the emitted photons back and forth through the material. In continuous lasers, one mirror is only partially reflecting and allows photons to leave the lasing region at a nearly constant rate.

FIGURE 28.20 Stimulated emission produces an increasing number of in-phase photons in the laser cavity.

Continual excitation of the atoms back to the metastable state provides a continuous source of stimulated emission photons.

Lasers can be used to transmit thousands of simultaneous telephone calls along thin glass fibers, read DVDs, scan bar codes, and remove certain kinds of tumors. Lasers are also used surgically to correct nearsightedness and farsightedness. All such procedures, including LASIK (laser-assisted in situ keratomileusis), PRK (photorefractive keratectomy), and other corneal ablation methods of vision correction, employ lasers to remove corneal tissue to reshape the eye so that it can focus an image directly on the retina.

Another important application is lidar (light detection and ranging). Lidar can be thought of as radar in which, instead of pulses of radio waves, pulses of laser light are used to measure the distance to an object. The differences in time of the return of the pulses and the changes in their wavelengths can be used to make three-dimensional images of the object. Lidar is used to make high-resolution maps.

REVIEW QUESTION 28.4 What is the main difference between spontaneous and stimulated emission of light?

28.5 Quantum numbers and Pauli's exclusion principle

Bohr's model was successful in explaining certain features of hydrogen's spectrum and structure, but it could not explain the structure of multi-electron atoms or some of the fine details of the hydrogen spectrum. We will describe one of these details next.

The Zeeman effect

In 1896 Dutch physicist Pieter Zeeman performed an experiment in which light from a known source passed through a region with sodium atoms in a strong magnetic field and then through a grating. He observed that when the magnetic field was on, a particular sodium spectral line was split into three closely spaced lines. The result of a similar experiment using mercury vapor is shown in Figure 28.21. The figure shows the split for the 404.7-nm spectral line in a 2.9-T magnetic field. The split is very small—about 1 nm. This phenomenon, which became known as the Zeeman effect, could only be explained after Bohr's model was extended.

Sommerfeld's extension of the Bohr model and additional quantum numbers

In 1915 Arnold Sommerfeld, a German theoretical physicist, modified Bohr's model to account for the Zeeman lines. Later, his modifications helped explain the spectral lines of more complex atoms. One of his innovations was the suggestion that electrons

FIGURE 28.21 Zeeman split of the 404.7-nm line of mercury in a 2.9-T magnetic field.

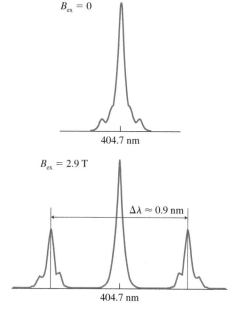

could have elliptical orbits. He proposed this through the introduction of two new quantum numbers related to the motion of the electron in the atom. (Prior to this, only one quantum number, n, was specified, and it was related to the energy and the size of the atom.) The new quantum numbers and the associated motion could explain the Zeeman effect.

In the Sommerfeld model, the n quantum number remains an indicator of the size and energy of the atom in a particular state, but not of the rotational (angular) momentum of the state. A new quantum number l, called the **orbital angular momentum quantum number**, identifies the angular momentum of a state whose value is $L = \sqrt{l(l + 1)}(h/2\pi)$, where $l = 0, 1, 2, \ldots, n - 1$ and $h/2\pi$ acts as a fundamental unit of angular momentum (see **Figure 28.22**).

FIGURE 28.22 The allowed electron $n = 1$ to 4 orbits in the Sommerfeld model.

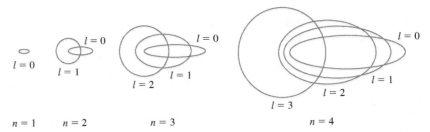

For the $n = 1$ energy state, l can only equal zero. Thus the angular momentum of that state could only be $L = \sqrt{0(0 + 1)}(h/2\pi) = 0$. For the $n = 2$ state, l can be zero or 1, resulting in two possible states $L = 0$ or $L = \sqrt{2}(h/2\pi)$. These two states correspond to electron orbits that look very different (see the shapes of the different n and l orbits in Figure 28.22). The orbit of the $l = 0$ state is that of an elongated, flat ellipse. Recall that the angular momentum of an object moving in an orbit is $L = mrv \sin \theta$, where θ is the angle between the r and v vectors. When the orbit is elongated, r and v are nearly parallel or antiparallel; therefore, θ is zero. Thus, $L = 0$. For the more circular higher L orbit, r and v are perpendicular and $\sin \theta$ is approximately 1.

According to the Sommerfeld model, in an $n = 2$ state, an electron zipping around the $l = 1$ orbit is the equivalent of an electric current loop, which therefore produces a magnetic dipole field. The elongated $l = 0$ orbit should not produce any magnetic field. This differs from Bohr's model, in which an electron in the $n = 1$ ground state has angular momentum and moves in a circular orbit, thus producing a circular current and a magnetic field. Sommerfeld's prediction is consistent with experimental results, which show that the ground state of hydrogen does not have this magnetic field. But how does the Sommerfeld model help explain the Zeeman effect?

Consider hydrogen. The electrons in most hydrogen atoms in a low-temperature gas are in the $n = 1$ ground state, the orbit nearest the nucleus. Suppose this cool gas is irradiated by photons of just the right frequency to excite an atom from the $n = 1$ state to one of the two possible $n = 2$ states of the atom. If the gas is placed in a strong magnetic field, the atoms in states with nonzero angular momentum would behave like tiny magnets (with their own dipole moments) and interact with the magnetic field. This would result in the slightly different energies of the two $n = 2$ states. The stronger the magnetic field, the greater the difference in energies of the states.

Zeeman's experiments on hydrogen gas using sensitive equipment and strong magnetic fields $(\vec{B}_{ex})$ produced three distinct spectral lines for transitions from the $n = 2$, $l = 1$ state to the $n = 1$, $l = 0$ state. This result could be explained if the $\vec{B}$ field due to the electron's motion in the $n = 2$, $l = 1$ state could be oriented in three discrete directions (see **Figure 28.23a**). This caused the $n = 2$, $l = 1$ state to split into three separate states (Figure 28.23b) due to different orientations of the electron orbit relative to the external magnetic field. The stronger the external magnetic field, the greater the splitting.

This reasoning led to the introduction of a third quantum number, the **magnetic quantum number** m_l. If this quantum number is allowed to have values

FIGURE 28.23 Explanation for the Zeeman effect.

(a)
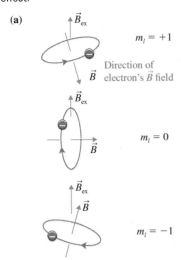

(b) Energy shifts due to the orientation of the magnetic dipole in the magnetic field

(c)
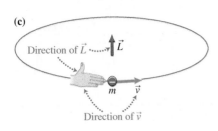

$m_l = 0, \pm 1, \pm 2, \ldots, \pm l$, then the Zeeman effect for the $n = 2$, $l = 1$ to $n = 1$, $l = 0$ transition is explained: three slightly different energy photons would be emitted, as shown in Figure 28.23b. Apparently, m_l is related to the orientation of the angular momentum vector of the electron in its orbit (the method for determining the orientation of the orbit's angular momentum is shown in Figure 28.23c). The component L_z of that angular momentum in the direction of $\vec{B}_{ex}$ is typically defined as the z-axis. L_z is then equal to m_l times the fundamental unit of angular momentum $h/2\pi$, or

$$L_z = m_l \frac{h}{2\pi} \qquad (28.10)$$

Understanding the splitting of spectral lines in a magnetic field led to many applications, specifically to studying different magnetic fields. By analyzing the wavelengths of the spectral lines in a magnetic field (like the split shown in Figure 28.21), scientists determined the magnitude of the $\vec{B}$ field of the Sun: about 4 T, 100,000 times stronger than the magnetic field of Earth.

Another difficulty with observed spectral lines

At about the same time that Zeeman made his observations, Thomas Preston performed Zeeman effect experiments with an even better magnet. He observed that some of the spectral lines for many elements did not split into three lines when a magnetic field was applied, but instead split into four, six, or more lines. This became known as the "anomalous" Zeeman effect. Sommerfeld's model could not explain it.

Around 1920 Sommerfeld and Alfred Landé suggested that the nucleus was affecting the angular momentum of the electron, thus producing this anomalous Zeeman effect. They incorporated this idea by assigning a quantum number to the nucleus, an ad hoc mathematical proposal with little physical model in mind.

Otto Stern and Walther Gerlach decided to test this idea by passing moving silver atoms through a nonhomogeneous magnetic field (**Figure 28.24**). A nonhomogeneous field was needed because the force that the magnetic field exerts on a stationary magnetic dipole differs from the force the field exerts on a single moving charged particle. This force is proportional to the rate of change per unit distance of the $\vec{B}$ field as opposed to just the magnitude of the $\vec{B}$ field. The combination of a pointed north pole of a magnet and a flat south pole produces a large rate of change in the vertical direction (Figure 28.24). Moving atoms were needed to make the split on the screen visible: the magnetic field would exert forces on two dipoles in opposite directions. If the Sommerfeld and Landé models were correct, then moving silver atoms should split into two streams, corresponding to the two values for the quantum number of the nucleus, rather than scatter randomly (the classical physics prediction). The experimental outcome matched the prediction based on the Sommerfeld and Landé model. However, even when a testing experiment matches the prediction, the model can still be wrong. Read on.

FIGURE 28.24 Stern-Gerlach experiment.

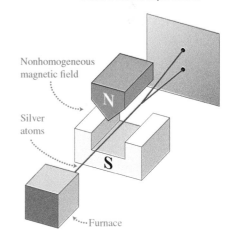

Nonhomogeneous magnetic field

Silver atoms

N

S

Furnace

A new fourth quantum number

In 1923–1924, Wolfgang Pauli disproved the Sommerfeld and Landé model using the theory of relativity. With relativistic corrections, the Landé and Sommerfeld model consequently made predictions that were contradicted by the Stern-Gerlach experiments.

In trying to explain the anomalous Zeeman effect, Pauli uncovered many unexplained patterns in the periodic table of the elements. Why did elements in the same column (group) have similar properties when they had very different numbers of electrons but very different properties from adjacent columns? Why didn't the properties of atoms change gradually with additional electrons? Why were atoms with an even number of electrons less chemically active than those with odd numbers? Why were elements in the periodic table with 2, 8, or 18 electrons almost chemically inert?

Finally, and perhaps most importantly, why at low temperatures weren't nearly all electrons in multi-electron atoms in the $n = 1$ ground state?

Pauli realized that these patterns in the periodic table and the anomalous Zeeman effect could be explained by a single, simple rule, namely, that each available state in the atom could hold at most *one* electron, an idea called the **Pauli exclusion principle**. This principle required the introduction of a fourth quantum number, but a different one from that of the Sommerfeld and Landé model. The fourth quantum number concerned the electron itself and was called m_s, the **spin quantum number**, because it was related to the electron's intrinsic angular momentum (called **spin**). Pauli reasoned that in a magnetic field, an electron's spin could orient along the direction of the external field $\vec{B}_{ex}$ or opposite to the external field $\vec{B}_{ex}$, but not in any other direction—like a compass needle that could point only precisely north or south. These two electron spin orientations correspond to values of $\pm 1/2$ for the quantum number m_s.

By suggesting this additional electron property, Pauli could explain the outcome of the Stern-Gerlach experiment and account for the anomalous Zeeman effect. Together with the exclusion principle, which states that the allowed states of an atom could be completely characterized by the four quantum numbers n, l, m_l, and m_s, and that only one electron could exist in each state, Pauli's ideas could account for many of the patterns observed in the periodic table (we will discuss the patterns in the periodic table in Section 28.7).

A failed attempt at a classical explanation for the spin quantum number

In 1925 Samuel Goudsmit and George Uhlenbeck attempted to explain the physical meaning of the m_s quantum number. For the electron to have angular momentum, it would have to be like a tiny spinning sphere, similar to Earth spinning about its axis (**Figure 28.25**). However, small charged parts Δq at the edge of this small spinning spherical shell of charged material would have to move faster than light speed to have the necessary angular momentum. This assumption contradicts special relativity. The spin of the electron seemed to be some sort of new property that could not be understood using classical physics.

Summary of quantum numbers

By 1925 scientists had four quantum numbers that could describe the state of an electron in an atom. Also, Pauli's exclusion principle accounted for the distribution of electrons in multiple-electron atoms. Let's summarize the four quantum numbers:

1. *Principal quantum number n* can be any positive integer:

$$n = 1, 2, 3, \ldots \tag{28.11}$$

and it indicates the energy of the electron in the atom. The larger the value of n, the larger the energy of the atom.

2. *Orbital angular momentum quantum number l* characterizes the orbital angular momentum of the electron. For a particular value of n, the allowed l values are

$$l = 0, 1, 2, \ldots, n - 1 \tag{28.12a}$$

The magnitude of the angular momentum of the electron is

$$L = \sqrt{l(l + 1)} \, \frac{h}{2\pi} \tag{28.12b}$$

3. *Magnetic quantum number m_l* describes the component of the electron's orbital angular momentum along an external magnetic field (Figure 28.23a). The energy of each l state in an external magnetic field splits according to the state's m_l value, which results in the additional lines in the spectra observed by Zeeman

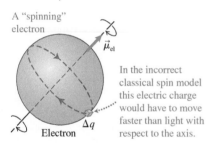

FIGURE 28.25 An incorrect classical model of electron spin.

A "spinning" electron

$\vec{\mu}_{el}$

In the incorrect classical spin model this electric charge would have to move faster than light with respect to the axis.

Δq

Electron

(Figure 28.23b). The magnetic quantum number m_l can have values that depend on the quantum number l:

$$m_l = 0, \pm 1, \pm 2, \ldots, \pm l \qquad (28.13a)$$

The component of the angular momentum in the z-direction is

$$L_z = m_l(h/2\pi) \qquad (28.13b)$$

4. *Spin magnetic quantum number m_s* Each electron has an intrinsic spin and corresponding magnetic dipole. The spin magnetic quantum number m_s indicates the two directions that this spin dipole can be oriented relative to an external magnetic field:

$$m_s = \pm\tfrac{1}{2} \qquad (28.14)$$

The spin quantum number doubles the number of states available for an electron in an atom.

Pauli's exclusion principle states that there is a relationship between these quantum numbers and the electrons in an atom, regardless of the type of atom.

Pauli exclusion principle Each electron in an atom must have a unique set of quantum numbers n, l, m_l, and m_s.

Consider the case of an $n = 2$ state for electrons in an atom. The possible values of l are 0 and 1. For $l = 0$, the only possible value of m_l is 0. For $l = 1$, the possible values of m_l are -1, 0, and 1. For each of these possibilities, m_s can be $\pm 1/2$. Thus, there are eight possible $n = 2$ states, as listed in Table 28.3.

An electron with $n = 2$ is said to be in the $n = 2$ shell of the atom. If a particular type of atom has eight $n = 2$ electrons, the atom is said to have a filled $n = 2$ shell—the shell can hold no more electrons.

You can use similar reasoning for other electron shells of atoms. You should find that the $n = 3$, $l = 2$ subshell of an atom can hold 10 electrons. The $n = 3$, $l = 1$ subshell can hold 6 electrons, and the $n = 3$, $l = 0$ subshell can hold 2 electrons. Thus, the $n = 3$ shell can hold a maximum of 18 electrons.

Pauli's ideas of atomic structure explain the possible states of electrons in atoms, and also the reason that not all electrons in multi-electron atoms are in the lowest energy state (the $n = 1$ state). However, these ideas do not explain how to calculate the values of the quantities that are based on these quantum numbers (energy, orbital angular momentum, etc.). Physicists needed to construct an entirely new theory for describing this subatomic world, the subject of the next section.

TABLE 28.3 The possible $n = 2$ states of an atom

$l = 0, m_l = 0, m_s = +1/2$
$l = 0, m_l = 0, m_s = -1/2$
$l = 1, m_l = -1, m_s = +1/2$
$l = 1, m_l = -1, m_s = -1/2$
$l = 1, m_l = 0, m_s = +1/2$
$l = 1, m_l = 0, m_s = -1/2$
$l = 1, m_l = +1, m_s = +1/2$
$l = 1, m_l = +1, m_s = -1/2$

REVIEW QUESTION 28.5 In what energy states are the 10 electrons in the atom neon? Label each by its set of four quantum numbers.

28.6 Particles are not just particles

While the work of Bohr, Sommerfeld, and Pauli explained many observed properties of atoms, it did not explain why quantum numbers existed. Why were the energies of atoms and the orbital and spin angular momenta of electrons quantized in the first place? Physicists began arriving at answers in the mid-1920s.

A proposal for the wave nature of particles

The theoretical foundation for this effort began in 1924 with a young French physicist, Louis de Broglie. Knowing that light had both wave and particle behaviors, he thought it seemed reasonable that the elementary constituents of matter might also have these behaviors. Given that the wavelength λ of a photon is $\lambda = h/p_{\text{photon}}$, de Broglie suggested that elementary particles also had a wavelength that could be determined in a similar way:

$$\lambda = \frac{h}{p} \qquad (28.15)$$

where h is Planck's constant and p is the particle's momentum (which could potentially be relativistic).

If matter has wave-like properties, then why don't macroscopic objects such as human beings have them as well? Using Eq. (28.15), the wavelength of a 50-kg person walking at speed 1 m/s would be 1.33×10^{-35} m. This wavelength is a hundred billion billion times smaller than the size of the nucleus of an atom. Thus, there is no real-world situation that would cause a macroscopic object to exhibit wave-like behavior. However, the smaller the mass of the object, the larger its wavelength. For example, for an electron (mass $m = 9.11 \times 10^{-31}$ kg) moving at a speed of 1000 m/s (a typical speed for electrons in cathode ray tubes), its wavelength would be

FIGURE 28.26 (a) Constructive and (b) destructive electron waves.

(a)

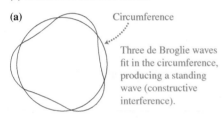

Circumference

Three de Broglie waves fit in the circumference, producing a standing wave (constructive interference).

(b)

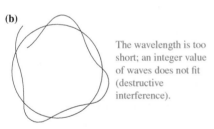

The wavelength is too short; an integer value of waves does not fit (destructive interference).

$$\lambda = \frac{6.63 \times 10^{-34} \text{ J} \cdot \text{s}}{(9.11 \times 10^{-31} \text{ kg})(1000 \text{ m/s})} = 7.3 \times 10^{-7} \text{ m} = 730 \text{ nm}$$

This is just a little longer than the wavelength of red light. Thus, the wave-like properties of atomic-sized particles might be observable, and might be very important for understanding their behavior.

De Broglie waves and Bohr's third postulate

De Broglie's radical idea contributed to building a new model of subatomic physics. Consider an electron moving around the nucleus of an atom in a stable Bohr orbit. Instead of modeling the electron as a point-like particle, de Broglie's ideas allow us to model it as a matter wave vibrating around the nucleus.

As the wave returns to the starting point in its orbit around the nucleus, it can interfere constructively with itself (**Figure 28.26a**); it can interfere destructively with itself; or some intermediate interference could occur (Figure 28.26b). If anything other than perfect constructive interference occurs, the wave will ultimately dampen and vanish as it circles many times around the nucleus. Thus, for the electron wave to be stable, an exact integer number n of electron wavelengths must be wrapped around the nucleus (**Figure 28.27**), with a circumference of $2\pi r$. Mathematically (and assuming the electron is not moving relativistically):

FIGURE 28.27 Electron standing waves.

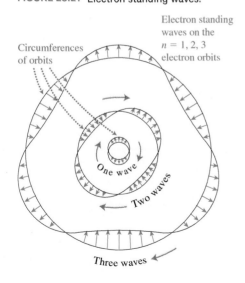

Electron standing waves on the $n = 1, 2, 3$ electron orbits

Circumferences of orbits

One wave

Two waves

Three waves

$$2\pi r = n\lambda = n\frac{h}{p} = n\frac{h}{mv}, \quad n = 1, 2, 3, \ldots$$

Because n must be a positive integer, this equation represents a quantization condition for the hydrogen atom. If we take this equation, multiply both sides by mv, and divide both sides by 2π, we get

$$mvr = n\frac{h}{2\pi}$$

This is precisely Bohr's third postulate [Eq. (28.2)]! Therefore, de Broglie's idea of the electron behaving like a standing wave explains Bohr's third postulate, which originally seemed rather arbitrary except that it led to the correct states of the hydrogen atom.

Testing de Broglie's matter wave idea

Recall our earlier experiment with a low-intensity photon beam irradiating a double slit (Section 27.4). The individual flashes of single photons on the screen gradually built up an interference pattern, suggesting that even individual photons had wave-like behavior. In Testing Experiment **Table 28.4**, we replicate that experiment using electrons instead of photons.

TESTING
EXPERIMENT TABLE 28.4   Do electrons behave like waves?

Testing experiment	Prediction	Outcome
We shoot electrons one by one toward a screen with two closely spaced narrow slits (similar to Young's double-slit experiment). The electrons that pass through the slits reach a screen that lights up (flashes) when hit. The electron speed is about 1000 m/s. 	If the electrons behave only like particles, there should be two images of the slits on the screen. Image of slits If the electrons possess wave-like properties, they should hit the screen like particles but gradually build up an alternating bright and dark interference pattern. Electron interference image 	An alternating bright and dark interference pattern gradually appears.

Conclusion

Electrons do indeed exhibit wave-like behavior. The experiment supports (that is, it does not disprove) de Broglie's hypothesis.

Another test of de Broglie waves

De Broglie's hypothesis was tested again a few years later. Shortly after de Broglie proposed his matter wave hypothesis, two scientists in New York, Clinton Davisson and his young colleague Lester Germer, were probing the structure of the atom by firing low-speed electrons at a nickel target made of small crystals and observing the resulting electron scattering (**Figures 28.28a and b**). The experiments did not give them any interesting results until they had a small accident in the laboratory in 1925: their

FIGURE 28.28 Davisson and Germer's experiment and accidental finding. (a) An electron beam excites a sample, and scattering is detected. (b) The scattering from a sample made up of small crystals. (c) The scattering from a sample accidentally turned into a large crystal.

equipment malfunctioned, and the nickel target previously consisting of many small crystals suddenly melted and resolidified into one large crystal.

When Davisson and Germer resumed the experiments with the large crystal, they saw a new pattern (compare Figure 28.28c with b). As they changed the position of the detector, they found alternating directions at which electrons were scattered and not scattered. The results made no sense to them until they heard about de Broglie's hypothesis. The large crystal was behaving like a three-dimensional interference grating, and they had observed the wave-like behavior of the electrons. This story is a perfect example of how accidental observations can turn into fundamental testing experiments, or how important serendipity is in science. Later, similar experiments performed with hydrogen nuclei and alpha particles supported the idea that all subatomic particles exhibit wave-like properties.

EXAMPLE 28.5 **Electron's wavelength**

An electron moves across a 1000-V potential difference in a cathode ray tube. What is its wavelength after crossing this potential difference?

Sketch and translate We've sketched the process below. The electron starts at the cathode (where the potential is 0 V) and accelerates to the anode (where the potential is +1000 V). We know the mass and the charge of the electron; thus we can determine its momentum at the end, and from its momentum, its wavelength.

Simplify and diagram We assume that the kinetic energy of the electron is zero at the cathode where it begins and that it moves nonrelativistically. We can represent the process with an energy bar chart. The system of interest is the electron and the electric field produced by the cathode ray tube. In the initial state, the system has zero kinetic and electric potential energy (since $V = 0$ at the cathode). In the final state, the system has positive kinetic energy and negative electric potential energy (since the electron's electric charge is negative).

Represent mathematically We can use the bar chart to apply energy constancy to the system ($-e$ represents the charge of the electron):

$$K_i + U_{qi} = K_f + U_{qf}$$

Inserting the appropriate energy expressions:

$$0 + 0 = \tfrac{1}{2}mv_f^2 + (-e)V_f$$

or

$$\tfrac{1}{2}mv_f^2 = eV_f$$

The electron's momentum when it reaches the anode is $p = mv_f$, and its wavelength, from Eq. (28.15), is $\lambda = h/p$.

Solve and evaluate Use the preceding energy equation to solve for the electron's final speed, v_f:

$$v_f = \sqrt{\frac{2eV_f}{m}}$$

We should check to see if the electron is nonrelativistic:

$$v_f = \sqrt{\frac{2(1.6 \times 10^{-19}\,\text{C})(1000\,\text{V})}{9.11 \times 10^{-31}\,\text{kg}}} = 1.9 \times 10^7\,\text{m/s}$$

This speed is extremely fast, but only about 6% of light speed, so using the low-speed expression for momentum was justified. Combining this with the expression for the momentum and wavelength of the electron, we get

$$\lambda = \frac{h}{p} = \frac{h}{mv_f} = \frac{h}{m\sqrt{\dfrac{2eV_f}{m}}} = \frac{h}{\sqrt{2meV_f}}$$

Inserting the appropriate values:

$$\lambda = \frac{6.63 \times 10^{-34}\,\text{J}\cdot\text{s}}{\sqrt{2(9.11 \times 10^{-31}\,\text{kg})(1.6 \times 10^{-19}\,\text{C})(1000\,\text{V})}}$$

$$= 3.9 \times 10^{-11}\,\text{m}$$

This is a tiny wavelength, the same order of magnitude as the wavelength of X-rays and the size of an atom! This knowledge of the value of the electron's wavelength led to the development of microscopes that use electrons to image individual atoms, a task that would be impossible with visible light.

Try it yourself Imagine that the electrons in this example were used to investigate the structure of a crystal. When the electrons reflect off the crystal, they form an interference pattern. What would happen to the distances between the peaks in the pattern if the electrons were accelerated by a potential difference that is half the value used in this example?

Answer

The electron wavelength would be longer, and the peaks in the pattern would be spread farther apart.

Wave functions

Louis de Broglie's revolutionary hypothesis changed the way scientists viewed elementary particles. However, it was not clear exactly what was "waving" in a de Broglie matter wave. In a water wave, the water "waves." In a light wave the $\vec{E}$ and $\vec{B}$ fields are "waving." But what is "waving" in an electron wave?

De Broglie himself thought of the wave as some sort of vibrating shroud surrounding the true electron (which he still thought of as a point-like particle). This idea is known as de Broglie's **pilot wave model**; the wave guides the actual particle within it. Others thought the electron wasn't point-like and was "smeared" over a certain volume; it was this smeared electron that was vibrating as it moved through space.

In 1926, German scientist Max Born suggested that a particle's matter wave is a mathematical distribution related to the likelihood of measuring the particle to be at various locations at a specific time. If the value (technically, the square of the value) of this matter wave at a particular location is near zero, then there is almost zero chance of detecting the particle at that location at that time. If the value is large, then there is a large chance of finding the electron there.

Erwin Schrödinger coined the term **wave function** for these matter waves and developed an equation for determining them. The field of **quantum mechanics** is based on understanding the particles of the micro-world in terms of wave functions and the Schrödinger equation (the mathematics of which is beyond the scope of this textbook). Based on all the above, now we can see electrons not as particles with a specific location, but as clouds smeared in three-dimensional space.

Atomic wave functions

Understanding matter in terms of wave functions lets us construct a better model of the hydrogen atom. When the Schrödinger equation is used to determine the electron wave functions, three of the quantum numbers (n, l, and m_l) appear naturally, without needing to be put in by hand. The fourth number, the spin magnetic quantum number m_s, appears naturally when quantum mechanics is extended so that it incorporates Einstein's theory of special relativity. This model resulted in a detailed understanding of the periodic table of the elements based on fundamental physics ideas. We will learn later in the book (Chapter 30) that this theory of relativistic quantum mechanics predicts the existence of antimatter and ultimately evolves into the standard model of particle physics, a theory that describes all the known elementary particles and their interactions.

REVIEW QUESTION 28.6 How did the Davisson and Germer experiment serve as a testing experiment for the idea that subatomic particles possess wave-like properties?

28.7 Multi-electron atoms and the periodic table

Together, the four quantum numbers n, l, m_l, and m_s specify the wave functions of possible electron states in atoms. Examples of the square of the $n = 2$ wave functions (that is, the values that generate the probability waves) are shown in **Figure 28.29**. The more darkly shaded regions indicate where the electron's probability wave has larger values; therefore, there is a greater probability of finding the electron at those locations. The $l = 0$ state actually has a greater probability than the $l = 1$ and 2 states of the electron's being at the nucleus—the dark region at the center of its probability distribution. This depiction provides an explanation for the lower energy in multi-electron atoms of the $l = 0$ electron states.

TIP In the previous paragraphs (and throughout the rest of this section) we use the word "particle" to refer to a fundamental constituent of matter (an electron, for example), even though we have learned that these constituents have both wave- and particle-like behaviors.

FIGURE 28.29 Probability distributions (wave functions squared) for $n = 2$ states.

(a)

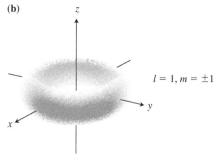

Dark areas have high probability of the electron being present ...

$l = 0, m = 0$

... while light areas have low probability of the electron being present.

(b)

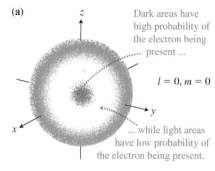

$l = 1, m = \pm 1$

(c)

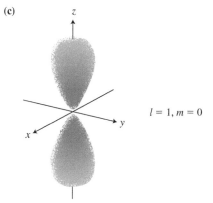

$l = 1, m = 0$

Subshells and number of electrons in them

Early in the study of the emission and absorption of light by atoms, scientists used letters to designate the different states of orbital angular momentum quantum number l. This notation is still widely used today. Table 28.5 summarizes the letter designations. A state with $n = 1$ and $l = 0$ is called a $1s$ state; an $n = 4$ and $l = 2$ state is called a $4d$ state.

TABLE 28.5 Letter designations of different l states and maximum number of electrons in that subshell

l value	0	1	2	3	4	5	6
Letter	s	p	d	f	g	h	i
Maximum number of electrons	2	6	10	14	18	22	26

Specifying the n and l quantum numbers specifies a group of states called a **subshell**. Specifying just the n quantum number specifies a group of subshells that is called a **shell**. The l quantum number also determines the maximum number of electrons that can populate that subshell.

How is the maximum number of electrons in the third row of Table 28.5 determined? Let's look at the $n = 1$, $l = 0$ s subshell. In this subshell, m_l can only be zero and m_s can be $+1/2$ or $-1/2$. Thus, there are two different states with $n = 1$ and $l = 0$. According to Pauli's exclusion principle, each state can hold just one electron; so the $1s$ shell can hold at most two electrons. This reasoning holds for any s state.

A subshell is said to be filled if each state in the subshell is occupied by an electron. The number of states in a subshell depends on the number of different allowed combinations of m_l and m_s for the given value of l. In the following exercise we consider the number of states in an $l = 2$ (or d) subshell.

CONCEPTUAL EXERCISE 28.6   **States in 3d subshell**

Determine the number of states and the quantum number designation of each state for the $3d$ subshell.

Sketch and translate For a $3d$ subshell, $n = 3$ and $l = 2$ (see Table 28.5). For an $l = 2$ state, the m_l quantum number can have the values $m_l = -2, -1, 0, 1,$ or 2. For each of these five values of m_l, the m_s quantum number can have values $m_s = -1/2$ or $+1/2$. Thus, there are 10 unique quantum states for the $3d$ subshell (5 m_l values times 2 m_s values).

Simplify and diagram The 10 states and their corresponding quantum numbers are summarized in Table 28.6. If one electron resides in each of these states, such as occurs in a copper atom, the atom is said to have a filled $3d$ subshell.

- -

Try it yourself How many electrons can be in a $4f$ subshell?

Answer

14 because $l = 3$, $m_l = -3, -2, -1, 0, 1, 2,$ or 3, and $m_s = -1/2$ or $+1/2$.

TABLE 28.6 The number of states and the quantum number designation of each state for the 3d subshell

n	l	m_l	m_s
3	2	-2	$+1/2$
3	2	-1	$+1/2$
3	2	0	$+1/2$
3	2	1	$+1/2$
3	2	2	$+1/2$
3	2	-2	$-1/2$
3	2	-1	$-1/2$
3	2	0	$-1/2$
3	2	1	$-1/2$
3	2	2	$-1/2$

Electron configuration for the ground state of an atom

The electrons in an atom normally occupy the states with lowest energy. In this case the atom is said to be in its ground state. An electron configuration for the ground state of an atom indicates the subshells occupied by its electrons and the number of electrons in each subshell. For example, a neutral sodium atom has 11 electrons distributed as follows: $1s^2 2s^2 2p^6 3s^1$. The superscript represents the number of electrons in each subshell: $1s^2$ indicates that two electrons occupy the $1s$ subshell, $2p^6$ indicates that six electrons occupy the $2p$ subshell, and so forth. The subshell $3s$ has only a single electron. The maximum numbers of electrons that can be in a subshell are listed in Table 28.7. The order of filling is shown in Figure 28.30. Note, for example, that electrons in the $4s$ subshell have lower energy than electrons in the $3d$ subshell. We see that for sodium, 10 electrons are needed to fill the $1s$, $2s$, and $2p$ subshells. The 11th electron in sodium occupies a state with the next available lowest energy subshell, the $3s$ subshell. When sodium is ionized to form Na^+, it loses the electron from the $3s$ subshell since that electron is most weakly bound to the nucleus.

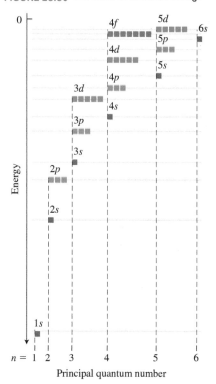

FIGURE 28.30 The order of subshell filling.

TABLE 28.7 Atomic subshells from lowest to highest energy (approximate)

	Quantum Numbers		Number of Quantum States	
n	l	m_l	In the Subshell	Total
1	0 (s)	0	2	2
2	0 (s)	0	2	8
	1 (p)	−1, 0, +1	6	
3	0 (s)	0	2	18
	1 (p)	−1, 0, +1	6	
	2 (d)	−2, −1, 0, +1, +2	10	
4	0 (s)	0	2	32
	1 (p)	−1, 0, +1	6	
	2 (d)	−2, −1, 0, +1, +2	10	
	3 (f)	−3, −2, −1, 0, +1, +2, +3	14	

The periodic table

Many of the properties of an atom, such as its tendency to form positive or negative ions and its tendency to combine with certain other atoms to form molecules, depend on the number of electrons in its outermost subshell and on the type of subshell it is. In addition, groups of atoms with the same number of outer electrons in subshells with the same letter designation (*s, p, d,* etc.) usually exhibit similar properties. This leads naturally to the classification of the atoms as presented in the **periodic table of the elements** (Table 28.8, on the next page). In the table, elements in the same column have similar electron configurations for their outer electrons and therefore similar chemical properties.

REVIEW QUESTION 28.7 What is the difference between a shell and a subshell?

TABLE 28.8 Periodic table of the elements*

I	II											III	IV	V	VI	VII	0
1 H 1.0080																	2 He 4.0026
3 Li 6.941	4 Be 9.0122											5 B 10.81	6 C 12.011	7 N 14.0067	8 O 15.9994	9 F 18.9984	10 Ne 20.179
11 Na 22.9898	12 Mg 24.305	transition										13 Al 26.9815	14 Si 28.086	15 P 30.9738	16 S 32.06	17 Cl 35.453	18 Ar 39.948
19 K 39.102	20 Ca 40.08	21 Sc 44.956	22 Ti 47.90	23 V 50.941	24 Cr 51.996	25 Mn 54.9380	26 Fe 55.847	27 Co 58.9332	28 Ni 58.71	29 Cu 63.54	30 Zn 65.37	31 Ga 69.72	32 Ge 72.59	33 As 74.9216	34 Se 78.96	35 Br 79.909	36 Kr 83.80
37 Rb 85.467	38 Sr 87.62	39 Y 88.906	40 Zr 91.22	41 Nb 92.906	42 Mo 95.94	43 Tc (99)	44 Ru 101.07	45 Rh 102.906	46 Pd 106.4	47 Ag 107.870	48 Cd 112.40	49 In 114.82	50 Sn 118.69	51 Sb 121.75	52 Te 127.60	53 I 126.9045	54 Xe 131.30
55 Cs 132.906	56 Ba 137.34	57 La 138.906	72 Hf 178.49	73 Ta 180.948	74 W 183.85	75 Re 186.2	76 Os 190.2	77 Ir 192.2	78 Pt 195.09	79 Au 196.967	80 Hg 200.59	81 Tl 204.37	82 Pb 207.2	83 Bi 208.981	84 Po (210)	85 At (210)	86 Rn (222)
87 Fr (223)	88 Ra 226.03	89 Ac 227.028	104 Rf (261)	105 Db (262)	106 Sg (266)	107 Bh (264)	108 Hs (269)	109 Mt (268)	110 Ds (271)	111 Rg (272)	112 Cn (285)	113 Nh (286)	114 Fl (289)	115 Mc (288)	116 Lv (293)	117 Ts (294)	118 Og (294)

Transition elements

Lanthanide series

58 Ce 140.12	59 Pr 140.908	60 Nd 144.24	61 Pm (147)	62 Sm 150.4	63 Eu 151.96	64 Gd 157.25	65 Tb 158.925	66 Dy 162.50	67 Ho 164.930	68 Er 167.26	69 Tm 168.934	70 Yb 173.04	71 Lu 174.97

Actinide series

90 Th 232.038	91 Pa 231.036	92 U 238.029	93 Np 237.048	94 Pu (242)	95 Am (243)	96 Cm (248)	97 Bk (249)	98 Cf (249)	99 Es (254)	100 Fm (257)	101 Md (258)	102 No (259)	103 Lr (260)

*Atomic masses of stable elements are those adopted in 1969 by the International Union of Pure and Applied Chemistry. For those elements having no stable isotope, the mass number of the "most stable" isotope is given in parentheses.

28.8 The uncertainty principle

Quantum mechanics provided a theoretical basis for observations of spectral lines. It also explained how different chemical elements could have similar chemical properties. Another profound prediction of quantum mechanics concerned the determinability of certain physical quantities.

Think of a classical particle. The ability to determine its velocity and its position would appear to be limited only by the precision of the measuring instruments. However, quantum mechanics predicts something quite different for subatomic particles.

Recall that if we send a laser beam through a single slit, it forms a pattern of bright and dark fringes on the screen behind the slit. What if you send a beam of electrons though the same horizontally oriented single slit and use a scintillating screen to register the electrons after they pass through the slit (**Figure 28.31**)? Electrons hit the screen at different locations. Instead of a sharp horizontal image of the slit as shown in Figure 28.31a, we see a set of "bright" and "dark" bands in the y-direction, parallel to the slit, as in Figure 28.31b. The pattern looks similar to the diffraction pattern that light produced when passing through a slit. Electron waves are being diffracted by the slit. The angular position θ_1 of the first dark region (where no electrons are detected) above or below the central maximum is determined by the light diffraction equation [Eq. (24.6) from Section 24.5]:

$$\sin \theta_1 = \frac{\lambda}{w}$$

Here w is the slit width and λ is the electron's wavelength (**Figure 28.32**).

Now, let's consider a single electron passing through the slit. Because the slit has width w, the electron's y-coordinate position can be determined only to within w. However, in order to hit the screen in any location other than directly across from the slit, the electron must have some y-component of momentum once it has passed through the slit. (We're considering the particle's momentum rather than velocity because the momentum determines its wavelength.)

TIP ▸ Remember that a significant diffraction pattern is being produced because of the wave-like properties of the electron combined with the narrowness of the slit.

A connection exists between the narrowness of the slit and the apparent y-component of the electron's momentum once it passes through. The narrower the slit, the larger the y-component of the electron's momentum might be. To put it another way, the more precisely we attempt to determine the y-coordinate of the electron's position y as it passes through the slit (by using a narrower slit), the less precisely we can determine the y-component of the electron's momentum p_y. It seems there is a limit on how precisely one can simultaneously determine the position y and momentum component p_y of the electron.

Let's see if we can make this idea more quantitative. The angular location of the first dark fringe is

$$\sin \theta_1 = \frac{\lambda}{w}$$

Inserting the expression for the wavelength of the electron, we have

$$\sin \theta_1 = \frac{(h/p)}{w}$$

Use Figure 28.32 to write $\sin \theta_1$ in terms of momentum:

$$\sin \theta_1 = \frac{p_y}{p} = \frac{(h/p)}{w} = \frac{h}{wp}$$

Combining and rearranging the second and fourth terms above, we get

$$wp_y = h$$

Note that w is the width of the slit and quantifies how determinable the y-position of the electron is (all we know is that the electron went through the slit). Assuming that nearly all of the electrons are detected between the first dark fringes on either side of the central maximum, p_y quantifies how determinable the y-component of the momentum of an electron is once it has passed through the slit (all we know is that it reaches the screen between the two dark fringes). We indicate this by saying $w = \Delta y$ and $p_y = \Delta p_y$. The relationship then becomes

$$\Delta y \, \Delta p_y \approx h$$

FIGURE 28.31 Electron diffraction patterns.

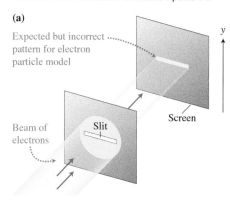

(a)

Expected but incorrect pattern for electron particle model

Beam of electrons

Slit

Screen

y

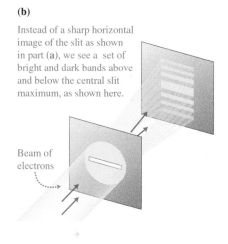

(b)

Instead of a sharp horizontal image of the slit as shown in part (**a**), we see a set of bright and dark bands above and below the central slit maximum, as shown here.

Beam of electrons

FIGURE 28.32 Single-slit diffraction with electrons.

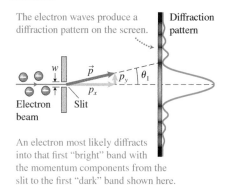

The electron waves produce a diffraction pattern on the screen.

Diffraction pattern

w

$\vec{p}$

p_y

θ_1

p_x

Electron beam

Slit

An electron most likely diffracts into that first "bright" band with the momentum components from the slit to the first "dark" band shown here.

Since the right-hand side of this equation is not zero, neither Δy nor Δp_y is exactly determinable. These limitations did not come from using a particular measuring device. They result solely from the electron having wave-like properties. Quantum mechanics sets a fundamental limit on how well the position and momentum of a particle in a particular direction can be determined simultaneously.

A more detailed analysis of the situation leads to an improved relationship (we switched to the x-direction, as it is the typical way to write this expression, but the same relations can be derived for the y- and z-directions):

$$\Delta x \, \Delta p_x \geq \frac{h}{4\pi} \qquad (28.16)$$

The factor 4π is a result of a more careful statistical analysis. Equation (28.16) is one example of what is known as the **uncertainty principle** of quantum mechanics, formulated by German physicist Werner Heisenberg in 1925.

Uncertainty principle for energy

Imagine that we have an atom that can potentially emit a photon of energy E. Remember that $E = pc$ for a photon. But because a fundamental limit exists on the determinability of the momentum of a particle, a fundamental limit should also exist on the determinability of the energy of the photon:

$$\Delta E = (\Delta p)c$$

Combining this with the uncertainty principle, we have

$$\Delta x \left(\frac{\Delta E}{c}\right) \geq \frac{h}{4\pi}$$

Photons travel at the speed of light, so they travel a distance Δx in a time interval $\Delta t = (\Delta x / c)$. Therefore, we can express Δx equivalently as $c\Delta t$. The equation then becomes

$$\Delta E \, \Delta t \geq \frac{h}{4\pi} \qquad (28.17)$$

In words, if a system (such as an atom in an excited state) exists for a certain time interval Δt before it emits a photon, then the energy of the state is only determinable within a range of ΔE. Thus, the energy states of an atom should be represented as "bands" whose thickness is determined by how long the atom can be in this state—the shorter-lived the state, the greater the energy uncertainty and the wider its energy spread (**Figure 28.33**).

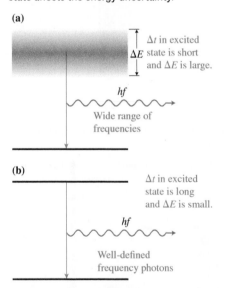

FIGURE 28.33 The lifetime Δt in an excited state affects the energy uncertainty.

(a)

Δt in excited state is short and ΔE is large.

hf

Wide range of frequencies

(b)

Δt in excited state is long and ΔE is small.

hf

Well-defined frequency photons

QUANTITATIVE EXERCISE 28.7 **Width of an energy state**

The lifetime of a hydrogen atom in the first excited state ($n = 2$) is about 10^{-9} s. Estimate the *width* of this energy state (the determinability of the energy of the state).

Represent mathematically Using Eq. (28.17) we can relate the uncertainty in the energy of the state to its lifetime:

$$\Delta E \, \Delta t \geq \frac{h}{4\pi}$$

Solve and evaluate Solving for ΔE, we have

$$\Delta E = \frac{h}{4\pi \, \Delta t}$$

We have changed the inequality to an equal sign, so we are calculating the minimum determinability of the energy of the state. Inserting the appropriate values gives

$$\Delta E = \frac{6.63 \times 10^{-34} \text{ J} \cdot \text{s}}{4\pi (10^{-9} \text{ s})} = 5.28 \times 10^{-26} \text{ J} \approx 5 \times 10^{-26} \text{ J}$$

Comparing this uncertainty to the value of the energy when in that energy state, 5.4×10^{-19} J, we see that the determinability of the energy is less than one-millionth of the value of the energy. This means that the energy of the $n = 2$ state is sharply defined, though not exactly determinable.

When we perform any measurements in physics we always encounter experimental uncertainty, although classical physics places no fundamental limits on how precise a measurement can be. Moreover, the uncertainty in one measurement generally is not related to the uncertainty in some other measurement. In the subatomic world the situation is different. The wavelike properties of particles make many pairs of physical quantities not simultaneously exactly determinable, for example, the components of position and momentum of a particle in a particular direction, or the energy and lifetime of a quantum state.

Tunneling

The energy-time interval uncertainty relationship, $\Delta E \Delta t \geq (h/4\pi)$, predicts that for a short time interval Δt the energy of a quantum system is only determinable to within $\Delta E \approx h/(4\pi \Delta t)$. In other words, energy conservation can be violated by an amount ΔE for a short time $\Delta t < h/(4\pi \Delta E)$. This means that the energy even of an isolated system is not constant in time but fluctuates. This result leads to an important phenomenon called **quantum tunneling**, which allows particles to have nonzero chances of existing in locations where classical physics forbids their existence.

In classical mechanics, the energy of a system is exactly determinable for all practical purposes because it is a macroscopic system. (Imagine a skier not having enough speed to go over a hill—he stops on the side of the hill before getting over the top.) However, a subatomic system's energy is not exactly determinable, so a particle that seems to be totally trapped actually isn't.

Consider a particle interacting with other particles so that the potential energy U of the particle-surroundings system as a function of position x looks as shown in **Figure 28.34** (the rectangular hump is the energy barrier). The horizontal line represents the system's energy E. The particle's kinetic energy at each position is then $K = E - U$. A classical particle on the left side of the barrier bounces back and forth between positions 0 and x_1. It cannot escape that region because its kinetic energy would be negative between positions x_1 and x_2 (the region where $U > E$), and that's impossible. In quantum mechanics, however, the square of a wave function determines the probability of finding the particle at different locations. Solving the Schrödinger equation for this situation reveals that the amplitude of the wave function decreases exponentially into the barrier region, rather than immediately becoming zero, meaning that after the barrier there is some nonzero wave function. This result means there is a nonzero probability that the particle will be detected on the right side of the barrier. It will have "tunneled through" the barrier.

FIGURE 28.34 A particle can pass through an energy barrier.

Quantum tunneling allows the particle to cross this barrier, where the kinetic energy is negative.

Hydrogen bonds and proton tunneling in DNA

A DNA molecule, which stores the genetic code, is made up of two complementary strands that coil about each other in a helical form. The two strands pair with each other through hydrogen bonding between adenine (A) and thymine (T) molecules and between cytosine (C) and guanine (G) molecules. A hydrogen bond forms when a hydrogen atom bonds with a more electronegative atom (such as oxygen or fluorine).

FIGURE 28.35 Hydrogen bonding between the two strands of a DNA molecule.

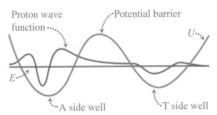

In each strand of the DNA molecule, an electron pair extends from either a nitrogen or an oxygen atom (consider the top bond in the A-T pair of molecules in **Figure 28.35**). The proton nucleus of a hydrogen atom, H in the figure, resides in a potential well created by the electron pair, represented by the two dots (⦙) on the A side of the figure. On the T side is another electron pair where the electric potential well is not as deep. The proton usually resides on the A side, but it can tunnel through the barrier between the wells to briefly reside on the T side (see **Figure 28.36**). In the next bond down in Figure 28.35, the well is deeper on the T side, and the proton spends more time there than on the A side.

FIGURE 28.36 Energy wells for a hydrogen nucleus associated with DNA strands. In a pairing between A and T, some small probability exists that a hydrogen nucleus can cross the kinetic energy barrier between the wells.

A new DNA molecule is formed when these H bonds are unzipped and new A, T, C, or G molecules bond with the complementary molecule on each side of the unzipped DNA to form two identical DNA molecules. However, there is a slight chance that a proton will have tunneled to the wrong side during the unzipping process. In that case, a different code is produced in the new DNA—a mutation occurs.

Let's consider the implications of this situation. Because the proton is a tiny particle that must be described using the principles of quantum mechanics, it is *impossible* for the genetic code to be 100% stable. A finite probability exists that because of proton tunneling, the proton will be on the wrong side of its bond when DNA is replicated. This approximately 1-nm displacement of protons through classically forbidden barriers plays a small part in the evolution of living organisms.

REVIEW QUESTION 28.8 What experimental evidence supports the uncertainty principle?

Summary

Nuclear model of atom The scattering of alpha particles at large angles from a thin foil indicated that the mass of an atom must be localized in a tiny nucleus (about 10^{-15} m in diameter) at the center of the atom. Low-mass electrons circle the nucleus. (Section 28.1)

Bohr's outdated model of one-electron atom
- A negatively charged electron moves in a stable circular orbit about a positively charged nucleus.
- In the stable orbits, the angular momentum of the electron is quantized and equal to a positive integer n times $h/2\pi$.
- When an electron transitions from one stable orbit to another, the atom's energy changes and a photon is emitted or absorbed. (Section 28.2)

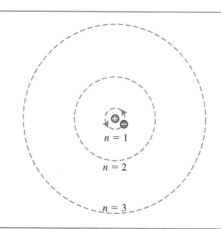

$$hf = E_i - E_f$$

Wave-particle duality EM radiation has a dual wave-particle (photon) behavior. Subatomic particles also have dual wave-particle behaviors. The wavelength $\lambda_{particle}$ depends on the momentum p of the particle. (Section 28.6)

$\lambda = h/p$

$$\lambda_{particle} = h/p \qquad \text{Eq. (28.15)}$$

Quantum numbers Electron states in atoms are described by four quantum numbers:

Principal quantum number n indicates the energy of the state, but *not* the angular momentum as in Bohr's atom.
Orbital angular momentum quantum number l indicates the orbital angular momentum L of the electron wave and its shape.
Magnetic quantum number m_l indicates the projection L_z of the electron's angular momentum on the direction of an external $\vec{B}_{ex}$ field.
Spin magnetic quantum number m_s indicates the two directions that the electron's spin can be oriented relative to an external $\vec{B}_{ex}$ field. (Section 28.5)

Direction of "electric current" due to electron's motion

$$n = 1, 2, 3, \ldots \qquad \text{Eq. (28.11)}$$

$$l = 0, 1, 2, \ldots, n-1 \qquad \text{Eq. (28.12a)}$$

$$L = \sqrt{l(l+1)}\,\frac{h}{2\pi} \qquad \text{Eq. (28.12b)}$$

$$m_l = 0, \pm 1, \pm 2, \ldots \pm l \qquad \text{Eq. (28.13a)}$$

$$L_z = m_l\,(h/2\pi) \qquad \text{Eq. (28.13b)}$$

$$m_s = \pm 1/2 \qquad \text{Eq. (28.14)}$$

Pauli exclusion principle Each available state in the atom can hold at most one electron. Each electron in an atom must have a unique set of quantum numbers n, l, m_l, and m_s. (Section 28.5)

Uncertainty principle The products of the uncertainties of position and momentum and of energy and lifetime of a state must be greater than or equal to $h/4\pi$. (Section 28.8)

$$\Delta x\, \Delta p_x \geq \frac{h}{4\pi} \qquad \text{Eq. (28.16)}$$

$$\Delta E\, \Delta t \geq \frac{h}{4\pi} \qquad \text{Eq. (28.17)}$$

Questions

Multiple Choice Questions

1. What could the plum pudding model of the atom account for (choose all answers that apply)?
   (a) The mass of the atom
   (b) The electric charge of the atom
   (c) Line spectra
   (d) The stability of the atom
   (e) Scattering experiments

2. What could the planetary model account for (choose all answers that apply)?
   (a) The mass of the atom
   (b) The electric charge of the atom
   (c) Line spectra
   (d) The stability of the atom
   (e) Scattering experiments

3. What was Bohr's model?
   (a) A revolutionary physics model that established modern physics
   (b) A compromise between classical and modern physics to account for experimental evidence
   (c) A classical model, as it relied on Newton's laws and Coulomb's law.

4. If you consider the energy of an electron-nucleus system to be zero when the electron is infinitely far from the nucleus and at rest, then
   (a) the energy of a hydrogen atom is positive.
   (b) the energy of a hydrogen atom is negative.
   (c) the sign of the energy depends on where in the atom the electron is located.

5. If you choose the electron to be the system (as opposed to an electron-nucleus system), its potential energy in the atom
   (a) is positive.   (b) is negative.
   (c) is zero.   (d) a single particle does not have potential energy.

6. Which of the spectra in **Figure Q28.6** correspond to each of the following three processes?
   (I) Light emitted by a gas placed in a strong electric field
   (II) Light emitted by a tungsten wire at a very high temperature
   (III) Light from the Sun reaching Earth

FIGURE Q28.6

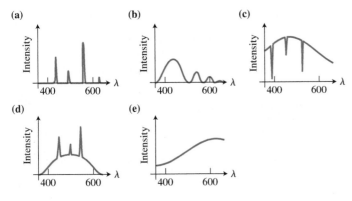

7. The Sun's emission spectrum is
   (a) continuous.   (b) line.
   (c) a combination of continuous and line.   (d) none of those.

8. The Sun's spectrum has
   (a) a continuous emission component and an emission line component.
   (b) a continuous emission component and an absorption line component.
   (c) a continuous absorption component and an emission line component.

9. What do the dark lines in the Sun's spectrum indicate?
   (a) Elements that are missing from the Sun
   (b) Elements that are present in the Sun's atmosphere
   (c) Elements that are present in Earth's atmosphere

10. What does wave-particle duality mean?
    (a) Waves and particles have the same properties.
    (b) Particles are simultaneously waves.
    (c) The behavior of elementary particles can be described simultaneously using physical quantities that describe the behavior of waves and of particles.

11. What does the uncertainty principle define?
    (a) The experimental uncertainty of our measurement
    (b) The uncertainty of our knowledge of some physical quantities that we try to determine simultaneously
    (c) The limits of our knowledge about nature

12. Why does the term "electron orbit" have no meaning in quantum mechanics?
    (a) The electron is considered to be at rest.
    (b) The electron does not follow a particular orbit.
    (c) The electron is a cloud smeared in a three-dimensional space.

Conceptual Questions

13. How do we know that the plum pudding model is not an accurate model of an atom?

14. How do we know that the planetary model is not an accurate model of the atom?

15. How do we know that Bohr's model is not an accurate model of the atom?

16. Why study atomic models that we do not think are accurate?

17. How is the contemporary model of the atom different from the planetary model? From Bohr's model?

18. Explain carefully why the scattering at large angles of alpha particles from atoms is inconsistent with the plum pudding model of the atom but is consistent with the nuclear model.

19. Describe how an energy state diagram helps explain the emission and absorption of photons by an atom.

20. Use available resources to learn what the Lyman and Paschen series are. Draw an energy diagram describing the series.

21. How can one determine if the Sun contains iron?

22. If you do not see helium absorption lines in the Sun's spectrum, does it mean that helium is missing from the Sun? Explain your answer.

23. Explain how an atom with only one electron can produce so many spectral lines.

24. A group of hydrogen atoms is contained in a tube at room temperature. Infrared radiation, visible light, and ultraviolet rays pass through the tube, but only the ultraviolet rays are absorbed at the wavelengths of the Lyman series. Explain.

25. If the principal quantum number n is 5, what are the values allowed for the orbital quantum number l? Explain how the maximum possible l quantum number is related to the n quantum number.

26. If the orbital quantum number is 3, what are possible values for the magnetic quantum number?

27. Explain why an electron cannot move in an orbit whose circumference is 1.5 times the electron's de Broglie wavelength.

28. If Planck's constant were approximately 50% bigger, would atoms be larger or smaller? Explain your answer.

29. Why do fluorine and chlorine exhibit similar properties?

30. Why do atoms in the first column of the periodic table exhibit similar properties?

31. How do astronomers know that atoms of particular elements are present in a star?

32. Explain how some elements can go undetected when they are actually present in a star.

33. What are the differences between spontaneous emission of light and stimulated emission?

34. What property of laser light makes it useful for (a) cutting metal and (b) carrying information?

Problems

Below, **BIO** indicates a problem with a biological or medical focus. Problems labeled **EST** ask you to estimate the answer to a quantitative problem rather than derive a specific answer. Asterisks indicate the level of difficulty of the problem. Problems with no * are considered to be the least difficult. A single * marks moderately difficult problems. Two ** indicate more difficult problems.

28.1 and 28.2 Early atomic models and Bohr's model of the atom: quantized orbits

1. * The electron in a hydrogen atom spends most of its time 0.53×10^{-10} m from the nucleus, whose radius is about 0.88×10^{-15} m. If each dimension of this atom was increased by the same factor and the radius of the nucleus was increased to the size of a tennis ball, how far from the nucleus would the electron be?

2. * **EST** A single layer of gold atoms lies on a table. The radius of each gold atom is about 1.5×10^{-10} m, and the radius of each gold nucleus is about 7×10^{-15} m. A particle much smaller than the nucleus is shot at the layer of gold atoms. Roughly, what is its chance of hitting a nucleus and being scattered? (The electrons around the atom have no effect.)

3. ** (a) Determine the mass of a gold foil that is 0.010 cm thick and whose area is 1 cm × 1 cm. The density of gold is 19,300 kg/m³. (b) Determine the number of gold atoms in the foil if the mass of each atom is 3.27×10^{-25} kg. (c) The radius of a gold nucleus is 7×10^{-15} m. Determine the area of a circle with this radius. (d) Determine the chance that an alpha particle passing through the gold foil will hit a gold nucleus. Ignore the alpha particle's size and assume that all gold nuclei are exposed to it; that is, no gold nuclei are hidden behind other nuclei.

4. * An object of mass M moving at speed v_0 has a direct elastic collision with a second object of mass m that is at rest. Using the energy and momentum conservation principles, it can be shown that the final velocity of the object of mass M is $v = (M - m)v_0/(M + m)$. Using this result, determine the final velocity of an alpha particle following a head-on collision with (a) an electron at rest and (b) a gold nucleus at rest. The alpha particle's velocity before the collision is $0.010c$; $m_{alpha} = 6.6 \times 10^{-27}$ kg; $m_{electron} = 9.11 \times 10^{-31}$ kg; and $m_{gold\ nucleus} = 3.3 \times 10^{-25}$ kg. (c) Based on your answers, could an alpha particle be deflected backward by hitting an electron in a gold atom?

5. * Describe what happens to the energy of the atom and represent your reasoning with an energy bar chart when (a) a hydrogen atom emits a photon and (b) a hydrogen atom absorbs a photon.

6. * How do we know that the energy of the hydrogen atom in the ground state is -13.6 eV?

7. * Determine the wavelength, frequency, and photon energies of the line with $n = 5$ in the Balmer series.

8. * Determine the wavelengths, frequencies, and photon energies (in electron volts) of the first two lines in the Balmer series. In what part of the electromagnetic spectrum do the lines appear?

9. * **Invent an equation** An imaginary atom is observed to emit electromagnetic radiation at the following wavelengths: 250 nm, 2250 nm, 6250 nm, 12,250 nm, ... Invent an empirical equation for calculating these wavelengths; that is, determine $\lambda = f(n)$, where f is an unknown function of an integer n, which can have the values 1, 2, 3, 4, ...

10. * Write three basic equations that are needed to derive the expressions for the allowed radii and the allowed energy of electron states in the Bohr model of the atom.

11. * If we know the value of n for the orbit of an electron in Bohr's model of a hydrogen atom, we can determine the values of three other quantities related to either the electron's motion or the atom as a whole. Briefly describe these quantities.

12. * **EST** Is it possible for a hydrogen atom to emit an X-ray? If so, describe the process and estimate the n value for the energy state. If not, indicate why not.

13. * A gas of hydrogen atoms in a tube is excited by collisions with free electrons. If the maximum excitation energy gained by an atom is 12.09 eV, determine all of the wavelengths of light emitted from the tube as atoms return to the ground state.

14. * Some of the energy states of a hypothetical atom, in units of electron volts, are $E_1 = -31.50$, $E_2 = -12.10$, $E_3 = -5.20$, and $E_4 = -3.60$. (a) Draw an energy diagram for this atom. (b) Determine the energy and wavelength of the least energetic photon that can be absorbed by these atoms when initially in their ground state.

15. ** Show that the frequency of revolution of an electron around the nucleus of a hydrogen atom is $f = (4\pi^2 k_C^2 e^4 m/h^3)(1/n^3)$.

16. * Are we justified in using nonrelativistic energy equations in the Bohr theory for hydrogen? (That is, is the electron's speed smaller than $0.1c$?)

17. * Determine the ratio of the electric force between the nucleus and an electron in the ground state of the hydrogen atom and the gravitational force between the two particles. Based on your answer, are we justified in ignoring the gravitational force in the Bohr theory?

28.3 Spectral analysis

18. * A group of hydrogen atoms in a discharge tube emit violet light of wavelength 410 nm. Determine the quantum numbers of the atom's initial and final states when undergoing this transition.

19. * Draw an energy state diagram for a hydrogen atom and explain (a) how an emission spectrum is formed and (b) how an absorption spectrum is formed.

20. * Explain what spectral lines could be emitted by hydrogen gas in a gas discharge tube with an 11.5-V potential difference across it. What assumptions did you make?

21. * The fractional population of an excited state of energy E_n compared to the population of the ground state of energy E_0 is $N_n/N_0 = e^{-(E_n - E_0)/k_B T}$. At approximately what temperature T are 20 % of hydrogen atoms in the first excited state?

22. * Draw an energy bar chart that describes the ionization process for a hydrogen atom due to collisions with other atoms.

28.4 Lasers

23. (a) A laser pulse emits 2.0 J of energy during 1.0×10^{-9} s. Determine the average power emitted during that short time interval. (b) Determine the average light intensity (power per unit area) in the laser beam if its cross-sectional area is 8.0×10^{-9} m².

24. A pulsed laser used for welding produces 100 W of power during 10 ms. Determine the energy delivered to the weld.

25. * **BIO** **Welding the retina** A pulsed argon laser of 476.5-nm wavelength emits 3.0×10^{-3} J of energy to produce a tiny weld to repair a detached retina. How many photons are in the laser pulse?

26. ** **BIO** **More welding the retina** A laser used to weld the damaged retina of an eye emits 20 mW of power for 100 ms. The light is focused on a spot 0.10 mm in diameter. Assume that the laser's energy is deposited in a small sheet of water of 0.10-mm diameter and 0.30-mm thickness. (a) Determine the energy deposited. (b) Determine the mass of this water. (c) Determine the increase in temperature of the water (assume that it does not boil and that its heat capacity is 4180 J/kg · C°).

27. ** Compare and contrast the energy state diagrams for a laser and for an atom of a hot gas that emits light.

28. * Explain why hot gases do not emit coherent light, but lasers do.

28.5 and 28.6 Quantum numbers and Pauli's exclusion principle and Particles are not just particles

29. * (a) An electron moves counterclockwise around a nucleus in a horizontal plane. Assuming that it behaves as a classical particle, what is the direction of the magnetic field produced by the electron? (b) The same one-electron atom is placed in an external magnetic field whose $\vec{B}$ field points horizontally from right to left. Draw a picture showing what happens to the atom.

30. * **EST** Estimate the wavelength of a tennis ball after a good serve.

31. ** What is the wavelength of the electron in a hydrogen atom at a distance of one Bohr radius from the nucleus?

32. * **EST** Estimate the average wavelength of hydrogen atoms at room temperature.

33. * Describe how you will determine the wavelength of an electron in a cathode ray tube if you know the potential difference between the electrodes.

34. * An electron first has an infinite wavelength and then after it travels through a potential difference has a de Broglie wavelength of 1.0×10^{-10} m. What is the potential difference that it traversed? Draw a picture of the situation and an energy bar chart, and explain why the electron's wavelength decreased.

35. * Describe two experiments whose outcomes can be explained using the concept of the de Broglie wavelength.

36. * How does the wavelength of an electron relate to the radius of its orbit in a hydrogen atom? Why?

37. * Discuss the similarities and differences in the way a hydrogen atom is pictured in Bohr's model and in quantum mechanics.

38. * A high-energy particle scattered from the nucleus of an atom helps determine the size and shape of the nucleus. For best results, the de Broglie wavelength of the particle should be the same size as the nucleus (approximately 10^{-14} m) or smaller. If the mass of the particle is 6.6×10^{-27} kg, at what speed must it travel to produce a wavelength of 10^{-14} m?

39. * (a) Use de Broglie's hypothesis to determine the speed of the electron in a hydrogen atom when in the $n = 1$ orbit. The radius of the orbit is 0.53×10^{-10} m. (b) Determine the electron's de Broglie wavelength. (c) Confirm that the circumference of the orbit equals one de Broglie wavelength.

40. * Repeat Problem 28.39 for the $n = 3$ orbit, whose radius is 4.77×10^{-10} m. Three de Broglie wavelengths should fit around the $n = 3$ orbit.

41. * A beam of electrons accelerated in an electric field is passing through two slits separated by a very small distance d and then hits a screen that glows when an electron hits it. What do you need to know about the electrons to be able to predict where on the screen, which is L meters from the slits, you will see the brightest and the darkest spots?

42. (a) An electron is in an $n = 4$ state. List the possible values of its l quantum number. (b) If the electron happens to be in an $l = 3$ state, list the possible values of its m_l quantum number. (c) List the possible values of its m_s quantum number.

43. (a) An electron in an atom is in a state with $m_l = 3$ and $m_s = +1/2$. What are the smallest possible values of l and n for that state? (b) Repeat for an $m_l = 2$ and $m_s = -1/2$ state.

44. (a) An atom is in the $n = 7$ state. List the possible values of the electron's l quantum number. (b) Of these different states, the electron occupies the $l = 4$ state. List the possible values of the m_l quantum number.

45. ** Draw schematic orbits and arrows representing the different m_l quantum states for an $l = 2$ atomic electron in a magnetic field.

28.7 Multi-electron atoms and the periodic table

46. List the n, l, m_l, and m_s states available for an electron in a $4p$ subshell.

47. List the n, l, m_l, and m_s states available for an electron in a $4f$ subshell.

48. Identify the values of n and l for each of the following subshell designations: $3s$, $2p$, $4d$, $5f$, and $6s$.

49. (a) Determine the electron configuration of sulfur (its atomic number is 16) and explain how you did it. (b) Why are sulfur and oxygen (atomic number 8) in the same group on the periodic table?

50. (a) Determine the electron configuration of silicon (atomic number 14) and explain how you did it. (b) Why are carbon and silicon in the same group of the periodic table?

51. Determine the electron configuration for iron (atomic number 26) and explain how you did it.

52. Manganese (atomic number 25) has two $4s$ electrons. How many $3d$ electrons does it have? Explain your answer.

53. * Determine the electron configurations of four elements of group I in the periodic table. Explain why these elements are likely to have similar properties. Note that a higher s shell fills before the next lower d shell—the electron in the s shell spends more time on average closer to the nucleus.

54. * Determine the electron configurations of three elements in group VI of the periodic table. Explain why these elements are likely to have similar properties.

28.8 The uncertainty principle

55. The uncertainty in the x-position of a 3×10^{-8}-g dust particle is about 1×10^{-6} m. Determine the uncertainty in the particle's velocity component v_x.

56. * Describe the experimental evidence that supports the concept of the uncertainty principle.

57. * The lifetime of the hydrogen atom in the $n = 3$ second excited state is 10^{-9} s. What is the uncertainty of that energy state of the atom? Compare this uncertainty with the magnitude of the -1.51-eV energy of the atom in that state.

58. * Use the uncertainty principle to discuss whether lasers can emit 100% monochromatic light. (Note that monochromatic light means light with a precisely defined frequency.)

General Problems

59. * (a) Determine the radii and energies of the $n = 1$, 2, and 3 states in the He^+ ion (it has two protons in its nucleus and one electron). (b) Construct an energy state diagram for this ion. Indicate any assumptions that you made.

60. * (a) Determine the radii and energies of the $n = 1$, 2, and 3 states of a sodium ion in which 10 of its electrons have been removed. (b) Construct an energy state diagram for the ion. Indicate any assumptions that you made.

61. * **EST** A uranium atom with $Z = 92$ has 92 protons in its nucleus. It has two electrons in an $n = 1$ orbit. Estimate the radius of this orbit. Indicate any assumptions that you made.

62. * **EST** Estimate the energy needed to remove an electron from (a) the $n = 1$ state of iron ($Z = 26$ is the number of protons in the nucleus) and (b) the $n = 1$ state of hydrogen ($Z = 1$).

63. An electron in a hydrogen atom changes from the $n = 4$ to the $n = 3$ state. Determine the wavelength of the emitted photon.

64. An electron in a He^+ ion changes its energy from the $n = 3$ to the $n = 1$ state. Determine the wavelength, frequency, and energy of the emitted photon.

65. Determine the energy, frequency, and wavelength of a photon whose absorption changes a He^+ ion from the $n = 1$ to the $n = 6$ state.

66. * A helium ion He^+ emits an ultraviolet photon of wavelength 164 nm. Determine the quantum numbers of the ion's initial and final states.

67. ** The average thermal energy due to the random translational motion of a hydrogen atom at room temperature is $(3/2)k_BT$. Here k_B is the Boltzmann constant. Would a typical collision between two hydrogen atoms be likely to transfer enough energy to one of the atoms to raise its energy from the $n = 1$ to the $n = 2$ energy state? Explain your answer. [Note: Earth's free hydrogen is in the molecular form H_2. However, the above reasoning is similar for atomic and molecular hydrogen.]

Reading Passage Problems

BIO **Electron microscope** The *scanning electron microscope* (*SEM*) uses high-energy electrons to form three-dimensional images of the surfaces of biological, geological, and integrated circuit samples. The electrons in the SEM have much shorter wavelengths than the light used in visible wavelength microscopes and consequently can produce high-resolution images of a surface, revealing details of about 1 to 5 nm in size with magnification ranging from 25 to 250,000.

A color-enhanced image of the inner surface of a lung is shown in Figure 28.37. The cavities are alveoli where inhaled oxygen is exchanged into the blood for carbon dioxide, a waste product removed in the lungs during exhalation.

A schematic diagram shows how a SEM works (see Figure 28.38). Electrons are emitted from a hot tungsten cathode (the electron gun) and accelerated

FIGURE 28.37 A scanning electron microscope photograph of lung tissue showing alveoli.

FIGURE 28.38 A schematic of a scanning electron microscope. Secondary electrons from the sample are recorded by the detector, producing a three-dimensional surface image.

Tungsten cathode

Electron beam

Condenser lenses

X-ray detector

Deflection coils

Objective lens

Sample

Secondary electron detector

across a potential difference of a few hundred volts to 40 kV. Two or more condenser lenses, each consisting of a coil of wire through which current flows, produce a magnetic field that bends and focuses the electron beam so that it irradiates a small 0.4-nm to 5-nm spot on the sample. The focused beam is then deflected so that it scans over a rectangular area of the sample surface.

The focused electron beam knocks electrons out of surface atoms. These so-called secondary electrons have low energy (< 50 eV) and are caught by a secondary electron detector. The number of secondary electrons produced depends on the orientation and nature of the exposed surface, thus forming a vivid three-dimensional image. The electron beam can also knock inner electrons out of the sample atoms. When outer electrons move into these vacant inner electron orbits, X-ray photons are emitted. The wavelengths of these photons are characteristic of the particular type of atom and thus provide information about the types of atoms in the sample.

68. Which answer below is closest to the speed of an electron accelerated from rest across a 100-V potential difference?
    (a) 600 m/s   (b) 6×10^4 m/s
    (c) 6×10^6 m/s   (d) 6×10^7 m/s

69. Which answer below is closest to the wavelength of an electron accelerated from rest across a 100-V potential difference?
    (a) 1×10^{-11} m   (b) 1×10^{-10} m
    (c) 1×10^{-9} m   (d) 1×10^{-8} m

70. Compared to the electron in Problem 28.69, the wavelength of an electron accelerated from rest across a 10,000-V potential difference would be
    (a) the same length.   (b) 10 times longer.
    (c) 1000 times longer.   (d) 1/10 times as long.
    (e) 1/1000 times as long.

71. An electron in the SEM electron beam is moving parallel to the paper and downward toward the bottom of the page. In which direction should a magnetic field point to deflect the electron toward the right as seen while looking at the page?
    (a) Right   (b) Left
    (c) Toward the top of the page   (d) Toward the bottom of the page
    (e) Into the paper   (f) Out of the paper

72. The SEM can also detect the types of atoms in the sample by measuring which of the following?
    (a) Energy of the secondary electrons
    (b) The wavelengths of X-rays emitted from the sample
    (c) Both (a) and (b)

73. The secondary electrons detected by the SEM detector are
    (a) outer electrons knocked out of atoms on or near the surface of the sample.
    (b) X-rays produced when outer electrons fall into vacant inner electron orbits.
    (c) electrons in the electron beam slowed by passing through the top layer of the sample.
    (d) none of the above.
    (e) (a)–(c) are correct.

BIO Electron transport chains in photosynthesis and metabolism Electron transport chains (ETCs), large proteins through which electrons move, play important roles in two of nature's fundamental processes: (1) the conversion of electromagnetic energy from the Sun into the energy in the chemical bonds in glucose molecules and (2) the conversion of the energy in these glucose molecules into useful forms for metabolic processes, such as muscle contraction, building proteins, and respiration.

Respiratory ETCs are located in the inner membranes of mitochondria, the power plants of eukaryotic cells. ETCs have molecular mass of 800,000 atomic mass units (u), are about 20 nm long, and vary in width from 3 to 6 nm. Each ETC has about 20 sites at which an energetic "free" electron traveling through the chain can make temporary stops. Most of the resting sites have metal ions at their centers. At three places along the chain the electron-protein system goes into a significantly lower energy state (≈ 0.4 eV). The released energy catalyzes a reaction that converts two other molecules into ATP, which carries the energy gained to other parts of the cell for different processes that require chemical energy.

How does a lone energetic electron traverse such a long distance along the bumpy electrical potential hills and valleys of a large protein and not have its energy transformed into thermal energy or light? The answer is still uncertain, but several possibilities have been proposed: (1) the ground state electron wave functions at resting sites are spread out (delocalized) over neighboring bonded atoms, which in turn are spread over and overlap with the next resting site; (2) an electron-atom system is excited into a higher energy, very delocalized state, after which the electron comes back down from this conduction state into a new location; or (3) the electron tunnels through the potential energy barrier between the resting sites. We'll examine briefly possibilities (2) and (3).

Explanation (2): Consider the visible spectra of metal atoms at the resting sites. If the excited state is delocalized over many atoms, as occurs in a semiconductor, this excited so-called conduction band is much broader than the excited state of a single atom. Thus, the visible spectra of a metal ion with a spread out excited conduction band should be somewhat broader than the excited state of a localized resting site with no conduction band. However, the observed visible spectra of the metal ion resting sites in ETCs look similar to metal ion sites found in other proteins not involved in electron conduction, with no apparent broader energy conduction band. Another problem with this explanation is the difficulty of exciting an electron at one site into the higher energy conduction band. The random kinetic energy at room temperature is about 0.025 eV, whereas visible spectra indicate that higher energy bands are 2–3 eV above the ground state. At room temperature, it is almost impossible to "promote" the electron to the conduction band.

Explanation (3): Tunneling can occur between the ground state of the atom at one site and an energy state at a nearby site—a state that is empty and slightly above the ground state. After the electron transfer, the electron-local atom system at the new site moves down to the slightly lower energy ground state, which prevents the electron from tunneling back. The system can also transition into a somewhat lower energy state and catalyze a reaction that converts the site's extra energy into the stable bond of some other molecule. Tunneling times seem consistent with measured electron transfer rates. But the actual transfer mechanism is still an open question.

74. If all of the electron resting sites were in a line along the length of an ETC, what would be the approximate distance from one site to the next?
    (a) 400 nm   (b) 20 nm
    (c) 1 nm   (d) $(1/20)$ nm

75. Electron tunneling involves electrons
    (a) jumping to a conduction band and then falling down at a different place.
    (b) diffusing with a small protein from one place to another.
    (c) converting to a photon and moving at light speed to another site.
    (d) passing through a potential barrier to a different location.
    (e) having an uncertain energy for a short time interval.
    (f) (d) and (e)

76. Suppose the barrier height above the electron energy $(U_{\text{barrier}} - E_{\text{electron}})$ is 1 eV and the barrier is 2 nm wide. According to the uncertainty principle, what is the minimum time interval that the electron's energy is in this classically forbidden region?
    (a) 3×10^{-16} s   (b) 3×10^{-15} s
    (c) 3×10^{-14} s   (d) 3×10^{-13} s
    (e) 3×10^{-12} s

77. Electron transport chains are fundamental parts of
    (a) the conversion of glucose into energetic molecules used in the body.
    (b) molecules used for fuel in trains and other vehicles.
    (c) the photosynthetic conversion of light into stable chemical compounds.
    (d) the passing of electric current in nerve cells.
    (e) (a) and (c)   (f) (a), (c), and (d)

78. The conduction band model of electron transport in electron transport chains may not work because
    (a) the electron transport chain is not entirely metal ions.
    (b) the conduction band is too localized.
    (c) the electron has little chance of being "promoted" into the conduction band.
    (d) the visible spectrum of the resting site is not a broad band as is expected.
    (e) (c) and (d)   (f) (a), (b), (c), and (d)

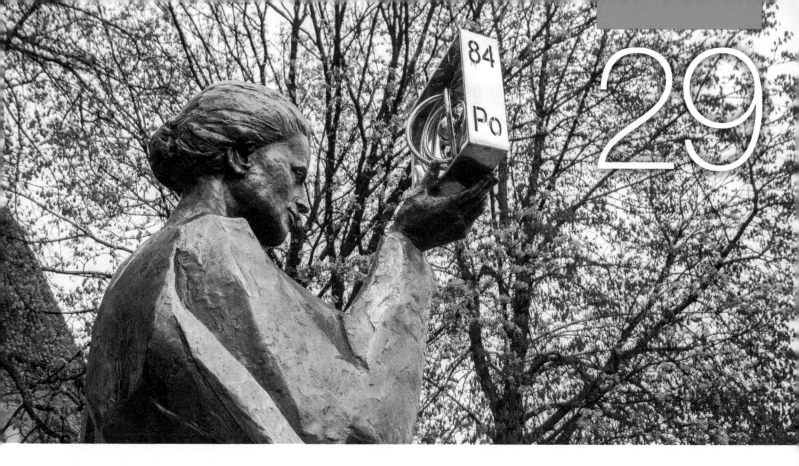

Nuclear Physics

In the late 1800s and early 1900s, Pierre and Marie Curie were investigating the radiation produced by salts of uranium. While extracting uranium from uranium ore, Marie made an accidental observation: the residual material produced more radiation than even the pure uranium. Could that material be a new element? Testing this hypothesis required processing several tons of the uranium ore. After 4 years of laborious, painstaking work, Curie isolated two new elements. She called them polonium (named after her native Poland) and radium. Marie Curie was awarded two Nobel Prizes for this work, in physics (1903) and in chemistry (1911), the former shared with her husband Pierre and Henri Becquerel. In this chapter, we will learn the mechanism by which uranium and other elements produce radiation.

- Who is Marie Curie, and why is her statue holding a model of the element polonium?

- What are the natural sources of ionizing radiation?

- How does carbon dating work?

BY 1911, the atomic nucleus was known to scientists from experiments with alpha particles scattering on thin gold foil. These scattering experiments indicated that the nucleus occupied about $10^{-12}\%$ of the atom's volume and yet contained 99.97% of the atom's mass. In this chapter we will investigate several questions about the nucleus: what is its structure, and what processes do nuclei undergo?

BE SURE YOU KNOW HOW TO:

- Use the right-hand rule for magnetic force to determine the direction of the force exerted by a magnetic field on a moving charged particle (Section 20.3).

- Relate mass to energy using the special theory of relativity (Section 26.9).

29.1 Radioactivity and an early nuclear model

Wilhelm Roentgen's discovery in 1895 of what became known as X-rays left open the question of what X-rays are. In this section we will review the findings that followed this discovery.

Becquerel and the emissions from uranyl crystals

Henri Becquerel, the third generation of a family of French scientists, was very interested in Roentgen's X-rays. In 1896, Becquerel was working with potassium uranyl sulfate crystals (crystals containing the element uranium). These crystals would glow in the dark after exposure to sunlight. In addition, Becquerel found that the crystals would produce images on photographic plates. At that time, scientists thought that photographic plates could only be exposed by light, ultraviolet rays, and X-rays. Consequently, Becquerel hypothesized that the uranium crystals absorbed the Sun's energy, which then caused them to emit X-rays. However, Becquerel also discovered that crystals that had not been exposed to sunlight still formed images on photographic plates. This meant that the uranium emitted radiation without an external source of energy.

Did the crystals emit electrically neutral high-energy X-rays—electromagnetic waves? If they did, then an external magnetic field would not deflect them. However, when Becquerel passed the rays through a region with a magnetic field, he found that the field deflected the rays. Therefore, they were electrically charged particles, not X-rays. Strangely, the magnetic field deflected the rays in two opposite directions, as if both negatively and positively charged particles were being emitted. He also found that the rays could discharge an electroscope independently of the sign of its initial charge, also consistent with the idea that the rays contained both positively and negatively charged particles.

This image of the uranium-laden cross that Becquerel used for his investigations appeared on a photographic plate.

Pierre and Marie Curie and the particles responsible for Becquerel's rays

In 1896—1898, Marie and Pierre Curie continued the work of Becquerel. They found a way to measure the rate of emission of "Becquerel rays" by using an electrometer, a device similar to an electroscope invented by Pierre. A charged electrometer maintains its charge in a dry room for a relatively long time (minutes) because dry air normally contains mostly electrically neutral particles. However, an electrometer discharges much more quickly when placed near a sample containing uranium salts because uranium rays ionize air molecules. By recording the time it takes the electrometer to discharge, one can infer the ion concentration in the air.

Marie Curie conducted several observational experiments on uranium salts using an electrometer.

Experiment 1: Curie measured the time of discharge of the electrometer for samples of uranium salt of different masses. Doubling the mass of the same salt caused a discharge in half the time.

Experiment 2: Curie measured the time of discharge of the electrometer for different uranium salts for which the mass of uranium was known. As long as the total uranium mass in the salt stayed the same, the discharge time did not change.

Experiment 3: Curie measured the time of discharge of the electrometer using the same sample of uranium salt, but varying the amount of light incident on it, its temperature, and its wetness. Despite all these variations, the discharge time did not change.

Based on her observations, Marie Curie came up with the following conclusions. Assuming that the ion concentration measured by the electrometer was due to the radiation from the uranium salts, then the amount of radiation depended only on

Uranium salts.

the amount of the uranium used and not on the chemical composition of the salt, its temperature, the presence of water, or the amount of light shining on it. These findings suggested that the electrons in the atoms were not responsible for the rays. Marie and Pierre Curie concluded that the Becquerel rays must come from the nuclei of atoms.

Rutherford and experiments investigating the charge of emitted particles

In 1899 Ernest Rutherford and his colleagues in England investigated the ability of uranium salts to ionize air. He set up two parallel metal plates with a potential difference between them. Because the air between the plates was a dielectric, there was no current in the circuit. However, when a uranium sample was placed between the plates, a considerable current was detected. The current was due to the ions and free electrons created by the uranium's radiation, which were then pulled to the plates by the electric field (Figure 29.1).

Rutherford then covered the uranium sample with thin aluminum sheets to investigate how metal layers affected the amount of radiation. He observed that the amount of current decreased as he added more sheets, but only up to a point. After this point, no further decrease in radioactivity was observed, even with the addition of more aluminum sheets. He proposed that the radiation consisted of at least two components, one of which was not significantly absorbed by the aluminum sheets. He reasoned that the component of radiation that was absorbed by the aluminum sheets consisted of charged particles. If true, then these particles should be deflected by a magnetic field. In contrast, if there were no charged particles in the rays, the magnetic field should not affect them.

In 1903, Rutherford tested his reasoning by passing rays from radium through a magnetic field. In Testing Experiment Table 29.1, we describe the experiment.

FIGURE 29.1 A schematic of Rutherford's uranium radiation detection device.

Radiation from uranium caused air molecules between the plates to ionize, causing an electric current.

TESTING EXPERIMENT TABLE 29.1   Testing the deflection of radiation by magnetic fields

Testing experiment	Prediction	Outcome
Radiation emanating from a radioactive sample passes through a $\vec{B}$ field pointing into the page. A scintillating screen (which glows when hit with radiation) is in the plane perpendicular to the beam.	If the radiation contains positively charged particles, then according to the right-hand rule, the magnetic field should deflect them upward with respect to the original direction.	The screen glowed in three places: straight ahead, deflected up, and deflected down. The location of the downward-deflected radiation was much farther from the original beam than the location of the upward-deflected radiation.
	If the radiation contains negatively charged particles, then the magnetic field should deflect them downward with respect to the original direction.	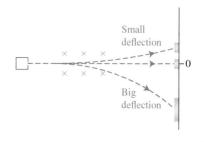

Conclusion

The radiation emanating from the radioactive sample consisted of positively charged particles, negatively charged particles, and particles with no electric charge.

Rutherford reasoned that the greater downward deflection could be the result of the downward-deflected negatively charged particles having a smaller mass-to-charge ratio than the upward-deflected positively charged particles. In the same year, using more powerful magnets, Rutherford found that the upward-bending particles were positively charged, with a mass-to-charge ratio twice that of a hydrogen ion. These positively charged massive particles were called **alpha particles**. These are the same alpha particles that Rutherford and his colleagues used later to probe the structure of the atom (Chapter 28). The downward-bending negatively charged particles had the same mass-to-charge ratio as that of electrons and were called **beta particles**. The radiation that was not deflected by the magnetic field was thought to be high-energy electromagnetic waves, called **gamma rays**.

Subsequent studies revealed that many elements with high atomic numbers were radioactive; all of them emitted alpha particles, beta particles, and/or gamma rays. Some elements with low atomic numbers were also radioactive, but generally these emitted only beta particles and gamma rays.

The early model of the nucleus

Based on these experiments, scientists developed a provisional explanation for radioactivity: the nucleus of an atom is made of positively charged alpha particles and negatively charged electrons. Their electrostatic attraction holds them together. When a nucleus contains a large number of alpha particles, they start repelling each other more strongly than the electrons can attract them, and the alpha particles leave the nucleus, a process called **alpha decay**. Alpha decay leaves behind a large number of electrons that also repel each other; thus beta particles are emitted (**beta decay**). After each transformation, the nucleus is left in an excited state and emits a high-energy photon, a gamma ray (**gamma decay**). This model provided a start for nuclear physics and was later significantly modified when new findings emerged.

A hydrogen atom, however, is lighter than an alpha particle. What, then, is the composition of its nucleus? In 1918, Rutherford noticed that when alpha particles were shot into nitrogen gas, particles that were knocked out of the nitrogen nuclei moved in curved paths that indicated they were positively charged. Further testing indicated that the particles had the same magnitude charge as an electron (only positive) and a mass that equaled that of a hydrogen nucleus. The particle was called a **proton**. Protons must be the nuclei of hydrogen atoms. The proton became an important part of a new model of the nucleus that began to emerge.

REVIEW QUESTION 29.1 How do we know that radioactive emission consists of three components: positively charged particles, negatively charged particles, and radiation with no electric charge?

29.2 A new particle and a new nuclear model

Werner Heisenberg's uncertainty principle (discussed in Chapter 28) had a profound effect on the nuclear model, in which protons and electrons were thought to be the primary nuclear constituents. However, it turned out that electrons were too light to reside inside the nucleus. Read on to follow the story.

Size of the nucleus: too small for an electron

Consider an electron confined to a carbon nucleus whose radius is approximately 2.7×10^{-15} m. The position determinability Δx of such an electron can be no greater than the size of the nucleus (**Figure 29.2**). According to the uncertainty principle, this electron must have a momentum determinability of

$$\Delta p \geq \frac{h}{4\pi\Delta x} = \frac{6.63 \times 10^{-34}\,\text{J}\cdot\text{s}}{4\pi(2.7 \times 10^{-15}\,\text{m})} = 1.95 \times 10^{-20}\,\text{kg}\cdot\text{m/s}$$

We can use this value as an estimate for the momentum of the electron confined to the nucleus. Using the equation $p = mv$, we would find that the speed of the electron is hundreds of times the speed of light, so we need to treat the electron relativistically. Using energy ideas, we find that the kinetic energy of the electron is the total energy of the electron minus its rest energy:

$$K = E_{\text{total}} - E_{\text{rest}}$$

Inserting the appropriate expressions:

$$K = \sqrt{(pc)^2 + (mc^2)^2} - mc^2$$

We can evaluate this expression using our estimate for the momentum of the electron, the mass of the electron, and the speed of light:

$$K = \sqrt{[(1.95 \times 10^{-20}\,\text{kg}\cdot\text{m/s})(3.00 \times 10^8\,\text{m/s})]^2 + [(9.11 \times 10^{-31}\,\text{kg})(3.00 \times 10^8\,\text{m/s})^2]^2}$$

$$- (9.11 \times 10^{-31}\,\text{kg})(3.00 \times 10^8\,\text{m/s})^2$$

$$= 5.77 \times 10^{-12}\,\text{J}$$

How does this result compare to the electric potential energy between the electron and the nucleus? When studying the Bohr model of the hydrogen atom, we found that the total energy of a system (potential plus kinetic) must be negative for the system to be bound (Chapter 28). Let's think, for example, about a carbon nucleus. Carbon is assigned atomic number 6, so its nucleus has a charge of $+6e$. If there are both protons and electrons in the nucleus, there must be 6 more protons than electrons. We'll make a rough estimate of the electric potential energy between an electron in a carbon nucleus and the rest of the nucleus by assuming that the electron is an average distance from the protons equal to half the diameter of the nucleus, or about 2.7×10^{-15} m. The electric potential energy between the electron and rest of the nucleus (total charge $+7$) is approximately

$$U_q = k_C \frac{q_{\text{nucleus}}q_e}{r} = \left(9 \times 10^9\,\frac{\text{N}\cdot\text{m}^2}{\text{C}^2}\right)\frac{(7 \cdot 1.6 \times 10^{-19}\,\text{C})(-1.6 \times 10^{-19}\,\text{C})}{2.7 \times 10^{-15}\,\text{m}}$$

$$\approx -6.0 \times 10^{-13}\,\text{J}$$

This value is slightly more than just one-tenth the kinetic energy of the electron. In other words, the total energy of the system (kinetic plus potential) is positive! This result means that an electron in the carbon nucleus would very rapidly escape the nucleus. This form of carbon is stable, however, and does not emit electrons. Similar reasoning applies to other nuclei.

Thus we have determined that, because of the uncertainty principle and their tiny mass, electrons cannot be components of nuclei.

The search for a neutral particle

Yet another mystery existed: an alpha particle had the charge of two protons ($+2e$) but had four times the mass of a proton. If there are no electrons to balance the two positive charges of the protons, what else is contributing to the alpha particle's mass? In 1920, Rutherford proposed that a neutral particle with the approximate mass of a proton, produced by the capture of an electron by a proton, should exist. This hypothesis stimulated a search for the particle. However, its electrical neutrality complicated the search because almost all experimental techniques of the time could detect only charged particles.

FIGURE 29.2 According to the uncertainty principle, if an electron is confined within a nucleus, its speed would be greater than light speed. Thus, it cannot be within the nucleus.

If an electron is in a nucleus, its position uncertainty is the size of the nucleus and its momentum can be estimated by the uncertainty principle.

TIP Notice that if you substitute the mass of a proton into the equation for kinetic energy, the kinetic energy goes to zero, making the proton a possible constituent of a nucleus due to its larger mass.

In 1928, German physicist Walther Bothe and his student Herbert Becker took an initial step in the search. They bombarded beryllium atoms with alpha particles and found that neutral radiation left the beryllium atoms. This radiation was initially thought to be high-energy gamma ray photons.

Detecting neutral radiation with paraffin

In 1932, Irène Joliot-Curie, one of Marie Curie's daughters, and her husband, Frédéric Joliot-Curie, used a stronger source of alpha particles to repeat and extend Bothe's experiment. They bombarded beryllium atoms with alpha particles, as Bothe had done. However, in their experiment they placed a block of paraffin (a wax-like substance) beyond the beryllium and a particle detector beyond the paraffin. They found that the neutral radiation leaving the beryllium ejected protons from the paraffin. The Joliot-Curies knew that ultraviolet and high-frequency visible photons could knock electrons from a metal surface (the photoelectric effect discussed in Chapter 27) and made the reasonable assumption that high-frequency gamma ray photons could knock protons out of paraffin.

The neutral radiation is not a gamma ray photon

In England, James Chadwick repeated the Joliot-Curie experiments. He placed a variety of materials, not just paraffin, beyond the beryllium (**Figure 29.3**). By comparing the energies and momenta of the particles knocked out of these different materials by the rays from the beryllium atoms, he found that the rays were not gamma ray photons but instead uncharged particles with a mass approximately equal to that of the proton. He called the new particle the **neutron**. In 1935, Chadwick received the Nobel Prize in physics for this work.

Revising ideas of the structure of the nucleus

The discovery of the neutron was a major factor in revising ideas of nuclear structure, as was the realization that electrons were definitely not constituents of the nucleus. A new model for a nucleus consisting of protons and neutrons evolved. The protons, each of charge $+e$, accounted for the electric charge of the nucleus, while the uncharged neutrons accounted for the extra mass. For example, two protons and two neutrons make a helium nucleus; five protons and five neutrons make a boron nucleus (**Figure 29.4**). The most common form of uranium is composed of 92 protons and 146 neutrons, with electric charge $+92e$, and 238 protons and neutrons.

Describing atomic nuclei

Nuclei are now represented by the notation $^A_Z X$, where X is the letter abbreviation for the element (e.g., Fe for iron, Na for sodium), and Z is the number of protons in the nucleus, called the **atomic number**. N (not shown in the notation) is the number of neutrons in the nucleus. $A = Z + N$ is the **mass number**, the total number of **nucleons** (protons plus neutrons). Thus, helium with two protons and two neutrons is shown as $^4_2 He$, and the most common form of uranium, with 92 protons and 146 neutrons (a total number of protons and neutrons of $A = 92 + 146 = 238$), is $^{238}_{92} U$.

The masses of atoms and nuclei are very small if given in kilograms. Thus, another useful mass unit, called the **atomic mass unit (u)**, was developed.

> **One atomic mass unit (u)** equals one-twelfth of the mass of a carbon atom with six protons and six neutrons in the nucleus ($^{12}_6 C$), including the six electrons in the atom. In terms of kilograms,
>
> $$1\ u = 1.660539 \times 10^{-27}\ kg$$

FIGURE 29.3 Chadwick's apparatus for detecting neutrons.

Alpha particles bombard beryllium.

Particles knocked out of beryllium hit the target material.

Vacuum

α

Alpha particle source

Be

Target material (such as paraffin)

Particle detector

FIGURE 29.4 The structure of a boron nucleus.

 = neutron

 = proton

Isotopes

In an earlier chapter we described the mass spectrometer—a device that allows scientists to measure the mass of an elementary particle by observing its motion in a magnetic field (Chapter 20). Scientists use mass spectrometers to measure the mass of ionized atoms, as described in Observational Experiment **Table 29.2**.

OBSERVATIONAL EXPERIMENT TABLE 29.2 Measurement of atomic mass

Observational experiment	Analysis

We accelerate carbon ions to high speeds by letting them pass through a region with high potential difference and then through a magnetic field, observing the paths they take. Ions of a single chemical element take different paths.

The force exerted by the magnetic field on the positive ions is perpendicular to their velocity. Thus, ions move in a semicircle. We assume that the force exerted on the ions by the magnetic field is the only force exerted on them. Newton's second law in radial component form becomes $(mv^2/R) = qvB$ or $R = (mv/qB)$. Ions entering the region with magnetic field have the same kinetic energy: $K = (mv^2/2) = q\,\Delta V$. Thus their speeds are

$$v = \sqrt{\frac{2q\,\Delta V}{m}}$$

Therefore, the radius of curvature for the ions of different mass is determined by

$$R = \frac{mv}{qB} = \frac{m}{qB}\sqrt{\frac{2q\,\Delta V}{m}} = \sqrt{\frac{m^2 2q\,\Delta V}{q^2 B^2 m}} = \sqrt{\frac{2m\,\Delta V}{qB^2}}$$

Pattern

Atoms of a single chemical element with the same atomic number have different masses.

Because atoms of a particular element have the same number of protons, the differences in mass among them can be explained only if we assume that the number of neutrons is different. Atoms with a different number of neutrons are called **isotopes** of that particular element. For example, carbon comes in three naturally occurring isotopic forms: $^{12}_{6}C$, $^{13}_{6}C$, and $^{14}_{6}C$, with six, seven, and eight neutrons, respectively. Each isotope has six protons and six electrons surrounding the nucleus.

The electron structure of an element's isotopes is the same, which means that their chemical behaviors are almost identical. However, the nuclei behave quite differently. Generally, only a small number (usually one or two) of the possible isotopes of an element are stable. The rest undergo radioactive decay, which we will consider shortly.

CONCEPTUAL EXERCISE 29.1 Reading symbolic representations of atomic nuclei

Determine the number of protons and neutrons in each of the following nuclei: $^{6}_{3}Li$, $^{16}_{8}O$, $^{27}_{13}Al$, $^{56}_{26}Fe$, $^{64}_{30}Zn$, and $^{107}_{47}Ag$.

Sketch and translate For lithium, $Z = 3$, so it has three protons in the nucleus. Lithium has $A = 6$ nucleons; thus there are three neutrons ($N = 3$).

Simplify and diagram We use the same strategy to determine the number of protons (Z) and the number of neutrons (N) in the other elements and summarize the results in the table shown at right.

$^{6}_{3}Li$		$^{16}_{8}O$		$^{27}_{13}Al$		$^{56}_{26}Fe$		$^{64}_{30}Zn$		$^{107}_{47}Ag$	
Z	N	Z	N	Z	N	Z	N	Z	N	Z	N
3	3	8	8	13	14	26	30	30	34	47	60

Try it yourself Determine Z and N for potassium-39, $^{39}_{19}K$.

Answer Z = 19 and N = 20.

Determine the chemical elements represented by the symbol X for each of the following nuclei: $^{20}_{10}$X, $^{52}_{24}$X, $^{59}_{27}$X, and $^{93}_{41}$X.

Sketch and translate To determine the element, we use the periodic table (Table 28.8) and the lower left number (Z) in the symbol.

Simplify and diagram The results for the four symbols are shown in the table at right.

$^{20}_{10}$X	$^{52}_{24}$X	$^{59}_{27}$X	$^{93}_{41}$X
Ne (neon)	Cr (chromium)	Co (cobalt)	Nb (niobium)

Try it yourself Determine the chemical element with the following Z and A: $^{63}_{29}$X.

Answer

An isotope of copper, that is, copper-63.

TIP   The mass number (A) in the periodic table might not match the number for the same element in the exercises above. In the periodic table, A is an average mass number over all isotopes, accounting for their relative abundances as well. As a result, it is generally not an exact integer. You can look up individual isotopes in Appendix B.

REVIEW QUESTION 29.2 Explain why the nucleus cannot contain any electrons.

29.3 Nuclear force and binding energy

So far, we have found that the nucleus consists of positively charged protons and neutral neutrons. The positively charged protons repel each other (**Figure 29.5a**). There are no electrons in the nucleus to cancel this repulsive force.

How can the protons stay bound together when they repel each other?

Nuclear force

Some attractive force must balance this electrical repulsive force and must attract neutrons as well—it has to be an attractive force for both protons and neutrons. (Figure 29.5b). We call this attractive force a **nuclear force**.

The nuclear force must also weaken to nearly zero extremely rapidly with increasing distance between nucleons. If it didn't, then nuclei of nearby atoms would be attracted to each other, clumping together into ever-larger nuclei. Since this does not occur, the nuclear force must be very short range—no greater than about 2×10^{-15} m. When nucleons are farther apart, the nuclear force they exert on each other is essentially zero (Figure 29.5c).

This 2×10^{-15}-m distance limit is important since 2×10^{-15} m is smaller than the size of many of the higher-Z nuclei. Therefore, not every nucleon is attracted to every other nucleon because some of them are farther away than this maximum range. If they were attracted equally strongly, then a massive uranium nucleus with 238 protons and neutrons should be exceedingly more stable than a helium nucleus with only 4 protons and neutrons simply because of the greater number. However, experiments show that these two nuclei are approximately equally stable. Thus, a proton in a small nucleus must be bound by the nuclear force to about the same number of protons and neutrons as a proton in a large nucleus. Evidently, the nuclear force involves only nearest neighbor nucleons (Figure 29.5b and c).

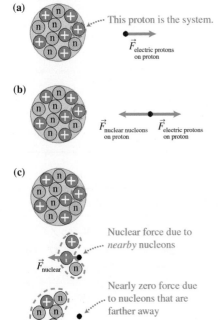

FIGURE 29.5 The nuclear force. (a) Protons exert an electric repulsion on each other. (b) This repulsion is balanced by the nuclear force. (c) Nearby nucleons exert a nuclear force; more distant nucleons do not.

However, in a nucleus with many protons, the repulsive force exerted on a particular proton by all the other protons must be strong; this repulsion might exceed the strong nuclear attraction between a proton and its neighbors. If this model of the nuclear force is correct, then we predict that such nuclei might be unstable. Could it be that we have found a reason for radioactive decay? Before we explore this possibility, let's consider another prediction based on the hypothesis of this attractive nuclear force.

Mass defect

Recall from our study of atoms (Chapter 28) that we must add energy to a system in order to separate the electrons from the nucleus, ionizing the atom. When the sum of added energy, electrical potential energy, and kinetic energy reaches zero, the electron becomes unbound from the nucleus. The magnitude of the energy that must be added to ionize the atom is known as the **binding energy** of the atom (also called the ionizing energy).

We could make similar statements about the nucleus itself, composed of protons and neutrons. The nucleus is a bound system, and therefore its **nuclear potential energy** plus electric potential energy (positive since the protons repel each other) plus kinetic energy of the protons and neutrons must be negative. We neglect the energy of the interaction between the nucleus and the electrons because electrons are 100,000 times farther from protons in the nucleus than protons are from neutrons. The binding energy of the nucleus is the energy that must be added to the nucleus to separate it into its component protons and neutrons. The energy can be added through a heating mechanism, by using a photon of light, or through a collision with a fast-moving particle.

Recall that all objects with mass have a corresponding rest energy given by $E_0 = mc^2$ (Chapter 26). According to Einstein's equation, the mass m of an object equals E_0/c^2, where E_0 is the corresponding rest energy of the object. Because the other three forms of energy of the nucleus add to a negative value, the rest energy of the nucleus should be less than the rest energy of the separated protons and neutrons. Consequently, the mass of the nucleus should be less than the total mass of its constituents. To check whether this prediction matches experimental evidence, we need to collect data on the masses of the nuclei and their constituents.

The data for the masses are below (see Appendix B). Note that we use the masses of atoms, not the nuclei, because in practice they are much easier to determine. The fact that the mass of the atom includes the mass of the electrons will not change the result, as you will see later.

$$m_{\text{helium}} = 4.002602 \text{ u}; \ m_{\text{hydrogen}} = 1.007825 \text{ u}; \ m_{\text{neutron}} = 1.008665 \text{ u}$$

Mass of helium constituents:

$$2m_{\text{hydrogen}} + 2m_{\text{neutron}} = 2(1.007825 \text{ u}) + 2(1.008665 \text{ u}) = 4.032980 \text{ u}$$

Missing mass:

$$(2m_{\text{hydrogen}} + 2m_{\text{neutron}}) - m_{\text{helium}} = 4.032980 \text{ u} - 4.002602 \text{ u} = 0.030378 \text{ u}$$

The missing mass is on the order of 1% of the mass of the nucleus. Our prediction matched the outcome of the experiment. The mass of the helium nucleus is less than the sum of the masses of the protons and neutrons that comprise it (**Figure 29.6**).

In Chapter 6 we learned that mass is constant in an isolated system. Now we have found that this is not true for an atomic nucleus because it is a bound system. This means that the mass of any bound system is less than the mass of its constituents, but at a macroscopic level the difference is tiny. Now we understand that mass is not a conserved quantity. However, energy is. We will use the idea of energy conservation to understand what happens to the energy of a system when a nucleus is formed.

Let us represent a hypothetical process of making a helium nucleus from its constituents using a bar chart (**Figure 29.7a** on the next page). The system is two protons and two neutrons. The initial state will be when the particles are far apart, and the final state is when they are all together in the form of a nucleus. Thus the protons and the neutrons

FIGURE 29.6 The mass of a helium nucleus is less than that of its constituents.

FIGURE 29.7 Energy bar charts representing the process of (a) making a nucleus out of several constituents; (b) splitting the nucleus into its constituents.

(a)

$$E_{0i} + U_{qi} + U_{ni} + E_{Add} = E_{0f} + U_{qf} + U_{nf} + \Delta U_{int}$$

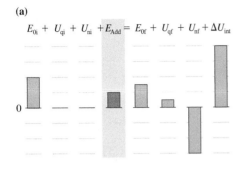

(b)

$$E_{0i} + U_{qi} + U_{ni} + E_{Add} = E_{0f} + U_{qf} + U_{nf} + \Delta U_{int}$$

have their initial total rest energy (E_{0i}), the sum of the rest energies of the individual particles. Because the protons are initially far away from each other, the electric potential energy of their repulsion is zero $(U_{qi} = 0)$, and we will assume that their kinetic energy is zero, too $(K_i = 0)$. The protons repel each other, so to bring them closer we need to add energy to the system (E_{Add}), for example, by doing work on the protons pushing them closer together. Once joined together, the rest energy of the four particles becomes less than it was before, the electric energy of repulsion increases because the positively charged protons are closer together, and now the system has negative nuclear potential energy of the resulting nucleus (U_{nf}). This U_{nf} energy is large enough to make the total energy of the nucleus, which consists of the rest energy, electric potential energy, and nuclear potential energy, negative.

To balance the energy bar chart, we need to assume that some energy is converted into the chaotic kinetic energy of the movement of the nuclei, which we will record as an increase in the internal energy ΔU_{int}. The bar chart predicts that when a helium nucleus is formed, some energy needs to be added to the system, but at the end of the process a much larger amount of energy is converted into internal energy—the process is exothermic (in exothermic processes, energy is released through the process of heating). We will see later how this prediction leads to the production of energy through fusion. We can also use a bar chart to represent the process of separating the helium nucleus into its constituents (Figure 29.7b). Notice that the initial and final states here are the flipped initial and final states in the first bar chart.

Helium is not unique in having a smaller mass than its constituents. For example, the nucleus of lithium $_3^7\text{Li}$ is made of three protons and four neutrons. The atom also has three electrons. Thus we can compare it to three hydrogen atoms $_1^1\text{H}$ and four neutrons. The total mass of the three hydrogen atoms and four neutrons is 7.058135 u, whereas the mass of a lithium atom is 7.016003 u, or 0.042132 u less than the mass of its constituents. On the basis of such findings, we can now define a new physical quantity called **mass defect**.

> **Mass defect** The mass defect of a nucleus is the difference in mass between the constituents of the atom (as hydrogen atoms and neutrons) and the mass of the atom itself. The mass defect Δm is
>
> $$\Delta m = \left[Z m_{\text{hydrogen atom}} + (A - Z) m_{\text{neutron}} \right] - m_{\text{atom}} \qquad (29.1)$$

As we mentioned above, the reason the mass of the hydrogen atom is used rather than the mass of the proton is to account for the mass of the electrons. When the mass of the atom is subtracted, the mass of the electrons cancels. In practice, it is also much easier to measure the masses of atoms than the masses of just their nuclei, which is why it's more useful to define mass defect in terms of the masses of atoms.

Binding energy *BE* of the nucleus

Earlier, we found the mass defect of a helium nucleus. This defect is related to the binding energy of the atomic nucleus.

> **Binding energy** The binding energy of a nucleus is the rest energy equivalent of its mass defect:
>
> $$BE = \Delta m \cdot c^2 \qquad (29.2)$$
>
> The binding energy represents the total energy needed to separate the nucleus into its component nucleons.

The binding energy of a nucleus is most easily expressed in units of million electron volts (MeV), where

$$1 \text{ MeV} = 10^6 \text{ eV} = 1.602 \times 10^{-13} \text{ J}$$

Recall that mass defect is easily expressed in atomic mass units (u). It is convenient to be able to quickly convert the mass defect of a nucleus into the corresponding binding energy. Because 1 u has units of mass, $1 \text{ u} \cdot c^2$ has units of energy. The goal is to convert $1 \text{ u} \cdot c^2$ into MeV:

$$1 \text{ u} \cdot c^2 = 1(1.660539 \times 10^{-27} \text{ kg})(2.9979 \times 10^8 \text{ m/s})^2$$
$$= 1.4924 \times 10^{-10} \text{ J}$$
$$= (1.4924 \times 10^{-10} \text{ J})\left(\frac{1 \text{ MeV}}{1.6022 \times 10^{-13} \text{ J}}\right) = 931.5 \text{ MeV}$$

Rearranging the last equation, we get the following conversion factor:

$$1 \text{ u} = 931.5 \text{ MeV}/c^2 \qquad (29.3)$$

As you can see, the mass defect allows us to calculate the binding energy of any nucleus and consequently the potential energy of attraction of the particles inside it.

Binding energy per nucleon

The binding energy of the nucleus of sodium-23 ($^{23}_{11}\text{Na}$) is 187 MeV; for lithium-7 ($^{7}_{3}\text{Li}$), the binding energy is 39.2 MeV; and earlier we found that the binding energy of helium-4 was 28.3 MeV. These differences means that more energy is needed to separate a sodium-23 nucleus into protons and neutrons than to separate a lithium-7 nucleus into protons and neutrons, and both require more energy than is needed to separate a helium-4 nucleus into protons and neutrons.

Does this mean that sodium is more stable than lithium, and that lithium is more stable than helium? Not necessarily. Sodium has an "unfair" advantage in total binding energy because it has more nucleons. Instead, a better measure of stability is the binding energy of a nucleus divided by its number of nucleons, A. For sodium-23, its binding energy per nucleon is $(187 \text{ MeV})/(23 \text{ nucleons}) = 8.1 \text{ MeV/nucleon}$. For lithium-7, it is $(39.2 \text{ MeV})/(7 \text{ nucleons}) = 5.6 \text{ MeV/nucleon}$, and for helium-4, it is $(28.3 \text{ MeV})/(4 \text{ nucleons}) = 7.1 \text{ MeV/nucleon}$. With this comparison, we can see that sodium is a more stable nucleus than helium, and that helium is more stable than lithium. Put another way, more energy is needed per nucleon to separate the sodium nucleus into its constituents than is needed for the helium nucleus, and the least stable is the lithium nucleus. Binding energy per nucleon is a much better indicator of nuclear stability than total binding energy.

Binding energy per nucleon Binding energy per nucleon is the binding energy BE of the atom divided by the number of nucleons A in the atom's nucleus:

$$\text{Binding energy per nucleon} = BE/A \qquad (29.4)$$

Binding energy per nucleon is a good indicator of the stability of a nucleus; the larger the binding energy per nucleon, the more stable the nucleus.

REVIEW QUESTIONS 29.3 What is the difference between nuclear binding energy and the energy needed to ionize an atom when in the ground state (atomic binding energy)?

29.4 Nuclear reactions

In 1919 Ernest Rutherford became the first person to transmute one element into another when he shot helium nuclei (alpha particles) at nitrogen and produced some oxygen-17 atoms (an isotope of oxygen that has nine neutrons and eight protons). This process, called a **nuclear reaction**, can be represented as follows:

$$^4_2\text{He} + {}^{14}_7\text{N} \rightarrow {}^{17}_8\text{O} + {}^1_1\text{H}$$

Rutherford's work was followed by many remarkable advances in the study of nuclear physics.

Representing nuclear reactions

Nuclear reactions, like chemical reactions, involve the transformation of reactants into different products. In a chemical reaction, molecules are transformed into other molecules, but the number of each type of atom remains the same. In a nuclear reaction, nuclei are transformed into different nuclei, resulting in different elements. In these reactions two nuclei may interact to form one or more new nuclei (e.g., $^1_1\text{H} + {}^7_3\text{Li} \rightarrow {}^4_2\text{He} + {}^4_2\text{He}$). Or a single nucleus may divide into two or more new nuclei, or perhaps emit a small particle, thus leaving behind a different nucleus (e.g., the radioactive decay experiments observed by Curie and others).

The advantage of writing nuclear reactions as shown above is that atomic masses found in atomic mass tables can be used to analyze the energy transformations that occur during the reactions. Even though atomic masses are used, the energy transformations are associated almost entirely with the rest energies of the reactant and product nuclei.

Rules for nuclear reactions

In balancing chemical reactions, the number of atoms of each type must be the same both before the reaction and after. Similar rules apply to nuclear reactions. Consider Observational Experiment **Table 29.3**, which describes reactions that have been observed to take place and reactions that have never been observed to occur.

OBSERVATIONAL EXPERIMENT TABLE 29.3 **Deducing rules for allowed nuclear reactions**

Observational experiment	Analysis
Nuclear reactions that have been observed: (a) $^1_1\text{H} + {}^7_3\text{Li} \rightarrow {}^4_2\text{He} + {}^4_2\text{He}$ (b) $^{226}_{88}\text{Ra} \rightarrow {}^{222}_{86}\text{Rn} + {}^4_2\text{He}$ (c) $^{14}_6\text{C} \rightarrow {}^{14}_7\text{N} + {}^0_{-1}e$ In the above reactions, the products are observed to have significantly more kinetic energy than the reactants. *Nuclear reactions that have never been observed:* (d) $^4_2\text{He} + {}^{27}_{13}\text{Al} \rightarrow {}^{32}_{15}\text{P} + {}^1_0\text{n}$ (e) $^2_1\text{H} + {}^3_1\text{H} \rightarrow {}^4_2\text{He} + {}^1_1\text{H}$	*Nuclear reactions that have been observed:* (a) A: $1 + 7 = 4 + 4$ Z: $1 + 3 = 2 + 2$ (b) A: $226 = 222 + 4$ Z: $88 = 86 + 2$ (c) A: $14 = 14 + 0$ Z: $6 = 7 - 1$ Both A and Z numbers are equal on both sides of the equations. *Nuclear reactions that have never been observed:* (d) A: $4 + 27 \neq 32 + 1$ Z: $2 + 13 = 15 + 0$ (e) A: $2 + 3 = 4 + 1$ Z: $1 + 1 \neq 2 + 1$ Either A or Z numbers are not equal on both sides of the equations.

Patterns

Two patterns appear in the observed nuclear reactions:

- The total numbers of nucleons A of the reactants and of the products are equal.
- The total Z numbers of the reactants and of the products are equal.
  (Z is the electric charge of the charged particles involved in the reaction.)

The second pattern, in which Z of the reactants is equal to Z of the products, is the result of electric charge conservation: in an isolated system, the total electric charge of the reactants equals that of the products; therefore, the charge is constant.

TIP The symbol for an electron is $_{-1}^{0}e$. It has charge $Z = -1$ and has zero nucleons, $A = 0$. It can be included in the reaction rules using this notation—see Reaction (c) in Table 29.3.

Why is the number of nucleons (protons plus neutrons) on both sides of the reactions equal? If a neutron vanished, it wouldn't violate charge conservation, but neutrons don't seem to do that. Evidently, they can transform into protons (see Reaction (c) in Table 29.3), but they cannot vanish. At the moment, we don't have a reason for why this pattern exists (but we will return to it in Chapter 30). For now, we will remember that in nuclear reactions the total charge of all participants and the total number of nucleons remain constant before and after the reaction. We summarize these two rules below.

Rule 1: Constant electric charge The total electric charge of the reacting nuclei and particles equals the total charge of the nuclei and particles produced by the reaction. This condition is satisfied if the total atomic number Z of the reactants equals the total atomic number of the products, including $Z = -1$ for free electrons.

Rule 2: Constant number of nucleons The total number of nucleons A (protons plus neutrons) of the reactants always equals the total number of nucleons of the products.

QUANTITATIVE EXERCISE 29.3 **Determine the missing product in the following reactions**

(a) $_{2}^{4}\text{He} + _{6}^{12}\text{C} \rightarrow _{7}^{15}\text{N} + ?$

(b) $_{1}^{2}\text{H} + _{1}^{3}\text{H} \rightarrow _{2}^{4}\text{He} + ?$

(c) $_{0}^{1}\text{n} + _{92}^{235}\text{U} \rightarrow _{54}^{140}\text{Xe} + ? + 2_{0}^{1}\text{n}$

(d) $_{55}^{137}\text{Cs} \rightarrow ? + _{-1}^{0}e$

Represent mathematically To determine the products, we need to use the rules stating that the A and Z numbers remain constant throughout the reaction.

(a) $4 + 12 = 15 + A$
$2 + 6 = 7 + Z$

(b) $2 + 3 = 4 + A$
$1 + 1 = 2 + Z$

(c) $1 + 235 = 140 + A + 2 \cdot 1$
$0 + 92 = 54 + Z + 2 \cdot 0$

(d) $137 = A + 0$
$55 = Z + (-1)$

Solve and evaluate Solving for A and Z in each case, we find the following:

(a) $A = 1$, $Z = 1$. The unknown product must be a hydrogen nucleus, $_{1}^{1}\text{H}$.

(b) $A = 1$, $Z = 0$. The unknown product must be a neutron, $_{0}^{1}\text{n}$.

(c) $A = 94$, $Z = 38$. The unknown product must be a strontium nucleus, $_{38}^{94}\text{Sr}$.

(d) $A = 137$, $Z = 56$. The unknown product must be a barium nucleus, $_{56}^{137}\text{Ba}$.

Try it yourself Determine the missing product in this reaction: $_{6}^{14}\text{C} \rightarrow ? + _{-1}^{0}e$.

Answer $_{7}^{14}\text{N}$.

Energy conversions in nuclear reactions

In the three observed reactions in Table 29.3, the products had significantly more kinetic energy than the reactants. How did the products get this kinetic energy? One hypothesis is that some of the rest mass energy of the reactants was converted to kinetic energy. If this is correct, then the mass of the reactants should be greater than the mass of the products. Let's test this hypothesis using the reaction described in Testing Experiment **Table 29.4**.

TESTING
EXPERIMENT TABLE 29.4

Accounting for the extra kinetic energy

Testing experiment	Prediction	Outcome
In the following reaction, the products have more kinetic energy than the reactants: $^1_1H + ^7_3Li \rightarrow ^4_2He + ^4_2He + $ kinetic energy Is this consistent with the idea that some rest mass energy of the reactants was transformed into kinetic energy of the products? The atomic masses of the atoms involved are $m(^1_1H) = 1.007825$ u $m(^7_3Li) = 7.016003$ u $m(^4_2He) = 4.002602$ u	Mass of reactants: $m(^1_1H) + m(^7_3Li) = 1.007825$ u $+ 7.016003$ u $= 8.023828$ u Mass of products: $m(^4_2He) + m(^4_2He) = 4.002602$ u $+ 4.002602$ u $= 8.005204$ u The energy equivalent of this mass difference is $\Delta mc^2 = (8.023828 \text{ u} - 8.005204 \text{ u})c^2 \left(\dfrac{931.5 \text{ MeV}}{\text{u} \cdot c^2} \right)$ $= 17.3$ MeV We predict that the products will have 17.3 MeV more kinetic energy than the reactants.	Through a complex process, the products are found to have 17.3 MeV more kinetic energy than the reactants, as predicted.

Conclusion

This experiment supports the idea that some of the rest energy of the reactants is converted into kinetic energy of the products.

These kinds of reactions, where rest energy is converted into kinetic energy, are called **exothermic**. The opposite can occur as well. In certain nuclear reactions the kinetic energy of the reactants is greater than the kinetic energy of the products. In this case, the rest energy of the products is greater than that of the reactants. In these **endothermic** reactions, some of the kinetic energy of the reactants is converted into rest energy of the products. Both types of reactions can be understood in terms of energy constancy in an isolated system.

> **Rule 3: Constant energy** The rest energy of the reactants equals the rest energy of the products plus any change in kinetic energy of the system:
>
> $$\sum_{\text{reactants}} mc^2 = \sum_{\text{products}} mc^2 + \Delta K$$
>
> ΔK can be either positive (if the products' rest energy is less that of the reactants) or negative (if the products' rest energy is greater). ΔK is also known as the reaction energy Q. Rearranging, we find an expression for Q:
>
> $$Q = \left(\sum_{\text{reactants}} m - \sum_{\text{products}} m \right) c^2 \tag{29.5}$$

We will have several opportunities to use this idea in the following sections.

REVIEW QUESTION 29.4 Is it possible for the nuclear reaction $^4_2He + ^6_3Li \rightarrow ^{11}_6C + 2 ^{~0}_{-1}e$ to occur? Explain your answer.

29.5 Nuclear sources of energy

Ernest Rutherford once said, "Anyone who expects a source of power from the transformation of the atom is talking moonshine."

Rutherford had an enormous impact on early atomic and nuclear physics, but he missed the mark with this statement. Let's consider more carefully why nuclear reactions can be a source of power.

Binding energy and energy release

Figure 29.8 is a graph of the binding energy per nucleon (BE/A) versus mass number A for atomic nuclei. Notice that nuclei with $A \approx 60$ have the greatest binding energy per nucleon, while very small and very large nuclei have less binding energy per nucleon. The higher the binding energy per nucleon, the more energy is needed to split the nucleus into its constituent protons and neutrons; therefore, the nuclei with the highest binding energy per nucleon—nuclei around $A \approx 60$—should be the most stable (in fact, iron, with $A = 56$, is the most stable element).

The graph in Figure 29.8 allows us to predict two new phenomena. First, note that the binding energy per nucleon for nuclei with small atomic numbers is less than that for nuclei with larger atomic numbers (up to about 60). Therefore, if two smaller nuclei were to combine to make a heavier one, the difference in their total binding energy should be released. Second, the graph shows that nuclei with very large atomic numbers have less binding energy per nucleon than those with atomic numbers near 60. Thus if there was a way to break one of these high-Z nuclei into two or more smaller-Z nuclei, the difference in binding energy could be released.

Now we have two predictions that are based on the patterns in the graph:

- When two small nuclei combine, energy should be released.
- When a large nucleus breaks apart into elements with atomic numbers below iron, energy should be released.

It turns out that both of these predicted processes do occur in nature. Note also that the first prediction is consistent with the analysis based on the bar chart in Figure 29.7a.

FIGURE 29.8 A graph of binding energy versus number of nucleons (BE/A).

Fusion and chemical elements

Small nuclei $(A < 60)$ do not spontaneously join to form heavier ones because the nuclei repel each other due to their positive electric charges. For the attractive nuclear force to be exerted, the nuclei must be very close to each other—less than 10^{-14} m apart. In solids and liquids the nuclei of neighboring atoms are about 10^{-10} m apart, far too distant from one another to have a chance to fuse. However, if a material is so hot that it ionizes completely (that is, it becomes a plasma), and the pressure is high enough, a small percentage of nuclei will come close enough so that the nuclear force pulls them together (**Figure 29.9a**) and a huge amount of energy is released (Figure 29.9b). This process, called **fusion**, occurs naturally in stars, which begin their lives with mostly hydrogen and a little helium. Fusion could potentially be a source of clean energy on Earth; however, the temperatures and pressures needed test the limits of current technology.

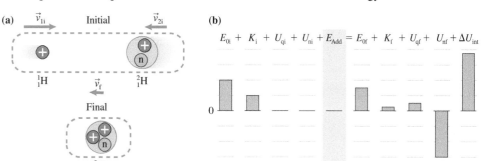

FIGURE 29.9 (a) A greatly simplified example of a fusion reaction. In plasma, ions are moving so fast that they can get close enough to each other during a collision that the nuclear force binds them together. (b) An energy bar chart representing the process.

The combining of light nuclei into heavier ones happens in the cores of stars where the temperature and pressure are very high. Stars with masses comparable to that of the Sun produce elements from helium to carbon through fusion. More massive stars produce elements up to iron. These stars finish their lives by cooling down, and the heavy elements produced in their cores remain there. In contrast, stars that are much more massive than the Sun explode at the end of their lives. These explosions, known as supernovae, produce elements heavier than iron.

Supernovae contribute to the chemical composition of the universe in two ways. First, the elements lighter than iron that are produced in the star's core before the explosion are ejected into space. (Lighter stars that do not explode as supernovae produce those elements, too, but those elements remain in their cores.) Second, elements heavier than iron that are produced during the explosion are also ejected into space. The Sun and Earth have a high abundance of elements heavier than helium; thus the Sun and Earth are made of elements produced long ago inside stars that exploded as supernovae. This means that many of the atoms in our bodies came from these supernovae.

QUANTITATIVE EXERCISE 29.4   **Fusion energy in the Sun**

The energy released by the Sun comes from several sources, including the so-called proton-proton chain of fusion reactions. This chain of reactions can be summarized as follows:

$$2{}_1^1\text{H} + 2{}_0^1\text{n} \rightarrow {}_2^4\text{He} + \text{energy}$$

Determine the amount of energy (in MeV) released in this chain of reactions. Use the masses ${}_1^1\text{H}$, 1.007825 u; ${}_0^1\text{n}$, 1.008665 u; and ${}_2^4\text{He}$, 4.002602 u to determine the rest energy converted to other forms.

Represent mathematically   The energy released, the reaction energy Q, is determined using Eq. (29.5):

$$Q = \left(\sum_{\text{reactants}} m - \sum_{\text{products}} m \right) c^2$$

Solve and evaluate   Inserting the appropriate values and converting to MeV gives

$$Q = [\, 2(1.007825 \text{ u}) + 2(1.008665 \text{ u}) - 4.002602 \text{ u}\,]$$
$$\times c^2 \left(\frac{931.5 \text{ MeV}}{\text{u} \cdot c^2} \right) = 28 \text{ MeV}$$

Compare this energy to the energy needed to remove an electron from a hydrogen atom, 13.6 eV. This energy has an order of magnitude typical of chemical reactions. Thus the energy released in nuclear chain reactions is about a million times greater than that released in chemical reactions.

- -

Try it yourself   What fraction of the total reactant rest energy in the proton-proton chain reactions is converted to other forms of energy during the above fusion reaction?

Answer

About (0.03 u)/(4.0 u) = 0.008, or 0.8%.

Just as in any reaction, once the reactants have been used up, the reaction stops. This means that eventually the Sun will run out of hydrogen and start burning helium into carbon. After helium is exhausted, fusion will cease (as the Sun is not massive enough to be able to fuse carbon), and our star will slowly go dark. How long will the Sun be able to shine?

EXAMPLE 29.5   **How long will the Sun shine?**

According to the present models of stellar structure, the central parts of stars, which contain about 10% of the star's mass, have conditions favorable for fusion. Assuming the Sun is made of pure hydrogen, how much time will it take the Sun to radiate the energy released when 10% of its hydrogen has fused into helium? The mass of the Sun is about 2.00×10^{30} kg. The total luminosity L (also power, or energy/second) emitted by the Sun is $L = 3.85 \times 10^{26}$ W.

Sketch and translate   Imagine the Sun as an object that simultaneously releases energy through fusion reactions and then radiates it as electromagnetic radiation.

Simplify and diagram   Assume that during a fusion reaction in the Sun's interior, two protons and two neutrons join together to form a helium nucleus. The amount of energy released is equal to

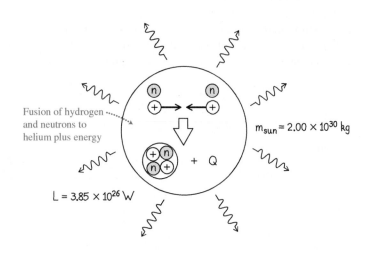

Fusion of hydrogen and neutrons to helium plus energy

$m_{\text{sun}} = 2.00 \times 10^{30}$ kg

$+ Q$

$L = 3.85 \times 10^{26}$ W

approximately 0.008 of the rest energy of the participating particles (see Quantitative Exercise 29.4). Further, assume that there are no other sources of energy, so that the rate at which energy is produced inside the Sun by fusion reactions is the same as the rate at which the energy is radiated. Finally, assume that the Sun emits the same amount of energy every second during its lifetime.

Represent mathematically The total energy released by fusion reactions throughout the lifetime of the Sun is equal to 0.8% of the rest energy of the 10% of the Sun's mass energy that is available for fusion:

$$E_{released} = (0.008)(0.10)m_{Sun}c^2$$

This equals the luminosity L of the Sun times the time interval Δt during which the Sun will shine, or

$$E_{released} = L\,\Delta t$$

Solve and evaluate Setting the above two expressions equal to each other, solving for Δt, and inserting the appropriate values, we estimate the expected lifetime of the Sun:

$$\Delta t = \frac{E_{released}}{L}$$

$$= \frac{(0.008)(0.10)(2.00 \times 10^{30}\ \text{kg})(3.00 \times 10^8\ \text{m/s})^2}{3.85 \times 10^{26}\ \text{W}}$$

$$= 3.7 \times 10^{17}\ \text{s}$$

$$= 3.7 \times 10^{17}\ \text{s}\left(\frac{1\ \text{year}}{3.16 \times 10^7\text{s}}\right) = 1.2 \times 10^{10}\ \text{years}$$

This result is about 2.5 times the age of Earth. Nuclear fusion is a possible mechanism for powering the Sun for billions of years.

Try it yourself Studying the energy emitted by stars of different masses, astronomers found that the energy that a typical star emits every second is proportional to its mass raised to the power of 3.5, that is, $L \propto m^{3.5}$. Will a 10-solar-mass star take more or less time than the Sun to exhaust its nuclear fuel, assuming that both the star and the Sun have 10% of their mass in hydrogen available for fusion?

Answer

$$= \frac{10L_{Sun}}{L_{star}} = \frac{10m_{Sun}^{3.5}}{m_{star}^{3.5}} = 10\left(\frac{10m_{Sun}}{m_{Sun}}\right)^{3.5} = 0.0032$$

$$\frac{\Delta t_{star}}{\Delta t_{Sun}} = \frac{\left(\frac{E_{released}}{L}\right)_{star}}{\left(\frac{E_{released}}{L}\right)_{Sun}} = \frac{\frac{(0.008)(0.10)(10m_{Sun})c^2}{L_{star}}}{\frac{(0.008)(0.10)m_{Sun}c^2}{L_{Sun}}}$$

The star will take significantly less time—about 0.003 times the time for the Sun:

FIGURE 29.10 Nuclear fission.

(a) Original heavy nucleus and a moving neutron

^{235}U

(b) The neutron enters the nucleus, increasing its energy.
$^{236}\text{U*}$

(c) The nucleus becomes unstable.
$^{236}\text{U*}$

(d) The excited nucleus divides into two smaller nuclei and several neutrons.

(e) The new neutrons can hit other heavy nuclei, producing a chain reaction.

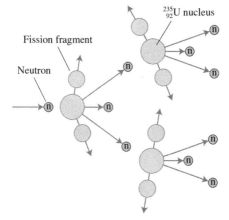

Fission and nuclear energy

We saw from the graph in Figure 29.8 that energy is released when very heavy nuclei split into nuclei of atoms that reside near the middle of the periodic table. However, the heavy nuclei do not just spontaneously split. One way the process can be initiated is the collision of a moving neutron with a heavy nucleus, which disturbs the equilibrium of the nucleus and causes it to split into two smaller nuclei. This process is called **fission** (**Figure 29.10a–d**). Heavy nuclei have a high neutron-to-proton ratio; however, nuclei near the middle of the periodic table (which are the typical product nuclei of these kinds of reactions) usually have a somewhat lower ratio. Thus splitting a heavy nucleus usually releases several neutrons, which can then catalyze the splitting of other

FIGURE 29.11 Energy bar chart representing fission.

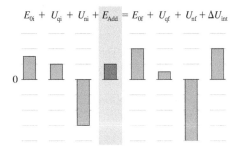

$$E_{0i} + U_{qi} + U_{ni} + E_{Add} = E_{0f} + U_{qf} + U_{nf} + \Delta U_{int}$$

nuclei, causing a chain reaction (Figure 29.10e). This reaction results in a huge release of energy (see the bar chart in **Figure 29.11**).

A common fission reaction is

$$^1_0n + {}^{235}_{92}U \rightarrow {}^{141}_{56}Ba + {}^{92}_{36}Kr + 3^1_0n + \text{energy}$$

Notice that one neutron absorbed by one uranium nucleus produces two smaller nuclei plus three new neutrons that can cause three new fission reactions. Let's calculate the energy released by the above fission reaction.

QUANTITATIVE EXERCISE 29.6 ⟩ **Energy from a fission reaction**

Determine the energy released in the above fission reaction. Use the following masses: ^1_0n, 1.008665 u; $^{235}_{92}U$, 235.043924 u; $^{141}_{56}Ba$, 140.914411 u; $^{92}_{36}Kr$, 91.926156 u. One neutron collides with and momentarily joins the U-235 atomic nucleus to form an excited U-236, which quickly splits into Ba-141, Kr-92, and three neutrons.

Represent mathematically The energy Q released in the reaction is the energy equivalent to the mass difference between the reactants and products.

$$Q = \left[m\left(^1_0n\right) + m\left(^{235}_{92}U\right) \right] c^2$$
$$\quad - \left[m\left(^{141}_{56}Ba\right) + m\left(^{92}_{36}Kr\right) + 3m\left(^1_0n\right) \right] c^2$$
$$\quad = \left[m\left(^{235}_{92}U\right) - m\left(^{141}_{56}Ba\right) - m\left(^{92}_{36}Kr\right) - 2m\left(^1_0n\right) \right] c^2$$

Solve and evaluate Inserting the appropriate values and converting to MeV, we get

$$Q = \left[235.043924 \text{ u} - 140.914411 \text{ u} - 91.926156 \text{ u} - 2(1.008665 \text{ u}) \right]$$
$$\times c^2 \left(\frac{931.5 \text{ MeV}}{u \cdot c^2} \right) = 173 \text{ MeV}$$

This is a typical energy release during fission reactions. This reaction produces over 300 million times more energy than burning one octane molecule.

- -

Try it yourself Your friend says that nuclear power plants could release energy by the fission of iron-56 (very abundant) instead of the much less abundant uranium-235. How do you respond?

Answer   Iron-56 has about the highest binding energy per nucleon of any nucleus. This means energy must be added to the nucleus to cause it to break into smaller nuclei with lower BE/A—exactly the opposite of what is desired in a nuclear power plant.

Lise Meitner.

Bohr's liquid drop model of the nucleus

Quantitative Exercise 29.6 represents an experiment performed by German scientists Lise Meitner, Otto Hahn, and Fritz Strassmann in the 1930s. Their goal was not to split uranium, but instead to produce elements heavier than uranium. They thought that bombarding uranium with neutrons would lead to neutron capture and subsequent beta decay that in turn would increase the number of protons in the nucleus, thus creating heavier elements. (You will find the explanation of how neutron capture leads to the production of heavier elements in Section 29.6.) To their surprise, the nuclei produced in the reaction behaved like isotopes of barium and other nuclei with about half the mass of uranium. They were at a complete loss to explain this result. At this point, Meitner, who was Jewish, had to flee Germany under Adolf Hitler's regime. She immigrated to Sweden. There, she asked her nephew Otto Robert Frisch (who was working in Denmark with Niels Bohr) to help interpret these results.

Meitner and Frisch decided that Bohr's **liquid drop model** of a nucleus could explain the strange experimental results (see Figure 29.10). In this model, the nucleus was compared to a drop of water: just as surface tension holds the water drop together, nuclear forces hold the nucleons together. However, in heavy nuclei, such as uranium, there are too many electrically charged protons present. The protons repel each other and can overwhelm the effect of the "surface tension," especially if the nucleus is not spherical. The model suggested that such a nucleus could stretch and divide into two smaller pieces. This meant that the uranium nucleus is very unstable, ready to split with the slightest provocation—such as being hit by a neutron. Meitner and Frisch had come up with a model for fission. Unfortunately, the Nobel Prize committee did not think that the work of Lise Meitner contributed enough to the discovery of fission, and so she did not share the Nobel Prize given to Otto Hahn in 1944.

REVIEW QUESTION 29.5 Explain how fusion can lead to a release of energy. Does this release mean that energy is not conserved in nuclear fusion reactions?

29.6 Mechanisms of radioactive decay

Now that we understand the stability of nuclei in terms of their binding energy per nucleon and the energy released in nuclear reactions (the Q of the reactions), we can explain why some nuclei undergo radioactive decay while others do not.

Alpha decay

Of the small nuclei, helium $\left(\frac{4}{2}\text{He}\right)$ is one of the most stable. Because of this, in a larger nucleus, groups of two protons and two neutrons tend to form helium nuclei within the larger nucleus (like a small group of friends joined together in a large crowd). In addition to the attractive forces that the two protons and two neutrons in a helium nucleus exert on each other, a repulsive electric force is exerted on the helium nuclei by the other protons in the larger nucleus. When the nucleus is large enough, the electric repulsion becomes significant compared to the nuclear attraction. Then there is a chance that one of the helium nuclei will leave the larger nucleus. This is the emission of an alpha particle.

When one of the alpha particles leaves, this emission reduces the number of protons in the original nucleus and also reduces the electric repulsion between the protons that remain. Because a helium nucleus also contains two neutrons, the total number of nucleons A in the original nucleus decreases by four. **Figure 29.12** illustrates the alpha decay of radium-226. When radium undergoes this decay, it becomes radon-222, and an alpha particle is released. The mass of the new radon nucleus (called the **daughter nucleus**) plus the alpha particle is less than the original radium nucleus (called the **parent nucleus**). Some of the rest energy of the radium nucleus is converted into the kinetic energy of the products.

$$^{226}_{88}\text{Ra} \rightarrow {}^{222}_{86}\text{Rn} + {}^{4}_{2}\text{He} + 4.9 \text{ MeV}$$

FIGURE 29.12 Alpha decay of radium-226. The result is a new element plus release of an alpha particle.

QUANTITATIVE EXERCISE 29.7 **Kinetic energy produced during alpha decay**

Determine the kinetic energy of the product nuclei when polonium-212 undergoes alpha decay. The masses of the nuclei involved in the decay are $m\left({}^{212}_{84}\text{Po}\right)$, 211.9889 u; $m\left({}^{208}_{82}\text{Pb}\right)$, 207.9766 u; and $m\left(\frac{4}{2}\text{He}\right)$, 4.0026 u.

Represent mathematically The alpha decay of polonium-212 is

$$^{212}_{84}\text{Po} \rightarrow {}^{208}_{82}\text{Pb} + {}^{4}_{2}\text{He} + \text{energy}$$

Use Eq. (29.5) to find the energy Q released during this reaction:

$$Q = \left[m\left({}^{212}_{84}\text{Po}\right) - m\left({}^{208}_{82}\text{Pb}\right) - m\left(\frac{4}{2}\text{He}\right) \right]c^2$$

Solve and evaluate Insert the appropriate values and convert to MeV:

$$Q = (211.9889 \text{ u} - 207.9766 \text{ u} - 4.0026 \text{ u})c^2$$
$$= (0.0097 \text{ u})c^2\left(\frac{931.5 \text{ MeV}}{\text{u} \cdot c^2}\right)$$
$$= 9.0 \text{ MeV}$$

Of this 9.0 MeV of released energy, 8.8 MeV is converted to the kinetic energy of the alpha particle $\left(\frac{4}{2}\text{He}\right)$. Most of the remaining 0.2 MeV is converted to the kinetic energy of the recoiling lead-208 nucleus. In addition, a small fraction of the energy may be released as an additional product in the reaction: a gamma ray (a high-energy photon).

Try it yourself Determine the product nucleus when thorium-232 $\left({}^{232}_{90}\text{Th}\right)$ undergoes alpha decay.

Answer Radium-228 $\left({}^{228}_{88}\text{Ra}\right)$.

Beta decay

The particle emitted from a nucleus during beta decay is an electron. But we know from the uncertainty principle that electrons cannot reside in the nucleus. How can we explain this apparent contradiction? One explanation is that a neutron inside a nucleus can spontaneously decay into a proton and an electron. The proton stays inside the nucleus, while the electron created during the decay leaves. This neutron decay reaction could be written as follows:

$$_0^1 n \rightarrow {}_1^1 p + {}_{-1}^0 e$$

where p represents a proton. Notice that both the total electric charge and the total nucleon number on both sides of the equation are equal, but, as we will see later, this equation is incomplete.

If this explanation is correct, then during beta decay, a particular element should transform to an element with a Z number that is larger by one. This is exactly what Ernest Rutherford and his colleague Frederick Soddy predicted while studying radioactive decay. Another prediction that follows from this explanation is that the heavier isotopes of a particular element should produce more beta radiation than the lighter isotopes of that element. This is exactly what is observed. For example, carbon-14 is a heavy isotope of carbon that has two more neutrons than protons, and it is unstable, undergoing beta decay, whereas carbon-12 and carbon-13 are stable. However, when scientists studied beta decay in greater detail, they encountered two difficulties.

The problem with conservation of angular momentum arising from spin You learned in Chapter 28 that electrons have an intrinsic spin quantum number. This spin quantum number was observed in many other experiments to be a conserved quantity. However, in beta decay experiments, it seemed not to be. For example, during the beta decay of a neutron, all three participating particles have spin values equal to $\pm 1/2$. But because there is only one particle on one side of the reaction but two particles on the other side, it is not possible that the total spin on both sides could be equal. Either spin is not a conserved quantity, or our understanding of beta decay is flawed.

The problem with energy conservation The second problem involved energy conservation. It, too, seemed to not be conserved in beta decay. The total energy of the products was always observed to be less than the total energy of the reactants, but it was not possible to detect where that missing energy went. This pattern was so persistent that originally Niels Bohr suggested that maybe energy was not always conserved in the subatomic world. However, this idea was too radical for most physicists to take seriously.

In 1930, Wolfgang Pauli proposed an explanation for beta decay that did not require abandoning energy conservation or spin number conservation. He hypothesized that some unknown particle carried away the missing energy and accounted for the discrepancy in spin number. This particle had zero electric charge, zero mass, and a spin number of either $+1/2$ or $-1/2$. Enrico Fermi called the particle a **neutrino**, meaning "little neutral one." Because the neutrino was thought to have zero mass, it was expected to travel at the speed of light.

Although this sounds like the description of a photon, there is a subtle difference between a photon and a neutrino. An atom can absorb a photon and as a result change its energy state. Allowed energy transitions occur only between states that differ by 1 in the angular momentum quantum number l. The photon's intrinsic spin must then be equal to the change in angular momentum of the atom, in this case, 1. A neutrino's intrinsic spin, however, must be $1/2$.

The existence of neutrinos was confirmed 25 years after Pauli's proposal. Large numbers of neutrinos produced by the Sun and other astronomical objects continually pass through our bodies. Yet they cause no damage and leave almost no trail because of the extremely small likelihood of them interacting with the atoms in their path.

We can now correct the equation for the decay of a neutron using our knowledge of neutrinos. The correct equation is

$$_0^1n \rightarrow {_1^1}p + {_{-1}^0}e + \bar{\nu}$$

where $\bar{\nu}$ is the antineutrino, the particle that is responsible for the conservation of angular momentum and energy in the decay. We will learn more about antineutrinos and why they are needed in this reaction as opposed to neutrinos in Chapter 30. Examples of other reactions involving beta decay, shown in **Figure 29.13** and represented mathematically below, always result in the production of a neutrino ν or an antineutrino $\bar{\nu}$ (we will discuss neutrinos and antimatter in more detail in Chapter 30). The first reaction is called **beta-minus decay**, as it produces a negative electron:

$$_6^{14}C \rightarrow {_7^{14}}N + {_{-1}^0}e + \bar{\nu}$$
$$_{11}^{22}Na \rightarrow {_{10}^{22}}Ne + {_{1}^0}e + \nu$$

The second beta decay is an example of a somewhat less common decay known as **beta-plus decay**, which produces a positron $\left({_1^0}e\right)$, which is otherwise identical to an electron but has positive charge e instead of the negative charge $-e$ of the electron. The positron is the antiparticle to the electron (we will discuss positrons further in Chapter 30).

What is the mechanism behind beta decay? Because neutrinos interact with atoms only extremely rarely, it's unlikely that the mechanism is electromagnetic in nature. For this and other technical reasons, Enrico Fermi proposed a new interaction in 1934 called the **weak nuclear interaction**, which explained beta decay. Like the nuclear interaction, the weak interaction is very short range and is much weaker (hence its name).

A free neutron (a neutron that is not inside a nucleus) is not a stable particle. It spontaneously decays rather quickly (in about 15 min on average), producing a proton, an electron, and an antineutrino. The weak interaction is responsible for this decay. When neutrons are bound inside nuclei they are very stable.

Gamma decay

We already know that alpha and beta decays are usually accompanied by rays that are not deflected by a magnetic field but that are easily detected by their interactions with atoms. These rays are gamma (γ) rays—photons that have higher energy and therefore shorter wavelength than X-rays.

One explanation for why gamma rays accompany alpha and beta decay is that the energy of the nucleus is quantized in a similar way to the energy in atoms. If this is true, then after alpha or beta decay the nucleus could be left in an excited state from which it then emits one or more photons to return to its ground state. Because the binding energy of a nucleus is much higher than that of the atom's electrons, the energy differences between states are also likely to be much larger. As a result, the energy of the emitted photons is much larger. Gamma decay happens, for example, when boron-12 undergoes beta decay to form carbon-12. The carbon-12 that is produced is in an excited state $\left({_6^{12}}C*\right)$ (the asterisk indicates that the nucleus is in an excited state). When it returns to its lowest energy state, a gamma ray photon is emitted (**Figure 29.14**):

$$_5^{12}B \rightarrow {_6^{12}}C* + {_{-1}^0}e$$
$$_6^{12}C* \rightarrow {_6^{12}}C + \gamma$$

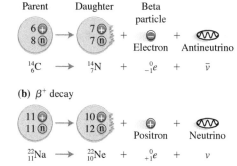

FIGURE 29.13 Two types of beta decay. (a) Beta-minus decay, releasing an electron and an antineutrino. (b) Beta-plus decay, releasing a positron and a neutrino.

(a) β^- decay

Parent   Daughter   Beta particle

$$_6^{14}C \rightarrow {_7^{14}}N + {_{-1}^0}e + \bar{\nu}$$

(b) β^+ decay

$$_{11}^{22}Na \rightarrow {_{10}^{22}}Ne + {_{1}^0}e + \nu$$

FIGURE 29.14 Gamma decay. (a) After beta decay, the daughter nucleus is left in an excited state and drops to the ground state, emitting a gamma ray photon. (b) An energy level diagram for the gamma ray emission process.

(a)

^{12}C in excited state

^{12}C returns to the ground state when it emits a γ ray.

Electron

Gamma ray photon

$$_5^{12}B \rightarrow {_6^{12}}C* + {_{-1}^0}e$$
$$_6^{12}C* \rightarrow {_6^{12}}C + \gamma$$

(b) E

Excited state

Ground state

γ ray photon

QUANTITATIVE EXERCISE 29.8   Alpha and beta decay

The following nuclei undergo different types of radioactive decay. Determine the daughter nucleus for each, and write an equation representing each decay reaction: (a) $^{239}_{94}$Pu alpha decay, (b) $^{144}_{58}$Ce beta-minus decay, (c) and $^{65}_{30}$Zn beta-plus decay. The latter produces a positron.

Represent mathematically

(a) We have to subtract $A = 4$ and $Z = 2$ (the $^4_2$He alpha particle) from the $^{239}_{94}$Pu. Thus, the daughter nucleus has $A = 239 - 4 = 235$ and $Z = 94 - 2 = 92$, which is uranium-235.

(b) We have to subtract $A = 0$ and $Z = -1$ (the $^{\ 0}_{-1}e$ beta-minus particle, an electron) from the $^{144}_{58}$Ce. Thus, the daughter nucleus has $A = 144 - 0 = 144$ and $Z = 58 - (-1) = 59$, which is praseodymium-144.

(c) We have to subtract $A = 0$ and $Z = +1$ (the ^0_1e beta-plus particle, a positron) from the $^{65}_{30}$Zn. Thus, the daughter nucleus has $A = 65 - 0 = 65$ and $Z = 30 - 1 = 29$, which is copper-65.

Solve and evaluate The reactions are then

(a) $^{239}_{94}$Pu $\rightarrow$ $^{235}_{92}$U + $^4_2$He + energy

(b) $^{144}_{58}$Ce $\rightarrow$ $^{144}_{59}$Pr + $^{\ 0}_{-1}e$ + $\bar{\nu}$ + energy

(c) $^{65}_{30}$Zn $\rightarrow$ $^{65}_{29}$Cu + ^0_1e + ν + energy

"Energy" here means the kinetic energy of the products and possibly one or more gamma rays.

Try it yourself Identify the daughter nucleus and write a reaction equation for the beta-minus decay of $^{131}_{53}$I.

Answer $^{131}_{53}$I $\rightarrow$ $^{131}_{54}$Xe + $^{\ 0}_{-1}e$ + $\bar{\nu}$ + energy.

EXAMPLE 29.9   Beta decay in our bodies

The body normally contains about 7 mg of radioactive potassium-40 $^{40}_{19}$K (it is present in some of the foods we eat, such as bananas). Each second, about 2×10^3 of these potassium nuclei undergo either beta-plus or beta-minus decay. Assume that all decays are beta-plus and that 40% of the energy released is absorbed by the body (the percent depends on the ratio of beta-minus decay to beta-plus decay, body thickness, the energy of the neutrinos produced in the decays, and other factors). Determine the energy in MeV transferred to the body each second by $^{40}_{19}$K beta-plus decay. What is the rate of energy transfer in watts?

Sketch and translate A single beta-plus decay is sketched below. If we determine the energy released by a single decay, we can then determine the rate of energy transfer to the body by multiplying the energy of one decay by 2000 (the number of decays that occur every second) and then taking 40% of that number.

$^{40}_{19}$K $\longrightarrow$ $^{40}_{18}$Ar + ^0_1e + ν

Positron   Neutrino

Simplify and diagram The system of interest is the potassium atom and its decay products.

Represent mathematically The Q of a single beta-plus decay is

$$Q = \left[m\left(^{40}_{19}\text{K}\right) - m\left(^{40}_{18}\text{Ar}\right) - m\left(^0_1e\right) \right]c^2$$

The rate of energy transferred to the body ($\Delta U/\Delta t$) is 40% of Q times the number ($\Delta N/\Delta t$) of decays per second, or

$$\frac{\Delta U}{\Delta t} = \frac{\Delta N}{\Delta t}(0.4Q)$$

Solve and evaluate From Appendix B, we find the masses of the atoms in atomic mass units (u):

$$Q = (39.964000\ \text{u} - 39.962384\ \text{u} - 5.4858 \times 10^{-4}\text{u})c^2$$

$$= (0.001068\ \text{u})c^2\left(\frac{931.5\ \text{MeV}}{\text{u} \cdot c^2}\right) = 0.995\ \text{MeV}$$

The energy transferred to the body per second is then

$$\frac{\Delta U}{\Delta t} = (2000\ \text{decays/s})(0.995\ \text{MeV})(0.4) = 800\ \text{MeV/s}$$

In watts this is

$$\frac{\Delta U}{\Delta t} = (800\ \text{MeV/s})\left(\frac{1.6 \times 10^{-13}\ \text{J}}{\text{MeV}}\right)$$

$$= 1.3 \times 10^{-10}\ \text{W} \approx 1 \times 10^{-10}\ \text{W}$$

This is extremely small compared to the average human metabolic rate of 100 W. We'll see in a later section whether these beta particles cause any negative health effects.

Try it yourself Is more or less energy released by potassium-40 beta-minus decay than beta-plus decay?

Answer   The product of the beta-minus decay is $^{40}_{20}$Ca, which has mass 39.962591 u. This is slightly greater than the 39.962383-u $^{40}_{18}$Ar product mass of the beta-plus decay. The beta-plus decay releases more energy.

REVIEW QUESTION 29.6 What was the observational evidence related to beta decay that made scientists think that a new particle was involved?

29.7 Half-life, decay rate, and exponential decay

Radioactive materials are used for medical diagnoses, and they are part of the world in which we live. Radioactive isotopes can also help determine the age of bones and other archeological artifacts through radioactive dating. For isotopes to serve these purposes, we need a quantitative description of how the number of radioactive nuclei in a sample changes and the rate at which radioactive decay occurs. These quantitative measures include half-life, the decay rate of nuclei, and the application of the exponential function to decay.

Half-life

Suppose a radioactive sample has a number N_0 of radioactive nuclei at time $t_0 = 0$. We can determine this number by dividing the mass of the sample by the mass of one atom. Using a particle detector such as a Geiger counter, we can estimate the number of nuclei that decay in a short time interval. By continually subtracting the number of decays in a short time interval from the initial number, we can determine the number N of radioactive nuclei that remain in the sample as a function of time. In Observational Experiment Table 29.5 we look for a pattern in the variation of N with time t.

OBSERVATIONAL EXPERIMENT TABLE 29.5 **The number N of radioactive nuclei versus time t**

Observational experiment	Analysis
We measure the decay rate in a radioactive sample and determine the number of radioactive nuclei in the sample as a function of time:	• The number decreased from N_0 to $0.50N_0$ from time 0 to 1 min.
	• The number decreased from $0.50N_0$ to $0.25N_0$ from time 1 min to 2 min.

Number N of radioactive nuclei	t (min)
N_0	0
$0.50N_0$	1
$0.25N_0$	2
$0.13N_0$	3
$0.06N_0$	4
$0.03N_0$	5
undetectable	10

• The number decreased from $0.25N_0$ to $0.13N_0$ from time 2 min to 3 min.

• The number decreased from $0.13N_0$ to $0.06N_0$ from time 3 min to 4 min.

• The number decreased from $0.06N_0$ to $0.03N_0$ from time 4 min to 5 min.

Pattern

For this sample, the number N of radioactive nuclei in the sample decreased by half during each 1-min time interval.

The data in Table 29.5 are plotted in Figure 29.15. As noted in the table, the number N of radioactive nuclei present in this particular radioactive sample decreased by half during each 1-min time interval. The reduction by one-half in a particular time interval, which varies depending on the type of nuclei, is a general property of radioactive samples called the **half-life** of the sample.

> **Half-life** The half-life T of a particular type of radioactive nucleus is the time interval during which the number of nuclei in a given sample is reduced by one-half. After n half-lives, the fraction of radioactive nuclei that remain is
>
> $$\frac{N}{N_0} = \frac{1}{2^n} \text{ at } t = nT \qquad (29.6)$$

FIGURE 29.15 A graph of radioactive decay.

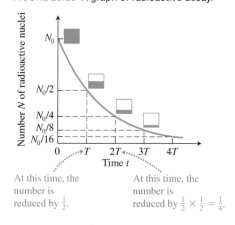

At this time, the number is reduced by $\frac{1}{2}$.

At this time, the number is reduced by $\frac{1}{2} \times \frac{1}{2} = \frac{1}{4}$.

Isotope	Half-life	Decay constant (s^{-1})
$^{87}_{37}Rb$	4.75×10^{10} years	4.62×10^{-19}
$^{238}_{92}U$	4.47×10^{9} years	4.91×10^{-18}
$^{40}_{19}K$	1.28×10^{9} years	1.72×10^{-17}
$^{239}_{94}Pu$	2.41×10^{4} years	9.11×10^{-13}
$^{14}_{6}C$	5730 years	3.83×10^{-12}
$^{226}_{88}Ra$	1600 years	1.37×10^{-11}
$^{137}_{55}Cs$	30.0 years	7.32×10^{-10}
$^{90}_{38}Sr$	28.9 years	7.60×10^{-10}
$^{3}_{1}H$	12.3 years	1.79×10^{-9}
$^{60}_{27}Co$	5.27 years	4.17×10^{-9}
$^{131}_{53}I$	8.03 day	9.99×10^{-7}
$^{11}_{6}C$	20.4 min	5.66×10^{-4}

Note that after one half-life, the ratio $N/N_0 = 1/2$. During the next half-life, there is another 50% reduction in the number of nuclei. Thus, after two half-lives, $N/N_0 = 1/2^2 = 1/4$, and so forth. For the radioactive sample in Table 29.5, the half-life is 1 min. According to Eq. (29.6), the fraction of radioactive nuclei that remains at $t = 4$ min $= 4T$ is $N/N_0 = 1/2^4 = 1/16 = 0.06$, in agreement with the data in the table.

This method can be used for fractional half-lives. For example, after 4.5 min (4.5 half-lives), the fraction of nuclei that remain is $N/N_0 = 1/2^{4.5} = 1/22.6 = 0.044$. The half-lives for a variety of nuclei are listed in Table 29.6.

Determining the source of carbon in plants

The complex photosynthesis process in plant growth can be summarized by the following chemical reaction:

$$6CO_2 + 6H_2O + \text{sunlight} \rightarrow C_6H_{12}O_6 + 6O_2$$

What is the source of the CO_2 in this process—is it part of the minerals found in the ground in which the plants grow, such as calcium carbonate $(CaCO_3)$, or is it the gaseous CO_2 in the air? A radioactive decay experiment was originally used to answer this question.

The carbon isotope carbon-11 $\left(^{11}_{6}C\right)$ is radioactive, with a half-life of 20 min. Carbon-11 was incorporated into the CO_2 in the air surrounding growing barley plants in a controlled environment. Investigators found that radioactive carbon-11 became part of the carbohydrate molecules produced by photosynthesis in the barley. This result is evidence that carbon in plants comes from CO_2 in the atmosphere. More evidence could be obtained by repeating the experiment with carbon-11 incorporated into soil minerals.

Decay rate

When working with radioactive samples, we seldom know or measure directly the number of radioactive nuclei in a sample. Instead, we measure the change in the number of nuclei that decay, ΔN, during a certain time interval Δt. From this measurement we can calculate the number of radioactive nuclei that still remain in the sample because the change in the number ΔN of nuclei that decay during time interval Δt is proportional to the starting number N of radioactive nuclei in the sample, and is proportional to Δt:

$$\Delta N = -\lambda N \Delta t \qquad (29.7)$$

This relationship makes sense if we assume that the absolute value of ΔN is the number of decayed nuclei. The number that decay is greater if the number N in the sample is greater. The number that decay during a certain time interval should also be proportional to that time interval Δt as long as the time interval is short compared to the half-life of the radioactive sample. λ is a proportionality constant, called the **decay constant**, that depends on the type of radioactive nucleus (see Table 29.6). The greater the value of λ, the greater the rate at which the radioactive nuclei decay. The ratio of ΔN and Δt is called the **decay rate**, or the **activity**, of a sample of radioactive material.

Decay rate (activity) A The ratio of the number of nuclei that decay ΔN during a certain time interval Δt and that time interval is called the decay rate, or activity, of a sample of that type of radioactive nucleus:

$$\text{Decay rate (activity) } A = -\frac{\Delta N}{\Delta t} = \lambda N \qquad (29.8)$$

where λ is the decay constant for that particular type of nucleus and ΔN is a negative number due to the decrease in the number of nuclei.

The unit of decay rate is the becquerel (Bq) and equals 1 decay/s. An older unit of decay rate is the curie (Ci), where $1 \text{ Ci} = 3.70 \times 10^{10} \text{ Bq}$. This unusual number for the curie was chosen because it represents roughly the activity of 1 g of pure radium, the radioactive element that Marie Curie isolated from tons of uranium ore. Radioactive tracers used in medicine usually have activities of tens of kilobecquerels or microcuries (μCi; $1 \mu\text{Ci} = 10^{-6} \text{ Ci}$). Radiation detection devices such as Geiger counters are used to measure the activity of radioactive samples.

Exponential function and decay

The N-versus-t data in Table 29.5 and in Figure 29.15 are characteristic of a variety of phenomena that can be represented mathematically by what is called the exponential function. We can use Eq. (29.8) to derive an expression for N as a function of time t. To start, we turn the Δt in Eq. (29.8) into a very short time interval $\Delta t'$ during which the number N changes by $\Delta N'$. We then rearrange the equation to

$$\frac{\Delta N'}{N} = -\lambda \Delta t'$$

These short time changes can be summed up for the time interval starting from time 0 when N_0 nuclei are present to some later time t when N nuclei remain. The result of this summation is the exponential function

$$N = N_0 e^{-\lambda t}$$

Because the activity A of a sample is proportional to N, the activity is also described by an exponential function:

$$A = -\frac{\Delta N}{\Delta t} = \lambda N = \lambda N_0 e^{-\lambda t}$$

The number of nuclei present in the sample decreases exponentially with time, as does the activity of the sample.

Exponential decay The number N of radioactive nuclei that remain at time t if the number at time zero was N_0 is given by an exponential function:

$$N = N_0 e^{-\lambda t} \tag{29.9}$$

where λ is the decay constant of the nucleus, and $e = 2.718\ldots$ is the base of the natural logarithms. The activity A of the sample also decreases exponentially:

$$A = -\frac{\Delta N}{\Delta t} = \lambda N_0 e^{-\lambda t}$$

N is an exponentially decreasing function of time and is plotted in the graph shown in Figure 29.15. We can now determine the number N of radioactive nuclei that remain in a sample at time t compared to the number N_0 at time zero by using either Eq. (29.6) or Eq. (29.9). The former equation, $(N/N_0) = (1/2^n)$, is often easier to use. However, rearrangement of the exponential function ($N = N_0 e^{-\lambda t}$) leads to an important equation for calculating the unknown age of a sample, which we will use when we discuss radioactive dating in the next section.

Equations for radioactive decay have numerous applications. For example, they can be used in medicine to measure the volume of blood in a patient. The process for this measurement is illustrated in the following example.

EXAMPLE 29.10 **Estimating blood volume**

When a small amount of radioactive material is placed in 1.0 cm^3 of water, it has an activity of 75,000 decays/min. Imagine that an identical amount of radioactive material is injected into an individual's bloodstream. Later, after the material has spread throughout the blood, a 1.0-cm^3 sample of blood taken from the individual has an activity of 16 decays/min. Determine the individual's total blood volume.

Sketch and translate The total blood volume can be thought of as a large container of unknown volume and the 1.0-cm^3 sample as a small portion of that. We know that a direct relationship exists between the decays/min in the sample and the decays/min in the total volume of blood. Therefore, we can use the value of one to determine the value of the other.

Simplify and diagram Assume that during this process the person does not drink any water (so that their blood volume doesn't change), that all the radioactive material is absorbed by the blood, and that none is filtered out by the kidneys. Assume also that the material disperses evenly throughout the blood. Lastly, assume that the half-life of that type of material is much longer than the time interval needed to complete the experiment so that the activity of the radioactive material can be considered constant.

Represent mathematically We use a ratio method to determine the blood volume by comparing the activity A and volume V of the two amounts of blood (the whole body and the sample):

$$\frac{V_{blood}}{V_{sample}} = \frac{A_{blood}}{A_{sample}}$$

Solve and evaluate Solve for V_{blood} and substitute the known quantities into the above to determine the blood volume:

$$V_{blood} = \frac{A_{blood}}{A_{sample}} V_{sample} = \left(\frac{75,000 \text{ decays/min}}{16 \text{ decays/min}}\right)(1.0 \text{ cm}^3)$$
$$= 4.7 \times 10^3 \text{ cm}^3$$

where $(4.7 \times 10^3 \text{ cm}^3)(1 \text{ L}/10^3 \text{ cm}^3) = 4.7$ L. This value is within the 4- to 6-L range that is considered a normal human blood volume.

Try it yourself Suppose the radioactive sample had a half-life of 1.0 h and you waited 2 h for the injected sample to distribute uniformly before measuring the activity of 1 cm^3 of blood. What would the activity of that blood sample be?

Answer

In two half-lives, the activity would be reduced to $(1/2^2)$ (16 decays/min) = 4 decays/min.

Decay rate and half-life

The values of the half-life and decay constant for several different radioactive nuclei were reported in Table 29.6. Because these two quantities are both related to how quickly a sample of radioactive material decays, we expect that a relationship exists between them. What is that relationship?

At time $t = T$ (one half-life), the number N of radioactive nuclei remaining is one-half the number N_0 at time zero. Using the equation $N = N_0 e^{-\lambda t}$, we can write

$$\tfrac{1}{2} N_0 = N_0 e^{-\lambda T}$$

or

$$\tfrac{1}{2} = e^{-\lambda T}$$

Take the natural logarithm of both sides of this equation and solve for the half-life T:

$$\ln\left(\tfrac{1}{2}\right) = -\lambda T \Rightarrow T = -\frac{1}{\lambda}\ln\left(\tfrac{1}{2}\right)$$

or

$$T = \frac{0.693}{\lambda} \tag{29.10}$$

This relationship between the half-life of the material and its decay constant makes sense. If the decay constant λ is large, then the material decays rapidly and consequently has a short half-life T.

REVIEW QUESTION 29.7 What does it mean that the half-life of a particular isotope is 200 years?

29.8 Radioactive dating

In the previous section, you saw how to calculate the fraction N/N_0 of radioactive nuclei that remain in a sample after a known time interval. Archeologists and geologists, in contrast, are interested in the reverse calculation—determining the age (that is, time interval that has passed) of a radioactive sample from the known fraction N/N_0 of radioactive nuclei that remain in a sample. Their method is based on a rearrangement of Eq. (29.9), $N = N_0 e^{-\lambda t}$.

First, divide both sides of the equation by N_0 and then take the natural logarithm of each side to get

$$\ln\left(\frac{N}{N_0}\right) = -\lambda t$$

Now substitute $\lambda = 0.693/T$ [Eq. (29.10)] and rearrange to get an expression for the age of a radioactive sample:

$$t = -\frac{\ln\left(N/N_0\right)}{0.693} T \tag{29.11}$$

In this equation T is the half-life of the radioactive material and t is the sample's age when N radioactive nuclei remain. The sample started at time zero with N_0 radioactive nuclei.

TIP Notice that in Eq. (29.11) you can use whatever units you want for t and T (seconds, years, thousands of years) provided that you are using the same units for t and T.

Carbon dating

An important type of radioactive dating, called carbon dating, is used to determine the age of objects that are less than 40,000 years old. Most of the carbon in our environment consists of the isotope carbon-12, but about 1 in 10^{12} carbon atoms has a radioactive carbon-14 nucleus with a half-life of about 5700 years. Carbon-14 is continuously produced in our atmosphere by collisions of neutrons from the solar wind with nitrogen nuclei in the atmosphere, resulting in the reaction

$$^{14}_{7}\text{N} + ^{1}_{0}\text{n} \rightarrow ^{14}_{6}\text{C} + ^{1}_{1}\text{H}$$

Atmospheric carbon-14 radioactively decays at the same rate as it is created by the above reaction: a stable equilibrium exists.

Any plant or animal that metabolizes carbon incorporates about one carbon-14 atom into its structure for every 10^{12} carbon-12 atoms it metabolizes. Because carbon is no longer absorbed and metabolized by the organism after death, the carbon-14 in the bones starts to transform by negative beta decay into nitrogen-14. After 5700 years the carbon-14 concentration decreases by one-half. After two half-lives, or 11,400 years after death, the concentration has dwindled to one-fourth what it was when the organism died. A measurement of the current carbon-14 concentration indicates the age of the remains.

EXAMPLE 29.11 ⟩ **Age of bone**

A bone found by an archeologist contains a small amount of radioactive carbon-14. The radioactive emissions from the bone produce a measured decay rate of 3.3 decays/s. The same mass of fresh cow bone produces 30.8 decays/s. Estimate the age of the sample.

Sketch and translate The situation is sketched below.

Fresh bone

Old bone

$\frac{\Delta N}{\Delta t}$ $(t_i = 0)$

$= 30.8 \frac{decays}{s}$

$\frac{\Delta N}{\Delta t}$ $(t_f = ?)$

$= 3.3 \frac{decays}{s}$

Simplify and diagram We assume that the concentration of radioactive carbon-14 in the atmosphere is the same today as it was at the time of death of the animal being studied. We assume the half-life of this isotope of carbon is 5730 years.

Represent mathematically The decay rate and number of radioactive nuclei are proportional to each other $(\Delta N/\Delta t = -\lambda N)$. Therefore, we can determine the ratio of carbon-14 in the old bone and what it was when the animal died.

$$\frac{(\Delta N/\Delta t)_{\text{now}}}{(\Delta N/\Delta t)_{\text{at death}}} = \frac{-\lambda N_{\text{now}}}{-\lambda N_{\text{at death}}} = \frac{N_{\text{now}}}{N_{\text{at death}}} = \frac{N}{N_0}$$

(CONTINUED)

From here we can use Eq. (29.11) to determine the age of the bone.

$$t = -\frac{\ln(N/N_0)}{0.693} T$$

Solve and evaluate The fraction of carbon-14 atoms remaining in the old bone is

$$\frac{N}{N_0} = \frac{3.3}{30.8} = 0.107$$

Substituting this value into Eq. (29.11), we find the age of the bone:

$$t = -\frac{\ln(0.107)}{0.693}(5730 \text{ years}) = 18,500 \text{ years}$$

The unit is correct and the order of magnitude is reasonable.

Try it yourself The ratio of carbon-14 in an old bone compared to the number in a fresh bone of the same mass is 0.050. Determine the age of the bone.

Answer   25,000 years.

Radioactive decay series

Many of the nuclei produced in radioactive decays are themselves radioactive. In fact, the decay of the original parent nucleus may start a chain of successive decays called a **decay series**. An example of a decay series is illustrated in **Figure 29.16**. First, the alpha decay of $^{238}_{92}U$ (top right) leads to the formation of $^{234}_{90}Th$. Thorium-234 then undergoes beta decay to form $^{234}_{91}Pa$. The series continues until $^{218}_{84}Po$ is formed. The series branches at this point, as polonium-218 can decay by either alpha or beta emission. Branching occurs at several other points as well. Eventually, the stable lead isotope $^{206}_{82}Pb$ is formed (lower left) and the series ends.

Radioactive series replenish our environment with nuclei that would normally have disappeared long ago. For example, radium ($^{226}_{88}Ra$) has a half-life of 1600 years.

FIGURE 29.16 A radioactive decay series. The sequence begins in the upper right with uranium-238.

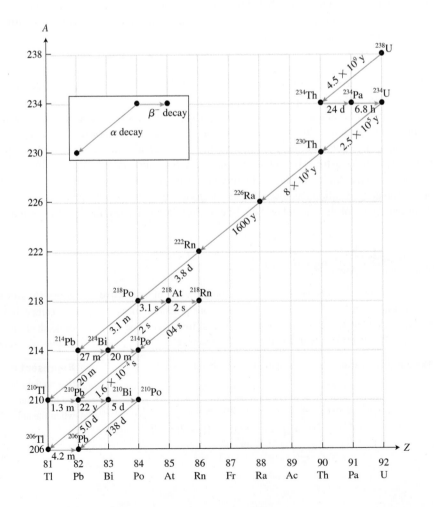

During the approximately 5×10^9 years of our solar system's existence, the original abundance of radium-226 would have long since been depleted by radioactive decay. However, the supply is continually replenished as uranium-238, with an extremely long 4.5×10^9-year half-life, decays and leads to the production of radium-226 via a series of reactions.

REVIEW QUESTION 29.8 How does the presence of radioactive carbon-14 help us determine the age of the bone of an animal that died thousands of years ago?

29.9 Ionizing radiation and its measurement

Ionizing radiation consists of photons and moving particles whose energy exceeds the 10 to 30 eV needed to ionize atoms and molecules. Ionizing radiation exists in many forms, such as ultraviolet, X-ray, and gamma ray photons, alpha and beta particles emitted during the radioactive decay of nuclei, and cosmic rays (high-energy particles) that reach Earth from space.

Life has evolved on Earth in the presence of a steady background of radiation, most of it coming from emissions of radioactive nuclei in Earth's crust and from cosmic rays and their collision products passing down through Earth's atmosphere. Using ionizing radiation for the purpose of diagnosing and treating health problems began more than 100 years ago. Today, our exposure from human-made sources of ionizing radiation accounts for about 40% of our total exposure.

Ionizing radiation's effects on living organisms are divided into two categories: genetic damage and somatic damage. Genetic damage occurs when DNA molecules in reproductive cells (those leading to eggs or sperm) are altered (mutated) by the radiation. These genetic changes are passed on to future generations. Somatic damage involves cellular changes to all other parts of the body except the reproductive cells.

Four different physical quantities are typically used to describe ionizing radiation and its effect on the matter that absorbs it. The first quantity, the decay rate or activity of a radioactive source, was defined in Section 29.7. The remaining three quantities are the absorbed dose, the relative biological effectiveness (RBE), and the dose equivalent.

The **absorbed dose** indicates the amount of ionizing radiation absorbed by biological material.

Absorbed dose The absorbed dose is the energy E of ionizing radiation absorbed by the exposed part of the body divided by the mass m of that exposed part:

$$\text{Absorbed dose} = \frac{\text{Energy absorbed by exposed material}}{\text{Mass of absorbing material}} \qquad (29.12)$$

The SI unit of absorbed dose is the gray (Gy), named after British physicist Louis Harold Gray, where 1 Gy = 1 J/kg. Another commonly used unit for absorbed dose is called the rad, short for radiation absorbed dose, where

$$1 \text{ rad} = 10^{-2} \text{ J/kg} = 10^{-2} \text{ Gy} \qquad (29.13)$$

The absorbed dose is not a completely satisfactory indicator of the damage we might expect when a living organism absorbs ionizing radiation. The reason is that when two different forms of ionizing radiation deposit the same amount of energy into organic material, the damage they cause is generally different. For example, a sample that absorbs 1 rad of alpha particles experiences about 20 times more damage than

TABLE 29.7 The approximate relative biological effectiveness (RBE) of different types of radiation

Type	RBE
X-rays and gamma rays	1
Electrons	1–2
Neutrons	4–20
Protons	10
Alpha particles	20
Heavy ions	2–5

when it absorbs 1 rad of X-rays. Alpha particles move through matter more slowly and slow down more gradually than other forms of radiation, and as a result, they interact with a larger number of atoms. X-rays and gamma rays, by contrast, deposit a large fraction of their energy within a small number of atoms.

Because of the variation in damage caused by different types of radiation, it is useful to define a quantity called **relative biological effectiveness (RBE)**. This measure indicates approximately the relative damage caused by different types of ionizing radiation. The RBE of several different types of radiation is listed in Table 29.7.

Relative biological effectiveness (RBE) Relative biological effectiveness is a dimensionless number that indicates the relative damage caused by a particular type of radiation compared with 2×10^5-eV X-rays. The approximate RBE for a particular type of radiation can be found in tables.

The last quantity is the **dose equivalent**, or simply **dose**—an indicator of the net biological effect of ionizing radiation.

Dose (dose equivalent) This quantity is a measure of the net biological effect (damage) of a particular exposure to ionizing radiation. The dose is the product of the absorbed dose and the relative biological effectiveness:

$$\text{Dose (dose equivalent)} = (\text{Absorbed dose})(\text{RBE})$$

$$= \left(\frac{\text{Energy absorbed}}{\text{Mass of absorbing material}}\right)(\text{RBE}) \quad (29.14)$$

The SI unit of dose is the sievert (Sv), named after Swedish medical physicist Rolf Sievert, where $1 \text{ Sv} = (1 \text{ Gy})(\text{RBE})$. Another commonly used unit is the rem (roentgen equivalent in man or in mammal), where $1 \text{ rem} = (1 \text{ rad})(\text{RBE})$.

QUANTITATIVE EXERCISE 29.12   **Ions produced by a chest X-ray**

In a typical chest X-ray, about 10 mrem (10×10^{-3} rem) of radiation is absorbed by about 5 kg of body tissue. The X-ray photons used in such exams each have energy of about 50,000 eV. Determine about how many ions are produced by this X-ray exam. Assume that each photon will produce approximately 100 ions because its energy is much larger than the ionization energy of atoms and molecules.

Represent mathematically We can estimate first the number of X-ray photons absorbed and then try to use that number to estimate the number of ions formed. We can accomplish this using Eq. (29.14):

$$\text{Dose} = (\text{Absorbed dose})(\text{RBE})$$

$$= \left(\frac{\text{Energy absorbed}}{\text{Mass of absorbing material}}\right)(\text{RBE})$$

Solve and evaluate The RBE of X-rays is 1. Thus, the dose of X-rays in rem equals the absorbed dose in rad. For an absorbed dose of 10 mrad, we can insert the known values and solve the equation to determine the energy absorbed per kilogram of exposed tissue:

$$10 \text{ mrad} = (10 \times 10^{-3} \text{ rad})\left(\frac{10^{-2} \text{ J/kg}}{1 \text{ rad}}\right) = 10^{-4} \text{ J/kg}$$

Because 5 kg of tissue receives this dose, the total absorbed energy is

$$(5 \text{ kg})(10^{-4} \text{ J/kg}) = 5 \times 10^{-4} \text{ J}$$

We can determine the number of X-ray photons by dividing this result by the energy of a single photon:

$$\frac{5 \times 10^{-4} \text{ J}}{\left(50,000 \dfrac{\text{eV}}{\text{photon}}\right)\left(\dfrac{1.6 \times 10^{-19} \text{ J}}{\text{eV}}\right)} = 6 \times 10^{10} \text{ photons}$$

or 60 billion photons. Because each photon produces 100 ions, the number of ions is approximately

$$6 \times 10^{10} \text{ photons} \times 100 \text{ ions/photon} = 6 \times 10^{12} \text{ ions}$$

Try it yourself How many ions may be produced by a bilateral mammogram, assuming that each breast has a mass of 0.5 kg, the absorbed dose for both breasts is 0.2 rad, and 50,000-eV X-ray photons are used?

Answer

$\approx 3 \times 10^{11}$ photons, or 3×10^{13} if each photon produces 100 ions.

Sources of ionizing radiation

The U.S. Environmental Protection Agency estimates that the average dose of ionizing radiation received by a person in the United States or in Canada is about 300–400 mrem/year. The sources of this radiation can be divided into two broad categories: natural and human-made.

Natural sources The three major natural sources of radiation are the following.

1. Radioactive elements in the Earth's crust, such as uranium-238, potassium-40, and radon-226. Small amounts of radioactive radon, an inert gaseous atom, diffuse out of the soil into buildings, exposing the occupants.

2. Foods that naturally contain radioactive isotopes, such as Brazil nuts, lima beans, and bananas.

3. Cosmic rays. A cosmic ray is an elementary particle moving at almost the speed of light. The original source of cosmic rays is primarily supernova explosions of stars in our galaxy.

Human-made sources Human-made radiation comes predominantly from medical applications such as X-rays used in diagnostic procedures, the medical use of radioactive nuclei as tracers, and gamma rays used in cancer treatment. Together, these sources account for a dose of about 60 mrem/year (see Table 29.8). Smoking and exposure to radon significantly increase the dose/year.

TABLE 29.8 Ionizing radiation sources and doses

Exposure mechanism	Typical dose
Having a smoke detector	0.008 mrem/year[1]
Living near nuclear plant	0.009 mrem/year[2]
Living near coal-fired plant	0.03 mrem/year[1]
5-hour airplane flight	2.5 mrem[1]
Natural gas stoves and heaters	6–9 mrem/year[3]
Dental X-ray	0.5 mrem/X-ray[1]
Chest X-ray	10 mrem/X-ray[1]
Smoking (one half-pack per day)	18 mrem/year[1]
Radon	200 mrem/year[2]

Sources: [1]American Nuclear Society; [2]PBS Frontline; [3]Lawrence Berkeley Lab.

REVIEW QUESTION 29.9 Why are X-rays, gamma rays, and fast-moving electrically charged particles potentially harmful to our bodies?

Summary

Terminology Nuclei are represented by the notation $^A_Z X$, where X is the letter abbreviation for that element, the atomic number Z is the number of protons in the nucleus, N is the number of neutrons in the nucleus, and the mass number $A = Z + N$ is the number of nucleons (protons plus neutrons) in the nucleus. (Section 29.2)

$^7_3 Li$

Isotopes Atoms with a common number of protons but differing numbers of neutrons are called isotopes. (Section 29.2)

Mass defect is the difference between the total mass of the proton and neutron components of a nucleus and its mass when the components are bound together.

Binding energy of a nucleus is the rest energy equivalent of its mass defect. The binding energy represents the total energy needed to separate a nucleus into protons and neutrons. (Section 29.3)

$$\Delta m_{Li} = [3m_H + 4m_n] - m_{Li}$$

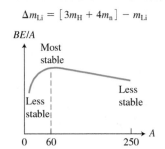

BE/A

Most stable

Less stable

Less stable

$0 \quad 60 \quad 250$ A

mass defect $= \Delta m$
$$= [Zm_{hydrogen\ atom} + (A - Z)m_{neutron}] - m_{atom} \quad \text{Eq. (29.1)}$$

$$BE = \Delta m \cdot c^2 \quad \text{Eq. (29.2)}$$

Binding energy per nucleon $= BE/A$ $\quad$ Eq. (28.4)

Types of radioactive decay

Alpha decay A heavy nucleus emits an alpha particle (α), the nucleus of a helium atom $^4_2 He$.

Parent — Daughter — Alpha particle

| 88 | 86 | 2 |
| 138 | 136 | 2 |

$^{226}_{88} Ra \longrightarrow\ ^{222}_{86} Rn + ^4_2 He + 4.9\ MeV$
(kinetic energy of products)

α decay:
$$^A_Z X \rightarrow ^{A-4}_{Z-2} X + ^4_2 He + \text{energy}$$

Beta decay The nucleus ejects a beta particle (β^- electron or β^+ positron) and an antineutrino $\bar{\nu}$ or neutrino ν.

Parent — Daughter — Beta particle

β^- decay

| 6 | 7 |
| 8 | 7 |

Electron +

$^{14}_6 C \longrightarrow\ ^{14}_7 N + ^{\ 0}_{-1} e + \bar{\nu}$

β^- decay:
$$^1_0 n \rightarrow ^1_1 p + ^{\ 0}_{-1} e + \bar{\nu}$$

Gamma decay A nucleus in an excited state ($^A_Z X^*$) emits a high-energy photon called a gamma ray. (Sections 29.1 and 29.6)

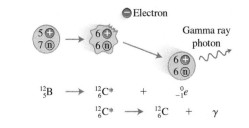

Electron

| 5 | 6 | 6 |
| 7 | 6 | 6 |

Gamma ray photon

$^{12}_5 B \longrightarrow\ ^{12}_6 C^* + ^{\ 0}_{-1} e$
$^{12}_6 C^* \longrightarrow\ ^{12}_6 C + \gamma$

γ decay:
$$^A_Z X^* \rightarrow ^A_Z X + \gamma$$

Half-life The half-life T of a particular type of radioactive nucleus is the time interval during which the number of these nuclei in a given sample is reduced by one-half. (Section 29.7)

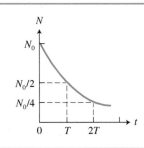

N

N_0

$N_0/2$

$N_0/4$

$0 \quad T \quad 2T$ t

$N/N_0 = 1/2^n$ at $t = nT$ $\quad$ Eq. (29.6)
where N_0 is the number at time $t = 0$, and N is the number at time t and n is the number of half-lives that have passed

Questions

Multiple Choice Questions

1. What were Becquerel rays?
   (a) Positively charged particles   (b) Negatively charged particles
   (c) High-energy photons   (d) All of the above
2. To decide whether Becquerel rays contained electrically charged particles, scientists could
   (a) let them pass through a magnetic field.
   (b) let them pass through an electric field.
   (c) study the existing literature.
   (d) both a and b
3. Why are there are no electrons in the nucleus?
   (a) The nucleus consists of protons and neutrons only.
   (b) Electrons do not exert a strong nuclear force.
   (c) Their kinetic energy is greater than the magnitude of the potential energy of their interaction with the nucleus.
   (d) Electrons are too big to fit into a nucleus.
4. What is radioactive decay?
   (a) A physical quantity   (b) An observable physical phenomenon
   (c) A law of physics   (d) A unit
5. Elements with higher decay constants decay
   (a) slower.   (b) faster.
   (c) The decay constant does not relate to how fast the substance decays.
6. One can speed up the rate of natural radioactive decay of a substance by
   (a) increasing its temperature.
   (b) illuminating it with high energy photons.
   (c) turning it into a powder.
   (d) No physical or chemical changes affect the rate of decay.
7. If the original number of radioactive nuclei in a sample is N_0, then after three half-life time intervals how many nuclei are left?
   (a) $N_0/3$ nuclei   (b) $N_0/6$ nuclei
   (c) $N_0/8$ nuclei   (d) $N_0/9$ nuclei
8. During alpha decay,
   (a) a neutron decays into an electron, a proton, and an antineutrino.
   (b) a photon is emitted by the nucleus changing its energy state.
   (c) a helium nucleus is emitted from the nucleus.
   (d) a hydrogen nucleus is emitted from the nucleus.
9. What happens during beta decay?
   (a) Electrons originally in the nucleus escape.
   (b) A neutron originally in the nucleus converts into a proton, an electron, and an antineutrino.
   (c) An alpha particle originally in the nucleus escapes.
   (d) Photons are emitted by the excited nucleus.

10. A sample of cesium-137 is placed in a lead container with a hole through which beta particles and gamma rays are emerging. A Geiger counter is placed next to the lead container as shown in **Figure Q29.10**. Placing the setup into which of the following *uniform* fields can result in an increase in the number of counts per unit time?
    (a) A magnetic field pointing perpendicularly into the page
    (b) A magnetic field pointing perpendicularly out of the page
    (c) An electric field pointing perpendicularly out of the page
    (d) An electric field pointing perpendicularly into the page
    (e) None of the above.

FIGURE Q29.10

Conceptual Questions

11. How did scientists figure out that radioactive materials emit radiation from the nucleus rather than from the electron structure of atoms and molecules?
12. Describe the experiments that helped scientists understand that uranium salts produce both fast-moving charged particles and high-energy photons.
13. Why did Becquerel have burns on his chest near the vest pocket in which he carried a test tube containing radium?
14. How did Marie Curie discover radium?
15. How did Rutherford determine that radioactivity consists of three types of particles, including a high-energy photon?
16. How did scientists devise the concept of half-life?
17. Explain why heavier isotopes of atoms with a small number of protons are unstable.
18. The half-life of radium-226 is 1600 years. Earth is about 4 billion years old, so essentially all of the radium-226 present originally should have decayed. How can you explain the fact that radium-226 still remains in our environment in moderate abundance?
19. Why is a high temperature usually needed for fusion to occur?
20. Compare and contrast isotopes and ions.
21. Compare and contrast fission and fusion.
22. Compare and contrast the binding energy of a nucleus and the ionization energy of an atom.

Problems

Below, **BIO** indicates a problem with a biological or medical focus. Problems labeled **EST** ask you to estimate the answer to a quantitative problem rather than derive a specific answer. Asterisks indicate the level of difficulty of the problem. Problems with no * are considered to be the least difficult. A single * marks moderately difficult problems. Two ** indicate more difficult problems.

29.1 and 29.2 Radioactivity and an early nuclear model and A new particle and a new nuclear model

1. **EST** Estimate the magnitude of the repulsive electrical force between two protons in a helium nucleus. How are they held together in the nucleus?
2. * **EST** Estimate the magnitude of the repulsive electrical force that the rest of a gold nucleus ($^{197}_{79}$Au) exerts on a proton near the edge of the nucleus. State any assumptions used in your estimation. This is an order-of-magnitude estimate, so do not become bogged down in details.
3. Determine the number of protons, neutrons, and nucleons in the following nuclei: $^9_4$Be, $^{16}_8$O, $^{27}_{13}$Al, $^{56}_{26}$Fe, ^{64}Zn, and ^{107}Ag.
4. * **EST** Suppose the radius of a copper nucleus is 10^{-14} m and the radius of the copper atom is about 0.1 nm. Estimate the atom's size if the nucleus were the size of a pin head.

5. ** **EST** **Electrons and nucleons in body** Estimate the total number of (a) nucleons and (b) electrons in your body. (c) Indicate roughly the volume in cubic centimeters occupied by these nucleons.

29.3 Nuclear force and binding energy

6. * Measurements with a mass spectrometer indicate the following particle masses: $^7_3$Li, 7.016003 u; $^1_1$H, 1.007825 u; and $^1_0$n, 1.008665 u. Compare the mass of the lithium atom to the mass of the particles of which it is made. What do you conclude? Note: 1 u = 1.660539×10^{-27} kg.
7. * A 5.5-g marble initially at rest is dropped in the Earth's gravitational field. How far must it fall before its decrease in gravitational potential energy is 938 MeV, the same as the rest mass energy of a proton?
8. * Determine the rest mass energies of an electron, a proton, and a neutron in units of mega-electron volts.
9. **EST** Use Figure 29.8 to estimate the total binding energy of $^{197}_{79}$Au.
10. * Determine the total binding energy and the binding energy per nucleon of $^{12}_6$C.
11. * Determine the binding energies per nucleon for $^{238}_{92}$U and $^{120}_{50}$Sn. Based on these numbers, which nucleus is more stable? Explain.

12. * Determine the energy that is needed to remove a neutron from ^7_3Li to produce ^6_3Li plus a free neutron. (Hint: Compare the mass of ^7_3Li with that of $^6_3\text{Li} + ^1_0\text{n}$.)

29.4 and 29.5 Nuclear reactions and Nuclear sources of energy

13. Insert the missing symbol in the following reactions.
     (a) $^4_2\text{He} + ^{12}_6\text{C} \rightarrow ^{15}_7\text{N} + ?$
     (b) $^2_1\text{H} + ^3_1\text{H} \rightarrow ^4_2\text{He} + ?$
     (c) $^1_0\text{n} + ^{235}_{92}\text{U} \rightarrow ^{140}_{54}\text{Xe} + ? + 2^1_0\text{n}$
     (d) $^3_1\text{H} \rightarrow ? + ^0_{-1}e$

14. Explain why the following reactions violate one or more of the rules for nuclear reactions.
     (a) $^4_2\text{He} + ^{27}_{13}\text{Al} \rightarrow ^{32}_{15}\text{P} + ^1_0\text{n}$
     (b) $^2_1\text{H} + ^3_1\text{H} \rightarrow ^4_2\text{He} + ^1_1\text{H}$
     (c) $^1_0\text{n} + ^{238}_{94}\text{Pu} \rightarrow ^{140}_{54}\text{Xe} + ^{96}_{40}\text{Zr} + 2^1_0\text{n}$
     (d) $^{14}_6\text{C} \rightarrow ^{14}_7\text{N} + ^0_1e$

15. * Explain why the following reaction does not occur spontaneously: $^4_2\text{He} \rightarrow ^3_1\text{H} + ^1_1\text{H}$.

16. ** The following reaction occurs in the Sun: $^3_2\text{He} + ^4_2\text{He} \rightarrow ^7_4\text{Be}$. How much energy is released?

17. * **Nuclear reaction in the Sun** Oxygen is produced in stars by the following reaction: $^{12}_6\text{C} + ^4_2\text{He} \rightarrow ^{16}_8\text{O}$. How much energy is absorbed or released by the reaction in units of mega-electron volts?

18. * **Another reaction in the Sun** One part of the carbon-nitrogen cycle that provides energy for the Sun is the reaction $^{12}_6\text{C} + ^1_1\text{H} \rightarrow ^{13}_7\text{N} + 1.943$ MeV. Using the known masses of ^{12}C and ^1H and the results of this reaction, determine the mass of ^{13}N.

19. * Determine the missing nucleus in the following reaction and calculate how much energy is released: $^{232}_{92}\text{U} \rightarrow ? + ^4_2\text{He} + \text{energy}$.

20. * Determine the missing nucleus in the following reaction and calculate its mass: $? \rightarrow ^{211}_{83}\text{Bi} + ^4_2\text{He} + 8.20$ MeV.

21. * Determine (a) the number of protons and neutrons in the missing fragment of the reaction shown below and (b) the mass of that fragment:
     $$^1_0\text{n} + ^{235}_{92}\text{U} \rightarrow ? + ^{136}_{54}\text{Xe} + 12^1_0\text{n} + 126.5 \text{ MeV}$$

22. * **More energy for the Sun** A series of reactions in the Sun leads to the fusion of three helium nuclei (^4_2He) to form one carbon nucleus ($^{12}_6\text{C}$). (a) Determine the net energy released by the reactions. (b) What fraction of the total mass of the three helium nuclei is converted to energy?

23. * **Another Sun process** A series of reactions that provides energy for the Sun and stars is summarized by the following equation: $6^1_1\text{H} \rightarrow 2^1_1\text{H} + 2^1_0\text{n} + 2^2_2\text{He}$. (a) Determine the net energy released by the reaction. (b) Convert this answer to units of joules per kilogram of deuterium (^2_1H).

24. * **Equation Jeopardy** Equations for determining the mass defect for two nuclear reactions are shown below. Represent each reaction in symbolic form (as in the previous problems). Decide whether each reaction results in energy release or energy absorption.
     (a) $235.0439 \text{ u} + 1.0087 \text{ u} \rightarrow 95.9343 \text{ u} + 137.9110 \text{ u}$
     $\qquad\qquad\qquad\qquad\qquad + 2(1.0087 \text{ u}) + \text{energy}$
     (b) $3(4.002602 \text{ u}) \rightarrow 12.000000 \text{ u} + \text{energy}$

29.6 Mechanisms of radioactive decay

25. * In 1913, Frederick Soddy collected the following data related to the radioactive transformation of uranium (**Figure P29.25**). The first product that appears in the sample is thorium, then protactinium, then another isotope of uranium, and so on. Examine the series of the transformation found by Soddy and explain using your knowledge of alpha and beta decays. Discuss what quantities are constant in each process.

FIGURE P29.25

26. * In the 1930s, Meitner, Hahn, and Strassmann did experiments irradiating uranium with neutrons. They predicted three possible outcomes:
     - Production of a new element (if the neutron undergoes beta decay in the nucleus).
     - Production of a heavier isotope of uranium (if the neutron stays in the nucleus).
     - Production of a slightly lighter nucleus (if the neutron captured by the nucleus causes one or two alpha particles to leave).

     Instead, the nuclei produced in the reaction included isotopes of barium and other nuclei with about half the mass of uranium. How could they explain their findings?

27. * The following nuclei produced in a nuclear reactor each undergo radioactive decay. Determine the daughter nucleus formed by each decay reaction: (a) $^{239}_{94}\text{Pu}$ alpha decay; (b) $^{144}_{58}\text{Ce}$ beta-minus decay; (c) $^{129}_{53}\text{I}$ beta-minus decay; and (d) $^{60}_{30}\text{Zn}$ beta-plus decay.

28. Potassium-40 ($^{40}_{19}\text{K}$) can undergo both beta-plus and beta-minus decay. Determine the daughter nucleus in each case.

29. * Radon-222 ($^{222}_{86}\text{Rn}$) is released into the air during uranium mining and undergoes alpha decay to form $^{218}_{84}\text{Po}$ of mass 218.0090 u. Determine the energy released by the decay reaction. Most of this energy is in the form of alpha particle kinetic energy.

30. * Carbon-11 ($^{11}_6\text{C}$) undergoes beta-plus decay. Determine the product of the decay and the energy released.

31. * (a) Determine the total binding energy of radium-226 ($^{226}_{88}\text{Ra}$). (b) Determine and add together the binding energies of a radon-222 ($^{222}_{86}\text{Rn}$) and an alpha particle. (c) Determine the difference of these numbers, which equals the energy released during alpha decay of ^{226}Ra.

32. ** **BIO** **Potassium decay in body** The body contains about 7 mg of radioactive ^{40}K that is absorbed with the foods we eat. Each second, about 2.0×10^3 of these potassium nuclei undergo beta decay (either beta-minus or beta-plus). About how many potassium nuclei are in the average body and what fraction decay each second?

33. * A radioactive ^{60}Co nucleus emits a gamma ray of wavelength 0.93×10^{-12} m. If the cobalt was initially at rest, use constancy of momentum to determine its speed following the gamma ray emission.

34. * A radioactive series different from those shown in Figures 29.16 and P29.25 begins with $^{232}_{90}\text{Th}$ and undergoes the following series of decays: $\alpha\beta^-\beta^-\alpha\alpha\alpha\beta^-\alpha\beta^-$. Determine each nucleus in the series.

29.7 Half-life, decay rate, and exponential decay

35. * **BIO** **O_2 emitted by plants** Propose a hypothetical experiment to determine whether O_2 emitted from plants comes from H_2O or from CO_2, the basic input molecules that lead to plant growth.

36. * A radioactive sample initially undergoes 4.8×10^4 decays/s. Twenty-four hours later, its activity is 1.2×10^4 decays/s. Determine the half-life of the radioactive species in the sample.

37. * Cesium-137, a waste product of nuclear reactors, has a half-life of 30 years. Determine the fraction of ^{137}Cs remaining in a reactor fuel rod (a) 120 years after it is removed from the reactor, (b) 240 years after, and (c) 1000 years after.

38. * A sample of radioactive technetium-99 of half-life 6 h is to be used in a clinical examination. The sample is delayed 15 h before arriving at the lab for use. Determine the fraction of radioactive technetium that remains.

39. ** **BIO** **Radiation therapy** If 120 mg of radioactive gold-198 with half-life 2.7 days is administered to a patient for radiation therapy, what is the gold-198 activity 3 weeks later if none is eliminated from the body by biological means?

40. **Fuel rod decay** How many years are required for the amount of krypton-85 ($^{85}_{36}\text{Kr}$) in a spent nuclear reactor fuel rod to be reduced to 1/8 of its original value? 1/32? 1/128? The half-life of ^{85}Kr is 10.8 years.

41. * **BIO** **Wisconsin glacier tree sample** A tree sample was uprooted and buried 60,000 years ago during part of the Wisconsin glaciation. How many years after it was buried was the radioactive carbon-14 (^{14}C) in the root reduced to (a) 1/2 of its original value? (b) 1/4? (c) 1/8? (d) What fraction remained after 60,000 years? The carbon-14 was not replenished after the tree stopped growing.

42. * How many years are required for the amount of strontium-90 (^{90}Sr) released from a nuclear explosion in the atmosphere to be reduced to (a) 1/16 of its original value? (b) 1/64? (c) 1/100?

43. ** [BIO] **Student swallows radioactive iodine** A student accidentally swallows 0.10 μg of iodine-131 while pipetting the radioactive material. (a) Determine the number of ^{131}I atoms swallowed (the atomic mass of ^{131}I is approximately 131 u). (b) Determine the activity of this material. The half-life of ^{131}I is 8.02 days. (c) What is the mass of radioactive iodine-131 that remains 21 days later if none leaves the body?

44. * An unlabeled container of radioactive material has an activity of 90 decays/min. Four days later the activity is 72 decays/min. Determine the half-life of the material. When will its activity reach 9 decays/min?

45. * One gram of pure, radioactive radium produces 130 J of energy per hour due to radium decay only and has an activity of 1.0 Ci. Determine the average energy in electron volts released by each radioactive decay.

29.8 Radioactive dating

46. **Age of mallet** A mallet found at an archeological excavation site has 1/16 the normal carbon-14 decay rate. Determine the mallet's age.

47. * The ^{235}U in a rock decays with a half-life of 7.04×10^8 years. A geologist finds that for each ^{235}U now remaining in the rock, 2.6 ^{235}U have already decayed to form daughter nuclei. Determine the age of the rock.

48. A sample of water from a deep, isolated well contains only 30% as much tritium as fresh rainwater. How old is the water in the well?

49. * The decay rate of ^{14}C from a bone uncovered at a burial site is 12.6 decays/min, whereas the decay rate from a fresh bone of the same mass is 1610 decays/min. Approximately how old is the uncovered bone?

50. ** [BIO] The tree sample described in Problem 29.41 contains 50 g of carbon when it is discovered. (a) If 1 in 10^{12} carbon atoms in a fresh tree sample are carbon-14, how many carbon-14 atoms would be in 50 g of carbon from a fresh tree? (b) Calculate the carbon-14 activity of the sample. (c) Determine the age of the buried tree if its 50 g of carbon has an activity of -2.2 s^{-1}.

29.9 Ionizing radiation and its measurement

51. Design an experiment in which radioactive nuclei are used to test the ability of different detergents to remove dirt from clothes.

52. * [BIO] **Radiation dose** A 70-kg person receives a 250-mrad whole-body absorbed dose of radiation. (a) How much energy does the person absorb? (b) Is it better to absorb 250 mrad of X-rays or 250 mrad of beta particles? Explain. (c) What is the dose in each case?

53. [BIO] **Dose due to different types of radiation** Determine the dose equivalent of a 70-mrad absorbed dose of the following types of radiation: (a) X-rays, (b) beta particles, (c) protons, (d) alpha particles, and (e) heavy ions.

54. * [BIO] **Potassium decay in body** The yearly whole-body dose caused by radioactive ^{40}K absorbed in our tissues is 17 mrem. (a) Assuming that ^{40}K undergoes beta decay with an RBE of 1.4, determine the absorbed dose in rads. (b) How much beta particle energy does an 80-kg person absorb in one year? Note: ^{40}K also emits gamma rays, many of which leave the body before being absorbed. Because fatty tissue has low potassium concentration and muscle has a higher concentration, gamma ray emissions indicate indirectly a person's fat content.

55. ** [BIO] **More potassium** Determine the number of ^{40}K nuclei in the body of an 80-kg person using the information provided in the previous problem and the fact that ^{40}K has a radioactive half-life of 1.28×10^9 years. Assume that each ^{40}K beta decay results in 1.4 MeV of energy that is deposited in the person's tissue.

56. * [BIO] **X-ray exam** During an X-ray examination a person receives a dose of 80 mrem in 4.0 kg of tissue. The RBE of X-rays is 1.0. (a) Determine the total energy absorbed by the 4.0 kg of body tissue. (b) Determine the energy in joules of each 40,000-eV X-ray photon. (c) Determine the number of photons absorbed by that tissue.

60. * [EST] Estimate the temperature at which two protons can come close enough together to form an isotope of a helium nucleus.

61. ** [EST] **Building a fusion reactor** The mass of a helium nucleus is less than the mass of the nucleons inside it. (a) Explain how this observation led scientists to the idea that it is possible to convert hydrogen into helium to produce thermal energy. (b) Will this process mean that energy is not conserved? Explain. (c) Why do you think scientists need very high temperatures and high pressures for this reaction? (d) Estimate the temperature at which two protons will join together due to their nuclear attraction. Remember that nuclear forces are effective at distances less than or equal to 10^{-15} m. (Hint: Use an energy approach, not a force approach.) (e) Suggest possible ways of containing ionized hydrogen to make the reaction possible (note that all solid containers will melt at this temperature).

62. ** **World energy use** World energy consumption in 2005 was about 4×10^{20} J. (a) Determine the number of deuterium nuclei ($^2_1$H) that would be needed to produce this energy. The fusion of two deuterium nuclei releases about 4 MeV of energy. (b) Determine the volume of water of density 1000 kg/m^3 needed to supply the energy. One mole of water has a mass of 18 g, and about one in every 6700 hydrogen atoms in water is a deuterium $^2_1$H.

63. * Convert the 200 MeV per nucleus energy that is released by ^{235}U fission to units of joules per kilogram. By comparison, the energy released by burning coal is 3.3×10^7 J/kg.

64. ** (a) Determine the energy used by a 1200-W hair dryer in 10 min. (b) Approximately how many ^{235}U fissions must occur in a nuclear power plant to provide this energy if 35% of the energy released by fission produces electrical energy? The energy released per fission is about 200 MeV.

65. * [EST] Estimate the number of ^{235}U nuclei that must undergo fission to provide the energy to lift your body 1 m. Indicate any assumptions made.

66. ** **Comparing nuclear and coal power plants** Suppose that a nuclear fission power plant and a coal-fired power plant both operate at 40% efficiency. Determine the ratio of the mass of coal that must be burned in 1 day of operating a 1000-MW plant compared to the mass of ^{235}U that must undergo fission in the same plant. The energy released by burning coal is 3.3×10^7 J/kg. The energy released per ^{235}U fission is 200 MeV.

67. ** **Uranium energy potential** The world's uranium supply is approximately 10^9 kg (10^6 tons), 0.7% of which is ^{235}U. (a) How much energy is available from the fission of this ^{235}U? (b) The world's energy consumption rate for production of electricity is about 10^{20} J/year. At this rate, how many years would the uranium last if it were used to provide all our electrical energy?

68. * A radioactive sample contains two different types of radioactive nuclei: A, with half-life 5.0 days, and B, with half-life 30.0 days. Initially, the decay rate of the A-type nucleus is 64 times that of the B-type nucleus. When will their decay rates be equal?

69. ** After a series of alpha and beta-minus decays, plutonium-239 ($^{239}_{94}$Pu) becomes lead-207 ($^{207}_{82}$Pb). Determine the number of alpha and beta-minus particles emitted in the complete decay process. Explain your method for determining these numbers.

70. * [BIO] [EST] **Estimate number of ants in nest** To estimate the number of ants in a nest, 100 ants are removed and fed sugar made from radioactive carbon of a long half-life. The ants are returned to the nest. Several days later, of 200 ants taken from the nest, only 5 are radioactive. Roughly how many ants are in the nest? Explain your calculation technique.

71. ** [BIO] **Ion production due to ionizing radiation** A body cell of 1.0×10^{-5}-m radius and density 1000 kg/m^3 receives a radiation absorbed dose of 1 rad. (a) Determine the mass of the cell (assume it is shaped like a sphere). (b) Determine the energy absorbed by the cell. (c) If the average energy needed to produce one positively charged ion is 100 eV, determine the number of positive ions produced in the cell. (d) Repeat the procedure for the 3.0×10^{-6}-m-radius nucleus of the cell, where its genetic information is stored. Indicate any assumptions that you made.

General Problems

57. Estimate the density of an atomic nucleus.

58. Estimate the radius of a sphere that would hold all of your mass if its density equaled the density of a nucleus.

59. Determine the chemical elements represented by the symbol X in each of the following: (a) $^{23}_{11}$X; (b) $^{56}_{26}$X; (c) $^{64}_{30}$X; and (d) $^{107}_{47}$X.

Reading Passage Problems

[BIO] **Nuclear accident and cancer risk** If 10,000 people are exposed to an average radiation dose of 1 rem, it is estimated that six cancer deaths will result. In March, 2011, a 9.0-magnitude earthquake and subsequent tsunami rocked Japan, causing hydrogen explosions in four reactors at the Fukushima Daiichi

Nuclear Power Station. As a result, radioactive materials such as cesium-137 and iodine-131 were released into the atmosphere. About 2 million people living within 80 km of the Fukushima reactors were each exposed to an average dose of approximately 1 rem.

72. Which answer below is closest to the Fukushima total person-rem exposure?
    (a) 2×10^4 person·rem   (b) 2×10^6 person·rem
    (c) 2×10^8 person·rem   (d) 2×10^{10} person·rem

73. Which answer below is closest to the number of statistically likely cancer deaths to be caused by this event?
    (a) 10   (b) 100   (c) 1000
    (d) 10,000   (e) 100,000

74. About 4000 of the 2 million residents die each year from cancer caused by other factors, such as smoking. If those Fukushima accident cancer deaths occurred during the 5 years following the accident, which number below is closest to the ratio of Fukushima-caused cancer deaths to normal cancer deaths for the exposed population?
    (a) 1/2000   (b) 1/20   (c) 1   (d) 20/1   (e) 200/1

75. Which answer below is closest to the number of cancer deaths in the United States due to an estimated 20 mrad/year average radon exposure to each person?
    (a) 50   (b) 500   (c) 5000   (d) 50,000   (e) 500,000

76. Which is the most important reason why it is difficult statistically to identify cancer deaths due to radiation exposure from the Fukushima accident?
    (a) An unknown number of people stayed in their homes following the accident.
    (b) The radiation detection devices were inaccurate.
    (c) The fraction of people who left the area was unknown.
    (d) The normal cancer rate significantly exceeded the cancer caused by the accident.
    (e) Authorities often distort information they provide about radiation exposure.

Was the sculpture of David from Notre Dame? Neutron activation analysis (NAA) is a sensitive analytical technique used in environmental studies, semiconductor quality control, forensic science, and archeological studies to identify trace and rare elements and determine the effect of rare elements on chronic diseases. In NAA a tiny specimen under investigation is irradiated with neutrons. The neutrons are absorbed by nuclei in the sample, thus creating artificially radioactive nuclei. These nuclei decay via the emission of alpha particles, beta particles, and gamma rays. The energy and wavelength of the gamma rays are characteristic of the element from which they were emitted.

NAA has been used to help identify the sources of art and sculpture by an elemental analysis of the paint or stone used. For example, the Metropolitan Museum of Art recently determined that stone used in the mid-13th-century *Head of King David* sculpture that they acquired in 1938 matched stone from

the cathedral of Notre Dame in Paris. Previously, it had been assumed that the sculpture came from Amiens Cathedral north of Paris. Figure 29.17 shows the elemental content of stone from Notre Dame (the red line) and from Amiens Cathedral (the blue line). Although the patterns on this logarithmic scale look similar, they are significantly different for some of the 16 elements tested.

FIGURE 29.17 NAA comparison showing that the relative amounts of different elements in stone from Cathedral of Notre Dame and from Amiens Cathedral are distinctively different.

77. In neutron activation analysis (NAA) a particular elemental concentration in a sample is determined by measuring which of the following?
    (a) Alpha particles   (b) Beta particles
    (c) Gamma rays   (d) Neutrons
    (e) Two of these choices

78. Which answer below is closest to the ratio of the relative amount of lutetium (Lu) in stone from Notre Dame compared to stone from Amiens Cathedral?
    (a) 0.5   (b) 1   (c) 2   (d) 10   (e) 1000

79. What does neutron absorption by sodium ($^{23}_{11}$Na) produce?
    (a) $^{22}_{11}$Na   (b) $^{24}_{11}$Na   (c) $^{24}_{12}$Mg   (d) $^{23}_{12}$Mg   (e) $^{22}_{10}$Ne

80. Neutron absorption by potassium produces an excited $^{40}_{19}$K nucleus, which emits a beta-minus and a gamma ray. Which nucleus below is a product of this decay?
    (a) $^{40}_{19}$K   (b) $^{40}_{18}$Ar   (c) $^{40}_{20}$Ca   (d) $^{41}_{19}$K   (e) $^{39}_{19}$K

81. Rank the elements Cr, Mn, and Na in stone samples from Notre Dame compared to stone from Amiens Cathedral as the best indicators for distinguishing the two types of stone (rated from best to worst).
    (a) Cr, Mn, Na   (b) Mn, Cr, Na
    (c) Na, Mn, Cr   (d) Mn, Na, Cr

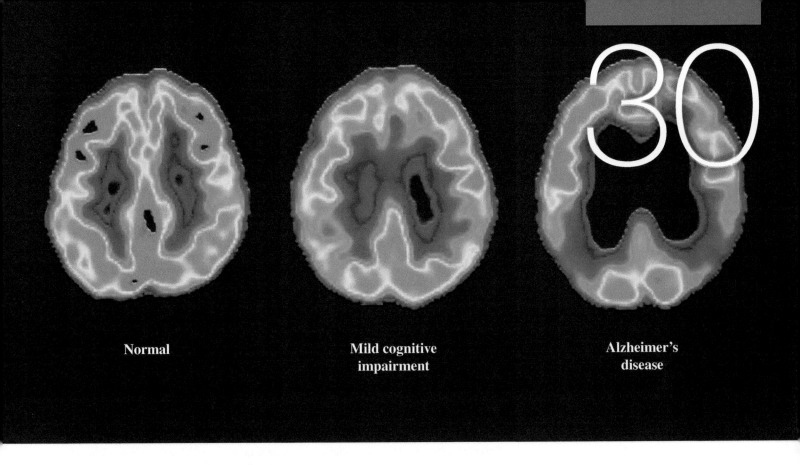

| Normal | Mild cognitive impairment | Alzheimer's disease |

Particle Physics

To obtain a three-dimensional image of the brain of a person who is suspected to have Alzheimer's disease, doctors use positron emission tomography (PET). Before the PET scan, the patient is given an injection of a glucose-like molecule containing fluorine-18, a radioactive isotope of fluorine. After an hour, the substance is absorbed into the patient's brain. Radioactive decay of the fluorine nuclei produces positrons, and by observing the photons that originate in the reactions of these positrons with nearby electrons, doctors can locate the active parts of the brain. A comparison of the active and inactive regions in the completed scan with scans of normal brains helps physicians determine whether Alzheimer's disease or other forms of dementia are present.

Beta decay produces antineutrinos, a form of antimatter. Every known particle has a corresponding antiparticle. In this chapter we investigate elementary particles such as the positron, which is an antielectron, and their fundamental interactions. By the end of this chapter, you should be able to understand the physics behind PET as well as the basic components of the universe, how they interact, and how they were discovered. This area of physics is called **particle physics**.

- ● **What is antimatter?**
- ● **What are the fundamental particles and interactions in nature?**
- ● **What was the Big Bang, and how has the universe evolved since?**

BE SURE YOU KNOW HOW TO:

- ● Use the right-hand rule to determine the direction of the magnetic force exerted by a magnetic field on a moving charged particle (Section 20.4).
- ● Explain beta decay (Section 29.6).
- ● Write an expression for the rest energy of a particle (Section 26.9).

30.1 Antiparticles

As you may recall, the word "atom" means indivisible. Physicists thought at one time that atoms were the smallest constituents of matter. However, in the late 19th century, physicists learned that atoms had an internal structure consisting of negatively charged electrons and a positively charged nucleus. The discovery of radioactivity and subsequent investigations indicated that the nucleus itself has a complex structure. In addition, the investigation of black body radiation and the photoelectric effect led scientists to conclude that light can be modeled as a photon—an object that has particle-like properties. By 1930, physicists had identified four particles—the electron, the proton, the neutron, and the photon. At that time, these were the only known truly **elementary particles**, a term used to indicate the simplest and most basic particles. This changed with the proposal and discovery of so-called antiparticles.

Antielectrons predicted

At the end of the 1920s, physicists solving the Schrödinger equation for the possible electron wave functions in the hydrogen atom produced solutions that were labeled by three quantum numbers: n, l, and m_l. However, the Schrödinger equation did not explain the existence of the fourth quantum number of the electron in the atom, its spin m_s.

In 1928 Paul Dirac extended quantum mechanics to incorporate Einstein's theory of special relativity. Dirac's model of relativistic quantum mechanics successfully predicted the existence of the spin quantum number for the electron. However, along with this prediction came an unexpected complication. Dirac's model also predicted that free electrons (electrons that are not interacting with other objects or fields) had an infinite number of possible quantum states with negative total energy. As a result, a free electron in a positive energy state should be able to transition to one of these negative energy states by emitting a photon whose energy equals the difference between those two states. All free electrons in the universe would transition to increasingly negative energy states, continually releasing electromagnetic energy. This model certainly was inconsistent with the observed behavior of electrons. How could a free electron have negative total energy? (Remember that only certain types of potential energies of systems, such as electric potential energy, can be negative.)

Dirac suggested that these negative energy states were occupied by an infinite number of so-called virtual electrons. Because only one electron can occupy an electron state according to the Pauli exclusion principle (Chapter 28), free electrons with positive energy could not then transition into these states. Over the next three years Dirac modified his model and eventually proposed that one of these virtual electrons in a negative energy state could be lifted out of its negative energy state to become a positive energy free electron. The empty state left behind would behave like a positively charged electron—like a particle with an electric charge of $+1.6 \times 10^{-19}$ C. Many quantum physicists disagreed with many aspects of his model. But Dirac argued that such particles should exist. He called them *antielectrons*, a new type of particle that had not yet been observed.

Antielectrons detected

To check Dirac's prediction, physicists had to search for a particle that behaved exactly like an electron but had a positive electric charge. At that time, to determine the sign of elementary particles, scientists used cloud chambers. A cloud chamber contained a gas that was supersaturated with vaporized water or alcohol. A charged particle like an electron or alpha particle passing through the vapor caused the vapor to condense into visible droplets, leaving a trail marking the particle's trajectory.

In 1932, American physicist Carl Anderson was studying cosmic rays, protons, and other particles coming to Earth from the Sun and from supernova explosions. He used a cloud chamber placed in a magnetic field. One of his photographs (Figure 30.1a) included the curved path of a particle passing through the chamber. The magnetic field pointed into the plane of the page.

The direction of travel of the particle and its charge were not immediately apparent from its path. If it entered the chamber from the top, then using the right-hand rule for magnetic force, it must be a negatively charged particle in order to curve as shown (Figure 30.1b). However, if it entered the chamber from the bottom, then it must be a positively charged particle (Figure 30.1c). How could Anderson distinguish between these two possibilities?

Notice that the lower part of the path is less curved than the upper part. Recall that a uniform magnetic field causes a charged particle to travel in a circular path (Chapter 20). The radius of its path can be expressed as

$$r = \frac{mv}{|q|B}$$

This means that the radius of the curved path is largest when the speed is largest—on the bottom half of the path. The particle passed through a thin lead plate in the middle of the chamber. This caused the particle's speed to decrease. Thus, the particle must have been traveling from the bottom of the chamber to the top, meaning it was a positively charged particle.

Anderson was able to determine the charge-to-mass ratio of the particle and found that it was the same as that for an electron. But it was positively charged! Anderson realized that the cloud chamber trace must have been produced by one of Dirac's antielectrons. Anderson found many more such tracks in other cloud chamber photographs. He called the new particle a **positron**.

Pair production

Electrons are abundant in our world, but there are few positrons. Under most circumstances, positrons exist for very short time periods. How are they produced? One way is during the radioactive decay of certain nuclear isotopes—for example, $^{11}_{6}C$, $^{13}_{7}N$, $^{15}_{8}O$, $^{22}_{11}Na$, and $^{40}_{19}K$ (we discussed this topic in Chapter 29).

Positrons can also be produced during the interaction of a high-energy gamma ray photon γ with matter. The photon has zero rest mass—you can't measure a photon's mass on a scale because the photon cannot exist at rest. However, under the right conditions this photon can simultaneously produce an electron $^{0}_{-1}e$ and a positron $^{0}_{+1}e$ (particles with nonzero rest mass). The process is called **pair production** and is represented by the reaction below:

$$\gamma \rightarrow {}^{0}_{-1}e + {}^{0}_{+1}e$$

For the photon to simultaneously produce both an electron and a positron, its energy must at least equal the combined rest energies of the electron and the positron. For the minimal energy calculation, we assume the kinetic energies of the electron and positron after their production are zero. The rest energy of the electron (and of the positron) is $E_0 = mc^2$. Thus, the photon must have double this energy:

$$E_{\text{photon}} = 2mc^2 = 2(9.11 \times 10^{-31} \text{ kg})(3.00 \times 10^8 \text{ m/s})^2$$

$$= 1.64 \times 10^{-13} \text{ J}$$

In electron volts, this is

$$E_{\text{photon}} = (1.64 \times 10^{-13} \text{ J})\left(\frac{1 \text{ eV}}{1.6 \times 10^{-19} \text{ J}}\right) = 1.02 \times 10^6 \text{ eV}$$

$$= 1.02 \text{ MeV}$$

FIGURE 30.1 (a) The cloud chamber photograph taken by Anderson. (b) The direction of the force that the magnetic field would exert on an electron if it were moving from the top to the bottom. (c) The direction of the force that the magnetic field would exert on a positron if it were moving from the bottom to the top.

(a)

$\vec{B}$ into the page

Path of particle

(b)

$\vec{B}$ into page

$\vec{F}_{\vec{B} \text{ on } \ominus}$ $\vec{v}$

(c)

$\vec{B}$ into page

$\vec{F}_{\vec{B} \text{ on } \oplus}$ $\vec{v}$

FIGURE 30.2 (a) A gamma ray photon converts to an electron and a positron. (b) If in a magnetic field, the magnetic force causes the particles to move apart in spiral circles. (c) A cloud chamber photograph of this pair production.

(a)

(b)

(c)

Positron   Electron

The frequency of such a photon is

$$f = \frac{E_{\text{photon}}}{h} = \frac{1.64 \times 10^{-13}\,\text{J}}{6.63 \times 10^{-34}\,\text{J} \cdot \text{s}} = 2.47 \times 10^{20}\,\text{Hz}$$

and its wavelength is

$$\lambda = \frac{c}{f} = \frac{3.00 \times 10^8\,\text{m/s}}{2.47 \times 10^{20}\,\text{Hz}} = 1.21 \times 10^{-12}\,\text{m}$$

This wavelength is about $1/100$ times the wavelength of an X-ray photon, making it a gamma ray photon.

If the positron and electron produced are to possess some kinetic energy, then a higher energy gamma ray is needed. If this pair production occurs in a magnetic field, the two particles should spiral in opposite directions.

To examine this idea, consider the photon as the system—it has momentum in the x-direction before producing the pair (**Figure 30.2a**). Immediately after pair production, the photon no longer exists, so the momentum of the two particles should be the same as the momentum of the original photon. The electron and positron should fly apart at an angle relative to each other, so that the y-components of momentum perpendicular to the original momentum of the photon cancel each other.

If we place the system in a magnetic field, the field exerts a force on the electron and on the positron in opposite directions because they have opposite electric charge (Figure 30.2b). This force will be perpendicular to their velocities, so they should curve in opposite directions. In addition, the radius of their circular paths should decrease as their speeds decrease, resulting in an inward spiraling motion. The photographs of pair production by gamma ray photons obtained by Anderson are consistent with this reasoning—see Figure 30.2c.

Pair annihilation

We have just learned that high-energy gamma ray photons can be used to create electron-positron pairs. Is the reverse process possible? If an electron and positron meet, will one or more photons be produced? The answer is yes, and the process is called **pair annihilation**.

CONCEPTUAL EXERCISE 30.1   **Pair annihilation**

Imagine that an electron and a positron meet and annihilate each other. Assume that they are moving directly toward each other at the same constant speed. Will one or two photons be produced? Write a reaction equation for this process. In what direction does the photon or photons move after the process?

Sketch and translate   If one photon is produced, the reaction for this process is

$$_{-1}^{0}e + {}_{+1}^{0}e \rightarrow \gamma$$

If two photons are produced, the reaction is

$$_{-1}^{0}e + {}_{+1}^{0}e \rightarrow \gamma + \gamma$$

Simplify and diagram   Assume that the process occurs in isolation, which means that the total electric charge and momentum of the system must remain constant. Electric charge is constant in both versions of the process. Because the two particles were moving directly toward each other at the same constant speed, the total initial momentum of the system is zero. In the first version of the process, however, momentum is not conserved. If only one photon (with momentum $p = E/c$) is produced, the system's final momentum after the annihilation cannot be zero. Therefore, for the momentum of the system to be constant in this process, two photons must be produced, and they must travel in opposite directions, as shown in the figure.

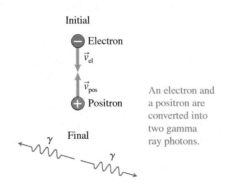

An electron and a positron are converted into two gamma ray photons.

(The details of the reaction are much more complicated than discussed here and are beyond the scope of this book.)

Try it yourself In this pair annihilation, two gamma rays were produced. But in pair production, an electron-positron pair is produced by only one gamma ray photon. How is this possible?

Answer

(see Figure 30.2).

The momentum of the single gamma ray photon can be converted to the combined momentum of the two particles

---

The previous exercise involved a positron. We have already discussed the creation of a positron and an electron by a high-energy gamma ray photon as well as the production of positrons by radioactive decay. Let's consider another possibility. Recall the explanation of the beta decay of a neutron in nuclear physics (Chapter 29). A free neutron is an unstable particle with half-life of about 10 min. A free neutron decays into a proton, an electron, and an antineutrino (the antiparticle of a neutrino):

$$\,^{1}_{0}n \rightarrow \,^{1}_{1}p + \,^{0}_{-1}e + \,^{0}_{0}\bar{\nu}$$

The electric charge of the system during this process is constant $(0 = +1 - 1 + 0)$, as is its nucleon number $(+1 = +1 + 0 + 0)$. The rest energy of the neutron is greater than the combined rest energies of the proton and electron; therefore, the system's energy can possibly be constant as well, provided the proton, electron, and neutrino have sufficient kinetic energy. Finally, an appropriate combination of the momenta of the three particles in the final state can equal the initial momentum of the neutron— the momentum of the system can be constant. This process could happen without interaction with the environment. Is there a similar process by which a positron is produced instead of an electron?

Beta-plus decay: transforming a proton into a neutron

Since the rest energy of the neutron is greater than the rest energy of the proton, it seems that transforming a proton into a neutron cannot occur. However, if a proton captures a gamma ray photon, it is possible for a neutron and the other particles to be produced:

$$\gamma + \,^{1}_{1}p \rightarrow \,^{1}_{0}n + \,^{0}_{+1}e + \,^{0}_{0}\nu$$

The proton absorbs the photon and then decays into a neutron, a positron, and a neutrino. This process is called **beta-plus decay** to indicate that a positron (not an electron) is produced as the result of the process.

Beta-plus decay can also occur without the proton absorbing a photon. This can happen inside a nucleus if some of the energy of the nucleus converts into the additional rest energy of the products. Nuclei that can undergo beta-plus decay include carbon-11, nitrogen-13, oxygen-15, fluorine-18, and potassium-40. In fact, about 20,000 positrons are produced each minute in your body from the beta-plus decay of potassium-40:

$$\,^{40}_{19}K \rightarrow \,^{40}_{18}Ar + \,^{0}_{+1}e + \,^{0}_{0}\nu$$

The potassium nucleus transforms into an argon nucleus and in the process emits a positron and a neutrino. The potassium-40 nuclei in your body come from the potassium in foods you eat, such as bananas.

Note that the positively charged positrons produced in your body by this radioactive decay travel infinitesimal distances. They are attracted to negatively charged electrons and abruptly undergo pair annihilation, producing high-energy gamma rays, most of which leave your body. This process is what makes positron emission tomography possible.

FIGURE 30.3 Positron emission tomography (PET).

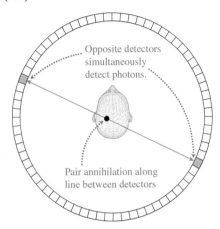

Opposite detectors
simultaneously
detect photons.

Pair annihilation along
line between detectors

Positron emission tomography (PET)

In the chapter opening story, we described positron emission tomography (PET), a process for imaging the brain using radioactive fluorine-18 injected into a person's body. The isotope undergoes beta-plus decay continually, producing positrons. The positrons immediately meet electrons and annihilate, producing a pair of gamma ray photons that move in opposite directions. The PET chamber surrounding the patient's head detects the pairs of simultaneously appearing photons (Figure 30.3) and determines where in the brain they were emitted. Combining many pairs of gamma rays produces a three-dimensional image of the active parts of the brain—the places where more photons are produced.

This technique not only allows doctors to find the active and inactive regions of the brain in sick patients, but also helps brain researchers identify parts of the brain that are active when a person performs a particular task, such as listening to a poem being read or thinking about a mathematical problem.

Other antiparticles

Other particles have antiparticles as well. The positively charged proton that is part of all nuclei has a negatively charged antiproton of the same mass but opposite electric charge. Even though the neutron has zero electric charge, it too has an antimatter counterpart; other properties besides charge differentiate the neutron from the antineutron (see Section 30.3 on quarks). Some particles are their own antiparticle; the photon is an example, and possibly the neutrino. Later in this chapter we will consider why our universe is composed almost entirely of ordinary particles and almost no antiparticles.

REVIEW QUESTION 30.1 Why does proton decay occur only inside nuclei and not when a proton is free?

30.2 Fundamental interactions

In our studies of physics we have learned about many different forces: gravitational, electric, elastic, buoyant, etc. Some of these forces characterize *fundamental* interactions, and some characterize *nonfundamental* interactions. Fundamental interactions are the most basic interactions known, such as the electromagnetic interaction between charged particles. Nonfundamental interactions, such as friction, can be understood in terms of fundamental interactions. Friction is a macroscopic manifestation of the electromagnetic interaction between the electrons of the two surfaces that are in contact. We use the term "interaction" rather than "force" when speaking in general terms because interactions can also be represented using energy ideas. The four fundamental interactions are the gravitational interaction, the electromagnetic interaction, the strong interaction, and the weak interaction.

Gravitational interaction All objects in the universe participate in **gravitational interactions** due to their mass (and, according to general relativity, also due to their energy content through $m = E/c^2$). The gravitational force that two objects exert on each other is proportional to the product of their masses and the inverse square of their separation $1/r^2$. This interaction is important for mega-objects (planets, stars, galaxies, etc.), much less important for objects in our daily lives (e.g., the gravitational force that one human exerts on another), and extremely insignificant for microscopic objects (atoms, electrons, etc.).

Electromagnetic interaction Electrically charged objects participate in **electromagnetic interactions**. The interaction is electric if the objects are at rest or in motion with respect to each other. The interaction is magnetic only if the objects are moving with respect to each other. Similar to the gravitational interaction, the electromagnetic interaction between two charged particles also decreases as the inverse square of the distance between them. The electromagnetic interaction between nuclei and electrons is important in understanding the structure of atoms.

The electromagnetic interaction is tremendously stronger than the gravitational interaction. For example, the electrostatic force that an electron and a proton exert on each other in an atom is about 10^{39} times greater than the gravitational force that they exert on each other (**Figure 30.4**).

Atoms are electrically neutral, but because they are composed of distributions of charged particles (protons and electrons), they do participate in electromagnetic interactions with each other. These interactions contribute to the formation of molecules and to holding liquids and solids together. Because these interactions occur between overall electrically neutral objects, they are called **residual interactions**. Interactions such as friction, tension, and buoyancy are nonfundamental interactions and also are macroscopic manifestations of residual electromagnetic interactions.

Strong interaction Atomic nuclei are made up of protons that repel each other via electromagnetic interactions and neutrons that do not significantly interact electromagnetically. The interaction that binds protons and neutrons together into a nucleus is a residual interaction of the **strong interaction**. The strong interaction is a very short range interaction, exerted by protons and neutrons only on their nearest neighbors within the nucleus (a range of $\sim 3 \times 10^{-15}$ m). This residual interaction of the strong interaction is called the nuclear interaction, and the force characterizing it is called the **nuclear force**.

Weak interaction The weak interaction is responsible for beta decay. Protons, neutrons, electrons, and neutrinos all participate in it. The **weak interaction** is (as the name implies) significantly weaker than the strong interaction and has a significantly shorter range, about 10^{-18} m or less.

Mechanisms of fundamental interactions

How do elementary particles interact with each other? Consider an analogy. Humans interact using speech. A vibration in a person's larynx produces a sound wave that travels through the air to another person. That sound wave vibrates the eardrums of the other person. This vibration is then converted into an electrical signal that travels to the brain, where it is interpreted. How can we use this idea to understand the microscopic mechanism behind the four fundamental interactions?

Earlier in this text we developed a field model for the interactions between electrically charged objects (Chapter 18). The idea behind this model was that electrically charged particles produce a disturbance in the region around them called an electric field. When a charged particle moves, the "signal" produced by that movement ripples through the electric field at a finite speed (light speed), and only when that signal reaches other charged particles does the force exerted on them by the field change.

We later used the field model to explain the mechanism behind the magnetic interaction, ultimately predicting the existence of electromagnetic waves, of which visible light is just one example. We also discussed how electromagnetic waves could be modeled as streams of photons and showed that this model was necessary to explain microscopic phenomena such as the photoelectric effect.

During the first half of the 20th century, physicists realized that the photon actually played a more central role in electromagnetic phenomena. Researchers learned that the mechanism behind all microscopic electromagnetic phenomena was the exchange of photons between electrically charged particles.

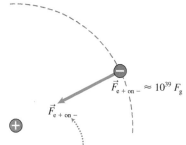

FIGURE 30.4 The electric force that atomic size objects exert on each other is much greater than the gravitational force they exert on each other.

$\vec{F}_{e+\text{ on }-} \approx 10^{39}\, F_g$

$\vec{F}_{e+\text{ on }-}$

The electrostatic force exerted by a proton on an electron in a hydrogen atom is about 10^{39} times greater than the gravitational force.

This particle exchange mechanism has since been successful in describing the weak and strong interactions, and it has had some success in describing the gravitational interaction (although there are still many unresolved difficulties). In each of the four fundamental interactions, when two objects interact, one of the two objects emits a particle that is then absorbed by the other. This emitted and absorbed particle is called a **mediator**. In the electromagnetic interaction, the mediator is the photon. Two electrons repel each other through the exchange of photons.

But the particle exchange mechanism of particle interactions appears to have a serious problem. Where does the energy required to create the mediating particle come from? To answer this question, we need to think about the uncertainty principle (Chapter 28):

$$\Delta E \Delta t \geq \frac{h}{4\pi}$$

The energy of an atomic-scale system is only determinable to within the range ΔE. When the system is involved in a process, the smaller the time interval Δt in which the process occurs, the larger the energy uncertainty of the system during the process:

$$\Delta E \approx \frac{h}{4\pi \Delta t}$$

This helps us understand how the mediator particle can exist. The energy of an atomic-scale system is not exactly determinable; it fluctuates from instant to instant. In the case of two electrons interacting, these fluctuations manifest as what are known as **virtual photons** being exchanged between the electrons. These mediator photons are called "virtual" because they exist only as extremely small energy fluctuations of the system. Virtual photons do not have independent energy of their own that could initiate a chemical change in your retina or in an electronic detector. As a result, virtual photons (and other virtual particles in general) cannot be experimentally detected in any direct way.

We see from the uncertainty principle that for interactions that are mediated over long distances and therefore require the mediator particles to exist for long time intervals Δt to travel between the particles, the energy uncertainty ΔE of the system must be small (large Δt, small ΔE). This means that photons mediating long-range electromagnetic interactions should have low total energy. This is possible because photons have zero rest energy and can have very low frequencies (recall $E_{\text{photon}} = hf$).

Interaction mediators

We can now think of the four fundamental interactions as exchange processes of four different mediators. The mediators of the electromagnetic, strong, and weak interactions have all been discovered. The hypothetical mediator for the gravitational interaction, the so-called **graviton**, has not. Below we summarize the four types of interaction mediators.

Photons (electromagnetic interaction) The electromagnetic interaction can be modeled as an exchange of photons. Photons are massless particles and always travel at the speed of light. They have zero electric charge and are their own antiparticles (the photon and antiphoton are the same particle).

Gluons (strong interaction) Gluons are massless and have zero electric charge but interact strongly with each other. As a result, the strong interaction has a very short range. There are eight different kinds of gluons. The exchange of gluons mediates the interaction of quarks and allows them to be bound together into protons, neutrons, and more exotic particles. (More on quarks below.)

Gluons were first hypothesized as the mediators of quark interactions in the 1960s, but over a decade later there was still no direct experimental evidence for them. In 1976,

Mary Gaillard, Graham Ross, and John Ellis proposed a search for gluons using specific collision events. The gluon was subsequently discovered at DESY (Deutsches Elektronen-Synchrotron) in 1979.

$W^{\pm}$ **and** Z^0 **(weak interaction)** Three particles mediate the weak interaction: the positively charged W^+, its antiparticle, the negatively charged W^-, and the neutral Z^0 (which is its own antiparticle). The existence of these particles was predicted in the 1960s by a theory that unified the electromagnetic and weak interactions. This electroweak theory of Sheldon Glashow, Abdus Salam, and Steven Weinberg remained largely theoretical until 1983, when the $W^{\pm}$ and Z^0 particles were discovered at the European Organization for Nuclear Research (CERN) in Geneva, Switzerland.

Electroweak theory predicted (correctly) that these mediators have a very large rest energy. Producing them, therefore, required the colliding particles to have extremely high total energy. When the $W^{\pm}$ and Z^0 particles were first predicted, the accelerator at CERN could accelerate protons and antiprotons only to energies of tens of giga-electron volts (1 GeV is one billion electron volts), which was well below the rest energy of the weak interaction mediators. This explains why it took so many years to confirm the mediators' existence—technology had to catch up.

Gravitons (gravitational interaction) Based on fairly widely accepted assumptions about what a quantum theory of gravity should be, the graviton is predicted to be electrically neutral and have zero mass and therefore to travel exclusively at the speed of light.

Our current understanding of the four fundamental interactions is summarized in Table 30.1. For reference, recall that the mass of a proton is 1.67×10^{-27} kg.

TIP Although the properties of the graviton seem identical to those of the photon, they are very different particles. In this text we won't discuss the properties that distinguish them.

TABLE 30.1 Fundamental interactions

Interaction (Force)	Range	Mediating particle		Relative strength
Electromagnetic	Infinity $\left(F \sim \dfrac{1}{r^2}\right)$	Photon		10^{-2}
		Mass 0 kg	Electric charge 0	
Strong	Short (about 10^{-15} m)	Gluon		1
		Mass 0 kg	Electric charge 0	
Weak	Short (about 10^{-18} m)	$W^{\pm}$		10^{-13}
		Mass 1.43×10^{-25} kg	Electric charge $+e$ for W^+ $-e$ for W^-	
		Z^0		
		Mass 1.62×10^{-25} kg	Electric charge 0	
Gravitational	Infinity $\left(F \sim \dfrac{1}{r^2}\right)$	Graviton		10^{-38}
		Mass 0 kg	Electric charge 0	

As you can see, the mediator particles are massless except for the $W^{\pm}$ and Z^0, which have masses of about 100 times that of a proton. This huge mass explains why the weak interaction has such a short range. When produced as virtual particles, the $W^{\pm}$ and Z^0 can only last very briefly, so briefly that even if they are traveling at near light speed they can travel only an average of 10^{-18} m.

In Table 30.1 we listed the masses of the interaction mediators in kilograms. This is not the common practice in particle physics because the values are so small in those units. In fact, mass is not generally used at all. Instead, the "mass" of a particle is given in terms of its rest energy $E_0 = mc^2$ measured in electron volts (eV). Typical particle "masses" are in the mega-electron volt (1 MeV $= 10^6$ eV) or giga-electron volt (1 GeV $= 10^9$ eV) range.

Units in particle physics

Convert the masses of the $W^\pm$ and Z^0 particles into electron volts.

Represent mathematically First we need to determine the rest energies of the particles in joules, then convert them into electron volts, where $1 \text{ eV} = 1.6 \times 10^{-19} \text{ J}$.

Solve and evaluate The rest energy of the $W^\pm$ particles is

$$E_0 = mc^2 = (1.43 \times 10^{-25} \text{ kg})(3.00 \times 10^8 \text{ m/s})^2$$
$$= 1.29 \times 10^{-8} \text{ J}$$

Converting to eV, we get

$$(1.29 \times 10^{-8} \text{ J})\left(\frac{1 \text{ eV}}{1.6 \times 10^{-19} \text{ J}}\right) = 80.6 \times 10^9 \text{ eV} = 80.6 \text{ GeV}$$

Using the same method, the rest energy of the Z^0 particle is 91.3 GeV.

Try it yourself Determine the mass in mega-electron volts of the 1.67×10^{-27}-kg proton and of the 9.11×10^{-31}-kg electron.

Answer   938 MeV and 0.51 MeV.

REVIEW QUESTION 30.2 Why do two protons repel each other—for example, in an ionized gas of hydrogen—but attract each other when they are inside a nucleus?

FIGURE 30.5 Inside the SLAC National Accelerator Laboratory.

FIGURE 30.6 Inside the Large Hadron Collider (LHC).

30.3 Elementary particles and the Standard Model

Since the early 20th century, physicists have been discovering new elementary particles using particle accelerators such as the electron-positron collider at the SLAC National Accelerator Laboratory (Figure 30.5), the proton-antiproton collider at Fermilab in Illinois, and the Large Hadron Collider (LHC; see Figure 30.6) at CERN in Switzerland. Because the total energy of accelerated particles is significantly greater than the electron and proton rest energies, it is possible for additional particles to be produced in the collisions. The properties of these additional particles can be determined using elaborate detectors. Most of the particles produced are not stable; they quickly decay into other particles until eventually only stable particles remain.

The interaction mediators we discussed in the previous section are the mechanism behind the four fundamental interactions. These form one category of elementary particles. The building blocks of matter form two more categories, **leptons** and **hadrons**, which are the topic of this section.

Leptons

Leptons interact only through the weak, electromagnetic (if electrically charged), and gravitational interactions (though very weakly), not through the strong interaction. The electron is an example of a lepton, as is the neutrino ν_e, which is produced in beta-plus decay (Chapter 29). Note the symbol e as the subscript for this neutrino, which we did not use before. The symbol indicates that it is an "electron neutrino," and you will see that there other types of neutrinos. The electron neutrino is electrically neutral. These two particles form what is called a generation (or family) of leptons.

In 1936 the first member of a second lepton generation was discovered by American physicists Carl Anderson and Seth Neddermeyer. The **muon** μ is similar to the electron in that it is negatively charged, but it is approximately 200 times more massive. It also is unstable, with a half-life of 1.5×10^{-6} s. The corresponding **muon neutrino** ν_μ was discovered in 1962. A third generation of leptons was later discovered: the **tau** τ (by Martin Lewis Perl and his collaborators in a series of experiments between 1974 and 1977 at the SLAC) and the **tau neutrino** ν_τ (in 2000 at Fermilab). The tau is 17 times more massive than the muon and has an even shorter half-life of 2.0×10^{-13} s.

Electrons, muons, and taus have their own distinct antiparticles, but it is not yet clear whether the antiparticles of neutrinos are distinct or whether neutrinos are their own antiparticles. The electron and the three neutrinos are stable particles and do not decay. Some of the properties of leptons are shown in Table 30.2.

TABLE 30.2 Lepton generations (families)

	Electron generation		Muon generation		Tau generation	
	Mass (GeV)	**Electric charge in e**	**Mass (GeV)**	**Charge 1.6×10^{-19} C**	**Mass (GeV)**	**Electric charge in e**
Particle	Electron e		Muon μ		Tau τ	
	5.1×10^{-4}	-1	0.106	-1	1.777	-1
Antiparticle	Positron e^+		Antimuon $\overline{\mu}$		Antitau $\overline{\tau}$	
	5.1×10^{-4}	$+1$	0.106	$+1$	1.777	$+1$
Paired neutrino	Electron neutrino ν_e		Muon neutrino ν_μ		Tau neutrino ν_τ	
	$<2 \times 10^{-9}$	0	$<1.9 \times 10^{-4}$	0	$<1.8 \times 10^{-2}$	0
Paired antineutrino	Electron antineutrino $\overline{\nu}_e$		Muon antineutrino $\overline{\nu}_\mu$		Tau antineutrino $\overline{\nu}_\tau$	
	$<2 \times 10^{-9}$	0	$<1.9 \times 10^{-4}$	0	$<1.8 \times 10^{-2}$	0

Until 1998, physicists had no evidence to suggest that the masses of the neutrinos were anything other than zero (though the possibility was considered as early as the 1950s). In 1998 the Super-Kamiokande neutrino detector in Japan (see Figure 30.7) established that the weak interaction allows for processes that can convert one type of neutrino into another, a process called **neutrino oscillation**. This process is only possible if one or more of the neutrinos have nonzero mass. Since then, additional evidence from both particle physics and astrophysics experiments suggests that all three neutrinos have a rest energy of about 1 eV. Experiments are under way to measure this more precisely. Because neutrinos do not interact via the electromagnetic or strong interaction, they are extremely difficult to detect.

FIGURE 30.7 Inside the Super-Kamiokande neutrino detector.

Hadrons

Two different types of hadrons can be distinguished—**baryons** and **mesons**. You already know two examples of baryons, the proton and the neutron, but there are many more. In 1949–1952, new higher mass baryons were discovered in interactions between cosmic rays and the atmosphere: the lambda particle Λ^0 and a set of four similar baryons known as the delta particles Δ^-, Δ^0, Δ^+, and Δ^{++}. Additional baryons were discovered in later years.

The first example of a meson was Hideki Yukawa's suggestion in 1935 of the existence of new particles that mediated the strong interaction. He called these particles mesons. The motivation for Yukawa's ideas came from combining relativity theory with quantum theory. The masses of Yukawa's mesons were predicted to be about 200 times the electron mass. In 1947 physicists discovered a meson in cosmic rays that participated in strong interactions and that had the correct properties to be Yukawa's meson (they called it a pi-meson, or pion for short). Later, additional mesons were discovered, such as the K^+ and K^0. Mesons and pions, however, were not the mediators of the strong interaction, as we will learn later.

The difference between baryons and mesons lies in their internal structure and is discussed in the next subsection. With the exception of the proton, all hadrons are unstable. Their half-lives range from 610 s for the neutron to 10^{-24} s for the shortest lived. A small sample of some hadrons and their properties is provided in **Table 30.3**. Hundreds of hadrons have been discovered since 1950.

TABLE 30.3 **Properties of hadrons**

	Particle, antiparticle	Mass (GeV)	Electric charge (in e)	Half-life (s)
Baryons				
Proton	$p, \bar{p}$	0.938	1, −1	Infinity
Neutron	$n, \bar{n}$	0.940	0, 0	610
Lambda	$\Lambda^0, \overline{\Lambda}^0$	1.116	0, 0	1.8×10^{-10}
Delta	$\Delta^{++}, \overline{\Delta}^{++}$	1.232	+2	3.9×10^{-24}
	$\Delta^{+}, \overline{\Delta}^{+}$	1.232	+1	3.9×10^{-24}
	$\Delta^{0}, \overline{\Delta}^{0}$	1.232	0	3.9×10^{-24}
	$\Delta^{-}, \overline{\Delta}^{-}$	1.232	−1	3.9×10^{-24}
Omega	Ω^-, Ω^+	1.672	−1, 1	5.7×10^{-11}
Mesons				
Pion	π^+, π^-	0.140	1, −1	1.8×10^{-8}
	π^0, self	0.135	0, 0	5.9×10^{-17}
Eta	η^0, self	0.548	0, 0	$<10^{-18}$
Kaon	K^+, K^-	0.498	1, −1	8.6×10^{-9}
	$K^0, \overline{K}^0$	0.498	0	No definite lifetime

Quarks

The large number of hadrons and the differences in their properties led to the belief that they were not truly elementary particles. In 1964 Murray Gell-Mann and his collaborators in the United States and Kazuhiko Nishijima in Japan independently proposed a new model of hadrons. In this model hadrons are complex objects made of a small number of more fundamental particles that combine in different combinations to make all of the existing hadrons. The idea is similar to combining protons, neutrons, and electrons in different combinations to make the elements of the periodic table.

The concept of hadrons having an internal structure helped explain the results of experimental investigations in which electron beams with energies of about 25 GeV (25 billion eV) were shot into a sample of liquid hydrogen. Electrons were scattered at various angles, similar to the alpha particles shot by Rutherford's colleagues at gold foil atoms. Electrons that were scattered at large angles had energy changes as if they had collided with tiny particles within the protons (**Figure 30.8**) rather than with the protons themselves.

These particles had to be electrically charged for the proton to be charged. Surprisingly, the experiments showed that the particles had *fractional* electric charge. These particles became known as **quarks**. When the quark model was first invented, only three quarks were needed to build all the hadrons that were known. These quarks were given the whimsical names **up, down,** and **strange**. Since then, three more quarks have been proposed (called **charm, top,** and **bottom**) to explain the structure of the ever more massive hadrons being produced in increasingly powerful accelerators. This brings the total number of quarks to six, all of which have since been discovered experimentally.

These different quark types are known in the physics community as **flavors** (having nothing to do with the sense of taste). Quarks interact through the strong (via gluons), weak (via the $W^{\pm}$ and Z^0), and electromagnetic (via photons) interactions. Because the quarks have nonzero mass, they also interact gravitationally, but only extremely

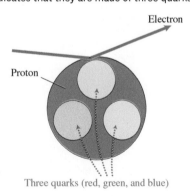

FIGURE 30.8 Electron scattering from protons indicates that they are made of three quarks.

Electron

Proton

Three quarks (red, green, and blue)

weakly. The difference between baryons and mesons comes from their quark content: baryons are bound states of three quarks, while mesons are bound states of one quark and one antiquark.

Recall that leptons come in three generations (electron, muon, or tau), each with a negatively charged member and a neutrino for a total of six leptons. The quarks also fall into three generations (up/down, charm/strange, top/bottom) for a total of six quarks. Is this a coincidence? In other words, is there a deep connection between leptons and quarks? Particle physicists hope to find a connection, and that it motivates investigations into grand unification, but no connection has yet been established.

Several properties of particles determine whether they participate in particular interactions. For example, objects with nonzero electric charge participate in the electromagnetic interaction. The property that determines whether a particle participates in the strong interaction is known as **color charge** (it is not related to the colors we can see, but rather is a technical term used for this property). Because quarks participate in the strong interaction, they have color charge. A particle with color charge can interact with another color-charged particle by exchanging gluons.

Unlike electric charge, color charge comes in three varieties, known as **red, green,** and **blue** (again, these do not correspond to the colors we see). Just as neutral atoms have a net electric charge of zero, composite particles such as the proton and neutron are made of three quarks with complementary colors, which results in an overall color-neutral object. In other words, a proton has one red quark, one blue quark, and one green quark. This neutrality is analogous to the effect of shining complementary beams of red, blue, and green light on a surface. That surface glows white, which corresponds to color neutral (sometimes called colorless).

We mentioned earlier that quarks have fractional electric charge. The up quark has charge $(2/3)e$, where e is the magnitude of the charge of the electron. The down quark has charge $(-1/3)e$. It is reasonable to suggest, then, that the proton is composed of two up quarks and one down quark, or *uud* for short (**Figure 30.9**). This results in a total electric charge of $(2/3)e + (2/3)e + (-1/3)e = e$, consistent with the charge of the proton. The three quarks also have color charge, one red, one green, and one blue, so that the proton is color neutral. In the following exercise we determine the quark content of a neutron.

FIGURE 30.9 A proton is made of three charged quarks, with colors red, green, and blue.

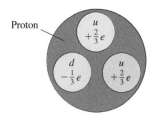

The proton's net charge is $+e$, and the proton is color neutral.

Making a neutron

What combination of quarks will have the correct properties to be a neutron?

Sketch and translate Because the electric charge of the neutron is zero, the charges of its component quarks must add to zero. The neutron is a baryon, so it is composed of three quarks (and no antiquarks).

Simplify and diagram This is accomplished if we include one up quark and two down quarks (*udd*).

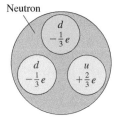

The neutron's net charge is zero, and the neutron is color neutral.

$$2(-1/3) + (2/3) = 0$$

Two *d* quarks   One *u* quark

Try it yourself How does the proton composition differ from the neutron composition? What would have to happen for a neutron to be converted into a proton?

Answer

The proton has two up quarks and one down quark. A neutron could be converted into a proton if one down quark were converted to an up quark. This sounds very similar to what occurs in beta decay.

We stated earlier that an antineutron has properties besides charge that differentiate it from a neutron. An antineutron is composed of three antiquarks, namely, one up antiquark and two down antiquarks.

Figure 30.10 summarizes our understanding of the structure of matter and the role of quarks in that structure. All of the matter that makes up our local world is composed of electrons and two types of quarks (up and down).

FIGURE 30.10 A summary of particles (matter) and their interactions.

Confinement

No experiment has ever produced a quark in isolation. Every quark and antiquark ever produced has always been part of a hadron (either a three-quark baryon or a quark-antiquark meson). Trying to split a proton into quarks by shooting a high-energy particle into it produces additional quark-antiquark pairs rather than separate quarks. These pairs then combine with the original proton's quarks to form new baryons and mesons. This phenomenon is called **confinement,** and it is an indication of a feature of the strong interaction that is very different from the other three interactions. The strong interaction between quarks is weakest when they are close together and gets stronger the farther apart the quarks are. The reasons why only the strong interaction exhibits confinement are beyond the scope of this book, but the feature is crucial to explaining the structure and stability of the protons and neutrons in every atomic nucleus in your body.

Development of the Standard Model

With this basic understanding of elementary particles (interaction mediators, leptons, and hadrons) we now can discuss the **Standard Model**, the name given to the combined theory of the building blocks of matter and their interactions. Its first versions were constructed during the first half of the 20th century, and it has continued to evolve ever since.

In the late 1940s physicists Richard Feynman, Julian Schwinger, and Sin-Itiro Tomonaga independently combined the ideas of special relativity and quantum mechanics into a single model of the electromagnetic interaction that explains all electromagnetic phenomena. (The three of them were jointly awarded the Nobel Prize in physics in 1965 for this work.) Their model is known as **quantum electrodynamics** (QED) and is a cornerstone of the Standard Model. In 1957–1959 Julian Schwinger, Sidney Bludman, and Sheldon Glashow, working separately, proposed a model in which the weak interaction was mediated by two massive charged particles, which were later called the W^+ and W^- particles. Chen-Ning Yang and Robert Mills developed the mathematical framework needed to describe this model.

In 1967 Sheldon Glashow, Abdus Salam, and Steven Weinberg independently put forth a model that unified the electromagnetic and weak interactions into a single interaction, which they called the **electroweak model**. This model made several striking predictions. First, it predicted that in the very distant past, when the universe was much smaller and very much hotter, all particles were massless. Second, it predicted the existence of a particle, which became known as the **Higgs particle** after physicist Peter Higgs. Third, it predicted that as the universe cooled, the Higgs particle began interacting significantly with other elementary particles, reconfiguring them into the familiar forms they have today (most having nonzero mass). This is known as the **Higgs mechanism**.

In 1973, using Yang's and Mills' mathematical framework, Harald Fritzsch and Murray Gell-Mann formulated **quantum chromodynamics** (QCD), a mathematical model of the strong interaction in terms of the exchange of gluons between quarks. By the end of 1974, results from the first experiments testing QCD began arriving. Among them was the discovery of the so-called J/ψ (J-psi) particle, a meson composed of one charm and one anticharm quark. The existence of the J-psi particle was a successful prediction of QCD. Between 1976 and 1979 scientists discovered the tau lepton and bottom quark and found direct evidence for gluons.

The 1980s brought the discoveries of the predicted weak interaction mediators $W^{\pm}$ and Z^0. The 1990s gave us the top quark, and in 2000 the tau neutrino was discovered. In July 2012 CERN announced the discovery of the long-sought-after Higgs particle, which was a triumph of the Standard Model (see **Figure 30.11**).

Although the predictions of the Standard Model of particle physics are very consistent with the outcomes of particle physics experiments, several unanswered questions remain and are the focus of active research:

1. Can the strong interaction be unified with the electroweak interaction, just as the weak and electromagnetic interactions were unified?
2. Why are there only three families of quarks/leptons?
3. Are the Standard Model particles truly fundamental, or do they have internal structure?
4. Are there additional particles beyond those predicted by the Standard Model?

FIGURE 30.11 A look inside ATLAS, one of the detectors of the Large Hadron Collider (LHC) at CERN, which was instrumental in the discovery of the Higgs particle. (a) The vertical tunnel leading underground to the accelerator part of the detector. (b) The front side of the heart of the ATLAS detector is 25 m in diameter and 46 m long and weighs about 7000 tons.

Summary of the Standard Model

The Standard Model is the currently accepted model of fundamental particles and their interactions (see Figure 30.10). It includes the following:

1. Quarks and leptons make up the matter of the universe. Quarks participate in all four fundamental interactions. Leptons participate in all but the strong interaction.
2. The model of strong interactions (QCD) mediated by gluons.
3. The model of electromagnetic (QED) and weak interactions mediated by photons and the $W^{\pm}$ and Z^0 particles.
4. The Higgs particle, which explains, through the Higgs mechanism, why some of the fundamental particles have nonzero mass.

The Standard Model does not include the gravitational interaction. How to combine this interaction with the Standard Model is a challenging, unsolved problem in physics. One framework for potentially doing this is *string theory*, and while much progress has been made in the past 40 years, no one knows yet whether string theory can achieve this goal.[1]

---

[1]The details of the discovery of the Higgs particle are beyond the scope of this book. If you wish to learn more about it, you can look up the following article: Lincoln, D. (2012). The Higgs boson: Is the end in sight? *The Physics Teacher* 50, 332.

REVIEW QUESTION 30.3 Using what you have learned about particle physics, describe as many differences as you can between a proton and an electron.

30.4 Cosmology

As we have seen, almost all elementary particles have distinct antiparticles. Furthermore, when particles are produced, they are always produced in matter-antimatter pairs. Why, then, is our universe not filled with equal numbers of particles and antiparticles? Why is there an imbalance? These questions are answered in part by particle physics and by **cosmology**—a branch of physics that studies the composition and evolution of the universe as a whole.

Recall that independently of which direction we look, distant galaxies are moving away from us—the universe is expanding (Chapter 26). If we could run this expansion in reverse, the universe would get smaller, denser, and hotter. About 13.7 billion years ago, the entire universe would have been in an unimaginably hot and dense state of nearly zero size, which then rapidly expanded. This initial expansion is called the **Big Bang** (see the timeline from the Big Bang to the present in **Figure 30.12**). The Big Bang model of cosmology was first proposed in 1927 by a Belgian Catholic priest, Georges Lemaître, and more fully developed by George Gamow, Ralph Alpher, and

FIGURE 30.12 The development of the universe since the Big Bang 13.7 billion years ago.

Robert Herman in the late 1940s. The model explained the observed red shifts of spiral nebulae, now called galaxies. It also accounted for the fact that the red shifts were greater for more distant nebulae than for closer nebulae. Below is a brief summary of what we know about the history of the universe.

Inflation

From $t = 0$ s to $t = 10^{-43}$ s, the average temperature of the universe was around 10^{32} K. At that time it was so hot that quarks and leptons could easily be converted into each other. At about $t = 10^{-36}$ s, the temperature of the universe had fallen to about 10^{28} K. For the first time, the universe became "cold" enough so that quarks and leptons became distinguishable particles. This caused a fundamental change in the structure of the universe, resulting in an extremely rapid exponential expansion known as cosmic **inflation**. During the time interval $t = 10^{-36}$ s to $t = 10^{-32}$ s, the volume of the universe increased by a factor of 10^{26} and then settled into a more gradual expansion. During inflation, small fluctuations in the density of the universe decreased. Areas where the density was slightly above average would later act as the seeds of galaxy formation. Even with these fluctuations, the density of the universe at a particular point differed from the average by only $1/1000$ of 1%. The universe was an extremely hot plasma of quarks, leptons, and interaction mediator particles.

From $t = 10^{-32}$ s to $t = 10^{-12}$ s, the average temperature was still so high that the random thermal motion of particles was at ultra-relativistic speeds (very close to light speed), and particles and antiparticles were continuously created and annihilated in collisions. However, for some reason (there are multiple explanations for this), the annihilations led to a small excess of quarks over antiquarks and leptons over antileptons. Understanding the details of these processes is an important goal of current research in physics.

Nucleosynthesis

At about $t = 10^{-6}$ s, the universe had cooled enough that quarks and gluons were able to combine to form baryons such as protons and neutrons. Because there was an excess of quarks over antiquarks, an excess of baryons over antibaryons was produced. The temperature was no longer high enough for particle collisions to create proton and antiproton pairs or neutron and antineutron pairs. Therefore, the baryons and antibaryons annihilated each other, destroying all antibaryons and leaving a small number of baryons. Only one in 10 billion protons survived this annihilation. These are the protons that we find in the universe today. After this annihilation, temperatures became low enough that the thermal motion of particles was no longer relativistic.

By a few minutes after the Big Bang, the temperature had dropped to about a billion degrees K, and the average density of the universe was close to the density of air at sea level on Earth today. For the first time, protons and neutrons were able to combine to form the simplest nuclei: deuterium (a heavy isotope of hydrogen with one proton and one neutron), helium, and trace amounts of lithium. This process is known as **Big Bang nucleosynthesis**. Some protons combined into helium nuclei, but most remained free as hydrogen nuclei. The ratio of hydrogen to helium nuclei was roughly 12 to 1.

Atoms, stars, galaxies

After 370,000 years the universe became cold enough for electrons to finally combine with nuclei to form the first neutral atoms. When this happened, the universe became transparent to electromagnetic radiation. Prior to this, the universe was a plasma, an ionized gas of nuclei and electrons. Plasmas do not allow photons to travel freely. Now that the universe was transparent, the photons that were produced when the neutral atoms formed were able to travel freely. These photons are still present in the universe today in the form of the **cosmic microwave background** (CMB), the afterglow of a process that took place more than 13 billion years ago. Due to the expansion of the universe, these photons have been red-shifted so that they have an effective temperature today of about 2.7 K. This can be thought of as the ambient temperature of the present universe.

The discovery of the CMB by Arno Penzias and Robert Wilson in 1965 was one of the most significant pieces of supporting evidence for the Big Bang model, as it had predicted this radiation. The model also made predictions about the relative abundances of hydrogen, helium, and lithium that resulted from nucleosynthesis. These predictions are consistent with experiments as well.

When the universe had cooled enough, the gravitational interaction became the dominant driver of its further evolution. Density fluctuations caused some regions to begin contracting. This process led to the formation of the first galaxies and stars just 500,000 years after the Big Bang. These early stars went through their life cycles, some ending in a violent collapse and explosion known as a **supernova**. Through nuclear fusion processes, heavier elements such as carbon, oxygen, iron, and gold were produced in supernovae and released into space to become parts of planets such as our own.

The present matter on Earth consists of quarks that were part of the early universe. These quarks combined to form protons and neutrons in the early universe, which then formed complex nuclei and atoms during the life cycles of stars. Thus, our bodies really are composed of matter that was created near the dawn of time and processed in stellar explosions before becoming part of us.

In the latter part of the 20th century two serious problems with this picture of the universe arose that have yet to be resolved. First, galaxies aren't nearly massive enough to explain the way stars move within them. Second, the universe is speeding up in its expansion rather than slowing down, as the gravitational interaction would predict. What possible explanations have physicists devised to explain these unexpected experimental results?

REVIEW QUESTION 30.4 Explain how the Big Bang model predicts the existence of the cosmic microwave background (CMB).

30.5 Dark matter and dark energy

Earlier (in Chapter 5), we learned that Earth's speed around the Sun is determined by the radius of Earth's orbit and the mass of the Sun. In a similar way, the solar system's speed around the Milky Way galaxy is determined by the radius of the solar system's orbit and the total mass of everything that lies between that orbit and the center of the galaxy (Figure 30.13).

When astronomers measure the mass of all the stars and gas that they can see, however, they find that the total mass is only about 5% of the mass needed to account for the speed of the solar system around the center of the galaxy. The pioneer of this work was American astronomer Vera Cooper Rubin. In the early 1980s, she found the discrepancy between the predicted rotational motion of galaxies and the observed motion by studying graphs of the orbital speeds of stars in a galaxy versus their distance to the galactic center. The same pattern was found for the motion of galaxies relative to each other; there is far too little visible mass to account for the galaxies' motion. Apparently, the universe is "missing" about 95% of the mass needed to account for the observed motion of stars and galaxies. How can this contradiction be resolved?

Dark matter

The first evidence of the problem came in 1933 when astrophysicist Fritz Zwicky looked at the galaxies in the Coma cluster. He found that the galaxies at the edge of the cluster were traveling much too fast to remain part of the cluster. He speculated

FIGURE 30.13 Our galaxy, with the position of the solar system marked.

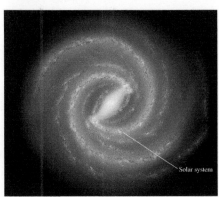

Solar system

that there must be some unseen **dark matter** present in the Coma cluster, but for about 40 years his observation was the only evidence for its existence.

In the 1970s Rubin presented further evidence. She looked at individual galaxies and found that stars near the edge of a galaxy were also traveling too fast to remain part of the galaxy. It was at this point that the dark matter explanation started to become more widely accepted.

Scientists suggested a new form of unseen matter and devised experiments to detect it as directly as possible. But as of 2017, direct detection of dark matter has not been accomplished. Astronomers are actually more certain about what this unseen dark matter is not than about what it is. First, it does not emit photons or otherwise participate in the electromagnetic interaction (this is why it is called "dark"). As a result, dark matter cannot be some sort of dark cloud of protons or gaseous atoms, because these could be detected by the scattering of radiation passing through them. Several hypotheses have been proposed, however, for what this dark matter might be. Two of these go by the names MACHOs (massive compact halo objects) and WIMPs (weakly interacting massive particles).

MACHOs (massive compact halo objects) MACHOs could be black holes, neutron stars, or brown dwarfs (objects that were not massive enough to achieve nuclear fusion and become stars). Astronomers have been trying to detect MACHOs using their gravitational effects on the light from distant objects (a phenomenon called gravitational microlensing). When one of these MACHOs passes in front of a distant object, such as a star or another galaxy, the light bends around the MACHO and is "focused" for a short time interval, making the light appear brighter. The MACHO Project has observed about 15 such lensing events over a span of 6 years of observations. This small number of events translates into MACHOs accounting for at most 20% of the dark matter in our galaxy. There must be another (or an additional) explanation.

WIMPs (weakly interacting massive particles) WIMPs are more exotic in nature. They are elementary particles, but not the quarks and leptons that make up ordinary matter. They are "weakly interacting" because they can pass through ordinary matter with almost no interaction, and light is not absorbed or emitted by them. They are "massive" in that their mass is not zero. Prime candidates for WIMPs include neutrinos, axions, and neutralinos, described next (the latter two are not Standard Model particles and hence require the Standard Model to be extended to accommodate their existence).

Cosmologists have a good idea of how many neutrinos there are in the universe. The Standard Model gives a zero mass to the neutrinos, but the model can be extended to include nonzero masses. Recent experiments have strongly suggested that neutrinos have a rest energy in the range of 0.1 to 1.0 eV, a million times smaller than that of the electron. For many years astronomers hoped that neutrinos could be the dark matter. However, it appears that neutrinos are too light for this. Some **grand unified theories** (theories that combine the strong, weak, and electromagnetic interactions into a single interaction) predict the existence of a so-called "sterile neutrino" that could be even less interactive than Standard Model neutrinos and also far more massive. Physicists do not know how to detect such a particle, but if it exists in sufficient abundance it could account for dark matter.

Physicists originally proposed the existence of **axions** to explain a mystery about the properties of the neutron. These proposed axions have very small mass (less than that of the electron), no electric charge, and very little interaction with Standard Model particles. They would have been produced abundantly in the Big Bang. Current searches for axions include Earth-based laboratory experiments and astronomical searches in the halo of our galaxy and in the Sun. However, axions have never been experimentally observed.

The **neutralino** is an example of another type of hypothetical particle known as a **superpartner**. Several outstanding problems in theoretical physics are understood better by suggesting the existence of what is known as **supersymmetry**. Supersymmetry is an extension of the Standard Model that effectively doubles the number of elementary particles, gives insight into the cosmological constant problem (to be described shortly), allows for a more precise understanding of the unification of interactions in grand unified theories, and gives a potential candidate for dark matter. None of the superpartners have ever been detected, but their detection is one of the primary design goals of the Large Hadron Collider.

Supersymmetry predicts that the lightest superpartner is stable and very weakly interacting. There are many versions of supersymmetry, but in some of them the lightest superpartner is the neutralino, predicted to have a mass of 100 to 1000 times the mass of a proton. Astronomers and physicists hope to detect neutralinos by using underground detectors, searching the universe for signs of their interactions, or producing them in particle accelerators.

Because none of the particles suggested as a solution to the dark matter problem have been detected, the mystery of the missing matter of the universe is still largely unsolved.

Although the dark matter problem has been around since the 1930s, one idea that seemed irrefutable through the 20th century was that the gravitational interaction between all the massive objects in the universe would gradually slow the expansion rate of the universe. Then in 1998, two independent experiments using the Hubble Space Telescope produced observations of very distant supernovae that showed that the universe is currently expanding more rapidly than it was 5–10 billion years ago. The expansion of the universe has not been slowing down; it has been speeding up! No one expected this, and no one at the time knew how to explain it. The 2011 Nobel Prize in physics was awarded to Saul Perlmutter, Brian P. Schmidt, and Adam G. Riess for the discovery of the accelerating universe. This led to extensions of the Big Bang model that could potentially explain this accelerated expansion. The most widely accepted of these is known as **dark energy**.

Dark energy

Physicists came up with many ideas to explain the accelerating expansion of the universe. Here are a few:

- A discarded feature of Einstein's general theory of relativity (our current best model of the gravitational interaction) known as the **cosmological constant**
- The existence of a strange kind of energy-fluid that fills space and has a repulsive gravitational effect
- A modified version of general relativity that includes a new kind of field that creates this cosmic acceleration

Let's look at each of these ideas.

The cosmological constant model An idea introduced and discarded by Einstein 80 years ago has been resurrected to explain recent observations. The cosmological constant is a term that Einstein included in general relativity to allow for a static universe. At the time he constructed general relativity, there was no experimental evidence for the Big Bang or the expansion of the universe. The dominant model of the universe was known as the **steady state model**, which asserted that the universe essentially didn't change in any major way as time passed. But general relativity predicted that a static universe was unstable. Einstein introduced the cosmological constant into general relativity in an attempt to allow the theory to accommodate a steady state universe.

Einstein could have predicted the expansion of the universe 10 years before it was discovered, but even he couldn't accept such a radical idea. It is said that he considered this his greatest blunder. However, because the cosmological constant introduces a repulsive effect into the equation to balance the attractive effect of gravity, it can also be used to describe the accelerated expansion.

The dark energy model The second idea is actually a suggestion about what the cosmological constant term means physically. It seems to represent a type of dark energy that is present at every point in space with equal density. Even as the universe expands, this density does not decrease because it is a property of space itself. What's even stranger is that this energy has a negative pressure. In general relativity this produces a gravitationally repulsive effect on space.

As the universe expands, the matter density decreases, but the dark energy density remains constant. This means that as time passes, the attractive gravitational effect of the matter decreases while the repulsive gravitational effect of the dark energy remains the same. In other words, as time goes on, the repulsion gets stronger relative to the attraction. This is precisely what astronomers have observed—the expansion is more rapid today than it was in the past.

Modified general relativity The third idea is in some sense the least radical. Just as general relativity is an improvement on Newton's view and is able to make more accurate predictions, perhaps the gravitational interaction theory can be revised so that it would make even better predictions than general relativity. The challenge has been to modify the theory in such a way that it doesn't contradict experiments that have already been done. Thus far physicists have been unsuccessful in doing this.

Mysteries Of these three ideas, dark energy has become the favored one. Various versions have been set forth, and the simplest of them represent the dark energy mathematically in general relativity as a cosmological constant. Dark energy models have a problem, though, and it comes from a basic feature of the combination of relativity and quantum mechanics.

This combined model predicts that each elementary particle in the Standard Model is actually an excitation of an associated quantum field, similar to how the photon is one quantum of excitation of the electromagnetic field. Each of these quantum fields has a certain minimum energy, called its **zero point energy**. In many dark energy models, the dark energy is the sum of the zero point energies of all the quantum fields in the universe. But when physicists make estimates of what value these models predict for the cosmological constant, they get a result that is 10^{120} times the observed value. It's been said that this is the largest disagreement between prediction and experiment in all of science. This so-called **cosmological constant problem** is a major unsolved problem in physics.

Supersymmetry, which predicts a doubling of all the types of elementary particles, suggests a possible solution. Each of these superpartner particles also has an associated quantum field. But the zero point energies of each of these additional quantum fields contribute oppositely to the dark energy compared to their nonsuper counterparts. This means the total dark energy density would be zero. That is not consistent with observations—we do observe an accelerated expansion—but the dark energy density is calculated to be very small. That's a much better prediction than 10^{120} times the observed value! None of the superpartner particles have been observed, however, which means that supersymmetry is not present in its full form in this universe. One of the goals of the Large Hadron Collider is to produce some of these superpartner particles and see whether supersymmetry can resolve the cosmological constant problem.

Explaining the accelerated expansion of the universe is an even greater puzzle than explaining dark matter. We know how much dark energy there is because we know how it affects the universe's expansion. Other than that, it is a complete mystery. It turns out that roughly 73% of the total energy of the universe is dark energy. The dark matter's rest energy makes up about 23%. The rest—everything on Earth, everything ever observed with all of our instruments, all normal matter—adds up to ~4% of the energy in the universe (see **Figure 30.14**). We understand only a few percent of what comprises the universe we live in. That is a humbling but highly motivating realization.

FIGURE 30.14 The proportion of matter, dark matter, and dark energy in the universe.

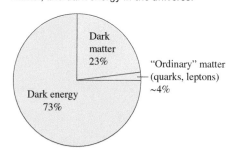

TIP Dark matter and dark energy are ideas invented by physicists to explain patterns observed in nature. Because neither of these has yet been detected, it is possible that one or both of these ideas could be disproven.

REVIEW QUESTION 30.5 How do we know that much more mass exists in the Milky Way galaxy than is present in the galaxy's stars and interstellar gas?

30.6 Is our pursuit of knowledge worthwhile?

In this textbook, we have been building models that describe the behavior of only 4% of the contents of our universe. The nature of the remaining 96% of our universe currently remains an unsolved problem. Is it worthwhile trying to learn about these other mysterious components of our universe?

In the late 1800s, the prime minister of England asked Michael Faraday what use there was for his idea of electromagnetic induction. Faraday could not say. But today we have electric power generators, microphones, credit card readers, electric guitar pickups, and electromagnetic braking systems for hybrid vehicles. All are based on electromagnetic induction.

In 1897, J. J. Thomson discovered and characterized the electron—he could not have imagined how this understanding would lead to the computing, communication, and entertainment devices that pervade our everyday lives.

In the 1930s, physicists at MIT were studying microwaves. What use would they have? Many historians believe that microwave radar saved England in World War II. Today, microwaves, a form of nonionizing radiation, are involved in satellite communications and GPS, and we use microwaves to cook our food.

Will our eventual knowledge of the other 96% of the universe someday make people's lives better? It's impossible to say for sure, but history suggests that it very likely will. Perhaps it will lead to new sustainable energy sources, the ability to easily travel to other planets, or ways to protect us from cosmic events that potentially threaten life on Earth. The most amazing future applications of understanding our universe are the ones no one has yet devised.

Summary

Antiparticles Each particle has an antiparticle with identical mass but opposite electric charge, color, etc. Some particles (such as the photon) are their own antiparticles. (Section 30.1)

Fundamental interactions All interactions between objects are gravitational, electromagnetic, weak, or strong. All other interactions are nonfundamental and can be understood in terms of these. (Section 30.2)

Elementary particles Every known particle is a quark (up, down, etc.), a lepton (electron, neutrino, etc.), or an interaction mediator (photon, gluon, etc.)

Standard Model The elementary particles together with the four fundamental interactions and the Higgs particle comprise the Standard Model of particle physics. (Section 30.3)

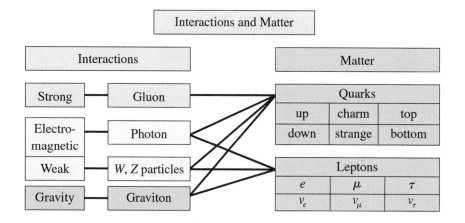

Cosmology The universe rapidly expanded from a hot, dense state 13.7 billion years ago, an event called the Big Bang. As it expanded and cooled, nuclei formed, followed by atoms, stars, and galaxies. (Section 30.4)

Dark matter and dark energy Most of the energy in the universe is thought to consist of two mysterious forms. Dark matter interacts only gravitationally and clumps around galaxies. Most of the mass of galaxies is proposed to be dark matter. Dark energy is thought to be spread uniformly throughout the universe and is responsible for its accelerated expansion. (Section 30.5)

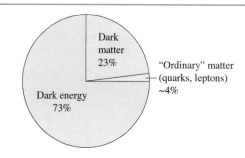

Questions

Multiple Choice Questions

1. What is the composition of all baryons? (a) Three different color quarks (b) Three quarks of any color (c) Two quarks of any color (d) Three gluons (e) Two quarks of complementary color
2. What is the composition of all mesons? (a) Three different color quarks (b) Three quarks of any color (c) Two quarks of any color (d) Three gluons (e) Two quarks of complementary color
3. What is the weak interaction mediated by? (a) Gluons (b) The W^+, W^-, or Z^0 particles (c) Photons (d) Gravitons (e) Mesons
4. What is the strong interaction mediated by? (a) Gluons (b) The W^+, W^-, or Z^0 particles (c) Photons (d) Gravitons (e) Mesons

Conceptual Questions

5. Give three examples of particles that are elementary (as far as is currently known) and three examples of those that are not.
6. What differs between a particle and its antiparticle?

7. Billions of neutrinos continually pass through each square centimeter of your body each second. Why don't you notice them?
8. How do we know that protons are very stable particles?
9. Free neutrons (those that are not part of a nucleus) decay into protons plus other particles with a half-life of about 10 min. Free protons, however, do not decay into neutrons plus other particles. Explain why.
10. What particle(s) could potentially not have an antiparticle or could be said to be its own antiparticle? Explain your reasoning.
11. Give an example of a long-range interaction and a short-range interaction. What is the difference between the mechanisms for each?
12. What are the components of the Standard Model?
13. Why do scientists need particle accelerators to discover new particles and to study the properties of known particles?
14. In 1930 the proton and neutron were thought to be elementary particles. We now know this is not true. How could physicists have been mistaken?

Problems

Below, **BIO** indicates a problem with a biological or medical focus. Asterisks indicate the level of difficulty of the problem. Problems with no * are considered to be the least difficult. A single * marks moderately difficult problems. Two ** indicate more difficult problems.

30.1 Antiparticles

1. Explain Dirac's reasoning for the existence of antielectrons.
2. How do scientists know that antiparticles exist in nature?
3. **BIO** Explain how positron emission tomography (PET) works.
4. A proton and an antiproton, both with negligible kinetic energy, annihilate each other to produce two photons. Determine the energy in electron volts of each photon and the frequency and wavelength of each photon.
5. An isolated slow-moving electron and positron annihilate each other. Why are two photons produced instead of just one?
6. Use Newtonian circular motion concepts to show that the radius r of the circle in which a charged particle spirals while moving perpendicular to a magnetic field is proportional to the particle's speed v.
7. A cosmic ray photon enters a cloud chamber from above. While inside the chamber the photon converts into an electron-positron pair. Draw a sketch of this situation. Label the photon, electron, and positron tracks and indicate the direction of the magnetic field within the cloud chamber.
8. A particle enters a cloud chamber from above traveling at $0.50c$ and leaves a curving track. The magnetic field within the cloud chamber has magnitude 0.120 T and points out of the page. The track curves to the right and has a radius of 0.020 m. Determine everything you can about this particle.

30.2 Fundamental interactions

9. Compare and contrast the fundamental interactions.
10. Even though the electromagnetic interaction is so much stronger than the gravitational interaction, the gravitational interaction clearly is much more relevant in everyday life. Do you agree with this statement? Explain either way.
11. Explain the mechanism behind elementary particle interactions.
12. * Explain how the energy to produce interaction mediators is provided.
13. What is the difference between a real particle and a virtual particle?
14. * Make an analogy between the interactions of elementary particles and some real-life interactions. Indicate which object in one system corresponds to each object in the other. Describe the limitations of your analogy.

30.3 Elementary particles and the Standard Model

15. Why are neutrinos difficult to detect?
16. Describe the evidence for neutrinos having nonzero mass.
17. What criteria do scientists use to classify elementary particles?
18. Compare and contrast leptons and hadrons.
19. How do we know that baryons are made of quarks? What is the experimental evidence that supports this idea?
20. Compare and contrast mesons and baryons.
21. How do quarks interact with each other inside a baryon?
22. * In what way is the interaction of electrically neutral atoms similar to the interactions of color-neutral protons and neutrons?
23. Describe the phenomenon of confinement.
24. What were four important steps in the building of the Standard Model?
25. ** What is the significance of the Higgs particle and the Higgs mechanism?
26. * What major piece of physics is not part of the Standard Model?
27. * Describe several open questions of the Standard Model.

30.4 Cosmology

28. * What is inflation, and what eventually happened as a result of the density fluctuations in the universe?
29. * What is the origin of the elementary particles?
30. * How would the universe be different if the number of particles in it had equaled the number of antiparticles?
31. * What process produced the cosmic microwave background?

32. * Our bodies contain significant amounts of carbon, oxygen, nitrogen, and other "heavy" elements. If only "light" elements such as hydrogen and helium were produced during Big Bang nucleosynthesis, how were the heavier elements produced?

30.5 Dark matter and dark energy

33. * What is the evidence that a large proportion of the mass of the universe is in the form of dark matter? Explain carefully.
34. * Describe as many hypotheses as you can about the nature of dark matter.
35. * What is the experimental evidence for dark energy? Explain carefully.
36. ** Describe as many hypotheses as you can that explain the accelerated expansion of the universe.
37. ** What is supersymmetry, and why is it a useful idea in physics?
38. ** What is the cosmological constant problem? Describe a possible resolution for it.

30.6 Is our pursuit of knowledge worthwhile?

39. What are the potential benefits of continued research in physics?

General Problems

40. * An electron and a positron are traveling directly toward each other at a speed of $0.90c$ (90% of light speed) with respect to the lab in which the experiment is being performed. The electron and positron collide, annihilate, and produce a pair of photons. Determine the wavelength of each photon. If you think they will have the same wavelength, explain why.
41. * In order to discover the W^+ and W^- particles, a large particle accelerator at CERN collided protons and antiprotons together at extremely high speeds. What is the minimum speed the proton and antiproton would need to produce a $W^\pm$ pair?

Reading Passage Problems

The solar neutrino problem Our Sun continually converts hydrogen into helium through nuclear fusion. The chain of reactions that occurs can be combined into the single reaction shown here:

$$4{}_1^1\text{H} \rightarrow {}_2^4\text{He} + 2e^+ + 2\nu_e$$

Four hydrogen nuclei fuse into one helium nucleus. In the process two positrons and two electron neutrinos are produced. These positrons interact immediately with electrons present in the Sun and annihilate with them, producing gamma ray photons. Because the Sun is a plasma, these photons interact strongly with it and push outward, supporting the Sun against gravitational collapse. The neutrinos, however, have only an extremely small chance of interacting with matter and therefore stream out of the Sun's core unhindered, making their way to Earth and beyond. So many neutrinos are produced in the fusion reactions that ~100 billion of them pass through each square centimeter of your skin each second.

Solar neutrinos act as a "window" into the fusion reactions occurring in the Sun that normally would be obscured from our view. For example, if we can measure the rate at which solar neutrinos reach Earth we can estimate the nuclear reaction rate in the Sun. We also have an independent way to measure the nuclear reaction rate—by measuring the intensity of the Sun's radiation. The higher the intensity of the Sun's radiation, the greater the nuclear reaction rate at its core. These two methods should produce consistent results. But they don't. The amount of electron neutrinos detected is less than half the amount expected. This is the **solar neutrino problem**.

The resolution of the problem comes from suggesting that some of the electron neutrinos convert into other types of neutrinos (muon or tau) on their way to Earth. This phenomenon is called **neutrino oscillation**. For this to be possible, the neutrinos cannot have zero mass. The experiment that first detected solar neutrinos could not detect muon or tau neutrinos. In 2001 a new detector was used that could detect all three types of neutrinos. About two-thirds of the electron neutrinos had in fact oscillated into muon and tau neutrinos. This resolved the solar neutrino problem.

42. Why don't you notice the solar neutrinos passing through your body?
    (a) The Sun is very far away so the neutrinos have very little energy by the time they get to Earth.
    (b) Nearly all the neutrinos oscillate into other types that can pass easily through your body.
    (c) Solar neutrinos only participate in nuclear fusion reactions and since nuclear fusion doesn't happen in your body you don't notice the neutrinos.
    (d) Neutrinos only interact very weakly with atoms.
    (e) Only a very small number of them pass through your body each second.

43. In 1987 a supernova was detected in the Large Magellanic Cloud. Neutrinos coming from the supernova were detected, but there were fewer than expected. What might be the reason for that?
    (a) The supernova actually occurred farther from Earth than astronomers thought.
    (b) Some of the neutrinos were converted into other types of neutrinos as they traveled to Earth.
    (c) The model used to predict the number of neutrinos produced in the supernova makes some unreasonable assumptions.
    (d) The detector used was not as efficient as the designers thought.
    (e) All of these are possible explanations.

44. Originally a different way to resolve the solar neutrino problem was suggested—the nuclear reaction rate wasn't as high as astrophysicists thought. What would have to be true for this suggestion to be reasonable?
    (a) The mass of the Sun would have to be significantly different than thought.
    (b) The brightness of the Sun would have to be significantly different than thought.
    (c) The size of the Sun would have to be significantly different than thought.
    (d) The orbits of the planets would have to be significantly different than thought.
    (e) Both (b) and (c) are correct.

APPENDIX A
Mathematics Review

A study of physics at the level of this textbook requires some basic math skills. The relevant math topics are summarized in this appendix. We strongly recommend that you review this material and become comfortable with it as quickly as possible so that, during your physics course, you can focus on the physics concepts and procedures that are being introduced without being distracted by unfamiliarity with the math that is being used.

A.1 Exponents

Exponents are used frequently in physics. When we write 3^4, the superscript 4 is called an **exponent** and the **base number** 3 is said to be raised to the fourth power. The quantity 3^4 is equal to $3 \times 3 \times 3 \times 3 = 81$. Algebraic symbols can also be raised to a power—for example, x^4. There are special names for the operation when the exponent is 2 or 3. When the exponent is 2, we say that the quantity is **squared**; thus, x^2 is x squared. When the exponent is 3, the quantity is **cubed**; hence, x^3 is x cubed.

Note that $x^1 = x$, and the exponent is typically not written. Any quantity raised to the zero power is defined to be unity (that is, 1). Negative exponents are used for reciprocals: $x^{-4} = 1/x^4$. The exponent can also be a fraction, as in $x^{1/4}$. The exponent $\frac{1}{2}$ is called a **square root**, and the exponent $\frac{1}{3}$ is called a **cube root**.

For example, $\sqrt{6}$ can also be written as $6^{1/2}$. Most calculators have special keys for calculating numbers raised to a power—for example, a key labeled y^x or one labeled x^2.

Exponents obey several simple rules, which follow directly from the meaning of raising a quantity to a power:

1. When two powers of the same quantity are multiplied, the exponents are added:

$$(x^n)(x^m) = x^{n+m}$$

   For example, $(3^2)(3^3) = 3^5 = 243$. To verify this result, note that $3^2 = 9$, $3^3 = 27$, and $(9)(27) = 243$.
   A special case of this rule is $(x^n)(x^{-n}) = x^{n+(-n)} = x^0 = 1$.

2. The product of two different base numbers raised to the same power is the product of the base numbers, raised to that power:

$$(x^n)(y^n) = (xy)^n$$

   For example, $(2^4)(3^4) = 6^4 = 1296$. To verify this result, note that $2^4 = 16$, $3^4 = 81$, and $(16)(81) = 1296$.

3. When a power is raised to another power, the exponents are multiplied:

$$(x^n)^m = x^{nm}$$

   For example, $(2^2)^3 = 2^6 = 64$. To verify this result, note that $2^2 = 4$, so $(2^2)^3 = (4)^3 = 64$.

If the base number is negative, it is helpful to know that $(-x)^n = (-1)^n x^n$, and $(-1)^n$ is $+1$ if n is even and -1 if n is odd.

QUANTITATIVE EXERCISE A.1 **Simplifying an exponential expression**

Simplify the expression $\dfrac{x^3y^{-3}xy^{4/3}}{x^{-4}y^{1/3}(x^2)^3}$ and calculate its numerical value when $x = 6$ and $y = 3$.

Represent mathematically, solve, and evaluate We simplify the expression as follows:

$$\frac{x^3x}{x^{-4}(x^2)^3} = x^3x^1x^4x^{-6} = x^{3+1+4-6} = x^2; \frac{y^{-3}y^{4/3}}{y^{1/3}} = y^{-3+\frac{4}{3}-\frac{1}{3}} = y^{-2}$$

$$\frac{x^3y^{-3}xy^{4/3}}{x^{-4}y^{1/3}(x^2)^3} = x^2y^{-2} = x^2\left(\frac{1}{y}\right)^2 = \left(\frac{x}{y}\right)^2$$

For $x = 6$ and $y = 3$, $\left(\dfrac{x}{y}\right)^2 = \left(\dfrac{6}{3}\right)^2 = 4$.

If we evaluate the original expression directly, we obtain

$$\frac{x^3y^{-3}xy^{4/3}}{x^{-4}y^{1/3}(x^2)^3} = \frac{(6^3)(3^{-3})(6)(3^{4/3})}{(6^{-4})(3^{1/3})([6^2]^3)}$$

$$= \frac{(216)(1/27)(6)(4.33)}{(1/1296)(1.44)(46,656)} = 4.00$$

which checks.

This example demonstrates the usefulness of the rules for manipulating exponents.

QUANTITATIVE EXERCISE A.2 **Solving an exponential expression for the base number**

If $x^4 = 81$, what is x?

Represent mathematically, solve, and evaluate We raise each side of the equation to the $\frac{1}{4}$ power: $(x^4)^{1/4} = (81)^{1/4}$. $(x^4)^{1/4} = x^1 = x$, so $x = (81)^{1/4}$ and $x = +3$ or $x = -3$. Either of these values of x gives $x^4 = 81$.

Notice that we raised *both sides* of the equation to the $\frac{1}{4}$ power. As will be explained in Section A.3, an operation performed on both sides of an equation does not affect the equation's validity.

A.2 Scientific notation and powers of 10

In physics, we frequently encounter very large and very small numbers, and it is important to use the proper number of significant digits when expressing a quantity. We can deal with both these issues by using **scientific notation**, in which a quantity is expressed as a decimal number with one digit to the left of the decimal point, multiplied by the appropriate power of 10. If the power of 10 is positive, it is the number of places the decimal point is moved to the right to obtain the fully written-out number. For example, $6.3 \times 10^4 = 63,000$. If the power of 10 is negative, it is the number of places the decimal point is moved to the left to obtain the fully written-out number. For example, $6.56 \times 10^{-3} = 0.00656$. In going from 6.56 to 0.00656, the decimal point is moved three places to the left, so 10^{-3} is the correct power of 10 to use when the number is written in scientific notation. Most calculators have keys for expressing a number in either decimal (floating-point) or scientific notation.

When two numbers written in scientific notation are multiplied (or divided), multiply (or divide) the decimal parts to get the decimal part of the result, and add (or subtract) the powers of 10 to get the power-of-10 portion of the result. You may have to adjust the location of the decimal point in the answer to express it in scientific notation. For example,

$$(8.43 \times 10^8)(2.21 \times 10^{-5}) = (8.43 \times 2.21)(10^8 \times 10^{-5})$$
$$= (18.6) \times (10^{8-5}) = 18.6 \times 10^3$$
$$= 1.86 \times 10^4$$

Similarly,

$$\frac{5.6 \times 10^{-3}}{2.8 \times 10^{-6}} = \left(\frac{5.6}{2.8}\right) \times \left(\frac{10^{-3}}{10^{-6}}\right) = 2.0 \times 10^{-3-(-6)} = 2.0 \times 10^3$$

Your calculator can handle these operations for you automatically, but it is important for you to develop good "number sense" for scientific notation manipulations.

A.3 Algebra

Solving equations

Equations written in terms of symbols that represent quantities are frequently used in physics. An **equation** consists of an equals sign and quantities to its left and to its right. Every equation tells us that the combination of quantities on the left of the equals sign has the same value as (is equal to) the combination on the right of the equals sign. For example, the equation $y + 4 = x^2 + 8$ tells us that $y + 4$ has the same value as $x^2 + 8$. If $x = 3$, then the equation $y + 4 = x^2 + 8$ says that $y = 13$.

Often, one of the symbols in an equation is considered to be the *unknown*, and we wish to solve for the unknown in terms of the other symbols or quantities. For example, we might wish to solve the equation $2x^2 + 4 = 22$ for the value of x. Or we might wish to solve the equation $x = v_0 t + \frac{1}{2} at^2$ for the unknown a in terms of x, t, and v_0.

An equation can be solved by using the following rule:

> **An equation remains true if any operation performed on one side of the equation is also performed on the other side.** The operations could be (a) adding or subtracting a number or symbol, (b) multiplying or dividing by a number or symbol, or (c) raising each side of the equation to the same power.

QUANTITATIVE EXERCISE A.3 ▸ **Solving a numerical equation**

Solve the equation $2x^2 + 4 = 22$ for x.

Represent mathematically, solve, and evaluate First we subtract 4 from both sides. This gives $2x^2 = 18$. Then we divide both sides by 2 to get $x^2 = 9$. Finally, we raise both sides of the equation to the $\frac{1}{2}$ power. (In other words, we take the square root of both sides of the equation.) This gives $x = \pm \sqrt{9} = \pm 3$. That is, $x = +3$ or $x = -3$. We can verify our answers by substituting our result back into the original equation: $2x^2 + 4 = 2(\pm 3)^2 + 4 = 2(9) + 4 = 18 + 4 = 22$,   so $x = \pm 3$ does satisfy the equation.

Notice that a square root always has *two* possible values, one positive and one negative. For instance, $\sqrt{4} = \pm 2$, because $(2)(2) = 4$ and $(-2)(-2) = 4$. Your calculator will give you only a positive root; it's up to you to remember that there are actually two. Both roots are correct mathematically, but in a physics problem only one may represent the answer. For instance, if you can get dressed in $\sqrt{4}$ minutes, the only physically meaningful root is 2 minutes!

QUANTITATIVE EXERCISE A.4 ▸ **Solving a symbolic equation**

Solve the equation $x = v_0 t + \frac{1}{2} at^2$ for a.

Represent mathematically, solve, and evaluate We subtract $v_0 t$ from both sides. This gives $x - v_0 t = \frac{1}{2} at^2$. Now we multiply both sides by 2 and divide both sides by t^2, giving $a = \dfrac{2(x - v_0 t)}{t^2}$.

As we've indicated, it makes no difference whether the quantities in an equation are represented by variables (such as x, v, and t) or by numerical values.

The quadratic formula

Using the methods of the previous subsection, we can easily solve the equation $ax^2 + c = 0$ for x:

$$x = \pm \sqrt{\frac{-c}{a}}$$

For example, if $a = 2$ and $c = -8$, the equation is $2x^2 - 8 = 0$ and the solution is

$$x = \pm\sqrt{\frac{-(-8)}{2}} = \pm\sqrt{4} = \pm 2$$

The equation $ax^2 + bx = 0$ is also easily solved by factoring out an x on the left side of the equation, giving $x(ax + b) = 0$. (To *factor out* a quantity means to isolate it so that the rest of the expression is either multiplied or divided by that quantity.) The equation $x(ax + b) = 0$ is true (that is, the left side equals zero) if either $x = 0$ or $x = -\frac{b}{a}$. Those are the equation's two solutions. For example, if $a = 2$ and $b = 8$, the equation is $2x^2 + 8x = 0$ and the solutions are $x = 0$ and $x = -\frac{8}{2} = -4$.

But if the equation is in the form $ax^2 + bx + c = 0$, with a, b, and c all nonzero, we cannot use our standard methods to solve for x. Such an equation is called a **quadratic equation**, and its solutions are expressed by the **quadratic formula**:

Quadratic formula For a quadratic equation in the form $ax^2 + bx + c = 0$, where a, b, and c are real numbers and $a \neq 0$, the solutions are given by the quadratic formula:

$$x = \frac{-b \pm \sqrt{b^2 - 4ac}}{2a}$$

In general, a quadratic equation has two roots (solutions). But if $b^2 - 4ac = 0$, then the two roots are equal. By contrast, if $b^2 - 4ac$ is negative, then both roots are complex numbers and cannot represent physical quantities. In such a case, the original quadratic equation has mathematical solutions, but no physical solutions.

QUANTITATIVE EXERCISE A.5 ⟩ **Solving a quadratic equation**

Find the values of x that satisfy the equation $2x^2 - 2x = 24$.

Represent mathematically, solve, and evaluate First we write the equation in the standard form $ax^2 + bx + c = 0$: $2x^2 - 2x - 24 = 0$. Then $a = 2$, $b = -2$, and $c = -24$. Next, the quadratic formula gives the two roots as

$$x = \frac{-(-2) \pm \sqrt{(-2)^2 - 4(2)(-24)}}{(2)(2)}$$

$$= \frac{+2 \pm \sqrt{4 + 192}}{4} = \frac{2 \pm 14}{4}$$

so $x = 4$ or $x = -3$. If x represents a physical quantity that takes only nonnegative values, then the negative root $x = -3$ is nonphysical and is discarded.

As we've mentioned, when an equation has more than one mathematical solution or root, it's up to *you* to decide whether one or the other or both represent the true physical answer. (If neither solution seems physically plausible, you should review your work.)

Simultaneous equations

If a problem has two unknowns—for example, x and y—then it takes two independent equations in x and y (that is, two equations for x and y, where one equation is not simply a multiple of the other) to determine their values uniquely. Such equations are called **simultaneous equations** (because you solve them together). A typical procedure is to solve one equation for x in terms of y and then substitute the result into the second equation to obtain an equation in which y is the only unknown. You then solve this equation for y and use the value of y in either of the original equations in order to solve for x. In general, to solve for n unknowns, we must have n independent equations.

| **QUANTITATIVE EXERCISE A.6** | Solving two equations in two unknowns |

Solve the following pair of equations for x and y:

$$x + 4y = 14$$
$$3x - 5y = -9$$

Represent mathematically, solve, and evaluate The first equation gives $x = 14 - 4y$. Substituting this for x in the second equation yields, successively, $3(14 - 4y) - 5y = -9$, $42 - 12y - 5y = -9$, and $-17y = -51$. Thus, $y = \frac{-51}{-17} = 3$. Then $x = 14 - 4y = 14 - 12 = 2$. We can verify that $x = 2$, $y = 3$ satisfies both equations.

An alternative approach is to multiply the first equation by -3, yielding $-3x - 12y = -42$. Adding this to the second equation gives, successively, $3x - 5y + (-3x) + (-12y) = -9 + (-42)$, $-17y = -51$, and $y = 3$, which agrees with our previous result.

As shown by the alternative approach, simultaneous equations can be solved in more than one way. The basic method we describe is easy to keep straight; other methods may be quicker, but may require more insight or forethought. Use the method you're comfortable with.

A pair of equations in which all quantities are symbols can be combined to eliminate one of the common unknowns.

| **QUANTITATIVE EXERCISE A.7** | Solving two symbolic equations in two unknowns |

Use the equations $v = v_0 + at$ and $x = v_0 t + \frac{1}{2}at^2$ to obtain an equation for x that does not contain a.

Represent mathematically, solve, and evaluate We solve the first equation for a:

$$a = \frac{v - v_0}{t}$$

We substitute this expression into the second equation:

$$x = v_0 t + \frac{1}{2}\left(\frac{v - v_0}{t}\right)t^2 = v_0 t + \frac{1}{2}vt - \frac{1}{2}v_0 t$$
$$= \frac{1}{2}v_0 t + \frac{1}{2}vt = \left(\frac{v_0 + v}{2}\right)t$$

When you solve a physics problem, it's often best to work with symbols for all but the final step of the problem. Once you've arrived at the final equation, you can plug in numerical values and solve for an answer.

A.4 Logarithmic and exponential functions

The base-10 logarithm, or **common logarithm** (log), of a number y is the power to which 10 must be raised to obtain y: $y = 10^{\log y}$. For example, $1000 = 10^3$, so $\log(1000) = 3$; you must raise 10 to the power 3 to obtain 1000. Most calculators have a key for calculating the log of a number.

Sometimes we are given the log of a number and are asked to find the number. That is, if $\log y = x$ and x is given, what is y? To solve for y, write an equation in which 10 is raised to the power equal to either side of the original equation: $10^{\log y} = 10^x$. But $10^{\log y} = y$, so $y = 10^x$. In this case, y is called the **antilog** of x. For example, if $\log y = -2.0$, then $y = 10^{-2.0} = 1.0 \times 10^{-2.0} = 0.010$.

The log of a number is positive if the number is greater than 1. The log of a number is negative if the number is less than 1, but greater than zero. The log of zero or of a negative number is not defined, and $\log 1 = 0$.

Another base that occurs frequently in physics is the quantity $e = 2.718....$ The **natural logarithm** (ln) of a number y is the power to which e must be raised to obtain y: $y = e^{\ln y}$. If $x = \ln y$, then $y = e^x$. Most calculators have keys for $\ln x$ and for e^x. For example, $\ln 10.0 = 2.30$, and if $\ln x = 3.00$, then $x = e^{3.00} = 20.1$. Note that $\ln 1 = 0$.

Logarithms with any choice of base, including base 10 or base e, obey several simple and useful rules:

1. $\log(ab) = \log a + \log b$

2. $\log\left(\dfrac{a}{b}\right) = \log a - \log b$

3. $\log(a^n) = n \log a$

A particular example of the second rule is

$$\log\left(\frac{1}{a}\right) = \log 1 - \log a = -\log a$$

since $\log 1 = 0$.

QUANTITATIVE EXERCISE A.8 ▶ **Solving a logarithmic equation**

If $\frac{1}{2} = e^{-\alpha T}$, solve for T in terms of α.

Represent mathematically, solve, and evaluate We take the natural logarithm of both sides of the equation: $\ln\left(\frac{1}{2}\right) = -\ln 2$ and $\ln\left(e^{-\alpha T}\right) = -\alpha T$. The equation thus becomes $-\alpha T = -\ln 2$, and it follows that $T = \dfrac{\ln 2}{\alpha}$.

The equation $y = e^{\alpha x}$ expresses y in terms of the exponential function $e^{\alpha x}$. The general rules for exponents in Appendix A.1 apply when the base is e, so $e^x e^y = e^{x+y}$, $e^x e^{-x} = e^{x+(-x)} = e^0 = 1$, and $\left(e^x\right)^2 = e^{2x}$.

FIGURE A.1

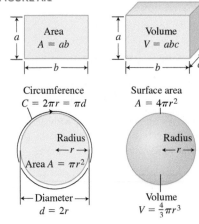

A.5 Areas and volumes

Figure A.1 illustrates the formulas for the areas and volumes of common geometric shapes:

- A rectangle with length a and width b has area $A = ab$.
- A rectangular solid (a box) with length a, width b, and height c has volume $V = abc$.
- A circle with radius r has diameter $d = 2r$, circumference $C = 2\pi r = \pi d$, and area $A = \pi r^2 = \pi d^2/4$.
- A sphere with radius r has surface area $A = 4\pi r^2$ and volume $V = \frac{4}{3}\pi r^3$.
- A cylinder with radius r and height h has volume $V = \pi r^2 h$.

A.6 Plane geometry and trigonometry

Following are some useful facts about angles:

1. Interior angles formed when two straight lines intersect are equal. For example, in **Figure A.2**, the two angles θ and ϕ are equal.
2. When two parallel lines are intersected by a diagonal straight line, the alternate interior angles are equal. For example, in **Figure A.3**, the two angles θ and ϕ are equal.
3. When the sides of one angle are each perpendicular to the corresponding sides of a second angle, then the two angles are equal. For example, in **Figure A.4**, the two angles θ and ϕ are equal.
4. The sum of the angles on one side of a straight line is 180°. In **Figure A.5**, $\theta + \phi = 180°$.
5. The sum of the angles in any triangle is 180°.

FIGURE A.2

Interior angles formed when two straight lines intersect are equal:
$\theta = \phi$

FIGURE A.3

When two parallel lines are intersected by a diagonal straight line, the alternate interior angles are equal:
$\theta = \phi$

FIGURE A.4

When the sides of one angle are each perpendicular to the corresponding sides of a second angle, then the two angles are equal:
$\theta = \phi$

FIGURE A.5

The sum of the angles on one side of a straight line is 180°:
$\theta + \phi = 180°$

Similar triangles

Triangles are **similar** if they have the same shape, but different sizes or orientations. Similar triangles have equal angles and equal ratios of corresponding sides. If the two triangles in **Figure A.6** are similar, then $\theta_1 = \theta_2$, $\phi_1 = \phi_2$, $\gamma_1 = \gamma_2$, and $\dfrac{a_1}{a_2} = \dfrac{b_1}{b_2} = \dfrac{c_1}{c_2}$.

If two similar triangles have the same size, they are said to be **congruent**. If triangles are congruent, one can be rotated to where it can be placed precisely on top of the other.

FIGURE A.6

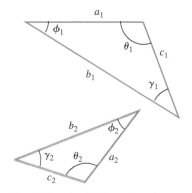

Two similar triangles: Same shape but not necessarily the same size

Right triangles and trig functions

In a right triangle, one angle is 90°. Therefore, the other two acute angles (*acute* means less than 90°) have a sum of 90°. In **Figure A.7**, $\theta + \phi = 90°$. The side opposite the right angle is called the **hypotenuse** (side c in the figure). In a right triangle, the square of the length of the hypotenuse equals the sum of the squares of the lengths of the other two sides. For the triangle in Figure A.7, $c^2 = a^2 + b^2$. This formula is called the **Pythagorean theorem**.

If two right triangles have the same value for one acute angle, then the two triangles are similar and have the same ratio of corresponding sides. This true statement allows us to define the functions **sine**, **cosine**, and **tangent**, which are ratios of a pair of sides. These functions, called **trigonometric functions** or **trig functions**, depend only on one of the angles in the right triangle. For an angle θ, these functions are written $\sin\theta$, $\cos\theta$, and $\tan\theta$.

In terms of the triangle in Figure A.7, the sine, cosine, and tangent of the angle θ are as follows:

FIGURE A.7

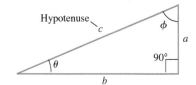

For a right triangle:
$\theta + \phi = 90°$
$c^2 = a^2 + b^2$ (Pythagorean theorem)

$$\sin\theta = \frac{\text{opposite side}}{\text{hypotenuse}} = \frac{a}{c}$$

$$\cos\theta = \frac{\text{adjacent side}}{\text{hypotenuse}} = \frac{b}{c}$$

$$\tan\theta = \frac{\text{opposite side}}{\text{adjacent side}} = \frac{a}{b}$$

Note that $\tan\theta = \dfrac{\sin\theta}{\cos\theta}$. For angle ϕ, $\sin\phi = \dfrac{b}{c}$, $\cos\phi = \dfrac{a}{c}$, and $\tan\phi = \dfrac{b}{a}$.

In physics, angles are expressed in either degrees or radians, where π radians $= 180°$. Most calculators have a key for switching between degrees and radians. Always be sure that your calculator is set to the appropriate angular measure.

Inverse trig functions, denoted, for example, by $\sin^{-1} x$ (or arcsin x), have a value equal to the angle that has the value x for the trig function. For example, $\sin 30° = 0.500$, so $\sin^{-1}(0.500) = \arcsin(0.500) = 30°$. Note that $\sin^{-1} x$ does *not* mean $\dfrac{1}{\sin x}$.

QUANTITATIVE EXERCISE A.9   **Using trigonometry I**

A right triangle has one angle of 30° and one side with length 8.0 cm, as shown in **Figure A.8**. What is the angle ϕ, and what are the lengths x and y of the other two sides of the triangle?

FIGURE A.8

8.0 cm

ϕ y

90°   30°

x

Represent mathematically, solve, and evaluate $\phi + 30° = 90°$, so $\phi = 60°$.

$$\tan 30° = \frac{8.0\ \text{cm}}{x}, \text{ so } x = \frac{8.0\ \text{cm}}{\tan 30°} = 13.9\ \text{cm}$$

To find y, we use the Pythagorean theorem: $y^2 = (8.0\ \text{cm})^2 + (13.9\ \text{cm})^2$, so $y = 16.0\ \text{cm}$.

Or we can say $\sin 30° = 8.0\ \text{cm}/y$, so $y = 8.0\ \text{cm}/\sin 30° = 16.0\ \text{cm}$, which agrees with the previous result.

Notice how we used the Pythagorean theorem in combination with a trig function. You will use these tools constantly in physics, so make sure that you can employ them with confidence.

QUANTITATIVE EXERCISE A.10   **Using trigonometry II**

A right triangle has two sides with lengths as specified in Figure A.9. What is the length x of the third side of the triangle, and what is the angle θ, in degrees?

FIGURE A.9

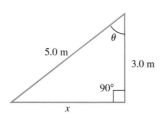

Represent mathematically, solve, and evaluate The Pythagorean theorem applied to this right triangle gives $(3.0 \text{ m})^2 + x^2 = (5.0 \text{ m})^2$, so $x = \sqrt{(5.0 \text{ m})^2 - (3.0 \text{ m})^2} = 4.0 \text{ m}$. (Since x is a length, we take the positive root of the equation.) We also have

$$\cos \theta = \frac{3.0 \text{ m}}{5.0 \text{ m}} = 0.600, \text{ so } \theta = \cos^{-1}(0.600) = 53.1°$$

In this case, we knew the lengths of two sides but none of the acute angles, so we used the Pythagorean theorem first and then an appropriate trig function.

FIGURE A.10

FIGURE A.11

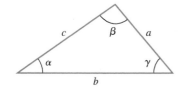

In a right triangle, all angles are in the range from 0° to 90°, and the sine, cosine, and tangent of the angles are all positive. This must be the case, since the trig functions are ratios of lengths. But for other applications, such as finding the components of vectors, calculating the oscillatory motion of a mass on a spring, or describing wave motion, it is useful to define the sine, cosine, and tangent for angles outside that range. Graphs of $\sin \theta$ and $\cos \theta$ are given in Figure A.10. The values of $\sin \theta$ and $\cos \theta$ vary between $+1$ and -1. Each function is periodic, with a period of 360°. Note the range of angles between 0° and 360° for which each function is positive and negative. The two functions $\sin \theta$ and $\cos \theta$ are 90° out of phase (that is, out of step). When one is zero, the other has its maximum magnitude.

For any triangle (see Figure A.11)—in other words, not necessarily a right triangle—the following two relations apply:

1. $\dfrac{\sin \alpha}{a} = \dfrac{\sin \beta}{b} = \dfrac{\sin \gamma}{c}$ (law of sines)

2. $c^2 = a^2 + b^2 - 2ab \cos \gamma$ (law of cosines)

Some of the relations among trig functions are called trig identities. The following table lists only a few, those most useful in introductory physics:

Useful trigonometric identities

$\sin(-\theta) = -\sin(\theta)$ ($\sin \theta$ is an odd function)

$\cos(-\theta) = \cos(\theta)$ ($\cos \theta$ is an even function)

$\sin 2\theta = 2 \sin \theta \cos \theta$

$\cos 2\theta = \cos^2 \theta - \sin^2 \theta = 2 \cos^2 \theta - 1 = 1 - 2 \sin^2 \theta$

$\sin(\theta \pm \phi) = \sin \theta \cos \phi \pm \cos \theta \sin \phi$

$\cos(\theta \pm \phi) = \cos \theta \cos \phi \mp \sin \theta \sin \phi$

$\sin(180° - \theta) = \sin \theta$

$\cos(180° - \theta) = -\cos \theta$

$\sin(90° - \theta) = \cos \theta$

$\cos(90° - \theta) = \sin \theta$

A.7 Examples of operations with vectors

EXAMPLE A.11   **Graphical addition of vectors**

A car travels 200 km west, 100 km south, and finally 150 km at an angle 60° south of east. Determine the net displacement of the car.

Reasoning Use the graphical addition technique with the three displacements drawn tail to head, as shown in Figure A.12a. To find the resultant displacement $\vec{R}$, draw an arrow from the tail of the first displacement to the head of the last (Figure A.12b). We measure the magnitude of the resultant with a ruler and find that its length is 5.2 cm. Since each centimeter represents 50 km, the magnitude of the resultant displacement is $(5.2 \text{ cm})(50 \text{ km/cm}) = 260$ km. Using a protractor, we confirm that the direction of the resultant is 60° south of west.

FIGURE A.12 (a) Three displacement vectors placed tail to head. (b) The resultant displacement.

EXAMPLE A.12   **Components of vectors**

Suppose we have a vector $\vec{N}$ of the magnitude of 10 units that is directed at 20° above the horizontal direction as shown in Figure A.13a. The hypotenuse and the x- and y-scalar components of $\vec{N}$ form a triangle for which:

$$\cos \theta = \frac{\text{adjacent side}}{\text{hypotenuse}} = \frac{N_x}{N} \quad \text{or} \quad N_x = N \cos \theta$$

$$\sin \theta = \frac{\text{opposite side}}{\text{hypotenuse}} = \frac{N_y}{N} \quad \text{or} \quad N_y = N \sin \theta$$

where N is the magnitude of the vector $\vec{N}$. Using the magnitude N and the angle, we can calculate the values of the scalar components of $\vec{N}$:

$$N_x = N \cos \theta = (10 \text{ units}) \cos 20° = 9.4 \text{ units}$$
$$N_y = N \sin \theta = (10 \text{ units}) \sin 20° = 3.4 \text{ units}$$

If the vector pointed in the opposite direction, as in Figure A.13b, the magnitude of the scalar stays the same but the signs are negative:

$$N_x = -N \cos \theta = -9.4 \text{ units}$$
$$N_y = -N \sin \theta = -3.4 \text{ units}$$

The signs depend on the orientation of the scalar components relative to the x- and y-axes.

FIGURE A.13 N_x and N_y can be calculated from known N and θ.

(a)

(b)

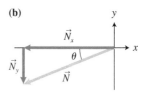

EXAMPLE A.13   Components of vectors

Determine the x- and y-components of each of the force vectors shown in Figure A.14.

Reasoning $F_{1x} = +24$ N   and $F_{1y} = +32$ N; $F_{2x} = -35$ N   and $F_{2y} = +35$ N. Note that the angle between $\vec{F}_3$ and the negative x-axis is $30°$ and not $60°$. Thus, $F_{3x} = -26$ N and $F_{3y} = -15$ N.

FIGURE A.14

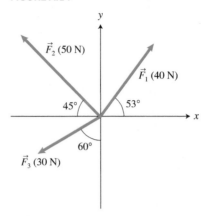

EXAMPLE A.14   Addition using components

Determine the sum of the three forces shown in Figure A.14. You can use the results of the component calculation in Example A.13.

Reasoning The x- and y-components of the resultant force are:

$$R_x = F_{1x} + F_{2x} + F_{3x} = +24 \text{ N} + (-35 \text{ N}) + (-26 \text{ N})$$
$$= -37 \text{ N}$$

$$R_y = F_{1y} + F_{2y} + F_{3y} = +32 \text{ N} + 35 \text{ N} + (-15 \text{ N})$$
$$= +52 \text{ N}$$

The magnitude of the resultant force is:

$$R = \sqrt{(-37 \text{ N})^2 + (52 \text{ N})^2} = 64 \text{ N}$$

The resultant vector is in the second quadrant (it has a negative x-component and a positive y-component) and makes the following angle above the negative x-axis:

$$\theta = \arctan \frac{52 \text{ N}}{37 \text{ N}} = \arctan 1.41 = 55°$$

Thus, the sum of the three forces has a magnitude of 64 N and points $55°$ above the negative x-axis.

Atomic and Nuclear Data

Atomic number (Z)	Element	Symbol	Mass number (A)	Atomic mass (u)	Percent abundance	Decay mode	Half-life $t_{1/2}$
0	(Neutron)	n	1	1.008 665		β^-	10.4 min
1	Hydrogen	H	1	1.007 825	99.985	stable	
	Deuterium	D	2	2.014 102	0.015	stable	
	Tritium	T	3	3.016 049		β^-	12.33 yr
2	Helium	He	3	3.016 029	0.000 1	stable	
			4	4.002 602	99.999 9	stable	
			6	6.018 886		β^-	0.81 s
3	Lithium	Li	6	6.015 121	7.50	stable	
			7	7.016 003	92.50	stable	
			8	8.022 486		β^-	0.84 s
4	Beryllium	Be	7	7.016 928		electron capture (EC)	53.3 days
			9	9.012 174	100	stable	
			10	10.013 534		β^-	1.5×10^6 yr
5	Boron	B	10	10.012 936	19.90	stable	
			11	11.009 305	80.10	stable	
			12	12.014 352		β^-	0.020 2 s
6	Carbon	C	10	10.016 854		β^+	19.3 s
			11	11.011 433		β^+	20.4 min
			12	12.000 000	98.90	stable	
			13	13.003 355	1.10	stable	
			14	14.003 242		β^-	5 730 yr
			15	15.010 599		β^-	2.45 s
7	Nitrogen	N	12	12.018 613		β^+	0.011 0 s
			13	13.005 738		β^+	9.96 min
			14	14.003 074	99.63	stable	
			15	15.000 108	0.37	stable	
			16	16.006 100		β^-	7.13 s
			17	17.008 450		β^-	4.17 s
8	Oxygen	O	14	14.008 595		EC	70.6 s
			15	15.003 065		β^+	122 s
			16	15.994 915	99.76	stable	
			17	16.999 132	0.04	stable	
			18	17.999 160	0.20	stable	
			19	19.003 577		β^-	26.9 s
9	Fluorine	F	17	17.002 094		EC	64.5 s
			18	18.000 937		β^+	109.8 min
			19	18.998 404	100	stable	
			20	19.999 982		β^-	11.0 s
10	Neon	Ne	19	19.001 880		β^+	17.2 s
			20	19.992 435	90.48	stable	
			21	20.993 841	0.27	stable	
			22	21.991 383	9.25	stable	

Atomic number (Z)	Element	Symbol	Mass number (A)	Atomic mass (u)	Percent abundance	Decay mode	Half-life $t_{1/2}$
11	Sodium	Na	22	21.994 434		β^+	2.61 yr
			23	22.989 770	100	stable	
			24	23.990 961		β^-	14.96 hr
12	Magnesium	Mg	24	23.985 042	78.99	stable	
			25	24.985 838	10.00	stable	
			26	25.982 594	11.01	stable	
13	Aluminum	Al	27	26.981 538	100	stable	
			28	27.981 910		β^-	2.24 min
14	Silicon	Si	28	27.976 927	92.23	stable	
			29	28.976 495	4.67	stable	
			30	29.973 770	3.10	stable	
			31	30.975 362		β^-	2.62 hr
15	Phosphorus	P	30	29.978 307		β^+	2.50 min
			31	30.973 762	100	stable	
			32	31.973 908		β^-	14.26 days
16	Sulfur	S	32	31.972 071	95.02	stable	
			33	32.971 459	0.75	stable	
			34	33.967 867	4.21	stable	
			35	34.969 033		β^-	87.5 days
			36	35.967 081	0.02	stable	
17	Chlorine	Cl	35	34.968 853	75.77	stable	
			36	35.968 307		β^-	3.0×10^5 yr
			37	36.965 903	24.23	stable	
18	Argon	Ar	36	35.967 547	0.34	stable	
			38	37.962 732	0.06	stable	
			39	38.964 314		β^-	269 yr
			40	39.962 384	99.60	stable	
			42	41.963 049		β^-	33 yr
19	Potassium	K	39	38.963 708	93.26	stable	
			40	39.964 000	0.01	β^+	1.28×10^9 yr
			41	40.961 827	6.73	stable	
20	Calcium	Ca	40	39.962 591	96.94	stable	
			42	41.958 618	0.64	stable	
			43	42.958 767	0.13	stable	
			44	43.955 481	2.08	stable	
			47	46.954 547		β^-	4.5 days
			48	47.952 534	0.18	stable	
24	Chromium	Cr	50	49.946 047	4.34	stable	
			52	51.940 511	83.79	stable	
			53	52.940 652	9.50	stable	
			54	53.938 883	2.36	stable	
26	Iron	Fe	54	53.939 613	5.9	stable	
			55	54.938 297		EC	2.7 yr
			56	55.934 940	91.72	stable	
			57	56.935 396	2.1	stable	
			58	57.933 278	0.28	stable	
27	Cobalt	Co	59	58.933 198	100	stable	
			60	59.933 820		β^-	5.27 yr

Atomic number (Z)	Element	Symbol	Mass number (A)	Atomic mass (u)	Percent abundance	Decay mode	Half-life $t_{1/2}$
28	Nickel	Ni	58	57.935 346	68.08	stable	
			60	59.930 789	26.22	stable	
			61	60.931 058	1.14	stable	
			62	61.928 346	3.63	stable	
			64	63.927 967	0.92	stable	
29	Copper	Cu	63	62.929 599	69.17	stable	
			65	64.927 791	30.83	stable	
37	Rubidium	Rb	96	95.934 27		stable	
38	Strontium	Sr	90	89.907 320		α	28.9 yr
47	Silver	Ag	107	106.905 091	51.84	stable	
			109	108.904 754	48.16	stable	
48	Cadmium	Cd	106	105.906 457	1.25	stable	
			109	108.904 984		EC	462 days
			110	109.903 004	12.49	stable	
			111	110.904 182	12.80	stable	
			112	111.902 760	24.13	stable	
			113	112.904 401	12.22	stable	
			114	113.903 359	28.73	stable	
			116	115.904 755	7.49	stable	
50	Tin	Sn	120	119.902 197	32.4		
53	Iodine	I	127	126.904 474	100	stable	
			129	128.904 984		β^-	1.6×10^7 yr
			131	130.906 124		β^-	8.03 days
54	Xenon	Xe	128	127.903 531	1.9	stable	
			129	128.904 779	26.4	stable	
			130	129.903 509	4.1	stable	
			131	130.905 069	21.2	stable	
			132	131.904 141	26.9	stable	
			133	132.905 906		β^-	5.4 days
			134	133.905 394	10.4	stable	
			136	135.907 215	8.9	stable	
55	Cesium	Cs	133	132.905 436	100	stable	
			137	136.907 078		β^-	30 yr
			138	137.911 017		β^-	32.2 min
56	Barium	Ba	131	130.906 931		EC	12 days
			133	132.905 990		EC	10.5 yr
			134	133.904 492	2.42	stable	
			135	134.905 671	6.59	stable	
			136	135.904 559	7.85	stable	
			137	136.905 816	11.23	stable	
			138	137.905 236	71.70	stable	
79	Gold	Au	197	196.966 543	100	stable	
			198	197.968 242		β^-	2.7 days
81	Thallium	Tl	203	202.972 320	29.524	stable	
			205	204.974 400	70.476	stable	
			207	206.977 403		β^-	4.77 min

Atomic number (Z)	Element	Symbol	Mass number (A)	Atomic mass (u)	Percent abundance	Decay mode	Half-life $t_{1/2}$
82	Lead	Pb	204	203.973 020	1.4	stable	
			205	204.974 457		EC	1.5×10^7 yr
			206	205.974 440	24.1	stable	
			207	206.975 871	22.1	stable	
			208	207.976 627	52.4	stable	
			210	209.984 163		α, β^-	22.3 yr
			211	210.988 734		β^-	36.1 min
83	Bismuth	Bi	208	207.979 717		EC	3.7×10^5 yr
			209	208.980 374	100	stable	
			211	210.987 254		α	2.14 min
			215	215.001 836		β^-	7.4 min
84	Polonium	Po	209	208.982 405		α	102 yr
			210	209.982 848		α	138.38 days
			215	214.999 418		α	0.001 8 s
			218	218.008 965		α, β^-	3.10 min
85	Astatine	At	218	218.008 685		α, β^-	1.6 s
			219	219.011 294		α, β^-	0.9 min
86	Radon	Rn	219	219.009 477		α	3.96 s
			220	220.011 369		α	55.6 s
			222	222.017 571		α, β^-	3.823 days
87	Francium	Fr	223	223.019 733		α, β^-	22 min
88	Radium	Ra	223	223.018 499		α	11.43 days
			224	224.020 187		α	3.66 days
			226	226.025 402		α	1 600 yr
			228	228.031 064		β^-	5.75 yr
89	Actinium	Ac	227	227.027 749		α, β^-	21.77 yr
			228	228.031 015		β^-	6.15 hr
90	Thorium	Th	227	227.027 701		α	18.72 days
			228	228.028 716		α	1.913 yr
			229	229.031 757		α	7 300 yr
			230	230.033 127		α	75.000 yr
			231	231.036 299		α, β^-	25.52 hr
			232	232.038 051	100	α	1.40×10^{10} yr
			234	234.043 593		β^-	24.1 days
91	Protactinium	Pa	231	231.035 880		α	32.760 yr
			234	234.043 300		β^-	6.7 hr
92	Uranium	U	232	232.03713		α	72 yr
			233	233.039 630		α	1.59×10^5 yr
			234	234.040 946		α	2.45×10^5 yr
			235	235.043 924	0.72	α	7.04×10^8 yr
			236	236.045 562		α	2.34×10^7 yr
			238	238.050 784	99.28	α	4.47×10^9 yr
93	Neptunium	Np	236	236.046 560		EC	1.15×10^5 yr
			237	237.048 168		α	2.14×10^6 yr
94	Plutonium	Pu	238	238.049 555		α	87.7 yr
			239	239.052 157		α	2.412×10^4 yr
			240	240.053 808		α	6 560 yr
			242	242.058 737		α	3.73×10^5 yr

Answers to Select Odd-Numbered Problems

Chapter 2

Multiple-Choice Questions

1. (c)   3. (b)   5. (b)   7. (a), (c)   9. (b)   11. (d)   13. (b)
15. (c), (e)

Problems

1.

5. (a) A, D, F, K, Q (b) A, G, H, K, L, M, Q (c) A, G, K, L, M, Q
(d) $\vec{M} + \vec{K} = \vec{F}, \vec{L} + \vec{Q} = \vec{D}, \vec{L} + \vec{M} = \vec{E}; \vec{L} - \vec{Q} = \vec{B}, \vec{Q} - \vec{K} = \vec{D},$
$\vec{M} - \vec{G} = \vec{F}$ (other solutions possible)
13. (a) 36 mph, 58 km/h, 16 m/s   (b) 349 km/h, 217 mph, 96.9 m/s
(c) 980 m/s, 3.53×10^3 km/h, 2.19×10^3 mph
15. (a) 60 mph
19. (1.4 ± 0.1) km
21. $(3.99 \pm 0.01) \times 10^{16}$ m
23. (a) Gabriele: 120 m, Xena: 160 m   (b) 30 s
(c) 60 m $-$ $(2.0$ m/s$)t$
27. (a) Gabriele: 3.0×10^3 m $-$ $(8.0$ m/s$)t$,
Xena 1.0×10^3 m $+$ $(6.0$ m/s$)t$
(b) 1.9 km, three ways   (c) -2.0×10^3 m $+$ $(14$ m/s$)t$
29. 3.6 mph
31. Impossible to reach this average speed
33. (a) $x_1(t) = 30$ m $+$ $(8.33$ m/s$)t$, $x_2(t) = -10$ m,
$x_3(t) = -10$ m $+$ $(5.0$ m/s$)t$, $x_4(t) = -10$ m $+$ $(-3.33$ m/s$)t$

39. (a)

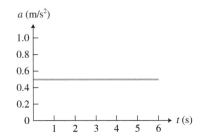

41. $a = 2 \times 10^2$ m/s^2 upward
43. 57.2 m/s^2, -9.2 m/s^2
45. The distance between the runners will increase.

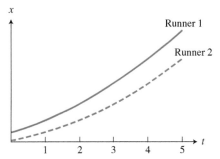

47. 4.0×10^3 m/s^2, 2.0 mm
49. 56 m/s^2, deceleration time: 0.12 s, deceleration distance: 1.4 m
51. (a) 5.6 m/s^2 (b) 11 m   (d) 7.9 s   (e) 9.9 ± 0.1 s
53. 42 m
57. (a) 1.8 s   (b) 17 m/s
61. 9.9 m/s, 2.0 s
63. 0.16 s
65. 5.3 m/s
67. 22 m/s
69. $x_A(t) = (200$ m$)$ $-$ $(20$ m/s$)t$, $x_B(t) = (-200$ m$)$ $+$ $(20$ m/s$)t$
71. The data support the hypothesis because the plot of d versus t^2 gives a straight line; -4.54 m/s^2, 5.08 m/s
73. 1.2 m, 2.2 s

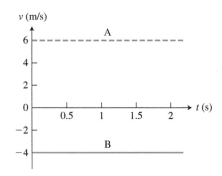

75. (a) Up   (b) 1.3 m, 0.7 m   (c) 2.4 m/s
77. Yes
79. 17 m, 32 m
85. No collision
87. (c)   89. (e)   91. (b)   93. (b)   95. (c)

Chapter 3

Multiple-Choice Questions

1. (c)   3. (a)   5. (c)   7. (b), (d)   9. (a)   11. (a)   13. (b)
15. (c)   17. (d)

Problems

1. (1) (d); (2) (a); (3) (d); (4) (b); 5(a)
3. (a)

(b)

(c)

5. (a)

(b)

(c)

7. (a) After the pulling stops, the cart continues to move with constant velocity, in accordance with Newton's first law.

(b)

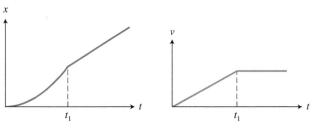

11. Both force diagrams are the same in each case.

17. 10 N to the right
25. (a) 4.9×10^2 N   (b) 5.9×10^2 N   (c) 3.9×10^2 N   (d) 0
27. 1.7×10^2 N
31. (b) 13 m/s^2
33. 2.4×10^3 N
41. (a)

(b) $\vec{F}_{\text{P on M}}$ and $\vec{F}_{\text{M on P}}$ (c) 0.24 N   (d) No; too many unknowns
43. (a) 2.4×10^5 N   (b) 0.67 m
45. (a) 0.9 N   (b) 30%
47. 1.6×10^3 N upward
49. 2400 N for both jumping and landing; ignore air resistance
51. $\sim 1 \times 10^{-12}$ m/s^2
53. (d)   55. (b)   57. (c)   59. (b)   61. (a)   63. (d)

Chapter 4

Multiple-Choice Questions

1. (c)   3. (b)   5. (b)   7. (a), (b)   9. (a)   11. (b)
13. I (a), (c); II (b), (c)

Problems

1. $F_{S \text{ on } C\,x} = 0$, $F_{S \text{ on } C\,y} = 250$ N, $F_{E \text{ on } C\,x} = 0$, $F_{E \text{ on } C\,y} = -150$ N,
$F_{P \text{ on } C\,x} = 173$ N, $F_{P \text{ on } C\,y} = -100$ N
3. $A_x = -7.1$ m, $A_y = 7.1$ m, $B_x = 5.0$ m, $B_y = 0$
5. (a) 224 N, 153°

(b) 500 N, 53°

(c) 500 N, 37° below +x-axis

9. (a) $a_x = \dfrac{\Sigma F_{\text{on W}\,x}}{m_W} = \dfrac{T_{R \text{ on W}}\cos\theta}{m_W}$,

$a_y = \dfrac{\Sigma F_{\text{on W}\,y}}{m_W} = \dfrac{T_{R \text{ on W}}\sin\theta + F_{S \text{ on W}} + (-F_{E \text{ on W}})}{m_W} = 0$

(b) $a_x = \dfrac{\Sigma F_{\text{on B}\,x}}{m_B} = \dfrac{-F_{R \text{ on B}}}{m_B}$, $a_y = \dfrac{\Sigma F_{\text{on B}\,y}}{m_B} = \dfrac{F_{S \text{ on B}} - F_{E \text{ on B}}}{m_B} = 0$

(c) $a_x = \dfrac{\Sigma F_{\text{on Y}\,x}}{m} = \dfrac{F_{E \text{ on Y}}\sin\theta}{m}$,

$a_y = \dfrac{\Sigma F_{\text{on Y}\,y}}{m_W} = \dfrac{F_{S \text{ on Y}} + (-F_{E \text{ on Y}}\cos\theta)}{m_W} = 0$

(d) $a_y = \dfrac{\Sigma F_{\text{on B}\,y}}{m_B} = \dfrac{F_{Y \text{ on B}} + (-F_{E \text{ on B}})}{m_B} = 0$

(e) For m_1: $a_x = \dfrac{\Sigma F_{\text{on } m_1\,x}}{m_1} = \dfrac{T_{R \text{ on } m_1} + (-T_{m_2 \text{ on } m_1})}{m_1}$,

$a_y = \dfrac{\Sigma F_{\text{on } m_1\,y}}{m_1} = \dfrac{F_{S \text{ on } m_1} - F_{E \text{ on } m_1}}{m_1} = 0$

For m_2: $a_x = \dfrac{\Sigma F_{\text{on } m_2\,x}}{m_2} = \dfrac{T_{m_1 \text{ on } m_2}}{m_2}$,

$a_y = \dfrac{\Sigma F_{\text{on } m_2\,y}}{m_2} = \dfrac{F_{S \text{ on } m_2} - F_{E \text{ on } m_2}}{m_2} = 0$

11. (a) $a_y = \dfrac{\Sigma F_{\text{on B}\,y}}{m_B} = \dfrac{F_{Y \text{ on B}} - F_{E \text{ on B}}}{m_B} = 0$

(b) $a_x = \dfrac{\Sigma F_{\text{on S}\,x}}{m_S} = \dfrac{T_{R \text{ on S}}\cos\theta}{m_S}$,

$a_y = \dfrac{\Sigma F_{\text{on S}\,y}}{m_S} = \dfrac{T_{R \text{ on S}}\sin\theta + F_{\text{Surface on S}} + (-F_{E \text{ on sled}})}{m_S} = 0$

(c) $a_x = \dfrac{\Sigma F_{\text{on S}\,x}}{m_S} = \dfrac{F_{R \text{ on S}} - F_{E \text{ on S}}\sin\theta}{m_S}$,

$a_y = \dfrac{\Sigma F_{\text{on S}\,y}}{m_S} = \dfrac{F_{\text{Surface on S}} + (-F_{E \text{ on S}}\cos\theta)}{m_S} = 0$

(d) $a_y = \dfrac{\Sigma F_{\text{on SD}\,y}}{m_{SD}} = \dfrac{F_{\text{air on SD}} - F_{E \text{ on SD}}}{m_{SD}} = 0$

13. $\Sigma F_{\text{on O}\,x} = F_{R2 \text{ on O}\,x} + F_{E \text{ on O}\,x} + F_{R1 \text{ on O}\,x} = 0 + 0 +$
$(30 \text{ N})\cos 245° = -13$ N, $\Sigma F_{\text{on O}\,y} = F_{R2 \text{ on O}\,y} + F_{E \text{ on O}\,y} + F_{R1 \text{ on O}\,y}$
$= 40$ N $+ (-10 \text{ N}) + (30 \text{ N})\sin 245° = 2.8$ N
17. 5.9 m/s
19. (a) A, horizontal   (b) 52°
21. 0.41
23. (a) 3.4×10^2 N   (b) 2.0 m/s²
25. (c)
27. g/μ_s
29. 99 N, 0.68
33. (a) 75 N   (b) 80 N
35. 4.1 m/s², 24 m/s
37. $\mu_s = 0.27$
43. 27 s, 22 m/s
45. 1.4×10^3 N, 1.7×10^3 N
47. 1.5×10^2 N, 60 N
51. 1.3 m/s²
53. $m_1 = 100$ g, $m_2 = 161$ g

57. On table   In air

65. 77 m/s
67. Both angles give the same range. Ignore air resistance.
69. (a) 12 m   (b) 41 m
75. (a) 0.28   (b) 23 N
77. ~30 N
79. (a) 1.6 m/s²   (b) 11 N
83. (a) $v_A = 0.5$ m/s, $v_B = 0.4$ m/s   (b) $d_A = 0.5$ m, $d_B = 0.6$ m
85. (a) 0.04   (b) 4 m/s
91. ~20 m
95. (c)   97. (a)   99. (b)   101. (e)   103. (a)

Chapter 5

Multiple-Choice Questions

1. (d)   3. (b)   5. (a)   7. (a)   9. (a)   11. (b)   13. (a)   15. (a)

Problems

1. $N_{\text{bottom}} = mg + mv^2/R$, $N_{\text{top}} = mg - mv^2/R$
5. ~6.6×10^{-3} m/s²
7. 92 m/s

11. (a) $d_A = 0$, $d_B = 2$ cm, $d_C = 4$ cm   (b) $v_A = 0$, $v_B = v$, $v_C = 2v$
(c) $v'_A = -v$, $v'_B = 0$, $v'_C = v$ (d) $v'_A = -2v$, $v'_B = -v$, $v'_C = 0$
13. The period is the same for all three people. $v_1 = 2\pi R/T$,
$v_2 = 0.6v_1$, $v_3 = 0.5v_1$, $a_1 = 4\pi^2 R/T^2$, $a_2 = 0.6a_1$, $a_3 = 0.5a_1$
15. (a)
17. ~ 60 m/s^2
19. No
21. 16 m/s, 36 mph
23. (a)

(b) $\vec{T}_{S \text{ on } B}$ increases, $\vec{F}_{E \text{ on } B}$ does not change, $\vec{N}_{I \text{ on } B}$ decreases

25. 2.1×10^2 N
27. 9.3 m/s
29. 90 N, radially inward
31. 9.0×10^3 N
35. (a) $v = \sqrt{rg\tan\theta}$
(b) Speed is independent of the mass of the object.
39. (a) 1.066   (b) Two (0.938)
41. (a) 3.0×10^4 m/s   (b) 2.0×10^{30} kg
43. 1.34
45. (a) 55 kg   (b) Mars: 204 N; Venus: 489 N; Saturn: 575 N
(c) Assume each planet is a perfect homogenous sphere.
47. 7.8 km/s
49. 7.9 km/s
51. 1.4 h
55. 1.8×10^3 N upward
57. $\sim 1.9 \times 10^3$ m/s^2
59. ~ 62 N
61. (b) At the equator: $N_{S \text{ on } P}^{EQ} = F_{E \text{ on } P} - ma_r^{EQ}$; at the North Pole,
$N_{S \text{ on } P}^{NP} = F_{E \text{ on } P}$ (c) Yes
67. 1.4 h
69. 1.7 km/s, yes
73. (c)   75. (b)   77. (c)   79. (a)   81. (a)   83. (b)

Chapter 6

Multiple-Choice Questions

1. (b)   3. (e)   5. (a)   7. (b)   9. (d)   11. (b)   13. (a), (e)

Problems

1. (a) 1.7 kg·m/s   (b) 5.3 m/s   (c) 3.0×10^2 kg·m/s
3. The ball that rebounds
7. $10v/7$
9. ~ 120 kg·m/s
11. 0.31 N·s, 0.13 N·s
15. (a) 4.0×10^5 N   (b) 100 s
17. 3.2×10^3 N
19. 7.2 N
21. (a) 25%   (b) -11%
23. $\sim 7 \times 10^1$ N
27. (c) For part (a), $p_{Bix} + J_{S \text{ on } Bx} = p_{Bfx}$; for part
(b), $p_{Aix} + p_{Bix} + J_x = p_{Afx} + p_{Bfx}$, where $J_x = J_{S \text{ on } Ax} + J_{S \text{ on } Bx}$

29.

$J_{N \text{ on } Bx} = mv$; Earth is the object of reference.
31. Cart will remain stationary
33. (a) 6.1 kg·m/s   (b) 57 N
35. 44 m/s
37. 1.3 m/s
39. (a) 0.080 s   (b) 0.64 m
41. 165 km/h
43. $\sim 1.8 \times 10^4$ N
45. (a) ~ 0.045 s   (b) ~ 0.34 m
47. (a) $v = \dfrac{Mu}{M + m}$ (b) $4.5m_0$, $0.83v_0$
49. Cart slows down
51. 8.0 m/s
53. 1.4×10^7 m/s
55. (a) 0.75 m/s   (b) 13 s
57. (a)

(b)

(c)

59. 56° north of east, 6.2 m/s
61. 15° north of east, 8.3 m/s
63. $\sim 1 \times 10^{-14}$ m/s
67. (a) 2.4 m/s   (b) 4×10^2 N
69. Large ball: 90°; small ball: $-270°$
71. ~ 200 N
73. 10 m/s
75. 13 m/s
79. (b)   81. (e)   83. (b)   85. (b)   87. (d)   89. (d)

Chapter 7

Multiple-Choice Questions

1. (b)   3. (b)   5. (a)   7. (a), (b), (c)   9. (d)   11. (a), (c), (d)   13. (b)

Problems

1. 5.4×10^2 J
3. (a) 7.5×10^2 J   (b) -7.5×10^2 J   (c) 7.5×10^2 J
5. Lifting 2.0×10^2 J; carrying 0; setting down -2.0×10^2 J; total 0
7. (a) 0.30 J   (b) 1.1 m/s
9. (a) 7.9×10^2 J   (b) 5.3×10^2 J
11. ~ 30 J
13. From 30 to 40 m/s
15. 3.8×10^6 J
17. (a) 26 m/s   (b) Ignore friction.   (c) Speed is independent of mass.
19. (a) 2.3×10^{11} J   (b) 0.07
23. 8.9×10^5 N/m
25. 5.7×10^2 m
27. $v_f = \sqrt{v_i^2 + 2gh - \dfrac{2fl}{m}}$

31. (a)

(b) 31 m/s   (c) Final speed is independent of mass.
33. 2.1 m/s
43. 22 m/s
47. (a) 1.0 m/s   (b) $K_{1i} = 8.0$ J, $K_{2i} = 4.5$ J; $K_{1f} = 2.0$ J, $K_{2f} = 0.5$ J
(c) The difference is converted to internal energy.
49. (a) 0.64 m/s   (b) 0.28 m
51. 2.2×10^2 m/s
55. (a) 9.8 W   (b) 2.0×10^2 W   (c) 2.0×10^2 W   (d) 2.0×10^2 W
57. (a) 1.2×10^7 J   (b) 47 s
61. (a) 1.1×10^7 J   (b) 0.24 kg   (c) 210 g
63. 12 m/s
67. 42 km/s
73. (a) $3\left(\frac{1}{2}mv_i^2\right)$ (b) mv
81. 6.9×10^2 N
83. (d)   85. (e)   87. (c)   89. (b)   91. (c)   93. (a)

Chapter 8

Multiple-Choice Questions

1. (b)   3. (a)   5. (d)   7. (d)   9. (c)   11. (c)

Problems

1. $\tau_1 = 0$, $\tau_2 = -1.8 \times 10^2$ N·m, $\tau_3 = +2.4 \times 10^2$ N·m,
$\tau_4 = -2.4 \times 10^2$ N·m
5. (b) $F_{A \text{ on } K\,x} + F_{J \text{ on } K\,x} + F_{L \text{ on } K\,x} = 0$, $F_{A \text{ on } K\,y} + F_{J \text{ on } K\,y} + F_{L \text{ on } K\,y} = 0$

(c) $F_{L \text{ on } K} = \sqrt{F_{L \text{ on } K\,x}^2 + F_{L \text{ on } K\,y}^2} = 32$ N,

$\theta = \tan^{-1}\left(\dfrac{F_{L \text{ on } K\,y}}{F_{L \text{ on } K\,x}}\right) = -83°$

7. (b) $T_{1 \text{ on } R\,x} + T_{2 \text{ on } R\,x} + T_{3 \text{ on } R\,x} + T_{4 \text{ on } R\,x} = 0$, $T_{1 \text{ on } R\,y} + T_{2 \text{ on } R\,y} + T_{3 \text{ on } R\,y} + T_{4 \text{ on } R\,y} = 0$
(c) 43 N, 9.9° above the x-axis
9. $T_1 = 1.2 \times 10^2$ N, $T_2 = 87$ N, $T_3 = 56$ N
11. $m_1 = 7.5$ kg, $m_2 = 3.8$ kg
13. $T_1 \cos \theta_1 - T_{Y \text{ on } 2} \cos \theta_2 = 0$, $T_1 \sin \theta_1 + T_{Y \text{ on } 2} \sin \theta_2 - T_1 = 0$,
$\theta_1 = 84°$, $\theta_2 = 2.9°$
15. 2.5°
19. Komila can sit anywhere from the left edge of the chair to 1.4 m to the right of the center of the chair.
21. 1.5 kg
25. 0.77 m
27. 0.84 m from head
29. 1.3 cm
31. 192 kg, 8.2×10^3 N
33. $m = \dfrac{F_{L \text{ on } P}}{g \sin \theta} = 36$ kg
35. $F = \frac{1}{2}mg$
37. No
39. $T = 37$ N, $F_{H \text{ on } B\,x} = 37$ N, $F_{H \text{ on } B\,y} = 39$ N, $F_{H \text{ on } B} = 54$ N,
$\theta = 46°$
41. 0.35 m from the clay
43. Yes
45. ~ 0.37 m
47. $\sim \pm 10$ N·m
51. $T_{H \text{ on } L} = 220$ lb, $F_{F \text{ on } B} = 200$ lb
53. 9.9×10^2 N
55. (a) 34 N   (b) 4.8×10^2 N
61. (a) 0.60 m from the 50-kg person   (b) No change
63. (a)   65. (c)   67. (d)   69. (e)

Chapter 9

Multiple-Choice Questions

1. (a)   3. (b)   5. (d)   7. (b), (d)   9. (b)   11. (c)

Problems

1. (a) 0.105 rad/s   (b) 0.021 m/s   (c) 0
3. (a) -0.058 rad/s^2 (b) 0.23 rad/s^2 (c) 0.63 rad/s^2 (d) 0.094 m/s^2
5. 8.4 s
7. 3.0×10^2 rad/s
9. (a) 2.0×10^2 m/s^2 (b) 1.3×10^2 rad/s^2
11. $a_t = 0.83$ m/s^2, $\alpha = 2.8$ rad/s^2, $\omega = 33$ rad/s, $\Delta t = 12$ s
13. 8.3×10^3
15. (a) $\omega^2/4\pi$ (b) $4\pi/\omega$ (c) $\dfrac{4\pi}{\omega}\left(\sqrt{2} - 1\right)$ (d) $4\pi l$
17. -1.2×10^{-14} rad/s^2
19. (a) 40 N·m   (b) 50 s
21. -6.9×10^{-4} N·m
23. (a) $T_2/T_1 = 1.33$ (b) $T_2 > \dfrac{4}{3}T_1$ (c) $T_2 < \dfrac{4}{3}T_1$
25. (a) 1.5 N   (b) $W_{H \text{ on } S} = +7.4$ J, $W_{S \text{ on } C} = +7.4$ J
29. $4m$ kg·m^2
31. 5.9×10^4 N·m
33. 27 N·m
35. -7.8×10^2 N

37. (a)

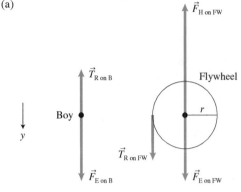

(b) $m_B g - T = m_B a$
(c) $T = I\alpha/r$
(d) 3.3 m/s^2, $2.0 \times 10^2 \text{ N}$

39. (a)

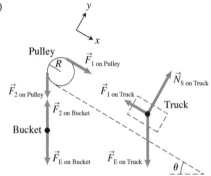

(b) $m_1 g \sin\theta - F_{1 \text{ on Truck}} = 0$, $F_{2 \text{ on Bucket}} - m_2 g = 0$
(c) $F_{2 \text{ on Pulley}} = F_{1 \text{ on Pulley}}$
41. (a) $13 \text{ kg} \cdot \text{m}^2/\text{s}$ (b) 400 rad/s
43. 13 rad/s, 2.5 m/s
47. 91 days
51. 96%
53. (a) $3.9 \times 10^7 \text{ J}$ (b) 230
57. (a) 12 rad/s (b) $-1.7 \times 10^{-3} \text{ J}$

59. (a) $\frac{1}{2} m_D r_D^2$ (b) $2\pi^2 f_i^2 m_B (d^2 - r_D^2) \left(\dfrac{\frac{1}{2} m_D r_D^2 + m_B d^2}{\frac{1}{2} m_D r_D^2 + m_B r_D^2} \right)$

65. $1.4 \times 10^2 \text{ N}$
67. $7.7 \times 10^3 \text{ N}$
73. 0.035 rad/s
75. (a)   77. (a)   79. (d)   81. (e)   83. (e)

Chapter 10

Multiple-Choice Questions

1. (c), (d)   3. (c)   5. (b)   7. (a)   9. (b)

Problems

3.

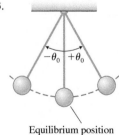

Equilibrium position

5. (a) x (cm)

(b) x (cm)

11. (a) 0.20 s (b) $x(t) = -(0.10 \text{ m})\cos(10\pi t)$ (c) 0 (d) $+0.10 \text{ m}$
13. 400 N/m, 0.25 m
15. 2.3 Hz
17. $3m$
21. (a) 0.10 m (b) 15 m/s (c) 13 m/s
23. (a) 11 J (b) 4.0 m/s (c) 0.052 m (d) 7.0 m/s
25. $\sqrt{2}$
27. (a) $3/4$ is kinetic energy, and $1/4$ is potential energy   (b) $A/\sqrt{2}$
33. Increase by 20%, decrease by 19%
37. 4.0 cm, 4.0 s, 0.25 Hz, 4.0 m
39. (b) 0.098 m
41. (a) $5.9 \times 10^3 \text{ N/m}$ (b) 7.4 J, 1.6 Hz
43. 3.7 m/s^2
47. 28 m/s
49. (a) 0.21 m (b) 0.53 J (c) 0.29 J (d) 0.36 m
53. (a) 2.2 m/s
55. 9.0 m/s
57. 0.41
63. (a) 0.44 s

(b)

$$U_{si} + K_i + W = U_{sf} + K_f$$

$$\vec{p}_i + \vec{J} = \vec{p}_f$$

67. 120 m/s
73. (d)   75. (a)   77. (a)   79. (a)   81. (c)

Chapter 11

Multiple-Choice Questions

1. (b)   3. (c)   5. (c)   7. (b)   9. (e)

Problems

11. (b) 820 m
13. (a) $v_2 > v_3 > v_1$ (b) $\lambda_2 > \lambda_1 > \lambda_3$ (c) $|y_2| > |y_1| > |y_3|$
15. Tightened
17. 0.7 m/s
19. (a) $v_1 = 11 \text{ m/s}$, $v_2 = 8.0 \text{ m/s}$ (b) 13 N
23. 40 m
27. $1 \times 10^2 \text{ m}$

29.

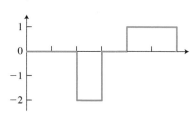

39. 700 m
43. (a) Yes   (b) No
(c) Beat frequency of 3.3 $\pm$ 0.1 Hz
45. The higher intensity is ten times greater than the lower intensity.
47. (a) 0.41 km/s   (b) 588 Hz, 882 Hz, 1.18 kHz
53. 4.9%
55. (a) 0.061 Hz   (b) 327
57. (a) Closed, 0.71 m   (b) 240 Hz, 480 Hz
59. $m_3 < m_2 < m_1 < m_4$
61. 5.7 Hz
63. Both pipes open-open: 0.85 m, 0.43 m; open-closed and open-open: 0.43 m, 0.43 m; open-open and open-closed: 0.85 m, 0.21 m; both pipes open-closed: 0.43 m, 0.21 m
65. (a) Yes   (b) No
67. (a) 337 Hz   (b) 266 Hz
69. 438 Hz, 367 Hz
71. 44 m/s
79. (b) 0.1 kg, 0.03 kg
81. (e)   83. (b)   85. (a)   87. (c)   89. (c)

Chapter 12

Multiple-Choice Questions

1. (c)   3. (c)   5. (d)   7. (c)   9. (c)   11. (b)   13. (b)   15. (d)

Problems

3. 3.3×10^{-21} kg/m^3
9. 6.4×10^4 N/m^2
13. (a) 2.9×10^3 N, 4.8 N   (b) 1.6 N/m^2
17. Yes
19. 5.65×10^{-21} J
21. 0.041 mol
23. $v_{1fx} = v_{2ix}$, $v_{2fx} = v_{1ix}$
25. 7.2 cm^3
27. 4×10^{-17} N/m^2, much less
29. Yes
31. 0.39 mol
37. 1.1 m
39. About three times more frequently

43. Decrease the temperature by a factor of 1.5.
45. 6.6×10^5 N/m^2
55. (a) $U = \dfrac{3mkT}{2M}$ (b) $P_2 = \frac{1}{2}P_1$, $T_2 = T_1$, $U_2 = U_1$
61. 2.7×10^5 m^{-3}, assume 100% hydrogen
67. $l_{H_2} = 93.3$ cm and $l_{N_2} = 6.67$ cm
71. ~3.5%
73. (d)   75. (a)   77. (e)   79. (b)   81. (b)

Chapter 13

Multiple-Choice Questions

1. (e)   3. (c)   5. (b)   7. (a)   9. (c)   11. (a)   13. (b)   15. (b)

Problems

1. 5.5×10^3 kg/m^3
3. 6.8×10^2 kg
5. 2.0×10^{14} kg
7. (a) $\rho_A = \rho_B = \rho_C$ (b) $V_B = 8V_A$, $V_C = 27V_A$
(c) $S_B = 4S_A$, $S_C = 9S_A$ (d) $A_B = 4A_A$, $A_C = 9A_A$
(e) $m_B = 8m_A$, $m_C = 27m_A$
11. 4600 kg/m^3, a solid material
15. 8.0×10^{-3} m^2
23. $P_A = P_B = P_C$, $F_A > F_C > F_B$
27. 0.14 m
33. 1.6×10^3 N/m^2, 3.2×10^3 N/m^2
35. 1.2 N directed from exterior toward interior of ear
39. (a)

(b)

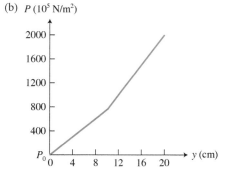

41. A, C, D, B
43. 1.0×10^5 Pa
47. (a) 1.1×10^4 N/m^2 (b) 1.1×10^4 N/m^2
53. $F_{\text{water on oak}} = F_{\text{mercury on oak}} > F_{\text{olive oil on oak}} > F_{\text{alcohol on oak}}$
55. 0.074 N
59. $\rho_A < \rho_B = \rho_D < \rho_C$
61. 2.3×10^3 kg/m^3
65. ρ_1/ρ_2

67. (a) 5.5 N   (b) 7.2×10^2 kg/m^3 (c) No change
69. 1.3 N, 1.3 m/s^2
77. 200 kg
79. 0.033 m
81. 0.016 m^3
85. 8.6×10^5 N
87. 1.96×10^4 N/m^2
91. 1031 kg/m^3, sink

93. $\rho_X = \rho\left(1 - \dfrac{T_{S \text{ on } O}}{mg}\right)$

95. (c)   97. (d)   99. (e)   101. (b)

Chapter 14

Multiple-Choice Questions

1. (b)   3. (c)   5. (b)   7. (a)   9. (d)   11. (b)

Problems

1. (a) 1.5×10^{-4} m^3/s   (b) 0.46 m/s
5. 6.6×10^{-3} cm/s

7. $\tilde{K}_1 + P_1 = P_2 + \tilde{K}_2 + \tilde{U}_{g2}$

17. $\tilde{K}_1 + \tilde{U}_{g1} + (P_1 - P_2) = \tilde{K}_2 + \tilde{U}_{g2}$

$P_1 + \frac{1}{2}\rho v_1^2 = P_2 + \frac{1}{2}\rho v_2^2$

19. (a) 5.2 m/s   (b) 1.4 cm   (c) No change in speed
21. (a) 4.4 m/s   (b) 3.7×10^3 N/m^2
23. 0.0041 m
25. (a) -19% (b) 23%
29. (a) 2.0×10^3 N/m^2 (b) 0.060 m
31. 2.3×10^{-2} m
33. 3.5×10^2 N/m^2
35. (a) 1.9 N   (b) 1.2×10^2 N
37. (a) 1.8 m   (b) 6.4×10^3 Pa
39. 17 N
41. (a) 5.3×10^{-12} m/s   (b) 5.2×10^6 hours
43. 0.25 m/s
45. Acetone, water, ethanol, blood
49. (a) 1.2×10^4 N/m^2 (b) 1.1 m
53. 2.44, 147 mm Hg
55. 6.7×10^{-4} kg
57. 6.4×10^{-2} m/s
59. (d)   61. (b)   63. (b)   65. (a)   67. (c)   69. (a)

Chapter 15

Multiple-Choice Questions

1. (b), (d)   3. (b)   5. (c)   7. (c)   9. (a)

Problems

3. (a) 0   (b) $\Delta T = 0$, $\rho_f = \rho_i/2$, $P_i = 2.4 \times 10^5$ N/m^2,
$P_f = 1.2 \times 10^5$ N/m^2, $\Delta \overline{K} = 0$, $\Delta U_{th} = 0$
5. (a) -1.3 kJ   (b) 1.2 mol   (c) 2.1×10^2 K
7. (b)

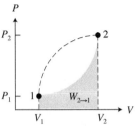

(c) < 0
17. 4.8 °C
19. 13.8 h
21. (a) 55
23. $m_{water} = 0.078$ kg, $\Delta U_{water} = 4.6 \times 10^3$ J, $\Delta U_{soup} = -4.6 \times 10^3$ J
25. (a) 7.3×10^{23} J   (b) 1.3×10^4 °C
27. $T_f = 38$ °C, $\Delta U_{Al} = -900$ J, $\Delta U_{water} = -2.5 \times 10^3$ J,
$\Delta U_{Fe} = +3.4 \times 10^3$ J
29. 2.1×10^5 J
31. (a) L_f
33. mL_f/H
35. (a) 1.8×10^5 J   (b) 64 °C
37. 6.8×10^4 J
39. 47 min
43. 2.1 kW
45. 16 J/s
47. 2.0 kg
49. -11 °C
51. 84 W
55. Wet air
57. (a) 850 J/s   (b) 1.8×10^{-4} kg   (c) 2 kg
59. -7.7 kW
61. -2.6 kW
63. (a) 1.0×10^3 K, 1.9×10^5 Pa
65. 13 min
71. (b)   73. (b)   75. (d)   77. (c)   79. (f)

Chapter 16

Multiple-Choice Questions

1. (a)   3. (c)   5. (b)   7. (d)   9. (a)   11. (c)   13. (a), (d), (e)

Problems

5. 4.0 °C, irreversible
7. (a) $(n_L, n_R) = (6, 0), (5, 1), (4, 2), (3, 3), (2, 4), (1, 5), (0, 6)$
(b) and (c)

Macrostate	Ω_{n_L}	S_i
(0, 6)	1	0
(1, 5)	6	$1.79k_B$
(2, 4)	15	$2.71k_B$
(3, 3)	20	$3.00k_B$
(4, 2)	15	$2.71k_B$
(5, 1)	6	$1.79k_B$
(6, 0)	1	0

9. (a) 1/8   (b) 1/64
11. (a) 972   (b) 25.2
13. (a)

n	2	3	4	5	6	7	8	9	10	11	12
Ω_n	1	2	3	4	5	6	5	4	3	2	1

(b) 7   (c) 2 and 12
17. (a) 53 °C   (b) +7.7 J/K, yes
19. $+1.94 \times 10^4$ J/K
21. +0.85 J/K
23. (a) 0.50   (b) 0.19   (c) 0.047
25. (a) 0.654   (b) 1.54×10^9 W
27. (a) $W_{12} = -3.6 \times 10^5$ J, $W_{23} = +2.7 \times 10^5$ J, $W_{31} = 0$
(b) -9.0×10^4 J   (c) $+9.0 \times 10^4$ J
(d) $T_1 = 1.4 \times 10^4$ K, $T_2 = 5.8 \times 10^4$ K, $T_3 = 7.2 \times 10^3$ K
(e) $U_{\text{th }1} = 1.8 \times 10^5$ J, $U_{\text{th }2} = 7.2 \times 10^5$ J, $U_{\text{th }3} = 9.0 \times 10^4$ J,
$\Delta U_{12} = 5.4 \times 10^5$ J, $\Delta U_{23} = -6.3 \times 10^5$ J, $\Delta U_{31} = 9.0 \times 10^4$ J
(f) $Q_{12} = 9.0 \times 10^5$ J, $Q_{23} = -9.0 \times 10^5$ J, $Q_{31} = +9.0 \times 10^4$ J
(g) 0.091
29. (a) 6.71×10^4 J   (b) 4.9×10^3 J   (c) 7.2×10^4 J
31. 0.36, 1800 J
33. 0.417, 700 MW
37. 1.3×10^3 K, 1.1×10^3 K
39. (b)   41. (a)   43. (d)

Chapter 17

Multiple-Choice Questions

1. (d)   3. (f)   5. (a), (c), (d), (f)   7. (b)   9. (c)   11. (c)

Problems

1. 1.25×10^{18}
3.

$m_1 \tan \varphi_1 = m_2 \tan \varphi_2$
5. 6.1×10^{-7} kg
7. 57 N, repulsive
9. 3.6×10^{-10} N
11. (a) 0   (b) 0   (c) 2×10^{-3} N, right
13. 2.2×10^{-9} N to the left
21. (a) Positive   (b) Increase their separation.
(c)

23. (a) Negative   (b) Decrease their separation.
(c)

25. (a) 4.3×10^{-9} N   (b) 3.6×10^{-9} N   (c) 3.9×10^{-13} J
(d) The work done is equal to the change in electrical potential energy.
31. $U_D < U_A < U_C < U_B$
43. (a) 8.5×10^{-3} m/s   (b) 7.4×10^{-3} m/s
45. 4×10^{-8} C
47. 9×10^{-6} C
49. (a) 1.9×10^6 m/s   (b) Initial kinetic energy is 2.9×10^{-15} J
51. 2.4×10^7 m/s
53. 3.2 C
57. (b)   59. (b)   61. (b)   63. (a)   65. (b)   67. (a)

Chapter 18

Multiple-Choice Questions

1. (c)   3. (e)   5. (d)   7. (a)

Problems

1.

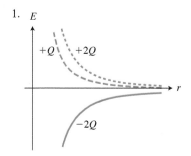

7. 0.27 m to the right of the 4.0-nC charge
9. 2.4×10^4 N/C at 242° counterclockwise from the x-axis
11. (a) $E_1 = E_0 + \dfrac{kq}{d^2}$ (b) $E_2 = E_0 - \dfrac{kq}{d^2}$
13. (a) 0.55 s   (b) 0.87 m
17. 9.8×10^5 N/C upward
21. -1.2×10^9 J
23. $\dfrac{2kq}{d}$
25. $E = 0, V = -\dfrac{4\sqrt{2}kq}{d}$
27. (a) -5.4×10^2 V   (b) -2.0×10^2 V
29. (a) -1.1×10^{-11} J   (b) 7.8 J   (c) 7.8%
33. (a) $E_F = E_G > E_A = E_B = E_C = E_D$
(b) $V_A > V_C = V_D > V_B > V_F > V_G$
(c) IV, I, III, II
35. 0
37.

39. (a) No   (b) Yes
41. (a) 9.3×10^6 V/m   (b) 1.5×10^{-12} N
43. (a) 2.4×10^4 V/m   (b) 4.0×10^{-5} m
45. (a) $q = C|\Delta V|, U_q = \frac{1}{2}C|\Delta V|^2$ (b) Same charge, U_q/κ

47. (a) 1.7×10^{-7} F   (b) 1.2×10^{-8} C   (c) 4.1×10^{-10} J
51. 9×10^3 V
55. (a) 7.5×10^6 J   (b) 1.7 h
57. (a) All charges are positive.   (b) $K_B > K_A = K_D > K_C$
(c) $v_A > v_B = v_C > v_D$
59. 9.5 g, 4.4×10^{-6} C
63. (b)   65. (b)   67. (b)   69. (b)   71. (d)   73. (a)

Chapter 19

Multiple-Choice Questions

1. (c), (d)   3. (f)   5. (c)   7. (a)   9. (a)   11. (a), (d), (g)   13. (a)

Problems

3. 3.4×10^{22}
11.

13. $R = 12\ \Omega$, $P = 12$ W
15. 1.2 A
17. (a) $I_{50\Omega} > I_{100\Omega} > I_{150\Omega}$ (b) $\Delta V_{50\Omega} = \Delta V_{100\Omega} = \Delta V_{150\Omega}$
19. (b) 2.7 kΩ
21. (a) 1.6 W   (b) More
23. 0.12 kW
29. (a) $\mathcal{E}_1 - IR_3 - \mathcal{E}_2 - IR_2 - IR_1 = 0$ (b) 0.50 A clockwise
(c) 11.5 V
31. $I_2 = 1.6$ A, $I_3 = 0.40$ A, 24 V
33. (a) $I_1 + I_3 = I_2, \mathcal{E}_2 - \mathcal{E}_1 - I_2 R_2 - I_1 R_1 = 0, \mathcal{E}_2 - I_3 R_3 - I_2 R_2 = 0,$
$I_1 = \dfrac{3}{110}$ A, $I_2 = \dfrac{15}{110}$ A, $I_3 = \dfrac{12}{110}$ A
(b) 3.3 V
35. (a) 40 Ω   (b) 0.25 A
37. 20 Ω
41. $q_1 = 0.0040$ C, $q_2 = 0.0060$ C
43. (b) $I_1 = 0.15$ A, $I_2 = 0.023$ A, $I_3 = 0.13$ A   (c) 2.5 V
45. 15.3 A
47. 13 W, 6.7 W
49. (a) $P = \dfrac{\mathcal{E}^2}{R(1 + r/R)^2}$
51. $I_1 > I_3 > I_4 > I_0 = I_2$
53. $R = 38\ \Omega$, $I = 0.040$ A, $P_{\text{batt}} = 0.060$ W
55. 2.0 m
57. 1.2
59. (a) 2   (b) $\frac{1}{4}$ (c) 0.61
61. (a) 30 V, 15 Ω
63. 2.2 W
65. (a) 3.3 V   (b) 0.13 W, 0.20 W
69.

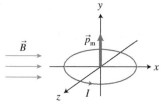

73. (a) $\sim 600\ \Omega$ (b) $\sim 2.0 \times 10^5\ \Omega$ (c) No

77. (a) Circuit 1   (b) Right terminal at higher potential   (c) 70 Ω
79. 10^9
81. (a) 1.1×10^{-5}A   (b) 7.16×10^{13} (c) New current is 5.7×10^{-6}A
83. (e)   85. (e)   87. (d)   89. (d)   91. (d)   93. (b)

Chapter 20

Multiple-Choice Questions

1. (d)   3. (d)   5. (b)   7. (a), (e), (f), (h)   9. (d)

Problems

3. No
7. 3.4×10^{-3} N west
15. (a) Segment a: 4.5×10^{-3} N out of page; segment b: 4.5×10^{-2} N upward; segment c: 4.5×10^{-3} N into page
(b) 0, 7.8×10^{-4} N·m, causing loop to rotate counterclockwise
19.

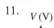

0.12 T
21. 2.5×10^{-11} N west
23. 1.4×10^7 m/s
29. 0.22 A
37. (a) 1.22×10^{-15} s   (b) 1.31×10^{-4} A   (c) 0.388 T, 1.85×10^{-23} A·m²
39. (a) 5.0×10^5 A   (b) Use coils with multiple loops.
41. (a) 1.2×10^6 m/s   (b) 1.7 m
45. (a) Into the page
47. (d)   49. (e)   51. (d)   53. (b)   55. (c)   57. (a)

Chapter 21

Multiple-Choice Questions

1. (b), (c), (d), (f), (g), (i)   3. (c)   5. (b)   7. (b)   9. (b)   11. (d)   13. (a)

Problems

13. (a) Clockwise   (b) Counterclockwise
19. 4.5×10^{-5} V
23. (b) a to b
27. 0.089 V
35. (a) 0.02 s
37. 7.6 cm²
39. (a) 25 cm²   (b) $(38\ \text{V})\sin[(2\pi)(80\ \text{s}^{-1})t]$ (c) 26 V
43. (a) 70.7 V   (b) 0.020 s   (c) 0.500 A   (d) 25.0 W
45. (a) 0.074 H   (b) 1.4 ms   (c) 0.011 A
47. (a) 2×10^{-9} F   (b) 2×10^7 Hz
49. 100
53. (a) 0.38 J   (b) 0.38 V   (c) 3.8 Ω
55. 4.1×10^{-5} V
65. 8 V
67. (a) Yes   (b) 4.8 s   (c) -23 m/s²
69. (d)   71. (e)   73. (c)   75. (b)

Chapter 22

Multiple-Choice Questions

1. (d)   3. (a), (b)   5. (b)   7. (c)   9. (b)   11. (c)

Problems

1. 17 m
3.

Moon

Shadow
of Earth

5. Inverted
9. 65° below the horizontal

Ray from Sun

$\theta_i = 25°$
$\theta_r = 25°$
Ray to wall
Mirror θ_M

11. (b) 120°   (c) 90° or 45°
13. 25°, 25°
15. 17°
17. $\theta_i = \theta_r = 45.6°$
19. 13 cm
21. 1.4
23. 64°
25. 58°
27. No
29. 24°
33. 1.25, yes
35. (a) 39.6°   (b) 1.6, 38°
37. (a) 39.3°   (b) No
43. (a) 33.9°   (b) 1.43
45. $1.44 \le n_g \le 1.88$
49. $\theta_1 = 60°, \theta_2 = 33°, \theta_3 = 27°, \theta_4 = 47°$

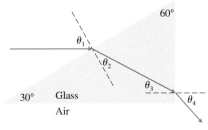

60°
θ_1
θ_2
θ_3
θ_4
30° Glass
Air

53. $x = \dfrac{Hl}{h + H}$
55. 1.1 m

30°
d | d
θ | θ
ϕ | ϕ
1.4 m

57. (a)   59. (e)   61. (e)   63. (c)   65. (c)   67. (d)

Chapter 23

Multiple-Choice Questions

1. (c)   3. (b)   5. (d)   7. (b)   9. (d)   11. (c)   13. (d)

Problems

17. (a) 0.86 m behind mirror   (b) 0.49 m
19. 6.7 cm in front of the mirror on the principal axis
27. (a) $s' \approx 15$ cm, real, inverted

Lens
Object
Image

(b) $s' \approx 38$ cm, real, inverted

Lens
Object
Image

(c) $s' \approx -10$ cm, virtual, upright

Lens
Image Object
Image

29. (a) $s' \approx -23$ cm, virtual, upright

Lens
Image
Object

(b) $s' \approx -4$ cm, virtual, upright

Object Image

33. (a) $s' \approx 12$ cm, real, inverted   (b) $s' \approx -2.7$ cm, virtual, upright

Lens
Object
Image
Lens
Object Image

37. 9.3 cm
39. ~3
41. 6.1 cm from the lens on the side opposite the object
43. (a) 252 cm   (b) −40 cm
45. 25 cm

47.

49.

51. (a) 2.10 cm   (b) 2.11 cm   (c) 2.29 cm
53. (a) −2.4 m   (b) 0.36 m
55. (a) Hyperopia, 0.50 m, infinity   (b) Myopia, 25 cm, 3.5 m
57. 5.8, increase, decrease
59. (a) 4.8 cm   (b) 31
61. (a) 20 cm to the right of the second lens   (b) Upright   (c) Real
63. (a)

(b) $m = 2$
65. −17 cm
67. (a) 1.1 m to the left of the second lens   (b) 1.5 mm   (c) 270
69. Distance between lenses is 1.24 m, 4.4
71. (a) 8.1 cm   (b) −20
73. 46 cm, 3.8 cm from eyepiece
79. 24 cm to 18 m
83. (a)   85. (a)   87. (a)   89. (b)   91. (d)   93. (b)

Chapter 24

Multiple-Choice Questions

1. (d)   3. (e)   5. (b)   7. (a), (b)   9. (b), (d), (g)

Problems

3. (a) 0.25°   (b) 13 mm
5. 39°

7.

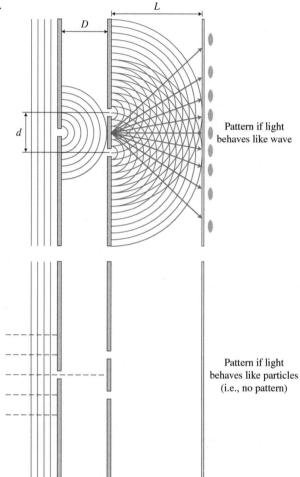

11. For $\lambda = 656$ nm, $\theta_1 = 20.3°$, $\theta_2 = 44.1°$; for $\lambda = 410$ nm, $\theta_1 = 12.6°$, $\theta_2 = 25.8°$
13. No change
15. 530 nm
17. (a) 1.5×10^{-6} m   (b) 4 Gb
23. (a) Yes   (b) No   (c) 480 nm   (d) 240 nm
25. (a) 600 nm   (b) 300 nm
33. 2.5°
35. 0.13 mm
37. 1.1×10^{-7} rad
41. 1.8 cm
45. (a) The magenta band appears between bands 2 and 3 for all gratings illuminated by white light.   (b) 4 m

47.

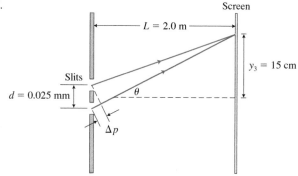

(b) Coherent light source, $\theta_3 = 4.29°$, $\lambda = 620$ nm
51. (a) 0.84°   (b) 3.1°
53. 130 nm, 389 nm
55. 50 μm
57. 0.30°
59. (a) 13 μm   (b) 9.1 μm
61. 5.2 mm
63. 0.65 cm
65. (a) 657 nm, 4.57×10^{14} Hz   (b) 993 nm, 3.02×10^{14} Hz
(c) The light may be Doppler shifted because the galaxy Hydra is receding from us.
67. ~0.04 mm
69. ~10 km
71. ~0.6 mm
73. 0.18 mm
75. 1×10^5 Hz
77. (d)   79. (d)   81. (d)   83. (b)   85. (d)

Chapter 25

Multiple-Choice Questions

1. (a)   3. (b)   5. (a)   7. (c)   9. (a), (c), (e)

Problems

1. 10 cm
3. (a) 0.40 W/m^2 (b) 0.17 W/m^2
9. 6300 years ago, yes
11. 1.8 to 10 km
13. 100 m
15. 80 km
17. (a) 6.67×10^{-8} T   (b) 9.00×10^5 N/C
19. 500 m
21. (a) 3.0×10^6 N/C

(b) $B_z = (1.0 \times 10^{-2} \text{ T}) \cos\left[2\pi\left(\dfrac{t}{1.0 \times 10^{-16} \text{ s}} - \dfrac{x}{3.0 \times 10^{-8} \text{ m}}\right)\right]$

(c) $E_y = (3.0 \times 10^6 \text{ N/C}) \cos\left[2\pi\left(\dfrac{t}{1.0 \times 10^{-16} \text{ s}} - \dfrac{x}{3.0 \times 10^{-8} \text{ m}}\right)\right]$

(d) Far field
23. UV-A: 7.5×10^{14} Hz $< f < 9.4 \times 10^{14}$ Hz;
UV-B: 9.4×10^{14} Hz $< f < 1.1 \times 10^{15}$ Hz
25. (a) 6.3×10^7 W   (b) 1.4×10^3 W/m^2
29. (a) 4.4×10^{13} W   (b) 3×10^{14} N/C
33. (a) 45°   (b) 72°
35. (a) 0.13   (b) 0.19
37. 39°
41. 6×10^{-23} kg/m^3
43. 4×10^{-18} W

47. $B_{max} = 6.7 \times 10^{-8}$ T, $\bar{u} = 1.8 \times 10^{-9}$ J/m^3,
$u_{max} = 3.6 \times 10^{-9}$ J/m^3, $I = 0.53$ W/m^2
49. $\dfrac{I_0}{8} \sin^2(2\theta)$

51. (c)   53. (d)   55. (e)   57. (b)   59. (d)

Chapter 26

Multiple-Choice Questions

1. (c)   3. (b)   5. (a)   7. (b)   9. (d)   11. (d)

Problems

1. 5 m/s
7. 1.3×10^{-10} s
9. $\Delta t_0 = 12$ s, $v_{spaceship} = 2.5 \times 10^8$ m/s
11. (a) 2.2×10^4 s   (b) Yes
13. 2.993×10^8 m/s
17. 12 cm
19. 2.9×10^8 m/s
21. Yes
23. (a) 4.5 y   (b) 1.1 y
(c)

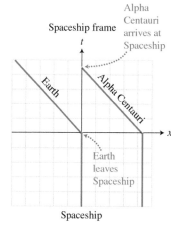

27. 50 m/s
33. (a) $1 + \frac{1}{2} \times 10^{-12}$ (b) 1.00005   (c) 1.00504   (d) 1.1
(e) 1.3   (f) 3.9
35. 4.2×10^7 m/s, 1.3×10^8 m/s, 2.6×10^8 m/s
37. 299,791,858 m/s
39. 1.7×10^{19} J, 5.4×10^{18} J, 1.2×10^{19} J

41. 59%
43. (a) 7.3 (b) 2.5
45. (a) 1.96×10^{-35} kg (b) 4.2×10^{-10}
47. 7.4×10^{-5} kg
49. (a) 3.3×10^{-8} kg/s (b) 1.1 kg
51. (a) 3.6×10^{26} J (b) 4.5×10^{-10}
53. 11.8 km/s
57. No
59. (a) 4.5 y (b) 1.1 y
61. (a) 10 s (b) 4.4 s (c) 390 m (d) Yes
63. (a) 2.8×10^8 m/s (b) 2.9×10^{20} J, 9.0×10^{19} J
65.

Person making measurement	Time interval for Bob's travel	Time interval for Charlie's travel
Alice	14 y	12 y
Bob	10 y	6.4 y
Charlie	11 y	5.8 y
Darien	14 y	12 y

67. (a) $E_{\text{therm}} = 1 \times 10^6$ J, $E_0/E_{\text{therm}} = 4 \times 10^{11}$ (b) 4×10^{-12}
69. (c)　71. (c)　73. (a)　75. (c)　77. (e)　79. (b)

Chapter 27

Multiple-Choice Questions

1. (c)　3. (e)　5. (e)　7. (c)　9. (c)　11. (d)

Problems

1. 9.35 μm
3. (a) 72.5 nm (b) 480 nm (c) 9.67 μm
5. 1.3×10^{-5} m^2
7. $P_{\text{Earth}}^{\text{received}} = 2 \times 10^{17}$ W, $P_{\text{Earth}}^{\text{emitted}} = 2 \times 10^{17}$ W
9. 0.0379 eV
11.

$$U_{qi} + hf = K_{ef}$$

13. (a) 5.1×10^{14} Hz (b) 1.0 eV
17. 1.2×10^{15} Hz, 250 nm
19. 4.1 eV
21. (a) 4×10^{-3} W (b) 30 nm
23. $(5.0 \times 10^{15})P\lambda$, where the λ must be given in nm and P in W
25. 1.4×10^{18}, 2 W
27. 2.6×10^{19}
29. (a) 7.0×10^{17} (b) 6.7×10^{-10} kg·m/s
33. (b) 9.7×10^{18} Hz, 3.1×10^{-11} m
35. 5.0×10^{11}
37. 2.0×10^{-12} W/m^2
39. (a) 8.0×10^{19} (b) 13 A
45. (a) 0.69 V (b) 1.4 μA
47. 1×10^2
49. (a) 9.1×10^{-22} W (b) 3.9×10^2 s
51. 1.0×10^{28} J, about 26 times greater
53. (a) 2.1×10^{11} W (b) 7.0×10^5 K
55. (c)　57. (d)　59. (c)　61. (d)　63. (e)

Chapter 28

Multiple-Choice Questions

1. (a), (b), (d)　3. (b)　5. (d)　7. (a)　9. (b)　11. (b)

Problems

1. 2 km
3. (a) 2×10^{-4} kg (b) 6×10^{20} (c) 1.5×10^{-28} m (d) 9×10^{-4}
7. 435 nm, 6.90×10^{14} Hz, 2.86 eV
9. $\lambda = 250(2n - 1)^2$ nm $= \left[1000(n - 1)n + 250\right]$ nm
11. $r_n = \dfrac{h^2}{4\pi k_C e^2 m_e} n^2$, $E_n = \dfrac{-13.6 \text{ eV}}{n^2}$, $L_n = n\dfrac{h}{2\pi}$
13. 658 nm, 122 nm, 103 nm
17. 2.3×10^{39}, yes
21. 4.6×10^{23} K
23. (a) 2.0×10^9 W (b) 2.5×10^{17} W/m^2
25. 7.2×10^{15}
29. (a) Upward

(b)

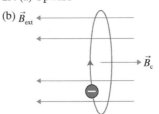

31. (a) 2.2×10^6 m/s (b) 3.3×10^{-10} m
41. The potential difference V through which the electrons are accelerated
43. (a) 3, 4 (b) 2, 3
45.

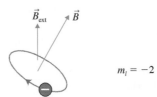

47. $(n, l, m_l, m_s) = (4, 3, \pm 3, \pm\frac{1}{2})$, $(4, 3, \pm 2, \pm\frac{1}{2})$, $(4, 3, \pm 1, \pm\frac{1}{2})$, $(4, 3, 0, \pm\frac{1}{2})$
49. (a) $1s^2, 2s^2, 2p^6, 3s^2, 3p^4$
(b) Because their outermost subshell is a p shell and they have the same number of electrons (4) in this subshell
51. $1s^2, 2s^2, 2p^6, 3s^2, 3p^6, 4s^2, 3d^6$

53. H: $1s^1$
Li: $1s^2, 2s^1$
Na: $1s^2, 2s^2, 2p^6, 3s^1$
K: $1s^2, 2s^2, 2p^6, 3s^2, 3p^6, 4s^1$
55. 2×10^{-18} m/s
57. 3.3×10^{-7} eV, $\Delta E_{min}/(1.51 \text{ eV}) = 2.2 \times 10^{-7}$
59. (a) 0.027 nm, 0.106 nm, 0.239 nm

(b)

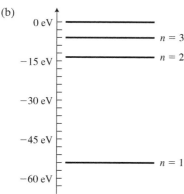

61. 5.8×10^{-13} m
63. 1.88 μm
65. 52.9 eV, 1.28×10^{16} Hz, 23.5 nm
67. No   69. (b)   71. (f)   73. (a)   75. (f)   77. (e)

Chapter 29

Multiple-Choice Questions

1. (d)   3. (c)   5. (b)   7. (c)   9. (b)

Problems

1. 58 N. The protons are held together by the strong nuclear interaction.
3.

	Protons	Neutrons	Nucleons
$^{9}_{4}$Be	4	5	9
$^{16}_{8}$O	8	8	16
$^{27}_{13}$Al	13	14	27
$^{56}_{26}$Fe	26	30	56
^{64}Zn	30	34	64
^{107}Ag	47	60	107

5. (a) 4×10^{28} (b) 2×10^{28} (c) 2×10^{-10} cm^3
7. 2.8 nm
9. 1.5 GeV
11. 7.57 MeV, 8.50 MeV, $^{120}_{50}$Sn
13. (a) $^1_1$H   (b) $^1_0 n$ (c) $^{94}_{38}$Sr   (d) $^3_2$He
17. 7.16 MeV released

19. $^{228}_{90}$Th, 5.41 MeV
21. (a) 38, 50   (b) 87.905586 u
23. (a) 43.2 MeV   (b) 3.45×10^{14} J/kg
27. (a) $^{235}_{92}$U   (b) $^{144}_{59}$Pr   (c) $^{129}_{54}$Xe   (d) $^{60}_{29}$Cu
29. 5.65 MeV
31. (a) 1732 MeV   (b) 1708 MeV, 28.30 MeV, 1737 MeV
(c) 4.87 MeV
33. 7.2×10^3 m/s
37. (a) 0.063   (b) 0.00391   (c) 9.24×10^{-11}
39. 4.94×10^{12} s^{-1}
41. (a) 5700 y   (b) 11,400 y   (c) 17,100 y   (d) 6.8×10^{-4}
43. (a) 4.6×10^{14} (b) 4.6×10^8 s^{-1} (c) 6.1×10^{-7} g
45. 6.09 MeV
47. 1.3×10^9 y
49. 40,000 y
53. (a) 7.0×10^{-4} Sv   (b) 1.2×10^{-3} Sv   (c) 7.0×10^{-3} Sv
(d) 7.0×10^{-3} (e) 2×10^{28}
55. 8.0×10^{19}
57. 2.4×10^{17} kg/m^3
59. (a) Na   (b) Fe   (c) Zn   (d) Ag
63. 8.2×10^{13} J/kg
65. 2×10^{13}
67. (a) 5.7×10^{20} J   (b) 5.7 y
69. 8, 4
71. (a) 4.2×10^{-12} kg   (b) 4.2×10^{-14} J   (c) 2.6×10^3
(d) 1.1×10^{-13} kg, 1.1×10^{-15} J, 71
73. (c)   75. (d)   77. (c)   79. (b)   81. (b)

Chapter 30

Multiple-Choice Questions

1. (a)   3. (b)

Problems

7.

Magnetic field points out of the page

17. Their fundamental interactions
21. Through all four fundamental interactions (gravity, electromagnetic, strong, and weak)
25. Higgs gives mass to the other particles.
31. The formation of neutral atoms 370,000 years after the Big Bang
41. 2.9977×10^8 m/s
43. (b)

Credits

All photos not listed below were provided by the authors.

Chapter 1

Opener: Muratart/Shutterstock; p. 2: John Reader/Science Source; p. 3: Gary Yim/Shutterstock; p. 5: Frank Polich/Reuters; p. 8 bottom: Passion Images/Shutterstock; p. 9: Vladischern/Fotolia.

Chapter 2

Opener: MadCircles/E+/Getty Images; Fig. 2.26: Jim Sugar/Corbis Documentary/Getty Images.

Chapter 3

Opener: FStop Images GmbH/Alamy Stock Photo; p. 61: Derek Bayes/Lebrecht Music and Arts Photo Library/Alamy Stock Photo; Fig. 3.7: Vereshchagin Dmitry/Shutterstock; p. 81: US Navy Permissions Dept./Academx Publishing; p. 82: Dotshock/Shutterstock.

Chapter 4

Opener: Jerry Wolford/Polaris/Newscom; p. 92: Henn Photography/Cultura/Getty Images; p. 710: Adam Rowley/Alamy Stock Photo; p. 922 bottom: Stefan Wermuth/Reuters.

Chapter 5

Opener: AM Corporation/Aflo Co. Ltd./Alamy Stock Photo.

Chapter 6

Opener: Joel Kowsky/NASA; Fig. 6.5: Ted Kinsman/Science Source; Fig. 6.7: Joel Kowsky/NASA; P6.52: Reinhard Dirscherl/Alamy Stock Photo.

Chapter 7

Opener: Christian Charisius/dpa picture alliance archive/Alamy Stock Photo; Fig. 7.3: Ted Foxx/Alamy Stock Photo; P7.77: Wikipedia User:BrandonR.

Chapter 8

Opener: Damir Sagolj/REUTERS/Alamy Stock Photo; Fig. 8.1: Alexander Y/Shutterstock; Fig. 8.13: Dylan Martinez/Reuters; Fig. 8.15: GIPhotoStock/Science Source; Fig. 8.22: Nmiskovic/Shutterstock; Fig. 8.24: Stringer Shanghai/Reuters; p. 241: Eric Schrader.

Chapter 9

Opener: Matt Tilghman/Shutterstock; p. 255: NASA.

Chapter 10

Opener: Afonskaya/iStock Editorial/Getty Images; Fig. 10.9: NASA; p. 299: David Hancock/Alamy Stock Photo; Fig.10.14: Cathy Keifer/Shutterstock; Fig.10.13: Joerg Reuther/imageBROKER/Alamy Stock Photo; Fig.10.16: AP Images; Fig.10.17: Eric Schrader/Fundamental Photographs, NYC.

Chapter 11

Opener: Merlin D. Tuttle/Science Source.

Chapter 12

Opener: MaxWebb/Shutterstock; Fig. 12.1: Richard Megna/Fundamental Photographs; Fig. 12.2: Egmont Strigl/ImageBROKER/Alamy Stock Photo.

Chapter 13

Opener: Topseller/Shutterstock; Fig.13.14: Akg-images/Superstock; Fig.13.15: Richard Megna/Fundamental Photographs, NYC.

Chapter 14

Opener: Biophoto Associates/Science Source; Fig.14.5: Thomas Otto/Fotolia; Fig.14.6: Dmitry Naumov/Fotolia; Fig.14.7: Ustyujanin/Shutterstock.

Chapter 15

Opener, left: William Osgood Field. 1941. Muir Glacier: From the Glacier Photograph Collection. Boulder, Colorado USA: National Snow and Ice Data Center. Digital media; right: Bruce F. Molnia. 2004. Muir Glacier: From the Glacier Photograph Collection. Boulder, Colorado USA: National Snow and Ice Data Center. Digital media; p. 464: Kirk Geisler/Shutterstock; Fig.15.17: Ted Kinsman/Science Source.

Chapter 16

Opener: Richard Megna/Fundamental Photographs, NYC; Fig.16.1a: iPics/Fotolia; p. 486: Michael Dalton/Fundamental Photographs, NYC; Fig.16.6: Misu/Fotolia; p. 491: Kenneth Sponsler/Shutterstock.

Chapter 17

Opener: Suksawad/Shutterstock; Fig. 17.2: GIPhotoStock/Science Source; Fig. 17.19: Sciencephotos/Alamy Stock Photo; p. 525: Ted Kinsman/Science Source; Fig. 17.21: Clive Streeter/Dorling Kindersley Ltd.; P17.15: Blickwinkel/Hartl/Alamy Stock Photo.

Chapter 18

Opener: Lyle Leduc/Photographer's Choice RF/Getty images; Fig.18.28: Matzsoca/Shutterstock.

Chapter 19

Opener: Bacho12345/123RF.

Chapter 20

Opener: CoolKengzz/Shutterstock; Fig. 20.7: Stephen Oliver/Dorling Kindersley Ltd.; p. 647: Thomas Dobner 2007/Alamy Stock Photo.

Chapter 21

Opener: Phanie/Alamy Stock Photo; Fig. 21.10: Sergei Kozak/Alamy Stock Photo; P21.50: Erik Charlton, Hammerhead, https://www.flickr.com/photos/erikcharlton/2935553971/in/photostream, © CC BY 2.0.

Chapter 22

Opener: Richard Megna/Fundamental Photographs, NYC; Fig. 22.1: Atith Jiranaphawiboon/123RF; Fig. 22.8: Artem Mazunov/Shutterstock; Fig. 22.9: Van Heuvelen, Alan; Fig. 22.14: Silkfactory/iStock/Getty Images; Fig. 22.17: Jeremy Woodhouse/Photodisc/Getty Images; P22.34: Dcwcreations/Shutterstock; p. 710: Pedrosala/Shutterstock.

Chapter 23

Opener: Juice Images242/Juice Images/Alamy Stock Photo; p. 713: I.Dr/Shutterstock; Fig. 23.3: Major Pix/Alamy Stock Photo; Fig. 23.20: Vladislav Lebedinski/Fotolia; Fig. 23.22: David Paul Morris/Bloomberg/Getty Images; P23.36: Anion/Fotolia.

Chapter 24

Opener: Richard Megna/Fundamental Photographs, NYC.; Fig. 24.4: Dietrich Zawischa, https://www.itp.unihannover.de/fileadmin/arbeitsgruppen/zawischa/static_html/multibeam.html; Fig. 24.7: Sciencephotos/Alamy Stock Photo; Fig. 24.9: Berenice Abbott/Science Source; Fig. 24.12a: GIPhotoStock/Science Source; Fig. 24.13: Langdu/Shutterstock; Fig. 24.15: Kelley/Fotolia; Fig. 24.16a: Andrew Lambert Photography/Science Source; Fig. 24.16b: Eric Schrader/Fundamental Photographs, NYC.; Fig. 24.19: Adam Filipowicz/Shutterstock; Fig. 24.20: Doug Schnurr/Fotolia; Fig. 24.21: Richard Megna/Fundamental Photographs, NYC.; Fig. 24.27: Images courtesy of Nick Strobel at www.astronomynotes.com; Fig. 24.29: Cphoto/Fotolia.

Chapter 25

Opener: Gorazd Planinsic; Fig. 25.9: National Oceanic and Atmospheric Administration (NOAA); P25.8: JPL-Caltech/R. Hurt/NASA; P25.41: P. Challis/R. Kirshner (CfA)/B. Su/NASA.

Chapter 26

Opener: Rafal Olechowski/123RF.

Chapter 27

Opener: NASA; Fig. 27.1: Dr. Arthur Tucker/Science Source; Fig. 27.10: Richard Megna/Fundamental Photographs, NYC; Fig. 27.11: Wilhelm Röntgen/Library of Congress; Fig. 27.18: The Physics Teacher 52, 212 (2014); doi: 10.1119/1.4868933; P27.62: NASA.

Chapter 28

Opener: Oxford Designers/Illustrators Ltd/Pearson Education, Inc.; Fig. 28.12: Wabash Instrument Corp/Fundamental Photographs, NYC; Fig. 28.37: Stefan Diller/Science Source.

Chapter 29

Opener: Joaquin Ossorio Castillo/Shutterstock; p. 922 bottom: Marcel Clemens/Shutterstock; p. 922 top: Science Source; p. 932: Chonrawit boonprakob/Shutterstock; p. 934: Chonrawit boonprakob/Shutterstock; p. 938: Photo Researchers/Science History Images/Alamy Stock Photo.

Chapter 30

Opener: Photo Researchers/Science History Images/Alamy Stock Photo; Fig. 30.1: Lawrence Livermore National Laboratory; Fig. 30.2: Lawrence Livermore National Laboratory; Fig. 30.5: Peter Ginter/RGB Ventures/SuperStock/Alamy Stock Photo; Fig. 30.6: Xenotar/iStock/Getty Images Plus; Fig. 30.7: Super Kamiokande Collaboration Japan/NASA; Fig. 30.13: JPL-Caltech/NASA.

Index of physics terms

For users of the two-volume edition, pages 1–499 are in Volume 1 and pages 500–981 are in Volume 2.